Data Mining and Knowledge Discovery Handbook

Second Edition

Oded Maimon · Lior Rokach
Editors

Data Mining and Knowledge Discovery Handbook

Second Edition

Editors
Prof. Oded Maimon
Tel Aviv University
Dept. Industrial Engineering
69978 Ramat Aviv
Israel
maimon@eng.tau.ac.il

Dr. Lior Rokach
Ben-Gurion University of the Negev
Dept. Information Systems
Engineering
84105 Beer-Sheva
Israel
liorrk@bgu.ac.il

ISBN 978-0-387-09822-7 e-ISBN 978-0-387-09823-4
DOI 10.1007/978-0-387-09823-4
Springer New York Dordrecht Heidelberg London

Library of Congress Control Number: 2010931143

Printed on acid-free paper

Springer is part of Springer Science+Business Media (www.springer.com)

To my family
– Oded Maimon

To my parents Ines and Avraham
– Lior Rokach

Preface

Knowledge Discovery demonstrates intelligent computing at its best, and is the most desirable and interesting end-product of Information Technology. To be able to discover and to extract knowledge from data is a task that many researchers and practitioners are endeavoring to accomplish. There is a lot of hidden knowledge waiting to be discovered – this is the challenge created by today's abundance of data.

Knowledge Discovery in Databases (KDD) is the process of identifying valid, novel, useful, and understandable patterns from large datasets. Data Mining (DM) is the mathematical core of the KDD process, involving the inferring algorithms that explore the data, develop mathematical models and discover significant patterns (implicit or explicit) -which are the essence of useful knowledge. This detailed guide book covers in a succinct and orderly manner the methods one needs to master in order to pursue this complex and fascinating area.

Given the fast growing interest in the field, it is not surprising that a variety of methods are now available to researchers and practitioners. This handbook aims to organize all major concepts, theories, methodologies, trends, challenges and applications of Data Mining into a coherent and unified repository. This handbook provides researchers, scholars, students and professionals with a comprehensive, yet concise source of reference to Data Mining (and additional selected references for further studies).

The handbook consists of eight parts, each part consists of several chapters. The first seven parts present a complete description of different methods used throughout the KDD process. Each part describes the classic methods, as well as the extensions and novel methods developed recently. Along with the algorithmic description of each method, the reader is provided with an explanation of the circumstances in which this method is applicable, and the consequences and trade-offs incurred by using that method. The last part surveys software and tools available today.

The first part describes preprocessing methods, such as cleansing, dimension reduction, and discretization. The second part covers supervised methods, such as regression, decision trees, Bayesian networks, rule induction and support vector machines. The third part discusses unsupervised methods, such as clustering, association rules, link analysis and visualization. The fourth part covers soft computing methods and their application to Data Mining. This part includes chapters about fuzzy logic, neural networks, and evolutionary algorithms.

Parts five and six present supporting and advanced methods in Data Mining, such as statistical methods for Data Mining, logics for Data Mining, DM query languages, text mining, web mining, causal discovery, ensemble methods, and a great deal more. Part seven provides an in-depth description of Data Mining applications in various interdisciplinary industries, such as finance, marketing, medicine, biology, engineering, telecommunications, software, and security.

The motivation: Over the past few years we have presented and written several scientific papers and research books in this fascinating field. We have also developed successful methods for very large complex applications in industry, which are in operation in several enterprises. Thus, we have first hand experience in the needs of the KDD/DM community in research and practice. This handbook evolved from these experiences.

The first edition of the handbook, which was published five years ago, was extremely well received by the data mining research and development communities. The field of data mining has evolved in several aspects since the first edition. Advances occurred in areas, such as Multimedia Data Mining, Data Stream Mining, Spatio-temporal Data Mining, Sequences Analysis, Swarm Intelligence, Multi-label classification and privacy in data mining. In addition new applications and software tools become available. We received many requests to include the new advances in the field in a second edition of the handbook. About half of the book is new in this edition. This second edition aims to refresh the previous material in the fundamental areas, and to present new findings in the field. The new advances occurred mainly in three dimensions: new methods, new applications and new data types, which can be handled by new and modified advanced data mining methods.

We would like to thank all authors for their valuable contributions. We would like to express our special thanks to Susan Lagerstrom-Fife of Springer for working closely with us during the production of this book.

Tel-Aviv, Israel *Oded Maimon*
Beer-Sheva, Israel *Lior Rokach*

April 2010

Contents

List of Contributors

Maria M. Abad
Software Engineering Department,
University of Granada, Spain

Ajith Abraham
Center of Excellence for Quantifiable
Quality of Service
Norwegian University of Science and
Technology,
Trondheim, Norway

Bruno Agard
Département de Mathmatiques et de
Génie Industriel,
École Polytechnique de Montréal,
Canada

Daniel Barbara
Department of Information and Software Engineering,
George Mason University, USA

Christopher D. Barko
Customer Analytics, Inc.

Irad Ben-Gal
Department of Industrial Engineering,
Tel-Aviv University, Israel

Moty Ben-Dov
School of Computing Science,
MDX University, London, UK.

Yoav Benjamini
Department of Statistics, Sackler
Faculty for Exact Sciences
Tel Aviv University, Israel

Richard A. Berk
Department of Statistics
UCLA, USA

Jean-Francois Boulicaut
INSA Lyon, France

Pavel Brazdil
Faculty of Economics,
University of Porto, Portugal

Abel Browarnik
Department of Industrial Engineering,
Tel-Aviv University, Israel

Christopher J.C. Burges
Microsoft Research, USA

Cory J. Butz
Department of Computer Science,
University of Regina, Canada

Nitesh V. Chawla
Department of Computer Science and
Engineering,
University of Notre Dame, USA

Ping Chen
Department of Computer and Mathematics Science,
University of Houston-Downtown, USA

Barak Chizi
Department of Industrial Engineering,
Tel-Aviv University, Israel

Shahar Cohen
Department of Industrial Engineering,
Tel-Aviv University, Israel

Antonio Congiusta
Dipartimento di Elettronica, Informatica e Sistemistica,
University of Calabria, Italy

Gautam Das
Computer Science and Engineering Department,
University of Texas, Arlington, USA

Swagatam Das
Department of Electronics and Telecommunication Engineering,
Jadavpur University, India.

Steve Donoho
Mantas, Inc. USA

Sašo Džeroski
Jožef Stefan Institute, Slovenia

Ronen Feldman
Department of Mathematics and Computer Science,
Bar-Ilan university, Israel

Eibe Frank
Department of Computer Science,
University of Waikato, New Zealand

Alex A. Freitas
Computing Laboratory,
University of Kent, UK

Johannes Fürnkranz
TU Darmstadt, Knowledge Engineering Group, Germany

Mohamed Medhat Gaber
Centre for Distributed Systems and Software Engineering
Monash University

Pierre Geurts
Department of Electrical Engineering and Computer Science,
University of Liège, Belgium

Christophe Giraud-Carrier
Department of Computer Science,
Brigham Young University, Utah, USA

Paolo Giudici
Faculty of Economics,
University of Pavia, Italy

Bart Goethals
Departement of Mathemati1cs and Computer Science,
University of Antwerp, Belgium

Jerzy W. Grzymala-Busse
Department of Electrical Engineering and Computer Science,
University of Kansas, USA

Witold J. Grzymala-Busse
FilterLogix Inc., USA

Dimitrios Gunopulos
Department of Computer Science and Engineering,
University of California at Riverside, USA

Petr Hájek
Institute of Computer Science,
Academy of Sciences of the Czech Republic

Maria Halkidi
Department of Computer Science and Engineering,
University of California at Riverside, USA

Mark Hall
Department of Computer Science,
University of Waikato, New Zealand

Howard J. Hamilton
Department of Computer Science,
University of Regina, Canada

Jiawei Han
Department of Computer Science,
University of Illinois, Urbana Champaign, USA

Geoffrey Holmes
Department of Computer Science,
University of Waikato, New Zealand

Frank Höppner
Department of Information Systems,
University of Applied Sciences Braunschweig/Wolfenbüttel, Germany

Yan Huang
Department of Computer Science,
University of Minnesota, USA

Sushil Jajodia
Center for Secure Information Systems,
George Mason University, USA

Ioannis Katakis
Dept. of Informatics, Aristotle University of Thessaloniki, 54124 Greece

Eamonn Keogh
Computer Science and Engineering Department,
University of California at Riverside, USA

Richard Kirkby
Department of Computer Science,
University of Waikato, New Zealand

Slava Kisilevich
University of Konstanz, Germany

Boris Kovalerchuk
Department of Computer Science,
Central Washington University, USA

Shonali Krishnaswamy
Centre for Distributed Systems and Software Engineering
Monash University

Andrew Kusiak
Department of Mechanical and Industrial Engineering,
The University of Iowa, USA

Nada Lavrač
Jožef Stefan Institute, Ljubljana, Slovenia
Nova Gorica Polytechnic, Nova Gorica, Slovenia

Moshe Leshno
Faculty of Management and Sackler Faculty of Medicine,
Tel Aviv University, Israel

Nissan Levin
Q-Ware Software Company, Israel

Tao Li
School of Computer Science,
Florida International University, USA

Churn-Jung Liau
Institute of Information Science,
Academia Sinica, Taiwan

Jessica Lin
Department of Computer Science and Engineering,
University of California at Riverside, USA

Tsau Y. Lin
Department of Computer Science,
San Jose State University, USA

Sheng Ma
Machine Learning for Systems
IBM T.J. Watson Research Center, USA

Oded Maimon
Department of Industrial Engineering,
Tel-Aviv University, Israel

Jonathan I. Maletic
Department of Computer Science,
Kent State University, USA

Florian Mansmann
University of Konstanz, Germany

Andrian Marcus
Department of Computer Science,
Wayne State University, USA

Cyrille Masson
INSA Lyon, France

Steve Moyle
Computing Laboratory,
Oxford University, UK

Mirco Nanni
University of Pisa,
Italy

Hamid R. Nemati
Information Systems and Operations Management Department
Bryan School of Business and Economics
The University of North Carolina at Greensboro, USA

Mitsunori Ogihara
Computer Science Department,
University of Rochester, USA

Nora Oikonomakou
Department of Informatics,
Athens University of Economics and Business (AUEB), Greece

Bernhard Pfahringer
Department of Computer Science,
University of Waikato, New Zealand

Marco F. Ramoni
Departments of Pediatrics and Medicine
Harvard University, USA

Chotirat Ann Ratanamahatana
Department of Computer Science and Engineering,
University of California at Riverside, USA

Yoram Reich
Center for Design Research,
Stanford University, Stanford, CA, USA

Salvatore Rinzivillo
Institute of Information Science and Technologies,
Italy

Lior Rokach
Department of Information Systems Engineering
Ben-Gurion University of the Negev,
Israel

Noa Ruschin Rimini
Department of Industrial Engineering,
Tel-Aviv University, Israel

Sigal Sahar
Department of Computer Science,
Tel-Aviv University, Israel

Paola Sebastiani
Department of Biostatistics,
Boston University, USA

Richard S. Segall
Arkansas State University,
Department of Computer and Info.
Tech., Jonesboro, AR
72467-0130,USA

Shashi Shekhar
Institute of Technology,
University of Minnesota, USA

Armin Shmilovici
Department of Information Systems
Engineering,
Ben-Gurion University of the Negev,
Israel

Gautam B. Singh
Department of Computer Science and
Engineering,
Center for Bioinformatics, Oakland
University, USA

Anoop Singhal
Center for Secure Information Systems,
George Mason University, USA

Domenico Talia
Dipartimento di Elettronica, Informatica
e Sistemistica,
University of Calabria, Italy

Kurt Thearling
Vertex Business Services
Richardson, Texas, USA

Vicenç Torra
Institut d'Investigació en Intel·ligència
Artificial, Spain

Paolo Trunfio
Dipartimento di Elettronica, Informatica
e Sistemistica,
University of Calabria, Italy

Grigorios Tsoumakas
Dept. of Informatics, Aristotle University of Thessaloniki, 54124 Greece

Jiong Yang
Department of Electronic Engineering
and Computer Science,
Case Western Reserve University, USA

Ying Yang
School of Computer Science and
Software Engineering,
Monash University, Melbourne, Australia

Hong Yao
Department of Computer Science,
University of Regina, Canada

Philip S. Yu
IBM T. J. Watson Research Center,
USA

Michalis Vazirgiannis
Department of Informatics,
Athens University of Economics and
Business, Greece

Ricardo Vilalta
Department of Computer Science,
University of Houston, USA

Evgenii Vityaev
Institute of Mathematics,
Russian Academy of Sciences, Russia

Michail Vlachos
IBM T. J. Watson Research Center,
USA

Ioannis Vlahavas
Dept. of Informatics, Aristotle University of Thessaloniki, 54124 Greece

Haixun Wang
IBM T. J. Watson Research Center,
USA

Wei Wang
Department of Computer Science,
University of North Carolina at Chapel Hill, USA

Geoffrey I. Webb
Faculty of Information Technology,
Monash University, Australia

Gary M. Weiss
Department of Computer and Information Science,
Fordham University, USA

Ian H. Witten
Department of Computer Science,
University of Waikato, New Zealand

Jacob Zahavi
The Wharton School,
University of Pennsylvania, USA

Arkady Zaslavsky
Centre for Distributed Systems and Software Engineering
Monash University

Peter G. Zhang
Department of Managerial Sciences,
Georgia State University, USA

Pusheng Zhang
Department of Computer Science and Engineering,
University of Minnesota, USA

Qingyu Zhang
Arkansas State University, Department of
Computer and Info. Tech.,
Jonesboro, AR 72467-0130,USA

Ruofei Zhang
Yahoo!, Inc. Sunnyvale, CA 94089

Zhongfei (Mark) Zhang
SUNY Binghamton, NY 13902-6000

Blaž Zupan
Faculty of Computer and Information Science,
University of Ljubljana, Slovenia

1

Introduction to Knowledge Discovery and Data Mining

Oded Maimon[1] and Lior Rokach[2]

[1] Department of Industrial Engineering, Tel-Aviv University, Ramat-Aviv 69978, Israel, maimon@eng.tau.ac.il

[2] Department of Information System Engineering, Ben-Gurion University, Beer-Sheba, Israel, liorrk@bgu.ac.il

Knowledge Discovery in Databases (KDD) is an automatic, exploratory analysis and modeling of large data repositories. KDD is the organized process of identifying valid, novel, useful, and understandable patterns from large and complex data sets. *Data Mining* (DM) is the core of the KDD process, involving the inferring of algorithms that explore the data, develop the model and discover previously unknown patterns. The model is used for understanding phenomena from the data, analysis and prediction.

The accessibility and abundance of data today makes Knowledge Discovery and Data Mining a matter of considerable importance and necessity. Given the recent growth of the field, it is not surprising that a wide variety of methods is now available to the researchers and practitioners. No one method is superior to others for all cases. The handbook of Data Mining and Knowledge Discovery from Data aims to organize all significant methods developed in the field into a coherent and unified catalog; presents performance evaluation approaches and techniques; and explains with cases and software tools the use of the different methods. The goals of this introductory chapter are to explain the KDD process, and to position DM within the information technology tiers. Research and development challenges for the next generation of the science of KDD and DM are also defined. The rationale, reasoning and organization of the handbook are presented in this chapter for helping the reader to navigate the extremely rich and detailed content provided in this handbook. In this chapter there are six sections followed by a brief discussion of the changes in the second edition.

1. The KDD Process 2. Taxonomy of Data Mining Methods 3. Data Mining within the Complete Decision Support System 4. KDD & DM Research Opportunities and Challenges 5. KDD & DM Trends 6. The Organization of the Handbook 7. New to This Edition

The special recent aspects of data availability that are promoting the rapid development of KDD and DM are the electronically readiness of data (though of different types and reliability). The internet and intranet fast development in particular pro-

O. Maimon, L. Rokach (eds.), *Data Mining and Knowledge Discovery Handbook*, 2nd ed.,
DOI 10.1007/978-0-387-09823-4_1, © Springer Science+Business Media, LLC 2010

mote data accessibility (as formatted or unformatted, voice or video, etc.). Methods that were developed before the Internet revolution considered smaller amounts of data with less variability in data types and reliability. Since the information age, the accumulation of data has become easier and less costly. It has been estimated that the amount of stored information doubles every twenty months. Unfortunately, as the amount of electronically stored information increases, the ability to understand and make use of it does not keep pace with its growth. Data Mining is a term coined to describe the process of sifting through large databases for interesting patterns and relationships. The studies today aim at evidence-based modeling and analysis, as is the leading practice in medicine, finance, security and many other fields. The data availability is increasing exponentially, while the human processing level is almost constant. Thus the potential gap increases exponentially. This gap is the opportunity for the KDD\DM field, which therefore becomes increasingly important and necessary.

1.1 The KDD Process

The knowledge discovery process (Figure 1.1) is iterative and interactive, consisting of nine steps. Note that the process is iterative at each step, meaning that moving back to adjust previous steps may be required. The process has many "artistic" aspects in the sense that one cannot present one formula or make a complete taxonomy for the right choices for each step and application type. Thus it is required to deeply understand the process and the different needs and possibilities in each step. Taxonomy for the Data Mining methods is helping in this process. It is presented in the next section.

The process starts with determining the KDD goals, and "ends" with the implementation of the discovered knowledge. As a result, changes would have to be made in the application domain (such as offering different features to mobile phone users in order to reduce churning). This closes the loop, and the effects are then measured on the new data repositories, and the KDD process is launched again. Following is a brief description of the nine-step KDD process, starting with a managerial step:

1. Developing an understanding of the application domain This is the initial preparatory step. It prepares the scene for understanding what should be done with the many decisions (about transformation, algorithms, representation, etc.). The people who are in charge of a KDD project need to understand and define the goals of the end-user and the environment in which the knowledge discovery process will take place (including relevant prior knowledge). As the KDD process proceeds, there may be even a revision and tuning of this step. Having understood the KDD goals, the preprocessing of the data starts, as defined in the next three steps (note that some of the methods here are similar to Data Mining algorithms, but are used in the preprocessing context):
2. Selecting and creating a data set on which discovery will be performed. Having defined the goals, the data that will be used for the knowledge discovery should

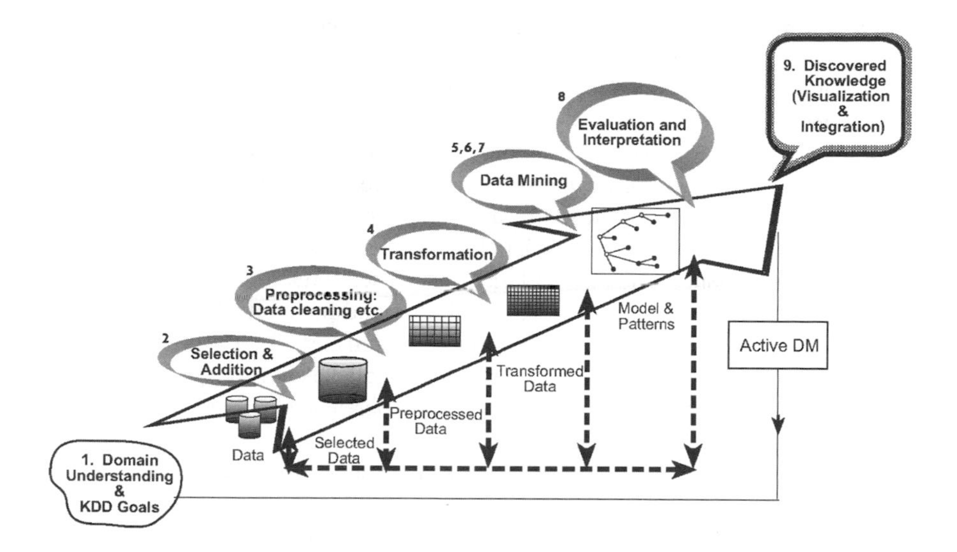

Fig. 1.1. The Process of Knowledge Discovery in Databases.

be determined. This includes finding out what data is available, obtaining additional necessary data, and then integrating all the data for the knowledge discovery into one data set, including the attributes that will be considered for the process. This process is very important because the Data Mining learns and discovers from the available data. This is the evidence base for constructing the models. If some important attributes are missing, then the entire study may fail. From success of the process it is good to consider as many as possible attribute at this stage. On the other hand, to collect, organize and operate complex data repositories is expensive, and there is a tradeoff with the opportunity for best understanding the phenomena. This tradeoff represents an aspect where the interactive and iterative aspect of the KDD is taking place. It starts with the best available data set and later expands and observes the effect in terms of knowledge discovery and modeling.

3. Preprocessing and cleansing. In this stage, data reliability is enhanced. It includes data clearing, such as handling missing values and removal of noise or outliers. Several methods are explained in the handbook, from doing nothing to becoming the major part (in terms of time consumed) of a KDD process in certain projects. It may involve complex statistical methods, or using specific Data Mining algorithm in this context. For example, if one suspects that a certain attribute is not reliable enough or has too many missing data, then this attribute could become the goal of a data mining supervised algorithm. A prediction model for this attribute will be developed, and then missing data can be predicted. The extension to which one pays attention to this level depends on many factors. In any case, studying these aspects is important and often revealing insight by itself, regarding enterprise information systems.

4. Data transformation. In this stage, the generation of better data for the data mining is prepared and developed. Methods here include dimension reduction (such as feature selection and extraction, and record sampling), and attribute transformation (such as discretization of numerical attributes and functional transformation). This step is often crucial for the success of the entire KDD project, but it is usually very project-specific. For example, in medical examinations, the quotient of attributes may often be the most important factor, and not each one by itself. In marketing, we may need to consider effects beyond our control as well as efforts and temporal issues (such as studying the effect of advertising accumulation). However, even if we do not use the right transformation at the beginning, we may obtain a surprising effect that hints to us about the transformation needed (in the next iteration). Thus the KDD process reflects upon itself and leads to an understanding of the transformation needed (like a concise knowledge of an expert in a certain field regarding key leading indicators). Having completed the above four steps, the following four steps are related to the Data Mining part, where the focus is on the algorithmic aspects employed for each project:
5. Choosing the appropriate Data Mining task. We are now ready to decide on which type of Data Mining to use, for example, classification, regression, or clustering. This mostly depends on the KDD goals, and also on the previous steps. There are two major goals in Data Mining: prediction and description. Prediction is often referred to as supervised Data Mining, while descriptive Data Mining includes the unsupervised and visualization aspects of Data Mining. Most data mining techniques are based on inductive learning, where a model is constructed explicitly or implicitly by generalizing from a sufficient number of training examples. The underlying assumption of the inductive approach is that the trained model is applicable to future cases. The strategy also takes into account the level of meta-learning for the particular set of available data.
6. Choosing the Data Mining algorithm. Having the strategy, we now decide on the tactics. This stage includes selecting the specific method to be used for searching patterns (including multiple inducers). For example, in considering precision versus understandability, the former is better with neural networks, while the latter is better with decision trees. For each strategy of meta-learning there are several possibilities of how it can be accomplished. Meta-learning focuses on explaining what causes a Data Mining algorithm to be successful or not in a particular problem. Thus, this approach attempts to understand the conditions under which a Data Mining algorithm is most appropriate. Each algorithm has parameters and tactics of learning (such as ten-fold cross-validation or another division for training and testing).
7. Employing the Data Mining algorithm. Finally the implementation of the Data Mining algorithm is reached. In this step we might need to employ the algorithm several times until a satisfied result is obtained, for instance by tuning the algorithm's control parameters, such as the minimum number of instances in a single leaf of a decision tree.
8. Evaluation. In this stage we evaluate and interpret the mined patterns (rules, reliability etc.), with respect to the goals defined in the first step. Here we consider

the preprocessing steps with respect to their effect on the Data Mining algorithm results (for example, adding features in Step 4, and repeating from there). This step focuses on the comprehensibility and usefulness of the induced model. In this step the discovered knowledge is also documented for further usage. The last step is the usage and overall feedback on the patterns and discovery results obtained by the Data Mining:

9. Using the discovered knowledge. We are now ready to incorporate the knowledge into another system for further action. The knowledge becomes active in the sense that we may make changes to the system and measure the effects. Actually the success of this step determines the effectiveness of the entire KDD process. There are many challenges in this step, such as loosing the "laboratory conditions" under which we have operated. For instance, the knowledge was discovered from a certain static snapshot (usually sample) of the data, but now the data becomes dynamic. Data structures may change (certain attributes become unavailable), and the data domain may be modified (such as, an attribute may have a value that was not assumed before).

1.2 Taxonomy of Data Mining Methods

There are many methods of Data Mining used for different purposes and goals. Taxonomy is called for to help in understanding the variety of methods, their interrelation and grouping. It is useful to distinguish between two main types of Data Mining: verification-oriented (the system verifies the user's hypothesis) and discovery-oriented (the system finds new rules and patterns autonomously). Figure 1.2 presents this taxonomy.

Discovery methods are those that automatically identify patterns in the data. The discovery method branch consists of prediction methods versus description methods. Descriptive methods are oriented to data interpretation, which focuses on understanding (by visualization for example) the way the underlying data relates to its parts. Prediction-oriented methods aim to automatically build a behavioral model, which obtains new and unseen samples and is able to predict values of one or more variables related to the sample. It also develops patterns, which form the discovered knowledge in a way which is understandable and easy to operate upon. Some prediction-oriented methods can also help provide understanding of the data.

Most of the discovery-oriented Data Mining techniques (quantitative in particular) are based on inductive learning, where a model is constructed, explicitly or implicitly, by generalizing from a sufficient number of training examples. The underlying assumption of the inductive approach is that the trained model is applicable to future unseen examples.

Verification methods, on the other hand, deal with the evaluation of a hypothesis proposed by an external source (like an expert etc.). These methods include the most common methods of traditional statistics, like goodness of fit test, tests of hypotheses (e.g., t-test of means), and analysis of variance (ANOVA). These methods are less associated with Data Mining than their discovery-oriented counterparts, because

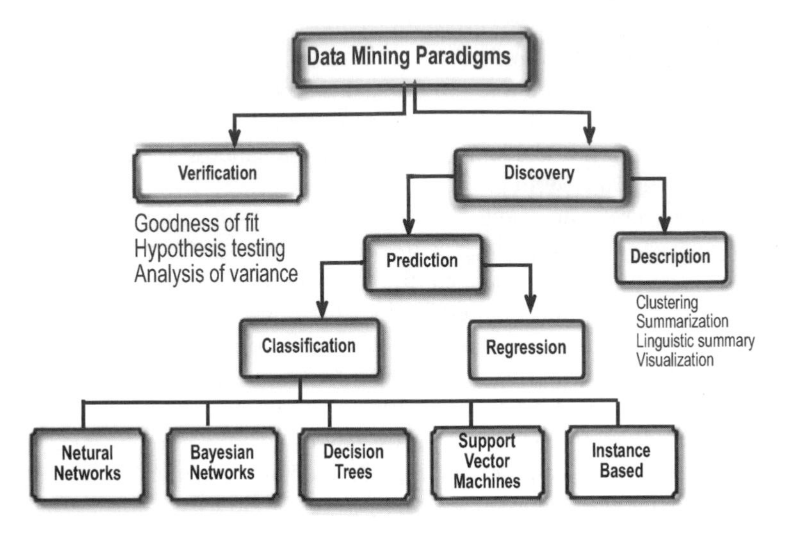

Fig. 1.2. Data Mining Taxonomy.

most Data Mining problems are concerned with discovering an hypothesis (out of a very large set of hypotheses), rather than testing a known one. Much of the focus of traditional statistical methods is on model estimation as opposed to one of the main objectives of Data Mining: model identification and construction, which is evidence based (though overlap occurs).

Another common terminology, used by the machine-learning community, refers to the prediction methods as supervised learning, as opposed to unsupervised learning. Unsupervised learning refers to modeling the distribution of instances in a typical, high-dimensional input space.

Unsupervised learning refers mostly to techniques that group instances without a prespecified, dependent attribute. Thus the term "unsupervised learning" covers only a portion of the description methods presented in Figure 1.2. For instance, it covers clustering methods but not visualization methods. Supervised methods are methods that attempt to discover the relationship between input attributes (sometimes called independent variables) and a target attribute sometimes referred to as a dependent variable). The relationship discovered is represented in a structure referred to as a model. Usually models describe and explain phenomena, which are hidden in the data set and can be used for predicting the value of the target attribute knowing the values of the input attributes. The supervised methods can be implemented on a variety of domains, such as marketing, finance and manufacturing. It is useful to distinguish between two main supervised models: classification models and regression models. The latter map the input space into a real-valued domain. For instance, a regressor can predict the demand for a certain product given its characteristics. On the other hand, classifiers map the input space into predefined classes. For example, classifiers can be used to classify mortgage consumers as good (fully payback the mortgage on time) and bad (delayed payback), or as many target classes as needed. There are many alternatives to represent classifiers. Typical examples include, support vector machines, decision trees, probabilistic summaries, or algebraic function.

1.3 Data Mining within the Complete Decision Support System

Data Mining methods are becoming part of integrated Information Technology (IT) software packages. Figure 1.3 illustrates the three tiers of the decision support aspect of IT. Starting from the data sources (such as operational databases, semi- and non-structured data and reports, Internet sites etc.), the first tier is the data warehouse, followed by OLAP (On Line Analytical Processing) servers and concluding with analysis tools, where Data Mining tools are the most advanced.

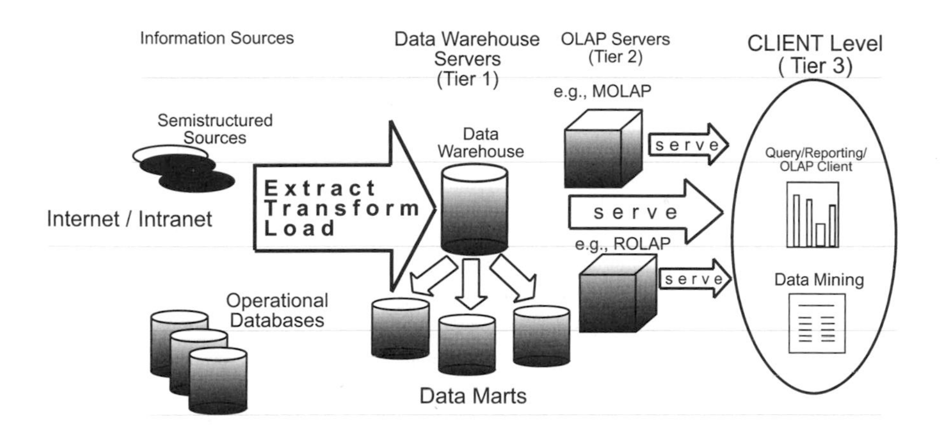

Fig. 1.3. The IT Decision Support Tiers.

The main advantage of the integrated approach is that the preprocessing steps are much easier and more convenient. Since this part is often the major burden for the KDD process (and can consumes most of the KDD project time), this industry trend is very important for expanding the use and utilization of Data Mining. However, the risk of the integrated IT approach comes from the fact that DM techniques are much more complex and intricate than OLAP, for example, so the users need to be trained appropriately.

This handbook shows the variety of strategies, techniques and evaluation measurements. We can naively distinguish among three levels of analysis. The simplest one is achieved by report generators (for example, presenting all claims that occurred because of a certain cause last year, such as car theft). We then proceed to OLAP multi-level analysis (for example presenting the ten towns where there was the highest increase of vehicle theft in the last month as compared to with the month before). Finally a complex analysis is carried out in discovering the patterns that predict car thefts in these cities, and what might occur if anti theft devices were installed. The latter is based on mathematical modeling of the phenomena, where the first two levels are ways of data aggregation and fast manipulation.

1.4 KDD and DM Research Opportunities and Challenges

Empirical comparison of the performance of different approaches and their variants in a wide range of application domains has shown that each performs best in some, but not all, domains. This phenomenon is known as the selective superiority problem, which means, in our case, that no induction algorithm can be the best in all possible domains. The reason is that each algorithm contains an explicit or implicit bias that leads it to prefer certain generalizations over others, and it will be successful only as long as this bias matches the characteristics of the application domain. Results have demonstrated the existence and correctness of this "no free lunch theorem". If one inducer is better than another in some domains, then there are necessarily other domains in which this relationship is reversed. This implies in KDD that for a given problem a certain approach can yield more knowledge from the same data than other approaches.

In many application domains, the generalization error (on the overall domain, not just the one spanned in the given data set) of even the best methods is far above the training set, and the question of whether it can be improved, and if so how, is an open and important one. Part of the answer to this question is to determine the minimum error achievable by any classifier in the application domain (known as the optimal Bayes error). If existing classifiers do not reach this level, new approaches are needed. Although this problem has received considerable attention, no generally reliable method has so far been demonstrated. This is one of the challenges of the DM research – not only to solve it, but even to quantify and understand it better. Heuristic methods can then be compared absolutely and not just against each other.

A subset of this generalized study is the question of which inducer to use for a given problem. To be even more specific, the performance measure needs to be defined appropriately for each problem. Though there are some commonly accepted measures it is not enough. For example, if the analyst is looking for accuracy only, one solution is to try each one in turn, and by estimating the generalization error, to choose the one that appears to perform best. Another approach, known as multi-strategy learning, attempts to combine two or more different paradigms in a single algorithm. The dilemma of which method to choose becomes even greater if other factors, such as comprehensibility are taken into consideration. For instance, for a specific domain, neural networks may outperform decision trees in accuracy. However from the comprehensibility aspect, decision trees are considered superior. In other words, in this case even if the researcher knows that neural network is more accurate, the dilemma of what methods to use still exists (or maybe to combine methods for their separate strength).

Induction is one of the central problems in many disciplines such as machine learning, pattern recognition, and statistics. However the feature that distinguishes Data Mining from traditional methods is its scalability to very large sets of varied types of input data. Scalability means working in an environment of high number of records, high dimensionality, and a high number of classes or heterogeneousness. Nevertheless, trying to discover knowledge in real life and large databases introduces time and memory problems. As large databases have become the norm in many fields (including astronomy, molecular biology, finance, marketing, health care, and many others), the use of Data Mining to discover patterns in them has become potentially very beneficial for the enterprise. Many companies are staking a large part of their future on these "Data Mining" applications, and turn to the research community for solutions to the fundamental problems they encounter. While a very large amount of available data used to be the dream of any data analyst, nowadays the synonym for "very large" has become "terabyte" or "pentabyte", a barely imaginable volume of information.

Information-intensive organizations (like telecom companies and financial institutions) are expected to accumulate several terabytes of raw data every one to two years. High dimensionality of the input (that is, the number of attributes) increases the size of the search space in an exponential manner (known as the "Curse of Dimensionality"), and thus increases the chance that the inducer will find spurious classifiers that in general are not valid. There are several approaches for dealing with a high number of records including: sampling methods, aggregation, massively parallel processing, and efficient storage methods.

1.5 KDD & DM Trends

This handbook covers the current state-of-the-art status of Data Mining. The field is still in its early stages in the sense that some basic methods are still being developed. The art expands but so does the understanding and the automation of the nine steps and their interrelation. For this to happen we need better characterization of the KDD problem spectrum and definition. The terms KDD and DM are not well-defined in terms of what methods they contain, what types of problem are best solved by these methods, and what results to expect. How are KDD\DM compared to statistics, machine learning, operations research, etc.? If subset or superset of the above fields? Or an extension\adaptation of them? Or a separate field by itself? In addition to the methods – which are the most promising fields of application and what is the vision KDD\DM brings to these fields? Certainly we already see the great results and achievements of KDD\DM, but we cannot estimate their results with respect to the potential of this field. All these basic analyses have to be studied and we see several trends for future research and implementation, including:

- Active DM – closing the loop, as in control theory, where changes to the system are made according to the KDD results and the full cycle starts again. Stability and controllability, which will be significantly different in these types of systems, need to be well-defined.

- Full taxonomy – for all the nine steps of the KDD process. We have shown a taxonomy for the DM methods, but a taxonomy is needed for each of the nine steps. Such a taxonomy will contain methods appropriate for each step (even the first one), and for the whole process as well.
- Meta-algorithms – algorithms that examine the characteristics of the data in order to determine the best methods, and parameters (including decompositions).
- Benefit analysis – to understand the effect of the potential KDD\DM results on the enterprise.
- Problem characteristics – analysis of the problem itself for its suitability to the KDD process.
- Mining complex objects of arbitrary type – Expanding Data Mining inference to include also data from pictures, voice, video, audio, etc. This will require adapting and developing new methods (for example, for comparing pictures using clustering and compression analysis).
- Temporal aspects - many data mining methods assume that discovered patterns are static. However, in practice patterns in the database evolve over time. This poses two important challenges. The first challenge is to detect when concept drift occurs. The second challenge is to keep the patterns up-to-date without inducing the patterns from scratch.
- Distributed Data Mining – The ability to seamlessly and effectively employ Data Mining methods on databases that are located in various sites. This problem is especially challenging when the data structures are heterogeneous rather than homogeneous.
- Expanding the knowledge base for the KDD process, including not only data but also extraction from known facts to principles (for example, extracting from a machine its principle, and thus being able to apply it in other situations).
- Expanding Data Mining reasoning to include creative solutions, not just the ones that appears in the data, but being able to combine solutions and generate another approach.

1.6 The Organization of the Handbook

This handbook is organized in eight parts. Starting with the KDD process, through to part six, the book presents a comprehensive but concise description of different methods used throughout the KDD process. Each part describes the classic methods as well as the extensions and novel methods developed recently. Along with the algorithmic description of each method, the reader is provided with an explanation of the circumstances in which this method is applicable and the consequences and the trade-offs of using the method including references for further readings. Part seven presents real-world case studies and how they can be solved. The last part surveys some software and tools available today. The first part is about preprocessing methods. This covers the preprocessing methods (Steps 3, 4 of the KDD process). The Data Mining methods are presented in the second part with the introduction and the very often-used supervised methods. The third part of the handbook considers the unsupervised methods. The fourth part is about methods termed soft computing, which include fuzzy logic, evolutionary algorithms, neural networks etc. Having established the foundation, we now proceed with supporting methods needed for Data Mining in the fifth part. The sixth part covers advanced methods like text mining and web mining. With all the methods described so far, the next section, the seventh, is concerned with applications for medicine, biology and manufacturing. The last and final part of this handbook deals with software tools. This part is not a complete survey of the software available, but rather a selected representative from different types of software packages that exist in today's market.

1.7 New to This Edition

Since the first edition that was published five years ago, the field of data mining has been evolved in the following aspects:

1.7.1 Mining Rich Data Formats

While in the past data mining methods could effectively analyze only flat tables, in recent years new mature techniques have been developed for mining rich data formats:

- Data Stream Mining - The conventional focus of data mining research was on mining resident data stored in large data repositories. The growth of technologies, such as wireless sensor networks, have contributed to the emergence of data streams. The distinctive characteristic of such data is that it is unbounded in terms of continuity of data generation. This form of data has been termed as data streams to express its owing nature. Mohamed Medhat Gaber, Arkady Zaslavsky, and Shonali Krishnaswamy present a review of the state of the art in mining data streams (Chapter 39). Clustering, classification, frequency counting, time series analysis techniques are been discussed. Different systems that use data stream mining techniques are also presented.
- Spatio-temporal - Spatio-temporal clustering is a process of grouping objects based on their spatial and temporal similarity. It is relatively new subfield of data mining, which gained high popularity especially in geographic information sciences due to the pervasiveness of all kinds of location-based or environmental devices that record position, time or/and environmental properties of an object or set of objects in real-time. As a consequence, different types and large amounts of spatio-temporal data became available and introduce new challenges to data analysis, which require novel approaches to knowledge discovery. Slava Kisilevich, Florian Mansmann, Mirco Nanni and Salvatore Rinzivillo provide a classification of different types of spatio-temporal data (Chapter 44). Then, they focus on one type of spatio-temporal clustering - trajectory clustering, provide an overview of the state-of-the-art approaches and methods of spatio-temporal clustering and finally present several scenarios in different application domains such as movement, cellular networks and environmental studies.

- Multimedia Data Mining - Zhongfei Mark Zhang and Ruofei Zhang present new methods for Multimedia Data Mining (Chapter 57). Multimedia data mining, as the name suggests, presumably is a combination of the two emerging areas: multimedia and data mining. Instead, the multimedia data mining research focuses on the theme of merging multimedia and data mining research together to exploit the synergy between the two areas to promote the understanding and to advance the development of the knowledge discovery multimedia data.

1.7.2 New Techniques

In this edition the following two new techniques are covered:

- In Chapter 23, Swagatam Das and Ajith Abraham present a family of bio-inspired algorithms, known as Swarm Intelligence (SI). SI has successfully been applied to a number of real world clustering problems. This chapter explores the role of SI in clustering different kinds of datasets. It also describes a new SI technique for partitioning a linearly non-separable dataset into an optimal number of clusters in the kernel- induced feature space. Computer simulations undertaken in this research have also been provided to demonstrate the effectiveness of the proposed algorithm.
- Multi-label classification - Most of the research in the field of supervised learning has been focused on single label tasks, where training instances are associated with a single label from a set of disjoint labels. However, Textual data, such as documents and web pages, are frequently annotated with more than a single label. In Chapter 34, Grigorios Tsoumakas, Loannis Katakis and Loannis Vlahavas review techniques for addressing multi-label classification task grouped into the two categories: i) *problem transformation*, and ii) *algorithm adaptation*. The first group of methods is algorithm independent. They transform the learning task into one or more single-label classification tasks, for which a large bibliography of learning algorithms exists. The second group of methods extends specific learning algorithms in order to handle multi-label data directly.
- Sequences Analysis - In Chapter 29, Noa Ruschin Rimini and Oded Maimon introduce a new visual analysis technique of sequences dataset using Iterated Function System (IFS). IFS produces a fractal representation of sequences. The proposed method offers an effective tool for visual detection of sequence patterns influencing a target attribute, and requires no understanding of mathematical or statistical algorithms. Moreover, it enables to detect sequence patterns of any length, without predefining the sequence pattern length.

1.7.3 New Application Domains

A new domain for KDD is the world of nanoparticles. Oded Maimon and Abel Browarnik present a smart repository system with text and data mining for this domain (Chapter 66). The impact of nanoparticles on health and the environment is a significant research subject, driving increasing interest from the scientific community, regulatory bodies and the general public. The growing body of knowledge in this area, consisting of scientific papers and other types of publications (such as surveys and whitepapers) emphasize the need for a methodology to alleviate the complexity of reviewing all the available information and discovering all the underlying facts, using data mining algorithms and methods. .

1.7.4 New Consideration

In Chapter 35, Vicenc Torra describes the main tools for privacy in data mining. He presents an overview of the tools for protecting data, and then focuses on protection procedures. Information loss and disclosure risk measures are also described.

1.7.5 Software

In Chapter 67, Zhang and Segall present selected commercial software for data mining, text mining, and web mining. The selected software are compared with their features and also applied to available data sets. Screen shots of each of the selected software are presented, as are conclusions and future directions.

1.7.6 Major Updates

Finally several chapters have been updated. Specifically, in Chapter 19, Alex Freitas presents a brief overview of EAs, focusing mainly on two kinds of EAs, viz. Genetic Algorithms (GAs) and Genetic Programming (GP). Then the chapter reviews the main concepts and principles used by EAs designed for solving several data mining tasks, namely: discovery of classification rules, clustering, attribute selection and attribute construction.

In Chapter 21, Peter Zhang provides an overview of neural network models and their applications to data mining tasks. He provides historical development of the field of neural networks and presents three important classes of neural models including feed forward multilayer networks, Hopfield networks, and Kohonen's self-organizing maps.

In Chapter 24, we discuss how fuzzy logic extends the envelope of the main data mining tasks: clustering, classification, regression and association rules. We begin by presenting a formulation of the data mining using fuzzy logic attributes. Then, for each task, we provide a survey of the main algorithms and a detailed description (i.e. pseudo-code) of the most popular algorithms.

References

Arbel, R. and Rokach, L., Classifier evaluation under limited resources, Pattern Recognition Letters, 27(14): 1619–1631, 2006, Elsevier.

Averbuch, M. and Karson, T. and Ben-Ami, B. and Maimon, O. and Rokach, L., Context-sensitive medical information retrieval, The 11th World Congress on Medical Informatics (MEDINFO 2004), San Francisco, CA, September 2004, IOS Press, pp. 282–286.

Cohen S., Rokach L., Maimon O., Decision Tree Instance Space Decomposition with Grouped Gain-Ratio, Information Science, Volume 177, Issue 17, pp. 3592-3612, 2007.

Hastie, T. and Tibshirani, R. and Friedman, J. and Franklin, J., The elements of statistical learning: data mining, inference and prediction, The Mathematical Intelligencer, 27(2): 83–85, 2005.

Han, J. and Kamber, M., Data mining: concepts and techniques, Morgan Kaufmann, 2006.

H. Kriege, K. M. Borgwardt, P. Krger, A. Pryakhin, M. Schubert and Arthur Zimek, Future trends in data mining, Data Mining and Knowledge Discovery, 15(1):87-97, 2007.

Larose, D.T., Discovering knowledge in data: an introduction to data mining, John Wiley and Sons, 2005.

Maimon O., and Rokach, L. Data Mining by Attribute Decomposition with semiconductors manufacturing case study, in Data Mining for Design and Manufacturing: Methods and Applications, D. Braha (ed.), Kluwer Academic Publishers, pp. 311–336, 2001.

Maimon O. and Rokach L., "Improving supervised learning by feature decomposition", Proceedings of the Second International Symposium on Foundations of Information and Knowledge Systems, Lecture Notes in Computer Science, Springer, pp. 178-196, 2002.

Maimon, O. and Rokach, L., Decomposition Methodology for Knowledge Discovery and Data Mining: Theory and Applications, Series in Machine Perception and Artificial Intelligence - Vol. 61, World Scientific Publishing, ISBN:981-256-079-3, 2005.

Rokach, L., Decomposition methodology for classification tasks: a meta decomposer framework, Pattern Analysis and Applications, 9(2006):257–271.

Rokach L., Genetic algorithm-based feature set partitioning for classification problems,Pattern Recognition, 41(5):1676–1700, 2008.

Rokach L., Mining manufacturing data using genetic algorithm-based feature set decomposition, Int. J. Intelligent Systems Technologies and Applications, 4(1):57-78, 2008.

Rokach L., Maimon O. and Lavi I., Space Decomposition In Data Mining: A Clustering Approach, Proceedings of the 14th International Symposium On Methodologies For Intelligent Systems, Maebashi, Japan, Lecture Notes in Computer Science, Springer-Verlag, 2003, pp. 24–31.

Rokach, L. and Maimon, O. and Averbuch, M., Information Retrieval System for Medical Narrative Reports, Lecture Notes in Artificial intelligence 3055, page 217-228 Springer-Verlag, 2004.

Rokach, L. and Maimon, O. and Arbel, R., Selective voting-getting more for less in sensor fusion, International Journal of Pattern Recognition and Artificial Intelligence 20 (3) (2006), pp. 329–350.

Rokach, L. and Maimon, O., Theory and applications of attribute decomposition, IEEE International Conference on Data Mining, IEEE Computer Society Press, pp. 473–480, 2001.

Rokach L. and Maimon O., Feature Set Decomposition for Decision Trees, Journal of Intelligent Data Analysis, Volume 9, Number 2, 2005b, pp 131–158.

Rokach, L. and Maimon, O., Clustering methods, Data Mining and Knowledge Discovery Handbook, pp. 321–352, 2005, Springer.

Rokach, L. and Maimon, O., Data mining for improving the quality of manufacturing: a feature set decomposition approach, Journal of Intelligent Manufacturing, 17(3):285–299, 2006, Springer.

Rokach, L., Maimon, O., Data Mining with Decision Trees: Theory and Applications, World Scientific Publishing, 2008.

Witten, I.H. and Frank, E., Data Mining: Practical machine learning tools and techniques, Morgan Kaufmann Pub, 2005.

Wu, X. and Kumar, V. and Ross Quinlan, J. and Ghosh, J. and Yang, Q. and Motoda, H. and McLachlan, G.J. and Ng, A. and Liu, B. and Yu, P.S. and others, Top 10 algorithms in data mining, Knowledge and Information Systems, 14(1): 1–37, 2008.

Part I

Preprocessing Methods

2

Data Cleansing: A Prelude to Knowledge Discovery

Jonathan I. Maletic[1] and Andrian Marcus[2]

[1] Kent State University
[2] Wayne State University

Summary. This chapter analyzes the problem of data cleansing and the identification of potential errors in data sets. The differing views of data cleansing are surveyed and reviewed and a brief overview of existing data cleansing tools is given. A general framework of the data cleansing process is presented as well as a set of general methods that can be used to address the problem. The applicable methods include statistical outlier detection, pattern matching, clustering, and Data Mining techniques. The experimental results of applying these methods to a real world data set are also given. Finally, research directions necessary to further address the data cleansing problem are discussed.

Key words: Data Cleansing, Data Cleaning, Data Mining, Ordinal Rules, Data Quality, Error Detection, Ordinal Association Rules

2.1 INTRODUCTION

The quality of a large real world data set depends on a number of issues (Wang *et al.*, 1995, Wang *et al.*, 1996), but the source of the data is the crucial factor. Data entry and acquisition is inherently prone to errors, both simple and complex. Much effort can be allocated to this front-end process with respect to reduction in entry error but the fact often remains that errors in a large data set are common. While one can establish an acquisition process to obtain high quality data sets, this does little to address the problem of existing or legacy data. The field errors rates in the data acquisition phase are typically around 5% or more (Orr, 1998, Redman, 1998) even when using the most sophisticated measures for error prevention available. Recent studies have shown that as much as 40% of the collected data is dirty in one way or another (Fayyad *et al.*, 2003).

For existing data sets the logical solution is to attempt to cleanse the data in some way. That is, explore the data set for possible problems and endeavor to correct the errors. Of course, for any real world data set, doing this task by hand is completely out of the question given the amount of person hours involved. Some organizations spend millions of dollars per year to detect data errors (Redman, 1998). A manual

O. Maimon, L. Rokach (eds.), *Data Mining and Knowledge Discovery Handbook*, 2nd ed.,
DOI 10.1007/978-0-387-09823-4_2,

process of data cleansing is also laborious, time consuming, and itself prone to errors. Useful and powerful tools that automate or greatly assist in the data cleansing process are necessary and may be the only practical and cost effective way to achieve a reasonable quality level in existing data.

While this may seem to be an obvious solution, little basic research has been directly aimed at methods to support such tools. Some related research addresses the issues of data quality (Ballou and Tayi, 1999, Redman, 1998, Wang *et al.*, 2001) and some tools exist to assist in manual data cleansing and/or relational data integrity analysis.

The serious need to store, analyze, and investigate such very large data sets has given rise to the fields of Data Mining (DM) and data warehousing (DW). Without clean and correct data the usefulness of Data Mining and data warehousing is mitigated. Thus, data cleansing is a necessary precondition for successful knowledge discovery in databases (KDD).

2.2 DATA CLEANSING BACKGROUND

There are many issues in data cleansing that researchers are attempting to tackle. Of particular interest here, is the search context for what is called in literature and the business world as "dirty data" (Fox *et al.*, 1994, Hernandez and Stolfo, 1998, Kimball, 1996). Recently, Kim (Kim *et al.*, 2003) proposed a taxonomy for dirty data. It is a very important issue that will attract the attention of the researchers and practitioners in the field. It is the first step in defining and understanding the data cleansing process.

There is no commonly agreed formal definition of data cleansing. Various definitions depend on the particular area in which the process is applied. The major areas that include data cleansing as part of their defining processes are: data warehousing, knowledge discovery in databases, and data/information quality management (e.g., Total Data Quality Management TDQM).

In the data warehouse user community, there is a growing confusion as to the difference between *data cleansing* and *data quality*. While many data cleansing products can help in transforming data, there is usually no persistence in this cleansing. Data quality processes ensure this persistence at the business level. Within the data warehousing field, data cleansing is typically applied when several databases are merged. Records referring to the same entity are often represented in different formats in different data sets. Thus, duplicate records will appear in the merged database. The issue is to identify and eliminate these duplicates. The problem is known as the *merge/purge problem* (Hernandez and Stolfo, 1998). In the literature instances of this problem are referred to as record linkage, semantic integration, instance identification, or the object identity problem. There are a variety of methods proposed to address this issue: knowledge bases (Lee *et al.*, 2001), regular expression matches and user-defined constraints (Cadot and di Martion, 2003), filtering (Sung *et al.*, 2002), and others (Feekin, 2000, Galhardas, 2001, Zhao *et al.*, 2002).

Data is deemed unclean for many different reasons. Various techniques have been developed to tackle the problem of data cleansing. Largely, data cleansing is an interactive approach, as different sets of data have different rules determining the validity of data. Many systems allow users to specify rules and transformations needed to clean the data. For example, Raman and Hellerstein (2001) propose the use of an interactive spreadsheet to allow users to perform transformations based on user-defined constraints, Galhardas (2001) allows users to specify rules and conditions on a SQL-like interface, Chaudhuri, Ganjam, Ganti and Motwani (2003) propose the definition of a *reference pattern* for records using fuzzy algorithms to match existing ones to the reference, and Dasu, Vesonder and Wright (2003) propose using business rules to define constraints on the data in the entry phase.

From this perspective data cleansing is defined in several (but similar) ways. In (Galhardas, 2001) data cleansing is the process of eliminating the errors and the inconsistencies in data and solving the object identity problem. Hernandez and Stolfo (1998) define the data cleansing problem as the merge/purge problem and proposes the basic sorted-neighborhood method to solve it.

Data cleansing is much more than simply updating a record with good data. Serious data cleansing involves decomposing and reassembling the data. According to (Kimball, 1996) one can break down the cleansing into six steps: elementizing, standardizing, verifying, matching, house holding, and documenting. Although data cleansing can take many forms, the current marketplace and technologies for data cleansing are heavily focused on customer lists (Kimball, 1996). A good description and design of a framework for assisted data cleansing within the merge/purge problem is available in (Galhardas, 2001).

Most industrial data cleansing tools that exist today address the duplicate detection problem. Table 2.1 lists a number of such tools. By comparison, there were few data cleansing tools available five years ago.

Table 2.1. Industrial data cleansing tools circa 2004

Tool	Company
Centrus Merge/Purge	*Qualitative Marketing Software*, http://www.qmsoft.com/
Data Tools Twins	*Data Tools*, http://www.datatools.com.au/
DataCleanser DataBlade	*Electronic Digital Documents*, http://www.informix.com
DataSet V	*iNTERCON http://www.ds-dataset.com*
DeDuce	*The Computing Group*
DeDupe	*International Software Publishing*
dfPower	*DataFlux Corporation*, http://www.dataflux.com/
DoubleTake	*Peoplesmith*, http://www.peoplesmith.com/
ETI Data Cleanse	*Evolutionary Technologies Intern*, http://www.evtech.com
Holmes	*Kimoce*, http://www.kimoce.com/
i.d.Centric	*firstLogic*, http://www.firstlogic.com/
Integrity	*Vality*, http://www.vality.com/
matchIT	*helpIT Systems Limited*, http://www.helpit.co.uk/
matchMaker	*Info Tech Ltd*, http://www.infotech.ie/
NADIS Merge/Purge Plus	*Group1 Software*, http://www.g1.com/
NoDupes	*Quess Inc*, http://www.quess.com/nodupes.html
PureIntegrate	*Carleton*, http://www.carleton.com/products/View/index.htm
PureName PureAddress	*Carleton*, http://www.carleton.com/products/View/index.htm
QuickAdress Batch	*QAS Systems*, http://207.158.205.110/
reUnion and MasterMerge	*PitneyBowes*, http://www.pitneysoft.com/
SSA-Name/Data Clustering Engine	*Search Software America* http://www.searchsoftware.co.uk/
Trillium Software System	*Trillium Software*, http://www.trilliumsoft.com/
TwinFinder	*Omikron*, http://www.deduplication.com/index.html
Ultra Address Management	*The Computing Group*

Total Data Quality Management (TDQM) is an area of interest both within the research and business communities. The data quality issue and its integration in the entire information business process are tackled from various points of view in the literature (Fox *et al.*, 1994, Levitin and Redman, 1995, Orr, 1998, Redman, 1998, Strong *et al.*, 1997, Svanks, 1984, Wang *et al.*, 1996). Other works refer to this as the enterprise data quality management problem. The most comprehensive survey of the research in this area is available in (Wang *et al.*, 2001).

Unfortunately, none of the mentioned literature explicitly refers to the data cleansing problem. A number of the papers deal strictly with the process management issues from data quality perspective, others with the definition of data quality. The later category is of interest here. In the proposed model of data life cycles with application to quality (Levitin and Redman, 1995) the data acquisition and data usage cycles contain a series of activities: assessment, analysis, adjustment, and discarding of data. Although it is not specifically addressed in the paper, if one integrated the data cleansing process with the data life cycles, this series of steps would define it in the proposed model from the data quality perspective. In the same framework of data quality, (Fox *et al.*, 1994) proposes four quality dimensions of the data: accuracy, current-ness, completeness, and consistency. The correctness of data is defined in terms of these dimensions. Again, a simplistic attempt to define the data cleansing process within this framework would be the process that assesses the correctness of data and improves its quality.

More recently, data cleansing is regarded as a first step, or a preprocessing step, in the KDD process (Brachman and Anand, 1996, Fayyad *et al.*, 1996) however no precise definition and perspective over the data cleansing process is given. Various KDD and Data Mining systems perform data cleansing activities in a very domain specific fashion. In (Guyon *et al.*, 1996) informative patterns are used to perform one kind of data cleansing by discovering *garbage patterns* – meaningless or mislabeled patterns. Machine learning techniques are used to apply the data cleansing process in the written characters classification problem. In (Simoudis *et al.*, 1995) data cleansing is defined as the process that implements computerized methods of examining databases, detecting missing and incorrect data, and correcting errors. Other recent work relating to data cleansing includes (Bochicchio and Longo, 2003, Li and Fang, 1989).

Data Mining emphasizes data cleansing with respect to the garbage-in-garbage-out principle. Furthermore, Data Mining specific techniques can be used in data cleansing. Of special interest is the problem of outlier detection where the goal is to find out exceptions in large data sets. These are often an indication of incorrect values. Different approaches have been proposed with many based on the notion of distance-based outliers (Knorr and Ng, 1998, Ramaswamy *et al.*, 2000). Other techniques such as *FindOut* (Yu *et al.*, 2002) combine clustering and outlier detection. Neural networks are also used in this task (Hawkins *et al.*, 2002), and outlier detection in multi-dimensional data sets is also addressed (Aggarwal and Yu, 2001).

2.3 GENERAL METHODS FOR DATA CLEANSING

With all the above in mind, data cleansing must be viewed as a process. This process is tied directly to data acquisition and definition or is applied after the fact, to improve data quality in an existing system. The following three phases define a data cleansing process:

- Define and determine error types
- Search and identify error instances
- Correct the uncovered errors

Each of these phases constitutes a complex problem in itself, and a wide variety of specialized methods and technologies can be applied to each. The focus here is on the first two aspects of this generic framework. The later aspect is very difficult to automate outside of a strict and well-defined domain. The intention here is to address and automate the data cleansing process outside domain knowledge and business rules.

While data integrity analysis can uncover a number of possible errors in a data set, it does not address more complex errors. Errors involving relationships between one or more fields are often very difficult to uncover. These types of errors require deeper inspection and analysis. One can view this as a problem in outlier detection. Simply put: if a large percentage (say 99.9%) of the data elements conform to a general form, then the remaining (0.1%) data elements are likely error candidates. These data elements are considered outliers. Two things are done here; identifying outliers or strange variations in a data set and identifying trends (or normality) in data. Knowing what data is supposed to look like allows errors to be uncovered. However, the fact of the matter is that real world data is often very diverse and rarely conforms to any standard statistical distribution. This fact is readily confirmed by any practitioner and supported by our own experiences. This problem is especially acute when viewing the data in several dimensions. Therefore, more than one method for outlier detection is often necessary to capture most of the outliers. Below is a set of general methods that can be utilized for error detection.

- **Statistical**: Identify outlier fields and records using the values such as mean, standard deviation, range, based on Chebyshev's theorem (Barnett and Lewis, 1994) and considering the confidence intervals for each field (Johnson and Wichern, 1998). While this approach may generate many false positives, it is simple and fast, and can be used in conjunction with other methods.
- **Clustering**: Identify outlier records using clustering techniques based on Euclidian (or other) distance (Rokach and Maimon, 2005). Some clustering algorithms provide support for identifying outliers (Knorr *et al.*, 2000, Murtagh, 1984). The main drawback of these methods is a high computational complexity.
- **Pattern-based**: Identify outlier fields and records that do not conform to existing patterns in the data. Combined techniques (partitioning, classification, and clustering) are used to identify patterns that apply to most records (Maimon and

Rokach, 2002). A pattern is defined by a group of records that have similar characteristics or behavior for $p\%$ of the fields in the data set, where p is a user-defined value (usually above 90).

- **Association rules:** Association rules with high confidence and support define a different kind of pattern. As before, records that do not follow these rules are considered outliers. The power of association rules is that they can deal with data of different types. However, Boolean association rules do not provide enough quantitative and qualitative information. Ordinal association rules, defined by (Maletic and Marcus, 2000, Marcus *et al.*, 2001), are used to find rules that give more information (e.g., ordinal relationships between data elements). The ordinal association rules yield special types of patterns, so this method is, in general, similar to the pattern-based method. This method can be extended to find other kind of associations between groups of data elements (e.g., statistical correlations).

2.4 APPLYING DATA CLEANSING

A version of each of the above-mentioned methods was implemented. Each method was tested using a data set comprised of real world data supplied by the Naval Personnel Research, Studies, and Technology (NPRST). The data set represents part of the Navy's officer personnel information system including midshipmen and officer candidates. Similar data sets are in use at personnel records division in companies all over the world. A subset of 5,000 records with 78 fields of the same type (dates) is used to demonstrate the methods. The size and type of the data elements allows fast and multiple runs without reducing the generality of the proposed methods.

The goal of this demonstration is to prove that these methods can be successfully used to identify outliers that constitute potential errors. The implementations are designed to work on larger data sets and without extensive amounts of domain knowledge.

2.4.1 Statistical Outlier Detection

Outlier values for particular fields are identified based on automatically computed statistics. For each field, the mean and standard deviation are utilized, and based on Chebyshev's theorem (Barnett and Lewis, 1994) those records that have values in a given field outside a number of standard deviations from the mean are identified. The number of standard deviations to be considered is customizable. Confidence intervals are taken into consideration for each field. A field f_i in a record r_j is considered an outlier if the value of $f_i > \mu_i + \varepsilon\sigma_i$ or the value of $f_i < \mu_i - \varepsilon\sigma_i$, where μ_i is the mean for the field f_i, σ_i is the standard deviation, and ε is a user defined factor. Regardless of the distribution of the field f_i, most values should be within a certain number ε of standard deviations from the mean. The value of ε can be user-defined, based on some domain or data knowledge.

In the experiments, several values were used for ε (i.e., 3, 4, 5, and 6), and the value 5 was found to generate the best results (i.e., less false positives and false negatives). Among the 5,000 records of the experimental data set, 164 contain outlier values detected using this method. A visualization tool was used to analyze the results. Trying to visualize the entire data set to identify the outliers by hand would be impossible.

2.4.2 Clustering

A combined clustering method was implemented based on the group-average clustering algorithm (Yang *et al.*, 2002) by considering the Euclidean distance between records. The clustering algorithm was run several times adjusting the maximum size of the clusters. Ultimately, the goal is to identify as outliers those records previously containing outlier values. However, computational time prohibits multiple runs in an every-day business application on larger data sets. After several executions on the same data set, it turned out that the larger the threshold value for the maximum distance allowed between clusters to be merged, the better the outlier detection. A faster clustering algorithm could be utilized that allows automated tuning of the maximum cluster size as well as scalability to larger data sets. Using domain knowledge, an important subspace could be selected to guide the clustering to reduce the size of the data. The method can be used to reduce the search space for other techniques.

The test data set has a particular characteristic: many of the data elements are empty. This particularity of the data set does not make the method less general, but allowed the definition of a new similarity measure that relies on this feature. Here, strings of zeros and ones, referred to as *Hamming value* (Hamming, 1980), are associated with each record. Each string has as many elements as the number of fields in the record. The Hamming distance (Hamming, 1980) is used to cluster the records into groups of similar records. Initially, clusters having zero Hamming distance between records were identified. Using the Hamming distance for clustering would not yield relevant outliers, but rather would produce clusters of records that can be used as search spaces for other methods and also help identify missing data.

2.4.3 Pattern-based detection

Patterns are identified in the data according to the distribution of the records per each field. For each field, the records are clustered using the Euclidian distance and the k-mean algorithm (Kaufman and Rousseauw, 1990), with k=6. The six starting elements are not randomly chosen, but at equal distances from the median. A pattern is defined by a large group of records (over $p\%$ of the entire data set) that cluster the same way for most of the fields. Each cluster is classified according to the number of records it contains (i.e., cluster number 1 has the largest size and so on). The following hypothesis is considered: if there is a pattern that is applicable to most of the fields in the records, then a record following that pattern should be part of the cluster with the same rank for each field.

This method was applied on the data set and a small number of records (0.3%) were identified that followed the pattern for more than 90% of the fields. The method can be adapted and applied on clusters of records generated using the Hamming distance, rather than the entire data set. Chances of identifying a pattern will increase since records in clusters will already have certain similarity and have approximately the same empty fields. Again, real-life data proved to be highly non-uniform.

2.4.4 Association Rules

The term *association rule* was first introduced by (Aggarwal *et al.*, 1993) in the context of market-basket analysis. Association rule of this type are also referred to in the literature as *classical* or *Boolean* association rules. The concept was extended in other studies and experiments. Of particular interest to this research are the *quantitative association rules* (Srikant *et al.*, 1996) and *ratio-rules* (Korn *et al.*, 1998) that can be used for the identification of possible erroneous data items with certain modifications. In previous work we argued that another extension of the association rule – *ordinal association rules* (Maletic and Marcus, 2000, Marcus *et al.*, 2001) – is more flexible, general, and very useful for identification of errors. Since this is a recently introduced concept, it is briefly defined.

Let $R = \{r_1, r_2, \ldots, r_n\}$ be a set of records, where each record is a set of k attributes $(a_1, \ldots, a_k)$. Each attribute a_i in a particular record r_j has a value $\phi(r_j, a_i)$ from a domain D. The value of the attribute may also be empty and is therefore included in D. The following relations (partial orderings) are defined over D, namely less or equal ($\leq$), equal (=) and, greater or equal ($\geq$) all having the standard meaning.

Then $(a_1, a_2, a_3, \ldots, a_m) \Rightarrow (a_1 \mu_1 a_2\ \mu_2 a_3\ \ldots\ \mu_{m-1} a_m)$, where each $\mu_i \in \{\leq, =, \geq\}$, is a an *ordinal association rule* if:

1. $a_1 \ldots a_m$ occur together (are non-empty) in at least $s\%$ of the n records, where s is the *support* of the rule;
2. and, in a subset of the records $R' \subseteq R$ where $a_1 \ldots a_m$ occur together and $\phi(r_j, a_1)$ $\mu_1 \ldots \mu_{m-1}$ $\phi(r_j, a_m)$ is true for each $r_j \in R'$. Thus $|R'|$ is the number of records that the rule holds for and the *confidence*, c, of the rule is the percentage of records that hold for the rule $c = |R'|/|R|$.

The process to identify potential errors in data sets using ordinal association rules is composed of the following steps:

1. Find ordinal rules with a minimum confidence c. This is done with a variation of *apriori* algorithm (Aggarwal *et al.*, 1993).
2. Identify data items that broke the rules and can be considered outliers (potential errors).

Here, the manner in which support of a rule is important differs from typical data-mining problem. We assume all the discovered rules that hold for more than two records represent valid possible partial orderings. Future work will investigate user-specified minimum support and rules involving multiple attributes.

The method first normalizes the data (if necessary) and then computes comparisons between each pair of attributes for every record. Only one scan of the data set is required. An array with the results of the comparisons is maintained in the memory. Figure 2.1 contains the algorithm for this step. The complexity of this step is only

$O(N * M^2)$ where N is the number of records in the data set, and M is the number of fields/attributes. Usually M is much smaller than N. The results of this algorithm are written to a temporary file for use in the next step of processing.

In the second step, the ordinal rules are identified based on the chosen minimum confidence. There are several researched methods to determine the strength including interestingness and statistical significance of a rule (e.g., minimum support and minimum confidence, chi-square test, etc.). Using confidence intervals to determine the minimum confidence is currently under investigation. However, previous work on the data set (Maletic and Marcus, 2000) used in our experiment showed that the distribution of the data was not normal. Therefore, the minimum confidence was chosen empirically, several values were considered and the algorithm was executed. The results indicated that a minimum confidence between 98.8 and 99.7 provide best results (less number of false negative and false positives).

```
Algorithm compare items.
    for each record in the data base (1...N)
        normalize or convert data
        for each attribute x in (1...M-1)
            for each attribute y in (x+1...M-1)
                compare the values in x and y
                update the comparisons array
            end for.
        end for.
        output the record with normalized data
    end for.
    output the comparisons array
end algorithm.
```

Fig. 2.1. The algorithm for the first step

The second component extracts the data associated with the rules from the temporary file and stores it in memory. This is done with a single scan (complexity $O(C(M,2))$. Then for each record in the data set, each pair of attributes that correspond to a pattern it is checked to see if the values in those fields are within the relationship indicated by the pattern. If they are not, each field is marked as possible error. Of course, in most cases only one of the two values will actually be an error. Once every pair of fields that correspond to a rule is analyzed, the average number of possible error marks for each marked field is computed. Only those fields that are marked as possible errors more times than the average are finally marked as having likely errors. Again, the average value was empirically chosen as a threshold to prune the possible errors set. Other methods to find such a threshold, without using domain knowledge or multiple experiments, are under investigation. The time complexity of this step is $O(N*C(M,2))$, and the analysis of each record is done entirely in the main memory. Figure 2.2 shows the algorithm used in the implementation of the second component. The results identify which records and fields are likely to have errors.

```
Algorithm analyze records.
    for each record in the data base (1N)
        for each rule in the pattern array
            determine rule type and pairs
            compare item pairs
            if pattern NOT holds
                then mark each item as possible
                     error
        end for.
        compute average number of marks
        select the high probability marked
              errors
    end for.
end algorithm.
```

Fig. 2.2. Algorithm for the second step

Using a 98% confidence, 9,064 records in 971 fields that had high probability errors were identified out of the extended data set of 30,000 records. These were compared with those outliers identified with statistical methods. These possible errors not only matched most of the previously discovered ones, but 173 were errors unidentified by the previous methods. The distribution of the data influenced dramatically the error identification of the data process in the previous utilized methods. This new method is proving to be more robust and is influenced less by the distribution of the data. Table 2.2 shows an error identified by ordinal association rules and missed with the previous methods. Here two patterns were identified with confidence higher than 98%: values in field 4 $\leq$ values in field 14, and values in field 4 $\leq$ values in field 15. In the record no. 199, both fields 14 and 15 were marked as high probability errors. Both values are in fact minimum values for their respective fields. The value in field 15 was identified previously as outlier but the value in field 14 was not because of the high value of the standard deviation for that field. It is obvious, even without consulting a domain expert, that both values are in fact wrong. The correct values (identified later) are 800704. Other values that did not lie at the edge of the distributions were identified as errors as well.

Table 2.2. A part of the data set. An error was identified in record 199, field 14, which was not identified previously. The data elements are dates in the format YYMMDD.

Record Number	Field 1	...	Field 4	...	Field 14	Field 15	...
199	600603	...	780709	...	**700804**	700804	...

2.5 CONCLUSIONS

Data cleansing is a very young field of research. This chapter presents some of the current research and practice in data cleansing. One missing aspect in the research is the definition of a solid theoretical foundation that would support many of the existing approaches used in an industrial setting. The philosophy promoted here is that a data cleansing framework must incorporate a variety of such methods to be used in conjunction. Each method can be used to identify a particular type of error in data. While not specifically addressed here, taxonomies like the one proposed in (Kim *et al.*, 2003) should be encouraged and extended by the research community. This will support the definition and construction of more general data cleansing frameworks.

Unfortunately, little basic research within the information systems and computer science communities has been conducted that directly relates to error detection and data cleansing. In-depth comparisons of data cleansing techniques and methods have not yet been published. Typically, much of the real data cleansing work is done in a customized, in-house, manner. This behind-the-scenes process often results in the use of undocumented and ad hoc methods. Data cleansing is still viewed by many as a "black art" being done "in the basement". Some concerted effort by the database and information systems groups is needed to address this problem.

Future research directions include the investigation and integration of various methods to address error detection. Combination of knowledge-based techniques with more general approaches should be pursued. In addition, a better integration of data cleansing in the data quality processes and frameworks should be achieved. The ultimate goal of data cleansing research is to devise a set of general operators and theory (much like relational algebra) that can be combined in well-formed statements to address data cleansing problems. This formal basis is necessary to design and construct high quality and useful software tools to support the data cleansing process.

References

Aggarwal, C. C. & Yu, P. S. Outlier detection for high dimensional data. Proceedings of ACM SIGMOD international Conference on Management of Data; 2001 May 21-24; Santa Barbara, CA. 37-46.

Agrawal, R., Imielinski, T., & Swami, A. Mining Association rules between Sets of Items in Large Databases. Proceedings of ACM SIGMOD International Conference on Management of Data; 1993 May; Washington D.C. 207-216.

Ballou, D. P. & Tayi, G. K. Enhancing Data Quality in Data Warehouse Environments, Communications of the ACM 1999; 42(1):73-78.

Barnett, V. & Lewis, T., Outliers in Statistical Data. John Wiley and Sons, 1994.

Bochicchio, M. A. & Longo, A. Data Cleansing for Fiscal Services: The Taviano Project. Proceedings of 5th International Conference on Enterprise Information Systems; 2003 April 22-26; Angers, France. 464-467.

Brachman, R. J., Anand, T., The Process of Knowledge Discovery in Databases — A Human–Centered Approach. In Advances in Knowledge Discovery and Data Mining, Fayyad, U. M., Piatetsky-Shapiro, G., Smyth, P., & Uth-urasamy, R., eds. MIT Press/AAAI Press, 1996.

Cadot, M. & di Martion, J. A data cleaning solution by Perl scripts for the KDD Cup 2003 task 2, ACM SIGKDD Explorations Newsletter 2003; 5(2):158-159.

Chaudhuri, S., Ganjam, K., Ganti, V., & Motwani, R. Robust and efficient fuzzy match for online data cleaning. Proceedings of ACM SIGMOD International Conference on Management of Data; 2003 june 9-12; San Diego, CA. 313-324.

Dasu, T., Vesonder, G. T., & Wright, J. R. Data quality through knowledge engineering. Proceedings of ACM SIGKDD International Conference on Knowledge Discovery and Data Mining; 2003 August 24-27; Washington, D.C. 705-710.

Fayyad, U. M., Piatetsky-Shapiro, G., & Smyth, P., From Data Mining to Knowledge Discovery: An Overview. In Advances in Knowledge Discovery and Data Mining, Fayyad, U. M., Piatetsky-Shapiro, G., Smyth, P., & Uthurasamy, R., eds. MIT Press/AAAI Press, 1996.

Fayyad, U. M., Piatetsky-Shapiro, G., & Uthurasamy, R. Summary from the KDD-03 Panel - Data Mining: The Next 10 Years, ACM SIGKDD Explorations Newsletter 2003; 5(2):191-196.

Feekin, A. & Chen, Z. Duplicate detection using k-way sorting method. Proceedings of ACM Symposium on Applied Computing; 2000 Como, Italy. 323-327.

Fox, C., Levitin, A., & Redman, T. The Notion of Data and Its Quality Dimensions, Information Processing and Management 1994; 30(1):9-19.

Galhardas, H. Data Cleaning: Model, Language and Algoritmes. University of Versailles, Saint-Quentin-En-Yvelines, Ph.D., 2001.

Guyon, I., Matic, N., & Vapnik, V., Discovering Information Patterns and Data Cleaning. In Advances in Knowledge Discovery and Data Mining, Fayyad, U. M., Piatetsky-Shapiro, G., Smyth, P., & Uthurasamy, R., eds. MIT Press/AAAI Press, 1996.

Hamming, R. W., Coding and Information Theory. New Jersey, Prentice-Hall, 1980.

Hawkins, S., He, H., Williams, G. J., & Baxter, R. A. Outlier Detection Using Replicator Neural Networks. Proceedings of 4th International Conference on Data Warehousing and Knowledge Discovery; 2002 September 04-06; 170-180.

Hernandez, M. & Stolfo, S. Real-world Data is Dirty: Data Cleansing and The Merge/Purge Problem, Data Mining and Knowledge Discovery 1998; 2(1):9-37.

Johnson, R. A. & Wichern, D. W., Applied Multivariate Statistical Analysis. Prentice Hall, 1998.

Kaufman, L. & Rousseauw, P. J., Finding Groups in Data: An Introduction to Cluster Analysis. John Wiley & Sons, 1990.

Kim, W., Choi, B.-J., Hong, E.-K., Kim, S.-K., & Lee, D. A taxonomy of dirty data, Data Mining and Knowledge Discovery 2003; 7(1):81-99.

Kimball, R. Dealing with Dirty Data, DBMS 1996; 9(10):55-60.

Knorr, E. M. & Ng, R. T. Algorithms for Mining Distance-Based Outliers in Large Datasets. Proceedings of 24th International Conference on Very Large Data Bases; 1998 New York. 392-403.

Knorr, E. M., Ng, R. T., & Tucakov, V. Distance-based outliers: algorithms and applications, The International Journal on Very Large Data Bases 2000; 8(3-4):237-253.

Korn, F., Labrinidis, A., Yannis, K., & Faloustsos, C. Ratio Rules: A New Paradigm for Fast, Quantifiable Data Mining. Proceedings of 24th VLDB Conference; 1998 New York. 582–593.

Lee, M. L., Ling, T. W., & Low, W. L. IntelliClean: a knowledge-based intelligent data cleaner. Proceedings of Sixth ACM SIGKDD International Conference on Knowledge Discovery and Data Mining; 2000 August 20-23; Boston, MA. 290-294.

Levitin, A. & Redman, T. A Model of the Data (Life) Cycles with Application to Quality, Information and Software Technology 1995; 35(4):217-223.

Li, Z., Sung, S. Y., Peng, S., & Ling, T. W. A New Efficient Data cleansing Method. Proceedings of Database and Expert Systems Applications (DEXA 2002); 2002 September 2-6; Aix-en-Provence, France. 484-493.

Maimon, O. and Rokach, L. Improving supervised learning by feature decomposition, Proceedings of the Second International Symposium on Foundations of Information and Knowledge Systems, Lecture Notes in Computer Science, Springer, 2002, 178-196

Maletic, J. I. & Marcus, A. Data Cleansing: Beynod Integrity Analysis. Proceedings of The Conference on Information Quality (IQ2000); 2000 October 20-22; Massachusetts Institute of Technology. 200-209.

Marcus, A., Maletic, J. I., & Lin, K. I. Ordinal Association Rules for Error Identification in Data Sets. Proceedings of Tenth International Conference on Information and Knowledge Management (CIKM 2001); 2001 November 3-5; Atlanta, GA. to appear.

Murtagh, F. A Survey of Recent Advances in Hierarchical Clustering Algorithms, The Computer Journal 1983; 26(4):354-359.

Orr, K. Data Quality and Systems Theory, Communications of the ACM 1998; 41(2):66-71.

Raman, V. & Hellerstein, J. M. Potter's wheel an interactive data cleaning system. Proceedings of 27th International Conference on Very Large Databases 2001 September 11-14; Rome, Italy. 381–391.

Ramaswamy, S., Rastogi, R., & Shim, K. Efficient Algorithms for Mining Outliers from Large Data Sets. Proceedings of ACM SIGMOD International Conference on Management of Data; 2000 Dallas. 427-438.

Redman, T. The Impact of Poor Data Quality on the Typical Enterprise, Communications of the ACM 1998; 41(2):79-82.

Rokach, L., Maimon, O. (2005), Clustering Methods, Data Mining and Knowledge Discovery Handbook, Springer, pp. 321-352.

Simoudis, E., Livezey, B., & Kerber, R., Using Recon for Data Cleaning. In Advances in Knowledge Discovery and Data Mining, Fayyad, U. M., Piatetsky-Shapiro, G., Smyth, P., & Uthurasamy, R., eds. MIT Press/AAAI Press, 1995.

Srikant, R., Vu, Q., & Agrawal, R. Mining Association Rules with Item Constraints. Proceedings of SIGMOD International Conference on Management of Data; 1996 June; Montreal, Canada. 1-12.

Strong, D., Yang, L., & Wang, R. Data Quality in Context, Communications of the ACM 1997; 40(5):103-110.

Sung, S. Y., Li, Z., & Sun, P. A fast filtering scheme for large database cleansing. Proceedings of Eleventh ACM International Conference on Information and Knowledge Management; 2002 November 04-09; McLean, VA. 76-83.

Svanks, M. Integrity Analysis: Methods for Automating Data Quality Assurance, EDP Auditors Foundation 1984; 30(10):595-605.

Wang, R., Storey, V., & Firth, C. A Framework for Analysis of Data Quality Research, IEEE Transactions on Knowledge and Data Engineering 1995; 7(4):623-639.

Wang, R., Strong, D., & Guarascio, L. Beyond Accuracy: What Data Quality Means to Data Consumers, Journal of Management Information Systems 1996; 12(4):5-34.

Wang, R., Ziad, M., & Lee, Y. W., Data Quality. Kluwer, 2001.

Yang, Y., Carbonell, J., Brown, R., Pierce, T., Archibald, B. T., & Liu, X. Learning Approaches for Detecting and Tracking News Events, IEEE Intelligent Systems 1999; 14(4).

Yu, D., Sheikholeslami, G., & Zhang, A. FindOut: Finding Outliers in Very Large Datasets, Knowledge and Information Systems 2002; 4(4):387-412.

Zhao, L., Yuan, S. S., Peng, S., & Ling, T. W. A new efficient data cleansing method. Proceedings of 13th International Conference on Database and Expert Systems Applications; 2002 September 02-06; 484-493.

3

Handling Missing Attribute Values

Jerzy W. Grzymala-Busse[1] and Witold J. Grzymala-Busse[2]

[1] University of Kansas
[2] FilterLogix Inc.

Summary. In this chapter methods of handling missing attribute values in Data Mining are described. These methods are categorized into sequential and parallel. In sequential methods, missing attribute values are replaced by known values first, as a preprocessing, then the knowledge is acquired for a data set with all known attribute values. In parallel methods, there is no preprocessing, i.e., knowledge is acquired directly from the original data sets. In this chapter the main emphasis is put on rule induction. Methods of handling attribute values for decision tree generation are only briefly summarized.

Key words: Missing attribute values, lost values, do not care conditions, incomplete data, imputation, decision tables.

3.1 Introduction

We assume that input data for Data Mining are presented in a form of a *decision table* (or *data set*) in which *cases* (or *records*) are described by *attributes* (independent variables) and a *decision* (dependent variable). A very simple example of such a table is presented in Table 3.1, with the attributes *Temperature*, *Headache*, and *Nausea* and with the decision *Flu*. However, many real-life data sets are incomplete, i.e., some attribute values are missing. In Table 3.1 missing attribute values are denoted by "?"s.

The set of all cases with the same decision value is called a *concept*. For Table 3.1, case set $\{1, 2, 4, 8\}$ is a concept of all cases such that the value of *Flu* is *yes*.

There is variety of reasons why data sets are affected by missing attribute values. Some attribute values are not recorded because they are irrelevant. For example, a doctor was able to diagnose a patient without some medical tests, or a home owner was asked to evaluate the quality of air conditioning while the home was not equipped with an air conditioner. Such missing attribute values will be called *"do not care" conditions*.

Another reason for missing attribute values is that the attribute value was not placed into the table because it was forgotten or it was placed into the table but later

O. Maimon, L. Rokach (eds.), *Data Mining and Knowledge Discovery Handbook*, 2nd ed.,
DOI 10.1007/978-0-387-09823-4_3, © Springer Science+Business Media, LLC 2010

Table 3.1. An Example of a Data Set with Missing Attribute Values.

Case	Attributes			Decision
	Temperature	Headache	Nausea	Flu
1	high	?	no	yes
2	very_high	yes	yes	yes
3	?	no	no	no
4	high	yes	yes	yes
5	high	?	yes	no
6	normal	yes	no	no
7	normal	no	yes	no
8	?	yes	?	yes

on was mistakenly erased. Sometimes a respondent refuse to answer a question. Such a value, that matters but that is missing, will be called *lost.*

The problem of missing attribute values is as important for data mining as it is for statistical reasoning. In both disciplines there are methods to deal with missing attribute values. Some theoretical properties of data sets with missing attribute values were studied in (Imielinski and Lipski, 1984, Lipski, 1979, Lipski, 1981).

In general, methods to handle missing attribute values belong either to *sequential methods* (called also *preprocessing methods*) or to *parallel methods* (methods in which missing attribute values are taken into account during the main process of acquiring knowledge).

Sequential methods include techniques based on deleting cases with missing attribute values, replacing a missing attribute value by the most common value of that attribute, assigning all possible values to the missing attribute value, replacing a missing attribute value by the mean for numerical attributes, assigning to a missing attribute value the corresponding value taken from the closest fit case, or replacing a missing attribute value by a new vale, computed from a new data set, considering the original attribute as a decision.

The second group of methods to handle missing attribute values, in which missing attribute values are taken into account during the main process of acquiring knowledge is represented, for example, by a modification of the LEM2 (Learning from Examples Module, version 2) rule induction algorithm in which rules are induced form the original data set, with missing attribute values considered to be "do not care" conditions or lost values. C4.5 (Quinlan, 1993) approach to missing attribute values is another example of a method from this group. C4.5 induces a decision tree during tree generation, splitting cases with missing attribute values into fractions and adding these fractions to new case subsets. A method of *surrogate splits* to handle missing attribute values was introduced in CART (Breiman *et al.*, 1984), yet another system to induce decision trees. Other methods of handling missing attribute values while generating decision trees were presented in (Brazdil and Bruha, 1992) and (Bruha, 2004)

In statistics, *pairwise deletion* (Allison, 2002) (Little and Rubin, 2002) is used to evaluate statistical parameters from available information.

In this chapter we assume that the main process is rule induction. Additionally for the rest of the chapter we will assume that all decision values are known, i.e., specified. Also, we will assume that for each case at least one attribute value is known.

3.2 Sequential Methods

In sequential methods to handle missing attribute values original incomplete data sets, with missing attribute values, are converted into complete data sets and then the main process, e.g., rule induction, is conducted.

3.2.1 Deleting Cases with Missing Attribute Values

This method is based on ignoring cases with missing attribute values. It is also called *listwise deletion* (or *casewise deletion*, or *complete case analysis*) in statistics. All cases with missing attribute values are deleted from the data set. For the example presented in Table 3.1, a new table, presented in Table 3.2, is created as a result of this method.

Table 3.2. Dataset with Deleted Cases with Missing Attribute Values.

Case	Attributes			Decision
	Temperature	Headache	Nausea	Flu
1	very_high	yes	yes	yes
2	high	yes	yes	yes
3	normal	yes	no	no
4	normal	no	yes	no

Obviously, a lot of information is missing in Table 3.2. However, there are some reasons (Allison, 2002), (Little and Rubin, 2002) to consider it a good method.

3.2.2 The Most Common Value of an Attribute

In this method, one of the simplest methods to handle missing attribute values, such values are replaced by the most common value of the attribute. In different words, a missing attribute value is replaced by the most probable known attribute value, where such probabilities are represented by relative frequencies of corresponding attribute values. This method of handling missing attribute values is implemented, e.g., in CN2 (Clark, 1989). In our example from Table 3.1, a result of using this method is presented in Table 3.3.

Table 3.3. Dataset with Missing Attribute Values replaced by the Most Common Values.

Case	Attributes			Decision
	Temperature	Headache	Nausea	Flu
1	high	yes	no	yes
2	very_high	yes	yes	yes
3	high	no	no	no
4	high	yes	yes	yes
5	high	yes	yes	no
6	normal	yes	no	no
7	normal	no	yes	no
8	high	yes	yes	yes

For case 1, the value of *Headache* in Table 3.3 is *yes* since in Table 3.1 the attribute *Headache* has four values *yes* and two values *no*. Similarly, for case 3, the value of *Temperature* in Table 3.3 is *high* since the attribute *Temperature* has the value *very_high* once, *normal* twice, and *high* three times.

3.2.3 The Most Common Value of an Attribute Restricted to a Concept

A modification of the method of replacing missing attribute values by the most common value is a method in which the most common value of the attribute restricted to the concept is used instead of the most common value for all cases. Such a concept is the same concept that contains the case with missing attribute value.

Let us say that attribute a has missing attribute value for case x from concept C and that the value of a for x is missing. This missing attribute value is exchanged by the known attribute value for which the conditional probability of a for case x given C is the largest. This method was implemented, e.g., in ASSISTANT (Kononenko *et al.*, 1984). In our example from Table 3.1, a result of using this method is presented in Table 3.4.

For example, in Table 3.1, case 1 belongs to the concept $\{1, 2, 4, 8\}$, all known values of *Headache*, restricted to $\{1, 2, 4, 8\}$, are *yes*, so the missing attribute value is replaced by *yes*. On the other hand, in Table 3.1, case 3 belongs to the concept $\{3, 5, 6, 7\}$, and the value of *Temperature* is missing. The known values of *Temperature*, restricted to $\{3, 5, 6, 7\}$ are: *high* (once) and *normal* (twice), so the missing attribute value is exchanged by *normal*.

3.2.4 Assigning All Possible Attribute Values to a Missing Attribute Value

This approach to missing attribute values was presented for the first time in (Grzymala-Busse, 1991) and implemented in LERS. Every case with missing attribute values is replaced by the set of cases in which every missing attribute value is replaced by all possible known values. In the example from Table 3.1, a result of using this method is presented in Table 3.5.

Table 3.4. Dataset with Missing Attribute Values Replaced by the Most Common Value of the Attribute Restricted to a Concept

Case	Attributes			Decision
	Temperature	Headache	Nausea	Flu
1	high	yes	no	yes
2	very_high	yes	yes	yes
3	normal	no	no	no
4	high	yes	yes	yes
5	high	no	yes	no
6	normal	yes	no	no
7	normal	no	yes	no
8	high	yes	yes	yes

Table 3.5. Dataset in Which All Possible Values are Assigned to Missing Attribute Values.

Case	Attributes			Decision
	Temperature	Headache	Nausea	Flu
1^i	high	yes	no	yes
1^{ii}	high	no	no	yes
2	very_high	yes	yes	yes
3^i	high	no	no	no
3^{ii}	very_high	no	no	no
3^{iii}	normal	no	no	no
4	high	yes	yes	yes
5^i	high	yes	yes	no
5^{ii}	high	no	yes	no
6	normal	yes	no	no
7	normal	no	yes	no
8^i	high	yes	yes	yes
8^{ii}	high	yes	no	yes
8^{iii}	very_high	yes	yes	yes
8^{iv}	very_high	yes	no	yes
8^v	normal	yes	yes	yes
8^{vi}	normal	yes	no	yes

In the example of Table 3.1, the first case from Table 3.1, with the missing attribute value for attribute *Headache*, is replaced by two cases, 1^i and 1^{ii}, where case 1^i has value *yes* for attribute *Headache*, and case 1^{ii} has values *no* for the same attribute, since attribute *Headache* has two possible known values, *yes* and *no*. Case 3 from Table 3.1, with the missing attribute value for the attribute *Temperature*, is replaced by three cases, 3^i, 3^{ii}, and 3^{iii}, with values *high*, *very_high*, and *normal*, since the attribute *Temperature* has three possible known values, *high*, *very_high*, and *normal*, respectively. Note that due to this method the new table, such as Table

3.5, may be inconsistent. In Table 3.5, case 1^{ii} conflicts with case 3^{i}, case 4 conflicts with case 5^{i}, etc. However, rule sets may be induced from inconsistent data sets using standard rough-set techniques, see, e.g., (Grzymala-Busse, 1988), (Grzymala-Busse, 1991), (Grzymala-Busse, 1992), (Grzymala-Busse, 1997), (Grzymala-Busse, 2002), (Polkowski and Skowron, 1998).

3.2.5 Assigning All Possible Attribute Values Restricted to a Concept

This method was described, e.g., in (Grzymala-Busse and Hu, 2000). Here, every case with missing attribute values is replaced by the set of cases in which every attribute *a* with the missing attribute value has its every possible known value restricted to the concept to which the case belongs. In the example from Table 3.1, a result of using this method is presented in Table 3.6.

Table 3.6. Dataset in which All Possible Values, Restricted to the Concept, are Assigned to Missing Attribute Values.

Case	Attributes			Decision
	Temperature	Headache	Nausea	Flu
1	high	yes	no	yes
2	very_high	yes	yes	yes
3^{i}	normal	no	no	no
3^{ii}	high	no	no	no
4	high	yes	yes	yes
5^{i}	high	yes	yes	no
5^{ii}	high	no	yes	no
6	normal	yes	no	no
7	normal	no	yes	no
8^{i}	high	yes	yes	yes
8^{ii}	high	yes	no	yes
8^{iii}	very_high	yes	yes	yes
8^{iv}	very_high	yes	no	yes

In the example of Table 3.1, the first case from Table 3.1, with the missing attribute value for attribute *Headache*, is replaced by one with value *yes* for attribute *Headache*, since attribute *Headache*, restricted to the concept $\{1, 2, 4, 8\}$ has one possible known value, *yes*. Case 3 from Table 3.1, with the missing attribute value for the attribute *Temperature*, is replaced by two cases, 3^{i} and 3^{ii}, with values *high* and *very_high*, since the attribute *Temperature*, restricted to the concept $\{3, 5, 6, 7\}$ has two possible known values, *normal* and *high*, respectively. Again, due to this method the new table, such as Table 3.6, may be inconsistent. In Table 3.6, case 4 conflicts with case 5^{i}, etc.

3.2.6 Replacing Missing Attribute Values by the Attribute Mean

This method is used for data sets with numerical attributes. An example of such a data set is presented in Table 3.7.

Table 3.7. An Example of a Dataset with a Numerical Attribute.

Case	Attributes			Decision
	Temperature	Headache	Nausea	Flu
1	100.2	?	no	yes
2	102.6	yes	yes	yes
3	?	no	no	no
4	99.6	yes	yes	yes
5	99.8	?	yes	no
6	96.4	yes	no	no
7	96.6	no	yes	no
8	?	yes	?	yes

In this method, every missing attribute value for a numerical attribute is replaced by the arithmetic mean of known attribute values. In Table 3.7, the mean of known attribute values for Temperature is 99.2, hence all missing attribute values for *Temperature* should be replaced by 99.2. The table with missing attribute values replaced by the mean is presented in Table 3.8. For symbolic attributes *Headache* and *Nausea*, missing attribute values were replaced using the most common value of the attribute.

Table 3.8. Data set in which missing attribute values are replaced by the attribute mean and the most common value

Case	Attributes			Decision
	Temperature	Headache	Nausea	Flu
1	100.2	yes	no	yes
2	102.6	yes	yes	yes
3	99.2	no	no	no
4	99.6	yes	yes	yes
5	99.8	yes	yes	no
6	96.4	yes	no	no
7	96.6	no	yes	no
8	99.2	yes	yes	yes

3.2.7 Replacing Missing Attribute Values by the Attribute Mean Restricted to a Concept

Similarly as the previous method, this method is restricted to numerical attributes. A missing attribute value of a numerical attribute is replaced by the arithmetic mean of all known values of the attribute restricted to the concept. For example from Table 3.7, case 3 has missing attribute value for *Temperature*. Case 3 belong to the concept $\{3, 5, 6, 7\}$. The arithmetic mean of known values of *Temperature* restricted to the concept, i.e., 99.8, 96.4, and 96.6 is 97.6, so the missing attribute value is replaced by 97.6. On the other hand, case 8 belongs to the concept $\{1, 2, 4, 8\}$, the arithmetic mean of 100.2, 102.6, and 99.6 is 100.8, so the missing attribute value for case 8 should be replaced by 100.8. The table with missing attribute values replaced by the mean restricted to the concept is presented in Table 3.9. For symbolic attributes *Headache* and *Nausea*, missing attribute values were replaced using the most common value of the attribute restricted to the concept.

Table 3.9. Data set in which missing attribute values are replaced by the attribute mean and the most common value, both restricted to the concept

Case	Attributes			Decision
	Temperature	Headache	Nausea	Flu
1	100.2	yes	no	yes
2	102.6	yes	yes	yes
3	97.6	no	no	no
4	99.6	yes	yes	yes
5	99.8	no	yes	no
6	96.4	yes	no	no
7	96.6	no	yes	no
8	100.8	yes	yes	yes

3.2.8 Global Closest Fit

The global closes fit method (Grzymala-Busse *et al.*, 2002) is based on replacing a missing attribute value by the known value in another case that resembles as much as possible the case with the missing attribute value. In searching for the closest fit case we compare two vectors of attribute values, one vector corresponds to the case with a missing attribute value, the other vector is a candidate for the closest fit. The search is conducted for all cases, hence the name global closest fit. For each case a distance is computed, the case for which the distance is the smallest is the closest fitting case that is used to determine the missing attribute value. Let x and y be two cases. The distance between cases x and y is computed as follows

$$distance(x,y) = \sum_{i=1}^{n} distance(x_i, y_i),$$

where

$$distance(x_i,y_i) = \begin{cases} 0 & \text{if } x_i = y_i\,, \\ 1 & \text{if } x \text{ and } y \text{ are symbolic and } x_i \neq y_i, \\ & \text{or } x_i = ? \text{ or } y_i = ?, \\ \frac{|x_i - y_i|}{r} & \text{if } x_i \text{ and } y_i \text{ are numbers and } x_i \neq y_i\,, \end{cases}$$

where r is the difference between the maximum and minimum of the known values of the numerical attribute with a missing value. If there is a tie for two cases with the same distance, a kind of heuristics is necessary, for example, select the first case. In general, using the global closest fit method may result in data sets in which some missing attribute values are not replaced by known values. Additional iterations of using this method may reduce the number of missing attribute values, but may not end up with all missing attribute values being replaced by known attribute values.

Table 3.10. Distance (1, x)

d(1, 2)	d(1, 3)	d(1, 4)	d(1, 5)	d(1, 6)	d(1, 7)	d (1, 8)
2.39	2.0	2.10	2.06	1.61	2.58	3.00

For the data set in Table 3.7, distances between case 1 and all remaining cases are presented in Table 3.10. For example, the distance $d(1,2) = \frac{|100.2-102.6|}{|102.6-96.4|} + 1 + 1 =$ 2.39. For case 1, the missing attribute value (for attribute *Headache*) should be the value of *Headache* for case 6, i.e., *yes*, since for this case the distance is the smallest. The table with missing attribute values replaced by values computed on the basis of the global closest fit is presented in Table 3.11. Table 3.11 is complete. However, in general, some missing attribute values may still be present in such a table. If so, it is recommended to use another method of handling missing attribute values to replace all remaining missing attribute values by some specified attribute values.

3.2.9 Concept Closest Fit

This method is similar to the global closest fit method. The difference is that the original data set, containing missing attribute values, is first split into smaller data sets, each smaller data set corresponds to a concept from the original data set. More precisely, every smaller data set is constructed from one of the original concepts, by restricting cases to the concept. For the data set from Table 3.7, two smaller data sets are created, presented in Tables 3.12 and 3.13.

Following the data set split, the same global closest fit method is applied to both tables separately. Eventually, both tables, processed by the global fit method, are merged into the same table. In our example from Table 3.7, the final, merged table is presented in Table 3.14.

Table 3.11. Data Set Processed by the Global Closest Fit Method.

Case	Attributes			Decision
	Temperature	Headache	Nausea	Flu
1	100.2	yes	no	yes
2	102.6	yes	yes	yes
3	100.2	no	no	no
4	99.6	yes	yes	yes
5	99.8	yes	yes	no
6	96.4	yes	no	no
7	96.6	no	yes	no
8	102.6	yes	yes	yes

Table 3.12. Dataset Restricted to the Concept $\{1, 2, 4, 8\}$.

Case	Attributes			Decision
	Temperature	Headache	Nausea	Flu
1	100.2	?	no	yes
2	102.6	yes	yes	yes
4	99.6	yes	yes	yes
8	?	yes	?	yes

Table 3.13. Dataset Restricted to the Concept $\{3, 5, 6, 7\}$.

Case	Attributes			Decision
	Temperature	Headache	Nausea	Flu
3	?	no	no	no
5	99.8	?	yes	no
6	96.4	yes	no	no
7	96.6	no	yes	no

3.2.10 Other Methods

There is a number of other methods to handle missing attribute values. One of them is *event-covering method* (Chiu and Wong, 1986), (Wong and Chiu, 1987), based on an interdependency between known and missing attribute values. The interdependency is computed from contingency tables. The outcome of this method is not necessarily a complete data set (with all attribute values known), just like in the case of closest fit methods.

Another method of handling missing attribute values, called D^3RJ was discussed in (Latkowski, 2003, Latkowski and Mikolajczyk, 2004). In this method a data set is decomposed into complete data subsets, rule sets are induced from such data subsets, and finally these rule sets are merged.

Table 3.14. Dataset Processed by the Concept Closest Fit Method.

Case	Attributes			Decision
	Temperature	Headache	Nausea	Flu
1	100.2	yes	no	yes
2	102.6	yes	yes	yes
3	96.4	no	no	no
4	99.6	yes	yes	yes
5	99.8	no	yes	no
6	96.4	yes	no	no
7	96.6	no	yes	no
8	102.6	yes	yes	yes

Yet another method of handling missing attribute values was refereed to as Shapiro's method in (Quinlan, 1989), where for each attribute with missing attribute values a new data set is created, such attributes take place of the decision and vice versa, the decision becomes one of the attributes. From such a table missing attribute values are learned using either a rule set or decision tree techniques. This method, identified as a *chase* algorithm, was also discussed in (Dardzinska and Ras, 2003A, Dardzinska and Ras, 2003B).

Learning missing attribute values from summary constraints was reported in (Wu and Barbara, 2002, Wu and Barbara, 2002). Yet another approach to handling missing attribute values was presented in (Greco *et al.*, 2000).

There is a number of statistical methods of handling missing attribute values, usually known under the name of *imputation* (Allison, 2002, Little and Rubin, 2002, Schikuta, 1996), such as maximum likelihood and the EM algorithm. Recently *multiple imputation* gained popularity. It is a Monte Carlo method of handling missing attribute values in which missing attribute values are replaced by many plausible values, then many complete data sets are analyzed and the results are combined.

3.3 Parallel Methods

In this section we will concentrate on handling missing attribute values in parallel with rule induction. We will distinguish two types of missing attribute values: *lost* and *do not care* conditions (for respective interpretation, see Introduction). First we will introduce some useful ideas, such as blocks of attribute-value pairs, characteristic sets, characteristic relations, lower and upper approximations. Later we will explain how to induce rules using the same blocks of attribute-value pairs that were used to compute lower and upper approximations. Input data sets are not preprocessed the same way as in sequential methods, instead, the rule learning algorithm is modified to learn rules directly from the original, incomplete data sets.

3.3.1 Blocks of Attribute-Value Pairs and Characteristic Sets

In this subsection we will quote some basic ideas of the rough set theory. Any decision table defines a function ρ that maps the direct product of the set U of all cases and the set A of all attributes into the set of all values. For example, in Table 3.1, $\rho(1, Temperature) = high$. In this section we will assume that all missing attribute values are denoted either by "?" or by "*", lost values will be denoted by "?", "do not care" conditions will be denoted by "*". Thus, we assume that all missing attribute values from Table 3.1 are lost. On the other hand, all attribute values from Table 3.15 are do not care conditions.

Table 3.15. An Example of a Dataset with Do Not Care Conditions.

Case	Attributes			Decision
	Temperature	Headache	Nausea	Flu
1	high	*	no	yes
2	very_high	yes	yes	yes
3	*	no	no	no
4	high	yes	yes	yes
5	high	*	yes	no
6	normal	yes	no	no
7	normal	no	yes	no
8	*	yes	*	yes

Let (a, v) be an attribute-value pair. For complete decision tables, a block of (a, v), denoted by $[(a, v)]$, is the set of all cases x for which $\rho(x, a) = v$. For incomplete decision tables the definition of a block of an attribute-value pair is modified. If for an attribute a there exists a case x such that $\rho(x, a) =?$, i.e., the corresponding value is lost, then the case x is not included in any block $[(a, v)]$ for every value v of attribute a. If for an attribute a there exists a case x such that the corresponding value is a "do not care" condition, i.e., $\rho(x, a) = *$, then the corresponding case x should be included in blocks $[(a, v)]$ for all known values v of attribute a. This modification of the attribute-value pair block definition is consistent with the interpretation of missing attribute values, lost and "do not care" conditions. Thus, for Table 3.1

[(Temperature, high)] = $\{1, 4, 5\}$,
[(Temperature, very_high)] = $\{2\}$,
[(Temperature, normal)] = $\{6, 7\}$,
[(Headache, yes)] = $\{2, 4, 6, 8\}$,
[(Headache, no)] = $\{3, 7\}$,
[(Nausea, no)] = $\{1, 3, 6\}$,
[(Nausea, yes)] = $\{2, 4, 5, 7\}$,

and for Table 3.15

[(Temperature, high)] = {1, 3, 4, 5, 8},
[(Temperature, very_high)] = {2, 3, 8},
[(Temperature, normal)] = {3, 6, 7, 8},
[(Headache, yes)] = {1, 2, 4, 5, 6, 8},
[(Headache, no)] = {1, 3, 5, 7},
[(Nausea, no)] = {1, 3, 6, 8},
[(Nausea, yes)] = {2, 4, 5, 7, 8}.

The *characteristic set* $K_B(x)$ is the intersection of blocks of attribute-value pairs (a, v) for all attributes a from B for which $\rho(x, a)$ is known and $\rho(x, a) = v$. For Table 3.1 and $B = A$,

$K_A(1) = \{1,4,5\} \cap \{1,3,6\} = \{1\}$,
$K_A(2) = \{2\} \cap \{2,4,6,8\} \cap \{2,4,5,7\} = \{2\}$,
$K_A(3) = \{3,7\} \cap \{1,3,6\} = \{3\}$,
$K_A(4) = \{1,4,5\} \cap \{2,4,6,8\} \cap \{2,4,5,7\} = \{4\}$,
$K_A(5) = \{1,4,5\} \cap \{2,4,5,7\} = \{4,5\}$,
$K_A(6) = \{6,7\} \cap \{2,4,6,8\} \cap \{1,3,6\} = \{6\}$,
$K_A(7) = \{6,7\} \cap \{3,7\} \cap \{2,4,5,7\} = \{7\}$, and
$K_A(8) = \{2,4,6,8\}$.

and for Table 3.15 and $B = A$,

$K_A(1) = \{1,3,4,5,8\} \cap \{1,3,6,8\} = \{1,3,8\}$,
$K_A(2) = \{2,3,8\} \cap \{1,2,4,5,6,8\} \cap \{2,4,5,7,8\} = \{2,8\}$,
$K_A(3) = \{1,3,5,7\} \cap \{1,3,6,8\} = \{1,3\}$,
$K_A(4) = \{1,3,4,5,8\} \cap \{1,2,4,5,6,8\} \cap \{2,4,5,7,8\} = \{4,5,8\}$,
$K_A(5) = \{1,3,4,5,8\} \cap \{2,4,5,7,8\} = \{4,5,8\}$,
$K_A(6) = \{3,6,7,8\} \cap \{1,2,4,5,6,8\} \cap \{1,3,6,8\} = \{6,8\}$,
$K_A(7) = \{3,6,7,8\} \cap \{1,3,5,7\} \cap \{2,4,5,7,8\} = \{7\}$, and
$K_A(8) = \{1,2,4,5,6,8\}$.

The characteristic set $K_B(x)$ may be interpreted as the smallest set of cases that are indistinguishable from x using all attributes from B, using a given interpretation of missing attribute values. Thus, $K_A(x)$ is the set of all cases that cannot be distinguished from x using all attributes. For further properties of characteristic sets see (Grzymala-Busse, 2003, Grzymala-Busse, 2004A, Grzymala-Busse, 2004B, Grzymala-Busse, 2004C). Incomplete decision tables in which all attribute values are lost, from the viewpoint of rough set theory, were studied for the first time in (Grzymala-Busse and Wang, 1997), where two algorithms for rule induction, modified to handle lost attribute values, were presented. This approach was studied later in (Stefanowski, 2001, Stefanowski and Tsoukias, 1999, Stefanowski and Tsoukias, 2001).

Incomplete decision tables in which all missing attribute values are "do not care" conditions, from the view point of rough set theory, were studied for the first time in (Grzymala-Busse, 1991), where a method for rule induction was introduced in which each missing attribute value was replaced by all values from the domain of the attribute. Originally such values were replaced by all values from the entire domain of the attribute, later, by attribute values restricted to the same concept to which a case with a missing attribute value belongs. Such incomplete decision tables, with all missing attribute values being "do not care conditions", were also studied in (Kryszkiewicz, 1995, Kryszkiewicz, 1999). Both approaches to missing attribute values were generalized in (Grzymala-Busse, 2003, Grzymala-Busse, 2004A, Grzymala-Busse, 2004B, Grzymala-Busse, 2004C).

3.3.2 Lower and Upper Approximations

Any finite union of characteristic sets of B is called a *B-definable* set. The lower approximation of the concept X is the largest definable sets that is contained in X and the upper approximation of X is the smallest definable set that contains X. In general, for incompletely specified decision tables lower and upper approximations may be defined in a few different ways (Grzymala-Busse, 2003, Grzymala-Busse, 2004A, Grzymala-Busse, 2004B, Grzymala-Busse, 2004C). Here we will quote the most useful definition of lower and upper approximations from the view point of Data Mining. A *concept* B-lower approximation of the concept X is defined as follows:

$$\underline{B}X = \cup\{K_B(x)|x \in X, K_B(x) \subseteq X\}.$$

A concept B-upper approximation of the concept X is defined as follows:

$$\overline{B}X = \cup\{K_B(x)|x \in X, K_B(x) \cap X \neq \emptyset\} = \cup\{K_B(x)|x \in X\}.$$

For the decision table presented in Table 3.1, the concept A-lower and A-upper approximations are

$$\underline{A}\{1,2,4,8\} = \{1,2,4\},$$

$$\underline{A}\{3,5,6,7\} = \{3,6,7\},$$

$$\overline{A}\{1,2,4,8\} = \{1,2,4,6,8\},$$

$$\overline{A}\{3,5,6,7\} = \{3,4,5,6,7\},$$

and for the decision table from Table 3.15, the concept A-lower and A-upper approximations are

$$\underline{A}\{1,2,4,8\} = \{2,8\},$$

$$\underline{A}\{3,5,6,7\} = \{7\},$$

$$\overline{A}\{1,2,4,8\} = \{1,2,3,4,5,6,8\},$$

$$\overline{A}\{3,5,6,7\} = \{1,3,4,5,6,7,8\}.$$

3.3.3 Rule Induction—MLEM2

The MLEM2 rule induction algorithm is a modified version of the algorithm LEM2, see chapter 12.6 in this volume. Rules induced from the lower approximation of the concept *certainly* describe the concept, so they are called *certain*. On the other hand, rules induced from the upper approximation of the concept describe the concept only *possibly* (or *plausibly*), so they are called *possible* (Grzymala-Busse, 1988). MLEM2 may induce both certain and possible rules from a decision table with some missing attribute values being lost and some missing attribute values being "do not care" conditions, while some attributes may be numerical. For rule induction from decision tables with numerical attributes see (Grzymala-Busse, 2004A). MLEM2 handles missing attribute values by computing (in a different way than in LEM2) blocks of attribute-value pairs, and then characteristic sets and lower and upper approximations. All these definitions are modified according to the two previous subsections, the algorithm itself remains the same.

Rule sets in the LERS format (every rule is equipped with three numbers, the total number of attribute-value pairs on the left-hand side of the rule, the total number of examples correctly classified by the rule during training, and the total number of training cases matching the left-hand side of the rule), induced from the decision table presented in Table 3.1 are:
certain rule set:

2, 1, 1
(Temperature, high) & (Nausea, no) -> (Flu, yes)
2, 2, 2
(Headache, yes) & (Nausea, yes) -> (Flu, yes)
1, 2, 2
(Temperature, normal) -> (Flu, no)
1, 2, 2
(Headache, no) -> (Flu, no)

and possible rule set:

1, 3, 4
(Headache, yes) -> (Flu, yes)
2, 1, 1
(Temperature, high) & (Nausea, no) -> (Flu, yes)
2, 1, 2
(Temperature, high) & (Nausea, yes) -> (Flu, no)
1, 2, 2
(Temperature, normal) -> (Flu, no)
1, 2, 2
(Headache, no) -> (Flu, no)

Rule sets induced from the decision table presented in Table 3.15 are:
certain rule set:

2, 2, 2

(Temperature, very_high) & (Nausea, yes) -> (Flu, yes)
3, 1, 1
(Temperature, normal) & (Headache, no) & (Nausea, yes) -> (Flu, no)

and possible rule set:

1, 4, 6
(Headache, yes) -> (Flu, yes)
1, 2, 3
(Temperature, very_high) -> (Flu, yes)
1, 2, 5
(Temperature, high) -> (Flu, no)
1, 3, 4
(Temperature, normal) -> (Flu, no)

3.3.4 Other Approaches to Missing Attribute Values

Through this section we assumed that the incomplete decision tables may only consist of lost values or do not care conditions. Note that the MLEM2 algorithm is able to handle not only these two types of tables but also decision tables with a mixture of these two cases, i.e., tables with some lost attribute values and with other missing attribute values being do not care conditions. Furthermore, other interpretations of missing attribute values are possible as well, see (Grzymala-Busse, 2003, Grzymala-Busse, 2004A).

3.4 Conclusions

In general, there is no best, universal method of handling missing attribute values. On the basis of existing research on comparison such methods (Grzymala-Busse and Hu, 2000, Grzymala-Busse and Siddhaye, 2004, Lakshminarayan *et al.*, 1999) we may conclude that for every specific data set the best method of handling missing attribute values should be chosen individually, using as the criterion of optimality the arithmetic mean of many multi-fold cross validation experiments (Weiss and Kulikowski, 1991). Similar conclusions may be drawn for decision tree generation (Quinlan, 1989).

References

Allison P.D. *Missing Data*. Sage Publications, 2002.

Brazdil P. and Bruha I. Processing unknown attribute values by ID3. Proceedings of the 4-th Int. Conference Computing and Information, Toronto, 1992, 227 – 230

Breiman L., Friedman J.H., Olshen R.A., Stone C.J. *Classification and Regression Trees*. Wadsworth & Brooks, Monterey, CA, 1984.

Bruha I. Meta-learner for unknown attribute values processing: Dealing with inconsistency of meta-databases. *Journal of Intelligent Information Systems* **22** 71–87, 2004.

Chiu, D. K. and Wong A. K. C. Synthesizing knowledge: A cluster analysis approach using event-covering. *IEEE Trans. Syst., Man, and Cybern.* **SMC-16** 251–259, 1986.

Clark P. and Niblett T. The CN2 induction algorithm. *Machine Learning* **3** 261–283, 1989.

Dardzinska A. and Ras Z.W. Chasing unknown values in incomplete information systems. Proceedings of the Workshop on Foundations and New Directions in Data Mining, associated with the third IEEE International Conference on Data Mining, Melbourne, FL, November 1922, 24–30, 2003A.

Dardzinska A. and Ras Z.W. On rule discovery from incomplete information systems. Proceedings of the Workshop on Foundations and New Directions in Data Mining, associated with the third IEEE International Conference on Data Mining, Melbourne, FL, November 1922, 31–35, 2003B.

Greco S., Matarazzo B., and Slowinski R. Dealing with missing data in rough set analysis of multi-attribute and multi-criteria decision problems. In *Decision Making: Recent developments and Worldwide Applications*, ed. by S. H. Zanakis, G. Doukidis, and Z. Zopounidis, Kluwer Academic Publishers, Dordrecht, Boston, London, 2000, 295–316.

Grzymala-Busse J.W. Knowledge acquisition under uncertainty—A rough set approach. *Journal of Intelligent & Robotic Systems* **1** (1988) 3–16.

Grzymala-Busse J.W. On the unknown attribute values in learning from examples. Proc. of the ISMIS-91, 6th International Symposium on Methodologies for Intelligent Systems, Charlotte, North Carolina, October 16–19, 1991. Lecture Notes in Artificial Intelligence, vol. 542, Springer-Verlag, Berlin, Heidelberg, New York, 1991, 368–377.

Grzymala-Busse J.W. LERS—A system for learning from examples based on rough sets. In Intelligent Decision Support. Handbook of Applications and Advances of the Rough Sets Theory, ed. by R. Slowinski, Kluwer Academic Publishers, Dordrecht, Boston, London, 1992, 3–18.

Grzymala-Busse J.W. A new version of the rule induction system LERS, *Fundamenta Informaticae* **31** (1997) 27–39.

Grzymala-Busse J.W. MLEM2: A new algorithm for rule induction from imperfect data. Proceedings of the 9th International Conference on Information Processing and Management of Uncertainty in Knowledge-Based Systems, IPMU 2002, Annecy, France, July 1–5, 2002, 243–250.

Grzymala-Busse J.W. Rough set strategies to data with missing attribute values. Proceedings of the Workshop on Foundations and New Directions in Data Mining, associated with the third IEEE International Conference on Data Mining, Melbourne, FL, November 1922, 2003, 56–63.

Grzymala-Busse J.W. Data with missing attribute values: Generalization of indiscernibility relation and rule induction. Transactions on Rough Sets, Lecture Notes in Computer Science Journal Subline, Springer-Verlag, vol. **1** 78–95, 2004A.

Grzymala-Busse J.W. Characteristic relations for incomplete data: A generalization of the indiscernibility relation. Proceedings of the RSCTC'2004, the Fourth International Conference on Rough Sets and Current Trends in Computing, Uppsala, Sweden, June 15, 2004. Lecture Notes in Artificial Intelligence 3066, Springer-Verlag pp.244–253, 2004B.

Grzymala-Busse J.W. Rough set approach to incomplete data. Proceedings of the ICAISC'2004, the Seventh International Conference on Artificial Intelligence and Soft Computing, Zakopane, Poland, June 711, 2004. Lecture Notes in Artificial Intelligence 3070, Springer-Verlag pp.50–55, 2004.

Grzymala-Busse J.W., Grzymala-Busse W.J., and Goodwin L.K. A comparison of three closest fit approaches to missing attribute values in preterm birth data. *International Journal of Intelligent Systems* **17** (2002) 125–134.

Grzymala-Busse, J.W. and Hu, M. A comparison of several approaches to missing attribute values in Data Mining. Proceedings of the Second International Conference on Rough Sets and Current Trends in Computing RSCTC'2000, Banff, Canada, October 16–19, 2000, 340–347.

Grzymala-Busse, J.W. and Wang A.Y. Modified algorithms LEM1 and LEM2 for rule induction from data with missing attribute values. Proc. of the Fifth International Workshop on Rough Sets and Soft Computing (RSSC'97) at the Third Joint Conference on Information Sciences (JCIS'97), Research Triangle Park, NC, March 2–5, 1997, 69–72.

Grzymala-Busse J.W. and Siddhaye S. Rough set approaches to rule induction from incomplete data. Proceedings of the IPMU'2004, the 10th International Conference on Information Processing and Management of Uncertainty in Knowledge-Based Systems, Perugia, Italy, July 49, 2004, vol. 2, 923930.

Imielinski T. and Lipski W. Jr. Incomplete information in relational databases, *Journal of the ACM* **31** (1984) 761–791.

Kononenko I., Bratko I., and Roskar E. Experiments in automatic learning of medical diagnostic rules. Technical Report, Jozef Stefan Institute, Lljubljana, Yugoslavia, 1984

Kryszkiewicz M. Rough set approach to incomplete information systems. Proceedings of the Second Annual Joint Conference on Information Sciences, Wrightsville Beach, NC, September 28–October 1, 1995, 194–197.

Kryszkiewicz M. Rules in incomplete information systems. *Information Sciences* **113** (1999) 271–292.

Lakshminarayan K., Harp S.A., and Samad T. Imputation of missing data in industrial databases. *Applied Intelligence* **11** (1999) 259 – 275.

Latkowski, R. On decomposition for incomplete data. *Fundamenta Informaticae* **54** (2003) 1-16.

Latkowski R. and Mikolajczyk M. Data decomposition and decision rule joining for classification of data with missing values. Proceedings of the RSCTC'2004, the Fourth International Conference on Rough Sets and Current Trends in Computing, Uppsala, Sweden, June 1–5, 2004. Lecture Notes in Artificial Intelligence 3066, Springer-Verlag 2004, 254–263.

Lipski W. Jr. On semantic issues connected with incomplete information databases. *ACM Transactions on Database Systems* **4** (1979), 262–296.

Lipski W. Jr. On databases with incomplete information. *Journal of the ACM* **28** (1981) 41–70.

Little R.J.A. and Rubin D.B. *Statistical Analysis with Missing Data*, Second Edition, J. Wiley & Sons, Inc., 2002.

Pawlak Z. Rough Sets. *International Journal of Computer and Information Sciences* **11** (1982) 341–356.

Pawlak Z. *Rough Sets. Theoretical Aspects of Reasoning about Data*. Kluwer Academic Publishers, Dordrecht, Boston, London, 1991.

Pawlak Z., Grzymala-Busse J.W., Slowinski R., and Ziarko, W. Rough sets. *Communications of the ACM* **38** (1995) 88–95.

Polkowski L. and Skowron A. (eds.) *Rough Sets in Knowledge Discovery, 2, Applications, Case Studies and Software Systems*, Appendix 2: Software Systems. Physica Verlag, Heidelberg New York (1998) 551–601.

Quinlan J.R. Unknown attribute values in induction. Proc. of the 6-th Int. Workshop on Machine Learning, Ithaca, NY, 1989, 164 – 168.

Quinlan J. R. *C4.5: Programs for Machine Learning*. Morgan Kaufmann Publishers, San Mateo CA (1993).

Schafer J.L. *Analysis of Incomplete Multivariate Data*. Chapman and Hall, London, 1997.

Slowinski R. and Vanderpooten D. A generalized definition of rough approximations based on similarity. *IEEE Transactions on Knowledge and Data Engineering* **12** (2000) 331–336.

Stefanowski J. Algorithms of Decision Rule Induction in Data Mining. Poznan University of Technology Prcss, Poznan, Poland (2001).

Stefanowski J. and Tsoukias A. On the extension of rough sets under incomplete information. Proceedings of the 7th International Workshop on New Directions in Rough Sets, Data Mining, and Granular-Soft Computing,
RSFDGrC'1999, Ube, Yamaguchi, Japan, November 8–10, 1999, 73–81.

Stefanowski J. and Tsoukias A. Incomplete information tables and rough classification. *Computational Intelligence* **17** (2001) 545–566.

Weiss S. and Kulikowski C.A. *Computer Systems That Learn: Classification and Prediction Methods from Statistics, Neural Nets, Machine Learning, and Expert Systems*, chapter *How to Estimate the True Performance of a Learning System*, pp. 17–49, San Mateo, CA: Morgan Kaufmann Publishers, Inc., 1991.

Wong K.C. and Chiu K.Y. Synthesizing statistical knowledge for incomplete mixed-mode data. IEEE Transactions on Pattern Analysis and Machine Intelligence **9** (1987) 796805.

Wu X. and Barbara D. Learning missing values from summary constraints. *ACM SIGKDD Explorations Newsletter* **4** (2002) 21 – 30.

Wu X. and Barbara D. Modeling and imputation of large incomplete multidimensional datasets. Proc. of the 4-th Int. Conference on Data Warehousing and Knowledge Discovery, Aix-en-Provence, France, 2002, 286 – 295

Yao Y.Y. On the generalizing rough set theory. Proc. of the 9th Int. Conference on Rough Sets, Fuzzy Sets, Data Mining and Granular Computing (RSFDGrC'2003), Chongqing, China, October 19–22, 2003, 44–51.

4

Geometric Methods for Feature Extraction and Dimensional Reduction - A Guided Tour

Christopher J.C. Burges

Microsoft Research

Summary. We give a tutorial overview of several geometric methods for feature extraction and dimensional reduction. We divide the methods into projective methods and methods that model the manifold on which the data lies. For projective methods, we review projection pursuit, principal component analysis (PCA), kernel PCA, probabilistic PCA, and oriented PCA; and for the manifold methods, we review multidimensional scaling (MDS), landmark MDS, Isomap, locally linear embedding, Laplacian eigenmaps and spectral clustering. The Nyström method, which links several of the algorithms, is also reviewed. The goal is to provide a self-contained review of the concepts and mathematics underlying these algorithms.

Key words: Feature Extraction, Dimensional Reduction, Principal Components Analysis, Distortion Discriminant Analysis, Nyström method, Projection Pursuit, Kernel PCA, Multidimensional Scaling, Landmark MDS, Locally Linear Embedding, Isomap

Introduction

Feature extraction can be viewed as a preprocessing step which removes distracting variance from a dataset, so that downstream classifiers or regression estimators perform better. The area where feature extraction ends and classification, or regression, begins is necessarily murky: an ideal feature extractor would simply map the data to its class labels, for the classification task. On the other hand, a character recognition neural net can take minimally preprocessed pixel values as input, in which case feature extraction is an inseparable part of the classification process (LeCun and Bengio, 1995). Dimensional reduction - the (usually non-invertible) mapping of data to a lower dimensional space - is closely related (often dimensional reduction is used as a step in feature extraction), but the goals can differ. Dimensional reduction has a long history as a method for data visualization, and for extracting key low dimensional features (for example, the 2-dimensional orientation of an object, from its high dimensional image representation). The need for dimensionality reduction

O. Maimon, L. Rokach (eds.), *Data Mining and Knowledge Discovery Handbook*, 2nd ed.,
DOI 10.1007/978-0-387-09823-4_4, © Springer Science+Business Media, LLC 2010

also arises for other pressing reasons. (Stone, 1982) showed that, under certain regularity assumptions, the optimal rate of convergence[1] for nonparametric regression varies as $m^{-p/(2p+d)}$, where m is the sample size, the data lies in $\mathscr{R}^d$, and where the regression function is assumed to be p times differentiable. Consider 10,000 sample points, for $p = 2$ and $d = 10$. If d is increased to 20, the number of sample points must be increased to approximately 10 million in order to achieve the same optimal rate of convergence. If our data lie (approximately) on a low dimensional manifold $\mathscr{L}$ that happens to be embedded in a high dimensional manifold $\mathscr{H}$, modeling the projected data in $\mathscr{L}$ rather than in $\mathscr{H}$ may turn an infeasible problem into a feasible one.

The purpose of this review is to describe the mathematics and ideas underlying the algorithms. Implementation details, although important, are not discussed. Some notes on notation: vectors are denoted by boldface, whereas components are denoted by x_a, or by $(\mathbf{x}_i)_a$ for the a'th component of the i'th vector. Following

(Horn and Johnson, 1985), the set of p by q matrices is denoted M_{pq}, the set of (square) p by p matrices by M_p, and the set of symmetric p by p matrices by S_p (all matrices considered are real). $\mathbf{e}$ with no subscript is used to denote the vector of all ones; on the other hand $\mathbf{e}_a$ denotes the a'th eigenvector. We denote sample size by m, and dimension usually by d or d', with typically $d' \ll d$. δ_{ij} is the Kronecker delta (the ij'th component of the unit matrix). We generally reserve indices i, j, to index vectors and a, b to index dimension.

We place feature extraction and dimensional reduction techniques into two broad categories: methods that rely on projections (Section 4.1) and methods that attempt to model the manifold on which the data lies (Section 4.2). Section 4.1 gives a detailed description of principal component analysis; apart from its intrinsic usefulness, PCA is interesting because it serves as a starting point for many modern algorithms, some of which (kernel PCA, probabilistic PCA, and oriented PCA) are also described. However it has clear limitations: it is easy to find even low dimensional examples where the PCA directions are far from optimal for feature extraction (Duda and Hart, 1973), and PCA ignores correlations in the data that are higher than second order. Section 4.2 starts with an overview of the Nyström method, which can be used to extend, and link, several of the algorithms described in this chapter. We then examine some methods for dimensionality reduction which assume that the data lie on a low dimensional manifold embedded in a high dimensional space $\mathscr{H}$, namely locally linear embedding, multidimensional scaling, Isomap, Laplacian eigenmaps, and spectral clustering.

[1] For convenience we reproduce Stone's definitions (Stone, 1982). Let θ be the unknown regression function, $\hat{T}_n$ an estimator of θ using n samples, and $\{b_n\}$ a sequence of positive constants. Then $\{b_n\}$ is called a lower rate of convergence if there exists $c > 0$ such that $\lim_n \inf_{\hat{T}_n} \sup_\theta P(\|\hat{T}_n - \theta\| \geq cb_n) = 1$, and it is called an achievable rate of convergence if there is a sequence of estimators $\{\hat{T}_n\}$ and $c > 0$ such that $\lim_n \sup_\theta P(\|\hat{T}_n - \theta\| \geq cb_n) = 0$; $\{b_n\}$ is called an optimal rate of convergence if it is both a lower rate of convergence and an achievable rate of convergence.

4.1 Projective Methods

If dimensional reduction is so desirable, how should we go about it? Perhaps the simplest approach is to attempt to find low dimensional *projections* that extract useful information from the data, by maximizing a suitable objective function. This is the idea of projection pursuit (Friedman and Tukey, 1974). The name 'pursuit' arises from the iterative version, where the currently optimal projection is found in light of previously found projections (in fact originally this was done manually[2]). Apart from handling high dimensional data, projection pursuit methods can be robust to noisy or irrelevant features (Huber, 1985), and have been applied to regression (Friedman and Stuetzle, 1981), where the regression is expressed as a sum of 'ridge functions' (functions of the one dimensional projections) and at each iteration the projection is chosen to minimize the residuals; to classification; and to density estimation (Friedman et al., 1984). How are the interesting directions found? One approach is to search for projections such that the projected data departs from normality (Huber, 1985). One might think that, since a distribution is normal if and only if all of its one dimensional projections are normal, if the least normal projection of some dataset is still approximately normal, then the dataset is also necessarily approximately normal, but this is not true; Diaconis and Freedman have shown that most projections of high dimensional data are approximately normal (Diaconis and Freedman, 1984) (see also below). Given this, finding projections along which the density departs from normality, if such projections exist, should be a good exploratory first step.

The sword of Diaconis and Freedman cuts both ways, however. If most projections of most high dimensional datasets are approximately normal, perhaps projections are not always the best way to find low dimensional representations. Let's review their results in a little more detail. The main result can be stated informally as follows: consider a model where the data, the dimension d, and the sample size m depend on some underlying parameter ν, such that as ν tends to infinity, so do m and d. Suppose that as ν tends to infinity, the fraction of vectors which are not approximately the same length tends to zero, and suppose further that under the same conditions, the fraction of pairs of vectors which are not approximately orthogonal to each other also tends to zero[3]. Then ((Diaconis and Freedman, 1984), theorem 1.1) the empirical distribution of the projections along any given unit direction tends to $N(0,\sigma^2)$ weakly in probability. However, if the conditions are not fulfilled, as for some long-tailed distributions, then the opposite result can hold - that is, most projections are *not* normal (for example, most projections of Cauchy distributed data[4] will be Cauchy (Diaconis and Freedman, 1984)).

---

[2] See J.H. Friedman's interesting response to (Huber, 1985) in the same issue.

[3] More formally, the conditions are: for σ^2 positive and finite, and for any positive ε, $(1/m)\text{card}\{j \leq m : |\|\mathbf{x}_j\|^2 - \sigma^2 d| > \varepsilon d\} \to 0$ and $(1/m^2)\text{card}\{1 \leq j,k \leq m : |\mathbf{x}_j \cdot \mathbf{x}_k| > \varepsilon d\} \to 0$ (Diaconis and Freedman, 1984).

[4] The Cauchy distribution in one dimension has density $c/(c^2+x^2)$ for constant c.

As a concrete example[5], consider data uniformly distributed over the unit $n+1$-sphere $\mathscr{S}^{n+1}$ for odd n. Let's compute the density projected along any line $\mathscr{I}$ passing through the origin. By symmetry, the result will be independent of

the direction we choose. If the distance along the projection is parameterized by $\xi \equiv \cos\theta$, where θ is the angle between $\mathscr{I}$ and the line from the origin to a point on the sphere, then the density at ξ is proportional to the volume of an n-sphere of radius $\sin\theta$: $\rho(\xi) = C(1-\xi^2)^{\frac{n-1}{2}}$. Requiring that $\int_{-1}^{1}\rho(\xi)d\xi = 1$ gives the constant C:

$$C = 2^{-\frac{1}{2}(n+1)}\frac{n!!}{(\frac{1}{2}(n-1))!} \tag{4.1}$$

Let's plot this density and compare against a one dimensional Gaussian density fitted using maximum likelihood. For that we just need the variance, which can be computed analytically: $\sigma^2 = \frac{1}{n+2}$, and the mean, which is zero. Figure 4.1 shows the result for the 20-sphere. Although data uniformly distributed on $\mathscr{S}^{20}$ is far from Gaussian, its projection along any direction is close to Gaussian for all such directions, and we cannot hope to uncover such structure using one dimensional projections.

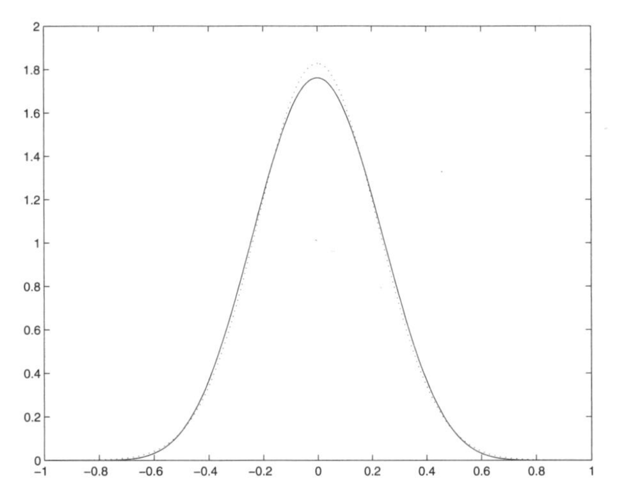

Fig. 4.1. Dotted line: a Gaussian with zero mean and variance 1/21. Solid line: the density projected from data distributed uniformly over the 20-sphere, to any line passing through the origin.

The notion of searching for non-normality, which is at the heart of projection pursuit (the goal of which is dimensional reduction), is also the key idea underlying independent component analysis (ICA) (the goal of which is source separation). ICA (Hyvärinen et al., 2001) searches for projections such that the probability distributions of the data along those projections are statistically independent: for example,

[5] The story for even n is similar but the formulae are slightly different

consider the problem of separating the source signals in a linear combinations of signals, where the sources consist of speech from two speakers who are recorded using two microphones (and where each microphone captures sound from both speakers). The signal is the sum of two statistically independent signals, and so finding those independent signals is required in order to decompose the signal back into the two original source signals, and at any given time, the separated signal values are related to the microphone signals by two (time independent) projections (forming an invertible 2 by 2 matrix). If the data is normally distributed, finding projections along which the data is uncorrelated is equivalent to finding projections along which it is independent, so although using principal component analysis (see below) will suffice to find independent projections, those projections will not be useful for the above task. For most other distributions, finding projections along which the data is statistically independent is a much stronger (and for ICA, useful) condition than finding projections along which the data is uncorrelated. Hence ICA concentrates on situations where the distribution of the data departs from normality, and in fact, finding the maximally non-Gaussian component (under the constraint of constant variance) will give you an independent component (Hyvärinen et al., 2001).

4.1.1 Principal Component Analysis (PCA)

PCA: Finding an Informative Direction

Given data $\mathbf{x}_i \in \mathscr{R}^d$, $i = 1, \cdots, m$, suppose you'd like to find a direction $\mathbf{v} \in \mathscr{R}^d$ for which the projection $\mathbf{x}_i \cdot \mathbf{v}$ gives a good one dimensional representation of your original data: that is, informally, the act of projecting loses as little information about your expensively-gathered data as possible (we will examine the information theoretic view of this below). Suppose that unbeknownst to you, your data in fact lies along a line $\mathscr{I}$ embedded in $\mathscr{R}^d$, that is, $\mathbf{x}_i = \mu + \theta_i \mathbf{n}$, where μ is the sample mean[6], $\theta_i \in \mathscr{R}$, and $\mathbf{n} \in \mathscr{R}^d$ has unit length. The sample variance of the projection along $\mathbf{n}$ is then

$$v_n \equiv \frac{1}{m} \sum_{i=1}^{m} ((\mathbf{x}_i - \mu) \cdot \mathbf{n})^2 = \frac{1}{m} \sum_{i=1}^{m} \theta_i^2 \tag{4.2}$$

and that along some other unit direction $\mathbf{n}'$ is

$$v_n' \equiv \frac{1}{m} \sum_{i=1}^{m} ((\mathbf{x}_i - \mu) \cdot \mathbf{n}')^2 = \frac{1}{m} \sum_{i=1}^{m} \theta_i^2 (\mathbf{n} \cdot \mathbf{n}')^2 \tag{4.3}$$

Since $(\mathbf{n} \cdot \mathbf{n}')^2 = \cos^2 \phi$, where ϕ is the angle between $\mathbf{n}$ and $\mathbf{n}'$, we see that the projected variance is maximized if and only if $\mathbf{n} = \pm \mathbf{n}'$. Hence in this case, finding the projection for which the projected variance is maximized gives you the direction you are looking for, namely $\mathbf{n}$, *regardless of the distribution of the data along* $\mathbf{n}$, as long as the data has finite variance. You would then quickly find that the variance along all directions orthogonal to $\mathbf{n}$ is zero, and conclude that your data in fact lies

[6] Note that if all x_i lie along a given line then so does μ.

along a one dimensional manifold embedded in $\mathscr{R}^d$. This is one of several basic results of PCA that hold for arbitrary distributions, as we shall see.

Even if the underlying physical process generates data that ideally lies along $\mathscr{I}$, noise will usually modify the data at various stages up to and including the measurements themselves, and so your data will very likely not lie exactly along $\mathscr{I}$. If the overall noise is much smaller than the signal, it makes sense to try to find $\mathscr{I}$ by searching for that projection along which the projected data has maximum variance. If in addition your data lies in a two (or higher) dimensional subspace, the above argument can be repeated, picking off the highest variance directions in turn. Let's see how that works.

PCA: Ordering by Variance

We've seen that directions of maximum variance can be interesting, but how can we find them? The variance along unit vector $\mathbf{n}$ (Eq. (4.2)) is $\mathbf{n}'C\mathbf{n}$ where C is the sample covariance matrix. Since C is positive semidefinite, its eigenvalues are positive or zero; let's choose the indexing such that the (unit normed) eigenvectors $\mathbf{e}_a$, $a = 1,\ldots,d$ are arranged in order of decreasing size of the corresponding eigenvalues λ_a. Since the $\{\mathbf{e}_a\}$ span the space, we can expand $\mathbf{n}$ in terms of them: $\mathbf{n} = \sum_{a=1}^d \alpha_a \mathbf{e}_a$, and we'd like to find the α_a that maximize $\mathbf{n}'C\mathbf{n} = \mathbf{n}'\sum_a \alpha_a C\mathbf{e}_a = \sum_a \lambda_a \alpha_a^2$, subject to $\sum_a \alpha_a^2 = 1$ (to give unit normed $\mathbf{n}$). This is just a convex combination of the λ's, and since a convex combination of any set of numbers is maximized by taking the largest, the optimal $\mathbf{n}$ is just $\mathbf{e}_1$, the principal eigenvector (or any one of the set of such eigenvectors, if multiple eigenvectors share the same largest eigenvalue), and furthermore, the variance of the projection of the data along $\mathbf{n}$ is just λ_1.

The above construction captures the variance of the data along the direction $\mathbf{n}$. To characterize the remaining variance of the data, let's find that direction $\mathbf{m}$ which is both orthogonal to $\mathbf{n}$, and along which the projected data again has maximum variance. Since the eigenvectors of C form an orthonormal basis (or can be so chosen), we can expand $\mathbf{m}$ in the subspace $\mathscr{R}^{d-1}$ orthogonal to $\mathbf{n}$ as $\mathbf{m} = \sum_{a=2}^d \beta_a \mathbf{e}_a$. Just as above, we wish to find the β_a that maximize $\mathbf{m}'C\mathbf{m} = \sum_{a=2}^d \lambda_a \beta_a^2$, subject to $\sum_{a=2}^d \beta_a^2 = 1$, and by the same argument, the desired direction is given by the (or any) remaining eigenvector with largest eigenvalue, and the corresponding variance is just that eigenvalue. Repeating this argument gives d orthogonal directions, in order of monotonically decreasing projected variance. Since the d directions are orthogonal, they also provide a complete basis. Thus if one uses all d directions, no information is lost, and as we'll see below, if one uses the $d' < d$ principal directions, then the mean squared error introduced by representing the data in this manner is minimized. Finally, PCA for feature extraction amounts to projecting the data to a lower dimensional space: given an input vector $\mathbf{x}$, the mapping consists of computing the projections of $\mathbf{x}$ along the $\mathbf{e}_a$, $a = 1,\ldots,d'$, thereby constructing the components of the projected d'-dimensional feature vectors.

PCA Decorrelates the Samples

Now suppose we've performed PCA on our samples, and instead of using it to construct low dimensional features, we simply use the full set of orthonormal eigenvectors as a choice of basis. In the old basis, a given input vector $\mathbf{x}$ is expanded as $\mathbf{x} = \sum_{a=1}^{d} x_a \mathbf{u}_a$ for some orthonormal set $\{\mathbf{u}_a\}$, and in the new basis, the same vector is expanded as $\mathbf{x} = \sum_{b=1}^{d} \tilde{x}_b \mathbf{e}_b$, so $\tilde{x}_a \equiv \mathbf{x} \cdot \mathbf{e}_a = \mathbf{e}_a \cdot \sum_b x_b \mathbf{u}_b$. The mean $\mu \equiv \frac{1}{m} \sum_i \mathbf{x}_i$ has components $\tilde{\mu}_a = \mu \cdot \mathbf{e}_a$ in the new basis. The sample covariance matrix depends on the choice of basis: if C is the covariance matrix in the old basis, then the corresponding covariance matrix in the new basis is $\tilde{C}_{ab} \equiv \frac{1}{m} \sum_i (\tilde{x}_{ia} - \tilde{\mu}_a)(\tilde{x}_{ib} - \tilde{\mu}_b) = \frac{1}{m} \sum_i \{\mathbf{e}_a \cdot (\sum_p x_{ip} \mathbf{u}_p - \mu)\}\{\sum_q x_{iq} \mathbf{u}_q - \mu) \cdot \mathbf{e}_b\} = \mathbf{e}_a' C \mathbf{e}_b = \lambda_b \delta_{ab}$. Hence in the new basis the covariance matrix is diagonal and the samples are uncorrelated. It's worth emphasizing two points: first, although the covariance matrix can be viewed as a geometric object in that it transforms as a tensor (since it is a summed outer product of vectors, which themselves have a meaning independent of coordinate system), nevertheless, the notion of correlation is basis-dependent (data can be correlated in one basis and uncorrelated in another). Second, PCA decorrelates the samples whatever their underlying distribution; it does not have to be Gaussian.

PCA: Reconstruction with Minimum Squared Error

The basis provided by the eigenvectors of the covariance matrix is also optimal for dimensional reduction in the following sense. Again consider some arbitrary orthonormal basis $\{\mathbf{u}_a,\ a = 1,\ldots,d\}$, and take the first d' of these to perform the dimensional reduction: $\tilde{\mathbf{x}} \equiv \sum_{a=1}^{d'} (\mathbf{x} \cdot \mathbf{u}_a)\mathbf{u}_a$. The chosen $\mathbf{u}_a$ form a basis for $\mathscr{R}^{d'}$, so we may take the components of the dimensionally reduced vectors to be $\mathbf{x} \cdot \mathbf{u}_a$, $a = 1,\ldots,d'$ (although here we leave $\tilde{\mathbf{x}}$ with dimension d). Define the reconstruction error summed over the dataset as $\sum_{i=1}^{m} \|\mathbf{x}_i - \tilde{\mathbf{x}}_i\|^2$. Again assuming that the eigenvectors $\{\mathbf{e}_a\}$ of the covariance matrix are ordered in order of non-increasing eigenvalues, choosing to use those eigenvectors as basis vectors will give minimal reconstruction error. If the data is not centered, then the mean should be subtracted first, the dimensional reduction performed, and the mean then added back[7]; thus in this case, the dimensionally reduced data will still lie in the subspace $\mathscr{R}^{d'}$, but that subspace will be offset from the origin by the mean. Bearing this caveat in mind, to prove the claim we can assume that the data is centered. Expanding $\mathbf{u}_a \equiv \sum_{p=1}^{d} \beta_{ap} \mathbf{e}_p$, we have

$$\frac{1}{m} \sum_i \|\mathbf{x}_i - \tilde{\mathbf{x}}_i\|^2 = \frac{1}{m} \sum_i \|\mathbf{x}_i\|^2 - \frac{1}{m} \sum_{a=1}^{d'} \sum_i (\mathbf{x}_i \cdot \mathbf{u}_a)^2 \tag{4.4}$$

with the constraints $\sum_{p=1}^{d} \beta_{ap} \beta_{bp} = \delta_{ab}$. The second term on the right is

[7] The principal eigenvectors are not necessarily the directions that give minimal reconstruction error if the data is not centered: imagine data whose mean is both orthogonal to the principal eigenvector and far from the origin. The single direction that gives minimal reconstruction error will be close to the mean.

$$-\sum_{a=1}^{d'} \mathbf{u}'_a C \mathbf{u}_a = -\sum_{a=1}^{d'} (\sum_{p=1}^{d} \beta_{ap} \mathbf{e}'_p) C (\sum_{q=1}^{d} \beta_{aq} \mathbf{e}_q) = -\sum_{a=1}^{d'} \sum_{p=1}^{d} \lambda_p \beta_{ap}^2$$

Introducing Lagrange multipliers ω_{ab} to enforce the orthogonality constraints (Burges, 2004), the objective function becomes

$$F = \sum_{a=1}^{d'} \sum_{p=1}^{d} \lambda_p \beta_{ap}^2 - \sum_{a,b=1}^{d'} \omega_{ab} \left(\sum_{p=1}^{d} \beta_{ap} \beta_{bp} - \delta_{ab} \right) \tag{4.5}$$

Choosing[8] $\omega_{ab} \equiv \omega_a \delta_{ab}$ and taking derivatives with respect to β_{cq} gives $\lambda_q \beta_{cq} = \omega_c \beta_{cq}$. Both this and the constraints can be satisfied by choosing $\beta_{cq} = 0 \ \forall q > c$ and $\beta_{cq} = \delta_{cq}$ otherwise; the objective function is then maximized if the first d' largest λ_p are chosen. Note that this also amounts to a proof that the 'greedy' approach to PCA dimensional reduction - solve for a single optimal direction (which gives the principal eigenvector as first basis vector), then project your data into the subspace orthogonal to that, then repeat - also results in the global optimal solution, found by solving for all directions at once. The same is true for the directions that maximize the variance. Again, note that this argument holds however your data is distributed.

PCA Maximizes Mutual Information on Gaussian Data

Now consider some proposed set of projections $W \in M_{d'd}$, where the rows of W are orthonormal, so that the projected data is $\mathbf{y} \equiv W\mathbf{x}$, $\mathbf{y} \in \mathscr{R}^{d'}$, $\mathbf{x} \in \mathscr{R}^d$, $d' \leq d$. Suppose that $\mathbf{x} \sim \mathscr{N}(0, C)$. Then since the $\mathbf{y}$'s are linear combinations of the $\mathbf{x}$'s, they are also normally distributed, with zero mean and covariance $C_y \equiv (1/m) \sum_i^m \mathbf{y}_i \mathbf{y}'_i = (1/m) W (\sum_i^m \mathbf{x}_i \mathbf{x}'_i) W' = WCW'$. It's interesting to ask how W can be chosen so that the mutual information between the distribution of the $\mathbf{x}$'s and that of the $\mathbf{y}$'s is maximized (Baldi and Hornik, 1995, Diamantaras and Kung, 1996). Since the mapping W is deterministic, the conditional entropy $H(\mathbf{y}|\mathbf{x})$ vanishes, and the mutual information is just $I(\mathbf{x}, \mathbf{y}) = H(\mathbf{y}) - H(\mathbf{y}|\mathbf{x}) = H(\mathbf{y})$. Using a small, fixed bin size, we can approximate this by the differential entropy,

$$H(\mathbf{y}) = -\int p(\mathbf{y}) \log_2 p(\mathbf{y}) d\mathbf{y} = \frac{1}{2} \log_2 (e(2\pi)^{d'}) + \frac{1}{2} \log_2 \det(C_y) \tag{4.6}$$

This is maximized by maximizing $\det(C_y) = \det(WCW')$ over choice of W, subject to the constraint that the rows of W are orthonormal. The general solution to this is $W = UE$, where U is an arbitrary d' by d' orthogonal matrix, and where the rows of $E \in M_{d'd}$ are formed from the first d' principal eigenvectors of C, and at the solution, $\det(C_y)$ is just the product of the first d' principal eigenvalues. Clearly, the choice of U does not affect the entropy, since $\det(UECE'U') = \det(U)\det(ECE')\det(U') = \det(ECE')$. In the special case where $d' = 1$, so that E consists of a single, unit length

[8] Recall that Lagrange multipliers can be chosen in any way that results in a solution satisfying the constraints.

vector $\mathbf{e}$, we have $\det(ECE') = \mathbf{e}'C\mathbf{e}$, which is maximized by choosing $\mathbf{e}$ to be the principal eigenvector of C, as shown above. (The other extreme case, where $d' = d$, is easy too, since then $\det(ECE') = \det(C)$ and E can be any orthogonal matrix). We refer the reader to (Wilks, 1962) for a proof for the general case $1 < d' < d$.

4.1.2 Probabilistic PCA (PPCA)

Suppose you've applied PCA to obtain low dimensional feature vectors for your data, but that you have also somehow found a partition of the data such that the PCA projections you obtain on each subset are quite different from those obtained on the other subsets. It would be tempting to perform PCA on each subset and use the relevant projections on new data, but how do you determine what is 'relevant'? That is, how would you construct a mixture of PCA models? While several approaches to such mixtures have been proposed, the first such probabilistic model was proposed by (Tipping and Bishop, 1999A, Tipping and Bishop, 1999B). The advantages of a probabilistic model are numerous: for example, the weight that each mixture component gives to the posterior probability of a given data point can be computed, solving the 'relevance' problem stated above. In this section we briefly review PPCA.

The approach is closely related to factor analysis, which itself is a classical dimensional reduction technique. Factor analysis first appeared in the behavioral sciences community a century ago, when Spearman hypothesised that intelligence could be reduced to a single underlying factor (Spearman, 1904). If, given an n by n correlation matrix between variables $x_i \in \mathscr{R}$, $i = 1, \cdots, n$, there is a single variable g such that the correlation between x_i and x_j vanishes for $i \neq j$ given the value of g, then g is the underlying 'factor' and the off-diagonal elements of the correlation matrix can be written as the corresponding off-diagonal elements of $\mathbf{z}\mathbf{z}'$ for some $\mathbf{z} \in \mathscr{R}^n$ (Darlington). Modern factor analysis usually considers a model where the underlying factors $\mathbf{x} \in \mathscr{R}^{d'}$ are Gaussian, and where a Gaussian noise term $\varepsilon \in \mathscr{R}^d$ is added:

$$\begin{aligned} \mathbf{y} &= W\mathbf{x} + \mu + \varepsilon \\ \mathbf{x} &\sim \mathscr{N}(0, \mathbf{1}) \\ \varepsilon &\sim \mathscr{N}(0, \Psi) \end{aligned} \tag{4.7}$$

Here $\mathbf{y} \in \mathscr{R}^d$ are the observations, the parameters of the model are $W \in M_{dd'}$ ($d' \leq d$), Ψ and μ, and Ψ is assumed to be diagonal. By construction, the $\mathbf{y}$'s have mean μ and 'model covariance' $WW' + \Psi$. For this model, given $\mathbf{x}$, the vectors $\mathbf{y} - \mu$ become uncorrelated. Since $\mathbf{x}$ and ε are Gaussian distributed, so is $\mathbf{y}$, and so the maximum likelihood estimate of $E[\mathbf{y}]$ is just μ. However, in general, W and Ψ must be estimated iteratively, using for example EM. There is an instructive exception to this (Basilevsky, 1994, Tipping and Bishop, 1999A). Suppose that $\Psi = \sigma^2\mathbf{1}$, that the $d - d'$ smallest eigenvalues of the model covariance are the same and are equal to σ^2, and that the sample covariance S is equal to the model covariance (so that σ^2 follows immediately from the eigendecomposition of S). Let $\mathbf{e}^{(j)}$ be the j'th orthonormal eigenvector of S with eigenvalue λ_j. Then by considering the spectral decomposition of S it is straightforward to show that $W_{ij} = \sqrt{(\lambda_j - \sigma^2)}\mathbf{e}_i^{(j)}$, $i = 1, \cdots, d$, $j = 1, \cdots, d'$, if the

$\mathbf{e}^{(j)}$ are in principal order. The model thus arrives at the PCA directions, but in a probabilistic way. *Probabilistic* PCA (PPCA) is a more general extension of factor analysis: it assumes a model of the form (4.7) with $\Psi = \sigma^2 \mathbf{1}$, but it drops the above assumption that the model and sample covariances are equal (which in turn means that σ^2 must now be estimated). The resulting maximum likelihood estimates of W and σ^2 can be written in closed form, as (Tipping and Bishop, 1999A)

$$W_{ML} = U(\Lambda - \sigma^2 \mathbf{1})R \tag{4.8}$$

$$\sigma^2_{ML} = \frac{1}{d-d'} \sum_{i=d'+1}^{d} \lambda_i \tag{4.9}$$

where $U \in M_{dd'}$ is the matrix of the d' principal column eigenvectors of S, Λ is the corresponding diagonal matrix of principal eigenvalues, and $R \in M_{d'}$ is an arbitrary orthogonal matrix. Thus σ^2 captures the variance lost in the discarded projections and the PCA directions appear in the maximum likelihood estimate of W (and in fact re-appear in the expression for the expectation of $\mathbf{x}$ given $\mathbf{y}$, in the limit $\sigma \to 0$, in which case the $\mathbf{x}$ become the PCA projections of the $\mathbf{y}$). This closed form result is rather striking in view of the fact that for general factor analysis we must resort to an iterative algorithm. The probabilistic formulation makes PCA amenable to a rich variety of probabilistic methods: for example, PPCA allows one to perform PCA when some of the data is missing components; and d' (which so far we've assumed known) can itself be estimated using Bayesian arguments (Bishop, 1999). Returning to the problem posed at the beginning of this Section, a mixture of PPCA models, each with weight $\pi_i \geq 0$, $\sum_i \pi_i = 1$, can be computed for the data using maximum likelihood and EM, thus giving a principled approach to combining several local PCA models (Tipping and Bishop, 1999B).

4.1.3 Kernel PCA

PCA is a linear method, in the sense that the reduced dimension representation is generated by linear projections (although the eigenvectors and eigenvalues depend non-linearly on the data), and this can severely limit the usefulness of the approach. Several versions of nonlinear PCA have been proposed (see e.g. (Diamantaras and Kung, 1996)) in the hope of overcoming this problem. In this section we describe a more recent algorithm called kernel PCA (Schölkopf et al., 1998). Kernel PCA relies on the "kernel trick", which is the following observation: suppose you have an algorithm (for example, k'th nearest neighbour) which depends only on dot products of the data. Consider using the same algorithm on transformed data: $\mathbf{x} \to \Phi(\mathbf{x}) \in \mathscr{F}$, where $\mathscr{F}$ is a (possibly infinite dimensional) vector space, which we will call feature space[9]. Operating in $\mathscr{F}$, your algorithm depends only on the dot products $\Phi(\mathbf{x}_i) \cdot \Phi(\mathbf{x}_j)$. Now suppose there exists a (symmetric) 'kernel' function $k(\mathbf{x}_i, \mathbf{x}_j)$

[9] In fact the method is more general: $\mathscr{F}$ can be any complete, normed vector space with inner product (i.e. any Hilbert space), in which case the dot product in the above argument is replaced by the inner product.

such that for all $\mathbf{x}_i,\ \mathbf{x}_j \in \mathscr{R}^d$, $k(\mathbf{x}_i,\mathbf{x}_j) = \Phi(\mathbf{x}_i)\cdot\Phi(\mathbf{x}_j)$. Then since your algorithm depends only on these dot products, you never have to compute $\Phi(\mathbf{x})$ explicitly; you can always just substitute in the kernel form. This was first used by (Aizerman et al., 1964) in the theory of potential functions, and burst onto the machine learning scene in (Boser et al., 1992), when it was applied to support vector machines. Kernel PCA applies the idea to performing PCA in $\mathscr{F}$. It's striking that, since projections are being performed in a space whose dimension can be much larger than d, the number of useful such projections can actually exceed d, so kernel PCA is aimed more at feature extraction than dimensional reduction.

It's not immediately obvious that PCA is eligible for the kernel trick, since in PCA the data appears in expectations over products of individual components of vectors, not over dot products between the vectors. However (Schölkopf et al., 1998) show how the problem can indeed be cast entirely in terms of dot products. They make two key observations: first, that the eigenvectors of the covariance matrix in $\mathscr{F}$ lie in the span of the (centered) mapped data, and second, that therefore no information in the eigenvalue equation is lost if the equation is replaced by m equations, formed by taking the dot product of each side of the eigenvalue equation with each (centered) mapped data point. Let's see how this works. The covariance matrix of the mapped data in feature space is

$$C \equiv \frac{1}{m}\sum_{i=1}^{m}(\Phi_i-\mu)(\Phi_i-\mu)^T \tag{4.10}$$

where $\Phi_i \equiv \Phi(\mathbf{x}_i)$ and $\mu \equiv \frac{1}{m}\sum_i \Phi_i$. We are looking for eigenvector solutions $\mathbf{v}$ of

$$C\mathbf{v} = \lambda\mathbf{v} \tag{4.11}$$

Since this can be written $\frac{1}{m}\sum_{i=1}^{m}(\Phi_i-\mu)[(\Phi_i-\mu)\cdot\mathbf{v}] = \lambda\mathbf{v}$, the eigenvectors $\mathbf{v}$ lie in the span of the $\Phi_i-\mu$'s, or

$$\mathbf{v} = \sum_i \alpha_i(\Phi_i-\mu) \tag{4.12}$$

for some α_i. Since (both sides of) Eq. (4.11) lie in the span of the $\Phi_i-\mu$, we can replace it with the m equations

$$(\Phi_i-\mu)^T C\mathbf{v} = \lambda(\Phi_i-\mu)^T\mathbf{v} \tag{4.13}$$

Now consider the 'kernel matrix' K_{ij}, the matrix of dot products in $\mathscr{F}$: $K_{ij} \equiv \Phi_i\cdot\Phi_j,\ \ i,j=1,\ldots,m$. We know how to calculate this, given a kernel function k, since $\Phi_i\cdot\Phi_j = k(\mathbf{x}_i,\mathbf{x}_j)$. However, what we need is the *centered* kernel matrix, $K_{ij}^C \equiv (\Phi_i-\mu)\cdot(\Phi_j-\mu)$. Happily, any m by m dot product matrix can be centered by left- and right multiplying by the projection matrix $P = \mathbf{1}-\frac{1}{m}\mathbf{e}\mathbf{e}'$, where $\mathbf{1}$ is the unit matrix in M_m and where $\mathbf{e}$ is the m-vector of all ones (see Section 4.2.2 for further discussion of centering). Hence we have $K^C = PKP$, and Eq. (4.13) becomes

$$K^C K^C \alpha = \bar{\lambda} K^C \alpha \tag{4.14}$$

where $\alpha \in \mathscr{R}^m$ and where $\bar{\lambda} \equiv m\lambda$. Now clearly any solution of

$$K^C \alpha = \bar{\lambda}\alpha \tag{4.15}$$

is also a solution of (4.14). It's straightforward to show that any solution of (4.14) can be written as a solution α to (4.15) plus a vector β which is orthogonal to α (and which satisfies $\sum_i \beta_i(\Phi_i - \mu) = 0$), and which therefore does not contribute to (4.12); therefore we need only consider Eq. (4.15). Finally, to use the eigenvectors $\mathbf{v}$ to compute principal components in $\mathscr{F}$, we need $\mathbf{v}$ to have unit length, that is, $\mathbf{v} \cdot \mathbf{v} = 1 = \bar{\lambda}\alpha \cdot \alpha$, so the α must be normalized to have length $1/\sqrt{\bar{\lambda}}$.

The recipe for extracting the i'th principal component in $\mathscr{F}$ using kernel PCA is therefore:

1. Compute the i'th principal eigenvector of K^C, with eigenvalue $\bar{\lambda}$.
2. Normalize the corresponding eigenvector, α, to have length $1/\sqrt{\bar{\lambda}}$.
3. For a training point $\mathbf{x}_k$, the principal component is then just

$$(\Phi(\mathbf{x}_k) - \mu) \cdot \mathbf{v} = \bar{\lambda}\alpha_k$$

4. For a general test point $\mathbf{x}$, the principal component is

$$(\Phi(\mathbf{x}) - \mu) \cdot \mathbf{v} = \sum_i \alpha_i k(\mathbf{x}, \mathbf{x}_i) - \frac{1}{m}\sum_{i,j} \alpha_i k(\mathbf{x}, \mathbf{x}_j)$$
$$- \frac{1}{m}\sum_{i,j} \alpha_i k(\mathbf{x}_i, \mathbf{x}_j) + \frac{1}{m^2}\sum_{i,j,n} \alpha_i k(\mathbf{x}_j, \mathbf{x}_n)$$

where the last two terms can be dropped since they don't depend on $\mathbf{x}$.

Kernel PCA may be viewed as a way of putting more effort into the up-front computation of features, rather than putting the onus on the classifier or regression algorithm. Kernel PCA followed by a linear SVM on a pattern recognition problem has been shown to give similar results to using a nonlinear SVM using the same kernel (Schölkopf et al., 1998). It shares with other Mercer kernel methods the attractive property of mathematical tractability and of having a clear geometrical interpretation: for example, this has led to using kernel PCA for de-noising data, by finding that vector $\mathbf{z} \in \mathscr{R}^d$ such that the Euclidean distance between $\Phi(\mathbf{z})$ and the vector computed from the first few PCA components in $\mathscr{F}$ is minimized (Mika et al., 1999). Classical PCA has the significant limitation that it depends only on first and second moments of the data, whereas kernel PCA does not (for example, a polynomial kernel $k(\mathbf{x}_i, \mathbf{x}_j) = (\mathbf{x}_i \cdot \mathbf{x}_j + b)^p$ contains powers up to order $2p$, which is particularly useful for e.g. image classification, where one expects that products of several pixel values will be informative as to the class). Kernel PCA has the computational limitation of having to compute eigenvectors for square matrices of side m, but again this can be addressed, for example by using a subset of the training data, or by using the Nyström method for approximating the eigenvectors of a large Gram matrix (see below).

4.1.4 Oriented PCA and Distortion Discriminant Analysis

Before leaving projective methods, we describe another extension of PCA, which has proven very effective at extracting robust features from audio (Burges et al., 2002, Burges et al., 2003). We first describe the method of oriented PCA (OPCA) (Diamantaras and Kung, 1996). Suppose we are given a set of 'signal' vectors $\mathbf{x}_i \in \mathscr{R}^d$, $i = 1,\ldots,m$, where each $\mathbf{x}_i$ represents an undistorted data point, and suppose that for each $\mathbf{x}_i$, we have a set of N distorted versions $\tilde{\mathbf{x}}_i^k$, $k = 1,\ldots,N$. Define the corresponding 'noise' difference vectors to be $\mathbf{z}_i^k \equiv \tilde{\mathbf{x}}_i^k - \mathbf{x}_i$. Roughly speaking, we wish to find linear projections which are as orthogonal as possible to the difference vectors, but along which the variance of the signal data is simultaneously maximized. Denote the unit vectors defining the desired projections by $\mathbf{n}_i$, $i = 1,\ldots,d'$, $\mathbf{n}_i \in \mathscr{R}^d$, where d' will be chosen by the user. By analogy with PCA, we could construct a feature extractor $\mathbf{n}$ which minimizes the mean squared reconstruction error $\frac{1}{mN}\sum_{i,k}(\mathbf{x}_i - \hat{\mathbf{x}}_i^k)^2$, where $\hat{\mathbf{x}}_i^k \equiv (\tilde{\mathbf{x}}_i^k \cdot \mathbf{n})\mathbf{n}$. The $\mathbf{n}$ that solves this problem is that eigenvector of $R_1 - R_2$ with largest eigenvalue, where R_1, R_2 are the correlation matrices of the $\mathbf{x}_i$ and $\mathbf{z}_i$ respectively. However this feature extractor has the undesirable property that the direction $\mathbf{n}$ will change if the noise and signal vectors are globally scaled with two different scale factors. OPCA (Diamantaras and Kung, 1996) solves this problem. The first OPCA direction is defined as that direction $\mathbf{n}$ that maximizes the generalized Rayleigh quotient (Duda and Hart, 1973, Diamantaras and Kung, 1996) $q_0 = \frac{\mathbf{n}'C_1\mathbf{n}}{\mathbf{n}'C_2\mathbf{n}}$, where C_1 is the covariance matrix of the signal and C_2 that of the noise. For d' directions collected into a column matrix $\mathscr{N} \in M_{dd'}$, we instead maximize $\frac{\det(\mathscr{N}'C_1\mathscr{N})}{\det(\mathscr{N}'C_2\mathscr{N})}$. For Gaussian data, this amounts to maximizing the ratio of the volume of the ellipsoid containing the data, to the volume of the ellipsoid containing the noise, where the volume is that lying inside an ellipsoidal surface of constant probability density. We in fact use the correlation matrix of the noise rather than the covariance matrix, since we wish to penalize the mean noise signal as well as its variance (consider the extreme case of noise that has zero variance but nonzero mean). Explicitly, we take

$$C \equiv \frac{1}{m}\sum_i (\mathbf{x}_i - E[\mathbf{x}])(\mathbf{x}_i - E[\mathbf{x}])' \tag{4.16}$$

$$R \equiv \frac{1}{mN}\sum_{i,k} \mathbf{z}_i^k(\mathbf{z}_i^k)' \tag{4.17}$$

and maximize $q = \frac{\mathbf{n}'C\mathbf{n}}{\mathbf{n}'R\mathbf{n}}$, whose numerator is the variance of the projection of the signal data along the unit vector $\mathbf{n}$, and whose denominator is the projected mean squared "error" (the mean squared modulus of all noise vectors $\mathbf{z}_i^k$ projected along $\mathbf{n}$). We can find the directions $\mathbf{n}_j$ by setting $\nabla q = 0$, which gives the generalized eigenvalue problem $C\mathbf{n} = qR\mathbf{n}$; those solutions are also the solutions to the problem of maximizing $\frac{\det(\mathscr{N}'C\mathscr{N})}{\det(\mathscr{N}'R\mathscr{N})}$. If R is not of full rank, it must be regularized for the problem to be well-posed. It is straightforward to show that, for positive semidefinite C, R, the generalized eigenvalues are positive, and that scaling either the signal or the noise leaves the OPCA directions unchanged, although the eigenvalues will change. Furthermore the $\mathbf{n}_i$ are, or may be chosen to be, linearly independent, and although the $\mathbf{n}_i$ are not necessarily orthogonal, they are conjugate with respect to both matrices C and R, that is, $\mathbf{n}_i'C\mathbf{n}_j \propto \delta_{ij}$, $\mathbf{n}_i'R\mathbf{n}_j \propto \delta_{ij}$. Finally, OPCA is similar to linear discriminant analysis (Duda and Hart, 1973), but where each signal point $\mathbf{x}_i$ is assigned its own class.

'Distortion discriminant analysis' (Burges et al., 2002, Burges et al., 2003) uses layers of OPCA projectors both to reduce dimensionality (a high priority for audio or video data) and to make the features more robust. The above features, computed by taking projections along the $\mathbf{n}$'s, are first translated and normalized so that the signal data has zero mean and the noise data has unit variance. For the audio application, for example, the OPCA features are collected over several audio frames into new 'signal' vectors, the corresponding 'noise' vectors are measured, and the OPCA directions for the next layer found. This has the further advantage of allowing different types of distortion to be penalized at different layers, since each layer corresponds to a different time scale in the original data (for example, a distortion that results from comparing audio whose frames are shifted in time to features extracted from the original data - 'alignment noise' - can be penalized at larger time scales).

4.2 Manifold Modeling

In Section 4.1 we gave an example of data with a particular geometric structure which would not be immediately revealed by examining one dimensional projections in input space[10]. How, then, can such underlying structure be found? This section outlines some methods designed to accomplish this. However we first describe the Nyström method (hereafter simply abbreviated 'Nyström'), which provides a thread linking several of the algorithms described in this review.

4.2.1 The Nyström method

Suppose that $K \in M_n$ and that the rank of K is $r \ll n$. Nyström gives a way of approximating the eigenvectors and eigenvalues of K using those of a small submatrix A. If A has rank r, then the decomposition is exact. This is a powerful method that can be used to speed up kernel algorithms (Williams and Seeger, 2001), to efficiently extend some algorithms (described below) to out-of-sample test points (Bengio et al., 2004), and in some cases, to make an otherwise infeasible algorithm feasible (Fowlkes et al., 2004). In this section only, we adopt the notation that matrix indices refer to sizes unless otherwise stated, so that e.g. A_{mm} means that $A \in M_m$.

[10] Although in that simple example, the astute investigator would notice that all her data vectors have the same length, and conclude from the fact that the projected density is independent of projection direction that the data must be uniformly distributed on the sphere.

Original Nyström

The Nyström method originated as a method for approximating the solution of Fredholm integral equations of the second kind (Press et al., 1992). Let's consider the homogeneous d-dimensional form with density $p(\mathbf{x})$, $\mathbf{x} \in \mathscr{R}^d$. This family of equations has the form

$$\int k(\mathbf{x},\mathbf{y})u(\mathbf{y})p(\mathbf{y})d\mathbf{y} = \lambda u(\mathbf{x}) \tag{4.18}$$

The integral is approximated using the quadrature rule (Press et al., 1992)

$$\lambda u(\mathbf{x}) \approx \frac{1}{m}\sum_{i=1}^{m} k(\mathbf{x},\mathbf{x}_i)u(\mathbf{x}_i) \tag{4.19}$$

which when applied to the sample points becomes a matrix equation K_{mm} $\mathbf{u}_m = m\lambda\mathbf{u}_m$ (with components $K_{ij} \equiv k(\mathbf{x}_i,\mathbf{x}_j)$ and $u_i \equiv u(\mathbf{x}_i)$). This eigensystem is solved, and the value of the integral at a new point $\mathbf{x}$ is approximated by using (4.19), which gives a much better approximation that using simple interpolation (Press et al., 1992). Thus, the original Nyström method provides a way to smoothly approximate an eigenfunction u, given its values on a sample set of points. If a different number m' of elements in the sum are used to approximate the same eigenfunction, the matrix equation becomes $K_{m'm'}\mathbf{u}_{m'} = m'\lambda\mathbf{u}_{m'}$ so the corresponding eigenvalues approximately scale with the number of points chosen. Note that we have not assumed that K is symmetric or positive semidefinite; however from now on we will assume that K is positive semidefinite.

Exact Nyström Eigendecomposition

Suppose that $\tilde{K}_{mm}$ has rank $r < m$. Since it's positive semidefinite it is a Gram matrix and can be written as $\tilde{K} = ZZ'$ where $Z \in M_{mr}$ and Z is also of rank r (Horn and Johnson, 1985). Order the row vectors in Z so that the first r are linearly independent: this just reorders rows and columns in $\tilde{K}$ to give K, but in such a way that K is still a (symmetric) Gram matrix. Then the principal submatrix $A \in S_r$ of K (which itself is the Gram matrix of the first r rows of Z) has full rank. Now letting $n \equiv m - r$, write the matrix K as

$$K_{mm} \equiv \begin{bmatrix} A_{rr} & B_{rn} \\ B'_{nr} & C_{nn} \end{bmatrix} \tag{4.20}$$

Since A is of full rank, the r rows $\begin{bmatrix} A_{rr} & B_{rn} \end{bmatrix}$ are linearly independent, and since K is of rank r, the n rows $\begin{bmatrix} B'_{nr} & C_{nn} \end{bmatrix}$ can be expanded in terms of them, that is, there exists H_{nr} such that

$$\begin{bmatrix} B'_{nr} & C_{nn} \end{bmatrix} = H_{nr}\begin{bmatrix} A_{rr} & B_{rn} \end{bmatrix} \tag{4.21}$$

The first r columns give $H = B'A^{-1}$, and the last n columns then give $C = B'A^{-1}B$. Thus K must be of the form[11]

[11] It's interesting that this can be used to perform 'kernel completion', that is, reconstruction of a kernel with missing values; for example, suppose K has rank 2 and that its first two

$$K_{mm} = \begin{bmatrix} A & B \\ B' & B'A^{-1}B \end{bmatrix} = \begin{bmatrix} A \\ B' \end{bmatrix}_{mr} A_{rr}^{-1} \begin{bmatrix} A & B \end{bmatrix}_{rm} \tag{4.22}$$

The fact that we've been able to write K in this 'bottleneck' form suggests that it may be possible to construct the *exact* eigendecomposition of K_{mm} (for its nonvanishing eigenvalues) using the eigendecomposition of a (possibly much smaller) matrix in M_r, and this is indeed the case (Fowlkes et al., 2004). First use the eigendecomposition of A, $A = U\Lambda U'$, where U is the matrix of column eigenvectors of A and Λ the corresponding diagonal matrix of eigenvalues, to rewrite this in the form

$$K_{mm} = \begin{bmatrix} U \\ B'U\Lambda^{-1} \end{bmatrix}_{mr} \Lambda_{rr} \begin{bmatrix} U \; \Lambda^{-1}U'B \end{bmatrix}_{rm} \equiv D\Lambda D' \tag{4.23}$$

This would be exactly what we want (dropping all eigenvectors whose eigenvalues vanish), if the columns of D were orthogonal, but in general they are not. It is straightforward to show that, if instead of diagonalizing A we diagonalize $Q_{rr} \equiv A + A^{-1/2}BB'A^{-1/2} \equiv U_Q\Lambda_Q U_Q'$, then the desired matrix of orthogonal column eigenvectors is

$$V_{mr} \equiv \begin{bmatrix} A \\ B' \end{bmatrix} A^{-1/2} U_Q \Lambda_Q^{-1/2} \tag{4.24}$$

(so that $K_{mm} = V\Lambda_Q V'$ and $V'V = \mathbf{1}_{rr}$) (Fowlkes et al., 2004).

Although this decomposition is exact, this last step comes at a price: to obtain the correct eigenvectors, we had to perform an eigendecomposition of the matrix Q which depends on B. If our intent is to use this decomposition in an algorithm in which B changes when new data is encountered (for example, an algorithm which requires the eigendecomposition of a kernel matrix constructed from both train and test data), then we must recompute the decomposition each time new test data is presented. If instead we'd like to compute the eigendecomposition just once, we must approximate.

Approximate Nyström Eigendecomposition

Two kinds of approximation naturally arise. The first occurs if K is only approximately low rank, that is, its spectrum decays rapidly, but not to exactly zero. In this case, $B'A^{-1}B$ will only approximately equal C above, and the approximation can be quantified as $\left\|C - B'A^{-1}B\right\|$ for some matrix norm $\|\cdot\|$, where the difference is known as the Schur complement of A for the matrix K (Golub and Van Loan, 1996).

The second kind of approximation addresses the need to compute the eigendecomposition just once, to speed up test phase. The idea is simply to take Equation (4.19), sum over d elements on the right hand side where $d \ll m$ and $d > r$, and approximate the eigenvector of the full kernel matrix K_{mm} by evaluating the left hand

rows (and hence columns) are linearly independent, and suppose that K has met with an unfortunate accident that has resulted in all of its elements, except those in the first two rows or columns, being set equal to zero. Then the original K is easily regrown using $C = B'A^{-1}B$.

side at all m points (Williams and Seeger, 2001). Empirically it has been observed that choosing d to be some small integer factor larger than r works well (Platt). How does using (4.19) correspond to the expansion in (4.23), in the case where the Schur complement vanishes? Expanding A, B in their definition in Eq. (4.20) to A_{dd}, B_{dn}, so that U_{dd} contains the column eigenvectors of A and U_{md} contains the approximated (high dimensional) column eigenvectors, (4.19) becomes

$$U_{md}\Lambda_{dd} \approx K_{md}U_{dd} = \begin{bmatrix} A \\ B' \end{bmatrix} U_{dd} = \begin{bmatrix} U\Lambda_{dd} \\ B'U_{dd} \end{bmatrix} \quad (4.25)$$

so multiplying by Λ_{dd}^{-1} from the right shows that the approximation amounts to taking the matrix D in (4.23) as the approximate column eigenvectors: in this sense, the approximation amounts to dropping the requirement that the eigenvectors be exactly orthogonal.

We end with the following observation (Williams and Seeger, 2001): the expression for computing the projections of a mapped test point along principal components in a kernel feature space is, apart from proportionality constants, exactly the expression for the approximate eigenfunctions evaluated at the new point, computed according to (4.19). Thus the computation of the kernel PCA features for a set of points can be viewed as using the Nyström method to approximate the full eigenfunctions at those points.

4.2.2 Multidimensional Scaling

We begin our look at manifold modeling algorithms with multidimensional scaling (MDS), which arose in the behavioral sciences (Borg and Groenen, 1997). MDS starts with a measure of dissimilarity between each pair of data points in the dataset (note that this measure can be very general, and in particular can allow for non-vectorial data). Given this, MDS searches for a mapping of the (possibly further transformed) dissimilarities to a low dimensional Euclidean space such that the (transformed) pair-wise dissimilarities become squared distances. The low dimensional data can then be used for visualization, or as low dimensional features.

We start with the fundamental theorem upon which 'classical MDS' is built (in classical MDS, the dissimilarities are taken to be squared distances and no further transformation is applied (Cox and Cox, 2001)). We give a detailed proof because it will serve to illustrate a recurring theme. Let $\mathbf{e}$ be the column vector of m ones. Consider the 'centering' matrix $P^e \equiv \mathbf{1} - \frac{1}{m}\mathbf{e}\mathbf{e}'$. Let X be the matrix whose rows are the datapoints $\mathbf{x} \in \mathscr{R}^n$, $X \in M_{mn}$. Since $\mathbf{e}\mathbf{e}' \in M_m$ is the matrix of all ones, $P^e X$ subtracts the mean vector from each row $\mathbf{x}$ in X (hence the name 'centering'), and in addition, $P^e\mathbf{e} = 0$. In fact $\mathbf{e}$ is the only eigenvector (up to scaling) with eigenvalue zero, for suppose $P^e\mathbf{f} = 0$ for some $\mathbf{f} \in \mathscr{R}^m$. Then each component of $\mathbf{f}$ must be equal to the mean of all the components of $\mathbf{f}$, so all components of $\mathbf{f}$ are equal. Hence P^e has rank $m-1$, and P^e projects onto the subspace $\mathscr{R}^{m-1}$ orthogonal to $\mathbf{e}$.

By a 'distance matrix' we will mean a matrix whose ij'th element is $\|\mathbf{x}_i - \mathbf{x}_j\|^2$ for some $\mathbf{x}_i$, $\mathbf{x}_j \in \mathscr{R}^d$, for some d, where $\|\cdot\|$ is the Euclidean norm.

Notice that the elements are squared distances, despite the name. P^e can also be used to center both Gram matrices and distance matrices. We can see this as follows. Let $[C(i,j)]$ be that matrix whose ij'th element is $C(i,j)$. Then $P^e[\mathbf{x}_i \cdot \mathbf{x}_j]P^e = P^e XX' P^e = (P^e X)(P^e X)' = [(\mathbf{x}_i - \mu)\cdot(\mathbf{x}_j - \mu)]$. In addition, using this result, $P^e[\|\mathbf{x}_i - \mathbf{x}_j\|^2]P^e = P^e[\|\mathbf{x}_i\|^2 e_i e_j + \|\mathbf{x}_j\|^2 e_i e_j - 2\mathbf{x}_i \cdot \mathbf{x}_j]P^e = -2P^e \mathbf{x}_i \cdot \mathbf{x}_j P^e = -2[(\mathbf{x}_i - \mu)\cdot(\mathbf{x}_j - \mu)]$.

For the following theorem, the earliest form of which is due to Schoenberg (Schoenberg, 1935), we first note that, for any $A \in M_m$, and letting $Q \equiv \frac{1}{m}\mathbf{e}\mathbf{e}'$,

$$P^e A P^e = \{(\mathbf{1} - Q)A(\mathbf{1} - Q)\}_{ij} = A_{ij} - A^R_{ij} - A^C_{ij} + A^{RC}_{ij} \tag{4.26}$$

where $A^C \equiv AQ$ is the matrix A with each column replaced by the column mean, $A^R \equiv QA$ is A with each row replaced by the row mean, and $A^{RC} \equiv QAQ$ is A with every element replaced by the mean of all the elements.

Theorem: Consider the class of symmetric matrices $A \in S_n$ such that $A_{ij} \geq 0$ and $A_{ii} = 0\ \forall i,j$. Then $\bar{A} \equiv -P^e A P^e$ is positive semidefinite if and only if A is a distance matrix (with embedding space $\mathscr{R}^d$ for some d). Given that A is a distance matrix, the minimal embedding dimension d is the rank of $\bar{A}$, and the embedding vectors are any set of Gram vectors of $\bar{A}$, scaled by a factor of $\frac{1}{\sqrt{2}}$.

Proof: Assume that $A \in S_m$, $A_{ij} \geq 0$ and $A_{ii} = 0\ \forall i$, and that $\bar{A}$ is positive semidefinite. Since $\bar{A}$ is positive semidefinite it is also a Gram matrix, that is, there exist vectors $\mathbf{x}_i \in \mathscr{R}^m$, $i = 1, \cdots, m$ such that $\bar{A}_{ij} = \mathbf{x}_i \cdot \mathbf{x}_j$. Introduce $\mathbf{y}_i = \frac{1}{\sqrt{2}}\mathbf{x}_i$. Then from Eq. (4.26),

$$\bar{A}_{ij} = (-P^e A P^e)_{ij} = \mathbf{x}_i \cdot \mathbf{x}_j = -A_{ij} + A^R_{ij} + A^C_{ij} - A^{RC}_{ij} \tag{4.27}$$

so that

$$\begin{aligned} 2(\mathbf{y}_i - \mathbf{y}_j)^2 \equiv (\mathbf{x}_i - \mathbf{x}_j)^2 &= A^R_{ii} + A^C_{ii} - A^{RC}_{ii} + A^R_{jj} + A^C_{jj} - A^{RC}_{jj} \\ &\quad -2(-A_{ij} + A^R_{ij} + A^C_{ij} - A^{RC}_{ij}) \\ &= 2A_{ij} \end{aligned} \tag{4.28}$$

using $A_{ii} = 0$, $A^R_{ij} = A^R_{jj}$, $A^C_{ij} = A^C_{ii}$, and from the symmetry of A, $A^R_{ij} = A^C_{ji}$. Thus A is a distance matrix with embedding vectors $\mathbf{y}_i$. Now consider a matrix $A \in S_n$ that is a distance matrix, so that $A_{ij} = (\mathbf{y}_i - \mathbf{y}_j)^2$ for some $\mathbf{y}_i \in \mathscr{R}^d$ for some d, and let Y be the matrix whose rows are the $\mathbf{y}_i$. Then since each row and column of P^e sums to zero, we have $\bar{A} = -(P^e A P^e) = 2(P^e Y)(P^e Y)'$, hence $\bar{A}$ is positive semidefinite. Finally, given a distance matrix $A_{ij} = (\mathbf{y}_i - \mathbf{y}_j)^2$, we wish to find the dimension of the minimal embedding Euclidean space. First note that we can assume that the $\mathbf{y}_i$ have zero mean ($\sum_i \mathbf{y}_i = 0$), since otherwise we can subtract the mean from each $\mathbf{y}_i$ without changing A. Then $\bar{A}_{ij} = \mathbf{x}_i \cdot \mathbf{x}_j$, again introducing $\mathbf{x}_i \equiv \sqrt{2}\mathbf{y}_i$, so the embedding vectors $\mathbf{y}_i$ are a set of Gram vectors of $\bar{A}$, scaled by a factor of $\frac{1}{\sqrt{2}}$. Now let r be the rank of $\bar{A}$. Since $\bar{A} = XX'$, and since $rank(XX') = rank(X)$ for any real matrix X (Horn and Johnson, 1985), and since $rank(X)$ is the number of linearly independent $\mathbf{x}_i$, the minimal embedding space for the $\mathbf{x}_i$ (and hence for the $\mathbf{y}_i$) has dimension r. □

General Centering

Is P^e the most general matrix that will convert a distance matrix into a matrix of dot products? Since the embedding vectors are not unique (given a set of Gram vectors, any global orthogonal matrix applied to that set gives another set that generates the same positive semidefinite matrix), it's perhaps not surprising that the answer is no. A distance matrix is an example of a conditionally negative definite (CND) matrix. A CND matrix $D \in S_m$ is a symmetric matrix that satisfies $\sum_{i,j} a_i a_j D_{ij} \leq 0 \, \forall \{a_i \in \mathscr{R} : \sum_i a_i = 0\}$; the class of CND matrices is a superset of the class of negative semidefinite matrices (Berg et al., 1984). Defining the projection matrix $P^c \equiv (\mathbf{1} - \mathbf{e}\mathbf{c}')$, for any $\mathbf{c} \in \mathscr{R}^m$ such that $\mathbf{e}'\mathbf{c} = 1$, then for any CND matrix D, the matrix $-P^c D P'^c$ is positive semidefinite (and hence a dot product matrix) (Schölkopf, 2001, Berg et al., 1984) (note that P^c is not necessarily symmetric). This is straightforward to prove: for any $\mathbf{z} \in \mathscr{R}^m$, $P'^c \mathbf{z} = (\mathbf{1} - \mathbf{c}\mathbf{e}')\mathbf{z} = \mathbf{z} - \mathbf{c}(\sum_a z_a)$, so $\sum_i (P'^c \mathbf{z})_i = 0$, hence $(P'^c \mathbf{z})' D (P'^c \mathbf{z}) \leq 0$ from the definition of CND. Hence we can map a distance matrix D to a dot product matrix K by using P^c in the above manner for any set of numbers c_i that sum to unity.

Constructing the Embedding

To actually find the embedding vectors for a given distance matrix, we need to know how to find a set of Gram vectors for a positive semidefinite matrix $\bar{A}$. Let E be the matrix of column eigenvectors $\mathbf{e}^{(\alpha)}$ (labeled by α), ordered by eigenvalue λ_α, so that the first column is the principal eigenvector, and $\bar{A}E = E\Lambda$, where Λ is the diagonal matrix of eigenvalues. Then $\bar{A}_{ij} = \sum_\alpha \lambda_\alpha e_i^{(\alpha)} e_j^{(\alpha)}$. The rows of E form the dual (orthonormal) basis to $e_i^{(\alpha)}$, which we denote $\tilde{e}_\alpha^{(i)}$. Then we can write $\bar{A}_{ij} = \sum_\alpha (\sqrt{\lambda_\alpha} \tilde{e}_\alpha^{(i)})(\sqrt{\lambda_\alpha} \tilde{e}_\alpha^{(i)})$. Hence the Gram vectors are just the dual eigenvectors with each component scaled by $\sqrt{\lambda_\alpha}$. Defining the matrix $\tilde{E} \equiv E\Lambda^{1/2}$, we see that the Gram vectors are just the rows of $\tilde{E}$.

If $\bar{A} \in S_n$ has rank $r \leq n$, then the final $n - r$ columns of $\tilde{E}$ will be zero, and we have directly found the r-dimensional embedding vectors that we are looking for. If $\bar{A} \in S_n$ is full rank, but the last $n - p$ eigenvalues are much smaller than the first p, then it's reasonable to approximate the i'th Gram vector by its first p components $\sqrt{\lambda_\alpha} \tilde{\mathbf{e}}_\alpha^{(i)}$, $\alpha = 1, \cdots, p$, and we have found a low dimensional approximation to the $\mathbf{y}$'s. This device - projecting to lower dimensions by lopping off the last few components of the dual vectors corresponding to the (possibly scaled) eigenvectors - is shared by MDS, Laplacian eigenmaps, and spectral clustering (see below). Just as for PCA, where the quality of the approximation can be characterized by the unexplained variance, we can characterize the quality of the approximation here by the squared residuals. Let $\bar{A}$ have rank r, and suppose we only keep the first $p \leq r$ components to form the approximate embedding vectors. Then denoting the approximation with a hat, the summed squared residuals are

$$\sum_{i=1}^{m} \|\hat{\mathbf{y}}_i - \mathbf{y}_i\|^2 = \frac{1}{2}\sum_{i=1}^{m} \|\hat{\mathbf{x}}_i - \mathbf{x}_i\|^2$$
$$= \frac{1}{2}\sum_{i=1}^{m}\sum_{a=1}^{p} \lambda_a \tilde{e}_a^{(i)2} + \frac{1}{2}\sum_{i=1}^{m}\sum_{a=1}^{r} \lambda_a \tilde{e}_a^{(i)2} - \sum_{i=1}^{m}\sum_{a=1}^{p} \lambda_a \tilde{e}_a^{(i)2}$$

but $\sum_{i=1}^{m} \tilde{e}_a^{(i)2} = \sum_{i=1}^{m} e_i^{(a)2} = 1$, so

$$\sum_{i=1}^{m} \|\hat{\mathbf{y}}_i - \mathbf{y}_i\|^2 = \frac{1}{2}\left(\sum_{a=1}^{r} \lambda_a - \sum_{a=1}^{p} \lambda_a\right) = \sum_{a=p+1}^{r} \lambda_a \tag{4.29}$$

Thus the fraction of 'unexplained residuals' is $\sum_{a=p+1}^{r} \lambda_a / \sum_{a=1}^{r} \lambda_a$, in analogy to the fraction of 'unexplained variance' in PCA.

If the original symmetric matrix A is such that $\bar{A}$ is not positive semidefinite, then by the above theorem there exist no embedding points such that the dissimilarities are distances between points in some Euclidean space. In that case, we can proceed by adding a sufficiently large positive constant to the diagonal of $\bar{A}$, or by using the closest positive semidefinite matrix, in Frobenius norm[12], to $\bar{A}$, which is $\hat{A} \equiv \sum_{\alpha:\lambda_\alpha>0} \lambda_\alpha \mathbf{e}^{(\alpha)} \mathbf{e}^{(\alpha)\prime}$. Methods such as classical MDS, that treat the dissimilarities themselves as (approximate) squared distances, are called metric scaling methods. A more general approach - 'non-metric scaling' - is to minimize a suitable cost function of the difference between the embedded squared distances, and some monotonic function of the dissimilarities (Cox and Cox, 2001); this allows for dissimilarities which do not arise from a metric space; the monotonic function, and other weights which are solved for, are used to allow the dissimilarities to nevertheless be represented approximately by low dimensional squared distances. An example of non-metric scaling is ordinal MDS, whose goal is to find points in the low dimensional space so that the distances there correctly reflect a given rank ordering of the original data points.

Landmark MDS

MDS is computationally expensive: since the distances matrix is not sparse, the computational complexity of the eigendecomposition is $O(m^3)$. This can be significantly reduced by using a method called Landmark MDS (LMDS) (Silva and Tenenbaum, 2002). In LMDS the idea is to choose q points, called 'landmarks', where $q > r$ (where r is the rank of the distance matrix), but $q \ll m$, and to perform MDS on landmarks, mapping them to $\mathscr{R}^d$. The remaining points are then mapped to $\mathscr{R}^d$ using only their distances to the landmark points (so in LMDS, the only distances considered are those to the set of landmark points). As first pointed out in (Bengio et al., 2004) and explained in more detail in (Platt, 2005), LMDS combines MDS with the Nyström algorithm. Let $E \in S_q$ be the matrix of landmark distances and U (Λ) the matrix of eigenvectors (eigenvalues) of the corresponding kernel matrix

[12] The only proof I have seen for this assertion is due to Frank McSherry, Microsoft Research.

$A \equiv -\frac{1}{2}P^c E P'^c$, so that the embedding vectors of the landmark points are the first d elements of the rows of $U\Lambda^{1/2}$. Now, extending E by an extra column and row to accommodate the squared distances from the landmark points to a test point, we write the extended distance matrix and corresponding kernel as

$$D = \begin{bmatrix} E & \mathbf{f} \\ \mathbf{f}' & g \end{bmatrix}, \quad K \equiv -\frac{1}{2}P^c D P'^c = \begin{bmatrix} A & \mathbf{b} \\ \mathbf{b}' & c \end{bmatrix} \tag{4.30}$$

Then from Eq. (4.23) we see that the Nyström method gives the approximate column eigenvectors for the extended system as

$$\begin{bmatrix} U \\ \mathbf{b}'U\Lambda^{-1} \end{bmatrix} \tag{4.31}$$

Thus the embedding coordinates of the test point are given by the first d elements of the row vector $\mathbf{b}'U\Lambda^{-1/2}$. However, we only want to compute U and Λ once - they must not depend on the test point. (Platt, 2005) has pointed out that this can be accomplished by choosing the centering coefficients c_i in $P^c \equiv \mathbf{1} - \mathbf{e}\mathbf{c}'$ such that $c_i = 1/q$ for $i \le q$ and $c_{q+1} = 0$: in that case, since

$$K_{ij} = -\frac{1}{2}\left(D_{ij} - e_i(\sum_{k=1}^{q+1} c_k D_{kj}) - e_j(\sum_{k=1}^{q+1} D_{ik}c_k) + e_i e_j(\sum_{k,m=1}^{q+1} c_k D_{km} c_m)\right)$$

the matrix A (found by limiting i, j to $1, \ldots, q$ above) depends only on the matrix E above. Finally, we need to relate $\mathbf{b}$ back to the measured quantities - the vector of squared distances from the test point to the landmark points. Using $b_i = (-\frac{1}{2}P^c D P'^c)_{q+1,i}$, $i = 1, \cdots, q$, we find that

$$b_k = -\frac{1}{2}\left[D_{q+1,k} - \frac{1}{q}\sum_{j=1}^{q} D_{q+1,j}e_k - \frac{1}{q}\sum_{i=1}^{q} D_{ik} + \frac{1}{q^2}\left(\sum_{i,j=1}^{q} D_{ij}\right)e_k\right]$$

The first term in the square brackets is the vector of squared distances from the test point to the landmarks, $\mathbf{f}$. The third term is the row mean of the landmark distance squared matrix, $\bar{E}$. The second and fourth terms are proportional to the vector of all ones $\mathbf{e}$, and can be dropped[13] since $U'\mathbf{e} = 0$. Hence, modulo terms which vanish when constructing the embedding coordinates, we have $\mathbf{b} \simeq -\frac{1}{2}(\mathbf{f} - \bar{E})$, and the coordinates of the embedded test point are $\frac{1}{2}\Lambda^{-1/2}U'(\bar{E} - \mathbf{f})$; this reproduces the form given in (Silva and Tenenbaum, 2002). Landmark MDS has two significant advantages: first, it reduces the computational complexity from $O(m^3)$ to $O(q^3 + q^2(m-q) = q^2 m)$; and second, it can be applied to any non-landmark point, and so gives a method of extending MDS (using Nyström) to out-of-sample data.

[13] The last term can also be viewed as an unimportant shift in origin; in the case of a single test point, so can the second term, but we cannot rely on this argument for multiple test points, since the summand in the second term depends on the test point.

4.2.3 Isomap

MDS is valuable for extracting low dimensional representations for some kinds of data, but it does not attempt to explicitly model the underlying manifold. Two methods that do directly model the manifold are Isomap and Locally Linear Embedding. Suppose that as in Section 4.1.1, again unbeknownst to you, your data lies on a curve, but in contrast to Section 4.1.1, the curve is not a straight line; in fact it is sufficiently complex that the minimal embedding space $\mathscr{R}^d$ that can contain it has high dimension d. PCA will fail to discover the one dimensional structure of your data; MDS will also, since it attempts to faithfully preserve all distances. Isomap (isometric feature map) (Tenenbaum, 1998), on the other hand, will succeed. The key assumption made by Isomap is that the quantity of interest, when comparing two points, is the distance along the curve between the two points; if that distance is large, it is to be taken, even if in fact the two points are close in $\mathscr{R}^d$ (this example also shows that noise must be handled carefully). The low dimensional space can have more than one dimension: (Tenenbaum, 1998) gives an example of a 5 dimensional manifold embedded in a 50 dimensional space. The basic idea is to construct a graph whose nodes are the data points, where a pair of nodes are adjacent only if the two points are close in $\mathscr{R}^d$, and then to approximate the geodesic distance along the manifold between any two points as the shortest path in the graph, computed using the Floyd algorithm (Gondran and Minoux, 1984); and finally to use MDS to extract the low dimensional representation (as vectors in $\mathscr{R}^{d'}$, $d' \ll d$) from the resulting matrix of squared distances (Tenenbaum (Tenenbaum, 1998) suggests using ordinal MDS, rather than metric MDS, for robustness).

Isomap shares with the other manifold mapping techniques we describe the property that it does not provide a direct functional form for the mapping $\mathscr{I} : \mathscr{R}^d \to \mathscr{R}^{d'}$ that can simply be applied to new data, so computational complexity of the algorithm is an issue in test phase. The eigenvector computation is $O(m^3)$, and the Floyd algorithm also $O(m^3)$, although the latter can be reduced to $O(hm^2 \log m)$ where h is a heap size (Silva and Tenenbaum, 2002). Landmark Isomap simply employs landmark MDS (Silva and Tenenbaum, 2002) to addresses this problem, computing all distances as geodesic distances to the landmarks. This reduces the computational complexity to $O(q^2 m)$ for the LMDS step, and to $O(hqm \log m)$ for the shortest path step.

4.2.4 Locally Linear Embedding

Locally linear embedding (LLE) (Roweis and Saul, 2000) models the manifold by treating it as a union of linear patches, in analogy to using coordinate charts to parameterize a manifold in differential geometry. Suppose that each point $\mathbf{x}_i \in \mathscr{R}^d$ has a small number of close neighbours indexed by the set $\mathscr{N}(i)$, and let $\mathbf{y}_i \in \mathscr{R}^{d'}$ be the low dimensional representation of $\mathbf{x}_i$. The idea is to express each $\mathbf{x}_i$ as a linear combination of its neighbours, and then construct the $\mathbf{y}_i$ so that they can be expressed as the same linear combination of their corresponding neighbours (the latter also indexed by $\mathcal{N}(i)$). To simplify the discussion let's assume that the number of the neighbours is fixed to n for all i. The condition on the $\mathbf{x}$'s can be expressed as finding that $W \in M_{mn}$ that minimizes the sum of the reconstruction errors, $\sum_i \|\mathbf{x}_i - \sum_{j\in\mathcal{N}(i)} W_{ij}\mathbf{x}_j\|^2$. Each reconstruction error $E_i \equiv \|\mathbf{x}_i - \sum_{j\in\mathcal{N}(i)} W_{ij}\mathbf{x}_j\|^2$ should be unaffected by any global translation $\mathbf{x}_i \to \mathbf{x}_i + \delta,\ \delta \in \mathscr{R}^d$, which gives the condition $\sum_{j\in\mathcal{N}(i)} W_{ij} = 1\ \forall i$. Note that each E_i is also invariant to global rotations and reflections of the coordinates. Thus the objective function we wish to minimize is

$$F \equiv \sum_i F_i \equiv \sum_i \left(\frac{1}{2}\|\mathbf{x}_i - \sum_{j\in\mathcal{N}(i)} W_{ij}\mathbf{x}_j\|^2 - \lambda_i \left(\sum_{j\in\mathcal{N}(i)} W_{ij} - 1 \right) \right)$$

where the constraints are enforced with Lagrange multipliers λ_i. Since the sum splits into independent terms we can minimize each F_i separately (Burges, 2004). Thus fixing i and letting $\mathbf{x} \equiv \mathbf{x}_i$, $\mathbf{v} \in \mathscr{R}^n$, $v_j \equiv W_{ij}$, and $\lambda \equiv \lambda_i$, and introducing the matrix $C \in S_n$, $C_{jk} \equiv \mathbf{x}_j \cdot \mathbf{x}_k$, $j,k \in \mathcal{N}(i)$, and the vector $\mathbf{b} \in \mathscr{R}^n$, $b_j \equiv \mathbf{x} \cdot \mathbf{x}_j$, $j \in \mathcal{N}(i)$, then requiring that the derivative of F_i with respect to v_j vanishes gives $\mathbf{v} = C^{-1}(\lambda\mathbf{e} + \mathbf{b})$. Imposing the constraint $\mathbf{e}'\mathbf{v} = 1$ then gives $\lambda = (1 - \mathbf{e}'C^{-1}\mathbf{b})/(\mathbf{e}'C^{-1}\mathbf{e})$. Thus W can be found by applying this for each i.

Given the W's, the second step is to find a set of $\mathbf{y}_i \in \mathscr{R}^{d'}$ that can be expressed in terms of each other in the same manner. Again no exact solution may exist and so $\sum_i \|\mathbf{y}_i - \sum_{j\in\mathcal{N}(i)} W_{ij}\mathbf{y}_j\|^2$ is minimized with respect to the $\mathbf{y}$'s, keeping the W's fixed. Let $Y \in M_{md'}$ be the matrix of row vectors of the points $\mathbf{y}$. (Roweis and Saul, 2000) enforce the condition that the $\mathbf{y}$'s span a space of dimension d' by requiring that $(1/m)Y'Y = \mathbf{1}$, although any condition of the form $Y'PY = Z$ where $P \in S_m$ and $Z \in S_{d'}$ is of full rank would suffice (see Section 4.2.5). The origin is arbitrary; the corresponding degree of freedom can be removed by requiring that the $\mathbf{y}$'s have zero mean, although in fact this need not be explicitly imposed as a constraint on the optimization, since the set of solutions can easily be chosen to have this property. The rank constraint requires that the $\mathbf{y}$'s have unit covariance; this links the variables so that the optimization no longer decomposes into m separate optimizations: introducing Lagrange multipliers $\lambda_{\alpha\beta}$ to enforce the constraints, the objective function to be minimized is

$$F = \frac{1}{2}\sum_i \|\mathbf{y}_i - \sum_j W_{ij}\mathbf{y}_j\|^2 - \frac{1}{2}\sum_{\alpha\beta} \lambda_{\alpha\beta} \left(\sum_i \frac{1}{m} Y_{i\alpha} Y_{i\beta} - \delta_{\alpha\beta} \right) \tag{4.32}$$

where for convenience we treat the W's as matrices in M_m, where $W_{ij} \equiv 0$ for $j \notin \mathcal{N}(i)$. Taking the derivative with respect to $Y_{k\delta}$ and choosing $\lambda_{\alpha\beta} = \lambda_\alpha \delta_{\alpha\beta} \equiv \Lambda_{\alpha\beta}$ gives[8] the matrix equation

$$(\mathbf{1} - W)'(\mathbf{1} - W)Y = \frac{1}{m} Y\Lambda \tag{4.33}$$

Since $(\mathbf{1} - W)'(\mathbf{1} - W) \in S_m$, its eigenvectors are, or can be chosen to be, orthogonal; and since $(\mathbf{1} - W)'(\mathbf{1} - W)\mathbf{e} = 0$, choosing the columns of Y to be the next d'

eigenvectors of $(1-W)'(1-W)$ with the smallest eigenvalues guarantees that the $\mathbf{y}$ are zero mean (since they are orthogonal to $\mathbf{e}$). We can also scale the $\mathbf{y}$ so that the columns of Y are orthonormal, thus satisfying the covariance constraint $Y'Y=\mathbf{1}$. Finally, the lowest-but-one weight eigenvectors are chosen because their corresponding eigenvalues sum to $m\sum_i \|\mathbf{y}_i - \sum_j W_{ij}\mathbf{y}_j\|^2$, as can be seen by applying Y' to the left of (4.33).

Thus, LLE requires a two-step procedure. The first step (finding the W's) has $O(n^3m)$ computational complexity; the second requires eigendecomposing the product of two sparse matrices in M_m. LLE has the desirable property that it will result in the same weights W if the data is scaled, rotated, translated and / or reflected.

4.2.5 Graphical Methods

In this section we review two interesting methods that connect with spectral graph theory. Let's start by defining a simple mapping from a dataset to an undirected graph G by forming a one-to-one correspondence between nodes in the graph and data points. If two nodes i, j are connected by an arc, associate with it a positive arc weight W_{ij}, $W \in S_m$, where W_{ij} is a similarity measure between points $\mathbf{x}_i$ and $\mathbf{x}_j$. The arcs can be defined, for example, by the minimum spanning tree, or by forming the N nearest neighbours, for N sufficiently large. The Laplacian matrix for any weighted, undirected graph is defined (Chung, 1997) by $\mathscr{L} \equiv D^{-1/2}LD^{-1/2}$, where $L_{ij} \equiv D_{ij} - W_{ij}$ and where $D_{ij} \equiv \delta_{ij}(\sum_k W_{ik})$. We can see that $\mathscr{L}$ is positive semidefinite as follows: for any vector $\mathbf{z} \in \mathscr{R}^m$, since $W_{ij} \geq 0$,

$$0 \leq \frac{1}{2}\sum_{i,j}(z_i - z_j)^2 W_{ij} = \sum_i z_i^2 D_{ii} - \sum_{i,j} z_i W_{ij} z_j = \mathbf{z}'L\mathbf{z}$$

and since L is positive semidefinite, so is the Laplacian. Note that L is never positive definite since the vector of all ones, $\mathbf{e}$, is always an eigenvector with eigenvalue zero (and similarly $\mathscr{L}D^{1/2}\mathbf{e} = 0$).

Let G be a graph and m its number of nodes. For $W_{ij} \in \{0,1\}$, the spectrum of G (defined as the set of eigenvalues of its Laplacian) characterizes its global properties (Chung, 1997): for example, a complete graph (that is, one for which every node is adjacent to every other node) has a single zero eigenvalue, and all other eigenvalues are equal to $\frac{m}{m-1}$; if G is connected but not complete, its smallest nonzero eigenvalue is bounded above by unity; the number of zero eigenvalues is equal to the number of connected components in the graph, and in fact the spectrum of a graph is the union of the spectra of its connected components; and the sum of the eigenvalues is bounded above by m, with equality iff G has no isolated nodes. In light of these results, it seems reasonable to expect that global properties of the data - how it clusters, or what dimension manifold it lies on - might be captured by properties of the Laplacian. The following two approaches leverage this idea. We note that using similarities in this manner results in local algorithms: since each node is only adjacent to a small set of similar nodes, the resulting matrices are sparse and can therefore be eigendecomposed efficiently.

Laplacian Eigenmaps

The Laplacian eigenmaps algorithm (Belkin and Niyogi, 2003) uses $W_{ij} = \exp^{-\|\mathbf{x}_i - \mathbf{x}_j\|^2/2\sigma^2}$. Let $\mathbf{y}(\mathbf{x}) \in \mathscr{R}^{d'}$ be the embedding of sample vector $\mathbf{x} \in \mathscr{R}^d$, and let $Y_{ij} \in M_{md'} \equiv (\mathbf{y}_i)_j$. We would like to find $\mathbf{y}$'s that minimize $\sum_{i,j} \left\|\mathbf{y}_i - \mathbf{y}_j\right\|^2 W_{ij}$, since then if two points are similar, their $\mathbf{y}$'s will be close, whereas if $W \approx 0$, no restriction is put on their $\mathbf{y}$'s. We have:

$$\sum_{i,j} \left\|\mathbf{y}_i - \mathbf{y}_j\right\|^2 W_{ij} = 2\sum_{i,j,a} (\mathbf{y}_i)_a (\mathbf{y}_j)_a (D_{ii}\delta_{ij} - W_{ij}) = 2\mathrm{Tr}(Y'LY) \tag{4.34}$$

In order to ensure that the target space has dimension d' (minimizing (4.34) alone has solution $Y = 0$), we require that Y have rank d. Any constraint of the form $Y'PY = Z$, where $P \in S_m$ and $m \geq d'$, will suffice, provided that $Z \in S_{d'}$ is of full rank. This can be seen as follows: since the rank of Z is d' and since the rank of a product of matrices is bounded above by the rank of each, we have that $d' = rank(Z) = rank(Y'PY) \leq \min(rank((Y'), rank(P),$
$rank(Y))$, and so $rank(Y) \geq d'$; but since $Y \in M_{md'}$ and $d' \leq m$, the rank of Y is at most d'; hence $rank(Y) = d'$. However, minimizing $\mathrm{Tr}(Y'LY)$ subject to the constraint $Y'DY = \mathbf{1}$ results in the simple generalized eigenvalue problem $L\mathbf{y} = \lambda D\mathbf{y}$ (Belkin and Niyogi, 2003). It's useful to see how this arises: we wish to minimize $\mathrm{Tr}(Y'LY)$ subject to the $d'(d'+1)/2$ constraints $Y'DY = \mathbf{1}$. Let $a, b = 1, \ldots, d$ and $i, j = 1, \ldots, m$. Introducing (symmetric) Lagrange multipliers λ_{ab} leads to the objective function $\sum_{i,j,a} y_{ia} L_{ij} y_{ja} - \sum_{i,j,a,b} \lambda_{ab}(y_{ia} D_{ij} y_{jb} - \delta_{ab})$, with extrema at $\sum_j L_{kj} y_{j\beta} = \sum_{\alpha,i} \lambda_{\alpha\beta} D_{ki} y_{i\alpha}$. We choose[8] $\lambda_{\alpha\beta} \equiv \lambda_\beta \delta_{\alpha\beta}$, giving $\sum_j L_{kj} y_{j\alpha} = \sum_i \lambda_\alpha D_{ki} y_{i\alpha}$. This is a generalized eigenvector problem with eigenvectors the columns of Y. Hence once again the low dimensional vectors are constructed from the first few components of the dual eigenvectors, except that in this case, the eigenvectors with lowest eigenvalues are chosen (omitting the eigenvector $\mathbf{e}$), and in contrast to MDS, they are not weighted by the square roots of the eigenvalues. Thus Laplacian eigenmaps must use some other criterion for deciding on what d' should be. Finally, note that the $\mathbf{y}$'s are conjugate with respect to D (as well as L), so we can scale them so that the constraints $Y'DY = \mathbf{1}$ are indeed met, and our drastic simplification of the Lagrange multipliers did no damage; and left multiplying the eigenvalue equation by $\mathbf{y}'_\alpha$ shows that $\lambda_\alpha = \mathbf{y}'_\alpha L \mathbf{y}_\alpha$, so choosing the smallest eigenvalues indeed gives the lowest values of the objective function, subject to the constraints.

Spectral Clustering

Although spectral clustering is a clustering method, it is very closely related to dimensional reduction. In fact, since clusters may be viewed as large scale structural features of the data, any dimensional reduction technique that maintains these structural features will be a good preprocessing step prior to clustering, to the point where very simple clustering algorithms (such as K-means) on the preprocessed data can work well (Shi and Malik, 2000, Meila and Shi, 2000, Ng et al., 2002). If a graph is partitioned into two disjoint sets by removing a set of arcs, the *cut* is defined as the sum of the weights of the removed arcs. Given the mapping of data to graph defined above, a cut defines a split of the data into two clusters, and the minimum cut encapsulates the notion of maximum dissimilarity between two clusters. However finding a minimum cut tends to just lop off outliers, so (Shi and Malik, 2000) define a normalized cut, which is now a function of all the weights in the graph, but which penalizes cuts which result in a subgraph g such that the cut divided by the sum of weights from g to G is large; this solves the outlier problem. Now suppose we wish to divide the data into two clusters. Define a scalar on each node, z_i, $i = 1, \ldots, m$, such that $z_i = 1$ for nodes in one cluster and $z_i = -1$ for nodes in the other. The solution to the normalized mincut problem is given by (Shi and Malik, 2000)

$$\min_{\mathbf{y}} \frac{\mathbf{y}' L \mathbf{y}}{\mathbf{y}' D \mathbf{y}} \text{ such that } y_i \in \{1, -b\} \text{ and } \mathbf{y}' D \mathbf{e} = 0 \tag{4.35}$$

where $\mathbf{y} \equiv (\mathbf{e} + \mathbf{z}) + b(\mathbf{e} - \mathbf{z})$, and b is a constant that depends on the partition. This problem is solved by relaxing $\mathbf{y}$ to take real values: the problem then becomes finding the second smallest eigenvector of the generalized eigenvalue problem $L\mathbf{y} = \lambda D\mathbf{y}$ (the constraint $\mathbf{y}' D\mathbf{e} = 0$ is automatically satisfied by the solutions), which is exactly the same problem found by Laplacian eigenmaps (in fact the objective function used by Laplacian eigenmaps was proposed as Eq. (10) in (Shi and Malik, 2000)). The algorithms differ in what they do next. The clustering is achieved by thresholding the element y_i so that the nodes are split into two disjoint sets. The dimensional reduction is achieved by treating the element y_i as the first component of a reduced dimension representation of the sample $\mathbf{x}_i$. There is also an interesting equivalent physical interpretation, where the arcs are springs, the nodes are masses, and the $\mathbf{y}$ are the fundamental modes of the resulting vibrating system (Shi and Malik, 2000). Meila and Shi (Meila and Shi, 2000) point out that that matrix $P \equiv D^{-1} L$ is stochastic, which motivates the interpretation of spectral clustering as the stationary distribution of a Markov random field: the intuition is that a random walk, once in one of the mincut clusters, tends to stay in it. The stochastic interpretation also provides tools to analyse the thresholding used in spectral clustering, and a method for learning the weights W_{ij} based on training data with known clusters (Meila and Shi, 2000). The dimensional reduction view also motivates a different approach to clustering, where instead of simply clustering by thresholding a single eigenvector, simple clustering algorithms are applied to the low dimensional representation of the data (Ng et al., 2002).

4.3 Pulling the Threads Together

At this point the reader is probably struck by how similar the mathematics underlying all these approaches is. We've used essentially the same Lagrange multiplier trick to enforce constraints three times; all of the methods in this review rely on an eigendecomposition. Isomap, LLE, Laplacian eigenmaps, and spectral clustering all share the property that in their original forms, they do not provide a direct functional form for the dimension-reducing mapping, so the extension to new data requires re-training. Landmark Isomap solves this problem; the other algorithms could also use Nyström to solve it (as pointed out by (Bengio et al., 2004)). Isomap is often called a 'global' dimensionality reduction algorithm, because it attempts to preserve all geodesic distances; by contrast, LLE, spectral clustering and Laplacian eigenmaps are local (for example, LLE attempts to preserve local translations, rotations and scalings of the data). Landmark Isomap is still global in this sense, but the landmark device brings the computational cost more in line with the other algorithms. Although they start from quite different geometrical considerations, LLE, Laplacian eigenmaps, spectral clustering and MDS all look quite similar under the hood: the first three use the dual eigenvectors of a symmetric matrix as their low dimensional representation, and MDS uses the dual eigenvectors with components scaled by square roots of eigenvalues. In light of this it's perhaps not surprising that relations linking these algorithms can be found: for example, given certain assumptions on the smoothness of the eigenfunctions and on the distribution of the data, the eigendecomposition performed by LLE can be shown to coincide with the eigendecomposition of the squared Laplacian (Belkin and Niyogi, 2003); and (Ham et al., 2004) show how Laplacian eigenmaps, LLE and Isomap can be viewed as variants of kernel PCA. (Platt, 2005) links several flavors of MDS by showing how landmark MDS and two other MDS algorithms (not described here) are in fact all Nyström algorithms. Despite the mathematical similarities of LLE, Isomap and Laplacian Eigenmaps, their different geometrical roots result in different properties: for example, for data which lies on a manifold of dimension d embedded in a higher dimensional space, the eigenvalue spectrum of the LLE and Laplacian Eigenmaps algorithms do not reveal anything about d, whereas the spectrum for Isomap (and MDS) does.

The connection between MDS and PCA goes further than the form taken by the 'unexplained residuals' in Eq. (4.29). If $X \in M_{md}$ is the matrix of m (zero-mean) sample vectors, then PCA diagonalizes the covariance matrix $X'X$, whereas MDS diagonalizes the kernel matrix XX'; but XX' has the same eigenvalues as $X'X$ (Horn and Johnson, 1985), and $m-d$ additional zero eigenvalues (if $m > d$). In fact if $\mathbf{v}$ is an eigenvector of the kernel matrix so that $XX'\mathbf{v} = \lambda\mathbf{v}$, then clearly $X'X(X'\mathbf{v}) = \lambda(X'\mathbf{v})$, so $X'\mathbf{v}$ is an eigenvector of the covariance matrix, and similarly if $\mathbf{u}$ is an eigenvector of the covariance matrix, then $X\mathbf{u}$ is an eigenvector of the kernel matrix. This provides one way to view how kernel PCA computes the eigenvectors of the (possibly infinite dimensional) covariance matrix in feature space in terms of the eigenvectors of the kernel matrix. There's a useful lesson here: given a covariance matrix (Gram matrix) for which you wish to compute those eigenvectors with nonvanishing eigenvalues, and if the corresponding Gram matrix (covariance matrix) is both available, and more easily eigendecomposed (has fewer elements), then compute the eigenvectors for the latter, and map to the eigenvectors of the former using the data matrix as above. Along these lines, Williams (Williams, 2001) has pointed out that kernel PCA can itself be viewed as performing MDS in feature space. Before kernel PCA is performed, the kernel is centered (i.e. $P^e K P^e$ is computed), and for kernels that depend on the data only through functions of squared distances between points (such as radial basis function kernels), this centering is equivalent to centering a distance matrix in feature space. (Williams, 2001) further points out that for these kernels, classical MDS in feature space is equivalent to a form of metric MDS in input space. Although ostensibly kernel PCA gives a function that can be applied to test points, while MDS does not, kernel PCA does so by using the Nyström approximation (see Section 4.2.1), and exactly the same can be done with MDS.

The subject of feature extraction and dimensional reduction is vast. In this review I've limited the discussion to mostly geometric methods, and even with that restriction it's far from complete, so I'd like to alert the reader to three other interesting leads. The first is the method of principal curves, where the idea is to find that smooth curve that passes through the data in such a way that the sum of shortest distances from each point to the curve is minimized, thus providing a nonlinear, one-dimensional summary of the data (Hastie and Stuetzle, 1989); the idea has since been extended by applying various regularization schemes (including kernel-based), and to manifolds of higher dimension (Schölkopf and Smola, 2002). Second, competitions have been held at recent NIPS workshops on feature extraction, and the reader can find a wealth of information there (Guyon, 2003). Finally, recent work on object detection has shown that boosting, where each weak learner uses a single feature, can be a very effective method for finding a small set of good (and mutually complementary) features from a large pool of possible features (Viola and Jones, 2001).

Acknowledgments

I thank John Platt for valuable discussions. Thanks also to Lawrence Saul, Bernhard Schölkopf, Jay Stokes and Mike Tipping for commenting on the manuscript.

References

M.A. Aizerman, E.M. Braverman, and L.I. Rozoner. Theoretical foundations of the potential function method in pattern recognition learning. *Automation and Remote Control*, 25:821–837, 1964.

P.F. Baldi and K. Hornik. Learning in linear neural networks: A survey. *IEEE Transactions on Neural Networks*, 6(4):837–858, July 1995.

A. Basilevsky. *Statistical Factor Analysis and Related Methods*. Wiley, New York, 1994.

M. Belkin and P. Niyogi. Laplacian eigenmaps for dimensionality reduction and data representation. *Neural Computation*, 15(6):1373–1396, 2003.

Y. Bengio, J. Paiement, and P. Vincent. Out-of-sample extensions for LLE, Isomap, MDS, Eigenmaps and spectral clustering. In *Advances in Neural Information Processing Systems 16*. MIT Press, 2004.

C. Berg, J.P.R. Christensen, and P. Ressel. *Harmonic Analysys on Semigroups*. Springer-Verlag, 1984.

C. M. Bishop. Bayesian PCA. In M. S. Kearns, S. A. Solla, and D. A. Cohn, editors, *Advances in Neural Information Processing Systems*, volume 11, pages 382–388, Cambridge, MA, 1999. The MIT Press.

I. Borg and P. Groenen. *Modern Multidimensional Scaling: Theory and Applications.* Springer, 1997.

B. E. Boser, I. M. Guyon, and V .Vapnik. A training algorithm for optimal margin classifiers. In *Fifth Annual Workshop on Computational Learning Theory*, pages 144–152, Pittsburgh, 1992. ACM.

C.J.C. Burges. Some Notes on Applied Mathematics for Machine Learning. In O. Bousquet, U. von Luxburg, and G. Rätsch, editors, *Advanced Lectures on Machine Learning*, pages 21–40. Springer Lecture Notes in Aritificial Intelligence, 2004.

C.J.C. Burges, J.C. Platt, and S. Jana. Extracting noise-robust features from audio. In *Proc. IEEE Conference on Acoustics, Speech and Signal Processing*, pages 1021–1024. IEEE Signal Processing Society, 2002.

C.J.C. Burges, J.C. Platt, and S. Jana. Distortion discriminant analysis for audio fingerprinting. *IEEE Transactions on Speech and Audio Processing*, 11(3):165–174, 2003.

F.R.K. Chung. *Spectral Graph Theory*. American Mathematical Society, 1997.

T.F. Cox and M.A.A. Cox., Multidimensional Scaling. Chapman and Hall, 2001.

R.B. Darlington. Factor analysis. Technical report, Cornell University, http://comp9.psych.cornell.edu/Darlington/factor.htm.

V. de Silva and J.B. Tenenbaum. Global versus local methods in nonlinear dimensionality reduction. In S. Becker, S. Thrun, and K. Obermayer, editors, *Advances in Neural Information Processing Systems 15*, pages 705–712. MIT Press, 2002.

P. Diaconis and D. Freedman. Asymptotics of graphical projection pursuit. *Annals of Statistics*, 12:793–815, 1984.

K.I. Diamantaras and S.Y. Kung. *Principal Component Neural Networks*. John Wiley, 1996.

R.O. Duda and P.E. Hart. *Pattern Classification and Scene Analysis*. John Wiley, 1973.

C. Fowlkes, S. Belongie, F. Chung, and J. Malik. Spectral grouping using the Nyström method. *IEEE Trans. Pattern Analysis and Machine Intelligence*, 26(2), 2004.

J.H. Friedman and W. Stuetzle. Projection pursuit regression. *Journal of the American Statistical Association*, 76(376):817–823, 1981.

J.H. Friedman, W. Stuetzle, and A. Schroeder. Projection pursuit density estimation. *J. Amer. Statistical Assoc.*, 79:599–608, 1984.

J.H. Friedman and J.W. Tukey. A projection pursuit algorithm for exploratory data analysis. *IEEE Transactions on Computers*, c-23(9):881–890, 1974.

G.H. Golub and C.F. Van Loan. *Matrix Computations*. Johns Hopkins, third edition, 1996.

M. Gondran and M. Minoux. *Graphs and Algorithms*. John Wiley and Sons, 1984.

I. Guyon. *NIPS 2003 workshop on feature extraction:* http://clopinet.com/isabelle/Projects/NIPS2003/.

J. Ham, D.D. Lee, S. Mika, and B. Schölkopf. A kernel view of dimensionality reduction of manifolds. In *Proceedings of the International Conference on Machine Learning*, 2004.

T.J. Hastie and W. Stuetzle. Principal curves. *Journal of the American Statistical Association*, 84(406):502–516, 1989.

R.A. Horn and C.R. Johnson. *Matrix Analysis*. Cambridge University Press, 1985.

P.J. Huber. Projection pursuit. *Annals of Statistics*, 13(2):435–475, 1985.

A. Hyvärinen, J. Karhunen, and E. Oja. *Independent Component Analysis*. Wiley, 2001.

Y. LeCun and Y. Bengio. Convolutional networks for images, speech and time-series. In M. Arbib, editor, *The Handbook of Brain Theory and Neural Networks*. MIT Press, 1995.

M. Meila and J. Shi. Learning segmentation by random walks.

In *Advances in Neural Information Processing Systems*, pages 873–879, 2000.

S. Mika, B. Schölkopf, A. J. Smola, K.-R. Müller, M. Scholz, and G. Rätsch. Kernel PCA and de–noising in feature spaces. In M. S. Kearns, S. A. Solla, and D. A. Cohn, editors,

Advances in Neural Information Processing Systems 11. MIT Press, 1999.
A. Y. Ng, M. I. Jordan, and Y. Weiss. On spectral clustering: analysis and an algorithm. In *Advances in Neural Information Processing Systems 14*. MIT Press, 2002.
J. Platt. *Private Communication.*
J. Platt. Fastmap, MetricMap, and Landmark MDS are all Nyström algorithms. In Z. Ghahramani and R. Cowell, editors, *Proc. 10th International Conference on Artificial Intelligence and Statistics*, 2005.
W.H. Press, B.P. Flannery, S.A. Teukolsky, and W.T. Vettering. *Numerical recipes in C: the art of scientific computing*. Cambridge University Press, 2nd edition, 1992.
S.T. Roweis and L.K. Saul. Nonlinear dimensionality reduction by locally linear embedding. *Science*, 290(22):2323–2326, 2000.
I.J. Schoenberg. Remarks to maurice frechet's article sur la définition axiomatique d'une classe d'espace distanciés vectoriellement applicable sur espace de hilbert. *Annals of Mathematics*, 36:724–732, 1935.
B. Schölkopf. The kernel trick for distances. In T.K. Leen, T.G. Dietterich, and V. Tresp, editors, *Advances in Neural Information Processing Systems 13*, pages 301–307. MIT Press, 2001.
B. Schölkopf and A. Smola. *Learning with Kernels*. MIT Press, 2002.
B. Schölkopf, A. Smola, and K-R. Muller. Nonlinear component analysis as a kernel eigenvalue problem. *Neural Computation*, 10(5):1299–1319, 1998.
J. Shi and J. Malik. Normalized cuts and image segmentation. *IEEE Transactions on Pattern Analysis and Machine Intelligence*, 22(8):888–905, 2000.
C.E. Spearman. 'General intelligence' objectively determined and measured. *American Journal of Psychology*, 5:201–293, 1904.
C.J. Stone. Optimal global rates of convergence for nonparametric regression. *Annals of Statistics*, 10(4):1040–1053, 1982.
J.B. Tenenbaum. Mapping a manifold of perceptual observations. In Michael I. Jordan, Michael J. Kearns, and Sara A. Solla, editors, *Advances in Neural Information Processing Systems*, volume 10. The MIT Press, 1998.
M.E. Tipping and C.M. Bishop. Probabilistic principal component analysis. *Journal of the Royal Statistical Society*, 61(3):611, 1999A.
M.E. Tipping and C.M. Bishop. Mixtures of probabilistic principal component analyzers. *Neural Computation*, 11(2):443–482, 1999B.
P. Viola and M. Jones. Robust real-time object detection. In *Second international workshop on statistical and computational theories of vision - modeling, learning, computing, and sampling*, 2001.
S. Wilks. *Mathematical Statistics*. John Wiley, 1962.
C.K.I. Williams. On a Connection between Kernel PCA and Metric Multidimensional Scaling. In T.K. Leen, T.G. Dietterich, and V. Tresp, editors, *Advances in Neural Information Processing Systems 13*, pages 675–681. MIT Press, 2001.
C.K.I. Williams and M. Seeger. Using the Nyström method to speed up kernel machines. In Leen, Dietterich, and Tresp, editors, *Advances in Neural Information Processing Systems 13*, pages 682–688. MIT Press, 2001.

5

Dimension Reduction and Feature Selection

Barak Chizi[1] and Oded Maimon[1]

Tel-Aviv University

Summary. Data Mining algorithms search for meaningful patterns in raw data sets. The Data Mining process requires high computational cost when dealing with large data sets. Reducing dimensionality (the number of attributed or the number of records) can effectively cut this cost. This chapter focuses a pre-processing step which removes dimension from a given data set before it is fed to a data mining algorithm. This work explains how it is often possible to reduce dimensionality with minimum loss of information. Clear dimension reduction taxonomy is described and techniques for dimension reduction are presented theoretically.

Key words: Dimension Reduction, Preprocessing

5.1 Introduction

Data Mining algorithms are used for searching meaningful patterns in raw data sets. Dimensionality (i.e., the number of data set attributes or groups of attributes) constitutes a serious obstacle to the efficiency of most Data Mining algorithms. This obstacle is sometimes known as the "curse of dimensionality" (Elder and Pregibon, 1996). Techniques quite efficient in low dimensions (e.g., nearest neighbors) cannot provide any meaningful results when the number of records goes beyond a 'modest' size of 10 attributes.

Data-mining algorithms are computationally intensive. Figure 5.1 describes the typical trade-off between the error rate of a Data Mining model and the cost of obtaining the model (in particular, the model may be a classification model). The cost is a function of the theoretical complexity of the Data Mining algorithm that derives the model, and is correlated with the time required for the algorithm to run, and the size of the data set. When discussing dimension reduction, given a set of records, the size of the data set is defined as the number of attributes, and is often used as an estimator to the mining cost.

Theoretically, knowing the exact functional relation between the cost and the error may point out the ideal classifier (i.e. a classifier that produces minimal error rate ε^* and costs h^* to be derived). On some occasions, one might prefer using an inferior

O. Maimon, L. Rokach (eds.), *Data Mining and Knowledge Discovery Handbook*, 2nd ed.,
DOI 10.1007/978-0-387-09823-4_5,

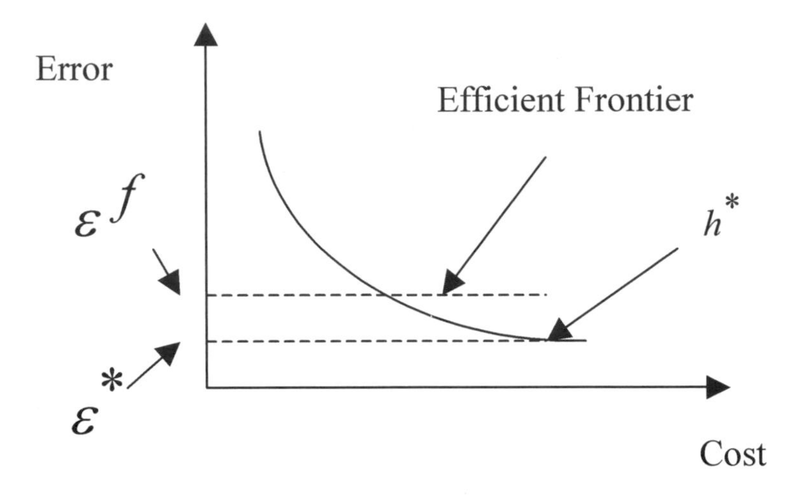

Fig. 5.1. typical cost-error relation in a classification models.

classifier that uses only a part of the data $h \leq h^*$ and produces an increased error rate. In practice, the exact tradeoff curve of Figure 5.1 is seldom known, and generating it might be computationally prohibitive. The objective of dimension reduction in Data Mining domains is to identify the smallest cost at which a Data Mining algorithm can keep the error rate below ε^f (this error rate is sometimes referred to efficiency frontier).

Feature selection, is a problem closely related to dimension reduction. The objective of feature selection is to identify features in the data-set as important, and discard any other feature as irrelevant and redundant information. Since feature selection reduces the dimensionality of the data, it holds out the possibility of more effective & rapid operation of data mining algorithms (i.e. Data Mining algorithms can be operated faster and more effectively by using feature selection). In some cases, as a result of feature selection, accuracy on future classification can be improved; in other instances, the result is a more compact, easily interpreted representation of the target concept (Hall, 1999).

On the other hand, feature selection is a costly process, and it also contradicts the initial assumption, that all information (i.e. attributes) is required in order to achieve maximum accuracy. While some attributes are less important, there are no attributes that are irrelevant or redundant. As described later on this work, feature selection problem is a sub problem of dimension reduction. Figure 5.2 is taxonomy of dimension reduction reasons.

It can be seen that there are four major reasons for performing dimension reduction. Each reason can be referred to as a distinctive sub-problem:

1. Decreasing the learning (model) cost;

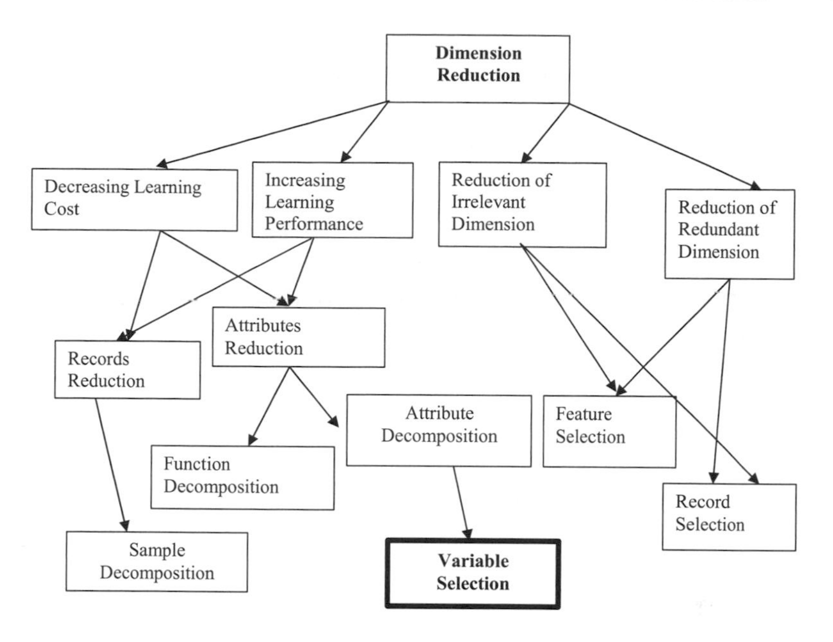

Fig. 5.2. Texonomy of dimension reduction problem.

2. Increasing the learning (model) performance;
3. Reducing irrelevant dimensions;
4. Reducing redundant dimensions.

Reduction of redundant dimensions and of irrelevant dimension can be further divided into two sub-problems:

Feature selection The objective of feature selection is to identify some features in the data-set as important, and discard any other feature as irrelevant and redundant information. The process of feature selection reduces the dimensionality of the data and enables learning algorithms to operate faster and more effectively. In some cases, the accuracy of future classifications can be improved; in others, the result is a more compact, easily interpreted model (Hall, 1999).

Record selection Just as some attributes are more useful than others, some records (examples) may better aid the learning process than others (Blum and Langley, 1997).

The other two sub-problems of dimension reduction, as described in Figure 5.2, are increasing learning performance and decreasing learning cost. Each of the these two sub-problems can also be divided into two further sub-problems: records reduction and attribute reduction. Record reduction is sometimes referred to as sample (or tuple) decomposition. Attribute reduction can be further divided into two sub problems: attribute decomposition and function decomposition. These decomposition problems embody an extensive methodology called decomposition methodology discussed in Chapter 50.7 of this volume.

A sub-problem of attribute decomposition, as seen in Figure 5.2, is variable selection. The solution to this problem is a pre-processing step which removes attributes from a given data set before feeding it to a Data Mining algorithm. The rationale for this step is the reduction of time required for running the Data Mining algorithm, since the running time depends both on the number of records as well as on the number of attributes in each record (the dimension). Variable selection may scarify some accuracy but saves time in the learning process.

This Chapter provides survey of feature selection techniques and variable selection techniques

5.2 Feature Selection Techniques

5.2.1 Feature Filters

The earliest approaches to feature selection within machine learning were filter methods. All filter methods use heuristics based on general characteristics of the data rather than a learning algorithm to evaluate the merit of feature subsets. As a consequence, filter methods are generally much faster than wrapper methods, and, as such, are more practical for use on data of high dimensionality.

FOCUS

Almuallim and Dietterich (1992) describe an algorithm originally designed for Boolean domains called FOCUS. FOCUS exhaustively searches the space of feature subsets until it finds the minimum combination of features that divides the training data into pure classes (that is, where every combination of feature values is associated with a single class). This is referred to as the "min- features bias". Following feature selection, the final feature subset is passed to ID3 (Quinlan, 1986), which constructs a decision tree.

There are two main difficulties with FOCUS, as pointed out by Caruanna and Freitag (1994) . Firstly, since FOCUS is driven to attain consistency on the training data, an exhaustive search may be difficult if many features are needed to attain consistency. Secondly, a strong bias towards consistency can be statistically unwarranted and may lead to over-fitting the training data— the algorithm will continue to add features to repair a single inconsistency.

The authors address the first of these problems in their paper (Almuallim and Dietterich, 1992). Three algorithms— each consisting of forward selection search coupled with a heuristic to approximate the min- features bias— are presented as methods to make FOCUS computationally feasible on domains with many features.

The first algorithm evaluates features using the following information theoretic formula:

$$Entropy(Q) = \sum_{i=0}^{2^{|Q|}-1} \frac{p_i+n_i}{|Sample|} \left[\frac{p_i}{p_i+n_i} \log_2 \frac{p_i}{p_i+n_i} + \frac{n_i}{p_i+n_i} \log_2 \frac{n_i}{p_i+n_i} \right] \tag{5.1}$$

For a given feature subset Q , there are $2^{|Q|}$ possible truth value assignments to the features. A given feature set divides the training data into groups of instances with the same truth value assignments to the features in Q. Equation 5.1 measures the overall entropy of the class values in these groups— p_i and n_i denote the number of positive and negative examples in the i-th group respectively. At each stage, the feature which minimizes Equation 5.1 is added to the current feature subset.

The second algorithm chooses the most discriminating feature to add to the current subset at each stage of the search. For a given pair of positive and negative examples, a feature is discriminating if its value differs between the two. At each stage, the feature is chosen which discriminates the greatest number of positive- negative pairs of examples— that have not yet been discriminated by any existing feature in the subset.

The third algorithm is like the second except that each positive- negative example pair contributes a weighted increment to the score of each feature that discriminates it. The increment depends on the total number of features that discriminate the pair.

LVF

Liu and Setiono (1996) describe an algorithm similar to FOCUS called LVF. Like FOCUS, LVF is consistency driven and, unlike FOCUS, can handle noisy domains if the approximate noise level is known a- priori.

LVF generates a random subset S from the feature subset space during each round of execution. If S contains fewer features than the current best subset, the inconsistency rate of the dimensionally reduced data described by S is compared with the inconsistency rate of the best subset. If S is at least as consistent as the best subset, S replaces the best subset.

The inconsistency rate of the training data prescribed by a given feature subset is defined over all groups of matching instances. Within a group of matching instances the inconsistency count is the number of instances in the group minus the number of instances in the group with the most frequent class value. The overall inconsistency rate is the sum of the inconsistency counts of all groups of matching instances divided by the total number of instances. Liu and Setiono report good results for LVF when applied to some artificial domains and mixed results when applied to commonly used natural domains. They also applied LVF to two "large" data sets— the first having 65,000 instances described by 59 attributes; the second having 5909 instances described by 81 attributes. They report that LVF was able to reduce the number of attributes on both data sets by more than half. They also note that due to the random nature of LVF, the longer it is allowed to execute, the better the results (as measured by the inconsistency criterion).

Filtering Features Through Discretization

Setiono and Liu (1996) note that discretization has the potential to perform feature selection among numeric features. If a numeric feature can justifiably be discretized to a single value, then it can safely be removed from the data.

The combined discretization and feature selection algorithm Chi2, uses a chi-square χ^2 statistic to perform discretization. Numeric attributes are initially sorted by placing each observed value into its own interval. Each numeric attribute is then repeatedly discretized by using theχ^2 test to determine when adjacent intervals should be merged.

The extent of the merging process is controlled by the use of an automatically set χ^2 threshold. The threshold is determined by attempting to maintain the original fidelity of the data— inconsistency (measured the same way as in the LVF algorithm described above) controls the process. The authors report results on three natural domains containing a mixture of numeric and nominal features, using C4.5 (Quinlan, 1993, Quinlan, 1986) before and after discretization. They conclude that Chi2 is effective at improving C4.5's performance and eliminating some features However, it is not clear whether C4.5's improvement is due entirely to some features having been removed or whether discretization plays a role as well.

Using One Learning Algorithm as a Filter for Another

Several researchers have explored the possibility of using a particular learning algorithm as a pre- processor to discover useful feature subsets for a primary learning algorithm. Cardie (1995) describes the application of decision tree algorithms to the task of selecting feature subsets for use by instance based learners. C4.5 was applied to three natural language data sets; only the features that appeared in the final decision trees were used with a *k*–nearest neighbor classifier. The use of this hybrid system resulted in significantly better performance than either C4.5 or the k nearest neighbor algorithm when used alone.

In a similar approach, Singh and Provan (1996) use a greedy oblivious decision tree algorithm to select features from which to construct a Bayesian network. Oblivious decision trees differ from those constructed by algorithms such as C4.5 in that all nodes at the same level of an oblivious decision tree test the same attribute. Feature subsets selected by three oblivious decision tree algorithms— each employing a different information splitting criterion— were evaluated with a Bayesian network classifier on several the-oretic machine learning datasets. Results showed that Bayesian networks using features selected by the oblivious decision tree algorithms outperformed Bayesian networks without feature selection.

Holmes and Nevill–Manning (1995) use Holte's 1R system (Holte, 1993) to estimate the predictive accuracy of individual features. 1R builds rules based on single features (called predictive 1- rules, 1- rules can be thought of as single level decision trees). If the data is split into training and test sets, it is possible to calculate a classification accuracy for each rule and hence each feature. From classification scores, a ranked list of features is obtained. Experiments with choosing a select number of the highest ranked features and using them with common machine learning algorithms showed that, on average, the top three or more features are as accurate as using the original set. This approach is unusual due to the fact that no search is conducted. Instead, it relies on the user to decide how many features to include from the ranked list in the final subset.

Pfahringer (1995) uses a program for inducing decision table majority classifiers to select features. DTM (Decision Table Majority) classifiers are a simple type of nearest neighbor classifier where the similarity function is restricted to returning stored instances that are exact matches with the instance to be classified. If no instances are returned, the most prevalent class in the training data is used as the predicted class; otherwise, the majority class of all matching instances is used. DTM works best when all features are nominal. Induction of a DTM is achieved by greedily searching the space of possible decision tables. Since a decision table is defined by the features it includes, induction is simply feature selection.

In Pfahringer's approach, the minimum description length principle (MDL) (Rissanen, 1978) guides the search by estimating the cost of encoding a decision table and the training examples it misclassifies with respect to a given feature subset. The features appearing in the final decision table are then used with other learning algorithms. Experiments on a small selection of machine learning datasets showed that feature selection by DTM induction can improve the accuracy of C4.5 in some cases. DTM classifiers induced using MDL were also compared with those induced using cross- validation (a wrapper approach) to estimate the accuracy of tables (and hence feature sets). The MDL approach was shown to be more efficient than, and perform as well as, as cross- validation.

An Information Theoretic Feature Filter

Koller and Sahami (1996) introduced a feature selection algorithm based on ideas from information theory and probabilistic reasoning. The rationale behind their approach is that, since the goal of an induction algorithm is to estimate the probability distributions over the class values, given the original feature set, feature subset selection should attempt to remain as close to these original distributions as possible.

More formally, let C be a set of classes, V a set of features, X a subset of V , v an assignment of values $(v_1, \ldots, v_n)$ to the features in V , and v_x the projection of the values in v onto the variables in X. The goal of the feature selector is to choose X so that $P(C|X = v_x)$ is as close as possible to $P(C|V = v)$.

To achieve this goal, the algorithm begins with all the original features and employs a backward elimination search to remove, at each stage, the feature that causes the least change between the two distributions. Because it is not reliable to estimate high order probability distributions from limited data, an approximate algorithm is given that uses pair wise combinations of features. Cross entropy is used to measure the difference between two distributions and the user must specify how many features are to be removed by the algorithm. The cross entropy of the class distribution given a pair of features is:

$$D(\Pr(C|V_i = v_i, V_j = v_j), \Pr(C|V_j = v_j)) = \sum_{c \in C} p(c|V_i = v_i, V_j = v_j) \log_2 \frac{p(c|V_i=v_i,V_j=v_j)}{p(c|V_j=v_j)} \quad (5.2)$$

For each feature i, the algorithm finds a set M_i, containing K attributes from those that remain, that is likely to include the information feature i has about the class values. M_i contains K features out of the remaining features for which the value of Equation 5.2 is smallest. The expected cross entropy between the distribution of the class values, given M_i , V_i , and the distribution of class values given just M_i , is calculated for each feature i. The feature for which this quantity is minimal is removed from the set. This process iterates until the user- specified number of features are removed from the original set.

Experiments on natural domains and two artificial domains using C4.5 and naïve Bayes as the final induction algorithm showed that the feature selector gives the best results when the size K of the conditioning set M is set to 2. In two domains containing over 1000 features the algorithm is able to reduce the number of features by more than half, while improving accuracy by one or two percent.

One problem with the algorithm is that it requires features with more than two values to be encoded as binary in order to avoid the bias that entropic measures have toward features with many values. This can greatly increase the number of features in the original data, as well as introducing further dependencies. Furthermore, the meaning of the original attributes is obscured, making the output of algorithms such as C4.5 hard to interpret.

An Instance Based Approach to Feature Selection – RELIEF

Kira and Rendell (1992) describe an algorithm called RELIEF that uses instance based learning to assign a relevance weight to each feature. Each feature's weight reflects its ability to distinguish among the class values. Features are ranked by weight and those that exceed a user- specified threshold are selected to form the final subset. The algorithm works by randomly sampling instances from the training data. For each instance sampled the nearest instance of the same class (nearest hit) and opposite class (nearest miss) is found. An attribute's weight is updated according to how well its values distinguish the sampled instance from its nearest hit and nearest miss. An attribute will receive a high weight if it differentiates between instances from different classes and has the same value for instances of the same class. Equation (5.3) shows the weight updating formula used by RELIEF:

$$W_X = W_X - \frac{diff(X,R,H)^2}{m} + \frac{diff(X,R,M)^2}{m} \quad (5.3)$$

where W_X is the weight for attribute X, R is a randomly sampled instance, H is the nearest hit, M is the nearest miss, and m is the number of randomly sampled instances.

The function diff calculates the difference between two instances for a given attribute. For nominal attributes it is defined as either 1 (the values are different) or 0 (the values are the same), while for continuous attributes the difference is the actual difference normalized to the interval [0; 1]. Dividing by m guarantees that all weights are in the interval [-1,1].

RELIEF operates on two- class domains. Kononenko (1994) describes enhancements to RELIEF that enable it to cope with multi- class, noisy and incomplete domains. Kira and Rendell provide experimental evidence that shows RELIEF to be effective at identifying relevant features even when they interact. Interacting features are those, whose values are dependent on the values of other features and the class, and as such, provide further information about the class. On the other hand, redundant features, are those whose values are dependent on the values of other features irrespective of the class, and as such, provide no further information about the class. (for example, in parity problems). However, RELIEF does not handle redundant features. The authors state: "If most of the given features are relevant to the concept, it (RELIEF) would select most of the given features even though only a small number of them are necessary for concept description."

Scherf and Brauer (1997) describe a similar instance based approach (EUBAFES) to assigning feature weights developed independently of RELIEF. Like RELIEF, EUBAFES strives to reinforce similarities between instances of the same class while simultaneously decrease similarities between instances of different classes. A gradient descent approach is employed to optimize feature weights with respect to this goal.

5.2.2 Feature Wrappers

Wrapper strategies for feature selection use an induction algorithm to estimate the merit of feature subsets. The rationale for wrapper approaches is that the induction method that will ultimately use the feature subset should provide a better estimate of accuracy than a separate measure that has an entirely different inductive bias (Langley and Sage, 1994).

Feature wrappers often achieve better results than filters due to the fact that they are tuned to the specific interaction between an induction algorithm and its training data. However, they tend to be much slower than feature filters because they must repeatedly call the induction algorithm and must be re- run when a different induction algorithm is used.

Since the wrapper is a well defined process, most of the variation in its application are due to the method used to estimate the off- sample accuracy of a target induction algorithm, the target induction algorithm itself, and the organization of the search. This section reviews work that has focused on the wrapper approach and methods to reduce its computational expense.

Wrappers for Decision Tree Learners

John, Kohavi, and Pfleger (1994) were the first to advocate the wrapper (Allen, 1974) as a general framework for feature selection in machine learning. They present formal for two degrees of feature relevance definitions, and claim that the wrapper is able to discover relevant features. A feature X_i is said to be strongly relevant to the target concept(s) if the probability distribution of the class values, given the full feature set, changes when X_i is removed. A feature X_i is said to be weakly relevant if it is not strongly relevant and the probability distribution of the class values, given some subset S (containing X_i) of the full feature set, changes when X_i is removed. All features that are not strongly or weakly relevant are irrelevant.

Experiments were conducted on three artificial and three natural domains using ID3 and C4.5 (Quinlan, 1993, Quinlan, 1986) as the induction algorithms. Accuracy was estimated by using 25- fold cross validation on the training data; a disjoint test set was used for reporting final accuracies. Both forward selection and backward elimination search were used. With the exception of one artificial domain, results showed that feature selection did not significantly change ID3 or C4.5's generalization performance. The main effect of feature selection was to reduce the size of the trees. Like John et al., Caruana and Freitag (1994) test a number of greedy search methods with ID3 on two calendar scheduling domains. As well as backward elimination and forward selection they also test two variants of stepwise bi- directional search— one starting with all features, the other with none.

Results showed that although the bi- directional searches slightly outperformed the forward and backward searches, on the whole there was very little difference between the various search strategies except with respect to computation time. Feature selection was able to improve the performance of ID3 on both calendar scheduling domains.

Vafaie and De Jong (1995) and Cherkauer and Shavlik (1996) have both applied genetic search strategies in a wrapper framework for improving the performance of decision tree learners. Vafaie and De Jong (1995) describe a system that has two genetic algorithm driven modules— the first performs feature selection, and the second performs constructive induction (Constructive induction is the process of creating new attributes by applying logical and mathematical operators to the original features (Michalski, 1983)). Both modules were able to significantly improve the performance of ID3 on a texture classification problem.

Cherkauer and Shavlik (1996) present an algorithm called SET- Gen which strives to improve the comprehensibility of decision trees as well as their accuracy. To achieve this, SET- Gen's genetic search uses a fitness function that is a linear combination of an accuracy term and a simplicity term:

$$Fitness(X) = \frac{3}{4}A + \frac{1}{4}(1 - \frac{S+F}{2}) \tag{5.4}$$

where X is a feature subset, A is the average cross- validation accuracy of C4.5, S is the average size of the trees produced by C4.5 (normalized by the number of training examples), and F is the number of features is the subset X (normalized by the total number of available features). Equation (5.4) ensures that the fittest population members are those feature subsets that lead C4.5 to induce small but accurate decision trees.

Wrappers for Instance-based Learning

The wrapper approach was proposed at approximately the same time and independently of John et al. (1994) by Langley and Sage (1994) during their investigation of the simple nearest neighbor algorithm's sensitivity to irrelevant attributes. Scaling experiments showed that the nearest neighbour's sample complexity (the number of training examples needed to reach a given accuracy) increases exponentially with the number of irrelevant attributes present in the data (Aha *et al.*, 1991, Langley and Sage, 1994). An algorithm called OBLIVION is presented which performs backward elimination of features using an oblivious decision tree (When all the original features are included in the tree and given a number of assumptions at classification time, Langley and Sage note that the structure is functionally equivalent to the simple nearest neighbor; in fact, this is how it is implemented in OBLIVION) as the induction algorithm. Experiments with OBLIVION using k- fold cross validation on several artificial domains showed that it was able to remove redundant features and learn faster than C4.5 on domains where features interact.

Moore and Lee (1994) take a similar approach to augmenting nearest neighbor algorithm but their system uses leave- one- out instead of k- fold cross- validation and concentrates on improving the prediction of numeric rather than discrete classes. Aha and Blankert (1994) also use leave- one- out cross validation, but pair it with a beam search (Beam search is a limited version of best first search that only remembers a portion of the search path for use in backtracking), instead of hill climbing. Their results show that feature selection can improve the performance of IB1 (a nearest neighbor classifier) on a sparse (very few instances) cloud pattern domain with many features. Moore, Hill, and Johnson (1992) encompass not only feature selection in the wrapper process, but also the number of nearest neighbors used in prediction and the space of combination functions. Using leave- one- out cross validation they achieve significant improvement on several control problems involving, the prediction of continuous classes.

In a similar vein, Skalak (1994) combines feature selection and prototype selection into a single wrapper process using random mutation hill climbing as the search strategy. Experimental results showed significant improvement in accuracy for nearest neighbor on two natural domains and a drastic reduction in the algorithm's storage requirement (number of instances retained during training).

Domingos (1997) describes a context sensitive wrapper approach to feature selection for instance based learners. The motivation for the approach is that there may be features that are either relevant in only a restricted area of the instance space and irrelevant elsewhere, or relevant given only certain values (weakly interacting) of other features and otherwise irrelevant. In either case, when features are estimated globally (over the instance space), the irrelevant aspects of these sorts of features may overwhelm their en-tire useful aspects for instance based learners. This is true even when using backward search strategies with the wrapper. In the wrapper approach, backward search strategies are generally more effective than forward search strategies in domains with feature interactions. Because backward search typically begins with all the features, the removal of a strongly interacting feature is usually detected by decreased accuracy during cross validation.

Domingos presents an algorithm called RC which can detect and make use of context sensitive features. RC works by selecting a (potentially) different set of features for each instance in the training set. It does this by using a search strategy and cross validation to estimate accuracy. For each instance in back-ward the training set, RC finds its nearest neighbour of the same class and removes those features in which the two differ. The accuracy of the entire training dataset is then estimated by cross validation. If the accuracy has not degraded, the modified instance in question is accepted; otherwise the instance is restored to its original state and deactivated (no further feature selection is attempted for it). The feature selection process continues until all instances are inactive.

Experiments on a selection of machine learning datasets showed that RC outperformed standard wrapper feature selectors using forward and backward search strategies with instance based learners. The effectiveness of the context sensitive approach was also shown on artificial domains engineered to exhibit restricted feature dependency. When features are globally relevant or irrelevant, RC has no advantage over standard wrapper feature selection. Furthermore, when few examples are available, or the data is noisy, standard wrapper approaches can detect globally irrelevant features more easily than RC.

Domingos also noted that wrappers that employ instance based learners (including RC) are unsuitable for use on databases containing many instances because they are quadratic in N (the number of instances).

Kohavi (1995) uses wrapper feature selection to explore the potential of decision table majority (DTM) classifiers. Appropriate data structures allow the use of fast incremental cross- validation with DTM classifiers. Experiments showed that DTM classifiers using appropriate feature subsets compared very favorably with sophisticated algorithms such as C4.5.

Wrappers for Bayes Classifiers

Due to the naive Bayes classifier's assumption that, within each class, probability distributions for attributes are independent of each other. Langley and Sage (1994) note that the classifier performance on domains with redundant features can be improved by removing such features. A forward search strategy is employed to select features for use with naïve Bayes, as opposed to the backward strategies that are used most often with decision tree algorithms and instance based learners. The rationale for a forward search is that it should immediately detect dependencies when harmful redundant attributes are added. Experiments showed overall improvement and increased learning rate on three out of six natural domains, with no change on the remaining three.

Pazzani (1995) combines feature selection and simple constructive induction in a wrapper framework for improving the performance of naive Bayes. Forward and backward hill climbing search strategies are compared. In the former case, the algorithm considers not only the addition of single features to the current subset, but also creating a new attribute by joining one of the as yet unselected features with each of the selected features in the subset. In the latter case, the algorithm considers both deleting individual features and replacing pairs of features with a joined feature. Results on a selection of machine learning datasets show that both approaches improve the performance of naïve Bayes.

The forward strategy does a better job at removing redundant attributes than the backward strategy. Because it starts with the full set of features, and considers all possible pairwise joined features, the backward strategy is more effective at identifying attribute interactions than the forward strategy. Improvement for naive Bayes using wrapper- based feature selection is also reported in (Kohavi and Sommerfield, 1995, Kohavi and John, 1996).

Provan and Singh (1996) have applied the wrapper to select features from which to construct Bayesian networks. Their results showed that while feature selection did not improve accuracy over networks which have been constructed from the full set of features, the networks created after feature selection were considerably smaller and faster to learn.

5.3 Variable Selection

This section aim to provide a survey of variable selection. Suppose is Y a variable of interest, and $X_1,\ldots,X_p$ is a set of potential explanatory variables or predictors, are vectors of n observations. The problem of variable selection, or subset selection as it is often called, arises when one wants to model the relationship between Y and a subset of $X_1,\ldots,X_p$, but there is uncertainty about which subset to use. Such a situation is particularly of interest when p is large and $X_1,\ldots,X_p$ is thought to contain many redundant or irrelevant variables The variable selection problem is most familiar in the linear regression context where attention is restricted to normal linear models. Letting $q\gamma$ index the subsets of $X_1,\ldots,X_p$ and letting q be the size of the γ–th subset, the problem is to select and fit a model of the form:

$$Y = X_\gamma \beta_\gamma + \varepsilon \tag{5.5}$$

where $X\gamma$ is an $nxq\gamma$ matrix whose columns correspond to the γth subset, β_γ is a $q\gamma x$ 1 vector of regression coefficients and $\varepsilon \approx \mathrm{N}(0,\sigma^2 I)$. More generally, the variable selection problem is a special case of the model selection problem, where each model under consideration corresponds to a distinct subset of $X_1,\ldots,X_p$. Typically, a single model class is simply applied to all possible subsets.

5.3.1 Mallows Cp (Mallows, 1973)

This method minimizes the mean square error of prediction:

$$C_p = \frac{RSS_\gamma}{\hat{\sigma}^2_{FULL}} + 2q_\gamma - n \tag{5.6}$$

where, RSS_γ is the residual sum of squares for the γth model and σ^2FULL is the usual unbiased estimate of σ^2 based on the full model.

The goal is to get a model with minimum C_p. By using C_p one can reduce dimension by finding the minimal subset which has minimum C_p,

5.3.2 AIC, BIC and F ratio

Two of the other most popular criteria, motivated from very different points of view, are AIC (for Akaike Information Criterion) and BIC (for Bayesian Information Criterion). Letting $\hat{l}_\gamma$ denote the maximum log likelihood of the γth model, AIC selects the model which maximizes $(\hat{l}_\gamma - q_\gamma)$ whereas BIC selects the model which maximizes ($\hat{l}_\gamma$- (logn)qγ/2).

For the linear model, many of the popular selection criteria are special cases of a penalized sum- of squares criterion, providing a unified framework for comparisons. Assuming σ^2 known to avoid complications, this general criterion selects the subset model that minimizes:

$$RSS_\gamma/\hat{\sigma}^2 + Fq_\gamma \tag{5.7}$$

where F is a preset "dimensionality penalty". Intuitively, the above penalizes $RSS\gamma/\sigma^2$ by F times $q\gamma$, the dimension of the γth model. AIC and minimum C_p are essentially equivalent, corresponding to $F = 2$, and BIC is obtained by setting $F = \log n$. By imposing a smaller penalty, AIC and minimum C_p will select larger models than BIC (unless n is very small).

5.3.3 Principal Component Analysis (PCA)

Principal component analysis (PCA) is the best, in the mean-square error sense, linear dimension reduction technique (Jackson, 1991, Jolliffe, 1986). Being based on the covariance matrix of the variables, it is a second-order method. In various fields, it is also known as the singular value decomposition (SVD), the Karhunen-Loeve transform, the Hotelling transform, and the empirical orthogonal function (EOF) method.In essence, PCA seeks to reduce the dimension of the data by finding a few orthogonal linear combinations (the PCs) of the original variables with the largest variance. The first PC, s_1, is the linear combination with the largest variance. We have s_1 = $x^T w_1$, where the p-dimensional coefficient vector w1 = $(w_1,1,\ldots,w_1,p)^T$ solves:

$$w_1 = \arg \max_{\|w=1\|} Var\left\{x^T w\right\} \tag{5.8}$$

The second PC is the linear combination with the second largest variance and orthogonal to the first PC, and so on. There are as many PCs as the number of the original variables. For many datasets, the first several PCs explain most of the variance, so that the rest can be disregarded with minimal loss of information.

5.3.4 Factor Analysis (FA)

Like PCA, factor analysis (FA) is also a linear method, based on the second-order data summaries. First suggested by psychologists, FA assumes that the measured variables depend on some unknown, and often unmeasurable, common factors. Typical examples include variables defined as various test scores of individuals, as such scores are thought to be related to a common "intelligence" factor. The goal of FA is to uncover such relations, and thus can be used to reduce the dimension of datasets following the factor model.

5.3.5 Projection Pursuit

Projection pursuit (PP) is a linear method that, unlike PCA and FA, can incorporate higher than second-order information, and thus is useful for non-Gaussian datasets. It is more computationally intensive than second-order methods. Given a projection index that defines the "interestingness" of a direction, PP looks for the directions that optimize that index. As the Gaussian distribution is the least interesting distribution (having the least structure), projection indices usually measure some aspect of non-Gaussianity. If, however, one uses the second-order maximum variance, subject that the projections be orthogonal, as the projection index, PP yields the familiar PCA.

5.3.6 Advanced Methods for Variable Selection

Chizi and Maimon (2002) describes in their work some new methods for variable selection. These methods based on simple algorithms and uses known evaluators like information gain, logistic regression coefficient and random selection. All the methods are presented with empirical results on benchmark datasets and with theoretical bounds on each method. Wider survey on variable selection can be found there provided with decomposition of the problem of dimension reduction.

In summary, features selection is useful for many application domains, such as: Manufacturing lr18,lr14, Security lr7,l10 and Medicine lr2,lr9, and for many data mining techniques, such as: decision trees lr6,lr12, lr15, clustering lr13,lr8, ensemble methods lr1,lr4,lr5,lr16 and genetic algorithms lr17,lr11.

References

Aha, D. W. and Blankert, R. L. Feature selection for case- based classification of cloud types. In Working Notes of th AAAI- 94 Workshop on Case- Based Reasoning, pages 106–112, 1994.

Aha, D. W. Kibler, and Albert, M. K. Instance based learning algorithms. Machine Learning, 6: 37–66, 1991.

Allen, D. The relationship between variable selection and data augmentation and a method for prediction. Technometrics, 16: 125– 127, 1974.

Almuallim, H. and Dietterich, T. G. Efficient algorithms for identifying relevant features. In.Proceedings of the Ninth Canadian Conference on Artificial Intelligence, pages 38–45. Morgan Kaufmann, 1992.

Almuallim, H. and Dietterich, T. G. Learning with many irrelevant features. In Proceedings of the Ninth National Conference on Artificial Intelligence, pages 547– 542. MIT Press, 1991.

Arbel, R. and Rokach, L., Classifier evaluation under limited resources, Pattern Recognition Letters, 27(14): 1619–1631, 2006, Elsevier.

Averbuch, M. and Karson, T. and Ben-Ami, B. and Maimon, O. and Rokach, L., Context-sensitive medical information retrieval, The 11th World Congress on Medical Informatics (MEDINFO 2004), San Francisco, CA, September 2004, IOS Press, pp. 282–286.

Blum P. and Langley, P. Selection Of Relevant Features And Examples In Machine Learning, Artificial Intelligence, 1997;97: 245-271

Cardie, C. Using decision trees to improve cased- based learning. In Proceedings of the First International Conference on Knowledge Discovery and Data Mining. AAAI Press, 1995.

Caruana, R. and Freitag, D. Greedy attribute selection. In Machine Learning: Proceedings of the Eleventh International Conference. Morgan Kaufmann, 1994.

Cherkauer, K. J. and Shavlik, J. W. Growing simpler decision trees to facilitate knowledge discovery. In Proceedings of the Second International Conference on Knowledge Discovery and Data Mining. AAAI Press, 1996.

Chizi, B. and Maimon, O. “On Dimensionality Reduction of High Dimensional Data Sets”, In “Frontiers in Artificial Intelligence and Applications”. IOS press, pp. 230-236, 2002.

Cohen S., Rokach L., Maimon O., Decision Tree Instance Space Decomposition with Grouped Gain-Ratio, Information Science, Volume 177, Issue 17, pp. 3592-3612, 2007.

Domingos, P. Context- sensitive feature selection for lazy learners. Artificial Intelligence Review, (11): 227– 253, 1997.

Elder, J.F. and Pregibon, D. “A Statistical perspective on knowledge discovery in databases” In Advances in Knowledge Discovery and Data Mining, Fayyad, U. Piatetsky-Shapiro, G. Smyth, P. & Uthurusamy, R. ed., AAAI/MIT Press., 1996.

George, E. and Foster. D. Empirical Bayes Variable Selection, Biometrika, 2000.

Hall, M. Correlation- based feature selection for machine learning, Ph.D. Thesis, Department of Computer Science, University of Waikato, 1999.

Holmes, G. and Nevill- Manning, C. G. . Feature selection via the discovery of simple classification rules. In Proceedings of the Symposium on Intelligent Data Analysis, Baden-Baden, Germany, 1995.

Holte, R. C. Very simple classification rules perform well on most commonly used datasets. Machine Learning, 11: 63– 91, 1993.

Jackson, J. A User’s Guide to Principal Components. New York: John Wiley and Sons, 1991

John, G. H. Kohavi, R. and Pfleger, P. Irrelevant features and the subset selection problem. In Machine Learning: Proceedings of the Eleventh International Conference. Morgan Kaufmann, 1994.

Jolliffe, I. Principal Component Analysis. Springer-Verlag, 1986

Kira, K. and Rendell, L. A.. A practical approach to feature selection. In Machine Learning: Proceedings of the Ninth International Conference, 1992.

Kohavi R. and John, G. Wrappers for feature subset selection. Artificial Intelligence, special issue on relevance, 97(1– 2): 273– 324, 1996

Kohavi, R. and Sommerfield, D. Feature subset selection using the wrapper method: Overfitting and dynamic search space topology. In Proceedings of the First International Conference on Knowledge Discovery and Data Mining. AAAI Press, 1995.

Kohavi, R. Wrappers for Performance Enhancement and Oblivious Decision Graphs. PhD thesis, Stanford University, 1995.

Koller, D. and Sahami, M. Towards optimal feature selection. In Machine Learning: Proceedings of the Thirteenth International Conference on machine Learning. Morgan Kaufmann, 1996.

Kononenko, I. Estimating attributes: Analysis and extensions of relief. In Proceedings of the European Conference on Machine Learning, 1994.

Langley, P. Selection of relevant features in machine learning. In Proceedings of the AAAI Fall Symposium on Relevance. AAAI Press, 1994.

Langley, P. and Sage, S. Scaling to domains with irrelevant features. In R. Greiner, editor, Computational Learning Theory and Natural Learning Systems, volume 4. MIT Press, 1994.

Liu, H. and Setiono, R. A probabilistic approach to feature selection: A filter solution. In Machine Learning: Proceedings of the Thirteenth International Conference on Machine Learning. Morgan Kaufmann, 1996.

Maimon O., and Rokach, L. Data Mining by Attribute Decomposition with semiconductors manufacturing case study, in Data Mining for Design and Manufacturing: Methods and Applications, D. Braha (ed.), Kluwer Academic Publishers, pp. 311–336, 2001.

Maimon O. and Rokach L., "Improving supervised learning by feature decomposition", Proceedings of the Second International Symposium on Foundations of Information and Knowledge Systems, Lecture Notes in Computer Science, Springer, pp. 178-196, 2002.

Maimon, O. and Rokach, L., Decomposition Methodology for Knowledge Discovery and Data Mining: Theory and Applications, Series in Machine Perception and Artificial Intelligence - Vol. 61, World Scientific Publishing, ISBN:981-256-079-3, 2005.

Mallows, C. L. Some comments on Cp . Technometrics 15, 661- 676, 1973

Michalski, R. S. A theory and methodology of inductive learning. Artificial Intelligence, 20(2): 111– 161, 1983.

Moore, A. W. and Lee, M. S. Efficient algorithms for minimizing cross validation error. In Machine Learning: Proceedings of the Eleventh International Conference. Morgan Kaufmann, 1994.

Moore, A. W. Hill, D. J. and Johnson, M. P. An empirical investigation of brute force to choose features, smoothers and function approximations. In S. Hanson, S. Judd, and T. Petsche, editors, Computational Learning Theory and Natural Learning Systems, volume 3. MIT Press, 1992.

Moskovitch R, Elovici Y, Rokach L, Detection of unknown computer worms based on behavioral classification of the host, Computational Statistics and Data Analysis, 52(9):4544–4566, 2008.

Pazzani, M. Searching for dependencies in Bayesian classifiers. In Proceedings of the Fifth International Workshop on AI and Statistics, 1995.

Pfahringer, B. Compression- based feature subset selection. In Proceeding of the IJCAI- 95 Workshop on Data Engineering for Inductive Learning, pages 109– 119, 1995.

Provan, G. M. and Singh, M. Learning Bayesian networks using feature selection. In D. Fisher and H. Lenz, editors, Learning from Data, Lecture Notes in Statistics, pages 291–300. Springer- Verlag, New York, 1996.

Quinlan, J.R. C4.5: Programs for machine learning. Morgan Kaufmann, Los Altos, California, 1993.

Quinlan, J.R. Induction of decision trees. Machine Learning, 1: 81– 106, 1986.

Rissanen, J. Modeling by shortest data description. Automatica, 14: 465–471, 1978.

Rokach, L., Decomposition methodology for classification tasks: a meta decomposer framework, Pattern Analysis and Applications, 9(2006):257–271.

Rokach L., Genetic algorithm-based feature set partitioning for classification problems,Pattern Recognition, 41(5):1676–1700, 2008.

Rokach L., Mining manufacturing data using genetic algorithm-based feature set decomposition, Int. J. Intelligent Systems Technologies and Applications, 4(1):57-78, 2008.

Rokach, L. and Maimon, O., Theory and applications of attribute decomposition, IEEE International Conference on Data Mining, IEEE Computer Society Press, pp. 473–480, 2001.

Rokach L. and Maimon O., Feature Set Decomposition for Decision Trees, Journal of Intelligent Data Analysis, Volume 9, Number 2, 2005b, pp 131–158.

Rokach, L. and Maimon, O., Clustering methods, Data Mining and Knowledge Discovery Handbook, pp. 321–352, 2005, Springer.

Rokach, L. and Maimon, O., Data mining for improving the quality of manufacturing: a feature set decomposition approach, Journal of Intelligent Manufacturing, 17(3):285–299, 2006, Springer.

Rokach, L., Maimon, O., Data Mining with Decision Trees: Theory and Applications, World Scientific Publishing, 2008.

Rokach L., Maimon O. and Lavi I., Space Decomposition In Data Mining: A Clustering Approach, Proceedings of the 14th International Symposium On Methodologies For Intelligent Systems, Maebashi, Japan, Lecture Notes in Computer Science, Springer-Verlag, 2003, pp. 24–31.

Rokach, L. and Maimon, O. and Averbuch, M., Information Retrieval System for Medical Narrative Reports, Lecture Notes in Artificial intelligence 3055, page 217-228 Springer-Verlag, 2004.

Rokach, L. and Maimon, O. and Arbel, R., Selective voting-getting more for less in sensor fusion, International Journal of Pattern Recognition and Artificial Intelligence 20 (3) (2006), pp. 329–350.

Scherf, M. and Brauer, W. Feature selection by means of a feature weighting approach. Technical Report FKI- 221- 97, Technische Universit at Munchen 1997.

Setiono, R. and Liu, H. Chi2: Feature selection and discretization of numeric attributes. In Proceedings of the Seventh IEEE International Conference on Tools with Artificial Intelligence, 1995

Singh, M. and Provan, G. M. Efficient learning of selective Bayesian classifiers. In Machine Learning: Proceedings of the Thirteenth International network Conference on Machine Learning. Morgan Kaufmann, 1996.

Skalak, B. Prototype and feature selection by sampling and random mutation hill climbing algorithms. In Machine Learning: Proceedings of the Eleventh International Conference. Morgan Kaufmann, 1994.

Vafaie, H. and De Jong, K. Genetic algorithms as a tool for restructuring feature space representations. In Proceedings of the International Conference on Tools with A. I. IEEE Computer Society Press, 1995.

Ward, B., What's Wrong with Economics. New York: Basic Books, 1972.

6

Discretization Methods

Ying Yang[1], Geoffrey I. Webb[2], and Xindong Wu[3]

[1] School of Computer Science and Software Engineering, Monash University, Melbourne, Australia yyang@mail.csse.monash.edu.au
[2] Faculty of Information Technology
Monash University, Australia
geoff.webb@infotech.monash.edu
[3] Department of Computer Science
University of Vermont, USA
xwu@cs.uvm.edu

Summary. Data-mining applications often involve quantitative data. However, learning from quantitative data is often less effective and less efficient than learning from qualitative data. Discretization addresses this issue by transforming quantitative data into qualitative data. This chapter presents a comprehensive introduction to discretization. It clarifies the definition of discretization. It provides a taxonomy of discretization methods together with a survey of major discretization methods. It also discusses issues that affect the design and application of discretization methods.

Key words: Discretization, quantitative data, qualitative data.

Introduction

Discretization is a data-processing procedure that transforms quantitative data into qualitative data.

Data Mining applications often involve quantitative data. However, there exist many learning algorithms that are primarily oriented to handle qualitative data (Kerber, 1992, Dougherty et al., 1995, Kohavi and Sahami, 1996). Even for algorithms that can directly deal with quantitative data, learning is often less efficient and less effective (Catlett, 1991, Kerber, 1992, Richeldi and Rossotto, 1995, Frank and Witten, 1999). Hence discretization has long been an active topic in Data Mining and knowledge discovery. Many discretization algorithms have been proposed. Evaluation of these algorithms has frequently shown that discretization helps improve the performance of learning and helps understand the learning results.

This chapter presents an overview of discretization. Section 6.1 explains the terminology involved in discretization. It clarifies the definition of discretization, which has been defined in many differing way in previous literature. Section 6.2 presents a

O. Maimon, L. Rokach (eds.), *Data Mining and Knowledge Discovery Handbook*, 2nd ed.,
DOI 10.1007/978-0-387-09823-4_6, © Springer Science+Business Media, LLC 2010

comprehensive taxonomy of discretization approaches. Section 6.3 introduces typical discretization algorithms corresponding to the taxonomy. Section 6.4 addresses the issue that different discretization strategies are appropriate for different learning problems. Hence designing or applying discretization should not be blind to its learning context. Section 6.5 provides a summary of this chapter.

6.1 Terminology

Discretization transforms one type of data to another type. In the large amount of existing literature that addresses discretization, there is considerable variation in the terminology used to describe these two data types, including 'quantitative' *vs.* 'qualitative', 'continuous' *vs.* 'discrete', 'ordinal' *vs.* 'nominal', and 'numeric' *vs.* 'categorical'. It is necessary to make clear the difference among the various terms and accordingly choose the most suitable terminology for discretization.

We adopt the terminology of statistics (Bluman, 1992, Samuels and Witmer, 1999), which provides two parallel ways to classify data into different types. Data can be classified into either *qualitative* or *quantitative*. Data can also be classified into different *levels of measurement scales*. Sections 6.1.1 and 6.1.2 summarize this terminology.

6.1.1 Qualitative *vs.* quantitative

Qualitative data, also often referred to as **categorical data**, are data that can be placed into distinct categories. Qualitative data sometimes can be arrayed in a meaningful order. But no arithmetic operations can be applied to them. Examples of qualitative data are: *blood type of a person: A, B, AB, O*; and *assignment evaluation: fail, pass, good, excellent*.

Quantitative data are numeric in nature. They can be ranked in order. They also admit to meaningful arithmetic operations. Quantitative data can be further classified into two groups, discrete or continuous.

Discrete data assume values that can be counted. The data cannot assume all values on the number line within their value range. An example is: *number of children in a family*.

Continuous data can assume all values on the number line within their value range. The values are obtained by measuring. An example is: *temperature*.

6.1.2 Levels of measurement scales

In addition to being classified into either qualitative or quantitative, data can also be classified by how they are categorized, counted or measured. This type of classification uses **measurement scales**, and four common levels of scales are: nominal, ordinal, interval and ratio.

The **nominal** level of measurement scales classifies data into mutually exclusive (non-overlapping), exhaustive categories in which no meaningful order or ranking can be imposed on the data. An example is: *blood type of a person: A, B, AB, O*

The **ordinal** level of measurement scales classifies data into categories that can be ranked. However, the differences between the ranks cannot be calculated by arithmetic. An example is: *assignment evaluation: fail, pass, good, excellent*. It is meaningful to say that the assignment evaluation of pass ranks higher than that of fail. It is not meaningful in the same way to say that the blood type of A ranks higher than that of B.

The **interval** level of measurement scales ranks data, and the differences between units of measure can be calculated by arithmetic. However, *zero* in the interval level of measurement does not mean 'nil' or 'nothing' as *zero* in arithmetic means. An example is: *Fahrenheit temperature*. It has a meaningful difference of one degree between each unit. But 0 degree Fahrenheit does not mean there is no heat. It is meaningful to say that 74 degree is two degrees higher than 72 degree. It is not meaningful in the same way to say that the evaluation of excellent is two degrees higher than the evaluation of good.

The **ratio** level of measurement scales possesses all the characteristics of interval measurement, and there exists a *zero* that, the same as arithmetic *zero*, means 'nil' or 'nothing'. In consequence, true ratios exist between different units of measure. An example is: *number of children in a family*. It is meaningful to say that family X has twice as many children as does family Y. It is not meaningful in the same way to say that 100 degree Fahrenheit is twice as hot as 50 degree Fahrenheit.

The nominal level is the lowest level of measurement scales. It is the least powerful in that it provides the least information about the data. The ordinal level is higher, followed by the interval level. The ratio level is the highest. Any data conversion from a higher level of measurement scales to a lower level of measurement scales will lose information. Table 6.1 gives a summary of the characteristics of different levels of measurement scales.

Table 6.1. Measurement Scales

Level	Ranking ?	Arithmetic operation ?	Arithmetic zero ?
Nominal	no	no	no
Ordinal	yes	no	no
Interval	yes	yes	no
Ratio	yes	yes	yes

6.1.3 Summary

In summary, the following classification of data types applies:

1. qualitative data:
   a) nominal;
   b) ordinal;

2. quantitative data:
   a) interval, either discrete or continuous;
   b) ratio, either discrete or continuous.

We believe that 'discretization' as it is usually applied in data mining is best defined as the transformation from *quantitative* data to *qualitative* data. In consequence, we will refer to data as either quantitative or qualitative throughout this chapter.

6.2 Taxonomy

There exist diverse taxonomies in the existing literature to classify discretization methods. Different taxonomies emphasize different aspects of the distinctions among discretization methods.

Typically, discretization methods can be either *primary* or *composite*. Primary methods accomplish discretization without reference to any other discretization method. Composite methods are built on top of some primary method(s).

Primary methods can be classified as per the following taxonomies.

1. **Supervised** *vs.* **Unsupervised** (Dougherty et al., 1995). Methods that use the class information of the training instances to select discretization cut points are supervised. Methods that do not use the class information are unsupervised. Supervised discretization can be further characterized as *error-based*, *entropy-based* or *statistics-based* according to whether intervals are selected using metrics based on error on the training data, entropy of the intervals, or some statistical measure.
2. **Parametric** *vs.* **Non-parametric**. Parametric discretization requires input from the user, such as the maximum number of discretized intervals. Non-parametric discretization only uses information from data and does not need input from the user.
3. **Hierarchical** *vs.* **Non-hierarchical**. Hierarchical discretization selects cut points in an incremental process, forming an implicit hierarchy over the value range. The procedure can be *split* or *merge* (Kerber, 1992). Split discretization initially has the whole value range as an interval, then continues splitting it into sub-intervals until some threshold is met. Merge discretization initially puts each value into an interval, then continues merging adjacent intervals until some threshold is met. Some discretization methods utilize both split and merge processes. For example, intervals are initially formed by splitting, and then a merge process is performed to post-process the formed intervals. Non-hierarchical discretization does not form any hierarchy during discretization. For example, many methods scan the ordered values only once, sequentially forming the intervals.
4. **Univariate** *vs.* **Multivariate** (Bay, 2000). Methods that discretize each attribute in isolation are univariate. Methods that take into consideration relationships among attributes during discretization are multivariate.
5. **Disjoint** *vs.* **Non-disjoint** (Yang and Webb, 2002). Disjoint methods discretize the value range of the attribute under discretization into disjoint intervals. No

intervals overlap. Non-disjoint methods discretize the value range into intervals that can overlap.

6. **Global** *vs.* **Local** (Dougherty et al., 1995). Global methods discretize with respect to the whole training data space. They perform discretization once only, using a single set of intervals throughout a single classification task. Local methods allow different sets of intervals to be formed for a single attribute, each set being applied in a different classification context. For example, different discretizations of a single attribute might be applied at different nodes of a decision tree (Quinlan, 1993).
7. **Eager** *vs.* **Lazy** (Hsu et al., 2000, Hsu et al., 2003). Eager methods perform discretization *prior* to classification time. Lazy methods perform discretization during the classification time.
8. **Time-sensitive** *vs.* **Time-insensitive**. Under time-sensitive discretization, the qualitative value associated with a quantitative value can change along the time. That is, the same quantitative value can be discretized into different values depending on the previous values observed in the time series. Time-insensitive discretization only uses the stationary pro-perties of the quantitative data.
9. **Ordinal** *vs.* **Nominal**. Ordinal discretization transforms quantitative data into ordinal qualitative data. It aims at taking advantage of the ordering information implicit in quantitative attributes, so as not to make values 1 and 2 as dissimilar as values 1 and 10. Nominal discretization transforms quantitative data into nominal qualitative data. The ordering information is hence discarded.
10. **Fuzzy** *vs.* **Non-fuzzy** (Wu, 1995, Wu, 1999, Ishibuchi et al., 2001). Fuzzy discretization first discretizes quantitative attribute values into intervals. It then places some kind of membership function at each cut point as fuzzy borders. The membership function measures the degree of each value belonging to each interval. With these fuzzy borders, a value can be discretized into a few different intervals at the same time, with varying degrees. Non-fuzzy discretization forms sharp borders without employing any membership function.

Composite methods first choose some primary discretization method to form the initial cut points. They then focus on how to adjust these initial cut points to achieve certain goals. The taxonomy of a composite method sometimes is flexible, depending on the taxonomy of its primary method.

6.3 Typical methods

Corresponding to our taxonomy in the previous section, we here enumerate some typical discretization methods. There are many other methods that are not reviewed due to the space limit. For a more comprehensive study on existing discretization algorithms, Yang (2003) and Wu (1995) offer good sources.

6.3.1 Background and terminology

A term often used for describing a discretization approach is 'cut point'. Discretization forms intervals according to the value range of the quantitative data. It then associates a qualitative value to each interval. A cut point is a value among the quantitative data where an interval boundary is located by a discretization method. Another commonly-mentioned term is 'boundary cut point', which are values between two instances with different classes in the sequence of instances sorted by a quantitative attribute. It has been proved that evaluating only the boundary cut points is sufficient for finding the minimum class information entropy (Fayyad and Irani, 1993).

We use the following terminology. Data comprises a set or sequence of *instances*. Each instance is described by a vector of *attribute* values. For classification learning, each instance is also labelled with a class. Each attribute is either qualitative or quantitative. Classes are qualitative. Instances from which one learns cut points or other knowledge are *training instances*. If a *test instance* is presented, a learning algorithm is asked to make a prediction about the test instance according to the evidence provided by the training instances.

6.3.2 Equal-width, equal-frequency and fixed-frequency discretization

We arrange to present these three methods together because they are seemingly similar but actually different. They all are typical of unsupervised discretization. They are also typical of parametric discretization.

When discretizing a quantitative attribute, equal width discretization (EWD) (Catlett, 1991, Kerber, 1992, Dougherty et al., 1995) predefines k, the number of intervals. It then divides the number line between v_{min} and v_{max} into k intervals of equal width, where v_{min} is the minimum observed value, v_{max} is the maximum observed value. Thus the intervals have width $w = (v_{max} - v_{min})/k$ and the cut points are at $v_{min} + w, v_{min} + 2w, \cdots, v_{min} + (k-1)w$.

When discretizing a quantitative attribute, equal-frequency discretization (EFD) (Catlett, 1991, Kerber, 1992, Dougherty et al., 1995) predefines k, the number of intervals. It then divides the sorted values into k intervals so that each interval contains approximately the same number of training instances. Suppose there are n training instances, each interval then contains n/k training instances with adjacent (possibly identical) values. Note that training instances with identical values must be placed in the same interval. In consequence it is not always possible to generate k equal-frequency intervals.

When discretizing a quantitative attribute, fixed-frequency discretization (FFD) (Yang and Webb, 2004) predefines a sufficient interval frequency k. Then it discretizes the sorted values into intervals so that each interval has approximately[4] the same number k of training instances with adjacent (possibly identical) values.

It is worthwhile contrasting EFD and FFD, both of which form intervals of equal frequency. EFD fixes the interval number that is usually arbitrarily chosen. FFD fixes

[4] Just as for EFD, because of the existence of identical values, some intervals can have instance frequency exceeding k.

the interval frequency that is not arbitrary but to ensure each interval contains sufficient instances to supply information such as for estimating probability.

6.3.3 Multi-interval-entropy-minimization discretization ((MIEMD)

Multi-interval-entropy-minimization discretization (Fayyad and Irani, 1993) is typical of supervised discretization. It is also typical of non-parametric discretization.

To discretize an attribute, MIEMD evaluates as a candidate cut point the midpoint between each successive pair of the sorted values. For evaluating each candidate cut point, the data are discretized into two intervals and the resulting class information entropy is calculated. A binary discretization is determined by selecting the cut point for which the entropy is minimal amongst all candidates. The binary discretization is applied recursively, always selecting the best cut point. A minimum description length criterion (MDL) is applied to decide when to stop discretization.

6.3.4 ChiMerge, StatDisc and InfoMerge discretization

EWD and EFD are non-hierarchical discretization. MIEMD involves a split procedure and hence is hierarchical discretization. A typical merge approach to hierarchical discretization is ChiMerge (Kerber, 1992). It uses the χ^2 (Chi square) statistic to determine if the relative class frequencies of adjacent intervals are distinctly different or if they are similar enough to justify merging them into a single interval. The ChiMerge algorithm consists of an initialization process and a bottom-up merging process. The initialization process contains two steps: (1) ascendingly sort the training instancċs according to their values for the attributes being discretized, (2) construct the initial discretization, in which each instance is put into its own interval. The interval merging process contains two steps, repeated continuously: (1) compute the χ^2 for each pair of adjacent intervals, (2) merge the pair of adjacent intervals with the lowest χ^2 value. Merging continues until all pairs of intervals have χ^2 values exceeding a predefined χ^2-*threshold*. That is, all intervals are considered significantly different by the χ^2 independence test. The recommended χ^2-*threshold* is at the 0.90, 0.95 or 0.99 significant level.

StatDisc discretization (Richeldi and Rossotto, 1995) extends ChiMerge to allow any number of intervals to be merged instead of only 2 as ChiMerge does. Both ChiMerge and StatDisc are based on a statistical measure of dependency. The statistical measures treat an attribute and a class symmetrically. A third merge discretization, InfoMerge (Freitas and Lavington, 1996) argues that an attribute and a class should be asymmetric since one wants to predict the value of the class attribute *given* the discretized attribute but not the reverse. Hence InfoMerge uses information loss, which is calculated as the amount of information necessary to identify the class of an instance after merging and the amount of information before merging, to direct the merge procedure.

6.3.5 Cluster-based discretization

The above mentioned methods are all univariate. A typical multivariate discretization technique is cluster-based discretization (Chmielewski and Grzymala-Busse, 1996).

This method consists of two steps. The first step is *cluster formation* to determine initial intervals for the quantitative attributes. The second step is *post-processing* to minimize the number of discretized intervals. Instances here are deemed as points in n-dimensional space which is defined by n attribute values. During cluster formation, the median cluster analysis method is used. Clusters are initialized by allowing each instance to be a cluster. New clusters are formed by merging two existing clusters that exhibit the greatest similarity between each other. The cluster formation continues as long as the level of consistency of the partition is not less than the level of consistency of the original data. Once this process is completed, instances that belong to the same cluster are indiscernible by the subset of quantitative attributes, thus a partition on the set of training instances is induced. Clusters can be analyzed in terms of all attributes to find out cut points for each attribute simultaneously. After discretized intervals are formed, post-processing picks a pair of adjacent intervals among all quantitative attributes for merging whose resulting class entropy is the smallest. If the consistency of the dataset after the merge is above a given threshold, the merge is performed. Otherwise this pair of intervals are marked as non-mergable and the next candidate is processed. The process stops when each possible pair of adjacent intervals are marked as non-mergable.

6.3.6 ID3 discretization

ID3 provides a typical example of local discretization. ID3 (Quinlan, 1986) is an inductive learning program that constructs classification rules in the form of a decision tree. It uses local discretization to deal with quantitative attributes. For each quantitative attribute, ID3 divides its sorted values into two intervals in all possible ways. For each division, the resulting information gain of the data is calculated. The attribute that obtains the maximum information gain is chosen to be the current tree node. And the data are divided into subsets corresponding to its two value intervals. In each subset, the same process is recursively conducted to grow the decision tree. The same attribute can be discretized differently if it appears in different branches of the decision tree.

6.3.7 Non-disjoint discretization

The above mentioned methods are all disjoint discretization. Non-disjoint discretization (NDD) (Yang and Webb, 2002), on the other hand, forms overlapping intervals for a quantitative attribute, always locating a value toward the middle of its discretized interval. This strategy is desirable since it can efficiently form for each single quantitative value a most appropriate interval.

When discretizing a quantitative attribute, suppose there are N instances. NDD identifies among the sorted values t' *atomic intervals*, $(a'_1, b'_1], (a'_2, b'_2], ..., (a'_{t'}, b'_{t'}]$, each containing s' instances, so that[5]

$$s' = \frac{s}{3}$$
$$s' \times t' = N. \tag{6.1}$$

One interval is formed for each set of three consecutive *atomic intervals*, such that the kth ($1 \leq k \leq t' - 2$) interval $(a_k, b_k]$ satisfies $a_k = a'_k$ and $b_k = b'_{k+2}$. Each value v is assigned to interval $(a'_{i-1}, b'_{i+1}]$ where i is the index of the *atomic interval* $(a'_i, b'_i]$ such that $a'_i < v \leq b'_i$, except when $i = 1$ in which case v is assigned to interval $(a'_1, b'_3]$ and when $i = t'$ in which case v is assigned to interval $(a'_{t'-2}, b'_{t'}]$. Figure 6.1 illustrates the procedure. As a result, except in the case of falling into the first or the last *atomic interval*, a numeric value is always toward the middle of its corresponding interval, and intervals can overlap with each other.

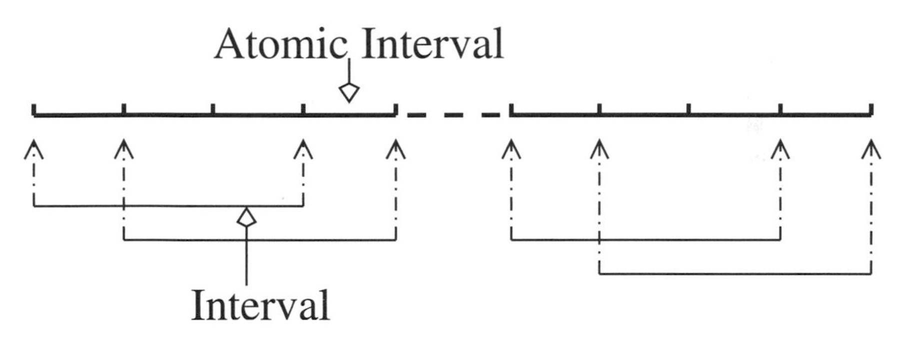

Fig. 6.1. *Atomic Intervals* Compose Actual Intervals

6.3.8 Lazy discretization

The above mentioned methods are all eager. In comparison, lazy discretization (LD) (Hsu et al., 2000,Hsu et al., 2003) defers discretization until classification time. It waits until a test instance is presented to determine the cut points for each quantitative attribute of this test instance. When classifying an instance, LD creates only one interval for each quantitative attribute containing its value from the instance, and leaves other value regions untouched. In particular, it selects a pair of cut points for each quantitative attribute such that the value is in the middle of its corresponding interval. Where the cut points locate is decided by LD's primary discretization method, such as EWD.

[5] Theoretically any odd number k besides 3 is acceptable in (6.1) as long as the same number k of atomic intervals are grouped together later for the probability estimation. For simplicity, we take $k = 3$ for demonstration.

6.3.9 Dynamic-qualitative discretization

The above mentioned methods are all time-insensitive while dynamic-qualitative discretization (Mora et al., 2000) is typically time-sensitive. Two approaches are individually proposed to implement dynamic-qualitative discretization. The first approach is to use statistical information about the preceding values observed from the time series to select the qualitative value which corresponds to a new quantitative value of the series. The new quantitative value will be associated to the same qualitative value as its preceding values if they belong to the same population. Otherwise, it will be assigned a new qualitative value. To decide if a new quantitative value belongs to the same population as the previous ones, a statistic with Student's t distribution is computed.

The second approach is to use distance functions. Two consecutive quantitative values correspond to the same qualitative value when the distance between them is smaller than a predefined threshold significant distance. The first quantitative value of the time series is used as reference value. The next values in the series are compared with this reference. When the distance between the reference and a specific value is greater than the threshold, the comparison process stops. For each value between the reference and the last value which has been compared, the following distances are computed: distance between the value and the first value of the interval, and distance between the value and the last value of the interval. If the former is lower than the latter, the qualitative value assigned is the one corresponding to the first value. Otherwise, the qualitative value assigned is the one corresponding to the last value.

6.3.10 Ordinal discretization

Ordinal discretization (Frank and Witten, 1999, Macskassy et al., 2001), as its name indicates, conducts a transformation of quantitative data that is able to preserve their ordering information. For a quantitative attribute, ordinal discretization first uses some primary discretization method to form a qualitative attribute with n values $(v_1, v_2, \cdots, v_n)$. Then it introduces $n-1$ boolean attributes. The ith boolean attribute represents the test $A^* \leq v_i$. These boolean attributes are substituted for the original A and are input to the learning process.

6.3.11 Fuzzy discretization

Fuzzy discretization (FD) (Ishibuchi et al., 2001) is employed for generating linguistic association rules, where many linguistic terms, such as 'short' and 'tall', can not be appropriately represented by intervals with sharp cut points. Hence, it employs a membership function, such as in (6.2), so that height 150 millimeter is of 0 degree to indicate 'tall'; height 175 millimeter is of 0.5 degree to indicate 'tall' and height 190 millimeter is of 1.0 degree to indicate 'tall'. The induction of rules will take those degrees into consideration.

$$Mem_{tall}(x) = \begin{cases} 0, & \text{if x <= 170;} \\ (x-170)/10 & \text{if 170<x<180;} \\ 1, & \text{if x>=180.} \end{cases} \tag{6.2}$$

FD uses the domain knowledge to define its linguistic membership functions. When dealing with data without such domain knowledge, fuzzy borders can still be set up with commonly used functions such as linear, polynomial and arctan, to fuzzify the sharp borders (Wu, 1999). Wu (1999) demonstrated that such fuzzy borders can be useful when applying rules produced by induction from training examples to a test example, no rules match the test example.

6.3.12 Iterative-improvement discretization

A typical composite discretization is iterative-improvement discretization (IID) (Pazzani, 1995). It initially forms a set of intervals using EWD or MIEMD, and then iteratively adjusts the intervals to minimize the classification error on the training data. It defines two operators: merge two contiguous intervals, or split an interval into two intervals by introducing a new cut point that is midway between each pair of contiguous values in that interval. In each loop of the iteration, for each quantitative attribute, IID applies both operators in all possible ways to the current set of intervals and estimates the classification error of each adjustment using leave-one-out cross validation. The adjustment with the lowest error is retained. The loop stops when no adjustment further reduces the error. IID can split as well as merge discretized intervals. How many intervals will be formed and where the cut points are located are decided by the error of the cross validation.

6.3.13 Summary

For each entry of our taxonomy presented in the previous section, we have reviewed a typical discretization method. Table 6.2 summarizes these methods by identifying their categories under each entry of our taxonomy.

6.4 Discretization and the learning context

Although various discretization methods are available, they are tuned to different types of learning, such as decision tree learning, decision rule learning, naive-Bayes learning, Bayes network learning, clustering, and association learning. Different types of learning have different characteristics and hence require different strategies of discretization. It is important to be aware of the leaning context whenever to design or employ discretization methods. It is unrealistic to pursue a universally optimal discretization approach that can be blind to its learning context.

For example, decision tree learners can suffer from the fragmentation problem, and hence they may benefit more than other learners from discretization that results in few intervals. Decision rule learners require pure intervals (containing instances dominated by a single class), while probabilistic learners such as naive-Bayes does not. Association rule learners value the relations between attributes, and thus they desire multivariate discretization that can capture the inter-dependencies among attributes. Lazy learners can further save training effort if coupled with lazy discretization. If a learning algorithm requires values of an attribute to be disjoint, such as decision tree learning, non-disjoint discretization is not applicable.

To explain this issue, we compare the discretization strategies of two popular learning algorithms, decision tree learning and naive-Bayes learning. Although both are widely used for inductive learning, decision trees and naive-Bayes classifiers have very different inductive biases and learning mechanisms. Correspondingly, their desirable discretization should take different approaches.

6.4.1 Discretization for decision tree learning

Decision tree learning represents the learned concept by a decision tree. Each non-leaf node tests an attribute. Each branch descending from that node corresponds to one of the attribute's values. Each leaf node assigns a class label. A decision tree classifies instances by sorting them down the tree from the root to some leaf node (Mitchell, 1997). ID3 (Quinlan, 1986) and its successor C4.5 (Quinlan, 1993) are well known exemplars of decision tree algorithms.

One popular discretization for decision tree learning is multi-interval-entropy-minimization discretization (MIEMD) (Fayyad and Irani, 1993), as we have reviewed in Section 6.3. MIEMD discretizes a quantitative attribute by calculating the class information entropy as if the classification only uses that *single* attribute after discretization. This can be suitable for the divide-and-conquer strategy of decision tree learning, but not necessarily appropriate for other learning mechanisms such as naive-Bayes learning (Yang and Webb, 2004).

Furthermore, MIEMD uses the minimum description length criterion (MDL) as the termination condition that decides when to stop further partitioning a quantitative attribute's value range. This has an effect to form qualitative attributes with few values (An and Cercone, 1999). This is only desirable for some learning contexts. For decision tree learning, it is important to minimize the number of values of an attribute, so as to avoid the fragmentation problem (Quinlan, 1993). If an attribute has many values, a split on this attribute will result in many branches, each of which receives relatively few training instances, making it difficult to select appropriate subsequent tests. However, minimizing the number of intervals has adverse impact on naive-Bayes learning as we will detail in the next section.

6.4.2 Discretization for naive-Bayes learning

When classifying an instance, naive-Bayes classifiers assume attributes conditionally independent of each other given the class[6]; and then apply Bayes' theorem to calculate the probability of each class given this instance. The class with the highest

[6] This assumption is often referred to as the *attribute independence assumption*.

probability is chosen as the class of this instance. Naive-Bayes classifiers are simple, effective[7], efficient, robust and support incremental training. These merits have seen them deployed in numerous classification tasks.

The appropriate discretization methods for naive-Bayes learning include fixed-frequency discretization (Yang, 2003) and non-disjoint discretization (Yang and Webb, 2002), which we have introduced in Section 6.3. Although it has demonstrated strong effectiveness for decision tree learning, MIEMD does not suit naive-Bayes learning. Naive-Bayes learning assumes that attributes are independent of one another given the class, and hence is not subject to the fragmentation problem of decision tree learning. MIEMD tends to minimize the number of discretized intervals, which has a strong potential to reduce the classification variance but increase the classification bias (Yang and Webb, 2004). As the data size becomes large, it is very likely that the loss through bias increase will soon overshadow the gain through variance reduction, resulting in inferior learning performance. However, naive-Bayes learning is particularly popular with learning from large data because of its efficiency. Hence, MIEMD is not a desirable approach for discretization in naive-Bayes learning.

The other way around, if we employ fixed-frequency discretization (FFD) for decision tree learning, the resulting learning performance can be inferior. FFD tends to maximize the number of discretized intervals as long as each interval contains sufficient instances for estimating the naive-Bayes probabilities. Hence FFD has a strong potential to cause a severe fragmentation problem for decision tree learning, especially when the data size is large.

6.5 Summary

Discretization is a process that transforms quantitative data to qualitative data. It builds a bridge between real-world data-mining applications where quantitative data flourish, and the learning algorithms many of which are more adept at learning from qualitative data. Hence, discretization has an important role in Data Mining and knowledge discovery. This chapter provides a high level overview of discretization. We have defined and presented terminology for discretization, clarifying the multiplicity of differing definitions among previous literature. We have introduced a comprehensive taxonomy of discretization. Corresponding to each entry of the taxonomy, we have demonstrated a typical discretization method. We have then illustrated the need to consider the requirements of a learning context before selecting a discretization technique. It is essential to be aware of the learning context where a discretization method is to be developed or employed. Different learning algorithms

[7] Although its assumption is suspicious to be often violated in real-world applications, naive-Bayes learning still achieves surprisingly good classification performance. Domingos and Pazzani (1997) suggested one reason is that the classification estimation under zero-one loss is only a function of the sign of the probability estimation. The classification accuracy can remain high even while the assumption violation causes poor probability estimation.

require different discretization strategies. It is unrealistic to pursue a universally optimal discretization approach.

References

An, A. and Cercone, N. (1999). Discretization of continuous attributes for learning classification rules. In *Proceedings of the 3rd Pacific-Asia Conference on Methodologies for Knowledge Discovery and Data Mining*, pages 509–514.

Bay, S. D. (2000). Multivariate discretization of continuous variables for set mining. In *Proceedings of the 6th ACM SIGKDD International Conference on Knowledge Discovery and Data Mining*, pages 315–319.

Bluman, A. G. (1992). *Elementary Statistics, A Step By Step Approach.* Wm.C.Brown Publishers. page5-8.

Catlett, J. (1991). On changing continuous attributes into ordered discrete attributes. In *Proceedings of the European Working Session on Learning*, pages 164–178.

Chmielewski, M. R. and Grzymala-Busse, J. W. (1996). Global discretization of continuous attributes as preprocessing for machine learning. *International Journal of Approximate Reasoning*, 15:319–331.

Dougherty, J., Kohavi, R., and Sahami, M. (1995). Supervised and unsupervised discretization of continuous features. In *Proceedings of the 12th International Conference on Machine Learning*, pages 194–202.

Fayyad, U. M. and Irani, K. B. (1993). Multi-interval discretization of continuous-valued attributes for classification learning. In *Proceedings of the 13th International Joint Conference on Artificial Intelligence*, pages 1022–1027.

Frank, E. and Witten, I. H. (1999). Making better use of global discretization. In *Proceedings of the 16th International Conference on Machine Learning*, pages 115–123. Morgan Kaufmann Publishers.

Freitas, A. A. and Lavington, S. H. (1996). Speeding up knowledge discovery in large relational databases by means of a new discretization algorithm. In *Advances in Databases, Proceedings of the 14th British National Conference on Databases*, pages 124–133.

Hsu, C.-N., Huang, H.-J., and Wong, T.-T. (2000). Why discretization works for naive Bayesian classifiers. In *Proceedings of the 17th International Conference on Machine Learning*, pages 309–406.

Hsu, C.-N., Huang, H.-J., and Wong, T.-T. (2003). Implications of the Dirichlet assumption for discretization of continuous variables in naive Bayesian classifiers. *Machine Learning*, 53(3):235–263.

Ishibuchi, H., Yamamoto, T., and Nakashima, T. (2001). Fuzzy Data Mining: Effect of fuzzy discretization. In *The 2001 IEEE International Conference on Data Mining.*

Kerber, R. (1992). Chimerge: Discretization for numeric attributes. In *National Conference on Artificial Intelligence*, pages 123–128. AAAI Press.

Kohavi, R. and Sahami, M. (1996). Error-based and entropy-based discretization of continuous features. In *Proceedings of the 2nd International Conference on Knowledge Discovery and Data Mining*, pages 114–119.

Macskassy, S. A., Hirsh, H., Banerjee, A., and Dayanik, A. A. (2001). Using text classifiers for numerical classification. In *Proceedings of the 17th International Joint Conference on Artificial Intelligence.*

Mitchell, T. M. (1997). *Machine Learning.* McGraw-Hill Companies.

Mora, L., Fortes, I., Morales, R., and Triguero, F. (2000). Dynamic discretization of continuous values from time series. In *Proceedings of the 11th European Conference on Machine Learning*, pages 280–291.

Pazzani, M. J. (1995). An iterative improvement approach for the discretization of numeric attributes in Bayesian classifiers. In *Proceedings of the 1st International Conference on Knowledge Discovery and Data Mining*, pages 228–233.

Quinlan, J. R. (1986). Induction of decision trees. *Machine Learning*, 1:81–106.

Quinlan, J. R. (1993). *C4.5: Programs for Machine Learning*. Morgan Kaufmann Publishers.

Rokach, L., Averbuch, M., and Maimon, O., Information retrieval system for medical narrative reports (pp. 217228). Lecture notes in artificial intelligence, 3055. Springer-Verlag (2004).

Richeldi, M. and Rossotto, M. (1995). Class-driven statistical discretization of continuous attributes (extended abstract). In *European Conference on Machine Learning*, 335-338. Springer.

Samuels, M. L. and Witmer, J. A. (1999). *Statistics For The Life Sciences, Second Edition*. Prentice-Hall. page10-11.

Wu, X. (1995). *Knowledge Acquisition from Databases*. Ablex Publishing Corp. Chapter 6.

Wu, X. (1996). A Bayesian discretizer for real-valued attributes. *The Computer Journal*, 39(8):688–691.

Wu, X. (1999). Fuzzy interpretation of discretized intervals. *IEEE Transactions on Fuzzy Systems*, 7(6):753–759.

Yang, Y. (2003). *Discretization for Naive-Bayes Learning*. PhD thesis, School of Computer Science and Software Engineering, Monash University, Melbourne, Australia.

Yang, Y. and Webb, G. I. (2001). Proportional k-interval discretization for naive-Bayes classifiers. In *Proceedings of the 12th European Conference on Machine Learning*, pages 564–575.

Yang, Y. and Webb, G. I. (2002). Non-disjoint discretization for naive-Bayes classifiers. In *Proceedings of the 19th International Conference on Machine Learning*, pages 666–673.

Yang, Y. and Webb, G. I. (2004). Discretization for naive-Bayes learning: Managing discretization bias and variance. *Submitted for publication.*

Table 6.2. Taxonomy of Discretization Methods

Method	Taxonomy (corresponding to Section 2)										
	0.	1.	2.	3.	4.	5.	6.	7.	8.	9.	10.
Equal-width											
Equal-frequency	primary	unsupervised	parametric	non-hierarchical	univariate	disjoint	global	eager	time-insensitive	nominal	non-fuzzy
Fixed-frequency											
Multi-interval-entropy-minimization	primary	supervised	non-parametric	hierarchical	univariate	disjoint	global	eager	time-insensitive	nominal	non-fuzzy
ChiMerge											
StatDisc	primary	supervised	non-parametric	hierarchical	univariate	disjoint	global	eager	time-insensitive	nominal	non-fuzzy
InfoMerge											
Cluster-based	primary	unsupervised	non-parametric	hierarchical	multivariate	disjoint	global	eager	time-insensitive	nominal	non-fuzzy
ID3	primary	supervised	parametric	hierarchical	univariate	disjoint	local	eager	time-insensitive	nominal	non-fuzzy
Non-disjoint	composite	unsupervised	*	non-hierarchical	univariate	non-disjoint	global	eager	time-insensitive	nominal	non-fuzzy
Lazy	composite	*	*	*	univariate	non-disjoint	global	lazy	time-insensitive	nominal	non-fuzzy
Dynamic-qualitative	primary	unsupervised	non-parametric	non-hierarchical	univariate	disjoint	local	lazy	time-sensitive	nominal	non-fuzzy
Ordinal	composite	*	*	*	univariate	disjoint	global	eager	time-insensitive	ordinal	non-fuzzy
Fuzzy	composite	*	*	*	univariate	non-disjoint	global	eager	time-insensitive	nominal	fuzzy
Iterative-improvement	composite	supervised	*	hierarchical	multivariate	disjoint	global	eager	time-insensitive	nominal	non-fuzzy

Note: each entry of the taxonomy is
0. primary *vs.* composite;
1. supervised *vs.* unsupervised;
2. parametric *vs.* non-parametric;
3. hierarchical *vs.* non-hierarchical;
4. univariate *vs.* multivariate;
5. disjoint *vs.* non-disjoint;
6. global *vs.* local;
7. eager *vs.* lazy;
8. time-sensitive *vs.* time-insensitive;
9. ordinal *vs.* nominal;
10. fuzzy *vs.* non-fuzzy.
An entry filled with '*' indicates that the corresponding method can be conducted in either way of the corresponding taxonomy entry.
This often happens for composite methods, whose taxonomy depends on their primary methods.

7

Outlier Detection

Irad Ben-Gal

Department of Industrial Engineering
Tel-Aviv University
Ramat-Aviv, Tel-Aviv 69978, Israel.
bengal@eng.tau.ac.il

Summary. Outlier detection is a primary step in many data-mining applications. We present several methods for outlier detection, while distinguishing between univariate vs. multivariate techniques and parametric vs. nonparametric procedures. In presence of outliers, special attention should be taken to assure the robustness of the used estimators. Outlier detection for Data Mining is often based on distance measures, clustering and spatial methods.

Key words: Outliers, Distance measures, Statistical Process Control, Spatial data

7.1 Introduction: Motivation, Definitions and Applications

In many data analysis tasks a large number of variables are being recorded or sampled. One of the first steps towards obtaining a coherent analysis is the detection of outlaying observations. Although outliers are often considered as an error or noise, they may carry important information. Detected outliers are candidates for aberrant data that may otherwise adversely lead to model misspecification, biased parameter estimation and incorrect results. It is therefore important to identify them prior to modeling and analysis (Williams *et al.*, 2002, Liu *et al.*, 2004).

An exact definition of an outlier often depends on hidden assumptions regarding the data structure and the applied detection method. Yet, some definitions are regarded general enough to cope with various types of data and methods. Hawkins (1980) defines an outlier *as an observation that deviates so much from other observations as to arouse suspicion that it was generated by a different mechanism.* Barnett and Lewis (1994) indicate that *an outlying observation, or outlier, is one that appears to deviate markedly from other members of the sample in which it occurs*, similarly, Johnson (1992) defines an outlier *as an observation in a data set which appears to be inconsistent with the remainder of that set of data.* Other case-specific definitions are given below.

Outlier detection methods have been suggested for numerous applications, such as credit card fraud detection, clinical trials, voting irregularity analysis, data cleansing, network intrusion, severe weather prediction, geographic information systems,

O. Maimon, L. Rokach (eds.), *Data Mining and Knowledge Discovery Handbook*, 2nd ed.,
DOI 10.1007/978-0-387-09823-4_7,

athlete performance analysis, and other data-mining tasks (Hawkins, 1980, Barnett and Lewis, 1994, Ruts and Rousseeuw, 1996, Fawcett and Provost, 1997, Johnson *et al.*, 1998, Penny and Jolliffe, 2001, Acuna and Rodriguez, 2004, Lu *et al.*, 2003).

7.2 Taxonomy of Outlier Detection Methods

Outlier detection methods can be divided between *univariate methods*, proposed in earlier works in this field, and *multivariate methods* that usually form most of the current body of research. Another fundamental taxonomy of outlier detection methods is between *parametric* (*statistical*) methods and *nonparametric* methods that are model-free (e.g., see (Williams *et al.*, 2002)). Statistical parametric methods either assume a known underlying distribution of the observations (e.g., (Hawkins, 1980, Rousseeuw and Leory, 1987, Barnett and Lewis, 1994)) or, at least, they are based on statistical estimates of unknown distribution parameters (Hadi, 1992, Caussinus and Roiz, 1990). These methods flag as outliers those observations that deviate from the model assumptions. They are often unsuitable for high-dimensional data sets and for arbitrary data sets without prior knowledge of the underlying data distribution (Papadimitriou *et al.*, 2002).

Within the class of non-parametric outlier detection methods one can set apart the data-mining methods, also called *distance-based methods*. These methods are usually based on local distance measures and are capable of handling large databases (Knorr and Ng, 1997, Knorr and Ng, 1998, Fawcett and Provost, 1997, Williams and Huang, 1997, Mouchel and Schonlau, 1998, Knorr *et al.*, 2000, Knorr *et al.*, 2001, Jin *et al.*, 2001, Breunig *et al.*, 2000, Williams *et al.*, 2002, Hawkins *et al.*, 2002, Bay and Schwabacher, 2003). Another class of outlier detection methods is founded on *clustering techniques*, where a cluster of small sizes can be considered as clustered outliers (Kaufman and Rousseeuw, 1990, Ng and Han, 1994, Ramaswamy *et al.*, 2000, Barbara and Chen, 2000, Shekhar and Chawla, 2002, Shekhar and Lu, 2001, Shekhar and Lu, 2002, Acuna and Rodriguez, 2004). Hu and Sung (2003) , whom proposed a method to identify both high and low density pattern clustering, further partition this class to *hard classifiers* and *soft classifiers*. The former partition the data into two non-overlapping sets: outliers and non-outliers. The latter offers a ranking by assigning each datum an outlier classification factor reflecting its degree of outlyingness. Another related class of methods consists of detection techniques for *spatial outliers*. These methods search for extreme observations or local instabilities with respect to neighboring values, although these observations may not be significantly different from the entire population (Schiffman *et al.*, 1981, Ng and Han, 1994, Shekhar and Chawla, 2002, Shekhar and Lu, 2001, Shekhar and Lu, 2002, Lu *et al.*, 2003).

Some of the above-mentioned classes are further discussed bellow. Other categorizations of outlier detection methods can be found in the following sources (Barnett and Lewis, 1994, Papadimitriou *et al.*, 2002, Acuna and Rodriguez, 2004, Hu and Sung, 2003).

7.3 Univariate Statistical Methods

Most of the earliest univariate methods for outlier detection rely on the assumption of an underlying known distribution of the data, which is assumed to be identically and independently distributed (i.i.d.). Moreover, many discordance tests for detecting univariate outliers further assume that the distribution parameters and the type of expected outliers are also known (Barnett and Lewis, 1994). Needless to say, in real world data-mining applications these assumptions are often violated.

A central assumption in statistical-based methods for outlier detection, is a generating model that allows a small number of observations to be randomly sampled from distributions $G_1,\ldots, G_k$, differing from the target distribution F, which is often taken to be a normal distribution $N\left(\mu,\sigma^2\right)$ (see (Ferguson, 1961, David, 1979, Barnett and Lewis, 1994, Gather, 1989, Davies and Gather, 1993)). The outlier identification problem is then translated to the problem of identifying those observations that lie in a so-called *outlier region*. This leads to the following definition (Davies and Gather, 1993):

For any *confidence coefficient* α, $0<\alpha<1$, the α-outlier region of the $N\left(\mu,\sigma^2\right)$ distribution is defined by

$$\operatorname{out}\left(\alpha,\mu,\sigma^2\right)=\left\{x:|x-\mu|>z_{1-\alpha/2}\sigma\right\}, \tag{7.1}$$

where z_q is the q quintile of the $N(0,1)$. A number x is an α-outlier with respect to F if $x\in\operatorname{out}\left(\alpha,\mu,\sigma^2\right)$. Although traditionally the normal distribution has been used as the target distribution, this definition can be easily extended to any unimodal symmetric distribution with positive density function, including the multivariate case.

Note that the outlier definition does not identify which of the observations are contaminated, i.e., resulting from distributions $G_1,\ldots, G_k$, but rather it indicates those observations that lie in the outlier region.

7.3.1 Single-step vs. Sequential Procedures

Davis and Gather (1993) make an important distinction between *single-step* and *sequential* procedures for outlier detection. Single-step procedures identify all outliers at once as opposed to successive elimination or addition of datum. In the sequential procedures, at each step, one observation is tested for being an outlier.

With respect to Equation 7.1, a common rule for finding the outlier region in a single-step identifier is given by

$$\operatorname{out}\left(\alpha_n,\hat{\mu}_n,\hat{\sigma}_n^2\right)=\left\{x:|x-\hat{\mu}_n|>g\left(n,\alpha_n\right)\hat{\sigma}_n\right\}, \tag{7.2}$$

where n is the size of the sample; $\hat{\mu}_n$ and $\hat{\sigma}_n$ are the estimated mean and standard deviation of the target distribution based on the sample; α_n denotes the confidence coefficient following the correction for multiple comparison tests; and $g\left(n,\alpha_n\right)$ defines the limits (critical number of standard deviations) of the outlier regions.

Traditionally, $\hat{\mu}_n, \hat{\sigma}_n$ are estimated respectively by the sample mean, $\bar{x}_n$, and the sample standard deviation, S_n. Since these estimates are highly affected by the presence of outliers, many procedures often replace them by other, more robust, estimates that are discussed in Section 7.3.3. The multiple-comparison correction is used when several statistical tests are being performed simultaneously. While a given α-value may be appropriate to decide whether a single observation lies in the outlier region (i.e., a single comparison), this is not the case for a set of several comparisons. In order to avoid spurious positives, the α-value needs to be lowered to account for the number of performed comparisons. The simplest and most conservative approach is the Bonferroni's correction, which sets the α-value for the entire set of n comparisons equal to α, by taking the α-value for each comparison equal to α/n. Another popular and simple correction uses $\alpha_n = 1-(1-\alpha)^{1/n}$. Note that the traditional Bonferroni's method is "quasi-optimal" when the observations are independent, which is in most cases unrealistic. The critical value $g(n, \alpha_n)$ is often specified by numerical procedures, such as Monte Carlo simulations for different sample sizes (e.g., (Davies and Gather, 1993)).

7.3.2 Inward and Outward Procedures

Sequential identifiers can be further classified to *inward* and *outward* procedures. In inward testing, or *forward selection* methods, at each step of the procedure the "most extreme observation", i.e., the one with the largest outlyingness measure, is tested for being an outlier. If it is declared as an outlier, it is deleted from the dataset and the procedure is repeated. If it is declared as a non-outlying observation, the procedure terminates. Some classical examples for inward procedures can be found in (Hawkins, 1980, Barnett and Lewis, 1994).

In outward testing procedures, the sample of observations is first reduced to a smaller sample (e.g., by a factor of two), while the removed observations are kept in a reservoir. The statistics are calculated on the basis of the reduced sample and then the removed observations in the reservoir are tested in reverse order to indicate whether they are outliers. If an observation is declared as an outlier, it is deleted from the reservoir. If an observation is declared as a non-outlying observation, it is deleted from the reservoir, added to the reduced sample, the statistics are recalculated and the procedure repeats itself with a new observation. The outward testing procedure is terminated when no more observations are left in the reservoir. Some classical examples for inward procedures can be found in (Rosner, 1975, Hawkins, 1980, Barnett and Lewis, 1994).

The classification to inward and outward procedures also applies to multivariate outlier detection methods.

7.3.3 Univariate Robust Measures

Traditionally, the sample mean and the sample variance give good estimation for data location and data shape if it is not contaminated by outliers. When the database is contaminated, those parameters may deviate and significantly affect the outlier-detection performance.

Hampel (1971, 1974) introduced the concept of the *breakdown point*, as a measure for the robustness of an estimator against outliers. The breakdown point is defined as the smallest percentage of outliers that can cause an estimator to take arbitrary large values. Thus, the larger breakdown point an estimator has, the more robust it is. For example, the sample mean has a breakdown point of $1/n$since a single large observation can make the sample mean and variance cross any bound. Accordingly, Hampel suggested the median and the median absolute deviation (MAD) as robust estimates of the location and the spread. The Hampel identifier is often found to be practically very effective (Perarson, 2002,Liu *et al.*, 2004). Another earlier work that addressed the problem of robust estimators was proposed by Tukey (1977) . Tukey introduced the Boxplot as a graphical display on which outliers can be indicated. The Boxplot, which is being extensively used up to date, is based on the distribution quadrants. The first and third quadrants, Q_1 and Q_3, are used to obtain the robust measures for the mean, $\hat{\mu}_n = (Q_1 + Q_3)/2$, and the standard deviation, $\hat{\sigma}_n = Q_3 - Q_1$. Another popular solution to obtain robust measures is to replace the mean by the median and compute the standard deviation based on (1–α) percents of the data points, where typically α¡5%.

Liu et al. (2004) proposed an outlier-resistant data filter-cleaner based on the earlier work of Martin and Thomson (1982) . The proposed data filter-cleaner includes an on-line outlier-resistant estimate of the process model and combines it with a modified Kalman filter to detect and "clean" outliers. The proposed method does not require an apriori knowledge of the process model. It detects and replaces outliers on-line while preserving all other information in the data. The authors demonstrated that the proposed filter-cleaner is efficient in outlier detection and data cleaning for autocorrelated and even non-stationary process data.

7.3.4 Statistical Process Control (SPC)

The field of Statistical Process Control (SPC) is closely-related to univariate outlier detection methods. It considers the case where the univariable stream of measures represents a stochastic process, and the detection of the outlier is required online. SPC methods are being applied for more than half a century and were extensively investigated in statistics literature.

Ben-Gal et al. (2003) categorize SPC methods by two major criteria: i) methods for *independent* data versus methods for *dependent* data; and ii) methods that are *model-specific*, versus methods that are *model-generic*. Model specific methods require a-priori assumptions on the process characteristics, usually defined by an underlying analytical distribution or a closed-form expression. Model-generic methods try to estimate the underlying model with minimum a-priori assumptions.

Traditional SPC methods, such as Shewhart, Cumulative Sum (CUSUM) and Exponential Weighted Moving Average (EWMA) are *model-specific* for *independent data*. Note that these methods are extensively implemented in industry, although the independence assumptions are frequently violated in practice.

The majority of *model-specific* methods for *dependent data* are based on time-series. Often, the underlying principle of these methods is as follows: find a time series model that can best capture the autocorrelation process, use this model to filter the data, and then apply traditional SPC schemes to the stream of residuals. In particular, the ARIMA (Auto Regressive Integrated Moving Average) family of models is widely implemented for the estimation and filtering of process autocorrelation. Under certain assumptions, the residuals of the ARIMA model are independent and approximately normally distributed, to which traditional SPC can be applied. Furthermore, it is commonly conceived that ARIMA models, mostly the simple ones such as AR(see Equation 7.1), can effectively describe a wide variety of industry processes (Box, 1976, Apley and Shi, 1999).

Model-specific methods for dependent data can be further partitioned to *parameter-dependent* methods that require explicit estimation of the model parameters (e.g., (Alwan and Roberts, 1988, Wardell *et al.*, 1994, Lu and Reynolds, 1999, Runger and Willemain, 1995, Apley and Shi, 1999)), and to *parameter-free* methods, where the model parameters are only implicitly derived, if at all (Montgomery and Mastrangelo, 1991, Zhang, 1998).

The Information Theoretic Process Control (ITPC) is an example for a *model-generic* SPC method for *independent data*, proposed in (Alwan *et al.*, 1998). Finally, a *model-generic* SPC method for *dependent* data is proposed in (Gal *et al.*, 2003).

7.4 Multivariate Outlier Detection

In many cases multivariable observations can not be detected as outliers when each variable is considered independently. Outlier detection is possible only when multivariate analysis is performed, and the interactions among different variables are compared within the class of data. A simple example can be seen in Figure 7.1, which presents data points having two measures on a two-dimensional space. The lower left observation is clearly a multivariate outlier but not a univariate one. When considering each measure separately with respect to the spread of values along the x and y axes, we can is seen that they fall close to the center of the univariate distributions. Thus, the test for outliers must take into account the relationships between the two variables, which in this case appear abnormal.

Data sets with multiple outliers or clusters of outliers are subject to *masking* and *swamping* effects. Although not mathematically rigorous, the following definitions from (Acuna and Rodriguez, 2004) give an intuitive understanding for these effects (for other definitions see (Hawkins, 1980, Iglewics and Martinez, 1982, Davies and Gather, 1993, Barnett and Lewis, 1994)):

Masking effect It is said that one outlier masks a second outlier, if the second outlier can be considered as an outlier only by itself, but not in the presence of the first outlier. Thus, after the deletion of the first outlier the second instance is emerged as an outlier. Masking occurs when a cluster of outlying observations skews the mean and the covariance estimates toward it, and the resulting distance of the outlying point from the mean is small.

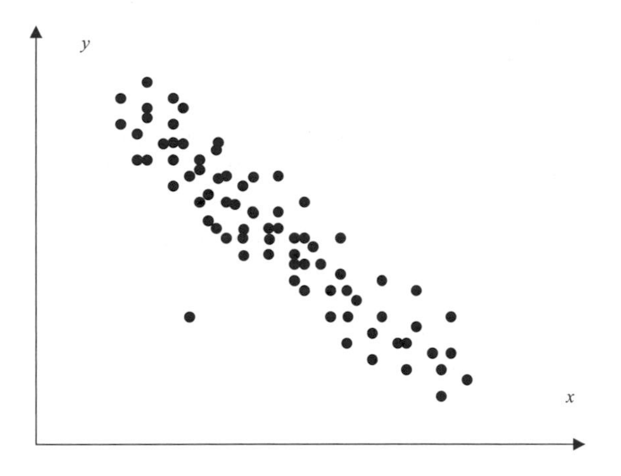

Fig. 7.1. A Two-Dimensional Space with one Outlying Observation (Lower Left Corner).

Swamping effect It is said that one outlier swamps a second observation, if the latter can be considered as an outlier only under the presence of the first one. In other words, after the deletion of the first outlier the second observation becomes a non-outlying observation. Swamping occurs when a group of outlying instances skews the mean and the covariance estimates toward it and away from other non-outlying instances, and the resulting distance from these instances to the mean is large, making them look like outliers. A single step procedure with low masking and swamping is given in (Iglewics and Martinez, 1982).

7.4.1 Statistical Methods for Multivariate Outlier Detection

Multivariate outlier detection procedures can be divided to statistical methods that are based on estimated distribution parameters, and data-mining related methods that are typically parameter-free.

Statistical methods for multivariate outlier detection often indicate those observations that are located relatively far from the center of the data distribution. Several distance measures can be implemented for such a task. The *Mahalanobis* distance is a well-known criterion which depends on estimated parameters of the multivariate distribution. Given n observations from a p-dimensional dataset (often$n¿¿ p$), denote the sample mean vector by $\bar{\mathbf{x}}_n$ and the sample covariance matrix by $\mathbf{V}_n$, where

$$\mathbf{V}_n = \frac{1}{n-1} \sum_{i=1}^{n} (\mathbf{x}_i - \bar{\mathbf{x}}_n)(\mathbf{x}_i - \bar{\mathbf{x}}_n)^T \tag{7.3}$$

The *Mahalanobis* distance for each multivariate data point $i, i = 1, \ldots, n$, is denoted by M_i and given by

$$M_i = \left(\sum_{i=1}^{n} (\mathbf{x}_i - \bar{\mathbf{x}}_n)^T \mathbf{V}_n^{-1} (\mathbf{x}_i - \bar{\mathbf{x}}_n) \right)^{1/2}. \quad (7.4)$$

Accordingly, those observations with a large *Mahalanobis* distance are indicated as outliers. Note that masking and swamping effects play an important rule in the adequacy of the Mahalanobis distance as a criterion for outlier detection. Namely, masking effects might decrease the Mahalanobis distance of an outlier. This might happen, for example, when a small cluster of outliers attracts $\bar{\mathbf{x}}_n$ and inflate $\mathbf{V}_n$ towards its direction. On the other hand, swamping effects might increase the Mahalanobis distance of non-outlying observations. For example, when a small cluster of outliers attracts $\bar{\mathbf{x}}_n$ and inflate $\mathbf{V}_n$ away from the pattern of the majority of the observations (see (Penny and Jolliffe, 2001)).

7.4.2 Multivariate Robust Measures

As in one-dimensional procedures, the distribution mean (measuring the location) and the variance-covariance (measuring the shape) are the two most commonly used statistics for data analysis in the presence of outliers (Rousseeuw and Leory, 1987). The use of robust estimates of the multidimensional distribution parameters can often improve the performance of the detection procedures in presence of outliers. Hadi (1992) addresses this problem and proposes to replace the mean vector by a vector of variable medians and to compute the covariance matrix for the subset of those observations with the smallest Mahalanobis distance. A modified version of Hadi's procedure is presented in (Penny and Jolliffe, 2001). Caussinus and Roiz (1990) propose a robust estimate for the covariance matrix, which is based on weighted observations according to their distance from the center. The authors also propose a method for a low dimensional projections of the dataset. They use the Generalized Principle Component Analysis (GPCA) to reveal those dimensions which display outliers. Other robust estimators of the location (centroid) and the shape (covariance matrix) include the minimum covariance determinant (MCD) and the minimum volume ellipsoid (MVE) (Rousseeuw, 1985, Rousseeuw and Leory, 1987, Acuna and Rodriguez, 2004).

7.4.3 Data-Mining Methods for Outlier Detection

In contrast to the above-mentioned statistical methods, data-mining related methods are often non-parametric, thus, do not assume an underlying generating model for the data. These methods are designed to manage large databases from high-dimensional spaces. We follow with a short discussion on three related classes in this category: distance-based methods, clustering methods and spatial methods.

Distance-based methods were originally proposed by Knorr and Ng (1997, 1998) . An observation is defined as a distance-based outlier if at least a fraction β of the observations in the dataset are further than r from it. Such a definition is based on a single, global criterion determined by the parameters r and β. As pointed out in Acuna and Rodriguez (2004), such definition raises certain difficulties, such as the determination of r and the lack of a ranking for the outliers. The time complexity of the algorithm is $O(pn^2)$, where p is the number of features and n is the sample size. Hence, it is not an adequate definition to use with very large datasets. Moreover, this definition can lead to problems when the data set has both dense and sparse regions (Breunig *et al.*, 2000, Ramaswamy *et al.*, 2000, Papadimitriou *et al.*, 2002). Alternatively, Ramaswamy et al. (2000) suggest the following definition: given two integers v and l (v¡ l), outliers are defined to be the top l sorted observations having the largest distance to their v-th nearest neighbor. One shortcoming of this definition is that it only considers the distance to the v-th neighbor and ignores information about closer observations. An alternative is to define outliers as those observations having a large *average distance* to the v-th nearest neighbors. The drawback of this alternative is that it takes longer to be calculated (Acuna and Rodriguez, 2004).

Clustering based methods consider a cluster of small sizes, including the size of one observation, as clustered outliers. Some examples for such methods are the *partitioning around medoids* (PAM) and the *clustering large applications* (CLARA) (Kaufman and Rousseeuw, 1990); a modified version of the latter for spatial outliers called CLARANS (Ng and Han, 1994); and a *fractal-dimension* based method (Barbara and Chen, 2000). Note that since their main objective is clustering, these methods are not always optimized for outlier detection. In most cases, the outlier detection criteria are implicit and cannot easily be inferred from the clustering procedures (Papadimitriou *et al.*, 2002).

Spatial methods are closely related to clustering methods. Lu et al. (2003) define a spatial outlier as a spatially referenced object whose non-spatial attribute values are significantly different from the values of its neighborhood. The authors indicate that the methods of spatial statistics can be generally classified into two sub categories: *quantitative tests* and *graphic approaches*. Quantitative methods provide tests to distinguish spatial outliers from the remainder of data. Two representative approaches in this category are the Scatterplot (Haining, 1993, Luc, 1994) and the Moran scatterplot (Luc, 1995). Graphic methods are based on visualization of spatial data which highlights spatial outliers. Variogram clouds and pocket plots are two examples for these methods (Haslett *et al.*, 1991, Panatier, 1996). Schiffman et al. (1981) suggest using a multidimensional scaling (MDS) that represents the similarities between objects spatially, as in a map. MDS seeks to find the best configuration of the observations in a low dimensional space. Both metric and non-metric forms of MDS are proposed in (Penny and Jolliffe, 2001). As indicated above, Ng and Han (1994) develop a clustering method for spatial data-mining called CLARANS which is based on randomized search. The authors suggest two spatial data-mining algorithms that use CLARANS. Shekhar et al. (2001, 2002) introduce a method for detecting spatial outliers in graph data set. The method is based on the distribution property of the difference between an attribute value and the average attribute value of its neighbors. Shekhar et al. (2003) propose a unified approach to evaluate spatial outlier-detection methods. Lu et al. (2003) propose a suite of spatial outlier detection algorithms to minimize false detection of spatial outliers when their neighborhood contains true spatial outliers.

Applications of spatial outliers can be found in fields where spatial information plays an important role, such as, ecology, geographic information systems, transportation, climatology, location-based services, public health and public safety (Ng and Han, 1994, Shekhar and Chawla, 2002, Lu *et al.*, 2003).

7.4.4 Preprocessing Procedures

Different paradigms were suggested to improve the efficiency of various data analysis tasks including outlier detection. One possibility is to reduce the size of the data set by assigning the variables to several representing groups. Another option is to eliminate some variables from the analyses by methods of *data reduction* (Barbara *et al.*, 1996), such as methods of *principal components* and *factor analysis* that are further discussed in Chapters 3.4 and 4.3 of this volume.

Another means to improve the accuracy and the computational tractability of multiple outlier detection methods is the use of biased sampling. Kollios et al. (2003) investigate the use of biased sampling according to the density of the data set to speed up the operation of general data-mining tasks, such as clustering and outlier detection.

7.5 Comparison of Outlier Detection Methods

Since different outlier detection algorithms are based on disjoints sets of assumption, a direct comparison between them is not always possible. In many cases, the data structure and the outlier generating mechanism on which the study is based dictate which method will outperform the others. There are few works that compare different classes of outlier detection methods.

Williams et al. (2002) , for example, suggest an outlier detection method based on *replicator neural networks* (RNNs). They provide a comparative study of RNNs with respect to two parametric (statistical) methods (one proposed in (Hadi, 1994), and the other proposed in (Knorr *et al.*, 2001)) and one data-mining non-parametric method (proposed in (Oliver *et al.*, 1996)). The authors find that RNNs perform adequately to the other methods in many cases, and particularly well on large datasets. Moreover, they find that some statistical outlier detection methods scale well for large dataset, despite claims to the contrary in the data-mining literature. They summaries the study by pointing out that in outlier detection problems simple performance criteria do not easily apply.

Shekhar et al. (2003) characterize the computation structure of spatial outlier detection methods and present scalable algorithms to which they also provide a cost model. The authors present some experimental evaluations of their algorithms using a traffic dataset. Their experimental results show that the *connectivity-clustered access model* (CCAM) achieves the highest clustering efficiency value with respect to a predefined performance measure. Lu et al. (2003) compare three spatial outlier detection algorithms. Two algorithms are sequential and one algorithm based on median as a robust measure for the mean. Their experimental results confirm the effectiveness of these approaches in reducing the risk of falsely negative outliers.

Finally, Penny and Jolliffe (2001) conduct a comparison study with six multivariate outlier detection methods. The methods' properties are investigated by means of a simulation study and the results indicate that no technique is superior to all others. The authors indicate several factors that affect the efficiency of the analyzed methods. In particular, the methods depend on: whether or not the data set is multivariate normal; the dimension of the data set; the type of the outliers; the proportion of outliers in the dataset; and the outliers' degree of contamination (outlyingness). The study motivated the authors to recommend the use of a "battery of multivariate methods" on the dataset in order to detect possible outliers. We fully adopt such a recommendation and argue that the battery of methods should depend, besides on the above-mentioned factors, but also on other factors such as, the data structure dimension and size; the time constraints in regard to single vs. sequential identifiers; and whether an online or an offline outlier detection is required.

References

Acuna E., Rodriguez C. A., "Meta analysis study of outlier detection methods in classification," Technical paper, Department of Mathematics, University of Puerto Rico at Mayaguez, Retrived from academic.uprm.edu/ eacuna/paperout.pdf. In proceedings IPSI 2004, Venice, 2004.

Alwan L.C., Ebrahimi N., Soofi E.S., "Information theoretic framework for process control," European Journal of Operations Research, 111, 526-542, 1998.

Alwan L.C., Roberts H.V., "Time-series modeling for statistical process control," Journal of Business and Economics Statistics, 6 (1), 87-95, 1988.

Apley D.W., Shi J., "The GLRT for statistical process control of autocorrelated processes," IIE Transactions, 31, 1123-1134, 1999.

Barbara D., Faloutsos C., Hellerstein J., Ioannidis Y., Jagadish H.V., Johnson T., Ng R., Poosala V., Ross K., Sevcik K.C., "The New Jersey Data Reduction Report," Data Eng. Bull., September, 1996.

Barbara D., Chen P., "Using the fractal dimension to cluster datasets," In Proc. ACM KDD 2000, 260-264, 2000.

Barnett V., Lewis T., Outliers in Statistical Data. John Wiley, 1994.

Bay S.D., Schwabacher M., "Mining distance-based outliers in near linear time with randomization and a simple pruning rule," In Proc. of the ninth ACM-SIGKDD Conference on Knowledge Discovery and Data Mining, Washington, DC, USA, 2003.

Ben-Gal I., Morag G., Shmilovici A., "CSPC: A Monitoring Procedure for State Dependent Processes," Technometrics, 45(4), 293-311, 2003.

Box G. E. P., Jenkins G. M., Times Series Analysis, Forecasting and Control, Oakland, CA: Holden Day, 1976.

Breunig M.M., Kriegel H.P., Ng R.T., Sander J., "Lof: Identifying density-based local outliers," In Proc. ACMSIGMOD Conf. 2000, 93–104, 2000.

Caussinus H., Roiz A., "Interesting projections of multidimensional data by means of generalized component analysis," In Compstat 90, 121-126, Heidelberg: Physica, 1990.

David H. A., "Robust estimation in the presence of outliers," In Robustness in Statistics, eds. R. L. Launer and G.N. Wilkinson, Academic Press, New York, 61-74, 1979.

Davies L., Gather U., "The identification of multiple outliers," Journal of the American Statistical Association, 88(423), 782-792, 1993.

DuMouchel W., Schonlau M., "A fast computer intrusion detection algorithm based on hypothesis testing of command transition probabilities," In Proceedings of the 4th International Conference on Knowledge Discovery and Data-mining (KDD98), 189–193, 1998.

Fawcett T., Provost F., "Adaptive fraud detection," Data-mining andKnowledge Discovery, 1(3), 291–316, 1997.

Ferguson T. S., "On the Rejection of outliers," In Proceedings of the Fourth Berkeley Symposium on Mathematical Statistics and Probability, vol. 1, 253-287, 1961.

Gather U., "Testing for multisource contamination in location / scale families," Communication in Statistics, Part A: Theory and Methods, 18, 1-34, 1989.

Grubbs F. E., "Proceadures for detecting outlying observations in Samples," Technometrics, 11, 1-21, 1969.

Hadi A. S., "Identifying multiple outliers in multivariate data," Journal of the Royal Statistical Society. Series B, 54, 761-771, 1992.

Hadi A. S., "A modification of a method for the detection of outliers in multivariate samples," Journal of the Royal Statistical Society, Series B, 56(2), 1994.

Hawkins D., Identification of Outliers, Chapman and Hall, 1980.

Hawkins S., He H. X., Williams G. J., Baxter R. A., "Outlier detection using replicator neural networks," In Proceedings of the Fifth International Conference and Data Warehousing and Knowledge Discovery (DaWaK02), Aix en Provence, France, 2002.

Haining R., Spatial Data Analysis in the Social and Environmental Sciences. Cambridge University Press, 1993.

Hampel F. R., "A general qualitative definition of robustness," Annals of Mathematics Statistics, 42, 1887–1896, 1971.

Hampel F. R., "The influence curve and its role in robust estimation," Journal of the American Statistical Association, 69, 382–393, 1974.

Haslett J., Brandley R., Craig P., Unwin A., Wills G., "Dynamic Graphics for Exploring Spatial Data With Application to Locating Global and Local Anomalies," The American Statistician, 45, 234–242, 1991.

Hu T., Sung S. Y., Detecting pattern-based outliers, Pattern Recognition Letters, 24, 3059-3068.

Iglewics B., Martinez J., Outlier Detection using robust measures of scale, Journal of Sattistical Computation and Simulation, 15, 285-293, 1982.

Jin W., Tung A., Han J., "Mining top-n local outliers in large databases," In Proceedings of the 7th International Conference on Knowledge Discovery and Data-mining (KDD01), San Francisco, CA, 2001.

Johnson R., Applied Multivariate Statistical Analysis. Prentice Hall, 1992.

Johnson T., Kwok I., Ng R., "Fast Computation of 2-Dimensional Depth Contours," In Proceedings of the Fourth International Conference on Knowledge Discovery and Data Mining, 224-228. AAAI Press, 1998.

Kaufman L., Rousseeuw P.J., Finding Groups in Data: An Introduction to Cluster Analysis. Wiley, New York, 1990.

Knorr E., Ng R., "A unified approach for mining outliers," In Proceedings Knowledge Discovery KDD, 219-222, 1997.

Knorr E., Ng. R., "Algorithms for mining distance-based outliers in large datasets," In Proc. 24th Int. Conf. Very Large Data Bases (VLDB), 392-403, 24-27, 1998.

Knorr, E., Ng R., Tucakov V., "Distance-based outliers: Algorithms and applications," VLDB Journal: Very Large Data Bases, 8(3-4):237-253, 2000.

Knorr E. M., Ng R. T., Zamar R. H., "Robust space transformations for distance based operations," In Proceedings of the 7th International Conference on Knowledge Discovery and Data-mining (KDD01), 126-135, San Francisco, CA, 2001.

Kollios G., Gunopulos D., Koudas N., Berchtold S., "Efficient biased sampling for approximate clustering and outlier detection in large data sets," IEEE Transactions on Knowledge and Data Engineering, 15 (5), 1170-1187, 2003.

Liu H., Shah S., Jiang W., "On-line outlier detection and data cleaning," Computers and Chemical Engineering, 28, 1635–1647, 2004.

Lu C., Chen D., Kou Y., "Algorithms for spatial outlier detection," In Proceedings of the 3rd IEEE International Conference on Data-mining (ICDM'03), Melbourne, FL, 2003.

Lu C.W., Reynolds M.R., "EWMA Control Charts for Monitoring the Mean of Autocorrelated Processes," Journal of Quality Technology, 31 (2), 166-188, 1999.

Luc A., "Local Indicators of Spatial Association: LISA," Geographical Analysis, 27(2), 93-115, 1995.

Luc A., "Exploratory Spatial Data Analysis and Geographic Information Systems," In M. Painho, editor, New Tools for Spatial Analysis, 45-54, 1994.

Martin R. D., Thomson D. J., "Robust-resistant spectrum estimation," In Proceeding of the IEEE, 70, 1097-1115, 1982.

Montgomery D.C., Mastrangelo C.M., "Some statistical process control methods for autocorrelated data," Journal of Quality Technology, 23 (3), 179-193, 1991.

Ng R.T., Han J., Efficient and Effective Clustering Methods for Spatial Data Mining, In Proceedings of Very Large Data Bases Conference, 144-155, 1994.

Oliver J. J., Baxter R. A., Wallace C. S., "Unsupervised Learning using MML," In Proceedings of the Thirteenth International Conference (ICML96), pages 364-372, Morgan Kaufmann Publishers, San Francisco, CA, 1996.

Panatier Y., Variowin. Software for Spatial Data Analysis in 2D., Springer-Verlag, New York, 1996.

Papadimitriou S., Kitawaga H., Gibbons P.G., Faloutsos C., "LOCI: Fast Outlier Detection Using the Local Correlation Integral," Intel research Laboratory Technical report no. IRP-TR-02-09, 2002.

Penny K. I., Jolliffe I. T., "A comparison of multivariate outlier detection methods for clinical laboratory safety data," The Statistician 50(3), 295-308, 2001.

Perarson R. K., "Outliers in process modeling and identification," IEEE Transactions on Control Systems Technology, 10, 55-63, 2002.

Ramaswamy S., Rastogi R., Shim K., "Efficient algorithms for mining outliers from large data sets," In Proceedings of the ACM SIGMOD International Conference on Management of Data, Dalas, TX, 2000.

Rokach, L., Averbuch, M., and Maimon, O., Information retrieval system for medical narrative reports. Lecture notes in artificial intelligence, 3055. pp. 217-228, Springer-Verlag (2004).

Rosner B., On the detection of many outliers, Technometrics, 17, 221-227, 1975.

Rousseeuw P., "Multivariate estimation with high breakdown point," In: W. Grossmann et al., editors, Mathematical Statistics and Applications, Vol. B, 283-297, Akademiai Kiado: Budapest, 1985.

Rousseeuw P., Leory A., Robust Regression and Outlier Detection, Wiley Series in Probability and Statistics, 1987.

Runger G., Willemain T., "Model-based and Model-free Control of Autocorrelated Processes," Journal of Quality Technology, 27 (4), 283-292, 1995.

Ruts I., Rousseeuw P., "Computing Depth Contours of Bivariate Point Clouds," In Computational Statistics and Data Analysis, 23,153-168, 1996.

Schiffman S. S., Reynolds M. L., Young F. W., Introduction to Multidimensional Scaling: Theory, Methods and Applications. New York: Academic Press, 1981.

Shekhar S., Chawla S., A Tour of Spatial Databases, Prentice Hall, 2002.

Shekhar S., Lu C. T., Zhang P., "Detecting Graph-Based Spatial Outlier: Algorithms and Applications (A Summary of Results)," In Proc. of the Seventh ACM-SIGKDD Conference on Knowledge Discovery and Data Mining, SF, CA, 2001.

Shekhar S., Lu C. T., Zhang P., "Detecting Graph-Based Spatial Outlier," Intelligent Data Analysis: An International Journal, 6(5), 451–468, 2002.

Shekhar S., Lu C. T., Zhang P., "A Unified Approach to Spatial Outliers Detection," GeoInformatica, an International Journal on Advances of Computer Science for Geographic Information System, 7(2), 2003.

Wardell D.G., Moskowitz H., Plante R.D., "Run-length distributions of special-cause control charts for correlated processes," Technometrics, 36 (1), 3–17, 1994.

Tukey J.W., Exploratory Data Analysis. Addison-Wesley, 1977.

Williams G. J., Baxter R. A., He H. X., Hawkins S., Gu L., "A Comparative Study of RNN for Outlier Detection in Data Mining," IEEE International Conference on Data-mining (ICDM'02), Maebashi City, Japan, CSIRO Technical Report CMIS-02/102, 2002.

Williams G. J., Huang Z., "Mining the knowledge mine: The hot spots methodology for mining large real world databases," In Abdul Sattar, editor, Advanced Topics in Artificial Intelligence, volume 1342 of Lecture Notes in Artificial Intelligence, 340–348, Springer, 1997.

Zhang N.F., "A Statistical Control Chart for Stationary Process Data," Technometrics, 40 (1), 24–38, 1998.

Part II

Supervised Methods

8

Supervised Learning

Lior Rokach[1] and Oded Maimon[2]

[1] Department of Information System Engineering, Ben-Gurion University, Beer-Sheba, Israel,
liorrk@bgu.ac.il
[2] Department of Industrial Engineering, Tel-Aviv University, Ramat-Aviv 69978, Israel,
maimon@eng.tau.ac.il

Summary. This chapter summarizes the fundamental aspects of supervised methods. The chapter provides an overview of concepts from various interrelated fields used in subsequent chapters. It presents basic definitions and arguments from the supervised machine learning literature and considers various issues, such as performance evaluation techniques and challenges for data mining tasks.

Key words: Attribute, Classifier, Inducer, Regression, Training Set, Supervised Methods, Instance Space, Sampling, Generalization Error

8.1 Introduction

Supervised methods are methods that attempt to discover the relationship between input attributes (sometimes called independent variables) and a target attribute (sometimes referred to as a dependent variable). The relationship discovered is represented in a structure referred to as a *model*. Usually models describe and explain phenomena, which are hidden in the dataset and can be used for predicting the value of the target attribute knowing the values of the input attributes. The supervised methods can be implemented in a variety of domains such as marketing, finance and manufacturing.

It is useful to distinguish between two main supervised models: *classification models* (*classifiers*) and *Regression Models*. Regression models map the input space into a real-value domain. For instance, a regressor can predict the demand for a certain product given its characteristics. On the other hand, classifiers map the input space into pre-defined classes. For instance, classifiers can be used to classify mortgage consumers as good (fully payback the mortgage on time) and bad (delayed payback). There are many alternatives for representing classifiers, for example, support vector machines, decision trees, probabilistic summaries, algebraic function, etc.

Along with regression and probability estimation, classification is one of the most studied models, possibly one with the greatest practical relevance. The potential ben-

O. Maimon, L. Rokach (eds.), *Data Mining and Knowledge Discovery Handbook*, 2nd ed.,
DOI 10.1007/978-0-387-09823-4_8,

efits of progress in classification are immense since the technique has great impact on other areas, both within Data Mining and in its applications.

8.2 Training Set

In a typical supervised learning scenario, a training set is given and the goal is to form a description that can be used to predict previously unseen examples.

The training set can be described in a variety of languages. Most frequently, it is described as a *bag instance* of a certain *bag schema*. A *bag instance* is a collection of tuples (also known as records, rows or instances) that may contain duplicates. Each tuple is described by a vector of attribute values. The bag schema provides the description of the attributes and their domains. A bag schema is denoted as $B(A \cup y)$. Where A denotes the set of input attributes containing n attributes: $A = \{a_1, \ldots, a_i, \ldots, a_n\}$ and y represents the class variable or the target attribute.

Attributes (sometimes called field, variable or feature) are typically one of two types: nominal (values are members of an unordered set), or numeric (values are real numbers). When the attribute a_i is nominal, it is useful to denote by $dom(a_i) = \{v_{i,1}, v_{i,2}, \ldots, v_{i,|dom(a_i)|}\}$ its domain values, where $|dom(a_i)|$ stands for its finite cardinality. In a similar way, $dom(y) = \{c_1, \ldots, c_{|dom(y)|}\}$ represents the domain of the target attribute. Numeric attributes have infinite cardinalities.

The instance space (the set of all possible examples) is defined as a Cartesian product of all the input attributes domains: $X = dom(a_1) \times dom(a_2) \times \ldots \times dom(a_n)$. The universal instance space (or the *labeled instance space)* U is defined as a Cartesian product of all input attribute domains and the target attribute domain, i.e.: $U = X \times dom(y)$.

The training set is a bag instance consisting of a set of m tuples. Formally the training set is denoted as $S(B) = (\langle x_1, y_1 \rangle, \ldots, \langle x_m, y_m \rangle)$ where $x_q \in X$ and $y_q \in dom(y)$.

It is usually assumed that the training set tuples are generated randomly and independently according to some fixed and unknown joint probability distribution D over U. Note that this is a generalization of the deterministic case when a supervisor classifies a tuple using a function $y = f(x)$.

We use the common notation of bag algebra to present projection (π) and selection (σ) of tuples (Grumbach and Milo, 1996).

8.3 Definition of the Classification Problem

Originally the machine learning community introduced the problem of *concept learning*. Concepts are mental categories for objects, events, or ideas that have a common set of features. According to Mitchell (1997): "each concept can be viewed as describing some subset of objects or events defined over a larger set" (e.g., the subset of a vehicle that constitues trucks). To learn a concept is to infer its general definition from a set of examples. This definition may be either explicitly formulated or left implicit, but either way it assigns each possible example to the concept or not. Thus, a concept can be regarded as a function from the Instance space to the Boolean set, namely: $c : X \rightarrow \{-1, 1\}$. Alternatively, one can refer a concept c as a subset of X, namely: $\{x \in X : c(x) = 1\}$. A *concept class* C is a set of concepts.

Other communities, such as the KDD community prefer to deal with a straightforward extension of *concept learning*, known as The *classification problem* or *multi-class classification problem*. In this case, we search for a function that maps the set of all possible examples into a pre-defined set of class labels which are not limited to the Boolean set. Most frequently the goal of the classifiers inducers is formally defined as follows.

Definition 1. *Given a training set S with input attributes set* $A = \{a_1, a_2, \ldots, a_n\}$ *and a nominal target attribute y from an unknown fixed distribution D over the labeled instance space, the goal is to induce an optimal classifier with minimum generalization error.*

The generalization error is defined as the misclassification rate over the distribution D. In case of the nominal attributes it can be expressed as:

$$\varepsilon(I(S), D) = \sum_{\langle x,y \rangle \in U} D(x,y) \cdot L(y, I(S)(x))$$

where $L(y, I(S)(x)$ is the zero-one loss function defined as:

$$L(y, I(S)(x)) = \begin{cases} 0 \; if \, y = I(S)(x) \\ 1 \; if \, y \neq I(S)(x) \end{cases} \tag{8.1}$$

In case of numeric attributes the sum operator is replaced with the integration operator.

8.4 Induction Algorithms

An *induction algorithm*, or more concisely an *inducer* (also known as learner), is an entity that obtains a training set and forms a model that generalizes the relationship between the input attributes and the target attribute. For example, an inducer may take as an input, specific training tuples with the corresponding class label, and produce a *classifier*.

The notation I represents an inducer and $I(S)$ represents a model which was induced by performing I on a training set S. Using $I(S)$ it is possible to predict the target value of a tuple x_q. This prediction is denoted as $I(S)(x_q)$.

Given the long history and recent growth of the field, it is not surprising that several mature approaches to induction are now available to the practitioner.

Classifiers may be represented differently from one inducer to another. For example, C4.5 (Quinlan, 1993) represents a model as a decision tree while Naïve Bayes (Duda and Hart, 1973) represents a model in the form of probabilistic summaries.

Furthermore, inducers can be deterministic (as in the case of C4.5) or stochastic (as in the case of back propagation)

The classifier generated by the inducer can be used to classify an unseen tuple either by explicitly assigning it to a certain class (crisp classifier) or by providing a vector of probabilities representing the conditional probability of the given instance to belong to each class (probabilistic classifier). Inducers that can construct probabilistic classifiers are known as probabilistic inducers. In this case it is possible to estimate the conditional probability $\hat{P}_{I(S)}(y = c_j \left| a_i = x_{q,i} \right. ; i = 1, \ldots, n)$ of an observation x_q. Note the addition of the "hat" - ^ - to the conditional probability estimation is to distinguish it from the actual conditional probability.

The following chapters review some of the major approaches to concept learning.

8.5 Performance Evaluation

Evaluating the performance of an inducer is a fundamental aspect of machine learning. As stated above, an inducer receives a training set as input and constructs a classification model that can classify an unseen instance . Both the classifier and the inducer can be evaluated using an evaluation criteria. The evaluation is important for understanding the quality of the model (or inducer); for refining parameters in the KDD iterative process; and for selecting the most acceptable model (or inducer) from a given set of models (or inducers).

There are several criteria for evaluating models and inducers. Naturally, classification models with high accuracy are considered better. However, there are other criteria that can be important as well, such as the computational complexity or the comprehensibility of the generated classifier.

8.5.1 Generalization Error

Let $I(S)$ represent a classifier generated by an inducer I on S. Recall that the generalization error of $I(S)$ is its probability to misclassify an instance selected according to the distribution D of the instance labeled space. The *classification accuracy* of a classifier is one minus the generalization error. The *training error* is defined as the percentage of examples in the training set correctly classified by the classifier, formally:

$$\hat{\varepsilon}(I(S), S) = \sum_{\langle x,y \rangle \in S} L(y, I(S)(x)) \tag{8.2}$$

where $L(y, I(S)(x))$ is the zero-one loss function defined in Equation 63.2.

Although generalization error is a natural criterion, its actual value is known only in rare cases (mainly synthetic cases). The reason for that is that the distribution D of the instance labeled space is not known.

One can take the training error as an estimation of the generalization error. However, using the training error as-is will typically provide an optimistically biased estimate, especially if the learning algorithm *over-fits* the training data.

There are two main approaches for estimating the generalization error: theoretical and empirical.

8.5.2 Theoretical Estimation of Generalization Error

A low training error does not guarantee low generalization error. There is often a trade-off between the training error and the confidence assigned to the training error as a predictor for the generalization error, measured by the difference between the generalization and training errors. The capacity of the inducer is a determining factor for this confidence in the training error. Indefinitely speaking, the capacity of an inducer indicates the variety of classifiers it can induce. The notion of VC-Dimension presented below can be used as a measure of the inducers capacity.

Inducers with a large capacity, e.g. a large number of free parameters, relative to the size of the training set are likely to obtain a low training error, but might just be memorizing or over-fitting the patterns and hence exhibit a poor generalization ability. In this regime, the low error is likely to be a poor predictor for the higher generalization error. In the opposite regime, when the capacity is too small for the given number of examples, inducers may under-fit the data, and exhibit both poor training and generalization error. For inducers with an insufficient number of free parameters, the training error may be poor, but it is a good predictor for the generalization error. In between these capacity extremes, there is an optimal capacity for which the best generalization error is obtained, given the character and amount of the available training data.

In "Mathematics of Generalization" Wolpert (1995) discuss four theoretical frameworks for estimating the generalization error, namely: PAC, VC and Bayesian, and statistical physics. All these frameworks combine the training error (which can be easily calculated) with some penalty function expressing the capacity of the inducers.

VC-Framework

Of all the major theoretical approaches to learning from examples, the Vapnik–Chervonenkis theory (Vapnik, 1995) is the most comprehensive, applicable to regression, as well as classification tasks. It provides general, necessary and sufficient conditions for the consistency of the induction procedure in terms of bounds on certain measures. Here we refer to the classical notion of consistency in statistics: both the training error and the generalization error of the induced classifier must converge to the same minimal error value as the training set size tends to infinity. Vapnik's theory also defines a capacity measure of an inducer, the VC-dimension, which is widely used.

VC-theory describes a worst-case scenario: the estimates of the difference between the training and generalization errors are bounds valid for any induction algorithm and probability distribution in the labeled space. The bounds are expressed in terms of the size of the training set and the VC-dimension of the inducer.

Theorem 1. *The bound on the generalization error of hypothesis space H with finite VC-Dimension d is given by:*

$$|\varepsilon(h,D) - \hat{\varepsilon}(h,S)| \leq \sqrt{\frac{d \cdot (\ln \frac{2m}{d} + 1) - \ln \frac{\delta}{4}}{m}} \quad \begin{array}{l} \forall h \in H \\ \forall \delta > 0 \end{array} \tag{8.3}$$

with probability of $1-\delta$ *where* $\hat{\varepsilon}(h,S)$ *represents the training error of classifier h measured on training set S of cardinality m and* $\varepsilon(h,D)$ *represents the generalization error of the classifier h over the distribution D.*

The VC-dimension is a property of a set of all classifiers, denoted by H, that have been examined by the inducer. For the sake of simplicity, we consider classifiers that correspond to the two-class pattern recognition case. In this case, the VC-dimension is defined as the maximum number of data points that can be shattered by the set of admissible classifiers. By definition, a set S of m points is shattered by H if and only if for every dichotomy of S there is some classifier in H that is consistent with this dichotomy. In other words, the set S is shattered by H if there are classifiers that split the points into two classes in all of the 2^m possible ways. Note that, if the VC-dimension of H is d, then there exists at least one set of d points that can be shattered by H. In general, however, it will not be true that every set of d points can be shattered by H.

A sufficient condition for consistency of an induction procedure is that the VC-dimension of the inducer is finite. The VC-dimension of a linear classifier is simply the dimension n of the input space, or the number of free parameters of the classifier. The VC-dimension of a general classifier may however be quite different from the number of free parameters and in many cases it might be very difficult to compute it accurately. In this case it is useful to calculate a lower and upper bound for the VC-Dimension. Schmitt (2002) have presented these VC bounds for neural networks.

PAC-Framework

The Probably Approximately Correct (PAC) learning model was introduced by Valiant (1984). This framework can be used to characterize the concept class "that can be reliably learned from a reasonable number of randomly drawn training examples and a reasonable amount of computation" (Mitchell, 1997). We use the following formal definition of PAC-learnable adapted from (Mitchell, 1997):

Definition 2. *Let C be a concept class defined over the input instance space X with n attributes. Let I be an inducer that considers hypothesis space H. C is said to be PAC-learnable by I using H if for all* $c \in C$*, distributions D over X,* ε *such that* $0 < \varepsilon < 1/2$ *and* δ *such that* $0 < \delta < 1/2$*, learner I with a probability of at least* $(1-\delta)$ *will output a hypothesis* $h \in H$ *such that* $\varepsilon(h,D) \leq \varepsilon$*, in time that is polynomial in* $1/\varepsilon$ *,* $1/\delta$*, n, and size(c), where size(c) represents the encoding length of c in C, assuming some representation for C.*

The PAC learning model provides a general bound on the number of training examples sufficient for any consistent learner I examining a finite hypothesis space H with probability at least (1 - δ) to output a hypothesis $h \in H$ within error ε of the target concept $c \in C \subseteq H$. More specifically, the size of the training set should be: $m \geq \frac{1}{\varepsilon}(\ln(1/\delta) + \ln|H|)$

8.5.3 Empirical Estimation of Generalization Error

Another approach for estimating the generalization error is to split the available examples into two groups: training and test sets. First, the training set is used by the inducer to construct a suitable classifier and then we measure the misclassification rate of this classifier on the test set. This test set error usually provides a better estimation of the generalization error than the training error. The reason for this is that the training error usually under-estimates the generalization error (due to the overfitting phenomena).

When data is limited, it is common practice to *re-sample* the data, that is, partition the data into training and test sets in different ways. An inducer is trained and tested for each partition and the accuracies averaged. By doing this, a more reliable estimate of the true generalization error of the inducer is provided.

Random sub-sampling and *n-fold cross-validation* are two common methods of re-sampling. In random subsampling, the data is randomly partitioned into disjoint training and test sets several times. Errors obtained from each partition are averaged. In n-fold cross-validation, the data is randomly split into n mutually exclusive subsets of approximately equal size. An inducer is trained and tested several times. Each time it is tested on one of the k folds and trained using the remaining $n-1$ folds.

The cross-validation estimate of the generalization error is the overall number of misclassifications, divided by the number of examples in the data. The random subsampling method has the advantage that it can be repeated an indefinite number of times. However, it has the disadvantage that the test sets are not independently drawn with respect to the underlying distribution of examples. Because of this, using a t-test for paired differences with random subsampling can lead to an increased chance of Type I error that is, identifying a significant difference when one does not actually exist. Using a t-test on the generalization error produced on each fold has a lower chance of Type I error but may not give a stable estimate of the generalization error. It is common practice to repeat n fold cross-validation n times in order to provide a stable estimate. However, this, of course, renders the test sets non-independent and increases the chance of Type I error. Unfortunately, there is no satisfactory solution to this problem. Alternative tests suggested by Dietterich (1998) have a low chance of Type I error but a high chance of Type II error - that is, failing to identify a significant difference when one does actually exist.

Stratification is a process often applied during random sub-sampling and n-fold cross-validation. Stratification ensures that the class distribution from the whole dataset is preserved in the training and test sets. Stratification has been shown to help reduce the variance of the estimated error especially for datasets with many classes. Stratified random subsampling with a paired t-test is used herein to evaluate accuracy.

8.5.4 Computational Complexity

Another useful criterion for comparing inducers and classifiers is their computational complexities. Strictly speaking, computational complexity is the amount of CPU consumed by each inducer. It is convenient to differentiate between three metrics of computational complexity:

- Computational complexity for generating a new classifier: This is the most important metric, especially when there is a need to scale the Data Mining algorithm to massive data sets. Because most of the algorithms have computational complexity, which is worse than linear in the numbers of tuples, mining massive data sets might be "prohibitively expensive".
- Computational complexity for updating a classifier: Given new data, what is the computational complexity required for updating the current classifier such that the new classifier reflects the new data?
- Computational Complexity for classifying a new instance: Generally this type is neglected because it is relatively small. However, in certain methods (like k-nearest neighborhood) or in certain real time applications (like anti-missiles applications), this type can be critical.

8.5.5 Comprehensibility

Comprehensibility criterion (also known as interpretability) refers to how well humans grasp the classifier induced. While the generalization error measures how the classifier fits the data, comprehensibility measures the "mental fit" of that classifier.

Many techniques, like neural networks or support vectors machines, are designed solely to achieve accuracy. However, as their classifiers are represented using large assemblages of real valued parameters, they are also difficult to understand and are referred to as black-box models.

It is often important for the researcher to be able to inspect an induced classifier. For domains such as medical diagnosis, the users must understand how the system makes its decisions in order to be confident of the outcome. Data mining can also play an important role in the process of scientific discovery. A system may discover salient features in the input data whose importance was not previously recognized. If the representations formed by the inducer are comprehensible, then these discoveries can be made accessible to human review (Hunter and Klein, 1993).

Comprehensibility can vary between different classifiers created by the same inducer. For instance, in the case of decision trees, the size (number of nodes) of the induced trees is also important. Smaller trees are preferred because they are easier to interpret. However, this is only a rule of thumb. In some pathologic cases, a large and unbalanced tree can still be easily interpreted (Buja and Lee, 2001).

As the reader can see, the accuracy and complexity factors can be quantitatively estimated, while comprehensibility is more subjective.

Another distinction is that the complexity and comprehensibility depend mainly on the induction method and much less on the specific domain considered. On the other hand, the dependence of error metrics on a specific domain cannot be neglected.

8.6 Scalability to Large Datasets

Obviously induction is one of the central problems in many disciplines such as machine learning, pattern recognition, and statistics. However the feature that distinguishes Data Mining from traditional methods is its scalability to very large sets of varied types of input data. The notion, "scalability" usually refers to datasets that fulfill at least one of the following properties: high number of records or high dimensionality.

"Classical" induction algorithms have been applied with practical success in many relatively simple and small-scale problems. However, trying to discover knowledge in real life and large databases, introduces time and memory problems.

As large databases have become the norm in many fields (including astronomy, molecular biology, finance, marketing, health care, and many others), the use of Data Mining to discover patterns in them has become a potentially very productive enterprise. Many companies are staking a large part of their future on these "Data Mining" applications, and looking to the research community for solutions to the fundamental problems they encounter.

While a very large amount of available data used to be the dream of any data analyst, nowadays the synonym for "very large" has become "terabyte", a hardly imaginable volume of information. Information-intensive organizations (like telecom companies and banks) are supposed to accumulate several terabytes of raw data every one to two years.

However, the availability of an electronic data repository (in its enhanced form known as a "data warehouse") has created a number of previously unknown problems, which, if ignored, may turn the task of efficient Data Mining into mission impossible. Managing and analyzing huge data warehouses requires special and very expensive hardware and software, which often causes a company to exploit only a small part of the stored data.

According to Fayyad *et al.* (1996) the explicit challenges for the data mining research community are to develop methods that facilitate the use of Data Mining algorithms for real-world databases. One of the characteristics of a real world databases is high volume data.

Huge databases pose several challenges:

- Computing complexity. Since most induction algorithms have a computational complexity that is greater than linear in the number of attributes or tuples, the execution time needed to process such databases might become an important issue.

- Poor classification accuracy due to difficulties in finding the correct classifier. Large databases increase the size of the search space, and the chance that the inducer will select an overfitted classifier that generally invalid.
- Storage problems: In most machine learning algorithms, the entire training set should be read from the secondary storage (such as magnetic storage) into the computer's primary storage (main memory) before the induction process begins. This causes problems since the main memory's capability is much smaller than the capability of magnetic disks.

The difficulties in implementing classification algorithms as is, on high volume databases, derives from the increase in the number of records/instances in the database and of attributes/features in each instance (high dimensionality). Approaches for dealing with a high number of records include:

- Sampling methods - statisticians are selecting records from a population by different sampling techniques.
- Aggregation - reduces the number of records either by treating a group of records as one, or by ignoring subsets of "unimportant" records.
- Massively parallel processing - exploiting parallel technology - to simultaneously solve various aspects of the problem.
- Efficient storage methods that enable the algorithm to handle many records. For instance, Shafer *et al.* (1996) presented the SPRINT which constructs an attribute list data structure.
- Reducing the algorithm's search space - For instance the PUBLIC algorithm (Rastogi and Shim, 2000) integrates the growing and pruning of decision trees by using MDL cost in order to reduce the computational complexity.

8.7 The "Curse of Dimensionality"

High dimensionality of the input (that is, the number of attributes) increases the size of the search space in an exponential manner, and thus increases the chance that the inducer will find spurious classifiers that are generally invalid. It is well-known that the required number of labeled samples for supervised classification increases as a function of dimensionality (Jimenez and Landgrebe, 1998). Fukunaga (1990) showed that the required number of training samples is linearly related to the dimensionality for a linear classifier and to the square of the dimensionality for a quadratic classifier. In terms of nonparametric classifiers like decision trees, the situation is even more severe. It has been estimated that as the number of dimensions increases, the sample size needs to increase exponentially in order to have an effective estimate of multivariate densities (Hwang *et al.*, 1994).

This phenomenon is usually called the "curse of dimensionality". Bellman (1961) was the first to coin this term, while working on complicated signal processing. Techniques, like decision trees inducers, that are efficient in low dimensions, fail to provide meaningful results when the number of dimensions increases beyond a "modest" size. Furthermore, smaller classifiers, involving fewer features (probably less than 10), are much more understandable by humans. Smaller classifiers are also more appropriate for user-driven Data Mining techniques such as visualization.

Most of the methods for dealing with high dimensionality focus on feature selection techniques, i.e. selecting a single subset of features upon which the inducer (induction algorithm) will run, while ignoring the rest. The selection of the subset can be done manually by using prior knowledge to identify irrelevant variables or by using proper algorithms.

In the last decade, feature selection has enjoyed increased interest by many researchers. Consequently many feature selection algorithms have been proposed, some of which have reported a remarkable improvement in accuracy. Please refer to Chapter 4.3 in this volume for further reading.

Despite its popularity, the usage of feature selection methodologies for overcoming the obstacles of high dimensionality has several drawbacks:

- The assumption that a large set of input features can be reduced to a small subset of relevant features is not always true. In some cases the target feature is actually affected by most of the input features, and removing features will cause a significant loss of important information.
- The outcome (i.e. the subset) of many algorithms for feature selection (for example almost any of the algorithms that are based upon the wrapper methodology) is strongly dependent on the training set size. That is, if the training set is small, then the size of the reduced subset will be also small. Consequently, relevant features might be lost. Accordingly, the induced classifiers might achieve lower accuracy compared to classifiers that have access to all relevant features.
- In some cases, even after eliminating a set of irrelevant features, the researcher is left with relatively large numbers of relevant features.
- The backward elimination strategy, used by some methods, is extremely inefficient for working with large-scale databases, where the number of original features is more than 100.

A number of linear dimension reducers have been developed over the years. The linear methods of dimensionality reduction include projection pursuit (Friedman and Tukey, 1973), factor analysis (Kim and Mueller, 1978), and principal components analysis (Dunteman, 1989). These methods are not aimed directly at eliminating irrelevant and redundant features, but are rather concerned with transforming the observed variables into a small number of "projections" or "dimensions". The underlying assumptions are that the variables are numeric and the dimensions can be expressed as linear combinations of the observed variables (and vice versa). Each discovered dimension is assumed to represent an unobserved factor and thus to provide a new way of understanding the data (similar to the curve equation in the regression models).

The linear dimension reducers have been enhanced by constructive induction systems that use a set of existing features and a set of pre-defined constructive operators to derive new features (Pfahringer, 1994, Ragavan and Rendell, 1993). These methods are effective for high dimensionality applications only if the original domain size of the input feature can be in fact decreased dramatically.

One way to deal with the above-mentioned disadvantages is to use a very large training set (which should increase in an exponential manner as the number of input features increases). However, the researcher rarely enjoys this privilege, and even if it does happen, the researcher will probably encounter the aforementioned difficulties derived from a high number of instances.

Practically most of the training sets are still considered "small" not due to their absolute size but rather due to the fact that they contain too few instances given the nature of the investigated problem, namely the instance space size, the space distribution and the intrinsic noise.

8.8 Classification Problem Extensions

In this section we survey a few extensions to the classical classification problem.

In classic supervised learning problems, classes are mutually exclusive by definition. In multi-label classification problems each training instance is given a set of candidate class labels but only one of the candidate labels is the correct one (Jin and Ghahramani, 2002). The reader should not be confused with multi-class classification problems which usually refer to simply having more than two possible disjoint classes for the classier to learn.

In practice, many real problems are formalized as a "Multiple Labels" problem. For example, this occurs when there is a disagreement regarding the label of a certain training instance. Another typical example of "multiple labels" occurs when there is a hierarchical structure over the class labels and some of the training instances are given the labels of the superclasses instead of the labels of the subclasses. For instance a certain training instance representing a course can be labeled as "engineering", while this class consists of more specific classes such as "electrical engineering", "industrial engineering", etc.

A closely-related problem is the "multi-label" classification problem. In this case, the classes are not mutually exclusive. One instance is actually associated with many labels, and all labels are correct. Such problems exist, for example, in text classifications. Texts may simultaneously belong to more than one genre (Schapire and Singer, 2000). In bioinformatics, genes may have multiple functions, yielding multiple labels (Clare and King, 2001). Boutella *et al.* (2004) presented a framework to handle multi-label classification problems. They present approaches for training and testing in this scenario and introduce new metrics for evaluating the results.

The difference between "multi-label" and "multiple Label" should be clarified. In "multi-label" each training instance can have multiple class labels, and all the assigned class labels are actually correct labels while in "Multiple Labels" problem only one of the assigned multiple labels is the target label.

Another closely-related problem is the fuzzy classification problem (Janikow, 1998), in which class boundaries are not clearly defined. Instead, each instance has a ceratin membership function for each class which represents the degree to which the instance belongs to this class.

Another related problem is "preference learning" (Furnkranz, 1997). The training set consists of a collection of training instances which are associated with a set of pairwise preferences between labels, expressing that one label is preferred over another. The goal of "preference learning" is to predict a ranking, of all possible labels for a new training example. Cohen *et al.* (1999) have investigated a more narrow version of the problem, the learning of one single preference function. The "constraint classification" problem (Har-Peled *et al.*, 2002) is a superset of the "preference learning" and "multi-label classification", in which each example is labeled according to some partial order.

In "multiple-instance" problems (Dietterich *et al.*, 1997), the instances are organized into bags of several instances, and a class label is tagged for every bag of instances. In the "multiple-instance" problem, at least one of the instances within each bag corresponds to the label of the bag and all other instances within the bag are just noises. Note that in "multiple-instance" problem the ambiguity comes from the instances within the bag.

Supervised learnig methods are useful for many application domains, such as: Manufacturing lr18,lr14,lr6, Security lr7,l10,lr12, Medicine lr2,lr9,lr15, and support many other data mining tasks, including unsupervised learning lr13,lr8,lr5,lr16 and genetic algorithms lr17,lr11,lr1,lr4.

References

Arbel, R. and Rokach, L., Classifier evaluation under limited resources, Pattern Recognition Letters, 27(14): 1619–1631, 2006, Elsevier.

Averbuch, M. and Karson, T. and Ben-Ami, B. and Maimon, O. and Rokach, L., Context-sensitive medical information retrieval, The 11th World Congress on Medical Informatics (MEDINFO 2004), San Francisco, CA, September 2004, IOS Press, pp. 282–286.

Boutella R. M., Luob J., Shena X., Browna C. M., Learning multi-label scene classification, Pattern Recognition, 37(9), pp. 1757-1771, 2004.

Buja, A. and Lee, Y.S., Data Mining criteria for tree based regression and classification, Proceedings of the 7th International Conference on Knowledge Discovery and Data Mining, (pp 27-36), San Diego, USA, 2001.

Clare, A., King R.D., Knowledge Discovery in Multi-label Phenotype Data, Lecture Notes in Computer Science, Vol. 2168, Springer, Berlin, 2001.

Cohen S., Rokach L., Maimon O., Decision Tree Instance Space Decomposition with Grouped Gain-Ratio, Information Science, Volume 177, Issue 17, pp. 3592-3612, 2007.

Cohen, W. W., Schapire R.E., and Singer Y., Learning to order things. Journal of Artificial Intelligence Research, 10:243270, 1999.

Dietterich, T. G., Approximate statistical tests for comparing supervised classification learning algorithms. Neural Computation, 10(7): 1895-1924, 1998.

Dietterich, T. G., Lathrop, R. H. , and Perez, T. L., Solving the multiple-instance problem with axis-parallel rectangles, Artificial Intelligence, 89(1-2), pp. 31-71, 1997.

Duda, R., and Hart, P., Pattern Classification and Scene Analysis, New-York, Wiley, 1973.

Dunteman, G.H., Principal Components Analysis, Sage Publications, 1989.

Fayyad, U., Piatesky-Shapiro, G. & Smyth P., From Data Mining to Knowledge Discovery: An Overview. In U. Fayyad, G. Piatetsky-Shapiro, P. Smyth, & R. Uthurusamy (Eds), Advances in Knowledge Discovery and Data Mining, pp 1-30, AAAI/MIT Press, 1996.

Friedman, J.H. & Tukey, J.W., A Projection Pursuit Algorithm for Exploratory Data Analysis, IEEE Transactions on Computers, 23: 9, 881-889, 1973.

Fukunaga, K., Introduction to Statistical Pattern Recognition. San Diego, CA: Academic, 1990.

Fürnkranz J. and Hüllermeier J., Pairwise preference learning and ranking. In Proc. ECML03, pages 145156, Cavtat, Croatia, 2003.

Grumbach S., Milo T., Towards Tractable Algebras for Bags. Journal of Computer and System Sciences 52(3): 570-588, 1996.

Har-Peled S., Roth D., and Zimak D., Constraint classification: A new approach to multiclass classification. In Proc. ALT02, pages 365379, Lubeck, Germany, 2002, Springer.

Hunter L., Klein T. E., Finding Relevant Biomolecular Features. ISMB 1993, pp. 190-197, 1993.

Hwang J., Lay S., and Lippman A., Nonparametric multivariate density estimation: A comparative study, IEEE Transaction on Signal Processing, 42(10): 2795-2810, 1994.

Janikow, C.Z., Fuzzy Decision Trees: Issues and Methods, IEEE Transactions on Systems, Man, and Cybernetics, Vol. 28, Issue 1, pp. 1-14. 1998.

Jimenez, L. O., & Landgrebe D. A., Supervised Classification in High- Dimensional Space: Geometrical, Statistical, and Asymptotical Properties of Multivariate Data. IEEE Transaction on Systems Man, and Cybernetics - Part C: Applications and Reviews, 28:39-54, 1998.

Jin, R. , & Ghahramani Z., Learning with Multiple Labels, The Sixteenth Annual Conference on Neural Information Processing Systems (NIPS 2002) Vancouver, Canada, pp. 897-904, December 9-14, 2002.

Kim J.O. & Mueller C.W., Factor Analysis: Statistical Methods and Practical Issues. Sage Publications, 1978.

Maimon O., and Rokach, L. Data Mining by Attribute Decomposition with semiconductors manufacturing case study, in Data Mining for Design and Manufacturing: Methods and Applications, D. Braha (ed.), Kluwer Academic Publishers, pp. 311–336, 2001.

Maimon O. and Rokach L., "Improving supervised learning by feature decomposition", Proceedings of the Second International Symposium on Foundations of Information and Knowledge Systems, Lecture Notes in Computer Science, Springer, pp. 178-196, 2002.

Maimon, O. and Rokach, L., Decomposition Methodology for Knowledge Discovery and Data Mining: Theory and Applications, Series in Machine Perception and Artificial Intelligence - Vol. 61, World Scientific Publishing, ISBN:981-256-079-3, 2005.

Mitchell, T., Machine Learning, McGraw-Hill, 1997.

Moskovitch R, Elovici Y, Rokach L, Detection of unknown computer worms based on behavioral classification of the host, Computational Statistics and Data Analysis, 52(9):4544–4566, 2008.

Pfahringer, B., Controlling constructive induction in CiPF, In Bergadano, F. and De Raedt, L. (Eds.), Proceedings of the seventh European Conference on Machine Learning, pp. 242-256, Springer-Verlag, 1994.

Quinlan, J. R., C4.5: Programs for Machine Learning, Morgan Kaufmann, Los Altos, 1993.

Ragavan, H. and Rendell, L., Look ahead feature construction for learning hard concepts. In Proceedings of the Tenth International Machine Learning Conference: pp. 252-259, Morgan Kaufman, 1993.

Rastogi, R., and Shim, K., PUBLIC: A Decision Tree Classifier that Integrates Building and Pruning,Data Mining and Knowledge Discovery, 4(4):315-344, 2000.
Rokach, L., Decomposition methodology for classification tasks: a meta decomposer framework, Pattern Analysis and Applications, 9(2006):257–271.
Rokach L., Genetic algorithm-based feature set partitioning for classification problems,Pattern Recognition, 41(5):1676–1700, 2008.
Rokach L., Mining manufacturing data using genetic algorithm-based feature set decomposition, Int. J. Intelligent Systems Technologies and Applications, 4(1):57-78, 2008.
Rokach, L. and Maimon, O., Theory and applications of attribute decomposition, IEEE International Conference on Data Mining, IEEE Computer Society Press, pp. 473–480, 2001.
Rokach L. and Maimon O., Feature Set Decomposition for Decision Trees, Journal of Intelligent Data Analysis, Volume 9, Number 2, 2005b, pp 131–158.
Rokach, L. and Maimon, O., Clustering methods, Data Mining and Knowledge Discovery Handbook, pp. 321–352, 2005, Springer.
Rokach, L. and Maimon, O., Data mining for improving the quality of manufacturing: a feature set decomposition approach, Journal of Intelligent Manufacturing, 17(3):285–299, 2006, Springer.
Rokach, L., Maimon, O., Data Mining with Decision Trees: Theory and Applications, World Scientific Publishing, 2008.
Rokach L., Maimon O. and Lavi I., Space Decomposition In Data Mining: A Clustering Approach, Proceedings of the 14th International Symposium On Methodologies For Intelligent Systems, Maebashi, Japan, Lecture Notes in Computer Science, Springer-Verlag, 2003, pp. 24–31.
Rokach, L. and Maimon, O. and Averbuch, M., Information Retrieval System for Medical Narrative Reports, Lecture Notes in Artificial intelligence 3055, page 217-228 Springer-Verlag, 2004.
Rokach, L. and Maimon, O. and Arbel, R., Selective voting-getting more for less in sensor fusion, International Journal of Pattern Recognition and Artificial Intelligence 20 (3) (2006), pp. 329–350.
Schapire R., Singer Y., Boostexter: a boosting-based system for text categorization, Machine Learning 39 (2/3):135168, 2000.
Schmitt , M., On the complexity of computing and learning with multiplicative neural networks, Neural Computation 14: 2, 241-301, 2002.
Shafer, J. C., Agrawal, R. and Mehta, M. , SPRINT: A Scalable Parallel Classifier for Data Mining, Proc. 22nd Int. Conf. Very Large Databases, T. M. Vijayaraman and Alejandro P. Buchmann and C. Mohan and Nandlal L. Sarda (eds), 544-555, Morgan Kaufmann, 1996.
Valiant, L. G. (1984). A theory of the learnable. Communications of the ACM 1984, pp. 1134-1142.
Vapnik, V.N., The Nature of Statistical Learning Theory. Springer-Verlag, New York, 1995.
Wolpert, D. H., The relationship between PAC, the statistical physics framework, the Bayesian framework, and the VC framework. In D. H. Wolpert, editor, The Mathematics of Generalization, The SFI Studies in the Sciences of Complexity, pages 117-214. AddisonWesley, 1995.

9

Classification Trees

Lior Rokach[1] and Oded Maimon[2]

[1] Department of Information System Engineering, Ben-Gurion University, Beer-Sheba, Israel,
liorrk@bgu.ac.il
[2] Department of Industrial Engineering, Tel-Aviv University, Ramat-Aviv 69978, Israel,
maimon@eng.tau.ac.il

Summary. Decision Trees are considered to be one of the most popular approaches for representing classifiers. Researchers from various disciplines such as statistics, machine learning, pattern recognition, and Data Mining have dealt with the issue of growing a decision tree from available data. This paper presents an updated survey of current methods for constructing decision tree classifiers in a top-down manner. The chapter suggests a unified algorithmic framework for presenting these algorithms and describes various splitting criteria and pruning methodologies.

Key words: Decision tree, Information Gain, Gini Index, Gain Ratio, Pruning, Minimum Description Length, C4.5, CART, Oblivious Decision Trees

9.1 Decision Trees

A decision tree is a classifier expressed as a recursive partition of the instance space. The decision tree consists of nodes that form a *rooted tree*, meaning it is a *directed tree* with a node called "root" that has no incoming edges. All other nodes have exactly one incoming edge. A node with outgoing edges is called an *internal* or test node. All other nodes are called leaves (also known as terminal or decision nodes). In a decision tree, each internal node splits the instance space into two or more subspaces according to a certain discrete function of the input attributes values. In the simplest and most frequent case, each test considers a single attribute, such that the instance space is partitioned according to the attribute's value. In the case of numeric attributes, the condition refers to a range.

Each leaf is assigned to one class representing the most appropriate target value. Alternatively, the leaf may hold a probability vector indicating the probability of the target attribute having a certain value. Instances are classified by navigating them from the root of the tree down to a leaf, according to the outcome of the tests along the path. Figure 9.1 describes a decision tree that reasons whether or not a potential customer will respond to a direct mailing. Internal nodes are represented as circles,

O. Maimon, L. Rokach (eds.), *Data Mining and Knowledge Discovery Handbook*, 2nd ed.,
DOI 10.1007/978-0-387-09823-4_9,

whereas leaves are denoted as triangles. Note that this decision tree incorporates both nominal and numeric attributes. Given this classifier, the analyst can predict the response of a potential customer (by sorting it down the tree), and understand the behavioral characteristics of the entire potential customers population regarding direct mailing. Each node is labeled with the attribute it tests, and its branches are labeled with its corresponding values.

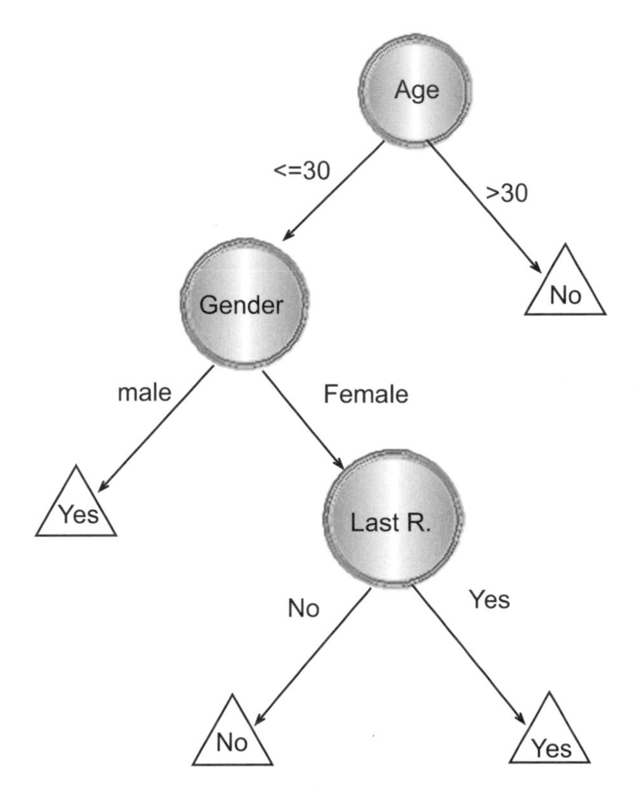

Fig. 9.1. Decision Tree Presenting Response to Direct Mailing.

In case of numeric attributes, decision trees can be geometrically interpreted as a collection of hyperplanes, each orthogonal to one of the axes. Naturally, decision-makers prefer less complex decision trees, since they may be considered more comprehensible. Furthermore, according to Breiman *et al.* (1984) the tree complexity has a crucial effect on its accuracy. The tree complexity is explicitly controlled by the stopping criteria used and the pruning method employed. Usually the tree complexity is measured by one of the following metrics: the total number of nodes, total number of leaves, tree depth and number of attributes used. Decision tree induction is closely related to rule induction. Each path from the root of a decision tree to one of its leaves can be transformed into a rule simply by conjoining the tests along the path to form the antecedent part, and taking the leaf's class prediction as the class value. For example, one of the paths in Figure 9.1 can be transformed into the rule: "If customer age is is less than or equal to or equal to 30, and the gender of the customer is "Male" – then the customer will respond to the mail". The resulting rule set can then be simplified to improve its comprehensibility to a human user, and possibly its accuracy (Quinlan, 1987).

9.2 Algorithmic Framework for Decision Trees

Decision tree inducers are algorithms that automatically construct a decision tree from a given dataset. Typically the goal is to find the optimal decision tree by minimizing the generalization error. However, other target functions can be also defined, for instance, minimizing the number of nodes or minimizing the average depth.

Induction of an optimal decision tree from a given data is considered to be a hard task. It has been shown that finding a minimal decision tree consistent with the training set is NP–hard (Hancock *et al.*, 1996). Moreover, it has been shown that constructing a minimal binary tree with respect to the expected number of tests required for classifying an unseen instance is NP–complete (Hyafil and Rivest, 1976). Even finding the minimal equivalent decision tree for a given decision tree (Zantema and Bodlaender, 2000) or building the optimal decision tree from decision tables is known to be NP–hard (Naumov, 1991).

The above results indicate that using optimal decision tree algorithms is feasible only in small problems. Consequently, heuristics methods are required for solving the problem. Roughly speaking, these methods can be divided into two groups: top–down and bottom–up with clear preference in the literature to the first group.

There are various top–down decision trees inducers such as ID3 (Quinlan, 1986)! C4.5 (Quinlan, 1993), CART (Breiman *et al.*, 1984). Some consist of two conceptual phases: growing and pruning (C4.5 and CART). Other inducers perform only the growing phase.

Figure 9.2 presents a typical algorithmic framework for top–down inducing of a decision tree using growing and pruning. Note that these algorithms are greedy by nature and construct the decision tree in a top–down, recursive manner (also known as "divide and conquer"). In each iteration, the algorithm considers the partition of the training set using the outcome of a discrete function of the input attributes. The selection of the most appropriate function is made according to some splitting measures. After the selection of an appropriate split, each node further subdivides the training set into smaller subsets, until no split gains sufficient splitting measure or a stopping criteria is satisfied.

9.3 Univariate Splitting Criteria

9.3.1 Overview

In most of the cases, the discrete splitting functions are univariate. Univariate means that an internal node is split according to the value of a single attribute. Consequently,

```
TreeGrowing (S,A,y,SplitCriterion,StoppingCriterion)
where:
S - Training Set
A - Input Feature Set
y - Target Feature
SplitCriterion - the method for evaluating a certain split
StoppingCriterion - the criteria to stop the growing process

Create a new tree T with a single root node.
IF StoppingCriterion(S) THEN
    Mark T as a leaf with the most
    common value of y in S as a label.
ELSE
    Find attribute a that obtains the best SplitCriterion(a_i,S).
    Label t with a
    FOR each outcome v_i of a:
        Set Subtree_i= TreeGrowing (σ_{a=v_i}S,A,y).
        Connect the root node of t_T to Subtree_i with
                an edge that is labelled as v_i
    END FOR
END IF RETURN TreePruning (S,T,y)
TreePruning (S,T,y) Where: S - Training Set
y - Target Feature T - The tree to be pruned
DO
    Select a node t in T such that pruning it
        maximally improve some evaluation criteria
    IF t≠Ø THEN T=pruned(T,t)
UNTIL t=Ø
RETURN T
```

Fig. 9.2. Top-Down Algorithmic Framework for Decision Trees Induction.

the inducer searches for the best attribute upon which to split. There are various univariate criteria. These criteria can be characterized in different ways, such as:

- According to the origin of the measure: information theory, dependence, and distance.
- According to the measure structure: impurity based criteria, normalized impurity based criteria and Binary criteria.

The following section describes the most common criteria in the literature.

9.3.2 Impurity-based Criteria

Given a random variable x with k discrete values, distributed according to $P = (p_1, p_2, \ldots, p_k)$, an impurity measure is a function ϕ:[0, 1]$^k \rightarrow R$ that satisfies the following conditions:

- ϕ (P)$\geq$0
- ϕ (P) is minimum if $\exists$i such that component $p_i = 1$.
- ϕ (P) is maximum if $\forall$i, $1 \leq i \leq k$, $p_i = 1/k$.
- ϕ (P) is symmetric with respect to components of P.
- ϕ (P) is smooth (differentiable everywhere) in its range.

Note that if the probability vector has a component of 1 (the variable x gets only one value), then the variable is defined as pure. On the other hand, if all components are equal, the level of impurity reaches maximum.

Given a training set S, the probability vector of the target attribute y is defined as:

$$P_y(S) = \left(\frac{\left|\sigma_{y=c_1} S\right|}{|S|}, \ldots, \frac{\left|\sigma_{y=c_{|dom(y)|}} S\right|}{|S|} \right)$$

The goodness–of–split due to discrete attribute a_i is defined as reduction in impurity of the target attribute after partitioning S according to the values $v_{i,j} \in dom(a_i)$:

$$\Delta\Phi(a_i, S) = \phi(P_y(S)) - \sum_{j=1}^{|dom(a_i)|} \frac{|\sigma_{a_i=v_{i,j}} S|}{|S|} \cdot \phi(P_y(\sigma_{a_i=v_{i,j}} S))$$

9.3.3 Information Gain

Information gain is an impurity-based criterion that uses the entropy measure (origin from information theory) as the impurity measure (Quinlan, 1987).

$$InformationGain(a_i, S) =$$
$$Entropy(y, S) - \sum_{v_{i,j} \in dom(a_i)} \frac{\left|\sigma_{a_i=v_{i,j}} S\right|}{|S|} \cdot Entropy(y, \sigma_{a_i=v_{i,j}} S)$$

where:

$$Entropy(y, S) = \sum_{c_j \in dom(y)} -\frac{\left|\sigma_{y=c_j} S\right|}{|S|} \cdot \log_2 \frac{\left|\sigma_{y=c_j} S\right|}{|S|}$$

9.3.4 Gini Index

Gini index is an impurity-based criterion that measures the divergences between the probability distributions of the target attribute's values. The Gini index has been used in various works such as (Breiman *et al.*, 1984) and (Gelfand *et al.*, 1991) and it is defined as:

$$Gini(y,S) = 1 - \sum_{c_j \in dom(y)} \left(\frac{\left|\sigma_{y=c_j} S\right|}{|S|} \right)^2$$

Consequently the evaluation criterion for selecting the attribute a_i is defined as:

$$GiniGain(a_i,S) = Gini(y,S) - \sum_{v_{i,j} \in dom(a_i)} \frac{\left|\sigma_{a_i=v_{i,j}} S\right|}{|S|} \cdot Gini(y, \sigma_{a_i=v_{i,j}} S)$$

9.3.5 Likelihood-Ratio Chi–Squared Statistics

The likelihood–ratio is defined as (Attneave, 1959)

$$G^2(a_i,S) = 2 \cdot \ln(2) \cdot |S| \cdot InformationGain(a_i,S)$$

This ratio is useful for measuring the statistical significance of the information gain criterion. The zero hypothesis (H_0) is that the input attribute and the target attribute are conditionally independent. If H_0 holds, the test statistic is distributed as χ^2 with degrees of freedom equal to: $(dom(a_i) - 1) \cdot (dom(y) - 1)$.

9.3.6 DKM Criterion

The DKM criterion is an impurity-based splitting criterion designed for binary class attributes (Dietterich *et al.*, 1996) and (Kearns and Mansour, 1999). The impurity-based function is defined as:

$$DKM(y,S) = 2 \cdot \sqrt{\left(\frac{\left|\sigma_{y=c_1} S\right|}{|S|} \right) \cdot \left(\frac{\left|\sigma_{y=c_2} S\right|}{|S|} \right)}$$

It has been theoretically proved (Kearns and Mansour, 1999) that this criterion requires smaller trees for obtaining a certain error than other impurity based criteria (information gain and Gini index).

9.3.7 Normalized Impurity Based Criteria

The impurity-based criterion described above is biased towards attributes with larger domain values. Namely, it prefers input attributes with many values over attributes with less values (Quinlan, 1986). For instance, an input attribute that represents the national security number will probably get the highest information gain. However, adding this attribute to a decision tree will result in a poor generalized accuracy. For that reason, it is useful to “normalize” the impurity based measures, as described in the following sections.

9.3.8 Gain Ratio

The gain ratio "normalizes" the information gain as follows (Quinlan, 1993):

$$GainRatio(a_i,S)=\frac{InformationGain(a_i,S)}{Entropy(a_i,S)}$$

Note that this ratio is not defined when the denominator is zero. Also the ratio may tend to favor attributes for which the denominator is very small. Consequently, it is suggested in two stages. First the information gain is calculated for all attributes. As a consequence, taking into consideration only attributes that have performed at least as good as the average information gain, the attribute that has obtained the best ratio gain is selected. It has been shown that the gain ratio tends to outperform simple information gain criteria, both from the accuracy aspect, as well as from classifier complexity aspects (Quinlan, 1988).

9.3.9 Distance Measure

The distance measure, like the gain ratio, normalizes the impurity measure. However, it suggests normalizing it in a different way (Lopez de Mantras, 1991):

$$\frac{\Delta\Phi(a_i,S)}{-\sum\limits_{v_{i,j}\in dom(a_i)}\ \sum\limits_{c_k\in dom(y)}\frac{|\sigma_{a_i=v_{i,j}AND\,y=c_k}S|}{|S|}\cdot\log_2\frac{|\sigma_{a_i=v_{i,j}AND\,y=c_k}S|}{|S|}}$$

9.3.10 Binary Criteria

The binary criteria are used for creating binary decision trees. These measures are based on division of the input attribute domain into two sub-domains.

Let $\beta(a_i,dom_1(a_i),dom_2(a_i),S)$ denote the binary criterion value for attribute a_i over sample S when $dom_1(a_i)$ and $dom_2(a_i)$ are its corresponding subdomains. The value obtained for the optimal division of the attribute domain into two mutually exclusive and exhaustive sub-domains is used for comparing attributes.

9.3.11 Twoing Criterion

The gini index may encounter problems when the domain of the target attribute is relatively wide (Breiman *et al.*, 1984). In this case it is possible to employ binary criterion called twoing criterion. This criterion is defined as:

$$twoing(a_i,dom_1(a_i),dom_2(a_i),S)=$$
$$0.25\cdot\frac{\left|\sigma_{a_i\in dom_1(a_i)}S\right|}{|S|}\cdot\frac{\left|\sigma_{a_i\in dom_2(a_i)}S\right|}{|S|}\cdot$$
$$\left(\sum_{c_i\in dom(y)}\left|\frac{\left|\sigma_{a_i\in dom_1(a_i)\ AND\ y=c_i}S\right|}{\left|\sigma_{a_i\in dom_1(a_i)}S\right|}-\frac{\left|\sigma_{a_i\in dom_2(a_i)AND\,y=c_i}S\right|}{\left|\sigma_{a_i\in dom_2(a_i)}S\right|}\right|\right)^2$$

When the target attribute is binary, the gini and twoing criteria are equivalent. For multi–class problems, the twoing criteria prefer attributes with evenly divided splits.

9.3.12 Orthogonal (ORT) Criterion

The ORT criterion was presented by Fayyad and Irani (1992). This binary criterion is defined as:

$$ORT(a_i, dom_1(a_i), dom_2(a_i), S) = 1 - cos\theta(P_{y,1}, P_{y,2})$$

where $\theta(P_{y,1}, P_{y,2})$ is the angle between two vectors $P_{y,1}$ and $P_{y,2}$. These vectors represent the probability distribution of the target attribute in the partitions $\sigma_{a_i \in dom_1(a_i)} S$ and $\sigma_{a_i \in dom_2(a_i)} S$ respectively.

It has been shown that this criterion performs better than the information gain and the Gini index for specific problem constellations.

9.3.13 Kolmogorov–Smirnov Criterion

A binary criterion that uses Kolmogorov–Smirnov distance has been proposed in Friedman (1977) and Rounds (1980). Assuming a binary target attribute, namely $dom(y) = \{c_1, c_2\}$, the criterion is defined as:

$$KS(a_i, dom_1(a_i), dom_2(a_i), S) =$$

$$\left| \frac{\left|\sigma_{a_i \in dom_1(a_i)\, AND\, y=c_1} S\right|}{\left|\sigma_{y=c_1} S\right|} - \frac{\left|\sigma_{a_i \in dom_1(a_i)\, AND\, y=c_2} S\right|}{\left|\sigma_{y=c_2} S\right|} \right|$$

This measure was extended in (Utgoff and Clouse, 1996) to handle target attributes with multiple classes and missing data values. Their results indicate that the suggested method outperforms the gain ratio criteria.

9.3.14 AUC–Splitting Criteria

The idea of using the AUC metric as a splitting criterion was recently proposed in (Ferri *et al.*, 2002). The attribute that obtains the maximal area under the convex hull of the ROC curve is selected. It has been shown that the AUC–based splitting criterion outperforms other splitting criteria both with respect to classification accuracy and area under the ROC curve. It is important to note that unlike impurity criteria, this criterion does not perform a comparison between the impurity of the parent node with the weighted impurity of the children after splitting.

9.3.15 Other Univariate Splitting Criteria

Additional univariate splitting criteria can be found in the literature, such as permutation statistics (Li and Dubes, 1986), mean posterior improvements (Taylor and Silverman, 1993) and hypergeometric distribution measures (Martin, 1997).

9.3.16 Comparison of Univariate Splitting Criteria

Comparative studies of the splitting criteria described above, and others, have been conducted by several researchers during the last thirty years, such as (Baker and Jain, 1976, BenBassat, 1978, Mingers, 1989, Fayyad and Irani, 1992, Buntine and Niblett, 1992, Loh and Shih, 1997, Loh and Shih, 1999, Lim *et al.*, 2000). Most of these comparisons are based on empirical results, although there are some theoretical conclusions.

Many of the researchers point out that in most of the cases, the choice of splitting criteria will not make much difference on the tree performance. Each criterion is superior in some cases and inferior in others, as the "No–Free–Lunch" theorem suggests.

9.4 Multivariate Splitting Criteria

In multivariate splitting criteria, several attributes may participate in a single node split test. Obviously, finding the best multivariate criteria is more complicated than finding the best univariate split. Furthermore, although this type of criteria may dramatically improve the tree's performance, these criteria are much less popular than the univariate criteria.

Most of the multivariate splitting criteria are based on the linear combination of the input attributes. Finding the best linear combination can be performed using a greedy search (Breiman *et al.*, 1984, Murthy, 1998), linear programming (Duda and Hart, 1973, Bennett and Mangasarian, 1994), linear discriminant analysis (Duda and Hart, 1973, Friedman, 1977, Sklansky and Wassel, 1981, Lin and Fu, 1983, Loh and Vanichsetakul, 1988, John, 1996) and others (Utgoff, 1989a, Lubinsky, 1993, Sethi and Yoo, 1994).

9.5 Stopping Criteria

The growing phase continues until a stopping criterion is triggered. The following conditions are common stopping rules:

1. All instances in the training set belong to a single value of y.
2. The maximum tree depth has been reached.
3. The number of cases in the terminal node is less than the minimum number of cases for parent nodes.
4. If the node were split, the number of cases in one or more child nodes would be less than the minimum number of cases for child nodes.
5. The best splitting criteria is not greater than a certain threshold.

9.6 Pruning Methods

9.6.1 Overview

Employing tightly stopping criteria tends to create small and under–fitted decision trees. On the other hand, using loosely stopping criteria tends to generate large decision trees that are over–fitted to the training set. Pruning methods originally suggested in (Breiman *et al.*, 1984) were developed for solving this dilemma. According to this methodology, a loosely stopping criterion is used, letting the decision tree to overfit the training set. Then the over-fitted tree is cut back into a smaller tree by removing sub–branches that are not contributing to the generalization accuracy. It has been shown in various studies that employing pruning methods can improve the generalization performance of a decision tree, especially in noisy domains.

Another key motivation of pruning is "trading accuracy for simplicity" as presented in (Bratko and Bohanec, 1994). When the goal is to produce a sufficiently accurate compact concept description, pruning is highly useful. Within this process, the initial decision tree is seen as a completely accurate one. Thus the accuracy of a pruned decision tree indicates how close it is to the initial tree.

There are various techniques for pruning decision trees. Most of them perform top-down or bottom-up traversal of the nodes. A node is pruned if this operation improves a certain criteria. The following subsections describe the most popular techniques.

9.6.2 Cost–Complexity Pruning

Cost-complexity pruning (also known as weakest link pruning or error-complexity pruning) proceeds in two stages (Breiman *et al.*, 1984). In the first stage, a sequence of trees $T_0, T_1, \ldots, T_k$ is built on the training data where T_0 is the original tree before pruning and T_k is the root tree.

In the second stage, one of these trees is chosen as the pruned tree, based on its generalization error estimation.

The tree T_{i+1} is obtained by replacing one or more of the sub–trees in the predecessor tree T_i with suitable leaves. The sub–trees that are pruned are those that obtain the lowest increase in apparent error rate per pruned leaf:

$$\alpha = \frac{\varepsilon(pruned(T,t),S) - \varepsilon(T,S)}{|leaves(T)| - |leaves(pruned(T,t))|}$$

where $\varepsilon(T,S)$ indicates the error rate of the tree T over the sample S and |leaves(T)| denotes the number of leaves in T. *pruned(T,t)* denotes the tree obtained by replacing the node t in T with a suitable leaf.

In the second phase the generalization error of each pruned tree $T_0, T_1, \ldots, T_k$ is estimated. The best pruned tree is then selected. If the given dataset is large enough, the authors suggest breaking it into a training set and a pruning set. The trees are constructed using the training set and evaluated on the pruning set. On the other hand, if the given dataset is not large enough, they propose to use cross–validation methodology, despite the computational complexity implications.

9.6.3 Reduced Error Pruning

A simple procedure for pruning decision trees, known as reduced error pruning, has been suggested by Quinlan (1987). While traversing over the internal nodes from the bottom to the top, the procedure checks for each internal node, whether replacing it with the most frequent class does not reduce the tree's accuracy. In this case, the node is pruned. The procedure continues until any further pruning would decrease the accuracy.

In order to estimate the accuracy, Quinlan (1987) proposes to use a pruning set. It can be shown that this procedure ends with the smallest accurate sub–tree with respect to a given pruning set.

9.6.4 Minimum Error Pruning (MEP)

The minimum error pruning has been proposed in (Olaru and Wehenkel, 2003). It performs bottom–up traversal of the internal nodes. In each node it compares the l-probability error rate estimation with and without pruning.

The l-probability error rate estimation is a correction to the simple probability estimation using frequencies. If S_t denotes the instances that have reached a leaf t, then the expected error rate in this leaf is:

$$\varepsilon'(t) = 1 - \max_{c_i \in dom(y)} \frac{\left|\sigma_{y=c_i} S_t\right| + l \cdot p_{apr}(y = c_i)}{|S_t| + l}$$

where $p_{apr}(y = c_i)$ is the *a–priori* probability of y getting the value c_i, and l denotes the weight given to the *a–priori* probability.

The error rate of an internal node is the weighted average of the error rate of its branches. The weight is determined according to the proportion of instances along each branch. The calculation is performed recursively up to the leaves.

If an internal node is pruned, then it becomes a leaf and its error rate is calculated directly using the last equation. Consequently, we can compare the error rate before and after pruning a certain internal node. If pruning this node does not increase the error rate, the pruning should be accepted.

9.6.5 Pessimistic Pruning

Pessimistic pruning avoids the need of pruning set or cross validation and uses the pessimistic statistical correlation test instead (Quinlan, 1993).

The basic idea is that the error ratio estimated using the training set is not reliable enough. Instead, a more realistic measure, known as the continuity correction for binomial distribution, should be used:

$$\varepsilon'(T,S) = \varepsilon(T,S) + \frac{|leaves(T)|}{2 \cdot |S|}$$

However, this correction still produces an optimistic error rate. Consequently, one should consider pruning an internal node t if its error rate is within one standard error from a reference tree, namely (Quinlan, 1993):

$$\varepsilon'(pruned(T,t),S) \leq \varepsilon'(T,S) + \sqrt{\frac{\varepsilon'(T,S)\cdot(1-\varepsilon'(T,S))}{|S|}}$$

The last condition is based on statistical confidence interval for proportions. Usually the last condition is used such that T refers to a sub–tree whose root is the internal node t and S denotes the portion of the training set that refers to the node t.

The pessimistic pruning procedure performs top–down traversing over the internal nodes. If an internal node is pruned, then all its descendants are removed from the pruning process, resulting in a relatively fast pruning.

9.6.6 Error–based Pruning (EBP)

Error–based pruning is an evolution of pessimistic pruning. It is implemented in the well–known C4.5 algorithm.

As in pessimistic pruning, the error rate is estimated using the upper bound of the statistical confidence interval for proportions.

$$\varepsilon_{UB}(T,S) = \varepsilon(T,S) + Z_{\alpha}\cdot\sqrt{\frac{\varepsilon(T,S)\cdot(1-\varepsilon(T,S))}{|S|}}$$

where $\varepsilon(T,S)$ denotes the misclassification rate of the tree T on the training set S. Z is the inverse of the standard normal cumulative distribution and α is the desired significance level.

Let *subtree(T,t)* denote the subtree rooted by the node t. Let *maxchild(T,t)* denote the most frequent child node of t (namely most of the instances in S reach this particular child) and let S_t denote all instances in S that reach the node t.

The procedure performs bottom–up traversal over all nodes and compares the following values:

1. $\varepsilon_{UB}(subtree(T,t),S_t)$
2. $\varepsilon_{UB}(pruned(subtree(T,t),t),S_t)$
3. $\varepsilon_{UB}(subtree(T,maxchild(T,t)),S_{maxchild(T,t)})$

According to the lowest value the procedure either leaves the tree as is, prune the node t, or replaces the node t with the subtree rooted by *maxchild(T,t)*.

9.6.7 Optimal Pruning

The issue of finding optimal pruning has been studied in (Bratko and Bohanec, 1994) and (Almuallim, 1996). The first research introduced an algorithm which guarantees optimality, known as OPT. This algorithm finds the optimal pruning based on dynamic programming, with the complexity of $\Theta(|leveas(T)|^2)$, where T is the initial decision tree. The second research introduced an improvement of OPT called OPT–2, which also performs optimal pruning using dynamic programming. However, the time and space complexities of OPT–2 are both $\Theta(|leveas(T^*)| \cdot |\text{internal}(T)|)$, where T^* is the target (pruned) decision tree and T is the initial decision tree.

Since the pruned tree is habitually much smaller than the initial tree and the number of internal nodes is smaller than the number of leaves, OPT–2 is usually more efficient than OPT in terms of computational complexity.

9.6.8 Minimum Description Length (MDL) Pruning

The minimum description length can be used for evaluating the generalized accuracy of a node (Rissanen, 1989, Quinlan and Rivest, 1989, Mehta *et al.*, 1995). This method measures the size of a decision tree by means of the number of bits required to encode the tree. The MDL method prefers decision trees that can be encoded with fewer bits. The cost of a split at a leaf t can be estimated as (Mehta *et al.*, 1995):

$$\text{Cost(t)} = \sum_{c_i \in dom(y)} \left|\sigma_{y=c_i} S_t\right| \cdot \ln \frac{|S_t|}{\left|\sigma_{y=c_i} S_t\right|} + \frac{|dom(y)|-1}{2} \ln \frac{|S_t|}{2} + \ln \frac{\pi^{\frac{|dom(y)|}{2}}}{\Gamma(\frac{|dom(y)|}{2})}$$

where S_t denotes the instances that have reached node t. The splitting cost of an internal node is calculated based on the cost aggregation of its children.

9.6.9 Other Pruning Methods

There are other pruning methods reported in the literature, such as the MML (Minimum Message Length) pruning method (Wallace and Patrick, 1993) and Critical Value Pruning (Mingers, 1989).

9.6.10 Comparison of Pruning Methods

Several studies aim to compare the performance of different pruning techniques (Quinlan, 1987, Mingers, 1989, Esposito *et al.*, 1997). The results indicate that some methods (such as cost–complexity pruning, reduced error pruning) tend to over–pruning, i.e. creating smaller but less accurate decision trees. Other methods (like error-based pruning, pessimistic error pruning and minimum error pruning) bias toward under–pruning. Most of the comparisons concluded that the "no free lunch" theorem applies in this case also, namely there is no pruning method that in any case outperforms other pruning methods.

9.7 Other Issues

9.7.1 Weighting Instances

Some decision trees inducers may give different treatments to different instances. This is performed by weighting the contribution of each instance in the analysis according to a provided weight (between 0 and 1).

9.7.2 Misclassification costs

Several decision trees inducers can be provided with numeric penalties for classifying an item into one class when it really belongs in another.

9.7.3 Handling Missing Values

Missing values are a common experience in real-world data sets. This situation can complicate both induction (a training set where some of its values are missing) as well as classification (a new instance that miss certain values).

This problem has been addressed by several researchers. One can handle missing values in the training set in the following way: let $\sigma_{a_i=?}S$ indicate the subset of instances in S whose a_i values are missing. When calculating the splitting criteria using attribute a_i, simply ignore all instances their values in attribute a_i are unknown, that is, instead of using the splitting criteria $\Delta\Phi(a_i,S)$ it uses $\Delta\Phi(a_i,S-\sigma_{a_i=?}S)$.

On the other hand, in case of missing values, the splitting criteria should be reduced proportionally as nothing has been learned from these instances (Quinlan, 1989). In other words, instead of using the splitting criteria $\Delta\Phi(a_i,S)$, it uses the following correction:

$$\frac{|S-\sigma_{a_i=?}S|}{|S|}\Delta\Phi(a_i,S-\sigma_{a_i=?}S).$$

In a case where the criterion value is normalized (as in the case of gain ratio), the denominator should be calculated as if the missing values represent an additional value in the attribute domain. For instance, the Gain Ratio with missing values should be calculated as follows:

$$GainRatio(a_i,S) = \frac{\frac{|S-\sigma_{a_i=?}S|}{|S|}InformationGain(a_i,S-\sigma_{a_i=?}S)}{-\frac{|\sigma_{a_i=?}S|}{|S|}\log(\frac{|\sigma_{a_i=?}S|}{|S|})-\sum\limits_{v_{i,j}\in dom(a_i)}\frac{|\sigma_{a_i=v_{i,j}}S|}{|S|}\log(\frac{|\sigma_{a_i=v_{i,j}}S|}{|S|})}$$

Once a node is split, it is required to add $\sigma_{a_i=?}S$ to each one of the outgoing edges with the following corresponding weight:

$$\left|\sigma_{a_i=v_{i,j}}S\right|/\left|S-\sigma_{a_i=?}S\right|$$

The same idea is used for classifying a new instance with missing attribute values. When an instance encounters a node where its splitting criteria can be evaluated due to a missing value, it is passed through to all outgoing edges. The predicted class will be the class with the highest probability in the weighted union of all the leaf nodes at which this instance ends up.

Another approach known as *surrogate splits* was presented by Breiman *et al.* (1984) and is implemented in the CART algorithm. The idea is to find for each split in the tree a surrogate split which uses a different input attribute and which most resembles the original split. If the value of the input attribute used in the original split is missing, then it is possible to use the surrogate split. The resemblance between two binary splits over sample S is formally defined as:

$$res(a_i,dom_1(a_i),dom_2(a_i),a_j,dom_1(a_j),dom_2(a_j),S) =$$
$$\frac{\left|\sigma_{a_i \in dom_1(a_i)\ AND\ a_j \in dom_1(a_j)} S\right|}{|S|} + \frac{\left|\sigma_{a_i \in dom_2(a_i)\ AND\ a_j \in dom_2(a_j)} S\right|}{|S|}$$

When the first split refers to attribute a_i and it splits $dom(a_i)$ into $dom_1(a_i)$ and $dom_2(a_i)$. The alternative split refers to attribute a_j and splits its domain to $dom_1(a_j)$ and $dom_2(a_j)$.

The missing value can be estimated based on other instances (Loh and Shih, 1997). On the learning phase, if the value of a nominal attribute a_i in tuple q is missing, then it is estimated by its mode over all instances having the same target attribute value. Formally,

$$estimate(a_i,y_q,S) = \underset{v_{i,j} \in dom(a_i)}{\text{argmax}} \left|\sigma_{a_i=v_{i,j}\ AND\ y=y_q} S\right|$$

where y_q denotes the value of the target attribute in the tuple q. If the missing attribute a_i is numeric, then instead of using mode of a_i it is more appropriate to use its mean.

9.8 Decision Trees Inducers

9.8.1 ID3

The ID3 algorithm is considered as a very simple decision tree algorithm (Quinlan, 1986). ID3 uses information gain as splitting criteria. The growing stops when all instances belong to a single value of target feature or when best information gain is not greater than zero. ID3 does not apply any pruning procedures nor does it handle numeric attributes or missing values.

9.8.2 C4.5

C4.5 is an evolution of ID3, presented by the same author (Quinlan, 1993). It uses gain ratio as splitting criteria. The splitting ceases when the number of instances to be split is below a certain threshold. Error–based pruning is performed after the growing phase. C4.5 can handle numeric attributes. It can induce from a training set that incorporates missing values by using corrected gain ratio criteria as presented above.

9.8.3 CART

CART stands for Classification and Regression Trees (Breiman *et al.*, 1984). It is characterized by the fact that it constructs binary trees, namely each internal node has exactly two outgoing edges. The splits are selected using the twoing criteria and the obtained tree is pruned by cost–complexity Pruning. When provided, CART can consider misclassification costs in the tree induction. It also enables users to provide prior probability distribution.

An important feature of CART is its ability to generate regression trees. Regression trees are trees where their leaves predict a real number and not a class. In case of regression, CART looks for splits that minimize the prediction squared error (the least–squared deviation). The prediction in each leaf is based on the weighted mean for node.

9.8.4 CHAID

Starting from the early seventies, researchers in applied statistics developed procedures for generating decision trees, such as: AID (Sonquist *et al.*, 1971), MAID (Gillo, 1972), THAID (Morgan and Messenger, 1973) and CHAID (Kass, 1980). CHAID (Chisquare–Automatic–Interaction–Detection) was originally designed to handle nominal attributes only. For each input attribute a_i, CHAID finds the pair of values in V_i that is least significantly different with respect to the target attribute. The significant difference is measured by the p value obtained from a statistical test. The statistical test used depends on the type of target attribute. If the target attribute is continuous, an F test is used. If it is nominal, then a Pearson chi–squared test is used. If it is ordinal, then a likelihood–ratio test is used.

For each selected pair, CHAID checks if the p value obtained is greater than a certain merge threshold. If the answer is positive, it merges the values and searches for an additional potential pair to be merged. The process is repeated until no significant pairs are found.

The best input attribute to be used for splitting the current node is then selected, such that each child node is made of a group of homogeneous values of the selected attribute. Note that no split is performed if the adjusted p value of the best input attribute is not less than a certain split threshold. This procedure also stops when one of the following conditions is fulfilled:

1. Maximum tree depth is reached.
2. Minimum number of cases in node for being a parent is reached, so it can not be split any further.
3. Minimum number of cases in node for being a child node is reached.

CHAID handles missing values by treating them all as a single valid category. CHAID does not perform pruning.

9.8.5 QUEST

The QUEST (Quick, Unbiased, Efficient, Statistical Tree) algorithm supports univariate and linear combination splits (Loh and Shih, 1997). For each split, the association between each input attribute and the target attribute is computed using the ANOVA F–test or Levene's test (for ordinal and continuous attributes) or Pearson's chi–square (for nominal attributes). If the target attribute is multinomial, two–means clustering is used to create two super–classes. The attribute that obtains the highest association with the target attribute is selected for splitting. Quadratic Discriminant Analysis (QDA) is applied to find the optimal splitting point for the input attribute. QUEST has negligible bias and it yields binary decision trees. Ten–fold cross–validation is used to prune the trees.

9.8.6 Reference to Other Algorithms

Table 9.1 describes other decision trees algorithms available in the literature. Obviously there are many other algorithms which are not included in this table. Nevertheless, most of these algorithms are a variation of the algorithmic framework presented above. A profound comparison of the above algorithms and many others has been conducted in (Lim *et al.*, 2000).

9.9 Advantages and Disadvantages of Decision Trees

Several advantages of the decision tree as a classification tool have been pointed out in the literature:

1. Decision trees are self–explanatory and when compacted they are also easy to follow. In other words if the decision tree has a reasonable number of leaves, it can be grasped by non–professional users. Furthermore decision trees can be converted to a set of rules. Thus, this representation is considered as comprehensible.
2. Decision trees can handle both nominal and numeric input attributes.
3. Decision tree representation is rich enough to represent any discrete–value classifier.
4. Decision trees are capable of handling datasets that may have errors.
5. Decision trees are capable of handling datasets that may have missing values.
6. Decision trees are considered to be a nonparametric method. This means that decision trees have no assumptions about the space distribution and the classifier structure.

On the other hand, decision trees have such disadvantages as:

1. Most of the algorithms (like ID3 and C4.5) require that the target attribute will have only discrete values.

Table 9.1. Additional Decision Tree Inducers.

Algorithm	Description	Reference
CAL5	Designed specifically for numerical–valued attributes	Muller and Wysotzki (1994)
FACT	An earlier version of QUEST. Uses statistical tests to select an attribute for splitting each node and then uses discriminant analysis to find the split point.	Loh and Vanichsetakul (1988)
LMDT	Constructs a decision tree based on multivariate tests are linear combinations of the attributes.	Brodley and Utgoff (1995)
T1	A one–level decision tree that classifies instances using only one attribute. Missing values are treated as a "special value". Support both continuous an nominal attributes.	Holte (1993)
PUBLIC	Integrates the growing and pruning by using MDL cost in order to reduce the computational complexity.	Rastogi and Shim (2000)
MARS	A multiple regression function is approximated using linear splines and their tensor products.	Friedman (1991)

2. As decision trees use the "divide and conquer" method, they tend to perform well if a few highly relevant attributes exist, but less so if many complex interactions are present. One of the reasons for this is that other classifiers can compactly describe a classifier that would be very challenging to represent using a decision tree. A simple illustration of this phenomenon is the replication problem of decision trees (Pagallo and Huassler, 1990). Since most decision trees divide the instance space into mutually exclusive regions to represent a concept, in some cases the tree should contain several duplications of the same sub-tree in order to represent the classifier. For instance if the concept follows the following binary function: $y = (A_1 \cap A_2) \cup (A_3 \cap A_4)$ then the minimal univariate decision tree that represents this function is illustrated in Figure 9.3. Note that the tree contains two copies of the same subt-ree.
3. The greedy characteristic of decision trees leads to another disadvantage that should be pointed out. This is its over–sensitivity to the training set, to irrelevant attributes and to noise (Quinlan, 1993).

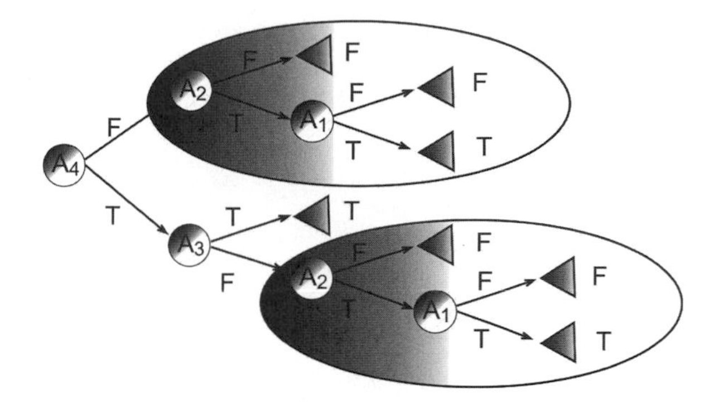

Fig. 9.3. Illustration of Decision Tree with Replication.

9.10 Decision Tree Extensions

In the following sub-sections, we discuss some of the most popular extensions to the classical decision tree induction paradigm.

9.10.1 Oblivious Decision Trees

Oblivious decision trees are decision trees for which all nodes at the same level test the same feature. Despite its restriction, oblivious decision trees are found to be effective for feature selection. Almuallim and Dietterich (1994) as well as Schlimmer (1993) have proposed forward feature selection procedure by constructing oblivious decision trees. Langley and Sage (1994) suggested backward selection using the same means. It has been shown that oblivious decision trees can be converted to a decision table (Kohavi and Sommerfield, 1998).

Figure 9.4 illustrates a typical oblivious decision tree with four input features: glucose level (G), age (A), hypertension (H) and pregnant (P) and the Boolean target feature representing whether that patient suffers from diabetes. Each layer is uniquely associated with an input feature by representing the interaction of that feature and the input features of the previous layers. The number that appears in the terminal nodes indicates the number of instances that fit this path. For example, regarding patients whose glucose level is less than 107 and their age is greater than 50, 10 of them are positively diagnosed with diabetes while 2 of them are not diagnosed with diabetes.

The principal difference between the oblivious decision tree and a regular decision tree structure is the constant ordering of input attributes at every terminal node of the oblivious decision tree, the property which is necessary for minimizing the overall subset of input attributes (resulting in dimensionality reduction). The arcs that connect the terminal nodes and the nodes of the target layer are labelled with the number of records that fit this path.

An oblivious decision tree is usually built by a greedy algorithm, which tries to maximize the mutual information measure in every layer. The recursive search for explaining attributes is terminated when there is no attribute that explains the target with statistical significance.

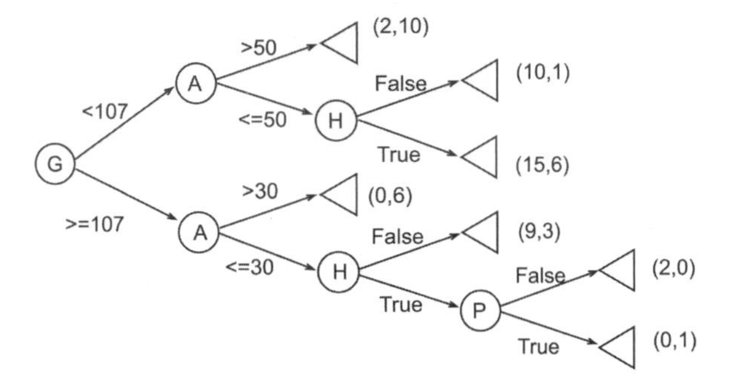

Fig. 9.4. Illustration of Oblivious Decision Tree.

9.10.2 Fuzzy Decision Trees

In classical decision trees, an instance can be associated with only one branch of the tree. Fuzzy decision trees (FDT) may simultaneously assign more than one branch to the same instance with gradual certainty.

FDTs preserve the symbolic structure of the tree and its comprehensibility. Nevertheless, FDT can represent concepts with graduated characteristics by producing real-valued outputs with gradual shifts

Janikow (1998) presented a complete framework for building a fuzzy tree including several inference procedures based on conflict resolution in rule-based systems and efficient approximate reasoning methods.

Olaru and Wehenkel (2003) presented a new fuzzy decision trees called soft decision trees (SDT). This approach combines tree-growing and pruning, to determine the structure of the soft decision tree, with refitting and backfitting, to improve its generalization capabilities. They empirically showed that soft decision trees are significantly more accurate than standard decision trees. Moreover, a global model variance study shows a much lower variance for soft decision trees than for standard trees as a direct cause of the improved accuracy.

Peng (2004) has used FDT to improve the performance of the classical inductive learning approach in manufacturing processes. Peng (2004) proposed to use soft discretization of continuous-valued attributes. It has been shown that FDT can deal with the noise or uncertainties existing in the data collected in industrial systems.

9.10.3 Decision Trees Inducers for Large Datasets

With the recent growth in the amount of data collected by information systems, there is a need for decision trees that can handle large datasets. Catlett (1991) has examined two methods for efficiently growing decision trees from a large database by reducing the computation complexity required for induction. However, the Catlett method requires that all data will be loaded into the main memory before induction. That is to say, the largest dataset that can be induced is bounded by the memory size. Fifield (1992) suggests parallel implementation of the ID3 Algorithm. However, like Catlett, it assumes that all dataset can fit in the main memory. Chan and Stolfo (1997) suggest partitioning the datasets into several disjointed datasets, so that each dataset is loaded separately into the memory and used to induce a decision tree. The decision trees are then combined to create a single classifier. However, the experimental results indicate that partition may reduce the classification performance, meaning that the classification accuracy of the combined decision trees is not as good as the accuracy of a single decision tree induced from the entire dataset.

The SLIQ algorithm (Mehta *et al.*, 1996) does not require loading the entire dataset into the main memory, instead it uses a secondary memory (disk). In other words, a certain instance is not necessarily resident in the main memory all the time. SLIQ creates a single decision tree from the entire dataset. However, this method also has an upper limit for the largest dataset that can be processed, because it uses a data structure that scales with the dataset size and this data structure must be resident in main memory all the time. The SPRINT algorithm uses a similar approach (Shafer *et al.*, 1996). This algorithm induces decision trees relatively quickly and removes all of the memory restrictions from decision tree induction. SPRINT scales any impurity based split criteria for large datasets. Gehrke *et al* (2000) introduced RainForest; a unifying framework for decision tree classifiers that are capable of scaling any specific algorithms from the literature (including C4.5, CART and CHAID). In addition to its generality, RainForest improves SPRINT by a factor of three. In contrast to SPRINT, however, RainForest requires a certain minimum amount of main memory, proportional to the set of distinct values in a column of the input relation. However, this requirement is considered modest and reasonable.

Other decision tree inducers for large datasets can be found in the literature (Alsabti *et al.*, 1998, Freitas and Lavington, 1998, Gehrke *et al.*, 1999).

9.10.4 Incremental Induction

Most of the decision trees inducers require rebuilding the tree from scratch for reflecting new data that has become available. Several researches have addressed the issue of updating decision trees incrementally. Utgoff (1989b, 1997) presents several methods for updating decision trees incrementally. An extension to the CART algorithm that is capable of inducing incrementally is described in (Crawford *et al.*, 2002).

Decision trees are useful for many application domains, such as: Manufacturing lr18,lr14, Security lr7,l10 and Medicine lr2,lr9, and for many data mining tasks, such as: supervised learning lr6,lr12, lr15, unsupervised learning lr13,lr8,lr5,lr16 and genetic algorithms lr17,lr11,lr1,lr4.

References

Almuallim H., An Efficient Algorithm for Optimal Pruning of Decision Trees. Artificial Intelligence 83(2): 347-362, 1996.

Almuallim H,. and Dietterich T.G., Learning Boolean concepts in the presence of many irrelevant features. Artificial Intelligence, 69: 1-2, 279-306, 1994.

Alsabti K., Ranka S. and Singh V., CLOUDS: A Decision Tree Classifier for Large Datasets, Conference on Knowledge Discovery and Data Mining (KDD-98), August 1998.

Attneave F., Applications of Information Theory to Psychology. Holt, Rinehart and Winston, 1959.

Arbel, R. and Rokach, L., Classifier evaluation under limited resources, Pattern Recognition Letters, 27(14): 1619–1631, 2006, Elsevier.

Averbuch, M. and Karson, T. and Ben-Ami, B. and Maimon, O. and Rokach, L., Context-sensitive medical information retrieval, The 11th World Congress on Medical Informatics (MEDINFO 2004), San Francisco, CA, September 2004, IOS Press, pp. 282–286.

Baker E., and Jain A. K., On feature ordering in practice and some finite sample effects. In Proceedings of the Third International Joint Conference on Pattern Recognition, pages 45-49, San Diego, CA, 1976.

BenBassat M., Myopic policies in sequential classification. IEEE Trans. on Computing, 27(2):170-174, February 1978.

Bennett X. and Mangasarian O.L., Multicategory discrimination via linear programming. Optimization Methods and Software, 3:29-39, 1994.

Bratko I., and Bohanec M., Trading accuracy for simplicity in decision trees, Machine Learning 15: 223-250, 1994.

Breiman L., Friedman J., Olshen R., and Stone C.. Classification and Regression Trees. Wadsworth Int. Group, 1984.

Brodley C. E. and Utgoff. P. E., Multivariate decision trees. Machine Learning, 19:45-77, 1995.

Buntine W., Niblett T., A Further Comparison of Splitting Rules for Decision-Tree Induction. Machine Learning, 8: 75-85, 1992.

Catlett J., Mega induction: Machine Learning on Vary Large Databases, PhD, University of Sydney, 1991.

Chan P.K. and Stolfo S.J, On the Accuracy of Meta-learning for Scalable Data Mining, J. Intelligent Information Systems, 8:5-28, 1997.

Cohen S., Rokach L., Maimon O., Decision Tree Instance Space Decomposition with Grouped Gain-Ratio, Information Science, Volume 177, Issue 17, pp. 3592-3612, 2007.

Crawford S. L., Extensions to the CART algorithm. Int. J. of ManMachine Studies, 31(2):197-217, August 1989.

Dietterich, T. G., Kearns, M., and Mansour, Y., Applying the weak learning framework to understand and improve C4.5. Proceedings of the Thirteenth International Conference on Machine Learning, pp. 96-104, San Francisco: Morgan Kaufmann, 1996.

Duda, R., and Hart, P., Pattern Classification and Scene Analysis, New-York, Wiley, 1973.

Esposito F., Malerba D. and Semeraro G., A Comparative Analysis of Methods for Pruning Decision Trees. EEE Transactions on Pattern Analysis and Machine Intelligence, 19(5):476-492, 1997.

Fayyad U., and Irani K. B., The attribute selection problem in decision tree generation. In proceedings of Tenth National Conference on Artificial Intelligence, pp. 104–110, Cambridge, MA: AAAI Press/MIT Press, 1992.

Ferri C., Flach P., and Hernández-Orallo J., Learning Decision Trees Using the Area Under the ROC Curve. In Claude Sammut and Achim Hoffmann, editors, Proceedings of the 19th International Conference on Machine Learning, pp. 139-146. Morgan Kaufmann, July 2002

Fifield D. J., Distributed Tree Construction From Large Datasets, Bachelor's Honor Thesis, Australian National University, 1992.

Freitas X., and Lavington S. H., Mining Very Large Databases With Parallel Processing, Kluwer Academic Publishers, 1998.

Friedman J. H., A recursive partitioning decision rule for nonparametric classifiers. IEEE Trans. on Comp., C26:404-408, 1977.

Friedman, J. H., "Multivariate Adaptive Regression Splines", The Annual Of Statistics, 19, 1-141, 1991.

Gehrke J., Ganti V., Ramakrishnan R., Loh W., BOAT-Optimistic Decision Tree Construction. SIGMOD Conference 1999: pp. 169-180, 1999.

Gehrke J., Ramakrishnan R., Ganti V., RainForest - A Framework for Fast Decision Tree Construction of Large Datasets,Data Mining and Knowledge Discovery, 4, 2/3) 127-162, 2000.

Gelfand S. B., Ravishankar C. S., and Delp E. J., An iterative growing and pruning algorithm for classification tree design. IEEE Transaction on Pattern Analysis and Machine Intelligence, 13(2):163-174, 1991.

Gillo M. W., MAID: A Honeywell 600 program for an automatised survey analysis. Behavioral Science 17: 251-252, 1972.

Hancock T. R., Jiang T., Li M., Tromp J., Lower Bounds on Learning Decision Lists and Trees. Information and Computation 126(2): 114-122, 1996.

Holte R. C., Very simple classification rules perform well on most commonly used datasets. Machine Learning, 11:63-90, 1993.

Hyafil L. and Rivest R.L., Constructing optimal binary decision trees is NP-complete. Information Processing Letters, 5(1):15-17, 1976

Janikow, C.Z., Fuzzy Decision Trees: Issues and Methods, IEEE Transactions on Systems, Man, and Cybernetics, Vol. 28, Issue 1, pp. 1-14. 1998.

John G. H., Robust linear discriminant trees. In D. Fisher and H. Lenz, editors, Learning From Data: Artificial Intelligence and Statistics V, Lecture Notes in Statistics, Chapter 36, pp. 375-385. Springer-Verlag, New York, 1996.

Kass G. V., An exploratory technique for investigating large quantities of categorical data. Applied Statistics, 29(2):119-127, 1980.

Kearns M. and Mansour Y., A fast, bottom-up decision tree pruning algorithm with near-optimal generalization, in J. Shavlik, ed., 'Machine Learning: Proceedings of the Fifteenth International Conference', Morgan Kaufmann Publishers, Inc., pp. 269-277, 1998.

Kearns M. and Mansour Y., On the boosting ability of top-down decision tree learning algorithms. Journal of Computer and Systems Sciences, 58(1): 109-128, 1999.

Kohavi R. and Sommerfield D., Targeting business users with decision table classifiers, in R. Agrawal, P. Stolorz & G. Piatetsky-Shapiro, eds, 'Proceedings of the Fourth International Conference on Knowledge Discovery and Data Mining', AAAI Press, pp. 249-253, 1998.

Langley, P. and Sage, S., Oblivious decision trees and abstract cases. in Working Notes of the AAAI-94 Workshop on Case-Based Reasoning, pp. 113-117, Seattle, WA: AAAI Press, 1994.

Li X. and Dubes R. C., Tree classifier design with a Permutation statistic, Pattern Recognition 19:229-235, 1986.

Lim X., Loh W.Y., and Shih X., A comparison of prediction accuracy, complexity, and training time of thirty-three old and new classification algorithms . Machine Learning 40:203-228, 2000.

Lin Y. K. and Fu K., Automatic classification of cervical cells using a binary tree classifier. Pattern Recognition, 16(1):69-80, 1983.

Loh W.Y.,and Shih X., Split selection methods for classification trees. Statistica Sinica, 7: 815-840, 1997.

Loh W.Y. and Shih X., Families of splitting criteria for classification trees. Statistics and Computing 9:309-315, 1999.

Loh W.Y. and Vanichsetakul N., Tree-structured classification via generalized discriminant Analysis. Journal of the American Statistical Association, 83: 715-728, 1988.

Lopez de Mantras R., A distance-based attribute selection measure for decision tree induction, Machine Learning 6:81-92, 1991.

Lubinsky D., Algorithmic speedups in growing classification trees by using an additive split criterion. Proc. AI&Statistics93, pp. 435-444, 1993.

Maimon O., and Rokach, L. Data Mining by Attribute Decomposition with semiconductors manufacturing case study, in Data Mining for Design and Manufacturing: Methods and Applications, D. Braha (ed.), Kluwer Academic Publishers, pp. 311–336, 2001.

Maimon O. and Rokach L., "Improving supervised learning by feature decomposition", Proceedings of the Second International Symposium on Foundations of Information and Knowledge Systems, Lecture Notes in Computer Science, Springer, pp. 178-196, 2002.

Maimon, O. and Rokach, L., Decomposition Methodology for Knowledge Discovery and Data Mining: Theory and Applications, Series in Machine Perception and Artificial Intelligence - Vol. 61, World Scientific Publishing, ISBN:981-256-079-3, 2005.

Martin J. K., An exact probability metric for decision tree splitting and stopping. An Exact Probability Metric for Decision Tree Splitting and Stopping, Machine Learning, 28, 2-3):257-291, 1997.

Mehta M., Rissanen J., Agrawal R., MDL-Based Decision Tree Pruning. KDD 1995: pp. 216-221, 1995.

Mehta M., Agrawal R. and Rissanen J., SLIQ: A fast scalable classifier for Data Mining: In Proc. If the fifth Int'l Conference on Extending Database Technology (EDBT), Avignon, France, March 1996.

Mingers J., An empirical comparison of pruning methods for decision tree induction. Machine Learning, 4(2):227-243, 1989.

Morgan J. N. and Messenger R. C., THAID: a sequential search program for the analysis of nominal scale dependent variables. Technical report, Institute for Social Research, Univ. of Michigan, Ann Arbor, MI, 1973.

Moskovitch R, Elovici Y, Rokach L, Detection of unknown computer worms based on behavioral classification of the host, Computational Statistics and Data Analysis, 52(9):4544–4566, 2008.

Muller W., and Wysotzki F., Automatic construction of decision trees for classification. Annals of Operations Research, 52:231-247, 1994.

Murthy S. K., Automatic Construction of Decision Trees from Data: A MultiDisciplinary Survey. Data Mining and Knowledge Discovery, 2(4):345-389, 1998.

Naumov G.E., NP-completeness of problems of construction of optimal decision trees. Soviet Physics: Doklady, 36(4):270-271, 1991.

Niblett T. and Bratko I., Learning Decision Rules in Noisy Domains, Proc. Expert Systems 86, Cambridge: Cambridge University Press, 1986.

Olaru C., Wehenkel L., A complete fuzzy decision tree technique, Fuzzy Sets and Systems, 138(2):221–254, 2003.

Pagallo, G. and Huassler, D., Boolean feature discovery in empirical learning, Machine Learning, 5(1): 71-99, 1990.

Peng Y., Intelligent condition monitoring using fuzzy inductive learning, Journal of Intelligent Manufacturing, 15 (3): 373-380, June 2004.

Quinlan, J.R., Induction of decision trees, Machine Learning 1, 81-106, 1986.

Quinlan, J.R., Simplifying decision trees, International Journal of Man-Machine Studies, 27, 221-234, 1987.

Quinlan, J.R., Decision Trees and Multivalued Attributes, J. Richards, ed., Machine Intelligence, V. 11, Oxford, England, Oxford Univ. Press, pp. 305-318, 1988.

Quinlan, J. R., Unknown attribute values in induction. In Segre, A. (Ed.), Proceedings of the Sixth International Machine Learning Workshop Cornell, New York. Morgan Kaufmann, 1989.

Quinlan, J. R., C4.5: Programs for Machine Learning, Morgan Kaufmann, Los Altos, 1993.

Quinlan, J. R. and Rivest, R. L., Inferring Decision Trees Using The Minimum Description Length Principle. Information and Computation, 80:227-248, 1989.

Rastogi, R., and Shim, K., PUBLIC: A Decision Tree Classifier that Integrates Building and Pruning,Data Mining and Knowledge Discovery, 4(4):315-344, 2000.

Rissanen, J., Stochastic complexity and statistical inquiry. World Scientific, 1989.

Rokach, L., Decomposition methodology for classification tasks: a meta decomposer framework, Pattern Analysis and Applications, 9(2006):257–271.

Rokach L., Genetic algorithm-based feature set partitioning for classification problems,Pattern Recognition, 41(5):1676–1700, 2008.

Rokach L., Mining manufacturing data using genetic algorithm-based feature set decomposition, Int. J. Intelligent Systems Technologies and Applications, 4(1):57-78, 2008.

Rokach, L. and Maimon, O., Theory and applications of attribute decomposition, IEEE International Conference on Data Mining, IEEE Computer Society Press, pp. 473–480, 2001.

Rokach L. and Maimon O., Feature Set Decomposition for Decision Trees, Journal of Intelligent Data Analysis, Volume 9, Number 2, 2005b, pp 131–158.

Rokach, L. and Maimon, O., Clustering methods, Data Mining and Knowledge Discovery Handbook, pp. 321–352, 2005, Springer.

Rokach, L. and Maimon, O., Data mining for improving the quality of manufacturing: a feature set decomposition approach, Journal of Intelligent Manufacturing, 17(3):285–299, 2006, Springer.

Rokach, L., Maimon, O., Data Mining with Decision Trees: Theory and Applications, World Scientific Publishing, 2008.

Rokach L., Maimon O. and Lavi I., Space Decomposition In Data Mining: A Clustering Approach, Proceedings of the 14th International Symposium On Methodologies For Intelligent Systems, Maebashi, Japan, Lecture Notes in Computer Science, Springer-Verlag, 2003, pp. 24–31.

Rokach, L. and Maimon, O. and Averbuch, M., Information Retrieval System for Medical Narrative Reports, Lecture Notes in Artificial intelligence 3055, page 217-228 Springer-Verlag, 2004.

Rokach, L. and Maimon, O. and Arbel, R., Selective voting-getting more for less in sensor fusion, International Journal of Pattern Recognition and Artificial Intelligence 20 (3) (2006), pp. 329–350.

Rounds, E., A combined non-parametric approach to feature selection and binary decision tree design, Pattern Recognition 12, 313-317, 1980.

Schlimmer, J. C. , Efficiently inducing determinations: A complete and systematic search algorithm that uses optimal pruning. In Proceedings of the 1993 International Conference on Machine Learning: pp 284-290, San Mateo, CA, Morgan Kaufmann, 1993.

Sethi, K., and Yoo, J. H., Design of multicategory, multifeature split decision trees using perceptron learning. Pattern Recognition, 27(7):939-947, 1994.

Shafer, J. C., Agrawal, R. and Mehta, M. , SPRINT: A Scalable Parallel Classifier for Data Mining, Proc. 22nd Int. Conf. Very Large Databases, T. M. Vijayaraman and Alejandro P. Buchmann and C. Mohan and Nandlal L. Sarda (eds), 544-555, Morgan Kaufmann, 1996.

Sklansky, J. and Wassel, G. N., Pattern classifiers and trainable machines. SpringerVerlag, New York, 1981.

Sonquist, J. A., Baker E. L., and Morgan, J. N., Searching for Structure. Institute for Social Research, Univ. of Michigan, Ann Arbor, MI, 1971.

Taylor P. C., and Silverman, B. W., Block diagrams and splitting criteria for classification trees. Statistics and Computing, 3(4):147-161, 1993.

Utgoff, P. E., Perceptron trees: A case study in hybrid concept representations. Connection Science, 1(4):377-391, 1989.

Utgoff, P. E., Incremental induction of decision trees. Machine Learning, 4: 161-186, 1989.

Utgoff, P. E., Decision tree induction based on efficient tree restructuring, Machine Learning 29, 1):5-44, 1997.

Utgoff, P. E., and Clouse, J. A., A Kolmogorov-Smirnoff Metric for Decision Tree Induction, Technical Report 96-3, University of Massachusetts, Department of Computer Science, Amherst, MA, 1996.

Wallace, C. S., and Patrick J., Coding decision trees, Machine Learning 11: 7-22, 1993.

Zantema, H., and Bodlaender H. L., Finding Small Equivalent Decision Trees is Hard, International Journal of Foundations of Computer Science, 11(2): 343-354, 2000.

10

Bayesian Networks

Paola Sebastiani[1], Maria M. Abad[2], and Marco F. Ramoni[3]

[1] Department of Biostatistics Boston University
sebas@bu.edu
[2] Software Engineering Department University of Granada, Spain
mabad@ugr.es
[3] Departments of Pediatrics and Medicine Harvard University
marco_ramoni@harvard.edu

Summary. Bayesian networks are today one of the most promising approaches to Data Mining and knowledge discovery in databases. This chapter reviews the fundamental aspects of Bayesian networks and some of their technical aspects, with a particular emphasis on the methods to induce Bayesian networks from different types of data. Basic notions are illustrated through the detailed descriptions of two Bayesian network applications: one to survey data and one to marketing data.

Key words: Bayesian networks, probabilistic graphical models, machine learning, statistics.

10.1 Introduction

Born at the intersection of Artificial Intelligence, statistics and probability, Bayesian networks (Pearl, 1988) are a representation formalism at the cutting edge of knowledge discovery and Data Mining (Heckerman, 1997, Madigan and Ridgeway, 2003, Madigan and York, 1995). Bayesian networks belong to a more general class of models called probabilistic graphical models (Whittaker, 1990,Lauritzen, 1996) that arise from the combination of graph theory and probability theory and their success rests on their ability to handle complex probabilistic models by decomposing them into smaller, amenable components. A probabilistic graphical model is defined by a graph where nodes represent stochastic variables and arcs represent dependencies among such variables. These arcs are annotated by probability distribution shaping the interaction between the linked variables. A probabilistic graphical model is called a Bayesian network when the graph connecting its variables is a directed acyclic graph (DAG). This graph represents conditional independence assumptions that are used to factorize the joint probability distribution of the network variables thus making the process of learning from large database amenable to computations. A Bayesian network induced from data can be used to investigate distant relationships between

O. Maimon, L. Rokach (eds.), *Data Mining and Knowledge Discovery Handbook*, 2nd ed.,
DOI 10.1007/978-0-387-09823-4_10,

variables, as well as making prediction and explanation, by computing the conditional probability distribution of one variable, given the values of some others.

The origins of Bayesian networks can be traced back as far as the early decades of the 20th century, when Sewell Wright developed path analysis to aid the study of genetic inheritance (Wright, 1923,Wright, 1934). In their current form, Bayesian networks were introduced in the early 80s as a knowledge representation formalism to encode and use the information acquired from human experts in automated reasoning systems to perform diagnostic, predictive, and explanatory tasks (Pearl, 1988, Charniak, 1991). Their intuitive graphical nature and their principled probabilistic foundations were very attractive features to acquire and represent information burdened by uncertainty. The development of amenable algorithms to propagate probabilistic information through the graph (Lauritzen and Spiegelhalter, 1988, Pearl, 1988) put Bayesian networks at the forefront of Artificial Intelligence research. Around same time, the machine learning community came to the realization that the sound probabilistic nature of Bayesian networks provided straightforward ways to learn them from data. As Bayesian networks encode assumptions of conditional independence, the first machine learning approaches to Bayesian networks consisted of searching for conditional independence structures in the data and encoding them as a Bayesian network (Glymour *et al.*, 1987, Pearl, 1988). Shortly thereafter, Cooper and Herskovitz (Cooper and Herskovitz, 1992) introduced a Bayesian method, further refined by (Heckerman *et al.*, 1995), to learn Bayesian networks from data. These results spurred the interest of the Data Mining and knowledge discovery community in the unique features of Bayesian networks (Heckerman, 1997): a highly symbolic formalism, originally developed to be used and understood by humans, well-grounded on the sound foundations of statistics and probability theory, able to capture complex interaction mechanisms and to perform prediction and classification.

10.2 Representation

A Bayesian network has two components: a directed acyclic graph and a probability distribution. Nodes in the directed acyclic graph represent stochastic variables and arcs represent directed dependencies among variables that are quantified by conditional probability distributions.

As an example, consider the simple scenario in which two variables control the value of a third. We denote the three variables with the letters A, B and C, and we assume that each is bearing two states: "True" and "False". The Bayesian network in Figure 10.1 describes the dependency of the three variables with a directed acyclic graph, in which the two arcs pointing to the node C represent the joint action of the two variables A and B. Also, the absence of any directed arc between A and B describes the *marginal independence* of the two variables that become dependent when we condition on the phenotype. Following the direction of the arrows, we call the node C a *child* of A and B, which become its *parents*. The Bayesian network in Figure 10.1 let us decompose the overall joint probability distribution of the three variables that would consist of $2^3 - 1 = 7$ parameters into three probability distri-

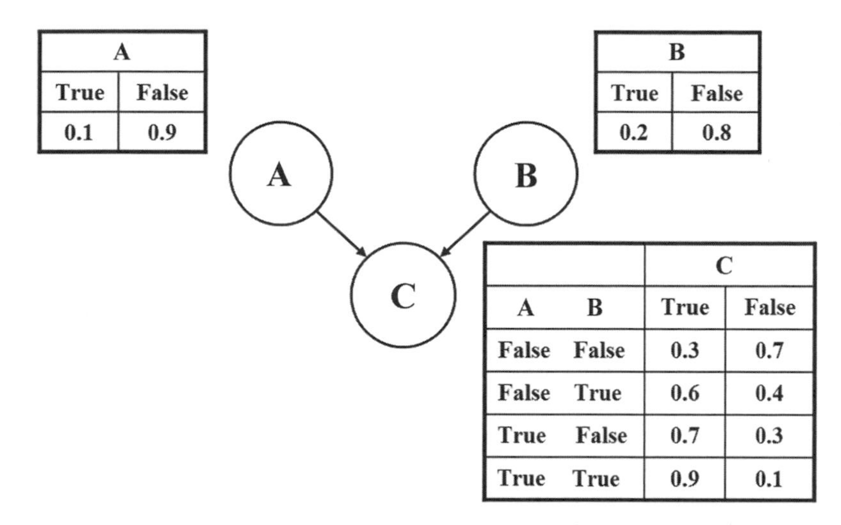

Fig. 10.1. A network describing the impact of two variables (nodes A and B) on a third one (node C). Each node in the network is associated with a probability table that describes the conditional distribution of the node, given its parents.

butions, one conditional distribution for the variable C given the parents, and two marginal distributions for the two parent variables A and B. These probabilities are specified by $1+1+4=6$ parameters. The decomposition is one of the key factors to provide both a verbal and a human understandable description of the system and to efficiently store and handle this distribution, which grows exponentially with the number of variables in the domain. The second key factor is the use of *conditional independence* between the network variables to break down their overall distribution into connected modules.

Suppose we have three random variables Y_1, Y_2, Y_3. Then Y_1 and Y_2 are independent given Y_3 if the conditional distribution of Y_1, given Y_2, Y_3 is only a function of Y_3. Formally:

$$p(y_1|y_2,y_3) = p(y_1|y_3)$$

where $p(y|x)$ denotes the conditional probability/density of Y, given $X=x$. We use capital letters to denote random variables, and small letters to denote their values. We also use the notation $Y_1 \perp Y_2 | Y_3$ to denote the conditional independence of Y_1 and Y_2 given Y_3.

Conditional and marginal independence are substantially different concepts. For example two variables can be marginally independent, but they may be dependent when we condition on a third variable. The directed acyclic graph in Figure 10.1 shows this property: the two parent variables are marginally independent, but they become dependent when we condition on their common child. A well known consequence of this fact is the Simpson's paradox (Whittaker, 1990) : two variables are independent but once a shared child variable is observed they become dependent.

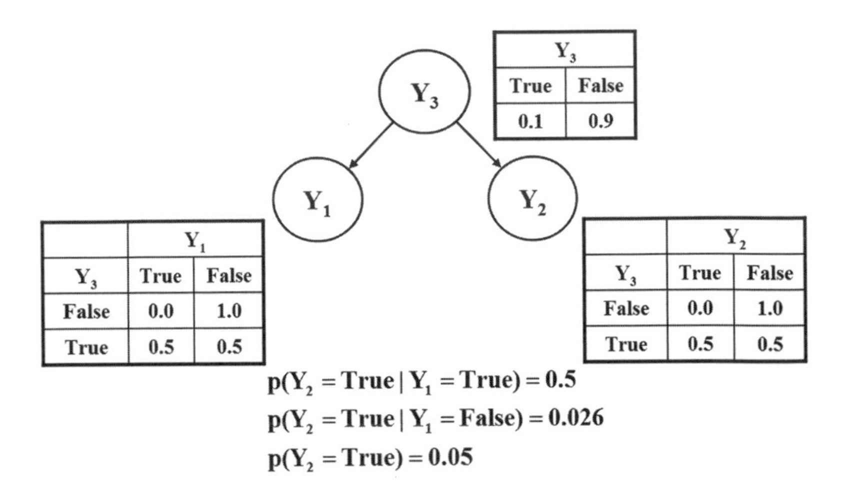

Fig. 10.2. A network encoding the conditional independence of Y_1, Y_2 given the common parent Y_3. The panel in the middle shows that the distribution of Y_2 changes with Y_1 and hence the two variables are conditionally dependent.

Conversely, two variables that are marginally dependent may be made conditionally independent by introducing a third variable. This situation is represented by the directed acyclic graph in Figure 10.2, which shows two children nodes (Y_1 and Y_2) with a common parent Y_3. In this case, the two children nodes are independent, given the common parent, but they may become dependent when we marginalize the common parent out.

The overall list of marginal and conditional independencies represented by the directed acyclic graph is summarized by the local and global Markov properties (Lauritzen, 1996) that are exemplified in Figure 10.3 using a network of seven variables. The *local Markov property* states that each node is independent of its non descendant given the parent nodes and leads to a direct factorization of the joint distribution of the network variables into the product of the conditional distribution of each variable Y_i given its parents $Pa(y_i)$. Therefore, the joint probability (or density) of the v network variables can be written as:

$$p(y_1, ..., y_v) = \prod_i p(y_i | pa(y_i)). \tag{10.1}$$

In this equation, $pa(y_i)$ denotes a set of values of $Pa(Y_i)$. This property is the core of many search algorithms for learning Bayesian networks from data. With this de-

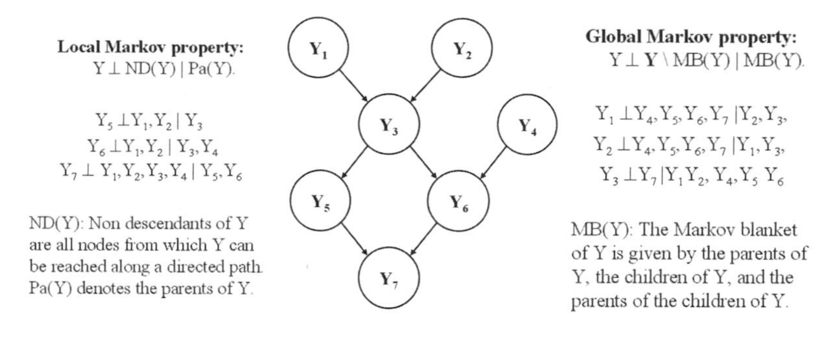

Local Markov property:
Y ⊥ ND(Y) | Pa(Y).
Y5 ⊥Y1,Y2 | Y3
Y6 ⊥Y1,Y2 | Y3,Y4
Y7 ⊥ Y1,Y2,Y3,Y4 | Y5,Y6
ND(Y): Non descendants of Y are all nodes from which Y can be reached along a directed path. Pa(Y) denotes the parents of Y.
Y1
Y2
Y3
Y4
Y5
Y6
Y7
Global Markov property:
Y ⊥ Y \ MB(Y) | MB(Y).
Y1 ⊥Y4,Y5,Y6,Y7 |Y2,Y3,
Y2 ⊥Y4,Y5,Y6,Y7 |Y1,Y3,
Y3 ⊥Y7 |Y1,Y2, Y4,Y5 Y6
MB(Y): The Markov blanket of Y is given by the parents of Y, the children of Y, and the parents of the children of Y.

Fig. 10.3. A Bayesian network with seven variables and some of the Markov properties represented by its directed acyclic graph. The panel on the left describes the local Markov property encoded by a directed acyclic graph and lists the three Markov properties that are represented by the graph in the middle. The panel on the right describes the global Markov property and lists three of the seven global Markov properties represented by the graph in the middle. The vector in bold denotes the set of variables represented by the nodes in the graph.

composition, the overall distribution is broken into modules that can be interrelated, and the network summarizes all significant dependencies without information disintegration. Suppose, for example, the variable in the network in Figure 10.3 are all categorical. Then the joint probability $p(y_1,\ldots,y_7)$ can be written as the product of seven conditional distributions:

$$p(y_1)p(y_2)p(y_3|y_1,y_2)p(y_4)p(y_5|y_3)p(y_6|y_3,y_4)p(y_7|y_5,y_6).$$

The *global Markov property*, on the other hand, summarizes all conditional independencies embedded by the directed acyclic graph by identifying the Markov Blanket of each node (Figure 10.3).

10.3 Reasoning

The modularity induced by the Markov properties encoded by the directed acyclic graph is the core of many search algorithms for learning Bayesian networks from data. By the Markov properties, the overall distribution is broken into modules that can be interrelated, and the network summarizes all significant dependencies without information disintegration. In the network in Figure 10.3, for example, we can compute the probability distribution of the variable Y_7, given that the variable Y_1 is observed to take a particular value (prediction) or, vice versa, we can compute the conditional distribution of Y_1 given the values of some other variables in the network (explanation). In this way, a Bayesian network becomes a complete simulation system able to forecast the value of unobserved variables under hypothetical conditions and, conversely, able to find the most probable set of initial conditions leading to observed situation.

Several exact algorithms exist to perform this inference when the network variables are all discrete, all continuous and modeled with Gaussian distributions, or the network topology is constrained to particular structures (Castillo *et al.*, 1997, Lauritzen and Spiegelhalter, 1988, Pearl, 1988). The most common approaches to evidence propagation in Bayesian networks can be summarized along four lines:

Polytrees When the topology of a Bayesian network is restricted to a *polytree structure* — a direct acyclic graph with only one path linking any two nodes in the graph — we can the fact that every node in the network divides the polytree into two disjoint sub-trees. In this way, propagation can be performed locally and very efficiently.

Conditioning The intuition underlying the Conditioning approach is that networks structures more complex than polytrees can be reduced to a set of polytrees when a subset of its nodes, known as *loop cutset*, are instantiated. In this way, we can efficiently propagate each polytree and then combine the results of these propagations. The source of complexity of these algorithms is the identification of the loop cutset (Cooper, 1990).

Clustering The algorithms developed following the Clustering approach (Lauritzen and Spiegelhalter, 1988) transforms the graphical structure of a Bayesian network into an alternative graph, called the *junction tree*, with a polytree structure by appropriately merging some variables in the network. This mapping consists first of transforming the directed graph into an undirected graph by joining the unlinked parents and triangulating the graph. The nodes in the junction tree cluster sets of nodes in the undirected graph into *cliques* that are defined as maximal and complete sets of nodes. The completeness ensures that there are links between every pair of nodes in the clique, while maximality guarantees that the set on nodes is not a proper subset of any other clique. The joint probability of the network variables can then be mapped into a probability distribution over the clique sets with some factorization properties.

Goal-Oriented This approach differs from the Conditioning and the Clustering approach in that it does not transform the entire network in an alternative structure to simultaneously compute the posterior probability of all variables but it rather *query* the probability distribution of a variable and targets the transformation of the network to the queried variable. The intuition is to identify the network variables that are irrelevant to compute the posterior probability of a particular variable (Shachter, 1986).

For general network topologies and non standard distributions, we need to resort to stochastic simulation (Cheng and Druzdzel, 2000). Among the several stochastic simulation methods currently available, Gibbs sampling (Geman and Geman, 1984,Thomas *et al.*, 1992) is particularly appropriate for Bayesian network reasoning because of its ability to leverage on the graphical decomposition of joint multivariate distributions to improve computational efficiency. Gibbs sampling is also useful for probabilistic reasoning in Gaussian networks, as it avoids computations with joint multivariate distributions. Gibbs sampling is a Markov Chain Monte Carlo) method that generates a sample from the joint distribution of the nodes in the network. The procedure works by generating an ergodic Markov chain

$$\begin{pmatrix} y_{10} \\ \vdots \\ y_{v0} \end{pmatrix} \rightarrow \begin{pmatrix} y_{11} \\ \vdots \\ y_{v1} \end{pmatrix} \rightarrow \begin{pmatrix} y_{12} \\ \vdots \\ y_{v2} \end{pmatrix} \rightarrow \cdots$$

that, under regularity conditions, converges to a stationary distribution. At each step of the chain, the algorithm generates y_{ik} from the conditional distribution of Y_i given all current values of the other nodes. To derive the marginal distribution of each node, the initial burns-in is removed, and the values simulated for each node are a sample generated from the marginal distribution. When one or more nodes in the networks are observed, they are fixed in the simulation so that the sample for each node is from the conditional distribution of the node given the observed nodes in the network.

Gibbs sampling in directed graphical models exploits the Global Markov property, so that to simulate from the conditional distribution of one node Y_i given the current values of the other nodes, the algorithm needs to simulate from the conditional probability/density

$$p(y_i|y \backslash y_i) \propto p(y_i|pa(y_i)) \prod_h p(c(y_i)_h|pa(c(y_i)_h))$$

where y denotes a set of values of all network variables, $pa(y_i)$ and $c(y_i)$ are values of the parents and children of Y_i, $pa(c(y_i)_h)$ are values of the parents of the hth child of Y_i, and the symbol $\backslash$ denotes the set difference.

10.4 Learning

Learning a Bayesian network from data consists of the induction of its two different components: 1) The graphical structure of conditional dependencies (*model selection*); 2) The conditional distributions quantifying the dependency structure (*parameter estimation*). While the process of parameter estimation follows quite standard statistical techniques (see (Ramoni and Sebastiani, 2003)), the automatic identification of the graphical model best fitting the data is a more challenging task. This automatic identification process requires two components: a scoring metric to select the best model and a search strategy to explore the space of possible, alternative models. This section will describe these two components — model selection and model search — and will also outline some methods to validate a graphical model once it has been induced from a data set.

10.4.1 Scoring Metrics

We describe the traditional Bayesian approach to model selection that solves the problem as hypothesis testing. Other approaches based on independence tests or variants of the Bayesian metric like the minimum description length (MDL) score or the

Bayesian information criterion (BIC) are described in (Lauritzen, 1996,Spirtes *et al.*, 1993,Whittaker, 1990). We suppose to have a set $\mathscr{M} = \{M_0, M_1, ..., M_g\}$ of Bayesian networks, each network describing an hypothesis on the dependency structure of the random variables $Y_1, ..., Y_v$. Our task is to choose one network after observing a sample of data $\mathscr{D} = \{y_{1k}, ..., y_{vk}\}$, for $k = 1, \ldots, n$. By Bayes' theorem, the data $\mathscr{D}$ are used to revise the prior probability $p(M_h)$ of each model into the posterior probability, which is calculated as

$$p(M_h|\mathscr{D}) \propto p(M_h)p(\mathscr{D}|M_h)$$

and the Bayesian solution consists of choosing the network with maximum posterior probability. The quantity $p(\mathscr{D}|M_h)$ is called the *marginal likelihood* and is computed by averaging out θ_h from the likelihood function $p(\mathscr{D}|\theta_h)$, where Θ_h is the vector parameterizing the distribution of $Y_1, ..., Y_v$, conditional on M_h. Note that, in a Bayesian setting, Θ_h is regarded as a random vector, with a prior density $p(\theta_h)$ that encodes any prior knowledge about the parameters of the model M_h. The likelihood function, on the other hand, encodes the knowledge about the mechanism underlying the data generation. In our framework, the data generation mechanism is represented by a network of dependencies and the parameters are usually a measure of the strength of these dependencies. By averaging out the parameters, the marginal likelihood provides an overall measure of the data generation mechanism that is independent of the values of the parameters. Formally, the marginal likelihood is the solution of the integral

$$p(\mathscr{D}|M_h) = \int p(\mathscr{D}|\theta_h)p(\theta_h)d\theta_h.$$

The computation of the marginal likelihood requires the specification of a parameterization of each model M_h that is used to compute the likelihood function $p(\mathscr{D}|\theta_h)$, and the elicitation of a prior distribution for Θ_h. The local Markov properties encoded by the network M_h imply that the joint density/probability of a case k in the data set can be written as

$$p(y_{1k}, ..., y_{vk}|\theta_h) = \prod_i p(y_{ik}|pa(y_i)_k, \theta_h). \tag{10.2}$$

Here, $y_{1k}, ..., y_{vk}$ is the set of values (*configuration*) of the variables for the kth case, and $pa(y_i)_k$ is the configuration of the parents of Y_i in case k. By assuming exchangeability of the data, that is, cases are independent given the model parameters, the overall likelihood is then given by the product

$$p(\mathscr{D}|\theta_h) = \prod_{ik} p(y_{ik}|pa(y_i)_k, \theta_h).$$

Computational efficiency is gained by using priors for Θ_h that obey the Directed Hyper-Markov law (Dawid and Lauritzen, 1993). Under this assumption, the prior density $p(\theta_h)$ admits the same factorization of the likelihood function, namely $p(\theta_h) = \prod_i p(\theta_{hi})$, where θ_{hi} is the subset of parameters used to describe the dependency of

Y_i on its parents. This parallel factorization of the likelihood function and the prior density allows us to write

$$p(\mathscr{D}|M_h) = \prod_{ik} \int p(y_{ik}|pa(y_i)_k, \theta_{hi}) p(\theta_{hi}) d\theta_{hi} = \prod_i p(D|M_{hi})$$

where $p(D|M_{hi}) = \prod_k \int p(y_{ik}|pa(y_i)_k, \theta_{hi}) p(\theta_{hi}) d\theta_{hi}$. By further assuming decomposable network prior probabilities that factorize as $p(M_h) = \prod_i p(M_{hi})$ (Heckerman *et al.*, 1995), the posterior probability of a model M_h is the product:

$$p(M_h|\mathscr{D}) = \prod_i p(M_{hi}|\mathscr{D}).$$

Here $p(M_{hi}|\mathscr{D})$ is the posterior probability weighting the dependency of Y_i on the set of parents specified by the model M_h. Decomposable network prior probabilities are encoded by exploiting the modularity of a Bayesian network, and are based on the assumption that the prior probability of a local structure M_{hi} is independent of the other local dependencies M_{hj} for $j \neq i$. By setting $p(M_{hi}) = (g+1)^{-1/v}$, where $g+1$ is the cardinality of the model space and v is the cardinality of the set of variables, there follows that uniform priors are also decomposable.

An important consequence of the likelihood modularity is that, in the comparison of models that differ for the parent structure of a variable Y_i, only the local marginal likelihood matters. Therefore, the comparison of two local network structures that specify different parents for the variable Y_i can be done by simply evaluating the product of the local *Bayes factor* $BF_{hk} = p(\mathscr{D}|M_{hi})/p(\mathscr{D}|M_{ki})$, and the prior odds $p(M_h)/p(M_k)$, to compute the posterior odds of one model versus the other:

$$p(M_{hi}|\mathscr{D})/p(M_{ki}|\mathscr{D}).$$

The posterior odds provide an intuitive and widespread measure of fitness. Another important consequence of the likelihood modularity is that, when the models are a priori equally likely, we can learn a model locally by maximizing the marginal likelihood node by node.

When there are no missing data, the marginal likelihood $p(\mathscr{D}|M_h)$ can be calculated in closed form under the assumptions that all variables are discrete, or all variables follow Gaussian distributions and the dependencies between children and parents are linear. These two cases are described in the next examples. We conclude by noting that the calculation of the marginal likelihood of the data is the essential component for the calculation of the Bayesian estimate of the parameter θ_h, which is given by the expected value of the posterior distribution:

$$p(\theta_h|\mathscr{D}) = \frac{p(\mathscr{D}|\theta_h)p(\theta_h)}{p(\mathscr{D}|M_h)} = \prod_i \frac{p(\mathscr{D}|\theta_{hi})p(\theta_{hi})}{p(\mathscr{D}|M_{hi})}.$$

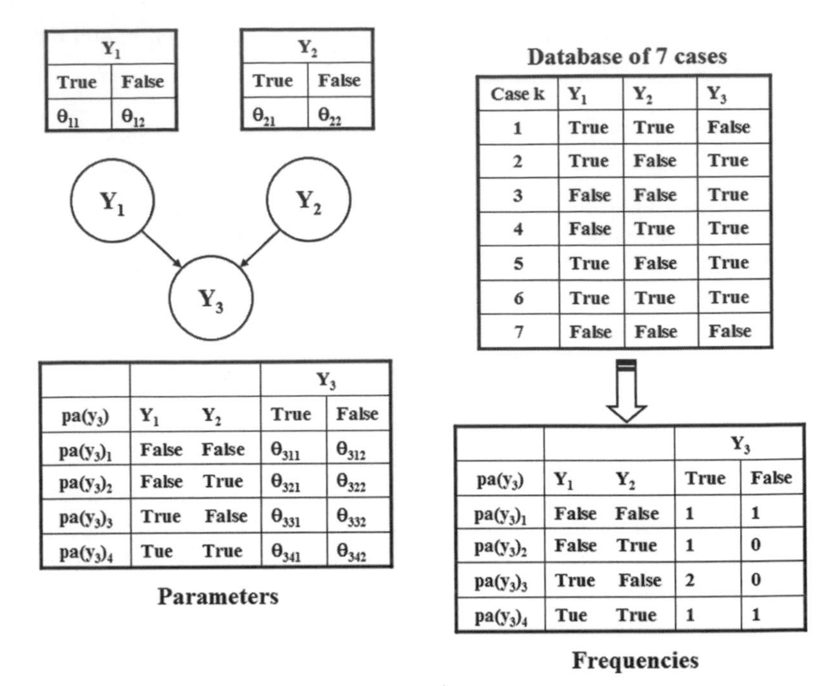

Fig. 10.4. A simple Bayesian network describing the dependency of Y_3 on Y_1 and Y_2 that are marginally independent. The table on the left describes the parameters θ_{3jk} ($j = 1,\ldots,4$ and $k = 1,2$) used to define the conditional distributions of $Y_3 = y_{3k}|pa(y_3)_j$, assuming all variables are binary. The two tables on the right describe a simple database of seven cases, and the frequencies n_{3jk}. The full joint distribution is defined by the parameters θ_{3jk}, and the parameters θ_{1k} and θ_{2k} that specify the marginal distributions of Y_1 and Y_2.

Discrete Variable Networks

Suppose the variables $Y_1,\ldots,Y_v$ are all discrete, and denote by c_i the number of categories of Y_i. The dependency of each variable Y_i on its parents is represented by a set of *multinomial distributions* that describe the conditional distribution of Y_i on the configuration j of the parent variables $Pa(Y_i)$. This representation leads to writing the likelihood function as:

$$p(\mathscr{D}|\theta_h) = \prod_{ijk} \theta_{ijk}^{n_{ijk}}$$

where the parameter θ_{ijk} denotes the conditional probability $p(y_{ik}|pa(y_i)_j)$; n_{ijk} is the sample frequency of $(y_{ik}, pa(y_i)_j)$, and $n_{ij} = \sum_k n_{ijk}$ is the marginal frequency of $pa(y_i)_j$. Figure 10.4 shows an example of the notation for a network with three variables. With the data in this example, the likelihood function is written as:

$$\{\theta_{11}^4\theta_{12}^3\}\{\theta_{21}^3\theta_{22}^4\}\{\theta_{311}^1\theta_{312}^1 \times \theta_{321}^1\theta_{322}^0 \times \theta_{331}^2\theta_{332}^0 \times \theta_{341}^1\theta_{342}^1\}.$$

The first two terms in the products are the contributions of nodes Y_1 and Y_2 to the likelihood, while the last product is the contribution of the node Y_3, with terms corresponding to the four conditional distributions of Y_3 given each of the four parent configurations.

The *hyper Dirichlet distribution* with parameters α_{ijk} is the conjugate Hyper Markov law (Dawid and Lauritzen, 1993) and it is defined by a density function proportional to the product $\prod_{ijk}\theta_{ijk}^{\alpha_{ijk}-1}$. This distribution encodes the assumption that the parameters θ_{ij} and $\theta_{i'j'}$ are independent for $i' \neq i$ and $j \neq j'$. These assumptions are known as *global and local parameter independence* (Spiegelhalter and Lauritzen, 1990), and are valid only under the assumption the hyper-parameters α_{ijk} satisfy the consistency rule $\sum_j \alpha_{ij} = \alpha$ for all i (Good, 1968, Geiger and Heckerman, 1997). Symmetric Dirichlet distributions satisfy easily this constraint by setting $\alpha_{ijk} = \alpha/(c_i q_i)$ where q_i is the number of states of the parents of Y_i. One advantage of adopting symmetric hyper Dirichlet priors in model selection is that, if we fix α constant for all models, then the comparison of posterior probabilities of different models is done conditionally on the same quantity α. With these parameterization and choice of prior distributions, the marginal likelihood is given by the equation

$$\prod_i p(\mathscr{D}|M_{hi}) = \prod_{ij} \frac{\Gamma(\alpha_{ij})}{\Gamma(\alpha_{ij}+n_{ij})} \prod_k \frac{\Gamma(\alpha_{ijk}+n_{ijk})}{\Gamma(\alpha_{ijk})}$$

where $\Gamma(\cdot)$ denotes the Gamma function, and the Bayesian estimate of the parameter θ_{ijk} is the posterior mean

$$E(\theta_{ijk}|\mathscr{D}) = \frac{\alpha_{ijk}+n_{ijk}}{\alpha_{ij}+n_{ij}}. \tag{10.3}$$

More details are in (Ramoni and Sebastiani, 2003).

Linear Gaussian Networks

Suppose now that the variables $Y_1,\ldots,Y_v$ are all continuous, and the conditional distribution of each variable Y_i given its parents $Pa(y_i) \equiv \{Y_{i1},\ldots,Y_{ip(i)}\}$ follows a *Gaussian distribution* with mean that is a linear function of the parent variables, and conditional variance $\sigma_i^2 = 1/\tau_i$. The parameter τ_i is called the precision. The dependency of each variable on its parents is represented by the linear regression equation:

$$\mu_i = \beta_{i0} + \sum_j \beta_{ij} y_{ij}$$

that models the conditional mean of Y_i given the parent values y_{ij}. Note that the regression equation is additive (there are no interactions between the parent variables) to ensure that the model is graphical (Lauritzen, 1996). In this way, the dependency of Y_i on a parent Y_{ij} is equivalent to having the regression coefficient $\beta_{ij} \neq 0$. Given a set of exchangeable observations $\mathscr{D}$, the likelihood function is:

$$p(\mathscr{D}|\theta_h) = \prod_i (\tau_i/(2\pi))^{n/2} \prod_k \exp[-\tau_i(y_{ik} - \mu_{ik})^2/2]$$

where μ_{ik} denotes the value of the conditional mean of Y_i, in case k, and the vector θ_h denotes the set of parameters τ_i, β_{ij}. It is usually more convenient to use a matrix notation and we use the $n \times (p(i)+1)$ matrix X_i to denote the matrix of regression coefficients, with kth row given by $(1, y_{i1k}, y_{i2k}, \ldots, y_{ip(i)k})$, β_i to denote the vector of parameters $(\beta_{i0}, \beta_{i1}, \ldots, \beta_{ip(i)})^T$ associated with Y_i and, in this example, y_i to denote the vector of observations $(y_{i1}, \ldots, y_{in})^T$. With this notation, the likelihood can be written in a more compact form:

$$p(\mathscr{D}|\theta_h) = \prod_i (\tau_i/(2\pi))^{n/2} \exp[-\tau_i(y_i - X_i\beta_i)^T(y_i - X_i\beta_i)/2]$$

There are several choices to model the prior distribution on the parameters τ_i and β_i. For example, the conditional variance can be further parameterized as:

$$\sigma_i^2 = V(Y_i) - cov(Y_i, Pa(y_i))V(Pa(y_i))^{-1}cov(Pa(y_i), Y_i)$$

where $V(Y_i)$ is the marginal variance of Y_i, $V(Pa(y_i))$ is the variance-covariance matrix of the parents of Y_i, and $cov(Y_i, Pa(y_i))$ $(cov(Pa(y_i), Y_i))$ is the row (column) vector of covariances between Y_i and each parent Y_{ij}. With this parameterization, the prior on τ_i is usually a hyper-Wishart distribution for the joint variance-covariance matrix of $Y_i, Pa(y_i)$ (Cowell *et al.*, 1999). The Wishart distribution is the multivariate generalization of a Gamma distribution. An alternative approach is to work directly with the conditional variance of Y_i. In this case, we estimate the conditional variances of each set of parents-child dependency and then the joint multivariate distribution that is needed for the reasoning algorithms is derived by multiplication. More details are described for example in (Whittaker, 1990) and (Geiger and Heckerman, 1994).

We focus on this second approach and again use the global parameter independence (Spiegelhalter and Lauritzen, 1990) to assign independent prior distributions to each set of parameters τ_i, β_i that quantify the dependency of the variable Y_i on its parents. In each set, we use the standard hierarchical prior distribution that consists of a marginal distribution for the precision parameter τ_i and a conditional distribution for the parameter vector β_i, given τ_i. The standard conjugate prior for τ_i is a *Gamma distribution*

$$\tau_i \sim \text{Gamma}(\alpha_{i1}, \alpha_{i2}) \;\; p(\tau_i) = \frac{1}{\alpha_{i2}^{\alpha_{i1}}\Gamma(\alpha_{i1})}\tau_i^{\alpha_{i1}-1}e^{-\tau_i/\alpha_{i2}}$$

where

$$\alpha_{i1} = \frac{\nu_{io}}{2}, \quad \alpha_{i2} = \frac{2}{\nu_{io}\sigma_{io}^2}.$$

This is the traditional Gamma prior for τ_i with hyper-parameters ν_{io} and σ_{io}^2 that can be given the following interpretation. The marginal expectation of τ_i is $E(\tau_i) = \alpha_{i1}\alpha_{i2} = 1/\sigma_{io}^2$ and

$$E(1/\tau_i) = \frac{1}{(\alpha_{i1}-1)\alpha_{i2}} = \frac{\nu_{io}\sigma_{io}^2}{\nu_{io}-2}$$

is the prior expectation of the population variance. Because the ratio $\nu_{io}\sigma_{io}^2/(\nu_{io}-2)$ is similar to the estimate of the variance in a sample of size ν_{io}, σ_{io}^2 is the prior population variance, based on ν_{io} cases seen in the past. Conditionally on τ_i, the prior density of the parameter vector β_i is supposed to be multivariate Gaussian:

$$\beta_i|\tau_i \sim N(\beta_{io}, (\tau_i R_{io})^{-1})$$

where $\beta_{io} = E(\beta_i|\tau_i)$. The matrix $(\tau_i R_{io})^{-1}$ is the prior variance-covariance matrix of $\beta_i|\tau_i$ and R_{io} is the identity matrix so that the regression coefficients are a priori independent, conditionally on τ_i. The density function of β_i is

$$p(\beta_i|\tau_i) = \frac{\tau_i^{(p(i)+1)/2}\det(R_{io})^{1/2}}{(2\pi)^{(p(i)+1)/2}} e^{-\tau_i/2(\beta_i-\beta_{io})^T R_{io}(\beta_i-\beta_{io})}.$$

With this prior specifications, it can be shown that the marginal likelihood $p(\mathscr{D}|M_h)$ can be written in product form $\prod_i p(\mathscr{D}|M_{hi})$, where each factor is given by the quantity:

$$p(\mathscr{D}|M_{hi}) = \frac{1}{(2\pi)^{n/2}} \frac{\det R_{io}^{1/2}}{\det R_{in}^{1/2}} \frac{\Gamma(\nu_{in}/2)}{\Gamma(\nu_{io}/2)} \frac{(\nu_{io}\sigma_{io}^2/2)^{\nu_{io}/2}}{(\nu_{in}\sigma_{in}^2/2)^{\nu_{in}/2}}$$

and the parameters are specified by the next updating rules:

$$\begin{aligned}
\alpha_{i1n} &= \nu_{io}/2 + n/2 \\
1/\alpha_{i2n} &= (-\beta_{in}^T R_{in}\beta_{in} + y_i^T y_i + \beta_{io}^T R_{io}\beta_{io})/2 + 1/\alpha_{i2} \\
\nu_{in} &= \nu_{io} + n \\
\sigma_{in} &= 2/(\nu_{in}\alpha_{i2n}) \\
R_{in} &= R_{io} + X_i^T X_i \\
\beta_{in} &= R_{in}^{-1}(R_{io}\beta_{io} + X_i^T y_i)
\end{aligned}$$

The Bayesian estimates of the parameters are given by the posterior expectations:

$$E(\tau_i|y_i) = \alpha_{i1n}\alpha_{i2n} = 1/\sigma_{in}^2, \quad E(\beta_i|y_i) = \beta_{in},$$

and the estimate of σ_i^2 is $\nu_{in}\sigma_{in}^2/(\nu_{in}-2)$. More controversial is the use of improper prior distributions that describe lack of prior knowledge about the network parameters by uniform distributions (Hagan, 1994). In this case, we set $p(\beta_i, \tau_i) \propto \tau_i^{-c}$, so that $\nu_{io} = 2(1-c)$ and $\beta_{io} = 0$. The updated hyper-parameters are:

$$\begin{aligned}
\nu_{in} &= \nu_{io} + n \\
R_{in} &= X_i^T X_i \\
\beta_{in} &= (X_i^T X_i)^{-1} X_i^T y_i \quad \text{least squares estimate of } \beta \\
\sigma_{in} &= RSS_i/\nu_{in} \\
RSS_i &= y_i^T y_i - y_i^T X_i (X_i^T X_i)^{-1} X_i^T y_i \quad \text{residual sum of squares}
\end{aligned}$$

and the marginal likelihood of each local dependency is

$$p(\mathscr{D}|M_{hi}) = \frac{\Gamma((n-p(i)-2c+1)/2)(RSS_i/2)^{-(n-p(i)-2c+1)/2}}{\det(X_i^T X_i)^{1/2}} \frac{1}{(2\pi)^{(n-p(i)-1)/2}}.$$

A very special case is $c = 1$ that corresponds to $\nu_{io} = 0$. In this case, the local marginal likelihood simplifies to

$$p(\mathscr{D}|M_{hi}) = \frac{1}{(2\pi)^{(n-p(i)-1)/2}} \frac{\Gamma((n-p(i)-1)/2)(RSS_i/2)^{-(n-p(i)-1)/2}}{\det(X_i^T X_i)^{1/2}}.$$

The estimates of the parameters σ_i and β_i become the traditional least squares estimates $RSS_i/(\nu_{in}-2)$ and β_{in}. This approach can be extended to model an unknown variance-covariance structure of the regression parameters, using Normal-Wishart priors (Geiger and Heckerman, 1994)

10.4.2 Model Search

The likelihood modularity allows local model selection and simplifies the complexity of model search. Still, the space of the possible sets of parents for each variable grows exponentially with the number of candidate parents and successful heuristic search procedures (both deterministic and stochastic) have been proposed to render the task feasible (Cooper and Herskovitz, 1992,Larranaga *et al.*, 1996,Singh and Valtorta, 1995, Zhou and Sakane, 2002). The aim of these heuristic search procedures is to impose some restrictions on the search space to capitalize on the decomposability of the posterior probability of each Bayesian network M_h. One suggestion, put forward by (Cooper and Herskovitz, 1992), is to restrict the model search to a subset of all possible networks that are consistent with an ordering relation $\succ$ on the variables $\{Y_1,...,Y_v\}$. This ordering relation $\succ$ is defined by $Y_j \succ Y_i$ if Y_i cannot be parent of Y_j. In other words, rather than exploring networks with arcs having all possible directions, this order limits the search to a subset of networks in which there is only a subset of directed associations. At first glance, the requirement for an order among the variables could appear to be a serious restriction on the applicability of this search strategy, and indeed this approach has been criticized in the artificial intelligence community because it limits the automation of model search. From a modeling point of view, specifying this order is equivalent to specifying the hypotheses that need to be tested, and some careful screening of the variables in the data set may avoid the effort to explore a set of not sensible models. For example, we have successfully applied this approach to model survey data (Sebastiani *et al.*, 2000, Sebastiani and Ramoni, 2001C) and more recently genotype data (1). Recent results have shown that restricting the search space by imposing an order among the variables yields a more regular space over the network structures (Friedman and Koller, 2003). Other search strategies based on genetic algorithms (Larranaga *et al.*, 1996), "ad hoc" stochastic methods (Singh and Valtorta, 1995) or Markov Chain

Monte Carlo methods (Friedman and Koller, 2003) can also be used. An alternative approach to limit the search space is to define classes of equivalent directed graphical models (Chickering, 2002).

The order imposed on the variables defines a set of candidate parents for each variable Y_i and one way to proceed is to implement an independent model selection for each variable Y_i and then link together the local models selected for each variable Y_i. A further reduction is obtained using the greedy search strategy deployed by the *K2 algorithm* (Cooper and Herskovitz, 1992). The K2 algorithm is a bottom-up strategy that starts by evaluating the marginal likelihood of the model in which Y_i has no parents. The next step is to evaluate the marginal likelihood of each model with one parent only and if the maximum marginal likelihood of these models is larger than the marginal likelihood of the independence model, the parent that increases the likelihood most is accepted and the algorithm proceeds to evaluate models with two parents. If none of the models has marginal likelihood that exceeds that of the independence model, the search stops. The K2 algorithm is implemented in Bayesware Discoverer (http://www.bayesware.com), and the R-package Deal (Bottcher and Dethlefsen, 2003). Greedy search can be trapped in local maxima and induce spurious dependency and a variant of this search to limit spurious dependency is stepwise regression (Madigan and Raftery, 1994). However, there is evidence that the K2 algorithm performs as well as other search algorithms (Yu *et al.*, 2002).

10.4.3 Validation

The automation of model selection is not without problems and both diagnostic and predictive tools are necessary to validate a multivariate dependency model extracted from data. There are two main approaches to model validation: one addresses the *goodness of fit* of the network selected from data and the other assesses the *predictive accuracy* of the network in some predictive/diagnostic tests.

The intuition underlying goodness of fit measures is to check the accuracy of the fitted model versus the data. In regression models in which there is only one dependent variable, the goodness of fit is typically based on some summary of the residuals that are defined by the difference between the observed data and the data reproduced by the fitted model. Because a Bayesian network describes a multivariate dependency model in which all nodes represent random variables, we developed *blanket residuals* (Sebastiani and Ramoni, 2003) as follows. Given the network induced from data, for each case k in the database we compute the values fitted for each node Y_i, given all the other values. Denote this fitted value by $\hat{y}_{ik}$ and note that, by the global Markov property, only the configuration in the Markov blanket of the node Y_i is used to compute the fitted value. For categorical variables, the fitted value $\hat{y}_{ik}$ is the most likely category of Y_i given the configuration of its Markov blanket, while for numerical variables the fitted value $\hat{y}_{ik}$ can be either the expected value of Y_i, given the Markov blanket, or the modal value. In both cases, the fitted values are computed by using one of the algorithms for probabilistic reasoning described in Section 10.2.

By repeating this procedure for each case in the database, we compute fitted values for each variable Y_i, and then define the blanket residuals by

$$r_{ik} = y_{ik} - \hat{y}_{ik}$$

for numerical variables, and by

$$c_{ik} = \delta(y_{ik}, \hat{y}_{ik})$$

for categorical variables, where the function $\delta(a,b)$ takes value $\delta = 0$ when $a = b$ and $\delta = 1$ when $a \neq b$. Lack of significant patterns in the residuals r_{ik} and approximate symmetry about 0 will provide evidence in favor of a good fit for the variable Y_i, while anomalies in the blanket residuals can help to identify weaknesses in the dependency structure that may be due to outliers or leverage points. Significance testing of the goodness of fit can be based on the standardized residuals:

$$R_{ik} = \frac{r_{ik}}{\sqrt{V(y_i)}}$$

where the variance $V(y_i)$ is computed from the fitted values. Under the hypothesis that the network fits the data well, we would expect to have approximately 95% of the standardized residuals within the limits [-2,2]. When the variable Y_i is categorical, the residuals c_{ik} identify the error in reproducing the data and can be summarized to compute the error rate for fit.

Because these residuals measure the difference between the observed and fitted values, anomalies in the residuals can identify inadequate dependencies in the networks. However, residuals that are on average not significantly different from 0 do not necessarily prove that the model is good. A better validation of the network should be done on an independent test set to show that the model induced from one particular data set is *reproducible* and gives good predictions. Measures of the predictive accuracy can be the monitors based on the *logarithmic scoring function* (Good, 1952). The basic intuition is to measure the degree of surprise in predicting that the variable Y_i will take a value y_{ih} in the hth case of an independent test set. The measure of surprise is defined by the score

$$s_{ih} = -\log p(y_{ih}|MB(y_i)_h)$$

where $MB(y_i)_h$ is the configuration of the Markov blanket of Y_i in the test case h, $p(y_{ih}|MB(y_i)_h)$ is the predictive probability computed with the model induced from data, and y_{ih} is the value of Y_i in the hth case of the test set. The score s_{ih} will be 0 when the model predicts y_{ih} with certainty, and increases as the probability of y_{ih} decreases. The scores can be summarized to derive *local and global monitors* and to define tests for predictive accuracy (Cowell *et al.*, 1999).

In absence of an independent test set, standard cross validation techniques are typically used to assess the predictive accuracy of one or more nodes (Hand, 1997). In K-fold cross validation, the data are divided into K non-overlapping sets of approximately the same size. Then $K-1$ sets are used for retraining (or inducing) the network from data that is then tested on the remaining set using monitors or other measures of the predictive accuracy (Hastie *et al.*, 2001). By repeating this process K times, we derive independent measures of the predictive accuracy of the network induced from data as well as measures of the robustness of the network to sampling variability. Note that the predictive accuracy based on cross-validation is usually an over-optimistic measure, and several authors have recently argued that cross-validation should be used with caution (Braga-Neto and Dougerthy, 2004), particularly with small sample sizes.

10.5 Bayesian Networks in Data Mining

This section describes the use of Bayesian networks to undertake other typical Data Mining tasks such as classification, and for modeling more complex models, such as nonlinear and temporal dependencies.

10.5.1 Bayesian Networks and Classification

The term "supervised classification" covers two complementary tasks: the first is to identify a function mapping a set of *attributes* onto a *class*, and the other is to assign a class label to a set of unclassified cases described by attribute values. We denote by C the variable whose states represent the class labels c_i, and by Y_i the attributes.

Classification is typically performed by first training a classifier on a set of labelled cases (*training set*) and then using it to label unclassified cases (*test set*). The supervisory component of this classifier resides in the training signal, which provides the classifier with a way to assess a dependency measure between attributes and classes. The classification of a case with attribute values $y_{1k},\dots,y_{vk}$ is then performed by computing the probability distribution $p(C \mid y_{1k},\dots,y_{vk})$ of the

class variable, given the attribute values, and by labelling the case with the most probable label. Most of the algorithms for learning classifiers described as Bayesian networks impose a restriction on the network structure, namely that there cannot be arcs pointing to the class variable. In this case, by the local Markov property, the joint probability $p(y_{1k},\dots,y_{vk},c_k)$ of class and attributes is factorized as $p(c_k)p(y_{1k},\dots,y_{vk} \mid c_k)$. The simplest example is known as a *Naïve Bayes classifier* (NBC) (Duda and Hart, 1973,Langley *et al.*, 1992), and makes the further simplification that the attributes Y_i are conditionally independent given the class C so that

$$p(y_{1k},\dots,y_{vk}|c_k) = \prod_i p(y_{ik}|c_k).$$

Figure 10.5 depicts the directed acyclic graph of a NBC. Because of the restriction on the network topology, the training step for a NBC classifier consists of estimating the conditional probability distributions of each attribute, given the class, from a training data set. When the attributes are discrete or continuous variables and follow Gaussian distributions, the parameters are learned by using the procedure described in Section 10.4. Once trained, the NBC classifies a case by computing the posterior probability distribution over the classes via Bayes' Theorem and assigns the case to the class with the highest posterior probability.

When the attributes are all continuous and modelled by Gaussian variables, and the class variable is binary, say $c_k = 0, 1$, the classification rule induced by a NBC is very similar to the Fisher discriminant rule and turns out to be a function of

$$r = \sum_i \{\log(\sigma_{i0}^2/\sigma_{i0}^2) - (y_i - \mu_{i1})^2/\sigma_{i1}^2 + (y_i - \mu_{i0})^2/\sigma_{i0}^2\}$$

where y_i is the value of attribute i in the new sample to classify and the parameters σ_{ik}^2 and μ_{ik} are the variance and mean of the attribute Gaussian distribution, conditional on the class membership, that are usually estimated by Maximum likelihood.

Other classifiers have been proposed to relax the assumption that attributes are conditionally independent given the class. Perhaps the most competitive one is the *Tree Augmented Naïve Bayes*(TAN) classifier (Friedman *et al.*, 1997) in which all the attributes have the class variable as a parent as well as another attribute. To avoid cycles, the attributes have to be ordered and the first attribute does not have other parents beside the class variable. Figure 10.6 shows an example of a TAN classifier with five attributes. An algorithm to infer a TAN classifier needs to choose both the dependency structure between attributes and the parameters that quantify this dependency. Due to the simplicity of its structure, the identification of a TAN classifiers does not require any search but rather the construction of a tree among the attributes. An "ad hoc" algorithm called *Construct-TAN* CTAN was proposed in (Friedman *et al.*, 1997). One limitation of the CTAN algorithm to build TAN classifiers is that it applies only to discrete attributes, and continuous attributes need to be discretized.

Other extensions of the NBC try to relax some of the assumptions made by the NBC or the TAN classifiers. Some examples are the *l-Limited Dependence Bayesian classifier* (l-LDB) in which the maximum number of parents that an attribute can have is l (Sahami, 1996). Another example is the *unrestricted Augmented Naïve Bayes classifier* (ANB) in which the number of parents is unlimited but the scoring metric used for learning, the minimum description length criterion, biases the search toward models with small number of parents per attribute (Friedman *et al.*, 1997). Due to the high dimensionality of the space of different ANB networks, algorithms that build

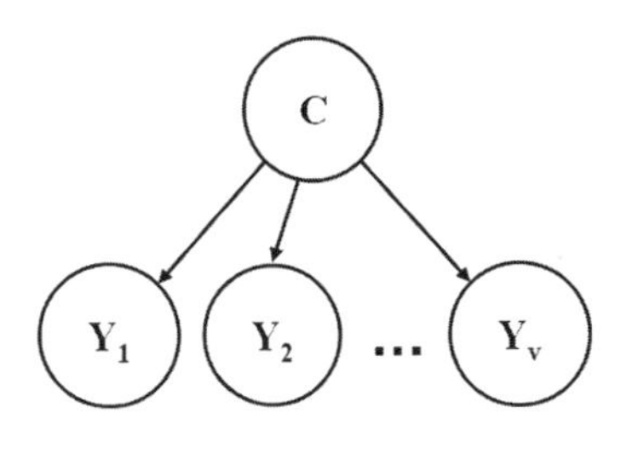

Fig. 10.5. The structure of the Naïve Bayes classifier.

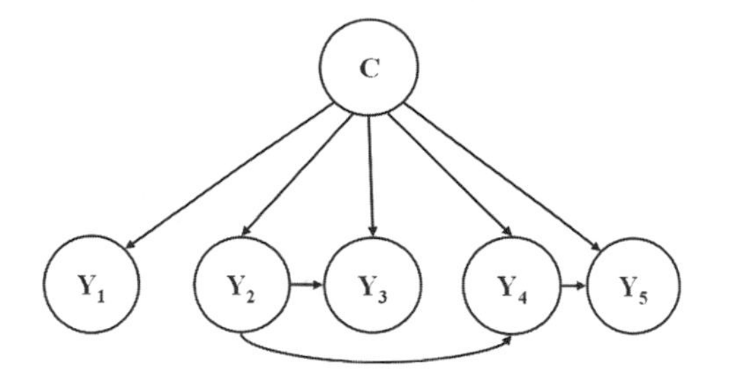

Fig. 10.6. The structure of a TAN classifier.

this type of classifiers must rely on heuristic searches. More examples are reported in (Friedman *et al.*, 1997).

10.5.2 Generalized Gamma Networks

Most of the work on learning Bayesian networks from data has focused on learning networks of categorical variables, or networks of continuous variables modeled by Gaussian distributions with linear dependencies. This section describes a new class of Bayesian networks, called Generalized Gamma networks (GGN), able to describe possibly nonlinear dependencies between variables with non-normal distributions (Sebastiani and Ramoni, 2003).
In a GGN the conditional distribution of each variable Y_i given the parents $Pa(y_i) = \{Y_{i1}, \ldots, Y_{ip(i)}\}$ follows a Gamma distribution $Y_i|pa(y_i), \theta_i \sim Gamma(\alpha_i, \mu_i(pa(y_i), \beta_i))$, where $\mu_i(pa(y_i), \beta_i)$ is the conditional mean of Y_i and $\mu_i(pa(y_i), \beta_i)^2/\alpha_i$ is the conditional variance. We use the standard parameterization of generalized linear models (McCullagh and Nelder, 1989), in which the mean $\mu_i(pa(y_i), \beta_i)$ is not restricted to be a linear function of the parameters β_{ij}, but the linearity in the parameters is enforced in the *linear predictor* η_i, which is itself related to the mean function by the *link function* $\mu_i = g(\eta_i)$. Therefore, we model the conditional density function as:

$$p(y_i|pa(y_i), \theta_i) = \frac{\alpha_i^{\alpha_i}}{\Gamma(\alpha_i)\mu_i^{\alpha_i}} y_i^{\alpha_i - 1} e^{-\alpha_i y_i/\mu_i}, \quad y_i \geq 0 \tag{10.4}$$

where $\mu_i = g(\eta_i)$ and the linear predictor η_i is parameterized as

$$\eta_i = \beta_{i0} + \sum_j \beta_{ij} f_j(pa(y_i))$$

and $f_j(pa(y_i))$ are possibly nonlinear functions. The linear predictor η_i is a function linear in the parameters β, but it is not restricted to be a linear function of the parent values, so that the generality of Gamma networks is in the ability to encode general non-linear stochastic dependency between the node variables. Table 10.1 shows example of non-linear mean functions. Figure 10.7 shows some examples of Gamma

Table 10.1. Link functions and parameterizations of the linear predictor.

LINK	$g(\cdot)$	LINEAR PREDICTOR η
IDENTITY	$\mu = \eta$	$\eta_i = \beta_{i0} + \sum_j \beta_{ij} y_{ij}$
INVERSE	$\mu = \eta^{-1}$	$\eta_i = \beta_{i0} + \sum_j \beta_{ij} y_{ij}^{-1}$
LOG	$\mu = e^{\eta}$	$\eta_i = \beta_{i0} + \sum_j \beta_{ij} log(y_{ij})$

density functions, for different shape parameters $\alpha = 1, 1.5, 5$ and mean $\mu = 400$. Note that approximately symmetrical distributions are obtained for particular values of the shape parameter α.

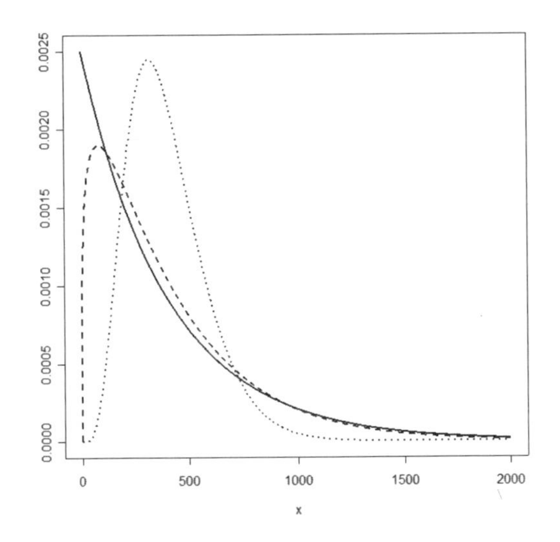

Fig. 10.7. Example of Gamma density functions for shape parameters $\alpha = 1$ (continuous line), $\alpha = 1.5$ (dashed line), and $\alpha = 5$ (dotted line) and mean $\mu = 400$. For fixed mean, the parameter α determines the shape of the distribution that is skewed to the left for small α and approaches symmetry as α increases.

Unfortunately, there is no closed form solution to learn the parameters of a GGN and we have therefore to resort to Markov Chain Monte Carlo methods to compute stochastic estimates (Madigan and Ridgeway, 2003), or to maximum likelihood to compute numerical approximation of the posterior modes (Kass and Raftery, 1995). A well know property of generalized linear models is that the parameters β_{ij} can be estimated independently of α_i, which is then estimated conditionally on β_{ij} (McCullagh and Nelder, 1989).

To compute the maximum likelihood estimates of the parameters β_{ij} within each family $(Y_i, Pa(y_i))$, we need to solve the system of equations

$\partial \log p(\mathscr{D}|\theta_i)/\partial \beta_{ij} = 0$. The Fisher Scoring method is the most efficient algorithm to find the solution of the system of equations. This iterative procedure is a generalization of the Newton Raphson procedure in which the Hessian matrix is replaced by its expected value. This modification speeds up the convergence rate of the iterative procedure that is known for being usually very efficient — it usually converges in 5 steps for appropriate initial values. Details can be found for example in (McCullagh and Nelder, 1989).

Once the ML estimates of β_{ij} are known, say $\hat{\beta}_i$, we compute the fitted means $\hat{\mu}_{ik} = g(\hat{\beta}_{i0} + \sum_j \hat{\beta}_{ij} f_j(pa(y_i)))$ and use these quantities to estimate the shape parameter α_i. Estimation of the shape parameter in Gamma distributions is an open issue, and authors have suggested several estimators (see for example (McCullagh and Nelder, 1989)). Popular choices are the deviance-based estimator that is defined as

$$\tilde{\alpha}_i = \frac{n-q}{\sum_k (y_{ik} - \hat{\mu}_{ik})^2 / \hat{\mu}_{ik}^2}$$

where q is the number of parameters β_{ij} that appear in the linear predictor. The maximum likelihood estimate $\hat{\alpha}_i$ of the shape parameter α_i would need the solution of the equation

$$n + n\log(\alpha_i) + n\frac{\Gamma(\alpha_i)'}{\Gamma(\alpha_i)} + -\sum_k \log(\hat{\mu}_{ik}) + \sum_k \log(y_{ik}) - \sum_i \frac{y_{ik}}{\hat{\mu}_{ik}} = 0$$

with respect to α_i. We have an approximate closed form solution to this equation based on a Taylor expansion that is discussed in (Sebastiani, Ramoni, and Kohane, 2003, Sebastiani *et al.*, 2004, Sebastiani, Yu, and Ramoni, 2003).

Also the model selection process requires the use of approximation methods. In this case, we use the Bayesian information criterion (BIC) (Kass and Raftery, 1995) to approximate the marginal likelihood by $2\log p(\mathscr{D}|\hat{\theta}) - n_p \log(n)$ where $\hat{\theta}$ is the maximum likelihood estimate of θ, and n_p is the overall number of parameters in the network. BIC is independent of the prior specification on the model space and trades off goodness of fit — measured by the term $2\log p(\mathscr{D}|\hat{\theta})$ — and model complexity — measured by the term $n_p \log(n)$. We note that BIC factorizes into a product of terms for each variable Y_i and makes it possible to conduct local structural learning.

While the general type of dependencies in Gamma networks makes it possible to model a variety of dependencies within the variables, exact probabilistic reasoning with the network becomes impossible and we need to resort to Gibbs sampling (see Section 10.2). Our simulation approach uses the adaptative rejection metropolis sampling (ARMS) of (Gilks and Roberts, 1996) when the conditional density $p(y_i|\mathscr{Y}\backslash y_i, \hat{\theta})$ is log-concave, and adaptive rejection with Metropolis sampling in the other cases (Sebastiani and Ramoni, 2003).

10.5.3 Bayesian Networks and Dynamic Data

One of the limitations of Bayesian networks is the inability to represents forward loops: by definition the directed graph that encodes the marginal and conditional independencies between the network variables cannot have cycles. This limitation makes traditional Bayesian networks unsuitable for the representation of many systems in which feedback controls are a critical aspect of many application domains, from control engineering to biomedical sciences. Dynamic Bayesian networks provide a general framework to integrate multivariate time series and to represent feedforward loops and feedback mechanisms.

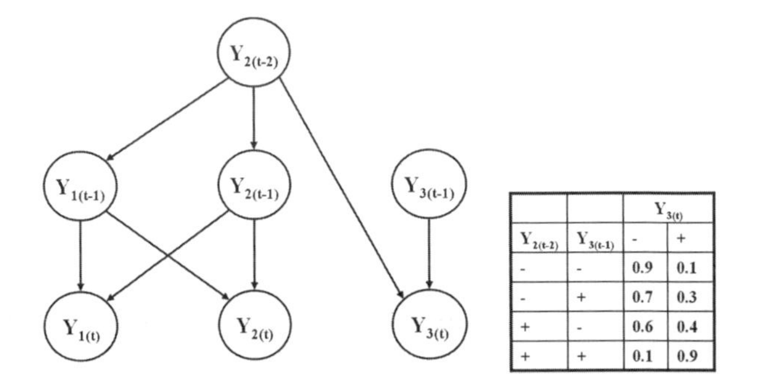

$Y_{2(t-2)}$	$Y_{3(t-1)}$	$Y_{3(t)}$ -	$Y_{3(t)}$ +
-	-	0.9	0.1
-	+	0.7	0.3
+	-	0.6	0.4
+	+	0.1	0.9

Fig. 10.8. A directed acyclic graph that represents the temporal dependency of three categorical variables describing positive (+) and negative (-) regulation.

A dynamic Bayesian network is defined by a directed acyclic graph in which nodes continue to represent stochastic variables and arrows represent temporal dependencies that are quantified by probability distributions. The crucial assumption is that the probability distributions of the temporal dependencies are time invariant, so that the directed acyclic graph of a dynamic Bayesian network represents only the necessary and sufficient time transitions to reconstruct the overall temporal process. Figure 10.8 shows the directed acyclic graph of a dynamic Bayesian network with three variables. The subscript of each node denotes the time lag, so that the arrows from the nodes $Y_{2(t-1)}$ and $Y_{1(t-1)}$ to the node $Y_{1(t)}$ describe the dependency of the probability distribution of the variable Y_1 at time t on the value of Y_1 and Y_2 at time $t-1$. Similarly, the directed acyclic graph shows that the probability distribution of the variable Y_2 at time t is a function of the value of Y_1 and Y_2 at time $t-1$. This symmetrical dependency allows us to represent feedback loops and we used it to describe the regulatory control of glucose in diabetic patients (Ramoni *et al.*, 1995). A dynamic Bayesian network is not restricted to represent temporal dependency of order 1. For example the probability distribution of the variable Y_3 at time t depends on the value of the variable at time $t-1$ as well as the value of the variable Y_2 at time $t-2$. The conditional probability table in Figure 10.8 shows an example when the variables Y_2, Y_3 are categorical.

By using the local Markov property, the joint probability distribution of the three variables at time t, given the past history

$$h_t := y_{1(t-1)}, \dots, y_{1(t-l)}, y_{2(t-1)}, \dots, y_{2(t-l)}, y_{3(t-1)}, \dots, y_{3(t-l)}$$

is given by the product of the three factors:

$$p(y_{1(t)}|h_t) = p(y_{1(t)}|y_{1(t-1)}, y_{2(t-1)})$$
$$p(y_{2(t)}|h_t) = p(y_{2(t)}|y_{1(t-1)}, y_{2(t-1)})$$
$$p(y_{3(t)}|h_t) = p(y_{3(t)}|y_{3(t-1)}, y_{2(t-2)})$$

that represent the probability of transition over time. By assuming that these probability distributions are time invariant, they are sufficient to compute the probability that a process that starts from known values $y_{1(1)}, y_{2(1)}, y_{3(0)}, y_{3(1)}$ evolves into $y_{1(T)}, y_{2(T)}, y_{3(T)}$, by using one of the algorithms for probabilistic reasoning described in Section 10.2. The same algorithms can be used to compute the probability that a process with values $y_{1(T)}, y_{2(T)}, y_{3(T)}$ at time T started from the initial states $y_{1(1)}, y_{2(1)}, y_{3(0)}, y_{3(1)}$.

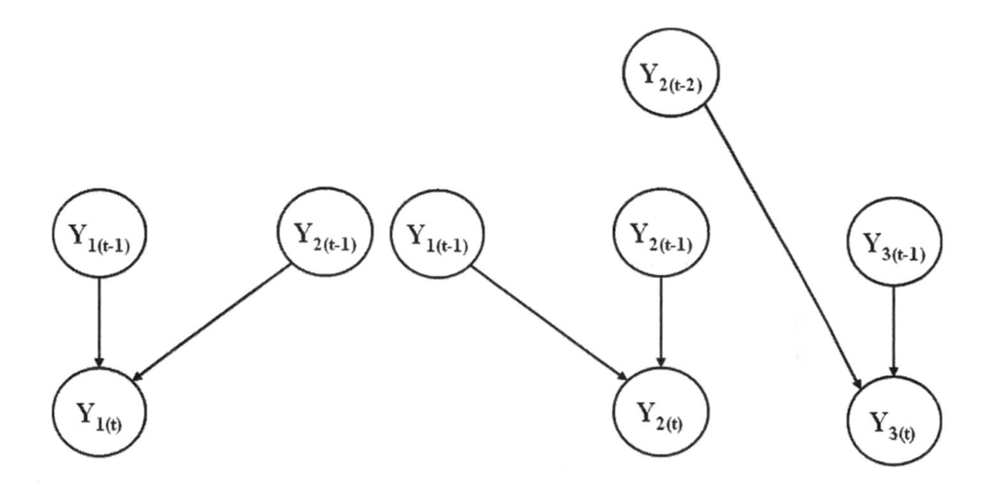

Fig. 10.9. Modular learning of the dynamic Bayesian network in Figure 10.8. First a regressive model is learned for each of the three variables at time t, and then the three models are joined by their common ancestors $Y_{1(t-1)}, Y_{2(t-2)}$ and $Y_{2(t-2)}$ to produce the directed acyclic graph in Figure 10.8.

Learning dynamic Bayesian networks when all the variables are observable is a straightforward parallel application of the structural learning described in Section

10.4. To build the network, we proceed by selecting the set of parents for each variable Y_i at time t, and then the models are joined by the common ancestors. An example is in Figure 10.9. The search of each local dependency structure is simplified by the natural ordering imposed on the variables by the temporal frame (Friedmanm *et al.*, 1998) that constrains the model space of each variable Y_i at time t: the set of candidate parents consists of the variables $Y_{i(t-1)},\ldots,Y_{i(t-p)}$ as well as the variables $Y_{h(t-j)}$ for all $h \neq i$, and $j = 1,\ldots,p$. The K2 algorithm (Cooper and Herskovitz, 1992) discussed in Section 10.4 appears to be particularly suitable for exploring the space of dependency for each variable $Y_{i(t)}$. The only critical issue is that the selection of the largest temporal order to explore depends on the sample size, because each temporal lag of order p leads to the loss of the first p temporal observations in the data set (Yu *et al.*, 2002).

10.6 Data Mining Applications

Bayesian networks have been used by us and others as knowledge discovery tools in a variety of fields, ranging from survey data analysis (Sebastiani and Ramoni, 2000, Sebastiani and Ramoni, 2001B) to customer profiling (Sebastiani *et al.*, 2000) and bioinformatics (Friedman, 2004, Sebastiani *et al.*, 2004,2). Here we describe two Data Mining and knowledge discovery applications based on Bayesian networks.

10.6.1 Survey Data

A major goal of surveys conducted by Federal Agencies is to provide citizens, consumers and decision makers with useful information in a compact and understandable format. Data are expected to improve the understanding of institutions, businesses, and citizens of the current state of affairs in the country and play a key role in political decisions. But the size and structure of this fast-growing databases pose the challenge of how effectively extracting and presenting this information to enhance planning, prediction, and decision making. An example of fast growing database is the Current Population Survey (CPS) database that collects monthly surveys of about 50,000 households conducted by the U.S. Bureau of the Census. These surveys are the primary source of information on the labor force characteristics of the U.S. population, they provide estimates for the nation as a whole and serve as part of model-based estimates for individual states and other geographic areas. Estimates obtained from the CPS include employment and unemployment, earnings, hours of work, and other indicators, and are often associated with a variety of demographic characteristics including age, sex, race, marital status, and education. CPS data are used by government policymakers and legislators as important indicators of the nations's economic situation and for planning and evaluating many government programs.

For most of the surveys conducted by the U.S Census Bureau, users can access both the microdata or summary tables. Summary tables provide easy access to findings of interest by relating a small number of preselected variables. In so doing, summary tables disintegrate the information contained in the original data into

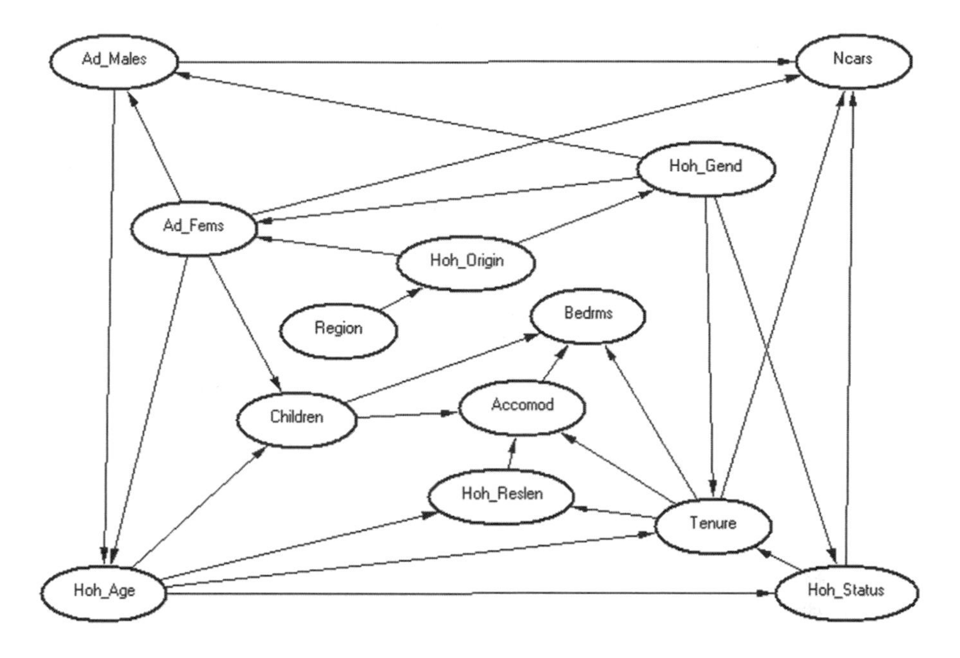

Fig. 10.10. Bayesian network induced from a portion of the 1996 General Household Survey, conducted between April 1996 and March 1997 by the British Office of National Statistics in Great Britain.

micro-components and fail to convey an overall picture of the process underlying the data. A different approach to the analysis of survey data would be to employ Data Mining tools to generate hypothesis and hence to make new discoveries in an automated way (Hand *et al.*, 2001, Hand *et al.*, 2002).

As an example, Figure 10.10 shows a Bayesian network learned from a data set of 13 variables extracted from the 1996 General Household Survey conducted between April 1996 and March 1997 by the British Office of National Statistics in Great Britain. Variables and their states are summarized in Table 10.2. The network structure shows interesting, directed dependencies and conditional independencies. For example, there is a dependency between the ethnic group of the heads of the households and the region of birth (variables *Region* and *HoH_origin*) and the conditional probability table that shapes this dependency reveals a more cosmopolitan society in England than Wales and Scotland, with a larger proportion of Blacks and Indians as household heads. The working status of the head of the household (*Hoh_status*) is independent of the ethnic group given gender and age. The conditional probability table that shapes this dependency shows that young female heads of household are much more likely to be inactive than male heads of household (40% compared to 6% when the age group is 17–36). This difference is attenuated as the age of the head

Variable	Description	Type	State description
Region	Hoh birth region	Nominal	England, Scotland and Wales
Ad_fems	No of adult females	Ordinal	0, 1, ≥ 2
Ad_males	No of adult males	Ordinal	0, 1, ≥ 2
Children	No of children	Ordinal	0, 1, 2, 3, ≥ 4
Hoh_age	Age of Hoh	Numeric	17-36; 36-50; 50-66; 66-98
Hoh_gend	Gender of Hoh	Nominal	M, F
Accomod	Accommodation	Nominal	Room, Flat, House, Other
Bedrms	No of bedrooms	Ordinal	1, 2, 3, ≥ 4
Ncars	No of cars	Ordinal	1, 2, 3, ≥ 4
Tenure	House status	Nominal	Rent, Owned, Soc-Sector
Hoh_reslen	Length of residence	Numeric	0-3; 3-9; 9-19; ≥ 19 (months)
Hoh_origin	Hoh ethnicity	Nominal	Caucas., Black, Chin., Indian, Other
Hoh_status	Status of Hoh	Nominal	Active, Inactive, Retired

Table 10.2. Description of the variables used in the analysis. Hoh denotes the Head of the Household. Numbers of adult males, females and children refer to the household.

of the household increases. The dependency of the gender of the household head on the ethnic group shows that Blacks have the smallest probability of having a male head of the household (64%) while Indians have the largest probability (89%). Other interesting discoveries are that the age of the head of the household depends directly on the number of adult males and females and shows that households with no females and two or more males are more likely to be headed by a young male, while on the other hand, households with no males and two or more females are headed by a mid age female. There appear to be more single households headed by an elder female than an elder male. Also the composition of the household changes in the ethnic groups and Indians have the smallest probability of living in a household with no adult males (10%), while Blacks have the largest probability (32%).

By propagating the network, one may investigate other undirected associations and discover that, for example, the typical Caucasian mid family with two children has 77% chance of being headed by a male who, with probability .57, is aged between 36 and 50 years. The probability that the head of the household is active is .84, and the probability that the household is in an owned house is .66. Results of these queries are displayed in Figure 10.11. These figures are slightly different if the head of the household is, for example, Black and the probability that the head of the household is male (given that there are two children in the household) is only .62 and the probability that he is active is .79. If the head of the household is Indian, then the probability that he is male is .90, and the probability that he is active is .88. On average, the ethnic group changes slightly the probability of the household being in an accommodation provided by the social service (26% for Blacks, 23% for Chinese, 20% Indians and 24% Caucasians). Similarly, Black household heads are more likely to be inactive than household heads from different ethnic groups (16% Blacks, 10% Indians, 14% Caucasians and Chinese) and to be living in a less wealthy household, as shown by the larger probability of living in accommodations with a smaller num-

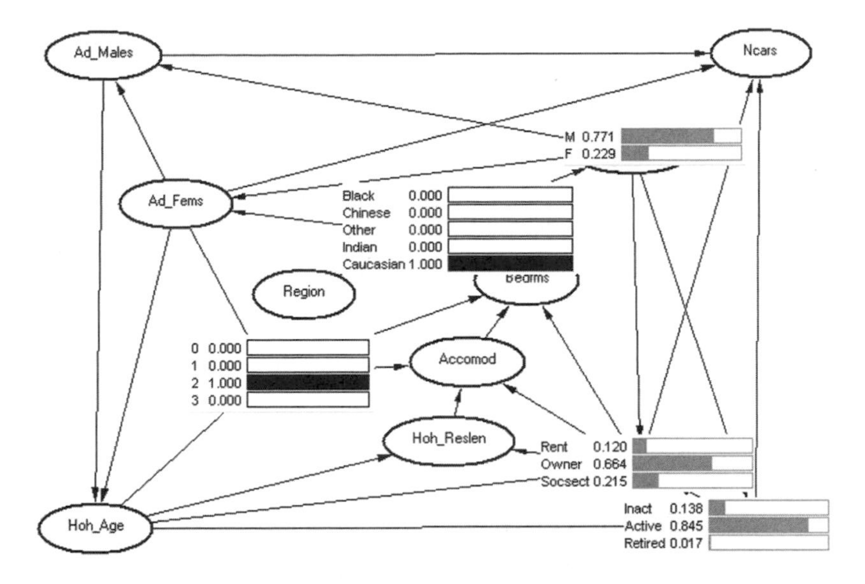

Fig. 10.11. An example of probabilistic reasoning using the Bayesian network induced from the 13 variables extracted from the 1996 General Household Survey.

ber of bedrooms and of having a smaller number of cars. The overall picture is that of households headed by a Black to be less wealthy than others, and this would be the conclusions one reaches if the gender of the head of the houschold is not taken into account. However, the dependency structure discovered shows that the gender of the head of the household and the number of adult females make all the other variables independent of the ethnic group. Thus, the extracted model supports the hypothesis that differences in the household wealth are more likely explained by the different household composition, and in particular by the gender of the head of the household, rather than racial factors.

10.6.2 Customer Profiling

A typical problem of direct mail fund raising campaigns is the low response rate. Recent studies have shown that adding incentives or gifts in the mailing can increase the response rate. This is the strategy implemented by an American Charity in the June '97 renewal campaign. The mailing included a gift of personalized name and address labels plus an assortment of 10 note cards and envelopes. Each mail cost the charity 0.68 dollars and resulted in a response rate of about 5% in the group of so called lapsed donors, that is, individuals who made their last donation more than a year before the '97 renewal mail. Since the donations received by the respondents ranged between 2 and 200 dollars, and the median donation was 13 dollars, the fund raiser needed to decide when it was worth sending the renewal mail to a donor, on the basis of the information available about him from the in-house database. Furthermore, the charity was interested in strategies to recapture Lapsed Donors and, therefore, in making a profile from which to understand motivations behind their lack of response.

We addressed these issues in (Sebastiani *et al.*, 2000) by building two causal models. The first model captured the dependency of the probability of response to the mailing campaign on the independent variables in the database. The second one modeled the dependency of the dollar amount of the gift and it was built by using only the 5% respondents to the '97 mailing campaign. We focused here on the first model, depicted in Figure 10.12, which shows that the probability of a donation (variable Target-B in the top-left corner) is directly affected by the wealth rating (variable Wealth1) and the donor's neighborhood (variable Domain1). The network shows that, marginally, only 5% of those who received the renewal mail are likely to respond. Persons living in suburbs, cities or towns have about 5% probability of responding, while donors living in rural or urban neighborhoods respond with probability 5%. The wealth rating of the donor neighborhood has a positive effect on the response rate of donors living in urban, suburban or city areas with donors living in wealthier neighborhoods being more likely to respond than donors living in poorer neighborhoods. The probability of responding raises up to about 6% for donors living in wealth city neighborhoods. The variable Domain1 is closely related to the variable Domain2 that represents an indicator of the socio-economic status of the donor neighborhood and it shows that donors living in suburbs or city are more likely to live in neighborhoods having a highly rated socio-economic status. Therefore, they may be more sensitive to political and social issues. The model also shows that donors living in neighborhoods with a high presence of males active in the Military (Malemili) are more likely to respond. Again, since the charity collects funds for military veterans, this fact supports the hypothesis that sensitivity to the problem for which funds are collected has a large effect on the probability of response. On the other hand, the wealth rating of donors living in rural neighborhood has the opposite effect: the higher the wealth rating, the smaller the probability that the donor responds, and the least likely to respond (3.8%) are donors living in wealth rural areas. A curiosity is that persons living in rural and poor neighborhood are more likely to respond positively to mail including a gift than donors living in wealthy city neighborhood.

By querying the network, we can profile respondents who are more likely to live in a wealth neighborhood, which is located in a suburb and they are less likely to have made a donation in the last 6 months than those who do not respond. One feature that discriminates respondents from nonrespondent is the household income, and respondents are 1.20 times more likely to be living in wealthy neighborhoods, and to be on higher income than nonrespondents.

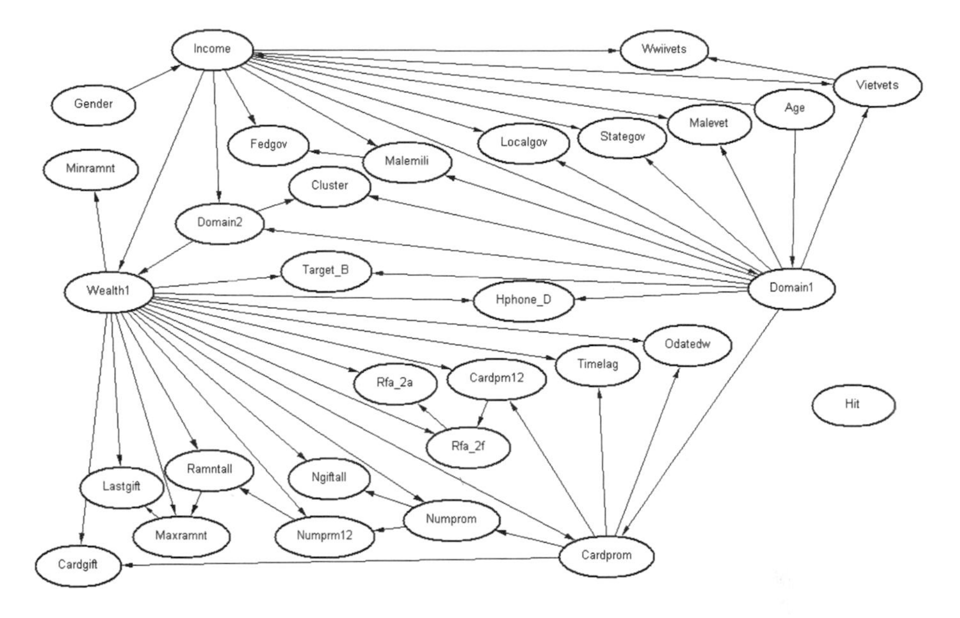

Fig. 10.12. The Bayesian network induced from the data warehouse to profile likely respondents to mail solicitations.

10.7 Conclusions and Future Research Directions

Bayesian networks are a representation formalism born at the intersection of statistics and Artificial Intelligence. Thanks to their solid statistical foundations, they have been successfully turned into a powerful Data Mining and knowledge discovery tool able to uncover complex models of interactions from large databases. Their high symbolic nature makes them easily understandable to human operators. Contrary to standard classification methods, Bayesian networks do not require the preliminary identification of an outcome variable of interest but they are able to draw probabilistic inferences on any variable in the database.

Notwithstanding these attractive properties, there are still several theoretical issues that limit the range of applicability of Bayesian networks to the practice of science and engineering. This chapter has described methods to learn Bayesian networks from databases with either discrete or continuous variables. How to induce Bayesian networks from databases containing both types of variables is still very much an open research issues. Imposing the assumption that discrete variables can only be parent nodes in the network, but cannot be children of any continuous Gaussian node leads to a closed form solution for the computation of the marginal likelihood (Lauritzen, 1992). This property has been applied, for example, to model-based clustering by (Ramoni *et al.*, 2002), and it is commonly used in classification problems (Cheeseman and Stutz, 1996). However, this restriction can quickly become unrealistic and greatly limit the set of models to explore. As a consequence, common practice is still to discretize continuous variables with possible loss of information, particularly when the continuous variables are highly skewed.

Another challenging research issue is how to learn Bayesian networks from incomplete data. The received view of the effect of missing data on statistical inference is based on the approach described by Rubin in (Rubin, 1987). This approach classifies the missing data mechanism as ignorable or not, according to whether the data are missing completely at random (MCAR), missing at random (MAR), or informatively missing (IM). According to this approach, data are MCAR if the probability that an entry is missing is independent of both observed and unobserved values. They are MAR if this probability is at most a function of the observed values in the database and, in all other cases, data are IM. The received view is that, when data are either MCAR or MAR, the missing data mechanism is ignorable for parameter estimation, but it is not when data are IM. An important but overlooked issue is whether the missing data mechanism generating data that are MAR is ignorable for model selection (Rubin, 1996, Sebastiani and Ramoni, 2001A). We have shown that this is not the case for regression type graphical models exemplified and introduced two approaches to model selection with partially ignorable missing data mechanisms: *ignorable imputation* and *model folding*. Contrary to standard imputation schemes (Geiger *et al.*, 1995, Little and Rubin, 1987, Schafer, 1997, Tanner, 1996, Thibaudeau and Winler, 2002), ignorable imputation accounts for the missing-data mechanism and produces, asymptotically, a proper imputation model as defined by Rubin (Rubin, 1987, Rubin *et al.*, 1995). However, the computation effort can be very demanding and model folding is a deterministic method to approximate the exact marginal likelihood that reaches high accuracy at a low computational cost, because the complexity of the model search is not affected by the presence of incomplete cases. Both ignorable imputation and model folding reconstruct a completion of the incomplete data by taking into account the variables responsible for the missing data. This property is in agreement with the suggestion put forward in (Heitjan and Rubin, 1991, Little and Rubin, 1987, Rubin, 1976) that the variables responsible for the missing data should be kept in the model. However, our approach allows us to also evaluate the likelihoods of models that do not depend explicitly on these variables.

Although this work provides the analytical foundations for a proper treatment of missing data when the inference task is model selection, it is limited to the very special situation in which only one variable is partially observed, data are supposed to be only MCAR or MAR, and the set of Bayesian networks is limited to those in which the partially observed variable is a child of the other variables. Research is needed to extend these results to the more general graphical structures, in which several variables can be partially observed and data can be MCAR, MAR or IM.

These two issues — learning mixed variables networks and handling incomplete databases — are still unsolved and they offer challenging research opportunities.

Acknowledgments

This work was supported in part by the National Science Foundation (ECS-0120309), the Spanish State Office of Education and Universities, the European Social Fund and the Fulbright Program of the US State Department.

References

S. G. Bottcher and C. Dethlefsen. Deal: A package for learning Bayesian networks. Available from http://www.jstatsoft.org/v08/i20/deal.pdf, 2003.

U. M. Braga-Neto and E. R. Dougerthy. Is cross-validation valid for small-sample microarray classification. *Bioinformatics*, 20:374–380, 2004.

E. Castillo, J. M. Gutierrez, and A. S. Hadi. *Expert Systems and Probabilistic Network Models*. Springer, New York, NY, 1997.

E. Charniak. Belief networks without tears. *AI Magazine*, pages 50–62, 1991.

P. Cheeseman and J. Stutz. Bayesian classification (AutoClass): Theory and results. In U. M. Fayyad, G. Piatetsky-Shapiro, P. Smyth, and R. Uthurusamy, editors, *Advances in Knowledge Discovery and Data Mining*, pages 153–180. MIT Press, Cambridge, MA, 1996.

J. Cheng and M. Druzdzel. AIS-BN: An adaptive importance sampling algorithm for evidential reasoning in large Bayesian networks. *J Artif Intell Res*, 13:155–188, 2000.

D. M. Chickering. Learning equivalence classes of Bayesian-network structures. *J Mach Learn Res*, 2:445–498, February 2002.

G. F. Cooper. The computational complexity of probabilistic inference using Bayesian belief networks. *aij*, 42:297–346, 1990.

G. F. Cooper and E. Herskovitz. A Bayesian method for the induction of probabilistic networks from data. *Mach Learn*, 9:309–347, 1992.

R. G. Cowell, A. P. Dawid, S. L. Lauritzen, and D. J. Spiegelhalter. *Probabilistic Networks and Expert Systems*. Springer, New York, NY, 1999.

A. P. Dawid and S. L. Lauritzen. Hyper Markov laws in the statistical analysis of decomposable graphical models. *Ann Stat*, 21:1272–1317, 1993. Correction ibidem, (1995), *23*, 1864.

R. O. Duda and P. E. Hart. *Pattern Classification and Scene Analysis*. Wiley, New York, NY, 1973.

N. Friedman. Inferring cellular networks using probabilistic graphical models. *Science*, 303:799–805, 2004.

N. Friedman, D. Geiger, and M. Goldszmidt. Bayesian network classifiers. *Mach Learn*, 29:131–163, 1997.

N. Friedman and D. Koller. Being Bayesian about network structure: A Bayesian approach to structure discovery in bayesian networks. *Machine Learning*, 50:95–125, 2003.

N. Friedman, K. Murphy, and S. Russell. Learning the structure of dynamic probabilistic networks. In *Proceedings of the 14th Annual Conference on Uncertainty in Artificial Intelligence (UAI-98)*, pages 139–147, San Francisco, CA, 1998. Morgan Kaufmann Publishers.

D. Geiger and D. Heckerman. Learning gaussian networks. In *Proceedings of the Tenth Annual Conference on Uncertainty in Artificial Intelligence (UAI-94)*, San Francisco, 1994. Morgan Kaufmann.

D. Geiger and D. Heckerman. A characterization of Dirichlet distributions through local and global independence. *Ann Stat*, 25:1344–1368, 1997.

A. Gelman, J. B. Carlin, H. S. Stern, and D. B. Rubin. *Bayesian Data Analysis*. Chapman and Hall, London, UK, 1995.

S. Geman and D. Geman. Stochastic relaxation, Gibbs distributions and the Bayesian restoration of images. *IEEE T Pattern Anal*, 6:721–741, 1984.

W. R. Gilks and G. O. Roberts. Strategies for improving MCMC. In W. R. Gilks, S. Richardson, and D. J. Spiegelhalter, editors, *Markov Chain Monte Carlo in Practice*, pages 89–114. Chapman and Hall, London, UK, 1996.

C. Glymour, R. Scheines, P. Spirtes, and K. Kelly. *Discovering Causal Structure: Artificial Intelligence, Philosophy of Science, and Statistical Modeling*. Academic Press, San Diego, CA, 1987.

I. J. Good. Rational decisions. *J Roy Stat Soc B*, 14:107–114, 1952.

I. J. Good. *The Estimation of Probability: An Essay on Modern Bayesian Methods*. MIT Press, Cambridge, MA, 1968.

D. J. Hand. *Construction and Assessment of Classification Rules*. Wiley, New York, NY, 1997.

D. J. Hand, N. M. Adams, and R. J. Bolton. *Pattern Detection and Discovery*. Springer, New York, 2002.

D. J. Hand, H. Mannila, and P. Smyth. *Principles of Data Mining*. MIT Press, Cambridge, 2001.

T. Hastie, R. Tibshirani, and J. Friedman. *The Elements of Statistical Learning*. Springer-Verlag, New York, 2001.

D. Heckerman. Bayesian networks for Data Mining. *Data Min Knowl Disc*, 1:79–119, 1997.

D. Heckerman, D. Geiger, and D. M. Chickering. Learning Bayesian networks: The combinations of knowledge and statistical data. *Mach Learn*, 20:197–243, 1995.

D. F. Heitjan and D. B. Rubin. Ignorability and coarse data. *Ann Stat*, 19:2244–2253, 1991.

R. E. Kass and A. Raftery. Bayes factors. *J Am Stat Assoc*, 90:773–795, 1995.

P. Langley, W. Iba, and K. Thompson. An analysis of Bayesian classifiers. In *Proceedings of the Tenth National Conference on Artificial Intelligence*, pages 223–228, Menlo Park, CA, 1992. AAAI Press.

P. Larranaga, C. Kuijpers, R. Murga, and Y. Yurramendi. Learning Bayesian network structures by searching for the best ordering with genetic algorithms. *IEEE T Pattern Anal*, 26:487–493, 1996.

S. L. Lauritzen. Propagation of probabilities, means and variances in mixed graphical association models. *J Am Stat Assoc*, 87(420):1098–108, 1992.

S. L. Lauritzen. *Graphical Models*. Oxford University Press, Oxford, UK, 1996.

S. L. Lauritzen and D. J. Spiegelhalter. Local computations with probabilities on graphical structures and their application to expert systems (with discussion). *J Roy Stat Soc B*, 50:157–224, 1988.

R. J. A. Little and D. B. Rubin. *Statistical Analysis with Missing Data*. Wiley, New York, NY, 1987.

D. Madigan and A. E. Raftery. Model selection and accounting for model uncertainty in graphical models using Occam's window. *J Am Stat Assoc*, 89:1535–1546, 1994.

D. Madigan and G. Ridgeway. Bayesian data analysis for Data Mining. In *Handbook of Data Mining*, pages 103–132. MIT Press, 2003.

D. Madigan and J. York. Bayesian graphical models for discrete data. *Int Stat Rev*, pages 215–232, 1995.

P. McCullagh and J. A. Nelder. *Generalized Linear Models*. Chapman and Hall, London, 2nd edition, 1989.

A. O'Hagan. *Bayesian Inference*. Kendall's Advanced Theory of Statistics. Arnold, London, UK, 1994.

J. Pearl. *Probabilistic Reasoning in Intelligent Systems: Networks of plausible inference*. Morgan Kaufmann, San Francisco, CA, 1988.

M. Ramoni, A. Riva, M. Stefanelli, and V. Patel. An ignorant belief network to forecast glucose concentration from clinical databases. *Artif Intell Med*, 7:541–559, 1995.

M. Ramoni and P. Sebastiani. Bayesian methods. In *Intelligent Data Analysis. An Introduction*, pages 131–168. Springer, New York, NY, 2nd edition, 2003.

M. Ramoni, P. Sebastiani, and I.S. Kohane. Cluster analysis of gene expression dynamics. *Proc Natl Acad Sci USA*, 99(14):9121–6, 2002.

L. Rokach, M. Averbuch, and O. Maimon, Information retrieval system for medical narrative reports. Lecture notes in artificial intelligence, 3055. pp. 217-228, Springer-Verlag (2004).

D. B. Rubin. Inference and missing data. *Biometrika*, 63:581–592, 1976.

D. B. Rubin. *Multiple Imputation for Nonresponse in Survey*. Wiley, New York, NY, 1987.

D. B. Rubin. Multiple imputation after 18 years. *J Am Stat Assoc*, 91:473–489, 1996.

D. B. Rubin, H. S. Stern, and V. Vehovar. Handling "don't know" survey responses: the case of the Slovenian plebiscite. *J Am Stat Assoc*, 90:822–828, 1995.

M. Sahami. Learning limited dependence Bayesian classifiers. In *Proceeding of the 2 Int. Conf. On Knowledge Discovery & Data Mining*, 1996.

J. L. Schafer. *Analysis of Incomplete Multivariate Data*. Chapman and Hall, London, UK, 1997.

P Sebastiani, M Abad, and M F Ramoni. Bayesian networks for genomic analysis. In E R Dougherty, I Shmulevich, J Chen, and Z J Wang, editors, *Genomic Signal Processing and Statistics*, Series on Signal Processing and Communications. EURASIP, 2004.

P. Sebastiani and M. Ramoni. Analysis of survey data with Bayesian networks. Technical Report, Knowledge Media Institute, The Open University, Walton Hall, Milton Keynes MK7 6AA, 2000. Available from authors.

P. Sebastiani and M. Ramoni. Bayesian selection of decomposable models with incomplete data. *J Am Stat Assoc*, 96(456):1375–1386, 2001A.

P. Sebastiani and M. Ramoni. Common trends in european school populations. *Res. Offic. Statist.*, 4(1):169–183, 2001B.

P. Sebastiani and M. F. Ramoni. On the use of Bayesian networks to analyze survey data. *Res. Offic. Statist.*, 4:54–64, 2001C.

P. Sebastiani and M. Ramoni. Generalized gamma networks. Technical report, University of Massachusetts, Department of Mathematics and Statistics, 2003.

P. Sebastiani, M. Ramoni, and A. Crea. Profiling customers from in-house data. *ACM SIGKDD Explorations*, 1:91–96, 2000.

P. Sebastiani, M. Ramoni, and I. Kohane. BADGE: Technical notes. Technical report, Department of Mathematics and Statistics, University of Massachusetts at Amherst, 2003.

P. Sebastiani, M. F. Ramoni, V. Nolan, C. Baldwin, and M. H. Steinberg. Discovery of complex traits associated with overt stroke in patients with sickle cell anemia by Bayesian network modeling. In *27th Annual Meeting of the National Sickle Cell Disease Program*, 2004. To appear.

P. Sebastiani, Y. H. Yu, and M. F. Ramoni. Bayesian machine learning and its potential applications to the genomic study of oral oncology. *Adv Dent Res*, 17:104–108, 2003.

R. D. Shachter. Evaluating influence diagrams. *Operation Research*, 34:871–882, 1986.

M. Singh and M. Valtorta. Construction of Bayesian network structures from data: A brief survey and an efficient algorithm. *Int J Approx Reason*, 12:111–131, 1995.

D. J. Spiegelhalter and S. L. Lauritzen. Sequential updating of conditional probabilities on directed graphical structures. *Networks*, 20:157–224, 1990.

P. Spirtes, C. Glymour, and R. Scheines. *Causation, prediction and search*. Springer, New York, 1993.

M. A. Tanner. *Tools for Statistical Inference*. Springer, New York, NY, third edition, 1996.

Y. Thibaudeau and W. E. Winler. Bayesian networks representations, generalized imputation, and synthetic microdata satisfying analytic restraints. Technical report, Statistical Research Division report RR 2002/09, 2002. http://www.census.gov/srd/www/byyear.html.

A. Thomas, D. J. Spiegelhalter, and W. R. Gilks. Bugs: A program to perform Bayesian inference using Gibbs Sampling. In J. Bernardo, J. Berger, A. P. Dawid, and A. F. M. Smith, editors, *Bayesian Statistics 4*, pages 837–42. Oxford University Press, Oxford, UK, 1992.

J. Whittaker. *Graphical Models in Applied Multivariate Statistics*. Wiley, New York, NY, 1990.

S. Wright. The theory of path coefficients: a reply to niles' criticism. *Genetics*, 8:239–255, 1923.

S. Wright. The method of path coefficients. *Annals of Mathematical Statistics*, 5:161–215, 1934.

J. Yu, V. Smith, P. Wang, A. Hartemink, and E. Jarvis. Using Bayesian network inference algorithms to recover molecular genetic regulatory networks. In *International Conference on Systems Biology 2002 (ICSB02)*, 2002.

H. Zhou and S. Sakane. Sensor planning for mobile robot localization using Bayesian network inference. *J. of Advanced Robotics*, 16, 2002. To appear.

11

Data Mining within a Regression Framework

Richard A. Berk

Department of Statistics
UCLA
berk@stat.ucla.edu

Summary. Regression analysis can imply a far wider range of statistical procedures than often appreciated. In this chapter, a number of common Data Mining procedures are discussed within a regression framework. These include non-parametric smoothers, classification and regression trees, bagging, and random forests. In each case, the goal is to characterize one or more of the distributional features of a response conditional on a set of predictors.

Key words: regression, smoothers, splines, CART, bagging, random forests

11.1 Introduction

Regression analysis can imply a broader range of techniques than ordinarily appreciated. Statisticians commonly define regression so that the goal is to understand "as far as possible with the available data how the the conditional distribution of some response y varies across subpopulations determined by the possible values of the predictor or predictors" (Cook and Weisberg, 1999). For example, if there is a single categorical predictor such as male or female, a legitimate regression analysis has been undertaken if one compares two income histograms, one for men and one for women. Or, one might compare summary statistics from the two income distributions: the mean incomes, the median incomes, the two standard deviations of income, and so on. One might also compare the shapes of the two distributions with a Q-Q plot.

There is no requirement in regression analysis for there to be a "model" by which the data were supposed to be generated. There is no need to address cause and effect. And there is no need to undertake statistical tests or construct confidence intervals. The definition of a regression analysis can be met by pure description alone. Construction of a "model," often coupled with causal and statistical inference, arc supplements to a regression analysis, not a necessary component (Berk, 2003).

Given such a definition of regression analysis, a wide variety of techniques and approaches can be applied. In this chapter, I will consider a range of procedures

O. Maimon, L. Rokach (eds.), *Data Mining and Knowledge Discovery Handbook*, 2nd ed.,
DOI 10.1007/978-0-387-09823-4_11, © Springer Sciencc+Busincss Media, LLC 2010

under the broad rubric of data mining. The coverage is intended to be broad rather than deep. Readers are encouraged to consult the references cited.

11.2 Some Definitions

There are almost as many definitions of Data Mining as there are treatises on the subject (Sutton and Barto, 1999, Cristianini and Shawe-Taylor, 2000, Witten and Frank, 2000, Hand et al., 2001, Hastie et al., 2001, Breiman, 2001b, Dasu and Johnson, 2003), and associated with Data Mining are a variety of names: statistical learning, machine learning, reinforcement learning, algorithmic modeling and others. By "Data Mining" I mean to emphasize the following.

The broad definition of regression analysis applies. Thus, the goal is to examine $\mathbf{y}|\mathbf{X}$ for a response $\mathbf{y}$ and a set of predictors $\mathbf{X}$, with the values of $\mathbf{X}$ treated as fixed. There is no need to commit to any particular feature of $\mathbf{y}|\mathbf{X}$, but emphasis will, nevertheless, be placed on the conditional mean, $\bar{\mathbf{y}}|\mathbf{X}$. This is the feature of $\mathbf{y}|\mathbf{X}$ that has to date drawn the most attention. [1]

Within the context of regression analysis, now consider a given a data set with N observations, a *single* predictor $\mathbf{x}$, and a *single* value of $\mathbf{x}$, x_0. The fitted value for $\hat{y}_0$ at x_0 can be written as

$$\hat{y}_0 = \sum_{j=1}^{N} \mathbf{S}_{0j} y_j, \tag{11.1}$$

where $\mathbf{S}$ is an N by N matrix of weights, the subscript 0 represents the row corresponding to the case whose value of $\mathbf{y}$ is to be constructed, and the subscript j represents the column in which the weight is found. That is, the fitted value $\hat{y}_0$ at x_0 is linear combination of all N values of $\mathbf{y}$, with the weights determined by $\mathbf{S}_{0j}$. If beyond description, estimation is the goal, one has a linear estimator of $\bar{\mathbf{y}}|\mathbf{x}$. In practice, the weights decline with distance from x_0, sometimes abruptly (as in a step function), so that many of the values in $\mathbf{S}_{0j}$ are often zero.[2]

In a regression context, $\mathbf{S}_{0j}$ is constructed from a function $f(\mathbf{x})$ that replaces $\mathbf{x}$ with transformations of $\mathbf{x}$. Then, we often require that

$$f(\mathbf{x}) = \sum_{m=1}^{M} \beta_m h_m(\mathbf{x}), \tag{11.2}$$

[1] In much of what follows I use the framework presented in (Hastie et al., 2001). Generally, matrices will be shown in capital letters in bold face type, vectors will be shown in small letters with bold face type, and scalars will be shown in small letter in italics. But by and large, the meaning will be clear from the context.

[2] It is the estimator that is linear. The function linking the response variable $\mathbf{y}$ to the predictor $\mathbf{x}$ can be highly non-linear. The role of $\mathbf{S}_{0j}$ has much in common the hat-matrix from conventional linear regression analysis: $\mathbf{H} = \mathbf{X}(\mathbf{X}^T\mathbf{X})^{-1}\mathbf{X}^T$. The hat-matrix transforms y_i in a linear fashion into $\hat{y}_i$. $\mathbf{S}_{0j}$ does the same thing but can be constructed in a more general manner.

where there are M transformation of $\mathbf{x}$ (which may include the $\mathbf{x}$ in its original form and a column of 1's for a constant), β_m is the weight given to the mth transformation, and $h_m(\mathbf{x})$ is the mth transformation of $\mathbf{x}$. Thus, one has a linear combination of transformed values of $\mathbf{x}$. The right hand side is sometime called a "linear basis expansion" in $\mathbf{x}$. Common transformations include polynomial terms, and indicator functions that break $\mathbf{x}$ up into several regions. For example, a cubic transformation of $\mathbf{x}$ might include three terms: $\mathbf{x}, \mathbf{x}^2, \mathbf{x}^3$. An indicator function might be defined so that it equals 1 if $\mathbf{x} < \mathbf{c}$ and $\mathbf{0}$ otherwise (where the vector $\mathbf{c}$ contains some constant). A key point is that this kind of formulation is both very flexible and computationally tractable.

Equation 11.2 can be generalized as follows so that more than one predictor may be included:

$$f(\mathbf{x}) = \sum_{j=1}^{p} \sum_{m=1}^{M_j} \beta_{jm} h_{jm}(\mathbf{x}), \tag{11.3}$$

where p is the number of predictors, and for each p, there are M_j transformations. Each predictor has its own set of transformations, and all of the transformations for all predictors, each with its own weight β_{jm}, are combined in a linear fashion.

Why the additive formulation when there is more than one predictor? As a practical matter, with each additional predictor the number of observations needed increases enormously; the volume to be filled with data goes up as a function of the power of the number of predictor dimensions. In addition, there can be very taxing computational demands. So, it is often necessary to restrict the class of functions of $\mathbf{x}$ examined. Equation 11.3 implies that one can consider the role of a large number of predictors within much the same additive framework used in conventional multiple regression.

To summarize, Data Mining within a regression framework will rely on regression analysis, broadly defined, so that there is no necessary commitment *a priori* to any particular function of the predictors. The relationships between the response and the predictors can be determined empirically from the data. We will be working within the spirit of procedures such as stepwise regression, but beyond allowing the data to determine which predictors are required, we allow the data to determine what function of each predictor is most appropriate. In practice, this will mean "subcontracting" a large part of one's data analysis to one or more computer algorithms. Attempting to proceed "by hand" typically is not be feasible.

In the pages ahead several specific Data Mining procedures will be briefly discussed. These are chosen because they are representative, widely used, and illustrate well how Data Mining can be undertaken within a regression framework. No claim is made that the review is exhaustive.

11.3 Regression Splines

A relatively small step beyond conventional parametric regression analysis is taken when regression splines are used in the fitting process. Suppose the goal is to fit the data with a broken line such that at each break the left hand edge meets the right hand edge. That is, the fit is a set of connected straight line segments. To illustrate, consider the three connected line segments as shown in Figure 11.1.

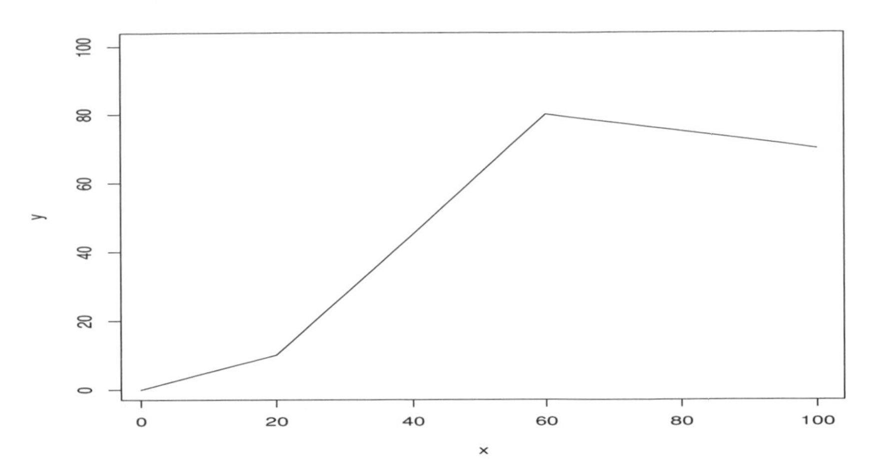

Fig. 11.1. An Illustration of Linear Regression Splines with Two Knots

Constructing such a fitting function for the conditional means is not difficult. To begin, one must decide where the break points on x will be. If there is a single predictor, as in this example, the break points might be chosen after examining a scatter plot of **y** on **x**. If there is subject-matter expertise to help determine the break points, all the better. For example, **x** might be years with the break points determined by specific historical events.

Suppose the break points are at $\mathbf{x} = \mathbf{a}$ and $\mathbf{x} = \mathbf{b}$ (with $\mathbf{b} > \mathbf{a}$). In Figure 11.1, $a = 20$ and $b = 60$. Now define two indicator variables. The first ($\mathbf{I}_a$) is equal to 1 if **x** is greater than the first break point and 0 otherwise. The second ($\mathbf{I}_b$) is equal to 1 if **x** is greater than the second break point and 0 otherwise. We let x_a be the value of **x** at the first break point and x_b be the value of **x** at the second break point.

The mean function is then[3]

$$\bar{\mathbf{y}}|\mathbf{x} = \beta_0 + \beta_1\mathbf{x} + \beta_2(\mathbf{x} - \mathbf{x}_a)\mathbf{I}_a + \beta_3(\mathbf{x} - \mathbf{x}_b)\mathbf{I}_b. \tag{11.4}$$

Looking back at equation 11.2, one can see that there are four $h_m(\mathbf{x})$'s, with the first function of **x** a constant. Now, the mean function for **x** less than **a** is,

$$\bar{\mathbf{y}}|\mathbf{x} = \beta_0 + \beta_1\mathbf{x}. \tag{11.5}$$

For values of **x** equal to or greater than **a** but less than **b**, the mean function is,

[3] To keep the equations consistent with the language of the text and to emphasize the descriptive nature of the enterprise, the conditional mean of **y** will be represented by $\bar{\mathbf{y}}$—x rather than by $E(\mathbf{y}|\mathbf{x})$. The latter implies, unnecessarily in this case, that **y** is a random variable.

$$\bar{\mathbf{y}}|\mathbf{x} = (\beta_0 - \beta_2\mathbf{x}_a) + (\beta_1 + \beta_2)\mathbf{x}. \quad (11.6)$$

If β_2 is positive, for $\mathbf{x} \geq \mathbf{a}$ the line is more steep with a slope of $(\beta_1 + \beta_2)$, and lower intercept of $(\beta_0 - \beta_2 x_a)$. If β_2 is negative, the reverse holds.

For values of $\mathbf{x}$ equal to or greater than $\mathbf{b}$, the mean function is,

$$\bar{\mathbf{y}}|\mathbf{x} = (\beta_0 - \beta_2\mathbf{x}_a - \beta_3\mathbf{x}_b) + (\beta_1 + \beta_2 + \beta_3)\mathbf{x}. \quad (11.7)$$

For values of $\mathbf{x}$ greater than $\mathbf{b}$, the slope is altered by adding β_3 to the slope of the previous line segment, and the intercept is altered by subtracting $\beta_3\mathbf{x}_b$. The sign of β_3 determines if the new line segment is steeper or flatter than the previous line segment and where the new intercept falls.

The process of fitting line segments to data is an example of "smoothing" a scatter plot, or applying a "smoother." Smoothers have the goal of constructing fitted values that are less variable than if each of the conditional means of $\mathbf{y}$ were connected by a series of broken lines. In this case, one might simply apply ordinary least squares using equation 11.4 as the mean function to compute of the regression parameters. These, in turn, would then be used to construct the fitted values. There would typically be little interpretative interest in the regression coefficients. The point of the exercise is to superimpose the fitted values on the a scatter plot of the data so that the relationship between $\mathbf{y}$ and $\mathbf{x}$ can be visualized. The relevant output is the picture. The regression coefficients are but a means to this end.

It is common to allow for somewhat more flexibility by fitting polynomials in $\mathbf{x}$ for each segment. Cubic functions of $\mathbf{x}$ are a popular choice because they balance well flexibility against complexity. These cubic line segments are known as "piecewise-cubic splines" when used in a regression format and are known as the "truncated power series basis" in spline parlance.

Unfortunately, simply joining polynomial line segments end to end will not produce an appealing fit where the polynomial segments meet. The slopes will often appear to change abruptly even if there is no reason in the data from them to do so. Visual continuity is achieved by requiring that the first derivative and the second derivative on either side of the break points are the same.[4]

Generalizing from the linear spline framework and keeping the continuity requirement, suppose there are a set of K interior break points, usually called "interior knots," at $\xi_1 < \cdots < \xi_K$ with two boundary knots at ξ_0 and ξ_{K+1}. Then, one can use piecewise cubic splines in the following regression formulation:

$$\bar{\mathbf{y}}|\mathbf{x} = \beta_0 + \beta_1\mathbf{x} + \beta_2\mathbf{x}^2 + \beta_3\mathbf{x}^3 + \sum_{j=1}^{K} \theta_j(\mathbf{x} - \xi_j)^3_+, \quad (11.8)$$

where the "+" indicates the positive values from the expression, and there are $K+4$ parameters to be estimated. This will lead to a conventional regression formulation with a matrix of predictor terms having $K+4$ columns and N rows. Each row will

[4] This is not a formal mathematical result. It stems from what seems to be the kind of smoothness the human eye can appreciate.

have the corresponding values of the piecewise-cubic spline function evaluated at the single value of **x** for that case. There is still only a single predictor, but now there are $K+4$ transformations.

Fitted values near the boundaries of **x** for piecewise-cubic splines can be unstable because they fall at the ends of polynomial line segments where there are no continuity constraints. Sometimes, constraints for behavior at the boundaries are added. One common constraint is that fitted values beyond the boundaries are linear in **x**. While this introduces a bit of bias, the added stability is often worth it. When these constraints are added, one has "natural cubic splines."

The option of including extra constraints to help stabilize the fit raises the well-known dilemma known as the variance-bias tradeoff. At a descriptive level, a smoother fit will usually be less responsive to the data, but easier to interpret. If one treats **y** as a random variable, a smoother fit implies more bias because the fitted values will typically be farther from the true conditional means of **y** ("in the population"), which are the values one wants to estimate from the data on hand. However, in repeated independent random samples (or random realizations of the data), the fitted values will vary less. Conversely, a rougher fit implies less bias but more variance over samples (or realizations), applying analogous reasoning.

For piecewise-cubic splines and natural cubic splines, the degree of smoothness is determined by the number of interior knots. The smaller the number of knots, the smoother the path of the fitted values. That number can be fixed *a priori* or more likely, determined through a model selection procedure that considers both goodness of fit and a penalty for the number of knots. The Akaike information criterion (AIC) is one popular measure, and the goal is to choose the number of knots that minimizes the AIC. Some software such as as R has procedures that can automate the model selection process.[5]

11.4 Smoothing Splines

There is a way to circumvent the need to determine the number of knots. Suppose that for a single predictor there is a fitting function $f(\mathbf{x})$ having two continuous derivatives. The goal is to minimize a "penalized" residual sum of squares

$$RSS(f,\lambda) = \sum_{i=1}^{N}[y_i - f(x_i)]^2 + \lambda \int [f''(t)]^2 dt, \tag{11.9}$$

where λ is a fixed smoothing parameter. The first term captures (as usual) how tight the fit is, while the second imposes a penalty for roughness. The integral quantifies

[5] In practice, the truncated power series basis is usually replaced by a B-spline basis. That is, the transformations of **x** required are constructed from another basis, not explicit cubic functions of **x**. In brief, all splines are linear combinations of B-splines; B-splines are a basis for the space of splines. They are also a well-conditioned basis, because they are fairly close to orthogonal, and they can be computed in a stable and efficient manner. Good discussions of B-splines can be found in (Gifi, 1990) and (Hastie et al., 2001).

how rough the function is, while λ determines how important that roughness will be in the fitting procedure. This is another instance of the variance-bias tradeoff. The larger the value of λ, the greater the penalty for roughness and the smoother the function. The value of λ is used in place of the number of knots to "tune" the variance-bias tradeoff.

Hastie and his colleagues (Hastie et al., 2001) explain that equation 11.9 has a unique minimizer based on a natural cubic spline with N knots.[6] While this might seem to imply that N degrees of freedom are used up, the impact of the N knots is altered because for $\lambda > 0$ there is shrinkage of the fitted values toward a linear fit. In practice, far fewer than N degrees of freedom are lost.

Like the number of knots, the value of λ can be determined *a priori* or through model selection procedures such as those based the generalized cross-validation (GCV). Thus, the value of λ can be chosen so that

$$GCV(\hat{f}_\lambda) = \frac{1}{N}\sum_{i=1}^{N}\left(\frac{y_i - \hat{f}_i(x_i)}{1 - \text{trace}(\mathbf{S}_\lambda)/\text{N}}\right) \tag{11.10}$$

is at small as possible. Using the GVC to select λ is one automated way to find a good compromise between the bias of the fit and its variance.

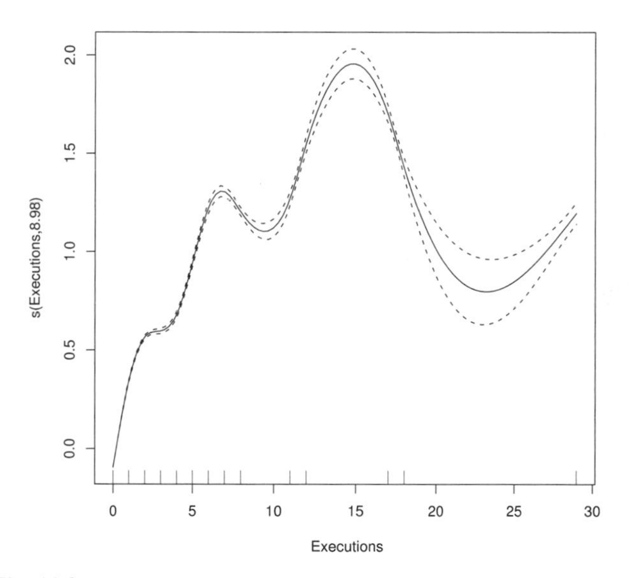

Fig. 11.2. An Illustration of Smoothing with Natural Cubic Splines

Figure 11.2 shows an application based on equations 11.9 and 11.10. The data come from states in the U.S. from 1977 to 1999. The response variable is the number of homicides in a state in a given year. The predictor is the number of inmates

[6] This assumes that there are N distinct values of $\mathbf{x}$. There will be fewer knots if there are less than N distinct values of $\mathbf{x}$.

executed 3 years earlier for capital crimes. Data such as these have been used to consider whether in the U.S. executions deter later homicides (e.g., (Mocan and Gittings, 2003)). Executions are on the horizontal axis (with a rug plot), and homicides are on the vertical axis, labeled as the smooth of executions using 8.98 as the effective degrees of freedom.[7] The solid line is for the fitted values, and the broken lines show the point-by-point 95% confidence interval around the fitted values.

The rug plot at the bottom of Figure 11.2 suggests that most states in most years have very few executions. A histogram would show that the mode is 0. But there are a handful of states that for a given year have a large number of executions (e.g., 18). These few observations are clear outliers.

The fitted values reveal a highly non-linear relationship that generally contradicts the deterrence hypotheses when the number of executions is 15 or less; with a larger number of executions, the number of homicides increases the following year. Only when the number of executions is greater than 15 do the fitted values seems consistent with deterrence. Yet, this is just where there is almost no data. Note that the confidence interval is much wider when the number of executions is between 18 and 28.[8]

The statistical message is that the relationship between the response and the predictor was derived directly from the data. No functional form was imposed *a priori*. And none of the usual regression parameters are reported. The story is Figure 11.2. Sometimes this form of regression analysis is called "nonparametric regression."

11.5 Locally Weighted Regression as a Smoother

Spline smoothers are popular, but there are other smoothers that are widely used as well. Lowess is one example (Cleveland, 1979). Lowess stands for "locally weighted linear regression smoother."

Consider again the one predictor case. The basic idea is that for any given value of the predictor x_0, a linear regression is constructed from observations with x-values near x_0. These data are weighted so that observations with x-values closer to x_0 are given more weight. Then, $\hat{y}_0$ is computed from the fitted regression line and used as the smoothed value of the response at x_0. This process is then repeated for all other x-values.

[7] The effective degrees of freedom is the degrees of freedom required by the smoother, and is calculated as the trace of $\mathbf{S}$ in equation 11.1. It is analogous to the degrees of freedom "used up" in a conventional linear regression analysis when the intercept and regression coefficients are computed. The smoother the fitted value, the greater the effective degrees of freedom used.

[8] Consider again equations 11.1 and 11.2. The natural cubic spline values for executions are the $h_m(\mathbf{x})$ in equation 11.2 which, in turn is the source of $\mathbf{S}$. From $\mathbf{S}$ and the number of homicides $\mathbf{y}$ ones obtains the fitted values $\hat{\mathbf{y}}$ shown in Figure 11.2.

The precise weight given to each observation depends on the weighting function employed; the normal distribution is one option.[9] The degree of smoothing depends on the proportion of the total number of observations used when each local regression line is constructed. The larger the "window" or "span," the larger the proportion of observations included, and the smoother the fit. Proportions between .25 and .75 are common because they seem to provide a good balance for the variance-bias tradeoff.

More formally, each local regression derives from minimizing the weighted sum of squares with respect to the intercept and slope for the $M \leq N$ observations included in the window. That is,

$$RSS^*(\beta) = (\mathbf{y}^* - \mathbf{X}^*\beta)^T \mathbf{W}^* (\mathbf{y}^* - \mathbf{X}^*\beta), \tag{11.11}$$

where the asterisk indicates that only the observations in the window are included, and $\mathbf{W}^*$ is an $M \times M$ diagonal matrix with diagonal elements w_i^*, which are a function of distance from x_0. The algorithm then operates as follows.

1. Choose the smoothing parameter f, which a proportion between 0 and 1.
2. Choose a point x_0 and from that the $(f \times N = M)$ nearest points on $\mathbf{x}$.
3. For these "nearest neighbor" points, compute a weighted least squares regression line for y on x.
4. Construct the fitted value $\hat{y}_0$ for that single x_0.
5. Repeat steps 2 through 4 for each value of $\mathbf{x}$.[10]
6. Connect these $\hat{y}$s with a line.

Lowess is a very popular smoother when there is a single predictor. With a judicious choice of the window size, Figure 11.2 could be effectively reproduced.

11.6 Smoothers for Multiple Predictors

In principle, it is easy to add more predictors and then smooth a multidimensional space. However, there are three major complications. First, there is the "curse of dimensionality." As the number of predictors increases, the space that needs to be filled with data goes up as a power function. So, the demand for data increases rapidly, and the risk is that the data will be far too sparse to get a meaningful fit.

Second, there are some difficult computational issues. For example, how is the neighborhood near $\mathbf{x}_0$ to be defined when predictors are correlated? Also, if the one predictor has much more variability than another, perhaps because of the units of measurement, that predictor can dominate the definition of the neighborhood.

---

[9] The tricube is another popular option. In practice, most of the common weighting functions give about the same results.

[10] As one approaches either tail of the distribution of $\mathbf{x}$, the window will tend to become asymmetrical. One implication is that the fitted values derived from x-values near the tails of $\mathbf{x}$ are typically less stable. Additional constraints are then sometimes imposed much like those imposed on cubic splines.

Third, there are interpretative difficulties. When there are more than two predictors one can no longer graph the fitted surface. How then does one make sense of a surface in more than three dimensions?

When there are only two predictors, there are some fairly straightforward extensions of conventional smoothers that can be instructive. For example, with smoother splines, the penalized sum of squares in equation 11.9 can be generalized. The solution is a set of "thin plate splines," and the results can be plotted. With more than two predictors, however, one generally need another strategy. The generalized additive model is one popular strategy that meshes well with the regression emphasis in this chapter.

11.6.1 The Generalized Additive Model

The mean function for generalized additive model (GAM) with p predictors can is written as

$$\bar{\mathbf{y}}|\mathbf{x} = \alpha + \sum_{j=1}^{p} f_j(\mathbf{x}_j). \tag{11.12}$$

Just as the generalized linear model (GLM), the generalized additive model allows for a number of "link functions" and disturbance distributions. For example, with logistic regression the link function is the log of the odds (the "logit") of the response, and disturbance distribution is logistic.

Each predictor is allowed to have its own functional relationship to the response, with the usual linear form as a special case. If the former, the functional form can be estimated from the data or specified by the researcher. If the latter, all of the usual regression options are available, including indicator variables. Functions of predictors that are estimated from the data rely on smoothers of the sort just discussed.[11]

With the additive form, one can use the same general conception of what it means to "hold constant" that applies to conventional linear regression. The fitting algorithm GAM removes linear dependence between predictors in a fashion that is analogous to the matrix operations behind conventional least squares estimates.

A GAM Fitting Algorithm

Many software packages use the backfitting algorithm to estimate the functions and constant in equation 11.12 (Hastie and Tibshirani, 1990). The basic idea is not difficult and proceeds in the following steps.

1. Initialize: $\alpha = \bar{y}_i, f_j = f_j^0, j = 1, \ldots, p$. Each predictor is given an initial functional relationship to the response such as a linear one. The intercept is given an initial value of the mean of $\mathbf{y}$.

[11] The functions constructed from the data are built so that they have a mean of zero. When all of the functions are estimated from the data, the generalized additive model is sometimes called"nonparametric." When some of the functions are estimated from the data and some are determined by the researcher, the generalized additive model is sometimes called "semiparametric."

2. Cycle: $j = 1, \ldots, p, 1, \ldots, p, \ldots$

$$f_k = \mathbf{S}_j(\mathbf{y} - \alpha - \sum_{j \neq k} f_j | \mathbf{x}_k) \tag{11.13}$$

   A single predictor is selected. Fitted values are constructed using all of the other predictors. These fitted values are subtracted from the response. A smoother $\mathbf{S}_j$ is applied to the resulting "residuals," taken to be a function of the single excluded predictor. The smoother updates the function for that predictor. Each of the other predictors is, in turn, subjected to the same process.
3. Continue 2 until the individual functions do not change.

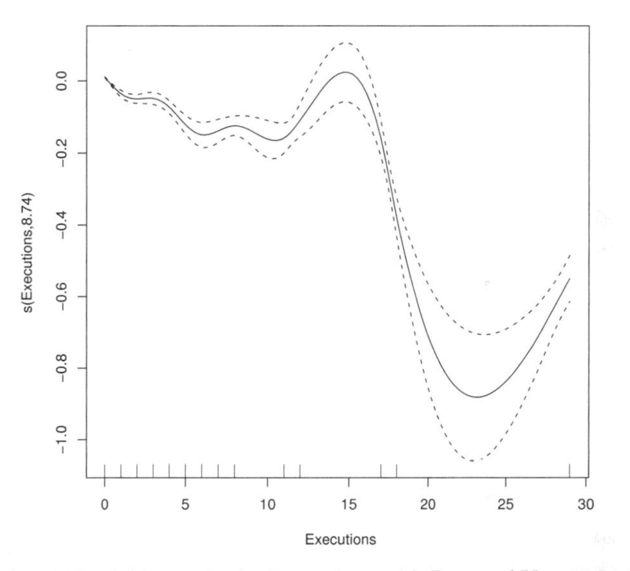

Fig. 11.3. GAM Homicide results for Executions with State and Year Held Constant

Some recent implementations of the generalized additive model do not rely on backfitting of this kind. Rather, they employ a form penalized regression much like in equation 11.9, implemented using B-splines (Wood, 2004). Initial experience suggests that this approach is computationally efficient and can produce more stable results that conventional backfitting.

There have also been a number of recent effort to allow for local determination of the smoothing window (Fan and Gijbels, 1996, Loader, 1999, Loader, 2004). The basic idea is to have the window size automatically shrink where the response function is changing more rapidly. These "adaptive" methods seem to be most useful when the data have a high signal to noise ration, when the response function is highly nonlinear, and when the variability in the response function changes dramatically from location to location. Experience to date suggests that data from the engineering and physical sciences are most likely to meet these criteria. Data from the social sciences are likely to be far too noisy.

Consider now an application of the generalized additive model. For data described earlier, Figure 11.3 shows the relationship between number of homicides and the number executions a year earlier, with state and year held constant. Indicator variables are included for each state to adjust for average differences over time in the number of homicides in each state. For example, states differ widely in population size, which is clearly factor in the raw number of homicides. Indicator variables for each state control for such differences. Indicator variables for year are included to adjust for average differences across states in the number of homicides each year. This controls for year to year trends for the country as a whole in the number of homicides.

There is now no apparent relationship between executions and homicides a year later except for the handful of states that in a very few years had a large number of executions. Again, any story is to be found in a few extreme outliers that are clearly atypical. The statistical point is that one can accommodate with GAM both smoother functions and conventional regression functions.

Figure 11.4 shows the relationship between number of homicides and 1) the number executions a year earlier and 2) the population of each state for each year. The two predictors were included in an additive fashion with their functions determined by smoothers.

The role of execution is about the same as in Figure 11.3, although at first glance the new vertical scale makes it looks a bit different. In addition, one can see that homicides increase monotonically with population size, as one would expect, but the rate of increase declines. The very largest states are not all that different from middle sized states.

11.7 Recursive Partitioning

Recall again equation 11.3 reproduced below for convenience as equation 11.14:

$$f(\mathbf{x}) = \sum_{j=1}^{p} \sum_{m=1}^{M_j} \beta_{jm} h_{jm}(\mathbf{x}), \tag{11.14}$$

An important special case sequentially includes basis functions that contribute to substantially to the fit. Commonly, this is done in much the same spirit as forward selection methods in stepwise regression. But, there are now two components to the fitting process. A function for each predictor is constructed. Then, only some of these functions are determined to be worthy and included in the final model. Classification and Regression Trees (Breiman *et al.*, 1984), commonly known as CART, is probably the earliest and most well known example of this approach.

11.7.1 Classification and Regression Trees and Extensions

CART can be applied to both categorical and quantitative response variables. We will consider first categorical response variables because they provide a better vehicle for explaining how CART functions.

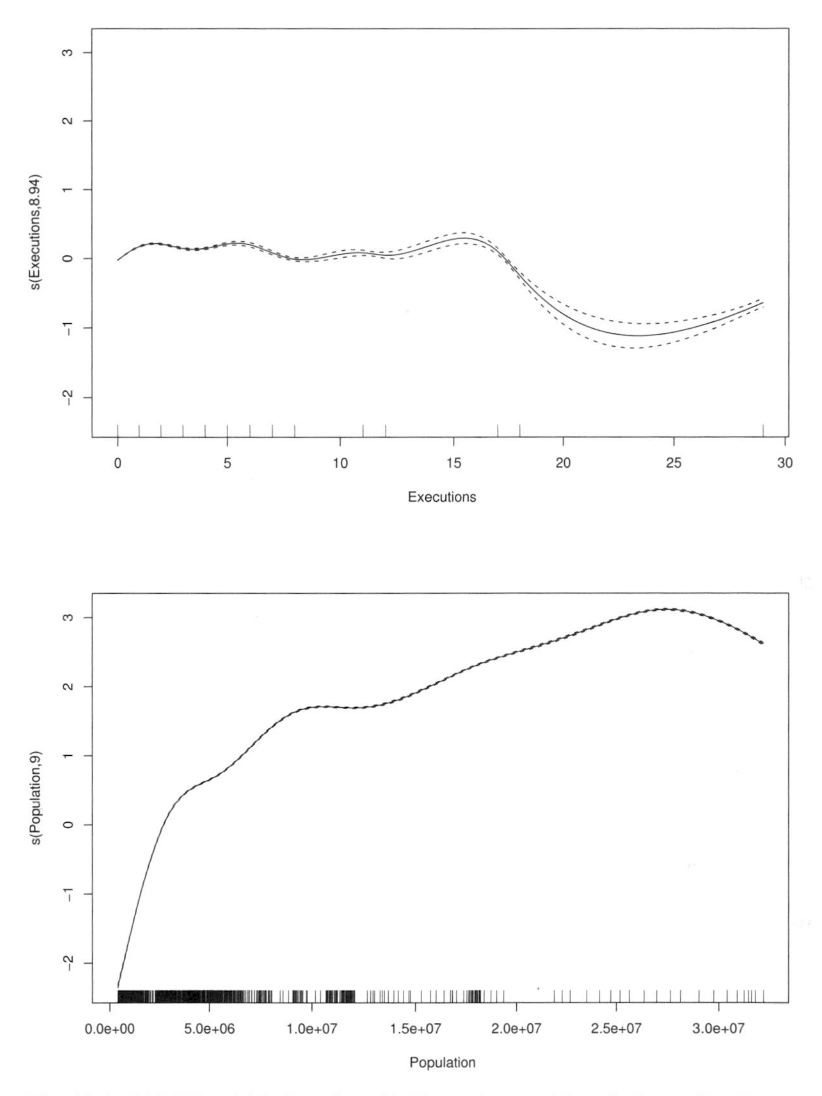

Fig. 11.4. GAM Homicide Results with Executions and Population as Predictors

CART uses a set of predictors to partition the data so that within each partition the values of the response variable are as homogeneous as possible. The data are partitioned one partition at a time. Once a partition is defined, it is unaffected by later partitions. The partitioning is accomplished with a series of straight-line boundaries, which define a break point for each selected predictor. Thus, the transformation for each predictor is an indicator variable.

Figure 11.5 illustrates a CART partitioning. There is a binary outcome coded "A" or "B" and in this simple illustration, just two predictors, **x** and **z**, are selected. The single vertical line defines the first partition. The double horizontal line defines the second partition. The triple horizontal line defines the third partition.

The data are first segmented left from right and then for the two resulting partitions, the data are further segmented separately into an upper and lower part. The upper left partition and the lower right partition are perfectly homogeneous. There remains considerable heterogeneity in the other two partitions and in principle, their partitioning could continue. Nevertheless, cases that are high on **z** and low on **x** are always "B." Cases that are low on **z** and high on **x** are always "A." In a real analysis, the terms "high" and "low" would be precisely defined by where the boundaries cross the x and z axes.

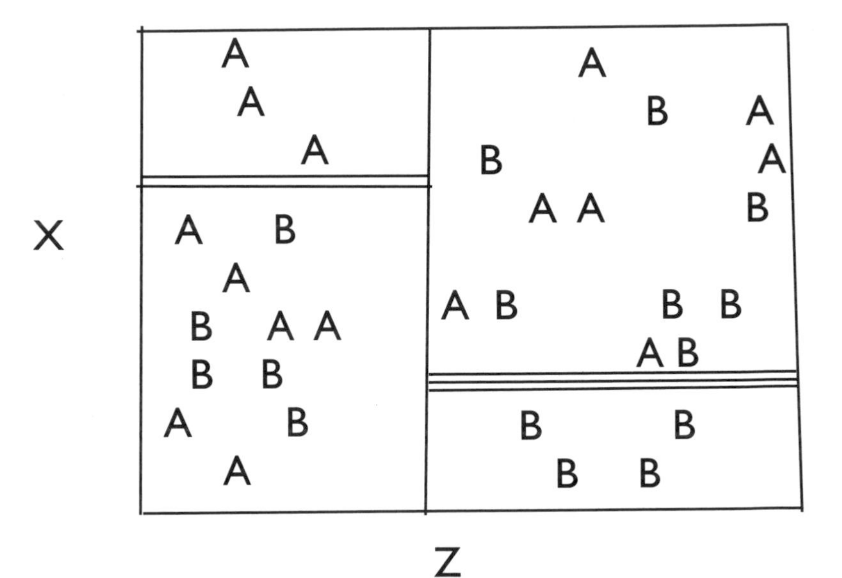

Fig. 11.5. Recursive Partitioning Logic in CART

The process by which each partition is constructed depends on two steps. First, each potential predictor individually is transformed into the indicator variable best able to split the data into two homogenous groups. All possible break points for each potential predictor are evaluated. Second, the predictor with the most effective indicator variable is selected for construction of the partition. For each partition, the process is repeated with all predictors, even ones used to construct earlier partitions. As a result, a given predictor can be used to construct more than one partition; some predictors will have more than one transformation selected.

Usually, CART output is displayed as an inverted tree. Figure 11.6 is a simple illustration. The full data set is contained in the root node. The final partitions are subsets of the data placed in the terminal nodes. The internal nodes contain subsets of data for intermediate steps.

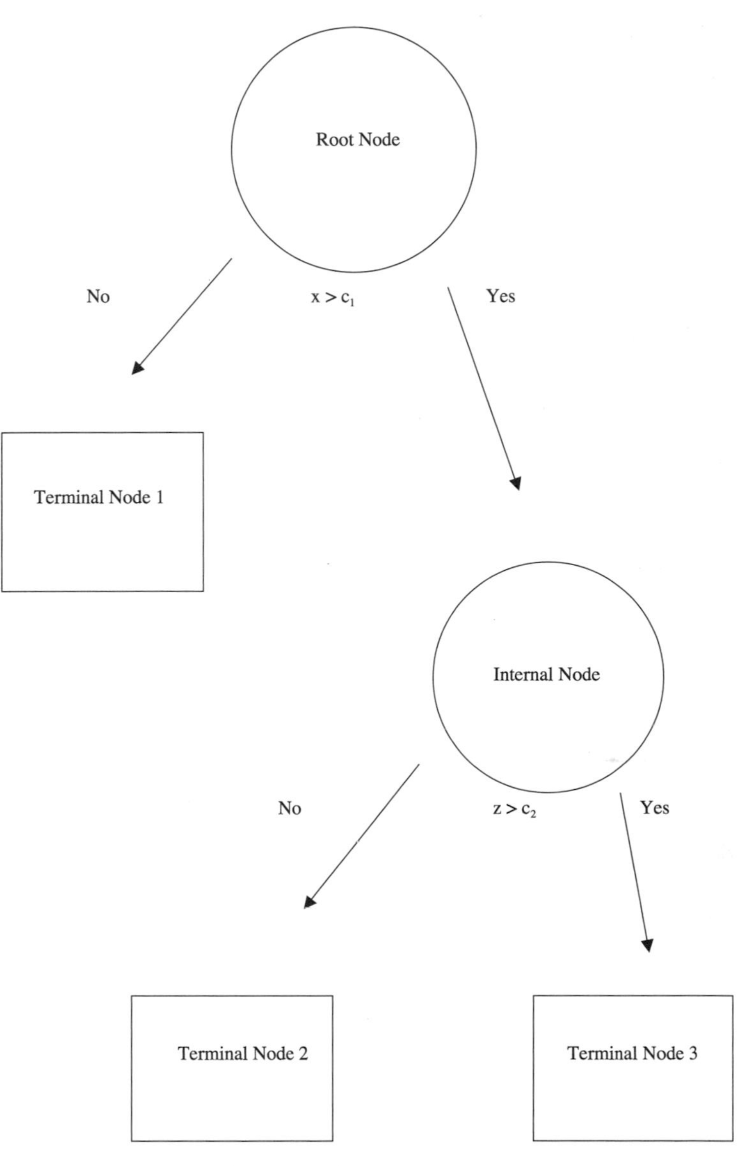

Fig. 11.6. CART Tree Structure

To achieve as much homogeneity as possible within data partitions, heterogeneity within data partitions is minimized. Two definitions of heterogeneity that are especially common. Consider a response that is a binary variable coded 1 or 0. Let the "impurity" i of node τ be a non-negative function of the probability that $\mathbf{y} = \mathbf{1}$. If τ is a node composed of cases that are all 1's or all 0's, its impurity is 0. If half the cases are 1's and half the cases are 0's, τ is the most impure it can be. Then, let

$$i(\tau) = \phi[p(\mathbf{y} = \mathbf{1}|\tau)], \tag{11.15}$$

where $\phi \geq 0$, $\phi(p) = \phi(1-p)$, and $\phi(0) = \phi(1) < \phi(p)$. Impurity is non-negative, symmetrical, and is at a minimum when all of the cases in τ are of one kind or another. The two most common options for the function ϕ are the entropy function shown in equation 11.16 and the Gini Index shown in equation 11.17:[12]

$$\phi(p) = -p\,log(p) - (1-p)\,log(1-p); \tag{11.16}$$

$$\phi(p) = p\,(1-p). \tag{11.17}$$

Both equations are concave with minimums at $p = 0$ and $p = 1$ and a maximum at $p = .5$. CART results from the two are often quite similar, but the Gini index seems to perform a bit better, especially when there are more than two categories in the response variable.

While it may not be immediately apparent, entropy and the Gini index are in much same spirit as the least squares criterion commonly use in regression, and the goal remains to estimate a set of conditional means. Because in classification problems the response can be coded as a 1 or a 0, the mean is a proportion.

Figure 11.7 shows a classification tree for an analysis of misconduct engaged in my inmates in prisons in California. The data are taken from a recent study of the California inmate classification system (Berk et al., 2003). The response variable is coded 1 for engaging in misconduct and 0 otherwise. Of the eight potential predictors, three were selected by CART: whether an inmate had a history of gang activity, the length of his prison term, and his age when he arrived at the prison reception center.

A node in the tree is classified as 1 if a majority of inmates in that node engaged in misconduct and 0 if a majority did not. The pair of numbers below each node classification show how the inmates are distributed with respect to misconduct. The right hand number is the count of inmates in the majority category. The left hand number is the count of inmates in the minority category. For example, in the terminal node at the far right side, there are 332 inmates. Because 183 of the 332 (55%) engaged in misconduct, the node is classified as a 1. The terminal nodes in Figure 11.7 are arranged so that the proportion of inmates engaging in misconduct increases from left to right.

In this application, one of the goals was to classify inmates by predictors of their proclivity to cause problems in prison. For inmates in the far right terminal node, if one claimed that all had engaged in misconduct, that claim would be incorrect 44% of the time. This is much better than one would do ignoring the predictors. In that

[12] Both can be generalized for nominal response variables with more than two categories (Hastie et al., 2001).

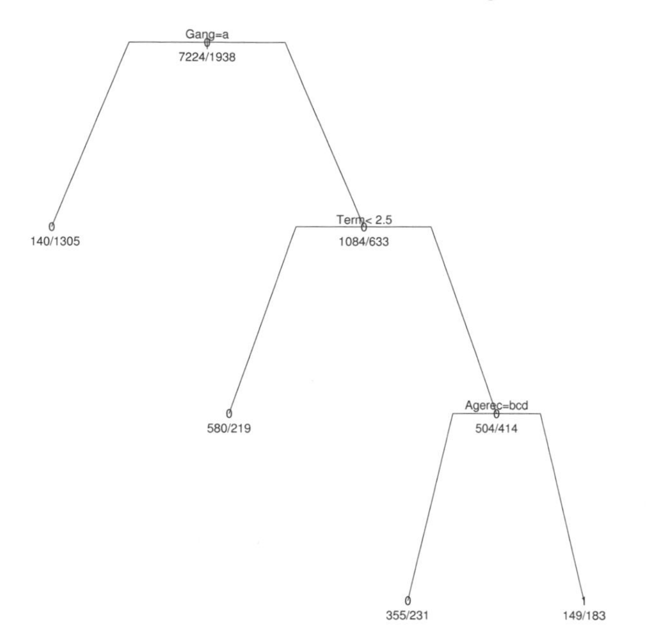

Fig. 11.7. Classification Tree for Inmate Misconduct

case, if one claimed that all inmates engaged in misconduct, that claim would be wrong 79% of the time.

The first predictor selected was gang activity. The "a" indicates that the inmates with a history of gang activity were placed in the right node, and inmates with a history of no gang activity were placed in the left node. The second predictor selected was only able meaningfully to improve the fit for inmates with a history of gang activity. Inmates with a sentence length ("Term") of less than 2.5 years were assigned to the left node, while inmates with a sentence length of 2.5 years or more were assigned to the right node. The final variable selected was only able meaningfully to improve the fit for the subset of inmates with a history of gang activity who were serving longer prison terms. That variable was the age of the inmate when he arrived at the prison reception center. Inmates with ages greater than 25 (age categories b, c, and d) were assigned to the left node, while inmates with ages less than 25 were assigned to the right node. In the end, this sorting makes good subject-matter sense. Prison officials often expect more trouble from younger inmates with a history of gang activity serving long terms.

When CART is applied with a quantitative response variable, the procedure is known as "Regression Trees." At each step, heterogeneity is now measured by the within-node sum of squares of the response:

$$i(\tau) = \sum(y_i - \bar{y}(\tau))^2, \tag{11.18}$$

where for node τ the summation is over all cases in that node, and $\bar{y}(\tau)$ is the mean of those cases. The heterogeneity for each potential split is the sum of the two sums of squares for the two nodes that would result. The split is chosen that reduces most this within-nodes sum of squares; the sum of squares of the parent node is compared to the combined sums of squares from each potential split into two offspring nodes. Generalization to Poisson regression (for count data) follows with the deviance used in place of the sum of squares.

11.7.2 Overfitting and Ensemble Methods

CART, like most Data Mining procedures, is vulnerable to overfitting. Because the fitting process is so flexible, the mean function tends to "over-respond" to idiosyncratic features of the data. If the data on hand are a random sample for a particular population, the mean function constructed from the sample can look very different from the mean function in the population (were it known). One implication is that a different random sample from the same population can lead to very different characterizations of how the response is related to the predictors. Conventional responses to overfitting (e.g., model selection based on the AIC) are a step in the right direction. However, they are often not nearly strong enough and usually provide few clues how a more appropriate model should be constructed.

It has been known for nearly a decade that one way to more effectively counteract overfitting is to construct average results over a number of random samples of the data (LeBlanc and Tibshirani, 1996, Mojirsheibani, 1999, Friedman et al., 2000). Cross-validaton can work on this principle. When the samples are bootstrap samples from a given data set, the procedures are sometimes called ensemble methods, with "bagging" as an early and important special case (Breiman, 1996).[13] Bagging can be applied to a wide variety of fitting procedures such as conventional regression, logistic regression and discriminant function analysis. Here the focus will be on bagging regression and classification trees.

The basic idea is that the various manifestations of overfitting cancel out in the aggregate over a large number of independent random samples from the same population. Bootstrap samples from the data on hand provide a surrogate for independent random samples from a well-defined population. However, the bootstrap sampling with replacement implies that bootstrap samples will share some observations with one another and that, therefore, the sets of fitted values across samples will not be independent. A bit of dependency is built it.

For recursive partitioning, the amount of dependence can be decreased substantially if in addition to random bootstrap samples, potential predictors are randomly sampled (with replacement) at each step. That is, one begins with a bootstrap sample of the data having the same number of observations as in the original data set. Then, when each decision is made about subdividing the data, only a random sample of predictors is considered. The random sample of predictors at each split may be relatively small (e.g., 5).

"Random forests" is one powerful approach exploiting these ideas. It builds on CART, and will generally fit the data better than standard regression models or CART

[13] "Bagging" stands for bootstrap aggregation.

itself (Breiman, 2001a). A large number of classification or regression trees is built (e.g., 500). Each tree is based on a bootstrap sample of the data on hand, and at each potential split, a random sample of predictors is considered. Then, average results are computed over the full set of trees. In the binary classification case, for example, a "vote" is taken over all of the trees to determine if a given case is assigned to one class or the other. So, if there are 500 trees and in 251 or more of these trees that case is classified as a "1," that case is treated as a "1."

One problem with random forests is that there is no longer a tree to interpret.[14] Partly in response to this defect, there are currently several methods under development that attempt to represent the importance of each predictor for the average fit. Many build on the following approach. The random forests procedure is applied to the data. For each tree, observations not included in the bootstrap sample are used as a "test" data set.[15] Some measure of the quality of the fit is computed with these data. Then, the values of a given explanatory variable in the test data are randomly shuffled, and a second measure of fit quality computed. Each of the measures is then averaged across the set of constructed trees. Any substantial decrease in the quality of the fit when the average of the first is compared to the average of the second must result from eliminating the impact of the shuffled variable.[16] The same process is repeated for each explanatory variable in turn.

There is no resolution to date of exactly what feature of the fit should be used to judge the importance of a predictor. Two that are commonly employed are the mean decline over trees in the overall measure of fit (e.g. the Gini Indix) and for classification problems, the mean decline over trees in how accurately cases are predicted. For example, suppose that for the full set explanatory variables an average of 75% of the cases are correctly predicted. If after the values of a given explanatory variable are randomly shuffled that figure drops to 65%, there is a reduction in predictive accuracy of 10%. Sometimes a standardized decline in predictive accuracy is used, which may be loosely interpreted as a z-score.

Figure 11.8 shows for the prison misconduct data how one can consider predictor importance using random forests. The number of explanatory variables included in the figure is truncated at four for ease of exposition. Term length is the most important explanatory variable by both the predictive accuracy and Gini measures. After that, the rankings from the two measures vary. Disagreements such as these are common because the Gini Index reflects the overall goodness of fit, while the predictive accuracy depends on how well the model actually predicts. The two are related, but they measure different things. Breiman argues that the decrease in predictive accuracy is the more direct, stable and meaningful indicator of variable importance (personal communication). If the point is to accurately predict cases, why not measure

[14] More generally, ensemble methods can lead to difficult interpretative problems if the links of inputs to outputs are important to describe.

[15] These are sometimes called "out-of-bag" observations. "Predicting" the values of the response for observations used to build the set of trees will lead to overly optimistic assessments of how well the procedure performs. Consequently, out-of-bag (OOB) observations are routinely used in random forests to determine how well random forests predicts.

[16] Small decreases could result from random sampling error.

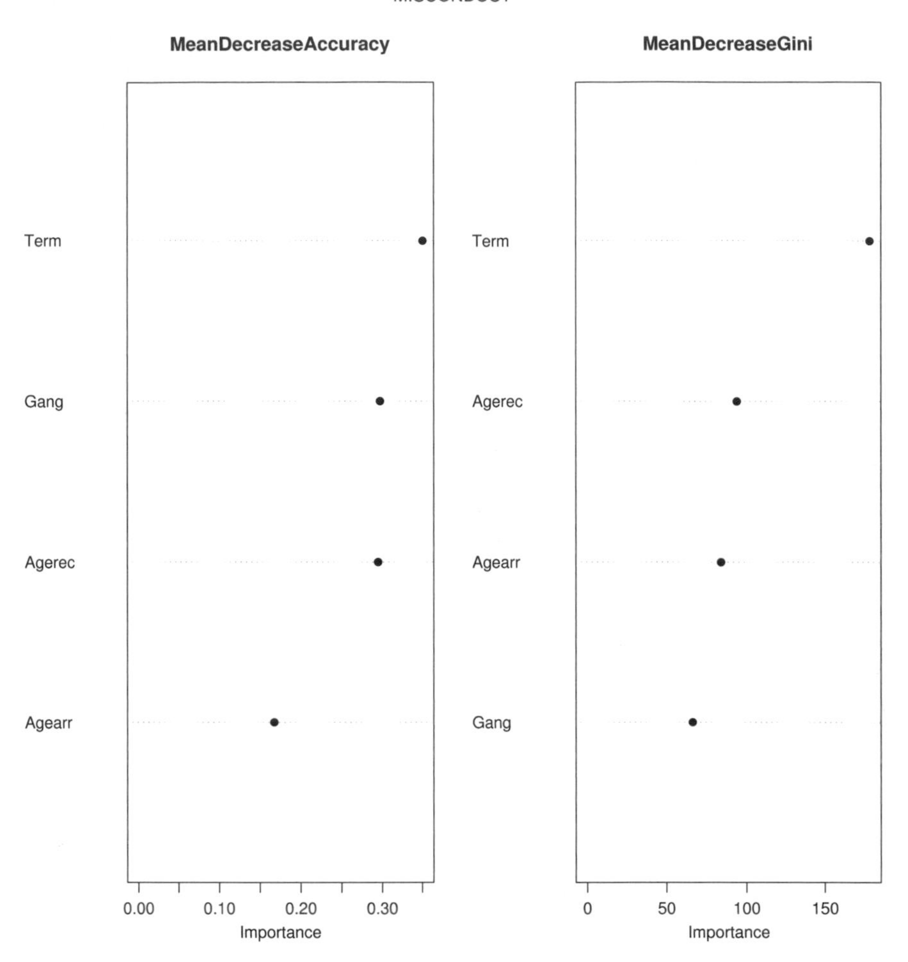

Fig. 11.8. Predictor Importance using Random Forests

importance by that criterion? In that case, the ranking of variables by importance is term length, gang activity, age at reception, and age when first arrested.

When the response variable is quantitative, importance is represented by the average increase in the within node sums of squares for the terminal nodes. The increase in this error sum of squares is related to how much the "explained variance" decreases when the values of a given predictor are randomly shuffled. There is no useful analogy in regression trees to correct or incorrect prediction.

11.8 Conclusions

A large number of Data Mining procedures can be considered within a regression framework. A representative sample of the most popular and powerful has been discussed in this chapter.[17] With more space, others could have been included: boosting (Freund and Schapire, 1995) and support vector machines (Vapnik, 1995) are two obvious candidates. Moreover, the development of new data mining methods is progressing very quickly, stimulated in part by relatively inexpensive computing power and in part by the Data Mining needs in a variety of disciplines. A revision of this chapter five years from now might look very different. Nevertheless, a key distinction between the more effective and the less effective Data Mining procedures is how overfitting is handled. Finding new and improved ways to fit data is often quite easy. Finding ways to avoid being seduced by the results is not (Svetnik et al., 2003, Reunanen, 2003).

Acknowledgments

The final draft of this chapter was funded in part by a grant from the National Science Foundation: (SES -0437169) "Ensemble methods for Data Analysis in the Behavioral, Social and Economic Sciences." This chapter was completed while visiting at the Department of Earth, Atmosphere, and Oceans, at the Ecole Normale Supérieur in Paris. Support from both is gratefully acknowledged.

References

Berk, R.A. (2003) *Regression Analysis: A Constructive Critique*. Newbury Park, CA.: Sage Publications.

Berk, R.A., Ladd, H., Graziano, H., and J. Baek (2003) "A Randomized Experiment Testing Inmate Classification Systems," *Journal of Criminology and Public Policy*, 2, No. 2: 215-242.

Breiman, L., Friedman, J.H., Olshen, R.A., and C.J. Stone, (1984) *Classification and Regression Trees*. Monterey, Ca: Wadsworth Press.

Breiman, L. (1996) "Bagging Predictors." *Machine Learning* 26:123-140.

Breiman, L. (2000) "Some Infinity Theory for Predictor Ensembles." *Technical Report 522*, Department of Statistics, University of California, Berkeley, California.

Breiman, L. (2001a) "Random Forests." *Machine Learning* 45: 5-32.

Breiman, L. (2001b) "Statistical Modeling: Two Cultures," (with discussion) *Statistical Science* 16: 199-231.

Cleveland, W. (1979) "Robust Locally Weighted Regression and Smoothing Scatterplots." *Journal of the American Statistical Association* 78: 829-836.

Cook, D.R. and Sanford Weisberg (1999) *Applied Regression Including Computing and Graphics*. New York: John Wiley and Sons.

[17] All of the procedures described in this chapter can be easily computed with procedures found in the programming language R.

Dasu, T., and T. Johnson (2003) *Exploratory Data Mining and Data Cleaning*. New York: John Wiley and Sons.

Christianini, N and J. Shawe-Taylor. (2000) *Support Vector Machines*. Cambridge, England: Cambridge University Press.

Fan, J., and I. Gijbels. (1996) *Local Polynomial Modeling and its Applications*. New York: Chapman & Hall.

Friedman, J., Hastie, T., and R. Tibsharini (2000). "Additive Logistic Regression: A Statistical View of Boosting" (with discussion). *Annals of Statistics* 28: 337-407.

Freund, Y., and R. Schapire. (1996) "Experiments with a New Boosting Algorithm," *Machine Learning: Proceedings of the Thirteenth International Conference*: 148-156. San Francisco: Morgan Freeman

Gigi, A. (1990) *Nonlinear Multivariate Analysis*. New York: John Wiley and Sons.

Hand, D., Manilla, H., and P Smyth (2001) *Principle of Data Mining*. Cambridge, Massachusetts: MIT Press.

Hastie, T.J. and R.J. Tibshirani. (1990) *Generalized Additive Models*. New York: Chapman & Hall.

Hastie, T., Tibshirani, R. and J. Friedman (2001) *The Elements of Statistical Learning*. New York: Springer-Verlag.

LeBlanc, M., and R. Tibshirani (1996) "Combining Estimates on Regression and Classification." *Journal of the American Statistical Association* 91: 1641–1650.

Loader, C. (1999) *Local Regression and Likelihood*. New York: Springer–Verlag.

Loader, C. (2004) "Smoothing: Local Regression Techniques," in J. Gentle, W. Härdle, and Y. Mori, *Handbook of Computational Statistics*. NewYork: Springer-Verlag.

Mocan, H.N. and K. Gittings (2003) "Getting off Death Row: Commuted Sentences and the Deterrent Effect of Capital Punishment." (Revised version of NBER Working Paper No. 8639) and forthcoming in the *Journal of Law and Economics*.

Mojirsheibani, M. (1999) "Combining Classifiers vis Discretization." *Journal of the American Statistical Association* 94: 600-609.

Reunanen, J. (2003) "Overfitting in Making Comparisons between Variable Selection Methods." *Journal of Machine Learning Research* 3: 1371-1382.

Sutton, R.S., and A.G. Barto. (1999). *Reinforcement Learning*. Cambridge, Massachusetts: MIT Press.

Svetnik, V., Liaw, A., and C.Tong. (2003) "Variable Selection in Random Forest with Application to Quantitative Structure-Activity Relationship." Working paper, Biometrics Research Group, Merck & Co., Inc.

Vapnik, V. (1995) *The Nature of Statistical Learning Theory*. New York: Springer-Verlag.

Witten, I.H. and E. Frank. (2000). *Data Mining*. New York: Morgan and Kaufmann.

Wood, S.N. (2004) "Stable and Eficient Multiple Smoothing Parameter Estimation for Generalized Additive Models," *Journal of the American Statistical Association*, Vol. 99, No. 467: 673-686.

12

Support Vector Machines

Armin Shmilovici

Ben-Gurion University

Summary. Support Vector Machines (SVMs) are a set of related methods for supervised learning, applicable to both classification and regression problems. A SVM classifiers creates a maximum-margin hyperplane that lies in a transformed input space and splits the example classes, while maximizing the distance to the nearest cleanly split examples. The parameters of the solution hyperplane are derived from a quadratic programming optimization problem. Here, we provide several formulations, and discuss some key concepts.

Key words: Support Vector Machines, Margin Classifier, Hyperplane Classifiers, Support Vector Regression, Kernel Methods

12.1 Introduction

Support Vector Machines (SVMs) are a set of related methods for supervised learning, applicable to both classification and regression problems. Since the introduction of the SVM classifier a decade ago (Vapnik, 1995), SVM gained popularity due to its solid theoretical foundation. The development of efficient implementations led to numerous applications (Isabelle, 2004).

The Support Vector learning machine was developed by Vapnik *et al.* (Scholkopf *et al.*, 1995, Scholkopf 1997) to constructively implement principles from *statistical learning* theory (Vapnik, 1998). In the statistical learning framework, learning means to estimate a function from a set of examples (the training sets). To do this, a learning machine must choose one function from a given set of functions, which minimizes a certain risk (the *empirical risk*) that the estimated function is different from the actual (yet unknown) function. The risk depends on the *complexity* of the set of functions chosen as well as on the training set. Thus, a learning machine must find the best set of functions - as determined by its complexity - and the best function in that set. Unfortunately, in practice, a bound on the risk is neither easily computable, nor very helpful for analyzing the quality of the solution (Vapnik and Chapelle, 2000).

O. Maimon, L. Rokach (eds.), *Data Mining and Knowledge Discovery Handbook*, 2nd ed., DOI 10.1007/978-0-387-09823-4_12,

Let us assume, for the moment, that the training set is separable by a *hyperplane*. It has been proved (Vapnik, 1995) that for the class of hyperplanes, the complexity of the hyperplane can be bounded in terms of another quantity, the *margin*. The margin is defined as the minimal distance of an example to a decision surface. Thus, if we bound the margin of a function class from below, we can control its complexity. Support vector learning implements this insight that the *risk is minimized when the margin is maximized*. A SVM chooses a maximum-margin hyperplane that lies in a transformed input space and splits the example classes, while maximizing the distance to the nearest cleanly split examples. The parameters of the solution hyperplane are derived from a quadratic programming optimization problem.

For example, consider a simple separable classification method in multi-dimensional space. Given two classes of examples clustered in feature space, any reasonable classifier hyperplane should pass *between* the means of the classes. *One* possible hyperplane is the decision surface that assigns a new point to the class whose mean is *closer* to it. This decision surface is geometrically equivalent to computing the class of a new point by checking the angle between two vectors - the vector connecting the two cluster means and the vector connecting the mid-point on that line with the new point. This angle can be formulated in terms of *a dot product* operation between vectors. The decision surface is implicitly defined in terms of the *similarity* between any new point and the cluster mean - *a kernel function*. This simple classifier is linear in the feature space while in the input domain it is represented by a kernel expansion in terms of the training examples. In the more sophisticated techniques presented in the next section, the selection of the examples that the kernels are centered on will no longer consider *all* training examples, and the weights that are put on each data point for the decision surface will no longer be uniform. For instance, we might want to remove the influence of examples that are far away from the decision boundary, either since we expect that they will not improve the generalization error of the decision function, or since we would like to reduce the computational cost of evaluating the decision function. Thus, the hyperplane will only depend on a subset of training examples, called *support vectors*.

There are numerous books and tutorial papers on the theory and practice of SVM (Scholkopf and Smola 2002, Cristianini and Shawe-Taylor 2000, Muller *et al.* 2001, Chen *et al.* 2003, Smola and Scholkopf 2004). The aim of this chapter is to introduce the main SVM models, and discuss their main attributes in the framework of supervised learning. The rest of this chapter is organized as follows: Section 12.2 describes the separable classifier case and the concept of kernels; Section 12.3 presents the non-separable case and some related SVM formulations; Section 12.4 discusses some practical computational aspects; Section 12.5 discusses some related concepts and applications; and Section 12.6 concludes with a discussion.

12.2 Hyperplane Classifiers

The task of classification is to find a rule, which based on external observations, assigns an object to one of several classes. In the simplest case, there are only two different classes. One possible formalization of this classification task is to estimate a function $f : \mathbf{R}^N \to \{-1, +1\}$ using input-output training data pairs generated identically and independently distributed (i.i.d.) according to an unknown probability distribution $P(\mathbf{x}, y)$ of the data $(\mathbf{x}_1, y_1), \ldots, (\mathbf{x}_n, y_n) \in \mathbf{R}^N \times Y$, $Y = \{-1, +1\}$ such that f will correctly classify unseen examples $(\mathbf{x}, y)$. The test examples are assumed to be generated from the same probability distribution as the training data. An example is assigned to class +1 if $f(x) \geq 0$ and to class -1 otherwise.

The best function f that one can obtain is the one minimizing the expected error (risk) - the integral of a certain *loss function* l according to the unknown probability distribution $P(\mathbf{x}, y)$ of the data. For classification problems, l is the so-called 0/1 loss function: $l(f(\mathbf{x}), y) = \theta(-yf(\mathbf{x}))$, where $\theta(z) = 0$ for $z < 0$ and $\theta(z) = 1$ otherwise. The loss framework can also be applied to regression problems where $y \in \mathbf{R}$, where the most common loss function is the squared loss: $l(f(\mathbf{x}), y) = (f(\mathbf{x}) - y)^2$.

Unfortunately, the risk cannot be minimized directly, since the underlying probability distribution $P(\mathbf{x}, y)$ is unknown. Therefore, we must try to estimate a function that is close to the optimal one based on the available information, i.e., the training sample and properties of the function class from which the solution f is chosen. To design a learning algorithm, one needs to come up with a class of functions whose capacity (to classify data) can be computed. The intuition, which is formalized in Vapnik (1995), is that a simple (e.g., linear) function that explains most of the data is preferable to a complex one (Occam's razor).

12.2.1 The Linear Classifier

Let us assume, for a moment that the training sample is separable by a hyperplane (see Figure 12.1) and we choose functions of the form

$$(\mathbf{w} \cdot \mathbf{x}) + b = 0 \quad \mathbf{w} \in \mathbf{R}^N, b \in \mathbf{R} \tag{12.1}$$

corresponding to decision functions

$$f(\mathbf{x}) = sign((\mathbf{w} \cdot \mathbf{x}) + b) \tag{12.2}$$

It has been shown (Vapnik, 1995) that, for the class of hyperplanes, the capacity of the function can be bounded in terms of another quantity, the margin (Figure 12.1). The margin is defined as the minimal distance of a sample to the decision surface. The margin, depends on the length of the weight vector $\mathbf{w}$ in Equation 12.1: since we assumed that the training sample is separable, we can rescale $\mathbf{w}$ and b such that the points closest to the hyperplane satisfy $|(\mathbf{w} \cdot \mathbf{x}_i) + b| = 1$ (i.e., obtain the so-called canonical representation of the hyperplane). Now consider two samples $\mathbf{x}_1$ and $\mathbf{x}_2$ from different classes with $|(\mathbf{w} \cdot \mathbf{x}_1) + b| = 1$ and $|(\mathbf{w} \cdot \mathbf{x}_2) + b| = 1$, respectively. Then, the margin is given by the distance of these two points, measured perpendicular to the hyperplane, i.e., $\left(\frac{\mathbf{w}}{\|\mathbf{w}\|} \cdot (\mathbf{x}_1 - \mathbf{x}_2)\right) = \frac{2}{\|\mathbf{w}\|}$.

Among all the hyperplanes separating the data, there exists a unique one yielding the maximum margin of separation between the classes:

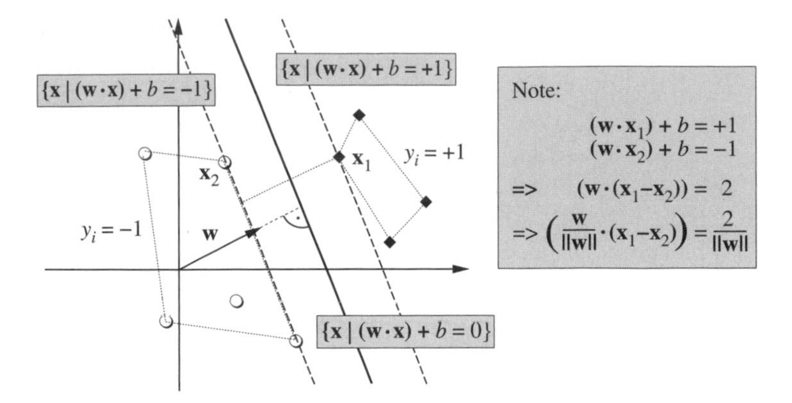

Fig. 12.1. A toy binary classification problem: separate balls from diamonds. The optimal hyperplane is orthogonal to the shortest line connecting the convex hull of the two classes (dotted), and itersects it half way between the two classes. In this case the margin is measured perpendicular to the hyperplane. Figure taken from Chen et al. (2001).

$$\underset{\{\mathbf{w},b\}}{\text{Max}} \ \min\left\{\|\mathbf{x}-\mathbf{x}_i\| : \mathbf{x}\in\mathbf{R}^N, (\mathbf{w}\cdot\mathbf{x})) + b = 0, i = 1,\ldots n\right\} \tag{12.3}$$

To construct this optimal hyperplane, one solves the following optimization problem:

$$\underset{\{\mathbf{w},b\}}{\text{Min}} \ \tfrac{1}{2}\|\mathbf{w}\|^2 \tag{12.4}$$

$$\text{Subject to } y_i\cdot((\mathbf{w}\cdot\mathbf{x}_i)+b)\geq 1, \ i=1,\ldots,n \tag{12.5}$$

This constraint optimization problem can be solved by introducing Lagrange multipliers $\alpha_i \geq 0$ and the Lagrangian function

$$L(\mathbf{w},b,\alpha) = \tfrac{1}{2}\|\mathbf{w}\|^2 - \sum_{i=1}^{n}\alpha_i\left(y_i\cdot((\mathbf{w}\cdot\mathbf{x}_i)+b)-1\right) \tag{12.6}$$

The Langrangian L has to be minimized with respect to the *primal* variables $\{\mathbf{w},b\}$ and maximized with respect to the *dual* variables α_i. The optimal point is a saddle point and we have the following equations for the primal variables:

$$\frac{\partial L}{\partial b} = 0; \quad \frac{\partial L}{\partial \mathbf{w}} = 0; \tag{12.7}$$

which translate into

$$\sum_{i=1}^{n}\alpha_i y_i = 0\,, \ \mathbf{w} = \sum_{i=1}^{n}\alpha_i y_i \mathbf{x}_i \tag{12.8}$$

The solution vector thus has an expansion in terms of a subset of the training patterns. The *Support Vectors* are those patterns corresponding with the non-zero α_i, and the non-zero α_i are called *Support Values*. By the Karush-Kuhn-Tucker (KKT) complimentary conditions of optimization, the α_i must be zero for all the constraints in Equation 12.5 which are not met as equality, thus

$$\alpha_i \left(y_i \cdot ((\mathbf{w} \cdot \mathbf{x}_i) + b) - 1\right) = 0\,,\; i = 1, \ldots, n \tag{12.9}$$

and all the Support Vectors lie on the margin (Figures 12.1,12.3) while the all remaining training examples are irrelevant to the solution. The hyperplane is completely captured by the patterns closest to it.

For a nonlinear problem like in the problem presented in Equations 12.4-12.5, called a primal problem, under certain conditions, the primal and dual problems have the same objective values. Therefor, we can solve the dual problem which may be easier than the primal problem. In particular, when working in feature space (Section 12.2.3) solving the dual may be the only way to train the SVM. By substituting Equation 12.8 into Equation 12.6, one eliminates the primal variables and arrives at the Wolfe dual (Wolfe, 1961) of the optimization problem for the multipliers α_i:

$$\max_{\alpha} \sum_{i=1}^{n} \alpha_i - \tfrac{1}{2} \sum_{i,j=1}^{n} \alpha_i \alpha_j y_i y_j \left(\mathbf{x}_i \cdot \mathbf{x}_j\right) \tag{12.10}$$

$$\text{Subject to } \alpha_i \geq 0,\;\; i = 1, \ldots, n \;,\; \sum_{i=1}^{n} \alpha_i y_i = 0 \tag{12.11}$$

The hyperplane decision function presented in Equation 12.2 can now be explicitly written as

$$f(\mathbf{x}) = sign(\sum_{i=1}^{n} \alpha_i y_i \left(\mathbf{x} \cdot \mathbf{x}_i\right) + b) \tag{12.12}$$

where b is computed from Equation 12.9 and from the set of support vectors $\mathbf{x}_i, i \in I \equiv \{i : \alpha_i \neq 0\}$.

$$b = \frac{1}{|I|} \sum_{i \in I} \left(y_i - \sum_{j=1}^{n} \alpha_j y_j \left(\mathbf{x}_i \cdot \mathbf{x}_j\right) \right) \tag{12.13}$$

12.2.2 The Kernel Trick

The choice of linear classifier functions seems to be very limited (i.e., likely to underfit the data). Fortunately, it is possible to have both linear models and a very rich set of nonlinear decision functions by using the kernel trick (Cortes and Vapnik, 1995) with maximum-margin hyperplanes. Using the kernel trick for SVM makes the maximum margin hyperplane be fit in a feature space F. The feature space F is a non-linear map $\Phi : R^N \rightarrow F$ from the original input space, usually of much higher dimensionality than the original input space. With the kernel trick, the same linear algorithm is worked on the transformed data $(\Phi(\mathbf{x}_1), y_1), \ldots, (\Phi(\mathbf{x}_n), y_n)$. In this way, non-linear SVMs can makes the maximum margin hyperplane be fit in a feature space. Figure 12.2 demonstrates such a case. In the original (linear) training algorithm (see Equations 12.10-12.12) the data appears in the form of dot products $\mathbf{x}_i \cdot \mathbf{x}_j$. Now, the training algorithm depends on the data through dot products in F, i.e., on functions of the form $\Phi(\mathbf{x}_i) \cdot \Phi(\mathbf{x}_j)$. If there exists a kernel function K such that $K(\mathbf{x}_i, \mathbf{x}_j) = \Phi(\mathbf{x}_i) \cdot \Phi(\mathbf{x}_j)$, we would only need to use K in the training algorithm and would never need to explicitly even know what Φ is.

Mercer's condition, (Vapnik, 1995) tells us the mathematical properties to check whether or not a prospective kernel is actually a dot product in some space, but it does not tell us how to construct Φ, or even what F is. Choosing the best kernel function is a subject of active research (Smola and Scholkopf 2002, Steinwart 2003). It was found that to a certain degree different choices of kernels give similar classification accuracy and similar sets of support vectors (Scholkopf *et al.* 1995), indicating that in some sense there exist "important" training points which characterize a given problem.

Some commonly used kernels are presented in Table 12.1. Note, however, that the Sigmoidal kernel only satisfies Mercer's condition for certain values of the parameters and the data. Hsu *et al.* (2003) advocate the use of the Radial Basis Function as a reasonable first choice.

Table 12.1. Commonly Used Kernel Functions.

Kernel	$K(\mathbf{x}, \mathbf{x}_i)$
Radial Basis Function	$\exp\left(-\gamma \lVert \mathbf{x} - \mathbf{x}_i \rVert^2\right)$, $\gamma > 0$
Inverse multiquadratic	$\frac{1}{\sqrt{\lVert \mathbf{x} - \mathbf{x}_i \rVert + \eta}}$
Polynomial of degree d	$\left(\left(\mathbf{x}^T \cdot \mathbf{x}_i\right) + \eta\right)^d$
Sigmoidal	$\tanh\left(\gamma\left(\mathbf{x}^T \cdot \mathbf{x}_i\right) + \eta\right)$, $\gamma > 0$
Linear	$\mathbf{x}^T \cdot \mathbf{x}_i$

12.2.3 The Optimal Margin Support Vector Machine

Using the kernel trick, replace every dot product $(\mathbf{x}_i \cdot \mathbf{x}_j)$ in terms of the kernel K evaluated on input patterns $\mathbf{x}_i, \mathbf{x}_j$. Thus, we obtain the more general form of Equation 12.12:

$$f(\mathbf{x}) = sign(\sum_{i=1}^{n} \alpha_i y_i K(\mathbf{x}, \mathbf{x}_i) + b) \tag{12.14}$$

and the following quadratic optimization problem

$$\max_{\alpha} \sum_{i=1}^{n} \alpha_i - \tfrac{1}{2} \sum_{i,j=1}^{n} \alpha_i \alpha_j y_i y_j K(\mathbf{x}_i, \mathbf{x}_j) \tag{12.15}$$

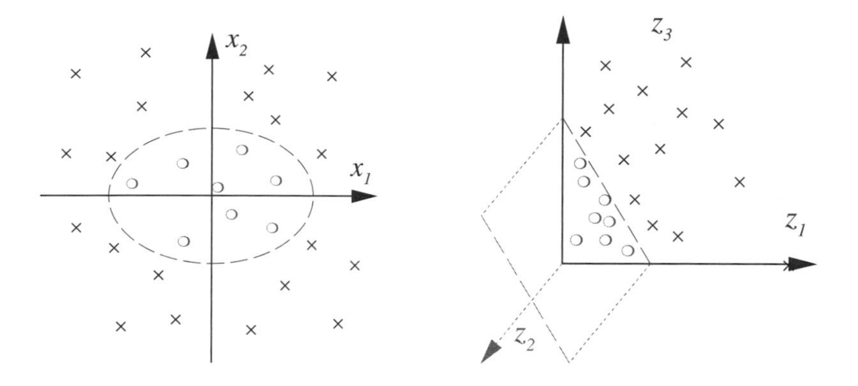

Fig. 12.2. The idea of SVM is to map the training data into a higher dimensional feature space via Φ, and construct a separating hyperplane with maximum margin there. This yields a nonlinear decision boundary in input space. In the following two-dimensional classification example, the transformation is $\Phi : \mathbf{R}^2 \to \mathbf{R}^3$, $(x_1,x_2) \to (z_1,z_2,z_3) \equiv \left(x_1^2, \sqrt{2}x_1x_2, x_2^2\right)$. The separating hyperplane is visible and the decision surface can be analytically found. Figure taken from Muller et al. (2001).

$$\text{Subject to } \alpha_i \geq 0,\ i = 1,\ldots,n\ ,\ \sum_{i=1}^{n} \alpha_i y_i = 0 \tag{12.16}$$

Formulation presented in Equations 12.15-12.16 is the standard SVM formulation. This dual problem has the same number of variables as the number of training variables, while the primal problem has a number of variables which depends on the dimensionality of the feature space, which could be infinite. Figure 12.3 presents an example of a decision function found with a SVM.

One of the most important properties of the SVM is that the solution is *sparse* in α, i.e. many patterns are outside the margin area and their optimal α_i is zero. Without this sparsity property, SVM learning would hardly be practical for large data sets.

12.3 Non-Separable SVM Models

The previous section considered the separable case. However, in practice, a separating hyperplane may not exits, e.g. if a high noise level causes some overlap of the classes. Using the previous SVM might not minimize the empirical risk. This section presents some SVM models that extend the capabilities of hyperplane classifiers to more practical problems.

12.3.1 Soft Margin Support Vector Classifiers

To allow for the possibility of examples violating constraint in Equation 12.5, Cortes and Vapnik (1995) introduced slack variables ξ_i that relax the hard margin constraints

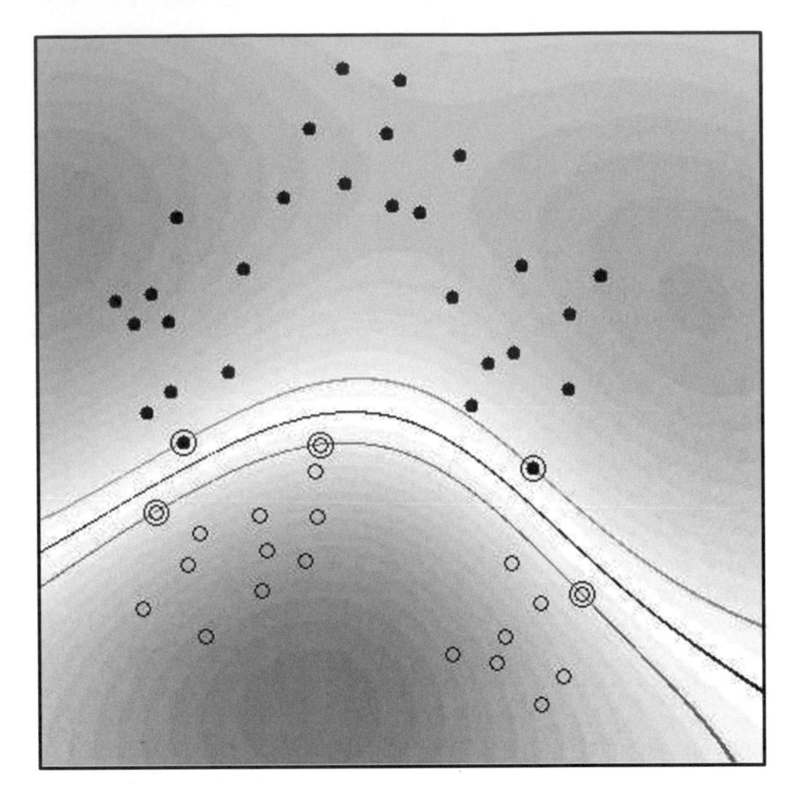

Fig. 12.3. Example of a Support Vector classifier found by using a radial basis function kernel. Circles and disks are two classes of training examples. Extra circles mark the Support Vectors found by the algorithm. The middle line is the decision surface. The outer lines precisely meet the constraint in Equation 12.16. The shades indicate the absolute value of the argument of the sign function in Equation 12.14. Figure taken from Chen et al. (2003).

$$y_i \cdot ((\mathbf{w} \cdot \Phi(\mathbf{x}_i)) + b) \geq 1 - \xi_i\,,\ \xi_i \geq 0,\ i = 1, \ldots, n \tag{12.17}$$

A classifier that generalizes well is then found by controlling both the classifier capacity (via $\|\mathbf{w}\|$) and the sum of the slacks $\sum_{i=1}^{n} \xi_i$, i.e. the number of training errors. One possible realization, called C-SVM, of a soft margin classifier is minimizing the following objective function

$$\min_{\mathbf{w},b,\xi} \ \frac{1}{2} \|\mathbf{w}\|^2 + C \sum_{i=1}^{n} \xi_i \tag{12.18}$$

The regularization constant $C > 0$ determines the trade-off between the empirical error and the complexity term. Incorporating Lagrange multipliers and solving, leads to the following dual problem:

$$\max_{\alpha} \sum_{i=1}^{n} \alpha_i - \tfrac{1}{2} \sum_{i,j=1}^{n} \alpha_i \alpha_j y_i y_j K(\mathbf{x}_i, \mathbf{x}_j) \tag{12.19}$$

$$\text{Subject to } 0 \leq \alpha_i \leq C, \; i = 1, \ldots, n \;, \; \sum_{i=1}^{n} \alpha_i y_i = 0 \tag{12.20}$$

The only difference from the separable case is the upper bound C on the Lagrange multipliers α_i. The solution remains sparse and the decision function retains the same form as Equation 12.14.

Another possible realization of a soft margin, called ν-SVM (Chen *et al.* 2003) was originally proposed for regression. The rather non-intuitive regularization constant C is replaced with another constant $\nu \in [0,1]$. The dual formulation of the ν-SVM is the following:

$$\max_{\alpha} \; -\tfrac{1}{2} \sum_{i,j=1}^{n} \alpha_i \alpha_j y_i y_j K(\mathbf{x}_i, \mathbf{x}_j) \tag{12.21}$$

$$\text{Subject to } 0 \leq \alpha_i \leq \tfrac{1}{n}, \; i = 1, \ldots, n \;, \; \sum_{i=1}^{n} \alpha_i y_i = 0 \;, \; \sum_{i=1}^{n} \alpha_i \geq \nu \tag{12.22}$$

For appropriate parameter choices, the ν-SVM yields exactly the same solutions as the C-SVM. The significance of ν is that under some mild assumptions about the data, ν is an upper bound on the fraction of margin errors (and hence also on the fraction of training errors); and ν is also a lower bound on the fraction of Support Vectors. Thus, controlling ν influences the tradeoff between the model's accuracy and the model's complexity

12.3.2 Support Vector Regression

One possible formalization of the regression task is to estimate a function $f : \mathbf{R}^N \rightarrow \mathbf{R}$ using input-output training data pairs generated identically and independently distributed (i.i.d.) according to an unknown probability distribution $P(\mathbf{x}, y)$ of the data. The concept of margin is specific to classification. However, we would still like to avoid too complex regression functions. The idea of SVR (Smola and Scholkopf, 2004) is that we find a function that has at most ε deviation from the actually obtained targets y_i for all the training data, and at the same time is as flat as possible. In other words, errors are unimportant as long as they are less then ε, but we do not tolerate deviations larger than this. An analogue of the margin is constructed in the space of the target values $y \subset \mathbf{R}$. By using Vapnik's ε-sensitive loss function (Figure 12.4).

$$|y - f(\mathbf{x})|_{\varepsilon} \equiv \max\{0, |y - f(\mathbf{x})| - \varepsilon\} \tag{12.23}$$

A tube with radius ε is fitted to the data, and a regression function that generalizes well is then found by controlling both the regression capacity (via $\|\mathbf{w}\|$) and the loss function. One possible realization, called C-SVR, of a is minimizing the following objective function

$$\min_{\mathbf{w},\mathbf{b},\xi} \ \frac{1}{2}\|\mathbf{w}\|^2 + C\sum_{i=1}^{n} |y_i - f(\mathbf{x})|_\varepsilon \tag{12.24}$$

The regularization constant $C > 0$ determines the trade-off between the empirical error and the complexity term.

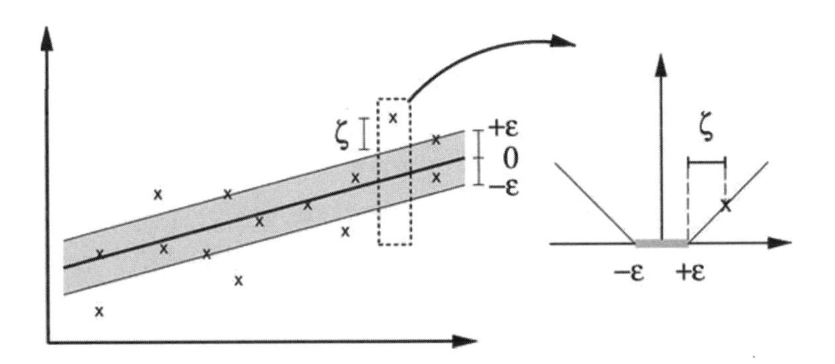

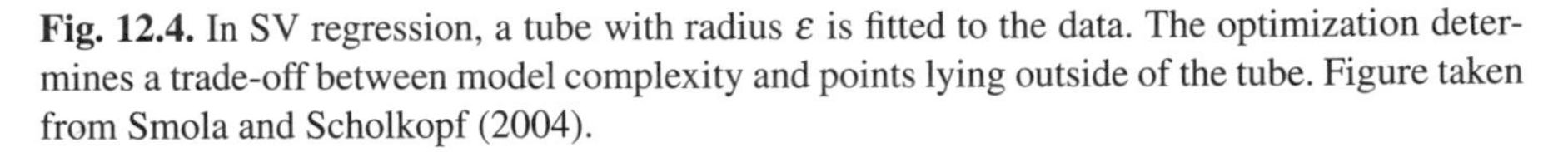

Fig. 12.4. In SV regression, a tube with radius ε is fitted to the data. The optimization determines a trade-off between model complexity and points lying outside of the tube. Figure taken from Smola and Scholkopf (2004).

Generalization to kernel-based regression estimation is carried out in complete analogy with the classification problem. Introducing Lagrange multipliers and choosing a-priory the regularization constants C, ε one arrives at a dual quadratic optimization problem. The support vectors and the support values of the solution define the following regression function

$$f(\mathbf{x}) = \sum_{i=1}^{n} \alpha_i K(\mathbf{x}, \mathbf{x}_i) + b \tag{12.25}$$

There are degrees of freedom for constructing SVR, such as how to penalize or regularize different parts of the vector, how to use the kernel trick, and the loss function to use. For example, in the ν-SVR algorithm implemented in LIBSVM (Chang and Lin 2001) one specifies an upper bound $0 \le \nu \le 1$ on the fraction of points allowed to be outside the tube (asymptotically, the number of Support Vectors). For a-priory chosen constants C, ν the dual quadratic optimization problem is as follows

$$\max_{\alpha,\alpha^*} \ \sum_{i=1}^{n} (\alpha_i^* - \alpha_i) y_i - \frac{1}{2} \sum_{i,j=1}^{n} (\alpha_i^* - \alpha_i)(\alpha_j^* - \alpha_j) K(\mathbf{x}_i, \mathbf{x}_j) \tag{12.26}$$

$$\text{Subject to } 0 \leq \alpha_i, \alpha_i^* \leq \tfrac{C}{n}, \quad \begin{array}{l} \sum\limits_{i=1}^{n} (\alpha_i^* + \alpha_i) \leq Cv \\ \sum\limits_{i=1}^{n} (\alpha_i^* - \alpha_i) \leq Cv \end{array} \quad i = 1, \ldots, n \tag{12.27}$$

and the regression solution is expressed as

$$f(\mathbf{x}) = \sum_{i=1}^{n} (\alpha_i^* - \alpha_i) K(\mathbf{x}, \mathbf{x}_i) + b \tag{12.28}$$

12.3.3 SVM-like Models

The power of SVM comes from the kernel representation that allows a non-linear mapping of input space to a higher dimensional feature space. However, the resulting quadratic programming equations may be computationally expensive for large problems. Smola *et al.* (1999) suggested an SVR like *linear programming* formulation that retains the form of the solution (Equation 12.25) while replacing the quadratic function in Equation 12.26 with a linear function subject to constraints on the error of kernel expansion (Equation 12.25).

Suykens *et al.* (2002) introduced the least squares SVM (LS-SVM) in which they modify the classifier of Equations 12.17-12.18 with the following equations:

$$\min_{\mathbf{w,b,e}} \; \tfrac{1}{2} \|\mathbf{w}\|^2 + \gamma \tfrac{1}{2} \sum_{i=1}^{n} e_i^2 \tag{12.29}$$

$$\text{Subject to } y_i \cdot ((\mathbf{w} \cdot \Phi(\mathbf{x}_i)) + b) = 1 - e_i, \quad i = 1, \ldots, n \tag{12.30}$$

Important differences with standard SVM are the equality constraint (see Equation 12.30) and the sum squared error terms, which greatly simplify the problem. Incorporating Lagrange multipliers and solving leads to the following dual *linear problem*:

$$\begin{bmatrix} \mathbf{0} & \mathbf{Y}^T \\ \mathbf{Y} & \blacksquare + \gamma^{-1}\mathbf{I} \end{bmatrix} \cdot \begin{bmatrix} b \\ \alpha \end{bmatrix} = \begin{bmatrix} \mathbf{0} \\ \mathbf{I} \end{bmatrix} \tag{12.31}$$

where the primal variables $\{\mathbf{w}, b\}$ define as before a decision surface like Equation 12.14, $Y = (y_1, \ldots, y_n)$, $(\Omega)_{i,j} = y_i y_j K(x_i, x_j)$, $\mathbf{I}, \mathbf{0}$ are appropriate size all ones (all zeros) matrices, and γ is a tuning parameter to be optimized. Equivalently, modifying the regression problem presented in Equations 12.26-12.27 also results in a linear system like (Equation 12.31) with an additional tuning parameter.

The LS-SVM can realize strongly nonlinear decision boundaries, and efficient matrix inversion methods can handle very large datasets. However, α is *not* sparse anymore (Suykens *et al.* 2002).

12.4 Implementation Issues with SVM

The purpose of this section is to overview some problems that face the application of SVM in machine learning.

12.4.1 Optimization Techniques

The solution of the SVM problem, is the solution of a constraint (convex) quadratic programming (QP) problem such as Equations 12.15-12.16. Equation 12.15 can be rewritten as maximizing $-\frac{1}{2}\alpha^T\hat{\mathbf{K}}\alpha + \mathbf{1}^T\alpha$, where $\mathbf{1}$ is a vector of all ones and $\hat{K}_{i,j} = y_i y_j k(x_i, x_j)$. When the Hessian matrix $\hat{\mathbf{K}}$ is positive definite, the objective function is convex and there is a *unique global solution*. If matrix $\hat{\mathbf{K}}$ is positive semi-definite, every maximum is also a global maximum, however, there can be several optimal solutions (different in their α) which might lead to different performance on the testing dataset.

In general, the support vector optimization can be solved analytically only when the number of training data is very small. The worst case computational complexity for the general analytic case results from the inversion of the Hessian matrix, thus is of order N_S^3, where N_S is the number of support vectors. There exists a vast literature on solving quadratic programs (Bertsekas 1995, Bazaraa *et al.* 1993) and several software packages are available. However, most quadratic programming algorithms are either only suitable for small problems or assume that the Hessian matrix $\hat{\mathbf{K}}$ is sparse, i.e., most elements of this matrix are zero. Unfortunately, this is not true for the SVM problem. Thus, using standard quadratic programming codes with more than a few hundred variables results in enormous training times and more demanding memory needs. Nevertheless, the structure of the SVM optimization problem allows the derivation of specially tailored algorithms, which allow for fast convergence with small memory requirements, even on large problems.

A key observation in solving large-scale SVM problems is the sparsity of the solution (Steinwart, 2004). Depending on the problem, many of the optimal α_i will either be zero or on the upper bound. If one could know beforehand which α_i were zero, the corresponding rows and columns could be removed from the matrix $\hat{\mathbf{K}}$ without changing the value of the quadratic form. Furthermore, a point can only be optimal if it fulfills the KKT conditions (such as Equation 12.5). SVM solvers decompose the quadratic optimization problem into a sequence of smaller quadratic optimization problems that are solved in sequence. Decomposition methods are based on the observations of Osuna *et al.* (1997) that each QP in a sequence of QPs always contains at least one sample violating the KKT conditions. The classifier built from solving the QP for part of the training data is used to test the rest of the training data. The next partial training set is generated from combining the support vectors already found (the "working set") with the points that most violate the KKT conditions, such that the partial Hessian matrix will fit the memory. The algorithm will eventually converge to the optimal solution. Decomposition methods differ in the strategies for generating the smaller problems and use sophisticated heuristics to select several patterns to add and remove from the sub-problem plus efficient caching methods. They usually achieve fast convergence even on large data sets with up to several thousands of support vectors. A quadratic optimizer is still required as part of the solver. Elements of the SVM solver can take advantage of parallel processing: such as simultaneous computing of the Hessian matrix, dot products, and the objective function. More details and tricks can be found in the literature (Platt, 1998,

Joachims 1999, Smola *et al.* 2000, Lin 2001, Chang and Lin 2001, Chew *et al.* 2003, Chung *et al.* 2004).

A fairly large selection of optimization codes for SVM classification and regression may be found on the Web (Kernel 2004), together with the appropriate references. They range from simple MATLAB implementation to sophisticated C, C++, or FORTRAN programs (e.g., LIBSVM: Chang and Lin 2001, SVMlight: Joachim 2004). Some solvers include integrated model selection and data rescaling procedures for improved speed and numerical stability. Hsu *et al.* (2003) advises about working with a SVM software on practical problems.

12.4.2 Model Selection

To obtain a high level of performance, some parameters of the SVM algorithm have to be tuned. These include 1) the selection of the kernel function; 2) the kernel parameter(s); 3) the regularization parameters (C, ν, ε) for the tradeoff between the model complexity and the model accuracy. Model selection techniques provide principled ways to select a proper kernel. Usually, a sequence of models is solved, and using some heuristic rules, next set of parameters is tested. The process is continued until a given criterion is obtained (e.g., 99% correct classification). For example, if we consider 3 alternative (single parameter) kernels, 5 partitions of the kernel parameters, and one regularization parameters with 5 partitions each, then we need to consider a total of 3x5x5=125 SVM evaluations.

The *cross validation* technique is widely used for a prediction of the generalization error, and is included in some SVM packages (such as LIBSVM: Chang and Lin 2001). Here, the training samples are divided into k subsets of equal size. Then, the classifier is trained k times: in the i-th iteration ($i = 1,...,k$), the classifier is trained on all subsets except the i-th one. Then, the classification error is computed for the i-th subset. It is known that the average of these k errors is a rather good estimate of the generalization error. k is typically 5 or 10. Thus, for the example above we need to consider at least 625 SVM evaluations to identify the model of the best SVM classifier.

In the *Bayesian evidence framework* the training of an SVM is interpreted as Bayesian inference, and the model selection is accomplished by maximizing the marginal likelihood (i.e., evidence). Law and Kwok (2000) and Chu (2003) provide iterative parameter updating formulas, and report a significantly smaller number of SVM evaluations.

12.4.3 Multi-Class SVM

Though SVM was originally designed for two-class problems, several approaches have been developed to extend SVM for multi-class data sets.

One approach to k-class pattern recognition is to consider the problem as a collection of binary classification problems. The technique of *one-against-the-rest* requires k binary classifiers to be constructed (when the label +1 is assigned to each class in its turn and the label -1 is assigned to the other $k-1$ classes). In the prediction stage, a voting scheme is applied to classify a new point. In the *winner-takes-all* voting scheme, one assigns the class with the largest real value. The *one-against-one* approach trains a binary SVM for any two classes of data and obtains a decision function. Thus, for a k-class problem, there are $k(k-1)/2$ decision functions where the voting scheme is designated to choose the class with the maximum number of votes. More elaborate voting schemes, such as *error-correcting-codes* consider the combined outputs from the n-parallel classifiers as a binary n-bit code word and selects the class with the closest (e.g. Hamming distance) code.

In Hsu and Lin (2002), it was experimentally shown that for general problems, using the C-SVM classifier, various multi-class approaches give similar accuracy. Rifkin and Klautau (2004) have similar observation, however, this may not always be the case. Multi-class methods must be considered together with parameter-selection strategies. That is, we search for appropriate regularization parameters and kernel parameters for constructing a better model. Chen, Lin and Scholkopf (2003) experimentally demonstrate inconsistent and marginal improvement in the accuracy when the parameters are trained differently for each classifier inside a multi-class C-SVM and ν-SVM classifiers.

12.5 Extensions and Application

Kernel algorithms have solid foundations in statistical learning theory and functional analysis, thus, kernel methods combine statistics and geometry. Kernels provide an elegant framework for studying fundamental issues of machine learning, such as similarity measures that can incorporate prior knowledge about the problem, and data representations. SVM have been one of the major kernel methods for supervised learning. It is not surprising that recent methods integrate SVM with kernel methods (Scholkopf *et al.* 1999, Scholkopf and Smola, 2002, Shawe-Taylor and Cristianini 2004) for unsupervised learning problems such as density estimation (Weston and Herbrich, 2000).

SVM has a strong analogy in *regularization theory* (Williamson *et al.*, 2001). Regularization is a method of solving problems by making some a-priori assumptions about the desired function. A penalty term that discourages over-fitting is added to the error function. A common choice of regularizer is given by the sum of the squares of the weight parameters and results in a functional similar to Equation 12.6. Like SVM, optimizing a functional of the learning function, such as its smoothness, leads to sparse solutions.

Boosting is a machine learning technique that attempts to improve a "weak" learning algorithm, by a convex combination of the original "weak" learning function, each one trained with a different distribution of the data in the training set. SVM can be translated to a corresponding boosting algorithm using the appropriate regularization norm (Ratsch *et al.*, 2001).

Successful applications of SVM algorithms have been reported for various fields, such as pattern recognition (Martin *et al.* 2002), text categorization (Dumais 1998,

Joachims 2002), time series prediction (Mukherjee, 1997), and bio-informatics (Zien *et al.* 2000). Historically, classification experiments with the U.S. Postal Service benchmark problem - the first real-world experiment of SVM (Cortes and Vapnik 1995, Scholkopf 1995) - demonstrated that plain SVMs give a performance very similar to other state-of-the-art methods. SVM has been achieving excellent results also on the Reuters-22173 text classification benchmark problem (Dumais, 1998). SVMs have been strongly improved by using prior knowledge about the problem to engineer the kernels and the support vectors with techniques such as virtual support vectors (Scholkopf 1997, Scholkopf *et al.* 1998). Isabelle (2004) and Kernel (2004) present many more applications.

12.6 Conclusion

Since the introduction of the SVM classifier a decade ago, SVM gained popularity due to its solid theoretical foundation in statistical learning theory. They differ radically from comparable approaches such as neural networks: they have a simple geometrical interpretation and SVM training always finds a global minimum. The development of efficient implementations led to numerous applications. Selected real-world applications served to exemplify that SVM learning algorithms are indeed highly competitive on a variety of problems.

SVM are a set of related methods for supervised learning, applicable to both classification and regression problems. This chapter provides an overview of the main SVM methods for the separable and non-separable case and for classification and regression problems. However, SVM methods are being extended to unsupervised learning problems.

A SVM is largely characterized by the choice of its kernel. The kernel can be viewed as a nonlinear similarity measure, and should ideally incorporate prior knowledge about the problem at hand. The best choice of kernel for a given problem is still an open research issue. A second limitation is the speed of training. Training for very large datasets (millions of support vectors) is still an unsolved problem.

References

Bazaraa M. S., Sherali H. D., and Shetty C. M. Nonlinear programming: theory and algorithms. Wiley, second edition, 1993.

Bertsekas D.P. Nonlinear Programming. Athena Scientific, MA, 1995.

Chang C.-C. and Lin C.-J. Training support vector classifiers: Theory and algorithms. Neural Computation 2001; 13(9):2119–2147.

Chang C.-C. and Lin C.-J. (2001). LIBSVM: a library for support vector machines. Software available at http://www.csie.ntu.edu.tw/~cjlin/libsvm.

Chen P.-H., Lin C. -J., and Scholkopf B. A tutorial on nu-support vector machines. 2003.

Chew H. G., Lim C. C., and Bogner R. E. An implementation of training dual-nu support vector machines. In Qi, Teo, and Yang, editors, Optimization and Control with Applications. Kluwer, 2003.

Chu W. Bayesian approach to support vector machines. PhD thesis, National University of Singapore , 2003; Available online http://citeseer.ist.psu.edu/chu03bayesian.html

Chung K.-M., Kao W.-C., Sun C.-L., and Lin C.-J. Decomposition methods for linear support vector machines. Neural Computation 2004; 16(8):1689-1704).

Cortes C. and Vapnik V. Support vector networks. Machine Learning 1995; 20:273–297.

Cristianini N. and Shawe-Taylor J. An Introduction to Support Vector Machines and other kernel-based learning methods. Cambridge Univ. Press, 2000.

Dumais S. Using SVMs for text categorization. IEEE Intelligent Systems 1998; 13(4).

Hsu C.-W. and Lin C.-J. A comparison of methods for multi-class support vector machines IEEE Transactions on Neural Networks 2002; 13(2); 415–425.

Hsu C.-W. Chang C.-C and Lin C.-J. A practical guide to support vector classification. 2003. Available Online: www.csie.ntu.edu.tw/~cjlin/papers/guide/guide.pdf

Isabelle 2004, (a collection of SVM applications) Available Online: http://www.clopinet.com/isabelle/Projects/SVWM/applist.html

Joachims T. Making large–scale SVM learning practical. In Scholkopf B., Burges C. J. C., and Smola A. J., editors, Advances in Kernel Methods — Support Vector Learning, pages 169–184, Cambridge, MA, MIT Press, 1999.

Joachims T. Learning to Classify Text using Support Vector Machines Methods, Theory, and Algorithms. Kluwer Academic Publishers, 2002.

Joachims T. 2004, SVMlight, available online http://www.cs.cornell.edu/People/tj/svm_light/

Kernel 2004, (a collection of literature, software and Web pointers dealing with SVM and Gaussian processes) Available Online http://www.kernel-machines.org.

Law M. H. and Kwok J. T. Bayesian support vector regression. Proceedings of the 8th International Workshop on Artificial Intelligence and Statistics (AISTATS) pages 239-244, Key-West, Florida, USA, January 2000.

Lin C.-J. Formulations of support vector machines: a note from an optimization point of view. Neural Computation 2001; 13(2):307–317.

Lin C.-J. On the convergence of the decomposition method for support vector machines. IEEE Transactions on Neural Networks 2001; 12(6):1288–1298.

Martin D. R., Fowlkes C. C., and Malik J. Learning to detect natural image boundaries using brightness and texture. In Advances in Neural Information Processing Systems, volume 14, 2002.

Mukherjee S., Osuna E., and Girosi F. Nonlinear prediction of chaotic time series using a support vector machine. In Principe J., Gile L., Morgan N. and Wilson E. editors, Neural Networks for Signal Processing VII - proceedings of the 1997 IEEE Workshop, pages 511–520, New-York, IEEE Press, 1997.

Muller K.-R., Mika S., Ratsch G., Tsuda K., and Scholkopf B., An introduction to kernel-based learning algorithms. IEEE Neural Networks 2001; 12(2):181-201.

Osuna E., Freund R., and Girosi F. An improved training algorithm for support vector machines. In Principe J., Gile L., Morgan N. and Wilson E. editors, Neural Networks for Signal Processing VII - proceedings of the 1997 IEEE Workshop, pages 276-285, New-York, IEEE Press, 1997.

Platt J. C. Fast training of support vector machines using sequential minimal optimization. In Scholkopf B., Burges C. J. C., and Smola A. J., editors, Advances in Kernel Methods - Support Vector Learning, Cambridge, MA, MIT Press, 1998.

Ratsch G., Onoda T., and Muller K.R. Soft margins for AdaBoost. Machine Learning 2001; 42(3):287–320.
Rifkin R. and Klautau A.. In Defense of One-vs-All Classification, Journal of Machine Learning Research 2004; 5:101-141.
Scholkopf B., Support Vector Learning. Oldenbourg Verlag, Munich, 1997.
Scholkopf B., Statistical learning and kernel methods, Technical Report MSR-TR-2000-23, Available Online http://research.microsoft.com/research/pubs /view.aspx?msr_tr_id= MSR-TR-2000-23
Scholkopf B., Burges C.J.C., and Vapnik V.N. Extracting support data for a given task. In Fayyad U.M. and Uthurusamy R., Editors, Proceedings, First International Conference on Knowledge Discovery and Data Mining. AAAI Press, Menlo Park, CA, 1995.
Scholkopf B., Simard P.Y., Smola A.J., and Vapnik V.N.. Prior knowledge in support vector kernels. In Jordan M., Kearns M., and Solla S., Editors, Advances in Neural Information Processing Systems 10, pages 640–646. MIT Press, Cambridge, MA, 1998.
Scholkopf B., Burges C. J. C., and Smola A. J., editors, Advances in Kernel Methods - Support Vector Learning, Cambridge, MA, MIT Press, 1999.
Scholkopf B. and Smola A. J. Learning with Kernels. MIT Press, Cambridge, MA, 2002.
Scholkopf B., Smola A. J., Williamson R. C., and Bartlett P. L. New support vector algorithms. Neural Computation 2000; 12:1207–1245.
Shawe-Taylor J. and Cristianini N. Kernel Methods for Pattern Analysis. Cambridge University Press, 2004.
Smola A. J., Bartlett P. L., Scholkopf B. and Schuurmans D. Advances in Large Margin Classifiers. MIT Press, Cambridge, MA, 2000.
Smola A.J. and Scholkopf B.. A tutorial on support vector regression. Statistics and Computing 2004; 14(13):199-222.
Smola A.J., Scholkopf B. and Ratsch G. Linear programs for automatic accuracy control in regression. Proceedings of International Conference on Artificial Neural Networks ICANN'99, Berlin, Springer 1999.
Steinwart I. On the optimal parameter choice for nu-support vector machines. IEEE Transactions on Pattern Analysis and Machine Intelligence 2003; 25: 1274-1284.
Steinwart I. Sparseness of support vector machines. Journal of Machine Learning Research 2004; 4(6):1071-1105.
Suykens J.A.K., Van Gestel T., De Brabanter J., De Moor B., and Vandewalle J. Least Squares Support Vector Machines. World Scientific Publishing, Singapore, 2002.
Vapnik V. The Nature of Statistical Learning Theory . Springer Verlag, New York, 1995.
Vapnik V. Statistical Learning Theory. Wiley, NY, 1998.
Vapnik V. and Chapelle O. Bounds on error expectation for support vector machines. Neural Computation 2000; 12(9):2013–2036.
Weston J. and Herbrich R., Adaptive margin support vector machines. In Smola A.J., Bartlett P.L., Scholkopf B., and Schuurmans D., Editors, Advances in Large Margin Classifiers, pages 281–296, MIT Press, Cambridge, MA, 2000,.
Williamson R. C., Smola A. J., and Scholkopf B., Generalization performance of regularization networks and support vector machines via entropy numbers of compact operators. IEEE Transactions on Information Theory 2001; 47(6):2516–2532.
Wolfe P. A duality theorem for non-linear programming. Quartely of Applied Mathematics 1961; 19:239–244.
Zien A., Ratsch G., Mika S., Scholkopf B., Lengauer T. and Muller K.R. Engineering support vector machine kernels that recognize translation initiation sites. Bio-Informatics 16(9):799–807.

13

Rule Induction

Jerzy W. Grzymala-Busse

University of Kansas

Summary. This chapter begins with a brief discussion of some problems associated with input data. Then different rule types are defined. Three representative rule induction methods: LEM1, LEM2, and AQ are presented. An idea of a classification system, where rule sets are utilized to classify new cases, is introduced. Methods to evaluate an error rate associated with classification of unseen cases using the rule set are described. Finally, some more advanced methods are listed.

Key words: Rule induction algorithms LEM1, LEM2, and AQ; LERS Data Mining system, LERS classification system, rule set types, discriminant rule sets, validation.

13.1 Introduction

Rule induction is one of the most important techniques of machine learning. Since regularities hidden in data are frequently expressed in terms of rules, rule induction is one of the fundamental tools of Data Mining at the same time. Usually rules are expressions of the form

$$\begin{gathered} if\ (attribute-1, value-1)\ and\ (attribute-2, value-2)\ and\ \cdots \\ and\ (attribute-n, value-n)\ then\ (decision, value). \end{gathered}$$

Some rule induction systems induce more complex rules, in which values of attributes may be expressed by negation of some values or by a value subset of the attribute domain.

Data from which rules are induced are usually presented in a form similar to a table in which *cases* (or *examples*) are *labels* (or *names*) for rows and variables are labeled as *attributes* and a *decision*. We will restrict our attention to rule induction which belongs to *supervised learning*: all cases are preclassified by an expert. In different words, the decision value is assigned by an expert to each case. Attributes are

O. Maimon, L. Rokach (eds.), *Data Mining and Knowledge Discovery Handbook*, 2nd ed.,
DOI 10.1007/978-0-387-09823-4_13, © Springer Science+Business Media, LLC 2010

independent variables and the decision is a dependent variable. A very simple example of such a table is presented as Table 13.1, in which attributes are: *Temperature*, *Headache*, *Weakness*, *Nausea*, and the decision is *Flu*. The set of all cases labeled by the same decision value is called a *concept*. For Table 13.1, case set $\{1, 2, 4, 5\}$ is a concept of all cases affected by flu (for each case from this set the corresponding value of *Flu* is *yes*).

Table 13.1. An Example of a Dataset.

Case	Attributes				Decision
	Temperature	Headache	Weakness	Nausea	Flu
1	very_high	yes	yes	no	yes
2	high	yes	no	yes	yes
3	normal	no	no	no	no
4	normal	yes	yes	yes	yes
5	high	no	yes	no	yes
6	high	no	no	no	no
7	normal	no	yes	no	no

Note that input data may be affected by errors. An example of such a data set is presented in Table 13.2. The case 7 has value 42.5 for Weakness, an obvious error, since the attribute Weakness is symbolic, with possible values yes and no. Such errors must be corrected before rule induction.

Table 13.2. An Example of an Erroneous Dataset

Case	Attributes				Decision
	Temperature	Headache	Weakness	Nausea	Flu
1	very_high	yes	yes	no	yes
2	high	yes	no	yes	yes
3	normal	no	no	no	no
4	normal	yes	yes	yes	yes
5	high	no	yes	no	yes
6	high	no	no	no	no
7	normal	no	42.5	no	no

Another problem is caused by numerical attributes, for example, Temperature may be represented by real numbers, as in Table 13.3.

Obviously, numerical attributes must be converted into symbolic attributes before or during rule induction. The process of converting numerical attributes into symbolic attributes is called *discretization* (or *quantization*).

Table 13.3. An Example of a Dataset with a Numerical Attribute.

Case	Attributes				Decision
	Temperature	Headache	Weakness	Nausea	Flu
1	41.6	yes	yes	no	yes
2	39.8	yes	no	yes	yes
3	36.8	no	no	no	no
4	37.0	yes	yes	yes	yes
5	38.8	no	yes	no	yes
6	40.2	no	no	no	no
7	36.6	no	yes	no	no

Input data may be incomplete, i.e., some attributes may have missing attribute values, as in Table 13.4, where ? denotes lack of the attribute value (for example, the original value was not recorded or was erased).

Table 13.4. An Example of a Dataset with Missing Attribute Values.

Case	Attributes				Decision
	Temperature	Headache	Weakness	Nausea	Flu
1	very_high	yes	yes	no	yes
2	?	yes	no	yes	yes
3	normal	no	?	no	no
4	normal	?	yes	yes	yes
5	high	no	yes	no	yes
6	high	no	no	no	no
7	normal	no	yes	no	no

Additionally, input data may be inconsistent, i.e., some cases may conflict with each other. Conflicting cases have the same attribute values yet different decision values. An example of an inconsistent data set is presented in Table 13.5. Cases 7 and 8 are conflicting.

In Section 13.2 a brief discussion of different rule types is presented. In the next section a few representative rule induction algorithms are discussed. Section 13.4 presents the main application of rule sets, classification systems, which are used to classify new cases on the basis of induced rule sets.

13.2 Types of Rules

A case x is *covered* by a rule r if and only if every condition (attribute-value pair) of r is satisfied by the corresponding attribute value for x. The concept C defined by

Table 13.5. An Example of an Inconsistent Dataset

Case	Attributes				Decision
	Temperature	Headache	Weakness	Nausea	Flu
1	very_high	yes	yes	no	yes
2	high	yes	no	yes	yes
3	normal	no	no	no	no
4	normal	yes	yes	yes	yes
5	high	no	yes	no	yes
6	high	no	no	no	no
7	normal	no	yes	no	no
8	normal	no	yes	no	yes

the right hand side of rule r is *indicated* by r. We say that a concept C is *completely* covered by a rule set R if and only if for every case x from C there exists a rule r from R such that r covers x. For a given data set, a rule set R is *complete* if and only if every concept from the data set is completely covered by R.

A rule r is *consistent* with the data set if and only if for every case x covered by r, x is a member of the concept C indicated by r. A rule set R is *consistent* with a data set if and only if every rule from R is consistent with the data set.

For example, case 1 from Table 13.1 is covered by the following rule r:

$$(Headache, yes) \rightarrow (Flu, yes).$$

The rule r indicates concept $\{1, 2, 4, 5\}$. Additionally, the concept $\{1, 2, 4, 5\}$ is not completely covered by the rule set consisting of r, since r covers only cases 1, 2, and 4, but the rule r is consistent with the data set from Table 13.1.

On the other hand, the single rule

$$(Headache, no) \rightarrow (Flu, no)$$

completely covers the concept $\{3, 6, 7\}$ in Table 13.1, though this rule is not consistent with the same data set. The above rule covers cases 3, 5, 6, and 7.

Any of the following two rules:

$$(Headache, yes) \ \& \ (Weakness, yes) \rightarrow (Flu, yes)$$

and

$$(Temperature, high) \ \& \ (Headache, yes) \rightarrow (Flu, yes)$$

is consistent with the data set from Table 13.1, but the concept $\{1, 2, 4, 5\}$ is not completely covered by the rule set consisting of the above two rules since case 5 is not covered by any rule. The first rule covers cases 1 and 4, the second rule covers case 2.

The most frequent task of rule induction is to induce a rule set R that is complete and consistent. Such a rule set R is called *discriminant* (Michalski, 1983). For Table 13.1, the rule set consisting of the following four rules:

$$(Headache, yes) \rightarrow (Flu, yes),$$

$$(Temperature, high) \ \& \ (Weakness, yes) \rightarrow (Flu, yes),$$

$$(Temperature, normal) \ \& \ (Headache, no) \rightarrow (Flu, no),$$

$$(Headache, no) \ \& \ (Weakness, no) \rightarrow (Flu, no).$$

is discriminant.

There are many other types of rules that are used. For example, some rule induction systems induce rule sets consisting of *strong* rules, i.e., rule sets in which every rule covers many cases. Another task is to induce *associative* rules, in which in both sides of a rule, left and right, involved variables are attributes. For Table 13.1, an example of such an associative rule is

$$(Nausea, yes) \rightarrow (Headache, yes).$$

13.3 Rule Induction Algorithms

In this section we will assume that input data sets are free of errors, numcrical attributes were already discretized, no missing attribute values are present in the input data sets, and that input data sets are consistent.

In general, rule induction algorithms may be categorized as *global* and *local*. In global rule induction algorithms the search space is the set of all attribute values, while in local rule induction algorithms the search space is the set of attribute-value pairs.

There exist many rule induction algorithms, we will discuss only three representative algorithms, all inducing discriminant rule sets. The first is an example of a global rule induction algorithm called LEM1 (Learning from Examples Module version 1).

13.3.1 LEM1 Algorithm

The algorithm LEM1, a component of the Data Mining system LERS (Learning from Examples using Rough Sets), is based on some rough set definitions (Pawlak, 1982), (Pawlak, 1991), (Pawlak *et al.*, 1995). Let B be a nonempty subset of the set A of all attributes. Let U denote the set of all cases. The indiscernibility relation $IND(B)$ is a relation on U defined for $x, y \in U$ by $(x, y) \in IND(B)$ if and only if for both x and y the values for all attributes from B are identical.

The indiscernibility relation $IND(B)$ is an equivalence relation. Equivalence classes of $IND(B)$ are called *elementary sets* of B. For example, for Table 13.1, and B = {Temperature, Headache}, elementary sets of $IND(B)$ are {1}, {2}, {3, 7}, {4}, {5, 6}.

The family of all B-elementary sets will be denoted B^*, for example, in Table 13.1,

$$\{Temperature, Headache\}^* = \{\{1\},\{2\},\{3,7\},\{4\},\{5,6\}\}.$$

For a decision d we say that $\{d\}$ depends on B if and only if $B^* \leq \{d\}^*$. A *global covering* (or *relative reduct*) of $\{d\}$ is a subset B of A such that $\{d\}$ depends on B and B is minimal in A. Thus, global coverings of $\{d\}$ are computed by comparing partitions B^* with $\{d\}^*$. The algorithm to compute a single global covering is presented below.

Algorithm to compute a single global covering
(**input**: the set $A$of all attributes, partition $\{d\}^*$ on U;
output: a single global covering R);
begin
compute partition A^*;
$P := A$;
$R := \emptyset$;
  if $A^* \leq \{d\}^*$
   then
    begin
    for each attribute a in A **do**
      begin
      $Q := P - \{a\}$;
      compute partition Q^*;
      if $Q^* \leq \{d\}^*$ **then** $P := Q$
    end {for}
    $R := P$
   end {then}
end {algorithm}.

On the basis of a global covering rules are computed using the *dropping conditions* technique (Michalski, 1983). For a rule of the form

$$C_1 \ \& \ C_2 \ \& \ ... \ \& \ C_n \ \rightarrow \ D$$

dropping conditions means scanning the list of all conditions, from the left to the right, with an attempt to drop any condition, checking against the decision table where the simplified rule does not violate consistency of the discriminant description.

For Table 13.1,

$$\{Temperature,\ Headache,\ Weakness,\ Nausea\}^* = \\ \{\{1\},\{2\},\{3\},\{4\},\{5\},\{6\},\{7\}\},$$

$$\{Flu\}^* = \{\{1,2,4,5\},\{3,6,7\}\},$$

and

$$\{Temperature,\ Headache,\ Weakness,\ Nausea\}^* \leq \{Flu\}^*.$$

Next we need to check whether

$$\{Headache,\ Weakness,\ Nausea\}^* \leq \{Flu\}^*.$$

This condition is false since

$$\{Headache,\ Weakness,\ Nausea\}^* = \\ \{\{1\},\{2\},\{3,6\},\{4\},\{5,7\}\}.$$

Then we compute

$$\{Temperature,\ Weakness,\ Nausea\}^* = \\ \{\{1\},\{2\},\{3\},\{4\},\{5\},\{6\},\{7\}\},$$

We observe that

$$\{Temperature,\ Weakness,\ Nausea\}^* \leq \{Flu\}^*.$$

The next partition to compute is

$$\{Temperature,\ Nausea\}^*,$$

equal to

$$\{\{1\},\{2\},\{3,7\},\{4\},\{5,6\}\},$$

and

$$\{Temperature,\ Nausea\}^* \not\leq \{Flu\}^*.$$

The last step is to compute

$$\{Temperature,\ Weakness\}^*,$$

equal to

$$\{\{1\},\{2,6\},\{3\},\{4,7\},\{5\}\},$$

since

$$\{Temperature,\ Weakness\}^* \not\leq \{Flu\}^*,$$

the total covering is

$$\{Temperature,\ Weakness,\ Nausea\}.$$

The first case from Table 13.1 implies the following preliminary rule

$$(Temperature, very\_high)\ \&\ (Weakness, yes)\ \&\ (Nausea, no) \\ \rightarrow\ (Flu, yes)$$

The above rule covers only the first case. The first condition,

$$(Temperature, very\_high),$$

cannot be dropped since the rule

$$(Weakness, yes)\ \&\ (Nausea, no)\ \rightarrow\ (Flu, yes)$$

covers cases 1 and 7 from different concepts. However, an attempt to drop the next condition, $(Weakness, yes)$ is successful since the rule

$$(Temperature, very\_high)\ \&\ (Nausea, no)\ \rightarrow\ (Flu, yes)$$

covers only case 1. The next possibility, to drop the last condition $(Weakness, yes)$ is successful as well, since the resulting rule

$$(Temperature, very\_high)\ \rightarrow\ (Flu, yes)$$

covers only case 1.

In a similar way the remaining rules are induced. The final rule set, induced by LEM1, is

$(Temperature, very\_high)\ \rightarrow\ (Flu, yes)$
$(Nausea, yes)\ \rightarrow\ (Flu, yes)$
$(Temperature, high)\ \&\ (Weakness, yes)\ \rightarrow\ (Flu, yes)$
$(Weakness, no)\ \&\ (Nausea, no)\ \rightarrow\ (Flu, no)$
$(Temperature, normal)\ \&\ (Nausea, no)\ \rightarrow\ (Flu, no)$

For Table 13.1, the second global covering is

$$\{Temperature,\ Headache,\ Weakness\}.$$

13.3.2 LEM2

An idea of blocks of attribute-value pairs is used in the rule induction algorithm LEM2 (Learning from Examples Module, version 2), another component of LERS. The option LEM2 of LERS is most frequently used since—in most cases—it gives better results. LEM2 explores the search space of attribute-value pairs. Its input data file is a lower or upper approximation of a concept (for definitions of lower and upper approximations of a concept see, e.g., (Grzymala-Busse, 1997)), so its input data file is always consistent. In general, LEM2 computes a local covering and then converts it into a rule set. We will quote a few definitions to describe the LEM2 algorithm (Chan and Grzymala-Busse, 1991), (Grzymala-Busse, 1992).

For an attribute-value pair $(a,\ v) = t$, a *block* of t, denoted by $[t]$, is a set of all cases from U such that for attribute a have value v. Let B be a nonempty lower or upper approximation of a concept represented by a decision-value pair (d,w). Set B *depends* on a set T of attribute-value pairs $t = (a,v)$ if and only if

$$\emptyset \neq [T] = \bigcap_{t \in T} [t] \subseteq B.$$

Set T is a *minimal complex* of B if and only if B depends on T and no proper subset T' of T exists such that B depends on T'. Let $\mathscr{T}$ be a nonempty collection of nonempty sets of attribute-value pairs. Then $\mathscr{T}$ is a *local covering* of B if and only if the following conditions are satisfied:

(1) each member T of $\mathscr{T}$ is a minimal complex of B,

(2) $\bigcup_{t \in \mathscr{T}} [T] = B$, and

$\mathscr{T}$ is minimal, i.e., $\mathscr{T}$ has the smallest possible number of members.

The procedure LEM2 is presented below.

Procedure LEM2
(**input**: a set B,
output: a single local covering $\mathscr{T}$ of set B);
begin
  $G := B$;
  $\mathscr{T} := \emptyset$;
  while $G \neq \emptyset$
    begin
    $T := \emptyset$;
    $T(G) := \{t | [t] \cap G \neq \emptyset\}$;
    while $T = \emptyset$ **or** $[T] \not\subseteq B$
      begin
        select a pair $t \in T(G)$ such that $|[t] \cap G|$ is
        maximum; if a tie occurs, select a pair $t \in T(G)$
        with the smallest cardinality of $[t]$;
        if another tie occurs, select first pair;

$T := T \cup \{t\}$;
$G := [t] \cap G$;
$T(G) := \{t | [t] \cap G \neq \emptyset\}$;
$T(G) := T(G) - T$;
end {while}
for each $t \in T$ **do**
if $[T - \{t\}] \subseteq$ B **then** $T := T - \{t\}$;
$\mathscr{T} := \mathscr{T} \cup \{T\}$;
$G := B - \cup_{T \in \mathscr{T}} [T]$;
end {while};
for each $T \in \mathscr{T}$ **do**
if $\bigcup_{S \in \mathscr{T} - \{T\}} [S] = B$ **then** $\mathscr{T} := \mathscr{T} - \{T\}$;
end {procedure}.

For a set X, $|X|$ denotes the cardinality of X.
The first step of the algorithm LEM2 is to compute all attribute-value pair blocks. For Table 13.1. these blocks are

$[(Temperature, very\_high)] = \{1\}$,
$[(Temperature, high)] = \{2,5,6\}$,
$[(Temperature, normal)] = \{3,4,7\}$,
$[(Headache, yes)] = \{1,2,4\}$,
$[(Headache, no)] = \{3,5,6,7\}$,
$[(Weakness, yes)] = \{1,4,5,7\}$,
$[(Weakness, no)] = \{2,3,6\}$,
$[(Nausea, no)] = \{1,3,5,6,7\}$,
$[(Nausea, yes)] = \{2,4\}$.

Let us induce rules for the concept {1,2,4,5}. Hence, $B = G = \{1,2,4,5\}$. The set $T(G)$ of all relevant attribute-value pairs is

$$\{(Temperature,\ very\_high), (Temperature,\ high), (Temperature,\ normal), (Headache,\ yes), (Headache,\ no), (Weakness,\ yes), (Weakness,\ no), (Nausea, no),\ (Nausea,\ yes)\}.$$

The next step is to identify attribute-value pairs (a, v) with the largest $|[(a,v)] \cap G|$. For two attribute-value pairs from $T(G)$, (*Headache*, *yes*) and (*Weakness, yes*), the cardinality of the set $|[(a,v)] \cap G|$ is equal to three. The next criterion is the size of the attribute-value pair block, this size is smaller for (*Headache*, *yes*) than for (*Weakness*, *yes*), so we select (*Headache*, *yes*). Besides, $[(Headache,\ yes)] \subseteq B$, so (*Headache*, *high*) is the first minimal complex of G.

The new set G is equal to $B[(Headache,\ yes)] = \{1,2,4,5\} - \{1,2,4\} = \{5\}$. A new set $T(G)$ is equal to

$$\{(Temperature,\ high), (Headache,\ no), (Weakness,\ yes), (Nausea, no)\}.$$

This time the first criterion, the largest $|[(a,v)] \cap G|$, identifies all four attribute-value pairs. The second criterion, the size of the attribute-value block, selects $(Temperature,\ high)$. However,

$$[(Temperature,\ high)] = \{2,5,6\} \not\subseteq B,$$

so we have to go through an additional iteration of the internal loop. The next candidates are $(Headache,\ no)$ and $(Weakness,\ yes)$, since for both of these attribute-value pairs the sizes of their blocks are equal to four. On the basis of heuristics, we will select $(Headache,\ no)$. But

$$[(Temperature,\ high)] \cap [(Headache, no)] = \{5,6\} \not\subseteq B = \{1,2,4,5\},$$

so we have to add $(Weakness, yes)]$ as well. This time

$$[(Temperature,\ high)] \cap [(Headache, no)] \cap [(Weakness, yes)] = \\ \{5\} \subseteq B = \{1,2,4,5\},$$

so our candidate for a minimal complex is the set

$$\{(Temperature,\ high), (Headache,\ no), (Weakness, yes)\}.$$

We have to run the following part of the LEM2 algorithm:

for each $t \in T$ **do**
  if $[T - \{t\}] \subseteq$ B **then** $T := T - \{t\}$;

As a result, the second minimal complex is identified:

$$\{(Temperature,\ high), (Weakness, yes)\}.$$

Eventually, the local covering of $B = \{1,2,4,5\}$ is the set

$$\{\{(Headache,\ yes)\}, \{(Temperature,\ high), (Weakness, yes)\}\}.$$

The complete rule set, induced by LEM2, is

$(Headache, yes) \rightarrow (Flu, yes)$
$(Temperature, high)\ \&\ (Weakness, yes) \rightarrow (Flu, yes)$
$(Temperature, normal)\ \&\ (Headache, no) \rightarrow (Flu, no)$
$(Headache, no)\ \&\ (Weakness, no) \rightarrow (Flu, no)$

Obviously, in general, rule sets induced by LEM1 differ from rule sets induced by LEM2 from the same data sets.

13.3.3 AQ

Another rule induction algorithm, developed by R. S. Michalski and his collaborators in the early seventies, is an algorithm called AQ. Many versions of the algorithm have been developed, under different names (Michalski *et al.*, 1986A), (Michalski *et al.*, 1986A).

Let us start by quoting some definitions from (Michalski *et al.*, 1986A), (Michalski *et al.*, 1986A). Let A be the set of all attributes, $A = \{A_1, A_2, ..., A_k\}$. A *seed* is a member of the concept, i.e., a positive case. A *selector* is an expression that associates a variable (attribute or decision) to a value of the variable, e.g., a negation of value, a disjunction of values, etc. A *complex* is a conjunction of selectors. A *partial star* $G(e|e_1)$ is a set of all complexes describing the seed $e = (x_1, x_2, ..., x_k)$ and not describing a negative case $e_1 = (y_1, y_2, ..., y_k)$. Thus, the complexes of $G(e|e1)$ are conjunctions of selectors of the form $(A_i, \neg y_i)$, for all i such that $x_i \neq y_i$. A *star* $G(e|F)$ is constructed from all partial stars $G(e|e_i)$, for all $e_i \in F$, and by conjuncting these partial stars by each other, using absorption law to eliminate redundancy. For a given concept C, a *cover* is a disjunction of complexes describing all positive cases from C and not describing any negative cases from $F = U - C$.

The main idea of the AQ algorithm is to generate a cover for each concept by computing stars and selecting from them single complexes to the cover.

For the example from Table 13.1, and concept $C = \{1,2,4,5\}$ described by (Flu, yes), set F of negative cases is equal to 3, 6, 7. A seed is any member of C, say that it is case 1. Then the partial star $G(1|3)$ is equal to

$$\{(Temperature, \neg normal), (Headache, \neg no), (Weakness, \neg no)\}.$$

Obviously, partial star $G(1|3)$ describes negative cases 6 and 7. The partial star $G(1|6)$ equals

$$\{(Temperature, \neg high), (Headache, \neg no), (Weakness, \neg no)\}$$

The conjunct of $G(1|3)$ and $G(1|6)$ is equal to

$$\begin{gathered}\{(Temperature, very\_high),\\ (Temperature, \neg normal) \;\&\; (Headache, \neg no),\\ (Temperature, \neg normal) \;\&\; (Weakness, \neg no),\\ (Temperature, \neg high) \;\&\; (Headache, \neg no),\\ (Headache, \neg no),\\ (Headache, \neg no) \;\&\; (Weakness, \neg no),\\ (Temperature, \neg high) \;\&\; (Weakness, \neg no),\\ (Headache, \neg no) \;\&\; Weakness, \neg no),\\ (Weakness, \neg no)\},\end{gathered}$$

after using the absorption law, this set is reduced to the following set $G(1|\{3,6\})$:

$$\{(Temperature, very\_high), (Headache \neg no), (Weakness, \neg no)\}.$$

The preceding set describes negative case 7. The partial star $G(1|7)$ is equal to

$$\{(Temperature, \neg normal), Headache, \neg no)\}.$$

The conjunct of $G(1|\{3,6\})$ and $G(1|7)$ is

$$\begin{gathered}\{(Temperature, very\_high),\\ (Temperature, very\_high) \ \& \ (Headache, \neg no),\\ (Temperature, \neg normal) \ \& \ Headache, \neg no),\\ (Headache, \neg no),\\ (Temperature, \neg normal) \ \& \ (Weakness, \neg no),\\ (Headache, \neg no) \ \& \ (Weakness, \neg no)\}.\end{gathered}$$

The above set, after using the absorption law, is already a star $G(1|F)$

$$\begin{gathered}\{(Temperature, very\_high),\\ (Headache, \neg no),\\ (Temperature, \neg normal) \ \& \ (Weakness, \neg no)\}.\end{gathered}$$

The first complex describes only one positive case 1, while the second complex describes three positive cases: 1, 2, and 4. The third complex describes two positive cases: 1 and 5. Therefore, the complex

$$(Headache, \neg no)$$

should be selected to be a member of the star of C. The corresponding rule is

$$(Headache, \neg no) \ \rightarrow \ (Flu, yes).$$

If rules without negation are preferred, the preceding rule may be replaced by the following rule

$$(Headache, yes) \rightarrow (Flu, yes).$$

The next seed is case 5, and the partial star $G(5|3)$ is the following set

$$\{(Temperature, \neg normal), (Weakness, \neg no)\}.$$

The partial star $G(5|3)$ covers cases 6 and 7. Therefore, we compute $G(5|6)$, equal to

$$\{(Weakness, \neg no)\}$$

A conjunct of $G(5|3)$ and $G(5|6)$ is the following set

$$\{(Temperature, \neg normal) \ \& \ (Weakness, \neg no), (Weakness, \neg no)\}$$

After simplification, the set $G(5|\{3,6\})$ equals

$$\{Weakness, \neg no)\}.$$

The above set covers case 7. The set $G(5|7)$ is equal to

$$\{(Temperature, \neg normal)\}$$

Finally, the partial star $G(5|\{3,6,7\})$ is equal to

$$\{(Temperature, \neg normal) \ \& \ (Weakness, \neg no)\},$$

so the second rule describing concept {1, 2, 4, 5} is

$$(Temperature, \neg normal) \ \& \ (Weakness, \neg no) \ \rightarrow \ (Flu, yes).$$

It is not difficult to see that the following rules describe the second concept from Table 13.1:

$$Temperature, \neg high) \ \& \ (Headache, \neg yes) \rightarrow (Flu, no),$$

$$(Headache, \neg yes) \ \& \ (Weakness, \neg yes) \rightarrow (Flu, no).$$

Note that the AQ algorithm demands computing conjuncts of partial stars. In the worst case, time complexity of this computation is $O(n^m)$, where n is the number of attributes and m is the number of cases. The authors of AQ suggest using the parameter MAXSTAR as a method of reducing the computational complexity. According to this suggestion, any set, computed by conjunction of partial stars, is reduced in size if the number of its members is greater than MAXSTAR. Obviously, the quality of the output of the algorithm is reduced as well.

13.4 Classification Systems

Rule sets, induced from data sets, are used mostly to classify new, unseen cases. Such rule sets may be used in rule-based expert systems.

There is a few existing classification systems, e.g., associated with rule induction systems LERS or AQ. A classification system used in LERS is a modification of the well-known bucket brigade algorithm (Booker *et al.*, 1990), (Holland *et al.*, 1986), (Stefanowski, 2001). In the rule induction system AQ, the classification system is based on a rule estimate of probability (Michalski *et al.*, 1986A), (Michalski *et al.*, 1986A). Some classification systems use a decision list, in which rules are ordered, the first rule that matches the case classifies it (Rivest, 1987). In this section we will concentrate on a classification system associated with LERS.

The decision to which concept a case belongs is made on the basis of three factors: *strength*, *specificity*, and *support*. These factors are defined as follows: *strength* is the total number of cases correctly classified by the rule during training. *Specificity* is the total number of attribute-value pairs on the left-hand side of the rule. The matching rules with a larger number of attribute-value pairs are considered more specific. The third factor, *support*, is defined as the sum of products of strength and specificity for all matching rules indicating the same concept. The concept C for which the support, i.e., the following expression

$$\sum_{matching\ rules\ r\ describing\ C} Strength(r) * Specificity(r)$$

is the largest is the winner and the case is classified as being a member of C.

In the classification system of LERS, if complete matching is impossible, all partially matching rules are identified. These are rules with at least one attribute-value pair matching the corresponding attribute-value pair of a case. For any partially matching rule r, the additional factor, called *Matching_factor* (r), is computed. Matching_factor (r) is defined as the ratio of the number of matched attribute-value pairs of r with a case to the total number of attribute-value pairs of r. In partial matching, the concept C for which the following expression is the largest

$$\sum_{\substack{partially\ matching\\ rules\ r\ describing\ C}} Matching\_factor(r) * Strength(r) * Specificity(r)$$

is the winner and the case is classified as being a member of C.

13.5 Validation

The most important performance criterion of rule induction methods

is the error rate. A complete discussion on how to evaluate the error rate from a data set is contained in (Weiss and Kulikowski, 1991). If the number of cases is less than 100, the *leaving-one-out* method is used to estimate the error rate of the rule set. In leaving-one-out, the number of learn-and-test experiments is equal to the number of cases in the data set. During the i-th experiment, the i-th case is removed from the data set, a rule set is induced by the rule induction system from the remaining cases, and the classification of the omitted case by rules produced is recorded. The error rate is computed as

$$\frac{total\ number\ of\ misclassifications}{number\ of\ cases}.$$

On the other hand, if the number of cases in the data set is greater than or equal to 100, the *ten-fold cross-validation* will be used. This technique is similar to leaving-one-out in that it follows the learn-and-test paradigm. In this case, however, all cases are randomly re-ordered, and then a set of all cases is divided into ten mutually disjoint subsets of approximately equal size. For each subset, all remaining cases are used for training, i.e., for rule induction, while the subset is used for testing. This method is used primarily to save time at the negligible expense of accuracy.

Ten-fold cross validation is commonly accepted as a standard way of validating rule sets. However, using this method twice, with different preliminary random re-ordering of all cases yields—in general—two different estimates for the error rate (Grzymala-Busse, 1997).

For large data sets (at least 1000 cases) a single application of the train-and-test paradigm may be used. This technique is also known as *holdout* (Weiss and Kulikowski, 1991). Two thirds of cases should be used for training, one third for testing.

13.6 Advanced Methodology

Some more advanced methods of machine learning in general and rule induction in particular were discussed in (Dietterich, 1997). Such methods include combining a few rule sets with associated classification systems, created independently, using different algorithms, to classify a new case by taking into account all individual decisions and using some mechanisms to resolve conflicts, e.g., voting. Another important problem is scaling up rule induction algorithms. Yet another important problem is learning from imbalanced data sets (Japkowicz, 2000), where some concepts are extremely small.

References

Booker L.B., Goldberg D.E., and Holland J.F. Classifier systems and genetic algorithms. In *Machine Learning. Paradigms and Methods*, Carbonell, J. G. (ed.), The MIT Press, Boston, MA, 1990, 235–282.

Chan C.C. and Grzymala-Busse J.W. On the attribute redundancy and the learning programs ID3, PRISM, and LEM2. Department of Computer Science, University of Kansas, TR-91-14, December 1991, 20 pp.

Dietterich T.G. Machine-learning research. *AI Magazine* 1997: 97–136.

Grzymala-Busse J.W. Knowledge acquisition under uncertainty—A rough set approach. *Journal of Intelligent & Robotic Systems* 1988; **1**: 3–16.

Grzymala-Busse J.W. LERS—A system for learning from examples based on rough sets. In *Intelligent Decision Support. Handbook of Applications and Advances of the Rough Sets Theory*, ed. by R. Slowinski, Kluwer Academic Publishers, Dordrecht, Boston, London, 1992, 3–18.

Grzymala-Busse J.W. A new version of the rule induction system LERS, *Fundamenta Informaticae* 1997; **31**: 27–39.

Holland J.H., Holyoak K.J., and Nisbett R.E. *Induction. Processes of Inference, Learning, and Discovery*, MIT Press, Boston, MA, 1986.

Japkowicz N. Learning from imbalanced data sets: a comparison of various strategies. Learning from Imbalanced Data Sets, AAAI Workshop at the 17th Conference on AI, AAAI-2000, Austin, TX, July 30–31, 2000, 10–17.

Michalski R.S. A Theory and Methodology of Inductive Learning. In *Machine Learning. An Artificial Intelligence Approach*, Michalski, R. S., J. G. Carbonell and T. M. Mitchell (eds.), Morgan Kauffman, San Mateo, CA, 1983, 83–134.

Michalski R.S., Mozetic I., Hong J., Lavrac N. The AQ15 inductive learning system: An overview and experiments, Report 1260, Department of Computer Science, University of Illinois at Urbana-Champaign, 1986A.

Michalski R.S., Mozetic I., Hong J., Lavrac N. The multi-purpose incremental learning system AQ 15 and its testing application to three medical domains. Proc. of the 5th Nat. Conf. on AI, 1986B, 1041–1045.

Pawlak Z.: Rough Sets. *International Journal of Computer and Information Sciences* 1982; **11**: 341–356.

Pawlak Z. *Rough Sets. Theoretical Aspects of Reasoning about Data.* Kluwer Academic Publishers, Dordrecht, Boston, London, 1991.

Pawlak Z., Grzymala-Busse J.W., Slowinski R. and Ziarko, W. Rough sets. *Communications of the ACM* 1995; **38**: 88–95.

Rivest R.L. Learning decision lists. *Machine Learning* 1987; **2**: 229–246.

Stefanowski J. *Algorithms of Decision Rule Induction in Data Mining*. Poznan University of Technology Press, Poznan, Poland, 2001.

Weiss S. and Kulikowski C.A. *Computer Systems That Learn: Classification and Prediction Methods from Statistics, Neural Nets, Machine Learning, and Expert Systems*, chapter *How to Estimate the True Performance of a Learning System*, pp. 17–49, San Mateo, CA: Morgan Kaufmann Publishers, Inc., 1991.

Part III

Unsupervised Methods

14

A survey of Clustering Algorithms

Lior Rokach

Department of Information Systems Engineering
Ben-Gurion University of the Negev
liorrk@bgu.ac.il

Summary. This chapter presents a tutorial overview of the main clustering methods used in Data Mining. The goal is to provide a self-contained review of the concepts and the mathematics underlying clustering techniques. The chapter begins by providing measures and criteria that are used for determining whether two objects are similar or dissimilar. Then the clustering methods are presented, divided into: hierarchical, partitioning, density-based, model-based, grid-based, and soft-computing methods. Following the methods, the challenges of performing clustering in large data sets are discussed. Finally, the chapter presents how to determine the number of clusters.

Key words: Clustering, K-means, Intra-cluster homogeneity, Inter-cluster separability,

14.1 Introduction

Clustering and classification are both fundamental tasks in Data Mining. Classification is used mostly as a supervised learning method, clustering for unsupervised learning (some clustering models are for both). The goal of clustering is descriptive, that of classification is predictive (Veyssieres and Plant, 1998). Since the goal of clustering is to discover a new set of categories, the new groups are of interest in themselves, and their assessment is intrinsic. In classification tasks, however, an important part of the assessment is extrinsic, since the groups must reflect some reference set of classes. *"Understanding our world requires conceptualizing the similarities and differences between the entities that compose it"* (Tyron and Bailey, 1970).

Clustering groups data instances into subsets in such a manner that similar instances are grouped together, while different instances belong to different groups. The instances are thereby organized into an efficient representation that characterizes the population being sampled. Formally, the clustering structure is represented as a set of subsets $C = C_1, \ldots, C_k$ of S, such that: $S = \bigcup_{i=1}^{k} C_i$ and $C_i \cap C_j = \emptyset$ for $i \neq j$. Consequently, any instance in S belongs to exactly one and only one subset.

O. Maimon, L. Rokach (eds.), *Data Mining and Knowledge Discovery Handbook*, 2nd ed.,
DOI 10.1007/978-0-387-09823-4_14,

Clustering of objects is as ancient as the human need for describing the salient characteristics of men and objects and identifying them with a type. Therefore, it embraces various scientific disciplines: from mathematics and statistics to biology and genetics, each of which uses different terms to describe the topologies formed using this analysis. From biological “taxonomies”, to medical “syndromes” and genetic “genotypes” to manufacturing ”group technology” — the problem is identical: forming categories of entities and assigning individuals to the proper groups within it.

14.2 Distance Measures

Since clustering is the grouping of similar instances/objects, some sort of measure that can determine whether two objects are similar or dissimilar is required. There are two main type of measures used to estimate this relation: distance measures and similarity measures.

Many clustering methods use distance measures to determine the similarity or dissimilarity between any pair of objects. It is useful to denote the distance between two instances x_i and x_j as: d(x_i,x_j). A valid distance measure should be symmetric and obtains its minimum value (usually zero) in case of identical vectors. The distance measure is called a metric distance measure if it also satisfies the following properties:

1. Triangle inequality $d(x_i,x_k) \leq d(x_i,x_j) + d(x_j,x_k) \quad \forall x_i,x_j,x_k \in S$.
2. $d(x_i,x_j)= 0 \Rightarrow x_i = x_j \quad \forall x_i ,x_j \in S$.

14.2.1 Minkowski: Distance Measures for Numeric Attributes

Given two p-dimensional instances, $x_i = (x_{i1}, x_{i2}, \ldots, x_{ip})$ and $x_j = (x_{j1}, x_{j2}, \ldots, x_{jp})$, The distance between the two data instances can be calculated using the Minkowski metric (Han and Kamber, 2001):

$$d(x_i,x_j) = (\left|x_{i1} - x_{j1}\right|^g + \left|x_{i2} - x_{j2}\right|^g + \ldots + \left|x_{ip} - x_{jp}\right|^g)^{1/g}$$

The commonly used Euclidean distance between two objects is achieved when $g = 2$. Given $g = 1$, the sum of absolute paraxial distances (Manhattan metric) is obtained, and with g=∞ one gets the greatest of the paraxial distances (Chebychev metric).

The measurement unit used can affect the clustering analysis. To avoid the dependence on the choice of measurement units, the data should be standardized. Standardizing measurements attempts to give all variables an equal weight. However, if each variable is assigned with a weight according to its importance, then the weighted distance can be computed as:

$$d(x_i,x_j) = (w_1 \left|x_{i1} - x_{j1}\right|^g + w_2 \left|x_{i2} - x_{j2}\right|^g + \ldots + w_p \left|x_{ip} - x_{jp}\right|^g)^{1/g}$$

where $w_i \in [0, \infty)$

14.2.2 Distance Measures for Binary Attributes

The distance measure described in the last section may be easily computed for continuous-valued attributes. In the case of instances described by categorical, binary, ordinal or mixed type attributes, the distance measure should be revised.

In the case of binary attributes, the distance between objects may be calculated based on a contingency table. A binary attribute is symmetric if both of its states are equally valuable. In that case, using the simple matching coefficient can assess dissimilarity between two objects:

$$d(x_i, x_j) = \frac{r+s}{q+r+s+t}$$

where q is the number of attributes that equal 1 for both objects; t is the number of attributes that equal 0 for both objects; and s and r are the number of attributes that are unequal for both objects.

A binary attribute is asymmetric, if its states are not equally important (usually the positive outcome is considered more important). In this case, the denominator ignores the unimportant negative matches (t). This is called the Jaccard coefficient:

$$d(x_i, x_j) = \frac{r+s}{q+r+s}$$

14.2.3 Distance Measures for Nominal Attributes

When the attributes are *nominal*, two main approaches may be used:

1. Simple matching:
$$d(x_i, x_j) = \frac{p-m}{p}$$
where p is the total number of attributes and m is the number of matches.
2. Creating a binary attribute for each state of each nominal attribute and computing their dissimilarity as described above.

14.2.4 Distance Metrics for Ordinal Attributes

When the attributes are *ordinal*, the sequence of the values is meaningful. In such cases, the attributes can be treated as numeric ones after mapping their range onto [0,1]. Such mapping may be carried out as follows:

$$z_{i,n} = \frac{r_{i,n} - 1}{M_n - 1}$$

where $z_{i,n}$ is the standardized value of attribute a_n of object i. $r_{i,n}$ is that value before standardization, and M_n is the upper limit of the domain of attribute a_n (assuming the lower limit is 1).

14.2.5 Distance Metrics for Mixed-Type Attributes

In the cases where the instances are characterized by attributes of *mixed-type*, one may calculate the distance by combining the methods mentioned above. For instance, when calculating the distance between instances i and j using a metric such as the Euclidean distance, one may calculate the difference between nominal and binary attributes as 0 or 1 ("match" or "mismatch", respectively), and the difference between numeric attributes as the difference between their normalized values. The square of each such difference will be added to the total distance. Such calculation is employed in many clustering algorithms presented below.

The dissimilarity $d(x_i,x_j)$ between two instances, containing p attributes of mixed types, is defined as:

$$d(x_i,x_j) = \frac{\sum\limits_{n=1}^{p} \delta_{ij}^{(n)} d_{ij}^{(n)}}{\sum\limits_{n=1}^{p} \delta_{ij}^{(n)}}$$

where the indicator $\delta_{ij}^{(n)}$ =0 if one of the values is missing. The contribution of attribute n to the distance between the two objects $d^{(n)}(x_i,x_j)$ is computed according to its type:

- If the attribute is binary or categorical, $d^{(n)}(x_i,x_j) = 0$ if $x_{in} = x_{jn}$, otherwise $d^{(n)}(x_i,x_j)$=1.
- If the attribute is continuous-valued, $d_{ij}^{(n)} = \frac{|x_{in} - x_{jn}|}{\max_h x_{hn} - \min_h x_{hn}}$, where h runs over all non-missing objects for attribute n.
- If the attribute is ordinal, the standardized values of the attribute are computed first and then, $z_{i,n}$ is treated as continuous-valued.

14.3 Similarity Functions

An alternative concept to that of the distance is the similarity function $s(x_i,x_j)$ that compares the two vectors x_i and x_j (Duda *et al.*, 2001). This function should be symmetrical (namely $s(x_i,x_j) = s(x_j,x_i)$) and have a large value when x_i and x_j are somehow "similar" and constitute the largest value for identical vectors.

A similarity function where the target range is [0,1] is called a dichotomous similarity function. In fact, the methods described in the previous sections for calculating the "distances" in the case of binary and nominal attributes may be considered as similarity functions, rather than distances.

14.3.1 Cosine Measure

When the angle between the two vectors is a meaningful measure of their similarity, the normalized inner product may be an appropriate similarity measure:

$$s(x_i, x_j) = \frac{x_i^T \cdot x_j}{||x_i|| \cdot ||x_j||}$$

14.3.2 Pearson Correlation Measure

The normalized Pearson correlation is defined as:

$$s(x_i, x_j) = \frac{(x_i - \bar{x}_i)^T \cdot (x_j - \bar{x}_j)}{||x_i - \bar{x}_i|| \cdot ||x_j - \bar{x}_j||}$$

where $\bar{x}_i$ denotes the average feature value of x over all dimensions.

14.3.3 Extended Jaccard Measure

The extended Jaccard measure was presented by (Strehl and Ghosh, 2000) and it is defined as:

$$s(x_i, x_j) = \frac{x_i^T \cdot x_j}{||x_i||^2 + ||x_j||^2 - x_i^T \cdot x_j}$$

14.3.4 Dice Coefficient Measure

The dice coefficient measure is similar to the extended Jaccard measure and it is defined as:

$$s(x_i, x_j) = \frac{2x_i^T \cdot x_j}{||x_i||^2 + ||x_j||^2}$$

14.4 Evaluation Criteria Measures

Evaluating if a certain clustering is good or not is a problematic and controversial issue. In fact Bonner (1964) was the first to argue that there is no universal definition for what is a good clustering. The evaluation remains mostly in the eye of the beholder. Nevertheless, several evaluation criteria have been developed in the literature. These criteria are usually divided into two categories: Internal and External.

14.4.1 Internal Quality Criteria

Internal quality metrics usually measure the compactness of the clusters using some similarity measure. It usually measures the intra-cluster homogeneity, the inter-cluster separability or a combination of these two. It does not use any external information beside the data itself.

Sum of Squared Error (SSE)

SSE is the simplest and most widely used criterion measure for clustering. It is calculated as:

$$SSE = \sum_{k=1}^{K} \sum_{\forall x_i \in C_k} \|x_i - \mu_k\|^2$$

where C_k is the set of instances in cluster k; μ_k is the vector mean of cluster k. The components of μ_k are calculated as:

$$\mu_{k,j} = \frac{1}{N_k} \sum_{\forall x_i \in C_k} x_{i,j}$$

where $N_k = |C_k|$ is the number of instances belonging to cluster k.

Clustering methods that minimize the SSE criterion are often called minimum variance partitions, since by simple algebraic manipulation the SSE criterion may be written as:

$$SSE = \frac{1}{2} \sum_{k=1}^{K} N_k \bar{S}_k$$

where:

$$\bar{S}_k = \frac{1}{N_k^2} \sum_{x_i, x_j \in C_k} \|x_i - x_j\|^2$$

(C_k=cluster k)

The SSE criterion function is suitable for cases in which the clusters form compact clouds that are well separated from one another (Duda *et al.*, 2001).

Other Minimum Variance Criteria

Additional minimum criteria to SSE may be produced by replacing the value of S_k with expressions such as:

$$\bar{S}_k = \frac{1}{N_k^2} \sum_{x_i, x_j \in C_k} s(x_i, x_j)$$

or:

$$\bar{S}_k = \min_{x_i, x_j \in C_k} s(x_i, x_j)$$

Scatter Criteria

The scalar scatter criteria are derived from the scatter matrices, reflecting the within-cluster scatter, the between-cluster scatter and their summation — the total scatter matrix. For the kth cluster, the scatter matrix may be calculated as:

$$S_k = \sum_{x \in C_k} (x - \mu_k)(x - \mu_k)^T$$

The within-cluster scatter matrix is calculated as the summation of the last definition over all clusters:

$$S_W = \sum_{k=1}^{K} S_k$$

The between-cluster scatter matrix may be calculated as:

$$S_B = \sum_{k=1}^{K} N_k (\mu_k - \mu)(\mu_k - \mu)^T$$

where μ is the total mean vector and is defined as:

$$\mu = \frac{1}{m} \sum_{k=1}^{K} N_k \mu_k$$

The total scatter matrix should be calculated as:

$$S_T = \sum_{x \in C_1, C_2, \ldots, C_K} (x - \mu)(x - \mu)^T$$

Three scalar criteria may be derived from S_W, S_B and S_T:

- **The trace criterion** — the sum of the diagonal elements of a matrix. Minimizing the trace of S_W is similar to minimizing SSE and is therefore acceptable. This criterion, representing the within-cluster scatter, is calculated as:

  $$J_e = tr[S_W] = \sum_{k=1}^{K} \sum_{x \in C_k} \|x - \mu_k\|^2$$

  Another criterion, which may be maximized, is the between cluster criterion:

  $$tr[S_B] = \sum_{k=1}^{K} N_k \|\mu_k - \mu\|^2$$

- **The determinant criterion** — the determinant of a scatter matrix roughly measures the square of the scattering volume. Since S_B will be singular if the number of clusters is less than or equal to the dimensionality, or if $m - c$ is less than the dimensionality, its determinant is not an appropriate criterion. If we assume that SW is nonsingular, the determinant criterion function using this matrix may be employed:

$$J_d = |S_W| = \left| \sum_{k=1}^{K} S_k \right|$$

- **The invariant criterion** — the eigenvalues $\lambda_1, \lambda_2, \ldots, \lambda_d$ of

$$S_W^{-1} S_B$$

are the basic linear invariants of the scatter matrices. Good partitions are ones for which the nonzero eigenvalues are large. As a result, several criteria may be derived including the eigenvalues. Three such criteria are:

1. $tr[S_W^{-1} S_B] = \sum_{i=1}^{d} \lambda_i$
2. $J_f = tr[S_T^{-1} S_W] = \sum_{i=1}^{d} \frac{1}{1+\lambda_i}$
3. $\frac{|S_W|}{|S_T|} = \prod_{i=1}^{d} \frac{1}{1+\lambda_i}$

Condorcet's Criterion

Another appropriate approach is to apply the Condorcet's solution (1785) to the ranking problem (Marcotorchino and Michaud, 1979). In this case the criterion is calculated as following:

$$\sum_{C_i \in C} \sum_{\substack{x_j, x_k \in C_i \\ x_j \neq x_k}} s(x_j, x_k) + \sum_{C_i \in C} \sum_{x_j \in C_i; x_k \notin C_i} d(x_j, x_k)$$

where $s(x_j, x_k)$ and $d(x_j, x_k)$ measure the similarity and distance of the vectors x_j and x_k.

The C-Criterion

The C-criterion (Fortier and Solomon, 1996) is an extension of Condorcet's criterion and is defined as:

$$\sum_{C_i \in C} \sum_{\substack{x_j, x_k \in C_i \\ x_j \neq x_k}} (s(x_j, x_k) - \gamma) + \sum_{C_i \in C} \sum_{x_j \in C_i; x_k \notin C_i} (\gamma - s(x_j, x_k))$$

where γ is a threshold value.

Category Utility Metric

The category utility (Gluck and Corter, 1985) is defined as the increase of the expected number of feature values that can be correctly predicted given a certain clustering. This metric is useful for problems that contain a relatively small number of nominal features each having small cardinality.

Edge Cut Metrics

In some cases it is useful to represent the clustering problem as an edge cut minimization problem. In such instances the quality is measured as the ratio of the remaining edge weights to the total precut edge weights. If there is no restriction on the size of the clusters, finding the optimal value is easy. Thus the min-cut measure is revised to penalize imbalanced structures.

14.4.2 External Quality Criteria

External measures can be useful for examining whether the structure of the clusters match to some predefined classification of the instances.

Mutual Information Based Measure

The mutual information criterion can be used as an external measure for clustering (Strehl *et al.*, 2000). The measure for m instances clustered using $C = \{C_1, \ldots, C_g\}$ and referring to the target attribute y whose domain is $dom(y) = \{c_1, \ldots, c_k\}$ is defined as follows:

$$C = \frac{2}{m} \sum_{l=1}^{g} \sum_{h=1}^{k} m_{l,h} \log_{g \cdot k} \left(\frac{m_{l,h} \cdot m}{m_{.,l} \cdot m_{l,.}} \right)$$

where $m_{l,h}$ indicate the number of instances that are in cluster C_l and also in class c_h. $m_{.,h}$ denotes the total number of instances in the class c_h. Similarly, $m_{l,.}$ indicates the number of instances in cluster C_l.

Precision-Recall Measure

The precision-recall measure from information retrieval can be used as an external measure for evaluating clusters. The cluster is viewed as the results of a query for a specific class. Precision is the fraction of correctly retrieved instances, while recall is the fraction of correctly retrieved instances out of all matching instances. A combined F-measure can be useful for evaluating a clustering structure (Larsen and Aone, 1999).

Rand Index

The Rand index (Rand, 1971) is a simple criterion used to compare an induced clustering structure (C_1) with a given clustering structure (C_2). Let a be the number of pairs of instances that are assigned to the same cluster in C_1 and in the same cluster in C_2; b be the number of pairs of instances that are in the same cluster in C_1, but not in the same cluster in C_2; c be the number of pairs of instances that are in the same cluster in C_2, but not in the same cluster in C_1; and d be the number of pairs of instances that are assigned to different clusters in C_1 and C_2. The quantities a and d

can be interpreted as agreements, and b and c as disagreements. The Rand index is defined as:

$$RAND = \frac{a+d}{a+b+c+d}$$

The Rand index lies between 0 and 1. When the two partitions agree perfectly, the Rand index is 1.

A problem with the Rand index is that its expected value of two random clustering does not take a constant value (such as zero). Hubert and Arabie (1985) suggest an adjusted Rand index that overcomes this disadvantage.

14.5 Clustering Methods

In this section we describe the most well-known clustering algorithms. The main reason for having many clustering methods is the fact that the notion of "cluster" is not precisely defined (Estivill-Castro, 2000). Consequently many clustering methods have been developed, each of which uses a different induction principle. Farley and Raftery (1998) suggest dividing the clustering methods into two main groups: hierarchical and partitioning methods. Han and Kamber (2001) suggest categorizing the methods into additional three main categories: *density-based methods*, *model-based clustering* and *grid-based methods*. An alternative categorization based on the induction principle of the various clustering methods is presented in (Estivill-Castro, 2000).

14.5.1 Hierarchical Methods

These methods construct the clusters by recursively partitioning the instances in either a top-down or bottom-up fashion. These methods can be sub-divided as following:

- Agglomerative hierarchical clustering — Each object initially represents a cluster of its own. Then clusters are successively merged until the desired cluster structure is obtained.
- Divisive hierarchical clustering — All objects initially belong to one cluster. Then the cluster is divided into sub-clusters, which are successively divided into their own sub-clusters. This process continues until the desired cluster structure is obtained.

The result of the hierarchical methods is a dendrogram, representing the nested grouping of objects and similarity levels at which groupings change. A clustering of the data objects is obtained by cutting the dendrogram at the desired similarity level.

The merging or division of clusters is performed according to some similarity measure, chosen so as to optimize some criterion (such as a sum of squares). The hierarchical clustering methods could be further divided according to the manner that the similarity measure is calculated (Jain *et al.*, 1999):

- **Single-link clustering** (also called the connectedness, the minimum method or the nearest neighbor method) — methods that consider the distance between two clusters to be equal to the shortest distance from any member of one cluster to any member of the other cluster. If the data consist of similarities, the similarity between a pair of clusters is considered to be equal to the greatest similarity from any member of one cluster to any member of the other cluster (Sneath and Sokal, 1973).
- **Complete-link clustering** (also called the diameter, the maximum method or the furthest neighbor method) - methods that consider the distance between two clusters to be equal to the longest distance from any member of one cluster to any member of the other cluster (King, 1967).
- **Average-link clustering** (also called minimum variance method) - methods that consider the distance between two clusters to be equal to the average distance from any member of one cluster to any member of the other cluster. Such clustering algorithms may be found in (Ward, 1963) and (Murtagh, 1984).

The disadvantages of the single-link clustering and the average-link clustering can be summarized as follows (Guha *et al.*, 1998):

- Single-link clustering has a drawback known as the "chaining effect": A few points that form a bridge between two clusters cause the single-link clustering to unify these two clusters into one.
- Average-link clustering may cause elongated clusters to split and for portions of neighboring elongated clusters to merge.

The complete-link clustering methods usually produce more compact clusters and more useful hierarchies than the single-link clustering methods, yet the single-link methods are more versatile. Generally, hierarchical methods are characterized with the following strengths:

- Versatility — The single-link methods, for example, maintain good performance on data sets containing non-isotropic clusters, including well-separated, chain-like and concentric clusters.
- Multiple partitions — hierarchical methods produce not one partition, but multiple nested partitions, which allow different users to choose different partitions, according to the desired similarity level. The hierarchical partition is presented using the dendrogram.

The main disadvantages of the hierarchical methods are:

- Inability to scale well — The time complexity of hierarchical algorithms is at least $O(m^2)$ (where m is the total number of instances), which is non-linear with the number of objects. Clustering a large number of objects using a hierarchical algorithm is also characterized by huge I/O costs.
- Hierarchical methods can never undo what was done previously. Namely there is no back-tracking capability.

14.5.2 Partitioning Methods

Partitioning methods relocate instances by moving them from one cluster to another, starting from an initial partitioning. Such methods typically require that the number of clusters will be pre-set by the user. To achieve global optimality in partitioned-based clustering, an exhaustive enumeration process of all possible partitions is required. Because this is not feasible, certain greedy heuristics are used in the form of iterative optimization. Namely, a relocation method iteratively relocates points between the k clusters. The following subsections present various types of partitioning methods.

Error Minimization Algorithms

These algorithms, which tend to work well with isolated and compact clusters, are the most intuitive and frequently used methods. The basic idea is to find a clustering structure that minimizes a certain error criterion which measures the "distance" of each instance to its representative value. The most well-known criterion is the Sum of Squared Error (SSE), which measures the total squared Euclidian distance of instances to their representative values. SSE may be globally optimized by exhaustively enumerating all partitions, which is very time-consuming, or by giving an approximate solution (not necessarily leading to a global minimum) using heuristics. The latter option is the most common alternative.

The simplest and most commonly used algorithm, employing a squared error criterion is the K-means algorithm. This algorithm partitions the data into K clusters $(C_1, C_2, \ldots, C_K)$, represented by their centers or means. The center of each cluster is calculated as the mean of all the instances belonging to that cluster.

Figure 14.1 presents the pseudo-code of the K-means algorithm. The algorithm starts with an initial set of cluster centers, chosen at random or according to some heuristic procedure. In each iteration, each instance is assigned to its nearest cluster center according to the Euclidean distance between the two. Then the cluster centers are re-calculated.

The center of each cluster is calculated as the mean of all the instances belonging to that cluster:

$$\mu_k = \frac{1}{N_k} \sum_{q=1}^{N_k} x_q$$

where N_k is the number of instances belonging to cluster k and μ_k is the mean of the cluster k.

A number of convergence conditions are possible. For example, the search may stop when the partitioning error is not reduced by the relocation of the centers. This indicates that the present partition is locally optimal. Other stopping criteria can be used also such as exceeding a pre-defined number of iterations.

The K-means algorithm may be viewed as a gradient-decent procedure, which begins with an initial set of K cluster-centers and iteratively updates it so as to decrease the error function.

Input: S (instance set), K (number of cluster)
Output: clusters
1: Initialize K cluster centers.
2: **while** termination condition is not satisfied **do**
3: Assign instances to the closest cluster center.
4: Update cluster centers based on the assignment.
5: **end while**

Fig. 14.1. K-means Algorithm.

A rigorous proof of the finite convergence of the K-means type algorithms is given in (Selim and Ismail, 1984). The complexity of T iterations of the K-means algorithm performed on a sample size of m instances, each characterized by N attributes, is: $O(T * K * m * N)$.

This linear complexity is one of the reasons for the popularity of the K-means algorithms. Even if the number of instances is substantially large (which often is the case nowadays), this algorithm is computationally attractive. Thus, the K-means algorithm has an advantage in comparison to other clustering methods (e.g. hierarchical clustering methods), which have non-linear complexity.

Other reasons for the algorithm's popularity are its ease of interpretation, simplicity of implementation, speed of convergence and adaptability to sparse data (Dhillon and Modha, 2001).

The Achilles heel of the K-means algorithm involves the selection of the initial partition. The algorithm is very sensitive to this selection, which may make the difference between global and local minimum.

Being a typical partitioning algorithm, the K-means algorithm works well only on data sets having isotropic clusters, and is not as versatile as single link algorithms, for instance.

In addition, this algorithm is sensitive to noisy data and outliers (a single outlier can increase the squared error dramatically); it is applicable only when mean is defined (namely, for numeric attributes);and it requires the number of clusters in advance, which is not trivial when no prior knowledge is available.

The use of the K-means algorithm is often limited to numeric attributes. Haung (1998) presented the K-prototypes algorithm, which is based on the K-means algorithm but removes numeric data limitations while preserving its efficiency. The algorithm clusters objects with numeric and categorical attributes in a way similar to the K-means algorithm. The similarity measure on numeric attributes is the square Euclidean distance; the similarity measure on the categorical attributes is the number of mismatches between objects and the cluster prototypes.

Another partitioning algorithm, which attempts to minimize the SSE is the K-medoids or PAM (partition around medoids — (Kaufmann and Rousseeuw, 1987)). This algorithm is very similar to the K-means algorithm. It differs from the latter mainly in its representation of the different clusters. Each cluster is represented by the most centric object in the cluster, rather than by the implicit mean that may not belong to the cluster.

The K-medoids method is more robust than the K-means algorithm in the presence of noise and outliers because a medoid is less influenced by outliers or other extreme values than a mean. However, its processing is more costly than the K-means method. Both methods require the user to specify K, the number of clusters.

Other error criteria can be used instead of the SSE. Estivill-Castro (2000) analyzed the total absolute error criterion. Namely, instead of summing up the squared error, he suggests to summing up the absolute error. While this criterion is superior in regard to robustness, it requires more computational effort.

Graph-Theoretic Clustering

Graph theoretic methods are methods that produce clusters via graphs. The edges of the graph connect the instances represented as nodes. A well-known graph-theoretic algorithm is based on the Minimal Spanning Tree — MST (Zahn, 1971). Inconsistent edges are edges whose weight (in the case of clustering-length) is significantly larger than the average of nearby edge lengths. Another graph-theoretic approach constructs graphs based on limited neighborhood sets (Urquhart, 1982).

There is also a relation between hierarchical methods and graph theoretic clustering:

- Single-link clusters are subgraphs of the MST of the data instances. Each subgraph is a *connected component*, namely a set of instances in which each instance is connected to at least one other member of the set, so that the set is maximal with respect to this property. These subgraphs are formed according to some similarity threshold.
- Complete-link clusters are *maximal complete subgraphs*, formed using a similarity threshold. A maximal complete subgraph is a subgraph such that each node is connected to every other node in the subgraph and the set is maximal with respect to this property.

14.5.3 Density-based Methods

Density-based methods assume that the points that belong to each cluster are drawn from a specific probability distribution (Banfield and Raftery, 1993). The overall distribution of the data is assumed to be a mixture of several distributions.

The aim of these methods is to identify the clusters and their distribution parameters. These methods are designed for discovering clusters of arbitrary shape which are not necessarily convex, namely:

$$x_i, x_j \in C_k$$

This does not necessarily imply that:

$$\alpha \cdot x_i + (1-\alpha) \cdot x_j \in C_k$$

The idea is to continue growing the given cluster as long as the density (number of objects or data points) in the neighborhood exceeds some threshold. Namely, the neighborhood of a given radius has to contain at least a minimum number of objects. When each cluster is characterized by local mode or maxima of the density function, these methods are called mode-seeking

Much work in this field has been based on the underlying assumption that the component densities are multivariate Gaussian (in case of numeric data) or multinominal (in case of nominal data).

An acceptable solution in this case is to use the maximum likelihood principle. According to this principle, one should choose the clustering structure and parameters such that the probability of the data being generated by such clustering structure and parameters is maximized. The expectation maximization algorithm — EM — (Dempster *et al.*, 1977), which is a general-purpose maximum likelihood algorithm for missing-data problems, has been applied to the problem of parameter estimation. This algorithm begins with an initial estimate of the parameter vector and then alternates between two steps (Farley and Raftery, 1998): an "E-step", in which the conditional expectation of the complete data likelihood given the observed data and the current parameter estimates is computed, and an "M-step", in which parameters that maximize the expected likelihood from the E-step are determined. This algorithm was shown to converge to a local maximum of the observed data likelihood.

The K-means algorithm may be viewed as a degenerate EM algorithm, in which:

$$p(k/x) = \begin{cases} 1 & k = \underset{k}{\operatorname{argmax}}\{\hat{p}(k/x)\} \\ 0 & \text{otherwise} \end{cases}$$

Assigning instances to clusters in the K-means may be considered as the E-step; computing new cluster centers may be regarded as the M-step.

The DBSCAN algorithm (density-based spatial clustering of applications with noise) discovers clusters of arbitrary shapes and is efficient for large spatial databases. The algorithm searches for clusters by searching the neighborhood of each object in the database and checks if it contains more than the minimum number of objects (Ester *et al.*, 1996).

AUTOCLASS is a widely-used algorithm that covers a broad variety of distributions, including Gaussian, Bernoulli, Poisson, and log-normal distributions (Cheeseman and Stutz, 1996). Other well-known density-based methods include: SNOB (Wallace and Dowe, 1994) and MCLUST (Farley and Raftery, 1998).

Density-based clustering may also employ nonparametric methods, such as searching for bins with large counts in a multidimensional histogram of the input instance space (Jain *et al.*, 1999).

14.5.4 Model-based Clustering Methods

These methods attempt to optimize the fit between the given data and some mathematical models. Unlike conventional clustering, which identifies groups of objects, model-based clustering methods also find characteristic descriptions for each group, where each group represents a concept or class. The most frequently used induction methods are decision trees and neural networks.

Decision Trees

In decision trees, the data is represented by a hierarchical tree, where each leaf refers to a concept and contains a probabilistic description of that concept. Several algorithms produce classification trees for representing the unlabelled data. The most well-known algorithms are:

COBWEB — This algorithm assumes that all attributes are independent (an often too naive assumption). Its aim is to achieve high predictability of nominal variable values, given a cluster. This algorithm is not suitable for clustering large database data (Fisher, 1987). CLASSIT, an extension of COBWEB for continuous-valued data, unfortunately has similar problems as the COBWEB algorithm.

Neural Networks

This type of algorithm represents each cluster by a neuron or "prototype". The input data is also represented by neurons, which are connected to the prototype neurons. Each such connection has a weight, which is learned adaptively during learning.

A very popular neural algorithm for clustering is the self-organizing map (SOM). This algorithm constructs a single-layered network. The learning process takes place in a "winner-takes-all" fashion:

- The prototype neurons compete for the current instance. The winner is the neuron whose weight vector is closest to the instance currently presented.
- The winner and its neighbors learn by having their weights adjusted.

The SOM algorithm is successfully used for vector quantization and speech recognition. It is useful for visualizing high-dimensional data in 2D or 3D space. However, it is sensitive to the initial selection of weight vector, as well as to its different parameters, such as the learning rate and neighborhood radius.

14.5.5 Grid-based Methods

These methods partition the space into a finite number of cells that form a grid structure on which all of the operations for clustering are performed. The main advantage of the approach is its fast processing time (Han and Kamber, 2001).

14.5.6 Soft-computing Methods

Section 14.5.4 described the usage of neural networks in clustering tasks. This section further discusses the important usefulness of other soft-computing methods in clustering tasks.

Fuzzy Clustering

Traditional clustering approaches generate partitions; in a partition, each instance belongs to one and only one cluster. Hence, the clusters in a hard clustering are disjointed. Fuzzy clustering (see for instance (Hoppner, 2005)) extends this notion and suggests a *soft clustering* schema. In this case, each pattern is associated with every cluster using some sort of membership function, namely, each cluster is a fuzzy set of all the patterns. Larger membership values indicate higher confidence in the assignment of the pattern to the cluster. A hard clustering can be obtained from a fuzzy partition by using a threshold of the membership value.

The most popular fuzzy clustering algorithm is the fuzzy c-means (FCM) algorithm. Even though it is better than the hard K-means algorithm at avoiding local minima, FCM can still converge to local minima of the squared error criterion. The design of membership functions is the most important problem in fuzzy clustering; different choices include those based on similarity decomposition and centroids of clusters. A generalization of the FCM algorithm has been proposed through a family of objective functions. A fuzzy c-shell algorithm and an adaptive variant for detecting circular and elliptical boundaries have been presented.

Evolutionary Approaches for Clustering

Evolutionary techniques are stochastic general purpose methods for solving optimization problems. Since clustering problem can be defined as an optimization problem, evolutionary approaches may be appropriate here. The idea is to use evolutionary operators and a population of clustering structures to converge into a globally optimal clustering. Candidate clustering are encoded as chromosomes. The most commonly used evolutionary operators are: selection, recombination, and mutation. A fitness function evaluated on a chromosome determines a chromosome's likelihood of surviving into the next generation. The most frequently used evolutionary technique in clustering problems is genetic algorithms (GAs). Figure 14.2 presents a high-level pseudo-code of a typical GA for clustering. A fitness value is associated with each clusters structure. A higher fitness value indicates a better cluster structure. A suitable fitness function is the inverse of the squared error value. Cluster structures with a small squared error will have a larger fitness value.

Input: S (instance set), K (number of clusters), n (population size)
Output: clusters

1: Randomly create a *population* of n structures, each corresponds to a valid K-clusters of the data.
2: **repeat**
3: Associate a fitness value $\forall structure \in population$.
4: Regenerate a new generation of structures.
5: **until** some termination condition is satisfied

Fig. 14.2. GA for Clustering.

The most obvious way to represent structures is to use strings of length m (where m is the number of instances in the given set). The i-th entry of the string denotes the cluster to which the i-th instance belongs. Consequently, each entry can have values from 1 to K. An improved representation scheme is proposed where an additional separator symbol is used along with the pattern labels to represent a partition. Using this representation permits them to map the clustering problem into a permutation problem such as the travelling salesman problem, which can be solved by using the permutation crossover operators. This solution also suffers from permutation redundancy.

In GAs, a selection operator propagates solutions from the current generation to the next generation based on their fitness. Selection employs a probabilistic scheme so that solutions with higher fitness have a higher probability of getting reproduced.

There are a variety of recombination operators in use; *crossover* is the most popular. Crossover takes as input a pair of chromosomes (called parents) and outputs a new pair of chromosomes (called children or offspring). In this way the GS explores the search space. Mutation is used to make sure that the algorithm is not trapped in local optimum.

More recently investigated is the use of edge-based crossover to solve the clustering problem. Here, all patterns in a cluster are assumed to form a complete graph by connecting them with edges. Offspring are generated from the parents so that they inherit the edges from their parents. In a hybrid approach that has been proposed, the GAs is used only to find good initial cluster centers and the K-means algorithm is applied to find the final partition. This hybrid approach performed better than the GAs.

A major problem with GAs is their sensitivity to the selection of various parameters such as population size, crossover and mutation probabilities, etc. Several researchers have studied this problem and suggested guidelines for selecting these control parameters. However, these guidelines may not yield good results on specific problems like pattern clustering. It was reported that hybrid genetic algorithms incorporating problem-specific heuristics are good for clustering. A similar claim is made about the applicability of GAs to other practical problems. Another issue with GAs is the selection of an appropriate representation which is low in order and short in defining length.

There are other evolutionary techniques such as evolution strategies (ESs), and evolutionary programming (EP). These techniques differ from the GAs in solution representation and the type of mutation operator used; EP does not use a recombination operator, but only selection and mutation. Each of these three approaches has been used to solve the clustering problem by viewing it as a minimization of the squared error criterion. Some of the theoretical issues, such as the convergence of these approaches, were studied. GAs perform a globalized search for solutions whereas most other clustering procedures perform a localized search. In a localized search, the solution obtained at the 'next iteration' of the procedure is in the vicinity of the current solution. In this sense, the K-means algorithm and fuzzy clustering algorithms are all localized search techniques. In the case of GAs, the crossover and mutation operators can produce new solutions that are completely different from the current ones.

It is possible to search for the optimal location of the centroids rather than finding the optimal partition. This idea permits the use of ESs and EP, because centroids can be coded easily in both these approaches, as they support the direct representation of a solution as a real-valued vector. ESs were used on both hard and fuzzy clustering problems and EP has been used to evolve fuzzy min-max clusters. It has been observed that they perform better than their classical counterparts, the K-means algorithm and the fuzzy c-means algorithm. However, all of these approaches are over sensitive to their parameters. Consequently, for each specific problem, the user is required to tune the parameter values to suit the application.

Simulated Annealing for Clustering

Another general-purpose stochastic search technique that can be used for clustering is simulated annealing (SA), which is a sequential stochastic search technique designed to avoid local optima. This is accomplished by accepting with some probability a new solution for the next iteration of lower quality (as measured by the criterion function). The probability of acceptance is governed by a critical parameter called the temperature (by analogy with annealing in metals), which is typically specified in terms of a starting (first iteration) and final temperature value. Selim and Al-Sultan (1991) studied the effects of control parameters on the performance of the algorithm. SA is statistically guaranteed to find the global optimal solution. Figure 14.3 presents a high-level pseudo-code of the SA algorithm for clustering.

Input: S (instance set), K (number of clusters), T_0 (initial temperature), T_f (final temperature), c (temperature reducing constant)
Output: clusters
1: Randomly select p_0 which is a K-partition of S. Compute the squared error value $E(p_0)$.
2: **while** $T_0 > T_f$ **do**
3: Select a neighbor p_1 of the last partition p_0.
4: **if** $E(p_1) > E(p_0)$ **then**
5: $p_0 \leftarrow p_1$ with a probability that depends on T_0
6: **else**
7: $p_0 \leftarrow p_1$
8: **end if**
9: $T_0 \leftarrow c * T_0$
10: **end while**

Fig. 14.3. Clustering Based on Simulated Annealing.

The SA algorithm can be slow in reaching the optimal solution, because optimal results require the temperature to be decreased very slowly from iteration to iteration. Tabu search, like SA, is a method designed to cross boundaries of feasibility or local optimality and to systematically impose and release constraints to permit exploration of otherwise forbidden regions. Al-Sultan (1995) suggests using Tabu search as an alternative to SA.

14.5.7 Which Technique To Use?

An empirical study of K-means, SA, TS, and GA was presented by Al-Sultan and Khan (1996). TS, GA and SA were judged comparable in terms of solution quality, and all were better than K-means. However, the K-means method is the most efficient in terms of execution time; other schemes took more time (by a factor of 500 to 2500) to partition a data set of size 60 into 5 clusters. Furthermore, GA obtained the best solution faster than TS and SA; SA took more time than TS to reach the best clustering. However, GA took the maximum time for convergence, that is, to obtain a population of only the best solutions, TS and SA followed.

An additional empirical study has compared the performance of the following clustering algorithms: SA, GA, TS, randomized branch-and-bound (RBA), and hybrid search (HS) (Mishra and Raghavan, 1994). The conclusion was that GA performs well in the case of one-dimensional data, while its performance on high dimensional data sets is unimpressive. The convergence pace of SA is too slow; RBA and TS performed best; and HS is good for high dimensional data. However, none of the methods was found to be superior to others by a significant margin.

It is important to note that both Mishra and Raghavan (1994) and Al-Sultan and Khan (1996) have used relatively small data sets in their experimental studies.

In summary, only the K-means algorithm and its ANN equivalent, the Kohonen net, have been applied on large data sets; other approaches have been tested, typically, on small data sets. This is because obtaining suitable learning/control parameters for ANNs, GAs, TS, and SA is difficult and their execution times are very high for large data sets. However, it has been shown that the K-means method converges to a locally optimal solution. This behavior is linked with the initial seed election in the K-means algorithm. Therefore, if a good initial partition can be obtained quickly using any of the other techniques, then K-means would work well, even on problems with large data sets. Even though various methods discussed in this section are comparatively weak, it was revealed, through experimental studies, that combining domain knowledge would improve their performance. For example, ANNs work better in classifying images represented using extracted features rather than with raw images, and hybrid classifiers work better than ANNs. Similarly, using domain knowledge to hybridize a GA improves its performance. Therefore it may be useful in general to use domain knowledge along with approaches like GA, SA, ANN, and TS. However, these approaches (specifically, the criteria functions used in them) have a tendency to generate a partition of hyperspherical clusters, and this could be a limitation. For example, in cluster-based document retrieval, it was observed that the hierarchical algorithms performed better than the partitioning algorithms.

14.6 Clustering Large Data Sets

There are several applications where it is necessary to cluster a large collection of patterns. The definition of 'large' is vague. In document retrieval, millions of instances with a dimensionality of more than 100 have to be clustered to achieve data abstraction. A majority of the approaches and algorithms proposed in the literature cannot handle such large data sets. Approaches based on genetic algorithms, tabu search and simulated annealing are optimization techniques and are restricted to reasonably small data sets. Implementations of conceptual clustering optimize some criterion functions and are typically computationally expensive.

The convergent K-means algorithm and its ANN equivalent, the Kohonen net, have been used to cluster large data sets. The reasons behind the popularity of the K-means algorithm are:

1. Its time complexity is $O(mkl)$, where m is the number of instances; k is the number of clusters; and l is the number of iterations taken by the algorithm to converge. Typically, k and l are fixed in advance and so the algorithm has linear time complexity in the size of the data set.
2. Its space complexity is $O(k+m)$. It requires additional space to store the data matrix. It is possible to store the data matrix in a secondary memory and access each pattern based on need. However, this scheme requires a huge access time because of the iterative nature of the algorithm. As a consequence, processing time increases enormously.
3. It is order-independent. For a given initial seed set of cluster centers, it generates the same partition of the data irrespective of the order in which the patterns are presented to the algorithm.

However, the K-means algorithm is sensitive to initial seed selection and even in the best case, it can produce only hyperspherical clusters. Hierarchical algorithms are more versatile. But they have the following disadvantages:

1. The time complexity of hierarchical agglomerative algorithms is $O(m^2 * \log m)$.
2. The space complexity of agglomerative algorithms is $O(m^2)$. This is because a similarity matrix of size m^2 has to be stored. It is possible to compute the entries of this matrix based on need instead of storing them.

A possible solution to the problem of clustering large data sets while only marginally sacrificing the versatility of clusters is to implement more efficient variants of clustering algorithms. A hybrid approach was used, where a set of reference points is chosen as in the K-means algorithm, and each of the remaining data points is assigned to one or more reference points or clusters. Minimal spanning trees (MST) are separately obtained for each group of points. These MSTs are merged to form an approximate global MST. This approach computes only similarities between a fraction of all possible pairs of points. It was shown that the number of similarities computed for 10,000 instances using this approach is the same as the total number of pairs of points in a collection of 2,000 points. Bentley and Friedman (1978) presents an algorithm that can compute an approximate MST in $O(m \log m)$ time. A scheme to generate an approximate dendrogram incrementally in $O(n \log n)$ time was presented.

CLARANS (Clustering Large Applications based on RANdom Search) have been developed by Ng and Han (1994). This method identifies candidate cluster centroids by using repeated random samples of the original data. Because of the use of random sampling, the time complexity is $O(n)$ for a pattern set of n elements.

The BIRCH algorithm (Balanced Iterative Reducing and Clustering) stores summary information about candidate clusters in a dynamic tree data structure. This tree hierarchically organizes the clusters represented at the leaf nodes. The tree can be rebuilt when a threshold specifying cluster size is updated manually, or when memory constraints force a change in this threshold. This algorithm has a time complexity linear in the number of instances.

All algorithms presented till this point assume that the entire dataset can be accommodated in the main memory. However, there are cases in which this assumption is untrue. The following sub-sections describe three current approaches to solve this problem.

14.6.1 Decomposition Approach

The dataset can be stored in a secondary memory (i.e. hard disk) and subsets of this data clustered independently, followed by a merging step to yield a clustering of the entire dataset.

Initially, the data is decomposed into number of subsets. Each subset is sent to the main memory in turn where it is clustered into k clusters using a standard algorithm.

In order to join the various clustering structures obtained from each subset, a representative sample from each cluster of each structure is stored in the main memory. Then these representative instances are further clustered into k clusters and the cluster labels of these representative instances are used to re-label the original dataset. It is possible to extend this algorithm to any number of iterations; more levels are required if the data set is very large and the main memory size is very small.

14.6.2 Incremental Clustering

Incremental clustering is based on the assumption that it is possible to consider instances one at a time and assign them to existing clusters. Here, a new instance is assigned to a cluster without significantly affecting the existing clusters. Only the cluster representations are stored in the main memory to alleviate the space limitations.

Figure 14.4 presents a high level pseudo-code of a typical incremental clustering algorithm.

The major advantage with incremental clustering algorithms is that it is not necessary to store the entire dataset in the memory. Therefore, the space and time requirements of incremental algorithms are very small. There are several incremental clustering algorithms:

Input: S (instances set), K (number of clusters), $Threshold$ (for assigning an instance to a cluster)
Output: clusters
1: $Clusters \leftarrow \emptyset$
2: **for all** $x_i \in S$ **do**
3: $As\_F = false$
4: **for all** $Cluster \in Clusters$ **do**
5: **if** $\|x_i - centroid(Cluster)\| < threshold$ **then**
6: $Update\, centroid(Cluster)$
7: $ins\_counter(Cluster) + +$
8: $As\_F = true$
9: Exit loop
10: **end if**
11: **end for**
12: **if** not(As_F) **then**
13: $centroid(newCluster) = x_i$
14: $ins\_counter(newCluster) = 1$
15: $Clusters \leftarrow Clusters \cup newCluster$
16: **end if**
17: **end for**

Fig. 14.4. An Incremental Clustering Algorithm.

1. The leading clustering algorithm is the simplest in terms of time complexity which is O(mk). It has gained popularity because of its neural network implementation, the ART network, and is very easy to implement as it requires only $O(k)$ space.
2. The shortest spanning path (SSP) algorithm, as originally proposed for data reorganization, was successfully used in automatic auditing of records. Here, the SSP algorithm was used to cluster 2000 patterns using 18 features. These clusters are used to estimate missing feature values in data items and to identify erroneous feature values.
3. The *COBWEB* system is an incremental conceptual clustering algorithm. It has been successfully used in engineering applications.
4. An incremental clustering algorithm for dynamic information processing was presented in (Can, 1993). The motivation behind this work is that in dynamic databases items might get added and deleted over time. These changes should be reflected in the partition generated without significantly affecting the current clusters. This algorithm was used to cluster incrementally an INSPEC database of 12,684 documents relating to computer science and electrical engineering.

Order-independence is an important property of clustering algorithms. An algorithm is *order-independent* if it generates the same partition for any order in which the data is presented, otherwise, it is *order-dependent*. Most of the incremental algorithms presented above are order-dependent. For instance the SSP algorithm and cobweb are order-dependent.

14.6.3 Parallel Implementation

Recent work demonstrates that a combination of algorithmic enhancements to a clustering algorithm and distribution of the computations over a network of workstations can allow a large dataset to be clustered in a few minutes. Depending on the clustering algorithm in use, parallelization of the code and replication of data for efficiency may yield large benefits. However, a global shared data structure, namely the cluster membership table, remains and must be managed centrally or replicated and synchronized periodically. The presence or absence of robust, efficient parallel clustering techniques will determine the success or failure of cluster analysis in large-scale data mining applications in the future.

14.7 Determining the Number of Clusters

As mentioned above, many clustering algorithms require that the number of clusters will be pre-set by the user. It is well-known that this parameter affects the performance of the algorithm significantly. This poses a serious question as to which K should be chosen when prior knowledge regarding the cluster quantity is unavailable.

Note that most of the criteria that have been used to lead the construction of the clusters (such as SSE) are monotonically decreasing in K. Therefore using these criteria for determining the number of clusters results with a trivial clustering, in which each cluster contains one instance. Consequently, different criteria must be applied here. Many methods have been presented to determine which K is preferable. These methods are usually heuristics, involving the calculation of clustering criteria measures for different values of K, thus making it possible to evaluate which K was preferable.

14.7.1 Methods Based on Intra-Cluster Scatter

Many of the methods for determining K are based on the intra-cluster (within-cluster) scatter. This category includes the within-cluster depression-decay (Tibshirani, 1996, Wang and Yu, 2001), which computes an error measure W_K, for each K chosen, as follows:

$$W_K = \sum_{k=1}^{K} \frac{1}{2N_k} D_k$$

where D_k is the sum of pairwise distances for all instances in cluster k:

$$D_k = \sum_{x_i, x_j \in Ck} \left\| x_i - x_j \right\|$$

In general, as the number of clusters increases, the within-cluster decay first declines rapidly. From a certain K, the curve flattens. This value is considered the appropriate K according to this method.

Other heuristics relate to the intra-cluster distance as the sum of squared Euclidean distances between the data instances and their cluster centers (the sum of square errors which the algorithm attempts to minimize). They range from simple methods, such as the PRE method, to more sophisticated, statistic-based methods.

An example of a simple method which works well in most databases is, as mentioned above, the proportional reduction in error (PRE) method. PRE is the ratio of reduction in the sum of squares to the previous sum of squares when comparing the results of using $K+1$ clusters to the results of using K clusters. Increasing the number of clusters by 1 is justified for PRE rates of about 0.4 or larger.

It is also possible to examine the SSE decay, which behaves similarly to the within cluster depression described above. The manner of determining K according to both measures is also similar.

An approximate F statistic can be used to test the significance of the reduction in the sum of squares as we increase the number of clusters (Hartigan, 1975). The method obtains this F statistic as follows:

Suppose that $P(m,k)$ is the partition of m instances into k clusters, and $P(m,k+1)$ is obtained from $P(m,k)$ by splitting one of the clusters. Also assume that the clusters are selected without regard to $x_{qi} \sim N(\mu_i, \sigma^2)$ independently over all q and i. Then the overall mean square ratio is calculated and distributed as follows:

$$R = \left(\frac{e(P(m,k)}{e(P(m,k+1)} - 1 \right) (m-k-1) \approx F_{N,N(m-k-1)}$$

where $e(P(m,k))$ is the sum of squared Euclidean distances between the data instances and their cluster centers.

In fact this F distribution is inaccurate since it is based on inaccurate assumptions:

- K-means is not a hierarchical clustering algorithm, but a relocation method. Therefore, the partition $P(m,k+1)$ is not necessarily obtained by splitting one of the clusters in $P(m,k)$.
- Each x_{qi} influences the partition.
- The assumptions as to the normal distribution and independence of x_{qi} are not valid in all databases.

Since the F statistic described above is imprecise, Hartigan offers a crude rule of thumb: only large values of the ratio (say, larger than 10) justify increasing the number of partitions from K to $K+1$.

14.7.2 Methods Based on both the Inter- and Intra-Cluster Scatter

All the methods described so far for estimating the number of clusters are quite reasonable. However, they all suffer the same deficiency: None of these methods examines the inter-cluster distances. Thus, if the K-means algorithm partitions an existing distinct cluster in the data into sub-clusters (which is undesired), it is possible that none of the above methods would indicate this situation.

In light of this observation, it may be preferable to minimize the intra-cluster scatter and at the same time maximize the inter-cluster scatter. Ray and Turi (1999), for example, strive for this goal by setting a measure that equals the ratio of intra-cluster scatter and inter-cluster scatter. Minimizing this measure is equivalent to both minimizing the intra-cluster scatter and maximizing the inter-cluster scatter.

Another method for evaluating the "optimal" K using both inter and intra cluster scatter is the validity index method (Kim *et al.*, 2001). There are two appropriate measures:

- MICD — mean intra-cluster distance; defined for the kth cluster as:

$$MD_k = \sum_{x_i \in C_k} \frac{\|x_i - \mu_k\|}{N_k}$$

- ICMD — inter-cluster minimum distance; defined as:

$$d_{\min} = \min_{i \neq j} \|\mu_i - \mu_j\|$$

In order to create cluster validity index, the behavior of these two measures around the real number of clusters (K^*) should be used.

When the data are under-partitioned ($K < K^*$), at least one cluster maintains large MICD. As the partition state moves towards over-partitioned ($K > K^*$), the large MICD abruptly decreases.

The ICMD is large when the data are under-partitioned or optimally partitioned. It becomes very small when the data enters the over-partitioned state, since at least one of the compact clusters is subdivided.

Two additional measure functions may be defined in order to find the under-partitioned and over-partitioned states. These functions depend, among other variables, on the vector of the clusters centers $\mu = [\mu_1, \mu_2, \ldots \mu_K]^T$:

1. Under-partition measure function:

$$v_u(K, \mu; X) = \frac{\sum_{k=1}^{K} MD_k}{K} \qquad 2 \leq K \leq K_{\max}$$

   This function has very small values for $K \geq K^*$ and relatively large values for $K < K^*$. Thus, it helps to determine whether the data is under-partitioned.
2. Over-partition measure function:

$$v_o(K, \mu) = \frac{K}{d_{\min}} \qquad 2 \leq K \leq K_{\max}$$

   This function has very large values for $K \geq K^*$, and relatively small values for $K < K^*$. Thus, it helps to determine whether the data is over-partitioned.

The validity index uses the fact that both functions have small values only at $K = K^*$. The vectors of both partition functions are defined as following:

$$V_u = [v_u(2,\mu;X),\ldots,v_u(K_{\max},\mu;X)]$$

$$V_o = [v_o(2,\mu),\ldots,v_o(K_{\max},\mu)]$$

Before finding the validity index, each element in each vector is normalized to the range [0,1], according to its minimum and maximum values. For instance, for the V_u vector:

$$v_u^*(K,\mu;X) = \frac{v_u(K,\mu;X)}{\max\limits_{K=2,\ldots,K_{\max}} \{v_u(K,\mu;X)\} - \min\limits_{K=2,\ldots,K_{\max}} \{v_u(K,\mu;X)\}}$$

The process of normalization is done the same way for the V_o vector. The validity index vector is calculated as the sum of the two normalized vectors:

$$v_{sv}(K,\mu;X) = v_u^*(K,\mu;X) + v_o^*(K,\mu)$$

Since both partition measure functions have small values only at $K = K^*$, the smallest value of v_{sv} is chosen as the optimal number of clusters.

14.7.3 Criteria Based on Probabilistic

When clustering is performed using a density-based method, the determination of the most suitable number of clusters K becomes a more tractable task as clear probabilistic foundation can be used. The question is whether adding new parameters results in a better way of fitting the data by the model. In Bayesian theory, the likelihood of a model is also affected by the number of parameters which are proportional to K. Suitable criteria that can used here include BIC (Bayesian Information Criterion)! MML (Minimum Message Length) and MDL (Minimum Description Length).

In summary, the methods presented in this chapetr are useful for many application domains, such as: Manufacturing lr18,lr14, Security lr7,l10 and Medicine lr2,lr9, and for many data mining tasks, such as: supervised learning lr6,lr12, lr15, unsupervised learning lr13,lr8,lr5,lr16 and genetic algorithms lr17,lr11,lr1,lr4.

References

A1-Sultan K. S., A tabu search approach to the clustering problem, Pattern Recognition, 28:1443-1451,1995.

Al-Sultan K. S. , Khan M. M. : Computational experience on four algorithms for the hard clustering problem. Pattern Recognition Letters 17(3): 295-308, 1996.

Arbel, R. and Rokach, L., Classifier evaluation under limited resources, Pattern Recognition Letters, 27(14): 1619–1631, 2006, Elsevier.

Averbuch, M. and Karson, T. and Ben-Ami, B. and Maimon, O. and Rokach, L., Context-sensitive medical information retrieval, The 11th World Congress on Medical Informatics (MEDINFO 2004), San Francisco, CA, September 2004, IOS Press, pp. 282–286.

Banfield J. D. and Raftery A. E. . Model-based Gaussian and non-Gaussian clustering. Biometrics, 49:803-821, 1993.

Bentley J. L. and Friedman J. H., Fast algorithms for constructing minimal spanning trees in coordinate spaces. IEEE Transactions on Computers, C-27(2):97-105, February 1978. 275

Bonner, R., On Some Clustering Techniques. IBM journal of research and development, 8:22-32, 1964.

Can F. , Incremental clustering for dynamic information processing, in ACM Transactions on Information Systems, no. 11, pp 143-164, 1993.

Cheeseman P., Stutz J.: Bayesian Classification (AutoClass): Theory and Results. Advances in Knowledge Discovery and Data Mining 1996: 153-180

Cohen S., Rokach L., Maimon O., Decision Tree Instance Space Decomposition with Grouped Gain-Ratio, Information Science, Volume 177, Issue 17, pp. 3592-3612, 2007.

Dhillon I. and Modha D., Concept Decomposition for Large Sparse Text Data Using Clustering. Machine Learning. 42, pp.143-175. (2001).

Dempster A.P., Laird N.M., and Rubin D.B., Maximum likelihood from incomplete data using the EM algorithm. Journal of the Royal Statistical Society, 39(B), 1977.

Duda, P. E. Hart and D. G. Stork, Pattern Classification, Wiley, New York, 2001.

Ester M., Kriegel H.P., Sander S., and Xu X., A density-based algorithm for discovering clusters in large spatial databases with noise. In E. Simoudis, J. Han, and U. Fayyad, editors, Proceedings of the 2nd International Conference on Knowledge Discovery and Data Mining (KDD-96), pages 226-231, Menlo Park, CA, 1996. AAAI, AAAI Press.

Estivill-Castro, V. and Yang, J. A Fast and robust general purpose clustering algorithm. Pacific Rim International Conference on Artificial Intelligence, pp. 208-218, 2000.

Fraley C. and Raftery A.E., "How Many Clusters? Which Clustering Method? Answers Via Model-Based Cluster Analysis", Technical Report No. 329. Department of Statistics University of Washington, 1998.

Fisher, D., 1987, Knowledge acquisition via incremental conceptual clustering, in machine learning 2, pp. 139-172.

Fortier, J.J. and Solomon, H. 1996. Clustering procedures. In proceedings of the Multivariate Analysis, '66, P.R. Krishnaiah (Ed.), pp. 493-506.

Gluck, M. and Corter, J., 1985. Information, uncertainty, and the utility of categories. Proceedings of the Seventh Annual Conference of the Cognitive Science Society (pp. 283-287). Irvine, California: Lawrence Erlbaum Associates.

Guha, S., Rastogi, R. and Shim, K. CURE: An efficient clustering algorithm for large databases. In Proceedings of ACM SIGMOD International Conference on Management of Data, pages 73-84, New York, 1998.

Han, J. and Kamber, M. Data Mining: Concepts and Techniques. Morgan Kaufmann Publishers, 2001.

Hartigan, J. A. Clustering algorithms. John Wiley and Sons., 1975.

Huang, Z., Extensions to the k-means algorithm for clustering large data sets with categorical values. Data Mining and Knowledge Discovery, 2(3), 1998.

Hoppner F. , Klawonn F., Kruse R., Runkler T., Fuzzy Cluster Analysis, Wiley, 2000.

Hubert, L. and Arabie, P., 1985 Comparing partitions. Journal of Classification, 5. 193-218.

Jain, A.K. Murty, M.N. and Flynn, P.J. Data Clustering: A Survey. ACM Computing Surveys, Vol. 31, No. 3, September 1999.

Kaufman, L. and Rousseeuw, P.J., 1987, Clustering by Means of Medoids, In Y. Dodge, editor, Statistical Data Analysis, based on the L1 Norm, pp. 405-416, Elsevier/North Holland, Amsterdam.

Kim, D.J., Park, Y.W. and Park,. A novel validity index for determination of the optimal number of clusters. IEICE Trans. Inf., Vol. E84-D, no.2, 2001, 281-285.

King, B. Step-wise Clustering Procedures, J. Am. Stat. Assoc. 69, pp. 86-101, 1967.

Larsen, B. and Aone, C. 1999. Fast and effective text mining using linear-time document clustering. In Proceedings of the 5th ACM SIGKDD, 16-22, San Diego, CA.

Maimon O., and Rokach, L. Data Mining by Attribute Decomposition with semiconductors manufacturing case study, in Data Mining for Design and Manufacturing: Methods and Applications, D. Braha (ed.), Kluwer Academic Publishers, pp. 311–336, 2001.

Maimon O. and Rokach L., "Improving supervised learning by feature decomposition", Proceedings of the Second International Symposium on Foundations of Information and Knowledge Systems, Lecture Notes in Computer Science, Springer, pp. 178-196, 2002.

Maimon, O. and Rokach, L., Decomposition Methodology for Knowledge Discovery and Data Mining: Theory and Applications, Series in Machine Perception and Artificial Intelligence - Vol. 61, World Scientific Publishing, ISBN:981-256-079-3, 2005.

Marcotorchino, J.F. and Michaud, P. Optimisation en Analyse Ordinale des Donns. Masson, Paris.

Mishra, S. K. and Raghavan, V. V., An empirical study of the performance of heuristic methods for clustering. In Pattern Recognition in Practice, E. S. Gelsema and L. N. Kanal, Eds. 425436, 1994.

Moskovitch R, Elovici Y, Rokach L, Detection of unknown computer worms based on behavioral classification of the host, Computational Statistics and Data Analysis, 52(9):4544–4566, 2008.

Murtagh, F. A survey of recent advances in hierarchical clustering algorithms which use cluster centers. Comput. J. 26 354-359, 1984.

Ng, R. and Han, J. 1994. Very large data bases. In Proceedings of the 20th International Conference on Very Large Data Bases (VLDB94, Santiago, Chile, Sept.), VLDB Endowment, Berkeley, CA, 144155.

Rand, W. M., Objective criteria for the evaluation of clustering methods. Journal of the American Statistical Association, 66: 846–850, 1971.

Ray, S., and Turi, R.H. Determination of Number of Clusters in K-Means Clustering and Application in Color Image Segmentation. Monash university, 1999.

Rokach, L., Decomposition methodology for classification tasks: a meta decomposer framework, Pattern Analysis and Applications, 9(2006):257–271.

Rokach L., Genetic algorithm-based feature set partitioning for classification problems,Pattern Recognition, 41(5):1676–1700, 2008.

Rokach L., Mining manufacturing data using genetic algorithm-based feature set decomposition, Int. J. Intelligent Systems Technologies and Applications, 4(1):57-78, 2008.

Rokach, L. and Maimon, O., Theory and applications of attribute decomposition, IEEE International Conference on Data Mining, IEEE Computer Society Press, pp. 473–480, 2001.

Rokach L. and Maimon O., Feature Set Decomposition for Decision Trees, Journal of Intelligent Data Analysis, Volume 9, Number 2, 2005b, pp 131–158.

Rokach, L. and Maimon, O., Clustering methods, Data Mining and Knowledge Discovery Handbook, pp. 321–352, 2005, Springer.

Rokach, L. and Maimon, O., Data mining for improving the quality of manufacturing: a feature set decomposition approach, Journal of Intelligent Manufacturing, 17(3):285–299, 2006, Springer.

Rokach, L., Maimon, O., Data Mining with Decision Trees: Theory and Applications, World Scientific Publishing, 2008.

Rokach L., Maimon O. and Lavi I., Space Decomposition In Data Mining: A Clustering Approach, Proceedings of the 14th International Symposium On Methodologies For Intelligent Systems, Maebashi, Japan, Lecture Notes in Computer Science, Springer-Verlag, 2003, pp. 24–31.

Rokach, L. and Maimon, O. and Averbuch, M., Information Retrieval System for Medical Narrative Reports, Lecture Notes in Artificial intelligence 3055, page 217-228 Springer-Verlag, 2004.

Rokach, L. and Maimon, O. and Arbel, R., Selective voting-getting more for less in sensor fusion, International Journal of Pattern Recognition and Artificial Intelligence 20 (3) (2006), pp. 329–350.

Selim, S.Z., and Ismail, M.A. K-means-type algorithms: a generalized convergence theorem and characterization of local optimality. In IEEE transactions on pattern analysis and machine learning, vol. PAMI-6, no. 1, January, 1984.

Selim, S. Z. AND Al-Sultan, K. 1991. A simulated annealing algorithm for the clustering problem. Pattern Recogn. 24, 10 (1991), 10031008.

Sneath, P., and Sokal, R. Numerical Taxonomy. W.H. Freeman Co., San Francisco, CA, 1973.

Strehl A. and Ghosh J., Clustering Guidance and Quality Evaluation Using Relationship-based Visualization, Proceedings of Intelligent Engineering Systems Through Artificial Neural Networks, 2000, St. Louis, Missouri, USA, pp 483-488.

Strehl, A., Ghosh, J., Mooney, R.: Impact of similarity measures on web-page clustering. In Proc. AAAI Workshop on AI for Web Search, pp 58–64, 2000.

Tibshirani, R., Walther, G. and Hastie, T., 2000. Estimating the number of clusters in a dataset via the gap statistic. Tech. Rep. 208, Dept. of Statistics, Stanford University.

Tyron R. C. and Bailey D.E. Cluster Analysis. McGraw-Hill, 1970.

Urquhart, R. Graph-theoretical clustering, based on limited neighborhood sets. Pattern recognition, vol. 15, pp. 173-187, 1982.

Veyssieres, M.P. and Plant, R.E. Identification of vegetation state and transition domains in California's hardwood rangelands. University of California, 1998.

Wallace C. S. and Dowe D. L., Intrinsic classification by mml – the snob program. In Proceedings of the 7th Australian Joint Conference on Artificial Intelligence, pages 37-44, 1994.

Wang, X. and Yu, Q. Estimate the number of clusters in web documents via gap statistic. May 2001.

Ward, J. H. Hierarchical grouping to optimize an objective function. Journal of the American Statistical Association, 58:236-244, 1963.

Zahn, C. T., Graph-theoretical methods for detecting and describing gestalt clusters. IEEE trans. Comput. C-20 (Apr.), 68-86, 1971.

15

Association Rules

Frank Höppner

University of Applied Sciences Braunschweig/Wolfenbüttel

Summary. Association rules are rules of the kind "70% of the customers who buy vine and cheese also buy grapes". While the traditional field of application is market basket analysis, association rule mining has been applied to various fields since then, which has led to a number of important modifications and extensions. We discuss the most frequently applied approach that is central to many extensions, the Apriori algorithm, and briefly review some applications to other data types, well-known problems of rule evaluation via support and confidence, and extensions of or alternatives to the standard framework.

Key words: Association Rules, Apriori

15.1 Introduction

To increase sales rates at retail a manager may want to offer some discount on certain products when bought in combination. Given the thousands of products in the store, how should they be selected (in order to maximize the profit)? Another possibility is to simply locate products which are often purchased in combination close to each other, to remind a customer, who just rushed into the store to buy product A, that she or he may also need product B. This may prevent the customer from visiting a – possibly different – store to buy B a short time after. The idea of "market basket analysis", the prototypical application of association rule mining, is to find such related products by analysing the content of the customer's market basket to find product associations like "70% of the customers who buy vine and cheese also buy grapes." The task is to find associated products within the set of offered products, as a support for marketing decisions in this case.

Thus, for the traditional form of association rule mining the database schema $S = \{A_1, ..., A_n\}$ consists of a large number of attributes (n is in the range of several hundred) and the attribute domains are binary, that is, $dom(A_i) = \{0, 1\}$. The attributes can be interpreted as properties an instance does have or does not have, such as a car may have an air conditioning system but no navigation system, or a cart in a supermarket may contain vine but no coffee. An alternative representation

O. Maimon, L. Rokach (eds.), *Data Mining and Knowledge Discovery Handbook*, 2nd ed.,
DOI 10.1007/978-0-387-09823-4_15,

of an instance with $A_7 = 1$, $A_{18} = 1$ and $A_i = 0$ for all other attributes, is therefore a set-oriented notation $\{A_7, A_{18}\}$. The task of association rule mining is, basically, to examine relationships between *all* possible subsets. For instance, a rule $\{A_9, A_{37}, A_{214}\} \rightarrow \{A_{189}, A_{165}\}$ would indicate that, whenever a database record possesses attributes values $A_9 = 1, A_{37} = 1$, and $A_{214} = 1$, also $A_{189} = 1$ and $A_{165} = 1$ will hold. The main difference to classification (see Chapter II in this book) is that association rules are usually not restricted to the prediction of a single attribute of interest. Even if we consider only $n = 100$ attributes and rules with two items in antecedent and consequent, we have already more than 23,500,000 possible rules. Every possible rule has to be verified against a very large database, where a single scan takes already considerable time. For this challenging task the technique of association rule mining has been developed (Agrawal and Shafer, 1996).

15.1.1 Formal Problem Definition

Let $I = \{a_1, \ldots, a_n\}$ be a set of literals, properties or *items*. A record $(t_1, \ldots, t_n) \in dom(A_1) \times \ldots \times dom(A_n)$ from our transaction database with schema $S = (A_1, \ldots, A_n)$, can be reformulated as an itemset T by $a_i \in T \Leftrightarrow t_i = 1$. We want to use the motivation of the introductory example to define an association explicitly. How can we characterize "associated products"?

Definition: We call a set $Z \subseteq I$ an *association,* if the frequency of occurrences of Z deviates from our expectation given the frequencies of individual $X \in Z$.

If the probability of having sausages (S) or mustard (M) in the shopping carts of our customers is 10% and 4%, resp., we expect 0.4% of the customers to buy both products at the same time. If we instead observe this behaviour for 1% of the customers, this deviates from our expectations and thus $\{S, M\}$ is an association. Do sausages require mustard ($S \rightarrow M$) or does mustard require sausages (M$\rightarrow$S)? If preference (or causality) induces a kind of direction in the association, it can be captured by rules:

Definition: We call $X \rightarrow Y$ an *association rule* with antecedent X and consequent Y, if $Z = X \cup Y \subseteq I$ and $X \cap Y = \emptyset$ holds.[1]

We say that an association rule $X \rightarrow Y$ *is supported* by database D, if there is a record T with $X \cup Y \subseteq T$. Naturally, a rule is the more reliable the more it is supported by records in the database:

Definition: Let $X \rightarrow Y$ be an association rule. The *support* of the rule is defined as $supp(X \rightarrow Y) := supp(X \cup Y)$, where the support of an arbitrary itemset S is defined as the probability $P(S)$ of observing S in a randomly selected record of D:

$$supp(S) = \frac{|\{T \in D | S \subseteq T\}|}{|D|}$$

If we interpret the binary values in the records as a truth values, an association rule can be considered as a logical implication. In this sense, an association rule holds

[1] From the foregoing definition it is clear that it would make sense to additionally require that Z must be an association. But this is not part of the "classical" problem statement.

if for every transaction T that supports X, Y is also supported. In our market basket scenario this means that, for a rule *butter* $\rightarrow$ *bread* to hold, nobody will ever buy butter without bread. This strict logical semantics is unrealistic in most applications (but an interesting special case, see Section 15.4.5), in general we allow for partially fulfilled association rules and measure their degree via confidence values:

Definition: Let $X \rightarrow Y$ be an association rule. The *confidence* of the rule is defined as the fraction of itemsets that support the rule among those that support the antecedent: $conf(X \rightarrow Y) := P(Y|X) = supp(X \cup Y) / supp(X)$.

Problem Statement: Given a database D and two thresholds min_{conf} , $min_{supp} \in [0,1]$, the problem of *association rule mining* is to generate all association rules $X \rightarrow Y$ with supp($X \rightarrow Y) > min_{supp}$ and $conf(X \rightarrow Y) > min_{conf}$ in D.

It has to be emphasized at this point, that these definitions are tailored towards the standard association rule framework; other definitions would have also been possible. For instance, it is commonly agreed that support and confidence are not particularly good as rule ranking measures in general. As an example, consider an everyday product, such as *bread*. It is very likely, that it is contained in many carts, giving it a high support, say 70%. If another product, say *batteries*, is completely independent from *bread*, then the conditional probability *P(bread|batteries)* will be identical to *P(bread)*. In such a case, where no relationship exists, a rule *batteries* $\rightarrow$ *bread* should not be flagged as interesting. However, given the high a priori probability of bread being in the cart, the confidence of this rule will be close to *P(bread)*=70%. Therefore, having found a rule that satisfies the constraints on support and confidence according to the problem definition *does not* mean that an association has been discovered (see footnote 1)! The reason why support and confidence are nevertheless so prominent is that their properties are massively exploited for the efficient enumeration of association rules (see Section 15.2). Their deficiencies (as sufficient conditions) for discovering associations can often be compensated by an additional postprocessing phase. There are, however, also competing approaches that use different measures from the beginning (often addressing the deficiencies of the minimum support as a necessary condition), which we will discuss in section 15.4.5.

In the next section, we review the most prominent association rule mining technique, which provides the central basis for many extensions and modifications, and illustrate its operation with an example. We then give a short impression on applications of the framework to other kinds of data than market basket. In section 15.4 we discuss several ways how to overcome or compensate the difficulties when support and confidence alone are used for rule evaluation.

15.2 Association Rule Mining

In this section we discuss the Apriori algorithm (Agrawal and Srikant, 1994,Agrawal and Shafer, 1996), which solves the association rule mining problem. In a first phase, it enumerates all itemsets found in the transaction database whose support exceeds min_{supp}. We call the itemsets enumerated in this way *frequent itemsets*, although the min_{supp} threshold may be only a few percent. In a second phase, the confidence threshold min_{conf} comes into play: The algorithm generates all rules from the frequent itemsets whose confidence exceeds this threshold. The enumeration will be discussed in section 15.2.1, the rule generation in section 60.2.1.

This subdivision into two independent phases is characteristic for many approaches to association rule mining. Depending on the properties of the database or problem at hand, the frequent itemset mining may be replaced by more efficient variants of Apriori, which will be discussed in detail in Chapter 15.5. Many different rule evaluation measures (see Section 60.2.2) can be calculated from the output of the first phase, such that rule generation may also be altered independently of the first phase. The two phase decomposition makes it easy to combine different itemset mining and rule generation algorithms, however, it also means that the individual properties of the rule evaluation measures used in phase two, which could potentially lead to pruning mechanism in phase one, are not utilized in the standard approach. In section 15.4.5 we will examine a few alternative methods from the literature, were both phases are more tightly coupled.

15.2.1 Association Mining Phase

We speak of a k-itemset X, if $|X|=k$. With the Apriori algorithm, the set of frequent itemsets is found iteratively by identifying all frequent 1-itemsets first, then searching for all frequent 2-itemsets, 3-itemsets, and so forth. Since the number of frequent itemsets is finite (bounded by $2^{|I|}$), for some k we will find no more frequent itemsets and stop the iteration. Can we stop already if we have found no frequent k-itemset, or is it possible that we miss some frequent $(k+j)$-itemsets, $j>0$, then? The following simple but important observation guarantees that this cannot happen:

Observation (Closure under Support): Given two itemsets X and Y with $X \subseteq Y$. Then $supp(X) \geq supp(Y)$.

Whenever a transaction contains the larger itemset Y, it also contains X, therefore this inequality is trivially true. Once we have found an itemset X to be infrequent, no superset Y of X can become frequent; therefore we can stop our iteration at the first k for which no frequent k-itemsets have been found.

The observation addresses the downward closure of frequent itemsets with respect to minimum support: Given a frequent itemset, all of its subsets must also be frequent. The Apriori algorithm explores the lattice over I (induced by set inclusion, see Figure 15.1) in a breadth-first search, starting from the empty set (which is certainly frequent), expanding every frequent itemset by a new item as long as the new itemset turns out to be frequent. This gives us already the main loop of the Apriori algorithm: Starting from a set C_k of *possibly* frequent itemsets of size k, so-called k-candidates, the database D is scanned to determine their support (function countSupport(C_k,D)). Afterwards, the set of frequent k-itemsets F_k is easily identified as the subset of C_k for which the minimum support is reached. From the frequent k-itemsets F_k we create candidates of size $k+1$ and the loop starts all over, until we finally obtain an empty set C_k and the iteration terminates. The frequent itemset mining step of the Apriori algorithm is illustrated in Figure 15.2.

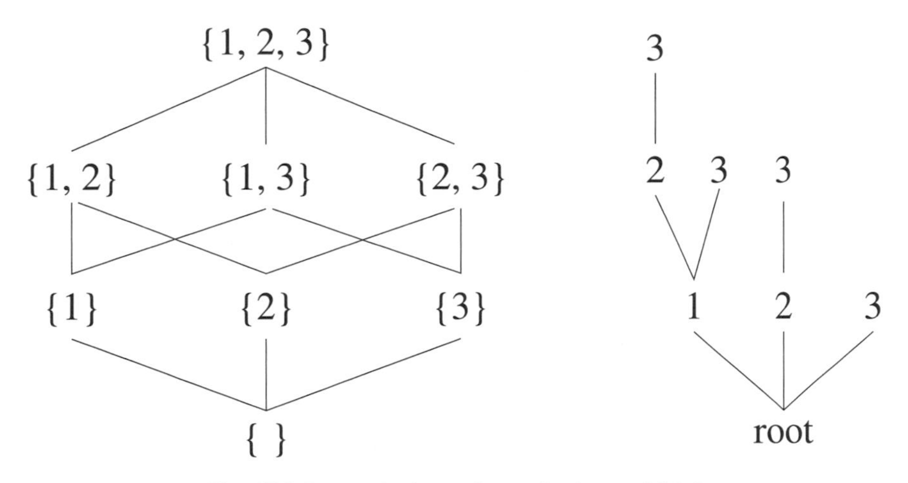

Fig. 15.1. Itemset lattice and tree of subsets of 1,2,3

```
C₁ := { {i} | i∈I };
k := 1;
while C_k ≠ ∅ do begin
    countSupport(C_k,D);
    F_k := { S∈C_k | supp(S) > min_supp };
    C_{k+1} := candidateGeneration(F_k);
    k := k+1;
end
F = F₁∪ F₂∪ ... ∪ F_{k-1}; // all freq. itemsets
```

Fig. 15.2. Apriori algorithm

Before examining some parts of the algorithm in greater detail, we introduce an appropriate tree data structure for the lattice, as shown in Figure 15.1. Let us fix an arbitrary order for items. Each node in the tree represents an itemset, which contains all items on the path from the node to the root node. If, at any node labeled x, we create the successors labeled y with y¿x (according to the item order), the tree encodes every itemset in the lattice exactly once.

Candidate Generation. The naïve approach for the candidate generation function would be to simply return all subsets of I of size $k+1$, which would lead to a brute force enumeration of all possible combinations. Given the size of I, this is of course prohibitive.

The key to escape from the combinatorial explosion is to exploit the closure under support for pruning: For a $(k+1)$-candidate set to be frequent, all of its subsets must be frequent. Any $(k+1)$-candidate has $k+1$ subsets of size k, and if one of them cannot be found in the set of frequent k-itemsets F_k, the candidate itself cannot be frequent. If we include only those $(k+1)$-sets in C_{k+1} that pass this test, the size of C_{k+1} is shrunk considerably. The candidateGeneration function therefore returns:

$$C_{k+1} = \{ C \subset I \mid \; |C|=k+1, \forall S \subset C: (|S|=k \Rightarrow S \in F_k) \}.$$

How can C_{k+1} be determined efficiently? Let us assume that the sets F_k are stored in the tree data structure according to Figure 15.1. Then the $k+1$ containment tests $(S \in F_k)$ can be done efficiently by descending the tree. Instead of enumerating all possible sets S of size $k+1$, it is more efficient to construct them from a union of two k-itemsets in F_k (as a positive effect, then only $k-1$ containment tests remain to be executed). This can be done as follows: All itemsets that have a common node at level $k-1$ are denoted as a block (they share $k-1$ items). From every two k-itemsets X and Y coming from the same block, we create a candidate $Z = X \cup Y$ (implying that $|Z|=k+1$). Candidates generated in this way can easily be used to extend the tree of k-itemsets into a tree of candidate $(k+1)$-itemsets.

Support Counting. Whenever the function countSupport(C_k,D) is invoked, a pass over the database is performed. At the beginning, the support counters for all candidate itemsets are set to zero. For each transaction T in the database, the counters for all subsets of T in C_k have to be incremented. If the candidates are represented as a tree, this can be done as follows: starting at the root node (empty set), follow all paths with an item contained in T. This procedure is repeated at every node. Whenever a node is reached at level k, a k-candidate contained in T has been found and its counter can be incremented. Note that for subsets of arbitrary size we would have to mark at each node whether it represents an itemset contained in the tree or it has been inserted just because it lies on a path to a larger itemset. Here, all itemsets in C_k or F_k are of the same length and therefore valid itemsets consist of all leafs at depth k only.

It is clear from Figure 15.1, that the leftmost subtree of the root node is always the largest and the rightmost subtree consists of a single node only. To make the subset search as fast as possible, the items should be sorted by their frequency in ascending order, such that the probability of descending the tree without finding a candidate is as small as possible.

Runtime complexity. Candidate generation is log-linear in the number of frequent patterns $|F_k|$. A single iteration of the main routine in Figure 15.2 depends linearly on the number of candidates and the database size. The most critical number is the number of candidate itemsets, which is usually tractable in the market basket data, because the transaction/item-table is sparse. However, if the support threshold is very small or the table is dense, the number of candidates (as well as frequent patterns) grows exponentially (see Section 15.4.4) and so does the runtime of this algorithm. A comparison of the Apriori runtime with other frequent itemset mining methods can be found in Chapter 15.5 of this volume.

Example. For illustration purposes, here is an example run of the Apriori algorithm. Consider $I = \{A, B, C, D, E\}$ and a database D consisting of the following transactions

Transaction #	A	B	C	D	E	Itemset
#1	1	1	1	0	1	A, B, C, E
#2	0	1	1	0	1	B, C, E
#3	1	0	1	1	1	A, C, D, E
#4	0	1	0	0	1	B, E
#5	1	1	1	0	0	A, B, C

For the 1-candidates $C_1 = \{\{A\}, \{B\}, \{C\}, \{D\}, \{E\}\}$ we obtain the support values 3/5, 4/5, 4/5, 1/5, 4/5. Suppose $min_{supp} = 2/5$, then $F_1 = \{\{A\}, \{B\}, \{C\}, \{E\}\}$. The 2-candidate generation always yields $\binom{|F_k|}{2}$ candidates, because the pruning has not yet any effect. By pairing all frequent 1-itemsets we obtain the following 2-candidates with respective support:

candidate 2-itemset	Support	frequent 2-itemset
{A, B}	2/5	{A, B}
{A, C}	3/5	{A, C}
{A, E}	2/5	{A, E}
{B, C}	3/5	{B, C}
{B, E}	3/5	{B, E}
{C, E}	3/5	{C, E}

Here, all 2-candidates are frequent. For the generation of 3-candidates, the alphabetic order is used and the blocks are indicated by double lines. From the first block (going from frequent 2-itemset {A,B} to {A,E}) we obtain the 3-candidates {A, B, C}, {A, B, E} (pairing {A,B} with its successors {A,C} and {A,E} in the same block), and {A, C, E} (pairing {A,C} with its successor {A,E}); from the second block {B, C, E} is created, and none from the third block.To qualify as candidate itemsets, we finally verify that all 2-subsets are frequent, e.g. for {A,B,C} (which was generated from {A,B} and {A,C}) we can see that third 2-subset {B,C} is also frequent.

cand. 3-itemset	Support	Freq. 3-itemset
{A, B, C}	2/5	{A, B, C}
{A, B, E}	1/5	
{A, C, E}	2/5	{A, C, E}
{B, C, E}	2/5	{B, C, E}

The 3-itemset $\{A, B, E\}$ turns out to be infrequent. Thus, three blocks with a single itemset remain and no candidate 4-itemset will be generated. As an example, $\{A, B, C, E\}$ would be considered as a candidate 4-itemset only if all of its subsets $\{B, C, E\}$, $\{A, C, E\}$, $\{A, B, E\}$ and $\{A, B, C\}$ are frequent, but $\{A, B, E\}$ is not.

15.2.2 Rule Generation Phase

Once frequent itemsets are available, rules can be extracted from them. The objective is to create for every frequent itemset Z and its subsets X a rule $X \rightarrow Z - X$ and include it in the result if $supp(Z)/supp(X) > min_{conf}$. The observation in section 15.2.1 states that for $X_1 \subset X_2 \subset Z$ we have $supp(X_1) \geq supp(X_2) \geq supp(Z)$, and therefore $supp(Z)/supp(X_1) \leq supp(Z)/supp(X_2) \leq 1$. Thus, the confidence value for rule $X_2 \rightarrow Z - X_2$ will be higher than that of $X_1 \rightarrow Z - X_1$. In general, the largest confidence value will be obtained for antecedent X_2 being as large as possible (or consequent $Z - X_2$ as small as possible). Therefore we start with all rules with a single item in the consequent. If such a rule does not reach the minimum confidence value, we do not need to evaluate a rule with a larger consequent.

Now, when we are about to consider a consequent $Y_1 = Z - X_1$ of a rule $X_1 \rightarrow Y_1$, to reach the minimum confidence, all rules $X_2 \rightarrow Y_2$ with $Y_2 \subset Y_1 \subset Z$ must also reach the minimum confidence. If for one Y_2 this is not the case, the rule $X_1 \rightarrow Y_1$ must also miss the minimum confidence. Thus, for rules $X \rightarrow Y$ with given $Z = X \cup Y$, we have the same downward closure of consequents (with respect to confidence) as we had with itemsets (with respect to support). This closure can be used for pruning in the same way as in the frequent itemset pruning in section 60.2.1: Before we look up the support value of a $(k+1)$-consequent Y, we check if all k-subsets of Y have led to rules with minimum confidence. If that is not the case, Y will not contribute to the output either. The following algorithm generates all rules from the set of frequent itemsets F. To underline the similarity to the algorithm in Figure 15.2, we use again the sets F_k and C_k, but this time they contain the candidate k-consequents (rather than candidate k-itemsets) and k-consequents of rules that satisfy min_{conf} (rather than frequent k-itemsets). The benefit of pruning, however, is much smaller here, because we only save a look up of the support values that have been determined in phase one.

```
forall Z∈F do begin
    R := ∅; // resulting set of rules
    C1 := { {i} | i∈Y }; // cand. 1-consequents
    k := 1;
    while Ck ≠ ∅ do begin
        Fk := { X∈Ck | conf(X→Z-X) > min_conf };
        R := R ∪ { X→Z-X | X∈Fk };
        Ck+1 := candidateGeneration(Fk);
        k := k+1;
    end
end
```

Fig. 15.3. Rule generation algorithm

As an example, consider the case of $Z = \{A, C, E\}$ in the outer loop. The set of candidate 1-consequents is $C_1 = \{\{A\}, \{C\}, \{E\}\}$. The confidence of the rules $AC \rightarrow E$ is 2/3, $AE \rightarrow C$ is 1, and $CE \rightarrow A$ is 2/3. With $min_{conf} = 0.7$, we obtain $F_1 = \{\{C\}\}$. Suppose $\{B, C\}$ is considered as a candidate, then both of its subsets $\{B\}$ and $\{C\}$ have to be found in F_1, which contains only $\{C\}$. Thus, the candidate generation yields only an empty set for C_2 and rule enumeration for Z is finished.

15.3 Application to Other Types of Data

The traditional application of association rules is market basket analysis, see for instance (Brijs *et al.*, 1999). Since then, the technique has been applied to other kinds of data, such as:

- Census data (Brin *et al.*, 1997A, Brin *et al.*, 1997B)
- Linguistic data for writer evaluation (Aumann and Lindell, 2003)
- Insurance data (Castelo and Giudici, 2003)
- Medical diagnosis (Gamberger *et al.*, 1999)

One of the first generalizations, which has still applications in the field of market basket analysis, is the consideration of temporal or sequential information, such as the date of purchase. Applications include:

- Market basket data (Agrawal and Srikant, 1995)
- Causes of plan failures (Zaki, 2001)
- Web personalization (Mobasher *et al.*, 2002)
- Text data (Brin *et al.*, 1997A, Delgado *et al.*, 2002)
- Publication databases (Lee *et al.*, 2001)

In the analysis of sequential data, very long sequences may easily occur (e.g. in text data), corresponding to frequent k-itemsets with large values of k. Given that any subsequence of such a sequence will also be frequent, the number of frequent sequences quickly becomes intractable (to discover a frequent sequence of length k, 2^k subsequences have to be found first). As a consequence, the reduced problem of enumerating only maximal sequences is considered. A sequence is *maximal*, if there is no frequent sequence that contains it as a subsequence. While in the Apriori algorithm a breadth first search is performed to find all frequent k-itemsets first, before $(k+1)$-itemsets are considered, for maximal sequences a depth first search may become computationally advantageous, because the exhaustive enumeration of subsequences is skipped. Techniques for mining maximal itemsets will be discussed in detail in the subsequent chapter.

In the analysis of sequential data, the traditional transactions are replaced by short sequences, such as the execution of a short plan, products bought by a single customer, or events caused by a single customer during website navigation. It is also possible to apply association mining when the observations consist of a few attributes observed over a long period of time. With such kind of data, the semantics of support counting has to be revised carefully to avoid misleading or uninterpretable results (Mannila *et al.*, 1997, Höppner and Klawonn, 2002). Usually a sliding window is shifted along the sequences and the probability of observing the pattern in a randomly selected position is defined as the support of the pattern. Examples for this kind of application can be found in:

- Telecommunication failures (Mannila *et al.*, 1997) (Patterns are sequences of parallel or consecutive events, e.g. "if alarm A is followed by simultaneous events B1 and B2, then event C is likely to occur".)

- Discovery of qualitative patterns in time series (Höppner and Klawonn, 2002) (Patterns consist of labeled intervals and their interval relationship to each other, e.g. "if an increase in time series A is finished by a decrease in series B, then it is likely that the decrease in B overlaps a decrease in C")

A number of extensions refer to the generalization of association rule mining to the case of numerical data. In the market basket scenario, these techniques may be applied to identify the most valuable customers (Webb, 2001). When broadening the range of scales for the attributes in the database (not only 0/1 data), the discovery of associations becomes very similar to cluster analysis and the discovery of association rules becomes closely related to machine learning approaches (Mitchell, 1997) to rule induction, classification and regression. Some approaches from the literature are:

- Mining of quantitative rules by discretizing numerical attributes into categorical attributes (Srikant and Agrawal, 1996)
- Mining of quantitative rules using mean values in the consequent (Aumann & Lindell, 1999) (Webb, 2001)
- Mining interval data (Miller and Yang, 1997)

Finally, here are a few examples for applications with more complex data structures:

- Discovery of spatial and spatio-temporal patterns (Koperski and Han, 1995,Tsoukatos and Gunopulos, 2001)
- Mining of molecule structure for drug discovery (Borgelt, 2002)
- Usage in inductive logic programming (ILP) (Dehaspe and Toivonen, 1999)

15.4 Extensions of the Basic Framework

Within the confidence/support framework for rule evaluation user's often complain that either the rules are well-known or even trivial, as it is often the case for rules with high confidence and large support, or that there are several hundreds of variants of the same rule with very similar confidence and support values. As we have seen already in the introduction, many rules may actually be incidental or represent nothing of interest (see batteries and bread examples in Section 15.1.1). In the literature, many different ways to tackle this problem have been proposed:

- The use of alternative rule measures
- Support rule browsing and interactive navigation through rules
- The use of a compressed representation of rules
- Additional limitations to further restrict the search space
- Alternatives to frequent itemset enumeration that allow for arbitrary rule support (highly interesting but rare patterns)

While the first few options may be implemented in a further postprocessing phase, this is not true for the last options. For instance, if no minimum support threshold is given, the Apriori-like enumeration of frequent itemsets has to be replaced completely. In the subsequent sections, we will briefly address all of these approaches.

15.4.1 Some other Rule Evaluation Measures

The set of rules returned to the user must be evaluated manually. This costly manual inspection should pay off, so only potentially interesting rules should be returned. Let us consider the example from the introduction again, where the support/confidence-framework flagged rules as worth investigating which were actually obtained from independent items. If items A and B are statistically independent, the empirical joint probability $P(A,B)$ is approximately equal to $P(A)P(B)$, therefore we could flag a rule as interesting only in case the rule support deviates from $P(A)P(B)$ (Piatetsky-Shapiro, 1991):

$$\text{rule-interest}(X \to Y) = supp(X \to Y) - supp(X)supp(Y)$$

If rule-interest < 0 (rule-interest > 0), then X is positively (negatively) correlated to Y. The same underlying idea is used for the lift measure, a well-known statistical measure, also known as interest or strength, where a division instead of a subtraction is used

$$\text{lift}(X \to Y) = supp(X \to Y)/(supp(X)supp(Y))$$

Lift can be interpreted as the number of times buyers of X are more likely to buy Y compared to all customers ($conf(X \to Y)/conf(\emptyset \to Y)$). For independent X and Y we obtain a value 1 (rather than 0 for rule-interest). The significance of the correlation between X and Y can be determined using a chi-square test for a 2 x 2 contingency table. A problem with this kind of test is, however, that statistics poses some conditions on the use of the chi-square test: (1) all cells in the contingency table should have expected value greater than 1, and (2) at least 80% of the cells should have expected value greater than 5 (Brin *et al.*, 1997A). These conditions are necessary since the binomial distribution (in case of items) is approximated by the normal distribution, but they do not necessarily hold for all associations to be tested.

The outcome of the statistical test partitions the set of rules into significant and non-significant, but cannot provide a ranking within the set of significant rules (the chi-square value itself could be used, but as it is not bounded, a comparison between rules is not that meaningful). Thus, chi-square alone is not sufficient (Tan and Kumar, 2002,Berzal *et al.*, 2001). On the other hand, a great advantage of using contingency tables is that also negative correlations are discovered (e.g. customers that buy coffee do *not* buy tea).

From an information-theoretical point of view, the ultimate rule evaluation measure is the J-measure (Smyth, 1992), which ranks rules by their information content. Consider X and Y as being random variables and $X = x \to Y = y$ being a rule. The

J-measure compares the a priori distribution of X (binary variable, either the antecedent holds $(X = x)$ or not $(X = \bar{x})$) with the a posteriori distribution of X given that $Y = y$. The relative information

$$j(X|Y = y) = \sum_{z \in \{x, \bar{x}\}} P(X = z|Y = y) \log_2 \left(\frac{P(X = z|Y = y)}{P(X = z)} \right)$$

yields the instantaneous information that $Y = y$ provides about X (j is also known as the Kullbach-Leibler distance). When applying the rule multiple times, on average we have the information $J(X|Y{=}y) = P(Y{=}y)j(X|Y{=}y)$, which is the J-value of the rule and is bounded by 0.53 bit. The drawback is, however, that highly infrequent rules do not carry much information *on average* (due to the factor $P(Y = y)$), such that highly interesting but rarely occurring associations may not appear under the top-ranked rules.

Other measures are conviction (a "directed", asymmetric lift) (Brin *et al.*, 1997B), certainty factors from MYCIN (Berzal *et al.*, 2001), correlation coefficients from statistics (Tan and Kumar, 2002), Laplace or Gini from rule induction (Clark and Boswell, 1991) or decision tree induction (Breiman, 1996). For a comparison of various measures of interestingness the reader is referred to (Hilderman and Hamilton, 2001), where also general properties rule measures should have are discussed. In (Bayardo and Agrawal, 1999) it is outlined that, *given a fixed consequent*, the ordering of rules obtained from confidence is identical to those obtained by lift or conviction (which is further generalized in (Bayardo *et al.*, 1999)).

15.4.2 Interactive or Knowledge-Based Filtering

Whatever the rule evaluation measure may propose, the final judgment about the interestingness and usefulness of a rule is made by the human expert or user. For instance, many measures consistently return those rules as most interesting that consists of a single item in the consequent, because in this case confidence is maximized (see Section 60.2.1). But the user may be interested in different items or item combinations in the consequent, therefore the *subjective* interestingness of these rules may be low in some applications. Indeed, all measures of interestingness rely on statistical properties of the items and do not take background information into account. But background knowledge about the presence or absence of correlation between items may alert a human expert at some rules, which are undistinguishable from others if looked at the interestingness rate provided by statistical measures alone.

Given the large number of rules, as a first step a visualization and navigation tool may help to quickly find interesting rules. In (Klemettinen *et al.*, 1994) rule templates are proposed, allowing the user to restrict the set of rules syntactically. Some visualization techniques are also presented, such as visualizing rules in a graph, where the items represent nodes and rules are represented by edges that lead from the antecedent to consequent attributes. The thickness of an edge may illustrate the rating of the rule.

The idea of (Dong and Li, 1998) is to compare the performance of a rule against the performance of similar rules and flag it only as interesting, if it deviates clearly from them. A distance measure for rules is used to define the neighborhood of a rule (containing all rules within a certain distance according to some distance measure). Several possibilities to flag a rule as interesting are discussed: it may qualify as interesting by an unexpectedly high confidence value, if its confidence deviates clearly from the average confidence in its neighborhood, or by an unexpectedly sparse neighborhood, if the number of mined rules is small compared to the number of possible rules in the neighborhood. Another rule filtering approach is proposed in (Liu *et al.*, 1999), where statistical correlation is used to define the direction of a rule: Basically, a rule has direction 1 / 0 / -1, if antecedent and consequent are positively correlated / uncorrelated / negatively correlated. From the set of predecessors of a rule, an expected direction can be derived (e.g. if all subrules have direction 0, we may expected an extended rule to have also direction 0), and a rule is flagged as interesting if its direction deviates from this expected direction (closely related to (Brin *et al.*, 1997A), see section 15.4.5).

A completely different approach is to let the expert formalize her or his domain knowledge in advance, which then can be used to test the newly discovered rules against the expert's belief. The Apriori algorithm is extended to find rules that contradict predefined rules in (Padmanabhan, 1998). Only contradictory rules are then reported, because they represent potentially new information to the expert.

The first interactive approaches were designed as post processing systems, which generate all rules first and then allow for a fast navigation through the rules. The idea of rule template matching can be extended to a rule query language that is supported by the mining algorithm (Srikant *et al.*, 1997). In contrast to the post-processing approach, such an integrated querying will be faster if only a few queries are posted and may succeed in occasions where a complete enumeration of all rules fails, e.g. if the minimum support threshold is very low but many other constraints can be exploited to limit the search space. The possibility of posing additional constraints is therefore crucial for successful interactive querying (see Section 15.4.4). In (Goethals and Van den Bussche, 2000) the user specifies items that must or must not appear in the antecedent or consequent of a rule. From this set of constraints, a reduced database (reduced in the size and number of transactions) is constructed, which is used for faster itemset enumeration. The generation of a working copy of the database is expensive, but may pay off for repeated queries, especially since intermediate results of earlier queries are reused. The technique is extended to process full-fledged Boolean expressions.

15.4.3 Compressed Representations

Possible motivations for using compressed or condensed representations of rules include

- Reduction of storage needs and, if possible, computational costs
- Derivation of new rules by condensing items to meta-items
- Reduction of the number of rules to be evaluated by an expert

Using a compressed rule set is motivated by problems occurring with dense databases (see Section 15.4.4): Suppose we have found a frequent itemset r containing an item i. If we include additional n items in our database that are perfectly correlated to item i, $2^n - 1$ variations of itemset r will be found. As another example, suppose the database consists of a single transaction $\{a_1, \ldots, a_n\}$ and $min_{supp} = 1$, then $2^n - 1$ frequent subsets will be generated to discover $\{a_1, \ldots, a_n\}$ as a frequent itemset. In such cases, a condensed representation (Mannila and Toivonen, 1996) can help to reduce the size of the rule set. An itemset X is called *closed*, if there is no superset X' that is contained in every transaction containing X. In the latter example, only a single closed frequent itemset would be discovered. The key to finding closed itemsets is the fact that the support value remains constant, regardless of which subset of the n items is contained in the itemset. Therefore, from the frequent closed itemsets, all frequent itemsets can be reconstructed including their support values, which makes closed itemsets a lossless representation. Algorithms that compute closed frequent itemsets are, to name just a few, *Close* (Pasquier *et al.*, 1999), *Charm* (Zaki, 2000), and *Closet* (Wang, 2003). An overview of lossless representations can be found in (Kryszkiewicz, 2001).

Secondly, for a rule to become more meaningful, it may be helpful to consider several items in combination. An example is the use of product taxonomies in market basket analysis, where many rules that differ only in one item, say *apple-juice*, *orange-juice*, and *cherry-juice*, may be outperformed (and thus replaced) by a single rule using the generalized meta-item *fruit-juice*. Such meta-items can be obtained from an a-priori known product taxonomy (Han *et al.*, 2000, Srikant and Agrawal, 1995). If such a taxonomy is not given, one may want to discover disjunctive combination of items that optimize the rule evaluation automatically (see (Zelenko, 1999) for disjunctive combination of items, or (Höppner, 2002) for generalization in the context of sequences).

Finally, the third motivation addresses the combinatorial explosion of rules that can be generated from a single (closed) frequent itemset. Since the expert herself knows best which items are interesting, why not present associations – as a compressed representation of many rules – rather than individual rules? In particular when sequences are mined, the order of items is fixed in the sequence, such that we have only two degrees of freedom: (1) Remove individual items before building a rule and (2) select the location which separates antecedent from consequent items. In (Höppner, 2002) every location in a frequent sequence is attributed with the J-value that can *at least* be achieved in predicting an arbitrary subset of the consequent. This series of J-values characterizes the predictive power of the whole sequence and thus represents a condensed representation of many associated rules. The tuple of J-values can also be used to define an order on sequences, allowing for a ranking of sequences rather than just rules.

15.4.4 Additional Constraints for Dense Databases

Suppose we have extracted a set of rules R from database D. Now we add a very frequent item f randomly to our transactions in $D$and derive rule set R'. Due to the randomness of f, the support and confidence values of the rules in R do not change substantially. Besides the rules in R, we will also obtain rules that contain f either in the antecedent or the consequent (with very similar support/confidence values). Even worse, since f is very frequent, almost every subset of I will be a good predictor of f, adding another $2^{|I|}$ rules to R. This shows that the inclusion of f increases the size of R' dramatically, while none of these rules are actually worth investigating. In the market basket application, such items f usually do not exist, but come easily into play if customer data is included in the analysis. For dense databases the bottom-up approach of enumerating *all* frequent itemsets quickly turns out to be infeasible. In this section some approaches that limit the search space further to reduce time and space requirements are briefly reviewed. (In contrast to the methods in the next section, the minimum support threshold is not necessarily released.)

In (Webb, 2000) it is argued that a user is not able to investigate many thousand rules and therefore will be satisfied if, say, only the top 1000 rules are returned. Then, optimistic pruning can be used to prune those parts of the search space that cannot contribute to the result since even under optimistic conditions there is no chance of getting better than the best 1000 rules found so far. This significantly limits the size of the space that has actually to be explored. The method requires, however, to run completely in main memory and may therefore be not applicable to very large databases. A modification to the case of numerical attributes (called impact rules) is proposed in (Webb, 2001).

In (Bayardo *et al.*, 1999) the set of rules is reduced by listing specializations of a rule only if they improve the confidence of a common base rule by more than a threshold min_{imp}. The improvement of a rule is defined as

$$imp(A \rightarrow C) = min \{ A' \subset A \mid conf(A \rightarrow C) - conf(A' \rightarrow C) \}$$

These additional constraints must be exploited during rule mining, since enumerating all frequent itemsets first and apply constraints thereafter is not feasible. The key to efficiency of *DenseMiner* is the derivation of upper bounds for confidence, improvement, and support, which are then used for optimistic pruning of itemset blocks. This saves the algorithm from determining the support of many frequent itemsets, whose rules will miss the confidence or improvement constraints anyway. This work is enhanced in (Bayardo and Agrawal, 1999) to mine only those rules that are the optimal rules according to a partial order of rules based on support and confidence. It is then shown that the optimal rules under various interestingness measures are included in this set. In practice, while the set of obtained rules is quite manageable, it characterizes only a specific subset of the database records (Bayardo and Agrawal, 1999). An alternative partial order is proposed (based on the subsumption of the transactions supported by rules) that better characterizes the database, but also leads to much larger rule sets.

One of the appealing properties of the Apriori algorithm is that it is complete in the sense that all association rules will be discovered. We have seen that, on the other hand, this is at the same time one of its drawbacks, since it leads to an unmanageably large rule set. If not completeness but quality of the rules has priority, many of the rule induction techniques from statistics and machine learning may be also applied to discover association rules. As an example, in (Friedman, 1997) local regions in the data (corresponding to associations) are sought where the distribution deviates from the distribution expected under the independence assumption. Starting with a box containing the whole data set, it is iteratively shrunk by limiting the set of admissible values for each variable (removing items from categorical attributes or shifting the borders of the interval of admissible values for numerical attributes). Among all possible ways to shrink the box, the one that exhibits the largest deviation from the expected distribution is selected. The refinement is stopped as soon as the support falls below a minimum support threshold. Having arrived at a box with a large deviation, a second phase tries to maximize the support of the box by enlarging the range of admissible values again, without compromising the high deviation found so far. From a box found in this way, association rules may be derived by exploring the correlation among the variables. The whole approach is not restricted to discovering associations, but finds local maxima of some objective function, such that it could also be applied to rule induction for numerical variables. Compared to partitioning approaches (such as decision or regression trees), the rules obtained from trees are much more complex and usually have much smaller support.

Many of the approaches in the next section can also be used for dense databases, because they do not enumerate *all* frequent itemsets either. Similarly, for some techniques in this section the minimum support threshold may be dropped with some databases, but in (Webb, 2000, Bayardo and Agrawal, 1999) and (Friedman, 1997) minimum support thresholds have been exploited.

15.4.5 Rules without Minimum Support

For some applications, the concept of a minimum support threshold is a rather artificial constraint primarily required for the efficient computation of large frequent itemsets. Even in market basket analysis one may miss interesting associations, e.g. between *caviar* and *vodka* (Cohen *et al.* (2007)). Therefore alternatives to Apriori have been developed that do not require a minimum support threshold for association rule enumeration. We discuss some of these approaches briefly in this section.

The observation in section 15.2.1 characterizes support as being *downward closed*: If an itemset has minimum support, all subsets also have minimum support. We have seen that the properties of a closure led to an efficient algorithm since it provided powerful pruning capabilities. In (Brin *et al.*, 1997A) the *upward closure* of correlation (measured by the chi-square test) is proven and exploited for mining. Since correlation is upward closed, to utilize it in a level-wise algorithm the inverse property is used: For a $(k+1)$-itemset to be *un*correlated, all of its k-subsets must be *un*correlated. The uncorrelated itemsets are then generated and verified in a similar fashion ("has minimum support" is substituted by "is uncorrelated") as in the Apriori algorithm (Figure 15.2). The interesting output, however, is not the set of uncorrelated items, but the set of correlated items. The border between both sets is identified during the database pass: whenever a candidate k-itemset has turned out to be correlated it is stored in a set of *minimally correlated itemsets* (otherwise it is stored in a set of uncorrelated k-itemsets and used for pruning in the next stage). From the set of minimally correlated itemsets we know that all supersets are also correlated (upward closure), so the partitioning into correlated and uncorrelated itemsets is complete. Note that in contrast to Apriori and derivatives, no rules but only sets of correlated attributes are provided, and that no ranking for the correlation can guide the user in manual investigation (see Section 60.2.2).

An interesting subproblem is that of finding all rules with a very high confidence (but no constraints on support), as it is proposed in (Li and Fang, 1989). The key idea is to, given a consequent, subdivide the database D into two parts, one, D_1, containing all transactions that contain the consequent and the other, D_2, containing those that do not. Itemsets that occur in D_1 but not at all in D_2 can be used as antecedents for 100%-confidence rules. A variation for rules with high confidence is also proposed.

Other proposals also address the enumeration of associations (or association rules), but have almost nothing in common with the Apriori algorithm. Such an approach is presented in (Cohen *et al.* (2007)), which heavily relies on the power of randomization. Associations are identified via correlation or *similarity* between attributes X and Y: $S(X,Y) = supp(X \cap Y) / supp(X \cup Y)$. The approach is restricted to the identification of pairs of similar attributes, from which groups of similar variables may be identified (via transitivity). Since only 2-itemsets are considered and due to the absence of a minimum support threshold, considerable effort is undertaken to prune the set of candidate 2-patterns of size $|I|^2$. (In (Brin *et al.*, 1997A) this problem is attacked by additionally introducing a support threshold guided by the chi-square test requirements.) The approach uses so-called *signatures* for each column, which are calculated in a first database scan. Several proposals for signatures are made, the most simple defines a random order on the rows and the signature itself is simply the first row index (under the ordering) in which the column has a 1. The probability that two columns X and Y have the same signature is proportional to their similarity $S(X,Y)$. To estimate this probability, k independently drawn orders are used to calculate k individual signatures. The estimated probabilities are used for pruning and exact similarities are calculated during a second pass.

The idea of *Mambo* (Castelo and Giudici, 2003) is that only those attributes should be considered in the antecedent of a rule, that *directly* influence the consequent attribute, which is reflected by conditional independencies: If X and Y are conditionally independent given Z (or $I(X,Y|Z)$), then once we know Z, Y (or X) does not tell us anything of importance about X (or Y):

$$I(X,Y|Z) \Leftrightarrow P(X,Y \mid Z) = P(X \mid Z)\, P(Y \mid Z)$$

As an example from automobile manufacturing (Rokach and Maimon, 2006), consider the variables “car model”, “top”, and “color”. If, for a certain car type, a removable top is available, the top’s color may be restricted to only a few possibilities, which consequently restricts the available colors for the car as a whole. The car model influences the color only indirectly, therefore we do not want to see rules leading from car model to color, but only from top to color. If conditional independencies *are known*, then they can be used to prevent a rule miner from listing such rules. The idea of *Mambo* is to find these independencies and use them for pruning. A minimal set of variables *MB* for a variable X such that for all variables $Y \notin MB \cup \{X\}$ we have *I(X,Y|MB),* is called a *Markov Blanket* of X. The blanket shields X from the remaining variables. The difficulty of this approach lies in the estimation of the Markov Blankets, which are obtained from a Markov Chain Monte Carlo method.

In (Aumann & Lindell, 1999) an approach for mining rules from one numerical attribute to another numerical attribute is proposed, that also does not require a minimum support threshold. The numerical attribute in the antecedent is restricted by an interval and the consequent characterizes the population mean in this case. A rule is generated only if the mean of the selected subset of the population differs significantly from the overall mean (according to a Z-test), and a "minimum difference" threshold is used to avoid enumerating rules which are significant but only marginally different. The idea is to sort the database according to the attribute in the antecedent (which is, however, a costly operation). Then any set of consecutive transactions in the sorted database, whose consequent values are above or below average, gives a new rule. They also make use of a closure property: if two neighboring intervals $[a,b]$ and $[b,c]$ lead to a significant change in the consequent variable, then so does the union $[a,c]$. This property is exploited to list only rule with maximal intervals in the antecedent.

15.5 Conclusions

We have discussed the common basis for many approaches to association rule mining, the Apriori algorithm, which gained its attractivity and popularity from its simplicity. Simplicity, on the other hand, always implies some insufficiencies and opens space for various (no longer simple) improvements. Since the first papers about association rule mining have been published, the number of papers in this area has exploded and it is almost impossible to keep track of all different proposals. Therefore, the overview provided here is necessarily incomplete.

Over time, the focus has shifted from sparse (market basket) data to (general purpose) dense data, and reasonably large itemsets (promoting less than 0.1% of the customers in a supermarket probably is not worth the effort) to small patterns, which represent deviations from the mean population (small set of most profitable customers, which shall be mailed directly). Then it may be desirable to get rid of constraints such as minimum support, which possibly hide some interesting patterns in the data. But on the other hand, the smaller the support the higher the probability of observing an incidental rather than a meaningful deviation from the average population, especially when taking the size of the database into account (Bolton *et al.*, 2002). A major issue for the upcoming research is therefore to limit the findings to substantive "real" patterns.

References

Agrawal R. and Srikant R. Fast Algorithms for mining association rules. Proc. Int. Conf. on Very Large Databases, 487-499, 1994

Agrawal R. and Srikant R. Mining Sequential Patterns. Proc. Int. Conf. on Data Engineering, 3-14, 1995.

Agrawal R., Mannila H., Srikant R., Toivonen H., Verkamo A.I. Fast Discovery of Association Rules. In: Advances in Knowledge Discovery and Data Mining, Fayyad U.M., Piatetsky-Shapiro G., Smyth P., Uthurusamy R. (eds)., AAAI Press / The MIT Press, 307-328, 1996

Aumann Y., Lindell, Y. A Statistical Theory for Quantitative Association Rules, Journal of Intelligent Information Systems, 20(3):255-283, 2003

Bayardo R.J., Agrawal R. Mining the Most Interesting Rules. Proc. ACM SIGKDD Int. Conf. on Knowledge Discovery and Data Mining, 145-154, 1999

Bayardo R.J., Agrawal R., Gunopulos D. Constrained-Based Rule Mining in Large, Dense Databases. Proc. 15th Int. Conf. on Data Engineering, 188-197, 1999

Berzal F., Blanco I., Sanchez D., Vila M.A. A New Framework to Assess Association Rules. Proc. Symp. on Intelligent Data Analysis, LNCS 2189, 95-104, Springer, 2001

Bolton R., Hand D.J., Adams, N.M. Determining Hit Rate in Pattern Search, Proc. Pattern Detection and Discovery, LNAI 2447, 36-48, Springer, 2002

Borgelt C., Berthold M. Mining Molecular Fragments: Finding Relevant Substructures of Molecules, Proc. Int. Conf. on Data Mining, 51-58, 2002

Breiman L., Friedman J., Olshen R., Stone C. Classification and Regression Trees. Chapman & Hall, New York, 1984.

Brijs T., Swinnen G., Vanhoof K. and Wets G. Using Association Rules for Product Assortment Decisions: A Case Study. Proc. ACM SIGKDD Int. Conf. on Knowledge Discovery and Data Mining, 254-260, 1999

Brin S., Motwani R., and Silverstein C. Beyond market baskets: Generalizing association rules to correlations. Proc. ACM SIGMOD Int. Conf. Management of Data, 265-276, 1997

Brin S., Motwani R., Ullman, J.D., Tsur, S. Dynamic itemset counting and implication rules for market basket data. SIGMOD Record 26(2):255-264, 1997

Castelo R., Feelders A., Siebes A. Mambo: Discovering Association Rules. Proc. Symp. on Intelligent Data Analysis. LNCS 2189, 289-298, Springer, 2001

Clark P., Boswell R. Rule Induction with CN2: Some recent improvements. In Proc. European Working Session on Learning EWSL-91, 151-163, 1991

Cohen E., Datar M., Fujiwara S., Gionis A., Indyk P., Motwani R., Ullman J., Yang C. Finding Interesting Associations without Support Pruning, IEEE Transaction on Knowledge Discovery 13(1):64-78, 2001

Dehaspe L., Toivonen H. Discovery of Frequent Datalog Patterns. Data Mining and Knowledge Discovery, 3(1):7-38,1999

Delgado M., Martin-Bautista M.J., Sanchez D., Vila M.A. Mining Text Data: Features and Patterns. Proc. Pattern Detection and Discovery, LNAI 2447, 140-153, Springer, 2002

Dong G., Li J. Interestingness of Discovered Association Rules in Terms of Neighbourhood-Based Unexpectedness. Proc. Pacific Asia Conf. on Knowledge Discovery in Databases, LNAI 1394, 72-86, 1998

Friedman J.H., Fisher N.I. Bump Hunting in High-Dimensional Data. Statistics and Computing 9(2), 123-143, 1999

Gamberger D., Lavrac N., Jovanoski, V. High Confidence Association Rules for Medical Diagnosis, Proc. of Intelligent Data Analysis in Medical Applications (IDAMAP), 42-51, 1999

Goethals B., Van den Bussche, J. On Supporting Interactive Association Rule Mining, Proc. Int. Conf. Data Warehousing and Knowledge Discovery, LNCS 1874, 307-316, Springer, 2000

Han J., Fu Y. Discovery of Multiple-Level Association Rules from Large Databases Proc. Int. Conf. on Very Large Databases, 420-431, 1995

Hilderman R.J. and Hamilton H.J. Knowledge Discovery and Measures of Interest. Kluwer Academic Publishers, 2001

Höppner F., Klawonn F. Finding Informative Rules in Interval Sequences. Intelligent Data Analysis, 6(3):237-256, 2002

Höppner F. Discovery of Core Episodes from Sequences. Proc. Pattern Detection and Discovery, LNAI 2447, 199-213, Springer, 2002

Klemettinen M., Mannila H., Ronkainen P., Toivonen, H. and Verkamo A.I. Finding Interesting Rules from Large Sets of Discovered Association Rules. Proc. Int. Conf. on Information and Knowledge Managament, 401-407, 1994.

Koperski K., Han J. Discovery of Spatial Association Rules in Geographic Information Databases. Proc. Int. Symp Advances in Spatial Databases, LNCS 951, 47-66, 1995

Kryszkiewicz M. Concise Representation of Frequent Patterns based on Disjunction-free Generators. Proc. Int. Conf. on Data Mining, 305-312, 2001.

Lee C.H., Lin C.R., Chen M.S. On Mining General Temporal Association Rules in a Publication Database. Proc. Int. Conf. on Data Mining, 337-344, 2001

Li J., Zhang X., Dong G., Ramamohanarao K., Sun Q. Efficient Mining of High Confidence Association Rules without Support Thresholds. Proc. Principles of Data Mining and Knowledge Discovery, 406-411, 1999

Liu B., Hsu W., Ma Y. Pruning and Summarizing the Discovered Associations. Proc. ACM SIGKDD Conf. Knowledge Discovery and Data Mining, 125-134, 1999

Mannila H., Toivonen H. Multiple uses of frequent sets and condensed representations. Proc. ACM SIGKDD Int. Conf. on Knowledge Discovery and Data Mining, 189-194, 1996

Mannila H., Toivonen H., Verkamo A.I. Discovery of Frequent Episodes in Event Sequences. Data Mining and Knowledge Discovery, 1(3):259-289, 1997

Miller R.J., Yang Y. Association Rules over Interval Data. Proc. Int. Conf. on Management of Data, 452-461, 1997

Mitchell T., Machine Learning, McGraw-Hill, 1997

Mobasher B., Dai H., Luo T., Nakagawa, M. Discovery and Evaluation of Aggregate Usage Profile for Web Personalization, Data Mining and Knowledge Discovery 6:61-82, 2002

Padmanabhan B., Tuzhilin A. A Belief-Driven Method for Discovering Unexpected Patterns. Int. Conf. on Knowledge Discovery and Data Mining, 94-100, 1998

Pasquier N., Bastide Y., Taouil R., Lakhal L. Efficient Mining of association rules using closed itemset lattices. Information Systems, 24(1):25-46, 1999

Piatetsky-Shapiro, G. Discovery, Analysis, and Presentation of Strong Rules. Proc. Knowledge Discovery in Databases, 229-248, 1991

Rokach, L., Averbuch, M. and Maimon, O., Information retrieval system for medical narrative reports. Lecture notes in artificial intelligence, 3055. pp. 217-228, Springer-Verlag (2004).

Rokach L. and Maimon O., Data mining for improving the quality of manufacturing: A feature set decomposition approach. Journal of Intelligent Manufacturing 17(3): 285299, 2006.

Smyth P. and Goodman, R.M. An Information Theoretic Approach to Rule Induction from Databases. IEEE Trans. Knowledge Discovery and Data Engineering, 4(4):301-316, 1992

Srikant R., Agrawal R. Mining Generalized Association Rules. Proc. Int. Conf. on Very Large Databases, 407-419, 1995

Srikant R., Agrawal R. Mining Quantitative Association Rules in Large Relational Tables. Proc. ACM SIGMOD Conf. on Management of Data, 1-12, 1996

Srikant R., Vu Q., Agrawal R. Mining Association Rules with Constraints. Proc. Int. Conf. Knowledge Discovery and Data Mining, 66-73, 1997

Tan P.N., Kumar V. Selecting the Right Interestingness Measure for Association Patterns, Proc. ACM SIGKDD Conf. Knowledge Discovery and Data Mining, 32-41, 2002

Tsoukatos, I. and Gunopulos, D. Efficient Mining of Spatiotemporal Patterns. Proc. Int. Symp. Spatial and Temporal Databases, LNCS 2121, pages 425-442, 2001

Wang J., Han J., Pei J. CLOSET+: Searching for the Best Strategies for mining Frequent Closed Itemsets, Proc. ACM SIGKDD Int. Conf. on Knowledge Discovery and Data Mining, 236-245, 2003

Webb G.I. Efficient Search for Association Rules. Proc. ACM SIGKDD Int. Conf. on Knowledge Discovery and Data Mining, 99-107, 2000

Webb G.I. Discovering Associations with numeric variables. Proc. ACM SIGKDD Int. Conf. on Knowledge Discovery and Data Mining, 383-388, 2001

Zaki M.J. Generating non-redundant Association Rules. In Proc. ACM SIGKDD Int. Conf. on Knowledge Discovery and Data Mining, 34-43, 2000

Zaki M.J. SPADE: An Efficient Algorithm for Mining Frequent Sequences. Machine Learning 42(1):31-60, 2001

Zelenko D. Optimizing Disjunctive Association Rules. Proc. of Int. Conf. on Principles of Data Mining and Knowledge Discovery, LNAI 1704, 204-213, 1999

16

Frequent Set Mining

Bart Goethals

Departement of Mathemati lcs and Computer Science, University of Antwerp, Belgium
bart.goethals@ua.ac.be

Summary. Frequent sets lie at the basis of many Data Mining algorithms. As a result, hundreds of algorithms have been proposed in order to solve the frequent set mining problem. In this chapter, we attempt to survey the most successful algorithms and techniques that try to solve this problem efficiently.

Key words: Frequent Set Mining, Association Rule, Support, Cover, Apriori

Introduction

Frequent sets play an essential role in many Data Mining tasks that try to find interesting patterns from databases, such as association rules, correlations, sequences, episodes, classifiers, clusters and many more of which the mining of association rules, as explained in Chapter 14.7.3 in this volume, is one of the most popular problems. The identification of sets of items, products, symptoms, characteristics, and so forth, that often occur together in the given database, can be seen as one of the most basic tasks in Data Mining.

Since its introduction in 1993 by Agrawal et al. (1993), the frequent set mining problem has received a great deal of attention. Hundreds of research papers have been published, presenting new algorithms or improvements to solve this mining problem more efficiently.

In this chapter, we explain the frequent set mining problem, some of its variations, and the main techniques to solve them. Obviously, given the huge amount of work on this topic, it is impossible to explain or even mention all proposed algorithms or optimizations. Instead, we attempt to give a comprehensive survey of the most influential algorithms and results.

16.1 Problem Description

The original motivation for searching frequent sets came from the need to analyze so called supermarket transaction data, that is, to examine customer behavior in terms

O. Maimon, L. Rokach (eds.), *Data Mining and Knowledge Discovery Handbook*, 2nd ed.,
DOI 10.1007/978-0-387-09823-4_16,

of the purchased products (Agrawal et al., 1993). Frequent sets of products describe how often items are purchased together.

Formally, let $\mathscr{I}$ be a set of items.

A *transaction* over $\mathscr{I}$ is a couple $T = (tid, I)$ where *tid* is the transaction identifier and I is a set of items from $\mathscr{I}$.

A *database* $\mathscr{D}$ over $\mathscr{I}$ is a set of transactions over $\mathscr{I}$ such that each transaction has a unique identifier. We omit $\mathscr{I}$ whenever it is clear from the context.

A transaction $T = (tid, I)$ is said to *support* a set X, if $X \subseteq I$. The *cover* of a set X in $\mathscr{D}$ consists of the set of transaction identifiers of transactions in $\mathscr{D}$ that support X. The *support* of a set X in $\mathscr{D}$ is the number of transactions in the cover of X in $\mathscr{D}$. The *frequency* of a set X in $\mathscr{D}$ is the probability that X occurs in a transaction, or in other words, the support of X divided by the total number of transactions in the database. We omit $\mathscr{D}$ whenever it is clear from the context.

A set is called *frequent* if its support is no less than a given absolute *minimal support threshold* σ_{abs}, with $0 \leq \sigma_{abs} \leq |\mathscr{D}|$. When working with frequencies of sets instead of their supports, we use a relative *minimal frequency threshold* σ_{rel}, with $0 \leq \sigma_{rel} \leq 1$. Obviously, $\sigma_{abs} = \lceil \sigma_{rel} \cdot |\mathscr{D}| \rceil$. In this chapter, we will mostly use the absolute minimal support threshold and omit the subscript *abs* unless explicitly stated otherwise.

Definition 1. *Let $\mathscr{D}$ be a database of transactions over a set of items $\mathscr{I}$, and σ a minimal support threshold. The collection of frequent sets in $\mathscr{D}$ with respect to σ is denoted by*

$$\mathscr{F}(\mathscr{D}, \sigma) := \{X \subseteq \mathscr{I} \mid support(X, \mathscr{D}) \geq \sigma\},$$

or simply $\mathscr{F}$ if $\mathscr{D}$ and σ are clear from the context.

Problem 1. (Frequent Set Mining) Given a set of items $\mathscr{I}$, a database of transactions $\mathscr{D}$ over $\mathscr{I}$, and minimal support threshold σ, find $\mathscr{F}(\mathscr{D}, \sigma)$.

In practice we are not only interested in the set of sets $\mathscr{F}$, but also in the actual supports of these sets.

For example, consider the database shown in Table 16.1 over the set of items $\mathscr{I} = \{\text{beer}, \text{chips}, \text{pizza}, \text{wine}\}$.

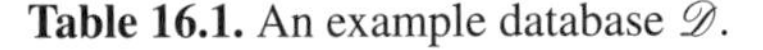

Table 16.1. An example database $\mathscr{D}$.

tid	set of items
100	{beer, chips, wine}
200	{beer, chips}
300	{pizza, wine}
400	{chips, pizza}

Table 16.2 shows all frequent sets in $\mathscr{D}$ with respect to a minimal support threshold equal to 1, their cover in $\mathscr{D}$, plus their support and frequency.

Note that the Set Mining problem is actually a special case of the Association Rule Mining problem explained in Chapter 14.7.3 in this volume. Indeed, if we are

Table 16.2. Frequent sets, their cover, support, and frequency in $\mathscr{D}$.

Set	Cover	Support	Frequency
{}	{100, 200, 300, 400}	4	100%
{beer}	{100,200}	2	50%
{chips}	{100,200,400}	3	75%
{pizza}	{300,400}	2	50%
{wine}	{100,300}	2	50%
{beer, chips}	{100,200}	2	50%
{beer, wine}	{100}	1	25%
{chips, pizza}	{400}	1	25%
{chips, wine}	{100}	1	25%
{pizza, wine}	{300}	1	25%
{beer, chips, wine}	{100}	1	25%

given the support threshold σ, then every frequent set X also represents the trivial rule $X \Rightarrow \{\}$ which holds with 100% confidence.

Nevertheless, the task of discovering all frequent sets is quite challenging. The search space is exponential in the number of items occurring in the database and the targeted databases tend to be massive, containing millions of transactions. Both these characteristics make it a worthwhile effort to seek the most efficient techniques to solve this task.

Search Space Issues

The search space of all sets contains exactly $2^{|\mathscr{I}|}$ different sets. If $\mathscr{I}$ is large enough, then the naive approach to generate and count the supports of all sets over the database can't be achieved within a reasonable period of time. For example, in many applications, $\mathscr{I}$ contains thousands of items, and then, the number of sets is more than the number of atoms in the universe ($\approx 10^{79}$).

Instead, we could limit ourselves to those sets that occur at least once in the database by generating only all subsets of all transactions in the database. Of course, for large transactions, this number could still be too large. As an optimization, we could generate only those subsets of at most a given maximum size. This technique, however, suffers from massive meory requirements for any but a database with only very small transactions (Amir et al., 1997). Most other efficient solutions perform a more directed search through the search space. During such a search, several collections of *candidate sets* are generated and their supports computed until all frequent sets have been generated. Obviously, the size of a collection of candidate sets must not exceed the size of available main memory. Moreover, it is important to generate as few candidate sets as possible, since computing the supports of a collection of sets is a time consuming procedure. In the best case, only the frequent sets are generated and counted. Unfortunately, this ideal is impossible in general, which will be shown later in this section.

The main underlying property exploited by most algorithms is that support is monotone decreasing with respect to extension of a set.

Property 1. **(Support monotonicity)** Given a database of transactions $\mathscr{D}$ over $\mathscr{I}$, and two sets $X, Y \subseteq \mathscr{I}$. Then,

$$X \subseteq Y \Rightarrow support(Y) \leq support(X).$$

Hence, if a set is infrequent, all of its supersets must be infrequent, and vice versa, if a set is frequent, all of its subsets must be frequent too. In the literature, this monotonicity property is also called the downward closure property, since the set of frequent sets is downward closed with respect to set inclusion. Similarly, the set of infrequent sets is upward closed.

Database Issues

To compute the supports of a collection of sets, we need to access the database. Since such databases tend to be very large, it is not always possible to store them into main memory.

An important consideration in most algorithms is the representation of the database. Conceptually, such a database can be represented by a two-dimensional binary matrix in which every row represents an individual transaction and the columns represent the items in $\mathcal{I}$. Such a matrix can be implemented in several ways. The most commonly used layout is the so called *horizontal layout*. That is, each transaction has a transaction identifier and a list of items occurring in that transaction. Another commonly used layout is the *vertical layout*, in which the database consists of a set of items, each followed by its cover (Holsheimer et al., 1995, Savasere et al., 1995, Zaki, 2000).

To count the support of a candidate set X using the horizontal layout, we need to scan the database completely and test for every transaction T whether $X \subseteq T$. Of course, this can be done for a large collection of sets at once. Although scanning the database is an I/O intensive operation, in most cases, this is not the major cost of such counting steps. Instead, updating the supports of all candidate sets contained in a transaction consumes considerably more time than reading that transaction from a file or from a database cursor. Indeed, for each transaction, we need to check for every candidate set whether it is included in that transaction, or otherwise, we need to check for every subset of that transaction whether it is in the set of candidate sets.

The vertical database layout on the other hand, has the major advantage that the support of a set X can be easily computed by simply intersecting the covers of any two subsets $Y, Z \subseteq X$, such that $Y \cup Z = X$ (Holsheimer et al., 1995, Savasere et al., 1995). Given a set of candidate sets, however, this technique requires that the covers of a lot of sets are available in main memory, which is evidently not always possible. Indeed, the covers of all singleton sets already represent the complete database.

In the next two sections, we will describe the standard algorithm for mining all frequent sets using the horizontal layout and the vertical database layout. After that, we consider several optimizations and variations of both approaches.

16.2 Apriori

Together with the introduction of the frequent set mining problem, also the first algorithm to solve it was proposed, later denoted as *AIS* (Agrawal et al., 1993). Shortly after that, the algorithm was improved and called *Apriori*. The main improvement was to exploit the monotonicity property of the support of sets (Agrawal and Srikant, 1994, Srikant and Agrawal, 1995). The same technique was independently proposed by Mannila et al. (1994). Both works were combined afterwards (Agrawal et al., 1996). Note that the Apriori algorithm actually solves the complete association rule mining problem, of which mining all frequent sets was only the first, but most difficult phase.

From now on, we assume for simplicity that items in transactions and sets are kept sorted in their lexicographic order, unless stated otherwise.

The set mining phase of the Apriori algorithm is given in Algorithm 16.1. We use the notation $X[i]$ to represent the ith item in X; the *k-prefix* of a set X is the k-set $\{X[1],\ldots,X[k]\}$, and $\mathcal{F}_k$ denotes the frequent k-sets.

Input: $\mathcal{D},\sigma$
Output: $\mathcal{F}(\mathcal{D},\sigma)$
1: $C_1 := \{\{i\} \mid i \in \mathcal{I}\}$
2: $k := 1$
3: **while** $C_k \neq \{\}$ **do**
4: **for all** transactions $(tid, I) \in \mathcal{D}$ **do**
5: **for all** candidate sets $X \in C_k$ **do**
6: **if** $X \subseteq I$ **then**
7: Increment $X.support$ by 1
8: **end if**
9: **end for**
10: **end for**
11: $\mathcal{F}_k := \{X \in C_k \mid X.support \geq \sigma\}$
12: $C_{k+1} := \{\}$
13: **for all** $X,Y \in \mathcal{F}_k$, such that $X[i] = Y[i]$
14: for $1 \leq i \leq k-1$, and $X[k] < Y[k]$ **do**
15: $I := X \cup \{Y[k]\}$
16: **if** $\forall J \subset I, |J| = k : J \in \mathcal{F}_k$ **then**
17: Add I to C_{k+1}
18: **end if**
19: **end for**
20: Increment k by 1
21: **end while**

Fig. 16.1. Apriori

The algorithm performs a breadth-first (levelwise) search through the search space of all sets by iteratively generating and counting a collection of candidate sets. More specifically, a set is candidate if all of its subsets are counted and frequent. In each iteration, the collection C_{k+1} of candidate sets of size $k+1$ is generated, starting with $k = 0$. Obviously, the initial set C_1 consists of all items in $\mathcal{I}$ (line 1). At a certain level k, all candidate sets of size $k+1$ are generated. This is done in two steps. First, in the *join* step, the union $X \cup Y$ of sets $X, Y \in \mathcal{F}_k$ is generated if they have the same $k-1$-prefix (lines 10–11). In the *prune* step, $X \cup Y$ is inserted into C_{k+1} only if all of its k-subsets are frequent and thus, must occur in $\mathscr{F}_k$ (lines 12–13).

To count the supports of all candidate k-sets, the database, which remains on secondary storage in the horizontal layout, is scanned one transaction at a time, and the supports of all candidate sets that are included in that transaction are incremented (lines 4–7). All sets that turn out to be frequent are inserted into $\mathscr{F}_k$ (line 8).

If the number of candidate sets is too large to remain into main memory, the algorithm can be easily modified as follows. The candidate generation procedure stops and the supports of all generated candidates is counted. In the next iteration, instead of generating candidate sets of size $k+2$, the remaining candidate $k+1$-sets are generated and counted repeatedly until all frequent sets of size $k+1$ are generated and counted.

Although this is a very efficient and robust algorithm, its main drawback lies in its inefficient support counting mechanism. As already explained, for each transaction, we need to check for every candidate set whether it is included in that transaction, or otherwise, we need to check for every subset of that transaction whether it is in the set of candidate sets.

When transactions are large, generating all k-subsets of a transaction and testing for each of them whether it is a candidate set, can take a prohibitive amount of time. For example, suppose we are counting candidate sets of size 5 in single transaction containing only 20 items. Then, we have to do already more than 15 000 set equality tests. Of course, this can be somewhat optimized since many of these sets have large intersections and hence, can be tested at the same time (Brin et al., 1997). Nevertheless, transactions can be much larger causing this method to become a significant bottleneck.

On the other hand, testing for each candidate set whether it is contained in the given transaction can also take to much time when the collection of candidate sets is large. For example, consider the case in which we have 1 000 frequent items. This means there are almost 500000 candidate 2-sets. Obviously, testing whether all of them occur in a single transaction, for every transaction, could take an immense amount of time. Fortunately, a lot of counting optimizations have been proposed for many different situations (Park et al., 1995, Srikant, 1996, Brin et al., 1997, Orlando et al., 2002). To reduce the number of iterations that are needed to go through the the database, it is also possible to combine the last few iterations of the algorithm. That is, generate every candidate set of size $k+\ell$ if all of its k-subsets are known to be frequent, for all possible $\ell > 1$. Of course, it is of crucial importance not to do this too early, since that could cause an exponential blowup in the number of generated candidate sets. It is possible, however, to bound the remaining number of candidate sets very accurately using a combinatorial technique proposed by Geerts et al. (2001). Given this bound, a combinatorial explosion can be avoided.

Another important aspect of the Apriori algorithm is the data structure used to store the candidate and frequent sets for the candidate generation and the support counting processes. Indeed, they both require an efficient data structure in which all candidate sets are stored since it is important to efficiently find the sets that are contained in a transaction or in another set. The two most successful data structures are the hash-tree and the trie. We refer the interested reader to other literature describing these data structures in more detail, e.g. (Srikant, 1996, Brin et al., 1997, Borgelt and Kruse, 2002).

16.3 Eclat

As explained earlier, when the database is stored in the vertical layout, the support of a set can be counted much easier by simply intersecting the covers of two of its subsets that together give the set itself. The original Eclat algorithm essentially used this technique inside the Apriori algorithm (Zaki, 2000). This is, however, not always possible since the total size of all covers at a certain iteration of the local set generation procedure could exceed main memory limits. Fortunately, it is possible to significantly reduce this total size by generating collections of candidate sets in a depth-first strategy. Also, in stead of using the intersection based technique already from the start, it is usually more efficient to first find the frequent items and frequent 2-sets separately and use the Eclat algorithm only for all larger sets (Zaki, 2000).

Given a database of transactions $\mathscr{D}$ and a minimal support threshold σ, denote the set of all frequent sets with the same prefix $I \subseteq \mathscr{I}$ by $\mathscr{F}[I](\mathscr{D},\sigma)$. (Note that $\mathscr{F}[\{\}](\mathscr{D},\sigma) = \mathscr{F}(\mathscr{D},\sigma)$.) The main idea of the search strategy is that all sets containing item $i \in \mathscr{I}$, but not containing any item smaller than i, can be found in the so called *i-conditional database* (Han et al., 2004), denoted by $\mathscr{D}^i$. That is, $\mathscr{D}^i$ consists of those transactions from $\mathscr{D}$ that contain i, and from which all items before i, and i itself are removed. In general, for a given set I, we can create the I-conditional database, $\mathscr{D}^I$, consisting of all transactions that contain I, but from which all items before the last item in I and that item itself have been removed. Then, for every frequent set found in $\mathscr{D}^I$, we add I to it, and thus, we found exactly all large tiles containing I, but not any item before the last item in I which is not in I, in the original database, $\mathscr{D}$. Finally, Eclat recursively generates for every item $i \in \mathscr{I}$ the set $\mathscr{F}[\{i\}](\mathscr{D}^i,\sigma)$.

For simplicity of presentation, we assume that all items that occur in the database are frequent. In practice, all frequent items can be computed during an initial scan over the database, after which all infrequent items will be ignored.

The final Eclat algorithm is given in Algorithm 16.2.

Note that a candidate set is now represented by each set $I \cup \{i, j\}$ of which the support is computed at line 6 and 7 of the algorithm. Since the algorithm doesn't fully exploit the monotonicity property, but generates a candidate set based on the frequency of only two of its subsets, the number of candidate sets that are generated is much larger as compared to Apriori's breadth-first approach. As a comparison, Eclat essentially generates candidate sets using only the join step from Apriori. The sets that are needed for the prune step are simply not available.

Recently, Zaki et al. proposed a significant improvement to this algorithm to reduce the amount of necessary memory and to compute the support of a set even faster using the vertical database layout (Zaki and Gouda, 2003). Instead of storing

Input: $\mathscr{D}, \sigma, I \subseteq \mathscr{I}$ (initially called with $I = \{\}$)
Output: $\mathscr{F}[I](\mathscr{D}, \sigma)$

```
 1: F[I] := {}
 2: for all i ∈ I occurring in D do
 3:    F[I] := F[I] ∪ {I ∪ {i}}
 4:    D^i := {}
 5:    for all j ∈ I occurring in D such that j > i do
 6:       C := cover({i}) ∩ cover({j})
 7:       if |C| ≥ σ then
 8:          D^i := D^i ∪ {(j,C)}
 9:       end if
10:    end for
11:    // Depth-first recursion
12:    Compute F[I ∪ {i}](D^i, σ) recursively
13:    F[I] := F[I] ∪ F[I ∪ {i}]
14: end for
```

Fig. 16.2. Eclat

the cover of a k-set I, the difference between the cover of I and the cover of the $k-1$-prefix of I is stored, called the *diffset* of I. To compute the support of I, we simply need to subtract the size of the diffset from the support of its $k-1$-prefix. This support can be provided as a parameter within the recursive function calls of the algorithm. The diffset of a set $I \cup \{i, j\}$, given the two diffsets of its subsets $I \cup \{i\}$ and $I \cup \{j\}$, with $i < j$, is computed as follows:

$$\mathit{diffset}(I \cup \{i, j\}) := \mathit{diffset}(I \cup \{j\}) \setminus \mathit{diffset}(I \cup \{i\}).$$

This technique has experimentally shown to result in significant performance improvements of the algorithm, now designated as *dEclat* (Zaki and Gouda, 2003). The original database is still stored in the original vertical database layout.

Observe an arbitrary recursion path of the algorithm starting from the set $\{i_1\}$, up to the k-set $I = \{i_1, \ldots, i_k\}$. The set $\{i_1\}$ has stored its cover and for each recursion step that generates a subset of I, we compute its diffset. Obviously, the total size of all diffsets generated on the recursion path can be at most $|cover(\{i_1\})|$. On the other hand, if we generate the cover of each generated set, the total size of all generated covers on that path is at least $(k-1) \cdot \sigma$ and can be at most $(k-1) \cdot |cover(\{i_1\})|$. This observation indicates that the total size of all diffsets that are stored in main memory at a certain point in the algorithm is much less than the total size of all covers. These predictions were supported by several experiments (Zaki and Gouda, 2003).

16.4 Optimizations

A lot of other algorithms proposed after the introduction of Apriori retain the same general structure, adding several techniques to optimize certain steps within the algorithm. Since the performance of the Apriori algorithm is almost completely dictated by its support counting procedure, most research has focused on that aspect of the Apriori algorithm.

The Eclat algorithm was not the first of its kind when considering the intersection based counting mechanism (Holsheimer et al., 1995, Savasere et al., 1995). Also, its original design did not pursue a depth-first traversal of the search space, although this is only a simple but effective change, which was later corrected in extensions of the algorithm (Zaki and Gouda, 2003, Zaki and Hsiao, 2002). The effectiveness of this change mainly shows in the amount of memory that is consumed. Indeed, the amount and total size of all covers or diffsets stored within a depth-first recursion is usually much smaller than compared to this amount during a breadth-first recursion (Goethals, 2004).

16.4.1 Item reordering

One of the most important optimizations which can be effectively exploited by almost any frequent set mining algorithm, is the reordering of items.

The underlying intuition is to assume statistical independence of all items. Then, items with high frequency tend to occur in more frequent sets, while low frequent items are more likely to occur in only very few sets.

For example, in the case of Apriori, sorting the items in support ascending order improves the distribution of the candidate sets within the used data structure (Borgelt and Kruse, 2002). Also, the number of candidate sets generated during the join step can be reduced in this way. Also in Eclat the number of candidate sets that is generated is reduced using this order, and hence, the number of intersections that need to be computed and the total size of the covers of all generated sets is reduced accordingly. In fact, in Eclat, such reordering can be performed at every recursion step of the algorithm.

Unfortunately, until now, no results have been presented on an optimal ordering of all items for any given algorithm and only vague intuitions and heuristics are given supported by practical experiments.

16.4.2 Partition

As the main drawback of Apriori is its slow and iterative support counting mechanism, Eclat has the drawback that it requires large parts of the (vertical) database to fit in main memory. To solve these issues, Savasere et al. proposed the Partition algorithm (Savasere et al., 1995). (Note, however, that this algorithm was already presented before Eclat and its relatives.)

The main difference in the Partition algorithm, compared to Apriori and Eclat, is that the database is partitioned into several disjoint parts and the algorithm generates for every part all sets that are relatively frequent within that part. This is can be done very efficiently by using the Eclat algorithm (originally, a slightly different algorithm was presented). The parts of the database are chosen in such a way that each part fits into main memory. Then, the algorithm merges all relatively frequent sets of every part together. This results in a superset of all frequent sets over the complete database, since a set that is frequent in the complete database must be relatively frequent in one of the parts. Finally, the actual supports of all sets are computed during a second scan through the database.

Although the covers of all items can be stored in main memory, during the generation of all local frequent sets for every part, it is still possible that the covers of all local candidate k-sets can not be stored in main memory. Also, the algorithm is highly dependent on the heterogeneity of the database and can generate too many local frequent sets, resulting in a significant decrease in performance. However, if the complete database fits into main memory and the total of all covers at any iteration also does not exceed main memory limits, then the database must not be partitioned at all and the algorithm essentially comes down to Eclat.

16.4.3 Sampling

Another technique to solve Apriori's slow counting and Eclat's large memory requirements is to use sampling as proposed by Toivonen (Toivonen, 1996).

The presented Sampling algorithm picks a random sample from the database, then finds all relatively frequent patterns in that sample, and then verifies the results with the rest of the database. In the cases where the sampling method does not produce all frequent sets, the missing sets can be found by generating all remaining potentially frequent sets and verifying their supports during a second pass through the database. The probability of such a failure can be kept small by decreasing the minimal support threshold. However, for a reasonably small probability of failure, the threshold must be drastically decreased, which can cause a combinatorial explosion of the number of candidate patterns. Nevertheless, in practice, finding all frequent patterns within a small sample of the database can be done very fast using Eclat or any other efficient frequent set mining algorithm. In the next step, all true supports of these patterns must be counted after which the standard levelwise algorithm could finish finding all other frequent patterns by generating and counting all candidate patterns iteratively. It has been shown that this technique usually needs only one more scan resulting in a significant performance improvement (Toivonen, 1996).

16.4.4 FP-tree

One of the most cited algorithms proposed after Apriori and Eclat is the FP-growth algorithm by Han et al. (2004). Like Eclat, it performs a depth-first search through all candidate sets and also recursively generates the so called i-conditional database $\mathscr{D}^i$, but in stead of counting the support of a candidate set using the intersection based approach, it uses a more advanced technique.

This technique is based on the so-called *FP-tree*. The main idea is to store all transactions in the database in a trie based structure. In this way, in stead of storing the cover of every frequent item, the transactions themselves are stored and each item has a linked list linking all transactions in which it occurs together. By using the trie structure, a prefix that is shared by several transactions is stored only once.

Nevertheless, the amount of consumed memory is usually much more as compared to Eclat (Goethals, 2004).

The main advantage of this technique is that it can exploit the so-called *single prefix path* case. That is, when it seems that all transactions in the currently observed conditional database share the same prefix, the prefix can be removed, and all subsets of that prefix can afterwards be added to all frequent sets that can still be found (Han et al., 2004), resulting in significant performance improvements. As we will see later, however, an almost equally effective technique can be used in Eclat, based on the notion of closure of a set.

16.5 Concise representations

If the number of frequent sets for a given database is large, it could become infeasible to generate them all. Moreover, if the database is dense, or the minimal support threshold is set too low, then there could exist a lot of very large frequent sets, which would make sending them all to the output infeasible to begin with. Indeed, a frequent set of size k includes the existence of at least $2^k - 1$ frequent sets, i.e. all of its subsets. To overcome this problem, several proposals have been made to generate only a concise representation of all frequent sets for a given database such that, if necessary, the frequency of a set, or the support of a set not in that representation can be efficiently determined or estimated (Gunopulos et al., 2003, Bayardo, 1998, Mannila, 1997, Pasquier et al., 1999, Boulicaut et al., 2003, Bykowski and Rigotti, 2001, Calders and Goethals, 2002, Calders and Goethals, 2003). In this section, we address the most popular.

16.5.1 Maximal Frequent Sets

Since the collection of all frequent sets is downward closed, it can be represented by its maximal elements, the so called *maximal frequent sets*. Most algorithms that have been proposed to find the maximal frequent sets rely on the same general structure as the Apriori and Eclat algorithm. The main additions are the use of several lookahead techniques and efficient subset checking.

The Max-Miner algorithm, proposed by Bayardo (1998), is an adapted version of the Apriori algorithm to which two lookahead techniques are added. Initially, all candidate $k+1$-sets are partitioned such that all sets sharing the same k-prefix are in a single part. Hence, in one such part, corresponding to a prefix set X, each candidate set adds exactly one item to X. Denote this set of 'added' items by I. When a superset of $X \cup I$ is already known to be frequent, this part of candidate sets can already be removed, since they can never belong to the maximal frequent sets anymore, and hence, also their supports don't need to be counted anymore. This subset checking procedure is done using a similar hash-tree as is used to store all frequent and candidate sets in Apriori.

First, during the support counting procedure, for each part, not only the support of all candidate sets is counted, but also the support of $X \cup I$. If it turns out that this set it frequent, again none of its subsets need to be generated anymore, since they can never belong to the maximal frequent sets. All other $k+1$-sets that turn out to be frequent are added to the collection of maximal sets unless a superset is already known to be frequent, and all subsets are removed from the collection, since, obviously, they are not maximal.

A second technique is the so called *support lower bounding* technique. That is, after counting the support of every candidate set $X \cup \{i\}$, it is possible to compute a lower bound on the support its supersets using the following inequality:

$$support(X \cup J) \geq support(X) - \sum_{i \in J} support(X) - support(X \cup \{i\}).$$

For every part with prefix set X, this bound is computed starting with J containing the most frequent item, after which items are added in frequency decreasing order as long as the total sum remains above the minimum support threshold. Finally, $X \cup J$ is added to the maximal frequent sets and all its subsets are removed.

Obviously, these techniques result in additional pruning power on top of the Apriori algorithm, when only maximal frequent sets are needed. Later, several other algorithms used similar lookahead techniques on top of depth-first algorithms such as Eclat. Among them, the most popular are GenMax (Gouda and Zaki, 2001) and MAFIA (Burdick *et al.*, 2001), which also use more advanced techniques to check whether a superset of a candidate set was already found to be frequent. Also the FP-tree approach has shown to be effective for maximal frequent set mining (G. Grahne, 2003, Liu et al., 2003).

A completely different approach, called *Dualize and Advance*, was proposed by Gunopulos et al. (2003). Here, a randomized algorithm finds a few maximal frequent sets by simply adding items to a frequent set until no extension is possible anymore. Then, all other maximal frequent sets can be found similarly by adding items to sets which are so called minimal hypergraph transversals of the complements of all already found maximal frequent sets. Although the algorithm has been theoretically shown to be better than all other proposed algorithms, until now, extensive experiments have only shown otherwise (Uno and Satoh, 2003, Goethals and Zaki, 2003).

16.5.2 Closed Frequent Sets

Another very popular concise representation of all frequent sets are the so called *closed frequent sets*, proposed by Pasquier et al (1999). A set is called closed if its support is different from the supports of its supersets. Although all frequent sets can essentially be closed, in practice, it shows that a lot of sets are not. Also here, several different algorithms, based on those described earlier, have been proposed to find only the closed frequent sets. The main added pruning technique simply checks for each set whether its support is the same as any of its subsets. If this is the case, the item can immediately be added to all frequent supersets of that subset, and does not need to be considered separately anymore as it can never result in a closed frequent set. Again, efficient subset checking techniques are necessary to make sure that a generated frequent has no closed superset with the same support that was generated earlier. Efficient algorithms include CHARM (Zaki and Hsiao, 2002) and CLOSET+ (Wang et al., 2003), and many of their improvements (G. Grahne, 2003, Liu et al., 2003).

16.5.3 Non Derivable Frequent Sets

Although the support monotonicity property is very simple and easy, it is possible to derive much better bounds on the support of a candidate set I, by using the inclusion-exclusion principle, given the supports of all subsets of I (Calders and Goethals, 2002). More specifically, for any subset $J \subseteq I$, we obtain a lower or an upper bound on the support of I using one of the following formulas.
If $|I \setminus J|$ is odd, then

$$support(I) \leq \sum_{J \subseteq X} (-1)^{|I \setminus X|+1} support(X). \quad (16.1)$$

If $|I \setminus J|$ is even, then

$$support(I) \geq \sum_{J \subseteq X} (-1)^{|I \setminus X|+1} support(X). \quad (16.2)$$

Then, when the smallest upper bound is less than the minimal support threshold, the set does not need to be counted anymore, but more interestingly, if the largest lower bound is equal to the smallest upper bound of the support of the set, then it also does not need to be counted anymore since these bounds are necessarily equal to support itself. Such a set is called *derivable* as its support can be derived from the supports of its subsets, or *non-derivable* otherwise. A nice property of the collection of non-derivable frequent sets is that it is downward closed. That is, every subset of a non-derivable set is non-derivable. An additional interesting property is that the size of the largest non-derivable set is at most $1 + \log |\mathscr{D}|$ where $|\mathscr{D}|$ denotes the total number of transactions in the database.

As a result, it makes sense to generate only the non-derivable frequent sets as its derivable counterparts essentially give no new information about the database. Also, the Apriori algorithm can easily be adapted to generate only the non-derivable frequent sets by implementing the inclusion-exclusion formulas as stated above. The resulting algorithm is called NDI (Calders and Goethals, 2002).

16.6 Theoretical Aspects

Already in the first section of this chapter, we made clear how hard the problem of frequent set mining is. More specifically, the search space of all possible frequent sets is exponential in the number of items and the number of transactions in the database tends to be huge such that the number of scans through it should be minimized. Of course, we can make it all sound as hard as we want, but fortunately, also some theoretical results have been presented, proving the hardness of the frequent set mining problems.

First, Gunupolos et al. studied the problem of counting the number of frequent sets and have proven it to be #P-hard (Gunopulos et al., 2003). Additionally, it was shown that deciding whether there is a maximal frequent set of size k, is NP-complete (Gunopulos et al., 2003). After that, Yang has shown that even counting the number of maximal frequent sets is #P-hard (Yang, 2004).

Ramesh et al. presented several results on the size distributions of frequent sets and their feasibility (G. Ramesh, 2003). Mielikäinen introduced and studied the *inverse frequent set mining problem*, i.e., given all frequent sets, what is the computational complexity of finding a database consistent with the collection of frequent sets (Mielikäinen, 2003). It is shown that this problem is NP-hard and its enumeration conterpart, counting the number of compatible databases, also #P-hard. Similarly, Calders introduced and studied the FREQSAT problem, i.e. given some set-interval pairs, does there exist a database such that for every pair, the support of the set falls in the interval? Again, it is shown that this problem is NP-complete (Calders, 2004).

16.7 Further Reading

During the first ten years after the proposal of the frequent set mining problem, several hundreds of scientific papers were written on the topic and it seems that this trend is keeping its pace. For a fair comparison of all these algorithms, a contest is organized to find the best implementations in order to to understand precisely why and under what conditions one algorithm would outperform another (Goethals and Zaki, 2003).

Of course, many articles also study variations of the frequent set mining problem. In this section, we list the most prominent, but refer the interested reader to the original articles.

Another interesting issue is how to effectively exploit more contraints next to the frequency constraint (Srikant et al., 1997). For example, find all sets contained in a specific set or containing a specific set, or boolean combinations of those (Goethals and den Bussche, 2000). Ng et al. have listed a large collection of constraints and classified them into several classes for which different optimization techniques could be used (Ng et al., 1998). The most studied classes or the class of so-called *anti-monotone* constraints, as is the minimal support threshold, and the *monotone constraints*, such as the minimum length constraint (Bonchi et al., 2003).

Combining the exploitation of constraints with the notion of concise representations for the collection of frequent sets has been widely studied within the *inductive database* framework (Mannila, 1997) as they are both crucial steps towards an effective optimization of so called *Data Mining queries*.

When databases contain only a small number of transactions, but a huge number of different items, then it is best to focus on only the closed frequent sets, and a slightly different approach might be benificial (Pan et al., 2003, Rioult et al., 2003). More specifically, as a closed set is essentially the intersection of transactions of the given database (while a non-closed set is not), these approaches perform a search traversal through all combinations of transactions in stead of all combinations of items.

Since privacy in Data Mining presents several important issues, also private frequent set mining has been studied (Vaidya and Clifton, 2002). Also from a theoretical point of view, several problems closely related to frequent set mining remain unsolved (Mannila, 2002).

References

Agrawal, R., Imielinski, T., and Swami, A. (1993). Mining association rules between sets of items in large databases. In Buneman, P. and Jajodia, S., editors, *Proceedings of the 1993 ACM SIGMOD International Conference on Management of Data*, volume 22(2) of *SIGMOD Record*, pages 207–216. ACM Press.

Agrawal, R., Mannila, H., Srikant, R., Toivonen, H., and Verkamo, A. (1996). Fast discovery of association rules. In Fayyad, U., Piatetsky-Shapiro, G., Smyth, P., and Uthurusamy, R., editors, *Advances in Knowledge Discovery and Data Mining*, pages 307–328. MIT Press.

Agrawal, R. and Srikant, R. (1994). Fast algorithms for mining association rules. In Bocca, J., Jarke, M., and Zaniolo, C., editors, *Proceedings 20th International Conference on Very Large Data Bases*, pages 487–499. Morgan Kaufmann.

Amir, A., Feldman, R., and Kashi, R. (1997). A new and versatile method for association generation. *Information Systems*, 2:333–347.

Bayardo, Jr., R. (1998). Efficiently mining long patterns from databases. In (Haas and Tiwary, 1998), pages 85–93.

Bonchi, F., Giannotti, F., Mazzanti, A., and Pedreschi, D. (2003). Exante: Anticipated data reduction in constrained pattern mining. In (Lavrac et al., 2003).

Borgelt, C. and Kruse, R. (2002). Induction of association rules: Apriori implementation. In Härdle, W. and Rönz, B., editors, *Proceedings of the 15th Conference on Computational Statistics*, pages 395–400. Physica-Verlag.

Boulicaut, J.-F., Bykowski, A., and Rigotti, C. (2003). Free-sets: A condensed representation of boolean data for the approximation of frequency queries. *Data Mining and Knowledge Discovery*, 7(1):5–22.

Brin, S., Motwani, R., Ullman, J., and Tsur, S. (1997). Dynamic itemset counting and implication rules for market basket data. In *Proceedings of the 1997 ACM SIGMOD International Conference on Management of Data*, volume 26(2) of *SIGMOD Record*, pages 255–264. ACM Press.

Burdick, D., Calimlim, M., and Gehrke, J. (2001). MAFIA: A maximal frequent itemset algorithm for transactional databases. In *Proceedings of the 17th International Conference on Data Engineering*, pages 443–452. IEEE Computer Society.

Bykowski, A. and Rigotti, C. (2001). A condensed representation to find frequent patterns. In *Proceedings of the Twentieth ACM SIGACT-SIGMOD-SIGART Symposium on Principles of Database Systems*, pages 267–273. ACM Press.

Calders, T. (2004). Computational complexity of itemset frequency satisfiability. In *Proceedings of the Twenty-third ACM SIGACT-SIGMOD-SIGART Symposium on Principles of Database Systems*, pages 143–154. ACM Press.

Calders, T. and Goethals, B. (2002). Mining all non-derivable frequent itemsets. In Elomaa, T., Mannila, H., and Toivonen, H., editors, *Proceedings of the 6th European Conference* on *Principles of Data Mining and Knowledge Discovery*, volume 2431 of *Lecture Notes in Computer Science*, pages 74–85. Springer.

Calders, T. and Goethals, B. (2003). Minimal k-free representations of frequent sets. In (Lavrac et al., 2003), pages 71–82.

Cercone, N., Lin, T., and Wu, X., editors (2001). *Proceedings of the 2001 IEEE International Conference on Data Mining*. IEEE Computer Society.

Dayal, U., Gray, P., and Nishio, S., editors (1995). *Proceedings 21th International Conference on Very Large Data Bases*. Morgan Kaufmann.

G. Grahne, J. Z. (2003). Efficiently using prefix-trees in mining frequent itemset. In (Goethals and Zaki, 2003).

G. Ramesh, W. Maniatty, M. Z. (2003). Feasible itemset distributions in Data Mining: theory and application. In *Proceedings of the Twenty-second ACM SIGACT-SIGMOD-SIGART Symposium on Principles of Database Systems*, pages 284–295. ACM Press.

Geerts, F., Goethals, B., and den Bussche, J. V. (2001). A tight upper bound on the number of candidate patterns. In (Cercone et al., 2001), pages 155–162.

Getoor, L., Senator, T., Domingos, P., and Faloutsos, C., editors (2003). *Proceedings of the Ninth ACM SIGKDD International Conference on Knowledge Discovery and Data Mining*. ACM Press.

Goethals, B. (2004). Memory issues in frequent itemset mining. In Haddad, H., Omicini, A., Wainwright, R., and Liebrock, L., editors, *Proceedings of the 2004 ACM symposium on Applied computing*, pages 530–534. ACM Press.

Goethals, B. and den Bussche, J. V. (2000). On supporting interactive association rule mining. In Kambayashi, Y., Mohania, M., and Tjoa, A., editors, *Proceedings of the Second International Conference on Data Warehousing and Knowledge Discovery*, volume 1874 of *Lecture Notes in Computer Science*, pages 307–316. Springer.

Goethals, B. and Zaki, M., editors (2003). *Proceedings of the ICDM 2003 Workshop on Frequent Itemset Mining Implementations*, volume 90 of *CEUR Workshop Proceedings*.

Gouda, K. and Zaki, M. (2001). Efficiently mining maximal frequent itemset. In (Cercone et al., 2001), pages 163–170.

Gunopulos, D., Khardon, R., Mannila, H., Saluja, S., Toivonen, H., and Sharma, R. (2003). Discovering all most specific sentences. *ACM Transactions on Database Systems*, 28(2):140–174.

Haas, L. and Tiwary, A., editors (1998). *Proceedings of the 1998 ACM SIGMOD International Conference on Management of Data*, volume 27(2) of *SIGMOD Record*. ACM Press.

Han, J., Pei, J., Yin, Y., and Mao, R. (2004). Mining frequent patterns without candidate generation: A frequent-pattern tree approach. *Data Mining and Knowledge Discovery*, 8(1):53–87.

Holsheimer, M., Kersten, M., Mannila, H., and Toivonen, H. (1995). A perspective on databases and Data Mining. In Fayyad, U. and Uthurusamy, R., editors, *Proceedings of the First International Conference on Knowledge Discovery and Data Mining*, pages 150–155. AAAI Press.

Lavrac, N., Gamberger, D., Blockeel, H., and Todorovski, L., editors (2003). *Proceedings of the 7th European Conference on Principles and Practice of Knowledge Discovery in Databases*, volume 2838 of *Lecture Notes in Computer Science*. Springer.

Liu, G., Lu, H., Yu, J., Wei, W., and Xiao, X. (2003). AFOPT: An efficient implementation of pattern growth approach. In (Goethals and Zaki, 2003).

Mannila, H. (1997). Inductive databases and condensed representations for Data Mining. In Maluszynski, J., editor, *Proceedings of the 1997 International Symposium on Logic*

Programming, pages 21–30. MIT Press.

Mannila, H. (2002). Local and global methods in Data Mining: Basic techniques and open problems. In Widmayer, P., Ruiz, F., Morales, R., Hennessy, M., Eidenbenz, S., and Conejo, R., editors, *Proceedings of the 29th International Colloquium on Automata, Languages and Programming*, volume 2380 of *Lecture Notes in Computer Science*, pages 57–68. Springer.

Mannila, H., Toivonen, H., and Verkamo, A. (1994). Efficient algorithms for discovering association rules. In Fayyad, U. and Uthurusamy, R., editors, *Proceedings of the AAAI Workshop on Knowledge Discovery in Databases*, pages 181–192. AAAI Press.

Mielikäinen, T. (2003). On inverse frequent set mining. In Du, W. and Clifton, C., editors, *2nd Workshop on Privacy Preserving Data Mining*, pages 18–23.

Ng, R., Lakshmanan, L., Han, J., and Pang, A. (1998). Exploratory mining and pruning optimizations of constrained association rules. In (Haas and Tiwary, 1998), pages 13–24.

Orlando, S., Palmerini, P., Perego, R., and Silvestri, F. (2002). Adaptive and resource-aware mining of frequent sets. In Kumar, V., Tsumoto, S., Yu, P., and N.Zhong, editors, *Proceedings of the 2002 IEEE International Conference on Data Mining*. IEEE Computer Society. To appear.

Pan, F., Cong, G., and A.K.H. Tung, J. Yang, M. Z. (2003). Carpenter: finding closed patterns in long biological datasets. In (Getoor et al., 2003), pages 637–642.

Park, J., Chen, M.-S., and Yu, P. (1995). An effective hash based algorithm for mining association rules. In *Proceedings of the 1995 ACM SIGMOD International Conference on Management of Data*, volume 24(2) of *SIGMOD Record*, pages 175–186. ACM Press.

Pasquier, N., Bastide, Y., Taouil, R., and Lakhal, L. (1999). Discovering frequent closed itemsets for association rules. In Beeri, C. and Buneman, P., editors, *Proceedings of the 7th International Conference on Database Theory*, volume 1540 of *Lecture Notes in Computer Science*, pages 398–416. Springer.

Rioult, F., Boulicaut, J.-F., and B. Crémilleux, J. B. (2003). Using transposition for pattern discovery from microarray data. In Zaki, M. and Aggarwal, C., editors, *ACM SIGMOD Workshop on Research Issues in Data Mining and Knowledge Discovery*, pages 73–79. ACM Press.

Rokach, L., Averbuch, M., and Maimon, O., Information retrieval system for medical narrative reports. Lecture notes in artificial intelligence, 3055. pp. 217-228, Springer-Verlag (2004).

Savasere, A., Omiecinski, E., and Navathe, S. (1995). An efficient algorithm for mining association rules in large databases. In (Dayal et al., 1995), pages 432–444.

Srikant, R. (1996). *Fast algorithms for mining association rules and sequential patterns*. PhD thesis, University of Wisconsin, Madison.

Srikant, R. and Agrawal, R. (1995). Mining generalized association rules. In (Dayal et al., 1995), pages 407–419.

Srikant, R., Vu, Q., and Agrawal, R. (1997). Mining association rules with item constraints. In Heckerman, D., Mannila, H., and Pregibon, D., editors, *Proceedings of the Third International Conference on Knowledge Discovery and Data Mining*, pages 66–73. AAAI Press.

Toivonen, H. (1996). Sampling large databases for association rules. In Vijayaraman, T., Buchmann, A., Mohan, C., and Sarda, N., editors, *Proceedings 22nd International Conference on Very Large Data Bases*, pages 134–145. Morgan Kaufmann.

Uno, T. and Satoh, K. (2003). Detailed description of an algorithm for enumeration of maximal frequent sets with irredundant dualization. In (Goethals and Zaki, 2003).

Vaidya, J. and Clifton, C. (2002). Privacy preserving association rule mining in vertically partitioned data. In Hand, D., Keim, D., and Ng, R., editors, *Proceedings of the Eight ACM SIGKDD International Conference on Knowledge Discovery and Data Mining*, pages 639–644. ACM Press.

Wang, J., Han, J., and Pei, J. (2003). CLOSET+: searching for the best strategies for mining frequent closed itemsets. In (Getoor et al., 2003), pages 236–245.

Yang, G. (2004). The complexity of mining maximal frequent itemsets and maximal frequent patterns. In DuMouchel, W., Gehrke, J., Ghosh, J., and Kohavi, R., editors, *Proceedings of the Tenth ACM SIGKDD International Conference on Knowledge Discovery and Data Mining*. ACM Press.

Zaki, M. (2000). Scalable algorithms for association mining. *IEEE Transactions on Knowledge and Data Engineering*, 12(3):372–390.

Zaki, M. and Gouda, K. (2003). Fast vertical mining using diffsets. In (Getoor et al., 2003), pages 326–335.

Zaki, M. and Hsiao, C.-J. (2002). CHARM: An efficient algorithm for closed itemset mining. In Grossman, R., Han, J., Kumar, V., Mannila, H., and Motwani, R., editors, *Proceedings of the Second SIAM International Conference on Data Mining*.

17

Constraint-based Data Mining

Jean-Francois Boulicaut[1] and Baptiste Jeudy[2]

[1] INSA Lyon, LIRIS CNRS FRE 2672
69621 Villeurbanne cedex, France. jean-francois.boulicaut@insa-lyon.fr
[2] University of Saint-Etienne, EURISE
42023 Saint-Etienne Cedex 2, France. baptiste.jeudy@univ-st-etienne.fr

Summary. Knowledge Discovery in Databases (KDD) is a complex interactive process. The promising theoretical framework of inductive databases considers this is essentially a querying process. It is enabled by a query language which can deal either with raw data or patterns which hold in the data. Mining patterns turns to be the so-called inductive query evaluation process for which constraint-based Data Mining techniques have to be designed. An inductive query specifies declaratively the desired constraints and algorithms are used to compute the patterns satisfying the constraints in the data. We survey important results of this active research domain. This chapter emphasizes a real breakthrough for hard problems concerning local pattern mining under various constraints and it points out the current directions of research as well.

Key words: Inductive querying, constraints, local patterns

17.1 Motivations

Knowledge Discovery in Databases (KDD) is a complex interactive and iterative process which involves many steps that must be done sequentially. Supporting the whole KDD process has enjoyed great popularity in recent years, with advances in both research and commercialization. We however still lack of a generally accepted underlying framework and this hinders the further development of the field. We believe that the quest for such a framework is a major research priority and that the *inductive database* approach (IDB) (Imielinski and Mannila, 1996, De Raedt, 2003) is one of the best candidates in this direction. IDBs contain not only data, but also patterns. Patterns can be either *local patterns* (e.g., itemsets, association rules, sequences) which are of descriptive nature, or *global patterns/models* (e.g., classifiers) which are generally of predictive nature. In an IDB, ordinary queries can be used to access and manipulate data, while *inductive queries* can be used to generate (mine), manipulate, and apply patterns. KDD becomes an extended querying process where the analyst can control the whole process since he/she specifies the data and/or patterns of interests.

O. Maimon, L. Rokach (eds.), *Data Mining and Knowledge Discovery Handbook*, 2nd ed.,
DOI 10.1007/978-0-387-09823-4_17,

The IDB framework is appealing because it employs *declarative* queries instead of ad-hoc *procedural constructs*. As declarative inductive queries are often formulated using constraints, inductive querying needs for *constraint-based Data Mining* techniques and is concerned with defining the necessary constraints.

It is useful to abstract the meaning of inductive queries. A simple model has been introduced in (Mannila and Toivonen, 1997). Given a language $\mathcal{L}$ of patterns (e.g., itemsets), the *theory* of a database $\mathcal{D}$ w.r.t. $\mathcal{L}$ and a selection predicate $\mathcal{C}$ is the set $\mathsf{Th}(\mathcal{D},\mathcal{L},\mathcal{C}) = \{\varphi \in \mathcal{L} \mid \mathcal{C}(\varphi,\mathcal{D}) = \mathsf{true}\}$. The predicate selection or *constraint* $\mathcal{C}$ indicates whether a pattern φ is interesting or not (e.g., φ is "frequent" in $\mathcal{D}$). We say that computing $\mathsf{Th}(\mathcal{D},\mathcal{L},\mathcal{C})$ is the evaluation for the inductive query $\mathcal{C}$ defined as a boolean expression over primitive constraints. Some of them can refer to the "behavior" of a pattern in the data (e.g., its "frequency" is above a threshold). Frequency is indeed the most studied case of *evaluation function*. Some others define syntactical restrictions (e.g., the "length" of the pattern is below a threshold) and checking them does not need any access to the data. Preprocessing concerns the definition of a mining context $\mathcal{D}$, the mining phase is generally the computation of a theory while post-processing is often considered as a querying activity on a materialized theory. To support the whole KDD process, it is important to support the specification and the computation of many different but correlated theories.

According to this formalization, solving an inductive query needs for the computation of every pattern which satisfies $\mathcal{C}$. We emphasized that the model is however quite general: beside the itemsets or sequences, $\mathcal{L}$ can denote, e.g., the language of partitions over a collection of objects or the language of decision trees on a collection of attributes. In these cases, classical constraints specify some function optimization. If the completeness assumption can be satisfied for most of the local pattern discovery tasks, it is generally impossible for optimization tasks like accuracy optimization during predictive model mining. In this case, heuristics or incomplete techniques are needed, which, e.g., compute sub-optimal decision trees. Very few techniques for constraint-based mining of models have been considered (see (Garofalakis and Rastogi, 2000) for an exception) and we believe that studying constraint-based clustering or constraint-based mining of classifiers will be a major topic for research in the near future. Starting from now, we focus on local pattern mining tasks.

It is well known that a "generate and test" approach that would enumerate the patterns of $\mathcal{L}$ and then test the constraint $\mathcal{C}$ is generally impossible. A huge effort has been made by data mining researchers to make an active use of the primitive constraints occurring in $\mathcal{C}$ (*solver* design) such that useful mining query evaluation is tractable. On one hand, researchers have designed solvers for important primitive constraints. A famous example is the one of frequent itemset mining (FIM) where the data is a set of transactions, the patterns are itemsets and the primitive constraint is a minimal frequency constraint. A second major line of research has been to consider specific, say ad-hoc, techniques for conjunctions of some primitives constraints. Examples of seminal work are (Srikant *et al.*, 1997) for syntactic constraints on frequent itemsets, (Pasquier *et al.*, 1999) for frequent and closed set mining, or (Garofalakis *et al.*, 1999) for mining sequences that are both frequent and satisfy a given regular expression in a sequence database. Last but not the least, a major progress has concerned the design of generic algorithms for mining under conjunctions or arbitrary boolean combination of primitive constraints. A pioneer contribution has been (Ng *et al.*, 1998) and this kind of work consists in a classification of constraint properties and the design of solving strategies according to these properties (e.g., anti-monotonicity, monotonicity, succinctness).

Along with constraint-based Data Mining, the concept of *condensed representation* has emerged as a key concept for inductive querying. The idea is to compute $\mathscr{CR} \subset \mathsf{Th}(\mathscr{D},\mathscr{L},\mathscr{C})$ while deriving $\mathsf{Th}(\mathscr{D},\mathscr{L},\mathscr{C})$ from $\mathscr{CR}$ can be performed efficiently. In the context of huge database mining, efficiently means without any further access to $\mathscr{D}$. Starting from (Mannila and Toivonen, 1996) and its concrete application to frequency queries in (Boulicaut and Bykowski, 2000), many useful condensed representations have been designed the last 5 years. Interestingly, we can consider condensed representation mining as a constraint-based Data Mining task (Jeudy and Boulicaut, 2002). It provides not only nice examples of constraint-based mining techniques but also important cross-fertilization possibilities (combining the both concepts) for optimizing inductive queries in very hard contexts.

Section 17.2 provides the needed notations and concepts. It introduces the pattern domains of itemsets and sequences for which most of the constraint-based Data Mining techniques have been designed. Section 17.3 recalls the principal results for solving anti-monotonic constraints. Section 17.4 concerns the introduction of non anti-monotonic constraints and the various strategies which have been proposed. Section 17.5 concludes and points out the actual directions of research.

17.2 Background and Notations

Given a database $\mathscr{D}$, a pattern language $\mathscr{L}$ and a constraint $\mathscr{C}$, let us first assume that we have to compute $\mathsf{Th}(\mathscr{D},\mathscr{L},\mathscr{C}) = \{\varphi \in \mathscr{L} \mid \mathscr{C}(\varphi,\mathscr{D}) = \mathsf{true}\}$. Our examples concern local pattern discovery tasks based on itemsets and sequences.

Itemsets have been studied a lot. Let $\mathscr{I} = \{\mathtt{A},\mathtt{B},\ldots\}$ be a set of *items*. A *transaction* is a subset of $\mathscr{I}$ and a database $\mathscr{D}$ is a multiset of transactions. An *itemset* is a set of items and a transaction t is said to *support* an itemset S if $S \subseteq t$. The *frequency* $\mathsf{freq}(S)$ of an itemset S is defined as the number of transactions that support S. $\mathscr{L}$ is the collection of all itemsets, i.e., $2^{\mathscr{I}}$. The most studied primitive constraint is the *minimum frequency constraint* $\mathscr{C}_{\sigma\text{-}\mathsf{freq}}$ which is satisfied by itemsets having a frequency greater than the threshold σ. Many other constraints have been studied such as syntactical constraints, e.g., $\mathtt{B} \in X$ whose testing does not need any access to the data. (Ng *et al.*, 1998) is a rather systematic study of many primitive constraints on itemsets (see also Section 17.4). (Boulicaut, 2004) surveys some new primitive constraints based on the closure evaluation function. The closure of an itemset S in $\mathscr{D}$, $f(S,\mathscr{D})$, is the maximal superset of S which has the same frequency than S in $\mathscr{D}$. Furthermore, a set S is closed in $\mathscr{D}$ if $S = f(S,\mathscr{D})$ in which case we say that it satisfies $\mathscr{C}_{\mathsf{clos}}$. Freeness is one of the first proposals for constraint-based mining of closed set generators: free itemsets (Boulicaut *et al.*, 2000) (also called key patterns in (Bastide *et al.*, 2000B)) are itemsets whose frequencies are different from the frequencies of all their subsets. We say that they satisfy the $\mathscr{C}_{\text{free}}$ constraint. An important result is that $\{f(S,\mathscr{D}) \in 2^{\mathscr{I}} \mid \mathscr{C}_{\text{free}}(S,\mathscr{D}) = \text{true}\} = \{S \in 2^{\mathscr{I}} \mid \mathscr{C}_{\text{clos}}(S,\mathscr{D}) = \text{true}\}$. For instance, in the toy data set of Figure 17.1, {A,C} is a free set and {A, C, D}, i.e., its closure, is a closed set.

Sequential pattern mining from sequence databases (i.e., $\mathscr{D}$ is a multiset of sequences) has been studied as well. Many different types of sequential patterns have been considered for which different *subpattern* relations can be defined. For instance, we could say that bc is a subpattern (substring) of abca but aa is not. In other proposals, aa would be considered as a subpattern of abca. Discussing this in details is not relevant for this chapter. The key point is that, a frequency evaluation function can be defined for sequential patterns (number of sequences in $\mathscr{D}$ for which the pattern is a subpattern). The pattern language $\mathscr{L}$ is then the infinite set of sequences which can be built on some alphabet. Many primitive constraints can be defined, e.g., minimal frequency or syntactical constraints specified by regular expressions. Interestingly, new constraints can exploit the spatial or temporal order, e.g., the min-gap and max-gap constraints (see, e.g., (Zaki, 2000) and (Pei *et al.*, 2002) for a recent survey).

Naive approaches that would compute $\text{Th}(\mathscr{D},\mathscr{L},\mathscr{C})$ by enumerating every pattern φ of the search space $\mathscr{L}$ and test the constraint $\mathscr{C}(\varphi,\mathscr{D})$ afterwards can not work. Even though checking $\mathscr{C}(\varphi,\mathscr{D})$ can be cheap, this strategy fails because of the size of the search space. For instance, we have $2^{|\mathscr{I}|}$ itemsets and we often have to cope with hundreds or thousands of items in practical applications. Moreover, for sequential pattern mining, the search space is infinite.

For a given constraint, the search space $\mathscr{L}$ is often structured by a specialization relation which provides a lattice structure. For important constraints, the specialization relation has an anti-monotonicity property. For instance, set inclusion for itemsets or substring for strings are anti-monotonic specialization relations w.r.t. a minimal frequency constraint. Anti-monotonicity means that when a pattern does not satisfy $\mathscr{C}$ (e.g., an itemset is not frequent) then none of its specializations can satisfy $\mathscr{C}$ (e.g., none of its supersets are frequent). It becomes possible to prune huge parts of the search space which can not contain interesting patterns. This has been studied within the "learning as search" framework (Mitchell, 1980) and the generic *level-wise algorithm* from (Mannila and Toivonen, 1997) has inspired many algorithmic developments (see Section 17.3). In this context where we say that the constraint $\mathscr{C}$ is anti-monotonic, the most specific patterns constitute the *positive border* of the theory (denoted $\mathscr{B}d^+(\mathscr{C})$) (Mannila and Toivonen, 1997) and $\mathscr{B}d^+(\mathscr{C})$ is a condensed representation of $\text{Th}(\mathscr{D},\mathscr{L},\mathscr{C})$. It corresponds to the S set in the terminology of versions spaces (Mitchell, 1980). For instance, the collection of the maximal frequent patterns $\mathscr{B}d^+(\mathscr{C}_{\sigma\text{-freq}})$ in $\mathscr{D}$ is generally several orders of magnitude smaller than the complete collection of the frequent patterns in $\mathscr{D}$. It is a condensed representation for $\text{Th}(\mathscr{D},2^{\mathscr{I}},\mathscr{C}_{\sigma\text{-freq}})$: deriving subsets (i.e., generalizations) of each maximal frequent set (i.e., each most specific pattern) enables to regenerate the whole collection of the frequent sets (i.e., the whole theory of interesting patterns w.r.t. the constraint).

In many applications, however, the user wants not only the collection of the patterns satisfying $\mathscr{C}$ but also the results of some evaluation functions for these patterns.

This is quite typical for the frequent pattern discovery problem: these patterns are generally exploited in a post-processing step to derive more useful statements about the data, e.g., the popular frequent association rules which have a high enough confidence (Agrawal *et al.*, 1996). This can be done efficiently if we compute not only the collection of frequent itemsets but also their frequencies. In fact, the semantics of an inductive query is better captured by the concept of *extended theories*. An extended theory w.r.t. an evaluation function f on a domain $\mathscr{V}$ is $\mathsf{Th}_x(\mathscr{D},\mathscr{L},\mathscr{C},f) = \{(\varphi, f(\varphi)) \in \mathscr{L} \otimes \mathscr{V} \mid \mathscr{C}(\varphi,\mathscr{D}) = \mathsf{true}\}$. The classical FIM problem turns to be the computation of $\mathsf{Th}_x(\mathscr{D}, 2^{\mathscr{I}}, \mathscr{C}_{\sigma\text{-freq}}, \mathsf{freq})$. Another example concerns the closure evaluation function.

For instance, $\{(\varphi, f(\varphi)) \in 2^{\mathscr{I}} \otimes 2^{\mathscr{I}} \mid \mathscr{C}_{\sigma\text{-freq}}(\varphi,\mathscr{D}) = \mathsf{true}\}$ is the collection of the frequent sets and their closures, i.e., the frequent closed sets.

An alternative and useful specification for the frequent closed sets is $\{(\varphi, f(\varphi)) \in 2^{\mathscr{I}} \otimes 2^{\mathscr{I}} \mid \mathscr{C}_{\sigma\text{-freq}}(\varphi,\mathscr{D}) \wedge \mathscr{C}_{\mathsf{free}}(\varphi,\mathscr{D}) = \mathsf{true}\}$.

Condensed representations can be designed for extended theories as well. Now, a condensed representation $\mathscr{CR}$ must enable to regenerate the patterns, but also the values of the evaluation function f on each pattern without any further access to the data. If the regenerated values for f are only approximated, the condensed representation is called *approximate*. Moreover, if the error on f can be bounded by ε, the approximate condensed representation is called an *ε-adequate representation* of the extended theory (Mannila and Toivonen, 1996). The idea is that we can trade off the precision on the evaluation function values with computational feasibility.

Most of condensed representations studied so far are condensed representations of the frequent itemsets. We have the maximal frequent itemsets (see, e.g., (Bayardo, 1998)), the frequent closed itemsets (see, e.g., (Pasquier *et al.*, 1999, Boulicaut and Bykowski, 2000)), the frequent free itemsets and the δ-free itemsets (Boulicaut *et al.*, 2000, Boulicaut *et al.*, 2003), the disjunction-free sets (Bykowski and Rigotti, 2003), the non-derivable itemsets (Calders and Goethals, 2002), the frequent pattern bases (Pei *et al.*, 2002), etc. Except for the maximal frequent itemsets from which it is not possible to get a useful approximation of the needed frequencies, these are condensed representations of the extended theory $\mathsf{Th}_x(\mathscr{D}, 2^{\mathscr{I}}, \mathscr{C}_{\sigma\text{-freq}}, \mathsf{freq})$ and δ-free itemsets and pattern bases are approximate representations.

Condensed representations have three main advantages. First, they contain (almost) the same information than the whole theory but are significantly smaller (generally by several orders of magnitude), which means that they are more easily stored or manipulated. Next, the computation of $\mathscr{CR}$ and the regeneration of the theory Th from $\mathscr{CR}$ is often less expensive than the direct computation of Th. One can even say that, as soon as a transactional data set is dense, mining condensed representations of the frequent itemsets is the only way to solve the FIM problem for practical applications. Last, many proposals emphasize the use of condensed representations for deriving directly useful patterns (i.e., skipping the regeneration phase). This is obvious for feature construction (see, e.g., (Kramer *et al.*, 2001)) but has been considered also for the generation of non redundant association rules (see, e.g., (Bastide *et al.*, 2000A)) or interesting classification rules (Crémilleux and Boulicaut, 2002)).

17.3 Solving Anti-Monotonic Constraints

In this section, we consider efficient solutions to compute (extended) theories for anti-monotonic constraints. We still focus on constraint-based mining of itemsets when the constraint is anti-monotonic. It is however straightforwardly extended to many other pattern domains.

An *anti-monotonic* constraint on itemsets is a constraint denoted $\mathscr{C}_{\mathsf{am}}$ such that for all itemsets $S, S' \in 2^{\mathscr{I}}$: $(S' \subseteq S \wedge S$ satisfies $\mathscr{C}_{\mathsf{am}}) \Rightarrow S'$ satisfies $\mathscr{C}_{\mathsf{am}}$. $\mathscr{C}_{\sigma\text{-freq}}$, $\mathscr{C}_{\mathsf{free}}$, $\mathtt{A} \notin S$, $S \subseteq \{\mathtt{A},\mathtt{B},\mathtt{C}\}$ and $S \cap \{\mathtt{A},\mathtt{B},\mathtt{C}\} = \emptyset$ are examples of anti-monotonic constraints. Furthermore, it is clear that a disjunction or a conjunction of anti-monotonic constraints is an anti-monotonic constraint.

Let us be more precise on the useful concept of border (Mannila and Toivonen, 1997). If $\mathscr{C}_{\mathsf{am}}$ denotes an anti-monotonic constraint and the goal is to compute $T = \mathsf{Th}(\mathscr{D}, 2^{\mathscr{I}}, \mathscr{C}_{\mathsf{am}})$, then $\mathscr{B}d^{+}(\mathscr{C}_{\mathsf{am}})$ is the collection of the maximal (w.r.t. the set inclusion) itemsets of T that satisfy $\mathscr{C}_{\mathsf{am}}$ and $\mathscr{B}d^{-}(\mathscr{C}_{\mathsf{am}})$ is the collection of the minimal (w.r.t. the set inclusion) itemsets that do not satisfy $\mathscr{C}_{\mathsf{am}}$.

Some algorithms have been designed for computing directly the positive borders, i.e., looking for the complete collection of the most specific patterns. A famous one is the Max-Miner algorithm which uses a clever enumeration technique for computing depth-first the maximal frequent sets (Bayardo, 1998). Other algorithms for computing maximal frequent sets are described in (Lin and Kedem, 2002, Burdick *et al.*, 2001, Goethals and Zaki, 2003). The computation of positive borders with applications to not only itemset mining but also dependency discovery, the generic “dualize and advance” framework, is studied in (Gunopulos *et al.*, 2003).

The levelwise algorithm by Mannila and Toivonen (Mannila and Toivonen, 1997) has influenced many research in data mining. It computes $\mathsf{Th}(\mathscr{D}, 2^{\mathscr{I}}, \mathscr{C}_{\mathsf{am}})$ levelwise in the lattice ($\mathscr{L}$ associated to its specialization relation) by considering first the most general patterns (e.g., the singleton in the FIM problem). Then, it alternates candidate evaluation (e.g., frequency counting or other checks for anti-monotonic constraints) and candidate generation (e.g., building larger itemsets from discovered interesting itemsets) phases. Candidate generation can be considered as the computation of the negative border of the previously computed collection. *Candidate pruning* is a major issue and it can be performed partly during the generation phase or just after: indeed, any candidate whose one generalization does not satisfy $\mathscr{C}_{\mathsf{am}}$ can be pruned safely (e.g., any itemset whose one of its subsets is not frequent can be removed). The algorithm stops when it can not generate new candidates or, in other terms, when the most specific patterns have been found (e.g., all the maximal frequent itemsets).

The Apriori algorithm (Agrawal *et al.*, 1996) is clearly the most famous instance of this levelwise algorithm. It computes $\mathsf{Th}(\mathscr{D}, 2^{\mathscr{I}}, \mathscr{C}_{\sigma\text{-freq}}, \mathsf{freq})$ and it uses a clever candidate generation technique. A lot of work has been done for efficient implementations of Apriori-like algorithms.

Pruning based on anti-monotonic constraints has been proved efficient on hard problems, i.e., huge volume and high dimensional data sets. The many experimental results which are available nowadays prove that the minimal frequency is often an extremely selective constraint in real data sets. Interestingly, an algorithm like AcMiner (Boulicaut *et al.*, 2000, Boulicaut *et al.*, 2003) which can compute frequent closed sets (closeness is not an anti-monotonic constraint) via the frequent free sets exploits these pruning possibilities. Indeed, the conjunction of freeness and minimal frequency is an anti-monotonic constraint which enables an efficient pruning in dense and/or highly correlated data sets.

The dual property of monotonicity is interesting as well. A *monotonic* constraint on itemsets is a constraint denoted $\mathscr{C}_{\mathsf{m}}$ such that for all itemsets $S, S' \in 2^{\mathscr{I}} : (S \subseteq S' \wedge S$ satisfies $\mathscr{C}_{\mathsf{m}}) \Rightarrow S'$ satisfies $\mathscr{C}_{\mathsf{m}}$. A constraint is monotonic when its negation is anti-monotonic (and vice-versa). In the itemset pattern domain, the maximal frequency constraint or a syntactic constraint like $\mathtt{A} \in S$ are examples of monotonic constraints.

The concept of border can be adapted to monotonic constraints. The positive border $\mathscr{B}d^{+}(\mathscr{C}_{\mathsf{m}})$ of a monotonic constraint $\mathscr{C}_{\mathsf{m}}$ is the collection of the most general patterns that satisfy the constraint. The theory $\mathsf{Th}(\mathscr{D}, \mathscr{L}, \mathscr{C}_{\mathsf{m}})$ is then the set of patterns that are more specific than the patterns of the border $\mathscr{B}d^{+}(\mathscr{C}_{\mathsf{m}})$. For instance, we have $\mathscr{B}d^{+}(\mathtt{A} \in S) = \{\mathtt{A}\}$ and the positive border of the monotonic maximal frequency constraint is the collection of the smallest itemsets which are not frequent in the data. In other terms, a monotonic constraint defines also a border in the search space which corresponds to the G set in the version space terminology (see Figure 17.1 for an example).

The recent work has indeed exploited this duality for solving conjunctions of monotonic and anti-monotonic constraints (see Section 17.4.2).

17.4 Introducing non Anti-Monotonic Constraints

Pushing anti-monotonic constraints in the levelwise algorithm always leads to less constraint checking. Of course, anti-monotonic constraints are exploited into alternative frameworks, like depth-first algorithms.

However, this is no longer the case when pushing non anti-monotonic constraints. For instance, if an itemset does not satisfy an anti-monotonic constraint $\mathscr{C}_{\mathsf{am}}$, then its supersets can be pruned. But if this itemset does not satisfy the non anti-monotonic constraint, then its supersets are not pruned since the algorithm does not test $\mathscr{C}_{\mathsf{am}}$ on it. Pushing non anti-monotonic constraint can therefore lead to less efficient pruning (Boulicaut and Jeudy, 2000, Garofalakis *et al.*, 1999). Clearly, we have here a trade-off between anti-monotonic pruning and monotonic pruning which can be decided if the selectivity of the various constraints is known in advance, which is obviously not the case in most of the applications. Nice contributions have considered boolean expressions over monotonic and anti-monotonic constraints. The problem is still quite open for optimization constraints.

17.4.1 The Seminal Work

MultipleJoins, Reorder and Direct

Srikant *et al.* (Srikant *et al.*, 1997) have been the first to address constraint-based mining of itemsets when the constraint $\mathscr{C}$ is not reduced to the minimum frequency constraint $\mathscr{C}_{\sigma\text{-freq}}$. They consider syntactical constraints built on two kinds of primitive constraints: $\mathscr{C}_i(S) = (i \in S)$, and $\mathscr{C}_{\neg i}(S) = (i \notin S)$ where $i \in \mathscr{I}$. They also introduce new constraints if a taxonomy on items is available. A taxonomy (also called a *is-a* relation) is an acyclic relation r on $\mathscr{I}$. For instance, if the items are products like Milk, Jackets... the relation can state that Milk *is-a* Beverages, Jackets *is-a* Outer-wear, ... The primitive constraints related to a taxonomy are: $\mathscr{C}_{\mathrm{a}(i)}(S) = (S \cap \mathrm{ancestor}(i) \neq \emptyset)$, $\mathscr{C}_{\mathrm{d}(i)}(S) = (S \cap \mathrm{descendant}(i) \neq \emptyset)$, and their negations. Functions ancestor and descendant are defined using the transitive closure r^* of r: we have $\mathrm{ancestor}(i) = \{i' \in \mathscr{I} \mid r^*(i',i)\}$ and $\mathrm{descendant}(i) = \{i' \in \mathscr{I} \mid r^*(i,i')\}$. These new constraints can be rewritten using the two primitive constraints $\mathscr{C}_i$ and $\mathscr{C}_{\neg i}$, e.g., $\mathscr{C}_{\mathrm{desc}(i)}(S) = \bigvee_{j \in \mathrm{descendant}(i)} \mathscr{C}_j(S)$.

It is now possible to specify syntactical constraints $\mathscr{C}_{\mathrm{synt}}$ as a boolean combination of the primitive constraints which is written in disjunctive normal form, i.e., $\mathscr{C}_{\mathrm{synt}} = D_1 \vee D_2 \vee \ldots \vee D_m$ where each D_k is $\mathscr{C}_{k1} \wedge \mathscr{C}_{k2} \wedge \ldots \wedge \mathscr{C}_{kn_k}$ and $\mathscr{C}_{kj}$ is either $\mathscr{C}_i$ or $\mathscr{C}_{\neg i}$ with $i \in \mathscr{I}$.

Srikant *et al.* (1997) provide three algorithms to compute $\mathsf{Th}_x(\mathscr{D}, 2^{\mathscr{I}}, \mathscr{C}, \mathsf{freq})$ where $\mathscr{C} = \mathscr{C}_{\sigma\text{-freq}} \wedge \mathscr{C}_{\mathrm{synt}}$. The first two algorithms (MultipleJoins and Reorder) use a relaxation of the syntactical constraint. They show how to compute from $\mathscr{C}_{\mathrm{synt}}$ an itemset T such that every itemset S satisfying the $\mathscr{C}_{\mathrm{synt}}$ also satisfies the constraint $S \cap T \neq \emptyset$. This constraint is pushed in an Apriori-like levelwise algorithm to obtain MultipleJoins and Reorder (Reorder is a simplification of MultipleJoins). The third algorithm, Direct, does not use a relaxation and pushes the whole syntactical constraint at the extended cost of a more complex candidate generation phase. Experimental results confirm that the behavior of the algorithms depends clearly of the selectivity of the constraints on the considered data sets.

CAP

The CAP algorithm (Ng *et al.*, 1998) computes the extended theory $\mathsf{Th}_x(\mathscr{D}, 2^{\mathscr{I}}, \mathscr{C}, \mathsf{freq})$ for $\mathscr{C} = \mathscr{C}_{\sigma\text{-freq}} \wedge \mathscr{C}_{\mathsf{am}} \wedge \mathscr{C}_{\mathrm{succ}}$ where $\mathscr{C}_{\mathsf{am}}$ is an anti-monotonic syntactical constraint and $\mathscr{C}_{\mathrm{succ}}$ is a *succinct* constraint. A constraint $\mathscr{C}$ is succinct (Ng *et al.*, 1998) if it is a syntactical constraint and if we have itemsets $I_1, I_2, \ldots I_k$ such that $\mathscr{C}(S) = S \subseteq I_1 \wedge S \not\subseteq I_2 \wedge \ldots \wedge S \not\subseteq I_k$. Efficient candidate generation techniques can be performed for such constraints which can be considered as special cases of conjunctions of anti-monotonic and monotonic syntactical constraints.

In (Ng *et al.*, 1998), the syntactical constraints are conjunctions of primitive constraints which are $\mathscr{C}_i$, $\mathscr{C}_{\neg i}$ and constraints based on aggregates. They indeed assume that a value v is associated with each item i and denoted $i.v$ such that several aggregate functions can be used:

$$\mathrm{MAX}(S) = \max\{i.v \mid i \in S\}, \qquad \mathrm{MIN}(S) = \min\{i.v \mid i \in S\},$$

$$\mathrm{SUM}(S) = \sum_{i \in S} i.v, \qquad \mathrm{AVG}(S) = \frac{\mathrm{SUM}(S)}{|S|}.$$

These aggregate functions enable to define new primitive constraints $\mathrm{AGG}(S)\theta n$ where AGG is an aggregation function, θ is in $\{=,<,>\}$ and n is a number. In a market basket analysis application, v can be the price of each item and we can define aggregate constraints to extract, e.g., itemsets whose average price of items is above a given threshold ($\mathrm{AVG}(S) > 10$). Among these constraints, some are anti-monotonic (e.g., $\mathrm{SUM}(S) < 100$ if all the values are positive, $\mathrm{MIN}(S) > 10$), some are succinct (e.g., $\mathrm{MAX}(S) > 10$, $|S| > 3$) and others have no special properties and must be relaxed to be used in the CAP algorithm (e.g., $\mathrm{SUM}(S) < 10$, $\mathrm{AVG}(S) < 10$).

The candidate generation function of CAP algorithm is an improvement over Direct algorithm. However, it can not use all syntactical constraints like Direct (only conjunction of anti-monotonic and succinct constraints can be used by CAP). The CAP algorithm can also use aggregate constraints. These constraints could also be used in Direct but they would need to be rewritten in disjunctive normal form using $\mathscr{C}_i$ and $\mathscr{C}_{\neg i}$. This rewriting stage can be computationally expensive such that, in practice, we can not push aggregate constraints into Direct.

SPIRIT

In (Garofalakis *et al.*, 1999), the authors present several version of the SPIRIT algorithm to extract frequent sequences satisfying a regular expression (such sequences are called *valid* w.r.t. the regular expression). For instance, if the sequences consist of letters, the valid sequences with respect to the regular expression `a*(bb|cc)e` are the sequences that start with several `a` followed by either `bbe` or `cce`. In the general case, such a syntactical constraint is not anti-monotonic. The different versions of SPIRIT use more and more selective relaxations of this regular expression constraint. The first algorithm, SPIRIT(N), uses an anti-monotonic relaxation of the syntactical constraint. This constraint $\mathscr{C}_N$ is satisfied by sequences s such that all the items appearing in s also appear in the regular expression. With our running example, $\mathscr{C}_N(s)$ is true if s is built on letters `a`, `b`, `c`, and `e` only. A constraint $\mathscr{C}_L$ is used by the second algorithm, SPIRIT(L). It is satisfied by a sequence s if s is a legal sequence w.r.t. the regular expression. A sequence s is legal if we can find a valid sequence s' such that s is a suffix of s'. For instance, `cce` is a legal sequence w.r.t. our running example. The SPIRIT(V) algorithm uses the constraint $\mathscr{C}_V$ which is satisfied by all contiguous sub-sequences of a valid sequence. Finally, the SPIRIT(R) algorithm uses the full constraint $\mathscr{C}_R$ which is satisfied only by valid sequences. For the three first algorithms, a final post-processing step is necessary to filter out non-valid sequences. There is a subset relationship between the theories computed by these four algorithms: $\mathsf{Th}(\mathscr{D},\mathscr{L},\mathscr{C}_R \wedge \mathscr{C}_{\sigma\text{-freq}}) \subseteq \mathsf{Th}(\mathscr{D},\mathscr{L},\mathscr{C}_V \wedge \mathscr{C}_{\sigma\text{-freq}}) \subseteq \mathsf{Th}(\mathscr{D},\mathscr{L},\mathscr{C}_L \wedge \mathscr{C}_{\sigma\text{-freq}}) \subseteq \mathsf{Th}(\mathscr{D},\mathscr{L},\mathscr{C}_N \wedge \mathscr{C}_{\sigma\text{-freq}})$. Clearly, the first two algorithms are based mostly on minimal frequency pruning while the two last ones exploit further regular expression pruning. Here again, only a prior knowledge on constraint selectivity enables to inform the choice of one of the algorithms, i.e., one of the pruning strategies.

17.4.2 Generic Algorithms

We now sketch some important results for the evaluation of quite general forms of inductive queries.

Conjunction of Monotonic and Anti-Monotonic Constraints

Let us assume that we use constraints that are conjunctions of a monotonic constraint and an anti-monotonic one denoted $\mathcal{C}_{\mathsf{am}} \wedge \mathcal{C}_{\mathsf{m}}$. The structure of $\mathsf{Th}(\mathcal{D}, \mathcal{L}, \mathcal{C}_{\mathsf{am}} \wedge \mathcal{C}_{\mathsf{m}})$ is well known. Given the positive borders $\mathcal{B}d^+(\mathcal{C}_{\mathsf{am}})$ and $\mathcal{B}d^+(\mathcal{C}_{\mathsf{m}})$, the patterns belonging to $\mathsf{Th}(\mathcal{D}, \mathcal{L}, \mathcal{C}_{\mathsf{am}} \wedge \mathcal{C}_{\mathsf{m}})$ are exactly the patterns that are more specific than a pattern of $\mathcal{B}d^+(\mathcal{C}_{\mathsf{m}})$ and more general than a pattern of $\mathcal{B}d^+(\mathcal{C}_{\mathsf{am}})$. This kind of convex pattern collection is called a *Version Space* and is illustrated on Fig. 17.1.

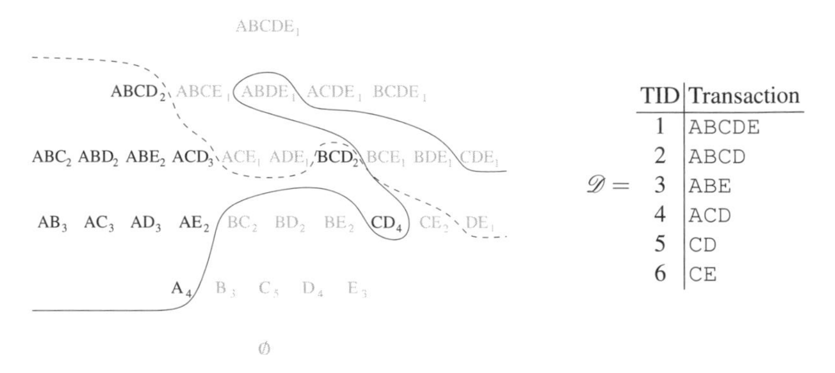

TID	Transaction
1	ABCDE
2	ABCD
3	ABE
4	ACD
5	CD
6	CE

Fig. 17.1. This figure shows the itemset lattice associated to $\mathcal{D}$ (the subscript number is the frequency of each itemset in $\mathcal{D}$). The itemsets above the black line satisfy the monotonic constraint $\mathcal{C}_{\mathsf{m}}(S) = (\mathtt{B} \in S) \vee (\mathtt{CD} \subseteq S)$ and the itemsets below the dashed line satisfy the anti-monotonic constraint $\mathcal{C}_{\mathsf{am}} = \mathcal{C}_{\mathsf{2\text{-}freq}}$. The black itemsets belong to the theory $\mathsf{Th}(\mathcal{D}, 2^{\mathcal{I}}, \mathcal{C}_{\mathsf{am}} \wedge \mathcal{C}_{\mathsf{m}})$. They are exactly the itemsets that are subsets of an element of $\mathcal{B}d^+(\mathcal{C}_{\mathsf{am}}) = \{\mathtt{ABCD}, \mathtt{ABE}, \mathtt{CE}\}$ and supersets of an element of $\mathcal{B}d^+(\mathcal{C}_{\mathsf{m}}) = \{\mathtt{A}, \mathtt{CD}\}$.

Several algorithms have been developed to deal with $\mathcal{C}_{\mathsf{am}} \wedge \mathcal{C}_{\mathsf{m}}$. The generic algorithm presented in (Boulicaut and Jeudy, 2000) computes the extended theory for a conjunction $\mathcal{C}_{\mathsf{am}} \wedge \mathcal{C}_{\mathsf{m}}$. It is a levelwise algorithm, but instead of starting the exploration with the most general patterns (as it is done for anti-monotonic constraints), it starts with the minimal itemsets (most general patterns) satisfying $\mathcal{C}_{\mathsf{m}}$, i.e., the itemsets of the border $\mathcal{B}d^+(\mathcal{C}_{\mathsf{m}})$. This is a generalization of MultipleJoins, Reorder and CAP: the constraint $T \cap S \neq \emptyset$ used in MultipleJoins and Reorder is indeed monotonic and succinct constraints used in CAP can be rewritten as the conjunction of a monotonic and an anti-monotonic constraints.

Since $\mathscr{B}d^+(\mathscr{C}_{\mathsf{am}})$ et $\mathscr{B}d^+(\mathscr{C}_{\mathsf{m}})$ characterize the theory of $\mathscr{C}_{\mathsf{am}} \wedge \mathscr{C}_{\mathsf{m}}$, these borders are a condensed representation of this theory. The Molfea algorithm and the Dualminer algorithms extract these two borders. They are interesting algorithms for feature extraction.

The Molfea algorithm presented in (Kramer *et al.*, 2001, De Raedt and Kramer, 2001) extract linear molecular fragments (i.e., strings) in a a partitioned database of molecules (say, active vs. inactive molecules). They consider conjunctions of a minimum frequency constraint (say in the active molecules), a maximum frequency constraint (say in the inactive ones) and syntactical constraints. The two borders are constructed in an incremental fashion, considering the constraints one after the other, using a level-wise algorithm for the frequency constraints and Mellish algorithm (Mellish, 1992) for the syntactical constraints. The Dualminer algorithm (Bucila *et al.*, 2003) uses a depth-first exploration similar to the one of Max-Miner whereas Dualminer deals with $\mathscr{C}_{\mathsf{am}} \wedge \mathscr{C}_{\mathsf{m}}$ instead of just $\mathscr{C}_{\mathsf{am}}$.

In (Bonchi *et al.*, 2003C), the authors consider the computation of not only borders but also the extended theory for $\mathscr{C}_{\mathsf{am}} \wedge \mathscr{C}_{\mathsf{m}}$. In this context, they show that the most efficient approach is not to reason on the search space only but both the search space and the transactions from the input data. They have a clever approach to data reduction based on the monotonic part. Not only it does not affect anti-monotonic pruning but also they demonstrate that the two pruning opportunities are mutually enhanced.

Arbitrary Expression over Monotonic and Anti Monotonic Constraints

The algorithms presented so far cannot deal with an arbitrary boolean expression consisting of monotonic and anti-monotonic constraints. These more general constraints are studied in (De Raedt *et al.*, 2002). Using the basic properties of monotonic and anti-monotonic constraints, the authors show that such a constraint can be rewritten as $(\mathscr{C}_{\mathsf{am}_1} \wedge \mathscr{C}_{\mathsf{m}_1}) \vee (\mathscr{C}_{\mathsf{am}_2} \wedge \mathscr{C}_{\mathsf{m}_2}) \vee \ldots \vee (\mathscr{C}_{\mathsf{am}_n} \wedge \mathscr{C}_{\mathsf{m}_n})$. The theory of each conjunction $(\mathscr{C}_{\mathsf{am}_i} \wedge \mathscr{C}_{\mathsf{m}_i})$ is a version space and the theory w.r.t. the whole constraint is a union of version spaces. The theory of each conjunction can be computed using any algorithm described in the previous sections. Since there are several ways to express the constraint as a disjunction of conjunctions, it is therefore desirable to find an expression in which the number of conjunction is minimal.

Conjunction of Arbitrary Constraints

When constraints are neither anti-monotonic nor monotonic, finding an efficient algorithm is difficult. The common approach is to design a specific strategy to deal with a particular class of constraints. Such algorithms are presented in the next section. A promising generic approach has been however presented recently. It is the concept of *witness* presented in (Kifer *et al.*, 2003) for itemset mining. This paper does not describe a mining algorithm but rather a pruning technique for non anti-monotonic and non monotonic constraints. Considering a sub-lattice $\AA$ of $2^{\mathscr{I}}$, the problem is to decide whether this sub-lattice can be pruned. A sub-lattice is characterized by its maximal element M and its minimal element m, i.e., the sub-lattice is the collection of all itemsets S such that $m \subseteq S \subseteq M$. To prune this sub-lattice, one must prove that none of its elements can satisfy the constraint $\mathscr{C}$. To check this, the authors introduce the concept of negative witness: a negative witness for $\mathscr{C}$ in the sub-lattice $\AA$ is an itemset W such that $\neg\mathscr{C}(W) \Rightarrow \forall X \in \AA,\ \neg\mathscr{C}(X)$. Therefore, if the constraint is not satisfied by the negative witness, then the whole sub-lattice can be pruned. Finding witnesses for anti-monotonic or monotonic constraints is easy : m is the witness for all anti-monotonic constraints and M for all monotonic ones. The authors then show how to compute efficiently witnesses for various tough constraints. For instance, for $\text{AVG}(S) > \sigma$, a witness is the set $m \cup \{i \in M \mid i.v > \sigma\}$. The authors also gives an algorithm (linear in the size of $\mathscr{I}$) to compute a witness for the difficult constraint $(\text{VAR}(S) > \sigma)$ where VAR denotes the variance.

17.4.3 Ad-hoc Strategies

Apart from generic algorithms, many algorithms have been designed to cope with specific classes of constraints. We select only two examples.

The FIC algorithm (Pei *et al.*, 2001) does a depth-first exploration of the itemset lattice. It is very efficient due to its clever data structure, a prefix-tree used to store the database. This algorithm can compute the extended theory for a conjunction $\mathscr{C}_{\text{am}} \wedge \mathscr{C}_{\text{m}} \wedge \mathscr{C}'$ where $\mathscr{C}'$ is *convertible anti-monotonic or monotonic*. A constraint $\mathscr{C}'$ is convertible anti-monotonic if there exists an order on the items such that, if itemsets are written using this order, every prefix of an itemset satisfying $\mathscr{C}'$ satisfies $\mathscr{C}'$. For instance, $\text{AVG}(S) > \sigma$ is convertible anti-monotonic if the items i are ordered by decreasing value $i.v$. The main problem with convertible constraints is that a conjunction of convertible constraints is generally not convertible.

Another example of an ad-hoc strategy is used in the c-Spade algorithm (Zaki, 2000). This algorithm is used to extract constrained sequences where each event in the sequences is dated. One of the constraints, the $max - gap$ constraint, states that two consecutive events occurring in a pattern must not be further apart than a given maximum gap. This constraint is neither anti-monotonic nor monotonic and a specific algorithm has been designed for it.

17.4.4 Other Directions of Research

Among others, let us introduce here three important directions of research.

Adaptive Pruning Strategies

We mentioned the trade-off between anti-monotonic pruning which is known to be quite efficient and pruning based on non anti-monotonic constraints. Since the selectivity of the various constraints is generally unknown, a quite exciting challenge is to look for adaptive strategies which can decide of the pruning strategy dynamically. (Bonchi *et al.*, 2003A, Bonchi *et al.*, 2003B) propose algorithms for frequent itemsets under syntactical monotonic constraints. (Albert-Lorincz and Boulicaut, 2003) considers frequent sequence mining under regular expression constraints. These are promising approaches to widen the applicability of constraint-based mining techniques in real contexts.

Combining Constraints and Condensed Representations

A few papers, e.g., (Boulicaut and Jeudy, 2000, Bonchi and Lucchese, 2004), deal with the problem of extracting constrained condensed representation. In these works, the aim is to compute a condensed representation of the extended theory $\mathsf{Th}_x(\mathscr{D}, 2^{\mathscr{I}}, \mathscr{C}_{\mathsf{am}} \wedge \mathscr{C}_{\mathsf{m}}, \mathsf{freq})$. In (Boulicaut and Jeudy, 2000), the authors use free itemsets, i.e., their algorithm computes the extended theory $\mathsf{Th}_x(\mathscr{D}, 2^{\mathscr{I}}, \mathscr{C}_{\mathsf{am}} \wedge \mathscr{C}_{\mathsf{m}} \wedge \mathscr{C}_{\mathsf{free}}, \mathsf{freq})$. In (Bonchi and Lucchese, 2004), the authors use closed itemsets, i.e., their algorithm computes the extended theory $\mathsf{Th}_x(\mathscr{D}, 2^{\mathscr{I}}, \mathscr{C}_{\mathsf{am}} \wedge \mathscr{C}_{\mathsf{m}} \wedge \mathscr{C}_{\mathsf{clos}}, \mathsf{freq})$. However, in these two works, the definition of free sets and closed sets have been modified to be able to regenerate the extended theory $\mathsf{Th}_x(\mathscr{D}, 2^{\mathscr{I}}, \mathscr{C}_{\mathsf{am}} \wedge \mathscr{C}_{\mathsf{m}}, \mathsf{freq})$ from the extracted theories. This kind of research combines the advantages of both condensed representations and constrained mining which result in very efficient algorithms.

Constraint-based Mining of more Complex Pattern Domains

Most of the recent results have concerned simple local pattern discovery tasks like the ones based on itemsets or sequences. We believe that inductive querying is much more general. Many open problems are however to be addressed. For instance, even constraint-based mining of association rules is already much harder than constraint-based mining of itemsets (Lakshmanan *et al.*, 1999, Jeudy and Boulicaut, 2002). The recent work on the `MINE RULE` query language (Meo *et al.*, 1998) is also typical of the difficulty to optimize constraint-based association rule mining (Meo, 2003). When considering model mining under constraints (e.g., classifier design or clustering), only very preliminary approaches are available (see, e.g., (Garofalakis and Rastogi, 2000)). We think that this will be a major issue for research in the next few years. For instance, for clustering, it seems important to go further than the classical similarity optimization constraints and enable to specify other constraints on clusters (e.g., enforcing that some objects are or are not within the same clusters).

17.5 Conclusion

In this chapter, we have considered constraint-based mining approaches, i.e., the core techniques for inductive querying.

This domain has been studied a lot for simple pattern domains like itemsets or sequences. Rather general forms of inductive queries on these domains (e.g., arbitrary boolean expressions over monotonic and anti-monotonic constraints) have been considered. Beside the many ad-hoc algorithms, an interesting effort has concerned generic algorithms. Many open problems are still there: how to solve tough constraints?, how to design relevant approximation or relaxation schemes? how to combine constraint-based mining with condensed representations, not only for simple pattern domains but also more complex ones?

Moreover, within the inductive database framework, the problem is to optimize sequences of queries and typically sequences of correlated inductive queries. It is crucial to consider that the optimization of a query and thus constraint-based mining must also take into account the previously solved queries. Looking for the formal properties between inductive queries, especially containment, is thus a major priority. Here again, we believe that condensed representations might play a major role.

Last but not the least, a quite challenging problem is to consider from where the constraints come. The analysts can think in terms of constraints or declarative specifications which are not supported by the available solvers: an obvious example could be unexpectedness or novelty w.r.t. some explicit background knowledge. To be able to derive appropriate inductive queries based on a limited number of primitives (and some associated solvers) from the constraints expressed by the analysts is challenging.

References

R. Agrawal, H. Mannila, R. Srikant, H. Toivonen, and A. I. Verkamo. Fast discovery of association rules. In *Advances in Knowledge Discovery and Data Mining*, pages 307–328. AAAI Press, 1996.

H. Albert-Lorincz and J.-F. Boulicaut. Mining frequent sequential patterns under regular expressions: a highly adaptative strategy for pushing constraints. In *Proc. SIAM DM'03*, pages 316–320, 2003.

Y. Bastide, N. Pasquier, R. Taouil, G. Stumme, and L. Lakhal. Mining minimal non-redundant association rules using frequent closed itemsets. In *Proc. CL 2000*, volume 1861 of *LNCS*, pages 972–986. Springer-Verlag, 2000.

Y. Bastide, R. Taouil, N. Pasquier, G. Stumme, and L. Lakhal. Mining frequent patterns with counting inference. *SIGKDD Explorations*, 2(2):66–75, 2000.

R. J. Bayardo. Efficiently mining long patterns from databases. In *Proc. ACM SIGMOD'98*, pages 85–93, 1998.

F. Bonchi, F. Giannotti, A. Mazzanti, and D. Pedreschi. Adaptive constraint pushing in frequent pattern mining. In *Proc. PKDD'03*, volume 2838 of *LNAI*, pages 47–58. Springer-Verlag, 2003A.

F. Bonchi, F. Giannotti, A. Mazzanti, and D. Pedreschi. Examiner: Optimized level-wise frequent pattern mining with monotone constraints. In *Proc. IEEE ICDM'03*, pages 11–18, 2003B.

F. Bonchi, F. Giannotti, A. Mazzanti, and D. Pedreschi. Exante: Anticipated data reduction in constrained pattern mining. In *Proc. PKDD'03*, volume 2838 of *LNAI*, pages 59–70. Springer-Verlag, 2003C.

F. Bonchi and C. Lucchese. On closed constrained frequent pattern mining. In *Proc. IEEE ICDM'04 (In Press)*, 2004.

J.-F. Boulicaut. Inductive databases and multiple uses of frequent itemsets: the cInQ approach. In *Database Technologies for Data Mining - Discovering Knowledge with Inductive Queries*, volume 2682 of *LNCS*, pages 1–23. Springer-Verlag, 2004.

J.-F. Boulicaut and A. Bykowski. Frequent closures as a concise representation for binary Data Mining. In *Proc. PAKDD'00*, volume 1805 of *LNAI*, pages 62–73. Springer-Verlag, 2000.

J.-F. Boulicaut, A. Bykowski, and C. Rigotti. Approximation of frequency queries by mean of free-sets. In *Proc. PKDD'00*, volume 1910 of *LNAI*, pages 75–85. Springer-Verlag, 2000.

J.-F. Boulicaut, A. Bykowski, and C. Rigotti. Free-sets : a condensed representation of boolean data for the approximation of frequency queries. *Data Mining and Knowledge Discovery*, 7(1):5–22, 2003.

J.-F. Boulicaut and B. Jeudy. Using constraint for itemset mining: should we prune or not? In *Proc. BDA'00*, pages 221–237, 2000.

J.-F. Boulicaut and B. Jeudy. Mining free-sets under constraints. In *Proc. IEEE IDEAS'01*, pages 322–329, 2001.

C. Bucila, J. E. Gehrke, D. Kifer, and W. White. Dualminer: A dual-pruning algorithm for itemsets with constraints. *Data Mining and Knowledge Discovery*, 7(4):241–272, 2003.

D. Burdick, M. Calimlim, and J. Gehrke. MAFIA: A maximal frequent itemset algorithm for transactional databases. In *Proc. IEEE ICDE'01*, pages 443–452, 2001.

A. Bykowski and C. Rigotti. DBC: a condensed representation of frequent patterns for efficient mining. *Information Systems*, 28(8):949–977, 2003.

T. Calders and B. Goethals. Mining all non-derivable frequent itemsets. In *Proc. PKDD'02*, volume 2431 of *LNAI*, pages 74–85. Springer-Verlag, 2002.

B. Crémilleux and J.-F. Boulicaut. Simplest rules characterizing classes generated by delta-free sets. In *Proc. ES 2002*, pages 33–46. Springer-Verlag, 2002.

L. De Raedt. A perspective on inductive databases. *SIGKDD Explorations*, 4(2):69–77, 2003.

L. De Raedt, M. Jaeger, S. Lee, and H. Mannila. A theory of inductive query answering. In *Proc. IEEE ICDM'02*, pages 123–130, 2002.

L. De Raedt and S. Kramer. The levelwise version space algorithm and its application to molecular fragment finding. In *Proc. IJCAI'01*, pages 853–862, 2001.

M. M. Garofalakis and R. Rastogi. Scalable Data Mining with model constraints. *SIGKDD Explorations*, 2(2):39–48, 2000.

M. M. Garofalakis, R. Rastogi, and K. Shim. SPIRIT: Sequential pattern mining with regular expression constraints. In *Proc. VLDB'99*, pages 223–234, 1999.

B. Goethals and M. J. Zaki, editors. *Proc. of the IEEE ICDM 2003 Workshop on Frequent Itemset Mining Implementations*, volume 90 of *CEUR Workshop Proceedings*, 2003.

D. Gunopulos, R. Khardon, H. Mannila, S. Saluja, H. Toivonen, and R. S. Sharm. Discovering all most specific sentences. *ACM Transactions on Database Systems*, 28(2):140–174, 2003.

T. Imielinski and H. Mannila. A database perspective on knowledge discovery. *Communications of the ACM*, 39(11):58–64, 1996.

B. Jeudy and J.-F. Boulicaut. Optimization of association rule mining queries. *Intelligent Data Analysis*, 6(4):341–357, 2002.

D. Kifer, J. E. Gehrke, C. Bucila, and W. White. How to quickly find a witness. In *Proc. ACM PODS'03*, pages 272–283, 2003.

S. Kramer, L. De Raedt, and C. Helma. Molecular feature mining in HIV data. In *Proc. ACM SIGKDD'01*, pages 136–143, 2001.

L. V. Lakshmanan, R. Ng, J. Han, and A. Pang. Optimization of constrained frequent set queries with 2-variable constraints. In *Proc. ACM SIGMOD'99*, pages 157–168, 1999.

D.-I. Lin and Z. M. Kedem. Pincer search: An efficient algorithm for discovering the maximum frequent sets. *IEEE Transactions on Knowledge and Data Engineering*, 14(3):553–566, 2002.

H. Mannila and H. Toivonen. Multiple uses of frequent sets and condensed representations. In *Proc. KDD'96*, pages 189–194. AAAI Press, 1996.

H. Mannila and H. Toivonen. Levelwise search and borders of theories in knowledge discovery. *Data Mining and Knowledge Discovery*, 1(3):241–258, 1997.

C. Mellish. The description identification problem. *Artificial Intelligence*, 52(2):151–168, 1992.

R. Meo. Optimization of a language for Data Mining. In *Proc. ACM SAC'03 - Data Mining Track*, pages 437–444, 2003.

R. Meo, G. Psaila, and S. Ceri. An extension to SQL for mining association rules. *Data Mining and Knowledge Discovery*, 2(2):195–224, 1998.

T. Mitchell. Generalization as search. *Artificial Intelligence*, 18(2):203–226, 1980.

R. Ng, L. V. Lakshmanan, J. Han, and A. Pang. Exploratory mining and pruning optimizations of constrained associations rules. In *Proc. ACM SIGMOD'98*, pages 13–24, 1998.

N. Pasquier, Y. Bastide, R. Taouil, and L. Lakhal. Efficient mining of association rules using closed itemset lattices. *Information Systems*, 24(1):25–46, 1999.

J. Pei, G. Dong, W. Zou, and J. Han. On computing condensed frequent pattern bases. In *Proc. IEEE ICDM'02*, pages 378–385, 2002.

J. Pei, J. Han, and L. V. S. Lakshmanan. Mining frequent itemsets with convertible constraints. In *Proc. IEEE ICDE'01*, pages 433–442, 2001.

R. Srikant, Q. Vu, and R. Agrawal. Mining association rules with item constraints. In *Proc. ACM SIGKDD'97*, pages 67–73, 1997.

M. J. Zaki. Sequence mining in categorical domains: incorporating constraints. In *Proc. ACM CIKM'00*, pages 422–429, 2000.

18

Link Analysis

Steve Donoho

Mantas, Inc.

Summary. Link analysis is a collection of techniques that operate on data that can be represented as nodes and links. This chapter surveys a variety of techniques including subgraph matching, finding cliques and K-plexes, maximizing spread of influence, visualization, finding hubs and authorities, and combining with traditional techniques (classification, clustering, etc). It also surveys applications including social network analysis, viral marketing, Internet search, fraud detection, and crime prevention.

Key words: Link analysis, Social network analysis, Graph theory

18.1 Introduction

The term "link analysis" does not refer to one specific technique or algorithm. Rather it refers to a collection of techniques that are bound together by the type of data they operate on. Link analysis techniques are applied to data that can be represented as nodes and links as in Figure 18.1.

A node represents an entity such as a person, a document, or a bank account. Nodes are sometimes referred to as "vertices." A link represents a relationship between two entities such as a parent/child relationship between two people, a reference relationship between two documents, or a transaction between two bank accounts. Links are sometimes referred to as "edges." Because links show relationships among entities, this type of data is often referred to as relational data.

This is as opposed to attribute vector data used by many other unsupervised and supervised Data Mining techniques. In most standard Data Mining techniques, data is represented as a set of tuples (a vector of attribute values). Each tuple represents an entity, but there is no explicit data about relationships among entities. In link analysis, information exists about the relationships among entities, and analysis of these relationships is the focus of the field.

The roots of link analysis predate the use of modern computers. Law enforcement officials have carried out manual link analysis for many years. When a crime is investigated, a network such as in Figure 18.1 is drawn where the nodes represent

O. Maimon, L. Rokach (eds.), *Data Mining and Knowledge Discovery Handbook*, 2nd ed.,
DOI 10.1007/978-0-387-09823-4_18,

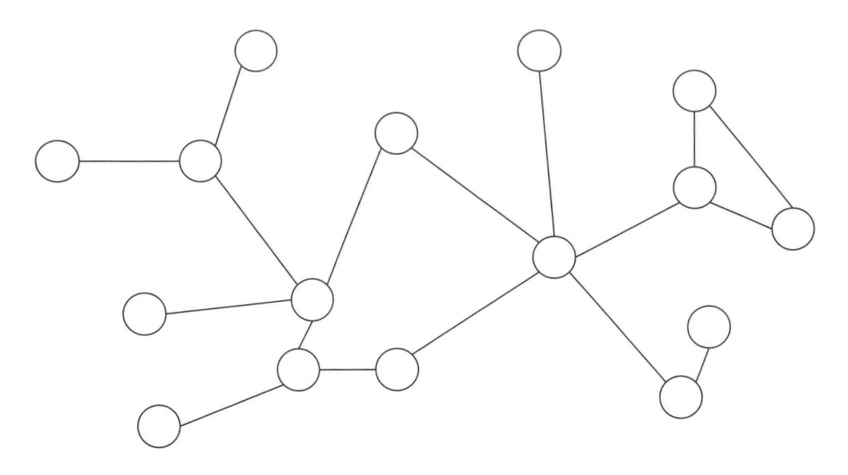

Fig. 18.1. Node and Link Data Used by Link Analysis Techniques.

people, weapons, crime scenes, etc. One person may be linked to another if they are family, friends, roommates, or business partners. A person may be linked to a weapon if it is registered in his name or if it was found at his home. Once a network of relationships is drawn out, the bigger picture of a crime emerges from the details. Holes in the network become apparent, and they are areas for further investigation. Hypotheses can be formed and tested.

Sociologists also performed manual link analysis long before there were computers. The structure of a clan or tribe would be mapped out with nodes representing people and links representing family, work, or social relationships. From this a sociologist could deduce who held powerful positions within the clan, who might influence who else, how information might spread within the clan, and what factions might arise.

The advent of computers allowed these techniques to become much more widespread and to be applied on a much larger scale. All 10 million of a bank's customers can be analysed for money laundering relationships. The hundreds of millions of documents on the Internet can be analysed to determine which are most respected and reliable. Large communities can be analysed to determine how information and opinions spread and who are the most influential individuals.

This chapter surveys the techniques that fall under the umbrella of link analysis and how these techniques are being applied. Section 18.2 presents some key concepts from the field of Social Network Analysis. Section 18.3 examines how link analysis techniques are used to improve search engine results. Section 18.4 looks at recent link analysis ideas emerging from the field of viral marketing. Section 18.5 shows how fraud detection and law enforcement have presented unique challenges and opportunities for link analysis. Finally, Section 18.6 surveys recent combinations of link analysis with traditional Data Mining techniques.

18.2 Social Network Analysis

The field of social network analysis (Wasserman, 1994, Hanneman, 2001) has developed over many years as sociologists developed formal methods of studying groups of people and their relationships. When studying a social network, there are many questions sociologists are interested in answering:

1. Which people are powerful?
2. Which people influence other people?
3. How does information spread within the network?
4. Who is relatively isolated, and who is well connected?
5. In a disagreement, who is likely to side with whom?
6. What roles do people play in an organization, and who has similar roles?

While concepts such as powerful, influential, isolated, and connected are somewhat subjective, social network analysis methods give us a baseline for measuring and making comparisons.

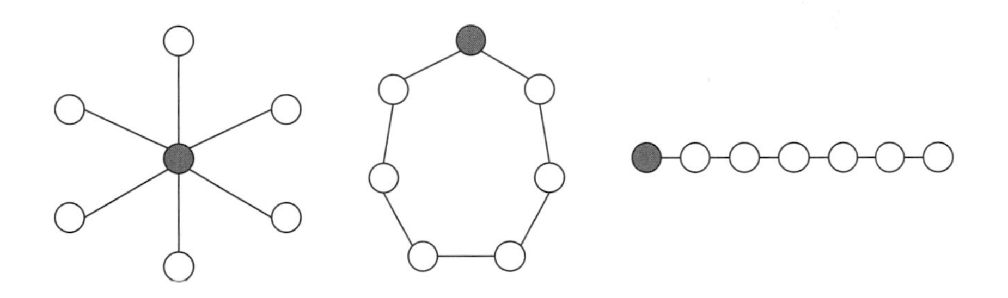

Fig. 18.2. Three Networks to Illustrate an Individual's Power within a Network.

Many things can make a person powerful within a group. Consider the shaded nodes in the three networks shown in Figure 18.2 (Hanneman, 2001). The person at the center of the star intuitively seems more powerful than the one in the circle or the one at the end of the line. If the people in the star want to communicate with each other they have to go through the center person, and that person has the power to either facilitate or hamper communication. If the people in the star want to engage in business, they have to go through the person in the center, and that person has the power to charge a fee as the middleman. In contrast, the shaded node in the circular network is the most convenient path of communication or trade for some nodes, but he is not the only path. Intuitively, he has less power than the center of the star. The shaded node at the end of the line is dependant on others for communication and trade but has no one who is dependant on him. Intuitively, he has little or no power.

The networks in Figure 18.2 illustrate how "centrality" is one measure of power. The node at the center of the star derives its power from being in the center of its network. The shaded nodes in the circle and line are less central to their networks and are therefore less powerful. Some quantitative methods of measuring centrality are:

1. Degree. The shaded node in the star network is linked to six other nodes and thus has a degree of six. All the other nodes in the star have a degree of one and are comparatively less central. All the nodes in the circle have the same degree: two. The shaded node in the line has a degree of one and is thus slightly less central than other nodes in the line with degree two.
2. Closeness. The average distance from the shaded node in the star to all other nodes is 1.0. This node has very direct access to everyone else. Other nodes in the star have an average distance of 1.8. All the nodes in the circle have an average distance of 2.0. The node at the end of the line has an average distance of 3.5 whereas the node in the center of the line has an average distance of 2.0.
3. Betweenness. The shaded node in the star is between all other 15 pairs of nodes. In the circle there are two paths between each pair of nodes. The shaded node in the circle is on a path between all other 15 pairs, but since there is an alternative path between each pair, the shaded node is on 50% of the paths between pairs. The node at the end of the line is between no pairs. The node one from the end of the line is on paths between 5 pairs (33% of 15 paths). The node at the center of the line is on paths between 9 pairs (60% of 15 paths).
4. Cutpoints. Related to betweenness, cutpoints are nodes that if removed divide the network into unconnected systems. These nodes hold particular power because they are the only point of contact between otherwise disconnected networks. If the center of the star is removed, six disconnected systems result. If a node in the circle is removed, the network is still connected. If a non-end node is removed from the line, two disconnected systems result.

A clique is a small, highly-interconnected group within a larger network. Cliques are of interest for several reasons. Ideas or information may spread extremely quickly within a clique because of the high connectivity. Members of a clique often act and behave as a cohesive unit. Disputes may form between cliques ("factions"). A person can be described with respect to the clique(s) they belong to. A person who is only connected to people in his clique is called a "local" and is strongly influenced by the clique. A person who belongs to many cliques is called a "cosmopolitan" and serves to bring outside ideas and information into a clique.

The most strict definition of a clique is a complete subgraph (all nodes in the clique must be linked to all other nodes). A couple more relaxed definitions are:

1. K-plexes. A group of N nodes is a K-plex if each of the nodes is connected to at least N-K other nodes in the group. Intuitively, if K=2 then every member of the clique has to be connected to all but two of the other members.
2. K-cores. The definition of a K-core is slightly more relaxed than that of a K-plex. A K-core is a maximal group of nodes all of which are connected to at least K other nodes in the group. For example, if K=4 then every member of the clique is connected to at least 4 other clique members.

The concept of "equivalence" is very important within social networks. It makes it possible to determine if a person is playing a particular role within a network. This allows both intra-network comparisons (one node has the same role as another node within one network) and inter-network comparison (two nodes in different networks are playing the same role). Two measures of equivalence are:

1. Structural Equivalence. This is a strict measure of equivalence between two nodes. Two nodes are exactly structurally equivalent if they are linked to exactly the same other nodes. If not exactly equivalent, the degree of partial structural equivalence can be measured using the degree of overlap in nodes they are linked to.
2. Regular equivalence. Regular equivalence is a less strict definition than structural equivalence. Two nodes have regular equivalence if the nodes they are linked to are regular equivalents. For example, Fred Flintstone is the regular equivalent of Barney Rubble because Fred is the husband of Wilma, and Barney is the husband of Betty, and Wilma and Betty are regular equivalents.

On a broader scale, equivalence of nodes lays the groundwork for measuring the similarity of one whole social network to another whole social network. This is useful for matching a network against a known template in order to identify the nature of the network as will be seen in Section 5 on Fraud Detection and Law Enforcement.

Many groups such as academic circles, fraud rings, business circles, shoppers with common interests, and professional societies can be represented as social networks. Because of this, Social Network Analysis lays the groundwork for many important real-world applications.

18.3 Search Engines

The Internet is rich in relational data by the simple fact that web pages are linked to other web pages. While traditional search techniques such as keyword searches focus exclusively on the content of a single page, newer techniques (Page *et al.*, 1998, Kleinberg, 1999) exploit relationships among pages. A user performing a search wants to find results that are not only relevant but are also authoritative and reliable. A keyword search on "stock market" will not only return authoritative sites such as the NASDAQ and NYSE pages, it will also return pages from thousands of self-proclaimed gurus selling books, software, and advice. The truly reliable sources of information are likely to be lost among the self-proclaimed gurus. Is there a way to separate the wheat from the chaff? This is where relational information contained in links comes into play.

An authoritative site such as the NASDAQ is likely to be recognized as authoritative by many people; therefore, many other sites are likely to point to the NASDAQ site. But a self-proclaimed stock market guru is less likely to have many other sites pointing to his site unless there truly is some merit to what he has to say. When one site references another site, it is in fact declaring that that site has some merit – it is casting a vote for the value and importance of the other site. Conceptually, this is similar to paper citations in academia. A paper that is cited often is considered to contain important ideas. A paper that is seldom or never cited is considered to be less important. The following paragraphs present two algorithms for incorporating link information into search engines: PageRank (Page *et al.*, 1998) and Kleinberg's Hubs and Authorities (Kleinberg, 1999).

The PageRank algorithm takes a set of interconnected pages and calculates a score for each. Intuitively, the score for a page is based on how many other pages point to that page and what their scores are. A page that is pointed to by a few other important pages is probably itself important. Similarly, a pages that is pointed to by numerous other marginally important pages is probably itself important. But a page that is not pointed to by anything probably isn't important.

A more formal definition taken from (Page *et al.*, 1998) is: Let u be a web page. Then let F_u be the set of pages u points to and B_u be the set of pages that point to u. Let $N_u = |F_u|$ be the number of links from u. Then let $E(u)$ be an *a priori* score assigned to u. Then $R(u)$, the score for u, is calculated:

$$R(u) = \sum_{v \in B_u} \frac{R(v)}{N_v} + E(u)$$

So the score for a page is some constant plus the sum of the scores of its incoming links. Each incoming link has the score of the page it is from divided by the number of outgoing links from that page (so a page's score is divided evenly among its outgoing links). The constant $E(u)$ serves a couple functions. First, it counterbalances the effect of "sinks" in the network. These are pages or groups of pages that are dead ends – they are pointed to, but they don't point out to any other pages. $E(u)$ provides a "source" of score that counterbalances the "sinks" in the network. Secondly, it provides a method of introducing *a priori* scores if certain pages are known to be authoritative.

The PageRank algorithm can be combined with other techniques to create a search engine. For example, PageRank is first used to assign a score to all pages in a population. Next a simple keyword search is used to find a list of relevant candidate pages. The candidate pages are ordered according to their PageRank score. The top ranked pages presented to the user are both relevant (based on the keyword match) and authoritative (based on the PageRank score). The Google search engine is based on PageRank combined with other factors including standard IR techniques and the text of incoming links.

Kleinberg's algorithm (Kleinberg, 1999) differs from the PageRank approach in two important respects:

1. Whereas the PageRank approach assigns a score to each page *before* applying a text search, Kleinberg assigns a score to a page within the context of a text search. For example, a page containing both the words "Microsoft" and "Oracle" may receive a high score if the search string is "Oracle" but a low score if the search string is "Microsoft." PageRank would assign a single score regardless of the text search.

2. Kleinberg's algorithm draws a distinction between "hubs" and "authorities." Hubs are pages that point out to many other pages on a topic (preferably many are authorities). Authorities are pages that are pointed to by many other pages (preferably many are hubs). Thus the two have a symbiotic relationship.

The first step in a search is to create a "focused subgraph." This is a small subset of the Internet that is rich in pages relevant to the search and also contains many of the strongest authorities on the searched topic. This is done by doing a pure text search and retrieving the top t pages (t about 200). This set is augmented by adding all the pages pointed to by pages in t and all pages that point to pages in t (for pages with a large in-degree, only a subset of the in-pointing pages are added). Note that adding these pages may add pages that do not contain the original search string! This is actually a good thing because often an authority on a topic may not contain the search string. For example, toyota.com may not contain the string "automobile manufacturer" (Rokach and Maimon, 2006) or a page may discuss several machine learning algorithms but not have the phrase "machine learning" because the authors always use the phrase "Data Mining." Adding the linked pages pulls in related pages whether they contain the search text or not.

The second step calculates two scores for each page: an authority score and a hub score. Intuitively, a page's authority score is the normalized sum of the hub scores of all pages that point to it. A page's hub score is the normalized sum of the authority scores of all the pages it points to. By iteratively recalculating each pages' hub and authority score, the scores converge to an equilibrium. The reinforcing relationship between hubs and authorities helps the algorithm differentiate true authorities on a topic from generally popular web sites such as amazon.com and yahoo.com.

In summary, both algorithms measure a web page's importance by its relationships with other web pages – an extension of the notion of importance in a social network being determined by position in the network.

18.4 Viral Marketing

Viral marketing relies heavily on "word-of-mouth" advertising where one individual who has bought a product tells their friends about the product (Domingos, 1996, Richardson, 2002, Kempe *et al.*, 2003). A famous example of viral marketing is the rapid spread of Hotmail as a free email service. Attached to each email was a short advertisement and Hotmail's URL, and customers spread the word about Hotmail simply by emailing their family and friends. Hotmail grew from zero to 12 million users in 18 months (Richardson, 2002). Word-of-mouth is a powerful form of advertising because if a friend tells me about a product, I don't believe that they have a hidden motive to sell the product. Instead I believe that they really believe in the inherent quality of the product enough to tell me about it.

Products are not the only things that spread by word of mouth. Fashion trends spread from person to person. Political ideas are transferred from one person to the next. Even technological innovations are transferred among a network of coworkers and peers. These problems can all be viewed as the diffusion of information, ideas, or influence among the members of a social network (Kempe *et al.*, 2003).

These social networks take on a variety of forms. The most easily understood is the old fashion "social network" where people are friends, neighbors, coworkers, etc. and their connections are personal and often face-to-face. For example, when a new medical procedure is invented, some physicians will be early adopters, but others will wait until close friends have tried the procedure and been successful. The Internet has also created social networks with virtual connections. In a collaborative filtering system, a recommendation for a book, movie, or musical CD may be made to Customer A based on N "similar" customers – customers who have bought similar items in the past. Customer A is influenced by these N other customer even though Customer A never meets them, communicates with them, or even knows their identities. Knowledge-sharing sites provide a second type of virtual connection with more explicit interaction. On these sites people provide reviews and ratings on things ranging from books to cars to restaurants. As an individual follows the advice offered by various "experts" they grow to trust the advice of some and not trust the advice of others.

Formally, this can be modeled as a social network where the nodes are people and node X_i is linked to node X_j if the person represented by X_i in some way influences X_j. From a marketing standpoint some natural questions emerge. "Which nodes should I market to to maximize my profit?" Or alternatively, "If I only have the budget to market to k of the n nodes in the network, which k should I choose to maximize the spread of influence?" Based on work from the field of social network analysis, two plausible approaches would be to pick nodes with the highest out-degree (nodes that influence a lot of other nodes) or nodes with good distance centrality (nodes that have a short average distance to the rest of the network). Two recent approaches to these questions are described in the following paragraphs.

The first approach proposed by (Domingos, 1996) and (Richardson, 2002) model the social network as a Markov random field. The probability that each node will purchase the product or adopt the idea is modeled as $P(X_i| \mathbf{N}_i, \mathbf{Y}, \mathbf{M})$ where $\mathbf{N}_i$ are the neighbors of X_i, the ones who directly influence X_i. $\mathbf{Y}$ is a vector of attributes describing the product. This reflects the fact that X_i is influenced not only by neighbors but also by the attributes of the product itself. A bald man probably won't buy a hairbrush even if all the people he trusts most do. M_i is the marketing action taken for X_i. This reflects the fact that a customer's decision to buy is influenced by whether he is marketed to such as if he receives a discount. This probability can be combined with other information (how much it costs to market to a customer, what the revenue is from a customer that was marketed to, and what the revenue is from a customer who was not marketed) to calculate the expected profit from a particular marketing plan. Various search techniques such as greedy search and hill-climbing can be employed to find local maxima for the profit.

A second approach proposed by (Kempe *et al.*, 2003) uses a more operational model of how ideas spread within a network. A set of nodes are initialized to be active (indicating they bought the product or adopted the idea) at time t= 0. If an inactive node X_i has active neighbors, those neighbors exert some influence on X_i

to become active. As more of X_i's neighbors become active, this may cause X_i to become active. Thus the process unfolds in a set of discrete steps where a set of nodes change their values at time t based on the set of active nodes at time $t-1$. Two models for how nodes become activated are:

1. Linear Threshold Model. Each link coming into X_i from its neighbors has a weight. When the sum of the weights of the links from X_i's active neighbors surpasses a threshold θ_i, then X_i becomes active at time $t+1$. The process runs until the network reaches an equilibrium state.
2. Independent Cascade Model. When the neighbor of an inactive node X_i first becomes active, it has one chance to activate X_i. It succeeds at activating X_i with a probability of $p_{i,j}$, and X_i becomes active at time $t+1$. The process runs until no new nodes become active.

Kempe presents a greedy hill-climbing algorithm and proves that its performance is within a factor of 63% of optimal. Empirical experiments show that the greedy algorithm performs better than picking nodes with the highest out-degree or the best distance centrality.

Areas of further research in viral marketing include dealing with the fact that network knowledge is often incomplete. Network knowledge can be acquired, but this involves a cost that must be factored into the overall marketing cost.

18.5 Law Enforcement & Fraud Detection

Link analysis was used in law enforcement long before the advent of computers. Police and detectives would manually create charts showing how people and pieces of evidence in a crime were connected. Computers greatly advanced these techniques in two key ways:

1. Visualization of crime/fraud networks. Charts that were previously manually drawn and static became automatically drawn and dynamic. A variety of link analysis visualization tools allowed users to perform such operations as:
   a) Automatically arranging networks to maximize clarity (e.g. minimize link crossings),
   b) Rearranging a network by dragging and dropping nodes,
   c) Filtering out links by weight or type,
   d) Grouping nodes by types.
2. Proliferation of databases containing information to link people, events, accounts, etc. Two people or accounts could be linked because:
   a) One sent a wire transfer to the other,
   b) They were in the same auto accident and were mentioned in the insurance claim together,
   c) They both owned the same house at different times,
   d) They share a phone number

All these pieces of information were gathered for non-law enforcement reasons and stored in databases, but they could be used to detect fraud and crime rings.

A pioneering work in automated link analysis for law enforcement is FAIS (Senator *et al.*, 1995), a system for investigating money laundering developed at FINCEN (Financial Crimes Enforcement Network). The data supporting FAIS was a database of Currency Transaction Reports (CTRs) and other forms filed by banks, brokerages, casinos, businesses, etc. when a customer conducts a cash transaction over $10,000. Entities were linked to each other because they appeared on the same CTR, shared an address, etc. The FAIS system provided leads to investigators on which people, businesses, accounts, or locations they should investigate. Starting with a suspicious entity, an investigator could branch out to everyone linked to that entity, then everyone linked to those entities, and so on. This information was then displayed in a link analysis visualization tool where it could be more easily manipulated and understood.

Insurance fraud is a crime that is usually carried out by rings of professionals. A ringleader orchestrates staged auto accidents and partners with fraudulent doctors, lawyers, and repair shops to file falsified claims. Over time, this manifests itself in the claim data as many claims involving an interlinked group of drivers, passengers, doctors, lawyers, and body shops. Each claim in isolation looks legitimate, but taken together they are extremely suspicious. The "NetMap for Claims" solution from NetMap Analytics like FAIS allows users to start with a person, find everyone directly linked to them (on a claim together), then find everyone two links away, then three links away, etc. These people and their interconnections are then displayed in a versatile visualization tool.

More recent work has focused on generating leads not based on any one node but instead on the relationships among nodes. This is because many nodes in isolation are perfectly legitimate or are only slightly suspicious but innocuous enough to stay "under the radar." But when several of these "slightly suspicious" nodes are linked together, it becomes very suspicious (Donoho and Johnstone, 1995). For example, if a bank account has cash deposits of between $2000 and $5000 in a month's time, this is only slightly suspicious. At a large bank there will be numerous accounts that meet this criteria – too many to investigate. But if it is found that 10 such accounts are linked by shared personal information or by transactions, this is suddenly very suspicious because it is highly unlikely this would happen by chance and this is exactly what money launderers do to hide their behavior.

Scenario-based approaches have instances of crimes represented as networks, and new situations are suspicious if they are sufficiently similar to a known crime instance. Fu *et al.* (2003) describes a system to detect contract murders by the Russian mafia. The system contains a library of known contract murders described as people linked by phone calls, meetings, wire transfers, etc. A new set of events may match a known instance if it has a similar network topology even if a phone call in one matches a meeting in the other (both are communication events) or if a wire transfer in one matches a cash payment in the other (both are payment events). The LAW system (Ruspini *et al.*, 2003) is a similar system to detect terrorist activities before a terrorist act occurs. LAW measures the similarity between two networks using edit distance – the number of edits needed to convert network #1 to be exactly link network #2.

In summary, crime rings and fraud rings are their own type of social network, and analyzing the relationships among entities is a powerful method of detecting those rings.

18.6 Combining with Traditional Methods

Many recent works focus on combining link analysis techniques with traditional knowledge discovery techniques such as inductive inference and clustering.

Jensen (1999) points out that certain challenges arise when traditional induction techniques are applied to linked data:

1. The linkages in the data may cause instances to no longer be statistically independent. If multiple instances are all linked to the same entity and draw some of their characteristics from that entity, then those instances are no long independent.
2. Sampling becomes very tricky. Because instances are interlinked with each other, sampling breaks many of these linkages. Relational attributes of an instance such as degree, closeness, betweenness can be drastically changed by sampling.
3. Attribute combinatorics are greatly increased in linked data. In addition to an instance's k intrinsic attributes, an algorithm can draw upon attributes from neighbors, neighbors' neighbors, etc. Yet more attributes arise from the combinations of neighbors and the topologies with which they are linked.

These and other challenges are discussed more extensively in (Jensen, 1999).

Neville and Jensen (2000) present an iterative method of classification using relational data. Some of an instance's attributes are static. These include intrinsic attributes (which contain information about an instance by itself regardless of linkages) and static relational attributes (which contain information about an instance's linked neighbors but are not dependent on the neighbors' classification). More interesting are the dynamic relational attributes. These attributes may change value as an instance's neighbors change classification. So the output of one instance (its class) is the input of a neighboring instance (in its dynamic relational attributes), and vice versa. The algorithm iteratively recalculates instances' classes. There are m iterations, and at iteration i it accepts class labels on the $N*(i/m)$ instances with the highest certainty. Classification proceeds from the instances with highest certainty to those with lowest certainty until all N instances are classified. In this way instances with the highest certainty have more opportunity to affect their neighbors.

Using linkage data in classification has also been applied to the classification of text (Chakrabarti *et al.*, 1998, Slattery, 2000, Oh *et al.*, 2000, Lu, 2003). Simply incorporating words from neighboring texts was not found to be helpful, but incorporating more targeted information such as hierarchical category information, predicted class, and anchor text was found to improve accuracy.

In order to cluster linked data, Neville *et al.* (2003) combines traditional clustering with graph partitioning techniques. Their work uses similarity metrics taken from traditional clustering to assign weights to linked graphs. Once this is done, several standard graph partitioning techniques can be used to partition the graph into clusters. Taskar *et al.* (2001) cluster linked data using probabilistic relational models.

While most work in link analysis assumes that the graph is complete and fairly correct, this is often far from the truth. Work in the area of link completion (Kubica *et al.*, 2002, Goldenberg *et al.*, 2003, Kubica *et al.*, 2003) induces missing links from previously observed data. This allows users to ask questions about a future state of the graph such as "Who is person XYZ likely to publish a paper with in the next year?"

There are many possible ways link analysis can be combined with traditional techniques, and many remain unexplored making this a fruitful area for future research.

18.7 Summary

Link analysis is a collection of techniques that operate on data that can be represented as nodes and links. A variety of applications rely on link analysis techniques or can be improved by link analysis techniques. Among these are Internet search, viral marketing, fraud detection, crime prevention, and sociologic study. To support these applications, a number of new link analysis techniques have emerged in recent years. This chapter has surveyed several of these including subgraph matching, finding cliques and K-plexes, maximizing spread of influence, visualization, and finding hubs and authorities. A fruitful area for future research is the combination of link analysis techniques with traditional Data Mining techniques.

References

Chakrabarti S, Dom B, Agrawal R, & Raghavan P. Scalable feature selection, classification and signature generation for organizing large text databases into hierarchical topic taxonomies. VLDB Journal: Very Large Data Bases 1998. 7:163 – 178.

Domingos P & Richardson M. Mining the network value of customers. Proceedings of the Seventh International Conference on Knowledge Discovery and Data Mining; 2001 August 26 – 29; San Francisco, CA. ACM Press, 2001.

Donoho S. & Lewis S. Understand behavior detection technology: Emerging approaches to dealing with three major consumer protection threats. April 2003.

Fu D, Remolina E, & Eilbert J. A CBR approach to asymmetric plan detection. Proceedings of Workshop on Link Analysis for Detecting Complex Behavior; 2003 August 27; Washington, DC.

Goldenberg A, Kubica J, & Komerak P. A comparison of statistical and machine learning algorithms on the task of link completion. Proceedings of Workshop on Link Analysis for Detecting Complex Behavior; 2003 August 27. Washington, DC.

Hanneman R. Introduction to social network methods. Univ of California, Riverside, 2001.

Jensen D. Statistical challenges to inductive inference in linked data. Preliminary papers of the 7th International Workshop on Artificial Intelligence and Statistics; 1999 Jan 4 – 6; Fort Lauderdale. FL.

Kempe D, Kleinberg J, & Tardos E. Maximizing the spread of influence through a social network. Proceedings of The Ninth ACM SIGKDD International Conference on Knowledge Discovery and Data Mining; 2003 August 24 – 27; Washington, DC. ACM Press, 2003.

Kleinberg J. Authoritative sources in a hyperlinked environment. Journal of the ACM, 1999, 46,5:604-632.

Kubica J, Moore A, Schneider J, & Yang Y. Stochastic link and group detection. Proceedings of The Eighth National Conference on Artificial Intelligence; 2002 July 28 – Aug 1; Edmonton, Alberta, Canada. ACM Press, 2002.

Kubica J, Moore A, Cohn D, & Schneider J. cGraph: A fast graph-based method for link analysis and queries. Proceedings of Text-Mining & Link-Analysis Workshop; 2003 August 9; Acapulco, Mexico.

Lu Q & Getoor L. Link-based text classification. Proceedings of Text-Mining & Link-Analysis Workshop; 2003 August 9; Acapulco, Mexico.

Neville J & Jensen D. Iterative classification in relational data. Proceeding of AAAI-2000 Workshop on Learning Statistical Models from Relational Data; 2000 August 3; Austin, TX. AAAI Press, 2000.

Neville J, Adler M, & Jensen D. Clustering relational data using attribute and link information. Proceedings of Text-Mining & Link-Analysis Workshop; 2003 August 9; Acapulco, Mexico.

Oh H, Myaeng S, & Lee M. A practical hypertext categorization method using links and incrementally available class information. Proceedings of the 23rd ACM International Conference on Research and Development in Information Retrieval (SIGIR-00); 2000 July; Athens, Greece.

Page L, Brin S, Motwani R, & Winograd T. The PageRank Citation Ranking: Bringing Order to the Web. Stanford Digital Library Technologies Project. 1998.

Richardson M & Domingos P. Mining knowledge-sharing sites for viral marketing. Proceedings of Eighth International Conference on Knowledge Discovery and Data Mining; 2002 July 28 – Aug 1; Edmonton, Alberta, Canada. ACM Press, 2002.

Rokach, L., Averbuch, M., and Maimon, O., Information retrieval system for medical narrative reports. Lecture notes in artificial intelligence, 3055. pp. 217-228, Springer-Verlag (2004).

Rokach, L., Averbuch, M., and Maimon, O., Information retrieval system for medical narrative reports. Lecture notes in artificial intelligence, 3055. pp. 217-228, Springer-Verlag (2004).

Rokach L. and Maimon O., Data mining for improving the quality of manufacturing: A feature set decomposition approach. Journal of Intelligent Manufacturing 17(3): 285299, 2006.

Ruspini E, Thomere J, & Wolverton M. Database-editing metrics for pattern matching. SRI Intern, March 2003.

Senator T, Goldberg H, Wooton J, Cottini A, Umar A, Klinger C, et al. The FinCEN Artificial Intelligence System: Identifying Potential Money Laundering from Reports of Large Cash Transactions. Proceedings of the 7th Conference on Innovative Applications of AI; 1995 August 21 – 23; Montreal, Quebec, Canada. AAAI Press, 1995.

Slattery S & Craven M. Combining statistical and relational methods for learning in hypertext domains. Proceedings of ILP-98, 8th International Conference on Inductive Logic

Programming; 1998 July 22 – 24; Madison, WI. Springer Verlag, 1998.
Taskar B, Segal E, & Koller D. Probabilistic clustering in relational data. Proceedings of Seventh International Joint Conference on Artificial Intelligence; 2001 August 4 – 10; Seattle, Washington.
Wasserman S & Faust K, Social Network Analysis. Cambridge University Press, 1994.

Part IV

Soft Computing Methods

19

A Review of Evolutionary Algorithms for Data Mining

Alex A. Freitas

University of Kent, UK, Computing Laboratory, A.A.Freitas@kent.ac.uk

Summary. Evolutionary Algorithms (EAs) are stochastic search algorithms inspired by the process of neo-Darwinian evolution. The motivation for applying EAs to data mining is that they are robust, adaptive search techniques that perform a global search in the solution space. This chapter first presents a brief overview of EAs, focusing mainly on two kinds of EAs, viz. Genetic Algorithms (GAs) and Genetic Programming (GP). Then the chapter reviews the main concepts and principles used by EAs designed for solving several data mining tasks, namely: discovery of classification rules, clustering, attribute selection and attribute construction. Finally, it discusses Multi-Objective EAs, based on the concept of Pareto dominance, and their use in several data mining tasks.

Key words: genetic algorithm, genetic programming, classification, clustering, attribute selection, attribute construction, multi-objective optimization

19.1 Introduction

The paradigm of Evolutionary Algorithms (EAs) consists of stochastic search algorithms inspired by the process of neo-Darwinian evolution (Back et al. 2000; De Jong 2006; Eiben & Smith 2003). EAs work with a population of individuals, each of them a candidate solution to a given problem, that "evolve" towards better and better solutions to that problem. It should be noted that this is a very generic search paradigm. EAs can be used to solve many different kinds of problems, by carefully specifying what kind of candidate solution an individual represents and how the quality of that solution is evaluated (by a "fitness" function).

In essence, the motivation for applying EAs to data mining is that EAs are robust, adaptive search methods that perform a global search in the space of candidate solutions. In contrast, several more conventional data mining methods perform a local, greedy search in the space of candidate solutions. As a result of their global search, EAs tend to cope better with attribute interactions than greedy data mining methods (Freitas 2002a; Dhar et al. 2000; Papagelis & Kalles 2001; Freitas 2001, 2002c).

O. Maimon, L. Rokach (eds.), *Data Mining and Knowledge Discovery Handbook*, 2nd ed.,
DOI 10.1007/978-0-387-09823-4_19, © Springer Science+Business Media, LLC 2010

Hence, intuitively EAs can discover interesting knowledge that would be missed by a greedy method.

The remainder of this chapter is organized as follows. Section 2 presents a brief overview of EAs. Section 3 discusses EAs for discovering classification rules. Section 4 discusses EAs for clustering. Section 5 discusses EAs for two data preprocessing tasks, namely attribute selection and attribute construction. Section 6 discusses multi-objective EAs. Finally, Section 7 concludes the chapter. This chapter is an updated version of (Freitas 2005).

19.2 An Overview of Evolutionary Algorithms

An Evolutionary Algorithm (EA) is essentially an algorithm inspired by the principle of natural selection and natural genetics. The basic idea is simple. In nature individuals are continuously evolving, getting more and more adapted to the environment. In EAs each "individual" corresponds to a candidate solution to the target problem, which could be considered a very simple "environment". Each individual is evaluated by a fitness function, which measures the quality of the candidate solution represented by the individual. At each generation (iteration), the best individuals (candidate solutions) have a higher probability of being selected for reproduction. The selected individuals undergo operations inspired by natural genetics, such as crossover (where part of the genetic material of two individuals are swapped) and mutation (where part of the generic material of an individual is replaced by randomly-generated genetic material), producing new offspring which will replace the parents, creating a new generation of individuals. This process is iteratively repeated until a stopping criterion is satisfied, such as until a fixed number of generations has been performed or until a satisfactory solution has been found.

There are several kinds of EAs, such as Genetic Algorithms, Genetic Programming, Classifier Systems, Evolution Strategies, Evolutionary Programming, Estimation of Distribution Algorithms, etc. (Back et al. 2000; De Jong 2006; Eiben & Smith 2003). This chapter will focus on Genetic Algorithms (GAs) and Genetic Programming (GP), which are probably the two kinds of EA that have been most used for data mining.

Both GA and GP can be described, at a high level of abstraction, by the pseudocode of Algorithm 1. Although GA and GP share this basic pseudocode, there are several important differences between these two kinds of algorithms. One of these differences involves the kind of solution represented by each of these kinds of algorithms. In GAs, in general a candidate solution consists mainly of values of variables – in essence, data. By contrast, in GP the candidate solution usually consists of both data and functions. Therefore, in GP one works with two sets of symbols that can be represented in an individual, namely the terminal set and the function set. The terminal set typically contains variables (or attributes) and constants; whereas the function set contains functions which are believed to be appropriate to represent good solutions for the target problem. In the context of data mining, the explicit use of a function set is interesting because it provides GP with potentially powerful means of changing the original data representation into a representation that is more suitable for knowledge discovery purposes, which is not so naturally done when using GAs or another EA where only attributes (but not functions) are represented by an individual. This ability of changing the data representation will be discussed particularly on the section about GP for attribute construction.

Note that in general there is no distinction between terminal set and function set in the case of GAs, because GAs' individuals usually consist only of data, not functions. As a result, the representation of GA individuals tend to be simpler than the representation of GP individuals. In particular, GA individuals are usually represented by a fixed-length linear genome, whereas the genome of GP individuals is often represented by a variable-size tree genome – where the internal nodes contain functions and the leaf nodes contain terminals.

Algorithm 1: Generic Pseudocode for GA and GP

```
1: Create initial population of individuals
2: Compute the fitness of each individual
3: repeat
4:    Select individuals based on fitness
5:    Apply genetic operators to selected individuals, creating new individuals
6:    Compute fitness of each of the new individuals
7:    Update the current population (new individuals replace old individuals)
8: until (stopping criteria)
```

When designing a GP algorithm, one must bear in mind two important properties that should be satisfied by the algorithm, namely closure and sufficiency (Banzhaf et al. 1998; Koza 1992). Closure means that every function in the function set must be able to accept, as input, the result of any other function or any terminal in the terminal set. Some approaches to satisfy the closure property in the context of attribute construction will be discussed in Subsection 5.2. Sufficiency means that the function set should be expressive enough to allow the representation of a good solution to the target problem. In practice it is difficult to know *a priori* which functions should be used to guarantee the sufficiency property, because in challenging real-world problems one often does not know the shape of a good solution for the problem. As a practical guideline, (Banzhaf et al. 1998) (p. 111) recommends:

"An approximate starting point for a function set might be the arithmetic and logic operations: PLUS, MINUS, TIMES, DIVIDE, OR, AND, XOR. ...Good solutions using only this function set have been obtained on several different classification problems,...,and symbolic regression problems."

We have previously mentioned some differences between GA and GP, involving their individual representation. Arguably, however, the most important difference between GAs and GP involves the fundamental nature of the solution that they represent. More precisely, in GAs (like in most other kinds of EA) each individual represents a solution to one particular instance of the problem being solved. In contrast, in GP a candidate solution should represent a generic solution – a program or an algorithm – to the kind of problem being solved; in the sense that the evolved program should be generic enough to be applied to any instance of the target kind of problem.

To quote (Banzhaf et al. 1998), p. 6:

> it is possible to define genetic programming as the direct evolution of *programs or algorithms* [our italics] for the purpose of inductive learning.

In practice, in the context of data mining, most GP algorithms evolve a solution (say, a classification model) *specific for a single data set*, rather than a *generic program* that can be applied to different data sets from different application domains. An exception is the work of (Pappa & Freitas 2006), proposing a grammar-based GP system that automatically evolves full rule induction algorithms, with loop statements, generic procedures for building and pruning classification rules, etc. Hence, in this system the output of a GP run is a *generic* rule induction algorithm (implemented in Java), which can be run on virtually any classification data set – in the same way that a manually-designed rule induction algorithm can be run on virtually any classification data set. An extended version of the work presented in (Pappa & Freitas 2006) is discussed in detail in another chapter of this book (Pappa & Freitas 2007).

19.3 Evolutionary Algorithms for Discovering Classification Rules

Most of the EAs discussed in this section are Genetic Algorithms, but it should be emphasized that classification rules can also be discovered by other kinds of EAs. In particular, for a review of Genetic Programming algorithms for classification-rule discovery, see (Freitas 2002a); and for a review of Learning Classifier Systems (a type of algorithm based on a combination of EA and reinforcement learning principles), see (Bull 2004; Bull & Kovacs 2005).

19.3.1 Individual Representation for Classification-Rule Discovery

This Subsection assumes that the EA discovers classification rules of the form "IF (conditions) THEN (class)" (Witten & Frank 2005). This kind of knowledge representation has the advantage of being intuitively comprehensible to the user – an important point in data mining (Fayyad et al. 1996). A crucial issue in the design of an individual representation is to decide whether the candidate solution represented by an individual will be a rule set or just a single classification rule (Freitas 2002a, 2002b).

The former approach is often called the "Pittsburgh approach", whereas the later approach is often called the "Michigan-style approach". This latter term is an extension of the term "Michigan approach", which was originally used to refer to one particular kind of EA called Learning Classifier Systems (Smith 2000; Goldberg 1989). In this chapter we use the extended term "Michigan-style approach" because, instead of discussing Learning Classifier Systems, we discuss conceptually simpler EAs sharing the basic characteristic that an individual represents a single classification rule, regardless of other aspects of the EA.

The difference between the two approaches is illustrated in Figure 19.1. Figure 19.1(a) shows the Pittsburgh approach. The number of rules, m, can be either variable, automatically evolved by the EA, or fixed by a user-specified parameter. Figure 19.1(b) shows the Michigan-style approach, with a single rule per individual. In both Figure 19.1(a) and 1(b) the rule antecedent (the "IF part" of the rule) consists of a conjunction of conditions. Each condition is typically of the form <Attribute, Operator, Value>, also known as attribute-value (or propositional logic) representation. Examples are the conditions: "Gender = Female" and "Age $<$ 25". In the case of continuous attributes it is also common to have rule conditions of the form <LowerBound, Operator, Attribute, Operator, UpperBound>, e.g.: "30K $\leq$ Salary $\leq$ 50K".

In some EAs the individuals can only represent rule conditions with categorical (nominal) attributes such as Gender, whose values (male, female) have no ordering – so that the only operator used in the rule conditions is "=", and sometimes "$\neq$". When using EAs with this limitation, if the data set contains continuous attributes – with ordered numerical values – those attributes have to be discretized in a pre-processing stage, before the EA is applied. In practice it is desirable to use an EA where individuals can represent rule conditions with both categorical and continuous attributes. In this case the EA is effectively doing a discretization of continuous values "on-the-fly", since by creating rule conditions such as "30K $\leq$ Salary $\leq$ 50K" the EA is effectively producing discrete intervals. The effectiveness of an EA that directly copes with continuous attributes can be improved by using operators that enlarge or shrink the intervals based on concepts and methods borrowed from the research area of discretization in data mining (Divina & Marchiori 2005).

It is also possible to have conditions of the form <Attribute, Operator, Attribute>, such as "Income $>$ Expenditure". Such conditions are associated with relational (or first-order logic) representations. This kind of relational representation has considerably more expressiveness power than the conventional attribute-value representation, but the former is associated with a much larger search space – which often requires a more complex EA and a longer processing time. Hence, most EAs for rule discovery use the attribute-value, propositional representation. EAs using the relational, first-order logic representation are described, for instance, in (Neri & Giordana 1995; Hekanaho 1995; Woung & Leung 2000; Divina & Marchiori 2002).

Note that in Figure 19.1 the individuals are representing only the rule antecedent, and not the rule consequent (predicted class). It would be possible to include the predicted class in each individual's genome and let that class be evolved along with its corresponding rule antecedent. However, this approach has one significant drawback, which can be illustrated with the following example. Suppose an EA has just generated an individual whose rule antecedent covers 100 examples, 97 of which have class c_1. Due to the stochastic nature of the evolutionary process and the

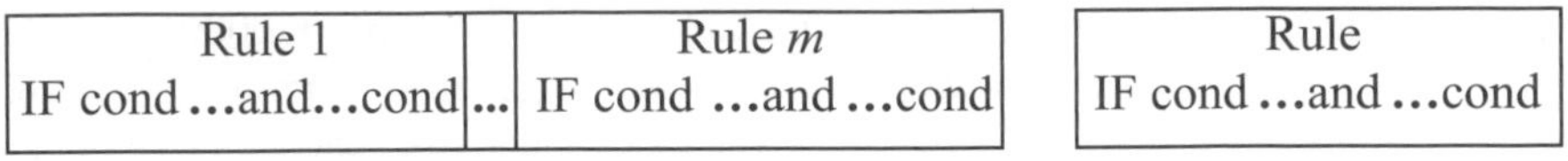

Rule 1 IF cond ...and...cond	...	Rule *m* IF cond ...and...cond

(a) Pittsburgh approach

Rule IF cond ...and...cond

(b) Michigan-style approach

Fig. 19.1. Pittsburgh vs. Michigan-style approach for individual representation

"blind-search" nature of the generic operators, the EA could associate that rule antecedent with class c_2, which would assign a very low fitness to that individual – a very undesirable result. This kind of problem can be avoided if, instead of evolving the rule consequent, the predicted class for each rule is determined by other (non-evolutionary) means. In particular, two such means are as follows.

First, one can simply assign to the individual the class of the majority of the examples covered by the rule antecedent (class c_1 in the above example), as a conventional, non-evolutionary rule induction algorithm would do. Second, one could use the "sequential covering" approach, which is often used by conventional rule induction algorithms (Witten & Frank 2005). In this approach, the EA discovers rules for one class at a time. For each class, the EA is run for as long as necessary to discover rules covering all examples of that class. During the evolutionary search for rules predicting that class, all individuals of the population will be representing rules predicting the same fixed class. Note that this avoids the problem of crossover mixing genetic material of rules predicting different classes, which is a potential problem in approaches where different individuals in the population represent rules predicting different classes. A more detailed discussion about how to represent the rule consequent in an EA can be found in (Freitas 2002a).

The main advantage of the Pittsburgh approach is that an individual represents a complete solution to a classification problem, i.e., an entire set of rules. Hence, the evaluation of an individual naturally takes into account rule interactions, assessing the quality of the rule *set*. In addition, the more complete information associated with each individual in the Pittsburgh approach can be used to design "intelligent", task-specific genetic operators. An example is the "smart" crossover operator proposed by (Bacardit & Krasnogor 2006), which heuristically selects, out of the N sets of rules in N parents (where $N \geq 2$), a good subset of rules to be included in a new child individual. The main disadvantage of the Pittsburgh approach is that it leads to long individuals and renders the design of genetic operators (that will act on selected individuals in order to produce new offspring) more difficult.

The main advantage of the Michigan-style approach is that the individual representation is simple, without the need for encoding multiple rules in an individual. This leads to relatively short individuals and simplifies the design of genetic operators. The main disadvantage of the Michigan-style approach is that, since each individual represents a single rule, a standard evaluation of the fitness of an individual ignores the problem of rule interaction. In the classification task, one usually wants to evolve a *good set* of rules, rather than a set of *good rules*. In other words, it is important to discover a rule set where the rules "cooperate" with each other. In particular, the rule set should cover the entire data space, so that each data instance should be covered by at least one rule. This requires a special mechanism to discover a diverse set of rules, since a standard EA would typically converge to a population where almost all the individuals would represent the same best rule found by the evolutionary process.

In general the previously discussed approaches perform a "direct" search for rules, consisting of initializing a population with a set of rules and then iteratively modifying those rules via the application of genetic operators. Due to a certain degree of randomness typically present in both initialization and genetic operations, some bad quality rules tend to be produced along the evolutionary process. Of course such bad rules are likely to be eliminated quickly by the selection process, but in any case an interesting alternative and "indirect" way of searching for rules has been proposed, in order to minimize the generation of bad rules. The basic idea of this new approach, proposed in (Jiao et al. 2006), is that the EA searches for good groups (clusters) of data instances, where each group consists of instances of the same class. A group is good to the extent that its data instances have similar attribute values and those attribute values are different from attribute values of the instances in other groups. After the EA run is over and good groups of instances have been discovered by the EA, the system extracts classification rules from the groups. This seems a promising new approach, although it should be noted that the version of the system described in (Jiao et al. 2006) has the limitation of coping only with categorical (not continuous) attributes.

In passing, it is worth mentioning that the above discussion on rule representation issues has focused on a generic classification problem. Specific kinds of classification problems may well be more effectively solved by EAs using rule representations "tailored" to the target kind of problem. For instance, (Hirsch et al. 2005) propose a rule representation tailored to document classification (i.e., a *text* mining problem), where strings of characters – in general fragments of words, rather than full words – are combined via Boolean operators to form classification rules.

19.3.2 Searching for a Diverse Set of Rules

This subsection discusses two mechanisms for discovering a diverse set of rules. It is assumed that each individual represents a single classification rule (Michigan-style approach). Note that the mechanisms for rule diversity discussed below are not normally used in the Pittsburgh approach, where an individual already represents a set of rules whose fitness implicitly depends on how well the rules in the set cooperate with each other.

First, one can use a niching method. The basic idea of niching is to avoid that the population converges to a single high peak in the search space and to foster the EA to create stable subpopulations of individuals clustered around each of the high peaks. In general the goal is to obtain a kind of "fitness-proportionate" convergence, where the size of the subpopulation around each peak is proportional to the height of that peak (i.e., to the quality of the corresponding candidate solution).

For instance, one of the most popular niching methods is fitness sharing (Goldberg & Richardson 1987; Deb & Goldberg 1989). In this method, the fitness of an individual is reduced in proportion to the number of similar individuals (neighbors), as measured by a given distance metric. In the context of rule discovery, this means that if there are many individuals in the current population representing the same rule or similar rules, the fitness of those individuals will be considerably reduced, and so they will have a considerably lower probability of being selected to produce new offspring. This effectively penalizes individuals which are in crowded regions of the search space, forcing the EA to discover a diverse set of rules.

Note that fitness sharing was designed as a generic niching method. By contrast, there are several niching methods designed specifically for the discovery of classification rules. An example is the "universal suffrage" selection method (Giordana et al. 1994; Divina 2005) where – using a political metaphor – individuals to be selected for reproduction are "elected" by the training data instances. The basic idea is that each data instance "votes" for a rule that covers it in a probabilistic fitness-based fashion. More precisely, let R be the set of rules (individuals) that cover a given data instance i, i.e., the set of rules whose antecedent is satisfied by data instance i. The better the fitness of a given rule r in the set R, the larger the probability that rule r will receive the vote of data instance i. Note that in general only rules covering the same data instances are competing with each other. Therefore, this selection method implements a form of niching, fostering the evolution of different rules covering different parts of the data space. For more information about niching methods in the context of discovering classification rules the reader is referred to (Hekanaho 1996; Dhar et al. 2000).

Another kind of mechanism that can be used to discover a diverse set of rules consists of using the previously-mentioned "sequential covering" approach – also known as "separate-and-conquer". The basic idea is that the EA discovers one rule at a time, so that in order to discover multiple rules the EA has to be run multiple times. In the first run the EA is initialized with the full training set and an empty set of rules. After each run of the EA, the best rule evolved by the EA is added to the set of discovered rules and the examples correctly covered by that rule are removed from the training set, so that the next run of the EA will consider a smaller training set. The process proceeds until all examples have been covered. Some examples of EAs using the sequential covering approach can be found in (Liu & Kwok 2000; Zhou et al. 2003; Carvalho & Freitas 2004). Note that the sequential covering approach is not specific to EAs. It is used by several non-evolutionary rule induction algorithms, and it is also discussed in data mining textbooks such as (Witten & Frank 2005).

19.3.3 Fitness Evaluation

One interesting characteristic of EAs is that they naturally allow the evaluation of a candidate solution, say a classification rule, as a whole, in a global fashion. This is in contrast with some data mining paradigms, which evaluate a partial solution. Consider, for instance, a conventional, greedy rule induction algorithm that incrementally builds a classification rule by adding one condition at a time to the rule. When the algorithm is evaluating several candidate conditions, the rule is still incomplete, being just a partial solution, so that the rule evaluation function is somewhat shortsighted (Freitas 2001, 2002a; Furnkranz & Flach 2003).

Another interesting characteristic of EAs is that they naturally allow the evaluation of a candidate solution by simultaneously considering different quality criteria. This is not so easily done in other data mining paradigms. To see this, consider again a conventional, greedy rule induction algorithm that adds one condition at a time to a candidate rule, and suppose one wants to favor the discovery of rules which are both accurate and simple (short). As mentioned earlier, when the algorithm is evaluating several candidate conditions, the rule is still incomplete, and so its size is not known yet. Hence, intuitively is better to choose the best candidate condition to be added to the rule based on a measure of accuracy only. The simplicity (size) criterion is better considered later, in a pruning procedure.

The fact that EAs evaluate a candidate solution as a whole and lend themselves naturally to simultaneously consider multiple criteria in the evaluation of the fitness of an individual gives the data miner a great flexibility in the design of the fitness function. Hence, not surprisingly, many different fitness functions have been proposed to evaluate classification rules. Classification accuracy is by far the criterion most used in fitness functions for evolving classification rules. This criterion is already extensively discussed in many good books or articles about classification, e.g. (Hand 1997; Caruana & Niculescu-Mizil 2004), and so it will not be discussed here – with the exception of a brief mention of overfitting issues, as follows. EAs can discover rules that overfit the training set – i.e. rules that represent very specific patterns in the training set that do not generalize well to the test set (which contains data instances unseen during training). One approach to try to mitigate the overfitting problem is to vary the training set at every generation, i.e., at each generation a subset of training instances is randomly selected, from the entire set of training instances, to be used as the (sub-)training or validation set from which the individuals' fitness values are computed (Bacardit et al. 2004; Pappa & Freitas 2006; Sharpe & Glover 1999; Bhattacharyya 1998). This approach introduces a selective pressure for evolving rules with a greater generalization power and tends to reduce the risk of overfitting, by comparison with the conventional approach of evolving rules for a training set which remains fixed throughout evolution. In passing, if the (sub)-training or validation set used for fitness computation is significantly smaller than the original training set, this approach also has the benefit of significantly reducing the processing time of the EA.

Hereafter this section will focus on two other rule-quality criteria (not based on accuracy) that represent different desirables properties of discovered rules in the context of data mining, namely: comprehensibility (Fayyad et al. 1996), or simplicity; and surprisingness, or unexpectedness (Liu et al. 1997; Romao et al. 2004; Freitas 2006).

The former means that ideally the discovered rule(s) should be comprehensible *to the user*. Intuitively, a measure of comprehensibility should have a strongly subjective, user-dependent component. However, in the literature this subjective component is typically ignored (Pazzani 2000; Freitas 2006), and comprehensibility is usually evaluated by a measure of the syntactic simplicity of the classifier, say the size of the rule set. The latter can be measured in an objective manner, for instance, by simply counting the total number of rule conditions in the rule set represented by an individual.

However, there is a natural way of incorporating a subjective measure of comprehensibility into the fitness function of an EA, namely by using an *interactive* fitness function. The basic idea of an interactive fitness function is that the user directly evaluates the fitness of individuals during the execution of the EA (Banzhaf 2000). The evaluation of the user is then used as the fitness measure for the purpose of selecting the best individuals of the current population, so that the EA evolves solutions that tend to maximize the subjective preference of the user.

An interactive EA for attribute selection is discussed e.g. in (Terano & Ishino 1998, 2002). In that work an individual represents a selected subset of attributes, which is then used by a classification algorithm to generate a set of rules. Then the user is shown the rules and selects good rules and rule sets according to her/his subjective preferences. Next the individuals having attributes that occur in the selected rules or rule sets are selected as parents to produce new offspring. The main advantage of interactive fitness functions is that intuitively they tend to favor the discovery of rules that are comprehensible and considered "good" by the user. The main disadvantage of this approach is that it makes the system considerably slower. To mitigate this problem one often has to use a small population size and a small number of generations.

Another kind of criterion that has been used to evaluate the quality of classification rules in the fitness function of EAs is the surprisingness of the discovered rules. First of all, it should be noted that accuracy and comprehensibility do not imply surprisingness. To show this point, consider the following classical hypothetical rule, which could be discovered from a hospital's database: IF (patient is pregnant) THEN (gender is female). This rule is very accurate and very comprehensible, but it is useless, because it represents an obvious pattern.

One approach to discover surprising rules consists of asking the user to specify a set of general impressions, specifying his/her previous knowledge and/or believes about the application domain (Liu et al. 1997). Then the EA can try to find rules that are surprising in the sense of contradicting some general impression specified by the user. Note that a rule should be reported to the user only if it is found to be both surprising and at least reasonably accurate (consistent with the training data). After all, it would be relatively easy to find rules which are surprising and inaccurate, but these rules would not be very useful to the user.

An EA for rule discovery taking this into account is described in (Romao et al. 2002, 2004). This EA uses a fitness function measuring both rule accuracy and rule surprisingness (based on general impressions). The two measures are multiplied to give the fitness value of an individual (a candidate prediction rule).

19.4 Evolutionary Algorithms for Clustering

There are several kinds of clustering algorithm, and two of the most popular kinds are iterative-partitioning and hierarchical clustering algorithms (Aldenderfer & Blashfield 1984; Krzanowski & Marriot 1995). In this section we focus mainly on EAs that can be categorized as iterative-partitioning algorithms, since most EAs for clustering seem to belong to this category.

19.4.1 Individual Representation for Clustering

A crucial issue in the design of an EA for clustering is to decide what kind of individual representation will be used to specify the clusters. There are at least three major kinds of individual representation for clustering (Freitas 2002a), as follows.

Cluster description-based representation – In this case each individual explicitly represents the parameters necessary to precisely specify each cluster. The exact nature of these parameters depends on the shape of clusters to be produced, which could be, e.g., boxes, spheres, ellipsoids, etc. In any case, each individual contains K sets of parameters, where K is the number of clusters, and each set of parameters determines the position, shape and size of its corresponding cluster. This kind of representation is illustrated, at a high level of abstraction, in Figure 19.2, for the case where an individual represents clusters of spherical shape. In this case each cluster is specified by its center coordinates and its radius. The cluster description-based representation is used, e.g., in (Srikanth et al. 1995), where an individual represents ellipsoid-based cluster descriptions; and in (Ghozeil and Fogel 1996; Sarafis 2005), where an individual represents hyperbox-shaped cluster descriptions. In (Sarafis 2005), for instance, the individuals represent rules containing conditions based on discrete numerical intervals, each interval being associated with a different attribute. Each clustering rule represents a region of the data space with homogeneous data distribution, and the EA was designed to be particularly effective when handling high-dimensional numerical datasets.

specification of cluster 1			specification of cluster K	
center 1 coordinates	radius 1	. . . . .	center K coordinates	radius K

Fig. 19.2. Structure of cluster description-based individual representation

Centroid/medoid-based representation – In this case each individual represents the coordinates of each cluster's centroid or medoid. A centroid is simply a point in the data space whose coordinates specify the centre of the cluster. Note that there may not be any data instance with the same coordinates as the centroid. By contrast, a medoid is the most "central" representative of the cluster, i.e., it is the data instance which is nearest to the cluster's centroid. The use of medoids tends to be more robust against outliers than the use of centroids (Krzanowski & Marriot 1995) (p. 83). This kind of representation is used, e.g., in (Hall et al. 1999; Estivill-Castro and Murray 1997) and other EAs for clustering reviewed in (Sarafis 2005). This representation is illustrated, at a high level of abstraction, in Figure 19.3. Each data instance is assigned to the cluster represented by the centroid or medoid that is nearest to that instance, according to a given distance measure. Therefore, the position of the centroids/medoids and the procedure used to assign instances to clusters implicitly determine the precise shape and size of the clusters.

cluster 1		cluster K
center 1 coordinates	. . . . .	center K coordinates

Fig. 19.3. Structure of centroid/medoid-based individual representation

Instance-based representation – In this case each individual consists of a string of n elements (genes), where n is the number of data instances. Each gene i, i=1,. . . ,n, represents the index (id) of the cluster to which the i-th data instance is assigned. Hence, each gene i can take one out of Kvalues, where K is the number of clusters. For instance, suppose that n = 10 and K= 3. The individual <2 1 2 3 3 2 1 1 2 3> corresponds to a candidate clustering where the second, seventh and eighth instances are assigned to cluster 1, the first, third, sixth and ninth instances are assigned to cluster 2 and the other instances are assigned to cluster 3. This kind of representation is used, for instance, in (Krishma and Murty 1999; Handl & Knowles 2004). A variation of this representation is used in (Korkmaz et al. 2006), where the value of a gene represents not the cluster id of a gene's associated data instance, but rather a link from the gene's instance to another instance which is considered to be in the same cluster. Hence, in this approach, two instances belong to the same cluster if there is a sequence of links from one of them to the other. This variation is more complex than the conventional instance-based representation, and it has been proposed together with repair operators that rectify the contents of an individual when it violates some pre-defined constraints.

Comparing different individual representations for clustering – In both the centroid/medoid-based representation and the instance-based representation, each instance is assigned to exactly one cluster. Hence, the set of clusters determine a partition of the data space into regions that are mutually exclusive and exhaustive. This is not the case in the cluster description-based representation. In the latter, the cluster descriptions may have some overlapping – so that an instance may be located within two or more clusters – and the cluster descriptions may not be exhaustive – so that some instance(s) may not be within any cluster.

Unlike the other two representations, the instance-based representation has the disadvantage that it does not scale very well for large data sets, since each individual's length is directly proportional to the number of instances being clustered. This representation also involves a considerable degree of redundancy, which may lead to problems in the application of conventional genetic operators (Falkenauer 1998). For instance, let $n = 4$ and $K = 2$, and consider the individuals <1 2 1 2> and <2 1 2 1>. These two individuals have different gene values in all the four genes, but they represent the same candidate clustering solution, i.e., assigning the first and third instances to one cluster and assigning the second and fourth instances to another cluster. As a result, a crossover between these two parent individuals can produce two children individuals representing solutions that are very different from the solutions represented by the parents, which is not normally the case in conventional crossover operators used by genetic algorithms. Some methods have been proposed to try to mitigate some redundancy-related problems associated with this kind of representation. For example, (Handl & Knowles 2004) proposed a mutation operator that is reported to work well with this representation, based on the idea that, when a gene has its value mutated – meaning that the gene's corresponding data instance is moved to another cluster – the system selects a number of "nearest neighbors" of that instance and moves all those nearest neighbors to the same cluster to which the mutated instance was moved. Hence, this approach effectively incorporates some knowledge of the clustering task to be solved in the mutation operator.

19.4.2 Fitness Evaluation for Clustering

In an EA for clustering, the fitness of an individual is a measure of the quality of the clustering represented by the individual. A large number of different measures have been proposed in the literature, but the basic ideas usually involve the following principles. First, the smaller the intra-cluster (within-cluster) distance, the better the fitness. The intra-cluster distance can be defined as the summation of the distance between each data instance and the centroid of its corresponding cluster – a summation computed over all instances of all the clusters. Second, the larger the inter-cluster (between-cluster) distance, the better the fitness. Hence, an algorithm can try to find optimal values for these two criteria, for a given fixed number of clusters. These and other clustering-quality criteria are extensively discussed in the clustering literature – see e.g. (Aldenderfer and Blashfield 1984; Backer 1995; Tan et al. 2006). A discussion of this topic in the context of EAs can be found in (Kim et al. 2000; Handl & Knowles 2004; Korkmaz et al. 2006; Krishma and Murty 1999; Hall et al. 1999).

In any case, it is important to note that, if the algorithm is allowed to vary the number of discovered clusters without any restriction, it would be possible to minimize intra-cluster distance and maximize inter-cluster distance in a trivial way, by assigning each example to its own singleton cluster. This would be clearly undesirable. To avoid this while still allowing the algorithm to vary the number of clusters, a common response is to incorporate in the fitness function a preference for a smaller number of clusters. It might also be desirable or necessary to incorporate in the fitness function a penalty term whose value is proportional to the number of empty clusters (i.e. clusters to which no data instance was assigned) (Hall et al. 1999).

19.5 Evolutionary Algorithms for Data Preprocessing

19.5.1 Genetic Algorithms for Attribute Selection

In the attribute selection task the goal is to select, out of the original set of attributes, a subset of attributes that are relevant for the target data mining task (Liu & Motoda 1998; Guyon and Elisseeff 2003). This Subsection assumes the target data mining task is classification – which is the most investigated task in the evolutionary attribute selection literature – unless mentioned otherwise.

The standard individual representation for attribute selection consists simply of a string of N bits, where N is the number of original attributes and the i-th bit, i=1,...,N, can take the value 1 or 0, indicating whether or not, respectively, the i-th attribute is selected. For instance, in a 10-attribute data set, the individual "1 0 1 0 1 0 0 0 0 1" represents a candidate solution where only the 1st, 3rd, 5th and 10th attributes are selected. This individual representation is simple, and traditional crossover and mutation operators can be easily applied. However, it has the disadvantage that it does not scale very well with the number of attributes. In applications with many thousands of attributes (such as text mining and some bioinformatics problems) an individual would have many thousands of genes, which would tend to lead to a slow execution of the GA.

An alternative individual representation, proposed by (Cherkauer & Shavlik 1996), consists of M genes (where M is a user-specified parameter), where each gene can contain either the index (id) of an attribute or a flag – say 0 – denoting no attribute. An attribute is considered selected if and only if it occurs in at least one of the M genes of the individual. For instance, the individual "3 0 8 3 0", where $M = 5$, represents a candidate solution where only the 3rd and the 8th attributes are selected. The fact that the 3rd attribute occurs twice in the previous individual is irrelevant for the purpose of decoding the individual into a selected attribute subset. One advantage of this representation is that it scales up better with respect to a large number of original attributes, since the value of M can be much smaller than the number of original attributes. One disadvantage is that it introduces a new parameter, M, which was not necessary in the case of the standard individual representation.

With respect to the fitness function, GAs for attribute selection can be roughly divided into two approaches – just like other kinds of algorithms for attribute selection – namely the wrapper approach and the filter approach. In essence, in the wrapper approach the GA uses the classification algorithm to compute the fitness of individuals, whereas in the filter approach the GA does not use the classification algorithm. The vast majority of GAs for attribute selection has followed the wrapper approach, and many of those GAs have used a fitness function involving two or more criteria to evaluate the quality of the classifier built from the selected attribute subset. This can be shown in Table 19.1, adapted from (Freitas 2002a), which lists the evaluation criteria used in the fitness function of a number of GAs following the wrapper approach. The columns of that table have the following meaning: *Acc* = accuracy; *Sens, Spec* = sensitivity, specificity; |Sel Attr| = number of selected attributes; |rule set| = number of discovered rules; *Info. Cont.* = information content of selected attributes;

Attr cost = attribute costs; *Subj eval* = subjective evaluation of the user; $|Sel\ ins|$ = number of selected instances.

Table 19.1. Diversity of criteria used in fitness function for attribute selection

<table>
<tr><th>Reference</th><th>Acc</th><th>Sens, Spec</th><th>|Sel Attr|</th><th>|rule set|</th><th>Info cont</th><th>Attr cost</th><th>Subj eval</th><th>|Sel ins|</th></tr>
<tr><td>(Bala et al. 1995)</td><td>yes</td><td></td><td>yes</td><td></td><td></td><td></td><td></td><td></td></tr>
<tr><td>(Bala et al. 1996)</td><td>yes</td><td></td><td>yes</td><td></td><td>yes</td><td></td><td></td><td></td></tr>
<tr><td>(Chen et al. 1999)</td><td>yes</td><td></td><td>yes</td><td></td><td></td><td></td><td></td><td></td></tr>
<tr><td>(Cherkauer & Shavlik 1996)</td><td>yes</td><td></td><td>yes</td><td>yes</td><td></td><td></td><td></td><td></td></tr>
<tr><td>(Emmanouilidis et al. 2000)</td><td>yes</td><td></td><td>yes</td><td></td><td></td><td></td><td></td><td></td></tr>
<tr><td>(Emmanouilidis et al. 2002)</td><td></td><td>yes</td><td>yes</td><td></td><td></td><td></td><td></td><td></td></tr>
<tr><td>(Guerra-Salcedo, Whitley 1998, 1999)</td><td>yes</td><td></td><td></td><td></td><td></td><td></td><td></td><td></td></tr>
<tr><td>(Ishibuchi & Nakashima 2000)</td><td>yes</td><td></td><td>yes</td><td></td><td></td><td></td><td></td><td>yes</td></tr>
<tr><td>(Llora & Garrell 2003)</td><td>yes</td><td></td><td></td><td></td><td></td><td></td><td></td><td></td></tr>
<tr><td>(Miller et al. 2003)</td><td></td><td>yes</td><td></td><td></td><td></td><td></td><td></td><td></td></tr>
<tr><td>(Moser & Murty 2000)</td><td>yes</td><td></td><td>yes</td><td></td><td></td><td></td><td></td><td></td></tr>
<tr><td>(Ni & Liu 2004)</td><td>yes</td><td></td><td></td><td></td><td></td><td></td><td></td><td></td></tr>
<tr><td>(Pappa et al. 2002)</td><td>yes</td><td></td><td></td><td>yes</td><td></td><td></td><td></td><td></td></tr>
<tr><td>(Rozsypal & Kubat 2003)</td><td>yes</td><td></td><td>yes</td><td></td><td></td><td></td><td></td><td>yes</td></tr>
<tr><td>(Terano & Ishino 1998)</td><td>yes</td><td></td><td></td><td>yes</td><td></td><td></td><td>yes</td><td></td></tr>
<tr><td>(Vafaie & DeJong 1998)</td><td>yes</td><td></td><td></td><td></td><td></td><td></td><td></td><td></td></tr>
<tr><td>(Yang & Honavar 1997, 1998)</td><td>yes</td><td></td><td></td><td></td><td></td><td>yes</td><td></td><td></td></tr>
<tr><td>(Zhang et al 2003)</td><td>yes</td><td></td><td></td><td></td><td></td><td></td><td></td><td></td></tr>
</table>

A precise definition of the terms used in the titles of the columns of Table 19.1 can be found in the corresponding references quoted in that table. The table refers to GAs that perform attribute selection for the classification task. GAs that perform attribute selection for the clustering task can be found, e.g., in (Kim et al. 2000; Jourdan 2003). In addition, in general Table 19.1 refers to GAs whose individuals directly represent candidate attribute subsets, but GAs can be used for attribute selection in other ways. For instance, in (Jong et al. 2004) a GA is used for attribute ranking. Once the ranking has been done, one can select a certain number of top-ranked attributes, where that number can be specified by the user or computed in a more automated way.

Empirical comparisons between GAs and other kinds of attribute selection methods can be found, for instance, in (Sharpe and Glover 1999; Kudo & Skalansky 2000). In general these empirical comparisons show that GAs, with their associated global search in the solution space, usually (though not always) obtain better results than local search-based attribute selection methods. In particular, (Kudo & Skalansky 2000) compared a GA with 14 non-evolutionary attribute selection methods (some of them variants of each other) across 8 different data sets. The authors concluded that the advantages of the global search associated with GAs over the local search associated with other algorithms is particularly important in data sets with a "large" number of attributes, where "large" was considered over 50 attributes in the context of their data sets.

19.5.2 Genetic Programming for Attribute Construction

In the attribute construction task the general goal is to construct new attributes out of the original attributes, so that the target data mining task becomes easier with the new attributes. This Subsection assumes the target data mining task is classification – which is the most investigated task in the evolutionary attribute construction literature.

Note that in general the problem of attribute construction is considerably more difficult than the problem of attribute selection. In the latter the problem consists just of deciding whether or not to select each attribute. By contrast, in attribute construction there is a potentially much larger search space, since there is a potentially large number of operations that can be applied to the original attributes in order to construct new attributes. Intuitively, the kind of EA that lends itself most naturally to attribute construction is GP. The reason is that, as mentioned earlier, GP was specifically designed to solve problems where candidate solutions are represented by both attributes and functions (operations) applied to those attributes. In particular, the explicit specification of both a terminal set and a function set is usually missing in other kinds of EAs.

Data Preprocessing vs. Interleaving Approach

In the data preprocessing approach, the attribute construction algorithm evaluates a constructed attribute without using the classification algorithm to be applied later. Examples of this approach are the GP algorithms for attribute construction proposed by (Otero et al. 2003; Hu 1998), whose attribute evaluation function (the fitness function) is the information gain ratio – a measure discussed in detail in (Quinlan 1993). In addition, (Muharram & Smith 2004) did experiments comparing the effectiveness of two different attribute-evaluation criteria in GP for attribute construction – viz. information gain ratio and gini index – and obtained results indicating that, overall, there was no significant difference in the results associated with those two criteria.

By contrast, in the interleaving approach the attribute construction algorithm evaluates the constructed attributes based on the performance of the classification algorithm with those attributes. Examples of this approach are the GP algorithms for attribute construction proposed by (Krawiec 2002; Smith and Bull 2003; Firpi et al. 2005), where the fitness functions are based on the accuracy of the classifier built with the constructed attributes.

Single-Attribute-per-Individual vs. Multiple-Attributes-per-Individual Representation

In several GPs for attribute construction, each individual represents a single constructed attribute. This approach is used for instance by CPGI (Hu 1998) and the GP algorithm proposed by (Otero et al. 2003). By default this approach returns to the user a single constructed attribute – the best evolved individual. However it can be extended to return to the user a set of constructed attributes, say returning a set of the best evolved individuals of a GP run or by running the GP multiple times and returning only the best evolved individual of each run. The main advantage of this approach is simplicity, but it has the disadvantage of ignoring interactions between the constructed attributes.

An alternative approach consists of associating with an individual a set of constructed attributes. The main advantage of this approach is that it takes into account interaction between the constructed attributes. In other words, it tries to construct the *best set* of attributes, rather than the set of *best attributes*. The main disadvantages are that the individuals' genomes become more complex and that it introduces the need for additional parameters such as the number of constructed attributes that should be encoded in one individual (a parameter that is usually specified in an ad-hoc fashion). In any case, the equivalent of this latter parameter would also have to be specified in the above-mentioned "extended version" of the single-attribute-per-individual approach when one wants the GP algorithm to return multiple constructed attributes.

Examples of this multiple-attributes-per-individual approach are the GP algorithms proposed by (Krawiec 2002; Smith & Bull 2003; Firpi et al. 2005). Here we briefly discuss the former two, as examples of this approach. In (Krawiec 2002) each individual encodes a fixed number K of constructed attributes, each of them represented by a tree, so that an individual consists of K trees – where K is a user-specified parameter. The algorithm also includes a method to split the constructed attributes encoded in an individual into two subsets, namely the subset of "evolving" attributes and the subset of "hidden" attributes. The basic idea is that high-quality constructed attributes are considered hidden (or "protected"), so that they cannot be manipulated by the genetic operators such as crossover and mutation. The choice of attributes to be hidden is based on an attribute quality measure. This measure evaluates the quality of each constructed attribute separately, and the best attributes of the individual are considered hidden.

Another example of the multiple-attributes-per-individual approach is the GAP (Genetic Algorithm and Programming) system proposed by (Smith & Bull 2003, 2004). GAP performs both attribute construction and attribute selection. The first stage consists of attribute construction, which is performed by a GP algorithm. As a result of this first stage, the system constructs an extended genotype containing both the constructed attributes represented in the best evolved individual of the GP run and original attributes that have not been used in those constructed attributes. This extended genotype is used as the basic representation for a GA that performs attribute selection, so that the GA searches for the best subset of attributes out of all (both constructed and original) attributes.

Satisfying the Closure Property

GP algorithms for attribute construction have used several different approaches to satisfy the closure property (briefly mentioned in Section 2). This is an important issue, because the chosen approach can have a significant impact on the types (e.g., continuous or nominal) of original attributes processed by the algorithm and on the types of attributes constructed by the algorithm. Let us see some examples.

A simple solution for the closure problem is used in the GAP algorithm (Smith and Bull 2003). Its terminal set contains only the continuous (real-valued) attributes of the data being mined. In addition, its function set consists only of arithmetic operators (+, –, *, %,) – where % denotes protected division, i.e. a division operator that handles zero denominator inputs by returning something different from an error (Banzhaf et al. 1998; Koza 1992) – so that the closure property is immediately satisfied. (Firpi et al. 2005) also uses the approach of having a function set consisting only of mathematical operators, but it uses a considerably larger set of mathematical operators than the set used by (Smith and Bull 2003).

The GP algorithm proposed by (Krawiec 2002) uses a terminal set including all original attributes (both continuous and nominal ones), and a function set consisting of arithmetical operators (+, –, *, %, log), comparison operators ($<$, $>$, =), an "IF (conditional expression)", and an "approximate equality operator" which compares its two arguments with tolerance given by the third argument. The algorithm did not enforce data type constraints, which means that expressions encoding the constructed attributes make no distinction between, for instance, continuous and nominal attributes. Values of nominal attributes, such as male and female, are treated as numbers. This helps to solve the closure problem, but at a high price: constructed attributes can contain expressions that make no sense from a semantical point of view. For instance, the algorithm could produce an expression such as "*Gender + Age*", because the value of the nominal attribute *Gender* would be interpreted as a number.

The GP proposed by (Otero et al. 2003) uses a terminal set including only the continuous attributes of the data being mined. Its function set consists of arithmetic operators (+, –, *, %,) and comparison operators ($\geq$, $\leq$). In order to satisfy the closure property, the algorithm enforces the data type restriction that the comparison operators can be used only at the root of the GP tree, i.e., they cannot be used as child nodes of other nodes in the tree. The reason is that comparison operators return a Boolean value, which cannot be processed by any operator in the function set (all operators accept only continuous values as input). Note that, although the algorithm can construct attributes only out of the continuous original attributes, the constructed attributes themselves can be either Boolean or continuous. A constructed attribute will be Boolean if its corresponding tree in the GP individual has a comparison operator at the root node; it will be continuous otherwise.

In order to satisfy the closure property, GPCI (Hu 1998) simply transforms all the original attributes into Boolean attributes and uses a function set containing only Boolean functions. For instance, if an attribute A is continuous (real-valued), such as the attribute *Salary*, it is transformed into two Boolean attributes, such as "Is *Salary* $> t$?" and "Is *Salary* $\leq t$?", where t is a threshold automatically chosen by the algorithm in order to maximize the ability of the two new attributes in discriminating between instances of different classes. The two new attributes are named "*positive-A*" and "*negative-A*", respectively. Once every original attribute has been transformed into two Boolean attributes, a GP algorithm is applied to the Boolean attributes. In this GP, the terminal set consists of all the pairs of attributes "*positive-A*" and "*negative-A*" for each original attribute A, whereas the function set consists of the Boolean operators $\{$AND, OR$\}$. Since all terminal symbols are Boolean, and all operators accept Boolean values as input and produce Boolean value as output, the closure property is satisfied.

Table 19.2 summarizes the main characteristics of the five GP algorithms for attribute construction discussed in this Section.

Table 19.2. Summary of GP Algorithms for Attribute Construction

Reference	Approach	Individual representation	Datatype of input attrib	Datatype of output attrib
(Hu 1998)	Data preprocessing	Single attribute	Any (attributes are booleanised)	Boolean
(Krawiec 2002)	Interleaving	Multiple attributes	Any (nominal attrib. values are interpreted as numbers)	Continuous
(Otero et al. 2003)	Data preprocessing	Single attribute	Continuous	Continuous or Boolean
(Smith & Bull 2003, 2004)	Interleaving	Multiple attributes	Continuous	Continuous
(Firpi et al. 2005)	Interleaving	Multiple attributes	Continuous	Continuous

19.6 Multi-Objective Optimization with Evolutionary Algorithms

There are many real-world optimization problems that are naturally expressed as the simultaneous optimization of two or more conflicting objectives (Coello Coello 2002; Deb 2001; Coello Coello & Lamont 2004). A generic example is to maximize the quality of a product and minimize its manufacturing cost in a factory. In the context of data mining, a typical example is, in the data preprocessing task of attribute selection, to minimize the error rate of a classifier trained with the selected attributes and to minimize the number of selected attributes.

The conventional approach to cope with such multi-objective optimization problems using evolutionary algorithms is to convert the problem into a single-optimization problem. This is typically done by using a weighted formula in the fitness function, where each objective has an associated weight reflecting its relative importance. For instance, in the above example of two-objective attribute selection, the fitness function could be defined as, say: "2/3 classification_error + 1/3 Number_of_selected_attributes".

However, this conventional approach has several problems. First, it mixes non-commensurable objectives (classification error and number of selected attributes in the previous example) into the same formula. This has at least the disadvantage that the value returned by the fitness function is not meaningful to the user. Second, note that different weights will lead to different selected attributes, since different weights represent different trade-offs between the two conflicting objectives. Unfortunately, the weights are usually defined in an ad-hoc fashion. Hence, when the EA returns the best attribute subset to the user, the user is presented with a solution that represents just one possible trade-off between the objectives. The user misses the opportunity to analyze different trade-offs.

Of course we could address this problem by running the EA multiple times, with different weights for the objectives in each run, and return the multiple solutions to the user. However, this would be very inefficient, and we would still have the problems of deciding which weights should be used in each run, how many runs we should perform (and so how many solutions should be returned to the user), etc.

A more principled approach consists of letting an EA answer these questions automatically, by performing a global search in the solution space and discovering as many good solutions, with as much diversity among them, as possible. This can be done by using a multi-objective EA, a kind of EA which has become quite popular in the EA community in the last few years (Deb 2001; Coello Coello 2002; Coello Coello & Lamont 2004). The basic idea involves the concept of Pareto dominance. A solution s_1 is said to dominate, in the Pareto sense, another solution s_2 if and only if solution s_1 is strictly better than s_2 in at least one of the objectives and solution s_1 is not worse than s_2 in any of the objectives. The concept of Pareto dominance is illustrated in Figure 19.4. This figure involves two objectives to be minimized, namely classification error and number of selected attributes (No_attrib). In that figure, solution D is dominated by solution B (which has both a smaller error and a smaller number of selected attributes than D), and solution E is dominated by solution C. Hence, solutions A, B and C are non-dominated solutions. They constitute the best "Pareto front" found by the algorithm. All these three solutions would be returned to the user.

The goal of a multi-objective EA is to find a Pareto front which is as close as possible to the true (unknown) Pareto front. This involves not only the minimization of the two objectives, but also finding a diverse set of non-dominated solutions, spread along the Pareto front. This allows the EA to return to the user a diverse set of good trade-offs between the conflicting objectives. With this rich information, the user can hopefully make a more intelligent decision, choosing the best solution to be used in practice.

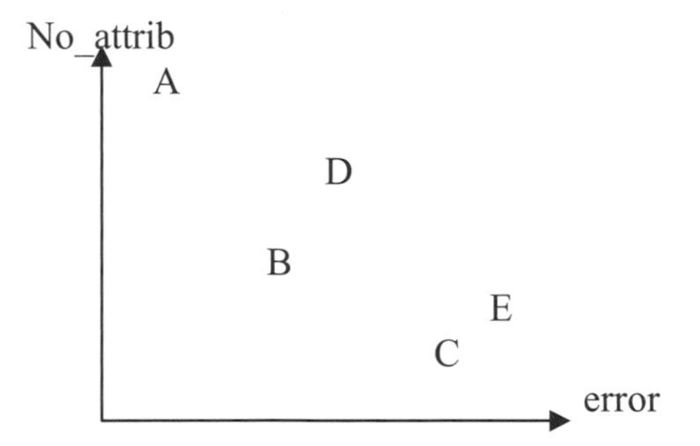

Fig. 19.4. Example of Pareto dominance

At this point the reader might argue that this approach has the disadvantage that the final choice of the solution to be used depends on the user, characterizing a subjective approach. The response to this is that the knowledge discovery process is interactive (Brachman & Anand 1996; Fayyad et al. 1996), and the participation of the user in this process is important to obtain useful results. The questions are *when and how* the user should participate (Deb 2001; Freitas 2004). In the above-described multi-objective approach, based on Pareto dominance, the user participates by choosing the best solution out of all the non-dominated solutions. This choice is made *a posteriori*, i.e., after the algorithm has run and has returned a rich source of information about the solution space: the discovered Pareto front. In the conventional approach – using an EA with a weighted formula and returning a single solution to the user – the user has to define the weights *a priori*, i.e., before running the algorithm, when the solution space was not explored yet. The multi-objective approach seems to put the user in the loop in a better moment, when valuable information about the solution space is available. The multi-objective approach also avoids the problems of ad-hoc choice of weights, mixing non-commensurable objectives into the same formula, etc.

Table 19.3 lists the main characteristics of multi-objective EAs for data mining. Most systems included in Table 19.3 consider only two objectives. The exceptions are the works of (Kim et al. 2000) and (Atkinson-Abutridy et al. 2003), considering 4 and 8 objectives, respectively. Out of the EAs considering only two objectives, the most popular choice of objectives – particularly for EAs addressing the classification task – has been some measure of classification accuracy (or its dual, error) and a measure of the size of the classification model (number of leaf nodes in a decision tree or total number of rule conditions – attribute-value pairs – in all rules). Note that the size of a model is typically used as a proxy for the concept of "simplicity" of that model, even though arguably this proxy leaves a lot to be desired as a measure of a model's simplicity (Pazzani 2000; Freitas 2006). (In practice, however, it seems no better proxy for a model's simplicity is known.) Note also that, when the task being solved is attribute selection for classification, the objective related to size can be the number of selected attributes, as in (Emmanouilidis et al. 2000), or the size of the classification model built from the set of selected attributes, as in (Pappa et al. 2002, 2004). Finally, when solving the clustering task a popular choice of objective has been some measure of intra-cluster distance, related to the total distance between each data instance and the centroid of its cluster, computed for all data instances in all the clusters. The number of clusters is also used as an objective in two out of the three EAs for clustering included in Table 19.3. A further discussion of multi-objective optimization in the context of data mining in general (not focusing on EAs) is presented in (Freitas 2004; Jin 2006).

Table 19.3. Main characteristics of multi-objective EAs for data mining

Reference	Data mining task	Objectives being Optimized
(Emmanouilidis et al. 2000)	attribute selection for classification	accuracy, number of selected attributes
(Pappa et al 2002, 2004)	attribute selection for classification	accuracy, number of leafs in decision tree
(Ishibuchi & Namba 2004)	selection of classification rules	error, number of rule conditions (in all rules)
(de la Iglesia 2007)	selection of classification rules	confidence, coverage
(Kim et al. 2004)	classification	error, number of leafs in decision tree
(Atkinson-Abutridy et al. 2003)	text mining	8 criteria for evaluating explanatory knowledge across text documents
(Kim et al. 2000)	attribute selection for clustering	Cluster cohesiveness, separation between clusters, number of clusters, number of selected attributes
(Handl & Knowles 2004)	clustering	Intra-cluster deviation and connectivity
(Korkmaz et al. 2006)	clustering	Intra-cluster variance and number of clusters

19.7 Conclusions

This chapter started with the remark that EAs are a very generic search paradigm. Indeed, the chapter discussed how EAs can be used to solve several different data mining tasks, namely the discovery of classification rules, clustering, attribute selection and attribute construction. The discussion focused mainly on the issues of individual representation and fitness function for each of these tasks, since these are the two EA-design issues that are more dependent of the task being solved. In any case, recall that the design of an EA also involves the issue of genetic operators. Ideally these three components – individual representation, fitness function and genetic operators – should be designed in a synergistic fashion and tailored to the data mining task being solved.

There are at least two motivations for using EAs in data mining, broadly speaking. First, as mentioned earlier, EAs are robust, adaptive search methods that perform a global search in the solution space. This is in contrast to other data mining paradigms that typically perform a greedy search. In the context of data mining, the global search of EAs is associated with a better ability to cope with attribute interactions. For instance, most "conventional", non-evolutionary rule induction algorithms are greedy, and therefore quite sensitive to the problem of attribute interaction. EAs can use the same knowledge representation (IF-THEN rules) as conventional rule induction algorithms, but their global search tends to cope better with attribute interaction and to discover interesting relationships that would be missed by a greedy search (Dhar et al. 2000; Papagelis & Kalles 2001; Freitas 2002a).

Second, EAs are a very flexible algorithmic paradigm. In particular, borrowing some terminology from programming languages, EAs have a certain "declarative" – rather than "procedural" – style. The quality of an individual (candidate solution) is evaluated, by a fitness function, in a way independent of how that solution was constructed. This gives the data miner a considerable freedom in the design of the individual representation, the fitness function and the genetic operators. This flexibility can be used to incorporate background knowledge into the EA and/or to hybridize EAs with local search methods that are specifically tailored to the data mining task being solved.

Note that declarativeness is a matter of degree, rather than a binary concept. In practice EAs are not 100% declarative, because as one changes the fitness function one might consider changing the individual representation and the genetic operators accordingly, in order to achieve the above-mentioned synergistic relationship between these three components of the EA. However, EAs still have a degree of declarativeness considerably higher than other data mining paradigms. For instance, as discussed in Subsection 3.3, the fact that EAs evaluate a complete (rather than partial) rule allows the fitness function to consider several different rule-quality criteria, such as comprehensibility, surprisingness and subjective interestingness to the user. In EAs these quality criteria can be directly considered during the search for rules. By contrast, in conventional, greedy rule induction algorithms – where the evaluation function typically evaluates a partial rule – those quality criteria would typically have to be considered in a post-processing phase of the knowledge discovery process, when it might be too late. After all, many rule set post-processing methods just try to select the most interesting rules out of all discovered rules, so that interesting rules that were missed by the rule induction method will remain missing after applying the post-processing method.

Like any other data mining paradigm, EAs also have some disadvantages. One of them is that conventional genetic operators – such as conventional crossover and mutation operators – are "blind" search operators in the sense that they modify individuals (candidate solutions) in a way independent from the individual's fitness (quality). This characteristic of conventional genetic operators increases the generality of EAs, but intuitively tends to reduce their effectiveness in solving a specific kind of problem. Hence, in general it is important to modify or extend EAs to use task specific-operators.

Another disadvantage of EAs is that they are computationally slow, by comparison with greedy search methods. The importance of this drawback depends on many factors, such as the kind of task being performed, the size of the data being mined, the requirements of the user, etc. Note that in some cases a relatively long processing time might be acceptable. In particular, several data mining tasks, such as classification, are typically an off-line task, and the time spent solving that task is usually less than 20% of the total time of the knowledge discovery process. In scenarios like this, even a processing time of hours or days might be acceptable to the user, at least in the sense that it is not the bottleneck of the knowledge discovery process.

In any case, if necessary the processing time of an EA can be significantly reduced by using special techniques. One possibility is to use parallel processing techniques, since EAs can be easily parallelized in an effective way (Cantu-Paz 2000; Freitas & Lavington 1998; Freitas 2002a). Another possibility is to compute the fitness of individuals by using only a subset of training instances – where that subset can be chosen either at random or using adaptive instance-selection techniques (Bhattacharyya 1998; Gathercole & Ross 1997; Sharpe & Glover 1999; Freitas 2002a).

An important research direction is to better exploit the power of Genetic Programming (GP) in data mining. Several GP algorithms for attribute construction were discussed in Subsection 5.2, and there are also several GP algorithms for discovering classification rules (Freitas 2002a; Wong & Leung 2000) or for classification in general (Muni et al. 2004; Song et al. 2005; Folino et al. 2006). However, the power of GP is still underexplored. Recall that the GP paradigm was designed to *automatically* discover computer *programs*, or *algorithms*, which should be *generic "recipes"* for solving a given kind of problem, and not to find the solution to one particular instance of that problem (like in most EAs). For instance, classification is a kind of problem, and most classification-rule induction algorithms are generic enough to be applied to different data sets (each data set can be considered just an instance of the kind of problem defined by the classification task). However, these generic rule induction algorithms have been *manually* designed by a human being. Almost all current GP algorithms for classification-rule induction are competing with conventional (greedy, non-evolutionary) rule induction algorithms, in the sense that both GP and conventional rule induction algorithms are discovering classification rules for a single data set at a time. Hence, the output of a GP for classification-rule induction is a set of rules for a given data set, which can be called a "program" or "algorithm" only in a very loose sense of these words.

A much more ambitious goal, which is more compatible with the general goal of GP, is to use GP to *automatically* discover a rule induction *algorithm*. That is, to perform *algorithm induction*, rather than rule induction. The first version of a GP algorithm addressing this ambitious task has been proposed in (Pappa & Freitas 2006), and an extended version of that work is described in detail in another chapter of this book (Pappa & Freitas 2007).

References

Aldenderfer MS & Blashfield RK (1984) *Cluster Analysis* (Sage University Paper Series on Quantitative Applications in the Social Sciences, No. 44) Sage Publications.

Atkinson-Abutridy J, Mellishm C, and Aitken S (2003) A semantically guided and domain-independent evolutionary model for knowledge discovery from texts. *IEEE Trans. Evolutionary Computation 7(6)*, 546-560.

Bacardit J, Goldberg DE, Butz MV, Llora X, Garrell JM (2004). Speeding-up Pittsburgh learning classifier systems: modeling time and accuracy. *Proc. Parallel Problem Solving From Nature (PPSN-2004), LNCS 3242,* 1021-1031, Springer.

Bacardit J and Krasnogor N (2006) Smart crossover operator with multiple parents for a Pittsburgh learning classifier system. *Proc. Genetic & Evolutionary Computation Conf. (GECCO-2006)*, 1441-1448. Morgan Kaufmann.

Backer E (1995) *Computer-Assisted Reasoning in Cluster Analysis*. Prentice-Hall.

Back T, Fogel DB and Michalewicz (Eds.) (2000) *Evolutionary Computation 1: Basic Algorithms and Operators*. Institute of Physics Publishing.

Bala J, De Jong K, Huang J, Vafaie H and Wechsler H (1995) Hybrid learning using genetic algorithms and decision trees for pattern classification. *Proc. Int. Joint Conf. on Artificial Intelligence (IJCAI-95)*, 719-724.

Bala J, De Jong K, Huang J, Vafaie H and Wechsler H (1996) Using learning to facilitate the evolution of features for recognizing visual concepts. *Evolutionary Computation 4(3)*: 297-312.

Banzhaf W (2000) Interactive evolution. In: T. Back, D.B. Fogel and T. Michalewicz (Eds.) *Evolutionary Computation 1*, 228-236. Institute of Physics Pub.

Banzhaf W, Nordin P, Keller RE, and Francone FD (1998) *Genetic Programming ~ an Introduction: On the Automatic Evolution of Computer Programs and Its Applications.* Morgan Kaufmann.

Bhattacharrya S (1998) Direct marketing response models using genetic algorithms. *Proceedings of the 4th Int. Conf. on Knowledge Discovery and Data Mining (KDD-98)*, 144-148. AAAI Press.

Brachman RJ and Anand T. (1996) The process of knowledge discovery in databases: a human-centered approach. In: U.M. Fayyad et al (Eds.) *Advances. in Knowledge Discovery and Data Mining*, 37-58. AAAI/MIT.

Bull L (Ed.) (2004) *Applications of Learning Classifier Systems*. Springer.

Bull L and Kovacs T (Eds.) (2005) *Foundations of Learning Classifier Systems*. Springer.

Cantu-Paz E (2000) *Efficient and Accurate Parallel Genetic Algorithms*. Kluwer.

Caruana R and Niculescu-Mizil A (2004) Data mining in metric space: an empirical analysis of supervised learning performance criteria. *Proc. 2004 ACM SIGKDD Int. Conf. on Knowledge Discovery and Data Mining (KDD-04)*, ACM.

Carvalho DR and Freitas AA (2004). A hybrid decision tree/genetic algorithm method for data mining. *Special issue on Soft Computing Data Mining, Information Sciences 163(1-3)*, pp. 13-35. 14 June 2004.

Chen S, Guerra-Salcedo C and Smith SF (1999) Non-standard crossover for a standard representation - commonality-based feature subset selection. *Proc. Genetic and Evolutionary Computation Conf. (GECCO-99)*, 129-134. Morgan Kaufmann.

Cherkauer KJ and Shavlik JW (1996). Growing simpler decision trees to facilitate knowledge discovery. *Proc. 2nd Int. Conf. on Knowledge Discovery and Data Mining (KDD-96)*, 315-318. AAAI Press.

Coello Coello CA, Van Veldhuizen DA and Lamont GB (2002) *Evolutionary Algorithms for Solving Multi-Objective Problems*. Kluwer.

Coello Coello CA and Lamont GB (Ed.) (2004) *Applications of Multi-objective Evolutionary Algorithms*. World Scientific.

Deb K (2001) *Multi-Objective Optimization Using Evolutionary Algorithms*. Wiley.

Deb K and Goldberg DE (1989). An investigation of niche and species formation in genetic function optimization. *Proc. 2nd Int. Conf. Genetic Algorithms (ICGA-89)*, 42-49.

De Jong K (2006) *Evolutionary Computation: a unified approach.* MIT.

De la Iglesia B (2007) Application of multi-objective metaheuristic algorithms in data mining. *Proc. 3rd UK Knowledge Discovery and Data Mining Symposium (UKKDD-2007)*, 39-44, University of Kent, UK, April 2007.

Dhar V, Chou D and Provost F (2000). Discovering interesting patterns for investment decision making with GLOWER – a genetic learner overlaid with entropy reduction. *Data Mining and Knowledge Discovery 4(4)*, 251-280.

Divina F (2005) Assessing the effectiveness of incorporating knowledge in an evolutionary concept learner. *Proc. EuroGP-2005 (European Conf. on Genetic Programming), LNCS 3447,* 13-24, Springer.

Divina F & Marchiori E (2002) Evolutionary Concept Learning. *Proc. Genetic & Evolutionary Computation Conf. (GECCO-2002)*, 343-350. Morgan Kaufmann.

Divina F & Marchiori E (2005) Handling continuous attributes in an evolutionary inductive learner. *IEEE Trans. Evolutionary Computation, 9(1),* 31-43, Feb. 2005.

Eiben AE and Smith JE (2003) *Introduction to Evolutionary Computing*. Springer.

Emmanouilidis C, Hunter A and J. MacIntyre J (2000) A multiobjective evolutionary setting for feature selection and a commonality-based crossover operator. *Proc. 2000 Congress on Evolutionary Computation (CEC-2000)*, 309-316. IEEE.

Emmanouilidis C (2002) Evolutionary multi-objective feature selection and ROC analysis with application to industrial machinery fault diagnosis. In: K. Giannakoglou et al. (Eds.) *Evolutionary Methods for Design, Optimisation and Control.* Barcelona: CIMNE.

Estivill-Castro V and Murray AT (1997) Spatial clustering for data mining with genetic algorithms. *Tech. Report FIT-TR-97-10*. Queensland University of Technology. Australia.

Falkenauer E (1998) *Genetic Algorithms and Grouping Problems.* John-Wiley & Sons.

Fayyad UM, Piatetsky-Shapiro G and Smyth P (1996) From data mining to knowledge discovery: an overview. In: U.M. Fayyad et al (Eds.) *Advances in Knowledge Discovery and Data Mining*, 1-34. AAAI/MIT.

Firpi H, Goodman E, Echauz J (2005) On prediction of epileptic seizures by computing multiple genetic programming artificial features. *Proc. 2005 European Conf. on Genetic Programming (EuroGP-2005), LNCS 3447,* 321-330. Springer.

Folino G, Pizzuti C and Spezzano G (2006) GP ensembles for large-scale data classification. *IEEE Trans. Evolutionary Computation 10(5)*, 604-616, Oct. 2006.

Freitas AA and. Lavington SH (1998) *Mining Very Large Databases with Parallel Processing*. Kluwer.

Freitas AA (2001) Understanding the crucial role of attribute interaction in data mining. *Artificial Intelligence Review 16(3)*, 177-199.

Freitas AA (2002a) *Data Mining and Knowledge Discovery with Evolutionary Algorithms.* Springer.

Freitas AA (2002b) A survey of evolutionary algorithms for data mining and knowledge discovery. In: A. Ghosh and S. Tsutsui. (Eds.) *Advances in Evolutionary Computation*, pp. 819-845. Springer-Verlag.

Freitas AA (2002c). Evolutionary Computation. In: W. Klosgen and J. Zytkow (Eds.) *Handbook of Data Mining and Knowledge Discovery*, pp. 698-706.Oxford Univ. Press.

Freitas AA (2004) A critical review of multi-objective optimization in data mining: a position paper. *ACM SIGKDD Explorations, 6(2),* 77-86, Dec. 2004.

Freitas AA (2005) Evolutionary Algorithms for Data Mining. In: O. Maimon and L. Rokach (Eds.) *The Data Mining and Knowledge Discovery Handbook*, pp. 435-467. Springer.

Freitas AA (2006) Are we really discovering "interesting" knowledge from data? *Expert Update, Vol. 9, No. 1,* 41-47, Autumn 2006.

Furnkranz J and Flach PA (2003). An analysis of rule evaluation metrics. *Proc.20th Int. Conf. Machine Learning (ICML-2003).* Morgan Kaufmann.

Gathercole C and Ross P (1997) Tackling the Boolean even N parity problem with genetic programming and limited-error fitness. *Genetic Programming 1997: Proc. 2nd Conf. (GP-97)*, 119-127. Morgan Kaufmann.

Ghozeil A and Fogel DB (1996) Discovering patterns in spatial data using evolutionary programming. *Genetic Programming 1996: Proceedings of the 1st Annual Conf.*, 521-527. MIT Press.

Giordana A, Saitta L, Zini F (2004) Learning disjunctive concepts by means of genetic algorithms. *Proc. 10th Int. Conf. Machine Learning (ML-94)*, 96-104. Morgan Kaufmann.

Goldberg DE (1989). *Genetic Algorithms in Search, Optimization and Machine Learning.* Addison-Wesley.

Goldberg DE and Richardson J (1987) Genetic algorithms with sharing for multimodal function optimization. *Proc. Int. Conf. Genetic Algorithms (ICGA-87)*, 41-49.

Guerra-Salcedo C and Whitley D (1998) Genetic search for feature subset selection: a comparison between CHC and GENESIS. *Genetic Programming 1998: Proc. 3rd Annual Conf.*, 504-509. Morgan Kaufmann.

Guerra-Salcedo C, Chen S, Whitley D, and Smith S (1999) Fast and accurate feature selection using hybrid genetic strategies. *Proc. Congress on Evolutionary Computation (CEC-99)*, 177-184. IEEE.

Guyon I and Elisseeff A (2003) An introduction to variable and feature selection. *Journal of Machine Learning Research 3*, 1157-1182.

Hall LO, Ozyurt IB, Bezdek JC (1999) Clustering with a genetically optimized approach. *IEEE Trans. on Evolutionary Computation 3(2)*, 103-112.

Hand DJ (1997) *Construction and Assessment of Classification Rules*. Wiley.

Handl J and Knowles J (2004) Evolutionary multiobjective clustering. *Proc. Parallel Problem Solving From Nature (PPSN-2004), LNCS 3242,* 1081-1091, Springer.

Hekanaho J (1995) Symbiosis in multimodal concept learning. *Proc. 1995 Int. Conf. on Machine Learning (ML-95)*, 278-285. Morgan Kaufmann.

Hekanaho J (1996) Testing different sharing methods in concept learning. *TUCS Technical Report No. 71*. Turku Centre for Computer Science, Finland.

Hirsch L, Saeedi M and Hirsch R (2005) Evolving rules for document classification. *Proc. 2005 European Conf. on Genetic Programming (EuroGP-2005), LNCS 3447,* 85-95, Springer.

Hu YJ (1998). A genetic programming approach to constructive induction. *Genetic Programming 1998: Proc. 3rd Annual Conf.*, 146-151. Morgan Kaufmann.

Ishibuchi H and Nakashima T (2000) Multi-objective pattern and feature selection by a genetic algorithm. *Proc. 2000 Genetic and Evolutionary Computation Conf. (GECCO-2000)*, 1069-1076. Morgan Kaufmann.

Ishibuchi H and Namba S (2004) Evolutionary multiobjective knowledge extraction for high-dimensional pattern classification problems. *Proc. Parallel Problem Solving From Nature (PPSN-2004), LNCS 3242,* 1123-1132, Springer.

Jiao L, Liu J and Zhong W (2006) An organizational coevolutionary algorithm for classification. *IEEE Trans. Evolutionary Computation, Vol. 10, No. 1,* 67-80, Feb. 2006.

Jin, Y (Ed.) (2006) *Multi-Objective Machine Learning*. Springer.

Jong K, Marchiori E and Sebag M (2004) Ensemble learning with evolutionary computation: application to feature ranking. *Proc. Parallel Problem Solving from Nature VIII (PPSN-2004), LNCS 3242*, 1133-1142. Springer, 2004.

Jourdan L, Dhaenens-Flipo C and Talbi EG (2003) Discovery of genetic and environmental interactions in disease data using evolutionary computation. In: G.B. Fogel and D.W. Corne (Eds.) *Evolutionary Computation in Bioinformatics*, 297-316. Morgan Kaufmann.

Kim Y, Street WN and Menczer F (2000) Feature selection in unsupervised learning via evolutionary search. *Proc. 6th ACM SIGKDD Int. Conf. on Knowledge Discovery and Data Mining (KDD-2000)*, 365-369. ACM.

Kim D (2004). Structural risk minimization on decision trees: using an evolutionary multiobjective algorithm. *Proc. 2004 European Conference on Genetic Programming (EuroGP-2004), LNCS 3003,* 338-348, Springer.

Korkmaz EE, Du J, Alhajj R and Barker (2006) Combining advantages of new chromosome representation scheme and multi-objective genetic algorithms for better clustering. *Intelligent Data Analysis 10* (2006),163-182.

Koza JR (1992) *Genetic Programming: on the programming g of computers by means of natural selection*. MIT Press.

Krawiec K (2002) Genetic programming-based construction of features for machine learning and knowledge discovery tasks. *Genetic Programming and Evolvable Machines 3(4)*, 329-344.

Krsihma K and Murty MN (1999) Genetic k-means algorithm. *IEEE Transactions on Systems, Man and Cyberneics - Part B: Cybernetics,* 29(3), 433-439.

Krzanowski WJ and Marriot FHC (1995) *Kendall's Library of Statistics 2: Multivariate Analysis - Part 2. Chapter 10 - Cluster Analysis*, pp. 61-94.London: Arnold.

Kudo M and Sklansky J (2000) Comparison of algorithms that select features for pattern classifiers. *Pattern Recognition 33(2000)*, 25-41.

Liu JJ and Kwok JTY (2000) An extended genetic rule induction algorithm. *Proc. 2000 Congress on Evolutionary Computation (CEC-2000)*. IEEE.

Liu H and Motoda H (1998) *Feature Selection for Knowledge Discovery and Data Mining*. Kluwer.

Liu B, Hsu W and Chen S (1997) Using general impressions to analyze discovered classification rules. *Proc. 3rd Int. Conf. on Knowledge Discovery and Data Mining (KDD-97)*, 31-36. AAAI Press.

Llora X and Garrell J (2003) Prototype induction and attribute selection via evolutionary algorithms. *Intelligent Data Analysis 7*, 193-208.

Miller MT, Jerebko AK, Malley JD, Summers RM (2003) Feature selection for computer-aided polyp detection using genetic algorithms. *Medical Imaging 2003: Physiology and Function: methods, systems and applications.* Proc. SPIE Vol. 5031.

Moser A and Murty MN (2000) On the scalability of genetic algorithms to very large-scale feature selection. *Proc. Real-World Applications of Evolutionary Computing (EvoWorkshops 2000). LNCS 1803*, 77-86. Springer.

Muharram MA and Smith GD (2004) Evolutionary feature construction using information gain and gene index. *Genetic Programming: Proc. 7th European Conf. (EuroGP-2003), LNCS 3003*, 379-388. Springer.

Muni DP, Pal NR and Das J (2004) A novel approach to design classifiers using genetic programming. *IEEE Trans. Evolutionary Computation 8(2)*, 183-196, April 2004.

Neri F and Giordana A (1995) Search-intensive concept induction. *Evolutionary Computation 3(4)*, 375-416.

Ni B and Liu J (2004) A novel method of searching the microarray data for the best gene subsets by using a genetic algorithms. *Proc. Parallel Problem Solving From Nature (PPSN-2004), LNCS 3242,* 1153-1162, Springer.

Otero FB, Silva MMS, Freitas AA and Nievola JC (2003) Genetic programming for attribute construction in data mining. *Genetic Programming: Proc. EuroGP-2003, LNCS 2610*, 384-393. Springer.

Papagelis A and Kalles D (2001) Breeding decision trees using evolutionary techniques. *Proc. 18th Int. Conf. Machine Learning (ICML-2001),* 393-400. Morgan Kaufmann.

Pappa GL and Freitas AA (2006) Automatically evolving rule induction algorithms. *Machine Learning: ECML 2006 – Proc. of the 17th European Conf. on Machine Learning, LNAI 4212,* 341-352. Springer.

Pappa GL and Freitas AA (2007) Discovering new rule induction algorithms with grammar-based genetic programming. *Maimon O and Rokach L (Eds.) Soft Computing for Knowledge Discovery and Data Mining*. Springer.

Pappa GL, Freitas AA and Kaestner CAA (2002) A multiobjective genetic algorithm for attribute selection. *Proc. 4th Int. Conf. On Recent Advances in Soft Computing (RASC-2002)*, 116-121. Nottingham Trent University, UK.

Pappa GL, Freitas AA and Kaestner CAA (2004) Multi-Objective Algorithms for Attribute Selection in Data Mining. In: Coello Coello CA and Lamont GB (Ed.) *Applications of Multi-objective Evolutionary Algorithms*, 603-626. World Scientific.

Pazzani MJ (2000) Knowledge discovery from data, *IEEE Intelligent Systems,* 10-13, Mar./Apr. 2000.

Quinlan JR. (1993) *C4.5: Programs for Machine Learning*. Morgan Kaufmann.

Romao W, Freitas AA and Pacheco RCS (2002) A Genetic Algorithm for Discovering Interesting Fuzzy Prediction Rules: applications to science and technology data. *Proc. Genetic and Evolutionary Computation Conf. (GECCO-2002)*, pp. 1188-1195. Morgan Kaufmann.

Romao W, Freitas AA, Gimenes IMS (2004) Discovering interesting knowledge from a science and technology database with a genetic algorithm. *Applied Soft Computing 4(2)*, pp. 121-137.

Rozsypal A and Kubat M (2003) Selecting representative examples and attributes by a genetic algorithm. *Intelligent Data Analysis 7*, 290-304.

Sarafis I (2005) *Data mining clustering of high dimensional databases with evolutionary algorithms*. PhD Thesis, School of Mathematical and Computer Sciences, Heriot-Watt

University, Edinburgh, UK.

Sharpe PK and Glover RP (1999) Efficient GA based techniques for classification. *Applied Intelligence 11*, 277-284.

Smith RE (2000) Learning classifier systems. In: T. Back, D.B. Fogel and T. Michalewicz (Eds.) *Evolutionary Computation 1: Basic Algorithms and Operators*, 114-123. Institute of Physics Publishing.

Smith MG and Bull L (2003) Feature construction and selection using genetic programming and a genetic algorithm. *Genetic Programming: Proc. EuroGP-2003, LNCS 2610*, 229-237. Springer.

Smith MG and Bull L (2004) Using genetic programming for feature creation with a genetic algorithm feature selector. *Proc. Parallel Problem Solving From Nature (PPSN-2004), LNCS 3242,* 1163-1171, Springer.

Song D, Heywood MI and Zincir-Heywood AN (2005) Training genetic programming on half a million patterns: an example from anomaly detection. *IEEE Trans. Evolutionary Computation 9(3)*, 225-239, June 2005.

Srikanth R, George R, Warsi N, Prabhu D, Petry FE, Buckles B (1995) A variable-length genetic algorithm for clustering and classification. *Pattern Recognition Letters 16(8)*, 789-800.

Tan PN, Steinbach M and Kumar V (2006) *Introduction to Data Mining*. Addison-Wesley.

Terano T and Ishino Y (1998) Interactive genetic algorithm based feature selection and its application to marketing data analysis. In: Liu H and Motoda H (Eds.) *Feature Extraction, Construction and Selection: a data mining perspective*, 393-406. Kluwer.

Terano T and Inada M (2002) Data mining from clinical data using interactive evolutionary computation. In: A. Ghosh and S. Tsutsui (Eds.) *Advances in Evolutionary Computing: theory and applications*, 847-861. Springer.

Vafaie H and De Jong K (1998) Evolutionary Feature Space Transformation. In: H. Liu and H. Motoda (Eds.) *Feature Extraction, Construction and Selection*, 307-323. Kluwer.

Witten IH and Frank E (2005) *Data Mining: practical machine learning tools and techniques* . 2nd Ed. Morgan Kaufmann.

Wong ML and Leung KS (2000) *Data Mining Using Grammar Based Genetic Programming and Applications*. Kluwer.

Yang J and Honavar V (1997) Feature subset selection using a genetic algorithm. *Genetic Programming 1997: Proc. 2nd Annual Conf. (GP-97)*, 380-385. Morgan Kaufmann.

Yang J and Honavar V (1998) Feature subset selection using a genetic algorithm. In: Liu, H. and Motoda, H (Eds.) *Feature Extraction, Construction and Selection*, 117-136. Kluwer.

Zhang P, Verma B, Kumar K (2003) Neural vs. Statistical classifier in conjunction with genetic algorithm feature selection in digital mammography. *Proc. Congress on Evolutionary Computation (CEC-2003)*. IEEE Press.

Zhou C, Xiao W, Tirpak TM and Nelson PC (2003) Evolving accurate and compact classification rules with gene expression programming. *IEEE Trans. on Evolutionary Computation 7(6)*, 519-531.

20

A Review of Reinforcement Learning Methods

Oded Maimon[1] and Shahar Cohen[1]

Department of Industrial Engineering, Tel-Aviv University, Ramat-Aviv 69978, Israel, maimon@eng.tau.ac.il

Summary. Reinforcement-Learning is learning how to best-react to situations, through trial and error. In the Machine-Learning community Reinforcement-Learning is researched with respect to artificial (machine) decision-makers, referred to as agents. The agents are assumed to be situated within an environment which behaves as a Markov Decision Process. This chapter provides a brief introduction to Reinforcement-Learning, and establishes its relation to Data-Mining. Specifically, the Reinforcement-Learning problem is defined; a few key ideas for solving it are described; the relevance to Data-Mining is explained; and an instructive example is presented.

Key words: Reinforcement-Learning

20.1 Introduction

Reinforcement-Learning (RL) is learning how to best-react to situations, through trial-and-error. The learning takes place as a decision-maker interacts with the environment she lives in. On a sequential basis, the decision-maker recognizes her *state* within the environment, and reacts by initiating an *action*. Consequently she obtains a *reward* signal, and enters another state. Both the reward and the next state are affected by the current state and the action taken. In the Machine Learning (ML) community, RL is researched with respect to artificial (machine) decision-makers, referred to as *agents*.

The mechanism that generates reward signals and introduces new states is referred to as the dynamics of the environment. As the RL agent begins learning, it is unfamiliar with that dynamics, and therefore initially it cannot correctly predict the outcome of actions. However as the agent interacts with the environment and observes the actual consequences of its decisions, it can gradually adapt its behavior accordingly. Through learning the agent chooses actions according to a *policy*. A policy is a means of deciding which action to choose when encountering a certain state. A policy is optimal if it maximizes an agreed-upon return function. A return function is usually some sort of expected weighted-sum over the sequence of rewards

O. Maimon, L. Rokach (eds.), *Data Mining and Knowledge Discovery Handbook*, 2nd ed.,
DOI 10.1007/978-0-387-09823-4_20,

obtained while following a specified policy. Typically the objective of the RL agent is to find an optimal policy.

RL research has been continually advancing over the past three decades. The aim of this chapter is to provide a brief introduction to this exciting research, and to establish its relation to Data-Mining (DM). For a more comprehensive RL survey, the reader is referred to Kaelbling *et al.* (1996). For a comprehensive introduction to RL, see Sutton and Barto (1998). A rigorous presentation of RL can be found in Bertsekas and Tsitsiklis (1996).

The rest of this chapter is organized as follows. Section 20.2 formally describes the basic mathematical model of RL, and reviews some key results for this model. Section 20.3 introduces some of the principles of computational methods in RL. Section 20.4 describes some extensions to the basic RL model and computation methods. Section 20.5 reviews several application of RL. Section 20.6 discusses RL in a DM perspective. Finally Section 20.7 presents an example of how RL is used to solve a typical problem.

20.2 The Reinforcement-Learning Model

RL is based on a well-known model called Markov Decision Process (MDP). An MDP is a tuple $\langle S, A, R, P \rangle$, where S is a set of states, A is a set of actions[1], $R: S \times A \rightarrow \Re$ is a mean-reward function and $P: S \times A \times S \rightarrow [0,1]$ is a state-transition function. An MDP evolves through discrete time stages. On stage t, the agent recognizes the state of its environment $s_t \in S$ and reacts by choosing an action $a_t \in A$. Consequently it obtains a reward r_t, whose mean-value is $R(s_t, a_t)$, and its environment is transited to a new state s_{t+1} with probability $P(s_t, a_t, s_{t+1})$. The two sets - of states and of actions - may be finite or infinite. In this chapter, unless specified otherwise, both sets are assumed to be finite. The RL agent begins interacting with the environment without any knowledge of the mean-reward function or the state-transition function.

Situated within its environment, the agent seeks an optimal policy. A policy $\pi : S \times A \rightarrow [0,1]$ is a mapping from state-action pairs to probabilities. Namely, an agent that observes the state s and follows the policy π will chose the action $a \in A$ in probability $\pi(s,a)$. Deterministic policies are of particular interest. A deterministic policy π_d is a policy in which for any state $s \in S$ there exists an action $a \in A$ so that $\pi_d(s,a) = 1$, and $\pi_d(s,a') = 0$ for all $a' \neq a$. A deterministic policy is therefore a mapping from states to actions. For a deterministic policy π_d the action $a = \pi_d(s)$ is the action for which $\pi_d(s,a) = 1$. The subscript "d" is added to differentiate deterministic from non-deterministic policies (when it is clear from the context that a policy is deterministic, the subscript is omitted). A policy (either deterministic or not) is optimal if it maximizes some agreed-upon return function. The most common return function is the expected geometrically-discounted infinite-sum of rewards. Considering this return function, the objective of the agent is defined as follows:

[1] It is possible to allow different sets of actions for different states (i.e. letting $A(s)$ be the set of allowable actions in state s for all $s \in S$). For ease of notation, it is assumed that all actions are allowed in all states.

$$E\left[\sum_{t=1}^{\infty}\gamma^{t-1}r_t\right]\rightarrow \max, \tag{20.1}$$

where $\gamma \in (0,1)$ is a discount factor representing the extent to which the agent is willing to compromise immediate rewards for the sake of future rewards. The discount factor can be interpreted either as a means of capturing some characteristics of the problem (for example an economic interest-rate) or as a mathematical trick that makes RL problems more tractable. Other useful return-functions are defined by expected finite-horizon sum of rewards, and expected long-run average reward (see Kaelbling *et al.*, 1996). This chapter assumes that the return function is the expected geometrically-discounted infinite-sum of rewards.

Given a policy π, the value of the state s is defined by:

$$V^{\pi}(s) = E_{\pi}\left[\sum_{t=1}^{\infty}\gamma^{t-1}r_t\,|s_1 = s\right]\ s \in S, \tag{20.2}$$

where the operator E_{π} represents expectation given that actions are chosen according to the policy π. The value of a state for a specific policy represents the return correlated with following the policy given a specific initial state. Similarly, given the policy π, the value of a state-action pair is defined by:

$$Q^{\pi}(s,a) = E_{\pi}\left[\sum_{t=1}^{\infty}\gamma^{t-1}r_t\,|s_1 = s, a_1 = a\right]\ s \in S; a \in A, \tag{20.3}$$

The value of a state-action pair for a specific policy is the return correlated with first choosing a specific action while being on a specific state, and thereafter choosing actions according to the policy. The optimal value of states is defined by:

$$V^{*}(s) = \max_{\pi} V^{\pi}(s),\ s \in S. \tag{20.4}$$

A policy π^* is optimal if it achieves the optimal values for all states, i.e. if:

$$V^{\pi^*}(s) = V^{*}(s)\ \forall s \in S. \tag{20.5}$$

If π^* is an optimal policy it also maximizes the value of all state-action pairs:

$$Q^{*}(s,a) = Q^{\pi^*}(s,a) = \max_{\pi} Q^{\pi}(s,a)\ \forall s \in S; a \in A. \tag{20.6}$$

A well-known result is that under the assumed return-function, any MDP has an optimal deterministic policy. This optimal policy, however, may not be unique. Any deterministic optimal policy must satisfy the following relation:

$$\pi^*(s) = \arg\max_{a} Q^{*}(s,a),\ \forall s \in S. \tag{20.7}$$

Finally, the relation between optimal values of states and of state-action pairs is established by the following set of equations:

$$\begin{array}{l} Q^*(s,a) = R(s,a) + \gamma \sum_{s' \in S} P(s,a,s') V^*(s') \\ V^*(s) = \max_a Q^*(s,a) \end{array} \quad , s \in S; a \in A. \tag{20.8}$$

For a more extensive discussion about MDPs, and related results, the reader is referred to Puterman (1994) or Ross (1983).

Some RL tasks are continuous while others are episodic. An episodic task is one that terminates after a (maybe random) number of stages. As a repeat of an episodic task terminates, another may begin, possibly at a different initial state. Continuous tasks, on the other hand, never terminate. The objective defined by Equation 20.1 considers an infinite horizon and therefore might be seen as inappropriate for episodic tasks (which are finite by definition). However, by introducing the concept of an absorbing state, episodic tasks can be viewed as infinite-horizon tasks (Sutton and Barto, 1998). An absorbing state is one from which all actions result in a transition to the same state and with zero reward.

20.3 Reinforcement-Learning Algorithms

The environment in RL problems is modeled as an MDP with unknown mean-reward and state transition-functions. Many RL algorithms are generalizations of dynamic-programming (DP) algorithms (Bellman, 1957; Howard, 1960) for finding optimal policies in MDPs given these functions. Sub-section 20.3.1 introduces a few key DP principles. The reader is referred to Puterman (1994), Bertsekas (1987) or Ross (1983) for a more comprehensive discussion. Sub-section 20.3.2 introduces several issues related to generalizing DP algorithms to RL problems. Please see Sutton and Barto (1998) for a comprehensive introduction to RL algorithms, and Bertsekas and Tsitsiklis (1996) for a more extensive treatment.

20.3.1 Dynamic-Programming

Typical DP algorithms begin with an arbitrary policy and proceed by evaluating the values of states or state-action pairs for this policy. These evaluations are used to derive a new, improved policy, for evaluating the values of states or state-action pairs for the new policy and so on. Given a deterministic policy π, the evaluation of values of states may take place by incorporating an iterative sequence of updates. The sequence begins with arbitrary initializations $V_1(s)$ for each s. On the k-th repeat of the sequence, the values $V_k(s)$ are used to derive $V_{k+1}(s)$ for all s:

$$V_{k+1}(s) = R(s,\pi(s)) + \gamma \sum_{s' \in S} P(s,\pi(s),s') V_k(s') \; \forall s \in S \, . \tag{20.9}$$

It can be shown that following this sequence, $V_k(s)$ converges to $V^{\pi}(s)$ for all s, as k increases.

Having the deterministic policy π and the values of states for this policy $V^{\pi}(s)$, the values of state-actions pairs are given by:

$$Q^{\pi}(s,a) = R(s,a) + \gamma \sum_{s' \in S} P(s,a,s') V^{\pi}(s) \;\; s \in S; a \in A \tag{20.10}$$

The values of state-action pairs for a given policy π_k, can be used to derive an improved deterministic policy π_{k+1}:

$$\pi_{k+1}(s) = \arg\max_{a \in A} Q^{\pi_k}(s,a) \;\; \forall s \in S \tag{20.11}$$

It can be shown that $V^{\pi_{k+1}}(s) \geq V^{\pi_k}(s)$ for all s, and that if the last relation is satisfied with equality for all states, then π_k is an optimal policy.

Improving the policy π may be done based on estimations of $V^{\pi}(s)$ instead of the exact values (i.e. if $V(s)$ estimate $V^{\pi}(s)$, a new policy can be derived by calculating $Q(s,a)$ according to Equation 20.10 where $V(s)$ replace $V^{\pi}(s)$, and by calculating the improved policy according to Equation 20.11, based on $Q(s,a)$. Estimations of $V^{\pi}(s)$ are usually the result of executing the sequence of updates defined by Equation 20.9, without waiting for $V_k(s)$ to converge to $V^{\pi}(s)$ for all $s \in S$. In particular, it is possible to repeatedly execute a single repeat of the sequence defined in Equation 20.9, to use the estimation results to derive a new policy as defined by Equations 20.10 and 20.11, to re-execute a single repeat of Equation 20.9 starting from the current estimation results, and so-on. This well-known approach, termed *value-iteration*, begins with arbitrary initialization $Q_1(s,a)$ for all s and a, and proceeds iteratively with the updates:

$$\begin{aligned} &Q_{t+1}(s,a) = R(s,a) + \gamma \sum_{s' \in S} P(s,a,s') V_t(s') \\ &\forall s \in S; a \in A; t = 1,2,\ldots \end{aligned} \tag{20.12}$$

where $V_t(s)$=$\max_a Q_t(s,a)$. It can be shown that using value-iteration, $Q_t(s,a)$ converges to $Q^*(s,a))$[2]. The algorithm terminates using some stopping conditions (e.g. the change in the values $Q_t(s,a)$ due to a single iteration is small enough)[3]. Let the termination occur at stage T. The output policy is calculated according to:

$$\pi(s) = \arg\max_{a \in A} Q_T(s,a) \;\; \forall s \in S \tag{20.13}$$

20.3.2 Generalization of Dynamic-Programming to Reinforcement-Learning

It should be noted that both the mean-reward and the state-transition functions are required in order to take on the computations described in the previous sub-section. In RL, these functions are initially unknown. Two different approaches, indirect and direct, may be used to generalize the discussion to the absence of these functions.

According to the indirect approach, samples from the consequences of choosing various actions at various states are gathered. These samples are used to approximate

[2] In addition, there are results concerning the rate of this convergence. See, for instance, the discussion in Puterman (1994).

[3] It is possible to establish a connection between the change in the values $Q_t(s,a)$ due to a single iteration and the distance between $Q_t(s,a)$ and $Q^*(s,a)$. See, for instance, the discussion in Puterman (1994).

the mean-reward and state-transition functions. Subsequently, the indirect approach uses the approximations to extract policies. During the extraction of policies, the approximated functions are used as if they were the exact ones.

The direct approach, the more common of the two, involves continuously maintaining estimations of the optimal values of state and state-action pairs without having any explicitly approximated mean-reward and state-transition functions. The overview in this sub-section focus on methods that take a direct approach.

In a typical direct method, the agent begins learning in a certain state, while having arbitrary estimations of the optimal values of states and state-action pairs. Subsequently the agent uses a while-learning policy to choose an action. Consequently a new state is encountered and an immediate reward is obtained (i.e. a new experience is gathered). The agent uses the new experience to update the estimations of optimal values for states and state-action pairs visited in previous stages.

The policy, that the agent uses while it learns, needs to solve a dilemma, known as the *exploration-exploitation dilemma*. Exploitation means using the knowledge gathered in order to obtain desired outcomes. In order to obtain desired outcomes on a certain stage, the agent needs to choose an action which corresponds with the maximal optimal value of state-action given the state. Since the exact optimal values of state-action pairs are unknown, at least the corresponding estimations are expected to be maximized. On the other hand, due to the random fluctuations of the reward signals and the random nature of the state-transition function, the agent's estimations are never accurate. In order to obtain better estimations, the agent must explore its possibilities. Exploration and exploitation are conflicting rationales, because by exploring possibilities the agent will sometimes choose actions that seem inferior at the time they are chosen.

In general, it is unknown how to best-solve the exploration-exploitation dilemma. There are, however, several helpful heuristics, which typically work as follows. During learning, the action chosen while being on state s is randomly chosen from the entire set of actions, but with a probability function that favors actions for which the current optimal value estimates are high. (See Sutton and Barto, 1998 for a discussion on the exploration-exploitation dilemma and its heuristic solutions).

Many RL algorithms are stochastic variations of DP algorithms. Instead of using an explicit mean-reward and state-transition functions, the agent uses the actual reward signals and state transitions while interacting with the environment. These actual outcomes implicitly estimate the real, unknown functions. There are several assumptions under which the estimates maintained by stochastic variations of DP algorithms converge to the optimal values of states and state-action pairs. Having the optimal values, a deterministic optimal policy may be derived according to Equation 20.7. The reader is referred to Bertsekas and Tsitsiklis (1996), Jaakkola *et al.* (1994) or Szepesvári and Littman (1999) for formal, general convergence results.

One of the most common RL algorithms is termed Q-Learning (Watkins, 1989; Watkins and Dayan 1992). Q-Learning takes the direct approach, and can be regarded as the stochastic version of value-iteration. At stage t of Q-Learning, the agent holds $Q_t(s,a)$ – estimations of the optimal values for state-action pairs $Q^*(s,a)$ for all state-action pairs. At this stage, the agent encounters state s_t and chooses the action

a_t. Following the execution of a_t from s_t, the agent obtains the actual reward r_t, and faces the new state s_{t+1}. The tuple $\langle s_t, a_t, r_t, s_{t+1}\rangle$ is referred to as the experience gathered on stage t. Given that experience, the agent updates its estimate as follows:

$$\begin{aligned} Q_{t+1}(s_t,a_t) &= (1-\alpha_t(s_t,a_t))\,Q_t(s_t,a_t)+\alpha_t(s_t,a_t)\,(r_t+\gamma V_t(s_{t+1})) \\ &= Q_t(s_t,a_t)+\alpha_t(s_t,a_t)\,(r_t+\gamma V_t(s_{t+1})-Q_t(s_t,a_t)) \end{aligned} \tag{20.14}$$

where $V_t(s) = \max_a Q_t(s,a)$, and $\alpha_t(s_t,a_t) \in (0,1)$ is a step size reflecting the extent to which the new experience needs to be blended into the current estimates[4]. It can be shown that as $t \to \infty$, $Q_t(s,a)$ converges to $Q^*(s,a)$ for all s and a and under several assumptions. (Convergence proofs can be found in Watkins and Dayan, 1992; Jaakkola *et al.* 1996; Szepesvári and Littman 1999 and Bertsekas and Tsitsiklis, 1996).

In order to understand the claim that Q-Learning is a stochastic version of value-iteration, it is helpful to address Equation 20.14 as an update of the estimated value of a certain state-action pair $Q_t(s_t,a_t)$ in the direction $r_t+\gamma V_t(s_{t+1})$, with a step size $\alpha_t(s_t,a_t)$. With this interpretation, Q-Learning can be compared to value-iteration (Equation 20.12). Referring to a certain state-action pair, noted by $\langle s_t, a_t\rangle$, and replacing the state space index s' with s_{t+1} Equation 20.12 can be re-phrased as:

$$Q_{t+1}(s_t,a_t) = R(s_t,a_t)+\gamma \sum_{s_{t+1}\in S} P(s_t,a_t,s_{t+1})\,V_t(s_{t+1}) \tag{20.15}$$

Rewriting Equation 20.14 with $\alpha_t(s_t,a_t)=1$ results in:

$$Q_{t+1}(s_t,a_t) = r_t+\gamma V_t(s_{t+1}) \tag{20.16}$$

The only difference between Equations 20.15 and 20.16 lies in the exchanged use of r_t instead of $R(s_t,a_t)$ and the exchanged use of $V_t(s_{t+1})$ instead of $\sum_{s_{t+1}\in S} P(s_t,a_t,s_{t+1})\,V_t(s_{t+1})$. It can be shown that:

$$E\left[r_t+\gamma V_t(s_{t+1})\right] = R(s_t,a_t)+\gamma \sum_{s_{t+1}\in S} P(s_t,a_t,s_{t+1})\,V_t(s_{t+1}), \tag{20.17}$$

namely Equation 20.16 is a stochastic version of Equation 15. It is appropriate to use a unit step-size when basing an update on exact values, but it is inappropriate to do so when basing the update on unbiased estimates, since the learning algorithm must be robust to the random fluctuations.

In order to converge, Q-Learning is assumed to infinitely update each state-action pair, but there is no explicit instruction as to which action to choose at each stage. However in order to boost the rate at which estimations converge to optimal values, heuristics that break the exploration-exploitation dilemma are usually used.

[4] In general, a unique step-size is defined for each state and action and for each stage. Usually step-sizes decrease with time.

20.4 Extensions to Basic Model and Algorithms

In general, the RL model is quite flexible and can be used to capture problems in a variety of domains. Applying an appropriate RL algorithm can lead to optimal solutions without requiring an explicit mean-reward or state-transition functions. There are some problems, however, that the RL model described in Section 20.2 cannot capture. There may also be some serious difficulties in applying RL algorithms described in Section 20.3. This section presents overviews of two extensions. The first extension involves, a multi-agent RL scenario, where learning agents co-exist in a single environment. This is followed by an overview of the problem of large (or even infinite) sets of states and actions.

20.4.1 Multi-Agent RL

In RL, as in real-life, the existence of one autonomous agent affects the outcomes obtained by other co-existing agents. An immediate approach for tackling multi-agent RL is to let each agent refer to its colleagues (or adversaries) as part of the environment. A learner that takes this approach is regarded as an *independent* learner (Claus and Boutilier, 1998). It is to be noticed that the environment of an independent learner consists of learning components (the other agents) and is therefore not stationary. The model described in Section 20.4, as well as the convergence results mentioned in Section 20.3, assumed a stationary environment (i.e. the mean-reward and state-transition functions do not change in time). Although convergence is not guaranteed when using independent learners in multi-agent problems, several authors have reported good empirical results for this approach. For example, Sen *et al.* (1994) used Q-Learning in multi-agent domains.

Littman (1994) proposed Markov Games (often referred to as Stochastic Games) as the theoretic model appropriate for multi-agent RL problems. A k-agents Stochastic Game (SG) is defined by a tuple $\langle S,\bar{A},\bar{R},P\rangle$, where S is a finite set of states (as in the case of MDPs); $\bar{A} = A_1 \times A_2 \times ... \times A_k$ is a Cartesian product of k action-sets available for the k agents; $\bar{R} : S \times \bar{A} \to \Re^k$ is a collection of k mean-reward functions for the k agents; and $P : S \times \bar{A} \times S \to [0,1]$ is a state-transition function. The evolution of an SG is controlled by k autonomous agents acting simultaneously rather than by a single agent. The notion of a game as a model for multi-agent RL problems raises the concept of *Nash-equilibrium* as an optimality criterion. Generally speaking, a joint policy is said to be in Nash-equilibrium if no agent can gain from being the only one to deviate from it.

Several algorithms that rely on the SG model appear in the literature (Littman, 1994; Hu and Wellman, 1998; Littman, 2001). In order to assure convergence, agents in these algorithms are programmed as *joint* learners (Claus and Boutilier, 1998). A joint learner is aware of the existence of other agents and in one way or another adapts its own behavior to the behavior of its colleagues. In general, the problem of multi-agent RL is still the subject of ongoing research. For a comprehensive introduction to SGs the reader is referred to Filar and Vrieze (1997).

20.4.2 Tackling Large Sets of States and Actions

The algorithms described in Section 20.3 assumed that the agent maintains a look-up table with a unique entry corresponding to each state or state-action pair. As the agent gathers new experience, it retrieves the entry corresponding with state or state-action pair for that experience, and updates the estimate stored within the entry. Representation of estimates to optimal values in the form of a look-up table is limited to problems with a reasonably small number of states and actions. Obviously, as the number of states and action increases, the memory required for the look-up table increases, and so does the time and experience needed to fill up this table with reliable estimates. That is to say, if there is a large number of states and actions, the agent cannot have the privilege of exploring them all, but must incorporate some sort of generalization.

Generalization takes place through function-approximation in which the agent maintains a single approximated value function from state or state-action pairs to mean-rewards. Function approximation is a central idea in Supervised-Learning (SL). The task in function-approximation is to find a function that will best approximate some unknown target-function based on a limited set of observations. The approximating function is located through a search over the space of parameters of a decided-upon family of parameterized functions. For example, the unknown function $Q^* : S \times A \rightarrow \Re$ may be approximated by an artificial neural network of pre-determined architecture. A certain network $f : S \times A \rightarrow \Re$ belongs to a family of parameterized networks Φ, where the parameters are the weights of the connections in the network.

RL with function approximation inherits the direct approach described in Section 20.3. That is, the agent repeatedly estimates the values of states or state-action pairs for its current policy; uses the estimates to derive an improved policy; estimates the values corresponding with the new policy; and so on. However, the function representation adds some more complications to the process. In particular, when the number of actions is large, the principle of policy improvement by finding the action that maximizes current estimates, does not scale well. Moreover, convergence results characterizing look-up table representations usually cannot be generalized to function-approximation representations. The reader is referred to Bertsekas and Tsitsiklis (1996) for an extensive discussion on RL with function-approximation and corresponding theoretic results.

20.5 Applications of Reinforcement-Learning

Using RL, an agent may learn how to best behave in a complex environment without any explicit knowledge regarding the nature or the dynamics of this environment. All that an agent needs in order to find an optimal policy is the opportunity to explore its options.

In some cases, RL occurs through interaction with the real environment under consideration. However, there are cases in which experience is expensive. For example, consider an agent that needs to learn a decision policy in a business environment.

In order to achieve good behavior, the agent must explore its environment. Exploration means trying different sort of actions in various situations. While exploring, some of the choices may be poor ones, which may lead to severe costs. In such cases, it is more appropriate to train the agent on a computer-simulated model of the environment. It is sometimes possible to simulate an environment without explicitly understanding it.

RL methods have been used to solve a variety of problems in a number of domains. Pednault *et al.* (2002) solved targeted marketing problems. Tesauro (1994, 1995) planned an artificial backgammon player with RL. Hong and Prabhu (2004) and Zhang and Dietterich (1996) used RL to solve manufacturing problems. Littman and Boyan (1993) have used RL for the solution of a networking routing problem. Using RL, Crites and Barto (1996) trained an elevator dispatching controller.

20.6 Reinforcement-Learning and Data-Mining

This chapter presents an overview of some of the ideas and computation methods in RL. In this section the relation and relevance of RL to DM is discussed.

Most DM learning methods are taken from ML. It is popular to distinguish between three categories of learning methods – Supervised Learning (SL), Unsupervised Learning and Reinforcement Learning. In SL, the learner is programmed to extract a model from a set of observations, where each observation consists of explaining variables and corresponding responses. In unsupervised learning there is a set of observations but no response, and the learner is expected to extract a helpful representation of the domain from which the observations were drawn. RL requires the learner to extract a model of response based on experience observations that include states, responses and the corresponding reinforcements.

SL methods are central in DM and a correlation may be established between SL and RL in the following manner. Consider a learner that needs to extract a model of response for different situations. A supervised learner will rely on a set of observations, each of which is labeled by an advisor (or an oracle). The label of each observation is regarded by the agent as the desired response for the situation introduced by the explanatory variables for this observation. In RL the privilege of having an advisor is not given. Instead, the learner views situations (in RL these are called states) chooses responses (in RL these are called actions) autonomously and obtains rewards that indicate how good the choices were. In this approach toward SL and RL, states and realizations of "explaining variables" are actually the same.

In some DM problems, cases arise in which the responses in one situation affect future outcomes. This is typically the case in *cost-sensitive* DM problems. Since SL relies on labeled observations and assumes no dependence between observations, it is sometimes inappropriate for such problems. The RL model, on the other hand, perfectly fits cost-sensitive DM problems[5]. For example, Pednault *et al.* (2002) used

[5] Despite this claim, there are several difficulties in applying RL methods to DM problems. A serious issue is that DM problems suggest batches of observation stored in a database, whereas RL methods require incremental accumulation of observations through interaction.

RL to solve a problem in targeted marketing – deciding on the optimal targeting of promotion efforts in order to maximize the benefits due to promotion. Targeted marketing is a classical DM problem in which the desired response is unknown, and responses taken at one point in time affect the future. (For example, deciding on an extensive campaign for a specific product this month may reduce the effectiveness of a similar campaign the following month).

Finally, DM may be defined as a process in which computer programs manipulate data in order to provide knowledge about the domain that produced the data. From the point of view implied by this definition, RL definitely needs to be considered as certain type of DM.

20.7 An Instructive Example

In this section, an example-problem from the area of supply-chain management is presented and solved through RL. Specifically, the modeling of the problem as an MDP with unknown reward and state-transition functions is shown; the sequence of Q-Learning is demonstrated; and the relations between RL and DM are discussed with respect to the problem.

The term "supply-chain management" refers to the attempts of an enterprise to optimize processes involved in purchasing, producing, shipping and distributing goods. Among other objectives, enterprises seek to formulate a cost-effective inventory policy. Consider the problem of an enterprise that purchases a single product from a manufacturer and sells it to end-customers. The enterprise may maintain a stock of the product in one or more warehouses. The stock help the enterprise respond to customer demand, which is usually stochastic. On the other hand, the enterprise has to invest in purchasing the stock and maintaining it. These activities lead to costs.

Consider an enterprise that has two warehouses in two different locations and behaves as follows. At the beginning of epoch t, the enterprise observes the stock levels $s_1(t)$ and $s_2(t)$, at the first and the second warehouses respectively. As a response, it may order from the manufacturer in quantities $a_1(t)$, $a_2(t)$ for the first and second warehouse respectively. The decision of how many units to order for each of the warehouses is taken centrally (i.e. simultaneously by a single decision-maker), but the actual orders are issued separately by the two warehouses. The manufacturer charges c_d for each unit ordered, and additional c_K for delivering an order to a warehouse (i.e. if the enterprise issues orders at both warehouses it is charged a fixed $2c_K$ in addition to direct costs of the units ordered). It is assumed that there is no lead-time (i.e. the units ordered become available immediately after issuing the orders). Subsequently, each of the warehouses observes a stochastic demand.

A warehouse that has enough units in stock sells the units and charges p for each sold unit. If one of the warehouses fails to respond to the demand, whereas the other warehouse, after delivering to its customers, can spare units, transshipment is initiated. Transshipment means transporting units between the warehouses in order to meet demand. Transshipment costs c_T for each unit transshipped. Any unit remaining in stock by the end of the epoch costs the enterprise c_i for that one epoch. The successive epoch begins with the number of units available at the end of the current epoch, and so-on.

The enterprise wants to formulate an optimal inventory policy (i.e. given the stock levels and in order to maximize its long-run expected profits the enterprise wants to know when to issue orders, and in what quantities). This problem can be modeled as an MDP (see the definition of MDP in Section 20.2). The stock levels $s_1(t)$ and $s_2(t)$ at the beginning of epochs are the states faced by the enterprise's decision-makers. The possible quantities for two orders are the possible actions given a state. As a consequence of choosing a certain action at a certain state, each warehouse obtains a deterministic quantity-on-hand. As the demand is observed and met (either directly or through transshipment), the actual, immediate profit r_t can be calculated as the revenue gained from selling products minus costs due to purchasing the products, delivering the orders, the transshipments and maintaining inventory. The stock levels at the end of the period, and thus the state for the successive epoch, are also determined. Since the demand is stochastic, both the reward (the profit) and the state-transition function are stochastic.

Assuming that the demand functions at the two warehouses are unknown, the problem of the enterprise is how to solve an MDP with unknown reward and state-transition functions. In order to solve the problem via RL, a large number of experience episodes needs to be presented to an agent. Gathering such experience is expensive, because in order to learn an optimal policy, the agent must *explore* its environment simultaneously to the *exploitation* of its current knowledge (see discussion on the exploration-exploitation dilemma on Section 20.3.2). However, in many cases learning may be based on simulated experience.

Consider using Q-Learning (see Section 20.3.2) for the solution of the enterprise's problem. Let this application be demonstrated for epoch $t = 158$, and the initial stock levels $s_1(158) = 4$, $s_2(158) = 2$. The agent constantly maintains a unique Q-value for each of the initial stock levels and the quantities ordered. Assumed that capacity at both warehouses is limited to 10 units of stock, the possible actions given the states are:

$$A(s_1(t), s_2(t)) = \{\langle a_1, a_2 \rangle : \; a_1 + s_1(t) \leq 10, \; a_2 + s_2(t) \leq 10\} \tag{20.18}$$

The agent chooses an action from the set of possible actions based on some heuristic that breaks the exploration-exploitation dilemma (see discussion in Section 20.3.2). Assume that the current Q-values for the state $s_1(158) = 4$, and $s_2(158) = 2$ are as described in Figure 20.1. The heuristic used should tend to choose actions for which the corresponding Q-value is high, while allowing each action to be chosen with a positive probability. Assume that the action chosen is $a_1(158) = 0$, $a_2(158) = 8$. This action means that the first warehouse does not issue an order while the second warehouse orders 8 units. Assume that the direct cost per unit is $c_d = 2$, and that the fixed cost for an order is $c_K = 10$. Since only the second warehouse issued an order, the enterprise's ordering costs are $10 + 8 \cdot 2 = 26$. The quantities-on-hand after receiving the order are 4 units in the first warehouse and 10 units in the second warehouse. Assume the demand realizations are 5 units from the first warehouse and a single unit from the second warehouse. Although the first warehouse can provide only 4 units directly, the second warehouse can spare a unit from its stock, transshipment occurs, and both warehouses meet demand. Assume the transshipment cost is $c_T = 1$ for each unit transshipped. Since only one unit needs to be transshipped, the total transshipment cost is 1. In epoch 158, six units were sold. Assuming the enterprise charges $p = 10$ for each unit sold, the revenue from selling products in this epoch is 60. At the end of the epoch, the stock levels are zero units for the first warehouse and 8 units for the second warehouse. Assuming the inventory costs are 0.5 per unit in stock for one period, the total inventory costs for epoch 158 are 4 ($= 8 \cdot 0.5$). The immediate reward for that epoch is 60-26-1-4=29. The state for the next epoch is $s_1(159) = 0$ and $s_2(159) = 8$. The agent can calculate $V_{158}(s_1(159), s_2(159))$ by maximizing the Q-values corresponding with $s_1(159)$ and $s_2(159)$, which it holds by the end of epoch 158. Assume that the result of this maximization is 25. Assume that the appropriate learning rate for $s_1 = 4$, $s_2 = 2$, $a_1 = 0$, $a_2 = 8$ and $t = 158$ is $\alpha_{158}(\langle 4,2 \rangle, \langle 0,8 \rangle) = 0.1$, and that the discount factor is 0.9. The agents update the appropriate entry according to the update rule in Equation 20.14 as follows.

$$\begin{aligned} Q_{159}(\langle 4,2\rangle, \langle 0,8\rangle) &= 0.9 \cdot Q_{158}(\langle 4,2\rangle, \langle 0,8\rangle) \\ &+ 0.1 \cdot [r_{158} + \gamma V_{158}(0,8)] \\ &= 0.9 \cdot 45 + 0.1 \cdot [29 + 0.9 \cdot 25] = 45.65 \end{aligned} \quad . \tag{20.19}$$

The consequence of this update results in a change in the corresponding Q-value as indicated in Figure 20.2. Figure 20.3 shows the learning curve of a Q-Learning agent that was trained to solve the enterprise's problem in accordance with the parameters assumed in this section. The agent was introduced to 200,000 simulated experience episodes, in each of which the demands were drawn from Poisson distributions with means 5 and 3 for the first and second warehouses respectively. The learning rates were set to 0.05 for all t, and a heuristic based on Boltzmann's distribution was used to break the exploration-exploitation dilemma (see Sutton and Barto, 1996). The figure shows a plot of the moving average reward (over 2000 episodes) against the experience of the agent while gaining these rewards.

This section shows how RL algorithms (specifically how Q-Learning) can be used to learn from data observation. As discussed in Section 20.6, this by itself makes RL, in this case, a DM tool. However, the term DM may imply the use of a SL algorithm. Within the scope of problem discussed here, SL is inappropriate. A supervised learner could induce an optimal (or at-least a near-optimal) policy based on examples of the form $\langle s_1, s_2, a_1, a_2 \rangle$ whereas s_1 and s_2 describe a certain state, and a_1 and a_2 are the optimal responses (orders quantity) for that state. However in the case discussed here, such examples are probably not available.

The methods presented in this chapter are useful for many application domains, such as: Manufacturing lr18,lr14, Security lr7,l10 and Medicine lr2,lr9, and for many data mining techniques, such as: decision trees lr6,lr12, lr15, clustering lr13,lr8, ensemble methods lr1,lr4,lr5,lr16 and genetic algorithms lr17,lr11.

Order Quantity by the First Warehouse

Order Quantity by the Second Warehouse	0	1	2	3	4	5	6
0	12	10	9	14	5	0	-3
1	16	18	5	12	4	-1	1
2	17	-4	2	19	0	7	12
3	3	18	4	22	31	21	4
4	8	17	12	21	3	14	24
5	15	43	8	36	12	10	11
6	46	34	36	37	33	31	28
7	29	17	36	48	41	7	23
8	**45**	39	44	25	35	24	12

Fig. 20.1. Q-values for the state encounter on epoch 158 before the update. The value corresponding with the action finally chosen is marked.

Order Quantity by the First Warehouse

Order Quantity by the Second Warehouse	0	1	2	3	4	5	6
0	12	10	9	14	5	0	-3
1	16	18	5	12	4	-1	1
2	17	-4	2	19	0	7	12
3	3	18	4	22	31	21	4
4	8	17	12	21	3	14	24
5	15	43	8	36	12	10	11
6	46	34	36	37	33	31	28
7	29	17	36	48	41	7	23
8	**45.65**	39	44	25	35	24	12

Fig. 20.2. Q-values for the state encounter on epoch 158 after the update. The value corresponding with the action finally chosen is marked.

References

Arbel, R. and Rokach, L., Classifier evaluation under limited resources, Pattern Recognition Letters, 27(14): 1619–1631, 2006, Elsevier.

Averbuch, M. and Karson, T. and Ben-Ami, B. and Maimon, O. and Rokach, L., Context-sensitive medical information retrieval, The 11th World Congress on Medical Informatics (MEDINFO 2004), San Francisco, CA, September 2004, IOS Press, pp. 282–286.

Bellman R. Dynamic Programming. Princeton University Press, 1957.

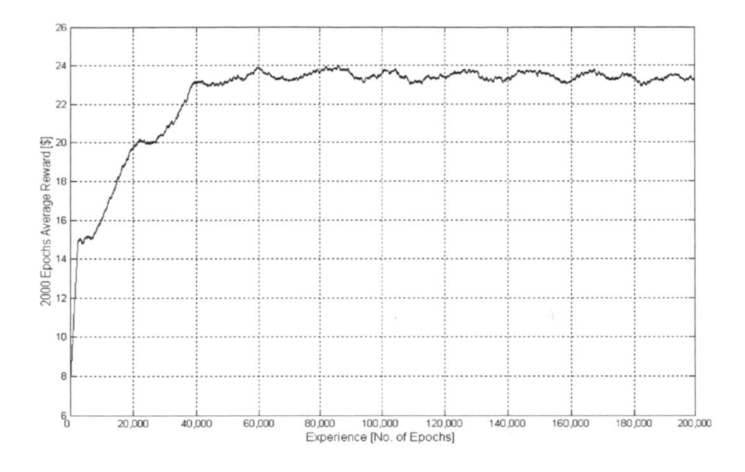

Fig. 20.3. The learning curve of a Q-Learning agent assigned to solve the enterprise's transshipment problem.

Bertsekas D.P. Dynamic Programming: Deterministic and Stochastic Models. Prentice-Hall, 1987.

Bertsekas D.P., Tsitsiklis J.N. Neuro-Dynamic Programming. Athena Scientific, 1996.

Claus C., Boutilier, C. The Dynamics of Reinforcement Learning in Cooperative Multiagent Systems. AAAI-97 Workshop on Multiagent Learning, 1998.

Cohen S., Rokach L., Maimon O., Decision Tree Instance Space Decomposition with Grouped Gain-Ratio, Information Science, Volume 177, Issue 17, pp. 3592-3612, 2007.

Crites R.H., Barto A.G. Improving Elevator Performance Using Reinforcement Learning. Advances in Neural Information Processing Systems: Proceedings of the 1995 Conference, 1996.

Filar J., Vriez K. Competitive Markov Decision Processes. Springer, 1997.

Hong J, Prabhu V.V. Distributed Reinforcement Learning for Batch Sequencing and Sizing in Just-In-Time Manufacturing Systems. Applied Intelligence, 2004; 20:71-87.

Howard, R.A. Dynamic Programming and Markov Processes, M.I.T Press, 1960.

Hu J., Wellman M.P. Multiagent Reinforcement Learning: Theoretical Framework and Algorithm. In Proceedings of the 15th International Conference on Machine Learning, 1998.

Jaakkola T., Jordan M.I.,Singh S.P. On the Convergence of Stochastic Iterative Dynamic Programming Algorithms. Neural Computation, 1994; 6:1185-201.

Kaelbling L.P., Littman L.M., Moore A.W. Reinforcement Learning: a Survey. Journal of Artificial Intelligence Research 1996; 4:237-85.

Littman M.L., Boyan J.A. A Distributed Reinforcement Learning Scheme for Network Routing. In Proceedings of the International Workshop on Applications of Neural Networks to Telecommunications, 1993.

Littman M.L. Markov Games as a Framework for Multi-Agent Reinforcement Learning. In Proceedings of the 7th International Conference on Machine Learning, 1994.

Littman M. L. Friend-or-Foe Q-Learning in General-Sum Games. Proceedings of the 18th International Conference on Machine Learning, 2001.

Maimon O., and Rokach, L. Data Mining by Attribute Decomposition with semiconductors manufacturing case study, in Data Mining for Design and Manufacturing: Methods and Applications, D. Braha (ed.), Kluwer Academic Publishers, pp. 311–336, 2001.

Maimon O. and Rokach L., "Improving supervised learning by feature decomposition", Proceedings of the Second International Symposium on Foundations of Information and Knowledge Systems, Lecture Notes in Computer Science, Springer, pp. 178-196, 2002.

Maimon, O. and Rokach, L., Decomposition Methodology for Knowledge Discovery and Data Mining: Theory and Applications, Series in Machine Perception and Artificial Intelligence - Vol. 61, World Scientific Publishing, ISBN:981-256-079-3, 2005.

Moskovitch R, Elovici Y, Rokach L, Detection of unknown computer worms based on behavioral classification of the host, Computational Statistics and Data Analysis, 52(9):4544–4566, 2008.

Pednault E., Abe N., Zadrozny B. Sequential Cost-Sensitive Decision making with Reinforcement-Learning. In Proceedings of the 8th ACM SIGKDD International Conference on Knowledge Discovery and Data Mining, 2002.

Puterman M.L. Markov Decision Processes. Wiley, 1994

Rokach, L., Decomposition methodology for classification tasks: a meta decomposer framework, Pattern Analysis and Applications, 9(2006):257–271.

Rokach L., Genetic algorithm-based feature set partitioning for classification problems,Pattern Recognition, 41(5):1676–1700, 2008.

Rokach L., Mining manufacturing data using genetic algorithm-based feature set decomposition, Int. J. Intelligent Systems Technologies and Applications, 4(1):57-78, 2008.

Rokach, L. and Maimon, O., Theory and applications of attribute decomposition, IEEE International Conference on Data Mining, IEEE Computer Society Press, pp. 473–480, 2001.

Rokach L. and Maimon O., Feature Set Decomposition for Decision Trees, Journal of Intelligent Data Analysis, Volume 9, Number 2, 2005b, pp 131–158.

Rokach, L. and Maimon, O., Clustering methods, Data Mining and Knowledge Discovery Handbook, pp. 321–352, 2005, Springer.

Rokach, L. and Maimon, O., Data mining for improving the quality of manufacturing: a feature set decomposition approach, Journal of Intelligent Manufacturing, 17(3):285–299, 2006, Springer.

Rokach, L., Maimon, O., Data Mining with Decision Trees: Theory and Applications, World Scientific Publishing, 2008.

Rokach L., Maimon O. and Lavi I., Space Decomposition In Data Mining: A Clustering Approach, Proceedings of the 14th International Symposium On Methodologies For Intelligent Systems, Maebashi, Japan, Lecture Notes in Computer Science, Springer-Verlag, 2003, pp. 24–31.

Rokach, L. and Maimon, O. and Averbuch, M., Information Retrieval System for Medical Narrative Reports, Lecture Notes in Artificial intelligence 3055, page 217-228 Springer-Verlag, 2004.

Rokach, L. and Maimon, O. and Arbel, R., Selective voting-getting more for less in sensor fusion, International Journal of Pattern Recognition and Artificial Intelligence 20 (3) (2006), pp. 329–350.

Ross S. Introduction to Stochastic Dynamic Programming. Academic Press. 1983.

Sen S., Sekaran M., Hale J. Learning to Coordinate Without Sharing Information. In Proceedings of the Twelfth National Conference on Artificial Intelligence, 1994.

Sutton R.S., Barto A.G. Reinforcement Learning, an Introduction. MIT Press, 1998.

Szepesvári C., Littman M.L. A Unified Analysis of Value-Function-Based Reinforcement-Learning Algorithms. Neural Computation, 1999; 11: 2017-60.

Tesauro G.T. TD-Gammon, a Self Teaching Backgammon Program, Achieves Master Level Play. Neural Computation, 1994; 6:215-19.

Tesauro G.T. Temporal Difference Learning and TD-Gammon. Communications of the ACM, 1995; 38:58-68.

Watkins C.J.C.H. Learning from Delayed Rewards. Ph.D. thesis; Cambridge University, 1989.

Watkins C.J.C.H., Dayan P. Technical Note: Q-Learning. Machine Learning, 1992; 8:279-92.

Zhang W., Dietterich T.G. High Performance Job-Shop Scheduling With a Time Delay TD(λ) Network. Advances in Neural Information Processing Systems, 1996; 8:1024-30.

21

Neural Networks For Data Mining

G. Peter Zhang

Georgia State University,
Department of Managerial Sciences,
gpzhang@gsu.edu

Summary. Neural networks have become standard and important tools for data mining. This chapter provides an overview of neural network models and their applications to data mining tasks. We provide historical development of the field of neural networks and present three important classes of neural models including feedforward multilayer networks, Hopfield networks, and Kohonen's self-organizing maps. Modeling issues and applications of these models for data mining are discussed.

Key words: neural networks, regression, classification, prediction, clustering

21.1 Introduction

Neural networks or artificial neural networks are an important class of tools for quantitative modeling. They have enjoyed considerable popularity among researchers and practitioners over the last 20 years and have been successfully applied to solve a variety of problems in almost all areas of business, industry, and science (Widrow, Rumelhart & Lehr, 1994). Today, neural networks are treated as a standard data mining tool and used for many data mining tasks such as pattern classification, time series analysis, prediction, and clustering. In fact, most commercial data mining software packages include neural networks as a core module.

Neural networks are computing models for information processing and are particularly useful for identifying the fundamental relationship among a set of variables or patterns in the data. They grew out of research in artificial intelligence; specifically, attempts to mimic the learning of the biological neural networks especially those in human brain which may contain more than 10^{11} highly interconnected neurons. Although the *artificial* neural networks discussed in this chapter are extremely simple abstractions of biological systems and are very limited in size, ability, and power comparing biological neural networks, they do share two very important characteristics: 1) parallel processing of information and 2) learning and generalizing from experience.

O. Maimon, L. Rokach (eds.), *Data Mining and Knowledge Discovery Handbook*, 2nd ed.,
DOI 10.1007/978-0-387-09823-4_21,

The popularity of neural networks is due to their powerful modeling capability for pattern recognition. Several important characteristics of neural networks make them suitable and valuable for data mining. First, as opposed to the traditional model-based methods, neural networks do not require several unrealistic *a priori* assumptions about the underlying data generating process and specific model structures. Rather, the modeling process is highly adaptive and the model is largely determined by the characteristics or patterns the network learned from data in the learning process. This data-driven approach is ideal for real world data mining problems where data are plentiful but the meaningful patterns or underlying data structure are yet to be discovered and impossible to be pre-specified.

Second, the mathematical property of the neural network in accurately approximating or representing various complex relationships has been well established and supported by theoretic work (Chen and Chen, 1995; Cybenko, 1989; Hornik, Stinchcombe, and White 1989). This universal approximation capability is powerful because it suggests that neural networks are more general and flexible in modeling the underlying data generating process than traditional fixed-form modeling approaches. As many data mining tasks such as pattern recognition, classification, and forecasting can be treated as function mapping or approximation problems, accurate identification of the underlying function is undoubtedly critical for uncovering the hidden relationships in the data.

Third, neural networks are nonlinear models. As real world data or relationships are inherently nonlinear, traditional linear tools may suffer from significant biases in data mining. Neural networks with their nonlinear and nonparametric nature are more cable for modeling complex data mining problems.

Finally, neural networks are able to solve problems that have imprecise patterns or data containing incomplete and noisy information with a large number of variables. This fault tolerance feature is appealing to data mining problems because real data are usually dirty and do not follow clear probability structures that typically required by statistical models.

This chapter aims to provide readers an overview of neural networks used for data mining tasks. First, we provide a short review of major historical developments in neural networks. Then several important neural network models are introduced and their applications to data mining problems are discussed.

21.2 A Brief History

Historically, the field of neural networks is benefited by many researchers in diverse areas such as biology, cognitive science, computer science, mathematics, neuroscience, physics, and psychology. The advancement of the filed, however, is not evolved steadily, but rather through periods of dramatic progress and enthusiasm and periods of skepticism and little progress.

The work of McCulloch and Pitts (1943) is the basis of modern view of neural networks and is often treated as the origin of neural network field. Their research is the first attempt to use mathematical model to describe how a neuron works. The main feature of their neuron model is that a weighted sum of input signals is compared to a threshold to determine the neuron output. They showed that simple neural networks can compute any arithmetic or logical function.

In 1949, Hebb (1949) published his book "The Organization of Behavior." The main premise of this book is that behavior can be explained by the action of neurons. He proposed one of the first learning laws that postulated a mechanism for learning in biological neurons.

In the 1950s, Rosenblatt and other researchers developed a class of neural networks called the perceptrons which are models of a biological neuron. The perceptron and its associated learning rule (Rosenblatt, 1958) had generated a great deal of interest in neural network research. At about the same time, Widrow and Hoff (1960) developed a new learning algorithm and applied it to their ADALINE (Adaptive Linear Neuron) networks which is very similar to perceptrons but with linear transfer function, instead of hard-limiting function typically used in perceptrons. The Widrow-Hoff learning rule is the basis of today's popular neural network learning methods. Although both perceptrons and ADALINE networks have achieved only limited success in pattern classification because they can only solve linearly-separable problems, they are still treated as important work in neural networks and an understanding of them provides the basis for understanding more complex networks.

The neural network research was hit by the book "Perceptrons" by Minsky and Papert (1969) who pointed out the limitation of the perceptrons and other related networks in solving a large class of nonlinearly separable problems. In addition, although Minsky and Papert proposed multilayer networks with hidden units to overcome the limitation, they were not able to find a way to train the network and stated that the problem of training may be unsolvable. This work causes much pessimism in neural network research and many researchers have left the filed. This is the reason that during the 1970s, the filed has been essentially dormant with very little research activity.

The renewed interest in neural network started in the 1980s when Hopfield (1982) used statistical mechanics to explain the operations of a certain class of recurrent network and demonstrated that neural networks could be trained as an associative memory. Hopfield networks have been used successfully in solving the Traveling Salesman Problem which is a constrained optimization problem (Hopfield and Tank, 1985). At about the same time, Kohonen (1982) developed a neural network based on self-organization whose key idea is to represent sensory signals as two-dimensional images or maps. Kohonen's networks, often called Kohonen's feature maps or self-organizing maps, organized neighborhoods of neurons such that similar inputs into the model are topologically close. Because of the usefulness of these two types of networks in solving real problems, more research was devoted to neural networks.

The most important development in the field was doubtlessly the invention of efficient training algorithms—called backpropagation—for multilayer perceptrons which have long been suspected to be capable of overcoming the linear separability limitation of the simple perceptron but have not been used due to lack of good training algorithms. The backpropagation algorithm, originated from Widrow and Hoff's learning rule, formalized by Werbos (1974), developed by Parker (1985), Rumelhart Hinton, and Williams (Rumelhart Hinton & Williams, 1986) and others, and popularized by Rumelhart, et al. (1986), is a systematic method for training multilayer neural networks. As a result of this algorithm, multilayer perceptrons are able to solve many important practical problems, which is the major reason that reinvigorated the filed of neural networks. It is by far the most popular learning paradigm in neural networks applications.

Since then and especially in the 1990s, there have been significant research activities devoted to neural networks. In the last 15 years or so, tens of thousands of papers have been published and numerous successful applications have been reported. It will not be surprising to see even greater advancement and success of neural networks in various data mining applications in the future.

21.3 Neural Network Models

As can be seen from the short historical review of development of the neural network field, many types of neural networks have been proposed. In fact, several dozens of different neural network models are regularly used for a variety of problems. In this section, we focus on three better known and most commonly used neural network models for data mining purposes: the multilayer feedforward network, the Hopfield network, and the Kohonen's map. It is important to point out that there are numerous variants of each of these networks and the discussions below are limited to the basic model formats.

21.3.1 Feedforward Neural Networks

The multilayer feedforward neural networks, also called multi-layer perceptrons (MLP), are the most widely studied and used neural network model in practice. According to Wong, Bodnovich, and Selvi (1997), about 95% of business applications of neural networks reported in the literature use this type of neural model. Feedforward neural networks are ideally suitable for modeling relationships between a set of predictor or input variables and one or more response or output variables. In other words, they are appropriate for any functional mapping problem where we want to know how a number of input variables affect the output variable(s). Since most prediction and classification tasks can be treated as function mapping problems, the MLP networks are very appealing to data mining. For this reason, we will focus more on feedforward networks and many issues discussed here can be extended to other types of neural networks.

Model Structure

An MLP is a network consisted of a number of highly interconnected simple computing units called neurons, nodes, or cells, which are organized in layers. Each neuron performs simple task of information processing by converting received inputs into processed outputs. Through the linking arcs among these neurons, knowledge can be generated and stored as arc weights regarding the strength of the relationship between different nodes. Although each neuron implements its function slowly and imperfectly, collectively a neural network is able to perform a variety of tasks efficiently and achieve remarkable results.

Figure 21.1 shows the architecture of a three-layer feedforward neural network that consists of neurons (circles) organized in three layers: input layer, hidden layer, and output layer. The neurons in the input nodes correspond to the independent or predictor variables that are believed to be useful for predicting the dependent variables which correspond to the output neurons. Neurons in the input layer are passive; they do not process information but are simply used to receive the data patterns and then pass them into the neurons into the next layer. Neurons in the hidden layer are connected to both input and output neurons and are key to learning the pattern in the data and mapping the relationship from input variables to the output variable. Although it is possible to have more than one hidden layer in a multilayer networks, most applications use only one layer. With nonlinear transfer functions, hidden neurons can process complex information received from input neurons and then send processed information to output layer for further processing to generate outputs. In feedforward neural networks, the information flow is one directional from the input to hidden then to output layer and there is no feedback from the output.

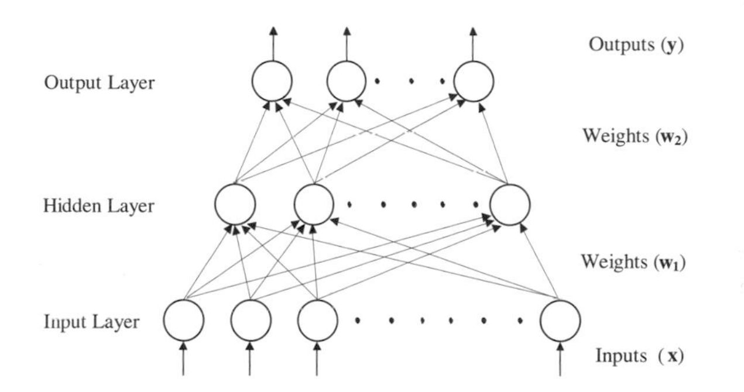

Fig. 21.1. Multi-layer feedforward neural network

Thus, a feedforward multilayer neural network is characterized by its architecture determined by the number of layers, the number of nodes in each layer, the transfer function used in each layer, as well as how the nodes in each layer connected to nodes in adjacent layers. Although partial connection between nodes in adjacent layers and direct connection from input layer to output layer are possible, the most commonly used neural network is so called fully connected one in that each node at one layer is fully connected only to all nodes in the adjacent layers.

To understand how the network in Figure 21.1 works, we need first understand the way neurons in the hidden and output layers process information. Figure 21.2 provides the mechanism that shows how a neuron processes information from several inputs and then converts it into an output. Each neuron processes information in two steps. In the first step, the inputs (x_i) are combined together to form a weighted sum of inputs and the weights (w_i) of connecting links. The 2nd step then performs a transformation that converts the sum to an output via a transfer function. In other words, the neuron in Figure 21.2 performs the following operations:

$$Out_n = f\left(\sum_i w_i x_i\right), \tag{21.1}$$

where Out_n is the output from this particular neuron and f is the transfer function. In general, the transfer function is a bounded nondecreasing function. Although there are many possible choices for transfer functions, only a few of them are commonly used in practice. These include

1. the sigmoid (logistic) function, $f(x) = (1+\exp(-x))^{-1}$,
2. the hyperbolic tangent function, $f(x) = \frac{\exp(x)-\exp(-x)}{\exp(x)+\exp(-x)}$,
3. the sine and cosine function, $f(x) = \sin(x)$, $f(x) = \cos(x)$, and
4. the linear or identity function, $f(x) = x$.

Among them, the logistic function is the most popular choice especially for the hidden layer nodes due to the fact that it is simple, has a number of good characteristics (bounded, nonlinear, and monotonically increasing), and bears a better resemblance to real neurons (Hinton, 1992).

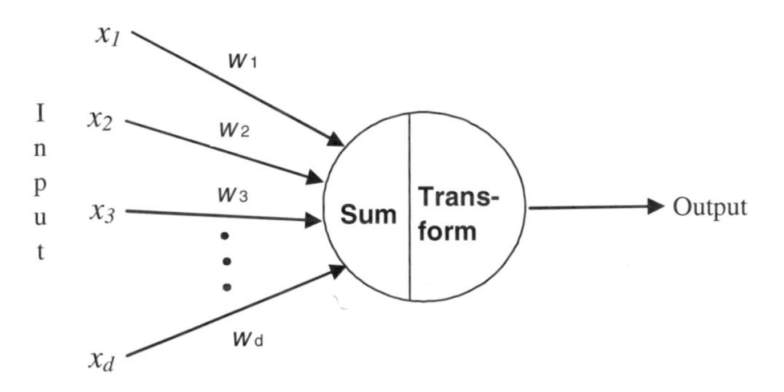

Fig. 21.2. Information processing in a single neuron

In Figure 21.1, let $\mathbf{x} = (x_1, x_2, ..., x_d)$ be a vector of d predictor or attribute variables, $\mathbf{y} = (y_1, y_2, ..., y_M)$be the M-dimensional output vector from the network, and $\mathbf{w_1}$ and $\mathbf{w_2}$ be the matrices of linking arc weights from input to hidden layer and from hidden to output layer, respectively. Then a three-layer neural network can be written as a nonlinear model of the form

$$\mathbf{y} = f_2(\mathbf{w_2} f_1(\mathbf{w_1}\mathbf{x})), \tag{21.2}$$

where f_1 and f_2 are the transfer functions for the hidden nodes and output nodes respectively. Many networks also contain node biases which are constants added to the hidden and/or output nodes to enhance the flexibility of neural network modeling. Bias terms act like the intercept term in linear regression.

In classification problems where desired outputs are binary or categorical, logistic function is often used in the output layer to limit the range of the network outputs. On the other hand, for prediction or forecasting purposes, since output variables are in general continuous, linear transfer function is a better choice for output nodes. Equation (63.2) can have many different specifications depending on the problem type, the transfer function, and numbers of input, hidden, and output nodes employed. For example, the neural network structure for a general univariate forecasting problem with logistic function for hidden nodes and identity function for the output node can be explicitly expressed as

$$y_t = w_{10} + \sum_{j=1}^{q} w_{1j} f(\sum_{i=1}^{p} w_{ij} x_{it} + w_{0j}) \tag{21.3}$$

where y_t is the observation of forecast variable and $\{x_{it},\ i = 1,\ 2,\ \ldots,\ p\}$ are p predictor variables at time t, p is also the number of input nodes, q is the number of hidden nodes, $\{w_{1j}, j = 0, 1, ..., n\}$ are weights from the hidden to output nodes and $\{w_{ij}, i = 0, 1, ..., p;\ j = 1, 2, ..., q\}$ are weights from the input to hidden nodes; α_0 and β_{0j} are bias terms, and f is the logistic function defined above.

Network Training

The arc weights are the parameters in a neural network model. Like in a statistical model, these parameters need to be estimated before the network can be adopted for further use. Neural network training refers to the process in which these weights are determined, and hence is the way the network learns. Network training for classification and prediction problems is performed via supervised learning in which known outputs and their associated inputs are both presented to the network.

The basic process to train a neural network is as follows. First, the network is fed with training examples, which consist of a set of input patterns and their desired outputs. Second, for each training pattern, the input values are weighted and summed at each hidden layer node and the weighted sum is then transmitted by an appropriate transfer function into the hidden node's output value, which becomes the input to the output layer nodes. Then, the network output values are calculated and compared to the desired or target values to determine how closely the actual network outputs match the desired outputs. Finally, the weights of the connection are changed so that the network can produce a better approximation to the desired output. This process typically repeats many times until differences between network output values and the known target values for all training patterns are as small as possible.

To facilitate training, some overall error measure such as the mean squared errors (MSE) or sum of squared errors (SSE) is often used to serve as an objective function or performance metric. For example, MSE can be defined as

$$\mathrm{MSE} = \frac{1}{M}\frac{1}{N}\sum_{m=1}^{M}\sum_{j=1}^{N}(d_{mj} - y_{mj})^2, \tag{21.4}$$

where d_{mj} and y_{mj} represent the desired (target) value and network output at the mth node for the jth training pattern respectively, M is the number of output nodes, and N is the number of training patterns. The goal of training is to find the set of weights that minimize the objective function. Thus, network training is actually an unconstrained nonlinear optimization problem. Numerical methods are usually needed to solve nonlinear optimization problems.

The most important and popular training method is the backpropagation algorithm which is essentially a gradient steepest descent method. The idea of steepest descent method is to find the best direction in the multi-dimension error space to move or change the weights so that the objective function is reduced most. This requires partial derivative of the objective function with respect to each weight to be calculated because the partial derivative represents the rate of change of the objective function. The weight updating therefore follows the following rule

$$\begin{aligned} w_{ij}^{new} &= w_{ij}^{old} + \Delta w_{ij} \\ \Delta w_{ij} &= -\eta \frac{\partial E}{\partial w_{ij}} \end{aligned} \tag{21.5}$$

where $\Delta w_{ij}$is the gradient of objective function E with respect to weight w_{ij}, and η is called the learning rate which controls the size of the gradient descent step. The algorithm requires an iterative process and there are two versions of weight updating schemes: batch mode and on-line mode. In the batch mode, weights are updated after all training patterns are evaluated, while in the on-line learning mode, the weights are updated after each pattern presentation. The basic steps with the batch mode training can be summarized as

initialize the weights to small random values from, say, a uniform distribution
choose a pattern and forward propagate it to obtain network outputs
calculate the pattern error and back-propagate it to obtain partial derivative of this error with respect to all weights
add up all the single-pattern terms to get the total derivative
update the weights with equation (63.6)
repeat steps 2-5 for next pattern until all patterns are passed through.

Note that each one pass of all patterns is called an epoch. In general, each weight update reduces the total error by only a small amount so many epochs are often needed to minimize the error. For information on further detail of the backpropagation algorithm, readers are referred to Rumelhart et al. (1986) and Bishop (1995).

It is important to note that there is no algorithm currently available which can guarantee global optimal solution for general nonlinear optimization problems such as those in neural network training. In fact, all algorithms in nonlinear optimization inevitably suffer from the local optima problems and the most we can do is to use the available optimization method which can give the "best" local optima if the true global solution is not available. It is also important to point out that the steepest descent method used in the basic backpropagation suffers the problems of slow convergence, inefficiency, and lack of robustness. Furthermore, it can be very sensitive to the choice of the learning rate. Smaller learning rates tend to slow the learning process while larger learning rates may cause network oscillation in the weight space. Common modifications to the basic backpropagation include adding in the weight updating formula (63.1) an additional momentum parameter proportional to the last weight change the to control the oscillation in weight changes and (63.2) a weight decay term that penalizes the overly complex network with large weights.

In light of the weakness of the standard backpropagation algorithm, the existence of many different optimization methods (Fletcher, 1987) provides various alternative choices for the neural network training. Among them, the second-order methods such as BFGS and Levenberg-Marquardt methods are more efficient nonlinear optimization methods and are used in most optimization packages. Their faster convergence, robustness, and the ability to find good local minima make them attractive in neural network training. For example, De Groot and Wurtz (1991) have tested several well-known optimization algorithms such as quasi-Newton, BFGS, Levenberg-Marquardt, and conjugate gradient methods and achieved significant improvements in training time and accuracy.

Modeling Issues

Developing a neural network model for a data mining application is not a trivial task. Although many good software packages exist to ease users' effort in building a neural network model, it is still critical for data miners to understand many important issues around the model building process. It is important to point out that building a successful neural network is a combination of art and science and software alone is not sufficient to solve all problems in the process. It is a pitfall to blindly throw data into a software package and then hope it will automatically identify the pattern or give a satisfactory solution. Other pitfalls readers need to be cautious can be found in Zhang (2007).

An important point in building an effective neural network model is the understanding of the issue of learning and generalization inherent in all neural network applications. This issue of learning and generalization can be understood with the concepts of model bias and variance (Geman, Bienenstock & Doursat, 1992). Bias and variance are important statistical properties associated with any empirical model. Model bias measures the systematic error of a model in *learning* the underlying relations among variables or observations. Model variance, on the other hand, relates to the stability of a model built on different data samples and therefore offers insights on *generalizability* of the model. A pre-specified or parametric model, which is less dependent on the data, may misrepresent the true functional relationship and hence cause a large bias. On the other hand, a flexible, data-driven model may be too dependent on the specific data set and hence have a large variance. Bias and variance are two important terms that impact a model's usefulness. Although it is desirable to have both low bias and low variance, we may not be able to reduce both terms at the same time for a given data set because these goals are conflicting. A model that is less dependent on the data tends to have low variance but high bias if the pre-specified model is incorrect. On the other hand, a model that fits the data well tends to have low bias but high variance when applied to new data sets. Hence a good predictive model should have an "appropriate" balance between model bias and model variance.

As a data-driven approach to data mining, neural networks often tend to fit the training data well and thus have low bias. But the potential price to pay is the overfitting effect that causes high variance. Therefore, attentions should be paid to address issues of overfitting and the balance of bias and variance in neural network model building.

The major decisions in building a neural network model include data preparation, input variable selection, choice of network type and architecture, transfer function, and training algorithm, as well as model validation, evaluation, and selection procedures. Some of these can be solved during the model building process while others must be considered before actual modeling starts.

Neural networks are data-driven techniques. Therefore, data preparation is a critical step in building a successful neural network model. Without an adequate and representative data set, it is impossible to develop a useful data mining model.

There are several practical issues around the data requirement for a neural network model. The first is the data quality. As data sets used for typical data mining tasks are massive and may be collected from multiple sources, they may suffer many quality problems such as noises, errors, heterogeneity, and missing observations. Results reported in Klein and Rossin (1999) suggest that data error rate and its magnitude can have substantial impact on neural network performance. Klein and Rossion believe that an understanding of errors in a dataset should be an important consideration to neural network users and efforts to lower error rates are well deserved. Appropriate treatment of these problems to clean the data is critical for successful application of any data mining technique including neural networks (Dasu and Johnson, 2003).

Another one is the size of the sample used to build a neural network. While there is no specific rule that can be followed for all situations, the advantage of having large samples should be clear because not only do neural networks have typically a large number of parameters to estimate, but also it is often necessary to split data into several portions for overfitting prevention, model selection, evaluation, and comparison. A larger sample provides better chance for neural networks to adequately approximate the underlying data structure.

The third issue is the data splitting. Typically for neural network applications, all available data are divided into an in-sample and an out-of-sample. The in-sample data are used for model fitting and selection, while the out-of-sample is used to evaluate the predictive ability of the model. The in-sample data often are further split into a training sample and a validation sample. The training sample is used for model parameter estimation while the validation sample is used to monitor the performance of neural networks and help stop training and select the final model. For a neural network to be useful, it is critical to test the model with an independent out-of-sample which is not used in the network training and model selection phase. Although there is no consensus on how to split the data, the general practice is to allocate more data for model building and selection although it is possible to allocate 50% vs. 50% for in-sample and out-of-sample if the data size is very large. Typical split in data mining applications reported in the literature uses convenient ratio varying from 70%:30% to 90%:10%.

Data preprocessing is another issue that is often recommended to highlight important relationships or to create more uniform data to facilitate neural network learning, meet algorithm requirements, and avoid computation problems. For time series forecasting, Azoff (1994) summarizes four methods typically used for input data normalization. They are along channel normalization, across channel normalization, mixed channel normalization, and external normalization. However, the necessity and effect of data normalization on network learning and forecasting are still not universally agreed upon. For example, in modeling and forecasting seasonal time series, some researchers (Gorr, 1994) believe that data preprocessing is not necessary because the neural network is a universal approximator and is able to capture all of the underlying patterns well. Recent empirical studies (Nelson, Hill, Remus & O'Connor, 1999; Zhang and Qi, 2002), however, find that pre-deseasonalization of the data is critical in improving forecasting performance.

Neural network design and architecture selection are important yet difficult tasks. Not only are there many ways to build a neural network model and a large number of choices to be made during the model building and selection process, but also numerous parameters and issues have to be estimated and experimented before a satisfactory model may emerge. Adding to the difficulty is the lack of standards in the process. Numerous rules of thumb are available but not all of them can be applied blindly to a new situation. In building an appropriate model, some experiments with different model structures are usually necessary. Therefore, a good experiment design is needed. For further discussions of many aspects of modeling issues for classification and forecasting tasks, readers may consult Bishop (1995), Zhang, Patuwo, and Hu (1998), and Remus and O'Connor (2001).

For network architecture selection, there are several decisions to be made. First, the size of output layer is usually determined by the nature of the problem. For example, in most time series forecasting problems, one output node is naturally used for one-step-ahead forecasting, although one output node can also be employed for multi-step-ahead forecasting in which case, iterative forecasting mode must be used. That is, forecasts for more than two-step ahead in the time horizon must be based on earlier forecasts. On the other hand, for classification problems, the number of output nodes is determined by the number of groups into which we classify objects. For a two-group classification problem, only one output node is needed while for a general M-group problem, M binary output nodes can be employed.

The number of input nodes is perhaps the most important parameter in an effective neural network model. For classification or causal forecasting problems, it corresponds to the number of feature (attribute) variables or independent (predictor) variables that data miners believe important in predicting the output or dependent variable. These input variables are usually pre-determined by the domain expert although variable selection procedures can be used to help identify the most important variables. For univariate forecasting problems, it is the number of past lagged observations. Determining an appropriate set of input variables is vital for neural networks to capture the essential relationship that can be used for successful prediction. How many and what variables to use in the input layer will directly affect the performance of neural network in both in-sample fitting and out-of-sample prediction.

Neural network model selection is typically done with the basic cross-validation process. That is the in-sample data is split into a training set and a validation set. The neural network parameters are estimated with the training sample, while the performance of the model is monitored and evaluated with the validation sample. The best model selected is the one that has the best performance on the validation sample. Of course, in choosing competing models, we must also apply the principle of parsimony. That is, a simpler model that has about the same performance as a more complex model should be preferred. Model selection can also be done with all of the in-sample data. This can be done with several in-sample selection criteria that modify the total error function to include a penalty term that penalizes for the complexity of the model. Some in-sample model selection approaches are based on criteria such as Akaike's information criterion (AIC) or Schwarz information criterion (SIC). However, it is important to note the limitation of these criteria as empirically demonstrated by Swanson and White (1995) and Qi and Zhang (2001). Other in-sample approaches are based on pruning methods such as node and weight pruning (see a review by Reed, 1993) as well as constructive methods such as the upstart and cascade correlation approaches (Fahlman and Lebiere, 1990; Frean, 1990).

After the modeling process, the finally selected model must be evaluated using data not used in the model building stage. In addition, as neural networks are often used as a nonlinear alternative to traditional statistical models, the performance of neural networks needs be compared to that of statistical methods. As Adya and Collopy (1998) point out, "if such a comparison is not conducted it is difficult to argue that the study has taught us much about the value of neural networks." They further propose three evaluation criteria to objectively evaluate the performance of a neural network: (63.1) comparing it to well-accepted (traditional) models; (63.2) using true out-of-samples; and (63.3) ensuring enough sample size in the out-of-sample (40 for classification problems and 75 for time series problems). It is important to note that the test sample served as out-of-sample should not in any way be used in the model building process. If the cross-validation is used for model selection and experimentation, the performance on the validation sample should not be treated as the true performance of the model.

Relationships with Statistical Methods

Neural networks especially the feedforward multilayer networks are closely related to statistical pattern recognition methods. Several articles that illustrate their link include Ripley (1993, 1994), Cheng and Titterington (1994), Sarle (1994), and Ciampi and Lechevallier (1997). This section provides a summary of the literature that links neural networks, particularly MLP networks to statistical data mining methods.

Bayesian decision theory is the basis for statistical classification methods. It provides the fundamental probability model for well known classification procedures.

It has been shown by many researchers that for classification problems, neural networks provide the direct estimation of the posterior probabilities under a variety of situations (Richard and Lippmann, 1991). Funahashi (1998) shows that for the two-group d-dimensional Gaussian classification problem, neural networks with at least $2d$ hidden nodes have the capability to approximate the posterior probability with arbitrary accuracy when infinite data is available and the training proceeds ideally. Miyake and Kanaya (1991) shows that neural networks trained with a generalized mean-squared error objective function can yield the optimal Bayes rule.

As the statistical counterpart of neural networks, discriminant analysis is a well-known supervised classifier. Gallinari, Thiria, Badran, and Fogelman-Soulie (1991) describe a general framework to establish the link between discriminant analysis and neural network models. They find that in quite general conditions the hidden layers of an MLP project the input data onto different clusters in a way that these clusters can be further aggregated into different classes. The discriminant feature extraction by the network with nonlinear hidden nodes has also been demonstrated in Webb and Lowe (1990) and Lim, Alder and Hadingham (1992).

Raudys (1998a, b) presents a detailed analysis of nonlinear single layer perceptron (SLP). He shows that by purposefully controlling the SLP classifier complexity during the adaptive training process, the decision boundaries of SLP classifiers are equivalent or close to those of seven statistical classifiers. These statistical classifiers include the Euclidean distance classifier, the Fisher linear discriminant function, the Fisher linear discriminant function with pseudo-inversion of the covariance matrix, the generalized Fisher linear discriminant function, the regularized linear discriminant analysis, the minimum empirical error classifier, and the maximum margin classifier.

Logistic regression is another important data mining tool. Schumacher, Robner and Vach (1996) make a detailed comparison between neural networks and logistic regression. They find that the added modeling flexibility of neural networks due to hidden layers does not automatically guarantee their superiority over logistic regression because of the possible overfitting and other inherent problems with neural networks (Vach Schumacher & Robner, 1996).

For time series forecasting problems, feedforward MLP are general nonlinear autoregressive models. For a discussion of the relationship between neural networks and general ARMA models, see Suykens, Vandewalle, and De Moor (1996).

21.3.2 Hopfield Neural Networks

Hopfield neural networks are a special type of neural networks which are able to store certain memories or patterns in a manner similar to the brain—the full pattern can be recovered if the network is presented with only partial or noisy information. This ability of brain is often called associative or content-addressable memory. Hopfield networks are quite different from the feedforward multilayer networks in several ways. From the model architecture perspective, Hopfield networks do not have a layer structure. Rather, a Hopfield network is a single layer of neurons with complete interconnectivity. That is, Hopfield networks are autonomous systems with all neurons being both inputs and outputs and no hidden neurons. In addition, unlike in feedforward networks where information is passed only in one direction, there are looping feedbacks among neurons.

Figure 21.3 shows a simple Hopfield network with only three neurons. Each neuron is connected to every other neuron and the connection strengths or weights are symmetric in that the weight from neuron i to neuron j (w_{ij}) is the same as that from neuron j to neuron $i(w_{ji})$. The flow of the information is not in a single direction as in the feedforward network. Rather it is possible for signals to flow from a neuron back to itself via other neurons. This feature is often called feedback or recurrent because neurons may be used repeatedly to process information.

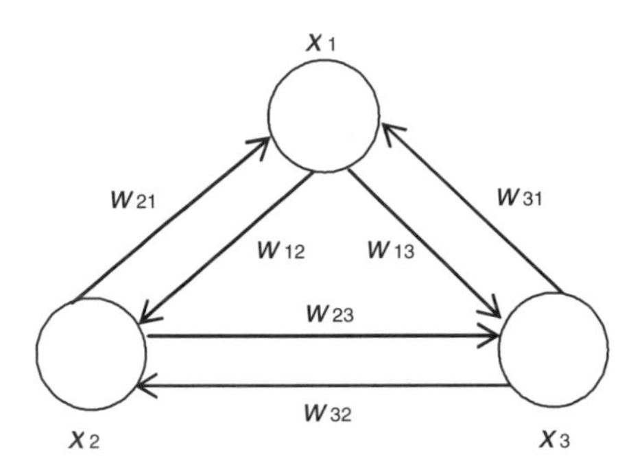

Fig. 21.3. A three-neuron Hopfield network

The network is completely described by a state vector which is a function of time t. Each node in the network contributes one component to the state vector and any or all of the node outputs can be treated as outputs of the network. The dynamics of neurons can be described mathematically as the following equations:

$$u_i(t) = \sum_{j=1}^{n} w_{ij}x_j(t) + v_i \tag{21.6}$$

where $u_i(t)$ is the internal state of the ith neuron, $x_i(t)$ is the output activation or output state of the ith neuron, v_i is the threshold to the ith neuron, n is the number of neurons, and *sign* is the sign function defined as $sign(x)$=1, if x >0 and -1 otherwise. Given a set of initial conditions $\mathbf{x}(0)$, and appropriate restrictions on the weights (such as symmetry), this network will converge to a fixed equilibrium point.

For each network state at any time, there is an energy associated with it. A common energy function is defined as

$$E(t) = -\frac{1}{2}\mathbf{x}(t)^T\mathbf{W}\mathbf{x}(t) - \mathbf{x}(t)^T\mathbf{v} \tag{21.7}$$

where $\mathbf{x}$(t) is the state vector, $\mathbf{W}$ is the weight matrix, $\mathbf{v}$ is the threshold vector, and T denote transpose. The basic idea of the energy function is that it always decreases or at least remains constant as the system evolves over time according to its dynamic rule in equations 6 and 7. It can be shown that the system will converge from an arbitrary initial energy to eventually a fixed point (a local minimum) on the surface of the energy function. These fixed points are stable states which correspond to the stored patterns or memories.

The main use of Hopfield's network is as associative memory. An associative memory is a device which accepts an input pattern and generates an output as the stored pattern which is most closely associated with the input. The function of the associate memory is to recall the corresponding stored pattern, and then produce a clear version of the pattern at the output. Hopfield networks are typically used for those problems with binary pattern vectors and the input pattern may be a noisy version of one of the stored patterns. In the Hopfield network, the stored patterns are encoded as the weights of the network.

There are several ways to determine the weights from a training set which is a set of known patterns. One way is to use a prescription approach given by Hopfield (1982). With this approach, the weights are given by

$$\mathbf{w} = \frac{1}{n}\sum_{i=1}^{p} z_i z_i^T \tag{21.8}$$

where $z_i, i = 1, 2, \ldots, p$ are $p$patterns that are to be stored in the network. Another way is to use an incremental, iterative process called Hebbian learning rule developed by Hebb (1949). It has the following learning process:
choose a pattern from the training set at random
present a pair of components of the pattern at the outputs of the corresponding nodes of the network
if two nodes have the same value then make a small positive increment to the interconnected weight. If they have opposite values then make a small negative decrement to the weight. The incremental size can be expressed as $\Delta w_{ij} = \alpha z_i^p z_j^p$, where α is a constant rate in between 0 and 1 and z_i^p is the ith component of pattern p.

Hopfield networks have two major limitations when used as a content addressable memory. First, the number of patterns that can be stored and accurately recalled is fairly limited. If too many patterns are stored, the network may converge to a spurious pattern different from all programmed patterns. Or, it may not converge at all. The second limitation is that the network may become unstable if the common patterns it shares are too similar. An example pattern is considered unstable if it is applied at time zero and the network converges to some other pattern from the training set.

21.3.3 Kohonen's Self-organizing Maps

Kohonen's self-organizing maps (SOM) are important neural network models for dimension reduction and data clustering. SOM can learn from complex, multidimensional data and transform them into a topological map of much fewer dimensions typically one or two dimensions. These low dimension plots provide much improved visualization capabilities to help data miners visualize the clusters or similarities between patterns.

SOM networks represent another neural network type that is markedly different from the feedforward multilayer networks. Unlike training in the feedforward MLP, the SOM training or learning is often called the unsupervised because there are no known target outputs associated with each input pattern in SOM and during the training process, the SOM processes the input patterns and learns to cluster or segment the data through adjustment of weights. A two-dimensional map is typically created in such a way that the orders of the interrelationships among inputs are preserved. The number and composition of clusters can be visually determined based on the output distribution generated by the training process. With only input variables in the training sample, SOM aims to learn or discover the underlying structure of the data.

A typical SOM network has two layers of nodes, an input layer and output layer (sometimes called the Kohonen layer). Each node in the input layer is fully connected to nodes in the two-dimensional output layer. Figure 21.4 shows an example of an SOM network with several input nodes in the input layer and a two dimension output layer with a 4x4 rectangular array of 16 neurons. It is also possible to use hexagonal array or higher dimensional grid in the Kohonen layer. The number of nodes in the input layer is corresponding to the number of input variables while the number of output nodes depends on the specific problem and is determined by the user. Usually, this number of neurons in the rectangular array should be large enough to allow a sufficient number of clusters to form. It has been recommended that this number is ten times the dimension of the input pattern (Deboeck and Kohonen, 1998)

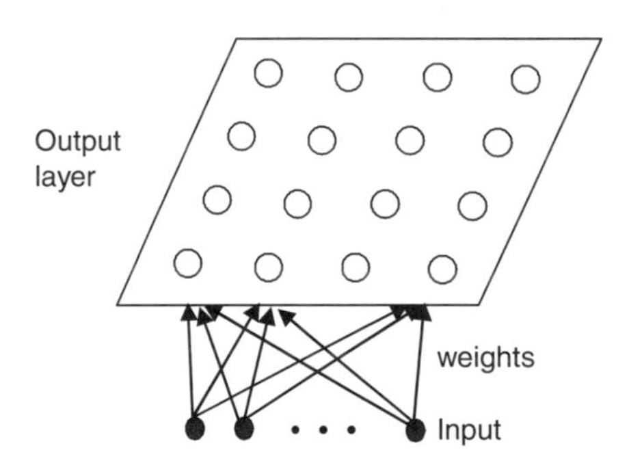

Fig. 21.4. A 4x4 SOM network

During the training process, input patterns are presented to the network. At each training step when an input pattern **x** randomly selected from the training set is presented, each neuron i in the output layer calculates how similar the input is to its weights $\mathbf{w}_i$. The similarity is often measured by some distance between **x** and $\mathbf{w}_i$. As the training proceeds, the neurons adjust their weights according to the topological relations in the input data. The neuron with the minimum distance is the winner and the weights of the winning node as well as its neighboring nodes are strengthened or adjusted to be closer to the value of input pattern. Therefore, the training with SOM is unsupervised and competitive with winner-take-all strategy.

A key concept in training SOM is the neighborhood N_k around a winning neuron, k, which is the collection of all nodes with the same radial distance. Figure 21.5 gives an example of neighborhood nodes for a 5x5 Kohonen layer at radius of 1 and 2.

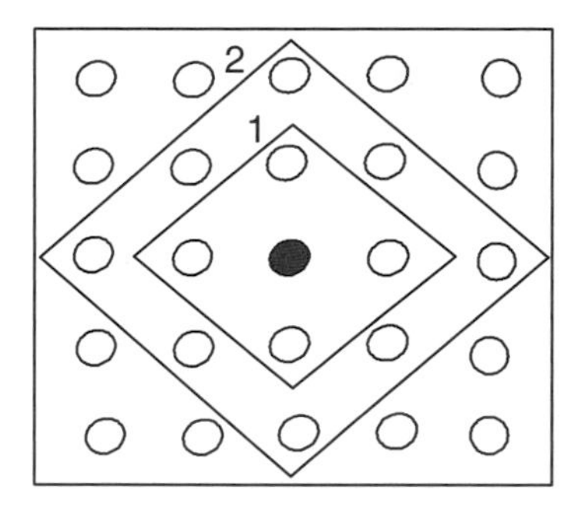

Fig. 21.5. A 5x5 Kohonen Layer with two neighborhood sizes

The basic procedure in training an SOM is as follows:
initialize the weights to small random values and the neighborhood size large enough to cover half the nodes
select an input pattern **x** randomly from the training set and present it to the network
find the best matching or "winning" node k whose weight vector w_k is closest to the current input vector **x** using the vector distance. That is:

$$\|x - w_k\| = \min_i \|x - w_i\|$$

where $\|.\|$ represents the Euclidean distance
update the weights of nodes in the neighborhood of k using the Kohonen learning rule:

$w_i^{new} = w_i^{old} + \alpha h_{ik}(x - w_i)$if i is in N_k
$w_i^{new} = w_i^{old}$if $i$is not in N_k(10)

where α is the learning rate between 0 and 1 and h_{ik} is a neighborhood kernel centered on the winning node and can take Gaussian form as

$$h_{ik} = \exp\left[-\frac{\|r_i - r_k\|^2}{2\sigma^2}\right] \tag{21.9}$$

where r_i and r_k are positions of neurons i and k on the SOM grid and σ is the neighborhood radius.
decrease the learning rate slightly
repeat Steps 1—5 with a number of cycles and then decrease the size of the neighborhood. Repeat until weights are stabilized.

As the number of cycles of training (epochs) increases, better formation of the clusters can be found. Eventually, the topological map is fine-tuned with finer distinctions of clusters within areas of the map. After the network has been trained, it can be used as a visualization tool to examine the data structure. Once clusters are identified, neurons in the map can be labeled to indicate their meaning. Assignment of meaning usually requires knowledge on the data and specific application area.

21.4 Data Mining Applications

Neural networks have been used extensively in data mining for a wide variety of problems in business, engineering, industry, medicine, and science. In general, neural networks are good at solving the following common data mining problems such as classification, prediction, association, and clustering. This section provides a short overview on the application areas.

Classification is one of the frequently encountered data mining tasks. A classification problem occurs when an object needs to be assigned into a predefined group or class based on a number of observed attributes related to that object. Many problems in business, industry, and medicine can be treated as classification problems. Examples include bankruptcy prediction, credit scoring, medical diagnosis, quality control, handwritten character recognition, and speech recognition. Feed-forward multilayer networks are most commonly used for these classification tasks although other types of neural networks can also be used.

Forecasting is central to effective planning and operations in all business organizations as well as government agencies. The ability to accurately predict the future is fundamental to many decision activities in finance, marketing, production, personnel, and many other business functional areas. Increasing forecasting accuracy could facilitate the saving of millions of dollars to a company. Prediction can be done with two approaches: causal and time series analysis, both of which are suitable for feed-forward networks. Successfully applications include predictions of sales, passenger volume, market share, exchange rate, futures price, stock return, electricity demand, environmental changes, and traffic volume.

Clustering involves categorizing or segmenting observations into groups or clusters such that each cluster is as homogeneous as possible. Unlike classification problems, the groups or clusters are usually unknown to or not predetermined by data miners. Clustering can simplify a complex large data set into a small number of groups based on the natural structure of data. Improved understanding of the data and subsequent decisions are major benefits of clustering. Kohonen or SOM networks are particularly useful for clustering tasks. Applications have been reported in market segmentation, customer targeting, business failure categorization, credit evaluation, document retrieval, and group technology.

With association techniques, we are interested in the correlation or relationship among a number variables or objects. Association is used in several ways. One use as in market basket analysis is to help identify the consequent items given a set of antecedent items. An association rule in this way is an implication of the form: IF $\mathbf{x}$, THEN Y, where $\mathbf{x}$ is a set of antecedent items and Y is the consequent items. This type of association rule has been used in a variety of data mining tasks including credit card purchase analysis, merchandise stocking, insurance fraud investigation,

Table 21.1. Data mining applications of neural networks

Data Mining Task	Application Area
Classification	bond rating (Dutta and shenkar, 1993)
	corporation failure (Zhang et al., 1999; Mckee and Greenstein, 2000)
	credit scoring (West, 2000)
	customer retention (Mozer and Wolniewics, 2000; Smith et al., 2000)
	customer satisfaction (Temponi et al., 1999)
	fraud detection (He et al., 1997)
	inventory (Partovi and Anandarajan, 2002)
	project (Thieme et al., 2000; Zhang et al., 2003)
	target marketing (Zahavi and Levin, 1997)
Prediction	air quality (Kolehmainen et al., 2001)
	business cycles and recessions (Qi, 2001)
	consumer expenditures (Church and Curram, 1996)
	consumer choice (West et al., 1997)
	earnings surprises (Dhar and Chou, 2001)
	economic crisis (Kim et al., 2004)
	exchange rate (Nag and Mitra, 2002)
	market share (Agrawal and Schorling, 1996)
	ozone concentration level (Prybutok et al., 2000)
	sales (Ansuj et al., 1996; Kuo, 2001; Zhang and Qi, 2002)
	stock market (Qi, 1999; Chen et al., 2003; Leung et al., 2000; Chun and Kim, 2004)
	tourist demand (Law, 2000)
	traffic (Dia, 2001; Qiao et al., 2001)
Clustering	bankruptcy prediction (Kiviluoto, 1998)
	document classification (Dittenbach et al., 2002)
	enterprise typology (Petersohn, 1998)
	fraud uncovering (Brockett et al., 1998)
	group technology (Kiang et al., 1995)
	market segmentation (Ha and Park, 1998; Vellido et al., 1999; Reutterer and Natter, 2000; Boone and Roehm, 2002)
	process control (Hu and Rose, 1995)
	property evaluation (Lewis et al., 1997)
	quality control (Chen and Liu, 2000)
	webpage usage (Smith and Ng, 2003)
Association/Pattern Recognition	defect recognition (Kim and Kumara, 1997)
	facial image recognition (Dai and Nakano, 1998)
	frequency assignment (Salcedo-Sanz et al., 2004)
	graph or image matching (Suganthan et al., 1995; Pajares et al., 1998)
	image restoration (Paik and Katsaggelos, 1992; Sun and Yu, 1995)
	imgage segmentation (Rout et al., 1998; Wang et al., 1992)
	landscape pattern prediction (Tatem et al., 2002)
	market basket analysis (Evans, 1997)
	object recognition (Huang and Liu, 1997; Young et al., 1997; Li and Lee, 2002)
	on-line marketing (Changchien and Lu, 2001)
	pattern sequence recognition (Lee, 2002)
	semantic indexing and searching (Chen et al., 1998)

market basket analysis, telephone calling pattern identification, and climate prediction. Another use is in pattern recognition. Here we train a neural network first to *remember* a number of patterns, so that when a distorted version of a stored pattern is presented, the network associates it with the closest one in its memory and returns the original version of the pattern. This is useful for restoring noisy data. Speech, image, and character recognitions are typical application areas. Hopfield networks are useful for this purpose.

Given an enormous amount of applications of neural networks in data mining, it is difficult if not impossible to give a detailed list. Table 21.1 provides a sample of several typical applications of neural networks for various data mining problems. It is important to note that studies given in Table 21.1 represent only a very small portion of all the applications reported in the literature, but we should still get an appreciation of the capability of neural networks in solving wide range of data mining problems. For real-world industrial or commercial applications, readers are referred to Widrow et al. (1994), Soulie and Gallinari (1998), Jain and Vemuri (1999), and Lisboa, Edisbury, and Vellido (2000).

21.5 Conclusions

Neural networks are standard and important tools for data mining. Many features of neural networks such as nonlinear, data-driven, universal function approximating, noise-tolerance, and parallel processing of large number of variables are especially desirable for data mining applications. In addition, many types of neural networks functionally are similar to traditional statistical pattern recognition methods in areas of cluster analysis, nonlinear regression, pattern classification, and time series forecasting. This chapter provides an overview of neural networks and their applications to data mining tasks. We present three important classes of neural network models: Feedforward multilayer networks, Hopfield networks, and Kohonen's self-organizing maps, which are suitable for a variety of problems in pattern association, pattern classification, prediction, and clustering.

Neural networks have already achieved significant progress and success in data mining. It is, however, important to point out that they also have limitations and may not be a panacea for every data mining problem in every situation. Using neural networks require thorough understanding of the data, prudent design of modeling strategy, and careful consideration of modeling issues. Although many rules of thumb exist in model building, they are not necessarily always useful for a new application. It is suggested that users should not blindly rely on a neural network package to "automatically" mine the data, but rather should study the problem and understand the network models and the issues in various stages of model building, evaluation, and interpretation.

References

Adya M., Collopy F. (1998), How effective are neural networks at forecasting and prediction? a review and evaluation. Journal of forecasting ; 17:481-495.

Agrawal D., Schorling C. (1996), Market share forecasting: an empirical comparison of artificial neural networks and multinomial logit model. Journal of Retailing ; 72:383-407.

Ahn H., Choi E., Han I. (2007), Extracting underlying meaningful features and canceling noise using independent component analysis for direct marketing Expert Systems with Applications; 33: 181-191.

Azoff E. M. (1994), Neural Network Time Series Forecasting of Financial Markets. Chichester: John Wiley & Sons.

Bishop M. (1995), Neural Networks for Pattern Recognition. Oxford: Oxford University Press.

Boone D., Roehm M. (2002), Retail segmentation using artificial neural networks. International Journal of Research in Marketing ; 19:287-301.

Brockett P.L., Xia X.H., Derrig R.A. (1998), Using Kohonen's self-organizing feature map to uncover automobile bodily injury claims fraud. The Journal of Risk and Insurance ; 65: 24

Changchien S.W., Lu T.C. (2001), Mining association rules procedure to support on-line recommendation by customers and products fragmentation. Expert Systems with Applications ; 20(4):325-335.

Chen T., Chen H. (1995), Universal approximation to nonlinear operators by neural networks with arbitrary activation functions and its application to dynamical systems, Neural Networks ; 6:911-917.

Chen F.L., Liu S.F. (2000), A neural-network approach to recognize defect spatial pattern in semiconductor fabrication. IEEE Transactions on Semiconductor Manufacturing; 13:366-37.

Chen S.K., Mangiameli P., West D. (1995), The comparative ability of self-organizing neural networks to define cluster structure. Omega ; 23:271-279.

Chen H., Zhang Y., Houston A.L. (1998), Semantic indexing and searching using a Hopfield net. Journal of Information Science ; 24:3-18.

Cheng B., Titterington D. (1994), Neural networks: a review from a statistical perspective. Statistical Sciences ; 9:2-54.

Chen K.Y., Wang, C.H. (2007), Support vector regression with genetic algorithms in forecasting tourism demand. Tourism Management ; 28:215-226.

Chiang W.K., Zhang D., Zhou L. (2006), Predicting and explaining patronage behavior toward web and traditional stores using neural networks: a comparative analysis with logistic regression. Decision Support Systems ; 41:514-531.

Church K. B., Curram S. P. (1996), Forecasting consumers' expenditure: A comparison between econometric and neural network models. International Journal of Forecasting ; 12:255-267.

Ciampi A., Lechevallier Y. (1997), Statistical models as building blocks of neural networks. Communications in Statistics: Theory and Methods ; 26:991-1009.

Crone S.F., Lessmann S., Stahlbock R. (2006), The impact of preprocessing on data mining: An evaluation of classifier sensitivity in direct marketing. European Journal of Operational Research ; 173:781-800.

Cybenko G. (1989), Approximation by superpositions of a sigmoidal function. Mathematical Control Signals Systems ; 2:303–314.

Dai Y., Nakano Y. (1998), Recognition of facial images with low resolution using a Hopfield memory model. Pattern Recognition ; 31:159-167.

Dasu T., Johnson T. (2003), Exploratory Data Mining and Data Cleaning. New Jersey: Wiley.

De Groot D., Wurtz D. (1991), Analysis of univariate time series with connectionist nets: A case study of two classical examples. Neurocomputing ;3:177-192.

Deboeck G., Kohonen T. (1998), Visual Explorations in Finance with Self-organizing Maps. London: Springer-Verlag.

Delen D., Sharda R., Bessonov M. (2006), Identifying significant predictors of injury severity in traffic accidents using a series of artificial neural networks Accident Analysis and Prevention ; 38:434-444.

Dhar V., Chou D. (2001), A comparison of nonlinear methods for predicting earnings surprises and returns. IEEE Transactions on Neural Networks ; 12:907-921.

Dia H. (2001), An object-oriented neural network approach to short-term traffic forecasting. European Journal of Operation Research ; 131:253-261.

Dittenbach M., Rauber A., Merkl, D. (2002), Uncovering hierarchical structure in data using the growing hierarchical self-organizing map. Neurocompuing ; 48:199-216.

Doganis P., Alexandridis A., Patrinos P., Sarimveis H. (2006), Time series sales forecasting for short shelf-life food products based on artificial neural networks and evolutionary computing. Journal of Food Engineering ; 75:196-204.

Dutot A.L., Rynkiewicz J., Steiner F.E., Rude J. (2007), A 24-h forecast of ozone peaks and exceedance levels using neural classifiers and weather predictions Modelling and Software; 22:1261-1269.

Dutta S., Shenkar S. (1993), "Bond rating: a non-conservative application of neural networks." In Neural Networks in Finance and Investing, Trippi, R., and Turban, E., eds. Chicago: Probus Publishing Company.

Enke D., Thawornwong S. (2005), The use of data mining and neural networks for forecasting stock market returns. Expert Systems with Applications ; 29:927-940.

Evans O.V.D. (1997), Discovering associations in retail transactions using neural networks. ICL Systems Journal ; 12:73-88.

Fahlman S., Lebiere C. (1990), "The cascade-correlation learning architecture." In Advances in Neural Information Processing Systems, Touretzky, D., ed. .

Fletcher R. (1987), Practical Methods of Optimization 2nd. Chichester: John Wiley & Sons.

Frean M. (1990), The Upstart algorithm: a method for constructing and training feed-forward networks. Neural Computations ; 2:198-209.

Funahashi K. (1998), Multilayer neural networks and Bayes decision theory. Neural Networks ; 11:209-213.

Gallinari P., Thiria S., Badran R., Fogelman-Soulie, F. (1991), On the relationships between discriminant analysis and multilayer perceptrons. Neural Networks ; 4:349-360.

Geman S., Bienenstock E., Doursat T. (1992), Neural networks and the bias/variance dilemma. Neural Computation ; 5:1-58.

Gorr L. (1994), Research prospective on neural network forecasting. International Journal of Forecasting ; 10:1-4.

He H., Wang J., Graco W., Hawkins S. (1997), Application of neural networks to detection of medical fraud. Expert Systems with Applications ; 13:329-336.

Hebb D.O. (1949), The Organization of Behavior. New York: Wiley.

Hinton G.E. (1992), How neural networks learn from experience. Scientific American; 9:145-151.

Hornik K., Stinchcombe M., White H. (1989), Multilayer feedforward networks are universal approximators. Neural Networks ; 2:359–366.

Hopfield J.J. (2558), (1982), Neural networks and physical systems with emergent collective computational abilities. Proceedings of National Academy of Sciences; 79(8):2554-2558.

Hopfield J.J., Tank D.W. (1985), Neural computation of decisions in optimization problems. Biological Cybernetics ; 52:141-152.

Hu J.Q., Rose, E. (1995), On-line fuzzy modeling by data clustering using a neural network. Advances in Process Control. , 4, 187-194.

Huang J.S., Liu H.C. (2004), Object recognition using genetic algorithms with a Huang Z. Chen, H., Hsu, C.J. Chen, W.H. and Wu, S., Credit rating analysis with support vector machines and neural networks: a market comparative study. Decision Support Systems ; 37:543-558.

Hopfield's neural model (1997). Expert Systems with Applications 1997; 13:191-199.

Jain L.C., Vemuri V.R. (1999), Industrial Applications of Neural Networks. Boca Raton: CRC Press.

Kiang M.Y., Hu, M.Y., Fisher D.M. (2006), An extended self-organizing map network for market segmentation—a telecommunication example Decision Support Systems ; 42:36-47.

Kiang M.Y., Kulkarni U.R., Tam K.Y. (1995), Self-organizing map network as an interactive clustering tool-An application to group technology. Decision Support Systems ; 15:351-374.

Kim T., Kumara S.R.T., (1997), Boundary defect recognition using neural networks. International Journal of Production Research; 35:2397-2412.

Kim T.Y., Oh K.J., Sohn K., Hwang C. (2004), Usefulness of artificial neural networks for early warning system of economic crisis. Expert Systems with Applications ; 26:583-590.

Kirkos E., Spathis C., Manolopoulos Y., (2007), Data Mining techniques for the detection of fraudulent financial statements. Expert Systems with Applications ; 32: 995-1003.

Kiviluoto K. (1998), Predicting bankruptcy with the self-organizing map. Neurocomputing ; 21:203-224.

Klein B.D., Rossin D. F. (1999), Data quality in neural network models: effect of error rate and magnitude of error on predictive accuracy. Omega ; 27:569-582.

Kohonen T. (1982), Self-organized formation of topologically correct feature maps. Biological Cybernetics ; 43:59-69.

Kolehmainen M., Martikainen H., Ruuskanen J. (2001), Neural networks and periodic components used in air quality forecasting. Atmospheric Environment ; 35:815-825.

Law R. (2000), Back-propagation learning in improving the accuracy of neural network-based tourism demand forecasting. Tourism Management ; 21:331-340.

Lee D.L. (2002), Pattern sequence recognition using a time-varying Hopfield network. IEEE Transactions on Neural Networks ; 13:330-343.

Lewis O.M., Ware J.A., Jenkins D. (1997), A novel neural network technique for the valuation of residential property. Neural Computing and Applications ; 5:224-229.

Li W.J., Lee T., (2002), Object recognition and articulated object learning by accumulative Hopfield matching. Pattern Recognition; 35:1933 1948.

Lim G.S., Alder M., Hadingham P. (1992), Adaptive quadratic neural nets. Pattern Recognition Letters ; 13: 325-329.

Lisboa P.J.G., Edisbury B., Vellido A. (2000), Business Applications of Neural Networks : The State-of-the-art of Real-world Applications. River Edge: World Scientific.

McCulloch W., Pitts W. (1943), A logical calculus of the ideas immanent in nervous activity. Bulletin of Mathematical Biophysics; 5:115-133.

Min S.H., Lee J., Han I. (2006), Hybrid genetic algorithms and support vector machines for bankruptcy prediction. Expert Systems with Applications ; 31: 652-660.

Minsky M. L., Papert S. A. (1969), Perceptrons. MA: MIT press.

Miyake S., Kanaya F. (1991), A neural network approach to a Bayesian statistical decision problem. IEEE Transactions on Neural Networks ; 2:538-540.

Mozer M.C., Wolniewics R. (2000), Predicting subscriber dissatisfaction and improving retention in the wireless telecommunication. IEEE Transactions on Neural Networks; 11:690-696.

Nag A.K., Mitra A. (2002), Forecasting daily foreign exchange rates using genetically optimized neural networks. Journal of Forecasting ; 21:501-512.

Nelson M., Hill T., Remus T., O'Connor, M. (1999), Time series forecasting using neural networks: Should the data be deseasonalized first? Journal of Forecasting ; 18:359-367.

O'Connor N., Madden M.G. (2006), A neural network approach to predicting stock exchange movements using external factors. Knowledge-Based Systems ; 19:371-378.

Paik J.K., Katsaggelos, A.K. (1992), Image restoration using a modified Hopfield neural network. IEEE Transactions on Image Processing ; 1:49-63.

Pajares G., Cruz J.M., Aranda, J. (1998), Relaxation by Hopfield network in stereo image matching. Pattern Recognition ; 31:561-574.

Panda C., Narasimhan V. (2007), Forecasting exchange rate better with artificial neural network. Journal of Policy Modeling ; 29:227-236.

Parker D.B. (1985), Learning-logic: Casting the cortex of the human brain in silicon, Technical Report TR-47, Center for Computational Research in Economics and Management Science, MIT.

Palmer A., Montaño J.J., Sesé, A. (2006), Designing an artificial neural network for forecasting tourism time series. Tourism Management ; 27: 781-790.

Partovi F.Y., Anandarajan M. (2002), Classifying inventory using an artificial neural network approach. Computers and Industrial Engineering ; 41:389-404.

Petersohn H. (1998), Assessment of cluster analysis and self-organizing maps. International Journal of Uncertainty Fuzziness and Knowledge-Based Systems. ; 6:139-149.

Prybutok V.R., Yi J., Mitchell D. (2000), Comparison of neural network models with ARIMA and regression models for prediction of Houston's daily maximum ozone concentrations. European Journal of Operational Research ; 122:31-40.

Qi M. (2001), Predicting US recessions with leading indicators via neural network models. International Journal of Forecasting ; 17:383-401.

Qi M., Zhang G.P. (2001), An investigation of model selection criteria for neural network time series forecasting. European Journal of Operational Research ; 132:666-680.

Qiao F., Yang H., Lam, W.H.K. (2001), Intelligent simulation and prediction of traffic flow dispersion. Transportation Research, Part B ; 35:843-863.

Raudys S. (1998), Evolution and generalization of a single neuron: I., Single-layer perceptron as seven statistical classifiers Neural Networks ; 11:283-296.

Raudys S. (1998), Evolution and generalization of a single neuron: II., Complexity of statistical classifiers and sample size considerations. Neural Networks ; 11:297-313.

Raviwongse R. Allada V., Sandidge T. (2000), Plastic manufacturing process selection methodology using self-organizing map (SOM)/fuzzy analysis. International Journal of Advanced Manufacturing Technology; 16:155-161.

Reed R. (1993), Pruning algorithms-a survey. IEEE Transactions on Neural Networks ; 4:740-747.

Remus W., O'Connor M. (2001), "Neural networks for time series forecasting." In Principles of Forecasting: A Handbook for Researchers and Practitioners, Armstrong, J. S. ed. Norwell:Kluwer Academic Publishers, 245-256.

Reutterer T., Natter M. (2000), Segmentation based competitive analysis with MULTICLUS and topology representing networks. Computers and Operations Research; 27:1227-1247.

Richard, M. (1991), D., Lippmann, R., Neural network classifiers estimate Bayesian *aposteriori* probabilities. Neural Computation ; 3:461-483.

Ripley A. (1993), "Statistical aspects of neural networks." In Networks and Chaos - Statistical and Probabilistic Aspects, Barndorff-Nielsen, O. E., Jensen J. L. and Kendall, W. S. eds. London: Chapman and Hall, 40-123.

Ripley A. (1994), Neural networks and related methods for classification. Journal of Royal Statistical Society, Series B ; 56:409-456.

Roh T. H. (2007), Forecasting the volatility of stock price index. Expert Systems with Applications ; 33:916-922.

Rosenblatt F. (1958), The perceptron: A probabilistic model for information storage and organization in the brain. Psychological Review ; 65:386-408.

Rout S., Srivastava, S.P., Majumdar, J. (1998), Multi-modal image segmentation using a modified Hopfield neural network. Pattern Recognition ; 31:743-750.

Rumelhart D.E., Hinton G.E., Williams R.J. (1986), "Learning internal representation by back-propagating errors." In Parallel Distributed Processing: Explorations in the Microstructure of Cognition Press, Rumelhart, D.E., McCleland, J.L. and the PDP Research Group, eds. MA: MIT.

Saad E.W., Prokhorov D.V., Wunsch, D.C. II. (1998), Comparative study of stock trend prediction using time delay, recurrent and probabilistic neural networks. IEEE Transactions on Neural Networks; 9:456-1470.

Salcedo-Sanz S., Santiago-Mozos R.,Bousono-Calzon, C. (2004), A hybrid Hopfield network-simulated annealing approach for frequency assignment in satellite communications systems. IEEE Transactions on System, Man and Cybernetics, Part B:108-116.

Sarle W.S. (1994), Neural networks and statistical models. Poceedings of the Nineteenth Annual SAS Users Group International Conference, Cary, NC: SAS Institute.

Schumacher M., Robner R., Vach W. (1996), Neural networks and logistic regression: Part I., Computational Statistics and Data Analysis ; 21:661-682.

Smith K.A., Ng, A. (2003), Web page clustering using a self-organizing map of user navigation patterns. Decision Support Systems ; 35:245-256.

Smith K.A., Willis R.J., Brooks M. (2000), An analysis of customer retention and insurance claim patterns using data mining: a case study. Journal of the Operational Research Society; 51:532-541.

Soulie F.F., Gallinari P. (1998), Industrial Applications of Neural Networks. River Edge, NJ: World Scientific.

Suganthan P.N., Teoh E.K., Mital D.P. (1995), Self-organizing Hopfield network for attributed relational graph matching. Image and Vision Computing; 13:61-71.

Sun Z.Z., Yu S. (1995), Improvement on performance of modified Hopfield neural network for image restoration. IEEE Transactions on Image processing; 4:683-692.

Suykens J.A.K., Vandewalle J.P.L., De Moor B.L.R. (1996), Artificial Neural Networks for Modeling and Control of Nonlinear Systems. Boston: Kluwer.

Swanson N.R., White H. (1995), A model-selection approach to assessing the information in the term structure using linear models and artificial neural networks. Journal of Business and Economic Statistics; 13;265-275.

Tatem A.J., Lewis H.G., Atkinson P.M., Nixon M.S. (2002), Supre-resolution land cover pattern prediction using a Hopfield neural network. Remote Sensing of Environment; 79:1-14.

Temponi C., Kuo Y.F., Corley H.W. (1999), A fuzzy neural architecture for customer satisfaction assessment. Journal of Intelligent & Fuzzy Systems; 7:173-183.

Thieme R.J., Song M., Calantone R.J. (2000), Artificial neural network decision support systems for new product developement project selection. Journal of Marketing Research; 37:543-558.

Vach W., Robner R., Schumacher M. (1996), Neural networks and logistic regression: Part I. Computational Statistics and Data Analysis; 21:683-701.

Wang T., Zhuang X., Xing X. (1992), Robust segmentation of noisy images using a neural network model. Image Vision Computing; 10:233-240.

Webb A.R., Lowe D., (1990), The optimized internal representation of multilayer classifier networks performs nonlinear discriminant analysis. Neural Networks; 3:367-375.

Werbos P.J., (1974), Beyond regression: New tools for prediction and analysis in the behavioral sciences. Ph.D. thesis, Harvard University, 1974.

West D., (2000), Neural network credit scoring models. Computers and Operations Research; 27:1131-1152.

West P.M., Brockett P.L., Golden L.L., (1997), A comparative analysis of neural networks and statistical methods for predicting consumer choice. Marketing Science; 16:370-391.

Widrow B., Hoff M.E., (1960), Adaptive switching circuits, 1960 IRE WESCON Convention Record, New York: IRE Part 4 1960:96-104.

Widrow B., Rumelhart D.E., Lehr M.A., (1994), Neural networks: applications in industry, business and science, Communications of the ACM; 37:93-105.

Wong B.K., Bodnovich T.A., Selvi Y., (1997), Neural network applications in business: A review and analysis of the literature (1988-1995). Decision Support Systems; 19:301-320.

Young S.S., Scott P.D., Nasrabadi, N.M., (1997), Object recognition using multilayer Hopfield neural network. IEEE Transactions on Image Processing; 6:357-372.

Zhang G.P., (2007), Avoiding Pitfalls in Neural Network Research. IEEE Transactions on Systems, Man, and Cybernetics; 37:3-16.

Zhang G.P., Hu M.Y., Patuwo B.E., Indro D.C., (1999), Artificial neural networks in bankruptcy prediction: general framework and cross-validation analysis. European Journal of Operational Research; 116:16-32.

Zhang G.P., Keil M., Rai A., Mann J., (2003), Predicting information technology project escalation: a neural network approach. European Journal of Operational Research 2003; 146:115–129.

Zhang G.P., Qi M. (2002), "Predicting consumer retail sales using neural networks." In Neural Networks in Business: Techniques and Applications, Smith, K. and Gupta, J.eds. Hershey: Idea Group Publishing, 26-40.

Zhang G.P., Patuwo E.P., Hu M.Y., (1998), Forecasting with artificial neural networks: the state of the art. International Journal of Forecasting; 14:35-62.

Zhang W., Cao Q., Schniederjans M.J., (2004), Neural Network Earnings per Share Forecasting Models: A Comparative Analysis of Alternative Methods. Decision Sciences; 35: 205–237.

Zhu Z., He H., Starzyk J.A., Tseng, C., (2007), Self-organizing learning array and its application to economic and financial problems. Information Sciences; 177:1180-1192.

Granular Computing and Rough Sets - An Incremental Development

Tsau Young ('T. Y.') Lin[1] and Churn-Jung Liau[2]

[1] Department of Computer Science
San Jose State University
San Jose, CA 95192
tylin@cs.sjsu.edu

[2] Institute of Information Science
Academia Sinica, Taipei 115, Taiwan
liaucj@iis.sinica.edu.tw

Summary. This chapter gives an overview and refinement of recent works on binary granular computing. For comparison and contrasting, granulation and partition are examined in parallel from the prospect of rough Set theory (RST).The key strength of RST is its capability in representing and processing knowledge in table formats. Even though such capabilities, for general granulation, are not available, this chapter illustrates and refines some such capability for binary granulation. In rough set theory, quotient sets, table representations, and concept hierarchy trees are all set theoretical, while in binary granulation, they are special kind of pretopological spaces, which is equivalent to a binary relation Here a pretopological space means a spacc that is equipped with a neighborhood system (NS). A NS is similar to the classical NS of a topological space, but without any axioms attached to it[3].

Key words: Granular computing, rough set, binary relation, equivalence relation

22.1 Introduction

Though the label, granular computing is relatively recent, the notion of granulation has in fact been appeared, under different names, in many related fields, such as programming, divide and conquer, fuzzy and rough set theories, pretopological spaces, interval computing, quantization, data compression, chunking, cluster analysis, belief functions, machine learning, databases, and many others. In the past few years, we have seen a renewed and fast growing interest in Granular Computing (GrC). Many applications of granular computing have appeared in fields, such as medicine, economics, finance, business, environment, electrical and computer engineering, a number of sciences, software engineering, and information science.

[3] This is an expansion of the article (Lin, 2005) in IEEE connections, the news letter of the IEEE Computational Intelligence Society

O. Maimon, L. Rokach (eds.), *Data Mining and Knowledge Discovery Handbook*, 2nd ed.,
DOI 10.1007/978-0-387-09823-4_22,

Granulation seems to be a natural problem-solving methodology deeply rooted in human thinking. Many daily "things" have been routinely granulated into sub"things;" human body has been granulated into head, neck, and so forth; geographic features into mountains, planes, and others. The notion is intrinsically fuzzy, vague and imprecise. Mathematicians idealized it into the notion of partitions, and developed it into a fundamental problem-solving methodology; it has played major roles throughout the entire history of mathematics.

Nevertheless, the notion of partitions, which absolutely does not permit any overlapping among its granules, seems to be too restrictive for real world problems. Even in natural science, classification *does* permit small degree of overlapping; there are beings that are both appropriate subjects of zoology and botany. A more general theory is needed.

Based on Zadeh's grand project on granular mathematics, during his sabbatical leave (1996/1997) at Berkeley, Lin focused on a subset of granular mathematics, which he called granular computing (Zadeh, 1998). To stimulate research on granular computing, a special interest group, with T. Y. Lin as its Chair, was formed within BISC (Berkeley Initiative in Soft Computing). Since then, granular computing has evolved into an active research area, generating many articles, books and presentations at conferences, workshops and special sessions. This chapter is devoted to present some of such development over the past few years.

There are two possible approaches: (1) One is starting from fuzzy side and moving down, and (2) the other one is from extreme crisp side and moving up. In this chapter, we take the second approach incrementally. Recall that algebraically a partition is an equivalence relation, so a natural next step is the binary granulation defined by a binary relation. For contrasting, we may call a partition A-granulation and the more general granulation B-granulation.

22.2 Naive Model for Problem Solving

An obvious approach to a large-scaled computing problem is: (1) To divide the problem into subtasks, might be point by point and level by level. (2) To elevate or abstract the problem into concept/knowledge spaces, could be in multilevels. (3) To integrate the solutions of subtasks and quotient tasks (knowledge spaces) of several levels

22.2.1 Information Granulations/Partitions

In the first step, we select an appropriate system of granulation/partition so that only the summaries of granules/equivalence classes may enter into the higher level computing. The information in data space is transformed to a concept space, possibly in levels, which may be locally at each point or globally at eh whole universe (Lin, 2003b). Classically, we granulate by partitioning (no overlapping on granules). Such examples are plentiful: in mathematics (quotient groups, quotient rings and etc. (Birkhoff and MacLane, 1977)), in theoretical computer science (divide-and-conquer (Aho *et al.*, 1974)), in software engineering (the structural, object oriented, and component based design and programming (Szyperski, 2002)), in artificial intelligence (Hobbs, 1985, Zhang and Zhang, 1992), in rough set theory (Pawlak, 1991) among others. However, these are all partition based, where no overlapping of granules is permitted. As we have observed, even in biology, classification does allow some overlapping. The focus of this presentation will be on *non-partition theory*, but only in an epsilon step away from partitioning method.

22.2.2 Knowledge Level Processing and Computing with Words

The information in each granule is summarized and the original problem is re-expressed in terms of symbols, words, predicates or linguistic variables. Such re-expressing is often referred to as knowledge representations. Its processing has been termed computing with symbols (table processing, computing with words, knowledge level processing, even precisiated natural language, depending on the complexity of the representations.

In this chapter, we are computing on the space of granules or "quotient space." in which each granule is represented by a word that carries different degree of semantics. For partition theory, the knowledge representation is in table format (Pawlak, 1991) and its computation is syntactic in nature. For binary granulation, that we have focused here, is semantic oriented. We expand and streamline the previous works (Lin, 1998a, Lin, 1998b, Lin, 2000); the main idea is to transfer the computing with words into computing with symbols.

Loosely speaking computing with symbols or symbolic computing is an "axiomatic" Computing: all rules of computing symbols are determined by the axioms. The computation follows the formal specifications. Such computing occurs only in an ideal situation. In many real world applications, unfortunately, such as non-linear computing, the formal specifications are often unavailable. So computing with words are needed; it can be processed informally. Semantics of words often may not be completely or precisely formalized. Their semantic computing is often carried out in the systems with human helps (the semantics of symbols are not implemented). Human enforced semantic computing are common in data processing environment.

22.2.3 Information Integration and Approximation Theory

Most applications require the solutions be presented in the same level as input data. So the solutions often need to be integrated from subtasks (solutions in granules) and quotient tasks (solutions in the spaces of granules). For some applications, such as Data Mining and some rough set theory, are aimed at high level information; in such cases this step can be skipped. In general, the integration is not easy. In partition world, many theories have been developed in mathematics; e. g., extension functors. The approximation theory of pretopological spaces and rough set theory can be regarded as in this step.

22.3 A Geometric Models of Information Granulations

For understanding the general idea, in this section, we recall and refine a previous formalization in (Lin, 1998a). The goal is to formalize Zadeh's informal notion of granulation mathematically.

As original thesis is informal, the best we could do is to present, hopefully, convincing arguments. We believe our formal theory is very close to the informal one. According to Zadeh (1996):

> Information granulation involves partitioning a class of objects(points) into granules, with a granule being a clump of objects (points) which are drawn together by indistinguishability, similarity or functionality.

We will literally take Zadeh's informal words as a formal definition of granulation. We observe that:

1. A granule is a group of objects that are draw together (by indistinguishability, similarity or functionality).
   The phrase "drawn together" implicitly implies certain level of symmetry among the objects in a granule. Namely, if p is drawn towards q, then q is also drawn towards p.
   Such symmetry, we believe, is imposed by imprecise-ness of natural language. To avoid such an implications, we will rephrase it to "drawn towards an object p," so that it is clear the reverse may or may not be true. So we have first revision:
2. A granule is a group $B(p)$ of objects that are draw toward an object p. Here p varies through every object in the universe.
3. Such an association between object p and a granule $B(p)$ induces a map from the object space to power set of object space. This map has been called a binary granulation (BG).
4. Geometric View:
   We may use geometric terminology and refer to the granule as a neighborhood of p, and the collection $\{B(p)\}$ a binary neighborhood system (BNS). It is possible that $B(p)$ is an empty set. In this case we will simply say p has no neighborhood (abuse of language; to be very correct, we should say p has an empty neighborhood). Also it is possible that different points may have the same neighborhood (granule) $B(p) = B(q)$. The set of all q, where $B(q)$ is equal to $B(p)$, is called the centers $C(p)$ of $B(p)$.
5. Algebraic View:
   Consider the set $R = \{(p,u)\}$, where u in $B(p)$ and p in U. It is clear that R is a subset of $U \times U$, hence defines a binary relation (BR), and vice versa.

Proposition 1 *A binary neighborhood system (BNS), A binary granulation (BG), and a binary relation (BR) are equivalent.*

From the analysis given above, we propose the following mathematical model for information granulation.

Definition 1 *By a (single level) information granulation defined on a set U we mean a binary granulation (binary neighborhood system, binary relation) defined on U.*

Let us goes a little bit further. Note that the binary relation is a mathematical expression of Zadeh's "indistinguishability, similarity or functionality." We abstract the three properties into a list of abstract binary relations $\{B_j \mid j$ run through some index set $\}$, where each B_j is a binary relation.

Note that at each point p, each B_j induces a neighborhood $B_j(p)$. Some may be empty, or identical. By removing empty set and duplications, the family have been we re-indexed $N_i(p)$. As in the single level case, we will define directly the granulation

$$N : U \to 2^{2^U}; p \mapsto \{B_i(p) \mid i \text{ run through some index set }\}.$$

The collection $\{B_i(p)\}$ is called a neighborhood system(NS)or (LNS); the latter one is used to distinguish itself from the neighborhood system (TNS) of a topological space (Lin, 1989a, Lin, 1992).

Definition 2 *By a local multi-level information granulation defined on U, we mean a neighborhood system (NS) is defined on U. By a global multi-level information granulation defined on U, we mean a set of BG is defined on U.*

All notions can be fuzzified. The right way to look at this section is to assume implicitly there is a modifier "crisp/fuzzy" to all notions presented above.

22.4 Information Granulations/Partitions

Technically, granular computing is actually computing with constraints. Especially in "infinite world", granulation is often given in terms of constraints. In this chapter, we concerns primarily with constraints that are mathematically represented as binary relations

22.4.1 Equivalence Relations(Partitions)

Partition is a decomposition of the universe into a family of disjoint subsets. They are called equivalence classes, because a partition induces an equivalence relation and vice versa. In this chapter, we will view the equivalence class in a special way. Let $A \subseteq U \times U$ be an equivalence relation (a reflexive, symmetric and transitive binary relation). For each p, let

$$A_p = \{v \in U : pAv\} \tag{22.1}$$

A_p is the equivalence class containing p, and will be called A-granule for the purpose of contrasting with general cases. Elements in A_p are equivalent to each other. Let us summarize the discussions in:

$$A : U \to 2^U : p \mapsto A_p \tag{22.2}$$

Proposition 2 *An equivalence relation on U $\Leftrightarrow$ a partition on U*

In RST, the pair (U,A) is called an approximation space and its topological properties are studied.

22.4.2 Binary Relation (Granulation) - Topological Partitions

In (Lin, 1998b), we observe that there is a derived partition for each BNS, that is, the map $B : V \to 2^U; p \mapsto B(p)$ induces a partition on V; the equivalence class $C(p) = B^{-1}(B(p))$ is the center of $B(p)$. In the case $V = U$, the $B(p)$ is the neighborhood of $C(p)$, and $C(p)$ consists of all the points that have the same neighborhood. So $B(p) = B((C(p))$. We observe that $\{C(p)\}$ is a partition. Since each $B(p)$ is a neighborhood of the set $C(p)$. The quotient set is a BNS (Lin, 1989a). We will call the collection of $C(p)$ topological partition with the understanding that there is a neighborhood $B(p)$ for each equivalence class $C(p)$. The neighborhoods capture the interaction among equivalence classes (Lin, 2000).

22.4.3 Fuzzy Binary Granulations (Fuzzy Binary Relations)

In (Lin, 1996), we have discussed various fuzzy sets. In this chapter, a fuzzy set is uniquely defined by its membership function. So a fuzzy set is a w-sofset, if we use the language of the cited paper.

A fuzzy binary relation is a fuzzification of a binary relation. Let I be the unit interval [0, 1]. Let FBR be a fuzzy binary relation, that is, there is a membership function: $FBR : V \times U \to I : (p,u) \mapsto r$. For each $p \in V$, there is a fuzzy set whose membership function $FM_p : U \to I$ is defined by $FM_p(u) = FBR(p,u)$, we call FM_p a fuzzy binary neighborhood/set.

Again, we can view the idea geometrically. We assume a fuzzy binary neighborhood system (FBNS) is imposed for V on U. For each object $p \in V$, we associate a fuzzy subset, denoted by $FB(p) \subseteq U$. In other words, we have a map $FB : V \to FZ(U) : p \mapsto FB(p)$, where $FZ(U)$ means all fuzzy subsets on U. $FB(p)$ is called a fuzzy binary neighborhood and FB a fuzzy binary granulation (FBG) and the collection $\{FB(p) | p \in V\}$ a fuzzy binary neighborhood system (FBNS).

It is clear that given a map FB, there is a binary relation FBR such that $FM_p = FB(p)$. So as in crisp cases, from now on we will use algebraic and geometric terms interchangeably. FB, FBNS, FBG, and FBR are synonyms.

22.5 Non-partition Application - Chinese Wall Security Policy Model

In 1989 IEEE Symposium on Security and Privacy, Brewer and Nash (BN) proposed a very intriguing security model, called Chinese Wall Security Policy (CWSP) model. Intuitively BN's idea was to build a family of impenetrable walls, called Chinese Walls, among the datasets of competing companies so that no datasets that are in conflict can be stored in the same side of Chinese Walls; this is BN's requirements and will be called *Aggressive (Strong) Chinese Wall Security Policy (ACWSP) Model.*

The methods are based on the formal analysis of the binary relations (CIR) of conflict of interests. Roughly, BN granulated the data sets by CIR and assumed the granulation was a partition. CIR is rarely an equivalence relation, for example, a company cannot be self conflicting; so reflexivity can never met by CIR. So a modified model, called an aggressive Chinese Wall Security Policy model (ACWSP) is proposed (Lin, 1989b). However, in that paper, the essential strength of **ACWSP** model had not brought out. With recent development in GrC, **ACWSP** model was refined (Lin, 2003a), and successfully captured the intuitive intention of BN "theory."

CWSP Model is essentially a Discretionary Access Control Model (DAC). The central notion of DAC is that owner of an object has discretionary authority on the access rights of that objects. The owner X of the dataset x may grant the read access of x to a user Y who owns a dataset y. The use Y may make a copy, Copy-of-x, in y. Even in the strict DAC model, this is permissible (Osborn*et al.*, 2000)). We have summarized the above grant access procedure, including making a copy, as a direct information flow (DIF) from X or x to Y or y respectively.

Let O be the set of all objects (corporate data),X and Y are typical objects in O. $CIR \subseteq O \times O$ represents the binary relation of conflict of interests. We will consider the following properties:

- CIR-1: CIR is symmetric.
- CIR-2: CIR is anti-reflexive.
- CIR-3: CIR is anti-transitive.

22.5.1 Simple Chinese Wall Security Policy

In (Brewer and Nash, 1988), Section "Simple Security", p. 207, BN asserted that "*people are only allowed access to information which is not held to conflict with any other information that they already possess.*" So if $(X,Y) \notin$ CIR, then X and Y could be assigned to one single agent. So we assume that information in X and Y have been disclosed to each other (since one agent knows both). So outside of CIR-class, there are direct information flows between any two objects.

Definition 3 *Simple CWSP : Direct Information Flow (DIF) may flow between X and Y if and only if* $(X,Y) \notin CIR$*,*

Simple CWSP is a requirement on DIF, it does not prevent information flow between X and Y indirectly. So we need composite information flow (CIF). By a CIF, we mean information flow between X and Y via a sequence of DIF's. An information flow from X to Y is called a *malicious Trojan horse*, if Simple CWSP is imposed on X and Y.

Definition 4 *(Strong) ACWSP: CIF may flow between X and Y if and only if* $(X,Y) \notin$ *CIR,*

Next, let us quote a theorem from (Lin, 2003a).

Theorem 1 Chinese Wall Security Theorem, *If CIR is symmetric, anti-reflexive and anti-transitive, then Simple CWSP implies (Strong) ACSWP.*

22.6 Knowledge Representations

At the current states, knowledge representations are mainly in table or tree formats. So the knowledge level processing is basically table processing. The main works, we will present here is the extension of the representation theory of equivalence relations to binary relations.

22.6.1 Relational Tables and Partitions

(Pawlak, 1982) and (Lee, 1983) observed that: A relational table is a knowledge representation of a universe of entities. Each column induces a partition on the universe; n columns induce n partitions. Here, we will explore the converse. How could we represent a finite set of partitions? The central idea is to assign meaningful name (a summary) to each equivalence class (Lin, 1998a, Lin, 1998b, Lin, 1999b).

We will illustrate the idea by example: Let $U = \{id_1, id_2, \ldots, id_9\}$ be a set of 9 balls with two partitions:

(1) $\{\{id_1, id_2, id_3\}, \{id_4, id_5\}, \{id_6, id_7, id_8, id_9\}\}$
(2) $\{\{id_1, id_2\}, \{id_3\}, \{id_4, id_5\}, \{id_6, id_7, id_8, id_9\}\}$

We name the first partition COLOR, (because it is the best summarization of the given partition from physical inspection).

COLOR = Name$(\{\{id_1, id_2, id_3\}, \{id_4, id_5\}, \{id_6, id_7, id_8, id_9\}\})$

Next, we will name each equivalence class to reflect its characteristic. We name the first equivalence class

Red = Name$(\{id_1, id_2, id_3\})$,

because each ball of this group has red color (appears to human). Note that this name reflects human's observation and meaningful to human only; its meaning (such as light spectrum) is not implemented or stored in the system. In AI, the term COLOR or Red are called semantic primitive (Barr and Feigenbaum, 1981). The same intent leads to the following names

Orange = Name$(\{id_4, id_5\})$
Yellow = Name$(\{id_6, id_7, id_8, id_9\})$

Next, we give names to the second partition, again by its characteristics (appear to human):

WEIGHT = Name($\{\{id_1, id_2\}, \{id_3\}, \{id_4, id_5\}, \{id_6, id_7, id_8, id_9\}\}$)
W1 = Name($\{id_1, id_2\}$)
W2 = Name($\{id_3\}$)
W3 = Name($\{id_4, id_5\}$)
W4 = Name($\{id_6, id_7, id_8, id_9\}$)

Base on these names, we have Table 22.1:

Table 22.1. Constructing an Information table by naming each partition and equivalence class

U	COLOR	WEIGHT
id_1	Red	W1
id_2	Red	W1
id_3	Red	W2
id_4	Orange	W3
id_5	Orange	W3
id_6	Yellow	W4
id_7	Yellow	W4
id_8	Yellow	W4
id_9	Yellow	W4

The first tuple can be interpreted as follows: the first ball belongs to the group that is labeled Red, and another group whose weight is labeled W1. We can do the same for rest of the tuples. This table is a classical bag relation.

The goal of this chapter is to generalize this naming methodology to general granulations. The word-representation of partitions is a very clean representation; each name (word) represents an equivalence class uniquely and independently. In next section, we will investigate the representations of binary relations, in which names have overlapping semantics.

22.6.2 Table Representations of Binary Relations

Real world granulation often cannot be expressed by equivalence relations. For example, the notions of "near","similar", and "conflict" are not equivalence relations. So there are intrinsic needs to generalize the theory of partition (RST) to the theory of more general granulation (granular computing). In this section, we will explain how to represent a finite set of binary granulations (binary relations) into a table format. So we can extend the relational theory from partitions to binary granulations. Most of the results are recall and refinements of the results observed in (Lin, 1998a, Lin, 1998b, Lin, 1999b, Lin, 2000).

The representation of a partition is rested on two properties:

(a) Each object p belongs to an equivalence class (the union of equivalence class covers the whole universe)
(b) No object belongs to two equivalence classes (equivalence class are pairwise disjoint)

The important question is: Does the family of binary granules have the same properties as equivalence classes? Obviously, a granulation does satisfy (a), but not (b), because granules may overlap each other. We need a different way to look at the problem: we restate the two properties into the following form:

- Each object belongs to one and only one equivalence class

If we assign each equivalence class a meaningful name, then each object is associated with a unique name (attribute value). Such an assignment construct one column of the table representation. Each equivalence relation get a column. So n equivalence relations construct a table of n columns.

With these observations, we can state a similar property for the binary granulation. Let B be a binary granulation

- Each object, $p \in V$, is assigned to one and only one B-granule $B_p \in 2^U$; $B : p \mapsto B_p$.

If we assign each B-granule a meaningful name, then each object is associated with a unique name (attribute value).

$$p(\in V) \xrightarrow{B} B_p(\in 2^U) \xrightarrow{\text{Name}} \text{Name}(B_p)(\in Dom(B)) \tag{22.3}$$

$$p \to \text{Name}(B_p)(\in Dom(B)) \tag{22.4}$$

Such an association allows us to represent

- a finite set of binary granulations by a "relational table", called granular table.

Note that we did not use the relationships "$\in$". Instead, we use the *assignment of neighborhoods (binary granules).*

We will illustrate the idea by modifying the last example. In binary granulation each p is associated with a unique binary neighborhood B_p. The following neighborhoods are given.

$B_{id_1} = B_{id_2} = B_{id_3} = \{id_1, id_2, id_3, id_4, id_5\}$
$B_{id_4} = B_{id_5} = \{id_1, id_2, id_3, id_4, id_5, id_6, id_7, id_8, id_9\}$
$B_{id_6} = B_{id_7} = B_{id_8} = B_{id_9} = \{id_4, id_5, id_6, id_7, id_8, id_9\}$.

By examining the characteristic of each binary neighborhood, we assign their names as follows:

Having-RED =Name(B_{id_1})=Name(B_{id_2})=Name(B_{id_3})
Having-RED+YELLOW =Name(B_{id_4})= Name(B_{id_5})
Having-YELLOW =Name(B_{id_6})=Name(B_{id_7})=Name(B_{id_8})= Name(B_{id_9})

For illustration, let us trace the journey of id_1: It is an object of V, and is moved to a subset, B_{id_1}, then stop at the name, Having-RED, in notation,

$$id_1 \ {}^{B}\!\!\rightarrow \ B_{id_1} \ {}^{\text{Name}}\!\!\rightarrow \ \text{Having-RED}.$$

By tracing every object of V, we get the second column of Table 22.2. For the third column, we use the same partition and naming scheme as in the previous section; so the third column is exactly the same as that in Table 22.1. The results are shown in Table 22.2.

Table 22.2. Granular table: Construct granular table by naming each binary granulations and binary granules

BALLs	Granulation 1	Granulation 2
id_1	Having-RED	W1
id_2	Having-RED	W1
id_3	Having-RED	W2
id_4	Having-RED+YELLOW	W3
id_5	Having-RED+YELLOW	W3
id_6	Having-YELLOW	W4
id_7	Having-YELLOW	W4
id_8	Having-YELLOW	W4
id_9	Having-YELLOW	W4

Perhaps, we should stress again that attribute values have overlapping semantics. The constraints among these words have to be properly handled. So, let us examine the "interactions" among attribute values of COLOR. Two attribute values, Having-RED and Having-RED+YELLOW, obviously have overlapping semantics. We need some preparations. We need one more concept, namely, the center

$$C_w = B^{-1}(B_p), \tag{22.5}$$

where w=Name(B_p). Verbally, C_w consists of all objects that have the same B-granule B_p. We use the granule's names to index the centers:

$$\begin{aligned} C_{\text{Having-RED}} &\equiv \text{Center of } B_{id_1} = \text{Center of } B_{id_2} \\ &= \text{Center of } B_{id_3} = \{id_1, id_2, id_3\} \\ C_{\text{Having-RED+YELLOW}} &= \text{Center of } B_{id_4} - \text{Center of } B_{id_5} \\ &= \{id_4, id_5\} \\ C_{\text{Having-YELLOW}} &\equiv \text{Center of } B_{id_6} = \text{Center of } B_{id_7} \\ &= \text{Center of } B_{id_8} = \text{Center of } B_{id_9} \\ &= \{id_6, id_7, id_8, id_9\} \end{aligned}$$

Now, we will define the binary relation B_{COLOR} in terms of BNS. First we observe that B_{COLOR} is reflexive, so we define the "other" points only. With a slight abuse of notation, we also denote B_{COLOR} by B. Let $w, u \in$ {Having-RED, Having-RED+YELLOW, Having-YELLOW}, then:

$$w \in B_u \Leftrightarrow \forall p \in C_u, B_p \cap C_w \neq \emptyset \Leftrightarrow \exists p \in C_u, B_p \cap C_w \neq \emptyset.$$

Thus, for example, we have: Having-RED+YELLOW $\in B_{\text{Having-RED}}$ since: $B_{id_1} \cap C_{\text{Having-RED+YELLOW}} \neq \emptyset$ and: $id_i \in C_{\text{Having-RED}}$. Analogously, we have: Having-RED $\in B_{\text{Having-RED+YELLOW}}$ etc.

Thus we have defined all B-granules. These B-granules defines a binary relation on the COLOR column, which is displayed in Table 22.3

Table 22.3. A Binary Relation on COLOR

Having-RED	Having-RED
Having-RED	Having-RED+YELLOW
Having-RED+YELLOW	Having-RED
Having-RED+YELLOW	Having-RED+YELLOW
Having-RED+YELLOW	Having-YELLOW
Having-YELLOW	Having-RED+YELLOW
Having-YELLOW	Having-YELLOW

Note that such a binary structure cannot be deduced from the table structure. We are ready to introduce the notion of semantic property.

Definition 5 *A property is said to be semantics if and only if it is not implied by the table structure. A property is said to be syntactic if and only if it is implied by the table structure.*

The binary relation (Table 22.3) is not derived from the table structure (of Table 22.2) so it is a semantic property. This type of tables has been studied in (Lin, 1988, Lin, 1989a) for approximate retrievals; and is called topological relations or tables. Formally,

Definition 6 *A table (e.g. Table 22.2) whose attributes are equipped with binary relations (e.g. Table 22.3 for COLOR attribute) is called a topological relation.*

22.6.3 New representations of topological relations

In (Lin, 2000), the granular table is transformed into topological information table. Here we will give a hew view and a refinement. By replacing the name of binary granule with centers in Table 22.2 and 22.3, we have Table 22.4 and Table 22.5; they are isomorphic. Table 22.5 provides the topology of Table 22.4. Table 22.4 and 22.5 provide a better interpretation than that of Table 22.2 and 22.3.

Table 22.4. Topological Table

BALLs	Granulation 1	Granulation 2
id_1	$C_{\text{Having-RED}}$	W1
id_2	$C_{\text{Having-RED}}$	W1
id_3	$C_{\text{Having-RED}}$	W2
id_4	$C_{\text{Having-RED+YELLOW}}$	W3
id_5	$C_{\text{Having-RED+YELLOW}}$	W3
id_6	$C_{\text{Having-YELLOW}}$	W4
id_7	$C_{\text{Having-YELLOW}}$	W4
id_8	$C_{\text{Having-YELLOW}}$	W4
id_9	$C_{\text{Having-YELLOW}}$	W4

Table 22.5. A Binary Relation on the Centers of COLOR

$C_{\text{Having-RED}}$	$C_{\text{Having-RED}}$
$C_{\text{Having-RED}}$	$C_{\text{Having-RED+YELLOW}}$
$C_{\text{Having-RED+YELLOW}}$	$C_{\text{Having-RED}}$
$C_{\text{Having-RED+YELLOW}}$	$C_{\text{Having-RED+YELLOW}}$
$C_{\text{Having-RED+YELLOW}}$	$C_{\text{Having-YELLOW}}$
$C_{\text{Having-YELLOW}}$	$C_{\text{Having-RED+YELLOW}}$
$C_{\text{Having-YELLOW}}$	$C_{\text{Having-YELLOW}}$

Theorem 2 *Given a finite binary relation B, a finite equivalence relation A can be induced. The knowledge representation of B is a topological representation of A.*

22.7 Topological Concept Hierarchy Lattices/Trees

We will examine a nested sequence of binary granulations; the essential ideas is in (Lin, 1998b, Lin, 2000). Each inner layer is strongly dependent on the immediate next outer layer (Section 22.8.2).

22.7.1 Granular Lattice

Let us continue on the same example: Each ball in U has a B-granule. Balls 1, 2, 3 have the same B-granule; it is labeled H-Red (abbreviation of Having-Red). Similarly, Balls 4, 5 have H-Red+Yellow, and Balls 6, 7 have H-Yellow.

The nested sequence (length) is display in Figure 22.1 as a tree:

The first generation children:

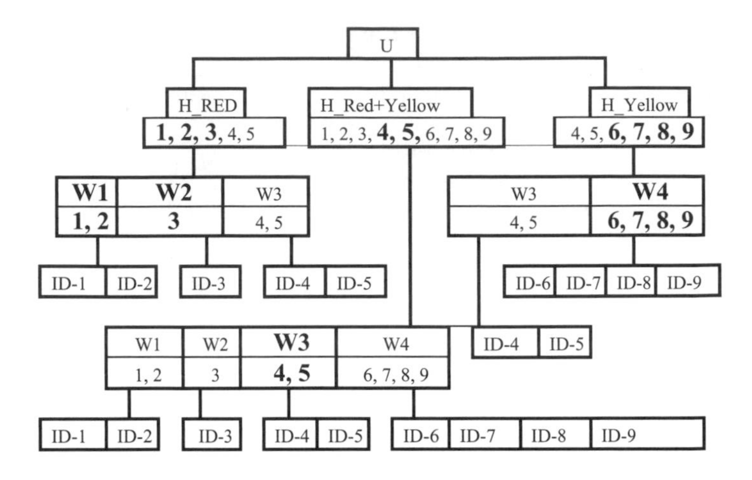

Fig. 22.1. In 2nd layer the bold print letters are in the centers.

1. *U* is granulated into three distinct children; they are named Having-Red Having-Red+Yellow, Having-Yellow; they are abbreviated to H-Red, H-Red+Yellow, and H-Yellow.
2. The three children are distinct, but *not* independent; their meanings have overlapping. Namely (1) there are interaction between H-Red+Yellow and H-Red+Yellow; (2) between H-Red+Yellow and H-Yellow; (3) there are NO interactions between H-Red and H-Yellow; The interactions are recorded in Table 22.3. This explains how the first level children are produced.
3. Every child has a center: the centers are $C_{\text{H-RED}}$ (abbreviation of $C_{\text{Having-RED}}$), $C_{\text{H-RED+Yellow}}$, $C_{\text{H-Yellow}}$. Centers are pairwise disjoint; they forms a partition.

The second generation children: Since COLOR-granulation strongly depends on WEIGHT-granulation, each COLOR-granule is a union of WEIGHT-granules. Thus one can regard that these WEIGHT-granules forms a granulation of this COLOR granule, so

1. H-Red (a COLOR-granule) is granulated into WEIGHT-granules, W1, W2, W3. Note that within each COLOR-granule the WEIGHT-granules are disjoint, so "granulated" is "partitioned."
2. H-Red+Yellow is granulated into W1, W2, W3, W4,
3. H-Yellow is granulated into W3, W4. This explains how the second level children are produced. We need information about the centers.
4. Since WEIGHT-granulation is a partition, the center is the same as granule.

Some Lattice Paths

1. $U \to$ H-Red $\to W1 \to id_1$
2. $U \to$ H-Red $\to W1 \to id_2$
3. $U \to$ H-Red $\to W2 \to id_3$
4. $U \to$ H-Red+YELLOW $\to W1 \to id_1$. This path has the same beginning and ending with Item 1; but the two paths are distinct.
5. $U \to$ H-Red+YELLOW $\to W1 \to id_2$;compare with Item 2.
6. $U \to$ H-Red+YELLOW $\to W2 \to id_3$; compare with Item 3.
7. $U \to$ H-Red+YELLOW $\to W3 \to id_4$
8. etc

22.7.2 Granulated/Quotient Sets

1. The children consists of three (overlapping) subsets, H-Red, H-Red+Yellow, H-Yellow. This collection is more than a classical set; there are interactions among them; It forms a BNS-space; see Table 22.3.
2. The grand children:
   a) Children of the first child $\{W1, W2, W3\}$ forms a classical set.
   b) Children of the second child $\{W1, W2, W3, W4\}$ forms a classical set.
   c) Children of the third child: $\{W3, W4\}$ forms a classical set.
3. Three distinct classical sets do have non-empty intersections.

Note that since WEIGHT-granulation is a partition, so the grand children under each individual child are disjoint. However, the grand children do overlap. The quotient set (of quotient set)

$$\begin{aligned}&\{\text{H-Red, H-Red+Yellow, H-Yellow}\}\\ =&\{\{\mathrm{W1,W2,W3}\},\{\mathrm{W2,W3,W4}\},\{\mathrm{W3,W4}\}\}\\ =&\{\{\{id_1,id_2\},\{id_3\},\{id_4,id_5\}\},\{\{id_3\},\{id_4,id_5\},\\ &\{id_6,id_7,id_8,id_9\}\},\{\{id_4,id_5\},\{id_6,id_7,id_8,id_9\}\}\}\end{aligned}$$

22.7.3 Tree of centers

In a granular lattice, children of every generation may overlap. Could we improve the situation? In deed, if we consider the centers only, then lattice becomes a tree (Figure 22.1a; observe the bold prints nodes).

1. The children consists of three (non-overlapping) subsets:
   a) $C_{H-Red} = \{id_1, id_2, id_3\}$,
   b) $C_{H-Red+Yellow} = \{id_4, id_5\}$,
   c) $C_{H-Yellow} = \{id_6, id_7, id_8.id_9\}$.
   They froms a classical set.
2. The grand children:
   a) Children of the first child: $W1 = \{id_1 id_2\}$, $W2 = \{id_3\}$.

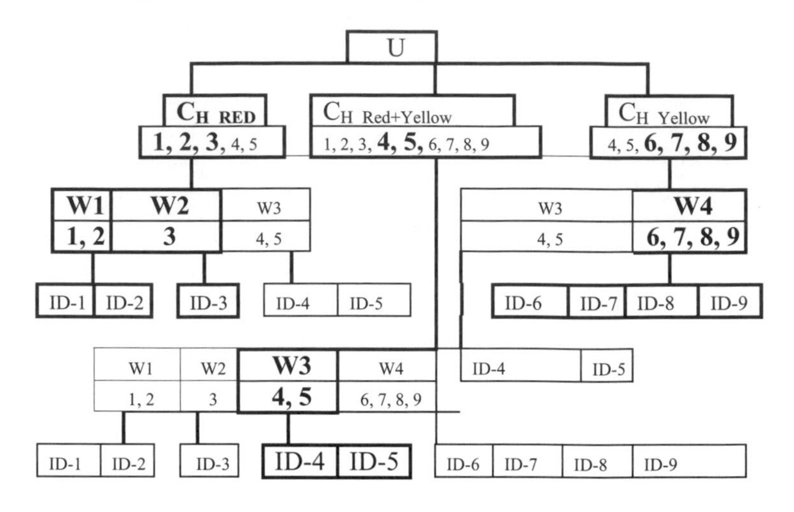

Fig. 22.2. A. Bold print letters are the centers (W*i* is its own center).

   b) Children of the second child $W3 = \{id_4, id_5\}$.
   c) Children of the third child: $W4 = \{id_6, id_7, id_8.id_9\}$.
3. The centers of each layers are disjoints; they forms a honest tree.

22.7.4 Topological tree

We will combine two trees in Figure 22.1 into one (with no information lost). We will take the tree of centers as the topological tree. Each node of the tree of centers is equipped with a *B*-granule (neighborhood), which is the corresponding node of the granular tree.

Here are the COLOR-neighborhoods of the centers of the first generation children:

- The neighborhood of $C_{\text{H-Red}}(= \{1,2,3\})$ is H-Red (= {1,2,3,4,5})
- The neighborhood of $C_{\text{H-Red+Yellow}}(= \{4,5\})$ is H-Red+Yellow (={1,2,3,4,5,6,7, 8,9})
- The neighborhood of $C_{\text{H-Yellow}}(= \{6,7,8,9\})$ is H-Yellow (={4,5,6,7,8,9})

For second generation, the WEIGHT-neighborhoods are:

- The neighborhood of $C_{W1} = W1 = \{W1, W2, W3\}$
- The neighborhood of = $C_{W2} = W2 = \{W1, W2, W3\}$
- The neighborhood of = C_{W3}= {W1, W2, W3, W4}
- The neighborhood of = C_{W4}= {W3, W4}

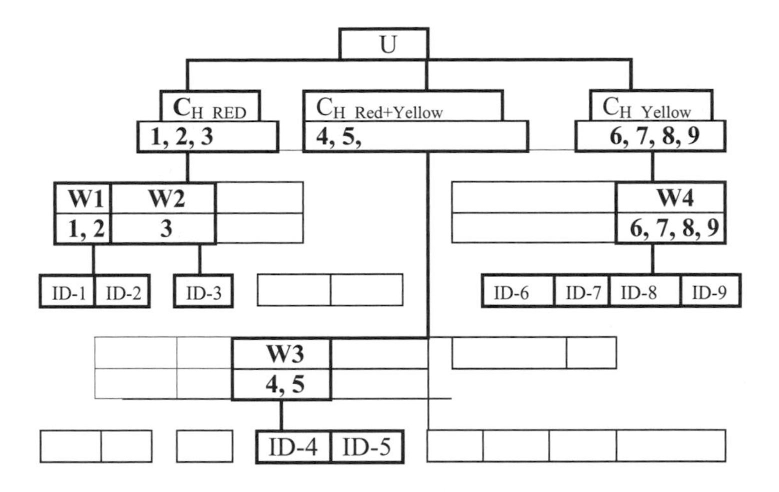

Fig. 22.3. B. The tree of centers.

22.7.5 Table Representation of Fuzzy Binary Relations

We will use a very common example to illustrate the idea. Let the universe be $V = \{0.1, 0.2, \ldots, 0.8, 0.9\}$. It contains 9 ordinary real numbers. Each number is associated with a special fuzzy set, called a fuzzy number (Zimmerman, 1991). For example, in Figure 22.4 the numbers, 01, 02, 03, and 0.4 are respectively associated with fuzzy numbers $N1, N2, N3$ and $N4$.

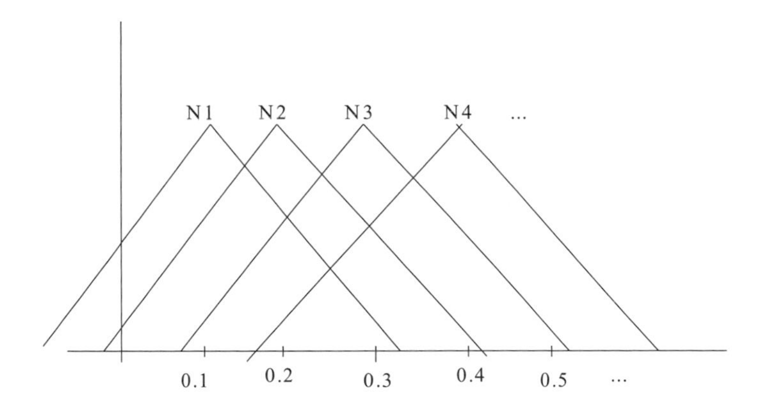

Fig. 22.4. Illustration of Fuzzy Numbers Association.

Table 22.6. Fuzzy Numbers

Points x	FB-granule	Name
0.1	$N1$	Fuzzy number 0.1=Name($N1$)
0.2	$N2$	Fuzzy number 0.2=Name($N2$)
0.3	$N3$	Fuzzy number 0.3=Name($N3$)
0.4	$N4$	Fuzzy number 0.4=Name($N4$)
...	...	...
0.9	$N9$	Fuzzy number 0.9=Name($N9$)

22.8 Knowledge Processing

Pawlak (Pawlak, 1991) interprets equivalent relations as knowledge and develop a theory. In this section, we will explain how to extend his view to binary relations (Lin, 1996,Lin, 1998a,Lin, 1998b,Lin, 1999a,Lin, 1999b,Lin, 2000,Lin and Hadjimichael, 1996, Lin *et al.*, 1998). To explain these concepts, we are tempted to use the same knowledge-oriented terminology. However, our results are not completely the same; after all, binary relations are not necessarily equivalence relations. We need to distinguish the differences, so mathematical terminology is used. Unless the intuitive support is needed, knowledge-oriented terms will not be employed.

22.8.1 The Notion of Knowledge

Pawlak views partitions (classification) as knowledge, and calls a finite set of equivalence relations on a given universe a knowledge base (Pawlak, 1991). He interprets refinements of equivalence relations as knowledge dependencies. We will take a stronger view: we regard the interpretations as the integral part of the knowledge. Here an interpretation means the naming of the mathematical structures based on real world characterization; the name is a summarization. Pawlak regards two isomorphic tables possess same knowledge (since they have the same knowledge base), however, we regard them as distinct knowledge. Let us summarize the discussions in a bullet:

- knowledge includes the knowledge representation (human interpretation) of a mathematical structure; it is a semantic notion.

For convenience, let us recall the notion of binary granular structures (Lin, 2000, Lin, 1998a, Lin, 1998b). It consists of 4-tuple

$$(V,U,B,C)$$

where V is called the object space, U the data space (V and U could be the same set), B is a set of finitely many crisp/fuzzy binary granulations, and C is the concept space which consists of all the names of B-granulations and granules. For us a piece of knowledge is a 4-tuple, while Pawlak only looks at the first three items (his definition of knowledge base).

22.8.2 Strong, Weak and Knowledge Dependence

Let B, P and Q be binary relations (binary granulations) for V on U (e.g. $B \subseteq V \times U$). Then we have the following:

Definition 7

1. *A subset $X \subseteq U$ is B-definable, if X is a union of B-granules B_p's. If the granulation is a partition, then a B-definable subset is definable in the sense of RST.*
2. *Q is strongly dependent on P, denoted by $P \Rightarrow Q$ if and only if every Q-granule is P-definable.*
3. *Q is weakly depends on P, denoted by $P \rightarrow Q$ if and only if every Q-granule contains some P-granule.*

We will adopt the language of partition theory to granulation. For $P \Rightarrow Q$, we will say P is finer than Q or Q is coarser than P. Write $Y_p = \text{Name}(Q_p)$ and $X_{p_i} = \text{Name}(P_{p_i})$. Since $Q_p = \cup_i P_{pi}$ for suitable choices of $p_i \in V$, we write informally

$$Y_p = X_{p_1} \vee X_{p_2} \vee \cdots$$

Note that Y_p and X_{p_i} are words and $\vee$ is the "logical" disjunction. So, this is a "formula" of informal logic. Formally, we have the following proposition.

Proposition 3 *If $P \Rightarrow Q$, then there is a map from the concept space of P to that of Q. The map f can be expressed by $Y_p = f(X_{p_1}, X_{p_2}, \ldots) = X_{p_1} \vee X_{p_2} \vee \cdots$; f will be termed knowledge dependence.*

This proposition is significant, since Name(P_p) is semantically interrelated. It implies that the semantic constraints among these words Name(P_p)'s are carried over to those words, Name(Q_p)'s consistently. Such semantic consistency among columns of granular tables allows us to extend the operations of classical information tables to granular tables.

22.8.3 Knowledge Views of Binary Granulations

Definition 8

1. *Knowledge P and Q are equivalent, denoted by $P \equiv Q$, if and only if $P \Rightarrow Q$ and $Q \Rightarrow P$*
2. *The intersection of P and Q, $P \wedge Q$, is a binary relation defined by*

$$(v,u) \in P \wedge Q \text{ if and only if } (v,u) \in P \text{ and } (v,u) \in Q$$

3. *Let $C = \{C_1, C_2, \ldots, C_m\}$ and $D = \{D_1, D_2, \ldots, D_n\}$ be two collections of binary relations. We write $C \Rightarrow D$, if and only if $C_1 \wedge C_2 \wedge \cdots \wedge C_m \Rightarrow D_1 \vee D_2 \vee \cdots \vee D_n$. By mimicking ((Pawlak, 1991), chapter 3), we write $IND(C) = C_1 \wedge C_2 \wedge \cdots \wedge C_m$; note that, all of them are binary relations, not necessarily equivalence relations.*

4. *C_j is dispensable in C if $IND(C) = IND(C - \{C_j\})$; otherwise C_j is indispensable.*
5. *C is independent if each $C_j \in C$ is indispensable; otherwise C is dependent.*
6. *S is a reduct of C if S is an independent subset of C such that $IND(S) = IND(C)$.*
7. *The set of all indispensable relations in C is called a core, and denoted by $CORE(C)$.*
8. *$CORE(C) = \cap RED(C)$, where RED is the set of all reducts in C.*

Corollary 1 *$P \wedge Q \Rightarrow P$ and $P \wedge Q \Rightarrow Q$.*

The fundamental procedures in table processing are to find cores and reducts of decision table. We hope readers are convinced that we have developed enough notions to extend these operations to granular tables.

22.9 Information Integration

Many applications would want the solutions be in the same level as input data. So this section is actually quite rich. There are many theories dedicated to this portion in mathematics. For example, suppose we know a normal subgroup and the quotient group of an unknown group, there is a theory to find this unknown group. For Data Mining and part of RST, the interests are on the high level information, so this step can be skipped. For RST, approximations are the only relevant part. In this section, we focus only on the approximation theory of granulations.

22.9.1 Extensions

Let $Z_4 = \{[0],[1],[2],[3]\}$ be the set of integers mod 4 and we will consider it as a commutative group (Birkhoff and MacLane, 1977). Next we consider a subgroup {[0],[2]} which is equivalent (isomorphic) to integer mod 2, Z_2, and its quotient group that consists of two elements, {[0],[2]} and {[1],[3]} and is also isomorphic to integer mod 2. The question is if we know the subgroup (subtasks) and the quotient group (quotient tasks), can we found the original universe. The answer is we have two universe, one is Z_4 and another is the Cartesian product of Z_2 by Z_2. So integration is not-trivial and is, outside of mathematics, unexplored teritory.

22.9.2 Approximations in Rough Set Theory (RST)

Let A be an equivalence relation on U. The pair (U,A) is called an approximation space.

1. $C(X) = \{x : A_x \cap X \neq \emptyset\}$ = Closure.
2. $I(X) = \{x : A_x \subseteq X\}$ = Interior,
3. $\overline{A}(X) = \cup\{A_x : A_x \cap X \neq \emptyset\}$ = Upper approximation.
4. $\underline{A}(X) = \cup\{A_x : A_x \subseteq X\}$ = Lower approximation.

5. $U(X) = \overline{A}(X)$ on (U,A)
6. $L(X) = \underline{A}(X)$ on (U,A)

Definition 9 *The pair* $(\overline{A}(X), \underline{A}(X))$ *is called a rough set.*

We should caution the readers that this is a technical definition of rough sets given by Pawlak (Pawlak, 1991). However, rough set theoreticians often use "rough set" as any subset X in the approximation space, where $\overline{A}(X)$ and $\underline{A}(X)$ are defined.

22.9.3 Binary Neighborhood System Spaces

We will be interested in the case $V = U$. Let B be a granulation. We will call (U,B) a NS-space(Section 22.3), which is a generalization of the RST and topological spaces. A subset X of U is open if for every object $p \in X$, there is a neighborhood $B(p) \subseteq X$. A subset X is closed if its complement is open. A BNS is open if every neighborhood is open. A BNS is topological, if BNS open and (U,B) is a usual topological space (Sierpenski and Krieger, 1956). So BNS-space is a generalization of topological space. Let X be a subset of U.

$I[X] = \{p : B(p) \subseteq X\}$ = Interior
$C[X] = \{p : X \cap B(p) \neq \emptyset\}$ = Closure

These are common notions in topological space; they were introduced to rough set community in (Lin, 1992), Subsequently re-defined and studied by (Yao, 1998, Grzymala-Busse, 2004). We should point out that $C[X]$ may not be closed; the closure in the sense of topology is transfinite C operations; see the notion of derived sets below. By porting the rough set style definitions to BNS-space, we have:

- $L[X] = \cup\{B(p) : B(p) \subseteq X\}$ = Lower approximation
- $H[X] = \cup\{B(p) : X \cap B(p) \neq \emptyset\}$ = Upper approximation

For BNS-space, these two definitions make sense. In fact, $H(X)$ is the neighborhood of a subset, that was used in (Lin, 1992) for defining the quotient set. In non-partition cases, upper and lower approximations do not equal to interior and closure. For NS-spaces (multilevel granulation), $H(X)$ defines a NS of subset X. The topological meaning of $L(X)$ is not clear. But we have used it in (Lin, 1998b) to compute belief functions, if all granules(neighborhoods) have basic probability assignments.

Note that in BNS, each object p has a unique neighborhood $B(p)$. In general neighborhood system (NS), each object is associated with a set of neighborhoods. In such NS, we have:

- An object p is a limit point of a set X, if every neighborhoods of p contains a point of X other than p. The set of all limit points of X is call derived set $D[X]$.
- Note that $C[X] = X \cup D[X]$ may not be closed. Some authors (e.g. (Sierpenski and Krieger, 1956)) define the closure as X together with repeated (transfinite) derived set. For such a closure it is a closed set.

22.10 Conclusions

Information granulation is a natural problem solving strategy since ancient time. Partition, the idealized form, has played a central role in the history of mathematics. Pawlak rough set theory has shown that the partition is also powerful notion in computer science; see (Pawlak, 1991) and a more recent survey in (Yao, 2004). Granulation, we believe, will play a similar role in real world problems. Some of its success has been demonstrated in fuzzy systems (Zadeh, 1973). Many ideas have been explored (Lin, 1988, Lin, 1989a, Chu and Chen, 1992, Raghavan, 1995, Miyamoto, 2004, Liu, 2004, Grzymala-Busse, 2004, Wang, 2004, Yao, 2004, Yao, 2004).

There are many strong applications in database, Data Mining, and security (Lin, 2004), (Lin, 2000) (Hu, 2004). The application to security may worth mention; it is a non-partition theory. It shares some light on the difficult problem of controlling of Trojan horses.

References

Aho, A., Hopcroft, J., and Ullman, J. (1974). *The Design and Analysis of Computer Algorithms*. Addison-Wesley.

Barr, A. and Feigenbaum, E. (1981). *The Handbook of Artificial Intelligence*. Addison-Wesley.

Birkhoff, G. and MacLane, S. (1977). *A Survey of Modern Algebra*. Macmillan.

David D. C. Brewer and Michael J. Nash: "The Chinese Wall Security Policy" IEEE Symposium on Security and Privacy, Oakland, May, 1988, pp 206-214,

Chu W. and Chen Q. (1992), Neighborhood and associative query answering, *Journal of Intelligent Information Systems*, **1**, 355-382, 1992.

Grzymala-Busse, J. W. (2004) Data with missing attribute values: Generalization of idiscernibility relation and rule induction. Transactions on Rough Sets, Lecture Notes in Computer Science Journal Subline, Springer-Verlag, vol. 1 (2004) 78-95.

Hobbs, J. (1985). Granularity. In *Proceedings of the Ninth Internation Joint Conference on Artificial Intelligence*, pages 432–435.

Hu X., Lin T.Y., Han J.,(2004) A New Rough Set Model Based on Database Systems, Journal of Fundamental Informatics, Vol. 59, Number 2,3,135-152

Lee, T. (1983). Algebraic theory of relational databases. *The Bell System Technical Journal*, 62(10):3159–3204.

Lin, T.Y. (1988). Neighborhood systems and relational database. In *Proceedings of CSC'88*, page 725.

Lin, T.Y. (1989). Neighborhood systems and approximation in database and knowledge base systems. In *Proceedings of the Fourth International Symposium on Methodologies of Intelligent Systems (Poster Session)*, pages 75–86.

Lin, T. Y. (1989), "Chinese Wall Security Policy–An Aggressive Model", Proceedings of the Fifth Aerospace Computer Security Application Conference, December 4-8, 1989, pp. 286-293.

Lin, T. Y.(1992) "Topological and Fuzzy Rough Sets," in: Decision Support by Experience - Application of the Rough Sets Theory, R. Slowinski (ed.), Kluwer Academic Publishers, 1992, 287-304.

Lin, T.Y. and Hadjimichael, M. (1996). Non-classificatory generalization in Data Mining. In *Proceedings of the 4th Workshop on Rough Sets, Fuzzy Sets, and Machine Discovery*, pages 404–412.

Lin, T.Y. (1996). A set theory for soft computing. In *Proceedings of 1996 IEEE International Conference on Fuzzy Systems*, pages 1140–1146.

Lin, T.Y. (1998a). Granular computing on binary relations i: Data Mining and neighborhood systems. In Skoworn, A. and Polkowski, L., editors, *Rough Sets In Knowledge Discovery*, pages 107–121. Physica-Verlag.

Lin, T.Y. (1998b). Rough set representations and belief functions ii. In Skoworn, A. and Polkowski, L., editors, *Rough Sets In Knowledge Discovery*, pages 121–140. Physica-Verlag.

Lin, T.Y., Zhong, N., Duong, J., and Ohsuga, S. (1998). Frameworks for mining binary relations in data. In Skoworn, A. and Polkowski, L., editors, *Rough sets and Current Trends in Computing*, LNCS 1424, pages 387–393. Springer-Verlag.

Lin, T.Y. (1999a). Data Mining: Granular computing approach. In *Methodologies for Knowledge Discovery and Data Mining: Proceedings of the 3rd Pacific-Asia Conference*, LNCS 1574, pages 24–33. Springer-Verlag.

Lin, T.Y. (1999b). Granular computing: Fuzzy logic and rough sets. In Zadeh, L. and Kacprzyk, J., editors, *Computing with Words in Information/Intelligent Systems*, pages 183–200. Physica-Verlag.

Lin, T.Y. (2000). Data Mining and machine oriented modeling: A granular computing approach. *Journal of Applied Intelligence*, 13(2):113–124.

Lin, T.Y. (2003a), "Chinese Wall Security Policy Models: Information Flows and Confining Trojan Horses." In: Data and Applications Security XVII: Status and Prospects,S. Vimercati, I. Ray & I. Ray 9eds) 2004, Kluwer Academic Publishers, 275-297 (Post conference proceedings of IFIP11.3 Working Conference on Database and Application Security, Aug 4-6, 2003, Estes Park, Co, USA

Lin, T.Y. (2003b), "Granular Computing: Structures, Representations, Applications and Future Directions." In: the Proceedings of 9th International Conference, RSFDGrC 2003, Chongqing, China, May 2003, Lecture Notes on Artificial Intelligence LNAI 2639, Springer-Verlag, 16-24.

Lin, T.Y. (2004), "A Theory of Derived Attributes and Attribute Completion," Proceedings of IEEE International Conference on Data Mining, Maebashi, Japan, Dec 9-12, 2002.

Lin, T.Y. (2005), Granular Computing - Rough Set Perspective, IEEE connections, The newsletter of the IEEE Computational Inelligence Society, Vol 2 Number 4, ISSN 1543-4281.

Liu, Q. (2004) Granular Language and Its Applications in Problem Solving, LNAI 3066,By Springer,127-132.

Miyamoto, S. (2004) Generalizations of multisets and rough approximations, International Journal of Intelligent Systems Volume 19, Issue 7, 639-652

Osborn S., Sanghu R. and Munawer Q.,"Configuring RoleBased Access Control to Enforce Mandatory and Discretionary Access Control Policies," ACM Transaction on Information and Systems Security, Vol 3, No 2, May 2002, Pages 85-106.

Pawlak, Z. (1982). Rough sets. *International Journal of Information and Computer Science*, 11(15):341–356.

Pawlak, Z. (1991). *Rough Sets–Theoretical Aspects of Reasoning about Data.* Kluwer Academic Publishers.

Raghavan, V. V.,Sever, H.,Deogun, J. S. (1995, August), Exploiting Upper Approximations in the Rough Set Model, Proceedings of the First International Conference on Knowledge Discovery and Data Mining (KDD'95), Spansored by AAAI in cooperation with IJCAI, Montreal, Quebec, Canada, August, 1995, pp. 69-74.

Rokach, L., Averbuch, M., and Maimon, O., Information retrieval system for medical narrative reports. Lecture notes in artificial intelligence, 3055. pp. 217-228, Springer-Verlag (2004).

Sierpenski, W. and Krieger, C. (1956). *General Topology*. University of Toronto Press.

Szyperski, C. (2002). *Component Software: Beyond Object-Oriented Programming*. Addison-Wesley.

Wang, D. W.,Liau, C. J.,Hsu, T.-S. (2004), Medical privacy protection based on granular computing, Artificial Intelligence in Medicine, 32(2), 137-149

Yao, Y. Y.: Relational interpretations of neighborhood operators and rough set approximation operators. *Information Sciences* **111** (1998) 239–259.

Yao, Y. Y. (2004) A partition model of granular computing to appear in LNCS Transactions on Rough Sets.

Yao Y.Y. , Zhao Y., Yao J.T., Level Construction of Decision Trees in a Partition-based Framework for Classification, Proceedings of the 16th International Conference on Software Engineering and Knowledge Engineering (SEKE'04), Banff, Alberta, Canada, June 20-24, 2004, pp199-204.

Zadeh. L. A. (1973) Outline of a New Approach to the Analysis of Complex Systems and Decision Process. IEEE Trans. Syst. Man.

Zadeh, L.A. (1979). Fuzzy sets and information granularity. In Gupta, N., Ragade, R., and Yager, R., editors, *Advances in Fuzzy Set Theory and Applications*, pages 3–18. North-Holland.

Zadeh, L.A. (1996). Fuzzy logic = computing with words. *IEEE Transactions on Fuzzy Systems*, 4(2):103–111.

Zadeh, L.A. (1997). Towards a theory of fuzzy information granulation and its centrality in human reasoning and fuzzy logic. *Fuzzy Sets and Systems*, 19:111–127.

Zadeh, L.A. (1998) Some reflections on soft computing, granular computing and their roles in the conception, design and utilization of information/ intelligent systems, Soft Computing, 2, 23-25.

Zhang, B. and Zhang, L. (1992). *Theory and Applications of Problem Solving*. North-Holland.

Zimmerman, H. (1991). *Fuzzy Set Theory –and its Applications*. Kluwer Acdamic Publisher.

23

Pattern Clustering Using a Swarm Intelligence Approach

Swagatam Das[1] and Ajith Abraham[2]

[1] Department of Electronics and Telecommunication Engineering, Jadavpur University, Kolkata 700032, India.
[2] Center of Excellence for Quantifiable Quality of Service Norwegian University of Science and Technology, Trondheim, Norway
ajith.abraham@ieee.org

Summary. Clustering aims at representing large datasets by a fewer number of prototypes or clusters. It brings simplicity in modeling data and thus plays a central role in the process of knowledge discovery and data mining. Data mining tasks, in these days, require fast and accurate partitioning of huge datasets, which may come with a variety of attributes or features. This, in turn, imposes severe computational requirements on the relevant clustering techniques. A family of bio-inspired algorithms, well-known as Swarm Intelligence (SI) has recently emerged that meets these requirements and has successfully been applied to a number of real world clustering problems. This chapter explores the role of SI in clustering different kinds of datasets. It finally describes a new SI technique for partitioning a linearly non-separable dataset into an optimal number of clusters in the kernel- induced feature space. Computer simulations undertaken in this research have also been provided to demonstrate the effectiveness of the proposed algorithm.

23.1 Introduction

Clustering means the act of partitioning an unlabeled dataset into groups of similar objects. Each group, called a 'cluster', consists of objects that are similar between themselves and dissimilar to objects of other groups. In the past few decades, cluster analysis has played a central role in a variety of fields ranging from engineering (machine learning, artificial intelligence, pattern recognition, mechanical engineering, electrical engineering), computer sciences (web mining, spatial database analysis, textual document collection, image segmentation), life and medical sciences (genetics, biology, microbiology, paleontology, psychiatry, pathology), to earth sciences (geography. geology, remote sensing), social sciences (sociology, psychology, archeology, education), and economics (marketing, business) (Evangelou *et al.*, 2001, Lillesand and Keifer, 1994, Rao, 1971, Duda and Hart, 1973, Everitt, 1993, Xu and Wunsch, 2008).

Human beings possess the natural ability of clustering objects. Given a box full of marbles of four different colors say red, green, blue, and yellow, even a child may separate these marbles into four clusters based on their colors. However, making a computer solve this type of problems is quite difficult and demands the attention of computer scientists and engineers all

O. Maimon, L. Rokach (eds.), *Data Mining and Knowledge Discovery Handbook*, 2nd ed.,
DOI 10.1007/978-0-387-09823-4_23,

over the world till date. The major hurdle in this task is that the functioning of the brain is much less understood. The mechanisms, with which it stores huge amounts of information, processes them at lightning speeds and infers meaningful rules, and retrieves information as and when necessary have till now eluded the scientists. A question that naturally comes up is: what is the point in making a computer perform clustering when people can do this so easily? The answer is far from trivial. The most important characteristic of this information age is the abundance of data. Advances in computer technology, in particular the Internet, have led to what some people call "data explosion": the amount of data available to any person has increased so much that it is more than he or she can handle. In reality the amount of data is vast and in addition, each data item (an abstraction of a real-life object) may be characterized by a large number of attributes (or *features*), which are based on certain measurements taken on the real-life objects and may be numerical or non-numerical. Mathematically we may think of a mapping of each data item into a point in the multi-dimensional feature space (each dimension corresponding to one feature) that is beyond our perception when number of features exceed just 3. Thus it is nearly impossible for human beings to partition tens of thousands of data items, each coming with several features (usually much greater than 3), into meaningful clusters within a short interval of time. Nonetheless, the task is of paramount importance for organizing and summarizing huge piles of data and discovering useful knowledge from them. So, can we devise some means to generalize to arbitrary dimensions of what humans perceive in two or three dimensions, as densely connected "patches" or "clouds" within data space? The entire research on cluster analysis may be considered as an effort to find satisfactory answers to this fundamental question.

The task of computerized data clustering has been approached from diverse domains of knowledge like graph theory, statistics (multivariate analysis), artificial neural networks, fuzzy set theory, and so on (Forgy, 1965, Zahn, 1971, Holeňa, 1996, Rauch, 1996, Rauch, 1997, Kohonen, 1995, Falkenauer, 1998, Paterlini and Minerva, 2003, Xu and Wunsch, 2005, Rokach and Maimon, 2005, Mitra *et al.*2002). One of the most popular approaches in this direction has been the formulation of clustering as an optimization problem, where the best partitioning of a given dataset is achieved by minimizing/maximizing one (single-objective clustering) or more (multi-objective clustering) objective functions. The objective functions are usually formed capturing certain statistical-mathematical relationship among the individual data items and the candidate set of representatives of each cluster (also known as cluster-centroids). The clusters are either hard, that is each sample point is unequivocally assigned to a cluster and is considered to bear no similarity to members of other clusters, or fuzzy, in which case a membership function expresses the degree of belongingness of a data item to each cluster.

Most of the classical optimization-based clustering algorithms (including the celebrated hard c-means and fuzzy c-means algorithms) rely on local search techniques (like iterative function optimization, Lagrange's multiplier, Picard's iterations etc.) for optimizing the clustering criterion functions. The local search methods, however, suffer from two great disadvantages. Firstly they are prone to getting trapped in some local optima of the multi-dimensional and usually multi-modal landscape of the objective function. Secondly performances of these methods are usually very sensitive to the initial values of the search variables.

Although many respected texts of pattern recognition describe clustering as an unsupervised learning method, most of the traditional clustering algorithms require a prior specification of the number of clusters in the data for guiding the partitioning process, thus making it not completely unsupervised. On the other hand, in many practical situations, it is impossible to provide even an estimation of the number of naturally occurring clusters in a previously unhandled dataset. For example, while attempting to classify a large database of handwritten characters in an unknown language; it is not possible to determine the correct number of distinct letters beforehand. Again, while clustering a set of documents arising from the query to a search engine, the number of classes can change for each set of documents that result from an interaction with the search engine. Data mining tools that predict future trends and behaviors for allowing businesses to make proactive and knowledge-driven decisions, demand fast and fully automatic clustering of very large datasets with minimal or no user intervention. Thus it is evident that the complexity of the data analysis tasks in recent times has posed severe challenges before the classical clustering techniques.

Recently a family of nature inspired algorithms, known as *Swarm Intelligence* (SI), has attracted several researchers from the field of pattern recognition and clustering. Clustering techniques based on the SI tools have reportedly outperformed many classical methods of partitioning a complex real world dataset. Algorithms belonging to the domain, draw inspiration from the collective intelligence emerging from the behavior of a group of social insects (like bees, termites and wasps). When acting as a community, these insects even with very limited individual capability can jointly (cooperatively) perform many complex tasks necessary for their survival. Problems like finding and storing foods, selecting and picking up materials for future usage require a detailed planning, and are solved by insect colonies without any kind of supervisor or controller. An example of particularly successful research direction in swarm intelligence is Ant Colony Optimization (ACO) (Dorigo *et al.*, 1996,Dorigo and Gambardella, 1997), which focuses on discrete optimization problems, and has been applied successfully to a large number of NP hard discrete optimization problems including the traveling salesman, the quadratic assignment, scheduling, vehicle routing, etc., as well as to routing in telecommunication networks. Particle Swarm Optimization (PSO) (Kennedy and Eberhart, 1995) is another very popular SI algorithm for global optimization over continuous search spaces. Since its advent in 1995, PSO has attracted the attention of several researchers all over the world resulting into a huge number of variants of the basic algorithm as well as many parameter automation strategies.

In this Chapter, we explore the applicability of these bio-inspired approaches to the development of self-organizing, evolving, adaptive and autonomous clustering techniques, which will meet the requirements of next-generation data mining systems, such as diversity, scalability, robustness, and resilience. The next section of the chapter provides an overview of the SI paradigm with a special emphasis on two SI algorithms well-known as Particle Swarm Optimization (PSO) and Ant Colony Systems (ACS). Section 3 outlines the data clustering problem and briefly reviews the present state of the art in this field. Section 4 describes the use of the SI algorithms in both crisp and fuzzy clustering of real world datasets. A new automatic clustering algorithm, based on PSO, is presented in Section 5. The algorithm requires no previous knowledge of the dataset to be partitioned, and can determine the optimal number of classes dynamically in a linearly non-separable dataset using a kernel-induced distance metric. The new method has been compared with two well-known, classical fuzzy clustering algorithms. The Chapter is concluded in Section 6 with discussions on possible directions for future research.

23.2 An Introduction to Swarm Intelligence

The behavior of a single ant, bee, termite and wasp often is too simple, but their collective and social behavior is of paramount significance. A look at National Geographic TV Channel reveals that advanced mammals including lions also enjoy social lives, perhaps for their self-existence at old age and in particular when they are wounded. The collective and social behavior of living creatures motivated researchers to undertake the study of today what is known as *Swarm Intelligence*. Historically, the phrase Swarm Intelligence (SI) was coined by Beny and Wang in late 1980s (Beni and Wang, 1989) in the context of cellular robotics. A group of researchers in different parts of the world started working almost at the same time to study the versatile behavior of different living creatures and especially the social insects. The efforts to mimic such behaviors through computer simulation finally resulted into the fascinating field of SI. SI systems are typically made up of a population of simple agents (an entity capable of performing/executing certain operations) interacting locally with one another and with their environment. Although there is normally no centralized control structure dictating how individual agents should behave, local interactions between such agents often lead to the emergence of global behavior. Many biological creatures such as fish schools and bird flocks clearly display structural order, with the behavior of the organisms so integrated that even though they may change shape and direction, they appear to move as a single coherent entity (Couzin *et al.*, 2002). The main properties of the collective behavior can be pointed out as follows and is summarized in Figure 1.

1. ***Homogeneity*:** every bird in flock has the same behavioral model. The flock moves without a leader, even though temporary leaders seem to appear.
2. ***Locality*:** its nearest flock-mates only influence the motion of each bird. Vision is considered to be the most important senses for flock organization.
3. ***Collision Avoidance*:** avoid colliding with nearby flock mates.
4. ***Velocity Matching:*** attempt to match velocity with nearby flock mates.
5. ***Flock Centering*:** attempt to stay close to nearby flock mates

Individuals attempt to maintain a minimum distance between themselves and others at all times. This rule is given the highest priority and corresponds to a frequently observed behavior of animals in nature (Rokach (2006)). If individuals are not performing an avoidance maneuver they tend to be attracted towards other individuals (to avoid being isolated) and to align themselves with neighbors (Partridge and Pitcher, 1980, Partridge, 1982).

Couzin *et al.* (2002) identified four collective dynamical behaviors as illustrated in Figure 2:

1. ***Swarm:*** an aggregate with cohesion, but a low level of polarization (parallel alignment) among members
2. ***Torus*:** individuals perpetually rotate around an empty core (milling). The direction of rotation is random.
3. ***Dynamic parallel group*:** the individuals are polarized and move as a coherent group, but individuals can move throughout the group and density and group form can fluctuate (Partridge and Pitcher, 1980, Major and Dill, 1978).
4. ***Highly parallel group*:** much more static in terms of exchange of spatial positions within the group than the dynamic parallel group and the variation in density and form is minimal.

As mentioned in (Grosan *et al.*, 2006) at a high-level, a swarm can be viewed as a group of agents cooperating to achieve some purposeful behavior and achieve some goal (Abraham *et al.*, 2006). This collective intelligence seems to emerge from what are often large groups: According to Milonas (1994), five basic principles define the SI paradigm. First is the proximity principle: the swarm should be able to carry out simple space and time computations. Second is the quality principle: the swarm should be able to respond to quality factors in the environment. Third is the principle of diverse response: the swarm should not commit its activities along excessively narrow channels. Fourth is the principle of stability: the swarm should

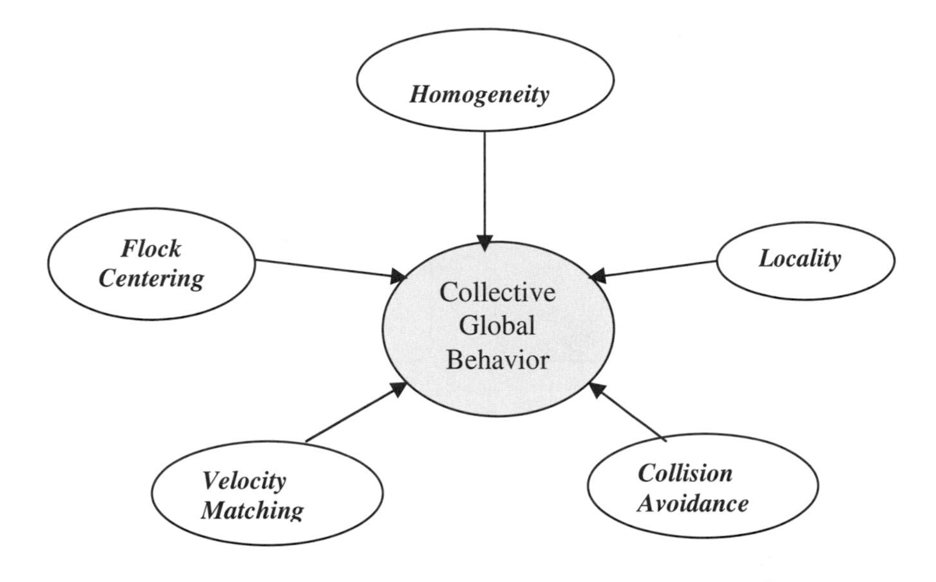

Fig. 23.1. Main traits of collective behavior.

not change its mode of behavior every time the environment changes. Fifth is the principle of adaptability: the swarm must be able to change behavior mote when it is worth the computational price. Note that principles four and five are the opposite sides of the same coin. Below we discuss in details two algorithms from SI domain, which have gained wide popularity in a relatively short span of time.

23.2.1 The Ant Colony Systems

The basic idea of a real ant system is illustrated in Figure 3. In the left picture, the ants move in a straight line to the food. The middle picture illustrates the situation soon after an obstacle is inserted between the nest and the food. To avoid the obstacle, initially each ant chooses to turn left or right at random. Let us assume that ants move at the same speed depositing pheromone in the trail uniformly. However, the ants that, by chance, choose to turn left will reach the food sooner, whereas the ants that go around the obstacle turning right will follow a longer path, and so will take longer time to circumvent the obstacle. As a result, pheromone accumulates faster in the shorter path around the obstacle. Since ants prefer to follow trails with larger amounts of pheromone, eventually all the ants converge to the shorter path around the obstacle, as shown in Figure 3.

An artificial Ant Colony System (ACS) is an agent-based system, which simulates the natural behavior of ants and develops mechanisms of cooperation and learning. ACS was proposed by Dorigo *et al.* (1997) as a new heuristic to solve combinatorial optimization problems. This new heuristic, called Ant Colony Optimization (ACO) has been found to be both robust and versatile in handling a wide range of combinatorial optimization problems.

The main idea of ACO is to model a problem as the search for a minimum cost path in a graph. Artificial ants as if walk on this graph, looking for cheaper paths. Each ant has a rather

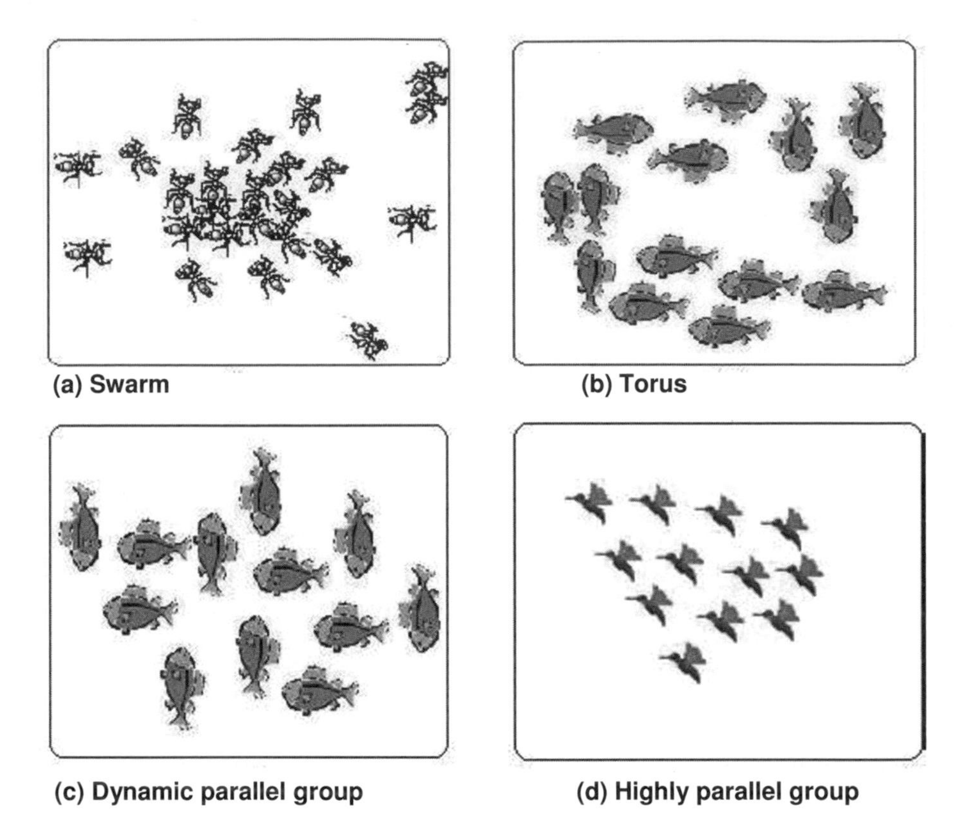

Fig. 23.2. Different models of collective behavior.

simple behavior capable of finding relatively costlier paths. Cheaper paths are found as the emergent result of the global cooperation among ants in the colony. The behavior of artificial ants is inspired from real ants: they lay pheromone trails (obviously in a mathematical form) on the graph edges and choose their path with respect to probabilities that depend on pheromone trails. These pheromone trails progressively decrease by evaporation. In addition, artificial ants have some extra features not seen in their counterpart in real ants. In particular, they live in a discrete world (a graph) and their moves consist of transitions from nodes to nodes.

Below we illustrate the use of ACO in finding the optimal tour in the classical Traveling Salesman Problem (TSP). Given a set of n cities and a set of distances between them, the problem is to determine a minimum traversal of the cities and return to the home-station at the end. It is indeed important to note that the traversal should in no way include a city more than once. Let $r\,(Cx, Cy)$ be a measure of cost for traversal from city Cx to Cy. Naturally, the total cost of traversing n cities indexed by $i1, i2, i3,\ldots, in$ in order is given by the following expression:

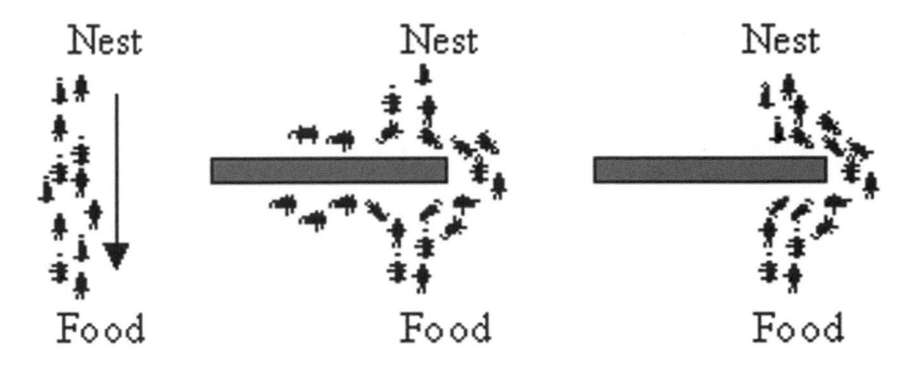

Fig. 23.3. Illustrating the behavior of real ant movements.

$$Cost(i_1, i_2, ..., i_n) = \sum_{j=1}^{n-1} r(Ci_j, Ci_{j+1}) + r(Ci_n, Ci_1) \tag{23.1}$$

The ACO algorithm is employed to find an optimal order of traversal of the cities. Let τ be a mathematical entity modeling the pheromone and $\eta ij = 1/r\ (i\ ,j)$ is a local heuristic. Also let *allowedk(t)* be the set of cities that are yet to be visited by ant q located in city i. Then according to the classical ant system (Xu and Wunsch, 2008) the probability that ant q in city i visits city j is given by

$$p_{ij}^{q}(t) = \frac{[\tau_{ij}(t)]^{\alpha} \cdot [\eta_{ij}]^{\beta}}{\sum_{h \in allowed_q(t)} [\tau_{ih}(t)]^{\alpha} \cdot [\eta_{ih}]^{\beta}}, \quad \text{if } j \in allowed_k(t)$$
$$= 0, \qquad \text{otherwise.} \tag{23.2}$$

In Equation 23.19 shorter edges with greater amount of pheromone are favored by multiplying the pheromone on edge $(i\ ,j\ )$ by the corresponding heuristic value $\eta(i,j\ )$. Parameters α (> 0) and β (> 0) determine the relative importance of pheromone versus cost. Now in ant system, pheromone trails are updated as follows. Let Dq be the length of the tour performed by ant q, $\Delta\tau_q(i,j) = 1/D_q$ if $(i, j) \in$ tour done by ant q and $\Delta\tau_q(i,j) = 0$ otherwise and finally let $\rho \in [0,1]$ be a pheromone decay parameter which takes care of the occasional evaporation of the pheromone from the visited edges. Then once all ants have built their tours, pheromone is updated on all the ages as,

$$\tau(i,j) = (1-\rho).\tau(i,j) + \sum_{p=1}^{m} \tau_k(i,j) \tag{23.3}$$

From equation 23.3, we can guess that pheromone updating attempts to accumulate greater amount of pheromone to shorter tours (which corresponds to high value of the second term in (3) so as to compensate for any loss of pheromone due to the first term). This conceptually resembles a reinforcement-learning scheme, where better solutions receive a higher reinforcement.

The ACO differs from the classical ant system in the sense that here the pheromone trails are updated in two ways. Firstly, when ants construct a tour they locally change the amount of pheromone on the visited edges by a local updating rule. Now if we let γ to be a decay parameter and $\Delta\tau(i,j) = \tau 0$ such that $\tau 0$ is the initial pheromone level, then the local rule may be stated as:

$$\tau(i,j) = (1-\gamma).\tau(i,j) + \gamma.\Delta\tau(i,j) \tag{23.4}$$

Secondly, after all the ants have built their individual tours, a global updating rule is applied to modify the pheromone level on the edges that belong to the best ant tour found so far. If κ be the usual pheromone evaporation constant, D_{gb} be the length of the globally best tour from the beginning of the trial and $\Delta\tau'/\ (i\ ,j) = 1/\ D_{gb}$ only when the edge (i, j) belongs to global-best-tour and zero otherwise, then we may express the global rule as follows:

$$\tau(i,j) = (1-\kappa).\tau(i,j) + \kappa.\Delta\tau'(i,j) \tag{23.5}$$

The main steps of ACO algorithm are presented below.

Procedure ACO
Begin
Initialize pheromone trails;
Repeat
Begin /* at this stage each loop is called an iteration */
Each ant is positioned on a starting node;
Repeat
Begin /* at this level each loop is called a step */
Each ant applies a *state transition rule like rule (2)* to incrementally build a solution and a *local pheromone-updating rule like rule (4);*
Until all ants have built a complete solution;
A *global pheromone-updating rule like rule (5)* is applied.
Until terminating condition is reached;
End

The concept of Particle Swarms, although initially introduced for simulating human social behaviors, has become very popular these days as an efficient search and optimization technique. The Particle Swarm Optimization (PSO) (Kennedy and Eberhart, 1995, Kennedy *et al.*, 2001), as it is called now, does not require any gradient information of the function to be optimized, uses only primitive mathematical operators and is conceptually very simple. In PSO, a population of conceptual 'particles' is initialized with random positions $\mathbf{X}_i$ and velocities $\mathbf{V}_i$, and a function, f, is evaluated, using the particle's positional coordinates as input values. In an D-dimensional search space, $\mathbf{X}_i = (x_{i1}, x_{i2}, ..., x_{iD})^T$ and $\mathbf{V}_i = (v_{i1}, v_{i2}, ..., v_{iD})^T$. In literature, the basic equations for updating the d-th dimension of the velocity and position of the i-th particle for PSO are presented most popularly in the following way:

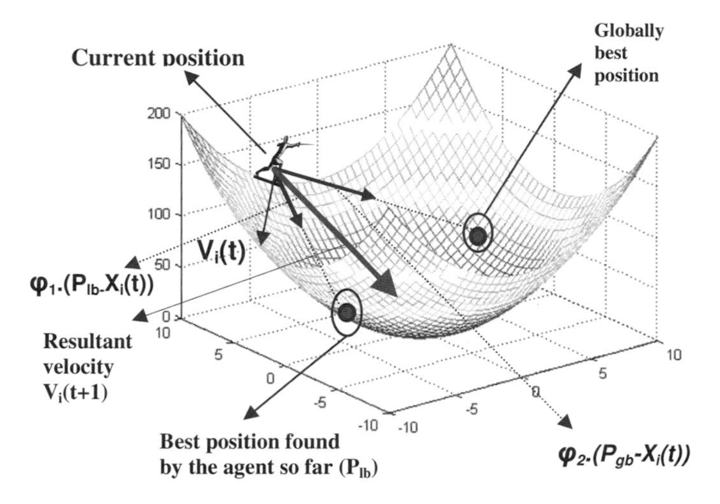

Fig. 23.4. Illustrating the velocity updating scheme of basic PSO.

$$v_{i,d}(t) = \omega.v_{i,d}(t-1) + \varphi_1.rand1_{i,d}(0,1).(p^l_{i,d} - x_{i,d}(t-1)) + \varphi_2.rand2_{i,d}(0,1).(p^g_d - x_{i,d}(t-1)) \tag{23.6}$$

$$x_{i,d}(t) = x_{i,d}(t-1) + v_{i,d}(t) \tag{23.7}$$

Please note that in 23.6 and 23.10, φ_1 and φ_2 are two positive numbers known as the *acceleration coefficients*. The positive constant $\omega$is known as *inertia factor*. $rand1_{i,d}(0,1)$ and $rand2_{i,d}(0,1)$ are the two uniformly distributed random numbers in the range of [0, 1]. While applying PSO, we define a maximum velocity $\mathbf{V}_{\max} = [v_{\max,1}, v_{\max,2}, \ldots, v_{\max,D}]^T$ of the particles in order to control their convergence behavior near optima. If $|v_{i,d}|$ exceeds a positive constant value $v_{\max,d}$ specified by the user, then the velocity of that dimension is assigned to $sgn(v_{i,d}).v_{\max,d}$ where *sgn* stands for the signum function and is defined as:

$$\begin{aligned} \text{sgn}(x) &= 1, & \text{if } x > 0 \\ &= 0, & \text{if } x = 0 \\ &= -1, & \text{if } x < 0 \end{aligned} \tag{23.8}$$

While updating the velocity of a particle, different dimensions will have different values for *rand*1 and *rand*2. Some researchers, however, prefer to use the same values of these random coefficients for all dimensions of a given particle. They use the following formula to update the velocities of the particles:

$$v_{i,d}(t) = \omega\, v_{i,d}(t-1) + \varphi_1.rand1_i(0,1).(p^l_{i,d}(t) - x_{i,d}(t-1)) + \varphi_2.rand2_i(0,1).(p^g_d(t) - x_{i,d}(t-1)) \tag{23.9}$$

Comparing the two variants in 23.6 and 23.12, the former can have a larger search space due to independent updating of each dimension, while the second is dimension-dependent and has a smaller search space due to the same random numbers being used for all dimensions The velocity updating scheme has been illustrated in Figure 4 with a humanoid particle.

A pseudo code for the PSO algorithm may be put forward as:

The PSO Algorithm

Input: Randomly initialized position and velocity of the particles: $\vec{X}_i(0)$ and $\vec{V}_i(0)$

Output: Position of the approximate global optima $\vec{X}*$

Begin
  While terminating condition is not reached **do**
    Begin
     for i = 1 to number of particles
        Evaluate the fitness: $= f(\vec{X}_i(t))$;
        Update $\vec{P}(t)$ and $\vec{g}(t)$;
        Adapt velocity of the particle using equation (6);
        Update the position of the particle;
        increase i;
  end while
end

23.3 Data Clustering – An Overview

In this Section, we first provide a brief and formal description of the clustering problem. We then discuss a few major classical clustering techniques.

23.3.1 Problem Definition

A *pattern* is a physical or abstract structure of objects. It is distinguished from others by a collective set of attributes called *features*, which together represent a pattern (Konar, 2005). Let P = {P1, P2... Pn} be a set of n patterns or data points, each having d features. These patterns can also be represented by a profile data matrix $\mathbf{X}n \times d$ having n d-dimensional row vectors. The i-th row vector$\mathbf{X}_i$characterizes the i-th object from the set P and each element Xi,j in $\mathbf{X}_i$corresponds to the j-th real value feature (j = 1, 2, ...,d) of the i-th pattern (i =1,2,...., n). Given such an $\mathbf{X}n \times d$, a partitional clustering algorithm tries to find a partition C = {C1, C2,......, Ck}of k classes, such that the similarity of the patterns in the same cluster is maximum and patterns from different clusters differ as far as possible. The partitions should maintain the following properties:

- Each cluster should have at least one pattern assigned i. e. $C_i \neq \Phi \forall i \in \{1, 2, ..., k\}$.
- Two different clusters should have no pattern in common. i.e. $C_i \bigcap C_j = \Phi, \forall i \neq j$ and $i, j \in \{1, 2, ..., k\}$. This property is required for crisp (hard) clustering. In Fuzzy clustering this property doesn't exist.
- Each pattern should definitely be attached to a cluster i.e. $\bigcup_{i=1}^{k} C_i = P$.

Since the given dataset can be partitioned in a number of ways maintaining all of the above properties, a fitness function (some measure of the adequacy of the partitioning) must be defined. The problem then turns out to be one of finding a partition **C*** of optimal or near-optimal adequacy as compared to all other feasible solutions **C** = { C1, C2,........, CN(n,*k*)} where,

$$N(n,k) = \frac{1}{k!}\sum_{i=1}^{k}(-1)^i \binom{k}{i}^i (k-i)^i \quad (23.10)$$

is the number of feasible partitions. This is same as,

$$Optimize_C f(X_{\mathrm{n\times d}}, C) \quad (23.11)$$

where C is a single partition from the set **C** and f is a statistical-mathematical function that quantifies the goodness of a partition on the basis of the similarity measure of the patterns. Defining an appropriate similarity measure plays fundamental role in clustering (Jain *et al.*, 1999). The most popular way to evaluate similarity between two patterns amounts to the use of *distance measure*. The most widely used distance measure is the Euclidean distance, which between any two d-dimensional patterns $\mathbf{X}_i$ and $\mathbf{X}_j$ is given by,

$$d(\mathbf{X}_i,\mathbf{X}_j) = \sqrt{\sum_{p=1}^{d}(X_{i,p}-X_{j,p})^2} = \|\mathbf{X}_i - \mathbf{X}_j\| \quad (23.12)$$

It has been shown in (Brucker, 1978) that the clustering problem is NP-hard when the number of clusters exceeds 3.

23.3.2 The Classical Clustering Algorithms

Data clustering is broadly based on two approaches: *hierarchical* and *partitional* (Frigui and Krishnapuram, 1999, Leung *et al.*, 2000). Within each of the types, there exists a wealth of subtypes and different algorithms for finding the clusters. In hierarchical clustering, the output is a tree showing a sequence of clustering with each cluster being a partition of the data set (Leung *et al.*, 2000). Hierarchical algorithms can be agglomerative (bottom-up) or divisive (top-down). Agglomerative algorithms begin with each element as a separate cluster and merge them in successively larger clusters. Divisive algorithms begin with the whole set and proceed to divide it into successively smaller clusters. Hierarchical algorithms have two basic advantages (Frigui and Krishnapuram, 1999). Firstly, the number of classes need not be specified a priori and secondly, they are independent of the initial conditions. However, the main drawback of hierarchical clustering techniques is they are static, i.e. data-points assigned to a cluster can not move to another cluster. In addition to that, they may fail to separate overlapping clusters due to lack of information about the global shape or size of the clusters (Jain *et al.*, 1999).

Partitional clustering algorithms, on the other hand, attempt to decompose the data set directly into a set of disjoint clusters. They try to optimize certain criteria. The criterion function may emphasize the local structure of the data, as by assigning clusters to peaks in the probability density function, or the global structure. Typically, the global criteria involve minimizing some measure of dissimilarity in the samples within each cluster, while maximizing the dissimilarity of different clusters. The advantages of the hierarchical algorithms are the disadvantages of the partitional algorithms and vice versa. An extensive survey of various clustering techniques can be found in (Jain *et al.*, 1999). The focus of this chapter is on the partitional clustering algorithms.

Clustering can also be performed in two different modes: crisp and fuzzy. In crisp clustering, the clusters are disjoint and non-overlapping in nature. Any pattern may belong to one and only one class in this case. In case of fuzzy clustering, a pattern may belong to all the classes with a certain fuzzy membership grade (Jain *et al.*, 1999).

The most widely used iterative k-means algorithm (MacQueen, 1967) for partitional clustering aims at minimizing the ICS (Intra-Cluster Spread) which for k cluster centers can be defined as

$$ICS(C_1, C_2, ..., C_k) = \sum_{i=1}^{k} \sum_{\mathbf{X}_i \in C_i} \|\mathbf{X}_i - \mathbf{m}_i\|^2 \tag{23.13}$$

The k-means (or hard c-means) algorithm starts with k cluster-centroids (these centroids are initially selected randomly or derived from some a priori information). Each pattern in the data set is then assigned to the closest cluster-centre. Centroids are updated by using the mean of the associated patterns. The process is repeated until some stopping criterion is met.

In the c-medoids algorithm (Kaufman and Rousseeuw, 1990), on the other hand, each cluster is represented by one of the representative objects in the cluster located near the center. Partitioning around medoids (PAM) (Kaufman and Rousseeuw, 1990) starts from an initial set of medoids, and iteratively replaces one of the medoids by one of the non-medoids if it improves the total distance of the resulting clustering. Although PAM works effectively for small data, it does not scale well for large datasets. Clustering large applications based on randomized search (CLARANS) (Ng and Han, 1994), using randomized sampling, is capable of dealing with the associated scalability issue.

The fuzzy c-means (FCM) (Bezdek, 1981) seems to be the most popular algorithm in the field of fuzzy clustering. In the classical FCM algorithm, a *within cluster sum* function Jm is minimized to evolve the proper cluster centers:

$$J_m = \sum_{j=1}^{n} \sum_{i=1}^{c} (u_{ij})^m \left\|\mathbf{X}_j - \mathbf{V}_i\right\|^2 \tag{23.14}$$

where $\mathbf{V}_i$ is the i-th cluster center, $\mathbf{X}_j$ is the j-th d-dimensional data vector and $||.||$is an inner product-induced norm in d dimensions. Given c classes, we can determine their cluster centers $\mathbf{V}_i$ for i=1 to c by means of the following expression:

$$\mathbf{V}_i = \frac{\sum_{j=1}^{n} (u_{ij})^m \mathbf{X}_j}{\sum_{j=1}^{n} (u_{ij})^m} \mathbf{V}_i = \frac{\sum_{j=1}^{n} (u_{ij})^m \mathbf{X}_j}{\sum_{j=1}^{n} (u_{ij})^m} \tag{23.15}$$

Here m (m>1) is any real number that influences the membership grade. Now differentiating the performance criterion with respect to $\mathbf{V}_i$ (treating uij as constants) and with respect to uij (treating $\mathbf{V}_i$ as constants) and setting them to zero the following relation can be obtained:

$$u_{ij} = \left[\sum_{k=1}^{c} \left(\frac{\left\|\mathbf{X}_j - \mathbf{V}_i\right\|^2}{\left\|\mathbf{X} - \mathbf{V}_i\right\|^2} \right)^{1/(m-1)} \right]^{-1} \tag{23.16}$$

Several modifications of the classical FCM algorithm can be found in (Hall *et al.*1999, Gath and Geva, 1989, Bensaid *et al.*, 1996, Clark *et al.*, 1994, Ahmed *et al.*, 2002, Wang X *et al.*, 2004).

23.3.3 Relevance of SI Algorithms in Clustering

From the discussion of the previous Section, we see that the SI algorithms are mainly stochastic search and optimization techniques, guided by the principles of collective behaviour and self organization of insect swarms. They are efficient, adaptive and robust search methods producing near optimal solutions and have a large amount of implicit parallelism. On the other hand, data clustering may be well formulated as a difficult global optimization problem; thereby making the application of SI tools more obvious and appropriate.

23.4 Clustering with the SI Algorithms

In this Section we first review the present state of the art clustering algorithms based on SI tools, especially the ACO and PSO. We then outline a new algorithm which employs the PSO model to automatically determine the number of clusters in a previously unhandled dataset. Computer simulations undertaken for this study have also been included to demonstrate the elegance of the new dynamic clustering technique.

23.4.1 The Ant Colony Based Clustering Algorithms

Ant colonies provide a means to formulate some powerful nature-inspired heuristics for solving the clustering problems. Among other social movements, researchers have simulated the way, ants work collaboratively in the task of grouping dead bodies so, as to keep the nest clean (Bonabeau *et al.*, 1999). It can be observed that, with time the ants tend to cluster all dead bodies in a specific region of the environment, thus forming piles of corpses.

Larval sorting and corpse cleaning by ant was first modeled by Deneubourg *et al.* (1991) for accomplishing certain tasks in robotics. This inspired the Ant-based clustering algorithm (Handl *et al.*, 2003). Lumer and Faieta modified the algorithm using a dissimilarity-based evaluation of the local density, in order to make it suitable for data clustering (Lumer and Faieta, 1994). This introduced standard Ant Clustering Algorithm (ACA). It has subsequently been used for numerical data analysis (Lumer and Faieta, 1994), data-mining (Lumer and Faieta, 1995), graph-partitioning (Kuntz and Snyers, 1994, Kuntz and Snyers, 1999, Kuntz *et al.*, 1998) and text-mining (Handl and Meyer B, 2002, Hoe *et al.*, 2002, Ramos and Merelo, 2002). Many authors (Handl and Meyer B, 2002, Ramos *et al.*, 2002) proposed a number of modifications to improve the convergence rate and to get optimal number of clusters. Monmarche *et al.* (1999) hybridized the Ant-based clustering algorithm with k-means algorithm and compared it to traditional k-means on various data sets, using the classification error for evaluation purposes. However, the results obtained with this method are not applicable to ordinary ant-based clustering since it differs significantly from the latter.

Like a standard ACO, ant-based clustering is a distributed process that employs positive feedback. Ants are modeled by simple agents that randomly move in their environment. The environment is considered to be a low dimensional space, more generally a two-dimensional plane with square grid. Initially, each data object that represents a multi-dimensional pattern is randomly distributed over the 2-D space. Data items that are scattered within this environment can be picked up, transported and dropped by the agents in a probabilistic way. The picking and dropping operation are influenced by the similarity and density of the data items within the ant's local neighborhood. Generally, the size of the neighborhood is 3×3. Probability of picking up data items is more when the object are either isolated or surrounded by dissimilar items. They trend to drop them in the vicinity of similar ones. In this way, a clustering of the elements on the grid is obtained.

The ants search for the feature space either through random walk or with jumping using a short term memory. Each ant picks up or drops objects according to the following local probability density measure:

$$f(\mathbf{X}_i) = \max\{0, \frac{1}{s^2} \sum_{\mathbf{X}_j \in N_{s \times s}(r)} [1 - \frac{d(\mathbf{X}_i, \mathbf{X}_j)}{\alpha(1 + \frac{v-1}{v_{\max}})} \tag{23.17}$$

In the above expression, $N_{s\times s}(r)$denotes the local area of perception surrounding the site of radius r, which the ant occupies in the two-dimensional grid. The threshold a scales the dissimilarity within each pair of objects, and the moving speed v controls the step-size of the ant searching in the space within one time unit. If an ant is not carrying an object and finds an object $\mathbf{X}_i$ in its neighborhood, it picks up this object with a probability that is inversely proportional to the number of similar objects in the neighborhood. It may be expressed as:

$$P_{pick-up}(\mathbf{X}_i) = [\frac{k_p}{k_p + f(\mathbf{X}_i)}]^2 \tag{23.18}$$

If however, the ant is carrying an object x and perceives a neighbor's cell in which there are other objects, then the ant drops off the object it is carrying with a probability that is directly proportional to the object's similarity with the perceived ones. This is given by:

$$\begin{aligned} P_{drop}(\vec{X}_i) &= 2.f(\vec{X}_i) \quad \text{if} \quad f(X_i) < k_d \\ &= 1 \qquad\qquad\; \text{if} \quad f(\vec{X}_i) \geq k_d \end{aligned} \tag{23.19}$$

The parameters k_p and k_d are the picking and dropping constants [41] respectively. Function $f(\mathbf{X}_i)$ provides an estimate of the density and similarity of elements in the neighborhood of object $\mathbf{X}_i$. The standard ACA algorithm is summarized in the following pseudo-code.

Kanade and Hall (2003) presented a hybridization of the ant systems with the classical FCM algorithm to determine the number of clusters in a given dataset automatically. In their fuzzy ant algorithm, at first the ant based clustering is used to create raw clusters and then these clusters are refined using the FCM algorithm. Initially the ants move the individual data objects to form heaps. The centroids of these heaps are taken as the initial cluster centers and the FCM algorithm is used to refine these clusters. In the second stage the objects obtained from the FCM algorithm are hardened according to the maximum membership criteria to form new heaps. These new heaps are then sometimes moved and merged by the ants. The final clusters formed are refined by using the FCM algorithm.

Procedure ACA

Place every item $\vec{X}_i$ on a random cell of the grid;
Place every ant k on a random cell of the grid unoccupied by ants;
iteration_count ← 1;
While iteration_count < maximum_iteration
  do
    for i = 1 to no_of_ants // for every ant
      do
        if unladen ant AND cell occupied by item $\vec{X}_i$,
        then compute $f(\vec{X}_i)$ and $P_{pick-up}(\vec{X}_i)$;
        pick up item $\vec{X}_i$ with probability $P_{pick-up}(\vec{X}_i)$
        else if ant carrying item xi AND cell empty,
          then compute $f(\vec{X}_i)$ and $P_{drop}(\vec{X}_i)$;
          drop item $\vec{X}_i$ with probability $P_{drop}(\vec{X}_i)$;
        end if
          move to a randomly selected, neighboring and unoccupied cell ;
    end for
    t ← t + 1
end while
print location of items;
end procedure

A number of modifications have been introduced to the basic ant based clustering scheme that improve the quality of the clustering, the speed of convergence and, in particular, the spatial separation between clusters on the grid, which is essential for the scheme of cluster retrieval. A detailed description of the variants and results on the qualitative performance gains afforded by these extensions are provided in (Tsang and Kwong, 2006).

23.4.2 The PSO-based Clustering Algorithms

Research efforts have made it possible to view data clustering as an optimization problem. This view offers us a chance to apply PSO algorithm for evolving a set of candidate cluster centroids and thus determining a near optimal partitioning of the dataset at hand. An important advantage of the PSO is its ability to cope with local optima by maintaining, recombining and comparing several candidate solutions simultaneously. In contrast, local search heuristics, such as the simulated annealing algorithm (Selim and Alsultan, 1991) only refine a single candidate solution and are notoriously weak in coping with local optima. Deterministic local search, which is used in algorithms like the k-means, always converges to the nearest local optimum from the starting position of the search.

PSO-based clustering algorithm was first introduced by Omran *et al.* (2002). The results of Omran *et al.* (2002, 2005) showed that PSO based method outperformed k-means, FCM and a few other state-of-the-art clustering algorithms. In their method, Omran *et al.* used a quantization error based fitness measure for judging the performance of a clustering algorithm. The quantization error is defined as:

$$J_e = \frac{\sum_{i=1}^{k} \sum_{\forall \mathbf{X}_j \in C_i} d(\mathbf{X}_j, \mathbf{V}_i)/n_i}{k} \tag{23.20}$$

where *Ci* is the *i*-th cluster center and *ni* is the number of data points belonging to the *i*-th cluster. Each particle in the PSO algorithm represents a possible set of *k* cluster centroids as:

$\vec{Z}_i(t)$ =

$\vec{V}_{i,1}$	$\vec{V}_{i,2}$	....	$\vec{V}_{i,k}$

where $\mathbf{V}_{i,p}$ refers to the *p*-th cluster centroid vector of the *i*-th particle. The quality of each particle is measured by the following fitness function:

$$f(\mathbf{Z}_i, M_i) = w_1 \bar{d}_{\max}(M_i, \mathbf{X}_i) + w_2(R_{\max} - d_{\min}(\mathbf{Z}_i)) + w_3 J_e \tag{23.21}$$

In the above expression, Rmax is the maximum feature value in the dataset and ***Mi*** is the matrix representing the assignment of the patterns to the clusters of the i-th particle. Each element mi, k, p indicates whether the pattern $\mathbf{X}_p$ belongs to cluster C_k of *i*-th particle. The user-defined constants $w1$, $w2$, and $w3$ are used to weigh the contributions from different sub-objectives. In addition,

$$\bar{d}_{\max} = \max_{j \in 1,2,\ldots,k} \{ \sum_{\forall \mathbf{X}_p \in C_{i,j}} d(\mathbf{X}_p, \mathbf{V}_{i,j})/n_{i,j} \} \tag{23.22}$$

and,

$$d_{\min}(\mathbf{Z}_i) = \min_{\forall p,q,p \neq q} \{ d(\mathbf{V}_{i,p}, \mathbf{V}_{i,q}) \} \tag{23.23}$$

is the minimum Euclidean distance between any pair of clusters. In the above, *ni,k* is the number of patterns that belong to cluster **C***i,k* of particle *i*. he fitness function is a multi-objective optimization problem, which minimizes the intra-cluster distance, maximizes inter-cluster separation, and reduces the quantization error. The PSO clustering algorithm is summarized below.

Step 1: Initialize each particle with k random cluster centers.
Step 2: repeat for iteration_count = 1 to maximum_iterations
(a) repeat for each particle i
(i) repeat for each pattern $\vec{X}_p$ in the dataset
- calculate Euclidean distance of $\vec{X}_p$ with all cluster centroids.
- assign $\vec{X}_p$ to the cluster that have nearest centroid to $\vec{X}_p$

(ii) calculate the fitness function $f(\vec{Z}_i, M_i)$
(b) find the personal best and global best position of each particle.
(c) Update the cluster centroids according to velocity updating and coordinate updating formula of PSO.

Van der Merwe and Engelbrecht hybridized this approach with the k-means algorithm for clustering general dataets (van der Merwe and Engelbrecht, 2003). A single particle of the swarm is initialized with the result of the k-means algorithm. The rest of the swarm is initialized randomly. In 2003, Xiao *et al* used a new approach based on the synergism of the PSO and the Self Organizing Maps (SOM) (Xiao *et al.*, 2003) for clustering gene expression data. They got promising results by applying the hybrid SOM-PSO algorithm over the gene expression data of Yeast and Rat Hepatocytes. Paterlini and Krink (2006) have compared the performance of k-means, GA (Holland, 1975, Goldberg, 1975), PSO and Differential Evolution (DE) (Storn and Price, 1997) for a representative point evaluation approach to partitional clustering. The results show that PSO and DE outperformed the k-means algorithm.

Cui et al. (2005) proposed a PSO based hybrid algorithm for classifying the text documents. They applied the PSO, k-means and a hybrid PSO clustering algorithm on four different text document datasets. The results illustrate that the hybrid PSO algorithm can generate more compact clustering results over a short span of time than the k-means algorithm.

23.5 Automatic Kernel-based Clustering with PSO

The Euclidean distance metric, employed by most of the exisiting partitional clustering algorithms, work well with datasets in which the natural clusters are nearly hyper-spherical and linearly seperable (like the artificial dataset 1 used in this paper). But it causes severe misclassifications when the dataset is complex, with linearly non-separable patterns (like the synthetic datasets 2, 3, and 4 described in Section 5.8.1 of the chapter). We would like to mention here that, most evolutionary algorithms could potentially work with an arbitrary distance function and are not limited to the Euclidean distance.

Moreover, very few works (Bandyopadhyay and Maulik, 2000, Rosenberger and Chehdi, 2000, Omran *et al.*, 2005, Sarkar *et al.*, 1997) have been undertaken to make an algorithm learn the correct number of clusters 'k' in a dataset, instead of accepting the same as a user input. Although, the problem of finding an optimal k is quite important from a practical point of view, the research outcome is still unsatisfactory even for some of the benchmark datasets (Rosenberger and Chehdi, 2000).

In this Section, we describe a new approach towards the problem of automatic clustering (without having any prior knowledge of k initially) in kernel space using a modified version of the PSO algorithm (Das *et al.*, 2008). Our procedure employs a kernel induced similarity measure instead of the conventional Euclidean distance metric. A kernel function measures the distance between two data points by implicitly mapping them into a high dimensional feature space where the data is linearly separable. Not only does it preserve the inherent structure of groups in the input space, but also simplifies the associated structure of the data patterns (Girolami, 2002). Several kernel-based learning methods, including the Support Vector Machine (SVM), have recently been shown to perform remarkably in supervised learning (Scholkopf and Smola, 2002, Vapnik, 1998, Zhang and Chen, 2003, Zhang and Rudnicky, 2002). The kernelized versions of the k-means and the fuzzy c-means (FCM) algorithms reported in (Zhang and Rudnicky, 2002) and (Zhang and Chen, 2003) respectively, have reportedly outperformed their original counterparts over several test cases.

Now, we may summarize the new contributions presented here in the following way:

1. Firstly, we develop an alternative framework for learning the number of partitions in a dataset besides the simultaneous refining of the clusters, through one shot of optimization.
2. We propose a new version of the PSO algorithm based on the multi-elitist strategy, well-known in the field of evolutionary algorithms. Our experiments indicate that the proposed MEPSO algorithm yields more accurate results at a faster pace than the classical PSO in context to the present problem.
3. We reformulate a recently proposed cluster validity index (known as the *CS measure* (Chou *et al.*, 2004)) using the kernelized distance metric. This reformulation eliminates the need to compute the cluster-centroids repeatedly for evaluating CS value, due to the implicit mapping via the kernel function. The new CS measure forms the objective function to be minimized for optimal clustering.

23.5.1 The Kernel Based Similarity Measure

Given a dataset **X** in the d-dimensional real space $\Re^d$, let us consider a non-linear mapping function from the input space to a high dimensional feature space H:

$$\varphi : \Re^d \rightarrow H, \mathbf{x}_i \rightarrow \varphi(\mathbf{x}_i) \tag{23.24}$$

where $\mathbf{x}_i = [x_{i,1}, x_{i,2}, \ldots.., x_{i,d}]^T$ and

$$\varphi(\mathbf{x}_i) = [\varphi_1(\mathbf{x}_i), \varphi_2(\mathbf{x}_i), \ldots.., \varphi_H(\mathbf{x}_i)]^T$$

By applying the mapping, a dot product $\mathbf{x}_i^T.\mathbf{x}_j$is transformed into $\varphi^T(\mathbf{x}_i).\varphi(\mathbf{x}_j)$. Now, the central idea in kernel-based learning is that the mapping function φneed not be explicitly specified. The dot product $\varphi^T(\mathbf{x}_i).\varphi(\mathbf{x}_j)$in the transformed space can be calculated through the kernel function $K(\mathbf{x}_i, \mathbf{x}_j)$ in the input space $\Re^d$. Consider the following simple example:

Example 1: let d = 2 and H = 3 and consider the following mapping:
$\varphi : \Re^2 \rightarrow H = \Re^3$, and $[x_{i,1}, x_{i,2}]^T \rightarrow [x_{i,1}^2, \sqrt{2}.x_{i,1}x_{i,2}, x_{i,2}^2]^T$
Now the dot product in feature space H:

$$\varphi^T(\mathbf{x}_i).\varphi(\mathbf{x}_j) = [x_{i,1}^2, \sqrt{2}.x_{i,1}x_{i,2}, x_{i,2}^2].[x_{j,1}^2, \sqrt{2}.x_{j,1}.x_{j,2}, x_{j,2}^2]^T.$$

$$= [x_{i,1}.x_{j,1} + x_{i,2}.x_{j,2}]^2$$

$$= [\mathbf{x}_i^T.\mathbf{x}_j]^2 = K(\mathbf{x}_i, \mathbf{x}_j)$$

Clearly the simple kernel function K is the square of the dot product of vectors $\mathbf{x}_i$ and $\mathbf{x}_j$in $\Re^d$. Hence, the kernelized distance measure between two patterns $\mathbf{x}_i$ and $\mathbf{x}_j$is given by:

$$\left\|\varphi(\mathbf{x}_i) - \varphi(\mathbf{x}_j)\right\|^2 = (\varphi(\mathbf{x}_i) - \varphi(\mathbf{x}_j))^T(\varphi(\mathbf{x}_i) - \varphi(\mathbf{x}_j))$$

$$= \varphi^T(\mathbf{x}_i).\varphi(\mathbf{x}_i) - 2.\varphi^T(\mathbf{x}_i).\varphi(\mathbf{x}_j) + \varphi^T(\mathbf{x}_j).\varphi(\mathbf{x}_j)$$

$$= K(\mathbf{x}_i, \mathbf{x}_i) - 2.K(\mathbf{x}_i, \mathbf{x}_j) + K(\mathbf{x}_j, \mathbf{x}_j) \tag{23.25}$$

Among the various kernel functions used in literature, in the present context, we have chosen the well-known Gaussian kernel (also referred to as the Radial Basis Function) owing to its better classification accuracy over the linear and polynomial kernels on many test problems (Pirooznia and Deng, 2006,Hertz *et al.*, 2006). The Gaussian Kernel may be represented as:

$$K(\mathbf{x}_i, \mathbf{x}_j) = \exp\left(-\frac{\left\|\mathbf{x}_i - \mathbf{x}_j\right\|^2}{2\sigma^2}\right) \tag{23.26}$$

where $\sigma > 0$. Clearly, for Gaussian kernel, $K(\mathbf{x}_i, \mathbf{x}_i)$= 1 and thus relation 23.25 reduces to:

$$\left\|\varphi(\mathbf{x}_i) - \varphi(\mathbf{x}_j)\right\|^2 = 2.(1 - K(\mathbf{x}_i, \mathbf{x}_j)) \tag{23.27}$$

23.5.2 Reformulation of CS Measure

Cluster validity indices correspond to the statistical-mathematical functions used to evaluate the results of a clustering algorithm on a quantitative basis. For crisp clustering, some of the well-known indices available in the literature are the Dunn's index (DI) (Dunn, 1974), Calinski-Harabasz index (Calinski and Harabasz, 1975), Davis-Bouldin (DB) index (Davies and Bouldin, 1979), PBM index (Pakhira *et al.*, 2004), and the CS measure (Chou *et al.*, 2004). In this work, we have based our fitness function on the CS measure as according to the authors, CS measure is more efficient in tackling clusters of different densities and/or sizes than the other popular validity measures, the price being paid in terms of high computational load with increasing k and n (Chou *et al.*, 2004). Before applying the CS measure, centroid of a cluster is computed by averaging the data vectors belonging to that cluster using the formula,

$$\mathbf{m}_i = \frac{1}{N_i} \sum_{x_j \in C_i} \mathbf{x}_j \tag{23.28}$$

A distance metric between any two data points $\mathbf{x}_i$ and $\mathbf{x}_j$ is denoted by $d(\mathbf{x}_i, \mathbf{x}_j)$. Then the CS measure can be defined as,

$$CS(k) = \frac{\frac{1}{k}\sum_{i=1}^{k}\left[\frac{1}{N_i}\sum_{\mathbf{X}_i \in C_i} \max_{\mathbf{X}_q \in C_i}\{d(\mathbf{x}_i, \mathbf{x}_q)\}\right]}{\frac{1}{k}\sum_{i=1}^{k}\left[\min_{j \in K, j \neq i}\{d(\mathbf{m}_i, \mathbf{m}_j)\}\right]} = \frac{\sum_{i=1}^{k}\left[\frac{1}{N_i}\sum_{\mathbf{X}_i \in C_i} \max_{\mathbf{X}_q \in C_i}\{d(\mathbf{x}_i, \mathbf{x}_q)\}\right]}{\sum_{i=1}^{k}\left[\min_{j \in K, j \neq i}\{d(\mathbf{m}_i, \mathbf{m}_j)\}\right]} \tag{23.29}$$

Now, using a Gaussian kernelized distance measure and transforming to the high dimensional feature space, the CS measure reduces to (using relation (23)):

$$CS_{\mathrm{ker}nel}(k) = \frac{\sum_{i=1}^{k}[\frac{1}{N_i}\sum_{\mathbf{X}_i \in C_i} \max_{\mathbf{X}_q \in C_i}\{||\varphi(\mathbf{x}_i) - \varphi(\mathbf{x}_q)||^2\}]}{\sum_{i=1}^{k}[\min_{j \in K, j \neq i}\{||\varphi(\mathbf{m}_i) - \varphi(\mathbf{m}_j)||\}]}$$
$$= \frac{\sum_{i=1}^{k}[\frac{1}{N_i}\sum_{\mathbf{X}_i \in C_i} \max_{\mathbf{X}_q \in C_i}\{2(1 - K(\mathbf{x}_i, \mathbf{x}_q))\}]}{\sum_{i=1}^{k}[\min_{j \in K, j \neq i}\{2(1 - K(\mathbf{m}_i, \mathbf{m}_j))\}]}$$

The minimum value of this CS measure indicates an optimal partition of the dataset. The value of 'k' which minimizes CSkernel(k) therefore gives the appropriate number of clusters in the dataset.

23.5.3 The Multi-Elitist PSO (MEPSO) Algorithm

The canonical PSO has been subjected to empirical and theoretical investigations by several researchers (Eberhart and Shi, 2001, Clerc and Kennedy, 2002). In many occasions, the convergence is premature, especially if the swarm uses a small inertia weight ω or constriction coefficient (Eberhart and Shi, 2001). As the global best found early in the searching process may be a poor local minima, we propose a multi-elitist strategy for searching the global best of the PSO. We call the new variant of PSO the MEPSO. The idea draws inspiration from the works reported in (Deb *et al.*, 2002). We define a growth rate β for each particle. When the fitness value of a particle at the t-th iteration is higher than that of a particle at the (t-1)-th iteration, the β will be increased. After the local best of all particles are decided in each generation, we move the local best, which has higher fitness value than the global best into the candidate area. Then the global best will be replaced by the local best with the highest growth rate β. The elitist concept can prevent the swarm from tending to the global best too early in the searching process. The MEPSO follows the *g_best* PSO topology in which the entire swarm is treated as a single neighborhood. The pseudo code about MEPSO is as follows:

Procedure MEPSO
For t =1 to t_{max}
For j =1 to N *// swarm size is N*
If (the fitness value of particle$_j$ in t-th time-step> that of particle$j$ in (t-1)-th time-step)
$\beta_j(t) = \beta_j(t\text{-}1) + 1;$
End
Update Local best$j$.
If (the fitness of Local bestj > that of Global best now)
Choose Local bestj put into candidate area.
End
End
Calculate β of every candidate, and record the candidate of β_{max} .
Update the Global best to become the candidate of
β_{max} .
Else
Update the Global best to become the particle of highest fitness value.
End
End

23.5.4 Particle Representation

In the proposed method, for n data points, each p-dimensional, and for a user-specified maximum number of clusters $k$max, a particle is a vector of real numbers of dimension $k$max + $k$max $\times$ p. The first $k$max entries are positive floating-point numbers in (0, 1), each of which determines whether the corresponding cluster is to be activated (i.e. to be really used for classifying the data) or not. The remaining entries are reserved for kmax cluster centers, each p-dimensional.

A single particle is illustrated as:

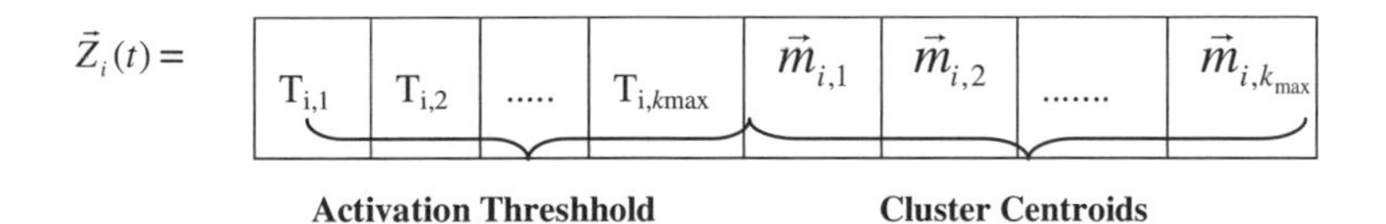

The j-th cluster center in the i-th particle is active or selected for partitioning the associated dataset if $T_{i,j} > 0.5$. On the other hand, if $T_{i,j} < 0.5$, the particular j-th cluster is inactive in the i-th particle. Thus the $T_{i,j}$ s behave like control genes (we call them *activation thresholds*) in the particle governing the selection of the active cluster centers. The rule for selecting the actual number of clusters specified by one chromosome is:

$$\textbf{IF } T_{i,j} > 0.5 \textbf{ THEN } \text{the } j\text{-th cluster center } \vec{m}_{i,j} \text{ is } \textbf{ACTIVE}$$
$$\textbf{ELSE } \vec{m}_{i,j} \text{ is } \textbf{INACTIVE} \qquad (23.30)$$

Consider the following example:
Example 2: Positional coordinates of one particular particle is illustrated below. There are at most five 3-dimensional cluster centers among which, according to the rule presented in Equation 23.30 the second (6, 4.4, 7), third (5.3, 4.2, 5) and fifth one (8, 4, 4) have been activated for partitioning the dataset and marked in bold. The quality of the partition yielded by such a particle can be judged by an appropriate cluster validity index.

0.3	**0.6**	**0.8**	0.1	**0.9**	6.1	3.2	2.1	**6**	**4.4**	**7**	**9.6**	**5.3**	**4.2**	5	8	4.6	**8**	**4**	**4**

During the PSO iterations, if some threshold T in a particle exceeds 1 or becomes negative, it is fixed to 1 or zero, respectively. However, if it is found that no flag could be set to one in a particle (all activation threshholds are smaller than 0.5), we randomly select 2 thresholds and re-initialize them to a random value between 0.5 and 1.0. Thus the minimum number of possible clusters is 2.

23.5.5 The Fitness Function

One advantage of the proposed algorithm is that it can use any suitable validity index as its fitness function. We have used the kernelized CS measure as the basis of our fitness function, which for i-th particle can be described as:

$$f_i = \frac{1}{CS_{\mathrm{ker}nel_i}(k) + eps} \tag{23.31}$$

where eps is a very small constant (we used 0.0002). Maximization of fi implies a minimization of the kernelized CS measure leading to the optimal partitioning of the dataset.

23.5.6 Avoiding Erroneous particles with Empty Clusters or Unreasonable Fitness Evaluation

There is a possibility that in our scheme, during computation of the kernelized CS index, a division by zero may be encountered. For example, the positive infinity (such as 4.0/0.0) or the not-a-number (such as 0.0/0.0) condition always occurs when one of the selected cluster centers is outside the boundary of distributions of data set as far as possible. To avoid this problem we first check to see if in any particle, any cluster has fewer than 2 data points in it. If so, the cluster center positions of this special particle are re-initialized by an average computation. If k clusters ($2 \leq k \leq k_{\max}$) are selected for this particle, we put n/k data points for every individual activated cluster center, such that a data point goes with a center that is nearest to it.

23.5.7 Putting It All Together

Step 1: Initialize each particle to contain k number of randomly selected cluster centers and k (randomly chosen) activation thresholds in [0, 1].

Step 2: Find out the active cluster centers in each particle with the help of the rule described in (10).

Step 3: For $t = 1$ to $t_{\max}$ **do**

i) For each data vector $\vec{X}_p$, calculate its distance metric $d(\vec{X}_p, \vec{m}_{i,j})$ from all active cluster centers of the i-th particle $\vec{V}_i$.

ii) Assign $\vec{X}_p$ to that particular cluster center $\vec{m}_{i,j}$ where

$$d(\vec{X}_p, \vec{m}_{i,j}) = \min_{\forall b \in \{1,2,\ldots,k\}} \left\{ d(\vec{X}_p, \vec{m}_{i,b}) \right\}$$

iii) Check if the number of data points belonging to any cluster center $\vec{m}_{i,j}$ is less than 2. If so, update the cluster centers of the particle using the concept of average described earlier.

iv) Change the population members according to the MEPSO algorithm. Use the fitness of the particles to guide the dynamics of the swarm.

Step 4: Report as the final solution the cluster centers and the partition obtained by the globally best particle (one yielding the highest value of the fitness function) at time $t = t_{\max}$.

23.5.8 Experimental Results

Comparison with Other Clustering Algorithms

To test the effectiveness of the proposed method, we compare its performance with six other clustering algorithms using a test-bed of five artificial and three real world datasets. Among the competitors, there are two recently developed automatic clustering algorithms well- known as the GCUK (Genetic Clustering with an Unknown number of clusters k) (Bandyopadhyay and Maulik, 2000) and the DCPSO (Dynamic Clustering PSO) (Omran *et al.*, 2005). Moreover, in order to investigate the effects of the changes made in the classical *g_best* PSO algorithm, we have compared MEPSO with an ordinary PSO based kernel-clustering method that uses the same particle representation scheme and fitness function as the MEPSO. Both the algorithms were let run on the same initial populations. The rest of the competitors are the kernel k-means algorithm (Zhang and Rudnicky, 2002) and a kernelized version of the subtractive clustering (Kim *et al.*, 2005). Both the algorithms were provided with the correct number of clusters as they are non-automatic.

We used datasets with a wide variety in the number and shape of clusters, number of datapoints and the count of features of each datapoint. The real life datasets used here are the Glass, the placeWisconsin breast cancer, the image segmentation, the Japanese Vowel and the automobile (Handl *et al.*, 2003). The synthetic datasets included here, comes with linearly non-separable clusters of different shapes (like elliptical, concentric circular dish and shell, rectangular etc). Brief details of the datasets have been provided in Table 1. Scatterplot of the synthetic datasets have also been shown in Figure 5.The clustering results were judged using Huang's accuracy measure (Huang and Ng, 1999):

$$r = \frac{\sum_{i=1}^{k} n_i}{n}, \tag{23.32}$$

where n_i is the number of data occurring in both the i-th cluster and its corresponding true cluster, and n is the total number of data points in the data set. According to this measure, a higher value of r indicates a better clustering result, with perfect clustering yielding a value of $r = 1$.

We used $\sigma = 1.1$ for all the artificial datasets, $\sigma = 0.9$ for breast cancer dataset and $\sigma = 2.0$ for the rest of the real life datasets for the RBF kernel following (Lumer and Faieta, 1995).In these experiments, the kernel k-means was run 100 times with the initial centroids randomly selected from the data set. A termination criterion of $\varepsilon = 0.001$. The parameters of the kernel-based subtractive methods were set to $\alpha = 5.4$ and $\beta = 1.5$ as suggested by Pal and Chakraborty (Kuntz and Snyers, 1994). For all the competitive algorithms, we have selected their best parameter settings as reported in the corresponding literatures. The control parameters for MEPSO where chosen after experimenting with several possible values. Some of the experiments focussing on the effects of parameter-tuning in MEPSO has been reported in the next subsection. The same set of parameters were used for all the test problems for all the algorithms. These parameter settings have been reported in Table 2.

Table 3 compares the algorithms on the quality of the optimum solution as judged by the Huang's measure. The mean and the standard deviation (within parentheses) for 40 independent runs (with different seeds for the random number generator) of each of the six algorithms are presented in Table 3. Missing values of standard deviation in this table indicate a zero standard deviation. The best solution in each case has been shown in bold. Table 4 and 5 present the mean and standard deviation of the number of classes found by the three automatic clustering algorithms. In Figure 2 we present the clustering results on the synthetic datasets by the new MEPSO algorithm (to save space we do not provide results for all the six algorithms).

For comparing the speed of the stochastic algorithms like GA, PSO etc. we choose *number of fitness function evaluations* (placeFEs) as a measure of computation time instead of generations or iterations. From the data provided in Table 3, we choose a threshold value of the classification accuracy for each dataset. This threshold value is somewhat larger than the minimum accuracy attained by each automatic clustering algorithm. Now we run an algorithm on each dataset and stop as soon as it achieves the proper number of clusters as well as the threshold accuracy. We then note down the number of fitness function evaluations the algorithm takes. A lower number of placeFEs corresponds to a faster algorithm. The speed comparison results are provided in Table 6. The kernel k-means and the subtractive clustering method are not included in this table, as they are non-automatic and do not employ evolutionary operators as in GCUK and PSO based methods.

Table 1: Description of the Datasets

Dateset	Number of Datapoints	Number of clusters	Data-dimension
Synthetic_1	500	2	2
Synthetic_2	52	2	2
Synthetic_3	400	4	3
Synthetic_4	250	5	2
Synthetic_5	600	2	2
Glass	214	6	9
Wine	178	3	13
Breast Cancer	683	2	9
Image Segmentation	2310	7	19
Japanese Vowel	640	9	12

Table 2: Parameter Settings for different algorithms

GCUK		DCPSO		PSO		MEPSO	
Pop_size	70	Pop_size	100	Pop_ size	40	Pop_ size	40
Cross-over Probability μ_c	0.85	Inertia Weight	0.72	Inertia Weight	0.75	Inertia Weight	0.794
Mutation probability μ_m	0.005	C_1, C_2	1.494	C_1, C_2	2.00	C_1, C_2	$0.35 \rightarrow 2.4$ $2.4 \rightarrow 0.35$
		P_{ini}	0.75				
K_{max} K_{min}	20 2	K_{max} K_{min}	20 2	K_{max} K_{min}	20 2	K_{max} K_{min}	20 2

A review of Tables 3, 4 and 5 reveals that the kernel based MEPSO algorithm performed markedly better as compared to the other considered clustering algorithms, in terms of both accuracy and convergence speed. We note that in general, the kernel based clustering methods outperform the GCUK or DCPSO algorithms (which do not use the kernelized fitness function) especially on linearly non-separable artificial datasets like synthetic_1, synthetic_2 and synthetic_5. Although the proposed method provided a better clustering result than the other methods for Synthetic_5 dataset, its accuracy for this data was lower than the seven other data sets considered. This indicates that the proposed approach is limited in its ability to classify non-spherical clusters.

The PSO based methods (especially MEPSO) on average took lesser computational time than the GCUK algorithm over most of the datasets. One possible reason of this may be the use of less complicated variation operators (like mutation) in PSO as compared to the operators used for GA. We also note that the MEPSO performs much better than the classical PSO based kernel-clustering scheme. Since both the algorithms use same particle representation and starts with the same initial population, difference in their performance must be due to the difference in their internal operators and parameter values. This demonstrates the effectiveness of the multi-elitist strategy incorporated in the MEPSO algorithm.

Table 3: Mean and standard deviation of the clustering accuracy (%) achieved by each clustering algorithm over 40 independent runs (Each run continued up to 50, 000 FEs for GCUK, DCPSO, Kernel_ PSO and Kernel_MEPSO)

Datasets	Algorithms					
	Kernel k-means	Kernel Sub_clust	GCUK	DC-PSO	Kernel_ PSO	Kernel_ MEPSO
Synthetic_1	83.45 (0.032)	87.28	54.98 (0.88)	57.84 (0.065)	90.56 (0.581)	**99.45 (0.005)**
Synthetic_2	71.32 (0.096)	75.73	65.82 (0.146)	59.91 (0.042)	61.41 (0.042)	**80.92 (0.0051)**
Synthetic_3	89.93 (0.88)	94.03	97.75 (0.632)	97.94 (0.093)	92.94 (0.193)	**99.31 (0.001)**
Synthetic_4	67.65 (0.104)	80.25	74.30 (0.239)	75.83 (0.033)	78.85 (0.638)	**87.84 (0.362)**
Synthetic_5	81.23 (0.127)	84.33	54.45 (0.348)	52.55 (0.209)	89.46 (0.472)	**99.75 (0.001)**
Glass	68.92 (0.032)	73.92	76.27 (0.327)	79.45 (0.221)	70.71 (0.832)	**92.01 (0.623)**
Wine	73.43 (0.234)	59.36	80.64 (0.621)	85.81 (0.362)	87.65 (0.903)	**93.17 (0.002)**
Breast Cancer	66.84 (0.321)	70.54	73.22 (0.437)	78.19 (0.336)	80.49 (0.342)	**86.35 (0.211)**
Image Segmentation	56.83 (0.641)	70.93	78.84 (0.336)	81.93 (1.933)	84.32 (0.483)	**87.72 (0.982)**
Japanese Vowel	44.89 (0.772)	61.83	70.23 (1.882)	82.57 (0.993)	79.32 (2.303)	**84.93 (2.292)**
Average	72.28	75.16	74.48	76.49	77.58	**91.65**

Table 4: Mean and standard deviation (in parenthesis) of the number of clusters found over the synthetic datasets for four automatic clustering algorithms over 40 independent runs.

Algorithms	Synthetic _1	Synthetic _2	Synthetic _3	Synthetic _4	Synthetic _5
GCUK	2.50 (0.021)	3.05 (0.118)	4.15 (0.039)	9.85 (0.241)	4.25 (0.921)
DCPSO	2.45 (0.121)	2.80 (0.036)	4.25 (0.051)	9.05 (0.726)	6.05 (0.223)
Ordinary PSO	2.50 (0.026)	2.65 (0.126)	4.10 (0.062)	9.35 (0.335)	2.25 (0.361)
Kernel_MEPSO	**2.10 (0.033)**	**2.15 (0.102)**	**4.00 (0.00)**	**10.05 (0.021)**	**2.05 (0.001)**

Table 5: Mean and standard deviation (in parenthesis) of the number of clusters found over the synthetic datasets for four automatic clustering algorithms over 40 independent runs.

Algorithms	Glass	Wine	Breast Cancer	Image Segmentation	Japanese Vowel
GCUK	5.85 (0.035)	4.05 (0.021)	2.25 (0.063)	7.05 (0.008)	9.50 (0.218)
DCPSO	5.60 (0.009)	3.75 (0.827)	2.25 (0.026)	7.50 (0.057)	10.25 (1.002)
Ordinary PSO	5.75 (0.075)	**3.00 (0.00)**	**2.00 (0.00)**	7.20 (0.025)	9.25 (0.822)
Kernel_MEPSO	**6.05 (0.015)**	**3.00 (0.00)**	**2.00 (0.00)**	**7.00 (0.00)**	**9.05 (0.021)**

Table 6: Mean and standard deviations of the number of fitness function evaluations (over 40 successful runs) required by each algorithm to reach a predefined cut-off value of the classification accuracy

Dateset	Threshold accuracy (in %)	GCUK	DCPSO	Ordinary PSO	Kernel_ MEPSO
Synthetic_1	50.00	48000.05 (21.43)	42451.15 (11.57)	43812.30 (2.60)	**37029.65 (17.48)**
Synthetic_2	55.00	41932.10 (12.66)	45460.25 (20.97)	40438.05 (18.52)	**36271.05 (10.41)**
Synthetic_3	85.00	40000.35 (4.52)	35621.05 (12.82)	37281.05 (7.91)	**32035.55 (4.87)**
Synthctic_4	65.00	46473.25 (7.38)	43827.65 (2.75)	42222.50 (2.33)	**36029.05 (6.38)**
Synthetic_5	50.00	43083.35 (5.30)	39392.05 (7.20)	42322.05 (2.33)	**35267.45 (9.11)**
Glass	65.00	47625.35 (6.75)	40382.15 (7.27)	38292.25 (10.32)	**37627.05 (12.36)**
Winc	55.00	44543.70 (44.89)	43425.00 (18.93)	3999.65 (45.90)	**35221.15 (67.92)**
Breast Cancer	65.00	40390.00 (11.45)	37262.65 (13.64)	35872.05 (8.32)	**32837.65 (4.26)**
Image Segmentation	55.00	39029.05 (62.09)	40023.25 (43.92)	35024.35 (101.26)	**34923.22 (24.28)**
Japanese Vowel	40.00	40293.75 (23.33)	28291.60 (121.72)	29014.85 (21.67)	**24023.95 (20.62)**

Choice of Parameters for MEPSO

The MEPSO has a number of control parameters that affect its performance on different clustering problems. In this Section we discuss the influence of parameters like swarm size, the inertia factor ω and the acceleration factors C1 and C2 on the Kernel_MEPSO algorithm.

1. **Swarm Size**: To investigate the effect of the swarm size, the MEPSO was executed separately with 10 to 80 particles (keeping all other parameter settings same as reported in

Table 2) on all the datasets. In Figure 6 we plot the convergence behavior of the algorithm (average of 40 runs) on the image segmentation dataset (with 2310 data points and 19 features, it is the most complicated synthetic dataset considered here) for different population sizes. We omit the other results here to save space. The results reveal that the number of particles more than 40 gives more or less identical accuracy of the final clustering results for MEPSO. This observation is in accordance with Van den Bergh and Engelbrecht, who in (van den Bergh and Engelbrecht, 2001) showed that though there is a slight improvement in the solution quality with increasing swarm sizes, a larger swarm increases the number of function evaluations to converge to an error limit. For most of the problems, it was found that keeping the swarm size 40 provides a reasonable trade-off between the quality of the final results and the convergence speed.

2. **The inertia factor ω:** Provided all other parameters are fixed at the values shown in Table 2, the MEPSO was run with several possible choices of the inertia factor ω. Specifically we used a time-varying ω (linearly decreasing from 0.9 to 0.4 following (Shi and Eberhart, 1999)), random ω (Eberhart and Shi, 2001),$\omega = 0.1$, $\omega = 0.5$ and finally $\omega = 0.794$ (Kennedy *et al.*, 2001). In Figure 7, we show how the fitness of the globally best particle (averaged over 40 runs) varies with no. of placeFEs for the image segmentation dataset over different values of ω. It was observed that for all the problems belonging to the current test suit, best convergence behavior of MEPSO is observed for $\omega = 0.794$.
3. **The acceleration coefficients $C1$ and $C2$:** Provided all other parameters are fixed at the values given in Table 2, we let MEPSO run for different settings of the acceleration coefficients $C1$ and $C2$ as reported in various literatures on PSO. We used $C1 = 0.7$ and $C2 = 1.4$, $C1 = 1.4$ and $C2 = 0.7$ (Shi and Eberhart, 1999), $C1 = C2 = 1.494$ (Kennedy *et al.*, 2001), $C1=C2=2.00$ (Kennedy *et al.*, 2001) and finally a time varying acceleration coefficients where $C1$ linearly increases from 0.35 to 2.4 and $C2$ linearly decreases from 2.4 to 0.35 (Ratnaweera and Halgamuge, 2004). We noted that the linearly varying acceleration coefficients gave the best clustering results over all the problems considered. This is perhaps due to the fact that an increasing $C1$ and gradually decreasing $C2$ boost the global search over the entire search space during the early part of the optimization and encourage the particles to converge to global optima at the end of the search. Figure 8 illustrates the convergence characteristics of MEPSO over the image segmentation dataset for different settings of $C1$ and $C2$.

23.6 Conclusion and Future Directions

In this Chapter, we introduced some of the preliminary concepts of Swarm Intelligence (SI) with an emphasis on particle swarm optimization and ant colony optimization algorithms. We then described the basic data clustering terminologies and also illustrated some of the past and ongoing works, which apply different SI tools to pattern clustering problems. We proposed a novel kernel-based clustering algorithm, which uses a deviant variety of the PSO. The proposed algorithm can automatically compute the optimal number of clusters in any dataset and thus requires minimal user intervention. Comparison with a state of the art GA based clustering strategy, reveals the superiority of the MEPSO-clustering algorithm both in terms of accuracy and speed.

Despite being an age old problem, clustering remains an active field of interdisciplinary research till date. No single algorithm is known, which can group all real world datasets efficiently and without error. To judge the quality of a clustering, we need some specially designed statistical-mathematical function called the clustering validity index. But a literature survey reveals that, most of these validity indices are designed empirically and there is no universally good index that can work equally well over any dataset. Since, majority of the PSO or ACO based clustering schemes rely on a validity index to judge the fitness of several possible partitioning of the data, research effort should be spent for defining a reasonably good index function and validating the same mathematically.

Feature extraction is an important preprocessing step for data clustering. Often we have a great number of features (especially for a high dimensional dataset like a collection of text documents) which are not all relevant for a given operation. Hence, future research may focus on integrating the automatic feature-subset selection scheme with the SI based clustering algorithm. The two-step process is expected to automatically project the data to a low dimensional feature subspace, determine the number of clusters and find out the appropriate cluster centers with the most relevant features at a faster pace.

Gene expression refers to a process through which the coded information of a gene is converted into structures operating in the cell. It provides the physical evidence that a gene has been "turned on" or activated for protein synthesis (Lewin, 1995). Proper selection, analysis and interpretation of the gene expression data can lead us to the answers of many important problems in experimental biology. Promising results have been reported in (Xiao *et al.*, 2003) regarding the application of PSO for clustering the expression levels of gene subsets. The research effort to integrate SI tools in the mechanism of gene expression clustering may in near future open up a new horizon in the field of bioinformatic data mining.

Hierarchical clustering plays an important role in fields like information retrieval and web mining. The self-assembly behavior of the real ants may be exploited to build up new hierarchical tree-structured partitioning of a data set according to the similarities between those data items. A description of the little but promising work already been undertaken in this direction can be found in (Azzag *et al.*, 2006). But a more extensive and systematic research effort is necessary to make the ant based hierarchical models superior to existing algorithms like Birch (Zhang *et al.*, 1997).

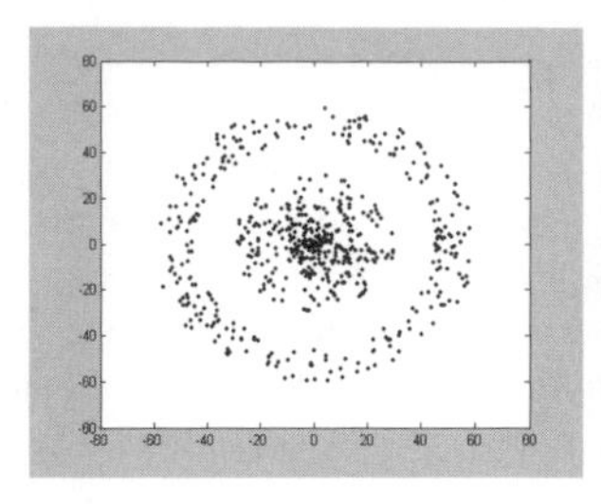

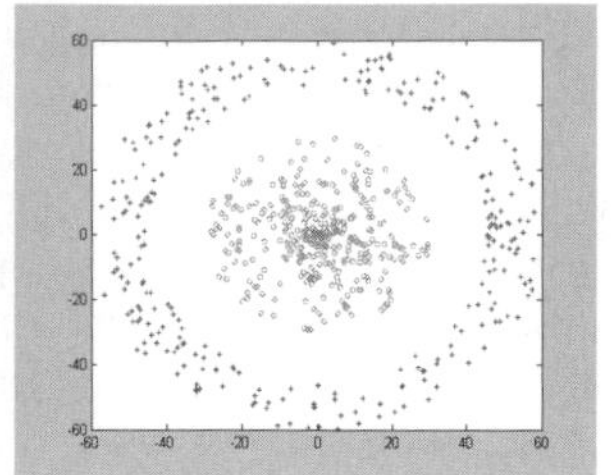

(a) Unlabeled Synthetic_1 Synthetic_1 Clustered with Kernel_MEPSO

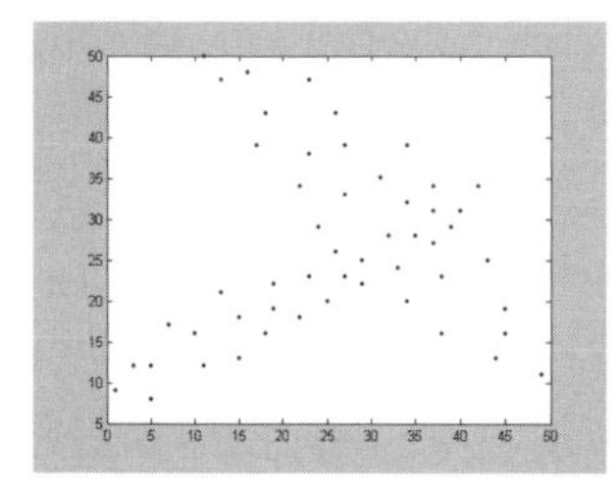

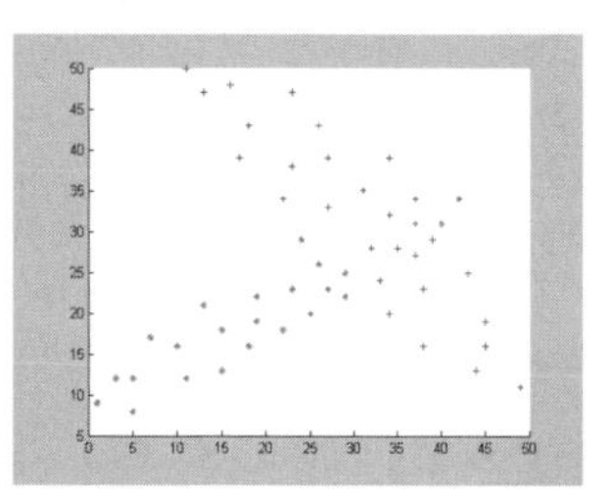

(b) Unlabeled Synthetic_2 Synthetic_2 Clustered with Kernel_MEPSO

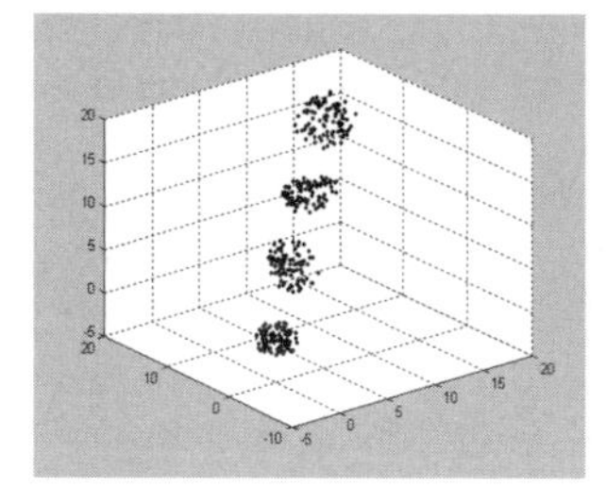

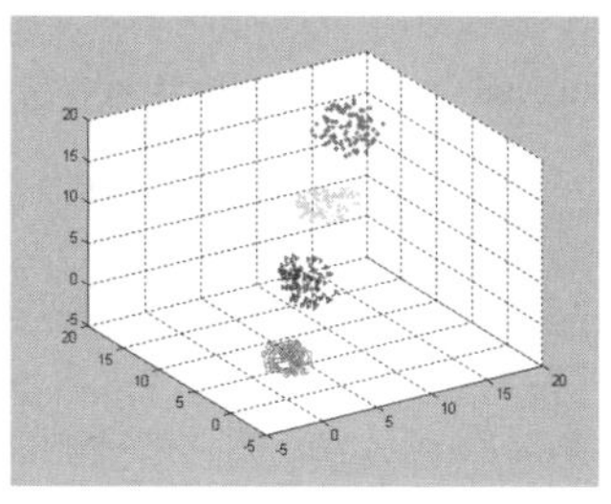

(c) Unlabeled Synthetic_3 Synthetic_3 Clustered with Kernel_MEPSO

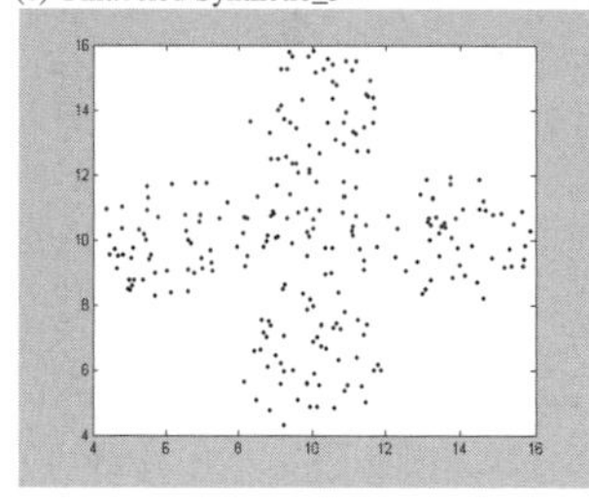

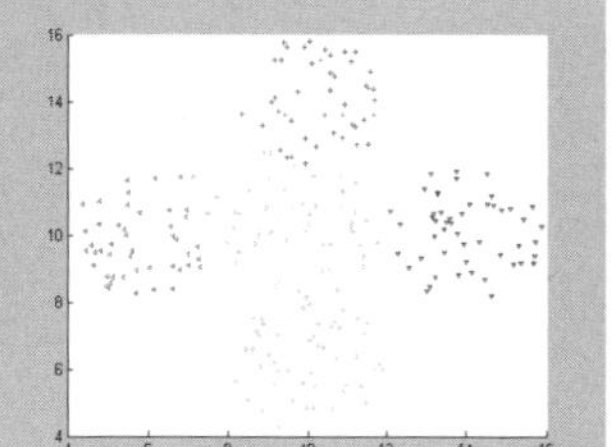

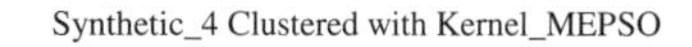

(d) Unlabeled Synthetic_4 Synthetic_4 Clustered with Kernel_MEPSO

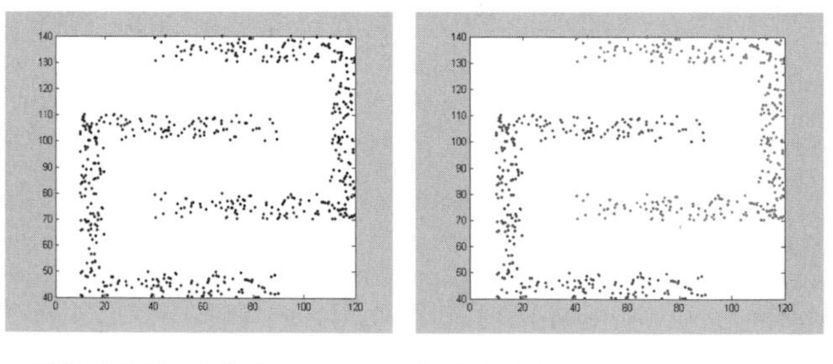

(e) Unlabeled Synthetic_5 Synthetic_5 Clustered with Kernel_MEPSO

Fig. 23.5. Two- and three-dimensional synthetic datasets clustered with MEPSO.

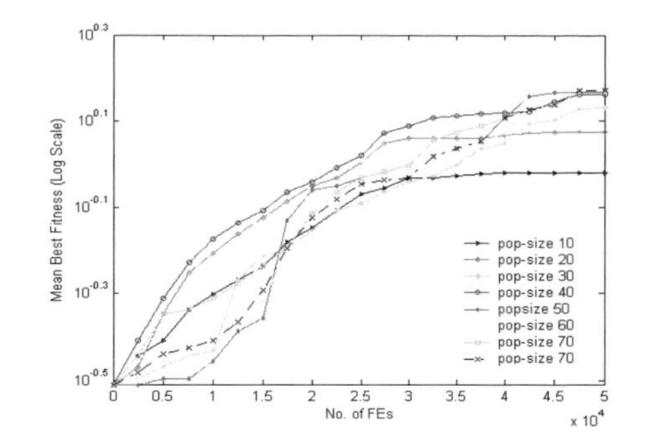

Fig. 23.6. The convergence characteristics of the MEPSO over the Image segmentation dataset for different population sizes.

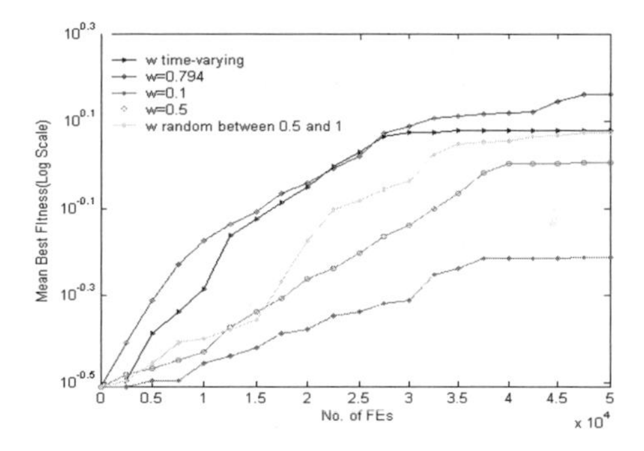

Fig. 23.7. The convergence characteristics of the MEPSO over the Image segmentation dataset for different inertia factors.

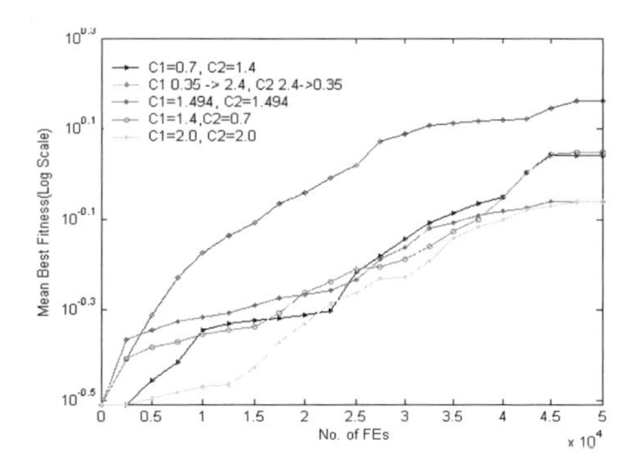

Fig. 23.8. The convergence characteristics of the MEPSO over the Image segmentation dataset for different acceleration coefficients.

References

A. Abraham, C. Grosan and V. Ramos (Eds.) (2006), Swarm Intelligence and Data Mining, Studies in Computational Intelligence, Springer Verlag, Germany, pages 270, ISBN: 3-540-34955-3.

Ahmed MN, Yaman SM, Mohamed N, Farag AA and Moriarty TA (2002) Modified fuzzy c-means algorithm for bias field estimation and segmentation of MRI data. IEEE Trans Med Imaging, 21, pp. 193199.

Azzag H, Guinot C and Venturini G (2006) Data and text mining with hierarchical clustering ants, in Swarm Intelligence in Data Mining, Abraham A, Grosan C and Ramos V (Eds), Springer, pp. 153-186.

Bandyopadhyay S and Maulik U (2000) Genetic clustering for automatic evolution of clusters and application to image classification, Pattern Recognition, 35, pp. 1197-1208.

Beni G and Wang U (1989) Swarm intelligence in cellular robotic systems. In NATO Advanced Workshop on Robots and Biological Systems, Il Ciocco, Tuscany, Italy.

Bensaid AM, Hall LO, Bezdek JC.and Clarke LP (1996) Partially supervised clustering for image segmentation. Pattern Recognition, vol. 29, pp. 859-871.

Bezdek JC (1981) Pattern recognition with fuzzy objective function algorithms. New York: Plenum.

Bonabeau E, Dorigo M and Theraulaz G (1999) Swarm Intelligence: From Natural to Artificial Systems. Oxford University Press, New York.

Brucker P (1978) On the complexity of clustering problems. Beckmenn M and Kunzi HP(Eds.), Optimization and Operations Research, Lecture Notes in Economics and Mathematical Systems, Berlin, Springer, vol.157, pp. 45-54.

Calinski RB and Harabasz J (1975) Adendrite method for cluster analysis, Commun. Statistics, 1 27.

Chou CH, Su MC, and Lai E (2004) A new cluster validity measure and its application to image compression, Pattern Analysis and Applications 7(2), 205-220.

Clark MC, Hall LO, Goldgof DB, Clarke LP, Velthuizen RP and Silbiger MS (1994) MRI segmentation using fuzzy clustering techniques. IEEE Eng Med Biol, 13, pp.730742.

Clerc M and Kennedy J. The particle swarm - explosion, stability, and convergence in a multidimensional complex space, In IEEE Transactions on Evolutionary Computation (2002) 6(1), pp. 58-73.

Couzin ID, Krause J, James R, Ruxton GD, Franks NR (2002) Collective Memory and Spatial Sorting in Animal Groups, Journal of Theoretical Biology, 218, pp. 1-11

Cui X and Potok TE (2005) Document Clustering Analysis Based on Hybrid PSO+Kmeans Algorithm, Journal of Computer Sciences (Special Issue), ISSN 1549-3636, pp. 27-33.

Das S, Abraham A, and Konar A (2008) Automatic Kernel Clustering with Multi-Elitist Particle Swarm Optimization Algorithm, Pattern Recognition Letters, Elsevier Science, Volume 29, pp. 688-699.

Davies DL and Bouldin DW (1979) A cluster separation measure, IEEE Transactions on Pattern Analysis and Machine Intelligence, 1, 224227.

Deb K, Pratap A, Agarwal S, and Meyarivan T (2002) A fast and elitist multiobjective genetic algorithm: NSGA-II, IEEE Trans. on Evolutionary Computation, Vol.6, No.2, April 2002.

Deneubourg JL, Goss S, Franks N, Sendova-Franks A, Detrain C and Chetien L (1991) The dynamics of collective sorting: Robot-like ants and ant-like robots. In Meyer JA and Wilson SW (Eds.) Proceedings of the First International Conference on Simulation of

Adaptive Behaviour: From Animals to Animats 1, pp. 356363. MIT Press, Cambridge, MA.

Dorigo M, Maniezzo V and Colorni A (1996), The ant system: Optimization by a colony of cooperating agents, IEEE Trans. Systems Man and Cybernetics Part B, vol. 26.

Dorigo M and Gambardella LM (1997) Ant colony system: A cooperative learning approach to the traveling salesman problem, IEEE Trans. Evolutionary Computing, vol. 1, pp. 5366.

Duda RO and Hart PE (1973) Pattern Classification and Scene Analysis. John Wiley and Sons, USA.

Dunn JC (1974) Well separated clusters and optimal fuzzy partitions. J. Cybern. 4, 95-104.

Eberhart RC and Shi Y (2001) Particle swarm optimization: Developments, applications and resources, In Proceedings of IEEE International Conference on Evolutionary Computation, vol. 1, pp. 81-86.

Evangelou IE, Hadjimitsis DG, Lazakidou AA, Clayton C (2001) Data Mining and Knowledge Discovery in Complex Image Data using Artificial Neural Networks, Workshop on Complex Reasoning an Geographical Data, Cyprus.

Everitt BS (1993) Cluster Analysis. Halsted Press, Third Edition.

Falkenauer E (1998) Genetic Algorithms and Grouping Problems, John Wiley and Son, Chichester.

Forgy EW (1965) Cluster Analysis of Multivariate Data: Efficiency versus Interpretability of classification, Biometrics, 21.

Frigui H and Krishnapuram R (1999) A Robust Competitive Clustering Algorithm with Applications in Computer Vision, IEEE Transactions on Pattern Analysis and Machine Intelligence 21 (5), pp. 450-465.

Gath I and Geva A (1989) Unsupervised optimal fuzzy clustering. IEEE Transactions on PAMI, 11, pp. 773-781.

Girolami M (2002) Mercer kernel-based clustering in feature space. IEEE Trans. Neural Nctworks 13(3), 780784.

Goldberg DE (1975) Genetic Algorithms in Search, Optimization and Machine Learning, Addison-Wesley, Reading, MA.

Grosan C, Abraham A and Monica C (2006) Swarm Intelligence in Data Mining, in Swarm Intelligence in Data Mining, Abraham A, Grosan C and Ramos V (Eds), Springer, pp. 1-16.

Hall LO, zyurt IB and Bezdek JC (1999) Clustering with a genetically optimized approach, IEEE Trans. Evolutionary Computing 3 (2) pp. 103112.

Handl J, Knowles J and Dorigo M (2003) Ant-based clustering: a comparative study of its relative performance with respect to k-means, average link and 1D-som. Technical Report TR/IRIDIA/2003-24. IRIDIA, Universite Libre de Bruxelles, Belgium

Handl J and Meyer B (2002) Improved ant-based clustering and sorting in a document retrieval interface. In Proceedings of the Seventh International Conference on Parallel Problem Solving from Nature (PPSN VII), volume 2439 of LNCS, pp. 913923. Springer-Verlag, Berlin, Germany.

Hertz T, Bar A, and Daphna Weinshall, H (2006) Learning a Kernel Function for Classification with Small Training Samples, Appearing in Proceedings of the 23rd International Conference on Machine Learning, Pittsburgh, PA.

Hoe K, Lai W, and Tai T (2002) Homogenous ants for web document similarity modeling and categorization. In Proceedings of the Third International Workshop on Ant Algorithms (ANTS 2002), volume 2463 of LNCS, pp. 256261. Springer-Verlag, Berlin, Germany.

Holland JH (1975) Adaptation in Natural and Artificial Systems, University of Michigan Press, Ann Arbor.

Huang Z and Ng MG (1999) A fuzzy k-modes algorithm for clustering categorical data. IEEE Trans. Fuzzy Systems 7 (4), 446452.

Jain AK, Murty MN and Flynn PJ (1999) Data clustering: a review, ACM Computing Surveys, vol. 31, no.3, pp. 264323.

Kanade PM and Hall LO (2003) Fuzzy Ants as a Clustering Concept. In Proceedings of the 22nd International Conference of the North American Fuzzy Information Processing Society (NAFIPS03), pp. 227-232.

Kaufman, L and Rousseeuw, PJ (1990) Finding Groups in Data: An Introduction to Cluster Analysis. John Wiley & Sons, New York.

Kennedy J, Eberhart R and Shi Y (2001) Swarm Intelligence, Morgan Kaufmann Academic Press.

Kennedy J and Eberhart R (1995) Particle swarm optimization, In Proceedings of IEEE International conference on Neural Networks, pp. 1942-1948.

Kim D W, Lee KY, Lee D, Lee KH (2005) A kernel-based subtractive clustering method. Pattern Recognition Letters 26(7), 879-891.

Kohonen T (1995) Self-Organizing Maps, Springer Series in Information Sciences, Vol 30, Springer-Verlag.

Konar A (2005) Computational Intelligence: Principles, Techniques and Applications, Springer.

Krause J and Ruxton GD (2002) Living in Groups. Oxford: Oxford University Press.

Kuntz P, Snyers D and Layzell P (1998) A stochastic heuristic for visualising graph clusters in a bi-dimensional space prior to partitioning. Journal of Heuristics, 5(3), pp. 327351.

Kuntz P and Snyers D (1994) Emergent colonization and graph partitioning. In Proceedings of the Third International Conference on Simulation of Adaptive Behaviour: From Animals to Animats 3, pp. 494 500. MIT Press, Cambridge, MA.

Kuntz P and Snyers D (1999) New results on an ant-based heuristic for highlighting the organization of large graphs. In Proceedings of the 1999 Congress on Evolutionary Computation, pp. 14511458. IEEE Press, Piscataway, NJ.

Leung Y, Zhang J and Xu Z (2000) Clustering by Space-Space Filtering, IEEE Transactions on Pattern Analysis and Machine Intelligence 22 (12), pp. 1396-1410.

Lewin B (1995) Genes VII. Oxford University Press, New York, NY.

Lillesand T and Keifer R (1994) Remote Sensing and Image Interpretation, John Wiley & Sons, USA.

Lumer E and Faieta B (1994) Diversity and Adaptation in Populations of Clustering Ants. In Proceedings Third International Conference on Simulation of Adaptive Behavior: from animals to animates 3, Cambridge, Massachusetts MIT press, pp. 499-508.

Lumer E and Faieta B (1995) Exploratory database analysis via self-organization, Unpublished manuscript.

MacQueen J (1967) Some methods for classification and analysis of multivariate observations, Proceedings of the Fifth Berkeley Symposium on Mathematical Statistics and Probability, pp. 281-297.

Major PF, Dill LM (1978) The three-dimensional structure of airborne bird flocks. Behavioral Ecology and Sociobiology, 4, pp. 111-122.

Mao J and Jain AK (1995) Artificial neural networks for feature extraction and multivariate data projection. IEEE Trans. Neural Networks. vol. 6, 296317.

Milonas MM (1994) Swarms, phase transitions, and collective intelligence, In Langton CG Ed., Artificial Life III, Addison Wesley, Reading, MA.

Mitchell T (1997) Machine Learning. McGraw-Hill, Inc., New York, NY.
Mitra S, Pal SK and Mitra P (2002) Data mining in soft computing framework: A survey, IEEE Transactions on Neural Networks, Vol. 13, pp. 3-14.
Monmarche N, Slimane M and Venturini G (1999) Ant Class: discovery of clusters in numeric data by a hybridization of an ant colony with the k means algorithm. Internal Report No. 213, E3i, Laboratoire dInformatique, Universite de Tours
Moskovitch R, Elovici Y, Rokach L (2008) Detection of unknown computer worms based on behavioral classification of the host, Computational Statistics and Data Analysis, 52(9):4544–4566.
Ng R and Han J (1994) Efficient and effective clustering method for spatial data mining. In: Proc. 1994 International Conf. Very Large Data Bases (VLDB94). Santiago, Chile, September pp. 144155.
Omran M, Salman A and Engelbrecht AP (2002) Image Classification using Particle Swarm Optimization. In Conference on Simulated Evolution and Learning, volume 1, pp. 370374.
Omran M, Engelbrecht AP and Salman A (2005) Particle Swarm Optimization Method for Image Clustering. International Journal of Pattern Recognition and Artificial Intelligence, 19(3), pp. 297322.
Omran M, Salman A and Engelbrecht AP (2005) Dynamic Clustering using Particle Swarm Optimization with Application in Unsupervised Image Classification. Fifth World Enformatika Conference (ICCI 2005), Prague, Czech Republic.
Pakhira MK, Bandyopadhyay S, and Maulik U (2004) Validity index for crisp and fuzzy clusters, Pattern Recognition Letters, 37, 487501.
Pal NR, Bezdek JC and Tsao ECK (1993) Generalized clustering networks and Kohonens self-organizing scheme. IEEE Trans. Neural Networks, vol 4, 549557.
Partridge BL, Pitcher TJ (1980) The sensory basis of fish schools: relative role of lateral line and vision. Journal of Comparative Physiology, 135, pp. 315-325.
Partridge BL (1982) The structure and function of fish schools. Science American, 245, pp. 90-99.
Paterlini S and Krink T (2006) Differential Evolution and Particle Swarm Optimization in Partitional Clustering. Computational Statistics and Data Analysis, vol. 50, pp. 1220 1247.
Paterlini S and Minerva T (2003) Evolutionary Approaches for Cluster Analysis. In Bonarini A, Masulli F and Pasi G (eds.) Soft Computing Applications. Springer-Verlag, Berlin. 167-178.
Pirooznia M and Deng Y: SVM Classifier a comprehensive java interface for support vector machine classification of microarray data, in Proc of Symposium of Computations in Bioinformatics and Bioscience (SCBB06), Hangzhou, China.
Ramos V, Muge F and Pina P (2002) Self-Organized Data and Image Retrieval as a Consequence of Inter-Dynamic Synergistic Relationships in Artificial Ant Colonies. Soft Computing Systems: Design, Management and Applications. 87, pp. 500509.
Ramos V and Merelo JJ (2002) Self-organized stigmergic document maps: Environments as a mechanism for context learning. In Proceedings of the First Spanish Conference on Evolutionary and Bio-Inspired Algorithms (AEB 2002), pp. 284293. Centro Univ. Merida, Merida, Spain.
Rao MR (1971) Cluster Analysis and Mathematical Programming,. Journal of the American Statistical Association, Vol. 22, pp 622-626.
Ratnaweera A and Halgamuge KS (2004) Self organizing hierarchical particle swarm optimizer with time-varying acceleration coefficients, In IEEE Trans. on Evolutionary Computation 8(3): 240-254.

Rokach L (2006), Decomposition methodology for classification tasks: a meta decomposer framework, Pattern Analysis and Applications, 9(2006):257–271.

Rokach L and Maimon O.(2001), Theory and applications of attribute decomposition, IEEE International Conference on Data Mining, IEEE Computer Society Press, pp. 473–480, 2001.

Rokach L and Maimon O (2005), Clustering Methods, Data Mining and Knowledge Discovery Handbook, Springer, pp. 321-352.

Rosenberger C and Chehdi K (2000) Unsupervised clustering method with optimal estimation of the number of clusters: Application to image segmentation, in Proc. IEEE International Conference on Pattern Recognition (ICPR), vol. 1, Barcelona, pp. 1656-1659.

Sarkar M, Yegnanarayana B and Khemani D (1997) A clustering algorithm using an evolutionary programming-based approach, Pattern Recognition Letters, 18, pp. 975986.

Scholkopf B and Smola AJ (2002) Learning with Kernels. The MIT Press, Cambridge.

Selim SZ and Alsultan K (1991) A simulated annealing algorithm for the clustering problem. Pattern recognition, 24(10), pp. 1003-1008.

Shi Y and Eberhart RCD (1999) Empirical Study of particle swarm optimization, In Proceedings of IEEE International Conference Evolutionary Computation, Vol. 3, 101-106.

Storn R and Price K (1997) Differential evolution A Simple and Efficient Heuristic for Global Optimization over Continuous Spaces, Journal of Global Optimization, 11(4), pp. 341359.

Tsang W and Kwong S (2006) Ant Colony Clustering and Feature Extraction for Anomaly Intrusion Detection, in Swarm Intelligence in Data Mining, Abraham A, Grosan C and Ramos V (Eds), Springer, pp. 101-121.

Vapnik VN (1998) Statistical Learning Theory. Wiley, New York.

Wang X, Wang Y and Wang L (2004) Improving fuzzy c-means clustering based on feature-weight learning. Pattern Recognition Letters, vol. 25, pp. 112332.

Xiao X, Dow ER, Eberhart RC, Miled ZB and Oppelt RJ (2003) Gene Clustering Using Self-Organizing Maps and Particle Swarm Optimization, Proc of the 17th International Symposium on Parallel and Distributed Processing (PDPS '03), IEEE Computer Society, Washington DC.

Xu, R., Wunsch, D.: (2005), Survey of Clustering Algorithms, IEEE Transactions on Neural Networks, Vol. 16(3): 645-678

Xu R and Wunsch D (2008) Clustering, IEEE Press Series on Computational Intelligence, USA.

Zahn CT (1971) Graph-theoretical methods for detecting and describing gestalt clusters, IEEE Transactions on Computers C-20, 6886.

Zhang T, Ramakrishnan R and Livny M (1997) BIRCH: A New Data Clustering Algorithm and Its Applications, Data Mining and Knowledge Discovery, vol. 1, no. 2, pp. 141-182.

Zhang DQ and Chen SC (2003) Clustering incomplete data using kernel-based fuzzy c-means algorithm. Neural Process Letters 18, 155162.

Zhang R and Rudnicky AI (2002) A large scale clustering scheme for kernel k-means. In: The Sixteenth International Conference on Pattern Recognition, p. 289292.

van den Bergh F and Engelbrecht AP (2001) Effects of swarm size on cooperative particle swarm optimizers, In Proceedings of GECCO-2001, San Francisco CA, 892-899.

van der Merwe DW and Engelbrecht AP (2003) Data clustering using particle swarm optimization. In: Proceedings of the 2003 IEEE Congress on Evolutionary Computation, pp. 215-220, Piscataway, NJ: IEEE Service Center

24

Using Fuzzy Logic in Data Mining

Lior Rokach[1]

Department of Information System Engineering, Ben-Gurion University, Israel
liorrk@bgu.ac.il

Summary. In this chapter we discuss how fuzzy logic extends the envelop of the main data mining tasks: clustering, classification, regression and association rules. We begin by presenting a formulation of the data mining using fuzzy logic attributes. Then, for each task, we provide a survey of the main algorithms and a detailed description (i.e. pseudo-code) of the most popular algorithms. However this chapter will not profoundly discuss neuro-fuzzy techniques, assuming that there will be a dedicated chapter for this issue.

24.1 Introduction

There are two main types of uncertainty in supervised learning: statistical and cognitive. Statistical uncertainty deals with the random behavior of nature and all existing data mining techniques can handle the uncertainty that arises (or is assumed to arise) in the natural world from statistical variations or randomness. While these techniques may be appropriate for measuring the likelihood of a hypothesis, they says nothing about the meaning of the hypothesis.

Cognitive uncertainty, on the other hand, deals with human cognition. Cognitive uncertainty can be further divided into two sub-types: vagueness and ambiguity.

Ambiguity arises in situations with two or more alternatives such that the choice between them is left unspecified. Vagueness arises when there is a difficulty in making a precise distinction in the world.

Fuzzy set theory, first introduced by Zadeh in 1965, deals with cognitive uncertainty and seeks to overcome many of the problems found in classical set theory.

For example, a major problem faced by researchers of control theory is that a small change in input results in a major change in output. This throws the whole control system into an unstable state. In addition there was also the problem that the representation of subjective knowledge was artificial and inaccurate. Fuzzy set theory is an attempt to confront these difficulties and in this chapter we show how it can be used in data mining tasks.

24.2 Basic Concepts of Fuzzy Set Theory

In this section we present some of the basic concepts of fuzzy logic. The main focus, however, is on those concepts used in the induction process when dealing with data mining. Since fuzzy

O. Maimon, L. Rokach (eds.), *Data Mining and Knowledge Discovery Handbook*, 2nd ed.,
DOI 10.1007/978-0-387-09823-4_24,

set theory and fuzzy logic are much broader than the narrow perspective presented here, the interested reader is encouraged to read (Zimmermann, 2005)).

24.2.1 Membership function

In classical set theory, a certain element either belongs or does not belong to a set. Fuzzy set theory, on the other hand, permits the gradual assessment of the membership of elements in relation to a set.

Definition 1. *Let U be a universe of discourse, representing a collection of objects denoted generically by u. A fuzzy set A in a universe of discourse U is characterized by a membership function μ_A which takes values in the interval [0, 1]. Where $\mu_A(u) = 0$ means that u is definitely not a member of A and $\mu_A(u) = 1$ means that u is definitely a member of A.*

The above definition can be illustrated on the vague set of *Young*. In this case the set U is the set of people. To each person in U, we define the degree of membership to the fuzzy set *Young*. The membership function answers the question "to what degree is person u young?". The easiest way to do this is with a membership function based on the person's age. For example Figure 24.1 presents the following membership function:

$$\mu_{Young}(u) = \begin{cases} 0 & age(u) > 32 \\ 1 & age(u) < 16 \\ \frac{32-age(u)}{16} & otherwise \end{cases} \tag{24.1}$$

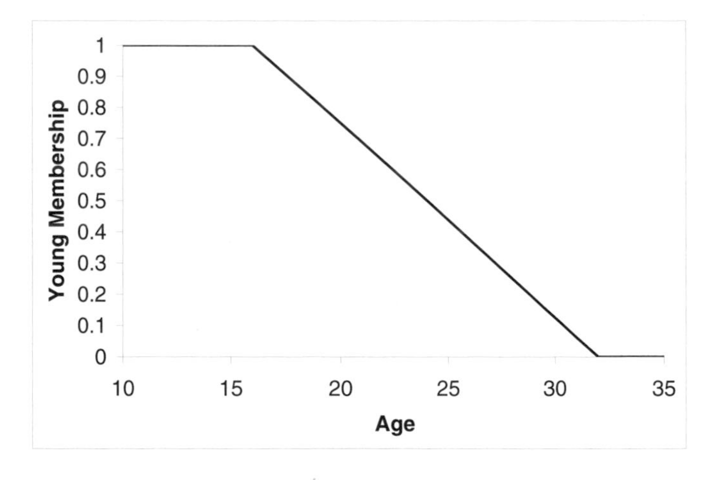

Fig. 24.1. Membership function for the young set.

Given this definition, John, who is 18 years old, has degree of youth of 0.875. Philip, 20 years old, has degree of youth of 0.75. Unlike probability theory, degrees of membership do not have to add up to 1 across all objects and therefore either many or few objects in the set may have high membership. However, an objects membership in a set (such as "young") and the sets complement ("not young") must still sum to 1.

The main difference between classical set theory and fuzzy set theory is that the latter admits to partial set membership. A classical or crisp set, then, is a fuzzy set that restricts its membership values to $\{0,1\}$, the endpoints of the unit interval. Membership functions can be used to represent a crisp set. For example, Figure 24.2 presents a crisp membership function defined as:

$$\mu_{CrispYoung}(u) = \begin{cases} 0 & age(u) > 22 \\ 1 & age(u) \leq 22 \end{cases} \tag{24.2}$$

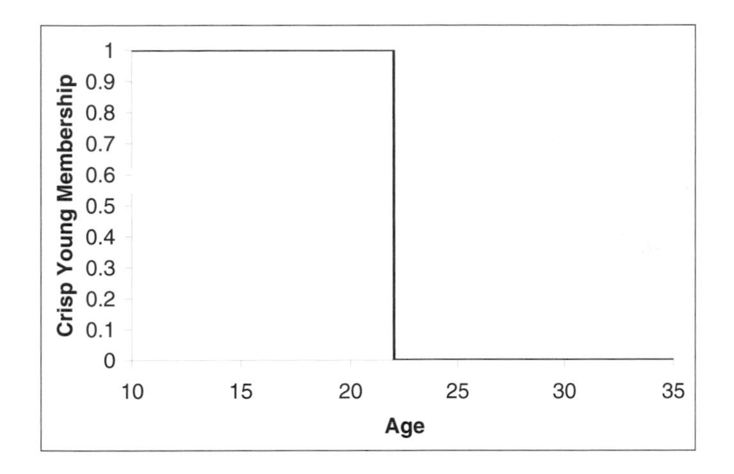

Fig. 24.2. Membership function for the crisp young set.

In regular classification problems, we assume that each instance takes one value for each attribute and that each instance is classified into only one of the mutually exclusive classes. To illustrate how fuzzy logic can help data mining tasks, we introduce the problem of modelling the preferences of TV viewers. In this problem there are 3 input attributes:

$A = \{\text{Time of Day}, \text{Age Group}, \text{Mood}\}$

and each attribute has the following values:

- $dom(\text{Time of Day}) = \{\text{Morning}, \text{Noon}, \text{Evening}, \text{Night}\}$
- $dom(\text{Age Group}) = \{\text{Young}, \text{Adult}\}$
- $dom(\text{Mood}) = \{\text{Happy}, \text{Indifferent}, \text{Sad}, \text{Sour}, \text{Grumpy}\}$

The classification can be the movie genre that the viewer would like to watch, such as $C = \{\text{Action}, \text{Comedy}, \text{Drama}\}$.

All the attributes are vague by definition. For example, peoples feelings of happiness, indifference, sadness, sourness and grumpiness are vague without any crisp boundaries between them. Although the vagueness of "Age Group" or "Time of Day" can be avoided by indicating the exact age or exact time, a rule induced with a crisp decision tree may then have an artificial crisp boundary, such as "IF Age $<$ 16 THEN action movie". But how about someone who is

17 years of age? Should this viewer definitely not watch an action movie? The viewer preferred genre may still be vague. For example, the viewer may be in a mood for both comedy and drama movies. Moreover, the association of movies into genres may also be vague. For instance the movie "Lethal Weapon" (starring Mel Gibson and Danny Glover) is considered to be both comedy and action movie.

Fuzzy concept can be introduced into a classical problem if at least one of the input attributes is fuzzy or if the target attribute is fuzzy. In the example described above , both input and target attributes are fuzzy. Formally the problem is defined as following (Yuan and Shaw, 1995):

Each class c_j is defined as a fuzzy set on the universe of objects U. The membership function $\mu_{c_j}(u)$ indicates the degree to which object u belongs to class c_j. Each attribute a_i is defined as a linguistic attribute which takes linguistic values from $dom(a_i) = \{v_{i,1}, v_{i,2}, \ldots, v_{i,|dom(a_i)|}\}$. Each linguistic value $v_{i,k}$ is also a fuzzy set defined on U. The membership $\mu_{v_{i,k}}(u)$ specifies the degree to which object u's attribute a_i is $v_{i,k}$. Recall that the membership of a linguistic value can be subjectively assigned or transferred from numerical values by a membership function defined on the range of the numerical value.

Typically, before one can incoporate fuzzy concepts into a data mining application, an expert is required to provide the fuzzy sets for the quantitative attributes, along with their corresponding membership functions. Alternatively the appropriate fuzzy sets are determined using fuzzy clustering.

24.2.2 Fuzzy Set Operations

Like classical set theory, fuzzy set theory includes operations union, intersection, complement, and inclusion, but also includes operations that have no classical counterpart, such as the modifiers concentration and dilation, and the connective fuzzy aggregation. Definitions of fuzzy set operations are provided in this section.

Definition 2. *The membership function of the union of two fuzzy sets A and B with membership functions μ_A and μ_B respectively is defined as the maximum of the two individual membership functions:*

$$\mu_{A \cup B}(u) = max\{\mu_A(u), \mu_B(u)\} \tag{24.3}$$

Definition 3. *The membership function of the intersection of two fuzzy sets A and B with membership functions μ_A and μ_B respectively is defined as the minimum of the two individual membership functions:*

$$\mu_{A \cap B}(u) = min\{\mu_A(u), \mu_B(u)\} \tag{24.4}$$

Definition 4. *The membership function of the complement of a fuzzy set A with membership function μ_A is defined as the negation of the specified membership function:*

$$\mu_{\overline{A}}(u) = 1 - \mu_A(u). \tag{24.5}$$

To illustrate these fuzzy operations, we elaborate on the previous example. Recall that John has a degree of youth of 0.875. Additionally John's happiness degree is 0.254. Thus, the membership of John in the set Young $\cup$ Happy would be $max(0.875, 0.254) = 0.875$, and its membership in Young $\cap$ Happy would be $min(0.875, 0.254) = 0.254$.

It is possible to chain operators together, thereby constructing quite complicated sets. It is also possible to derive many interesting sets from chains of rules built up from simple operators. For example John's membership in the set $\overline{Young} \cup$ Happy would be $max(1-0.875, 0.254) = 0.254$

The usage of the max and min operators for defining fuzzy union and fuzzy intersection, respectively is very common. However, it is important to note that these are not the only definitions of union and intersection suited to fuzzy set theory.

Definition 5. *The fuzzy subsethood $S(A,B)$ measures the degree to which A is a subset of B.*

$$S(A,B) = \frac{M(A \cap B)}{M(A)} \tag{24.6}$$

where $M(A)$ is the *cardinality* measure of a fuzzy set A and is defined as

$$M(A) = \sum_{u \in U} \mu_A(u) \tag{24.7}$$

The subsethood can be used to measure the truth level of the rule of classification rules. For example given a classification rule such as "IF Age is Young AND Mood is Happy THEN Comedy" we have to calculate $S(Hot \cap Sunny, Swimming)$ in order to measure the truth level of the classification rule.

24.3 Fuzzy Supervised Learning

In this section we survey supervised methods that incoporate fuzzy sets. Supervised methods are methods that attempt to discover the relationship between input attributes and a target attribute (sometimes referred to as a dependent variable). The relationship discovered is represented in a structure referred to as a model. Usually models describe and explain phenomena, which are hidden in the dataset and can be used for predicting the value of the target attribute knowing the values of the input attributes.

It is useful to distinguish between two main supervised models: classification models (classifiers) and Regression Models. Regression models map the input space into a real-value domain. For instance, a regressor can predict the demand for a certain product given its characteristics. On the other hand, classifiers map the input space into pre-defined classes. For instance, classifiers can be used to classify mortgage consumers as good (fully payback the mortgage on time) and bad (delayed payback).

Fuzzy set theoretic concepts can be incorporated at the input, output, or into to backbone of the classifier. The data can be presented in fuzzy terms and the output decision may be provided as fuzzy membership values. In this chapter we will concentrate on fuzzy decision trees.

24.3.1 Growing Fuzzy Decision Tree

Decision tree is a predictive model which can be used to represent classifiers. Decision trees are frequently used in applied fields such as finance, marketing, engineering, medicine and security (Moskovitch *et al.* (2008)). In the opinion of many researchers decision trees gained popularity mainly due to their simplicity and transparency. Decision tree are self-explained. There is no need to be an expert in data mining in order to follow a certain decision tree.

There are several algorithms for induction of fuzzy decision trees, most of them extend existing decision trees methods. The UR-ID3 algorithm (Maher and Clair, 1993)) starts by building a strict decision tree, and subsequently fuzzifies the conditions of the tree. Tani and Sakoda (1992) use the ID3 algorithm to select effective numerical attributes. The obtained splitting intervals are used as fuzzy boundaries. Regression is then used in each subspace to form fuzzy rules. Cios and Sztandera (1992) use the ID3 algorithm to convert a decision tree into a layer of a feedforward neural network. Each neuron is represented as a hyperplane with a fuzzy boundary. The nodes within the hidden layer are generated until some fuzzy entropy is reduced to zero. New hidden layers are generated until there is only one node at the output layer.

Fuzzy-CART (Jang (1994)) is a method which uses the CART algorithm to build a tree. However, the tree, which is the first step, is only used to propose fuzzy sets of the continuous domains (using the generated thresholds). Then, a layered network algorithm is employed to learn fuzzy rules. This produces more comprehensible fuzzy rules and improves the CART's initial results.

Another complete framework for building a fuzzy tree including several inference procedures based on conflict resolution in rule-based systems and efficient approximate reasoning methods was presented in (Janikow, 1998).

Olaru and Wehenkel (2003) presented a new type of fuzzy decision trees called soft decision trees (SDT). This approach combines tree-growing and pruning, to determine the structure of the soft decision tree. Refitting and backfitting are used to improve its generalization capabilities. The researchers empirically showed that soft decision trees are significantly more accurate than standard decision trees. Moreover, a global model variance study shows a much lower variance for soft decision trees than for standard trees as a direct cause of the improved accuracy.

Peng (2004) has used FDT to improve the performance of the classical inductive learning approach in manufacturing processes. Peng proposed using soft discretization of continuous-valued attributes. It has been shown that FDT can deal with the noise or uncertainties existing in the data collected in industrial systems.

In this chapter we will focus on the algorithm proposed in (Yuan and Shaw, 1995). This algorithm can handle the classification problems with both fuzzy attributes and fuzzy classes represented in linguistic fuzzy terms. It can also handle other situations in a uniform way where numerical values can be fuzzified to fuzzy terms and crisp categories can be treated as a special case of fuzzy terms with zero fuzziness. The algorithm uses classification ambiguity as fuzzy entropy. The classification ambiguity directly measures the quality of classification rules at the decision node. It can be calculated under fuzzy partitioning and multiple fuzzy classes.

The fuzzy decision tree induction consists of the following steps:

- Fuzzifying numeric attributes in the training set.
- Inducing a fuzzy decision tree.
- Simplifying the decision tree.
- Applying fuzzy rules for classification.

Fuzzifying numeric attributes

When a certain attribute is numerical, it needs to be fuzzified into linguistic terms before it can be used in the algorithm. The fuzzification process can be performed manually by experts or can be derived automatically using some sort of clustering algorithm. Clustering groups the data instances into subsets in such a manner that similar instances are grouped together; different instances belong to different groups. The instances are thereby organized into an efficient representation that characterizes the population being sampled.

Yuan and Shaw (1995) suggest a simple algorithm to generate a set of membership functions on numerical data. Assume attribute a_i has numerical value x from the domain X. We can cluster X to k linguistic terms $v_{i,j}, j = 1, \ldots, k$. The size of k is manually predefined. For the first linguistic term $v_{i,1}$, the following membership function is used:

$$\mu_{v_{i,1}}(x) = \begin{cases} 1 & x \leq m_1 \\ \frac{m_2 - x}{m_2 - m_1} & m_1 < x < m_2 \\ 0 & x \geq m_2 \end{cases} \tag{24.8}$$

For each $v_{i,j}$ when $j = 2, \ldots, k-1$ has a triangular membership function as follows:

$$\mu_{v_{i,j}}(x) = \begin{cases} 0 & x \leq m_{j-1} \\ \frac{x - m_{j-1}}{m_j - m_{j-1}} & m_{j-1} < x \leq m_j \\ \frac{m_{j+1} - x}{m_{j+1} - m_j} & m_j < x < m_{j+1} \\ 0 & x \geq m_{j+1} \end{cases} \tag{24.9}$$

Finally the membership function of the last linguistic term $v_{i,k}$ is:

$$\mu_{v_{i,k}}(x) = \begin{cases} 0 & x \leq m_{k-1} \\ \frac{x - m_{k-1}}{m_k - m_{k-1}} & m_{k-1} < x \leq m_k \\ 1 & x \geq m_k \end{cases} \tag{24.10}$$

Figure 24.3 illustrates the creation of four groups defined on the age attribute: "young", "early adulthood", "middle-aged" and "old age". Note that the first set ("young") and the last set ("old age") have a trapezoidal form which can be uniquely described by the four corners. For example, the "young" set could be represented as $(0,0,16,32)$. In between, all other sets ("early adulthood" and "middle-aged") have a triangular form which can be uniquely described by the three corners. For example, the set "early adulthood" is represented as $(16,32,48)$.

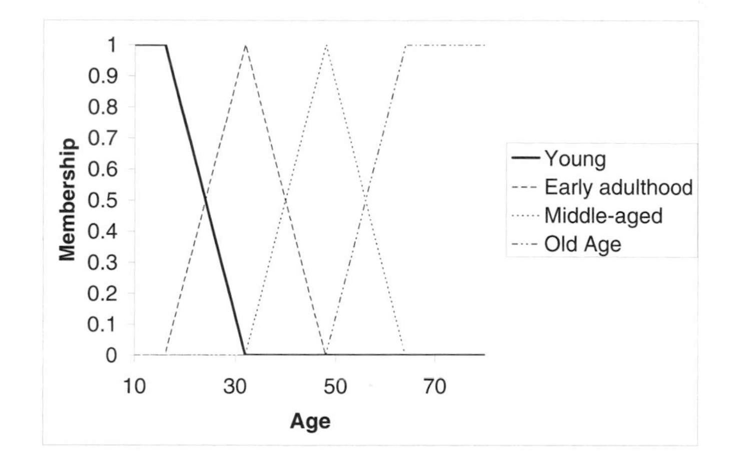

Fig. 24.3. Membership function for various groups in the age attribute.

The only parameters that need to be determined are the set of k centers $M = \{m_1, \ldots, m_k\}$. The centers can be found using the algorithm presented in Algorithm 1. Note that in order to use the algorithm, a monotonic decreasing learning rate function should be provided.

Algorithm 1: Algorithm for fuzzifying numeric attributes

Input: X - a set of values, $\eta(t)$ - some monotonic decreasing scalar function representing the learning rate.
Output: $M = \{m_1, \ldots, m_k\}$
1: Initially set m_i to be evenly distributed on the range of X.
2: $t \leftarrow 1$
3: **repeat**
4: Randomly draw one sample x from X
5: Find the closest center m_c to x.
6: $m_c \leftarrow m_c + \eta(t) \cdot (x - m_c)$
7: $t \leftarrow t + 1$
8: $D(X,M) \leftarrow \sum_{x \in X} \min_i \|x - m_i\|$
9: **until** $D(X,M)$ converges

The Induction Phase

The induction algorithm of fuzzy decision tree is presented in Algorithm 2. The algorithm measures the classification ambiguity associated with each attribute and split the data using the attribute with the smallest classification ambiguity. The classification ambiguity of attribute a_i with linguistic terms $v_{i,j}, j = 1, \ldots, k$ on fuzzy evidence S, denoted as $G(a_i|S)$, is the weighted average of classification ambiguity calculated as:

$$G(a_i|S) = \sum_{j=`1}^{k} w(v_{i,j}|S) \cdot G(v_{i,j}|S) \tag{24.11}$$

where $w(v_{i,j}|S)$ is the weight which represents the relative size of $v_{i,j}$ and is defined as:

$$w(v_{i,j}|S) = \frac{M(v_{i,j}|S)}{\sum_k M(v_{i,k}|S)} \tag{24.12}$$

The classification ambiguity of $v_{i,j}$ is defined as $G(v_{i,j}|S) = g\left(\mathbf{p}\left(C\left|v_{i,j}\right.\right)\right)$, which is measured based on the possibility distribution vector $\mathbf{p}\left(C\left|v_{i,j}\right.\right) = \left(p\left(c_1\left|v_{i,j}\right.\right), \ldots, p\left(c_{|k|}\left|v_{i,j}\right.\right)\right)$.

Given $v_{i,j}$, the possibility of classifying an object to class c_l can be defined as:

$$p\left(c_l\left|v_{i,j}\right.\right) = \frac{S(v_{i,j}, c_l)}{\max_k S(v_{i,j}, c_k)} \tag{24.13}$$

where $S(A,B)$ is the fuzzy subsethood that was defined in Definition 5. The function $g(\mathbf{p})$ is the possibilistic measure of ambiguity or nonspecificity and is defined as:

$$g(\mathbf{p}) = \sum_{i=1}^{|\mathbf{p}|} \left(p_i^* - p_{i+1}^*\right) \cdot \ln(i) \tag{24.14}$$

where $\mathbf{p}^* = \left(p_1^*, \ldots, p_{|\mathbf{p}|}^*\right)$ is the permutation of the possibility distribution $\mathbf{p}$ sorted such that $p_i^* \geq p_{i+1}^*$.

All the above calculations are carried out at a predefined significant level α. An instance will take into consideration of a certain branch $v_{i,j}$ only if its corresponding membership is greater than α. This parameter is used to filter out insignificant branches.

After partitioning the data using the attribute with the smallest classification ambiguity, the algorithm looks for nonempty branches. For each nonempty branch, the algorithm calculates the truth level of classifying all instances within the branch into each class. The truth level is caluclated using the fuzzy subsethood measure $S(A,B)$.

If the truth level of one of the classes is above a predefined threshold β then no additional partitioning is needed and the node become a leaf in which all instance will be labeled to the class with the highest truth level. Otherwise the procedure continues in a recursive manner. Note that small values of β will lead to smaller trees with the risk of underfitting. A higher β may lead to a larger tree with higher classification accuracy. However, at a certain point, higher values β may lead to overfitting.

Algorithm 2: Fuzzy decision tree induction

Input: S - Training Set A - Input Feature Set y - Target Feature
Output: Fuzzy Decision Tree

1: Create a new fuzzy tree FT with a single root node.
2: **if** S is empty OR Truth level of one of the classes $\geq \beta$ **then**
3: Mark FT as a leaf with the most common value of y in S as a label.
4: Return FT.
5: **end if**
6: $\forall a_i \in A$ find a with the smallest classification ambiguity.
7: **for** each outcome v_i of a **do**
8: Recursively call procedure with corresponding partition v_i.
9: Connect the root to the subtree with an edge that is labeled as v_i.
10: **end for**
11: Return FT

Simplifying the decision tree

Each path of branches from root to leaf can be converted into a rule with the condition part representing the attributes on the passing branches from the root to the leaf and the conclusion part representing the class at the leaf with the highest truth level classification. The corresponding classification rules can be further simplified by removing one input attribute term at a time for each rule we try to simplify . Select the term to remove with the highest truth level of the simplified rule. If the truth level of this new rule is not lower than the threshold β or the truth level of the original rule, the simplification is successful. The process will continue until no further simplification is possible for all the rules.

Using the Fuzzy Decision Tree

In a regular decision tree, only one path (rule) can be applied for every instance. In a fuzzy decision tree, several paths (rules) can be applied for one instance. In order to classify an unlabeled instance, the following steps should be performed (Yuan and Shaw, 1995):

- Step 1: Calculate the membership of the instance for the condition part of each path (rule). This membership will be associated with the label (class) of the path.
- Step 2: For each class calculate the maximum membership obtained from all applied rules.
- Step 3: An instance may be classified into several classes with different degrees based on the membership calculated in Step 2.

24.3.2 Soft Regression

Regressions are used to compute correlations among data sets. The "classical" approach uses statistical methods to find these correlations. Soft regression is used when we want to compare data sets that are temporal and interdependent. The use of fuzzy logic can overcome many of the difficulties associated with the classical approach. The fuzzy techniques can achieve greater flexibility, greater accuracy and generate more information in comparison to econometric modeling based on (statistical) regression techniques. In particular, the fuzzy method can potentially be more successful than conventional regression methods, especially under circumstances that severely violate the fundamental conditions required for the reliable use of conventional methods.

Soft regression techniques have been proposed in (Shnaider et al., 1991, Shnaider and Schneider, 1988).

24.3.3 Neuro-fuzzy

Neuro-fuzzy refers to hybrids of artificial neural networks and fuzzy logic. Neuro-fuzzy is the most visible hybrid paradigm and has been adequately investigated (Mitra and Pal, 2005)

Neuro-fuzzy hybridization can be done in two ways (Mitra, 2000): fuzzy-neural network (FNN) which is a neural network equipped with the capability of handling fuzzy information and a neural-fuzzy system (NFS) which is a fuzzy system augmented by neural networks to enhance some of its characteristics like flexibility, speed, and adaptability.

A neurofuzzy system can be viewed as a special 3layer neural network (Nauck, 1997). The first layer represents input variables, the hidden layer represents fuzzy rules and the third layer represents output variables. Fuzzy sets are encoded as (fuzzy) connection weights. Usually after learning the obtained model is interpreted as a system of fuzzy rules.

24.4 Fuzzy Clustering

The goal of clustering is descriptive, that of classification is predictive. Since the goal of clustering is to discover a new set of categories, the new groups are of interest in themselves, and their assessment is intrinsic. In classification tasks, however, an important part of the assessment is extrinsic, since the groups must reflect some reference set of classes.

Clustering of objects is as ancient as the human need for describing the salient characteristics of men and objects and identifying them with a type. Therefore, it embraces various scientific disciplines: from mathematics and statistics to biology and genetics, each of which uses different terms to describe the topologies formed using this analysis. From biological “taxonomies”, to medical “syndromes” and genetic “genotypes” to manufacturing ”group technology” — the problem is identical: forming categories of entities and assigning individuals to the proper groups within it.

Clustering groups data instances into subsets in such a manner that similar instances are grouped together, while different instances belong to different groups. The instances are thereby organized into an efficient representation that characterizes the population being sampled. Formally, the clustering structure is represented as a set of subsets $C = C_1, \ldots, C_k$ of S, such that: $S = \bigcup_{i=1}^{k} C_i$ and $C_i \cap C_j = \emptyset$ for $i \neq j$. Consequently, any instance in S belongs to exactly one and only one subset.

Traditional clustering approaches generate partitions; in a partition, each instance belongs to one and only one cluster. Hence, the clusters in a hard clustering are disjointed. Fuzzy clustering extends this notion and suggests a *soft clustering* schema. In this case, each pattern is associated with every cluster using some sort of membership function, namely, each cluster is a fuzzy set of all the patterns. Larger membership values indicate higher confidence in the assignment of the pattern to the cluster. A hard clustering can be obtained from a fuzzy partition by using a threshold of the membership value.

The most popular fuzzy clustering algorithm is the fuzzy c-means (FCM) algorithm. Even though it is better than the hard K-means algorithm at avoiding local minima, FCM can still converge to local minima of the squared error criterion. The design of membership functions is the most important problem in fuzzy clustering; different choices include those based on similarity decomposition and centroids of clusters. A generalization of the FCM algorithm has been proposed through a family of objective functions. A fuzzy c-shell algorithm and an adaptive variant for detecting circular and elliptical boundaries have been presented.

FCM is an iterative algorithm. The aim of FCM is to find cluster centers (centroids) that minimize a dissimilarity function. To accommodate the introduction of fuzzy partitioning, the membership matrix(U) is randomly initialized according to Equation 24.15.

$$\sum_{i=1}^{c} u_{ij} = 1, \forall j = 1, \ldots, n \tag{24.15}$$

The algorithm minimizes a dissimilarity (or distance) function which is given in Equation 24.16:

$$J(U, c_1, c_2, \ldots, c_c) = \sum_{i=1}^{c} J_i = \sum_{i=1}^{c} \sum_{j=1}^{n} u_{ij}^m d_{ij}^2 \tag{24.16}$$

where, u_{ij} is between 0 and 1; c_i is the centroid of cluster i; d_{ij} is the Euclidian distance between i-th centroid and j-th data point; m is a weighting exponent.

To reach a minimum of dissimilarity function there are two conditions. These are given in Equation 24.17 and Equation 24.18.

$$c_i = \frac{\sum_{j=1}^{n} u_{ij}^m x_j}{\sum_{j=1}^{n} u_{ij}^m} \tag{24.17}$$

$$u_{ij} = \frac{1}{\sum_{k=1}^{c} \left(\frac{d_{ij}}{d_{kj}} \right)^{2/(m-1)}} \tag{24.18}$$

Algorithm 3 presents the fuzzy c-means that was originally proposed in (Bezdek, 1973).

By iteratively updating the cluster centers and the membership grades for each data point, FCM iteratively moves the cluster centers to the ”right” location within a data set. However,

Algorithm 3: FCM Algorithm

Input: X - Data Set
c - number of clusters
t - convergence threshold (termination criterion)
m - exponential weight
Output: U - membership matrix
1: Randomly initialize matrix U with c clusters and fulfils Eq. 24.15
2: **repeat**
3: Calculate c_i by using Equation 24.17.
4: Compute dissimilarity between centroids and data points using Eq. 24.16.
5: Compute a new U using Eq. 24.18
6: **until** The improvement over previous iteration is below t.

FCM does not ensure that it converges to an optimal solution. The random initilization of U might have uncancelled effect on the final performance.

There are several extensions to the basic FCM algorithm, The Fuzzy Trimmed C Prototype (FTCP) algorithm (Kim et al., 1996) increases the robustness of the clusters by trimming away observations with large residuals. The Fuzzy C Least Median of Squares (FCLMedS) algorithm (Nasraoui and Krishnapuram, 1997) replaces the summation presented in Equation 24.16 with the median.

24.5 Fuzzy Association Rules

Association rules are rules of the kind "70% of the customers who buy vine and cheese also buy grapes". While the traditional field of application is market basket analysis, association rule mining has been applied to various fields since then, which has led to a number of important modifications and extensions.

In this section, an algorithm based on the *apriori* data mining algorithm is described to discover large itemsets. Fuzzy sets are used to handle quantitative values, as described in (Hong et al., 1999). Our algorithm is applied with some differences. We will use the following notation:

- n – number of transactions in the database.
- m – number of items (attributes) in the database.
- d_i – the i-th transaction.
- I_j – the j-th attribute.
- I_{ij} – the value of $I_j ford_i$.
- μ_{ijk} – the membership grade of I_{ij} in the region k.
- R_{jk} – the k-th fuzzy region of the attribute I_j.
- $num(R_{jk})$ – number of occurrences of the attribute region R_{jk} in the whole database, where $\mu_{ijk} > 0$.
- C_r – the set of candidate itemsets with r attributes.
- c_r – candidate itemset with r attributes.
- f_j^i – the membership value of d_i in region s_j.
- f_{cr}^i – the fuzzy value of the itemset c_r in the transaction d_i.
- L_r – the set of large itemsets with r items.

Algorithm 4: Fuzzy Association Rules Algorithm

1: **for all** transaction i **do**
2: **for all** attribute j **do**
3: I_{ij}^{f}=(μ_{ij1}/R$_{j1}$ + μ_{ij2}/R$_{j2}$+...+μ_{ijk}/R$_{jk}$) // where the superscript f denotes fuzzy set
4: **end for**
5: **end for**
6: For each attribute region R$_{jk}$, count the number of occurrences, where $\mu_{ijk} > 0$, in the whole database. The output is $num(R_{jk})$. $num(R_{jk}) = \sum_{i=1}^{n} 1\{\mu_{ijk}/R_{jk} \neq 0\}$
7: L_1={R$_{jk}$|num(R$_{jk}$) ≥minnum, 1≤j≤m, 1≤k≤numR(I$_j$)}.
8: r=1 (r is the number of items that composed the large itemsets in the current stage).
9: Generate the candidate set C$_{r+1}$ from L$_r$
10: **for all** newly formed candidate itemset c$_{r+1}$ in C$_{r+1}$, that is composed of the items (s$_1$, s$_2$,...,s$_{r+1}$) **do**
11: For each transaction d_i calculate its intersection fuzzy value as: $f^i_{(cr+1)} = f^i_1 \cap \mathrm{f}^i_2 \cap \ldots \cap \mathrm{f}^i_{r+1}$.
12: Calculate the frequency of c$_{r+1}$ on the transactions, where $f^i_{(cr+1)} > 0$. $num(c_{r+1})$ is output.
13: If the frequency of the itemset is larger than or equal to the predefined number of occurrences minnum, put it in the set of large r+1-itemsets L_{r+1}.
14: **end for**
15: **if** L_{r+1} is not empty **then**
16: $r = r + 1$
17: go to Step 9.
18: **end if**
19: **for all** large itemset l_r, r≥ 2 **do**
20: Calculate its support as: sup(l$_r$) = $\Sigma \mathrm{f}^i_{(lr)}$.
21: Calculate its strength as: str(l$_r$)= sup(l$_r$)/num(l$_r$).
22: **end for**
23: For each large itemset l$_r$, r≥2, generate the possible association rules as in (Agrawal et al., 1993).
24: For each association rule $s_1, s_2, \ldots, s_n \geq s_{n+1}, \ldots, s_r$, calculate its confidence as: num(s$_1$, s$_2$...s$_n$,s$_{n+1}$...s$_r$)/num(s$_1$, s$_2$...s$_n$).
25: **if** the confidence is higher than the predefined threshold minconf **then**
26: output the rule as an association rule.
27: **end if**
28: For each association rule $s_1, s_2, \ldots, s_n \geq s_{n+1}, \ldots, s_r$, record its strength as str(s$_1$, s$_2$...s$_n$,s$_{n+1}$...s$_r$), and its support as sup(l$_r$).

- l_r – a large itemset with r items.
- $num(I_1,\ldots,I_s)$ – the occurrences number of the itemset $(I_1,\ldots,I_s)$.
- $numR(I_j)$ – the number of the membership function regions for the attribute I_j.

Algorithm 4 presents the fuzzy association algorithm proposed in (Komem and Schneider, 2005). The quantitative values are first transformed into a set of membership grades, by using predefined membership functions. Every membership grade represents the agreement of a quantitative value with a linguistic term. In order to avoid discriminating the importance level of data, each point must have membership grade of 1 in one membership function; Thus, the membership functions of each attribute produce a continuous line of $\mu = 1$. Additionally, in order to diagnose the bias direction of an item from the center of a membership function region, almost each point get another membership grade which is lower than 1 in other membership functions region. Thus, each end of membership function region is touching, close to, or slightly overlapping an end of another membership function (except the outside regions, of course).

By this mechanism, as point "a" moves right, further from the center of the region "middle", it gets a higher value of the label "middle-high", additionally to the value 1 of the label "middle".

24.6 Conclusion

This chapter discussed how fuzzy logic can be used to solve several different data mining tasks, namely classification clustering, and discovery of association rules. The discussion focused mainly one representative algorithm for each of these tasks.

There are at least two motivations for using fuzzy logic in data mining, broadly speaking. First, as mentioned earlier, fuzzy logic can produce more abstract and flexible patterns, since many quantitative features are involved in data mining tasks. Second, the crisp usage of metrics is better replaced by fuzzy sets that can reflect, in a more natural manner, the degree of belongingness/membership to a class or a cluster.

References

R. Agrawal, T. Imielinski and A. Swami: Mining Association Rules between Sets of Items in Large Databases. Proceeding of ACM SIGMOD, 207-216. Washington, D.C, 1993.

Arbel, R. and Rokach, L., Classifier evaluation under limited resources, Pattern Recognition Letters, 27(14): 1619–1631, 2006, Elsevier.

Averbuch, M. and Karson, T. and Ben-Ami, B. and Maimon, O. and Rokach, L., Context-sensitive medical information retrieval, The 11th World Congress on Medical Informatics (MEDINFO 2004), San Francisco, CA, September 2004, IOS Press, pp. 282–286.

J. C. Bezdek. Fuzzy Mathematics in Pattern Classification. PhD Thesis, Applied Math. Center, Cornell University, Ithaca, 1973.

Cios K. J. and Sztandera L. M., Continuous ID3 algorithm with fuzzy entropy measures, Proc. IEEE lnternat. Con/i on Fuzz)' Systems,1992, pp. 469-476.

Cohen S., Rokach L., Maimon O., Decision Tree Instance Space Decomposition with Grouped Gain-Ratio, Information Science, Volume 177, Issue 17, pp. 3592-3612, 2007.

T.P. Hong, C.S. Kuo and S.C. Chi: A Fuzzy Data Mining Algorithm for Quantitative Values. 1999 Third International Conference on Knowledge-Based Intelligent Information Engineering Systems. Proceedings. IEEE 1999, pp. 480-3.

T.P. Hong, C.S. Kuo and S.C. Chi: Mining Association Rules from Quantitative Data. Intelligent Data Analysis, vol.3, no.5, nov. 1999, pp363-376.

Jang J., "Structure determination in fuzzy modeling: A fuzzy CART approach," in Proc. IEEE Conf. Fuzzy Systems, 1994, pp. 480485.

Janikow, C.Z., Fuzzy Decision Trees: Issues and Methods, IEEE Transactions on Systems, Man, and Cybernetics, Vol. 28, Issue 1, pp. 1-14. 1998.

Kim, J., Krishnapuram, R. and Dav, R. (1996). Application of the Least Trimmed Squares Technique to Prototype-Based Clustering, Pattern Recognition Letters, 17, 633-641.

Joseph Komem and Moti Schneider, On the Use of Fuzzy Logic in Data Mining, in The Data Mining and Knowledge Discovery Handbook, O. Maimon, L. Rokach (Eds.), pp. 517-533, Springer, 2005.

Maher P. E. and Clair D. C, Uncertain reasoning in an ID3 machine learning framework, in Proc. 2nd IEEE Int. Conf. Fuzzy Systems, 1993, pp. 712.

Maimon O., and Rokach, L. Data Mining by Attribute Decomposition with semiconductors manufacturing case study, in Data Mining for Design and Manufacturing: Methods and Applications, D. Braha (ed.), Kluwer Academic Publishers, pp. 311–336, 2001.

Maimon O. and Rokach L., “Improving supervised learning by feature decomposition”, Proceedings of the Second International Symposium on Foundations of Information and Knowledge Systems, Lecture Notes in Computer Science, Springer, pp. 178-196, 2002.

Maimon, O. and Rokach, L., Decomposition Methodology for Knowledge Discovery and Data Mining: Theory and Applications, Series in Machine Perception and Artificial Intelligence - Vol. 61, World Scientific Publishing, ISBN:981-256-079-3, 2005.

S. Mitra, Y. Hayashi, "Neuro-fuzzy Rule Generation: Survey in Soft Computing Framework." IEEE Trans. Neural Networks, Vol. 11, N. 3, pp. 748-768, 2000.

S. Mitra and S. K. Pal, Fuzzy sets in pattern recognition and machine intelligence, Fuzzy Sets and Systems 156 (2005) 381386

Moskovitch R, Elovici Y, Rokach L, Detection of unknown computer worms based on behavioral classification of the host, Computational Statistics and Data Analysis, 52(9):4544–4566, 2008.

Nasraoui, O. and Krishnapuram, R. (1997). A Genetic Algorithm for Robust Clustering Based on a Fuzzy Least Median of Squares Criterion, Proceedings of NAFIPS, Syracuse NY, 217-221.

Nauck D., Neuro-Fuzzy Systems: Review and Prospects Paper appears in Proc. Fifth European Congress on Intelligent Techniques and Soft Computing (EUFIT'97), Aachen, Sep. 8-11, 1997, pp. 1044-1053

Olaru C., Wehenkel L., A complete fuzzy decision tree technique, Fuzzy Sets and Systems, 138(2):221–254, 2003.

Peng Y., Intelligent condition monitoring using fuzzy inductive learning, Journal of Intelligent Manufacturing, 15 (3): 373-380, June 2004.

Rokach, L., Decomposition methodology for classification tasks: a meta decomposer framework, Pattern Analysis and Applications, 9(2006):257–271.

Rokach L., Genetic algorithm-based feature set partitioning for classification problems,Pattern Recognition, 41(5):1676–1700, 2008.

Rokach L., Mining manufacturing data using genetic algorithm-based feature set decomposition, Int. J. Intelligent Systems Technologies and Applications, 4(1):57-78, 2008.

Rokach L., Maimon O. and Lavi I., Space Decomposition In Data Mining: A Clustering Approach, Proceedings of the 14th International Symposium On Methodologies For Intelligent Systems, Maebashi, Japan, Lecture Notes in Computer Science, Springer-Verlag, 2003, pp. 24–31.

Rokach, L. and Maimon, O. and Averbuch, M., Information Retrieval System for Medical Narrative Reports, Lecture Notes in Artificial intelligence 3055, page 217-228 Springer-Verlag, 2004.

Rokach, L. and Maimon, O. and Arbel, R., Selective voting-getting more for less in sensor fusion, International Journal of Pattern Recognition and Artificial Intelligence 20 (3) (2006), pp. 329–350.

Rokach, L. and Maimon, O., Theory and applications of attribute decomposition, IEEE International Conference on Data Mining, IEEE Computer Society Press, pp. 473–480, 2001.

Rokach L. and Maimon O., Feature Set Decomposition for Decision Trees, Journal of Intelligent Data Analysis, Volume 9, Number 2, 2005b, pp 131–158.

Rokach, L. and Maimon, O., Clustering methods, Data Mining and Knowledge Discovery Handbook, pp. 321–352, 2005, Springer.

Rokach, L. and Maimon, O., Data mining for improving the quality of manufacturing: a feature set decomposition approach, Journal of Intelligent Manufacturing, 17(3):285–299, 2006, Springer.

Rokach, L., Maimon, O., Data Mining with Decision Trees: Theory and Applications, World Scientific Publishing, 2008.

E. Shnaider and M. Schneider, Fuzzy Tools for Economic Modeling. In: Uncertainty Logics: Applications in Economics and Management. Proceedings of SIGEF'98 Congress, 1988.

Shnaider E., M. Schneider and A. Kandel, 1997, A Fuzzy Measure for Similarity of Numerical Vectors, Fuzzy Economic Review, Vol. II, No. 1, 1997, pp. 17 -

Tani T. and Sakoda M., Fuzzy modeling by ID3 algorithm and its application to prediction of heater outlet temperature, Proc. IEEE lnternat. Conf. on Fuzzy Systems, March 1992, pp. 923-930.

Yuan Y., Shaw M., Induction of fuzzy decision trees, Fuzzy Sets and Systems 69(1995):125-139.

Zimmermann H. J., Fuzzy Set Theory and its Applications, Springer, 4th edition, 2005.

Part V

Supporting Methods

25

Statistical Methods for Data Mining

Yoav Benjamini[1] and Moshe Leshno[2]

[1] Department of Statistics, School of Mathematical Sciences, Sackler Faculty for Exact Sciences Tel Aviv University
ybenja@post.tau.ac.il
[2] Faculty of Management and Sackler Faculty of Medicine
Tel Aviv University
leshnom@post.tau.ac.il

Summary. The aim of this chapter is to present the main statistical issues in Data Mining (DM) and Knowledge Data Discovery (KDD) and to examine whether traditional statistics approach and methods substantially differ from the new trend of KDD and DM. We address and emphasize some central issues of statistics which are highly relevant to DM and have much to offer to DM.

Key words: Statistics, Regression Models, False Discovery Rate (FDR), Model selection and False Discovery Rate (FDR)

25.1 Introduction

In the words of anonymous saying there are two problems in modern science: too many people using different terminology to solve the same problems and even more people using the same terminology to address completely different issues. This is particularly relevant to the relationship between traditional statistics and the new emerging field of knowledge data discovery (KDD) and Data Mining (DM). The explosive growth of interest and research in the domain of KDD and DM of recent years is not surprising given the proliferation of low-cost computers and the requisite software, low-cost database technology (for collecting and storing data) and the ample data that has been and continues to be collected and organized in databases and on the web. Indeed, the implementation of KDD and DM in business and industrial organizations has increased dramatically, although their impact on these organizations is not clear. The aim of this chapter is mainly to present the main statistical issues in DM and KDD and to examine the role of traditional statistics approach and methods in the new trend of KDD and DM. We argue that data miners should be familiar with statistical themes and models and statisticians should be aware of the capabilities and limitation of Data Mining and the ways in which Data Mining differs from traditional statistics.

Statistics is the traditional field that deals with the quantification, collection, analysis, interpretation, and drawing conclusions from data. Data Mining is an interdisciplinary field that draws on computer sciences (data base, artificial intelligence, machine learning, graphical and

O. Maimon, L. Rokach (eds.), *Data Mining and Knowledge Discovery Handbook*, 2nd ed.,
DOI 10.1007/978-0-387-09823-4_25,

visualization models), statistics and engineering (pattern recognition, neural networks). DM involves the analysis of large existing data bases in order to discover patterns and relationships in the data, and other findings (unexpected, surprising, and useful). Typically, it differs from traditional statistics on two issues: the size of the data set and the fact that the data were initially collected for purpose other than the that of the DM analysis. Thus, *experimental design*, a very important topic in traditional statistics, is usually irrelevant to DM. On the other hand asymptotic analysis, sometimes criticized in statistics as being irrelevant, becomes very relevant in DM.

While in traditional statistics a data set of 100 to 10^4 entries is considered large, in DM even 10^4 may be considered a small set set fit to be used as an example, rather than a problem encountered in practice. Problem sizes of 10^7 to 10^{10} are more typical. It is important to emphasize, though, that data set sizes are not all created equal. One needs to distinguish between the number of cases (observations) in a large data set (n), and the number of features (variables) available for each case (m). In a large data set,n ,m or both can be large, and it does matter which, a point on which we will elaborate in the continuation. Moreover these definitions may change when the same data set is being used for two different purposes. A nice demonstration of such an instance can be found in the 2001 KDD competition, where in one task the number of cases was the number of purchasing customers, the click information being a subset of the features, and in the other task the clicks were the cases.

Our aim in this chapter is to indicate certain focal areas where statistical thinking and practice have much to offer to DM. Some of them are well known, whereas others are not. We will cover some of them in depth, and touch upon others only marginally. We will address the following issues:

- Size
- Curse of Dimensionality
- Assessing uncertainty
- Automated analysis
- Algorithms for data analysis in Statistics
- Visualization
- Scalability
- Sampling
- Modelling relationships
- Model selection

We briefly discuss these issues in the next section and then devote special sections to three of them. In section 25.3 we explain and present how the most basic of statistical methodologies, namely regression analysis, has developed over the years to create a very flexible tool to model relationships, in the form of Generalized Linear Models (GLMs). In section 25.4 we discuss the False Discovery Rate (FDR) as a scalable approach to hypothesis testing. In section 25.5 we discuss how FDR ideas contribute to flexible model selection in GLM. We conclude the chapter by asking whether the concepts and methods of KDD and DM differ from those of traditional statistical, and how statistics and DM should act together.

25.2 Statistical Issues in DM

25.2.1 Size of the Data and Statistical Theory

Traditional statistics emphasizes the mathematical formulation and validation of a methodology, and views simulations and empirical or practical evidence as a lesser form of validation.

The emphasis on rigor has required proof that a proposed method will work prior to its use. In contrast, computer science and machine learning use experimental validation methods. In many cases mathematical analysis of the performance of a statistical algorithm is not feasible in a specific setting, but becomes so when analyzed asymptotically. At the same time, when size becomes extremely large, studying performance by simulations is also not feasible. It is therefore in settings typical of DM problems that asymptotic analysis becomes both feasible and appropriate. Interestingly, in classical asymptotic analysis the number of cases n tends to infinity. In more contemporary literature there is a shift of emphasis to asymptotic analysis where the number of variables m tends to infinity. It is a shift that has occurred because of the interest of statisticians and applied mathematicians in wavelet analysis (see Chapter 26.3 in this volume), where the number of parameters (wavelet coefficients) equals the number of cases, and has proved highly successful in areas such as the analysis of gene expression data from microarrays.

25.2.2 The Curse of Dimensionality and Approaches to Address It

The curse of dimensionality is a well documented and often cited fundamental problem. Not only do algorithms face more difficulties as the the data increases in dimension, but the structure of the data itself changes. Take, for example, data uniformly distributed in a high-dimensional ball. It turns out that (in some precise way, see (Meilijson, 1991)) most of the data points are very close to the surface of the ball. This phenomenon becomes very evident when looking for the k-Nearest Neighbors of a point in high-dimensional space. The points are so far away from each other that the radius of the neighborhood becomes extremely large.

The main remedy offered for the curse of dimensionality is to use only part of the available variables per case, or to combine variables in the data set in a way that will summarize the relevant information with fewer variables. This dimension reduction is the essence of what goes on in the data warehousing stage of the DM process, along with the cleansing of the data. It is an important and time-consuming stage of the DM operations, accounting for 80-90% of the time devoted to the analysis.

The dimension reduction comprises two types of activities: the first is quantifying and summarizing information into a number of variables, and the second is further reducing the variables thus constructed into a workable number of combined variables. Consider, for instance, a phone company that has at its disposal the entire history of calls made by a customer. How should this history be reflected in just a few variables? Should it be by monthly summaries of the number of calls per month for each of the last 12 months, such as their means (or medians), their maximal number, and a certain percentile? Maybe we should use the mean, standard deviation and the number of calls below two standard deviations from the mean? Or maybe we should use none of these but rather variables capturing the monetary values of the activity? If we take this last approach, should we work with the cost itself or will it be more useful to transfer the cost data to the log scale? Statistical theory and practice have much to offer in this respect, both in measurement theory, and in data analysis practices and tools. The variables thus constructed now have to be further reduced into a workable number of combined variables. This stage may still involve judgmental combination of previously defined variables, such as cost per number of customers using a phone lines, but more often will require more automatic methods such as principal components or independent components analysis (for a further discussion of principle component analysis see (Roberts and Everson, 2001)).

We cannot conclude the discussion on this topic without noting that occasionally we also start getting the *blessing of dimensionality*, a term coined by David Donoho (Donoho, 2000) to describe the phenomenon of the high dimension helping rather than hurting the analysis, that we often encounter as we proceed up the scale in working with very high dimensional data. For example, for large m if the data we study is pure noise, the i-th largest observation is very close to its expectations under the model for the noise! Another case in point is microarray analysis, where the many non-relevant genes analyzed give ample information about the distribution of the noise, making it easier to identify real discoveries. We shall see a third case below.

25.2.3 Assessing Uncertainty

Assessing the uncertainty surrounding knowledge derived from data is recognized as a the central theme in statistics. The concern about the uncertainty is down-weighted in KDD, often because of the myth that all relevant data is available in DM. Thus, standard errors of averages, for example, will be ridiculously low, as will prediction errors. On the other hand experienced users of DM tools are aware of the variability and uncertainty involved. They simply tend to rely on seemingly "non-statistical" technologies such as the use of a training sample and a test sample. Interestingly the latter is a methodology widely used in statistics, with origins going back to the 1950s. The use of such validation methods, in the form of cross-validation for smaller data sets, has been a common practice in exploratory data analysis when dealing with medium size data sets.

Some of the insights gained over the years in statistics regarding the use of these tools have not yet found their way into DM. Take, for example, data on food store baskets, available for the last four years, where the goal is to develop a prediction model. A typical analysis will involve taking a random training and validation samples from the data, then testing the model on the validation sample, with the results guiding us as to the choice of the most appropriate model. However, the model will be used next year, not last year. The main uncertainty surrounding its conclusions may not stem from the person to person variability captured by the differences between the values in the validation sample, but rather follow from the year to year variability. If this is the case, we have all the data, but only four observations. The choice of the data for validation and training samples should reflect the higher sources of variability in the data, by each time setting the data of one year aside to serve as the source for the test sample (for an illustrated yet profound discussion of these issues in exploratory data analysis (see (Mosteller and Tukey, 1977), Ch. 7,8).

25.2.4 Automated Analysis

The inherent dangers of the necessity to rely on automatic strategies for analyzing the data, another main theme in DM, have been demonstrated again and again. There are many examples where trivial non-relevant variables, such as case number, turned out to be the best predictors in automated analysis. Similarly, variables displaying a major role in predicting a variable of interest in the past, may turn out to be useless because they reflect some strong phenomenon not expected to occur in the future (see for example the conclusions using the onion metaphor from the 2002 KDD competition). In spite of these warnings, it is clear that large parts of the analysis should be automated, especially at the warehousing stage of the DM.

This may raise new dangers. It is well known in statistics that having even a small proportion of outliers in the data can seriously distort its numerical summary. Such unreasonable values, deviating from the main structure of the data, can usually be identified by a careful human data analyst, and excluded from the analysis. But once we have to warehouse information about millions of customers, summarizing the information about each customer by a few numbers has to be automated and the analysis should rather deal automatically with the possible impact of a few outliers.

Statistical theory and methodology supply the framework and the tools for this endeavor. A numerical summary of the data that is not unboundedly influenced by a negligible proportion of the data is called a resistant summary. According to this definition the average is not resistant, for even one straying data value can have an unbounded effect on it. In contrast, the median is resistant. A resistant summary that retains its good properties under less than ideal situations is called a robust summary, the α-trimmed mean (rather than the median) being an example of such. The concepts of robustness and resistance, and the development of robust statistical tools for summarizing location, scale, and relationships, were developed during the 1970's and the 1980's, and resulting theory is quite mature (see, for instance, (Ronchetti et al., 1986,Dell'Aquila and Ronchetti, 2004)), even though robustness remains an active area of contemporary research in statistics. Robust summaries, rather than merely averages, standard deviations, and simple regression coefficients, are indispensable in DM. Here too, some adaptation of the computations to size may be needed, but efforts in this direction are being made in the statistical literature.

25.2.5 Algorithms for Data Analysis in Statistics

Computing has always been a fundamental to statistic, and it remained so even in times when mathematical rigorousity was perceived as the most highly valued quality of a data analytic tool. Some of the important computational tools for data analysis, rooted in classical statistics, can be found in the following list: efficient estimation by maximum likelihood, least squares and least absolute deviation estimation, and the EM algorithm; analysis of variance (ANOVA, MANOVA, ANCOVA), and the analysis of repeated measurements; nonparametric statistics; log-linear analysis of categorial data; linear regression analysis, generalized additive and linear models, logistic regression, survival analysis, and discriminant analysis; frequency domain (spectrum) and time domain (ARIMA) methods for the analysis of time series; multivariate analysis tools such as factor analysis, principal component and later independent component analyses, and cluster analysis; density estimation, smoothing and de-noising, and classification and regression trees (decision trees); Bayesian networks and the Markov Chain Monte Carlo (MCMC)) algorithm for Bayesian inference.

For an overview of most of these topics, with an eye to the DM community (Hastie *et al.*, 2001). Some of the algorithms used in DM which were not included in classical statistic, are considered by some statisticians to be part of statistics (Friedman, 1998). For example, rule induction (AQ, CN2, Recon, etc.), associate rules, neural networks, genetic algorithms and self-organization maps may be attributed to classical statistics.

25.2.6 Visualization

Visualization of the data and its structure, as well as visualization of the conclusions drawn from the data, are another central theme in DM. Visualization of quantitative data as a major activity flourished in the statistics of the 19th century, faded out of favor through most of the 20th century, and began to regain importance in the early 1980s. This importance in reflected in the development of the Journal of Computational and Graphical Statistics of the American Statistical Association. Both the theory of visualizing quantitative data and the practice have dramatically changed in recent years. Spinning data to gain a 3-dimensional understanding of pointclouds, or the use of projection pursuit are just two examples of visualization technologies that emerged from statistics.

It is therefore quite frustrating to see how much KDD software deviates from known principles of good visualization practices. Thus, for instance, the fundamental principle that the retinal variable in a graphical display (length of line, or the position of a point on a scale) should be proportional to the quantitative variable it represents is often violated by introducing a dramatic perspective. Add colors to the display and the result is even harder to understand.

Much can be gained in DM by mining the knowledge about visualization available in statistics, though the visualization tools of statistics are usually not calibrated for the size of the data sets commonly dealt within DM. Take for example the extremely effective Boxplots display, used for the visual comparisons of batches of data. A well-known rule determines two fences for each batch, and points outside the fences are individually displayed. There is a traditional default value in most statistical software, even though the rule was developed with batches of very small size in mind (in DM terms). In order to adapt the visualization technique for routine use in DM, some other rule which will probably be adaptive to the size of the batch should be developed. As this small example demonstrates, visualization is an area where joint work may prove to be extremely fruitful.

25.2.7 Scalability

In machine learning and Data Mining *scalability* relates to the ability of an algorithm to scale up with size, an essential condition being that the storage requirement and running time should not become infeasible as the size of the problem increases. Even simple problems like multivariate histograms become a serious task, and may benefit from complex algorithms that scale up with size. Designing scalable algorithms for more complex tasks, such as decision tree modeling, optimization algorithms, and the mining of association rules, has been the most active research area in DM. Altogether, scalability is clearly a fundamental problem in DM mostly viewed with regard to its algorithmic aspects. We want to highlight the duality of the problem by suggesting that concepts should be scalable as well. In this respect, consider the general belief that hypothesis testing is a statistical concept that has nothing to offer in DM. The usual argument is that data sets are so large that every hypothesis tested will turn out to be statistically significant - even if differences or relationships are minuscule. Using association rules as an example, one may wonder whether an observed lift for a given rule is "really different from 1", but then find that at the traditional level of significance used (the mythological 0.05) an extremely large number of rules are indeed significant. Such findings brought David Hand (Hand, 1998) to ask "what should replace hypothesis testing?" in DM. We shall discuss two such important scalable concepts in the continuation: the testing of multiple hypotheses using the False Discovery Rate and the penalty concept in model selection.

25.2.8 Sampling

Sampling is the ultimate scalable statistical tool: if the number of cases n is very large the conclusions drawn from the sample depend only on the size of the sample and not on the size of the data set. It is often used to get a first impression of the data, visualize its main features, and reach decisions as to the strategy of analysis. In spite of its scalability and usefulness sampling has been attacked in the KDD community for its inability to find very rare yet extremely interesting pieces of knowledge.

Sampling is a very well developed area of statistics (see for example (Cochran, 1977)), but is usually used in DM at the very basic level. Stratified sampling, where the probability of picking a case changes from one stratum to another, is hardly ever used. But the questions are relevant even in the simplest settings: should we sample from the few positive responses at the same rate that we sample from the negative ones? When studying faulty loans, should we sample larger loans at a higher rate? A thorough investigations of such questions, phrased in the realm of particular DM applications may prove to be very beneficial.

Even greater benefits might be realized when more advanced sampling models, especially those related to super populations, are utilized in DM. The idea here is that the population of customers we view each year, and from which we sample, can itself be viewed as a sample of the same super population. Hence next year's customers will again be a population sampled from the super population. We leave this issue wide open.

25.3 Modeling Relationships using Regression Models

Demonstrating that statistics, like Data Mining, is concerned with turning data into information and knowledge, even though the terminology may differ, in this section we present a major statistical approach being used in Data Mining, namely regression analysis. In the late 1990s, statistical methodologies such as regression analysis were not included in commercial Data Mining packages. Nowadays, most commercial Data Mining software includes many statistical tools and in particular regression analysis. Although regression analysis may seem simple and anachronistic, it is a very powerful tool in DM with large data sets, especially in the form of the generalized linear models (GLMs). We emphasize the assumptions of the models being used and how the underlying approach differs from that of machine learning. The reader is referred to (McCullagh and Nelder, 1991) and Chapter 10.7 in this volume for more detailed information on the specific statistical methods.

25.3.1 Linear Regression Analysis

Regression analysis is the process of determining how a variable y is related to one, or more, other variables $x_1,\ldots,x_k$. The y is usually called the *dependent* variable and the x_i's are called the *independent* or *explanatory* variables. In a linear regression model we assume that

$$y_i = \beta_0 + \sum_{j=1}^{k} \beta_j x_{ji} + \varepsilon_i \quad i = 1,\ldots,M \tag{25.1}$$

and that the ε_i's are independent and are identically distributed as $\mathcal{N}(0,\sigma^2)$ and M is the number of data points. The expected value of y_i is given by

$$E(y_i) = \beta_0 + \sum_{j=1}^{k} \beta_j x_{ji} \tag{25.2}$$

To estimate the coefficients of the linear regression model we use the least square estimation which gives results equivalent to the estimators obtained by the maximum likelihood method. Note that for the linear regression model there is an explicit formula of the β's. We can write (63.1) in matrix form by $Y = X \cdot \beta^t + \varepsilon^t$ where β^t is the transpose of the vector $[\beta_0, \beta_1, \ldots, \beta_k]$, ε^t is the transpose of the vector $\varepsilon = [\varepsilon_1, \ldots, \varepsilon_M]$ and the matrix X is given by

$$X = \begin{pmatrix} 1 & x_{11} & \cdots & x_{1k} \\ 1 & x_{21} & \cdots & x_{2k} \\ 1 & \vdots & \ddots & \vdots \\ 1 & x_{k1} & \cdots & x_{Mk} \end{pmatrix} \quad (25.3)$$

The estimates of the β's are given (in matrix form) by $\hat{\beta}=(X^tX)^{-1}X^tY$. Note that in linear regression analysis we assume that for a given $x_1,\ldots,x_k$ y_i is distributed as $\mathcal{N}(\beta_0 + \sum_{j=1}^k \beta_j x_{ji}, \sigma^2)$. There is a large class of general regression models where the relationship between the y_is and the vector x is not assumed to be linear, that can be converted to a linear model.

Machine learning approach, when compared to regression analysis, aims to select a function $f \in \mathcal{F}$ from a given set of functions $\mathcal{F}$, that best approximates or fits the given data. Machine learning assumes that the given data $(\mathbf{x}_i, y_i)$, $(i = 1,\ldots,M)$ is obtained by a data generator, producing the data according to an unknown distribution $p(\mathbf{x},y) = p(\mathbf{x})p(y|\mathbf{x})$. Given a loss function $\Psi(y - f(\mathbf{x}))$, the quality of an approximation produced by the machine learning is measured by the expected loss, the expectation being under the unknown distribution $p(\mathbf{x},y)$. The subject of statistical machine learning is the following optimization problem:

$$\min_{f \in \mathcal{F}} \int \Psi(y - f(\mathbf{x}))dp(\mathbf{x},y) \quad (25.4)$$

when the density function $p(\mathbf{x},y)$ is unknown but a random independent sample of $(\mathbf{x}_i, y_i)$ is given. If $\mathcal{F}$ is the set of all linear function of $\mathbf{x}$ and $\Psi(y - f(\mathbf{x})) = (y - f(\mathbf{x}))^2$ then if $p(y|\mathbf{x})$ is normally distributed then the minimization of (25.4) is equivalent to linear regression analysis.

25.3.2 Generalized Linear Models

Although in many cases the set of linear function is good enough to model the relationship between the stochastic response y as a function of $\mathbf{x}$ it may not always suffice to represent the relationship. The generalized linear model increases the family of functions $\mathcal{F}$ that may represent the relationship between the response y and $\mathbf{x}$. The tradeoff is between having a simple model and a more complex model representing the relationship between y and $\mathbf{x}$. In the general linear model the distribution of y given $\mathbf{x}$ does not have to be normal, but can be any of the distributions in the exponential family (McCullagh and Nelder, 1991). Instead of the expected value of $y|\mathbf{x}$ being a linear function, we have

$$g(E(y_i)) = \beta_0 + \sum_{j=1}^{k} \beta_j x_{ji} \quad (25.5)$$

where $g(\cdot)$ is a monotone differentiable function.

In the *generalized additive models*, $g(E(y_i))$ need not to be a linear function of $\mathbf{x}$ but has the form:

$$g(E(y_i)) = \beta_0 + \sum_{j=1}^{k} \sigma_j(x_{ji}) \quad (25.6)$$

where $\sigma(\cdot)$'s are smooth functions. Note that neural networks are a special case of the generalized additive linear models. For example the function that a multilayer feedforward neural network with one hidden layer computes is (see Chapter 21 in this volume for detailed information):

$$y_i = f(\mathbf{x}) = \sum_{l=1}^{m} \beta_j \cdot \sigma\left(\sum_{j=1}^{k} \mathbf{w}_{jl}\mathbf{x}_{ji} - \theta_j\right) \tag{25.7}$$

where m is the number of processing-units in the hidden layer. The family of functions that can be computed depends on the number of neurons in the hidden layer and the activation function σ. Note that a standard multilayer feedforward network with a smooth activation function σ can approximate any continuous function on a compact set to any degree of accuracy if and only if the network's activation function σ is not a polynomial (Leshno et al., 1993).

There are methods for fitting generalized additive models. However, unlike linear models for which there exits a methodology of statistical inference, for machine learning algorithms as well as generalized additive methods, no such methodology have yet been developed. For example, using a statistical inference framework in linear regression one can test the hypothesis that all or part of the coefficients are zero.

The total sum of squares (SST) is equal to the sum of squares due to regression (SSR) plus the residual sum of square (RSS_k), i.e.

$$\underbrace{\sum_{i=1}^{M}(y_i - \bar{y})^2}_{SST} = \underbrace{\sum_{i=1}^{M}(\hat{y}_i - \bar{y})^2}_{SSR} + \underbrace{\sum_{i=1}^{M}(y_i - \hat{y}_i)^2}_{RSS_k} \tag{25.8}$$

The percentage of variance explained by the regression is a very popular method to measure the goodness-of-fit of the model. More specifically R^2 and the adjusted R^2 defined below are used to measure the goodness of fit.

$$R^2 = \frac{\sum_{i=1}^{M}(\hat{y}_i - \bar{y})^2}{\sum_{i=1}^{M}(y_i - \bar{y})^2} = 1 - \frac{RSS_k}{SST} \tag{25.9}$$

$$\text{Adjusted-}R^2 = 1 - (1 - R^2)\frac{M-1}{M-k-1} \tag{25.10}$$

We next turn to a special case of the general additive model that is very popular and powerful tool in cases where the responses are binary values.

25.3.3 Logistic Regression

In logistic regression the y_is are binary variables and thus not normally distributed. The distribution of y_i given $\mathbf{x}$ is assumed to follow a Bernoulli distribution such that:

$$log\left(\frac{p(y_i = 1|\mathbf{x})}{1 - p(y_i = 1|\mathbf{x})}\right) = \beta_0 + \sum_{j=1}^{k} \beta_j x_{ji} \tag{25.11}$$

If we denote $\pi(\mathbf{x}) = p(y = 1|\mathbf{x})$ and the real valued function $g(t) = \frac{t}{1-t}$ then $g(\pi(\mathbf{x}))$ is a linear function of $\mathbf{x}$. Note that we can write $y = \pi(\mathbf{x}) + \varepsilon$ such that if $y = 1$ then $\varepsilon = 1 - \pi(\mathbf{x})$ with probability $\pi(\mathbf{x})$, and if $y = 0$ then $\varepsilon = -\pi(\mathbf{x})$ with probability $1 - \pi(\mathbf{x})$. Thus, $\pi(\mathbf{x}) = E(y|\mathbf{x})$ and

$$\pi(\mathbf{x}) = \frac{e^{\beta_0+\sum_{j=1}^{k}\beta_j x_j}}{1+e^{\beta_0+\sum_{j=1}^{k}\beta_j x_j}} \tag{25.12}$$

Of the several methods to estimates the β's, the method of maximum likelihood is one most commonly used in the logistic regression routine of the major software packages.

In linear regression, interest focuses on the size of R^2 or adjusted-R^2. The guiding principle in logistic regression is similar: the comparison of observed to predicted values is based on the log likelihood function. To compare two models - a full model and a reduced model, one uses the following likelihood ratio:

$$D = -2\ln\left(\frac{\text{likelihod of the reduced model}}{\text{likelihod of the full model}}\right) \tag{25.13}$$

The statistic D in equation (25.13), is called the deviance (McCullagh and Nelder, 1991). Logistic regression is a very powerful tool for classification problems in discriminant analysis and is applied in many medical and clinical research studies.

25.3.4 Survival Analysis

Survival analysis addresses the question of how long it takes for a particular event to happen. In many medical applications the most important response variable often involves time; the event is some hazard or death and thus we analyze the patient's survival time. In business application the event may be a failure of a machine or market entry of a competitor. There are two main characteristics of survival analysis that make it different from regression analysis. The first is that the presence of *censored observation*, where the event (e.g. death) has not necessarily occurred by the end of the study. Censored observation may also occur when patients are lost to follow-up for one reason or another. If the output is censored, we do not have the value of the output, but we do have some information about it. The second, is that the distribution of survival times is often skewed or far from normality. These features require special methods of analysis of survival data, two functions describing the distribution of survival times being of central importance: the *hazard* function and the *survival* function. Using T to represent survival time, the *survival* function denoted by $S(t)$, is defined as the probability of survival time to be greater than t, i.e. $S(t) = \Pr(T > t) = 1 - F(t)$, where $F(t)$ is the cumulative distribution function of the output. The hazard function, $h(t)$, is defined as the probability density of the output at time t conditional upon survival to time t, that is $h(t) = f(t)/S(t)$, where $f(t)$ is the probability density of the output. Is is also known as the instantaneous failure rate and presents the probability that an event will happen in a small time interval Δt, given that the individual has survived up to the beginning of this interval, i.e. $h(t) = lim_{\Delta t \downarrow 0}\frac{\Pr(t \le T < t+\Delta t | t \le T)}{\Delta t} = f(t)/S(t)$. The hazard function may remain constant, increase, decrease or take some more complex shape. Most modeling of survival data is done using a proportional hazard model. A proportional-hazard model, which assumes that the hazard function is of the form

$$h(t) = \alpha(t)\exp\left(\beta_0 + \sum_{i=1}^{n}\beta_i x_i\right) \tag{25.14}$$

$\alpha(t)$ is a hazard function on its own, called the baseline hazard function, corresponding to that for the average value of all the covariates $x_1,\ldots,x_n$. This is called a proportional-hazard model, because the hazard function for two different patients have a constant ratio. The interpretation of the β's in this model is that the effect is multiplicative.

There are several approaches to survival data analysis. The simplest it to assume that the baseline hazard function is constant which is equivalent to assuming exponential distribution for the time to event. Another simple approach would be to assume that the baseline hazard function is of the two-parameter family of function, like the Weibull distribution. In these cases the standard methods such as maximum likelihood can be used. In other cases one may restrict $\alpha(t)$ for example by assuming it to be monotonic. In business application, the baseline hazard function can be determined by experimentation, but in medical situations it is not practical to carry out an experiment to determine the shape of the baseline hazard function. The Cox proportional hazards model (Cox, 1972), introduced to overcome this problem, has become the most commonly used procedure for modelling the relationship of covariates to a survival outcome and it is used in almost all medical analyses of survival data. Estimation of the β's is based on the partial likelihood function introduced by Cox (Cox, 1972, Therneau and Grambsch, 2000).

There are many other important statistical themes that are highly relevant to DM, among them: statistical classification methods, spline and wavelets, decision trees and others (see Chapters 8.8 and 26.3 in this volume for more detailed information on these issues). In the next section we elaborate on the False Discovery Rate (FDR) approach (Benjamini and Hochberg, 1995), a most salient feature of DM.

25.4 False Discovery Rate (FDR) Control in Hypotheses Testing

As noted before there is a feeling that the testing of a hypothesis is irrelevant in DM. However the problem of separating a real phenomenon from its background noise is just as fundamental a concern in DM as in statistics. Take for example an association rule, with an observed lift which is bigger than 1, as desired. Is it also significantly bigger than 1 in the statistical sense, that is beyond what is expected to happen as a result of noise? The answer to this question is given by the testing of the hypothesis that the lift is 1. However, in DM a hypothesis is rarely tested alone, as the above point demonstrates. The tested hypothesis is always a member of a larger family of similar hypotheses, all association rules of at least a given support and confidence being tested simultaneously. Thus, the testing of hypotheses in DM always invokes the "Multiple Comparisons Problem" so often discussed in statistics. Interestingly it is the first demonstration of a DM problem in the statistics of 50 years ago: when a feature of interest (a variable) is measured on 10 subgroups (treatments), and the mean values are compared to some reference value (such as 0), the problem is a small one, but take these same means and search among all pairwise comparisons between the treatments to find a significant difference, and the number of comparisons increases to 10*(10-1)/2=45 - which is in general quadratic in the number of treatments. It becomes clear that if we allow an .05 probability of deciding that a difference exists in a single comparison even if it really does not, thereby making a false discovery (or a type I error in statistical terms), we can expect to find on the average 2.25 such errors in our pool of discoveries. No wonder this DM activity is sometimes described in statistics as "post hoc analysis" - a nice definition for DM with a traditional flavor.

The attitude that has been taken during 45 years of statistical research is that in such problems the probability of making even one false discovery should be controlled, that is controlling the Family Wise Error rate (FWE) as it is called. The simplest way to address the multiple comparisons problem, offering FWE control at some desired level α with no further assumptions, is to use the Bonferroni procedure: conduct each of the m tests at level α/m. In problems where m becomes very large the penalty to the researcher from the extra caution becomes heavy, in the sense that the probability of making any discovery becomes very small, and so it is not uncommon to observe researchers avoiding the need to adjust for multiplicity.

The False Discovery Rate (FDR), namely the expectation of the proportion of false discoveries (rejected true null hypotheses) among the discoveries (the rejected hypotheses), was developed by (Benjamini and Hochberg, 1995) to bridge these two extremes. When the null hypothesis is true for all hypotheses - the FDR and FWE criteria are equivalent. However, when there are some hypotheses for which the null hypotheses are false, an FDR controlling procedure may yield many more discoveries at the expense of having a small proportion of false discoveries.

Formally, let H_{0i}, $i = 1, \ldots m$ be the tested null hypotheses. For $i = 1, \ldots m_0$ the null hypotheses are true, and for the remaining $m_1 = m - m_0$ hypotheses they are not. Thus, any discovery about a hypothesis from the first set is a false discovery, while a discover about a hypothesis from the second set is a true discovery. Let V denote the number of false discoveries and R the total number of discoveries. Let the proportion of false discoveries be

$$Q = \begin{cases} V/R & \text{if } R > 0 \\ 0 & \text{if } R = 0 \end{cases},$$

and define $FDR = E(Q)$.

Benjamini and Hochberg advocated that the FDR should be controlled at some desirable level q, while maximizing the number of discoveries made. They offered the linear step-up procedure as a simple and general procedure that controls the FDR. The linear step-up procedure makes use of the m p-values, $\mathbf{P} = (P_1, \ldots P_m)$ so in a sense it is very general. It compares the ordered values $P_{(1)} \leq \ldots \leq P_{(m)}$ to the set of constants linearly interpolated between q and q/m.

Definition 25.4.1 *The Linear step-up Procedure: Let* $k = \max\{i : P_{(i)} \leq iq/m\}$, *and reject the* k *hypotheses associated with* $P_{(1)}, \ldots P_{(k)}$. *If no such a* k *exists reject none.*

The procedure was first suggested by Eklund (Seeger, 1968) and forgotten, then independently suggested by Simes (Simes, 1986). At both points in time it went out of favor because it does not control the FWE. (Benjamini and Hochberg, 1995), showed that the procedure does control the FDR, raising the interest in this procedure. Hence it is now referred to also as the Benjamini and Hochberg procedure (BH procedure), or (unfortunately) the FDR procedure (e.g. in SAS) (for a detailed historical review see (Benjamini and Hochberg, 2000)).

For the purpose of practical interpretation and flexibility in use, the results of the linear step-up procedure can also be reported in terms of the FDR adjusted p-values. Formally, the FDR adjusted p-value of $H_{(i)}$ is $p^{LSU}_{(i)} = \min\{\frac{mp_{(j)}}{j} \mid j \geq i\}$. Thus the linear step-up procedure at level q is equivalent to rejecting all hypotheses whose FDR adjusted p-value is $\leq q$.

It should also be noted that the dual linear step-down procedure, which uses the same constants but starts with the smallest p-value and stops at the last $\{P_{(i)} \leq iq/m\}$, also controls the FDR (Sarkar, 2002). Even though it is obviously less powerful, it is sometimes easier to calculate in very large problems.

The linear step-up procedure is quite striking in its ability to control the FDR at precisely $q \cdot m_0/m$, regardless of the distributions of the test statistics corresponding to false null hypotheses (when the distributions under the simple null hypotheses are independent and continuous).

Benjamini and Yekutieli (Benjamini and Yekutieli, 2001) studied the procedure under dependency. For some type of positive dependency they showed that the above remains an upper bound. Even under the most general dependence structure, where the FDR is controlled merely at level $q(1+1/2+1/3+\cdots+1/m)$, it is again conservative by the same factor m_0/m (Benjamini and Yekutieli, 2001).

Knowledge of m_0 can therefore be very useful in this setting to improve upon the performance of the FDR controlling procedure. Were this information to be given to us by an "oracle", the linear step-up procedure with $q' = q \cdot m/m_0$ would control the FDR at precisely the desired level q in the independent and continuous case. It would then be more powerful in rejecting many of the hypotheses for which the alternative holds. In some precise asymptotic sense, Genovese and Wasserman (Genovese and Wasserman, 2002a) showed it to be the best possible procedure.

Schweder and Spjotvoll (Schweder and Spjotvoll, 1982) were the first to try and estimate this factor, albeit informally. Hochberg and Benjamini (Hochberg and Benjamini, 1990) formalized the approach. Benjamini and Hochberg (Benjamini and Hochberg, 2000) incorporated it into the linear step-up procedure, and other adaptive FDR controlling procedures make use of other estimators (Efron and Tibshirani, 1993, Storey, 2002, Storey, Taylor and Siegmund, 2004). (Benjamini, Krieger and Yekutieli, 2001), offer a very simple and intuitive two-stage procedure based on the idea that the value of m_0 can be estimated from the results of the linear step-up procedure itself, and prove it controls the FDR at level q.

Definition 25.4.2 *Two-Stage Linear Step-Up Procedure (TST):*

1. *Use the linear step-up procedure at level* $q' = \frac{q}{1+q}$. *Let* r_1 *be the number of rejected hypotheses. If* $r_1 = 0$ *reject no hypotheses and stop; if* $r_1 = m$ *reject all* m *hypotheses and stop; or otherwise*
2. *Let* $\hat{m}_0 = (m - r_1)$.
3. *Use the linear step-up procedure with* $q^* = q' \cdot m/\hat{m}_0$

Recent papers have illuminated the FDR from many different points of view: asymptotic, Bayesian, empirical Bayes, as the limit of empirical processes, and in the context of penalized model selection (Efron and Tibshirani, 1993, Storey, 2002, Genovese and Wasserman, 2002a, Abramovich et al., 2001). Some of the studies have emphasized variants of the FDR, such as its conditional value given some discovery is made (the positive FDR in (Storey, 2002)), or the distribution of the proportion of false discoveries itself (the FDR in (Genovese and Wasserman, 2002a, Genovese and Wasserman, 2002b)).

Studies on FDR methodologies have become a very active area of research in statistics, many of them making use of the large dimension of the problems faced, and in that respect relying on the blessing of dimensionality. FDR methodologies have not yet found their way into the practice and theory of DM, though it is our opinion that they have a lot to offer there, as the following example shows

Example 1: (Zytkov and Zembowicz, 1997) and (Zembowicz and Zytkov, 1996), developed the 49er software to mine association rules using chi-square tests of significance for the independence assumption, i.e. by testing whether the lift is significantly > 1. Finding that too many of the m potential rules are usually significant, they used $1/m$ as a threshold for significance, comparing each p-value to the threshold, and choosing only the rules that pass the threshold. Note that this is a Bonferroni-like treatment of the multiplicity problem, controlling the FWE at $\alpha = 1$. Still, they further suggest increasing the threshold if a few hypotheses are rejected. In particular they note that the performance of the threshold is especially good if the largest p-value of the selected k rules is smaller than k times the original $1/m$ threshold. This is exactly the BH procedure used at level $q = 1$, and they arrived at it by merely checking the actual performance on a specific problem. In spite of this remarkable success, theory further tells us that it is important to use $q < 1/2$, and not 1, to always get good performance. The preferable values for q are, as far as we know, between 0.05 and 0.2. Such values for q further allow us to conclude that only approximately q of the discovered association rules are not real ones. With $q = 1$ such a statement is meaningless.

25.5 Model (Variables or Features) Selection using FDR Penalization in GLM

Most of the commonly used variable selection procedures in linear models choose the appropriate subset by minimizing a model selection criterion of the form: $RSS_k + \sigma^2 k\lambda$, where RSS_k is the residual sum of squares for a model with k parameters as defined in section 25.3, and λ is the penalization parameter. For the generalized linear models discussed above twice the logarithm of the likelihood of the model takes on the role of RSS_k, but for simplicity of exposition we shall continue with the simple linear model. This penalized sum of squares might ideally be minimized over all k and all subsets of variables of size k, but practically in larger problems it is usually minimized either by forward selection or backward elimination, adding or dropping one variable at a time. The different selection criteria can be identified by the value of λ they use. Most traditional model selection criteria make use of a fixed λ and can also be described as fixed level testing. The Akaike Information Criterion (AIC) and the C_p criterion of Mallows both make use of $\lambda = 2$, and are equivalent to testing at level 0.16 whether the coefficient of each newly included variable in the model is different than 0. Usual backward and forward algorithms use similar testing at the .05 level, which is approximately equivalent to using $\lambda = 4$.

Note that when the selection of the model is conducted over a large number of potential variables m, the implications of the above approach can be disastrous. Take for example $m =$ 500 variables, not an unlikely situation in DM. Even if there is no connection whatsoever between the predicted variable and the potential set of predicting variables, you should expect to get 65 variables into the selected model - an unacceptable situation.

Model selection approaches have been recently examined in the statistical literature in settings where the number of variables is large, even tending to infinity. Such studies, usually held under an assumption of orthogonality of the variables, have brought new insight into the choice of λ. Donoho and Jonhstone (Donoho and Johnstone, 1995) suggested using λ, where $\lambda = 2log(m)$, whose square root is called the "universal threshold" in wavelet analysis. Note that the larger the pool over which the model is searched, the larger is the penalty per variable included. This threshold can also be viewed as a multiple testing Bonferroni procedure at the level α_m, with $.2 \leq \alpha_m \leq .4$ for $10 \leq m \leq 10000$. More recent studies have emphasized that the penalty should also depend on the size of the already selected model k, $\lambda = \lambda_{k,m}$, increasing in m and decreasing in k. They include (Abramovich and Benjamini, 1996, Birge and Massart, 2001, Abramovich et al., 2001, Tibshirani, 1996, George and Foster, 2000), and (Foster *et al.*, 2002). As a full review is beyond our scope, we shall focus on the suggestion that is directly related to FDR testing.

In the context of wavelet analysis (Abramovich and Benjamini, 1996) suggested using FDR testing, thereby introducing a threshold that increases in m and decreases with k. (Abramovich et al., 2001), were able to prove in an asymptotic setup, where m tends to infinity and the model is sparse, that using FDR testing is asymptotically minimax in a very wide sense. Their argument hinges on expressing the FDR testing as a penalized RSS as follows:

$$RSS_k + \sigma^2 \sum_{i=1}^{i=k} z^2_{\frac{i}{m} \cdot \frac{q}{2}}, \qquad (25.15)$$

where z_α is the $1-\alpha$ percentile of a standard normal distribution. This is equivalent to using $\lambda_{k,m} = \frac{1}{k}\sum_{i=1}^{i=k} z^2_{\frac{i}{m}\cdot\frac{q}{2}}$ in the general form of penalty. When the models considered are sparse, the penalty is approximately $2\sigma^2 log(\frac{m}{k}\cdot\frac{2}{q})$. The FDR level controlled is q, which should be kept at a level strictly less than 1/2.

In a followup study (Gavrilov, 2003), investigated the properties of such penalty functions using simulations, in setups where the number of variables is large but finite, and where the potential variables are correlated rather than orthogonal. The results show the dramatic failure of all traditional "fixed penalty per-parameter" approaches. She found the FDR-penalized selection procedure to have the best performance in terms of minimax behavior over a large number of situations likely to arise in practice, when the number of potential variables was more than 32 (and a close second in smaller cases). Interestingly she recommends using $q = .05$, which turned out to be well calibrated value for q for problems with up to 200 variables (the largest investigated).

Example 2: (Foster *et al.*, 2002), developed these ideas for the case when the predicted variable is 0-1, demonstrating their usefulness in DM, in developing a prediction model for loan default. They started with approximately 200 potential variables for the model, but then added all pairwise interactions to reach a set of some 50,000 potential variables. Their article discusses in detail some of the issues reviewed above, and has a very nice and useful discussion of important computational aspects of the application of the ideas in real a large DM problem.

25.6 Concluding Remarks

KDD and DM are a vaguely defined field in the sense that the definition largely depends on the background and views of the definer. Fayyad defined DM as the nontrivial process of identifying valid, novel, potentially useful, and ultimately understandable patterns in data. Some definitions of DM emphasize the connection of DM to databases containing ample of data. Another definitions of KDD and DM is the following definition: "Nontrivial extraction of implicit, previously unknown and potentially useful information from data, or the search for relationships and global patterns that exist in databases". Although mathematics, like computing is a tool for statistics, statistics has developed over a long time as a subdiscipline of mathematics. Statisticians have developed mathematical theories to support their methods and a mathematical formulation based on probability theory to quantify the uncertainty. Traditional statistics emphasizes a mathematical formulation and validation of its methodology rather than empirical or practical validation. The emphasis on rigor has required a proof that a proposed method will work prior to the use of the method. In contrast, computer science and machine learning use experimental validation methods. Statistics has developed into a closed discipline, with its own scientific jargon and academic objectives that favor analytic proofs rather than practical methods for learning from data. We need to distinguish between the theoretical mathematical background of statistics and its use as a tool in many experimental scientific research studies. We believe that computing methodology and many of the other related issues in DM should be incorporated into traditional statistics. An effort has to be made to correct the negative connotations that have long surrounded Data Mining in the statistics literature (Chatfield, 1995) and the statistical community will have to recognize that empirical validation does constitute a form of validation (Friedman, 1998).

Although the terminology used in DM and statistics may differ, in many cases the concepts are the same. For example, in neural networks we use terms like "learning", "weights" and "knowledge" while in statistics we use "estimation", "parameters" and "value of parameters", respectively. Not all statistical themes are relevant to DM. For example, as DM analyzes existing databases, experimental design is not relevant to DM. However, many of them, including those covered in this chapter, are highly relevant to DM and any data miner should be familiar with them.

In summary, there is a need to increase the interaction and collaboration between data miners and statisticians. This can be done by overcoming the terminology barriers and by working jointly on problems stemming from large databases. A question that has often been raised among statisticians is whether DM is not merely part of statistics. The point of this chapter was to show how each can benefit from the other, making the inquiry from data a more successful endeavor, rather than dwelling on where the disciplinary boundaries should pass.

References

Abramovich F. and Benjamini Y., (1996). Adaptive thresholding of wavelet coefficients. *Computational Statistics & Data Analysis*, 22:351–361.

Abramovich F., Bailey T .C. and Sapatinas T., (2000). Wavelet analysis and its statistical applications. *Journal of the Royal Statistical Society Series D-The Statistician*, 49:1–29.

Abramovich F., Benjamini Y., Donoho D. and Johnstone I., (2000). Adapting to unknown sparsity by controlling the false discovery rate. Technical Report 2000-19, Department of Statistics, Stanford University.

Benjamini Y. and Hochberg Y., (1995). Controlling the false discover rate: A practical and powerful approach to multiple testing. *J. R. Statist. Soc. B*, 57:289–300.

Benjamini Y. and Hochberg Y., (2000). On the adaptive control of the false discovery fate in multiple testing with independent statistics. *Journal of Educational and Behavioral Statistics*, 25:60–83.

Benjamini Y., Krieger A.M. and Yekutieli D., (2001). Two staged linear step up for controlling procedure. Technical report, Department of Statistics and O.R., Tel Aviv University.

Benjamini Y. and Yekutieli D., (2001). The control of the false discovery rate in multiple testing under dependency. *Annals of Statistics*, 29:1165–1188.

Berthold M. and Hand D., (1999). *Intelligent Data Analysis: An Introduction*. Springer.

Birge L. and Massart P., (2001). Gaussian model selection. *Journal of the European Mathematical Society*, 3:203–268.

Chatfield C., (1995). Model uncertainty, Data Mining and statistical inference. *Journal of the Royal Statistical Society A*, 158:419–466.

Cochran W.G., (1977). *Sampling Techniques*. Wiley.

Cox D.R., (1972). Regressio models and life-tables. *Journal of the Royal Statistical Society B*, 34:187–220.

Dell'Aquila R. and Ronchetti E.M., (2004). *Introduction to Robust Statistics with Economic and Financial Applications*. Wiley.

Donoho D.L. and Johnstone I.M., (1995). Adapting to unknown smoothness via wavelet shrinkage. *Journal of the American Statistical Association*, 90:1200–1224.

Donoho D., (2000). American math. society: Math challenges of the 21st century: *High-dimensional data analysis: The curses and blessings of dimensionality.*

Efron B., Tibshirani R.J., Storey J.D. and Tusher V., (2001). Empirical Bayes analysis of a microarray experiment. *Journal of the American Statistical Association*, 96:1151–1160.

Friedman J.H., (1998). *Data Mining and Statistics: What's the connections?*, Proc. 29th Symposium on the Interface (D. Scott, editor).

Foster D.P. and Stine R.A., (2004). Variable selection in Data Mining: Building a predictive model for bankruptcy. *Journal of the American Statistical Association*, 99:303–313.

Gavrilov Y., (2003). Using the falls discovery rate criteria for model selection in linear regression. M.Sc. Thesis, Department of Statistics, Tel Aviv University.

Genovese C. and Wasserman L., (2002a). Operating characteristics and extensions of the false discovery rate procedure. *Journal of the Royal Statistical Society Series B*, 64:499–517.

Genovese C. and Wasserman L., (2002b). A stochastic process approach to false discovery rates. Technical Report 762, Department of Statistics, Carnegie Mellon University.

George E.I. and Foster D.P., (2000). Calibration and empirical Bayes variable selection. *Biometrika*, 87:731–748.

Hand D., (1998). Data Mining: Statistics and more? *The American Statistician*, 52:112–118.

Hand D., Mannila H. and Smyth P., (2001). *Principles of Data Mining*. MIT Press.

Han J. and Kamber M., (2001). *Data Mining: Concepts and Techniques*. Morgan Kaufmann Publisher.

Hastie T., Tibshirani R. and Friedman J., (2001). *The Elements of Statistical Learning: Data Mining, Inference, and Prediction*. Springer.

Hochberg Y. and Benjamini Y., (1990). More powerful procedures for multiple significance testing. *Statistics in Medicine*, 9:811–818.

Leshno M., Lin V.Y., Pinkus A. and Schocken S., (1993). Multilayer feedforward networks with a non polynomial activation function can approximate any function. *Neural Networks*, 6:861–867.

McCullagh P. and Nelder J.A., (1991). *Generalized Linear Model*. Chapman & Hall.

Meilijson I., (1991). The expected value of some functions of the convex hull of a random set of points sampled in r^d. *Isr. J. of Math.*, 72:341–352.

Mosteller F. and Tukey J.W., (1977). *Data Analysis and Regression : A Second Course in Statistics*. Wiley.

Roberts S. and Everson R. (editors), (2001). *Independent Component Analysis : Principles and Practice*. Cambridge University Press.

Ronchetti E.M., Hampel F.R., Rousseeuw P.J. and Stahel W.A., (1986). *Robust Statistics : The Approach Based on Influence Functions*. Wiley.

Sarkar S.K., (2002). Some results on false discovery rate in stepwise multiple testing procedures. *Annals of Statistics*, 30:239–257.

Schweder T. and Spjotvoll E., (1982). Plots of p-values to evaluate many tests simultaneously. *Biometrika*, 69:493–502.

Seeger P., (1968). A note on a method for the analysis of significances en mass. *Technometrics*, 10:586–593.

Simes R.J., (1986). An improved Bonferroni procedure for multiple tests of significance. *Biometrika*, 73:751–754.

Storey J.D., (2002). A direct approach to false discovery rates. *Journal of the Royal Statistical Society Series B*, 64:479–498.

Storey J.D., Taylor J.E. and Siegmund D., (2004). Strong control, conservative point estimation, and simultaneous conservative consistency of false discovery rates: A unified approach. *Journal of the Royal Statistical Society Series B*, 66:187–205.

Therneau T.M. and Grambsch P.M., (2000). *Modeling Survival Data, Extending the Cox Model*. Springer.

Tibshirani R. and Knight K., (1999). The covariance inflation criterion for adaptive model selection. *Journal of the Royal Statistical Society Series B*, 61:Part 3 529–546.

Zembowicz R. and Zytkov J.M., (1996). From contingency tables to various froms of knowledge in databases. In U.M. Fayyad, R. Uthurusamy, G. Piatetsky-Shapiro and P. Smyth (editors) *Advances in Knowledge Discovery and Data Mining* (pp. 329-349). MIT Press.

Zytkov J.M. and Zembowicz R., (1997). Contingency tables as the foundation for concepts, concept hierarchies and rules: The 49er system approach. *Fundamenta Informaticae*, 30:383–399.

26

Logics for Data Mining

Petr Hájek

Institute of Computer Science
Academy of Sciences of the Czech Republic
182 07 Prague, Czech Republic
hajek@cs.cas.cz

Summary. Systems of formal (symbolic) logic suitable for Data Mining are presented, main stress being put to various kinds of generalized quantifiers.

Key words: logic, Data Mining, generalized quantifiers, GUHA method

Introduction

Data Mining, as presently understood, is a broad term, including search for "association rules", classification, regression, clustering and similar. Here we shall restrict ourselves to search for "rules" in a rather general sense, namely general dependencies valid in given data and expressed by formulas of a formal logical language. The present theoretical approach is the result of a long development of the GUHA method of automated generation of hypotheses (General Unary Hypotheses Automaton, see a paragraph in Section 26.2) but is believed to be fully relevant for contemporary mining of association rules and its possible generalization. See (Agrawal *et al.*, 1996, Hoppner, 2005, Adamo, 2001) for association rules.

Data are assumed to have the form of one or more tables, matrices or relations. A rectangular matrix may be understood as giving data on *objects* (corresponding to rows of the matrix) and their *attributes* (columns). Or rows may correspond to objects from one set, columns to objects of the same or a different set and the whole matrix is understood as one binary attribute. In the former case we have *variables* x, y for objects and a unary predicate for each column (P_i for i-th column, say); $P_i(x)$ denotes the value of P_i for the object x. In the latter case we have variables for objects from the first set (x, say), other variables for objects from the other set (y, say) and one binary predicate P; then $P(x,y)$ denotes the value of the attribute for the pair (x,y) of objects.

For example, rows correspond to patients, the first column corresponds to having feaver (yes – value 1, no – value 0). Then patient Novák satisfies $P_1(x)$ if he has feaver. Secondly, the matrix describes the relation "being a married couple" and we take MC for the predicate. Then the couple (Novák, Nováková) satisfies $P(x,y)$ if they are a married couple, thus the corresponding field in the matrix has value 1.

These were examples of Boolean (0-1-valued) data; more generally, they may take values from some set of values (reals, colours, ...).

O. Maimon, L. Rokach (eds.), *Data Mining and Knowledge Discovery Handbook*, 2nd ed.,
DOI 10.1007/978-0-387-09823-4_26, © Springer Science+Business Media, LLC 2010

It should be clear what is the value of $P(x)$ for an object in the former case and value of $P(x,y)$ for a pair of objects in the latter.

Logic enables us to construct composed formulas from atomic formulas as above using some *connectives* (as conjunction, disjunction, implication, negation in the Boolean logic, notation: $\wedge, \vee, \rightarrow, \neg$) and quantifiers (universal $\forall$ and existential $\exists$ in classical Boolean logic). Our *main message* is that in Data Mining we have to deal with *generalized quantifiers* of some particular kind and that their logical properties are very important. For simplicity we restrict ourselves to two-valued (0-1-valued) data. Our approach generalizes easily to categorical (finitely valued) data when we work with atomic formulas of the form $(X)P(x)$ (or $(X)P(x,y)$ etc.) where X is a subset of the domain of values of the attribute P and an object o satisfies $(X)P(x)$ iff the P-value of o is in X, similarly for $(X)P(x,y)$. (For example let P denote age in years $0,\ldots,100$, let X be the set of numbers $0 \leq n \leq 30$.)

The reader having some knowledge of classical propositional and predicate calculus will have no problems with the mentioned notions; the reader having difficulties is recommended to have a look to a textbook of mathematical logic, e. g. (Ebbinghaus *et al.*, 1984).

26.1 Generalized quantifiers

We shall present the notion of a generalized quantifier and supply several examples. (In the next section we shall study various classes of quantifiers.) For simplicity we shall work with data having the form of a rectangular boolean matrix, rows corresponding to object and columns to (yes-no) attributes. Predicates $P_1,\ldots,P_n$ are names of the attributes; we have an object variable x, and $P_1(x),\ldots,P_n(x)$ are atomic formulas. For each formula φ, we have its *negation* $\neg\varphi$; an object satisfies $\neg\varphi$ if it does not satisfy φ. For each pair φ,ψ of formulas we have their *conjunction* $\varphi \wedge \psi$ and *disjunction* $\varphi \vee \psi$. An object satisfies $\varphi \wedge \psi$ if it satisfies both φ and ψ; it satisfies at least one of φ,ψ. Similarly for conjunction/disjunction of three, four... formulas. An object satisfies implication $\varphi \rightarrow \psi$ if it satisfies ψ or does not satisfy φ. Formulas built from atomic formulas using the connectives $\neg, \wedge\vee, \rightarrow$ are *open:* each open formula φ defines an attribute: each object of our data either satisfies φ or does not satisfy it. This is uniquely determined by the data. Thus φ defines two numbers: r – the number of objects satisfying φ and s – the number of objects satisfying $\neg\varphi$. The pair (r,s) may be called the *two-fold table* of φ (given by the data); $r+s=m$ is the number of objects in the data (rows of the matrix).

A (one-dimensional) *quantifier* Q applied to an open formula φ describes the behaviour of the attribute defined by φ in the data as a whole, i.e. gives a global characterization of the attribute (in the data). The reader surely knows the *classical* quantifiers $\forall$ (universal) and $\exists$ (existential). The formula $(\forall x)\varphi$ is *true* in the data if each object satisfies φ (thus $s=0$); the formula $(\exists x)\varphi$ is true in the data of at least one object satisfies φ (thus $r \geq 1$). You see that truth of such a quantified formula does not depend on any ordering of the rows of the data matrix, just it is given by the two-fold table of φ.

A (one-dimensional) quantifier Q is determined by its truth function Tr_Q assigning to each two-fold table (r,s) either 1 (true) or 0 (false). For each open formula φ, the *closed formula* $(Qx)\varphi$ is true in the data iff the two-fold table (r,s) of φ (given by the data) satisfies $Tr_Q(r,s)=1$. Clearly, $Tr_\forall(r,s)=1$ iff $s=0$ and $Tr_\exists(r,s)=1$ iff $r>0$.
Some other examples:
Majority: $Tr_{\mathrm{Maj}}(r,s)=1$ iff $r>s$; $(\mathrm{Maj}\,x)\varphi$ says that the majority of objects satisfy φ.
Many: Let $0<p<1$; $Tr_{\mathrm{Many}_p}(r,s)=1$ iff $r/(r+s) \geq p$; $(\mathrm{Many}_p x)\varphi$ says that the relative

frequence of objects satisfying φ is at least p.
At least: $Tr_{\exists \geq n}(r,s) = 1$ iff $r \geq n$ (at least n objects satisfy φ).
Odd: $Tr_{\mathrm{Odd}}(r,s) = 1$ iff r is an odd number.
The reader may produce many more examples. We shall be more general:

A *two-dimensional quantifier* Q, applied to a pair φ, ψ of open formulas describes the behaviour of the pair of attributes defined by φ and ψ in the data as a whole, thus gives a global characterization of the mutual relation of φ, ψ (in the data). The *closed formula* given by Q, φ, ψ is written as $(Qx)(\varphi, \psi)$. Its truth/falsity in the data is determined by the *four-fold table* (a,b,c,d) where a,b,c,d denotes the number of objects in the data satisfying $\varphi \wedge \psi$, $\varphi \wedge \neg\psi$, $\neg\varphi \wedge \psi$, $\neg\varphi \wedge \neg\psi$ respectively. This is often displayed as

	ψ	$\neg\psi$	
φ	a	b	r
$\neg\varphi$	c	d	s
	k	l	m

where $r = a+b$ (number of objects satisfying φ), $s = c+d$, $k = a+c$, $l = b+d$ (marginal sums), $m = a+b+c+d = r+s = k+l$. Thus the *truth function* of a two-dimensional quantifier Q assigns to each four-fold table (a,b,c,d) the value $Tr_Q(a,b,c,d) \in \{0,1\}$.

All (two-dimensional) $\varphi \overset{(x)}{\Rightarrow} \psi$ says "all φ's are ψ's)"; $Tr_{\Rightarrow}(a,b,c,d) = 1$ iff $b = 0$. This is definable by one-dimensional $\forall$ and the connective $\rightarrow$, namely $\varphi \overset{(x)}{\Rightarrow} \psi$ says the same as $(\forall x)(\varphi \rightarrow \psi)$.

Many: for $0 < p \leq 1$, $\varphi \overset{(x)}{\Rightarrow}_p \psi$ says "p-many φ's are ψ's", i.e. the relative frequence of objects satisfying ψ among those satisfying φ is $> p$, thus $Tr_{\Rightarrow_p}(a,b,c,d) = 1$ iff $a/(a+b) \geq p$. Caution: this is *not* the same as $(\mathrm{Many}_p x)(\varphi \rightarrow \psi)$: for example if (a,b,c,d) is $(2,2,5,5)$ then $a/(a+b) = 1/2$ but the number of objects satisfying $\varphi \rightarrow \psi$ is $a+c+d = 12$, thus $(a+c+d)/m = 12/14 = 6/7$. Thus if $p = 0.8$ then $\varphi \overset{(x)}{\Rightarrow} \psi$ is false but $(\forall x)(\varphi \rightarrow \psi)$ is true. But $\varphi \overset{(x)}{\Rightarrow} \psi$ can be also written as $(\mathrm{Many}_p x)\varphi/\psi$, understood as saying that the formula $(\mathrm{Many}_p x)\varphi$ is true in the subtable consisting of rows satisfying ψ.

p-equivalence: $\varphi \overset{(x)}{\Leftrightarrow} \psi$ (φ is p-equivalent to ψ) is true in the data if both $\varphi \overset{(x)}{\Rightarrow}_p \psi$ and $\neg\varphi \overset{(x)}{\Rightarrow}_p \neg\psi$ is true, thus $a/(a+b) \geq p$ and $d/(c+d) \geq p$.

Foundedness: Let t be a natural number. $(\mathrm{Fdd}\, x)(\varphi, \psi)$ says that at least t objects satisfy $\varphi \wedge \psi$, i.e. $a \geq t$. Similarly:

Support: Let $0 < \sigma < 1$. $(\mathrm{Supp}_\sigma x)(\varphi, \psi)$ is $\mathrm{Many}_\sigma(\varphi \wedge \psi)$, thus says that the relative frequence of $\varphi \wedge \psi$ in the data is at least σ.

Founded implication: $(\mathrm{FIMPL}_{p,t} x)(\varphi, \psi)$ (or just $\varphi \Rightarrow_{p,t} \psi$) is $(\mathrm{Many}_p x)(\varphi, \psi)$ and $\mathrm{Fdd}_t s)(\varphi \wedge \psi)$, hence $Tr_{\Rightarrow_{p,t}}(a,b,c,d) = 1$ iff $a/(a+b) \geq p$ and $a \geq t$.

Agrawal: $\varphi \Rightarrow^{Agr}_{p,\sigma} \psi$ is $\mathrm{Many}_p(\varphi, \psi)$ and $\mathrm{Supp}_\sigma(\varphi, \psi)$, hence $Tr_{\Rightarrow^{Agr}_{p,\sigma}}(a,b,c,d) = 1$ iff $a/(a+b) \geq p$ and $a/(a,b,c,d) \geq \sigma$.

Clearly the last two quantifiers differ only very little: $\varphi \Rightarrow^{*}_{p,t} \psi$ is equivalent to $\varphi \Rightarrow^{Agr}_{p,\sigma}$ for $\sigma = t/(a+b+c+d)$. Now $\Rightarrow^{Agr}$ is the quantifier of the association rules of Agrawal; it is little known and has to be stressed that the "almost the same" quantifier of founded implication was used in GUHA to generate "association rules" in the presently common sense as soon as in mid-sixties of the past century (Hájek *et al.*, 1966).

The reader may play by defining more and more two-dimensional quantifiers; clearly not all of them are relevant for Data Mining. We close this section by two important remarks.

Closed formulas. Each formula (of the present formalism, with unary predicates and just one object variable) beginning with a quantifier is *closed*, i.e. does not refer to any particular object but expresses some global pattern found in the data. Further closed formulas result from those beginning by a quantifier using connectives (e.g. we have seen that $\varphi \overset{(x)}{\Leftrightarrow}_p \psi$ is equivalent to $(\varphi \overset{(x)}{\Rightarrow}_p \psi) \wedge (\neg\varphi \overset{(x)}{\Rightarrow}_p \neg\psi)$, etc.). A closed formula is a *tautology* (logical truth) if it is true in each data. To give a trivial example, observe that if $p_1 \leq p_2$ then the formula $(\varphi \overset{(x)}{\Rightarrow}_{p_2} \psi) \to (\varphi \overset{(x)}{\Rightarrow}_{p_1} \psi)$ is a tautology.

Predicates of higher arity. If our data contain information on relations of higher arity (binary, ternary,...) we have to use predicates of higher arity and more than one object variable. A quantifier always *binds* (quantifies) a variable. This leads to the logical notion of free and bound variables of a formula, free variables varying over arbitrary objects of the data. For example, take a binary predicate P; $P(x,y)$ is a formula in which x,y are free, $(\mathrm{Many}_p y)P(x,y)$ is a formula in which only x is free. An object o satisfies the last formula iff it is P-related with p-many objects. (Let $P(x,y)$ say "x knows y", let p be 0.8. An object o satisfies $(\mathrm{Many}_{0.8} y)P(x,y)$ if he knows at least 80% objects (from the data). We may form composed formulas using several quantifiers binding different variables, e.g. $(\forall x)((\mathrm{Many}_{0.8} y)P(x,y) \to R(x))$ (saying "each object knowing at least 80% objects has the property R") etc. This sort of formulas is used in *relational Data Mining* (Džeroski and Lavrač, 2001). We shall not go into details.

26.2 Some important classes of quantifiers

26.2.1 One-dimensional

Let us call a one-dimensional quantifier Q *multitudinal* if its truth function Tr_Q is not decreasing in its first argument and non-increasing in the second, i.e. for any two-fold tables (r_1,s_1), (r_2,s_2) whenever
$r_1 \leq r_2$, $s_1 \geq s_2$ and $Tr_Q(r,s_1) = 1$ then $Tr_Q(r_1,s_2) = 1$. This means that the formula $(Qx)\varphi$ says, in some sense given by Tr_Q, that *sufficiently many* objects satisfy φ. "Sufficiently many" may mean "all" $(\forall)$, at least one $(\exists)$, at least 7 $(\exists_7)$, at least $100p\%$ (Many_p) etc.

Very important: The quantifier may correspond to a *statistical test* of high probability. Telegraphically: Our hypothesis is that the probability of the attribute φ is bigger than p (under frame assumptions saying that all objects have the same probability of having φ and are mutually independent). Take a small α (e.g. 0.05 – significance level). The number $\sum_{i=r}^{r+s}\binom{r+s}{i}p^i(1-p)^{r+s-i}$ is the probability that at least r objects (form our $r+s$ objects) will have φ, assuming that the probability of φ is p. If this sum is $\leq \alpha$ then we can reject the (null) hypothesis saying that the probability of φ is *p or less* (since if it were then what we have observed would be improbable). This is the (simplified) idea of statistical hypothesis testing. We get a *quantifier* of testing high probability, $HProb_{p,\alpha}$.

$$Tr_{HProb_{p,\alpha}}(r,s) = 1 \text{ iff } \sum_{i=r}^{r+s}\binom{r+s}{i}p^i(1-p)^{r+s-i} \leq \alpha.$$

This is an example of a statistically motivated one-dimensional quantifier; it can be proved to be multitudinal). See Chapter 31.4.6 for statistical hypothesis testing and (Hájek and Havránek, 1978) for its logical foundations.

26.2.2 Two-dimensional

Recall that a two-dimensional quantifier is given by its truth function assigning to each four-fold table (a,b,c,d) a truth value (1 or 0). In Data Mining we are especially interested in two-dimensional quantifiers expressing in some sense a kind of association of two attributes (described by two open formulas). In some sense, the formula $(Qx)(\varphi,\psi)$ should say that there are sufficiently many coincidences in (the truth values of) φ,ψ and not too many differences. This leads to the following definition (Hájek and Havránek, 1978):

A two-dimensional quantifier is *associational* if it satisfies the following for each pair (a_1,b_1,c_1,d_1), (a_2,b_2,c_2,d_2) of four-fold tables: $a_2 \geq a_1, b_2 \leq b_1, c_2 \leq c_1, d_2 \geq d_1$ and $Tr_Q(a_1,b_1,c_1,d_1) = 1$ implies $Tr_Q(a_2,b_2,c_2,d_2) = 1$. In other words: if

	ψ_1	$\neg\psi_1$
φ_1	a_1	b_1
$\neg\varphi_1$	c_1	d_1

	ψ_2	$\neg\psi_2$
φ_2	a_2	b_2
$\neg\varphi_2$	c_2	d_2

are four-fold tables of the pairs (φ_1,ψ_1), (φ_2,ψ_2) of open formulas in given data, if $(Qx)(\varphi_1,\psi_1)$ is true in the data and the above inequalities hold $(a_2 \geq a_1, b_2 \leq b_1, c_2 \leq c_1, d_2 \geq d_1)$, then $(Qx)(\varphi_2,\psi_2)$ is also true. The second table has more coincidences $(a_2 \geq a_1, d_2 \geq d_1)$ and less differences $(b_2 \leq b_1, c_2 \leq c_1)$.

A quantifier Q is *locally associational* if the above condition holds for all (a_1,b_1,c_1,d_1), (a_2,b_2,c_2,d_2) satisfying the additional assumption $a_1+b_1+c_1+d_1 = a_2+b_2+c_2+d_2$ (i.e. the tables correspond to two data matrices of the same cardinality; in particular, think of $\varphi_1,\psi_1,\varphi_2,\psi_2$ evaluated in *the same* data matrix).

We shall deal with (locally) associational quantifiers of two important kinds: *implicational* and *comparative*. We give examples and state general (deductive) properties of quantifiers in these classes.

Implicational quantifiers formalize the association formulated as "many φ's are ψ's". (They could be also called two-dimensional multitudinal quantifiers.) The definition reads as follows:

A two-dimensional quantifier Q is *implicational* if each pair (a_1,b_1, c_1,d_1), (a_2,b_2,c_2,d_2) of four fold tables satisfies the following condition: If $a_2 \geq a_1, b_2 \leq b_1$ and $Tr_Q(a_1,b_1,c_1,d_1) = 1$ then $Tr_Q(a_2,b_2,c_2,d_2) = 1$ Q is *locally implicational* if this condition is satisfied for each pair of four-fold tables with the same sum $(a_1+b_1+c_1+d_1 = c_2+b_2+c_2+d_2)$.

Clearly, the quantifier $\Rightarrow_p$ (p-many) is implicational: $a_2 \geq a_1$ and $b_2 \leq b_1$ imply $a_2/(a_2+b_2) \geq a_1/(a_1+b_1)$. The quantifier $\Rightarrow^*_{p,t}$ of founded implication is also implicational: if $a_2 \geq a_1$ and $a_1 \geq t$ then trivially $a_2 \geq t$. The "almost the same" Agrawal's quantifier $\Rightarrow^{Agr}_{p,\sigma}$ is locally implicational: if the tables have equal sum and $a_2 \geq a_1$ then trivially $a_2/(a_2+b_2+c_2+d_2) \geq a_1/(a_1+b_1+c_1+d_1)$.

Note the statistical parallel of $\Rightarrow_p$: The hypothesis of $P(\psi|\varphi) \geq p$ (conditional probability of ψ, given φ) is tested using the statistic $\sum_{i=a}^{a+b}\binom{a+b}{i} p_i(1-p)^{a+b-i}$. The corresponding quantifier $\Rightarrow^!_{p,\alpha}$ of likely p-implication (with significance level α) is

$$Tr_{\Rightarrow^!_{p,\alpha}}(a,b,c,d) = 1 \text{ iff } \sum_{i=a}^{a+b}\binom{a+b}{i} p^i(1-p)^{a+b-i} \leq \alpha.$$

This is also an implicational quantifier (see (Hájek and Havránek, 1978), where also another statistically motivated implicational quantifier is discussed). For each (locally) implicational quantifier (denote it $\Rightarrow^\#$) the following two deduction rules are sound (in the sense that whenever the assumption is true in your data, the consequence is also true):

$$\frac{(\varphi_1 \wedge \varphi_2) \Rightarrow^\# \psi}{\varphi_1 \Rightarrow^\# \psi \vee \neg\varphi_2}, \qquad \frac{\varphi \Rightarrow^\# \psi_1}{\varphi \Rightarrow^\# (\psi_1 \vee \psi_2)}$$

For example, if the formula "p-many probands being smokers and older then 50 have cancer" is true in your data then the following is true too: "p-many probands being smokers have cancer or are not older than 50". Second: If the "association rule" "x buys Lidové noviny $\Rightarrow x$ is Czech" is 90%-true with support 1000 then also

$$x \text{ buys Lidové noviny } \Rightarrow (x \text{ is Czech or } x \text{ is Slovak})$$

is 90% true with the same support. These deduction rules are extremely useful for optimizing search for formulas ("rules") of the form $\varphi \Rightarrow^\# \psi$ where $\Rightarrow^\#$ is an implicational quantifier, φ is an elementary conjunction (conjunction of atomic open formulas and negated atomic formulas containing each predicate at most once, e.g. $P_1(x) \wedge \neg P_3(x) \wedge P_7(x)$) and ψ is an elementary disjunction (similar definition, e.g. $\neg P_2(x) \vee \neg P_{10}(x)$).

Caution: For the "classical" quantifier $\Rightarrow_1$ ($\varphi \Rightarrow_1 \psi$ saying "all φ's are ψ's") the first rule can be converted, thus truth of $\varphi_1 \Rightarrow_1 \psi \vee \neg\varphi_2$ implies truth of $(\varphi_1 \wedge \varphi_2) \Rightarrow_1 \psi$. But this is *not* true for $\Rightarrow_p$ and other mentioned implicational (locally implicational) quantifiers.

Let us stress once more that implicational quantifiers formalize, in various possible ways, what we mean saying "many φ's are ψ's". Agrawal's association rules are a particular case, with one particular implicational quantifier and also with specific open formulas (no negation allowed, just conjunction of atoms). Even if this may be the most used case, the reader is invited to consider broader, more general and more powerful possibilities.

We now turn our attention to a very important class of quantifiers that we shall call *comparative.* The intuitive meaning of association expressed by a comparative quantifier is that the formula $(Qx)(\varphi,\psi)$ should say that presence of φ *positively contributes to the presence of* ψ. This does not mean that many φ's are ψ's, thus that the relative frequence of ψ among φ (denoted $Freq(\psi|\varphi)$) is big but that $Freq(\psi|\varphi)$ is (sufficiently) bigger than $Freq(\psi|\neg\varphi)$. For example, imagine that 30% of smokers have an illness and only 5% of non-smokers have the same illness. The simplest quantifier of this kind is called the *simple* associational quantifier, denoted SIMPLE or $\sim_0$ (see (Hájek and Havránek, 1978, Hájek *et al.*, 1995) or other GUHA papers); the truth function is $Tr_{\sim_0}(a,b,c,d) = 1$ if $ad > bc$. A trivial computation shows that $ad > bc$ is equivalent both to $\frac{a}{a+b} > \frac{a+c}{a+b+c+d}$ (if (a,b,c,d) is the fourfold table of φ,ψ then this says that, in the data, $Freq(\psi|\varphi) > Freq(\psi)$) and to $\frac{a}{a+b} > \frac{c}{c+d}$ ($Freq(\psi|\varphi) > Freq(\psi|\neg\varphi)$). You may make this quantifier parametric, demanding $ad > h.bc$, for some $h \geq 1$.

Thus let us accept the following definition: A two-dimensional quantifier Q is *comparative* if $Tr_Q(a,b,c,d) = 1$ implies $ad > bc$. The statistical counterpart is *Fisher quantifier* $\sim^F_\alpha$ based on the test of the hypothesis $P(\psi|\varphi) > P(\psi)$ (against the null hypothesis $P(\psi|\varphi) \leq P(\psi)$), with significance α. The formula is:

$$Tr_{\sim_\alpha}(a,b,c,d) = 1 \text{ if } ad > bc \text{ and}$$
$$\sum_{i=a}^{\min(a+b,a+c)} \binom{a+b}{i}\binom{b+d}{a+b-i}\binom{a+b+c+d}{a+b} \leq \alpha.$$

If we adopt the usual notation $a+b=r$, $a+c=k$, $b+d=l$, $a+b+c+d=m$ then the last formula becomes

$$\sum_{i=a}^{\min(r,k)} \binom{k}{i}\binom{l}{r-i} / \binom{m}{r} \leq \alpha.$$

This is a rather complicated formula; there are non-trivial algorithms for computing the sum in question. Fisher quantifier can be proved to be associational (Hájek and Havránek, 1978).

Let us mention that another comparative associational statistically motivated quantifier is based on the statistical *chi-square* test.

Indeed,

$$Tr_{\sim_\alpha^{CHISQ}}(a,b,c,d) = 1 \text{ if } ad > bc \text{ and } \frac{m(ad-bc)^2}{rskl} \geq \chi^2_\alpha,$$

where χ^2_α is a constant (the (1-α)-quantile of the χ^2 distribution function)

Now let us present three deduction rules and ask if our quantifiers obey them. Once more, it means that whenever the assumption (above the line) is true in the data then the conclusion (below the line) is true. Here $\sim$ stands for a quantifier; we write $\varphi \sim \psi$ instead of $(\sim x)(\varphi, \psi)$.

Rule of symmetry: (SYM)

$$\frac{\varphi \sim \psi}{\psi \sim \varphi}$$

Rule of negation: (NEG)

$$\frac{\varphi \sim \psi}{\neg\varphi \sim \neg\psi}$$

Rule of conversion: (CNVS)

$$\frac{\varphi \sim \psi}{\neg\psi \sim \neg\varphi}$$

Fact

The simple quantifier $\sim_0$, the Fisher quantifier $\sim_\alpha^F$ as well as the chi-square quantifier $\sim_\alpha^{CHI}$ obey all the rules (SYM), (NEG), (CNVS).

For a proof see again (Hájek and Havránek, 1978), observing that if any quantifier obeys (SYM) and (NEG) then it automatically obeys (CNVS). Now we present three more quantifiers occuring in the literature, each obeying just one of our present rules.

The quantifier of *pure p-equivalence* $\equiv_p$ (Rauch, see e.g. (Rauch, 1998A)). The formula $\varphi \equiv_p \psi$ is true if both $\varphi \Rightarrow_p \psi$ and $\neg\varphi \Rightarrow_p \neg\psi$ are true, thus $Tr_{\equiv_p}(a,b,c,d) = 1$ if $a/(a+b) \geq p$ and $d/(c+d) \geq p$. For $p > \frac{1}{2}$ this quantifier is comparative. (Indeed, if $a/(a+b) > \frac{1}{2}$ and $d/(c+d) > \frac{1}{2}$ then $c/(c+d) < \frac{1}{2} < a/(a+b)$, which gives $bc < ad$.)

The quantifier of *conviction* (Adamo, 2001). $\varphi \sim_h^{conv} \psi$ is true if
$(a+b)(b+d) > h.b(a+b+c+d)$, or equivalently, $(rl)/(bm) > h$, where h is a parameter, $h \geq 1$. An elementary computation gives that the last inequality for $h = 1$ (and hence for each $h \geq 1$) implies $ad < bc$; the quantifier is comparative.

The quantifier "*above average*" is a variant of SIMPLE (used in the program 4FT-miner (lispminer)). $\varphi \sim_h^{AA} \psi$ is true if
$a/(a+b) > h.(a+c)/(a+b+c+d)$ (thus $a/r > h.k/m$), which means that is $Fr(\psi/\varphi) > h.Fr(\psi)$. For $h = 1$ this is equivalent to the simple quantifier with $h = 1$; evidently, for each $h \geq 1$, the *AA* quantifier is comparative.

But these last three quantifiers differ as far as our deduction rules are concerned:

Fact

(1) The quantifier *AA* obeys symmetry but for $h > 1$ neither negation nor conversion. (2) The quantifier of pure p-equivalence obeys negation but for $p < 1$ neither symmetry nor conversion. (3) The quantifier of conviction obeys conversion but for $h > 1$ neither symmetry nor negation.

The positive claims are verified by easy computations; the negative claims can be all witnessed e.g. by the table $(9,1,10,80)$.

Let us also mention the quantifier of *double p-implication* $\Leftrightarrow_p$ (Rauch): $\varphi \Leftrightarrow_p \psi$ is true if both $\varphi \Rightarrow_p \psi$ and $\psi \Rightarrow_p \varphi$ is true. Show that this quantifier is *not* comparative (consider e.g. $(9,1,1,0)$); it obeys symmetry but (for $p < 1$) neither negation nor conversion.

The study of deductive rules is important for interpretation of results of Data Mining as well as for optimization of mining algorithms.

To close this section let us mention that each two-dimensional quantifier $\sim$ can be used to define a three-dimensional quantifier by partializing: the formula $(\varphi \sim \psi)/\chi$ is true in the data matrix in question iff $\varphi \sim \psi$ is true in the submatrix of objects satisfying χ. Cf. (Hájek, 2003).

26.3 Some comments and conclusion

Using four-fold tables

Even is we have dealt with *logical* aspects of Data Mining we feel obliged to stress once more the importance of the statistical side of the game. We already referred to (Giudici, 2003) ; let us make some further references. Glymour's (Glymour *et al.*, 1996) is a good reading on the prehistory of Data Mining, namely *exploratory data analysis* and of some dangers of using statistically motivated notions in Data Mining. (Zytkow and Zembowicz, 1997) deal with generating knowledge from four-fold tables (and describe their database discovery system "49-er"). Recently, the chi-square statistic was used to define "generalized association rules" (we would say: using a comparative quantifier) by Hegland (Hegland, 2001) and Brin (Brin *et al.*, 1998). Papers by Rauch (et al.) discuss several further classes of quantifiers, see references.

Two generalizations

First, the logical approach to mining generalized association rules can be and has been generalized to *fuzzy logic*. We refer to Holeňa's papers ((Holeňa, 1996) – (Holeňa, 1996)) and also to Chen et. al. (Chen *et al.*, 2003). Second, we have only mentioned *relational Data Mining* and its techniques of *inductive logic* programming. Besides Džeroski and Lavrač (Džeroski and Lavrač, 2001) the reader may consult e.g. Dehaspe and Toivonen (Dehaspe and Toivonen, 1999).

The GUHA method

The reader should be informed on the more then 30 years old story of the GUHA method of automated generation of hypotheses (General Unary Hypotheses Automaton) which is undoubtely one of the oldest methods of computerized exploratory data analysis (or, if you want, mining of association rules) starting with (Hájek *et al.*, 1966) from 1966. We already mentioned above that formulas almost identical with Agrawal's association rules were considered and algorithms for their generation were discussed in that paper. This was followed by a long period of research culminating in 1978 by the monograph (Hájek and Havránek, 1978) by Hájek and Havránek, presenting a logical and statistical fundations that are still relevant for contemporary Data Mining. (Note that the book is presently available on web, see references.) The research has continued; see (Hájek *et al.*, 2003) and (Hájek and Holeňa, 2003, Hájek, 2001) for a survey of the present state and relation to other Data Mining methods. The GUHA approach offers observational logical calculi (based on generated quantifiers as presented here), logical foundations of statistical inference (theoretical logical calculi), theory of some auxiliary (helpful) quantifiers good for compression of results, three semantics of missing information and several other facts, notions and techniques. There have been several implementations; for two presently available see (GUHA+-, lispminer).

It is regrettable that the mainstream of Data Mining has neglected the GUHA approach ((Liu *et al.*, 2000) being one of few exceptions); this subsection is a small attempt to change this.

Conclusion

The study of logical aspects of Data Mining is interesting and useful: it gives an exact abstract approach to "association rules" based on the notion of (generalized) quantifiers, important classes of quantifiers, deductive properties of associations expressed using such quantifiers as well as other results not mentioned here (as e.g. results on computational complexity). Hopefully the present chapter will help the reader to enjoy this.

Acknowledgments

Partial support of the COST Action 274 (TARSKI) is recognized.

References

Adamo, J. M. *Data Mining for association rules and sequential patterns.* Springer 2001.

Agrawal, R., Mannila, H., Srikant, R., Toivonen, H., and A. I. Verkamo. "Fast discovery of association rules." In: Advances in knowledge discovery and Data Mining. Fayyad U. M. et al., ed., AAAI Press/MIT Press,1996

Brin, S., Motwani, R., and C. Silverstein. "Beyond market baskets: Generalizing association rules to correlations". http://citeseer.ist.psu.edu/brin97beyond.html

Chen, G., Wei, Q., and E. E. Kerre. "Fuzzy logic approaches for the mining of association rules: an overview". In: Data Mining and knowledge discovery approaches based on rule induction techniques (Triantaphyllou E. et al., ed.) Kluwer, 2003

Dehaspe, L., and H. Toivonen. Discovery of frequent Datalog patterns. Data Mining and knowledge discovery 1999; 3:7-36.

Džeroski, S., and N. Lavrač. *Relational data mining.* Springer, 2001

Ebbinghaus, H. D., Flum, J., and W. Thomas. *Mathematical logic.* Springer 1984.

Giudici, P. "Data Mining model comparison (Statistical models for Data Mining)". Chapter 31.4.6, This volume.

Glymour, C., Madigan, D., Pregibon, D., and P. Smyth. "Statistical themes and lessons for Data Mining." Data Mining and knowledge discovery 1996; 1:25-42.

Hájek, P. "The GUHA method and mining association rules." Proc. CIMA'2001 (Bangor, Wales) 533-539.

Hájek, P. "The new version of the GUHA procedure ASSOC", COMPSTAT 1984, 360-365.

Hájek, P. "Generalized quantifiers, finite sets and Data Mining". In: (Klopotek et al. ed.) Intelligent Information Processing and Web Mining. Springer 2003, 489-496.

Hájek, P., Havel, I. and M. Chytil. "The GUHA method of automatic hypotheses determination", Computing 1966; 1:293-308.

Hájek, P. and T. Havránek. *Mechanizing Hypothesis Formation (Mathematical Foundations for a General Theory)*, Springer-Verlag 1978, 396 pp.

Hájek, P. and T. Havránek. *Mechanizing Hypothesis Formation (Mathematical Foundations for a General Theory).* Internet edition (freely accessible) http://www.cs.cas.cz/~hajek/guhabook/

Hájek, P. and M. Holeňa. "Formal logics of discovery and hypotheses formation by machine". Theor. Comp. Sci. 2003; 299:245-357.

Hájek, P., Holeňa, M. and J. Rauch. "The GUHA method and foundations of (relational) data mining." In: (de Swart et al., ed.) Theory and applications of relational structures as knowledge instruments. Lecture Notes in Computer Science vol. 2929, Springer 2003, 17-37.

Hájek, P., Sochorová, A. and J. Zvárová. "GUHA for personal computers", Comp. Stat. and Data Anal. 1995; 19:149-153.

Hegland, M. "Data Mining techniques". Acta numerica 2001; 10:313-355.

Holeňa, M. "Exploratory data processing using a fuzzy generalization of the Guha approach". In: J. F. Baldwin, editor, Fuzzy Logic, John Wiley and Sons, New York 1996, 213-229.

Holeňa, M. "Fuzzy hypotheses for Guha implications". Fuzzy Sets and Systems, 1998; 98:101–125.

Holeňa, M. "A fuzzy logic framework for testing vague hypotheses with empirical data". In: Proceedings of the Fourth International ICSC Symposium on Soft Computing and Intelligent Systems for Industry, ICSC Academic Press 2001, 401–407.

Holeňa, M. "A fuzzy logic generalization of a Data Mining approach." Neural Network World 2001; 11:595–610.

Holeňa, M. "Exploratory data processing using a fuzzy generalization of the GUHA approach", Fuzzy Logic, Baldwin et al., ed. Willey et Sons New York 1996, 213-229.

Höppner, F. "Association rules". Chapter 14.7.3, This volume.

Liu, W., Alvarez, S. A., and C. Ruiz. "Collaborative recommendation via adaptive association rule mining". KDD-2000 Workshop on Web Mining for E-Commerce, Boston, MA.
Rauch, J. "Logical problems of statistical data analysis in databases". Proc. Eleventh Int. Seminar on Database Management Systems 1988, 53-63.
Rauch, J. "GUHA as a Data Mining Tool, Practical Aspects of Knowledge management". Schweizer Informatiker Gesellshaft Basel 1996, 10 pp.
Rauch, J. "Logical Calculi for Knowledge Discovery". Red. Komorowski, J. – Żytkow, J., Berlin, Springer Verlag 1997, 47-57.
Rauch, J.: Classes of Four-Fold Table Quantifiers. In Principles of Data Mining and Knowledge Discovery, (J. Zytkow, M. Quafafou, eds.), Springer-Verlag, 203-211, 1998.
Rauch, J. "Four-fold Table Calculi and Missing Information". In: JCIS'98 Proceedings, (Paul P. Wang, editor), Association for Intelligent Machinery, 375-378.
Rauch, J., and M. Šimůnek. "Mining for 4ft association rules". Proc. Discovery Science 2000 Kyoto, Springer Verlag, 268-272.
Rauch, J. and M. Šimůnek. "Mining for statistical association rules". Proc. PAKDD 2001 Hong Kong, 149-158.
Rauch, J. "Association Rules and Mechanizing Hypothesis Formation". Working notes of ECML'2001 Workshop: Machine Learning as Experimental Philosophy of Science. See also
http://www.informatik.uni-freiburg.de/ ml/ecmlpkdd/.
Rauch, J. and M. Šimůnek. "Mining for 4ft Association Rules by 4ft-Miner". In: INAP 2001, The Proceeding of the International Rule-Based Data Mining – in conjunction with INAP 2001, Tokyo.
Rauch, J. "Interesting Association Rules and Multi-relational Association Rules". Communications of Institute of Information and Computing Machinery, Taiwan, 2002; 5, 2:77-82.
Żytkow, J. M. and R. Zembowicz. "Contingency tables as the foundation for concepts, concept hierarchies and rules: the 49er approach". Fundamenta informaticae 1997; 30:383-399.
GUHA+– project web site http://www.cs.cas.cz/ click Research, Software.
http://lispminer.vse.cz/overview/4ftminer.html

27

Wavelet Methods in Data Mining

Tao Li[1], Sheng Ma[2], and Mitsunori Ogihara[3]

[1] School of Computer Science Florida International University
Miami, FL 33199
taoli@cs.fiu.edu
[2] Machine Learning for Systems, IBM T.J. Watson Research Center
19 Skyline Drive, Hawthorne, NY 10532
shengma@us.ibm.com
[3] Computer Science Department, University of Rochester
Rochester, NY 14627-0226
ogihara@cs.rochester.edu

Summary. Recently there has been significant development in the use of wavelet methods in various Data Mining processes. This article presents general overview of their applications in Data Mining. It first presents a high-level data-mining framework in which the overall process is divided into smaller components. It reviews applications of wavelets for each component. It discusses the impact of wavelets on Data Mining research and outlines potential future research directions and applications.

Key words:
Wavelet Transform, Data Management, Short Time Fourier Transform, Heisenberg's Uncertainty Principle, Discrete Wavelet Transform, Multiresolution Analysis, Harr Wavelet Transform, Trend and Surprise Abstraction, Preprocessing, Denoising, Data Transformation, Dimensionality Reduction, Distributed Data Mining

27.1 Introduction

The wavelet transform is a synthesis of ideas that emerged over many years from different fields. Generally speaking, the wavelet transform is a tool that partitions data, functions, or operators into different frequency components and then studies each component with a resolution matched to its scale (Daubechies, 1992). Therefore, it can provide economical and informative mathematical representation of many objects of interest (Abramovich *et al.*, 2000). Nowadays many software packages contain fast and efficient programs that perform wavelet transforms. Due to such easy accessibility wavelets have quickly gained popularity among scientists and engineers, both in theoretical research and in applications.

Data Mining is a process of automatically extracting novel, useful, and understandable patterns from a large collection of data. Over the past decade this area has become significant both in academia and in industry. Wavelet theory could naturally play an important role in Data

O. Maimon, L. Rokach (eds.), *Data Mining and Knowledge Discovery Handbook*, 2nd ed.,
DOI 10.1007/978-0-387-09823-4_27,

Mining because wavelets could provide data presentations that enable efficient and accurate mining process and they can also could be incorporated at the kernel for many algorithms. Although standard wavelet applications are mainly on data with temporal/spatial localities (e.g., time series data, stream data, and image data), wavelets have also been successfully applied to various Data Mining domains.

In this chapter we present a general overview of wavelet methods in Data Mining with relevant mathematical foundations and of research in wavelets applications. An interested reader is encouraged to consult with other chapters for further reading (for references, see (Li, Li, Zhu, and Ogihara, 2003)). This chapter is organized as follows: Section 27.2 presents a high-level Data Mining framework, which reduces Data Mining process into four components. Section 27.3 introduces some necessary mathematical background. Sections 27.4, 27.5, and 27.6 review wavelet applications in each of the components. Finally, Section 27.7 concludes.

27.2 A Framework for Data Mining Process

Here we view Data Mining as an iterative process consisting of: **data management**, **data preprocessing**, **core mining process** and **post-processing**. In **data management**, the mechanism and structures for accessing and storing data are specified. The subsequent **data preprocessing** is an important step, which ensures the data quality and improves the efficiency and ease of the mining process. Real-world data tend to be incomplete, noisy, inconsistent, high dimensional and multi-sensory etc. and hence are not directly suitable for mining. **Data preprocessing** includes data cleaning to remove noise and outliers, data integration to integrate data from multiple information sources, data reduction to reduce the dimensionality and complexity of the data, and data transformation to convert the data into suitable forms for mining. **Core mining** refers to the essential process where various algorithms are applied to perform the Data Mining tasks. The discovered knowledge is refined and evaluated in **post-processing** stage.

The four-component framework above provides us with a simple systematic language for understanding the steps that make up the data mining process. Of the four, **post-processing** mainly concerns the non-technical work such as documentation and evaluation, we will focus our attention on the first three components.

27.3 Wavelet Background

27.3.1 Basics of Wavelet in $L^2(R)$

So, first, **what is a wavelet?** Simply speaking, a mother wavelet is a function $\psi(x)$ such that $\{\psi(2^j x-k), i,k \in Z\}$ is an orthonormal basis of $L^2(R)$. The basis functions are usually referred to wavelets [4]. The term wavelet means a small wave. The smallness refers to the condition that we desire that the function is of finite length or compactly supported. The wave refers to the condition that the function is oscillatory. The term mother implies that the functions with different regions of support that are used in the transformation process are derived by dilation and translation of the mother wavelet.

[4] Note that this orthogonality is not an essential property of wavelets. We include it in the definition because we discuss wavelet in the context of Daubechies wavelet and orthogonality is a good property in many applications.

At first glance, wavelet transforms are very much the same as Fourier transforms except they have different bases. So **why bother to have wavelets? What are the real differences between them**? The simple answer is that wavelet transform is capable of providing time and frequency localizations simultaneously while Fourier transforms could only provide frequency representations. Fourier transforms are designed for stationary signals because they are expanded as sine and cosine waves which extend in time forever, if the representation has a certain frequency content at one time, it will have the same content for all time. Hence Fourier transform is not suitable for non-stationary signal where the signal has time varying frequency (Polikar, 2005). Since FT doesn't work for non-stationary signal, researchers have developed a revised version of Fourier transform, The Short Time Fourier Transform (STFT). In STFT, the signal is divided into small segments where the signal on each of these segments could be assumed as stationary. Although STFT could provide a time-frequency representation of the signal, Heisenberg's Uncertainty Principle makes the choice of the segment length a big problem for STFT. The principle states that one cannot know the exact time-frequency representation of a signal and one can only know the time intervals in which certain bands of frequencies exist. So for STFT, longer length of the segments gives better frequency resolution and poorer time resolution while shorter segments lead to better time resolution but poorer frequency resolution. Another serious problem with STFT is that there is no inverse, i.e., the original signal can not be reconstructed from the time-frequency map or the spectrogram.

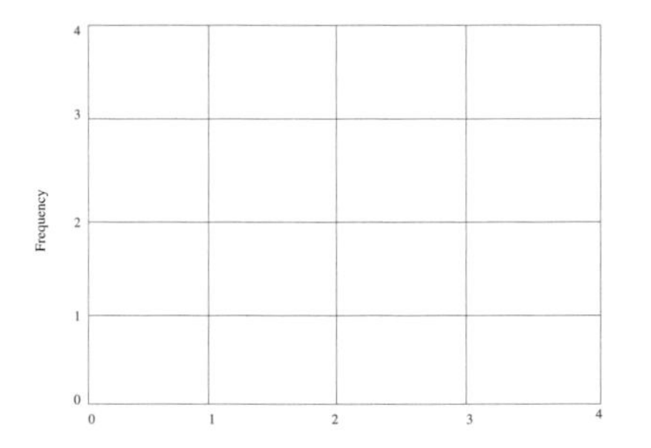

Fig. 27.1. Time-Frequency Structure of STFT. The graph shows that time and frequency localizations are independent. The cells are always square.

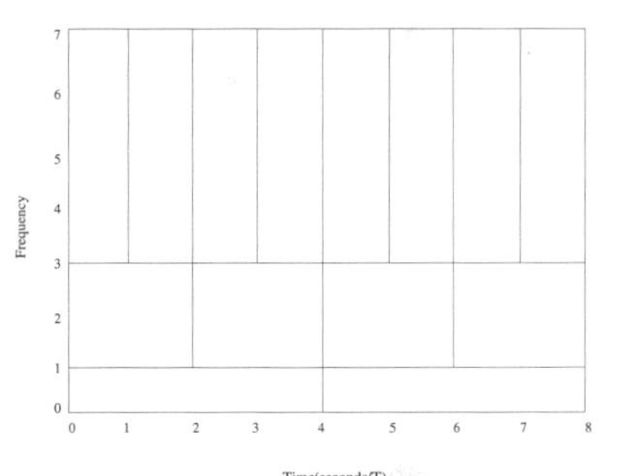

Fig. 27.2. Time Frequency structure of WT. The graph shows that frequency resolution is good for low frequency and time resolution is good at high frequencies.

Wavelet is designed to give good time resolution and poor frequency resolution at high frequencies and good frequency resolution and poor time resolution at low frequencies (Polikar, 2005). This is useful for many practical signals since they usually have high frequency components for a short durations (bursts) and low frequency components for long durations (trends). The time-frequency cell structures for STFT and WT are shown in Figure 27.1 and Figure 27.2 , respectively. In Data Mining practice, the key concept in use of wavelets is the discrete wavelet transform (DWT). Our discussions will focus on DWT.

27.3.2 Dilation Equation

How to find the wavelets? The key idea is self-similarity. Start with a function $\phi(x)$ that is made up of smaller version of itself. This is the refinement (or 2-scale, dilation) equation $\phi(x) = \sum_{k=-\infty}^{\infty} a_k \phi(2x-k)$, where $a_k's$ are called filter coefficients or masks. The function $\phi(x)$ is called the scaling function (or father wavelet). Under certain conditions,

$$\psi(x) = \sum_{k=-\infty}^{\infty} (-1)^k b_k \phi(2x-k) = \sum_{k=-\infty}^{\infty} (-1)^k \bar{a}_{1-k} \phi(2x-k) \tag{27.1}$$

gives a wavelet [5]. Figure 27.3 shows Haar wavelet [6] and Figure 27.4 shows Daubechies-2(db_2) wavelet that is supported on intervals $[0,3]$. In general, db_n represents the family of Daubechies Wavelets and n is the order. Generally it can be shown that: (1) The support for db_n is on the interval $[0, 2n-1]$, (2) The wavelet db_n has n vanishing moments, and (3) The regularity increases with the order. db_n has rn continuous derivatives (r is about 0.2).

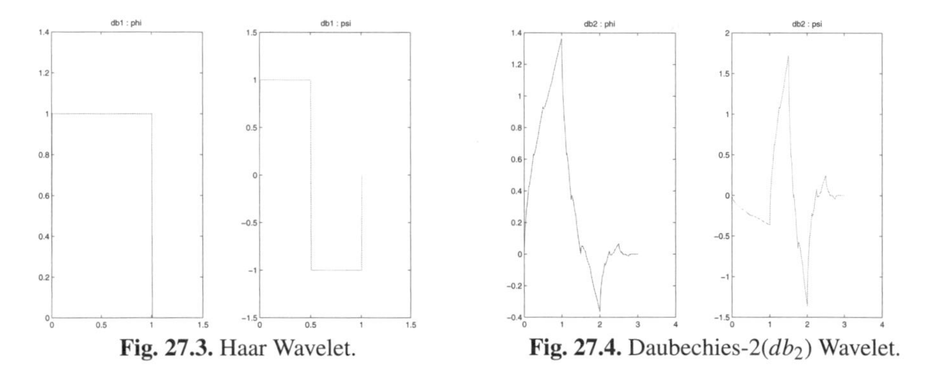

Fig. 27.3. Haar Wavelet.

Fig. 27.4. Daubechies-2(db_2) Wavelet.

27.3.3 Multiresolution Analysis (MRA) and Fast DWT Algorithm

How to efficiently compute wavelet transforms? To answer the question, we need to touch on some material of Multiresolution Analysis (MRA). MRA was first introduced in (Mallat, 1989) and there is a fast family of algorithms based on it. The motivation of MRA is to use a sequence of embedded subspaces to approximate $L^2(R)$ so that a proper subspace for a specific application task can be chosen to get a balance between accuracy and efficiency. Mathematically, MRA studies the property of a sequence of closed subspaces $V_j, j \in Z$ which approximate $L^2(R)$ and satisfy $\cdots V_{-2} \subset V_{-1} \subset V_0 \subset V_1 \subset V_2 \subset \cdots$, $\overline{\bigcup_{j \in Z} V_j} = L^2(R)$ ($L^2(R)$ space is the closure of the union of all V_j), and $\bigcap_{j \in Z} V_j = \emptyset$ (the intersection of all V_j is empty). So **what does multiresolution mean?** The multiresolution is reflected by the additional requirement $f \in V_j \Longleftrightarrow f(2x) \in V_{j+1}, j \in Z$ (This is equivalent to $f(x) \in V_0 \Longleftrightarrow f(2^j x) \in V_j$), i.e., all the spaces are scaled versions of the central space V_0.

So **how does this related to wavelets**? Because the scaling function ϕ easily generates a sequence of subspaces which can provide a simple multiresolution analysis. First, the translations of $\phi(x)$, i.e., $\phi(x-k), k \in Z$, span a subspace, say V_0 (Actually, $\phi(x-k), k \in Z$ constitutes an orthonormal basis of the subspace V_0). Similarly $2^{-1/2}\phi(2x-k), k \in Z$ span another subspace, say V_1. The dilation equation tells us that ϕ can be represented by a basis of V_1. It implies that ϕ falls into subspace V_1 and so the translations $\phi(x-k), k \in Z$ also fall into subspace V_1. Thus V_0 is embedded into V_1. With different dyadic, it is straightforward to obtain a sequence of embedded subspaces of $L^2(R)$ from only one function. It can be shown that the closure of the union of these subspaces is exactly $L^2(R)$ and their intersections are

[5] $\bar{a}$ means the conjugate of a.

[6] Haar wavelet represents the same wavelet as Daubechies wavelets with support at $[0,1]$, called db_1.

empty sets (Daubechies, 1992). here, j controls the observation resolution while k controls the observation location. Formal proof of wavelets' spanning complement spaces can be found in (Daubechies, 1992).

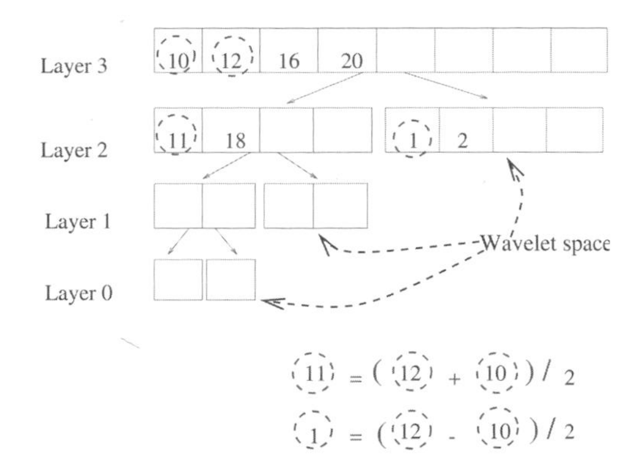

Fig. 27.5. Fast Discrete Wavelet Transform.

A direct application of multiresolution analysis is the fast discrete wavelet transform algorithm, called the *pyramid* algorithm (Mallat, 1989). The core idea is to progressively smooth the data using an iterative procedure and keep the detail along the way, i.e., analyze projections of f to W_j. We use Haar wavelets to illustrate the idea through the following example. In Figure 27.5, the raw data is in resolution 3 (also called layer 3). After the first decomposition, the data are divided into two parts: one is of average information (projection in the scaling space V_2 and the other is of detail information (projection in the wavelet space W_2). We then repeat the similar decomposition on the data in V_2, and get the projection data in V_1 and W_1, etc. The fact that $L^2(R)$ is decomposed into an infinite wavelet subspace is equivalent to the statement that $\psi_{j,k}, j,k \in Z$ span an orthonormal basis of $L^2(R)$. An arbitrary function $f \in L^2(R)$ then can be expressed as $f(x) = \sum_{j,k \in Z} d_{j,k} \psi_{j,k}(x)$, where $d_{j,k} = \langle f, \psi_{j,k} \rangle$ is called the *wavelet coefficients*. Note that j controls the observation resolution and k controls the observation location. If data in some location are relatively smooth (it can be represented by low-degree polynomials), then its corresponding wavelet coefficients will be fairly small by the vanishing moment property of wavelets.

27.3.4 Illustrations of Harr Wavelet Transform

We demonstrate the Harr wavelet transform using a discrete time series $x(t)$, where $0 \leq t \leq 2^K$. In $L^2(R)$, discrete wavelets can be represented as $\phi_j^m(t) = 2^{-j/2}\phi(2^{-j}t - m)$, where j and m are positive integers. j represents the dilation, which characterizes the function $\phi(t)$ at different time-scales. m represents the translation in time. Because $\phi_j^m(t)$ are obtained by dilating and translating a mother function $\phi(t)$, they have the same shape as the mother wavelet and therefore self-similar to each other.

A discrete-time process $x(t)$ can be represented through its inverse wavelet transform $x(t) = \sum_{j=1}^{K} \sum_{m=0}^{2^{K-j}-1} d_j^m \phi_j^m(t) + \phi_0$, where $0 \leq t < 2^K$. ϕ_0 is equal to the average value of $x(t)$ over $t \in [0, 2^K - 1]$. Without loss of generality, ϕ_0 is assumed to be zero. d_j^m's are wavelet coefficients and can be obtained through the wavelet transform $d_j^m = \sum_{t=0}^{2^K-1} x(t)\phi_j^m(t)$. To explore the relationships among wavelets, a tree diagram and the corresponding one-dimensional indices of wavelet coefficients were defined (Luettgen, 1993). The left picture of Figure 27.6 shows an example of Haar wavelets for $K = 3$, and the right figure shows the corresponding tree diagram. The circled numbers represent the one-dimensional indices of the wavelet basis functions, and are assigned sequentially to wavelet coefficients from the top to the bottom down and the left to the right. The one-dimensional index s is thus a one-to-one mapping to the two dimensional index $(j(s), m(s))$, where $j(s)$ and $m(s)$ represent the scale and the shift indices of the s-th wavelet. The equivalent notation [7] of d_s is then $d_{j(s)}^{m(s)}$. In addition, we denote the parent and the neighboring wavelets of a wavelet through the tree diagram. As shown in Figure 27.6, $\gamma(s)$ and $\nu(s)$ are the parent and the left neighbor of node s, respectively.

27.3.5 Properties of Wavelets

In this section, we summarize and highlight the properties of wavelets which make they are useful tools for Data Mining and many other applications.

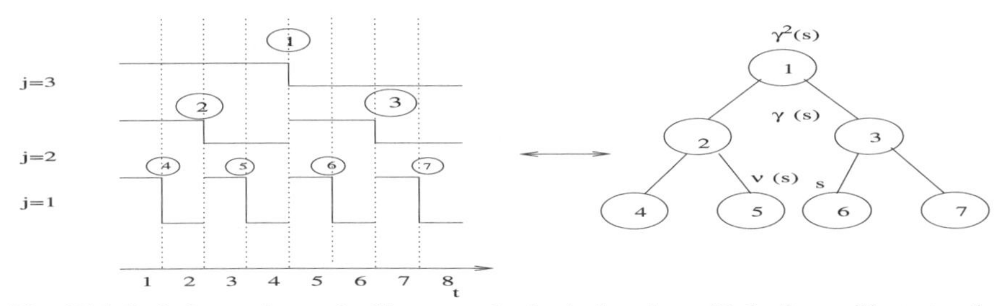

Fig. 27.6. Left figure shows the Haar wavelet basis functions. Right figure illustrates the corresponding tree diagram and two types of operations. The number in the circle represents the one dimension index of the wavelet basis functions. For example, the equivalent notation of d_1^2 is d_6. s, $\nu(s)$ and $\gamma(s)$ represent the one dimension index of wavelet coefficients. $\gamma(s)$ is defined to be the parent node of node s. $\nu(s)$ is defined to be the left neighbor of node s.

Computational Complexity: First, the computation of wavelet transform can be very efficient. Discrete Fourier transform(DFT) requires $O(N^2)$ multiplications and fast Fourier transform also needs $O(N \log N)$ multiplications. However fast wavelet transform based on Mallat's pyramidal algorithm) only needs $O(N)$ multiplications. The space complexity is also linear.

Vanishing Moments: Another important property of wavelets is vanishing moments. A function $f(x)$ which is supported in bounded region ω is called to have n-vanishing moments if it satisfies the following equation: $\int_{\omega} f(x)x^j dx = 0, j = 0, 1, \ldots, n$. For example, Haar wavelet has 1-vanishing moment and db_2 has 2-vanishing moment. The intuition of vanishing moments of wavelets is the oscillatory nature which can thought to be the characterization of difference or details between a datum with the data in its neighborhood. Note that the filter [1, -1] corresponding to Haar wavelet is exactly a difference operator. With higher vanishing moments, if data can be represented by low-degree polynomials, their wavelet coefficients are equal to zero.

Compact Support: Each wavelet basis function is supported on a finite interval. Compact support guarantees the localization of wavelets. In other words, processing a region of data with wavelet does not affect the the data out of this region.

[7] For example, d_6 is d_1^2 (The shift index, m, starts from 0.) in the given example.

Decorrelated Coefficients: Another important aspect of wavelets is their ability to reduce temporal correlation so that the correlation of wavelet coefficients are much smaller than the correlation of the corresponding temporal process (Flandrin, 1992). Hence, the wavelet transform could be able used to reduce the complex process in the time domain into a much simpler process in the wavelet domain.

Parseval's Theorem: Assume that $e \in L^2$ and ψ_i be the orthonormal basis of L^2. Parseval's theorem states that $\|e\|_2^2 = \sum_i | < e, \psi_i > |^2$. In other words, the energy, which is defined to be the square of its L_2 norm, is preserved under the orthonormal wavelet transform.

In addition, the multi-resolution property of scaling and wavelet functions leads to hierarchical representations and manipulations of the objects and has widespread applications. There are also some other favorable properties of wavelets such as the symmetry of scaling and wavelet functions, smoothness and the availability of many different wavelet basis functions etc.

27.4 Data Management

One of the features that distinguish Data Mining from other types of data analytic tasks is the huge amount of data. The purpose of data management is to find methods for storing data to facilitate fast and efficient access. The wavelet transformation provides a natural hierarchy structure and multidimensional data representation and hence could be applied to data management. A novel wavelet based tree structures was introduced in (Shahabi *et al.*, 2001, Shahabi *et al.*, 2000): TSA-tree and 2D TSA-tree, to improve the efficiency of multilevel trends and surprise queries on time sequence data. Frequent queries on time series data are to identify rising and falling trends and abrupt changes at multiple level of abstractions. To support such multi-level queries, a large amount of raw data usually needs to be retrieved and processed. TSA (Trend and Surprise Abstraction) tree is designed to expedite the query process. It is constructed based on the procedure of discrete wavelet transform. The root is the original time series data. Each level of the tree corresponds to a step in wavelet decomposition. At the first decomposition level, the original data is decomposed into a low frequency part (trend) and a high frequency part (surprise). The left child of the root records the trend and the right child records the surprise. At the second decomposition level, the low frequency part obtained in the first level is further divided into a trend part and a surprise part. This process is repeated until the last level of the decomposition. The structure of the TSA tree is described in Figure 27.7. The 2D TSA tree is just the two dimensional extensions of the TSA tree using two dimensional discrete wavelet transform.

27.5 Preprocessing

Real world data sets are usually not directly suitable for performing Data Mining algorithms. They contain noise, missing values and may be inconsistent. In addition, real world data sets tend to be too large and high-dimensional. Wavelets provide a way to estimate the underlying function from the data. With the vanishing moment property of wavelets, we know that only some wavelet coefficients are significant in most cases. By retaining selective wavelet coefficients, wavelet transform could then be applied to denoising and dimensionality reduction. Moreover, since wavelet coefficients are generally decorrelated, we could transform the original data into wavelet domain and then carry out Data Mining tasks.

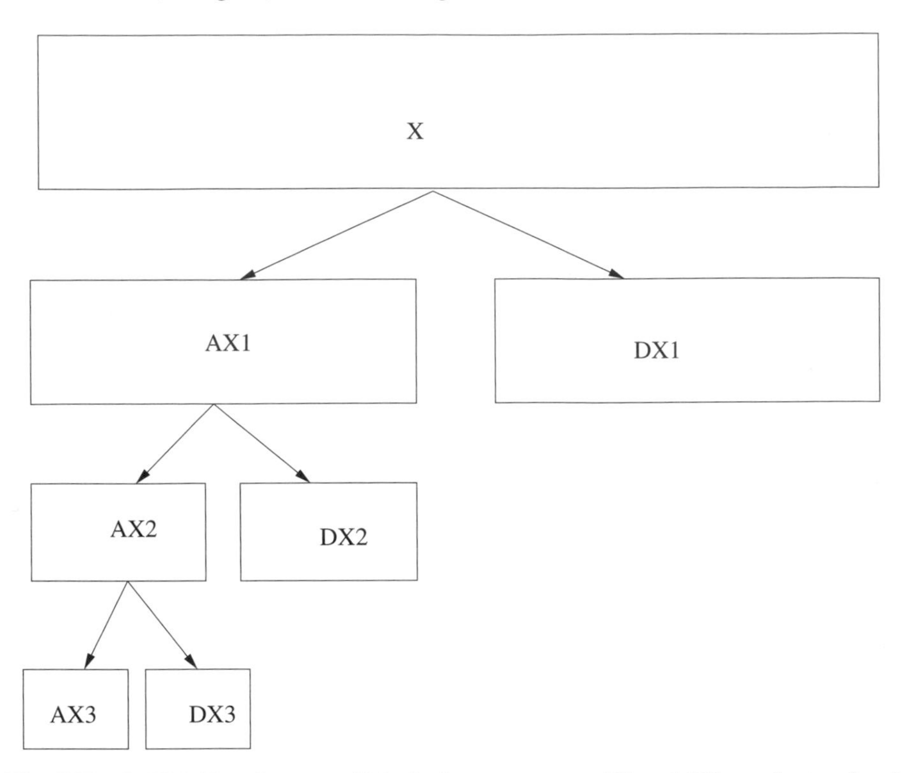

Fig. 27.7. 1D TSA Tree Structure: X is the input sequence. AX_i and DX_i are the trend and surprise sequence at level i.

27.5.1 Denoising

Noise is a random error or variance of a measured variable. Removing noise from data can be considered as a process of identifying outliers or constructing optimal estimates of unknown data from available noisy data. Wavelet techniques provide an effective way to denoise and have been successfully applied in various areas especially in image research. Formally, Suppose observation data $y = (y_1, \dots, y_n)$ is a noisy realization of the signal $x = (x_1, \dots, x_n)$: $y_i = x_i + \varepsilon_i, \quad i = 1, \dots, n$, where ε_i is noise. It is commonly assumed that ε_i are independent from the signal and are independent and identically distributed (*iid*) Gaussian random variables. A usual way to denoise is to find $\hat{x}$ such that it minimizes the mean square error (MSE), $MSE(\hat{x}) = \frac{1}{n}\sum_{i=1}^{n}(\hat{x}_i - x_i)^2$. The main idea of wavelet denoising is to transform the data into a different basis, the wavelet basis, where the *large* coefficients are mainly the useful information and the *smaller* ones represent noise. By suitably modifying the coefficients in the new basis, noise can be directly removed from the data.

A methodology called *WaveShrink* for estimating x was developed in (Donoho and Johnstone, 1998). WaveShrink includes three steps: (1) Transform data y to the wavelet domain, (2) Shrink the empirical wavelet coefficients towards zero, and (3) Transform the shrunk coefficients back to the data domain. There are three commonly used shrinkage functions: the hard, the soft, and the non-negative garrote shrinkage functions:

$$\delta_{\lambda}^{H}(x)=\begin{cases}0 & |x|\leq\lambda \\ x & |x|>\lambda\end{cases},\delta_{\lambda}^{S}(x)=\begin{cases}0 & |x|\leq\lambda \\ x-\lambda & x>\lambda \\ \lambda-x & x<-\lambda\end{cases},$$

$$\delta_{\lambda}^{H}(x)=\begin{cases}0 & |x|\leq\lambda \\ x-\lambda^2/x & |x|>\lambda\end{cases}$$

where $\lambda\in[0,\infty)$ is the threshold. Wavelet denoising is generally different from traditional filtering approaches and it is nonlinear, due to a thresholding step. Determining threshold λ is the key issue in WaveShrink denoising. Minimax threshold is one of commonly used thresholds. The *minimax* (Minimize Maximal Risk) threshold λ^* is defined as threshold λ which minimizes expression $\inf_{\lambda}\sup_{\theta}\left\{\frac{R_{\lambda}(\theta)}{n^{-1}+\min(\theta^2,1)}\right\}$, where $R_{\lambda}(\theta)=E(\delta_{\lambda}(x)-\theta)^2, x\sim N(\theta,1)$.

The use of wavelet preprocessing to alleviate the effect of noisy data for biological data classification was investigated in (Li *et al.*, 2002). The research has showed that, if the localities of the attributes are strong enough, wavelet denoising is able to improve the performance
.

27.5.2 Data Transformation

A wide class of operations can be performed directly in the wavelet domain by operating on coefficients of the wavelet transforms of original data sets. Operating in the wavelet domain enables to perform these operations progressively in a coarse-to-fine fashion, to operate on different resolutions, manipulate features at different scales, and localize the operation in both spatial and frequency domains. Performing such operations in the wavelet domain and then reconstructing the result is more efficient than performing the same operation in the standard direct fashion and reduces the memory footprint. In addition, wavelet transformations have the ability to reduce temporal correlation so that the correlation of wavelet coefficients are much smaller than the correlation of corresponding temporal process. Hence simple models that are insufficient in the original domain may be quite accurate in the wavelet domain. Researchers have proposed a new approach of applying Principal Component Analysis (PCA) on the wavelet subband (Feng *et al.*, 2000). Wavelet transform is used to decompose an image into different frequency subbands and a mid-range frequency subband is used for PCA representation. A probability model for natural images, based on empirical observation of their statistics in the wavelet transform domain was developed in (Buccigrossi and Simoncelli, 1997). The researchers noted that pairs of wavelet coefficients, corresponding to basis functions at adjacent spatial locations, orientations, and scales, generally to be non-Gaussian in both their marginal and joint statistical properties and specifically, their marginals are heavy-tailed, and although they are typically decorrelated, their magnitudes are highly correlated.

27.5.3 Dimensionality Reduction

The goal of dimension reduction is to express the original data set using some smaller set of data with or without a loss of information. Dimensionality reduction can be thought as an extension of the data transformation presented in Section 27.5.2: while data transformation just transforms original data into wavelet domain without discarding any coefficients, dimensionality reduction only keeps a collection of selective wavelet coefficients. More formally, the dimensionality reduction problem is to project the n-dimensional tuples that represent the data in a k-dimensional space so that $k<<n$ and the distances are preserved as well as possible.

Based on the different choices of wavelet coefficients, there are two different ways for dimensionality reduction using wavelet: (a) Keeping the largest k coefficients and approximate the rest with 0, and (b) Keeping the first k coefficients and approximate the rest with 0. Keeping the largest k coefficients achieves more accurate representation while keeping the first k coefficients is useful for indexing. Keeping the first k coefficients implicitly assumes a priori the significance of all wavelet coefficients in the first k coarsest levels and that all wavelet coefficients at a higher resolution levels are negligible. Such a strong prior assumption heavily depends on a suitable choice of k and essentially denies the possibility of local singularities in the underlying function (Abramovich *et al.*, 2000).

It has been shown that (Stonllnitz *et al.*, 1996), if the basis is orthonormal, in terms of L_2 loss, maintaining the largest k wavelet coefficients provides the optimal k-term Haar approximation. Using the largest k wavelet coefficients, given a predefined precision ε, Dimension reduction procedure can be summarized in the following steps: (1) Compute the wavelet coefficients of the original data set, (2) Sort the coefficients in order of decreasing magnitude to produce the sequence $c_0, c_1, \ldots, c_{M-1}$, and (3) Starting with $M' = M$, find the best M' such that $\sum_{i=M'}^{M-1} ||c_i|| \leq \varepsilon$. $||c_i||$ is the norm of c_i. In general, the norm can be chosen as L_2 norm or L_1 norm or other norms.

27.6 Core Mining Process

Core mining refers to the essential procedure where intelligent methods are applied to extract useful information patterns. There are many Data Mining tasks such as clustering and classification etc. Each task can be thought as a particular kind of problem to be solved by a Data Mining algorithm. Generally there are many different algorithms could serve the purpose of the same task. Meanwhile, some algorithms can be applied to different tasks. In this section, we review the wavelet applications in Data Mining tasks and algorithms.

27.6.1 Clustering

Intuitively, the clustering problem can be described as follows: Let W be a set of n data points in a multi-dimensional space. Find a partition of W into classes such that the points within each class are *similar* to each other. The multi-resolution property of wavelet transforms inspires the researchers to consider algorithms that could identify clusters at different scales. WaveCluster (Sheikholeslami *et al.*, 1998) is a multi-resolution clustering approach for very large spatial databases. Spatial data objects can be represented in an n-dimensional feature space and the numerical attributes of a spatial object can be represented by a feature vector where each element of the vector corresponds to one numerical attribute (feature). Partitioning the data space by a grid reduces the number of data objects while inducing only small errors. From a signal processing perspective, if the collection of objects in the feature space is viewed as an n-dimensional signal, the high frequency parts of the signal correspond to the regions of the feature space where there is a rapid change in the distribution of objects (i.e., the boundaries of clusters) and the low frequency parts of the n-dimensional signal which have high amplitude correspond to the areas of the feature space where the objects are concentrated (i.e., the clusters). Applying wavelet transform on a signal decomposes it into different frequency sub-bands. Hence identifying clusters is converted to finding connected components in the transformed feature space. Moreover, application of wavelet transformation provides multiresolution data representation and hence finding the connected components could be carried out at different resolution levels.

27.6.2 Classification

Classification problems aim to identify the characteristics that indicate the group to which each instance belongs. Classification can be used both to understand the existing data and to predict how new instances will behave. Wavelets can be very useful for classification tasks. First, classification methods can be applied on the wavelet domain of the original data as discussed in Section 27.5.2 or selective dimensions of the wavelet domain as we will discussed in this section. Second, the multi-resolution property of wavelets can be incorporated into classification procedures to facilitate the process.

A wavelet-based classification algorithm on large two-dimensional data sets typically large digital images was developed by (Castelli *et al.*, 1996). The image is viewed as a real-valued configuration on a rectangular subset of the integer lattice Z^2 and each point on the lattice (i.e. pixel) is associated with a vector denoting as pixel-values and a label denoting its class. The classification problem here consists of observing an image with known pixel-values but unknown labels and assigning a label to each point and it was motivated primarily by the need to classify quickly and efficiently large images in digital libraries. Traditional pixel-by-pixel analysis (Cromp and Campbell, 1993) is fairly computationally expensive and does not consider the correlation between the labels of adjacent pixels. The wavelet-based classification method is based on the progressive classification (Castelli *et al.*, 1996) framework: It uses generic classifiers on a low-resolution representation of the data obtained using discrete wavelet transform. The wavelet transformation produces a multiresolution pyramid representation of the data. In this representation, at each level each coefficient corresponds to a $k \times k$ pixel block in the original image. At each step of the classification, the algorithm decides whether each coefficient corresponds to a homogeneous block of pixels and assigns the same class label to the whole block or to re-examine the data at a higher resolution level. And the same process is repeated iteratively. The wavelet-based classification method achieves a significant speedup over traditional pixel-wise classification methods. For images with pixel values that are highly correlated, the method will give more accurate results than the corresponding non-progressive classifier because DWT produces a weight average of the values for a $k \times k$ block and the algorithm tend to assume more uniformity in the image than may appear when we look at individual pixels. A paper by (Mojsilovic and Popovic, 1997) also proposed a wavelet-based approach for classification of texture samples with small dimensions. The idea is first to decompose the given image with a filter bank derived from an orthonormal wavelet basis and to form an image approximation with nigher resolution. Texture energy measures calculated at each output of the filter bank as well as energies if synthesized images are used as texture features for a classification procedure based on modified statistical t-test. A paper by (Aggarwal, 2002) exploited the multi-resolution property of wavelet decomposition to create a scheme which can mine classification characteristics at different levels of granularity.

27.6.3 Regression

Regression uses existing values to forecast what other values will be and it is one of the fundamental tasks of Data Mining. Consider the standard univariate nonparametric regression setting: $y_i = g(t_i) + \varepsilon_i, i = 1, \ldots, n$ where ε_i are independent $N(0, \sigma^2)$ random variables. The goal is to recover the underlying function g from the noisy data y_i, without assuming any particular parametric structure for g. The basic approach of using wavelets for nonparametric regression is to consider the unknown function g expanded as a generalized wavelet series and then to estimate the wavelet coefficients from the data. Hence the original nonparametric problem is thus transformed to a parametric one (Abramovich *et al.*, 2000). Note that the denoise problem we discussed in Section 27.5.1 can be regarded as a subtask of the regression problem since the estimation of the underlying function involves the noise removal from the observed data.

Linear Regression. For linear regression, we can express $g(t) = c_0\phi(t) + \sum_{j=0}^{\infty}\sum_{k=0}^{2^j-1} w_{jk}\psi_{jk}(t)$, where $c_0 =< g,\phi >, w_{jk} =< g,\psi_{jk} >$. If we assume g belongs to a class of functions with certain regularity, then the corresponding norm of the sequence of w_{jk} is finite and w_{jk}'s decay to zero. So $g(t) = c_0\phi(t) + \sum_{j=0}^{M}\sum_{k=0}^{2^j-1} w_{jk}\psi_{jk}(t)$, for some M and a corresponding truncated wavelet estimator is (Abramovich *et al.*, 2000) $\hat{g}_M(t) = \hat{c}_0\phi(t) + \sum_{j=0}^{M}\sum_{k=0}^{2^j-1} \hat{w}_{jk}\psi_{jk}(t)$. Thus the original nonparametric problem reduces to linear regression and the sample estimates of the coefficients are given by: $\hat{c}_0 = \frac{1}{n}\sum_{i=1}^{n}\phi(t_i)y_i, \hat{w}_{jk} = \frac{1}{n}\sum_{i=1}^{n}\psi_{jk}(t_i)y_i$. The performance of the truncated wavelet estimator clearly depends on an appropriate choice of M. Various methods such as Akaike's Information Criterion and cross-validation can be used for choosing M. We should point out that: the linear regression approach here is similar to the dimensionality reduction by keeping the first several wavelet coefficients discussed in section 27.5.3. There is an implicit strong assumption underlying the approach: all wavelet coefficients in the first M coarsest levels are significant while all wavelet coefficients at a higher resolution levels are negligible. Such a strong assumption clearly would not hold for many functions. It has been shown that no linear estimator will be optimal in minimax sense for estimating inhomogeneous functions with local singularities (Donoho and Johnstone, 1998).

Nonlinear Regression. A nonlinear wavelet estimator of g based on reconstruction from a more judicious selection of the empirical wavelet coefficients was proposed in (Donoho and Johnstone, 1998). The vanishing moments property of wavelets makes it reasonable to assume that essentially only a few 'large' $\hat{w}_{jk}$ contain information about the underlying function g, while 'small' $\hat{w}_{jk}$ can be attributed to noise. If we can decide which are the 'significant' large wavelet coefficients, then we can retain them and set all the others equal to zero, so obtaining an approximate wavelet representation of underlying function g. The key concept here is thresholding. Thresholding allows the data itself to decide which wavelet coefficients are significant. Clearly an appropriate choice of the threshold value λ is fundamental to the effectiveness of the estimation procedure. Too large threshold might "cut off" important parts of the true function underlying the data while too small a threshold retains noise in the selective reconstruction. As described in Section 27.5.1, there are three commonly used thresholding functions. It has been shown that hard thresholding results in larger variance in the function estimate while soft thresholding has large bias. To comprise the trade-off between bias and variance, Bruce and Gao (Bruce and Gao, 1996) suggested a firm thresholding that combines the hard and soft thresholding. More discussion on nonlinear regression can be found in (Antoniadis, 1999).

27.6.4 Distributed Data Mining

Over the years, data set sizes have grown rapidly. Moreover, many of these data sets are, in nature, geographically distributed across multiple sites. To mine such large and distributed data sets, it is important to investigate efficient distributed algorithms to reduce the communication overhead, central storage requirements, and computation times. The orthogonal property of wavelet basis could play an important role in distributed data mining since the orthogonality guarantees correct and independent local analysis that can be used as a building-block for a global model. In addition, the compact support property of wavelets could be used to design parallel algorithms since the compact support guarantees the localization of wavelet and processing a region of data with wavelet does not affect the the data out of this region.

The idea of performing distributed data analysis using wavelet-based *Collective Data Mining*(CDM) from heterogeneous sites was introduced by (Kargupta *et al.*, 2000). The main steps for the approach can be summarized as follows: (1) choose an orthonormal representation that is appropriate for the type of data model to be constructed, (2) generate approximate orthonormal basis coefficients at each local site, (3) if necessary, move an approximately chosen sample of the datasets from each site to a single site and generate the approximate basis coefficients corresponding to non-linear cross terms, and (4) combine the local models, transform the model into the user described canonical representation and output the model. The foundation of CDM is based on the fact that any function can be represented in a distributed fashion using an appropriate basis. If we use wavelet basis, The orthogonality guarantees correct and independent local analysis that can be used as a building-block for a global model.

27.6.5 Similarity Search/Indexing

The problem of similarity search in Data Mining is: given a pattern of interest, try to find similar patterns in the data set based on some similarity measures. This task is most commonly used for time series, image and text data sets. Formally, A dataset is a set denoted $DB = \{X_1, X_2, \ldots, X_i, \ldots, X_N\}$, where $X_i = [x_0^i, x_1^i, \ldots, x_n^i]$ and a given pattern is a sequence of data points $Q = [q_0, q_1, \ldots, q_n]$. Given a pattern Q, the result set R from the data set is $R = \{X_{i_1}, X_{i_2}, \ldots, X_{i_j}, \ldots, X_{i_m}\}$, where $\{i_1, i_2, \cdots, i_m\} \subseteq \{1, \cdots, N\}$, such that $D(X_{i_j}, Q) < d$. If we use Euclidean distance between X and Y as the distance function $D(X,Y)$, then $D(X,Y) = \sqrt{\sum_j |x_j - y_j|^2}$, which is the aggregation of the point to point distance of two patterns. Wavelets could be applied into similarity search in several different ways. First, wavelets could transform the original data into the wavelet domain as described in Section 27.5.2 and we may also only keep selective wavelet coefficients to achieve dimensionality reduction as in Section 27.5.3. The similarity search are then conducted in the transformed domain and could be more efficient. Although the idea here is similar to that reviewed in Section 27.5.2 and Section 27.5.3: both involves transforming the original data into wavelet domain and may also selecting some wavelet coefficients. However, it should be noted that here for the data set: to project the n-dimensional space into a k-dimensional space using wavelets, the same k-wavelet coefficients should be stored for objects in the data set. Obviously, this is not optimal for all objects. To find the k optimal coefficients for the data set, we need to compute the average energy for each coefficient. Second, wavelet transforms could be used to extract compact feature vectors and define new similarity measure to facilitate search. Third, wavelet transforms are able to support similarity search at different scales. The similarity measure could then be defined in an adaptive and interactive way.

Wavelets have been extensively used in similarity search in time series and excellent can be found in (Chiann and Morettin, 1999). Chan and Fu (Chan and Fu, 1999) proposed efficient time series matching strategy by wavelets. Haar transform wavelet transform is first applied and the first few coefficients of the transformed sequences are indexed in an R-Tree for similarity search. The method provides efficient for range and nearest neighborhood queries. A comprehensive comparison between DFT and DWT in time series matching was presented in (Wu *et al.*, 2000). The experimental results show that although DWT does not reduce relative matching error and does not increase query precision in similarity search, DWT based techniques have several advantage such as DWT has multi-resolution property and DWT has complexity of $O(N)$ while DFT has complexity of $O(N \log N)$. A new similarity measures based on the special presentations derived from Haar wavelet transform was presented by (Struzik and Siebes, 1999). Instead of keeping selective wavelet coefficients, the special representations keep only the sign of the wavelet coefficients (sign representation) or keep the difference of the logarithms (DOL) of the values of the wavelet coefficient at highest scale and the working scale (DOL representation).

Wavelets also have widespread applications in content-based similarity search in image/audio databases. A method of using *image querying metric* for fast and efficient content-based image querying was presented by (Jacobs *et al.*, 1995). The image querying metric is computed on the *wavelet signatures* which are obtained by truncated and quantized wavelet decomposition. *WALRUS* (WAveLet-based Retrieval of User-specified Scenes) (Natsev *et al.*, 1999) is an algorithm for similarity retrieval in image diastases. WALRUS first uses dynamic programming to compute wavelet signatures for sliding windows of varying size, then clusters the signatures in wavelet space and finally the similarity measure between a pair of images is calculated to be the fraction of the area the two images covered by matching signatures. *Windsurf* (Wavelet-Based Indexing of Images Using Region Fragmentation) (Ardizzoni *et al.*, 1999) is a new approach for image retrieval. Windsurf uses Haar wavelet transform to extract color and texture features and applies clustering techniques to partition the image into regions. Similarity is then computed as the Bhattcharyya metric (Campbell, 1997) between matching regions. Brambilla (Brambilla, 1999) defined an effective strategy which exploits multi-resolution wavelet transform to effectively describe image content and is capable of interactive learning of the similarity measure. *WBIIS* (Wavelet-Based Image Indexing and Searching) (Wang, Wiederhold, Firschein and Wei, 1997) is a new image indexing and retrieval algorithm with partial sketch image searching capability for large image databases. WBIIS applies Daubechies-8 wavelets for each color component and low frequency wavelet coefficients and their variance are stored as feature vectors. Wang, Wiederhold and Firschein (Wang, Wiederhold and Firschein, 1997) described *WIPETM* (Wavelet Image Pornography Elimination) for image retrieval. WIPETM uses Daubechies-3 wavelets, normalized central moments and color histograms to provide feature vector for similarity matching. Subramanya and Youssef (Subramanya and Youssef, 1998) presented a scalable content-based image indexing and retrieval system based on vector coefficients of color images where highly decorrelated wavelet coefficient planes are used to acquire a search efficient feature space. A fast wavelet histogram techniques for image indexing was proposed in (Mandal *et al.*, 1999). There are also lots of applications of wavelets in audio/music information processing such as (Tzanetakis and Cook, 2002, Li, Ogihara, and Li, 2003).

27.6.6 Approximate Query Processing

Query processing is a general task in Data Mining and similarity search discussed in Section 27.6.5 is one of the specific form of query processing. Approximate query processing has recently emerged as a viable solution for large-scale decision support. Due to the exploratory nature of many decision support applications, there are a number of scenarios where an exact answer may not be required and a user may in fact prefer a fast approximate answer. Wavelet-based techniques can be applied as a data reduction mechanism to obtain *wavelet synopses* of the data on which the approximate query could then operate. The wavelet synopses are compact sets of wavelet coefficients obtained by the wavelet decomposition. Note that some of wavelet methods described here might overlap with those described in Section 27.5.3. The wavelet synopses reduce large amount of data to compact sets and hence could provide fast and reasonably approximate answers to queries.

A wavelet-based technique to build histograms on the underlying data distributions for selectivity estimation was presented in (Matias *et al.*, 1998, Matias *et al.*, 2000). Moreover s wavelet-based techniques for the approximation of range-sum queries over OLAP data cubes was proposed in (Vitter and Wang, 1999, Vitter *et al.*, 1998). Generally, the central idea is to apply multidimensional wavelet decomposition on the input data collection (attribute columns or OLAP cube) to obtain a compact data synopsis by keeping a selective small collection of wavelet coefficients.

An extension (Chakrabarti *et al.*, 2001) to previous work on wavelet techniques in approximate query answering demonstrates that wavelets could be used as a generic and effective tool for decision support applications. The generic approach consists of three steps: the wavelet-coefficient synopses are first computed and then using novel query processing algorithms SQL operators such as select, project and join can be executed entirely in the wavelet-coefficient domain. Finally the results is mapped from the wavelet domain to relational tuples (*Rendering*).

Techniques for computing small space representations of massive data streams by keeping a small number of wavelet coefficients and using the representations for approximate aggregate queries was presented in (Gilbert *et al.*, 2001). Garofalakis and Gibbons (Garofalakis and Gibbons, 2002) introduced probabilistic wavelet synopses that provably enabled unbiased data reconstruction with guarantees on the accuracy of individual approximate answers. The probabilistic technique is based on probabilistic thresholding scheme to assign each coefficient a probability of being retained instead of deterministic thresholding.

27.6.7 Traffic Modeling

Traffic modeling and understanding is imperative to providing Quality of Service (QoS) to network management and control, and to network design and simulation. It is therefore important to accurately model key statistical properties of traffic; while achieving computational efficiency when the model is used for generating synthetic traffic or for prediction. In early 90s, a surge of work have shown that Internet IP traffic, web traffic and multimedia stream (e.g. MPEG traffic) have the complex temporal correlation structure best characterized by both the short-range dependence (SRD) and long-range dependence (LRD) (Garrett and Willinger, 1994). SRD traffic has a short memory; its auto-correlation function decays exponentially. In contrast, LRD has a long memory and a self-similar property; its auto-correlation function decays hyperbolically. Figure 27.8 shows the correlation function of FARIMA (fractional differentiation of an auto-regressive moving average) model that possesses both SRD (the correlation function drops exponentially when the lag is small) and LRD (the correlation function drops polynomially when the lag is large). Its correlation is still non-negligible when lag is 400.

Wavelet models have been demonstrated as a good candidate to characterize traffic (Abry and Darryl, 1998) and further model traffic (Ma and Ji, 2001, Ribeiro *et al.*, 1999) with a complex temporal correlation structure as its base functions match the self-similar property of traffic. Let $x(t)$ be a time series representing traffic, $t \in [0, 2^K - 1]$. Through wavelet transformation, we have $x(t) = \sum_{j=1}^{K} \sum_{m=0}^{2^{K-j}-1} d_j^m \phi_j^m(t) + \phi_0$, where ϕ_0 is equal to the average value of $x(t)$ over $t \in [0, 2^K - 1]$. When $x(t)$ is a random process, the corresponding wavelet coefficients d_j^m's define a two-dimensional random processes in terms of j and m. Due to the one-to-one correspondence between $x(t)$ and its wavelet coefficients, the statistical properties of the wavelet coefficients are completely determined by those of $x(t)$. Likewise, if the statistical properties of the wavelet coefficients are well specified, they can be used to characterize the original random process. This motivates the approach of traffic modeling by characterizing statistical properties of waveletcoefficients.

Why modeling in wavelet domain can be more effective than that in time domain? Figure 27.9 shows the correlation matrix of wavelet coefficients of the FARIMA model, in which a pixel (i,k) in an image represents the correlation between the i-th and the k-th wavelet coefficients, where i and k result in the one dimension index. The gray level is proportional to the magnitude of the correlation. The higher the magnitude of the correlation, the whiter the pixel in the image. The diagonal line is corresponding to the variance of each wavelet. The bright lines (i.e. significant correlations) from diagonal correspond to the correlations between $\gamma^k(s)$ and s, where $\gamma(s)$ represents the parent of the node s, and $\gamma^k(s)$ denotes the parent of the node $\gamma^{k-1}(s)$ with k being $1,2,3,4$ from the diagonal line. Comparing Figure 27.8 and Figure 27.9, we can see that the complicated temporal correlation concentrates on only a few key correlation patterns in the wavelet domain. Thus, we can use a compact model in the wavelet domain to represent the original traffic. Ma and Ji (Ma and Ji, 2001) have demonstrated an independent wavelet model can have a very satisfactory accuracy in matching the possible complex temporal correlation and matching the queuing loss behaviors. Further, a low order Markov model in the wavelet domain that model father-son relationships can boost modeling accuracy. A further development of multiscale wavelet model and its theoretical inside was presented in (Ribeiro *et al.*, 1999).

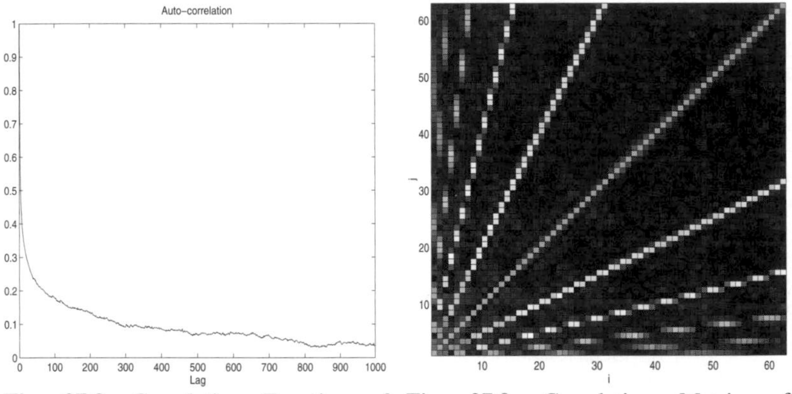

Fig. 27.8. Correlation Function of "FARIMA(1,0.4,0)".

Fig. 27.9. Correlation Matrix of "FARIMA(1,0.4,0)".

27.7 Conclusion

Wavelet techniques have a lot of advantages and there already exists numerous successful applications in Data Mining. It goes without saying that wavelet approaches will be of growing importance in Data Mining.

It should also be mentioned that most of current works on wavelet applications in Data Mining are based orthonormal wavelet basis. However, we argue that orthonormal basis may not be the best representation for noisy data even though the vanishing moments can help them achieve denoising and dimensionality reduction purpose. Intuitively, orthogonality is the most economical representation. In other words, in each direction, it contains equally important information. Therefore, it is usually likely that thresholding wavelet coefficients remove useful information when they try to remove the noise or redundant information (noise can also be regarded as one kind of redundant information). To represent redundant information, it might be good to use redundant wavelet representation – wavelet frames. Except orthogonality, wavelet frames preserve all other properties that an orthonormal wavelet basis owns, such as vanishing moment, compact support, multiresolution. The redundancy of a wavelet frame means that the frame functions are not independent anymore. For example, vectors $[0,1]$ and $[1,0]$ is an orthonormal basis of a plane R^2, while vectors $[1/2,1/2]$, $[-1/2,1/2]$, and $[0,-1]$ are not independent, and consist a frame for R^2. So when data contain noise, frames may provide some specific directions to record the noise. Our work will be the establishment of criteria to recognize the direction of noise or redundant information.

Wavelets could also potentially enable many other new researches and applications such as conventional database compression, multiresolution data analysis and fast approximate Data Mining etc. Finally we eagerly await many future developments and applications of wavelet approaches in Data Mining.

References

F. Abramovich, T. Bailey, and T. Sapatinas. Wavelet analysis and its statistical applications. *JRSSD*, (48):1–30, 2000.

P. Abry and V. Darryl. Wavelet analysis of long-range-dependent traffic. *IEEE Transactions on Information Theory*, 44(1):2–16, 1998.

C. C. Aggarwal. On effective classification of strings with wavelets. In *Proceedings of the eighth ACM SIGKDD international conference on Knowledge discovery and Data Mining*, pages 163–172, 2002.

A. Antoniadis. Wavelets in statistics: a review. *J. It. Statist. Soc.*, 1999.

S. Ardizzoni, I. Bartolini, and M. Patella. Windsurf: Region-based image retrieval using wavelets. In *DEXA Workshop*, pages 167–173, 1999.

C. Brambilla, A. D. Ventura, I. Gagliardi, and R. Schettini. Multiresolution wavelet transform and supervised learning for content-based image retrieval. In *Proceedings of the IEEE International Conference on Multimedia Computing and Systems*, pages 9183–9188, 1999.

A. Bruce and H.-Y. Gao. Waveshrink with firm shrinkage. *Statistica Sinica*, (4):855–874, 1996.

R. W. Buccigrossi and E. P. Simoncelli. Image compression via joint statistical characterization in the wavelet domain. In *Proceedings ICASSP-97 (IEEE International Conference on Acoustics, Speech and Signal Processing)*, number 414, pages 1688–1701, 1997.

J. P. Campbell. Speaker recognition: A tutorial. In *Proceedings of the IEEE*, volume 85, pages 1437–1461, Sept. 1997.

V. Castelli, C. Li, J. Turek, and I. Kontoyiannis. Progressive classification in the compressed domain for large EOS satellite databases, April 1996.

K. Chakrabarti, M. Garofalakis, R. Rastogi, and K. Shim. Approximate query processing using wavelets. *VLDB Journal*, 10(2-3):199–223, 2001.

K. P. Chan and A. W.-C. Fu. Efficient time series matching by wavelets. In *ICDE*, pages 126–133, 1999.

C. Chiann and P. A. Morettin. A wavelet analysis for time series. *Journal of Nonparametric Statistics*, 10(1):1–46, 1999.

R. F. Cromp and W. J. Campbell. Data Mining of multidimensional remotely sensed images. In *Proc. 2nd International Conference of Information and Knowledge Management,*, pages 471–480, 1993.

I. Daubechies. *Ten Lectures on Wavelets*. Capital City Press, Montpelier, Vermont, 1992.

D. L. Donoho and I. M. Johnstone. Minimax estimation via wavelet shrinkage. *Annals of Statistics*, 26(3):879–921, 1998.

G. C. Feng, P. C. Yuen, and D. Q. Dai. Human face recognition using PCA on wavelet subband. *SPIE Journal of Electronic Imaging*, 9(2):226–233, 2000.

P. Flandrin. Wavelet analysis and synthesis of fractional Brownian motion. *IEEE Transactions on Information Theory*, 38(2):910–917, 1992.

M. Garofalakis and P. B. Gibbons. Wavelet synopses with erro guarantee. In *Proceedings of 2002 ACM SIGMOD*, pages 476–487, 2002.

M. W. Garrett and W. Willinger. Analysis, modeling and generation of self-similar VBR video traffic. In *Proceedings of SIGCOM*, pages 269–279, 1994.

A. C. Gilbert, Y. Kotidis, S. Muthukrishnan, and M. Strauss. Surfing wavelets on streams: One-pass summaries for approximate aggregate queries. In *The VLDB Journal*, pages 79–88, 2001.

C. E. Jacobs, A. Finkelstein, and D. H. Salesin. Fast multiresolution image querying. *Computer Graphics*, 29:277–286, 1995.

J.S.Vitter, M. Wang, and B. Iyer. Data cube approximation and histograms via wavelets. In *Proc. of the 7th Intl. Conf. On Infomration and Knowledge Management*, pages 96–104, 1998.

H. Kargupta, B. Park, D. Hershbereger, and E. Johnson. Collective Data Mining: A new perspective toward distributed data mining. In *Advances in Distributed Data Mining*, pages 133–184. 2000.

Q. Li, T. Li, and S. Zhu. Improving medical/biological data classification performance by wavelet pre-processing. In *ICDM*, pages 657–660, 2002.

T. Li, Q. Li, S. Zhu, and M. Ogihara. A survey on wavelet applications in Data Mining. *SIGKDD Explorations*, 4(2):49–68, 2003.

T. Li, M. Ogihara, and Q. Li. A comparative study on content-based music genre classification. In *Proceedings of 26th Annual ACM Conference on Research and Development in Information Retrieval (SIGIR 2003)*, pages 282–289, 2003.

M. Luettgen, W. C. Karl, and A. S. Willsky. Multiscale representations of markov random fields. *IEEE Trans. Signal Processing*, 41:3377–3396, 1993.

S. Ma and C. Ji. Modeling heterogeneous network traffic in wavelet domain. *IEEE/ACM Transactions on Networking*, 9(5):634–649, 2001.

S. Mallat. A theory for multiresolution signal decomposition: the wavelet representation. *IEEE Transactions on Pattern Analysis and Machine Intelligence*, 11(7):674–693, 1989.

M. K. Mandal, T. Aboulnasr, and S. Panchanathan. Fast wavelet histogram techniques for image indexing. *Computer Vision and Image Understanding: CVIU*, 75(1–2):99–110, 1999.

Y. Matias, J. S. Vitter, and M. Wang. Wavelet-based histograms for selectivity estimation. In *ACM SIGMOD*, pages 448–459. ACM Press, 1998.

Y. Matias, J. S. Vitter, and M. Wang. Dynamic maintenance of wavelet-based histograms. In *Proceedings of 26th International Conference on Very Large Data Bases*, pages 101–110, 2000.

A. Mojsilovic and M. V. Popovic. Wavelet image extension for analysis and classification of infarcted myocardial tissue. *IEEE Transactions on Biomedical Engineering*, 44(9):856–866, 1997.

A. Natsev, R. Rastogi, and K. Shim. Walrus:a similarity retrieval algorithm for image databases. In *Proceedings of ACM SIGMOD International Conference on Management of Data*, pages 395–406. ACM Press, 1999.

R. Polikar. The wavelet tutorial. Internet Resources:http://engineering.rowan.edu/ polikar/WAVELETS/WTtutorial.html.

V. Ribeiro, R. Riedi, M. Crouse, and R. Baraniuk. Simulation of non-gaussian long-range-dependent traffic using wavelets. In *Proc. ACM SIGMETRICS'99*, pages 1–12, 1999.

C. Shahabi, S. Chung, M. Safar, and G. Hajj. 2d TSA-tree: A wavelet-based approach to improve the efficiency of multi-level spatial Data Mining. In *Statistical and Scientific Database Management*, pages 59–68, 2001.

C. Shahabi, X. Tian, and W. Zhao. TSA-tree: A wavelet-based approach to improve the efficiency of multi-level surprise and trend queries on time-series data. In *Statistical and Scientific Database Management*, pages 55–68, 2000.

G. Sheikholeslami, S. Chatterjee, and A. Zhang. WaveCluster: A multi-resolution clustering approach for very large spatial databases. In *Proc. 24th Int. Conf. Very Large Data Bases, VLDB*, pages 428–439, 1998.

E. J. Stonllnitz, T. D. DeRose, and D. H. Salesin. *Wavelets for computer graphics, theory and applications*. Morgan Kaufman Publishers, San Francisco, CA, USA, 1996.

Z. R. Struzik and A. Siebes. The haar wavelet transform in the time series similarity paradigm. In *Proceedings of PKDD'99*, pages 12–22, 1999.

S. R. Subramanya and A. Youssef. Wavelet-based indexing of audio data in audio/multimedia databases. In *IW-MMDBMS*, pages 46–53, 1998.

G. Tzanetakis and P. Cook. Musical genre classification of audio signals. *IEEE Transactions on Speech and Audio Processing*, 10(5):293–302, July 2002.

J. S. Vitter and M. Wang. Approximate computation of multidimensional aggregates of sparse data using wavelets. In *Proceedings of the 1999 ACM SIGMOD International Conference on Management of Data*, pages 193–204, 1999.

J. Z. Wang, G. Wiederhold, and O. Firschein. System for screening objectionable images using daubechies' wavelets and color histograms. In *Interactive Distributed Multimedia Systems and Telecommunication Services*, pages 20–30, 1997.

J. Z. Wang, G. Wiederhold, O. Firschein, and S. X. Wei. Content-based image indexing and searching using daubechies' wavelets. *International Journal on Digital Libraries*, 1(4):311–328, 1997.

Y.-L. Wu, D. Agrawal, and A. E. Abbadi. A comparison of DFT and DWT based similarity search in time-series databases. In *CIKM*, pages 488–495, 2000.

28

Fractal Mining - Self Similarity-based Clustering and its Applications

Daniel Barbara[1] and Ping Chen[2]

[1] George Mason University
Fairfax, VA 22030
dbarbara@gmu.edu

[2] University of Houston-Downtown
Houston, TX 77002
chenp@uhd.edu

Summary. Self-similarity is the property of being invariant with respect to the scale used to look at the data set. Self-similarity can be measured using the fractal dimension. Fractal dimension is an important charactaristics for many complex systems and can serve as a powerful representation technique. In this chapter, we present a new clustering algorithm, based on self-similarity properties of the data sets, and also its applications to other fields in Data Mining, such as projected clustering and trend analysis. Clustering is a widely used knowledge discovery technique. The new algorithm which we call Fractal Clustering (FC) places points incrementally in the cluster for which the change in the fractal dimension after adding the point is the least. This is a very natural way of clustering points, since points in the same clusterhave a great degree of self-similarity among them (and much less self-similarity with respect to points in other clusters). FC requires one scan of the data, is suspendable at will, providing the best answer possible at that point, and is incremental. We show via experiments that FC effectively deals with large data sets, high-dimensionality and noise and is capable of recognizing clusters of arbitrary shape.

Key words: self-similarity, clustering, projected clustering, trend analysis

28.1 Introduction

Clustering is one of the most widely used techniques in Data Mining. It is used to reveal structure in data that can be extremely useful to the analyst. The problem of clustering is to partition a data set consisting of n points embedded in a d-dimensional space into k sets or clusters, in such a way that the data points within a cluster are more similar among them than to data points in other clusters. A precise definition of clusters does not exist. Rather, a set of functional definitions have been adopted. A cluster has been defined (Backer, 1995) as a set of entities which are alike (and different from entities in other clusters), an aggregation of points such that the distance between any point in the cluster is less than the distance to points in other clusters, and as a connected region with a relatively high density of points. Our method

O. Maimon, L. Rokach (eds.), *Data Mining and Knowledge Discovery Handbook*, 2nd ed.,
DOI 10.1007/978-0-387-09823-4_28, © Springer Science+Business Media, LLC 2010

adopts the first definition (likeness of points) and uses a fractal property to define similarity between points.

The area of clustering has received an enormous attention as of late in the database community. The latest techniques try to address pitfalls in the traditional clustering algorithms (for a good coverage of traditional algorithms see (Jain and Dubes, 1988)). These pitfalls range from the fact that traditional algorithms favor clusters with spherical shapes (as in the case of the clustering techniques that use centroid-based approaches), are very sensitive to outliers (as in the case of all-points approach to clustering, where all the points within a cluster are used as representative of the cluster), or are not scalable to large data sets (as is the case with all traditional approaches).

New approaches need to satisfy the Data Mining desiderata (Bradley *et al.*, 1998):

- Require at most one scan of the data.
- Have on-line behavior: provide the best answer possible at any given time and be suspendable at will.
- Be incremental by incorporating additional data efficiently.

In this chapter we present a clustering algorithm that follows this desiderata, while providing a very natural way of defining clusters that is not restricted to spherical shapes (or any other type of shape). This algorithm is based on self-similarity (namely, a property exhibited by self-similar data sets, i.e., the fractal dimension) and clusters points in such a way that data points in the same cluster are more *self-affine* among themselves than to points in other clusters.

This chapter is organized as follows. Section 28.2 offers a brief introduction to the fractal concepts we need to explain the algorithm. Section 28.3 describes our clustering technique and experimental results. Section 28.4 discusses its application on projected clustering, and section 28.5 shows its application on trend analysis. Finally, Section 28.6 offers conclusions and future work.

28.2 Fractal Dimension

Nature is filled with examples of phenomena that exhibit seemingly chaotic behavior, such as air turbulence, forest fires and the like. However, under this behavior it is almost always possible to find *self-similarity*, i.e. an invariance with respect to the scale used. The structures that exhibit self-similarity over every scale are known as *fractals* (Mandelbrot). On the other hand, many data sets, that are not fractal, exhibit self-similarity over a range of scales.

Fractals have been used in numerous disciplines (for a good coverage of the topic of fractals and their applications see (Schroeder, 1991)). In the database area, fractals have been successfully used to analyze R-trees (Faloutsos and Kamel, 1997), Quadtrees (Faloutsos and Gaede, 1996), model distributions of data (Faloutsos *et al.*, 1996) and selectivity estimation (Belussi and Faloutsos, 1995).

Self-similarity can be measured using the *fractal dimension*. Loosely speaking, the fractal dimension measures the number of dimensions "filled" by the object represented by the data set. In truth, there exists an infinite family of fractal dimensions. By embedding the data set in an n-dimensional grid which cells have sides of size r, we can count the frequency with which data points fall into the i-th cell, p_i, and compute D_q, the generalized fractal dimension (Grassberger, 1983, Grassberger and Procaccia, 1983), as shown in Equation 63.1.

$$D_q = \begin{cases} \frac{\partial \log \sum_i p_i \log p_i}{\partial \log r} & \text{for } q = 1 \\ \frac{1}{q-1} \frac{\partial \log \sum_i p_i^q}{\partial \log r} & \text{otherwise} \end{cases} \tag{28.1}$$

Among the dimensions described by Equation 63.1, the *Hausdorff fractal dimension* ($q = 0$), the *Information Dimension* ($\lim_{q \to 1} D_q$), and the *Correlation dimension* ($q = 2$) are widely used. The Information and Correlation dimensions are particularly useful for Data Mining, since the numerator of D_1 is Shannon's entropy, and D_2 measures the probability that two points chosen at random will be within a certain distance of each other. Changes in the Information dimension mean changes in the entropy and therefore point to changes in trends. Equally, changes in the Correlation dimension mean changes in the distribution of points in the data set.

The traditional way to compute fractal dimensions is by means of the box-counting plot. For a set of N points, each of D dimensions, one divides the space in grid cells of size r (hypercubes of dimension D). If $N(r)$ is the number of cells occupied by points in the data set, the plot of $N(r)$ versus r in log-log scales is called the *box-counting plot*. The negative value of the slope of that plot corresponds to the Hausdorff fractal dimension D_0. Similar procedures are followed to compute other dimensions, as described in (Liebovitch and Toth, 1989).

To clarify the concept of box-counting, let us consider the famous example of George Cantor's dust, constructed in the following manner. Starting with the closed unit interval [0,1] (a straight-line segment of length 1), we erase the open middle third interval $(\frac{1}{3}, \frac{2}{3})$ and repeat the process on the remaining two segments, recursively. Figure 28.1 illustrates the procedure. The "dust" has a length measure of zero and yet contains an uncountable number of points. The Hausdorff dimension can be computed the following way: it is easy to see that for the set obtained after n iterations, we are left with $N = 2^n$ pieces, each of length $r = (\frac{1}{3})^n$. So, using a unidimensional box size with $r = (\frac{1}{3})^n$, we find 2^n of the boxes populated with points. If, instead, we use a box size twice as big, i.e., $r = 2(\frac{1}{3})^n$, we get 2^{n-1} populated boxes and so on. The log-log plot of box population vs. r renders a line with slope $D_0 = -log2/log3 = -0.63...$. The value 0.63 is precisely the fractal dimension of the Cantor's dust data set.

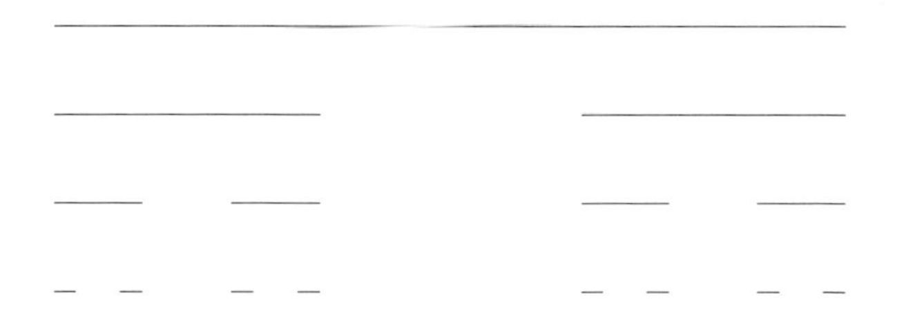

Fig. 28.1. The construction of the Cantor dust. The final set has fractal (Hausdorff) dimension 0.63.

In what follows of this section we present a motivating example that illustrates how the fractal dimension can be a powerful way for driving a clustering algorithm. Figure 28.2 shows the effect of superimposing two different Cantor dust sets. After erasing the open middle interval which results of dividing the original line in three intervals, the left-most interval gets divided in 9 intervals, and only the alternative ones survive (5 in total).

The rightmost interval gets divided in three, as before, erasing the open middle interval. The result is that if one considers grid cells of size $\frac{1}{3\times 9^n}$ at the n-th iteration, the number of occupied cells turns out to be $5^n + 6^n$. The slope of the log-log plot for this set is $D_0^l = lim_{n\to\infty}(log(5^n+6^n))/log(3\times 9^n)$. It is easy to show that $D_0^l > D_0^r$, where $D_0^r = log2/log3$ is the fractal dimension of the rightmost part of the data set (the Cantor dust of Figure 28.1). Therefore, one could say that the inclusion of the leftmost part of the data set produces a change in the fractal dimension and this subset is therefore "anomalous" with respect to the rightmost subset (or vice-versa). From the clustering point of view, for a human being it is easy to recognize the two Cantor sets as two different clusters. And, in fact, an algorithm that exploits the fractal dimension (as the one presented in this paper) will indeed separate these two sets as different clusters. Any point in the right Cantor set would change the fractal dimension of the left Cantor set if included in the left cluster (and viceversa). This fact is exploited by our algorithm (as we shall explain later) to place the points accordingly.

Fig. 28.2. A "hybrid" Cantor dust set. The final set has fractal (Hausdorff) dimension larger than that of the the rightmost set (which is the Cantor dust set of Figure 28.1

To further motivate the algorithm, let us consider two of the clusters in Figure 28.7: the right-top ring and the left-bottom (square-like) ring. Figure 28.3 shows two log-log plots of number of occupied boxes against grid size. The first is obtained by using the points of the left-bottom ring (except one point). The slope of the plot (in its linear region) is equal to 1.57981, which is the fractal dimension of this object. The second plot, obtained by adding to the data set of points on the left-bottom ring the point (93.285928,71.373638) – which naturally corresponds to this cluster– almost coincides with the first plot, with a slope (in its linear part) of 1.57919. Figure 28.4 on the other hand, shows one plot obtained by the data set of points in the right-top ring, and another one obtained by adding to that data set the point (93.285928,71.373638). The first plot exhibits a slope in its linear portion of 1.08081 (the fractal dimension of the data set of points in the right-top ring); the second plot has a slope of 1.18069 (the fractal dimension after adding the above-mentioned point). While the change in the fractal dimension brought about the point (93.285928,71.373638) in the bottom-left cluster is 0.00062, the change in the right-top ring data set is 0.09988, more than 3 orders of magnitude bigger than the first change. Our algorithm would proceed to place point (93.285928, 71.373638) in the left-bottom ring, based on these changes.

Figures 28.3 and 28.4 also illustrate another important point. The "ring" used for the box counting algorithm is not a pure mathematical fractal set, as the Cantor Dust (Figure 28.1), or the Sierpinski Triangle (Mandelbrot) are. Yet, this data set exhibits a fractal dimension (or more precisely a linear behavior in the log-log box counting plot) through a (relatively) large range of grid sizes. This fact serves to illustrate the point that our algorithm does not depend on the clusters being "pure" fractals, but rather to have a measurable dimension (i.e., their box count plot has to exhibit linearity over a range of grid sizes). Since we base our definition of cluster in the self-similarity of points within the cluster, this is an easy constraint to meet.

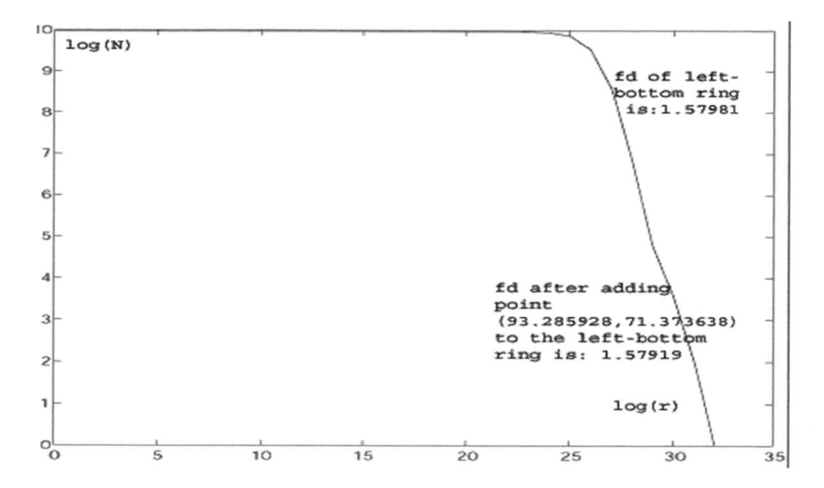

Fig. 28.3. The box-counting plots of the bottom-left ring data set of Figure 28.7, before and after the point (93.285928,71.373638) has been added to the data set. The difference in the slopes of the linear region of the plots is the "fractal impact" (0.00062). (The two plots are so similar that lie almost on top of each other.)

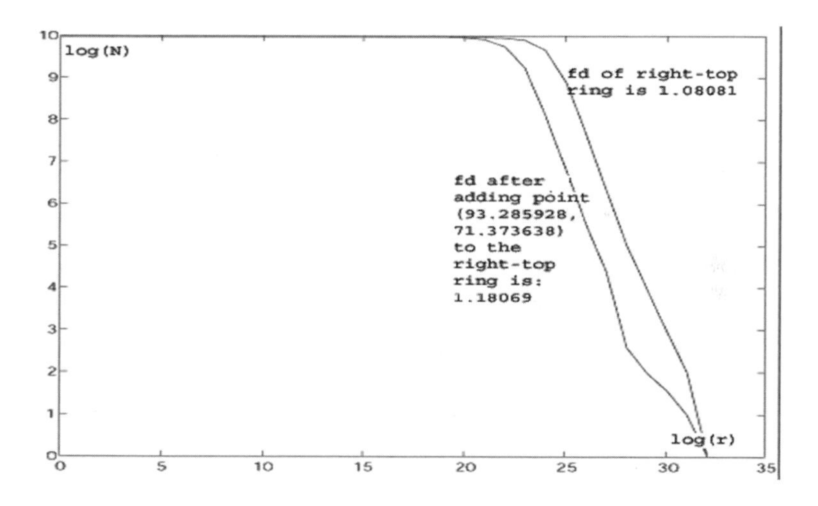

Fig. 28.4. TThe box-counting plots of the top-right ring data set of Figure 28.7, before and after the point (93.285928,71.373638) has been added to the data set. The difference in the slopes of the linear region of the plots is the "fractal impact" (0.09988), much bigger than the corresponding impact shown in Figure 28.3

28.3 Clustering Using the Fractal Dimension

Incremental clustering using the fractal dimension, abbreviated as Fractal Clustering, or FC, is a form of grid-based clustering (where the space is divided in cells by a grid; other techniques that use grid-based clustering are STING (Wang *et al.*, 1997), WaveCluster (Sheikholeslami *et al.*, 1998) and Hierarchical Grid Clustering (Schikuta, 1996)). The main idea behind FC is to group points in a cluster in such a way that none of the points in the cluster changes the cluster's fractal dimension radically. FC also combines connectness, closeness and data points position information to pursue high clustering quality.

Our algorithm takes a first step of initializing a set of clusters, and then, incrementally adds points to that set. In what follows, we describe the initialization and incremental steps.

28.3.1 FC Initialization Step

In clustering algorithms the quality of initial clusters is extremely important, and has direct effect on the final clustering quality. Obviously, before we can apply the main concept of our technique, i.e., adding points incrementally to existing clusters, based on how they affect the clusters' fractal dimension, some initial clusters are needed. In other words, we need to "bootstrap" our algorithm via an initialization procedure that finds a set of clusters, each with sufficient points so its fractal dimension can be computed. If the wrong decisions are made at this step, we will be able to correct them later by reshaping the clusters dynamically.

Initialization Algorithm

The process of initialization is made easy by the fact that we are able to convert a problem of clustering a set of multidimensional data points (which is a subset of the original data set) into a much simpler problem of clustering 1-dimensional points. The problem is further simplified by the fact that the set of data points that we use for the initialization step fits in memory. Figure 28.3.1 shows the pseudo-code of the initialization step. Notice that lines 3 and 4 of the code map the points of the initial set into unidimensional values, by computing the effect that each point has in the fractal dimension of the rest of the set (we could have computed the difference between the fractal dimension of S and that of S minus a point, but the result would have been the same). Line 5 of the code deserves further explanation: in order to cluster the set of Fd_i values, we can use any known algorithm. For instance, we could feed the fractal dimension values Fd_i, and a value k to a K-means implementation (Selim and Ismail, 1984, Fukunaga, 1990). Alternatively, we can let a hierarchical clustering algorithm (e.g., CURE (Guha *et al.*, 1998)) cluster the sequence of Fd_i values.

Although, in principle, any of the dimensions in the family described by Equation 63.1 can be used in line 4 of the initialization step, we have found that the best results are achieved by using D_2, i.e., the correlation dimension.

28.3.2 Incremental Step

After we get the initial clusters, we can proceed to cluster the rest of the data set. Each cluster found by the initialization step is represented by a set of boxes (cells in a grid). Each box in the set records its population of points. Let k be the number of clusters found in the initialization step, and $C = \{C_1, C_2, \ldots, C_k\}$ where C_i is the set of boxes that represent cluster i. Let $F_d(C_i)$ be the fractal dimension of cluster i.

1: Given an initial set S of points $\{p_1, \cdots, p_M\}$ that fit in main memory (obtained by sampling the data set).
2: **for** $i = 1, \cdots, M$ **do**
3: Define group $G_i = S - \{p_i\}$
4: Calculate the fractal dimension of the set G_i, Fd_i.
5: **end for**
6: Cluster the set of Fd_i values,(The resulting clusters are the initial clusters.)

Fig. 28.5. Initialization Algorithm for FC.

The incremental step brings a new set of points to main memory and proceeds to take each point and add it to each cluster, computing its new fractal dimension. The pseudo-code of this step is shown in Figure 28.6. Line 5 computes the fractal dimension for each modified cluster (adding the point to it). Line 6 finds the proper cluster to place the point (the one for which the change in fractal dimension is minimal). We call the value $|F_d(C_i' - F_d(C_i)|$ the *Fractal Impact* of the point being clustered over cluster i. The quantity $min_i|F_d(C_i' - F_d(C_i)|$ is the *Minimum Fractal Impact* of the point. Line 7 is used to discriminate "noise." If the Minimum Fractal Impact of the point is bigger than a threshold τ, then the point is simply rejected as noise (Line 8). Otherwise, it is included in that cluster. We choose to use the Hausdorff dimension, D_0, for the fractal dimension computation of Line 5 in the incremental step. We chose D_0 since it can be computed faster than the other dimensions and it proves robust enough for the task.

1: Given a batch S of points brought to main memory:
2: **for** each point $p \in S$ **do**
3: **for** $i = 1, \cdots, k$ **do**
4: Let $C_i' = C_i \bigcup \{p\}$
5: Compute $F_d(C_i')$
6: Find $\hat{i} = min_i(|F_d(C_i' - F_d(C_i)|)$
7: **if** $|F_d(C_{\hat{i}}') - F_d(C_{\hat{i}})| > \tau$ **then**
8: Discard p as noise
9: **else**
10: place p in cluster $C_{\hat{i}}$
11: **end if**
12: **end for**
13: **end for**

Fig. 28.6. The Incremental Step for FC.

To compute the fractal dimension of the clusters every time a new point is added to them, we keep the cluster information using a series of grid representations, or layers. In each layer, boxes (i.e., grids) have a size that is smaller than in the previous layer. The sizes of the boxes are computed in the following way. For the first layer (largest boxes), we divide the cardinality of each dimension in the data set by 2, for the next layer, we divide the cardinality of each dimension by 4 and so on. Accordingly, we get $2^D, 2^{2D}, \cdots, 2^{LD}$ D-dimensional boxes in each layer, where D is the dimensionality of the data set, and L the maximum layer we will store. Then, the information kept is not the actual location of points in the boxes, but rather, the number of points in each box. It is important to remark that the number of boxes in layer L

can grow considerably, specially for high-dimensionality data sets. However, we need only to save boxes for which there is any population of points, i.e., empty boxes are not needed. The number of populated boxes at that level is, in practical data sets, considerably smaller (that is precisely why clusters are formed, in the first place). Let us denote by B the number of populated boxes in level L. Notice that, B is likely to remain very stable throughout passes over the incremental step.

Every time a point is assigned to a cluster, we register that fact in a table, adding a row that maps the cluster membership to the point identifier (rows of this table are periodically saved to disk, each cluster into a file, freeing the space for new rows). The array of layers is used to drive the computation of the fractal dimension of the cluster, using a box-counting algorithm. In particular, we chose to use FD3 (Sarraille and DiFalco, 2004), an implementation of a box counting algorithm based on the ideas described in (Liebovitch and Toth, 1989).

28.3.3 Reshaping Clusters in Mid-Flight

It is possible that the number and form of the clusters may change after having processed a set of data points using the step of Figure 28.6. This may occur because the data used in the initialization step does not accurately reflect the true distribution of the overall data set or because we are clustering an incoming stream of data, whose distribution changes over time. There are two basic operations that can be performed: splitting a cluster and merging two or more clusters into one.

A good indication that a cluster may need to be split is given by how much the fractal dimension of the cluster has changed since its inception during the initialization step. (This information is easy to keep and does not occupy much space.) A large change may indicate that the points inside the cluster do not belong together. (Notice that these points were included in that cluster because it was the *best choice* at the time, i.e., it was the cluster for which the points caused the least amount of change on the fractal dimension; but this does not mean this cluster is an ideal choice for the points.)

Once the decision of splitting a cluster has been made, the actual procedure is simple. Using the box (finest resolution layer, i.e., the lst layer of boxes) population we can run the initialization step. That will define how many clusters (if more than one) are needed to represent the set of points. Notice that up to that point, there is no need to re-process the actual points that compose the splitting cluster (i.e., no need to bring them to memory). This is true since the initialization step can be run over the box descriptions directly (the box populations represent an approximation of the real set of points, but this approximation is good enough for the purpose). On the other hand, after the new set of clusters has been decided upon, we need to relabel the points and a pass over that portion of the data set is needed (we assume that the points belonging to the splitting cluster can be retrieved from disk without looking at the entire data set: this can be easily accomplish by keeping each cluster in a separate file).

Merging clusters is even simpler. As an indication of the need to merge two clusters, we keep the minimum distance between clusters, defined by the distance between two points P_1 and P_2, such that P_1 belongs to the first cluster and P_2 to the second, and P_1 and P_2 are the closest pair of such points. When this minimum distance is smaller than a threshold, it is time to consider merging the two clusters. The threshold used is the minimum of the $\kappa = \kappa_0 \times \hat{d}$ for each of the two clusters. (Recall that $\hat{d}$ is the average pairwise distance in the cluster.) The merging can be done by using box population at the highest level of resolution (smallest box size), for all the clusters that are deemed as too close. To actually decide whether the clusters ought to be merged or not, we perform the initialization algorithm 2, using the center of the populated boxes (at the highest resolution layer) as "points." Notice that it is not necessary to bring previously examined points back to memory, since the relabeling can simply be done by equating the labels of the merged clusters at the end. In this sense, merging does not affect the "one-pass" property of fractal clustering (as splitting does, although only for the points belonging to the splitting cluster).

28.3.4 Complexity of the Algorithm

We assume that the cost of computing the fractal dimension of a set of n points is $O(n\log(n))$, as it is the case for the software (FD3 (Sarraille and DiFalco, 2004)) that we have chosen for our experiments.

For the initialization algorithm, the complexity is $O(M^2\log(M))$, where M is the size of the sample of points. This follows from the fact that for each point in the sample, we need to compute the fractal dimension of the rest of the sample set (minus the point), incurring a cost of $O(M\log(M))$ per point. The incremental step is executed $O(N)$ times, where N is the size of the data set. The complexity of the incremental step is $O(n\log(n))$ where n is the number of points involved in the computation of the fractal dimension. Now, since we do not use the point information, but rather the box population to drive the computation of the fractal dimension, we can claim that n is $O(B)$ (the number of populated boxes in the highest layer). Now, since $B << N$, it follows that the incremental part of FC will take time linear with respect to the size of the data set.

For small data sets, the first initialization algorithm time becomes dominant in FC. However, for large data sets, i.e., when $M << N$, the cost of the incremental step dominates, making FC linear in the size of the data set.

28.3.5 Confidence Bounds

One question we need to settle is how to determine if we are placing points as outliers correctly. A point is deemed an outlier in the test of Line 7, in Figure 28.6, when the Minimum Fractal Impact of the point exceeds a threshold τ. To add confidence to the stability of the clusters that are defined by this step, we can use the Chernoff bound (Chernoff, 1952) and the concept of adaptive sampling (Lipton *et al.*, 1993, Lipton and Naughton, 1995, Domingo *et al.*, 1998, Domingo *et al.*, 2000, Domingos and Hulten, 2000), to find the minimum number of points that must be successfully clustered after the initialization algorithm in order to guarantee with a high probability that our clustering decisions are correct. We present these bounds in this section.

Consider the situation immediately after the initial clusters have been found, and we start clustering points using FC. Let us define a random variable X_i, whose value is 1 if the i-th point to be clustered by FC has a Minimum Fractal Impact which is less than τ, and 0 otherwise. Using Chernoff's inequality one can bound the expectation of the sum of the X_i's, $X = \sum_i^n X_i$, which is another random variable whose expected value is np, where $p = Pr[X_i = 1$, and n is the number of points clustered. The bound is shown in Equation 28.2, where ε is a small constant.

$$Pr[\,X/n > (1+\varepsilon)p\,] \leq exp(-pn\varepsilon^2/3) \tag{28.2}$$

Notice that we really do not know p, but rather have an estimated value of it, namely $\hat{p}$, given by the number of times that X_i is 1 divided by n. (I.e., the number of times we can successfully cluster a point divided by the total number of times we try.) In order that the estimated value of p, $\hat{p}$ obeys Equation 28.3, which bounds the estimate close to the real value with an arbitrarily large probability (controlled by δ), one needs to use a sample of n points, with n satisfying the inequality shown in Equation 28.4.

$$Pr[\ \|\hat{p}-p\|\ ] > 1-\delta \tag{28.3}$$

$$n > \frac{3}{p\varepsilon^2} ln(\frac{2}{\delta}) \tag{28.4}$$

By using adaptive sampling, one can keep bringing points to cluster until obtaining at least a number of successful events (points whose minimum fractal impact is less than τ) equal to s. It can be proven that in adaptive sampling (Watanabe, 2000), one needs to have s bound by the inequality shown in Equation 28.5, in order for Equation 28.3 to hold. Moreover, with probability greater than $1-\delta/2$, the sample size (number of points processed) n, would be bound by the inequality of Equation 28.6. (Notice that the bound of Equation 28.6 and that of Equation 28.4 are very close; The difference is that the bound of Equation 28.6 is achieved without *knowing* p in advance.)

$$s > \frac{3(1+\varepsilon)}{\varepsilon^2} ln(\frac{2}{\delta}) \tag{28.5}$$

$$n \leq \frac{3(1+\varepsilon)}{(1-\varepsilon)\varepsilon^2 p} ln(\frac{2}{\delta}) \tag{28.6}$$

Therefore, after seeing s positive results, while processing n points where n is bounded by Equation 28.6 one can be confident that the clusters will be stable and the probability of successfully clustering a point is the expected value of the random variable X divided by n (the total number of points that we attempted to cluster).

28.3.6 Memory Management

Our algorithm is very space-efficient, by the virtue of requiring memory just to hold the boxes population at any given time during its execution. This fact makes FC scale very well with the size of the set. Notice that if the initialization sample is a good representative of the rest of the data, the initial clusters are going to remain intact (just containing large populations in the boxes). In that case, the memory used during the entire clustering task remains stable.

However, there are cases in which we will have demands beyond the available memory. Mainly, there are two cases where this can happen. If the sample is not a good representative (or the data changes with time in an incoming stream) we will be forced to change the number and structure of the clusters (as explained in Section 28.3.3), possibly requiring more space. The other case arises when we deal with high dimensional sets, where the number of boxes needed to describe the space may exceed the available memory.

For these cases, we have devised a series of memory reduction techniques that aim to achieve reasonable trade-offs between the memory used and the performance of the algorithm, both in terms of its running time and the quality of the uncovered clusters.

Memory Reduction Technique 1:

In this technique, we cache boxes in memory, while keeping others swapped out to the disk, replacing the ones in memory on demand. Our experience shows that the boxes of smallest size consume 75% of all memory. So, we share the cache only amongst the smallest boxes, keeping the other layers always in memory. Of course, we cluster the boxes in pages, and use the pages as a caching unit. This reduction technique affects the running time but not the clustering quality.

Memory Reduction Technique 2:

A way of requiring less memory is to ignore boxes with very few points. While this method can, in principle, affect the quality of clusters, it may actually be a good way to eliminate noise from the data set.

28.3.7 Experimental Results

In this section we will show the results of using FC to cluster a series of data sets. Each data set aims to test how well FC does in each of the issues we have discussed in the Section 28.3. For each one of the experiments we have used a value of $\tau = 0.03$ (the threshold used to decide if a point is noise or it really belongs to a cluster). We performed the experiments in a Sun Ultra2 with 500 Mb. of RAM, running Solaris 2.5. When using the first initialization algorithm, we have used K-means to cluster the unidimensional vector of effects. In each of the experiments, the points are distributed equally among the clusters (i.e., each cluster has the same number of points). After we run FC, for each cluster found, we count the number of points that were placed in that cluster and that also belonged there. The accuracy of FC is then measured for each cluster as the percentage of points correctly placed there. (We know, for each data set, the membership of each point; in one of the data sets we spread the space with outliers: in that case, the outliers are considered as belonging to an extra "cluster.")

Scalability

In this subsection we show experimental results of running time and cluster quality using a range of data sets of increasing sizes and a high-dimensional data set.

First, we use data sets whose distribution follows the one shown in Figure 28.7 for scalability experiments. We use a complex set of clusters in this experiment, in order to show how FC can deal with arbitrarily-shaped clusters. (Not only do we have a square-shaped cluster, but also one of the clusters resides inside of another one.) We vary the total number of points in the data set to measure the performance of our clustering algorithm. In every case, we pick a sample of 600 points to run the initialization step. The results are summarized in Table 28.1.

Experiment on a Real Dataset

We performed an experiment using our fractal clustering algorithm to cluster points in a real data set. The data set used was a picture of the world map in black and white (see Figure 28.8), where the black pixels represent land and the white pixels water. The data set contains 3,319,530 pixels or points. With the second initialization algorithm the running time was 589 sec The quality of the clusters is extremely good, totally five clusters were found. Cluster 0 spans the European, Asian and African continents (these continents are very close, so the algorithm did not separate them and we did not run the split technique for the cluster), Cluster 1 corresponds to the North American continent, Cluster 2 corresponds to the South American continent; Cluster 3 corresponds to Australia, and finally Cluster 4 shows Antarctica.

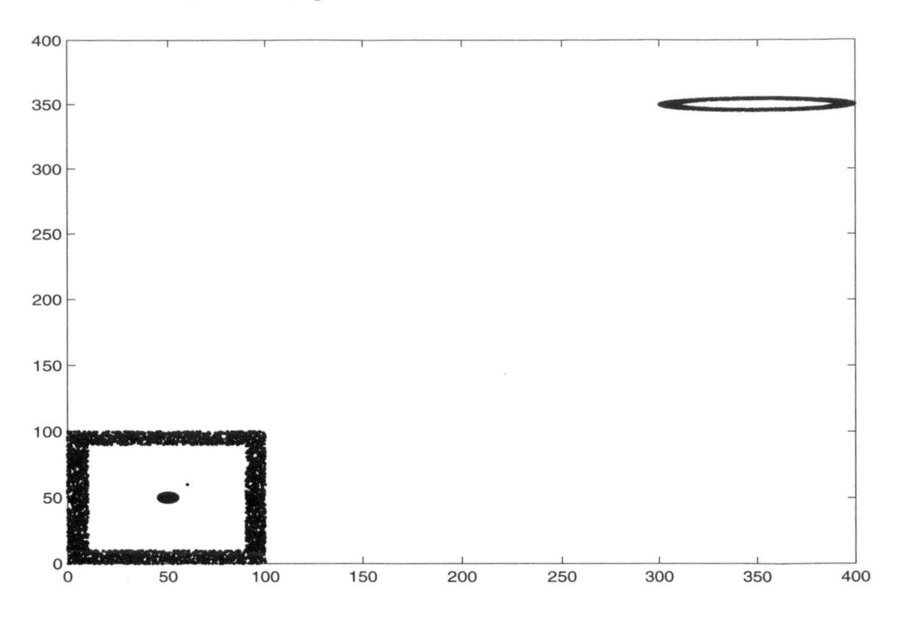

Fig. 28.7. Three-cluster Dataset for Scalability Experiments.

Table 28.1. Results of using FC in a data set (of several sizes) whose composition is shown in Figure 28.7. The table shows the data set size (*N*), the running time for FC (time), memory used is 64KB, and the composition for each cluster found (column C) in terms of points assigned to the cluster (points in cluster) and their provenance, i.e., whether they actually belong to cluster1, cluster2 or cluster3. Finally, the accuracy column shows the percentage of points that were correctly put in each cluster.

			Assigned	Coming from			accuracy
N	time	C	to	cluster1	cluster2	cluster3	%
30K	12s.	1	10,326	9,972	0	354	99.72
		2	11,751	0	10,000	1,751	100
		3	7,923	28	0	7,895	78.95
300K	56s.	1	103,331	99,868	0	3,463	99.86
		2	117,297	0	100,000	17,297	100
		3	79,372	132	0	79,240	79.24
3M	485s.	1	1,033,795	998,632	0	35,163	99.86
		2	1,172,895	0	999,999	173,896	99.99
		3	793,310	1,368	0	791,942	79.19
30M	4,987s.	1	10,335,024	9,986,110	22	348,897	99.86
		2	11,722,887	0	9,999,970	1,722,917	99.99
		3	7,942,084	13,890	8	7,928,186	79.28

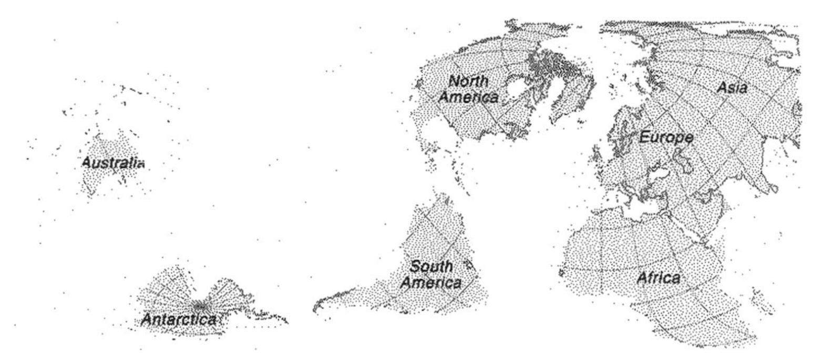

Fig. 28.8. A World Map Picture as a Real Dataset.

28.4 Projected Fractal Clustering

Fractal clustering is a grid-based clustering algorithm, whose memory usage is increased exponentially with the number of dimensions. Although we develop some memory reduction techniques, fractal clustering can not work on a dataset with hundreds of dimensions. To make fractal clustering useful on a very high dimensional dataset we develop a new algorithm called projected fractal clustering (PFC).

Figure 28.9 shows our projected fractal clustering algorithm. First we sample the dataset, run the initialization algorithm on the sample set, and get initial clusters. Then compute the fractal dimension for each cluster. After running SVD on each cluster we get an "importance" index of dimension for each cluster. We prune off unimportant dimensions for each cluster according to its fractal dimension, and only use the remaining dimensions for the following incremental clustering step. In the incremental step we perform fractal clustering and get all clusters in the end.

1: sample the original dataset D and get a sample set S
2: run FC initialization algorithm shown above on S and get initial clusters C_i(i=1,k, k is the number clusters found)
3: compute C_i's fractal dimension f_i
4: run SVD analysis on C_i, and keep only n_i dimensions of C_i(n_i is decided by f_i), prune off unimportant dimensions, these n_i dimensions of C_i is stored in FD_i
5: **for all** points in D **do**
6:    input a point p
7:    **for** i=1,k **do**
8:       prune p according to FD_i, put p into C_i
9:       compute C_i's fractal dimension change fdc_i
10:    **end for**
11:    compare fdc_i(i=1,k), put p into C_i with the smallest fdc_i
12: **end for**

Fig. 28.9. Projected Fractal Clustering Algorithm.

28.5 Tracking Clusters

Organizations today accumulate data at a astonishing rate. This fact brings new challenges for Data Mining. For instance, finding out when patterns change in the data opens the possibility of making better decisions and discovering new interesting facts. The challenge is to design algorithms that can track changes in an incremental way and without making growing demands on memory.

In this section we present a technique to track changes in cluster models. Clustering is a widely used technique that helps uncovering structures in data that were previously not known. Our technique helps in discovering the points in the data stream in which the cluster structure is changing drastically from the current structure. Finding changes in clusters as new data is collected can prove fruitful in scenarios like the following:

- Tracking the evolution of the spread of illnesses. As new cases are reported, finding out how clusters evolve can prove crucial in identifying sources responsible for the spread of the illness.
- Tracking the evolution of workload in an e-commerce server (clustering has already been successfully used to characterize e-commerce workloads (Menascé *et al.*, 1999)), which can help in dynamically fine tune the server to obtain better performance.
- Tracking meteorological data, such as temperatures registered throughout a region, by observing how clusters of spatial-meteorological points evolve in time.

Our idea is to track the number of outliers that the next batch of points produce with respect to the current clusters, and with the help of analytical bounds decide if we are in the presence of data that does not follow the patterns (clusters) found so far. If that is the case, we proceed to re-cluster the points to find the new model.

As we get a new batch of points to be clustered, we can ask ourselves if these points can be adequately clustered using the models we have so far. The key to answer this question is to count the number of outliers in this batch of points. A point is deemed an outlier in the test of Line 7, in Figure 28.6, when the MFI of the point exceeds a threshold τ. We can use the Chernoff bound (Chernoff, 1952) and the concept of adaptive sampling (Lipton *et al.*, 1993, Lipton and Naughton, 1995, Domingo *et al.*, 1998, Domingo *et al.*, 2000, Domingos and Hulten, 2000), to find the minimum number of points that must be successfully clustered after the initialization algorithm in order to guarantee with a high probability that our clustering decisions are correct.

These bounds can be used to drive our tracking algorithm *Tracking*, described in Figure 28.10. Essentially, the algorithm takes n new points (where n is given by the lower bound of Equation 28.6) and checks how many of them can be successfully clustered by FC, using the current set of clusters. (Recall that if a point has a MFI bigger than τ, it is deemed an outlier.) If after attempting to cluster the n points, one finds too many outliers (tested in Line 9, by comparing the successful count r, with the computed bound s, given by Equation 28.5), then we call this a *turning point* and proceed to redefine the clusters. This is done by throwing away all the information of the previous clusters and clustering the n points of the current batch. Notice that after each iteration, the value of p is re-estimated as the ratio of successfully clustered points divided by the total number of points tried.

28.5.1 Experiment on a Real Dataset

We describe in this section the result of two experiments using our *tracking* algorithm. We performed the experiments in a Sun Ultra2 with 500 Mb. of RAM, running Solaris 2.5.

1: Initialize the count of successfully clustered points, i.e., $r = 0$
2: Given a batch S of n points, where n is computed as the lower bound of Equation 28.6, using the estimated p from the previous round of points
3: **for** each point in S **do**
4: Use FC to cluster the point.
5: **if** the point is not an outlier **then**
6: Increase the count of successfully clustered points, i.e., $r = r + 1$
7: **end if**
8: **end for**
9: Compute s as the lower bound of Equation 28.5
10: **if** $r < s$ **then**
11: flag this batch of points S as a turning point and use S to find the new clusters.
12: **else**
13: re-estimate $p = r/n$
14: **end if**

Fig. 28.10. Algorithm to Track Cluster Changes.

The experiment used data from the U.S. Historical Climatology Network (CDIA, 2004), which contains (among other types of data) data sets with the average temperature per month, for several years measured in many meteorological stations throughout the United States. We chose the data for the years 1990 to 1994 for the state of Virginia for this experiment (the data comes from 19 stations throughout the state). We organized the data as follows. First we feed the algorithm with the data of the month of January for all the years 1990-1994, since (we were interested in finding how the average temperature changes throughout the months of the year, during those 5 years. Our clustering algorithm found initially a single cluster for points throughout the region in the month of January. This cluster contained 1,716 data points. Using $\delta = 0.15$, and $\varepsilon = 0.1$, and with the estimate of $p = 0.9$ (given by the number of initial points that were successfully clustered), we get a window $n = 1055$, and a value of s, the minimum number of points that need to be clustered successfully, of 855. (Which means that if we find more than 1055-855 = 200 outliers, we will declare the need to re-cluster.) We proceeded to feed the data corresponding to the next month (February for the years 1990-1994) in chunks of 1055 points, always finding less than 200 outliers per window. With the March data, we found a window with more than 2000 outliers and decided to re-cluster the data points (using only that window of data). After that, with the data corresponding to April, fed to the algorithm in chunks of n points (p stays roughly the same, so n and s remain stable at 1055 and 255, respectively) we did not find any window with more than 200 outliers. The next window that prompts re-clustering comes within the May data (for which we reclustered). After that, re-clustering became necessary for windows in the months of July, October and December. The τ used throughout the algorithm was 0.001. The total running time was 1 second, and the total number of data points processed was 20,000.

28.6 Conclusions

In this chapter we presented a new clustering algorithm based on the usage of the fractal dimension. This algorithm clusters points according to the effect they have on the fractal dimension of the clusters that have been found so far. The algorithm is, by design, incremental and its complexity is $O(N)$. Our experiments have proven that the algorithm has very desirable properties. It is resistant to noise, capable of finding clusters of arbitrary shape and capable of dealing with points of high dimensionality. Also We applied FC to projected clustering and tracking changes in cluster models for evolving data sets.

References

E. Backer. *Computer-Assisted Reasoning in Cluster Analysis*. Prentice Hall, 1995.

A. Belussi and C. Faloutsos. Estimating the Selectivity of Spatial Queries Using the 'Correlation' Fractal Dimension. In *Proceedings of the International Conference on Very Large Data Bases*, pages 299–310, September 1995.

P.S. Bradley, U. Fayyad, and C. Reina. Scaling Clustering Algorithms to Large Databases (Extended Abstract). In *Proceedings of the ACM SIGMOD Workshop on Research Issues in Data Mining and Knowledge Discovery*, June 1998.

CDIA. U.S. Historical Climatology Network Data. http://cdiac.esd.ornl.gov /epubs/ndp019/ ushcn_r3.html.

H. Chernoff. A Measure of Asymptotic Efficiency for Tests of a Hypothesis Based on the Sum of Observations. *Annals of Mathematical Statistics*, pages 493–509, 1952.

C. Domingo, R. Gavaldá, and O. Watanabe. Practical Algorithms for Online Selection. In *Proceedings of the first International Conference on Discovery Science*, 1998.

C. Domingo, R. Gavaldá, and O. Watanabe. Adaptive Sampling Algorithms for Scaling Up Knowledge Discovery Algorithms. In *Proceedings of the second International Conference on Discovery Science*, 2000.

P. Domingos and G. Hulten. Mining High-Speed Data Streams. In *Proceedings of the Sixth ACM-SIGKDD International Conference on Knowledge Discovery and Data Mining, Boston, MA*, 2000.

C. Faloutsos and V. Gaede. Analysis of the Z-ordering Method Using the hausdorff Fractal Dimension. In *Proceedings of the International Conference on Very Large Data Bases*, pages 40–50, September 1996.

C. Faloutsos and I. Kamel. Relaxing the Uniformity and Independence Assumptions, Using the Concept of Fractal Dimensions. *Journal of Computer and System Sciences*, 55(2):229–240, 1997.

C. Faloutsos, Y. Matias, and A. Silberschatz. Modeling Skewed Distributions Using Multifractals and the '80-20 law'. In *Proceedings of the International Conference on Very Large Data Bases*, pages 307–317, September 1996.

K. Fukunaga. *Introduction to Statistical Pattern Recognition*. Academic Press, San Diego, California, 1990.

P. Grassberger. Generalized Dimensions of Strange Attractors. *Physics Letters*, 97A:227–230, 1983.

P. Grassberger and I. Procaccia. Characterization of Strange Attractors. *Physical Review Letters*, 50(5):346–349, 1983.

S. Guha, R. Rastogi, and K. Shim. CURE: An Efficient Clustering Algorithm for Large Databases. In *Proceedings of the ACM SIGMOD Conference on Management of Data, Seattle, Washington*, pages 73–84, 1998.

A. Jain and R. C. Dubes. *Algorithms for Clustering Data*. Prentice Hall, Englewood Cliffs, New Jersey, 1988.

L.S. Liebovitch and T. Toth. A Fast Algorithm to Determine Fractal Dimensions by Box Countig. *Physics Letters*, 141A(8), 1989.

R.J. Lipton and J.F Naughton. Query Size Estimation by Adaptive Sampling. *Journal of Computer Systems Science*, pages 18–25, 1995.

R.J. Lipton, J.F. Naughton, D.A. Schneider, and S. Seshadri. Efficient Sampling Strategies for Relational Database Operations. *Theoretical Computer Science*, pages 195–226, 1993.

B.B. Mandelbrot. *The Fractal Geometry of Nature*. W.H. Freeman, New York, 1983.

D.A. Menascé, V.A. Almeida, R.C. Fonseca, and M.A. Mendes. A Methodology for Workload Characterizatoin for E-commerce Servers. In *Proceedings of the ACM Conference in Electronic Commerce, Denver, CO*, November 1999.

J. Sarraille and P. DiFalco. FD3. http://tori.postech.ac.kr/softwares/.

E. Schikuta. Grid clustering: An efficient hierarchical method for very large data sets. In *Proceedings of the 13th Conference on Pattern Recognition, IEEE Computer Society Press*, pages 101–105, 1996.

M. Schroeder. *Fractals, Chaos, Power Laws: Minutes from an Infinite Paradise*. W.H. Freeman, New York, 1991.

S.Z. Selim and M.A. Ismail. K-Means-Type Algorithms: A Generalized Convergence Theorem and Characterization of Local Optimality. *IEEE Transactions on Pattern Analysis and Machine Intelligence*, PAMI-6(1), 1984.

G. Sheikholeslami, S. Chatterjee, and A. Zhang. WaveCluster: A Multi-Resolution Clustering Approach for Very Large Spatial Databases. In *Proceedings of the 24th Very Large Data Bases Conference*, pages 428–439, 1998.

W. Wang, J. Yand, and R. Muntz. STING: A statistical information grid approach to spatial data mining. In *Proceedings of the 23rd Very Large Data Bases Conference*, pages 186–195, 1997.

O. Watanabe. Simple Sampling Techniques for Discovery Science. *IEICE Transactions on Information and Systems*, January 2000.

29

Visual Analysis of Sequences Using Fractal Geometry

Noa Ruschin Rimini and Oded Maimon

Department of Industrial Engineering, Tel-Aviv University, Tel-Aviv, Israel

Summary. Sequence analysis is a challenging task in the data mining arena, relevant for many practical domains. We propose a novel method for visual analysis and classification of sequences based on Iterated Function System (IFS). IFS is utilized to produce a fractal representation of sequences. The proposed method offers an effective tool for visual detection of sequence patterns influencing a target attribute, and requires no understanding of mathematical or statistical algorithms. Moreover, it enables to detect sequence patterns of any length, without predefining the sequence pattern length. It also enables to visually distinguish between different sequence patterns in cases of reoccurrence of categories within a sequence. Our proposed method provides another significant added value by enabling the visual detection of rare and missing sequences per target class.

29.1 Introduction

Mining sequential data is an important challenge relevant for many practical domains, such as analysis of the impact of operation sequence on product quality (See Da Cunha *et al.*, 2006, Rokach *et al.*, 2008 and Ruschin-Rimini *et al.*, 2009) analysis of customers purchase history for determining the next best offer, analysis of products failure history for the purpose of root cause analysis, security (Moskovitch *et al.*, 2008) and more.

This chapter presents a novel approach for detecting sequence patterns that influence a target attribute and therefore act as sequence classifiers. It extends existing methods by providing a visual application, enabling domain experts such as production engineers, sales and customer service managers, to visually detect sequence patterns that affect a target attribute.

Moreover, the proposed method overcomes limitations of existing methods such as the *n*-gram approach utilized by Da Cunha *et al.* (2006), by enabling the detection of sequence patterns of any length, without predefining the pattern length, and by enabling to visually distinguish between different sequence patterns, even in cases of reoccurrences of categories in a sequence. The proposed method provides another significant added value by enabling the visual detection of rare and missing sequences per target attribute value.

The proposed approach is based on Iterated Function System (IFS) for producing a Fractal representation of sequences.

In particular, we developed a unique software application for visual detection of sequence patterns. Our application comprises such features as color codes and zoom functions, in order

O. Maimon, L. Rokach (eds.), *Data Mining and Knowledge Discovery Handbook*, 2nd ed.,
DOI 10.1007/978-0-387-09823-4_29,

to facilitate the process of visual detection of sequence patterns affecting a target attribute. Visual data analysis usually allows faster data exploration and often provides better results than automatic data mining techniques from statistics or machine learning, especially in cases where automatic algorithms fail (Keim, 2002). Since the proposed method and application are aimed at domain experts, it is of great importance that the sequence analysis process and results be intuitive and easy to understand.

29.2 Iterated Function System (IFS)

Iterated Function System (IFS) was originally developed as a method for constructing fractals as discussed in detail in Barnsley (1988). Other main applications of IFS are image compression (Barnsley and Hurd, 1993) and analysis of genomic sequences (Jeffery, 1990)

We utilize IFS in order to represent sequences as fractals. For this purpose, IFS is used as an iterative contractive mapping technique that represents a sequence as vectors in $\Re$. This type of transformation of a sequence is also known as the 'Chaos Game Representation'. It produces a self similar fractal formed graph, and has the following significant properties:

1. An IFS of a sequence provides a unique representation of it. It can be seen as the 'finger-print' of a sequence. Every point on the graph achieved via IFS represents uniquely all sequence history up to this point, hence an IFS representation comprises all information regarding all subsequences existing in a sequence.
2. The source of the sequence can be inversed from the graph.

Barnsley (1988) provides the following definition for IFS: An Iterated Function System consists of a complete metric space *(X, d)* together with a finite set of contraction mappings $w_n : X \to X$, with respective contractivity factors s_n, for $n=1,2,\ldots,m$.

A mapping $w_i(x)$ is contractive in *(X, d)*, if $d(w_i(y), w_i(z)) \leq s_i \cdot d(y,z)\ \forall y,z \in X$ for some contractivity factor $0 < s_i < 1$.

Barnsley (1988) uses the following general notation for IFS transformation in $\Re^2$: Varying the IFS parameters produces a variety of IFS transformations.

As an example, IFS transformation is utilized for visual analysis of genomic sequences since Jeffrey's foundation essay in 1990. This can be accomplished by using the following IFS: In order to apply this IFS transformation for visual analysis of genomic sequences, one follows the following procedure:

1. Associate each nucleotide (A,C,G,T) with one of the contractive mappings $wi(x)$, $i \in \{1,2,3,4\}$. For example: A = $w1(x)$, C = $w2(x)$, G = $w3(x)$, T = $w4(x)$.
2. Accordingly, represent the genomic sequence of length N as a sequence of N corresponding contractive mappings $\{wi(n)(x)$: $i \in \{1,2,3,4\}$and $n= 1,2,\ldots,N\}$.
3. Plot $x0$, an arbitrary point in $\Re^2$.
4. Recursively apply each of the N contractive mappings $w_{i(23.1)}(x), w_{i(23.19)}(x), \ldots, w_{i(N)}(x)$ by their sequence order as follows: apply contractive mapping$w_{i(23.1)}(x)$to point $x0$ in order to achieve point $x1$, then apply contractive mapping$w_{i(23.19)}(x)$to point $x1$ in order to achieve point $x2$, etc. More generally, $x_n = w_{i(n)}(x_{n-1})$for $n = 1,2,3,\ldots,N$ and $i \in \{1,2,3,4\}$. This results in a sequence of N points in$\Re^2$ $\{x_n : n = 1,2,3,\ldots,N\}$.

The sequence of points produces a square-formed fractal-like graph that enables visual analysis of genomic sequences. It also enables visual comparison between different genomic sequences. If the original sequence of four category types would be uniformly random and long enough, this IFS transformation would produce an equally filled in square. Since the transformed sequence is not uniformly random, the graph reveals its underlying correlations by varying densities of points in different zones. For illustration purposes, Figure 29.1 shows the results for implementing the above IFS transformation on amylase enzyme. This program was coded in $MATLAB^{TM}$ language.

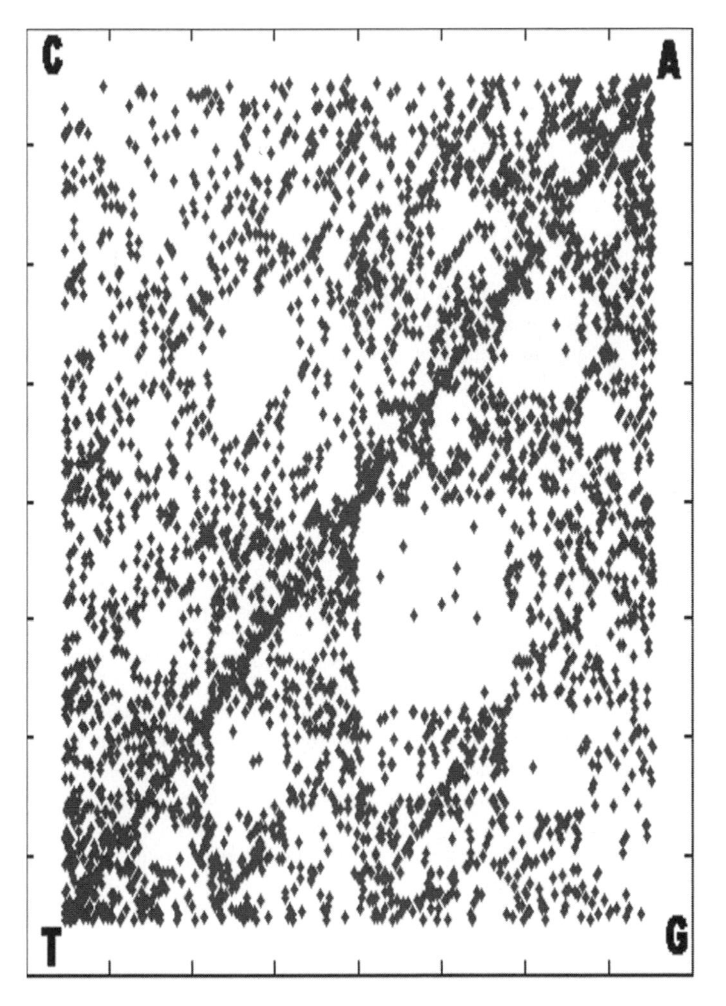

Fig. 29.1. IFS transformation of amylase enzyme

The interpretation of the graph is done by the addresses of points on a fractal (Barnsley, 1988). Figure 29.2 shows the addresses for this specific IFS. The addresses are utilized for the purpose of inversing and analyzing the subsequences that led to a specific point. The length of the inversed subsequence depends on the address resolution. In theory, we can display infinite address resolutions or subsequence lengths. In practice this is limited by display resolution.

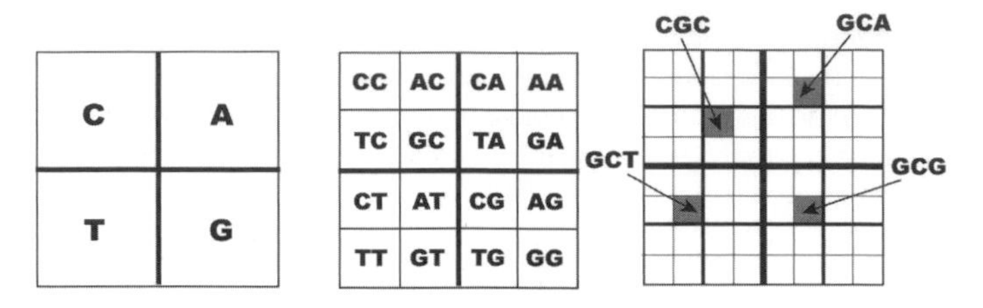

Fig. 29.2. IFS addresses

Based on the IFS addresses, one is able to provide the following analysis of the amylase enzyme IFS transformation presented in Figure 1: The largest empty square formed area right and below the center diagonal line and other empty areas indicate that the subsequence 'CG' is rare in amylase enzyme. Moreover, the A to T diagonal indicates that various combinations of A and T are frequent in amylase enzyme.

This approach is restricted to visual analysis of sequences consisting of four category types. The proposed method extends this approach, by enabling visual analysis and classification of sequences consisting of any number of category types, yet keeping representation in$\Re^2$.

29.3 Algorithmic Framework

29.3.1 Overview

In order to enable visual analysis of sequences for the purpose of sequences classification, we suggest using the framework presented in Figure 3. The process consists of four phases:

1. **Sequence Representation:** A domain-specific task designed to represent each set of sequences that share a common target class as one long string.
2. **Sequence Transformation:** Applying an IFS scheme with circle transformation as developed by Weiss (2008) to each established string in order to receive a fractal graph in $\Re^2$ that enables visual analysis of sequences.
3. **Sequence Patterns Detection**: Visual detection of sequence patterns for each group of sequences that share a common target class. The process is based on interpretation of the graph via addresses of points on fractals, and such features as color codes and zoom function provided by our developed application.
4. **Classifiers Selection:** A heuristic for filtering the detected sequence patterns in order to select the ones that affect the target class. The sequence patterns that influence the target class are selected as classifiers.

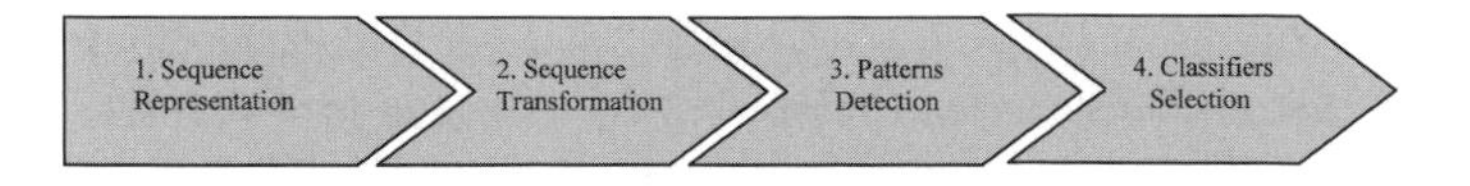

Fig. 29.3. The process overview

The following subsections describe in detail each of the above phases.

29.3.2 Sequence Representation

Each sequence is represented as a string of tokens, each token representing a different category type. We group sequences that share a common target class. We concatenate the sequences of each group to form one string, while adding a delimiter between sequences in order to differentiate between them. The delimiter is also added at the beginning and at the end of the concatenated string.

29.3.3 Sequence Transformation

We utilize an IFS with circle transformation, based on the IFS developed by Weiss (2008). This unique IFS transformation provides the flexibility of analyzing sequences consisting of any number of categories, yet keeping representation in $\Re^2$ in order to enable visual analysis (Weiss, 2008).

Following is a description of the suggested IFS transformation for a sequence consisting of *m* types of categories:

$$\begin{aligned} w_i(x) &= \begin{bmatrix} \alpha_i & 0 \\ 0 & \alpha_i \end{bmatrix} \begin{bmatrix} x_1 \\ x_2 \end{bmatrix} + \begin{bmatrix} \beta_i \\ \delta_i \end{bmatrix} && \text{for } i = 1,2,...,m \\ \text{When } \beta_i &= \cos(i \cdot \tfrac{2\pi}{m}) && \text{for } i = 1,2,...,m \\ \delta_i &= \sin(i \cdot \tfrac{2\pi}{m}) && \text{for } i = 1,2,...,m \\ \alpha_i &= \alpha \,\forall\, i && \text{for } i = 1,2,...,m \end{aligned}$$

It is also required that a fulfills: $\frac{\alpha}{1-\alpha} < \sin(\frac{\pi}{m})$. For proof see Ruschin-Rimini et al. (2009). We introduce an explanation of the steps for applying IFS with circle transformation to a sequence of length *N* consisting of *m* category types *C* = *c1, c2,. . . ,cm.*

1. Associate each category type *c1, c2,. . . ,cm* with one of the contractive mappings *wi*(*x*), $i \in \{1,2,...,m\}$.
2. Accordingly, represent the sequence of length *N* consisting of *m* category types as a sequence of *N* corresponding contractive mappings

   $\{wi(n)(x)$: $i \in \{1,2,...,m\}$and *n*= *1,2,. . . ,N*$\}$.

1. Plot *x0*, an arbitrary point in$\Re^2$.
2. Recursively apply each of the *N* contractive mappings $w_{i(23.1)}(x), w_{i(23.19)}(x), ..., w_{i(N)}(x)$ by their sequence order as follows: apply contractive mapping$w_{i(23.1)}(x)$to point *x0* in order to achieve point *x1*, then apply contractive mapping$w_{i(23.19)}(x)$ to point *x1* in order to achieve point *x2*, etc. More generally, $x_n = w_{i(n)}(x_{n-1})$for $n = 1,2,3,...,N$ and $i \in \{1,2,3,...,m\}$. This results in a sequence of *N* points in$\Re^2$ $\{x_n : n = 1,2,3,...,N\}$.

The sequence of points produces a circle-formed fractal graph that enables visual analysis and classification of sequences. If the original sequence of *m* category types is uniformly random and long enough, this IFS transformation results in a graph of a self-similar fractal consisting of *m* equally filled and disconnected circles, each circle at each resolution also comprising *m* circles. If the sequence is not uniformly random, the graph will reveal its underlying correlations by varying densities of points in different zones.

For example, Figure 29.4 shows the result of transforming a uniformly random sequence of length *N*=20,000 consisting of *m* =26 category types. In this example, we chose a=0.08. Note that a fulfills the requirement mentioned above. The Figure presents the 1st and 2nd

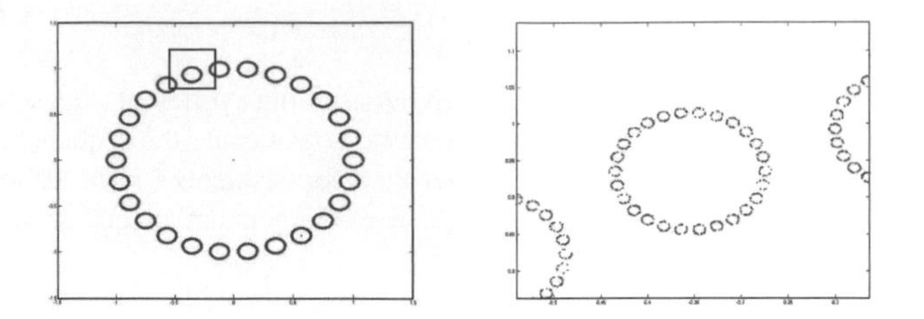

Fig. 29.4. IFS circle transformation of a random sequence of m=26 categories, 1st and 2nd resolutions

resolutions of the graph. By examining the 2nd resolution (zooming into the upper left area) we can conclude that the sequence was only pseudorandom, since the circles are not equally filled. This program was coded in $MATLAB^{TM}$ language.

During this stage we therefore apply the IFS of circle transformation separately to each of the concatenated strings of tokens established in the first stage of the process (See Section 3.2)

29.3.4 Sequence Pattern Detection

We refer to the term sequence pattern as a frequent subsequence existing in a set of sequences sharing the same target class. The frequency level is defined by some threshold. The fractal graph is interpreted by utilizing the address of points on a fractal (Barnsley, 1988). The location of every point on the graph holds the information of the whole sequence up to this point; hence theoretically the full sequence from the beginning until a specific element can be recovered using the address of the point. This allows us to translate areas of the graph such as empty areas, areas of relatively low density, and areas of relatively high density, into missing subsequences, rare subsequences, and frequent subsequences, respectively. As mentioned in Section 3.3, the IFS of circle transformation results in a fractal-like graph consisting of m disconnected circles, each circle at each resolution also comprising m circles. Since we associated every category type to a certain contractive mapping, the address of every circle represents a category type. More specifically, category ci, which was associated with mapping$w_i(x)$, is the address of the circle centered at $\begin{bmatrix} \beta_i \\ \delta_i \end{bmatrix}$. The address length is determined by the graph resolution.

Figure 29.5 demonstrates addresses of 1st, 2nd, and 3rd resolutions for IFS of circle transformation. In the example we show the result of transforming a uniformly random sequence of length N=15,000 consisting of m =9 categories (c1=1, c2=2,...,c8=8, c9=0). The IFS parameter a=0.08 was chosen.

Defining the addresses of points on the graph enables us to suggest the following algorithm for the detection of sequence patterns.

29.3.5 Sequence Pattern Detection Algorithm Description:

1. Detect an area of relatively high density on the 1st resolution of the fractal graph, i.e. one of the m circles comprising a high percentage of points. A circle of relatively high density is defined by a certain percentage threshold.

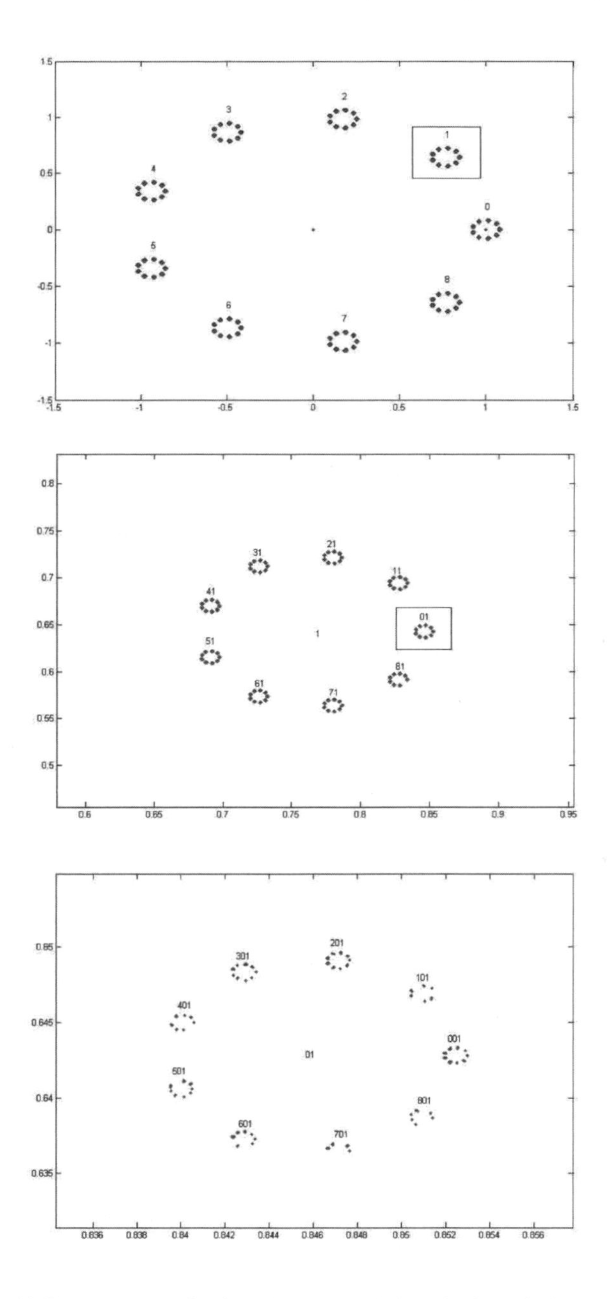

Fig. 29.5. Addresses of points for 1st resolution, 2nd resolution (zooming into circle address 1) and 3rd resolution (zooming into circle address 01) for IFS circle transformation

2. Drill into the relevant circle of m circles
3. Detect an area of relatively high density on the next resolution.
4. Drill into the relevant circle of m circles
5. Repeat steps 3 and 4 until the relevant circle contains points almost uniformly distributed between approximately m circles, with no circle of ***relatively*** high density. This is where the sequence pattern ends.
6. Compute the address of the relevant circle location in order to recover the sequence pattern. For example, if the sequence pattern ends at the 4th resolution, we conclude the sequence pattern's length is 3. We will therefore recover the 3 length address of the relevant circle from the 3rd resolution.
7. Repeat steps 1 – 6 for a different area of relatively high density in order to reveal another pattern.
8. End after exploring all areas of relatively high density, i.e. after all sequence patterns have been revealed.

During this stage we therefore apply the Sequence Pattern Detection Algorithm separately to each of the fractal graphs established in the prior stage of the process (See Section 3.3). It should be noted, that we could use the same algorithm to explore *empty areas* and *areas of relatively low density*, for the detection of missing and rare operation sequences per product quality measure, respectively.

29.3.6 Classifiers Selection

Applying stages 1-3 of the algorithm presented above results in a set of sequence patterns detected for each target class. Next, we filter the sequence patterns that have been detected by using a simple heuristic comprising the following steps.

1. Select one of the detected sequence patterns per target class.
2. Check whether the sequence pattern was detected per other target classes.
   a) If the sequence pattern was detected exclusively per the specific target class, go to step 3.
   b) If the sequence pattern was detected per other target classes, go to step 4.
3. Conclude that the selected sequence pattern affects the specific target class and therefore can be selected as a classifier.
4. Conclude that the selected sequence pattern does not affect the target class and therefore cannot be used as a classifier.
5. Repeat steps 1-4 for other detected sequence patterns.
6. End after exploring all sequence patterns detected per all target classes.

It should be emphasized, that a more accurate approach would be to apply any data mining classification algorithm at this stage. For this purpose, each detected sequence pattern represents a different feature. The feature value is 1 if the corresponding sequence pattern exists in a sequence, otherwise it is 0. A classification algorithm such as C4.5, which creates a classification decision tree (Quinlan, 1993), could be utilized at this stage. Nevertheless, since we aim to provide domain experts a practical tool that requires no data mining or statistical background, our simple heuristic is sufficient.

29.4 Fault Sequence Detection Application

We develop an efficient, user-friendly software application aimed at domain experts, based on the proposed algorithmic framework. The software application is developed in $MATLAB^{TM}$ language.

Following are the main features of the proposed application:

1. *Color code*: The detection of areas of relatively high density should be obvious to the end user. We therefore add a color code for relatively high density areas.
2. *Zoom function*: In order to explore areas of relatively high density until the full sequence pattern is revealed, it is necessary to zoom into the relevant circles. Our application offers this functionality.
3. *Method for recovering the sequence pattern address*: In order to ease the process of circle address computation, our application presents the corresponding addresses near each circle as a part of the graph.

A demonstration of a complete process of visual analysis of sequences affecting a target attribute using the proposed method and application can be found in Ruschin-Rimini *et al.* (2009). A comparison of the proposed method to existing approaches such as n-gram and Hidden Markov Model (HMM) is also presented in Ruschin-Rimini *et al.* (2009), including demonstration of cases in which the proposed method was the only method capable of detecting the complete sequence patterns influencing a target class.

29.5 Conclusions and Future Research

This paper presents a novel method for visual analysis of sequential data based on fractal geometry, demonstrated by a unique application for visual detection of sequence patterns affecting a target attribute.

Compared to existing methods, the proposed method has the following advantages:

1. The proposed method enables visual data analysis. The main advantages of visual data exploration techniques over automatic data mining techniques from statistics or machine learning are as follows (Keim, 2002): *i)* visual data exploration can easily deal with noisy data. *ii)* Visual data exploration is intuitive and requires no understanding of mathematical or statistical algorithms or parameters. Accordingly, we provide an analytical tool that can be easily utilized by domain experts such as engineering staff in a real-world manufacturing environment.
2. The proposed method can detect sequence patterns of any length without predefining the length of the pattern.
3. In case a category appears more than once in a sequence, such as might occur in realistic conditions, the proposed method is able to distinguish between the different occurrences of the same category. For example, if the following two sequence patterns: 8504 and 472 affect a target class, our method will detect these two different patterns and clearly distinguish between the different occurrences of category 4.
4. The proposed method provides a significant added value by enabling the visual detection of rare and missing sequences per target class.

On the next phase of the research we plan to further develop the proposed scheme which is based on fractal representation, to account for online changes in monitored processes. We plan to suggest a novel type of online interactive SPC chart that enables a dynamic inspection of non-linear state dependant processes.

The presented algorithmic framework is applicable for many practical domains, for example visual analysis of the affect of operation sequence on product quality (See Ruschin-Rimini *et al.*, 2009), visual analysis of customers action history, visual analysis of products defect codes history, and more.

The developed application was utilized by the General Motors research labs located in Bangalore, India, for visual analysis of vehicle failure history.

References

Arbel, R. and Rokach, L., Classifier evaluation under limited resources, Pattern Recognition Letters, 27(14): 1619–1631, 2006, Elsevier.

Barnsley M., *Fractals Everywhere*, Academic Press, Boston, 1988

Barnsley, M., Hurd L. P., *Fractal Image Compression*, A. K. Peters, Boston, 1993

Cohen S., Rokach L., Maimon O., Decision Tree Instance Space Decomposition with Grouped Gain-Ratio, Information Science, Volume 177, Issue 17, pp. 3592-3612, 2007.

Da Cunha C., Agard B., and Kusiak A., Data mining for improvement of product quality, *International Journal of Production Research*, 44(18-19), pp. 4027-4041, 2006

Falconer K., *Techniques in Fractal geometry*, John Wiley & Sons, 1997

Jeffrey H. J., Chaos game representation of genetic sequences, *Nucleic Acids Res.*, vol. 18, pp. 2163 – 2170, 1990

Keim D. A., Information Visualization and Visual Data mining, *IEEE Transactions of Visualization and Computer Graphics*, Vol. 7, No. 1, pp. 100-107, 2002

Maimon O., and Rokach, L. Data Mining by Attribute Decomposition with semiconductors manufacturing case study, in Data Mining for Design and Manufacturing: Methods and Applications, D. Braha (ed.), Kluwer Academic Publishers, pp. 311–336, 2001.

Moskovitch R, Elovici Y, Rokach L, Detection of unknown computer worms based on behavioral classification of the host, Computational Statistics and Data Analysis, 52(9):4544–4566, 2008.

Quinlan, J. R., C4.5: Programs for Machine Learning, *Morgan Kaufmann*, 1993

Rokach L., Mining manufacturing data using genetic algorithm-based feature set decomposition, Int. J. Intelligent Systems Technologies and Applications, 4(1):57-78, 2008.

Rokach L., Genetic algorithm-based feature set partitioning for classification problems,Pattern Recognition, 41(5):1676–1700, 2008.

Rokach, L., Decomposition methodology for classification tasks: a meta decomposer framework, Pattern Analysis and Applications, 9(2006):257–271.

Rokach, L. and Maimon, O., Theory and applications of attribute decomposition, IEEE International Conference on Data Mining, IEEE Computer Society Press, pp. 473–480, 2001.

Rokach L. and Maimon O., Feature Set Decomposition for Decision Trees, Journal of Intelligent Data Analysis, Volume 9, Number 2, 2005b, pp 131–158.

Rokach L., and Maimon O., Data mining for improving the quality of manufacturing: A feature set decomposition approach. *Journal of Intelligent Manufacturing,* 17(23.3), pp. 285-299, 2006

Rokach, L. and Maimon, O. and Arbel, R., Selective voting-getting more for less in sensor fusion, International Journal of Pattern Recognition and Artificial Intelligence 20 (3) (2006), pp. 329–350.
Rokach, L. and Maimon, O. and Averbuch, M., Information Retrieval System for Medical Narrative Reports, Lecture Notes in Artificial intelligence 3055, page 217-228 Springer-Verlag, 2004.
Rokach L., Maimon O. and Lavi I., Space Decomposition In Data Mining: A Clustering Approach, Proceedings of the 14th International Symposium On Methodologies For Intelligent Systems, Maebashi, Japan, Lecture Notes in Computer Science, Springer-Verlag, 2003, pp. 24–31.
Rokach L., Romano R. and Maimon O., Mining manufacturing databases to discover the effect of operation sequence on the product quality, *Journal of Intelligent Manufacturing*, 2008
Ruschin-Rimini N., Maimon O. and Romano R., Visual Analysis of Quality-related Manufacturing Data Using Fractal Geometry, working paper submitted for publication, 2009.
Weiss C. H., Visual Analysis of Categorical Time Series, *Statistical Methodology* 5, pp. 56-71, 2008

30

Interestingness Measures - On Determining What Is Interesting

Sigal Sahar

Department of Computer Science,
Tel-Aviv University, Israel
gales@post.tau.ac.il

Summary. As the size of databases increases, the sheer number of mined from them can easily overwhelm users of the KDD process. Users run the KDD process because they are overloaded by data. To be successful, the KDD process needs to extract *interesting* patterns from large masses of data. In this chapter we examine methods of tackling this challenge: how to identify *interesting* patterns.

Key words: Interestingness Measures, Association Rules

Introduction

According to (Fayyad et al., 1996) "Knowledge Discovery in Databases (KDD) is the non-trivial process of identifying valid, novel, potentially useful, and ultimately understandable patterns in data." Mining algorithms primarily focus on discovering patterns in data, for example, the Apriori algorithm (Agrawal and Shafer, 1996) outputs the exhaustive list of association rules that have at least the predefined support and confidence thresholds. Interestingness differentiates between the "valid, novel, potentially useful and ultimately understandable" mined association rules and those that are not—differentiating the interesting patterns from those that are not interesting. Thus, determining what is interesting, or interestingness, is a critical part of the KDD process. In this chapter we review the main approaches to determining what is interesting.

Figure 30.1 summarizes the three main types of interestingness measures, or approaches to determining what is interesting. **Subjective interestingness** explicitly relies on users' specific needs and prior knowledge. Since what is interesting to any user is ultimately subjective, these subjective interestingness measures will have to be used to reach any complete solution of determining what is interesting. (Silberschatz and Tuzhilin, 1996) differentiate between subjective and objective interestingness. **Objective interestingness** refers to measures of interest "where interestingness of a pattern is measured in terms of its structure and the underlying data used in the discovery process" (Silberschatz and Tuzhilin, 1996) but requires user intervention to select which of these measures to use and to initialize it. **Impartial interestingness**, introduced in (Sahar, 2001), refers to measures of interest that can be applied automatically

O. Maimon, L. Rokach (eds.), *Data Mining and Knowledge Discovery Handbook*, 2nd ed.,
DOI 10.1007/978-0-387-09823-4_30,

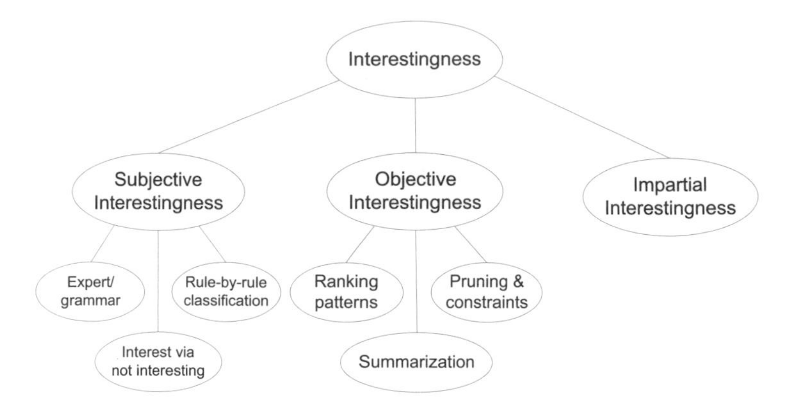

Fig. 30.1. Types of Interestingness Approaches.

to the output of any association rule mining algorithm to reduce the number of not-interesting rules independently of the domain, task and users.

30.1 Definitions and Notations

Let Λ be a set of attributes over the boolean domain. Λ is the superset of all attributes we discuss in this chapter. An **itemset** I is a set of attributes: $I \subseteq \Lambda$. A **transaction** is a subset of attributes of Λ that have the boolean value TRUE. We will refer to the set of transactions over Λ as a **database**. If exactly $s\%$ of the transactions in the database contain an itemset I then we say that I has **support** s, and express the support of I as $P(I)$. Given a support threshold s we will call itemsets that have at least support s **large** or **frequent**.

Let A and B be two sets of attributes such that $A, B \subseteq \Lambda$ and $A \cap B = \emptyset$. Let $\mathscr{D}$ be a set of transactions over Λ. Following the definition in (Agrawal and Shafer, 1996), an **association rule** $A \rightarrow B$ is defined to have support $s\%$ and confidence $c\%$ in $\mathscr{D}$ if $s\%$ of the transactions in $\mathscr{D}$ contain $A \cup B$ and $c\%$ of the transactions that contain A also contain B. For convenience, in an association rule $A \rightarrow B$ we will refer to A as the **assumption** and B as the **consequent** of the rule. We will express the support of $A \rightarrow B$ as $P(A \cup B)$. We will express the confidence of $A \rightarrow B$ as $P(B|A)$ and denote it with *confidence*$(A \rightarrow B)$. (Agrawal and Shafer, 1996) presents an elegantly simple algorithm to mine the exhaustive list of association rules that have at least predefined support and confidence thresholds from a boolean database.

30.2 Subjective Interestingness

What is interesting to users is ultimately subjective; what is interesting to one user may be known or irrelevant, and therefore not interesting, to another user. To determine what is subjectively interesting, users' domain knowledge—or at least the portion of it that pertains to the data at hand—needs to be incorporated into the solution. In this section we review the three main approaches to this problem.

30.2.1 The Expert-Driven Grammatical Approach

In the first and most popular approach, the domain knowledge required to subjectively determine which rules are interesting is explicitly described through a predefined grammar. In this approach a domain expert is expected to express, using the predefined grammar what is, or what is not, interesting. This approach was introduced by (Klemettinen et al., 1994), who were the first to apply subjective interestingness, and many other applications followed. (Klemettinen et al., 1994) define pattern templates that describe the structure of interesting association rules through inclusive templates and the structure of not-interesting rules using restrictive templates. (Liu et al., 1997) present a formal grammar that allows the expression of imprecise or vague domain knowledge, the General Impressions. (Srikant et al., 1997) introduce into the mining process user defined constraints, including taxonomical constraints, in the form of boolean expressions, and (Ng et al., 1998) introduce user constraints as part of an architecture that supports exploratory association rule mining. (Padmanabhan and Tuzhilin, 2000) use a set of predefined user beliefs in the mining process to output a minimal set of unexpected association rules with respect to that set of beliefs. (Adomavicius and Tuzhilin, 1997) define an action hierarchy to determine which which association rules are actionable; actionability is an aspect of being subjectively interesting. (Adomavicius and Tuzhilin, 2001, Tuzhilin and Adomavicius, 2002) iteratively apply expert-driven validation operators to incorporate subjective interestingness in the personalization and bioinformatics domains.

In some cases the required domain knowledge can be obtained from a pre-existing knowledge base, thus eliminating the need to engage directly with a domain expert to acquire it. For example, in (Basu et al., 2001) the WordNet lexical knowledge-base is used to measure the novelty—an indicator of interest—of an association rule by assessing the dissimilarity between the assumption and the consequent of the rule. An example of a domain where such a knowledge base exists naturally is when detecting rule changes over time, as in (Liu et al., 2001a). In many domains, these knowledge-bases are not readily available. In those cases the success of this approach is conditioned on the availability of a domain expert willing and able to complete the task of defining all the required domain knowledge. This is no easy task: the domain expert may unintentionally neglect to define some of the required domain knowledge, some of it may not be applicable across all cases, and could change over time. Acquiring such a domain expert for the duration of the task is often costly and sometimes unfeasible. But given the domain knowledge required, this approach can output the small set of subjectively interesting rules.

30.2.2 The Rule-By-Rule Classification Approach

In the second approach, taken in (Subramonian, 1998), the required domain knowledge base is constructed by classifying rules from prior mining sessions. This approach does not depend on the availability of domain experts to define the domain knowledge, but does require very intensive user interaction of a mundane nature. Although the knowledge base can be constructed incrementally, this, as the author says, can be a tedious process.

30.2.3 Interestingness Via What Is Not Interesting Approach

The third approach, introduced by (Sahar, 1999), capitalizes on an inherent aspect in the interestingness task: the majority of the mined association rules are not interesting. In this approach a user is iteratively presented with simple rules, with only one attribute in their assumption and one attribute in the consequent, for classification. These rules are selected so that a single user classification of a rule can imply that a large number of the mined association rules are also not-interesting. The advantages of this approach are that it is simple so that a naive user can use it without depending on a domain expert to provide input, that it very quickly, with only a few questions, can eliminate a significant portion of the not-interesting rules, and that it circumvents the need to define why a rule is interesting. However, this approach is used only to reduce the size of the interestingness problem by substantially decreasing the number of potentially interesting association rules, rather than pinpointing the exact set of interesting rules. This approach has been integrated into the mining process in (Sahar, 2002b).

30.3 Objective Interestingness

The domain knowledge needed in order to apply subjective interestingness criteria is difficult to obtain. Although subjective interestingness is needed to reach the short list of interesting patterns, much can be done without explicitly using domain knowledge. The application of objective interestingness measures depends only the structure of the data and the patterns extracted from it; some user intervention will still be required to select the measure to be used, etc. In this section we review the three main types of objective interestingness measures.

30.3.1 Ranking Patterns

To rank association rules according to their interestingness, a mapping, f, is introduced from the set of mined rules, Ω, to the domain of real numbers:

$$f : \Omega \rightarrow \Re. \tag{30.1}$$

The number an association rule is mapped to is an indication of how interesting this rule is; the larger the number a rule is mapped to, the more interesting the rule is assumed to be. Thus, the mapping imposes an order, or ranking, of interest on a set of association rules.

Ranking rules according to their interest has been suggested in the literature as early as (Piatetsky-Shapiro, 1991). (Piatetsky-Shapiro, 1991) introduced the first three principles of interestingness evaluation criteria, as well as a simple mapping that could satisfy them: P-S$(A \rightarrow B) = P(A \cup B) - P(B) \cdot P(A)$. Since then many different mappings, or rankings, have been proposed as measures of interest. Many definitions of such mappings, as well as their empirical and theoretical evaluations, can be found in (Klösgen, 1996, Bayardo Jr. and Agrawal, 1999, Sahar and Mansour, 1999, Hilderman and Hamilton, 2000, Hilderman and Hamilton, 2001, Tan et al., 2002). The work on the principles introduced by (Piatetsky-Shapiro, 1991) has been expanded by (Major and Mangano, 1995, Kamber and Shinghal, 1996). (Tan et al., 2002) extends the studies of the properties and principles of the ranking criteria. (Hilderman and Hamilton, 2001) provide a very thorough review and study of these criteria, and introduce an interestingness theory for them.

30.3.2 Pruning and Application of Constraints

The mapping in Equation 30.1 can also be used as a pruning technique: prune as not-interesting all the association rules that are mapped to an interest score lower than a user-defined threshold. Note that in this section we only refer to pruning and application of constraints performed using objective interestingness measures, and not subjective ones, such as removing rules if they contain, or do not contain, certain attributes.

Additional methods can be used to prune association rules without requiring the use of the an interest mapping. Statistical tests such as the χ^2 test are used for pruning in (Brin et al., 1997,Liu et al., 1999,Liu et al., 2001b). These tests have parameters that need to be initialized. A collection of pruning methods is described in (Shah et al., 1999).

Another type of pruning is the constraint based approach of (Bayardo Jr. et al., 1999). To output a more concise list of rules as the output of the mining process, the algorithm of (Bayardo Jr. et al., 1999) only mines rules that comply with the usual constraints of minimum support and confidence thresholds as well as with two new constraints. The first constraint is a user-specified consequent (subjective interestingness). The second, unprecedented, constraint is of a user-specified minimum confidence improvement threshold. Only rules whose confidence is at least the minimum confidence improvement threshold greater than the confidence of any of their simplifications are outputted; a simplification of a rule is formed by removing one or more attributes from its assumption.

30.3.3 Summarization of Patterns

Several distinct methods fall under the summarization approach. (Aggarwal and Yu, 1998) introduce a redundancy measure that summarizes all the rules at the predefined support and confidence levels very compactly by using more "complex" rules. The preference to complex rules is formally defined as follows: a rule $C \rightarrow D$ is redundant with respect to $A \rightarrow B$ if (1) $A \cup B = C \cup D$ and $A \subset C$, or (2) $C \cup D \subset A \cup B$ and $A \subseteq C$. A different type of summary that favors less "complex" rules was introduced by (Liu et al., 1999). (Liu et al., 1999) provide a summary of association rules with a single attributed consequent using a subset of "direction-setting" rules, rules that represent the direction a group of non-direction-setting rules follows. The direction is calculated using the χ^2 test, which is also used to prune the mined rules prior to the discovery of direction-setting rules. (Liu et al., 2000) present a summary that simplifies the discovered rules by providing an overall picture of the relationships in the data and their exceptions. (Zaki, 2000) introduces an approach to mining only the non-redundant association rules from which all the other rules can be inferred. (Zaki, 2000) also favors "less-complex" rules, defining a rule $C \rightarrow D$ to be redundant if there exists another rule $A \rightarrow B$ such that $A \subseteq C$ and $B \subseteq D$ and both rules have the same confidence.

(Adomavicius and Tuzhilin, 2001) introduce summarization through similarity based rule grouping. The similarity measure is specified via an attribute hierarchy, organized by a domain expert who also specifies a level of rule aggregation in the hierarchy, called a cut. The association rules are then mapped to aggregated rules by mapping to the cut, and the aggregated rules form the summary of all the mined rules.

(Toivonen et al., 1995) suggest clustering rules "that make statements about the same database rows [...]" using a simple distance measure, and introduce an algorithm to compute rule covers as short descriptions of large sets of rules. For this approach to work without losing any information, (Toivonen et al., 1995) make a monotonicity assumption, restricting the databases on which the algorithm can be used. (Sahar, 2002a) introduce a general clustering framework for association rules to facilitate the exploration of masses of mined rules by automatically organizing them into groups according to similarity. To simplify interpretation of the resulting clusters, (Sahar, 2002a) also introduces a data-inferred, concise representation of the clusters, the ancestor coverage.

30.4 Impartial Interestingness

To determine what is interesting, users need to first determine which interestingness measures to use for the task. Determining interestingness according to different measures can result in different sets of rules outputted as interesting. This dependence of the output of the interestingness analysis on the interestingness measure used is clear when domain knowledge is applied explicitly, in the case of the subjective interestingness measures (Section 30.2). When domain knowledge is applied implicitly, this dependence may not be as clear, but it still exists. As (Sahar, 2001) shows, objective interestingness measures depend implicitly on domain knowledge. This dependence is manifested during the *selection* of the objective interestingness measure to be used, and, when applicable, during its initialization (for pruning and constraints) and the interpretation of the results (for summarization).

(Sahar, 2001) introduces a new type of interestingness measure, as part of an interestingness framework, that can be applied automatically to eliminate a portion of the rules that is not interesting, as in Figure 30.2. This type of interestingness is called impartial interestingness because it is domain-independent, task-independent, and user-independent, making it impartial to all considerations affecting other interestingness measures. Since the impartial interestingness measures do not require any user intervention, they can be applied sequentially and automatically, directly following the Data Mining process, as depicted in Figure 30.2. The impartial interestingness measure preprocess the mined rules to eliminate those rules that are not interesting regardless of the domain, task and user, and so they form the Interestingness PreProcessing Step. This step is followed by Interestingness Processing, which includes the application of objective (when needed) and subjective interestingness criteria.

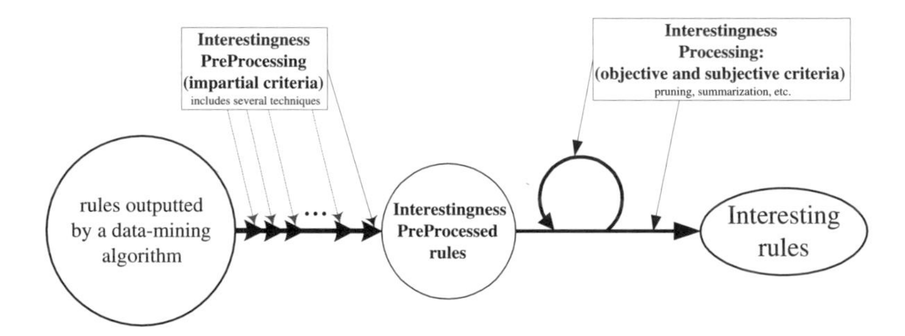

Fig. 30.2. Framework for Determining Interestingness.

To be able to define impartial measures, (Sahar, 2001) assume that the goal of the interestingness analysis on a set of mined rules is to find a subset of interesting rules, rather than to infer from the set of mined rules rules that have not been mined that could potentially be interesting. An example of an impartial measure is (Overfitting, (Sahar, 2001)) the deletion of all rules $r = A \cup C \to B$ if there exists another mined rule $\widehat{r} = A \to B$ such that $\mathit{confidence}(\widehat{r}) \geq \mathit{confidence}(r)$.

30.5 Concluding Remarks

Characterizing what is interesting is a difficult problem, primarily because what is interesting is ultimately subjective. Numerous attempts have been made to formulate these qualities, ranging from evidence and simplicity to novelty and actionability, with no formal definition for "interestingness" emerging so far. In this chapter we reviewed the three main approaches to tackling the challenge of discovering which rules are interesting under certain assumptions.

Some of the interestingness measures reviewed have been incorporated into the mining process as opposed to being applied after the mining process. (Spiliopoulou and Roddick, 2000) discuss the advantages of processing the set of rules after the mining process, and introduce the concept of higher order mining, showing that rules with higher order semantics can be extracted by processing the mined results. (Hipp and Günter, 2002) argue that pushing constraints into the mining process "[...] is based on an understanding of KDD that is no longer up-to-date" as KDD is an iterative discovery process rather than "pure hypothesis investigation". There is no consensus on whether it is advisable to push constraints into the mining process. An optimal solution is likely to be produced through a balanced combination of these approaches; some interestingness measures (such as the impartial ones) can be pushed into the mining process without overfitting its output to match the subjective interests of only a small audience, permitting further interestingness analysis that will tailor it to each user's subjective needs.

Data Mining algorithms output patterns. Interestingness discovers the potentially interesting patterns. To be successful, the KDD process needs to extract the interesting patterns from large masses of data. That makes interestingness a very important capability in the extremely data-rich environment in which we live. It is likely that our environment will continue to inundate us with data, making determining interestingness critical for success.

References

Adomavicius, G. and Tuzhilin, A. (1997). Discovery of actionable patterns in databases: The action hierarchy approach. In *Proceedings of the Third International Conference on Knowledge Discovery and Data Mining*, pages 111–114, Newport Beach, CA, USA. AAAI Press.

Adomavicius, G. and Tuzhilin, A. (2001). Expert-driven validation of rule-based user models in personalization applications. *Data Mining and Knowledge Discovery*, 5(1/2):33–58.

Aggarwal, C. C. and Yu, P. S. (1998). A new approach to online generation of association rules. Technical Report Research Report RC 20899, IBM T J Watson Research Center.

Agrawal, R., Heikki, M., Srikant, R., Toivonen, H., and Verkamo, A. I. (1996). *Advances in Knowledge Discovery and Data Mining*, chapter 12: *Fast Discovery of Association Rules*, pages 307–328. AAAI Press/The MIT Press, Menlo Park, California.

Basu, S., Mooney, R. J., Pasupuleti, K. V., and Ghosh, J. (2001). Evaluating the novelty of text-mined rules using lexical knowledge. In *Proceedings of the Seventh ACM SIGKDD International Conference on Knowledge Discovery and Data Mining*, pages 233–238, San Francisco, CA, USA.

Bayardo Jr., R. J. and Agrawal, R. (1999). Mining the most interesting rules. In *Proceedings of the Fifth ACM SIGKDD International Conference on Knowledge Discovery and Data Mining*, pages 145–154, San Diego, CA.

Bayardo Jr., R. J., Agrawal, R., and Gunopulos, D. (1999). Constraint-based rule mining in large, dense databases. In *Proceedings of the Fifteenth IEEE ICDE International Conference on Data Engineering*, pages 188–197, Sydney, Australia.

Brin, S., Motwani, R., and Silverstein, C. (1997). Beyond market baskets: Generalizing association rules to correlations. In *Proceedings of ACM SIGMOD International Conference on Management of Data*, pages 265–276, Tucson, AZ, USA.

Fayyad, U. M., Piatetsky-Shapiro, G., and Smyth, P. (1996). *Advances in Knowledge Discovery and Data Mining*, chapter 1: *From Data Mining to Knowledge Discovery: An Overview*, pages 1–34. AAAI Press.

Hilderman, R. J. and Hamilton, H. J. (2000). Principles for mining summaries using objective measures of interestingness. In *Proceedings of the Twelfth IEEE International Conference on Tools with Artificial Intelligence (ICTAI)*, pages 72–81, Vancouver, Canada.

Hilderman, R. J. and Hamilton, H. J. (2001). *Knowledge Discovery and Measures of Interest*. Kluwer Academic Publishers.

Hipp, J. and Günter, U. (2002). Is pushing constraints deeply into the mining algorithms really what we want? *SIGKDD Explorations*, 4(1): 50–55.

Kamber, M. and Shinghal, R. (1996). Evaluating the interestingness of characteristic rules. In *Proceedings of the Second International Conference on Knowledge Discovery and Data Mining*, pages 263–266, Portland, OR, USA.

Klemettinen, M., Mannila, H., Ronkainen, P., Toivonen, H., and Verkam, A. I. (1994). Finding interesting rules from large sets of discovered association rules. In *Proceedings of the Third ACM CIKM International Conference on Information and Knowledge Management*, pages 401–407, Orlando, FL, USA. ACM Press.

Klösgen, W. (1996). *Advances in Knowledge Discovery and Data Mining*, chapter 10: *Explora: a Multipattern and Multistrategy Discovery Assistant*, pages 249–271. AAAI Press.

Liu, B., Hsu, W., and Chen, S. (1997). Using general impressions to analyze discovered classification rules. In *Proceedings of the Third International Conference on Knowledge Discovery and Data Mining*, pages 31–36, Newport Beach, CA, USA. AAAI Press.

Liu, B., Hsu, W., and Ma, Y. (1999). Pruning and summarizing the discovered associations. In *Proceedings of the Fifth ACM SIGKDD International Conference on Knowledge Discovery and Data Mining*, pages 125–134, San Diego, CA, USA.

Liu, B., Hsu, W., and Ma, Y. (2001a). Discovery the set of fundamental rule changes. In *Proceedings of the Seventh ACM SIGKDD International Conference on Knowledge Discovery and Data Mining*, pages 335–340, San Francisco, CA, USA.

Liu, B., Hsu, W., and Ma, Y. (2001b). Identifying non-actionable association rules. In *Proceedings of the Seventh ACM SIGKDD International Conference on Knowledge Discovery and Data Mining*, pages 329–334, San Francisco, CA, USA.

Liu, B., Hu, M., and Hsu, W. (2000). Multi-level organization and summarization of the discovered rules. In *Proceedings of the Sixth ACM SIGKDD International Conference on Knowledge Discovery and Data Mining*, pages 208–217, Boston, MA, USA.

Major, J. A. and Mangano, J. J. (1995). Selecting among rules induced from a hurricane databases. *Journal of Intelligent Information Systems*, 4:39–52.

Ng, R. T., Lakshmanan, L. V. S., Han, J., and Pang, A. (1998). Exploratory mining and pruning optimizations of constrained association rules. In *Proceedings of ACM SIGMOD International Conference on Management of Data*, pages 13–24.

Padmanabhan, B. and Tuzhilin, A. (2000). Small is beautiful: Discovering the minimal set of unexpected patterns. In *Proceedings of the Sixth ACM SIGKDD International Conference on Knowledge Discovery and Data Mining*, pages 54–63, Boston, MA, USA.

Piatetsky-Shapiro, G. (1991). *Knowledge Discovery in Databases*, chapter 13: *Discovery, Analysis, and Presentation of Strong Rules*, pages 248–292. AAAI/MIT Press.

Rokach, L., Averbuch, M., and Maimon, O., Information retrieval system for medical narrative reports. Lecture notes in artificial intelligence, 3055. pp. 217-228, Springer-Verlag (2004).

Sahar, S. (1999). Interestingness via what is not interesting. In *Proceedings of the Fifth ACM SIGKDD International Conference on Knowledge Discovery and Data Mining*, pages 332–336, San Diego, CA, USA.

Sahar, S. (2001). Interestingness preprocessing. In *Proceedings of the IEEE ICDM International Conference on Data Mining*, pages 489–496, San Jose, CA, USA.

Sahar, S. (2002a). Exploring interestingness through clustering: A framework. In *Proceedings of the IEEE ICDM International Conference on Data Mining*, pages 677–680, Maebashi City, Japan.

Sahar, S. (2002b). On incorporating subjective interestingness into the mining process. In *Proceedings of the IEEE ICDM International Conference on Data Mining*, pages 681–684, Maebashi City, Japan.

Sahar, S. and Mansour, Y. (1999). An empirical evaluation of objective interestingness criteria. In *SPIE Conference on Data Mining and Knowledge Discovery*, pages 63–74, Orlando, FL, USA.

Shah, D., Lakshmanan, L. V. S., Ramamritham, K., and Sudarshan, S. (1999). Interestingness and pruning of mined patterns. In *Proceedings of the ACM SIGMOD Workshop on Research Issues in Data Mining and Knowledge Discovery (DMKD)*, Philadelphia, PA, USA.

Silberschatz, A. and Tuzhilin, A. (1996). What makes patterns interesting in knowledge discovery systems. *IEEE Transactions on Knowledge and Data Engineering (TKDE)*, 8(6):970–974.

Spiliopoulou, M. and Roddick, J. F. (2000). Higher order mining: Modeling and mining the results of knowledge discovery. In *Proceedings of the Second Conference on Data Mining Methods and Databases*, pages 309–320, Cambridge, UK. WIT Press.

Srikant, R., Vu, Q., and Agrawal, R. (1997). Mining association rules with item constraints. In *Proceedings of the Third International Conference on Knowledge Discovery and Data Mining*, pages 67–73, Newport Beach, CA, USA. AAAI Press.

Subramonian, R. (1998). Defining diff as a Data Mining primitive. In *Proceedings of the Fourth International Conference on Knowledge Discovery and Data Mining*, pages 334–338, New York City, NY, USA. AAAI Press.

Tan, P.-N., Kumar, V., and Srivastava, J. (2002). Selecting the right interestingness measure for association patterns. In *Proceedings of the Eight ACM SIGKDD International Conference on Knowledge Discovery and Data Mining*, pages 32–41, Edmonton, Alberta, Canada.

Toivonen, H., Klemettinen, M., Ronkainen, P., Hätönen, K., and Mannila, H. (1995). Pruning and grouping discovered association rules. In *Proceedings of the MLnet Familiarization Workshop on Statistics, Machine Learning and Knowledge Discovery in Databases*, pages 47–52, Heraklion, Crete, Greece.

Tuzhilin, A. and Adomavicius, G. (2002). Handling very large numbers of association rules in the analysis of microarray data. In *Proceedings of the Eight ACM SIGKDD International Conference on Knowledge Discovery and Data Mining*, pages 396–404, Edmonton, Alberta, Canada.

Zaki, M. J. (2000). Generating non-redundant association rules. In *Proceedings of the Sixth ACM SIGKDD International Conference on Knowledge Discovery and Data Mining*, pages 34–43, Boston, MA, USA.

31

Quality Assessment Approaches in Data Mining

Maria Halkidi[1] and Michalis Vazirgiannis[2]

[1] Department of Computer Science and Engineering, University of California at Riverside, USA,
Department of Informatics, Athens University of Economics and Business, Greece
mhalkidi@cs.ucr.edu

[2] Department of Informatics, Athens University of Economics and Business, Greece
mvazirg@aueb.gr

Summary. The *Data Mining process* encompasses many different specific techniques and algorithms that can be used to analyze the data and derive the discovered knowledge. An important problem regarding the results of the Data Mining process is the development of efficient indicators of assessing the quality of the results of the analysis. This, the quality assessment problem, is a cornerstone issue of the whole process because: i) *The analyzed data may hide interesting patterns* that the Data Mining methods are called to reveal. Due to the size of the data, the requirement for automatically evaluating the validity of the extracted patterns is stronger than ever.
ii)*A number of algorithms and techniques have been proposed* which under different assumptions can lead to different results. iii)*The number of patterns generated during the Data Mining process* is very large but only a few of these patterns are likely to be of any interest to the domain expert who is analyzing the data. In this chapter we will introduce the main concepts and quality criteria in Data Mining. Also we will present an overview of approaches that have been proposed in the literature for evaluating the Data Mining results .

Key words: cluster validity, quality assessment, unsupervised learning, clustering

Introduction

Data Mining is mainly concerned with methodologies for extracting patterns from large data repositories. There are many Data Mining methods which accomplishing a limited set of tasks produces a particular enumeration of patterns over data sets. The main tasks of Data Mining which have already been discussed in previous sections are: i) *Clustering*, ii) *Classification*, iii) *Association Rule Extraction*, iv)*Time Series*, v) *Regression*, and vi) *Summarization*.

Since a Data Mining system could generate under different conditions thousands or million of patterns, questions arise for the quality of the Data Mining results, such as which of the extracted patterns are interesting and which of them represent knowledge.

In general terms, a pattern is interesting if it is easily understood, valid, potentially useful and novel. A pattern also is considered as interesting if it validates a hypothesis that a user

O. Maimon, L. Rokach (eds.), *Data Mining and Knowledge Discovery Handbook*, 2nd ed.,
DOI 10.1007/978-0-387-09823-4_31,

seeks to confirm. An interesting pattern represents knowledge. The quality of patterns depends both on the quality of the analysed data and the quality of the Data Mining results. Thus several techniques have been developed aiming at evaluating and preparing the data used as input to the Data Mining process. Also a number of techniques and measures have been developed aiming at evaluating and interpreting the extracted patterns.

Generally, the term 'Quality' in Data Mining corresponds to the following issues:

- *Representation of the 'real' knowledge* included in the analyzed data. The analyzed data hides interesting information that the Data Mining methods are called to reveal. The requirement for evaluating the validity of the extracted knowledge and representing it to be exploitable by the experts of domain is stronger than ever.
- *Algorithms tuning*. A number of algorithms and techniques have been proposed which under different assumptions could lead to different results. Also, there are Data Mining approaches considered as more suitable for specific application domains (e.g. spatial data, business, marketing etc.). The selection of a suitable method for a specific data analysis task in terms of their performance and the quality of its results is one of the major problems in Data Mining.
- *Selection of the most interesting and representative patterns for the data*. The number of patterns generated during the Data Mining process is very large but only a few of these patterns are likely to be of any interest to the domain expert analyzing the data. Many of the patterns are either irrelevant or obvious and do not provide new knowledge. The selection of the most representative patterns for a data set is another important issue in terms of the quality assessment.

Depending on the data-mining task the quality assessment approaches aim at estimating different aspects of quality. Thus, in the case of classification the quality refers to i) the ability of the designed classification model to correctly classify new data samples, ii) the ability of an algorithm to define classification models with high accuracy, and iii) the interestingness of the patterns extracted during the classification process. In clustering, the quality of extracted patterns is estimated in terms of their validity and their fitness to the analyzed data. The number of groups into which the analyzed data can be partitioned is another important problem in the clustering process. On the other hand the quality in association rules corresponds to the significance and interestingness of the extracted rules. Another quality criterion for association rules is the proportion of data that the extracted rules represent. Since the quality assessment is widely recognized as a major issue in Data Mining, techniques for evaluating the relevance and usefulness of discovered patterns attract the interests of researchers. These techniques are broadly referred to as:

- *Interestingness measures* in case of classification or association rules applications.
- *Cluster validity* indices (or measures) in case of clustering.

In the following section, there is a brief discussion about the role of pre-processing in the quality assessment. Then we proceed with the presentation of quality assessment techniques related to the Data Mining tasks. These techniques, depending on the Data Mining tasks refer to, are organized into the following categories: i) *Classifiers accuracy techniques* and related measures, ii) *Classification rules interestingness measures*, iii) *Association Rules Interestingness Measures*, and iv) *Cluster Validity* approaches.

31.1 Data Pre-processing and Quality Assessment

Data in the real world tends to be 'dirty'. Database users frequently report errors, unusual values, and inconsistencies in the stored data. Then, it is usual in the real world for the analyzed data to be:

- *incomplete* (i.e. lacking attribute values, lacking certain attributes of interest, or containing only aggregate data),
- *noisy*: containing errors or outliers,
- *inconsistent*: containing discrepancies in codes used to categorize items or the names used to refer to the same data items.

Based on a set of data that lacks quality the results of the mining process unavoidably tend to be inaccurate and lack any interest for the expert of domain. In other words quality decisions must be based on quality data. Data pre-processing is a major step in the knowledge discovery process. Data pre-processing techniques applied prior to Data Mining step could help to improve the quality of analyzed data and consequently the accuracy and efficiency of the subsequent mining processes.

There are a number of data pre-processing techniques aimed at substantially improving the overall quality of the extracted patterns (i.e. information included in analyzed data). The most widely used are summarized below (Han and Kamber, 2001):

- *Data cleaning* which can be applied to remove noise and correct inconsistencies in the data.
- *Data transformation.* A common transformation technique is the normalization. It is applied to improve the accuracy and efficiency of mining algorithms involving distance measurements.
- *Data reduction.* It is applied to reduce the data size by aggregating or eliminating redundant features.

31.2 Evaluation of Classification Methods

Classification is one of the most commonly applied Data Mining tasks and a number of classification approaches has been proposed in literature. These approaches can be compared and evaluated based on the following criteria (Han and Kamber, 2001):

- *Classification model accuracy*: The ability of the classification model to correctly predict the class into which new or previously unseen data are classified.
- *Speed:* It refers to the computation costs in building and using a classification model.
- *Robustness*: The ability of the model to handle noise or data with missing values and make correct predictions.
- *Scalability*: The method ability to construct the classification model efficiently given large amounts of data.
- *Interpretability*: It refers to the level of understanding that the constructed model provides.

31.2.1 Classification Model Accuracy

The accuracy of a classification model designed according to a set of training data is one of the most important and widely used criteria in the classification process. It allows one to evaluate how accurately the designed model (classifier) will classify future data (i.e. data on which the model has not been trained). Accuracy also helps in the comparison of different classifiers. The most common techniques for assessing the accuracy of a classifier are:

1. *Hold-out method.* The given data set is randomly partitioned into two independent sets, a training set and a test set. Usually, two thirds of the data are considered for the training set and the remaining data are allocated to the test set. The training data are used to define the classification model (classifier). Then the classifiers accuracy is estimated based on the test data. Since only a proportion of the data is used to derive the model the estimate of accuracy tends to be pessimistic. A variation of the hold-out method is the random subsampling technique. In this case the hold-out method is repeated k times and the overall accuracy is estimated as the average of the accuracies obtained from each iteration.
2. *k-fold cross-validation.* The initial data set is partitioned into k subsets, called 'folds', let $S = \{S_1, \ldots !S_k\}$. These subsets are mutually exclusive and have approximately equal size. The classifier is iteratively trained and tested k times. In iteration i, the S_i subset is reserved as the test set while the remaining subsets are used to train the classifier. Then the accuracy is estimated as the overall number of correct classifications from the k iterations, divided by the total number of samples in the initial data. A variation of this method is the stratified cross-validation in which the subsets are stratified so that the class distribution of the samples in each subset is approximately the same as that in the initial data set.
3. *Bootstraping.* This method is k-fold cross validation, with k set to the number of initial samples. It samples the training instances uniformly with replacement and leave-one-out. In each iteration, the classifier is trained on the set of $k-1$ samples that is randomly selected from the set of initial samples, S. The testing is performed using the remaining subset.

Though the use of the above-discussed techniques for estimating classification model accuracy increases the overall computation time, they are useful for assessing the quality of classification models and/or selecting among the several classifiers.

Alternatives to the Accuracy Measure

There are cases that the estimation of an accuracy rate may mislead one about the quality of a derived classifier. For instance, assume a classifier is trained to classify a set of data as 'positive' or 'negative'. A high accuracy rate may not be acceptable since the classifier could correctly classify only the negative samples giving no indication about the ability of the classifier to recognize positive and negative samples. In this case, the sensitivity and specificity measures can be used as an alternative to the accuracy measures (Han and Kamber, 2001).

Sensitivity assesses how well the classifier can recognize positive samples and is defined as

$$Sensitivity = \frac{true\_positive}{positive} \tag{31.1}$$

where $true\_positive$ corresponds to the number of the true positive samples and positive is the number of positive samples.

Specificity measures how well the classifier can recognize negative samples. It is defined as

$$Specificity = \frac{true\_negative}{negative} \tag{31.2}$$

where $true\_negative$ corresponds to the number of the true negative examples and negative the number of samples that is negative.

The measure that assesses the percentage of samples classified as positive that are actually positive is known as *precision*. That is,

$$Precision = \frac{true\_positive}{true\_positive + false\_positive} \tag{31.3}$$

Based on above definitions the *accuracy* can be defined as a function of *sensitivity* and *specificity*:

$$Accuracy = Sensitivity \cdot \frac{positive}{positive + negative} + Specificity \cdot \frac{negative}{positive + negative} \tag{31.4}$$

In the classification problem discussed above it is considered that each training sample belongs to only one class, i.e. the data are uniquely classified. However, there are cases where it is more reasonable to assume that a sample may belong to more than one class. It is then necessary to derive models that assign data to classes with an attached degree of belief. Thus classifiers return a probability class distribution rather than a class label. The accuracy measure is not appropriate in this case since it assumes the unique classification of samples for its definition. An alternative is to use heuristics where a class prediction is considered correct if it agrees with the first or second most probable class.

31.2.2 Evaluating the Accuracy of Classification Algorithms

A classification (learning) algorithm is a function that given a set of examples and their classes constructs a classifier. On the other hand a classifier is a function that given an example assigns it to one of the predefined classes. A variety of classification methods have already been developed (Han and Kamber, 2001). The main question that arises in the development and application of these algorithms is about the accuracy of the classifiers they produce.

Below we shall discuss some of the most common statistical methods proposed (Dietterich, 1998) for answering the following question: *Given two classification algorithms A and B and a data set S, which algorithm will produce more accurate classifiers when trained on data sets of the same size?*

McNemar's Test

Let S be the available set of data, which is divided into a training set R, and a test set T. Then we consider two algorithms A and B trained on the training set and the result is the definition of two classifiers and . These classifiers are tested on T and for each example $x \in T$ we record how it was classified. Thus the contingency table presented in Table 31.1 is constructed.

Table 31.1. McNemar 's test:Contingency table

Number of examples misclassified by both classifiers (n_{00})	Number of examples misclassified by $\hat{f}_A$ but not by $\hat{f}_B$(n_{01})
Number of examples misclassified by $\hat{f}_B$ but not by $\hat{f}_A$(n_{10})	Number of examples misclassified neither by $\hat{f}_A$ nor $\hat{f}_B$(n_{11})

The two algorithms should have the same error rate under the null hypothesis, Ho. McNemar's test is based on a χ^2 test for goodness-of-fit that compares the distribution of counts

Table 31.2. Expected counts under Ho

n_{00}	$(n_{01}+n_{10})/2)$
$(n_{01}+n_{10})/2)$	$n_{11})$

expected under null hypothesis to the observed counts. The expected counts under Ho are presented in Table 31.2.

The following statistic, s, is distributed as χ^2 with 1 degree of freedom. It incorporates a "continuity correction" term (of -1 in the numerator) to account for the fact that the statistic is discrete while the χ^2 distribution is continuous:

$$s = \frac{(|n_{10} - n_{01}| - 1)^2}{n_{10} + n_{01}}$$

According to the probabilistic theory (Athanasopoulos, 1991), if the null hypothesis is correct, the probability that the value of the statistic, s, is greater than $\chi^2_{1,0.95}$ is less than 0.05, i.e. $P(|s| > \chi^2_{1,0.95}) < 0.05$. Then to compare the algorithms A and the definied classifiers $\hat{f}_A$ and $\hat{f}_B$ are tested on T and the value of s is estimated as described above. Then if $|s| > \chi^2_{1,0.95}$, the null hypothesis could be rejected in favor of the hypothesis that the two algorithms have different performance when trained on the particular training set R.

The shortcomings of this test are:

1. It does not directly measure variability due to the choice of the training set or the internal randomness of the learning algorithm. The algorithms are compared using a single training set R. Thus McNemar's test should be only applied if we consider that the sources of variability are small.
2. It compares the performance of the algorithms on training sets, which are substantially smaller than the size of the whole data set. Hence we must assume that the relative difference observed on training sets will still hold for training sets of size equal to the whole data set.

A Test for the Difference of Two Proportions

This statistical test is based on measuring the difference between the error rate of algorithm A and the error rate of algorithm B (Snedecor and Cochran, 1989). More specifically, let $p_A = (n_{00} + n_{01})/n$ be the proportion of test examples incorrectly classified by algorithm A and let $p_B = (n_{00} + n_{10})/n$ be the proportion of test examples incorrectly classified by algorithm B. The assumption underlying this statistical test is that when algorithm A classifies an example x from the test set T, the probability of misclassification is p_A. Then the number of misclassifications of n test examples is a binomial random variable with mean np_A and variance $p_A(1 - p_A)n$.

The binomial distribution can be well approximated by a normal distribution for reasonable values of n. The difference between two independent normally distributed random variables is itself normally distributed. Thus, the quantity $p_A - p_B$ can be viewed as normally distributed if we assume that the measured error rates p_A and p_B are independent. Under the null hypothesis, Ho, it will have a mean of zero and a standard deviation error of

$$se = \sqrt{2p \cdot \left(1 - \frac{p_A + p_B}{2}\right)/n}$$

where n is the number of test examples.

Based on the above analysis, we obtain the statistic

$$z = \frac{p_A - p_B}{\sqrt{2p(1-p)/n}}$$

which has a standard normal distribution. According to the probabilistic theory if the z value is greater than $Z_{0.975}$ the probability of incorrectly rejecting the null hypothesis is less than 0.05. Thus the null hypothesis could be rejected if $|z| > Z_{0.975} = 1.96$ in favor of the hypothesis that the two algorithms have different performances. There are several problems with this statistic, two of the most important being:

1. The probabilities p_A and p_B are measured on the same test set and thus they are not independent.
2. The test does not measure variation due to the choice of the training set or the internal variation of the learning algorithm. Also it measures the performance of the algorithms on training sets of size significantly smaller than the whole data set.

The Resampled Paired t Test

The resampled paired t test is the most popular in machine learning. Usually, the test conducts a series of 30 trials. In each trial, the available sample S is randomly divided into a training set R (it is typically two thirds of the data) and a test set T. The algorithms A and B are both trained on R and the resulting classifiers are tested on T. Let $p_A^{(i)}$ and $p_B^{(i)}$ be the observed proportions of test examples misclassified by algorithm A and B respectively during the i-th trial. If we assume that the 30 differences $p^{(i)} = p_A^{(i)} - p_B^{(i)}$ were drawn independently from a normal distribution, then we can apply Student's t test by computing the statistic

$$t = \frac{\bar{p} \cdot \sqrt{n}}{\sqrt{\frac{\sum_{i=1}^{n}(p^{(i)} - \bar{p})^2}{n-1}}}$$

where $\bar{p} = \frac{1}{n} \cdot \sum_{i=1}^{n} p^{(i)}$. Under null hypothesis this statistic has a t distribution with $n-1$ degrees of freedom. Then for 30 trials, the null hypothesis could be rejected if $|t| > t_{29,0.975} = 2.045$. The main drawbacks of this approach are:

1. Since $p_A^{(i)}$ and $p_B^{(i)}$ are not independent, the difference $p^{(i)}$ will not have a normal distribution.
2. The $p^{(i)}$'s are not independent, because the test and training sets in the trials overlap.

The k-fold Cross-validated Paired$t$ Test

This approach is similar with the resampled paired t test except that instead of constructing each pair of training and test sets by randomly dividing S, the data set is randomly divided into k disjoint sets of equal size, $T_1, T_2, \ldots, T_k$. Then k trials are conducted. In each trial, the test set is T_i and the training set is the union of all of the others T_j, $j \neq i$. The t statistic is computed as described in Section 31.2.2. The advantage of this approach is that each test set is independent of the others. However, there is the problem that the training sets overlap. This overlap may prevent this statistical test from obtaining a good estimation of the amount of variation that would be observed if each training set were completely independent of the others training sets.

31.2.3 Interestingness Measures of Classification Rules

The number of classification patterns generated could be very large and it is possible that different approaches do result in different sets of patterns. The patterns extracted during the classification process could be represented in the form of rules, known as classification rules. It is important to evaluate the discovered patterns identifying these ones that are valid and provide new knowledge. Techniques that aim at this goal are broadly referred to as *interestingness measures*. The *interestingness* of the patterns that discovered by a classification approach could also be considered as another quality criterion. Some representative measures (Hilderman and Hamilton, 1999) for ranking the usefulness and utility of discovered classification patterns (classification rules) are:

- *Rule-Interest Function*. Piatetsky-Shapiro introduced the rule-interest (Piatetsky-Shapiro, 1991) that is used to quantify the correlation between attributes in a classification rule. It is suitable only for the single classification rules, i.e. the rules whose both the left- and right-hand sides correspond to a single attribute.
- *Smyth and Goodman's J-Measure*. The J-measure (Smyth and Goodman, 1991) is a measure for probabilistic classification rules and is used to find the best rules relating discrete-valued attributes. A probabilistic classification rule is a logical implication, $X \rightarrow Y$, satisfied with some probability p. The left- and right-hand sides of this implication correspond to a single attribute. The right-hand side is restricted to simple single-valued assignment expression while the left-hand-side may be a conjunction of simple expressions.
- *General Impressions*. In (Liu et al., 1997) general impression is proposed as an approach for evaluating the importance of classification rules. It compares discovered rules to an approximate or vague description of what is considered to be interesting. Thus a general impression can be considered as a kind of specification language.
- *Gago and Bento's Distance Metric*. The distance metric (Gago and Bentos, 1998) measures the distance between classification rules and is used to determine the rules that provide the highest coverage for the given data. The rules with the highest average distance to the other rules are considered to be most interesting.

For additional discussion regarding interestingness measures please refer to Chapter 29.5 in this volume.

31.3 Association Rules

Mining rules is one of the main tasks in the Data Mining process. It has attracted considerable interest because the rule provides a concise statement of potentially useful information that is easily understood by the end-users.

There is a lot of research in the field of association rule extraction, resulting in a variety of algorithms that efficiently analyzing data and extract rules from them. The extracted rules have to satisfy some user-defined thresholds related with association rule measures (such as support, confidence, leverage, lift).

These measures give an indication of the association rules' importance and confidence. They may represent the predictive advantage of a rule and help to identify interesting patterns of knowledge in data and make decisions. Below we shall briefly summarize these measures.

31.3.1 Association Rules Interestingness Measures

Let $LHS \rightarrow RHS$ be an association rule. Further we refer to the left hand side and the right hand side of the rule as LHS and RHS respectively. Below some of the most known measures of the rule interestingness are presented (Han and Kamber, 2001, Berry and Linoff, 1996).

Coverage

The coverage of an association rule is the proportion of cases in the data that have the attribute values or items specified on the Left Hand Side of the rule:

$$Coverage = \frac{n(LHS)}{N} \tag{31.5}$$

or

$$Coverage = P(LHS)$$

where N is the total number of cases under consideration and $n(LHS)$ denotes the number of cases covered by the Left Hand Side. *Coverage* takes values in $[0,1]$. An association rule with *coverage* value near 1 can be considered as an interesting association rule.

Support

The *support* of an association rule is the proportion of all cases in the data set that satisfy a rule, i.e. both LHS and RHS of the rule. More specifically, *support* is defined as

$$Support = \frac{n(LHS \cap RHS)}{N} \tag{31.6}$$

or

$$Support = P(LHS \cap RHS)$$

where N is the total number of cases under consideration and $n(LHS)$ denotes the number of cases covered by the Left Hand Side.

Support can be considered as an indication of how often a rule occurs in a data set and as a consequence how significant is a rule.

Confidence

The *confidence* of an association rule is the proportion of the cases covered by LHS of the rule that are also covered by RHS:

$$Confidence = \frac{n(LHS \cap RHS)}{n(LHS)} \tag{31.7}$$

or

$$Confidence = \frac{P(LHS \cap RHS)}{P(LHS)}$$

where $n(LHS)$ denotes the number of cases covered by the Left Hand Side. Confidence takes values in $[0,1]$. A value of *confidence* near to 1 is an indication of an important association rule.

The above discussed interestingness measures, *support* and *confidence*, are widely used in the association rule extraction process and are also known as Agrawal and Srikant's Itemset measures. From their definitions, we could say that *confidence* corresponds to the strength while *support* to the statistical significance of a rule.

Leverage

The *leverage* (MAGOpus) of an association rule is the proportion of additional cases covered by both the LHS and RHS above those expected if the LHS and RHS were independent of each other. This is a measure of the importance of the association that includes both the *confidence* and the *coverage* of the rule. More specifically, it is defined as

$$Leverage = p(RHS|LHS) - (p(LHS) \cdot p(RHS)) \tag{31.8}$$

Leverage takes values in $[-1,1]$. Values of *leverage* equal or under 0, indicate a strong independence between LHS and RHS. On the other hand, values of *leverage* near to 1 are an indication of an important association rule.

Lift

The *lift* of an association rule is the *confidence* divided by the proportion of all cases that are covered by the RHS. This is a measure of the importance of the association and it is independent of coverage.

$$Lift = \frac{p(LHS \cap RHS)}{p(LHS) \cdot p(RHS)} \tag{31.9}$$

or

$$Lift = \frac{Confidence}{p(RHS)}$$

It takes values in R^+ (the space of the real positive numbers). Based on the values of *lift* we get the following inferences for the rules interestingness:

1. $lift \to 1$ means that RHS and LHS are independent, which indicates that the rule is not interesting.
2. *lift* values close to $+\infty$. Here, we have the following sub-cases:
   - $RHS \subseteq LHS$ **or** $LHS \subseteq RHS$. If any of these cases is satisfied, we may conclude that the rule is not interesting.
   - $P(RHS)$ **is close to** 0 **or** $P(RHS|LHS)$ **is close to 1**. The first case indicates that the rule is not important. On the other hand, the second case is a good indication that the rule is an interesting one.
3. $lift = 0$ means that $P(RHS|LHS) = 0 \Leftrightarrow P(RHS \cap LHS) = 0$, which indicates that the rule is not important.

Further discussion on interestigness measures

Based on the definition of the association rules and the related measures it is obvious that *support* is an indication of the rule's importance based on the amount of data that support it. For instance, assuming the rule $A \to B$, a high support of the rule is an indication that a high number of tuples contains both the left hand side and right hand side of this rule and thus it can be considered as a representative rule of our data set. Moreover! *confidence* expresses our confidence based on the available data that when the left hand side of the rule happens, the right hand side also happens.

Though support and confidence are useful to mine association rules in many applications, they could mislead us in some cases. Based on support-confidence framework a rule can be identified as interesting even though the occurrence of A does not imply the occurrence of B. In this case lift and *leverage* could be considered as alternative interestingness measures giving also an indication about the correlation of LHS and RHS.

Also *lift* is another measure, which may give an indication of rule significance, or how interesting is the rule. *Lift* represents the predictive advantage a rule offers over simply guessing based on the frequency of the rule consequence (RHS). Thus, *lift* may be an indication whether a rule could be considered as representative of the data so as to use it in the process of decision-making (Roberto et al., 1999). For instance, let a rule $A + B \rightarrow G$ with confidence 85% and $support(G) = 90\%$. Due to the high confidence of the rule we may conclude that it is a significant rule. On the other hand, the right hand side of the rule represents the 90% of the studied data, that is, a high proportion of the data contains G. Then, the rule may not be very interesting since there is a high probability the right hand side of the rule (G) to be satisfied by our data. More specifically, the rule may be satisfied by a high percentage of the data under consideration but at the same time the consequence of the rule (RHS) is high supported. As a consequence this rule may not make sense in making decisions or extracting general rule as regards the behavior of the data. Finally, the leverage expresses the hyper-representation of the rule in relation with its representation in data set if there is no interaction between LHS and RHS.

A similar measure, *conviction*! was proposed by Brin etal. (Brin et al., 1997). The formal definition is (n is the number of transactions in the database):

$$Conviction = \frac{n - p(RHS)}{(1 - Confidence)} \tag{31.10}$$

Both the lift and the conviction are monotone in the confidence.

31.3.2 Other approaches for evaluating association rules

There are also some other well-known approaches and measures for evaluating association rules are:

- *Rule templates* are used to describe a pattern for those attributes that can appear in the left- or right-hand side of an association rule. A rule template may be either inclusive or restrictive. An inclusive rule template specifies desirable rules that are considered to be interesting. On the other hand a restrictive rule template specifies undesirable rules that are considered to be uninteresting. Rule pruning can be done by setting support, confidence and rule size thresholds.
- *Dong and Li's interestingness measure* (Dong and Li, 1998) is used to evaluate the importance of an association rule by considering its unexpectedness in terms of other association rules in its neighborhood. The neighborhood of an association rule consists of association rules within a given distance.
- *Gray and Orlowska's Interestingness* (Gray and Orlowka, 1998) to evaluate the confidence of associations between sets of items in the extracted association rules. Though *suppor* and *confidence* have been shown to be useful for characterizing association rules, *interestingness* contains a discriminator component that gives an indication of the independence of the antecedent and consequent.
- *Peculiarity* (Zhong et al., 1999) is a distance-based measure of rules interestingness. It is used to determine the extent to which one data object differs from other similar data objects.
- *Closed Association Rules Mining*. It is widely recognized that the larger the set of frequent itemsets, the more association rules are presented to the user, many of which turn out to be redundant. However it is not necessary to mine all frequent itemsets to guarantee that all non-redundant association rules will be found. It is sufficient to consider only the *closed*

frequent itemsets (Zaki and Hsiao, 2002, Pasquier et al., 1999, Pei et al., 2000). The set of closed frequent itemsets can guarantee completeness even in dense domains and *all non-redundant association rules* can be defined on it. **CHARM** is an efficient algorithm for closed association rules mining.

31.4 Cluster Validity

Clustering is a major task in the Data Mining process for discovering groups and identifying interesting distributions and patterns in the underlying data (Fayyad et al., 1996). Thus, the main problem in the clustering process is to reveal the organization of patterns into 'sensible' groups, which allow us to discover similarities and differences, as well as to derive useful inferences about them (Guha et al., 1999).

In the literature a wide variety of algorithms have been proposed for different applications and sizes of data sets (Han and Kamber, 2001), (Jain et al., 1999). The application of an algorithm to a data set aims at, assuming that the data set offers a clustering tendency, discovering its inherent partitions. However, the clustering process is perceived as an unsupervised process, since there are no predefined classes and no examples that would show what kind of desirable relations should be valid among the data (Berry and Linoff, 1996). Then, the various clustering algorithms are based on some assumptions in order to define a partitioning of a data set. As a consequence, they may behave in a different way depending on: i) the features of the data set (geometry and density distribution of clusters) and ii) the input parameter values.

A problem that we face in clustering is to decide the optimal number of clusters into which our data can be partitioned. In most algorithms' experimental evaluations 2D-data sets are used in order that the reader is able to visually verify the validity of the results (i.e. how well the clustering algorithm discovered the clusters of the data set). It is clear that visualization of the data set is a crucial verification of the clustering results. In the case of large multidimensional data sets (e.g. more than three dimensions) effective visualization of the data set would be difficult. Moreover the perception of clusters using available visualization tools is a difficult task for humans that are not accustomed to higher dimensional spaces.

As a consequence, if the clustering algorithm parameters are assigned an improper value, the clustering method may result in a partitioning scheme that is not optimal for the specific data set leading to wrong decisions. The problems of deciding the number of clusters (i.e. partitioning) better fitting a data set as well as the evaluation of the clustering results has been the subject of several research efforts (Dave, 1996, Gath and Geva, 1989, Theodoridis and Koutroubas, 1999, Xie and Beni, 1991).

We shall now discuss the fundamental concepts of clustering validity. Furthermoref we present the external and internal criteria in the context of clustering validity assessment while the relative criteria will be discussed in Section 31.4.4.

31.4.1 Fundamental Concepts of Cluster Validity

The procedure of evaluating the results of a clustering algorithm is known under the term *cluster validity*. In general terms, there are three approaches to investigating cluster validity (Theodoridis and Koutroubas, 1999). The first is based on *external criteria*. This implies that we evaluate the results of a clustering algorithm based on a pre-specified structure, which is imposed on a data set and reflects our intuition about the clustering structure of the data set. The second approach is based on *internal criteria*. The results of a clustering algorithm are

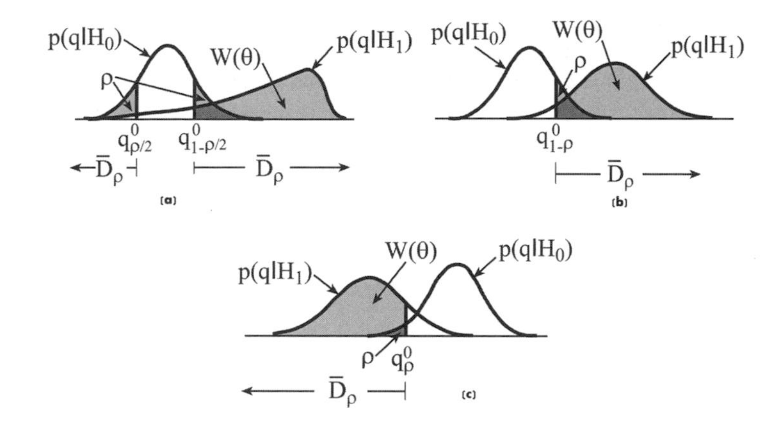

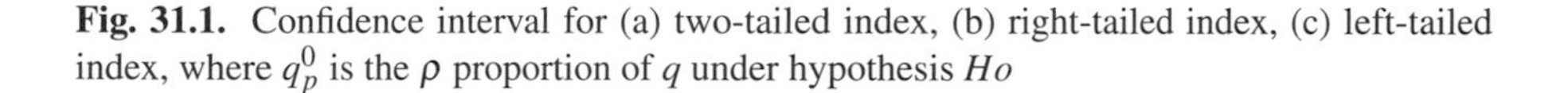
Fig. 31.1. Confidence interval for (a) two-tailed index, (b) right-tailed index, (c) left-tailed index, where q_p^0 is the ρ proportion of q under hypothesis Ho

evaluated in terms of quantities that involve the data themselves (e.g. proximity matrix). The third approach of clustering validity is based on *relative criteria*. Here the basic idea is the evaluation of a clustering structure by comparing it to other clustering schemes, resulting with the same algorithm but with different parameter values.

The first two approaches are based on statistical tests and their major drawback is their high computational cost. Moreover, the indices related to these approaches aim at measuring the degree to which a data set confirms an a priori specified scheme. On the other hand, the third approach aims at finding the best clustering scheme that a clustering algorithm can define under certain assumptions and parameters.

External and Internal Validity Indices

In this section, we discuss methods suitable for the quantitative evaluation of the clustering results, known as cluster validity methods. However, these methods give an indication of the quality of the resulting partitioning and thus they can only be considered as a tool at the disposal of the experts in order to evaluate the clustering results.

The cluster validity approaches based on external and internal criteria rely on statistical hypothesis testing. In the following section, an introduction to the fundamental concepts of hypothesis testing in cluster validity is presented.

Hypothesis Testing in Cluster Validity

In cluster validity the basic idea is to test whether the points of a data set are randomly structured or not. This analysis is based on the *Null Hypothesis*, denoted as Ho, expressed as a statement of random structure of a data set X. To test this hypothesis we use statistical tests, which lead to a computationally complex procedure. Monte Carlo techniques, discussed below, are used as a solution to high computational problems (Theodoridis and Koutroubas, 1999).

How Monte Carlo is used in Cluster Validity
The goal of using Monte Carlo techniques is the computation of the probability density function (*pdf*) of the validity indices. They rely on simulating the process of estimating the *pdf* of a validity index using a sufficient number of computer-generated data. First, a large amount of synthetic data sets is generated by normal distribution. For each one of these synthetic data sets, called X_i, the value of the defined index, denoted q_i, is computed. Then based on the respective values of q_i for each of the data sets X_i, we create a scatter-plot. This scatter-plot is an approximation of the probability density function of the index. Figure 31.1 depicts the three possible cases of probability density function's shape of an index q. There are three different possible shapes depending on the critical interval $\bar{D_\rho}$, corresponding to *significant level* ρ (statistic constant). The probability density function of a statistic index q, under Ho, has a single maximum and the region is either a half line, or a union of two half lines (Theodoridis and Koutroubas, 1999). Assuming that the scatter-plot has been generated using r-values of the index q, called q_i, in order to accept or reject the *Null Hypothesis* Ho we examine the following conditions:

---

if the shape is right-tailed (Figure 31.1b) **if** (q's value of our data set, is greater than $(1-\rho)\cdot r$ of q_i values) **then** Reject Ho **else** Accept Ho **endif else if** the shape is left-tailed (Figure 31.1c), **if** (q's value for our data set, is smaller than $\rho \cdot r$ r of q_i values) **then** Reject Ho **else** Accept Ho **endif else if** the shape is two-tailed (Figure 31.1a) **if** (q is greater than $(\rho/2)\cdot r$ number of q_i values and smaller than $(1-\rho/2)\cdot r$ of q_i values) **then** Accept Ho **endif endif**

31.4.2 External Criteria

Based on external criteria we can work in two different ways. Firstly, we can evaluate the resulting clustering structure **C**, by comparing it to an independent partition of the data P built according to our intuition about the clustering structure of the data set. Secondly, we can compare the proximity matrix P to the partition P.

Comparison of C with partition P (non- hierarchical clustering)
Let $C = \{C_1 \ldots C_m\}$ be a clustering structure of a data set X and $P = \{P_1, \ldots, P_s\}$ be a defined partition of the data. We refer to a pair of points (x_v, x_u) from the data set using the following terms:

- **SS**: if both points belong to the same cluster of the clustering structure C and to the same group of partition P.
- **SD**: if points belong to the same cluster of C and to different groups of P.
- **DS**: if points belong to different clusters of C and to the same group of P.
- **DD**: if both points belong to different clusters of C and to different groups of P.

Assuming now that **a, b, c** and **d** are the number of SS, SD, DS and DD pairs respectively, then $a+b+c+d = M$ which is the maximum number of all pairs in the data set (meaning, $M = N\cdot(N-1)/2$ where N is the total number of points in the data set).

Now we can define the following indices to measure the degree of similarity between C and P:

1. *Rand Statistic*: $R = (a+d)/M$
2. *Jaccard Coefficient*: $J = a/(a+b+c)$ The above two indices range between 0 and 1, and are maximized when m=s. Another index is the:
3. *Folkes and Mallows index:*

$$FM = a/\sqrt{m_1 \cdot m_2} = \sqrt{\frac{a}{a+b} \cdot \frac{a}{a+c}} \tag{31.11}$$

where $m_1 = (a+b)$, $m_2 = (a+c)$. For the previous three indices it has been proven that high values of indices indicate great similarity between C and P. The higher the values of these indices are the more similar C and P are. Other indices are:
4. *Huberts Γ statistic*:

$$\Gamma = (1/M) \sum_{i=1}^{N-1} \sum_{j=i+1}^{N} X(i,j) \cdot Y(i,j) \tag{31.12}$$

High values of this index indicate a strong similarity between the matrices X and Y.
5. *Normalized Γ statistic:*

$$\hat{\Gamma} = \frac{\left[(1/M)\sum_{i=1}^{N-1}\sum_{j=i+1}^{N}(X(i,j)-\mu_X)(Y(i,j)-\mu_Y)\right]}{\sigma_X \cdot \sigma_Y} \tag{31.13}$$

where $X(i,j)$ and $Y(i,j)$ are the (i,j) element of the matrices X, Y respectively that we wish to compare. Also μ_x, μ_y, σ_x, σ_y are the respective means and variances of X, Y matrices. This index takes values between -1 and 1.

All these statistics have right-tailed probability density functions, under the random hypothesis. In order to use these indices in statistical tests we must know their respective probability density function under the Null Hypothesis, *Ho*, which is the hypothesis of random structure of our data set. This means that using statistical tests, if we accept the Null Hypothesis then our data are randomly distributed. However, the computation of the probability density function of these indices is computationally expensive. A solution to this problem is to use Monte Carlo techniques. The procedure is as follows:

Algorithm 1: Monte Carlo Algorithm

1: **for** $i = 1$ to r **do**
2: Generate a data set X_i with N vectors (points) in the area of X (i.e. having the same dimension with those of the data set X).
3: Assign each vector $y_{j,i}$ of X_i to the group that $x_j \in X$ belongs, according to the partition P.
4: Run the same clustering algorithm used to produce structure C, for each X_i, and let Ci the resulting clustering structure.
5: Compute $q(C_i)$ value of the defined index q for P and C_i.
6: **end for**
7: Create scatter-plot of the r validity index values, $q(C_i)$ (that computed into the for loop).

After having plotted the approximation of the probability density function of the defined statistic index, its value, denoted by q, is compared to the $q(C_i)$ values, further referred to as q_i. The indices R, J, FM, G defined previously are used as the q index mentioned in the above procedure.

Comparison of P (proximity matrix) with partition P
Let P be the proximity matrix of a data set X and **P** be its partitioning. Partition **P** can be considered as a mapping

$$g : X \rightarrow 1, \ldots, n_c$$

where n_c is the number of clusters.

Assuming the matrix Y defined as:

$$Y(i,j) = \begin{cases} 1 \text{ if } g(x_i) \neq g(x_j), \\ 0 \text{ otherwise} \end{cases}$$

The Γ (or normalized Γ) statistic index can be computed using the proximity matrix P and the matrix Y. Based on the index value, we may have an indication of the two matrices' similarity.

To proceed with the evaluation procedure we use the Monte Carlo techniques as mentioned above. In the *'Generate'* step of the procedure the corresponding mappings gi is generated for every generated X_i data set. So in the *'Compute'* step the matrix Y_i is computed for each X_i in order to find the Γ_i corresponding statistic index.

31.4.3 Internal Criteria

Using this approach of cluster validity the goal is to evaluate the clustering result of an algorithm using only quantities and features inherited from the data set. There are two cases in which we apply internal criteria of cluster validity depending on the clustering structure: a) hierarchy of clustering schemes, and b) single clustering scheme.

Validating hierarchy of clustering schemes
A matrix called cophenetic matrix, P_c, can represent the hierarchy diagram that is produced by a hierarchical algorithm. The element $P_c(i,j)$, of cophenetic matrix represents the proximity level at which the two vectors x_i and x_j are found in the same cluster for the first time. We may define a statistical index to measure the degree of similarity between P_c and P (proximity matrix) matrices. This index is called *Cophenetic Correlation Coefficient* and defined as:

$$CPCC = \frac{(1/M) \cdot \sum_{i=1}^{N-1} \sum_{j=i+1}^{N} d_{ij} \cdot c_{ij} - \mu_P \cdot \mu_C}{\sqrt{\left[(1/M) \sum_{i=1}^{N-1} \sum_{j=i+1}^{N} (d_{ij}^2 - \mu_P^2)\right] \cdot \left[(1/M) \sum_{i=1}^{N-1} \sum_{j=i+1}^{N} (c_{ij}^2 - \mu_C^2)\right]}} \quad (31.14)$$

where $M = N \cdot (N-1)/2$ and N is the number of points in a data set. Also, μ_p and μ_c are the means of matrices P and P_c respectively, and are defined as follows:

$$\mu_P = (1/M) \sum_{i=1}^{N-1} \sum_{j=i+1}^{N} P(i,j), \mu_C = (1/M) \sum_{i=1}^{N-1} \sum_{j=i+1}^{N} P_c(i,j) \quad (31.15)$$

Moreover, d_{ij}, c_{ij} are the (i,j) elements of P and P_c matrices respectively. The CPCC are between -1 and 1. A value of the index close to 1 is an indication of a significant similarity between the two matrices. The procedure of the Monte Carlo techniques described above is also used in this case of validation.

Validating a single clustering scheme
The goal here is to find the degree of match between a given clustering scheme C, consisting of n_c clusters, and the proximity matrix P. The defined index for this approach is Hubert's G statistic (or normalized G statistic). An additional matrix for the computation of the index is used, that is

$$Y = \begin{cases} 1 \text{ if } x_i \text{ and } x_j \text{ belong to different clusters} \\ 0 \text{ otherwise} \end{cases}$$

where $i, j = 1, \ldots, N$.

The application of Monte Carlo techniques is also the way to test the random hypothesis in a given data set.

31.4.4 Relative Criteria

The basis of the above described validation methods is statistical testing. Thus, the major drawback of techniques based on internal or external criteria is their high computational demands. A different validation approach is discussed in this section. It is based on relative criteria and does not involve statistical tests. The fundamental idea of this approach is to choose the best clustering scheme of a set of defined schemes according to a pre-specified criterion. More specifically, the problem can be stated as follows:

Let P_{alg} be the set of parameters associated with a specific clustering algorithm (e.g. the number of clusters n_c). Among the clustering schemes C_i, $i = 1, \ldots, n_c$, is defined by a specific algorithm. For different values of the parameters in P_{alg}, choose the one that best fits the data set.

Then, we can consider the following cases of the problem:

1. P_{alg} **does not contain the number of clusters, n_c, as a parameter**. In this case, the choice of the optimal parameter values are described as follows: The algorithm runs for a wide range of its parameters' values and the largest range for which n_c remains constant is selected (usually $n_c << N$ (number of tuples)). Then the values that correspond to the middle of this range are chosen as appropriate values of the P_{alg} parameters. Also, this procedure identifies the number of clusters that underlie our data set.
2. P_{alg} **contains n_c as a parameter**. The procedure of identifying the best clustering scheme is based on a validity index. Selecting a suitable performance index, q, we proceed with the following steps:
   - The clustering algorithm runs for all values of nc between a minimum $n_{c_{min}}$ and a maximum $n_{c_{max}}$. The minimum and maximum values have been defined a priori by the user.
   - For each of the values of n_c, the algorithm runs r times, using different sets of values for the other parameters of the algorithm (e.g. different initial conditions).
   - The best values of the index q obtained by each n_c is plotted as the function of n_c.

Based on this plot we may identify the best clustering scheme. We have to stress that there are two approaches for defining the best clustering depending on the behavior of q with respect to nc. Thus, if the validity index does not exhibit an increasing or decreasing trend as n_c increases we seek the maximum (minimum) of the plot. On the other hand, for indices that increase (or decrease) as the number of clusters increase we search for the values of nc at which a significant local change in value of the index occurs. This change appears as a 'knee' in the plot and it is an indication of the number of clusters underlying the data set. Moreover, the absence of a knee may be an indication that the data set possesses no clustering structure. Below, some representative validity indices for crisp and fuzzy clustering are presented.

Crisp Clustering

Crisp clustering considers non overlapping partitions meaning that a data point either belongs to a class or not. In this section we discuss validity indices suitable for crisp clustering.

The modified Hubert Γ statistic
The definition of the modified Hubert Γ (Theodoridis and Koutroubas, 1999) statistic is given by the equation

$$\Gamma = (1/M) \sum_{i=1}^{N-1} \sum_{j=i+1}^{N} P(i,j) \cdot Q(i,j) \tag{31.16}$$

where N is the number of objects in a data set, $M = N \cdot (N-1)/2$, P is the proximity matrix of the data set and Q is an $N \times N$ matrix whose (i,j) element is equal to the distance between the representative points (v_{c_i}, v_{c_j}) of the clusters where the objects x_i and x_j belong.

Similarly, we can define the normalized Hubert σ statistic, given by equation

$$\hat{\Gamma} = \frac{\left[(1/M)\sum_{i=1}^{N-1}\sum_{j=i+1}^{N}(P(i,j)-\mu_P)(Q(i,j)-\mu_Q)\right]}{\sigma_P \cdot \sigma_Q} \tag{31.17}$$

where μ_P, μ_Q, σ_P, σ_Q are the respective means and variances of P, Q matrices.

If the $d(v_{c_i}, v_{c_j})$ is close to $d(x_i, x_j)$ for $i, j = 1, 2, \ldots, N$, P and Q will be in close agreement and the values of Γ and $\hat{\Gamma}$(normalized Γ) will be high. Conversely, a high value of Γ ($\hat{\Gamma}$) indicates the existence of compact clusters. Thus, in the plot of normalized Γ versus n_c, we seek a significant knee that corresponds to a significant increase of normalized G. The number of clusters at which the knee occurs is an indication of the number of clusters that occurs in the data. We note, that for $n_c = 1$ and $n_c = N$ the index is not defined.

Dunn family of indices
A cluster validity index for crisp clustering proposed in (Dunn, 1974), aims at the identification of 'compact and well separated clusters'. The index is defined in the following equation for a specific number of clusters

$$D_{n_c} = min_{i=1,\ldots,n_c}\left\{min_{j=i+1,\ldots,n_c}\left(\frac{d(c_i,c_j)}{max_{k=1,\ldots,n_c}(diam(c_k))}\right)\right\} \tag{31.18}$$

where $d(c_i, c_j)$ is the dissimilarity function between two clusters c_i and c_j defined as $d(c_i, c_j) = min_{x \in C_i, y \in C_j} d(x,y)$, and $diam(c)$ is the diameter of a cluster, which may be considered as a measure of clusters' dispersion. The diameter of a cluster C can be defined as follows:

$$diam(C) = max_{x,y \in C}\{d(x,y)\} \tag{31.19}$$

If the data set contains compact and well-separated clusters, the distance between the clusters is expected to be large and the diameter of the clusters is expected to be small. Thus, based on the Dunn's index definition, we may conclude that large values of the index indicate the presence of compact and well-separated clusters.

Index D_{n_c} does not exhibit any trend with respect to number of clusters. Then the maximum in the plot of D_{n_c} versus the number of clusters can be an indication of the number of clusters that fits the data.

The problems of the *Dunn* index are: i) its considerable time complexity, and ii) its sensitivity to the presence of noise in data sets, since these are likely to increase the values of $diam(c)$ (i.e. dominator of equation 31.18).

Three indices, are proposed in (Pal and Biswas, 1997) that are more robust to the presence of noise. They are known as Dunn-like indices since they are based on the Dunn index. Moreover, the three indices use for their definition the concepts of Minimum Spanning Tree (MST), the relative neighborhood graph (RNG) and the Gabriel graph respectively (Theodoridis and Koutroubas, 1999). Consider the index based on MST. Let a cluster ci and the complete graph G_i whose vertices correspond to the vectors of c_i. The weight, we, of an edge, e, of this graph equals the distance between its two end points, x, y. Let E_i^{MST} be the set of edges of the *MST* of the graph G_i, and e_i^{MST} the edge in E_i^{MST} with the maximum weight. Then the diameter of C_i is defined as the weight of e_i^{MST}. Dunn-like index based on the concept of the *MST* is given by equation

$$D_{n_c} = min_{i=1,\ldots,n_c} \left\{ min_{j=i+1,\ldots,n_c} \left(\frac{d(c_i, c_j)}{max_{k=1,\ldots,n_c}(diam_k^{MST})} \right) \right\} \tag{31.20}$$

The number of clusters at which D_m^{MST} takes its maximum value indicates the number of clusters in the underlying data. Based on similar arguments we may define the *Dunn-like* indices for GG and RGN graphs.

The Davies-Bouldin (DB) index

A similarity measure R_{ij} between the clusters C_i and C_j is defined based on a measure of dispersion of a cluster C_i, denoted by s_i, and a dissimilarity measure between two clusters, d_{ij}. The R_{ij} index is defined to satisfy the following conditions (Davies and Bouldin, 1979):

1. $R_{ij} = 0$
2. $R_{ij} = R_{ji}$
3. if $s_i = 0$ and $s_j = 0$ then $R_{ij} = 0$
4. if $s_j > s_k$ and $d_{ij} = d_{ik}$ then $R_{ij} > R_{ik}$
5. if $s_j = s_k$ and $d_{ij} < d_{ik}$ then $R_{ij} > R_{ik}$.

These conditions state that R_{ij} is non-negative and symmetric. A simple choice for R_{ij} that satisfies the above conditions is (Davies and Bouldin, 1979):

$$R_{ij} = (s_i + s_j)/d_{ij} \tag{31.21}$$

Then the *DB* index is defined as

$$DB_{n_c} = \frac{1}{n_c} \cdot \sum_{i=1}^{n_c} R_i, and R_i = max_{j=1,\ldots,n_c, j \neq i}\{R_{ij}\}, i = 1, \ldots, n_c \tag{31.22}$$

It is clear for the above definition that DB_{n_c} is the average similarity between each cluster c_i, $i = 1, \ldots, n_c$ and its most similar one. It is desirable for the clusters to have the minimum possible similarity to each other; therefore we seek partitionings that minimize DB_{n_c}. The DB_{n_c} index exhibits no trends with respect to the number of clusters and thus we seek the minimum value of DB_{n_c} in its plot versus the number of clusters.

Some alternative definitions of the dissimilarity between two clusters as well as the dispersion of a cluster, c_i, is defined in (Davies and Bouldin, 1979).

Three variants of the DB_{n_c} index are proposed in (Pal and Biswas, 1997). They are based on the MST, RNG and GG concepts, similar to the cases of the Dunn-like indices.

Other validity indices for crisp clustering have been proposed in (Dave, 1996) and (Milligan and Cooper, 1985). The implementation of most of these indices is computationally very expensive, especially when the number of clusters and objects in the data set grows very large (Xie and Beni, 1991). In (Milligan and Cooper, 1985), an evaluation study of 30 validity indices proposed in literature is presented. It is based on tiny data sets (about 50 points each) with well-separated clusters. The results of this study (Milligan and Cooper, 1985) place Caliski and Harabasz (1974), Je(2)/Je(1) (1984), C-index (1976), Gamma and Beale among the six best indices. However, it is noted that although the results concerning these methods are encouraging they are likely to be data dependent. Thus, the behavior of indices may change if different data structures are used. Also, some indices are based on a sample of clustering results. A representative example is Je(2)/Je(1) whose computations based only on the information provided by the items involved in the last cluster merge.

RMSSDT, SPR, RS, CD

This family of validity indices is applicable in the cases that hierarchical algorithms are used to cluster the data sets. Below we present the definitions of four validity indices, which have to be used simultaneously to determine the number of clusters existing in the data set. These four indices are applied to each step of a hierarchical clustering algorithm (Sharma, 1996).

- $RMSSTD$ (root mean square standard deviation) of a new clustering scheme defined at a level of a clustering hierarchy is the square root of the variance of all the variables (attributes used in the clustering process). This index measures the homogeneity of the formed clusters at each step of the hierarchical algorithm. Since the objective of cluster analysis is to form homogeneous groups the $RMSSTD$ of a cluster should be as small as possible. Where the values of RMSSTD are higher than the ones of the previous step, we have an indication that the new clustering scheme is worse.
  In the following definitions we shall use the term SS, which means Sum of Squares and refers to the equation:
  $$SS = \sum_{i=1}^{n} (X_i - \bar{X})^2 \qquad (31.23)$$
  Along with this we shall use some additional symbolism like:
  1. SSw referring to the sum of squares within group,
  2. SSb referring to the sum of squares between groups,
  3. SSt referring to the total sum of squares, of the whole data set.
- *SPR (Semi-Partial R-squared)* for the new cluster is the difference between SSw of the new cluster and the sum of the SSw's values of clusters joined to obtain the new cluster (loss of homogeneity), divided by the SSt for the whole data set. This index measures the loss of homogeneity after merging the two clusters of a single algorithm step. If the index value is zero then the new cluster is obtained by merging two perfectly homogeneous clusters. If its value is high then the new cluster is obtained by merging two heterogeneous clusters.
- *RS (R-Squared)* of the new cluster is the ratio of SSb over SSt. SSb is a measure of difference between groups. Since $SSt = SSb + SSw$, the greater the SSb the smaller the SSw and vice versa. As a result, the greater the differences between groups, the more homogenous

each group is and vice versa. Thus, *RS* may be considered as a measure of dissimilarity between clusters. Furthermore, it measures the degree of homogeneity between groups. The values of *RS* range between 0 and 1. Where the value of *RS* is zero, there is an indication that no difference exists among groups. On the other hand, when *RS* equals 1 there is an indication of significant difference among groups.

- The *CD index* measures the distance between the two clusters that are merged in a given step of the hierarchical clustering. This distance depends on the selected representatives for the hierarchical clustering we perform. For instance, in the case of Centroid hierarchical clustering the representatives of the formed clusters are the centers of each cluster, so *CD* is the distance between the centers of the clusters. Where we use single linkage, CD measures the minimum Euclidean distance between all possible pairs of points. In case of complete linkage, *CD* is the maximum Euclidean distance between all pairs of data points, and so on.

Using these four indices we determine the number of clusters that exist in a data set, plotting a graph of all these indices values for a number of different stages of the clustering algorithm. In this graph we search for the steepest knee, or in other words, the greatest jump of these indices' values from the higher to the smaller number of clusters.

The SD validity index

Another clustering validity approach is proposed in (Halkidi et al., 2000). The *SD* validity index definition is based on the concepts of *average scattering for clusters* and *total separation between clusters*. Below, we give the fundamental definition for this index.

Average scattering for clusters. It evaluates scattering of the points in the clusters comparing the variance of the considered clustering scheme with the variance of the whole data set. The average scattering for clusters is defined as follows:

$$Scat(n_c) = \frac{1}{n_c} \cdot \frac{\sum_{i=1}^{n_c} \|\sigma(v_i)\|}{\|\sigma(S)\|} \tag{31.24}$$

The term $\sigma(S)$ is the variance of a data set; and its p-th dimension is defined as follows:

$$\sigma^p = \frac{1}{n} \cdot \sum_{k=1}^{n} \left(x_k^p - \bar{x}^p\right) \tag{31.25}$$

where $\bar{x}^p$ is the p-th dimension of $\bar{X} = \frac{1}{n} \cdot \sum_{k=1}^{n} x_k, \forall x_k \in S$.

The term $\sigma(v_i)$ is the variance of cluster c_i and its p-th dimension is given by the equation

$$\sigma(v_i)^p = \sum_{k=1}^{n_i} (x_k^p - v_i^p)^2 / n_i \tag{31.26}$$

Further the term $\|Y\|$ is defined as: $\|Y\| = \left(Y^T Y\right)^{1/2}$, where $Y = (y_1, \ldots, y_k)$ is a vector (e.g. $\sigma(v_i)$).

Total separation between clusters. The definition of total scattering (separation) between clusters is given in the following equation:

$$Dis(n_c) = \frac{D_{max}}{D_{min}} \cdot \sum_{k=1}^{n_c} \left(\sum_{z=1}^{n_c} \|v_k - v_z\| \right)^{-1} \tag{31.27}$$

where $D_{max} = max(\|v_i - v_j\|)$ $i, j \in 1,2,3,,n_c$ is the maximum distance between cluster centers. The $D_{min} = min(\|v_i - v_j\|)$, $\forall i, j \in \{1,2,\dots,n_c\}$ is the minimum distance between cluster centers.

Now, we can define a validity index based on 31.24 and 31.27 as follows

$$SD(n_c) = a \cdot Scat(n_c) + Dis(n_c) \tag{31.28}$$

where a is a weighting factor equal to $Dis(c_{max})$ and where c_{max} is the maximum number of input clusters.

The first term (i.e. $Scat(n_c)$) is defined in Eq. 24, indicating the average compactness of clusters (i.e. intra-cluster distance). A small value for this term indicates compact clusters and as the scattering within clusters increases (i.e. they become less compact) the value of $Scat(n_c)$ also increases. The second term $Dis(n_c)$ indicates the total separation between the nc clusters (i.e. an indication of inter-cluster distance). Contrary to the first term the second one, $Dis(n_c)$, is influenced by the geometry of the clusters and increases with the number of clusters. The two terms of *SD* are of the different range, thus a weighting factor is needed in order to incorporate both terms in a balanced way. The number of clusters, n_c, that minimizes the above index is an optimal value. Also, the influence of the maximum number of clusters c_{max}, related to the weighting factor, in the selection of the optimal clustering scheme, is discussed in (Halkidi et al., 2000). It is proved that *SD* proposes an optimal number of clusters almost irrespectively of the c_{max} value.

The *S_Dbw* validity index

A recent validity index is proposed in (Halkidi and Vazirgiannis, 2001a). It exploits the inherent features of clusters to assess the validity of results and select the optimal partitioning for the data under concern. Similarly with the SD index, its definition is based on the compactness and separation of clusters. The average scattering for clusters is defined as above in 31.24.

Inter-cluster Density (ID) - It evaluates the average density in the region among clusters in relation with the density of the clusters. The goal is the density among clusters to be significantly low in comparison with the density in the considered clusters. Then, considering a partitioning of the data set into more than two clusters (i.e. $n_c > 1$) the inter-cluster density is defined as follows:

$$Dens\_bw(c) = \frac{1}{n_c \cdot (n_c - 1)} \sum_{i=1}^{n_c} \left(\sum_{j=1, j \neq i}^{n_c} \frac{density(u_{ij})}{max\{density(v_i), density(v_j)\}} \right), c > 1 \tag{31.29}$$

where v_i, v_j are the centers of clusters c_i, c_j respectively, and u_{ij} the middle point of the line segment defined by the clusters' centers v_i, v_j.

The term $density(u)$ is defined in the following equation:

$$density(u) = \sum_{l=1}^{n_{ij}} f(x_l, u) \tag{31.30}$$

where x_l is a point of data set S, n_{ij} is the number of points (tuples) that belong to the clusters c_i and c_j, i.e. $x_l \in c_i \cup c_j \subseteq S$. It represents the number of points in the neighborhood of u. In our work, the neighborhood of a data point, u, is defined to be a hyper-sphere with center u and radius the average standard deviation of the clusters, $stdev$. The standard deviation of the clusters is given by the following equation:

$$stdev = \frac{1}{n_c}\sqrt{\sum_{i=1}^{n_c}\|\sigma(v_i)\|}$$

where c is the number of clusters and $s(v_i)$ is the variance of cluster C_i.

More specifically, the function $f(x,u)$ is defined as:

$$Y = \begin{cases} 0 \text{ if } d(x,u) > stdev\ , \\ 1 \text{ otherwise} \end{cases} \tag{31.31}$$

It is obvious that a point belongs to the neighborhood of u if its distance from u is smaller than the average standard deviation of clusters. Here we assume that the data has been scaled to consider all dimensions (bringing them into comparable ranges), as is equally important during the process of finding the neighbors of a multidimensional point (Berry and Linoff, 1996).

Then the validity index *S_Dbw* is defined as:

$$S\_Dbw(n_c) = Scat(n_c) + Dens\_bw(n_c) \tag{31.32}$$

The above definitions refer to the case that a cluster presents clustering tendency, i.e. it can be partitioned into at least two clusters. The index is not defined for $n_c = 1$.

The definition of *S_Dbw* indicates that both criteria of 'good' clustering (i.e. compactness and separation) are properly combined, enabling reliable evaluation of clustering results. Also, the density variations among clusters are taken into account to achieve more reliable results. The number of clusters, n_c, that minimizes the above index is an optimal value indicating the number of clusters present in the data set.

Moreover, an approach based on the *S_Dbw* index is proposed in (Halkidi and Vazirgiannis, 2001b). It evaluates the clustering schemes of a data set as defined by different clustering algorithms and selects the algorithm resulting in optimal partitioning of the data.

In general terms, *S_Dbw* enables the selection both of the algorithm and its parameter values for which the optimal partitioning of a data set is defined (assuming that the data set presents clustering tendency). However, the index cannot properly handle arbitrarily shaped clusters. The same applies to all the aforementioned indices.

There are a number of applications where it is important to identify non-convex clusters such as medical or spatial data applications. An approach to handle arbitrarily shaped clusters in the cluster validity process is presented in (Halkidi and Vazirgiannis, 2002).

31.4.5 Fuzzy Clustering

In this section, we present validity indices suitable for fuzzy clustering. The objective is to seek clustering schemes where most of the vectors of the data set exhibit a high degree of membership in one cluster. Fuzzy clustering is defined by a matrix $U = \left[u_{ij}\right]$, where u_{ij} denotes the degree of membership of the vector x_i in cluster j. Also, a set of cluster representatives is defined. Similar to a crisp clustering case a validity index, $q!$ is defined and we search for the minimum or maximum in the plot of q versus n_c.

Also, where q exhibits a trend with respect to the number of clusters, we seek a significant knee of decrease (or increase) in the plot of q.

Below two categories of fuzzy validity indices are discussed. The first category uses only the membership values, u_{ij}, of a fuzzy partition of data. The second involves both the U matrix and the data set itself.

Validity Indices involving only the membership values

Bezdek proposed in (Bezdeck et al., 1984) the *partition coefficient*, which is defined as

$$PC = \frac{1}{N}\sum_{i=1}^{N}\sum_{j=1}^{n_c} u_{ij}^2 \tag{31.33}$$

The PC index values range in $[1/n_c, 1]$, where n_c is the number of clusters. The closer the index is to unity the "crisper" the clustering is. In case that all membership values to a fuzzy partition are equal, that is, $u_{ij} = 1/n_c$, the PC obtains its lower value. Thus, the closer the value of PC is to $1/n_c$, the fuzzier the clustering is. Furthermore, a value close to $1/n_c$ indicates that there is no clustering tendency in the considered data set or the clustering algorithm failed to reveal it.

The *partition entropy coefficient* is another index of this category. It is defined as follows

$$PE = -\frac{1}{N}\sum_{i=1}^{N}\sum_{j=1}^{n_c} u_{ij} \cdot log_a(u_{ij}) \tag{31.34}$$

where a is the base of the logarithm. The index is computed for values of n_c greater than 1 and its value ranges in $[0, log_a n_c]$. The closer the value of PE to 0, the 'crisper' the clustering is. As in the previous case, index values close to the upper bound (i.e. $log_a n_c$), indicate absence of any clustering structure in the data set or inability of the algorithm to extract it.

The drawbacks of these indices are:

- their monotonous dependency on the number of clusters. Thus, we seek significant knees of increase (for PC) or decrease (for PE) in the plots of the indices versus the number of clusters,
- their sensitivity to the fuzzifier, m. More specifically, as $m \to 1$ the indices give the same values for all values of n_c. On the other hand when $m \to \infty$, both PC and PE exhibit significant knee at $n_c = 2$,
- the lack of direct connection to the geometry of the data (Dave, 1996), since they do not use the data itself.

Indices involving the membership values and the data set

The *Xie-Beni index* (Xie and Beni, 1991), XB, also called the compactness and separation validity function, is a representative index of this category. Consider a fuzzy partition of the data set $X = \{x_j; j = 1,\ldots,n\}$ with v_i, $(i = 1,\ldots,n_c)$ the centers of each cluster and u_{ij} the membership of the jth data point belonging to the ith cluster. The *fuzzy deviation* of x_j form cluster i, d_{ij}, is defined as the distance between x_j and the center of cluster weighted by the fuzzy membership of data point j with regards to cluster i. It is given by the following equation:

$$d_{ij} = u_{ij}\left\|x_j - v_i\right\| \tag{31.35}$$

Also, for a cluster i, the sum of the squares of fuzzy deviation of the data point in X, denoted σ_i, is called variation of cluster i.

The term $\pi_i = (\sigma_i/n_i)$, is called compactness of cluster i. Since n_i is the number of point in cluster belonging to cluster i, π_i is the average variation in cluster i. Then the compactness of a partitioning of n_c clusters is defined as the average compactness of the defined clusters, given by the equation:

$$\pi = \frac{\sum_{i=1}^{n_c} \pi_i}{n_c} \tag{31.36}$$

Also, the separation of the fuzzy partitions is defined as the minimum distance between cluster centers, that is

$$d_{min} = min \left\| v_i - v_j \right\| \tag{31.37}$$

Then *XB index* is defined as

$$XB = \frac{\pi}{N \cdot (d_{min})^2} \tag{31.38}$$

where N is the number of points in the data set.

It is clear that small values of XB are expected for compact and well-separated clusters. We note, however, that XB is monotonically decreasing when the number of clusters n_c gets very large and close to n. One way to eliminate this decreasing tendency of the index is to determine a starting point, c_{max}, of the monotonic behavior and to search for the minimum value of XB in the range [2, c_{max}]. Moreover, the values of the index XB depend on the fuzzifier values, so as if $m \to \infty$ then $XB \to \infty$.

Another index of this category is the *Fukuyama-Sugeno index*, which is defined as

$$FS_m = \sum_{i=1}^{N} \sum_{j=1}^{n_c} u_{ij}^m \left(\left\| x_i - v_j \right\|_A^2 - \left\| v_j - v \right\|_A^2 \right) \tag{31.39}$$

where v is the mean vector of X and A is an lxl positive definite, symmetric matrix. When $A = I$, the above distance becomes the squared Euclidean distance. It is clear that for compact and well-separated clusters we expect small values for FS_m. The first term in brackets measures the compactness of the clusters while the second one measures the distances of the clusters representatives.

Also some other fuzzy validity indices are proposed in (Gath and Geva, 1989), which are based on the concepts of hyper volume and density.

31.4.6 Other Approaches for Cluster Validity

Another approach for finding the optimal number of clusters of a data set was proposed in (Smyth, 1996). It introduces a practical clustering algorithm based on Monte Carlo cross-validation. More specifically, the algorithm consists of M cross-validation runs over M chosen train/test partitions of a data set, D. For each partition u, the EM algorithm is used to define n_c clusters to the training data, while n_c is varied from 1 to c_{max}. Then, the log-likelihood $L_c^u(D)$ is calculated for each model with n_c clusters. It is defined using the probability density function of the data as

$$L_k(D) = \sum_{i=1}^{N} log f_k(x_i / \Phi_k) \tag{31.40}$$

where f_k is the probability density function for the data and Φ_k denotes parameters that have been estimated from data. This is repeated M times and the M cross-validated estimates are averaged for each of n_c. Based on these estimates we may define the posterior probabilities for each value of the number of clusters n_c, $p(n_c/D)$. If one of $p(n_c/D)$ is near 1, there is strong evidence that the particular number of clusters is the best for our data set. The evaluation approach proposed in (Smyth, 1996) is based on density functions considered for the data set. Thus, it is based on concepts related to probabilistic models in order to estimate the number of clusters, better fitting a data set, and it does not use concepts directly related to the data, (i.e. inter-cluster and intra-cluster distances).

References

Athanasopoulos, D. (1991). *Probabilistic Theory*. Stamoulis, Piraeus.

Berry, M. and Linoff, G. (1996). *Data Mining Techniques for Marketing, Sales and Customer Support.* John Wiley and Sons, Inc.

Bezdeck, J., Ehrlich, R., and Full, W. (1984). Fcm:fuzzy c-means algorithm. *Computers and Geoscience.*

Brin, S., Motwani, R., Ullman, J., and Tsur, S. (1997). Dynamic itemset counting and implication rules for market basket data. *CSIGMOD Record (ACM Special Interest Group on Management of Data)*, 26(2).

Dave, R. (1996). Validating fuzzy partitions obtained through c-shells clustering. *Pattern Recognition Letters*, 10:613–623.

Davies, D. and Bouldin, D. (1979). A cluster separation measure. *PIEEE Transactions on Pattern Analysis and Machine Intelligence*, 1(2).

Dietterich, T. (1998). Approximate statistical tests for comparing supervised classification learning algorithms. *Neural Computation*, 10(7):6.

Dong, G. and Li, J. (1998). Research and development in knowledge discovery and Data Mining. In *Proc. 2nd Pacific-Asia Conf. Knowledge Discovery and Data Mining(PAKDD).*

Dunn, J. (1974). Well separated clusters and optimal fuzzy partitions. *Cybernetics*, 4:95–104.

Fayyad, M., Piatesky-Shapiro, G., Smuth, P., and Uthurusamy, R. (1996). *Advances in Knowledge Discovery and Data Mining*. AAAI Press.

Gago, P. and Bentos, C. (1998). A metric for selection of the most promising rules. In *Proceedings of the 2nd European Conference on The Pronciples of Data Mining and Knowledge Discovery (PKDD'98).*

Gath, I. and Geva, A. (1989). Unsupervised optimal fuzzy clustering. *IEEE Transactions on Pattern Analysis and Machine Intelligence*, 11(7).

Gray, B. and Orlowka, M. (1998). Ccaiia: Clustering categorial attributed into interseting accociation rules. In *Proceedings of the 2nd Pacific-Asia Conference on Knowledge Discovery and Data Mining (PAKDD '98).*

Guha, S., Rastogi, R., and Shim, K. (1999). Rock: A robust clustering algorithm for categorical attributes. In *Proceedings of the IEEE Conference on Data Engineering*.

Halkidi, M. and Vazirgiannis, M. (2001a). Clustering validity assessment: Finding the optimal partitioning of a data set. In *Proceedings of ICDM*. California, USA.

Halkidi, M. and Vazirgiannis, M. (2001b). A data set oriented approach for clustering algorithm selection. In *Proceedings of PKDD*. Freiburg, Germany.

Halkidi, M. and Vazirgiannis, M. (2002). Clustering validity assessment: Finding the optimal partitioning of a data set. In *Poster paper in the Proceedings of SETN Conference*. April, Thessaloniki, Greece.

Halkidi, M., Vazirgiannis, M., and Batistakis, I. (2000). Quality scheme assessement in the clustering process. In *Proceedings of PKDD*. Lyon, France.

Han, J. and Kamber, M. (2001). *Data Mining: Concepts and Techniques*. Morgan Kaufmann Publishers.

Hilderman, R. and Hamilton, H. (1999). Knowledge discovery and interestingness measures: A survey. In *Technical Report CS 99-04*. Department of Computer Science, University of Regina.

Jain, A., Murty, M., and Flyn, P. (1999). Data clustering: A review. *ACM Computing Surveys*, 31(3).

Liu, H., Hsu, W., and Chen, S. (1997). Using general impressions to analyze discovered classification rules. In *Proceedings of the Third International Conference on Knowledge Discovery and Data Mining (KDD'97)*. Newport Beach, California.

MAGOpus. V1.1 software. g.i. webb and assoc. In *RuleQuest Research Pty Ltd, 30 Athena Avenue, St Ives NSW 2075, Australia.*

Milligan, G. and Cooper, M. (1985). An examination of procedures for determining the number of clusters in a data set. *Psychometrika*, 50(3):159–179.

Pal, N. and Biswas, J. (1997). Cluster validation using graph theoretic concepts. *Pattern Recognition*, 30(6).

Pasquier, N., Bastide, Y., Taouil, R., and Lakhal, L. (1999). Discovering frequent closed itemsets for association rules. In *Proceedings of the 7th International Conference on Database Theory.*

Pei, J., Han, J., and Mao, R. (2000). Dcloset: An efficient algorithm for mining frequent closed itemsets. In *Proceedings of ACM-SIGMOD International Workshop on Data Mining and Knowledge Discovery (DMKD'00).*

Piatetsky-Shapiro, G. (1991). *Discovery analysis and presentation of strong rules.* Knowledge Discovery in Databases, AAAI/MIT Press.

Roberto, J., Bayardo, J., Agrawal, R., and Gunopulos, D. (1999). Constraint-based rule mining in large, dense databases. In *Proceedings of the 15th ICDE.*

Rokach, L., Averbuch, M., and Maimon, O., Information retrieval system for medical narrative reports. Lecture notes in artificial intelligence, 3055. pp. 217-228, Springer-Verlag (2004).

Sharma, S. (1996). *Applied Multivariate Techniques.* John Wiley and Sons.

Smyth, P. (1996). Clustering using monte carlo cross-validation. In *Proceedings of KDD Conference.*

Smyth, P. and Goodman, R. (1991). *Rule induction using information theory.* Knowledge Discovery in Databases, AAAI/MIT Press.

Snedecor, G. and Cochran, W. (1989). *Statistical Methods.* owa State University Press, Ames, IA, 8th Edition.

Theodoridis, S. and Koutroubas, K. (1999). *Pattern recognition.* Knowledge Discovery in Databases, Academic Press.

Xie, X. and Beni, G. (1991). A validity measure for fuzzy clustering. *IEEE Transactions on Pattern Analysis and machine Intelligence*, 13(4).

Zaki, M. and Hsiao, C. (2002). Charm: An efficient algorithm for closed itemset mining. In *Proceedings of the 2nd SIAM International Conference on Data Mining.*

Zhong, N., Yao, Y., and Ohsuga, S. (1999). Peculiarity-oriented multi-database mining. In *Proceedings of the 3rd European Conference on the Principles of Data Mining and Knowledge Discovery.*

32

Data Mining Model Comparison

Paolo Giudici

University of Pavia

Summary. The aim of this contribution is to illustrate the role of statistical models and, more generally, of statistics, in choosing a Data Mining model. After a preliminary introduction on the distinction between Data Mining and statistics, we will focus on the issue of how to choose a Data Mining methodology. This well illustrates how statistical thinking can bring real added value to a Data Mining analysis, as otherwise it becomes rather difficult to make a reasoned choice. In the third part of the paper we will present, by means of a case study in credit risk management, how Data Mining and statistics can profitably interact.

Key words: Model choice, statistical hypotheses testing, cross-validation, loss functions, credit risk management, logistic regression models.

32.1 Data Mining and Statistics

Statistics has always been involved with creating methods to analyse data. The main difference compared to the methods developed in Data Mining is that statistical methods are usually developed in relation to the data being analyzed but also according to a conceptual reference paradigm. Although this has made the various statistical methods available coherent and rigorous at the same time, it has also limited their ability to adapt quickly to the methodological requests put forward by the developments in the field of information technology.

There are at least four aspects that distinguish the statistical analysis of data from Data Mining.

First, while statistical analysis traditionally concerns itself with analyzing primary data that has been collected to check specific research hypotheses, Data Mining can also concern itself with secondary data collected for other reasons. This is the norm, for example, when analyzing company data that comes from a data warehouse. Furthermore, while in the statistical field the data can be of an experimental nature (the data could be the result of an experiment which randomly allocates all the statistical units to different kinds of treatment) in Data Mining the data is typically of an observational nature.

Second, Data Mining is concerned with analyzing great masses of data. This implies new considerations for statistical analysis. For example, for many applications it is impossible to analyst or even access the whole database for reasons of computer efficiency. Therefore

O. Maimon, L. Rokach (eds.), *Data Mining and Knowledge Discovery Handbook*, 2nd ed.,
DOI 10.1007/978-0-387-09823-4_32,

it becomes necessary to have a sample of the data from the database being examined. This sampling must be carried out bearing in mind the Data Mining aims and, therefore, it cannot be analyzed with the traditional statistical sampling theory tools.

Third, many databases do not lead to the classic forms of statistical data organization. This is true, for example, of data that comes from the Internet. This creates the need for appropriate analytical methods to be developed, which are not available in the statistics field.

One last but very important difference that we have already mentioned is that Data Mining results must be of some consequence. This means that constant attention must be given to business results achieved with the data analysis models.

32.2 Data Mining Model Comparison

Several classes of computational and statistical methods for data mining are available. Once a class of models has been established the problem is to choose the "best" model from it. In this chapter, summarized from chapter 6 in (Giudici, 2003) we present a systematic comparison of them.

Comparison criteria for Data Mining models can be classified schematically into: criteria based on statistical tests, based on scoring functions, Bayesian criteria, computational criteria, and business criteria.

The first are based on the theory of statistical hypothesis testing and, therefore, there is a lot of detailed literature related to this topic. See for example a text about statistical inference, such as (Mood *et al.*, 1991) or (Bickel and Doksum, 1977). A statistical model can be specified by a discrete probability function or by a probability density function, $f(x)$ Such model is usually left unspecified, up to unknown quantities that have to be estimated on the basis of the data at hand. Typically, the observed sample it is not sufficient to reconstruct each detail of $f(x)$, but can indeed be used to approximate $f(x)$ with a certain accuracy. Often a density function is parametric so that it is defined by a vector of parameters $\Theta=(\theta_1,\ldots,\theta_I)$, such that each value θ of Θ corresponds to a particular density function, $p_\theta(x)$. In order to measure the accuracy of a parametric model, one can resort to the notion of distance between a model f, which underlies the data, and an approximating model g (see, for instance, (Zucchini, 2000)).

Notable examples of distance functions are, for categorical variables: the entropic distance, which describes the proportional reduction of the heterogeneity of the dependent variable; the chi-squared distance, based on the distance from the case of independence; the 0-1 distance, which leads to misclassification rates.

The entropic distance of a distribution g from a target distribution f, is:

$$_E d = \sum_i f_i \log \frac{f_i}{g_i} \tag{32.1}$$

The chi-squared distance of a distribution g from a target distribution f is instead:

$$\chi^2 d = \sum_i \frac{(f_i - g_i)^2}{g_i} \tag{32.2}$$

The 0-1 distance between a vector of predicted values, X_{gr}, and a vector of observed values, X_{fr}, is:

$$_{0-1}d = \sum_{r=1}^{n} 1\left(X_{fr} - X_{gr}\right) \tag{32.3}$$

where $1(w,z) = 1$ if $w = z$ and 0 otherwise.

For quantitative variables, the typical choice is the Euclidean distance, representing the distance between two vectors in the Cartesian plane. Another possible choice is the uniform distance, applied when nonparametric models are being used.

The Euclidean distance between a distribution g and a target f is expressed by the equation:

$$_2d\left(X_f, X_g\right) = \sqrt{\sum_{r=1}^{n}\left(X_{fr} - X_{gr}\right)^2} \tag{32.4}$$

Given two distribution functions F and G with values in [0, 1] it is defined uniform distance the quantity:

$$\sup_{0 \leq t \leq 1} \left|F(t) - G(t)\right| \tag{32.5}$$

Any of the previous distances can be employed to define the notion of discrepancy of a statistical model. The discrepancy of a model, g, can be obtained as the discrepancy between the unknown probabilistic model, f, and the best (closest) parametric statistical model. Since f is unknown, closeness can be measured with respect to a sample estimate of the unknown density f.

Assume that f represents the unknown density of the population, and let g= p_θ be a family of density functions (indexed by a vector of I parameters, θ) that approximates it. Using, to exemplify, the Euclidean distance, the discrepancy of a model g, with respect to a target model f is:

$$\Delta\left(f, p_\vartheta\right) = \sum_{i=1}^{n}\left(f\left(x_i\right) - p_\vartheta\left(x_i\right)\right)^2 \tag{32.6}$$

A common choice of discrepancy function is the Kullback-Leibler divergence, that derives from the entropic distance, and can be applied to any type of observations. In such context, the best model can be interpreted as that with a minimal loss of information from the true unknown distribution.

The Kullback-Leibler divergence of a parametric model p_θ with respect to an unknown density f is defined by:

$$\Delta_{K-L}(f, p_\vartheta) = \sum_i f(x_i) \log \frac{f(x_i)}{p_{\hat{\theta}}(x_i)} \tag{32.7}$$

where the parametric density in the denominator has been evaluated in terms of the values of the parameters which minimizes the distance with respect to f.

It can be shown that the statistical tests used for model comparison are generally based on estimators of the total Kullback-Leibler discrepancy. The most used of such estimators is the log-likelihood score. Statistical hypothesis testing is based on subsequent pairwise comparisons between pairs of alternative models. The idea is to compare the log-likelihood score of two alternative models.

The log-likelihood score is then defined by:

$$-2\sum_{i=1}^{n} \log\left[p_{\hat{\theta}}\left(x_i\right)\right] \tag{32.8}$$

Hypothesis testing theory allows to derive a threshold below which the difference between two models is not significant and, therefore, the simpler models can be chosen. To summarize, using statistical tests it is possible to make an accurate choice among the models, based on the observed data. The defect of this procedure is that it allows only a partial ordering of models, requiring a comparison between model pairs and, therefore, with a large number of alternatives it is necessary to make heuristic choices regarding the comparison strategy (such as choosing among forward, backward and stepwise criteria, whose results may diverge). Furthermore, a probabilistic model must be assumed to hold, and this may not always be a valid assumption.

A less structured approach has been developed in the field of information theory, giving rise to criteria based on score functions. These criteria give each model a score, which puts them into some kind of complete order. We have seen how the Kullback-Leibler discrepancy can be used to derive statistical tests to compare models. In many cases, however, a formal test cannot be derived. For this reason, it is important to develop scoring functions, that attach a score to each model. The Kullback-Leibler discrepancy estimator is an example of such a scoring function that, for complex models, can be often be approximated asymptotically. A problem with the Kullback-Leibler score is that it depends on the complexity of a model as described, for instance, by the number of parameters. It is thus necessary to employ score functions that penalise model complexity.

The most important of such functions is the AIC (Akaike Information Criterion, see (Akaike, 1974)). The AIC criterion is defined by the following equation:

$$AIC = -2\log L(\hat{\vartheta};x_1,...,x_n) + 2q \tag{32.9}$$

where the first term is minus twice the the logarithm of the likelihood function calculated in the maximum likelihood parameter estimate and q is the number of parameters of the model.

From its definition notice that the AIC score essentially penalises the log-likelihood score with a term that increases linearly with model complexity. The AIC criterion is based on the implicit assumption that q remains constant when the size of the sample increases. However this assumption is not always valid and therefore the AIC criterion does not lead to a consistent estimate of the dimension of the unknown model. An alternative, and consistent, scoring function is the BIC criterion (Bayesian Information Criterion), also called SBC, formulated in (Schwarz, 1978). The BIC criterion is defined by the following expression:

$$BIC = -2\log L\left(\hat{\vartheta};x_1,...,x_n\right) + q\log(n) \tag{32.10}$$

As can be seen from its definition the BIC differs from the AIC only in the second part which now also depends on the sample size n. Compared to the AIC, when n increases the BIC favours simpler models. As n gets large, the first term (linear in n) will dominate the second term (logarithmic in n). This corresponds to the fact that, for a large n, the variance term in the mean squared error expression tends to be negligible. We also point out that, despite the superficial similarity between the AIC and the BIC, the first is usually justified by resorting to classical asymptotic arguments, while the second by appealing to the Bayesian framework.

To conclude, the scoring function criteria for selecting models are easy to calculate and lead to a total ordering of the models. From most statistical packages we can get the AIC and BIC scores for all the models considered. A further advantage of these criteria is that they can be used also to compare non-nested models and, more generally, models that do not belong to the same class (for instance a probabilistic neural network and a linear regression model).

However, the limit of these criteria is the lack of a threshold, as well the difficult interpretability of their measurement scale. In other words, it is not easy to determine if the difference between two models is significant or not, and how it compares to another difference. These criteria are indeed useful in a preliminary exploration phase. To examine this criteria and to compare it with the previous ones see, for instance, (Zucchini, 2000) or (Hand *et al.*, 2001).

A possible "compromise" between the previous two criteria is the Bayesian criteria which could be developed in a rather coherent way (see e.g. (Bernardo and Smith, 1994)). It appears to combine the advantages of the two previous approaches: a coherent decision threshold and a complete ordering. One of the problems that may arise is connected to the absence of a general purpose software. For Data Mining works using Bayesian criteria the reader could see, for instance, (Giudici, 2003) and (Giudici and Castelo, 2001).

The intensive wide spread use of computational methods has led to the development of computationally intensive model comparison criteria. These criteria are usually based on using dataset different than the one being analyzed (external validation) and are applicable to all the models considered, even when they belong to different classes (for example in the comparison between logistic regression, decision trees and neural networks, even when the latter two are non probabilistic). A possible problem with these criteria is that they take a long time to be designed and implemented, although general purpose softwares have made this task easier.

The most common of such criterion is based on cross-validation. The idea of the cross-validation method is to divide the sample into two sub-samples, a "training" sample, with $n-m$ observations, and a "validation" sample, with m observations. The first sample is used to fit a model and the second is used to estimate the expected discrepancy or to assess a distance. Using this criterion the choice between two or more models is made by evaluating an appropriate discrepancy function on the validation sample. Notice that the cross-validation idea can be applied to the calculation of any distance function.

One problem regarding the cross-validation criterion is in deciding how to select m, that is, the number of the observations contained in the "validation sample". For example, if we select $m = n/2$ then only $n/2$ observations would be available to fit a model. We could reduce m but this would mean having few observations for the validation sampling group and therefore reducing the accuracy with which the choice between models is made. In practice proportions of 75% and 25% are usually used, respectively for the training and the validation samples.

To summarize these criteria have the advantage of being generally applicable but have the disadvantage of taking a long time to be calculated and of being sensitive to the characteristics of the data being examined. A way to overcome this problem is to consider model combination methods, such as bagging and boosting. For a thorough description of these recent methodologies, see (Hastie *et al.*, 2001).

One last group of criteria seem specifically tailored for the data mining field. These are criteria that compare the performance of the models in terms of their relative losses, connected to the errors of approximation made by fitting Data Mining models. Criteria based on loss functions have appeared recently, although related ideas are known since longtime in Bayesian decision theory (see for instance (Bernardo and Smith, 1994)) . They are of great interest and have great application potential although at present they are mainly concerned with solving problems regarding classification. For a more detailed examination of these criteria the reader can see for example (Hand , 1997,Hand *et al.*, 2001) or the reference manuals on Data Mining software, such as that of SAS Enterprise Miner.

The idea behind these methods is that it is important to focus the attention, in the choice among alternative models, to compare the utility of the results obtained from the models and not just to look exclusively at the statistical comparison between the models themselves. Since the main problem dealt with by data analysis is to reduce uncertainties on the risk factors or "loss" factors, reference is often made to developing criteria that minimize the loss connected to the problem being examined. In other words, the best model is the one that leads to the least loss.

Most of the loss function based criteria apply to predictive classification problems, where the concept of a confusion matrix arises. The confusion matrix is used as an indication of the properties of a classification (discriminant) rule. It contains the number of elements that have been correctly or incorrectly classified for each class. On its main diagonal we can see the number of observations that have been correctly classified for each class while the off-diagonal elements indicate the number of observations that have been incorrectly classified. If it is (explicitly or implicitly) assumed that each incorrect classification has the same cost, the proportion of incorrect classifications over the total number of classifications is called rate of error, or misclassification error, and it is the quantity which must be minimized. Of course the assumption of equal costs can be replaced by weighting errors with their relative costs.

The confusion matrix gives rise to a number of graphs that can be used to assess the relative utility of a model, such as the Lift Chart, and the ROC Curve. For a detailed illustration of these graphs we refer to (Hand , 1997) or (Giudici, 2003). The lift chart puts the validation set observations, in increasing or decreasing order, on the basis of their score, which is the probability of the response event (success), as estimated on the basis of the training set. Subsequently, it subdivides such scores in deciles. It then calculates and graphs the observed probability of success for each of the decile classes in the validation set. A model is valid if the observed success probabilities follow the same order (increasing or decreasing) as the estimated ones. Notice that, in order to be better interpreted, the lift chart of a model is usually compared with a baseline curve, for which the probability estimates are drawn in the absence of a model, that is, taking the mean of the observed success probabilities.

The ROC (Receiver Operating Characteristic) curve is a graph that also measures predictive accuracy of a model. It is based on four conditional frequencies that can be derived from a model, and the choice of a cut-off points for its scores:

- the observations predicted as events and effectively such (sensitivity)
- the observations predicted as events and effectively non events
- the observations predicted as non events and effectively events;
- the observations predicted as non events and effectively such (specificity)

The ROC curve is obtained representing, for any fixed cut-off value, a point in the Cartesian plane having as x-value the false positive value (1-specificity) and as y-value the sensitivity value. Each point in the curve corresponds therefore to a particular cut-off. In terms of model comparison, the best curve is the one that is leftmost, the ideal one coinciding with the y-axis. To summarize, criteria based on loss functions have the advantage of being easy to interpret and, therefore, well suited for Data Mining applications but, on the other hand, they still need formal improvements and mathematical refinements. In the next section we give an example of how this can be done, and show that statistics and Data Mining applications can fruitfully interact.

32.3 Application to Credit Risk Management

We now apply the previous considerations to a case-study that concerns credit risk management. The objective of the analysis is the evaluation of the credit reliability of small and medium enterprises (SMEs) that demand financing for their development.

In order to assess credit reliability each applicant for credit is associated with a score, usually expressed in terms of probability of repayment (default probability). Data Mining methods are used to estimate such score and, on the basis of it, to classify applicants as being reliable (worth of credit) or not.

Data Mining models for credit scoring are of the predictive (or supervised) kind: they use explanatory variables obtained from information available on the applicant in order to get an estimate of the probability of repayment (target or response variable). The methods most used in practical credit scoring applications are: linear and logistic regression models, neural networks and classification tress. Often, in banking practice, the resulting scores are called "statistical" and supplemented with subjective, judgemental evaluations.

In this section we consider the analysis of a database that includes 7134 SMEs belonging to the retail segment of an important Italian bank. The retail segment contains companies with total annual sales less than 2,5 million per year. On each of this companies the bank has calculated a score, in order to evaluate their financing (or refinancing) in the period from April 1st, 1999 to April 30th, 2000. After data cleaning, 13 variables are included in the analysis database, of which one binary variable that expresses credit reliability (BAD =0 for the reliables, BAD=1 for the non reliables) can be considered as the response or target variable. The sample contains about 361 companies with BAD=1 (about 5%) and 6773 observed with BAD=0 (about 95%). The objective of the analysis is to build a statistical rule that explains the target variable as a function of the explanatory one. Once built on the observed data, such rule will be extrapolated to assess and predict future applicants for credit. Notice the unbalancedness of the distribution of the target response: this situation, typical in predictive Data Mining problems, poses serious challenges to the performance of a model.

The remaining 12 available variables are retained to influence reliability, and can be considered as explanatory predictors. Among them we have: the age of the company, its legal status, the number of employees, the total sales and variation of the sales in the last period, the region of residence, the specific business, the duration of the relationship of the managers of the company with the bank. Most of them can be considered as "demographic" information on the company, stable in time but indeed not very powerful to build a statistical model. However, it must be said that, being the companies considered all SMEs, it is rather difficult to rely on other, such as balance sheet, information.

A preliminary exploratory analysis can give indications on how to code the explanatory variables, in order to maximize their predictive power. In order to reach this objective we have employed statistical measures of association between pairs of variables, such as chi-squared based measures and statistical measures of dependence, such as Goodman and Kruskal's (see (Giudici, 2003) for a systematic comparison of such measures). We remark that the use of such tools is very much beneficial for the analysis, and can considerably improve the final performance results. As a result of our analysis, all explanatory variables have been discretised, with a number of levels ranging from 2 to 26.

In order to focus on the issue of model comparison we now concentrate on the comparison of three different logistic regression models on the data. This model is the most used in credit scoring applications; other models that are employed are classification trees, linear discriminant analysis and neural networks. Here we prefer to compare models belonging to the same class, to better illustrate our issue; for a detailed comparison of credit scoring methods, on a different data set, see (Giudici, 2003). Our analysis have been conducted using SAS and SAS Enterprise Miner softwares, available at the bank subject of the analysis.

We have chosen, in agreement with the bank's experts, three logistic regression models: a saturated model, that contains all explanatory variables, with the levels obtained from the explanatory analysis; a statistically selected model, using pairwise statistical hypotheses testing; and a model that minimizes the loss function. In the following, the saturated model will be named "RegA (model A)"; the chosen model, according to a statistical selection strategy "RegB (model B)", the model chosen minimizing the loss function "RegC (model C)". Statistical model comparison has been carried out using a stepwise model selection approach, with a reference value of 0,05 to compare p-values with. On the other hand, the loss function has been expressed by the bank's experts, as a function of the classification errors. Table 32.1 below describes such a loss function.

Table 32.1. The chosen loss function

Predicted Actual	BAD	GOOD
BAD	0	20
GOOD	-1	0

The table contains the estimated losses (in scale free values) corresponding to the combinations of actual and predicted values of the target variable. The specified loss function means that it is retained that giving credit to a non reliable (bad) enterprise is 20 times more costly that not giving credit to a reliable (good) enterprise. In statistical terms, the type I error costs 20 times the type II error. As each of the four scenarios in Table 32.1 has an occurrence probability, it is possible to calculate the expected loss of each considered statistical model. The best one will be that minimizing such expected loss.

In the SAS Enterprise Miner tool the Assessment node provides a common framework to compare models, in terms of their predictions. This requires that data has been partitioned in two or more datasets, according to computational criteria of model comparison. The Assessment node produces a table view of the model results that lists relevant statistics and model adequacy and several different charts/reports depending on whether the target variable is continuous or categorical and whether a profit/loss function has been specified.

In the case under examination, the initial dataset (5351 observations) has been split in two, using a sampling mechanism stratified with respect to the target variable. The training dataset contains about 70% of the observations (about 3712) and the validation dataset the remaining 30% (about1639 observations). As the samples are stratified, in both the resulting datasets the percentages of "bad" and "good" enterprises remain the same as those in the combined dataset (5 % e il 95%).

The first model comparison tool we consider is the lift chart. For a binary target, the lift (also called gains chart) is built as follows. The scored data set is sorted by the probabilities of the target event in descending order; observations are then grouped into deciles. For each decile, a lift chart can calculate either: the percentage of target responses (Bad repayers here) or the ratio between the percentage and the corresponding one for the baseline (random) model, called the lift. Lift charts show the percent of positive response or the lift value on the vertical axis. Table 54.1 show the calculations that give rise to the lift chart, for the credit scoring problem considered here. Figure 32.3 shows the corresponding curves.

Table 32.2. Calculations for the lift chart

Number of observations in each group	percentile	% of captured responses (BASELINE)	% di of captured responses % (REG A)	% di of captured responses % (REG B)	% di of captured responses % (REG C)
163.90	10	5.064	20.134	22.575	22.679
163.90	20	5.064	12.813	12.813	14.033
163.90	30	5.064	9.762	10.103	10.293
163.90	40	5.064	8.237	8.237	8.542
163.90	50	5.064	7.322	7.383	7.445
163.90	60	5.064	6.508	6.913	6.624
163.90	70	5.064	5.753	6.237	6.096
163.90	80	5.064	5.567	5.567	5.644
163.90	90	5.064	5.288	5.220	5.185
163.90	100	5.064	5.064	5.064	5.064

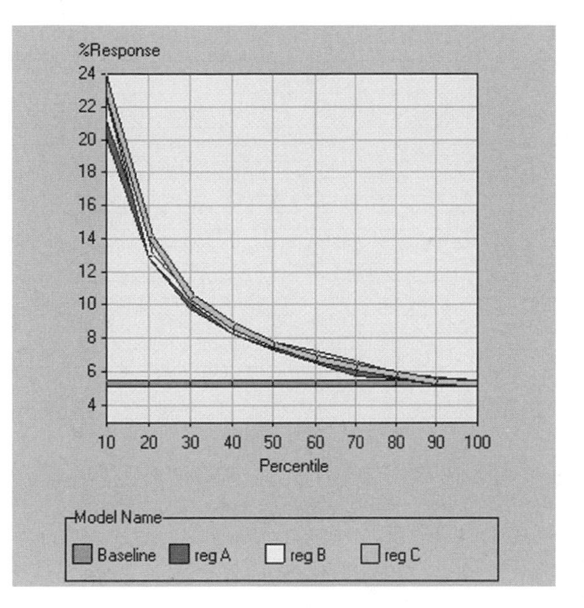

Fig. 32.1. Lift charts for the best model

Comparing the results in Table 54.1 and Figure 39.1 it emerges that the performances of the three models being compared are rather similar; however the best model seem to be model C (the model that minimises the losses) as it is the model that, in the first deciles, is able to effectively capture more bad enterprises, a difficult task in the given problem. Recalling that the actual percentage of bad enterprises observed is equal to 5%, the previous graph can be normalized by dividing the percentage of bads in each decile by the overall 5% percentage. The result is the actual lift of a model, that is, the actual improvement with respect to the baseline situation of absence of a model (as if each company were estimated good/bad according to a purely random mechanism). In terms of model C, in the first decile (with about 164 enterprises) the lift is equal to 4,46 (i.e. $22,7\%/5,1\%$); this means that, using model C it is expected to obtain, in the first decile, a number of enterprises 4,5 times higher with respect to a random sample of the considered enterprises.

The second Assessment tool we consider is the threshold chart. Threshold-based charts enable to display the agreement between the predicted and actual target values across a range of threshold levels. The threshold level is the cutoff that is used to classify an observation that is based on the event level posterior probabilities. The default threshold level is 0.50. For the credit scoring case the calculations leading to the threshold chart are in Table 32.3 and the corresponding figure in Figure 32.3 below.

In order to interpret correctly the previous table and figure, let us consider some numerical examples. First we remark that the results refer to the validation dataset, with 1629 enterprises

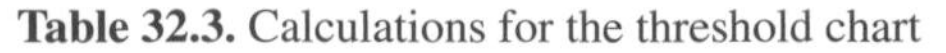

Table 32.3. Calculations for the threshold chart

cutoff	%accuracy (model A)	Freq.	% accuracy (model B)	Freq	% accuracy (model C)	Freq.
95	0	1	0	1	0	1
90	0	1	0	1	0	1
85	0	1	0	1	0	1
80	0	1	0	1	0	1
75	0	1	0	1	0	1
70	0	1	0	1	0	1
65	0	1	0	1	0	1
60	0	1	0	1	0	2
55	0	2	0	1	0	2
50	0.6666666667	6	0	1	0	2
45	0.5714285714	7	0	2	0	2
40	0.6666666667	9	0	4	0	2
35	0.6111111111	18	0	8	0	2
30	0.4642857143	28	0.4230769231	26	0	8
25	0.3902439024	41	0.3673469388	49	0	18
20	0.298245614	57	0.3529411765	51	0.3513513514	37
15	0.2352941176	102	0.2871287129	101	0.2857142857	56
10	0.1833333333	180	0.2402597403	154	0.2364864865	148
5	0.1136363636	396	0.1076555024	418	0.1415384615	325

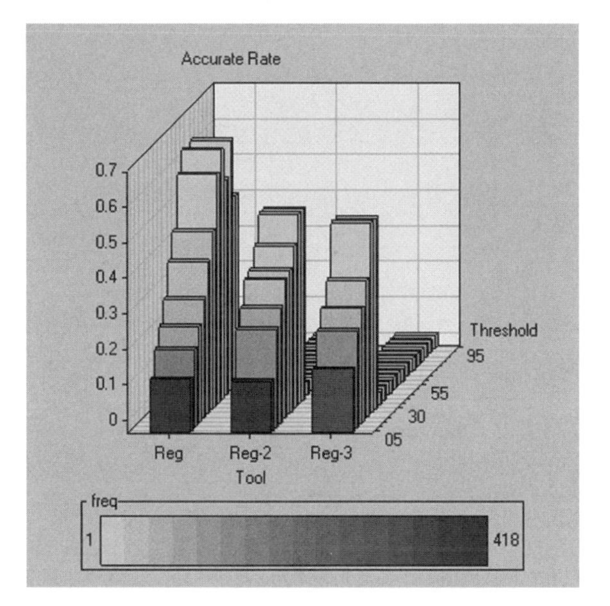

Fig. 32.2. Threshold charts of the models

of which 5% (i.e. 83) are "bad" and 95% (i.e. 1556) are "good". Looking at model A and considering a cut-off level of 5% notice that the model classifies as "bad" 396 enterprises. Clearly this figure is higher than the actual number of bad enterprises and, consequently, the accuracy rate of the model will be low. Indeed, of the 396 enterprises estimated as "bad" only 45 are effectively such, and this leads to an accuracy rate of 11.36% for the model. Model A reaches its maximum accuracy for cut off equal to 40% and 50%. Similar conclusions can be drawn for the other two models.

To summarize, from the *Response Threshold Chart* we can state that, for the examined dataset:

For low levels of the cut-off (i.e. until 15%) the highest accuracy rates are those of Reg-3 (Model C);

For higher levels of the cut-off (between 20% and 55%) model A shows a greater accuracy in predicting the occurrence of default (bad) situations.

In the light of the previous considerations it seems natural to ask which of the three is actually the "best" model. Indeed this question does not have a unique answer; the solution depends on the cut-off level retained more opportune to fix in relationship with the business problem at hand. In our case, being the default a "rare event" a low cut-off is typically chosen, for instance equal to the observed bad rate. Under this setting, model C (Reg-3) turns out to be the best choice.

We also remark that, from our discussion, it seems appropriate to employ the threshold chart not only as a tool to choose a model, rather as a support to individuate and choose, for each built model, the cut off level which corresponds to the highest accuracy in predicting the target event (here the default in repaying). For instance, for model A, the cut-off levels that give rise to the highest accuracy rates are 40% and 50%. Instead, for model C, 25% or 30%.

The third assessment tool we consider is the receiver operating characteristic (ROC) chart. The ROC chart is a graphical display that gives the measure of the predictive accuracy of a model. It displays the sensitivity (a measure of accuracy for predicting events that is equal to the ratio between the true positives and the total actual positive) and specificity (a measure of accuracy for predicting nonevents that is equal to the ratio between true negative and total actual negative) of a classifier for a range of cutoffs. In order to better comprehend the ROC curve it is important to define precisely the quantities contained in it. Table 32.4 below is helpful in determining the elements involved in the ROC curve. For each combination of observed and predicted events and non events it reports a symbol that corresponds to a frequency.

Table 32.4. Elements of the ROC curve

predicted *observed*	EVENTS	NON EVENTS	TOTAL
EVENTS	a	b	a + b
NON EVENTS	c	d	c + d
TOTAL	a + c	b + d	a+b+c+d

The ROC curve is built on the basis of the frequencies contained in Table 32.4. More precisely, let us define the following conditional frequencies (probabilities in the limit):

- *Sensitivity* $\left(a/(a+b)\right)$: proportion of *events* that a model correctly predicts as such (true positives);
- *specificity* $\left(d/(c+d)\right)$: proportion of *non events* that the model correclt predicts as such (true negatives);
- *false positives rate* $\left(c/(c+d)\right)$ = *1-specificity:* proportion of non events that the model predicts as events (type II error);
- *false negatives rate* $\left(b/(a+b)\right)$ = *1-sensitivity:* proportion of events that the model predicts as non events (type I error).

Each of the previous quantities is, evidently, function of the cut-off chosen to classify observations in the validation dataset. Notice also that the accuracy, defined about the threshold curve, is different from the sensitivity. Accuracy can be indeed obtained as $\left(a/(a+c)\right)$: it is a different conditional frequency.

The ROC curve is obtained representing, for each given cut-off point, a point in the plane having as x-value the false positives rate and as y-value the sensitivity. In this way a monotone non decreasing function is obtained. Each point on the curve corresponds to a particular cut-off point. Points closer to the upper right corner correspond to lower cut-offs; points closer to the lower left corner correspond to higher cut-offs.

The choice of the cut-off thus represents a trade-off between sensitivity and specificity. Ideally one wants high values of both, so the model can well predict both events and non events. Usually a low cut-off increases the frequencies (a,c) and decreases (b,d) and, therefore, gives a higher false positives rate, indeed with a higher sensitivity. Conversely, a high cut-off gives a lower false positives rate, at the price of a lower sensitivity

For the examined case study the ROC curves of the three models are represented in Figure 32.3. From Figure 32.3 it emerges that, among the three considered models, the best one is model C ("Reg-3"). Focusing on such model it can be noticed, for example, that, if one wanted to predict correctly 45,6% of "bad" enterprises, it had to allow a type II error equal to 10%.

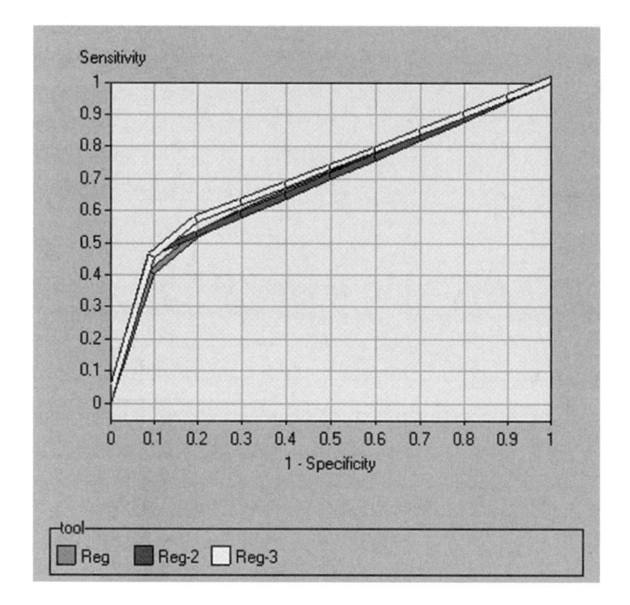

Fig. 32.3. ROC curves for the models

It appears that model choice depends on the chosen cut-off. In the case being examined, involving predicting company defaults, it seems reasonable to have the highest possible values of the sensitivity, yet with acceptable levels of false positives. This because type I errors (predicting as "good" and "bad" enterprises) are typically more costly than type II errors (as the choice of the loss function previously introduced shows). In conclusion, what mostly matters is the maximization of the sensitivity or, equivalently, the minimization of type I errors.

Therefore, in order to compare the entertained models, it can be opportune to compare, for given levels of false positives, the sensitivity of the considered models, so to maximize it. We remark that, in this case, cut-offs can vary and, therefore, they can differ, for the same level of 1-specificity, differently from what occurs with the ROC curve. Table 32.5 below gives the results of such comparison for our case, fixing low levels for the false positives rate.

Table 32.5. Comparison of the sensitivities

1-specificity	Sensitivity (model A)	sensitivity (model B)	sensitivity (model C)
0	0	0	0
0.01	0.4036853296	0.4603162651	0.4556974888
0.02	0.5139617293	0.5189006024	0.5654574445
0.03	0.5861660751	0.5784700934	0.6197752639
0.04	0.6452852072	0.6386886515	0.6740930834
0.05	0.7044043393	0.6989072096	0.7284109028
0.06	0.7635234715	0.7591257677	0.7827287223
0.07	0.8226426036	0.8193443257	0.8370465417
0.08	0.8817617357	0.8795628838	0.8913643611
0.09	0.9408808679	0.9397814419	0.9456821806
1	1	1	1

From Table 32.5 it turns out a substantial similarity of the models with a slight advantage, indeed, for model C.

To summarize our analysis, on the basis of the model comparison criteria being presented, it is possible to conclude that, although the three compared models have similar performances, the model with the best predictive performance results to be model C, not surprisingly, as the model was chosen in terms of minimization of the loss function.

32.4 Conclusions

We have presented a collection of model assessment measures for Data Mining models. We indeed remark that their application depends on the specific problem at hand. It is well known that Data Mining methods can be classified into exploratory, descriptive (or unsupervised), predictive (or supervised) and local (see e.g. (Hand *et al.*, 2001)). Exploratory methods are preliminary to others and, therefore, do not need a performance measure. Predictive problems, on the other hand, are the setting where model comparison methods are most needed, mainly because of the abundance of the models available. All presented criteria can be applied to predictive models: this is a rather important aid for model choice. For descriptive and local methods, which are simpler to implement and interpret, it is not easy to find model assessment tools. Some of the methods described before can be applied; however a great deal of attention is needed to arrive at valid choice solutions.

In particular, it is quite difficult to assess local models, such as association rules, for the bare fact that a global measure of evaluation of such model contradicts with the very notion of a local model. The idea that prevails in the literature is to measure the utility of patterns in terms of how interesting or unexpected they are to the analyst. As it is quite difficult to model an analyst's opinion, it is usually assumed a situation of a completely uninformed opinion. As measures of interest one can consider, for instance, the support, the confidence and the lift. Which of the three measures of interestingness is ideal for selecting a set of rules depends on the user's needs. The former is to be used to assess the importance of a rule, in terms of its frequency in the database; the second can be used to investigate possible dependencies between variables; finally the lift can be employed to measure the distance from the situation of independence.

For descriptive models aimed at summarizing variables, such as clustering methods, the evaluation of the results typically proceeds on the basis of the Euclidean distance, leading at the R^2 index. We remark that is important to examine the ratio between the "between" and "total" sums of squares, that leads to R^2 separately for each variable in the dataset. This can give a variable-specific measure of the goodness of the cluster representation.

In conclusion, we believe more research is needed in the area of statistical methods for Data Mining model comparison. Our contribution shows, both theoretically and at the applied level, that good statistical thinking, as well as subject-matter experience, is crucial to achieve a good performance for Data Mining models.

References

Akaike, H. A new look at statistical model identification. IEEE Transactions on Automatic Control 1974; 19: 716-723

Bernardo, J.M. and Smith, A.F.M., Bayesian Theory. New York: Wiley, 1994.

Bickel, P.J. and Doksum, K.A., Mathematical Statistics. New Jersey: Prentice and Hall, 1977.

Castelo, R. and Giudici, P., Improving Markov chain model search for Data Mining. Machine Learning,50:127-158,2003.

Giudici, P., Applied Data Mining. London: Wiley, 2003.

Giudici P., Castelo R.. Association models for web mining, Data mining and knowledge discovery, 5, 183-196, 2001.

Hand, D.J.,Mannila, H. and Smyth, P., Principles of Data Mining. New York: MIT press, 2001.

Hand, D. Construction and assessment of classification rules. London: Wiley, 1997.

Hastie, T., Tibshirani, R., Friedman, J. The elements of statistical learning: Data Mining, inference and prediction. New York: Springer-Verlag, 2001.

Mood, A.M., Graybill, F.A. and Boes, D.C. Introduction to the theory of Statistics. Tokyo: McGraw Hill, 1991.

Rokach, L., Averbuch, M., and Maimon, O., Information retrieval system for medical narrative reports. Lecture notes in artificial intelligence, 3055. pp. 217-228, Springer-Verlag (2004).

Schwarz, G. Estimating the dimension of a model. Annals of Statistics 1978; 62: 461-464.

Zucchini, W. An Introduction to Model Selection. Journal of Mathematical Psychology 2000; 44: 41-61

33

Data Mining Query Languages

Jean-Francois Boulicaut[1] and Cyrille Masson[1]

INSA Lyon, LIRIS CNRS FRE 2672
69621 Villeurbanne cedex, France.
jean-francois.boulicaut,Cyrille.Masson@insa-lyon.fr

Summary. Many Data Mining algorithms enable to extract different types of patterns from data (e.g., local patterns like itemsets and association rules, models like classifiers). To support the whole knowledge discovery process, we need for integrated systems which can deal either with patterns and data. The inductive database approach has emerged as an unifying framework for such systems. Following this database perspective, knowledge discovery processes become querying processes for which query languages have to be designed. In the prolific field of association rule mining, different proposals of query languages have been made to support the more or less declarative specification of both data and pattern manipulations. In this chapter, we survey some of these proposals. It enables to identify nowadays shortcomings and to point out some promising directions of research in this area.

Key words: Query languages, Association Rules, Inductive Databases.

33.1 The Need for Data Mining Query Languages

Since the first definition of the Knowledge Discovery in Databases (KDD) domain in (Piatetsky-Shapiro and Frawley, 1991), many techniques have been proposed to support these "From Data to Knowledge" complex interactive and iterative processes. In practice, knowledge elicitation is based on some extracted and materialized (collections of) patterns which can be global (e.g., decision trees) or local (e.g., itemsets, association rules). Real life KDD processes imply complex pre-processing manipulations (e.g., to clean the data), several extraction steps with different parameters and types of patterns (e.g., feature construction by means of constrained itemsets followed by a classifying phase, association rule mining for different thresholds values and different objective measures of interestingness), and post-processing manipulations (e.g., elimination of redundancy in extracted patterns, crossing-over operations between patterns and data like the search of transactions which are exceptions to frequent and valid association rules or the selection of misclassified examples with a decision tree). Looking for a tighter integration between data and patterns which hold in the data, Imielinski and Mannila have proposed in (Imielinski and Mannila, 1996) the concept of inductive database (IDB). In an IDB, ordinary queries can be used to access and manipulate data, while *inductive queries* can be used to generate (mine), manipulate, and apply patterns. KDD becomes

O. Maimon, L. Rokach (eds.), *Data Mining and Knowledge Discovery Handbook*, 2nd ed.,
DOI 10.1007/978-0-387-09823-4_33,

an extended querying process where the analyst can control the whole process since he/she specifies the data and/or patterns of interests. Therefore, the quest for query languages for IDBs is an interesting goal. It is actually a long-term goal since we still do not know which are the relevant primitives for Data Mining. In some sense, we still lack from a well-accepted set of primitives. It might recall the context at the end of the 60's before the Codd's relational algebra proposal.

In some limited contexts, researchers have, however, designed data mining query languages. Data Mining query languages can be used for specifying inductive queries on some pattern domains. They can be more or less coupled to standard query languages for data manipulation or pattern postprocessing manipulations. More precisely, a Data Mining query language, should provide primitives to (1) select the data to be mined and pre-process these data, (2) specify the kind of patterns to be mined, (3) specify the needed background knowledge (as item hierarchies when mining generalized association rules), (4) define the constraints on the desired patterns, and (5) post-process extracted patterns.

Furthermore, it is important that Data Mining query languages satisfy the closure property, i.e., the fact that the result of a query can be queried. Following a classical approach in database theory, it is also needed that the language is based on a well-defined (operational or even better declarative) semantics. It is the only way to make query languages that are not only "syntactical sugar" on top of some algorithms but true query languages for which query optimization strategies can be designed. Again, if we consider the analogy with SQL, relational algebra has paved the way towards query processing optimizers that are widely used today. Ideally, we would like to study containment or equivalence between mining queries as well.

Last but not the least, the evaluation of Data Mining queries is in general very expensive. It needs for efficient constraint-based data mining algorithms, the so-called solvers (De Raedt, 2003,Boulicaut and Jeudy, 2005). In other terms, data mining query languages are often based on primitives for which some more or less ad-hoc solvers are available. It is again typical of a situation where a consensus on the needed primitives is yet missing.

So far, no language proposal is generic enough to provide support for a broad kind applications during the whole KDD process. However, in the active field of association rule mining, some interesting query languages have been proposed. In Section 33.2, we recall the main steps of a KDD process based on association rule mining and thus the need for querying support. In Section 33.3, we introduce several relevant proposals for association rule mining query languages. It contains a short critical evaluation (see (Botta *et al.*, 2004) for a detailed one). Section 33.4 concludes.

33.2 Supporting Association Rule Mining Processes

We assume that the reader is familiar with association rule mining (see, e.g., (Agrawal *et al.*, 1996) for an introduction). In this context, data is considered as a multiset of transactions, i.e., sets of items. Frequent associations rules are built on frequent itemsets (itemsets which are subsets of a certain percentage of the transactions). Many objective interestingness measures can inform about the quality of the extracted rules, the confidence measure being one of the most used. Importantly, many objective measures appear to be complementary: they enable to rank the rules according to different points of view. Therefore, it seems important to provide support for various measures, including the definition of new ones, e.g., application specific ones.

When a KDD process is based on itemsets or association rules, many operations have to be performed by means of queries. First, the language should allow to manipulate and extract source data. Typically, the raw data is not always available as transactional data. One of the typical problems concerns the transformation of numerical attributes into items (or boolean properties). More generally, deriving the transactional context to be mined from raw data can be a quite tedious task (e.g., deriving a transactional data set about WWW resources loading per session from raw WWW logs in a WWW Usage Mining application). Some of these preprocessing are supported by SQL but a programming extension like PL/SQL is obviously needed.

Then, the language should allow the user to specify a broad kind of constraints on the desired patterns (e.g., thresholds for the objective measures of interestingness, syntactical constraints on items which must appear or not in rule components). So far, the primitive constraints and the way to combine them is tightly linked with the kinds of constraints the underlying evaluation engine or solvers can process efficiently (typically anti-monotonic or succinct constraints). One can expect that minimal frequency and minimal confidence constraints are available. However, many other primitive constraints can be useful, including the ones based on aggregates (Ng *et al.*, 1998) or closures (Jeudy and Boulicaut, 2002, Boulicaut, 2004).

Once rules have been extracted and materialized (e.g., in relational tables), it is important that the query language provides techniques to manipulate them. We can wish, for instance, to find a cover of a set of extracted rules (i.e., non redundant association rules based on closed sets (Bastide *et al.*, 2000)), which requires to have subset operators, primitives to access bodies and heads of rules, and primitives to manipulate closed sets or other condensed representations of frequent sets (Boulicaut, 2004) and (Calders and Goethals, 2002). Another important issue is the need for crossing-over primitives. It means that, for instance, we need simple way to select transactions that satisfy or do not satisfy a given rule.

The so-called closure property is important. It enables to combine queries, to support the reuse of KDD scenarios, and it gives rise to opportunities for compiling schemes over sequences of queries (Boulicaut *et al.*, 1999). Finally, we could also ask for a support to pattern uses. In other terms, once relevant patterns have been stored, they are generally used by some software component. To the best of our knowledge, very few tools have been designed for this purpose (see (Imielinski *et al.*, 1999) for an exception).

We can distinguish two major approaches in the design of Data Mining query languages. The first one assumes that all the required objects (data and pattern storage systems and solvers) are already embedded into a common system. The motivation for the query language is to provide more understandable primitives: the risk is that the query

language provides mainly "syntactic sugar" on top of solvers. In that framework, if data are stored using a classical relational DBMS, it means that source tables are views or relations and that extracted patterns are stored using the relational technology as well. `MSQL`, `DMQL` and `MINE RULE` can be considered as representative of this approach. A second approach assumes that we have no predefined integrated systems and that storage systems are loosely coupled with solvers which can be available from different providers. In that case, the language is not only an interface for the analyst but also a facilitator between the DBMS and the solvers. It is the approach followed by `OLE DB` for DM (Microsoft). It is an API between different components that also provides a language for creating and filling extraction contexts, and then access them for manipulations and tests. It is primarily designed to work on top of SQL Server and can be plugged with different solvers provided that they comply the API standard.

33.3 A Few Proposals for Association Rule Mining

33.3.1 MSQL

MSQL (Imielinski and Virmani, 1999) has been designed at the Rutgers University. It extracts rules that are based on descriptors, each descriptor being an expression of the type $(A_i = a_{ij})$, where A_i is an attribute and a_{ij} is a value or a range of values in the domain of A_i. We define a *conjunctset* as the conjunction of an arbitrary number of descriptors such that there are no couple of descriptors built on the same attribute. MSQL extracts propositional rules of the form $\mathscr{A} \Rightarrow \mathscr{B}$, where $\mathscr{A}$ is a conjunctset and $\mathscr{B}$ is a descriptor. As a consequence, only one attribute can appear in the consequent of a rule. Notice that MSQL defines the support of an association rule $\mathscr{A} \Rightarrow \mathscr{B}$ as the number of tuples containing $\mathscr{A}$ in the original table and its confidence as the ratio between the number of tuples containing $\mathscr{A}$ et $\mathscr{B}$ and the support of the rule.

From a practical point of view, MSQL can be seen as an extension of SQL with some primitives tailored for association rule mining (given their semantics of association rules). Specific queries are used to mine rules (inductive queries starting with GetRules) while other queries are post-processing queries over a materialized collection of rules (queries starting with SelectRules). The global syntax of the language for rule extraction is the following one:

```
GetRules(C) [INTO <rulebase name>]
    [WHERE <rule constraints>]
    [SQL-group-by clause]
    [USING encoding-clause]
```

C is the source table and rule_constraints are conditions on the desired rules, e.g., the kind of descriptors which must appear in rule components, the minimal frequency or confidence of the rules or some mutual exclusion constraints on attributes which can appear in a rule. The USING part enables to discretize numerical values. rulebase_name is the name of the object in which rules will be stored. Indeed, using MSQL, the analyst can explicitly materialize a collection of rules and then query it with the following generic statement where <conditions> can specify constraints on the body, the head, the support or the confidence of the rule:

```
SelectRules(rulebase name)
        [where <conditions>]
```

Finally, MSQL provides a few primitives for post-processing. Indeed, it is possible to use Satisfy and Violate clauses to select rules which are supported (or not) in a given table.

33.3.2 MINE RULE

MINE RULE (Meo *et al.*, 1998) has been designed at the University of Torino and the Politecnico di Milano. It is an extension of SQL which is coupled with a relational DBMS. Data can be selected using the full power of SQL. Mined association rules are materialized into relational tables as well. MINE RULE extracts association rule between values of attributes in a relational table. However, it is up to the user to specify the form of the rules to be extracted. More precisely, the user can specify the cardinality of body and head of the desired rules and the attributes on which rule components can be built. An interesting aspect of MINE RULE is that it is possible to work on different levels on grouping during the extraction (in a similar way as the GROUP BY clause of SQL). If there is one level of grouping, rule support will be computed w.r.t. the number of groups in the table. Defining a second level of grouping leads to the definition of clusters (sub-groups). In that case, rules components can be taken in two different clusters, eventually ordered, inside a same group. It is thus possible to extract some elementary sequential patterns (by clustering on a time-related attribute). For instance, grouping purchases by customers and then clustering them by date, we can obtain rules like $Butter \wedge Milk \Rightarrow Oil$ to say that customers who buy first *Butter* and *Milk* tend to buy *Oil* after. Concerning interestingness measures, MINE RULE enables to specify minimal frequency and confidence thresholds. The general syntax of a MINE RULE query for extracting rules is:

```
MINE RULE <TableName> AS
 SELECT DISTINCT [<Cardinality>]  <Attributes>
                 AS BODY,
                 [<Cardinality>]  <Attributes>
                 AS HEAD
                 [,SUPPORT] [,CONFIDENCE]
 FROM <Table> [ WHERE <WhereClause> ]
 GROUP BY <Attributes> [ HAVING <HavingClause> ]
 [ CLUSTER BY <Attributes>
              [ HAVING <HavingClause> ]]
 EXTRACTING RULES WITH
              SUPPORT:<real>, CONFIDENCE:<real>
```

33.3.3 DMQL

DMQL (Han *et al.*, 1996) has been designed at the Simon Fraser University, Canada. It has been designed to support various rule mining extractions (e.g., classification rules, comparison rules, association rules). In this language, an association rule is a relation between the values of two sets of predicates that are evaluated on the relations of a database. These predicates are of the form $P(X,c)$ where P is a predicate taking the name of an attribute of a relation, X is a variable and c is a value in the domain of the attribute. A typical example of association rule that can be extracted by DMQL is $buy(X,milk) \wedge town(X,Berlin) \Rightarrow buy(X,beer)$. An important possibility in DMQL is the definition of meta-patterns, i.e., a powerful way to restrict the syntactic aspect of the extracted rules (expressive syntactic constraints). For instance, the meta-pattern $buy^{+}(X,Y) \wedge town(X,Berlin) \Rightarrow buy(X,Z)$ restricts the search to association rules concerning implication between bought products for customers living in Berlin. Symbol $+$ denotes that the predicate *buy* can appear several times in the left part of the rule. Moreover, beside the classical frequency and confidence, DMQL also enables to define thresholds on the noise or novelty of extracted rules. Finally, DMQL enables to define a hierarchy on attributes such that generalized association rules can be extracted. The general syntax of DMQL for the extraction of association rules is the following one:

```
Use database ⟨database_name⟩

{Use hierarchy ⟨hierarchy_name⟩
For ⟨attribute⟩ }

Mine associations [as ⟨pattern_name⟩]

[ Matching ⟨metapattern⟩]

From ⟨relation(s)⟩ [ Where ⟨condition⟩]

[ Order by ⟨order_list⟩]
[ Group by ⟨grouping_list⟩] [ Having ⟨condition⟩]

With ⟨interest_measure⟩
Threshold = value
```

33.3.4 OLE DB for DM

OLE DB for DM has been designed by Microsoft Corporation (Netz *et al.*, 2000). It is an extension of the OLE DB API to access database systems. More precisely, it aims at supporting the communication between the data sources and the solvers that are not necessarily implemented inside the query evaluation system. It can thus work with many different solvers and types of patterns. To support the manipulation of the objects of the API during a KDD process, OLE DB for DM proposes a language as an extension to SQL. The concept of OLE DB for DM relies on the definition of Data Mining Models (DMM), i.e. object that correspond to extraction contexts in KDD. Indeed, whereas the other language proposals made the assumption that the data almost have a suitable format for the extraction, OLE DB for DM considers it is not always the case and let the user defines a virtual object that will have a suitable format for the extraction and that will be populated with the needed data. Once the extraction algorithm has been applied on this DMM, the DMM will become an object containing patterns or models. It will then be possible to query this DMM as a rule base or to use it as a classifier. The global syntax for creating a DMM is the following:

```
CREATE MINING MODEL <DMM name>
              (<columns definition>)
              USING <algorithm>
              [(<algorithm parameters>)]
```

For each column, it is possible to specify the data type and if it is the target attribute of the model to be learnt in case of classification. Moreover, a column can correspond to a nested table, which is useful when populating the mining model with data taken in tables linked by a one-to-many relationship. For the moment, OLE DB for DM is implemented in the SQL

Server 2000 software and it provides only two mining algorithms: one for decision trees and one for clustering. However, the 2005 version of SQL server should provide neural network and association rule extractors. This latter one will enable to define minimal and maximal rule support, minimal confidence,and minimal and maximal sizes of itemsets on which the rules are based.

33.3.5 A Critical Evaluation

Let us now emphasize the main advantages and drawbacks of the different proposals. A detailed evaluation of these four languages has been performed on a simple but realistic association rule mining scenario (Botta *et al.*, 2004). We summarize the results of this study and it enables to point some important problems that must be addressed on our way to query languages for inductive databases.

The advantages of the proposed languages is that they are all designed as extensions of SQL. It facilitates the work for database experts and it is useful for data manipulation (or the needed standard queries). They all satisfy the closure property. Indeed, even if all the languages do not systematically provide operators for manipulating extracted rules, it is always possible to access materialized collections of rules using SQL queries. Notice, however, that most of the needed pre-processing or post-processing techniques will need not only SQL queries but also PL/SQL statements. Some languages provide primitives to simplify some typical preprocessing, e.g., the discretization of numerical values. Even if is quite preliminary, it is an important support for the practical use of the association rule mining technique. Finally, the concept of `OLE DB` for `DM` is quite relevant as it enables external providers to plug-in new solvers to the existing systems.

The first major limitation of the proposed languages is the poor support to pre- and post-postprocessing operations. Indeed, they are essentially designed around the extraction step and mainly provide primitives for rule extractions, these primitives being generally fixed, e.g., the possibilities to specify minimal thresholds for a few selected objective measures of interestingness or to define syntactical constraints on the rules. Only `MSQL` and `OLE DB` for `DM` propose restricted mechanisms for discretization. Typical preprocessing techniques for, e.g., sampling or boosting, are not supported. It has been shown that pre-processing processes for KDD are tedious phases for which the use of integrated tools and operators is needed (see, e.g., the MINING MART "Enabling End-User Datawarehouse Mining" EU funded project IST-1999-11993 (Morik and Scholz, 2004)). The lack of primitives for post-processing is also obvious. Only `MSQL` provides a `SelectRules` operator which enables to query rule databases and primitives for crossing-over operations between rules and data. The others rely on SQL and its programming extensions for accessing and manipulating the rules. For instance, using `MINE RULE`, extracted rules are stored in relational tables that have to be queried with SQL. In that case, writing a query which simply returns tuples of a table which satisfy a given rule can be very complex because of SQL mechanisms for handling subset relationships (see (Botta *et al.*, 2004) for examples). Not only the SQL post-processing queries are hard to write but also difficult to optimize given the current state of the art for SQL optimization. A solution can come from query languages dedicated to pattern database manipulations. It is the case of `RULE-QL` (Tuzhilin and Liu, 2002) which extends SQL with operators allowing to access rules components and to specify subset relationships. It is thus easier to write queries that, for instance, select rules that have a left part contained in the consequent of another rule. `RULE-QL` can be seen as a good complement to languages like `MINE RULE`. More generally, some basic research is needed on pattern database querying where patterns can be rules, clusters, classifiers, etc. An interesting work in this direction is done by the PANDA

"Patterns for Next-Generation Database Systems" EU funded Working Group IST/FET-2001-33058 (Theodoridis and Vassiliadis, 2004, Catania *et al.*, 2004).

The second main drawback of the proposed languages is that they appear to be quite *ad hoc* proposals. By this term, we mean that they have been proposed on top of some specific algorithms or solvers. The available constraints or conjunction of constraints are the one for which solvers were available at the time of design. When considering the evaluation architecture (described, e.g., for `MINE RULE`), we can see that different solvers cope with specific conjunctions of constraints on the association rules. This is also the case for `DMQL` and `OLE DB` for `DM` proposals, i.e. languages that can extract several types of patterns. For instance, with `DMQL`, each type of rule that

can be extracted is indeed related to a particular solver.

To summarize, primitives are missing and the integration of new primitives by the analyst is not possible. This is obviously due to the lack of consensus on a good collection of primitives. This is true for simple pattern domains like association rules but also for more complex ones. It is interesting to note that the semantics of the association rules for the different query language proposals is not the same. When looking at the details, we can see that even simple evaluation functions like frequency can be defined differently. In other terms, we still lack from a consensus on what is an association rule and what is the semantics of a constrained association rule. The situation is the same for other kinds of patterns, e.g., see the many different semantics for constrained sequential patterns which have been proposed the last 10 years.

We believe that looking for a formal semantics of Data Mining query languages is crucial for the development of the field. Indeed, if we draw a parallel with the development of standard database query languages, we know that (extended) relational algebra have played a major role for their design but also the implementation of efficient query optimizers. The same goal should be taken if we wish to develop Data Mining query languages that are not just "syntactic sugar" on top of solvers. For instance, based on the `MINE RULE` formal semantics, it has been possible to analyze how to optimize queries and also to exploit properties on the relationship between queries. Thanks to data dependencies in the source tables, (Meo, 2003) shows that containment and dominance relations between queries can be used to speed-up the evaluation of new mining queries.

It was one of the main goals of the cINQ "**c**onsortium on knowledge discovery by **In**ductive **Q**ueries" EU funded project IST/FET-2000-26469 to make a breakthrough in this direction. Considering several pattern domains (e.g., association rules, sequences, molecular fragments), they have been looking for useful primitives, new ways to combine them, and not only ad-hoc but also generic solvers for complex inductive queries (e.g., arbitrary boolean expressions over monotonic and anti-monotonic constraints (De Raedt *et al.*, 2002)). A simple formal language is sketched in (De Raedt, 2003) to describe both data and pattern manipulations via inductive queries. Some recent contributions to database support for Data Mining are collected in (Meo *et al.*, 2004). It contains, among others, extended contributions of the first two workshops organized by the cINQ project.

33.4 Conclusion

In this chapter, we have considered Data Mining query languages issues. To support the whole knowledge discovery process, we need for integrated systems which can deal either with patterns and data. Designing such systems is the goal of the emerging inductive database approach. Following this database perspective, knowledge discovery processes become querying processes for which query languages have to be designed. On one hand, interesting conceptual, or say abstract, proposals have been made like (Giannotti and Manco, 1999, De Raedt, 2003, Catania *et al.*, 2004). On another hand, concrete query languages have been designed and implemented for specific pattern domains, mainly association rules (Han *et al.*, 1996, Meo *et al.*, 1998, Imielinski and Virmani, 1999, Netz *et al.*, 2000). The first approach emphasizes the need for general-purpose primitives and is looking for generic approaches in combining these primitives and designing generic solvers. The second approach is pragmatic: providing an immediate support to practitioners by means of better Data Mining tools. Doing so, the primitives are often tailored to some specific pattern domain, or even some application domain. Ad-hoc solvers are designed for an efficient evaluation of concrete queries. Standards like PMML ((http://www.dmg.org) are also immediately useful for practitioners and software companies. This XML-based language provides a standard format for representing various patterns and this is important to support interoperability between various tools. Let us notice however that it does not provide primitives for pattern manipulation. We strongly believe that both directions are useful on our road towards inductive databases and inductive database management systems.

Acknowledgments

The authors want to thank the colleagues of the cInQ IST-2000-26469 (consortium on knowledge discovery by inductive queries) for interesting discussions on Data Mining query languages. A special thank goes to Rosa Meo for her contribution to this domain and the critical evaluation (Botta *et al.*, 2004).

References

R. Agrawal, H. Mannila, R. Srikant, H. Toivonen, and A. I. Verkamo. Fast discovery of association rules. In *Advances in Knowledge Discovery and Data Mining*, pages 307–328. AAAI Press, 1996.

Y. Bastide, N. Pasquier, R. Taouil, G. Stumme, and L. Lakhal. Mining minimal non-redundant association rules using frequent closed itemsets. In *Proc. CL 2000*, volume 1861 of *LNCS*, pages 972–986. Springer-Verlag, 2000.

M. Botta, J.-F. Boulicaut, C. Masson, and R. Meo. Query languages supporting descriptive rule mining: a comparative study. In *Database Technologies for Data Mining - Discovering Knowledge with Inductive Queries*, volume 2682 of *LNCS*, pages 27–54. Springer-Verlag, 2004.

J.-F. Boulicaut. Inductive databases and multiple uses of frequent itemsets: the cInQ approach. In *Database Technologies for Data Mining - Discovering Knowledge with Inductive Queries*, volume 2682 of *LNCS*, pages 3–26. Springer-Verlag, 2004.

J.-F. Boulicaut and B. Jeudy. Constraint-based Data Mining. In *Data Mining and Knowledge Discovery Handbook*. Chapter 16.7, this volume, Kluwer, 2005.

J.-F. Boulicaut, M. Klemettinen, and H. Mannila. Modeling KDD processes within the inductive database framework. In *Proc. DaWaK'99*, volume 1676 of *LNCS*, pages 293–302. Springer-Verlag, 1999.

T. Calders and B. Goethals. Mining all non-derivable frequent itemsets. In *Proc. PKDD*, volume 2431 of *LNCS*, pages 74–85. Springer-Verlag, 2002.

B. Catania, A. Maddalena, M. Mazza, E. Bertino, and S. Rizzi. A framework for Data Mining pattern management. In *Proc. PKDD'04*, volume 3202 of *LNAI*, pages 87–98. Springer-Verlag, 2004.

L. De Raedt. A perspective on inductive databases. *SIGKDD Explorations*, 4(2):69–77, 2003.

L. De Raedt, M. Jaeger, S. Lee, and H. Mannila. A theory of inductive query answering. In *Proc. IEEE ICDM'02*, pages 123–130, 2002.

F. Giannotti and G. Manco. Querying inductive databases via logic-based user-defined aggregates. In *Proc. PKDD'99*, volume 1704 of *LNCS*, pages 125–135. Springer-Verlag, 1999.

J. Han, Y. Fu, W. Wang, K. Koperski, and O. Zaiane. DMQL: a Data Mining query language for relational databases. In R. Ng, editor, *Proc. ACM SIGMOD Workshop DMKD'96*, Montreal, Canada, 1996.

T. Imielinski and H. Mannila. A database perspective on knowledge discovery. *Communications of the ACM*, 39(11):58–64, November 1996.

T. Imielinski and A. Virmani. MSQL: A query langugage for database mining. *Data Mining and Knowledge Discovery*, 3(4):373–408, 1999.

T. Imielinski, A. Virmani, and A. Abdulghani. DMajor-application programming interface for database mining. *Data Mining and Knowledge Discovery*, 3(4):347–372, 1999.

B. Jeudy and J.-F. Boulicaut. Optimization of association rule mining queries. *Intelligent Data Analysis*, 6(4):341–357, 2002.

R. Meo. Optimization of a language for Data Mining. In *Proc. ACM SAC'03 - Data Mining track*, pages 437–444, 2003.

R. Meo, P. L. Lanzi, and M. Klemettinen, editors. *Database Technologies for Data Mining - Discovering Knowledge with Inductive Queries*, volume 2682 of *LNCS*. Springer-Verlag, 2004.

R. Meo, G. Psaila, and S. Ceri. An extension to SQL for mining association rules. *Data Mining and Knowledge Discovery*, 2(2):195–224, 1998.

K. Morik and M. Scholz. The Mining Mart approach to knowledge discovery in databases. In *Intelligent Technologies for Information Analysis*. Springer-Verlag, 2004.

A. Netz, S. Chaudhuri, J. Bernhardt, and U. Fayyad. Integration of Data Mining and relational databases. In *Proc. VLDB'00*, pages 719–722, Cairo, Egypt, 2000. Morgan Kaufmann.

R. Ng, L. V. Lakshmanan, J. Han, and A. Pang. Exploratory mining and pruning optimizations of constrained associations rules. In *Proc. ACM SIGMOD'98*, pages 13–24, 1998.

G. Piatetsky-Shapiro and W. J. Frawley. *Knowledge Discovery in Databases*. AAAI/MIT Press, 1991.

Y. Theodoridis and P. Vassiliadis, editors. *Proc. of Pattern Representation and Management PaRMa 2004 co-located with EDBT 2004*. CEUR Workshop Proceedings 96 Technical University of Aachen (RWTH), 2004.

A. Tuzhilin and B. Liu. Querying multiple sets of discovered rules. In *Proc. ACM SIGKDD'02*, pages 52–60, 2002.

Part VI

Advanced Methods

34

Mining Multi-label Data

Grigorios Tsoumakas, Ioannis Katakis, and Ioannis Vlahavas

Dept. of Informatics, Aristotle University of Thessaloniki, 54124 Greece
{greg,katak,vlahavas}@csd.auth.gr

34.1 Introduction

A large body of research in supervised learning deals with the analysis of *single-label* data, where training examples are associated with a single label λ from a set of disjoint labels L. However, training examples in several application domains are often associated with a *set* of labels $Y \subseteq L$. Such data are called *multi-label*.

Textual data, such as documents and web pages, are frequently annotated with more than a single label. For example, a news article concerning the reactions of the Christian church to the release of the "Da Vinci Code" film can be labeled as both *religion* and *movies*. The categorization of textual data is perhaps the dominant multi-label application.

Recently, the issue of learning from multi-label data has attracted significant attention from a lot of researchers, motivated from an increasing number of new applications, such as semantic annotation of images (Boutell et al., 2004, Zhang & Zhou, 2007a, Yang et al., 2007) and video (Qi et al., 2007, Snoek et al., 2006), functional genomics (Clare & King, 2001, Elisseeff & Weston, 2002, Blockeel et al., 2006, Cesa-Bianchi et al., 2006a, Barutcuoglu et al., 2006), music categorization into emotions (Li & Ogihara, 2003, Li & Ogihara, 2006, Wieczorkowska et al., 2006, Trohidis et al., 2008) and directed marketing (Zhang et al., 2006). Table 34.1 presents a variety of applications that are discussed in the literature.

This chapter reviews past and recent work on the rapidly evolving research area of multi-label data mining. Section 2 defines the two major tasks in learning from multi-label data and presents a significant number of learning methods. Section 3 discusses dimensionality reduction methods for multi-label data. Sections 4 and 5 discuss two important research challenges, which, if successfully met, can significantly expand the real-world applications of multi-label learning methods: a) exploiting label structure and b) scaling up to domains with large number of labels. Section 6 introduces benchmark multi-label datasets and their statistics, while Section 7 presents the most frequently used evaluation measures for multi-label learning. We conclude this chapter by discussing related tasks to multi-label learning in Section 8 and multi-label data mining software in Section 9.

34.2 Learning

There exist two major tasks in supervised learning from multi-label data: *multi-label classification* (MLC) and *label ranking* (LR). MLC is concerned with learning a model that outputs

O. Maimon, L. Rokach (eds.), *Data Mining and Knowledge Discovery Handbook*, 2nd ed.,
DOI 10.1007/978-0-387-09823-4_34,

Data type	Application	Resource	Labels Description (Examples)	References
text	categorization	news article	Reuters topics (agriculture, fishing)	(Schapire, 2000)
		web page	Yahoo! directory (health, science)	(Ueda & Saito, 2003)
		patent	WIPO (paper-making, fibreboard)	(Godbole & Sarawagi, 2004, Rousu et al., 2006)
		email	R&D activities (delegation)	(Zhu et al., 2005)
		legal document	Eurovoc (software, copyright)	(Mencia & Fürnkranz, 2008)
		medical report	MeSH (disorders, therapies)	(Moskovitch et al., 2006)
		radiology report	ICD-9-CM (diseases, injuries)	(Pestian et al., 2007)
		research article	Heart conditions (myocarditis)	(Ghamrawi & McCallum, 2005)
		research article	ACM classification (algorithms)	(Veloso et al., 2007)
		bookmark	Bibsonomy tags (sports, science)	(Katakis et al., 2008)
		reference	Bibsonomy tags (ai, kdd)	(Katakis et al., 2008)
		adjectives	semantics (object-related)	(Boleda et al., 2007)
image	semantic annotation	pictures	concepts (trees, sunset)	(Boutell et al., 2004, Zhang & Zhou, 2007a, Yang et al., 2007)
video	semantic annotation	news clip	concepts (crowd, desert)	(Qi et al., 2007)
audio	noise detection	sound clip	type (speech, noise)	(Streich & Buhmann, 2008)
	emotion detection	music clip	emotions (relaxing-calm)	(Li & Ogihara, 2003, Trohidis et al., 2008)
structured	functional genomics	gene	functions (energy, metabolism)	(Elisseeff & Weston, 2002, Clare & King, 2001, Blockeel et al., 2006)
	proteomics	protein	enzyme classes (ligases)	(Rousu et al., 2006)
	directed marketing	person	product categories	(Zhang et al., 2006)

Table 34.1. Applications of multi-label Learning

a bipartition of the set of labels into relevant and irrelevant with respect to a query instance. LR on the other hand is concerned with learning a model that outputs an ordering of the class labels according to their relevance to a query instance. Note that LR models can also be learned from training data containing single labels, total rankings of labels, as well as pairwise preferences over the set of labels (Vembu & Gärtner, 2009).

Both MLC and LR are important in mining multi-label data. In a news filtering application for example, the user must be presented with interesting articles only, but it is also important to see the most interesting ones in the top of the list. Ideally, we would like to develop methods that are able to mine both an ordering and a bipartition of the set of labels from multi-label data. Such a task has been recently called *multi-label ranking* (MLR) (Brinker et al., 2006) and poses a very interesting and useful generalization of MLC and LR.

In the following subsections we present MLC, LR and MLR methods grouped into the two categories proposed in (Tsoumakas & Katakis, 2007): i) *problem transformation*, and ii) *algorithm adaptation*. The first group of methods are algorithm independent. They transform the learning task into one or more single-label classification tasks, for which a large bibliography of learning algorithms exists. The second group of methods extend specific learning algorithms in order to handle multi-label data directly.

For the formal description of these methods, we will use $L = \{\lambda_j : j = 1 \dots q\}$ to denote the finite set of labels in a multi-label learning task and $D = \{(\mathbf{x_i}, Y_i), i = 1 \dots m\}$ to denote a set of multi-label training examples, where $\mathbf{x_i}$ is the feature vector and $Y_i \subseteq L$ the set of labels of the *i*-th example.

34.2.1 Problem Transformation

Problem transformation methods will be exemplified through the multi-label data set of Figure 34.1. It consists of four examples that are annotated with one or more out of four labels: λ_1, λ_2, λ_3, λ_4. As the transformations only affect the label space, in the rest of the figures of this section, we will omit the attribute space for simplicity of presentation.

Example	Attributes	Label set
1	$\mathbf{x_1}$	$\{\lambda_1, \lambda_4\}$
2	$\mathbf{x_2}$	$\{\lambda_3, \lambda_4\}$
3	$\mathbf{x_3}$	$\{\lambda_1\}$
4	$\mathbf{x_4}$	$\{\lambda_2, \lambda_3, \lambda_4\}$

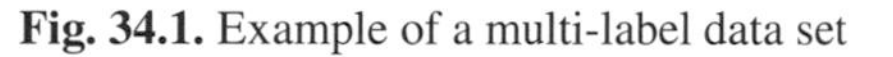

Fig. 34.1. Example of a multi-label data set

There exist several simple transformations that can be used to convert a multi-label data set to a single-label data set with the same set of labels (Boutell et al., 2004, Chen et al., 2007). A single-label classifier that outputs probability distributions over all classes can then be used to learn a ranking. The class with the highest probability will be ranked first, the class with the second best probability will be ranked second, and so on. The *copy* transformation replaces each multi-label example (x_i, Y_i) with $|Y_i|$ examples (x_i, λ_j), for every $\lambda_j \in Y_i$. A variation of this transformation, dubbed *copy-weight*, associates a weight of $\frac{1}{|Y_i|}$ to each of the produced examples. The *select* family of transformations replaces Y_i with one of its members. This label could be the most (*select-max*) or least (*select-min*) frequent among all examples. It could also be randomly selected (*select-random*). Finally, the *ignore* transformation simply discards every multi-label example. Figure 34.2 shows the transformed data set using these simple transformations.

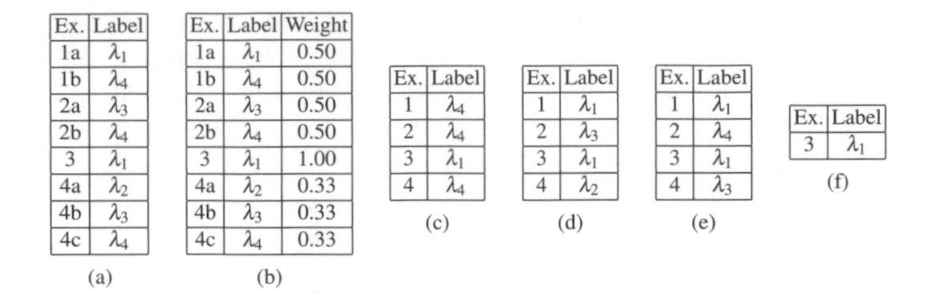

(a)

Ex.	Label
1a	λ_1
1b	λ_4
2a	λ_3
2b	λ_4
3	λ_1
4a	λ_2
4b	λ_3
4c	λ_4

(b)

Ex.	Label	Weight
1a	λ_1	0.50
1b	λ_4	0.50
2a	λ_3	0.50
2b	λ_4	0.50
3	λ_1	1.00
4a	λ_2	0.33
4b	λ_3	0.33
4c	λ_4	0.33

(c)

Ex.	Label
1	λ_4
2	λ_4
3	λ_1
4	λ_4

(d)

Ex.	Label
1	λ_1
2	λ_3
3	λ_1
4	λ_2

(e)

Ex.	Label
1	λ_1
2	λ_4
3	λ_1
4	λ_3

(f)

Ex.	Label
3	λ_1

Fig. 34.2. Transformation of the data set in Figure 34.1 using (a) *copy*, (b) *copy-weight*, (c) *select-max*, (d) *select-min*, (e) *select-random* (one of the possible) and (f) *ignore*

Label powerset (LP) is a simple but effective problem transformation method that works as follows: It considers each unique set of labels that exists in a multi-label training set as one of the classes of a new single-label classification task. Figure 34.3 shows the result of transforming the data set of Figure 34.1 using LP.

Ex.	Label
1	$\lambda_{1,4}$
2	$\lambda_{3,4}$
3	λ_1
4	$\lambda_{2,3,4}$

Fig. 34.3. Transformed data set using the label powerset method

Given a new instance, the single-label classifier of LP outputs the most probable class, which is actually a set of labels. If this classifier can output a probability distribution over all classes, then LP can also rank the labels following the approach in (Read, 2008). Table 34.2 shows an example of a probability distribution that could be produced by LP, trained on the data of Figure 34.3, given a new instance **x** with unknown label set. To obtain a label ranking we calculate for each label the sum of the probabilities of the classes that contain it. This way LP can solve the complete MLR task.

Table 34.2. Example of obtaining a ranking from LP

c	$p(c\|\mathbf{x})$	λ_1	λ_2	λ_3	λ_4
$\lambda_{1,4}$	0.7	1	0	0	1
$\lambda_{3,4}$	0.2	0	0	1	1
λ_1	0.1	1	0	0	0
$\lambda_{2,3,4}$	0.0	0	1	1	1
	$\sum_c p(c\|\mathbf{x})\lambda_j$	0.8	0.0	0.2	0.9

The computational complexity of LP with respect to q depends on the complexity of the base classifier with respect to the number of classes, which is equal to the number of distinct label sets in the training set. This number is upper bounded by $\min(m, 2^q)$ and despite that it typically is much smaller, it still poses an important complexity problem, especially for large values of m and q. The large number of classes, many of which are associated with very few examples, makes the learning process difficult as well.

The pruned problem transformation (PPT) method (Read, 2008) extends LP in an attempt to deal with the aforementioned problems. It prunes away label sets that occur less times than a small user-defined threshold (e.g. 2 or 3) and optionally replaces their information by introducing disjoint subsets of these label sets that do exist more times than the threshold.

The random k-labelsets (RA*k*EL) method (Tsoumakas & Vlahavas, 2007) constructs an ensemble of LP classifiers. Each LP classifier is trained using a different small random subset of the set of labels. This way RA*k*EL manages to take label correlations into account, while avoiding LP's problems. A ranking of the labels is produced by averaging the zero-one predictions of each model per considered label. Thresholding is then used to produce a bipartition as well.

Binary relevance (BR) is a popular problem transformation method that learns q binary classifiers, one for each different label in L. It transforms the original data set into q data sets $D_{\lambda_j}, j = 1 \dots q$ that contain all examples of the original data set, labeled positively if the label set of the original example contained λ_j and negatively otherwise. For the classification of a new instance, BR outputs the union of the labels λ_j that are positively predicted by the q classifiers. Figure 34.4 shows the four data sets that are constructed by BR when applied to the data set of Figure 34.1.

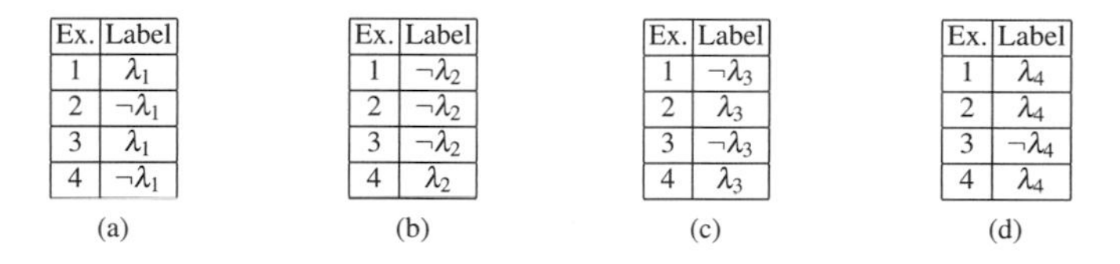

(a)

Ex.	Label
1	λ_1
2	$\neg\lambda_1$
3	λ_1
4	$\neg\lambda_1$

(b)

Ex.	Label
1	$\neg\lambda_2$
2	$\neg\lambda_2$
3	$\neg\lambda_2$
4	λ_2

(c)

Ex.	Label
1	$\neg\lambda_3$
2	λ_3
3	$\neg\lambda_3$
4	λ_3

(d)

Ex.	Label
1	λ_4
2	λ_4
3	$\neg\lambda_4$
4	λ_4

Fig. 34.4. Data sets produced by the BR method

Ranking by pairwise comparison (RPC) (Hüllermeier et al., 2008) transforms the multi-label dataset into $\frac{q(q-1)}{2}$ binary label datasets, one for each pair of labels $(\lambda_i, \lambda_j), 1 \leq i < j \leq q$. Each dataset contains those examples of D that are annotated by at least one of the two corresponding labels, but not both. A binary classifier that learns to discriminate between the two labels, is trained from each of these data sets. Given a new instance, all binary classifiers are invoked, and a ranking is obtained by counting the votes received by each label. Figure 34.5 shows the data sets that are constructed by RPC when applied to the data set of Figure 34.1. The multi-label pairwise perceptron (MLPP) algorithm (Loza Mencia & Fürnkranz, 2008a) is an instantiation of RPC using perceptrons for the binary classification tasks.

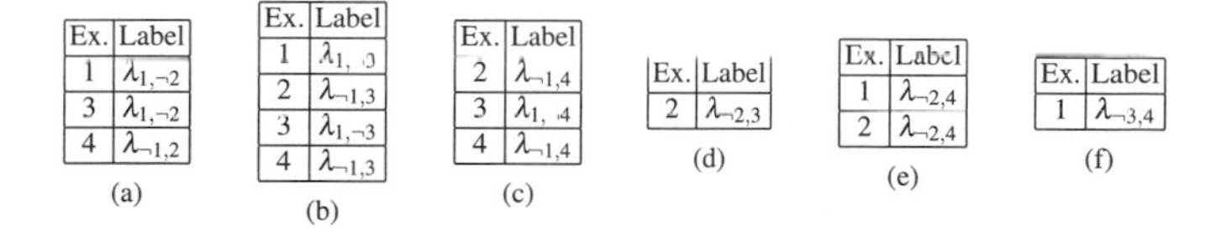

(a)

Ex.	Label
1	$\lambda_{1,\neg 2}$
3	$\lambda_{1,\neg 2}$
4	$\lambda_{\neg 1,2}$

(b)

Ex.	Label
1	$\lambda_{1,\neg 3}$
2	$\lambda_{\neg 1,3}$
3	$\lambda_{1,\neg 3}$
4	$\lambda_{\neg 1,3}$

(c)

Ex.	Label
2	$\lambda_{\neg 1,4}$
3	$\lambda_{1,\neg 4}$
4	$\lambda_{\neg 1,4}$

(d)

Ex.	Label
2	$\lambda_{\neg 2,3}$

(e)

Ex.	Label
1	$\lambda_{\neg 2,4}$
2	$\lambda_{\neg 2,4}$

(f)

Ex.	Label
1	$\lambda_{\neg 3,4}$

Fig. 34.5. Data sets produced by the RPC method

Calibrated label ranking (CLR) (Fürnkranz et al., 2008) extends RPC by introducing an additional virtual label, which acts as a natural breaking point of the ranking into relevant and irrelevant sets of labels. This way, CLR manages to solve the complete MLR task. The binary models that learn to discriminate between the virtual label and each of the other labels, correspond to the models of BR. This occurs, because each example that is annotated with a given label is considered as positive for this label and negative for the virtual label, while each example that is not annotated with a label is considered negative for it and positive for the virtual label. When applied to the data set of Figure 34.1, CLR would construct both the datasets of Figure 34.5 and those of Figure 34.4.

The INSDIF algorithm (Zhang & Zhou, 2007b) computes a prototype vector for each label, by averaging all instances of the training set that belong to this label. After that, every instance is transformed to a bag of q instances, each equal to the difference between the initial instance and one of the prototype vectors. A two level classification strategy is then employed to learn form the transformed data set.

34.2.2 Algorithm Adaptation

The C4.5 algorithm was adapted in (Clare & King, 2001) for the handling of multi-label data. In specific, multiple labels were allowed at the leaves of the tree and the formula of entropy calculation was modified as follows:

$$\text{Entropy}(D) = -\sum_{j=1}^{q} \left(p(\lambda_j) log p(\lambda_j) + q(\lambda_j) log q(\lambda_j) \right) \quad (34.1)$$

where $p(\lambda_j)$ = relative frequency of class λ_j and $q(\lambda_j) = 1 - p(\lambda_j)$.

AdaBoost.MH and AdaBoost.MR (Schapire, 2000) are two extensions of AdaBoost for multi-label data. While AdaBoost.MH is designed to minimize Hamming loss, AdaBoost.MR is designed to find a hypothesis which places the correct labels at the top of the ranking.

A combination of AdaBoost.MH with an algorithm for producing alternating decision trees was presented in (de Comite et al., 2003). The main motivation was the production of multi-label models that can be understood by humans.

A probabilistic generative model is proposed in (McCallum, 1999), according to which, each label generates different words. Based on this model a multi-label document is produced by a mixture of the word distributions of its labels. A similar word-based mixture model for multi-label text classification is presented in (Ueda & Saito, 2003). A deconvolution approach is proposed in (Streich & Buhmann, 2008), in order to estimate the individual contribution of each label to a given item.

The use of conditional random fields is explored in (Ghamrawi & McCallum, 2005), where two graphical models that parameterize label co-occurrences are proposed. The first one, *collective multi-label*, captures co-occurrence patterns among labels, whereas the second one, *collective multi-label with features*, tries to capture the impact that an individual feature has on the co-occurrence probability of a pair of labels.

BP-MLL (Zhang & Zhou, 2006) is an adaptation of the popular back-propagation algorithm for multi-label learning. The main modification to the algorithm is the introduction of a new error function that takes multiple labels into account.

The multi-class multi-label perceptron (MMP) (Crammer & Singer, 2003) is a family of online algorithms for label ranking from multi-label data based on the perceptron algorithm. MMP maintains one perceptron for each label, but weight updates for each perceptron are performed so as to achieve a perfect ranking of all labels.

An SVM algorithm that minimizes the ranking loss (see Section 34.7.2) is proposed in (Elisseeff & Weston, 2002). Three improvements to instantiating the BR method with SVM classifiers are given in (Godbole & Sarawagi, 2004). The first two could easily be abstracted in order to be used with any classification algorithm and could thus be considered an extension to BR itself, while the third is specific to SVMs.

The main idea in the first improvement is to extend the original data set with q additional features containing the predictions of each binary classifier. Then a second round of training q new binary classifiers takes place, this time using the extended data sets. For the classification of a new example, the binary classifiers of the first round are initially used and their output is appended to the features of the example to form a meta-example. This meta-example is then classified by the binary classifiers of the second round. Through this extension, the approach takes into consideration the potential dependencies among the different labels. Note here that this improvement is actually a specialized case of applying Stacking (Wolpert, 1992), a method for the combination of multiple classifiers, on top of BR.

The second improvement, *ConfMat*, consists in removing negative training instances of a complete label if it is very similar to the positive label, based on a confusion matrix that is estimated using any fast and moderately accurate classifier on a held out validation set. The third improvement *BandSVM*, consists in removing very similar negative training instances that are within a threshold distance from the learned hyperplane.

A number of methods (Luo & Zincir-Heywood, 2005, Wieczorkowska et al., 2006, Brinker & Hüllermeier, 2007, Zhang & Zhou, 2007a, Spyromitros et al., 2008) are based on the popular k Nearest Neighbors (kNN) lazy learning algorithm. The first step in all these approaches is the same as in kNN, i.e. retrieving the k nearest examples. What differentiates them is the aggregation of the label sets of these examples.

For example, ML-kNN (Zhang & Zhou, 2007a), uses the maximum a posteriori principle in order to determine the label set of the test instance, based on prior and posterior probabilities for the frequency of each label within the k nearest neighbors.

MMAC (Thabtah et al., 2004) is an algorithm that follows the paradigm of associative classification, which deals with the construction of classification rule sets using association rule mining. MMAC learns an initial set of classification rules through association rule mining, removes the examples associated with this rule set and recursively learns a new rule set from the remaining examples until no further frequent items are left. These multiple rule sets might contain rules with similar preconditions but different labels on the right hand side. Such rules are merged into a single multi-label rule. The labels are ranked according to the support of the corresponding individual rules.

Finally, an approach that combines lazy and associative learning is proposed in (Veloso et al., 2007), where the inductive process is delayed until an instance is given for classification.

34.3 Dimensionality Reduction

Several application domains of multi-label learning (e.g. text, bioinformatics) involve data with large number of features. Dimensionality reduction has been extensively studied in the case of single-label data. Some of the existing approaches are directly applicable to multi-label data, while others have been extended for handling them appropriately. We present past and very recent approaches to multi-label dimensionality reduction, organized into two categories: i) *feature selection* and ii) *feature extraction*.

34.3.1 Feature Selection

The *wrapper* approach to feature selection (Kohavi & John, 1997) is directly applicable to multi-label data. Given a multi-label learning algorithm, we can search for the subset of features that optimizes a multi-label loss function (see Section 34.7) on an evaluation data set.

A different line of attacking the multi-label feature selection problem is to transform the multi-label data set into one or more single-label data sets and use existing feature selection methods, particularly those that follow the *filter* paradigm. One of the most popular approaches, especially in text categorization, uses the BR transformation in order to evaluate the discriminative power of each feature with respect to each of the labels independently of the rest of the labels. Subsequently the obtained scores are aggregated in order to obtain an overall ranking. Common aggregation strategies include taking the maximum or a weighted average of the obtained scores (Yang & Pedersen, 1997). The LP transformation was used in (Trohidis et al., 2008), while the *copy*, *copy-weight*, *select-max*, *select-min* and *ignore* transformations are used in (Chen et al., 2007).

34.3.2 Feature Extraction

Feature extraction methods construct new features out of the original ones either using class information (supervised) or not (unsupervised).

Unsupervised methods, such as principal component analysis and latent semantic indexing (LSI) are obviously directly applicable to multi-label data. For example, in (Gao et al., 2004), the authors directly apply LSI based on singular value decomposition in order to reduce the dimensionality of the text categorization problem.

Supervised feature extraction methods for single-label data, such as linear discriminant analysis (LDA), require modification prior to their application to multi-label data. LDA has been modified to handle multi-label data in (Park & Lee, 2008). A version of the LSI method that takes into consideration label information (MLSI) was proposed in (Yu et al., 2005), while a supervised multi-label feature extraction algorithm based on the Hilbert-Schmidt independence criterion was proposed in (Zhang & Zhou, 2008). In (Ji et al., 2008) a framework for extracting a subspace of features is proposed. Finally, a hypergraph is employed in (Sun et al., 2008) for modeling higher-order relations among instances sharing the same label. A spectral learning method is then used for computing a low-dimensional embedding that preserves these relations.

34.4 Exploiting Label Structure

In certain multi-label domains, such as text mining and bioinformatics, labels are organized into a tree-shaped general-to-specific hierarchical structure. An example of such a structure, called functional catalogue (FunCat) (Ruepp et al., 2004), is an annotation scheme for the functional description of proteins from several living organisms. The 1362 functional categories in version 2.1 of FunCat are organized in a tree like structure with up to six levels of increasing specificity. Many more hierarchical structures exist for textual data, such as the MeSH[1] for medical articles and the ACM computing classification system[2] for computer

[1] www.nlm.nih.gov/mesh/

[2] www.acm.org/class/

science articles. Taking into account such structures when learning from multi-label data is important, because it can lead to improved predictive performance and time complexity.

A general-to-specific tree structure of labels implies that an example cannot be associated with a label λ if it isn't associated with its parent label par(λ). In other words, the set of labels associated with an example must be a union of the labels found along zero or more paths starting at the root of the hierarchy. Some applications may require such paths to end at a leaf, but in the general case they can be partial.

Given a label hierarchy, a straightforward approach to learning a multi-label classifier is to train a binary classifier for each non-root label λ of this hierarchy, using as training data those examples of the full training set that are annotated with par(λ). During testing, these classifiers are called in a top-down manner, calling a classifier for λ only if the classifier for par(λ) has given a positive output. We call this the *hierarchical binary relevance* (HBR) method.

An online learning algorithm that follows the HBR approach, using a regularized least squares estimator at each node, is presented in (Cesa-Bianchi et al., 2006b). Better results were found compared to an instantiation of HBR using perceptrons. Other important contributions of (Cesa-Bianchi et al., 2006b) are the definition of a hierarchical loss function (see Section 34.7.1) and a thorough theoretical analysis of the proposed algorithm. An approach that follows the training process of HBR but uses a bottom-up procedure during testing is presented in (Cesa-Bianchi et al., 2006a).

The HBR approach can be reformulated in a more generalized fashion as the training of a multi-label (instead of binary) classifier in all non-leaf (instead of non-root) nodes (Esuli et al., 2008, Tsoumakas et al., 2008). TreeBoost.MH (Esuli et al., 2008) uses Adaboost.MH (see Section 34.2.2) at each non-leaf node. Experimental results indicate that not only is TreeBoost.MH more efficient in training and testing than Adaboost.MH, but that it also improves predictive accuracy.

Two different approaches for exploiting tree-shaped hierarchies are (Blockeel et al., 2006, Rousu et al., 2006). Predictive clustering trees are used in (Blockeel et al., 2006), while a large margin method for structured output prediction is used in (Rousu et al., 2006).

The directed acyclic graph (DAG) is a more general type of structure, where a node can have multiple parents. This is the case for the Gene Ontology (GO) (Harris et al., 2004), which covers several domains of molecular and cellular biology. A Bayesian framework for combining a hierarchy of support vector machines based on the GO is proposed in (Barutcuoglu et al., 2006). An extension of the work in (Blockeel et al., 2006) for handling DAG label structures is presented in (Vens et al., 2008).

34.5 Scaling Up

Problems with large number of labels can be found in several domains. For example, the *Eurovoc*[3] taxonomy contains approximately 4000 descriptors European for documents, while in collaborative tagging systems such as *delicious*[4], the user assigned tags can be hundreds of thousands.

The high dimensionality of the label space may challenge a multi-label learning algorithm in many ways. Firstly, the number of training examples annotated with each particular label will be significantly less than the total number of examples. This is similar to the class imbalance problem in single-label data (Chawla et al., 2004). Secondly, the computational cost of

[3] europa.eu/eurovoc/

[4] delicious.com

training a multi-label model may be strongly affected by the number of labels. There are simple algorithms, such as BR with linear complexity with respect to q, but there are others, such as LP, whose complexity is worse. Thirdly, although the complexity of using a multi-label model for prediction is linear with respect to q in the best case, this may still be inefficient for applications requiring fast response times. Finally, methods that need to maintain a large number of models in memory, may fail to scale up to such domains.

HOMER (Tsoumakas et al., 2008) constructs a Hierarchy Of Multilabel classifiERs each one dealing with a much smaller set of labels compared to q and a more balanced example distribution. This leads to improved predictive performance along with linear training and logarithmic testing complexities withs respect to q. At a first step, HOMER automatically organizes labels into a tree-shaped hierarchy. This is accomplished by recursively partitioning the set of labels into a number of nodes using a balance clustering algorithm. It then builds one multi-label classifier at each node apart from the leafs, following the HBR approach described in the previous Section. The multi-label classifiers predict one or more meta-labels μ, each one corresponding to the disjunction of a child node's labels. Figure 34.6 presents a sample tree of multi-label classifiers constructed by HOMER for a domain with 8 labels.

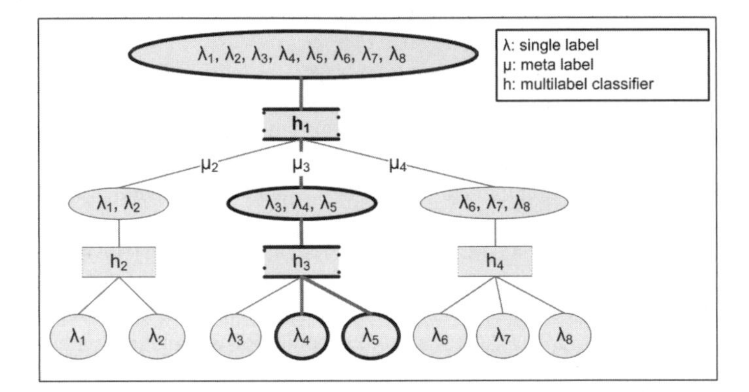

Fig. 34.6. Sample hierarchy for a multi-label domain with 8 labels.

To deal with the memory problem of RPC, an extension of MLPP with reduced space complexity in the presence of large number of labels is described in (Loza Mencia & Fürnkranz, 2008b).

34.6 Statistics and Datasets

In some applications the number of labels of each example is small compared to q, while in others it is large. This could be a parameter that influences the performance of the different multi-label methods. We here introduce the concepts of label cardinality and label density of a data set.

Label cardinality of a dataset D is the average number of labels of the examples in D:

$$\text{Label-Cardinality} = \frac{1}{m}\sum_{i=1}^{m}|Y_i|$$

Label density of D is the average number of labels of the examples in D divided by q:

$$\text{Label-Density} = \frac{1}{m}\sum_{i=1}^{m}\frac{|Y_i|}{q} \tag{34.2}$$

Label cardinality is independent of the number of labels q in the classification problem, and is used to quantify the number of alternative labels that characterize the examples of a multi-label training data set. Label density takes into consideration the number of labels in the domain. Two data sets with the same label cardinality but with a great difference in the number of labels (different label density) might not exhibit the same properties and cause different behavior to the multi-label learning methods. The number of distinct label sets is also important for many algorithm transformation methods that operate on subsets of labels.

Table 34.3 presents some benchmark datasets[5] from various domains among with their corresponding statistics and source reference. The statistics of the Reuters (rcv1v2) dataset are averages over the 5 subsets.

34.7 Evaluation Measures

The evaluation of methods that learn from multi-label data requires different measures than those used in the case of single-label data. This section presents the various measures that have been proposed in the past for the evaluation of i) bipartitions and ii) rankings with respect to the ground truth of multi-label data. It concludes with a subsection on measures that take into account an existing label hierarchy.

For the definitions of these measures we will consider an evaluation data set of multi-label examples $(\mathbf{x_i}, Y_i)$, $i = 1 \dots m$, where $Y_i \subseteq L$ is the set of true labels and $L = \{\lambda_j : j = 1 \dots q\}$ is the set of all labels. Given instance $\mathbf{x_i}$, the set of labels that are predicted by an MLC method is denoted as Z_i, while the rank predicted by an LR method for a label λ is denoted as $r_i(\lambda)$. The most relevant label, receives the highest rank (1), while the least relevant one, receives the lowest rank (q).

34.7.1 Bipartitions

Some of the measures that evaluate bipartitions are calculated based on the average differences of the actual and the predicted sets of labels over all examples of the evaluation data set. Others decompose the evaluation process into separate evaluations for each label, which they subsequently average over all labels. We call the former *example-based* and the latter *label-based* evaluation measures.

Example-based

The Hamming loss (Schapire, 2000) is defined as follows:

$$\text{Hamming-Loss} = \frac{1}{m}\sum_{i=1}^{m}\frac{|Y_i \triangle Z_i|}{M}$$

[5] All datasets are available for download at http://mlkd.csd.auth.gr/multilabel.html

Table 34.3. Multilabel datasets and their statistics

name	domain	instances	nominal	numeric	labels	cardinality	density	distinct	source
delicious	text (web)	16105	500	0	983	19.020	0.019	15806	(Tsoumakas & Katakis, 2007)
emotions	music	593	0	72	6	1.869	0.311	27	(Trohidis et al., 2008)
genbase	biology	662	1186	0	27	1.252	0.046	32	(Diplaris et al., 2005)
mediamill	multimedia	43907	0	120	101	4.376	0.043	6555	(Snoek et al., 2006)
rcv1v2 (avg)	text	6000	0	47234	101	2.6508	0.026	937	(Lewis et al., 2004)
scene	multimedia	2407	0	294	6	1.074	0.179	15	(Boutell et al., 2004)
yeast	biology	2417	0	103	14	4.237	0.303	198	(Elisseeff & Weston, 2002)
tmc2007	text	28596	49060	0	22	2.158	0.098	1341	(Srivastava & Zane-Ulman, 2005)

where $\triangle$ stands for the symmetric difference of two sets, which is the set-theoretic equivalent of the exclusive disjunction (XOR operation) in Boolean logic.

Classification accuracy (Zhu et al., 2005) or subset accuracy (Ghamrawi & McCallum, 2005) is defined as follows:

$$\text{ClassificationAccuracy} = \frac{1}{m}\sum_{i=1}^{m} I(Z_i = Y_i)$$

where $I(\text{true}) = 1$ and $I(\text{false}) = 0$. This is a very strict evaluation measure as it requires the predicted set of labels to be an exact match of the true set of labels.

The following measures are used in (Godbole & Sarawagi, 2004):

$$\text{Precision} = \frac{1}{m}\sum_{i=1}^{m} \frac{|Y_i \cap Z_i|}{|Z_i|} \qquad \text{Recall} = \frac{1}{m}\sum_{i=1}^{m} \frac{|Y_i \cap Z_i|}{|Y_i|}$$

$$F_1 = \frac{1}{m}\sum_{i=1}^{m} \frac{2|Y_i \cap Z_i|}{|Z_i| + |Y_i|} \qquad \text{Accuracy} = \frac{1}{m}\sum_{i=1}^{m} \frac{|Y_i \cap Z_i|}{|Y_i \cup Z_i|}$$

Label-based

Any known measure for binary evaluation can be used here, such as accuracy, area under the ROC curve, precision and recall. The calculation of these measures for all labels can be achieved using two averaging operations, called *macro-averaging* and *micro-averaging* (Yang, 1999). These operations are usually considered for averaging precision, recall and their harmonic mean (F-measure) in Information Retrieval tasks.

Consider a binary evaluation measure $B(tp, tn, fp, fn)$ that is calculated based on the number of true positives (tp), true negatives (tn), false positives (fp) and false negatives (fn). Let tp_λ, fp_λ, tn_λ and fn_λ be the number of true positives, false positives, true negatives and false negatives after binary evaluation for a label λ. The macro-averaged and micro-averaged versions of B, are calculated as follows:

$$B_{\text{macro}} = \frac{1}{q}\sum_{\lambda=1}^{q} B(tp_\lambda, fp_\lambda, tn_\lambda, fn_\lambda)$$

$$B_{\text{micro}} = B\left(\sum_{\lambda=1}^{q} tp_\lambda, \sum_{\lambda=1}^{q} fp_\lambda, \sum_{\lambda=1}^{q} tn_\lambda, \sum_{\lambda=1}^{q} fn_\lambda\right)$$

Note that micro-averaging has the same result as macro-averaging for some measures, such as accuracy, while it differs for other measures, such as precision, recall and area under the ROC curve. Note also that the average (macro/micro) accuracy and Hamming loss sum up to 1, as Hamming loss is actually the average binary classification error.

34.7.2 Ranking

One-error evaluates how many times the top-ranked label is not in the set of relevant labels of the instance:

$$\text{1-Error} = \frac{1}{m}\sum_{i=1}^{m} \delta(\arg\min_{\lambda \in L} r_i(\lambda))$$

where

$$\delta(\lambda) = \begin{cases} 1 & \text{if } \lambda \notin Y_i \\ 0 & \text{otherwise} \end{cases}$$

Coverage evaluates how far we need, on average, to go down the ranked list of labels in order to cover all the relevant labels of the example.

$$\text{Cov} = \frac{1}{m} \sum_{i=1}^{m} \max_{\lambda \in Y_i} r_i(\lambda) - 1$$

Ranking loss expresses the number of times that irrelevant labels are ranked higher than relevant labels:

$$\text{R-Loss} = \frac{1}{m} \sum_{i=1}^{m} \frac{1}{|Y_i||\overline{Y_i}|} |\{(\lambda_a, \lambda_b) : r_i(\lambda_a) > r_i(\lambda_b), (\lambda_a, \lambda_b) \in Y_i \times \overline{Y_i}\}|$$

where $\overline{Y_i}$ is the complementary set of Y_i with respect to L.

Average precision evaluates the average fraction of labels ranked above a particular label $\lambda \in Y_i$ which actually are in Y_i.

$$\text{AvgPrec} = \frac{1}{m} \sum_{i=1}^{m} \frac{1}{|Y_i|} \sum_{\lambda \in Y_i} \frac{|\{\lambda' \in Y_i : r_i(\lambda') \leq r_i(\lambda)\}|}{r_i(\lambda)}$$

34.7.3 Hierarchical

The hierarchical loss (Cesa-Bianchi et al., 2006b) is a modified version of the Hamming loss that takes into account an existing hierarchical structure of the labels. It examines the predicted labels in a top-down manner according to the hierarchy and whenever the prediction for a label is wrong, the subtree rooted at that node is not considered further in the calculation of the loss. Let $\text{anc}(\lambda)$ be the set of all the ancestor nodes of λ. The hierarchical loss is defined as follows:

$$\text{H-Loss} = \frac{1}{m} \sum_{i=1}^{m} |\{\lambda : \lambda \in Y_i \triangle Z_i, \text{anc}(\lambda) \cap (Y_i \triangle Z_i) = \emptyset\}|$$

Several other measures for hierarchical (multi-label) classification are examined in (Moskovitch et al., 2006, Sun & Lim, 2001).

34.8 Related Tasks

One of the most popular supervised learning tasks is *multi-class* classification, which involves a set of labels L, where $|L| > 2$. The critical difference with respect to multi-label classification is that each instance is associated with only one element of L, instead of a subset of L.

Jin and Ghahramani (Jin & Ghahramani, 2002) call *multiple-label problems*, the semi-supervised classification problems where each example is associated with more than one classes, but only one of those classes is the true class of the example. This task is not that common in real-world applications as the one we are studying.

Multiple-instance or *multi-instance* learning is a variation of supervised learning, where labels are assigned to bags of instances (Maron & p Erez, 1998). In certain applications, the training data can be considered as both multi-instance and multi-label (Zhou, 2007). In image classification for example, the different regions of an image can be considered as multiple-instances, each of which can be labeled with a different concept, such as *sunset* and *sea*.

Several methods have been recently proposed for addressing such data (Zhou & Zhang, 2006, Zha et al., 2008).

In *Multitask learning* (Caruana, 1997) we try to solve many similar tasks in parallel usually using a shared representation. Taking advantage of the common characteristics of these tasks a better generalization can be achieved. A typical example is to learn to identify hand written text for different writers in parallel. Training data from one writer can aid the construction of better predictive models for other authors.

34.9 Multi-Label Data Mining Software

There exists a number of implementations of specific algorithms for mining multi-label data, most of which have been discussed in Section 34.2.2. The BoosTexter system[6], implements the boosting-based approaches proposed in (Schapire, 2000). There also exist Matlab implementations for ML*k*NN[7] and BPMLL[8].

There are also more general-purpose software that handle multi-label data as part of their functionality. LibSVM (Chang & Lin, 2001) is a library for support vector machines that can learn from multi-label data using the binary relevance transformation. Clus[9] is a predictive clustering system that is based on decision tree learning. Its capabilities include (hierarchical) multi-label classification.

Finally, Mulan[10] is an open-source software devoted to multi-label data mining. It includes implementations of a large number of learning algorithms, basic capabilities for dimensionality reduction and hierarchical multi-label classification and an extensive evaluation framework.

References

Barutcuoglu, Z., Schapire, R. E. & Troyanskaya, O. G. (2006). Bioinformatics 22, 830–836.

Blockeel, H., Schietgat, L., Struyf, J., Dz?eroski, S. & Clare, A. (2006). Lecture Notes in Computer Science (including subseries Lecture Notes in Artificial Intelligence and Lecture Notes in Bioinformatics) 4213 LNAI, 18–29.

Boleda, G., im Walde, S. S. & Badia, T. (2007). In Proceedings of the 2007 Joint Conference on Empirical Methods in Natural Language Processing and Computational Natural Language Learning pp. 171–180,, Prague.

Boutell, M., Luo, J., Shen, X. & Brown, C. (2004). Pattern Recognition 37, 1757–1771.

Brinker, K., Fürnkranz, J. & Hüllermeier, E. (2006). In Proceedings of the 17th European Conference on Artificial Intelligence (ECAI '06) pp. 489–493,, Riva del Garda, Italy.

Brinker, K. & Hüllermeier, E. (2007). In Proceedings of the 20th International Conference on Artificial Intelligence (IJCAI '07) pp. 702–707,, Hyderabad, India.

Caruana, R. (1997). Machine Learning 28, 41–75.

---

[6] http://www.cs.princeton.edu/ schapire/boostexter.html

[7] http://lamda.nju.edu.cn/datacode/MLkNN.htm

[8] http://lamda.nju.edu.cn/datacode/BPMLL.htm

[9] http://www.cs.kuleuven.be/ dtai/clus/

[10] http://sourceforge.net/projects/mulan/

Cesa-Bianchi, N., Gentile, C. & Zaniboni, L. (2006a). In ICML '06: Proceedings of the 23rd international conference on Machine learning pp. 177–184,.

Cesa-Bianchi, N., Gentile, C. & Zaniboni, L. (2006b). Journal of Machine Learning Research 7, 31–54.

Chang, C.-C. & Lin, C.-J. (2001). LIBSVM: a library for support vector machines. Software available at http://www.csie.ntu.edu.tw/~cjlin/libsvm.

Chawla, N. V., Japkowicz, N. & Kotcz, A. (2004). SIGKDD Explorations 6, 1–6.

Chen, W., Yan, J., Zhang, B., Chen, Z. & Yang, Q. (2007). In Proc. 7th IEEE International Conference on Data Mining pp. 451–456, IEEE Computer Society, Los Alamitos, CA, USA.

Clare, A. & King, R. (2001). In Proceedings of the 5th European Conference on Principles of Data Mining and Knowledge Discovery (PKDD 2001) pp. 42–53,, Freiburg, Germany.

Crammer, K. & Singer, Y. (2003). Journal of Machine Learning Research 3, 1025–1058.

de Comite, F., Gilleron, R. & Tommasi, M. (2003). In Proceedings of the 3rd International Conference on Machine Learning and Data Mining in Pattern Recognition (MLDM 2003) pp. 35–49,, Leipzig, Germany.

Diplaris, S., Tsoumakas, G., Mitkas, P. & Vlahavas, I. (2005). In Proceedings of the 10th Panhellenic Conference on Informatics (PCI 2005) pp. 448–456,, Volos, Greece.

Elisseeff, A. & Weston, J. (2002). In Advances in Neural Information Processing Systems 14.

Esuli, A., Fagni, T. & Sebastiani, F. (2008). Information Retrieval 11, 287–313.

Fürnkranz, J., Hüllermeier, E., Mencia, E. L. & Brinker, K. (2008). Machine Learning .

Gao, S., Wu, W., Lee, C.-H. & Chua, T.-S. (2004). In Proceedings of the 21st international conference on Machine learning (ICML '04) p. 42,, Banff, Alberta, Canada.

Ghamrawi, N. & McCallum, A. (2005). In Proceedings of the 2005 ACM Conference on Information and Knowledge Management (CIKM '05) pp. 195–200,, Bremen, Germany.

Godbole, S. & Sarawagi, S. (2004). In Proceedings of the 8th Pacific-Asia Conference on Knowledge Discovery and Data Mining (PAKDD 2004) pp. 22–30,.

Harris, M. A., Clark, J., Ireland, A., Lomax, J., Ashburner, M., Foulger, R., Eilbeck, K., Lewis, S., Marshall, B., Mungall, C., Richter, J., Rubin, G. M., Blake, J. A., Bult, C., Dolan, M., Drabkin, H., Eppig, J. T., Hill, D. P., Ni, L., Ringwald, M., Balakrishnan, R., Cherry, J. M., Christie, K. R., Costanzo, M. C., Dwight, S. S., Engel, S., Fisk, D. G., Hirschman, J. E., Hong, E. L., Nash, R. S., Sethuraman, A., Theesfeld, C. L., Botstein, D., Dolinski, K., Feierbach, B., Berardini, T., Mundodi, S., Rhee, S. Y., Apweiler, R., Barrell, D., Camon, E., Dimmer, E., Lee, V., Chisholm, R., Gaudet, P., Kibbe, W., Kishore, R., Schwarz, E. M., Sternberg, P., Gwinn, M., Hannick, L., Wortman, J., Berriman, M., Wood, V., de La, Tonellato, P., Jaiswal, P., Seigfried, T. & White, R. (2004). Nucleic Acids Res 32.

Hüllermeier, E., Fürnkranz, J., Cheng, W. & Brinker, K. (2008). Artificial Intelligence 172, 1897–1916.

Ji, S., Tang, L., Yu, S. & Ye, J. (2008). In Proceedings of the 14th SIGKDD International Conferece on Knowledge Discovery and Data Mining, Las Vegas, USA.

Jin, R. & Ghahramani, Z. (2002). In Proceedings of Neural Information Processing Systems 2002 (NIPS 2002), Vancouver, Canada.

Katakis, I., Tsoumakas, G. & Vlahavas, I. (2008). In Proceedings of the ECML/PKDD 2008 Discovery Challenge, Antwerp, Belgium.

Kohavi, R. & John, G. H. (1997). Artificial Intelligence 97, 273–324.

Lewis, D. D., Yang, Y., Rose, T. G. & Li, F. (2004). J. Mach. Learn. Res. 5, 361–397.

Li, T. & Ogihara, M. (2003). In Proceedings of the International Symposium on Music Information Retrieval pp. 239–240,, Washington D.C., USA.

Li, T. & Ogihara, M. (2006). IEEE Transactions on Multimedia 8, 564–574.

Loza Mencia, E. & Fürnkranz, J. (2008a). In 2008 IEEE International Joint Conference on Neural Networks (IJCNN-08) pp. 2900–2907,, Hong Kong.

Loza Mencia, E. & Fürnkranz, J. (2008b). In 12th European Conference on Principles and Practice of Knowledge Discovery in Databases, PKDD 2008 pp. 50–65,, Antwerp, Belgium.

Luo, X. & Zincir-Heywood, A. (2005). In Proceedings of the 15th International Symposium on Methodologies for Intelligent Systems pp. 161–169,.

Maron, O. & p Erez, T. A. L. (1998). In Advances in Neural Information Processing Systems 10 pp. 570–576, MIT Press.

McCallum, A. (1999). In Proceedings of the AAAI' 99 Workshop on Text Learning.

Mencia, E. L. & Fürnkranz, J. (2008). In 12th European Conference on Principles and Practice of Knowledge Discovery in Databases, PKDD 2008, Antwerp, Belgium.

Moskovitch, R., Cohenkashi, S., Dror, U., Levy, I., Maimon, A. & Shahar, Y. (2006). Artificial Intelligence in Medicine 37, 177–190.

Park, C. H. & Lee, M. (2008). Pattern Recogn. Lett. 29, 878–887.

Pestian, J. P., Brew, C., Matykiewicz, P., Hovermale, D. J., Johnson, N., Cohen, K. B. & Duch, W. (2007). In BioNLP '07: Proceedings of the Workshop on BioNLP 2007 pp. 97–104, Association for Computational Linguistics, Morristown, NJ, USA.

Qi, G.-J., Hua, X.-S., Rui, Y., Tang, J., Mei, T. & Zhang, H.-J. (2007). In MULTIMEDIA '07: Proceedings of the 15th international conference on Multimedia pp. 17–26, ACM, New York, NY, USA.

Read, J. (2008). In Proc. 2008 New Zealand Computer Science Research Student Conference (NZCSRS 2008) pp. 143–150,.

Rokach L., Genetic algorithm-based feature set partitioning for classification problems,Pattern Recognition, 41(5):1676–1700, 2008.

Rokach L., Mining manufacturing data using genetic algorithm-based feature set decomposition, Int. J. Intelligent Systems Technologies and Applications, 4(1):57-78, 2008.

Rokach L., Maimon O. and Lavi I., Space Decomposition In Data Mining: A Clustering Approach, Proceedings of the 14th International Symposium On Methodologies For Intelligent Systems, Maebashi, Japan, Lecture Notes in Computer Science, Springer-Verlag, 2003, pp. 24–31.

Rousu, J., Saunders, C., Szedmak, S. & Shawe-Taylor, J. (2006). Journal of Machine Learning Research 7, 1601–1626.

Ruepp, A., Zollner, A., Maier, D., Albermann, K., Hani, J., Mokrejs, M., Tetko, I., Güldener, U., Mannhaupt, G., Münsterkötter, M. & Mewes, H. W. (2004). Nucleic Acids Res 32, 5539–5545.

Schapire, R.E. Singer, Y. (2000). Machine Learning 39, 135–168.

Snoek, C. G. M., Worring, M., van Gemert, J. C., Geusebroek, J.-M. & Smeulders, A. W. M. (2006). In MULTIMEDIA '06: Proceedings of the 14th annual ACM international conference on Multimedia pp. 421–430, ACM, New York, NY, USA.

Spyromitros, E., Tsoumakas, G. & Vlahavas, I. (2008). In Proc. 5th Hellenic Conference on Artificial Intelligence (SETN 2008).

Srivastava, A. & Zane-Ulman, B. (2005). In IEEE Aerospace Conference.

Streich, A. P. & Buhmann, J. M. (2008). In 12th European Conference on Principles and Practice of Knowledge Discovery in Databases, PKDD 2008, Antwerp, Belgium.

Sun, A. & Lim, E.-P. (2001). In ICDM '01: Proceedings of the 2001 IEEE International Conference on Data Mining pp. 521–528, IEEE Computer Society, Washington, DC, USA.

Sun, L., Ji, S. & Ye, J. (2008). In Proceedings of the 14th SIGKDD International Conferece on Knowledge Discovery and Data Mining, Las Vegas, USA.

Thabtah, F., Cowling, P. & Peng, Y. (2004). In Proceedings of the 4th IEEE International Conference on Data Mining, ICDM '04 pp. 217–224,.

Trohidis, K., Tsoumakas, G., Kalliris, G. & Vlahavas, I. (2008). In Proc. 9th International Conference on Music Information Retrieval (ISMIR 2008), Philadelphia, PA, USA, 2008.

Tsoumakas, G. & Katakis, I. (2007). International Journal of Data Warehousing and Mining 3, 1–13.

Tsoumakas, G., Katakis, I. & Vlahavas, I. (2008). In Proc. ECML/PKDD 2008 Workshop on Mining Multidimensional Data (MMD'08) pp. 30–44,.

Tsoumakas, G. & Vlahavas, I. (2007). In Proceedings of the 18th European Conference on Machine Learning (ECML 2007) pp. 406–417,, Warsaw, Poland.

Ueda, N. & Saito, K. (2003). Advances in Neural Information Processing Systems 15 , 721–728.

Veloso, A., Wagner, M. J., Goncalves, M. & Zaki, M. (2007). In Proceedings of the 11th European Conference on Principles and Practice of Knowledge Discovery in Databases (PKDD 2007) vol. LNAI 4702, pp. 605–612, Springer, Warsaw, Poland.

Vembu, S. & Gärtner, T. (2009). In Preference Learning, (Fürnkranz, J. & Hüllermeier, E., eds),. Springer.

Vens, C., Struyf, J., Schietgat, L., Džeroski, S. & Blockeel, H. (2008). Machine Learning 73, 185–214.

Wieczorkowska, A., Synak, P. & Ras, Z. (2006). In Proceedings of the 2006 International Conference on Intelligent Information Processing and Web Mining (IIPWM'06) pp. 307–315,.

Wolpert, D. (1992). Neural Networks 5, 241–259.

Yang, S., Kim, S.-K. & Ro, Y. M. (2007). Circuits and Systems for Video Technology, IEEE Transactions on 17, 324–335.

Yang, Y. (1999). Journal of Information Retrieval 1, 67–88.

Yang, Y. & Pedersen, J. O. (1997). In Proceedings of ICML-97, 14th International Conference on Machine Learning, (Fisher, D. H., ed.), pp. 412–420, Morgan Kaufmann Publishers, San Francisco, US, Nashville, US.

Yu, K., Yu, S. & Tresp, V. (2005). In SIGIR '05: Proceedings of the 28th annual international ACM SIGIR conference on Research and development in information retrieval pp. 258–265, ACM Press, Salvador, Brazil.

Zha, Z.-J., Hua, X.-S., Mei, T., Wang, J., Qi, G.-J. & Wang, Z. (2008). In Computer Vision and Pattern Recognition, 2008. CVPR 2008. IEEE Conference on pp. 1–8,.

Zhang, M.-L. & Zhou, Z.-H. (2006). IEEE Transactions on Knowledge and Data Engineering 18, 1338–1351.

Zhang, M.-L. & Zhou, Z.-H. (2007a). Pattern Recognition 40, 2038–2048.

Zhang, M.-L. & Zhou, Z.-H. (2007b). In Proceedings of the Twenty-Second AAAI Conference on Artificial Intelligence pp. 669–674, AAAI Press, Vancouver, Britiths Columbia, Canada.

Zhang, Y., Burer, S. & Street, W. N. (2006). Journal of Machine Learning Research 7, 1315–1338.

Zhang, Y. & Zhou, Z.-H. (2008). In Proceedings of the Twenty-Third AAAI Conference on Artificial Intelligence, AAAI 2008 pp. 1503–1505, AAAI Press, Chicago, Illinois, USA.
Zhou, Z.-H. (2007). In Proceedings of the 3rd International Conference on Advanced Data Mining and Applications (ADMA'07) p. 1. Springer.
Zhou, Z. H. & Zhang, M. L. (2006). In NIPS, (Schölkopf, B., Platt, J. C. & Hoffman, T., eds), pp. 1609–1616, MIT Press.
Zhu, S., Ji, X., Xu, W. & Gong, Y. (2005). In Proceedings of the 28th annual international ACM SIGIR conference on Research and development in Information Retrieval pp. 274–281.

35

Privacy in Data Mining

Vicenç Torra

IIIA - CSIC, Campus UAB s/n, 08193 Bellaterra Catalonia, Spain
vtorra@iiia.csic.es

Summary. In this chapter we describe the main tools for privacy in data mining. We present an overview of the tools for protecting data, and then we focus on protection procedures. Information loss and disclosure risk measures are also described.

35.1 Introduction

Data is nowadays gathered in large amounts by companies and national offices. This data is often analyzed either using statistical methods or data mining ones. When such methods are applied within the walls of the company that has gathered them, the danger of disclosure of sensitive information might be limited. In contrast, when the analysis have to be performed by third parties, privacy becomes a much more relevant issue.

To make matters worst, it is not uncommon the scenario where an analysis does not only require data from a single data source, but from several data sources. This is the case of banks looking for fraud detection and hospitals analyzing deseases and treatments. In the first case, data from several banks might help on fraud detection. Similarly, data from different hospitals might help on the process of finding the causes of a bad response to a given treatment, or the causes of a given desease.

Privacy-Preserving Data Mining (Aggarwal and Yu, 2008) (PPDM) and Statistical Disclosure Control (Willenborg, 2001, Domingo-Ferrer and Torra, 2001a) (SDC) are two related fields with a similar interest on ensuring data privacy. Their goal is to avoid the disclosure of sensitive or proprietary information to third parties.

Within these fields, several methods have been proposed for processing and analysing data without compromising privacy, for releasing data ensuring some levels of data privacy; measures and indices have been defined for evaluating disclosure risk (that is, in what extent data satisfy the privacy constraints), and data utility or information loss (that is, in what extent the protected data is still useful for applications). In addition, tools have been proposed to visualize and compare different approaches for data protection.

In this chapter we will review some of the existing methods and give an overview of the measures. The structure of the chapter is as follows. In Section 35.2, we present a classification of protection procedures. In Section 35.3, we review different interpretations for risk and give an overview of disclosure risk measures. In Section 35.4, we present major protection procedures. Also in this section we review k-anonymity. Then, Section 35.5 is focused on how

O. Maimon, L. Rokach (eds.), *Data Mining and Knowledge Discovery Handbook*, 2nd ed.,
DOI 10.1007/978-0-387-09823-4_35,

to measure data utility and information loss. A few information loss measures are reviewed there. The chapter finishes in Section 35.6 presenting different approaches for visualizing the trade-off between risk and utility, or risk and information loss. Some conclusions close the chapter.

35.2 On the Classification of Protection Procedures

The literature on Privacy Preserving Data Mining (PPDM) and on Statistical Disclosure Control (SDC) is vast, and a large number of procedures for ensuring privacy have been proposed. We classify them in two categories according to the prior knowledge the data owner has about the usage of the data.

Data-driven or general purpose protection procedures. In this case, no specific analysis or usage is foreseen for the data. The data owner does not know what kind of analysis will be performed by the third party.
This is the case when data is released for public use, as there is no way to know what kind of study a potential user will perform. This situation is common in National Statistical Offices, where data obtained from census and questionnaires can be e.g. downloaded from internet (census.gov). A similar case can occur for other public offices that publish regularly data obtained from questionnaires. Another case is when data are transferred to e.g. researchers so that they can analyse them. Hospitals and other healthcare institutions can also be the target of such protection procedures, as they can be interested in protection procedures that permit different researchers to apply different data analysis tools (e.g., regression, clustering, association rules).
Within data-driven procedures, subcategories can be distinguished according to the type of data used. The main distinction about data types is between original datafiles (e.g., individuals described in terms of attributes) and aggregates of the data (e.g., contingency tables). In the statistical disclosure control community, the former type corresponds to microdata and the later to tabular data.
With respect to the type or structure of the original files, most of the research has been focused on standard files with numerical or categorical data (ordinal or nominal categorical data). Nevertheless, other more complex types of data have also been considered in the literature, as, e.g., multirelational databases, logs, and social networks. Another aspect to be considered in relation to the structure of the files is about the constraints that the protected data needs to satisfy (e.g., when there is a linear combination of some variables). Data protection methods need to consider such constraints so that the protected data also satisfies them (see e.g. (Torra, 2008) for details on a classification of the constraints and a study of microaggregation under this light).

Computation-driven or specific purpose protection procedures. In this case it is known beforehand which type of analysis has to be applied to the data. As the data uses are known, protection procedures are defined according to the intented subsequent computation. Thus, protection procedures are tailored to a specific purpose.
This will be the case of a retailer with a commercial database with information on customers having a fidelity card, when such data has to be transferred to a third party for market basket analysis. For example, there exist tailored procedures for data protection for association rules. They can be applied in this context of market basket analysis.

Results-driven protection procedures. In this case, privacy concerns to the result of applying a particular data mining method to some particular data (Atallah *et al.*, 1999, Atzori *et al.*,

2008). For example, the association rules obtained from a commercial database should not permit the disclosure of sensitive information about particular customers.
Although this class of procedures can be seen as computation-driven, they are important enough to deserve their own class. This class of methods are also known by *anonymity preserving pattern discovery* (Atzori *et al.*, 2008), *result privacy* (Bertino *et al.*, 2008), and *output secrecy* (Haritsa, 2008).

Other dimensions have been considered in the literature for classifying protection procedures. One of them concerns the number of data sources.

Single data source. The data analysis only requires data from a single source.
Multiple data sources. Data from different sources have to be combined in order to compute a certain analysis.
The analysis of data protection procedures for multiple data sources usually falls within the computation-driven approach. A typical scenario in this setting is when a few companies collaborate in a certain analysis, each one providing its own data base. In the typical scenario within data privacy, data owners want to compute such analysis without disclosing their own data to the other data owners. So, the goal is that at the end of the analysis the only additional information obtained by each of the data owners is the result of the analysis itself. That is, no extra knowledge should be acquired while computing the analysis.
A *trivial* approach for solving this problem is to consider a trusted third party (TTP) that computes the analysis. This is the centralized approach. In this case, data is just transferred using a completely secure channel (i.e., using cryptographic protocols). In contrast, in distributed privacy preserving data mining, data owners compute the analysis in a collaborative manner. In this way, the trusted third party is not needed. For such computation, cryptographic tools are also used.
Multiple data sources for data-driven protection procedures has limited interest. Each data owner can publish its own data protected using general purpose protection procedures, and then data can be linked (using e.g. record linkage algorithms) and finally analysed. So, this roughly corresponds to multidatabase mining.

The literature often classifies protection procedures using another dimension concerning the type of tools used. That is, methods are classified either as following the perturbative or the cryptographic approach. Our classification given above encompasses these two approaches. General purpose protection procedures follow the so-called perturbative approach, while computation-driven protection procedures mainly follow the cryptographic approach.

Note, however, that there are some papers on perturbative approaches as e.g. noise addition for specific uses as e.g. association rules (see (Atallah *et al.*, 1999)). Nevertheless, such methods are general enough to be used in other applications. So, they are general purpose protection procedures.

In addition, it is important to underline that, in this chapter, we will not use the term *perturbative approach* with the interpretation above. Instead, we will use the term perturbative methods/approaches in a more restricteed way (see Section 35.4), as it is usual in the statistical disclosure control community.

In the rest of this section we further discuss both computation-driven and data-driven procedures.

35.2.1 Computation-Driven Protection Procedures: the Cryptographic Approach

As stated above, cryptographic protocols are often applied in applications where the analysis (or function) to be computed from the data is known. In fact, it is usually applied to scenarios with multiple data sources. We illustrate below this scenario with an example.

Example 1. Parties $P_1,\ldots,P_n$ own databases $DB_1,\ldots,DB_n$. The parties want to compute a function, say f, of these databases (i.e., $f(DB_1,\ldots,DB_n)$) without revealing unnecessary information. In other words, after computing $f(DB_1,\ldots,DB_n)$ and delivering this result to all P_i, what P_i knows is nothing more than what can be deduced from his DB_i and the function f. So, the computation of f has not given P_i any extra knowledge.

Distributed privacy preserving data mining is based on the secure multiparty computation, which was introduced by A. C. Yao in 1982 (Yao, 1982). For example, (Lindell and Pinkas, 2000) and (Lindell and Pinkas, 2002) defined a method based on cryptographic tools for computing a decision tree from two data sets owned by two different parties. (Bunn and Ostrovsky, 2007) discusses clustering data from different parties.

When data is represented in terms of records and attributes, two typical scenarios are considered in the literature: vertical partitioning of the data and horizontal partitioning. They are as follows.

- Vertically partitioned data. All data owners share the same records, but different data owners have information about different attributes (i.e., different data owners have different views of the same records or individuals).
- Horizontally partitioned data. All data owners have information about the same attributes, nevertheless the records or individuals included in their data bases are different.

As stated above, for both centralized and distributed PPDM the only information that should be learnt by the data owners is the one that can be inferred from his original data and the final computed analysis. In this setting, the centralized approach is considered as a reference result when analyzing the privacy of the distributed approach. Privacy leakage for the distributed approach is usually analyzed considering two types of adversaries.

- **Semi-honest adversaries.** Data owners follow the cryptographic protocol but they analyse all the information they get during its execution to discover as much information as they can.
- **Malicious adversaries.** Data owners try to fool the protocol (e.g. aborting it or sending incorrect messages on purpose) so that they can infer confidential information.

Computation-driven protection procedures using cryptographic approaches present some clear advantges with respect to general purpose ones. The first one is the good quality of the computed function (analysis). That is, the function we compute is exactly the one the users want to compute. This is not so, as we will see later, when other general purpose protection methods are used. In this latter case, the resulting function is just an approximation of the function we would compute from the original data. At the same time, cryptographic tools ensure an optimal level of privacy.

Nevertheless, this approach has some limitations. The first one is that we need to know beforehand the function (or analysis) to be computed. As different functions lead to different cryptographic protocols, any change on the function to be computed (even small ones) requires a redefinition of the protocol. A second disadvantage is that the computational costs of the protocols are very high. In addition, it is even harder when malicious adversaries are considered. (Kantarcioglu, 2008) discusses other limitations. One is that most literature only considers the types of adversaries described above (honest, semi-honest and malicious). No other types are studied. Another one is the fact that in these methods no trade-off can be found between privacy and information loss (they use the term accuracy). As we will see later in Sections 35.5 and 35.6, most general purpose protection procedures permit the user to select an appropriate trade-off between these two contradictory issues. When using cyptographic protocols, the only trade-off that can be implemented easily is the one between privacy and efficiency.

35.2.2 Data-driven Protection Procedures

Given a data set, data-driven protection procedures construct a new data set so that the new one does not permit a third party to infer confidential information present in the original data. Different methods have been developed for this purpose. We will focus on the case where the data set is a standard file defined in terms of records and attributes (microdata following the jargon of statistical disclosure control). As stated above, we can also consider other types of data sets as e.g. aggregate data (tabular data following the jargon of SDC).

All data-driven methods are similar in the sense that they construct the new data set reducing the quality of the original one. As quality reduction might cause data to be unsuitable for a particular analysis, measures have been developed to evaluate in what extent the protected data set is still valid. These measures are known as information loss measures or utility measures.

Data-driven procedures are much more efficient with respect to computational cost than the ones using cryptographic protocols. Nevertheless, this major efficiency is at the cost of not ensuring complete privacy. Due to this, some measures of risk have been developed to determine in which extent a protected data set ensures privacy. These measures are known as disclosure risk measures.

These two families of measures, information loss and disclosure risk, are in contradiction and, thus, methods should look for an appropriate trade-off between risk and utility. Tools have been developed to visualize this trade-off and also to quantify this trade-off, so that protection methods can be compared.

We will present some of the data protection procedures in Section 35.4, information loss measures in Section 35.5, and visualization methods in Section 35.6. Section 35.3 includes a description of the standard scenario for evaluating risk before reviewing disclosure risk measures.

35.3 Disclosure Risk Measures

Disclosure risk is defined in terms of the additional confidential information (in general, additional knowledge) that an intruder can acquire from the protected data set. According to (Lambert, 1993, Paass, 1985), disclosure risk can be studied from two perspectives:

- **Identity disclosure.** This disclosure takes place when a respondent is linked to a particular record in the protected data set. This process of linking is known as re-identification (of the respondent).
- **Attribute disclosure.** In this case, defining disclosure as the disclosure of the identity of the individual is considered too strong. Disclosure takes place when the intruder can

learn something new about an attribute of a respondent, even when no relationship can be established between the individual and the data. That is, disclosure takes place when the published data set permits the intruder to increase his accuracy on an attribute of the respondent. This approach was first formulated in (Dalenius, 1977) (see also (Duncan and Lambert, 1986) and (Duncan and Lambert, 1989)).

Interval disclosure is a measure, proposed in (Domingo-Ferrer *et al.*, 2001) and (Domingo-Ferrer and Torra, 2001b), for attribute disclosure. It is defined according to the following procedure. Each attribute is independently ranked and a rank interval is defined around the value the attribute takes on each record. The ranks of values within the interval for an attribute around record r should differ less than p percent of the total number of records and the rank in the center of the interval should correspond to the value of the attribute in record r. Then, the proportion of original values that fall into the interval centered around their corresponding protected value is a measure of disclosure risk. A 100 percent proportion means that an attacker is completely sure that the original value lies in the interval around the protected value (interval disclosure).

Identity disclosure has received much attention in the last years and has been used to evaluate different protection methods. Its formulation needs a concrete scenario. We present it in the next section. Some identity disclosure risk measures will be reviewed later using this scenario.

35.3.1 An Scenario for Identity Disclosure

The typical scenario is to consider the protected data set and an intruder having some partial information about the individuals in the published data set. The protected data set is assumed to be a data file, and it is usual to consider that intruder's information can be represented in the same way. See e.g. (Sweeney, 2002, Torra *et al.*, 2006). Formally, we consider data sets X with the usual structure of r rows (*records*) and k columns (*attributes*). Naturally, each row contains the values of the attributes for an individual.

Then, the attributes in X can be classified (Dalenius, 1986, Samarati, 2001, Torra *et al.*, 2006) in three non-disjoint categories.

- **Identifiers.** These are attributes that *unambiguously* identify the respondent. Examples are passport number, social security number, full name, etc.
- **Quasi-identifiers.** These are attributes that, in combination, can be linked with external information to re-identify some of the respondents. Examples are age, birth date, gender, job, zipcode, etc. Although a single attribute cannot identify an individual, a subset of them can.
- **Confidential.** These are attributes which contain sensitive information on the respondent. For example, salary, religion, political affiliation, health condition, etc.

Using these three categories, an original data set X is defined as $X = id||X_{nc}||X_c$, where id are the identifiers, X_{nc} are the non-confidential quasi-identifier attributes, and X_c are the confidential attributes. Let us consider the protected data set X'. X' is obtained from the application of a protection procedure to X. This process takes into account the type of the attributes. It is usual to proceed as follows.

- **Identifiers.** To avoid disclosure, identifiers are usually removed or encrypted in a preprocessing step. In this way, information cannot be linked to specific respondents.
- **Confidential.** These attributes X_c are usually not modified. So, we have $X'_c = X_c$.

- **Quasi-identifiers.** They cannot be removed as almost all attributes can be quasi-identifiers. The usual approach to preserve the privacy of the individuals is to apply protection procedures to these attributes. We will use ρ to denote the protection procedure. Therefore, we have $X'_{nc} = \rho(X_{nc})$.

Therefore, we have $X' = \rho(X_{nc})||X_c$. Proceeding in this way, we allow third parties to have precise information on confidential data without revealing to whom the confidential data belongs to.

In this scenario we have identity disclosure when an intruder, having some information described in terms of a set of records and some quasi-identifiers, can link his information with the published data set. That is, he is able to link his records with the ones in the protected data set. Then, if the links between records are correct, he will be able to obtain the right values for the confidential attributes.

Figure 35.1 represents this situation. A represents the file with data from the protected data set (i.e., containing records from X') and B represents the file with the records of the intruder. B is usually defined in terms of the original data set X, because it is assumed that the intruder has a subset of X. In general, the number of records owned by the intruder and the number of records in the protected data file will differ.

Reidentification is achieved using some common quasi-identifiers on both X and X'. They permit to link pairs of records (using record linkage algorithms) from both files, and, then, the confidential attribute is linked to the identifiers. At this point reidentification is achieved.

Formally, following (Torra *et al.*, 2006, Nin *et al.*, 2007, Sweeney, 2002) and the notation in Figure 35.1, the intruder is assumed to know the non-confidential quasi-identifiers $X_{nc} = \{a_1, \ldots, a_n\}$ together with the identifiers $Id = \{i_1, i_2, \ldots\}$. Then, the linkage is between identifiers $(a_1, \ldots, a_n)$ from the protected data (X'_{nc}) and the same attributes from the intruder (X_{nc}).

35.3.2 Measures for Identity Disclosure

Two main approaches exists for measuring identity disclosure risk. They are known by uniqueness and re-identification. We describe them below.

- **Re-identification.** Risk is defined as an estimation of the number of re-identifications that might be obtained by an intruder. This estimation is obtained empirically through record linkage algorithms. This approach for measuring disclosure risk goes back, at least, to (Spruill, 1983) and (Paass, 1985) (using *e.g.* the algorithm described in (Paass and Wauschkuhn, 1985)). (Torra *et al.*, 2006, Nin *et al.*, 2007, Sweeney, 2002) are more recent papers using this approach. This approach is general enough to be applied in different contexts. It can be applied under different assumptions of intruder's knowledge, and under different assumptions on protection procedures. It can even be applied when protected data has been generated using a synthetic data generator (i.e., data is constructed using a particular data model – see Section 35.4.3 for details). For example, (Torra *et al.*, 2006) describes empirical results about using record linkage algorithms on synthetic data. The performance of different algorithms is discussed. (Winkler, 2004) considers a similar problem.
- **Uniqueness.** Informally, the risk of identity disclosure is measured as the probability that rare combinations of attribute values in the protected data set are indeed rare in the original population.
  This approach is typically used when data is protected using sampling (Willenborg, 2001) (i.e., X' is just a subset of X). Note that with perturbative methods it makes no sense to

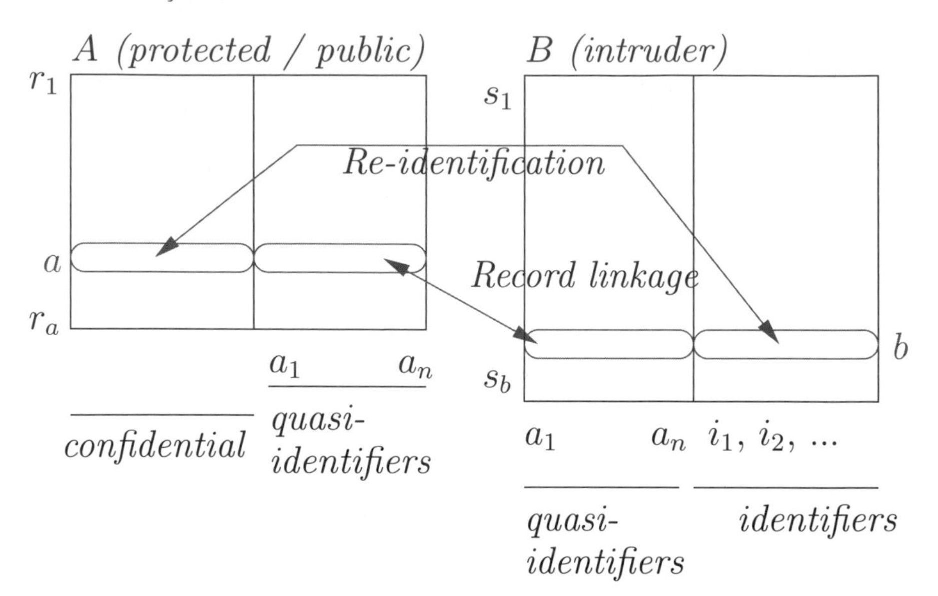

Fig. 35.1. Disclosure Risk Scenario.

investigate the probability that a rare combination of protected values is rare in the original data set, because *that* combination is most probably *not found* in the original data set.

In the next sections we describe these two approaches in more detail.

Uniqueness

Two types of disclosure risk measures based on uniqueness can be distinguished: file-level and record-level. We describe them below.

- **File-level uniqueness.** Disclosure risk is defined as the probability that a sample unique (SU) is a population unique (PU) (Elliot *et al.*, 1998). According to (Elamir, 2004), this probability can be computed as

$$P(PU|SU) = \frac{P(PU,SU)}{P(SU)} = \frac{\sum_j I(F_j = 1, f_j = 1)}{\sum_j I(f_j = 1)}$$

  where $j = 1, \ldots, J$ denotes possible values in the sample, F_j is the number of individuals in the population with key value j (frequency of j in the population), f_j is the same frequency for the sample and I stands for the cardinality of the selection. Unless the sample size is much smaller than the population size, $P(PU|SU)$ can be dangerously high; in that case, an intruder who locates a unique value in the released sample can be almost certain that there is a single individual in the population with that value, which is very likely to lead to that individual's identification.

- **Record-level risk uniqueness.** They are also known as individual risk measures. Disclsoure risk is defined as the probability that a particular sample record is re-identified, *i.e.* recognized as corresponding to a particular individual in the population. As (Elliot, 2002) points out, the main rationale behind this approach is that risk is not homogeneous within a data file. We summarize next the description given in (Franconi.Polettini.2004) of the record-level risk estimation.
  Assume that there are K possible combinations of key attributes. These combinations induce a partition both in the population and in the released sample. If the frequency of the k-th combination in the population was known to be F_k, then the individual disclosure risk of a record in the sample with the k-th combination of key attributes would be $1/F_k$. Since the population frequencies F_k are generally unknown but the sample frequencies f_k of the combinations are known, the distribution of frequencies F_k given f_k is considered. Under reasonable assumptions, the distribution of $F_k|f_k$ can be modeled as a negative binomial. The per-record risk of disclosure is then measured as the posterior mean of $1/F_k$ with respect to the distribution of $F_k|f_k$.

Record Linkage

This approach for measuring disclosure risk directly follows the scenario in Figure 35.1. That is, record linkage consists of linking each record b of the intruder (file B) to a record a in the original file A. The pair (a,b) is a match if b turns out to be the original record corresponding to a. For applying record linkage, the common approach is to use the shared attributes (some quasi-identifiers). As the number of matches is an estimation of the number of re-identifications that an intruder can achieve, disclosure risk is defined as the proportion of matches among the total number of records in B.

Two main types of record linkage algorithms are described in the literature: distance-based and probabilistic. They are outlined below. For details on these methods see (Torra and Domingo-Ferrer, 2003).

- **Distance-based record linkage.** Each record b in B is linked to its nearest record a in A. An appropriate definition of a record-level distance has to be supplied to the algorithm to express *nearness*. This distance is usually constructed from distance functions defined at the level of attributes. In addition, we need to standardize attributes as well as assign weights to them.
  (Pagliuca *et al.*, 1999) proposed distance-based record linkage to assess the disclosure risk for microaggregation. They used Euclidean distance and equal weight for all attributes. Later, in (Domingo-Ferrer and Torra, 2001b), distance-based record linkage (also with Euclidean distance and equal weights) was used for evaluating other masking methods as well. In their empirical work, distance-based record linkage outperforms probabilistic record linkage (See Section 35.3.2 below).
  The main advantages of using distances for record linkage are simplicity for the implementer and intuitiveness for the user. Another strong point is that subjective information (about individuals or attributes) can be included in the re-identification process by means of appropriate distances.
  The main difficulties for distance-based record linkage are (i) the selection of the appropriate distance function, and (ii) the determination of the weights. In relation to the distance function, for numerical data, the Euclidean distance is the most used distance. Nevertheless, other distances have also been used as e.g. Mahalanobis (Torra *et al.*, 2006), and some Kernel-based ones (Torra *et al.*, 2006). The difficulty of choosing a distance is

especially thorny in the cases of categorical attributes and of masking methods such as local recoding where the masked file contains new labels with respect to the original data set. The determination of the weights is also a rellevant problem that is difficult to solve. In the case of the Euclidean distance, it is common to assign equal weights to all attributes, and in the case of the Mahalanobis distance, this problem is avoided because weights are extracted from the covariance matrix.

- **Probabilistic record linkage**
  Probabilistic record linkage also links pairs of records (a,b) in data sets A and B, respectively. For each pair, an index is computed. Then, two thresholds LT and NLT in the index range are used to label the pair as linked, clerical or non-linked pair: if the index is above LT, the pair is linked; if it is below NLT, the pair is non-linked; a clerical pair is one that cannot be automatically classified as linked or non-linked. When independence between attributes is assumed, the index can be computed from the following two conditional probabilities for each attribute: the probability $P(1|M)$ of coincidence between the values of the attribute in two records a and b given that these records are a real match, and the probability $P(0|U)$ of non-coincidence between the values of the attribute given that a and b are a real unmatch.
  To use probabilistic record linkage in an effective way, we need to set the thresholds LT and NLT and estimate the conditional probabilities $P(1|M)$ and $P(0|U)$ used in the computation of the indices. In plain words, thresholds are computed from: (i) the probability $P(LP|U)$ of linking a pair that is an unmatched pair (a *false positive* or *false linkage*) and (ii) the probability $P(NP|M)$ of not linking a pair that is a match (a *false negative* or *false unlinkage*). Conditional probabilities $P(1|M)$ and $P(0|U)$ are usually estimated using the EM algorithm (Dempster *et al.*, 1977).
  The original description of probabilistic record linkage can be found in (Fellegi and Sunter, 1969) and (Jaro, 1989). (Torra and Domingo-Ferrer, 2003) describe the method in detail (with examples) and (Winkler, 1993) presents a review of the state of the art on probabilistic record linkage. In particular, this latter paper includes a discussion concerning non-independent attributes. A (hierarchical) graphical model has recently been proposed (Ravikumar and Cohen, 2004) that compares favorably with previous approaches.
  Probabilistic record linkage methods are less simple than distance-based ones. However, they do not require rescaling or weighting of attributes. The user only needs to provide the two probabilities $P(LP|U)$ (false positives) and $P(NP|M)$ (false negatives).

The literature presents some other record linkage algorithms, some of which are variations of the ones presented here. For example, (Bacher *et al.*, 2002) presents a method based on cluster analysis. The results are similar to the ones of distance-based record linkage as cluster analysis assigns objects (in this case, records) that are similar (in this case, near) to each other, to the same cluster. The algorithms presented here permit two records of the intruder b_1 and b_2 to be assigned to the same record a. There are algorithms that force different records in B to be linked to different records in A.

The approaches described so far for record linkage do not use any information about the data protection process. That is, they use files A and B and try to re-identify as much records as possible. In this sense, they are general purpose record linkage algorithms.

In the last years, specific record linkage algorithms have been developed. They take advantage of any information available about the data protection procedure. That is, protection procedures are analyzed in detail to find flaws that can be used for computing more efficient, with larger matching rates, record linkage algorithms. Attacks tailored for two protection procedures are reported in the literature. (Torra and Miyamoto, 2004) was the first specific record linkage approach for microaggregation. More effective algorithms have been proposed in (Nin and Torra, 2009, Nin *et al.*, 2008b) (for either univariate and multivariate microaggregation). (Nin *et al.*, 2007) describes an algorithm for data protection using rank swapping.

The scenario described above can be relaxed so that the published file and the one of the intruder do not share the set of variables. I.e., there are no common quasi-identifiers in the two files. A few record linkage algorithms have been developed under this premise. In this case, some structural information is assumed to be common in both files. (Torra, 2004) follows this approach. Its use for disclosure risk assessment is described in (Domingo-Ferrer and Torra, 2003).

35.4 Data Protection Procedures

Protection methods can be classified into three different categories depending on how they manipulate the original data to define the protected data set.

- **Perturbative.** The original data set is distorted in some way, and the new data set might contain some erroneous information. E.g. noise is added to an attribute following a $N(0,a)$ for a given a. In this way, some combinations of values disappear, and, new combinations appear in the protected data set. At the same time, combinations in the protected data set no longer correspond to the ones in the original data set. This obfuscation makes disclosure difficult for intruders.
- **Non-perturbative.** Protection is achieved through replacing an original value by another one that is not incorrect but less specific. For example, we replace a real number by an interval. In general, non-perturbative methods reduce the level of detail of the data set. This detail reduction causes different records to have the same combinations of values, which makes disclosure difficult to intruders.
- **Synthetic Data Generators.** In this case, instead of distorting the original data, new artificial data is generated and used to substitute the original values. Formally, synthetic data generators build a data model from the original data set and, subsequently, a new (protected) data set is randomly generated constrained by the model computed.

An alternative dimension to classify protection methods is based on the type of data. Basic distinction is about numerical and categorical data, although other types of data (as e.g. time series (Nin and Torra, 2006)), sequences of events for location privacy, logs, etc. have also been considered in the literature.

- **Numerical.** As usual, an attribute is numerical when arithmetic operations as e.g. substraction can be performed with it. Income and age are typical examples of such attributes. With respect to disclosure risk, numerical values are likely to be unique in a database and, therefore, leading to disclosure if no action is taken.
- **Categorical.** In this case, the attribute takes values over a finite set and standard numerical operations do not make sense. Ordinal and nominal scales are typically distinguished among categorical attributes. In ordinal scales the order between values is relevant (*e.g.* academic degree), whereas in nominal scales it is not (*e.g.* hair color). Therefore, max and min operations are meaningful in ordinal scales but not on nominal scales.
  Structured attributes is a subclass of categorical attributes. In this case, different categories are related in terms of subclasses or *member of* relationships. In some cases, a hierarchy between categories can be inferred from these relationships. Cities, counties, and provinces are typical examples of these hierarchical attributes. For some attributes, the hierarchy is given but for others not but constructed by the protection procedure.

In the next sections we review some of the existing protection methods following the classification above. Some good reviews on data protection procedures are (Adam and Wortmann, 1989, Domingo-Ferrer and Torra, 2001a, Willenborg, 2001). In addition, we have a section about k-anonymity. As we will see later, k-anonymity is not a protection method but a general approach for avoiding disclosure up to a certain extent. Different instantiations exist, some using perturbative and some using non-perturbative procedures.

In this section we will use X to denote the original data, X' to denote the protected data set, and $x_{i,V}$ to represent the value of the attribute V in the ith record.

35.4.1 Perturbative Methods

In this section we review some of the perturbative methods. Among them, the ones that are most used by the statistical agencies are rank swapping and microaggregation (Felso *et al.*, 2001), but the literature on privacy preserving data mining, more oriented to business-related applications, largely focus on additive noise and microaggregation. Microaggregation and rank swapping are simple and have a low computational cost. Most of the methods described in this section, with some of their variants are implemented in the sdcMicro package in R (Templ, 2008) and in the μ-Argus software (Hundepool *et al.*, 2003).

Rank Swapping

Rank swapping was originally proposed for ordinal attributes in (Moore, 1996), but also applied to numerical data in (Domingo-Ferrer and Torra, 2001b). It was classified in (Domingo-Ferrer and Torra, 2001b) among the best microdata protection methods for numerical attributes and in (Torra, 2004) among the best for categorical attributes.

Rank swapping is defined for a single attribute V as described below. The application of this method to a data file with several attributes is done attribute-wise, in a sequential way. The algorithm depends on a parameter p that permits the user to control the amount of disclosure risk. Normally, p corresponds to a percent of the total number of records in X.

- records of X (for the considered attribute V) are sorted in increasing order.
- Let us assume, for simplicity, that the records are already sorted and that $(a_1, \dots, a_n)$ are the sorted values in X. That is, $a_i \leq a_\ell$ for all $1 \leq i < \ell \leq n$.
- Each value a_i is swapped with another value a_ℓ, randomly and uniformly chosen from the limited range $i < \ell \leq i + p$.
- The sorting step is undone.

The algorithm shows that the smaller the p, the larger the risk. Note that when p increases the difference between x_i and x_ℓ may increase accordingly. Therefore, the risk decreases. Nevertheless, in this case the differences between the original and the protected data set are higher, so information loss increases.

(Nin *et al.*, 2007) proves that specific attacks can be designed for this kind of rank swapping and proposed two alternative algorithms where the swapping is not constrained to a specific interval. In this way, the range for swapping includes the whole set $(a_1, \dots, a_n)$, although farther data have small probability of being swapped. In this way, the intruder cannot take advantage of the closed interals in the attack. p-buckets and p-distribution rank swapping are the names of such algorithms. Other variants of rank swapping include (Carlson and Salabasis, 2002) and (Takemura, 2002).

Microaggregation

Microaggregation was originally (Defays and Nanopoulos, 1993) defined for numerical atributes (see also (Domingo and Mateo, 2002)) and later extended to categorical data (Torra, 2004) (see also (Domingo-Ferrer and Torra, 2005)) and to time series (Nin and Torra, 2006).

(Felso *et al.*, 2001) shows that microaggregation is a method used by many statistical agencies, and (Domingo-Ferrer and Torra, 2001b) shows that, for numerical data, is one of the methods with a better trade-off between information loss and disclosure risk. (Torra, 2004) describes its good performance in comparison with other methods for categorical data.

Microaggregation is operationally defined in terms of two steps: partition and aggregation.

- **Partition.** Records are partitioned into several clusters, each of them consisting of at least k records.
- **Aggregation.** For each of the clusters a representative (the centroid) is computed, and then original records are replaced by the representative of the cluster to which they belong to.

This approach permits protected data to satisfy privacy constraints, as all k records in the cluster are replaced by the same value. In this way, k controls the privacy in the protected data.

We can formalize microaggregation using u_{ij} to describe the partition of the records in X. That is, $u_{ij} = 1$ if record j is assigned to the ith cluster. Let v_i be the representative of the ith cluster, then a general formulation of microaggregation with g clusters and a given k is as follows:

$$\begin{aligned} &\text{Minimize} \quad SSE = \textstyle\sum_{i=1}^{g} \sum_{j=1}^{n} u_{ij} (d(x_j, v_i))^2 \\ &\text{Subject to} \; \textstyle\sum_{i=1}^{g} u_{ij} = 1 \text{ for all } j = 1, \ldots, n \\ &\qquad 2k \geq \textstyle\sum_{j=1}^{n} u_{ij} \geq k \text{ for all } i = 1, \ldots, g \\ &\qquad u_{ij} \in \{0, 1\} \end{aligned}$$

For numerical data it is usual to require that $d(x,v)$ is the Euclidean distance. In the general case, when attributes $\mathbf{V} = (V_1, \ldots, V_s)$ are considered, x and v are vectors, and d becomes $d^2(x,v) = \sum_{V_i \in V} (x_v - v_{V_i})^2$. In addition, it is also common to require for numerical data that v_i is defined as the arithmetic mean of the records in the cluster. I.e., $v_i = \sum_{j=1}^{n} u_{ij} x_i / \sum_{j=1}^{n} u_{ij}$. In the case of univariate microaggregation (for Euclidean distance and arithmetic mean), there exists algorithms to find an optimal solution in polynomial time (Hansen and Mukherjee, 2003) (Algorithm 1 describes such method). In contrast, for multivariate data sets, the problem becomes an NP-Hard (Oganian and Domingo-Ferrer, 2000). For this reason, heuristic methods have been proposed in the literature.

The general formulation given above permits us to apply microaggregation to multidimensional data. Nevertheless, when the number of attributes is large, it is usual to apply microaggregation to subsets of attributes; otherwise, the information loss is very high (Aggarwal, 2005). Individual ranking is a multivariate approach that consists of applying microaggregation to each of the attributes in an independent way. Alternatively, a partition of the attributes is constructed and microaggregation is applied to each subset.

Applying microaggregation to subsets of attributes decrease information loss but at the cost of increasing disclosure risk. See (Nin *et al.*, 2008a) for an analysis of how to build these partitions (i.e., whether it is preferable to select correlated or uncorrelated attributes when defining the partition) and their effect on information loss and disclsoure risk. (Nin *et al.*, 2008a) shows that the selection of uncorrelated attributes decrease disclosure risk and can lead to a better trade-off between disclosure risk and information loss.

Algorithm 1: Optimal Univariate Microaggregation

Data: X: original data set, k: integer
Result: X': protected data set

1 **begin** Let $X = (a_1 \dots a_n)$ be a vector of size n containing all the values for the attribute being protected. Sort the values of X in ascending order so that if $i < j$ then $a_i \leq a_j$.
2 Given A and k, a graph $G_{k,n}$ is defined as follows.
3 **begin** Define the nodes of G as the elements a_i in A plus one additional node g_0 (this node is later needed to apply the Dijkstra algorithm).
4 For each node g_i, add to the graph the directed edges (g_i, g_j) for all j such that $i+k \leq j < i+2k$. The edge (g_i, g_j) means that the values $(a_i, \dots, a_j)$ might define one of the possible clusters.
5 The cost of the edge (g_i, g_j) is defined as the within-group sum of squared error for such cluster. That is, $SSE = \Sigma_{l=i}^{j}(a_l - \bar{a})^2$, where $\bar{a}$ is the average record of the cluster.
6 The optimal univariate microaggregation is defined by the shortest path algorithm between the nodes g_0 and g_n. This shortest path can be computed using the Dijkstra algorithm.

Algorithm 2: Projected Microaggregation

Data: X: original data set, k: integer
Result: X': protected data set

1 **begin** Split the data set X into r sub-data sets $\{X_i\}_{1 \leq i \leq r}$, each one with a_i attributes of the n records, such that $\sum_{i=1}^{r} a_i = A$
2 **foreach** $(X_i \in X)$ **do**
3 Apply a projection algorithm to the attributes in X_i, which results in an univariate vector z_i with n components (one for each record)
4 Sort the components of z_i in increasing order
5 Apply to the sorted vector z_i the following variant of the univariate optimal microaggregation method: use the algorithm defining the cost of the edges $\langle z_{i,s}, z_{i,t} \rangle$, with $s < t$, as the within-group sum of square error for the a_i-dimensional cluster in X_i which contains the original attributes of the records whose projected values are in the set $\{z_{i,s}, z_{i,s+1}, \dots, z_{i,t}\}$
6 For each cluster resulting from the previous step, compute the v_i-dimensional centroid and replace all the records in the cluster by the centroid

Heuristic approaches for sets of attributes can be classified into two categories. One approach consists of projecting the records (using e.g. Principal Components or Zscores) and then applying optimal microaggregation on the projected dimension (Algorithm 2 describes this approach). The other is to develop adhoc algorithms. The MDAV (Domingo and Mateo, 2002) (Maximum Distance to Average Vector) algorithm follows this second approach. It is explained in detail in Algorithm 3, when applied to a data set X with n records and A attributes. The implementation of MDAV for categorical data is given in (Domingo-Ferrer and Torra, 2005).

Microaggregation has been used as a way to implement k-anonymity. Section 35.4.4 discusses the relation between both microaggregation and k-anonymity.

Algorithm 3: MDAV

Data: X: original data set, k: integer
Result: X': protected data set

1. **begin while**($|X| > k$) Compute the average record $\bar{x}$ of all records in X
2. Consider the most distant record x_r to the average record $\bar{x}$
3. Form a cluster around x_r. The cluster contains x_r together with the $k-1$ closest records to x_r
4. Remove these records from data set X
5. **if**($|X| > k$) Find the most distant record x_s from record x_r
6. Form a cluster around x_s. The cluster contains x_s together with the $k-1$ closest records to x_s
7. Remove these records from data set X
8. Form a cluster with the remaining records

Additive noise

This method protects data adding noise into the original file. That is,

$$X' = X + \varepsilon,$$

where ε is the noise. The simplest approach is to require ε to be such that $E(\varepsilon) = 0$ and $Var(\varepsilon) = kVar(X)$ for a given constant k.

Uncorrelated noise addition corresponds to the case that for variables V_i and V_j, noise is such that $Cov(\varepsilon_i, \varepsilon_j) = 0$ for $i \neq j$. In this case, additive noise preserves means and covariances, but neither variances $Var(X') = Var(X) + kVar(X)$ nor correlation coefficients. Note that,

$$E(X') = E(X) + E(\varepsilon) = E(X)$$
$$Cov(X'_i, X'_j) = Cov(X_i, X_j) \text{ for } i \neq j$$
$$Var(X') = Var(X) + kVar(X) = (1+k)Var(X)$$
$$\rho_{X'_i,X'_j} = \frac{Cov(X'_i, X'_j)}{\sqrt{Var(X'_i)Var(X'_j)}} = \frac{Cov(X_i, X_j)}{(1+k)\sqrt{Var(X_i)Var(X_j)}} = \frac{1}{1+k}\rho_{X_i,X_j}$$

Correlated noise addition preserves correlation coefficients and means. In this case, however, neither variance nor covariance is preserved: they are proportional to the variance and covariance of the original data set.

In correlated noise addition, ε follows a normal distribution $N(0, k\Sigma)$ where Σ is the covariance matrix of X.

$$E(X') = E(X) + E(\varepsilon) = E(X)$$
$$Cov(X'_i, X'_j) = (1+k)Cov(X_i, X_j) \text{ for } i \neq j$$
$$Var(X') = Var(X) + kVar(X) = (1+k)Var(X)$$
$$\rho_{X'_i,X'_j} = \frac{Cov(X'_i, X'_j)}{\sqrt{Var(X'_i)Var(X'_j)}} = \frac{(1+k)Cov(X_i, X_j)}{(1+k)\sqrt{Var(X_i)Var(X_j)}} = \rho_{X_i,X_j}$$

First extensive testing of noise addition was due to Spruill (Spruill, 1983). (Brand, 2002) gives an overview of these approaches for noise addition as well as more sophisticated techniques. (Domingo-Ferrer *et al.*, 2004) also describes some of the existing methods as well as the difficulties for its application in privacy. In addition to that, there exists a related approach known as multiplicative noise (see e.g. (Kim and Winkler, 2003, Liu *et al.*, 2006) for details).

PRAM

PRAM, Post-RAndomization Method (Gouweleeuw *et al.*, 1998), is a method for categorical data where categories are replaced according to a given probability.

Formally, it is based on a Markov matrix on the set of categories. Let $C = \{c_1, \ldots, c_c\}$ be the set of categories, then P is the Markov matrix on C when $P : C \times C \rightarrow [0,1]$ such that $\sum_{c_j \in C} P(c_i, c_j) = 1$. Then, X' is constructed from X replacing, with probability $P(c_i, c_j)$, each c_i in X by a c_j.

The application of PRAM requires an adequate definition of the probabilities $P(c_i, c_j)$. (Gouweleeuw *et al.*, 1998) proposes the Invariant PRAM. Given $T = (T(c_1) \ldots T(c_c))$ the vector of frequencies of categories in C, it consists of defining P such that frequencies are kept after PRAM. That is, $\sum_{i=1}^{c} T(c_i) p_{ij} = T(c_j)$ for all j. Then, assuming without loss of generality $T(c_k) \geq T(c_i)$ for all i, and given a parameter θ such that $0 < \theta < 1$, p_{ij} is defined as follows:

$$p_{ij} = \begin{cases} 1 - (\theta T(c_k)/T(c_i)) & \text{if } i = j \\ \theta T(c_k)/((k-1)T(c_i)) & \text{if } i \neq j \end{cases}$$

Note that a θ equal to zero implies no perturbation, and θ equal to 1 implies total perturbation. So, θ permits the user to control the degree of distortion suffered by the data set.

(Gross *et al.*, 2004) proposes the computation of matrix P from a preference matrix $W = \{w_{ij}\}$ where w_{ij} is our degree of preference about replacing category c_i by category c_j. Formally, given W the probabilities P are determined from the following optimization function:

Minimize $\sum_{i,j} w_{ij} p_{ij}$
Subject to
$p_{ij} \geq 0$
$\sum_j p_{ij} = 1$
$\sum_{i=1}^{c} T(c_i) p_{ij} = T(c_j)$ for all j

(Gross *et al.*, 2004) use integers to express preferences, and $w_{ij} = 1$ is the most preferred change, $w_{ij} = 2$ is the second most preferred changes, and so on.

Lossy Compression

This approach, first proposed in (Domingo-Ferrer and Torra, 2001a), consists of viewing a numerical data file as a grey-level image. Rows are records and columns are attributes. Then, a lossy compression method is applied to the *image*, obtaining a *compressed image*. This *image* is then decompressed and the *decompressed image* corresponds to the masked file.

Different compression rates lead to files with different degrees of distortion. I.e., the more compression, the more distortion. (Domingo-Ferrer and Torra, 2001a) used JPEG, which is based on DCT, for the compression. (Jimenez and Torra, 2009) uses JPEG 2000, which is based on wavelets.

The transformation of the original data file into an *image* requires, in general, a quantization. As JPEG 2000 allows a higher dynamic range (65536 levels of grey) than JPEG (only 256), this quantization step is more accurate in JPEG 2000.

35.4.2 Non-perturbative Methods

In this section we review some of the non-perturbative algorithms. We review generalization (also known as recoding), top and bottom coding, and suppression. Sampling (Willenborg, 2001), which is not discussed here, can also be seen as a non-perturbative method.

Generalization and Recoding

This method is mainly applied to categorical attributes. Protection is achieved by means of combining a few categories into a more general one. Local and global recoding can be distinguished.

Global recoding (Willenborg, 2001, LeFevre *et al.*, 2005) corresponds to the case that the same recoding is applied to all the categories in the original data file. Formally, if Π is a partition of the categories in C, then each c in the original data file is replaced by the partition element in Π that contains c.

In contrast, in local recoding (Takemura, 2002) different categories might be replaced by different generalizations. Constrained local recoding is when the data space is partitioned and within the region the same recoding is used, but different regions use different recodings.

In general, global recoding has a larger information loss (as changes are applied to all records without taking into account whether they need them to ensure privacy) than local recoding. Nevertheless, local recoding generates a data set that has a larger diversity on the terms used to describe the records. This situation might cause difficulties when using the protected data set in analysis.

While most of the literature considers the recoding as functions of a single variable (i.e., single-dimension recoding), it is also possible to consider recoding of several variables at once. This is multidimensional recoding (LeFevre *et al.*, 2005). Formally, when n variables are considered, recoding is understood as a function of n values.

One of the main difficulties for applying recoding is the need for a hierarchy of the categories. In some applications such hierarchy is assumed to be known, while in others it is constructed by the protection procedure. In this later case, difficulties appear to determine the optimal generalization structure. In general, to find an optimal generalization for data protection is a hard problem and besides of that the constructed hierarchy might have small semantic interpretation. E.g. generalizing Zip codes 08192 and 09195 into 0**9*.

Top and Bottom Coding

Top and bottom coding are two methods that consist of replacing the lowest and largest values (given a threshold) by a generalized category. These two methods can be considered as particular cases of global recoding.

These methods are applied in data protection when there are only a few records that have extreme values. This kind of generalization permits us to reduce the disclosure risk as records assigned to the same category are indistinguishable.

Local Suppression

Suppression consists of replacing some values by a special label denoting that the value has been suppressed. Suppression is often applied (Samarati and Sweeney, 1998, Sweeney, 2002, Kisilevich *et al.*, 2010) in combination with recoding, and mainly for categorical data.

35.4.3 Synthetic Data Generators

In recent years, a new research trend has emerged for protecting data. It consists of publishing synthetic data instead of the original one. The rationale is that synthetic data cannot compromise the privacy of the respondents as the data is not *"real"*. Synthetic data generation methods consist first of constructing a model of the data, and then generating artificial data from this model.

The difficulties of the approach are twofold. On the one hand, although data is synthetic some reidentification is possible and, therefore, disclosure risk should be studied for this type of data. See e.g. (Torra *et al.*, 2006) that presents the results of a series of reidentification experiments. Due to this, synthetic data do not avoid disclosure risk and still represents a threat to privacy. On the other hand, as the synthetic data is generated from a particular data model built from the original data, all those aspects that are not explicitly included in the model, are often not included in the data. Due to this, unforseen analysis on the protected data might lead to results rather different than the same analysis on the original data.

Several methods for synthetic data generation have been developed in the last few years. Data distortion by probability distribution (Liew *et al.*, 1985) is one of the first protection procedures that can be classified as such. This procedure is defined by the following three steps:

- identify an underlying density function and determine its parameters
- generate distorted series using the estimated density function
- the distorted series are put in place of the original ones

The procedure was originally defined for univariate density functions, although it can be applied to multivariate density functions.

Information Preserving Statistical Obfuscation (IPSO) (Burridge, 2003) is another family of methods for synthetic data generation. It includes three different methods with names IPSO-A, IPSO-B, and IPSO-C. In this family of methods, IPSO-A is the simplest method and IPSO-C is the most elaborated one. They satisfy the property that the data generated by IPSO-C is more similar to the original data than the data generated by IPSO-A. The larger the complexity, the more the protected data resembles in terms of statistics to the original one.

All three methods presume that the variables of the original data file can be divided into two sets of variables X and Y, X contains the confidential outcome variables and Y the quasi-identifier variables.

IPSO-A is defined as follows. Take X as independent and Y as dependent variables, and then fit a multiple regression model of Y on X. Compute Y'_A with this model. IPSO-A releases the variables X and Y'_A in place of X and Y.

The following can be stated with respect to IPSO-A. In the above setting, conditional on the specific confidential attributes x_i, the quasi-identifier attributes Y_i are assumed to follow a multivariate normal distribution with covariance matrix $\Sigma = \{\sigma_{jk}\}$ and a mean vector $x_i B$, where B is the matrix of regression coefficients. Let $\hat{B}$ and $\hat{\Sigma}$ be the maximum likelihood estimates of B and Σ derived from the complete data set (y, x). If a user fits a multiple regression model to (y'_A, x), she will get estimates $\hat{B}_A$ and $\hat{\Sigma}_A$ which, in general, are different from the estimates $\hat{B}$ and $\hat{\Sigma}$ obtained when fitting the model to the original data (y, x).

The goal of IPSO-B is to modify y'_A into y'_B in such a way that the estimate $\hat{B}_B$ obtained by multiple linear regression from (y'_B, x) satisfies $\hat{B}_B = \hat{B}$.

Finally, the goal of IPSO-C is to obtain a matrix y'_C such that when a multivariate multiple regression model is fitted to (y'_C, x), *both* sufficient statistics $\hat{B}$ and $\hat{\Sigma}$ obtained on the original data (y, x) are preserved.

(Muralidhar and Sarathy, 2008) is another example of a data generator for synthetic data.

35.4.4 *k*-Anonymity

k-Anonymity is not a protection procedure in itself but a condition to be satisfied by the protected data set. Nevertheless, in most of the cases, *k*-anonymity is achieved via a combination of recoding and suppression. As we will discuss later, it can also be achieved via microaggregation.

Formally, a data set X satisfies *k*-anonymity (Samarati and Sweeney, 1998,Samarati, 2001, Sweeney, 2002, Sweeney, 2002) when X is partitioned into sets of at least k indistinguishable records.

In the literature there exist different approaches for achieving *k*-anonymity. As stated above, most of them are based on generalization and suppression. Nevertheless, optimal *k*-anonymity with generalization and suppression is an NP-Hard problem. Due to this, heuristic algorithms have been defined.

As *k*-anonymity pressumes no disclosure, only information loss measures are of interest here.

An alternative approach for achieving *k*-anonymity is the use of microaggregation methods. In this case, data homogeneity is ensured. Microaggregation has to be applied considering all the variables at once, otherwise, *k*-anonymity would not be guaranteed. (Nin *et al.*, 2008a) measures real *k*-anonymity when not all variables are microaggregated at once.

When *k*-anonymity is satisfied for quasi-identifiers, disclosure might occur if all the values of a confidential attribute are the same for a particular combination of quasi-identifiers. *p*-sensitive *k*-anonymity was defined to avoid this type of problems.

Definition 1. *(Truta and Vinay, 2006) A data set is said to satisfy p-sensitive k-anonymity for $k > 1$ and $p \leq k$ if it satisfies k-anonymity and, for each group of records with the same combination of values for quasi-identifiers, the number of distinct values for each confidential value is at least p (within the same group).*

To satisfy *p*-sensitive *k*-anonymity we require p different values in all sets. This requirement might be difficult to achieve in some databases.

l-diversity (Machanavajjhala *et al.*, 2006) is an alternative approach for *k*-anonymity. Similar to the case of *p*-sensitivity, *l*-diversity forces l different categories in each set. However, in this case, categories should have to be *well-represented*. Different meanings have been given to what *well-represented* means.

t-closeness (Li *et al.*, 2007) is another alternative approach to standard *k*-anonymity. In this case it is required that the distribution of the attribute in any *k*-anonymous subset of the database is similar to the one of the full database. Similarity is defined in terms of the distance between the two distributions and such distance should be below a given threshold t.

This approach permits the data protecter to limit the disclosure. Nevertheless, a low threshold forces all the sets to have the same distribution as the full data set. This might cause a large information loss: any correlation between the confidential attributes and the one used for l-diversity might be lost when forcing all the sets to have the same distribution.

35.5 Information Loss Measures

As stated in the introduction, information loss measures are to quantify in which extent the distortion embedded in the protected data set distorts the results that would be obtained with the original data. These measures are defined for general-purpose protection procedures.

When it is known which type of analysis the user will perform on the data, the analysis of the distortion of a particular protection procedure can be done in detail. That is, measures can be developed, and protection procedures can be compared and ranked using such measures. *Specific information loss measures* are the indices that permits us to quantify such distortion.

Nevertheless, when the type of analysis to be performed is not known, only generic indices can be computed. *Generic information loss measures* are the indices to be applied in this case. They have been defined to evaluate the utility of the protected data but not for a specific application but for *any* of them. Naturally, as these indices aggregate a few components, it might be the case that a protection procedure with a good *average* performance behaves badly in a specific application.

Nevertheless, information loss (either specific or generic) should not be considered in isolation but disclosure risk should also be considered. In Section 35.3, we discussed disclosure risk measures and in Section 35.6 we will describe the tools for taking into account both elements.

In this section we describe some of the existing methods for measuring the utility of the protected data. We will start with some generic information loss measures and then point out some specific ones. We focus on methods for numerical data.

35.5.1 Generic Information Loss Measures

First steps on the definition of a family of information loss measures were presented in (Domingo-Ferrer *et al.*, 2001) and (Domingo-Ferrer and Torra, 2001b). There, information loss was defined as the discrepancy between a few matrices obtained on the original data and on the masked one. Covariance matrices, correlation matrices and a few other matrices (as well as the original file X and the masked file X') were used. Mean square error, mean absolute error and mean variation were used to compute matrix discrepancy. E.g., mean square error of the difference between the original data and the protected one is defined as $(\sum_{i=1}^{n}\sum_{i=1}^{p}(x_{ij}-x'_{ij})^2)/np$ (here p is the number of attributes and n is the number of records).

Nevertheless, these information loss measures are unbounded, and problems appear e.g. when the original values are close to zero. To solve these problems, (Yancey *et al.*, 2002) proposed to replace the mean variation of X and X' by a measure more stable when the original values are close to zero

Trottini (Trottini, 2003) detected that using such information loss measures in combination to disclosure risk measures for comparing masking methods was not well defined as information loss measures were unbounded. That is, when an overall score is computed for a method (see (Domingo-Ferrer *et al.*, 2001) and (Yancey *et al.*, 2002)) as *e.g.* the average of information loss and disclosure risk, both measures should be defined in the same commensurable range. To solve this problem, Trottini proposed to settle a predefined maximum value of error. More recently, in (Mateo-Sanz *et al.*, 2005), probabilistic information loss measures were introduced to solve the same problem avoiding the need of such predefined values.

Definitions in (Mateo-Sanz *et al.*, 2005) start considering the discrepancy between a population parameter θ on X and a sample statistic Θ on X'. Let $\hat{\Theta}$ be the value of this statistic for a specific sample. Then, the standardized sample discrepancy corresponds to

$$Z = \frac{\hat{\Theta} - \theta}{\sqrt{Var(\hat{\Theta})}}$$

This discrepancy can be assumed to follow a $N(0,1)$ (see (Mateo-Sanz *et al.*, 2005) for details). Then, we define the probabilistic information loss measure for $\hat{\Theta}$ as follows:

$$pil(\hat{\Theta}) := 2 \cdot P\left(0 \leq Z \leq \frac{\hat{\theta} - \theta}{\sqrt{Var(\hat{\Theta})}}\right) \tag{35.1}$$

Following (Mateo-Sanz *et al.*, 2005, Domingo-Ferrer and Torra, 2001b), we consider five measures for analysis. They are based on the following statistics:

- Mean for attribute V ($Mean(V)$):

$$\sum_{i=1}^{n} x_{i,V}/n$$

- Variance for attribute V ($Var(V)$):

$$\sum_{i=1}^{n} (x_{i,V} - Mean(V))^2/n$$

- Covariance for attribute V and V' ($Cov(V'V')$):

$$\frac{\sum_{i=1}^{n'} \left(x_{i,V} - Mean(V)\right)\left(x_{i,V'} - Mean(V')\right)}{n'}$$

- Correlation coefficient for V and V' ($\rho(V,V')$):

$$\frac{\sum_{i=1}^{n} \left(x_{i,V} - Mean(V)\right)\left(x_{i,V'} - Mean(V')\right)}{\sum_{i=1}^{n} \left(x_{i,V'} - Mean(V')\right)^2 \sum_{i=1}^{n} \left(x_{i,V} - Mean(V)\right)}$$

- Quantiles for attribute V: That is, the values that divide the distribution in such a way that a given proportion of the observations are below the quantile. (Mateo-Sanz *et al.*, 2005) uses the quantiles for i from 5% to 95% with increments of 5%.

$pil(Mean)$, $pil(Var)$, $pil(Cov)$, $pil(\rho)$ and $pil(Q)$ are computed using Expression 35.1 above. In fact, for a given data file with several attributes V_i, these measures are computed for each V_i (or pair V_i, V_j) and then the corresponding *pil* averaged. In the particular case of the quantile, $pil(Q(V))$ is first defined as the average of the set of measures $pil(Q_i(V))$ for $i = 5\%$ to 95% with increments of 5%.

Then, the average of all these can be defined as a generic information loss. We denote it by $aPil$. That is,

$$aPil := (pil(Mean) + pil(Var) + pil(Cov) + pil(\rho) + pil(Q))/5.$$

This measure has been used in (Mateo-Sanz *et al.*, 2005) and (Ladra and Torra, 2008) for evaluating and ranking different protection methods.

35.5.2 Specific Information Loss Measures

These measures depend on the type of application. In general, these measures compare the performance of an analysis applied to the original data and the same analysis to the protected data. When parametric protection procedures are considered, the more distortion is applied to the data, the larger the difference between the analysis on the original data and the analysis on the protected data.

The comparison of the analysis depends on what type of analysis is performed. In some cases, the comparison of the results is not trivial. For example, when the analysis corresponds to the application of a (crisp) clustering algorithm, measures should compare the partition obtained from the original data and the one obtained from the protected data. That is, we need functions to measure the similarity of two partitions. Similarly, in the case of fuzzy clustering, functions to measure the similarity between two fuzzy partitions are required.

In the case that protected data has to be used for classification, the comparison should be done on the resulting classifiers themselves or on the performance of the classifiers. As an example, (Agrawal and Srikant, 2000) analyses the effects of noise addition on data classifiers (decision trees).

In general, the analysis of the results will show the decline of the performance of the classifiers with respect to the distortion included in the data. The best protection procedures are those that permit a large distortion (and a low disclosure risk) but with a low decline of the performance of classifiers (or clustering).

35.6 Trade-off and Visualization

In this chapter we have described a few protection methods and described a few ways to analyse them. We have seen that the distortion caused by protection methods permits us to reduce the risk of disclosure but at the same time this causes some information loss. In the last section we have described a few measures for this purpose.

To compare data protection procedures and to visualize the trade-off between information loss and disclosure risk, a few tools have been developed. We describe some of them below.

35.6.1 The Score

In (Domingo-Ferrer and Torra, 2001b), the average between disclosure risk and information loss was proposed. That is,

$$Score(method) = \frac{IL(method) + DR(method)}{2}.$$

where IL corresponds to the information loss and DR the disclosure risk. Different measures will, of course, result into different scores.

(Domingo-Ferrer and Torra, 2001b) used generic information loss measures and generic disclosure risk. Disclosure risk was defined as an average of interval disclosure ID (one of the measures described in Section 35.3 for attribute disclosure) and reidentification RD (a measure for individual disclosure). That is, $DR(method) = (ID(method) + RD(method))/2$.

Then, in relation to the measure about reidentification, as we can pressume that the intruder will use the most effective method for disclosure risk, we define RD as the maximum percentage of the reidentification obtained by a set of reidentification algorithms (RDA). That is,

method	aPIL	DR	Score
Rank19	34.36	15.55220303	24.95390177
Rank20	34.46	18.856305	26.6562812
Rank17	35.88	17.66314794	26.77372912
Rank15	33.41	20.60728858	27.00680666
Rank18	34.69	20.728565	27.70810968
Rank16	33.25	22.996878	28.12175372
Rank13	31.29	25.26048534	28.27610286
Rank14	34.12	26.5717775	30.34832651
Rank11	29.79	32.09452819	30.94388101
Rank12	31.08	31.800747	31.43809107
Rank09	24.13	40.95472341	32.54038346
Rank10	28.39	41.191185	34.79133316
Rank07	20.95	52.05937146	36.5030496
Micmul03	38.36	35.61935825	36.98824959
Micmul04	42.80	31.30605884	37.05470347
Micmul05	46.47	27.98280127	37.2245081
Micmul08	51.00	23.4813759	37.24028758
Micmul06	48.60	25.94948057	37.27647059
Rank08	22.42	52.32661	37.37232476
Micmul09	52.71	22.06353532	37.38619949
Micmul07	50.55	24.37663265	37.46127816
Micmul10	53.53	21.82407737	37.6762343
Rank05	16.27	64.60353016	40.43547191
Mic4mul03	20.44	60.58282339	40.51198937
Mic4mul04	27.09	54.33719341	40.71608119

Table 35.1. Methods with the best performace according to the score.

$$RD(method) = \max_{rda \in RDA} RD_{rda}(method).$$

For illustration, Table 35.1 displays the 25 best scores from the selection of methods included in (Domingo-Ferrer and Torra, 2001b). The best methods are rank swapping (standard implementation) and microaggregation (individual ranking and MDAV). The analysis included other methods as lossy compression (using JPEG) and noise addition.

In Table 35.1 both information loss and disclosure risk are represented in the scale [0,100]. Information loss, disclosure risk and the score are defined as in (Nin *et al.*, 2008c). That is, they are defined as follows. Disclosure risk corresponds to the average between interval disclosure and the maximum percentage of reidentification using a few reidentification algorithms (including probabilistic record linkage and distance-based record linkage, using Euclidean, Mahalanobis, and also another distance based on kernel functions). Information loss was computed using $aPil$. For both measures, 0 is the optimal value (no information loss and no risk) and 100 is the worst value (maximum information loss and 100% of reidentifications). Note that using the identity function ($X' = X$) as a protection method, we obtain a score of 50 (full risk and no loss). In the same way, a complete distortion (e.g., completely random data) has also a score equal to 50 (full loss and no risk). So, only scores lower than 50 are of interest.

In comparison with (Domingo-Ferrer and Torra, 2001b), were the average was used to combine only two reidentification algorithms (probabilistic record linkage and distance-based one), the results are similar. Some parameterizations of rank swapping (*Rank* with parameter p in the Table) and microaggregation (*Micmul* with parameter k in the Table) are ranked in both (Domingo-Ferrer and Torra, 2001b) and here among the best algorithms.

The comparison can be extended evaluating new masking methods and comparing them with the existing scores. For example, results from (Jimenez and Torra, 2009) would permit to include in this table (with a score lower than 40) some parameterizations of lossy compression using JPEG 2000.

35.6.2 R-U Maps

(Duncan *et al.*, 2001, Duncan *et al.*, 2004) propose the R-U maps, for Risk-Utility maps. This is a graphical representation of the two measures. R for risk and U for utility.

Figure 35.2 represents an R-U map for the methods listed in the previous section each with several parameterizations. Namely, RankXXX corresponds to Rank Swapping, MicXXX are variations of Microaggregation, JPEGXXX corresponds to Lossy Compression using JPEG, and RemuestX is resampling (not described in this chapter). In the figure, DR corresponds to the Disclosure Risk (R following the standard jargon of R-U maps), and IL to information loss (in our case computed as $aPIL$). Formally, IL and utility U are related as follows: $1 - U = IL$.

Note that in addition to the protection procedures represented in Table 35.1, the figure includes all the other methods analyzed in (Domingo-Ferrer and Torra, 2001b) but with the new measures DR and $aPIL$ described above. In this figure, the lines represent scores of 50, 40, 30, and 20. Naturally, the nearer a method to $(0,0)$, the better.

35.7 Conclusions

In this chapter we have reviewed the major topics concerning privacy in data mining. We have rewiewed major protection methods, and discussed how to measure disclosure risk and information loss. Finally, some tools for visualizing such measures and for comparing the methods have been described.

Acknowledgements

Part of the research described in this chapter is supported by the Spanish MEC (projects ARES – CONSOLIDER INGENIO 2010 CSD2007-00004 – and eAEGIS – TSI2007-65406-C03-02).

References

Adam, N. R., Wortmann, J. C. (1989) Security-control for statistical databases: a comparative study, ACM Computing Surveys, Volume: 21, 515-556.

Aggarwal, C. (2005) On k-anonymity and the curse of dimensionality, Proceedings of the 31st International Conference on Very Large Databases, pages 901-909.

Aggarwal, C. C., Yu, P. S. (2008) Privacy-Preserving Data Mining: Models and Algorithms, Springer.

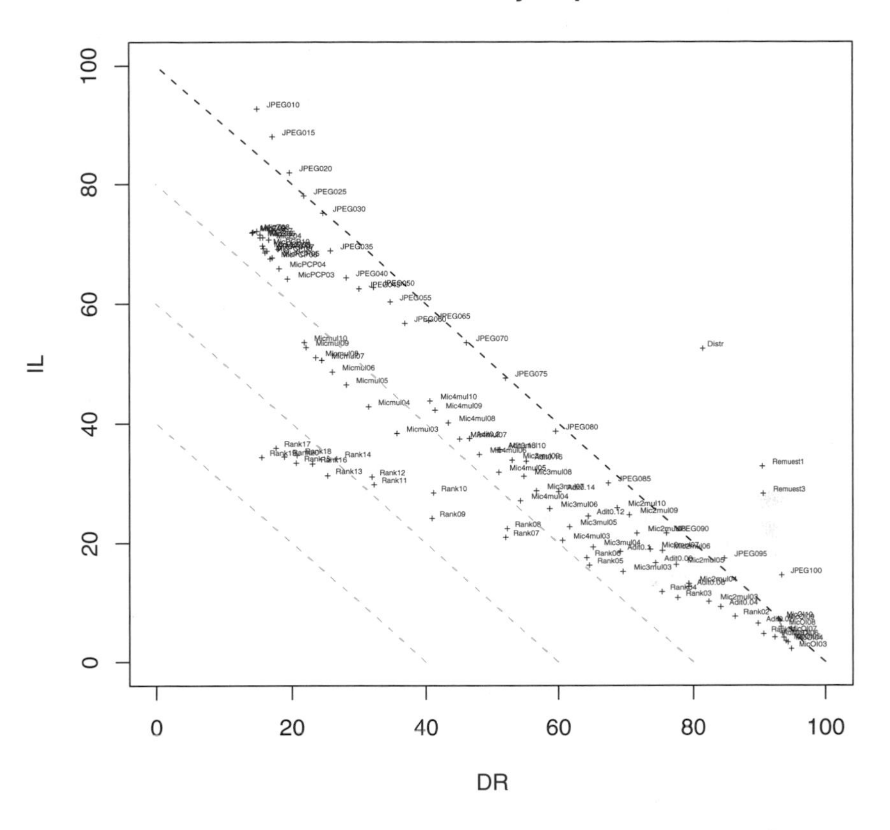

Fig. 35.2. R-U Maps for some protection methods. IL computed with PIL.

Agrawal, R., Srikant, R. (2000) Privacy Preserving Data Mining, Proc. of the ACM SIGMOD Conference on Management of Data, 439-450.

Atallah, M., Bertino, E., Elmagarmid, A., Ibrahim, M., Verykios, V. (1999) Disclosure limitation of sensitive rules, Proc. of IEEE Knowledge and Data Engineering Exchange Workshop (KDEX).

Atzori, M., Bonchi, F., Giannotti, F., Pedreschi, D. (2008) Anonymity preserving pattern discovery, The VLDB Journal 17 703-727.

Bacher, J., Brand, R., Bender, S. (2002) Re-identifying register data by survey data using cluster analysis: an empirical study, Int. J. of Unc., Fuzz. and Knowledge Based Systems 10:5 589-607.

Bertino, E., Lin, D., Jiang, W. (2008) A survey of quantification of privacy preserving data mining algorithms, in C. C. Aggarwal, P. S. Yu (eds.) Privacy-Preserving Data Mining:

Models and Algorithms, Springer, 183-205.

Brand, R. (2002) Microdata protection through noise addition, in J. Domingo-Ferrer (ed.) Inference Control in Statistical Databases, Lecture Notes in Computer Science 2316 97-116.

Bunn, P., Ostrovsky, R. (2007) Secure two-party k-means clustering, Proc. of CCS'07, ACM Press, 486-497.

Burridge, J. (2003) Information preserving statistical obfuscation, Statistics and Computing, 13:321–327.

Carlson, M., Salabasis, M. (2002) A data swapping technique using ranks: a method for disclosure control, Research on Official Statistics 5:2 35-64.

Dalenius, T. (1977) Towards a methodology for statistical disclosure control, Statistisk Tidskrift 5 429-444.

Dalenius, T. (1986) Finding a needle in a haystack - or identifying anonymous census records, Journal of Official Statistics 2:3 329-336.

Defays, D., Nanopoulos, P. (1993) Panels of enterprises and confidentiality: the small aggregates method, Proc. of 92 Symposium on Design and Analysis of Longitudinal Surveys, Statistics Canada, 195-204.

Dempster, A. P., Laird, N. M., Rubin, D. B. (1977) Maximum Likelihood From Incomplete Data Via the EM Algorithm, Journal of the Royal Statistical Society 39 1-38.

Domingo-Ferrer, J., Mateo-Sanz, J. M. (2002) Practical data-oriented microaggregation for statistical disclosure control, IEEE Trans. on Knowledge and Data Engineering 14:1 189-201.

Domingo-Ferrer, J., Mateo-Sanz, J. M., Torra, V. (2001) Comparing SDC methods for microdata on the basis of information loss and disclosure risk, Pre-proceedings of ETK-NTTS'2001, (Eurostat, ISBN 92-894-1176-5), Vol. 2, 807-826, Creta, Greece.

Domingo-Ferrer, J., Sebe, F., Castella-Roca, J. (2004) On the security of noise addition for privacy in statistical databases, PSD 2004, Lecture Notes in Computer Science 3050 149-161.

Domingo-Ferrer, J., Torra, V. (2001) Disclosure Control Methods and Information Loss for Microdata, in P. Doyle, J. I. Lane, J. J. M. Theeuwes, L. Zayatz (eds.) Confidentiality, Disclosure, and Data Access: Theory and Practical Applications for Statistical Agencies, Elsevier Science, 91-110.

Domingo-Ferrer, J., Torra, V. (2001) A quantitative comparison of disclosure control methods for microdata, in P. Doyle, J. I. Lane, J. J. M. Theeuwes, L. Zayatz (eds.) Confidentiality, Disclosure and Data Access: Theory and Practical Applications for Statistical Agencies, North-Holland, 111-134.

Domingo-Ferrer, J., Torra, V. (2003) Disclosure Risk Assessment in Statistical Microdata Protection via advanced record linkage, Statistics and Computing, 13 343-354.

Domingo-Ferrer, J., Torra, V. (2005) Ordinal, Continuous and Heterogeneous k-Anonymity Through Microaggregation, Data Mining and Knowledge Discovery 11:2 195-212.

Duncan, G. T., Keller-McNulty, S. A., Stokes, S. L. (2001) Disclosure risk vs. data utility: The R-U confidentiality map, Technical Report 121, National Institute of Statistical Sciences.

Duncan, G. T., Keller-McNulty, S. A., Stokes, S. L. (2001) Database security and confidentiality: examining disclosure risk vs. data utility through the R-U confidentiality map, Technical Report 142, National Institute of Statistical Sciences.

Duncan, G. T., Lambert, D. (1986) Disclosure-limited data dissemination, Journal of the American Statistical Association, 81 10-18.

Duncan, G. T., Lambert, D. (1989) The risk disclosure for microdata, Journal of Business and Economic Statistics 7 207-217.

Elamir, E. A. H. (2004) Analysis of re-identification risk based on log-linear models, PSD 2004, Lecture Notes in Computer Science 3050 273-281.

Elliot, M. (2002) Integrating file and record level disclosure risk assessment, in J. Domingo-Ferrer, Inference Control in Statistical Databases, Lecture Notes in Computer Science 2316 126-134.

Elliot, M. J. Skinner, C. J., Dale, A. (1998) Special Uniqueness, Random Uniques and Sticky Populations: Some Counterintuitive Effects of Geographical Detail on Disclosure Risk, Research in Official Statistics 1:2 53-67.

Fellegi, I. P., Sunter, A. B. (1969) A theory for record linkage, Journal of the American Statistical Association 64:328 1183-1210.

Felsö, F., Theeuwes, J., Wagner, G., (2001) Disclosure Limitation in Use: Results of a Survey, in P. Doyle, J. I. Lane, J. J. M. Theeuwes, L. Zayatz (eds.) Confidentiality, Disclosure, and Data Access: Theory and Practical Applications for Statistical Agencies, Elsevier Science, 17-42.

Franconi, L., Polettini, S. (2004) Individual risk estimation in μ-Argus: a review, PSD 2004, Lecture Notes in Computer Science 3050 262-272.

Gouweleeuw, J. M., Kooiman, P., Willenborg, L. C. R. J., De Wolf, P.-P. (1998) Post Randomisation for Statistical Disclosure Control: Theory and Implementation', Journal of Official Statistics 14:4 463-478. Also as Research Paper No. 9731, Voorburg: Statistics Netherlands (1997).

Gross, B., Guiblin, P., Merrett, K. (2004) Implementing the Post Randomisation method to the individual sample of anonymised records (SAR) from the 2001 Census, paper presented at "The Samples of Anonymised Records, An Open Meeting on the Samples of Anonymised Records from the 2001 Census". http://www.ccsr.ac.uk/sars/events/2004-09-30/gross.pdf

Hansen, S., Mukherjee, S. (2003) A Polynomial Algorithm for Optimal Univariate Microaggregation, IEEE Trans. on Knowledge and Data Engineering 15:4 1043-1044.

Haritsa, J. R. (2008) Mining association rules under privacy constraints, in C. C. Aggarwal, P. S. Yu (eds.) Privacy-Preserving Data Mining: Models and Algorithms, Springer, 239-266.

Hundepool, A., van de Wetering, A., Ramaswamy, R., Franconi, L., Capobianchi, C., de Wolf, P.-P., Domingo-Ferrer, J., Torra, V., Brand, R., Giessing, S. (2003) μ-ARGUS version 3.2 Software and User's Manual, Voorburg NL,Statistics Netherlands, February, 2003; version 4.0 published on may 2005. http://neon.vb.cbs.nl/casc.

Jaro, M. A. (1989) Advances in record-linkage methodology as applied to matching the 1985 Census of Tampa, Florida, Journal of the American Statistical Association 84:406 414-420.

Jiménez, J., Torra, V. (2009) Utility and risk of JPEG-based continuous microdata protection methods, Proc. Int. Conf. on Availability, Reliability and Security (ARES 2009), 929-934.

Kantarcioglu, M. (2008) A survey of privacy-preserving methods across horizontally partitioned data, in C. C. Aggarwal, P. S. Yu (eds.) Privacy-Preserving Data Mining: Models and Algorithms, Springer, 313-335.

Kim, J., Winkler, W. (2003) Multiplicative noise for masking continuous data, Research Report Series (Statistics 2003-01), U. S. Bureau of the Census.

Kisilevich S., Rokach L., Elovici Y., Shapira B., Efficient Multidimensional Suppression for K-Anonymity, IEEE Transactions on Knowledge and Data Engineering, vol. 22, no. 3, pp. 334-347, Mar. 2010

Ladra, S., Torra, V. (2008) On the comparison of generic information loss measures and cluster-specific ones, Intl. J. of Unc., Fuzz. and Knowledge-Based Systems, 16:1 107-120.

Lambert, D. (1993) Measures of Disclosure Risk and Harm, Journal of Official Statistics 9 313-331.

LeFevre, K., DeWitt, D. J., Ramakrishnan, R. (2005) Multidimensional *k*-anonymity, Technical Report 1521, University of Wisconsin.

LeFevre, K., DeWitt, D. J., Ramakrishnan, R. (2005) Incognito: Efficient Full-Domain K-Anonymity, SIGMOD 2005.

Li, N., Li, T., Venkatasubramanian, S. (2007) T-closeness: privacy beyond k-anonymity and l-diversity, Proc. of the IEEE ICDE 2007.

Liew, C. K., Choi, U. J., Liew, C. J. (1985) A data distortion by probability distribution, ACM Transactions on Database Systems 10 395-411.

Lindell, Y., Pinkas, B. (2002) Privacy Preserving Data Mining, Journal of Cryptology, 15:3.

Lindell, Y., Pinkas, B. (2000) Privacy Preserving Data Mining, Crypto'00, Lecture Notes in Computer Science 1880 20-24.

Liu, K., Kargupta, H., Ryan, J. (2006) Random projection based multiplicative data perturbation for privacy preserving data mining, IEEE Trans. on Knowledge and Data Engineering 18:1 92-106.

Machanavajjhala, A., Gehrke, J., Kiefer, D., Venkitasubramanian, M. (2006) L-diversity: privacy beyond k-anonymity, Proc. of the IEEE ICDE.

Mateo-Sanz, J. M., Domingo-Ferrer, J. Sebé, F. (2005) Probabilistic information loss measures in confidentiality protection of continuous microdata, Data Mining and Knowledge Discovery, 11:2 181-193.

Moore, R. (1996) Controlled data swapping techniques for masking public use microdata sets, U. S. Bureau of the Census (unpublished manuscript).

Muralidhar, K., Sarathy, R. (2008) Generating Sufficiency-based Non-Synthetic Perturbed Data, Transactions on Data Privacy 1:1 17 - 33

Nin, J., Herranz, J., Torra, V. (2007) Rethinking Rank Swapping to Decrease Disclosure Risk, Data and Knowledge Engineering, 64:1 346-364.

Nin, J., Herranz, J., Torra, V. (2008) How to Group Attributes in Multivariate Microaggregation, Intl. J. of Unc., Fuzz. and Knowledge-Based Systems, 16:1 121-138.

Nin, J., Herranz, J., Torra, V. (2008) On the Disclosure Risk of Multivariate Microaggregation, Data and Knowledge Engineering, 67:3 399-412.

Nin, J., Herranz, J., Torra, V. (2008) Towards a More Realistic Disclosure Risk Assessment, Lecture Notes in Computer Science, 5262 152-165.

Nin, J. Torra, V. (2006) Extending microaggregation procedures for time series protection, Lecture Notes in Artificial Intelligence, 4259 899-908.

Nin, J., Torra, V. (2009) Analysis of the Univariate Microaggregation Disclosure Risk, New Generation Computing, 27 177-194.

Oganian, A., Domingo-Ferrer, J. (2000) On the Complexity of Optimal Microaggregation for Statistical Disclosure Control, Statistical J. United Nations Economic Commission for Europe, 18, 4, 345-354.

Paass, G. (1985) Disclosure risk and disclosure avoidance for microdata, Journal of Business and Economic Statistics 6 487-500.

Paass, G., Wauschkuhn, U. (1985) Datenzugang, Datenschutz und Anonymisierung - Analysepotential und Identifizierbarkeit von Anonymisierten Individualdaten, Oldenbourg Verlag.

Pagliuca, D., Seri, G. (1999) Some results of individual ranking method on the system of enterprise accounts annual survey, Esprit SDC Project, Deliverable MI-3/D2.

Pinkas, B. (2002) Cryptographic techniques for privacy-preserving data mining, ACM SIGKDD Explorations 4:2.

Ravikumar, P., Cohen, W. W. (2004) A hierarchical graphical model for record linkage, Proc. of UAI 2004.

Rokach L., Genetic algorithm-based feature set partitioning for classification problems,Pattern Recognition, 41(5):1676–1700, 2008.

Rokach L., Maimon O. and Lavi I., Space Decomposition In Data Mining: A Clustering Approach, Proceedings of the 14th International Symposium On Methodologies For Intelligent Systems, Maebashi, Japan, Lecture Notes in Computer Science, Springer-Verlag, 2003, pp. 24–31.

Samarati, P. (2001) Protecting Respondents' Identities in Microdata Release, IEEE Trans. on Knowledge and Data Engineering, 13:6 1010-1027.

Samarati, P., Sweeney, L. (1998) Protecting privacy when disclosing information: *k*-anonymity and its enforcement through generalization and suppression, SRI Intl. Tech. Rep.

Spruill, N. L. (1983) The confidentiality and analytic usefulness of masked business microdata, Proc. of the Section on Survery Research Methods 1983, American Statistical Association, 602-610.

Sweeney, L. (2002) Achieving *k*-anonymity privacy protection using generalization and suppression, Int. J. of Unc., Fuzz. and Knowledge Based Systems 10:5 571-588.

Sweeney, L. (2002) *k*-anonymity: a model for protecting privacy, Int. J. of Unc., Fuzz. and Knowledge Based Systems 10:5 557-570.

Takemura, A. (2002) Local recoding and record swapping by maximum weight matching for disclosure control of microdata sets, Journal of Official Statistics 18 275-289. Preprint (1999) Local recoding by maximum weight matching for disclosure control of microdata sets.

Templ, M. (2008) Statistical Disclosure Control for Microdata Using the R-Package sdcMicro, Transactions on Data Privacy 1 67-85.

Torra, V. (2004) Microaggregation for categorical variables: a median based approach, Proc. Privacy in Statistical Databases (PSD 2004), Lecture Notes in Computer Science 3050 162-174.

Torra, V. (2004) OWA operators in data modeling and reidentification, IEEE Trans. on Fuzzy Systems 12:5 652-660.

Torra, V. (2008) Constrained Microaggregation: Adding Constraints for Data Editing, Transactions on Data Privacy 1:2 86-104.

Torra, V., Abowd, J. M., Domingo-Ferrer, J. (2006) Using Mahalanobis Distance-Based Record Linkage for Disclosure Risk Assessment, Lecture Notes in Computer Science 4302 233-242.

Torra, V., Domingo-Ferrer, J. (2003) Record linkage methods for multidatabase data mining, in V. Torra (ed.) Information Fusion in Data Mining, Springer, 101-132.

Torra, V., Miyamoto, S. (2004) Evaluating fuzzy clustering algorithms for microdata protection, PSD 2004, Lecture Notes in Computer Science 3050 175-186.

Trottini, M. (2003) Decision models for data disclosure limitation, PhD Dissertation, Carnegie Mellon University. http://www.niss.org/dgii/TR/Thesis-Trottini-final.pdf

Truta, T. M., Vinay, B. (2006) Privacy protection: p-sensitive k-anonymity property. Proc. 2nd Int. Workshop on Privacy Data management (PDM 2006) p. 94.

Willenborg, L., de Waal, T. (2001) *Elements of Statistical Disclosure Control*, Lecture Notes in Statistics, Springer-Verlag.

Winkler, W. E. (1993) Matching and record linkage, Statistical Research Division, U. S. Bureau of the Census (USA), RR93/08.

Winkler, W. E. (2004) Re-identification methods for masked microdata, PSD 2004, Lecture Notes in Computer Science 3050 216-230.

Yancey, W. E., Winkler, W. E., Creecy, R. H. (2002) Disclosure risk assessment in perturbative microdata protection, in J. Domingo-Ferrer (ed.) Inference Control in Statistical Databases, Lecture Notes in Computer Science 2316 135-152.

Yao, A. C. (1982) Protocols for Secure Computations, Proc. of 23rd IEEE Symposium on Foundations of Computer Science, Chicago, Illinois, 160-164.

http://www.census.gov

36

Meta-Learning - Concepts and Techniques

Ricardo Vilalta[1], Christophe Giraud-Carrier[2], and Pavel Brazdil[3]

[1] University of Houston
[2] Brigham Young University
[3] University of Porto

Summary. The field of meta-learning has as one of its primary goals the understanding of the interaction between the mechanism of learning and the concrete contexts in which that mechanism is applicable. The field has seen a continuous growth in the past years with interesting new developments in the construction of practical model-selection assistants, task-adaptive learners, and a solid conceptual framework. In this chapter we give an overview of different techniques necessary to build meta-learning systems. We begin by describing an idealized meta-learning architecture comprising a variety of relevant component techniques. We then look at how each technique has been studied and implemented by previous research. In addition we show how meta-learning has already been identified as an important component in real-world applications.

Key words: Meta-learning

36.1 Introduction

We are used to thinking of a learning system as a rational agent capable of adapting to a specific environment by exploiting knowledge gained through experience; encountering multiple and diverse scenarios sharpens the ability of the learning system to predict the effect produced from selecting a particular course of action. In this case, learning is made manifest because the quality of the predictions normally improves with an increasing number of scenarios or examples. Nevertheless, if the predictive mechanism were to start afresh on different tasks, the learning system would find itself at a considerable disadvantage; learning systems capable of modifying their own predictive mechanism would soon outperform our base learner by being able to change their learning strategy according to the characteristics of the task under analysis.

Meta-learning differs from *base-learning* in the scope of the level of adaptation; whereas learning at the base-level is based on accumulating experience on a specific learning task (e.g., credit rating, medical diagnosis, mine-rock discrimination, fraud detection, etc.), learning at the meta-level is based on accumulating experience on the performance of multiple applications of a learning system. If a base-learner fails to perform efficiently, one would expect the

O. Maimon, L. Rokach (eds.), *Data Mining and Knowledge Discovery Handbook*, 2nd ed.,
DOI 10.1007/978-0-387-09823-4_36,

learning mechanism itself to adapt in case the same task is presented again. Meta-learning is then important in understanding the interaction between the mechanism of learning and the concrete contexts in which that mechanism is applicable. Briefly stated, the field of meta-learning is focused on the relation between tasks or domains and learning strategies. In that sense, by learning or explaining what causes a learning system to be successful or not on a particular task or domain, we go beyond the goal of producing more accurate learners to the additional goal of understanding the conditions (e.g., types of example distributions) under which a learning strategy is most appropriate.

From a practical stance, meta-learning can solve important problems in the application of machine learning and Data Mining tools, particularly in the area of classification and regression. First, the successful use of these tools outside the boundaries of research (e.g., industry, commerce, government) is conditioned on the appropriate selection of a suitable predictive model (or combinations of models) according to the domain of application. Without any kind of assistance, model selection and combination can turn into stumbling blocks to the end-user who wishes to access the technology more directly and cost-effectively. End-users often lack not only the expertise necessary to select a suitable model, but also the availability of many models to proceed on a trial-and-error basis (e.g., by measuring accuracy via some re-sampling technique such as n-fold cross-validation). A solution to this problem is attainable through the construction of meta-learning systems. These systems can provide automatic and systematic user guidance by mapping a particular task to a suitable model (or combination of models).

Second, a problem commonly observed in the practical use of ML and DM tools is how to profit from the repetitive use of a predictive model over similar tasks. The successful application of models in real-world scenarios requires a continuous adaptation to new needs. Rather than starting afresh on new tasks, we expect the learning mechanism itself to re-learn, taking into account previous experience (Thrun, 1998, Pratt *et al.*, 1991, Caruana, 1997, Vilalta and Drissi, 2002). Again, meta-learning systems can help control the process of exploiting cumulative expertise by searching for patterns across tasks.

Our goal in this chapter is to give an overview of different techniques necessary to build meta-learning systems. To impose some structure, we begin by describing an idealized meta-learning architecture comprising a variety of relevant component techniques. We then look at how each technique has been studied and implemented by previous research. We hope that by proceeding in this way the reader can not only learn from past work, but in addition gain some insight on how to construct meta-learning systems.

We also hope to show how recent advances in meta-learning are increasingly filling the gaps in the construction of practical model-selection assistants and task-adaptive learners, as well as in the development of a solid conceptual framework (Baxter, 1998, Baxter, 2000, Giraud-Carrier *et al.*, 2004).

This chapter is organized as follows. In the next section we illustrate an idealized meta-learning architecture and detail on its constituent parts. In Section 65.3.3 we describe previous research in meta-learning and its relation to our architecture. Section 65.3.4 describes a meta-learning tool that has been instrumental as a decision support tool in real applications. Lastly, section 65.3.5 discusses future directions and provides our conclusions.

36.2 A Meta-Learning Architecture

In this section we provide a general view of a software architecture that will be used as a reference to describe many of the principles and current techniques in meta-learning. Though not every technique in meta-learning fits into this architecture, such a general view helps us understand the challenges we need to overcome before we can turn the technology into a set of useful and practical tools.

36.2.1 Knowledge-Acquisition Mode

To begin, we propose a meta-learning system that divides into two modes of operation. During the first mode, also known as the *knowledge-acquisition mode*, the main goal is to learn about the learning process itself. Figure 36.1 illustrates this mode of operation. We assume the input to the system is made of more than one dataset of examples (e.g., more than one set of pairs of feature vectors and classes; Figure 36.1A). Upon arrival of each dataset, the meta-learning system invokes a component responsible for extracting dataset characteristics or meta-features (Figure 36.1B). The goal of this component is to gather information that transcends the particular domain of application. We look for information that can be used to generalize to other example distributions. Section 36.3.1 details current research pointing in this direction.

During the knowledge acquisition mode, the learning technique (Figure 36.1C) does not exploit knowledge across different datasets or tasks. Each dataset is considered independently of the rest; the output to the system is a learning strategy (e.g., a classifier or combination of classifiers, Figure 36.1D). Statistics derived from the output model or its performance (Figure 36.1E) may also serve as a form of characterizing the task under analysis (Sections 36.3.1 and 36.3.1).

Information derived from the meta-feature generator and the performance evaluation module can be combined into a *meta-knowledge base* (Figure 36.1F). This knowledge base is the main result of the knowledge–acquisition phase; it reflects experience accumulated across different tasks. Meta-learning is tightly linked to the process of acquiring and exploiting meta-knowledge. One can even say that advances in the field of meta-learning hinge around one specific question: how can we acquire and exploit knowledge about learning systems (i.e., meta-knowledge) to understand and improve their performance? As we describe current research in meta-learning we will be pointing out to different forms of meta-knowledge.

36.2.2 Advisory Mode

The efficiency of the meta-learner increases as it accumulates meta-knowledge. We assume the lack of experience at the beginning of the learner's life compels the meta-learner to use one or more learning strategies without a clear preference for one of them; experimenting with many different strategies becomes time consuming. However, as more training sets have been examined, we expect the expertise of the meta-learner to dominate in deciding which learning strategy best suits the characteristics of the training set.

In the *advisory mode*, meta-knowledge acquired in the exploratory mode is used to configure the learning system in a manner that exploits the characteristics of the new data distribution. Meta-features extracted from the dataset (Figure 36.2B) are matched with the meta-knowledge base (Figure 36.2F) to produce a recommendation regarding the best available learning strategy. At this point we move away from the use of static base learners to the ability to do model selection or combining base learners (Figure 36.2C).

Two observations are worth considering at this point. First, the nature of the match between the set of meta-features and the meta-knowledge base can have several interpretations. The traditional view poses this problem as a learning problem itself where a meta-learner is invoked to output an approximating function mapping meta-features to learning strategies

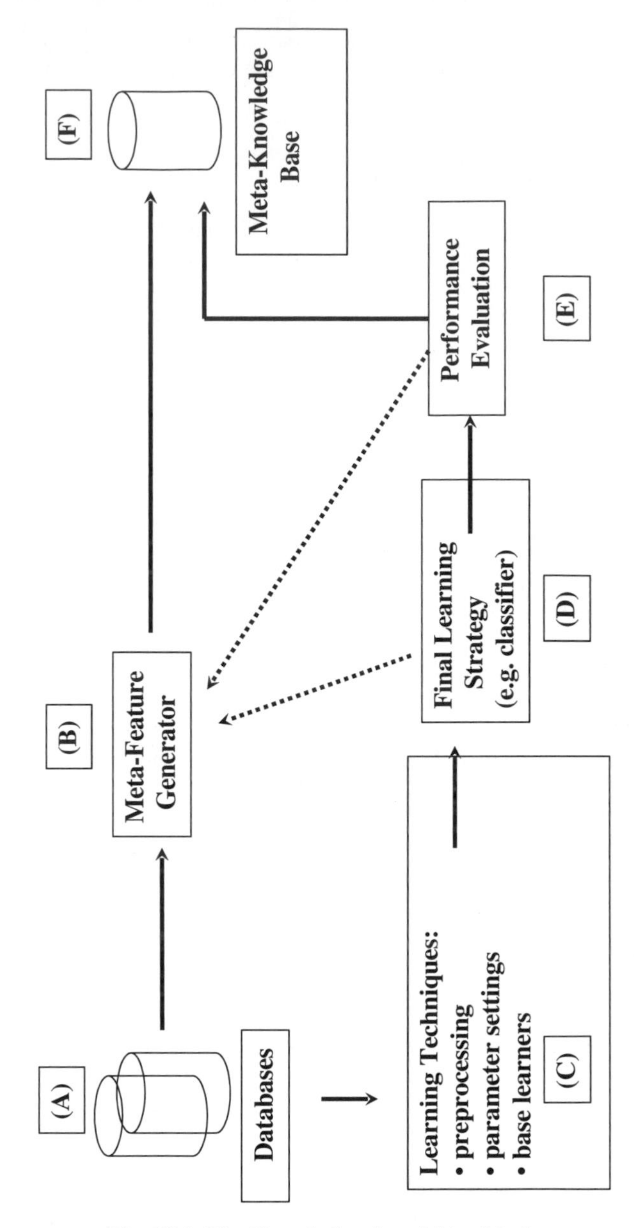

Fig. 36.1. The Knowledge-Acquisition Mode

(e.g., learning model). This view is problematic as the meta-learner is now a learning system subject to improvement through meta-learning (Schmidhuber, 1995, Vilalta, 2001). Second, the matching process is not intended to modify our set of available learning techniques, but simply enables us to select one or more strategies that seem effective given the characteristics of the dataset under analysis.

The final classifier (or combination of classifiers; Figure 36.2D) is selected based not only on its generalization performance over the current dataset, but also on information derived from exploiting past experience. In this case, the system has moved from using a single learning strategy to the ability of selecting one dynamically from among a variety of different strategies.

We will show how the constituent components conforming our two-mode meta-learning architecture can be studied and utilized through a variety of different methodologies:

1. The characterization of datasets can be performed under a variety of statistical, information-theoretic, and model-based approaches (Section 36.3.1).
2. Matching meta-features to predictive model(s) can be used for model selection or model ranking (Section 36.3.1).
3. Information collected from the performance of a set of learning algorithms at the base level can be combined through a meta-learner (Section 36.3.1).
4. Within the learning-to-learn paradigm, a continuous learner can extract knowledge across domains or tasks to accelerate the rate of learning convergence (Section 36.3.1).
5. The learning strategy can be modified in an attempt to shift this strategy dynamically (Section 36.3.2). A meta-learner in effect explores not only the space of hypotheses within a fixed family set, but the space of families of hypotheses.

36.3 Techniques in Meta-Learning

In this section we describe how previous research has tackled the implementation and application of various methodologies in meta-learning.

36.3.1 Dataset Characterization

First, a critical component of any meta-learning system is in charge of extracting relevant information about the task under analysis (Figure 36.1B). The central idea is that high-quality dataset characteristics or meta-features provide some information to differentiate the performance of a set of given learning strategies. We describe a representative set of techniques in this area.

Statistical and Information-Theoretic Characterization

Much work in dataset characterization has concentrated on extracting statistical and information-theoretic parameters estimated from the training set (Aha, 1992, Michie *et al.*, 1994, Gama and Brazdil, 1995, Brazdil, 1998) (Engels and Theusinger, 1998, Sohn, 1999). Measures include number of classes, number of features, ratio of examples to features, degree of correlation between features and target concept, average class entropy and class-conditional entropy, skewness, kurtosis, signal to noise ratio, etc. This work has produced a number of research projects with positive and tangible results (e.g., ESPRIT Statlog and METAL).

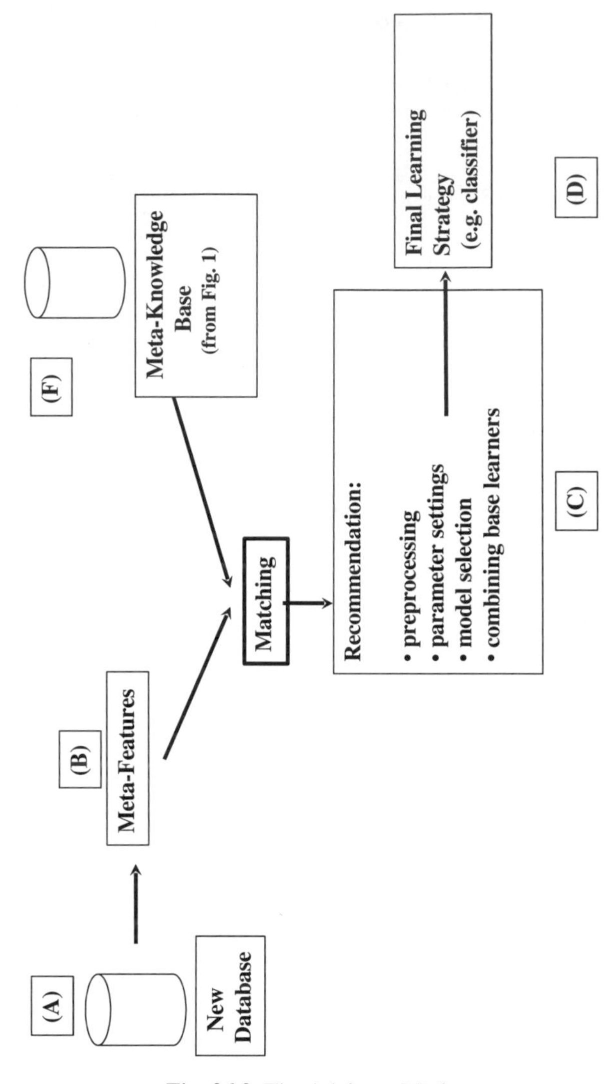

Fig. 36.2. The Advisory Mode

Model-Based Characterization

In addition to statistical measures, a different form of dataset characterization exploits properties of the induced hypothesis as a form of representing the dataset itself. This has several advantages: 1) the dataset is summarized into a data structure that can embed the complexity and performance of the induced hypothesis (and thus is not limited to the example distribution); 2) the resulting representation can serve as a basis to explain the reasons behind the performance of the learning algorithm. As an example, one can build a decision tree from a dataset and collect properties of the tree (e.g., nodes per feature, maximum tree depth, shape, tree imbalance, etc.), as a means to characterize the dataset (Bensusan, 1998, Bensusan and Giraud-Carrier, 2000b, Hilario and Kalousis, 2000, Peng *et al.*, 1995).

Landmarking

Another source of characterization falls within the concept of landmarking (Bensusan and Giraud-Carrier, 2000a, Pfahringer *et al.*, 2000). The idea is to exploit information obtained from the performance of a set of simple learners (i.e., learning systems with low capacity) that exhibit significant differences in their learning mechanism. The accuracy (or error rate) of these *landmarkers* is used to characterize a dataset. The goal is to identify areas in the input space where each of the simple learners can be regarded as an expert. This meta-knowledge can be subsequently exploited to produce more accurate learners.

Another idea related to landmarking is to exploit information obtained on simplified versions of the data (e.g. small samples). Accuracy results on these samples serve to characterise individual datasets and are referred to as *sampling landmarks*. This information is subsequently used to select a learning algorithm (Furnkranz, 1997, Soares *et al.*, 2001).

36.3.2 Mapping Datasets to Predictive Models

An important and practical use of meta-learning is the construction of an engine that maps an input space composed of datasets or applications to an output space composed of predictive models. Criteria such as accuracy, storage space, and running time can be used for performance assessment (Giraud-Carrier, 1998). Several approaches have been developed in this area.

Hand-Crafting Meta Rules

First, using human expertise and empirical evidence, a number of meta-rules matching domain characteristics with learning techniques may be crafted manually (Brodley, 1993, Brodley, 1994). For example, in decision tree learning, a heuristic rule can be used to switch from univariate tests to linear tests if there is a need to construct non-orthogonal partitions over the input space. Crafting rules manually has the disadvantage of failing to identify many important rules. As a result most research has focused on learning these meta-rules automatically as explained next.

Learning at the Meta-Level

The characterization of a dataset is a form of meta-knowledge (Figure 36.1F) that is commonly embedded in a meta-dataset as follows. After learning from several tasks, one can construct a meta-dataset where each element pair is made up of the characterization of a dataset (meta-feature vector) and a class label corresponding to the model with best performance on that dataset. A learning algorithm can then be applied to this well-defined learning task to induce a hypothesis mapping datasets to predictive models.

As in base-learning, the hand-crafting and the learning approach can be combined; in this case the hand-crafted rules can serve as background knowledge to the meta-learner.

Mapping Query Examples to Models

Instead of mapping a task or dataset to a predictive model, a different approach consists of selecting a model for each individual query example. The idea is similar to the nearest-neighbour approach: select the model displaying best performance around the neighbourhood of the query example (Merz, 1995A, Merz, 1995B). Model selection is done according to best-accuracy performance using a re-sampling technique (e.g., cross-validation).

A variation to the approach above is to look at the neighbourhood of a query example in the space of meta-features. When a new training set arrives, the k-nearest neighbour instances (i.e., datasets) around the query example (i.e., query dataset) are gathered to select the model with best average performance (Keller *et al.*, 2000).

Ranking

Rather than mapping a dataset to a single predictive model, one may also produce a ranking over a set of different models. One can argue that such rankings are more flexible and informative for users. In a practical scenario, users should not be limited to a single kind of advice; this is important if the suggested final model turns unsatisfactory. Rankings provide alternative solutions to users who may wish to incorporate their own expertise or any other criterion (e.g., financial constraints) on their decision-making process. Multiple approaches have been suggested attacking the problem of ranking predictive models (Gama and Brazdil, 1995, Nakhaeizadeh *et al.*, 2002, Berrer *et al.*, 2000, Brazdil and Soares, 2000, Keller *et al.*, 2000, Soares and Brazdil, 2000, Brazdil and Soares, 2003).

36.3.3 Learning from Base-Learners

Another approach to meta-learning consists of learning from base learners. The idea is to make explicit use of information collected from the performance a set of learning algorithms at the base level; such information is then incorporated into the meta-learning process.

Stacked Generalization

Meta-knowledge (Figure 36.1F) can incorporate predictions of base learners, a process known as stacked generalization (Wolpert, 1997). The process works under a layered architecture as follows. Each of a set of base-classifiers is trained on a dataset; the original feature representation is then extended to include the predictions of these classifiers. Successive layers receive as input the predictions of the immediately preceding layer and the output is passed on to the next layer. A single classifier at the topmost level produces the final prediction. Most research in this area focuses on a two-layer architecture (Wolpert, 1997, Breiman, 1996, Chan and Stolfo, 1998, Ting, 1994).

Stacked generalization is considered a form of meta-learning because the transformation of the training set conveys information about the predictions of the base-learners (i.e., conveys meta-knowledge). Research in this area investigates what base-learners and meta-learners produce best empirical results (Chan and Stolfo, 1993, Chan and Stolfo, 1996, Gama and Brazdil, 2000); how to represent class predictions (class labels versus class-posterior probabilities; (Ting, 1994); what higher-level learners can be invoked (Gama and Brazdil, 2000, Dzeroski, 2002); and what are novel definitions of meta-features (Brodley, 1996, Ali and Pazzani, 1995).

Boosting

A popular approach to combining base learners is called boosting (Freund and Schapire, 1995, Friedman, 1997, Hastie *et al.*, 2001). The basic idea is to generate a set of base learners by generating variants of the training set. Each variant is generated by sampling with replacement under a weighted distribution. This distribution is modified for every new variant by giving more attention to those examples incorrectly classified by the most recent hypothesis.

Boosting is considered a form of meta-learning because it takes into consideration the predictions of each hypothesis over the original training set to progressively improve the classification of those examples for which the last hypothesis failed.

Landmarking Meta-Learning

We mentioned before how landmarking can be used as a form of dataset characterization by exploiting the accuracy (or error rate) of a set of base (simple) learners called landmarkers. Meta-learning based on landmarking may be viewed as a form of learning from base learners; these base learners provide a new representation of the dataset that can be used in finding areas of learning expertise. Here we assume there is a second set of advanced learners (i.e., learning systems with high capacity), one of which must be selected for the current task under analysis. Under this framework, meta-learning is the process of correlating areas of expertise as dictated by simple learners, with the performance of other -more advanced- learners.

Meta-Decision Trees

Another approach in the field of learning from base learners consists of combining several inductive models by means of induction of meta-decision trees (Todorovski and Dzeroski, 1999, Todorovski and Dzeroski, 2000, Todorovski and Dzeroski, 2003). The general idea is to build a decision tree where each internal node is a meta-feature that measures a property of the class probability distributions predicted for a given example by a set of given models. Each leaf node corresponds to a predictive model. Given a new example, a meta-decision tree indicates the model that appears most suitable in predicting its class label.

36.3.4 Inductive Transfer and Learning to Learn

We have mentioned above how learning is not an isolated task that starts from scratch on every new task. As experience accumulates, a learning mechanism is expected to perform increasingly better. One approach to simulate the accumulation of experience is by transferring meta-knowledge across domains or tasks; a process known as inductive transfer (Pratt *et al.*, 1991). The goal here is not to match meta-features with a meta-knowledge base (Figure 36.2), but simply to incorporate the meta-knowledge into the new learning task.

A review of how neural networks can learn from related tasks is provided by (Pratt *et al.*, 1991). Caruana (1997) shows the reasons explaining why learning works well in the context of neural networks using backpropagation. In essence, training with many domains in parallel on a single neural network induces information that accumulates in the training signals; a new domain can then benefit from such past experience. Thrun (1998) proposes a learning algorithm that groups similar tasks into clusters. A new task is assigned to the most related cluster; inductive transfer takes place when generalization exploits information about the selected cluster.

A Theoretical Framework of Learning-to-Learn

Several studies have provided a theoretical analysis of the learning-to-learn paradigm within a Bayesian view (Baxter, 1998), and within a Probably Approximately Correct or PAC view (Baxter, 2000). In the PAC view, meta-learning takes place because the learner is not only looking for the right hypothesis in a hypothesis space, but in addition is searching for the right hypothesis space in a family of hypothesis spaces. Both the VC dimension and the size of the family of hypothesis spaces can be used to derive bounds on the number of tasks, and the number of examples on each task, required to ensure with high probability that we will find a solution having low error on new training tasks.

36.3.5 Dynamic-Bias Selection

A field related to the idea of learning-to-learn is that of dynamic-bias selection. This can be understood as the search for the right hypothesis space or concept representation as the learning system encounters new tasks. The idea, however, departs slightly from our architecture; meta-learning is not divided into two modes (i.e., knowledge-acquisition and advisory), but rather occurs on a single step. In essence, the performance of a base learner (Figure 36.1E) can trigger the need to explore additional hypothesis spaces, normally through small variations of the current hypothesis space.

As an example, DesJardins and Gordon (1995) develop a framework for the study of dynamic bias as a search in different tiers. Whereas the first tier refers to a search over a hypothesis space, additional tiers search over families of hypothesis spaces. Other approaches to dynamic-bias selection are based on changing the representation of the feature space by adding or removing features (Utgoff, 1986,Gordon, 1989,Gordon, 1990). Alternatively, Baltes (1992) describes a framework for dynamic selection of bias as a case-based meta-learning system; concepts displaying some similarity to the target concept are retrieved from memory and used to define the hypothesis space.

A slightly different approach is to look at dynamic-bias selection as a form of data variation, but as a time-dependent feature (Widmer, 1996A, Widmer, 1996B, Widmer, 1997). The idea is to perform online detection of concept drift with a single base-level classifier. The meta-learning task consists of identifying contextual clues, which are used to make the base-level classifier more selective with respect to training instances for prediction. Features that are characteristic of a specific context are identified and contextual features are used to focus on relevant examples (i.e., only those instances that match the context of the incoming training example are used as a basis for prediction).

36.4 Tools and Applications

36.4.1 METAL DM Assistant

The METAL DM Assistant (DMA) is the result of an ambitious European Research and Development project broadly aimed at the development of methods and tools for providing support to users of machine learning and Data Mining technology. DMA is a web-enabled prototype assistant system that supports users with model selection and model combination. The project has as its main goal improving the utility of Data Mining tools and in particular to provide significant savings in experimentation time.

DMA follows a ranking strategy as the basis for its advice in model selection (Section 36.3.1). Instead of delivering a single model candidate, the software assistant produces an ordered list of models, sorted from best to worst, based on a weighted combination of parameters such as accuracy and training time. The task characterisation is based on statistical and information-theoretic measures (Section 36.3.1). DMA incorporates more than one ranking method. One of them exploits a ratio of accuracies and times (Brazdil and Soares, 2003). Another, referred to as DCRanker (Keller *et al.*, 1999), is based on a technique known as Data Envelopment Analysis (Andersen and Petersen, 1993, Paterson, 2000).

DMA is the result of a long and consistent effort in providing a practical and effective tool to users in need for assistance in model selection and guidance (Metal, 1998). In addition to a large number of controlled experiments on synthetic datasets and real-world datasets, DMA has been instrumental as a decision support tool within DaimlerChrysler and in the field of Computer-Aided Engineering Design (Keller *et al.*, 2000).

36.5 Future Directions and Conclusions

One important research direction in meta-learning consists of searching for alternative meta-features in the characterization of datasets (Section 36.3.1). A proper characterization of datasets can elucidate the interaction between the learning mechanism and the task under analysis. Current work has only started to unveil relevant meta-features; clearly much work lies ahead. For example, many statistical and information-theoretic measures adopt a global view of the example distribution under analysis; meta-features are obtained by averaging results over the entire training set, implicitly smoothing the actual example distribution (e.g., class-conditional entropy is estimated by projecting all training examples over a single feature dimension.). There is a need for alternative -more detailed- descriptors of the example distribution in a form that can be related to learning performance.

Another interesting path for future work is to understand the difference between the nature of the meta-learner and that of the base-learners. In particular, our general architecture assumes a meta-learner (i.e., a high-level generalization method) performing a form of model selection, mapping a training set into a learning strategy (Figure 36.2). Commonly we look at the problem as a learning problem itself where a meta-learner is invoked to output an approximating function mapping meta-features to learning strategies (e.g., learning model). This opens many questions, such as how can we improve the meta-learner which can now be regarded as a base learner? (Schmidhuber, 1995, Vilalta, 2001). Future research should investigate how the nature of the meta-learner can differ from the base-learners to improve the learning performance as we extract knowledge across domains or tasks.

We conclude this chapter by emphasizing the important role of meta-learning as an assistant tool in the tasks of model selection and combination (Section 65.3.4). Classification and regression tasks are common in daily business practice across a number of sectors. Hence, any form of decision support offered by a meta-learning assistant has the potential of bearing a strong impact for Data Mining practitioners. In particular, since prior expert knowledge is often expensive, not always readily available, and subject to bias and personal preferences, meta-learning can serve as a promising complement to this form of advice through the automatic accumulation of experience based on the performance of multiple applications of a learning system.

References

Aha D. W. Generalizing from Case Studies: A Case Study. Proceedings of the Ninth International Workshop on Machine Learning; 1-10, Morgan Kaufman, 1992.

Ali K., Pazzani M. J. Error Reduction Through Learning Model Descriptions. Machine Learning, 24, 173-202, 1996.

Andersen, P., Petersen, N.C. A Procedure for Ranking Efficient Units in Data Envelopment Analysis. *Management Science*, **39**(10):1261-1264, 1993.

Baltes J. Case-Based Meta Learning: Sustained Learning Supported by a Dynamically Biased Version Space. Proceedings of the Machine Learning Workshop on Biases in Inductive Learning, 1992.

Baxter, J. Theoretical Models of Learning to Learn. In Learning to Learn, Chapter 4, 71-94, MA: Kluwer Academic Publishers, 1998.

Baxter, J. A Model of Inductive Learning Bias. Journal of Artificial Intelligence Research, 12: 149-198, 2000.

Bensusan, H. God Doesn't Always Shave with Occam's Razor – Learning When and How to Prune. In Proceedings of the Tenth European Conference on Machine Learning, 1998.

Bensusan, H., Giraud-Carrier, C. Discovering Task Neighbourhoods Through Landmark Learning Performances. In Proceedings of the Fourth European Conference on Principles and Practice of Knowledge Discovery in Databases, 2000.

Bensusan H., Giraud-Carrier C., Kennedy C. J. A Higher-Order Approach to Meta-Learning. Eleventh European Conference on Machine Learning, Workshop on Meta-Learning: Building Automatic Advice Strategies for Model Selection and Method Combination, Barcelona, Spain. 2000.

Berrer, H., Paterson, I., Keller, J. Evaluation of Machine-learning Algorithm Ranking Advisors. In Proceedings of the PKDD-2000 Workshop on Data-Mining, Decision Support, Meta-Learning and ILP: Forum for Practical Problem Presentation and Prospective Solutions, 2000.

Brazdil P. Data Transformation and Model Selection by Experimentation and Meta-Learning. Proceedings of the ECML-98 Workshop on Upgrading Learning to Meta-Level: Model Selection and Data Transformation, 11-17, Technical University of Chemnitz, 1998.

Brazdil, P., Soares, C. A Comparison of Ranking Methods for Classification Algorithm Selection. In Proceedings of the Twelfth European Conference on Machine Learning, 2000.

Brazdil, P., Soares, C., Pinto da Costa, J. Ranking Learning Algorithms: Using IBL and Meta-Learning on Accuracy and Time Results. Machine Learning, 50(3): 251-277, 2003.

Breiman, L. Stacked Regressions. *Machine Learning, 24:49-64,* 1996.

Brodley, C. Addressing the Selective Superiority Problem: Automatic Algorithm/Model Class Selection. Proceedings of the Tenth International Conference on Machine Learning, 17-24, San Mateo, CA, Morgan Kaufman, 1993.

Brodley, C. Recursive Automatic Bias Selection for Classifier Construction. Machine Learning, 20, 1994.

Brodley C., Lane T. Creating and Exploiting Coverage and Diversity. Proceedings of the AAAI-96 Workshop on Integrating Multiple Learned Models, 8-14, Portland, Oregon, 1996.

Caruana, R. Multitask Learning. Second Special Issue on Inductive Transfer. Machine Learning, 28: 41-75, 1997.

Chan P., Stolfo S. Experiments on Multistrategy Learning by Meta-Learning. Proceedings of the International Conference on Information Knowledge Management, 314-323, 1993.

Chan, P., Stolfo, S. On the Accuracy of Meta-Learning for Scalable Data Mining. Journal of Intelligent Information Systems, 8:3-28, 1996.

Chan P., Stolfo S. On the Accuracy of Meta-Learning for Scalable Data Mining. Journal of Intelligent Integration of Information, Ed. L. Kerschberg, 1998.

DesJardins M., Gordon D. F. Evaluation and Selection of Biases in Machine Learning. Machine Learning, 20, 5-22, 1995.

Dzeroski, Z. Is Combining Classifiers Better than Selecting the Best One? Proceedings of the Nineteenth International Conference on Machine Learning, pp 123-130, San Francisco, CA, Morgan Kaufmann, 2002.

Engels, R., Theusinger, C. Using a Data Metric for Offering Preprocessing Advice in Data-mining Applications. In Proceedings of the Thirteenth European Conference on Artificial Intelligence, 1998.

Freund, Y., Schapire, R. E. Experiments with a New Boosting Algorithm. In Proceedings of the 13th International Conference on Machine Learning, 148-156, Morgan Kaufmann, 1996.

Friedman, J., Hastie, T., Tibshirani, R. Additive Logistic Regression: A Statistical View of Boosting. Annals of Statistics 28: 337-387, 2000.

Fürnkranz, J., Petrak J. An Evaluation of Landmarking Variants, in C. Giraud-Carrier, N. Lavrac, Steve Moyle, and B. Kavsek, editors, Working Notes of the ECML/PKDD 2000 Workshop on Integrating Aspects of Data Mining, Decision Support and Meta-Learning, 2001.

Gama, J., Brazdil, P. A Characterization of Classification Algorithms. Proceedings of the Seventh Portuguese Conference on Artificial Intelligence, EPIA, 189-200, Funchal, Madeira Island, Portugal, 1995.

Gama, J., Brazdil P. Cascade Generalization, Machine Learning, 41(3), Kluwer, 2000.

Giraud-Carrier, C. Beyond Predictive Accuracy: What? Proceedings of the ECML-98 Workshop on Upgrading Learning to Meta-Level: Model Selection and Data Transformation, 78-85, Technical University of Chemnitz, 1998.

Giraud-Carrier, C., Vilalta, R., Brazdil, P. Introduction to the Special Issue on Meta-Learning. Machine Learning, 54: 187-193, 2004.

Gordon D. Perlis D. Explicitly Biased Generalization. Computational Intelligence, 5, 67-81, 1989.

Gordon D. F. Active Bias Adjustment for Incremental, Supervised Concept Learning. PhD Thesis, University of Maryland, 1990.

Hastie, T., Tibshirani, R., Friedman, J. The Elements of Statistical Learning: Data Mining, Inference, and Prediction. Springer Series, 2001.

Hilario, M., Kalousis, A. Building Algorithm Profiles for Prior Model Selection in Knowledge Discovery Systems. Engineering Intelligent Systems, 8(2), 2000.

Keller, J., Holzer, I., Silvery, S. Using Data Envelopment Analysis and Cased-based Reasoning Techniques for Knowledge-based Engine-intake Port Design. In Proceedings of the Twelfth International Conference on Engineering Design, 1999.

Keller, J., Paterson, I., Berrer, H. An Integrated Concept for Multi-Criteria-Ranking of Data-Mining Algorithms. Eleventh European Conference on Machine Learning, Workshop on Meta-Learning: Building Automatic Advice Strategies for Model Selection and Method Combination, Barcelona, Spain, 2000.

Merz C. Dynamic Learning Bias Selection. Preliminary papers of the Fifth International Workshop on Artificial Intelligence and Statistics, 386-395, Florida, 1995A.

Merz C. Dynamical Selection of Learning Algorithms. Learning from Data: Artificial Intelligence and Statistics, D. Fisher and H. J. Lenz (Eds.), Springer-Verlag, 1995B.

Metal. A Meta-Learning Assistant for Providing User Support in Machine Learning and Data Mining, 1998.

Michie, D., Spiegelhalter, D. J., Taylor, C.C. Machine Learning, Neural and Statistical Classification. England: Ellis Horwood, 1994.

Nakhaeizadeh, G., Schnabel, A. Development of Multi-criteria Metrics for Evaluation of Data-mining Algorithms. In Proceedings of the Third International Conference on Knowledge Discovery and Data-Mining, 1997.

Paterson, I. New Models for Data Envelopment Analysis, Measuring Efficiency with the VRS Frontier. Economics Series No. 84, Institute for Advanced Studies, Vienna, 2000.

Peng, Y., Flach, P., Brazdil, P., Soares, C. Decision Tree-Based Characterization for Meta-Learning. In: ECML/PKDD'02 Workshop on Integration and Collaboration Aspects of Data Mining, Decision Support and Meta-Learning, 111-122. University of Helsinki, 2002.

Pfahringer, B., Bensusan, H., Giraud-Carrier, C. Meta-learning by Landmarking Various Learning Algorithms. In Proceedings of the Seventeenth International Conference on Machine Learning, 2000.

Pratt, L., Thrun, S. Second Special Issue on Inductive Transfer. Machine Learning, 28, 1997.

Pratt S., Jennings B. A Survey of Connectionist Network Reuse Through Transfer. In Learning to Learn, Chapter 2, 19-43, Kluwer Academic Publishers, MA, 1998.

Rokach, L., Averbuch, M., and Maimon, O., Information retrieval system for medical narrative reports. Lecture notes in artificial intelligence, 3055. pp. 217-228, Springer-Verlag (2004).

Schmidhuber J. Discovering Solutions with Low Kolmogorov Complexity and High Generalization Capability. Proceedings of the Twelve International Conference on Machine Learning, 488-49, Morgan Kaufman, 1995.

Skalak, D. Prototype Selection for Composite Nearest Neighbor Classifiers. PhD thesis, University of Massachusetts, Amherst, 1997.

Soares, C., Brazdil, P. Zoomed Ranking: Selection of Classification Algorithms Based on Relevant Performance Information. In Proceedings of the Fourth European Conference on Principles and Practice of Knowledge Discovery in Databases, 2000.

Soares, C., Petrak, J., Brazdil, P. Sampling-Based Relative Landmarks: Systematically Test-Driving Algorithms Before Choosing. Proceedings of the 10th Portuguese Conference on Artificial Intelligence, Springer, 2001.

Sohn, S.Y. Meta Analysis of Classification Algorithms for Pattern Recognition. IEEE Transactions on Pattern Analysis and Machine Intelligence, 21(11): 1137-1144, 1999.

Thrun, S. Lifelong Learning Algorithms. In Learning to Learn, Chapter 8, 181-209, MA: Kluwer Academic Publishers, 1998.

Ting, K. M., Witten I. H. Stacked generalization: When does it work?. In Proceedings of the 15th International Joint Conference on Artificial Intelligence, pp 866-873, Nagoya, Japan, Morgan Kaufmann, 1997.

Todorovski, L., Dzeroski, S. Experiments in Meta-level Learning with ILP. In Proceedings of the Third European Conference on Principles and Practice of Knowledge Discovery in Databases, 1999.

Todorovski, L., Dzeroski, S. Combining Multiple Models with Meta Decision Trees. In Proceedings of the Fourth European Conference on Principles and Practice of Knowledge Discovery in Databases, 2000.

Todorovski, L., Dzeroski, S. Combining Classifiers with Meta Decision Trees. Machine Learning 50 (3), 223-250, 2003.

Utgoff P. Shift of Bias for Inductive Concept Learning. In Michalski, R.S. et al (Ed), Machine Learning: An Artificial Intelligence Approach Vol. II, 107-148, Morgan Kaufman, California, 1986.

Vilalta, R. Research Directions in Meta-Learning: Building Self-Adaptive Learners. International Conference on Artificial Intelligence, Las Vegas, Nevada, 2001.

Vilalta, R., Drissi, Y. A Perspective View and Survey of Meta-Learning. Journal of Artificial Intelligence Review, 18 (2): 77-95, 2002.

Widmer, G. On-line Metalearning in Changing Contexts. MetaL(B) and MetaL(IB). In Proceedings of the Third International Workshop on Multistrategy Learning (MSL-96), 1996A.

Widmer, G. Recognition and Exploitation of Contextual Clues via Incremental Meta-Learning. In Proceedings of the Thirteenth International Conference on Machine Learning (ICML-96), 1996B.

Widmer, G. Tracking Context Changes through Meta-Learning. Machine Learning, **27**(3): 259-286, 1997.

Wolpert D. Stacked Generalization. Neural Networks, 5: 241-259, 1992.

37

Bias vs Variance Decomposition For Regression and Classification

Pierre Geurts

Department of Electrical Engineering and Computer Science, University of Liège, Belgium.
Postdoctoral Researcher, F.N.R.S., Belgium

Summary. In this chapter, the important concepts of bias and variance are introduced. After an intuitive introduction to the bias/variance tradeoff, we discuss the bias/variance decompositions of the mean square error (in the context of regression problems) and of the mean misclassification error (in the context of classification problems). Then, we carry out a small empirical study providing some insight about how the parameters of a learning algorithm influence bias and variance.

Key words: bias, variance, supervised learning, overfitting

37.1 Introduction

The general problem of supervised learning is often formulated as an optimization problem. An error measure is defined that evaluates the quality of a model and the goal of learning is to find, in a family of models (the hypothesis space), a model that minimizes this error estimated on the learning sample (or dataset) S. So, at first sight, if no good enough model is found in this family, it should be sufficient to extend the family or to exchange it for a more powerful one in terms of model flexibility. However, we are often interested in a model that generalizes well on unseen data rather than on a model that perfectly predicts the output for the learning sample cases. And, unfortunately, in practice, good results on the learning set do not necessarily imply good generalization performance on unseen data, especially if the "size" of the hypothesis space is large in comparison to the sample size.

Let us use a simple one-dimensional regression problem to explain intuitively why larger hypothesis spaces do not necessarily lead to better models. In this synthetic problem, learning outputs are generated according to $y = f_b(x)+\varepsilon$, where f_b is represented by the dashed curves in Figure 39.1 and ε is distributed according to a Gaussian $N(0,\sigma)$ distribution. With squared error loss, we will see below that the best possible model for this problem is f_b and its average squared error is σ^2. Let us consider two extreme situations of a bad model structure choice.

- A too simple model: using a linear model $y = w.x + b$ and minimizing squared error on the learning set, we obtain the estimations given in the left part of Figure 39.1 for two

O. Maimon, L. Rokach (eds.), *Data Mining and Knowledge Discovery Handbook*, 2nd ed.,
DOI 10.1007/978-0-387-09823-4_37,

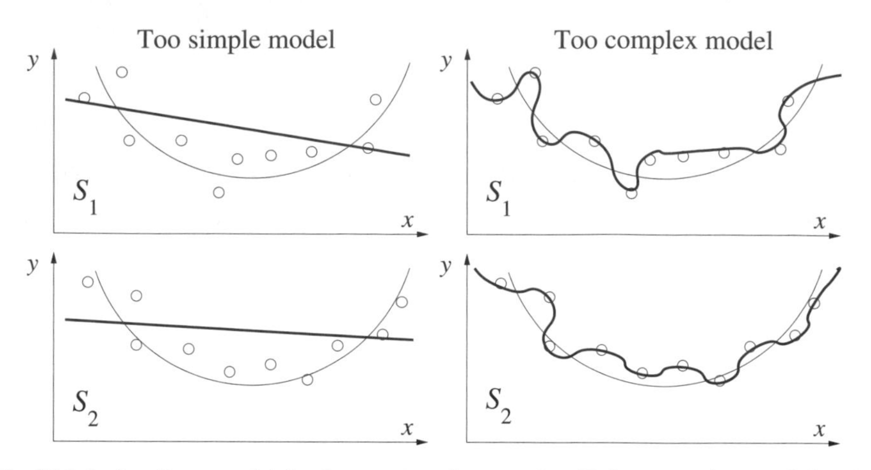

Fig. 37.1. Left, a linear model fitted to two learning samples. Right, a neural network fitted to the same samples

different learning set choices. These models are not very good, neither on their learning sets, nor in generalization. Whatever the learning set, there will always remain an error due to the fact that the model is too simple with respect to the complexity of f_b.

- A too complex model: by using a very complex model like a neural network with two hidden layers of ten neurons each, we get the functions at the right part of Figure 39.1 for the same learning sets. This time, models receive an almost perfect score on the learning set. However, their generalization errors are still not very good because of two phenomena. First, the learning algorithm is able to match perfectly the learning set and hence also the noise term. We say in this case that the learning algorithm "overfits" the data. Second, even if there is no noise, there will still remain some errors due to the high complexity of the model. Indeed, the learning algorithm has many different models at its disposal and if the learning set size is relatively small, several of them will realize a perfect match of the learning set. As at most one of them is a perfect image of the best model, any other choice by the learning algorithm will result in suboptimality.

The main source of error is very different in both cases. In the first case, the error is essentially independent of the particular learning set and must be attributed to the lack of complexity of the model. This source of error is called *bias*. In the second case, on the other hand, the error may be attributed to the variability of the model from one learning set to another (which is due on one hand to overfitting and on the other hand to the sparse nature of the learning set with respect to the complexity of the model). This source of error is called *variance*. Note that in the first case there is also a dependence of the model on the learning set and thus some variability of the predictions. However the resulting variance is negligible with respect to bias. In general, bias and variance both depend on the complexity of the model but in opposite direction and thus there must exist an optimal tradeoff between these two sources of error. As a matter of fact, this optimal tradeoff depends also on the smoothness of the best model and on the sample size. An important consequence of this is that, because of variance, we should always take care of not increasing too much the complexity of the model structure with respect to the complexity of the problem and the size of the learning sample.

In the next section, we give a formal additive decomposition of the mean (over all learning set choices) squared error into two terms which represent the bias and the variance effect.

Some propositions of similar decompositions in the context of 0-1 loss-functions are also discussed. They show some fundamental differences between the two types of problems although bias and variance concepts are always useful. Section 3 discusses procedures to estimate bias and variance terms for practical problems. In Section 4, we give some experiments and applications of bias/variance decompositions.

37.2 Bias/Variance Decompositions

Let us introduce some notations. A learning sample S is a collection of m input/output pairs $(<\mathbf{x}_1, y_1>, \ldots, <\mathbf{x}_m, y_m>)$, each one randomly and independently drawn from a probability distribution $P_D(\mathbf{x}, y)$. A learning algorithm I produces a model $I(S)$ from S, i.e. a function of inputs $\mathbf{x}$ to the domain of y. The error of this model is computed as the expectation:

$$Error(I(S)) = E_{\mathbf{x},y}[L(y, I(S)(\mathbf{x}))],$$

where L is some loss function that measures the discrepancy between its two arguments. Since the learning sample S is randomly drawn from some distribution D, the model $I(S)$ and its prediction $I(S)(\mathbf{x})$ at $\mathbf{x}$ are also random. Hence, $Error(I(S))$ is again a random variable and we are interested in studying the expected value of this error (over the set of all learning sets of size m) $E_S[Error(I(S))]$. This error can be decomposed into:

$$\begin{aligned} E_S[Error(I(S))] &= E_{\mathbf{x}}[E_S[E_{y|\mathbf{x}}[L(y, I(S)(\mathbf{x}))]]] \\ &= E_{\mathbf{x}}[E_S[Error(I(S)(\mathbf{x}))]], \end{aligned}$$

where $Error(I(S)(\mathbf{x}))$ denotes the local error at point $\mathbf{x}$.

Bias/variance decompositions usually try to decompose this error into three terms: the residual or minimal attainable error, the systematic error, and the effect of the variance. The exact decomposition depends on the loss function L. The next two subsections are devoted to the most common loss functions, i.e. the squared loss for regression problems and the 0-1 loss for classification problems. Notice however that these loss functions are not the only plausible loss functions and several authors have studied bias/variance decompositions for other loss functions (Wolpert, 1997, Hansen, 2000). Actually, several of the decompositions for 0-1 loss presented below are derived as special cases of more general bias/variance decompositions (Tibshirani, 1996, Wolpert, 1997, Heskes, 1998, Domingos, 1996, James, 2003). The interested reader may refer to these references for more details.

37.2.1 Bias/Variance Decomposition of the Squared Loss

When the output y is numerical, the usual loss function is the squared loss $L_2(y_1,y_2)=(y_1 - y_2)^2$. With this loss function, it is easy to show that the best possible model is $f_b(\mathbf{x}) = E_{y|\mathbf{x}}[y]$, which takes the expectation of the target y at each point $\mathbf{x}$. The best model according to a given loss function is often called the Bayes model in statistical pattern recognition. Introducing this model in the mean local error, we get with some elementary calculations:

$$E_S[Error(I(S)(\mathbf{x}))] = E_{y|\mathbf{x}}[(y - f_b(\mathbf{x}))^2] + E_S[(f_b(\mathbf{x}) - I(S)(\mathbf{x}))^2]. \tag{37.1}$$

Symmetrically to the Bayes model, let us define the average model, $f_{avg}(\mathbf{x}) = E_S[I(S)(\mathbf{x})]$ which outputs the average prediction among all learning sets. Introducing this model in the second term of Equation (63.1), we obtain:

$$E_S[(f_b(\mathbf{x}) - I(S)(\mathbf{x}))^2] = (f_b(\mathbf{x}) - f_{avg}(\mathbf{x}))^2 + E_S[(I(S)(\mathbf{x}) - f_{avg}(\mathbf{x}))^2].$$

In summary, we have the following well-known decomposition of the mean square error at a point $\mathbf{x}$:

$$E_S\left[Error(I(S)(\mathbf{x}))\right] = \sigma_R^2(\mathbf{x}) + \text{bias}_R^2(\mathbf{x}) + \text{var}_R(\mathbf{x})$$

by defining:

$$\sigma_R^2(\mathbf{x}) = E_{y|\mathbf{x}}\left[(y - f_b(\mathbf{x}))^2\right], \quad (37.2)$$

$$\text{bias}_R^2(\mathbf{x}) = (f_b(\mathbf{x}) - f_{avg}(\mathbf{x}))^2, \quad (37.3)$$

$$\text{var}_R^2(\mathbf{x}) = E_S[(I(S)(\mathbf{x}) - f_{avg}(\mathbf{x}))^2]. \quad (37.4)$$

This error decomposition is well known in estimation theory and has been introduced in the automatic learning community by (Geman *et al.*, 1995).

The residual squared error, $\sigma^2(\mathbf{x})$, is the error obtained by the best possible model. It provides a theoretical lower bound that is independent of the learning algorithm. Thus, the suboptimality of a particular learning algorithm is composed of two terms: the (squared) bias measures the discrepancy between the best and the average model. It measures how well is the estimate in average. The variance measures the variability of the predictions with respect to the learning set randomness.

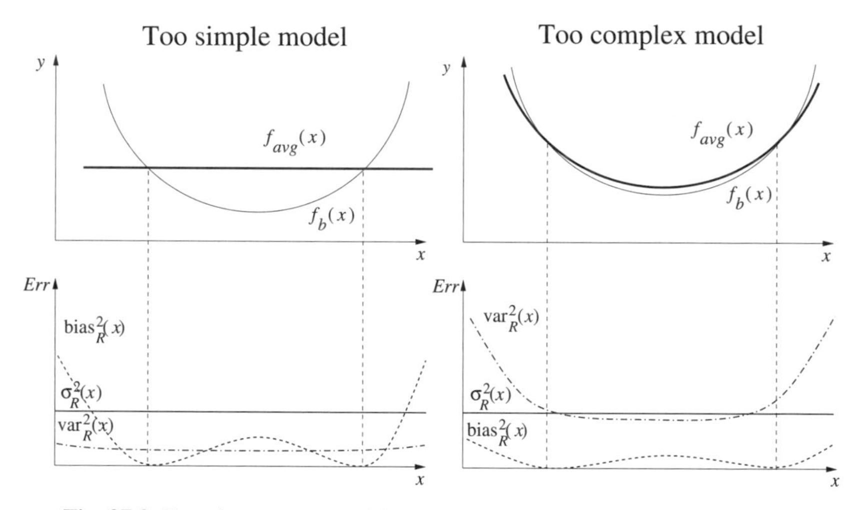

Fig. 37.2. Top: the average models; bottom: residual error, bias, and variance

To explain why these two terms are indeed the consequence of the two phenomena discussed in the introduction of this chapter, let us come back to our simple regression problem. The average model is depicted in the top of Figure 39.2 for the two cases of bad model choice. Residual error, bias and variance for each position x are drawn in the bottom of the same figure. The residual error is entirely specified by the problem and loss criterion and hence independent of the algorithm and learning set used. When the model is too simple, the average model is far from the Bayes model almost everywhere and thus the bias is large. On the other hand, the variance is small as the model does not match very strongly the learning set and thus the prediction at each point does not vary too much from one learning set to another. Bias is thus the dominant term of error. When the model is too complex, the distribution of predictions matches very strongly the distribution of outputs at each point. The average prediction is thus close to the Bayes model and the bias is small. However, because of the noise and the small learning set size, predictions are highly variable at each point. In this case, variance is the dominant term of error.

37.2.2 Bias/variance decompositions of the 0-1 loss

The usual loss function for classification problems (i.e. a discrete target variable) is the 0-1 loss function, $L_c(y_1,y_2)$=1 if $y_1 = y_2$, 0 otherwise, which yields the mean misclassification error at x:

$$\begin{aligned} E_S[Error(I(S)(\mathbf{x}))] &= E_S[E_{y|\mathbf{x}}[L_c(y,I(S)(\mathbf{x}))]] \\ &= P_{D,S}(y \neq I(S)(\mathbf{x})|\mathbf{x}). \end{aligned}$$

The Bayes model in this case is the model that outputs the most probable class at $\mathbf{x}$, i.e. $f_b(\mathbf{x}) = \arg\max_c P_D(y = c\,|\mathbf{x})$. The corresponding residual error is:

$$\sigma_C(\mathbf{x}) = 1 - P_D(y = f_b(\mathbf{x})\,|\mathbf{x})\;. \tag{37.5}$$

By analogy with the decomposition of the square error, it is possible to define what we call "natural" bias and variance terms for the 0-1 loss function. First, by symmetry with the Bayes model and by analogy with the square loss decomposition, the equivalent in classification of the average model is the majority vote classifier defined by:

$$f_{avg}(\mathbf{x}) = \arg\max_c P_S(I(S)(\mathbf{x}) = c),$$

which outputs at each point the class receiving the majority of votes among the distribution of classifiers induced from the distribution of learning sets. The square bias is the error of the average model with respect to the best possible model. This definition yields here:

$$\text{bias}_C(\mathbf{x}) - L_c(f_b(\mathbf{x}), f_{maj}(\mathbf{x})).$$

So, biased points are those for which the majority vote classifier disagrees with the Bayes classifier. On the other hand, variance can be naturally defined as:

$$\text{var}_C(\mathbf{x}) = E_S\left[L_c(I(S)(\mathbf{x}), f_{maj}(\mathbf{x}))\right] = P_S(I(S)(\mathbf{x}) \neq f_{maj}(\mathbf{x})),$$

which is the average error of the models induced from random learning samples S with respect to the majority vote classifier. This definition is indeed a measure of the variability of the predictions at $\mathbf{x}$: when $\text{var}_C(\mathbf{x}) = 0$, every model outputs the same class whatever the learning set from which it is induced and $\text{var}_C(\mathbf{x})$ is maximal when the probability of the class given by the majority vote classifier is equal to $1/z$ (with z the number of classes), which corresponds to the most uncertain distribution of predictions.

Unfortunately, these natural bias and variance terms do not sum up with the residual error to give the local misclassification error. In other words:

$$E_S\left[Error(I(S)(\mathbf{x}))\right] \neq \sigma_C(\mathbf{x}) + \text{bias}_C(\mathbf{x}) + \text{var}_C(\mathbf{x}).$$

Let us illustrate on a simple example how increased variance may decrease the average classification error in some situations. Let us suppose that we have a 3 classes problem such that the true class probability distribution is given by $(P_D(y=c_1|\mathbf{x}), P_D(y=c_2|\mathbf{x}), P_D(y=c_3|\mathbf{x}))$=(0.7, 0.2, 0.1). The best possible prediction at $\mathbf{x}$ is thus the class c_1 and the corresponding minimal error is 0.3. Let us suppose that we have two learning algorithms I_1 and I_2 and that the distribution of predictions of the models built by these algorithms are given by:

$$(P_S(I_1(S)(\mathbf{x})=c_1), P_S(I_1(S)(\mathbf{x})=c_2), P_S(I_1(S)(\mathbf{x})=c_3)) = (.1,.8,.1)$$
$$(P_S(I_2(S)(\mathbf{x})=c_1), P_S(I_2(S)(\mathbf{x})=c_2), P_S(I_2(S)(\mathbf{x})=c_3)) = (.4,.5,.1)$$

So, we observe that both algorithms produce models that most probably will decide class c_2 (respectively with probability 0.8 and 0.5). Thus, the two methods are biased ($\text{bias}_C(\mathbf{x})=1$). On the other hand, the variances of the two methods are obtained in the following way:

$$\text{var}_C^1(\mathbf{x}) = 1-0.8 = 0.2 \text{ and } \text{var}_C^2(\mathbf{x}) = 1-0.5 = 0.5,$$

and their mean misclassification errors are found to be

$$E_S[Error(I_1(S)(\mathbf{x}))] = 0.76 \text{ and } E_S[Error(I_2(S)(\mathbf{x}))] = 0.61.$$

Thus between these two methods with identical bias, it is the one having the largest variance that has the smallest average error rate.

It is easy to see that this happens here because of the existence of a bias. Indeed, with 0-1 loss, an algorithm that has small variance and high bias is an algorithm that systematically (i.e. whatever the learning sample) produces a wrong answer, whereas an algorithm that has a high bias but also a high variance is only wrong for a majority of learning samples, but not necessarily systematically. So, this latter may be better than the former. In other words, with 0-1 loss, much variance can be beneficial because it can lead the system closer to the Bayes classification.

As a result of this counter-intuitive interaction between bias and variance terms with 0-1 loss, several authors have proposed their own decompositions. We briefly describe below the most representative of them. For a more detailed discussion of these decompositions, see for example (Geurts, 2002) or (James, 2003). In the following sections, we present a very different approach to study bias and variance of 0-1 loss due to Friedman (1997), which relates the mean error to the squared bias and variance terms of the class probability estimates.

Some decompositions

Tibshirani (1996) defines the bias as the difference between the probability of the Bayes class and the probability of the majority vote class:

$$\text{bias}_T(\mathbf{x}) = P_D(y=f_b(\mathbf{x})|\mathbf{x}) - P_D(y=f_{maj}(\mathbf{x})|\mathbf{x}). \quad (37.6)$$

Thus, the sum of this bias and the residual error is actually the misclassification error of the majority vote classifier:

$$\sigma_C(\mathbf{x}) + \text{bias}_T(\mathbf{x}) = 1 - P_D(y=f_{maj}(\mathbf{x})|\mathbf{x}) = Error(f_{maj}(\mathbf{x})).$$

This is exactly the part of the error that would remain if we could completely cancel the variability of the predictions. The variance is then defined as the difference between the mean misclassification error and the error of the majority vote classifier:

$$\mathrm{var}_T(\mathbf{x}) = E_S[Error(I(S)(\mathbf{x}))] - Error(f_{maj}(\mathbf{x})). \tag{37.7}$$

Tibshirani (1996) denotes this variance term the aggregation effect. Indeed, this is the variation of error that results from the aggregation of the predictions over all learning sets. Note that this variance term is not necessarily positive. From different considerations, James (2003) has proposed exactly the same decomposition. To distinguish (63.3) and (63.5) from the natural bias and variance terms, he calls them systematic and variance effect respectively. Dietterich and Kong (1995) have proposed a decomposition that applies only to the noise-free case but that exactly reduces to Tibshirani's decomposition in this latter case.

Domingos (2000) agrees with the natural definition of bias, variance given in the introduction of this section and he combines them into a non-additive expression like:

$$E_S[Error(I(S)(\mathbf{x}))] = b_1(\mathbf{x}).\sigma_C(\mathbf{x}) + \mathrm{bias}_C(\mathbf{x}) + b_2(\mathbf{x}).\mathrm{var}_C(\mathbf{x}),$$

where b_1 and b_2 are two factors that are in fact functions of the true class distribution and of the distribution of predictions.

Kohavi and Wolpert (1996) have proposed a very different decomposition which is closer in spirit to the decomposition of the squared loss. Their decomposition makes use of quadratic functions of the probabilities $P_S(I(S)(\mathbf{x})|\mathbf{x})$ and $P(y|\mathbf{x})$. Heskes (1998) adopts the natural variance term var_C and, ignoring the residual error, defines bias as the difference between the mean misclassification error and his variance. As a consequence, it can happen that his bias is smaller than the residual error. Breiman (1996a, 2000) has successively proposed two decompositions. In the first one, bias and variance are defined globally instead of locally. Bias is the part of the error due to biased points (i.e. such that $\mathrm{bias}_C(\mathbf{x}) = 1$) and variance is defined as the part of the error due to unbiased points.

This multitude of decompositions translates well the complexity of the interaction between bias and variance in classification. Each decomposition has its pros and cons. Notably, we may observe in some case counterintuitive behavior with respect to what would be observed with the classical decomposition of the squared error (e.g. a negative variance). This makes the choice, both in theoretical and empirical studies, of a particular decomposition difficult. Nevertheless, all decompositions have proven to be useful to analyze classification algorithms, each one at least in the context of its introduction.

Bias and variance of class probability estimates

Many classification algorithms work by first computing an estimate $I^c(S)(\mathbf{x})$ of the conditional probability of each class c at $\mathbf{x}$ and then deriving their classification model by:

$$I(S)(\mathbf{x}) = \arg\max_c I^c(S)(\mathbf{x}).$$

Obviously, if these (numerical) estimates have a small variance and bias the corresponding classifier is stable with respect to random variations of the learning set and close to the Bayes classifier. Thus, a complementary approach to study bias and variance of a classification algorithm is to connect, in a quantitative way, the bias and variance terms of these estimates to the mean misclassification error of the resulting classification rule.

Friedman (1997) has done this connection in the particular case of a two-class problem and assuming that the distribution of $I^c(S)(\mathbf{x})$ with respect to S is close to Gaussian. In this case, the mean misclassification error at some point $\mathbf{x}$ may be written (see (Friedman, 1997)):

$$E_S\left[Error(I(S)(\mathbf{x}))\right] = \sigma_C(\mathbf{x}) + \Phi\left(\frac{E_S[I^{c_b}(S)(\mathbf{x})]-0.5}{var_S[I^{c_b}(S)(\mathbf{x})]}\right).(2.P(y=c_b|\mathbf{x})-1)$$

where c_b is the Bayes class at $\mathbf{x}$ and $\Phi(.)$ is the upper tail of the standard normal distribution which is a positive and monotonically decreasing function of its argument and such that $\Phi(+\infty)=0$. The numerator in Φ is called the "boundary bias" and the denominator is exactly the variance (37.4) of the regression model $I^{c_b}(S)(\mathbf{x})$. There are two possible situations depending on the sign of the boundary bias:

- When the average probability estimates of the Bayes prediction is greater than 0.5 (a majority of models are right), a decrease of the variance of these estimates will decrease the error.
- On the other hand, when the average probability estimates is lower than 0.5 (a majority of models are wrong), a decrease of variance will yield an increase of the error.

Hence, the conclusions are similar to what we found in our illustrative problem above: *in classification, more variance is beneficial for biased points and detrimental for unbiased ones*.

Another important conclusion can be drawn from this decomposition: whatever the regression bias on the approximation of $I^c(S)(\mathbf{x})$, the classification error can be driven to its minimum value by reducing solely the variance, under the assumption that $E_S[I^{c_b}(S)(\mathbf{x})]$ remains greater than 0.5. This means that perfect classification rules can be induced from very bad (rather biased, but of small variance) probability estimators. In this sense, we can say that reducing variance is more important than reducing bias in classification.

This decomposition is certainly complementary to the decompositions of the previous section. One of its main advantages over these decompositions is that the behavior of bias and variance of probability estimates is predictable in the usual (squared loss) way, while some of the bias and variance terms introduced above are less interpretable.

37.3 Estimation of Bias and Variance

Bias and variance are useful tools to understand the behavior of learning algorithms. So, it is very desirable to be able to compute them in practice for a given learning algorithm and problem. However, bias and variance definitions make intensive use of the knowledge of the distribution of learning samples and, usually, the only knowledge we have about this distribution is a data set of randomly drawn samples. So, in practice, true values of bias and variance terms have to be exchanged for some estimates. One way to obtain these estimates, assuming that we have enough data, is to split the available dataset into two disjoint parts, *PS* (*P* for "pool") and *TS* (*T* for "test") and use them in the following way:

- *PS* is used to approximate the learning set generation mechanism. A good candidate for this is to replace sampling from $P_D(\mathbf{x}, y)$ by sampling with replacement from *PS*. This is called "bootstrap" sampling in the statistical literature (Efron and Tibshirani, 1993) and the idea behind it is to use the empirical distribution of the finite sample as an approximation of the true distribution. The bigger is *PS* with respect to the learning set size, the better will be the approximation. For example, denoting *PS* by $(< \mathbf{x}_1, y_1 >, \ldots, < \mathbf{x}_{Mp}, y_{Mp} >)$, we can estimate the average regression model by the following procedure: (i) draw T learning sets of size m (with $m \leq M_p$) with replacement from *PS*, $(S_1, ..., S_T)$, (ii) build a model from each S_i using the inducer I, and (iii) compute:

$$f_{avg}(\mathbf{x}) \cong \frac{1}{T} \sum_{i=1}^{T} I(S_i)(\mathbf{x}). \tag{37.8}$$

- The set *TS*, which is independent of *PS*, is used as a test sample to estimate errors and bias and variance terms. For example, denoting by $(< \mathbf{x}_1, y_1 >, \ldots, < \mathbf{x}_{Mv}, y_{Mv} >)$ this set, the mean error of a model is estimated by:

$$Error(f) = E_{\mathbf{x},y}\left[L(f(\mathbf{x}), y)\right] \cong \frac{1}{M_v} \sum_{i=1}^{M_v} L(f(\mathbf{x}_i), y_i).$$

However, some of the previously defined terms are difficult to estimate without any knowledge of the problem other than a dataset *TS*. Indeed, the estimation of the Bayes model $f_b(\mathbf{x})$ from data is nothing but the final goal of supervised learning. So, the bias and variance terms that make explicit use of these latter will be mostly impossible to estimate for real datasets (from which we do not have any knowledge of the underlying distribution). For example, the regression noise and bias terms depend on the Bayes model and thus are impossible to estimate separately only from data. A common solution to circumvent this problem is to assume that there is no noise, i.e. $f_b(\mathbf{x}_i) = y_i, i = 1, \ldots, M_v$, and to estimate the bias term from *TS* by:

$$E_{\mathbf{x}}\left[\text{bias}_R^2(\mathbf{x})\right] = E_{\mathbf{x}}\left[(f_b(\mathbf{x}) - f_{avg}(\mathbf{x}))^2\right] \cong \frac{1}{M_v} \sum_{i=1}^{M_v} (y_i - f_{avg}(\mathbf{x}_i))^2, \tag{37.9}$$

using in this latter expression the estimation of the average model given by (63.6). If it happens that actually there is noise, this expression is an estimation of the error of the average model and hence, this amounts at estimating the sum of the residual error and bias terms. The fact that we can not distinguish errors which are due to noise or bias is not very dramatic, since usually we are mainly interested in studying relative variations of bias and variance more than their absolute values and the part of (63.7) which is due to residual error is constant. The regression variance may then be estimated by the following expression:

$$E_{\mathbf{x}}[\text{var}_R(\mathbf{x})] \cong \frac{1}{M_v}\sum_{i=1}^{M_v}\frac{1}{T}\sum_{j=1}^{T}(I(S_j)(\mathbf{x}_i) - f_{avg}(\mathbf{x}_i))^2,$$

or equivalently by the difference between the mean error and the sum of noise and bias as estimated by (63.7).

Of course, this estimation procedure also works for estimating the different bias and variance terms for the 0-1 loss function and the discussion about the estimation of the Bayes model still applies.

The preceding procedure may yield very unstable estimators (suffering of a high variance) especially if the available data is not sufficiently large. Several techniques are possible to stabilize the estimations. For example, a simple method is to use several random divisions of the data set into *PS* and *TS* and to average the estimates found for each separation. More complex estimates may be constructed to further reduce the variance of the estimation (see for example (Tibshirani, 1996), (Wolpert, 1997) or (Webb, 2000)).

37.4 Experiments and Applications

In this section, we carry out some experiments to illustrate the interest of a bias/variance analysis. These experiments are restricted to a regression problem but most of the discussion can be directly applied to classification problems as well. The illustrative problem is an artificial problem introduced in (Friedman, 1997), which has 10 input attributes all independently and uniformly distributed in [0,1]. The regression output variable is obtained by $y = f_b(\mathbf{x}) + \varepsilon$ where $f_b(\mathbf{x}) = 10.\sin(\pi x_1 x_2) + 20(x_3 - 0.5)^2 + 10x_4 + 5x_5$ depends only on the first five inputs and ε is a noise term distributed according to a Gaussian distribution of zero mean and unit variance. Bias and variance are estimated using the protocol of the previous section with *PS* and *TS* of respectively 8000 and 2000 cases. The learning set size m is 300 and T=50 models are constructed.

37.4.1 Bias/variance tradeoff

As bias and variance are both positive and contribute directly to the error, they should both be minimized as much as possible. Unfortunately, there is a compromise, called the bias/variance tradeoff, between these two types of error. Indeed, usually, the more you fit your model to the data, the lower is the bias but at the same time the higher the variance since the dependence of the model to the learning sample increases. On the opposite, if you reduce the dependence of the model to the learning sample, usually, you will increase the bias. The goodness of the fit mainly depends on the model complexity (the size of the hypothesis space) but also on the amount of optimization carried out by the machine learning method.

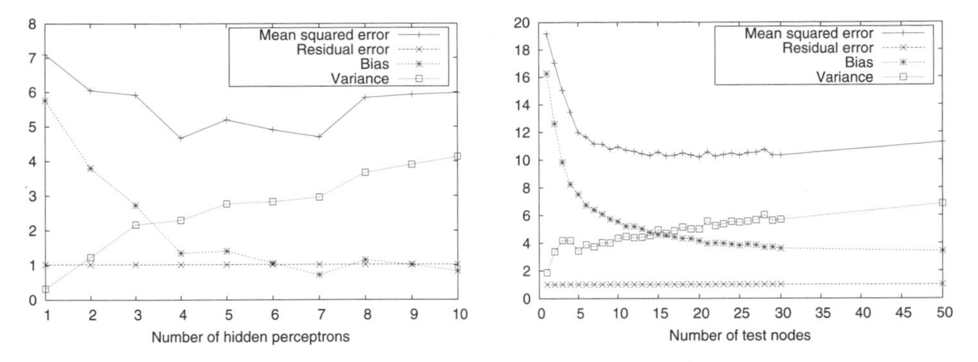

Fig. 37.3. Evolution of bias and variance with respect to the number of perceptrons in the hidden layer of a neural network (top) and the number of tests in a regression tree (bottom)

Figure 39.3 shows on our illustrative problem the evolution of bias and variance, left when we increase the number of hidden neurons in a neural network with one layer (Bishop, 1995) and right when we increase the number of test nodes in a regression tree (Breiman *et al.*, 1984). These curves clearly show the expected evolution of bias and variance. In both cases, the mean squared error goes through a minimum that corresponds to the best tradeoff between bias and variance. In the context of a particular machine learning, there are often many parameters that regulate a different bias/variance tradeoff. It is thus necessary to control these parameters in order to find the optimal tradeoff. There exist many techniques to do this in the context of a given learning algorithm. Examples of such techniques are early stopping or weight decay for neural networks (Bishop, 1995) and pruning in the context of decision/regression trees (Breiman *et al.*, 1984).

37.4.2 Comparison of some learning algorithms

The bias/variance tradeoff is different from one learning algorithm to another. Some algorithms intrinsically present high variance but low bias and other algorithms present high bias but low variance. For example, linear models and Naive Bayes method because of their strong hypothesis often suffer from a high bias. On the other hand, because of their small number of parameters, their variance is small and, on some problems, they may therefore be competitive with more complex algorithms, even if their underlying hypothesis is clearly violated. While the bias of the nearest neighbor method (1-NN), neural networks, and regression/decision trees are generally smaller, the increase in flexibility of their model is paid by an increase of the variance.

Table 54.1 provides a comparison of various algorithms on our illustrative problem. The variance of the linear model is negligible and hence, its error is mainly due

Table 37.1. A comparison of bias/variance decompositions for several algorithms

Method	Mean squared error	Noise	Bias	Variance
Linear regression	7.0	1.0	5.8	0.2
k-NN (*k*=1)	15.4	1.0	4.0	10.4
k-NN (*k*=10)	8.5	1.0	6.2	1.3
MLP (10)	2.0	1.0	0.2	0.8
MLP (10-10)	4.6	1.0	0.4	3.2
Regression tree	10.2	1.0	2.5	6.7
Tree bagging	5.3	1.0	2.8	1.5

to bias. For the k nearest neighbors, the smaller value of k gives a very high variance and also a rather high bias. Increasing k to 10 allows reducing variance significantly, but at the price of bias. All in all, this method is less accurate than linear regression in spite of the fact that it may in principle handle the non-linearity. Multilayer perceptrons provide overall the best results on this problem, having both negligible bias and small variance. The two simulations correspond respectively to one hidden layer with 10 neurons (10) and two hidden layers of 10 neurons each (10-10). The variance of the more complex structure is significantly more important. Finally, (un-pruned) regression trees present a very high variance on this problem. So, although its bias is small, this method is less accurate than linear regression and the 10-NN method.

37.4.3 Ensemble methods: bagging

Bias/variance analyses are especially useful to understand the behavior of ensemble methods. The idea of ensemble methods is to generate by some means a set of models using any learning algorithm and then to aggregate the predictions of these models to yield a more stable final prediction. By doing so, ensemble methods usually change the intrinsic bias/variance tradeoff of the learning algorithm they are applied to. For example, Breiman (1996b) has introduced bagging as a procedure to approximate the average model f_{avg} or the majority vote classifier f_{maj}, which both have zero variance by definition. To this end, Bagging replaces sampling from the distribution $P_D(\mathbf{x},y)$ by bootstrap sampling from the learning sample. Usually, it reduces mainly the variance and only slightly increases the bias (Bauer and Kohavi, 1999). For example, on our illustrative problem, Bagging with 25 bootstrap samples reduces the variance of regression trees to 1.5 and slightly increases their bias to 2.8. All in all, it reduces the mean squared error from 10.2 to 5.3. Another well-known ensemble method is boosting which has been shown to reduce both bias and variance when applied to decision trees (Bauer and Kohavi, 1999).

37.5 Discussion

In automatic learning, bias and variance both contribute to prediction error, whatever the learning problem, algorithm and sample size. The error decomposition allows us to better understand the way an automatic learning algorithm will respond to changing conditions. It allows us also to compare different methods in terms of their weaknesses. This understanding can then be exploited in order to select methods in practice, to study their performances in research, and to guide us in order to find appropriate ways to improve automatic learning methods.

References

Bauer, E., Kohavi, R. An Empirical Comparison of Voting Classification Algorithms: Bagging, Boosting, and Variants. Machine Learning 1999; 36:105-139.

Bishop, C.M. Neural Networks for Pattern Recognition. Oxford: Oxford University Press, 1995.

Breiman, L., Friedman, J.H., Olsen, R.A., Stone, C.J. Classification and Regression Trees. California: Wadsworth International, 1984.

Breiman, L. Bias, Variance, and Arcing Classifiers. Technical Report 460, Statistics Department, University of California Berkeley, 1996.

Breiman, L. Bagging Predictors. Machine Learning 1996; 24(2):123-140.

Breiman, L. Randomizing Outputs to Increase Prediction Accuracy. Machine Learning 2000; 40(3):229-242.

Dietterich, T.G. and Kong, E.B. Machine Learning Bias, Statistical Bias, and Statistical Variance of Decision Tree Algorithms. Technical Report. Department of Computer Science, Oregon State University, 1995.

Domingos, P. An unified Bias-Variance Decomposition for Zero-One and Squa-red Loss. Proceedings of the 17th International Conference on Machine Learning, Morgan Kaufman, San Francisco, CA, 2000.

Efron, B., Tibshirani, R.J. An Introduction to the Bootstrap. Chapman & Hall, 1993.

Freund, Y., Schapire, R.E. A Decision-Theoretic Generalization of Online Learning and an Application to Boosting. Proceedings of the second European Conference on Computational Learning Theory, 1995.

Friedman, J.H. On Bias, Variance, 0/1-Loss, and the Curse-of-Dimensionality. Data Mining and Knowledge Discovery 1997; 1:55-77.

Geman, S., Bienenstock, E. and Doursat, R. Neural Networks and the Bias/Vari-ance Dilemna. Neural computation 1992; 4:1-58.

Geurts, P. Contribution to Decision Tree Induction: Bias/Variance Tradeoff and Time Series Classification. Phd thesis. Department of Electrical Engineering and Computer Science, University of Liège, 2002.

Hansen, J.V. Combining Predictors: Meta Machine Learning Methods and Bias/Variance & Ambiguity Decompositions. PhD thesis. Department of Computer Science, University of Aarhus, 2000.

Heskes, T. Bias/Variance Decompositions for Likelihood-Based Estimators. Neural Computation 1998; 10(6):1425-1433.

James, G.M. Variance and Bias for General Loss Functions. Machine Learning 2003; 51:115-135.

Kohavi, R. and Wolpert, D. H. Bias Plus Variance Decomposition for Zero-One Loss Functions. Proceedings of the 13th International Conference on Machine Learning, Morgan Kaufman, 1996.

Tibshirani, R. Bias, Variance and Prediction Error for Classification Rules. Technical Report, Department of Statistics, University of Toronto, 1996.

Wolpert, D.H. On Bias plus Variance. Neural Computation 1997; 1211-1243.

Webb, G. MultiBoosting: A Technique for Combining Boosting and Wagging. Machine Learning 2000; 40(2):159-196.

38

Mining with Rare Cases

Gary M. Weiss

Department of Computer and Information Science
Fordham University
441 East Fordham Road
Bronx, NY 10458
gweiss@cis.fordham.edu

Summary. Rare cases are often the most interesting cases. For example, in medical diagnosis one is typically interested in identifying relatively rare diseases, such as cancer, rather than more frequently occurring ones, such as the common cold. In this chapter we discuss the role of rare cases in Data Mining. Specific problems associated with mining rare cases are discussed, followed by a description of methods for addressing these problems.

Key words: Rare cases, small disjuncts, inductive bias, sampling

38.1 Introduction

Rare cases are often of special interest. This is especially true in the context of Data Mining, where one often wants to uncover subtle patterns that may be hidden in massive amounts of data. Examples of mining rare cases include learning word pronunciations (Van den Bosch *et al.*, 1997), detecting oil spills from satellite images (Kubat *et al.*, 1998), predicting telecommunication equipment failures (Weiss and Hirsh, 1998) and finding associations between infrequently purchased supermarket items (Liu *et al.*, 1999). Rare cases warrant special attention because they pose significant problems for Data Mining algorithms.

We begin by discussing what is meant by a rare case. Informally, a *case* corresponds to a region in the instance space that is meaningful with respect to the domain under study and a *rare case* is a case that covers a small region of the instance space and covers relatively few training examples. As a concrete example, with respect to the class *bird*, *non-flying bird* is a rare case since very few birds (e.g., ostriches) do not fly. Figure 38.1 shows rare cases and common cases for unlabeled data (Figure 38.1A) and for labeled data (Figure 38.1B). In each situation the regions associated with each case are outlined. Unfortunately, except for artificial domains, the borders for rare and common cases are not known and can only be approximated.

O. Maimon, L. Rokach (eds.), *Data Mining and Knowledge Discovery Handbook*, 2nd ed.,
DOI 10.1007/978-0-387-09823-4_38, © Springer Science+Business Media, LLC 2010

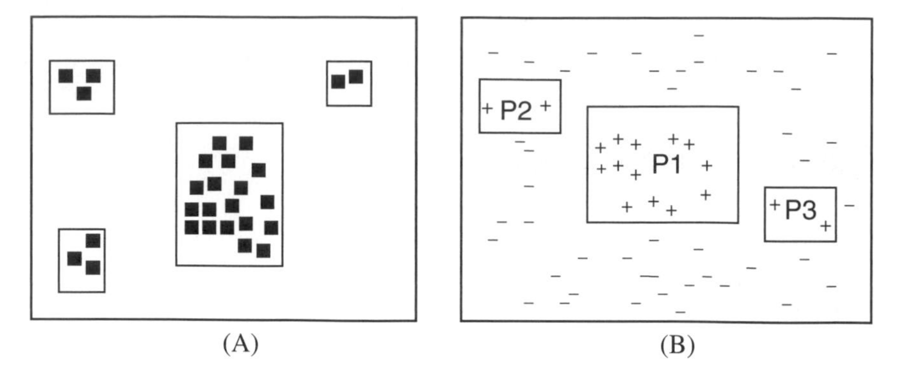

Fig. 38.1. Rare and common cases in unlabeled (A) and labeled (B) data

One important Data Mining task associated with unsupervised learning is *clustering*, which involves the grouping of entities into categories. Based on the data in Figure 38.1A, a clustering algorithm might identify four clusters. In this situation we could say that the algorithm has identified one common case and three rare cases. The three rare cases will be more difficult to detect and generalize from because they contain fewer data points. A second important unsupervised learning task is *association rule mining*, which looks for associations between items (Agarwal *et al.*, 1993). Groupings of items that co-occur frequently, such as *milk* and *cookies*, will be considered common cases, while other associations may be extremely rare. For example, *mop* and *broom* will be a rare association (i.e., case) in the context of supermarket sales, not because the items are unlikely to be purchased together, but because neither item is frequently purchased in a supermarket (Liu *et al.*, 1999).

Figure 38.1B shows a *classification problem* with two classes: a positive class P and a negative class N. The positive class contains one common case, $P1$, and two rare cases, $P2$ and $P3$. For classification tasks the rare cases may manifest themselves as *small disjuncts*. Small disjuncts are those disjuncts in the *learned* classifier that cover few training examples (Holte *et al.*, 1989). If a decision tree learner were to form a leaf node to cover case $P2$, the disjunct (i.e., leaf node) will be a small disjunct because it covers only two training examples. Because rare cases are not easily identified, most research focuses on their learned counterparts–small disjuncts.

Existing research indicates that rare cases and small disjuncts pose difficulties for Data Mining. Experiments using artificial domains show that rare cases have a much higher misclassification rate than common cases (Weiss, 1995, Japkowicz, 2001), a problem we refer to as the problem with rare cases. A large number of studies demonstrate a similar problem with small disjuncts. These studies show that small disjuncts consistently have a much higher error rate than large disjuncts (Ali and Pazzani, 1995, Weiss, 1995, Holte *et al.*, 1989, Ting, 1994, Weiss and Hirsh, 2000). Most of these studies also show that small disjuncts collectively cover a substantial fraction of all examples and cannot simply be eliminated—doing so will substantially degrade the performance of a classifier. The most thorough empirical study of small disjuncts showed that, in the classifiers induced from thirty real-world data sets, most errors are contributed by the smaller disjuncts (Weiss and Hirsh, 2000).

One important question to consider is whether the rarity of a case should be determined with respect to some absolute threshold number of training examples ("absolute rarity") or with respect to the relative frequency of occurrence in the underlying distribution of data ("relative rarity"). If we use absolute rarity, then if a rare case covers only three examples from a training set, then it should be considered rare. However, if additional training data are obtained so that the training set increases by factor of 100, so that this case now covers 300 examples, then absolute rarity says this case is no longer a rare case. However, if the case covers only 1% of the training data in both situations, then relative rarity would say it is rare in both situations. From a practical perspective we are concerned with both absolute and relative rarity since, as we shall see, both forms of rarity pose problems for virtually all Data Mining systems.

This chapter focuses on rare cases. In the remainder of this chapter we discuss problems associated with mining rare cases and techniques to address these problems. Rare *classes* pose similar problems to those posed by rare cases and for this reason we comment on the connection between the two at the end of this chapter.

38.2 Why Rare Cases are Problematic

Rare cases pose difficulties for Data Mining systems for a variety of reasons. The most obvious and fundamental problem is the associated lack of data—rare cases tend to cover only a few training examples (i.e., absolute rarity). This lack of data makes it difficult to detect rare cases and, even if the rare case is detected, makes generalization difficult since it is hard to identify regularities from only a few data points.

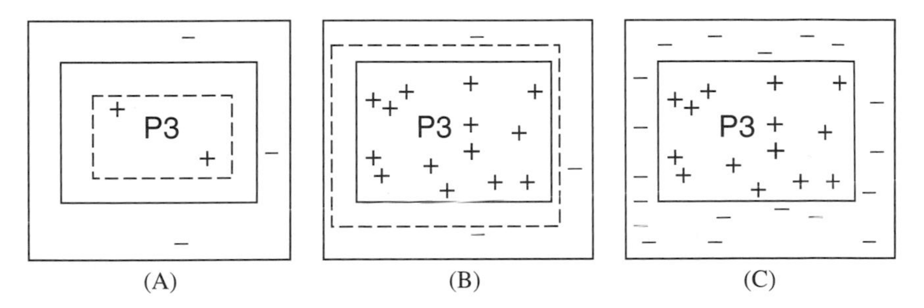

Fig. 38.2. The problem with absolute rarity

The learned decision boundaries are displayed in Figure 38.2A and Figure 38.2B using dashed lines. The learned boundary in Figure 38.2A is far off from the "true" boundary and excludes a substantial portion of *P*3. The inclusion of additional positive examples in Figure 38.2B addresses the problem with absolute rarity and causes all of *P*3 to be covered/learned—although some examples not belonging to *P*3 will be mistakenly assigned a positive label. Figure 38.2C, which includes additional positive and negative examples, corrects this last problem (the learned decision boundary nearly overlaps the true boundary and hence is not shown). Figures 38.2B and 38.2C demonstrate that additional data can address the problem with absolute rarity. Of course, in practice it is not always possible to obtain additional training data.

Another problem associated with mining rare cases is reflected by the phrase: *like a needle in a haystack*. The difficulty is not so much due to the needle being small—or there being only one needle—but by the fact that the needle is obscured by a huge number of strands of hay. Similarly, in Data Mining, rare cases may be obscured by common cases (relative rarity). This is especially a problem when Data Mining algorithms rely on greedy search heuristics that examine one variable at a time, since rare cases may depend on the conjunction of many conditions and any single condition in isolation may not provide much guidance. As a specific example of the problem with relative rarity, consider the association rule mining problem described earlier, where we want to be able to detect the association between *mop* and *broom*. Because this association occurs rarely, this association can only be found if the minimum support (*minsup*) threshold, the number of times the association is found in the data, is set very low. However, setting this threshold low would cause a combinatorial explosion because frequently occurring items will be associated with one another in an enormous number of ways (most of which will be random and/or meaningless). This is called the *rare item problem* (Liu *et al.*, 1999).

The metrics used during Data Mining and to evaluate the results of Data Mining can also make it difficult to mine rare cases. For example, because a common case covers more examples than a rare case, classification accuracy will cause classifier induction programs to focus their attention more on common cases than on rare cases. As a consequence, rare cases may be totally ignored. Furthermore, consider the manner in which decision trees are induced. Most decision trees are grown in a top-down manner, where test conditions are repeatedly evaluated and the best one selected. The metrics (e.g., information gain) used to select the best test generally prefer tests that result in a balanced tree where purity is increased for most of the examples, over a test that yields high purity for a relatively small subset of the data but low purity for the rest (Riddle *et al.*, 1994). Thus, rare cases, which correspond to high purity branches covering few examples will often not be included in the decision tree. The problem is even easier to understand for association rule mining, since rules that do not cover at least *minsup* examples will never be considered.

The bias of a Data Mining system is critical to its performance. This extra-evidentiary bias makes it possible to generalize from specific examples. Unfortunately, the bias used by most data mining systems impacts their ability to mine rare cases. This is because many Data Mining systems, especially those used to induce classifiers, employ a maximum-generality bias (Holte *et al.*, 1989). This means that when a disjunct that covers some set of training examples is formed, only the most general set of conditions that satisfy those examples are selected. This can be contrasted with a maximum-specificity bias, which would add all possible, shared, conditions. The maximum-generality bias will work well for common cases/large disjuncts but does not work well for rare cases/small disjuncts. This leads to the problem with small disjuncts described earlier. Attempts to address the problem of small disjuncts by carefully selecting the bias of the learner are described in Section 38.3.2.

Noisy data may also make it difficult to mine rare cases, since, given a sufficiently high level of background noise, a learner may not be able to distinguish between true exceptions (i.e., rare cases) and noise-induced ones (Weiss, 1995). To see this, consider the rare case, *P*3, in Figure 38.1B. Because *P*3 contains so few training examples, if attribute noise causes even a few negative examples to appear within *P*3, this would prevent *P*3 from being learned correctly. However, common cases such as *P*1 are not nearly as susceptible to noise. Unfortunately, there is not much that can be done to minimize the impact of noise on rare cases. Pruning and other overfitting avoidance techniques—as well as inductive biases that foster generalization—can minimize the overall impact of noise, but, because these methods tend to remove both the rare cases and "noise-generated" ones, they do so at the expense of the rare cases.

38.3 Techniques for Handling Rare Cases

A number of techniques are available to address the issues with rare cases described in the previous section. We describe only the most popular techniques.

38.3.1 Obtain Additional Training Data

Obtaining additional training data is the most direct way of addressing the problems associated with mining rare cases. However, if one obtains additional training data from the original distribution, then most of the new data will be associated with the common cases. Nonetheless, because some of the data will be associated with the rare cases, this approach may help with the problem of "absolute rarity". However, this approach does not address the problem of relative rarity at all, since the same proportion of the training data will cover common cases. Only by *selectively* obtaining additional training data for the rare cases can one address the issues with relative rarity (such a sampling scheme would also be quite efficient at dealing with absolute rarity). Japkowicz (2001) applied this non-uniform sampling approach to artificial domains and demonstrated that it can be very beneficial. Unfortunately, since one can only identify rare cases for artificial domains, this approach generally cannot be implemented and has not been used in practice.[1] However, based on the assumption

[1] Because rare *classes* are trivial to identify, it is straightforward to increase the proportion of rare classes in the training data. Thus this approach is routinely used to address the problem with relative rarity for rare classes.

that small disjuncts are the manifestation of the rare cases in the learned classifier, this approach can be approximated by preferentially sampling examples that fall into the small disjuncts of some initial classifier. This approach warrants additional research.

38.3.2 Use a More Appropriate Inductive Bias

Rare cases tend to cause error-prone small disjuncts to be formed in a classifier induced from labeled data (Weiss, 1995). As discussed earlier, the error prone nature of small disjuncts is at least partly due to the bias used by most learners. Simple strategies that eliminate all small disjuncts or use statistical significance testing to prevent small disjuncts from being formed perform poorly (Holte *et al.*, 1989). A number of studies have investigated more sophisticated approaches for adjusting the bias of a learner in order to minimize the problem with small disjuncts.

Holte *et al.* (1989) modified CN2 so that its maximum generality bias is used only for large disjuncts. A maximum specificity bias was then used for small disjuncts. This was shown to improve the performance of the small disjuncts but degrade the performance of the large disjuncts, yielding poorer overall performance. This occurred because the "emancipated" examples–those that would previously have been classified by small disjuncts–were then misclassified at an even higher rate by the large disjuncts. Going on the assumption that this change in bias was too extreme, a selective specificity bias was then evaluated. This yielded further improvements, but not enough to improve overall classification accuracy.

This approach was subsequently refined to ensure that the more specific bias used to induce the small disjuncts does not affect—and therefore cannot degrade—the performance of the large disjuncts. This was accomplished by using different learners for examples that fall into large disjuncts and examples that fall into small disjuncts (Ting, 1994). While the results of this study are encouraging and show that this hybrid approach can improve the accuracy of the small disjuncts, the results were not conclusive. Carvalho & Freitas (2002A, 2002B) use essentially the same approach, except that the set of training examples falling into each individual small disjunct are used to generate a separate classifier.

A final study advocates the use of instance-based learning for domains with many rare cases/small disjuncts, because of the highly specific bias associated with this learning method (Van den Bosch *et al.*, 1997). The authors of this study were mainly interested in learning word pronunciations, which, by their very nature, have "pockets of exceptions" (i.e., rare cases) that cause many small disjuncts to be formed during learning. Results are not provided to demonstrate that instance-based learning outperforms others learning methods in this situation. Instead the authors argue that instance-based learning methods should be used because they store all examples in memory, while other approaches ignore examples when they fall below some utility threshold (e.g., due to pruning).

In summary, several attempts have been made to perform better on rare cases by using a highly specific bias for the induced small disjuncts. These methods have shown only mixed success. We view this approach to addressing rarity to be promising and worthy of future investigation.

38.3.3 Using More Appropriate Metrics

Data Mining can better handle rare cases by using evaluation metrics that, unlike accuracy, do not discount the importance of rare cases. These metrics can then better guide the Data Mining process and better evaluate the results of Data Mining. Precision and recall are metrics from the information retrieval community that have been used to mine rare cases. Given a classification rule R that predicts target class C, the recall of R is the percentage of examples belonging to C that are correctly identified while the precision of R is the percentage of times the rule is correct. Rare cases can be given more prominence by increasing the importance of precision over recall. Timeweaver (Weiss, 1999), a genetic-algorithm based classification system, searches for rare cases by carefully altering the relative importance of precision versus recall. This ensures that a diverse population of classification rules is developed, which leads to rules that perform well with respect to precision, recall, or both. Thus, precise rules that cover rare cases will be generated.

Two-phase rule induction is another approach that utilizes precision and recall. This approach is motivated by the observation that it is very difficult to optimize precision and recall simultaneously—and trying to do so will miss rare cases. PNrule (Joshi, Agarwal and Kumar, 2001) uses two-phase rule induction to focus on each measure separately. In the first phase, if high precision rules cannot be found then lower precision rules are accepted, as long as they have relatively high recall. So, the first phase focuses on recall. In the second phase precision is optimized. This is accomplished by learning to identify false positives within the rules from phase 1. Returning to the needle and haystack analogy, this approach identifies regions likely to contain needles in the first phase and then learns to discard the hay strands within these regions in the second phase. Two-phase rule induction deals with rare cases because the first phase is sensitive to the problem of small disjuncts while the second phase allows the false positives to all be grouped together, making it easier to identify the false positives. Experimental results indicate that PNrule performs competitively with other disjunctive learners on easy problems and is able to maintain its high performance as more complex concepts with many rare cases are introduced—something the other learners cannot do.

38.3.4 Employ Non-Greedy Search Techniques

Most Data Mining algorithms are greedy in the sense that they make locally optimal decisions without regard to what may be best globally. This is done to ensure that the Data Mining algorithms are tractable. However, because rare cases may depend on the conjunction of many conditions and any single condition in isolation may not provide much guidance, such greedy methods are often ineffective when dealing with rare cases. Thus, one approach for handling rare cases is to use more powerful, global, search methods. Genetic algorithms, which operate on a population of candidate solutions rather than a single solution, fit this description and cope well with attribute interactions (Goldberg *et al.*, 1992). For this reason genetic algorithms are being increasingly used for Data Mining (Freitas and Lavington, 1998) and several systems have used genetic algorithms to handle rare cases. In particular, Timeweaver (Weiss, 1999) uses a genetic algorithm to predict very rare events and Carvalho and Freitas (2002A, 2002B) use a genetic algorithm to "discover small disjunct rules".

More conventional learning methods can also be adapted to better handle rare cases. For example, Brute (Riddle, et al., 1994) is a rule-learning algorithm that performs an exhaustive depth-bounded search for accurate conjunctive rules. The goal is to find accurate rules, even if they cover relatively few training examples. Brute performs quite well when compared to other algorithms, although the lengths of the rules needs to be limited to make the algorithm tractable. Brute is capable of locating "nuggets" of information that other algorithms may not be able to find.

Association-rule mining systems generally employ an exhaustive search algorithm (Agarwal *et al.*, 1993). However, while these algorithms are in theory capable of finding rare associations, they become intractable if the minimum level of support, *minsup*, is set small enough to find rare associations. Thus, such algorithms are heuristically inadequate for finding rare associations and suffer from the rare item problem described earlier. This problem has been addressed by modifying the standard Apriori algorithm so that it can handle multiple minimum levels of support (Liu et al., 1999). Using this approach, the user specifies a different minimum support for each item, based on the frequency of the item in the distribution. The minimum support for an association rule is then the lowest *minsup* value amongst the items in the rule. Empirical results indicate that these enhancements permit the modified algorithm to find meaningful associations involving rare items, without producing a huge number of meaningless rules involving common items.

38.3.5 Utilize Knowledge/Human Interaction

Knowledge, while generally useful when Data Mining, is especially useful when rare cases are present. Knowledge can take many forms. For example, an expert's domain knowledge, including knowledge of how variables interact, can be used to generate sophisticated features capable of identifying rare cases (most experts naturally tend to identify features that are useful for predicting rare, but important, cases). Knowledge can also be applied interactively during Data Mining to help identify rare cases. For example, in association rule mining a human expert may indicate which preliminary results are interesting and warrant further mining and which are uninteresting and should not be pursued. At the end of the Data Mining process the expert can also help distinguish between meaningful rare cases and spurious correlations.

38.3.6 Employ Boosting

Boosting algorithms, such as AdaBoost, are iterative algorithms that place different weights on the training distribution each iteration (Schapire, 1999). Following each iteration boosting increases the weights associated with the incorrectly classified examples and decreases the weights associated with the correctly classified examples. This forces the learner to focus more on the incorrectly classified examples in the next iteration. Because rare cases are difficult to predict, it is reasonable to believe that boosting will improve their classification performance. A recent study showed that boosting can help with rarity if the base learner can effectively trade-off precision and recall (Joshi *et al.*, 2002). An algorithm, RareBoost (Joshi, Kumar and Agarwal, 2001), has been developed that modifies the standard boosting weight-update mechanism to improve the performance of rare classes and rare cases.

38.3.7 Place Rare Cases Into Separate Classes

Rare cases complicate classification tasks because different rare cases may have little in common between them, making it difficult to assign the same class value to all of them. One possible solution is to reformulate the original problem so that the rare cases are viewed as separate classes. The general approach is to 1) separate each class into subclasses using clustering (an unsupervised learning technique) and then 2) learn after re-labeling the training examples with the new classes (Japkowicz, 2002). Because multiple clustering experiments were used in step 1, step 2 involves learning multiple models, which are subsequently combined using voting. The performance results from this study are promising, but not conclusive, and additional research is needed.

38.4 Conclusion

This chapter describes various problems associated with mining rare cases and methods for addressing these problems. While a significant amount of research on rare cases is available, much of this work is still in its infancy. That is, there are no well-established, proven, methods for generally handling rare cases. We expect research on this topic to continue, and accelerate, as increasingly more difficult Data Mining tasks are tackled.

This chapter covers rare cases. Rare classes, which result from highly skewed class distributions, share many of the problems associated with rare cases. Furthermore, rare cases and rare classes are connected. First, while rare cases can occur within both rare classes and common classes, we expect rare cases to be more of an issue for rare classes (e.g., rare classes will never have any very common cases). A study by Weiss & Provost (2003) confirms this connection by showing that rare classes tend to have smaller disjuncts than common classes (small disjuncts are assumed to indicate the presence of rare cases). Other research shows that rare cases and rare classes can also be viewed from a common perspective. Japkowicz (2001) views rare classes as a consequence of between-class imbalance and rare cases as a consequence of within-class imbalances. Thus, both forms of rarity are a type of data imbalance. Recent work further demonstrates the similarity between rare cases and rare classes by showing that they introduce the same set of problems and that these problems can be addressed using the same set of techniques (Weiss, 2004). More intriguing still, some research indicates that rare classes per se are not a problem, but rather it is the rare cases within the rare classes that are the fundamental problem (Japkowicz, 2001).

References

Agarwal, R., Imielinski, T., Swami, A. Mining association rules between sets of items in large databases. Proceedings of the 1993 ACM SIGMOD International Conference on Management of Data; 1993.

Ali, K., Pazzani, M. HYDRA-MM: learning multiple descriptions to improve classification accuracy. International Journal of Artificial Intelligence Tools 1995; 4.

Carvalho, D. R., Freitas, A. A. A genetic algorithm for discovering small-disjunct rules in Data Mining. Applied Soft Computing 2002, 2(2):75-88.

Carvalho, D. R., Freitas, A. A. New results for a hybrid decision tree/genetic algorithm for Data Mining. Proceedings of the Fourth International Conference on Recent Advances in Soft Computing; 2002.

Freitas, A. A. Evolutionary computation. In Handbook of Data Mining and Knowledge Discovery; Oxford University Press, 2002.

Goldberg, D. E. Genetic Algorithms in Search, Optimization and Machine Learning. Addison-Wesley, 1989.

Holte, R. C., Acker, L. E., Porter, B. W. Concept learning and the problem of small disjuncts. In Proceedings of the Eleventh International Joint Conference on Artificial Intelligence; 1989.

Japkowicz, N. Concept learning in the presence of between-class and within-class imbalances. Proceedings of the Fourteenth Conference of the Canadian Society for Computational Studies of Intelligence, Springer-Verlag; 2001.

Japkowicz, N. Supervised learning with unsupervised output separation. International Conference on Artificial Intelligence and Soft Computing; 2002.

Japkowicz, N., Stephen, S. The class imbalance problem: a systematic study. Intelligent Data Analysis 2002; 6(5):429-450.

Joshi, M. V., Agarwal, R. C., Kumar, V. Mining needles in a haystack: classifying rare classes via two-phase rule induction. SIGMOD '01 Conference on Management of Data; 2001.

Joshi, M. V., Kumar, V., Agarwal, R. C. Evaluating boosting algorithms to classify rare cases: comparison and improvements. First IEEE International Conference on Data Mining; 2001.

Joshi, M. V., Agarwal, R. C., Kumar, V. Predicting rare classes: can boosting make any weak learner strong? Proceedings of the Eighth ACM SIGKDD International Conference on Knowledge Discovery and Data Mining; 2002.

Kubat, M., Holte, R. C., Matwin, S. Machine learning for the detection of oil spills in satellite radar images. Machine Learning 1998; 30(2):195-215.

Liu, B., Hsu, W., Ma, Y. Mining association rules with multiple minimum supports. Proceedings of the Fifth ACM SIGKDD International Conference on Knowledge Discovery and Data Mining; 1999.

Riddle, P., Segal, R., Etzioni, O. Representation design and brute-force induction in a Boeing manufacturing design. Applied Artificial Intelligence 1994; 8:125-147.

Rokach L. and Maimon O., Data mining for improving the quality of manufacturing: A feature set decomposition approach. Journal of Intelligent Manufacturing 17(3): 285299, 2006.

Schapire, R. E. A brief introduction to boosting. Proceedings of the Sixteenth International Joint Conference on Artificial Intelligence, 1999.

Ting, K. M. The problem of small disjuncts: its remedy in decision trees. Proceeding of the Tenth Canadian Conference on Artificial Intelligence; 1994.

Van den Bosch, A., Weijters, T., Van den Herik, H. J., Daelemans, W. When small disjuncts abound, try lazy learning: A case study. Proceedings of the Seventh Belgian-Dutch Conference on Machine Learning; 1997.

Weiss, G. M. Learning with rare cases and small disjuncts. Proceedings of the Twelfth International Conference on Machine Learning; Morgan Kaufmann, 1995.

Weiss, G. M., Hirsh, H. Learning to predict rare events in event sequences. Proceedings of the Fourth International Conference on Knowledge Discovery and Data Mining; 1998.

Weiss, G. M. Timeweaver: a genetic algorithm for identifying predictive patterns in sequences of events. Proceedings of the Genetic and Evolutionary Computation Conference; Morgan Kaufmann, 1999.

Weiss, G. M., Hirsh, H. A quantitative study of small disjuncts. Proceedings of the Seventeenth National Conference on Artificial Intelligence; AAAI Press, 2000.

Weiss, G. M. Mining with Rarity—Problems and Solutions: A Unifying Framework. SIGKDD Explorations 2004: 6(1):7-19.

39

Data Stream Mining

Mohamed Medhat Gaber, Arkady Zaslavsky, and Shonali Krishnaswamy

Centre for Distributed Systems and Software Engineering
Monash University
{Shonali.Krishnaswamy}@infotech.monash.edu.au
Shonali.Krishnaswamy@infotech.monash.edu.au

39.1 Introduction

Data mining is concerned with the process of computationally extracting hidden knowledge structures represented in models and patterns from large data repositories. It is an interdisciplinary field of study that has its roots in databases, statistics, machine learning, and data visualization. Data mining has emerged as a direct outcome of the *data explosion* that resulted from the success in database and data warehousing technologies over the past two decades (Fayyad, 1997,Fayyad, 1998,Kantardzic, 2003).

The conventional focus of data mining research was on mining resident data stored in large data repositories. The growth of technologies such as wireless sensor networks (Akyildiz *et al.*, 2002) have contributed to the emergence of data streams (Muthukrishnan, 2003). The distinctive characteristic of such data is that it is unbounded in terms of continuity of data generation. This form of data has been termed as data streams to express its flowing nature (Henzinger *et al.*, 1998). Examples of such streams of data and their characteristics are:

- a pair of Landsat 7 and Terra spacecraft generates 350 GB of data per day in NASA Earth Observation System *EOS* (Park and Kargupta, 2002);
- an oil drill transmits data about its current drilling conditions at 1 Mb/Second (Muthukrishnan, 2003);
- NASA satellites generate around 1.5 TB/day (Coughlan, 2004); and
- AT&T collects a total of 100 GB/day of NetFlow data (Coughlan, 2004).

The widespread dissemination and rapid increase of data stream generators coupled with high demand to utilize these streams of data in critical real time data analysis tasks have led to the emerging focus on stream processing (Muthukrishnan, 2003,Babcock *et al.*, 2002,Henzinger *et al.*, 1998). Data stream processing is broadly classified into two main categories according to the type of processing namely:

- *Data stream management*: this represents querying and summarization of data streams for further processing (Babcock *et al.*, 2002,Abadi *et al.*, 2003).

O. Maimon, L. Rokach (eds.), *Data Mining and Knowledge Discovery Handbook*, 2nd ed.,
DOI 10.1007/978-0-387-09823-4_39,

- *Data stream mining*: performing traditional data mining techniques with linear/sublinear time and space complexity (Muthukrishnan, 2003).

Applications of data stream mining can vary from time-critical astronomical and geophysical applications to real-time decision support in business applications. Data stream mining has been used in many applications including:

- Analyzing biosensor measurements around a city for security reasons (Cormode and Muthukrishnan, 2004);
- Analysis of simulation results and on-board sensors in scientific laboratories and spacecrafts has its potential in changing the mission plan or the experimental settings in real time (Burl *et al.*, 1999,Castano *et al.*, 2003,Srivastava and Stroeve, 2003, Tanner *et al.*, 2002);
- Analysis of web logs and web clickstreams (Nasraoui *et al.*, 2003);
- Real-time analysis of data streams generated from stock markets (Kargupta *et al.*, 2002);
- A traveling salesperson performing customer profiling (Grossman, 1998);
- Continuous monitoring and analyzing of status information received for intrusion detection or laboratory experiments (Gaber *et al.*, 2004, Moskovitch *et al.* (2008));
- Analysis of data from sensors in moving vehicles to prevent fatal accidents through early detection by monitoring and analysis of status information (Kargupta, 2004); and
- Performing preliminary mining of data generated in a sensor network (Krishnamachari and Iyengar, 2003, Krishnamachari and Iyengar, 2004).

Table 39.1 shows the major differences between data stream processing and traditional data processing. The objective of this table is to clearly differentiate between traditional stored data processing and stream processing as a step towards focusing on the data mining aspects of data stream processing systems.

Knowledge extraction from data streams has attracted attention in recent years (Gaber *et al.*, 2005). The continuous high-speed generation of data from sensors, web clickstreams, and stock market information has created new challenges for the data mining community (Muthukrishnan, 2003). Thus, data mining researchers have developed linear/sublinear techniques that can produce acceptable approximate mining results. Different statistical and algorithmic methods have been proposed for data stream mining. Sampling and projection cope with the high data rate of data streams. Sampling refers to the process of selecting data records from a data stream according to a bounding statistical measure that guarantees a minimum acceptable accuracy of output. Projection is used for dimensionality reduction using sketching techniques (Muthukrishnan, 2003). Group testing, tree method, robust approximation and exponential histograms have been used as algorithmic techniques for reducing space and time complexity (Babcock *et al.*, 2002, Muthukrishnan, 2003). These techniques address a number of research issues in order to realize effective data stream mining systems. The following highlights of these research issues (Babcock *et al.*, 2002, Gaber *et al.*, 2004, Kargupta *et al.*, 2002) include:

Table 39.1. Streaming and Traditional Processing

Stream Processing	Traditional Processing
Real-time processing.	Offline processing.
Rapid data generation relative to the available computational resources.	Normal or slow data generation relative to the available computational resources.
Storage of data is not feasible.	Storage of data is feasible.
Approximate results are acceptable.	Accurate results are required.
Processing of samples of data is the usual task.	Processing of every data item/record is the usual task.
Storage of aggregated and summarized data only.	Storage of the raw data.
Spatial and temporal contexts are particularly important.	Spatial and temporal contexts are considered for certain classes of applications such as Geographical Information Systems (GIS).
Linear and sublinear computational techniques are widely used.	Techniques with high space and time complexity are used if necessary.

- *Handling the continuous flow of data streams*: the flow of data records in data streams is characterized by its continuity. This feature requires the development of novel management and analysis techniques that can cope with the continuous, rapid flow of data elements.
- *Unbounded memory requirements*: This is due to the continuous flow of data. Many data stream sources require data processing on small computational devices. Sensors and/or handheld devices lack sufficient memory to run traditional data mining techniques which require the results to be resident in memory over the time of data processing.
- *Change detection and modeling of mining results over time*: due to the evolving nature of data streams, some researchers have pointed out that capturing the change of data mining results is more important in this area than the data mining results. Modeling this change is a real challenge given the high speed of streaming data.
- *Minimizing energy consumption in wireless environments*: data stream mining in sensor networks or wireless environments faces the problem of short battery life-time of sensors or mobile devices. Therefore energy efficient design of data mining algorithms will positively impact on power consumption.
- *Transferring data mining results over a wireless network with a limited bandwidth*: wireless environments are characterized by unreliable connections and limited bandwidth. If the number of mobile devices involved in a data mining process is high such as multiple sensors in a sensor network, the process of transferring the results back to a processing site represents a challenging research issue

- *Visualization of data mining results on the small screen of a mobile device*: the user interface of a handheld device to visualize data mining results is a real challenging issue in applications that involve handheld devices such as analyzing stock market data (Kargupta *et al.*, 2002), given that the visualization of data mining results on a large screen is still an issue due to the high dimensionality of data (Fayyad *et al.*, 2001). Novel visualization techniques that are concerned with the size of image should be investigated.
- *Interactive data mining environment to satisfy user requirements*: the user should be able to change the process settings in real-time. The research problem is represented in how the mining technique can use the generated results to be integrated with the new results after the change in the settings.

The rest of the chapter is organized as follows. Section 39.2 reviews the area of data stream clustering. Classification techniques developed for streaming applications are discussed in Section 39.3. Frequent pattern mining for data streams is detailed in Section 39.4. Section 39.5 reviews time series analysis techniques in data streams. Systems and applications that use data stream mining are discussed in Section 39.6. Following the review of different mining techniques and systems, a classification of different approaches used in the area is given in Section 39.7. Related work in data stream management systems is given in Section 39.8. Future directions in the area are briefly discussed in Section 39.9. Finally, the chapter is concluded with a summary in Section 39.10.

39.2 Clustering Techniques

Clustering (Kantardzic, 2003) is the process of grouping similar objects/records. It is the main strategy for unsupervised learning. Clustering is used in data set description and as a preliminary step in some predictive learning tasks for better understanding the data to be analyzed.

Guha et al (Guha *et al.*, 2000, Guha *et al.*, 2003) have studied clustering data streams using K-median technique. The *K-median* clustering technique identifies cluster centers from the records observed in the dataset. The proposed algorithm makes a single pass over the data and uses small space and is classified as a "divide and conquer" algorithm. It requires $O(nk)$ time and $O(n\varepsilon)$ space where k is the number of centers, n is the number of points and $\varepsilon < 1$. The algorithm uses constant-factor approximation. The algorithm has not been implemented nor validated, but the analysis of space and time requirements has been studied analytically. It has been proven that any *K-median* algorithm that achieves a constant-factor approximation can not achieve a better run time than $O(nk)$. The algorithm starts by calculating the size of a sample bounded by the error rate ε. The sample is then clustered into $2k$ (twice the number of the user specified number of centers), and then at a second level, the algorithm continues clustering a number of samples into $2k$. This process is repeated to a number of a statistically calculated number of levels that guarantee the required accuracy of the output. Finally, the algorithm clusters the already calculated approximate $2k$ clusters into the required k clusters.

Babcock et al (Babcock *et al.*, 2003) have used Exponential Histogram *EH* data structure to enhance Guha et al (Guha *et al.*, 2000, Guha *et al.*, 2003) algorithm described above. However they have addressed the problem of merging clusters by maintaining the EH data structure when the two sets of cluster centers are far apart. This algorithm also has been studied analytically to enhance the performance of the previous algorithm.

Charikar et al (Charikar *et al.*, 2003) have proposed a K-median algorithm that improves accuracy of Guha et al (Guha *et al.*, 2000, Guha *et al.*, 2003) algorithm by increasing the number of levels of the previously described "divide and conquer algorithm". This technique also has been studied analytically.

Domingos and Hulten (Domingos and Hulten, 2000, Domingos and Hulten, 2001, Domingos and Hulten, 2001) have proposed a generic method for scaling up machine learning algorithms termed Very Fast Machine Learning (VFML). This method depends on determining an upper bound for the learner's loss as a function in the number of examples (data records) in each step of the algorithm. Hoeffding bound (Hoeffding, 1963) has been used for the development of the VFML techniques. It states that with probability $1-\delta$, the true mean (r) is at least $(r'-\varepsilon)$ where r' is the estimated mean value with the following equation.

$$\varepsilon = \sqrt{\frac{R^2 \ln 1/\delta}{2n}} \tag{39.1}$$

Where R is the range of the estimated number and n is the number of points. This generic method has been applied to an extension of the conventional *K-means* clustering algorithm *VFKM* and decision tree classification *VFDT* techniques. *VFKM* and *VFDT* algorithms have been implemented and tested on synthetic data sets as well as real web data. Unlike K-median, the K-means algorithm computes the cluster centers by using the mean values of the data records assigned to the cluster under examination. *VFKM* uses Hoeffding bound to determine the number of examples needed in each step of K-means algorithm. *VFKM* runs as a sequence of K-means executions with each run uses more data records than the previous one until the calculated statistical Hoeffding bound is satisfied.

Ordonez (Ordonez, 2003) has proposed several improvements to k-means algorithm for clustering of binary data streams. The improved algorithm is an incremental version of the K-means technique. The experimental studies were conducted on real data sets as well as synthetic ones. It has been demonstrated experimentally that the proposed algorithm outperforms the scalable K-means in most of the cases. The proposed algorithm is a one pass algorithm in $O(Tkn)$ complexity, where T is the average transaction size, n is number of transactions and k is number of centers. The use of binary data simplifies the manipulation of categorical data and eliminates the need for data normalization. The main idea behind the proposed algorithm is that it updates the centers and cluster weights after reading a batch of transactions which is equal to the square root of the number of transactions rather than updating them in real-time. Thus, it solves the problem of high data rate by batch processing of the streaming information.

O'Callaghan et al (O'Callaghan *et al.*, 2002) have proposed *STREAM* and *LOCALSEARCH* algorithms for high performance data stream clustering. The STREAM algorithm starts by determining the size of the sample and then applies the *LOCALSEARCH* algorithm if the sample size is larger than a pre-specified statistical bound that guarantees the accuracy of the output. This process is repeated for each data chunk. Finally, the *LOCALSEARCH* algorithm is applied to the cluster centers generated in the previous iterations to find the final clustering model.

Aggarwal et al (Aggarwal *et al.*, 2003) have proposed a framework for clustering data streams termed as *CluStream* algorithm. The proposed technique divides the clustering process into two components. The online component stores summary statistics about the data streams and the offline one performs clustering on the summarized data according to a number of user preferences such as the time frame and the number of clusters. A number of experiments on real datasets have been conducted to prove the accuracy and efficiency of the proposed algorithm. Furthermore, they have proposed an extension to *CluStream* termed as *HPStream* (Aggarwal *et al.*, 2004); a projected clustering for high dimensional data streams. *HPStream* has outperformed *CluStream* in a number of case studies. The main motivation behind the development of *HPStream* is that *CluStream* has not performed effectively with high dimensionality streaming information.

Gaber et al. (Gaber *et al.*, 2005) have developed Lightweight Clustering *LWC*. It is an AOG-based algorithm. The algorithm adjusts a threshold that represents the minimum distance measure between data items in different clusters. This adjustment is done regularly according to a pre-specified time frame. It is done according to the available resources by monitoring the input-output rate. This process is followed by merging clusters when the memory is full.

Gaber and Yu (Gaber and Yu, 2006) have developed *RA-Cluster*. It is an incremental online clustering algorithm that has all the required parameters to enable resource-awareness. Memory adaptation is done through threshold adaptation and outlier and inactive cluster elimination. CPU adaptation is done through randomized assignment. Battery adaptation is done through the change in sampling rate.

39.3 Classification Techniques

Classification methods (Hand *et al.*, 2001, Hastie *et al.*, 2001) have been studied thoroughly as a major category of the data analysis in machine learning, statistical inference (Hand, 1999) and data mining. Classification methods represent the set of supervised learning techniques where a set of dependent variables needs to be predicted based on another set of input attributes (Kantardzic, 2003). There are two main distinctive approaches under the supervised learning category: classification and prediction. Classification is mainly concerned with categorical attributes as dependent variables. Major classification techniques are decision trees, rule induction, nearest neighbor, and artificial neural networks. Prediction is concerned with numerical attributes as its output. Linear regression is the main prediction technique. The classification process is divided into two phases: model building and model testing.

In model building, a learning algorithm runs over a dataset to induce a model that could be used in estimating output attributes. The quality of this estimation is assessed in the model testing phase. The model building is referred to as training as well.

Wang et al. (Wang *et al.*, 2003) have proposed a generic framework for mining concept drifting data streams. The framework is based on the observation that data stream mining algorithms proposed so far have not addressed the concept of drifting in the evolving data. The idea is based on using a set of classification models such as decision trees using C4.5, RIPPER, naive Bayesian and others to vote for the classification output to increase the accuracy of the predicted output. The framework was developed to address three research challenges in data stream classification:

- *Accuracy*: accuracy of the output of the existing classifiers is very sensitive to concept drifts in the incoming high speed evolving data streams.
- *Efficiency*: Building classifiers is a complex computational task and the update of the model due to concept drifts is a complicated computational process that may not be efficient with data streams.
- *Ease of use*: the classifiers should be used easily in the context of data stream applications which is not the case with the existing techniques.

The main motivation behind the framework is to deal with the expiration of old data streams. The idea of using the most recent data streams to build and use the developed classifiers may not be valid for most applications. Although the old data streams can affect the accuracy of the classification model in a negative way, it is still important to keep track of this data in the current model. Wang and his colleagues (Wang *et al.*, 2003) have proposed using weighted classifier ensembles according to the current accuracy of each classifier used in the ensemble. The weight of every classifier is calculated and taken onto account to predict the final output. Experimentally, the framework has outperformed single classifiers. Figure 39.1 shows the proposed ensemble-based classification framework.

Domingos and Hulten (Domingos and Hulten, 2000) have developed Very Fast Decision Trees *VFDT* which is a decision tree learning system based on *Hoeffding* trees. It splits the tree using the current best attribute taking into consideration that the number of examples used satisfies the *Hoeffding* bound. *VFDT* is an extended version of *Hoeffding* tree algorithm that addresses the research issues of data streams. These research issues are:

- *Ties of attributes*: occur when two or more attributes have close values of the splitting criteria such as information gain.
- High speed nature of data streams: represents an inherent feature of data streams.
- *Bounded memory*: the tree can grow till the algorithm runs out of memory.
- *Accuracy of the output*: is an issue in all data stream mining algorithms.

The extension of *Hoeffding* trees in *VFDT* has been done using the following techniques:

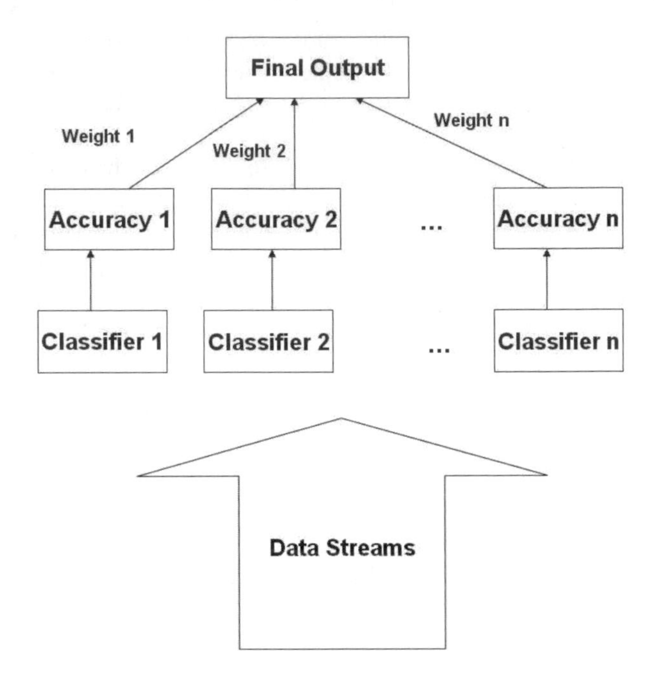

Fig. 39.1. Ensemble-based Classification of Data Streams

- Ties of attributes have been overcome using a user-specified threshold of acceptable error measure for the output. That way the algorithm running time will be reduced and it overcomes the risk of infinite running time of the algorithm.
- The high speed nature of the streaming information has been addressed using batch processing. The computation of the splitting criteria is done in a batch processing rather than online processing. This significantly saves the time of recalculating the criteria for all the attributes with each incoming record of the stream.
- Bounded memory has been addressed by deactivating the least promising leaves and ignoring the poor attributes. The calculation of these poor attributes is done through the difference between the splitting criteria of the highest and lowest attributes. If the difference is greater than a pre-specified value, the attribute with the lowest splitting measure will be freed from memory saving the memory of the data stream computing environment.
- The accuracy of the output has been taken into consideration using multiple scans over the data streams in the case of low data rates, and by using an accurate initialization of the tree using a different more accurate technique to build an initial decision tree.

All of the above improvements have been tested using special synthetic data sets. The experiments have proved efficiency of these improvements. Figure 39.2 shows the VFDT learning system. The VFDT has been extended to address the problem of concept drift in evolving data streams by Hulten et al (Hulten *et al.*, 2001). The new framework has been termed as CVFDT. It is mainly running VFDT over fixed sliding windows in order to have the most updated classifier. The change occurs when the splitting criteria change significantly across the input attributes.

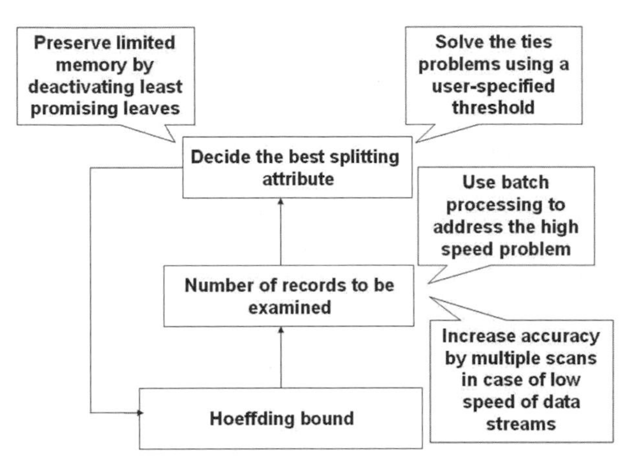

Fig. 39.2. VFDT Learning System

Aggarwal et al (Aggarwal *et al.*, 2004) have adopted the idea of micro-clusters introduced in CluStream (Aggarwal *et al.*, 2003) in On-Demand classification. CluStream as described earlier divides the clustering process into two components: offline and online. The online component stores summarized statistics about the data streams and the offline one performs clustering on the summarized data according to a number of user preferences such as the time frame and the number of clusters.

On-Demand Classification (Aggarwal *et al.*, 2004) uses clustering results to classify data using statistics of class distribution. The main motivation behind the technique is that the classification model should be used over a time period according to the application. The technique uses micro-clustering for each class in the data stream. This initialization is followed by a nearest neighbor classification of the unlabeled data. The micro-clusters have the key of the proposed technique which is the "subtractive property". This property enables the extraction of the needed micro-clusters over the required time period. Figure 39.3 depicts the On-Demand classification technique.

Last (Last, 2002) has proposed an online classification system that can adapt to concept drifting in data streams. The system re-builds the classification model with the most recent examples. According to the error rate as a guide to concept drifting, the frequency of model building and the window size (number of examples used in model building) can change over time.

The *OLIN* system uses info-fuzzy techniques for building a tree-like classification model termed Info-Fuzzy Network *IFN*. The tree is different from conventional decision trees in that each level of the tree represents only one attribute except the

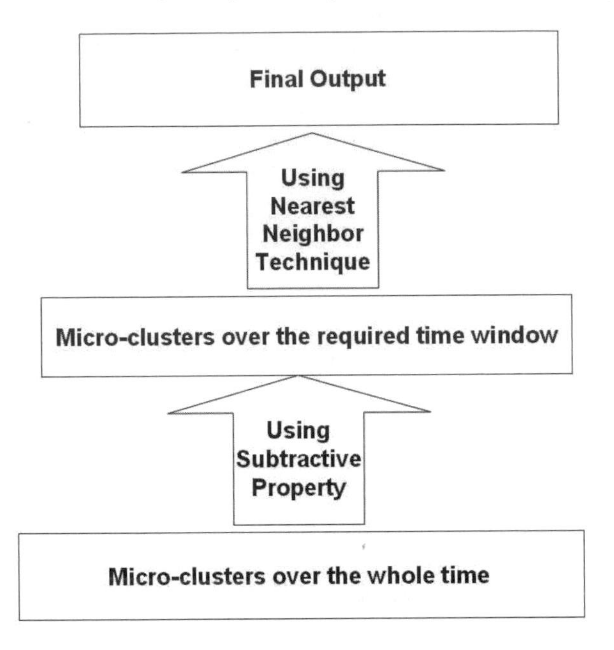

Fig. 39.3. On-Demand Classification

root node level. The nodes represent different values of a sigle attribute. The process of inducing the class label is similar to the one of conventional decision trees. The process of constructing this tree has been termed as Information Network *IN*. *IN* technique uses a similar procedure of building conventional decision trees by determining if the split of an attribute would decrease the entropy or not. The measure used is mutual conditional information that assesses the dependency between the current input attribute under examination and the output attribute. At each iteration, the algorithm chooses the attribute with the maximum mutual information and adds a level with each node representing a different value of this attribute. The iterations stop once there is no increase in the mutual information measure for any of the remaining attributes that have not been considered in the tree.

OLIN system repeatedly uses the IN algorithm for building a new classification model. The system uses the information theory to calculate the window size (refers to number of examples). It uses a less conservative measure than Hoeffding bound used in VFDT (Domingos and Hulten, 2000) reviewed earlier in this section. This measure is derived from the mutual conditional information in the IN algorithm by applying the likelihood ratio test to assess the statistical significance of the mutual information. The main idea behind the system is to change the window size of the model reconstruction according to the classification error rate. The error rate is calculated by measuring the difference between the error rate during the training on one hand and the error rate during the model validation on the other hand. A significant increase in the error rate indicates a high probability of a concept drifting. The window size changes according to the value of this increase. The OLIN is depicted in Figure 39.4.

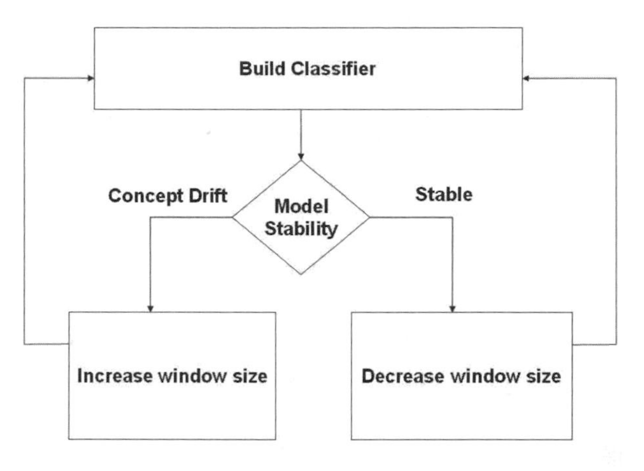

Fig. 39.4. OLIN System

Law et al (Law and Zaniolo, 2005) have proposed an incremental classification algorithm termed Adaptive Nearest Neighbor Classification for Data-streams (ANNCAD). The algorithm uses Haar Wavelets Transformation for multi-resolution data representation. A grid-based representation at each level is used. The process of classification starts with classifying the incoming data record according to the majority nearest neighbors at finer levels. If the finer levels are unable to differentiate between the classes with a pre-specified threshold, the coarser levels are used in a hierarchical way. To address the concept drifting problem of the evolving data streams, an exponential forgetting factor is used to decrease the weight of old data in the classification process. Ensemble classifiers are used to overcome the errors of initial quantization of data. Experimental results over real data sets have the better accuracy compared to the VFDT and CVFDT discussed earlier in this section. The drawback of ANNCAD related to inability of dealing with sudden concept drifting as the exponential forgetting factor takes time to reach the specified accuracy level. Figure 39.5 shows the ANNCAD framework.

Ganti et al (Ganti *et al.*, 2002a) have described an algorithm for model maintenance under insertion and deletion of blocks of data records called *GEMM*. This algorithm can be applied to any incremental data mining model. They have also described a generic framework for change detection between two data sets in terms of the data mining results they induce called FOCUS. The algorithms are not implemented, but are applied analytically to decision tree models and the frequent itemset model. *GEMM* algorithm accepts a class of models and an incremental model maintenance algorithm for the unrestricted window option, and outputs a model maintenance algorithm for both window-independent and window-dependent block selec-

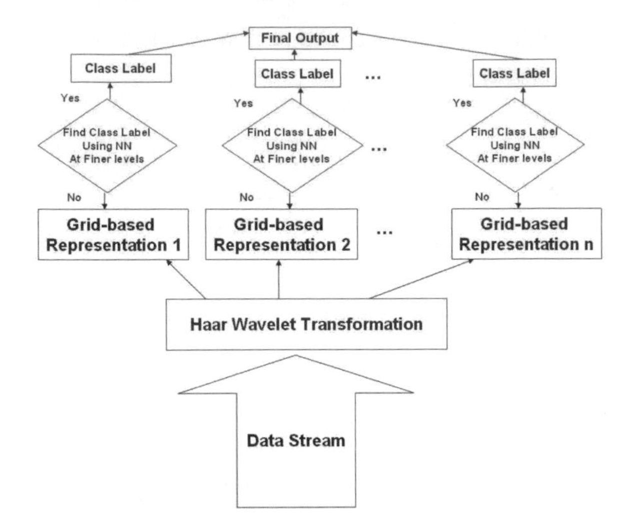

Fig. 39.5. ANNCAD Framework

tion sequence. *FOCUS* framework uses the difference between data mining models as the deviation in data sets.

Ferrer-Troyano et al (Ferrer-Troyano *et al.*,2004) have proposed a scalable classification algorithm for numerical data streams. The algorithm has been termed as Scalable Classification Algorithm by Learning decisiOn Patterns *SCALLOP*. The algorithm starts by reading a number of user-specified labeled records. A number of rules are created for each class from these records. For each record read after creating these rules, there are three cases:
a) Positive covering: a new record that strengthens a current discovered rule.
b) Possible expansion: a new record that is associated with at least one rule however is not covered by any discovered rule.
c) Negative covering: a new record that weakens a current discovered rule.
For each of the above cases, a different procedure is used as follows:
a) Positive covering: an update of the positive support and confidence of the rule is calculated and assigned to the existing rule.
b) Possible expansion: the rule is extended if it satisfies two conditions:
1. It is bounded within a user-specified growth bounds to avoid a possible wrong expansion of the rule.
2. There is no intersection between the expanded rule and any already discovered rule associated with the same class label.
c) Negative covering: an update of the negative support and confidence is calculated. If the confidence is less than a minimum user-specified threshold, a new rule is added.

Having read a user-defined number of records, a rule refining process takes place. Merge of rules in the same class and within a user-defined acceptable distance measure is used in this process with a condition non-intersecting with rules associated with other class labels. The resulting hypercube should also be within the growth bounds of the rules. The second step of the refining stage release the uninteresting rules from the current model. The rules that have less than the minimum positive support are released from the model. Also the rules that are not covered by at least one of the records of the last user-defined number of received records are also released from the classifier. Figure 39.6 shows an illustration of the basic process of using *SCALLOP* to build a data stream classifier.

Finally a voting-based classification technique is used to classify the unlabelled records for model use. If there is a rule covers the current record, the label associated with that rule is used as the classifier output; otherwise a voting over the current rules within the growth bounds is used to infer the class label.

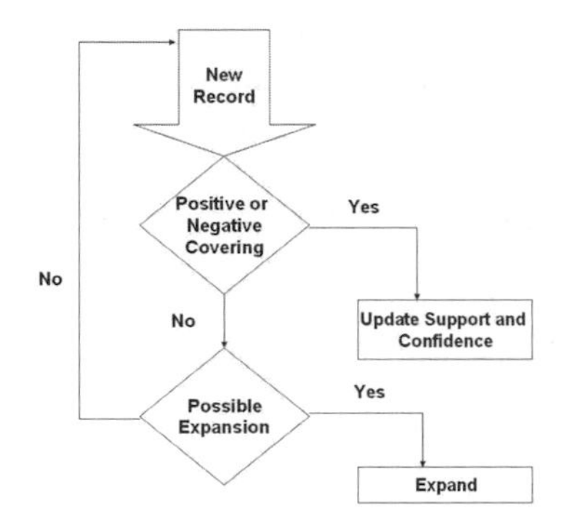

Fig. 39.6. Basic SCALLOP Process

Papadimitriou et al (Papadimitriou *et al.*, 2003) have proposed AWSOM (Arbitrary Window Stream mOdeling Method) for discovering interesting patterns from sensor data. They developed a one-pass algorithm to incrementally update the patterns. Their method requires only $O(logN)$ memory where N is the length of the sequence. They conducted experiments with real and synthetic data sets. They use wavelet coefficients as compact information representation and correlation structure detection, and then apply a linear regression model in the wavelet domain. The system depends on creating compact representation to address the high speed streaming problem. The experimental results show the efficiency in detecting correlation.

Gaber et al. (Gaber *et al.*, 2005) have developed Lightweight Classification LW-Class. It is a variation of LWC. It is also an AOG-based technique. The idea is to use Knearest neighbors with updating the frequency of class occurrence given the data stream features. In case of contradiction between the incoming stream and the stored summary of the cases, the frequency is reduced. In case of the frequency is equalized to zero, all the cases represented by this class is released from the memory.

39.4 Frequent Pattern Mining Techniques

Frequency counting is the process of identifying the highest frequent items. It could be used as a stand alone technique to discover the heavy hitters (Cormode and Muthukrishnan, 2003). It could also be used as a step towards finding association rules. The main idea is to find data items with a probability greater than or equal to a pre-specified minimum threshold known in the context of frequent items as the item support (Dunham, 2003). The item support is calculated by dividing the number of times the observed item appears to the total number of records.

Giannella et al (Giannella *et al.*, 2003) have proposed and implemented a frequent itemsets mining algorithm over data stream. They have used tilted windows to calculate the frequent patterns for the most recent transactions based on the fact that users are more interested in the most recent streaming information rather than older data streams. They have developed an incremental algorithm to maintain the FP-stream, which is a tree data structure to represent and discover frequent itemsets in data streams. FP-stream has been developed based on FP-tree, which has been first introduced by Han et al (Han *et al.*, 2000) as a graphical representation for discovering frequent itemsets. A number of experiments have been conducted to prove the algorithm efficiency. The results show that with limited memory, the algorithm can discover the frequent itemsets with approximate support.

Manku and Motwani (Manku and Motwani, 2002) have proposed and implemented an approximate frequency counting algorithm in data streams. The implemented algorithm uses all the previous historical data to calculate the frequent patterns incrementally. Two algorithms have been introduced: sticky sampling and lossy counting algorithms. Although the first algorithm analytically should have a better performance because it has better worst-case bound, the experimental studies have proved the lossy count algorithm has a better practical performance. The sticky sampling algorithm uses sampling that attracts the new records with already existing entries to have a higher probability to be sampled. The other algorithm uses that idea of group testing using buckets for counting items within the same group by maintaining one counter only.

Cormode and Muthukrishnan (Cormode and Muthukrishnan, 2003) have developed an algorithm for counting frequent items. The algorithm uses group testing to find the hottest k items. The algorithm can process turnstile data stream model which allows addition as well as deletion of data records. An approximation randomized algorithm has been used to approximately discover the most frequent items. The algorithm can recall the frequent items with given item support and probability. It is worth mentioning that the turnstile data stream model is the hardest to analyze. Time series and cash register models are easier. The former does not allow increments and decrements and the later one allows only increments.

Jin et al (Jin *et al.*, 2003) have proposed hCount algorithm to discovering frequent items in data streams. This algorithm also deals with the turnstile data stream model where insertion and deletion from the data are allowed. The algorithm dynamically works with any range of data and does not need any prior knowledge about the data. The algorithm is classified as an approximation technique that keeps the number of counters that can guarantees a minimum acceptable error. The algorithm simply keeps the number of counters that analytically can result in the final approximated output deviated with a user given threshold of error.

Gaber et al. (Gaber *et al.*, 2005) have developed one more AOG-based algorithm: Lightweight frequency counting LWF. It has the ability to find an approximate solution to the most frequent items in the incoming stream using adaptation and releasing the least frequent items regularly in order to count the more frequent ones.

39.5 Time Series Analysis

Time series analysis is concerned with discovering patterns in attribute values that vary over temporal basis. Three main functions are performed in time series mining: clustering of similar time series, predicting future values in a time series, and classifying the behavior of a time series (Dunham, 2003).

Indyk et al (Indyk *et al.*, 2000) have proposed approximate solutions with probabilistic error bounding to two problems in time series analysis: relaxed periods and average trends. The algorithms use dimensionality reduction sketching techniques. The process starts with computing the sketches over an arbitrarily chosen time window. This creates what so called sketch pool. Sketching is the process of random projection over a number of attributes. Using this pool of sketches, relaxed periods and average trends are computed. Relaxed periods refer to those periods in time series that are repeated over time. Since exact repetition is rare, similar ones using distance functions are acceptable. Average trend is the mean values of a subsequence of observation of a pre-specified length in a time series. The algorithms have shown experimentally efficiency in running time and accuracy.

Perlman and Java (Perlman and Java, 2003) have proposed an approach to mine astronomical time series streams. The technique starts with handling missing data using interpolation. A normalization process then takes place for a two-phase preprocessing step. A process of finding frequently occurring shapes in times series using time windows represents the first processing step. Then, clustering the discovered patterns of shapes is the second step. Rule extraction and filtering over the created clusters represent final step in the approach. The limitation of the implemented system is that it can process only one time series at any time. Figure 39.7 shows a simple flow chart of the approach.

Zhu and Shasha (Zhu and Shasha, 2003) have proposed techniques to compute a set of statistical measures over time series data streams. The proposed techniques use discrete Fourier transform to create synopsis data structure. The system is called StatStream and is able to compute approximate error bounded correlations and inner products. The system works over an arbitrarily chosen sliding window.

Keogh et al (Keogh *et al.*, 2003) have proved empirically that most cited clustering time series data streams algorithms proposed so far in the literature result in meaningless results in subsequence clustering. They have proposed a solution using k-motif to choose the subsequences that the algorithm can work on. The 1-motif is the subsequence that has the highest count of not-trivial matches in a time series.

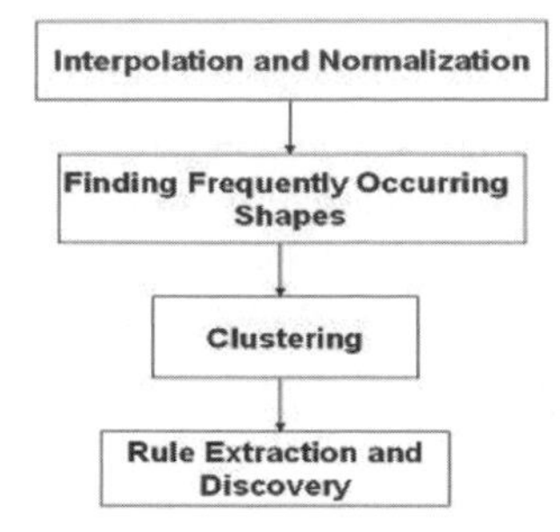

Fig. 39.7. Astronomical Time Series Analysis

Thus, the k-motif is the highest k subsequences that satisfy the condition of highest count of matches. Experimental results show the success of the techniques in extracting meaningful time series clustering results.

Lin et al (Lin *et al.*, 2003) have proposed the use of symbolic representation of time series data streams that has been termed Symbolic Aggregate approXimation (SAX). This representation allows dimensionality/numerosity reduction. Numerosity reduction refers to reducing the number of records. They have demonstrated the applicability of the proposed representation by applying it to clustering, classification, indexing and anomaly detection mining techniques. The approach has two main stages. The first one is the transformation of time series data to Piecewise Aggregate Approximation followed by transforming the output to discrete string symbols in the second stage.

Chen et al (Chen *et al.*, 2002) have proposed the application of what so called regression cubes for data streams. Due to the success of OnLine Analytical Processing *OLAP* technology in the application of static stored data, it has been proposed to use multidimensional regression analysis to create a compact cube that could be used for answering aggregate queries over the incoming data streams. This research has been extended to be adopted in the undergoing project Mining Alarming Incidents in Data Streams MAIDS. The technique has shown experimentally efficiency in analyzing time series data streams.

39.6 Systems and Applications

Recently systems and applications that deal with mining data streams have been developed. The systems are application-oriented except for MAIDS developed by Cai et al (Cai *et al.*, 2004) which represents the first attempt to develop a generic data stream mining system. The following list introduces these systems and applications with short descriptions.

Burl et al (Burl *et al.*, 1999) have developed Diamond Eye for NASA and JPL. The aim of the project is to enable remote systems as well as scientists to extract patterns from spatial objects in real time image streams. The success of this project will enable "a new era of exploration using highly autonomous spacecraft, rovers, and sensors" (Burl *et al.*, 1999). The system uses a high performance computational facility for processing the data mining request. The scientist uses a web interface that uses java applets to connect to the server that requests that images to perform the image mining process.

Kargupta et al (Kargupta *et al.*, 2002) have developed the first ubiquitous data stream mining system termed MobiMine. It is a client/server PDA-based distributed data mining application for financial data streams. The system prototype has been developed using a single data source and multiple mobile clients; however the system is designed to handle multiple data sources. The server functionalities in the proposed system are data collection from different financial web sites and storage, selection of active stocks using common statistics methods, and applying online data mining techniques to the stock data. The client functionalities are portfolio management using a mobile micro-database to store portfolio data and information about user's preferences, and construction of the WatchList and this is the first point of interaction between the client and the server. The server computes the most active stocks in the market, and the client in turn selects a subset of this list to construct the personalized WatchList according to an optimization module. The second point of interaction between the client and the server is that the server performs online data mining and then transforms the results using Fourier transformation and finally sends this to the client. The client in turn visualizes the results on the PDA screen. It is worth pointing out that the data mining process in MobiMine has been performed at the server side given the resource constraints of a mobile device. With the increase need for onboard data mining in resource-constrained computing environments, Kargupta et al (Kargupta, 2004) have developed onboard mining techniques for a different application in mining vehicle sensory data streams.

Kargupta et al (Kargupta, 2004) have developed Vehicle Data Stream Mining System *VEDAS*. It is a ubiquitous data stream mining system that allows continuous monitoring and pattern extraction from data streams generated on-board a moving vehicle. The mining component is located on the PDA. VEDAS uses online incremental clustering for modeling of driving behavior.

Tanner et al (Tanner *et al.*, 2002) have developed EnVironment for On-Board Processing (EVE) for astronomical data streams. The system analyzes data streams continuously generated from measurements of different on-board sensors. Only interesting patterns are sent to the ground stations for further analysis preserving the limited bandwidth.

Srivastava and Stroeve (Srivastava and Stroeve, 2003) work in a NASA project for onboard detection of geophysical processes such as snow, ice and clouds using kernel clustering methods for data compression preserving limited bandwidth needed to send image streams to the ground centers. The kernel methods have been chosen due to its low computational complexity.

Cai et al (Cai *et al.*, 2004) have developed an integrated mining and querying system. The system can classify, cluster, count frequency and query over data streams. Mining Alarming Incidents of Data Streams MAIDS is currently under development and recently the project team has demonstrated its prototype implementation.

Sequential pattern mining and hidden network mining are currently under development.

Pirttikangas et al (Pirttikangas *et al.*, 2001) have implemented a mobile agent-based ubiquitous data mining for a context-aware health club for cyclists. The system is called Genie of the Net. The process starts by collecting information from sensors and databases in order to recognize the needed information for the specific application. This information includes user's context and other needed information collected by mobile agents. The main scenario for the health club system is that the user has a plan for an exercise. All the needed information about the health such as heart rate is recorded during the exercise. This information is analyzed using data mining techniques to advise the user after each exercise.

Having discussed the state-of-the-art in mining data streams in terms of developed techniques as well as systems used in different applications, we can use this review as a base for classifying these techniques into generic categories

39.7 Taxonomy of Data Stream Mining Approaches

Research problems and challenges that have been discussed earlier in mining data streams have its solutions using well-established statistical and computational approaches. We can categorize these solutions to data-based and task-based ones. In data-based solutions, the idea is to examine only a subset of the whole dataset or to transform the data vertically or horizontally to an approximate smaller size data

representation. On the other hand, in task-based solutions, techniques from computational theory have been adopted to achieve time and space efficient solutions. In this section we review these theoretical foundations.

39.7.1 Data-based Techniques

Data-based techniques refer to summarizing the whole dataset or choosing a subset of the incoming stream to be analyzed. Sampling, load shedding and sketching techniques represent the former one. Synopsis data structures and aggregation represent the later one. The following subsections represent an outline of the basics of these techniques with pointers to its applications in the context of data stream mining.

Sampling

Sampling refers to the process of probabilistic choice of a data item to be processed (Toivonen, 1996). Sampling is an old statistical technique that has been used for a long time in the context of conventional data mining for large databases. In the context of data stream mining, boundaries of error rate of the computation are given as a function in the sampling rate or size. Very Fast Machine Learning techniques (Domingos and Hulten, 2000) have used Hoeffding bound (Hoeffding, 1963) to measure the sample size according to a derived loss function according to the running mining algorithm. The problem with using sampling in the context of data stream analysis is the unknown dataset size. Thus the treatment of data stream should follow a special analysis to find the error bounds. Another problem with sampling is that it is important to check for anomalies for surveillance analysis as an application in mining data streams. Sampling is not the right choice for such an application. Sampling also does not address the problem of fluctuating data rates. It would be worth investigating the relationship among the three parameters: data rate, sampling rate and error bounds.

Load Shedding

Load shedding refers (Babcock *et al.*, 2003, Tatbul *et al.*, 2003, Tatbul *et al.*, 2003) to the process of dropping a sequence of data streams. Load shedding has been used successfully in querying data streams. It has the same problems of sampling. Load shedding is difficult to be used with mining algorithms because it drops chunks of data streams that could be used in the structuring of the generated models or it might represent a pattern of interest in time series analysis. However recently it has been used in the classification problem with an acceptable accuracy in an algorithm developed by Chi et al (Chi *et al.*, 2005). The algorithm has been termed as Loadstar. It represents the first attempt for using load shedding in high speed data stream classification problems.

Sketching

Sketching (Babcock *et al.*, 2002, Muthukrishnan, 2003) is the process of randomly project a subset of the features. It is the process of vertically sample the incoming stream. Sketching has been applied in comparing different data streams and in aggregate queries. The major drawback of sketching is that of accuracy. It is hard to use it in the context of data stream mining. Principal Component Analysis (PCA) would be a better solution that has been applied in streaming applications (Kargupta, 2004).

Synopsis Data Structures

Creating synopsis of data refers to the process of applying summarization techniques that are capable of summarizing the incoming stream for further analysis. Wavelet analysis (Gilbert *et al.*, 2003), histograms, quantiles and frequency moments (Babcock *et al.*, 2002) have been proposed as synopsis data structures. Since synopsis of data does not represent all the characteristics of the dataset, approximate answers are produced when using such data structures.

Aggregation

Aggregation is the process of computing statistical measures such as means and variance that summarize the incoming data stream. Using this aggregated data could then be used by the data mining algorithm. The problem with aggregation is that it does not perform well with highly fluctuating data distributions. Merging online aggregation with offline mining has been studies in (Aggarwal *et al.*, 2003, Aggarwal *et al.*, 2004, Aggarwal *et al.*, 2004) for clustering and classification of data streams.

Definitions, advantages and disadvantages of all of the above data-based approaches are given in Table 39.2.

39.7.2 Task-based Techniques

Task-based techniques are those methods that modify existing techniques or develop new ones in order to address the computational challenges of data stream processing. Approximation algorithms, sliding window techniques represent this category. In the following subsections, we examine each of these techniques and its application in the context of data stream analysis.

Approximation algorithms

Approximation algorithms (Muthukrishnan, 2003) have their roots in algorithm design. It is concerned with design algorithms for computationally hard problems. These algorithms can result in an approximate solution with error bounds. The idea is that data stream mining algorithms are considered hard computational problems given its features of continuity and speed and the resource-constrained computational environment. Approximation algorithms have attracted researchers as a direct solution to data stream mining problems. However, the problem of data rates with regard to the available resources could not be solved using approximation algorithms. Other tools should be used along with these algorithms in order to adapt to the available resources. Approximation algorithms have been used in (Cormode and Muthukrishnan, 2003, Jin *et al.*, 2003) for discovering frequent items.

Sliding Window

The inspiration behind sliding window techniques is that the user is more concerned with the analysis of most recent data streams. Thus, the detailed analysis is done over the most recent data items and summarized versions of the old ones. This idea has been adopted in many techniques in the undergoing comprehensive data stream mining system MAIDS (Dong *et al.*, 2003). The main issue of the sliding window techniques is how to remove the expired results from the current created model.

Algorithm Output Granularity

The algorithm output granularity (AOG) (Gaber *et al.*, 2005, Gaber *et al.*, 2004) introduces the first resource-aware data analysis approach that can cope with fluctuating very high data rates according to the available memory and the processing speed represented in time constraints. The AOG performs the local data analysis on a resource

Table 39.2. Data-based Techniques

Technique	Definition	Pros	Cons
Sampling	The process of choosing a subset of a dataset for the sake of analysis using probability theory.	• Well established techniques. • Error boundaries guaranteed	• Poor for anomaly detection.
Load Shedding	The process of ignoring a continuous chunk of streaming data	• Proved efficiency with data stream querying. • Used recently with success in data stream mining	• Very poor for anomaly detection.
Sketching	Randomly projection of a set of features to be analyzed	• Considerably improve the running time.	• Some unselected features might be of great importance.
Synopsis Data Structure	Quick transformation of the incoming stream into a summarized compressed form.	• Analysis task independent.	• might not be sufficient with high data rates.
Aggregation	Calculating statistical measures that capture the features of data.	• Analysis task independent.	• Aggregation measures do not capture all the required features of data.

constrained device that generates or receive streams of information. AOG has three main stages. Mining followed by adaptation to resources and data stream rates represent the first two stages. Merging the generated knowledge structures when running out of memory represents the last stage. AOG has been used in clustering, classification and frequency counting (Gaber *et al.*, 2005).

Figure 39.8 shows a flowchart of AOG-mining process. It shows the sequence of the three stages of AOG.

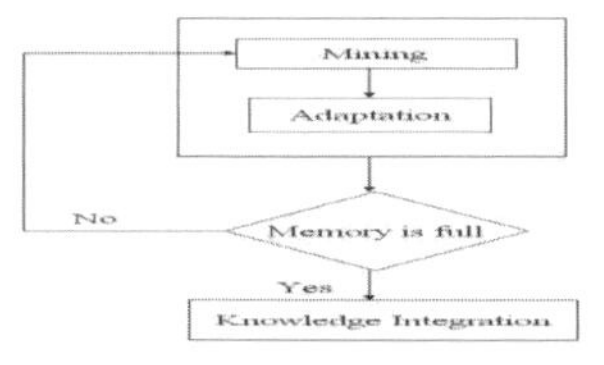

Fig. 39.8. AOG Approach

Definitions, advantages and disadvantages of all of the above task-based approaches are given in Table 39.3.

39.8 Related Work

The last few years have witnessed the emergence of data management strategies focusing on data stream issues (Babcock *et al.*, 2002). Querying and summarizing data that could be stored for further analysis are the main processing tasks studied in data stream management systems. Extension of query languages, query planning, scheduling, and optimization are the major research activities conducted in this area. Aurora (Abadi *et al.*, 2003), COUGAR (Yao and Gehrke, 2002), Gigascope (Cranor *et al.*, 2003), STREAM (Arasu *et al.*, 2003), TelegraphCQ (Krishnamurthy *et al.*, 2003) represent the first generation of data stream management systems. In this section, a brief description of each one is given as follows:

- *STREAM:* STanford stREam datA Manager (STREAM) (Arasu *et al.*, 2003) is a data stream management system that handles multiple continuous data streams and supports long-running continuous queries. The intermediate results of a continuous query are stored in a data structure termed Scratch Store. The results of a query could be a data stream transferred to the user or it could be a relation that also could be stored for re-processing. To support continuous queries over data streams, a continuous query language termed as CQL has been developed as part of the system. The language supports relation-to-relation, stream-to-relation, and relation-to-stream operators.
- *Gigascope:* is a specialized data stream management system (Cranor *et al.*, 2003) for the application of network monitoring. It has its own SQL-like query language termed as GSQL. Unlike CQL, the input and output of this language are only

Table 39.3. Task-based Techniques

Technique	Definition	Pros	Cons
Approximation Algorithms	Design algorithms that approximate mining results with error bounds.	• Efficiency in running time.	• the problem of data rates with regard to the available resources could not be solved using approximation algorithms.
Sliding Window	Analyzing the most recent data streams	• Applicable to most of data stream applications.	• don't provide a model for the whole data stream.
Algorithm Output Granularity	Adapting the algorithm parameters according to data stream rate and memory consumption	• Generic approach that could be used with any mining technique with no or minor modifications	• It has an overhead when running for long period of time

data streams. GSQL supports merge, selection, join and aggregation operations on data streams. Query optimization and performance considerations have been addressed in developing the language. The system serves a number of network related applications including intrusion detection and traffic analysis.

- *TelegraphCQ:* is a continuous query processing system (Krishnamurthy *et al.*, 2003) built on the basis of PostgreSQL open source query language. The system supports creating data streams, sources, wrappers and queries.
- *COUGAR:* is a data stream management system (Yao and Gehrke, 2002) designed for sensor networks. Motivated by the fact that local computation in sensor networks is cheaper than transferring data generated from sensors over wireless connections, a loosely coupled distributed architecture has been proposed to answer in-network queries.
- *Aurora:* is a data stream management system (Abadi *et al.*, 2003) that has the optimization features for load shedding, real-time query scheduling and QoS assessment. It is mainly designed to deal with very large numbers of data streams.

Queries over data streams have some similarities with data stream mining in terms of research issues and challenges. The two main constraints for querying data streams are the unbounded memory requirement and the high data rate. Thus, the computation time per data element/record should be less than the data rate or the sampling rate. Furthermore, the unbounded memory requirement compounds the challenge by necessitating approximate rather than exact results. Significant research efforts have been conducted to approximate the query results (Babcock *et al.*, 2002, Garofalakis *et al.*, 2002b).

The data stream mining algorithms have used some of the techniques introduced in the data stream management research. Sampling and load shedding (Muthukrishnan, 2003) are among the basic techniques that have been introduced in querying data streams and extended to the data mining process.

39.9 Future Directions

The field of data stream mining is in a nascent stage of evolution. The last few years have witnessed increased attention to this area of research due to the dissemination of data stream sources. Based on the state-of-the-art in the area and demands of data streaming applications, we can identify the future directions of research as follows:

- Developing data mining algorithms for wireless sensor networks to serve a number of real-time critical applications.
- Online medical, scientific and biological data stream mining using data generated from medical, biological instruments and various tools employed in scientific laboratories.
- Hardware solutions to small devices emitting or receiving data streams in order to enable high performance computation on small devices.
- Developing software architectures that serve data streaming applications.

39.10 Summary

In this chapter, a review of the state of the art in mining data streams has been presented. Clustering, classification, frequency counting, time series analysis techniques have been discussed. Different systems that use data stream mining techniques have been also presented. Generalization of the approaches used in developing data stream mining techniques is given. The approaches have been broadly classified into data-based and task-based strategies. Sampling, load shedding, sketching, synopsis data structure creation and aggregation represent the data-based approaches. Approximation algorithms, sliding window and algorithm output granularity are the two approaches that form the task-based approaches. The chapter is concluded with pointers to future research directions in the area.

References

A. Arasu, B. Babcock. S. Babu, M. Datar, K. Ito, I. Nishizawa, J. Rosenstein, and J. Widom. STREAM: The Stanford Stream Data Manager Demonstration description - short overview of system status and plans, in Proc. of the ACM Intl Conf. on Management of Data (SIGMOD 2003), June 2003, pp. 665 - 665.

D. Abadi, D. Carney, U. Cetintemel, M. Cherniack, C. Convey, C. Erwin, E. Galvez, M. Hatoun, J. Hwang, A. Maskey, A. Rasin, A. Singer, M. Stonebraker, N. Tatbul, Y. Xing, R.Yan, S. Zdonik. Aurora: A Data Stream Management System (Demonstration). Proceedings of the ACM SIGMOD International Conference on Management of Data (SIGMOD'03), San Diego, CA, June 2003.

C. Aggarwal, J. Han, J. Wang, P. S. Yu, A Framework for Clustering Evolving Data Streams, Proc. 2003 Int. Conf. on Very Large Data Bases (VLDB'03), Berlin, Germany, Sept. 2003, pp 81-92.

C. Aggarwal, J. Han, J. Wang, and P. S. Yu, A Framework for Projected Clustering of High Dimensional Data Streams, Proc. 2004 Int. Conf. on Very Large Data Bases (VLDB'04), Toronto, Canada, Aug. 2004, pp. 852-863.

C. Aggarwal, J. Han, J. Wang, and P. S. Yu, On Demand Classification of Data Streams, Proc. 2004 Int. Conf. on Knowledge Discovery and Data Mining (KDD'04), Seattle, WA, Aug. 2004, pp. 503-508.

I.F. Akyildiz, W. Su, Y. Sankarasubramaniam, and E. Cayirci. A Survey on Sensor Networks, IEEE Communication Magazine, August, 2002, pp. 102-114.

B. Babcock, S. Babu, M. Datar, R. Motwani, and J. Widom. Models and issues in data stream systems, Proceedings of PODS, 2002, pp. 1-16.

B. Babcock, M. Datar, and R. Motwani. Load Shedding Techniques for Data Stream Systems (short paper), Proc. of the 2003 Workshop on Management and Processing of Data Streams (MPDS 2003), June 2003

B. Babcock, M. Datar, R. Motwani, L. O'Callaghan, Maintaining Variance and k-Medians over Data Stream Windows, Proceedings of the 22nd Symposium on Principles of Database Systems (PODS 2003), pp. 234 - 243.

M. Burl, Ch. Fowlkes, J. Roden, A. Stechert, and S. Mukhtar, Diamond Eye: A distributed architecture for image data mining, in SPIE DMKD, Orlando, April 1999, pp. 197-206.

M. Charikar, L. O'Callaghan, and R. Panigrahy, Better streaming algorithms for clustering problems, Proc. of 35th ACM Symposium on Theory of Computing (STOC), 2003, pp. 30-39.

Y.D. Cai, D. Clutter, G. Pape, J. Han, M. Welge, and L. Auvil, MAIDS: Mining Alarming Incidents from Data Streams, (system demonstration), Proc. 2004 ACM-SIGMOD Int. Conf. Management of Data (SIGMOD'04), Paris, France, June 2004, pp. 919 - 920.

Y. Chen, G. Dong, J. Han, B. W. Wah, and J. Wang, Multi-Dimensional Regression Analysis of Time-Series Data Streams, Proceedings of VLDB Conference, 2002, pp. 323-334.

B. Castano, M. Judd, R. C. Anderson, and T. Estlin, Machine Learning Challenges in Mars Rover Traverse Science, Proc. of the ICML 2003 workshop on Machine Learning Technologies for Autonomous Space Applications.

C. Cranor , Johnson, T., Spataschek, O., and Shkapenyuk, V., Gigascope: a stream database for network applications, In Proceedings of the 2003 ACM SIGMOD international Conference on Management of Data (San Diego, California, June 09 - 12, 2003). SIGMOD '03. ACM, New York, NY, 647-651

L. O'Callaghan, Nina Mishra, Adam Meyerson, Sudipto Guha, and Rajeev Motwani, Streaming-data algorithms for high-quality clustering, Proceedings of IEEE International Conference on Data Engineering, March 2002, pp. 685-697.

G. Cormode, S. Muthukrishnan, What's hot and what's not: tracking most frequent items dynamically, PODS 2003, pp. 296-306

J. Coughlan, Accelerating Scientific Discovery at NASA, SIAM SDM 2004, Florida USA.

G. Cormode and S. Muthukrishnan., What is new: Finding significant differences in network data streams, INFOCOM 2004.

Y. Chi, Philip S. Yu, Haixun Wang, Richard R. Muntz, Loadstar: A Load Shedding Scheme for Classifying Data Streams, The 2005 SIAM International Conference on Data Mining (SIAM SDM'05), 2005.

G. Dong, J. Han, L.V.S. Lakshmanan, J. Pei, H. Wang and P.S. Yu. Online mining of changes from data streams: Research problems and preliminary results, Proceedings of the 2003 ACM SIGMOD Workshop on Management and Processing of Data Streams. In cooperation with the 2003 ACM-SIGMOD International Conference on Management of Data (SIGMOD'03), San Diego, CA, June 8, 2003.

P. Domingos and G. Hulten, Mining High-Speed Data Streams, In Proceedings of the Association for Computing Machinery Sixth International Conference on Knowledge Discovery and Data Mining, 2000, pp. 71-80

P. Domingos and G. Hulten. Catching Up with the Data: Research Issues in Mining Data Streams, Workshop on Research Issues in Data Mining and Knowledge Discovery, 2001. Santa Barbara, CA

P. Domingos and G. Hulten, A General Method for Scaling Up Machine Learning Algorithms and its Application to Clustering, Proceedings of the Eighteenth International Conference on Machine Learning, 2001, Williamstown, MA, Morgan Kaufmann, pp. 106-113.

M. Dunham. Data Mining: Introductory and Advanced Topics. Pearson Education, 2003.

F.J. Ferrer-Troyano, J.S. Aguilar-Ruiz and J.C. Riquelme, Discovering Decision Rules from Numerical Data Streams, ACM Symposium on Applied Computing - SAC04, 2004, ACM Press, pp. 649-653.

U.M. Fayyad: Knowledge Discovery in Databases: An Overview. ILP 1997, pp. 3-16

U.M. Fayyad: Mining Databases: Towards Algorithms for Knowledge Discovery. IEEE Data Eng. Bull. 21(1), 1998 pp. 39-48.

U.M. Fayyad, Georges G. Grinstein, Andreas Wierse: Information Visualization in Data Mining and Knowledge Discovery Morgan Kaufmann 2001.

M.M. Gaber , Yu P. S., A Holistic Approach for Resource-aware Adaptive Data Stream Mining, Journal of New Generation Computing, Special Issue on Knowledge Discovery from Data Streams, 2006.

V. Ganti, Johannes Gehrke, Raghu Ramakrishnan: Mining Data Streams under Block Evolution. SIGKDD Explorations 3(2), 1002 pp. 1-10.

M. Garofalakis, Johannes Gehrke, Rajeev Rastogi: Querying and mining data streams: you only get one look a tutorial. SIGMOD Conference 2002: 635

C. Giannella, J. Han, J. Pei, X. Yan, and P.S. Yu, Mining Frequent Patterns in Data Streams at Multiple Time Granularities, in H. Kargupta, A. Joshi, K. Sivakumar, and Y. Yesha (eds.), Next Generation Data Mining, AAAI/MIT, 2003.

A.C. Gilbert, Yannis Kotidis, S. Muthukrishnan, Martin Strauss: One-Pass Wavelet Decompositions of Data Streams. TKDE 15(3), 2003, pp. 541-554.

M.M. Gaber, Krishnaswamy, S., and Zaslavsky, A., On-board Mining of Data Streams in Sensor Networks, a book chapter in Advanced Methods of Knowledge Discovery from Complex Data, (Eds.) Sanghamitra Badhyopadhyay, Ujjwal Maulik, Lawrence Holder and Diane Cook, Springer Verlag,.2005.

R. Grossman, Supporting the Data Mining Process with Next Generation DataMining Systems, Enterprise Systems, August 1998

M.M. Gaber, Zaslavsky, A., and Krishnaswamy, S., Towards an Adaptive Approach for Mining Data Streams in Resource Constrained Environments, Proceedings of Sixth International Conference on Data Warehousing and Knowledge Discovery - Industry Track (DaWaK 2004), Zaragoza, Spain, 30 August - 3 September, Lecture Notes in Computer Science (LNCS), Springer Verlag.

S. Guha, N. Mishra, R. Motwani, and L. O'Callaghan, Clustering data streams, Proceedings of the Annual Symposium on Foundations of Computer Science. IEEE, November 2000, pp. 359-366.

S. Guha, Adam Meyerson, Nina Mishra, Rajeev Motwani, and Liadan O'Callaghan, Clustering Data Streams: Theory and Practice TKDE special issue on clustering, vol. 15, 2003, pp. 515-528.

D.J. Hand, Statistics and Data Mining: Intersecting Disciplines, ACM SIGKDD Explorations, 1, 1, June 1999, pp. 16-19.

D.J. Hand, Mannila H., and Smyth P. Principles of data mining, MIT Press, 2001.

W. Hoeffding. Probability inequalities for sums of bounded random variables, Journal of the American Statistical Association (58), 1963, pp. 13-30.

J. Han, Pei, J., and Yin, Y, Mining frequent patterns without candidate generation, In Proc. 2000 ACM-SIGMOD Int. Conf. Management of Data (SIGMOD'00), pp. 1-12.

G. Hulten, L. Spencer, and P. Domingos. Mining Time-Changing Data Streams. ACM SIGKDD 2001, pp. 97-106.

M. Henzinger, P. Raghavan and S. Rajagopalan, Computing on data streams , Technical Note 1998-011, Digital Systems Research Center, Palo Alto, CA, May 1998

T. Hastie, R. Tibshirani, J. Friedman, The elements of statistical learning: data mining, inference, and prediction, New York: Springer, 2001

P. Indyk, N. Koudas, and S. Muthukrishnan, Identifying Representative Trends in Massive Time Series Data Sets Using Sketches. In Proc. of the 26th Int. Conf. on Very Large Data Bases, Cairo, Egypt, September 2000, pp. 363 - 372.

C. Jin, Weining Qian, Chaofeng Sha, Jeffrey X. Yu, and Aoying Zhou, Dynamically Maintaining Frequent Items over a Data Stream, In Proceedings of the 12th ACM Conference on Information and Knowledge Management (CIKM'2003), pp. 287-294

M. Kantardzic, Data mining : concepts, models, methods and algorithms, Piscataway, NJ: IEEE Pr. Wiley Interscience, 2003.

H. Kargupta, Ruchita Bhargava, Kun Liu, Michael Powers, Patrick Blair, Samuel Bushra, James Dull, Kakali Sarkar, Martin Klein, Mitesh Vasa, and David Handy, VEDAS: A Mobile and Distributed Data Stream Mining System for Real-Time Vehicle Monitoring, Proceedings of SIAM International Conference on Data Mining 2004.

S. Krishnamurthy, S. Chandrasekaran, O. Cooper, A. Deshpande, M. Franklin, J. Hellerstein, W. Hong, S. Madden, V. Raman, F. Reiss, and M. Shah. TelegraphCQ: An Architectural Status Report. IEEE Data Engineering Bulletin, Vol 26(1), March 2003.

E. Keogh, J. Lin, and W. Truppel. Clustering of Time Series Subsequences is Meaningless: Implications for Past and Future Research. In proceedings of the 3rd IEEE International Conference on Data Mining. Melbourne, FL. Nov 19-22, 2003, pp. 115-122.

H. Kargupta, Park, B., Pittie, S., Liu, L., Kushraj, D. and Sarkar, K. (2002). MobiMine: Monitoring the Stock Market from a PDA. ACM SIGKDD Explorations. January 2002. Volume 3, Issue 2, ACM Press, pp. 37-46.

B. Krishnamachari and S.S. Iyengar. Efficient and Fault-tolerant Feature Extraction in Sensor Networks. In Proceedings of the 2nd International Workshop on Information Processing in Sensor Networks (IPSN '03), Palo Alto, California, April 2003.

B. Krishnamachari and S. Iyengar. Distributed Bayesian Algorithms for Fault-tolerant Event Region Detection in Wireless Sensor Networks. IEEE Transactions on Computers, vol. 53, No. 3, March 2004.

M. Last, Online Classification of Nonstationary Data Streams, Intelligent Data Analysis, Vol. 6, No. 2, 2002, pp. 129-147.

Y. Law, C. Zaniolo, An Adaptive Nearest Neighbor Classification Algorithm for Data Streams, Proceedings of the 9th European Conference on the Principals and Practice of Knowledge Discovery in Databases (PKDD 2005), Springer Verlag, Porto, Portugal, October 3-7, 2005, pp. 108-120.

J. Lin, E. Keogh, S. Lonardi, and B. Chiu, A Symbolic Representation of Time Series, with Implications for Streaming Algorithms, In proceedings of the 8th ACM SIGMOD Workshop on Research Issues in Data Mining and Knowledge Discovery. San Diego, CA. June 13, 2003, pp. 2-11.

G.S. Manku and R. Motwani. Approximate frequency counts over data streams. In Proceedings of the 28th International Conference on Very Large Data Bases, Hong Kong, China, August 2002, pp. 346-357.

R. Moskovitch, Y. Elovici, L. Rokach, Detection of unknown computer worms based on behavioral classification of the host, Computational Statistics and Data Analysis, 52(9):4544–4566, 2008.

S. Muthukrishnan, Data streams: algorithms and applications. Proceedings of the fourteenth annual ACM-SIAM symposium on discrete algorithms, 2003.

O. Nasraoui , Cardona C., Rojas C., and Gonzalez F., Mining Evolving User Profiles in Noisy Web Clickstream Data with a Scalable Immune System Clustering Algorithm, in Proc. of WebKDD 2003 - KDD Workshop on Web mining as a Premise to Effective and Intelligent Web Applications, Washington DC, August 2003, p. 71

C. Ordonez. Clustering Binary Data Streams with K-means ACM DMKD 2003.

B. Park and H. Kargupta. Distributed Data Mining: Algorithms, Systems, and Applications, Data Mining Handbook. Editor: Nong Ye. 2002.

E. Perlman and A. Java, Predictive Mining of Time Series Data in Astronomy. In ASP Conf. Ser. 295: Astronomical Data Analysis Software and Systems XII, 2003.

S. Papadimitriou, C. Faloutsos, and A. Brockwell, Adaptive, Hands-Off Stream Mining, 29th International Conference on Very Large Data Bases VLDB, 2003.

S. Pirttikangas, J. Riekki, J. Kaartinen, J. Miettinen, S. Nissila, J. Roning. Genie Of The Net: A New Approach For A Context-Aware Health Club. In Proceedings of Joint 12th ECML'01 and 5th European Conference on PKDD'01. September 3-7, 2001, Freiburg, Germany.

L. Rokach, Decomposition methodology for classification tasks: a meta decomposer framework, Pattern Analysis and Applications, 9(2006):257–271.

L. Rokach, O. Maimon and R. Arbel, Selective voting-getting more for less in sensor fusion, International Journal of Pattern Recognition and Artificial Intelligence 20 (3) (2006), pp. 329–350.

A. Srivastava and J. Stroeve, Onboard Detection of Snow, Ice, Clouds and Other Geophysical Processes Using Kernel Methods, Proceedings of the ICML'03 workshop on Machine Learning Technologies for Autonomous Space Applications.

S. Tanner, M. Alshayeb, E. Criswell, M. Iyer, A. McDowell, M. McEniry, K. Regner, EVE: On-Board Process Planning and Execution, Earth Science Technology Conference, Pasadena, CA, Jun. 11 - 14, 2002.

N. Tatbul, U. Cetintemel, S. Zdonik, M. Cherniack and M. Stonebraker, Load Shedding in a Data Stream Manager Proceedings of the 29th International Conference on Very Large Data Bases (VLDB), September, 2003.

N. Tatbul, U. Cetintemel, S. Zdonik, M. Cherniack, M. Stonebraker. Load Shedding on Data Streams, In Proceedings of the Workshop on Management and Processing of Data Streams (MPDS 03), San Diego, CA, USA, June 8, 2003.

H. Toivonen, Sampling large databases for association rules, Proceeding of VLDB Conference, 1996

Y. Yao, J. E. Gehrke, The Cougar Approach to In-Network Query Processing in Sensor Networks, SIGMOD Record, Volume 31, Number 3. September 2002, pp. 9-18.

H. Wang, W. Fan, P. Yu and J. Han, Mining Concept-Drifting Data Streams using Ensemble Classifiers, in the 9th ACM International Conference on Knowledge Discovery and Data Mining (SIGKDD), Aug. 2003, Washington DC, USA.

Y. Zhu and D. Shasha, Efficient Elastic Burst Detection in Data Streams, The Ninth ACM SIGKDD International Conference on Knowledge Discovery and Data Mining KDD-2003 24 August 2003 - 27 August 2003, pp 336 - 345.

40

Mining Concept-Drifting Data Streams

Haixun Wang[1], Philip S. Yu[2], and Jiawei Han[3]

[1] IBM T. J. Watson Research Center
haixun@us.ibm.com
[2] IBM T. J. Watson Research Center
psyu@us.ibm.com
[3] University of Illinois, Urbana Champaign
hanj@cs.uiuc.edu

Summary. Knowledge discovery from infinite data streams is an important and difficult task. We are facing two challenges, the overwhelming volume and the concept drifts of the streaming data. In this chapter, we introduce a general framework for mining concept-drifting data streams using weighted ensemble classifiers. We train an ensemble of classification models, such as C4.5, RIPPER, naive Bayesian, etc., from sequential chunks of the data stream. The classifiers in the ensemble are judiciously weighted based on their expected classification accuracy on the test data under the time-evolving environment. Thus, the ensemble approach improves both the efficiency in learning the model and the accuracy in performing classification. Our empirical study shows that the proposed methods have substantial advantage over single-classifier approaches in prediction accuracy, and the ensemble framework is effective for a variety of classification models.

Key words: Data Mining, concept learning, classifier design and evaluation

40.1 Introduction

Knowledge discovery on streaming data is a research topic of growing interest (Babcock *et al.*, 2002, Chen *et al.*, 2002, Domingos and Hulten, 2000, Hulten *et al.*, 2001). The fundamental problem we need to solve is the following: given an infinite amount of continuous measurements, how do we model them in order to capture time-evolving trends and patterns in the stream, and make time-critical predictions?

Huge data volume and drifting concepts are not unfamiliar to the Data Mining community. One of the goals of traditional Data Mining algorithms is to learn models from large databases with bounded-memory. It has been achieved by several classification methods, including Sprint (Shafer *et al.*, 1996), BOAT (Gehrke *et al.*, 1999), etc. Nevertheless, the fact that these algorithms require multiple scans of the training data makes them inappropriate in the streaming environment where examples are coming in at a higher rate than they can be repeatedly analyzed.

O. Maimon, L. Rokach (eds.), *Data Mining and Knowledge Discovery Handbook*, 2nd ed.,
DOI 10.1007/978-0-387-09823-4_40,

Incremental or online Data Mining methods (Utgoff, 1989, Gehrke *et al.*, 1999) are another option for mining data streams. These methods continuously revise and refine a model by incorporating new data as they arrive. However, in order to guarantee that the model trained incrementally is identical to the model trained in the batch mode, most online algorithms rely on a costly model updating procedure, which sometimes makes the learning even slower than it is in batch mode. Recently, an efficient incremental decision tree algorithm called VFDT is introduced by Domingos et al (Domingos and Hulten, 2000). For streams made up of discrete type of data, Hoeffding bounds guarantee that the output model of VFDT is asymptotically nearly identical to that of a batch learner.

The above mentioned algorithms, including incremental and online methods such as VFDT, all produce a single model that represents the entire data stream. It suffers in prediction accuracy in the presence of concept drifts. This is because the streaming data are not generated by a stationary stochastic process, indeed, the future examples we need to classify may have a very different distribution from the historical data.

In order to make time-critical predictions, the model learned from the streaming data must be able to capture transient patterns in the stream. To do this, as we revise the model by incorporating new examples, we must also eliminate the effects of examples representing outdated concepts. This is a non-trivial task. The challenge of maintaining an accurate and up-to-date classifier for infinite data streams with concept drifts including the following:

- ACCURACY. It is difficult to decide what are the examples that represent outdated concepts, and hence their effects should be excluded from the model. A commonly used approach is to 'forget' examples at a constant rate. However, a higher rate would lower the accuracy of the 'up-to-date' model as it is supported by a less amount of training data and a lower rate would make the model less sensitive to the current trend and prevent it from discovering transient patterns.
- EFFICIENCY. Decision trees are constructed in a greedy divide-and-conquer manner, and they are non-stable. Even a slight drift of the underlying concepts may trigger substantial changes (e.g., replacing old branches with new branches, re-growing or building alternative subbranches) in the tree, and severely compromise learning efficiency.
- EASE OF USE. Substantial implementation efforts are required to adapt classification methods such as decision trees to handle data streams with drifting concepts in an incremental manner (Hulten *et al.*, 2001). The usability of this approach is limited as state-of-the-art learning methods cannot be applied directly.

In light of these challenges, we propose using *weighted classifier ensembles* to mine streaming data with concept drifts. Instead of continuously revising a single model, we train an ensemble of classifiers from sequential data chunks in the stream. Maintaining a most up-to-date classifier is not necessarily the ideal choice, because potentially valuable information may be wasted by discarding

results of previously-trained less-accurate classifiers. We show that, in order to avoid overfitting and the problems of conflicting concepts, the expiration of old data must rely on data's distribution instead of only their arrival time. The ensemble approach offers this capability by giving each classifier a weight based on its expected prediction accuracy on the current test examples. Another benefit of the ensemble approach is its efficiency and ease-of-use. Our method also works in a cost-sensitive senario, where *instance-based ensemble pruning* method (Wang *et al.*,2003) can be applied so that a pruned ensemble delivers the same level of benefits as the entire set of classifiers.

40.2 The Data Expiration Problem

The fundamental problem in learning drifting concepts is how to identify in a timely manner those data in the training set that are no longer consistent with the current concepts. These data must be discarded. A straightforward solution, which is used in many current approaches, discards data indiscriminately after they become old, that is, after a fixed period of time T has passed since their arrival. Although this solution is conceptually simple, it tends to complicate the logic of the learning algorithm. More importantly, it creates the following dilemma which makes it vulnerable to unpredictable conceptual changes in the data: if T is large, the training set is likely to contain outdated concepts, which reduces classification accuracy; if T is small, the training set may not have enough data, and as a result, the learned model will likely carry a large variance due to overfitting.

We use a simple example to illustrate the problem. Assume a stream of 2-dimensional data is partitioned into sequential chunks based on their arrival time. Let S_i be the data that came in between time t_i and t_{i+1}. Figure 40.1 shows the distribution of the data and the optimum decision boundary during each time interval.

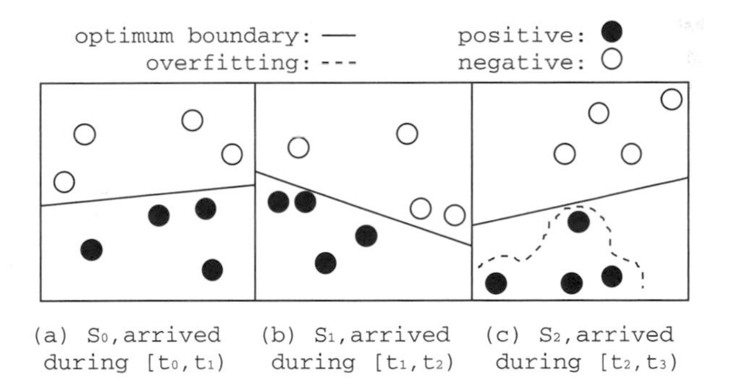

Fig. 40.1. Data Distributions and Optimum Boundaries.

The problem is: after the arrival of S_2 at time t_3, what part of the training data should still remain influential in the current model so that the data arriving after t_3 can be most accurately classified?

On one hand, in order to reduce the influence of old data that *may* represent a different concept, we shall use nothing but the most recent data in the stream as the training set. For instance, use the training set consisting of S_2 only (i.e., $T = t_3 - t_2$, data S_1, S_0 are discarded). However, as shown in Figure 40.1(c), the learned model may carry a significant variance since S_2's insufficient amount of data are very likely to be overfitted.

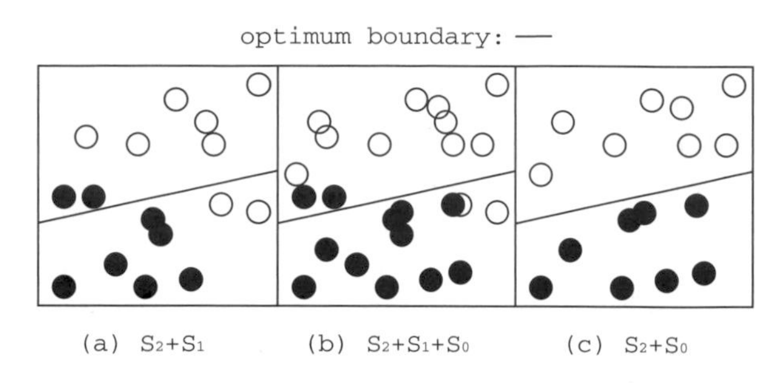

Fig. 40.2. Which Training Dataset to Use?

The inclusion of more historical data in training, on the other hand, may also reduce classification accuracy. In Figure 40.2(a), where $S_2 \cup S_1$ (i.e., $T = t_3 - t_1$) is used as the training set, we can see that the discrepancy between the underlying concepts of S_1 and S_2 becomes the cause of the problem. Using a training set consisting of $S_2 \cup S_1 \cup S_0$ (i.e., $T = t_3 - t_0$) will not solve the problem either. Thus, there may not exists an optimum T to avoid problems arising from overfitting and conflicting concepts.

We should not discard data that may still provide useful information to classify the current test examples. Figure 40.2(c) shows that the combination of S_2 and S_0 creates a classifier with less overfitting or conflicting-concept concerns. The reason is that S_2 and S_0 have similar class distribution. Thus, instead of discarding data using the criteria based solely on their arrival time, we shall make decisions based on their class distribution. Historical data whose class distributions are similar to that of current data can reduce the variance of the current model and increase classification accuracy.

However, it is a non-trivial task to select training examples based on their class distribution. We argue that a carefully weighted classifier ensemble built on a set of data partitions $S_1, S_2, \cdots, S_n$ is more accurate than a single classifier built on $S_1 \cup S_2 \cup \cdots \cup S_n$. Due to space limitation, we refer readers to (Wang *et al.*,2003) for the proof.

40.3 Classifier Ensemble for Drifting Concepts

A weighted classifier ensemble can outperform a single classifier in the presence of concept drifts (Wang *et al.*,2003). To apply it to real-world problems we need to assign an actual weight to each classifier that reflects its predictive accuracy on the current testing data.

40.3.1 Accuracy-Weighted Ensembles

The incoming data stream is partitioned into sequential chunks, $S_1, S_2, \cdots$, S_n, with S_n being the most up-to-date chunk, and each chunk is of the same size, or `ChunkSize`. We learn a classifier $\mathscr{C}_i$ for each S_i, $i \geq 1$.

According to the error reduction property, given test examples T, we should give each classifier $\mathscr{C}_i$ a weight reversely proportional to the expected error of $\mathscr{C}_i$ in classifying T. To do this, we need to know the actual function being learned, which is unavailable.

We derive the weight of classifier $\mathscr{C}_i$ by *estimating* its expected prediction error on the test examples. We assume the class distribution of S_n, the most recent training data, is closest to the class distribution of the current test data. Thus, the weights of the classifiers can be approximated by computing their classification error on S_n.

More specifically, assume that S_n consists of records in the form of (x, c), where c is the true label of the record. $\mathscr{C}_i$'s classification error of example (x, c) is $1 - f_c^i(x)$, where $f_c^i(x)$ is the probability given by $\mathscr{C}_i$ that x is an instance of class c. Thus, the mean square error of classifier $\mathscr{C}_i$ can be expressed by:

$$\text{MSE}_i = \frac{1}{|S_n|} \sum_{(x,c) \in S_n} (1 - f_c^i(x))^2$$

The weight of classifier $\mathscr{C}_i$ should be reversely proportional to MSE_i. On the other hand, a classifier predicts randomly (that is, the probability of x being classified as class c equals to c's class distributions $p(c)$) will have mean square error:

$$\text{MSE}_r = \sum_c p(c)(1 - p(c))^2$$

For instance, if $c \in \{0, 1\}$ and the class distribution is uniform, we have $\text{MSE}_r =$ .25. Since a random model does not contain useful knowledge about the data, we use MSE_r, the error rate of the random classifier as a threshold in weighting the classifiers. That is, we discard classifiers whose error is equal to or larger than MSE_r. Furthermore, to make computation easy, we use the following weight w_i for classifier $\mathscr{C}_i$:

$$w_i = \text{MSE}_r - \text{MSE}_i \tag{40.1}$$

For cost-sensitive applications such as credit card fraud detection, we use the benefits (e.g., total fraud amount detected) achieved by classifier C_i on the most recent training data S_n as its weight.

Table 40.1. Benefit Matrix $b_{c,c'}$.

	predict $fraud$	predict $\neg fraud$
actual $fraud$	$t(x) - cost$	0
actual $\neg fraud$	$-cost$	0

Assume the benefit of classifying transaction x of actual class c as a case of class c' is $b_{c,c'}(x)$. Based on the benefit matrix shown in Table 40.1 (where $t(x)$ is the transaction amount, and *cost* is the fraud investigation cost), the total benefits achieved by $\mathscr{C}_i$ is:

$$b_i = \sum_{(x,c) \in S_n} \sum_{c'} b_{c,c'}(x) \cdot f^i_{c'}(x)$$

and we assign the following weight to $\mathscr{C}_i$:

$$w_i = b_i - b_r \tag{40.2}$$

where b_r is the benefits achieved by a classifier that predicts randomly. Also, we discard classifiers with 0 or negative weights.

Since we are handling infinite incoming data flows, we will learn an infinite number of classifiers over the time. It is impossible and unnecessary to keep and use all the classifiers for prediction. Instead, we only keep the top K classifiers with the highest prediction accuracy on the current training data. In (Wang *et al.*,2003), we studied ensemble pruning in more detail and presented a technique for instance-based pruning.

Figure 40.3 gives an outline of the classifier ensemble approach for mining concept-drifting data streams. Whenever a new chunk of data has arrived, we build a classifier from the data, and use the data to tune the weights of the previous classifiers. Usually, `ChunkSize` is small (our experiments use chunks of size ranging from 1,000 to 25,000 records), and the entire chunk can be held in memory with ease.

The algorithm for classification is straightforward, and it is omitted here. Basically, given a test case y, each of the K classifiers is applied on y, and their outputs are combined through weighted averaging.

Input S: a dataset of `ChunkSize` from the incoming stream
    K: the total number of classifiers
    $\mathscr{C}$: a set of K previously trained classifiers
Output $\mathscr{C}$: a set of K classifiers with updated weights

train classifier $\mathscr{C}'$ from S
compute error rate / benefits of $\mathscr{C}'$ via cross validation on S
derive weight w' for $\mathscr{C}'$ using (40.1) or (40.2)
for each classifier $\mathscr{C}_i \in \mathscr{C}$ **do**
  apply $\mathscr{C}_i$ on S to derive MSE_i or b_i
  compute w_i based on (40.1) and (40.2)
end for
$\mathscr{C} \leftarrow K$ of the top weighted classifiers in $C \cup \{C'\}$
return $\mathscr{C}$

Fig. 40.3. A Classifier Ensemble Approach for Mining Concept-Drifting Data Streams.

40.4 Experiments

We conducted extensive experiments on both synthetic and real life data streams. Our goals are to demonstrate the error reduction effects of weighted classifier ensembles, to evaluate the impact of the frequency and magnitude of the concept drifts on prediction accuracy, and to analyze the advantage of our approach over alternative methods such as incremental learning. The base models used in our tests are C4.5 (Quinlan, 1993), the RIPPER rule learner (Cohen, 1995), and the Naive Bayesian method. The tests are conducted on a Linux machine with a 770 MHz CPU and 256 MB main memory.

40.4.1 Algorithms used in Comparison

We denote a classifier ensemble with a capacity of K classifiers as E_K. Each classifier is trained by a data set of size `ChunkSize`. We compare with algorithms that rely on a single classifier for mining streaming data. We assume the classifier is continuously being revised by the data that have just arrived and the data being faded out. We call it a window classifier, since only the data in the most recent window have influence on the model. We denote such a classifier by G_K, where K is the number of data chunks in the window, and the total number of the records in the window is $K \cdot$ `ChunkSize`. Thus, ensemble E_K and G_K are trained from the same amount of data. Particularly, we have $E_1 = G_1$. We also use G_0 to denote the classifier built on the entire historical data starting from the beginning of the data stream up to now. For instance, BOAT (Gehrke *et al.*, 1999) and VFDT (Domingos and Hulten, 2000) are G_0 classifiers, while CVFDT (Hulten *et al.*, 2001) is a G_K classifier.

40.4.2 Streaming Data

Synthetic Data

We create synthetic data with drifting concepts based on a moving hyperplane. A hyperplane in d-dimensional space is denoted by equation:

$$\sum_{i=1}^{d} a_i x_i = a_0 \tag{40.3}$$

We label examples satisfying $\sum_{i=1}^{d} a_i x_i \geq a_0$ as positive, and examples satisfying $\sum_{i=1}^{d} a_i x_i < a_0$ as negative. Hyperplanes have been used to simulate time-changing concepts because the orientation and the position of the hyperplane can be changed in a smooth manner by changing the magnitude of the weights (Hulten *et al.*, 2001).

We generate random examples uniformly distributed in multi-dimensional space $[0,1]^d$. Weights a_i $(1 \leq i \leq d)$ in (40.3) are initialized randomly in the range of $[0,1]$. We choose the value of a_0 so that the hyperplane cuts the multi-dimensional space in two parts of the same volume, that is, $a_0 = \frac{1}{2}\sum_{i=1}^{d} a_i$. Thus, roughly half of the examples are positive, and the other half negative. Noise is introduced by randomly switching the labels of $p\%$ of the examples. In our experiments, the noise level $p\%$ is set to 5%.

We simulate concept drifts by a series of parameters. Parameter k specifies the total number of dimensions whose weights are changing. Parameter $t \in \mathscr{R}$ specifies the magnitude of the change (every N examples) for weights $a_1, \cdots, a_k$, and $s_i \in \{-1, 1\}$ specifies the direction of change for each weight a_i, $1 \leq i \leq k$. Weights change continuously, i.e., a_i is adjusted by $s_i \cdot t/N$ after each example is generated. Furthermore, there is a possibility of 10% that the change would reverse direction after every N examples are generated, that is, s_i is replaced by $-s_i$ with probability 10%. Also, each time the weights are updated, we recompute $a_0 = \frac{1}{2}\sum_{i=1}^{d} a_i$ so that the class distribution is not disturbed.

Credit Card Fraud Data

We use real life credit card transaction flows for cost-sensitive mining. The data set is sampled from credit card transaction records within a one year period and contains a total of 5 million transactions. Features of the data include the time of the transaction, the merchant type, the merchant location, past payments, the summary of transaction history, etc. A detailed description of this data set can be found in (Stolfo *et al.*, 1997). We use the benefit matrix shown in Table 40.1 with the cost of disputing and investigating a fraud transaction fixed at $cost = \$90$.

The total benefit is the sum of recovered amount of fraudulent transactions less the investigation cost. To study the impact of concept drifts on the benefits, we derive two streams from the dataset. Records in the 1st stream are ordered by transaction time, and records in the 2nd stream by transaction amount.

40.4.3 Experimental Results

Time Analysis

We study the time complexity of the ensemble approach. We generate synthetic data streams and train single decision tree classifiers and ensembles with varied `ChunkSize`. Consider a window of $K = 100$ chunks in the data stream. Figure 40.4 shows that the ensemble approach E_K is much more efficient than the corresponding single-classifier G_K in training.

Smaller `ChunkSize` offers better training performance. However, `ChunkSize` also affects classification error. Figure 40.4 shows the relationship between error rate (of E_{10}, e.g.) and `ChunkSize`. The dataset is generated with certain concept drifts (weights of 20% of the dimensions change $t = 0.1$ per $N = 1000$ records), large chunks produce higher error rates because the ensemble cannot detect the concept drifts occurring inside the chunk. Small chunks can also drive up error rates if the number of classifiers in an ensemble is not large enough. This is because when `ChunkSize` is small, each individual classifier in the ensemble is not supported by enough amount of training data.

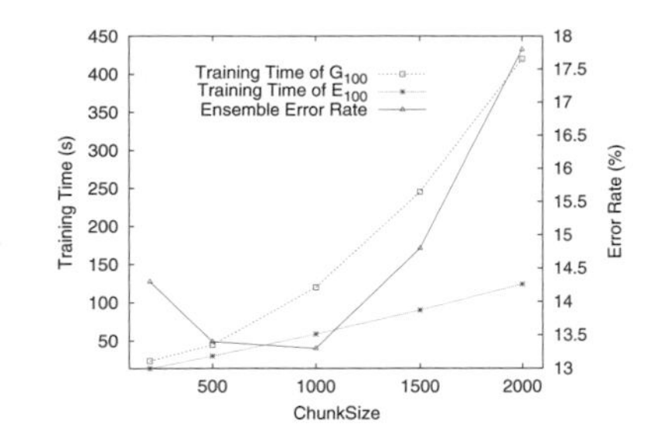

Fig. 40.4. Training Time, `ChunkSize`, and Error Rate

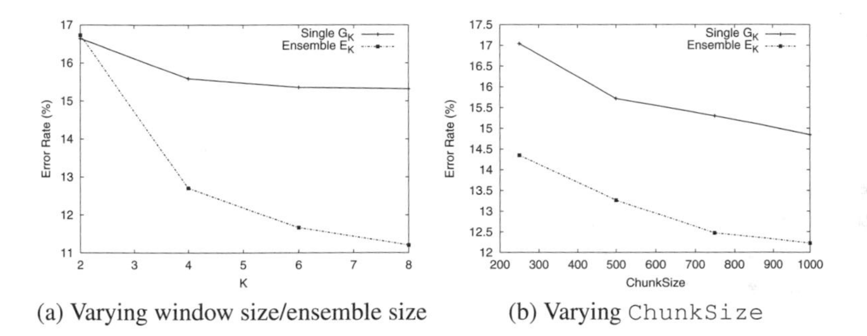

(a) Varying window size/ensemble size (b) Varying `ChunkSize`

Fig. 40.5. Average Error Rate of Single and Ensemble Decision Tree Classifiers.

Table 40.2. Error Rate (%) of Single and Ensemble Decision Tree Classifiers.

ChunkSize	G_0	$G_1 = E_1$	G_2	E_2	G_4	E_4	G_8	E_8
250	18.09	18.76	**18.00**	18.37	16.70	**14.02**	16.76	**12.19**
500	17.65	17.59	**16.39**	17.16	16.19	**12.91**	14.97	**11.25**
750	17.18	16.47	16.29	**15.77**	15.07	**12.09**	14.86	**10.84**
1000	16.49	16.00	15.89	**15.62**	14.40	**11.82**	14.68	**10.54**

Table 40.3. Error Rate (%) of Single and Ensemble Naive Bayesian Classifiers.

ChunkSize	G_0	G_1=E_1	G_2	E_2	G_4	E_4	G_6	E_6	G_8	E_8
250	11.94	8.09	7.91	**7.48**	8.04	**7.35**	8.42	**7.49**	8.70	**7.55**
500	12.11	7.51	7.61	**7.14**	7.94	**7.17**	8.34	**7.33**	8.69	**7.50**
750	12.07	7.22	7.52	**6.99**	7.87	**7.09**	8.41	**7.28**	8.69	**7.45**
1000	15.26	7.02	7.79	**6.84**	8.62	**6.98**	9.57	**7.16**	10.53	**7.35**

Table 40.4. Error Rate (%) of Single and Ensemble RIPPER Classifiers.

ChunkSize	G_0	G_1=E_1	G_2	E_2	G_4	E_4	G_8	E_8
50	27.05	24.05	22.85	**22.51**	21.55	**19.34**	19.34	**17.84**
100	25.09	21.97	**19.85**	20.66	**17.48**	17.50	17.50	**15.91**
150	24.19	20.39	**18.28**	19.11	17.22	**16.39**	16.39	**15.03**

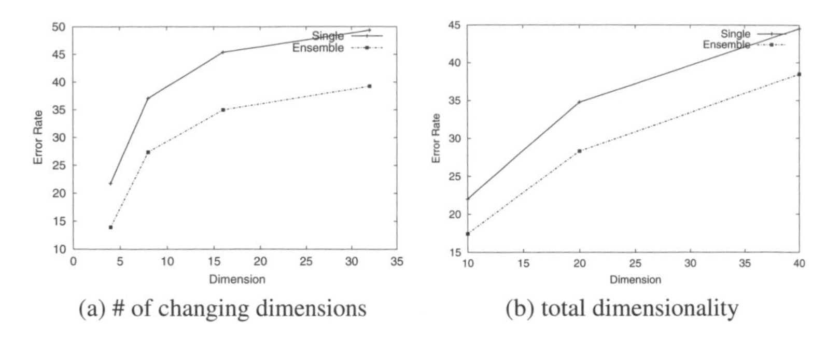

(a) # of changing dimensions (b) total dimensionality

Fig. 40.6. Magnitude of Concept Drifts.

Error Analysis

We use C4.5 as our base model, and compare the error rates of the single classifier approach and the ensemble approach. The results are shown in Figure 40.5 and Table 40.2. The synthetic datasets used in this study have 10 dimensions ($d = 10$). Figure 40.5 shows the averaged outcome of tests on data streams generated with varied concept drifts (the number of dimensions with changing weights ranges from 2 to 8, and the magnitude of the change t ranges from 0.10 to 1.00 for every 1000 records).

First, we study the impact of ensemble size (total number of classifiers in the ensemble) on classification accuracy. Each classifier is trained from a dataset of size ranging from 250 records to 1000 records, and their averaged error rates are shown in Figure 40.5(a). Apparently, when the number of classifiers increases, due to the increase of diversity of the ensemble, the error rate of E_k drops significantly. The single classifier, G_k, trained from the same amount of the data, has a much higher error rate due to the changing concepts in the data stream. In Figure 40.5(b), we vary the chunk size and average the error rates on different K ranging from 2 to 8. It shows that the error rate of the ensemble approach is about 20% lower than the single-classifier approach in all the cases. A detailed comparison between single- and ensemble-classifiers is given in Table 40.2, where G_0 represents the global classifier trained by the entire history data, and we use **bold** font to indicate the better result of G_k and E_k for $K = 2,4,6,8$.

We also tested the Naive Bayesian and the RIPPER classifier under the same setting. The results are shown in Table 40.3 and Table 40.4. Although C4.5, Naive Bayesian, and RIPPER deliver different accuracy rates, they confirmed that, with a reasonable amount of classifiers (K) in the ensemble, the ensemble approach outperforms the single classifier approach.

Concept Drifts

Figure 40.6 studies the impact of the magnitude of the concept drifts on classification error. Concept drifts are controlled by two parameters in the synthetic data: i) the number of dimensions whose weights are changing, and ii) the magnitude of

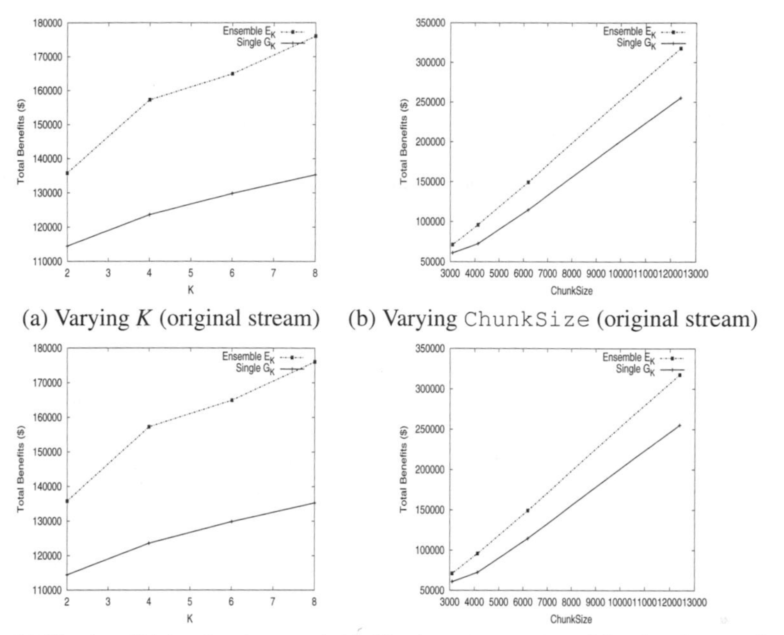

(a) Varying K (original stream) (b) Varying `ChunkSize` (original stream)

(c) Varying K (simulated stream) (d) Varying `ChunkSize` (simulated stream)

Fig. 40.7. Averaged Benefits using Single Classifiers and Classifier Ensembles.

weight change per dimension. Figure 40.6 shows that the ensemble approach outperform the single-classifier approach under all circumstances. Figure 40.6(a) shows the classification error of G_k and E_k (averaged over different K) when 4, 8, 16, and 32 dimensions' weights are changing (the change per dimension is fixed at $t = 0.10$). Figure 40.6(b) shows the increase of classification error when the dimensionality of dataset increases. In the datasets, 40% dimensions' weights are changing at ± 0.10 per 1000 records. An interesting phenomenon arises when the weights change monotonically (weights of some dimensions are constantly increasing, and others constantly decreasing).

Table 40.5. Benefits (US $) using Single Classifiers and Classifier Ensembles (Simulated Stream).

`Chunk`	G_0	G_1=E_1	G_2	E_2	G_4	E_4	G_8	E_8
12000	296144	207392	233098	**268838**	248783	**313936**	275707	**360486**
6000	146848	102099	102330	**129917**	113810	**148818**	123170	**162381**
4000	96879	62181	66581	**82663**	72402	**95792**	76079	**103501**
3000	65470	51943	55788	**61793**	59344	**70403**	66184	**77735**

Table 40.6. Benefits (US $) using Single Classifiers and Classifier Ensembles (Original Stream).

Chunk	G_0	G_1=E_1	G_2	E_2	G_4	E_4	G_8	E_8
12000	201717	203211	197946	**253473**	211768	**269290**	215692	**289129**
6000	103763	98777	101176	**121057**	102447	**138565**	106576	**143620**
4000	69447	65024	68081	**80996**	69346	**90815**	70325	**96153**
3000	43312	41212	42917	**59293**	44977	**67222**	46139	**71660**

Cost-sensitive Learning

For cost-sensitive applications, we aim at maximizing benefits. In Figure 40.7(a), we compare the single classifier approach with the ensemble approach using the credit card transaction stream. The benefits are averaged from multiple runs with different chunk size (ranging from 3000 to 12000 transactions per chunk). Starting from $K = 2$, the advantage of the ensemble approach becomes obvious.

In Figure 40.7(b), we average the benefits of E_k and G_k ($K = 2, \cdots, 8$) for each fixed chunk size. The benefits increase as the chunk size does, as more fraudulent transactions are discovered in the chunk. Again, the ensemble approach outperforms the single classifier approach.

To study the impact of concept drifts of different magnitude, we derive data streams from the credit card transactions. The simulated stream is obtained by sorting the original 5 million transactions by their transaction amount. We perform the same test on the simulated stream, and the results are shown in Figure 40.7(c) and 40.7(d).

Detailed results of the above tests are given in Table 40.6 and 40.5.

40.5 Discussion and Related Work

Data stream processing has recently become a very important research domain. Much work has been done on modeling (Babcock *et al.*, 2002), querying (Babu and Widom, 2001, Gao and Wang, 2002, Greenwald and Khanna, 2001), and mining data streams, for instance, several papers have been published on classification (Domingos and Hulten, 2000, Hulten *et al.*, 2001, Street and Kim, 2001), regression analysis (Chen *et al.*, 2002), and clustering (Guha *et al.*, 2000).

Traditional Data Mining algorithms are challenged by two characteristic features of data streams: the infinite data flow and the drifting concepts. As methods that require multiple scans of the datasets (Shafer *et al.*, 1996) can not handle infinite data flows, several incremental algorithms (Gehrke *et al.*, 1999, Domingos and Hulten, 2000) that refine models by continuously incorporating new data from the stream have been proposed. In order to handle drifting concepts, these methods are revised again to achieve the goal that effects of old examples are eliminated at a certain rate. In terms of an incremental decision tree classifier, this means we have to discard, re-grow sub trees, or build alternative subtrees under a node (Hulten *et al.*, 2001). The resulting algorithm is often complicated, which indicates substantial efforts are required to adapt state-of-the-art learning methods to the infinite, concept-drifting streaming environment. Aside from this undesirable aspect, incremental methods are also hindered by their prediction accuracy. Since old examples are discarded at a fixed rate (no matter if they represent the changed concept or not), the learned model is supported only by the current snapshot – a relatively small amount of data. This usually results in larger prediction variances.

Classifier ensembles are increasingly gaining acceptance in the data mining community. The popular approaches to creating ensembles include changing the instances used for training through techniques such as Bagging (Bauer and Kohavi, 1999) and Boosting (Freund and Schapire, 1996). The classifier ensembles have several advantages over single model classifiers. First, classifier ensembles offer a significant improvement in prediction accuracy (Freund and Schapire, 1996, Tumer and Ghosh, 1996). Second, building a classifier ensemble is more efficient than building a single model, since most model construction algorithms have super-linear complexity. Third, the nature of classifier ensembles lend themselves to scalable parallelization (Hall *et al.*, 2000) and on-line classification of large databases. Previously, we used averaging ensemble for scalable learning over very-large datasets (Fan, Wang , Yu, and Stolfo, 2003). We show that a model's performance can be estimated before it is completely learned (Fan, Wang , Yu, and Lo, 2002, Fan, Wang , Yu, and Lo, 2003). In this work, we use weighted ensemble classifiers on concept-drifting data streams. It combines multiple classifiers weighted by their expected prediction accuracy on the current test data. Compared with incremental models trained by data in the most recent window, our approach combines talents of set of experts based on their credibility and adjusts much nicely to the underlying concept drifts. Also, we introduced the dynamic classification technique (Fan, Chu, Wang, and Yu, 2002) to the concept-drifting streaming environment, and our results show that it enables us to dynamically select a subset of classifiers in the ensemble for prediction without loss in accuracy.

Ackowledgement

We thank Wei Fan of IBM T. J. Watson Research Center for providing us with a revised version of the C4.5 decision tree classifier and running some experiments.

References

Babcock B., Babu S. , Datar M. , Motawani R. , and Widom J., Models and issues in data stream systems, In *ACM Symposium on Principles of Database Systems (PODS)*, 2002.

Babu S. and Widom J., Continuous queries over data streams. *SIGMOD Record*, 30:109–120, 2001.

Bauer, E. and Kohavi, R., An empirical comparison of voting classification algorithms: Bagging, boosting, and variants. *Machine Learning*, 36(1-2):105–139, 1999.

Chen Y., Dong G., Han J., Wah B. W., and Wang B. W., Multi-dimensional regression analysis of time-series data streams. In *Proc. of Very Large Database (VLDB)*, Hongkong, China, 2002.

Cohen W., Fast effective rule induction. In *Int'l Conf. on Machine Learning (ICML)*, pages 115–123, 1995.

Domingos P., and Hulten G., Mining high-speed data streams. In *Int'l Conf. on Knowledge Discovery and Data Mining (SIGKDD)*, pages 71–80, Boston, MA, 2000. ACM Press.

Fan W., Wang H., Yu P., and Lo S. , Progressive modeling. In *Int'l Conf. Data Mining (ICDM)*, 2002.

Fan W., Wang H., Yu P., and Lo S. , Inductive learning in less than one sequential scan, In *Int'l Joint Conf. on Artificial Intelligence*, 2003.

Fan W., Wang H., Yu P., and Stolfo S., A framework for scalable cost-sensitive learning based on combining probabilities and benefits. In *SIAM Int'l Conf. on Data Mining (SDM)*, 2002.

Fan W., Chu F., Wang H., and Yu P. S., Pruning and dynamic scheduling of cost-sensitive ensembles, In *Proceedings of the 18th National Conference on Artificial Intelligence (AAAI)*, 2002.

Freund Y., and Schapire R. E., Experiments with a new boosting algorithm, In *Int'l Conf. on Machine Learning (ICML)*, pages 148–156, 1996.

Gao L. and Wang X., Continually evaluating similarity-based pattern queries on a streaming time series, In *Int'l Conf. Management of Data (SIGMOD)*, Madison, Wisconsin, June 2002.

Gehrke J., Ganti V., Ramakrishnan R., and Loh W., BOAT– optimistic decision tree construction, In *Int'l Conf. Management of Data (SIGMOD)*, 1999.

Greenwald M., and Khanna S., Space-efficient online computation of quantile summaries, In *Int'l Conf. Management of Data (SIGMOD)*, pages 58–66, Santa Barbara, CA, May 2001.

Guha S., Milshra N., Motwani R., and O'Callaghan L., Clustering data streams, In *IEEE Symposium on Foundations of Computer Science (FOCS)*, pages 359–366, 2000.

Hall L., Bowyer K., Kegelmeyer W., Moore T., and Chao C., Distributed learning on very large data sets, In *Workshop on Distributed and Parallel Knowledge Discover*, 2000.

Hulten G., Spencer L., and Domingos P., Mining time-changing data streams, In *Int'l Conf. on Knowledge Discovery and Data Mining (SIGKDD)*, pages 97–106, San Francisco, CA, 2001. ACM Press.

Quinlan J. R., *C4.5: Programs for Machine Learning*. Morgan Kaufmann, 1993.

Rokach L., Mining manufacturing data using genetic algorithm-based feature set decomposition, Int. J. Intelligent Systems Technologies and Applications, 4(1):57-78, 2008.

Shafer C., Agrawal R., and Mehta M., Sprint: A scalable parallel classifier for Data Mining, In *Proc. of Very Large Database (VLDB)*, 1996.

Stolfo S., Fan W., Lee W., Prodromidis A., and Chan P., Credit card fraud detection using meta-learning: Issues and initial results. In *AAAI-97 Workshop on Fraud Detection and Risk Management*, 1997.

Street W. N. and Kim Y. S., A streaming ensemble algorithm (SEA) for large-scale classification. In *Int'l Conf. on Knowledge Discovery and Data Mining (SIGKDD)*, 2001.

Tumer K. and Ghosh J., Error correlation and error reduction in ensemble classifiers, *Connection Science*, 8(3-4):385–403, 1996.

Utgoff, P. E., Incremental induction of decision trees, *Machine Learning*, 4:161–186, 1989.

Wang H., Fan W., Yu P. S., and Han J., Mining concept-drifting data streams using ensemble classifiers, In *Int'l Conf. on Knowledge Discovery and Data Mining (SIGKDD)*, 2003.

41

Mining High-Dimensional Data

Wei Wang[1] and Jiong Yang[2]

[1] Department of Computer Science, University of North Carolina at Chapel Hill
[2] Department of Electronic Engineering and Computer Science, Case Western Reserve University

Summary. With the rapid growth of computational biology and e-commerce applications, high-dimensional data becomes very common. Thus, mining high-dimensional data is an urgent problem of great practical importance. However, there are some unique challenges for mining data of high dimensions, including (1) the curse of dimensionality and more crucial (2) the meaningfulness of the similarity measure in the high dimension space. In this chapter, we present several state-of-art techniques for analyzing high-dimensional data, e.g., frequent pattern mining, clustering, and classification. We will discuss how these methods deal with the challenges of high dimensionality.

Key words: High-dimensional Data Mining, frequent pattern, clustering high-dimensional data, classifying high-dimensional data

41.1 Introduction

The emergence of various new application domains, such as bioinformatics and e-commerce, underscores the need for analyzing high dimensional data. In a gene expression microarray data set, there could be tens or hundreds of dimensions, each of which corresponds to an experimental condition. In a customer purchase behavior data set, there may be up to hundreds of thousands of merchandizes, each of which is mapped to a dimension. Researchers and practitioners are very eager in analyzing these data sets.

Various Data Mining models have been proven to be very successful for analyzing very large data sets. Among them, frequent patterns, clusters, and classifiers are three widely studied models to represent, analyze, and summarize large data sets. In this chapter, we focus on the state-of-art techniques for constructing these three Data Mining models on massive high-dimensional data sets.

O. Maimon, L. Rokach (eds.), *Data Mining and Knowledge Discovery Handbook*, 2nd ed., DOI 10.1007/978-0-387-09823-4_41,

41.2 Chanllenges

Before presenting any algorithm for building individual Data Mining models, we first discuss two common challenges for analyzing high-dimensional data. The first one is the curse of dimensionality. The complexity of many existing Data Mining algorithms is exponential with respect to the number of dimensions. With increasing dimensionality, these algorithms soon become computationally intractable and therefore inapplicable in many real applications.

Secondly, the specificity of similarities between points in a high dimensional space diminishes. It was proven in (Beyer *et al.*, 1999) that, for any point in a high dimensional space, the expected gap between the Euclidean distance to the closest neighbor and that to the farthest point shrinks as the dimensionality grows. This phenomenon may render many Data Mining tasks (e.g., clustering) ineffective and fragile because the model becomes vulnerable to the presence of noise. In the remainder of this chapter, we present several state-of-art algorithms for mining high-dimensional data sets.

41.3 Frequent Pattern

Frequent pattern is a useful model for extracting salient features of the data. It was originally proposed for analyzing market basket data (Agrawal, 1994). A market basket data set is typically represented as a set of transactions. Each transaction contains a set of items from a finite vocabulary. In principle, we can represent the data as a matrix, each row represents a transaction and each column represents an item. The goal is to find the collection of itemsets appearing in a large number of transactions, defined by a support threshold t. Most algorithms for mining frequent patterns utilize the Apriori property stated as follows. If an itemset A is frequent (i.e., present in more than t transactions), then every subset of A must be frequent. On the other hand, if an itemset A is infrequent (i.e, present in less than t transactions), then any superset of A is also infrequent. This property is the basis of all level-wise search algorithms. The general procedure consists of a series of iterations beginning with counting item occurrences and identifying the set of frequent items (or equivalently, frequent 1-itemsets). During each subsequent iteration, candidates for frequent k-itemsets are proposed from frequent (k-1)-itemsets using the Apriori property. These candidates are then validated by explicitly counting their actual occurrences. The value of k is incremented before the next iteration starts. The process terminates when no more frequent itemset can be generated. We often refer to this level-wise approach as the breadth-first approach because it evaluates the itemsets residing at the same depth in the lattice formed by imposing the partial order of subset-superset relationship between itemsets.

It is a well-known problem that the full set of frequent patterns contains significant redundant information and consequently the number of frequent patterns is often too large. To address this issue, Pasquier et al. (1999) proposed to mine a selective subset of frequent patterns, called *closed frequent patterns*. If the number of occurrences of a pattern is the same to all its immediate subpatterns, then the pattern is considered as a closed pattern. The CLOSET algorithm (Pei *et al.*, 2000) is proposed to expedite the mining of closed frequent patterns. CLOSET uses a novel frequent pattern tree (FP structure) as a compact representation to organize the data set. It performs a depth-first search, that is, after discovering a frequent itemset A, it searches for superpatterns of A before checking A's siblings.

A more recent algorithm for mining frequent closed pattern is CHARM (Zaki and Hsiao, 2002). Similar to CLOSET, CHARM searches for patterns in a depth-first manner. The difference between CHARM and CLOSET is that CHARM stores the data set in a vertical format where a list of row IDs is maintained for each dimension. These row ID lists are then merged during a "column enumeration" procedure that generates row ID lists for other nodes in the enumeration tree. In addition, a technique called *diffset* is used to reduce the length of the row ID lists as well as the computational complexity of merging them.

All previous algorithms can find frequent closed patterns when the dimensionality is low to moderate. When the number of dimensions is very high, e.g., greater than 100, the efficiency of these algorithms could be significantly impacted. CARPENTER (Pan *et al.*, 2003) is therefore proposed to solve this problem. It first transposes the matrix representing the data set. Next, CARPENTER performs a depth-first rowwise enumeration on the transposed matrix. It has been shown that this algorithm can greatly reduce the computation time especially when the dimensionality is high.

41.4 Clustering

Clustering is a widely adopted Data Mining model that partitions data points into a set of groups, each of which is called a *cluster*. A data point has a shorter distance to points within the cluster than those outside the cluster. In a high dimensional space, for any point, its distance to its closest point and that to the farthest point tend to be similar. This phenomenon may render the clustering result sensitive to any small perturbation to the data due to noise and make the exercise of clustering useless. To solve this problem, Agrawal et. al. proposed a subspace clustering model (Agrawal *et al.*, 1998). A subspace cluster consists of a subset of objects and a subset of dimensions such that the distance among these objects is small within the given set of dimensions. The CLIQUE algorithm (Agrawal *et al.*, 1998) is proposed to find the subspace clusters.

In many applications, users are more interested in the objects that exhibit a consistent trend (rather than points having similar values) within a subset of dimensions. One such example is the bicluster model (Cheng and Church, 2000) proposed for analyzing gene expression profiles. A bicluster is a subset of objects (U) and a subset dimensions (D) such that objects in U have the same trend (i.e., fluctuating simultaneously) across dimensions in D. This is particular useful in analyzing gene expression levels in a microarray experiment since the expression levels of some genes may be inflated/deflated systematically in some experiments. Thus, the absolute value is not as important as the trend. If two genes have similar trends across a large set of experiments, they are likely to be co-regulated. In the bicluster model, the mean squared error residue is used to qualify a bicluster. Cheng and Church (2000) used a heuristic randomized algorithm to find biclusters. It consists of a series of iterations, each of which locates one bicluster. To prevent the same bicluster from being reported again in subsequent iterations, each time when a bicluster is found, the values in the bicluster are replaced by uniform noise before the next iteration starts. This procedure continues until a desired number of biclusters are discovered.

Although the bicluster model and algorithm have been used in several applications in bioinformatics, it has two major drawbacks: (1) the mean squared error residue may not be the best measure to qualify a bicluster. A big cluster may have small mean squared error residue even if it includes a small number of objects whose trends are vastly different in the selected dimensions; (2) the heuristic algorithm may be interfered by the noise artificially injected after each iteration and hence may not discover overlapped clusters properly. To solve these two problems, the authors of (Wang *et al.*, 2002) proposed the *p-cluster* model. A p-cluster consists of a subset of objects U and a subset of dimensions D where for each pair of objects u_1 and u_2 in U and each pair of dimension d_1 and d_2 in D, the change of u_1 from d_1 to d_2 should be similar to that of u_2 from $d_1$to d_2. A threshold is used to evaluate the dissimilarity between two objects on two dimensions. Given a subset of objects and a subset of dimensions, if the dissimilarity between every pair of objects on every pair of dimensions is less than the threshold, then these objects constitute a p-cluster in the given dimensions. A novel deterministic algorithm is developed in (Wang *et al.*, 2002)to find all maximal p-clusters, which utilizes the Apriori property held on p-clusters.

41.5 Classification

The classification is also a very powerful data analysis tool. In a classification problem, the dimensions of an object can be divided into two types. One dimension records the class type of the object and the rest dimensions are attributes. The goal of classification is to build a model that captures the intrinsic associations between the class type and the attributes so that an (unknown) class type can be accurately predicted from the attribute values. For this purpose, the data is usually divided into a training set and a test set, where the training set is used to build the classifier which is validated by the test set. There are several models developed for classifying high dimensional data, e.g., naïve Bayesian, neural networks, decision trees (Mitchell, 1997), SVMs, rule-based classifiers, and so on.

Supporting vector machine (SVM) (Vapnik, 1998) is one of the newly developed classification models. The success of SVM in practice is drawn by its solid mathematical foundation that conveys the following two salient properties. (1) The classification boundary functions of SVMs maximize the margin, which equivalently optimize the general performance given a training data set. (2) SVMs handle a nonlinear classification efficiently using the kernel trick that implicitly transforms the input space into another higher dimensional feature space. However, SVM suffers from two problems. First, the complexity of training an SVM is at least $O(N^2)$ where

N is the number of objects in the training data set. It could be too costly when the training data set is large. Second, since an SVM essentially draws a hyper-plain in a transformed high dimensional space, it is very difficult to identify the principal (original) dimensions that are most responsible for the classification.

Rule-based classifiers (Liu *et al.*, 2000) offer some potential to address the above two problems. A rule-based classifier consists of a set of rules in the following form: $A_1[l_1,u_1] \cap A_2[l_2,u_2] \cap \ldots \cap A_m[l_m,u_m] \rightarrow C$, where $A_i[l_i, u_i]$ is the range of attribute A_i's value and C is the class type. The above rule can be interpreted as that, if an object whose attributes' values fall in the ranges in the left hand side, then its class type is likely to be C (with some high probability). Each rule is also associated with a confidence level that depicts the probability that such a rule holds. When an object satisfies several rules, either the rule with the highest confidence (e.g., CBA (Liu *et al.*, 2000)) or a weighted voting of all valid rules (e.g., CPAR (Yin and Han, 2003)) may be used for class prediction. However, neither CBA nor CPAR are targeted for high dimensional data. An algorithm called FARMER (Cong *et al.*, 2004) is proposed to generate rule-based classifiers for high dimensional data set. It first quantizes the attributes into a set of bins. Each bin is treated as an item subsequently. FARMER then generates the closed frequent itemsets using a method similar to CARPENTER. These closed frequent itemsets are the basis to generate rules. Since the dimensionality is high, the number of possible rules in the classifier could be very large. FARMER finally organizes all rules into compact rule groups.

References

Agrawal R., Gehrke J., Gunopulos D., Raghavan P.: "Automatic Subspace Clustering of High Dimensional Data for Data Mining Applications", Proc. ACM SIGMOD Int. Conf. on Management of Data, Seattle, WA, 1998, pp. 94-105.

Agrawal R., and Srikant R., Fast Algorithms for Mining Association Rules in Large Databases. In Proc. of the 20th VLDB Conf., pages 487-499, 1994.

Beyer K.S., Goldstein J., Ramakrishnan R. and Shaft U.: "When Is 'Nearest Neighbor' Meaningful?", Proceedings 7th International Conference on Database Theory (ICDT'99), pp. 217-235, Jerusalem, Israel, 1999.

Cheng Y., and Church, G., Biclustering of expression data. In Proceedings of the Eighth International Conference on Intelligent Systems for Molecular Biology, pp. 93-103. San Diego, CA, August 2000.

Cong G., Tung Anthony K. H., Xu X., Pan F., and Yang J., Farmer: Finding interesting rule groups in microarray datasets. In the 23rd ACM SIGMOD International Conference on Management of Data, 2004.

Liu B., Ma Y., Wong C. K., Improving an Association Rule Based Classifier, Proceedings of the 4th European Conference on Principles of Data Mining and Knowledge Discovery, p.504-509, September 13-16, 2000.

Mitchell T., Machine Learning. WCB McGraw Hill, 1997.

Pan F., Cong G., Tung A. K. H., Yang J., and Zaki M. J., CARPENTER: finding closed patterns in long biological data sets. Proceedings of ACM SIGKDD International Conference on Knowledge Discovery and Data Mining, 2003.

Pasquier, N., Bastide, Y., Taouil, R., Lakhal, L.: Discovering frequent closed itemsets for association rules. In Beeri, C., Buneman, P., eds., Proc. of the 7th Int'l Conf. on Database Theory (ICDT'99), Jerusalem, Israel, Volume 1540 of Lecture Notes in Computer Science., pp. 398-416, Springer-Verlag, January 1999.

Pei, J., Han, J., and Mao, R., CLOSET: an efficient Algorithm for mining frequent closed itemsets. In D. Gunopulos and R. Rastogi, eds., ACM SIGMOD Workshop on Research Issues in Data Mining and Knowledge Discovery, pp 21-30, 2000.

Vapnik, V.N., Statistical Learning Theory. John Wiley and Sons, 1998.

Wang H., Wang W., Yang J. and Yu P., Clustering by pattern similarity in large data sets. Proceedings of the ACM SIGMOD International Conference on Management of Data (SIGMOD), pp. 394-405, 2002.

Yin X., Han J., CPAR: classification based on predictive association rules. Proceedings of SIAM International Conference on Data Mining, San Fransisco, CA, pp. 331-335, 2003.

Zaki M. J. and Hsiao C., CHARM: An efficient algorithm for closed itemset mining. In Proceedings of the Second SIAM International Conference on Data Mining, Arlington, VA, 2002. SIAM

42

Text Mining and Information Extraction

Moty Ben-Dov[1] and Ronen Feldman[2]

[1] MDX University, London
[2] Hebrew university, Israel

Summary. Text Mining is the automatic discovery of new, previously unknown information, by automatic analysis of various textual resources. Text mining starts by extracting facts and events from textual sources and then enables forming new hypotheses that are further explored by traditional Data Mining and data analysis methods. In this chapter we will define text mining and describe the three main approaches for performing information extraction. In addition, we will describe how we can visually display and analyze the outcome of the information extraction process.

Key words: text mining, content mining, structure mining, text classification, information extraction, Rules Based Systems.

42.1 Introduction

The information age has made it easy for us to store large amounts of texts. The proliferation of documents available on the Web, on corporate intranets, on news wires, and elsewhere is overwhelming. However, while the amount of information available to us is constantly increasing, our ability to absorb and process this information remains constant. Search engines only exacerbate the problem by making more and more documents available in a matter of a few key strokes; So-called "push" technology makes the problem even worse by constantly reminding us that we are failing to track news, events, and trends everywhere. We experience information overload, and miss important patterns and relationships even as they unfold before us. As the old adage goes, "we can't see the forest for the trees."

Text-mining (TM), also known as Knowledge discovery from text (KDT), refers to the process of extracting interesting patterns from very large text database for the purposes of discovering knowledge. Text-mining applies the same analytical functions of data-mining but also applies analytic functions from natural language (NL) and information retrieval (IR) techniques (Dorre *et al.*, 1999).

The text-mining tools are used for:

O. Maimon, L. Rokach (eds.), *Data Mining and Knowledge Discovery Handbook*, 2nd ed.,
DOI 10.1007/978-0-387-09823-4_42, © Springer Science+Business Media, LLC 2010

- Extracting relevant information from a document – extract the features (entities) from a document by using NL, IR and association metrics algorithms (Feldman *et al.*, 1998) or pattern matching (Averbuch *et al.*, 2004).
- Finding trend or relations between people/places/organizations etc. by aggregating and comparing information extracted from the documents.
- Classifying and organizing documents according to their content (Tkach, 1998)
- Retrieving documents based on the various sorts of information about the document content.
- Clustering documents according to their content (Wai-chiu and Fu 2000).

A Text Mining system is composed of 3 major components (See Figure 42.1):

Information Feeders enable the connection between various textual collections and the tagging modules. This component connects to any web site, streamed source (such a news feed), internal document collections and any other types of textual collections.

Intelligent Tagging A component responsible for reading the text and distilling (tagging) the relevant information. This component can perform any type of tagging on the documents such as statistical tagging (categorization and term extraction), semantic tagging (information extraction) and structural tagging (extraction from the visual layout of documents).

Business Intelligence Suite A component responsible for consolidating the information from disparate sources, allowing for simultaneous analysis of the entire information landscape.

The TM task can be separated into two major categories according to their task and according to the algorithms and formal frameworks that they are using.

The first is the Task-oriented preprocessing approaches that envision the process of creating a structured document representation in terms of tasks and sub-tasks and usually involve some sort of preparatory goal or problem that needs to be solved.

The second is the preprocessing approaches that rely on techniques that derive from formal methods for analyzing complex phenomena and can be also applied to natural language texts. Such approaches include classification schemes, probabilistic models, rule-based systems approaches and other methodologies.

In this chapter we will first talk about the differences between TM and Text retrieval. Second we will describe the two approaches in TM - the task and the formal frameworks process.

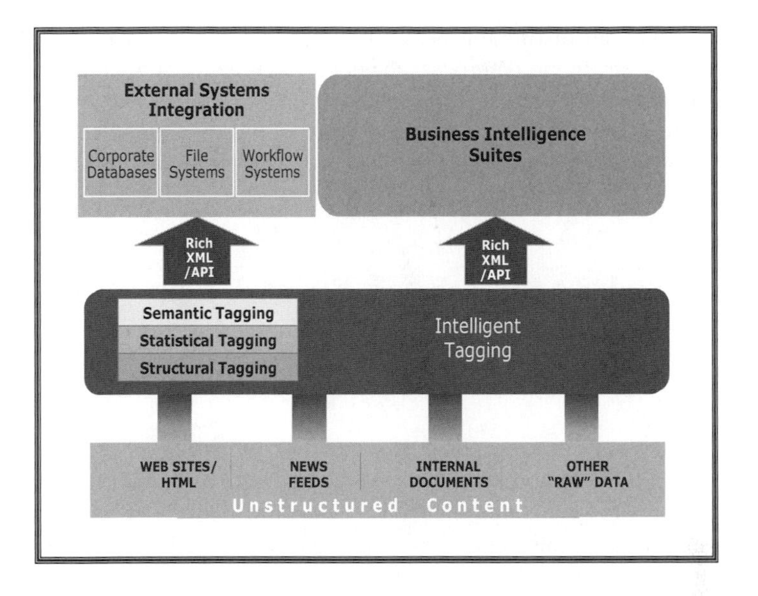

Fig. 42.1. Architecture of Text-Mining systems

42.2 Text Mining vs. Text Retrieval

It is important to differentiate between Text Mining (TM) and Text Retrieval (or information retrieval, as it is more widely known).

The goal of information retrieval is to help users find documents that satisfy their information needs (Baeza-Yates and Ribeiro-Neto, 1999). The standard procedure is analogous to looking for needles in a needle stack - the problem isn't so much that the desired information is not known, but rather that the desired information coexists with many other valid pieces of information (Hearst, 1999). The outcome of information retrieval process is documents.

The goal of TM is to discover or derive new information from text, finding patterns across datasets, and/or separating signal from noise. The fact that an information retrieval system can return a document that contains the information a user requested does not imply that a new discovery has been made: the information had to have already been known to the author of the text; otherwise the author could not have written it down.

On the other hand, the results of certain types of text processing can yield tools that indirectly *aid* in the information access process. Examples include text clustering to create thematic overviews of text collections (Cutting *et al.*, 1992), automatically categorizing search results (Chen and Dumais, 2000), automatically generating term associations to aid in query expansion (Xu and Croft., 1996), and using co-citation analysis to find general topics within a collection or identify central web pages (Kleinberg, 1999).

The most important distinction between TM and Information Retrieval is the output of each process. In the IR process the out is documents, some it's clustered or ordered or scored but at the end to get the information we have to read the documents. In contrast the results of TM process can be features, patterns, connections, profiles or trends and to find the information we need we don't necessary have to read the documents.

42.3 Task-Oriented Approaches vs. Formal Frameworks

Two clear ways of categorizing the totality of preparatory document structuring techniques are according to their task and according to the algorithms and formal frameworks that they use.

Task-oriented preprocessing approaches envision the process of creating a structured document representation in terms of tasks and sub-tasks and usually involve some sort of preparatory goal or problem that needs to be solved. Other preprocessing approaches rely on techniques that derive from formal methods for analyzing complex phenomena that can be also applied to natural language texts. Such approaches include classification schemes, probabilistic models, rule-based systems approaches and other methodologies.

42.4 Task-Oriented Approaches

A document is an abstract which has a variety of possible actual representations. The task of the document structuring process is to take the most "raw" representation and convert it to the representation where the meaning of the document is understandable.

In order to cope with this extremely difficult problem, a "divide-and-conquer" strategy is typically employed. The problem is separated into a set of smaller subtasks, each of which is solved separately. The subtasks can broadly be divided into three classes (see Figure 42.2) – preparatory processing, general-purpose natural language processing ("NLP") tasks, and problem-dependent tasks.

- Preparatory processing converts the raw representation into a structure suitable for further linguistic processing. For example, the raw input may be a PDF document, a scanned page, or even recorded speech. The task of the preparatory processing is to convert the raw input into a stream of text, possibly labeling the internal text zones, such as paragraphs, columns, or tables. It is sometimes also possible for the preparatory processing to extract some document-level fields, such as ¡Author¿ or ¡Title¿, in case where the visual position of the fields allows their identification.

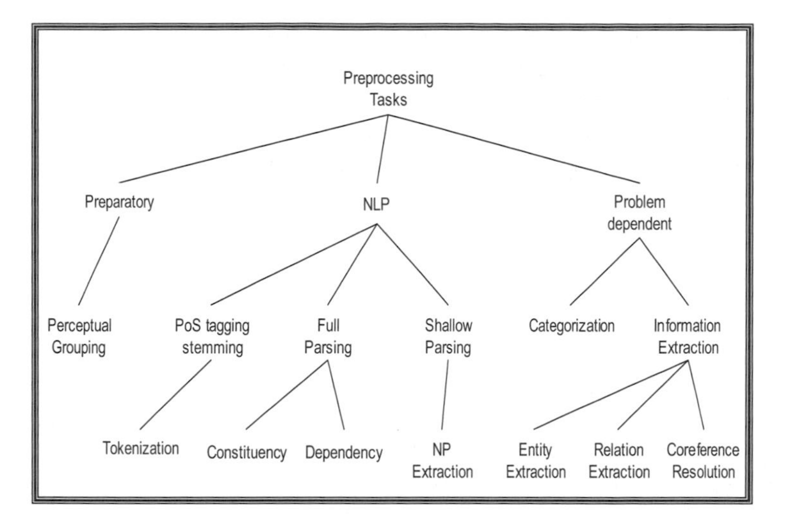

Fig. 42.2. Text Preprocessing Tasks.

- NLP process - The general-purpose NLP tasks process text documents using the general knowledge about natural language. The tasks may include tokenization, morphological analysis, part-of-speech tagging, and syntactic parsing, either shallow or deep.
  - Tokenization - The first step in information extraction from text is to identify the words in sentences in the text. The tokenizer performs this function. In English text, this process is fairly simple since white spaces and punctuation marks separate words. In other languages (e.g., Chinese, Japanese), in which spaces do not separate words, this process is more complex. Also, in some languages, use of hyphens to compound words (e.g., German, Dutch), this is a crucial step.
  - Part-of-speech Tagging - Part-of-speech tagging is the process of identifying a word's part of speech in a sentence (e.g., noun, verb, etc.) by its context. Tagging serves as the basis for the information retrieval system to perform syntax-sensitive filtering and analysis. Usually, PoS taggers at some stage of their processing perform morphological analysis of words. Thus, an additional output of a PoS tagger is a sequence of stems (also known as "lemmas") of the input words (Brill, 1992; Kupiec, 1992; Brill, 1995).
  - Syntactical parsing components perform a full syntactical analysis of sentences, according to a certain grammar theory. The basic division is between the constituency and dependency grammars (Keller, 1992; Pollard and Sag, 1994; Rambow and Joshi, 1994; Neuhaus and Broker, 1997).
    Shallow Parsing (Hammerton *et al.*, 2002) is the task of diving documents into non overlapping word sequences or phrases such that syntactically related words are grouped together. Each phrase is then tagged by one of a set

of predefined tags such as: Noun Phrase, Verb Phrase, Prepositional Phrase, Adverb Phrase, Subordinated clause, Adjective Phrase, Conjunction Phrase, and List Marker. Shallow parsing is generally useful as a preprocessing step, either for bootstrapping, extracting information from corpora for use by more sophisticated parsers, or for end-user applications such as information extraction. Shallow parsing allows morphological analysis and the identification of relationships between the object, subject and/or spatial/temporal location within a sentence.

42.4.1 Problem Dependant Task - Information Extraction in Text Mining

Information Extraction (Hobbs *et al.*, 1992; Riloff, 1993; Riloff, 1994; Riloff and Lehnert, 1994; Huffman, 1995; Grishman, 1996; Cardie, 1997; Grishman, 1997; Leek, 1997; Wilks, 1997; Freitag, 1998), is perhaps the most prominent technique currently used in text mining pre-processing operations. Without IE (Information Extraction) techniques, text mining systems would have much more limited knowledge discovery capabilities.

IE technology would allow one to rapidly create extraction systems for new tasks whose performance was on par with human performance. However, even systems that do not have anything near perfect recall and precision can be of real value. In such cases, the results of the IE system would need to be fed into an auditing environment that would allow auditors to fix the system's precision (easy) and recall (much harder) errors. These types of systems would also be of value in cases when the vast amount of information does not enable the users to read all of it, and hence even a partially correct IE system would do much better than the option of not the getting any potentially relevant information at all.

In general, IE systems are useful if the following conditions are met:

- The information to be extracted is specified explicitly and no further inference is needed.
- A small number templates are sufficient to summarize the relevant parts of the document.
- The needed information is expressed relatively locally in the text.

As a first step in tagging documents for text mining systems, each document is processed to find (i.e., extract) entities and relationships that are likely to be meaningful and content-bearing. With respect to relationships, what are referred to here are facts or events involving certain entities.

By way of example, a possible **event** may be that a company has entered into a joint venture to develop a new drug. A **fact** may be that a gene causes a certain disease. **Facts** are static in nature and usually do not change; **events** are more dynamic in nature and generally have a specific time stamp associated with them. The extracted information provides more concise and precise data for the mining process than the more naive word-based approaches such as those used for text categorization, and tends to represent concepts and relationships that are more meaningful and relate directly to the examined document's domain.

From text we can extract four basic elements:

Entities Entities are the basic building blocks that can be found in text documents. Examples include people, companies, locations, genes, drugs, etc.

Attributes Attributes are features of the extracted entities. Examples of attributes might include the title of a person, the age of a person, the type of an organization, etc.

Facts Facts are the relations that exist between entities. Examples could include an employment relationship between a person and a company, Phosphorylation between two proteins, etc.

Event An event is an activity or occurrence of interest in which entities participate. Examples could include a terrorist act, a merger between two companies, a birthday, etc.

Figure 42.4.1 demonstrates of tagged entities and relations from a document:

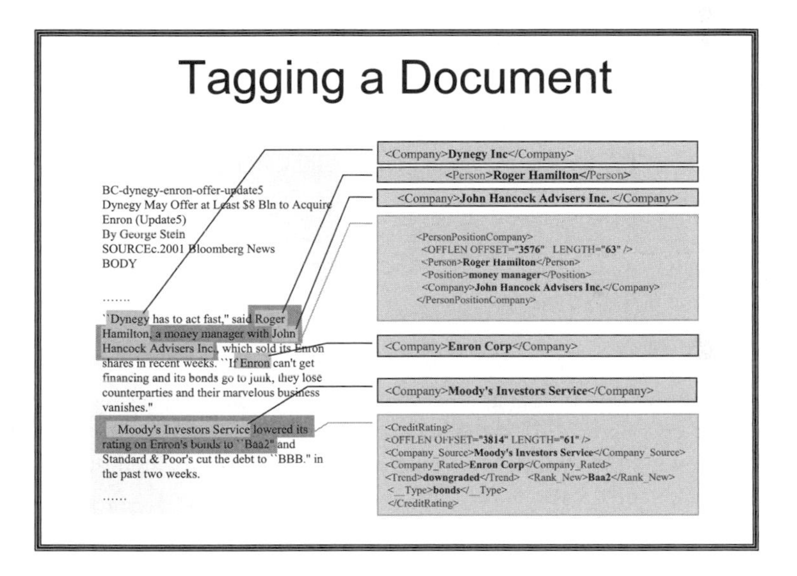

Fig. 42.3. Tagged Document.

42.5 Formal Frameworks And Algorithm-Based Techniques

42.5.1 Text Categorization

There are two main approaches to the categorization problem. The first approach is the *knowledge engineering* approach where the user is defining manually a set of rules encoding expert knowledge how to classify documents under given categories. The other approach is the *machine learning* approach where a general inductive process automatically builds an automatic text classifier by learning from a set of pre classified documents.

Knowledge Engineering Approach

An example of knowledge engineering approach is the CONSTRUE system (Hayes *et al.*, 1988; Hayes *et al.*, 1990; Hayes and Weinstein 1990; Hayes 1992) built by the Carnegie group for Reuters. A typical rule in the CONSTRUE system:

1 **if** DNF (disjunction of conjunctive clauses) formula **then** category **else** ¬ category

An example of this rule being applied might look like the following:

1 **If** ((wheat & farm) or (wheat & commodity) or (bushels & export) or (wheat & tonnes) or (wheat & winter & ¬ soft)) **then** Wheat **else** ¬ Wheat

The main drawback of this approach is what might be referred to as the knowledge acquisition bottleneck. The rules must be manually defined by a knowledge engineer interviewing a domain expert. If the set of categories is modified, then these two professionals must intervene again.

The Machine Learning Approach

The machine learning approach, on the other hand, is based on the existence of a training set of document that are already pre-tagged using the predefined set of categories.

There are two main methods for performing ML based categorization. One method is to perform "Hard" (fully automated) classification where for each pair of category and document we assign a truth value (either TRUE if the document belongs to the category or FALSE otherwise). The other approach is to perform a ranking (semi-automated) based classification. In this approach rather than returning a truth value the classifier return a Categorization Status Value (CSV), i.e. a number between 0 and 1 that represents the evidence for the fact that the document belongs to the category. Documents are then ranked according to their CSV value. Specific text categorization algorithms are discussed below. We will use the following definitions:

- $D = \{d_1, d_2. \ldots., d_n\}$, the training document collection
- $C = \{c_1, c_2. \ldots., c_k\}$, the set of possible categories to be assigned to the documents
- $T = \{t_1, t_2. \ldots., t_m\}$, the set of terms appearing in the documents
- w_{ij}: The weight of the jth term of the ith document
- $CSV_i(d_j)$: A number between 0 and 1 that represents the certainty that a category c_i should be assigned to document d_j
- $DisD_i, D_j)$: The distance between document D_i and D_j. This number represents the similarity between the documents.
- Probabilistic Classifiers view $CSV_i(d_j)$ in terms of $P(c_i|d_j)$, i.e. the probability that a document represented by a vector $\overrightarrow{d_j}$=¡$w_{1j}, \ldots, w_{mj}$¿ of (binary or weighted) terms belongs to c_i, and compute this probability by an application of Bayes' theorem:
  $P(c_i|\overrightarrow{d_j}) = \frac{P(c_i)P(\overrightarrow{d_j}|c_i)}{P(\overrightarrow{d_j})}$.
  In order to compute $P(d_j)$ and $P(d_j|\mathrm{C}_i)$ we need to make the assumption that any two coordinates of the document vector are, when viewed as random variables, statistically independent of each other; this independence assumption is encoded by the equation:
  $P(\overrightarrow{d_j}|c_i) = \prod_{k=1}^{|T|} p(w_{kj}|c_i)$.
- Example-based Classifiers do not build an explicit, declarative representation of the category c_i, but rely on the category labels attached to the training documents similar to the test document. These methods have thus been called lazy learners, since they defer the decision on how to generalize beyond the training data until each new query instance is encountered. The most prominent example of example-based classifier is KNN (K-Nearest-Neighbor).
  For deciding whether $d_j \in c_i$, k-NN looks at whether the k training documents most similar to d_j also are in c_i; if the answer is positive for a large enough proportion of them, a positive decision is taken, and a negative decision is taken otherwise. Distance-weighted version of k-NN is a variation of K-NN such that we weight the contribution of each neighbor by its similarity with the test document. Classifying d_j by means of k-NN thus comes down to computing
  $CSV_i(d_j) = \sum_{d_z \in Tr_k(d_j)} Dis(d_j, d_z) \cdot C_i(d_z)$.
  One interesting problem is how to pick the best value for k. Larkey and Croft (1996) use k = 20, while Yang (2001) has found $30 \leq$ k ≤ 45 to yield the best effectiveness. Various experiments have shown that increasing the value of k does not significantly degrade the performance.
- Propositional Rules Learners – There is a family of algorithms that try to learn the propositional definition of the category. One of the prominent examples of

this family of algorithms is Ripper (Cohen, 1995a; Cohen, 1995b, Cohen and Singer, 1996). Ripper learns rules that are disjunctions of conjunctions.

- Support Vector Machines – The support vector machine (SVM) algorithm was proven to be very fast and effective for text classification problems (Drucker *et al.*, 1999; Taira and Haruno, 1999; Joachims, 2000; Takamura and Matsumoto, 2001). SVMs were introduced by Vapnik in his work on structural risk minimization (Vapnik, 1995).
  A linear SVM is a hyperplane that separates with the maximum margin a set of positive examples from a set of negative examples. The margin is the distance from the hyperplane to the nearest example from the positive and negative sets.

Further Reading: Text Categorization

Sebastiani (2002) provides an excellent tutorial on text categorization.

- Papers that discuss various algorithms to text categorization include (Apte *et al.*, 1994; Cavnar and Trenkle, 1994; Iwayama and Tokunaga, 1994; Lewis and Ringuette, 1994; Yang and Chute, 1994; Cohen, 1995a; Goldberg, 1995; Cohen and Singer, 1996; Lam *et al.*, 1997; Ruiz and Srinivasan, 1997; Attardi *et al.*, 1998; Dumais *et al.*, 1998; Joachims, 1998; Kwok, 1998; Lam and Ho, 1998; Yavuz and Guvenir, 1998; Jo, 1999; Ruiz and Srinivasan, 1999; Taira and Haruno, 1999; Weigend *et al.*, 1999; Yang and Liu, 1999; Chen and Ho, 2000; D'Alessio *et al.*, 2000; Frank *et al.*, 2000; Junker *et al.*, 2000; Ko and Seo, 2000; Lewis, 2000; Siolas and d'Alche-Buc, 2000; Bao *et al.*, 2001; Ferilli *et al.*, 2001; Sable and Church, 2001; Soucy and Mineau, 2001; Tan, 2001; Vert, 2001; Yang, 2001; Zhang and Oles, 2001; Ko *et al.*, 2002; Ko and Seo, 2002; Leopold and Kindermann, 2002; Tan *et al.*, 2002; Bigi, 2003; Zhang *et al.*, 2003; Zhang and Yang, 2003).
- Approaches that combine several algorithms by using committees of algorithms or by using boosting are described in (Larkey and Croft, 1996; Liere and Tadepalli, 1997; Liere and Tadepalli, 1998; Forsyth, 1999; Ruiz and Srinivasan, 1999; Schapire and Singer, 2000; Sebastiani *et al.*, 2000; Al-Kofahi *et al.*, 2001; Bao *et al.*, 2001; Lam and Lai, 2001; Taira and Haruno, 2001; Nardiello *et al.*, 2003)
- Approaches that integrate linguistic knowledge and background knowledge into the categorization process can be found in ..(Jacobs 1992; Rodriguez, Gomez-Hidalgo *et al.*, 1997; Aizawa, 2001; Benkhalifa *et al.*, 2001a; Benkhalifa *et al.*, 2001b)
- Applications of text categorization are described in (Hayes *et al.*, 1988; Ittner *et al.*, 1995; Larkey, 1998; Lima *et al.*, 1998; Attardi *et al.*, 1999; Drucker *et al.*, 1999; Moens and Dumortier, 2000; Yang *et al.*, 2000; Gentili *et al.*, 2001; Krier and Zacc'a, 2002; Fall *et al.*, 2003; Giorgetti and Sebastiani, 2003; Giorgetti and Sebastiani, 2003)

42.5.2 Probabilistic models for Information Extraction

Probabilistic models often show better accuracy and robustness against the noise than categorical models. The ultimate reason for this is not quite clear, and can be an excellent subject for a philosophical debate.

Nevertheless, several probabilistic models have turned out to be especially useful for the different tasks in extracting meaning from natural language texts. Most prominent among these probabilistic approaches are Hidden Markov Models ("HMMs"), Stochastic Context-Free Grammars ("SCFG"), and Maximal Entropy ("ME").

Hidden Markov Models

HMM is a finite state automaton with stochastic state transitions and symbol emissions (Rabiner 1989). The automaton models a probabilistic generative process. In this process a sequences of symbols is produced by starting in an initial state, transitioning to a new state, emitting a symbol selected by the state and repeating this transition/emission cycle until a designated final state is reached.

Leek used the HMM for IE of gene names and location from scientific abstracts (Leek, 1997). In his work, leek build a HMM that classifies and parses natural language assertions about genes being located at a particular position on chromosomes. The HMM was trained on a small set of sentences fragments chosen from the collected scientific abstracts. Leek got precision of 80%. The HMM approach, in contrast with the traditional NLP methods, make no use of part-of-speech taggers or dictionaries, just using non-emitting states to assemble modules roughly corresponding to noun, verb and prepositional phrase.

The NYMBLE system, which was build for the MUC-6 task, (Bikel *et al.*, 1997) used the HMM approach for extracting names out of text. They created a HMM with only eight internal states (the name classes, including the NOT-A-NAME class), with two special state, the START-OF-SENTENCE and the END-OF-SENTENCE state. They trained the model for extracting names in two languages – English and Spanish. They results was the F-measure between 90% and 93%.

Another use of HMM for extracting the names of genes was done by Collier *et al.* (2000). Their research was to automatically extract facts from scientific abstracts and full papers in the molecular-biology domain. They trained the HMM entirely with bigrams based on lexical and character features in a relatively small corpus of 100 MEDLINE abstracts that were marked up by domain experts with term classes such as proteins and DNA. They get 0.73 f-score.

Seymore *et al.* (1999) explore the use of HMM models for IE tasks. They used their model for extracting important fields from the headers of computer science research papers. Their experiments show that HMM models do well at extracting important information from the headers of research papers. They achieved an accuracy of 90.1% over all fields of the header, and 98.3% for titles and 93.2% for authors. They found that the use of distantly-labeled data improved the results of the model by 10.7% accuracy in extracting for headlines.

The above are examples of the researches which has been done to implement the HMM for IE tasks. The results we get for IE by using the HMM are good comparing to other techniques but there are few problems in using HMM.

The main disadvantage of using an HMM for Information extraction is the need for a large amount of training data the more training data we have the better results we get. To build such training data it a time consuming task. We need to do lot of manually tagging which must to be done by experts of the specific domain we are working with.

The second one is that the HMM model is a flat model, so the most it can do is assign a tag to each token in a sentence. This is suitable for the tasks where the tagged sequences do not nest and where there are no explicit relations between the sequences. Part-of-speech tagging and entity extraction belong to this category, and indeed the HMM-based PoS taggers and entity extractors are state-of-the-art. Extracting relationships is different, because the tagged sequences can (and must) nest, and there are relations between them which must be explicitly recognized.

Stochastic Context-Free Grammars

A stochastic context-free grammar (SCFG) (Lari and Young, 1990; Collins, 1996; Kammeyer and Belew, 1996; Keller and Lutz, 1997a; Keller and Lutz, 1997b; Osborne and Briscoe, 1998) is a quintuple $G = (T,N,S,R,P)$, where T is the alphabet of terminal symbols (tokens), N is the set of nonterminals, S is the starting nonterminal, R is the set of rules, and $P : R \to [0..1]$ defines their probabilities. The rules have the form $n \to s_1 s_2 \ldots s_k$, where n is a nonterminal and each s_i either token or another nonterminal. As can be seen, SCFG is a usual context-free grammar with the addition of the P function.

Similarly to a canonical (non-stochastic) grammar, SCFG is said to *generate* (or *accept*) a given string (sequence of tokens) if the string can be produced starting from a sequence containing just the starting symbol S, and one by one expanding nonterminals in the sequence using the rules from the grammar. The particular way the string was generated can be naturally represented by a *parse tree* with the starting symbol as a root, nonterminals as internal nodes and the tokens as leaves.

The semantics of the probability function P is straightforward. If r is the rule $n \to s_1 s_2 \ldots s_k$, then $P(r)$ is the frequency of expanding n using this rule. Or, in Bayesian terms, if it is known that a given sequence of tokens was generated by expanding n, then $P(r)$ is the apriori likelihood that n was expanded using the rule r. Thus, it follows that for every nonterminal n the sum $\sum P(r)$ of probabilities of all rules r headed by n must equal to one.

Maximal Entropy Modelling

Consider a random process of an unknown nature which produces a single output value y, a member of a finite set Y of possible output values. The process of generating y may be influenced by some contextual information x, a member of the set X

of possible contexts. The task is to construct a statistical model that accurately represents the behavior of the random process. Such a model is a method of estimating the conditional probability of generating y given the context x.

Let $P(x,y)$ be denoted as the unknown true joint probability distribution of the random process, and $p(y|x)$ the model we are trying to build, taken from the class $\wp$ of all possible models. In order to build the model we are given a set of training samples, generated by observing the random process for some time. The training data consists of a sequence of pairs (x_i, y_i) of different outputs produced in different contexts.

In many interesting cases the set X is too large and underspecified to be directly used. For instance, X may be the set of all dots "." in all possible English texts. For contrast, the Y may be extremely simple, while remaining interesting. In the above case, the Y may contain just two outcomes: "SentenceEnd" and "NotSentenceEnd". The target model $p(y|x)$ would in this case solve the problem of finding sentence boundaries.

In cases like that it is impossible to directly use the context x to generate the output y. However, there are usually many regularities and correlations, which can be exploited. Different contexts are usually similar to each other in all manner of ways, and similar contexts tend to produce similar output distributions (Berger *et al.*, 1996; Ratnaparkhim, 1996; Rosenfeld, 1997; McCallum *et al.*, 2000; Hopkins and Cui, 2004).

42.6 Hybrid Approaches - TEG

The knowledge engineering (mostly rule based) systems traditionally were the top performers in most IE benchmarks, such as MUC (Chinchor *et al.*, 1994), ACE (ACE, 2002) and the KDD CUP (Yeh *et al.*, 2002). Recently though, the machine learning systems became state-of-the-art, especially for simpler tagging problems, such as named entity recognition (Bikel, *et al.*, 1999; Chieu and Ng, 2002), or field extraction (McCallum *et al.*, 2000).

Still, the knowledge engineering approach retains some of its advantages. It is focused around manually writing patterns to extract the entities and relations. The patterns are naturally accessible to human understanding, and can be improved in a controllable way. Whereas, improving the results of a pure machine learning system, would require providing it with additional training data. However, the impact of adding more data soon becomes infinitesimal while the cost of manually annotating the data grows linearly.

TEG (Rosenfeld *et al.*, 2004) is a hybrid entities and relations extraction system, which combines the power of knowledge-based and statistical machine learning approaches. The system is based upon SCFGs. The rules for the extraction grammar are written manually, while the probabilities are trained from an annotated corpus. The powerful disambiguation ability of PCFGs allows the knowledge engineer to write very simple and naive rules while retaining their power, thus greatly reducing the required labor.

In addition, the size of the needed training data is considerably smaller than the size of the training data needed for pure machine learning system (for achieving comparable accuracy results). Furthermore, the tasks of rule writing and corpus annotation can be balanced against each other.

Although the formalisms based upon probabilistic finite-state automata are quite successful for entity extraction, they have shortcomings, which make them harder to use for the more difficult task of extracting relationships.

One problem is that a finite-state automaton model is flat, so its natural task is assignment of a tag (state label) to each token in a sequence. This is suitable for the tasks where the tagged sequences do not nest and where there are no explicit relations between the sequences. Part-of-speech tagging and entity extraction tasks belong to this category, and indeed the HMM-based PoS taggers and entity extractors are state-of-the-art.

Extracting relationships is different in that the tagged sequences can and must nest, and there are relations between them, which must be explicitly recognized. While it is possible to use nested automata to cope with this problem, we felt that using more general context-free grammar formalism would allow for a greater generality and extendibility without incurring any significant performance loss.

42.7 Text Mining – Visualization and Analytics

One of the crucial needs in text mining process is the ability enables the user to visualize relationships between entities that were extracted from the documents. This type of interactive exploration enables one to identify new types of entities and relationships that can be extracted and, better explore the results of the information extraction phase. There are tools that can do the analytic and visualization task, the first is Clear Research (Aumann *et al.*, 1999; Feldman*et al.*, 2001; Feldman *et al.*, 2002).

42.7.1 Clear Research

Clear Research has five different visualization tools to analyze the entities and relationships. The following subsections present each one of them.

Category Connection Map

Category Connection Maps provide a means for concise visual representation of connections between different categories, e.g. between companies and technologies, countries and people, or drugs and diseases. The system finds all the connections between the terms in the different categories. To visualize the output, all the terms in the chosen categories are depicted on a circle, with each category placed on a separate part on the circle. A line is depicted between terms of different categories which are related. A color coding scheme represents stronger links with darker colors. An example of a Category Connection Map is presented in Figure 42.4. In this chapter we used a text collection (1354 documents) from yahoo-news about Bin Laden organization. In Figure 42.4 we can see the connection between Persons and Organizations.

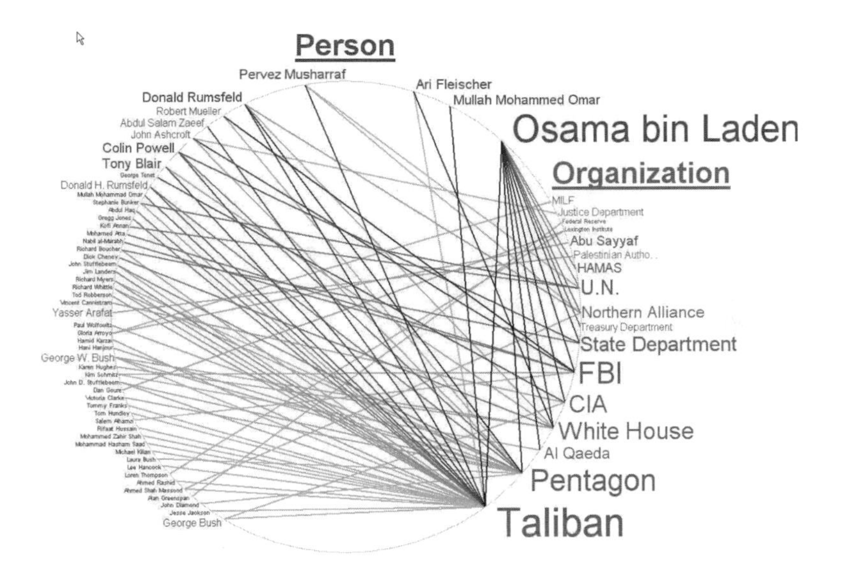

Fig. 42.4. Category map – connections between Persons and Organizations

Relationship Maps

Relationship maps provide a visual means for concise representation of the relationship between many terms in a given context. In order to define a relationship map the user defines:

- A taxonomy category (e.g. "companies"), which determines the nodes of the circle graph (e.g. companies)
- An optional context node (e.g. "joint venture"): which will determine the type of connection we wish to find among the graph nodes.

In Figure 42.5 we can see an example of relations map between Persons. The graph gives the user a summary of the entire collection in one view. The user can appreciate the overall structure of the connections between persons in this context, even before reading a single document!

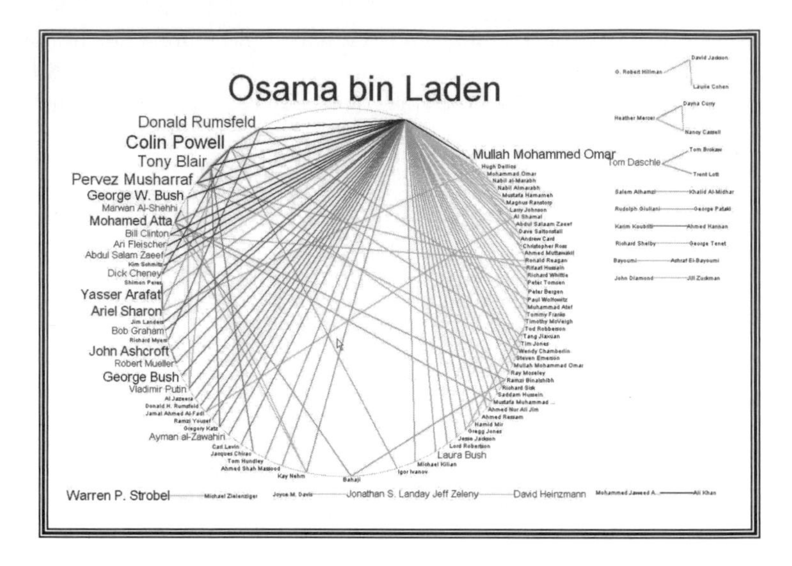

Fig. 42.5. Relationship map– relations between Persons

Spring Graph

A spring graph is a 2D graph where the distance between 2 elements should reflect the strength of the relationships between the elements. The stronger the relationship the closer the two elements should be. An example of a spring graph is shown in Figure 42.6. The graph represents the relationships between the people in a document collection. We can see that Osama Bin Laden is at the center connected to many of the other key players related to the tragic events.

Link Analysis

This query enables users to find interesting but previously unknown implicit information within the data. The Links Analysis query automatically organizes links (associations) between entities that are not present in individual documents. The results of a link analysis query can give new insight into the data and interprets the relevant interconnections between entities.

The Links Analysis query results graphically illustrate the links that indicate the associations among the selected entities. The results screen arranges the source and

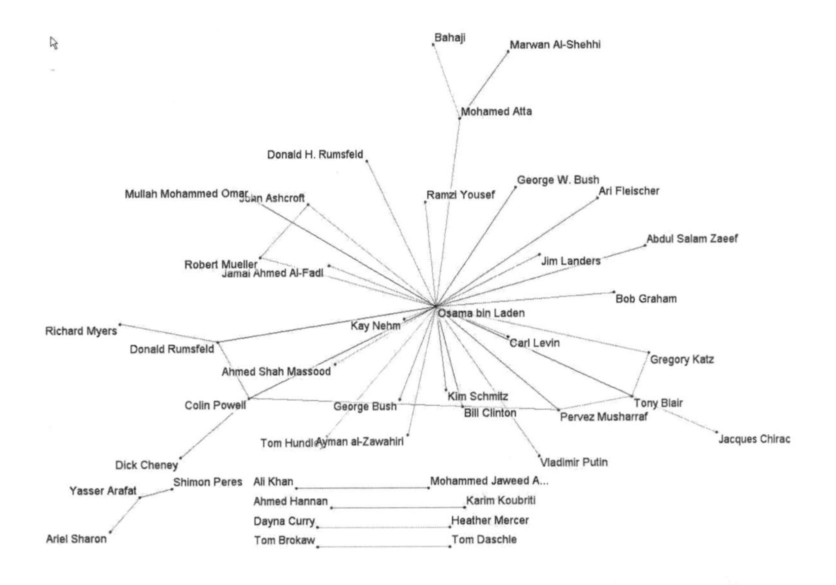

Fig. 42.6. Spring Graph

destination nodes at opposite ends and places the connecting nodes between them enabling users to follow the path that links the nodes together. The Links Analysis query is useful to users that require a graphical analysis that charts the interconnections among entities through implicit channels.

The Link Analysis query implicitly illustrates inter-relationships between entities. Users define the query criterion by defining the: source, destination and connection through entities. In this manner - the results, if any relations are found, will display the defined entities and the paths that show how they connect to one another, e.g. through third party or more entities.

In Figure 42.7 we can see a link analysis query about relation between Osama Bin Laden and John Paul II. We can see that there is no direct connection between the two but we can find indirect connection between them.

For more information regarding Link Analysis please refer to Chapter 17.5 in this volume.

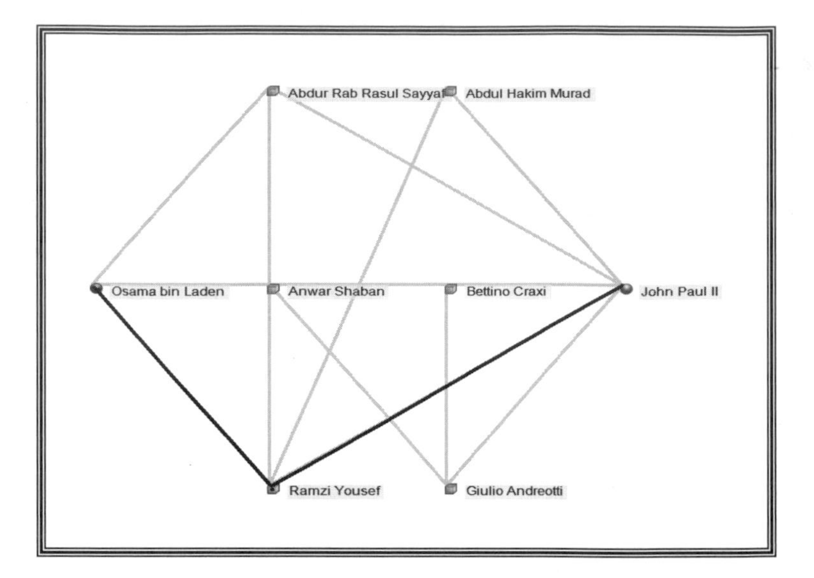

Fig. 42.7. Link Analysis – relations between Bin Laden and John Paul II.

42.7.2 Other Visualization and Analytical Approaches

The BioTeKS is an IBM prototype system for text analysis, search, and text-mining methods to support problem solving in life science, which was build by several groups in the IBM Research Division. The system is called "BioTeKS" ("Biological Text Knowledge Services"), and it integrates research technologies from multiple IBM Research labs (Mack *et al.*, 2004)

The SPIRE text visualization system, which images information from free text documents as natural terrains, serves as an example of the "ecological approach" in its visual metaphor, its text analysis, and its specializing procedures (Wise, 1999).

The ThemeRiver visualization depicts thematic variations over time within a large collection of documents. The thematic changes are shown in the context of a time line and corresponding external events. The focus on temporal thematic change within a context framework allows a user to discern patterns that suggest relationships or trends. For example, the sudden change of thematic strength following an external event may indicate a causal relationship. Such patterns are not readily accessible in other visualizations of the data (Havre *et al.*, 2002).

An approach for visualization technique of association rules is described in the following article (Wong *et al.*, 1999). We can find a technique for visualizing Sequential Patterns was describe in the work done by the Pacific Northwest National Laboratory (Wong *et al.*, 2000).

References

ACE (2002). http://www.itl.nist.gov/iad/894.01/tests/ace/. ACE - Automatic Content Extraction.

Aizawa, A. (2001). Linguistic Techniques to Improve the Performance of Automatic Text Categorization. Proceedings of NLPRS-01, 6th Natural Language Processing Pacific Rim Symposium. Tokyo, JP: 307-314.

Al-Kofahi, K., A. Tyrrell, A., Vachher, A., Travers, T., and Jackson (2001). Combining Multiple Classifiers for Text Categorization. Proceedings of CIKM-01, 10th ACM International Conference on Information and Knowledge Management. H. P. a. L. L. a. D. Grossman. Atlanta, US, ACM Press, New York, US: 97-104.

Apte, C., Damerau, F. J., and Weiss, S. M. (1994). Automated learning of decision rules for text categorization. ACM Transactions on Information Systems, 12(3): 233-251.

Attardi, G., Gulli, A., and Sebastiani, F. (1999). Automatic Web Page Categorization by Link and Context Analysis. In C. H. a. G. Lanzarone (Ed.), Proceedings of THAI-99, 1st European Symposium on Telematics, Hypermedia and Artificial Intelligence: 105-119. Varese,

Attardi, G., Marco, S. D., and Salvi, D. (1998). Categorization by context. Journal of Universal Computer Science, 4(9): 719-736.

Aumann Y., Feldman R., Ben Yehuda Y., Landau D., Lipshtat O., and Y, S. (1999). Circle Graphs: New Visualization Tools for Text-Mining. Paper presented at the PKDD.

Averbuch, M., Karson, T., Ben-Ami, B., Maimon, O., and Rokach, L. (2004). Context-sensitive medical information retrieval, MEDINFO-2004, San Francisco, CA, September. IOS Press, pp. 282-262.

Bao, Y., Aoyama, S., Du, X., Yamada, K., and Ishii, N. (2001). A Rough Set-Based Hybrid Method to Text Categorization. In M. T. O. a. H.-J. S. a. K. T. a. Y. Z. a. Y. Kambayashi (Ed.), Proceedings of WISE-01, 2nd International Conference on Web Information Systems Engineering: 254-261. Kyoto, JP: IEEE Computer Society Press, Los Alamitos, US.

Baeza-Yates, R. and Ribeiro-Neto, B. (1999). Modern Information Retrieval, Addison-Wesley.

Benkhalifa, M., Mouradi, A., and Bouyakhf, H. (2001a). Integrating External Knowledge to Supplement Training Data in Semi-Supervised Learning for Text Categorization. Information Retrieval, 4(2): 91-113.

Benkhalifa, M., Mouradi, A., and Bouyakhf, H. (2001b). Integrating WordNet knowledge to supplement training data in semi-supervised agglomerative hierarchical clustering for text categorization. International Journal of Intelligent Systems, 16(8): 929-947.

Berger, A. L., Della Pietra, S. A., and Della Pietra, V. J. (1996). A maximum entropy approach to natural language processing. Computational Linguistics, 22.

Bigi, B. (2003). Using Kullback-Leibler distance for text categorization. Proceedings of ECIR-03, 25th European Conference on Information Retrieval. F. Sebastiani. Pisa, IT, Springer Verlag: 305-319.

Bikel, D. M., S. Miller, *et al.* (1997). Nymble: a high-performance learning name-finder. Proceedings of ANLP-97: 194-201.

Bikel, D. M., Miller, S., Schwartz, R., and Weischedel, R. (1997). Nymble: a high-performance learning name-finder, Proceedings of ANLP-97: 194-201.

Brill, E. (1992). A simple rule-based part of speech tagger. Third Annual Conference on Applied Natural Language Processing, ACL.

Brill, E. (1995). "Transformation-based Error-driven Learning and Natural Language Processing: A Case Study in Part-Of-Speech Tagging." Computational Linguistics, 21(4): 543-565.

Cardie, C. (1997). "Empirical Methods in Information Extraction." AI Magazine, 18(4): 65-80.

Cavnar, W. B. and J. M. Trenkle (1994). N-Gram-Based Text Categorization. Proceedings of SDAIR-94, 3rd Annual Symposium on Document Analysis and Information Retrieval. Las Vegas, US: 161-175.

Chen, H. and S. T. Dumais (2000). Bringing order to the Web: automatically categorizing search results. Proceedings of CHI-00, ACM International Conference on Human Factors in Computing Systems. Den Haag, NL, ACM Press, New York, US: 145-152.

Chen, H. and T. K. Ho (2000). Evaluation of Decision Forests on Text Categorization. Proceedings of the 7th SPIE Conference on Document Recognition and Retrieval. San Jose, US, SPIE - The International Society for Optical Engineering: 191-199.

Chieu, H. L. and H. T. Ng (2002). Named Entity Recognition: A Maximum Entropy Approach Using Global Information. Proceedings of the 17th International Conference on Computational Linguistics.

Chinchor, N., Hirschman, L., and Lewis, D. (1994). Evaluating Message Understanding Systems: An Analysis of the Third Message Understanding Conference (MUC-3). Computational Linguistics, 3(19): 409-449.

Cohen, W. and Y. Singer (1996). Context Sensitive Learning Methods for Text categorization. SIGIR'96.

Cohen, W. W. (1995a). Learning to classify English text with ILP methods. Advances in inductive logic programming. L. D. Raedt. Amsterdam, NL, IOS Press: 124-143.

Cohen, W. W. (1995b). Text categorization and relational learning. Proceedings of ICML-95, 12th International Conference on Machine Learning. Lake Tahoe, US, Morgan Kaufmann Publishers, San Francisco, US: 124-132.

Collier, N., Nobata, C., and Tsujii, J. (2000). Extracting the names of genes and gene products with a Hidden Markov Model.

Collins, M. J. (1996). A neew statistical parser based on bigram lexical dependencies. 34 th Annual Meeting of the Association for Computational Linguistics., university of California, Santa Cruz USA.

Cutting, D. R., Pedersen, J. O., Karger, D., and Tukey., J. W. 1992. Scatter/Gather: A cluster-based approach to browsing large document collections. Paper presented at the In Proceedings of the 15th Annual International ACM/SIGIR Conference, pages 318-329, Copenhagen, Denmark.

D'Alessio, S., Murray, K., Schiaffino, R., and Kershenbaum, A. 2000. The effect of using Hierarchical classifiers in Text Categorization, Proceeding of RIAO-00, 6th International Conference "Recherche d'Information Assistee par Ordinateur": 302-313

Dorre, J., Gerstl, P., and Seiffert, R. (1999). Text mining: finding nuggets in mountains of textual data, Proceedings of KDD-99, 5th ACM International Conference on Knowledge Discovery and Data Mining: 398-401. San Diego, US: ACM Press, New York, US.

Drucker, H., Vapnik, V., and Wu, D. (1999). Support vector machines for spam categorization. IEEE Transactions on Neural Networks, 10(5): 1048-1054.

Dumais, S. T., Platt, J., Heckerman, D., and Sahami, M. (1998). Inductive learning algorithms and representations for text categorization. Paper presented at the Seventh International Conference on Information and Knowledge Management (CIKM'98).

Fall, C. J., Torcsvari, A., Benzineb, K., and Karetka, G. (2003). Automated Categorization in the International Patent Classification. SIGIR Forum, 37(1).

Feldman, R., Aumann, Y., Finkelstein-Landau, M., Hurvitz, E., Regev, Y., and Yaroshevich, A. (2002). A Comparative Study of Information Extraction Strategies, CICLing: 349-359.

Feldman, R., Aumann, Y., Liberzon, Y., Ankori, K., Schler, J., and Rosenfeld, B. (2001). A Domain Independent Environment for Creating Information Extraction Modules., CIKM: 586-588.

Feldman, R., Fresko, M., Kinar, Y., Lindell, Y., Liphstar, O., Rajman, M., Schler, Y., and Zamir, O. (1998). Text Mining at the Term Level. Paper presented at the In Proceedings of the 2nd European Symposium on Principles of Data Mining and Knowledge Discovery, Nantes, France.

Ferilli, S., Fanizzi, N., and Semeraro, G. (2001). Learning logic models for automated text categorization. In F. Esposito (Ed.), Proceedings of AI*IA-01, 7th Congress of the Italian Association for Artificial Intelligence: 81-86. Bari, IT: Springer Verlag, Heidelberg, DE.

Forsyth, R. S. (1999). New directions in text categorization. Causal models and intelligent data management. A. Gammerman. Heidelberg, DE, Springer Verlag: 151-185.

Frank, E., Chui, C., and Witten, I. H. (2000). Text Categorization Using Compression Models. In J. A. S. a. M. Cohn (Ed.), Proceedings of DCC-00, IEEE Data Compression Conference: 200-209.

Freitag, D. (1998). Machine Learning for Information Extraction in Informal Domains. Computer Science Department. Pittsburgh, PA, Carnegie Mellon University: 188.

Gentili, G. L., Marinilli, M., Micarelli, A., and Sciarrone, F. 2001. Text categorization in an intelligent agent for filtering information on the Web. International Journal of Pattern Recognition and Artificial Intelligence, 15(3): 527-549.

Giorgetti, D. and F. Sebastiani (2003). "Automating Survey Coding by Multiclass Text Categorization Techniques." Journal of the American Society for Information Science and Technology, 54(12): 1269-1277.

Giorgetti, D. and F. Sebastiani (2003). Multiclass Text Categorization for Automated Survey Coding. Proceedings of SAC-03, 18th ACM Symposium on Applied Computing. Melbourne, US, ACM Press, New York, US: 798-802.

Goldberg, J. L. (1995). CDM: an approach to learning in text categorization. Proceedings of ICTAI-95, 7th International Conference on Tools with Artificial Intelligence. Herndon, US, IEEE Computer Society Press, Los Alamitos, US: 258-265.

Grishman, R. (1996). The role of syntax in Information Extraction. Advances in Text Processing: Tipster Program Phase II, Morgan Kaufmann.

Grishman, R. (1997). Information Extraction: Techniques and Challenges. SCIE: 10-27.

Hammerton, J., Miles Osborne, Susan Armstrong, and Daelemans, W. 2002. Introduction to the Special issue on Machine Learning Approaches to Shallow Parsing. Journal of Machine Learning Research, 2(Special Issue Website): 551-558.

Havre S., Hetzler E., Whitney P., and Nowell L., (2002). "ThemeRiver: Visualizing Thematic Changes in Large Document Collections." IEEE Transactions on Visualization and Computer Graphics, 8(1): 9-20.

Hayes, P. (1992). Intelligent High-Volume Processing Using Shallow, Domain-Specific Techniques. Text-Based Intelligent Systems: Current Research and Practice in Information Extraction and Retrieval: 227-242.

Hayes, P. J., Andersen, P. M., Nirenburg, I. B., and Schmandt, L. M. (1990). Tcs: a shell for content-based text categorization, Proceedings of CAIA-90, 6th IEEE Conference on Artificial Intelligence Applications: 320-326. Santa Barbara, US: IEEE Computer Society Press, Los Alamitos, US..

Hayes, P. J., Knecht, L. E., and Cellio, M. J. (1988). A news story categorization system, Proceedings of ANLP-88, 2nd Conference on Applied Natural Language Processing: 9-17. Austin, US: Association for Computational Linguistics, Morristown, US.

Hayes, P. J. and S. P. Weinstein (1990). Construe/Tis: a system for content-based indexing of a database of news stories. Proceedings of IAAI-90, 2nd Conference on Innovative Applications of Artificial Intelligence. AAAI Press, Menlo Park, US: 49-66.

Hearst, M. A. (1999). Untangling Text Data Mining. Proceedings of ACL'99: the 37th Annual Meeting of the Association for Computational Linguistics, University of Maryland.

Hobbs, J. R., Appelt, D. E., John Bear, D. I., Kameyama, M., and Tyson, M. (1992). FASTUS: A System for Extracting Information from Text. Paper presented at the Human Language Technology.

Hopkins, J. and J. Cui (2004). Maximum Entropy Modeling in Sparse Semantic Tagging, NSF grant numbers IIS- 0121285.

Huffman, S. B. (1995). Learning information extraction patterns from examples. Learning for Natural Language Processing: 246-260.

Ittner, D. J., Lewis, D. D., and Ahn, D. D. (1995). Text categorization of low quality images, Proceedings of SDAIR-95, 4th Annual Symposium on Document Analysis and Information Retrieval: 301-315. Las Vegas, US.

Iwayama, M. and T. Tokunaga (1994). A Probabilistic Model for Text Categorization Based on a Single Random Variable with Multiple Values. In Proceedings of the 4th Conference on Applied Natural Language Processing.

Jacobs, P. (1992). Joining Statistics with NLP for Text Categorization. In Proceedings of the 3rd Conference on Applied Natural Language Processing.

Jo, T. C. (1999). Text categorization with the concept of fuzzy set of informative keywords. Proceedings of FUZZ-IEEE'99, IEEE International Conference on Fuzzy Systems. Seoul, KR, IEEE Computer Society Press, Los Alamitos, US: 609-614.

Joachims, T. (1998). Text categorization with support vector machines: learning with many relevant features. Proceedings of ECML-98, 10th European Conference on Machine Learning. Chemnitz, DE, Springer Verlag, Heidelberg, DE: 137-142.

Joachims, T. (2000). Estimating the Generalization Performance of a SVM Efficiently. Proceedings of ICML-00, 17th International Conference on Machine Learning. P. Langley. Stanford, US, Morgan Kaufmann Publishers, San Francisco, US: 431-438.

Junker, M., Sintek, M., and Rinck, M. (2000). Learning for text categorization and information extraction with ILP. In J. C. a. S. Dzeroski (Ed.), Proceedings of the 1st Workshop on Learning Language in Logic: 247-258. Bled, SL: Springer Verlag, Heidelberg, DE.

Kammeyer, T. and Belew, R. K. (1996). Stochastic Context-Free Grammar Induction with a Genetic Algorithm Using Local Search. Foundations of Genetic Algorithms, Morgan Kaufmann.

Keller, B. (1992). A Logic for Representing Grammatical Knowledge. European Conference on Artificial Intelligence: 538-542, European Conference on Artificial Intelligence.

Keller B. and Lutz R. (1997a). Evolving stochastic context-free grammars from examples using a minimum description length principle. Workshop on Automata Induction Grammatical Inference and Language Acquisition, ICML-97, Nashville, Tennessee.

Keller, B. and R. Lutz (1997b). Learning stochastic context-free grammars from corpora using a genetic algorithm. International conference on Artificial Neural Networks and Genetic Algorithms.

Kleinberg, J. M. (1999). "Authoritative sources in a hyperlinked environment." Journal of the ACM, 46(5): 604-632.

Ko, Y., Park, J., and Seo, J. (2002). Automatic Text Categorization using the Importance of Sentences, Proceedings of COLING-02, the 19th International Conference on Computational Linguistics. Taipei, TW.

Ko, Y. and J. Seo (2000). Automatic Text Categorization by Unsupervised Learning. Proceedings of COLING-00, the 18th International Conference on Computational Linguistics. Saarbrucken, DE.

Ko, Y. and J. Seo (2002). Text Categorization using Feature Projections. Proceedings of COLING-02, the 19th International Conference on Computational Linguistics. Taipei, TW.

Krier, M. and F. Zacc'a (2002). "Automatic categorization applications at the European Patent Office." World Patent Information, 24: 187-196.

Kupiec, J. (1992). "Robust Part-of-speech tagging using a hidden Markov model." Computer Speech and Language, 6.

Kwok, J. T. (1998). Automated text categorization using support vector machine. Proceedings of ICONIP'98, 5th International Conference on Neural Information Processing. Kitakyushu, JP: 347-351.

Lam, W. and C. Y. Ho (1998). Using a generalized instance set for automatic text categorization. Proceedings of SIGIR-98, 21st ACM International Conference on Research and Development in Information Retrieval. Melbourne, AU, ACM Press, New York, US: 81-89.

Lam, W. and K.-Y. Lai (2001). A Meta-Learning Approach for Text Categorization. Proceedings of SIGIR-01, 24th ACM International Conference on Research and Development in Information Retrieval. New Orleans, US, ACM Press, New York, US: 303-309.

Lam, W., Low, K. F., and Ho, C. Y. (1997). Using a Bayesian Network Induction Approach for Text Categorization. In M. E. Pollack (Ed.), Proceedings of IJCAI-97, 15th International Joint Conference on Artificial Intelligence: 745-750. Nagoya, JP: Morgan Kaufmann Publishers, San Francisco, US.

Lari, K. and Young, S. J. (1990). "The estimation of stochastic context-free grammars using the Inside-Outside algorithm." Computer Speech and Language, 4: 35–56.

Larkey, L. S. (1998). Automatic essay grading using text categorization techniques. Proceedings of SIGIR-98, 21st ACM International Conference on Research and Development in Information Retrieval. Melbourne, AU, ACM Press, New York, US: 90-95.

Larkey, L. S. and W. B. Croft (1996). Combining classifiers in text categorization. Proceedings of SIGIR-96, 19th ACM International Conference on Research and Development in Information Retrieval. Zurich, CH, ACM Press, New York, US: 289-297.

Leek, T. R. (1997). "Information extraction using hidden Markov models."

Leopold, E. and J. Kindermann (2002). "Text Categorization with Support Vector Machines: How to Represent Texts in Input Space?" Machine Learning, 46(1/3): 423-444.

Lewis, D. D. (2000). Machine learning for text categorization: background and characteristics. Proceedings of the 21st Annual National Online Meeting. M. E. Williams. New York, US, Information Today, Medford, USA: 221-226.

Lewis, D. D. and M. Ringuette (1994). A comparison of two learning algorithms for text categorization. Proceedings of SDAIR-94, 3rd Annual Symposium on Document Analysis and Information Retrieval. Las Vegas, US: 81-93.

Liere, R. and P. Tadepalli (1997). Active learning with committees for text categorization. Proceedings of AAAI-97, 14th Conference of the American Association for Artificial Intelligence. Providence, US, AAAI Press, Menlo Park, US: 591-596.

Liere, R. and P. Tadepalli (1998). Active Learning with Committees: Preliminary Results in Comparing Winnow and Perceptron in Text Categorization. Proceedings of CONALD-

98, 1st Conference on Automated Learning and Discovery. Pittsburgh, US, AAAI Press, Menlo Park, US.

Lima, L. R. D., Laender, A. H., and Ribeiro-Neto, B. A. (1998). A hierarchical approach to the automatic categorization of medical documents. In L. Bouganim (Ed.), Proceedings of CIKM-98, 7th ACM International Conference on Information and Knowledge Management: 132-139. Bethesda, US: ACM Press, New York, US.

Mack R., Mukherjea S., A. Soffer, N. Uramoto, E. Brown, A. Coden, J. Cooper, A. Inokuchi, B. Iyer, Y. Mass, H. Matsuzawa, and Subramaniam, L. V (2004). "Text analytics for life science using the Unstructured Information Management Architecture." IBN systems journal, 43.

McCallum, A., Freitag, D., and Pereira, F. (2000a). Maximum Entropy Markov Models for Information Extraction and Segmentation, Proc. 17th International Conf. on Machine Learning: 591-598: Morgan Kaufmann, San Francisco, CA.

McCallum, A., Freitag, D., and Pereira, F. (2000b). Maximum Entropy Markov Models for Information Extraction and Segmentation. Paper presented at the Proceedings of the 17th International Conference on Machine Learning.

Moens, M.-F. and J. Dumortier (2000). "Text categorization: the assignment of subject descriptors to magazine articles." Information Processing and Management, 36(6): 841-861.

Nardiello, P., Sebastiani, F., and Sperduti, A. (2003). Discretizing continuous attributes in AdaBoost for text categorization. In F. Sebastiani (Ed.), Proceedings of ECIR-03, 25th European Conference on Information Retrieval: 320-334. Pisa, IT: Springer Verlag.

Neuhaus, P. and N. Broker (1997). The Complexity of Recognition of Linguistically Adequate Dependency Grammars. Proceedings of the Thirty-Fifth Annual Meeting of the Association for Computational Linguistics and Eighth Conference of the European Chapter of the Association for Computational Linguistics., New Jersey.

Osborne M. and Briscoe T. (1998). Learning Stochastic Categorial Grammars. Computational Natural Language Learning, Association for Computational Linguistics: 80-87.

Pollard, C. and I. A. Sag (1994). "Head-Driven Phrase Structure Grammar." Chicago, Illinois, University of Chicago Press and CSLI Publications.

Rambow, O. and A. K. Joshi (1994). " A Formal Look at Dependency Grammars and Phrase-Structure Grammars, with Special Consideration of Word-Order Phenomena." Current Issues in Meaning-Text Theory. L. Wanner. London, UK, Pinter.

Ratnaparkhi, A. (1996). A maximum entropy model for part-of-speech tagging. Proc. EMNLP: Association for Computational Linguistics, New Brunswick, New Jersey.

Riloff, E. (1993a). Automatically Constructing a Dictionary for Information Extraction Tasks. In Proceedings of the Eleventh National Congress on Artificial Intelligence, AAAI Press / MIT Press.

Riloff, E. (1993b). Automatically Constructing a Dictionary for Information Extraction Tasks. National Conference on Artificial Intelligence: 811-816.

Riloff, E. (1994). Information Extraction as a Basis for Portable Text Classification Systems. Amherst, US, Department of Computer Science, University of Massachusetts.

Riloff, E. and W. Lehnert (1994). "Information extraction as a basis for high-precision text classification." ACM Transactions on Information Systems, 12(3): 296-333.

Rodriguez, M. D. B., Gomez-Hidalgo J. M., and Diaz-Agudo, B. (1997). Using WordNet to Complement Training Information in Text Categorization. Proceedings of RANLP-97, 2nd International Conference on Recent Advances in Natural Language Processing. Tzigov Chark, BL.

Rosenfeld, R. (1997). A whole sentence maximum entropy language model. Proceedings of the IEEE Workshop on Speech Recognition and Understanding., Santa Barbara, California.

Rosenfeld B., Feldman R., *et al.* (2004). TEG: a hybrid approach to information extraction. Conference on Information and Knowledge Management, Washington, D.C., USA.

Ruiz, M. E. and P. Srinivasan (1997). Automatic Text Categorization Using Neural Networks. Proceedings of the 8th ASIS/SIGCR Workshop on Classification Research. E. Efthimiadis. Washington, US, American Society for Information Science, Washington, US: 59-72.

Ruiz, M. E. and P. Srinivasan (1999). Combining Machine Learning and Hierarchical Indexing Structures for Text Categorization. Proceedings of the 10th ASIS/SIGCR Workshop on Classification Research. Washington, US, American Society for Information Science, Washington, US.

Ruiz, M. E. and P. Srinivasan (1999). Hierarchical neural networks for text categorization. Proceedings of SIGIR-99, 22nd ACM International Conference on Research and Development in Information Retrieval. Berkeley, US, ACM Press, New York, US: 281-282.

Sable, C. and K. Church (2001). Using Bins to Empirically Estimate Term Weights for Text Categorization. Proceedings of EMNLP-01, 6th Conference on Empirical Methods in Natural Language Processing. Pittsburgh, US, Association for Computational Linguistics, Morristown, US: 58-66.

Schapire, R. E. and Y. Singer (2000). "BoosTexter: a boosting-based system for text categorization." Machine Learning, 39(2/3): 135-168.

Sebastiani, F. (2002). "Machine learning in automated text categorization." ACM Computing Surveys, 34(1): 1-47.

Sebastiani, F., Sperduti A., and Valdambrini, N. (2000). An improved boosting algorithm and its application to automated text categorization. Proceedings of CIKM-00, 9th ACM International Conference on Information and Knowledge Management, US, ACM Press, New York, US: 78-85.

Seymore, K., McCallum A., and Rosenfeld, R. (1999). Learning Hidden Markov Model Structure for Information Extraction. AAAI 99 Workshop on Machine Learning for Information Extraction.

Siolas, G. and F. d'Alche-Buc (2000). Support Vector Machines based on a semantic kernel for text categorization. Proceedings of IJCNN-00, 11th International Joint Conference on Neural Networks. Como, IT, IEEE Computer Society Press, Los Alamitos, US. 5: 205-209.

Soucy, P. and G. W. Mineau (2001). A Simple KNN Algorithm for Text Categorization. Proceedings of ICDM-01, IEEE International Conference on Data Mining. San Jose, CA, IEEE Computer Society Press, Los Alamitos, US: 647-648.

Taira, H. and M. Haruno (1999). Feature selection in SVM text categorization. Proceedings of AAAI-99, 16th Conference of the American Association for Artificial Intelligence. Orlando, US, AAAI Press, Menlo Park, US: 480-486.

Taira, H. and M. Haruno (2001). Text Categorization Using Transductive Boosting. Proceedings of ECML-01, 12th European Conference on Machine Learning. Freiburg, DE, Springer Verlag, Heidelberg, DE: 454-465.

Takamura, H. and Y. Matsumoto (2001). Feature Space Restructuring for SVMs with Application to Text Categorization. Proceedings of EMNLP-01, 6th Conference on Empirical Methods in Natural Language Processing. Pittsburgh, US, Association for Computational Linguistics, Morristown, US: 51-57.

Tan, A.-H. (2001). Predictive Self-Organizing Networks for Text Categorization. Proceedings of PAKDD-01, 5th Pacific-Asia Conferenece on Knowledge Discovery and Data Mining. Hong Kong, CN, Springer Verlag, Heidelberg, DE: 66-77.

Tan, C.-M., Wang, Y.-F., and Lee, C. D. (2002). "The use of bigrams to enhance text categorization." Information Processing and Management, 38(4): 529-546.

Tkach, D. (1998). "Turning information into knowledge." a white paper from IBM.

Vapnik, V. (1995). The Nature of Statistical Learning Theory, Springer-Verlag.

Vert, J.-P. (2001). Text Categorization Using Adaptive Context Trees. Proceedings of CICLING-01, 2nd International Conference on Computational Linguistics and Intelligent Text Processing. A. Gelbukh. Mexico City, ME, Springer Verlag, Heidelberg, DE: 423-436.

Wai-chiu, W. and A. W.-c. Fu (2000). Incremental Document Clustering for Web Page Classification. In Proceedings of 2000 International Conference on Information Society in the 21st Century: Emerging Technologies and New Challenges (IS2000), Aizu-Wakameatsu City, Fukushima, Japan.

Weigend, A. S., Wiener, E. D., and Pedersen, J. O. (1999). "Exploiting hierarchy in text categorization." Information Retrieval, 1(3): 193-216.

Wilks, Y. (1997). Information Extraction as a Core Language Technology. SCIE: 1-9.

Wise, J. A. (1999). "The ecological approach to text visualization." Journal of the American Society for Information Science, 50(13): 1224-1233.

Wong P., Cowley W., Foote H., Jurrus E., Thomas J. (2000), "Visualizing sequential patterns for text mining," Proc. IEEE Information Visualization.

Wong P., Whitney P., Thomas J. (1999), Visualizing Association Rules for Text Mining. Proceedings of the 1999 IEEE Symposium on Information Visualization.

Xu, J. and W. B. Croft. (1996). Query expansion using local and global document analysis. In SIGIR '96: Proceedings of the 19th Annual International ACM SIGIR Conference on Research and Development in Information Retrieval, pages 4-11, Zurich.

Yang, Y. (2001). A Study on Thresholding Strategies for Text Categorization. Proceedings of SIGIR-01, 24th ACM International Conference on Research and Development in Information Retrieval. New Orleans, US, ACM Press, New York, US: 137-145.

Yang, Y., Ault, T., Pierce, T., and Lattimer, C. W. (2000). Improving text categorization methods for event tracking. Proceedings of SIGIR-00, 23rd ACM International Conference on Research and Development in Information Retrieval. Athens, GR, ACM Press, New York, US: 65-72.

Yang, Y. and C. G. Chute (1994). "An example-based mapping method for text categorization and retrieval." ACM Transactions on Information Systems, 2(3): 252-277.

Yang, Y. and X. Liu (1999). A re-examination of text categorization methods. Proceedings of SIGIR-99, 22nd ACM International Conference on Research and Development in Information Retrieval. Berkeley, US, ACM Press, New York, US: 42-49.

Yavuz, T. and H. A. Guvenir (1998). Application of k-nearest neighbor on feature projections classifier to text categorization. Proceedings of ISCIS-98, 13th International Symposium on Computer and Information Sciences, Ankara, TR, IOS Press, Amsterdam, NL: 135-142.

Yeh, A., Hirschman, L., and Morgan, A. (2002). "Background and Overview for KDD Cup 2002 Task 1: Information Extraction from Biomedical Articles." KDD Explorarions, 4(2): 87-89.

Zhang, J., R. Jin, Yang Y., and Hauptmann, A. (2003). Modified Logistic Regression: An Approximation to SVM and Its Applications in Large-Scale Text Categorization. Proceedings of ICML-03, 20th International Conference on Machine Learning. Washington,

DC, Morgan Kaufmann Publishers, San Francisco, US.
Zhang, J. and Y. Yang (2003). Robustness of regularized linear classification methods in text categorization. Proceedings of SIGIR-03, 26th ACM International Conference on Research and Development in Information Retrieval, Smeaton. Toronto, CA, ACM Press, New York, US: 190-197.
Zhang, T. and F. J. Oles (2001). "Text Categorization Based on Regularized Linear Classification Methods." Information Retrieval, 4(1): 5-31.

43

Spatial Data Mining

Shashi Shekhar[1], Pusheng Zhang[1], and Yan Huang[1]

University of Minnesota

Summary. Spatial Data Mining is the process of discovering interesting and previously unknown, but potentially useful patterns from large spatial datasets. Extracting interesting and useful patterns from spatial datasets is more difficult than extracting the corresponding patterns from traditional numeric and categorical data due to the complexity of spatial data types, spatial relationships, and spatial autocorrelation. This chapter provides an overview on the unique features that distinguish spatial data mining from classical Data Mining, and presents major accomplishments of spatial Data Mining research.

Key words: Spatial Data Mining, Spatial Autocorrelation, Location Prediction, Spatial Outliers, Co-location, Spatial Clustering

43.1 Introduction

The explosive growth of spatial data and widespread use of spatial databases emphasize the need for the automated discovery of spatial knowledge. Spatial Data Mining (Roddick and Spiliopoulou, 1999, Shekhar and Chawla, 2003) is the process of discovering interesting and previously unknown, but potentially useful patterns from spatial databases. The complexity of spatial data and intrinsic spatial relationships limits the usefulness of conventional Data Mining techniques for extracting spatial patterns. Efficient tools for extracting information from geo-spatial data are crucial to organizations which make decisions based on large spatial datasets, including the National Aeronautics and Space Administration (NASA), the National Geospatial-Intelligence Agency (NGA), the National Cancer Institute (NCI), and the United States Department of Transportation (USDOT). These organizations are spread across many application domains including ecology and environmental management, public safety, transportation, Earth science, epidemiology, and climatology.

General purpose Data Mining tools, such as SPSS Clementine, Statistica Data Miner, IBM Intelligent Miner, and SAS Enterprise Miner, are designed to analyze

O. Maimon, L. Rokach (eds.), *Data Mining and Knowledge Discovery Handbook*, 2nd ed., DOI 10.1007/978-0-387-09823-4_43,

large commercial databases. However, extracting interesting and useful patterns from spatial data sets is more difficult than extracting corresponding patterns from traditional numeric and categorical data due to the complexity of spatial data types, spatial relationships, and spatial autocorrelation.

Specific features of spatial data that preclude the use of general purpose Data Mining algorithms are: rich data types(e.g., extended spatial objects), implicit spatial relationships among the variables, observations that are not independent, and spatial autocorrelation among the features. In this chapter we focus on the unique features that distinguish spatial Data Mining from classical Data Mining, and present major accomplishments of spatial data mining research, especially regarding predictive modeling, spatial outlier detection, spatial co-location rule mining, and spatial clustering.

43.2 Spatial Data

The data inputs of spatial Data Mining are more complex than the inputs of classical Data Mining because they include extended objects such as points, lines, and polygons. The data inputs of spatial Data Mining have two distinct types of attributes: non-spatial attribute and spatial attribute. Non-spatial attributes are used to characterize non-spatial features of objects, such as name, population, and unemployment rate for a city. They are the same as the attributes used in the data inputs of classical Data Mining. Spatial attributes are used to define the spatial location and extent of spatial objects (Bolstad, 2002). The spatial attributes of a spatial object most often include information related to spatial locations, e.g., longitude, latitude and elevation, as well as shape.

Relationships among non-spatial objects are explicit in data inputs, e.g., arithmetic relation, ordering, is_instance_of, subclass_of, and membership_of. In contrast, relationships among spatial objects are **often implicit**, such as overlap, intersect, and behind. One possible way to deal with implicit spatial relationships is to materialize the relationships into traditional data input columns and then apply classical Data Mining techniques (Quinlan, 1993, Barnett and Lewis, 1994, Agrawal and Srikant, 1994, Jain and Dubes, 1988). However, the materialization can result in loss of information. Another way to capture implicit spatial relationships is to develop models or techniques to incorporate spatial information into the spatial data mining process.

Statistical models (Cressie, 1993) are often used to represent observations in terms of random variables. These models can then be used for estimation, description, and prediction based on probability theory. Spatial data can be thought of as resulting from observations on the stochastic process $Z(s)$: $s \in D$, where s is a spatial location and D is possibly a random set in a spatial framework. Here we present three spatial statistical problems one might encounter: point process, lattice, and geostatistics.

Point process: A point process is a model for the spatial distribution of the points in a point pattern. Several natural processes can be modeled as spatial point patterns, e.g., positions of trees in a forest and locations of bird habitats in a wetland. Spatial

Table 43.1. Relationships among Non-spatial Data and Spatial Data

Non-spatial Relationship	*Spatial Relationship*
Arithmetic	Set-oriented: union, intersection, membership, ···
Ordering	Topological: meet, within, overlap, ···
Is_instance_of	Directional: North, NE, left, above, behind, ···
Subclass_of	Metric: e.g., distance, area, perimeter, ···
Part_of	Dynamic: update, create, destroy, ···
Membership_of	Shape-based and visibility

point patterns can be broadly grouped into random or non-random processes. Real point patterns are often compared with a random pattern(generated by a Poisson process) using the average distance between a point and its nearest neighbor. For a random pattern, this average distance is expected to be $\frac{1}{2*\sqrt{density}}$, where density is the average number of points per unit area. If for a real process, the computed distance falls within a certain limit, then we conclude that the pattern is generated by a random process; otherwise it is a non-random process.

Lattice: A lattice is a model for a gridded space in a spatial framework. Here the lattice refers to a countable collection of regular or irregular spatial sites related to each other via a neighborhood relationship. Several spatial statistical analyses, e.g., the spatial autoregressive model and Markov random fields, can be applied on lattice data.

Geostatistics: Geostatistics deals with the analysis of spatial continuity and weak stationarity (Cressie, 1993), which is an inherent characteristics of spatial data sets. Geostatistics provides a set of statistics tools, such as kriging (Cressie, 1993) to the interpolation of attributes at unsampled locations.

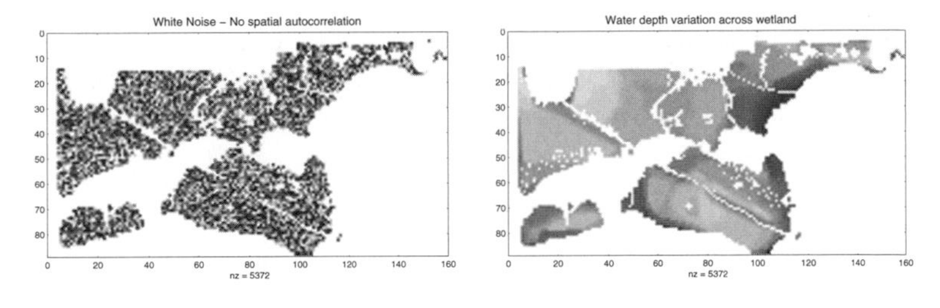

(a) Attribute with an Independent Identical Distribution

(b) Attribute with Spatial Autocorrelation

Fig. 43.1. Attribute Values in Space with Independent Identical Distribution and Spatial Autocorrelation

One of the fundamental assumptions of statistical analysis is that the data samples are independently generated: like successive tosses of coin, or the rolling of a die. However, in the analysis of spatial data, the assumption about the independence of samples is generally false. In fact, spatial data tends to be highly self correlated. For example, people with similar characteristics, occupation and background tend to cluster together in the same neighborhoods. The economies of a region tend to be similar. Changes in natural resources, wildlife, and temperature vary gradually over space. The property of like things to cluster in space is so fundamental that geographers have elevated it to the status of the first law of geography: *"Everything is related to everything else but nearby things are more related than distant things"* (Tobler, 1979). In spatial statistics, an area within statistics devoted to the analysis of spatial data, this property is called **spatial autocorrelation**. For example, Figure 43.1 shows the value distributions of an attribute in a spatial framework for an independent identical distribution and a distribution with spatial autocorrelation.

Knowledge discovery techniques which ignore spatial autocorrelation typically perform poorly in the presence of spatial data. Often the spatial dependencies arise due to the inherent characteristics of the phenomena under study, but in particular they arise due to the fact that the spatial resolution of imaging sensors are finer than the size of the object being observed. For example, remote sensing satellites have resolutions ranging from 30 meters (e.g., the Enhanced Thematic Mapper of the Landsat 7 satellite of NASA) to one meter (e.g., the IKONOS satellite from SpaceImaging), while the objects under study (e.g., Urban, Forest, Water) are often much larger than 30 meters. As a result, per-pixel-based classifiers, which do not take spatial context into account, often produce classified images with *salt and pepper* noise. These classifiers also suffer in terms of classification accuracy.

The spatial relationship among locations in a spatial framework is often modeled via a contiguity matrix. A simple contiguity matrix may represent a neighborhood relationship defined using adjacency, Euclidean distance, etc. Example definitions of neighborhood using adjacency include a four-neighborhood and an eight-neighborhood. Given a gridded spatial framework, a four-neighborhood assumes that a pair of locations influence each other if they share an edge. An eight-neighborhood assumes that a pair of locations influence each other if they share either an edge or a vertex.

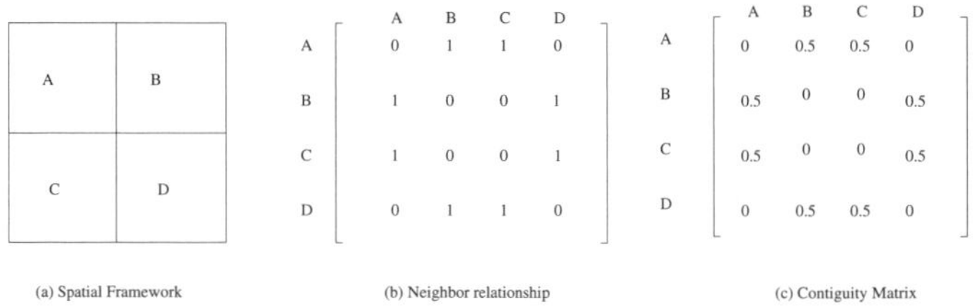

A	B
C	D

(a) Spatial Framework

	A	B	C	D
A	0	1	1	0
B	1	0	0	1
C	1	0	0	1
D	0	1	1	0

(b) Neighbor relationship

	A	B	C	D
A	0	0.5	0.5	0
B	0.5	0	0	0.5
C	0.5	0	0	0.5
D	0	0.5	0.5	0

(c) Contiguity Matrix

Fig. 43.2. A Spatial Framework and Its Four-neighborhood Contiguity Matrix.

Figure 43.2(a) shows a gridded spatial framework with four locations, A, B, C, and D. A binary matrix representation of a four-neighborhood relationship is shown in Figure 43.2(b). The row-normalized representation of this matrix is called a contiguity matrix, as shown in Figure 43.2(c). Other contiguity matrices can be designed to model neighborhood relationships based on distance. The essential idea is to specify the pairs of locations that influence each other along with the relative intensity of interaction. More general models of spatial relationships using cliques and hypergraphs are available in the literature (Warrender and Augusteijn, 1999). In spatial statistics, spatial autocorrelation is quantified using measures such as Ripley's K-function and Moran's I (Cressie, 1993).

In the rest of the chapter, we present case studies of the discovering four important patterns for spatial Data Mining: spatial outliers, spatial co-location rules, predictive models, and spatial clusters.

43.3 Spatial Outliers

Outliers have been informally defined as observations in a dataset which appear to be inconsistent with the remainder of that set of data (Barnett and Lewis, 1994), or which deviate so much from other observations so as to arouse suspicions that they were generated by a different mechanism (Hawkins, 1980). The identification of global outliers can lead to the discovery of unexpected knowledge and has a number of practical applications in areas such as credit card fraud, athlete performance analysis, voting irregularity, and severe weather prediction. This section focuses on spatial outliers, i.e., observations which appear to be inconsistent with their neighborhoods. Detecting spatial outliers is useful in many applications of geographic information systems and spatial databases, including transportation, ecology, public safety, public health, climatology, and location-based services.

A spatial outlier is a spatially referenced object whose non-spatial attribute values differ significantly from those of other spatially referenced objects in its spatial neighborhood. Informally, a spatial outlier is a local instability (in values of non-spatial attributes) or a spatially referenced object whose non-spatial attributes are extreme relative to its neighbors, even though the attributes may not be significantly different from the entire population. For example, a new house in an old neighborhood of a growing metropolitan area is a spatial outlier based on the non-spatial attribute house age.

Illustrative Examples We use an example to illustrate the differences among global and spatial outlier detection methods. In Figure 43.3(a), the X-axis is the location of data points in one-dimensional space; the Y-axis is the attribute value for each data point. Global outlier detection methods ignore the spatial location of each data point and fit the distribution model to the values of the non-spatial attribute. The outlier detected using this approach is the data point G, which has an extremely high attribute value 7.9, exceeding the threshold of $\mu + 2\sigma = 4.49 + 2 * 1.61 = 7.71$, as shown in Figure 43.3(b). This test assumes a normal distribution for attribute values. On the other hand, S is a spatial outlier whose observed value is significantly different than its neighbors P and Q.

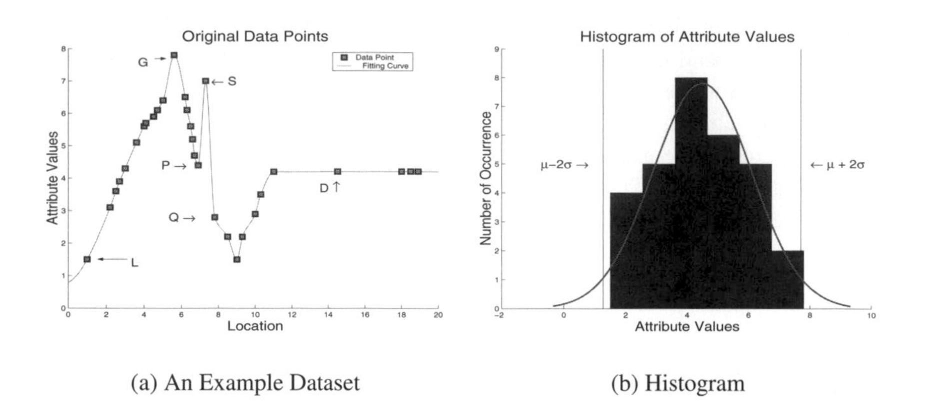

(a) An Example Dataset (b) Histogram

Fig. 43.3. A Dataset for Outlier Detection.

Tests for Detecting Spatial Outliers Tests to detect spatial outliers separate spatial attributes from non-spatial attributes. Spatial attributes are used to characterize location, neighborhood, and distance. Non-spatial attribute dimensions are used to compare a spatially referenced object to its neighbors. Spatial statistics literature provides two kinds of bi-partite multidimensional tests, namely graphical tests and quantitative tests. Graphical tests, which are based on the visualization of spatial data, highlight spatial outliers. Example methods include variogram clouds and Moran scatterplots. Quantitative methods provide a precise test to distinguish spatial outliers from the remainder of data. Scatterplots (Anselin, 1994) are a representative technique from the quantitative family.

A variogram-cloud (Cressie, 1993) displays data points related by neighborhood relationships. For each pair of locations, the square-root of the absolute difference between attribute values at the locations versus the Euclidean distance between the locations are plotted. In datasets exhibiting strong spatial dependence, the variance in the attribute differences will increase with increasing distance between locations. Locations that are near to one another, but with large attribute differences, might indicate a spatial outlier, even though the values at both locations may appear to be reasonable when examining the dataset non-spatially. Figure 43.4(a) shows a variogram cloud for the example dataset shown in Figure 43.3(a). This plot shows that two pairs (P,S) and (Q,S) on the left hand side lie above the main group of pairs, and are possibly related to spatial outliers. The point S may be identified as a spatial outlier since it occurs in both pairs (Q,S) and (P,S). However, graphical tests of spatial outlier detection are limited by the lack of precise criteria to distinguish spatial outliers. In addition, a variogram cloud requires non-trivial post-processing of highlighted pairs to separate spatial outliers from their neighbors, particularly when multiple outliers are present, or density varies greatly.

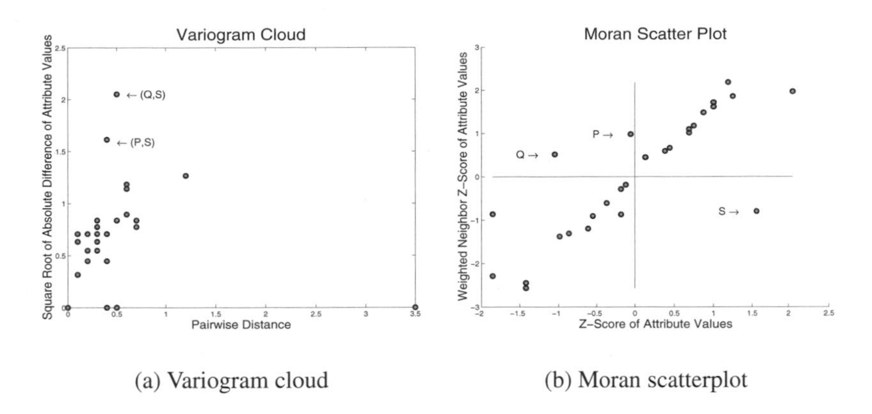

(a) Variogram cloud

(b) Moran scatterplot

Fig. 43.4. Variogram Cloud and Moran Scatterplot to Detect Spatial Outliers.

A Moran scatterplot (Anselin, 1995) is a plot of normalized attribute value ($Z[f(i)] = \frac{f(i)-\mu_f}{\sigma_f}$) against the neighborhood average of normalized attribute values ($W \cdot Z$), where W is the row-normalized (i.e., $\sum_j W_{ij} = 1$) neighborhood matrix, (i.e., $W_{ij} > 0$ iff neighbor(i, j)). The upper left and lower right quadrants of Figure 43.4(b) indicate a spatial association of dissimilar values: low values surrounded by high value neighbors(e.g., points P and Q), and high values surrounded by low values (e.g,. point S). Thus we can identify points(nodes) that are surrounded by unusually high or low value neighbors. These points can be treated as spatial outliers.

A scatterplot (Anselin, 1994) shows attribute values on the X-axis and the average of the attribute values in the neighborhood on the Y-axis. A least square regression line is used to identify spatial outliers. A scatter sloping upward to the right indicates a positive spatial autocorrelation (adjacent values tend to be similar); a scatter sloping upward to the left indicates a negative spatial autocorrelation. The residual is defined as the vertical distance (Y-axis) between a point P with location (X_p, Y_p) to the regression line $Y = mX + b$, that is, residual $\varepsilon = Y_p - (mX_p + b)$. Cases with standardized residuals, $\varepsilon_{standard} = \frac{\varepsilon - \mu_\varepsilon}{\sigma_\varepsilon}$, greater than 3.0 or less than -3.0 are flagged as possible spatial outliers, where μ_ε and σ_ε are the mean and standard deviation of the distribution of the error term ε. In Figure 43.5(a), a scatterplot shows the attribute values plotted against the average of the attribute values in neighboring areas for the dataset in Figure 43.3(a). The point S turns out to be the farthest from the regression line and may be identified as a spatial outlier.

A location (sensor) is compared to its neighborhood using the function $S(x) = [f(x) - E_{y \in N(x)}(f(y))]$, where $f(x)$ is the attribute value for a location x, $N(x)$ is the set of neighbors of x, and $E_{y\in N(x)}(f(y))$ is the average attribute value for the neighbors of x (Shekhar et al., 2003). The statistic function $S(x)$ denotes the difference of the attribute value of a sensor located at x and the average attribute value of $x's$ neighbors.

Spatial statistic $S(x)$ is normally distributed if the attribute value $f(x)$ is normally distributed. A popular test for detecting spatial outliers for normally distributed $f(x)$ can be described as follows: Spatial statistic $Z_{s(x)} = |\frac{S(x)-\mu_s}{\sigma_s}| > \theta$. For each location x with an attribute value $f(x)$, the $S(x)$ is the difference between the attribute value at location x and the average attribute value of $x's$ neighbors, μ_s is the mean value of $S(x)$, and σ_s is the value of the standard deviation of $S(x)$ over all stations. The choice of θ depends on a specified confidence level. For example, a confidence level of 95 percent will lead to $\theta \approx 2$.

Figure 43.5(b) shows the visualization of the spatial statistic method described above. The X-axis is the location of data points in one-dimensional space; the Y-axis is the value of spatial statistic $Z_{s(x)}$ for each data point. We can easily observe that point S has a $Z_{s(x)}$ value exceeding 3, and will be detected as a spatial outlier. Note that the two neighboring points P and Q of S have $Z_{s(x)}$ values close to -2 due to the presence of spatial outliers in their neighborhoods.

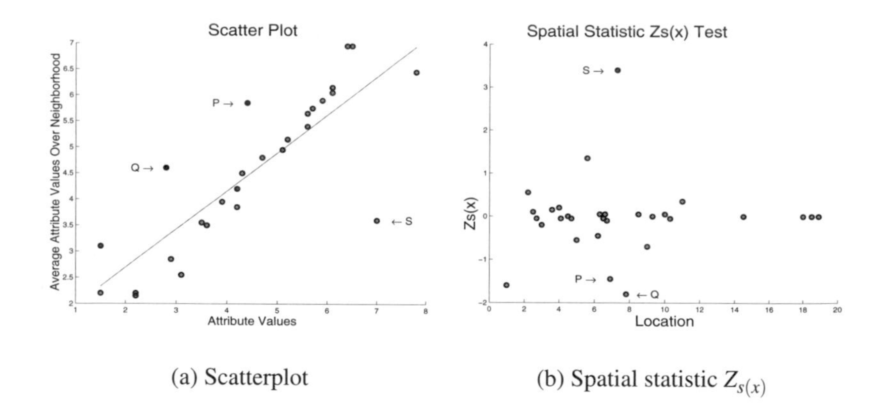

(a) Scatterplot (b) Spatial statistic $Z_{s(x)}$

Fig. 43.5. Scatterplot and Spatial Statistic $Z_{s(x)}$ to Detect Spatial Outliers.

43.4 Spatial Co-location Rules

Spatial co-location patterns represent subsets of boolean spatial features whose instances are often located in close geographic proximity. Examples include symbiotic species, e.g., the Nile Crocodile and Egyptian Plover in ecology and frontage-roads and highways in metropolitan road maps. Boolean spatial features describe the presence or absence of geographic object types at different locations in a two dimensional or three dimensional metric space, e.g., surface of the Earth. Examples of boolean spatial features include plant species, animal species, disease, crime, business types, climate disturbances, etc.

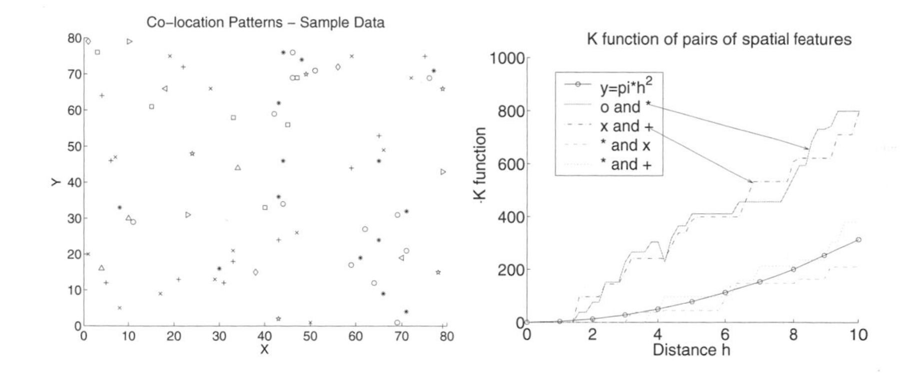

Fig. 43.6. a) A spatial dataset. Shapes represent different spatial feature types. b) Spatial features in sets {'+', '×'} and {'o', '*'} are co-located in (a) as shown by Ripley's K function

Spatial co-location rules are models to infer the presence of boolean spatial features in the neighborhood of instances of other boolean spatial features. For example, "Nile Crocodiles → Egyptian Plover" predicts the presence of Egyptian Plover birds in areas with Nile Crocodiles. Figure 43.6(a) shows a dataset consisting of instances of several boolean spatial features, each represented by a distinct shape. A careful visual review reveals two prevalent co-location patterns, i.e., ('+','×') and ('o','*'). These co-location patterns are also identified via a spatial statistical interest measure, namely Ripley's K function (Ripley, 1977). This interest measure has a value of πh^2 for a co-location pattern with a pair of spatial independent features for a given distance h. The co-location patterns ('+','×') and ('o','*') have much higher values of this interest measure relative to that of an independent pair illustrated in Figure 43.6 (b). Also note that we will refer to Ripley's K function as the K function in the rest of the chapter for simplicity.

Spatial co-location rule discovery is a process to identify co-location patterns from spatial datasets of instances of a number of boolean features. It is not trivial to adapt association rule mining algorithms to mine co-location patterns since instances of spatial features are embedded in a continuous space and share a variety of spatial relations. Reusing association rule algorithm may require transactionizing spatial datasets, which is challenging due to the risk of transaction boundaries splitting co-location pattern instances across distinct transactions as illustrated in Figure 43.7, which uses cells of a rectangular grid to define transactions. Transaction boundaries split many instances of ('+','×') and ('o','*'), which are highlighted using ellipses. Transaction-based association rule mining algorithms need to be extended to correctly and completely identify co-locations defined by interest measures, such as the K function, whose values may be adversely affected by the split instances.

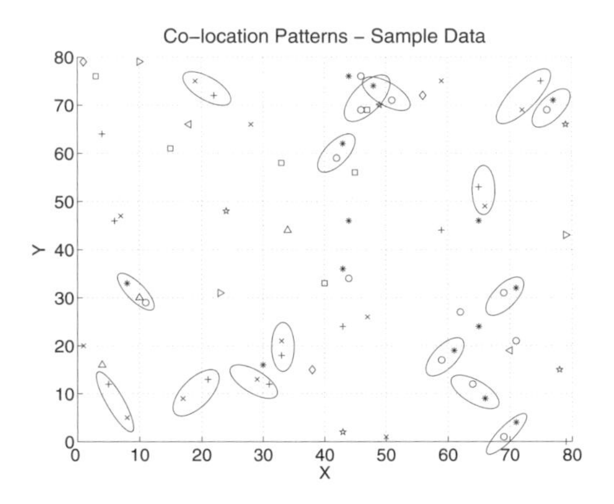

Fig. 43.7. Transactions split circled instances of co-location patterns

Approaches to discovering spatial co-location rules in the literature can be categorized into two classes, namely spatial statistics and association rules. In spatial statistics, interest measures such as the K function (Ripley, 1977) (and variations such as the L function (Cressie, 1993) and G function (Cressie, 1993)), mean nearest-neighbor distance, and quadrat count analysis (Cressie, 1993) are used to identify co-located spatial feature types. The K function for a pair of spatial features is defined as follows: $K_{ij}(h) = \lambda_j^{-1}$ E [number of type j event within distance h of a randomly chosen type i event], where λ_j is the density (number per unit area) of event j and h is the distance. Without edge effects, the K-function could be estimated by: $\hat{K}_{ij}(h) = \frac{1}{\lambda_i \lambda_j W} \sum_k \sum_l I_h(d(i_k, j_l))$, where $d(i_k, j_l)$ is the distance between the k'th location of type i and the l'th location of type j, I_h is the indicator function assuming value 1 if $d(i, j) \leq h$, and value 0 otherwise, and W is the area of the study region. $\lambda_j \times \hat{K}_{ij}(h)$ estimates the expected number of type j event instances within distance h of a type i event. The value of πh^2 is expected for a pair of independent pair of spatial features. The variance of the K function can be estimated by Monte Carlo simulation (Cressie, 1993) in general and by a close form equation under special circumstances (Cressie, 1993). Pointwise confidence intervals, e.g., 95%, can be estimated by simulating many realizations of the spatial patterns. The critical values for a test of independence could be calculated accordingly. In Figure 43.6 (b), the K functions of the two pairs of spatial features, i.e., {'+','x'} and {'o','*'}, are well above the $y = \pi * h^2$ while the K functions of the other random two pairs of spatial features, i.e., {'*','x'} and {'*','+'}, are very close to complete spatial independence. This figure does not show the confidence band. We are not aware of the definition of the K function for subsets with 3 or more spatial features. Even if the definition is generalized, computing spatial correlation measures for all possible co-location patterns can be computationally expensive due to the exponential number of candidate subsets given a large collection of spatial boolean features.

Data Mining approaches to spatial co-location mining can be broadly divided into transaction-based and spatial join-based approaches. The transaction based approaches focus on the creation of transactions over space so that an association rule mining algorithm (Agrawal and Srikant, 1994) can be used. Transactions over space have been defined by a reference-feature centric model (Koperski and Han, 1995) or a data-partition (Morimoto, 2001) approach. In the reference feature centric model (Koperski and Han, 1995), transactions are created around instances of a special user-specified spatial feature. The association rules are derived using the *apriori* (Agrawal and Srikant, 1994) algorithm. The rules found are all related to the reference feature. Generalizing this paradigm to the case where no reference feature is specified is non-trivial. Defining transactions around locations of instances of all features may yield duplicate counts for many candidate associations. Transactions in the data-partition approach (Morimoto, 2001) are formulated via grouping the spatial instances into disjoint partitions using different partitioning methods, which may yield distinct sets of transactions, which in turn yields different values of support of the co-location. Occasionally, imposing artificial disjoint transactions via space partitioning may undercount instances of tuples intersecting the boundaries of artificial transactions. In addition, to the best of our knowledge, no previous study has identified the relationship between transaction-based interest measures(e.g., support and confidence) (Agrawal and Srikant, 1994) and commonly used spatial interest measures(e.g., K function).

Spatial join-based approaches work directly with spatial data and include the cluster-then-overlay approaches (Estivill-Castro and Murray, 1998, Estivill-Castro and Lee, 2001) and instance join-based approach (Shekhar and Huang, 2001). The former treats every spatial attribute as a map layer and first identifies spatial clusters of instance data in each layer. Given X and Y as sets of layers, a clustered spatial association rule is defined as $X \Rightarrow Y(CS, CC\%)$, for $X \bigcap Y = \emptyset$, where CS is the clustered support, defined as the ratio of the area of the cluster (region) that satisfies both X and Y to the total area of the study region S, and $CC\%$ is the clustered confidence, which can be interpreted as $CC\%$ of areas of clusters (regions) of X intersect with areas of clusters(regions) of Y. The value of interest measures, e.g., clustered support and clustered confidence, depend on the choice of clustering algorithms from a large collection of choices (Han et al., 2001). To our knowledge, the relationship between these interest measures and commonly used spatial statistical measures(e.g., K function) is not yet established. In recent work (Huang et al., 2004), an instance join-based approach was proposed that uses join selectivity as the prevalence interest measures and provided interpretation models by relating those to other interest measures, e.g., K function.

43.5 Predictive Models

The prediction of events occurring at particular geographic locations is very important in several application domains. Examples of problems which require location prediction include crime analysis, cellular networking, and natural disasters such as fires, floods, droughts, vegetation diseases, and earthquakes. In this section we provide two spatial Data Mining techniques for predicting locations, namely the Spatial Autoregressive Model (SAR) and Markov Random Fields (MRF).

An Application Domain We begin by introducing an example to illustrate the different concepts related to location prediction in spatial Data Mining. We are given data about two wetlands, named Darr and Stubble, on the shores of Lake Erie in Ohio USA in order to *predict* the spatial distribution of a marsh-breeding bird, the red-winged blackbird (*Agelaius phoeniceus*). The data was collected from April to June in two successive years, 1995 and 1996.

A uniform grid was imposed on the two wetlands and different types of measurements were recorded at each cell or pixel. In total, the values of seven attributes were recorded at each cell. Domain knowledge is crucial in deciding which attributes are important and which are not. For example, *Vegetation Durability* was chosen over *Vegetation Species* because specialized knowledge about the bird-nesting habits of the red-winged blackbird suggested that the choice of nest location is more dependent on plant structure, plant resistance to wind, and wave action than on the plant species.

An important goal is to build a model for predicting the location of bird nests in the wetlands. Typically, the model is built using a portion of the data, called the learning or training data, and then tested on the remainder of the data, called the testing data. In this study we build a model using the 1995 Darr wetland data and then tested it 1995 Stubble wetland data. In the learning data, all the attributes are used to build the model and in the training data, one value is hidden, in our case the location of the nests. Using knowledge gained from the 1995 Darr data and the value of the independent attributes in the test data, we want to predict the location of the nests in 1995 Stubble data.

Modeling Spatial Dependencies Using the SAR and MRF Models Several previous studies (Jhung and Swain, 1996), (Solberg et al., 1996) have shown that the modeling of spatial dependency (often called context) during the classification process improves overall classification accuracy. Spatial context can be defined by the relationships between spatially adjacent pixels in a small neighborhood. In this section, we present two models to model spatial dependency: the spatial autoregressive model(SAR) and Markov random field(MRF)-based Bayesian classifiers.

Spatial Autoregressive Model The spatial autoregressive model decomposes a classifier $\hat{f}_C$ into two parts, namely spatial autoregression and logistic transformation. We first show how spatial dependencies are modeled using the framework of logistic

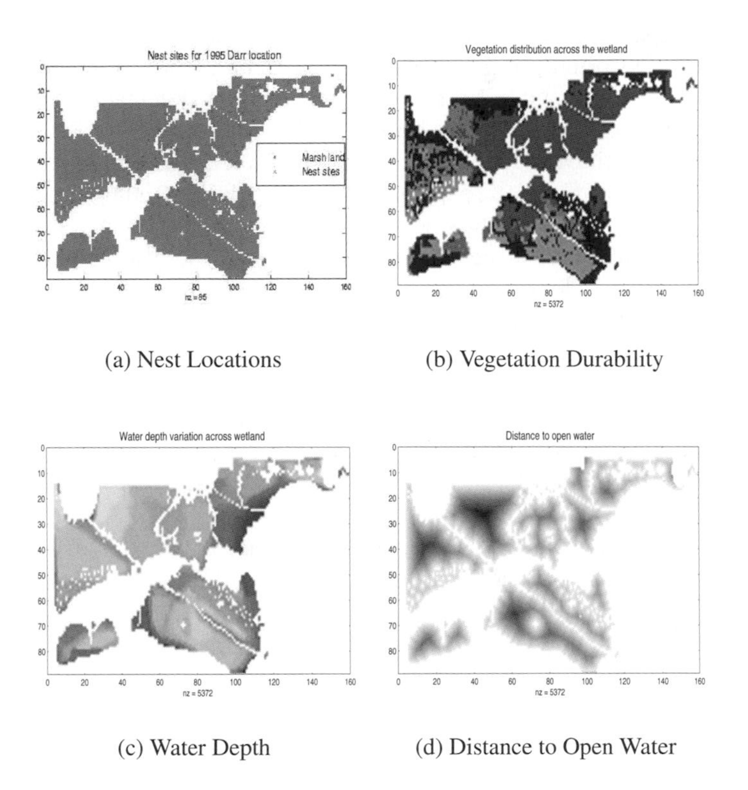

(a) Nest Locations (b) Vegetation Durability

(c) Water Depth (d) Distance to Open Water

Fig. 43.8. (a) Learning dataset: The geometry of the Darr wetland and the locations of the nests, (b) The spatial distribution of *vegetation durability* over the marshland, (c) The spatial distribution of *water depth*, and (d) The spatial distribution of *distance to open water*.

regression analysis. In the spatial autoregression model, the spatial dependencies of the error term, or, the dependent variable, are directly modeled in the regression equation (Anselin, 1988). If the dependent values y_i are related to each other, then the regression equation can be modified as

$$y = \rho W y + X\beta + \varepsilon. \tag{43.1}$$

Here W is the neighborhood relationship contiguity matrix and ρ is a parameter that reflects the strength of the spatial dependencies between the elements of the dependent variable. After the correction term $\rho W y$ is introduced, the components of the residual error vector ε are then assumed to be generated from independent and identical standard normal distributions. As in the case of classical regression, the SAR equation has to be transformed via the logistic function for binary dependent variables.

We refer to this equation as the Spatial Autoregressive Model (SAR). Notice that when $\rho = 0$, this equation collapses to the classical regression model. The benefits of modeling spatial autocorrelation are many: The residual error will have much lower spatial autocorrelation (i.e., systematic variation). With the proper choice of W, the residual error should, at least theoretically, have no systematic variation. If the spatial autocorrelation coefficient is statistically significant, then SAR will quantify the presence of spatial autocorrelation. It will indicate the extent to which variations in the dependent variable (y) are explained by the average of neighboring observation values. Finally, the model will have a better fit, (i.e., a higher R-squared statistic).

Markov Random Field-based Bayesian Classifiers Markov random field-based Bayesian classifiers estimate the classification model $\hat{f}_C$ using MRF and Bayes' rule. A set of random variables whose interdependency relationship is represented by an undirected graph (i.e., a symmetric neighborhood matrix) is called a Markov Random Field (Li, 1995). The Markov property specifies that a variable depends only on its neighbors and is independent of all other variables. The location prediction problem can be modeled in this framework by assuming that the class label, $l_i = f_C(s_i)$, of different locations, s_i, constitutes an MRF. In other words, random variable l_i is independent of l_j if $W(s_i, s_j) = 0$.

The Bayesian rule can be used to predict l_i from feature value vector X and neighborhood class label vector L_i as follows:

$$Pr(l_i|X, L_i) = \frac{Pr(X|l_i, L_i)Pr(l_i|L_i)}{Pr(X)} \tag{43.2}$$

The solution procedure can estimate $Pr(l_i|L_i)$ from the training data, where L_i denotes a set of labels in the neighborhood of s_i excluding the label at s_i, by examining the ratios of the frequencies of class labels to the total number of locations in the spatial framework. $Pr(X|l_i, L_i)$ can be estimated using kernel functions from the observed values in the training dataset. For reliable estimates, even larger training datasets are needed relative to those needed for the Bayesian classifiers without spatial context, since we are estimating a more complex distribution. An assumption on $Pr(X|l_i, L_i)$ may be useful if the training dataset available is not large enough. A common assumption is the uniformity of influence from all neighbors of a location. For computational efficiency it can be assumed that only local explanatory data $X(s_i)$ and neighborhood label L_i are relevant in predicting class label $l_i = f_C(s_i)$. It is common to assume that all interaction between neighbors is captured via the interaction in the class label variable. Many domains also use specific parametric probability distribution forms, leading to simpler solution procedures. In addition, it is frequently easier to work with a Gibbs distribution specialized by the locally defined MRF through the Hammersley-Clifford theorem (Besag, 1974).

A more detailed theoretical and experimental comparison of these methods can be found in (Shekhar et al., 2002). Although MRF and SAR classification have different formulations, they share a common goal, estimating the posterior probability distribution: $p(l_i|X)$. However, the posterior for the two models is computed differently with different assumptions. For MRF the posterior is computed using Bayes' rule. On the other hand, in logistic regression, the posterior distribution is directly fit to the data. One important difference between logistic regression and MRF is that logistic regression assumes no dependence on neighboring classes. Logistic regression and logistic SAR models belong to a more general exponential family. The exponential family is given by $Pr(u|v) = e^{A(\theta_v)+B(u,\pi)+\theta_v^T u}$ where u, v are location and label respectively. This exponential family includes many of the common distributions such as Gaussian, Binomial, Bernoulli, and Poisson as special cases.

Experiments were carried out on the Darr and Stubble wetlands to compare classical regression, SAR, and the MRF-based Bayesian classifiers. The results showed that the MRF models yield better spatial and classification accuracies over SAR in the prediction of the locations of bird nests. We also observed that SAR predictions are extremely localized, missing actual nests over a large part of the marsh lands.

43.6 Spatial Clusters

Spatial clustering is a process of grouping a set of spatial objects into clusters so that objects within a cluster have high similarity in comparison to one another, but are dissimilar to objects in other clusters. For example, clustering is used to determine the "hot spots" in crime analysis and disease tracking. Hot spot analysis is the process of finding unusually dense event clusters across time and space. Many criminal justice agencies are exploring the benefits provided by computer technologies to identify crime hot spots in order to take preventive strategies such as deploying saturation patrols in hot spot areas.

Spatial clustering can be applied to group similar spatial objects together; the implicit assumption is that patterns in space tend to be grouped rather than randomly located. However, the statistical significance of spatial clusters should be measured by testing the assumption in the data. The test is critical before proceeding with any serious clustering analyses.

Complete Spatial Randomness, Cluster, and Decluster In spatial statistics, the standard against which spatial point patterns are often compared is a completely spatially random point process, and departures indicate that the pattern is not distributed randomly in space. *Complete spatial randomness (CSR)* (Cressie, 1993) is synonymous with a homogeneous Poisson process. The patterns of the process are independently and uniformly distributed over space, i.e., the patterns are equally likely to occur anywhere and do not interact with each other. However, patterns generated by a non-random process can be either cluster patterns(aggregated patterns) or decluster patterns(uniformly spaced patterns).

To illustrate, Figure 43.9 shows realizations from a completely spatially random process, a spatial cluster process, and a spatial decluster process (each conditioned to have 80 points) in a square. Notice in Figure 43.9 (a) that the complete spatial randomness pattern seems to exhibit some clustering. This is not an unrepresentive realization, but illustrates a well-known property of homogeneous Poisson processes: event-to-nearest-event distances are proportional to χ_2^2 random variables, whose densities have a substantial amount of probability near zero (Cressie, 1993). Spatial clustering is more statistically significant when the data exhibit a cluster pattern rather than a CSR pattern or decluster pattern.

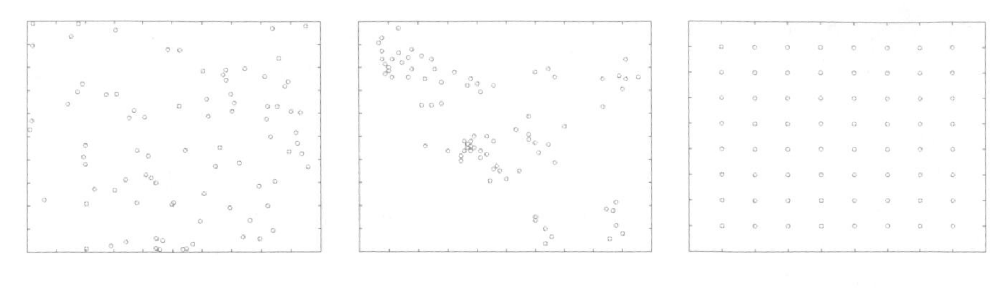

(a) CSR Pattern (b) Cluster Pattern (c) Decluster Pattern

Fig. 43.9. Illustration of CSR, Cluster, and Decluster Patterns

Several statistical methods can be applied to quantify deviations of patterns from a complete spatial randomness point pattern (Cressie, 1993). One type of descriptive statistics is based on quadrats (i.e., well defined area, often rectangle in shape). Usually quadrats of random location and orientations in the quadrats are counted, and statistics derived from the counters are computed. Another type of statistics is based on distances between patterns; one such type is Ripley's K-function (Cressie, 1993).

After the verification of the statistical significance of the spatial clustering, classical clustering algorithms (Han et al., 2001) can be used to discover interesting clusters.

43.7 Summary

In this chapter, we have focused on the features of spatial data mining that distinguish it from classical Data Mining. We have discussed major research accomplishments and techniques in spatial Data Mining, especially those related to four important output patterns: predictive models, spatial outliers, spatial co-location rules, and spatial clusters.

Acknowledgments

This work was supported in part by the Army High Performance Computing Research Center under the auspices of the Department of the Army, Army Research Laboratory cooperative agreement number DAAD19-01-2-0014, the content of which does not necessarily reflect the position or the policy of the government, and no official endorsement should be inferred.

We are particularly grateful to our collaborators Prof. Vipin Kumar, Prof. Paul Schrater, Dr. Sanjay Chawla, Dr. Chang-Tien Lu, Dr. Weili Wu, and Prof. Uygar Ozesmi for their various contributions. We also thank Xiaobin Ma, Hui Xiong, Jin Soung Yoo, Qingsong Lu, Baris Kazar, and anonymous reviewers for their valuable feedbacks on early versions of this chapter.

References

Agrawal, R. and Srikant, R. (1994). Fast Algorithms for Mining Association Rules. In *Proc. of Very Large Databases.*

Anselin, L. (1988). *Spatial Econometrics: Methods and Models.* Kluwer, Dordrecht, Netherlands.

Anselin, L. (1994). Exploratory Spatial Data Analysis and Geographic Information Systems. In Painho, M., editor, *New Tools for Spatial Analysis*, pages 45–54.

Anselin, L. (1995). Local Indicators of Spatial Association: LISA. *Geographical Analysis*, 27(2):93–115.

Barnett, V. and Lewis, T. (1994). *Outliers in Statistical Data.* John Wiley, 3rd edition edition.

Besag, J. (1974). Spatial Interaction and Statistical Analysis of Lattice Systems. *Journal of Royal Statistical Society: Series B*, 36:192–236.

Bolstad, P. (2002). *GIS Foundamentals: A Fisrt Text on GIS.* Eider Press.

Cressie, N. (1993). *Statistics for Spatial Data (Revised Edition).* Wiley, New York.

Estivill-Castro, V. and Lee, I. (2001). Data Mining Techniques for Autonomous Exploration of Large Volumes of Geo-referenced Crime Data. In *Proc. of the 6th International Conference on Geocomputation.*

Estivill-Castro, V. and Murray, A. (1998). Discovering Associations in Spatial Data - An Efficient Medoid Based Approach. In *Proc. of the Second Pacific-Asia Conference on Knowledge Discovery and Data Mining.*

Han, J., Kamber, M., and Tung, A. (2001). Spatial Clustering Methods in Data Mining: A Survey. In Miller, H. and Han, J., editors, *Geographic Data Mining and Knowledge Discovery.* Taylor and Francis.

Hawkins, D. (1980). *Identification of Outliers.* Chapman and Hall.

Huang, Y., Shekhar, S., and Xiong, H. (2004). Discovering Co-location Patterns from Spatial Datasets:A General Approach. *IEEE Transactions on Knowledge and Data Engineering*, 16(12).

Jain, A. and Dubes, R. (1988). *Algorithms for Clustering Data.* Prentice Hall.

Jhung, Y. and Swain, P. H. (1996). Bayesian Contextual Classification Based on Modified M-Estimates and Markov Random Fields. *IEEE Transaction on Pattern Analysis and Machine Intelligence*, 34(1):67–75.

Koperski, K. and Han, J. (1995). Discovery of Spatial Association Rules in Geographic Information Databases. In *Proc. Fourth International Symposium on Large Spatial Databases, Maine. 47-66.*

Li, S. (1995). A Markov Random Field Modeling. *Computer Vision.*

Morimoto, Y. (2001). Mining Frequent Neighboring Class Sets in Spatial Databases. In *Proc. ACM SIGKDD International Conference on Knowledge Discovery and Data Mining.*

Quinlan, J. (1993). *C4.5: Programs for Machine Learning.* Morgan Kaufmann Publishers.

Ripley, B. (1977). Modelling spatial patterns. *Journal of the Royal Statistical Society*, Series B 39:172–192.

Roddick, J.-F. and Spiliopoulou, M. (1999). A Bibliography of Temporal, Spatial and Spatio-Temporal Data Mining Research. *SIGKDD Explorations 1(1): 34-38 (1999).*

Shekhar, S. and Chawla, S. (2003). Spatial Databases: A Tour. *Prentice Hall (ISBN 0-7484-0064-6).*

Shekhar, S. and Huang, Y. (2001). Co-location Rules Mining: A Summary of Results. In *Proc. of the 7th Int'l Symp. on Spatial and Temporal Databases.*

Shekhar, S., Lu, C., and Zhang, P. (2003). A Unified Approach to Detecting Spatial Outliers. *GeoInformatica*, 7(2).

Shekhar, S., Schrater, P. R., Vatsavai, R. R., Wu, W., and Chawla, S. (2002). Spatial Contextual Classification and Prediction Models for Mining Geospatial Data. *IEEE Transaction on Multimedia*, 4(2).

Solberg, A. H., Taxt, T., and Jain, A. K. (1996). A Markov Random Field Model for Classification of Multisource Satellite Imagery. *IEEE Transaction on Geoscience and Remote Sensing*, 34(1):100–113.

Tobler, W. (1979). *Cellular Geography, Philosophy in Geography*. Gale and Olsson, Eds., Dordrecht, Reidel.

Warrender, C. E. and Augusteijn, M. F. (1999). Fusion of image classifications using Bayesian techniques with Markov rand fields. *International Journal of Remote Sensing*, 20(10):1987–2002.

44

Spatio-temporal clustering

Slava Kisilevich, Florian Mansmann, Mirco Nanni, Salvatore Rinzivillo

[1] University of Konstanz, Germany, slaks@dbvis.inf.uni-konstanz.de
[2] University of Konstanz, Germany Florian.Mansmann@uni-konstanz.de
[3] University of Pisa, Italy, rinziv@di.unipi.it
[4] Institute of Information Science and Technologies, Italy, mirco.nanni@isti.cnr.it

Summary. Spatio-temporal clustering is a process of grouping objects based on their spatial and temporal similarity. It is relatively new subfield of data mining which gained high popularity especially in geographic information sciences due to the pervasiveness of all kinds of location-based or environmental devices that record position, time or/and environmental properties of an object or set of objects in real-time. As a consequence, different types and large amounts of spatio-temporal data became available that introduce new challenges to data analysis and require novel approaches to knowledge discovery. In this chapter we concentrate on the spatio-temporal clustering in geographic space. First, we provide a classification of different types of spatio-temporal data. Then, we focus on one type of spatio-temporal clustering - trajectory clustering, provide an overview of the state-of-the-art approaches and methods of spatio-temporal clustering and finally present several scenarios in different application domains such as movement, cellular networks and environmental studies.

44.1 Introduction

Geographic and temporal properties are a key aspect of many data analysis problems in business, government, and science. Through the availability of cheap sensor devices we have witnessed an exponential growth of geo-tagged data in the last few years resulting in the availability of fine-grained geographic data at small temporal sampling intervals. Therefore, the actual challenge in geo-temporal analysis is moving from acquiring the right data towards large-scale analysis of the available data.

Clustering is one approach to analyze geo-temporal data at a higher level of abstraction by grouping the data according to its similarity into meaningful clusters. While the two dimensional geographic dimensions are relatively manageable, their combination with time results in a number of challenges. It is mostly application dependent how the weight of the time dimension should be considered in a distance metric. When tracking pedestrians, for example, two geographically close sample points co-occurring within a minute interval could belong to the same cluster, whereas two sample points at near distance within a time interval of a few nanoseconds in a physics experiment might belong to different clusters. In addition to this, representing temporal information on a map becomes extremely challenging.

When considering a group of points in time as a single entity, more complex data types such as trajectories emerge. Analysis questions might then deal with the correlation of these

O. Maimon, L. Rokach (eds.), *Data Mining and Knowledge Discovery Handbook*, 2nd ed.,
DOI 10.1007/978-0-387-09823-4_44,

trajectories among each others, resulting in extraction of patterns such as important places from trajectories or clustering of trajectories with common features.

Yet on a higher level, the problem of moving clusters arises. An exemplary analysis question might therefore be if there are groups of commuters within a city that move from one area of the city to another one within a particular time frame. This kind of analysis can give meaningful hints to city planners in order to avoid regular traffic jams.

The rest of this chapter first details basic concepts of spatio-temporal clustering and then lists a number of applications for spatio-temporal clustering found in the literature. Afterwards, we identify open issues in spatio-temporal clustering with a high need for future research. Finally, the last section summarizes our view on spatio-temporal clustering.

44.2 Spatio-temporal clustering

Whatever the analysis objective or the computational schema adopted, the clustering task heavily depends on the specific characteristics of the data considered. In particular, the spatio-temporal context is a large container, which includes several kinds of data types that exhibit extremely different properties and offer sensibly different opportunities of extracting useful knowledge. In this section we provide a taxonomy of the data types that are available in the spatio-temporal domain, briefly describe each class of data with a few examples taken from the spatio-temporal clustering literature, and finally report in detail the state-of-art of clustering methods for a particular kind of data – trajectories – that constitute the main focus of this chapter.

44.2.1 A classification of spatio-temporal data types

Several different forms of spatio-temporal data types are available in real applications. While they all share the availability of some kind of spatial and temporal aspects, the extent of such information and the way they are related can combine to several different kinds of data objects. Figure 44.1 visually depicts a possible classification of such data types, based on two dimensions:

- the *temporal dimension* describes to which extent the evolution of the object is captured by the data. The very basic case consists of objects that do not evolve at all, in which case only a static snapshot view of each object is available. In slightly more complex contexts, each object can change its status, yet only its most recent value (i.e., an updated snapshot) is known, therefore without any knowledge about its past history. Finally, we can have the extreme case where the full history of the object is kept, thus forming a time series of the status it traversed;
- the *spatial dimension* describes whether the objects considered are associated to a fixed location (e.g., the information collected by sensors fixed to the ground) or they can move, i.e., their location is dynamic and can change in time.

In addition to these two dimensions, a third, auxiliary one is mentioned in our classification, which is related to the spatial extension of the objects involved. The simplest case, which is also the most popular in real world case studies, considers point-wise objects, while more complex cases can take into consideration objects with an extension, such as lines and areas. In particular, Figure 44.1 focuses on point-wise objects, while their counterparts with spatial extension are omitted for the sake of presentation.

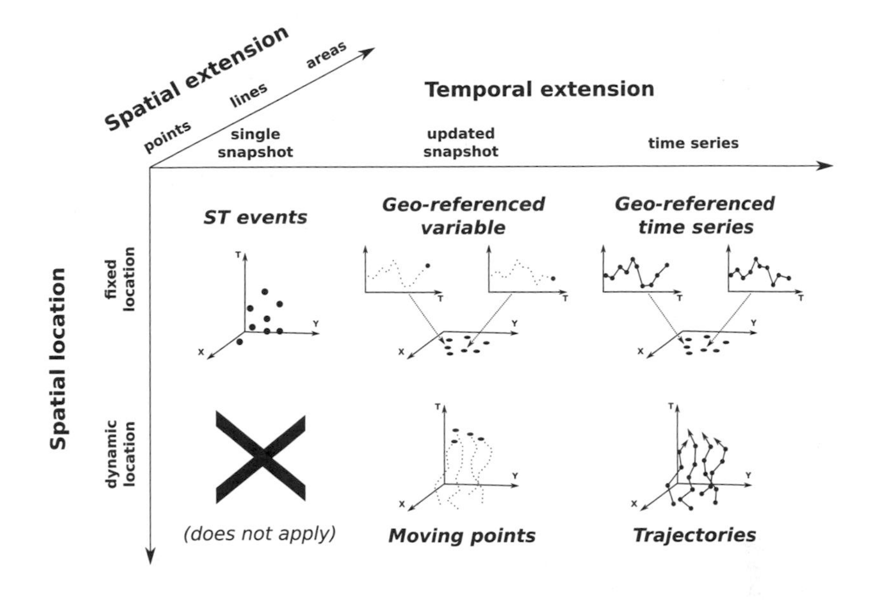

Fig. 44.1. Context for ST Clustering

In the following we briefly describe the main classes of data types we obtain for point-wise objects.

ST events. A very basic example of spatio-temporal information are spatio-temporal events, such as earth tremors captured by sensors or geo-referenced records of an epidemic. Each event is usually associated with the location where it was recorded and the corresponding timestamp. Both the spatial and the temporal information associated with the events are static, since no movement or any other kind of evolution is possible. Finding clusters among events means to discover groups that lie close both in time and in space, and possibly share other non-spatial properties. A classical example of that is (Kulldorff(1997))'s spatial scan statistics, that searches spatio-temporal cylinders (i.e., circular regions considered within a time interval) where the density of events of the same type is higher than outside, essentially representing areas where the events occurred consistently for a significant amount of time. In some applications, such as epidemiology, such area is expected to change in size and location, therefore extensions of the basic scan statistics have been proposed that consider shapes different from simple cylinders. For instance, (Iyengar(2004)) introduces (reversed) pyramid shapes, representing a small region (the pinpoint of the pyramid, e.g. the origin of an epidemic) that grows in time (the enlarging section of the pyramid, e.g. the progressive outbreak) till reaching its maximal extension (the base of the pyramid). From another viewpoint, (Wang et al(2006)Wang, Wang, and Li) proposed two spatio-temporal clustering algorithms (ST-GRID and ST-DBSCAN) for analysis of sequences of seismic events. ST-GRID is based on partitioning of the spatial and temporal dimensions into cells. ST-DBSCAN is an extension of the DBSCAN algorithm to handle spatio-temporal clustering. The k-dist graph proposed in (Ester et al(1996)Ester,

Kriegel, Sander, and Xu) as a heuristic for determination of the input parameters was used in both approaches. Hence, in the first step, the k-dist graph was created using spatial and temporal dimensions. By means of the graph, the analyst could infer the suitable thresholds for the spatial and temporal cell lengths. In the second step, the inferred cell lengths are provided to ST-GRID algorithm as an input and the dense clusters are extracted. ST-DBSCAN introduced the second parameter of the neighborhood radius in addition to the spatial neighborhood radius ε, namely temporal neighborhood radius ε_t. These two parameters were determined using k-dist graph and provided to ST-DBSCAN as an input. Thus, point p is considered as *core* when the number of points in the neighborhood is greater or equal to the threshold $MinPts$ within spatial and temporal thresholds.

Geo-referenced variables. When it is possible to observe the evolution in time of some phenomena in a fixed location, we have what is usually called a geo-referenced variable, i.e., the time-changing value of some observed property. In particular, the basic settings might allow only to remember the most recent value of such variable. In this case, the clustering task can be seen as very similar to the case of events discussed above, with the exception that the objects compared refer to the same time instant (the actual time) and their non-spatial features (variables) are not constant. A typical problem in this context consists in efficiently computing a clustering that (i) takes into account both the spatial and non-spatial features, and (ii) exploits the clusters found at the previous time stamp, therefore trying to detect the relevant changes in the data and incrementally update the clusters, rather than computing them from scratch.

Geo-referenced time series. In a more sophisticated situation, it might be possible to store the whole history of the evolving object, therefore providing a (geo-referenced) time-series for the measured variables. When several variables are available, they are usually seen as a single, multidimensional time series. In this case, clustering a set of objects requires to compare the way their time series evolve and to relate that to their spatial position. A classical problem consists in detecting the correlations (and therefore forming clusters) among different time series trying to filter out the effects of spatial autocorrelation, i.e., the mutual interference between objects due to their spatial proximity, e.g., (Zhang et al(2003)Zhang, Huang, Shekhar, and Kumar). Moreover, spatio-temporal data in the form of sequences of images (e.g., fields describing pressure and ground temperature, remotely sensed from satellites) can be seen as a particular case where location points are regularly distributed in space along a grid.

Moving objects. When (also) the spatial location of the data object is time-changing, we are dealing with moving objects. In the simplest case, the available information about such objects consists in their most recent position, as in the context of real-time monitoring of vehicles for security applications, and no trace of the past locations is kept. As in the case of geo-referenced variables, a typical clustering problem in this context consists in keeping an up-to-date set of clusters through incremental update from previous results, trying to detect the recent changes in the data (in particular, their recent movements) that were significant or that are likely to be followed by large changes in the close future, e.g., due to a change of heading of the object. An example is provided by the work in (Li et al(2004a)Li, Han, and Yang), where a *micro-clustering* technique based on direction and speed of objects is applied to achieve a large scalability.

Trajectories. When the whole history of a moving object is stored and available for analysis, the sequence of spatial locations visited by the object, together with the time-stamps of such visits, form what is called a *trajectory*. Trajectories describe the movement behavior of objects, and therefore clustering can be used to detect groups of objects that behaved in a similar way, for instance by following similar paths (maybe in different time periods), by moving consistently together (i.e., keeping close to each other for long time intervals) or by sharing other properties of movement. Recent literature is relatively rich of examples in this area, which will be the focus of this chapter and will be described in detail in the following sections.

Analogous classes of data types can be obtained through similar combination of the temporal and spatial properties on objects that possess a spatial extension, such as lines (e.g., road segments) and areas (e.g., extension of a tornado). In these cases, a dynamic spatial attribute can result not only to movement, but also to a change of shape and size. Due to the limited availability of this form of information in real scenarios and the absence of studies of specific analysis methods – especially for the dynamic cases – these contexts will not be further examined in this chapter, which instead will focus on point-wise objects in the richest setting, i.e., trajectories of moving objects.

44.2.2 Clustering Methods for Trajectory Data

Here we will focus on the context of moving objects that can be traced along the time, resulting in trajectories that describe their movements. On one hand, trajectories represent the most complex and promising (from a knowledge extraction viewpoint) form of data among those based on point-wise information. On the other hand, point-wise information is becoming nowadays largely available and usable in real contexts, while spatio-temporal data with more complex forms of spatial components are still rarely seen in real world problems – exception made for a few, very specific contexts, such as climate monitoring.

Clustering is one of the general approaches to a descriptive modeling of a large amount of data, allowing the analyst to focus on a higher level representation of the data. Clustering methods analyze and explore a dataset to associate objects in groups, such that the objects in each groups have common characteristics. These characteristics may be expressed in different ways: for example, one may describe the objects in a cluster as the population generated by a joint distribution, or as the set of objects that minimize the distances from the centroid of the group.

Descriptive and generative model-based clustering

The objective of this kind of methods is to derive a global model capable of describing the whole dataset. Some of these methods rely on a definition of multivariate density distribution and look for a set of fitting parameters for the model. In (Gaffney and Smyth(1999)) it is proposed a clustering method based on a mixture model for continuous trajectories. The trajectories are represented as functional data, i.e. each individual is modeled as a sequence of measurement given by a function of time depending on a set of parameters that models the interaction of the different distributions. The objects that are likely to be generated from a core trajectory plus gaussian noise are grouped together by means of the EM algorithm. In a successive work (Chudova et al(2003)Chudova, Gaffney, Mjolsness, and Smyth), spatial and temporal shift of trajectories within each cluster is also considered. Another approach based on a model-based technique is presented in (Alon et al(2003)Alon, Sclaroff, Kollios, and Pavlovic), where the representative of a cluster is expressed by means of a Markov model that estimates the transition between successive positions. The parameter estimation task for the model is performed by means of EM algorithm.

Distance-based clustering methods

Another approach to cluster complex form of data, like trajectories, is to transform the complex objects into features vectors, i.e. a set of multidimensional vectors where each dimension represents a single characteristic of the original object, and then to cluster them using generic clustering algorithms, like, for example, k-means. However, the complex structure of the trajectories not alway allows an approach of this kind, since most of these methods require that all the vectors are of equal length. In contrast to this, one of the largely adopted approach to the clustering of trajectories consists in defining distance functions that encapsulate the concept of similarity among the data items.

Using this approach, the problem of clustering a set of trajectories can be reduced to the problem of choosing a generic clustering algorithm, that determines how the trajectories are joined together in a cluster, and a distance function, that determines which trajectories are candidate to be in the same group. The chosen method determines also the “shape” of the resulting clusters: center-based clustering methods, like *k-means*, produce compact, spherical clusters around a set of centroids and are very sensitive to noisy outliers; hierarchical clusters organize the data items in a multi-level structure; density-based clustering methods form maximal, dense clusters, not limiting the groups number, the groups size and shape.

The concepts of similarities of spatio-temporal trajectories may vary depending on the considered application scenario. For example, two objects may be considered similar if they have followed the same spatio-temporal trajectory within a given interval, i.e. they have been in the same places at the same times. However, the granularity of the observed movements (i.e. the number of sampled spatio-temporal points for each trajectory), the uncertainty on the measured points, and, in general, other variations of the availability of the locations of the two compared objects have required the definition of several similarity measures for spatio-temporal trajectories. The definition of these measures is not only tailored to the cluster analysis task, but it is strongly used in the field of Moving Object Databases for the similarity search problem (Theodoridis(2003)), and it is influenced also by the work on time-series analysis (Agrawal et al(1993)Agrawal, Faloutsos, and Swami, Berndt and Clifford(1996), Chan and chee Fu(1999)) and Longest Common Sub Sequence (LCSS) model (Vlachos et al(2002)Vlachos, Kollios, and Gunopulos, Vlachos et al(2003)Vlachos, Hadjieleftheriou, Gunopulos, and Keogh, Chen et al(2005)Chen, Özsu, and Oria). The distance functions defined in (Nanni and Pedreschi(2006), Pelekis et al(2007)Pelekis, Kopanakis, Marketos, Ntoutsi, Andrienko, and Theodoridis) are explicitly defined on the trajectory domain and take into account several spatio-temporal characteristics of the trajectories, like direction, velocity and co-location in space and time.

Density-based methods and the DBSCAN family

The density-based clustering methods use a density threshold around each object to distinguish the relevant data items from noise. DBSCAN (Ester et al(1996)Ester, Kriegel, Sander, and Xu), one of the first example of density-based clustering, visits the whole dataset and tags each object either as *core object* (i.e. an object that is definitively within a cluster), *border object* (i.e. objects at the border of a cluster), or *noise* (i.e. objects definitively outside any cluster). After this first step, the core objects that are close each other are joined in a cluster. In this method, the density threshold is espressed by means of two parameters: a maximum radius ε around each object, and a minimum number of objects, say *MinPts*, within this interval. An object p is defined a *core object* if its neighborhood of radius ε (denoted as $N_\varepsilon(p)$) contains at least *MinPts* objects. Using the core object condition, the input dataset is scanned and the status of each object is determined. A cluster is determined both by its core objects and the objects that are reachable from a core object, i.e. the objects that do not satisfy the core object condition but that are contained in the Eps-neighborhood of a core object. The concept of "reachable" is express in terms of the *reachability distance*. It is possible to define two measures of distances for a core object c and an object in its ε-neighborhood: the *core distance*, which is the distance of the *MinPts*-th object in the neighborhood of c in order of distance ascending from c, and the *reachability distance*, i.e. the distance of an object p from c except for the case when p's distance is less than the *core distance*; in this case the distance is normalized to the *core distance*. Given a set of *core* and *border* object for a dataset, the clusters are formed by visiting all the objects, starting from a core point: the cluster formed by the single point is extended by including other objects that are within a reachability distance; the process is repeated by including all the objects reachable by the new included items, and so on. The growth of the cluster stops when all the border points of the cluster have been visited and there are no more reachable items. The visit may continue from another core object, if avaiable.

The OPTICS method (Ankerst et al(1999)Ankerst, Breunig, Kriegel, and Sander) proceeds by exploring the dataset and enumerating all the objects. For each object p it checks if the core object conditions are satisfied and, in the positive case, starts to enlarge the potential cluster by checking the condition for all neighbors of p. If the object p is not a core object, the scanning process continues with the next unvisited object of D. The results are summarized in a reachability plot: the objects are represented along the horizontal axis in the order of visiting them and the vertical dimension represents their reachability distances. Intuitively, the reachability distance of an object p_i corresponds to the minimum distance from the set of its predecessors $p_j, 0 < j < i$. As a consequence, a high value of the reachability distance roughly means a high distance from the other objects, i.e. indicates that the object is in a sparse area. The actual clusters may be determined by defining a reachability distance threshold and grouping together the consecutive items that are below the chosen threshold in the plot. The result of the OPTICS algorithm is insensitive to the original order of the objects in the dataset. The objects are visited in this order only until a core object is found. After that, the neighborhood of the core object is expanded by adding all density-connected objects. The order of visiting these objects depends on the distances between them and not on their order in the dataset. It is also not important which of density-connected objects will be chosen as the first core object since the algorithm guarantees that all the objects will be put close together in the resulting ordering. A formal proof of this property of the algorithm is given in (Ester et al(1996)Ester, Kriegel, Sander, and Xu).

It is clear that the density methods strongly rely on an efficient implementation of the neighborhood query. In order to improve the performances of such algorithms it is necessary to have the availability of valid index data structure. The density based algorithms are largely used in different context and they take advantages of many indices like R-tree, kd-tree, etc. When dealing with spatio-temporal data, it is necessary to adapt the existing approaches also for the spatio-temporal domain (Frentzos et al(2007)Frentzos, Gratsias, and Theodoridis) or use a general distance based index (e.g. M-tree, (Ciaccia et al(1997)Ciaccia, Patella, and Zezula))

The approach of choosing a clustering method and a distance function is just a starting point for a more evolute approach to mining. For example, in (Nanni and Pedreschi(2006)) the basic notion of the distance function is exploited to stress the importance of the temporal characteristics of trajectories. The authors propose a new approach called *temporal focusing* to better exploit the temporal aspect and improve the quality of trajectory clustering. For example, two trajectories may be very different if the whole time interval is considered. However, if only a small sub-interval is considered, these trajectories may be found very similar. Hence, it is very crucial for the algorithm to efficiently work on different spatial and temporal granularities. As mentioned by the authors, usually some parts of trajectories are more important than others. For example, in rush hours it can be expected that many people moving from home to work and viceversa form movement patterns that can be grouped together. On weekends, people's activity can be less ordered where the local distribution of people is more influential than collective movement behavior. Hence, there is a need for discovering the most interesting time intervals in which movement behavior can be organized into meaningful clusters. The general idea of the time focusing approach is to cluster trajectories using all possible time intervals (time windows), evaluate the results and find the best clustering. Since the time focusing method is based on OPTICS, the problem of finding the best clusters converges to finding the best input parameters. The authors proposed several quality functions based on density notion of clusters that measures the quality of the produced clustering and are expressed in terms of average reachability (Ankerst et al(1999)Ankerst, Breunig, Kriegel, and Sander) with respect to a time interval I and reachability threshold ε'. In addition, ways of finding optimal values of ε' for every time interval I were provided.

Visual-aided approaches

Analysis of movement behavior is a complex process that requires understanding of the nature of the movement and phenomena it incurs. Automatic methods may discover interesting behavioral patterns with respect to the optimization function but it may happen that these patterns are trivial or wrong from the point of view of the phenomena that is under investigation. The visual analytics field tries to overcome the issues of automatic algorithms introducing frameworks implementing various visualization approaches of spatio-temporal data and proposing different methods of analysis including trajectory aggregation, generalization and clustering (Andrienko and Andrienko(2006), Andrienko et al(2007)Andrienko, Andrienko, and Wrobel, Andrienko and Andrienko(2008), Andrienko et al(2009)Andrienko, Andrienko, Rinzivillo, Nanni, Pedreschi, and Giannotti, Andrienko and Andrienko(2009)). These tools often target different application domains (movement of people, animals, vehicles) and support many types of movement data (Andrienko et al(2007)Andrienko, Andrienko, and Wrobel). The advantages of visual analytics in analysis of movement data is clear. The analyst can control the computational process by setting different input parameters, interpret the results and direct the algorithm towards the solution that better describes the underlying phenomena.

In (Rinzivillo et al(2008)Rinzivillo, Pedreschi, Nanni, Giannotti, Andrienko, and Andrienko) the authors propose progressive clustering approach to analyze the movement behavior of objects. The main idea of the approach is the following. The analyst or domain expert progressively applies different distance functions that work with spatial, temporal, numerical or categorical variables on the spatio-temporal data to gain understanding of the underlying data in a stepwise manner. This approach is orthogonal to commonly used approaches in machine learning and data mining where the distance functions are combined together to optimize the outcome of the algorithm.

Micro clustering methods

In (Hwang et al(2005)Hwang, Liu, Chiu, and Lim) a different approach is proposed, where trajectories are represented as piece-wise segments, possibly with missing intervals. The proposed method tries to determine a *close time interval*, i.e. a maximal time interval where all the trajectories are pair-wise close to each other. The similarity of trajectories is based on the amount of time in which trajectories are close and the mining problem is to find all the trajectory groups that are close within a given threshold.

A similar approach based on an extension of *micro-clustering* is proposed in (Li et al(2004b)Li, Han, and Yang). In this case, the segments of different trajectories within a given rectangle are grouped together if they occur in similar time intervals. The objective of the method is to determine the maximal group size and temporal dimension within the threshold rectangle.

In (Lee et al(2007)Lee, Han, and Whang), the trajectories are represented as sequences of points without explicit temporal information and they are partitioned into a set of quasi-linear segments. All the segments are grouped by means of a density based clustering method and a representative trajectory for each cluster is determined.

Flocks and convoy

In some application domains there is a need in discovering group of objects that move together during a given period of time. For example, migrating animals, flocks of birds or convoys of vehicles. (Kalnis et al(2005)Kalnis, Mamoulis, and Bakiras) proposed the notion of *moving clusters* to describe the problem of discovery of sequence of clusters in which objects may leave or enter the cluster during some time interval but having the portion of common objects higher than a predefined threshold. Other patterns of moving clusters were proposed in the literature: (Gudmundsson and van Kreveld(2006), Vieira et al(2009)Vieira, Bakalov, and Tsotras) define a flock pattern, in which the same set of objects stay together in a circular region of a predefined radius, while (Jeung et al(2008)Jeung, Yiu, Zhou, Jensen, and Shen) defines a convoy pattern, in which the same set of objects stay together in a region of arbitrary shape and extent.

(Kalnis et al(2005)Kalnis, Mamoulis, and Bakiras) proposed three algorithms for discovery of moving clusters. The basic idea of these algorithms is the following. Assuming that the locations of each object were sampled at every timestamp during the lifetime of the object, a snapshot $S_{t=i}$ of objects' positions is taken at every timestamp $t = i$. Then, DBSCAN (Ester et al(1996)Ester, Kriegel, Sander, and Xu), a density-based clustering algorithm, is applied on the snapshot forming clusters $c_{t=i}$ using density constraints of *MinPts* (minimum points in the neighborhood) and ε (radius of the neighborhood). Having two snapshots clusters $c_{t=i}$ and $c_{t=i+1}$, the moving cluster $c_{t=i}c_{t=i+1}$ is formed if $\frac{|c_{t=i} \cap c_{t=i+1}|}{|c_{t=i} \cup c_{t=i+1}|} > \theta$, where θ is an integrity threshold between 0 and 1.

(Jeung et al(2008)Jeung, Yiu, Zhou, Jensen, and Shen) adopts DBSCAN algorithm to find candidate convoy patterns. The authors proposed three algorithms that incorporate trajectory simplification techniques in the first step. The distance measures are performed on the segments of trajectories as opposed to commonly used point based distance measures. They show that the clustering of trajectories at every timestamp as it is performed in moving clusters is not applicable to the problem of convoy patterns because the global integrity threshold θ may be not known in advance and time constraint (lifetime) is not taken into account, which is important in convoy patterns. Another problem is related to the trajectory representation: Some trajectories may have missing timestamps or be measured at different time intervals. Therefore, the density measures cannot be applied between trajectories with different timestamps. To handle the problem of missing timestamps, the authors proposed to interpolate the trajectories creating virtual time points and apply density measures on segments of the trajectories. Additionally, the convoy was defined as candidate when it had at least k clusters during k consequent timestamps.

Five on-line algorithms for discovery flock patterns in spatio-temporal databases were presented in (Vieira et al(2009)Vieira, Bakalov, and Tsotras). The flock pattern Φ is defined as the maximal number of trajectories and greater or equal to density threshold μ that move together during minimum time period δ. Additionally, the disc with radius $\varepsilon/2$ with the center $c_k^{t_i}$ of the flock k at time t_i should cover all the points of flock trajectories at time t_i. All the algorithms employ the grid-based structure. The input space is divided into cells with edge size ε. Every trajectory location sampled at time t_i is placed in one of the cells. After processing all the trajectories at time t_i, a range query with radius ε is performed on every point p to find neighbor points whose distance from p is at most ε and the number of neighbor points is not less than μ. Then, for every pairs of points found, density of neighbor points with minimum radius $\varepsilon/2$ is determined. If the density of a disk is less than μ, the disk is discarded otherwise the common points of two valid disks are found. If the number of common points is above the threshold then the disk is added to a list of candidate disks. In the basic algorithm that generate flock patterns, the candidate disk at time t_i is compared to the candidate disk at time t_{i-1} and augmented together if they have the common number of points above the threshold. The flock is generated if the augmented clusters satisfy the time constraint δ. In other four proposed algorithms, different heuristics were applied to speed up the performance by improving generation of candidate disks. In one of the approaches called *Cluster Filtering Evaluation*, DBSCAN with parameters μ as a density threshold and ε for neighborhood radius is used to generate candidate disks. Once candidate disks are obtained, the basic algorithm for finding flocks is applied. This approach works particularly well when trajectory dataset is relatively small and many trajectories have similar moving patterns.

Important places

In the work of (Kang et al(2004)Kang, Welbourne, Stewart, and Borriello), the authors proposed an incremental clustering for identification of important places in a single trajectory. Several factors for the algorithm were defined: arbitrary number of clusters, exclusion of as much unimportant places as possible and being not computationally expensive to allow running on mobile devices. The algorithm is based on finding important places where many location measurements are clustered together. Two parameters controlled the cluster creation - distance between positions and time spent in a cluster. The basic idea is the following. Every new location measurement provided by a location-based device (Place Lab, in this case) is compared to the previous location. If the distance between previous location is less than a threshold, the new location is added to the previously created cluster. Otherwise, the new candidate cluster is created with the new location. The candidate cluster becomes a cluster of important places when the time difference between first point in a cluster and the last point is greater than the threshold. Similar ideas of finding interesting places in trajectories were used in later works (Alvares et al(2007)Alvares, Bogorny, Kuijpers, de Macedo, Moelans, and Vaisman, Zheng et al(2009)Zheng, Zhang, Xie, and Ma).

A similar task was performed in (Palma et al(2008)Palma, Bogorny, Kuijpers, and Alvares), this time by using speed characteristics. For this, the original definition of DBSCAN was altered to accommodate the temporal aspect. Specifically, the point p of a trajectory called *core* point if the time difference between first and last neighbor points of p was greater or equal to some predefined threshold $MinTime$ (minimum time). This definition corresponds to the maximum average speed condition $\varepsilon/MinTime$ in the neighborhood of point p. Since original DBSCAN requires two parameters to be provided for clustering: ε - radius of the neighborhood and $MinPts$ - minimum number of points in the neighborhood of p, similarly, the adopted version required providing two parameters: ε and $MinTime$. However, without knowing the characteristic of the trajectory it is difficult for the user to provide meaningful parameters. The authors proposed to regard the trajectory as a list of distances between two consecutive points and obtain means and standard deviations of these distances. Then, Gaussian curve can be plotted using these parameters that should give some information about the properties of the trajectory and inverse cumulative distribution function can be constructed expressed in terms of mean and standard deviation. In order to obtain ε, the user should provide a value between 0 and 1 that reflects the proportion of points that can be expected in a cluster.

Borderline cases: patterns

Patterns that are mined from trajectories are called *trajectory patterns* and characterize interesting behaviors of single object or group of moving objects (Fosca and Dino(2008)). Different approach exist in mining trajectory patterns. We present two examples. The first one is based on grid-based clustering and finding dense regions (Giannotti et al(2007)Giannotti, Nanni, Pinelli, and Pedreschi), the second is based on partitioning of trajectories and clustering of trajectories' segments (Kang and Yong(2009)).

(Giannotti et al(2007)Giannotti, Nanni, Pinelli, and Pedreschi) presented an algorithm to find frequent movement patterns that represent cumulative behavior of moving objects where a pattern, called *T-pattern*, was defined as a sequence of points with temporal transitions between consecutive points. A *T-pattern* is discovered if its spatial and temporal components approximately correspond to the input sequences (trajectories). The meaning of these patterns is that different objects visit the same places with similar time intervals. Once the patterns are discovered, the classical sequence mining algorithms can be applied to find frequent patterns. Crucial to the determination of *T-patterns* is the definition of the visiting regions. For this, the *Region-of-Interest (RoI)* notion was proposed. A *RoI* is defined as a place visited by many objects. Additionally, the duration of stay can be taken into account. The idea behind *Roi* is to divide the working region into cells and count the number of trajectories that intersect the cell. The algorithm for finding popular regions was proposed, which accepted the grid with cell densities and a density threshold δ as input. The algorithm scans the cells and tries to expand the region in four directions (left, right, up, down). The direction that maximizes the average cell density is selected and the cells are merged. After the regions of interest are obtained, the sequences can be created by following every trajectory and matching the regions of interest they intersect. The timestamps are assigned to the regions in two ways: (1) Using the time when the trajectory entered the region or (2) Using the starting time if the trajectory started in that region. Consequently, the sequences are used in mining frequent *T-patterns*. The proposed approach was evaluated on the trajectories of 273 trucks in Athens, Greece having 112,203 points in total.

(Kang and Yong(2009)) argues that methods based on partition of the working space into grids may lose some patterns if the cell lengths are too large. In addition, some methods require trajectory discretization according to its recorded timestamps which can lead to creation of redundant and repeating sequences in which temporal aspects are contained in the sequentially ordered region ids. As a workaround to these issues, the authors proposed two refinements: (1) Partitioning trajectories into disjoint segments, which represent meaningful spatio-temporal changes of the movement of the object. The segment is defined as an area having start and end points as well as the time duration within the area. (2) Applying clustering algorithm to group similar segments. A *ST-pattern* (Spatio-temporal pattern) was defined as a sequences of segments (areas) with time duration described as a height of 3-dimensional cube. Thus, the sequences of *ST-patterns* are formed by clustering similar cubes. A four-step approach was proposed to mine frequent *ST-patterns*. In the first step, the trajectories are simplified using the DP (Douglas-Peucker) algorithm dividing the trajectories into segments. The segments are then normalized using linear transformation to allow comparison between segments having different offsets. In the next step, the spatio-temporal segments are clustered using the BIRCH (Zhang et al(1996)Zhang, Ramakrishnan, and Livny) algorithm. In the final step, a DFS-based (depth-first search) method is applied on the clustered regions to find frequent patterns.

44.3 Applications

The literature on spatio-temporal clustering is usually centered around concrete methods rather than application contexts. Nevertheless, in this chapter, we would like to bring examples of several possible scenarios where spatio-temporal clustering can be used along with other data mining methods.

For the sake of simplicity, we divide spatio-temporal data into three main categories according to the way these data are collected: *movement, cellular networks* and *environmental.* Movement data are often obtained by location based devices such as GPS and contain id of an object, its coordinates and timestamp. Cellular network data are obtained from mobile operators at the level of network bandwidth. Environmental data are usually obtained by censor networks and RFID technology.

The specificity of properties of these data require different approaches for analysis and also result in unique tasks. For example, in movement data the possible analysis tasks could be analysis of animal movement, their behavior in time, people's mobility and tracking of group objects. Phone calls that people make in a city can be used in the analysis of urban activity. Such information will be valuable for local authorities, service providers, decision makers, etc. Environmental processes are analyzed using information about locations and times of specific events. This information is of high importance to ecologists and geographers.

Table 44.1 summarizes the categories of spatio-temporal data, tasks considered in these categories, examples of applications, the basic methods used for solving tasks and the selected literature.

44.3.1 Movement data

Trajectory data obtained from location-aware devices usually comes as a sequence of points annotated by coordinate and time. However, not all of these points are equally important. Many application domains require identification of important parts from the trajectories. A single trajectory or a group of trajectories can be used for finding important parts. For example, in analysis of people's daily activities, some places like home or work could be identified as important while movement from one place to another would be considered as not important. Knowledge of such places can be used in analysis of activity of an object or group of objects (people, animals). Moreover, the information can be used in personalized applications (Kang et al(2004)Kang, Welbourne, Stewart, and Borriello). Usually, the place can be considered as important if the object spends in it considerable amount of time or the place is visited frequently by one or many objects. GPS-based devices are the main source of movement trajectories. However, the main disadvantage is that they loose signal indoor. A new approach to data collection using Place Lab (Schilit et al(2003)Schilit, LaMarca, Borriello, Griswold, McDonald, Lazowska, Balachandran, Hong, and Iverson) was proposed in which a WiFi enabled

Table 44.1. Overview of spatio-temporal methods and applications

Category	Problem	Application	Method based on	Selected literature
Movement	Trajectory clustering	Cars, evacuation traces, landings and interdictions of migrant boats		(Nanni and Pedreschi(2006)) (Rinzivillo et al(2008)Rinzivillo, Pedreschi, Nanni, Giannotti, Andrienko, and Andrienko) (Andrienko and Andrienko(2008))
	Trajectory aggregation		OPTICS	(Andrienko and Andrienko(2009))
	Trajectory generalization			(Andrienko et al(2009)Andrienko, Andrienko, Rinzivillo, Nanni, Pedreschi, and Giannotti)
	Moving clusters	Migrating animals, flocks, convoys of vehicles	DBSCAN	(Kalnis et al(2005)Kalnis, Mamoulis, and Bakiras)
			Gridding, DBSCAN	(Jeung et al(2008)Jeung, Yiu, Zhou, Jensen, and Shen)
			DBSCAN	(Vieira et al(2009)Vieira, Bakalov, and Tsotras)
	Extracting important places from trajectories	People's trajectories	DBSCAN,	(Palma et al(2008)Palma, Bogorny, Kuijpers, and Alvares)
			Incremental clustering	(Kang et al(2004)Kang, Welbourne, Stewart, and Borriello)
	Trajectory patterns	Fleet of trucks	Density of spatial regions	(Giannotti et al(2007)Giannotti, Nanni, Pinelli, and Pedreschi)
		Synthetic data	BIRCH	(Kang and Yong(2009))
Cellular networks	Urban activity	Phone calls	k-means	(Reades et al(2007)Reades, Calabrese, Sevtsuk, and Ratti)
Environmental	Oceanography	Seawater distribution	DBSCAN	(Birant and Kut(2006), Birant and Kut(2007))
	Seismology	Seismic activity	Gridding, DBSCAN	(Wang et al(2006)Wang, Wang, and Li)

device can get location positions from various wireless access points installed in cities. This approach can be used by mobile devices in real time applications even when the person is inside a building. In the example presented by (Kang et al(2004)Kang, Welbourne, Stewart, and Borriello), the mobile device should identify the important place and act according to some scenario. For example it can switch to a silent mode when the person enters a public place. For this, incremental spatio-temporal clustering was used to identify important places.

Two fictitious but possible scenarios of analysis of movement were proposed at VAST 2008 mini challenge (Grinstein et al(2008)Grinstein, Plaisant, Laskowski, OConnell, Scholtz, and Whiting) and addressed in (Andrienko and Andrienko(2009)). In the first scenario called *Evacuation traces*, a bomb, set up by a religious group, exploded in the building. All employees and visitors in the building wore RFID badges that enabled recording location of every person. Five analytical questions were asked: *Where was the device set off, Identify potential suspects and/or witnesses to the event, Identify any suspects and/or witnesses who managed to escape the building, Identify any casualties, Describe the evacuation*. Clustering of trajectories comes in handy for answering the second and third questions. In order to find suspects of the event, the place of the explosion epicenter was identified and people's trajectories were separated into *normal* (trajectories not passing through the place of explosion) and *suspected*. In order to answer the third question, trajectories were clustered according to the common destination. This enabled to find people who managed to escape the building and those who didn't. The second scenario called *Migrant boats*, described a problem of illegal immigration of people by boats to the US. The data consisted, among the others, of the following fields: location and date where the migrant boat left the place and where the boat was intercepted or landed. The questions were to *characterize the choice of landing sites and their evolution over the years* and *characterize the geographical patterns of interdiction*. Spatio-temporal clustering with different distance functions was applied on the data and the following patterns were found: landings at the Mexican coast and period of migration started from 2006 and increased towards 2007, while the number of landings at the coast of Florida and nearby areas was significantly smaller during 2006-2007 than on 2005. It was shown that the strategy of migration changed over the years. The migration routes increased and included new destinations. Consequently, the patrolling extended over larger areas and the rate of successful landings increased.

44.3.2 Cellular networks

Until recently, surveys were the only data collection method for analysis of various urban activities. With the rapid development of mobile networks and their global coverage, new opportunities for analysis of urban systems using phone call data have emerged. (Reades et al(2007)Reades, Calabrese, Sevtsuk, and Ratti) were one of the first who attempted to analyze urban dynamics on a city level using *Erlang* data. Erlang data is a measure of network bandwidth and indicate the load of cellular antenna as an average number of calls made over specific time period (usually hour). As such, these data are considered spatio-temporal, where the spatial component relates to the location of a transmitting antenna and temporal aspect is an aggregation of phone calls by time interval. Since the data do not contain object identifiers, only group activity can be learned from it.

The city of Rome was divided into cells of $1,600m^2$ each, $262,144$ cells in total. The Erlang value was computed for every cell taking into consideration the signal decay and positions of antennas. For each cell, the average Erlang value was obtained using 15 minutes interval during 90 day period. Thus, every cell contained seven (for every day of the week) observations of phone call activities during 90 days and 96 measurements for each day (using 15 minutes interval). Initially, six cells corresponding to different parts of the city and types of activities (residential areas, touristic places, nighttime spots) with significantly different Erlang values were selected. The analysis of these places revealed six patterns in the daily activity when there were rapid changes in cellular network usage: $1a.m.$, $7a.m.$, $11a.m.$, $2p.m.$, $5p.m.$, and $9p.m.$. To check this hypothesis, k-means was applied on all $262,144$ cells using 24-dimensional feature vector of six daily periods averaged for Monday through Thursday and separate six daily periods for Friday, Saturday and Sunday. The result of clustering suggested that the phone call activity is divided into eight separate clusters. The visual interpretation of these clusters revealed the correspondence of places to expected types of people's activity over time.

44.3.3 Environmental data

Very early examples of spatio-temporal analysis of environmental data, including clustering, are given in (Stolorz et al(1995)Stolorz, Nakamura, Mesrobian, Muntz, Santos, Yi, and Ng) as applications of an exploratory data analysis environment called CONQUEST. The system is specifically devoted to deal with sequences of remotely-sensed images that describe the evolution of some geophysical measures in some spatial areas. A most relevant application example is cyclones detection, i.e., extracting locations of cyclones and the tracks (trajectories) they follow. Since cyclones are events rather than physical objects, and there is not a straightforward way to locate them, cyclone detection requires a multi-step analysis process, where spatio-temporal data is subject to transformations from a data type to another one. First, for each time instant all candidate cyclone occurrences are located by means of a local minima heuristics based on sea level pressure, i.e. spatial locations where the sea level pressure is lower than their neighbourhood (namely, a circle of given radius) are selected. The result is essentially a set of spatio-temporal events, so far considered as independent from each other. Then, the second step consists in spatio-temporally clustering such cyclone occurrences, by iteratively merging occurrences that are temporally close and have a small spatial distance. The latter condition is relaxed when the instantaneous wind direction and magnitude are coherent with the relative positions of the occurrences – i.e., the cyclone can move fast, if wind conditions allow that. The output of this second phase is a set of trajectories, each describing the movement in time of a cyclone, which can be visually inspected and compared against geographical and geophysical features of the territory. In summary, this application shows an interesting analysis process where original geo-referenced time series are collectively analyzed to locate complex spatio-temporal events, and such events are later connected – i.e., associated to the same entity – to form trajectories.

More recently (Birant and Kut(2006), Birant and Kut(2007)) studied spatio-temporal marine data with the following attributes: sea surface temperature, the sea surface height residual, the significant wave height and wind speed values of four seas (the Black Sea, the Marmara Sea, the Aegean Sea, and the eastern part of the Mediterranean). The authors proposed ST-DBSCAN algorithm as an extension of classical DBSCAN to find seawater regions that have similar physical characteristics. In particular, the authors pursued three goals: (1) to discover regions with a similar sea surface temperature (2) to discover regions with similar sea surface height residual values and (3) to find regions with significant wave height. The database that was used for analysis contained measurements of sea surface temperature from 5340 stations obtained between 2001 and 2004, sea surface height collected over five-day periods between 1992 and 2002 and significant wave height collected over ten-day periods between 1992 and 2002 from 1707 stations. The ST-DBSCAN algorithm was integrated into the interactive system to facilitate the analysis.

44.4 Open Issues

Spatio-temporal properties of the data introduce additional complexity to the data mining process and to the clustering in particular. We can differentiate between two types of issues that the analyst should deal with or take into consideration during analysis: general and application dependent. The general issues involve such aspects as data quality, precision and uncertainty (Miller and Han(2009)). Scalability, spatial resolution and time granularity can be related to application dependent issues.

Data quality (spatial and temporal) and precision depends on the way the data is generated. Movement data is usually collected using GPS-enabled devices attached to an object. For example, when a person enters a building a GPS signal can be lost or the positioning may be inaccurate due to a weak connection to satellites. As in the general data preprocessing step, the analyst should decide how to handle missing or inaccurate parts of the data - should it be ignored, tolerated or interpolated.

The computational power does not go in line with the pace at which large amounts of data are being generated and stored. Thus, the scalability becomes a significant issue for the analysis and demand new algorithmic solutions or approaches to handle the data.

Spatial resolution and time granularity can be regarded as most crucial in spatio-temporal clustering since change in the size of the area over which the attribute is distributed or change in time interval can lead to discovery of completely different clusters and therefore, can lead to the improper explanation of the phenomena under investigation. There are still no general guidelines for proper selection of spatial and temporal resolution and it is rather unlikely that such guidelines will be proposed. Instead, ad hoc approaches are proposed to handle the problem in specific domains (see for example (Nanni and Pedreschi(2006))). Due to this, the involvement of the domain expert in every step of spatio-temporal clustering becomes essential. The geospatial visual analytics field has recently emerged as the discipline that combines automatic data mining approaches including spatio-temporal clustering with visual reasoning supported by the knowledge of domain experts and has been successfully applied at different geographical spatio-temporal phenomena ((Andrienko and Andrienko(2006), Andrienko et al(2007)Andrienko, Andrienko, and Wrobel, Andrienko and Andrienko(2010))).

A class of application-dependent issues that is quickly emerging in the spatio-temporal clustering field is related to exploitation of available background knowledge. Indeed, most of the methods and solutions surveyed in this chapter work on an abstract space where locations have no specific meanings and the analysis process extracts information from scratch, instead of starting from (and integrating to) possible a priori knowledge of the phenomena under consideration. On the opposite, a priori knowledge about such phenomena and about the context they take place in is commonly available in real applications, and integrating them in the mining process might improve the output quality (Alvares et al(2007)Alvares, Bogorny, Kuijpers, de Macedo, Moelans, and Vaisman, Baglioni et al(2009)Baglioni, Antonio Fernandes de Macedo, Renso, Trasarti, and Wachowicz, Kisilevich et al(2010)Kisilevich, Keim, and Rokach). Examples of that include the very basic knowledge of the street network and land usage, that can help in understanding which aspects of the behavior of our objects (e.g., which parts of the trajectory of a moving object) are most discriminant and better suited to form homogeneous clusters; or the existence of recurring events, such as rush hours and planned road maintenance in a urban mobility setting, that are known to interfere with our phenomena in predictable ways.

Recently, the spatio-temporal data mining literature has also pointed out that the relevant context for the analysis mobile objects includes not only geographic features and other physical constraints, but also the population of objects themselves, since in most application scenarios objects can interact and mutually interfere with each other's activity. Classical examples include traffic jams – an entity that emerges from the interaction of vehicles and, in turn, dominates their behavior. Considering interactions in the clustering process is expected to improve the reliability of clusters, yet a systematic taxonomy of relevant interaction types is still not available (neither a general one, nor any application-specific one), it is still not known how to detect such interactions automatically, and understanding the most suitable way to integrate them in a clustering process is still an open problem.

44.5 Conclusions

In this chapter we focused on geographical spatio-temporal clustering. We presented a classification of main spatio-temporal types of data: *ST events, Geo-referenced variables, Moving objects* and *Trajectories*. We described in detail how spatio-temporal clustering is applied on trajectories, provided an overview of recent research developments and presented possible scenarios in several application domains such as movement, cellular networks and environmental studies.

References

Agrawal R, Faloutsos C, Swami AN (1993) Efficient Similarity Search In Sequence Databases. In: Lomet D (ed) Proceedings of the 4th International Conference of Foundations of Data Organization and Algorithms (FODO), Springer Verlag, Chicago, Illinois, pp 69–84

Alon J, Sclaroff S, Kollios G, Pavlovic V (2003) Discovering clusters in motion time-series data. In: CVPR (1), pp 375–381

Alvares LO, Bogorny V, Kuijpers B, de Macedo JAF, Moelans B, Vaisman A (2007) A model for enriching trajectories with semantic geographical information. In: GIS '07: Proceedings of the 15th annual ACM international symposium on Advances in geographic information systems, pp 1–8

Andrienko G, Andrienko N (2008) Spatio-temporal aggregation for visual analysis of movements. In: Proceedings of IEEE Symposium on Visual Analytics Science and Technology (VAST 2008), IEEE Computer Society Press, pp 51–58

Andrienko G, Andrienko N (2009) Interactive cluster analysis of diverse types of spatiotemporal data. ACM SIGKDD Explorations

Andrienko G, Andrienko N (2010) Spatial generalization and aggregation of massive movement data. IEEE Transactions on Visualization and Computer Graphics (TVCG) Accepted

Andrienko G, Andrienko N, Wrobel S (2007) Visual analytics tools for analysis of movement data. SIGKDD Explorations Newsletter 9(2):38–46

Andrienko G, Andrienko N, Rinzivillo S, Nanni M, Pedreschi D, Giannotti F (2009) Interactive Visual Clustering of Large Collections of Trajectories. VAST 2009

Andrienko N, Andrienko G (2006) Exploratory analysis of spatial and temporal data: a systematic approach. Springer Verlag

Ankerst M, Breunig MM, Kriegel HP, Sander J (1999) Optics: ordering points to identify the clustering structure. SIGMOD Rec 28(2):49–60

Baglioni M, Antonio Fernandes de Macedo J, Renso C, Trasarti R, Wachowicz M (2009) Towards semantic interpretation of movement behavior. Advances in GIScience pp 271–288

Berndt DJ, Clifford J (1996) Finding patterns in time series: a dynamic programming approach. Advances in knowledge discovery and data mining pp 229–248

Birant D, Kut A (2006) An algorithm to discover spatialtemporal distributions of physical seawater characteristics and a case study in turkish seas. Journal of Marine Science and Technology pp 183–192

Birant D, Kut A (2007) St-dbscan: An algorithm for clustering spatial-temporal data. Data Knowl Eng 60(1):208–221

Chan KP, chee Fu AW (1999) Efficient time series matching by wavelets. In: In ICDE, pp 126–133

Chen L, Özsu MT, Oria V (2005) Robust and fast similarity search for moving object trajectories. In: SIGMOD '05: Proceedings of the 2005 ACM SIGMOD international conference on Management of data, ACM, New York, NY, USA, pp 491–502

Chudova D, Gaffney S, Mjolsness E, Smyth P (2003) Translation-invariant mixture models for curve clustering. In: KDD '03: Proceedings of the ninth ACM SIGKDD international conference on Knowledge discovery and data mining, ACM, New York, NY, USA, pp 79–88

Ciaccia P, Patella M, Zezula P (1997) M-tree: An efficient access method for similarity search in metric spaces. In: Jarke M, Carey M, Dittrich KR, Lochovsky F, Loucopoulos P, Jeusfeld MA (eds) Proceedings of the 23rd International Conference on Very Large Data Bases (VLDB'97), Morgan Kaufmann Publishers, Inc., Athens, Greece, pp 426–435

Cohen S., Rokach L., Maimon O., Decision Tree Instance Space Decomposition with Grouped Gain-Ratio, Information Science, Volume 177, Issue 17, pp. 3592-3612, 2007.

Ester M, Kriegel HP, Sander J, Xu X (1996) A density-based algorithm for discovering clusters in large spatial databases with noise. Data Mining and Knowledge Discovery pp 226–231

Fosca G, Dino P (2008) Mobility, Data Mining and Privacy: Geographic Knowledge Discovery. Springer

Frentzos E, Gratsias K, Theodoridis Y (2007) Index-based most similar trajectory search. In: ICDE, pp 816–825

Gaffney S, Smyth P (1999) Trajectory clustering with mixtures of regression models. In: KDD '99: Proceedings of the fifth ACM SIGKDD international conference on Knowledge discovery and data mining, ACM, New York, NY, USA, pp 63–72

Giannotti F, Nanni M, Pinelli F, Pedreschi D (2007) Trajectory pattern mining. In: Proceedings of the 13th ACM SIGKDD international conference on Knowledge discovery and data mining, ACM, p 339

Grinstein G, Plaisant C, Laskowski S, OConnell T, Scholtz J, Whiting M (2008) VAST 2008 Challenge: Introducing mini-challenges. In: Proceedings of IEEE Symposium, vol 1, pp 195–196

Gudmundsson J, van Kreveld M (2006) Computing longest duration flocks in trajectory data. In: GIS '06: Proceedings of the 14th annual ACM international symposium on Advances in geographic information systems, ACM, New York, NY, USA, pp 35–42

Hwang SY, Liu YH, Chiu JK, Lim EP (2005) Mining mobile group patterns: A trajectory-based approach. In: PAKDD, pp 713–718

Iyengar VS (2004) On detecting space-time clusters. In: Proceedings of the 10th International Conference on Knowledge Discovery and Data Mining (KDD'04), ACM, pp 587–592

Jeung H, Yiu ML, Zhou X, Jensen CS, Shen HT (2008) Discovery of convoys in trajectory databases. Proc VLDB Endow 1(1):1068–1080

Kalnis P, Mamoulis N, Bakiras S (2005) On discovering moving clusters in spatio-temporal data. Advances in Spatial and Temporal Databases pp 364–381

Kang J, Yong HS (2009) Mining Trajectory Patterns by Incorporating Temporal Properties. Proceedings of the 1st International Conference on Emerging Databases

Kang JH, Welbourne W, Stewart B, Borriello G (2004) Extracting places from traces of locations. In: WMASH '04: Proceedings of the 2nd ACM international workshop on Wireless mobile applications and services on WLAN hotspots, ACM, New York, NY, USA, pp 110–118

Kisilevich S, Keim D, Rokach L (2010) A novel approach to mining travel sequences using collections of geo-tagged photos. In: The 13th AGILE International Conference on Geographic Information Science

Kulldorff M (1997) A spatial scan statistic. Communications in Statistics: Theory and Methods 26(6):1481–1496

Lee JG, Han J, Whang KY (2007) Trajectory clustering: a partition-and-group framework. In: SIGMOD Conference, pp 593–604

Li Y, Han J, Yang J (2004a) Clustering moving objects. In: Proceedings of the 10th International Conference on Knowledge Discovery and Data Mining (KDD'04), ACM, pp 617–622

Li Y, Han J, Yang J (2004b) Clustering moving objects. In: KDD, pp 617–622

Maimon O., and Rokach, L. Data Mining by Attribute Decomposition with semiconductors manufacturing case study, in Data Mining for Design and Manufacturing: Methods and Applications, D. Braha (ed.), Kluwer Academic Publishers, pp. 311–336, 2001.

Miller HJ, Han J (2009) Geographic data mining and knowledge discovery. Chapman & Hall/CRC

Nanni M, Pedreschi D (2006) Time-focused clustering of trajectories of moving objects. Journal of Intelligent Information Systems 27(3):267–289

Palma AT, Bogorny V, Kuijpers B, Alvares LO (2008) A clustering-based approach for discovering interesting places in trajectories. In: SAC '08: Proceedings of the 2008 ACM symposium on Applied computing, pp 863–868

Pelekis N, Kopanakis I, Marketos G, Ntoutsi I, Andrienko G, Theodoridis Y (2007) Similarity search in trajectory databases. In: TIME '07: Proceedings of the 14th International Symposium on Temporal Representation and Reasoning, IEEE Computer Society, Washington, DC, USA, pp 129–140

Reades J, Calabrese F, Sevtsuk A, Ratti C (2007) Cellular census: Explorations in urban data collection. IEEE Pervasive Computing 6(3):30–38

Rinzivillo S, Pedreschi D, Nanni M, Giannotti F, Andrienko N, Andrienko G (2008) Visually driven analysis of movement data by progressive clustering. Information Visualization 7(3):225–239

Rokach L. and Maimon O., Feature Set Decomposition for Decision Trees, Journal of Intelligent Data Analysis, Volume 9, Number 2, 2005b, pp 131–158.

Rokach L., Genetic algorithm-based feature set partitioning for classification problems,Pattern Recognition, 41(5):1676–1700, 2008.

Rokach L., Maimon O. and Lavi I., Space Decomposition In Data Mining: A Clustering Approach, Proceedings of the 14th International Symposium On Methodologies For Intelligent Systems, Maebashi, Japan, Lecture Notes in Computer Science, Springer-Verlag, 2003, pp. 24–31.

Schilit BN, LaMarca A, Borriello G, Griswold WG, McDonald D, Lazowska E, Balachandran A, Hong J, Iverson V (2003) Challenge: ubiquitous location-aware computing and the "place lab" initiative. In: WMASH '03: Proceedings of the 1st ACM international workshop on Wireless mobile applications and services on WLAN hotspots, ACM, New York, NY, USA, pp 29–35

Stolorz P, Nakamura H, Mesrobian E, Muntz RR, Santos JR, Yi J, Ng K (1995) Fast spatio-temporal data mining of large geophysical datasets. In: Proceedings of the First International Conference on Knowledge Discovery and Data Mining (KDD'95), AAAI Press, pp 300–305

Theodoridis Y (2003) Ten benchmark database queries for location-based services. The Computer Journal 46(6):713–725

Vieira MR, Bakalov P, Tsotras VJ (2009) On-line discovery of flock patterns in spatio-temporal data. In: GIS '09: Proceedings of the 17th ACM SIGSPATIAL International Conference on Advances in Geographic Information Systems, ACM, New York, NY, USA, pp 286–295

Vlachos M, Kollios G, Gunopulos D (2002) Discovering similar multidimensional trajectories. In: Proceedings of the International Conference on Data Engineering, pp 673–684

Vlachos M, Hadjieleftheriou M, Gunopulos D, Keogh E (2003) Indexing multi-dimensional time-series with support for multiple distance measures. In: KDD '03: Proceedings of the ninth ACM SIGKDD international conference on Knowledge discovery and data mining, ACM, New York, NY, USA, pp 216–225

Wang M, Wang A, Li A (2006) Mining Spatial-temporal Clusters from Geo-databases. Lecture Notes in Computer Science 4093:263

Zhang P, Huang Y, Shekhar S, Kumar V (2003) Correlation analysis of spatial time series datasets: A filter-and-refine approach. In: In the Proc. of the 7th PAKDD

Zhang T, Ramakrishnan R, Livny M (1996) BIRCH: an efficient data clustering method for very large databases. ACM SIGMOD Record 25(2):103–114

Zheng Y, Zhang L, Xie X, Ma WY (2009) Mining interesting locations and travel sequences from gps trajectories. In: WWW '09: Proceedings of the 18th international conference on World wide web, pp 791–800

45

Data Mining for Imbalanced Datasets: An Overview

Nitesh V. Chawla

Department of Computer Science and Engineering
University of Notre Dame
IN 46530, USA
nchawla@cse.nd.edu

Summary. A dataset is imbalanced if the classification categories are not approximately equally represented. Recent years brought increased interest in applying machine learning techniques to difficult "real-world" problems, many of which are characterized by imbalanced data. Additionally the distribution of the testing data may differ from that of the training data, and the true misclassification costs may be unknown at learning time. Predictive accuracy, a popular choice for evaluating performance of a classifier, might not be appropriate when the data is imbalanced and/or the costs of different errors vary markedly. In this Chapter, we discuss some of the sampling techniques used for balancing the datasets, and the performance measures more appropriate for mining imbalanced datasets.

Key words: imbalanced datasets, classification, sampling, ROC, cost-sensitive measures, precision and recall

45.1 Introduction

The issue with imbalance in the class distribution became more pronounced with the applications of the machine learning algorithms to the real world. These applications range from telecommunications management (Ezawa et al., 1996), bioinformatics (Radivojac et al., 2004), text classification (Lewis and Catlett, 1994, Dumais et al., 1998, Mladenić and Grobelnik, 1999, Cohen, 1995b), speech recognition (Liu et al., 2004), to detection of oil spills in satellite images (Kubat et al., 1998). The imbalance can be an artifact of class distribution and/or different costs of errors or examples. It has received attention from machine learning and Data Mining community in form of Workshops (Japkowicz, 2000b, Chawla et al., 2003a, Dietterich et al., 2003, Ferri et al., 2004) and Special Issues (Chawla et al., 2004a). The range of papers in these venues exhibited the pervasive and ubiquitous nature of the class imbalance issues faced by the Data Mining community. Sampling methodologies continue to be popular in the research work. However, the research continues to evolve with different applications, as each application provides a compelling problem. One focus of the initial workshops was primarily

O. Maimon, L. Rokach (eds.), *Data Mining and Knowledge Discovery Handbook*, 2nd ed.,
DOI 10.1007/978-0-387-09823-4_45,

the performance evaluation criteria for mining imbalanced datasets. The limitation of the accuracy as the performance measure was quickly established. ROC curves soon emerged as a popular choice (Ferri et al., 2004).

The compelling question, given the different class distributions is: *What is the correct distribution for a learning algorithm?* Weiss and Provost presented a detailed analysis on the effect of class distribution on classifier learning (Weiss and Provost, 2003). Our observations agree with their work that the natural distribution is often not the best distribution for learning a classifier (Chawla, 2003). Also, the imbalance in the data can be more characteristic of "sparseness" in feature space than the class imbalance. Various re-sampling strategies have been used such as random oversampling with replacement, random undersampling, focused oversampling, focused undersampling, oversampling with synthetic generation of new samples based on the known information, and combinations of the above techniques (Chawla et al., 2004b).

In addition to the issue of inter-class distribution, another important probem arising due to the sparsity in data is the distribution of data within each class (Japkowicz, 2001a). This problem was also linked to the issue of small disjuncts in the decision tree learning. Yet another, school of thought is a recognition based approach in the form of a one-class learner. The one-class learners provide an interesting alternative to the traditional discriminative approach, where in the classifier is learned on the target class alone (Japkowicz, 2001b, Juszczak and Duin, 2003, Raskutti and Kowalczyk, 2004, Tax, 2001).

In this chapter[1], we present a liberal overview of the problem of mining imbalanced datasets with particular focus on performance measures and sampling methodologies. We will present our novel oversampling technique, SMOTE, and its extension in the boosting procedure — SMOTEBoost.

45.2 Performance Measure

A classifier is, typically, evaluated by a confusion matrix as illustrated in Figure 45.1 (Chawla et al., 2002). The columns are the Predicted class and the rows are the Actual class. In the confusion matrix, TN is the number of negative examples correctly classified (True Negatives), FP is the number of negative examples incorrectly classified as positive (False Positives), FN is the number of positive examples incorrectly classified as negative (False Negatives) and TP is the number of positive examples correctly classified (True Positives). Predictive accuracy is defined as $Accuracy = (TP+TN)/(TP+FP+TN+FN)$.

However, predictive accuracy might not be appropriate when the data is imbalanced and/or the costs of different errors vary markedly. As an example, consider the classification of pixels in mammogram images as possibly cancerous (Woods et al., 1993). A typical mammography dataset might contain 98% normal pixels and 2% abnormal pixels. A simple default strategy of guessing the majority class would give a predictive accuracy of 98%. The nature of the application requires a fairly high rate of correct detection in the minority class and allows for a small error rate in the majority class in order to achieve this (Chawla et al., 2002). Simple predictive accuracy is clearly not appropriate in such situations.

[1] The chapter will utilize excerpts from our published work in various Journals and Conferences. Please see the references for the original publications.

	Predicted Negative	Predicted Positive
Actual Negative	TN	FP
Actual Positive	FN	TP

Fig. 45.1. Confusion Matrix

45.2.1 ROC Curves

The Receiver Operating Characteristic (ROC) curve is a standard technique for summarizing classifier performance over a range of tradeoffs between true positive and false positive error rates (Swets, 1988). The Area Under the Curve (AUC) is an accepted performance metric for a ROC curve (Bradley, 1997).

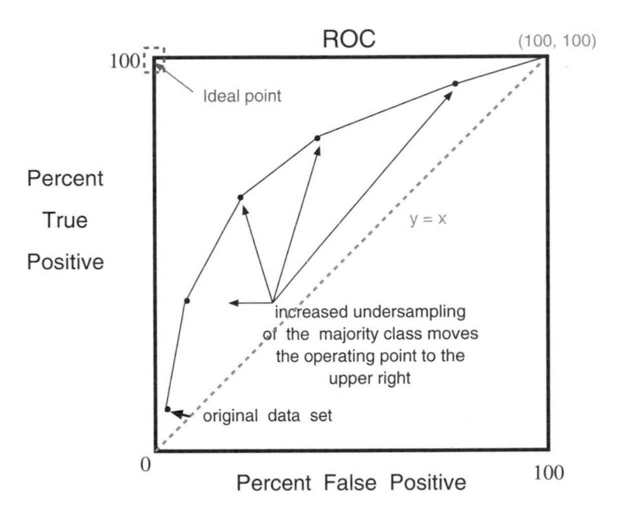

Fig. 45.2. Illustration of Sweeping out an ROC Curve through under-sampling. Increased under-sampling of the majority (negative) class will move the performance from the lower left point to the upper right.

ROC curves can be thought of as representing the family of best decision boundaries for relative costs of TP and FP. On an ROC curve the X-axis represents $\%FP = FP/(TN+FP)$ and the Y-axis represents $\%TP = TP/(TP+FN)$. The ideal point on the ROC curve would be (0,100), that is all positive examples are classified correctly and no negative examples are misclassified as positive. One way an ROC curve can be swept out is by manipulating the balance of training samples for each class in the training set. Figure 45.2 shows an illustration (Chawla et al., 2002). The line y = x represents the scenario of randomly guessing the class. A single operating point of a classifier can be chosen from the trade-off between the

%TP and %FP, that is, one can choose the classifier giving the best %TP for an acceptable %FP (Neyman-Pearson method) (Egan, 1975). Area Under the ROC Curve (AUC) is a useful metric for classifier performance as it is independent of the decision criterion selected and prior probabilities. The AUC comparison can establish a dominance relationship between classifiers. If the ROC curves are intersecting, the total AUC is an average comparison between models (Lee, 2000).

The ROC convex hull can also be used as a robust method of identifying potentially optimal classifiers (Provost and Fawcett, 2001). Given a family of ROC curves, the ROC convex hull can include points that are more towards the north-west frontier of the ROC space. If a line passes through a point on the convex hull, then there is no other line with the same slope passing through another point with a larger true positive (TP) intercept. Thus, the classifier at that point is optimal under any distribution assumptions in tandem with that slope (Provost and Fawcett, 2001).

Moreover, distribution/cost sensitive applications can require a ranking or a probabilistic estimate of the instances. For instance, revisiting our mammography data example, a probabilistic estimate or ranking of cancerous cases can be decisive for the practitioner (Chawla, 2003, Maloof, 2003). The cost of further tests can be decreased by thresholding the patients at a particular rank. Secondly, probabilistic estimates can allow one to threshold ranking for class membership at values < 0.5. The ROC methodology by (Hand, 1997) allows for ranking of examples based on their class memberships — whether a randomly chosen majority class example has a higher majority class membership than a randomly chosen minority class example. It is equivalent to the Wilcoxon test statistic.

45.2.2 Precision and Recall

From the confusion matrix in Figure 45.1, we can derive the expression for *precision* and *recall* (Buckland and Gey, 1994).

$$precision = \frac{TP}{TP+FP}$$

$$recall = \frac{TP}{TP+FN}$$

The main goal for learning from imbalanced datasets is to improve the *recall* without hurting the *precision*. However, *recall* and *precision* goals can be often conflicting, since when increasing the true positive for the minority class, the number of false positives can also be increased; this will reduce the precision. The *F-value* metric is one measure that combines the trade-offs of *precision* and *recall*, and outputs a single number reflecting the "goodness" of a classifier in the presence of rare classes. While ROC curves represent the trade-off between values of TP and FP, the *F-value* represents the trade-off among different values of TP, FP, and FN (Buckland and Gey, 1994). The expression for the *F-value* is as follows:

$$F - value = \frac{(1+\beta^2) * recall * precision}{\beta^2 * recall + precision}$$

where β corresponds to the relative importance of *precision* vs *recall*. It is usually set to 1.

45.2.3 Cost-sensitive Measures

Cost Matrix

Cost-sensitive measures usually assume that the costs of making an error are known (Turney, 2000, Domingos, 1999, Elkan, 2001). That is one has a cost-matrix, which defines the costs incurred in false positives and false negatives. Each example, x, can be associated with a cost $C(i,j,x)$, which defines the cost of predicting class i for x when the "true" class is j. The goal is to take a decision to minimize the expected cost. The optimal prediction for x can be defined as

$$\sum_j P(j|x)C(i,j,x) \qquad (45.1)$$

The aforementioned equation requires a computation of conditional probablities of class j given feature vector or example x. While the cost equation is straightforward, we don't always have a cost attached to making an error. The costs can be different for every example and not only for every type of error. Thus, $C(i,j)$ is not always $\equiv$ to $C(i,j,x)$.

Cost Curves

(Drummond and Holte, 2000) propose cost-curves, where the x-axis represents of the fraction of the positive class in the training set, and the y-axis represents the expected error rate grown on each of the training sets. The training sets for a data set is generated by under (or over) sampling. The error rates for class distributions not represented are construed by interpolation. They define two cost-sensitive components for a machine learning algorithm: 1) producing a variety of classifiers applicable for different distributions and 2) selecting the appropriate classifier for the right distribution. However, when the misclassification costs are known, the x-axis can represent the "probability cost function", which is the normalized product of $C(-\mid +) * P(+)$; the y-axis represents the expected cost.

45.3 Sampling Strategies

Over and under-sampling methodologies have received significant attention to counter the effect of imbalanced data sets (Solberg and Solberg, 1996, Japkowicz, 2000a, Chawla et al., 2002, Weiss and Provost, 2003, Kubat and Matwin, 1997, Jo and Japkowicz, 2004, Batista et al., 2004, Phua and Alahakoon, 2004, Laurikkala, 2001, Ling and Li, 1998). Various studies in imbalanced datasets have used different variants of over and under sampling, and have presented (sometimes conflicting) viewpoints on usefulness of oversampling versus undersampling (Chawla, 2003, Maloof, 2003, Drummond and Holte, 2003, Batista et al., 2004).

The random under and over sampling methods have their various short-comings. The random undersampling method can potentially remove certain important examples, and random oversampling can lead to overfitting. However, there has been progression in both the under and over sampling methods. (Kubat and Matwin, 1997) used one-sided selection to selectively undersample the original population. They used Tomek Links (Tomek, 1976) to identify the noisy and borderline examples. They also used the Condensed Nearest Neighbor (CNN) rule (Hart, 1968) to remove examples from the majority class that are far away from the decision border. (Laurikkala, 2001) proposed Neighborhood

Cleaning Rule (NCL) to remove the majority class examples. The author computes three nearest neighbors for each of the (E_i) examples in the training set. If E_i belongs to the majority class, and it is misclassified by its three nearest neighbors, then E_i is removed. If E_i belongs to the minority class, and it is misclassified by its three nearest neighbors then the majority class examples among the three nearest neighbors are removed. This approach can reach a computational bottleneck for very large datasets, with a large majority class.

(Japkowicz, 2000a) discussed the effect of imbalance in a dataset. She evaluated three strategies: under-sampling, resampling and a recognition-based induction scheme. She considered two sampling methods for both over and undersampling. Random resampling consisted of oversampling the smaller class at random until it consisted of as many samples as the majority class and "focused resampling" consisted of oversampling only those minority examples that occurred on the boundary between the minority and majority classes. Random under-sampling involved under-sampling the majority class samples at random until their numbers matched the number of minority class samples; focused under-sampling involved under-sampling the majority class samples lying further away. She noted that both the sampling approaches were effective, and she also observed that using the sophisticated sampling techniques did not give any clear advantage in the domain considered. However, her oversampling methodologies did not construct any new examples.

(Ling and Li, 1998) also combined over-sampling of the minority class with under-sampling of the majority class. They used lift analysis instead of accuracy to measure a classifier's performance. They proposed that the test examples be ranked by a confidence measure and then lift be used as the evaluation criteria. In one experiment, they under-sampled the majority class and noted that the best lift index is obtained when the classes are equally represented. In another experiment, they over-sampled the positive (minority) examples with replacement to match the number of negative (majority) examples to the number of positive examples. The over-sampling and under-sampling combination did not provide significant improvement in the lift index.

We developed a novel oversampling technique called SMOTE (Synthetic Minority Oversampling TEchnique). It can be essential to provide new related information on the positive class to the learning algorithm, in addition to undersampling the majority class. This was the first attempt to introduce new examples in the training data to enrich the data space and counter the sparsity in the distribution. We will discuss SMOTE in more detail in the subsequent section. We combined SMOTE with undersampling. We used ROC analyses to present the results of our findings.

Batista et al. (Batista et al., 2004) evaluated various sampling methodologies on a variety of datasets with different class distributions. They included various methods in both oversampling and undersampling. They conclude that SMOTE+Tomek and SMOTE+ENN are more applicable and give very good results for datasets with a small number of positive class examples. They also noted that the decision trees constructed from the oversampled datasets are usually very large and complex. This is similar to the observation by (Chawla et al., 2002).

45.3.1 Synthetic Minority Oversampling TEchnique: SMOTE

Over-sampling by replication can lead to similar but more specific regions in the feature space as the decision region for the minority class. This can potentially lead to overfitting on the multiple copies of minority class examples. To overcome the overfitting and broaden the decision region of minority class examples, we introduced a novel technique to generate synthetic examples by operating in "feature space" rather than "data space" (Chawla et al., 2002). The minority class is over-sampled by taking each minority class sample and introducing synthetic examples along the line segments joining any/all of the k minority class nearest neighbors. Depending upon the amount of over-sampling required, neighbors from the k nearest neighbors are randomly chosen. Synthetic samples are generated in the following way: Take the difference between the feature vector (sample) under consideration and its nearest neighbor. Multiply this difference by a random number between 0 and 1, and add it to the feature vector under consideration. This causes the selection of a random point along the line segment between two specific features. This approach effectively forces the decision region of the minority class to become more general. For the nominal cases, we take the majority vote for the nominal value amongst the nearest neighbors. We use the modification of Value Distance Metric (VDM) (Cost and Salzberg, 1993) to compute the nearest neighbors for the nominal valued features.

The synthetic examples cause the classifier to create larger and less specific decision regions, rather than smaller and more specific regions, as typically caused by over-sampling with replication. More general regions are now learned for the minority class rather than being subsumed by the majority class samples around them. The effect is that decision trees generalize better.

SMOTE was tested on a variety of datasets, with varying degrees of imbalance and varying amounts of data in the training set, thus providing a diverse testbed. SMOTE forces focused learning and introduces a bias towards the minority class. On most of the experiments, SMOTE using C4.5 (Quinlan, 1992) and Ripper (Cohen, 1995a) as underlying classifiers outperformed other methods including sampling strategies, Ripper's Loss Ratio, and even Naive Bayes by varying the class priors.

45.4 Ensemble-based Methods

Combination of classifiers can be an effective technique for improving prediction accuracy. As one of the most popular combining techniques, boosting (Freund and Schapire, 1996), uses adaptive sampling of instances to generate a highly accurate ensemble of classifiers whose individual global accuracy is only moderate. In boosting, the classifiers in the ensemble are trained serially, with the weights on the training instances adjusted adaptively according to the performance of the previous classifiers. The main idea is that the classification algorithm should concentrate on the instances that are difficult to learn. Boosting has received extensive empirical study (Dietterich, 2000, Bauer and Kohavi, 1999), but most of the published work focuses on improving the accuracy of a weak classifier on datasets with well-balanced class distributions. There has been significant interest in the recent literature for embedding cost-sensitivities in the boosting algorithm. We proposed SMOTEBoost that embeds the SMOTE procedure during boosting iterations. CSB (Ting, 2000) and AdaCost boosting algorithms (Fan et al., 1999) update the weights of examples according to the misclassification costs. On the other side, Rare-Boost (Joshi et al., 2001) updates the weights of the examples differently for all four entries shown in Figure 45.1. Guo and Viktor (Guo and Viktor, 2004) propose another technique that modifies the boosting procedure — DataBoost. As compared to SMOTEBoost, which only focuses on the hard minority class cases, this technique employs a synthetic data generation process for both minority and majority class cases.

In addition to boosting, popular sampling techniques have also been deployed to construct ensembles. Radivojac et al. (Radivojac et al., 2004) combined bagging with oversampling methodlogies for the bioinformatics domain. Liu et al. (Liu et al., 2004) also applied a variant of bagging by bootstrapping at equal proportions from both the minority and majority classes. They applied this technique to the problem of sentence boundary detection. Phua et. al (Phua and Alahakoon, 2004) combine bagging and stacking to identify the best mix of classifiers. In their insurance fraud detection domain, they note that stacking-bagging achieves the best cost-savings

45.4.1 SMOTEBoost

SMOTEBoost algorithm combines SMOTE and the standard boosting procedure (Chawla et al., 2003b). We want to utilize SMOTE for improving the accuracy over the minority classes, and we want to utilize boosting to maintain accuracy over the entire data set. The major goal is to better model the minority class in the data set, by providing the learner not only with the minority class instances that were misclassified in previous boosting iterations, but also with a broader representation of those instances.

The standard boosting procedure gives equal weights to all misclassified examples. Since boosting samples from a pool of data that predominantly consists of the majority class, subsequent samplings of the training set may still be skewed towards the majority class. Although boosting reduces the variance and the bias in the final ensemble (Freund and Schapire, 1996), it might not hold for datasets with skewed class distributions. There is a very strong learning bias towards the majority class cases in a skewed data set, and subsequent iterations of boosting can lead to a broader sampling from the majority class. Boosting (Adaboost) treats both kinds of errors (FP and FN) in a similar fashion. Our goal is to reduce the bias inherent in the learning procedure due to the class imbalance, and increase the sampling weights for the minority class. Introducing SMOTE in each round of boosting will enable each learner to be able to sample more of the minority class cases, and also learn better and broader decision regions for the minority class. SMOTEBoost approach outperformed boosting, Ripper (Cohen, 1995a), and AdaCost on a variety of datasets (Chawla et al., 2003b).

45.5 Discussion

Mining from imbalanced datasets is indeed a very important problem from both the algorithmic and performance perspective. Not choosing the right distribution or the objective function while developing a classification model can introduce bias towards majority (potentially uninteresting) class. Furthermore, predictive accuracy is not a useful measure when evaluating classifiers learned on imbalance data sets. Some of the measures discussed in Section 45.2 can be more appropriate.

Sampling methods are very popular in balancing the class distribution before learning a classifier, which uses an error based objective function to search the hypothesis space. We focused on SMOTE in the chapter. Consider the effect on the decision regions in feature space when minority over-sampling is done by replication (sampling with replacement) versus the introduction of synthetic examples. With replication, the decision region that results in a classification decision for the minority class can actually become smaller and more specific as the minority samples in the region are replicated. This is the opposite of the desired effect. Our method of synthetic over-sampling works to cause the classifier to build larger decision regions that contain nearby minority class points. The same reasons may be applicable to why SMOTE performs better than Ripper's loss ratio and Naive Bayes; these methods, nonetheless, are still learning from the information provided in the dataset, albeit with different cost information. SMOTE provides more related minority class samples to learn from, thus allowing a learner to carve broader decision regions, leading to more coverage of the minority class. The

SMOTEBoost methodology that embeds SMOTE within the Adaboost procedure provided further improvements to the minority class prediction.

One compelling problem arising from sampling methodologies is: *Can we identify the right distribution?* Is balanced the best distribution? It is not straightforward. This is very domain and classifier dependent, and is usually driven by empirical observations. (Weiss and Provost, 2003) present a budgeted sampling approach, which represents a heuristic for searching for the right distribution. Another compelling issue is :*What if the test distribuion remarkably differs from the training distribution?* If we train a classifier on a distribution tuned on the discovered distribution, will it generalize enough on the testing set. In such cases, one can assume that the natural distribution holds, and apply a form of cost-sensitive learning. If a cost-matrix is known and is static across the training and testing sets, learn from the original or natural distribution, and then apply the cost-matrix at the time of classification. It can also be the case that the majority class is of an equal interest as the minority class — the imbalance here is a mere artifiact of class distribution and not of different types of errors (Liu et al., 2004). In such a scenario, it is important to model both the majority and minority classes without a particular bias towards any one class.

We believe mining imbalanced datasets opens a front of interesting problems and research directions. Given that Data Mining is becoming pervasive and ubiquitous in various applications, it is important to investigate along the lines of imbalance both in class distribution and costs.

Acknowledgements

I would like to thank Larry Hall, Kevin Bowyer and Philip Kegelmeyer for their valuable input during my Ph.D. research in this field. I am also extremely grateful to all my collaborators and co-authors in the area of learning from imbalanced datasets. I have enjoyed working with them and contributing to this field.

References

Batista, G. E. A. P. A., Prati, R. C., and Monard, M. C. (2004). A study of the behavior of several methods for balancing machine learning training data. *SIGKDD Explorations*, 6(1).

Bauer, E. and Kohavi, R. (1999). An empirical comparison of voting classification algorithms: Bagging, boosting and variants. *Machine Learning*, 36(1,2).

Bradley, A. P. (1997). The Use of the Area Under the ROC Curve in the Evaluation of Machine Learning Algorithms. *Pattern Recognition*, 30(6):1145–1159.

Buckland, M. and Gey, F. (1994). The Relationship Between Recall and Precision. *Journal of the American Society for Information Science*, 45(1):12–19.

Chawla, N. V. (2003). C4.5 and Imbalanced Data sets: Investigating the Effect of Sampling Method, Probabilistic Estimate, and Decision Tree Structure. In *ICML Workshop on Learning from Imbalanced Data sets*, Washington, DC.

Chawla, N. V., Bowyer, K. W., Hall, L. O., and Kegelmeyer, W. P. (2002). SMOTE: Synthetic Minority Oversampling TEchnique. *Journal of Artificial Intelligence Research*, 16:321–357.

Chawla, N. V., Japkowicz, N., and Kołcz, A., editors (2003a). *Proceedings of the ICML'2003 Workshop on Learning from Imbalanced Data Sets II.*

Chawla, N. V., Japkowicz, N., and Kołcz, A., editors (2004a). *SIGKDD Special Issue on Learning from Imbalanced Datasets.*

Chawla, N. V., Japkowicz, N., Kolcz, A. (2004b), Editorial: Learning form Imbalanced Datasets, SIGKDD Explorations, 6(1).

Chawla, N. V., Lazarevic, A., Hall, L. O., and Bowyer, K. W. (2003b). Smoteboost: Improving Prediction of the Minority Class in Boosting. In *Seventh European Conference on Principles and Practice of Knowledge Discovery in Databases*, pages 107–119, Dubrovnik, Croatia.

Cohen, W. (1995a). Fast Effective Rule Induction. In *Proceedings of the Twelfth International Conference on Machine Learning*, pages 115–123. Department of Computer Science, Katholieke Universiteit Leuven.

Cohen, W. (1995b). Learning to Classify English Text with ILP Methods. In *Proceedings of the 5th International Workshop on Inductive Logic Programming*, pages 3–24. Department of Computer Science, Katholieke Universiteit Leuven.

Cost, S. and Salzberg, S. (1993). A Weighted Nearest Neighbor Algorithm for Learning with Symbolic Features. *Machine Learning*, 10(1):57–78.

Dietterich, T. (2000). An Empirical Comparison of Three Methods for Constructing Ensembles of Decision Trees: Bagging, Boosting and Randomization. *Machine Learning*, 40(2):139 – 157.

Dietterich, T., Margineantu, D., Provost, F., and Turney, P., editors (2003). *Proceedings of the ICML'2000 Workshop on COST-SENSITIVE LEARNING.*

Domingos, P. (1999). Metacost: A General Method for Making Classifiers Cost-sensitive. In *Proceedings of the Fifth ACM SIGKDD International Conference on Knowledge Discovery and Data Mining*, pages 155–164, San Diego, CA. ACM Press.

Drummond, C. and Holte, R. (2003). C4.5, class imbalance, and cost sensitivity: Why under-sampling beats over-sampling. In *Proceedings of the ICML'03 Workshop on Learning from Imbalanced Data Sets.*

Drummond, C. and Holte, R. C. (2000). Explicitly Representing Expected Cost: An Alternative to ROC Representation. In *Proceedings of the Sixth ACM SIGKDD International Conference on Knowledge Discovery and Data Mining*, pages 198–207, Boston. ACM.

Dumais, S., Platt, J., Heckerman, D., and Sahami, M. (1998). Inductive Learning Algorithms and Representations for Text Categorization. In *Proceedings of the Seventh International Conference on Information and Knowledge Management.*, pages 148–155.

Egan, J. P. (1975). Signal Detection Theory and ROC Analysis. In *Series in Cognition and Perception*. Academic Press, New York.

Elkan, C. (2001). The Foundations of Cost-sensitive Learning. In *Proceedings of the Seventeenth International Joint Conference on Artificial Intelligence*, pages 973–978, Seattle, WA.

Ezawa, K., J., Singh, M., and Norton, S., W. (1996). Learning Goal Oriented Bayesian Networks for Telecommunications Risk Management. In *Proceedings of the International Conference on Machine Learning, ICML-96*, pages 139–147, Bari, Italy. Morgan Kauffman.

Fan, W., Stolfo, S., Zhang, J., and Chan, P. (1999). Adacost: Misclassification Cost-sensitive Boosting. In *Proceedings of Sixteenth International Conference on Machine Learning*, pages 983–990, Slovenia.

Ferri, C., Flach, P., Orallo, J., and Lachice, N., editors (2004). *ECAI' 2004 First Workshop on ROC Analysis in AI.* ECAI.

Freund, Y. and Schapire, R. (1996). Experiments with a New Boosting Algorithm. In *Thirteenth International Conference on Machine Learning*, Bari, Italy.

Guo, H. and Viktor, H. L. (2004). Learning from imbalanced data sets with boosting and data generation: The DataBoost-IM approach. *SIGKDD Explorations*, 6(1).

Hand, D. J. (1997). *Construction and Assessment of Classification Rules*. John Wiley and Sons.

Hart, P. E. (1968). The Condensed Nearest Neighbor Rule. *IEEE Transactions on Information Theory*, 14:515–516.

Japkowicz, N. (2000a). The Class Imbalance Problem: Significance and Strategies. In *Proceedings of the 2000 International Conference on Artificial Intelligence (IC-AI'2000): Special Track on Inductive Learning*, Las Vegas, Nevada.

Japkowicz, N. (2000b). Learning from Imbalanced Data sets: A Comparison of Various Strategies. In *Proceedings of the AAAI'2000 Workshop on Learning from Imbalanced Data Sets*, Austin, TX.

Japkowicz, N. (2001a). Concept-learning in the presence of between-class and within-class imbalances. In *Proceedings of the Fourteenth Conference of the Canadian Society for Computational Studies of Intelligence*, pages 67–77.

Japkowicz, N. (2001b). Supervised versus unsupervised binary-learning by feedforward neural networks. *Machine Learning*, 42(1/2):97–122.

Jo, T. and Japkowicz, N. (2004). Class imbalances versus small disjuncts. *SIGKDD Explorations*, 6(1).

Joshi, M., Kumar, V., and Agarwal, R. (2001). Evaluating Boosting Algorithms to Classify Rare Classes: Comparison and Improvements. In *Proceedings of the First IEEE International Conference on Data Mining*, pages 257–264, San Jose, CA.

Juszczak, P. and Duin, R. P. W. (2003). Uncertainty sampling methods for one-class classifiers. In *Proceedings of the ICML'03 Workshop on Learning from Imbalanced Data Sets*.

Kubat, M., Holte, R., and Matwin, S. (1998). Machine Learning for the Detection of Oil Spills in Satellite Radar Images. *Machine Learning*, 30:195–215.

Kubat, M. and Matwin, S. (1997). Addressing the Curse of Imbalanced Training Sets: One Sided Selection. In *Proceedings of the Fourteenth International Conference on Machine Learning*, pages 179–186, Nashville, Tennesse. Morgan Kaufmann.

Laurikkala, J. (2001). Improving Identification of Difficult Small Classes by Balancing Class Distribution. Technical Report A-2001-2, University of Tampere.

Lee, S. S. (2000). Noisy Replication in Skewed Binary Classification. *Computational Statistics and Data Analysis*, 34.

Lewis, D. and Catlett, J. (1994). Heterogeneous Uncertainity Sampling for Supervised Learning. In *Proceedings of the Eleventh International Conference of Machine Learning*, pages 148–156, San Francisco, CA. Morgan Kaufmann.

Ling, C. and Li, C. (1998). Data Mining for Direct Marketing Problems and Solutions. In *Proceedings of the Fourth International Conference on Knowledge Discovery and Data Mining (KDD-98)*, New York, NY. AAAI Press.

Liu, Y., Chawla, N. V., Shriberg, E., Stolcke, A., and Harper, M. (2004). Resampling Techniques for Sentence Boundary Detection: A Case Study in Machine Learning from Imbalanced Data for Spoken Language Processing. *Under Review*.

Maloof, M. (2003). Learning when data sets are imbalanced and when costs are unequal and unknown. In *Proceedings of the ICML'03 Workshop on Learning from Imbalanced Data Sets*.

Mladenić, D. and Grobelnik, M. (1999). Feature Selection for Unbalanced Class Distribution and Naive Bayes. In *Proceedings of the 16th International Conference on Machine Learning.*, pages 258–267. Morgan Kaufmann.

Phua, C. and Alahakoon, D. (2004). Minority report in fraud detection: Classification of skewed data. *SIGKDD Explorations*, 6(1).

Provost, F. and Fawcett, T. (2001). Robust Classification for Imprecise Environments. *Machine Learning*, 42/3:203–231.

Quinlan, J. R. (1992). *C4. 5: Programs for Machine Learning.* Morgan Kaufmann, San Mateo, CA.

Radivojac, P., Chawla, N. V., Dunker, K., and Obradovic, Z. (2004). Classification and Knowledge Discovery in Protein Databases. *Journal of Biomedical Informatics*, 37(4):224–239.

Raskutti, B. and Kowalczyk, A. (2004). Extreme rebalancing for svms: a case study. *SIGKDD Explorations*, 6(1).

Solberg, A. H. and Solberg, R. (1996). A Large-Scale Evaluation of Features for Automatic Detection of Oil Spills in ERS SAR Images. In *International Geoscience and Remote Sensing Symposium*, pages 1484–1486, Lincoln, NE.

Swets, J. (1988). Measuring the Accuracy of Diagnostic Systems. *Science*, 240:1285–1293.

Tax, D. (2001). *One-class classification.* PhD thesis, Delft University of Technology.

Ting, K. (2000). A comparative study of cost-sensitive boosting algorithms. In *Proceedings of Seventeenth International Conference on Machine Learning*, pages 983–990, Stanford, CA.

Tomek, I. (1976). Two Modifications of CNN. *IEEE Transactions on Systems, Man and Cybernetics*, 6:769–772.

Turney, P. (2000). Types of Cost in Inductive Concept Learning. In *Workshop on Cost-Sensitive Learning at the Seventeenth International Conference on Machine Learning*, pages 15–21, Stanford, CA.

Weiss, G. and Provost, F. (2003). Learning when Training Data are Costly: The Effect of Class Distribution on Tree Induction. *Journal of Artificial Intelligence Research*, 19:315–354.

Woods, K., Doss, C., Bowyer, K., Solka, J., Priebe, C., and Kegelmeyer, P. (1993). Comparative Evaluation of Pattern Recognition Techniques for Detection of Microcalcifications in Mammography. *International Journal of Pattern Recognition and Artificial Intelligence*, 7(6):1417–1436.

46

Relational Data Mining

Sašo Džeroski

Jožef Stefan Institute
Jamova 39, SI-1000 Ljubljana, Slovenia
saso.dzeroski@ijs.si

Summary. Data Mining algorithms look for patterns in data. While most existing Data Mining approaches look for patterns in a single data table, relational Data Mining (RDM) approaches look for patterns that involve multiple tables (relations) from a relational database. In recent years, the most common types of patterns and approaches considered in Data Mining have been extended to the relational case and RDM now encompasses relational association rule discovery and relational decision tree induction, among others. RDM approaches have been successfully applied to a number of problems in a variety of areas, most notably in the area of bioinformatics. This chapter provides a brief introduction to RDM.

Key words: relational Data Mining, inductive logic programming, relational association rules, relational decision trees

46.1 In a Nutshell

Data Mining algorithms look for patterns in data. Most existing Data Mining approaches are propositional and look for patterns in a single data table. Relational Data Mining (RDM) approaches (Džeroski and Lavrač, 2001), many of which are based on inductive logic programming (Muggleton, 1992), look for patterns that involve multiple tables (relations) from a relational database. To emphasize this fact, RDM is often referred to as multi-relational data mining (Džeroski *et al.*, 2002). In this chapter, we will use the terms RDM and MRDM interchangeably. In this introductory section, we take a look at data, patterns, and algorithms in RDM, and mention some application areas.

46.1.1 Relational Data

A relational database typically consists of several tables (relations) and not just one table. The example database in Table 46.1 has two relations: `Customer` and `MarriedTo`. Note that relations can be defined extensionally (by tables, as in our example) or intensionally through

O. Maimon, L. Rokach (eds.), *Data Mining and Knowledge Discovery Handbook*, 2nd ed.,
DOI 10.1007/978-0-387-09823-4_46, © Springer Science+Business Media, LLC 2010

database views (as explicit logical rules). The latter typically represent relationships that can be inferred from other relationships. For example, having extensional representations of the relations mother and father, we can intensionally define the relations grandparent, grandmother, sibling, and ancestor, among others.

Intensional definitions of relations typically represent general knowledge about the domain of discourse. For example, if we have extensional relations listing the atoms that make a compound molecule and the bonds between them, functional groups of atoms can be defined intensionally. Such general knowledge is called domain knowledge or background knowledge.

Table 46.1. A relational database with two tables and two classification rules: a propositional and a relational.

Customer table

ID	*Gender*	*Age*	*Income*	*TotalSpent*	*BigS*
c1	Male	30	214000	18800	Yes
c2	Female	19	139000	15100	Yes
c3	Male	55	50000	12400	No
c4	Female	48	26000	8600	No
c5	Male	63	191000	28100	Yes
c6	Male	63	114000	20400	Yes
c7	Male	58	38000	11800	No
c8	Male	22	39000	5700	No
...	...	...	...	...	...

MarriedTo table

Spouse1	*Spouse2*
c1	c2
c2	c1
c3	c4
c4	c3
c5	c12
c6	c14
...	...

Propositional rule

IF Income > 108000
THEN BigSpender = Yes

Relational rule

`big_spender`(C1,Age1,Income1,TotalSpent1) $\leftarrow$
`married_to`(C1,C2) $\wedge$
`customer`(C2,Age2,Income2,TotalSpent2,BS2) $\wedge$
Income2 $\geq$ 108000.

46.1.2 Relational Patterns

Relational patterns involve multiple relations from a relational database. They are typically stated in a more expressive language than patterns defined on a single data table. The major types of relational patterns extend the types of propositional patterns considered in single table Data Mining. We can thus have relational classification rules, relational regression trees, and relational association rules, among others.

An example relational classification rule is given in Table 46.1, which involves the relations `Customer` and `MarriedTo`. It predicts a person to be a big spender if the person is married to somebody with high income (compare this to the propositional rule that states a person is a big spender if she has high income). Note that the two persons *C*1 and *C*2 are connected through the relation `MarriedTo`.

Relational patterns are typically expressed in subsets of first-order logic (also called predicate or relational logic). Essentials of predicate logic include predicates (`MarriedTo`) and variables ($C1, C2$), which are not present in propositional logic. Relational patterns are thus more expressive than propositional ones (Rokach *et al.*, 2004).

Most commonly, the logic programming subset of first-order logic, which is strongly related to deductive databases, is used as the formalism for expressing relational patterns. E.g., the relational rule in Table 46.1 is a logic program clause. Note that a relation in a relational database corresponds to a predicate in first-order logic (and logic programming).

46.1.3 Relational to propositional

RDM tools can be applied directly to multi-relational data to find relational patterns that involve multiple relations. Most other Data Mining approaches assume that the data resides in a single table and require preprocessing to integrate data from multiple tables (e.g., through joins or aggregation) into a single table before they can be applied. Integrating data from multiple tables through joins or aggregation, however, can cause loss of meaning or information.

Suppose we are given relations $customer(CustID, Name, Age, SpendsALot)$ and $purchase(CustID, ProductID, Date, Value, PaymentMode)$, where each customer can make multiple purchases, and we are interested in characterizing customers that spend a lot. Integrating the two relations via a natural join will give rise to a relation $purchase1$ where each row corresponds to a purchase and not to a customer. One possible aggregation would give rise to the relation $customer1(CustID, Age, NofPurchases, TotalValue, SpendsALot)$. In this case, however, some information has been clearly lost during aggregation.

The following pattern can be discovered if the relations *customer* and *purchase* are considered together.

$$\begin{aligned}
&customer(CID, Name, Age, SpendsALot) \leftarrow SpendsALot = yes \wedge \\
&\qquad Age > 30 \wedge purchase(CID, PID, D, Value, PM) \wedge \\
&\qquad PM = credit_card \wedge Value > 100.
\end{aligned}$$

This pattern says: “a customer spends a lot if she is older than 30, has purchased a product of value more than 100 and paid for it by credit card.” It would not be possible to induce such a pattern from either of the relations $purchase1$ and $customer1$ considered on their own.

Besides the ability to deal with data stored in multiple tables directly, RDM systems are usually able to take into account generally valid background (domain) knowledge given as a logic program. The ability to take into account background knowledge and the expressive power of the language of discovered patterns are distinctive for RDM.

Note that Data Mining approaches that find patterns in a given single table are referred to as attribute-value or propositional learning approaches, as the patterns they find can be expressed in propositional logic. RDM approaches are also referred to as first-order learning approaches, or relational learning approaches, as the patterns they find are expressed in the relational formalism of first-order logic. A more detailed discussion of the single table assumption, the problems resulting from it and how a relational representation alleviates these problems is given by Wrobel (Wrobel, 2001, Džeroski and Lavrač, 2001).

46.1.4 Algorithms for relational Data Mining

A RDM algorithm searches a language of relational patterns to find patterns valid in a given database. The search algorithms used here are very similar to those used in single table Data

Mining: one can search exhaustively or heuristically (greedy search, best-first search, etc.). Just as for the single table case, the space of patterns considered is typically lattice-structured and exploiting this structure is essential for achieving efficiency. The lattice structure is traversed by using refinement operators (Shapiro, 1983), which are more complicated in the relational case. In the propositional case, a refinement operator may add a condition to a rule antecedent or an item to an item set. In the relational case, a link to a new relation (table) can be introduced as well.

Just as many Data Mining algorithms come from the field of machine learning, many RDM algorithms come form the field of inductive logic programming (Muggleton, 1992, Lavrač and Džeroski, 1994). Situated at the intersection of machine learning and logic programming, ILP has been concerned with finding patterns expressed as logic programs. Initially, ILP focussed on automated program synthesis from examples, formulated as a binary classification task. In recent years, however, the scope of ILP has broadened to cover the whole spectrum of Data Mining tasks (classification, regression, clustering, association analysis). The most common types of patterns have been extended to their relational versions (relational classification rules, relational regression trees, relational association rules) and so have the major Data Mining algorithms (decision tree induction, distance-based clustering and prediction, etc.).

Van Laer and De Raedt (Van Laer and De Raedt, 2001,Džeroski and Lavrač, 2001))present a generic approach of upgrading single table Data Mining algorithms (propositional learners) to relational ones (first-order learners). Note that it is not trivial to extend a single table Data Mining algorithm to a relational one. Extending the key notions to (e.g., defining distance measures for) multi-relational data requires considerable insight and creativity. Efficiency concerns are also very important, as it is often the case that even testing a given relational pattern for validity is computationally expensive, let alone searching a space of such patterns for valid ones. An alternative approach to RDM (called propositionalization) is to create a single table from a multi-relational database in a systematic fashion (Kramer *et al.*, 2001, Džeroski and Lavrač, 2001): this approach shares some efficiency concerns and in addition can have limited expressiveness.

A pattern language typically contains a very large number of possible patterns even in the single table case: this number is in practice limited by setting some parameters (e.g., the largest size of frequent itemsets for association rule discovery). For relational pattern languages, the number of possible patterns is even larger and it becomes necessary to limit the space of possible patterns by providing more explicit constraints. These typically specify what relations should be involved in the patterns, how the relations can be interconnected, and what other syntactic constraints the patterns have to obey. The explicit specification of the pattern language (or constraints imposed upon it) is known under the name of declarative bias (Nedellec *et al.*, 1996).

46.1.5 Applications of relational Data Mining

The use of RDM has enabled applications in areas rich with structured data and domain knowledge, which would be difficult to address with single table approaches. RDM has been used in different areas, ranging from analysis of business data, through environmental and traffic engineering to web mining, but has been especially successful in bioinformatics (including drug design and functional genomics). For a comprehensive survey of RDM applications we refer the reader to Džeroski (Džeroski and Lavrač, 2001).

46.1.6 What's in this chapter

The remainder of this chapter first gives a brief introduction to inductive logic programming, which (from the viewpoint of RDM) is mainly concerned with the induction of relational classification rules for two-class problems. It then proceeds to introduce the basic RDM techniques of discovery of relational association rules and induction of relational decision trees. The chapter concludes with an overview of the RDM literature and Internet resources.

46.2 Inductive logic programming

From a KDD perspective, we can say that inductive logic programming (ILP) is concerned with the development of techniques and tools for relational Data Mining. Patterns discovered by ILP systems are typically expressed as logic programs, an important subset of first-order (predicate) logic, also called relational logic. In this section, we first briefly discuss the language of logic programs, then proceed with a discussion of the major task of ILP and some approaches to solving it.

46.2.1 Logic programs and databases

Logic programs consist of clauses. We can think of clauses as first-order rules, where the conclusion part is termed the head and the condition part the body of the clause. The head and body of a clause consist of atoms, an atom being a predicate applied to some arguments, which are called terms. In Datalog, terms are variables and constants, while in general they may consist of function symbols applied to other terms. Ground clauses have no variables.

Consider the clause $father(X,Y) \vee mother(X,Y) \leftarrow parent(X,Y)$. It reads: "if X is a parent of Y then X is the father of Y or X is the mother of Y" ($\vee$ stands for logical or). $parent(X,Y)$ is the body of the clause and $father(X,Y) \vee mother(X,Y)$ is the head. *parent*, *father* and *mother* are predicates, X and Y are variables, and $parent(X,Y)$, $father(X,Y)$, $mother(X,Y)$ are atoms. We adopt the Prolog (Bratko, 2001) syntax and start variable names with capital letters. Variables in clauses are implicitly universally quantified. The above clause thus stands for the logical formula $\forall X \forall Y : father(X,Y) \vee mother(X,Y) \vee \neg parent(X,Y)$. Clauses are also viewed as sets of literals, where a literal is an atom or its negation. The above clause is then the set $\{father(X,Y), mother(X,Y), \neg parent(X,Y)\}$.

As opposed to full clauses, definite clauses contain exactly one atom in the head. As compared to definite clauses, program clauses can also contain negated atoms in the body. While the clause in the paragraph above is a full clause, the clause $ancestor(X,Y) \leftarrow parent(Z,Y) \wedge ancestor(X,Z)$ is a definite clause ($\wedge$ stands for logical and). It is also a recursive clause, since it defines the relation *ancestor* in terms of itself and the relation *parent*. The clause $mother(X,Y) \leftarrow parent(X,Y) \wedge not\ male(X)$ is a program clause.

A set of clauses is called a clausal theory. Logic programs are sets of program clauses. A set of program clauses with the same predicate in the head is called a predicate definition. Most ILP approaches learn predicate definitions.

A predicate in logic programming corresponds to a relation in a relational database. A n-ary relation p is formally defined as a set of tuples (Ullman, 1988), i.e., a subset of the Cartesian product of n domains $D_1 \times D_2 \times \ldots \times D_n$, where a domain (or a type) is a set of values. It is assumed that a relation is finite unless stated otherwise. A relational database (RDB) is a set of relations.

Table 46.2. Database and logic programming terms.

DB terminology	*LP terminology*
relation name p	predicate symbol p
attribute of relation p	argument of predicate p
tuple $\langle a_1, \ldots, a_n \rangle$	ground fact $p(a_1, \ldots, a_n)$
relation p - a set of tuples	predicate p - defined extensionally by a set of ground facts
relation q defined as a view	predicate q defined intensionally by a set of rules (clauses)

Thus, a predicate corresponds to a relation, and the arguments of a predicate correspond to the attributes of a relation. The major difference is that the attributes of a relation are typed (i.e., a domain is associated with each attribute). For example, in the relation $lives\_in(X,Y)$, we may want to specify that X is of type *person* and Y is of type *city*. Database clauses are typed program clauses.

A deductive database (DDB) is a set of database clauses. In deductive databases, relations can be defined extensionally as sets of tuples (as in RDBs) or intensionally as sets of database clauses. Database clauses use variables and function symbols in predicate arguments and the language of DDBs is substantially more expressive than the language of RDBs (Lloyd, 1987, Ullman, 1988). A deductive Datalog database consists of definite database clauses with no function symbols.

Table 46.2 relates basic database and logic programming terms. For a full treatment of logic programming, RDBs, and deductive databases, we refer the reader to (Lloyd, 1987) and (Ullman, 1988).

46.2.2 The ILP task of relational rule induction

Logic programming as a subset of first-order logic is mostly concerned with deductive inference. Inductive logic programming, on the other hand, is concerned with inductive inference. It generalizes from individual instances/observations in the presence of background knowledge, finding regularities or hypotheses about yet unseen instances.

The most commonly addressed task in ILP is the task of learning logical definitions of relations (Quinlan, 1990), where tuples that belong or do not belong to the target relation are given as examples. From training examples ILP then induces a logic program (predicate definition) corresponding to a view that defines the target relation in terms of other relations that are given as background knowledge. This classical ILP task is addressed, for instance, by the seminal MIS system (Shapiro, 1983) (rightfully considered as one of the most influential ancestors of ILP) and one of the best known ILP systems FOIL (Quinlan, 1990).

Given is a set of examples, i.e., tuples that belong to the target relation p (positive examples) and tuples that do not belong to p (negative examples). Given are also background relations (or background predicates) q_i that constitute the background knowledge and can be used in the learned definition of p. Finally, a hypothesis language, specifying syntactic restrictions on the definition of p is also given (either explicitly or implicitly). The task is to find a definition of the target relation p that is consistent and complete, i.e., explains all the positive and none of the negative tuples.

Formally, given is a set of examples $E = P \cup N$, where P contains positive and N negative examples, and background knowledge B. The task is to find a hypothesis H such that $\forall e \in P : B \wedge H \models e$ (H is complete) and $\forall e \in N : B \wedge H \not\models e$ (H is consistent), where $\models$ stands for logical implication or entailment. This setting, introduced by Muggleton (Muggleton, 1991), is thus also called learning from entailment. In an alternative setting proposed by De Raedt and Džeroski (De Raedt and Dzeroski, 1994), the requirement that $B \wedge H \models e$ is replaced by the requirement that H be true in the minimal Herbrand model of $B \wedge e$: this setting is called learning from interpretations.

In the most general formulation, each e, as well as B and H can be a clausal theory. In practice, each e is most often a ground example (tuple), B is a relational database (which may or may not contain views) and H is a definite logic program. The semantic entailment ($\models$) is in practice replaced with syntactic entailment ($\vdash$) or provability, where the resolution inference rule (as implemented in Prolog) is most often used to prove examples from a hypothesis and the background knowledge. In learning from entailment, a positive fact is explained if it can be found among the answer substitutions for h produced by a query $?-b$ on database B, where $h \leftarrow b$ is a clause in H. In learning from interpretations, a clause $h \leftarrow b$ from H is true in the minimal Herbrand model of B if the query $b \wedge \neg h$ fails on B.

As an illustration, consider the task of defining relation $daughter(X,Y)$, which states that person X is a daughter of person Y, in terms of the background knowledge relations $female$ and $parent$. These relations are given in Table 46.3. There are two positive and two negative examples of the target relation $daughter$. In the hypothesis language of definite program clauses it is possible to formulate the following definition of the target relation,

$$daughter(X,Y) \leftarrow female(X), parent(Y,X).$$

which is consistent and complete with respect to the background knowledge and the training examples.

Table 46.3. A simple ILP problem: learning the *daughter* relation. Positive examples are denoted by $\oplus$ and negative by $\ominus$.

Training examples		Background knowledge	
$daughter(mary,ann).$	$\oplus$	$parent(ann,mary).$	$female(ann).$
$daughter(eve,tom).$	$\oplus$	$parent(ann,tom).$	$female(mary).$
$daughter(tom,ann).$	$\ominus$	$parent(tom,eve).$	$female(eve).$
$daughter(eve,ann).$	$\ominus$	$parent(tom,ian).$	

In general, depending on the background knowledge, the hypothesis language and the complexity of the target concept, the target predicate definition may consist of a set of clauses, such as

$$daughter(X,Y) \leftarrow female(X), mother(Y,X). \; daughter(X,Y) \leftarrow female(X), father(Y,X).$$

if the relations *mother* and *father* were given in the background knowledge instead of the *parent* relation.

The hypothesis language is typically a subset of the language of program clauses. As the complexity of learning grows with the expressiveness of the hypothesis language, restrictions have to be imposed on hypothesized clauses. Typical restrictions are the exclusion of recursion and restrictions on variables that appear in the body of the clause but not in its head (so-called new variables).

From a Data Mining perspective, the task described above is a binary classification task, where one of two classes is assigned to the examples (tuples): $\oplus$ (positive) or $\ominus$ (negative). Classification is one of the most commonly addressed tasks within the Data Mining community and includes approaches for rule induction. Rules can be generated from decision trees (Quinlan, 1993) or induced directly (Michalski *et al.*, 1986, Clark and Boswel, 1991).

ILP systems dealing with the classification task typically adopt the covering approach of rule induction systems. In a main loop, a covering algorithm constructs a set of clauses. Starting from an empty set of clauses, it constructs a clause explaining some of the positive examples, adds this clause to the hypothesis, and removes the positive examples explained. These steps are repeated until all positive examples have been explained (the hypothesis is complete).

In the inner loop of the covering algorithm, individual clauses are constructed by (heuristically) searching the space of possible clauses, structured by a specialization or generalization operator. Typically, search starts with a very general rule (clause with no conditions in the body), then proceeds to add literals (conditions) to this clause until it only covers (explains) positive examples (the clause is consistent).

When dealing with incomplete or noisy data, which is most often the case, the criteria of consistency and completeness are relaxed. Statistical criteria are typically used instead. These are based on the number of positive and negative examples explained by the definition and the individual constituent clauses.

46.2.3 Structuring the space of clauses

Having described how to learn sets of clauses by using the covering algorithm for clause/rule set induction, let us now look at some of the mechanisms underlying single clause/rule induction. In order to search the space of relational rules (program clauses) systematically, it is useful to impose some structure upon it, e.g., an ordering. One such ordering is based on θ-subsumption, defined below.

A substitution $\theta = \{V_1/t_1, ..., V_n/t_n\}$ is an assignment of terms t_i to variables V_i. Applying a substitution θ to a term, atom, or clause F yields the instantiated term, atom, or clause $F\theta$ where all occurrences of the variables V_i are simultaneously replaced by the term t_i. Let c and

c' be two program clauses. Clause c θ-subsumes c' if there exists a substitution θ, such that $c\theta \subseteq c'$ (Plotkin, 1969).

To illustrate the above notions, consider the clause $c = daughter(X,Y) \leftarrow parent(Y,X)$. Applying the substitution $\theta = \{X/mary, Y/ann\}$ to clause c yields

$$c\theta = daughter(mary, ann) \leftarrow parent(ann, mary).$$

Clauses can be viewed as sets of literals: the clausal notation $daughter(X,Y) \leftarrow parent(Y,X)$ thus stands for $\{daughter(X,Y), \neg parent(Y,X)\}$ where all variables are assumed to be universally quantified, $\neg$ denotes logical negation, and the commas denote disjunction. According to the definition, clause c θ-subsumes c' if there is a substitution θ that can be applied to c such that every literal in the resulting clause occurs in c'. Clause c θ-subsumes $c' = daughter(X,Y) \leftarrow female(X), parent(Y,X)$ under the empty substitution $\theta = \emptyset$, since $\{daughter(X,Y), \neg parent(Y,X)\}$ is a proper subset of $\{daughter(X,Y), \neg female(X), \neg parent(Y,X)\}$. Furthermore, under the substitution $\theta = \{X/mary, Y/ann\}$, clause c θ-subsumes the clause:

$c' = daughter(mary, ann) \leftarrow female(mary), parent(ann, mary), parent(ann, tom)$.

θ-subsumption introduces a syntactic notion of generality. Clause c is at least as general as clause c' ($c \leq c'$) if c θ-subsumes c'. Clause c is more general than c' ($c < c'$) if $c \leq c'$ holds and $c' \leq c$ does not. In this case, we say that c' is a specialization of c and c is a generalization of c'. If the clause c' is a specialization of c then c' is also called a refinement of c.

Under a semantic notion of generality, c is more general than c' if c logically entails c' ($c \models c'$). If c θ-subsumes c' then $c \models c'$. The reverse is not always true. The syntactic, θ-subsumption based, generality is computationally more feasible. Namely, semantic generality is in general undecidable. Thus, syntactic generality is frequently used in ILP systems.

The relation $\leq$ defined by θ-subsumption introduces a lattice on the set of reduced clauses (Plotkin, 1969): This enables ILP systems to prune large parts of the search space. θ-subsumption also provides the basis for clause construction by top-down searching of refinement graphs and bounding the search of refinement graphs from below by using a bottom clause (which can be constructed as least general generalizations, i.e., least upper bounds of example clauses in the θ-subsumption lattice).

46.2.4 Searching the space of clauses

Most ILP approaches search the hypothesis space of program clauses in a top-down manner, from general to specific hypotheses, using a θ-subsumption-based specialization operator. A specialization operator is usually called a refinement operator (Shapiro, 1983). Given a hypothesis language $\mathscr{L}$, a refinement operator ρ maps a clause c to a set of clauses $\rho(c)$ which are specializations (refinements) of c: $\rho(c) = \{c' \mid c' \in \mathscr{L}, c < c'\}$.

A refinement operator typically computes only the set of minimal (most general) specializations of a clause under θ-subsumption. It employs two basic syntactic operations:

- apply a substitution to the clause, and
- add a literal to the body of the clause.

The hypothesis space of program clauses is a lattice, structured by the θ-subsumption generality ordering. In this lattice, a refinement graph can be defined as a directed, acyclic graph in which nodes are program clauses and arcs correspond to the basic refinement operations: substituting a variable with a term, and adding a literal to the body of a clause.

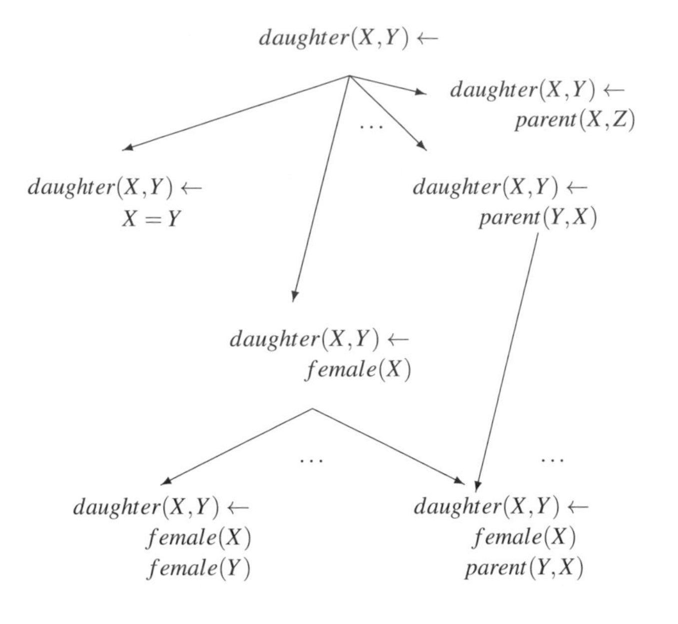

Fig. 46.1. Part of the refinement graph for the family relations problem.

Figure 46.1 depicts a part of the refinement graph for the family relations problem defined in Table 46.3, where the task is to learn a definition of the *daughter* relation in terms of the relations *female* and *parent*.

At the top of the refinement graph (lattice) is the clause with an empty body $c = daughter(X,Y) \leftarrow$. The refinement operator ρ generates the refinements of c, which are of the form $\rho(c) = \{daughter(X,Y) \leftarrow L\}$, where L is one of following literals:

- literals having as arguments the variables from the head of the clause: $X = Y$ (applying a substitution X/Y), $female(X)$, $female(Y)$, $parent(X,X)$, $parent(X,Y)$, $parent(Y,X)$, and $parent(Y,Y)$, and
- literals that introduce a new distinct variable Z ($Z \neq X$ and $Z \neq Y$) in the clause body: $parent(X,Z)$, $parent(Z,X)$, $parent(Y,Z)$, and $parent(Z,Y)$.

This assumes that the language is restricted to definite clauses, hence literals of the form *not* L are not considered, and non-recursive clauses, hence literals with the predicate symbol *daughter* are not considered.

The search for a clause starts at the top of the lattice, with the clause $d(X,Y)$ that covers all example (positive and negative). Its refinements are then considered, then their refinements in turn, an this is repeated until a clause is found which covers only positive examples. In the example above, the clause $daughter(X,Y) \leftarrow female(X), parent(Y,X)$ is such a clause. Note that this clause can be reached in several ways from the top of the lattice, e.g., by first adding $female(X)$, then $parent(Y,X)$ or vice versa.

The refinement graph is typically searched heuristically level-wise, using heuristics based on the number of positive and negative examples covered by a clause. As the branching factor is very large, greedy search methods are typically applied which only consider a limited number of alternatives at each level. Hill-climbing considers only one best alternative at each level, while beam search considers n best alternatives, where n is the beam width. Occasionally, complete search is used, e.g., A^* best-first search or breadth-first search. This search can be bound from below by using so-called bottom clauses, which can be constructed by least general generalization (Muggleton and Feng, 1990) or inverse resolution/entailment (Muggleton, 1995).

46.2.5 Transforming ILP problems to propositional form

One of the early approaches to ILP, implemented in the ILP system LINUS (Lavrač *et al.*, 1991), is based on the idea that the use of background knowledge can introduce new attributes for learning. The learning problem is transformed from relational to attribute-value form and solved by an attribute-value learner. An advantage of this approach is that Data Mining algorithms that work on a single table (and this is the majority of existing Data Mining algorithms) become applicable after the transformation.

This approach, however, is feasible only for a restricted class of ILP problems. Thus, the hypothesis language of LINUS is restricted to function-free program clauses which are typed (each variable is associated with a predetermined set of values), constrained (all variables in the body of a clause also appear in the head) and nonrecursive (the predicate symbol the head does not appear in any of the literals in the body).

The LINUS algorithm which solves ILP problems by transforming them into propositional form consists of the following three steps:

- The learning problem is transformed from relational to attribute-value form.
- The transformed learning problem is solved by an attribute-value learner.
- The induced hypothesis is transformed back into relational form.

The above algorithm allows for a variety of approaches developed for propositional problems, including noise-handling techniques in attribute-value algorithms, such as CN2 (Clark and Niblett, 1989), to be used for learning relations. It is illustrated on the simple ILP problem of learning family relations. The task is to define the target relation $daughter(X,Y)$, which states that person X is a daughter of person Y, in terms of the background knowledge relations $female$, $male$ and $parent$.

Table 46.4. Non-ground background knowledge for learning the *daughter* relation.

Training examples		Background knowledge		
$daughter(mary,ann).$	$\oplus$	$parent(X,Y) \leftarrow$	$mother(ann,mary).$	$female(ann).$
$daughter(eve,tom).$	$\oplus$	$mother(X,Y).$	$mother(ann,tom).$	$female(mary).$
$daughter(tom,ann).$	$\ominus$	$parent(X,Y) \leftarrow$	$father(tom,eve).$	$female(eve).$
$daughter(eve,ann).$	$\ominus$	$father(X,Y).$	$father(tom,ian).$	

Table 46.5. Propositional form of the *daughter* relation problem.

Variables			Propositional features							
C	X	Y	$f(X)$	$f(Y)$	$m(X)$	$m(Y)$	$p(X,X)$	$p(X,Y)$	$p(Y,X)$	$p(Y,Y)$
$\oplus$	*mary*	*ann*	*true*	*true*	*false*	*false*	*false*	*false*	*true*	*false*
$\oplus$	*eve*	*tom*	*true*	*false*	*false*	*true*	*false*	*false*	*true*	*false*
$\ominus$	*tom*	*ann*	*false*	*true*	*true*	*false*	*false*	*false*	*true*	*false*
$\ominus$	*eve*	*ann*	*true*	*true*	*false*	*false*	*false*	*false*	*false*	*false*

All the variables are of the type *person*, defined as *person* $= \{ann,\ eve,\ ian,\ mary,\ tom\}$. There are two positive and two negative examples of the target relation. The training examples and the relations from the background knowledge are given in Table 46.3. However, since the LINUS approach can use non-ground background knowledge, let us assume that the background knowledge from Table 46.4 is given.

The first step of the algorithm, i.e., the transformation of the ILP problem into attribute-value form, is performed as follows. The possible applications of the background predicates on the arguments of the target relation are determined, taking into account argument types. Each such application introduces a new attribute. In our example, all variables are of the same type *person*. The corresponding attribute-value learning problem is given in Table 46.5, where f stands for $female$, m for *male* and p for *parent*. The attribute-value tuples are generalizations (relative to the given background knowledge) of the individual facts about the target relation.

In Table 46.5, *variables* stand for the arguments of the target relation, and *propositional features* denote the newly constructed attributes of the propositional learning task. When learning function-free clauses, only the new attributes (propositional features) are considered for learning.

In the second step, an attribute-value learning program induces the following if-then rule from the tuples in Table 46.5:

$Class = \oplus$ **if** $[female(X) = true] \wedge [parent(Y,X) = true]$

In the last step, the induced if-then rules are transformed into clauses. In our example, we get the following clause:

$daughter(X,Y) \leftarrow female(X), parent(Y,X).$

The LINUS approach has been extended to handle determinate clauses (Džeroski *et al.*, 1992, Lavrač and Džeroski, 1994), which allow the introduction of determinate new variables (which have a unique value for each training example). There also exist a number of other approaches to propositionalization, some of them very recent: an overview is given by Kramer et al. (Kramer *et al.*, 2001).

Let us emphasize again, however, that it is in general not possible to transform an ILP problem into a propositional (attribute-value) form efficiently. De Raedt (De Raedt, 1998) treats the relation between attribute-value learning and ILP in detail, showing that propositionalization of some more complex ILP problems is possible, but results in attribute-value problems that are exponentially large. This has also been the main reason for the development of a variety of new RDM/ILP techniques by upgrading propositional approaches.

46.2.6 Upgrading propositional approaches

ILP/RDM algorithms have many things in common with propositional learning algorithms. In particular, they share the learning as search paradigm, i.e., they search for patterns valid in the given data. The key differences lie in the representation of data and patterns, refinement operators/generality relationships, and testing coverage (i.e., whether a rule explains an example).

Van Laer and De Raedt (Van Laer and De Raedt, 2001) explicitly formulate a recipe for upgrading propositional algorithms to deal with relational data and patterns. The key idea is to keep as much of the propositional algorithm as possible and upgrade only the key notions. For rule induction, the key notions are the refinement operator and coverage relationship. For distance-based approaches, the notion of distance is the key one. By carefully upgrading the key notions of a propositional algorithm, a RDM/ILP algorithm can be developed that has the original propositional algorithm as a special case.

The recipe has been followed (more or less exactly) to develop ILP systems for rule induction, well before it was formulated explicitly. The well known FOIL (Quinlan, 1990) system can be seen as an upgrade of the propositional rule induction program CN2 (Clark and Niblett, 1989). Another well known ILP system, PROGOL (Muggleton, 1995) can be viewed as upgrading the AQ approach (Michalski *et al.*, 1986) to rule induction.

More recently, the upgrading approach has been used to develop a number of RDM approaches that address Data Mining tasks other than binary classification. These include the discovery of frequent Datalog patterns and relational association rules (Dehaspe and Toivonen, 2001, Džeroski and Lavrač, 2001), (Dehaspe, 1999), the induction of relational decision trees (structural classification and regression trees (Kramer and Widmer, 2001) and first-order logical decision trees (Blockeel and De Raedt, 1998)), and relational distance-based approaches to classification and clustering (Kirsten *et al.*, 2001, Džeroski and Lavrač, 2001, Emde and Wettschereck, 1996). The algorithms developed have as special cases well known propositional algorithms, such as the APRIORI algorithm for finding frequent patterns; the CART and C4.5 algorithms for learning decision trees; k- nearest neighbor classification, hierarchical and k-medoids clustering. In the following two sections, we briefly review how the propositional approaches for association rule discovery and decision tree inducion have been lifted to a relational framework, highlighting the key differences between the relational algorithms and their propositional counterparts.

46.3 Relational Association Rules

The discovery of frequent patterns and association rules is one of the most commonly studied tasks in Data Mining. Here we first describe frequent relational patterns (frequent Datalog patterns) and relational association rules (query extensions). We then look into how a well-known algorithm for finding frequent itemsets has been upgraded do discover frequent relational patterns.

46.3.1 Frequent Datalog queries and query extensions

Dehaspe and Toivonen (Dehaspe, 1999, Dehaspe and Toivonen, 2001, Džeroski and Lavrač, 2001) consider patterns in the form of Datalog queries, which reduce to SQL queries. A Datalog query has the form $? - A_1, A_2, \ldots A_n$, where the A_i's are logical atoms.

An example Datalog query is

$$? - person(X), parent(X,Y), hasPet(Y,Z)$$

This query on a Prolog database containing predicates *person*, *parent*, and *hasPet* is equivalent to the SQL query

```
SELECT PERSON.ID, PARENT.KID, HASPET.AID
FROM PERSON, PARENT, HASPET
WHERE PERSON.ID = PARENT.PID
AND PARENT.KID = HASPET.PID
```

on a database containing relations PERSON with argument ID, PARENT with arguments PID and KID, and HASPET with arguments PID and AID. This query finds triples (x, y, z), where child y of person x has pet z.

Datalog queries can be viewed as a relational version of itemsets (which are sets of items occurring together). Consider the itemset $\{person, parent, child, pet\}$. The market-basket interpretation of this pattern is that a person, a parent, a child, and a pet occur together. This is also partly the meaning of the above query. However, the variables X, Y, and Z add extra information: the person and the parent are the same, the parent and the child belong to the same family, and the pet belongs to the child. This illustrates the fact that queries are a more expressive variant of itemsets.

To discover frequent patterns, we need to have a notion of frequency. Given that we consider queries as patterns and that queries can have variables, it is not immediately obvious what the frequency of a given query is. This is resolved by specifying an additional parameter of the pattern discovery task, called the key. The key is an atom which has to be present in all queries considered during the discovery process. It determines what is actually counted. In the above query, if $person(X)$ is the key, we count persons, if $parent(X,Y)$ is the key, we count (parent,child) pairs, and if $hasPet(Y,Z)$ is the key, we count (owner,pet) pairs. This is described more precisely below.

Submitting a query $Q = ? - A_1, A_2, \ldots A_n$ with variables $\{X_1, \ldots X_m\}$ to a Datalog database $\mathbf{r}$ corresponds to asking whether a grounding substitution exists (which replaces each of the variables in Q with a constant), such that the conjunction $A_1, A_2, \ldots A_n$ holds in $\mathbf{r}$. The answer to the query produces answering substitutions $\theta = \{X_1/a_1, \ldots X_m/a_m\}$ such that $Q\theta$ succeeds. The set of all answering substitutions obtained by submitting a query Q to a Datalog database $\mathbf{r}$ is denoted $answerset(Q, \mathbf{r})$.

The absolute frequency of a query Q is the number of answer substitutions θ for the variables in the key atom for which the query $Q\theta$ succeeds in the given database, i.e., $a(Q, \mathbf{r}, key) = |\{\theta \in answerset(key, \mathbf{r}) | Q\theta$ succeeds w.r.t. $\mathbf{r}\}|$. The relative frequency (support) can be calculated as $f(Q, \mathbf{r}, key) = a(Q, \mathbf{r}, key) / |\{\theta \in answerset(key, \mathbf{r})\}|$. Assuming the key is $person(X)$, the absolute frequency for our query involving parents, children and pets can be calculated by the following SQL statement:

```
SELECT count(distinct *)
FROM SELECT PERSON.ID
    FROM PERSON, PARENT, HASPET
    WHERE PERSON.ID = PARENT.PID
    AND PARENT.KID = HASPET.PID
```

Association rules have the form $A \rightarrow C$ and the intuitive market-basket interpretation "customers that buy A typically also buy C". If itemsets A and C have supports f_A and f_C, respectively, the confidence of the association rule is defined to be $c_{A \rightarrow C} = f_C / f_A$. The task of association rule discovery is to find all association rules $A \rightarrow C$, where f_C and $c_{A \rightarrow C}$ exceed prespecified thresholds (minsup and minconf).

Association rules are typically obtained from frequent itemsets. Suppose we have two frequent itemsets A and C, such that $A \subset C$, where $C = A \cup B$. If the support of A is f_A and the support of C is f_C, we can derive an association rule $A \rightarrow B$, which has confidence f_C / f_A. Treating the arrow as implication, note that we can derive $A \rightarrow C$ from $A \rightarrow B$ ($A \rightarrow A$ and $A \rightarrow B$ implies $A \rightarrow A \cup B$, i.e., $A \rightarrow C$).

Relational association rules can be derived in a similar manner from frequent Datalog queries. From two frequent queries $Q_1 = ? - l_1, \ldots l_m$ and $Q_2 = ? - l_1, \ldots l_m, l_{m+1}, \ldots l_n$, where Q_2 θ-subsumes Q_1, we can derive a relational association rule $Q_1 \rightarrow Q_2$. Since Q_2 extends Q_1, such a relational association rule is named a query extension.

A query extension is thus an existentially quantified implication of the form $? - l_1, \ldots l_m \rightarrow ? - l_1, \ldots l_m, l_{m+1}, \ldots l_n$ (since variables in queries are existentially quantified). A shorthand notation for the above query extension is $? - l_1, \ldots l_m \rightsquigarrow l_{m+1}, \ldots l_n$. We call the query $? - l_1, \ldots l_m$ the body and the sub-query $l_{m+1}, \ldots l_n$ the head of the query extension. Note, however, that the head of the query extension does not correspond to its conclusion (which is $? - l_1, \ldots l_m, l_{m+1}, \ldots l_n$).

Assume the queries $Q_1 = ? - person(X), parent(X,Y)$ and $Q_2 = ? - person(X), parent(X,Y), hasPet(Y,Z)$ are frequent, with absolute frequencies of 40 and 30, respectively. The query extension E, where E is defined as $E = ? - person(X), parent(X,Y) \rightsquigarrow hasPet(Y,Z)$, can be considered a relational association rule with a support of 30 and confidence of $30/40 = 75\%$. Note the difference in meaning between the query extension E and two obvious, but incorrect, attempts at defining relational association rules. The clause $person(X), parent(X,Y) \rightarrow hasPet(Y,Z)$ (which stands for the logical formula $\forall XYZ : person(X) \wedge parent(X,Y) \rightarrow hasPet(Y,Z)$) would be interpreted as follows: "if a person has a child, then this child has a pet". The implication $? - person(X), parent(X,Y) \rightarrow ? - hasPet(Y,Z)$, which stands for $(\exists XY : person(X) \wedge parent(X,Y)) \rightarrow (\exists YZ : hasPet(Y,Z))$ is trivially true if at least one person in the database has a pet. The correct interpretation of the query extension E is: "if a person has a child, then this person also has a child that has a pet."

46.3.2 Discovering frequent queries: WARMR

The task of discovering frequent queries is addressed by the RDM system WARMR (Dehaspe, 1999). WARMR takes as input a database **r**, a frequency threshold *minfreq*, and declarative language bias $\mathscr{L}$. The latter specifies a *key* atom and input-output modes for predicates/relations, discussed below.

WARMR upgrades the well-known APRIORI algorithm for discovering frequent patterns, which performs levelwise search (Agrawal *et al.*, 1996) through the lattice of itemsets. APRIORI starts with the empty set of items and at each level l considers sets of items of cardinality l. The key to the efficiency of APRIORI lies in the fact that a large frequent itemset can only be generated by adding an item to a frequent itemset. Candidates at level $l+1$ are thus generated by adding items to frequent itemsets obtained at level l. Further efficiency is achieved using the fact that all subsets of a frequent itemset have to be frequent: only candidates that pass this tests get their frequency to be determined by scanning the database.

In analogy to APRIORI, WARMR searches the lattice of Datalog queries for queries that are frequent in the given database **r**. In analogy to itemsets, a more complex (specific) frequent query Q_2 can only be generated from a simpler (more general) frequent query Q_1 (where Q_1 is more general than Q_2 if Q_1 θ-subsumes Q_2; see Section 46.2.3 for a definition of θ-subsumption). WARMR thus starts with the query $? - key$ at level 1 and generates candidates for frequent queries at level $l+1$ by refining (adding literals to) frequent queries obtained at level l.

Table 46.6. An example specification of declarative language bias settings for WARMR.

```
warmode_key(person(-)).
warmode(parent(+, -)).
warmode(hasPet(+, cat)).
warmode(hasPet(+, dog)).
warmode(hasPet(+, lizard)).
```

Suppose we are given a Prolog database containing the predicates *person*, *parent*, and *hasPet*, and the declarative bias in Table 46.6. The latter contains the key atom $parent(X)$ and input-output modes for the relations *parent* and *hasPet*. Input-output modes specify whether a variable argument of an atom in a query has to appear earlier in the query (+), must not (−) or may, but need not to (±). Input-output modes thus place constraints on how queries can be refined, i.e., what atoms may be added to a given query.

Given the above, WARMR starts the search of the refinement graph of queries at level 1 with the query $? - person(X)$. At level 2, the literals $parent(X,Y)$, $hasPet(X,cat)$, $hasPet(X,dog)$ and $hasPet(X,lizard)$ can be added to this query, yielding the queries $? - person(X), parent(X,Y)$, $? - person(X), hasPet(X,cat)$, $? - person(X), hasPet(X,dog)$, and $? - person(X), hasPet(X,lizard)$. Taking the first of the level 2 queries, the following literals are added to obtain level 3 queries: $parent(Y,Z)$ (note that $parent(Y,X)$ cannot be added, because X already appears in the query being refined), $hasPet(Y,cat)$, $hasPet(Y,dog)$ and $hasPet(Y,lizard)$.

While all subsets of a frequent itemset must be frequent in APRIORI, not all sub-queries of a frequent query need be frequent queries in WARMR. Consider the query $? - person(X), parent(X,Y), hasPet(Y,cat)$ and assume it is frequent. The sub-query $? - person(X), hasPet(Y,cat)$ is not allowed, as it violates the declarative bias constraint that the first argument of *hasPet* has to appear earlier in the query. This causes some complications in pruning the generated candidates for frequent queries: WARMR keeps a list of infrequent queries and checks whether the generated candidates are subsumed by a query in this list. The WARMR algorithm is given in Table 46.7.

WARMR upgrades APRIORI to a multi-relational setting following the upgrading recipe (see Section 46.2.6). The major differences are in finding the frequency of queries (where we have to count answer substitutions for the key atom) and the candidate query generation (by using a refinement operator and declarative bias). WARMR has APRIORI as a special case: if we only have predicates of zero arity (with no arguments), which correspond to items, WARMR can be used to discover frequent itemsets.

More importantly, WARMR has as special cases a number of approaches that extend the discovery of frequent itemsets with, e.g., hierarchies on items (Srikant and Agrawal, 1995), as well as approaches to discovering sequential patterns (Agrawal and Srikant, 1995), including general episodes (Mannila and Toivonen, 1996). The individual approaches mentioned make

Table 46.7. The WARMR algorithm for discovering frequent Datalog queries.

Algorithm WARMR(**r**, $\mathscr{L}$, *key*, *minfreq*; Q)
Input: Database **r**; Declarative language bias $\mathscr{L}$ and *key* ;
threshold *minfreq*;
Output: All queries $Q \in \mathscr{L}$ with frequency $\geq$ *minfreq*

1. Initialize level $d := 1$
2. Initialize the set of candidate queries $\mathscr{Q}_1 := \{$*?- key*$\}$
3. Initialize the set of (in)frequent queries $\mathscr{F} := \emptyset$; $\mathscr{I} := \emptyset$
4. While $\mathscr{Q}_d$ not empty
5. Find frequency of all queries $Q \in \mathscr{Q}_d$
6. Move those with frequency below *minfreq* to $\mathscr{I}$
7. Update $\mathscr{F} := \mathscr{F} \cup \mathscr{Q}_d$
8. Compute new candidates:
$\mathscr{Q}_{d+1}$ = WARMRgen($\mathscr{L}$; $\mathscr{I}$; $\mathscr{F}$; $\mathscr{Q}_d$))
9. Increment d
10. Return $\mathscr{F}$

Function WARMRgen($\mathscr{L}$; $\mathscr{I}$; $\mathscr{F}$; $\mathscr{Q}_d$);

1. Initialize $\mathscr{Q}_{d+1} := \emptyset$
2. For each $Q_j \in \mathscr{Q}_d$, and for each refinement $Q'_j \in \mathscr{L}$ of Q_j:
Add Q'_j to $\mathscr{Q}_{d+1}$, unless:
(i) Q'_j is more specific than some query $\in \mathscr{I}$, or
(ii) Q'_j is equivalent to some query $\in \mathscr{Q}_{d+1} \cup \mathscr{F}$
3. Return $\mathscr{Q}_{d+1}$

use of the specific properties of the patterns considered (very limited use of variables) and are more efficient than WARMR for the particular tasks they address. The high expressive power of the language of patterns considered has its computational costs, but it also has the important advantage that a variety of different pattern types can be explored without any changes in the implementation.

WARMR can be (and has been) used to perform propositionalization, i.e., to transform MRDM problems to propositional (single table) form. WARMR is first used to discover frequent queries. In the propositional form, examples correspond to answer substitutions for the key atom and the binary attributes are the frequent queries discovered. An attribute is true for an example if the corresponding query succeeds for the corresponding answer substitution. This approach has been applied with considerable success to the tasks of predictive toxicology (Dehaspe *et al.*, 1998) and genome-wide prediction of protein functional class (King *et al.*, 2000).

46.4 Relational Decision Trees

Decision tree induction is one of the major approaches to Data Mining. Upgrading this approach to a relational setting has thus been of great importance. In this section, we first look into what relational decision trees are, i.e., how they are defined, then discuss how such trees can be induced from multi-relational data.

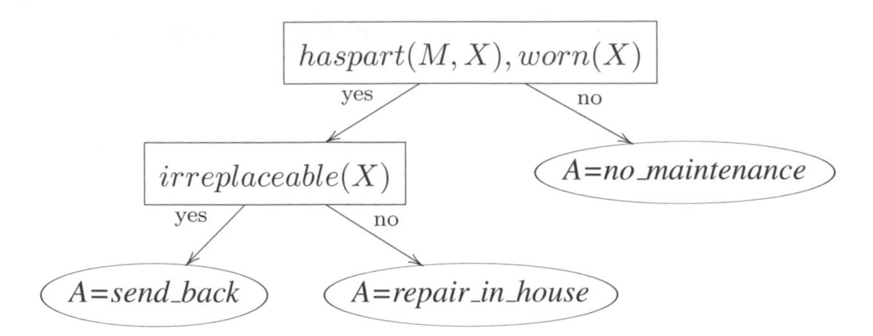

Fig. 46.2. A relational decision tree, predicting the class variable A in the target predicate $maintenance(M,A)$.

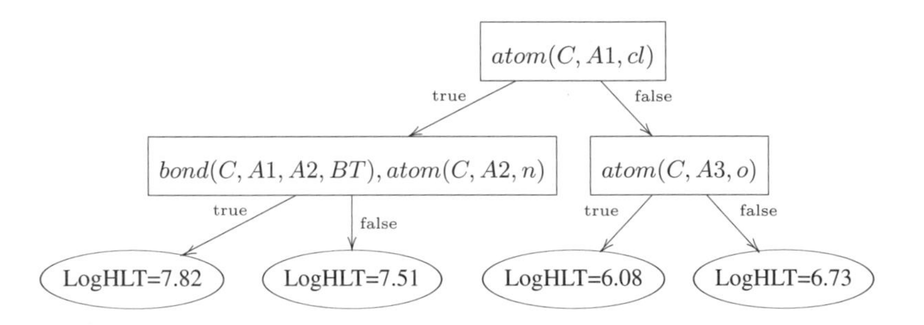

Fig. 46.3. A relational regression tree for predicting the degradation time *LogHLT* of a chemical compound C (target predicate $degrades(C,LogHLT)$).

46.4.1 Relational Classification, Regression, and Model Trees

Without loss of generality, we can say the task of relational prediction is defined by a two-place target predicate $target(ExampleID,ClassVar)$, which has as arguments an example ID and the class variable, and a set of background knowledge predicates/relations. Depending on whether the class variable is discrete or continuous, we talk about relational classification or regression. Relational decision trees are one approach to solving this task.

An example relational decision tree is given in Figure 46.2. It predicts the maintenance action A to be taken on machine M ($maintenance(M,A)$), based on parts the machine contains ($haspart(M,X)$), their condition ($worn(X)$) and ease of replacement ($irreplaceable(X)$). The target predicate here is $maintenance(M,A)$, the class variable is A, and background knowledge predicates are $haspart(M,X)$, $worn(X)$ and $irreplaceable(X)$.

Relational decision trees have much the same structure as propositional decision trees. Internal nodes contain tests, while leaves contain predictions for the class value. If the class variable is discrete/continuous, we talk about relational classification/regression trees. For regression, linear equations may be allowed in the leaves instead of constant class-value predictions: in this case we talk about relational model trees.

The tree in Figure 46.2 is a relational classification tree, while the tree in Figure 46.3 is a relational regression tree. The latter predicts the degradation time (the logarithm of the mean half-life time in water (Džeroski *et al.*, 1999)) of a chemical compound from its chemical structure, where the latter is represented by the atoms in the compound and the bonds between them. The target predicate is $degrades(C,LogHLT)$, the class variable $LogHLT$, and the background knowledge predicates are $atom(C,AtomID,Element)$ and $bond(C,A_1,A_2, BondType)$. The test at the root of the tree $atom(C,A1,cl)$ asks if the compound C has a chlorine atom $A1$ and the test along the left branch checks whether the chlorine atom $A1$ is connected to a nitrogen atom $A2$.

As can be seen from the above examples, the major difference between propositional and relational decision trees is in the tests that can appear in internal nodes. In the relational case, tests are queries, i.e., conjunctions of literals with existentially quantified variables, e.g., $atom(C,A1,cl)$ and $haspart(M,X),worn(X)$. Relational trees are binary: each internal node has a left (yes) and a right (no) branch. If the query succeeds, i.e., if there exists an answer substitution that makes it true, the yes branch is taken.

It is important to note that variables can be shared among nodes, i.e., a variable introduced in a node can be referred to in the left (yes) subtree of that node. For example, the X in $irreplaceable(X)$ refers to the machine part X introduced in the root node test $haspart(M,X),worn(X)$. Similarly, the $A1$ in $bond(C,A1,A2,BT)$ refers to the chlorine atom introduced in the root node $atom(C,A1,cl)$. One cannot refer to variables introduced in a node in the right (no) subtree of that node. For example, referring to the chlorine atom $A1$ in the right subtree of the tree in Figure 46.3 makes no sense, as going along the right (no) branch means that the compound contains no chlorine atoms.

The actual test that has to be executed in a node is the conjunction of the literals in the node itself and the literals on the path from the root of the tree to the node in question. For example, the test in the node $irreplaceable(X)$ in Figure 46.2 is actually $haspart(M,X),worn(X), irreplaceable(X)$. In other words, we need to send the machine back to the manufacturer for maintenance only if it has a part which is both worn and irreplaceable (Rokach and Maimon, 2006). Similarly, the test in the node $bond(C,A1,A2,BT)$, $atom(C,A2,n)$ in Figure 46.3 is in fact $atom(C,A1,cl),bond(C,A1,A2,BT)$, $atom(C,A2,n)$. As a consequence, one cannot transform relational decision trees to logic programs in the fashion "one clause per leaf" (unlike propositional decision trees, where a transformation "one rule per leaf" is possible).

Table 46.8. A decision list representation of the relational decision tree in Figure 46.2.

$maintenance(M,A) \leftarrow haspart(M,X),worn(X),$
  $irreplaceable(X)$!, $A = send\_back$
$maintenance(M,A) \leftarrow haspart(M,X),worn(X)$, !,
  $A = repair\_in\_house$
$maintenance(M,A) \leftarrow A = no\_maintenance$

Relational decision trees can be easily transformed into first-order decision lists, which are ordered sets of clauses (clauses in logic programs are unordered). When applying a decision list to an example, we always take the first clause that applies and return the answer produced. When applying a logic program, all applicable clauses are used and a set of answers can be produced. First-order decision lists can be represented by Prolog programs with cuts (!) (Bratko, 2001): cuts ensure that only the first applicable clause is used.

Table 46.9. A decision list representation of the relational regression tree for predicting the biodegradability of a compound, given in Figure 46.3.

$degrades(C, LogHLT) \leftarrow atom(C, A1, cl),$
$\quad bond(C, A1, A2, BT), atom(C, A2, n), LogHLT = 7.82, !$
$degrades(C, LogHLT) \leftarrow atom(C, A1, cl),$
$\quad LogHLT = 7.51, !$
$degrades(C, LogHLT) \leftarrow atom(C, A3, o),$
$\quad LogHLT = 6.08, !$
$degrades(C, LogHLT) \leftarrow LogHLT = 6.73.$

Table 46.10. A logic program representation of the relational decision tree in Figure 46.2.

$a(M) \leftarrow haspart(M, X), worn(X), irreplaceable(X)$
$b(M) \leftarrow haspart(M, X), worn(X)$
$maintenance(M, A) \leftarrow not\ a(M), A = no\_aintenance$
$maintenance(M, A) \leftarrow b(M), A = repair\_in\_house$
$maintenance(M, A) \leftarrow a(M), not\ b(M), A = send\_back$

A decision list is produced by traversing the relational regression tree in a depth-first fashion, going down left branches first. At each leaf, a clause is output that contains the prediction of the leaf and all the conditions along the left (yes) branches leading to that leaf. A decision list obtained from the tree in Figure 46.2 is given in Table 46.8. For the first clause (*send_back*), the conditions in both internal nodes are output, as the left branches out of both nodes have been followed to reach the corresponding leaf. For the second clause, only the condition in the root is output: to reach the *repair_in_house* leaf, the left (yes) branch out of the root has been

Table 46.11. The TDIDT part of the SCART algorithm for inducing relational decision trees.

```
procedure DIVIDEANDCONQUER(TestsOnYesBranchesSofar, DeclarativeBias, Examples)

if TERMINATIONCONDITION(Examples)
then
    NewLeaf = CREATENEWLEAF(Examples)
    return NewLeaf
else
    PossibleTestsNow = GENERATETESTS(TestsOnYesBranchesSofar, DeclarativeBias)
    BestTest = FINDBESTTEST(PossibleTestsNow, Examples)
    (Split_1, Split_2) = SPLITEXAMPLES(Examples, TestsOnYesBranchesSofar, BestTest)
    LeftSubtree = DIVIDEANDCONQUER(TestsOnYesBranchesSofar ∧ BestTest, Split_1)
    RightSubtree = DIVIDEANDCONQUER(TestsOnYesBranchesSofar, Split_2)
    return [BestTest, LeftSubtree, RightSubtree]
```

followed, but the right (no) branch out of the *irreplaceable*(X) node has been followed. A decision list produced from the relational regression tree in Figure 46.3 is given in Table 46.9.

Generating a logic program from a relational decision tree is more complicated. It requires the introduction of new predicates. We will not describe the transformation process in detail, but rather give an example. A logic program, corresponding to the tree in Figure 46.2 is given in Table 46.10.

46.4.2 Induction of Relational Decision Trees

The two major algorithms for inducing relational decision trees are upgrades of the two most famous algorithms for inducting propositional decision trees. SCART (Kramer, 1996, Kramer and Widmer, 2001) is an upgrade of CART (Breiman *et al.*, 1984), while TILDE (Blockeel and De Raedt, 1998, De Raedt *et al.*, 2001) is an upgrade of C4.5 (Quinlan, 1993). According to the upgrading recipe, both SCART and TILDE have their propositional counterparts as special cases. The actual algorithms thus closely follow CART and C4.5. Here we illustrate the differences between SCART and CART by looking at the TDIDT (top-down induction of decision trees) algorithm of SCART (Table 46.11).

Given a set of examples, the TDID algorithm first checks if a termination condition is satisfied, e.g., if all examples belong to the same class c. If yes, a leaf is constructed with an appropriate prediction, e.g., assigning the value c to the class variable. Otherwise a test is selected among the possible tests for the node at hand, examples are split into subsets according to the outcome of the test, and tree construction proceeds recursively on each of the subsets. A tree is thus constructed with the selected test at the root and the subtrees resulting from the recursive calls attached to the respective branches.

The major difference in comparison to the propositional case is in the possible tests that can be used in a node. While in CART these remain (more or less) the same regardless of where the node is in the tree (e.g., $A = v$ or $A < v$ for each attribute and attribute value), in SCART the set of possible tests crucially depend on the position of the node in the tree. In particular, it depends on the tests along the path from the root to the current node, more precisely on the variables appearing in those tests and the declarative bias. To emphasize this, we can think of a GENERATETESTS procedure being separately employed before evaluating the tests. The inputs to this procedure are the tests on positive branches from the root to the current node and the declarative bias. These are also inputs to the top level TDIDT procedure.

The declarative bias in SCART contains statements of the form *schema(CofL,TandM)*, where *CofL* is a conjunction of literals and *TandM* is a list of type and mode declarations for the variables in those literals. Two such statements, used in the induction of the regression tree in Figure 46.3 are as follows: *schema((bond(V, W, X, Y), atom(V, X, Z)), [V:chemical:'+', W:atomid:'+', X:atomid:'-', Y:bondtype:'-', Z:element: '=']*) and *schema(bond (V, W, X, Y), [V: chemical:'+', W:atomid:'+', X:atomid:'-', Y:bondtype: '=']*). In the lists, each variable in the conjunction is followed by its type and mode declaration: '+' denotes that the variable must be bound (i.e., appear in *TestsOnYesBranchesSofar*), - that it must not be bound, and = that it must be replaced by a constant value.

Assuming we have taken the left branch out of the root in Figure 46.3, *TestsOnYesBranchesSofar* $= atom(C,A1,cl)$. Taking the declarative bias with the two schema statements above, the only choice for replacing the variables V and W in the schemata are the variables C and $A1$, respectively. The possible tests at this stage are thus of the form $bond(C,A1,A2,BT)$, $atom(C,A2,E)$, where E is replaced with an element (such as cl - chlorine, s - sulphur, or n - nitrogen), or of the form $bond(C,A1,A2,BT)$, where BT is replaced with a bond type (such as *single*, *double*, or *aromatic*). Among the possible tests, the test $bond(C,A1,A2,BT)$, $atom(C,A2,n)$ is chosen.

The approaches to relational decision tree induction are among the fastest MRDM approaches. They have been successfully applied to a number of practical problems. These include learning to predict the biodegradability of chemical compounds (Džeroski *et al.*, 1999) and learning to predict the structure of diterpene compounds from their NMR spectra (Džeroski *et al.*, 1998).

46.5 RDM Literature and Internet Resources

The book *Relational Data Mining*, edited by Džeroski and Lavrač (Džeroski and Lavrač, 2001) provides a cross-section of the state-of-the-art in this area at the turn of the millennium. This introductory chapter is largely based on material from that book.

The RDM book originated from the *International Summer School on Inductive Logic Programming and Knowledge Discovery in Databases* (ILP&KDD-97), held 15–17 September 1997 in Prague, Czech Republic, organized in conjunction with the *Seventh International Workshop on Inductive Logic Programming* (ILP-97). The teaching materials from this event are available on-line at `http://www-ai.ijs.si/SasoDzeroski/ILP2/ilpkdd/`.

A special issue of *SIGKDD Explorations* (vol. 5(1)) was recently devoted to the topic of multi-relational Data Mining. This chapter is a shortened version of the introductory article of that issue. Two journal special issues address the related topic of using ILP for KDD: *Applied Artificial Intelligence* (vol. 12(5), 1998), and *Data Mining and Knowledge Discovery* (vol. 3(1), 1999).

Many papers related to RDM appear in the ILP literature. For an overview of the ILP literature, see Chapter 3 of the RDM book (Džeroski and Lavrač, 2001). ILP-related bibliographic information can be found at ILPnet2's on-line library.

The major publication venue for ILP-related papers is the annual ILP workshop. The first *International Workshop on Inductive Logic Programming (ILP-91)* was organized in 1991. Since 1996, the proceedings of the ILP workshops are published by Springer within the Lecture Notes in Artificial Intelligence/Lecture Notes in Computer Science series.

Papers on ILP appear regularly at major Data Mining, machine learning and artificial intelligence conferences. The same goes for a number of journals, including *Journal of Logic Programming*, *Machine Learning*, and *New Generation Computing*. Each of these has published several special issues on ILP. Special issues on ILP containing extended versions of selected papers from ILP workshops appear regularly in the *Machine Learning* journal.

Selected papers from the ILP-91 workshop appeared as a book *Inductive Logic Programming*, edited by Muggleton (Muggleton, 1992), while selected papers from ILP-95 appeared as a book *Advances in Inductive Logic Programming*, edited by De Raedt (De Raedt, 1996). Authored books on ILP include *Inductive Logic Programming: Techniques and Applications* by Lavrač and Džeroski (Lavrač and Džeroski, 1994) and *Foundations of Inductive Logic Programming* by Nienhuys-Cheng and de Wolf (Nienhuys-Cheng and de Wolf, 1997). The first provides a practically oriented introduction to ILP, but is dated now, given the fast development of ILP in the recent years. The other deals with ILP from a theoretical perspective.

Besides the Web sites mentioned so far, the ILPnet2 site @ IJS (`http://www-ai.ijs.si/~ilpnet2/`) is of special interest. It contains an overview of ILP related resources in several categories. These include a list of and pointers to ILP-related educational materials, ILP applications and datasets, as well as ILP systems. It also contains a list of ILP-related events and an electronic newsletter. For a detailed overview of ILP-related Web resources we refer the reader to Chapter 16 of the RDM book (Džeroski and Lavrač, 2001).

References

Agrawal R. and Srikant R. , Mining sequential patterns. In Proceedings of the Eleventh International Conference on Data Engineering, pages 3–14. IEEE Computer Society Press, Los Alamitos, CA, 1995.

Agrawal R., Mannila H., Srikant R., Toivonen H., and Verkamo A. I., Fast discovery of association rules. In U. Fayyad, G. Piatetsky-Shapiro, P. Smyth, and R. Uthurusamy, editors, Advances in Knowledge Discovery and Data Mining, pages 307-328. AAAI Press, Menlo Park, CA, 1996.

Blockeel H. and De Raedt L., Top-down induction of first order logical decision trees. Artificial Intelligence, 101: 285–297, 1998.

Bratko I., Prolog Programming for Artificial Intelligence, 3rd edition. Addison Wesley, Harlow, England, 2001.

Breiman L., Friedman J. H., Olshen R. A., and Stone C. J., Classification and Regression Trees. Wadsworth, Belmont, 1984.

Clark P. and Boswel, R., Rule induction with CN2: Some recent improvements. In Proceedings of the Fifth European Working Session on Learning, pages 151–163. Springer, Berlin, 1991.

Clark P. and Niblett T., The CN2 induction algorithm. Machine Learning, 3(4): 261–283, 1989.

Dehaspe L., Toivonen H., and King R. D., Finding frequent substructures in chemical compounds. In Proceedings of the Fourth International Conference on Knowledge Discovery and Data Mining, pages 30–36. AAAI Press, Menlo Park, CA, 1998.

Dehaspe L. and Toivonen H., Discovery of frequent datalog patterns. Data Mining and Knowledge Discovery, 3(1): 7–36, 1999.

Dehaspe L. and Toivonen H., Discovery of Relational Association Rules. In (Džeroski and Lavrač, 2001), pages 189–212, 2001.

De Raedt L., editor. Advances in Inductive Logic Programming. IOS Press, Amsterdam, 1996.

De Raedt L., Attribute-value learning versus inductive logic programming: the missing links (extended abstract). In Proceedings of the Eighth International Conference on Inductive Logic Programming, pages 1–8. Springer, Berlin, 1998.

De Raedt L., Blockeel H., Dehaspe L., and Van Laer W., Three Companions for Data Mining in First Order Logic. In (Džeroski and Lavrač, 2001), pages 105–139, 2001.

De Raedt L. and Džeroski S., First order *jk*-clausal theories are PAC-learnable. Artificial Intelligence, 70: 375–392, 1994.

Džeroski S. and Lavrač N., editors. Relational Data Mining. Springer, Berlin, 2001.

Džeroski S., Muggleton S., and Russell S., PAC-learnability of determinate logic programs. In Proceedings of the Fifth ACM Workshop on Computational Learning Theory, pages 128–135. ACM Press, New York, 1992.

Džeroski S., Schulze-Kremer S., Heidtke K., Siems K., Wettschereck D., and Blockeel H., Diterpene structure elucidation from ^{13}C NMR spectra with Inductive Logic Programming. Applied Artificial Intelligence, 12: 363–383, 1998.

Džeroski S., Blockeel H., Kompare B., Kramer S., Pfahringer B., and Van Laer W., Experiments in Predicting Biodegradability. In Proceedings of the Ninth International Workshop on Inductive Logic Programming, pages 80–91. Springer, Berlin, 1999.

Džeroski S., Relational Data Mining Applications: An Overview. In (Džeroski and Lavrač, 2001), pages 339–364, 2001.

Džeroski S., De Raedt L., and Wrobel S., editors. Proceedings of the First International Workshop on Multi-Relational Data Mining. KDD-2002: Eighth ACM SIGKDD International Conference on Knowledge Discovery and Data Mining, Edmonton, Canada, 2002.

Emde W. and Wettschereck D., Relational instance-based learning. In Proceedings of the Thirteenth International Conference on Machine Learning, pages 122–130. Morgan Kaufmann, San Mateo, CA, 1996.

King R.D., Karwath A., Clare A., and Dehaspe L., Genome scale prediction of protein functional class from sequence using Data Mining. In Proceedings of the Sixth International Conference on Knowledge Discovery and Data Mining, pages 384–389. ACM Press, New York, 2000.

Kirsten M., Wrobel S., and Horváth T., Distance Based Approaches to Relational Learning and Clustering. In (Džeroski and Lavrač, 2001), pages 213–232, 2001.

Kramer S., Structural regression trees. In Proceedings of the Thirteenth National Conference on Artificial Intelligence, pages 812–819. MIT Press, Cambridge, MA, 1996.

Kramer S. and Widmer G., Inducing Classification and Regression Trees in First Order Logic. In (Džeroski and Lavrač, 2001), pages 140–159, 2001.

Kramer S., Lavrač N., and Flach P., Propositionalization Approaches to Relational Data Mining. In (Džeroski and Lavrač, 2001), pages 262–291, 2001.

Lavrač N., Džeroski S., and Grobelnik M., Learning nonrecursive definitions of relations with LINUS. In Proceedings of the Fifth European Working Session on Learning, pages 265–281. Springer, Berlin, 1991.

Lavrač N. and Džeroski S., Inductive Logic Programming: Techniques and Applications. Ellis Horwood, Chichester, 1994.

Lloyd J., Foundations of Logic Programming, 2nd edition. Springer, Berlin, 1987.

Mannila H. and Toivonen H., Discovering generalized episodes using minimal occurrences. In Proceedings of the Second International Conference on Knowledge Discovery and Data Mining, pages 146–151. AAAI Press, Menlo Park, CA, 1996.

Michalski R., Mozetič I., Hong J., and Lavrač N., The multi-purpose incremental learning system AQ15 and its testing application on three medical domains. In Proceedings of the Fifth National Conference on Artificial Intelligence, pages 1041–1045. Morgan Kaufmann, San Mateo, CA, 1986.

Muggleton S., Inductive logic programming. New Generation Computing, 8 (4) : 295–318, 1991.

Muggleton S., editor. Inductive Logic Programming. Academic Press, London, 1992.

Muggleton S., Inverse entailment and Progol. New Generation Computing, 13: 245–286, 1995.

Muggleton S. and Feng C., Efficient induction of logic programs. In Proceedings of the First Conference on Algorithmic Learning Theory, pages 368–381. Ohmsha, Tokyo, 1990.

Nedellec C., Rouveirol C., Ade H., Bergadano F., and Tausend B., Declarative bias in inductive logic programming. In L. De Raedt, editor, Advances in Inductive Logic Programming, pages 82–103. IOS Press, Amsterdam, 1996.

Nienhuys-Cheng S.-H. and de Wolf R., Foundations of Inductive Logic Programming. Springer, Berlin, 1997.

Plotkin G., A note on inductive generalization. In B. Meltzer and D. Michie, editors, Machine Intelligence 5, pages 153–163. Edinburgh Univ. Press, 1969.
Quinlan J. R., Learning logical definitions from relations. Machine Learning, 5(3): 239–266, 1990.
Quinlan J. R., C4.5: Programs for Machine Learning. Morgan Kaufmann, San Mateo, CA, 1993.
Rokach, L., Averbuch, M., and Maimon, O., Information retrieval system for medical narrative reports (pp. 217228). Lecture notes in artificial intelligence, 3055. Springer-Verlag (2004).
Rokach L. and Maimon O., Data mining for improving the quality of manufacturing: A feature set decomposition approach. Journal of Intelligent Manufacturing 17(3): 285299, 2006.
Shapiro E., Algorithmic Program Debugging. MIT Press, Cambridge, MA, 1983.
Srikant R. and Agrawal R., Mining generalized association rules. In Proceedings of the Twenty-first International Conference on Very Large Data Bases, pages 407–419. Morgan Kaufmann, San Mateo, CA, 1995.
Ullman J., Principles of Database and Knowledge Base Systems, volume 1. Computer Science Press, Rockville, MA, 1988.
Van Laer V. and De Raedt L., How to Upgrade Propositional Learners to First Order Logic: A Case Study. In (Džeroski and Lavrač, 2001), pages 235-261, 2001.
Wrobel S., Inductive Logic Programming for Knowledge Discovery in Databases. In (Džeroski and Lavrač, 2001), pages 74–101, 2001.

47

Web Mining

Johannes Fürnkranz

TU Darmstadt, Knowledge Engineering Group

Summary. The World-Wide Web provides every internet citizen with access to an abundance of information, but it becomes increasingly difficult to identify the relevant pieces of information. Research in web mining tries to address this problem by applying techniques from data mining and machine learning to Web data and documents. This chapter provides a brief overview of web mining techniques and research areas, most notably hypertext classification, wrapper induction, recommender systems and web usage mining.

Key words: web mining, content mining, structure mining, usage mining, text classification, hypertext classification, information extraction, wrapper induction, collaborative filtering, recommender systems, Semantic Web

47.1 Introduction

The advent of the World-Wide Web (WWW) (Berners-Lee, Cailliau, Loutonen, Nielsen & Secret, 1994) has overwhelmed home computer users with an enormous flood of information. To almost any topic one can think of, one can find pieces of information that are made available by other internet citizens, ranging from individual users that post an inventory of their record collection, to major companies that do business over the Web.

To be able to cope with the abundance of available information, users of the Web need assistance of intelligent software agents (often called *softbots*) for finding, sorting, and filtering the available information (Etzioni, 1996, Kozierok and Maes, 1993). Beyond search engines, which are already commonly used, research concentrates on the development of agents that are general, high-level interfaces to the Web (Etzioni, 1994, Fürnkranz *et al.*, 2002), programs for filtering and sorting e-mail messages (Maes, 1994, Payne and Edwards, 1997) or Usenet netnews articles (Lashkari *et al.*, 1994, Sheth, 1993, Lang, 1995, Mock, 1996), recommender systems for suggesting Web sites (Armstrong *et al.*, 1995, Pazzani *et al.*, 1996, Balabanovi and Shoham, 1995) or products (Doorenbos *et al.*, 1997, Burke *et al.*, 1996), automated answering systems (Burke *et al.*, 1997, Scheffer, 2004) and many more.

O. Maimon, L. Rokach (eds.), *Data Mining and Knowledge Discovery Handbook*, 2nd ed.,
DOI 10.1007/978-0-387-09823-4_47, © Springer Science+Business Media, LLC 2010

Many of these systems are based on machine learning and Data Mining techniques. Just as Data Mining aims at discovering valuable information that is hidden in conventional databases, the emerging field of *web mining* aims at finding and extracting relevant information that is hidden in Web-related data, in particular in (hyper-)text documents published on the Web. Like Data Mining, web mining is a multi-disciplinary effort that draws techniques from fields like information retrieval, statistics, machine learning, natural language processing, and others.

Web mining is commonly divided into the following three sub-areas:

Web Content Mining: application of Data Mining techniques to unstructured or semi-structured text, typically HTML-documents

Web Structure Mining: use of the hyperlink structure of the Web as an (additional) information source

Web Usage Mining: analysis of user interactions with a Web server

An excellent textbook for the field is (Chakrabarti, 2002), an earlier effort (Chang *et al.*, 2001). Brief surveys can be found in (Chakrabarti, 2000, Kosala and Blockeel, 2000). For surveys of content mining, we refer to (Sebastiani, 2002), while a survey of usage mining can be found in (Srivastava *et al.*, 2000). We are not aware of a previous survey on structure mining.

In this chapter, we will organize the material somewhat differently. We start with a brief introduction on the Web, in particular on its unique properties as a graph (Section 47.2), and subsequently discuss how these properties are exploited for improved retrieval performance in search engines (Section 47.3). After a brief recapitulation of text classification (Section 47.4), we discuss approaches that attempt to use the link structure of the Web for improving hypertext classification (Section 47.5). Subsequently, we summarize important research in the areas information extraction and wrapper induction (Section 47.6), and briefly discuss the web mining opportunities of the Semantic Web (Section 47.7). Finally, we present research in web usage mining (Section 47.8) and recommender systems (Section 47.9).

47.2 Graph Properties of the Web

While conventional information retrieval focuses primarily on information that is provided by the text of Web documents, the Web provides additional information through the way in which different documents are connected to each other via *hyperlinks*. The Web may be viewed as a (directed) graph with documents as nodes and hyperlinks as edges.

Several authors have tried to analyze the properties of this graph. The most comprehensive study is due to (Broder *et al.*, 2000). They used data from an AltaVista crawl (May 1999) with 203 million URLs and 1466 million links, and stored the underlying graph structure in a connectivity server (Bharat *et al.*, 1998), which implements an efficient document indexing technique that allows fast access to both outgoing and incoming hyperlinks of a page. The entire graph fitted in 9.5 GB of storage, and a breadth-first search that reached 100M nodes took only about 4 minutes. Their main result is an analysis of the structure of the web graph, which, according to them, looks like a giant bow tie, with a strongly connected core component (SCC) of 56 million pages in the middle, and two components with 44 million pages each on the sides, one containing pages from which the SCC can be reached (the IN set), and the other containing pages that can be reached from the SCC (the OUT set). In addition, there are "tubes" that allow to reach the OUT set from the IN set without passing through the SCC, and many "tendrils", that lead out of the IN set or into the OUT set without connecting to other components. Finally, there are also several smaller components that cannot be reached from any point in this structure. Broder *et al.* (2000) also sketch a diagram of this structure, which is somewhat deceptive because the prominent role of the IN, OUT, and SCC sets is based on size only, and there are other structures with a similar shape, but of somewhat smaller size (e.g., the tubes may contain other strongly connected components that differ from the SCC only in size). The main result is that there are several disjoint components. In fact, the probability that a path between two randomly selected pages exists is only about 0.24.

Based on the analysis of this structure, Broder *et al.* (2000) estimated that the diameter (i.e., the maximum of the lengths of the shortest paths between two nodes) of the SCC is larger than 27, that the diameter of the entire graph is larger than 500, and that the average length of such a path is about 16. This is, of course only for cases where a path between two pages exists. These results correct earlier estimates obtained by Albert, Jeong, and Barabási (1999) who estimated the average length at about 19. Their analysis was based on a probabilistic argument using estimates for the in-degrees and out-degrees, thereby ignoring the possibility of disjoint components.

Albert *et al.* (1999) base their analysis on the observation that the in-degrees (number of incoming links) and out-degrees (number of outgoing links) follow a power law distribution $P(d) \approx d^{-\gamma}$. They estimated values of y=2.45 and y=2.1. for the in-degrees and out-degrees respectively. They also note that these power law distributions imply a much higher probability of encountering documents with large in- or out-degrees than would be the case for random networks or random graphs. The power-law results have been confirmed by Broder *et. al.* (2000) who also observed a power law distribution for the sizes of strongly connected components in the web graph. Faloutsos, Faloutsos & Faloutsos (1999) observed a Zipf distribution $P(d) \approx r(d)^{-\gamma}$ for the out-degree of nodes ($r(d)$ is the rank of the degree in a sorted list of out-degree values). Similarly, a model of the behavior of web surfers was shown to follow a Zipf distribution (Levene *et al.*, 2001).

Finally, another interesting property is the size of the Web. Lawrence and Giles (1998) propose to estimate the size of the Web from the overlap that different search engines return for identical queries. Their method is based on the assumption that the probability that a page is indexed by search engine *A* is independent of the probability that this page is indexed by search engine *B*. In this case, the percentage of pages in the result set of a query for search engine *B* that are also indexed by search engine *A* could be used as an estimate for the overall percentage of pages indexed by *A*. Obviously, the independence assumption on which this argument is based does not hold in practice, so that the estimated percentage is larger than the real percentage (and the obtained estimates of the web size are more like lower bounds). Lawrence and Giles (1998) used the results of several queries to estimate that the largest search engine indexes only about one third of the indexable Web (the portion of the Web that is accessible to crawlers, i.e., not hidden behind query interfaces). Similar arguments were used by Bharat and Broder (1998) to estimate the relative size of search engines.

47.3 Web Search

Whereas conventional query interfaces concentrate on indexing documents by the words that appear in them (Salton, 1989), the potential of utilizing the information contained in the hyperlinks pointing to a page has been recognized early on. Anchor texts (texts on hyperlinks in an HTML document) of predecessor pages were already indexed by the World-Wide Web Worm, one of the first search engines and web crawlers (McBryan, 1994). Spertus (1997) introduced a taxonomy of different types of (hyper-)links that can be found on the Web, and discussed how the links can be exploited for various information retrieval tasks on the Web.

However, the main break-through was the realization that the popularity and hence the importance of a page is—to some extent—correlated to the number of incoming links, and that this information can be advantageously used for sorting the query results of a search engine. The in-degree alone, however, is a poor measure of importance because many pages are frequently pointed to without being connected to the contents of the referring page (think, e.g., of the numerous "best viewed with..." hyperlinks that point to browser home-pages). More sophisticated measures are needed.

Kleinberg (1999) suggests that are two types of pages that could be relevant for a query: *authorities* are pages that contain useful information about the query topic, while *hubs* contain pointers to good information sources. Obviously, both types of pages are typically connected: good hubs contain pointers to many good authorities, and good authorities are pointed to by many good hubs. Kleinberg (1999) suggests to make practical use of this relationship by associating each page x with a hub score $H(x)$ and an authority score $A(x)$, which are computed iteratively:

$$H_{i+1}(x) = \sum_{(x,s)} A_i(s) \quad A_{i+1}(x) = \sum_{(p,x)} H_i(s)$$

where (x,y) denotes that there is a hyperlink from page x to page y. This computation is conducted on a so-called *focused subgraph* of the Web, which is obtained by enhancing the search result of a conventional query (or a bounded subset of the result) with all predecessor and successor pages (or, again, a bounded subset of them). The hub and authority scores are initialized uniformly with $A_0(x) = H_0(x) = 1.0$. and normalized so that they sum up to one before each iteration. It can be proved that this algorithm (called HITS) will always converge (Kleinberg, 1999), and practical experience shows that it will typically do so within a few (about 5) iterations (Chakrabarti *et al.*, 1998b). Variants of the HITS algorithm have been used for identifying relevant documents for topics in web catalogues (Chakrabarti *et al.*, 1998b, Bharat and Henzinger, 1998) and for implementing a "Related Pages" functionality (Dean and Henzinger, 1999).

The main drawback of this algorithm is that the hubs and authority score must be computed iteratively from the query result, which does not meet the real-time constraints of an on-line search engine. However, the implementation of a similar idea in the Google search engine resulted in a major break-through in search engine technology (Brin *et al.*, 1998). The key idea is to use the probability that a page is visited by a random surfer on the Web as an important factor for ranking search results. This probability is approximated by the so-called *page rank*, which is again computed iteratively:

$$PR_{i+1}(x) = (1-l)\frac{1}{N} + l \sum_{(p,x)} \frac{PR_i(p)}{|(p,y)|}$$

The first term of this sum models the behavior that a surfer gets bored and jumps to a randomly selected page of the entire set of N pages (with probability $(1-l)$, where l is typically set to 0.85). The second term uniformly distributes the current page rank of a page to all its successor pages. Thus, a page receives a high page rank if it is linked by many pages, which in turn have a high page rank and/or only few successor pages. The main advantage of the page rank over the hubs and authority scores is that it can be computed off-line, i.e., it can be precomputed for all pages in the index of a search engine. Its clever (but secret) integration with other information that is typically used by search engines (number of matching query terms, location of matches, proximity of matches, etc.) promoted Google from a student project to the main player in search engine technology.

47.4 Text Classification

Text classification is the task of sorting documents into a given set of categories. One of the most common web mining tasks is the automated induction of such text classifiers from a set of training documents for which the category is known. A detailed overview of this field can be found in (Sebastiani, 2002), as well as in the corresponding Chapter of this book. The main problem, in comparison to conventional classification tasks, is the additional degree of freedom that results from the need to extract a suitable feature set for the classification task. Typically, each word is considered as a separate feature with either a Boolean value indicating whether the word occurs or does not occur in the document (*set-of-words* representation) or a numeric value that indicates the frequency (*bag-of-words* representation). A comparison of these two basic models can be found in (McCallum and Nigam, 1998). Advanced approaches use different weights for terms (Salton and Buckley, 1988), more elaborate feature sets like *n*-grams (Mladenić and Grobelnik, 1998, Fürnkranz, 1998) or linguistic features (Lewis, 1992, Fürnkranz *et al.*, 1998, Scott and Matwin, 1999), linear combinations of features (Deerwester *et al.*, 1990) or rely on automated feature selection techniques (Yang and Pedersen, 1997, Mladenić, 1998a).

There are numerous application areas for this type of learning task (Mladenić, 1999). For example, the generation of web catalogues such as `http://www.dmoz.org/` is basically a classification task that assigns documents to labels in a structured hierarchy of classes. Typically, this task is performed manually by a large user community or employees of companies that specialize in such efforts, like Yahoo!. Automating this assignment is a rewarding task for text categorization and text classification (Mladenić, 1998b).

Similarly, the sorting of one's personal E-mail messages into a flat or structured hierarchy of mail folders is a text categorization task that is mostly performed manually, sometimes supported with manually defined classification rules. Again, there have been numerous attempts in augmenting this procedure with automatically induced content-based classification rules (Cohen, 1996, Payne and Edwards, 1997, Crawford *et al.*, 2002). Recently, a related task has received increased attention, namely automated filtering of spam mail. Training classifiers for recognizing spam mail is a particularly challenging problem for machine learning, involving skewed example distributions, misclassification costs, concept drift, undefined feature sets, and more (Fawcett, 2003). Most algorithms, such as the built-in spam filter of the Mozilla open source browser (Graham, 2003), rely on Bayesian learning for tackling this problem. A comparison of different learning algorithms for this problem can be found in (Androutsopoulos *et al.*, 2004).

47.5 Hypertext Classification

Not surprisingly, recent research has also looked at the potential of hyperlinks as an additional information source for hypertext categorization tasks. Many authors addressed this problem in one way or another by merging (parts of) the text of the predecessor pages with the text of the page to classify, or by keeping a separate feature set for the predecessor pages. For example, Chakrabarti, Dom, and Indyk (1998a) evaluate two variants: (1) appending the text of the neighboring (predecessor and successor) pages to the text of the target page, and (2) using two different sets of features, one for the target page and one for a concatenation of the neighboring pages. The results were negative: in two domains both approaches performed worse than the conventional technique that uses only features of the target document. Chakrabarti *et al.* (1998a) concluded that the text from the neighbors is too unreliable to help classification. Consequently, a different technique was proposed that included predictions for the class labels of the neighboring pages into the model. Unless the labels for the neighbors are known a priori, the implementation of this approach requires an iterative technique for assigning the labels, because changing the class of a page may potentially change the class assignments for all neighboring pages as well. The authors implemented a relaxation labeling technique, and showed that it improves performance over the standard text-based approach that ignores the hyperlink structure. The utility of class predictions for neighboring pages was confirmed by the results of Oh, Myaeng, and Lee (2000) and Yang, Slattery, and Ghani (2002).

A different line of research concentrates on explicitly encoding the relational structure of the Web in first-order logic. For example, a binary predicate link_to(page1,page2) can be used to represent the fact that there is a hyperlink on page1 that points to page2. In order to be able to deal with such a representation, one has to go beyond traditional attribute-value learning algorithms and resort to inductive logic programming, aka relational Data Mining (Džeroski and Lavrač, 2001). Craven, Slattery & Nigam (1998) use a variant of Foil (Quinlan, 1990) to learn classification rules that can incorporate features from neighboring pages. The algorithm uses a deterministic version of relational path-finding (Richards and Mooney, 1992), which overcomes Foil's restriction to determinate literals (Quinlan, 1991), to construct chains of `link_to/2` predicates that allow the learner to access the words on a page via a predicate of the type `has_word(page,word)`. For example, the conjunction `link_to(P1,P), has_word(P1,word)` means "there exists a predecessor page `P1` that contains the word `word`. Slattery and Mitchell (2000) improve the basic Foil-like learning algorithm by integrating it with ideas originating from the HITS algorithm for computing hub and authority scores of pages, while Craven and Slattery (2001) combine it favorably with a Naive Bayes classifier.

At its core, using features of pages that are linked via a `link_to/2` predicate is quite similar to the approach evaluated in (Chakrabarti *et al.*, 1998a) where words of neighboring documents are added as a separate feature set: in both cases, the learner has access to all the features in the neighboring documents. The main difference lies in the fact that in the relational representation, the learner may control the depth of the chains of `link_to/2` predicates, i.e., it may incorporate features from pages that are several clicks apart. From a practical point of view, the main difference lies in the characteristics of the used learning algorithms: while inductive logic programming typically relies on rule learning algorithms which classify pages with "hard" classification rules that predict a class by looking only at a few selected features, Chakrabarti *et al.* (1998a) used learning algorithms that always take all available features into account (such as a Naive Bayes classifier). Yang *et al.* (2002) discuss both approaches and relate them to a taxonomy of five possible regularities that may be present in the neighborhood of a target page. They also experimentally compare these approaches under different conditions.

However, the above-mentioned approaches still suffer from several short-comings, most notably that only portions of the predecessor pages are relevant, and that not all predecessor pages are equally relevant. A solution attempt is provided by the use of *hyperlink ensembles* for classification of hypertext pages (Fürnkranz, 2002). The idea is quite simple: instead of training a classifier that classifies *pages* based on the words that appear in their text, a classifier is trained that classifies *hyperlinks* according to the class of the pages they point to, based on the words that occur in their neighborhood of the link (in the simplest case the anchor text of the link). Consequently, each page will be assigned multiple predictions for its class membership, one for each incoming hyperlink. These individual predictions are then combined to a final prediction by some voting procedure. Thus, the technique is a member of the family of ensemble learning methods (Dietterich, 2000a). In a preliminary empirical evaluation in the Web→KB domain (where the task is to recognize typical entities in Computer Science departments, such as faculty, student, course, and project pages.), hyperlink ensembles outperformed a conventional full-text classifier in a study that employed a variety of voting schemes for combining the individual classifiers and a variety of feature extraction techniques for representing the information around an incoming hyperlink (e.g., the anchor text on a hyperlink, the text in the sentence that contains the hyperlink, or the text of an entire paragraph). The overall classifier improved the full-text classifier from about 70% accuracy to about 85% accuracy in this domain. It remains to be seen whether this generalizes to other domains.

47.6 Information Extraction and Wrapper Induction

Information extraction is concerned with the extraction of certain information items from unstructured text. For example, you might want to extract the title, show times, and prices from web pages of movie theaters near you. While web search can be used to find the relevant pages, information extraction is needed to identify these particular items on each page. An excellent survey of the field can be found in (Eikvil, 1999). Premier events in this field include the *Message Understanding Conferences (MUC)*, and numerous workshops devoted to special aspects of this topic (Califf, 1999, Pazienza, 2003).

Information extraction has a long history. There are numerous algorithms that work with unstructured textual documents, mostly employing natural language processing. A typical system is AutoSlog (Riloff, 1996b), which was developed as a method for automatically constructing domain-specific extraction patterns from an annotated training corpus. As input, AutoSlog requires a set of noun phrases that constitute the information that should be extracted from the training documents. AutoSlog then uses syntactic heuristics to create linguistic patterns that can extract the desired information from the training documents (and from unseen documents). The extracted patterns typically represent subject–verb or verb–direct-object relationships (e.g., *<subject> teaches* or *teaches <direct-object>*) as well as prepositional phrase attachments (e.g., *teaches at <noun-phrase>* or *teacher at <noun-phrase>*). An extension, AutoSlog-TS (Riloff, 1996a), removes the need for an annotated training corpus by generating extraction patterns for *all* noun phrases in the training corpus whose syntactic role matches one of the syntactic heuristics.

Other systems that work with unstructured text are based on inductive rule learning algorithms that can make use of a multitude of features, including linguistic tags, HTML tags, font size, etc., and learn a set of extraction rules that specify which combination of features indicates an appearance of the target information. WHISK (Soderland, 1999) and SRV (Freitag, 1998) employ a top-down, general-to-specific search for finding a rule that covers a subset of the target patterns, whereas RAPIER (Califf, 2003) employs a bottom-up search that successively generalizes a pair of target patterns.

While the above-mentioned systems typically work on unstructured or semi-structured text, a new direction focused on the extraction of items from structured HTML-pages. Such *wrappers* identify their content primarily via a sequence of HTML tags (or an XPath in a

DOM-tree). Kushmerick (2000) first studied the problem of inducing such wrappers from a set of training examples where the information to extract is marked. He studies a variety of types of wrapper algorithms with different expressiveness. The simplest class, LR wrappers, assume a highly regular source page that allows to map its content into a database table by learning delimiters for each attribute. LR wrappers were able to wrap 53% of the pages in an experimental study, more expressive classes were able to wrap up to 70%. Moreover, it was shown that all studied wrapper classes are PAC-learnable. Grieser, Jantke, Lange & Thomas (2000) extend this work with a study of theoretical properties and learnability results for island wrappers, a generalization of the wrapper types studied by Kushmerick (2000). SoftMealy (Hsu and Dung, 1998) addresses several of the short-comings of the framework of Kushmerick (2000), most notably the restriction to single sequences of features, by learning a finite-state transducer that allows to encode all occurring sequences of features. Lerman, Minton, and Knoblock (2003) discuss learning approaches for supporting the maintenance of existing wrappers.

The field has also seen numerous commercial efforts, such as the Lixto project (Gottlob *et al.*, 2004) or IBM's Andes project (Myllymaki, 2001). The most notable application of information extraction techniques are comparison shopping agents (Doorenbos *et al.*, 1997).

47.7 The Semantic Web

The Semantic Web is a term coined by Tim Berner-Lee for the vision of making the information on the Web machine-processable (Berners-Lee *et al.*, 2001). The basic idea is to enrich web pages with machine-processable knowledge that is represented in the form of *ontologies* (Staab and Studer, 2004, Fensel, 2001). Ontologies define certain types of objects and the relations between them. As ontologies are readily accessible (like other web documents), a computer program can use them to draw inferences about the information provided on web pages.

One of the research challenges in that area is to annotate the information that is currently available on the Web with semantic tags. Typically, techniques from text classification, hypertext classification and information extraction are used for that purpose. A landmark application in this area was the WebKB project at Carnegie-Mellon University (Craven *et al.*, 2000). Its goal was to assign web pages or parts of web pages to entities in an ontology. A simple test ontology modeled knowledge about computer science departments: there are entities like students (graduate and undergraduate), faculty members (professors, researchers, lecturers, post-docs, ...), courses, projects, etc., and relations between these entities, such as "courses are taught by one lecturer and attended by several students" or "every graduate student is advised by a professor". Many applications could be imagined for such an ontology. For example, it could enhance the capabilities of search engines by enabling them to answer queries like "Who teaches course X at university Y? " or "How many students are in department Z? ", or serve as a backbone for web catalogues (Staab and Maedche, 2001). A description of the first prototype system can be found in (Craven *et al.*, 2000).

Semantic Web Mining emerged as research field that focuses on the interactions of web mining and the Semantic Web (Berendt *et al.*, 2002). On the one hand, web mining can support the learning of ontologies in various ways (Maedche and Staab, 2001, Maedche *et al.*, 2003, Doan *et al.*, 2003). On the other hand, background knowledge in the form of ontologies may be used for supporting web mining tasks. Several workshops have been devoted to these topics (Staab *et al.*, 2000, Maedche *et al.*, 2001, Stumme *et al.*, 2001, Stumme *et al.*, 2002).

47.8 Web Usage Mining

Most of the previous approaches are concerned with the analysis of the contents of web documents (content mining) or the graph structure of the web (structure mining). Additional information can be inferred from data sources that capture the interaction of users with a web site, e.g., from server-side web logs or from client-side applets that observe a single user's browsing patterns. Such information may, e.g., provide important clues for restructuring web sites (Perkowitz and Etzioni, 2000, Berendt, 2002), personalizing web services (Mobasher *et al.*, 2000, Mobasher *et al.*, 2002, Pierrakos *et al.*, 2003), optimizing search engines (Joachims, 2002), recognizing web spiders (Tan and Kumar, 2002) and many more. An excellent overview and taxonomy of this research area can be found in (Srivastava *et al.*, 2000).

As an example, let us consider systems that make user-specific browsing recommendations (Armstrong *et al.*, 1995, Pazzani *et al.*, 1996, Balabanovi and Shoham, 1995). For example, the WebWatcher system (Armstrong *et al.*, 1995) predicts which links on the currently viewed page are most interesting to the user's search goal, which has to be specified in advance, and recommends the user to follow these links. However, these early systems rely on user intervention by specification of a search goal (Armstrong *et al.*, 1995) or explicit feedback about interesting or not interesting pages (Pazzani *et al.*, 1996). More advanced systems try to infer this information from web logs, thereby removing the need for user feedback. For example, Personal WebWatcher (Mladenić, 1996) is an early attempt that replaces WebWatcher's requirement for an explicitly specified search goal with a user model that has been inferred by a text classification system trained on pages that the user has been observed to visit (positive examples) or not to visit (negative examples). These pages have been obtained by a client-side applet that logs the user's browsing behavior.

More recently, it was tried to infer this information from server-side web logs (Mobasher *et al.*, 2000). The information contained in a web log includes the IP-address of the client, the page that has been retrieved, the time at which the request was initiated, the page from which the link originated, the browsing agent used, etc. However, unless additional information is used (e.g., session cookies), there is no way to reliably determine the browsing path that a user takes. Problems include missing page requests because of client-side caches or merged sessions because of multiple users operating from the same IP-addresses. Special techniques have to be used to infer the browsing paths (so-called *click streams*) of individual users (Cooley *et al.*, 1999). These click-streams can then be mined using clustering and association rule finding techniques, and the resulting models be used for making page recommendations. The WUM Web Utilization Miner (Spiliopoulou, 1999) is a publicly available, prototypical system that allows to mine web logs using advanced association rule discovery algorithms.

47.9 Collaborative Filtering

Collaborative filtering (Goldberg *et al.*, 1992) may be considered a special case of usage mining, which relies on previous recommendations by other users in order to predict which among a set of items are most interesting for the current user. Such systems are also known as *recommender systems* (Resnick, 1997). Naturally, recommender systems have many applications, most notably in E-commerce (Schafer *et al.*, 2000), but also in science (e.g., assigning papers to reviewers) (Basu *et al.*, 2001).

Recommender systems typically store a data table that records for each user/item pair whether the user made a recommendation for the item or not and possibly also the strength of this recommendation. Such recommendations can either be made explicitly by giving some sort of feedback (e.g., by assigning a rating to a movie) or implicitly (e.g., by buying a video of the movie). The elegant idea of collaborative filtering systems is that recommendations can be based on user similarity, and that user similarity can in turn be defined by the similarity of their recommendations. Alternatively, recommender systems can also be based on item similarities, which are defined via the recommendations of the users that recommended the items in question (Sarwar *et al.*, 2001).

Early recommender systems followed a *memory-based* approach, which means that they directly computed this similarity for each new query. For example, the GroupLens system (Konstan *et al.*, 1997) required readers of Usenet news articles to rate an article on a scale with five values. From that, similarities between users are cached by computing a correlation coefficient over their votes for individual items.

In a landmark paper, Breese, Heckerman, and Kadie (1998) compare memory-based approaches to *model-based* approaches, which use the stored data for inducing an explicit model for the recommendations of the users. The results show that a Bayesian network outperforms alternative approaches, in particular memory-based approaches. Other types of models that have been studied include clustering (Ungar and Foster, 1998), latent semantic models (Hofmann and Puzicha, 1999) and association rules (Lin *et al.*, 2002).

An active research area is to combine integrate collaborative filtering with content-based approaches to recommender systems, i.e., approaches that make predictions based on background knowledge of characteristics of users and/or items. An interesting approach is followed by Cohen and Fan (2000), who propose to model content-based similarities in the form of artificial users. For example, an artificial user could represent a certain musical genre and comment positively on all representatives of that genre. Melville, Mooney, and Nagarajan (2002) propose a similar approach by suggesting the use of content-based predictions for replacing missing recommendations. Popescul, Ungar, Pennock, and Lawrence (2001) extend the approach taken by Hofmann and Puzicha (1999), who associate users and items with a hidden layer of emerging concepts, by merging word occurrence information into the latent models.

47.10 Conclusion

Web mining is a very active research area. A survey like this can only scratch on the surface. We tried to include references to the most important works in this area, but we necessarily had to be selective. Nevertheless, we hope to have provided the reader with a good starting point for her own explorations into this rapidly expanding and exciting research field.

References

R. Albert, H. Jeong, and A.-L. Barabási. Diameter of the world-wide web. *Nature*, 401:130–131, September 1999.

I. Androutsopoulos, G. Paliouras, and E. Michelakis. Learning to filter unsolicited commercial e-mail. Technical Report 2004/2, NCSR Demokritos, March 2004.

R. Armstrong, D. Freitag, T. Joachims, and T. Mitchell. WebWatcher: A learning apprentice for the world wide web. In C. Knoblock and A. Levy, editors, *Proceedings of AAAI Spring Symposium on Information Gathering from Heterogeneous, Distributed Environments*, pages 6–12. AAAI Press, 1995. Technical Report SS-95-08.

M. Balabanovi and Y. Shoham. Learning information retrieval agents: Experiments with automated web browsing. In C. Knoblock and A. Levy, editors, *Proceedings of AAAI Spring Symposium on Information Gathering from Heterogeneous, Distributed Environments*, pages 13–18. AAAI Press, 1995. Technical Report SS-95-08.

C. Basu, H. Hirsh, W. W. Cohen, and C. Nevill-Manning. Technical paper recommendation: A study in combining multiple information sources. *Journal of Artificial Intelligence Research*, 14: 231–252, 2001.

B. Berendt. Using site semantics to analyze, visualize, and support navigation. *Data Mining and Knowledge Discovery*, 6(1): 37–59, 2002.

B. Berendt, A. Hotho, and G. Stumme. Towards semantic web mining. In I. Horrocks and J. Hendler, editors, *Proceedings of the 1st International Semantic Web Conference (ISWC-02)*, pages 264–278. Springer-Verlag, 2002.

T. Berners-Lee, R. Cailliau, A. Loutonen, H. Nielsen, and A. Secret. The World Wide Web. *Communications of the ACM*, 37(8):76–82, 1994.

T. Berners-Lee, J. Hendler, and O. Lassila. The Semantic Web. *Scientific American*, May 2001.

K. Bharat and A. Broder. A technique for measuring the relative size and overlap of public web search engines. *Computer Networks*, 30(1–7):107–117, 1998. Proceedings of the 7th International World Wide Web Conference (WWW-7), Brisbane, Australia.

K. Bharat, A. Broder, M. R. Henzinger, P. Kumar, and S. Venkatasubramanian. The connectivity server: Fast access to linkage information on the Web. *Computer Networks*, 30(1–7):469–477, 1998. Proceedings of the 7th International World Wide Web Conference (WWW-7), Brisbane, Australia.

K. Bharat and M. R. Henzinger. Improved algorithms for topic distillation in a hyperlinked environment. In *Proceedings of the 21st ACM SIGIR Conference on Research and Development in Information Retrieval (SIGIR-98)*, pages 104–111, 1998.

J. S. Breese, D. Heckerman, and C. Kadie. Empirical analysis of predictive algorithms for collaborative filtering. In G. F. Cooper and S. Moral, editors, *Proceedings of the 14th Conference on Uncertainty in Artificial Intelligence (UAI-98)*, pages 43–52, Madison, WI, 1998. Morgan Kaufmann.

S. Brin and L. Page. The anatomy of a large-scale hypertextual Web search engine. *Computer Networks*, 30(1–7):107–117, 1998. Proceedings of the 7th International World Wide Web Conference (WWW-7), Brisbane, Australia.

A. Broder, R. Kumar, F. Maghoul, P. Raghavan, S. Rajagopalan, R. Stata, A. Tomkins, and J. Wiener. Graph structure in the Web. *Computer Networks*, 33(1–6):309–320, 2000. Proceedings of the 9th International World Wide Web Conference (WWW-9).

R. D. Burke, K. J. Hammond, V. Kulyukin, S. L. Lytinen, N. Tomuro, and S. Scott Schoenberg. Frequently-asked question files: Experiences with the FAQ finder system. *AI Magazine*, 18(2):57–66, 1997.

R. D. Burke, K. J. Hammond, and B. C. Young. Knowledge-based navigation of complex information spaces. In *Proceedings of 13th National Conference on Artificial Intelligence (AAAI-96)*, pages 462–468. AAAI Press, 1996.

M. E. Califf, editor. Machine Learning for Information Extraction: Proceedings of the AAAI-99 Workshop, 1999. AAAI Press. Technical Report WS-99-11.

M. E. Califf. Bottom-up relational learning of pattern matching rules for information extraction. *Journal of Machine Learning Research*, 4:177–210, 2003.

S. Chakrabarti. Data Mining for hypertext: A tutorial survey. *SIGKDD explorations*, 1(2):1–11, January 2000.

S. Chakrabarti. Mining the Web: Analysis of Hypertext and Semi Structured Data. Morgan Kaufmann, 2002.

S. Chakrabarti, B. Dom, and P. Indyk. Enhanced hypertext categorization using hyperlinks. In *Proceedings of the ACM SIGMOD International Conference on Management on Data*, pages 307–318, Seattle, WA, 1998a. ACM Press.

S. Chakrabarti, B. Dom, P. Raghavan, S. Rajagopalan, D. Gibson, and J. Kleinberg. Automatic resource compilation by analyzing hyperlink structure and associated text. *Computer Networks*, 30(1–7):65–74, 1998b. Proceedings of the 7th International World Wide Web Conference (WWW-7), Brisbane, Australia.

G. Chang, M. J. Healy, J. A. M. McHugh, and J. T. L. Wang. *Mining the World Wide Web: An Information Search Approach*. Kluwer Academic Publishers, 2001.

W. W. Cohen. Learning rules that classify e-mail. In M. Hearst and H. Hirsh, editors, *Proceedings of the AAAI Spring Symposium on Machine Learning in Information Access*, pages 18–25. AAAI Press, 1996. Technical Report SS-96-05.

W. W. Cohen and W. Fan. Web-collaborative filtering: Recommending music by crawling the web. In *Proceedings of the 9th International World Wide Web Conference (WWW-9)*, 2000.

R. Cooley, B. Mobasher, and J. Srivastava. Data preparation for mining world wide web browsing patterns. *Knowledge and Information Systems*, 1(1): 5–32, 1999.

M. Craven, D. DiPasquo, D. Freitag, A. McCallum, T. Mitchell, K. Nigam, and S. Slattery. Learning to construct knowledge bases from the World Wide Web. *Artificial Intelligence*, 118(1-2):69–114, 2000.

M. Craven and S. Slattery. Relational learning with statistical predicate invention: Better models for hypertext. *Machine Learning*, 43(1-2):97–119, 2001.

M. Craven, S. Slattery, and K. Nigam. First-order learning for Web mining. In C. Nédellec and C. Rouveirol, editors, *Proceedings of the 10th European Conference on Machine Learning (ECML-98)*, pages 250–255, Chemnitz, Germany, 1998. Springer-Verlag.

E. Crawford, J. Kay, and E. McCreath. IEMS – The Intelligent Email Sorter. In C. Sammut and A. G. Hoffmann, editors, *Proceedings of the 19th International Conference on Machine Learning (ICML-02)*, pages 263–272, Sydney, Australia, 2002. Morgan Kaufmann.

J. Dean and M. R. Henzinger. Finding related pages in the World Wide Web. In A. Mendelzon, editor, *Proceedings of the 8th International World Wide Web Conference (WWW-8)*, pages 389–401, Toronto, Canada, 1999.

S. C. Deerwester, S. T. Dumais, T. K. Landauer, G. W. Furnas, and R. A. Harshman. Indexing by latent semantic analysis. *Journal of the American Society of Information Science*, 41(6):391–407, 1990.

T. G. Dietterich. Ensemble methods in machine learning. In J. Kittler and F. Roli, editors, *First International Workshop on Multiple Classifier Systems*, pages 1–15. Springer-Verlag, 2000.

A. Doan, J. Madhavan, R. Dhamankar, P. Domingos, and A. Y. Halevy. Learning to match ontologies. *VLDB Journal*, 12(4):303–319, 2003. Special Issue on the Semantic Web.

R. B. Doorenbos, O. Etzioni, and D. S. Weld. A scalable comparison-shopping agent for the World-Wide Web. In *Proceedings of the 1st International Conference on Autonomous Agents*, pages 39–48, Marina del Rey, CA, 1997.

S. Džeroski and N. Lavrač, editors. Relational Data Mining: Inductive Logic Programming for Knowledge Discovery in Databases. Springer-Verlag, 2001.

L. Eikvil. Information extraction from world wide web – a survey. Technical Report 945, Norwegian Computing Center, 1999.

O. Etzioni and D. Weld. A softbot-based interface to the internet. *Communications of the ACM*, 37(7):72–76, July 1994. Special Issue on *Intelligent Agents*.

O. Etzioni. Moving up the information food chain: Deploying softbots on the world wide web. In *Proceedings of the 13th National Conference on Artificial Intelligence (AAAI-96)*, pages 1322–1326. AAAI Press, 1996.

M. Faloutsos, P. Faloutsos, and C. Faloutsos. On power-law relationships of the internet topology. In Proceedings of the ACM Conference on Applications, Technologies, Architectures, and Protocols for Computer Communication (SIGCOMM-99), pages 251–262, Cambridge, MA, 1999. ACM Press.

T. Fawcett. "In vivo" spam filtering: A challenge problem for Data Mining. *SIGKDD explorations*, 5(2), December 2003.

D. Fensel. Ontologies: Silver Bullet for Knowledge Management and Electronic Commerce. Springer-Verlag, Berlin, 2001.

D. Freitag. Information extraction from HTML: Application of a general machine learning approach. In *Proceedings of the 15th National Conference on Artificial Intelligence (AAAI-98)*. AAAI Press, 1998.

J. Fürnkranz. A study using *n*-gram features for text categorization. Technical Report OEFAI-TR-98-30, Austrian Research Institute for Artificial Intelligence, Wien, Austria, 1998.

J. Fürnkranz. Hyperlink ensembles: A case study in hypertext classification. *Information Fusion*, 3(4):299–312, December 2002. Special Issue on Fusion of Multiple Classifiers.

J. Fürnkranz, C. Holzbaur, and R. Temel. User profiling for the Melvil knowledge retrieval system. *Applied Artificial Intelligence*, 16(4): 243–281, 2002.

J. Fürnkranz, T. Mitchell, and E. Riloff. A case study in using linguistic phrases for text categorization on the WWW. In M. Sahami, editor, *Learning for Text Categorization: Proceedings of the 1998 AAAI/ICML Workshop*, pages 5–12, Madison, WI, 1998. AAAI Press. Technical Report WS-98-05.

D. Goldberg, D. Nichols, B. M. Oki, and D. Terry. Using collaborative filtering to weave and information tapestry. *Communications of the ACM*, 35(12):61–70, December 1992.

G. Gottlob, C. Koch, R. Baumgartner, M. Herzog, and S. Flesca. The Lixto data extraction project — Back and forth between theory and practice. In *Proceedings of the Symposium on Principles of Database Systems (PODS-04)*, 2004.

P. Graham. Better bayesian filtering. In *Proceedings of the 2003 Spam Conference*, Cambridge, MA, 2003

G. Grieser, K. P. Jantke, S. Lange, and B. Thomas. A unifying approach to HTML wrapper representation and learning. In S. Arikawa and S. Morishita, editors, *Proc. 3rd International Conference on Discovery Science*, pages 50–64. Springer–Verlag, 2000.

T. Hofmann and J. Puzicha. Latent class models for collaborative filtering. In *Proceedings of the 16th International Joint Conference on Artificial Intelligence (IJCAI-99)*, pages 688–693, 1999.

C. N. Hsu and M. T. Dung. Generating finite-state transducers for semistructured data extraction from the web. *Information Systems*, 23(8):521–538, 1998. Special Issue on Semistructured Data.

T. Joachims. Optimizing search engines using clickthrough data. In *Proceedings of the 8th ACM SIGKDD International Conference on Knowledge Discovery and Data Mining (KDD-02)*, pages 133–142. ACM Press, 2002.

J. M. Kleinberg. Authoritative sources in a hyperlinked environment. *Journal of the ACM*, 46(5):604–632, September 1999. ISSN 0004-5411.

J. A. Konstan, B. N. Miller, D. Maltz, J. L. Herlocker, L. R. Gordon, and J. Riedl. Grouplens: Applying collaborative filtering to usenet news. *Communications of the ACM*, 40(3):77–87, 1997. Special Issue on Recommender Systems.

R. Kosala and H. Blockeel. Web mining research: A survey. *SIGKDD explorations*, 2(1):1–15, 2000

R. Kozierok and P. Maes. Learning interface agents. In *Proceedings of the 11th National Conference on Artificial Intelligence (AAAI-93)*, pages 459–465. AAAI Press, 1993.

N. Kushmerick. Wrapper induction: Efficiency and expressiveness. *Artificial Intelligence*, 118:15–68, 2000.

K. Lang. NewsWeeder: Learning to filter netnews. In A. Prieditis and S. Russell, editors, *Proceedings of the 12th International Conference on Machine Learning (ML-95)*, pages 331–339. Morgan Kaufmann, 1995.

Y. Lashkari, M. Metral, and P. Maes. Collaborative interface agents. In *Proceedings of the 12th National Conference on Artificial Intelligence (AAAI-94)*, pages 444–450, Seattle, WA, 1994. AAAI Press.

S. Lawrence and C. L. Giles. Searching the world wide web. *Science*, 280:98–100, 1998.

K. Lerman, S. N. Minton, and C. A. Knoblock. Wrapper maintenance: A machine learning approach. *Journal of Artificial Intelligence Research*, 18: 149–181, 2003.

M. Levene, J. Borges, and G. Louizou. Zipf's law for Web surfers. *Knowledge and Information Systems*, 3(1): 120–129, 2001.

D. D. Lewis. An evaluation of phrasal and clustered representations on a text categorization task. In *Proceedings of the 15th Annual International ACM SIGIR Conference on Research and Devlopment in Information Retrieval*, pages 37–50, 1992.

W. Lin, S. A. Alvarez, and C. Ruiz. Efficient adaptive-support association rule mining for recommender systems. *Data Mining and Knowledge Discovery*, 6(1): 83–105, 2002.

A. Maedche, C. Nédellec, S. Staab, and E. Hovy, editors. *Proceedings of the 2nd Workshop on Ontology Learning (OL-2001)*, volume 38 of *CEUR Workshop Proceedings*, Seattle, WA, 2001. IJCAI-01.

A. Maedche, V. Pekar, and S. Staab. Ontology learning part one — on discovering taxonomic relations from the web. In N.Zhong, J. Liu, and Y. Y. Yao, editors, *Web Intelligence*, pages 301–321. Springer-Verlag, 2003.

A. Maedche and S. Staab. Learning ontologies for the semantic web. *IEEE Intelligent Systems*, 16(2), 2001.

P. Maes. Agents that reduce work and information overload. *Communications of the ACM*, 37(7):30–40, July 1994. Special Issue on *Intelligent Agents*.

O. A. McBryan. GENVL and WWWW: Tools for taming the Web. In *Proceedings of the 1st World-Wide Web Conference (WWW-1)*, pages 58–67, Geneva, Switzerland, 1994. Elsevier.

A. McCallum and K. Nigam. A comparison of event models for naive bayes text classification. In M. Sahami, editor, *Learning for Text Categorization: Proceedings of the 1998 AAAI/ICML Workshop*, pages 41–48, Madison, WI, 1998. AAAI Press.

P. Melville, R. J. Mooney, and R. Nagarajan. Content-boosted collaborative filtering for improved recommendations. In *Proceedings of the 18th National Conference on Artificial Intelligence (AAAI-2002)*, pages 187–192, Edmonton, Canada, 2002.

D. Mladenić. Personal WebWatcher: Implementation and design. Technical Report IJS-DP-7472, Department of Intelligent Systems, Jožef Stefan Institute, 1996.

D. Mladenić. Feature subset selection in text-learning. In C. Nédellec and C. Rouveirol, editors, *Proceedings of the 10th European Conference on Machine Learning (ECML-98)*, pages 95–100, Chemnitz, Germany, 1998a. Springer-Verlag.

D. Mladenić. Turning Yahoo into an automatic web-page classifier. In H. Prade, editor, *Proceedings of the 13th European Conference on Artificial Intelligence (ECAI-98)*, pages 473–474, Brighton, U.K., 1998b. Wiley.

D. Mladenić. Text-learning and related intelligent agents: A survey. *IEEE Intelligent Systems*, 14(4):44–54, July/August 1999.

D. Mladenić and M. Grobelnik. Word sequences as features in text learning. In *Proceedings of the 17th Electrotechnical and Computer Science Conference (ERK-98)*, Ljubljana, Slovenia, 1998. IEEE section.

B. Mobasher, R. Cooley, and J. Srivastava. Automatic personalization based on web usage mining. *Communications of the ACM*, 43(8):142–151, 2000.

B. Mobasher, H. Dai, T. Luo, and M. Nakagawa. Discovery and evaluation of aggregate usage profiles for web personalization. *Data Mining and Knowledge Discovery*, 6(1): 61–82, 2002.

K. J. Mock. Hybrid hill-climbing and knowledge-based methods for intelligent news filtering. In *Proceedings of the 13th National Conference on Artificial Intelligence (AAAI-96)*, pages 48–53. AAAI Press, 1996.

J. Myllymaki. Effective web data extraction with standard XML technologies (HTML). In *Proceedings of the 10th International World Wide Web Conference (WWW-01)*, Hong Kong, May 2001.

H. J. Oh, S. H. Myaeng, and M.-H. Lee. A practical hypertext categorization method using links and incrementally available class information. In *Proceedings of the 23rd ACM International Conference on Research and Development in Information Retrieval (SIGIR-00)*, pages 264–271, Athens, Greece, 2000.

T. R. Payne and P. Edwards. Interface agents that learn: An investigation of learning issues in a mail agent interface. *Applied Artificial Intelligence*, 11(1): 1–32, 1997.

M. T. Pazienza, editor. Information Extraction in the Web Era: Natural Language Communication for Knowledge Acquisition and Intelligent Information Agents (SCIE-02), Rome, Italy, 2003. Springer-Verlag.

M. Pazzani, J. Muramatsu, and D. Billsus. Syskill & Webert: Identifying interesting web sites. In *Proceedings of the 13th National Conference on Artificial Intelligence (AAAI-96)*, pages 54–61. AAAI Press, 1996.

M. Perkowitz and O. Etzioni. Towards adaptive web sites: Conceptual framework and case study. *Artificial Intelligence*, 118:245–275, 2000.

D. Pierrakos, G. Paliouras, C. Papatheodorou, and C. D. Spyropoulos. Web usage mining as a tool for personalization: A survey. *User Modeling and User-Adapted Interaction*, 13 (4):311–372, 2003.

A. Popescul, L. Ungar, D. Pennock, and S. Lawrence. Probabilistic models for unified collaborative and content-based recommendation in sparse-data environments. In *Proceedings of the 17th Conference on Uncertainty in Artificial Intelligence (UAI-2001)*, pages 437–444. Morgan Kaufmann, 2001.

J. R. Quinlan. Learning logical definitions from relations. *Machine Learning*, 5:239–266, 1990.

J. R. Quinlan. Determinate literals in inductive logic programming. In *Proceedings of the 8th International Workshop on Machine Learning (ML-91)*, pages 442–446, 1991.

P. Resnick and H. R. Varian. Special issue on recommender systems. *Communications of the ACM*, 40(3), 1997.

B. L. Richards and R. J. Mooney. Learning relations by pathfinding. In *Proceedings of the 10th National Conference on Artificial Intelligence (AAAI-92)*, pages 50–55, San Jose, CA, 1992. AAAI Press.

E. Riloff. Automatically generating extraction patterns from untagged text. In *Proceedings of the 13th National Conference on Artificial Intelligence (AAAI-96)*, pages 1044–1049. AAAI Press, 1996a.

E. Riloff. An empirical study of automated dictionary construction for information extraction in three domains. *Artificial Intelligence*, 85:101–134, 1996b.

G. Salton. Automatic Text Processing: The Transformation, Analysis, and Retrieval of Information by Computer. Addison-Wesley, Reading, MA, 1989.

G. Salton and C. Buckley. Term-weighting approaches in automatic text retrieval. *Information Processing and Management*, 24 (5):513–523, 1988.

G. Salton, A. Wong, and C. S. Yang. A vector space model for automatic indexing. *Communications of the ACM*, 18(11):613–620, November 1975.

B. M. Sarwar, G. Karypis, J. A. Konstan, and J. Riedl. Item-based collaborative filtering recommendation algorithms. In *Proceedings of the 10th International World Wide Web Conference (WWW-10)*, Hong Kong, May 2001.

J. B. Schafer, J. A. Konstan, and J. Riedl. Electronic commerce recommender applications. *Data Mining and Knowledge Discovery*, 5(1/2): 115–152, 2000.

T. Scheffer. Email answering assistance by semi-supervised text classification. *Intelligent Data Analysis*, 8(5), 2004.

S. Scott and S. Matwin. Feature engineering for text classification. In I. Bratko and S. Džeroski, editors, *Proceedings of 16th International Conference on Machine Learning (ICML-99)*, pages 379–388, Bled, SL, 1999. Morgan Kaufmann Publishers, San Francisco, US.

F. Sebastiani. Machine learning in automated text categorization. *ACM Computing Surveys*, 34(1):1–47, March 2002.

B. Sheth and P. Maes. Evolving agents for personalized information filtering. In *Proceedings of the 9th Conference on Artificial Intelligence for Applications (CAIA-93)*, pages 345–352. IEEE Press, 1993.

S. Slattery and T. Mitchell. Discovering test set regularities in relational domains. In P. Langley, editor, *Proceedings of the 17th International Conference on Machine Learning (ICML-00)*, pages 895–902, Stanford, CA, 2000. Morgan Kaufmann.

S. Soderland. Learning information extraction rules for semi-structured and free text. *Machine Learning*, 34(1–3):233–272, 1999.

E. Spertus. ParaSite: Mining structural information on the Web. *Computer Networks and ISDN Systems*, 29 (8-13):1205–1215, September 1997. Proceedings of the 6th International World Wide Web Conference (WWW-6).

M. Spiliopoulou. The laborious way from Data Mining to web log mining. *Journal of Computer Systems Science and Engineering*, 14:113–126, 1999. Special Issue on Semantics of the Web.

J. Srivastava, R. Cooley, M. Deshpande, and P.-N. Tan. Web usage mining: Discovery and applications of usage patterns from web data. *SIGKDD explorations*, 1(2):12–23, 2000.

S. Staab and A. Maedche. Knowledge portals — ontologies at work. *AI Magazine*, 21(2):63–75, Summer 2001.

S. Staab, A. Maedche, C. Nédellec, and P. Wiemer-Hastings, editors. *Proceedings of the 1st Workshop on Ontology Learning (OL-2000)*, volume 31 of *CEUR Workshop Proceedings*, Berlin, 2000. ECAI-00.

S. Staab and R. Studer, editors. *Handbook on Ontologies.* International Handbooks on Information Systems. Springer-Verlag, 2004.

G. Stumme, A. Hotho, and B. Berendt, editors. *Proceedings of the ECML PKDD 2001 Workshop on Semantic Web Mining*, Freiburg, Germany, 2001.

G. Stumme, A. Hotho, and B. Berendt, editors. *Proceedings of the ECML PKDD 2002 Workshop on Semantic Web Mining*, Helsinki, Finland, 2002.

P. N. Tan and V. Kumar. Discovery of web robot sessions based on their navigational patterns. *Data Mining and Knowledge Discovery*, 6(1): 9–35, 2002.

L. H. Ungar and D. P. Foster. Clustering methods for collaborative filtering. In H. Kautz, editor, *Proceedings of the AAAI-98 Workshop on Recommender Systems*, page 112, Madison, Wisconsin, 1998. AAAI Press. Technical Report WS-98-08.

Y. Yang and J. O. Pedersen. A comparative study on feature selection in text categorization. In D. Fisher, editor, *Proceedings of the 14th International Conference on Machine Learning (ICML-97)*, pages 412–420, Nashville, TN, 1997. Morgan Kaufmann.

Y. Yang, S. Slattery, and R. Ghani. A study of approaches to hypertext categorization. *Journal of Intelligent Information Systems*, 18 (2–3):219–241, March 2002. Special Issue on Automatic Text Categorization.

48

A Review of Web Document Clustering Approaches

Nora Oikonomakou[1] and Michalis Vazirgiannis[2]

[1] Department of Informatics
Athens University of Economics and Business (AUEB)
Patision 76, 10434, Greece
oikonomn@aueb.gr

[2] Department of Informatics
Athens University of Economics and Business (AUEB)
Patision 76, 10434, Greece
mvazirg@aueb.gr

Summary. Nowadays, the Internet has become the largest data repository, facing the problem of information overload. Though, the web search environment is not ideal. The existence of an abundance of information, in combination with the dynamic and heterogeneous nature of the Web, makes information retrieval a difficult process for the average user. It is a valid requirement then the development of techniques that can help the users effectively organize and browse the available information, with the ultimate goal of satisfying their information need. Cluster analysis, which deals with the organization of a collection of objects into cohesive groups, can play a very important role towards the achievement of this objective. In this chapter, we present an exhaustive survey of web document clustering approaches available on the literature, classified into three main categories: text-based, link-based and hybrid. Furthermore, we present a thorough comparison of the algorithms based on the various facets of their features and functionality. Finally, based on the review of the different approaches we conclude that although clustering has been a topic for the scientific community for three decades, there are still many open issues that call for more research.

Key words: Clustering, World Wide Web, Web-Mining, Text-Mining

48.1 Introduction

Nowadays, the internet has become the largest data repository, facing the problem of information overload. In the same time, more and more people use the World Wide Web as their main source of information. The existence of an abundance of information, in combination with the dynamic and heterogeneous nature of the Web, makes information retrieval a tedious process for the average user. Search engines, meta-search engines and Web Directories have been developed in order to help the users quickly and easily satisfy their information need.

O. Maimon, L. Rokach (eds.), *Data Mining and Knowledge Discovery Handbook*, 2nd ed.,
DOI 10.1007/978-0-387-09823-4_48,

Usually, a user searching for information submits a query composed by a few keywords to a search engine (such as Google (http://www.google.com) or Lycos (http://www.lycos.com)). The search engine performs exact matching between the query terms and the keywords that characterize each web page and presents the results to the user. These results are long lists of URLs, which are very hard to search. Furthermore, users without domain expertise are not familiar with the appropriate terminology thus not submitting the right (in terms of relevance or specialization) query terms, leading to the retrieval of more irrelevant pages.

This has led to the need for the development of new techniques to assist users effectively navigate, trace and organize the available web documents, with the ultimate goal of finding those best matching their needs. One of the techniques that can play an important role towards the achievement of this objective is *document clustering*. The increasing importance of document clustering and the variety of its applications has led to the development of a wide range of algorithms with different quality/complexity tradeoffs.

The contribution of this chapter is a review and a comparison of the existing web document clustering approaches. A comparative description of the different approaches is important in order to understand the needs that led to the development of each approach (i.e. the problems that it intended to solve) and the various issues related to web document clustering. Finally, we determine problems and open issues that call for more research in this context.

48.2 Motivation for Document Clustering

Clustering (or cluster analysis) is one of the main data analysis techniques and deals with the organization of a set of objects in a multidimensional space into cohesive groups, called clusters. Each cluster contains objects that are very similar to each other and very dissimilar to objects in other clusters (Rasmussen, 1992). An example of a clustering is depicted in figure 48.1. The input objects are shown in figure 48.1a and the existing clusters are shown in 48.1b. Objects belonging to the same cluster are depicted with the same symbol. Cluster analysis aims at discovering objects that have some representative behavior in the collection. The basic idea is that if a rule is valid for one object, it is very possible that the rule also applies to all the objects that are very similar to it. With this technique one can trace dense and sparse regions in the data space and, thus, discover hidden similarities, relationships and concepts and to group large datasets with regard to the common characteristics of their objects. Clustering is a form of *unsupervised classification*, which means that the categories into which the collection must be partitioned are not known, and so the clustering process involves the discovering of these categories.

In order to cluster documents, one must first choose the type of the characteristics or attributes (e.g. words, phrases or links) of the documents on which the clustering algorithm will be based and their representation. The most commonly used model is the Vector Space Model (Salton et al., 1975). Each document is represented as a feature vector whose length is equal to the number of unique document attributes in the collection. Each component of that vector has a weight associated to it, which indicates the degree of importance of the particular attribute for the characterization of the document. The weight can be either 0 or 1, depending on if the attribute characterizes or not the document respectively (binary representation). It can also be a function of the frequency of occurrence of the attribute in the document (tf) and the frequency of occurrence of the attribute in the entire collection (tf-idf). Then, an appropriate similarity measure must be chosen for the calculation of the similarity between two documents (or clusters). Some widely used similarity measures are the Cosine Coefficient, which gives the cosine of the angle between the two feature vectors, the Jaccard Coefficient and the Dice

Coefficient (all normalized versions of the simple matching coefficient). More on the similarity measures can be found in Van Rijsbergen (1979), Willet (1988) and Strehl et al. (2000).

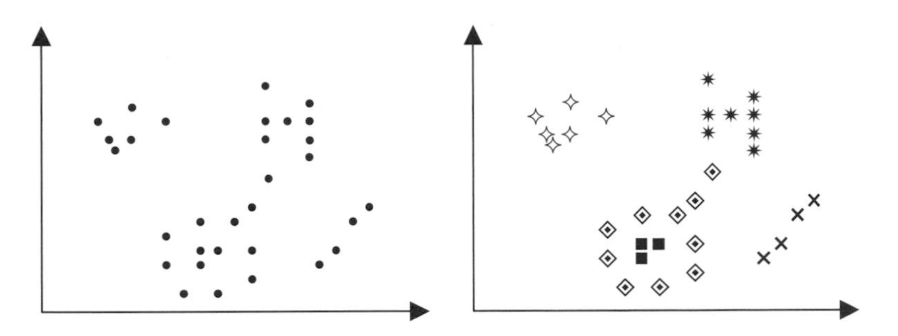

Fig. 48.1. Clustering example: a) input and b) clusters.

Many uses of clustering as part of the Web Information Retrieval process have been proposed in the literature. Firstly, based on the cluster hypothesis, clustering can increase the efficiency and the effectiveness of the retrieval (Van Rijsbergen, 1979). The fact that the users query is not matched against each document separately, but *against each cluster* can lead to an increase in the effectiveness, as well as the efficiency, by returning more relevant and less non relevant documents. Furthermore, clustering can be used as a very powerful mechanism for browsing a collection of documents or for presenting the results of the retrieval (e.g. suffix tree clustering (Zamir and Etzioni,1998), Scatter/Gather (Cutting et al., 1992)). A typical retrieval on the Internet will return a long list of web pages. The organization and presentation of the pages in small and meaningful groups (usually followed by short descriptions or summaries of the contents of each group) gives the user the possibility to focus exactly on the subject of his interest and find the desired documents more quickly. Furthermore, the presentation of the search results in clusters can provide an overview of the major subject areas related to the users topic of interest. Finally, other applications of clustering include query refinement (automatic inclusion or exclusion of terms from the users query in order to increase the effectiveness of the retrieval), tracing of similar documents and the ranking of the retrieval results (Kleinberg, 1997 & Page et al.,1998).

48.3 Web Document Clustering Approaches

There are many document clustering approaches proposed in the literature. They differ in many parts, such as the types of attributes they use to characterize the documents, the similarity measure used, the representation of the clusters etc. Based on the characteristics or attributes of the documents that are used by the clustering algorithm, the different approaches can be categorized into *i. text-based*, in which the clustering is based on the content of the document, *ii. link-based*, based on the link structure of the pages in the collection and *iii. hybrid* ones, which take into account both the content and the links of the document.

Most algorithms in the first category were developed for use in static collections of documents that were stored and could be retrieved from a database and not for collections of web pages, although they are used for the later case too. But, contrary to traditional document retrieval systems, the World Wide Web is a *directed graph*. This means that apart from its content, a web page contains other characteristics that can be very useful to clustering. The most important among these are the hyperlinks that play the role of citations between the web pages. The basic idea is that when two documents are cited together by many other documents (i.e. have many common incoming links) or cite the same documents (i.e. have many common outgoing links) there exists a semantic relationship between them. Consequently, traditional algorithms, developed for text retrieval, need to be refitted to incorporate these new sources of information about documents associations. In the Web Information Retrieval literature there are many applications based on the use of hyperlinks in the clustering process and the calculation of the similarity based on the link structure of the documents has proven to produce high quality clusters.

In the following sections we consider n to be the number of documents in the document collection under consideration.

48.3.1 Text-based Clustering

The text-based web document clustering approaches characterize each document according to its content, i.e. the words (or sometimes phrases) contained in it. The basic idea is that if two documents contain many common words then it is likely that the two documents are very similar.

The text-based approaches can be further classified according to the clustering method used into the following categories: *partitional*, *hierarchical*, *graph-based*, *neural network-based* and *probabilistic*. Furthermore, according to the way a clustering algorithm handles uncertainty in terms of cluster overlapping, an algorithm can be either *crisp* (or hard), which considers non-overlapping partitions, or *fuzzy* or soft) with which a document can be classified to more than one cluster. Most of the existing algorithms are crisp, meaning that a document either belongs to a cluster or not. It must also be noted that most of the mentioned approaches in this category are general clustering algorithms that can be applied to any kind of data. In this chapter though, we are interested in their application to documents. In the following paragraphs we present the main text-based document clustering approaches, their characteristics and the representative algorithms of each category. We also present a rather new approach to document clustering, which relies on the use of ontologies in order to calculate the similarity between the words that characterize the documents.

Partitional Clustering

The partitional or non-hierarchical document clustering approaches attempt a flat partitioning of a collection of documents into a *predefined* number of *disjoint* clusters. Partitional clustering algorithms are divided into iterative or reallocation methods and single pass methods. Most of them are iterative and the single pass methods are usually used in the beginning of a reallocation method, in order to produce the first partitioning of the data.

The partitional clustering algorithms use a feature vector matrix[3] and produce the clusters by optimizing a criterion function. Such criterion functions are the following: maximize the sum of the average pairwise cosine similarities between the documents assigned to a cluster, minimize the cosine similarity of each cluster centroid to the centroid of the entire collection etc. Zhao and Karypis (2001) compared eight criterion functions and concluded that the selection of a criterion function can affect the clustering solution and that the overall quality

[3] Each row of the feature vector matrix corresponds to a document and each column to a term. The ij-th entry has a value equal to the weight of the term j in document i

depends on the degree to which they can correctly operate when the dataset contains clusters of different densities and the degree to which they can produce balanced clusters.

The most common partitional clustering algorithm is k-means, which relies on the idea that the center of the cluster, called *centroid*, can be a good representation of the cluster. The algorithm starts by selecting k cluster centroids. Then the cosine distance[4] between each document in the collection and the centroids is calculated and the document is assigned to the cluster with the nearest centroid. After all documents have been assigned to clusters, the new cluster centroids are recalculated and the procedure runs iteratively until some criterion is met. Many variations of the k-means algorithm are proposed, e.g. ISODATA (Jain et al., 1999) and bisecting k-means (Steinbach et al., 2000). Another approach to partitional clustering is used in the Scatter/Gather system.

Scatter/Gather uses two linear-time partitional algorithms, Buckshot and Fractionation, which also apply HAC logic[5]. The idea is to use these algorithms to find the initial cluster centers and then find the clusters using the assign-to-nearest approach. Finally, the single pass method (Rasmussen, 1992) is another approach to partitional clustering which is based on the assignment of each document to the cluster with the most similar representative is above a threshold. The clusters are formed after only one pass of the data and no iteration takes place. Consequently, the order in which the documents are processed influences the clustering.

The advantages of these algorithms consist in their simplicity and their low computational complexity. The disadvantage is that the clustering is rather arbitrary since it depends on many parameters, like the values of the target number of clusters, the selection of the initial cluster centroids and the order of processing the documents.

Hierarchical Clustering

Hierarchical clustering algorithms produce a sequence of nested partitions. Usually the similarity between each pair of documents is stored in a nxn similarity matrix. At each stage, the algorithm either merges two clusters (agglomerative methods) or splits a cluster in two (divisive methods). The result of the clustering can be displayed in a tree-like structure, called a *dendrogram*, with one cluster at the top containing all the documents of the collection and many clusters at the bottom with one document each. By choosing the appropriate level of the dendrogram we get a partitioning into as many clusters as we wish. The dendrogram is a useful representation when considering retrieval from a clustered set of documents, since it indicates the paths that the retrieval precess may follow (Rasmussen, 1992).

Almost all the hierarchical algorithms used for document clustering are agglomerative (HAC). The steps of the typical HAC algorithm are the following:

1. Assign each document to a single cluster
2. Compute the similarity between all pairs of clusters and store the result in a similarity matrix, in which the ij-th entry stores the similarity between the i-th and j-th cluster
3. Merge the two most similar (closest) clusters

[4] K-means does not generally use the cosine similarity measure, but when applying k-means to documents it seems to be more appropriate

[5] Buckshot and Fractionation both use a cluster subroutine that applies the group average hierarchical clustering method.

4. Update the similarity matrix with the similarity between the new cluster and the original clusters
5. Repeat steps 3 and 4 until only one cluster remains or until a threshold[6] is reached.

The hierarchical agglomerative clustering methods differ in the way they calculate the similarity between two clusters. The existing methods are the following (Rasmussen, 1992; El. Handouchi and Willet, 1989; Willet, 1988):

- *Single link*: The similarity between a pair of clusters is calculated as the similarity between the two most similar documents, one of which is in each cluster. This method tends to produce long, loosely bound clusters with little internal cohesion (chaining effect). The single link method incorporates useful mathematical properties and can have small computational complexity. There are many algorithms based on this method. Their complexities vary from $O(nlogn)$ to $O(n^5)$. Single link algorithms include van Rijsbergen's algorithm (Van Rijsbergen, 1979), SLINK (Sibson, 1973), Minimal Spanning Tree (Rasmussen, 1992) and Voorhees's algorithm (Voorhees, 1986).
- *Complete link*: The similarity between a pair of clusters is taken to be the similarity between the least similar documents, one of which is in each cluster. This definition is much stricter than that of the single link method and, thus, the clusters are small and tightly bound. Implementations of this method are the CLINK algorithm (Defays, 1977), which is a variation of the SLINK algorithm, and the algorithm proposed by Voorhees (Voorhees, 1986).
- *Group average:* This method produces clusters such that each document in a cluster has greater average similarity with the other documents in the cluster than with the documents in any other cluster. All the documents in the cluster contribute in the calculation of the pairwise similarity and, thus, this method is a mid-point between the above two methods. Usually the complexity of the group average algorithm is higher than $O(n^2)$. Voorhees proposed an algorithm for the group average method that calculates the pairwise similarity as the inner product of two vectors with appropriate weights (Voorhees, 1986). Steinbach et al. (2000) used UPGMA for the implementation of the group average method and obtained very good results.
- *Ward's method*: In this method the cluster pair to be merged is the one whose merger minimizes the increase in the total within-group error sum of squares based on the distance between the cluster centroids (i.e. the sum of the distances from each document to the centroid of the cluster containing it). This method tends to result in spherical, tightly bound clusters and is less sensitive to outliers. Wards method can be implemented using the reciprocal-nearest neighbor (RNN) algorithm (Murtagh, 1983), which was modified for document clustering by Handouchi and Willett (1986).
- *Centroid/Median Methods*: Each cluster as is it formed is represented by the group centroid/median. At each stage of the clustering the pair of clusters with the most similar mean centroid/median is merged. The difference between the centroid and the median is that the second is not weighted proportionally to the size of the cluster.

The HAC approaches produce high quality clusters but have very high computational requirements (at least $O(n^2)$). They are typically greedy. This means that the pair of clusters that is chosen for agglomeration at each time is the one, which is considered the best at that time, without regard to future consequences. Also, if a merge that has taken place is not appropriate, there is no backtracking to correct the mistake.

---

[6] Some examples of such threshold are the desired number of clusters, the maximum number of documents in a cluster or the maximum similarity value below which mo merge is done

There are many experiments in the literature comparing the different HAC methods. Most of them conclude that the single link method, although the only method applicable for large document sets, does not give high quality results (El-Hamdouchi and Willett, 1989; Willett, 1988; Steinbach et al., 2000). As for the best HAC method, the group average method seems to work slightly better than the complete link and Ward's method (El-Hamdouchi and Willett, 1989; Steinbach et al., 2000; Zhao and Karypis, 2002). This may be because the single link method decides using very little information and complete link considers the clusters to be very dissimilar. The group average method overcomes these problems by calculating the mean distance between the clusters (Steinbach et al., 2000).

Graph based clustering

In this case the documents to be clustered can be viewed as a set of nodes and the edges between the nodes represent the relationship between them. The edges bare a weight, which denotes the strength of that relationship. Graph based algorithms rely on *graph partitioning*, that is, they identify the clusters by cutting edges from the graph such that the edge-cut, i.e. the sum of the weights of the edges that are cut, is minimized. Since each edge in the graph represents the similarity between the documents, by cutting the edges with the minimum sum of weights the algorithm minimizes the similarity between documents in different clusters. The basic idea is that the weights of the edges in the same cluster will be greater than the weights of the edges across clusters. Hence, the resulting cluster will contain highly related documents.

The different graph based algorithms may differ in the way they produce the graph and in the graph partitioning algorithm that they use. Chameleon's (Karypis et al., 1999) graph representation of the document set is based on the knearest neighbor graph approach. Each node represents a document and there exists an edge between two nodes if the document corresponding to either of the nodes is among the k most similar documents of the document corresponding to the other node. The resulting k-nearest neighbor graph is sparse and captures the neighborhood of each document. Chameleon then applies a graph partitioning algorithm, hMETIS (Karypis and Kumar, 1999) to identify the clusters. These clusters are further clustered using a hierarchical agglomerative clustering algorithm and based on a dynamic model (Relative Interconnectivity and Relative Closeness) to determine the similarity between two clusters. So, Chameleon is actually a *hybrid* (graph based and HAC) text-based algorithm.

Association Rule Hypergraph Partitioning (ARHP) (Boley et al., 1999) is another graph based approach which is based on hypergraphs. A hypergraph is an extension of a graph in the sense that each hyperedge can connect more than two nodes. In ARHP the hyperedges connect a set of nodes that consist a frequent item set. A frequent item set captures the relationship between two or more documents and it consists of documents with many common terms characterizing them. In order to determine these sets in the document collection and to weight the hyperedge, the algorithm uses an association rule discovery algorithm (Apriori). Then the hypergraph is partitioned using a hypergraph partitioning algorithm to get the clusters. This algorithm is used in the WebACE project (Han et al., 1997) to cluster web pages that have been returned by a search engine in response to a user's query. It can also be used for term clustering.

Another graph based approach is the algorithm proposed by Dhillon (2001) which uses iterative bipartite graph partitioning to co-cluster documents and words. The advantages of these approaches are that can capture the structure of the data and that they work effectively in high dimensional spaces. The disadvantage is that the graph must fit the memory.

Neural Network based Clustering

The Kohonen's Self-Organizing feature Maps (SOM) (Kohonen, 1995) is a widely used unsupervised neural network model. It consists of two layers: the input layer with n input nodes, which correspond to the n documents, and an output layer with k output nodes, which correspond to k decision regions (i.e. clusters). The input units receive the input data and propagate them onto the output units. Each of the k output units is assigned a weight vector. During each learning step, a document from the collection is associated with the output node, which has the most similar weight vector. The weight vector of that 'winner' node is then adapted in such a way that it will become even more similar to the vector that represents that document, i.e. the weight vector of the output node 'moves closer' to the feature vector of the document. This process runs iteratively until there are no more changes in the weight vectors of the output nodes. The output of the algorithm is the arrangement of the input documents in a 2-dimensional space in such a way that the similarity between the input documents is mirrored in terms of topographic distance between the k decision regions.

Another approach proposed in the literature is the *hierarchical feature map* (Merkl, 1998) model, which is based on a hierarchical organization of more than one self-organizing feature maps. The aim of this approach is to overcome the limitations imposed by the 2-dimensional output grid of the SOM model, by arranging a number of SOMs in a hierarchy, such that for each unit on one level of the hierarchy a 2-dimensional self-organizing map is added to the next level.

Neural networks are usually useful in environments where there is a lot of noise, and when dealing with data with complex internal structure and frequent changes. The advantage of this approach is the ability to give high quality results without having high computational complexity. The disadvantages are the difficulty to explain the results and the fact that the 2-dimensional output grid may restrict the mirroring and result in loss of information. Furthermore, the selection of the initial weights may influence the result (Jain et al., 1999).

Fuzzy Clustering

All the aforementioned approaches produce clusters in such a way that each document is assigned to one and only one cluster. Fuzzy clustering approaches, on the other hand, are non-exclusive, in the sense that each document can belong to more than one clusters. Fuzzy algorithms usually try to find the best clustering by optimizing a certain criterion function. The fact that a document can belong to more than one clusters is described by a *membership function*. The membership function computes for each document a membership vector, in which the i-th element indicates the degree of membership of the document in the i-th cluster.

The most widely used fuzzy clustering algorithm is Fuzzy c-means (Bezdek, 1984), a variation of the partitional k-means algorithm. In fuzzy c-means each cluster is represented by a *cluster prototype* (the center of the cluster) and the membership degree of a document to each cluster depends on the distance between the document and each cluster prototype. The closest the document is to a cluster prototype, the greater is the membership degree of the document in the cluster. Another fuzzy approach, that tries to overcome the fact that fuzzy c-means does not take into account the distribution of the document vectors in each cluster, is the Fuzzy Clustering and Fuzzy Merging algorithm (FCFM) (Looney, 1999). The FCFM uses Gaussian weighted feature vectors to represent the cluster prototypes. If a document vector is equally close to two prototypes, then it belongs more to the widely distributed cluster than to the narrowly distributed cluster.

Probabilistic Clustering

Another way of dealing with uncertainty is to use probabilistic clustering algorithms. These algorithms use statistical models to calculate the similarity between the data instead of some predefined measures. The basic idea is the assignment of probabilities for the membership of a document in a cluster. Each document can belong to more than one cluster according to the probability of belonging to each cluster. Probabilistic clustering approaches are based on finite mixture modeling (Everitt and Hand, 1981). They assume that the data can be partitioned into clusters that are characterized by a probability distribution function (p.d.f.). The p.d.f. of a cluster gives the probability of observing a document with particular weight values on its feature vector in that cluster. Since the membership of a document in each cluster is not known a priori, the data are characterised by a distribution, which is the mixture of all the cluster distributions. Two widely used probabilistic algorithms are Expectation Maximization (EM) and AutoClass (Cheeseman and Stutz, 1996). The output of the probabilistic algorithms is the set of distribution function parameter values and the probability of membership of each document to each cluster.

Using Ontologies

The algorithms described above, most often rely on *exact* keyword matching, and do not take into account the fact that the keywords may have some *semantic proximity* between each other. This is, for example, the case with synonyms or words that are part of other words (whole-part relationship). For instance a document might be characterized by the words 'camel, desert' and another with the word 'animal, Sahara'. By using traditional techniques these documents would be judged unrelated. Using an ontology can help capture this semantic proximity of the documents. An ontology, in our context, is a structure (a lexicon) that organizes words in a net connected according to the semantic relationship that exists between them. More on ontologies can be found in Ding (2001).

THESUS (Varlamis et al.) is a system that clusters web documents that are characterized by weighted keywords of an ontology. The ontology used is a tree of terms connected according to the IS-A relationship. Given this ontology and a set of document characterized by keywords the algorithm proposes a clustering scheme based on a novel similarity measure between sets of terms that are hierarchically related. Firstly, the keywords that characterize each document are mapped onto terms in the ontology. Then, the similarity between the documents is calculated based on the proximity of their terms in the ontology. In order to do that, an extension of the Wu and Palmer similarity measure is used (Wu and Palmer, 1994). Finally, a modified version of the DBSCAN clustering algorithm is used to provide the clusters. The advantage of using an ontology in clustering is that it provides a very useful structure not only for the calculation of document similarity, but also for dimensionality reduction by abstracting the keywords that characterize the documents to terms in the ontology.

48.3.2 Link-based Clustering

Text-based clustering approaches were developed for use in small, static and homogeneous collections of documents. On the contrary, the www is a huge collection of heterogeneous and interconnected web pages. Moreover, the web pages have additional information attached to them (web document metadata, hyperlinks) that can be very useful to clustering. According to Kleinberg (1997), *the link structure of a hypermedia environment can be a rich source of information about the content of the environment*. The link-based document clustering approaches take into account information extracted by the link structure of the collection. The underlying idea is that when two documents are connected via a link there exists a semantic relationship between them, which can be the basis for the partitioning of the collection into clusters.

The use of the link structure for clustering a collection is based on citation analysis from the field of bibliometrics (White and McCain, 1989). Citation analysis assumes that if a person creating a document cites two other documents then these documents must be somehow related in the mind of that person. In this way, the clustering algorithm tries to incorporate the human judgement when characterizing the documents. Two measures of similarity between two documents p and q based on citation analysis that are widely used are: *co-citation*, which is the number of documents that co-cite p and q and *bibliographic coupling*, which is the number of documents that are cited by both p and q. The greater the value of these measures the stronger the relationship between the documents p and q is. Also, the length of the path that connects two documents is sometimes considered when calculating the document similarity.

There are many uses of the link structure of a web page collection in web IR. Crofts Inference Network Model (Croft, 1993) uses the links that connect two web pages to enhance the word representation of a web page by the words contained in the pages linked to it. Frei & Stieger (1995) characterise a hyperlink by the common words contained in the documents that it connects. This method is proposed for the ranking of the results returned to a user's query. Page et al.(1998) also proposed an algorithm for the ranking of the search results. Their approach, PageRank, assigns at each web page a score, which denotes the importance of that page and depends on the number and importance of pages that point to it. Finally, Kleinberg proposed the HITS algorithm (Kleinberg, 1997) for the identification of mutually reinforcing communities, called hubs and authorities. Pages with many incoming links are called authorities and are considered very important. The hubs are pages that point to many important pages.

As far as clustering is concerned, one of the first link-based algorithms was proposed by Botafogo & Shneiderman (1991). Their approach is based on a graph theoretic algorithm that found strongly connected components in a hypertexts graph structure. The algorithm uses a *compactness* measure, which indicates the interconnectedness of the hypertext, and is a function of the average link distance between the hypertext nodes. The higher that compactness the more relevant the nodes are. The algorithm identifies clusters as highly connected subgraphs of the hypertext graph. Later, Botafogo (1993) extended his idea to include also the number of the different paths that connect two nodes in the calculation of the compactness. This extended algorithm produces more discriminative clusters, with reasonable size and with highly related nodes.

Another link-based algorithm was proposed by Larson (1996), who applied cocitation analysis to a collection of web documents. Co-citation analysis begins with the construction of a co-citation frequency matrix, whose ij-th entry contains the number of documents citing both documents i and j. Then, correlation analysis is applied to convert the raw frequencies into correlation coefficients. The last step is the multivariate analysis of the correlation matrix using multidimensional scaling techniques (SAS MDS), which mirrors the data onto a 2-dimensional map. The interpretation of the 'map' can reveal interesting relationships and groupings of the documents. The complexity of the algorithm is $O(n^2/2-n)$.

Finally, another interesting approach to clustering of web pages is trawling (Kumar et al., 1999), which clusters related web pages in order to discover new emerging cyber-communities that have not yet been identified by large web directories. The underlying idea in trawling is that these relevant pages are very frequently cited together even before their creators realise that they have created a community. Furthermore, based on Kleinberg's idea, trawling assumes that these communities consist of mutually reinforcing hubs and authorities. So, trawling combines the idea of co-citation and HITS to discover clusters. Based on the above assumptions, Web communities are characterized by dense directed bipartite subgraphs[7]. These graphs, that are the signatures of web communities, contain at least one core, which are complete directed bipartite graphs with a minimum number of nodes. Trawling aims at discovering these cores and then applies graph-based algorithms to discover the clusters.

48.3.3 Hybrid Approaches

The link-based document clustering approaches described above characterize the document solely by the information extracted from the link structure of the collection, just as the text-based approaches characterize the documents only by the words they contain. Although the links can be seen as a recommendation of the creator of one page to another page, they do not intend to indicate the similarity. Furthermore, these algorithms may suffer from poor or too dense link structures. On the other hand, text-based algorithms have problems when dealing with different languages or with particularities of the language (synonyms, homonyms etc.). Also, web pages contain other forms of information except text, such as images or multimedia. As a consequence, hybrid document clustering approaches have been proposed in order to combine the advantages and limit the disadvantages of the two approaches.

Pirolli et al. (1996) described a method that represents the pages as vectors containing information from the content, the linkage, the usage data and the meta-information attached to each document. The method uses spreading activation techniques to cluster the collection. These techniques start by 'activating' a node in the graph (giving a starting value to it) and 'spreading' the value across the graph through its links. In the end, the nodes with the highest values are considered very related to the starting node. The problem with the algorithm proposed by Pirolli et al. is that there is no scheme for combining the different information about the documents. Instead, there is a different graph for each attribute (text, links etc.) and the algorithm is applied to each one, leading to many different clustering solutions.

The 'content-link clustering' algorithm, which was proposed by Weiss et al. (1996), is a hierarchical agglomerative clustering algorithm that uses the complete link method and a hybrid similarity measure. The similarity between two documents is taken to be the maximum between the text similarity and the link similarity:

$$S_{ij} = max(S_{ij}{}^{terms}, S_{ij}{}^{links}) \tag{48.1}$$

The text similarity is computed as the normalized dot product of the term vectors representing the documents. The link similarity is a linear combination of three parameters: the number of Common Ancestors (i.e. common incoming links), the number of Common Descendants (i.e. common outgoing links) and the number of Direct Paths between the two documents. The strength of the relationship between the documents is also proportional to the length of the shortest paths between the two documents and between the documents and their common ancestors and common descendants. This algorithm is used in the HyPursuit system to provide a set of services such as query routing, clustering of the retrieval results, query refinement, cluster-based browsing and result set expansion. The system also provides summaries of the cluster contents, called content labels, in order to support the system operations.

Finally, another hybrid text- and link-based clustering approach is the toric k-means algorithm, proposed by Modha and Spangler (2000). The algorithm starts by gathering the results

[7] A bipartite graph is a graph whose node set can be partitioned into two sets N_1 and N_2. Each directed edge in the graph is directed from a node in N_1 to a node in N_2

returned to a user's query from a search engine and expands the set by including the web pages that are linked to the pages in the original set. Each document is represented as a triplet of unit vectors (D, F, B). The components D, F and B capture the information about the words contained in the document, the out-links originating at the document and the in-links terminating at the document, respectively. The representation follows the Vector Space Model, mentioned earlier. The document similarity is a weighted sum of the inner products of the individual components. Each disjoint cluster is represented by a vector called 'concept triplet' (like the centroid in k-means). Then, the k-means algorithm is applied to produce the clusters. Finally, Modha & Spangler also provide a scheme for presenting the contents of each cluster to the users by describing various aspects of the cluster.

48.4 Comparison

The choice of the best clustering methods is a tedious problem, firstly, because each method has its advantages and disadvantages, and also because the effectiveness of each method depends on the particular data collection and the application domain (Jain et al., 1999; Steinbach et al., 2000).

There are many studies in the literature that try to evaluate and compare the different clustering methods. Most of them concentrate on the two most widely used approaches to text-based clustering: partitional and HAC algorithms. As mentioned earlier, among the HAC methods, the single link method has the lowest complexity but gives the worst results whereas group average gives the best. In comparison to the partitional methods, the general conclusion is that the partitional algorithms have lower complexities than the HAC, but they dont produce high quality clusters. HAC, on the other hand, are much more effective but their computational requirements forbid them from being used in large document collections (Steinbach et al. 2000; Zhao et Karypis, 2002; Cutting et al., 1992). Indeed, the complexity of the partitional algorithms is linear to the number of documents in the collection, whereas the HAC take at least $O(n^2)$ time. But, as far as the quality of the clustering is concerned, the HAC are ranked higher. This may be due to the fact that the output of the partitional algorithms depends on many parameters (predefined number of clusters, initial cluster centers, criterion function, processing order of documents). Hierarchical algorithms are more efficient in handling noise and outliers. Another advantage of the HAC algorithms is the tree-like structure, which allows the examination of different abstraction levels. Steinbach et al. (2000), on the other hand, compared these two categories of text-based algorithms and drove to slightly different conclusions. They implemented k-means and UPGMA in 8 different test data and found that k-means produces better clusters. According to them, this was because they used an incremental variation of the k-means algorithm and because they run the algorithm many times. When k-means is run more than one times it may give better clusters than the HAC. Finally, a disadvantage of the HAC algorithms, compared to partitional, is that they cannot correct the mistakes in the merges. This leads to the development of hybrid partitional HAC methods, in order to overcome the problems of each method. This is the case with Scatter/Gather (Cutting et al., 1992), where a HAC algorithm (Buckshot or Fractionation) is used to select the initial cluster centers and then an iterative partitional algorithm is used for the refinement of the clusters, and with bisecting k-means (Steinbach et al., 2000), which is a divisive hierarchical algorithm that uses k-means for the division of a cluster in two. Chameleon, on the other hand, is useful when dealing with clusters of arbitrary shapes and sizes. ARHP has the advantage that the hypergraphs can include information about the relationship between more than two documents. Finally, fuzzy approaches can be very useful for representing the human experience and because it is very frequent that a web page deals with more than one topic. The table that follows the reference section presents the main text-based document clustering approaches according to various aspects of their features and functionality, as well as their most important advantages and disadvantages.

The link-based document clustering approaches exploit a very useful source of information: the link structure of the document collection. As mentioned earlier, compared to most text-based approaches, they are developed for use in large, heterogeneous, dynamic and linked collections of web pages. Furthermore, they can include pages that contain pictures, multimedia and other types of data and they overcome problems with the particularities of each language. Although the links can be seen as a recommendation of a page's author to another page, they do not always intend to indicate the similarity. In addition, these algorithms may suffer from poor or dense link structures, in which case no clusters can be found because the algorithm cannot trace dense and sparse regions in the graph. The hybrid document clustering approaches try to use both the content and the links of a web page in order to use as much information as possible for the clustering. It is expected that, as in most cases, the hybrid approaches will be more effective.

48.5 Conclusions and Open Issues

The conclusion derived from the literature review of the document clustering algorithms is that clustering is a very useful technique and an issue that prompts for new solutions in order to deal more efficiently and effectively with the large, heterogeneous and dynamic web page collections. Clustering, of course, is a very complex procedure as it depends on the *collection* on which it is applied as well as the choice of the various *parameter values*. Hence, a careful selection of these is very crucial to the success of the clustering. Furthermore, the development of link-based clustering approaches has proven that the links can be a very useful source of information for the clustering process.

Although there is already much research conducted on the field of web document clustering, it is clear that there are still some open issues that call for more research. These include the achievement of better *quality-complexity tradeoffs*, as well as effort to deal with each methods disadvantages. In addition, another very important issue is *incrementality*, because the web pages change very frequently and because new pages are always added to the web. Also, the fact that very often a web page relates to more than one subject should also be considered and lead to algorithms that allow for *overlapping clusters*. Finally, more attention should also be given to the description of the clusters' contents to the users, the *labelling issue*.

References

Bezdek, J.C., Ehrlich, R., Full, W. *FCM: Fuzzy C-Means Algorithm*. Computers and Geosciences, 1984.

Boley, D., Gini, M., Gross, R., Han, E.H., Hastings, K., Karypis, G., Kumar, V., Mobasher, B., Moore, J. *Partitioning-based clustering for web document categorization*. Decision Support Systems, 27(3):329-341, 1999.

Botafogo, R.A., Shneiderman, B. *Identifying aggregates in hypertext structures*. Proc. 3rd ACM Conference on Hypertext, pp.63-74, 1991.

Botafogo, R.A. *Cluster analysis for hypertext systems.* Proc. ACM SIGIR Conference on Research and Development in Information Retrieval, pp.116- 125, 1993.

Cheeseman, P., Stutz, J. *Bayesian Classification (AutoClass): Theory and Results.* Advances in Knowledge Discovery and Data Mining, AAAI/MIT Press, pp. 153-180, 1996.

Croft, W. B. *Retrieval strategies for hypertext.* Information Processing and Management, 29:313-324, 1993.

Cutting, D.R., Karger, D.R., Pedersen, J.O., Tukey, J.W. *Scatter/Gather: A Cluster-based Approach to Browsing Large Document Collections.* Proc. ACM SIGIR Conference on Research and Development in Information Retrieval, pp.318-329, 1992.

Defays, D. *An efficient algorithm for the complete link method.* The Computer Journal, 20:364-366, 1977.

Dhillon, I.S. *Co-clustering documents and words using Bipartite Spectral Graph Partitioning.* UT CS Technical Report TR2001-05 20, 2001, (http://www.cs.texas.edu/users/inderjit/public_papers/kdd_bipartite.pdf).

Ding, Y. *IR and AI: The role of ontology.* Proc. 4th International Conference of Asian Digital Libraries, Bangalore, India, 2001.

El-Hamdouchi, A., Willett, P. *Hierarchic document clustering using Ward's method.* Proceedings of the Ninth International Conference on Research and Development in Information Retrieval. ACM, Washington, pp.149-156, 1986.

El-Hamdouchi, A., Willett, P. *Comparison of hierarchic agglomerative clustering methods for document retrieval.* The Computer Journal 32, 1989.

Everitt, B. S., Hand, D. J. *Finite Mixture Distributions.* London: Chapman and Hall, 1981.

Frei, H. P., Stieger, D. *The Use of Semantic Links in Hypertext Information Retrieval.* Information Processing and Management, 31(1):1-13, 1995.

Han, E.H., Boley, D., Gini, M., Gross, R., Hastings, K., Karypis, G., Kumar, V., Mobasher, B., Moore, J. *WebACE: a web agent for document categorization and exploration.* Technical Report TR-97-049, Department of Computer Science, University of Minnesota, Minneapolis, 1997, (http://www.users.cs.umn.edu/ karypis/publications/ir.html).

Jain, A.K., Murty, M.N., Flyn, P.J. *Data Clustering: A Review.* ACM Computing Surveys, Vol. 31, No. 2, 1999.

Karypis, G., Han, E.H, Kumar, V. *CHAMELEON: A Hierarchical Clustering Algorithm Using Dynamic Modelling.* IEEE Computer, 32(8):68- 75, 1999.

Karypis, G., Kumar, V. *A fast and highly quality multilevel scheme for partitioning irregular graphs.* SIAM Journal on Scientific Computing, 20(1), 1999.

Kleinberg, J. *Authoritative sources in a hyperlinked environment.* Proc. of the 9th ACM-SIAM Symposium on Discrete Algorithms, 1997.

Kohonen, T. *Self-organizing maps.* Springer-Verlag, Berlin, 1995.

Kumar, S.R., Raghavan, P., Rajagopalan, S., Tomkins, A. *Trawling the Web for Emerging Cyber-Communities.* Proc. 8th WWW Conference, 1999.

Larson, R.R. *Bibliometrics of the World Wide Web: An Exploratory Analysis of the Intellectual Structure of Cyberspace.* Proc. 1996 American Society for Information Science Annual Meeting, 1996.

Looney, C. *A Fuzzy Clustering and Fuzzy Merging Algorithm.* Technical Report, CS-UNR-101-1999, 1999.

Merkl, D. *Text Data Mining.* Dale, R., Moisl, H., Somers, H. (eds.), A handbook of natural language processing: techniques and applications for the processing of language as text, Marcel Dekker, New York

Modha, D., Spangler, W.S. *Clustering hypertext with applications to web searching*. Proc. ACM Conference on Hypertext and Hypermedia, 2000.

Murtagh, F. *A survey of recent advances in hierarchical clustering algorithms*. The Computer Journal, 26:354-359

Page, L., Brin, S., Motwani, R., Winograd, T. *The PageRank citation ranking: Bringing order to the Web*. Technical report, Stanford, 1998, (http://www.stanford.edu/ backrub/pageranksub.ps)

Pirolli, P., Pitkow, J., Rao, R. *Silk from a sow's ear: Extracting usable structures from the Web* Proc. ACM SIGCHI Conference on Human Factors in Computing, 1996.

Rasmussen, E. *Clustering Algorithms*. Information Retrieval, W.B. Frakes & R. Baeza-Yates, Prentice Hall PTR, New Jersey, 1992.

Salton, G., Wang, A., Yang, C. *A vector space model for information retrieval*. Journal of the American Society for Information Science, 18:613–620, 1975.

Sibson, R. *SLINK: an optimally efficient algorithm for the single link cluster method*. The Computer Journal 16:30-34, 1973

Steinbach, M., G. Karypis, G., Kumar, V. *A Comparison of Document Clustering Techniques*. KDD Workshop on Text Mining, 2000.

Strehl, A., Joydeep, G., Mooney, R. *Impact of Similarity Measures on Web-page Clustering*. Proc. 17th National Conference on Artificial Intelligence: Workshop of Artificial Intelligence for Web Search, pp.30-31, 2000.

Van Rijsbergen, C. J. *Information Retrieval*. Butterworths, 1979.

Varlamis, I., Vazirgiannis, M., Halkidi, M., Nguyen, B. *THESUS: Effective Thematic Selection And Organization Of Web Document Collections based on Link Semantics*. To appear in the IEEE Transactions on Knowledge And Data Engineering Journal

Voorhees, E. M. *Implementing agglomerative hierarchic clustering algorithms for use in document retrieval*. Information Processing & Management, 22: 465-476, 1986.

Weiss, R., Velez, B., Sheldon, M., Nemprempre, C., Szilagyi, P., Gifford, D.K. *HyPursuit: A Hierarchical Network Search Engine that Exploits Content-Link Hypertext Clustering*. Proc. Seventh ACM Conference on Hypertext, 1996.

White, D.H., McCain, K.W. *Bibliometrics*. Annual Review of Information Science Technology, 24:119-165, 1989.

Willett, P. *Recent Trends in Hierarchic document Clustering: a critical review*. Information & Management, 24(5):577-597, 1988.

Wu, Z., Palmer, M. *Verb Semantics and Lexical Selection*. 32nd Annual Meetings of the Associations for Computational Linguistics, pp.133-138, 1994.

Zamir, O., Etzioni, O. *Web document clustering: a feasibility demonstration*. Proc. of SIGIR '98, Melbourne, Appendix-Questionnaire, pp.46-54, 1998.

Zhao, Y., Karypis, G. *Criterion Functions for Document Clustering: Experiments and Analysis*. Technical Report 01-40. University of Minnesota, Computer Science Department. Minneapolis, MN, 2001 (http://wwwusers. cs.umn.edu/ karypis/publications/ir.html)

Zhao, Y., Karypis, G. *Evaluation of Hierarchical Clustering Algorithms for Document Datasets*. ACM Press, 16:515-524, 2002.

Name	Complexity Time Space	Input	Output	Similarity Criterion	Type of clusters	Overlap	Handling Outliers	Advantages	Disadvantages
Single linkage	$O(n^2)$ $O(n)$ (Time: $O(nlogn)-O(n^5)$)	Similarity Matrix	Assign documents to clusters, dendrogram	Join clusters with most similar pair of documents	Few, long, ellipsoidal loosely bound, chaining effect	Crisp clusters	No	- Sound theoretical properties - Efficient implementations	- Not suitable for poorly separated clusters - Poor quality
Group Average	$O(n^2)$ $O(n)$	Similarity Matrix	Assign documents to clusters, dendrogram	Average pairwise similarity between all objects in the 2 clusters	Intermediate in tightness between single and complete linkeage	Crisp clusters	No	- High quality results	- Expensive in large collections
Complete linkage	$O(n^3)$ $O(n^2)$ (worst case) in sparse matrix less	Similarity Matrix	Assign documents to clusters, dendrogram	Join clusters with least similar pair of documents	Small, tightly bound	Crisp clusters	No	- Good results (Voorhees alg.)	- Not applicable in large datasets
Ward's Method	$O(n^2)$ $O(n)$	Similarity Matrix	Assign documents to clusters, dendrogram	Join clusters whose merge minimizes the increase in the total error sum of squares	Homogeneous clusters, symmetric hierarchy	Crisp clusters	No	- Good at discovering cluster structure	- Very sensitive to outliers - Poor at recovering elongated clusters
Centroid/ Median HAC	$O(n^2)$ $O(n)$	Similarity Matrix	Assign documents to clusters	Join clusters with most similar centroids/medians	-	Crisp clusters	No		- Small changes may cause large changes in the hierarchy
K-means	$O(nkt)$ $O(n+k)$ (k:initial clusters, t: iterations)	K, iter Feature vector matrix	Assign documents to clusters, refinement of initial clusters	Euclidean or cosine metric	Arbitrary sizes	Crisp clusters	No	- Efficient (no sim matrix required) - Suitable for large databases	- Very sensitive to input parameters

Name	Complexity Time Space	Input	Output	Similarity Criterion	Type of clusters	Overlap	Handling Outliers	Advantages	Disadvantages
Single-Pass	$O(nlogn)$ $O(n)$	Similarity threshold, Feature vector matrix	Assign documents to clusters	If distance to closest centroid > threshold assign, else create new cluster	Large	Crisp clusters	No	- Efficient - Simple	- Results depend on the order of document presentation to the algorithm
Cheme-leon	$O(nm + nlogn + m^2logm)$ m:sub-clusters	k (for knn graph), MINSIZE, scheme for combining RI, RC	Assign documents to clusters, dendrogram	Relative Interconnectivity, Relative Closeness	Natural, homogeneous, arbitrary sizes	Crisp clusters	Yes	- Dynamic modelling	- Very sensitive to parameters - Graph must fit memory - Cannot correct merges
ARHP	$O(n)$ $O(n)$	Apriori, HMETIS parameters, confidence threshold	Assign documents to clusters	Min-cut of hyperedges	-	Crisp clusters	Yes	- Efficient - No centroid / similarity measure	- Sensitive to the choice of Apriori parameters
Fuzzy C-Means	$O(n)$	Initial c prototypes	Membership values for each document (u_{ik})	Minimize $\sum\sum u_{ik}d^2(x_k, u_i)$	hyperspherical, same sizes	Fuzzy clusters	No	- Handles uncertainty - Reflects the human experience	- Sensitive to initial parameters - Poor at recovering clusters with different densities
SOM	$O(k^2n)$ (k: input units)	Weights (m_i)	Topological ordering of input patterns	$m_i(t+1) = m_i(t) + a(t) * h_{ci}(t) * [x(t) - m_i(t)]$	hyperspherical	-	Yes	- Suitable for collections that change frequently	- Fixed number of output nodes limits interpretation of results

Name	**Complexity Time Space**	**Input**	**Output**	**Similarity Criterion**	**Type of clusters**	**Overlap**	**Handling Outliers**	**Advantages**	**Disadvantages**
Scatter/ Gather	Buckshot: $O(kn)$ Fractionation: $O(nm)$	k: number of clusters	Assign documents to clusters with short summary	Hybrid:first partitional then HAC	-	Crisp clusters	No	- Dynamic clustering - Clusters presented with summaries - Fast	- Must have a very quick clustering algorithm - Focus on speed but not on accuracy
Suffix Tree Clustering	$O(n)$	Similarity threshold for the merge of the base clusters	Assign documents to clusters	- Sim = 1 if $\lvert B_m \cap B_n\rvert/\lvert B_m\rvert >$ threshold and $\lvert B_m \cap B_n\rvert/\lvert B_n\rvert >$ threshold, else - Sim = 0	-	Fuzzy clusters	No	- Incremental - Captures the word sequence	- Snippets usually introduce noise - Snippets may not be a good description of a web page

49

Causal Discovery

Hong Yao[1], Cory J. Butz[1], and Howard J. Hamilton[1]

Department of Computer Science, University of Regina
Regina, SK, S4S 0A2, Canada
{yao2hong, butz,hamilton}@cs.uregina.ca

Summary. Many algorithms have been proposed for learning a causal network from data. It has been shown, however, that learning all the conditional independencies in a probability distribution is a NP-hard problem. In this chapter, we present an alternative method for learning a causal network from data. Our approach is novel in that it learns functional dependencies in the sample distribution rather than probabilistic independencies. Our method is based on the fact that functional dependency logically implies probabilistic conditional independency. The effectiveness of the proposed approach is explicitly demonstrated using fifteen real-world datasets.

Key words: Causal networks, functional dependency, conditional independency

49.1 Introduction

Causal networks (CNs) (Pearl, 1988) have been successfully established as a framework for uncertainty reasoning. A CN is a *directed acyclic graph* (DAG) together with a corresponding set of *conditional probability distributions* (CPDs). Each node in the DAG represents a variable of interest, while an edge can be interpreted as direct casual influence. CNs facilitate knowledge acquisition as the *conditional independencies* (CIs) (Wong *et al.*, 2000) encoded in the DAG indicate that the product of the CPDs is a joint probability distribution.

Numerous algorithms have been proposed for learning a CN from data (Neapolitan, 2003). Developing a method for learning a CN from data is tantamount to obtaining an effective graphical representation of the CIs holding in the data. It has been shown, however, that discovering all the CIs in a probability distribution is a NP-hard problem (Bouckaert, 1994). In addition, choosing an initial DAG is important for reducing the search space, as many learning algorithms use greedy search techniques.

In this chapter, we present a method, called FD2CN, for learning a CN from data using *functional dependencies* (FDs) (Maier, 1983). We have recently developed a method for learning FDs from data (Yao *et al.*, 2002). Learning FDs from data is useful, since it has been

O. Maimon, L. Rokach (eds.), *Data Mining and Knowledge Discovery Handbook*, 2nd ed.,
DOI 10.1007/978-0-387-09823-4_49,

shown that FD logically implies CI (Butz *et al.*, 1999). We show how to combine the obtained FDs with the chain rule of probability to construct a DAG of a CN. Given a set of FDs obtained from data, an ordering of variables is obtained such that the Markov boundaries of some variables are determined. Representing joint probability distribution of variables in the resulting ordering by the chain rule, the Markov boundaries of the other variables are determined. A DAG of a CN is constructed by designating each Markov boundary of variable as the parent set of the variable. During this process, we take full advantage of known results in both CNs (Pearl, 1988) and relational databases (Maier, 1983). We demonstrate the effectiveness of our approach using fifteen real-world datasets. The DAG constructed in our approach can also be used as an initial DAG for previous approaches. The work here further illustrates the intrinsic relationship between CNs and relational databases (Wong *et al.*, 2000, Wong and Butz, 2001).

The remainder of this chapter is organized as follows. Background knowledge is given in Section 49.2. In Section 49.3, the theoretical foundation of our approach is provided. The algorithm to construct a CN is developed in Section 49.4. In Section 49.5, the experimental results are presented. Conclusions are drawn in Section 49.6.

49.2 Background Knowledge

Let U be a finite set of discrete variables, each with a finite domain. Let V be the Cartesian product of the variable domains. A *joint probability distribution* (Pearl, 1988) $p(U)$ is a function p on V such that $0 \leq p(v) \leq 1$ for each configuration $v \in V$ and $\sum_{v \in V} p(v) = 1.0$. The *marginal distribution* $p(X)$ for $X \subseteq U$ is defined as $\sum_{U-X} p(U)$. If $p(X) > 0$, then the *conditional probability distribution* $p(Y|X)$ for $X,Y \subseteq U$ is defined as $p(XY)/p(X)$. In this chapter, we may write a_i for the singleton set $\{a_i\}$, and we use the terms attribute and variable interchangeably. Similarly, for the terms tuple and configuration.

Definition 1. A *causal network* (CN) is a *directed acyclic graph* (DAG) $\mathscr{D}$ together with a *conditional probability distribution* (CPD) $p(a_i|P_i)$ for each variable a_i in $\mathscr{D}$, where P_i denotes the parent set of a_i in $\mathscr{D}$.

The DAG $\mathscr{D}$ graphically encodes CIs regarding the variables in U.

Definition 2. *(Wong et al., 2000).* Let X, Y, and Z be three disjoint sets of variables. X is said to be *conditionally independent* of Y given Z, denoted $I(X,Z,Y)$, if $p(X|Y,Z) = p(X|Z)$.

As previously mentioned, the CIs encoded in the DAG $\mathscr{D}$ indicate that the product of the given CPDs is a joint probability distribution $p(U)$.

Example 1. One CN on the set $U = \{a_1,a_2,a_3,a_4,a_5,a_6\}$ is the DAG in Figure 49.1(i) together with the CPDs $p(a_1)$, $p(a_2|a_1)$, $p(a_3|a_1)$, $p(a_4|a_2)$, $p(a_5|a_3)$, and $p(a_6|a_4,a_5)$. This DAG encodes, in particular, $I(a_3,a_1,a_2)$, $I(a_4,a_2,a_1a_3)$, $I(a_5,a_3,a_1a_2a_4)$ and $I(a_6,a_4a_5,a_1a_2a_3)$. By the chain rule, the joint probability distribution $p(U)$ can be expressed as:

$$p(U) = p(a_1)p(a_2|a_1)p(a_3|a_1,a_2)p(a_4|a_1,a_2,a_3)p(a_5|a_1,a_2,a_3,a_4)$$
$$p(a_6|a_1,a_2,a_3,a_4,a_5).$$

The above CIs can be used to rewrite $p(U)$ as:

$$p(U) = p(a_1)p(a_2|a_1)p(a_3|a_1)p(a_4|a_2)p(a_5|a_3)p(a_6|a_4,a_5).$$

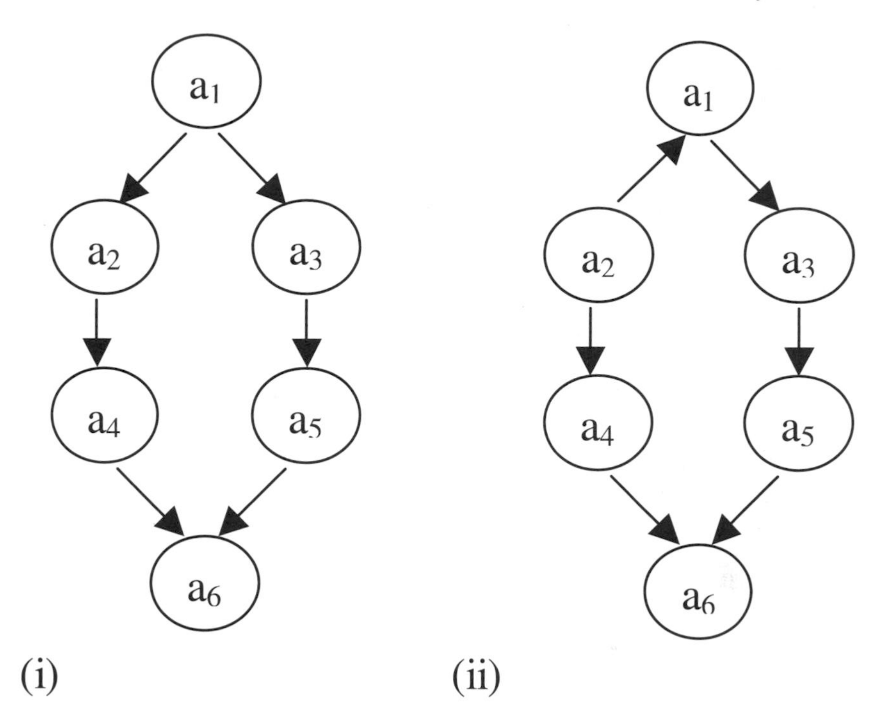

Fig. 49.1. Two causal networks.

The term CN is somewhat misleading because it may be possible to reverse a directed edge without disturbing the encoded CI information. For example, the two CNs in Figure 49.1 encode the same independency information. Thus, it is perhaps better to view a CN as encoding independency information rather than causal relationships.

Using CI information encoded in the DAG, the Markov boundary of a variable can be defined.

Definition 3. Let $U = \{a_1, \ldots, a_n\}$, and $\mathscr{O}$ be an ordering $\langle a_1, \ldots, a_n \rangle$ of variables of U. Let $U_i = \{a_1, \ldots, a_{i-1}\}$ be a subset of U with respect to the ordering $\mathscr{O}$. A *Markov boundary* of a variable a_i over U_i, denoted B_i, is any subset X of U_i such that $p(U_i)$ satisfies $I(a_i, X, U_i - X - a_i)$ and $a_i \notin X$, but $p(U_i)$ does not satisfy $I(a_i, X', U_i - X' - a_i)$ for any $X' \subset X$.

Example 2. Recall the DAG in Figure 49.1(i). Let the ordering $\mathscr{O}$ be $\langle a_1, a_2, a_3, a_4, a_5, a_6 \rangle$. Then $U_1 = \{\}$, $U_2 = \{a_1\}$, $U_3 = \{a_1, a_2\}$, $U_4 = \{a_1, a_2, a_3\}$, $U_5 = \{a_1, a_2, a_3, a_4\}$, and $U_6 = \{a_1, a_2, a_3, a_4, a_5\}$. The Markov boundary of variable a_3 over U_3 is $B_3 = \{a_1\}$, since $p(U_3)$ satisfies $I(a_3,\ a_1,\ a_2)$, but does not satisfy $I(a_3, \phi, a_1 a_2)$.

The Markov boundary B_i of a variable a_i over U_i encodes CI $I(a_i, B_i, U_i - B_i - a_i)$ over $p(U_i)$. Using the Markov boundary of each variable, a boundary DAG is defined as follows.

Definition 4. Let $p(U)$ be a joint probability distribution over U, $\mathscr{O}$ be an ordering $\langle a_1,\ldots,a_n\rangle$ of variables of U, and $U_i = \{a_1,\ldots,a_{i-1}\}$ be a subset of U with respect to the ordering $\mathscr{O}$. $\{B_1,\ldots,B_n\}$ is an ordered set of subsets of U such that each B_i is a Markov boundary of a_i over the U_i. The DAG created by designating each B_i as a parent set of variable a_i is called a *boundary DAG* of $p(U)$ relative to $\mathscr{O}$.

The next theorem (Pearl, 1988) indicates that the boundary DAG of $p(U)$ relative to an ordering $\mathscr{O}$ is a DAG of CN of $p(U)$.

Theorem 1. Let $p(U)$ be a joint probability distribution and $\mathscr{O}$ be an ordering of variables of U. If $\mathscr{D}$ is a boundary DAG of $p(U)$ relative to $\mathscr{O}$, then $\mathscr{D}$ is a DAG of CN of $p(U)$.

It is important to realize that we can construct a DAG of a CN after the Markov boundary of each variable has been obtained according to Definition 4 and Theorem 1.

Example 3. Let $U = \{a_1,\ldots,a_6\}$ *and* $\mathscr{O} = \langle a_2,a_1,a_3,a_4,a_5,a_6\rangle$ *be an ordering of the variables of* U. *With respect to the ordering* $\mathscr{O}$, $U_1 = \{a_2\}$, $U_2 = \{\}$, $U_3 = \{a_1,a_2\}$, $U_4 = \{a_1,a_2,a_3\}$, $U_5 = \{a_1,a_2,a_3,a_4\}$, *and* $U_6 = \{a_1,a_2,a_3,a_4,a_5\}$. *Supposing we assign* $B_1 = \{a_2\}$, $B_2 = \{\}$, $B_3 = \{a_1\}$, $B_4 = \{a_2\}$, $B_5 = \{a_3\}$ *and* $B_6 = \{a_4,a_5\}$, *then the DAG shown in Figure 49.1 (ii) is the learned DAG of CN of* $p(U)$.

49.3 Theoretical Foundation

In this section, several theorems relevant to our approach are provided. We define a relation $r(U)$ as a finite set of tuples over U. We begin with functional dependency (Maier, 1983).

Definition 5. Let $r(U)$ be a relation over U and $X,Y \subseteq U$. The *functional dependency* (FD) $X \to Y$ is satisfied by $r(U)$ if every two tuples t_1 and t_2 of $r(U)$ that agree on X also agree on Y.

If a relation $r(U)$ satisfies the FD $X \to Y$, but not $X' \to Y$ for every $X' \subset X$, then $X \to Y$ is called *left-reduced* (Maier, 1983).

The next theorem shows that FD logically implies CI.

Theorem 2. (Butz *et al.*, 1999). Let $r(U)$ be a relation over U. Let $p(U)$ be a joint distribution over $r(U)$. Let $X,Y \subseteq U$ and $Z = U - XY$. Having FD $X \to Y$ satisfied by $r(U)$ is a sufficient condition for the CI $I(Y,X,Z)$ to be satisfied by $p(U)$.

By exploiting the implication relationship between functional dependency and conditional independency, we can relate a left-reduced FD $X \to a_i$ to the Markov boundary of variable a_i.

Theorem 3. Let $U = \{a_1,\ldots,a_n\}$, U_i be a subset of U, and $X \subseteq U_i$. If FD $X \to a_i$ is left-reduced, then X is the Markov boundary of variable a_i over U_i.

Proof: Since $X \to a_i$ is a FD and $X \subseteq U_i$, according to Definition 5, $X \to a_i$ holds over $U_i \cup \{a_i\}$. Since FD $X \to a_i$ is left-reduced, by Theorem 2, $p(U_i)$ satisfies $I(a_i,X,U_i - X - a_i)$ but not $I(a_i,X',U_i - X' - a_i)$ for any $X' \subset X$.

Theorem 3 indicates that the Markov boundary of variable a_i over U_i can be learned from a left-reduced FD $X \to a_i$, if X is a subset of U_i. We define those variables that can be learned from a set of left-reduced FDs as follows.

Definition 6. Let $U = \{a_1, \ldots, a_n\}$, $X \subset U$, $a_i \notin X$, and F be a set of left-reduced FDs over U. If exists $X \to a_i \in F$, then a_i is a *decided* variable. Otherwise, a_i is an *undecided* variable.

Considering a variable as decided indicates that its Markov boundary can be learned from FDs implied in data.

49.4 Learning a DAG of CN by FDs

In this section, we use learned FDs to construct a CN. We illustrate our algorithm using the heart disease dataset, which contains 13 attributes and 230 rows, from the UCI Machine Learning Repository (Blake and Merz, 1998).

Example 4. The heart disease dataset has $U = \{a_1, \ldots, a_{13}\}$. Using FD_Mine, the discovered set of left-reduced FDs is $F = \{a_1a_5 \to a_3$, $a_1a_5 \to a_6$, $a_1a_5 \to a_{11}$, $a_1a_5 \to a_{13}$, $a_1a_8 \to a_7$, $a_4a_5a_9 \to a_2$, $a_1a_5a_{10} \to a_4$, $a_1a_2a_5 \to a_8$, $a_1a_5a_{10} \to a_9$, $a_1a_5a_{10} \to a_{12}\}$.

As indicated by Theorem 1, a DAG of a CN is constructed when the Markov boundary of each variable relative to an ordering $\mathscr{O}$ is obtained. We obtain the Markov boundary of each variable in two steps. First, in Section 49.4.1, we show how to obtain an ordering $\mathscr{O}$ such that the Markov boundary of each decided variable with respect to $\mathscr{O}$ can be obtained from the given FDs. Second, we determine the Markov boundary of each undecided variable by the chain rule in Section 49.4.2.

49.4.1 Learning an Ordering of Variables from FDs

Given a set F of left-reduced FDs, the algorithm in Figure 49.4.1 will determine an ordering $\mathscr{O}$ of the variables of U such that the Markov boundary of each decided variable with respect to $\mathscr{O}$ can be obtained from F.

We use the FDs in Example 4 to demonstrate how Algorithm 1 works.

Example 5. First, the FD $a_1a_5 \to a_3$ is selected in line 2 by Algorithm 1. In line 3, since a_3 is not in Y for any $Y \to a_i$ $(a_i \neq a_3)$ in F, variable a_3 is removed for U in line 4, thus $U = \{a_1, a_2, a_4, \ldots, a_{13}\}$. We obtain $\mathscr{O} = \langle a_3 \rangle$ in line 5. In line 6, FD $a_1a_5 \to a_3$ is removed from F. Because F is not empty, Algorithm 1 continues performing lines $2-8$. At this time, the FD $a_1a_5 \to a_6$ is selected in line 2. Variable a_6 is removed from U in line 4, $U = \{a_1, a_2, a_4, a_5, a_7, \ldots, a_{13}\}$. We obtain $\mathscr{O} = \langle a_6, a_3 \rangle$ in line 5. By recursively performing lines 2 to 8 until F is empty, we obtain an ordering $\mathscr{O} = \langle a_{12}, a_9, a_4, a_2, a_8, a_7, a_{13}, a_{11}, a_6, a_3 \rangle$ and $U = \{a_1, a_5, a_{10}\}$. In line 9, we prepend U to the head of $\mathscr{O}$. Thus, for the variables in U, $\mathscr{O} = \langle a_1, a_5, a_{10}, a_{12}, a_9, a_4, a_2, a_8, a_7, a_{13}, a_{11}, a_6, a_3 \rangle$.

The next theorem guarantees that the Markov boundary of each decided variable is determined by $\mathscr{O}$ as obtained by Algorithm 1.

Theorem 4. Let $U = \{a_1, \ldots, a_n\}$, $\mathscr{O} = \langle a_1, \ldots, a_n \rangle$ be an ordering obtained by the Algorithm 1, and $U_i = \{a_1, \ldots, a_{i-1}\}$ be a subset of U with respect to $\mathscr{O}$. Let B_i be the Markov boundary of a_i over U_i. If a_i is a decided variable, then $B_i \subseteq U_i$.

Proof: Since a_i is a decided variable, then there exists a FD $X \to a_i$. When Algorithm 1 performs lines 2-8, then a_i is always deleted from U before any variable of X according to lines 3-5 of Algorithm 1. Thus, for any $a_j \in X$, a_j is always before a_i in $\mathscr{O}$, so we have $X \subseteq U_i$. Since $B_i = X$, according to Theorem 3, then $B_i \subseteq U_i$.

Algorithm 1.
Input: $U = \{a_1, \ldots, a_n\}$, and a set F of left-reduced FDs.
Output: an ordered list $\mathcal{O}$ of the variables of U.
Begin
1. $\mathcal{O} = \langle\rangle$.
2. for each $X \rightarrow a_i \in F$
3. if $a_i \notin Y$ for all $Y \rightarrow y \in F$ $(y \neq a_i)$
4. $U = U - \{a_i\}$.
5. prepend a_i to the head of $\mathcal{O}$.
6. $F = F - \{Y \rightarrow a_i | Y \rightarrow a_i \in F\}$.
7. end if
8. end for
9. prepend U to the head of $\mathcal{O}$.
10. return($\mathcal{O}$)
End

Fig. 49.2. An algorithm to obtain an ordering $\mathcal{O}$ of the variables of U.

49.4.2 Learning the Markov Boundaries of Undecided Variables

Once an ordering $\mathcal{O} = \langle a_1, \ldots, a_n \rangle$ of variables of U is obtained by Algorithm 1, the joint probability distribution $p(U)$ can be expressed using the chain rule as follows.

$$p(U) = p(a_1) \ldots p(a_j | a_1, \ldots, a_{j-1}) \ldots p(a_n | a_1, \ldots, a_{n-1}). \quad (49.1)$$

If a_i is an undecided variable, then there is no FD $X \rightarrow a_i \in F$. According to Algorithm 1, a_i is not deleted from U and is prepended to the head of $\mathcal{O}$ at the line 9 of Algorithm 1. This indicates that all undecided variables appear before all decided variables in $\mathcal{O}$.

Suppose $a_1, \ldots, a_j$ are all the undecided variables, and $B_{j+1}, \ldots, B_n$ are the Markov boundaries of all the decided variables. By Definition 3, CI $I(a_i, B_i, U_i - B_i - a_i)$ holds for each variable a_i, $j+1 \leq i \leq n$. Thus, each $p(a_i | a_1, \ldots, a_{i-1}) = p(a_i | B_i)$. Equation 63.1 can be rewritten as:

$$p(U) = p(a_1) \ldots p(a_j | a_1, \ldots, a_{j-1}) p(a_{j+1} | B_{j+1}) \ldots p(a_n | B_n). \quad (49.2)$$

By assigning $B_k = \{a_1, \ldots, a_{k-1}\}$ as the Markov boundary of each undecided variable a_k, $1 \leq k \leq j$, Equation 63.2 can be expressed as:

$$p(U) = p(a_1 | B_1) \ldots p(a_j | B_j) \ldots p(a_n | B_n) = \prod_{a_i \in U} p(a_i | B_i). \quad (49.3)$$

Equation 63.5 indicated that a joint probability distribution $p(U)$ can be represented by the Markov boundary of all variables of U. Thus, a boundary DAG relative to $\mathcal{D}$ can be constructed. Based on the above analysis, we developed the algorithm shown in Figure 49.4.2, called FD2CN, which learns a DAG $\mathcal{D}$ of a CN from a dataset $r(U)$.

Algorithm 2. FD2CN
Input: A dataset $r(U)$ over variable set U.
Output: A DAG $\mathscr{D}\langle U,E\rangle$ of a CN learned from $r(U)$.
Begin
1. F= FD_Mine($r(U)$). //return a set of left-reduced FDs
2. Obtaining an ordering $\mathscr{O}$ using Algorithm 1.
3. $U_i = \{\}$.
4. while $\mathscr{O}$ is not empty
5. a_i= pophead($\mathscr{O}$).
6. if there exists a FD $X \to a_i \in F$ then
7. $B_i = X$
8. else
9. $B_i = U_i$.
10. $U_i = U_i \cup \{a_i\}$.
11. end while
12. Constructing a DAG $\mathscr{D}\langle U,E\rangle$ such that $\{(b,a_i) \in E | b \in B_i,\ a_i, b \in U\}$.
End

Fig. 49.3. The algorithm, FD2CN, to learn a DAG of a CN from data.

In line 6 of Algorithm FD2BN, we determined whether or not a variable is decided. If so, in the line 7, we obtain its Markov boundary. If not, in line 9, we obtain its Markov boundary. In the line 12, a DAG of a CN is constructed by making each B_i as the parent set of a_i of U. In other words, if $b_i \in B_i$, then we add a edge $b \to a_i$ in DAG $\mathscr{D}$. According to Definition 4 and Theorem 1, we know the constructed DAG $\mathscr{D}$ is a DAG of a CN.

Example 6. Applying Algorithm FD2CN on the heart disease dataset, F in line 1 is obtained as in Example 4. $\mathscr{O}$ in line 2 is obtained as in Example 5. According to obtained $\mathscr{O}$, we obtain $B_1 = \{\}$, $B_5 = \{a_1\}$, $B_{10} = \{a_1, a_5\}$, $B_{12} = \{a_1, a_5, a_{10}\}$, $B_9 = \{a_1, a_5, a_{10}\}$, $B_4 = \{a_1, a_5, a_{10}\}$, $B_2 = \{a_4, a_5, a_9\}$, $B_8 = \{a_1, a_5, a_2\}$, $B_7 = \{a_1, a_8\}$, $B_{13} = \{a_1, a_5\}$, $B_{11} = \{a_1, a_5\}$, $B_6 = \{a_1, a_5\}$, and $B_3 = \{a_1, a_5\}$. By making each B_i as the parent set of a_i of U, a $\mathscr{D}$ is constructed. For example, since $B_2 = \{a_4, a_5, a_9\}$, then $\mathscr{D}$ have edges $a_4 \to a_2$, $a_5 \to a_2$, and $a_9 \to a_2$. the DAG of a CN learned from the heart disease dataset in Figure 49.5 is depicted in Figure 49.4.

49.5 Experimental Results

Experiments were carried out on fifteen real-world datasets obtained from the UCI Machine Learning Repositories (Blake and Merz, 1998). The results are shown in Figure 49.5. The last column gives the elapsed time to construct a CN, measured on a 1GHz Pentium III PC with 256 MB RAM. The results show that the processing time is mainly determined by the

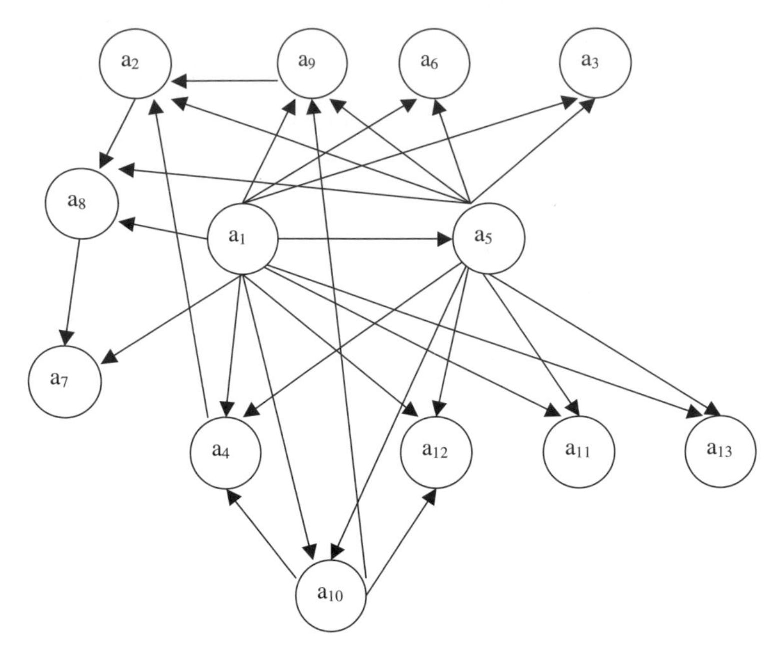

Fig. 49.4. The learned DAG of a CN from the heart disease dataset.

number of attributes. Since it also indicates that many FDs hold in some datasets, our proposed approach is a feasible way to learn a CN.

Dataset Name	# of attributes	# of rows	# of FDs	Time (seconds)
Abalone	8	4,177	60	1
Breast-cancer	10	191	3	0
Bridge	13	108	62	2
Cancer-Wisconsin	10	699	19	1
Chess	7	28,056	1	3
Crx	16	690	1,099	10
Echocardiogram	13	132	583	0
Glass	10	142	119	0
Heart disease	13	270	10	2
Hepatitis	20	155	8,250	1,327
Imports-85	26	205	4,176	8,322
Iris	5	150	4	0
Led	8	50	11	0
Nursery	9	12,960	1	16
Pendigits	17	7,494	29,934	920

Fig. 49.5. Experimental results using fifteen real-world datasets.

49.6 Conclusion

In this chapter, we presented a novel method for learning a CN. Although a CN encodes probabilistic conditional independencies, our method is based on learning FDs (Yao *et al.*, 2002). Since functional dependency logically implies conditional independency (Butz *et al.*, 1999), we described how to construct a CN from data dependencies. We implemented our approach and encouraging experimental results have been obtained.

Since functional dependency is a special case of conditional independency, we acknowledge that our approach may not utilize all the independency information encoded in the sample data. However, previous methods also suffer from this disadvantage as learning all CIs from sample data is a NP-hard problem (Bouckaert, 1994).

References

Bouckaert, R. (1994). Properties of learning algorithms for Bayesian belief networks. In *Proceedings of the 10th Conference on Uncertainty in Artificial Intelligence*, 102–109.

Butz, C. J., Wong, S. K. M., and Yao, Y. Y. (1999) On Data and Probabilistic Dependencies, *IEEE Canadian Conference on Electrical and Computer Engineering*, 1692-1697.

Maier, D. (1983). *The Theory of Relational Databases*, Computer Science Press.

Neapolitan, R. E. (2003) *Learning Bayesian Networks*, Prentice Hall.

Pearl, J. (1988) *Probabilistic Reasoning in Intelligent Systems: Networks of Plausible Inference*, Morgan Kaufmann Publishers.

Blake, C.L. and Merz, C.J. (1998). UCI Repository of machine learning databases. Irvine, CA: University of California, Department of Information and Computer Science.

Wong, S. K. M., Butz, C. J., and Wu, D. (2000). On the Implication Problem for Probabilistic Conditional Independency, *IEEE Transactions on Systems, Man, and Cybernetics, Part A: Systems and Humans*, 30(6), 785-805.

Wong, S. K. M., and Butz, C. J. (2001), Constructing the Dependency Structure of a Multi-Agent Probabilistic Network, *IEEE Transactions on Knowledge and Data Engineering*, 13(3), 395-415.

Yao, H., Hamilton, H. J., and Butz, C. J. (2002). FD_Mine: Discovering Functional Dependencies in a Database Using Equivalences, In *Proceedings of the Second IEEE International Conference on Data Mining*, 729-732.

50

Ensemble Methods in Supervised Learning

Lior Rokach

Department of Information Systems Engineering
Ben-Gurion University of the Negev
liorrk@bgu.ac.il

Summary. The idea of ensemble methodology is to build a predictive model by integrating multiple models. It is well-known that ensemble methods can be used for improving prediction performance. In this chapter we provide an overview of ensemble methods in classification tasks. We present all important types of ensemble methods including boosting and bagging. Combining methods and modeling issues such as ensemble diversity and ensemble size are discussed.

Key words: Ensemble, Boosting, AdaBoost, Windowing, Bagging, Grading, Arbiter Tree, Combiner Tree

50.1 Introduction

The main idea of ensemble methodology is to combine a set of models, each of which solves the same original task, in order to obtain a better composite global model, with more accurate and reliable estimates or decisions than can be obtained from using a single model. The idea of building a predictive model by integrating multiple models has been under investigation for a long time. Bühlmann and Yu (2003) pointed out that the history of ensemble methods starts as early as 1977 with Tukeys Twicing, an ensemble of two linear regression models. Ensemble methods can be also used for improving the quality and robustness of clustering algorithms (Dimitriadou *et al.*, 2003). Nevertheless, in this chapter we focus on classifier ensembles.

In the past few years, experimental studies conducted by the machine-learning community show that combining the outputs of multiple classifiers reduces the generalization error (Domingos, 1996, Quinlan, 1996, Bauer and Kohavi, 1999, Opitz and Maclin, 1999). Ensemble methods are very effective, mainly due to the phenomenon that various types of classifiers have different "inductive biases" (Geman *et al.*, 1995, Mitchell, 1997). Indeed, ensemble methods can effectively make use of such diversity to reduce the variance-error (Tumer and Ghosh, 1999, Ali and Pazzani, 1996) without increasing the bias-error. In certain situations, an ensemble can also reduce bias-error, as shown by the theory of large margin classifiers (Bartlett and Shawe-Taylor, 1998).

O. Maimon, L. Rokach (eds.), *Data Mining and Knowledge Discovery Handbook*, 2nd ed.,
DOI 10.1007/978-0-387-09823-4_50,

The ensemble methodology is applicable in many fields such as: finance (Leigh *et al.*, 2002), bioinformatics (Tan *et al.*, 2003), healthcare (Mangiameli *et al.*, 2004), manufacturing (Maimon and Rokach, 2004), geography (Bruzzone *et al.*, 2004) etc.

Given the potential usefulness of ensemble methods, it is not surprising that a vast number of methods is now available to researchers and practitioners. This chapter aims to organize all significant methods developed in this field into a coherent and unified catalog. There are several factors that differentiate between the various ensembles methods. The main factors are:

1. Inter-classifiers relationship — How does each classifier affect the other classifiers? The ensemble methods can be divided into two main types: sequential and concurrent.
2. Combining method — The strategy of combining the classifiers generated by an induction algorithm. The simplest combiner determines the output solely from the outputs of the individual inducers. Ali and Pazzani (1996) have compared several combination methods: uniform voting, Bayesian combination, distribution summation and likelihood combination. Moreover, theoretical analysis has been developed for estimating the classification improvement (Tumer and Ghosh, 1999). Along with simple combiners there are other more sophisticated methods, such as stacking (Wolpert, 1992) and arbitration (Chan and Stolfo, 1995).
3. Diversity generator — In order to make the ensemble efficient, there should be some sort of diversity between the classifiers. Diversity may be obtained through different presentations of the input data, as in bagging, variations in learner design, or by adding a penalty to the outputs to encourage diversity.
4. Ensemble size — The number of classifiers in the ensemble.

The following sections discuss and describe each one of these factors.

50.2 Sequential Methodology

In sequential approaches for learning ensembles, there is an interaction between the learning runs. Thus it is possible to take advantage of knowledge generated in previous iterations to guide the learning in the next iterations. We distinguish between two main approaches for sequential learning, as described in the following sections (Provost and Kolluri, 1997).

50.2.1 Model-guided Instance Selection

In this sequential approach, the classifiers that were constructed in previous iterations are used for manipulating the training set for the following iteration. One can embed this process within the basic learning algorithm. These methods, which are also known as constructive or conservative methods, usually ignore all data instances on which their initial classifier is correct and only learn from misclassified instances.

The following sections describe several methods which embed the sample selection at each run of the learning algorithm.

Uncertainty Sampling

This method is useful in scenarios where unlabeled data is plentiful and the labeling process is expensive. We can define uncertainty sampling as an iterative process of manual labeling of examples, classifier fitting from those examples, and the use of the classifier to select new examples whose class membership is unclear (Lewis and Gale, 1994). A teacher or an expert is asked to label unlabeled instances whose class membership is uncertain. The pseudo-code is described in Figure 50.1.

Input: I (a method for building the classifier), b (the selected bulk size), U (a set on unlabled instances), E (an Expert capable to label instances)
Output: C
1: $X_{new} \leftarrow Random\, set\, of\, size\, b\, selected\, from\, U$
2: $Y_{new} \leftarrow E(X_{new})$
3: $S \leftarrow (X_{new}, Y_{new})$
4: $C \leftarrow I(S)$
5: $U \leftarrow U - X_{new}$
6: **while** E is willing to label instances **do**
7: $X_{new} \leftarrow$ Select a subset of U of size b such that C is least certain of its classification.
8: $Y_{new} \leftarrow E(X_{new})$
9: $S \leftarrow S \cup (X_{new}, Y_{new})$
10: $C \leftarrow I(S)$
11: $U \leftarrow U - X_{new}$
12: **end while**

Fig. 50.1. Pseudo-Code for Uncertainty Sampling.

It has been shown that using uncertainty sampling method in text categorization tasks can reduce by a factor of up to 500 the amount of data that had to be labeled to obtain a given accuracy level (Lewis and Gale, 1994).

Simple uncertainty sampling requires the construction of many classifiers. The necessity of a cheap classifier now emerges. The cheap classifier selects instances "in the loop" and then uses those instances for training another, more expensive inducer. The *Heterogeneous Uncertainty Sampling* method achieves a given error rate by using a cheaper kind of classifier (both to build and run) which leads to reducted computational cost and run time (Lewis and Catlett, 1994).

Unfortunately, an uncertainty sampling tends to create a training set that contains a disproportionately large number of instances from rare classes. In order to balance this effect, a modified version of a C4.5 decision tree was developed (Lewis and Catlett, 1994). This algorithm accepts a parameter called loss ratio (LR). The parameter specifies the relative cost of two types of errors: false positives (where negative instance is classified positive) and false negatives (where positive instance is classified negative). Choosing a loss ratio greater than 1 indicates that false positives errors are more costly than the false negative. Therefore, setting the LR above 1 will counterbalance the over-representation of positive instances. Choosing the exact value of LR requires sensitivity analysis of the effect of the specific value on the accuracy of the classifier produced.

The original C4.5 determines the class value in the leaves by checking whether the split decreases the error rate. The final class value is determined by majority vote. In a modified C4.5, the leaf's class is determined by comparison with a probability threshold of LR/(LR+1) (or its appropriate reciprocal). Lewis and Catlett (1994) show that their method leads to significantly higher accuracy than in the case of using random samples ten times larger.

Boosting

Boosting (also known as arcing — Adaptive Resampling and Combining) is a general method for improving the performance of any learning algorithm. The method works by repeatedly running a weak learner (such as classification rules or decision trees), on various distributed training data. The classifiers produced by the weak learners are then combined into a single composite strong classifier in order to achieve a higher accuracy than the weak learner's classifiers would have had.

Schapire introduced the first boosting algorithm in 1990. In 1995 Freund and Schapire introduced the AdaBoost algorithm. The main idea of this algorithm is to assign a weight in each example in the training set. In the beginning, all weights are equal, but in every round, the weights of all misclassified instances are increased while the weights of correctly classified instances are decreased. As a consequence, the weak learner is forced to focus on the difficult instances of the training set. This procedure provides a series of classifiers that complement one another.

The pseudo-code of the AdaBoost algorithm is described in Figure 50.2. The algorithm assumes that the training set consists of m instances, labeled as -1 or +1. The classification of a new instance is made by voting on all classifiers $\{C_t\}$, each having a weight of α_t. Mathematically, it can be written as:

$$H(x) = sign(\sum_{t=1}^{T} \alpha_t \cdot C_t(x))$$

Input: I (a weak inducer), T (the number of iterations), S (training set)
Output: $C_t, \alpha_t; t = 1, \ldots, T$
1: $t \leftarrow 1$
2: $D_1(i) \leftarrow 1/m; i = 1, \ldots, m$
3: **repeat**
4: Build Classifier C_t using I and distribution D_t
5: $\varepsilon_t \leftarrow \sum_{i:C_t(x_i) \neq y_i} D_t(i)$
6: **if** $\varepsilon_t > 0.5$ **then**
7: $T \leftarrow t - 1$
8: exit Loop.
9: **end if**
10: $\alpha_t \leftarrow \frac{1}{2} \ln(\frac{1-\varepsilon_t}{\varepsilon_t})$
11: $D_{t+1}(i) = D_t(i) \cdot e^{-\alpha_t y_t C_t(x_i)}$
12: Normalize D_{t+1} to be a proper distribution.
13: $t{+}{+}$
14: **until** $t > T$

Fig. 50.2. The AdaBoost Algorithm.

The basic AdaBoost algorithm! described in Figure 50.2, deals with binary classification. Freund and Schapire (1996) describe two versions of the AdaBoost algorithm (AdaBoost.M1, AdaBoost.M2), which are equivalent for binary classification and differ in their handling of multiclass classification problems. Figure 50.3 describes the pseudo-code of AdaBoost.M1. The classification of a new instance is performed according to the following equation:

$$H(x) = \underset{y \in dom(y)}{\arg\max} \left(\sum_{t:C_t(x)=y} \log \frac{1}{\beta_t} \right)$$

Input: I (a weak inducer), T (the number of iterations), S (the training set)
Output: $C_t, \beta_t; t = 1, \ldots, T$

1: $t \leftarrow 1$
2: $D_1(i) \leftarrow 1/m; i = 1, \ldots, m$
3: **repeat**
4: Build Classifier C_t using I and distribution D_t
5: $\varepsilon_t \leftarrow \sum_{i:C_t(x_i) \neq y_i} D_t(i)$
6: **if** $\varepsilon_t > 0.5$ **then**
7: $T \leftarrow t - 1$
8: exit Loop.
9: **end if**
10: $\beta_t \leftarrow \frac{\varepsilon_t}{1-\varepsilon_t}$
11: $D_{t+1}(i) = D_t(i) \cdot \begin{cases} \beta_t & C_t(x_i) = y_i \\ 1 & Otherwise \end{cases}$
12: Normalize D_{t+1} to be a proper distribution.
13: $t{+}{+}$
14: **until** $t > T$

Fig. 50.3. The AdaBoost.M.1 Algorithm.

All boosting algorithms presented here assume that the weak inducers which are provided can cope with weighted instances. If this is not the case, an unweighted dataset is generated from the weighted data by a resampling technique. Namely, instances are chosen with probability according to their weights (until the dataset becomes as large as the original training set).

Boosting seems to improve performances for two main reasons:

1. It generates a final classifier whose error on the training set is small by combining many hypotheses whose error may be large.
2. It produces a combined classifier whose variance is significantly lower than those produced by the weak learner.

On the other hand, boosting sometimes leads to deterioration in generalization performance. According to Quinlan (1996) the main reason for boosting's failure is overfitting. The objective of boosting is to construct a composite classifier that performs well on the data, but a large number of iterations may create a very complex composite classifier, that is significantly less accurate than a single classifier. A possible way to avoid overfitting is by keeping the number of iterations as small as possible.

Another important drawback of boosting is that it is difficult to understand. The resulted ensemble is considered to be less comprehensible since the user is required to capture several classifiers instead of a single classifier. Despite the above drawbacks, Breiman (1996) refers to the boosting idea as the most significant development in classifier design of the nineties.

Windowing

Windowing is a general method aiming to improve the efficiency of inducers by reducing the complexity of the problem. It was initially proposed as a supplement to the ID3 decision tree in order to address complex classification tasks that might have exceeded the memory capacity of computers. Windowing is performed by using a sub-sampling procedure. The method may be summarized as follows: a random subset of the training instances is selected (a window). The subset is used for training a classifier, which is tested on the remaining training data. If the accuracy of the induced classifier is insufficient, the misclassified test instances are removed from the test set and added to the training set of the next iteration. Quinlan (1993) mentions two different ways of forming a window: in the first, the current window is extended up to some specified limit. In the second, several "key" instances in the current window are identified and the rest are replaced. Thus the size of the window stays constant. The process continues until sufficient accuracy is obtained, and the classifier constructed at the last iteration is chosen as the final classifier. Figure 50.4 presents the pseudo-code of the windowing procedure.

Input: I (an inducer), S (the training set), r (the initial window size), t (the maximum allowed windows size increase for sequential iterations).
Output: C

```
1: Window ← Select randomly r instances from S.
2: Test ← S-Window
3: repeat
4:    C ← I(Window)
5:    Inc ← 0
6:    for all (x_i, y_i) ∈ Test do
7:       if C(x_i) ≠ y_i then
8:          Test ← Test − (x_i, y_i)
9:          Window = Window ∪ (x_i, y_i)
10:         Inc++
11:      end if
12:      if Inc = t then
13:         exit Loop
14:      end if
15:   end for
16: until Inc = 0
```

Fig. 50.4. The Windowing Procedure.

The windowing method has been examined also for separate-and-conquer rule induction algorithms (Furnkranz, 1997). This research has shown that for this type of algorithm, significant improvement in efficiency is possible in noise-free domains. Contrary to the basic windowing algorithm, this one removes all instances that have been classified by consistent rules from this window, in addition to adding all instances that have been misclassified. Removal of instances from the window keeps its size small and thus decreases induction time.

In conclusion, both windowing and uncertainty sampling build a sequence of classifiers only for obtaining an ultimate sample. The difference between them lies in the fact that in windowing the instances are labeled in advance, while in uncertainty, this is not so. Therefore, new training instances are chosen differently. Boosting also builds a sequence of classifiers, but combines them in order to gain knowledge from them all. Windowing and uncertainty sampling do not combine the classifiers, but use the best classifier.

50.2.2 Incremental Batch Learning

In this method the classifier produced in one iteration is given as "prior knowledge" to the learning algorithm in the following iteration (along with the subsample of that iteration). The learning algorithm uses the current subsample to evaluate the former classifier, and uses the former one for building the next classifier. The classifier constructed at the last iteration is chosen as the final classifier.

50.3 Concurrent Methodology

In the concurrent ensemble methodology, the original dataset is partitioned into several subsets from which multiple classifiers are induced concurrently. The subsets created from the original training set may be disjoint (mutually exclusive) or overlapping. A combining procedure is then applied in order to produce a single classification for a given instance. Since the method for combining the results of induced classifiers is usually independent of the induction algorithms, it can be used with different inducers at each subset. These concurrent methods aim either at improving the predictive power of classifiers or decreasing the total execution time. The following sections describe several algorithms that implement this methodology.

Bagging

The most well-known method that processes samples concurrently is bagging (bootstrap aggregating). The method aims to improve the accuracy by creating an improved composite classifier, I^*, by amalgamating the various outputs of learned classifiers into a single prediction.

Figure 50.5 presents the pseudo-code of the bagging algorithm (Breiman, 1996). Each classifier is trained on a sample of instances taken with replacement from the training set. Usually each sample size is equal to the size of the original training set.

Input: I (an inducer), T (the number of iterations), S (the training set), N (the subsample size).
Output: $C_t; t = 1, \ldots, T$
1: $t \leftarrow 1$
2: **repeat**
3: $S_t \leftarrow$ Sample N instances from S with replacment.
4: Build classifier C_t using I on S_t
5: $t{+}{+}$
6: **until** $t > T$

Fig. 50.5. The Bagging Algorithm.

Note that since sampling with replacement is used, some of the original instances of S may appear more than once in S_t and some may not be included at all. So the training sets S_t

are different from each other, but are certainly not independent. To classify a new instance, each classifier returns the class prediction for the unknown instance. The composite bagged classifier, I^*, returns the class that has been predicted most often (voting method). The result is that bagging produces a combined model that often performs better than the single model built from the original single data. Breiman (1996) notes that this is true especially for unstable inducers because bagging can eliminate their instability. In this context, an inducer is considered unstable if perturbing the learning set can cause significant changes in the constructed classifier. However, the bagging method is rather hard to analyze and it is not easy to understand by intuition what are the factors and reasons for the improved decisions.

Bagging, like boosting, is a technique for improving the accuracy of a classifier by producing different classifiers and combining multiple models. They both use a kind of voting for classification in order to combine the outputs of the different classifiers of the same type. In boosting, unlike bagging, each classifier is influenced by the performance of those built before, so the new classifier tries to pay more attention to errors that were made in the previous ones and to their performances. In bagging, each instance is chosen with equal probability, while in boosting, instances are chosen with probability proportional to their weight. Furthermore, according to Quinlan (1996), as mentioned above, bagging requires that the learning system should not be stable, where boosting does not preclude the use of unstable learning systems, provided that their error rate can be kept below 0.5.

Cross-validated Committees

This procedure creates k classifiers by partitioning the training set into k-equal-sized sets and in turn, training on all but the i-th set. This method, first used by Gams (1989), employed 10-fold partitioning. Parmanto *et al.* (1996) have also used this idea for creating an ensemble of neural networks. Domingos (1996) has used cross-validated committees to speed up his own rule induction algorithm RISE, whose complexity is $O(n^2)$, making it unsuitable for processing large databases. In this case, partitioning is applied by predetermining a maximum number of examples to which the algorithm can be applied at once. The full training set is randomly divided into approximately equal-sized partitions. RISE is then run on each partition separately. Each set of rules grown from the examples in partition p is tested on the examples in partition $p+1$, in order to reduce overfitting and improve accuracy.

50.4 Combining Classifiers

The way of combining the classifiers may be divided into two main groups: simple multiple classifier combinations and meta-combiners. The simple combining methods are best suited for problems where the individual classifiers perform the same task and have comparable success. However, such combiners are more vulnerable to outliers and to unevenly performing classifiers. On the other hand, the meta-combiners are theoretically more powerful but are susceptible to all the problems associated with the added learning (such as over-fitting, long training time).

50.4.1 Simple Combining Methods

Uniform Voting

In this combining schema, each classifier has the same weight. A classification of an unlabeled instance is performed according to the class that obtains the highest number of votes.

Mathematically it can be written as:

$$Class(x) = \underset{c_i \in dom(y)}{\operatorname{argmax}} \sum_{\forall k c_i = \underset{c_j \in dom(y)}{\operatorname{argmax}} \hat{P}_{M_k}(y=c_j|x)} 1$$

where M_k denotes classifier k and $\hat{P}_{M_k}(y = c|x)$ denotes the probability of y obtaining the value c given an instance x.

Distribution Summation

This combining method was presented by Clark and Boswell (1991). The idea is to sum up the conditional probability vector obtained from each classifier. The selected class is chosen according to the highest value in the total vector. Mathematically, it can be written as:

$$Class(x) = \underset{c_i \in dom(y)}{\operatorname{argmax}} \sum_k \hat{P}_{M_k}(y = c_i|x)$$

Bayesian Combination

This combining method was investigated by Buntine (1990). The idea is that the weight associated with each classifier is the posterior probability of the classifier given the training set.

$$Class(x) = \underset{c_i \in dom(y)}{\operatorname{argmax}} \sum_k P(M_k|S) \cdot \hat{P}_{M_k}(y = c_i|x)$$

where $P(M_k|S)$ denotes the probability that the classifier M_k is correct given the training set S. The estimation of $P(M_k|S)$ depends on the classifier's representation. Buntine (1990) demonstrates how to estimate this value for decision trees.

Dempster–Shafer

The idea of using the Dempster–Shafer theory of evidence (Buchanan and Shortliffe, 1984) for combining models has been suggested by Shilen (1990; 1992). This method uses the notion of basic probability assignment defined for a certain class c_i given the instance x:

$$bpa(c_i,x) = 1 - \prod_k (1 - \hat{P}_{M_k}(y = c_i|x))$$

Consequently, the selected class is the one that maximizes the value of the belief function:

$$Bel(c_i,x) = \frac{1}{A} \cdot \frac{bpa(c_i,x)}{1 - bpa(c_i,x)}$$

where A is a normalization factor defined as:

$$A = \sum_{\forall c_i \in dom(y)} \frac{bpa(c_i,x)}{1 - bpa(c_i,x)} + 1$$

Naïve Bayes

Using Bayes' rule, one can extend the Naïve Bayes idea for combining various classifiers:

$$class(x) = \underset{\substack{c_j \in dom(y) \\ \hat{P}(y=c_j)>0}}{\operatorname{argmax}} \hat{P}(y=c_j) \cdot \prod_{k=1} \frac{\hat{P}_{M_k}(y=c_j\,|x)}{\hat{P}(y=c_j)}$$

Entropy Weighting

The idea in this combining method is to give each classifier a weight that is inversely proportional to the entropy of its classification vector.

$$Class(x) = \underset{c_i \in dom(y)}{\operatorname{argmax}} \sum_{k:c_i = \underset{c_j \in dom(y)}{\operatorname{argmax}} \hat{P}_{M_k}(y=c_j|x)} Ent(M_k, x)$$

where:

$$Ent(M_k, x) = -\sum_{c_j \in dom(y)} \hat{P}_{M_k}(y=c_j\,|x) \log\left(\hat{P}_{M_k}(y=c_j\,|x)\right)$$

Density-based Weighting

If the various classifiers were trained using datasets obtained from different regions of the instance space, it might be useful to weight the classifiers according to the probability of sampling x by classifier M_k, namely:

$$Class(x) = \underset{c_i \in dom(y)}{\operatorname{argmax}} \sum_{k:c_i = \underset{c_j \in dom(y)}{\operatorname{argmax}} \hat{P}_{M_k}(y=c_j|x)} \hat{P}_{M_k}(x)$$

The estimation of $\hat{P}_{M_k}(x)$ depend on the classifier representation and can not always be estimated.

DEA Weighting Method

Recently there has been attempt to use the DEA (Data Envelop Analysis) methodology (Charnes *et al.*, 1978) in order to assign weight to different classifiers (Sohn and Choi, 2001). They argue that the weights should not be specified based on a single performance measure, but on several performance measures. Because there is a trade-off among the various performance measures, the DEA is employed in order to figure out the set of efficient classifiers. In addition, DEA provides inefficient classifiers with the benchmarking point.

Logarithmic Opinion Pool

According to the logarithmic opinion pool (Hansen, 2000) the selection of the preferred class is performed according to:

$$Class(x) = \underset{c_j \in dom(y)}{\operatorname{argmax}} e^{\sum_k \alpha_k \cdot \log(\hat{P}_{M_k}(y=c_j|x))}$$

where α_k denotes the weight of the k-th classifier, such that:

$$\alpha_k \geq 0; \sum \alpha_k = 1$$

Order Statistics

Order statistics can be used to combine classifiers (Tumer and Ghosh, 2000). These combiners have the simplicity of a simple weighted combining method with the generality of meta-combining methods (see the following section). The robustness of this method is helpful when there are significant variations among classifiers in some part of the instance space.

50.4.2 Meta-combining Methods

Meta-learning means learning from the classifiers produced by the inducers and from the classifications of these classifiers on training data. The following sections describe the most well-known meta-combining methods.

Stacking

Stacking is a technique whose purpose is to achieve the highest generalization accuracy. By using a meta-learner, this method tries to induce which classifiers are reliable and which are not. Stacking is usually employed to combine models built by different inducers. The idea is to create a meta-dataset containing a tuple for each tuple in the original dataset. However, instead of using the original input attributes, it uses the predicted classification of the classifiers as the input attributes. The target attribute remains as in the original training set.

Test instance is first classified by each of the base classifiers. These classifications are fed into a meta-level training set from which a meta-classifier is produced. This classifier combines the different predictions into a final one. It is recommended that the original dataset will be partitioned into two subsets. The first subset is reserved to form the meta-dataset and the second subset is used to build the base-level classifiers. Consequently the meta-classifier predications reflect the true performance of base-level learning algorithms. Stacking performances could be improved by using output probabilities for every class label from the base-level classifiers. In such cases, the number of input attributes in the meta-dataset is multiplied by the number of classes.

Džeroski and Ženko (2004) have evaluated several algorithms for constructing ensembles of classifiers with stacking and show that the ensemble performs (at best) comparably to selecting the best classifier from the ensemble by cross validation. In order to improve the existing stacking approach, they propose to employ a new multi-response model tree to learn at the meta-level and empirically showed that it performs better than existing stacking approaches and better than selecting the best classifier by cross-validation.

Arbiter Trees

This approach builds an arbiter tree in a bottom-up fashion (Chan and Stolfo, 1993). Initially the training set is randomly partitioned into k disjoint subsets. The arbiter is induced from a pair of classifiers and recursively a new arbiter is induced from the output of two arbiters. Consequently for k classifiers, there are $\log_2(k)$ levels in the generated arbiter tree.

The creation of the arbiter is performed as follows. For each pair of classifiers, the union of their training dataset is classified by the two classifiers. A selection rule compares the classifications of the two classifiers and selects instances from the union set to form the training set for the arbiter. The arbiter is induced from this set with the same learning algorithm used in the base level. The purpose of the arbiter is to provide an alternate classification when the base classifiers present diverse classifications. This arbiter, together with an arbitration rule, decides on a final classification outcome, based upon the base predictions. Figure 50.6 shows how the final classification is selected based on the classification of two base classifiers and a single arbiter.

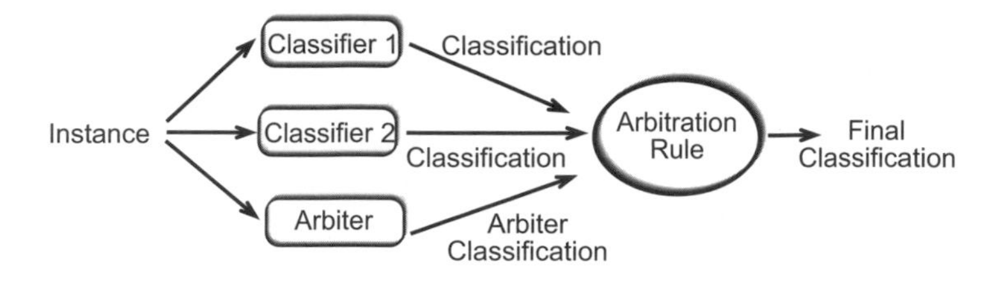

Fig. 50.6. A Prediction from Two Base Classifiers and a Single Arbiter.

The process of forming the union of data subsets; classifying it using a pair of arbiter trees; comparing the classifications; forming a training set; training the arbiter; and picking one of the predictions, is recursively performed until the root arbiter is formed. Figure 50.7 illustrate an arbiter tree created for $k = 4$. $T_1 - T_4$ are the initial four training datasets from which four classifiers $C_1 - C_4$ are generated concurrently. T_{12} and T_{34} are the training sets generated by the rule selection from which arbiters are produced. A_{12} and A_{34} are the two arbiters. Similarly, T_{14} and A_{14} (root arbiter) are generated and the arbiter tree is completed.

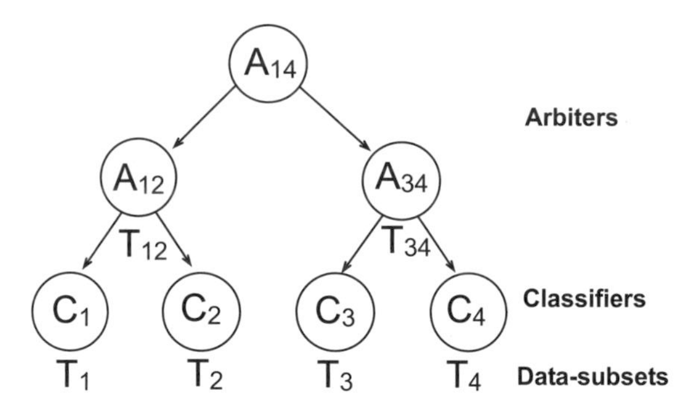

Fig. 50.7. Sample Arbiter Tree.

Several schemes for arbiter trees were examined and differentiated from each other by the selection rule used. Here are three versions of rule selection:

- Only instances with classifications that disagree are chosen (group 1).
- Like group 1 defined above, plus instances that their classifications agree but are incorrect (group 2).
- Like groups 1 and 2 defined above, plus instances that have the same correct classifications (group 3).

Two versions of arbitration rules have been implemented; each one corresponds to the selection rule used for generating the training data at that level:

- For selection rule 1 and 2, a final classification is made by a majority vote of the classifications of the two lower levels and the arbiter's own classification, with preference given to the latter.
- For selection rule 3, if the classifications of the two lower levels are not equal, the classification made by the sub-arbiter based on the first group is chosen. In case this is not true and the classification of the sub-arbiter constructed on the third group equals those of the lower levels — then this is the chosen classification. In any other case, the classification of the sub-arbiter constructed on the second group is chosen. Chan and Stolfo (1993) achieved the same accuracy level as in the single mode applied to the entire dataset but with less time and memory requirements. It has been shown that this meta-learning strategy required only around 30% of the memory used by the single model case. This last fact, combined with the independent nature of the various learning processes, make this method robust and effective for massive amounts of data. Nevertheless, the accuracy level depends on several factors such as the distribution of the data among the subsets and the pairing scheme of learned classifiers and arbiters in each level. The decision in any of these issues may influence performance, but the optimal decisions are not necessarily known in advance, nor initially set by the algorithm.

Combiner Trees

The way combiner trees are generated is very similar to arbiter trees. A combiner tree is trained bottom-up. However, a combiner, instead of an arbiter, is placed in each non-leaf node of a combiner tree (Chan and Stolfo, 1997). In the combiner strategy, the classifications of the learned base classifiers form the basis of the meta-learner's training set. A composition rule determines the content of training examples from which a combiner (meta-classifier) will be generated. In classifying an instance, the base classifiers first generate their classifications and based on the composition rule, a new instance is generated. The aim of this strategy is to combine the classifications from the base classifiers by learning the relationship between these classifications and the correct classification. Figure 50.8 illustrates the result obtained from two base classifiers and a single combiner.

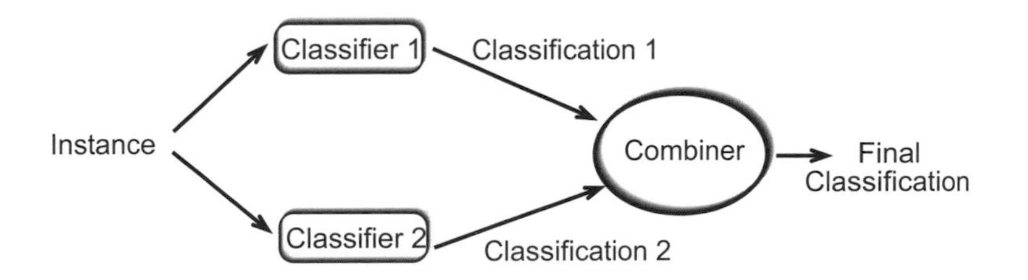

Fig. 50.8. A Prediction from Two Base Classifiers and a Single Combiner.

Two schemes of composition rule were proposed. The first one is the stacking schema. The second is like stacking with the addition of the instance input attributes. Chan and Stolfo (1995) showed that the stacking schema per se does not perform as well as the second schema. Although there is information loss due to data partitioning, combiner trees can sustain the accuracy level achieved by a single classifier. In a few cases, the single classifier's accuracy was consistently exceeded.

Grading

This technique uses "graded" classifications as meta-level classes (Seewald and Furnkranz, 2001). The term graded is used in the sense of classifications that have been marked as correct or incorrect. The method transforms the classification made by the k different classifiers into k training sets by using the instances k times and attaching them to a new binary class in each occurrence. This class indicates whether the k–th classifier yielded a correct or incorrect classification, compared to the real class of the instance.

For each base classifier, one meta-classifier is learned whose task is to classify when the base classifier will misclassify. At classification time, each base classifier classifies the unlabeled instance. The final classification is derived from the classifications of those base classifiers that are classified to be correct by the meta-classification schemes. In case several base classifiers with different classification results are classified as correct, voting, or a combination considering the confidence estimates of the base classifiers, is performed. Grading may be considered as a generalization of cross-validation selection (Schaffer, 1993), which divides the training data into k subsets, builds $k-1$ classifiers by dropping one subset at a time and then using it to find a misclassification rate. Finally, the procedure simply chooses the classifier corresponding to the subset with the smallest misclassification. Grading tries to make this decision separately for each and every instance by using only those classifiers that are predicted to classify that instance correctly. The main difference between grading and combiners (or stacking) are that the former does not change the instance attributes by replacing them with class predictions or class probabilities (or adding them to it). Instead it modifies the class values. Furthermore, in grading several sets of meta-data are created, one for each base classifier. Several meta-level classifiers are learned from those sets.

The main difference between grading and arbiters is that arbiters use information about the disagreements of classifiers for selecting a training set, while grading uses disagreement with the target function to produce a new training set.

50.5 Ensemble Diversity

In an ensemble, the combination of the output of several classifiers is only useful if they disagree about some inputs (Tumer and Ghosh, 1996). According to Hu (2001) diversified classifiers lead to uncorrelated errors, which in turn improve classification accuracy.

50.5.1 Manipulating the Inducer

A simple method for gaining diversity is to manipulate the inducer used for creating the classifiers. Ali and Pazzani (1996) propose to change the rule learning HYDRA algorithm in the following way: Instead of selecting the best literal in each stage (using, for instance, information gain measure), the literal is selected randomly such that its probability of being selected is proportional to its measure value. Dietterich (2000a) has implemented a similar idea for C4.5 decision trees. Instead of selecting the best attribute in each stage, it selects randomly (with equal probability) an attribute from the set of the best 20 attributes. The simplest way to manipulate the back-propagation inducer is to assign different initial weights to the network (Kolen and Pollack, 1991). MCMC (Markov Chain Monte Carlo) methods can also be used for introducing randomness in the induction process (Neal, 1993).

50.5.2 Manipulating the Training Set

Most ensemble methods construct the set of classifiers by manipulating the training instances. Dietterich (2000b) distinguishes between three main methods for manipulating the dataset.

Manipulating the Tuples

In this method, each classifier is trained on a different subset of the original dataset. This method is useful for inducers whose variance-error factor is relatively large (such as decision trees and neural networks), namely, small changes in the training set may cause a major change in the obtained classifier. This category contains procedures such as bagging, boosting and cross-validated committees.

The distribution of tuples among the different subsets could be random as in the bagging algorithm or in the arbiter trees. Other methods distribute the tuples based on the class distribution such that the class distribution in each subset is approximately the same as that in the entire dataset. Proportional distribution was used in combiner trees (Chan and Stolfo, 1993). It has been shown that proportional distribution can achieve higher accuracy than random distribution.

Recently Christensen et al. (2004) suggest a novel framework for construction of an ensemble in which each instance contributes to the committee formation with a fixed weight, while contributing with different individual weights to the derivation of the different constituent models. This approach encourages model diversity whilst not biasing the ensemble inadvertently towards any particular instance.

Manipulating the Input Feature Set

Another less common strategy for manipulating the training set is to manipulate the input attribute set. The idea is to simply give each classifier a different projection of the training set.

50.5.3 Measuring the Diversity

For regression problems *variance* is usually used to measure diversity (Krogh and Vedelsby, 1995). In such cases it can be easily shown that the ensemble error can be reduced by increasing ensemble diversity while maintaining the average error of a single model.

In classification problems, a more complicated measure is required to evaluate the diversity. Kuncheva and Whitaker (2003) compared several measures of diversity and concluded that most of them are correlated. Furthermore, it is usually assumed that increasing diversity may decrease ensemble error (Zenobi and Cunningham, 2001).

50.6 Ensemble Size

50.6.1 Selecting the Ensemble Size

An important aspect of ensemble methods is to define how many component classifiers should be used. This number is usually defined according to the following issues:

- Desired accuracy — Hansen (1990) argues that ensembles containing ten classifiers is sufficient for reducing the error rate. Nevertheless, there is empirical evidence indicating that in the case of AdaBoost using decision trees, error reduction is observed in even relatively large ensembles containing 25 classifiers (Opitz and Maclin, 1999). In the disjoint partitioning approaches, there may be a tradeoff between the number of subsets and the final accuracy. The size of each subset cannot be too small because sufficient data must be available for each learning process to produce an effective classifier. Chan and Stolfo (1993) varied the number of subsets in the arbiter trees from 2 to 64 and examined the effect of the predetermined number of subsets on the accuracy level.
- User preferences — Increasing the number of classifiers usually increase computational complexity and decreases the comprehensibility. For that reason, users may set their preferences by predefining the ensemble size limit.
- Number of processors available — In concurrent approaches, the number of processors available for parallel learning could be put as an upper bound on the number of classifiers that are treated in paralleled process.

Caruana et al. (2004) presented a method for constructing ensembles from libraries of thousands of models. They suggest using forward stepwise selection in order to select the models that maximize the ensemble's performance. Ensemble selection allows ensembles to be optimized to performance metrics such as accuracy, cross entropy, mean precision, or ROC Area.

50.6.2 Pruning Ensembles

As in decision trees induction, it is sometime useful to let the ensemble grow freely and then prune the ensemble in order to get more effective and more compact ensembles. Empirical examinations indicate that pruned ensembles may obtain a similar accuracy performance as the original ensemble (Margineantu and Dietterich, 1997).

The efficiency of pruning methods when meta-combining methods are used have been examined in (Prodromidis *et al.*, 2000). In such cases the pruning methods can be divided into two groups: pre-training pruning methods and post-training pruning methods. Pre-training pruning is performed before combining the classifiers. Classifiers that seem to be attractive are included in the meta-classifier. On the other hand, post-training pruning methods, remove classifiers based on their effect on the meta-classifier. Three methods for pre-training pruning (based on an individual classification performance on a separate validation set, diversity metrics, the ability of classifiers to classify correctly specific classes) and two methods for post-training pruning (based on decision tree pruning and the correlation of the base classifier to the unpruned meta-classifier) have been examined in (Prodromidis *et al.*, 2000). As in (Margineantu and Dietterich, 1997), it has been shown that by using pruning, one can obtain similar or better accuracy performance, while compacting the ensemble.

The GASEN algorithm was developed for selecting the most appropriate classifiers in a given ensemble (Zhou *et al.*, 2002). In the initialization phase, GASEN assigns a random weight to each of the classifiers. Consequently, it uses genetic algorithms to evolve those weights so that they can characterize to some extent the fitness of the classifiers in joining the ensemble. Finally, it removes from the ensemble those classifiers whose weight is less than a predefined threshold value. Recently a revised version of the GASEN algorithm called GASEN-b has been suggested (Zhou and Tang, 2003). In this algorithm, instead of assigning a weight to each classifier, a bit is assigned to each classifier indicating whether it will be used in the final ensemble. They show that the obtained ensemble is not only smaller in size, but in some cases has better generalization performance.

Liu et al. (2004) conducted an empirical study of the relationship of ensemble sizes with ensemble accuracy and diversity, respectively. They show that it is feasible to keep a small ensemble while maintaining accuracy and diversity similar to those of a full ensemble. They proposed an algorithm called $LVFd$ that selects diverse classifiers to form a compact ensemble.

50.7 Cluster Ensemble

This chapter focused mainly on ensembles of classifiers. However ensemble methodology can be used for other Data Mining tasks such as regression and clustering.

The cluster ensemble problem refers to the problem of combining multiple partitionings of a set of instances into a single consolidated clustering. Usually this problem is formalized as a combinatorial optimization problem in terms of shared mutual information.

Dimitriadou et al. (2003) have used ensemble methodology for improving the quality and robustness of clustering algorithms. In fact they employ the same ensemble idea that has been used for many years in classification and regression tasks. More specifically they suggested various aggregation strategies and studied a greedy forward aggregation.

Hu and Yoo (2004) have used ensemble for clustering gene expression data. In this research the clustering results of individual clustering algorithms are converted into a distance matrix. These distance matrices are combined and a weighted graph is constructed according to the combined matrix. Then a graph partitioning approach is used to cluster the graph to generate the final clusters.

Strehl and Ghosh (2003) propose three techniques for obtaining high-quality cluster combiners. The first combiner induces a similarity measure from the partitionings and then reclusters the objects. The second combiner is based on hypergraph partitioning. The third one collapses groups of clusters into meta-clusters, which then compete for each object to determine the combined clustering. Moreover, it is possible to use supra-combiners that evaluate all three approaches against the objective function and pick the best solution for a given situation.

In summary, the methods presented in this chapetr are useful for many application domains, such as: Manufacturing lr18,lr14, Security lr7,l10 and Medicine lr2,lr9, and for many data mining techniques, such as: decision trees lr6,lr12, lr15, clustering lr13,lr8,lr5,lr16 and genetic algorithms lr17,lr11,lr1,lr4.

References

Ali K. M., Pazzani M. J., Error Reduction through Learning Multiple Descriptions, Machine Learning, 24: 3, 173-202, 1996.

Arbel, R. and Rokach, L., Classifier evaluation under limited resources, Pattern Recognition Letters, 27(14): 1619–1631, 2006, Elsevier.

Averbuch, M. and Karson, T. and Ben-Ami, B. and Maimon, O. and Rokach, L., Context-sensitive medical information retrieval, The 11th World Congress on Medical Informatics (MEDINFO 2004), San Francisco, CA, September 2004, IOS Press, pp. 282–286.

Bartlett P. and Shawe-Taylor J., Generalization Performance of Support Vector Machines and Other Pattern Classifiers, In "Advances in Kernel Methods, Support Vector Learning", Bernhard Scholkopf, Christopher J. C. Burges, and Alexander J. Smola (eds.), MIT Press, Cambridge, USA, 1998.

Bauer, E. and Kohavi, R., "An Empirical Comparison of Voting Classification Algorithms: Bagging, Boosting, and Variants". Machine Learning, 35: 1-38, 1999.

Breiman L., Bagging predictors, Machine Learning, 24(2):123-140, 1996.

Bruzzone L., Cossu R., Vernazza G., Detection of land-cover transitions by combining multidate classifiers, Pattern Recognition Letters, 25(13): 1491–1500, 2004.

Buchanan, B.G. and Shortliffe, E.H., Rule Based Expert Systems, 272-292, Addison-Wesley, 1984.

Buhlmann, P. and Yu, B., Boosting with L_2 loss: Regression and classification, Journal of the American Statistical Association, 98, 324338. 2003.

Buntine, W., A Theory of Learning Classification Rules. Doctoral dissertation. School of Computing Science, University of Technology. Sydney. Australia, 1990.

Caruana R., Niculescu-Mizil A. , Crew G. , Ksikes A., Ensemble selection from libraries of models, Twenty-first international conference on Machine learning, July 04-08, 2004, Banff, Alberta, Canada.

Chan P. K. and Stolfo, S. J., Toward parallel and distributed learning by meta-learning, In AAAI Workshop in Knowledge Discovery in Databases, pp. 227-240, 1993.

Chan P.K. and Stolfo, S.J., A Comparative Evaluation of Voting and Meta-learning on Partitioned Data, Proc. 12th Intl. Conf. On Machine Learning ICML-95, 1995.

Chan P.K. and Stolfo S.J, On the Accuracy of Meta-learning for Scalable Data Mining, J. Intelligent Information Systems, 8:5-28, 1997.

Charnes, A., Cooper, W. W., and Rhodes, E., Measuring the efficiency of decision making units, European Journal of Operational Research, 2(6):429-444, 1978.

Christensen S. W. , Sinclair I., Reed P. A. S., Designing committees of models through deliberate weighting of data points, The Journal of Machine Learning Research, 4(1):39–66, 2004.

Clark, P. and Boswell, R., "Rule induction with CN2: Some recent improvements." In Proceedings of the European Working Session on Learning, pp. 151-163, Pitman, 1991.

Cohen S., Rokach L., Maimon O., Decision Tree Instance Space Decomposition with Grouped Gain-Ratio, Information Science, Volume 177, Issue 17, pp. 3592-3612, 2007.

Džeroski S., Ženko B., Is Combining Classifiers with Stacking Better than Selecting the Best One?, Machine Learning, 54(3): 255–273, 2004.

Dietterich, T. G., An Experimental Comparison of Three Methods for Constructing Ensembles of Decision Trees: Bagging, Boosting and Randomization. 40(2):139-157, 2000.

Dietterich T., Ensemble methods in machine learning. In J. Kittler and F. Roll, editors, First International Workshop on Multiple Classifier Systems, Lecture Notes in Computer Science, pages 1-15. Springer-Verlag, 2000

Dimitriadou E., Weingessel A., Hornik K., A cluster ensembles framework, Design and application of hybrid intelligent systems, IOS Press, Amsterdam, The Netherlands, 2003.

Domingos, P., Using Partitioning to Speed Up Specific-to-General Rule Induction. In Proceedings of the AAAI-96 Workshop on Integrating Multiple Learned Models, pp. 29-34, AAAI Press, 1996.

Freund Y. and Schapire R. E., Experiments with a new boosting algorithm. In Machine Learning: Proceedings of the Thirteenth International Conference, pages 325-332, 1996.

Fürnkranz, J., More efficient windowing, In Proceeding of The 14th national Conference on Artificial Intelegence (AAAI-97), pp. 509-514, Providence, RI. AAAI Press, 1997.

Gams, M., New Measurements Highlight the Importance of Redundant Knowledge. In European Working Session on Learning, Montpeiller, France, Pitman, 1989.

Geman S., Bienenstock, E., and Doursat, R., Neural networks and the bias variance dilemma. Neural Computation, 4:1-58, 1995.

Hansen J., Combining Predictors. Meta Machine Learning Methods and Bias Variance & Ambiguity Decompositions. PhD dissertation. Aurhus University. 2000.

Hansen, L. K., and Salamon, P., Neural network ensembles. IEEE Transactions on Pattern Analysis and Machine Intelligence, 12(10), 993–1001, 1990.

Hu, X., Using Rough Sets Theory and Database Operations to Construct a Good Ensemble of Classifiers for Data Mining Applications. ICDM01. pp. 233-240, 2001.

Hu X., Yoo I., Cluster ensemble and its applications in gene expression analysis, Proceedings of the second conference on Asia-Pacific bioinformatics, pp. 297–302, Dunedin, New Zealand, 2004.

Kolen, J. F., and Pollack, J. B., Back propagation is sesitive to initial conditions. In Advances in Neural Information Processing Systems, Vol. 3, pp. 860-867 San Francisco, CA. Morgan Kaufmann, 1991.

Krogh, A., and Vedelsby, J., Neural network ensembles, cross validation and active learning. In Advances in Neural Information Processing Systems 7, pp. 231-238 1995.

Kuncheva, L., & Whitaker, C., Measures of diversity in classifier ensembles and their relationship with ensemble accuracy. Machine Learning, pp. 181–207, 2003.

Leigh W., Purvis R., Ragusa J. M., Forecasting the NYSE composite index with technical analysis, pattern recognizer, neural networks, and genetic algorithm: a case study in romantic decision support, Decision Support Systems 32(4): 361–377, 2002.

Lewis D., and Catlett J., Heterogeneous uncertainty sampling for supervised learning. In Machine Learning: Proceedings of the Eleventh Annual Conference, pp. 148-156 , New Brunswick, New Jersey, Morgan Kaufmann, 1994.

Lewis, D., and Gale, W., Training text classifiers by uncertainty sampling, In seventeenth annual international ACM SIGIR conference on research and development in information retrieval, pp. 3-12, 1994.

Liu H., Mandvikar A., Mody J., An Empirical Study of Building Compact Ensembles. WAIM 2004: pp. 622-627.

Maimon O., and Rokach, L. Data Mining by Attribute Decomposition with semiconductors manufacturing case study, in Data Mining for Design and Manufacturing: Methods and Applications, D. Braha (ed.), Kluwer Academic Publishers, pp. 311–336, 2001.

Maimon O. and Rokach L., "Improving supervised learning by feature decomposition", Proceedings of the Second International Symposium on Foundations of Information and Knowledge Systems, Lecture Notes in Computer Science, Springer, pp. 178-196, 2002.

Maimon O. Rokach L., Ensemble of Decision Trees for Mining Manufacturing Data Sets, Machine Engineering, vol. 4 No1-2, 2004.

Maimon, O. and Rokach, L., Decomposition Methodology for Knowledge Discovery and Data Mining: Theory and Applications, Series in Machine Perception and Artificial Intelligence - Vol. 61, World Scientific Publishing, ISBN:981-256-079-3, 2005.

Mangiameli P., West D., Rampal R., Model selection for medical diagnosis decision support systems, Decision Support Systems, 36(3): 247–259, 2004.

Margineantu D. and Dietterich T., Pruning adaptive boosting. In Proc. Fourteenth Intl. Conf. Machine Learning, pages 211–218, 1997.

Mitchell, T., Machine Learning, McGraw-Hill, 1997.

Moskovitch R, Elovici Y, Rokach L, Detection of unknown computer worms based on behavioral classification of the host, Computational Statistics and Data Analysis, 52(9):4544–4566, 2008.

Neal R., Probabilistic inference using Markov Chain Monte Carlo methods. Tech. Rep. CRG-TR-93-1, Department of Computer Science, University of Toronto, Toronto, CA, 1993.

Opitz, D. and Maclin, R., Popular Ensemble Methods: An Empirical Study, Journal of Artificial Research, 11: 169-198, 1999.

Parmanto, B., Munro, P. W., and Doyle, H. R., Improving committee diagnosis with resampling techinques. In Touretzky, D. S., Mozer, M. C., and Hesselmo, M. E. (Eds). Advances in Neural Information Processing Systems, Vol. 8, pp. 882-888 Cambridge, MA. MIT Press, 1996.

Prodromidis, A. L., Stolfo, S. J. and Chan, P. K., Effective and efficient pruning of metaclassifiers in a distributed Data Mining system. Technical report CUCS-017-99, Columbia Univ., 1999.

Provost, F.J. and Kolluri, V., A Survey of Methods for Scaling Up Inductive Learning Algorithms, Proc. 3rd International Conference on Knowledge Discovery and Data Mining, 1997.

Quinlan, J. R., C4.5: Programs for Machine Learning, Morgan Kaufmann, Los Altos, 1993.

Quinlan, J. R., Bagging, Boosting, and C4.5. In Proceedings of the Thirteenth National Conference on Artificial Intelligence, pages 725-730, 1996.

Rokach, L., Decomposition methodology for classification tasks: a meta decomposer framework, Pattern Analysis and Applications, 9(2006):257–271.

Rokach L., Genetic algorithm-based feature set partitioning for classification problems,Pattern Recognition, 41(5):1676–1700, 2008.

Rokach L., Mining manufacturing data using genetic algorithm-based feature set decomposition, Int. J. Intelligent Systems Technologies and Applications, 4(1):57-78, 2008.

Rokach, L. and Maimon, O., Theory and applications of attribute decomposition, IEEE International Conference on Data Mining, IEEE Computer Society Press, pp. 473–480, 2001.

Rokach L. and Maimon O., Feature Set Decomposition for Decision Trees, Journal of Intelligent Data Analysis, Volume 9, Number 2, 2005b, pp 131–158.

Rokach, L. and Maimon, O., Clustering methods, Data Mining and Knowledge Discovery Handbook, pp. 321–352, 2005, Springer.

Rokach, L. and Maimon, O., Data mining for improving the quality of manufacturing: a feature set decomposition approach, Journal of Intelligent Manufacturing, 17(3):285–299, 2006, Springer.

Rokach, L., Maimon, O., Data Mining with Decision Trees: Theory and Applications, World Scientific Publishing, 2008.

Rokach L., Maimon O. and Lavi I., Space Decomposition In Data Mining: A Clustering Approach, Proceedings of the 14th International Symposium On Methodologies For Intelligent Systems, Maebashi, Japan, Lecture Notes in Computer Science, Springer-Verlag, 2003, pp. 24–31.

Rokach, L. and Maimon, O. and Averbuch, M., Information Retrieval System for Medical Narrative Reports, Lecture Notes in Artificial intelligence 3055, page 217-228 Springer-Verlag, 2004.

Rokach, L. and Maimon, O. and Arbel, R., Selective voting-getting more for less in sensor fusion, International Journal of Pattern Recognition and Artificial Intelligence 20 (3) (2006), pp. 329–350.

Schaffer, C., Selecting a classification method by cross-validation. Machine Learning 13(1):135-143, 1993.

Seewald, A.K. and Fürnkranz, J., Grading classifiers, Austrian research institute for Artificial intelligence, 2001.

Sharkey, A., On combining artificial neural nets, Connection Science, Vol. 8, pp.299-313, 1996.

Shilen, S., Multiple binary tree classifiers. Pattern Recognition 23(7): 757-763, 1990.

Shilen, S., Nonparametric classification using matched binary decision trees. Pattern Recognition Letters 13: 83-87, 1992.

Sohn S. Y., Choi, H., Ensemble based on Data Envelopment Analysis, ECML Meta Learning workshop, Sep. 4, 2001.

Strehl A., Ghosh J. (2003), Cluster ensembles - a knowledge reuse framework for combining multiple partitions, The Journal of Machine Learning Research, 3: 583-617, 2003.

Tan A. C., Gilbert D., Deville Y., Multi-class Protein Fold Classification using a New Ensemble Machine Learning Approach. Genome Informatics, 14:206–217, 2003.

Tukey J.W., Exploratory data analysis, Addison-Wesley, Reading, Mass, 1977.

Tumer, K. and Ghosh J., Error Correlation and Error Reduction in Ensemble Classifiers, Connection Science, Special issue on combining artificial neural networks: ensemble approaches, 8 (3-4): 385-404, 1996.

Tumer, K., and Ghosh J., Linear and Order Statistics Combiners for Pattern Classification, in Combining Articial Neural Nets, A. Sharkey (Ed.), pp. 127-162, Springer-Verlag, 1999.

Tumer, K., and Ghosh J., Robust Order Statistics based Ensembles for Distributed Data Mining. In Kargupta, H. and Chan P., eds, Advances in Distributed and Parallel Knowledge Discovery , pp. 185-210, AAAI/MIT Press, 2000.

Wolpert, D.H., Stacked Generalization, Neural Networks, Vol. 5, pp. 241-259, Pergamon Press, 1992.

Zenobi, G., and Cunningham, P. Using diversity in preparing ensembles of classifiers based on different feature subsets to minimize generalization error. In Proceedings of the European Conference on Machine Learning, 2001.

Zhou, Z. H., and Tang, W., Selective Ensemble of Decision Trees, in Guoyin Wang, Qing Liu, Yiyu Yao, Andrzej Skowron (Eds.): Rough Sets, Fuzzy Sets, Data Mining, and Granular Computing, 9th International Conference, RSFDGrC, Chongqing, China, Proceedings. Lecture Notes in Computer Science 2639, pp.476-483, 2003.

Zhou, Z. H., Wu J., Tang W., Ensembling neural networks: many could be better than all. Artificial Intelligence 137: 239-263, 2002.

51

Data Mining using Decomposition Methods

Lior Rokach[1] and Oded Maimon[2]

[1] Department of Information System Engineering, Ben-Gurion University, Beer-Sheba, Israel,
liorrk@bgu.ac.il
[2] Department of Industrial Engineering, Tel-Aviv University, Ramat-Aviv 69978, Israel,
maimon@eng.tau.ac.il

Summary. The idea of decomposition methodology is to break down a complex Data Mining task into several smaller, less complex and more manageable, sub-tasks that are solvable by using existing tools, then joining their solutions together in order to solve the original problem. In this chapter we provide an overview of decomposition methods in classification tasks with emphasis on elementary decomposition methods. We present the main properties that characterize various decomposition frameworks and the advantages of using these framework. Finally we discuss the uniqueness of decomposition methodology as opposed to other closely related fields, such as ensemble methods and distributed data mining.

Key words: Decomposition, Mixture-of-Experts, Elementary Decomposition Methodology, Function Decomposition, Distributed Data Mining, Parallel Data Mining

51.1 Introduction

One of the explicit challenges in Data Mining is to develop methods that will be feasible for complicated real-world problems. In many disciplines, when a problem becomes more complex, there is a natural tendency to try to break it down into smaller, distinct but connected pieces. The concept of breaking down a system into smaller pieces is generally referred to as *decomposition*. The purpose of decomposition methodology is to break down a complex problem into smaller, less complex and more manageable, sub-problems that are solvable by using existing tools, then joining them together to solve the initial problem. Decomposition methodology can be considered as an effective strategy for changing the representation of a classification problem. Indeed, Kusiak (2000) considers decomposition as the "most useful form of transformation of data sets".

The decomposition approach is frequently used in statistics, operations research and engineering. For instance, decomposition of time series is considered to be a practical way to improve forecasting. The usual decomposition into trend, cycle, seasonal and irregular components was motivated mainly by business analysts, who wanted to get a clearer picture of

O. Maimon, L. Rokach (eds.), *Data Mining and Knowledge Discovery Handbook*, 2nd ed.,
DOI 10.1007/978-0-387-09823-4_51, © Springer Science+Business Media, LLC 2010

the state of the economy (Fisher, 1995). Although the operations research community has extensively studied decomposition methods to improve computational efficiency and robustness, identification of the partitioned problem model has largely remained an ad hoc task (He *et al.*, 2000).

In engineering design, problem decomposition has received considerable attention as a means of reducing multidisciplinary design cycle time and of streamlining the design process by adequate arrangement of the tasks (Kusiak *et al.*, 1991). Decomposition methods are also used in decision-making theory. A typical example is the AHP method (Saaty, 1993). In artificial intelligence finding a good decomposition is a major tactic, both for ensuring the transparent end-product and for avoiding a combinatorial explosion (Michie, 1995).

Research has shown that no single learning approach is clearly superior for all cases. In fact, the task of discovering regularities can be made easier and less time consuming by decomposition of the task. However, decomposition methodology has not attracted as much attention in the KDD and machine learning community (Buntine, 1996).

Although decomposition is a promising technique and presents an obviously natural direction to follow, there are hardly any works in the Data Mining literature that consider the subject directly. Instead, there are abundant practical attempts to apply decomposition methodology to specific, real life applications (Buntine, 1996). There are also many discussions on closely related problems, largely in the context of distributed and parallel learning (Zaki and Ho, 2000) or ensembles classifiers (see Chapter 49.6 in this volume). Nevertheless, there are a few important works that consider decomposition methodology directly. Various decomposition methods have been presented (Kusiak, 2000). There was also suggestion to decompose the exploratory data analysis process into 3 parts: *model search*, *pattern search*, and *attribute search* (Bhargava, 1999). However, in this case the notion of "decomposition" refers to the entire KDD process, while this chapter focuses on decomposition of the model search.

In the neural network community, several researchers have examined the decomposition methodology (Hansen, 2000). The *"mixture-of-experts"* (ME) method decomposes the input space, such that each expert examines a different part of the space (Nowlan and Hinton, 1991). However, the sub-spaces have soft "boundaries", namely sub-spaces are allowed to overlap. Figure 51.1 illustrates an n-expert structure. Each expert outputs the conditional probability of the target attribute given the input instance. A gating network is responsible for combining the various experts by assigning a weight to each network. These weights are not constant but are functions of the input instance x.

An extension to the basic mixture of experts, known as hierarchical mixtures of experts (HME), has been proposed by Jordan and Jacobs (1994). This extension decomposes the space into sub-spaces, and then recursively decomposes each sub-space to sub-spaces.

Variation of the basic mixtures of experts methods have been developed to accommodate specific domain problems. A specialized modular network called the Meta-p_i network has been used to solve the vowel-speaker problem (Hampshire and Waibel, 1992, Peng *et al.*, 1995). There have been other extensions to the ME such as nonlinear gated experts for time-series (Weigend *et al.*, 1995); revised modular network for predicting the survival of AIDS patients (Ohno-Machado and Musen, 1997); and a new approach for combining multiple experts for improving handwritten numerals recognition (Rahman and Fairhurst, 1997).

However, none of these works presents a complete framework that considers the coexistence of different decomposition methods, namely: when we should prefer a specific method and whether it is possible to solve a given problem using a hybridization of several decomposition methods.

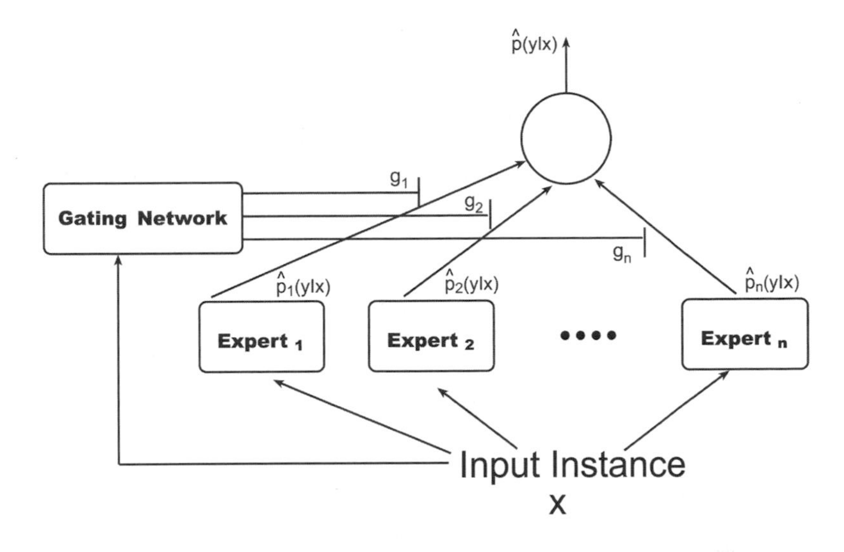

Fig. 51.1. Illustration of n-Expert Structure.

51.2 Decomposition Advantages

51.2.1 Increasing Classification Performance (Classification Accuracy)

Decomposition methods can improve the predictive accuracy of regular methods. In fact Sharkey (1999) argues that improving performance is the main motivation for decomposition. Although this might look surprising at first, it can be explained by the bias-variance tradeoff. Since decomposition methodology constructs several simpler sub-models instead a single complicated model, we might gain better performance by choosing the appropriate sub-models' complexities (i.e. finding the best bias-variance tradeoff). For instance, a single decision tree that attempts to model the entire instance space usually has high variance and small bias. On the other hand, Naïve Bayes can be seen as a composite of single-attribute decision trees (each one of these trees contains only one unique input attribute). The bias of the Naïve Bayes is large (as it can not represent a complicated classifier); on the other hand, its variance is small. Decomposition can potentially obtain a set of decision trees, such that each one of the trees is more complicated than a single-attribute tree (thus it can represent a more complicated classifier and it has lower bias than the Naïve Bayes) but not complicated enough to have high variance.

There are other justifications for the performance improvement of decomposition methods, such as the ability to exploit the specialized capabilities of each component, and consequently achieve results which would not be possible in a single model. An excellent example to the contributions of the decomposition methodology can be found in Baxt (1990). In this research, the main goal was to identify a certain clinical diagnosis. Decomposing the problem and building two neural networks significantly increased the correct classification rate.

51.2.2 Scalability to Large Databases

One of the explicit challenges for the KDD research community is to develop methods that facilitate the use of Data Mining algorithms for real-world databases. In the information age, data is automatically collected and therefore the database available for mining can be quite large, as a result of an increase in the number of records in the database and the number of fields/attributes in each record (high dimensionality).

There are many approaches for dealing with huge databases including: sampling methods; massively parallel processing; efficient storage methods; and dimension reduction. Decomposition methodology suggests an alternative way to deal with the aforementioned problems by reducing the volume of data to be processed at a time. Decomposition methods break the original problem into several sub-problems, each one with relatively small dimensionality. In this way, decomposition reduces training time and makes it possible to apply standard machine-learning algorithms to large databases (Sharkey, 1999).

51.2.3 Increasing Comprehensibility

Decomposition methods suggest a conceptual simplification of the original complex problem. Instead of getting a single and complicated model, decomposition methods create several sub-models, which are more comprehensible. This motivation has often been noted in the literature (Pratt *et al.*, 1991, Hrycej, 1992, Sharkey, 1999). Smaller models are also more appropriate for user-driven Data Mining that is based on *visualization techniques*. Furthermore, if the decomposition structure is induced by automatic means, it can provide new insights about the explored domain.

51.2.4 Modularity

Modularity eases the maintenance of the classification model. Since new data is being collected all the time, it is essential once in a while to execute a rebuild process to the entire model. However, if the model is built from several sub-models, and the new data collected affects only part of the sub-models, a more simple re-building process may be sufficient. This justification has often been noted (Kusiak, 2000).

51.2.5 Suitability for Parallel Computation

If there are no dependencies between the various sub-components, then parallel techniques can be applied. By using parallel computation, the time needed to solve a mining problem can be shortened.

51.2.6 Flexibility in Techniques Selection

Decomposition methodology suggests the ability to use different inducers for individual sub-problems or even to use the same inducer but with a different setup. For instance, it is possible to use neural networks having different topologies (different number of hidden nodes). The researcher can exploit this freedom of choice to boost classifier performance.

The first three advantages are of particular importance in commercial and industrial Data Mining. However, as it will be demonstrated later, not all decomposition methods display the same advantages.

51.3 The Elementary Decomposition Methodology

Finding an optimal or quasi-optimal decomposition for a certain supervised learning problem might be hard or impossible. For that reason Rokach and Maimon (2002) proposed *elementary decomposition methodology*. The basic idea is to develop a meta-algorithm that recursively decomposes a classification problem using elementary decomposition methods. We use the term "elementary decomposition" to describe a type of simple decomposition that can be used to build up a more complicated decomposition. Given a certain problem, we first select the most appropriate elementary decomposition to that problem. A suitable decomposer then decomposes the problem, and finally a similar procedure is performed on each sub-problem. This approach agrees with the "no free lunch theorem", namely if one decomposition is better than another in some domains, then there are necessarily other domains in which this relationship is reversed.

For implementing this decomposition methodology, one might consider the following issues:

- What type of elementary decomposition methods exist for classification inducers?
- Which elementary decomposition type performs best for which problem? What factors should one take into account when choosing the appropriate decomposition type?
- Given an elementary type, how should we infer the best decomposition structure automatically?
- How should the sub-problems be re-composed to represent the original concept learning?
- How can we utilize prior knowledge for improving decomposing methodology?

Figure 51.2 suggests an answer to the first issue. This figure illustrates a novel approach for arranging the different elementary types of decomposition in supervised learning (Maimon and Rokach, 2002).

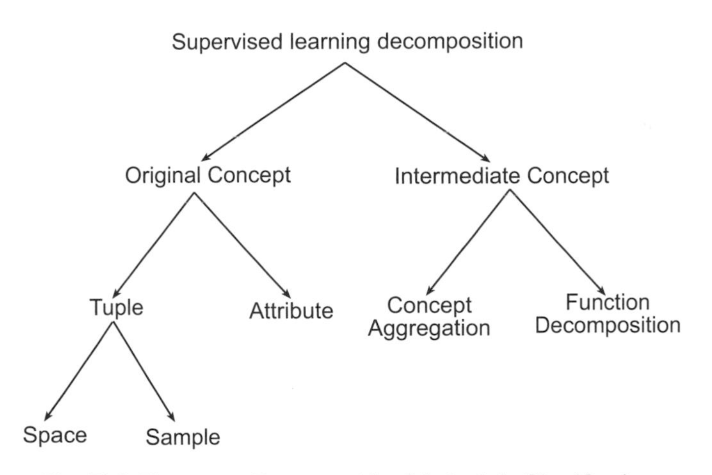

Fig. 51.2. Elementary Decomposition Methods in Classification.

In *intermediate concept* decomposition, instead of inducing a single complicated classifier, several sub-problems with different and more simple concepts are defined. The intermediate concepts can be based on an aggregation of the original concept's values (*concept aggregation*) or not (*function decomposition*).

Classical concept aggregation replaces the original target attribute with a function, such that the domain of the new target attribute is smaller than the original one.

Concept aggregation has been used to classify free text documents into predefined topics (Buntine, 1996). This paper suggests breaking the topics up into groups (co-topics). Instead of predicting the document's topic directly, the document is first classified into one of the co-topics. Another model is then used to predict the actual topic in that co-topic.

A general concept aggregation algorithm called *Error-Correcting Output Coding* (ECOC) which decomposes multi-class problems into multiple, two-class problems has been suggested by Dietterich and Bakiri (1995). A classifier is built for each possible binary partition of the classes. Experiments show that ECOC improves the accuracy of neural networks and decision trees on several multi-class problems from the UCI repository.

The idea to decompose a K class classification problems into K two class classification problems has been proposed by Anand *et al.* (1995). Each problem considers the discrimination of one class to the other classes. Lu and Ito (1999) extend the last method and propose a new method for manipulating the data based on the class relations among training data. By using this method, they divide a K class classification problem into a series of $K(K-1)/2$ two-class problems where each problem considers the discrimination of one class to each one of the other classes. They have examined this idea using neural networks.

Fürnkranz (2002) studied the round-robin classification problem (pairwise classification), a technique for handling multi-class problems, in which one classifier is constructed for each pair of classes. Empirical study has showed that this method can potentially improve classification accuracy.

Function decomposition was originally developed in the Fifties and Sixties for designing switching circuits. It was even used as an evaluation mechanism for checker playing programs (Samuel, 1967). This approach was later improved by Biermann *et al.* (1982). Recently, the machine-learning community has adopted this approach. Michie (1995) used a manual decomposition of the problem and an expert-assisted selection of examples to construct rules for the concepts in the hierarchy. In comparison with standard decision tree induction techniques, structured induction exhibits about the same degree of classification accuracy with the increased transparency and lower complexity of the developed models. Zupan *et al.* (1998) presented a general-purpose function decomposition approach for machine-learning. According to this approach, attributes are transformed into new concepts in an iterative manner and create a hierarchy of concepts. Recently, Long (2003) has suggested using a different function decomposition known as bi-decomposition and shows it applicability in data mining.

Original Concept decomposition means dividing the original problem into several subproblems by partitioning the training set into smaller training sets. A classifier is trained on each sub-sample seeking to solve the original problem. Note that this resembles ensemble methodology but with the following distinction: each inducer uses only a portion of the original training set and ignores the rest. After a classifier is constructed for each portion separately, the models are combined in some fashion, either at learning or classification time.

There are two obvious ways to break up the original dataset: tuple-oriented or attribute (feature) oriented. Tuple decomposition by itself can be divided into two different types: sample and space. In sample decomposition (also known as partitioning), the goal is to partition the training set into several sample sets, such that each sub-learning task considers the entire space.

In space decomposition, on the other hand, the original instance space is divided into several sub-spaces. Each sub-space is considered independently and the total model is a (possibly soft) union of such simpler models.

Space decomposition also includes the divide and conquer approaches such as mixtures of experts, local linear regression, CART/MARS, adaptive subspace models, etc., (Johansen and Foss, 1992, Jordan and Jacobs, 1994, Ramamurti and Ghosh, 1999, Holmstrom *et al.*, 1997).

Feature set decomposition (also known as attribute set decomposition) generalizes the task of feature selection which is extensively used in Data Mining. Feature selection aims to provide a representative set of features from which a classifier is constructed. On the other hand, in feature set decomposition, the original feature set is decomposed into several subsets. An inducer is trained upon the training data for each subset independently, and generates a classifier for each one. Subsequently, an unlabeled instance is classified by combining the classifications of all classifiers. This method potentially facilitates the creation of a classifier for high dimensionality data sets because each sub-classifier copes with only a projection of the original space.

In the literature there are several works that fit the feature set decomposition framework. However, in most of the papers the decomposition structure was obtained ad-hoc using prior knowledge. Moreover, as a result of a literature review, Ronco *et al.* (1996) have concluded that "*There exists no algorithm or method susceptible to perform a vertical self-decomposition without a-priori knowledge of the task!*". Bay (1999) presented a feature set decomposition algorithm known as MFS which combines multiple nearest neighbor classifiers, each using only a subset of random features. Experiments show MFS can improve the standard nearest neighbor classifiers. This procedure resembles the well-known bagging algorithm (Breiman, 1996). However, instead of sampling instances with replacement, it samples features without replacement.

Another feature set decomposition was proposed by Kusiak (2000). In this case, the features are grouped according to the attribute type: nominal value features, numeric value features and text value features. A similar approach was used by Gama (2000) for developing the linear-bayes classifier. The basic idea consists of aggregating the features into two subsets: the first subset containing only the nominal features and the second subset only the continuous features.

An approach for constructing an ensemble of classifiers using rough set theory was presented by Hu (2001). Although Hu's work refers to ensemble methodology and not decomposition methodology, it is still relevant for this case, especially as the declared goal was to construct an ensemble such that different classifiers use different attributes as much as possible. According to Hu, diversified classifiers lead to uncorrelated errors, which in turn improve classification accuracy. The method searches for a set of reducts, which include all the indispensable attributes. A reduct represents the minimal set of attributes which has the same classification power as the entire attribute set.

In another research, Tumer and Ghosh (1996) propose decomposing the feature set according to the target class. For each class, the features with low correlation relating to that class have been removed. This method has been applied on a feature set of 25 sonar signals where the target was to identify the meaning of the sound (whale, cracking ice, etc.). Cherkauer (1996) used feature set decomposition for radar volcanoes recognition. Cherkauer manually decomposed a feature set of 119 into 8 subsets. Features that are based on different image processing operations were grouped together. As a consequence, for each subset, four neural networks with different sizes were built. Chen *et al.* (1997) proposed a new combining framework for feature set decomposition and demonstrate its applicability in text-independent speaker identification. Jenkins and Yuhas (1993) manually decomposed the features set of a certain truck backer-upper problem and reported that this strategy has important advantages.

A paradigm, termed co-training, for learning with labeled and unlabeled data was proposed in Blum and Mitchell (1998). This paradigm can be considered as a feature set decomposition for classifying Web pages, which is useful when there is a large data sample, of which only a small part is labeled. In many applications, unlabeled examples are significantly easier to collect than labeled ones. This is especially true when the labeling process is time-consuming or expensive, such as in medical applications. According to the co-training paradigm, the input space is divided into two different views (i.e. two independent and redundant sets of features). For each view, Blum and Mitchell built a different classifier to classify unlabeled data. The newly labeled data of each classifier is then used to retrain the other classifier. Blum and Mitchell have shown, both empirically and theoretically, that unlabeled data can be used to augment labeled data.

More recently, Liao and Moody (2000) presented another option to a decomposition technique whereby all input features are initially grouped by using a hierarchical clustering algorithm based on pairwise mutual information, with statistically similar features assigned to the same group. As a consequence, several feature subsets are constructed by selecting one feature from each group. A neural network is subsequently constructed for each subset. All netwroks are then combined.

In the statistics literature, the most well-known decomposition algorithm is the MARS algorithm (Friedman, 1991). In this algorithm, a multiple regression function is approximated using linear splines and their tensor products. It has been shown that the algorithm performs an ANOVA decomposition, namely the regression function is represented as a grand total of several sums. The first sum is of all basic functions that involve only a single attribute. The second sum is of all basic functions that involve exactly two attributes, representing (if present) two-variable interactions. Similarly, the third sum represents (if present) the contributions from three-variable interactions, and so on.

Other works on feature set decomposition have been developed by extending the Naïve Bayes classifier. The Naïve Bayes classifier (Domingos and Pazzani, 1997) uses the Bayes' rule to compute the conditional probability of each possible class, assuming the input features are conditionally independent given the target feature. Due to the conditional independence assumption, this method is called "Naïve". Nevertheless, a variety of empirical researches show surprisingly that the Naïve Bayes classifier can perform quite well compared to other methods, even in domains where clear feature dependencies exist (Domingos and Pazzani, 1997). Furthermore, Naïve Bayes classifiers are also very simple and easy to understand (Kononenko, 1990).

Both Kononenko (1991) and Domingos and Pazzani (1997), suggested extending the Naïve Bayes classifier by finding the single best pair of features to join by considering all possible joins. Kononenko (1991) described the semi-Naïve Bayes classifier that uses a conditional independence test for joining features. Domingos and Pazzani (1997) used estimated accuracy (as determined by leave–one–out cross-validation on the training set). Friedman *et al.* (1997) have suggested the tree augmented Naïve Bayes classifier (TAN) which extends the Naïve Bayes, taking into account dependencies among input features. The selective Bayes Classifier (Langley and Sage, 1994) preprocesses data using a form of feature selection to delete redundant features. Meretakis and Wthrich (1999) introduced the large Bayes algorithm. This algorithm employs an *a-priori*-like frequent pattern-mining algorithm to discover frequent and interesting features in subsets of arbitrary size, together with their class probability estimation.

Recently Maimon and Rokach (2005) suggested a general framework that searches for helpful feature set decomposition structures. This framework nests many algorithms, two of which are tested empirically over a set of benchmark datasets. The first algorithm performs a serial search while using a new Vapnik-Chervonenkis dimension bound for multiple oblivious trees as an evaluating schema. The second algorithm performs a multi-search while using wrapper evaluating schema. This work indicates that feature set decomposition can increase the accuracy of decision trees.

It should be noted that some researchers prefer the terms "horizontal decomposition" and "vertical decomposition" for describing "space decomposition" and "attribute decomposition" respectively (Ronco *et al.*, 1996).

51.4 The Decomposer's Characteristics

51.4.1 Overview

The following sub-sections present the main properties that characterize decomposers. These properties can be useful for differentiating between various decomposition frameworks.

51.4.2 The Structure Acquiring Method

This important property indicates how the decomposition structure is obtained:

- Manually (explicitly) based on an expert's knowledge in a specific domain (Blum and Mitchell, 1998, Michie, 1995). If the origin of the dataset is a relational database, then the schema's structure may imply the decomposition structure.
- Predefined due to some restrictions (as in the case of distributed Data Mining)
- Arbitrarily (Domingos, 1996, Chan and Stolfo, 1995) - The decomposition is performed without any profound thought. Usually, after setting the size of the subsets, members are randomly assigned to the different subsets.
- Induced without human interaction by a suitable algorithm (Zupan *et al.*, 1998).

Some may justifiably claim that searching for the best decomposition might be time-consuming, namely prolonging the Data Mining process. In order to avoid this disadvantage, the complexity of the decomposition algorithms should be kept as small as possible. However, even if this cannot be accomplished, there are still important advantages, such as better comprehensibility and better performance that makes decomposition worth the additional computational complexity.

Furthermore, it should be noted that in an ongoing Data Mining effort (like in a churning application) searching for the best decomposition structure might be performed in wider time buckets (for instance, once a year) than when training the classifiers (for instance once a week). Moreover, for acquiring decomposition structure, only a relatively small sample of the training set may be required. Consequently, the execution time of the decomposer will be relatively small compared to the time needed to train the classifiers.

Ronco *et al.* (1996) suggest a different categorization in which the first two categories are referred as "ad-hoc decomposition" and the last two categories as "self-decomposition".

Usually in real-life applications the decomposition is performed manually by incorporating business information into the modeling process. For instance Berry and Linoff (2000) provide a practical example in their book saying:

> It may be known that platinum cardholders behave differently from gold cardholders. Instead of having a Data Mining technique figure this out, give it the hint by building separate models for the platinum and gold cardholders.

Berry and Linoff (2000) state that decomposition can be also useful for handling missing data. In this case they do not refer to sporadic missing data but to the case where several attribute values are available for some tuples but not for all of them. For instance: "Historical data, such as billing information, is available only for customers who have been around for a sufficiently long time" or "Outside data, such as demographics, is available only for the subset of the customer base that matches"). In this case, one classifier can be trained for customers having all the information and a second classifier for the remaining customers.

51.4.3 The Mutually Exclusive Property

This property indicates whether the decomposition is mutually exclusive (*disjointed decomposition*) or partially overlapping (i.e. a certain value of a certain attribute in a certain tuple is utilized more than once). For instance, in the case of sample decomposition, "mutually exclusive" means that a certain tuple cannot belong to more than one subset (Domingos, 1996,Chan and Stolfo, 1995). Bay (1999), on the other hand, has used non-exclusive feature decomposition.

Similarly CART and MARS perform mutually exclusive decomposition of the input space, while HME allows sub-spaces to overlap.

Mutually exclusive decomposition can be deemed as a *pure* decomposition. While pure decomposition forms a restriction on the problem space, it has some important and helpful properties:

- A greater tendency in reduction of execution time than non-exclusive approaches. Since most learning algorithms have computational complexity that is greater than linear in the number of attributes or tuples, partitioning the problem dimensionality in a mutually exclusive manner means a decrease in computational complexity (Provost and Kolluri, 1997).
- Since mutual exclusiveness entails using smaller datasets, the models obtained for each sub-problem are smaller in size. Without the mutually exclusive restriction, each model can be as complicated as the model obtained for the original problem. Smaller models contribute to comprehensibility and ease in maintaining the solution.
- According to Bay (1999), mutually exclusive decomposition may help avoid some error correlation problems that characterize non-mutually exclusive decompositions. However, Sharkey (1999) argues that mutually exclusive training sets do not necessarily result in low error correlation. This point is true when each sub-problem is representative (i.e. represent the entire problem, as in sample decomposition).
- Reduced tendency to contradiction between sub-models. When a mutually exclusive restriction is unenforced, different models might generate contradictive classifications using the same input. Reducing inter-models contraindications help us to grasp the results and to combine the sub-models into one model. Ridgeway *et al.* (1999), for instance, claim that the resulting predictions of ensemble methods are usually inscrutable to end-users, mainly due to the complexity of the generated models, as well as the obstacles in transforming theses models into a single model. Moreover, since these methods do not attempt to use all relevant features, the researcher will not obtain a complete picture of which attribute actually affects the target attribute, especially when, in some cases, there are many relevant attributes.
- Since the mutually exclusive approach encourages smaller datasets, they are more feasible. Some Data Mining tools can process only limited dataset size (for instance when the program requires that the entire dataset will be stored in the main memory). The mutually

exclusive approach can make certain that Data Mining tools are fairly scalable to large data sets (Chan and Stolfo, 1997, Provost and Kolluri, 1997).

- We claim that end-users can grasp mutually exclusive decomposition much easier than many other methods currently in use. For instance, boosting, which is a well-known ensemble method, distorts the original distribution of instance space, a fact that non-professional users find hard to grasp or understand.

51.4.4 The Inducer Usage

This property indicates the relation between the decomposer and the inducer used. Some decomposition implementations are "inducer-free", namely they do not use intrinsic inducers at all. Usually the decomposition procedure needs to choose the best decomposition structure among several structures that it considers. In order to measure the performance of a certain decomposition structure, there is a need to realize the structure by building a classifier for each component. However since "inducer-free" decomposition does not use any induction algorithm, it uses a frequency table of the Cartesian product of the feature values instead. Consider the following example. The training set consists of four binary input attributes (a_1, a_2, a_3, a_4) and one target attribute (y). Assume that an "inducer-free" decomposition procedure examines the following feature set decomposition: (a_1, a_3) and (a_2, a_4). In order to measure the classification performance of this structure, it is required to build two classifiers; one classifier for each subset. In the absence of an induction algorithm, two frequency tables are built; each table has $2^2 = 4$ entries representing the Cartesian product of the attributes in each subset. For each entry in the table, we measure the frequency of the target attribute. Each one of the tables can be separately used to classify a new instance x: we search for the entry that corresponds to the instance x and select the target value with the highest frequency in that entry. This "inducer-free" strategy has been used in several places. For instance the extension of Naïve Bayes suggested by Domingos and Pazzani (1997), can be considered as a feature set decomposition with no intrinsic inducer. Zupan *et al.* (1998) have developed the function decomposition by using sparse frequency tables.

Other implementations are considered as an "inducer-dependent" type, namely these decomposition methods use intrinsic inducers, and they have been developed specifically for a certain inducer. They do not guarantee effectiveness in any other induction method. For instance, the work of Lu and Ito (1999) was developed specifically for neural networks.

The third type of decomposition method is the "inducer-independent" type. These implementations can be performed on any given inducer, however, the same inducer is used in all subsets. As opposed to the "inducer-free" implementation, which does not use any inducer for its execution, "inducer-independent" requires the use of an inducer. Nevertheless, it is not limited to a specific inducer like the "inducer-dependent".

The last type is the "inducer-chooser" type, which, given a set of inducers, the system uses the most appropriate inducer on each sub-problem.

51.4.5 Exhaustiveness

This property indicates whether all data elements should be used in the decomposition. For instance, an exhaustive feature set decomposition refers to the situation in which each feature participates in at least one subset.

51.4.6 Combiner Usage

This property specifies the relation between the decomposer and the combiner. Some decomposers are combiner-dependent. That is to say they have been developed specifically for a certain combination method like voting or Naïve Bayes. For additional combining methods see Chapter 49.6 in this volume. Other decomposers are combiner-independent; the combination method is provided as input to the framework. Potentially there could be decomposers that, given a set of combiners, would be capable of choosing the best combiner in the current case.

51.4.7 Sequentially or Concurrently

This property indicates whether the various sub-classifiers are built sequentially or concurrently. In sequential framework the outcome of a certain classifier may effect the creation of the next classifier. On the other hand, in concurrent framework each classifier is built independently and their results are combined in some fashion. Sharkey (1996) refers to this property as "The relationship between modules" and distinguishes between three different types: successive, cooperative and supervisory. Roughly speaking the "successive" refers to "sequential" while "cooperative" refers to "concurrent". The last type applies to the case in which one model controls the other model. Sharkey (1996) provides an example in which one neural network is used to tune another neural network.

The original problem in *intermediate concept decomposition* is usually converted to a sequential list of problems, where the last problem aims to solve the original one. On the other hand, in *original concept decomposition* the problem is usually divided into several subproblems which exist on their own. Nevertheless, there are some exceptions. For instance, Quinlan (1993) proposed an original concept framework known as "windowing" that is considered to be sequential. For other examples the reader is referred to Chapter 49.6 in this volume.

Naturally there might be other important properties which can be used to differentiate a decomposition scheme. Table 51.1 summarizes the most relevant research performed on each decomposition type.

Table 51.1. Summary of Decomposition Methods in the Literature.

Paper	Decomposition Type	Mutually Exclusive	Structure Acquiring Method
(Anand *et al.*, 1995)	Concept	No	Arbitrarily
(Buntine, 1996)	Concept	Yes	Manually
(Michie, 1995)	Function	Yes	Manually
(Zupan *et al.*, 1998)	Function	Yes	Induced
(Ali and Pazzani, 1996)	Sample	No	Arbitrarily
(Domingos, 1996)	Sample	Yes	Arbitrarily
(Ramamurti and Ghosh, 1999)	Space	No	Induced
(Kohavi *et al.*, 1997)	Space	Yes	Induced
(Bay, 1999)	Attribute	No	Arbitrarily
(Kusiak, 2000)	Attribute	Yes	Manually

51.5 The Relation to Other Methodologies

The main distinction between existing approaches, such as ensemble methods and distributed Data Mining to decomposition methodology, focuses on the following fact: the assumption that each model has access to a comparable quality of data is not valid in the decomposition approach (Tumer and Ghosh, 2000):

> A fundamental assumption in all the multi-classifier approaches is that the designer has access to the entire data set, which can be used in its entirety, resampled in a random (bagging) or weighted (boosting) way, or randomly partitioned and distributed. Thus, except for boosting situations, each classifier sees training data of comparable quality. If the individual classifiers are then appropriately chosen and trained properly, their performances will be (relatively) comparable in any region of the problem space. So gains from combining are derived from the diversity among classifiers rather that by compensating for weak members of the pool.

This assumption is clearly invalid for decomposition methodology, where classifiers may have significant variations in their overall performance. Furthermore when individual classifiers have substantially different performances over different parts of the input space, combining is still desirable (Tumer and Ghosh, 2000). Nevertheless neither simple combiners nor more sophisticated combiners are particularly well-suited for the type of problems that arise (Tumer and Ghosh, 2000):

> The simplicity of averaging the classifier outputs is appealing, but the prospect of one poor classifier corrupting the combiner makes this a risky choice. Weighted averaging of classifier outputs appears to provide some flexibility. Unfortunately, the weights are still assigned on a per classifier basis rather than a per tuple basis. If a classifier is accurate only in certain areas of the input space, this scheme fails to take advantage of the variable accuracy of the classifier in question. Using a combiner that provides different weights for different patterns can potentially solve this problem, but at a considerable cost.

The ensemble methodology is closely related to the decomposition methodology (see Chapter 49.6 in this volume). In both cases the final model is a composite of multiple models combined in some fashion. However, Sharkey (1996) distinguishes between these methodologies in the following way: the main idea of ensemble methodology is to combine a set of models, each of which solves the same original task. The purpose of ensemble methodology is to obtain a more accurate and reliable performance than when using a single model. On the other hand, the purpose of decomposition methodology is to break down a complex problem into several manageable problems, enabling each inducer to solve a different task. Therefore, in ensemble methodology, any model can provide a sufficient solution to the original task. On the other hand, in decomposition methodology, a combination of all models is mandatory for obtaining a reliable solution.

Distributed Data Mining (DDM) deals with mining data that might be inherently distributed among different, loosely coupled sites with slow connectivity, such as geographically distributed sites connected over the Internet (Kargupta and Chan, 2000). Usually DDM is categorized according to data distribution:

Homogeneous. In this case, the datasets in all the sites are built from the same common set of attributes. This state is equivalent to the sample decomposition discussed above, when the decomposition structure is set by the environment.

Heterogeneous. In this case, the quality and quantity of data available to each site may vary substantially. Since each specific site may contain data for different attributes, leading to large discrepancies in their performance, integrating classification models derived from distinct and distributed databases is complex.

DDM can be useful also in the case of "mergers and acquisitions" of corporations. In such cases, since each company involved may have its own IT legacy systems, different sets of data are available.

In DDM the different sources are given, namely the instances are pre-decomposed. As a result, DDM is mainly focused on combining the various methods. Several researchers discuss ways of leveraging distributed techniques in knowledge discovery, such as data cleaning and preprocessing, transformation, and learning.

Prodromidis *et al.* (1999) proposed the JAM system a meta-learning approach for DDM. The meta-learning approach is about combining several models (describing several sets of data from several sources of data) into one high-level model. Guo and Sutiwaraphun (1998) describe a meta-learning concept know-as *knowledge probing*. In knowledge probing, supervised learning is organized into two stages. In the first stage, a set of base classifiers is constructed using the distributed data sets. In the second stage, the relationship between an attribute vector and the class predictions from all of the base classifiers is determined. Grossman *et al.* (1999) outline fundamental challenges for mining large-scale databases, one of them being the need to develop DDM algorithms.

A closely related field is *Parallel Data Mining* (PDM). PDM deals with mining data by using several tightly-coupled systems with fast interconnection, as in the case of a cluster of shared memory workstations (Zaki and Ho, 2000).

The main goal of PDM techniques is to scale-up the speed of the Data Mining on large datasets. It addresses the issue by using high performance, multi-processor computers. The increasing availability of such computers calls for extensive development of data analysis algorithms that can scale up as we attempt to analyze data sets measured in terabytes on parallel machines with thousands of processors. This technology is particularly suitable for applications that typically deal with large amounts of data, e.g. company transaction data, scientific simulation and observation data. Another important example of PDM is the SPIDER project that uses shared-memory multiprocessors systems (SMPs) to accomplish PDM on distributed data sets (Zaki, 1999). Please refer to Chapter 52.5 for more information.

51.6 Summary

In this chapter we have reviewed the necessity of decomposition methodology in Data Mining and knowledge discovery. We have suggested an approach to categorize elementary decomposition methods. We also discussed the main characteristics of decomposition methods and showed its suitability to the current research in the literature.

The methods presented in this chapetr are useful for many application domains, such as: Manufacturing lr18,lr14, Security lr7,l10 and Medicine lr2,lr9, and for many data mining techniques, such as: decision trees lr6,lr12, lr15, clustering lr13,lr8,lr5,lr16 and genetic algorithms lr17,lr11,lr1,lr4.

References

Ali K. M., Pazzani M. J., Error Reduction through Learning Multiple Descriptions, Machine Learning, 24: 3, 173-202, 1996.

Anand R, Methrotra K, Mohan CK, Ranka S. Efficient classification for multiclass problems using modular neural networks. IEEE Trans Neural Networks, 6(1): 117-125, 1995.

Arbel, R. and Rokach, L., Classifier evaluation under limited resources, Pattern Recognition Letters, 27(14): 1619–1631, 2006, Elsevier.

Averbuch, M. and Karson, T. and Ben-Ami, B. and Maimon, O. and Rokach, L., Context-sensitive medical information retrieval, The 11th World Congress on Medical Informatics (MEDINFO 2004), San Francisco, CA, September 2004, IOS Press, pp. 282–286.

Baxt, W. G., Use of an artificial neural network for data analysis in clinical decision making: The diagnosis of acute coronary occlusion. Neural Computation, 2(4):480-489, 1990.

Bay, S., Nearest neighbor classification from multiple feature subsets. Intelligent Data Analysis, 3(3): 191-209, 1999.

Bhargava H. K., Data Mining by Decomposition: Adaptive Search for Hypothesis Generation, INFORMS Journal on Computing Vol. 11, Iss. 3, pp. 239-47, 1999.

Biermann, A. W., Faireld, J., and Beres, T., 1982. Signature table systems and learning. IEEE Trans. Syst. Man Cybern., 12(5):635-648.

Blum A., and Mitchell T., Combining Labeled and Unlabeled Data with CoTraining. In Proc. of the 11th Annual Conference on Computational Learning Theory, pages 92-100, 1998.

Breiman L., Bagging predictors, Machine Learning, 24(2):123-140, 1996.

Buntine, W., "Graphical Models for Discovering Knowledge", in U. Fayyad, G. Piatetsky-Shapiro, P. Smyth, and R. Uthurusamy, editors, Advances in Knowledge Discovery and Data Mining, pp 59-82. AAAI/MIT Press, 1996.

Chan P.K. and Stolfo S.J, On the Accuracy of Meta-learning for Scalable Data Mining, J. Intelligent Information Systems, 8:5-28, 1997.

Chen K., Wang L. and Chi H., Methods of Combining Multiple Classifiers with Different Features and Their Applications to Text-Independent Speaker Identification, International Journal of Pattern Recognition and Artificial Intelligence, 11(3): 417-445, 1997.

Cherkauer, K.J., Human Expert-Level Performance on a Scientific Image Analysis Task by a System Using Combined Artificial Neural Networks. In
Working Notes, Integrating Multiple Learned Models for Improving and Scaling Machine Learning Algorithms Workshop, Thirteenth National Conference on Artificial Intelligence. Portland, OR: AAAI Press, 1996.

Cohen S., Rokach L., Maimon O., Decision Tree Instance Space Decomposition with Grouped Gain-Ratio, Information Science, Volume 177, Issue 17, pp. 3592-3612, 2007.

Dietterich, T. G., and Ghulum Bakiri. Solving multiclass learning problems via error-correcting output codes. Journal of Artificial Intelligence Research, 2:263-286, 1995.

Domingos, P., Using Partitioning to Speed Up Specific-to-General Rule Induction. In Proceedings of the AAAI-96 Workshop on Integrating Multiple Learned Models, pp. 29-34, AAAI Press, 1996.

Domingos, P., & Pazzani, M., On the Optimality of the Naive Bayes Classifier under Zero-One Loss, Machine Learning, 29: 2, 103-130, 1997.

Fischer, B., "Decomposition of Time Series - Comparing Different Methods in Theory and Practice", Eurostat Working Paper, 1995.

Friedman, J. H., "Multivariate Adaptive Regression Splines", The Annual Of Statistics, 19, 1-141, 1991.

Friedman N., Geiger D., and Goldszmidt M., Bayesian Network Classifiers, Machine Learning 29: 2-3, 131-163, 1997.

Gama J., A Linear-Bayes Classifier. In C. Monard, editor, Advances on Artificial Intelligence – SBIA2000. LNAI 1952, pp 269-279, Springer Verlag, 2000

Grossman R., Kasif S., Moore R., Rocke D., and Ullman J., Data Mining research: Opportunities and challenges. Report of three NSF workshops on mining large, massive, and distributed data, 1999.

Guo Y. and Sutiwaraphun J., Knowledge probing in distributed Data Mining, in Proc. 4h Int. Conf. Knowledge Discovery Data Mining, pp 61-69, 1998.

Hansen J., Combining Predictors. Meta Machine Learning Methods and Bias, Variance & Ambiguity Decompositions. PhD dissertation. Aurhus University. 2000.

Hampshire, J. B., and Waibel, A. The meta-Pi network - building distributed knowledge representations for robust multisource pattern-recognition. Pattern Analyses and Machine Intelligence 14(7): 751-769, 1992.

He D. W., Strege B., Tolle H., and Kusiak A., Decomposition in Automatic Generation of Petri Nets for Manufacturing System Control and Scheduling, International Journal of Production Research, 38(6): 1437-1457, 2000.

Holmstrom, L., Koistinen, P., Laaksonen, J., and Oja, E., Neural and statistical classifiers - taxonomy and a case study. IEEE Trans. on Neural Networks, 8,:5–17, 1997.

Hrycej T., Modular Learning in Neural Networks. New York: Wiley, 1992.

Hu, X., Using Rough Sets Theory and Database Operations to Construct a Good Ensemble of Classifiers for Data Mining Applications. ICDM01. pp 233-240, 2001.

Jenkins R. and Yuhas, B. P. A simplified neural network solution through problem decomposition: The case of Truck backer-upper, IEEE Transactions on Neural Networks 4(4):718-722, 1993.

Johansen T. A. and Foss B. A., A narmax model representation for adaptive control based on local model -Modeling, Identification and Control, 13(1):25-39, 1992.

Jordan, M. I., and Jacobs, R. A., Hierarchical mixtures of experts and the EM algorithm. Neural Computation, 6, 181-214, 1994.

Kargupta, H. and Chan P., eds, Advances in Distributed and Parallel Knowledge Discovery , pp. 185-210, AAAI/MIT Press, 2000.

Kohavi R., Becker B., and Sommerfield D., Improving simple Bayes. In Proceedings of the European Conference on Machine Learning, 1997.

Kononenko, I., Comparison of inductive and Naive Bayes learning approaches to automatic knowledge acquisition. In B. Wielinga (Ed.), Current Trends in Knowledge Acquisition, Amsterdam, The Netherlands IOS Press, 1990.

Kononenko, I., SemiNaive Bayes classifier, Proceedings of the Sixth European Working Session on Learning, pp. 206-219, Porto, Portugal: SpringerVerlag, 1991.

Kusiak, A., Decomposition in Data Mining: An Industrial Case Study, IEEE Transactions on Electronics Packaging Manufacturing, Vol. 23, No. 4, pp. 345-353, 2000.

Kusiak, E. Szczerbicki, and K. Park, A Novel Approach to Decomposition of Design Specifications and Search for Solutions, International Journal of Production Research, 29(7): 1391-1406, 1991.

Langley, P. and Sage, S., Oblivious decision trees and abstract cases. in Working Notes of the AAAI-94 Workshop on Case-Based Reasoning, pp. 113-117, Seattle, WA: AAAI Press, 1994.

Liao Y., and Moody J., Constructing Heterogeneous Committees via Input Feature Grouping, in Advances in Neural Information Processing Systems, Vol.12, S.A. Solla, T.K. Leen and K.-R. Muller (eds.),MIT Press, 2000.

Long C., Bi-Decomposition of Function Sets Using Multi-Valued Logic, Eng. Doc. Dissertation, Technischen Universitat Bergakademie Freiberg 2003.

Lu B.L., Ito M., Task Decomposition and Module Combination Based on Class Relations: A Modular Neural Network for Pattern Classification, IEEE Trans. on Neural Networks, 10(5):1244-1256, 1999.

Maimon O., and Rokach, L. Data Mining by Attribute Decomposition with semiconductors manufacturing case study, in Data Mining for Design and Manufacturing: Methods and Applications, D. Braha (ed.), Kluwer Academic Publishers, pp. 311–336, 2001.

Maimon O. and Rokach L., "Improving supervised learning by feature decomposition", Proceedings of the Second International Symposium on Foundations of Information and Knowledge Systems, Lecture Notes in Computer Science, Springer, pp. 178-196, 2002.

Maimon, O. and Rokach, L., Decomposition Methodology for Knowledge Discovery and Data Mining: Theory and Applications, Series in Machine Perception and Artificial Intelligence - Vol. 61, World Scientific Publishing, ISBN:981-256-079-3, 2005.

Meretakis, D. and Wthrich, B., Extending Nave Bayes Classifiers Using Long Itemsets, in Proceedings of the Fifth International Conference on Knowledge Discovery and Data Mining, pp. 165-174, San Diego, USA, 1999.

Michie, D., Problem decomposition and the learning of skills, in Proceedings of the European Conference on Machine Learning, pp. 17-31, Springer-Verlag, 1995.

Moskovitch R, Elovici Y, Rokach L, Detection of unknown computer worms based on behavioral classification of the host, Computational Statistics and Data Analysis, 52(9):4544–4566, 2008.

Nowlan S. J., and Hinton G. E. Evaluation of adaptive mixtures of competing experts. In Advances in Neural Information Processing Systems, R. P. Lippmann, J. E. Moody, and D. S. Touretzky, Eds., vol. 3, pp. 774-780, Morgan Kaufmann Publishers Inc., 1991.

Ohno-Machado, L., and Musen, M. A. Modular neural networks for medical prognosis: Quantifying the benefits of combining neural networks for survival prediction. Connection Science 9, 1, 1997, 71-86.

Peng, F. and Jacobs R. A., and Tanner M. A., Bayesian Inference in Mixtures-of-Experts and Hierarchical Mixtures-of-Experts Models With an Application to Speech Recognition, Journal of the American Statistical Association, 1995.

Pratt, L. Y., Mostow, J., and Kamm C. A., Direct Transfer of Learned Information Among Neural Networks, in: Proceedings of the Ninth National Confcrence on Artificial Intelligence, Anaheim, CA, 584-589, 1991.

Provost, F.J. and Kolluri, V., A Survey of Methods for Scaling Up Inductive Learning Algorithms, Proc. 3rd International Conference on Knowledge Discovery and Data Mining, 1997.

Quinlan, J. R., C4.5: Programs for Machine Learning, Morgan Kaufmann, Los Altos, 1993.

Rahman, A. F. R., and Fairhurst, M. C. A new hybrid approach in combining multiple experts to recognize handwritten numerals. Pattern Recognition Letters, 18: 781-790,1997.

Ramamurti, V., and Ghosh, J., Structurally Adaptive Modular Networks for Non-Stationary Environments, IEEE Transactions on Neural Networks, 10 (1):152-160, 1999.

Ridgeway, G., Madigan, D., Richardson, T. and O'Kane, J., Interpretable Boosted Naive Bayes Classification, Proceedings of the Fourth International Conference on Knowledge Discovery and Data Mining, pp 101-104, 1998.

Rokach, L., Decomposition methodology for classification tasks: a meta decomposer framework, Pattern Analysis and Applications, 9(2006):257–271.

Rokach L., Genetic algorithm-based feature set partitioning for classification problems,Pattern Recognition, 41(5):1676–1700, 2008.

Rokach L., Mining manufacturing data using genetic algorithm-based feature set decomposition, Int. J. Intelligent Systems Technologies and Applications, 4(1):57-78, 2008.

Rokach, L. and Maimon, O., Theory and applications of attribute decomposition, IEEE International Conference on Data Mining, IEEE Computer Society Press, pp. 473–480, 2001.

Rokach L. and Maimon O., Feature Set Decomposition for Decision Trees, Journal of Intelligent Data Analysis, Volume 9, Number 2, 2005b, pp 131–158.

Rokach, L. and Maimon, O., Clustering methods, Data Mining and Knowledge Discovery Handbook, pp. 321–352, 2005, Springer.

Rokach, L. and Maimon, O., Data mining for improving the quality of manufacturing: a feature set decomposition approach, Journal of Intelligent Manufacturing, 17(3):285–299, 2006, Springer.

Rokach, L., Maimon, O., Data Mining with Decision Trees: Theory and Applications, World Scientific Publishing, 2008.

Rokach L., Maimon O. and Lavi I., Space Decomposition In Data Mining: A Clustering Approach, Proceedings of the 14th International Symposium On Methodologies For Intelligent Systems, Maebashi, Japan, Lecture Notes in Computer Science, Springer-Verlag, 2003, pp. 24–31.

Rokach, L. and Maimon, O. and Averbuch, M., Information Retrieval System for Medical Narrative Reports, Lecture Notes in Artificial intelligence 3055, page 217-228 Springer-Verlag, 2004.

Rokach, L. and Maimon, O. and Arbel, R., Selective voting-getting more for less in sensor fusion, International Journal of Pattern Recognition and Artificial Intelligence 20 (3) (2006), pp. 329–350.

Ronco, E., Gollee, H., and Gawthrop, P. J., Modular neural network and self-decomposition. CSC Research Report CSC-96012, Centre for Systems and Control, University of Glasgow, 1996.

Saaty, X., The analytic hierarchy process: A 1993 overview. Central European Journal for Operations Research and Economics, Vol. 2, No. 2, p. 119-137, 1993.

Samuel, A., Some studies in machine learning using the game of checkers II: Recent progress. IBM J. Res. Develop., 11:601-617, 1967.

Sharkey, A., On combining artificial neural nets, Connection Science, Vol. 8, pp.299-313, 1996.

Sharkey, A., Multi-Net Iystems, In Sharkey A. (Ed.) Combining Artificial Neural Networks: Ensemble and Modular Multi-Net Systems. pp. 1-30, Springer
-Verlag, 1999.

Tumer, K. and Ghosh J., Error Correlation and Error Reduction in Ensemble Classifiers, Connection Science, Special issue on combining artificial neural networks: ensemble approaches, 8 (3-4): 385-404, 1996.

Tumer, K., and Ghosh J., Linear and Order Statistics Combiners for Pattern Classification, in Combining Articial Neural Nets, A. Sharkey (Ed.), pp. 127-162, Springer-Verlag, 1999.

Weigend, A. S., Mangeas, M., and Srivastava, A. N. Nonlinear gated experts for time-series - discovering regimes and avoiding overfitting. International Journal of Neural Systems 6(5):373-399, 1995.

Zaki, M. J., Ho C. T., and Agrawal, R., Scalable parallel classification for Data Mining on shared- memory multiprocessors, in Proc. IEEE Int. Conf. Data Eng., Sydney, Australia, WKDD99, pp. 198– 205, 1999.

Zaki, M. J., Ho C. T., Eds., Large- Scale Parallel Data Mining. New York: Springer- Verlag, 2000.

Zupan, B., Bohanec, M., Demsar J., and Bratko, I., Feature transformation by function decomposition, IEEE intelligent systems & their applications, 13: 38-43, 1998.

52

Information Fusion - Methods and Aggregation Operators

Vicenç Torra

Institut d'Investigació en Intel·ligència Artificial

Summary. Information fusion techniques are commonly applied in Data Mining and Knowledge Discovery. In this chapter, we will give an overview of such applications considering their three main uses. This is, we consider fusion methods for data preprocessing, model building and information extraction. Some aggregation operators (i.e. particular fusion methods) and their properties are briefly described as well.

Key words: Information fusion, aggregation operators, preprocessing, multi-database Data Mining, re-identification algorithms, ensemble methods, information summarization

52.1 Introduction

Data, in any of their possible shapes, is the basic *material* for knowledge discovery. However, this *material* is often not polished and, therefore, it has to be prepared before Data Mining methods are applied. Information fusion offers some basic methods that are useful in this initial step of data preprocessing. This is, to improve the quality of the data prior to subsequent analysis and to the application of Data Mining methods.

This is not the only situation in which information fusion can be applied. In fact, fusion techniques are known to be also used for building data models and to extract information. For example, they are used in ensemble methods to build composite models or for computing representatives of the data.

In this chapter we will describe the main uses of information fusion in knowledge discovery. The structure of the chapter is as follows. In Section 52.2, we will give an overview of information fusion techniques for data preprocessing. Then, in Section 52.3, we will review their use for building models (for both building composite models and for defining data models). Section 52.4 is devoted to information extraction and summarization. The chapter finishes in Section 52.5 with some conclusions.

O. Maimon, L. Rokach (eds.), *Data Mining and Knowledge Discovery Handbook*, 2nd ed.,
DOI 10.1007/978-0-387-09823-4_52,

52.2 Preprocessing Data

Collected data usually contains errors (either introduced on purpose *e.g.* for protecting confidentiality as in privacy preserving Data Mining, or due to incorrect data handling). Such errors make data processing a difficult task as incorrect models might be inferred from the erroneous data.

This situation is even more noticeable in multi database data mining (Wrobel, 1997,Zhong *et al.*, 1999). In such framework, models have to be extracted from data distributed among several databases. Then, data is usually non consistent, attributes in different databases are not codified in a unified way (they might have different names) and the domain of the attributes is not the same.

Information fusion techniques permit to deal with some of these difficulties. We describe below some of the current techniques in use for dealing with these problems. Namely, re-identification algorithms in multi database Data Mining; fusion and aggregation operators for improving the quality of data (for both multi-database and single source database data mining).

52.2.1 Re-identification Algorithms

In the construction of models from multiple databases, re-identification methods play a central role. They are to link those data descriptions that while distributed in different data files belong to the same object. To formalize such methods, let us consider a database A and a database B, both containing information about the same individuals but being the former described in terms of attributes $A_1, \ldots, A_n$ and the latter in terms of attributes $B_1, \ldots, B_m$.

In this setting, we can distinguish two groups of algorithms. They are the following ones:

Record Linkage (or Record Matching) Methods

Given a record r in A, such methods consist on finding all records in B that correspond to the same individual than r. Different methods rely on different assumptions on the attributes A_i and B_i and on the underlying model for the data.

Classical methods assume that both files share a large enough set of attributes and that such attributes have the same domain. Difficulties on re-identifying the records are solely caused by the errors in the data. In this setting, different algorithms have been developed: probabilistic record linkage, distance-based record linkage. Recently, a cluster-based approach was introduced, similar in spirit to distance-based record linkage.

Probabilistic record linkage is described in detail in (Jaro, 1989). This work describes the application of such method to the 1985 Census of Tampa (Florida). A more up-to-date description of the field is given in (Winkler, 1995a) and (Winkler, 1995b). Distance-based record linkage was proposed by (Pagliuca and Seri, 1999). Both methods are reviewed and compared in (Torra and Domingo-Ferrer, 2003). Cluster-based approach is given in (Bacher *et al.*, 2002).

New methods (Torra, 2000b, Domingo-Ferrer and Torra, 2003) have been proposed recently that weaken the conditions required for the variables. Variables are no longer assumed to be the same for both files but only *similar*. In this case, re-identification is based on the assumption that there is some structural information that is present in both files and that can be extracted from the relationships between objects. Differences on the methods correspond to differences on the way relationships are expressed.

For example, Torra (2000c) expresses such similarities considering partitions. This is, similar objects are clustered together while dissimilar objects are distributed in different clusters; and Torra (2000b, 2004) computes aggregates for records and use such aggregates to compute similarities between records.

Scheme Matching Methods

These methods establish proper correspondence between two or more database schemes.

The most typical situation is to find correspondences between attributes. Methods to deal with this situation are the so-called *attribute correspondence identification methods*.

Given the attributes $A_1,\ldots,A_n$ of a database A, such methods are to find the attributes in $B_1,\ldots,B_m$ of database B that describe the same concept. Such relationships can be one-to-one, many-to-one or many-to-many. The first case is when an attribute in one file corresponds to an attribute in the second file (although they can use different domains, or the same domain with different granularities); the second case is when one attribute in one file is represented by many attributes in the second file (e.g. address in one file, vs. street and number in the second file) and the third case, when only complex relations can be established between attributes. For examples of the first case see e.g. (Bergamaschi *et al.*, 2001) and (Do and Rahm, 2002). See e.g. (Borkar *et al.*, 2000) for an example of the second case. The third case is not presently studied in the literature.

Structure-level matching methods are an alternative to attribute correspondence identification methods that deal with more general situations not directly corresponding to attribute correspondence. See (Rahm and Bernstein, 2001) for a review.

Although we have divided re-identification algorithms in two classes, there are algorithms that can be applied to both situations. This is so because the attribute correspondence identification problem is similar to the record re-identification one. On this basis, a similar algorithm was applied in (Torra, 2000c) and (Domingo-Ferrer and Torra, 2003) to problems of both classes.

52.2.2 Fusion to Improve the Quality of Data

Once data from several sources is known to refer to the same individual, they can be fused to obtain data of better quality (in some applications this is known by data consolidation) or to obtain data having a broader scope (data permitting to cope with aspects that data supplied from a single source do not permit).

Fusion methods are also applied in situations in which data is supplied by a single source but at different time instants. This is the case, for example, of the moving average on a window for temporal variables. See e.g. (Tsumoto, 2003).

Aggregation operators are the particular functions used to combine data into a new and more reliable datum. Aggregation operators (Calvo, Mayor and Mesiar, 2002) exist for data of several types. For example, there are operators for:

- Numerical data: e.g. The arithmetic or the weighted mean and fuzzy integrals: Choquet and Sugeno integral (Choquet, 1954, Sugeno, 1974).
- Linguistic or categorical data: e.g. the median and the majority or plurality rule (Roberts, 1991)

Other operators exist for combining more complex data as partitions, or dendrograms (hierarchical structures). See (Day, 1986) on methods for partitions and dendrograms.

Differences on the operators rely on the different assumptions that apply for sources and domains. The family of operators encompassed by the Choquet integral (an operator for numerical data) illustrates this situation. See (Torra, 2003) or (Calvo, Kolesárová, Komorníková and Mesiar, 2002) for a more detailed account of the Choquet integral family and a detailed description of aggregation operators for numerical data.

The arithmetic mean ($\Sigma a_i/N$) is the simplest aggregation operator for numerical scales. However, its definition implies that all sources have the same importance. Instead, the weighted mean ($\Sigma p_i a_i$ with $p_i \geq 0$ and $\Sigma p_i = 1$) overcomes this assumption permitting the user to assign weights to the sources (e.g. a sensor is twice as reliable as another sensor). Alternatively, the OWA operator (Yager, 1988) permits to assign importance to the values but not to the sources. That is, the OWA operator permits to model situations in which smaller values are more important than larger ones, or, in other words, compensation is allowed between values. Also, the OWA operator permits to diminish the importance of outliers.

The WOWA operator (Torra, 1997) (or the OWA with importances defined by Yager in 1996) permits both the importance of the values and the one of the sources. This is, the operator allows for compensation and at the same time some sources can be said to be more important than some other ones.

Finally, Choquet integrals (Choquet, 1954) permit to overcome some implicit assumptions of the weighted mean: all sources are independent. The Choquet integral, which is defined as an integral of a function (the values to be aggregated) with respect to a fuzzy measure, permits to consider sources that interact.

Aggregation Operators: Formalization

In ordered scales D (ordinal or numerical scales), aggregation operators are typically functions that combine N values and that satisfy idempotency or unanimity (i.e., when all inputs are the same, the output is also the same), monotonicity (i.e., when any of the input data is increased, the output is also increased) and the output is a compromise value (i.e., the output is between the minimum and the maximum of the input values). Formally speaking, aggregation operators are functions C (from Consensus) from $D^N \to D$ that satisfy:

1. $C(a, a, \ldots, a) = a$
2. if $a_i \leq b_i$ then $C(a_1, a_2, \ldots, a_N) \leq C(b_1, b_2, \ldots, b_N)$
3. $\min a_i \leq C(a_1, a_2, \ldots, a_N) \leq \max a_i$

In some applications, the role of the information sources is important and some prior knowledge about their relevance or reliability is given. In this case, the set of sources $X = \{x_1, \ldots, x_N\}$ is considered in the aggregation process and the value that source x_i supplies is expressed by means of a function f from X into D. In this way, $f(x_i) = a_i$. Then, knowledge on the sources is given as functions on X (or on subsets of X). For example, the weighted mean corresponds to: Σ_i p(x$_i$) f(x$_i$). In the same way, a Choquet integral is an integral of the function f with respect to a fuzzy measure μ defined on the parts of X. Each μ(A) for $A \subseteq X$ measures the reliability or importance of the sources of A.

52.3 Building Data Models

The main goal of Data Mining is the extraction of knowledge from data. Information fusion methods also play a role in this process.

In fact, Data Mining techniques are usually rooted on a particular kind of knowledge representation formalism, or a particular data model. Information fusion can be used on the one hand as one of such data models and on the other hand to combine different data models for building a more accurate one. We review both situations below.

52.3.1 Data Models Using Aggregation Operators

Aggregation operators, as any mathematical function, can be used for data modeling. They are appropriate to model situations that are compatible with the properties of these operators. This is, in particular, the variable to be modeled is monotonic with respect to the inputs, the operator satisfies idempotency and the output is between the minimum and the maximum of the input values. Additional constraints should be considered when a particular aggregation operator is selected. For example, when the weighted mean is used as the model, independence of the attributes is required.

Even though aggregation operators are restricted to be used when data satisfy these properties, they can be used in composite models (e.g. hierarchical models) when data do not follow such constraints.

The validity of aggregation operators in such more complex situations is rooted on the following two results:

1. Two step Choquet integrals with constant permit the approximation of any monotonic function at the desired level of accuracy. See (Murofushi and Narukawa, 2002) for details.
2. Hierarchies of quasi-weighted means and hierarchies of Choquet integrals with constant can be used to approximate any arbitrary function at the desired level of detail. This result is given in (Narukawa and Torra, 2003).

The construction of an appropriate model, based on experimental data or expert's experience, requires the selection of an appropriate aggregation function as well as of its parameters. In the case that composite models are used, the architecture of the model is also required. Research has been carried out on operator and parameter selection. In the next section, we concentrate on methods for learning parameters from examples. This is, we consider the case where there exists a set of (input x output) pairs that the model is intended to reproduce. Other research is based on the existence of an expert that gives relevant information for the selection of the operator or for the parameters of a given operator. See e.g. Saaty's Analytical Hierarchy Process (Saaty, 1980) or O'Hagan's selection of OWA operator's weight from the degree of compensation (Hagan, 1988).

Learning Methods for Aggregation Operators

At present there exists a plethora of methods to learn parameters from examples once the aggregation operator is selected.

Filev and Yager (1998) developed a method for the case of the OWA operator that is also appropriate when the operator selected is the weighted mean. A more effective method was presented in (Torra, 1999) and extended in (Torra, 2002) to be applied to the quasi-weighted means.

The first approach in the literature to consider the learning of fuzzy measures for Choquet integrals is (Tanaka and Murofushi, 1989). The same problem, but considering alternative assumptions, has been considered, among others, in (Marichal and Roubens, 2000, Imai *et al.*,

2000, Wang *et al.*, 1999, Torra, 2000a). While the first two papers correspond to general methods the last two are for constrained fuzzy measures, namely, Sugeno λ measures and distorted probabilities. A recent survey of learning methods for the Choquet integral is given in (Grabisch, 2003).

For more details on learning methods for aggregation operators see the excellent work by Beliakov (Beliakov, 2003).

52.3.2 Aggregation Operators to Fuse Data Models

A second use of aggregation operators for data modeling corresponds to the so-called ensemble methods (Dietterich, 1997). In this case, aggregation is used to combine several different models constructed on the basis of different subsets of objects (as in Bagging (Breiman, 1996) and Boosting (Schapire, 1990)) or based on different machine learning approaches.

The operators used in the literature depend on the kind of problem considered and on the models built. Two different cases can be underlined:

Regression Problems

The output of individual modules is a numerical value. Accordingly, aggregation operators for numerical data are used. Arithmetic and weighted means are the most commonly used operators (see e.g. (Merz, 1999)). Nevertheless, fuzzy integrals, as the Choquet or Sugeno integrals, can also be used.

Classification Problems

The output of individual modules typically correspond to a categorical value (the class for a particular instance problem), therefore, aggregation operators for categorical values (in nominal scales: non-ordered categorical domains) are considered. The plurality rule (voting or weighted voting procedures) is the most usual aggregation operator used (see e.g. (Bauer and Kohavi, 1999) and (Merz, 1999)). In a recent study (Kuncheva, 2003) several fuzzy combination methods have been compared in classification problems against some nonfuzzy techniques as the (weighted) majority voting.

52.4 Information Extraction

Information fusion can be used in Data Mining as a method for information extraction. In particular, the following applications can be envisioned:

52.4.1 Summarization

Fusion methods are used to build summaries so that the quantity of available data is reduced. In this way, data can be represented in a more compact way. For example, Detyniecki (2000) uses aggregation operators to build summaries of video sequences and Kacprzyk, Yager and Zadrozny (2000) and Yager (2003) use aggregation operators to build linguistic summaries (fuzzy rules) of databases. Methods for dimensionality reduction or multidimensional scaling (Cox and Cox, 1994) can also be seen from the perspective of information fusion and summarization. Namely, the reduction of the dimensionality of records (reduction on the number of variables) and the reduction on the number of records (e.g. clustering to build prototypes from data).

52.4.2 Knowledge from Aggregation Operators

For some information fusion methods, their parameters can be used to extract information from the data. This is the case of Choquet integral. It has been said that this operator can be used to aggregate situations in which the sources are not independent. Accordingly, when a data model is built using a Choquet integral, the corresponding fuzzy measure can be interpreted and interactions between variables arise from the analysis of the measure. Grabisch (2000) describes the use of Choquet integrals models to extract features from data.

1. Preprocessing data
   a. Re-identification algorithms:
      Record and scheme matching methods
   b. Fusion to improve the quality of data:
      Aggregation operators
2. Building data models
   a. Models based on aggregation operators:
      Learning methods for aggregation operators
   b. Aggregation operators to fuse data models:
      Ensemble methods
3. Information extraction
   a. Summarization:
      Dimensionality reduction and linguistic summaries
   b. Knowledge from aggregation operators:
      Interpreting operators parameters

Fig. 52.1. Main topics and methods for Information Fusion in Data Mining and Knowledge Discovery.

52.5 Conclusions

In this chapter we have reviewed the application of information fusion in Data Mining and knowledge discovery. We have seen that aggregation operators and information fusion methods can be applied for three main uses: preprocessing data, building data models and information extraction. Figure 52.4.2 gives a more detailed account of such applications.

Acknowledgments

Partial support of the MCyT under the contract STREAMOBILE (TIC2001-0633-C03-01/02) is acknowledged.

References

Bacher, J., Brand, R., Bender, S., Re-identifying register data by survey data using cluster analysis: an empirical study, Int. J. of Unc., Fuzziness and KBS, 10:5 589-607, 2002.

Beliakov, G., How to Build Aggregation Operators from Data, Int. J. of Intel. Syst., 18 903-923, 2003.

Bergamaschi, S., Castano, S., Vincini, M., Beneventano, D., Semantic integration of heterogeneous information sources, Data and Knowledge Engineering 36 215-249, 2001.

Bauer, E., Kohavi, R., An Empirical Comparison of Voting Classification Algorithms: Bagging, Boosting and Variants, Machine Learning 36 105–139, 1999.

Borkar, V. R., Deshmukh, K., Sarawagi, S., Automatically extracting structure from free text addresses, Bulletin of the Technical Committee on Data Engineering, 23: 4, 27-32, 2000.

Breiman, L., Bagging predictors, Machine Learning, 24 123–140, 1996.

Calvo, T., Mayor, G., Mesiar, R., (Eds), Aggregation operators: New trends and applications, Physica-Verlag: Springer, 2002.

Calvo, T., Kolesárová, A., Komorníková, M., Mesiar, R., Aggregation Operators: Properties, Classes and Construction Methods, in T. Calvo, G. Mayor, R. Mesiar (Eds.), Aggregation operators: New Trends and Applications, Physica-Verlag, 3-123, 2002.

Choquet, G., Theory of Capacities, Ann. Inst. Fourier 5 131-296, 1954.

Cox, T. F., Cox, M. A. A., Multidimensional scaling, Chapman and Hall, 1994.

Day, W.H.E., Special issue on comparison and consensus of classifications, Journal of Classification, 3, 1986.

Detyniecki, M., Mathematical Aggregation Operators and their Application to Video Querying, PhD Dissertation, University of Paris VI, Paris, France, 2000.

Dietterich, T.G., Machine-Learning Research: Four Current Directions, AI Magazine, Winter 1997, 97-136.

Do, H.-H., Rahm, E., COMA - A system for flexible combination of schema matching approaches, Proc. of the 28th VLDB Conference, Hong-Kong, China, 2002.

Domingo-Ferrer, J., Torra, V., Disclosure Risk Assessment in Statistical Microdata Protection via advanced record linkage, Statistics and Computing, 13 343-354, 2003.

Filev, D., Yager, R.R., On the issue of obtaining OWA operator weights, Fuzzy Sets and Systems, 94, 157-169, 1998.

Grabisch, M., Fuzzy integral for classification and feature extraction, in M. Grabisch, T. Murofushi, M. Sugeno (Eds.), Fuzzy Measures and Integrals, Physica-Verlag, 415–434, 2000.

Grabisch, M., Modelling data by the Choquet integral, in V. Torra, Information Fusion in Data Mining, Springer, 135–148, 2003.

Imai, H., Miyamori, M., Miyakosi, M., Sato, Y., An algorithm based on alternative projections for a fuzzy measures identification problem, Proc. of the Iizuka conference, Iizuka, Japan (CD-ROM), 2000.

Jaro, M. A., Advances in record-linkage methodology as applied to matching the 1985 Census of Tampa, Florida, J. of the American Stat. Assoc., 84:406 414-420, 1989.

Kacprzyk, J., Yager, R. R., Zadrozny, S., "A fuzzy logic based approach to linguistic summaries in databases, Int. J. of Applied Mathematical Computer Science 10 813-834, 2000.

Kuncheva, L. I., "Fuzzy" Versus "Nonfuzzy" in Combining Classifiers Designed by Boosting, IEEE Trans. on Fuzzy Systems, 11:6 729-741, 2003.

Marichal, J.-L., Roubens, M., Determination of weights of interacting criteria from a reference set, European Journal of Operational Research, 124:3 641-650, 2000.

Merz, C. J., Using Correspondence Analysis to Combine Classifiers, Machine Learning, 36 33-58, 1999.

Merz, C. J., Pazzani, M. J., Combining regression estimates, Machine Learning, 36 9–32, 1999.

Murofushi, T., Narukawa, Y., A Characterization of multi-level discrete Choquet integral over a finite set (in Japanese). Proc. of the 7th Workshop on Evaluation of Heart and Mind 33-36, 2002.

Narukawa, Y., Torra, V., Choquet integral based models for general approximation, in I. Aguiló, L. Valverde, M. T. Escrig (Eds.), Artificial Intelligence Research and Development, IOS Press, 39-50, 2003.

O'Hagan, M., Aggregating template or rule antecedents in real-time expert systems with fuzzy set logic, Proceedings of the 22nd Annual IEEE Asilomar Conference on Signals, Systems and Computers, Pacific Grove, CA, 1988, 681-689, 1988.

Pagliuca, D., Seri, G., Some Results of Individual Ranking Method on the System of Enterprise Accounts Annual Survey, Esprit SDC Project, Deliverable MI-3/D2, 1999.

Rahm, E., Bernstein, P. A., A survey of approaches to automatic schema matching, The VLDB Journal, 10 334-350, 2001.

Roberts, F. S., On the indicator function of the plurality function, Mathematical Social Sciences, 22 163-174, 1991.

Saaty, T. L., The Analytic Hierarchy Process (McGraw-Hill, New York), 1980.

Schapire, R. E., The strength of weak learnability, Machine Learning, 5: 2 197–227, 1990.

Sugeno, M., Theory of Fuzzy Integrals and its Applications. (PhD Dissertation). Tokyo Institute of Technology, Tokyo, Japan, 1974.

Tanaka, A., Murofushi, T., A learning model using fuzzy measure and the Choquet integral, Proc. of the 5th Fuzzy System Symposium, 213-217, Kobe, Japan (in Japanese), 1989.

Torra, V., The Weighted OWA operator, Int. J. of Intel. Systems, 12 153-166, 1997.

Torra, V., On the learning of weights in some aggregation operators: The weighted mean and the OWA operators, Mathware and Soft Computing, 6 249-265, 1999.

Torra, V., Learning weights for Weighted OWA operators, Proc. IEEE Int. Conf. on Industrial Electr. Control and Instrumentation, 2000a.

Torra, V., Re-identifying Individuals using OWA operators, Proc. of the 6th Int. Conference on Soft Computing, (CD Rom), Iizuka, Fukuoka, Japan, 2000b.

Torra, V., Towards the re-identification of individuals in data files with Non-common variables, Proc. of the 14th ECAI, 326-330, Berlin, 2000c.

Torra, V., Learning weights for the quasi-weighted means, IEEE Trans. on Fuzzy Systems, 10:5 (2002) 653-666, 2002.

Torra, V., On some aggregation operators for numerical information, in V. Torra (Ed.), Information Fusion in Data Mining, 9-26, Springer, 2003.

Torra, V., OWA operators in data modeling and re-identification, IEEE Trans. on Fuzzy Systems, 12:5 2004.

Torra, V., Domingo-Ferrer, J., Record linkage methods for multidatabase Data Mining, in V. Torra, Information Fusion in Data Mining, 101-132, Springer, 2003.

Tsumoto, S., Discovery of Temporal Knowledge in Medical Time-Series Databases using Moving Average, Multiscale Matching and Rule Induction, in Torra V., Information Fusion in Data Mining, Springer, 79–100, 2003.

Wang, Z., Leung, K. S., Wang, J., A genetic algorithm for determining nonadditive set functions in information fusion, *Fuzzy sets and systems*, 102 462-469, 1999.

Winkler, W. E., Matching and record linkage, in B. G. Cox (Ed) Business Survey Methods, Wiley, 355–384, 1995.

Winkler, W. E., Advanced methods for record linkage, American Statistical Association, Proceedings of the Section on Survey Research Methods, 467-472, 1995.

Wrobel, S., An Algorithm for Multi-relational Discovery of Subgroups, J. Komorowski et al. (eds.), Principles of Data Mining and Knowledge Discovery, Lecture Notes in Artificial Intelligence 1263, Springer, 367-375, 1997.

Yager, R. R., On ordered weighted averaging aggregation operators in multi-criteria decision making, IEEE Trans. on SMC, 18 183-190, 1988.

Yager, R. R., Quantifier Guided Aggregation Using OWA operators, Int. J. of Int. Systems, 11 49-73, 1996.

Yager, R. R., Data Mining Using Granular Linguistic Summaries, in V. Torra, Information Fusion in Data Mining, Springer, 211–229, 2003.

Zhong, N., Yao, Y. Y., Ohsuga, S., Pecularity Oriented Multi-Database Mining", J. Zytkow, J. Rauch, (eds.), Principles of Data Mining and Knowledge Discovery, Lecture Notes in Artificial Intelligence 1704, Springer, 136-146, 1999.

53

Parallel And Grid-Based Data Mining – Algorithms, Models and Systems for High-Performance KDD

Antonio Congiusta[1], Domenico Talia[1], and Paolo Trunfio[1]

DEIS – University of Calabria
acongiusta,talia,trunfio@deis.unical.it

Summary. Data Mining often is a computing intensive and time requiring process. For this reason, several Data Mining systems have been implemented on parallel computing platforms to achieve high performance in the analysis of large data sets. Moreover, when large data repositories are coupled with geographical distribution of data, users and systems, more sophisticated technologies are needed to implement high-performance distributed KDD systems. Since computational Grids emerged as privileged platforms for distributed computing, a growing number of Grid-based KDD systems has been proposed. In this chapter we first discuss different ways to exploit parallelism in the main Data Mining techniques and algorithms, then we discuss Grid-based KDD systems. Finally, we introduce the Knowledge Grid, an environment which makes use of standard Grid middleware to support the development of parallel and distributed knowledge discovery applications.

Key words: Parallel Data Mining, Grid-based Data Mining, Knowledge Grid, Distributed Knowledge Discovery

53.1 Introduction

Today the information overload is a problem like the shortage of information is. In our daily activities we often deal with flows of data much larger than we can understand and use. Thus we need a way to sift those data for extracting what is interesting and relevant for our activities. Knowledge discovery in large data repositories can find what is interesting in them representing it in an understandable way (Berry and Linoff, 1997).

Mining large data sets requires powerful computational resources. In fact, Data Mining algorithms working on very large data sets take a very long time on conventional computers to get results. One approach to reduce response time is sampling. But, in some cases reducing data might result in inaccurate models, in some other cases it is not useful (e.g., outliers identification). Another approach is parallel computing. High–performance computers and parallel Data Mining algorithms can offer a very efficient way to mine very large data sets (Freitas and Lavington, 1998, Skillicorn, 1999) by analyzing them in parallel.

O. Maimon, L. Rokach (eds.), *Data Mining and Knowledge Discovery Handbook*, 2nd ed.,
DOI 10.1007/978-0-387-09823-4_53, © Springer Science+Business Media, LLC 2010

It is not uncommon to have sequential Data Mining applications that require several days or weeks to complete their task. Parallel computing systems can bring significant benefits in the implementation of Data Mining and knowledge discovery applications by means of the exploitation of inherent parallelism of Data Mining algorithms. Data Mining and knowledge discovery on large amounts of data can benefit of the use of parallel computers both to improve performance and quality of data selection. When Data Mining tools are implemented on high-performance parallel computers, they can analyze massive databases in a reasonable time. Faster processing also means that users can experiment with more models to understand complex data.

Beyond the development of KDD systems based on parallel computing platforms to achieve high performance in the analysis of large data sets stored in a single site, a lot of work has been devoted to design KDD systems able to handle and analyze multi-site data repositories. The combination of large-sized data sets, geographical distribution of data, users, resources, and computationally intensive analysis, demands for an advanced infrastructure for parallel and distributed knowledge discovery (PDKD).

Advances in networking technology and computational infrastructures made it possible to construct large-scale high-performance distributed computing environments, or computational Grids, that enable the integrated use of remote high-end computers, databases, scientific instruments, networks, and other resources. The Grid has the potential to fundamentally change the way we think about computing, as our ability to compute will no longer be limited to the resources we currently have on our desktop or office. Grid applications often involve large amounts of computing and/or data. For these reasons, Grids can offer an effective support to the implementation and use of PDKD systems.

This chapter discusses state of the art and recent advances in both parallel and Grid-based Data Mining.

Section 53.2 analyzes different forms of parallelism that can be exploited in Data Mining techniques and algorithms. The main goal is to introduce Data Mining techniques on parallel architectures and to show how large–scale Data Mining and knowledge discovery applications can achieve scalability by using systems, tools and performance offered by parallel processing systems. For several Data Mining techniques, such as rule induction, clustering algorithms, decision trees, genetic algorithms! and neural networks, different strategies to exploit parallelism are presented and discussed. Furthermore, some experiences and results in parallelizing Data Mining algorithms according to different approaches are examined.

Section 53.3 analyzes the Grid-based Data Mining approach. First, the main benefits coming from the use of Grid models and platforms in developing distributed knowledge discovery systems are discussed. Secondly, we review and compare Grid-based Data Mining systems.

Section 53.4 introduces a reference software architecture for geographically distributed PDKD systems called Knowledge Grid. Its architecture is built on top of Grid infrastructures providing dependable, consistent, and pervasive access to high-end computational resources. The Knowledge Grid uses the basic Grid services and defines a set of additional layers to implement specialized services for PDKD on world-wide connected sites, where each node may be a sequential or a parallel machine. The Knowledge Grid enables the collaboration of scientists that need to mine data stored in different research centers, as well as executive managers that necessitate a knowledge management system operating on several data warehouses located in different company establishments.

53.2 Parallel Data Mining

Main goals of the use of parallel computing technologies in the Data Mining field are:

- performance improvements of existing techniques,
- implementation of new (parallel) techniques and algorithms, and
- concurrent analysis using different Data Mining techniques in parallel and result integration to get a better model (i.e., more accurate results).

We identify three main strategies in the exploitation of parallelism in data mining algorithms:

- *independent parallelism*,
- *task parallelism*,
- *SPMD parallelism*.

Independent parallelism is exploited when processes are executed in parallel in an independent way. Generally, each process accesses to the whole data set and does not communicate or synchronize with other processes. According to task parallelism (or control parallelism) each process executes different operations on (a different partition of) the data set. Finally, in Single Program Multiple Data (SPMD) parallelism a set of processes execute in parallel the same algorithm on different partitions of a data set, and processes cooperate to exchange partial results. These three strategies for parallelizing Data Mining algorithms are not necessarily alternative. They can be combined to improve both performance and accuracy of results. In combination with strategies for parallelization, different data partition strategies may be used:

- *sequential partitioning*: separate partitions are defined without overlapping among them;
- *cover-based partitioning*: some data can be replicated on different partitions;
- *range-based query partitioning*: partitions are defined on the basis of some queries that select data according to attribute values.

53.2.1 Parallelism in Data Mining Techniques

This section presents different parallelization strategies for each data mining technique and describes some parallel Data Mining tools, algorithms, and systems.

Table 53.1 contains the main Data Mining tasks, and for each task the main techniques used to solve them are listed. In the following we describe different approaches for parallel implementation of some techniques listed in Table 53.1.

Parallel Decision Trees (Parallel Induction)

Classification is the process of assigning new objects to predefined categories or classes. Decision trees are an effective technique for classification. They are tree-shaped structures that represent sets of decisions. These decisions generate rules for the classification of a data set. The tree leaves represent the classes and the tree nodes represent attribute values. The path from the root to a leaf gives the features of a class in terms of attribute–value pairs.

Task parallel approach. According to the task parallelism approach one process is associated to each sub-tree of the decision tree that is built to represent a classification model. The search occurs in parallel in each sub-tree, thus the degree of parallelism P is equal to the number of active processes at a given time. A possible implementation of this approach is based

Table 53.1. Data Mining tasks and used techniques

Data Mining Tasks	Data Mining Techniques
Classification	induction, neural networks, genetic algorithms
Association	Apriori, statistics, genetic algorithms
Clustering	neural networks, induction, statistics
Regression	induction, neural networks, statistics
Episode discovery	induction, neural networks, genetic algorithms
Summarization	induction, statistics

on farm parallelism in which there is a master process that controls the computation and a set of P workers that are assigned to the sub-trees.

SPMD approach. In the exploitation of SPDM parallelism each process classifies the items of a subset of data. The P processes search in parallel in the whole tree using a partition D/P of the data set D. The global result is obtained by exchanging partial results. The data set partitioning may be operated in two main different ways:

- by partitioning the D tuples of the data set: D/P per processor.
- by partitioning the n attributes of each tuple: D tuples of n/P attributes per processor.

In (Kufrin, 1997) a parallel implementation of the C4.5 algorithm that uses the independent parallelism approach is discussed. Other significant examples of parallel algorithms that use decision trees are SPRINT (Shafer *et al.*, 1996), and Top-Down Induction of Decision Trees (Pearson, 2000).

Discovery of Association Rules in Parallel

Association rules algorithms, such as Apriori, allow automatic discovery of complex associations in a data set. The task is to find all frequent itemsets, i.e., to list all combinations of items that are found in a sufficient number of examples. Given a set of transactions D, the problem of mining association rules is to generate all association rules that have support (how often a combination occurred overall) and confidence (how often the association rule holds true in the data set) greater than the user-specified minimum support and minimum confidence respectively. An example of such a rule might be that "*% of customers that purchase tires and auto accessories also get automotive services done*".

SPMD approach. In the SPMD strategy the data set D is partitioned among the P processors but candidate itemsets I are replicated on each processor. Each process p counts in parallel the partial support S_p of the global itemsets on its local partition of the data set of size D/P. At the end of this phase the global support S is obtained by collecting all local supports S_p. The replication of the candidate itemsets minimizes communication, but does not use memory efficiently. Due to low communication overhead, scalability is good.

Task parallel approach. In this case both the data set D and the candidate itemsets I are partitioned on each processor. Each process p counts the global support S_i of its candidate itemset I_p on the entire data set D. After scanning its local data set partition D/P, a process must scan all remote partitions for each iteration. The partitioning of data set and candidate itemsets minimizes the use of memory but requires high communication overhead in distributed memory architectures. Due to communication overhead this approach is less scalable than the previous one.

Hybrid approaches. Combination of different parallelism approaches can be designed. For example, SPMD and task parallelism can be combined by defining C clusters of processors composed of the same number of processing nodes. The data set is partitioned among the C clusters, thus each cluster is responsible to compute the partial support S_c of the candidate itemsets I according to the SPMD approach. Each processor in a cluster uses the task parallel approach to compute the support of its disjoint set of candidates I_p by scanning the data set stored on the processors of its cluster. At the end of each iteration the clusters cooperate each other to compute the global support S.

The Apriori algorithm (Agrawal and Srikant, 1994) is the most known algorithm for association rules discovery. Several parallel implementations have been proposed for this algorithm. In (Agrawal and Shafer, 1996) two different parallel algorithms called Count Distribution (CD) and Data Distribution (DD) are presented. The first one is based on independent parallelism and the second one is based on task parallelism. In (Han *et al.*, 2000) two different parallel approaches to Apriori called Intelligent Data Distribution (IDD) and Hybrid Distribution (HD) are presented. A complete review of parallel algorithms for association rules can be found in (Zaki, 1999).

Parallel Neural Networks

Neural networks (NN) are a biology-inspired model of parallel computing that can be used in knowledge discovery. Supervised NN are used to implement classification algorithms and unsupervised NN are used to implement clustering algorithms. A lot of work on parallel implementation of neural networks has been done in the past. Theoretically, each neuron can be executed in parallel, but in practice the grain of processors is generally larger then the grain of neurons. Moreover, the processor interconnection degree is restricted in comparison with neuron interconnection. Hence a subset of neurons is generally mapped on each processor. There are several different ways to exploit parallelism in a neural network:

- *parallelism among training sessions*: it is based on simultaneous execution of different training sessions;
- *parallelism among training examples*: each processor trains the same network on a subset of $1/P$ examples;
- *layer parallelism*: each layer of a neural network is mapped on a different processor;
- *column parallelism*: the neurons that belong to a column are executed on a different processor;
- *weight parallelism*: weight summation for connections of each neuron is executed in parallel.

These parallel approaches may be combined to form different hybrid parallelization strategies. Different combinations can raise up different issues to be faced for efficient implementation such as interconnection topology, mapping strategies, load balancing among the processors, and communication latency.

Typical parallelism approaches used for the implementation of neural networks on parallel architectures are task parallelism, SPMD parallelism, and farm parallelism.

A parallel Data Mining system based on neural networks is Clementine. Several task-parallel implementations of back-propagation networks and parallel implementations of Self-organizing maps have been implemented for Data Mining tasks. Finally, Neural Network Utility (Bigus, 1996) is a neural network-based Data Mining environment that has been also implemented on a IBM SP2 parallel machine.

Parallel Genetic Algorithms

Genetic algorithms are used today for several Data Mining tasks such as classification, association rules, and episode discovery. Parallelism can be exploited in three main phases of a genetic algorithm:

- population initialization,
- fitness computation, and
- execution of the mutation operator,

without modifying the behavior of the algorithm in comparison to the sequential version. On the other hand, the parallel execution of selection and crossover operations requires the definition of new strategies that modify the behavior (and results) of a genetic algorithm in comparison to the sequential version. The most used approach is called *global parallelization*. It is based on the parallel execution of the fitness function and mutation operator while the other operations are executed sequentially. However, there are two possible SMPD variants:

- each processor receives a subset of elements and evaluates their fitness using the entire data set D;
- each processor receives a subset D/P of the data set and evaluates the fitness of every population element (data item) on its local subset.

Global parallelization can be effective when very large data sets are to be mined. This approach is simple and has the same behavior of its sequential version, however its implementations did not achieve very good performance and scalability on distributed memory machines because of communication overhead.

Two different parallelization strategies that can change the behavior of the genetic algorithm are the *island model* (coarse grained), where each processor executes the genetic algorithm on a subset N/P of elements (sub-demes) and periodically the best elements of a sub-population are migrated towards the other processors, and the *diffusion model* (fine grained), where population is divided into a large number of sub-populations composed of few individuals (D/n where $n \gg P$) that evolve in parallel. Several subsets are mapped on one processor. Typically, elements are arranged in a regular topology (e.g., a grid). Each element evolves in parallel and executes the selection and crossover operations with the neighbor elements.

A very simple strategy is the independent parallel execution of P independent copies of a genetic algorithm on P processors. The final result is selected as the best one among the P results. Different parameters and initial populations should be used for each copy. In this approach there is no communication overhead. The main goal here is not getting a higher performance but a better accuracy. Some significant examples of Data Mining systems based on the parallel execution of genetic algorithms are GA-MINER, REGAL (Neri and Giordana, 1995), and G-NET.

Parallel Cluster Analysis

Clustering algorithms arrange data items into several groups, called *clusters* so that similar items fall into the same group. This is done without any suggestion from an external supervisor, so classes are not given a priori but they must be discovered by the algorithm. When used to classify large data sets, clustering algorithms are very computing demanding.

Clustering algorithms can roughly be classified into two groups: hierarchical and partitioning models. Hierarchical methods generate a hierarchical decomposition of a set of N items represented by a *dendogram*. Each level of a dendogram identifies a possible set of clusters. Dendograms can be built starting from one cluster and iteratively splitting this cluster until N clusters are obtained (*divisive methods*), or starting with the N clusters and merging at each step a couple of clusters until only one is left (*agglomerative methods*).

Partitioning methods divide a set of objects into K clusters using a distance measure. Most of these approaches assume that the number K of groups has been given a priori. Usually these methods generate clusters by optimizing a criterion function. The K-means clustering is a well-known and effective method for many practical applications that employs the squared error criterion.

Parallelism in clustering algorithms can be exploited both in the clustering strategy and in the computation of the similarity or distance among the data items, by computing on each processor the distance/similarity of a different partition of items. In the parallel implementation of clustering algorithms the three main parallel strategies described in section 53.2 can be exploited.

Independent parallel approach. Each processor uses the whole data set D and performs a different classification based on a different number of clusters K_p. To get the load among the processors balanced, until the clustering task is complete a new classification is assigned to a processor that completed its assigned classification.

Task parallel approach. Each processor executes a different task that composes the clustering algorithm and cooperates with other processors exchanging partial results. For example, in partitioning methods processors can work on disjoint regions of the search space using the whole data set. In hierarchical methods a processor can be responsible of one or more clusters. It finds the nearest neighbor cluster by computing the distance among its cluster and the others. Then all the local shortest distances are exchanged to find the global shortest distance between two clusters that must be merged. The new cluster will be assigned to one of the two processors that handled the merged clusters.

SPMD approach. Each processor executes the same algorithm on a different partition D/P of the data set to compute partial clustering results. Local results are then exchanged among all the processors to get global values on every processor. The global values are used in all processors to start the next clustering step until a convergence is reached or a given number of steps are executed. The SPMD strategy can be also used to implement clustering algorithms where each processor generates a local approximation of a model (classification), which at each iteration can be passed to the other processors that in turn use it to improve their clustering model.

In (Olson, 1995) a set of hierarchical clustering algorithms and an analysis of time complexity on different parallel architectures can be found. An example of parallel implementation of a clustering algorithm is P-CLUSTER (Judd *et al.*, 1996). Other parallel algorithms are discussed in (Bruynooghe, 1989,Li and Fang, 1989,Foti *et al.*, 2000). In particular, in (Foti *et al.*, 2000) an SPDM implementation of the AutoClass algorithm, named P-AutoClass is described. That paper shows interesting performance results on distributed memory MIMD machines. Table 53.2 shows experimental performance results we obtained by running P-AutoClass on a parallel machine using up to 10 processors for clustering a data set composed of 100,000 tuples with two real valued attributes. In particular, Table 53.2 contains execution times and absolute speedup on 2, 4, 6, 8 and 10 processors. We can observe how the system behavior is scalable; speedup on 10 processors is about 8 and execution time significantly decreases from 245 to 31 minutes.

Table 53.2. Execution time and speedup of P-AutoClass

Processors	Execution Time (secs)	Speedup
1	14683	1.0
2	7372	2.0
4	3598	4.1
6	2528	5.8
8	2248	6.5
10	1865	7.9

53.2.2 Architectural and Research Issues

In presenting the different strategies for the parallel implementation of Data Mining techniques we did not address architectural issues such as:

- distributed memory versus shared memory implementation,
- interconnection topology of processors,
- optimal communication strategies,
- load balancing of parallel Data Mining algorithms,
- memory usage and optimization, and
- I/O impact on algorithm performance.

These issues (and others) must be taken into account in the parallel implementation of Data Mining techniques. The architectural issues are strongly related to the parallelization strategies and there is a mutual influence between knowledge extraction strategies and architectural features. For instance, increasing the parallelism degree in some cases corresponds to an increment of the communication overhead among the processors. However, communication costs can be also balanced by the improved knowledge that a Data Mining algorithm can get from parallelization. At each iteration the processors share the approximated models produced by each of them. Thus each processor executes a next iteration using its own previous work and also the knowledge produced by the other processors. This approach can improve the rate at which a Data Mining algorithm finds a model for data (knowledge) and can make up for lost time in communication.

Parallel execution of different Data Mining algorithms and techniques can be integrated not just to get high–performance but also high accuracy. Here we list some promising research issues in the parallel Data Mining area:

- it is necessary to develop environments and tools for interactive high performance Data Mining and knowledge discovery;

- the use of parallel knowledge discovery techniques in text mining must be extensively investigated;
- parallel and distributed Web mining is a very promising area for exploiting high-performance computing techniques;
- the integration of parallel Data Mining techniques with parallel databases and data warehouses is a crucial aspect for private enterprises and public organizations.

53.3 Grid-Based Data Mining

Although the use of parallel techniques on dedicated parallel machines or clusters of computers is convenient to extract knowledge from large data sets stored in a single site, an advanced computing infrastructure is needed to build distributed KDD systems able to address a wide-area distribution of data, algorithms, and users.

The Grid emerged as a privileged computing infrastructure to develop applications over geographically distributed sites, providing for protocols and services enabling the integrated and seamless use of remote computing power, storage, software, and data, managed and shared by different organizations.

Basic Grid protocols and services are provided by toolkits and environments such as Globus Toolkit (www.globus.org/toolkit), Condor (www.cs.wisc.edu/condor), Legion (legion.virginia.edu), and Unicore (www.unicore.org). In particular, the Globus Toolkit is the most widely used middleware in scientific and data-intensive Grid applications, and is becoming a de facto standard for implementing Grid systems. The toolkit addresses security, information discovery, resource and data management, communication, fault-detection, and portability issues. A wide set of applications is being developed for the exploitation of Grid platforms. Since application areas range from scientific computing to industry and business, specialized services are required to meet needs in different application contexts. In particular, *data Grids* have been designed to easily store, move, and manage large data sets in distributed data-intensive applications.

Besides core data management services, *knowledge-based Grids,* built on top of computational and data Grid environments, are needed to offer higher-level services for data analysis, inference, and discovery in scientific and business areas (Moore, 2001).

Berman (2001), Johnston (2002), and some of us (Cannataro *et al.*, 2001) claimed that the creation of knowledge Grids is the enabling condition for developing high-performance knowledge discovery processes and meeting the challenges posed by the increasing demand of power and abstractness coming from complex problem solving environments.

53.3.1 Grid-Based Data Mining Systems

Whereas some high-performance PDKD systems have been proposed (Kargupta and Chan, 2000) - see also (Cannataro *et al.*, 2001) - there are few projects attempting to implement and/or support knowledge discovery processes over computational Grids. A main issue here is the integration of two main demands: synthesizing useful and usable knowledge from data, and performing sophisticated large-scale computations leveraging the Grid infrastructure. Such integration must pass through a clear representation of the knowledge base used in order to translate moderately abstract domain-specific queries into computations and data analysis operations able to answer such queries by operating on the underlying systems (Berman, 2001).

In the remainder of this section we shortly review the most significant systems oriented at supporting knowledge discovery processes over distributed or Grid infrastructures. The systems discussed here provide different approaches in supporting knowledge discovery on Grids. We discuss them starting from general frameworks, such as the TeraGrid infrastructure, then outlining data-intensive oriented systems, such as DataCutter and InfoGrid, and, finally, describing KDD systems such as Discovery Net, and some significant Data Mining testbed experiences.

The *TeraGrid* project is building a powerful Grid infrastructure, called *Distributed TeraScale Facility* (*DTF*), connecting four main sites in USA (the San Diego Supercomputer Center, the National Center for Supercomputing Applications, Caltech and Argonne National Laboratory). Recently, the NSF funded the integration into the DTF of the *TeraScale Computing System* (*TCS-1*) at the Pittsburgh Supercomputer Center; the resulting Grid environment will provide, besides tera-scale data storage, 21 TFLOPS of computational capacity (Catlett, 2002). Furthermore, the TeraGrid network connections, whose bandwidth is in the order of tenths of Gbps, have been designed in such a way that all resources appear as a single physical site. The connections have also been optimized in order to support peak requirements rather than an average load, as it is natural in Grid environments. The TeraGrid adopts Grid software technologies and, from this point of view, appears as a "virtual system" in which each resource describes its own capabilities and behavior through *Service Specifications*. The basic software components are called *Grid Services*, and are organized into three distinct layers. The *Basic* layer comprises authentication, resource allocation, data access and resource information services; the *Core* layer comprises services such as advanced data management, single job scheduling and monitoring; the *Advanced* layer comprises superschedulers, resource discovery services, repositories etc. Finally, *TeraGrid Application Services* are built using Grid Services.

The most challenging application on the TeraGrid will be the synthesis of knowledge from very large scientific data sets. The development of knowledge synthesis tools and services will enable the TeraGrid to operate as a knowledge Grid. A first application is the establishment of the Biomedical Informatics Research Network to allow brain researchers at geographically distributed advanced imaging centers to share data acquired from different subjects and using different techniques. Such applications make a full use of a distributed data Grid with hundreds of terabytes of data online, enabling the TeraGrid to be used as a knowledge Grid in the biomedical domain.

InfoGrid is a service-based data integration middleware engine designed to operate on Grids. Its main objective is to provide information access and querying services to knowledge discovery applications (Giannadakis *et al.*, 2003). The information integration approach of InfoGrid is not based on the classical idea of providing a "universal" query system: instead of abstracting everything for users, it gives a personalized view of the resources for each particular application domain. The assumption here is that users have enough knowledge and expertise to handle with the absence of "transparency". In InfoGrid the main entity is the *Wrapper*; wrappers are distributed on a Grid and each node publishes a directory of the wrappers it owns. A wrapper can wrap information sources and programs, or can be built by composing other wrappers (*Composite Wrapper*). Each wrapper provides: (*i*) a set of query construction interfaces, that can be used to query the underlying information sources in their native language; (*ii*) a set of administration interfaces, that can be used to configure its properties (access metadata, linkage metadata, configuration files). In summary, InfoGrid puts the emphasis on delivering metadata describing resources and providing an extensible framework for composing queries.

DataCutter is another middleware infrastructure that aims to provide specific services for the support of multi-dimensional range querying, data aggregation and user-defined filtering over large scientific datasets in shared distributed environments (Beynon *et al.*, 2001). DataCutter has been developed in the context of the *Chaos* project at the University of Maryland; it uses and extends features of the *Active Data Repository* (*ADR*), that is a set of tools for the optimization of storage, retrieval and processing of very large multi-dimensional datasets. In ADR, data processing is performed at the site where data is stored.

In the DataCutter framework, an application is decomposed into a set of processes, called *filters*, that are able to perform a rich set of queries and data transformation operations. Filters can execute anywhere but are intended to run on a machine close (in terms of connectivity) to the storage server. DataCutter supports efficient indexing. In order to avoid the construction of a huge single index that would result very costly to use and keep updated, the system adopts a multi-level hierarchical indexing scheme, specifically targeted at the multi-dimensional data model adopted.

Differently from the two environments discussed above, the *Datacentric Grid* is a system directed at knowledge discovery on Grids designed for mainly dealing with immovable data (Skillicorn and Talia, 2002). The system consists of four kinds of entities. The nodes at which computations happen are called *Data/Compute Servers* (*DCS*). Besides a compute engine and a data repository, each DCS comprises a *metadata tree*, that is a structure for maintaining relationships among raw datasets and models extracted from them. Furthermore, extracted models become new datasets, potentially useful at subsequent steps and/or for other applications.

The *Grid Support Nodes* (*GSNs*) maintain information about the whole Grid. Each GSN contains a directory of DCSs with static and dynamic information about them (e.g. properties and usage), and an execution plan cache containing recent plans along with their achieved performance. Since a computation in the Datacentric Grid is always executed on a single node, execution plans are simple. However, they can start at different places in the model hierarchy because, when they reach a node, they could find or not already computed models. The *User Support Nodes* (*USNs*) carry out execution planning and maintain results. USNs are basically proxies for user interface nodes (called *User Access Points*, *UAPs*). This is because user requests (i.e. task descriptions) and their results can be small in size, so in principle UAPs could be simple devices not always online, and USNs could interact with the Datacentric Grid when users are not connected.

An agent-based Data Mining framework, called *ADaM* (*Algorithm Development and Mining*), has been developed at the University of Alabama (datamining.itsc.uah.edu/adam). Initially, this framework was adopted for processing large datasets for geophysical phenomena. More recently, it has been ported to the NASA's *Information Power Grid* (*IPG*) environment, for the mining of satellite data (Hinke and Novonty, 2000). In this system, the user specifies *what* is to be mined (datasets names and locations), *how* and *where* to perform mining (sequence of operations, required parameters and IPG processors to be used). Initially, "thin" agents are associated to the sequence of mining operations; such agents acquire and combine the needed mining operations from repositories that can be public or private, i.e. provided by mining users or private companies. ADaM comprises a moderately rich set of interoperable operation modules, comprising *data readers* and *writers* for a variety of formats, *preprocessing modules*, e.g. for data sub-setting, and *analysis modules* providing Data Mining algorithms.

The InfoGrid system mentioned before has been designed as an application specific layer for constructing and publishing knowledge discovery services. In particular, it is intended to be used in the *Discovery Net* (*D-NET*) system (Curcin *et al.*, 2002). D-NET is a project of the Engineering and Physical Sciences Research Council, at the Imperial College (ex.doc.ic.ac.uk/new) whose main goal is to design, develop and implement an infrastructure to effectively support scientific knowledge discovery processes from high-throughput informatics. In this context, a series of testbeds and demonstrations are being carried out, for using the technology in the areas of life sciences, environmental modeling and geo-hazard prediction.

The building blocks in Discovery Net are the so-called *Knowledge Discovery Services* (*KDS*), distinguished in *Computation Services* and *Data Services*. The former typically comprise algorithms, e.g. data preparation and Data Mining, while the latter define relational tables (as queries) and other data sources. Both kinds of services are described (and registered) by means of *Adapters*, providing information such as input and output types, parameters, location and/or platform/operating system constraints, *factories* (objects allowing to retrieve references to services and to download them), keywords and a human-readable description. KDS are used to compose moderately complex data-pipelined processes. The composition may be carried out by means of a GUI which provides access to a library of services. The XML-based language used to describe processes is called *Discovery Process Markup Language* (*DPML*). Each composed process can be deployed and published as a new process. Typically, process descriptions are not bound to specific servers since the actual resources are later resolved by lookup servers (see below).

Discovery Net is based on an open architecture using common protocols and infrastructures such as Globus Toolkit. Servers are distinguished into (*i*) *Knowledge Servers*, allowing storage and retrieval of knowledge (meant as raw data and knowledge models) and processes; (*ii*) *Resource Discovery Servers*, providing a knowledge base of service definitions and performing resource resolution; (*iii*) *Discovery Meta-Information Servers*, used to store information about the *Knowledge Schema*, i.e. the sets of features of known databases, their types, and how they can be composed with each other.

Finally, we outline here some interesting Data Mining testbeds developed at the National Center for Data Mining (NCDM) at the University of Illinois at Chicago (UIC) (www.ncdm.uic.edu/testbeds.htm):

- *The Terra Wide Data Mining Testbed* (*TWDM*). TWDM is an infrastructure for the remote analysis, distributed mining, and real-time exploration of scientific, engineering, business, and other complex data. It consists of five geographically distributed nodes linked by optical networks through *StarLight* (an advanced optical infrastructure) in Chicago. These sites include StarLight itself, the Laboratory for Advanced Computing at UIC, SARA in Amsterdam, and the Dalhousie University in Halifax. In 2003 new sites will be connected, including the Imperial College in London. A central idea in TWDM is to keep generated predictive models up-to-date with respect to newly available data, in order to achieve better predictions (as this is an important aspect in many "critical" domains, such as infectious disease tracking). TWDM is based on *DataSpace*, another NCDM project for supporting real-time streaming data; in DataSpace the *Data Tranformation Markup Language* (*DTML*) is used to describe how to update "profiles", i.e. aggregate data which are inputs of predictive models, on the basis of new "events", i.e. new bits of information.
- *The Terabyte Challenge Testbed.* The Terabyte Challenge Testbed is an open, distributed testbed for DataSpace tools, services, and protocols. It involves a number of organizations, including the University of Illinois at Chicago, the University of Pennsylvania, the University of California at Davis and the Imperial College. The testbed consists of ten sites distributed over three continents connected by high–performance links. Each site provides a number of local clusters of workstations which are connected to form wide area *meta-clusters* maintained by the *National Scalable Cluster Project*. So far, meta-clusters have been used by applications in high energy physics, computational chemistry, nonlinear simulation, bioinformatics, medical imaging, network traffic analysis, digital libraries

of video data, etc. Currently, the Terabyte Challenge Testbed consists of approximately 100 nodes and 2 terabytes of disk storage.

- *The Global Discovery Network* (*GDN*). The GDN is a collaboration between the Laboratory for Advanced Computing of the National Center for Data Mining and the Discovery Net project (see above). It will link the Discovery Net to the Terra Wide Data Mining Testbed to create a combined global testbed with a critical mass of data.

The *GridMiner* project at the University of Vienna aims to cover the main aspects of knowledge discovery on Grids. GridMiner is a model based on the OGSA framework (Foster *et al.*, 2002), and embraces an open architecture in which a set of services are defined for handling data distribution and heterogeneity, supporting different types of analysis strategies, as well as tools and algorithms, and providing for OLAP support. Key components in GridMiner are the *Data Access* service, the *Data Mediation* service, and the *Data Mining* service. Data Access implements the data access to databases and data repositories; Data Mediation provides for a view of distributed data by logically integrating them into *virtual data sources* (*VDS*) and allowing to send queries to them and combine and deliver back the results. The Data Mining layer comprises a set of specific services useful to prepare and execute a Data Mining application, as well as present its results. The system has not been yet implemented on a Grid; a preliminary fully centralized version of the system is currently available.

GATES (*Grid-based AdapTive Execution on Streams*) is an OGSA based system that provides support for processing data streams in a Grid environment (Agrawal, 2003). GATES aims to support the distributed analysis of data streams arising from distributed sources (e.g., data from large–scale experiments/simulations), providing automatic resource discovery, and an interface for enabling self-adaptation to meet real-time constraints.

Some of the systems discussed above support specific domains applications, others support a more general class of problems. Moreover, some of such systems are mainly advanced interfaces for integrating, accessing, and elaborating large datasets, whereas others provide more specific functionalities for the support of typical knowledge discovery processes.

In the next section we present a Grid-based environment, named Knowledge Grid, whose aim is to support general PDKD applications, providing an interface both to manage and access large remote data sets, and to execute high-performance data analysis on them.

53.4 The Knowledge Grid

The Knowledge Grid (Cannataro and Talia, 2003) is an environment providing knowledge discovery services for a wide range of high–performance distributed applications. Data sets and Data Mining and analysis tools used in such applications are increasingly becoming available as stand-alone packages and as remote services on the Internet. Examples include gene and DNA databases, network access and intrusion data, drug features and effects data repositories, astronomy data files, and data about web usage, content, and structure.

Knowledge discovery procedures in all these applications typically require the creation and management of complex, dynamic, multi-step workflows. At each step, data from various sources can be moved, filtered, and integrated and fed into a Data Mining tool. Based on the output results, the analyst chooses which other data sets and mining components should be integrated in the workflow, or how to iterate the process to get a knowledge model. Workflows are mapped on a Grid by assigning its nodes to the Grid hosts and using interconnections for implementing communication among the workflow nodes.

The Knowledge Grid supports such activities by providing mechanisms and high–level services for searching resources, representing, creating, and managing knowledge discovery processes, and for composing existing data services and data mining services in a structured manner, allowing designers to plan, store, document, verify, share and re-execute their workflows as well as manage their output results.

The Knowledge Grid architecture is composed of a set of services divided in two layers: the *Core K-Grid layer* that interfaces the basic and generic Grid middleware services and the *High-level K-Grid layer* that interfaces the user by offering a set of services for the design and execution of knowledge discovery applications. Both layers make use of repositories that provide information about resource metadata, execution plans, and knowledge obtained as result of knowledge discovery applications.

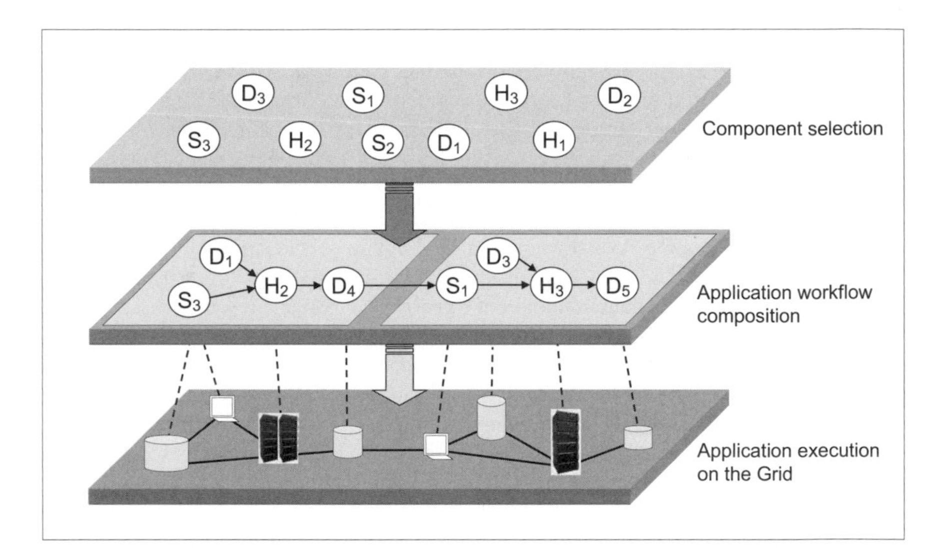

Fig. 53.1. Main steps of application composition and execution in the Knowledge Grid

In the Knowledge Grid environment, discovery processes are represented as workflows that a user may compose using both concrete and abstract Grid resources. Knowledge discovery workflows are defined using a visual interface that shows resources (data, tools, and hosts) to the user and offers mechanisms for integrating them in a workflow. Information about single resources and workflows are stored using an XML-based notation that represents a workflow (called *execution plan* in the Knowledge Grid terminology) as a data-flow graph of nodes, each representing either a Data Mining service or a data transfer service. The XML representation allows the workflows for discovery processes to be easily validated, shared, translated in executable scripts, and stored for future executions. Figure 53.1 shows the main steps of the composition and execution processes of a knowledge discovery application on the Knowledge Grid.

53.4.1 Knowledge Grid Components and Tools

Figure 53.2 shows the general structure of the Knowledge Grid system and its main components and interaction patterns.

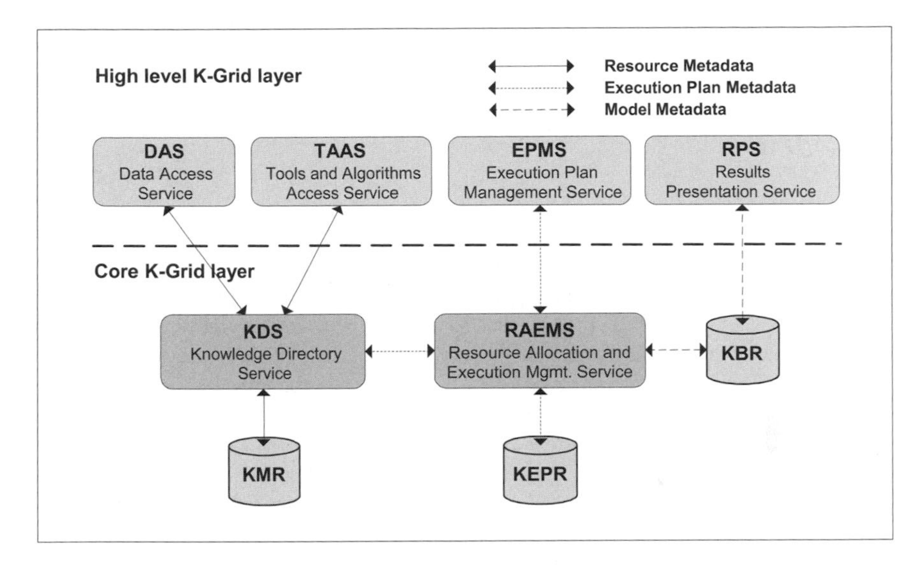

Fig. 53.2. The Knowledge Grid general structure and components

The High-level K-Grid layer includes services used to compose, validate, and execute a parallel and distributed knowledge discovery computation. Moreover, the layer offers services to store and analyze the discovered knowledge. Main services of the High-level K-Grid layer are:

- The *Data Access Service* (*DAS*) allows for the search, selection, transfer, transformation, and delivery of data to be mined.
- The *Tools and Algorithms Access Service* (*TAAS*) is responsible for searching, selecting and downloading Data Mining tools and algorithms.
- The *Execution Plan Management Service* (*EPMS*). An execution plan is represented by a graph describing interactions and data flows among data sources, extraction tools, Data Mining tools, and visualization tools. The Execution Plan Management Service allows for defining the structure of an application by building the corresponding graph and adding a set of constraints about resources. Generated execution plans are stored, through the RAEMS, in the *Knowledge Execution Plan Repository* (*KEP*
R).
- The *Results Presentation Service* (*RPS*) offers facilities for presenting and visualizing the extracted knowledge models (e.g., association rules, clustering models, classifications).

The Core K-Grid layer includes two main services:

- The *Knowledge Directory Service* (*KDS*) that manages metadata describing Knowledge Grid resources. Such resources comprise hosts, repositories of data to be mined, tools and algorithms used to extract, analyze, and manipulate data, distributed knowledge discovery execution plans and knowledge obtained as result of the mining process. The metadata information is represented by XML documents stored in a Knowledge Metadata Repository (KMR).

- The *Resource Allocation and Execution Management Service* (*RAEMS*) is used to find a suitable mapping between an "abstract" execution plan (formalized in XML) and available resources, with the goal of satisfying the constraints (computing power, storage, memory, database, network performance) imposed by the execution plan. After the execution plan activation, this service manages and coordinates the application execution and the storing of knowledge results in the *Knowledge Base Repository* (*KBR*).

An Application Scenario

We discuss here a simple meta-learning process over the Knowledge Grid, to show how the execution of a distributed Data Mining application can benefit from the Knowledge Grid services (Cannataro *et al.*, 2002B). Meta-learning aims to generate a number of independent classifiers by applying learning programs to a collection of distributed data sets in parallel. The classifiers computed by learning programs are then collected and combined to obtain a global classifier (Prodromidis *et al.*, 2000).

Figure 53.3 shows a distributed meta-learning scenario, in which a global classifier *GC* is obtained on *NodeZ* starting from the original data set *DS* stored on *NodeA*.

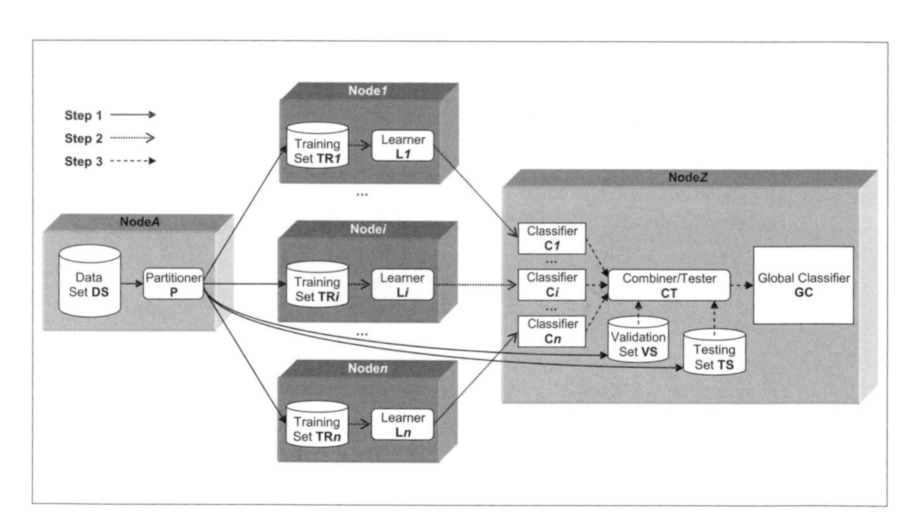

Fig. 53.3. A distributed meta-learning scenario

This process can be described through the following steps:

1. On *NodeA*, training sets $TR_1,\ldots,TR_n$, testing set *TS* and validation set *VS* are extracted from *DS* by the partitioner *P*. Then $TR_1,\ldots,TR_n$, *TS* and *VS* are respectively moved from *NodeA* to $Node_1,\ldots,Node_n$, and to *NodeZ*.
2. On each $Node_i$ (i=1,...,n) the classifier C_i is trained from TR_i by the learner L_i. Then each C_i is moved from $Node_i$ to *NodeZ*.

3. On *NodeZ*, the $C_1,\ldots,C_n$ classifiers are combined and tested on *TS* and validated on *VS* by the combiner/tester *CT* to produce the global classifier *GC*.

To design such an application, a Knowledge Grid user interacts with the EPMS service, which provides a visual interface - see below - to compose a workflow (execution plan) describing at a high level the needed activities involved in the overall Data Mining computation.

Through the execution plan, computing, software and data resources are specified along with a set of requirements on them. In our example the user requires a set of n nodes providing the *Learner* software and a node providing the *Combiner/Tester* software, all of them satisfying given platform constraints and performance requirements. In addition, the execution plan includes information about how to coordinate the execution of all the steps, as outlined above.

The execution plan is then processed by the RAEMS, which takes care of its allocation. In particular, it first finds appropriate resources matching user requirements (i.e., a set of concrete hosts $Node_1,\ldots,Node_n$ offering the software *L*, and a host $Node_Z$ providing the *CT* software), using the KDS services. Next, it manages the execution of the overall application, enforcing dependencies among data extraction, transfer, and mining steps, as specified in the execution plan. The operations of data extraction and transfer are performed at a lower level by invoking the DAS services. We observe here that, where needed, the RAEMS may perform software staging by means of the TAAS service.

Finally, the RAEMS manages results retrieving (i.e., transferring of the global classifier *GC* to the user host), and visualizes them using the RPS facilities.

Implementation

A software environment that implements the main components of the Knowledge Grid, comprising services and functionalities ranging from information and discovery services to visual design and execution facilities is *VEGA - Visual Environment for Grid Applications* - (Cannataro *et al.*, 2002A, Cannataro *et al.*, 2002A).

The main goal of VEGA is to offer a set of visual functionalities that give users the possibility to design applications starting from a view of the present Grid status (i.e., available nodes and resources), and composing the different stages constituting them inside a structured environment. The high-level features offered by VEGA are intended to provide the user with easy access to Grid facilities with a high level of abstraction, in order to leave her free to concentrate on the application design process. To fulfill this aim VEGA builds a visual environment based on the component framework concept, by using and enhancing basic services offered by the Knowledge Grid and the Globus Toolkit.

Key concepts in the VEGA approach to the design of a Grid application are the *visual language* used to describe in a component-like manner, and through a graphical representation, the jobs constituting an application, and the possibility to group these jobs in *workspaces* to form specific interdependent stages. A consistency checking module parses the model of the computation both while the design is in progress and prior to execute it, monitoring and driving user actions so as to obtain a correct and consistent graphical representation of the application. Together with the workspace concept, VEGA makes available also the *virtual resource* abstraction, thanks to these entities it is possible to compose applications working on data processed/generated in previous phases even if the execution has not been performed yet. VEGA includes an *execution service*, which gives the user the possibility to execute the designed application, monitor its status, and visualize results.

53.5 Summary

Parallel and Grid-based Data Mining are key technologies to enhance performance of knowledge discovery processes on large amount of data. Parallel Data Mining is a mature area that produced algorithms and techniques broadly integrated in data mining systems and suites. Today parallel Data Mining systems and algorithms can be integrated as components of Grid-based systems to develop high-performance knowledge discovery applications.

This chapter introduced Data Mining techniques on parallel architectures, showing how large-scale Data Mining and knowledge discovery applications can achieve scalability by using systems, tools and performance offered by parallel processing systems. Some experiences and results in parallelizing Data Mining algorithms according to different approaches have been also reported.

To perform Data Mining on massive data sets, distributed across multiple sites, knowledge discovery systems based on Grid infrastructures are emerging. The chapter discussed the main benefits coming from the use of Grid models and platforms in developing distributed knowledge discovery systems, analyzing some emerging Grid-based Data Mining systems.

Parallel and Grid-based Data Mining will play a more and more important role for data analysis and knowledge extraction in several application contexts. The Knowledge Grid environment, we shortly described here, is a representative effort to build a Grid-based parallel and distributed knowledge discovery system for a wide set of high–performance distributed applications.

References

Agrawal G. High-level Interfaces and Abstractions for Grid-based Data Mining. Workshop on Data Mining and Exploration Middleware for Distributed and Grid Computing; 2003 September 18–19; Minneapolis, MI.

Agrawal R., Shafer J.C. Parallel Mining of Association Rules. IEEE Transactions on Knowledge and Data Engineering 1996; 8: 962-969.

Agrawal R, Srikant R. Fast Algorithms for Mining Association Rules. Proceedings of the 20th International Conference on Very Large Databases; 1994; Santiago, Chile.

Berman F. From TeraGrid to Knowledge Grid. Communications of the ACM 2001; 44(11): 27-28.

Berry, M. JA, Linoff, G., Data Mining Techniques for Marketing, Sales, and Customer Support. New York: Wiley Computer Publishing, 1997.

Beynon M, Kurc T, Catalyurek U, Chang C, Sussman A, Saltz J. Distributed Processing of Very Large Datasets with DataCutter. Parallel Computing 2001. 27(11):1457-1478.

Bigus, J. P., Data Mining with Neural Networks. New York: McGraw-Hill, 1996.

Bruynooghe M., Parallel Implementation of Fast Clustering Algorithms. Proceedings of the International Symposium on High Performance Computing; 1989 March 22-24; Montpellier, France. Elsevier Science, 1989; 65-78.

Cannataro M, Congiusta A, Talia D, Trunfio P. A Data Mining Toolset for Distributed High-performance Platforms. Proceedings of the International Conference on Data Mining Methods and Databases for Engineering; 2002 September 25-27; Bologna, Italy. Wessex Institute Press, 2002; 41-50.

Cannataro M., Talia D. The Knowledge Grid. Communications of the ACM 2003; 46(1):89-93.

Cannataro M, Talia D, Trunfio P. KNOWLEDGE GRID: High Performance Knowledge Discovery Services on the Grid. Proceedings of the 2nd International Workshop GRID 2001; 2001 November; Denver, CO. Springer-Verlag, 2001; LNCS 2242:38-50.

Cannataro M., Talia D., Trunfio P. Distributed Data Mining on the Grid. Future Generation Computer Systems 2002. 18(8):1101-1112.

Congiusta A, Talia D, Trunfio P. VEGA: A Visual Environment for Developing Complex Grid Applications. Proceedings of the First International Workshop on Knowledge Grid and Grid Intelligence (KGGI); 2003 October 13; Halifax, Canada.

Catlett C. The TeraGrid: a Primer, 2002.

Curcin V, Ghanem M, Guo Y, Kohler M, Rowe A, Syed J, Wendel P. Discovery Net: Towards a Grid of Knowledge Discovery. Proceedings of the 8th International Conference on Knowledge Discovery and Data Mining; 2002 July 23-26; Edmonton, Canada.

Foster I, Kesselman C, Nick J, Tuecke S (2002). The Physiology of the Grid: An Open Grid Services Architecture for Distributed Systems Integration.

Foti D, Lipari D, Pizzuti C, Talia D. Scalable Parallel Clustering for Data Mining on Multicomputers. Proceedings of the 3rd International Workshop on High Performance Data Mining; 2000; Cancun. Springer-Verlag, 2000; LNCS 1800:390-398.

Freitas, A. A., Lavington, S. H, Mining Very Large Database with Parallel Processing. Boston: Kluwer Academic Publishers, 1998.

Giannadakis N., Rowe A., Ghanem M., Guo Y. InfoGrid: Providing Information Integration for Knowledge Discovery. Information Sciences 2003; 155:199-226.

Han E. H., Karypis G., Kumar V. Scalable Parallel Data Mining for Association Rules. IEEE Transactions on Knowledge and Data Engineering 2000; 12(2):337-352

Hinke T., Novonty J. Data Mining on NASA's Information Power Grid. Proceedings 9th International Symposium on High Performance Distributed Computing; 2000 August 1-4; Pittsburgh, PA.

Johnston W. E. Computational and Data Grids in Large-Scale Science and Engineering. Future Generation Computer Systems 2002; 18(8):1085-1100.

Judd D, McKinley K, Jain AK. Large-Scale Parallel Data Clustering. Proceedings of the International Conference On Pattern Recognition; 1996; Wien.

Kargupta, H., Chan, P. (Eds.), Advances in Distributed and Parallel Knowledge Discovery. Boston: AAAI/MIT Press, 2000.

Kufrin R. Generating C4.5 Production Rules in Parallel. Proceedings of the 14th National Conference on Artificial Intelligence; AAAI Press, 1997.

Li X., Fang Z. Parallel Clustering Algorithms. Parallel Computing 1989; 11: 275–290.

Moore R.W. (2001). Knowledge-Based Grids: Two Use Cases. GGF-3 Meeting.

Neri F, Giordana A. A Parallel Genetic Algorithm for Concept Learning. Proceedings of the 6th International Conference on Genetic Algorithms; 1995 July 15-19; Pittsburgh, PA. Morgan Kaufmann, 1995; 436-443.

Olson C.F. Parallel Algorithms for Hierarchical Clustering. Parallel Computing 1995; 21:1313-1325.

Pearson, R. A. "A Coarse-grained Parallel Induction Heuristic." In Parallel Processing for Artificial Intelligence 2, H. Kitano, V. Kumar, C.B. Suttner, ed. Elsevier Science, 1994.

Prodromidis, A. L., Chan, P. K., Stolfo, S. J. "Meta-Learning in Distributed Data Mining Systems: Issues and Approaches", In Advances in Distributed and Parallel Knowledge Discovery, H. Kargupta, P. Chan, ed. AAAI Press, 2000.

Shafer J, Agrawal R, Mehta M. SPRINT: A Scalable Parallel Classifier for Data Mining. Proceedings of the 22nd International Conference Very Large Databases; 1996; Bombay.

Skillicorn D. Strategies for Parallel Data Mining. IEEE Concurrency 1999; 7(4):26-35.

Skillicorn D., Talia D. Mining Large Data Sets on Grids: Issues and Prospects. Computing and Informatics 2002; 21:347-362.

Witten, I. H., Frank, E., Data Mining: Practical Machine Learning Tools and Techniques with Java Implementations. San Francisco: Morgan Kaufmann, 2000.

Zaki M.J. Parallel and Distributed Association Mining: A Survey. IEEE Concurrency 1999; 7(4):14-25.

54

Collaborative Data Mining

Steve Moyle

Oxford University Computing Laboratory

Summary. Collaborative Data Mining is a setting where the Data Mining effort is distributed to multiple collaborating agents – human or software. The objective of the collaborative Data Mining effort is to produce solutions to the tackled Data Mining problem which are considered better by some metric, with respect to those solutions that would have been achieved by individual, non-collaborating agents. The solutions require evaluation, comparison, and approaches for combination. Collaboration requires communication, and implies some form of community. The human form of collaboration is a social task. Organizing communities in an effective manner is non-trivial and often requires well defined roles and processes. Data Mining, too, benefits from a standard process. This chapter explores the standard Data Mining process CRISP-DM utilized in a collaborative setting.

Key words: Collaborative Data Mining, CRISP-DM, ROC

54.1 Introduction

Data Mining is about solving problems using data (Witten and Frank, 2000), and such it is normally a creative activity leveraging human intelligence. This is similar to the spirit and practices of scientific discovery (Bacon, 1994, Popper, 1977, Kuhn, 1970) which utilize many techniques including induction, abduction, hunches, and clever guessing to propose hypotheses that aid in understanding the problem and finally lead to a solution. Collaboration is the act of working together with one or more people in order to achieve something (Soukhanov, 2001). Collaboration in intelligence-intensive activities may lead to increased results. However, collaboration brings its own difficulties including communication, coordination, as well as cultural and social difficulties. Some of these difficulties can be analyzed by the *e-Collaboration Space* model (McKenzie and van Winkelen, 2001).

Data Mining projects benefit from a rigorous process and methodology (Adriaans and Zantinge, 1996, Fayyad *et al.*, 1996, Chapman *et al.*, 2000). For collaborative Data Mining, such processes need to be embedded in a broader set of processes that support the collaboration

O. Maimon, L. Rokach (eds.), *Data Mining and Knowledge Discovery Handbook*, 2nd ed.,
DOI 10.1007/978-0-387-09823-4_54,

setting. Such well defined and strong processes include, for instance, clear model evaluation procedures (Blockeel and Moyle, 2002).

Different perspectives exist on what collaborative Data Mining is (this is discussed further in section 54.5). Three interpretations are: 1) multiple software agents applying Data Mining algorithms to solve the same problem; 2) humans using modern collaboration techniques to apply Data Mining to a single, defined problem; 3) Data Mining the artifacts of human collaboration. This chapter will focus solely on the second item – that of humans using collaboration techniques to apply data mining to a single task. With sufficient definition of a *particular* Data Mining problem, this is similar to a multiple software agent Data Mining framework (the first item), although this is not the aim of the chapter. Many of the difficulties encountered in human collaboration will also be encountered in designing a system for software agent collaboration.

Collaborative Data Mining aims to combine the results generated by isolated experts, by enabling the collaboration of geographically dispersed laboratories and companies. For each Data Mining problem, a virtual team of experts is selected on the basis of adequacy and availability. Experts apply their methods to solving the problem – but also communicate with each other to share their growing understanding of the problem. It is here that collaboration is key.

The process of analyzing data through models has many similarities to experimental research. Like the process of scientific discovery, Data Mining can benefit from different techniques used by multiple researchers who collaborate, compete, and compare results to improve their combined understanding. The rest of this chapter is organized as follows. The potential difficulties in (remote) collaboration and a framework for analyzing such difficulties are outlined. A standard Data Mining process is reviewed, and studied for the likely contributions that can be achieved collaboratively. A collaboration process for Data Mining is presented, with clear guidelines for the practitioner so that they may avoid the potential pitfalls related to collaborative Data Mining. A brief summary of real examples of the application of collaborative Data Mining are presented. The chapter concludes with a discussion.

54.2 Remote Collaboration

This section considers the motivations behind (remote) collaboration[1], and types of collaboration it enables. It then reviews the framework proposed by McKenzie and Van Winkelen (McKenzie and van Winkelen, 2001) for working within *e-Collaboration Space*. The term *e-Collaboration* will be used as shorthand for *remote* collaboration, but many of the principles can be applied to local collaboration also.

54.2.1 E-Collaboration:Motivations and Forms

The main motivation for collaboration (Moyle *et al.*, 2003) is to harness dispersed expertise and to enable knowledge sharing and learning in a manner that builds intellectual capital (Edvinsson and Malone, 1997). This offers tantalizing potential rewards including boosting innovation, flexible resource management, and reduced risk (Amara, 1990, Mowshowitz, 1997, Nohria and Eccles, 1993, Snow *et al.*, 1996), but these rewards are offset by numerous difficulties mainly due to the increased complexity of a virtual environment.

In (McKenzie and van Winkelen, 2001) seven distinct forms of e-collaborating organizations that can be distinguished either by their structure or the intent behind their formation are

[1] The term “remote” is removed in the sequel.

identified. These are: 1) virtual/smart organizations; 2) a community of interest and practice; 3) a virtual enterprise; 4) virtual teams; 5) a community of creation; 6) collaborative product commerce or customer communities; and 7) virtual sourcing and resource coalitions. For collaborative data mining forms 4, and 5 are most relevant. These forms are summarized below.

- Virtual Teams are temporary culturally diverse geographically dispersed work groups that communicate electronically. These can be smaller entities within virtual enterprises, or within a transnational organization. They can be categorized by changing membership and multiple organizational contexts.
- A Community of creation is revolves around a central firm and shares its knowledge for the purpose of innovation. This structure consists of individuals and organizations with ever changing boundaries.

Having recognized the collaboration form makes it possible to analyze the difficulties that might be encountered. Such an analysis can be performed with respect to the e-collaboration space model described in the next section.

54.2.2 E-Collaboration Space

Each type of e-collaboration form can be usefully analyzed with respect to McKenzie and Van Winkelen's *e-Collaboration Space* model (McKenzie and van Winkelen, 2001). This model casts each form into the space by studying their location on the three dimensions of: number of *boundaries crossed*, *task*, and *relationships*.

- **Boundaries crossed**: The more boundaries that are crossed in e-collabo ration, the more barriers to a successful outcome are present. All communication takes place across some boundary (Wilson, 2002). Fewer boundaries between agents lead to a lower risk of misunderstanding. In e-collaboration the number of boundaries is automatically increased. Influential boundaries to successful e-collaboration are: technological, temporal, organizational, and cultural.
- **Task**: The nature of the tasks involved in the collaborative project is influenced by the complexity of the processes, uncertainty of the available information and outcomes, and interdependence of the various stages of the task. The complexity can be broadly classified into linear – step-by-step processes; or non-linear. The interdependence of a task relates to whether it can be decomposed into subtasks which can be worked on independently by different participants.
- **Relationships**: Relationships are key to any successful collaboration. When electronic communication is the only mode of interaction it is harder for relationships to form, because the instinctive reading of signals that establish trust and mutual understanding are less accessible to participants.

For the remainder of the chapter only the dimension of *task* will be highlighted within the e-collaboration space model. As will be described in the next sub-section, task complexity makes collaborative Data Mining risk prone.

54.2.3 Collaborative Data Mining in E-Collaboration Space

Different forms of e-collaboration – as measured relative to the dimensions of task, boundaries, and relationships – can be viewed as locations in a three dimensional e-collaboration space. The location of a collaborative Data Mining project depends on the actual setting of such a project. The most well defined dimension with respect to the Data Mining process (refer back to section 60.2.1) is that of *task*.

The ***task*** complexity of Data Mining is high. Not only is there a high level of expertise involved in a Data Mining project, but also there is the risk that in reaching the final solution(s), much effort will appear – in hindsight – to have been wasted. Data miners have long understood the need for a methodology to support the Data Mining process (Adriaans and Zantinge, 1996, Fayyad *et al.*, 1996, Chapman *et al.*, 2000). All these methodologies are explicit that the Data Mining process is non-linear, and warns that information uncovered in later phases can invalidate assumptions made in earlier phases. As a result the previous phases may need to be re-visited. To exacerbate the situation, Data Mining is by its very nature a speculative process – there may be no valuable information contained in the data sources at all, or the techniques being used may not have sufficient power to uncover it. A typical Data Mining project at the start of the collaboration is summarized with respect to the e-collaboration model in Table 54.1.

Table 54.1. The position of a disperse collaborative Data Mining project in E-collaboration space († potential boundary depending on situation).

Task	**Boundaries Crossed**	**Relationships**
High	*High*	*Medium High*
- Complex non-linear interdependencies - Uncertainty	- Medium technological - temporal† - geographical - large organizational† - cultural†	- Medium commonality of view - Medium duration of existing relationship - Medium duration of collaboration

54.3 The Data Mining Process

Data Mining processes broadly consist of a number of phases. These phases, however, are interrelated and are not necessarily executed in a linear manner. For example, the results of one phase may uncover more detail relating to an earlier phase and may force more effort to be expended on a phase previously thought complete. The CRISP-DM methodology — CRoss Industry Standard Process for Data Mining (Chapman *et al.*, 2000), is an attempt to standardise the process of Data Mining. In CRISP-DM, six interrelated phases are used to describe the Data Mining process: *business understanding*, *data understanding*, *data preparation*, *modelling*, *evaluation*, and *deployment* (Figure 54.1). The main outputs of the *business understanding* phase are the definition of business and data mining objectives as well as business and Data Mining evaluation criteria. In this phase an assessment of resource requirements and estimation of risk is performed. In the *data understanding* phase data collected and characterized. Data quality is also assessed.

During *data preparation*, tables, records and attributes are selected and transformed for modelling. *Modelling* is the process of extracting input/output patterns from given data and deriving models — typically mathematical or logical models. In the modelling phase, various techniques (e.g. association rules, decision trees, logistic regression, k-means clustering)

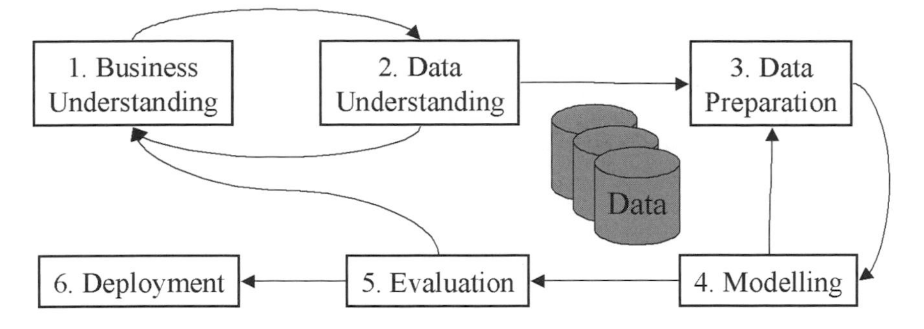

Fig. 54.1. The CRISP-DM cycle

are selected and applied and their parameters are calibrated – or *tuned* – to optimal values. Different models are compared, and possibly combined.

In the *evaluation* phase models are selected and reviewed according to the business criteria. The whole Data Mining process is reviewed and a list of possible actions is elaborated. In the last phase, *deployment* is planned, implemented, and monitored. The entire project is typically documented and summarized in a report.

The CRISP-DM handbook (Chapman *et al.*, 2000) describes in detail how each of the main phases is subdivided into specific tasks, with clearly defined predecessors/successors, and inputs/outputs.

54.4 Collaborative Data Mining Guidelines

The CRISP-DM Data Mining process described in the preceding section can be adopted by Data Mining agents collaborating remotely on a particular Data Mining project (SolEuNet, 2002, Flach *et al.*, 2003). Not all of the CRISP-DM methodology can be entirely performed in a collaboartive setting. *Business understanding* for instance, requires intense close contact with the business environment for which the Data Mining is being performed. The phases that can most easily be performed in a remote-collaborative fashion are *data preparation* and *modelling*. The other phases can nevertheless benefit from a collaborative approach. Although many of the specific tasks can be carried out independently, care must be taken by the participants to ensure that efforts are not wasted. Principles to guide the process of collaboration should be established in advance of a collaborative Data Mining project. For instance, individual agents must communicate or share any intermediate results – or improvements in the *current best understanding* of the Data Mining problem – so that all agents have the new knowledge. Providing a catalogue of up-to-date knowledge about the problem assists new agents entering the Data Mining project. Furthermore, procedures are required for how results from different agents are compared, and ultimately combined, so that the value of efforts is greater than the sum of the individual components.

54.4.1 Collaboration Principles

(Moyle *et al.*, 2003) present a framework for collaborative Data Mining, involving both principles and technological support. Collaborative groupware technology, with specific functionality to support data mining are described (Vo *et al.*, 2001). Principles for collaborative data mining are outlined as follows (Moyle *et al.*, 2003).

1. *Requisite management.* Sufficient management processes should be established. In particular the definition and objectives of the Data Mining problem should be clear from the start of the project to all participants. An infrastructure ensuring information flows within the network of agents should be provided.
2. *Problem Solving Freedom.* Agents should use their expertise and tools to execute Data Mining tasks to solve problem in the manner they find best.
3. *Start any time.* All the necessary information about the Data Mining problem should be captured and made available to participants at all times. This includes problem definition, data, evaluation criteria, and any knowledge produced.
4. *Stop any time.* Participants should work on their solutions so that a working solution – however crude – is available whenever a stop signal is issued. These solutions will typically be Data Mining models. One approach is to try simpler modeling techniques first (Holte, 1993).
5. *Online knowledge sharing.* The knowledge about the Data Mining problem gained by each participant at each phase should be shared with all participants in a timely manner.
6. *Security.* Data and information about the Data Mining problem may contain sensitive information and must not to be revealed outside the project. Access to information must be controlled.

Having established a collaborative Data Mining project with appropriate principles and support, how can the results of the Data Mining efforts be compared and combined so that the results are maximized? This is the question that the next section deals with.

54.4.2 Data Mining model evaluation and combination

One of the main outputs from the Data Mining process (Chapman *et al.*, 2000) are the Data Mining models. These may take many forms including decision trees, rules, artificial neural-networks, regression equations (see (Mitchell, 1997) as an introduction to machine learning, and (Hair *et al.*, 1998) as an introduction to statistics text). Different agents may produce models in the different forms, which requires methods for both evaluating them and combining them.

When multiple agents produce multiple models as the result of data mining effort a process for evaluating their relative merits must be established. Such processes are well defined in Data Mining challenge problems (e.g. (Srinivasan *et al.*, 1999,Page and Hatzis, 2001)). For example a challenge recipe for the production of classificatory models can be found in (Moyle and Srinivasan, 2001). To ensure accurate comparisons, models built by different agents must be evaluated in exactly the same way, on the same data. This sounds like an obvious statement, but agents can easily make adjustments to their copy of the data to suit their particular approaches, without making the changes available to the other agents. This makes any model evaluation ad comparison extremely difficult.

Furthermore, the evaluation criterion or criteria (there may be several) deemed most appropriate may change during the knowledge discovery process. For instance, at some point one may wish to redefine the data set on which models are evaluated (e.g. because it is found that it contains outliers that make the evaluation procedure inaccurate) and re-evaluate previously built models. In (Blockeel and Moyle, 2002) it is discussed how this evaluation and re-evaluation leads to significant extra efforts for the different agents and consequently is a barrier to the knowledge discovery process, unless adequate software support is provided.

One approach to control model evaluation is to centralize the process. Consider an abstracted Data Mining process where agents first *tune* their modeling algorithm (which outputs the algorithm and its parameter settings, I), before *building* a final model (which is output as M). The agent then uses the model to *predict* the labels on a test set (producing predictions, P), from which an overall *evaluation* of the model (resulting in a score S) is determined. The point at which these outputs are published for all agents to access depend on the architecture of the evaluation system as shown in Figure 54.2. A single evaluation agent provides the evaluation procedures; different agents submit information on their models to this agent, which stores this information and automatically evaluates it according to all relevant criteria. If criteria change, the evaluation agent automatically re-evaluates previously submitted models.

In such a framework information about produced models can be submitted at several levels, as illustrated in Figure 54.2. Agents can run their own models on a test set and send only predictions to the evaluation agent (assuming evaluation is based on predictions only), they can submit descriptions of the models themselves, or even just send a complete description on the model producing algorithm and the used parameters to the evaluation agent which has been augmented with modeling algorithms. These respective options offer increased centralization and increasingly flexible evaluation possibilities, but also involve increasingly sophisticated software support (Blockeel and Moyle, 2002).

Communicating Data Mining models to the evaluation agent can be performed using a standard format. For instance in (Flach *et al.*, 2003) models from multiple agents were submitted in a standard, XML style, format (using the standard Predictive Markup Modeling Language (PMML) (The Data Mining Group, 2003)). Such a procedure has been adopted for a real-world collaborative Data Mining project (Flach *et al.*, 2003).

Model combination is not always possible. However, when restricted to binary-classificatory models it is possible to utilize Receiver Operating Characteristic (Provost and Fawcett, 2001) curves to assist both model comparison, and model combination. ROC analysis plots different binary-classification models on a two dimensional space with respect to the type of errors the models make – false positive errors, and false negative errors[2]. The actual performance of a model at run-time depends on the costs of errors at run-time, and the distribution of the classes at run-time. The values of these run-time parameters – or operating characteristics – determine the optimal model(s) for use in prediction. ROC analysis enables models to be compared, which may result in some models *never* being optimal under any operating conditions and can be discarded. The remaining models are those that are located on the ROC convex hull (ROCCH).

As well as determining non-optimal models, ROC analysis can be used to combine models. One method is to use more two adjacent models on the ROCCH that are located either side of the operating condition in combination to make run-time predictions. Another approach to using ROCCH is to modify a single model into multiple models, that then can be plotted in ROC space (Flach *et al.*, 2001) resulting in models that fit a broader range of operating conditions. (Wettschereck *et al.*, 2003) describe a support system that performs model evaluation, model visualization, and model comparison, which has been applied in a collaborative Data Mining setting (Flach *et al.*, 2003).

[2] The axes on an ROC curve are actually the true positive rate versus the false positive rate.

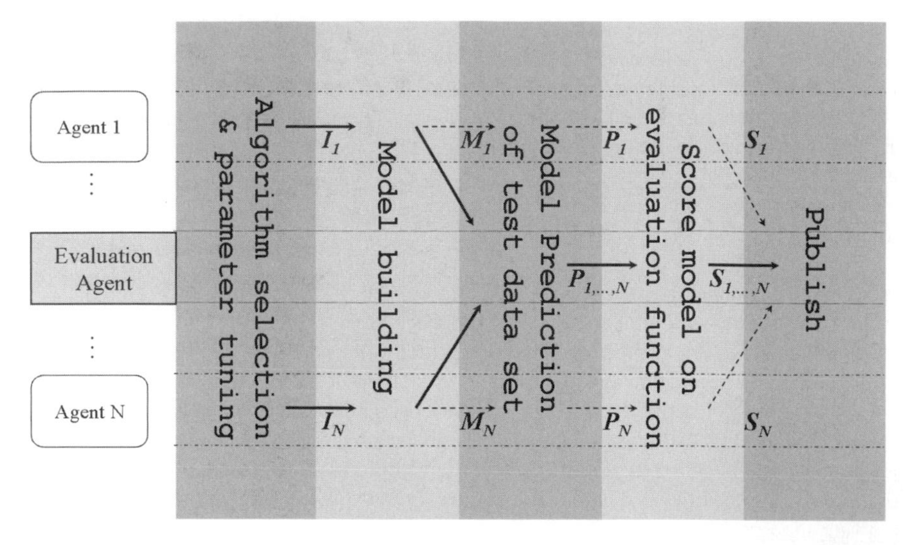

Fig. 54.2. Two different architectures for model evaluation. The path finishing in dashed arrows depicts agents in charge of building and evaluating their own models before publishing their results centrally. The path of solid arrows depicts Data Mining agents submitting their models to a centralized evaluation agent which provides the services of executing submitted models on a test set, evaluating the predictions to produce scores, and then publishing the results. The information submitted to the central evaluation agent is: I=algorithm and parameter settings to produce models; M=models; P=predictions made by the models on a test set; S=scores of the value of the models.

54.5 Discussion

References containing the keywords: *collaborative Data Mining collaboration* partition naturally into the following categories.

- *Multiple software agents applying Data Mining algorithms to solve the same problem:* (e.g. (Ramakrishnan, 2001)) this presupposes that the Data Mining task and its associated data are well defined *a priori*.
- *Humans using modern collaboration techniques to apply Data Mining to a single, defined problem* (e.g. (Mladenic *et al.*, 2003)).
- *Data Mining the artifacts of human collaboration:* (e.g. (Biuk-Aghai and Simoff, 2001)) these artifacts are typically the conversations and associated documents collected via some electronic based discussion forum.
- *The collaboration process itself resulting in increased knowledge:* a form of knowledge growth by collection within a context.
- *Grid style computing facilities collaborating to provide resources for Data Mining:* (e.g. (Singh *et al.*, 2003)) these resources are typically providing either federated data or distributed computing power.
- *Integrating Data Mining techniques into business process software:* (e.g. (McDougall, 2001)) for example Enterprise Resource Planning systems, and groupware. Note that this, too, implies *a priori* knowledge of what the Data Mining problems are to be solved.

This chapter focused mainly on the second item – that of humans using collaboration techniques to apply Data Mining to a single task. With sufficient definition of a *particular* Data Mining problem, this can lead to a multiple software agent Data Mining framework (the first item), although this is not the aim of this chapter.

Many Data Mining challenges have been issued, which by their nature always result in "winners" and "losers". However, in collaborative approaches, much can be learned from the losers as the Data Mining projects proceed. Much initial effort is required to establish a Data Mining challenge (e.g. problem specification, data collection and preprocessing, specification of evaluation criteria) – even before the participants register. This effort also needs to be expended in a collaborative setting so that the objectives of the Data Mining project are clearly articulated in advance.

The application of the collaborative methodology and techniques described here has been performed with mixed success in the data mining projects (Flach *et al.*, 2003, Stepnkov *et al.*, 2003, Jorge *et al.*, 2003). More development of collaborative Data Mining processes and supporting tools and communication environments are likely to improve the results of harnessing dispersed Data Mining expertise.

54.6 Conclusions

Collaborative Data Mining is more difficult the single team setting. Data mining benefits from adhering to established processes. One key notion in Data Mining methodologies is that of *understanding* (e.g. CRISP-DM contains the phases, business understanding and data understanding). How are such understandings produced, articulated, maintained, and communicated to all collaborating agents? What happens when understandings change – how much of the data mining process will need re-work? How does one agent's understanding differ from another, simply due to communication, language and cultural differences?

Practitioners embarking on collaborative Data Mining might wish to heed some of the lessons learned from other collaborative Data Mining projects:

- Analyze the form of collaboration proposed and understand how difficult it is likely to be.
- Establish a methodology that all participants can utilize along with support tools and technologies.
- Ensure that all results – intermediate or otherwise – are recorded, and shared in a timely manner.
- Encourage competition among participants.
- Define metrics for success at all stages.
- Define model evaluation and combination procedures.

References

Adriaans, P., and Zantinge, D., *Data Mining*. Addison-Wesley, New York, 1996.

Amara, R., *New directions for innovations*. Futures **53-22**(2): p. 142 - 152, 1990.

Bacon, F., *Novum Organum*, eds. P. Urbach and J. Gibson. Open Court Publishing Company, 1994.

Biuk-Aghai, R.P. and S.J. Simoff. *An integrative framework for knowledge extraction in collaborative virtual environments*. In *The 2001 International ACM SIGGROUP Conference on Supporting Group Work*. Boulder, Colorado, USA, 2001.

Blockeel, H. and S.A. Moyle. *Collaborative Data Mining needs centralised model evaluation*. In *Proceedings of the ICML-2002 Workshop on Data Mining Lessons Learned*. The University of New South Wales, Sydney, 2002.

Chapman, P., Clinton, J., Kerber, R., Khabaza, T., Reinartz, T., Shearer, C., and Wirth, R. *CRISP-DM 1.0: Step-by-step data mining guide*. The CRISP-DM consortium, 2000.

Edvinsson, L. and Malone, M.S. *Intellectual Capital: Realizing Your Company's True Value by Finding Its Hidden Brainpower*. HarperBusiness, New York, USA, 1997.

Fayyad, U., et al., eds. *Advances in Knowledge Discovery and Data Mining*. MIT Press, 1996.

Flach, P.A., et al., *Decision support for Data Mining: introduction to ROC analysis and its application*. In Data Mining and Decision Support: Integration and Collaboration, D. Mladenic, et al., editors. Kluwer Academic Publishers, 2003.

Flach, P., Blockeel, H., Gaertner, T., Grobelnik, M., Kavsek, B., Kejkula, M., Krzywania, D., Lavrac, N., Mladenic, D., Moyle, S., Raeymaekers, S., Rauch, J., Ribeiro, R., Sclep, G., Struyf, J., Todorovski, L., Torgo, L., Wettsc -hereck, D., and Wu, S. *On the road to knowledge: mining 21 years of UK traffic accident reports*, In Data Mining and Decision Support: Integration and Collaboration, D. Mladenic, et al., editors. Kluwer Academic Publishers, 2003.

Hair, J.F., Anderson, R.E., Tatham, R.L., and Black, W.C. *Multivariate Data Analysis*. Prentice Hall, 1998.

Holte, R.C., *Very Simple Classification Rules Perform Well on Most Commonly Used Datasets*. Machine Learning, 1993.

53-3: p. 63-91.

Jorge, J., Alves, M.A., Grobelnik, M., Mladenic, D., and Petrak, J. *Web site access analysis for a national statistical agency*. In Data Mining and Decision Support: Integration and Collaboration, D. Mladenic, et al., editors, p. 157 – 166. Kluwer Academic Publishers, 2003.

Kuhn, T.S., *The structure of scientific revolutions*. 2nd, enlarged ed. 1962, University of Chicago Press, Chicago, 1970.

McDougall, P., *Companies that dare to share information are cashing in on new opportunities*. InformationWeek, May 7, 2001.

McKenzie, J. and C. van Winkelen. *Exploring E-collaboration Space*. In the proceedings of *The first annual Knowledge Management Forum Conference*. Henley Management College, 2001.

Mitchell, T. *Machine Learning*. Department of Computer Science, Carnegie Mellon University. McGraw-Hill Book Company, Pittsburgh, 1997.

Mladenic, D., Lavrac, N., Bohanec, M., and Moyle, S. editors. *Data Mining and Decision Support: Integration and Collaboration*. Kluwer Academic Publishers, 2003.

Mowshowitz, A., *Virtual Organization*. Communications of ACM, **53-40**(9): p. 30 - 37. 1997.

Moyle, S. A., Srinivasan A., *Classificatory challenge-Data Mining: a recipe*. Informatica **53-25**(3): p. 343–347. 2001.

Moyle, S., J. McKenzie, and A. Jorge, *Collaboration in a Data Mining virtual organization*. In *Data Mining and Decision Support: Integration and Collaboration*, D. Mladenic, et al., editors. Kluwer Academic Publishers, 2003.

Nohria, N. and R.G. Eccles, eds. *Network and organizations; structure form and action*. Harvard Business School Press, Boston, 1993.

Page, C.D. and C. Hatzis, *KDD Cup 2001*. University of Wisconsin, `http://www.cs.wisc.edu/~dpage/kddcup2001/`, 2001.

Popper, K. *The Logic of Scientific Discovery*. Routledge, 1977.

Provost, F. and T. Fawcett. *Robust Classification for Imprecise Environments.* Machine Learning **53-42**: p. 203-231, 2001.

Ramakrishnan., R. *Mass Collaboration and Data Mining (keynote address).* In *The Seventh ACM SIGKDD International Conference on Knowledge Discovery and Data Mining (KDD-2001).* San Francisco, California, 2001.

Singh, R., Leigh, J., DeFanti, T.A., and Karayannis F. *TeraVision: a High Resolution Graphics Streaming Device for Amplified Collaboration Environments.* Journal of Future Generation Computer Systems (FGCS). **53-19**(6): p. 957-972, 2003.

Snow, C.C., S.A. Snell, and S.C. Davison. *Using transnational teams to globalize your company.* Organizational Dynamics **53-24**(4): p. 50 - 67, 1996.

SolEuNet. *The Solomon European Netowrk – Data Mining and Decision Support for Business Competitiveness: A European Virtual Enterprise.* `http://soleunet.ijs.si/`, 2002.

Soukhanov, A., ed. *Microsoft Encarta College Dictionary: The First Dictionary for the Internet Age.* St. Martin's Press, 2001.

A. Srinivasan, R.D. King, and D.W. Bristol. *An assessment of submissions made to the Predictive Toxicology Evaluation Challenge.* In *Proceedings of the Sixteenth International Conference on Artificial Intelligence (IJCAI-99).* Morgan Kaufmann, Los Angeles, CA, 1999.

Stepnkov, O., J. Klma, and P. Mikovsk. *Collaborative Data Mining with RAMSYS and Sumatra TT: Prediction of resources for a health farm.* In *Data Mining and Decision Support: Integration and Collaboration*, D. Mladenic, et al., editors. p. 215 – 227. Kluwer Academic Publishers, 2003.

The Data Mining Group, *The Predictive Model Markup Language (PMML).* `http://www.dmg.org/`, 2003.

Vo, A., Richter, G., Moyle, S., Jorge, A. *Collaboration support for virtual data mining enterprises.* In *3rd International Workshop on Learning Software Organizations (LSO'01).* Springer-Verlag, 2001.

Wettschereck, D., A. Jorge, and S. Moyle. *Visaulisation and Evaluation Support of Knowledge Discovery through the Predictive Model Markup Language.* In *7th International Knowledge-Based Intelligent Information and Engineering Systems (KES 2003)*, Oxford. Springer-Verlag, 2003.

Wilson, T.D. *The nonsense of knowledge management.* Information Research **53-8**(1), 2002.

Witten, I.H. and E. Frank. *Data Mining: Practical Machine Learning Tools and Techniques with Java Implementations.* Morgan Kaufmann, San Francisco, 2000.

55

Organizational Data Mining

Hamid R. Nemati[1] and Christopher D. Barko[2]

[1] Information Systems and Operations Management Department
Bryan School of Business and Economics
The University of North Carolina at Greensboro
nemati@uncg.edu
[2] Customer Analytics, Inc.
7009 Austin Creek Drive
Summerfield, NC 27358
chris.barko@customer-analytics.com

Summary. Many organizations today possess substantial quantities of business information but have very little real business knowledge. A recent survey of 450 business executives reported that managerial intuition and instinct are more prevalent than hard facts in driving organizational decisions. To reverse this trend, businesses of all sizes would be well advised to adopt Organizational Data Mining (ODM). ODM is defined as leveraging Data Mining tools and technologies to enhance the decision-making process by transforming data into valuable and actionable knowledge to gain a competitive advantage. ODM has helped many organizations optimize internal resource allocations while better understanding and responding to the needs of their customers. The fundamental aspects of ODM can be categorized into Artificial Intelligence (AI), Information Technology (IT), and Organizational Theory (OT), with OT being the key distinction between ODM and Data Mining. In this chapter, we introduce ODM, explain its unique characteristics, and report on the current status of ODM research. Next we illustrate how several leading organizations have adopted ODM and are benefiting from it. Then we examine the evolution of ODM to the present day and conclude our chapter by contemplating ODM's challenging yet opportunistic future.

Key words: Organizational Data Mining, Customer Relationship Management

55.1 Introduction

Data experts estimate that in 2002 the world generated 5 exabytes of information. This amount of data is more than all the words ever spoken by human beings. And the rate of growth is just as staggering – the amount of data produced in 2002 was up 68% from just two years earlier. The size of the typical business database has grown a hundred-fold during the past five years as a result of Internet commerce, ever-expanding computer systems and mandated record keeping

O. Maimon, L. Rokach (eds.), *Data Mining and Knowledge Discovery Handbook*, 2nd ed.,
DOI 10.1007/978-0-387-09823-4_55,

by government regulations. To better grasp how much data this is, consider the following: if one byte of data is the equivalent of this dot ⊙, the amount of data produced globally in 2002 would equal the diameter of 4,000 suns. And that amount has probably doubled since then (Hardy, 2004).

In spite of this enormous growth in enterprise databases, research from IBM reveals that organizations use less than 1 percent of their data for analysis (Brown, 2002). This is the fundamental irony of the Information Age we live in: organizations possess enormous amounts of business information, yet have so little real business knowledge. And to magnify the problem further, a leading business intelligence firm recently surveyed executives at 450 companies and discovered that 90 percent of these organizations rely on gut instinct rather than hard facts for most of their decisions because they lack the necessary information when they need it (Brown, 2002). And in cases where sufficient business information is available, those organizations are only able to utilize less than 7 percent of it (The Economist, 2001).

This proclamation about data volume growth is no longer surprising, but continues to amaze even the experts. Although for businesses, more data isn't always better. Organizations must assess what data they need to collect and how to best leverage it. Collecting, storing and managing business data and associated databases can be costly, and expending scarce resources to acquire and manage extraneous data fuels inefficiency and hinders optimal performance. The generation and management of business data also loses much of its potential organizational value unless important conclusions can be extracted from it quickly enough to influence decision making while the business opportunity is still present. Managers must rapidly and thoroughly understand the factors driving their business in order to sustain a competitive advantage. Organizational speed and agility supported by fact-based decision making are critical to ensure an organization remains at least one step ahead of its competitors.

In the past, companies have struggled to make decisions because of the lack of data. But in the current environment, more and more organizations are struggling to overcome "information paralysis" – there is so much data available that it is difficult to determine what is relevant and how to extract meaningful knowledge. Organizations today routinely collect and manage terabytes of data in their databases, thereby making information paralysis a key challenge in enterprise decision-making. Once the essential data elements are identified, the data must be reformatted, pre-processed and analyzed to generate knowledge. The resulting knowledge is then delivered to the decision-makers for collaboration, review and action. Once decided upon, the final decision must be communicated to the appropriate parties in a rapid, efficient and cost-effective manner.

55.2 Organizational Data Mining

The manner in which organizations execute this intricate decision-making process is critical to their well-being and industry competitiveness. Those organizations making swift, fact-based decisions by optimally leveraging their data resources will outperform those organizations that do not. A robust technology that facilitates this process of optimal decision-making is known as Organizational Data Mining (ODM). ODM is defined as leveraging Data Mining tools and technologies to enhance the decision-making process by transforming data into valuable and actionable knowledge to gain a competitive advantage (Nemati and Barko, 2001). ODM eliminates the guesswork that permeates so much of corporate decision making. By adopting ODM, an organization's managers and employees are able to act sooner rather than later, be proactive rather than reactive and know rather than guess. ODM technology has helped many organizations optimize internal resource allocations while better understanding and responding to the needs of their customers.

ODM spans a wide array of technologies, including, but not limited to, e-business intelligence, data analysis, online analytical processing (OLAP), customer relationship management (CRM), electronic CRM (e-CRM), executive information systems (EIS), digital dashboards and information portals. ODM enables organizations to answer questions about the past (what has happened?), the present (what is happening?), and the future (what might happen?). Armed with this capability, organizations can generate valuable knowledge from their data, which in turn enhances enterprise decisions. This decision-enhancing technology enables many advantages in operations (faster product development, increased market share with quicker time to market, optimal supply chain management), marketing (higher profitability and increased customer loyalty through more effective marketing campaigns and customer profitability analyses) finance (improved performance through financial analytics and economic evaluation of business units and products) and strategy implementation (business performance management (BPM), the Balanced Scorecard, and related strategy alignment and measurement systems). The result of this enhanced decision making at all levels of the organization is optimal resource allocation and improved business performance.

Profitability in business today relies on speed, agility and efficiency at quality levels thought unobtainable just a few years ago. The slightest imbalance along the supply chain can increase costs, lengthen internal cycle times and delay new product introductions. These imbalances can eventually lead to a loss in both market share and competitive advantage. Meanwhile, organizations are also forging closer relationships with their customers and suppliers by defining tighter agreements in terms of shared processes and risks. As a result, many businesses are deeply immersed in continuously reengineering their processes to improve quality. Six sigma and Balanced Scorecard type efforts are increasingly prevalent. ODM enables organizations to remove supply chain imbalances while improving the speed, flexibility and efficiency of their business processes. This leads to stronger customer and partner relationships and a sustainable competitive advantage.

55.3 ODM versus Data Mining

Data Mining is the process of discovering and interpreting previously unknown patterns in databases. It is a powerful technology that converts data into information and potentially actionable knowledge. However, obtaining new knowledge in an organizational vacuum does not facilitate optimal decision making in a business setting. The unique organizational challenge of understanding and leveraging ODM to engineer actionable knowledge requires assimilating insights from a variety of organizational and technical fields and developing a comprehensive framework that supports an organization's quest for a sustainable competitive advantage. These multidisciplinary fields include Data Mining, business strategy, organizational learning and behavior, organizational culture, organizational politics, business ethics and privacy, knowledge management, information sciences and decision support systems. These fundamental elements of ODM can be summarized into three main groups: Artificial Intelligence (AI), Information Technology (IT), and Organizational Theory (OT). Our research and industry experience suggest that successfully leveraging ODM requires integrating insights from all three categories in an organizational setting typically characterized by complexity and uncertainty. This is the essence and uniqueness of ODM. Obtaining maximum value from ODM involves a cross-department team effort that includes statisticians/data miners, software engineers, business analysts, line-of-business managers, subject matter experts, and upper management support.

55.3.1 Organizational Theory and ODM

Organizations are primarily concerned with studying how operating efficiencies and profitability can be achieved through the effective management of customers, suppliers, partners, and employees. To achieve these goals, research in Organizational Theory (OT) suggests that organizations use data in three vital knowledge creation activities. This organizational knowledge creation and management is a learned ability that can only be achieved via an organized and deliberate methodology. This methodology is a foundation for successfully leveraging ODM within the organization. The three knowledge creation activities (Choo, 1997) are:

- **Sense making** is the ability to interpret and understand information about the environment and events happening both inside and outside the organization.
- **Knowledge making** is the ability to create new knowledge by combining the expertise of members to learn and innovate.
- **Decision making** is the ability to process and analyze information and knowledge in order to select and implement the appropriate course of action.

First, organizations use data to make sense of changes and developments in the external environments – a process called sense making. This is a vital activity wherein managers discern the most significant changes, interpret their meaning, and develop appropriate responses. Secondly, organizations create, organize, and process data to generate new knowledge through organizational learning. This knowledge creation activity enables the organization to develop new capabilities, design new products and services, enhance existing offerings, and improve organizational processes. Third, organizations search for and evaluate data in order to make decisions. This data is critical since all organizational actions are initiated by decisions and all decisions are commitments to actions, the consequences of which will, in turn, lead to the creation of new data. Adopting an OT methodology enables an enterprise to enhance the knowledge engineering and management process.

In another OT study, researchers and academic scholars have observed that there is no direct correlation between information technology (IT) investments and organizational performance. Research has confirmed that identical IT investments in two different companies may give a competitive advantage to one company but not the other. Therefore, a key factor for the competitive advantage in an organization is not the IT investment but the effective utilization of information as it relates to organizational performance (Brynjolfsson and Hitt, 1996). This finding emphasizes the necessity of integrating OT practices with robust information technology and artificial intelligence techniques in successfully leveraging ODM.

55.4 Ongoing ODM Research

Given the scarcity of past research in ODM along with its growing acceptance and importance in organizations, we conducted empirical research during the past several years that explored the utilization of ODM in organizations along with project implementation factors critical for success. We surveyed ODM professionals from multiple industries in both domestic and international organizations. Our initial research examined the ODM industry status and best practices, identified both technical and business issues related to ODM projects, and elaborated on how organizations are benefiting through enhanced enterprise decision-making (Nemati and Barko, 2001). The results of our research suggest that ODM can improve the quality and accuracy of decisions for any organization willing to make the investment.

After exploring the status and utilization of ODM in organizations, we decided to focus subsequent research on how organizations implement ODM projects and the factors critical for its success. Similar to our initial research, this was pursued in response to the scarcity of empirical research investigating the implementation of ODM projects. To that end, we developed a new ODM Implementation Framework based on data, technology, organizations, and the Iron Triangle (Nemati and Barko, 2003). Our research demonstrated that selected organizational Data Mining project factors, when modeled under this new framework, have a significant influence on the successful implementation of ODM projects.

Our latest research has focused on a specific ODM technology known as Electronic Customer Relationship Management (e-CRM) and its data integration role within organizations. We developed a new e-CRM Value Framework to better examine the significance of integrating data from all customer touch-points with the goal of improving customer relationships and creating additional value for the firm. Our research findings suggest that despite the cost and complexity, data integration for e-CRM projects contributes to a better understanding of the customer and leads to higher return on investment (ROI), a greater number of benefits, improved user satisfaction and a higher probability of attaining a competitive advantage (Nemati *et al.*, 2003).

55.5 ODM Advantages

A 2002 Strategic Decision Making study conducted by Hackett Best Practices determined that "world-class" companies have adopted ODM technologies at more than twice the rate of "average" companies (Hoblitzell, 2002). ODM technologies provide these world-class organizations greater opportunities to understand their business and make informed decisions. ODM also enables world-class organizations to leverage their internal resources more efficiently and effectively than their "average" counterparts who have not fully embraced ODM.

Many of today's leading organizations credit their success to the development of an integrated, enterprise-level ODM system. For example, Harrah's Entertainment has saved over $20 million per year since implementing its Total Rewards CRM program. This ODM system has given Harrah's a better understanding of its customers and enabled the company to create targeted marketing campaigns that almost doubled the profit per customer and delivered same-store sales growth of 14 percent after only the first year. In another notable case, Travelocity.com, an Internet-based travel agency, implemented an ODM system and improved total bookings and earnings by 100 percent in 2000. Gross profit margins improved 150 percent, and booker conversion rates rose to 8.9 percent, the highest in the online travel services industry.

In another significant study, executives from twenty-four leading companies in customer-knowledge management, including FedEx, Frito-Lay, Harley-Davidson, Procter & Gamble and 3M, all realized that in order to succeed, they must go beyond simply collecting customer data and translate it into meaningful knowledge about existing and potential customers (Davenport *et al.*, 2001). This study revealed that several objectives were common to all of the leading companies, and these objectives can be facilitated by ODM. A few of these objectives are segmenting the customer base, prioritizing customers, understanding online customer behavior, engendering customer loyalty, and increasing cross-selling opportunities.

55.6 ODM Evolution

55.6.1 Past

Initially, IT systems were developed to automate expensive manual systems. This automation provided cost savings through labor reductions and more accurate, faster processes. Over the last three decades, the organizational role of information technology has evolved from efficiently processing large amounts of batch transactions to providing information in support of tactical and strategic decision-making activities. This evolution from automating expensive manual systems to providing strategic organizational value led to the birth of Decision Support Systems (DSS) such as data warehousing and Data Mining. Operational and decision support systems are now a vital part of many organizations. The organizational need to combine data from multiple stand-alone systems (e.g. financial, manufacturing and distribution) grew as corporations began to acknowledge the power of combining these data sources for reporting. This spurred the growth of data warehousing where multiple data sources were stored in a format that supported advanced data analysis.

The slowness in adoption of ODM techniques in the 1990s was partly due to an organizational and cultural resistance. Business management has always been reluctant to trust something it doesn't fully understand. Until recently, most businesses were managed by instinct, intuition and "gut feel". The transition over the past twenty years to a method of managing by the numbers is both the result of technology advances as well as a generational shift in the business world as younger managers arrive with information technology training and experience.

55.6.2 Present

Many current ODM techniques trace their origins to traditional statistics and artificial intelligence research from the 1980s. Today, there are extensive vertical Data Mining applications providing analysis in the domains of banking and credit, bioinformatics, CRM, e-CRM, healthcare, human resources, e-commerce, insurance, investment, manufacturing, marketing, retail, entertainment, and telecommunications. Our latest survey findings indicate that the banking, accounting/financial, e-commerce, and retail industries display the highest ODM maturity level to date. The need for service organizations (banking, financial, healthcare and insurance) to build a holistic view of their customers through a mass customization marketing strategy is critical to remaining competitive. And organizations in the e-commerce industry are continuing to improve online customer relationships and overall profitability via e-CRM technologies (Nemati and Barko, 2001). Continuous technological innovations now enable the affordable exploration of enormous volumes of data. It is the combination of technological innovation, creation of new advanced pattern-recognition and data-analysis techniques, ongoing research in organizational theory, and the availability of large quantities of data that have guided ODM to where it is today.

55.6.3 Future

The number of ODM projects is projected to grow more than 300 percent in the next decade (Linden, 1999). As the collection, organization and storage of data rapidly increases, ODM will be the only means of extracting timely and relevant knowledge from large corporate databases. The growing mountains of business data coupled with recent advances in Organizational Theory and technological innovations provide organizations with a framework to effectively use their data to gain a competitive advantage. An organization's future success will depend largely on whether or not they adopt and leverage this ODM framework. ODM will continue to expand and mature as the corporate demand for one-to-one marketing, CRM, e-CRM, Web personalization, and related interactive media increases.

As information technology advances, organizations are able to collect, store, process, analyze and distribute an ever-increasing amount of data. Data and information are rampant, but knowledge is scarce. As a result, most organizations today are governed by managerial intuition and historical reporting. This is the byproduct of years of system automation. However, we believe organizations are slowly moving from the Information Age to the Knowledge Age where decision-makers will leverage ODM and Internet technologies to augment intuition in order to allocate scarce enterprise resources for optimal performance.

As organizations set a strategic course into the Knowledge Age, there are a number of difficulties awaiting them. As its name suggests, ODM is part technological and part organizational. Organizations are comprised of individuals, management, politics, culture, hierarchies, teams, processes, customers, partners, suppliers, and shareholders. The never-ending challenge is to successfully integrate Data Mining technologies with organizations to enhance decision-making with the objective of optimally allocating scarce enterprise resources. As many consultants, professionals, industry leaders and authors of this chapter can attest, this is not an easy task. The media can oversimplify the effort, but successfully implementing ODM is not accomplished without political battles, project management struggles, cultural shocks, business process reengineering, personnel changes, short-term financial and budgetary shortages, and overall disarray. ODM is a journey, not a destination, so there must be a continual effort in revising existing knowledge bases and generating new ones. But the benefits far outweigh both the technical and organizational costs, and the enhanced decision-making capabilities can lead to a sustainable competitive advantage.

Recent ODM research has revealed a number of industry predictions that are expected to be key ODM issues in the future (Nemati and Barko, 2001). About 80 percent of survey respondents expect web farming/mining and consumer privacy to be significant issues, while over 90 percent predict ODM integration with external data sources to be important. We also foresee the development of widely accepted standards for ODM processes and techniques to be an influential factor for knowledge seekers in the 21st century. One attempt at ODM standardization is the creation of the Cross Industry Standard Process for Data Mining (CRISP-DM) project that developed an industry and tool neutral data-mining process model to solve business problems. Another attempt at industry standardization is the work of the Data Mining Group in developing and advocating the Predictive Model Markup Language (PMML), which is an XML-based language that provides a quick and easy way for companies to define predictive models and share models between compliant vendors' applications. Lastly, Microsoft's OLE DB for Data Mining is a further attempt at industry standardization and integration. This specification offers a common interface for Data Mining that will enable developers to embed data-mining capabilities into their existing applications. One only has to consider Microsoft's industry-wide dominance of the office productivity (Microsoft Office), software development (Visual Basic and .Net) and database (SQL Server) markets to envision the potential impact this could have on the ODM market and its future direction.

55.7 Summary

Although many improvements have materialized over the last decade, the knowledge gap in many organizations is still prevalent. Industry professionals have suggested that many corporations could maintain current revenues at half the current costs if they optimized their use of corporate data. Whether this finding is true or not, it sheds light on an important issue. Leading corporations in the next decade will adopt and weave these ODM technologies into the fabric of their organizations at all levels, from upper management all the way down to the lowest organizational level. Those enterprises that see the strategic value of evolving into knowledge organizations by leveraging ODM will benefit directly in the form of improved profitability, increased efficiency, and a sustainable competitive advantage. Once the first organization within an industry realizes a competitive advantage through ODM, it is only a matter of time before one of three events transpires: its industry competitors adopt ODM, change industries, or vanish. By adopting ODM, an organization's managers and employees are able to act sooner rather than later, anticipate rather than react, know rather than guess, and ultimately, succeed rather than fail.

References

Anonymous (2001), "The slow progress of fast wires", The Economist, London, Vol. 358, No. 8209, February 17.

Brown, E. (2002), "Analyze This", Forbes, Vol. 169, No. 8, April 1, pp. 96-98.

Brynjolfsson, E. and Hitt, L. (1996), "The Customer Counts", InformationWeek, September 9,www.informationweek.com/596/96mit.htm.

Choo, C. W. (1997), The Knowing Organization: How Organizations Use Information to Construct Meaning, Create Knowledge, and Make Decisions, Oxford University Press, www.choo.fis.utoronto.ca/fis/ko/default.html.

Davenport, T. H., Harris, J. G. and Kohli, A. K. (2001), "How Do They Know Their Customers So Well?", Sloan Management Review, Vol. 42, No. 2, Winter, pp. 63-73.

Hardy, Q. (2004), "Data of Reckoning", Forbes, Vol. 173, No. 10, May 10, pp 151-154.

Hoblitzell, T. (2002), "Disconnects in Today's BI Systems", DM Review, Vol. 12, No. 6, July, pp. 56-59.

Linden, A. (1999), CIO Update: Data Mining Applications of the Next Decade, Inside Gartner Group, Gartner Inc., July 7.

Nemati, H. R. and Barko, C. D. (2001), "Issues in Organizational Data Mining: A Survey of Current Practices", Journal of Data Warehousing, Vol. 6, No. 1, Winter, pp. 25-36.

Nemati, H. R. and Barko, C. D. (2003), "Key Factors for Achieving Organizational Data Mining Success", Industrial Management and Data Systems, Vol. 103, No. 4, pp. 282-292.

Nemati, H. R., Barko, C. D. and Moosa, A. (2003), "E-CRM Analytics: The Role of Data Integration", Journal of Electronic Commerce in Organizations, Vol. 1, No. 3, July-Sept, pp. 73-89.

56

Mining Time Series Data

Chotirat Ann Ratanamahatana[1], Jessica Lin[1], Dimitrios Gunopulos[1], Eamonn Keogh[1], Michail Vlachos[2], and Gautam Das[3]

[1] University of California, Riverside
[2] IBM T.J. Watson Research Center
[3] University of Texas, Arlington

Summary. Much of the world's supply of data is in the form of time series. In the last decade, there has been an explosion of interest in mining time series data. A number of new algorithms have been introduced to classify, cluster, segment, index, discover rules, and detect anomalies/novelties in time series. While these many different techniques used to solve these problems use a multitude of different techniques, they all have one common factor; they require some high level representation of the data, rather than the original raw data. These high level representations are necessary as a feature extraction step, or simply to make the storage, transmission, and computation of massive dataset feasible. A multitude of representations have been proposed in the literature, including spectral transforms, wavelets transforms, piecewise polynomials, eigenfunctions, and symbolic mappings. This chapter gives a high-level survey of time series Data Mining tasks, with an emphasis on time series representations.

Key words: Data Mining, Time Series, Representations, Classification, Clustering, Time Series Similarity Measures

56.1 Introduction

Time series data accounts for an increasingly large fraction of the world's supply of data. A random sample of 4,000 graphics from 15 of the world's newspapers published from 1974 to 1989 found that more than 75% of all graphics were time series (Tufte, 1983). Given the ubiquity of time series data, and the exponentially growing sizes of databases, there has been recently been an explosion of interest in time series Data Mining. In the medical domain alone, large volumes of data as diverse as gene expression data (Aach and Church, 2001), electrocardiograms, electroencephalograms, gait analysis and growth development charts are routinely created. Similar remarks apply to industry, entertainment, finance, meteorology and virtually every other field of human endeavour. Although statisticians have worked with time series for more than a century, many of their techniques hold little utility for researchers working with massive time series databases (for reasons discussed below).

Below are the major task considered by the time series Data Mining community.

O. Maimon, L. Rokach (eds.), *Data Mining and Knowledge Discovery Handbook*, 2nd ed., DOI 10.1007/978-0-387-09823-4_56,

- **Indexing** (Query by Content): Given a query time series Q, and some similarity/dissimilarity measure $D(Q,C)$, find the most similar time series in database DB (Chakrabarti *et al.*, 2002, Faloutsos *et al.*, 1994, Kahveci and Singh, 2001, Popivanov *et al.*, 2002).
- **Clustering**: Find natural groupings of the time series in database DB under some similarity/dissimilarity measure $D(Q,C)$ (Aach and Church, 2001, Debregeas and Hebrail, 1998, Kalpakis *et al.*, 2001, Keogh and Pazzani, 1998).
- **Classification**: Given an unlabeled time series Q, assign it to one of two or more predefined classes (Geurts, 2001, Keogh and Pazzani, 1998).
- **Prediction** (Forecasting): Given a time series Q containing n data points, predict the value at time $n+1$.
- **Summarization**: Given a time series Q containing n data points where n is an extremely large number, create a (possibly graphic) approximation of Q which retains its essential features but fits on a single page, computer screen, etc. (Indyk *et al.*, 2000, Wijk and Selow, 1999).
- **Anomaly Detection** (Interestingness Detection): Given a time series Q, assumed to be normal, and an unannotated time series R, find all sections of R which contain anomalies or "surprising/interesting/unexpected" occurrences (Guralnik and Srivastava, 1999, Keogh *et al.*, 2002, Shahabi *et al.*, 2000).
- **Segmentation**: (**a**) Given a time series Q containing n data points, construct a model $\bar{Q}$, from K piecewise segments ($K << n$), such that $\bar{Q}$ closely approximates Q (Keogh and Pazzani, 1998). (**b**) Given a time series Q, partition it into K internally homogenous sections (also known as change detection (Guralnik and Srivastava, 1999)).

Note that indexing and clustering make *explicit* use of a distance measure, and many approaches to classification, prediction, association detection, summarization, and anomaly detection make *implicit* use of a distance measure. We will therefore take the time to consider time series similarity in detail.

56.2 Time Series Similarity Measures

56.2.1 Euclidean Distances and L_p Norms

One of the simplest similarity measures for time series is the Euclidean distance measure. Assume that both time sequences are of the same length n, we can view each sequence as a point in n-dimensional Euclidean space, and define the dissimilarity between sequences C and Q and $D(C,Q) = L_p(C,Q)$, i.e. the distance between the two points measured by the L_p norm (when $p = 2$, it reduces to the familiar Euclidean distance). Figure 56.1 shows a visual intuition behind the Euclidean distance metric.

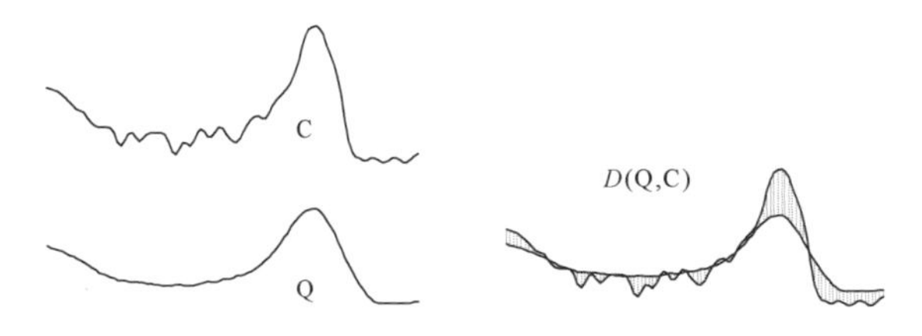

Fig. 56.1. The intuition behind the Euclidean distance metric

Such a measure is simple to understand and easy to compute, which has ensured that the Euclidean distance is the most widely used distance measure for similarity search (Agrawal *et al.*, 1993, Chan and Fu, 1999, Faloutsos *et al.*, 1994). However, one major disadvantage is that it is very brittle; it does not allow for a situation where two sequences are alike, but one has been "stretched" or "compressed" in the Y-axis. For example, a time series may fluctuate with small amplitude between 10 and 20, while another may fluctuate in a similar manner with larger amplitude between 20 and 40. The Euclidean distance between the two time series will be large. This problem can be dealt with easily with offset translation and amplitude scaling, which requires normalizing the sequences before applying the distance operator[4].

In Goldin and Kanellakis (1995) , the authors describe a method where the sequences are normalized in an effort to address the disadvantages of the L_p as a similarity measure. Figure 56.2 illustrates the idea.

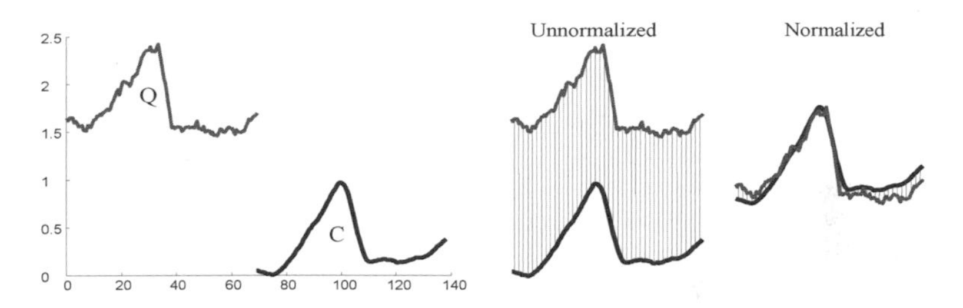

Fig. 56.2. A visual intuition of the necessity to normalize time series before measuring the distance between them. The two sequences Q and C appear to have approximately the same shape, but have different offsets in Y-axis. The unnormalized data greatly overstate the subjective dissimilarity distance. Normalizing the data reveals the true similarity of the two time series.

More formally, let $\mu(C)$ and $\sigma(C)$ be the mean and standard deviation of sequence $C = \{c_1, \ldots, c_n\}$. The sequence C is replaced by the normalized sequences C', where

$$c_i' = \frac{c_i - \mu(C)}{\sigma(C)}$$

Even after normalization, the Euclidean distance measure may still be unsuitable for some time series domains since it does not allow for acceleration and deceleration along the time axis. For example, consider the two subjectively very similar sequences shown in Figure 56.3A. Even with normalization, the Euclidean distance will fail to detect the similarity between the two signals. This problem can generally be handled by Dynamic Time Warping distance measure, which will be discussed in the next section.

56.2.2 Dynamic Time Warping

In some time series domains, a very simple distance measure such as the Euclidean distance will suffice. However, it is often the case that the two sequences have approximately the same

[4] In unusual situations, it might be more appropriate not to normalize the data, e.g. when offset and amplitude changes are important.

overall component shapes, but these shapes do not line up in X-axis. Figure 56.3 shows this with a simple example. In order to find the similarity between such sequences or as a preprocessing step before averaging them, we must "warp" the time axis of one (or both) sequences to achieve a better alignment. Dynamic Time Warping (DTW) is a technique for effectively achieving this warping.

In Berndt and Clifford (1996) , the authors introduce the technique of dynamic time warping to the Data Mining community. Dynamic time warping is an extensively used technique in speech recognition, and allows acceleration-deceleration of signals along the time dimension. We describe the basic idea below.

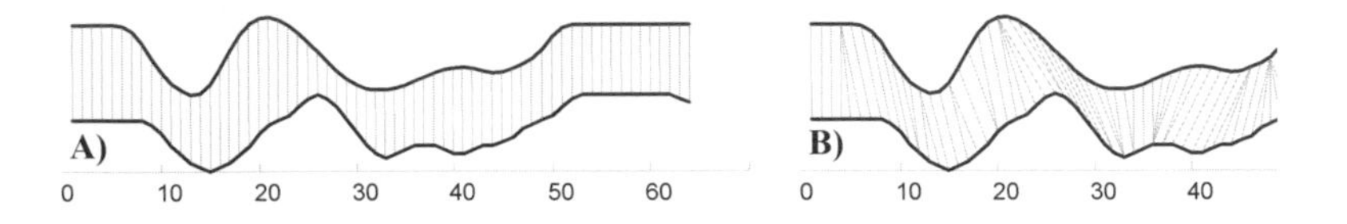

Fig. 56.3. Two time series which require a warping measure. Note that while the sequences have an overall similar shape, they are not aligned in the time axis. Euclidean distance, which assumes the ith point on one sequence is aligned with i^{th} point on the other (A), will produce a pessimistic dissimilarity measure. A nonlinear alignment (B) allows a more sophisticated distance measure to be calculated.

Consider two sequence (of possibly different lengths), $C = \{c_1, \ldots, c_m\}$ and $Q = \{q_1, \ldots, q_n\}$. When computing the similarity of the two time series using Dynamic Time Warping, we are allowed to extend each sequence by repeating elements.

A straightforward algorithm for computing the Dynamic Time Warping distance between two sequences uses a bottom-up dynamic programming approach, where the smaller sub-problems $D(i, j)$ are first determined, and then used to solve the larger sub-problems, until $D(m,n)$ is finally achieved, as illustrated in Figure 56.4 below.

Although this dynamic programming technique is impressive in its ability to discover the optimal of an exponential number alignments, a basic implementation runs in $O(mn)$ time. If a warping window w is specified, as shown in Figure 56.4B, then the running time reduces to $O(nw)$, which is still too slow for most large scale application. In (Ratanamahatana and Keogh, 2004), the authors introduce a novel framework based on a learned warping window constraint to further improve the classification accuracy, as well as to speed up the DTW calculation by utilizing the lower bounding technique introduced in (Keogh, 2002).

56.2.3 Longest Common Subsequence Similarity

The longest common subsequence similarity measure, or LCSS, is a variation of edit distance used in speech recognition and text pattern matching. The basic idea is to match two sequences by allowing some elements to be unmatched. The advantage of the LCSS method is that some elements may be unmatched or left out (e.g. outliers), where as in Euclidean and DTW, all elements from both sequences must be used, even the outliers. For a general discussion of string edit distances, see (Kruskal and Sankoff, 1983).

For example, consider two sequences: $C = \{1,2,3,4,5,1,7\}$ and $Q = \{2,5,4,5,3,1,8\}$. The longest common subsequence is $\{2,4,5,1\}$.

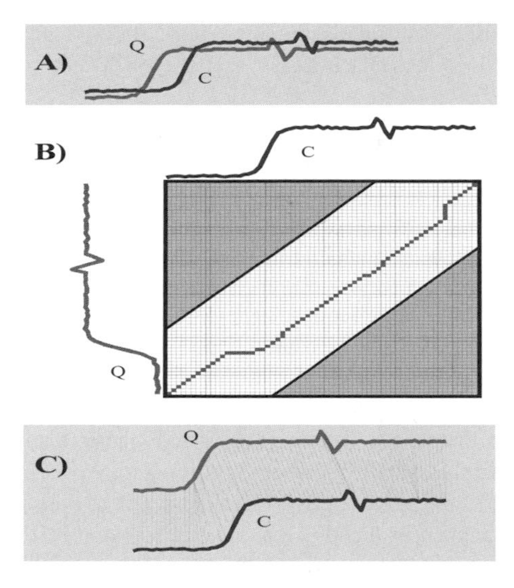

Fig. 56.4. A) Two similar sequences Q and C, but out of phase. B) To align the sequences, we construct a warping matrix, and search for the optimal warping path, shown with solid squares. Note that the "corners" of the matrix (shown in dark gray) are excluded from the search path (specified by a warping window of size w) as part of an Adjustment Window condition. C) The resulting alignment

More formally, let C and Q be two sequences of length m and n, respectively. As was done with dynamic time warping, we give a recursive definition of the length of the longest common subsequence of C and Q. Let $L(i,j)$ denote the longest common subsequences $\{c_1, \ldots, c_i\}$ and $\{q_1, \ldots, q_j\}$. $L(i, j)$ may be recursively defined as follows:

IF $a_i = b_j$ THEN

$L(i,j) = 1 + L(i-1,j-1)$

ELSE

$L(i,j) = \max \{D(i-1,j), D(i,j-1)\}$

We define the dissimilarity between C and Q as

$$LCSS(C,Q) = \frac{m+n-2l}{m+n}$$

where l is the length of the longest common subsequence. Intuitively, this quantity determines the minimum (normalized) number of elements that should be removed from and inserted into C to transform C to Q. As with dynamic time warping, the LCSS measure can be computed by dynamic programming in $O(mn)$ time. This can be improved to $O((n+m)w)$ time if a matching window of length w is specified (i.e. where $|i-j|$ is allowed to be at most w).

With time series data, the requirement that the corresponding elements in the common subsequence should match exactly is rather rigid. This problem is addressed by allowing some tolerance (say $\varepsilon > 0$) when comparing elements. Thus, two elements a and b are said to match if $a(1-\varepsilon) < b < a(1+\varepsilon)$.

In the next two subsections, we discuss approaches that try to incorporate local scaling and global scaling functions in the basic LCSS similarity measure.

Using local Scaling Functions

In (Agrawal *et al.*, 1995), the authors develop a similarity measure that resembles LCSS-like similarity with local scaling functions. Here, we only give an intuitive outline of the complex algorithm; further details may be found in this work.

The basic idea is that two sequences are similar if they have enough non-overlapping time-ordered pairs of contiguous subsequences that are similar. Two contiguous subsequences are similar if one can be scaled and translated appropriately to approximately resemble the other. The scaling and translation function is local, i.e. it may be different for other pairs of subsequences.

The algorithmic challenge is to determine how and where to cut the original sequences into subsequences so that the overall similarity is minimized. We describe it briefly here (refer to (Agrawal *et al.*, 1995) for further details). The first step is to find all pairs of atomic subsequences in the original sequences A and Q that are similar (atomic implies subsequences of a certain small size, say a parameter w). This step is done by a spatial self-join (using a spatial access structure such as an R-tree) over the set of all atomic subsequences. The next step is to "stitch" similar atomic subsequences to form pairs of larger similar subsequences. The last step is to find a non-overlapping ordering of subsequence matches having the longest match length. The stitching and subsequence ordering steps can be reduced to finding longest paths in a directed acyclic graph, where vertices are pairs of similar subsequences, and a directed edge denotes their ordering along the original sequences.

Using a global scaling function

Instead of different local scaling functions that apply to different portions of the sequences, a simpler approach is to try and incorporate a single global scaling function with the LCSS similarity measure. An obvious method is to first normalize both sequences and then apply LCSS similarity to the normalized sequences. However, the disadvantage of this approach is that the normalization function is derived from all data points, including outliers. This defeats the very objective of the LCSS approach which is to ignore outliers in the similarity calculations.

In (Bollobas *et al.*, 2001), an LCSS-like similarity measure is described that derives a global scaling and translation function that is independent of outliers in the data. The basic idea is that two sequences C and Q are similar if there exists constants a and b, and long common subsequences C' and Q' such that Q' is approximately equal to $aC' + b$. The scale+translation linear function (i.e. the constants a and b) is derived from the subsequences, and not from the original sequences. Thus, outliers cannot taint the scale+translation function.

Although it appears that the number of all linear transformations is infinite, Bollobas *et al.* (2001) shows that the number of different unique linear transformations is $O(n^2)$. A naive implementation would be to compute LCSS on all transformations, which would lead to an algorithm that takes $O(n^3)$ time. Instead, in (Bollobas *et al.*, 2001), an efficient randomized approximation algorithm is proposed to compute this similarity.

56.2.4 Probabilistic methods

A different approach to time-series similarity is the use of a probabilistic similarity measure. Such measures have been studied in (Ge and Smyth, 2000, Keogh and Smyth, 1997). While previous methods were "distance" based, some of these methods are "model" based. Since time series similarity is inherently a fuzzy problem, probabilistic methods are well suited for handling noise and uncertainty. They are also suitable for handling scaling and offset translations. Finally, they provide the ability to incorporate prior knowledge into the similarity measure. However, it is not clear whether other problems such as time-series indexing, retrieval and clustering can be efficiently accomplished under probabilistic similarity measures.

Here, we briefly describe the approach in (Ge and Smyth, 2000). Given a sequence C, the basic idea is to construct a probabilistic generative model M_C, i.e. a probability distribution on waveforms. Once a model M_C has been constructed for a sequence C, we can compute similarity as follows. Given a new sequence pattern Q, similarity is measured by computing $p(Q|M_C)$, i.e. the likelihood that M_C generates Q.

56.2.5 General Transformations

Recognizing the importance of the notion of "shape" in similarity computations, an alternate approach was undertaken by Jagadish *et al.* (1995) . In this paper, the authors describe a general similarity framework involving a transformation rules language. Each rule in the transformation language takes an input sequence and produces an output sequence, at a cost that is associated with the rule. The similarity of sequence C to sequence Q is the minimum cost of transforming C to Q by applying a sequence of such rules. The actual rules language is application specific.

56.3 Time Series Data Mining

The last decade has seen the introduction of hundreds of algorithms to classify, cluster, segment and index time series. In addition, there has been much work on novel problems such as rule extraction, novelty discovery, and dependency detection. This body of work draws on the fields of statistics, machine learning, signal processing, information retrieval, and mathematics. It is interesting to note that with the exception of indexing, researches in the tasks enumerated above predate not only the decade old interest in Data Mining, but in computing itself. What then, are the essential differences between the classic and the Data Mining versions of these problems? The key difference is simply one of size and scalability; time series data miners routinely encounter datasets that are gigabytes in size. As a simple motivating example, consider hierarchical clustering. The technique has a long history and well-documented utility. If however, we wish to hierarchically cluster a mere million items, we would need to construct a matrix with 10^{12} cells, well beyond the abilities of the average computer for many years to come. A Data Mining approach to clustering time series, in contrast, must explicitly consider the scalability of the algorithm (Kalpakis *et al.*, 2001).

In addition to the large volume of data, most classic machine learning and Data Mining algorithms do not work well on time series data due to their unique structure; it is often the case that each individual time series has a very high dimensionality, high feature correlation, and large amount of noise (Chakrabarti *et al.*, 2002), which present a difficult challenge in time series Data Mining tasks. Whereas classic algorithms assume relatively low dimensionality (for example, a few measurements such as "height, weight, blood sugar, etc."), time series Data Mining algorithms must be able to deal with dimensionalities in the hundreds or thousands. The problems created by high dimensional data are more than mere computation time considerations; the very meanings of normally intuitive terms such as "similar to" and "cluster forming" become unclear in high dimensional space. The reason is that as dimensionality increases, all objects become essentially equidistant to each other, and thus classification and clustering lose their meaning. This surprising result is known as the "curse of dimensionality" and has been the subject of extensive research (Aggarwal *et al.*, 2001). The key insight that allows meaningful time series Data Mining is that although the actual dimensionality may be high, the *intrinsic* dimensionality is typically much lower. For this reason, virtually all time series Data Mining algorithms avoid operating on the original "raw" data; instead, they consider some higher-level representation or abstraction of the data.

Before giving a full detail on time series representations, we first briefly explore some of the classic time series Data Mining tasks. While these individual tasks may be combined to obtain more sophisticated Data Mining applications, we only illustrate their main basic ideas here.

56.3.1 Classification

Classification is perhaps the most familiar and most popular Data Mining technique. Examples of classification applications include image and pattern recognition, spam filtering, medical diagnosis, and detecting malfunctions in industry applications. Classification maps input data into predefined groups. It is often referred to as supervised learning, as the classes are determined prior to examining the data; a set of predefined data is used in training process and learn to recognize patterns of interest. Pattern recognition is a type of classification where an input pattern is classified into one of several classes based on its similarity to these predefined classes. Two most popular methods in time series classification include the Nearest Neighbor classifier and Decision trees. Nearest Neighbor method applies the similarity measures to the object to be classified to determine its best classification based on the existing data that has already been classified. For decision tree, a set of rules are inferred from the training data, and this set of rules is then applied to any new data to be classified. Note that even though decision trees are defined for real data, attempting to apply raw time series data could be a mistake due to its high dimensionality and noise level that would result in deep, bushy tree. Instead, some researchers suggest representing time series as Regression Tree to be used in Decision Tree training (Geurts, 2001).

The performance of classification algorithms is usually evaluated by measuring the accuracy of the classification, by determining the percentage of objects identified as the correct class.

56.3.2 Indexing (Query by Content)

Query by content in time series databases has emerged as an area of active interest since the classic first paper by Agrawal et al. (1993) . This also includes a sequence matching task which has long been divided into two categories: whole matching and subsequence matching (Faloutsos *et al.*, 1994, Keogh *et al.*, 2001).

Whole Matching: a query time series is matched against a database of individual time series to identify the ones similar to the query

Subsequence Matching: a short query subsequence time series is matched against longer time series by sliding it along the longer sequence, looking for the best matching location.

While there are literally hundreds of methods proposed for whole sequence matching (See, e.g. (Keogh and Kasetty, 2002) and references therein), in practice, its application is limited to cases where some information about the data is known *a priori*.

Subsequence matching can be generalized to whole matching by dividing sequences into non-overlapping sections by either a specific period or, more arbitrarily, by its shape. For example, we may wish to take a long electrocardiogram and extract the individual heartbeats. This informal idea has been used by many researchers.

Most of the indexing approaches so far use the original GEMINI framework (Faloutsos *et al.*, 1994) but suggest a different approach to the dimensionality reduction stage. There is increasing awareness that for many Data Mining and information retrieval tasks, very fast approximate search is preferable to slower exact search (Chang *et al.*, 2002). This is particularly true for exploratory purposes and hypotheses testing. Consider the stock market data. While it makes sense to look for approximate patterns, for example, "*a pattern that rapidly decreases after a long plateau*", it seems pedantic to insist on *exact* matches. Next we would like to discuss similarity search in some more detail.

Given a database of sequences, the simplest way to find the closest match to a given query sequence Q, is to perform a *linear* or *sequential* scan of the data. Each sequence is retrieved from disk and its distance to the query Q is calculated according to the pre-selected distance measure. After the query sequence is compared to all the sequences in the database, the one with the smallest distance is returned to the user as the closest match.

This brute-force technique is costly to implement, first because it requires many accesses to the disk and second because it operates or the raw sequences, which can be quite long. Therefore, the performance of linear scan on the raw data is typically very costly.

A more efficient implementation of the linear scan would be to store two levels of approximation of the data; the raw data and their compressed version. Now the linear scan is performed on the compressed sequences and a *lower bound* to the original distance is calculated for all the sequences. The raw data are retrieved in the order suggested by the lower bound approximation of their distance to the query. The smallest distance to the query is updated after each raw sequence is retrieved. The search can be terminated when the lower bound of the currently examined object exceeds the smallest distance discovered so far.

A more efficient way to perform similarity search is to utilize an *index structure* that will cluster similar sequences into the same group, hence providing faster access to the most promising sequences. Using various pruning techniques, indexing structures can avoid examining large parts of the dataset, while still guaranteeing that the results will be identical with the outcome of linear scan. Indexing structures can be divided into two major categories: vector based and metric based.

Vector Based Indexing Structures

Vector based indices work on the compressed data dimensionality. The original sequences are compacted using a dimensionality reduction method, and the resulting multi-dimensional vectors can be grouped into similar clusters using some vector-based indexing technique, as shown in Figure 56.5.

Vector-based indexing structures can also appear in two flavors; hierarchical or non-hierarchical. The most common hierarchical vector based index is the R-tree or some variant. The R-tree consists of multi-dimensional vectors on the leaf levels, which are organized in the tree fashion using hyper-rectangles that can potentially overlap, as illustrated in Figure 56.6.

In order to perform the search using an index structure, the query is also projected in the compressed dimensionality and then probed on the index. Using the R-tree, only neighboring hyper-rectangles to the query's projected location need to be examined.

Other commonly used hierarchical vector-based indices are the kd-B-trees (Robinson, 1981) and the quad-trees (Tzouramanis *et al.*, 1998). Non-

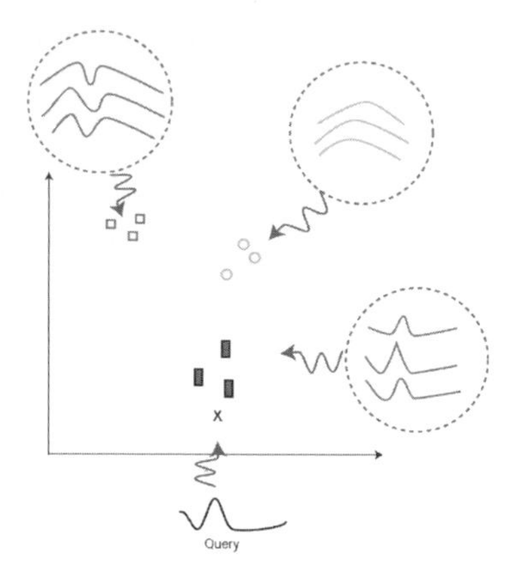

Fig. 56.5. Dimensionality reduction of time-series into two dimensions

hierarchical vector based structures are less common and are typically known as grid files (Nievergelt *et al.*, 1984). For example, grid files have been used in (Zhu and Shasha, 2002) for the discovery of the most correlated data sequences.

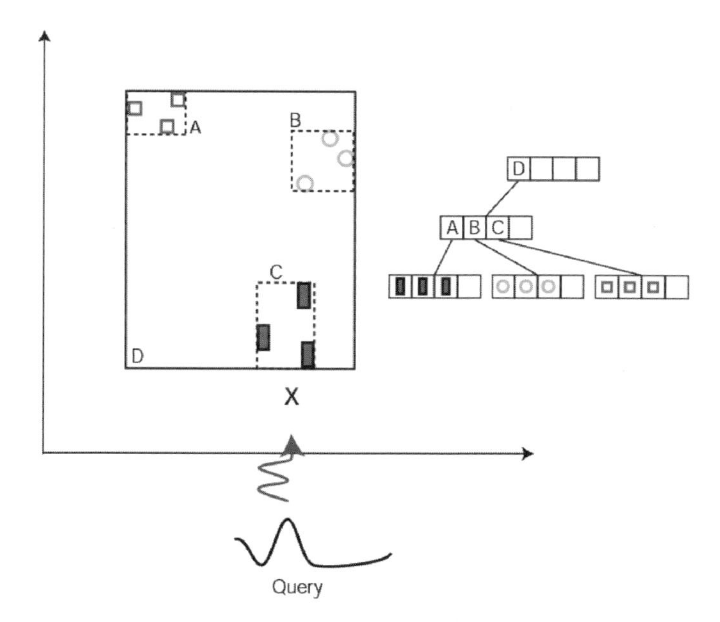

Fig. 56.6. Hierarchical organization using an R-tree

However, such types of indexing structures work well only for low compressed dimensionalities (typically$<$5). For higher dimensionalities, the pruning power of vector-based indices diminishes exponentially. This can be experimentally and analytically shown and it is coined under the term 'dimensionality curse' (Agrawal *et al.*, 1993). This inescapable fact suggests that even when using an index structure, the complete dataset would have to be retrieved from disk for higher compressed dimensionalities.

Metric Based Indexing Structures

Metric based structures can typically perform much better than vector based indices, even for higher dimensionalities (up to 20 or 30). They are more flexible because they require only distances between objects. Thus, they do not cluster objects based on their compressed features but based on relative object distances. The choice of reference objects, from which all object distances will be calculated, can vary in different approaches. Examples of metric trees include the Vantage Point (VP) tree (Yianilos, 1992), M-tree (Ciaccia *et al.*, 1997) and GNAT (Brin, 1995). All variations of such trees, exploit the distances to the reference points in conjunction with the triangle inequality to prune parts of the tree, where no closer matches (to the ones already discovered) can be found. A recent use of VP-trees for time-series search under Euclidean distance using compressed Fourier descriptors can be found in (Vlachos *et al.*, 2004).

56.3.3 Clustering

Clustering is similar to classification that categorizes data into groups; however, these groups are not predefined, but rather defined by the data itself, based on the similarity between time series. It is often referred to as unsupervised learning. The clustering is usually accomplished by determining the similarity among the data on predefined attributes. The most similar data are grouped into clusters, but the clusters themselves should be very dissimilar. And since the clusters are not predefined, a domain expert is often required to interpret the meaning of the created clusters. The two general methods of time series clustering are Partitional Clustering and Hierarchical Clustering. Hierarchical Clustering computes pairwise distance, and then merges similar clusters in a bottom-up fashion, without the need of providing the number of clusters. We believe that this is one of the best (subjective) tools to data evaluation, by creating a dendrogram of several time series from the domain of interest (Keogh and Pazzani, 1998), as shown in Figure 56.7. However, its application is limited to only small datasets due to its quadratic computational complexity.

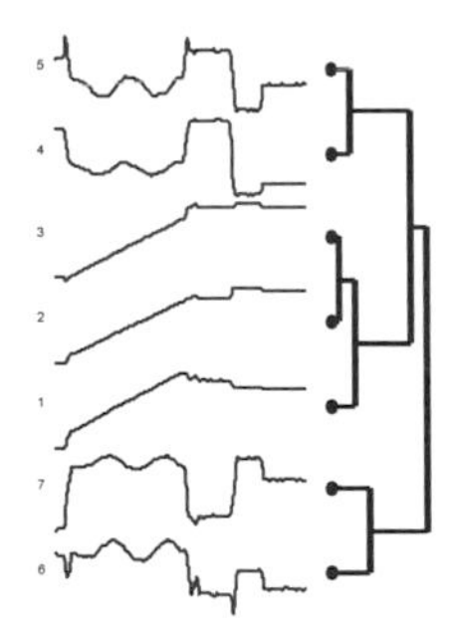

Fig. 56.7. A hierarchical clustering of time series

On the other hand, Paritional Clustering typically uses the K-means algorithm (or some variant) to optimize the objective function by minimizing the sum of squared intra-cluster errors. While the algorithm is perhaps the most commonly used clustering algorithm in the literature, one of its shortcomings is the fact that the number of clusters, K, must be pre-specified.

Clustering has been used in many application domains including biology, medicine, anthropology, marketing, and economics. It is also a vital process for condensing and summarizing information, since it can provide a synopsis of the stored data. Similar to query by content, there are two types of time series clustering: whole clustering and subsequence clustering. The notion of whole clustering is similar to that of conventional clustering of discrete objects. Given a set of individual time series data, the objective is to group similar time series into the same cluster. On the other hand, given a single (typically long) time series, subsequence clustering is performed on each individual time series (subsequence) extracted from the long time series with a sliding window. Subsequence clustering is a common pre-processing step for many pattern discovery algorithms, of which the most well-known being the one proposed for time series rule discovery. Recent empirical and theoretical results suggest that subsequence clustering may not be meaningful on an entire dataset (Keogh *et al.*, 2003), and that clustering should only be applied to a subset of the data. Some feature extraction algorithm must choose the subset of data, but we cannot use clustering as the feature extraction algorithm, as this would open the possibility of a chicken and egg paradox. Several researchers have suggested using time series motifs (see below) as the feature extraction algorithm (Chiu *et al.*, 2003).

56.3.4 Prediction (Forecasting)

Prediction can be viewed as a type of clustering or classification. The difference is that prediction is predicting a future state, rather than a current one. Its applications include obtaining forewarning of natural disasters (flooding, hurricane, snowstorm, etc), epidemics, stock crashes, etc. Many time series prediction applications can be seen in economic domains, where a prediction algorithm typically involves regression analysis. It uses known values of data to predict future values based on historical trends and statistics. For example, with the rise of competitive energy markets, forecasting of electricity has become an essential part of an efficient power system planning and operation. This includes predicting future electricity demands based on historical data and other information, e.g. temperature, pricing, etc. As another example, the sales volume of cellular phone accessories can be forecasted based on the number of cellular phones sold in the past few months. Many techniques have been proposed to increase the accuracy of time series forecast, including the use of neural network and dimensionality reduction techniques.

56.3.5 Summarization

Since time series data can be massively long, a summarization of the data may be useful and necessary. A statistic summarization of the data, such as the mean or other statistical properties can be easily computed even though it might not be particularly valuable or intuitive information. Rather, we can often utilize natural language, visualization, or graphical summarization to extract useful or meaningful information from the data. Anomaly detection and motif discovery (see the next section below) are special cases of summarization where only anomalous/repeating patterns are of interest and reported. Summarization can also be viewed as a special type of clustering problem that maps data into subsets with associated simple (text or graphical) descriptions and provides a higher-level view of the data. This new simpler description of the data is then used in place of the entire dataset. The summarization may be done at multiple granularities and for different dimensions.

Some of popular approaches for visualizing massive time series datasets include *Time-Searcher*, *Calendar-Based Visualization*, *Spiral* and *VizTree*.

TimeSearcher (Hochheiser and Shneiderman, 2001) is a query-by-example time series exploratory and visualization tool that allows user to retrieve time series by creating queries, so called TimeBoxes. Figure 56.8 shows three TimeBoxes being drawn to specify time series that start low, increase, then fall once more. However, some knowledge about the datasets may be needed in advance and users need to have a general idea of what to look for or what is interesting.

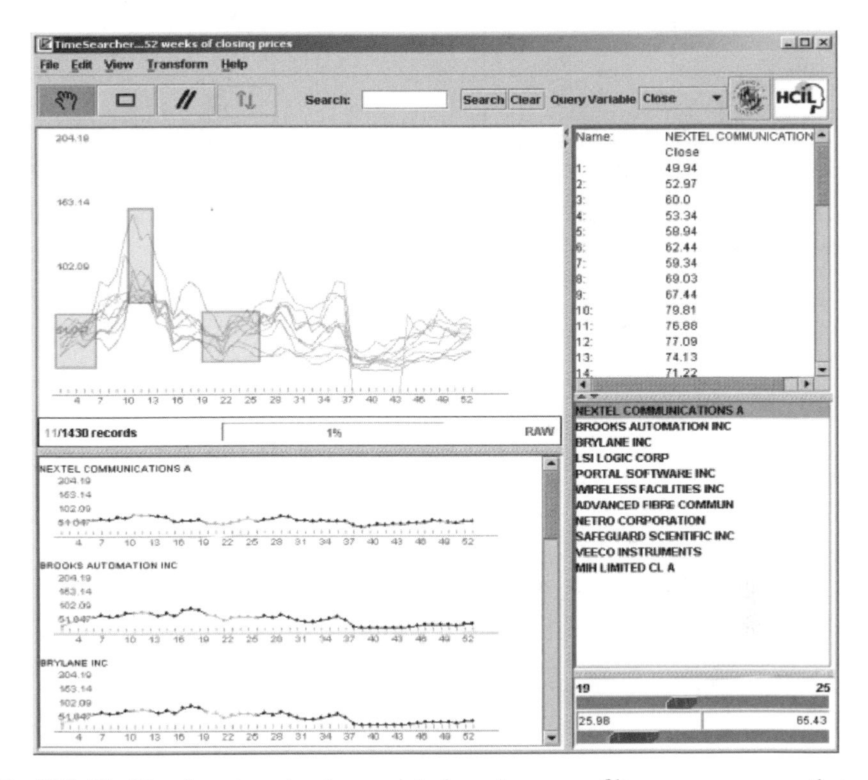

Fig. 56.8. The TimeSearcher visual query interface. A user can filter away sequences that are not interesting by insisting that all sequences have at least one data point within the query boxes

Cluster and Calendar-Based Visualization (Wijk and Selow, 1999) is a visualization system that 'chunks' time series data into sequences of day patterns, and these day patterns are clustered using a bottom-up clustering algorithm. The system displays patterns represented by cluster average, along with a calendar with each day color-coded by the cluster it belongs to. Figure 56.9 shows an example view of this visualization scheme. From viewing patterns which are linked to a calendar we can potentially discover simple rules such as: "*In the winter months the power consumption is greater than in summer months*".

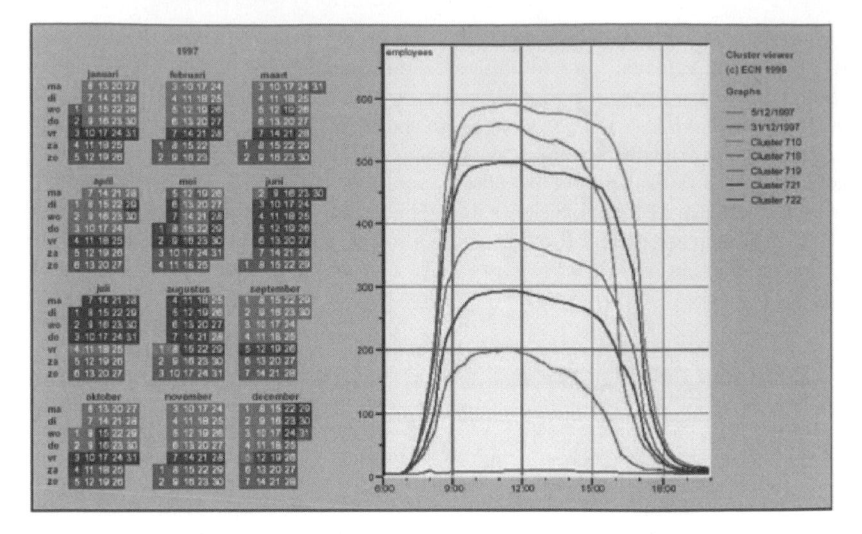

Fig. 56.9. The cluster and calendar-based visualization on employee working hours data. It shows six clusters, representing different working-day pattern

Spiral (Weber *et al.*, 2000) maps each periodic section of time series onto one "ring" and attributes such as color and line thickness are used to characterize the data values. The main use of the approach is the identification of periodic structures in the data. Figure 56.10 displays the annual power usage that characterizes the normal "9-to-5" working week pattern. However, the utility of this tool is limited for time series that do not exhibit periodic behaviors, or when the period is unknown.

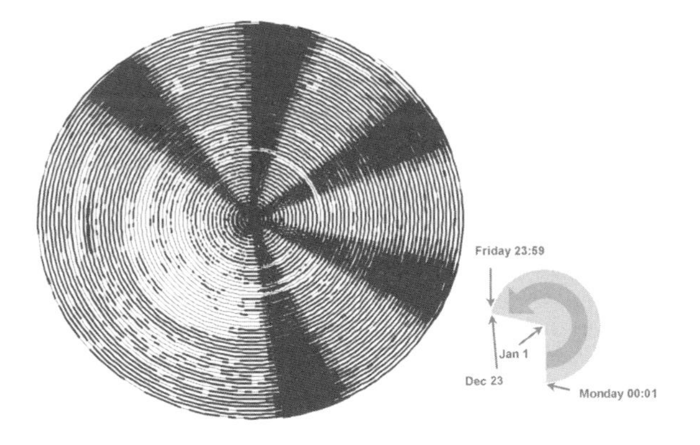

Fig. 56.10. The Spiral visualization approach applied to the power usage dataset

VizTree (Lin *et al.*, 2004) is recently introduced with the aim to discover previously *unknown* patterns with little or no knowledge about the data; it provides an overall visual summary, and potentially reveal hidden structures in the data. This approach first transforms the time series into a symbolic representation, and encodes the data in a modified suffix tree in which the frequency and other properties of patterns are mapped onto colors and other visual properties. Note that even though the tree structure needs the data to be discrete, the original time series data is not. Using a time-series discretization introduced in (Lin *et al.*, 2003), continuous data can be transformed into discrete domain, with certain desirable properties such as lower-bounding distance, dimensionality reduction, etc. While frequently occurring patterns can be detected by thick branches in VizTree, simple anomalous patterns can be detected by unusually thin branches. Figure 56.11 demonstrates both motif discovery and simple anomaly detection on ECG data.

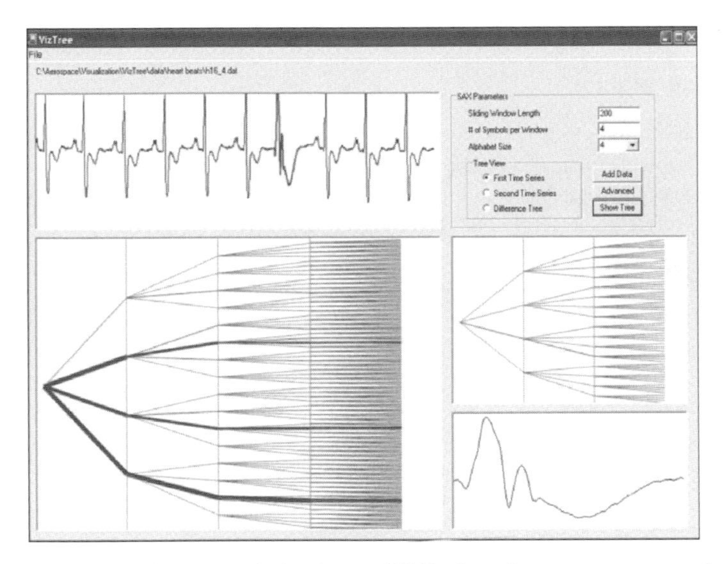

Fig. 56.11. ECG data with anomaly is shown. While the subsequence tree can be used to identify motifs, it can be used for simple anomaly detection as well

56.3.6 Anomaly Detection

In time series Data Mining and monitoring, the problem of detecting anomalous/surprising/novel patterns has attracted much attention (Dasgupta and Forrest, 1999, Ma and Perkins, 2003, Shahabi *et al.*, 2000). In contrast to subsequence matching, anomaly detection is identification of previously *unknown* patterns. The problem is particularly difficult because what constitutes an anomaly can greatly differ depending on the task at hand. In a general sense, an anomalous behavior is one that deviates from "normal" behavior. While there have been numerous definitions given for anomalous or surprising behaviors, the one given by (Keogh *et al.*, 2002) is unique in that it requires no explicit formulation of what is anomalous. Instead, the authors simply define an anomalous pattern as on "*whose frequency of occurrences differs substantially from that expected, given previously seen data*". The problem of anomaly detection in time series has been generalized to include the detection of surprising or interesting patterns (which are not necessarily anomalies). Anomaly detection is closely related to Summarization, as discussed in the previous section. Figure 56.12 illustrates the idea.

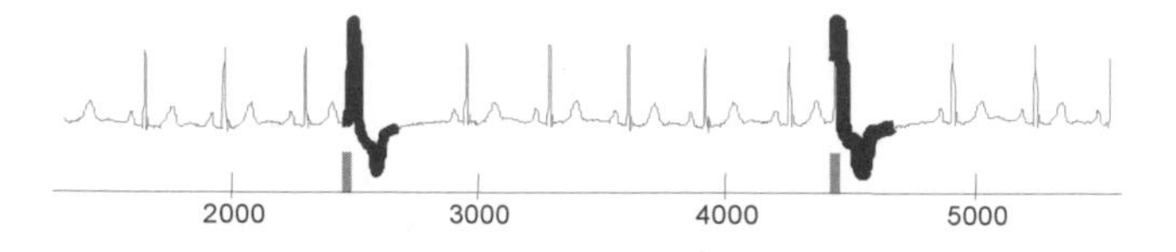

Fig. 56.12. An example of anomaly detection from the MIT-BIH Noise Stress Test Database. Here, we show only a subsection containing the two most interesting events detected by the compression-based algorithm (Keogh et al., 2004) (the thicker the line, the more interesting the subsequence). The gray markers are independent annotations by a cardiologist indicating Premature Ventricular Contractions.

56.3.7 Segmentation

Segmentation in time series is often referred to as a dimensionality reduction algorithm. Although the segments created could be polynomials of an arbitrary degree, the most common representation of the segments is of linear functions. Intuitively, a Piecewise Linear Representation (PLR) refers to the approximation of a time series Q, of length n, with K straight lines. Figure 56.13 contains an example.

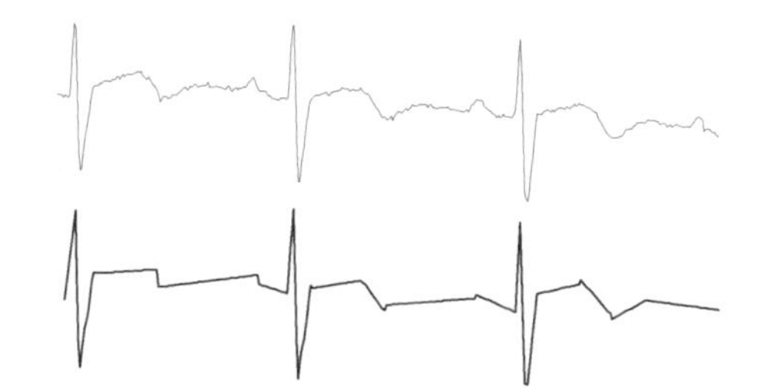

Fig. 56.13. An example of a time series segmentation with its piecewise linear representation

Because K is typically much smaller than n, this representation makes the storage, transmission, and computation of the data more efficient.

Although appearing under different names and with slightly different implementation details, most time series segmentation algorithms can be grouped into one of the following three categories.

- **Sliding-Windows (SW)**: A segment is grown until it exceeds some error bound. The process repeats with the next data point not included in the newly approximated segment.
- **Top-Down (TD)**: The time series is recursively partitioned until some stopping criteria is met.
- **Bottom-Up (BU)**: Starting from the finest possible approximation, segments are merged until some stopping criteria are met.

We can measure the quality of a segmentation algorithm in several ways, the most obvious of which is to measure the reconstruction error for a fixed number of segments. The reconstruction error is simply the Euclidean distance between the original data and the segmented representation. While most work in this area has consider static cases, recently researchers have consider obtaining and maintaining segmentations on streaming data sources (Palpanas *et al.*, 2004)

56.4 Time Series Representations

As noted in the previous section, time series datasets are typically very large, for example, just eight hours of electroencephalogram data can require in excess of a gigabyte of storage. Rather than analyzing or finding statistical properties on time series data, time series data miners' goal is more towards discovering useful information from the massive amount of data efficiently. This is a problem because for almost all Data Mining tasks, most of the execution time spent by algorithm is used simply to move data from disk into main memory. This is acknowledged as the major bottleneck in Data Mining because many naïve algorithms require multiple accesses of the data. As a simple example, imagine we are attempting to do k-means clustering of a dataset that does not fit into main memory. In this case, every iteration of the algorithm will require that data in main memory to be swapped. This will result in an algorithm that is thousands of times slower than the main memory case.

With this in mind, a generic framework for time series Data Mining has emerged. The basic idea (similar to GEMINI framework) can be summarized in Table 56.1.

Table 56.1. A generic time series Data Mining approach.

1)	Create an approximation of the data, which will fit in main memory, yet retains the essential features of interest.
2)	Approximately solve the problem at hand in main memory.
3)	Make (hopefully very few) accesses to the original data on disk to confirm the solution obtained in Step 2, or to modify the solution so it agrees with the solution we would have obtained on the original data.

As with most problems in computer science, the suitable choice of representation/approximation greatly affects the ease and efficiency of time series Data Mining. It should be clear that the utility of this framework depends heavily on the quality of the approximation created in Step 1). If the approximation is very faithful to the original data, then the solution obtained in main memory is likely to be the same as, or very close to, the solution we would have obtained on the original data. The handful of disk accesses made in Step 2) to confirm or slightly modify the solution will be inconsequential, compared to the number of disks accesses required if we had worked on the original data. With this in mind, there has been a huge interest in approximate representation of time series, and various solutions to the diverse set of problems frequently operate on high-level abstraction of the data, instead of the original data. This includes the Discrete Fourier Transform (DFT) (Agrawal *et al.*, 1993), the Discrete Wavelet Transform (DWT) (Chan and Fu, 1999, Kahveci and Singh, 2001, Wu *et al.*, 2000), Piecewise Linear, and Piecewise Constant models (PAA) (Keogh *et al.*, 2001, Yi and Faloutsos, 2000), Adaptive Piecewise Constant Approximation (APCA) (Keogh *et al.*, 2001), and Singular Value Decomposition (SVD) (Kanth *et al.*, 1998, Keogh *et al.*, 2001, Korn *et al.*, 1997).

Figure 56.14 illustrates a hierarchy of the representations proposed in the literature.

It may seem paradoxical that, after all the effort to collect and store the precise values of a time series, the exact values are abandoned for some high level approximation. However, there are two important reasons why this is so.

We are typically not interested in the exact values of each time series data point. Rather, we are interested in the trends, shapes and patterns contained within the data. These may best be captured in some appropriate high-level representation.

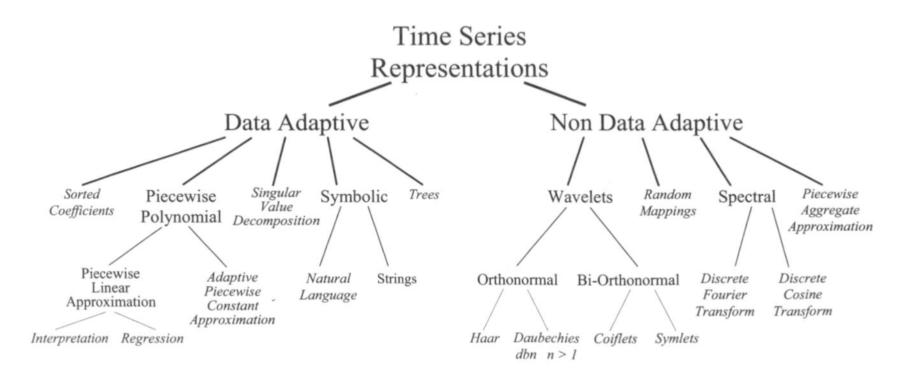

Fig. 56.14. A hierarchy of time series representations

As a practical matter, the size of the database may be much larger than we can effectively deal with. In such instances, some transformation to a lower dimensionality representation of the data may allow more efficient storage, transmission, visualization, and computation of the data.

While it is clear no one representation can be superior for all tasks, the plethora of work on mining time series has not produced any insight into how one should choose the best representation for the problem at hand and data of interest. Indeed the literature is not even consistent on nomenclature. For example, one time series representation appears under the names Piecewise Flat Approximation (Faloutsos *et al.*, 1997), Piecewise Constant Approximation (Keogh *et al.*, 2001) and Segmented Means (Yi and Faloutsos, 2000).

To develop the reader's intuition about the various time series representations, we have discussed and illustrated some of the well-known representations in the following subsections below.

56.4.1 Discrete Fourier Transform

The first technique suggested for dimensionality reduction of time series was the Discrete Fourier Transform (DFT) (Agrawal *et al.*, 1993). The basic idea of spectral decomposition is that any signal, no matter how complex, can be represented by the super position of a finite number of sine/cosine waves, where each wave is represented by a single complex number known as a Fourier coefficient. A time series represented in this way is said to be in the frequency domain. A signal of length n can be decomposed into $n/2$ sine/cosine waves that can be recombined into the original signal. However, many of the Fourier coefficients have very low amplitude and thus contribute little to reconstructed signal. These low amplitude coefficients can be discarded without much loss of information thereby saving storage space.

To perform the dimensionality reduction of a time series C of length n into a reduced feature space of dimensionality N, the Discrete Fourier Transform of C is calculated. The transformed vector of coefficients is truncated at $N/2$. The reason the truncation takes place at $N/2$ and not at N is that each coefficient is a complex number, and therefore we need one dimension each for the imaginary and real parts of the coefficients.

Given this technique to reduce the dimensionality of data from n to N, and the existence of the lower bounding distance measure, we can simply "slot in" the DFT into the GEMINI

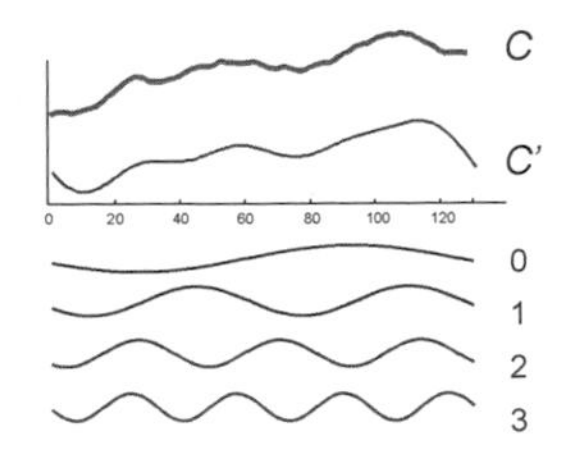

Fig. 56.15. A visualization of the DFT dimensionality reduction technique

framework. The time taken to build the entire index depends on the length of the queries for which the index is built. When the length is an integral power of two, an efficient algorithm can be employed.

This approach, while initially appealing, does have several drawbacks. None of the implementations presented thus far can guarantee no false dismissals. Also, the user is required to input several parameters, including the size of the alphabet, but it is not obvious how to choose the best (or even reasonable) values for these parameters. Finally, none of the approaches suggested will scale very well to massive data since they require clustering all data objects prior to the discretizing step.

56.4.2 Discrete Wavelet Transform

Wavelets are mathematical functions that represent data or other functions in terms of the sum and difference of a prototype function, so called the "analyzing" or "mother" wavelet. In this sense, they are similar to DFT. However, one important difference is that wavelets are localized in time, i.e. some of the wavelet coefficients represent small, local subsections of the data being studied. This is in contrast to Fourier coefficients that always represent global contribution to the data. This property is very useful for Multiresolution Analysis (MRA) of the data. The first few coefficients contain an overall, coarse approximation of the data; addition coefficients can be imagined as "zooming-in" to areas of high detail, as illustrated in Figure 56.16.

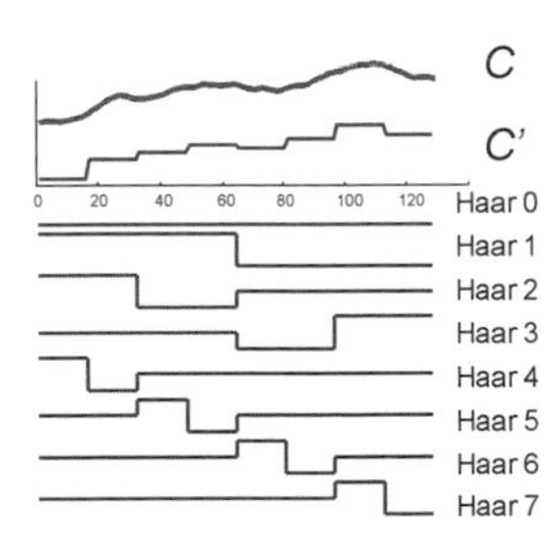

Fig. 56.16. A visualization of the DWT dimensionality reduction technique

Recently, there has been an explosion of interest in using wavelets for data compression, filtering, analysis, and other areas where Fourier methods have previously been used. Chan and Fu (1999) produced a breakthrough for time series indexing with wavelets by producing a distance measure defined on wavelet coefficients which provably satisfies the lower bounding requirement. The work is based on a simple, but powerful type of wavelet known as the Haar Wavelet. The Discrete Haar Wavelet Transform (DWT) can be calculated efficiently and an entire dataset can be indexed in $O(mn)$.

DTW does have some drawbacks, however. It is only defined for sequence whose length is an integral power of two. Although much work has been undertaken on more flexible distance measures using Haar wavelet (Huhtala *et al.*, 1995, Struzik and Siebes, 1999), none of those techniques are indexable.

56.4.3 Singular Value Decomposition

Singular Value Decomposition (SVD) has been successfully used for indexing images and other multimedia objects (Kanth *et al.*, 1998, Wu *et al.*, 1996) and has been proposed for time series indexing (Chan and Fu, 1999, Korn *et al.*, 1997).

Singular Value Decomposition is similar to DFT and DWT in that it represents the shape in terms of a linear combination of basis shapes, as shown in 56.17. However, SVD differs from DFT and DWT in one very important aspect. SVD and DWT are local; they examine one data object at a time and apply a transformation. These transformations are completely independent of the rest of the data. In contrast, SVD is a global transformation. The entire dataset is examined and is then rotated such that the first axis has the maximum possible variance, the second axis has the maximum possible variance orthogonal to the first, the third axis has the maximum possible variance orthogonal to the first two, etc. The global nature of the transformation is both a weakness and strength from an indexing point of view.

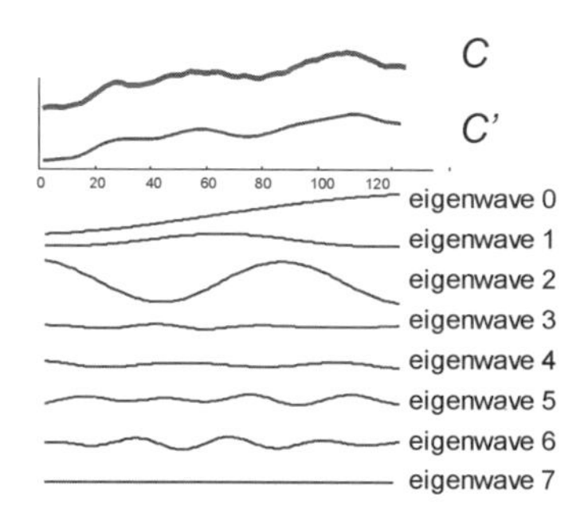

Fig. 56.17. A visualization of the SVD dimensionality reduction technique.

SVD is the optimal transform in several senses, including the following: if we take the SVD of some dataset, then attempt to reconstruct the data, SVD is the optimal (linear) transform that minimizes reconstruction error (Ripley, 1996). Given this, we should expect SVD to perform very well for the indexing task.

56.4.4 Piecewise Linear Approximation

The idea of using piecewise linear segments to approximate time series dates back to 1970s (Pavlidis and Horowitz, 1974). This representation has numerous advantages, including data compression and noise filtering. There are numerous algorithms available for segmenting time series, many of which were pioneered by (Pavlidis and Horowitz, 1974). Figure 56.18 shows an example of a time series represented by piecewise linear segments.

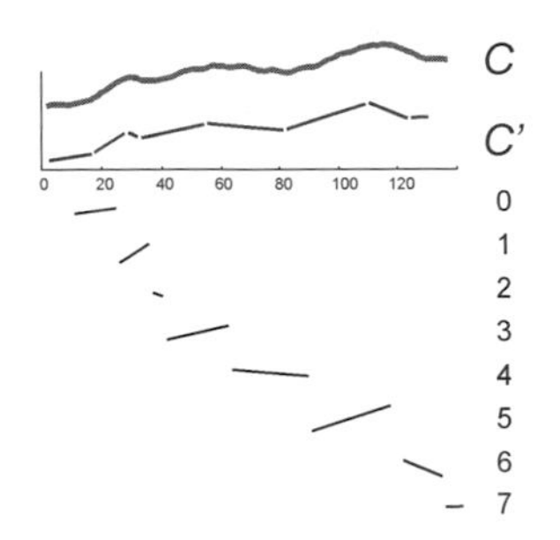

Fig. 56.18. A visualization of the PLA dimensionality reduction technique

An open question is how to best choose K, the "optimal" number of segments used to represent a particular time series. This problem involves a trade-off between accuracy and compactness, and clearly has no general solution.

56.4.5 Piecewise Aggregate Approximation

The recent work (Keogh *et al.*, 2001, Yi and Faloutsos, 2000) (independently) suggest approximating a time series by dividing it into equal-length segments and recording the mean value of the data points that fall within the segment. The authors use different names for this representation. For clarity here, we refer to it as Piecewise Aggregate Approximation (PAA). This representation reduces the data from n dimensions to N dimensions by dividing the time series into N equi-sized 'frames'. The mean value of the data falling within a frame is calculated, and a vector of these values becomes the data reduced representation. When $N = n$, the transformed representation is identical to the original representation. When $N = 1$, the transformed representation is simply the mean of the original sequence. More generally, the transformation produces a piecewise constant approximation of the original sequence, hence the name, Piecewise Aggregate Approximation (PAA). This representation is also capable of handling queries of variable lengths.

In order to facilitate comparison of PAA with other dimensionality reduction techniques discussed earlier, it is useful to visualize it as approximating a sequence with a linear combination of box functions. Figure 56.19 illustrates this idea.

This simple technique is surprisingly competitive with the more sophisticated transform. In addition, the fact that each segment in PAA is of the same length facilitates indexing of this representation.

56.4.6 Adaptive Piecewise Constant Approximation

As an extension to the PAA representation, Adaptive Piecewise Constant Approximation (APCA) is introduced (Keogh *et al.*, 2001). This representation allows the segments to have arbitrary lengths, which in turn needs two numbers per segment. The first number records the

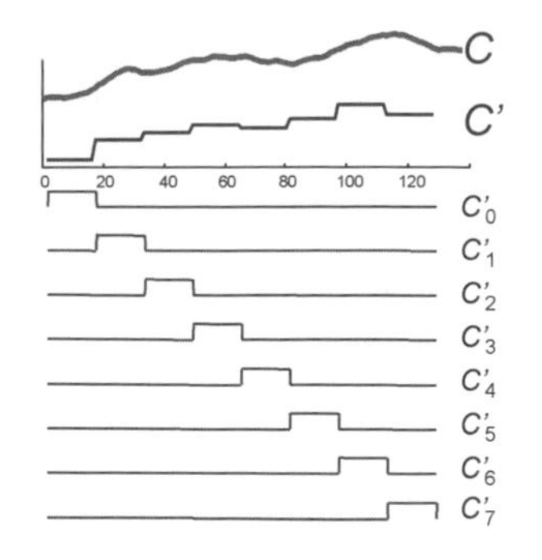

Fig. 56.19. A visualization of the PAA dimensionality reduction technique

mean value of all the data points in segment, and the second number records the length of the segment.

It is difficult to make any intuitive guess about the relative performance of this technique. On one hand, PAA has the advantage of having twice as many approximating segments. On the other hand, APCA has the advantage of being able to place a single segment in an area of low activity and many segments in areas of high activity. In addition, one has to consider the structure of the data in question. It is possible to construct artificial datasets, where one approach has an arbitrarily large reconstruction error, while the other approach has reconstruction error of zero.

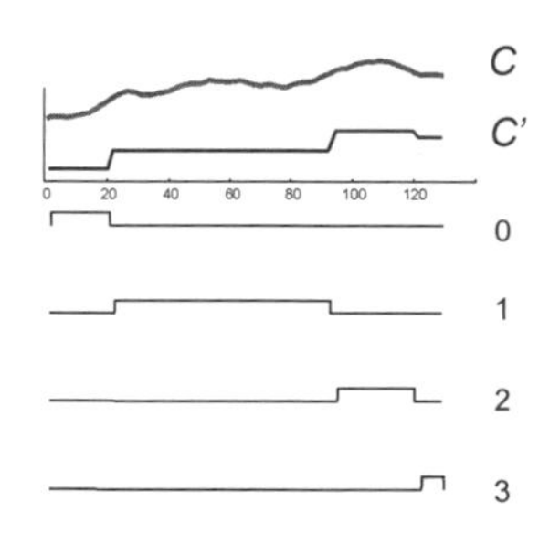

Fig. 56.20. A visualization of the APCA dimensionality reduction technique

In general, finding the optimal piecewise polynomial representation of a time series requires a $O(Nn^2)$ dynamic programming algorithm (Faloutsos *et al.*, 1997). For most purposed, however, an optimal representation is not required. Most researchers, therefore, use a greedy suboptimal approach instead (Keogh and Smyth, 1997). In (Keogh *et al.*, 2001), the authors utilize an original algorithm which produces high quality approximations in $O(nlog(n))$. The algorithm works by first converting the problem into a wavelet compression problem, for which there are well-known optimal solutions, then converting the solution back to the APCA representation and (possible) making minor modification.

56.4.7 Symbolic Aggregate Approximation (SAX)

Symbolic Aggregate Approximation is a novel symbolic representation for time series recently introduced by (Lin *et al.*, 2003), which has been shown to preserve meaningful information from the original data and produce competitive results for classifying and clustering time series.

The basic idea of SAX is to convert the data into a discrete format, with a small alphabet size. In this case, every part of the representation contributes about the same amount of information about the shape of the time series. To convert a time series into symbols, it is first normalized, and two steps of discretization will be performed. First, a time series T of length n is divided into w equal-sized segments; the values in each segment are then approximated and replaced by a single coefficient, which is their average. Aggregating these w coefficients form the Piecewise Aggregate Approximation (PAA) representation of T. Next, to convert the PAA coefficients to symbols, we determine the breakpoints that divide the distribution space into α equiprobable regions, where α is the alphabet size specified by the user (or it could be determined from the Minimum Description Length). In other words, the breakpoints are determined such that the probability of a segment falling into any of the regions is approximately the same. If the symbols are not equi-probable, some of the substrings would be more probable than others. Consequently, we would inject a probabilistic bias in the process. In (Crochemore *et al.*, 1994), Crochemore et al. show that a suffix tree automation algorithm is optimal if the letters are equiprobable.

Once the breakpoints are determined, each region is assigned a symbol. The PAA coefficients can then be easily mapped to the symbols corresponding to the regions in which they reside. The symbols are assigned in a bottom-up fashion, i.e. the PAA coefficient that falls in the lowest region is converted to "*a*", in the one above to "*b*", and so forth. Figure 56.21 shows an example of a time series being converted to string *baabccbc*. Note that the general shape of the time series is still preserved, in spite of the massive amount of dimensionality reduction, and the symbols are equiprobable.

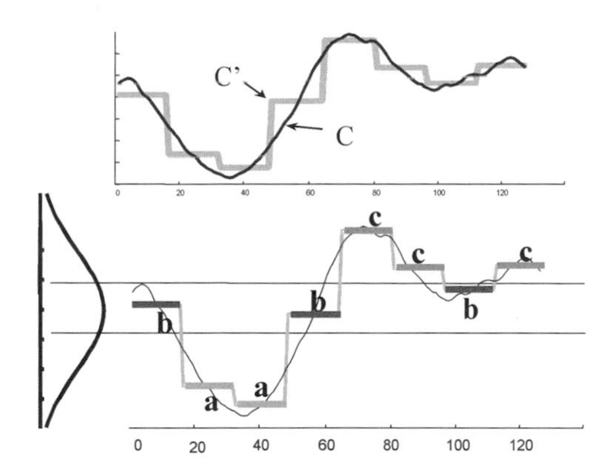

Fig. 56.21. A visualization of the SAX dimensionality reduction technique

To reiterate the significance of time series representation, Figure 56.22 illustrates four of the most popular representations.

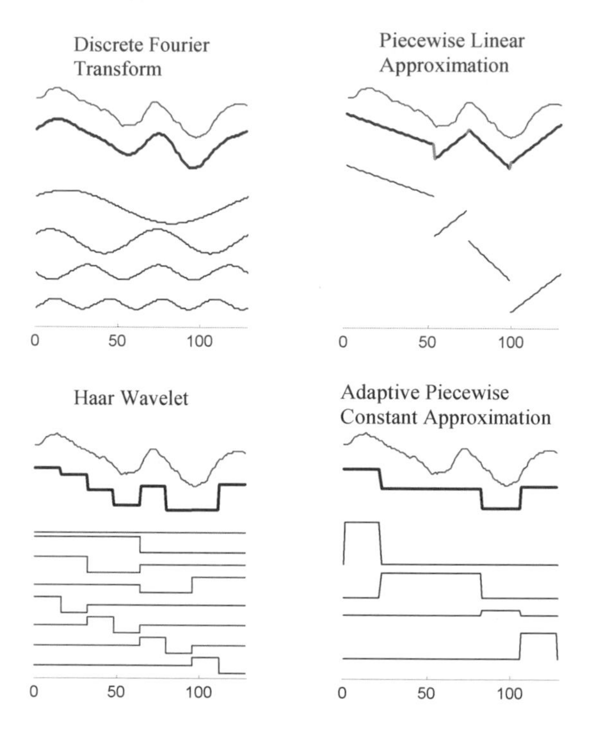

Fig. 56.22. Four popular representations of time series. For each graphic, we see a raw time series of length 128. Below it, we see an approximation using 1/8 of the original space. In each case, the representation can be seen as a linear combination of basis functions. For example, the Discrete Fourier representation can be seen as a linear combination of the four sine/cosine waves shown in the bottom of the graphics.

Given the plethora of different representations, it is natural to ask which is best. Recall that the more faithful the approximation, the less clarification disks accesses we will need to make in Step 3 of Table 56.1. In the example shown in Figure 56.22, the discrete Fourier approach seems to model the original data the best. However, it is easy to imagine other time series where another approach might work better. There have been many attempts to answer the question of which is the best representation, with proponents advocating their favorite technique (Chakrabarti *et al.*, 2002, Faloutsos *et al.*, 1994, Popivanov *et al.*, 2002, Rafiei *et al.*, 1998). The literature abounds with mutually contradictory statements such as "*Several wavelets outperform the ... DFT*" (Popivanov *et al.*, 2002), "*DFT-base and DWT-based techniques yield comparable results*" (Wu *et al.*, 2000), "*Haar wavelets perform ... better than DFT*" (Kahveci and Singh, 2001). However, an extensive empirical comparison on 50 diverse datasets suggests that while some datasets favor a particular approach, overall, there is little difference between the various approaches in terms of their ability to approximate the data (Keogh and Kasetty, 2002). There are however, other important differences in the usability of each approach (Chakrabarti *et al.*, 2002). We will consider some representative examples of strengths and weaknesses below.

The wavelet transform is often touted as an ideal representation for time series Data Mining, because the first few wavelet coefficients contain information about the overall shape of the sequence while the higher order coefficients contain information about localized trends (Popivanov *et al.*, 2002, Shahabi *et al.*, 2000). This multiresolution property can be exploited by some algorithms, and contrasts with the Fourier representation in which every coefficient represents a contribution to the global trend (Faloutsos *et al.*, 1994, Rafiei *et al.*, 1998). However, wavelets do have several drawbacks as a Data Mining representation. They are only defined for data whose length is an integer power of two. In contrast, the Piecewise Constant Approximation suggested by (Yi and Faloutsos, 2000), has exactly the fidelity of resolution of as the Haar wavelet, but is defined for arbitrary length time series. In addition, it has several other useful properties such as the ability to support several different distance measures (Yi and Faloutsos, 2000), and the ability to be calculated in an incremental fashion as the data arrives (Chakrabarti *et al.*, 2002). One important feature of all the above representations is that they are real valued. This somewhat limits the algorithms, data structures, and definitions available for them. For example, in anomaly detection, we cannot meaningfully define the probability of observing any particular set of wavelet coefficients, since the probability of observing any real number is zero. Such limitations have lead researchers to consider using a symbolic representation of time series (Lin *et al.*, 2003).

56.5 Summary

In this chapter, we have reviewed some major tasks in time series data mining. Since time series data are typically very large, discovering information from these massive data becomes a challenge, which leads to the enormous research interests in approximating the data in reduced representation. The dimensionality reduction of the data has now become the heart of time series Data Mining and is the primary step to efficiently deal with Data Mining tasks for massive data. We review some of important time series representations proposed in the literature. We would like to emphasize that the key step in any successful time series Data Mining endeavor always lies in choosing the right representation for the task at hand.

References

Aach, J. and Church, G. Aligning gene expression time series with time warping algorithms. Bioinformatics; 2001, Volume 17, pp. 495-508.

Aggarwal, C., Hinneburg, A., Keim, D. A. On the surprising behavior of distance metrics in high dimensional space. In proceedings of the 8th International Conference on Database Theory; 2001 Jan 4-6; London, UK, pp 420-434.

Agrawal, R., Faloutsos, C., Swami, A. Efficient Similarity Search in Sequence Data bases. International Conference on Foundations of Data Organization (FODO); 1993.

Agrawal, R., Lin, K.-I., Sawhney, H.S., Shim, K. Fast Similarity Search in the Presence of Noise, Scaling, and Translation in Trime-Series Databases. Proceedings of 21st International Conference on Very Large Databases; 1995 Sep; Zurich, Switzerland, pp. 490-500.

Berndt, D.J., Clifford, J. Finding Patterns in Time Series: A Dynamic Programming Approach. In Advances in Knowledge Discovery and Data Mining AAAI/MIT Press, Menlo Park, CA, 1996, pp. 229-248.

Bollobas, B., Das, G., Gunopulos, D., Mannila, H. Time-Series Similarity Problems and Well-Separated Geometric Sets. Nordic Jour. of Computing 2001; 4.

Brin, S. Near neighbor search in large metric spaces. Proceedings of 21st VLDB; 1995.

Chakrabarti, K., Keogh, E., Pazzani, M., Mehrotra, S. Locally adaptive dimensionality reduction for indexing large time series databases. ACM Transactions on Database Systems. Volume 27, Issue 2, (June 2002). pp 188-228.

Chan, K., Fu, A.W. Efficient time series matching by wavelets. Proceedings of 15th IEEE International Conference on Data Engineering; 1999 Mar 23-26; Sydney, Australia, pp. 126-133.

Chang, C.L.E., Garcia-Molina, H., Wiederhold, G. Clustering for Approximate Similarity Search in High-Dimensional Spaces. IEEE Transactions on Knowledge and Data Engineering 2002; Jul – Aug, 14(4): 792-808.

Chiu, B.Y., Keogh, E., Lonardi, S. Probabilistic discovery of time series motifs. Proceedings of ACM SIGKDD; 2003, pp. 493-498.

Ciaccia, P., Patella, M., Zezula, P. M-tree: An efficient access method for similarity search in metric spaces. Proceedings of 23rd VLDB; 1997, pp. 426-435.

Crochemore, M., Czumaj, A., Gasjeniec, L, Jarominek, S., Lecroq, T., Plandowski, W., Rytter, W. Speeding up two string-matching algorithms. Algorithmica; 1994; Vol. 12(4/5), pp. 247-267.

Dasgupta, D., Forrest, S. Novelty Detection in Time Series Data Using Ideas from Immunology. Proceedings of 8th International conference on Intelligent Systems; 1999 Jun 24-26; Denver, CO.

Debregeas, A., Hebrail, G. Interactive interpretation of kohonen maps applied to curves. In proceedings of the 4th Int'l Conference of Knowledge Discovery and Data Mining; 1998 Aug 27-31; New York, NY, pp 179-183.

Faloutsos, C., Jagadish, H., Mendelzon, A., Milo, T. A signature technique for similarity-based queries. Proceedings of the International Conference on Compression and Complexity of Sequences; 1997 Jun 11-13; Positano-Salerno, Italy.

Faloutsos, C., Ranganathan, M., Manolopoulos, Y. Fast subsequence matching in time-series databases. In proceedings of the ACM SIGMOD Int'l Conference on Management of Data; 1994 May 25-27; Minneapolis, MN, pp 419-429.

Ge, X., Smyth, P. Deformable Markov Model Templates for Time-Series Pattern Matching. Proceedings of 6th ACM SIGKDD International Conference on Knowledge Discovery and Data Mining; 2000 Aug 20-23; Boston , MA, pp. 81-90.

Geurts, P. Pattern extraction for time series classification. Proceedings of Principles of Data Mining and Knowledge Discovery, 5th European Conference; 2001 Sep 3-5; Freiburg, Germany, pp 115-127.

Goldin, D.Q., Kanellakis, P.C. On Similarity Queries for Time-Series Data: Constraint Specification and Implementation. Proceedings of the 1st International Conference on the Principles and Practice of Constraint Programming; 1995 Sep 19-22; Cassis, France, pp. 137-153.

Guralnik, V., Srivastava, J. Event detection from time series data. In proceedings of the 5th ACM SIGKDD Int'l Conference on Knowledge Discovery and Data Mining; 1999 Aug 15-18; San Diego, CA, pp 33-42.

Huhtala, Y., Karkkainen, J, Toivonen, H. Mining for similarities in aligned time series using wavelet. Data Mining and Knowledge Discovery: Theory, Tools, and Technology, SPIE Proceedings Series 1995; Orlando, FL, Vol. 3695, pp. 150-160.

Hochheiser, H., Shneiderman,, B. Interactive Exploration of Time-Sereis Data. Proceedings of 4th International conference on Discovery Science; 2001 Nov 25-28; Washington, DC, pp. 441-446.

Indyk, P., Koudas, N., Muthukrishnan, S. Identifying representative trends in massive time series data sets using sketches. In proceedings of the 26th Int'l Conference on Very Large Data Bases; 2000 Sept 10-14; Cairo, Egypt, pp 363-372.

Jagadish, H.V., Mendelzon, A.O., and Milo, T. Similarity-Based Queries. Proceedings of ACM PODS; 1995 May; San Jose, CA, pp. 36-45.

Kahveci, T., Singh, A. Variable length queries for time series data. In proceedings of the 17th Int'l Conference on Data Engineering; 2001 Apr 2-6; Heidelberg, Germany, pp 273-282.

Kalpakis, K., Gada, D., Puttagunta, V. Distance measures for effective clustering of ARIMA time-series. Proceedings of the IEEE Int'l Conference on Data Mining; 2001 Nov 29-Dec 2; San Jose, CA, pp 273-280.

Kanth, K.V., Agrawal, D., Singh, A. Dimensionality reduction for similarity searching in dynamic databases. Proceedings of ACM SIGMOD International Conference; 1998, pp. 166-176.

Keogh, E. Exact indexing of dynamic time warping. Proceedings of 28th Internation Conference on Very Large Databases; 2002; Hong Kong, pp. 406-417.

Keogh, E., Chakrabarti, K., Mehrotra, S., Pazzani, M. Locally adaptive dimensionality reduction for indexing large time series databases. Proceedings of ACM SIGMOD International Conference; 2001.

Keogh, E., Chakrabarti, K., Pazzani, M., Mehrotra, S. Dimensionality reduction for fast similarity search in large time series databases. Knowledge and Information Systems 2001; 3: 263-286.

Keogh, E., Lin, J., Truppel, W. Clustering of Time Series Subsequences is Meaningless: Implications for Previous and Future Research. Proceedings of ICDM; 2003, pp. 115-122.

Keogh, E., Lonardi, S., Chiu, W. Finding Surprising Patterns in a Time Series Database In Linear Time and Space. In the 8th ACM SIGKDD International Conference on Knowledge Discovery and Data Mining; 2002 Jul 23 – 26; Edmonton, Alberta, Canada, pp 550-556.

Keogh, E., Lonardi, S., Ratanamahatana, C.A. Towards Parameter-Free Data Mining. Proceedings of 10th ACM SIGKDD International Conference on Knowledge Discovery and Data Mining; 2004 Aug 22-25; Seattle, WA.

Keogh, E., Pazzani, M. An enhanced representation of time series which allows fast and accurate classification, clustering and relevance feedback. Proceedings of the 4th Int'l Conference on Knowledge Discovery and Data Mining; 1998 Aug 27-31; New York, NY, pp 239-241.

Keogh, E. and Kasetty, S. On the Need for Time Series Data Mining Benchmarks: A Survey and Empirical Demonstration. In the 8th ACM SIGKDD International Conference on Knowledge Discovery and Data Mining; 2002 Jul 23 – 26; Edmonton, Alberta, Canada, pp 102-111.

Keogh, E., Smyth, P. A Probabilistic Approach to Fast Pattern matching in Time Series Databases. Proceedings of 3rd International conference on Knowledge Discovery and Data Mining; 1997 Aug 14-17; Newport Beach, CA, pp. 24-30.

Korn, F., Jagadish, H., Faloutsos, C. Efficiently supporting ad hoc queries in large datasets of time sequences. Proceedings of SIGMOD International Conferences 1997; Tucson, AZ, pp. 289-300.

Kruskal, J.B., Sankoff, D., Editors. Time Warps, String Edits, and Macromolecules: The Theory and Practice of Sequence Comparison. Addison-Wesley, 1983.

Lin, J., Keogh, E., Lonardi, S., Chiu, B. A Symbolic Representation of Time Series, with Implications for Streaming Algorithms. Workshop on Research Issues in Data Mining and Knowledge Discovery, 8th ACM SIGMOD; 2003 Jun 13; San Diego, CA.

Lin, J., Keogh, E., Lonardi, S., Lankford, J. P., Nystrom, D. M. Visually Mining and Monitoring Massive Time Series. Proceedings of the 10th ACM SIGKDD International Conference on Knowledge Discovery and Data Mining; 2004 Aug 22-25; Seattle, WA.

Ma, J., Perkins, S. Online Novelty Detection on Temporal Sequences. Proceedings of 9th International Conference on Knowledge Discovery and Data Mining; 2003 Aug 24-27; Washington DC.

Nievergelt, H., Hinterberger, H., Sevcik, K.C. The grid file: An adaptable, symmetricmultikey file structure. ACM Trans. Database Systems; 1984; 9(1): 38-71.

Palpanas, T., Vlachos, M., Keogh, E., Gunopulos, D., Truppel, W. Online Amnestic Approximation of Streaming Time Series. Proceedings of 20th International Conference on Data Engineering; 2004, Boston, MA.

Pavlidis, T., Horowitz, S. Segmentation of plane curves. IEEE Transactions on Computers; 1974 August; Vol. C-23(8), pp. 860-870.

Popivanov, I., Miller, R. J. Similarity search over time series data using wave -lets. In proceedings of the 18th Int'l Conference on Data Engineering; 2002 Feb 26-Mar 1; San Jose, CA, pp 212-221.

Rafiei, D., Mendelzon, A. O. Efficient retrieval of similar time sequences using DFT. In proceedings of the 5th Int'l Conference on Foundations of Data Organization and Algorithms; 1998 Nov 12-13; Kobe, Japan.

Ratanamahatana, C.A., Keogh, E. Making Time-Series Classification More Accurate Using Learned Constrints. Proceedings of SIAM International Conference on Data Mining; 2004 Apr 22-24; Lake Buena Vista, FL, pp.11-22.

Ripley, B.D. Pattern recognition and neural networks. Cambridge University Press, Cambridge, UK, 1996.

Robinson, J.T. The K-d-b-tree: A search structure for large multidimensional dynamic indexes. Proceedings of ACM SIGMOD; 1981.

Shahabi, C., Tian, X., Zhao, W. TSA-tree: a wavelet based approach to improve the efficiency of multi-level surprise and trend queries. In proceedings of the 12th Int'l Conference on Scientific and Statistical Database Management; 2000 Jul 26-28; Berlin, Germany, pp 55-68.

Struzik, Z., Siebes, A. The Haar wavelet transform in the time series similarity paradigm. Proceedings of 3rd European Conference on Principles and Practice of Knowledge Discovery in Databases; 1999; Prague, Czech Republic, pp. 12-22.

Tufte, E. The visual display of quantitative information. Graphics Press, Cheshire, Connecticut, 1983.

Tzouramanis, T., Vassilakopoulos, M., Manolopoulos, Y. Overlapping Linear Quadtrees: A Spatio-Temporal Access Method. ACM-GIS; 1998, pp. 1-7.

Guralnik, V., Srivastava, J. Event Detection from Time Series Data. Proceedings of ACM SIGKDD; 1999, pp 33-42.

Vlachos, M., Gunopulos, D., Das, G. Rotation Invariant Distance Measures for Trajectories. Proceedings of 10th International Conference on Knowledge Discovery and Data Mining; 2004 Aug 22-25; Seattle, WA.

Vlachos, M., Meek, C., Vagena, Z., Gunopulos, D. Identification of Similarities, Periodicities & Bursts for Online Search Queries. Proceedings of International Conference on Management of Data; 2004; Paris, France.

Weber, M., lexa, M., Muller, W. Visualizing Time Series on Spirals. Proceedings of IEEE Symposium on Information Visualization; 2000 Oct 21-26; San Diego, CA, pp. 7-14.

Wijk, J.J. van, E. van Selow. Cluster and calendar-based visualization of time series data. Proceedings of IEEE Symposium on Information Visualization; 1999 Oct 25-26, IEEE Computer Society, pp 4-9.

Wu, D., Agrawal, D., El Abbadi, A., Singh, A, Smith, T.R. Efficient retrieval for browsing large image databases. Proceedings of 5th International Conference on Knowledge Information; 1996; Rockville, MD, pp. 11-18.

Wu, Y., Agrawal, D., El Abbadi, A. A comparison of DFT and DWT based similarity search in time-series databases. In proceedings of the 9th ACM CIKM Int'l Conference on Information and Knowledge Management; 2000 Nov 6-11; McLean, VA, pp 488-495.

Yi, B., Faloutsos, C. Fast time sequence indexing for arbitrary lp norms. Proceedings of the 26th Int'l Conference on Very Large Databases; 2000 Sep 10-14; Cairo, Egypt, pp 385-394.

Yianilos, P. Data structures and algorithms for nearest neighbor search in general metric spaces. Proceedings of 3rd SIAM on Discrete Algorithms; 1992.

Zhu, Y., Shasha, D. StatStream: Statistical Monitoring of Thousands of Data Streams in Real Time, Proceedings of VLDB; 2002, pp. 358-369.

Part VII

Applications

57

Multimedia Data Mining

Zhongfei (Mark) Zhang and Ruofei Zhang

[1] SUNY at Binghamton, NY 13902-6000, zhongfei@cs.binghamton.edu
[2] Yahoo!, Inc., Sunnyvale, CA 94089 rzhang@yahoo-inc.com

Summary. *Each chapter should be preceded by an abstract (10–15 lines long) that summarizes the content. The abstract will appear *online* at www.SpringerLink.com and be available with unrestricted access. This allows unregistered users to read the abstract as a teaser for the complete chapter. As a general rule the abstracts will not appear in the printed version of your book unless it is the style of your particular book or that of the series to which your book belongs. Please use the 'starred' version of the new Springer abstract command for typesetting the text of the online abstracts (cf. source file of this chapter template abstract) and include them with the source files of your manuscript. Use the plain abstract command if the abstract is also to appear in the printed version of the book.

57.1 Introduction

Multimedia data mining, as the name suggests, presumably is a combination of the two emerging areas: *multimedia* and *data mining*. However, multimedia data mining is *not* a research area that just simply combines the research of multimedia and data mining together. Instead, the multimedia data mining research focuses on the theme of merging multimedia and data mining research together to exploit the synergy between the two areas to promote the understanding and to advance the development of the knowledge discovery in multimedia data. Consequently, multimedia data mining exhibits itself as a unique and distinct research area that synergistically relies on the state-of-the-art research in multimedia and data mining but at the same time fundamentally differs from either multimedia or data mining or a simple combination of the two areas.

Multimedia and data mining are two very interdisciplinary and multidisciplinary areas. Both areas started in early 1990s with only a very short history. Therefore, both areas are relatively young areas (in comparison, for example, with many well established areas in computer science such as operating systems, programming languages, and artificial intelligence). On the other hand, with substantial application demands, both areas have undergone independently and simultaneously rapid developments in recent years.

Multimedia is a very diverse, interdisciplinary, and multidisciplinary research area[3]. The word *multimedia* refers to a combination of multiple media types together. Due to the advanced

[3] Here we are only concerned with a research area; multimedia may also be referred to industries and even social or societal activities.

O. Maimon, L. Rokach (eds.), *Data Mining and Knowledge Discovery Handbook*, 2nd ed.,
DOI 10.1007/978-0-387-09823-4_57,

development of the computer and digital technologies in early 1990s, multimedia began to emerge as a research area (Furht, 1996, Steinmetz & Nahrstedt, 2002). As a research area, multimedia refers to the study and development of an effective and efficient multimedia system targeting a specific application. In this regard, the research in multimedia covers a very wide spectrum of subjects, ranging from multimedia indexing and retrieval, multimedia databases, multimedia networks, multimedia presentation, multimedia quality of services, multimedia usage and user study, to multimedia standards, just to name a few.

While the area of multimedia is so diverse with many different subjects, those that are related to multimedia data mining mainly include multimedia indexing and retrieval, multimedia databases, and multimedia presentation (Faloutsos et al., 1994, Jain, 1996, Subrahmanian, 1998). Today, it is well known that multimedia information is ubiquitous and is often required, if not necessarily essential, in many applications. This phenomenon has made multimedia repositories widespread and extremely large. There are tools for managing and searching within these collections, but the need for tools to extract hidden useful knowledge embedded within multimedia collections is becoming pressing and central for many decision-making applications. For example, it is highly desirable for developing the tools needed today for discovering relationships between objects or segments within images, classifying images based on their content, extracting patterns in sound, categorizing speech and music, and recognizing and tracking objects in video streams.

At the same time, researchers in multimedia information systems, in the search for techniques for improving the indexing and retrieval of multimedia information, are looking for new methods for discovering indexing information. A variety of techniques, from machine learning, statistics, databases, knowledge acquisition, data visualization, image analysis, high performance computing, and knowledge-based systems, have been used mainly as research handcraft activities. The development of multimedia databases and their query interfaces recalls again the idea of incorporating multimedia data mining methods for dynamic indexing.

On the other hand, data mining is also a very diverse, interdisciplinary, and multidisciplinary research area. The terminology *data mining* refers to knowledge discovery. Originally, this area began with knowledge discovery in databases. However, data mining research today has been advanced far beyond the area of databases (Faloutsos, 1996, Han & Kamber, 2006). This is due to the following two reasons. First, today's knowledge discovery research requires more than ever the advanced tools and theory beyond the traditional database area, noticeably mathematics, statistics, machine learning, and pattern recognition. Second, with the fast explosion of the data storage scale and the presence of multimedia data almost everywhere, it is not enough for today's knowledge discovery research to just focus on the structured data in the traditional databases; instead, it is common to see that the traditional databases have evolved into data warehouses, and the traditional structured data have evolved into more non-structured data such as imagery data, time-series data, spatial data, video data, audio data, and more general multimedia data. Adding into this complexity is the fact that in many applications these non-structured data do not even exist in a more traditional "database" anymore; they are just simply a collection of the data, even though many times people still call them databases (e.g., image database, video database).

Examples are the data collected in fields such as art, design, hypermedia and digital media production, case-based reasoning and computational modeling of creativity, including evolutionary computation, and medical multimedia data. These exotic fields use a variety of data sources and structures, interrelated by the nature of the phenomenon that these structures describe. As a result there is an increasing interest in new techniques and tools that can detect and discover patterns that lead to new knowledge in the problem domain where the data have been collected. There is also an increasing interest in the analysis of multimedia data generated by different distributed applications, such as collaborative virtual environments, virtual communities, and multi-agent systems. The data collected from such environments include a record of the actions in them, a variety of documents that are part of the business process, asynchronous threaded discussions, transcripts from synchronous communications, and other data records. These heterogeneous multimedia data records require sophisticated preprocessing, synchronization, and other transformation procedures before even moving to the analysis stage.

Consequently, with the independent and advanced developments of the two areas of multimedia and data mining, with today's explosion of the data scale and the existence of the pluralism of the data media types, it is natural to evolve into this new area called *multimedia data mining*. While it is presumably true that multimedia data mining is a combination of the research between multimedia and data mining, the research in multimedia data mining refers to the synergistic application of knowledge discovery theory and techniques in a multimedia database or collection. As a result, "inherited" from its two parent areas of multimedia and data mining, multimedia data mining by nature is also an interdisciplinary and multidisciplinary area; in addition to the two parent areas, multimedia data mining also relies on the research from many other areas, noticeably from mathematics, statistics, machine learning, computer vision, and pattern recognition. Figure 57.1 illustrates the relationships among these interconnected areas.

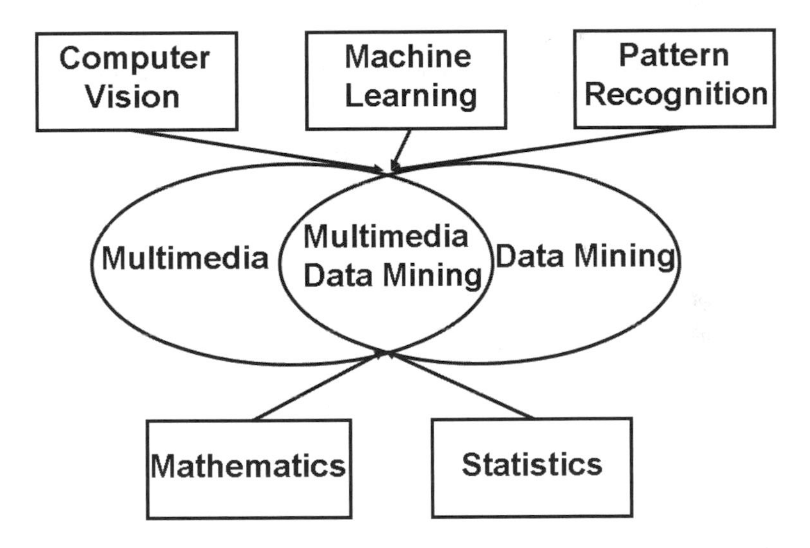

Fig. 57.1. Relationships among the interconnected areas to multimedia data mining.

While we have clearly given the working definition of multimedia data mining as an emerging, active research area, due to historic reasons, it is helpful to clarify several misconceptions and to point out several pitfalls at the beginning.

- *Multimedia Indexing and Retrieval* vs. *Multimedia Data Mining*: It is well-known that in the classic data mining research, the pure text retrieval or the classic information retrieval is *not* considered as part of data mining, as there is no knowledge discovery involved.

However, in multimedia data mining, when it comes to the scenarios of multimedia indexing and retrieval, this boundary becomes vague. The reason is that a typical multimedia indexing and/or retrieval system reported in the recent literature often contains a certain level of knowledge discovery such as feature selection, dimensionality reduction, concept discovery, as well as mapping discovery between different modalities (e.g., imagery annotation where a mapping from an image to textual words is discovered and word-to-image retrieval where a mapping from a textual word to images is discovered). In this case, multimedia information indexing and/or retrieval is considered as part of multimedia data mining. On the other hand, if a multimedia indexing or retrieval system uses a "pure" indexing system such as the text-based indexing technology employed in many commercial imagery/video/audio retrieval systems on the Web, this system is not considered as a multimedia data mining system.

- *Database* vs. *Data Collection*: In a classic database system, there is always a database management system to govern all the data in the database. This is true for the classic, structured data in the traditional databases. However, when the data become non-structured data, in particular, multimedia data, often we do not have such a management system to "govern" all the data in the collection. Typically, we simply just have a whole collection of multimedia data, and we expect to develop an indexing/retrieval system or other data mining system on top of this data collection. For historic reasons, in many literature references, we still use the terminology of "database" to refer to such a multimedia data collection, even though this is different from the traditional, structured database in concept.
- *Multimedia Data* vs. *Single Modality Data*: Although "multimedia" refers to the multiple modalities and/or multiple media types of data, conventionally in the area of multimedia, multimedia indexing and retrieval also includes the indexing and retrieval of a single, non-text modality of data, such as image indexing and retrieval, video indexing and retrieval, and audio indexing and retrieval. Consequently, in multimedia data mining, we follow this convention to include the study of any knowledge discovery dedicated to any single modality of data as part of the multimedia data mining research. Therefore, studies in image data mining, video data mining, and audio data mining alone are considered as part of the multimedia data mining area.

Multimedia data mining, although still in its early booming stage as an area that is expected to have further development, has already found enormous application potential in a wide spectrum covering almost all the sectors of society, ranging from people's daily lives to economic development to government services. This is due to the fact that in today's society almost all the real-world applications often have data with multiple modalities, from multiple sources, and in multiple formats. For example, in homeland security applications, we may need to mine data from an air traveler's credit history, traveling patterns, photo pictures, and video data from surveillance cameras in the airport. In the manufacturing domains, business processes can be improved if, for example, part drawings, part descriptions, and part flow can be mined in an integrated way instead of separately. In medicine, a disease might be predicted more accurately if the MRI (magnetic resonance imaging) imagery is mined together with other information about the patient's condition. Similarly, in bioinformatics, data are available in multiple formats.

The rest of the chapter is organized as follows. In the next section, we give the architecture for a typical multimedia data mining system or methodology in the literature. Then in order to showcase a specific multimedia data mining system and how it works, we present an example of a specific method on concept discovery in an imagery database in the following section. Finally, the chapter is concluded in Sec. 57.4.

57.2 A Typical Architecture of a Multimedia Data Mining System

A typical multimedia data mining system, or framework, or method always consists of the following three key components. Given the raw multimedia data, the very first step for mining the multimedia data is to convert a specific raw data collection (or a database) into a representation in an abstract space which is called the feature space. This process is called feature extraction. Consequently, we need a feature representation method to convert the raw multimedia data to the features in the feature space, before any mining activities are able to be conducted. This component is very important as the success of a multimedia data mining system to a large degree depends upon how good the feature representation method is. The typical feature representation methods or techniques are taken from the classic computer vision research, pattern recognition research, as well as multimedia information indexing and retrieval research in multimedia area.

Since knowledge discovery is an intelligent activity, like other types of intelligent activities, multimedia data mining requires the support of a certain level of knowledge. Therefore, the second key component is the knowledge representation, i.e., how to effectively represent the required knowledge to support the expected knowledge discovery activities in a multimedia database. The typical knowledge representation methods used in the multimedia data mining literature are directly taken from the general knowledge representation research in artificial intelligence area with the possible special consideration in the multimedia data mining problems such as spatial constraints based reasoning.

Finally, we come to the last key component — the actual mining or learning theory and/or technique to be used for the knowledge discovery in a multimedia database. In the current literature of multimedia data mining, there are mainly two paradigms of the learning or mining theory/techniques that can be used separately or jointly in a specific multimedia data mining application. They are *statistical learning theory* and *soft computing theory*, respectively. The former is based on the recent literature on machine learning and in particular statistical machine learning, whereas the latter is based on the recent literature on soft computing such as fuzzy logic theory. This component typically is the core of the multimedia data mining system.

In addition to the three key components, in many multimedia data mining systems, there are user interfaces to facilitate the communications between the users and the mining systems. Like the general data mining systems, for a typical multimedia data mining system, the quality of the final mining results can only be judged by the users. Hence, it is necessary in many cases to have a user interface to allow the communications between the users and the mining systems and the evaluations of the final mining quality; if the quality is not acceptable, the users may need to use the interface to tune different parameter values of a specific component used in the system, or even to change different components, in order to achieve better mining results, which may go into an iterative process until the users are happy with the mining results.

Figure 57.2 illustrates this typical architecture of a multimedia data mining system.

57.3 An Example — Concept Discovery in Imagery Data

In this section, as an example to showcase the research as well as the technologies developed in multimedia data mining, we address the image database modeling problem in general and, in particular, focuses on developing a hidden semantic concept discovery methodology to address effective semantics-intensive image data mining and retrieval. In the approach proposed in this section, each image in the database is segmented into regions associated with homogenous

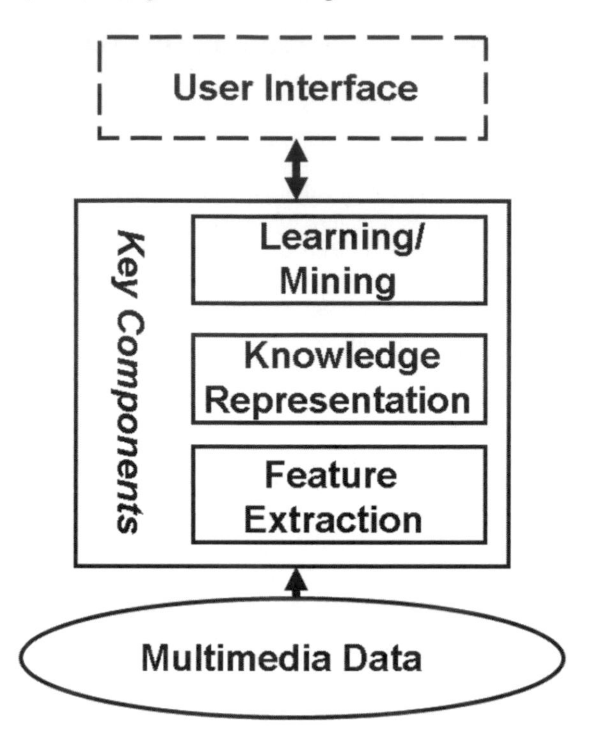

Fig. 57.2. The typical architecture of a multimedia data mining system.

color, texture, and shape features. By exploiting regional statistical information in each image and employing a vector quantization method, a uniform and sparse region-based representation is achieved. With this representation a probabilistic model based on the statistical-hidden-class assumptions of the image database is obtained, to which the Expectation-Maximization (EM) technique is applied to discover and analyze semantic concepts hidden in the database. An elaborated mining and retrieval algorithm is designed to support the probabilistic model. The semantic similarity is measured through integrating the posterior probabilities of the transformed query image, as well as a constructed negative example, to the discovered semantic concepts. The proposed approach has a solid statistical foundation; the experimental evaluations on a database of 10,000 general-purpose images demonstrate the promise and the effectiveness of the proposed approach.

57.3.1 Background and Related Work

As is obvious, large collections of images have become available to the public, from photo collections to Web pages or even video databases. To effectively mine or retrieve such a large collection of imagery data is a huge challenge. After more than a decade of research, it has been found that content based image data mining and retrieval are a practical and satisfactory solution to this challenge. At the same time, it is also well known that the performance of the existing approaches in the literature is mainly limited by the *semantic gap* between low-level features and high-level semantic concepts (Smeulders et al., 2000). In order to reduce this gap, region based features (describing object level features), instead of raw features of the whole image, to represent the visual content of an image are widely used (Carson et al., 2002, Wang et al., 2001, Jing et al., 2004, Chen & Wang, 2002).

In contrast to traditional approaches (Huang & et al., 1997, Flickner et al., 1995, Pentland et al., 1994), which compute global features of images, the region based methods extract features of the segmented regions and perform similarity comparisons at the granularity of regions. The main objective of using region features is to enhance the ability to capture and represent the focus of users' perception of the image content.

One important issue significantly affecting the success of an image data mining methodology is how to compare two images, i.e., the definition of the image similarity measurement. A straightforward solution adopted by most early systems (Carson et al., 2002, Ma & Manjunath, 1997, Wood et al., 1998) is to use individual region-to-region similarity as the basis of the comparisons. When using such schemes, the users are forced to select a limited number of regions from a query image in order to start a query session. As discussed in (Wang et al., 2001), due to the uncontrolled nature of the visual content in an image, automatically and precisely extracting image objects is still beyond the reach of the state-of-the-art in computer vision. Therefore, these systems tend to partition one object into several regions, with none of them being representative for the object. Consequently, it is often difficult for users to determine which regions should be used for their interest.

To provide users a simpler querying interface and to reduce the influence of inaccurate segmentation, several image-to-image similarity measurements that combine information from all of the regions have been proposed (Greenspan et al., 2004, Wang et al., 2001, Chen & Wang, 2002). Such systems only require users to impose a query image and therefore relieve the users from making the puzzling decisions. For example, the SIMPLIcity system (Wang et al., 2001) uses integrated region matching as its image similarity measure. By allowing a many-to-many relationship of the regions, the approach is robust to inaccurate segmentation. Greenspan et al (Greenspan et al., 2001) propose a continuous probabilistic framework for image matching. In this framework, each image is represented as a Gaussian mixture distribution, and images are compared and matched via a probabilistic measure of similarity between distributions. Improved image matching results are reported.

Ideally, what we strive to measure is the *semantic similarity*, which physically is very difficult to define, or even to describe. The majority of the existing methodologies do not explicitly connect the extracted features with the pursued semantics reflected in the visual content. They define region-to-region and/or image-to-image similarities to attempt to approximate the semantic similarity. However, the approximation is typically heuristic and consequently not reliable and effective. Thus, the retrieval and mining accuracies are rather limited.

To deal with the inaccurate approximation problem, several research efforts have been attempted to link regions to semantic concepts by supervised learning. Barnard et al proposed several statistical models (Barnard et al., 2003, Duygulu et al., 2002, Barnard & Forsyth, 2001) which connect *image blobs* and linguistic words. The objective is to predict words associated with whole images (auto-annotation) and corresponding to particular image regions (region naming). In their approaches, a number of models are developed for the joint distribution of image regions and words. The models are multi-modal and correspondence extensions to Hofmann's hierarchical clustering aspect model (Hofmann & Puzicha, 1998, Hofmann et al., 1996, Hofmann, 2001), a translation model adapted from statistical machine translation, and a multi-modal extension to the mixture of latent Dirichlet allocation models (Blei et al., 2001). The models are used to automatically annotate testing images, and the reported performance is promising. Recognizing that these models fail to exploit spatial context in the images and words, Carbonetto et al augmented the models such that spatial relationships between regions are learned. The model proposed is more expressive in the sense that the spatial correspondences are incorporated into the joint probability learning (Carbonetto et al., 2004, Carbonetto et al., 2003), which improves the accuracy of object recognition in image annotation. Recently, Feng et al proposed a Multiple Bernoulli Relevance Model (MBRM) (Feng et al., 2004) for image-word association, which is based on the Continuous-space Relevance Model (CRM) proposed by (Jeon et al., 2003). In the MBRM model, the word probabilities are estimated using a multiple Bernoulli model and the image feature probabilities using a non-parametric kernel density estimate.

We argue that for all the feature based image mining and retrieval methods, the semantic concepts related to the content of the images are always hidden. By hidden, we mean (1) objectively, there is no direct mapping from the numerical image features to the semantic meanings in the images, and (2) subjectively, given the same region, there are different corresponding semantic concepts, depending on different context and/or different user interpretations. This observation justifies the need to discover the hidden semantic concepts that is a key step toward effective image retrieval.

In this chapter, we propose a probabilistic approach to addressing the hidden semantic concept discovery. A region-based sparse but uniform image representation scheme is developed (unlike the block-based uniform representation in (Zhu et al., 2002), region-based representation is more effective for image mining and retrieval due to the fact that humans pay more attention to objects than blocks in an image), which facilitates the indexing scheme based on a region-image-concept probabilistic model with validated assumptions. This model has a solid statistical foundation and is intended for the objective of semantics-intensive image retrieval. To describe the semantic concepts hidden in the region and image distributions of a database, the Expectation-Maximization (EM) technique is used. With a derived iterative procedure, the posterior probabilities of each region in an image for the hidden semantic concepts are quantitatively obtained, which act as the basis for the *semantic similarity* measure for image mining and retrieval. Therefore, the effectiveness is improved as the similarity measure is based on the discovered semantic concepts, which are more reliable than the region features used in most of the existing systems in the literature. Figure 57.3 shows the architecture of the proposed approach.

Different from the models reviewed above, the model and the approach we propose and present here do not require training data; we formulate a generative model to discover the clusterings in a probabilistic scheme by unsupervised learning. In this model, the regions and images are connected through a hidden layer — the concept layer, which constitutes the basis of the image similarity measures. In addition, users' relevance feedback is incorporated into the model fitting procedure such that the subjectivity in image mining and retrieval is addressed explicitly and the model fitting is customized toward users' querying needs.

57.3.2 Region Based Image Representation

In the proposed approach, the query image and images in a database are first segmented into homogeneous color-texture regions. Then representative properties are extracted for every region by incorporating multiple features, specifically, color, texture, and shape properties. Based on the extracted regions, a visual token catalog is generated to explore and exploit the content similarities of the regions, which facilitates the indexing and mining scheme based on the region-image-concept probabilistic model elaborated in Section 57.3.3.

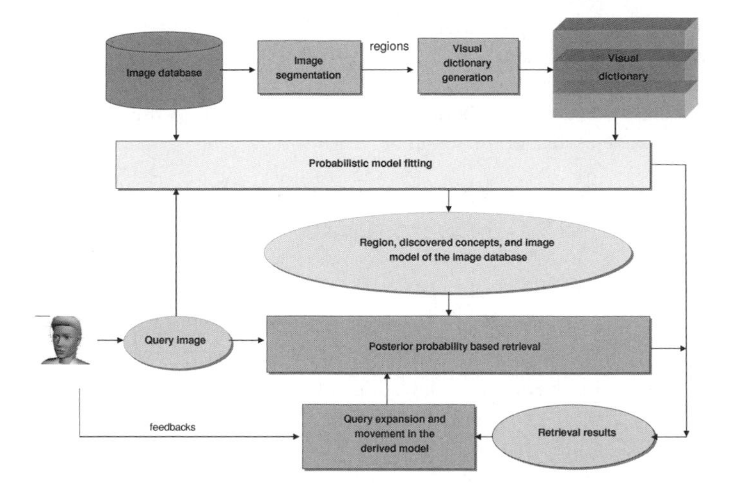

Fig. 57.3. The architecture of the latent semantic concept discovery based image data mining and retrieval approach. Reprint from (Zhang & Zhang, 2007) ©2007 IEEE Signal Processing Society Press.

Image Segmentation

To segment an image, the system first partitions the image into blocks of 4 by 4 pixels to compromise between the texture effectiveness and the computation time. Then a feature vector consisting of nine features from each block is extracted. Three of the features are average color components in the 4 by 4 pixel size block; we use the *LAB* color space due to its desired property that the perceptual color difference is proportional to the numerical difference. The other six features are the texture features extracted using wavelet analysis.

To extract texture information of each block, we apply a set of Gabor filters (Manjunath & Ma, 1996), which are shown to be effective for image indexing and retrieval (Ma & Manjunath, 1995), to the block to measure the response. The Gabor filters measure the two-dimensional wavelets. The discretization of a two-dimensional wavelet applied to the blocks is given by

$$W_{mlpq} = \int\int I(x,y)\psi_{ml}(x - p\triangle x, y - q\triangle y)dxdy \tag{57.1}$$

where I denotes the processed block; $\triangle x$ and $\triangle y$ denote the spatial sampling rectangle; p, q are image positions; and m, l specify the scale and orientation of the wavelets. The base function $\psi_{ml}(x,y)$ is given by

$$\psi_{ml}(x,y) = a^{-m}\psi(\widetilde{x},\widetilde{y}) \tag{57.2}$$

where

$$\widetilde{x} = a^{-m}(x\cos\theta + y\sin\theta)$$
$$\widetilde{y} = a^{-m}(-x\sin\theta + y\cos\theta)$$

denote a dilation of the mother wavelet (x,y) by a^{-m}, where a is the scale parameter, and a rotation by $\theta = l \times \triangle\theta$, where $\triangle\theta = 2\pi/V$ is the orientation sampling period; V is the number of orientation sampling intervals.

In the frequency domain, with the following Gabor function as the mother wavelet, we use this family of wavelets as our filter bank:

$$\begin{aligned} \Psi(u,v) &= \exp\{-2\pi^2(\sigma_x^2 u^2 + \sigma_y^2 v^2)\} \otimes \delta(u-W) \\ &= \exp\{-2\pi^2(\sigma_x^2 (u-W)^2 + \sigma_y^2 v^2)\} \\ &= \exp\{-\frac{1}{2}(\frac{(u-W)^2}{\sigma_u^2} + \frac{v^2}{\sigma_v^2})\} \end{aligned} \tag{57.3}$$

where $\otimes$ is a convolution symbol, $\delta(\cdot)$ is the impulse function, $\sigma_u = (2\pi\sigma_x)^{-1}$, and $\sigma_v = (2\pi\sigma_y)^{-1}$; σ_x and σ_y are the standard deviations of the filter along the x and y directions, respectively. The constant W determines the frequency bandwidth of the filters.

Applying the Gabor filter bank to the blocks, for every image pixel (p,q), in U (the number of scales in the filter bank) by V array of responses to the filter bank, we only need to retain the magnitudes of the responses:

$$F_{mlpq} \doteq |W_{mlpq}| \quad m = 0,\ldots,U-1,\ l = 0,\ldots V-1 \tag{57.4}$$

Hence, a texture feature is represented by a vector, with each element of the vector corresponding to the energy in a specified scale and orientation sub-band w.r.t. a Gabor filter. In the implementation, a Gabor filter bank of 3 orientations and 2 scales is used for each image in the database, resulting in a 6-dimensional feature vector (i.e., 6 means for $|W_{ml}|$) for the texture representation.

After we obtain feature vectors for all blocks, we perform normalization on both color and texture features such that the effects of different feature ranges are eliminated. Then a k-means based segmentation algorithm, similar to that used in (Chen & Wang, 2002), is applied to clustering the feature vectors into several classes, with each class corresponding to one region in the segmented image.

Figure 57.4 gives four examples of the segmentation results of images in the database, which show the effectiveness of the segmentation algorithm employed.

After the segmentation, the edge map is used with the water-filling algorithm (Zhou et al., 1999) to describe the shape feature for each region due to its reported effectiveness and efficiency for image mining and retrieval (Moghaddam et al., 2001). A 6-dimensional shape feature vector is obtained for each region by incorporating the statistics defined in (Zhou et al., 1999), such as the filling time histogram and the fork count histogram. The mean of the color-texture features of all the blocks in each region is determined to combine with the corresponding shape feature as the extracted feature vector of the region.

Visual Token Catalog

Since the region features $f \in \mathbb{R}^n$, it is necessary to perform regularization on the region property set such that they can be indexed and mined efficiently. Considering that many regions from different images are very similar in terms of the features, vector quantization (VQ) techniques are required to group similar regions together. In the proposed approach, we create a visual token catalog for region properties to represent the visual content of the regions. There are three advantages to creating such a visual token catalog. First, it improves mining and retrieval robustness by tolerating minor variations among visual properties. Without the visual

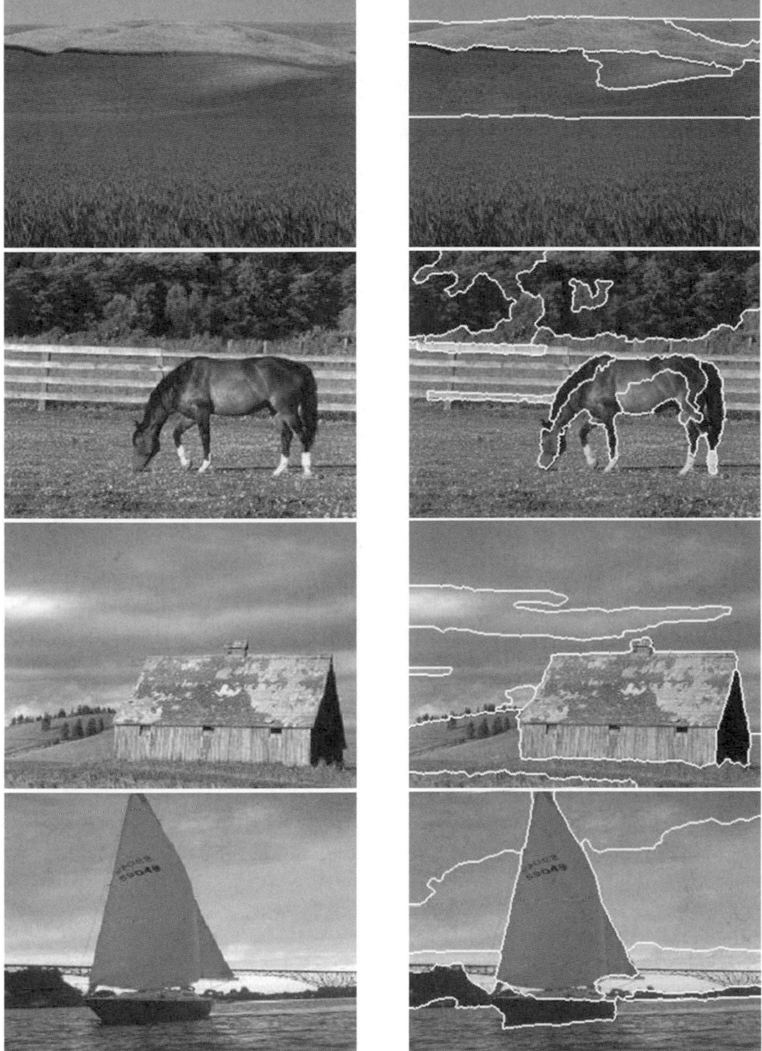

Fig. 57.4. The segmentation results. Left column shows the original images; right column shows the corresponding segmented images with the region boundary highlighted.

token catalog, since very few feature values are exactly shared by different regions, we would have to consider feature vectors of all the regions in the database. This makes it not effective to compare the similarity among regions. However, based on the visual token catalog created, low-level features of regions are quantized such that images can be represented in a way resistant to perception uncertainties (Chen & Wang, 2002). Second, the region-comparison efficiency is significantly improved by mapping the expensive numerical computation of the distances between region features to the inexpensive symbolic computation of the differences between "code words" in the visual token catalog. Third, the utilization of the visual token catalog reduces the storage space without sacrificing the accuracy.

We create the visual token catalog for region properties by applying the Self-Organization Map (SOM) (Kohonen et al., 2000) learning strategy. SOM is ideal for this problem, as it projects the high-dimensional feature vectors to a 2-dimensional plane through mapping similar features together while separating different features at the same time. The SOM learning algorithm we have used is competitive and unsupervised. The nodes in a 2-dimensional array become specifically tuned to various classes of input feature patterns in an orderly fashion.

A procedure is designed to create "code words" in the dictionary. Each "code word" represents a set of visually similar regions. The procedure follows 4 steps:

1. Performing the Batch SOM learning (Kohonen et al., 2000) algorithm on the region feature set to obtain the visualized model (node status) displayed on a 2-dimensional plane map. The distance metric used is Euclidean for its simplicity.
2. Regarding each node as a "pixel" in the 2-dimensional plane map such that the map becomes a binary lattice with the value of each pixel i defined as follows:

$$p(i) = \begin{cases} 0 \text{ if } count(i) \geq t \\ 1 \text{ else} \end{cases}$$

   where $count(i)$ is the number of features mapped to node i and the constant t is a preset threshold. Pixel value 0 denotes the objects, while pixel value 1 denotes the background.
3. Performing the morphological erosion operation (Castleman, 1996) on the resulting lattice to make sparse connected objects in the image disjointed. The size of the erosion mask is determined to be the minimum to make two sparsely connected objects separated.
4. With connected component labeling (Castleman, 1996), we assign each separated object a unique ID, a "code word". For each "code word", the mean of all the features associated with it is determined and stored. All "code words" constitute the visual token catalog to be used to represent the visual properties of the regions.

Figure 57.5 illustrates this procedure on a portion of the map we have obtained.

Simple yet effective Euclidean distance is used in the SOM learning to determine the "code word" to which each region belongs. The proof of the convergence of the SOM learning process in the 2-dimensional plane map is given in (Kohonen, 2001). The details about the selection of the parameters are also covered in (Kohonen, 2001). Each labeled component represents a region feature set among which the intra-distance is low. The extent of similarity in each "code word" is controlled by the parameters in the SOM algorithm and the threshold t. With this procedure, the number of the "code words" is adaptively determined and the similarity-based feature grouping is achieved. The experiments reported in Section 57.3.6 show that the visual token catalog created captures the clustering characteristics existing in the feature set well. We note that the threshold t is highly correlated to the number of the "code words" generated; it is determined empirically by balancing the efficiency and the accuracy.

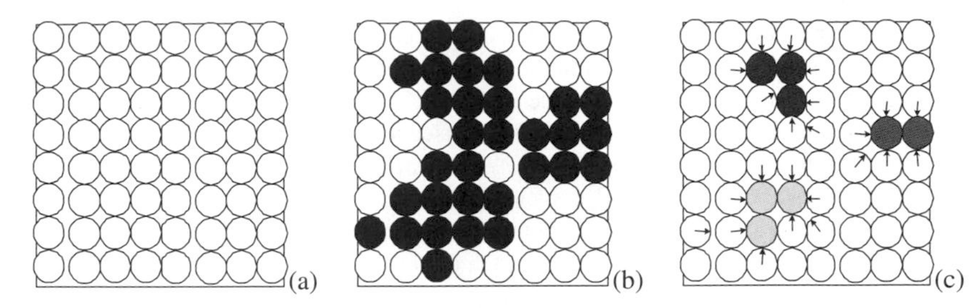

Fig. 57.5. Illustration of the procedure: (a) the initial map; (b) the binary lattice obtained after the SOM learning is converged; (c) the labeled object on the final lattice. The arrows indicate the objects that the corresponding nodes belong to. Reprint from (Zhang & Zhang, 2007) ©2007 IEEE Signal Processing Society Press.

We discuss the issue of choosing the appropriate number of the "code words" in the visual token catalog in Section 57.3.6. Figure 57.6 shows the process of the generation of the visual token catalog. Each rounded rectangle in the third column of the figure is one "code word" in the dictionary.

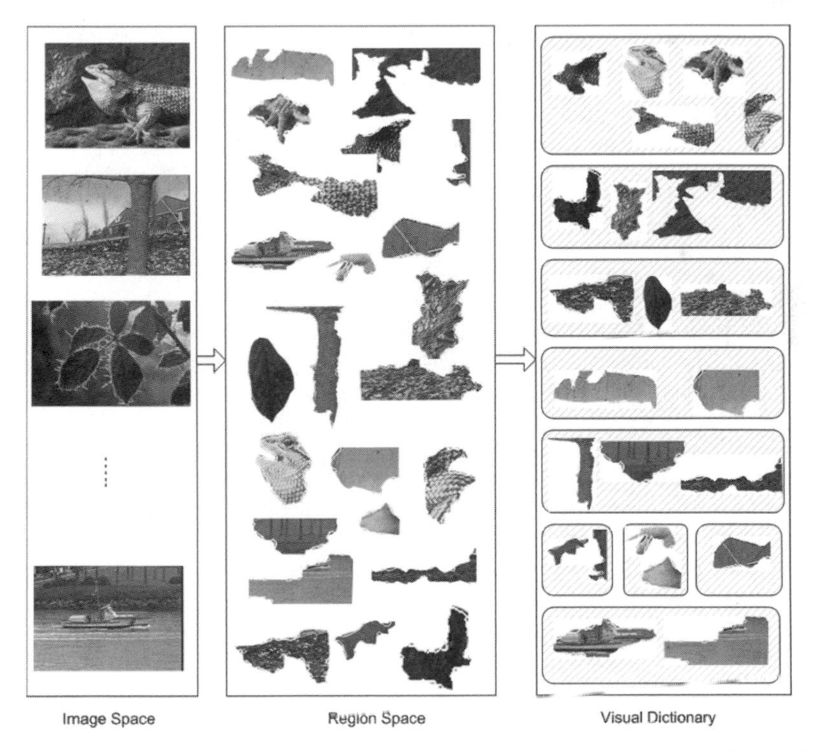

Fig. 57.6. The process of the generation of the visual token catalog. Reprint from (Zhang & Zhang, 2007) ©2007 IEEE Signal Processing Society Press and from (Zhang & Zhang, 2004a) ©2004 IEEE Computer Society Press.

For each region of an image in the database, the "code word" that it is associated with is identified and the corresponding index in the visual token catalog is stored, while the original feature of this region is discarded. For the region of a new image, the closest entry in the dictionary is found and the corresponding index is used to replace its feature. In the rest of this chapter, we use the terminologies *region* and "*code word*" interchangeably; they both denote an entry in the visual token catalog equivalently.

Based on the visual token catalog, each image is represented in a uniform vector model. In this representation, an image is a vector with each dimension corresponding to a "code word". More formally, the uniform representation $\mathbf{I}_u$ of an image I is a vector $\mathbf{I}_u = \{w_1, w_2, \ldots, w_M\}$, where M is the number of the "code words" in the visual token catalog. For a "code word" $C_i, 1 \leq i \leq M$, if there exists a region R_j of I that corresponds to it, then $w_i = W_{Rj}$ for $\mathbf{I}_u$, where W_{Rj} is the number of the occurrences of R_j in the image I; otherwise, $w_i = 0$. This uniform representation is sparse, for an image usually contains a few regions compared with the number of the "code words" in the visual token catalog. Based on this representation of all the images, the database is modeled as a $M \times N$ "code word"-image matrix which records the occurrences of every "code word" in each image, where N is the number of the images in the database.

57.3.3 Probabilistic Hidden Semantic Model

To achieve the automatic semantic concept discovery, a region-based probabilistic model is constructed for the image database with the representation of the "code word"-image matrix. The probabilistic model is analyzed by the Expectation-Maximization (EM) technique (Dempster et al., 1977) to discover the latent semantic concepts, which act as a basis for effective image mining and retrieval via the concept similarities among images.

Probabilistic Database Model

With a uniform "code word" vector representation for each image in the database, we propose a probabilistic model. In this model, we assume that the specific (region, image) pairs are known i.i.d. samples from an unknown distribution. We also assume that these samples are associated with an unobserved *semantic concept* variable $z \in Z = \{z_1, \ldots, z_K\}$, where K is the number of concepts to be discovered. Each observation of one region ("code word") $r \in R = \{r_1, \ldots, r_M\}$ in an image $g \in G = \{g_1, \ldots, g_N\}$ belongs to one concept class z_k. To simplify the model, we have two further assumptions. First, the observation pairs (r_i, g_j) are generated independently. Second, the pairs of random variables (r_i, g_j) are conditionally independent given the respective hidden concept z_k, i.e., $P(r_i, g_j|z_k) = P(r_i|z_k)P(g_j|z_k)$. Intuitively, these two assumptions are reasonable, which are further validated by the experimental evaluations. The region and image distribution may be treated as a randomized data generation process, described as follows:

- Choose a concept with probability $P(z_k)$;
- Select a region $r_i \in R$ with probability $P(r_i|z_k)$; and
- Select an image $g_j \in G$ with probability $P(g_j|z_k)$.

As a result, one obtains an observed pair (r_i, g_j), while the concept variable z_k is discarded.

Based on the theory of the generative model (Mclachlan & Basford, 1988), the above process is equivalent to the following:

- Select an image g_j with probability $P(g_j)$;

- Select a concept z_k with probability $P(z_k|g_j)$;
- Generate a region r_i with probability $P(r_i|z_k)$.

Translating this process into a joint probability model results in the expression

$$P(r_i,g_j) = P(g_j)P(r_i|g_j) = P(g_j)\sum_{k=1}^{K} P(r_i|z_k)P(z_k|g_j) \quad (57.5)$$

Inverting the conditional probability $P(z_k|g_j)$ in Equation 57.5 with the application of Bayes' rule results in

$$P(r_i,g_j) = \sum_{k=1}^{K} P(z_k)P(r_i|z_k)P(g_j|z_k) \quad (57.6)$$

Following the likelihood principle, one determines $P(z_k)$, $P(r_i|z_k)$, and $P(g_j|z_k)$ by the maximization of the log-likelihood function

$$\mathscr{L} = \log P(R,G) = \sum_{i=1}^{M}\sum_{j=1}^{N} n(r_i,g_j)\log P(r_i,g_j) \quad (57.7)$$

where $n(r_i,g_j)$ denotes the number of the regions r_i that occurred in image g_j. From Equations 57.7 and 57.5 we derive that the model is a statistical mixture model (Mclachlan & Basford, 1988), which can be resolved by applying the EM technique (Dempster et al., 1977).

Model Fitting with EM

One powerful procedure for maximum likelihood estimation in hidden variable models is the EM method (Dempster et al., 1977). EM alternates in two steps iteratively: (i) an expectation (E) step where posterior probabilities are computed for the hidden variable z_k, based on the current estimates of the parameters, and (ii) a maximization (M) step, where parameters are updated to maximize the expectation of the complete-data likelihood $\log P(R,G,Z)$ for the given posterior probabilities computed in the previous E-step.

Applying Bayes' rule with Equation 57.5, we determine the posterior probability for z_k under (r_i,g_j):

$$P(z_k|r_i,g_j) = \frac{P(z_k)P(g_j|z_k)P(r_i|z_k)}{\sum_{k'=1}^{K} P(z_{k'})P(g_j|z_{k'})P(r_i|z_{k'})} \quad (57.8)$$

The expectation of the complete-data likelihood $\log P(R,G,Z)$ for the estimated $P(Z|R,G)$ derived from Equation 57.8 is

$$E\{\log P(R,G,Z)\} = \sum_{(i,j)=1}^{K}\sum_{i=1}^{M}\sum_{j=1}^{N} n(r_i,g_j)\log[P(z_{i,j})P(g_j|z_{i,j})P(r_i|z_{i,j})]P(Z|R,G) \quad (57.9)$$

where

$$P(Z|R,G) = \prod_{m=1}^{M}\prod_{n=1}^{N} P(z_{m,n}|r_m;g_n)$$

In Equation 57.9 the notation $z_{i,j}$ is the concept variable that is associated with the region-image pair (r_i,g_j). In other words, (r_i,g_j) belongs to concept z_t where $t = (i,j)$.

With the normalization constraint $\sum_{(i,j)=1}^{K} P(z_{i,j}|r_i,g_j) = 1$, Equation 57.9 further becomes:

$$E\{\log P(R,G,Z)\} = \sum_{l=1}^{K}\sum_{i=1}^{M}\sum_{j=1}^{N} n(r_i,g_j)\log[P(r_i|z_l)P(g_j|z_l)]P(z_l|r_i,g_j) + \\ + \sum_{l=1}^{K}\sum_{i=1}^{M}\sum_{j=1}^{N} n(r_i,g_j)\log[P(z_l)]P(z_l|r_i,g_j) \tag{57.10}$$

Maximizing Equation 57.10 with Lagrange multipliers to $P(z_l)$, $P(r_u|z_l)$, and $P(g_v|z_l)$, respectively, under the following normalization constraints

$$\sum_{k=1}^{K} P(z_k) = 1 \tag{57.11}$$

$$\sum_{k=1}^{K} P(z_k|r_i,g_j) = 1 \tag{57.12}$$

$$\sum_{i=1}^{M} P(r_i|z_l) = 1 \tag{57.13}$$

for any r_i, g_j, and z_l, the parameters are determined as

$$P(z_k) = \frac{\sum_{i=1}^{M}\sum_{j=1}^{N} n(r_i,g_j)P(z_k|r_i,g_j)}{\sum_{i=1}^{M}\sum_{j=1}^{N} u(r_i,g_j)} \tag{57.14}$$

$$P(r_u|z_l) = \frac{\sum_{j=1}^{N} n(r_u,g_j)P(z_l|r_u,g_j)}{\sum_{i=1}^{M}\sum_{j=1}^{N} u(r_i,g_j)P(z_l|r_i,g_j)} \tag{57.15}$$

$$P(g_v|z_l) = \frac{\sum_{i=1}^{M} n(r_i,g_v)P(z_l|r_i,g_v)}{\sum_{i=1}^{M}\sum_{j=1}^{N} u(r_i,g_j)P(z_l|r_i,g_j)} \tag{57.16}$$

Alternating Equation 57.8 with Equations 57.14–57.16 defines a convergent procedure that approaches a local maximum of the expectation in Equation 57.10. The initial values for $P(z_k)$, $P(g_j|z_k)$, and $P(r_i|z_k)$ are set to be the same as if the distributions of $P(Z)$, $P(G|Z)$, and $P(R|Z)$ are the uniform distributions; in other words, $P(z_k) = 1/K$, $P(r_i|z_k) = 1/M$, and $P(g_j|z_k) = 1/N$. We have found in the experiments that different initial values only affect the number of iterative steps to the convergence but have no effects on the converged values of them.

Estimating the Number of Concepts

The number of concepts, K, must be determined in advance to initiate the EM model fitting. Ideally, we would like to select the value of K that best represents the number of the semantic classes in the database. One readily available notion of the goodness of the fitting is the log-likelihood. Given this indicator, we apply the Minimum Description Length (MDL) principle (Rissanen, 1978, Rissanen, 1989) to select the best value of K. This can be operationalized as follows (Rissanen, 1989): choose K to maximize

$$\log(P(R,G)) - \frac{m_K}{2}\log(MN) \tag{57.17}$$

where the first term is expressed in Equation 57.7 and m_K is the number of the free parameters needed for a model with K mixture components. In the case of the proposed probabilistic model, we have

$$m_K = (K-1)+K(M-1)+K(N-1) = K(M+N-1)-1$$

As a consequence of this principle, when models using two values of K fit the data equally well, the simpler model is selected. In the database used in the experiments reported in Section 57.3.6, K is determined through maximizing Equation 57.17.

57.3.4 Posterior Probability Based Image Mining and Retrieval

Based on the probabilistic model, we can derive the posterior probability of each image in the database for every discovered concept by applying Bayes' rule as

$$P(z_k|g_j) = \frac{P(g_j|z_k)P(z_k)}{P(g_j)} \tag{57.18}$$

which can be determined using the estimations in Equations 57.14–57.16. The posterior probability vector $P(Z|g_j) = [P(z_1|g_j), P(z_2|g_j), \ldots, P(z_K|g_j)]^T$ is used to quantitatively describe the semantic concepts associated with the image g_j. This vector can be treated as a representation of g_j (which originally has a representation in the M-dimensional "code word" space) in the K-dimensional *concept space* determined using the estimated $P(z_k|r_i, g_j)$ in Equation 57.8.

For each query image, after obtaining the corresponding "code words" as described in Section 57.3.2, we attain its representation in the discovered concept space by substituting it in the EM iteration derived in Section 57.3.3. The only difference is that $P(r_i|z_k)$ and $P(z_k)$ are fixed to be the values we have obtained for the whole database modeling (which are obtained in the indexing phase, i.e., to determine the concept space representation of every image in the database).

In designing a region-based image mining and retrieval methodology, there are two characteristics of the region representation that must be taken into consideration:

1. The number of the segmented regions in one image is normally small.
2. Not all regions in one image are semantically relevant to a given image; some are unrelated or even non-relevant; which regions are relevant or irrelevant depends on the user's querying subjectivity.

Incorporating the "code words" corresponding to unrelated or non-relevant regions would hurt the mining or retrieval accuracy because the occurrences of these regions in one image tend to "fool" the probabilistic model such that erroneous concept representations would be generated. To address the two characteristics in image mining and retrieval explicitly, we employ the relevance feedback for the similarity measurement in the concept space. Relevance feedback has been demonstrated as great potential to capture users' querying subjectivity both in text retrieval and in image retrieval (Vasconcelos & Lippman, 2000,Rui et al., 1997). Consequently, a mining and retrieval algorithm based on the relevance feedback strategy is designed to integrate the probabilistic model to deliver a more effective mining and retrieval performance.

In the algorithm, we move the query point in the "code word" token space toward the good example points (the relevant images labeled by the user) and away from the bad example points (the irrelevant images labeled by the user) such that the region representation has more supports to the probabilistic model. At the same time, the query point is expanded with the "code words" of the labeled relevant images. On the other hand, we construct a negative example "code word" vector by applying a similar vector moving strategy such that the constructed negative vector lies near the bad example points and away from the good example points. The vector moving strategy uses a form of Rocchio's formula (Rocchio, 1971). Rocchio's formula for relevance feedback and feature expansion has proven to be one of the best iterative optimization techniques in the field of information retrieval. It is frequently used to estimate the "optimal query" in relevance feedback for sets of relevant documents D_R and irrelevant documents D_I given by the user. The formula is

$$Q' = \alpha Q + \beta(\frac{1}{N_R} \sum_{j \in D_R} D_j) - \gamma(\frac{1}{N_I} \sum_{j \in D_I} D_j) \tag{57.19}$$

where α, β, and γ are suitable constants; N_R and N_I are the number of documents in D_R and D_I, respectively; and Q' is the updated query of the previous query Q.

In the algorithm, based on the vector moving strategy and Rocchio's formula, in each iteration a modified query vector *pos* and a constructed negative example *neg* are computed; their representations in the discovered concept space are obtained and their similarities to each image in the database are measured through the cosine metric (Baeza-Yates & Ribeiro-Neto, 1999) of the corresponding vectors in the concept space, respectively. The retrieved images are ranked based on the similarity to *pos* as well as the dissimilarity to *neg*. The algorithm is described in Algorithm 3.

Algorithm 1: A semantic concept mining based retrieval algorithm

1 **Input:** q, "code word" vector of the query image
2 **Output:** Images retrieved for the query image q
3 **Method:**

1: Plug q to the model to compute the vector $P(Z|q)$;
2: Retrieve and rank images based on the cosine similarity measure of the vectors $P(Z|q)$ and $P(Z|g)$ of each image in the database;
3: $rs = \{rel_1, rel_2, \ldots, rel_a\}$, where rel_i is a "code word" vector of each image labeled as relevant by the user on the retrieved result;
4: $is = \{ire_1, ire_2, \ldots, ire_b\}$, where ire_j is a "code word" vector of each image labeled as irrelevant by the user on the retrieved result;
5: $pos = \alpha q + \beta(\frac{1}{a}\sum_{i=1}^{a} rel_i) - \gamma(\frac{1}{b}\sum_{j=1}^{b} ire_j)$;
6: $neg = \alpha(\frac{1}{b}\sum_{j=1}^{b} ire_j) - \gamma(\frac{1}{a}\sum_{i=1}^{a} rel_i)$;
7: **for** $k = 1$ to K **do**
8: Determine $P(z_k|pos)$ and $P(z_k|neg)$ with EM and Equation 57.18;
9: **end for**
10: $n = 1$;
11: **while** $n <= N$ **do**
12: $sim1(g_n) = \frac{P(Z|pos) \bullet P(Z|g_n)}{\|P(Z|pos)\| \|P(Z|g_n)\|}$;
13: $sim2(g_n) = \frac{P(Z|neg) \bullet P(Z|g_n)}{\|P(Z|neg)\| \|P(Z|g_n)\|}$;
14: **if** $(sim1(g_n) > sim2(g_n))$ **then**
15: $sim(g_n) = sim1(g_n) - sim2(g_n)$;
16: **else**
17: $sim(g_n) = 0$;
18: **end if**
19: Rank the images in the database based on $sim(g_n)$;
20: **end while**

We use the cosine metric to compute $sim1(\bullet)$ and $sim2(\bullet)$ in Algorithm 3 because the posterior probability vectors are the basis for the similarity measure in this proposed approach. The vectors are uniform, and the value of each component in the vectors is between 0 and 1. The cosine similarity is effective and ideal for measuring the similarity for the space composed of these kinds of vectors. The experiments reported in Section 57.3.6 show the effectiveness of the cosine similarity measure. At the same time, we note that Algorithm 3 itself is orthogonal to the selections of similarity measure metrics. The parameters α, β, and γ in Algorithm 3 are assigned a value of 1.0 in the current implementation of the prototype system for the sake of simplicity. However, other values may be used to emphasize the different weights between good sample points and bad sample points.

57.3.5 Approach Analysis

It is worth comparing the proposed probabilistic model and the fitting methodology with the existing region based statistical clustering methods in the image mining and retrieval literature, such as (Zhang & Zhang, 2004b, Chen et al., 2003). In the clustering methods, one typically associates a class variable with each image or each region in the database based on specific similarity metrics cast. One fundamental problem overlooked in such methods is that the semantic concepts of a region are typically not entirely determined by the features of the region itself; rather, they are dependent upon and affected by the contextual environment around the region in the image. In other words, a region in a different context in an image may convey a different concept. It is also noticeable that the degree of a specific region associated with several semantic concepts varies with different contextual region co-occurrences in an image. For example, it is likely that the *sand* "code word" conveys the concept of *beach* when it co-occurs in the context of the *water*, *sky*, and *people* "code words"; on the other hand, it becomes likely that the same *sand* "code word" conveys the concept of *African* with a high probability when it co-occurs in the context of the *plant* and *black* "code words". Wang et al (Wang et al., 2001) attempted to alleviate the effect caused by this problem by using integrated region matching to incorporate similarity between two images for all their region pairs; this matching scheme, however, is heuristic such that it is impossible for a more rigorous analysis.

The probabilistic model we have described addresses these problems quantitatively and analytically in an optimal framework. Given a region in an image the conditional probability of each concept and the conditional probability of each image in a concept are iteratively determined to fit the model representing the database as formulated in Equations 57.8 and 57.16. Since the EM technique always converges to a local optimality, from the experiments reported in Section 57.3.6, we have found that the local optimum is satisfactory for typical image data mining and retrieval applications. The effectiveness of this methodology in real image databases is demonstrated in the experimental analysis presented in Section 57.3.6. To find the global maximum is computationally intractable for a large-scale database, and the advantage of such model fitting compared to the model fitting obtained through this proposed approach is not obvious and is under further investigation.

With the proposed probabilistic model, we are able to concurrently obtain $P(z_k|r_i)$ and $P(z_k|g_j)$ such that both regions and images have an interpretation in the concept space simultaneously, while typical image clustering based approaches, such as (Jing et al., 2004), do not have this flexibility. Since in the proposed scheme, every region and/or image may be represented as a weighted sum of the components along the discovered concept axes, the proposed model acts as a factoring analysis (Mclachlan & Basford, 1988), yet the same model offers important advantages, such as that each weight has a clear probabilistic meaning and the factoring is two-fold, i.e., both regions and images in the database have probabilistic representations with the discovered concepts.

Another advantage of the proposed methodology is its capability to reduce the dimensionality. The image similarity comparison is performed in a derived K-dimensional concept space Z instead of in the original M-dimensional "code word" token space R. Note that typically $K << M$, as has been demonstrated in the experiments reported in Section 57.3.6. The derived subspace represents the hidden semantic concepts conveyed by the regions and the images, while the noise and all the non-intrinsic information are discarded in the dimensionality reduction, which makes the semantic comparison of regions and images more effective and efficient. The coordinates in the concept space for each image as well as for each region are determined by automatic model fitting. The computation requirement in the lower-dimensional concept space is reduced as compared with that required in the original "code word" space.

Algorithm 3 integrates the posterior probability of the discovered concepts with the query expansion and the query vector moving strategy in the "code word" token space. Consequently, the accuracy of the representation of the semantic concepts of a user's query is enhanced in the "code word" token space, which also improves the accuracy of the position obtained for the query image in the concept space. Moreover, the constructed negative example *neg* improves the discriminative power of the probabilistic model. Both the similarity to the modified query representation and the dissimilarity to the constructed negative example in the concept space are employed.

57.3.6 Experimental Results

We have implemented the approach in a prototype system on a platform of a Pentium IV 2.0 GHz CPU and 256 MB memory. The interface of the system is shown in Figure 57.13. The following reported evaluations are performed on a general-purpose color image database containing 10,000 images from the COREL collection with 96 semantic categories. Each semantic category consists of 85–120 images. In Table 57.1, exemplar categories in the database are provided. We note that the category information in the COREL collection is only used to ground-truth the evaluation, and we do not make use of this information in the indexing, mining, and retrieval procedures. Figure 57.7 shows a few examples of the images in the database.

To evaluate the image retrieval performance, 1,500 images are randomly selected from all the categories as the query set. The relevancy of the retrieved images is subjectively examined by users. The ground truth used in the mining and retrieval experiments is the COREL category label if the query image is in the database. If the query image is a new image outside the database, users' specified relevant images in the mining and retrieval results are used to calculate the mining and retrieval accuracy statistics. Unless otherwise noted, the default results of the experiments are the averages of the top 30 returned images for each of the 1,500 queries.

In the experiments, the parameters of the image segmentation algorithm (Wang et al., 2001) are adjusted with the consideration of the balance of the depiction detail and the computation complexity such that there is an average of 8.3207 regions in each image. To determine the size of the visual token catalog, different numbers of the "code words" are selected and evaluated. The average precisions (without the query expansion and movement) within the top 20, 30, and 50 images, denoted as P(20), P(30), and P(50), respectively, are shown in Figure 57.8. It indicates that the general trend is that the larger the visual token catalog size, the higher the mining and retrieval accuracy. However, a larger visual token catalog size means a larger number of image feature vectors, which implies a higher computation complexity in the process of the hidden semantic concept discovery. Also, a larger visual token catalog leads to a larger storage space. Therefore, we use 800 as the number of the "code words", which

Table 57.1. Examples of the 96 categories and their descriptions. Reprint from (Zhang & Zhang, 2007) ©2007 IEEE Signal Processing Society Press.

ID	Category description
1	reptile, animal, rock
2	Britain, royal events, queen, prince, princess
3	Africa, people, landscape, animal
4	European, historical building, church
5	woman, fashion, model, face, cloth
6	hawk, sky
7	New York City, skyscrapers, skyline
8	mountain, landscape
9	antique, craft
10	Easter egg, decoration, indoor, man-made
11	waterfall, river, outdoor
12	poker cards
13	beach, vacation, sea shore, people
14	castle, grass, sky
15	cuisine, food, indoor
16	architecture, building, historical building
..	......

Fig. 57.7. Sample images in the database. The images in each column are assigned to one category. From left to right, the categories are Africa rural area, historical building, waterfalls, British royal event, and model portrait, respectively.

corresponds to the first turning point in Figure 57.8. Since there are a total of 83,307 regions in the database, on average each "code word" represents 104.13 regions.

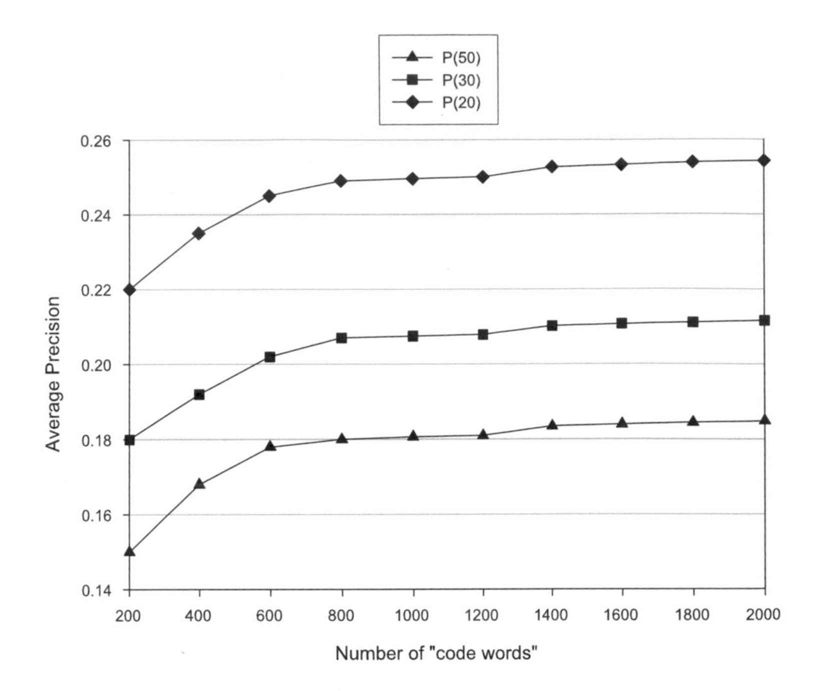

Fig. 57.8. Average precision (without the query expansion and movement) for different sizes of the visual token catalog. Reprint from (Zhang & Zhang, 2007) ©2007 IEEE Signal Processing Society Press and from (Zhang & Zhang, 2004a) ©2004 IEEE Computer Society Press.

Applying the method of estimating the number of the hidden concepts described in Section 57.3.3, the number of the concepts is determined to be 132. Performing the EM model fitting, we have obtained the conditional probability of each "code word" to every concept, i.e., $P(r_i|z_k)$. Manual examination of the visual content of the region sets corresponding to the top 10 highest "code words" in every semantic concept reveals that these discovered concepts indicate semantic interpretations, such as "people", "building", "outdoor scenery", "plant", and "automotive race". Figure 57.9 shows several exemplar concepts discovered and the top regions corresponding to $P(r_i|z_k)$ obtained.

In terms of the computational complexity, despite the iterative nature of EM, the computing time for the model fitting at $K = 132$ is acceptable (less than 1 second). The average number of iterations upon convergence for one image is less than 5.

We give an example for discussion. Figure 57.10 shows one image, *Im*, belonging to the "medieval building" category in the database. *Im* (i.e., Figure 57.10(a)) has 6 "code words" associated. Each "code word" is presented using a unique color graphically in Figure 57.10(b). For the sake of discussion, the indices for these "code words" are assigned to be 1–6, respectively.

Figure 57.11 shows the $P(z_k|r_i,Im)$ for each "code word" r_i (represented as a different color) and the posterior probability $P(z_k|Im)$ after the first iteration and the last iteration in the

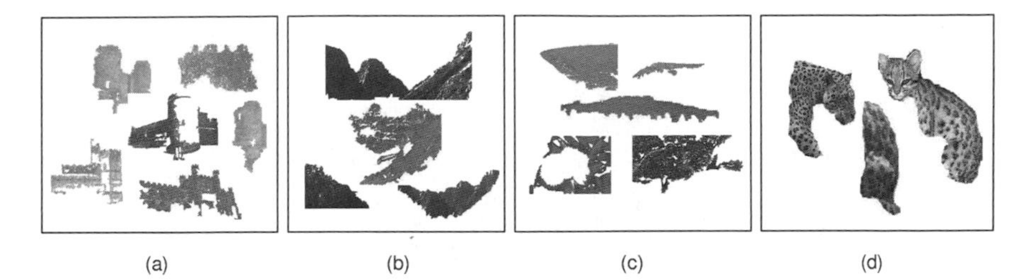

Fig. 57.9. The regions with the top $P(r_i|z_k)$ to the different concepts discovered. (a) "castle"; (b) "mountain"; (c) "meadow and plant"; (d) "cat". Reprint from (Zhang & Zhang, 2007) ©2007 IEEE Signal Processing Society Press.

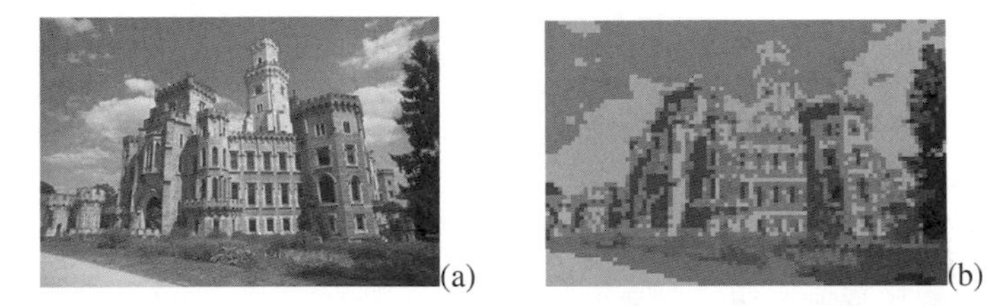

Fig. 57.10. Illustration of one query image in the "code word" space. (a) Image Im; (b) "code word" representation. Reprint from (Zhang & Zhang, 2007) ©2007 IEEE Signal Processing Society Press.

course of the EM model fitting. Here the 4 concepts with highest $P(z_k|Im)$ are shown. From left to right in Figure 57.11, they represent "plant", "castle", "cat", and "mountain", respectively, interpreted through manual examination. As is seen in the figure, the "castle" concept has indeed the highest weight after the first iteration; nevertheless, the other three concepts still account for more than half of the probability. The probability distribution changes after several EM iterations, since the proposed probabilistic model incorporates co-occurrence patterns between the "code words"; i.e., $P(z_k|r_i)$ is not only related to one "code word" (r_i) but is also related to all the co-occurring "code words" in the image. For example, although "code word" 2, which accounts for "meadow", has higher fitness in the concept "plant" after the first iteration, the context of the other regions in image Im increases the probability that this "code word" is related to the concept "castle" and decreases its probability related to "plant" as well.

Figure 57.12 shows the similar plot to Figure 57.11 except that we apply the relevance feedback based query expansion and moving strategy to image Im as described in the Algorithm 3. The "code word" vector of image Im is expanded to contain 10 "code words". Compared with Figure 57.11, it is clear that with the expansion of the relevant "code words" to Im and the query moving strategy toward the relevant image set, the posterior probabilities favoring the concept "castle" increase while the posterior probabilities favoring other concepts decrease substantially, resulting in an improved mining and retrieval precision, accordingly.

To show the effectiveness of the probabilistic model in image mining and retrieval, we have compared the accuracy of this methodology with that of UFM (Chen & Wang, 2002) proposed by Chen and Wang. UFM is a method based on the fuzzified region representation to build region-to-region similarity measures for image retrieval; it is an improvement of their early work SIMPLIcity (Wang et al., 2001). The reasons why we compare this proposed approach with UFM are: (1) the UFM system is available to us; and (2) UFM reflects the

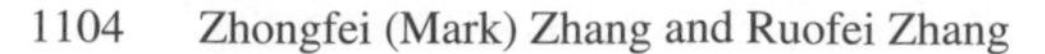

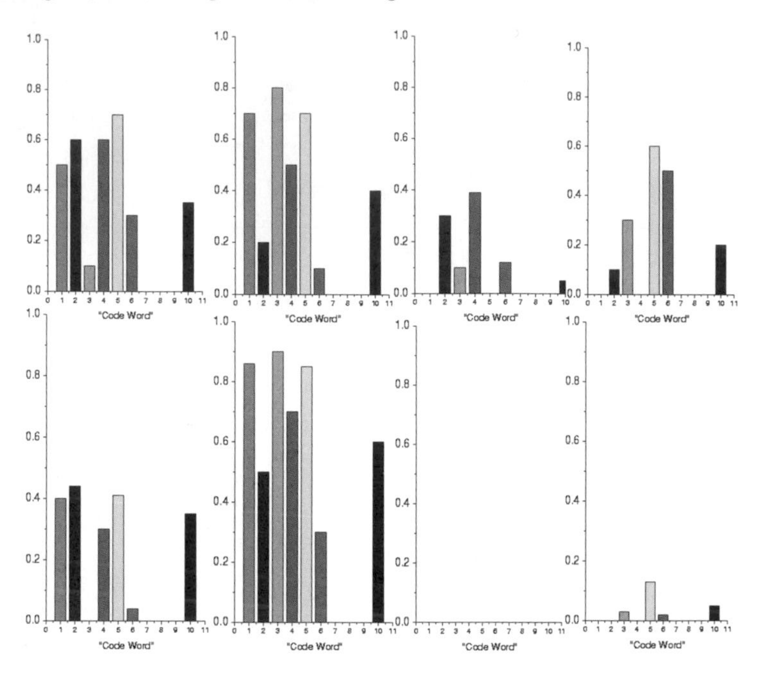

Fig. 57.11. $P(z_k|r_i, Im)$ (each color column for a "code word") and $P(z_k|Im)$ (rightmost column in each bar plot) for image Im for the four concept classes (semantically related to "plant", "castle", "cat", and "mountain", from left to right, respectively) after the first iteration (first row) and the last iteration (second row). Reprint from (Zhang & Zhang, 2007) ©2007 IEEE Signal Processing Society Press.

performance of the state-of-the-art image mining and retrieval performance. In addition, the same image segmentation and feature extraction methods are used in UFM such that a fair comparison on the performance between the two systems is ensured. Figure 57.13 shows the top 16 retrieved images by the prototype system and as well as by UFM, respectively, using image Im as a query.

More systematic comparison results on the 1,500 query image set are reported in Figure 57.14. Two versions of the prototype (one with the query expansion and moving strategy and the other without) and UFM are evaluated. It is demonstrated that the performances of the probabilistic model in both versions of the prototype have higher overall precisions than that of UFM, and the query expansion and moving strategy with the interaction of the constructed negative examples boost the mining and retrieval accuracy significantly.

57.4 Summary

In this chapter we have introduced the new, emerging area called multimedia data mining. We have given a working definition of what this area is about; we have corrected a few misconceptions that typically exist in the related research communities; and we have given a typical

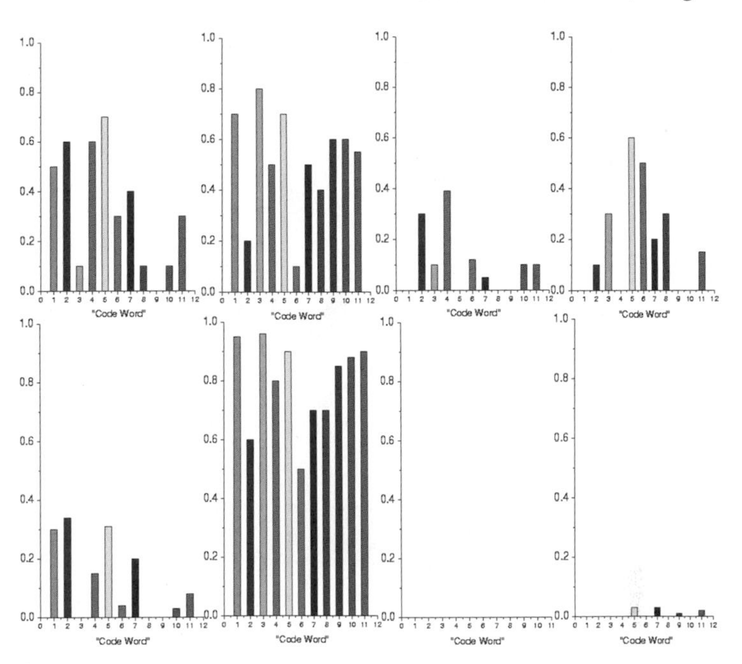

Fig. 57.12. The similar plot to Figure 57.11 with the application of the query expansion and moving strategy. Reprint from (Zhang & Zhang, 2007) ©2007 IEEE Signal Processing Society Press.

architecture for a multimedia data mining sytem or methodology. Finally, in order to showcase what a typical multimedia data mining system does and how it works, we have given an example of a specific method for semantic concept discovery in an imagery database.

Multimedia data mining, though it is a new and emerging area, has undergone an independent and rapid development over the last few years. A systematic introduction to this area may be found in (Zhang & Zhang, 2008) as well as the further readings contained in the book.

Ackonwledgments

This work is supported in part by the National Science Foundation through grants IIS-0535162 and IIS-0812114. Any opinions, findings, and conclusions or recommendations expressed in this material are those of the authors and do not necessarily reflect the views of the National Science Foundation.

References

Baeza-Yates, R. & Ribeiro-Neto, B. (1999). Modern Information Retrieval. Addison-Wesley.

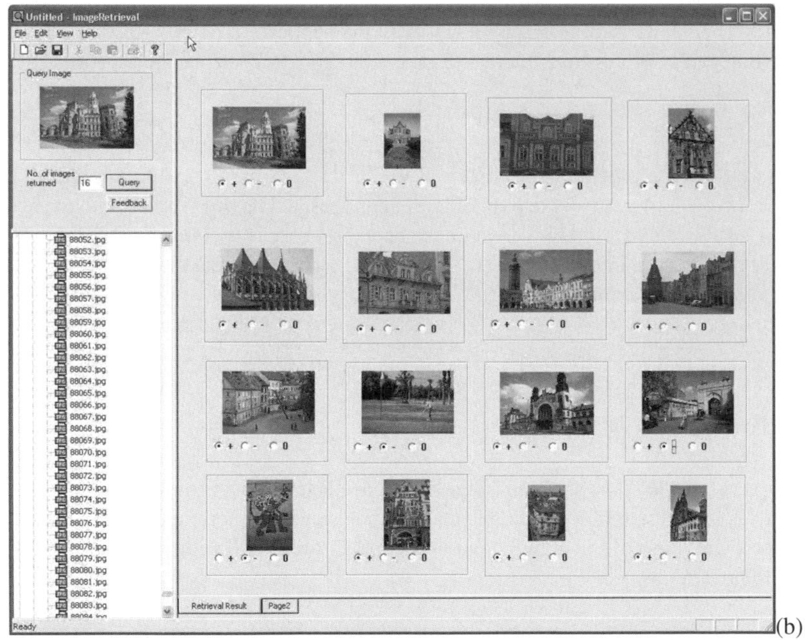

Fig. 57.13. Retrieval performance comparisons between UFM and the prototype system using image *Im* in Figure 57.10 as the query. (a) Images returned by UFM (9 of the 16 images are relevant). (b) Images returned by the prototype system (14 of the 16 images are relevant).

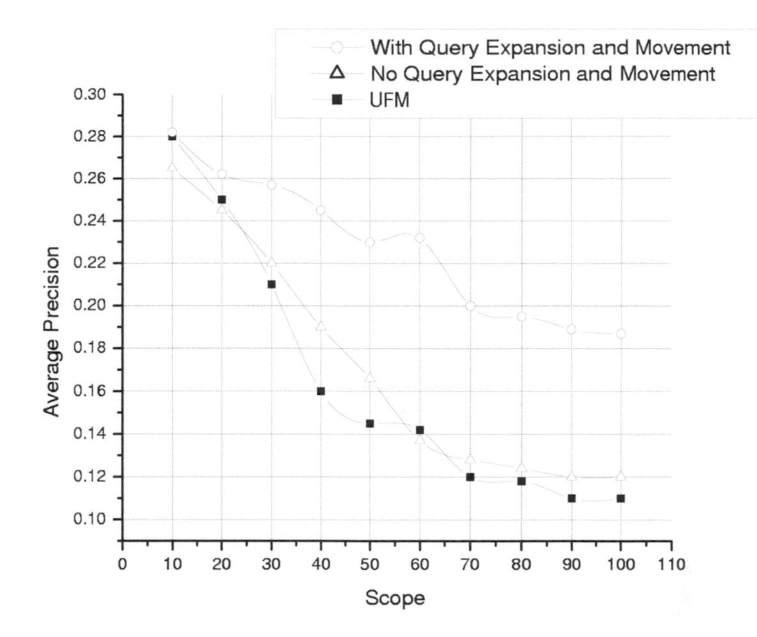

Fig. 57.14. Average precision comparisons between the two versions of the prototype and UFM. Reprint from (Zhang & Zhang, 2007) ©2007 IEEE Signal Processing Society Press and from (Zhang & Zhang, 2004a) ©2004 IEEE Computer Society Press.

Barnard, K., Duygulu, P., d.Freitas, N., Blei, D. & Jordan, M. I. (2003). Journal of Machine Learning Research 3, 1107–1135.

Barnard, K. & Forsyth, D. (2001). In The International Conference on Computer Vision vol. II, pp. 408–415,.

Blei, D., Ng, A. & Jordan, M. (2001). In The International Conference on Neural Information Processing Systems.

Carbonetto, P., d. Freitas, N. & Barnard, K. (2004). In The 8th European Conference on Computer Vision.

Carbonetto, P., d. Freitas, N., Gustafson, P. & Thompson, N. (2003). In The 9th International Workshop on Artificial Intelligence and Statistics.

Carson, C., Belongie, S., Greenspan, H. & Malik, J. (2002). IEEE Trans. on PAMI 24, 1026–1038.

Castleman, K. (1996). Digital Image Processing. Prentice Hall, Upper Saddle River, NJ.

Chen, Y. & Wang, J. (2002). IEEE Trans. on PAMI 24, 1252–1267.

Chen, Y., Wang, J. & Krovetz, R. (2003). In the 5th ACM SIGMM International Workshop on Multimedia Information Retrieval pp. 193–200,, Berkeley, CA.

Dempster, A., Laird, N. & Rubin, D. (1977). Journal of the Royal Statistical Society, Series B 39, 1C38.

Duygulu, P., Barnard, K., d. Freitas, J. F. G. & Forsyth, D. A. (2002). In The 7th European Conference on Computer Vision vol. IV, pp. 97–112,, Copenhagon, Denmark.

Faloutsos, C. (1996). Searching Multimedia Databases by Content. Kluwer Academic Publishers.

Faloutsos, C., Barber, R., Flickner, M., Hafner, J., Niblack, W., Petkovic, D. & Equitz, W. (1994). Journal of Intelligent Information Systems 3, 231–262.
Feng, S. L., Manmatha, R. & Lavrenko, V. (June, 2004). In The International Conference on Computer Vision and Pattern Recognition, Washington, DC.
Flickner, M., Sawhney, H., Ashley, J., Huang, Q., Dom, B., Gorkani, M., Hafner, J., Lee, D., Petkovic, D., Steele, D. & Yanker, P. (1995). IEEE Computer 28, 23–32.
Furht, B., ed. (1996). Multimedia Systems and Techniques. Kluwer Academic Publishers.
Greenspan, H., Dvir, G. & Rubner, Y. (2004). Journal of Computer Vision and Image Understanding 93, 86–109.
Greenspan, H., Goldberger, J. & Ridel, L. (2001). Journal of Computer Vision and Image Understanding 84, 384–406.
Han, J. & Kamber, M. (2006). Data Mining — Concepts and Techniques. 2 edition, Morgan Kaufmann.
Hofmann, T. (2001). Machine Learning 42, 177C196.
Hofmann, T. & Puzicha, J. (1998). AI Memo 1625.
Hofmann, T., Puzicha, J. & Jordan, M. I. (1996). In The International Conference on Neural Information Processing Systems.
Huang, J. & et al., S. R. K. (1997). In IEEE Int'l Conf. Computer Vision and Pattern Recognition Proceedings, Puerto Rico.
Jain, R. (1996). In Multimedia Systems and Techniques, (Furht, B., ed.),. Kluwer Academic Publishers.
Jeon, J., Lavrenko, V. & Manmatha, R. (2003). In the 26th Annual International ACM SIGIR Conference on Research and Development in Information Retrieval.
Jing, F., Li, M., Zhang, H.-J. & Zhang, B. (2004). IEEE Trans. on Image Processing 13.
Kohonen, T. (2001). Self-Organizing Maps. Springer, Berlin, Germany.
Kohonen, T., Kaski, S., Lagus, K., Salojärvi, J., Honkela, J., Paatero, V. & Saarela, A. (2000). IEEE Trans. on Neural Networks 11, 1025–1048.
Ma, W. & Manjunath, B. S. (1995). In Internation Conference on Image Processing pp. 2256–2259,.
Ma, W. Y. & Manjunath, B. (1997). In IEEE Int'l Conf. on Image Processing Proceedings pp. 568–571,, Santa Barbara, CA.
Maimon O., and Rokach, L. Data Mining by Attribute Decomposition with semiconductors manufacturing case study, in Data Mining for Design and Manufacturing: Methods and Applications, D. Braha (ed.), Kluwer Academic Publishers, pp. 311–336, 2001.
Manjunath, B. S. & Ma, W. Y. (1996). IEEE Trans. on Pattern Analysis and Machine Intelligence 18.
Mclachlan, G. & Basford, K. E. (1988). Mixture Models. Marcel Dekker, Inc., Basel, NY.
Moghaddam, B., Tian, Q. & Huang, T. (2001). In The International Conference on Multimedia and Expo 2001.
Pentland, A., Picard, R. W. & Sclaroff, S. (1994). In SPIE-94 Proceedings pp. 34–47,.
Rissanen, J. (1978). Automatica 14, 465–471.
Rissanen, J. (1989). Stochastic Complexity in Statistical Inquiry. World Scientific.
Rocchio, J. J. J. (1971). In The SMART Retreival System — Experiments in Automatic Document Processing pp. 313–323. Prentice Hall, Inc Englewood Cliffs, NJ.
Rokach L., Mining manufacturing data using genetic algorithm-based feature set decomposition, Int. J. Intelligent Systems Technologies and Applications, 4(1):57-78, 2008.
Rokach, L. and Maimon, O. and Averbuch, M., Information Retrieval System for Medical Narrative Reports, Lecture Notes in Artificial intelligence 3055, page 217-228 Springer-Verlag, 2004.

Rui, Y., Huang, T. S., Mehrotra, S. & Ortega, M. (1997). In IEEE Workshop on Content-based Access of Image and Video Libraries, in conjunction with CVPR'97 pp. 82–89,.

Smeulders, A. W. M., Worring, M., Santini, S., Gupta, A. & Jain, R. (2000). IEEE Trans. on Pattern Analysis and Machine Intelligence 22, 1349–1380.

Steinmetz, R. & Nahrstedt, K. (2002). Multimedia Fundamentals — Media Coding and Content Processing. Prentice-Hall PTR.

Subrahmanian, V. (1998). Principles of Multimedia Database Systems. Morgan Kaufmann.

Vasconcelos, N. & Lippman, A. (2000). In IEEE Workshop on Content-based Access of Image and Video Libraries (CBAIVL'00), Hilton Head, South Carolina.

Wang, J., Li, J. & Wiederhold, G. (2001). IEEE Trans. on PAMI 23.

Wood, M. E. J., Campbell, N. W. & Thomas, B. T. (1998). In ACM Multimedia 98 Proceedings, Bristol, UK.

Zhang, R. & Zhang, Z. (2004a). In IEEE International Conference on Computer Vision and Pattern Recogntion (CVPR) 2004, Washington, DC.

Zhang, R. & Zhang, Z. (2004b). EURASIP Journal on Applied Signal Processing 2004, 871–885.

Zhang, R. & Zhang, Z. (2007). IEEE Transactions on Image Processing 16, 562–572.

Zhang, Z. & Zhang, R. (2008). Multimedia Data Mining — A Systematic Introduction to Concepts and Theory. Taylor & Francis.

Zhou, X. S., Rui, Y. & Huang, T. S. (1999). In IEEE Conf. on Image Processing Proceedings.

Zhu, L., Rao, A. & Zhang, A. (2002). ACM Transaction on Information Systems 20, 224–257.

58

Data Mining in Medicine

Nada Lavrač[1] and Blaž Zupan[2]

[1] Jožef Stefan Institute, Jamova 39, 1000 Ljubljana, Slovenia,
Nova Gorica Polytechnic, Vipavska 13, 5000 Nova Gorica, Slovenia
[2] Faculty of Computer and Information Science, University of Ljubljana, Tržaška 25, 1000 Ljubljana, Slovenia
Department of Molecular and Human Genetics, Baylor College of Medicine, 1 Baylor Plaza, Houston, TX 77030, USA

Summary. Extensive amounts of data stored in medical databases require the development of specialized tools for accessing the data, data analysis, knowledge discovery, and effective use of stored knowledge and data. This chapter focuses on Data Mining methods and tools for knowledge discovery. The chapter sketches the selected Data Mining techniques, and illustrates their applicability to medical diagnostic and prognostic problems.

Key words: Data Mining in Medicine, Inductive Logic Programming, Decision Trees, Rule Induction, Case-based Reasoning, Instance-based Learning, Supervised Learning, Neural Networks

58.1 Introduction

Extensive amounts of knowledge and data stored in medical databases require the development of specialized tools for accessing the data, data analysis, knowledge discovery, and effective use of stored knowledge and data, since the increase in data volume causes difficulties in extracting useful information for decision support. The traditional manual data analysis has become insufficient, and methods for efficient computer-based analysis indispensable, such as the technologies developed in the area of *Data Mining* and *knowledge discovery in databases* (Frawley, 1991).

Knowledge discovery in databases is frequently defined as a *process* (Fayyad, 1996) consisting of the following steps: understanding the domain, forming the data set and cleaning the data, extracting of regularities hidden in the data thus formulating knowledge in the form of patterns or models (this step is referred to as *Data Mining* (DM)), postprocessing of discovered knowledge, and exploiting the results.

Important issues that arise from the rapidly emerging globality of data and information are:

O. Maimon, L. Rokach (eds.), *Data Mining and Knowledge Discovery Handbook*, 2nd ed.,
DOI 10.1007/978-0-387-09823-4_58,

- the provision of standards in terminology, vocabularies and formats to support multilinguality and sharing of data,
- standards for the abstraction and visualization of data,
- standards for interfaces between different sources of data,
- integration of heterogeneous types of data, including images and signals, and
- reusability of data, knowledge, and tools.

Many environments still lack standards, which hinders the use of data analysis tools on large global data sets, limiting their application to data sets collected for specific diagnostic, screening, prognostic, monitoring, therapy support or other patient management purposes. The emerging standards that relate to Data Mining are CRISP-DM and PMML. CRISP-DM is a Data Mining process standard that was crafted by Cross-Industry Standard Process for Data Mining Interest Group (www.crisp-dm.org). PMML (Predictive Data Mining Markup Language, www.dmg.org), on the other hand, is a standard that defines how to use XML markup language to store predictive Data Mining models, such as classification trees and classification rule sets.

Modern hospitals are well equipped with monitoring and other data collection devices which provide relatively inexpensive means to collect and store the data in inter- and intra-hospital information systems. Large collections of medical data are a valuable resource from which potentially new and useful knowledge can be discovered through Data Mining. Data Mining is increasingly popular as it is aimed at gaining an insight into the relationships and patterns hidden in the data.

Patient records collected for diagnosis and prognosis typically encompass values of anamnestic, clinical and laboratory parameters, as well as results of particular investigations, specific to the given task. Such data sets are characterized by their incompleteness (missing parameter values), incorrectness (systematic or random noise in the data), sparseness (few and/or non-representable patient records available), and inexactness (inappropriate selection of parameters for the given task). The development of Data Mining tools for medical diagnosis and prediction was frequently motivated by the requirements for dealing with these characteristics of medical data sets (Bratko and Kononenko, 1987, Cestnik *et al.*, 1987).

Data sets collected in monitoring (either acute monitoring of a particular patient in an intensive care unit, or discrete monitoring over long periods of time in the case of patients with chronic diseases) have additional characteristics: they involve the measurements of a set of parameters at different times, requesting the temporal component to be taken into account in data analysis. These data characteristics need to be considered in the design of analysis tools for prediction, intelligent alarming and therapy support.

In medicine, Data Mining can be used for solving descriptive and predictive Data Mining tasks. *Descriptive* Data Mining tasks are concerned with finding interesting patterns in the data, as well as interesting clusters and subgroups of data, where typical methods include association rule learning, and (hierarchical or k-means) clustering, respectively. In contrast, *predictive* Data Mining starts from the entire data set and aims at inducing a predictive model that holds on the data and can be used for prediction or classification of yet unseen instances. Learning in the predictive Data Mining setting requires labelled data items. Class labels can be either categorical or continuous; accordingly, predictive tasks concern building classification models or regression models, respectively.

Data Mining in medicine is most often used for building classification models, these being used for either diagnosis, prognosis or treatment planning. Predictive Data Mining, which is the focus of this chapter, is concerned with the analysis of classificatory properties of data tables. Data represented in the tables may be collected from measurements or acquired from experts. Rows in the table usually correspond to individuals (training examples) to be analyzed in terms of their properties (attributes) and the class (concept) to which they belong. In a medical setting, a concept of interest can be a disease or a medical outcome. Supervised learning assumes that training examples are classified whereas unsupervised learning concerns the analysis of unclassified examples.

This chapter is organized as follows. Section 58.2 presents a selection of symbolic classification methods. Section 58.3 complements it by outlining selected subsymbolic classification methods. Finally, Section 58.4 concludes with a brief outline of other methods for supporting medical knowledge discovery.

58.2 Symbolic Classification Methods

In medical data analysis it is very important that the results of data mining can be communicated to humans in an understandable way. In this respect, the analysis tools have to deliver transparent results and preferably facilitate human intervention in the analysis process. A good example of such methods are symbolic machine learning algorithms that, as a result of data analysis, aim to derive a symbolic model (e.g., a decision tree or a set of rules) of preferably low complexity but high transparency and accuracy.

58.2.1 Rule Induction

If-then Rules

Given a set of classified examples, a rule induction system constructs a set of rules. An if-then rule has the form:

```
IF Condition THEN Conclusion.
```

The condition of a rule contains one or more attribute tests of the form $A_i = v_{i_k}$ for discrete attributes, and $A_i < v$ or $A_i > v$ for continuous attributes. The condition of a rule is a conjunction of attribute tests (or a disjunction of conjunctions of attribute tests). The conclusion has the form $C = c_i$, assigning a particular value c_i to class C. An example is *covered* by a rule if the attribute values of the example satisfy the condition in the antecedent of the rule.

An example rule below, induced in the domain of early diagnosis of rheumatic diseases (Lavrač *et al.*, 1993, Džeroski and Lavrač, 1996), assigns the diagnosis crystal-induced synovitis to male patients older than 46 who have more than three painful joints and psoriasis as a skin manifestation.

```
IF     Sex = male
   AND Age > 46
   AND Number_of_painful_joints > 3
   AND Skin_manifestations = psoriasis
THEN   Diagnosis = crystal_induced_synovitis
```

If-then rule induction, studied already in the eighties (Michalski, 1986), resulted in a series of AQ algorithms, including the AQ15 system which was applied also to the analysis of medical data (Michalski et al. 1986).

Here we describe the rule induction system CN2 (Clark and Niblett, 1989, Clark and Boswell, 1991) which is among the best known if-then rule learners capable of handling imperfect/noisy data. Like the AQ algorithms, CN2 also uses the covering approach to construct a set of rules for each possible class c_i in turn: when rules for class c_i are being constructed, examples of this class are treated as positive, and all other examples as negative. The covering approach works as follows: CN2 constructs a rule that correctly classifies some positive examples, removes the positive examples covered by the rule from the training set and repeats the process until no more positive examples remain uncovered. To construct a single rule that classifies examples into class c_i, CN2 starts with a rule with an empty condition (IF part) and the selected class c_i as the conclusion (THEN part). The antecedent of this rule is satisfied by all examples in the training set, and not only those of the selected class. CN2 then progressively refines the antecedent by adding conditions to it, until only examples of class c_i satisfy the antecedent. To allow for the handling imperfect data, CN2 may construct a set of rules which is imprecise, i.e., does not classify all examples in the training set correctly.

Consider a partially built rule. The conclusion part is fixed to c_i and there are some (possibly none) conditions in the IF part. The examples covered by this rule form the current training set. For discrete attributes, all conditions of the form $A_i = v_{i_k}$, where v_{i_k} is a possible value for A_i, are considered for inclusion in the condition part. For continuous attributes, all conditions of the form $A_i \leq \frac{v_{i_k}+v_{i_{k+1}}}{2}$ and $A_i > \frac{v_{i_k}+v_{i_{k+1}}}{2}$ are considered, where v_{i_k} and $v_{i_{k+1}}$ are two consecutive values of attribute A_i that actually appear in the current training set. For example, if the values 4.0, 1.0, and 2.0 for attribute A_i appear in the current training set, the conditions $A_i \leq 1.5$, $A_i > 1.5$, $A_i \leq 3.0$, and $A_i > 3.0$ will be considered.

Note that both the structure (set of attributes to be included) and the parameters (values of the attributes for discrete ones and boundaries for the continuous ones) of the rule are determined by CN2. Which condition will be included in the partially built rule depends on the number of examples of each class covered by the refined rule and the heuristic estimate of the quality of the rule.

The heuristic estimates used in rule induction are mainly designed to estimate the performance of the rule on unseen examples in terms of classification accuracy. This is in accordance with the task of achieving high classification accuracy on unseen cases. Suppose a rule covers p positive and n negative examples of class c_j. Its accuracy an be estimated by the relative frequency of positive examples of class c_j covered, computed as $p/(p+n)$. This heuristic, used in early rule induction algorithms, prefers rules which cover examples of only one class. The problem with this metric is that it tends to select very specific rules supported by few examples. In the extreme case, a maximally specific rule will cover one example and hence have an unbeatable score using the metrics of apparent accuracy (scoring 100% accuracy). Apparent accuracy on the training data, however, does not necessarily reflect true predictive accuracy, i.e., accuracy on new test data. It has been shown (Holte *et al.*, 1989) that rules supported by few examples have very high error rates on new test instances.

The problem lies in the estimation of the probabilities involved, i.e., the estimate of the probability that a new instance is correctly classified by a given rule. If we use relative frequency, the estimate is only good if the rule covers many examples. In practice, however, not enough examples are available to estimate these probabilities reliably at each step. Therefore, probability estimates that are more reliable when few examples are given should be used, such as the Laplace estimate which, in two-class problems, estimates the accuracy as $(p+1)/(p+n+2)$ (Niblett and Bratko, 1986). This is the search heuristic used in CN2. The m-estimate (Cestnik, 1990) is a further upgrade of the Laplace estimate, taking also into account the prior distribution of classes.

Rule induction can be used for early diagnosis of rheumatic diseases (Lavrač *et al.*, 1993, Džeroski and Lavrač, 1996), for the evaluation of EDSS in multiple sclerosis (Gaspari *et al.*, 2001) and in numerous other medical domains.

Rough Sets

If-then rules can be also induced using the theory of *rough sets* (Pawlak, 1981, Pawlak, 1991). Rough sets (RS) are concerned with the analysis of classificatory properties of data aimed at approximations of concepts. RS can be used both for supervised and unsupervised learning.

Let us introduce the main concepts of the rough set theory. Let U denote a non-empty finite set of *objects* called the *universe* and A a non-empty finite set of *attributes*. Each object $x \in U$ is assumed to be described by a subset of attributes B, $B \subseteq A$. The basic concept of RS is an *indiscernibility* relation. Two objects x and y are indiscernible on the basis of the available attribute subset B if they have the same values of attributes B. It is usually assumed that this relation is reflexive, symmetric and transitive. The set of objects indiscernible from x using attributes B forms an equivalence class and is denoted by $[x]_B$. There are extensions of RS theory that do not require transitivity to hold.

Let $X \subseteq U$, and let $Ind_B(X)$ denote a set of equivalence classes of examples that are indiscernible, i.e., a set of subsets of examples that cannot be distinguished on the basis of attributes in B. The subset of attributes B is sufficient for classification if for every $[x]_B \in Ind_B(X)$ all the examples in $[x]_B$ belong to the same decision class. In this case crisp definitions of classes can be induced; otherwise, only 'rough' concept definitions can be induced since some examples can not be decisively classified.

The goal of RS analysis is to induce approximations of concepts c_i. Let X consist of training examples of class c_i. X may be approximated using only the information contained in B by constructing the *B-lower* and *B-upper approximations of* X, denoted $\underline{B}X$ and $\overline{B}X$ respectively, where $\underline{B}X = \{x \mid x \in X,\ [x]_B \subseteq X\}$ and $\overline{B}X = \{x \mid x \in U,\ [x]_B \cap X \neq \emptyset\}$. On the basis of knowledge in B the objects in $\underline{B}X$ can be classified with certainty as members of X, while the objects in $\overline{B}X$ can be only classified as possible members of X. The set $BN_B(X) = \overline{B}X - \underline{B}X$ is called the *B-boundary region of* X thus consisting of those objects that on the basis of knowledge in B cannot be unambiguously classified into X or its complement. The set $U - \overline{B}X$ is called the *B-outside region of* X and consists of those objects which can be with certainty classified as not belonging to X. A set is said to be *rough* (respectively *crisp*) if the boundary region is non-empty (respectively empty). The boundary region consists of examples that are indiscernible from some examples in X and therefore can not be decisively classified into c_i; this region consists of the union of equivalence classes each of which contains some examples from X and some examples not in X.

The main task of RS analysis is to find minimal subsets of attributes that preserve the indiscernibility relation. This is called the *reduct* computation. Note that there are usually many reducts. Several types of reducts exist. Decision rules are generated from reducts by reading off the values of the attributes in each reduct. The main challenge in inducing rules lies in determining which attributes should be included in the condition of the rule. Rules induced from the (standard) reducts will usually result in large sets of rules and are likely to overfit the data. Instead of standard reducts, attribute sets that "almost" preserve the indiscernibility relation are generated. Good results have been achieved with *dynamic reducts* (Skowron, 1995) that use a combination of reduct computation and statistical resampling. Many RS approaches to discretization, feature selection, symbolic attribute grouping, have also been designed (Polkowski and Skowron, 1998a, Polkowski and Skowron, 1998b). There exist also several software tools for RS, such as the Rosetta system (Rumelhart, 1986).

The list of applications of RS in medicine is significant. It includes extracting diagnostic rules, image analysis and classification of histological pictures, modelling set residuals, EEG signal analysis, etc (Averbuch *et al.*, 2004, Rokach *et al.*, 2004). Examples of RS analysis in medicine include (Grzymala-Busse, 1998, Komorowski and Øhrn, 1998, Tsumoto, 1998). For references that include medical applications, see (Polkowski and Skowron, 1998a, Polkowski and Skowron, 1998b, Lin and Cercone, 1997).

Ripple Down Rules

The knowledge representation of the form of ripple down rules allows incremental learning by including exceptions to the current rule set. Ripple down rules (RDR) (Compton and Jansen, 1988, Compton *et al.*, 1989) have the following form:

```
IF Conditions THEN Conclusion BECAUSE Case EXCEPT
    IF ...
ELSE IF ...
```

For the domain of lens prescription (Cendrowka, 1987) an example RDR (Sammut, 1998) is shown below.

```
IF true THEN no_lenses BECAUSE case0
    EXCEPT
        IF astigmatism = not_astigmatic and
           tear_production = normal
        THEN
           soft_lenses BECAUSE case2
ELSE
    IF prescription = myope and
       tear_production = normal
    THEN
       hard_lenses BECAUSE case4
```

The contact lenses RDR is interpreted as follows: The default rule is that a person does not use lenses, stored in the rule base together with a ‘dummy’ case0. No update of the system is needed after entering the data on the first patient who needs no lenses. But the second patient (case2) needs soft lenses and the rule is updated according to the conditions that hold for case2. Case3 is again a patient who does not need lenses, but the rule needs to be updated w.r.t. the conditions of the fourth patient (case4) who needs hard lenses.

The above example illustrates also the incremental learning of ripple down rules in which EXCEPT IF THEN and ELSE IF THEN statements are added to the RDRs to make them consistent with the current database of patients.

If the RDR from example above were rewritten as an IF-THEN-ELSE statement it would look as follows:

```
IF true THEN
     IF astigmatism = not_astigmatic and
        tear_production = normal
     THEN
        soft_lenses ELSE no_lenses
ELSE
     IF prescription = myope and
        tear_production = normal
     THEN
        hard_lenses
```

There were many successful medical applications of the RDR approach, including the system PEIRS (Edwards *et al.*, 1993) which is an RDR reconstruction of the hand-built GARVAN expert system knowledge base on thyroid function tests (Horn *et al.*, 1985).

58.2.2 Learning of Classification and Regression Trees

Systems for Top-Down Induction of Decision Trees (Quinlan, 1986) generate a decision tree from a given set of examples. Each of the interior nodes of the tree is labelled by an attribute, while branches that lead from the node are labelled by the values of the attribute.

The tree construction process is heuristically guided by choosing the 'most informative' attribute at each step, aimed at minimizing the expected number of tests needed for classification. Let E be the current (initially entire) set of training examples, and $c_1, \ldots, c_N$ the decision classes. A decision tree is constructed by repeatedly calling a tree construction algorithm in each generated node of the tree. Tree construction stops when all examples in a node are of the same class (or if some other stopping criterion is satisfied). This node, called a leaf, is labelled by class value. Otherwise the 'most informative' attribute, say A_i, is selected as the root of the (sub)tree, and the current training set E is split into subsets E_i according to the values of the most informative attribute. Recursively, a subtree T_i is built for each E_i.

Ideally, each leaf is labelled by exactly one class value. However, leaves can also be empty, if there are no training examples having attribute values that would lead to a leaf, or can be labelled by more than one class value (if there are training examples with same attribute values and different class values).

One of the most important features is tree pruning, used as a mechanism for handling noisy data (Quinlan, 1993). Tree pruning is aimed at producing trees which do not overfit possibly erroneous data. In tree pruning, the unreliable parts of a tree are eliminated in order to increase the classification accuracy of the tree on unseen instances.

An early decision tree learner, ASSISTANT (Cestnik *et al.*, 1987), that was developed specifically to deal with the particular characteristics of medical data sets, supports the handling of incompletely specified training examples (missing attribute values), binarization of continuous attributes, binary construction of decision trees, pruning of unreliable parts of the tree and plausible classification based on the 'naive' Bayesian principle to calculate the classification in the leaves for which no evidence is available. An example decision tree that can be used to predict outcome of patients after severe head injury (Pilih, 1997) is shown in Figure 58.1. The two attributes in the nodes of the tree are CT score (number of abnormalities detected by Computer axial Tomography) and GCS (evaluation of coma according to the Glasgow Coma Scale).

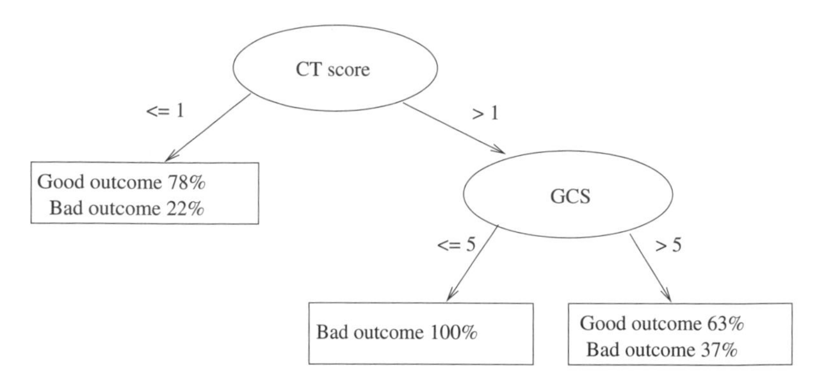

Fig. 58.1. Decision tree for outcome prediction after severe head injury. In the leaves, the percentages indicate the probabilities of class assignment.

Implementations of the ASSISTANT algorithm include ASSISTANT-R and ASSISTANT-R2 (Kononenko and Šimec, 1995). Instead of the standardly used informativity search heuristic, ASSISTANT-R employs ReliefF as a heuristic for attribute selection (Kononenko, 1994, Kira and Rendell, 1992b). This heuristic is an extension of RELIEF (Kira and Rendell, 1992a, Kira and Rendell, 1992b) which is a non-myopic heuristic measure that is able to estimate the quality of attributes even if there are strong conditional dependencies between attributes. In addition, wherever appropriate, instead of the relative frequency, ASSISTANT-R uses the m-estimate of probabilities (Cestnik, 1990).

The best known decision tree learner is C4.5 (Quinlan, 1993) (See5 and J48 are its more recent upgrades) which is widely used and has been incorporated into commercial Data Mining tools as well as in the publicly available WEKA Data Mining toolbox (Witten and Frank, 1999). The system is reliable, efficient and capable of dealing with large sets of training examples.

Learning of regression trees is similar to decision tree learning: it also uses a top-down greedy approach to tree construction. The main difference is that decision tree construction involves the classification into a finite set of discrete classes whereas in regression tree learning the decision variable is continuous and the leaves of the tree either consist of a prediction into a numeric value or a linear combination of variables (attributes). An early learning system CART (Breiman *et al.*, 1984) featured both classification and regression tree learning.

There are many applications of decision trees for analysis of medical data sets. For instance, CART has been applied to the problem of mining a diabetic data warehouse composed of a complex relational database with time series and sequencing information (Breault and Goodall, 2002). Decision tree learning has been applied to the diagnosis of sport injuries (Zelic *et al.*, 1997), patient recovery prediction after traumatic brain injury (Andrews *et al.*, 2002), prediction of recurrent falling in community-dwelling older persons (Stel *et al.*, 2003), and numerous other medical domains.

58.2.3 Inductive Logic Programming

Inductive logic programming (ILP) systems learn relational concept descriptions from relational data. Well known ILP systems include FOIL (Quinlan, 1990), Progol (Muggleton, 1995) and Claudien (De Raedt and Dehaspe, 1997). LINUS is an ILP environment (Lavrač and Džeroski, 1994), enabling the transformation of relational learning problems into the form appropriate for standard attribute-value learners, while in general ILP systems learn relational descriptions without such a transformation to propositional learning.

In ILP, induced rules typically have the form of Prolog clauses. The output of an ILP system is illustrated by a rule of ocular fundus image classification for glaucoma diagnosis, induced by an ILP system GKS (Mizoguchi *et al.*, 1997) specially designed to deal with low-level measurement data including images.

```
class(Image, Segment, undermining) :-
    clockwise(Segment, Adjacent, 1),
    class_confirmed(Image, Adjacent, undermining).
```

Compared to rules induced by a rule learning algorithm of the form IF *Condition* THEN *Conclusion*, Prolog rules have the form *Conclusion* :- *Condition*. For example, the rule for glaucoma diagnosis means that Segment of Image is classified as undermining (i.e., not normal) if the conditions of the right-hand side of the clause are fulfilled. Notice that the conditions consist of a conjunction of predicate clockwise/3 defined in the background knowledge, and predicate class_confirmed/3, added to the background knowledge in one of the previous iterative runs of the GKS algorithm. This shows one of the features of ILP learning, namely that learning can be done in several cycles of the learning algorithm in which definitions of new background knowledge predicates are learned and used in the subsequent runs of the learner; this may improve the performance of the learner.

ILP has been successfully applied to carcinogenesis prediction in the predictive toxicology evaluation challenge (Srinivasan *et al.*, 1997) and to the recognition of arrhythmia from electrocardiograms (Carrault *et al.*, 2003).

58.2.4 Discovery of Concept Hierarchies and Constructive Induction

The data can be decomposed into equivalent but smaller, more manageable and potentially easier to comprehend data sets. A method that uses such an approach is called *function decomposition* (Zupan and Bohanec, 1998). Besides the discovery of appropriate data sets, function decomposition arranges them into a concept hierarchy. Function decomposition views classification data (example set) with attributes $X = \{x_1, \ldots, x_n\}$ and an output concept (class) y defined as a partially specified function $y = F(X)$. The core of the method is a single step decomposition of F into $y = G(A, c)$ and $c = H(B)$, where A and B are proper subsets of input attributes such that $A \cup B = X$. Single step decomposition constructs the example sets that partially specify new functions G and H. Functions G and H are determined in the decomposition process and are not predefined in any way. Their joint complexity (determined by some complexity measure) should be lower than the complexity of F. Obviously, there are many candidates for partitioning X into A and B; the decomposition chooses the partition that yields functions G and H of lowest complexity. In this way, single step decomposition also discovers a new intermediate concept $c = H(B)$. Since the decomposition can be applied recursively on H and G, the result in general is a hierarchy of concepts. For each concept in the hierarchy, there is a corresponding function (such as $H(B)$) that determines the dependency of that concept on its immediate descendants in the hierarchy.

In terms of data analysis, the benefits of function decompositions are:

- Discovery of new data sets that use fewer attributes than the original one and include fewer instances as well. Because of lower complexity, such data sets may then be easier to analyze.
- Each data set represents some concept. Function decomposition organizes discovered concepts in a hierarchy, which may itself be interpretable and can help to gain insight into the data relationships and underlying attribute groups.

Consider for example a concept hierarchy in Figure 58.2 that was discovered for a data set that describes a nerve fiber conduction-block (Zupan *et al.*, 1997). The original data set used 2543 instances of six attributes (aff, nl, k-conc, na-conc, scm, leak) and a single class variable (block) determining nerve fiber conducts or not. Function decomposition found three intermediate concepts, c1, c2, and c3. When interpreted by the domain expert, it was found that the discovered intermediate concepts are physiologically meaningful and constitute useful intermediate biophysical properties. Intermediate concept c1, for example, couples the concentration of ion channels (na-conc and k-conc) and ion leakage (leak) that are all the axonal properties and together influence the combined current source/sink capacity of the axon which is the driving force for all propagated action potentials. Moreover, new concepts use fewer attributes and instances: c1, c2, c3, and the output concept block described 125, 25, 184, and 65 instances, respectively.

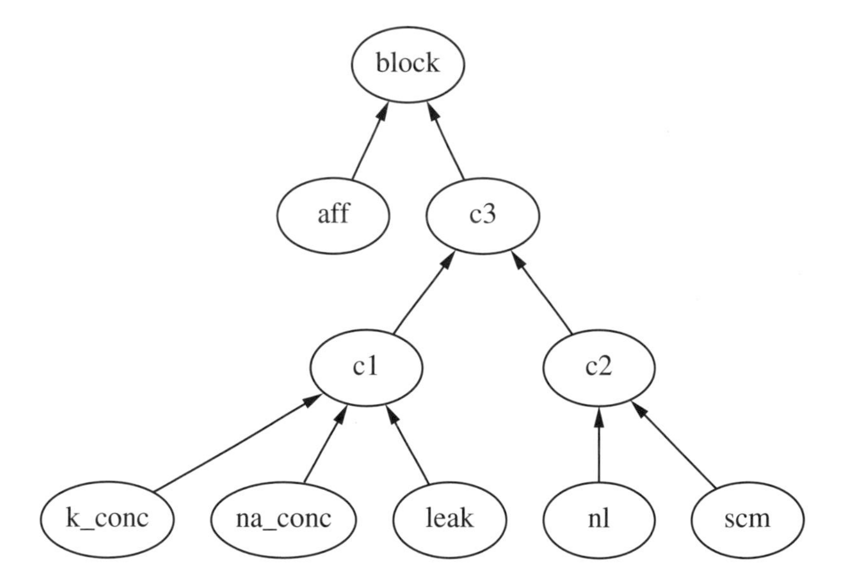

Fig. 58.2. Discovered concept hierarchy for the conduction-block domain.

Intermediate concepts discovered by decomposition can also be regarded as new features that can, for example, be added to the original example set, which can then be examined by some other data analysis method. Feature discovery and constructive induction, first investigated in (Michalski, 1986), are defined as an ability of the system to derive and use new attributes in the process of learning. Besides pure performance benefits in terms of classification accuracy, constructive induction is useful for data analysis as it may help to induce simpler and more comprehensible models and to identify interesting inter-attribute relationships. New attributes may be constructed based on available background knowledge of the domain: an example of how this facilitated learning of more accurate and comprehensible rules in the domain of early diagnosis of rheumatic diseases is given in (Džeroski and Lavrač, 1996). Function decomposition, on the other hand, may help to discover attributes from classified instances alone. For the same rheumatic domain, this is illustrated in (Zupan and Džeroski, 1998). Although such discovery may be carried out automatically, the benefits of the involvement of experts in new attribute selection are typically significant (Zupan *et al.*, 2001).

58.2.5 Case-Based Reasoning

Case-based reasoning (CBR) uses the knowledge of past experience when dealing with new cases (Aamodt and Plaza, 1994, Macura and Macura, 1997). A "case" refers to a problem situation. Although, as in instance-based learning (Aha *et al.*, 1991), cases (examples) can be described by a simple attribute-value vector, CBR most often uses a richer, often hierarchical data structure. CBR relies on a database of past cases that has to be designed in the way to facilitate the retrieval of similar cases. CBR is a four stage process:

1. Given a new case to solve, a set of similar cases is retrieved from the database.
2. The retrieved cases are reused in order to obtain a solution for a new case. This may be simply achieved by selecting the most frequent solution used with similar past cases, or, if appropriate background knowledge or a domain model exist, retrieved solutions may be adapted for a new case.
3. The solution for the new case is then checked by the domain expert, and, if not correct, repaired using domain-specific knowledge or expert's input. The specific revision may be saved and used when solving other new cases.
4. The new case, its solution, and any additional information used for this case that may be potentially useful when solving new cases are then integrated in the case database.

CBR offers a variety of tools for data analysis. The similar past cases are not just retrieved, but are also inspected for most relevant features that are similar or different to the case in question. Because of the hierarchical data organization, CBR may incorporate additional explanation mechanisms. The use of symbolic domain knowledge for solution adaptation may further reveal specifics and interesting case's features. When applying CBR to medical data analysis, however, one has to address several non-trivial questions, including the appropriateness of similarity measures used, the actuality of old cases (as the medical knowledge is rapidly changing), how to handle different solutions (treatment actions) by different physicians, etc.

Several CBR systems were used, adapted for, or implemented to support reasoning and data analysis in medicine. Some are described in the special issue of *Artificial Intelligence in Medicine* (Macura and Macura, 1997) and include CBR systems for reasoning in cardiology by Reategui et al., learning of plans and goal states in medical diagnosis by López and Plaza, detection of coronary heart disease from myocardial scintigrams by Haddad et al., and treatment advice in nursing by Yearwood and Wilkinson. Others include a system that uses CBR to assist in the prognosis of breast cancer (Mariuzzi *et al.*, 1997), case classification in the domain of ultrasonography and body computed tomography (Kahn and Anderson, 1994), and a CBR-based expert system that advises on the identification of nursing diagnoses in a new client (Bradburn *et al.*, 1993). There is also an application of case-based distance measurements in coronary interventions (Gyöngyösi, 2002).

58.3 Subsymbolic Classification Methods

In medical problem solving it is important that a decision support system is able to explain and justify its decisions. Especially when faced with an unexpected solution of a new problem, the user requires substantial justification and explanation. Hence the interpretability of induced knowledge is an important property of systems that induce solutions from data about past solved cases. Symbolic Data Mining methods have this property since they induce symbolic representations (such as decision trees) from data. On the other hand, subsymbolic Data Mining methods typically lack this property which hinders their use in situations for which explanations are required. Nevertheless, when classification accuracy is the main applicability criterion subsymbolic methods may turn out to be very appropriate since they typically achieve accuracies that are at least as good as those of symbolic classifiers.

58.3.1 Instance-Based Learning

Instance-based learning (IBL) algorithms (Aha *et al.*, 1991) use specific instances to perform classification, rather than generalizations induced from examples, such as induced if-then rules. IBL algorithms are also called lazy learning algorithms, as they simply save some or all of the training examples and postpone all the inductive generalization effort until classification time. They assume that similar instances have similar classifications: novel instances are classified according to the classifications of their most similar neighbors.

IBL algorithms are derived from the nearest neighbor pattern classifier (Fix and Hodges, 1957, Cover and Hart, 1968). The nearest neighbor (NN) algorithm is one of the best known classification algorithms; an enormous body of research exists on the subject (Dasarathy, 1990). In essence, the NN algorithm treats attributes as dimensions of an Euclidean space and examples as points in this space. In the training phase, the classified examples are stored without any processing. When classifying a new example, the Euclidean distance between this example and all training examples is calculated and the class of the closest training example is assigned to the new example.

The more general k-NN method takes the k nearest training examples and determines the class of the new example by majority vote. In improved versions of k-NN, the votes of each of the k nearest neighbors are weighted by the respective proximity to the new example (Dudani, 1975). An optimal value of k may be determined automatically from the training set by using leave-one-out cross-validation (Weiss and Kulikowski, 1991). In the k-NN algorithm implementation described in (Wettschereck, 1994), the best k from the range [1,75] was selected in this manner. This implementation also incorporates feature weights determined from the training set. Namely, the contribution of each attribute to the distance may be weighted, in order to avoid problems caused by irrelevant features (Wolpert, 1989).

Let $n = N_{at}$. Given two examples $x = (x_1, \ldots, x_n)$ and $y = (y_1, \ldots, y_n)$, the distance between them is calculated as

$$\text{distance}(x,y) = \sqrt{\sum_{i=1}^{n} w_i \cdot \text{difference}(x_i, y_i)^2} \tag{58.1}$$

where w_i is a non-negative weight value assigned to feature (attribute) A_i and the difference between attribute values is defined as follows

$$\text{difference}(x_i, y_i) = \begin{cases} |x_i - y_i| & \text{if } A_i \text{ is continuous} \\ 0 & \text{if } A_i \text{ is discrete and } x_i = y_i \\ 1 & \text{otherwise} \end{cases} \quad (58.2)$$

When classifying a new instance z, k-NN selects the set K of k-nearest neighbors according to the distance defined above. The vote of each of the k nearest neighbors is weighted by its proximity (inverse distance) to the new example. The probability $p(z, c_j, K)$ that instance z belongs to class c_j is estimated as

$$p(z, c_j, K) = \frac{\sum_{x \in K} x_{c_j} / \text{distance}(z, x)}{\sum_{x \in K} 1 / \text{distance}(z, x)} \quad (58.3)$$

where x is one of the k nearest neighbors of z and x_{c_j} is 1 if x belongs to class c_j. Class c_j with largest value of $p(z, c_j, K)$ is assigned to the unseen example z.

Before training (respectively before classification), the continuous features are normalized by subtracting the mean and dividing by the standard deviation so as to ensure that the values output by the difference function are in the range $[0,1]$. All features have then equal maximum and minimum potential effect on distance computations. However, this bias handicaps k-NN as it allows redundant, irrelevant, interacting or noisy features to have as much effect on distance computation as other features, thus causing k-NN to perform poorly. This observation has motivated the creation of many methods for computing feature weights.

The purpose of a feature weight mechanism is to give low weight to features that provide no information for classification (e.g., very noisy or irrelevant features), and to give high weight to features that provide reliable information. In the k-NN implementation of Wettschereck (Wettschereck, 1994), feature A_i is weighted according to the mutual information (Shannon, 1948) $I(c_j, A_i)$ between class c_j and attribute A_i.

Instance-based learning was applied to the problem of early diagnosis of rheumatic diseases (Džeroski and Lavrač, 1996).

58.3.2 Neural Networks

Artificial neural networks can be used for both supervised and unsupervised learning. For each learning type, we briefly describe the most frequently used approaches.

Supervised Learning

For supervised learning and among different neural network paradigm, feed-forward multi-layered neural networks (Rumelhart and McClelland, 1986,Fausett, 1994) are most frequently used for modeling medical data. They are computational structures consisting of a interconnected processing elements (PE) or nodes arranged on a multi-layered hierarchical architecture. In general, a PE computes the weighted sum of its inputs and filters it through some sigmoid function to obtain the output (Figure 58.3.a). Outputs of PEs of one layer serve as inputs to PEs of the next layer (Figure 58.3.b). To obtain the output value for selected instance, its attribute values are stored in input nodes of the network (the network's lowest layer). Next, in each step, the outputs of the higher-level processing elements are computed (hence the name feed-forward), until the result is obtained and stored in PEs at the output layer.

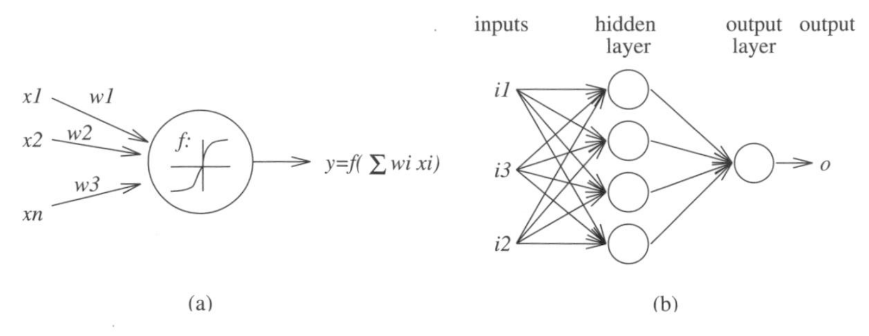

Fig. 58.3. Processing element (a) and an example of the typical structure of the feed-forward multi-layered neural network with four processing elements at hidden layer and one at output layer (b).

A typical architecture of multi-layered neural network comprising an input, a hidden and and output layer of nodes is given in Figure 58.3.b. The number of nodes in the input and output layers is domain-dependent and, respectively, is related to number and type of attributes and a type of classification task. For example, for a two-class classification problem, a neural net may have two output PEs, each modelling the probability of a distinct class, or a single PE, if a problem is coded properly.

Weights that are associated with each node are determined from training instances. The most popular learning algorithm for this is backpropagation (Rumelhart and McClelland, 1986, Fausett, 1994). Backpropagation initially sets the weights to some arbitrary value, and then considering one or several training instances at the time adjusts the weights so that the error (difference between the expected and the obtained value of nodes at the output level) is minimized. Such a training step is repeated until the overall classification error across all of the training instances falls below some specified threshold.

Most often, a single hidden layer is used and the number of nodes has to be either defined by the user or determined through learning. Increasing the number of nodes in a hidden layer allows more modeling flexibility but may cause overfitting of the data. The problem of determining the "right architecture", together with the high complexity of learning, are two of the limitations of feed-forward multi-layered neural networks. Another is the need for proper preparation of the data (Kattan and Beck, 1995): a common recommendation is that all inputs are scaled over the range from 0 to 1, which may require normalization and encoding of input attributes.

For data analysis tasks, however, the most serious limitation is the lack of explanational capabilities: the induced weights together with the network's architecture do not usually have an obvious interpretation and it is usually difficult or even impossible to explain "why" a certain decision was reached. Recently, several approaches for alleviating this limitation have been proposed. A first approach is based on pruning of the connections between nodes to obtain sufficiently accurate, but in terms of architecture significantly less complex, neural networks (Chung and Lee, 1992). A second approach, which is often preceded by the first one to reduce the complexity, is to represent a learned neural network with a set of symbolic rules (Andrews *et al.*, 1995, Craven and Shavlik, 1997, Setiono, 1997, Setiono, 1999).

Despite the above-mentioned limitations, multi-layered neural networks often have equal or superior predictive accuracy when compared to symbolic learners or statistical approaches (Kattan and Beck, 1995, Shawlik *et al.*, 1991). They have been extensively used to model medical data. Example applications areas include survival analysis (Liestøl *et al.*, 1994), clinical medicine (Baxt, 1995), pathology and laboratory medicine (Astion and Wilding, 1992), molecular sequence analysis (Wu, 1997), pneumonia risk assessment (Caruana *et al.*, 1995), and prostate cancer survival (Kattan *et al.*, 1997). There are fewer applications where rules were extracted from neural networks: an example of such data analysis is finding rules for breast cancer diagnosis (Setiono, 1996).

Different types of neural networks for supervised learning include Hopfield's recurrent networks and neural networks based on adaptive resonance theory mapping (ARTMAP). For the first, an example application is tumor boundary detection (Zhu and Yan, 1997). Example studies of application of ARTMAP in medicine include classification of cardiac arrhythmias (Ham and Han, 1996) and treatment selection for schizophrenic and unipolar depressed in-patients (Modai *et al.*, 1996). Learned ARTMAP networks can also be used to extract symbolic rules (Carpenter and Tan, 1993, Downs *et al.*, 1996). There are numerous medical applications of neural networks, including brain volumes characterization (Bona *et al.*, 2003).

Unsupervised Learning

For unsupervised learning — learning which is presented with unclassified instances and aims at identifying groups of instances with similar attribute values — the most frequently used neural network approach is that of Kohonen's self organizing maps (SOM) (Kohonen, 1988). Typically, SOM consist of a single layer of output nodes. An output node is fully connected with nodes at the input layer. Each such link has an associated weight. There are no explicit connections between nodes of the output layer.

The learning algorithm initially sets the weights to some arbitrary value. At each learning step, an instance is presented to the network, and a winning output node is chosen based on instance's attribute values and node's present weights. The weights of the winning node and of the topologically neighboring nodes are then updated according to their present weights and instance's attribute values. The learning results in the internal organization of SOM such that when two similar instances are presented, they yield a similar "pattern" of networks output node values. Hence, data analysis based on SOM may be additionally supported by proper visualization methods that show how the patterns of output nodes depend on input data (Kohonen, 1988). As such, SOM may not only be used to identify similar instances, but can, for example, also help to detect and analyze time changes of input data. Example applications of SOM include analysis of ophthalmic field data (Henson *et al.*, 1997), classification of lung sounds (Malmberg *et al.*, 1996), clinical gait analysis (Koehle *et al.*, 1997), analysis of molecular similarity (Barlow, 1995), and analysis of a breast cancer database (Markey *et al.*, 2002).

58.3.3 Bayesian Classifier

The Bayesian classifier uses the naive Bayesian formula to calculate the probability of each class c_j given the values v_{i_k} of all the attributes for a given instance to be classified (Kononenko, 1993, 1). For simplicity, let $(v_1,\ldots,v_n)$ denote the n-tuple of values of example e_k to be classified. Assuming the conditional independence of the attributes given the class, i.e., assuming $p(v_1..v_n|c_j) = \prod_i p(v_i|c_j)$, then $p(c_j|v_1..v_n)$ is calculated as follows:

$$p(c_j|v_1..v_n) = \frac{p(c_j.v_1..v_n)}{p(v_1..v_n)} = \frac{p(v_1..v_n|c_j)\cdot p(c_j)}{p(v_1..v_n)} = \qquad (58.4)$$

$$\frac{\prod_i p(v_i|c_j)\cdot p(c_j)}{p(v_1..v_n)} = \frac{p(c_j)}{p(v_1..v_n)}\prod_i \frac{p(c_j|v_i)\cdot p(v_i)}{p(c_j)} =$$

$$p(c_j)\frac{\prod_i p(v_i)}{p(v_1..v_n)}\prod_i \frac{p(c_j|v_i)}{p(c_j)}$$

A new instance will be classified into the class with maximal probability.

In the above equation, $\frac{\prod_i p(v_i)}{p(v_1..v_n)}$ is a normalizing factor, independent of the class; it can therefore be ignored when comparing values of $p(c_j|v_1..v_n)$ for different classes c_j. Hence, $p(c_j|v_1..v_n)$ is proportional to:

$$p(c_j)\prod_i \frac{p(c_j|v_i)}{p(c_j)} \tag{58.5}$$

Different probability estimates can be used for computing the probabilities, i.e., the relative frequency, the Laplace estimate (Niblett and Bratko, 1986), and the *m*-estimate (Cestnik, 1990, Kononenko, 1993, 1).

Continuous attributes have to be pre-discretized in order to be used by the naive Bayesian classifier. The task of discretization is the selection of a set of boundary values that split the range of a continuous attribute into a number of intervals which are then considered as discrete values of the attribute. Discretization can be done manually by the domain expert or by applying a discretization algorithm (Richeldi and Rossotto, 1995).

The problem of (strict) discretization is that minor changes in the values of continuous attributes (or, equivalently, minor changes in boundaries) may have a drastic effect on the probability distribution and therefore on the classification. Fuzzy discretization may be used to overcome this problem by considering the values of the continuous attribute (or, equivalently, the boundaries of intervals) as fuzzy values instead of point values (Kononenko, 1993). The effect of fuzzy discretization is that the probability distribution is smoother and the estimation of probabilities more reliable, which in turn results in more reliable classification.

Bayesian computation can also be used to support decisions in different stages of a diagnostic process (McSherry, 1997) in which doctors use *hypothetico-deductive reasoning* for gathering evidence which may help to confirm a diagnostic hypothesis, eliminate an alternative hypothesis, or discriminate between two alternative hypotheses. In particular, Bayesian computation can help in identifying and selecting the most useful tests, aimed at confirming the target hypothesis, eliminating the likeliest alternative hypothesis, increase the probability of the target hypothesis, decrease the probability of the likeliest alternative hypothesis or increase the probability of the target hypothesis relative to the likeliest alternative hypothesis. Bayesian classification has been applied to different medical domains, including the diagnosis of sport injuries (Zelic *et al.*, 1997).

58.4 Other Methods Supporting Medical Knowledge Discovery

There is a variety of other methods and tools that can support medical data analysis and can be used separately or in combination with the classification methods introduced above. We here mention only several most frequently used techniques.

The problem of discovering *association rules* has recently received much attention in the Data Mining community. The problem of inducing association rules (Agrawal *et al.*, 1996) is defined as follows: Given a set of transactions, where each transaction is a set of items (i.e., literals of the form $Attribute = value$), an *association rule* is an expression of the form $X \rightarrow Y$ where X and Y are sets of items. The intuitive meaning of such a rule is that transactions in a database which contain X tend to contain Y. Consider a sample association rule: "80% of patients with pneumonia also have high fever. 10% of all transactions contain both of these items." Here 80% is called *confidence* of the rule, and 10% support of the rule. Confidence of the rule is calculated as the ratio of the number of records having true values for all items in X and Y to the number of records having true values for all items in X. Support of the rule is the ratio of the number of records having true values for all items in X and Y to the number of all records in the database. The problem of association rule learning is to find all rules that satisfy the minimum support and minimum confidence constraints.

Association rule learning was applied in medicine, for example, to identify new and interesting patterns in surveillance data, in particular in the analysis of the *Pseudomonas aeruginosa* infection control data (Brossette *et al.*, 1998). An algorithm for finding a more expressive variant of association rules, where data and patterns are represented in first-order logic, was successfully applied to the problem of predicting whether chemical compounds are carcinogenic or not (Toivonen and King, 1998).

Subgroup discovery (Wrobel, 1997, Gamberger and Lavrač, 2002, Lavrač *et al.*, 2004) has the goal to uncover characteristic properties of population subgroups by building short rules which are highly significant (assuring that the distribution of classes of covered instances are statistically significantly different from the distribution in the training set) and have a large coverage (covering many target class instances). The approach, using a beam search rule learning algorithm aimed at inducing short rules with large coverage, was successfully applied to the problem of coronary heart disease risk group detection (Gamberger *et al.*, 2003).

Genetic algorithms (Goldberg, 1989) are optimization procedures that maintain candidate solutions encoded as strings (or chromosomes). A fitness function is defined that can assess the quality of a solution represented by some chromosome. A genetic algorithm iteratively selects best chromosomes (i.e., those of highest fitness) for reproduction, and applies crossover and mutation operators to search in the problem space. Most often, genetic algorithms are used in combination with some classifier induction technique or some schema for classification rules in order to optimize their performance in terms of accuracy and complexity (e.g., (Larranaga *et al.*, 1997) and (Dybowski *et al.*, 1996)). They can also be used alone, e.g., for the estimation of Doppler signals (Gonzalez *et al.*, 1999) or for multi-disorder diagnosis (Vinterbo and Ohno-Machado 1999). For more information please refer to Chapter 19 in this book.

Data analysis approaches reviewed so far in this chapter mostly use crisp logic: the attributes take a single value and when evaluated, decision rules return a single class value. *Fuzzy logic* (Zadeh, 1965) provides an enhancement compared to classical AI approaches (Steinmann, 1997): rather than assigning an attribute a single value, several values can be assigned, each with its own degree or grade. Classically, for example, "body temperature" of 37.2°C can be represented by a discrete value "high", while in fuzzy logic the same value can be represented by two values: "normal" with degree 0.3 and "high" with degree 0.7. Each value in a fuzzy set (like "normal" and "high") has a corresponding membership function that determines how the degree is computed from the actual continuous value of an attribute. Fuzzy systems may thus formalize a gradation and may allow handling of vague concepts—both being natural characteristics of medicine (Steinmann, 1997)—while still supporting comprehensibility and transparency by computationally relying on a fuzzy rules. In medical data analysis, the best developed approaches are those that use data to induce a straightforward tabular rule-based mapping from input to control variables and to find the corresponding membership functions. Example applications studies include design of patient monitoring and alarm system (Becker and Thull, 1997), support system for breast cancer diagnosis (Kovalerchuk *et al.*, 1997), design of a rule-based visuomotor control (Prochazka, 1996). Fuzzy logic control applications in medicine are discussed in (Rau *et al.*, 1995).

Support vector machines (SVM) are a classification technique originated from statistical learning theory (Cristianini, 2000, Vapnik, 1998). Depending on the chosen kernel, SVM selects a set of data examples (support vectors) that define the decision boundary between classes. SVM have been proven for excellent classification performance, while it is arguable whether support vectors can be effectively used in communication of medical knowledge to the domain experts.

Bayesian networks (Pearl, 1988) are probabilistic models that can be represented by a directed graph with vertices encoding the variables in the model and edges encoding their dependency. Given a Bayesian network, one can compute any joint or conditional probability of interest. In terms of intelligent data analysis, however, it is the learning of the Bayesian network from data that is of major importance. This includes learning of the structure of the network, identification and inclusion of hidden nodes, and learning of conditional probabilities that govern the networks (Szolovits, 1995, Lam, 1998). The data analysis then reasons about the structure of the network (examining the inter-variable dependencies) and the conditional probabilities (the strength and types of such dependencies). Examples of Bayesian network learning for medical data analysis include a genetic algorithm-based construction of a Bayesian network for predicting the survival in malignant skin melanoma (Larranaga *et al.*, 1997), learning temporal probabilistic causal models from longitudinal data (Riva and Bellazzi, 1996), learning conditional probabilities in modeling of the clinical outcome after bone marrow transplantation (Quaglini *et al.*, 1994), cerebral modeling (Labatut *et al.*, 2003) and cardiac SPECT image interpretation (Sacha *et al.*, 2002).

There are also different forms of unsupervised learning, where the input to the learner is a set of unclassified instances. Besides unsupervised learning using neural networks described in Section 58.3.2 and learning of association rules described in Section 58.4, other forms of unsupervised learning include conceptual clustering (Fisher, 1987, Michalski and Stepp, 1983) and qualitative modeling (Bratko, 1989).

The *data visualization techniques* may either complement or additionally support other data analysis techniques. They can be used in the preprocessing stage (e.g., initial data analysis and feature selection) and the postprocessing stage (e.g., visualization of results, tests of performance of classifiers, etc.). Visualization may support the analysis of the classifier and thus increase the comprehensibility of discovered relationships. For example, visualization of results of naive Bayesian classification may help to identify which are the important factors that speak for and against a diagnosis (Zelic *et al.*, 1997), and a 3D visualization of a decision tree may assist in tree exploration and increase its transparency (Kohavi *et al.*, 1997).

58.5 Conclusions

There are many Data Mining methods from which one can chose for mining the emerging medical data bases and repositories. In this chapter, we have reviewed most popular ones, and gave some pointers where they have been applied. Despite the potential and promising approaches, the utility of Data Mining methods to analyze medical data sets is still sparse, especially when compared to classical statistical approaches. It is gaining ground, however, in the areas where data is accompanied with knowledge bases, and where data repositories storing heterogenous data from different sources took ground.

Acknowledgments

This work was supported by the Slovenian Ministry of Education, Science and Sport. Thanks to Elpida Keravnou, Riccardo Bellazzi, Peter Flach, Peter Hammond, Jan Komorowski, Ramon M. Lopez de Mantaras, Silvia Miksch, Enric Plaza and Claude Sammut for their comments on individual parts of this chapter.

References

Aamodt, A. and Plaza, E., Case-based reasoning: Foundational issues, methodological variations, and system approaches, *AI Communications*, 7(1): 39–59 (1994).

Agrawal, R., Manilla, H., Srikant, R., Toivonen, H. and Verkamo A.I., "Fast discovery of association rules." In: Advances in Knowledge Discovery and Data Mining (Fayyad, U.M., Piatetsky-Shapiro, G., Smyth, P. and Uthurusamy, R., eds.), AAAI Press, 1996, pp. 307–328 (1996).

Aha, D., Kibler, D., Albert, M., "Instance-based learning algorithms," *Machine Learning*, 6(1): 37–66 (1991).

Andrews, R., Diederich, J. and Tickle, A.B., "A survey and critique of techniques for extracting rules from trained artificial neural networks," *Knowledge Based Systems*, 8(6): 373–389 (1995).

Andrews, P.J., Sleeman, D.H., Statham, P.F., *et al.* "Predicting recovery in patients suffering from traumatic brain injury by using admission variables and physiological data: a comparison between decision tree analysis and logistic regression." *J Neurosurg.* 97(2): 326-336 (2002).

Astion, M.L. and Wilding, P., "The application of backpropagation neural networks to problems in pathology and laboratory medicine," *Arch Pathol Lab Med*, 116(10): 995–1001 (1992).

Averbuch, M., Karson, T., Ben-Ami, B., Maimon, O., and Rokach, L. (2004). Context-sensitive medical information retrieval, MEDINFO-2004, San Francisco, CA, September. IOS Press, pp. 282-262.

Barlow, T.W., "Self-organizing maps and molecular similarity," *Journal of Molecular Graphics*, 13(1): 53–55 (1995).

Baxt, W.G. "Application of artificial neural networks to clinical medicine," *Lancet*, 364(8983) 1135–1138 (1995).

Becker, K., Thull, B., Kasmacher-Leidinger, H., Stemmer, J., Rau, G., Kalff, G. and Zimmermann, H.J. "Design and validation of an intelligent patient monitoring and alarm system based on a fuzzy logic process model," *Artificial Intelligence in Medicine*, 11(1): 33–54 (1997).

Bradburn, C., Zeleznikow, J. and Adams, A., "Florence: synthesis of case-based and model-based reasoning in a nursing care planning system," *Computers in Nursing*, 11(1): 20–24 (1993).

Bratko, I., Kononenko, I. Learning diagnostic rules from incomplete and noisy data. In Phelps, B. (ed.) *AI Methods in Statistics*. Gower Technical Press, 1987.

Bratko, I., Mozetič, I. and Lavrač, N., *KARDIO: A Study in Deep and Qualitative Knowledge for Expert Systems*, The MIT Press, 1989.

Breiman, L., Friedman, J.H., Olshen, R.A. and Stone, C.J., *Classification and Regression Trees*. Wadsworth, Belmont, 1984.

Brossette, S.E., Sprague, A.P., Hardin, J.M., Waites, K.B., Jones, W.T., Moser, S.A. "Association rules and Data Mining in hospital infection control and public health surveillance." *Journal of the Americal Medical Inform. Assoc.* 5(4): 373–81 (1998).

Breault, J.L., Goodall, C.R., Fos P.J., *Data Mining a Diabetic Data Warehouse*. Artificial Intelligence in Medicine 26(1-2): 37–54 2002.

Carpenter, G.A. and Tan, A.H., "Rule extraction, fuzzy artmap and medical databases." In: *Proc. World Cong. Neural Networks*, pp. 501–506 (1993).

Carrault, G., Cordier, M., Quiniou, R., Wang, F., *Temporal Abstraction and Inductive Logic Programming for Arrhythmia Recognition from Electrocardiograms*. Artificial Intelligence in Medicine 28(3): 231–236 (2003).

Caruana, R., Baluja, S., and Mitchell, T., "Using the Future to Sort Out the Present: Rankprop and Multitask Learning for Medical Risk Analysis," *Advances in Neural Information Processing Systems (NIPS*95)* 8: 959–965 (1995).

Cendrowka, J. "PRISM: An algorithm for inducing modular rules," *Int. J. Man-Machine Studies* 27: 349–370 (1987).

Cestnik B., "Estimating Probabilities: A Crucial Task in Machine Learning," In: *Proc. European Conf. on Artificial Intelligence*, pp. 147-149 (1990).

Cestnik B., Kononenko I., Bratko I., "ASSISTANT 86: A knowledge elicitation tool for sophisticated users." In: *Progress in Machine learning* (Bratko, I., Lavrač, N., eds.), Wilmslow: Sigma Press (1987).

Chung, F.L. and Lee, L. "A node prunning algorithm for backpropagation network," *Int. J. Neural Systems*, 3: 301–314 (1992).

Clark, P., Boswell, R., "Rule induction with CN2: Some recent improvements." In: *Proc. Fifth European Working Session on Learning*, Springer, pp. 151–163 (1991).

Clark, P., Niblett, T. The CN2 induction algorithm. *Machine Learning*, 3(4): 261–283 (1989).

Compton, P. and Jansen, R., "Knowledge in context: A strategy for expert system maintenance." In: *Proc. 2nd Australian Joint Artificial Intelligence Conference*, Springer LNAI 406, pp. 292–306 (1988).

Compton, P., Horn, R., Quinlan, R. and Lazarus, L., "Maintaining an expert system." In: *Applications of Expert Systems* (Quinlan, R., ed.), Addison Wesley, pp. 366–385 (1989).

Cover, T.M., Hart, P.E., "Nearest neighbor pattern classification," *IEEE Transactions on Information Theory*, 13: 21–27 (1968).

Craven, M.W., and Shavlik, J.W. (1997) "Using neural networks for Data Mining," *Future generation computer systems*, 13(2–3): 211–229 (1997).

Cristianini, N., and Shawe-Taylor, J. *An introduction to support vector machines and other kernel-based learning methods*. Cambridge Univ. Press (2000).

Dasarathy, B.V. (ed.) *Nearest Neighbor (NN) Norms: NN Pattern Classification Techniques.* IEEE Computer Society Press, Los Alamitos, CA (1990).

Dehaspe, L, Toivonen, H. and King, R.D. "Finding frequent substructures in chemical compounds." In: *Proc. 4th International Conference on Knowledge Discovery and Data Mining, (KDD-98)* (Agrawal, R., Stolorz, P. and Piatetsky-Shapiro, G., eds.), AAAI Press, pp. 30–37 (1998)..

De Raedt, L. and Dehaspe, L., "Clausal discovery." *Machine Learning*, 26: 99–146 (1997).

Di Bona, S., Niemann, H., Pieri, G., Salvetti, O., *Brain Volumes Characterisation Using Hierarchical Neural Networks*. AI in Medicine 28 (2003).

Downs, J., Harrison, R.F., Kennedy, R.L., and Cross, S.C., "Application of the fuzzy artmap neural network model to medical pattern classification tasks," *Artificial Intelligence in Medicine*, 8(4): 403–428 (1996).

Dudani, S.A., "The distance-weighted *k*-nearest neighbor rule," *IEEE Transactions on Systems, Man and Cybernetics*, 6(4): 325–327 (1975).

Dybowski R., Weller P., Chang R., Gant V. "Prediction of outcome in the critically ill using an artificial neural network synthesised by a genetic algorithm," *Lancet*, 347: 1146-1150 (1996).

Džeroski, S., Lavrač, N., "Rule induction and instance-based learning applied in medical diagnosis," *Technology and Health Care*, 4(2): 203–221 (1996).

Edwards, G., Compton, P., Malor, R., Srinivasan, A. and Lazarus, L., "PEIRS: A pathologist maintained expert system for the interpretation of chamical pathology reports," *Pathology* 25: 27–34 (1993).

Fayyad, U.M., Piatetsky-Shapiro, G., Smyth, P., "The KDD process for extracting useful knowledge from volumes of data," *Communications of the ACM*, 39(11):27–41 (1996).

Fausett, L.V., *Fundamentals of neural networks: Architectures, algorithms and applications*, Prentice Hall, Upper Saddle River, NJ (1994).

Fisher, D.H. (1987) Knowledge acquisition via incremental conceptual clustering. *Machine Learning* 2: 139–172 (1987).

Fix, E., Hodges, J.L., "Discriminatory analysis. Nonparametric discrimination. Consistency properties." Technical Report 4, US Air Force School of Aviation Medicine. Randolph Field, TX (1957).

Frawley, W., Piatetsky-Shapiro, G., Matheus, C. "Knowledge discovery in databases: An overview." In: *Knowledge discovery in databases* (Piatetsky-Shapiro, G., Frawley, W., eds.), The AAAI Press, Menlo Park, CA (1991).

Gamberger D. and Lavrač N., Expert-guided subgroup discovery: Methodology and application. *Journal of Artficial Intelligence Research* 17: 501–527 (2002).

Gamberger, D., Lavrač, N., & Krstačić, G. Active subgroup mining: A case study in coronary heart disease risk group detection. *Artificial Intelligence in Medicine* 28(1): 27–57 (2003).

Gaspari, M., Roveda, G., Scandellari, C., Stecchi, S., *An Expert System for the Evaluation of EDSS in Multiple Sclerosis*. Artificial Intelligence in Medicine 25(2): 187–210, (2001).

Goldberg, D.E. *Genetic Algorithms in Search, Optimization and Machine Learning*, pp. 68–74, Addison-Wesley (1989).

Gonzalez, J.S., Rodrigez, K., Garcia Nocetti, D.F., *Model-based Spectral Estimation of Doppler Signals Using Parallel Genetic Algorithms*. Artificial Intelligence in Medicine 19(1): 75–89 (1999).

Grzymała-Busse, J., "Applications of the rule induction systems LERS," In: (Polkowski and Skowron, 1998a), pp. 366–375 (1998).

Gyöngyösi, M., Ploner, M., Porenta, G., Sperker, W., Wexberg, P., Strehblow, C., Glogar, D., *Case-based Distance Measurements for the Selection of Controls in Case-matched studies: Application in Coronary Interventions*. Artificial Intelligence in Medicine 26(3): 237–53 (2002).

Ham, F.M. and Han, S. "Classification of cardiac arrhythmias using fuzzy artmap," *IEEE Transactions on Biomedical Engineering*, 43(4): 425–430 (1996).

Henson, D.B. , Spenceley, S.E., and Bull, D.R. "Artificial neural network analysis of noisy visual field data in glaucoma," *Artificial Intelligence in Medicine*, 10(2): 99–113 (1997).

Holte, R., Acker, L., Porter, B. "Concept learning and the problem of small disjuncts." In: *Proc. Tenth International Joint Conference on Artificial Intelligence*. Morgan Kaufmann, San Mateo, CA, pp. 813–818 (1989).

Horn, K., Compton, P.J., Lazarus, L. and Quinlan, J.R. "An expert system for the interpretation of thyroid assays in a clnical laboratory," *Austr. Comput. Journal* 17(1): 7–11 (1985).
Kahn, C.E. Jr. and Anderson, G. M., "Case-based reasoning and imaging procedure selection," *Invest Radiol.*, 29(6): 643-647 (1994).
Kattan, M.W. and Beck, J.R., "Artificial neural networks for medical classification decisions," *Arch Pathol Lab Med*, 119: 672–677 (1995).
Kattan, M.W., Ishida, H., Scardino, P.T. and Beck, J.R., "Applying a neural network to prostate cancer survival data." In: *Intelligent data analysis in medicine and pharmacology* (Lavrač, N. Keravnou, E. and Zupan, B., eds.), Kluwer, pp. 295–306 (1997).
Kira, K., Rendell, L. "A practical approach to feature selection." In: *Proc. Intern. Conf. on Machine Learning* (Sleeman, D., Edwards, P., eds.), Aberdeen, Morgan Kaufmann, pp. 249–256 (1992).
Kira, K., Rendell, L. "The feature selection problem: traditional methods and new algorithm." In: *Proc. AAAI'92*, San Jose, pp. 129–134 (1992).
Koehle, M., Merkl, D., Kastner, J. "Clinical Gait Analysis by Neural Networks - Issues and Experiences." In: *Proc. IEEE Symposium on Computer-Based Medical Systems (CBMS'97)* (Kokol, P., Štiglič, B., eds.), Maribor, IEEE Press, pp. 138–143 (1997).
Kohavi, R., Sommerfield, D. and Dougherty, J., "Data Mining using MLC++, a machine learning library in C++," *International Journal of Artificial Intelligence Tools*, 6(4): 537–566 (1997).
Kohonen, T., *Self-organization and associative memory*, Springer-Verlag, New York (1988).
Komorowski, J. and Øhrn, A., "Modelling prognostic power of cardiac tests using rough sets," *Artificial Intelligence in Medicine*, 15(2): 167–91 (1998).
Kononenko, I., "Semi-naive Bayesian classifier." In: *Proc. European Working Session on Learning* (Kodratoff, Y., ed.), Porto, Springer, pp. 206–219 (1991).
Kononenko, I., "Inductive and Bayesian learning in medical diagnosis," *Applied Artificial Intelligence*, 7: 317–337 (1993).
Kononenko, I. (1994) "Estimating attributes: Analysis and extensions of Relief." In: *Proc. European Conf. on Machine Learning* (De Raedt, L., Bergadano, F., eds.), Catania, Springer, pp. 171–182 (1994).
Kononenko, I., Šimec, E. (1995) "Induction of decision trees using RELIEFF." In: *Proc. of ISSEK Workshop on Mathematical and Statistical Methods in Artificial Intelligence* (Della Riccia, G., Kruse, R., Viertl, R., eds.), Udine, September 1994, Springer, pp. 199–220 (1995).
Kovalerchuk, B., Triantaphyllou, E., Ruiz, J.F. and Clayton, J., "Fuzzy logic in computer-aided breast cancer diagnosis: analysis of lonulation," *Artificial Intelligence in Medicine*, 11(1): 75–87 (1997).
Labatut, V., Pastor, J., Ruff, S., Demonet, J.F., Celsis, P., *Cerebral Modeling and Dynamic Bayesian Networks. Artificial Intelligence in Medicine* 30(1): 119–39 (2003).
Lam, W. (1998), Bayesian network refinement via machine learning approach, *IEEE Transactions on Pattern Analysis and Machine Intelligence*, 20(3): 240–251 (1998).
Larranaga, P., Sierra, B., Gallego, M.J., Michelena, M.J., Picaza, J.M., Learning Bayesian networks by genetic algorithms: a case study in the prediction of survival in malignant skin melanoma. In *Proc. Artificial Intelligence in Medicine Europe* (E. Keravnou, C. Garbay, R. Baud, J. Wyatt, eds.), pp. 261–272 (1997).
Lavrač, N., Džeroski, S., Pirnat, V., Križman, V. "The utility of background knowledge in learning medical diagnostic rules," *Applied Artificial Intelligence*, 7: 273–293 (1993).
Lavrač, N., Džeroski, S., *Inductive Logic Programming: Techniques and Applications.* Ellis Horwood, Chichester (1994).

Lavrač, N., Kavšek, B., Flach, P., and Todorovski, L. Subgroup discovery with CN2-SD. *Journal of Machine Learning Research*, 5: 153–188, 2004.

Liestøl, K., Andersen, P.K. and Andersen, U. "Survival analysis and neural nets," *Statist. Med.*, 13(2): 1189–1200 (1994).

Lin, T.Y and Cercone, N., eds., "Rough Sets and Data Mining", Kluwer (1997).

Lubsen, J., Pool, J., van der Does, E. A practical device for the application of a diagnostic or prognostic function. *Methods Inf. Med.* 17(2): 127–129 (1978).

Macura, R.T. and Macura, K., eds., "Case-based reasoning: opportunities and applications in health care," *Artificial Intelligence in Medicine*, 9(1): 1–4 (1997).

Macura, R.T. and Macura, K., eds., *Artificial Intelligence in Medicine: Special Issue on Case-Based Reasoning*, 9(1) (1997).

Malmberg, L.P., Kallio, K., Haltsonen, S., Katila, T. and Sovijarvi, A.R., "Classification of lung sounds in patients with asthma, emphysema, fibrosing alveolitis and healthy lungs by using self-organizing maps," *Clinical Physiology*, 16(2): 115–129 (1996).

Mariuzzi, G., Mombello, A., Mariuzzi, L., Hamilton, P.W., Weber, J.E., Thompson D. and Bartels, P.H., "Quantitative study of ductal breast cancer–patient targeted prognosis: an exploration of case base reasoning," *Pathology, Research & Practice*, 193(8): 535–542 (1997).

Markey, M.K., Lo, J.Y., Tourassi, G.D. AND Floyd Jr., C.E., "Self-organizing map for cluster analysis of a breast cancer database," *Artificial Intelligence in Medicine*, 27(2): 113-127 (2002).

McSherry, D., "Hypothesist: A development environment for intelligent diagnostic systems." In: *Proc. Sixth Conference on Artificial Intelligence in Medicine* (AIME'97), Springer, pp. 223–234 (1997).

Michalski, R.S. and Stepp, R.E., "Learning from observation: Conceptual clustering." In: *Machine Learning: An AI Approach* (Michalski, R.S., Carbonell, J. and Mitchell, T.M., eds.), volume I, Palo Alto, CA. Tioga., pp. 331–363 (1983).

Michalski, R.S. (1986) "Understanding the nature of learning: Issues and research directions." In: *Machine Learning: An AI Approach* (Michalski, R.S., Carbonnel, J. and Mitchell, T.M., eds.) Morgan Kaufmann, pp. 3–25 (1986).

Michalski, R.S., Mozetič, I., Hong, J. and Lavrač, N., "The multi-purpose incremental learning system AQ15 and its testing application on three medical domains." In *Proc. Fifth National Conference on Artificial Intelligence*, Morgan Kaufmann, 1986, pp. 1041–1045.

Mizoguchi, F., Ohwada, H., Daidoji, M., Shirato, S., "Using Inductive Logic Programming to learn classification rules that identify glaucomatous eyes." In: *Intelligent Data Analysis in Medicine and Pharmacology* (Lavrač, N., Keravnou, E., Zupan, B., eds.), Kluwer, pp. 227–242 (1997).

Modai, I., Israel, A., Mendel, S., Hines, E.L. and Weizman, R., "Neural network based on adaptive resonance theory as compared to experts in suggesting treatment for schizophrenic and unipolar depressed in-patients," *Journal of Medical Systems*, 20(6): 403–412 (1996).

Muggleton, S., "Inverse entailment and Progol," *New Generation Computing, Special Issue on Inductive Logic Programming*, 13(3–4): 245–286 (1995).

Niblett, T. and Bratko, I., "Learning decision rules in noisy domains." In: *Research and Development in Expert Systems III* (Bramer, M., ed.), Cambridge University Press, pp. 24–25 (1986).

Pawlak, Z., Information systems – theoretical foundations. *Information Systems*, 6: 205–218 (1981).

Pawlak, Z., *Rough Sets: Theoretical Aspects of Reasoning about Data*, volume 9 of *Series D: System Theory, Knowledge Engineering and Problem Solving*. Kluwer (1991).

Pearl, J., *Probabilistic Reasoning in Intelligent Systems: Networks of Plausible Inference*. Morgan Kaufmann, San Mateo, CA (1988).

Pilih, I.A., Mladenič, D., Lavrač, N., Prevec, T.S., "Data analysis of patients with severe head injury." In: *Intelligent Data Analysis in Medicine and Pharmacology* (Lavrač, N., Keravnou, E., Zupan, B., eds.), Kluwer, pp. 131–148 (1997).

Polkowski, L. and Skowron, A., eds., *Rough Sets in Knowledge Discovery 1: Methodology and Applications*, volume 18 of *Studies in Fuzziness and Soft Computing*. Physica-Verlag (1998).

Polkowski, L. and Skowron, A., eds. (1998) *Rough Sets in Knowledge Discovery 2: Applications, Case Studies and Software Systems*, volume 18 of *Studies in Fuzziness and Soft Computing*. Physica-Verlag (1998).

Prochazka, A., "The fuzzy logic of visuomotor control," *Canadian Journal of Physiology & Pharmacology*, 74(4): 456–462 (1996).

Quaglini, S., Bellazzi, R., Locatelli, F., Stefanelli, M., Salvaneschi, C., "An Influence Diagram for Assessing GVHD Prophylaxis after Bone Marrow Transplantation in Children." *Medical Decision Making*, 14:223-235 (1994).

Quinlan, J.R., "Induction of decision trees." *Machine Learning*, 1(1): 81–106 (1986).

Quinlan, J.R., "Learning logical definitions from relations," *Machine Learning*, 5(3): 239–266 (1990).

Quinlan, J.R., *C4.5: Programs for Machine Learning*, San Mateo, CA, Morgan Kaufmann (1993).

Rau, G., Becker, K., Kaufmann, R. and Zimmermann, H.J., Fuzzy logic and control: principal approach and potential applications in medicine, *Artificial Organs*, 19(1): 105–112 (1995).

Richeldi, M., Rossotto, M., "Class-driven statistical discretization of continuous attributes." In: *Machine Learning: Proc. ECML-95* (Lavrač, N., Wrobel, S., eds.), Springer, pp. 335-342 (1995).

Riva, A. and Bellazzi, R., "Learning Temporal Probabilistic Causal Models from Longitudinal Data." *Artificial Intelligence in Medicine*, 8(3): 217–234 (1996).

Rokach, L., Averbuch, M., and Maimon, O., Information retrieval system for medical narrative reports (pp. 217228). Lecture notes in artificial intelligence, 3055. Springer-Verlag (2004).

Rosetta: A rough set toolkit for the analysis of data.

Rumelhart, D.E. and McClelland, J.L., eds., *Parallel Distributed Processing, Vol. 1: Foundations*. MIT Press, Cambridge, MA (1986).

Sacha, J.P., Goodenday, L.S. and Cios, K.J., "Bayesian Learning for Cardiac SPECT Image Interpretation". AI in Medicine 26(1): 109–143 (2002).

Sammut, C., "Introduction to Ripple Down Rules." `http://www.cse.unsw.edu.au/~claude/teaching/AI/notes/prolog/Extenions/rdr.lens.html` (1998).

Setiono, R., "Extracting rules from pruned networks for breast cancer diagnosis," *Artificial Intelligence in Medicine*, 8(1): 37–51 (1996).

Setiono, R. "Extracting rules from neural networks by pruning and hidden-unit splitting," *Neural Computation*, 9(1): 205–225 (1997).

Setiono, R. "Generating Concise and Accurate Classification Rules for Breast Cancer Diagnosis" *Artificial Intelligence in Medicine*, 18(3): 205–219, (1999).

Shannon, C.E., "A mathematical theory of communication." *Bell. Syst. Techn. J.*, 27: 379–423 (1948).

Shawlik, J.W., Mooney, R.J. and Towell, G.G., "Symbolic and neural learning algorithms: An experimental comparison," *Machine Learning*, 6(2): 111–143 (1991).

Skowron, A., "Synthesis of adaptive decision systems from experimantal data (invited talk)." In: *Proc. of the Fifth Scandinavian Conference on Artificial Intelligence SCAI-95* (A. Aamodt and J. Komorowski, eds.), IOS Press Ohmsa, Amsterdam, pp. 220–238 (1995).

Stel, V.S., Pluijm, S.M., Deeg, D.J., Smit, J.H., Bouter, L.M., Lips, P. "A classification tree for predicting recurrent falling in community-dwelling older persons." *J Am Geriatr Soc.* 51(10): 1356–1364 (2003).

Srinivasan, A., King, R.D., Muggleton, S.H. and Sternberg, M.J.E., "Carcinogenesis predictions using inductive logic programming." In *Intelligent Data Analysis in Medicine and Pharmacology* (Lavrač, N. Keravnou, E. and Zupan, B., eds.), Kluwer, pp. 243–260 (1997)..

Steinmann, F., "Fuzzy set theory in medicine," *Artificial Intelligence in Medicine*, 11(1) 1–7 (1997).

Szolovits, P., "Uncertainty and Decision in Medical Informatics," *Methods of Information in Medicine*, 34: 111–121 (1995).

Tsumoto, S., "Modelling medical diagnostic rules based on rough sets", In: *Proc. First International Conference on Rough Sets and Soft Computing – RSCTC'98* (Polkowski, L. and Skowron, A., eds.), volume 1424 of *Lecture Notes in Artificial Intelligence, Springer Verlag*. Springer, pp. 475–482 (1998).

Vapnik, V.N. *Statistical Learning Theory*. Wiley (1998).

Vinterbo, S. and Ohno-Machado, L., *A Genetic algorithm approach to multi-disorder diagnosis. AI in Medicine*, 18(2): 117–32 (1999).

Weiss, S.M., Kulikowski, C.A., *Computer Systems that Learn*. Morgan Kaufmann, San Mateo, CA, 1991.

Wettschereck, D., "A study of distance-based machine learning algorithms," PhD Thesis, Department of Computer Science, Oregon State University, Corvallis, OR (1994).

Witten, I. H. and E. Frank, *Data Mining: practical machine learning tools and techniques with Java implementations*. Morgan Kaufmann, San Francisco, CA (1999).

Wolpert, D., "Constructing a generalizer superior to NETtalk via mathematical theory of generalization," *Neural Networks*, 3: 445–452 (1989).

Wrobel, S., "An algorithm for multi-relational discovery of subgroups." In: *Proc. First European Symposium on Principles of Data Mining and Knowledge Discovery*, Springer, pp. 78–87 (1997).

Wu, C.H., Artificial neural networks for molecular sequence analysis, *Computers & Chemistry*, 21(4): 237–56 (1997).

Zadeh, L.A., Fuzzy sets. *Information and Control*, vol. 8, pp. 338–353 (1965).

Zelič, I., Kononenko, I., Lavrač, N., Vuga, V., "Induction of decision trees and Bayesian classification applied to diagnosis of sport injuries," *Journal of Medical Systems*, 21(6): 429–444 (1997).

Zhu, Y. and Yan, H., "Computerized tumor boundary detection using a hopfield neural network," *IEEE Transactions on Medical Imaging*, 16(1): 55–67 (1997).

Zupan, B. and Džeroski, S., "Acquiring and validating background knowledge for machine learning using function decomposition," *Artificial Intelligence in Medicine*, 14(1–2): 101–118 (1998).

Zupan, B., Halter, J.A. and Bohanec, M. (1997) "Concept discovery by decision table decomposition and its application in neurophysiology." In *Intelligent Data Analysis in Medicine and Pharmacology* (Lavrač, N., Keravnou, E. and Zupan, B., eds.), Kluwer, pp. 261–277 (1997).

Zupan, B., Bohanec, M., Demšar, J. and Bratko, I., "Feature transformation by function decomposition," *IEEE Intelligent Systems*, 13(2): 38–43 (1998).

Zupan, B., Demsar, J., Smrke, D., Bozikov, K., Stankovski, V., Bratko, I., Beck, J.R., "Predicting patient's long-term clinical status after hip arthroplasty using hierarchical decision modelling and Data Mining." *Methods Inf Med.* 40(1): 25–31 (2001).

59

Learning Information Patterns in Biological Databases - Stochastic Data Mining

Gautam B. Singh

Department of Computer Science and Engineering, Center for Bioinformatics, Oakland University, Rochester, MI 48309, USA.

Summary. This chapter aims at developing the computational theory for modeling patterns and their hierarchical coordination within biological sequences. With the exception of the promoters and enhancers, the functional significance of the non-coding DNA is not well understood. Scientists are now discovering that specific regions of non-coding DNA interact with the cellular machinery and help bring about the expression of genes. Our premise is that it is possible to study the arrangements of patterns in biological sequences through machine learning algorithms. As the biological database continue their exponential growth, it becomes feasible to apply *in-silico* Data Mining algorithms to discover interesting patterns of motif arrangements and the frequency of their re-iteration. A systematic procedure for achieving this goal is presented.

Key words: DNA Pattern Model, Hidden Markov Model, Matrix Attachment Regions, Transcription Potential, Promoter Models

59.1 Background

Every cell in an organism contains a genetic program that forms the basis of life. This program is stored inside the nucleus in eukaryotes (multi-cellular organisms, such as humans, primates, etc.) as a linear macromolecule comprised of nitrogenous bases. This large macromolecule is also known as DNA or deoxy-ribonucleic acid. A DNA molecule is comprised of four *bases*, namely, Adenosine (A), Cytosine (C), Thymidine (T) and Guanine (G). The DNA bases occur in pairs, with the base A always pairing with T, and the base C pairing with G. Accordingly, the length of a DNA macromolecule, collectively referred to as the *genome*, is measured as *base-pairs* or *bp*. The size of the human genome is 3×10^9 bp and stores the blueprints for the synthesis of a variety of proteins – the macromolecules that enable an organism to be structurally and functionally viable.

The blueprint or the program for the synthesis of a single protein is called a *gene*, a unit of the DNA sequence that is generally between 1×10^3 to 1×10^6 in length based upon the

O. Maimon, L. Rokach (eds.), *Data Mining and Knowledge Discovery Handbook*, 2nd ed.,
DOI 10.1007/978-0-387-09823-4_59,

complexity of the protein that it *codes* for. The process of synthesizing a protein using its genetically coded blueprint is known as *gene expression*. A higher level eukaryote may contain as many as 30,000-40,000 genes. It has also been estimated that only about 2000-10,000 genes may be expressed by a given cell. Although each cell in an organism contains the same genetic program, the subset of genes expressed in one cell type is different from another and is determined by the functional role of that cell. Apart from the gene *coding* regions that account for ~2% of the genome, the majority of nuclear DNA is *non-coding*. However, non-coding appears to be important as researchers believe that the genetic program is able to regulate the expression of genes based upon the biologically significant DNA sequence patterns that are primarily observed in the non-coding regions within the neighborhood of a gene (Kliensmith and Kish, 1995).

DNA is not a homogeneous string of characters, but is comprised of a mosaic of sequence level patterns that come together in a synergistic manner to coordinate and regulate in synthesis of proteins. Special sequences of regulatory importance such as introns, promoters, enhancers, Matrix Association Regions (MARs), and repeats are found in the non-coding DNA. These regions contain DNA sequence patterns that represent functional control points for cell specific or *differential* gene expression (Roeder, 1996, Kadonaga, 1998), while others such as the *repetitive* DNA patterns serve as a biological clock (Hartwell and Kasten, 1994). These and numerous other examples suggest that non-coding sequence motifs often characterized by complex patterns play a vital role for the organism's viability; and that deviations from patterns are deleterious to the organism.

In this chapter our focus is to begin with the *known* set of DNA-level sequences or patterns and find a higher level organization that is represented in large number sequences present in the genomic databases. In this regard, our present state of knowledge is like that of a blind man feeling different parts of an elephant. He does not see the whole elephant and thus the abstraction that the parts belong to the whole is missing from his cognition. His focus is on individual parts of the body and his efforts directed towards describing their shape, size, texture and other properties.

The field of biological sciences is in a similar predicament. Experimental biologists often aim to find any higher-level organizations and relationships between lower level patterns. As an evidence of our limited understanding in this regard, consider the case of the two commonly appearing patterns on the DNA, represented as the sequence patterns TATAAA (TATA-box) and CCAATCT (CAAT-box). Their relative locations, when both are present near a gene, are shown in Figure 59.1. We know that most genes exhibit the presence of a TATA-box, and some genes exhibit the presence of a CAAT-box within a 80-100 bp vicinity of each other (Penotti, 1990,Nussinov, 1991,Bucher and Trifonov, 1231). However, what we don't know is the answer to this simple question: *Does the presence of a CAAT-box imply the presence of a TATA-box?*

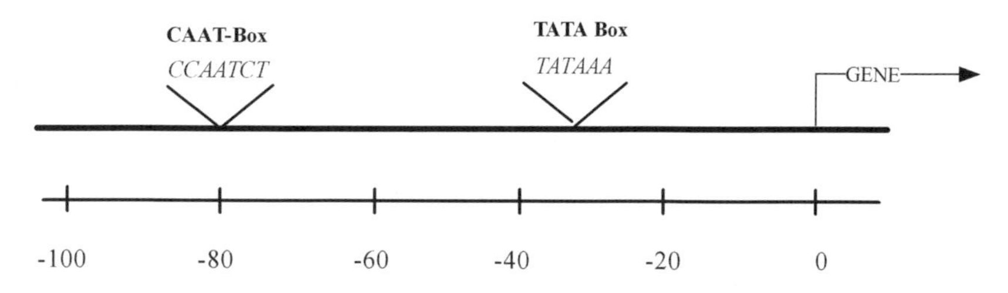

Fig. 59.1. Promoter samples from genomic databases can yield the frequency with which TATA-Box and CAAT-Box co-occur.

Common belief is that these organizational structures represent the solution of a jigsaw puzzle for which we are continually discovering newer and newer pieces. The question before us is whether we can discover an organization of these patterns using an *unsupervised learning* approach. The answer to this question is *yes* and is based upon the following premise. Any higher level organization of these patterns should be evident with sufficient support in an exhaustive analysis of the vast amounts of DNA sequence data that we have accumulated. Such *meta-pattern* should include both the existential and structural pattern organizations. In an existential organization, the existence of one pattern implies the presence or the absence of another. The structural organization further imposes a distance constraint by requiring the dependent pattern to occur within a pre-specified distance.

59.2 Learning Stochastic Pattern Models

The sequence level patterns observed on a DNA are the anchor points where the class of molecules, called the *Transcription Factors* or TFs, bind and initiate gene expression. For example, when the cell is in a state of heat shock, one of the heat-shock transcription factor is produced that binds to the DNA string where the pattern CTACCAAAATAACG is located. One can expect to find this pattern close to a gene that needs to be expressed when the cell is experiencing a heat-shock. Generally, the binding footprints for most transcription factors are about 5–50 bp (Nussinov, 1991, Faisst and Meyer, 1992, Berg and Hippel, 1987, Ghosh, 1990) in length. Due to the continued research in transcription factors and their functional implications, the information specifying their chemical properties as well as their DNA-binding footprint (which we refer to as the *pattern, or the DNA-level pattern*) has been steadily accumulating. Repositories containing this and a variety of other information on transcription factors are maintained by Genome Centers in the US, Europe and Japan (Ghosh, 1998, Wingender *et al.*, 2001, Margoulis *et al.*, 2002, Matys *et al.*, 2003). Moreover, as researchers continue to deposit DNA sequences into international databanks, such as GenBank, GSDB, EMBL and DDBJ, their size also continues to grow at an exponential rate.

The processing flow for learning the mathematical models for biological patterns is shown in Figure 59.2. Pattern data is assimilated from a variety of pattern databases and processed to remove redundancy and reduced to eliminate information not necessary for model generation. The patterns thus assimilated are next clustered into groups based on their sequence level similarity. The clustering process will thus partition the input patterns into groups of similar DNA-patterns. A stochastic model such as Position Specific Score Matrix (PSSM) or a Hidden Markov Model (HMM) is next associated with each of the pattern clusters. These steps are described in more detail below.

59.2.1 Assimilating the Pattern Sets

This step requires us to filter the appropriate fields from the pattern databases such as TFD, TRANSFAC, SMART DB, etc. The databases such as the Transcription Factor Database (TFD) incorporate a *relational schema* that enable them to efficiently manage the relationships between sequence motifs, the factors that bind there, the domains that they belong to, and the gene/organism that they are specific to. TFD is updated 4 times every year (Ghosh, 1998). The other database on transcription factors, TRANSFAC, is based in Heidelberg, Germany (Wingender *et al.*, 2001, Margoulis *et al.*, 2002) and records information similar to that found in TFD. The additional information contained in the TRANSFAC, as described below, describes

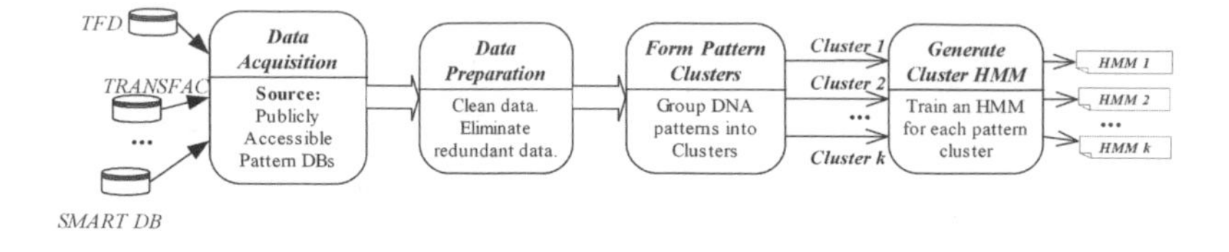

Fig. 59.2. The processing stages required for generating the Markovian models for DNA patterns

these elements and defines consensus and matrices for elements of certain function, and thus to provide means of identifying regulatory signals in anonymous genomic sequences.

TRANSFAC: Transcription Factors and Regulation

The development path for the TRANSFAC database has been geared by the objective to provide a *biological context* for understanding the function of regulatory signals found in genomic sequences. The aim of this compilation of signals was meant to provide all relevant data about the regulating proteins and allow researchers to trace back transcriptional control cascades to their origin (Wingender *et al.*, 2001, Matys *et al.*, 2003). The TRANSFAC database contains information about regulatory DNA sequences and the transcription factors binding to and acting through them. At the core of this database are its components describing the transcription factor (FACTOR) and its corresponding binding site (SITE) and the regulation of the corresponding gene (GENE). The GENE table is one of the central tables in this database. It is linked to several other databases including S/MARtDB (Liebich *et al.*, 2002), TransCOMPEL (Margoulis *et al.*, 2002), LocusLink, OMIM, and RefSeq (Wheeler *et al.*, 2004).

Sites are experimentally proven for their inclusion in the database. The experimental evidence of the transcription factor and the DNA-binding site is described, and the cell type from which the factor is derived is linked to the respective entry in the CELL table. A set of weight matrices are derived from the collection of binding sites. These matrices are recorded in the MATRIX table. Moreover, as determined by their DNA-binding domain, the transcription factors are assigned to a certain class, and hence link to the CLASS table is established.

The starting point for accessing these databases is the following web-site: www.gene-regulation.com. As an example, consider the following somewhat edited entry from the SITES table in the TRANSFAC database shown in Figure 59.3. The entry provides a wide variety of information about the transcription factor, such as the binding sequence motif (SQ) and the first (SF) and the last position (ST) of the factor binding site. The accession number of the binding factor itself is provided (BF) – this is in fact a key for the FACTOR table in TRANSFAC. The source of the factor is identified (SO). The specific type of cells where the factor was found to be active are identified, 3T3, C2 myoblasts, and F9 in this case. Additional information about these cells is accessible under the CELL table with the accession numbers of 0003, 0042 and 0069 respectively. External database references and their corresponding accession numbers are provided under the (DR) field, as well as publication titles (RT) and citation information (RL).

```
AC   R00026
TY   D
DE   CA-ACT (cardiac alpha-actin); G000193.
SQ   CCAAATAAGG.
SF   -113
ST   -84
BF   T00765 SRF (504 AA); Quality: 4; Species: mouse, Mus musculus.
OS   human, Homo sapiens
OC   eukaryota; animalia; metazoa; chordata; vertebrata; tetrapo da; mammalia;
OC   eutheria; primates
SO   0003 3T3
SO   0042 C2 myoblasts
SO   0069 F9
MM   gel retardation
DR   EMBL: M13483; HSACTCA (377:386).
DR   EPD: EP16033; HS_ACTC.
RN   [1]
RX   MEDLINE; 89093119.
RA   Boxer L. M., Miwa T., Gustafson T. A., Kede s L.
RT   Identification and Characterization of a Factor That Binds to Two Human
RT   Sarcomeric Actin Promoters
RL   J. Biol. Chem. 264:1284-1292 (1989).
//
```

Fig. 59.3. A sample record from TRANSFAC database

Another set of patterns are significant for bringing about the *structural modifications* to the DNA. It is necessary for the DNA to be in a structurally *open* conformation[1] for the gene expression to successfully occur. The MARS or Matrix Attachment regions are relatively short (100-1000 bp long) sequences that anchor the chromatin loops to the nuclear matrix and enable it to adopt the open conformation needed for gene expression (Bode, 1996). Approximately 100,000 matrix attachment sites are believed to exist in the mammalian nucleus, a number that roughly equals the number of genes. MARs have been observed to flank the ends of genic domains encompassing various transcriptional units (Bode, 1996,Nikolaev *et al.*, 1996). A list of structural motifs that are responsible for attaching DNA to the nuclear matrix is shown in Table 59.1.

59.2.2 Clustering Biological Patterns

Clustering is an important step as it directly impacts the success of the downstream model generation process. Given a set of sequence patterns, *S*, the objective of the clustering process is to partition these into groups such that each group represents patterns that are either related due to sequence level, functional or structural similarity. The pattern similarity measured purely at the sequence level can be measured by the string-edit or *Levenstein's* distance. In most cases, the sequence level similarity implies functional and/or structural similarity. However, sometimes known similarity in function (for example, the categorization of MAR specific patterns above) may be used to form clusters regardless of the sequence level similarity.

Consider a given sequence pair, $\vec{a} = a_1 a_2 ... a_n$ and $\vec{b} = b_1 b_2 ... b_m$, where both the sequences are defined over the alphabet, $A=\{A,C,T,G\}$. Let $d(a_i, b_j)$ denote the distance be-

[1] Within a cell, the DNA can be in a loosely packed, open conformation by adopting a 11 nm fiber structure, or in a tightly packed, closed conformation by adopting a 30 nm fiber structure.

Table 59.1. Polymorphism is commonly observed in biological patterns. A stochastic basis for pattern representation is thus justifiable. The list of motifs that are functionally related to MARs was generated by studying related literature.

Index	Motif Name	DNA Signature
m_1	ORI Signal	ATTA
m_2	ORI Signal	ATTTA
m_3	ORI Signal	ATTTTA
m_4	TG-Rich Signal	TGTTTTG
m_5	TG-Rich Signal	TGTTTTTTG
m_6	TG-Rich Signal	TTTTGGGG
m_7	Curved DNA Signal	AAAANNNNNNNAAAANNNNNNNAAAA
m_8	Curved DNA Signal	TTTTNNNNNNNTTTTNNNNNNNTTTT
m_9	Curved DNA Signal	TTTAAA
m_{10}	Kinked DNA Signal	TANNNTGNNNCA
m_{11}	Kinked DNA Signal	TANNNCANNNTG
m_{12}	Kinked DNA Signal	TGNNNTANNNCA
m_{13}	Kinked DNA Signal	TGNNNCANNNTA
m_{14}	Kinked DNA Signal	CANNNTANNNTG
m_{15}	Kinked DNA Signal	CANNNTGNNNTA
m_{16}	mtopo-II Signal	RNYNNCNNGYNGKTNYNY
m_{17}	dtopo-II Signal	GTNWAYATTNATNNR
m_{18}	AT-Rich Signal	WWWWWW

tween the ith symbol of sequence $\vec{a}$ and j^{th} symbol of sequence $\vec{b}$. $d(a_i, b_j)$ is defined as $|g_i - g_j|$. Also, let $g(k)$ be the cost of inserting (or deleting) an additional gap of size k. If the distance between these two pattern sequences of lengths m and n is denoted as $D_{m,n}$, the recursive formulation of *Levenstein's* distance is defined by

$$D_{i,j} = Min \begin{cases} D_{i-1,j-1} + d(a_i, b_j), \\ Min_{1 \le k \le j}\{D_{i,j-k} + g(k)\}, \\ Min_{1 \le l \le i}\{D_{i-l,j} + g(l)\} \end{cases} \tag{59.1}$$

Having computed the similarity between all pattern pairs, a clustering algorithm described below is applied for grouping these into *pattern-clusters*. This clustering approach is based upon the work described in (Zahn, 1971, Page, 1974). In this graph-theoretic approach, each vertex v_x represents a pattern $x \in S$, belonging to the set of patterns being clustered. The normalized *Levenstein's* distance between two patterns x and y, denoted as δ_{xy}, is the weight of an edge e_{xy} connecting vertices v_x and v_y. The clustering process proceeds as follows:

- **Construct a Minimum Spanning Tree (MST)**. The MST covers the entire set S of patterns. The MST is build using Prim's algorithm. Since the MST covers the entire set of training sequences, it is considered to be the *root-cluster* that is iteratively sub-divided into smaller child clusters, by repeated applications of steps and below.
- **Identify an Inconsistent Edge in the MST**. This process is based on the value of mean μ_i and standard deviation σ_i of distance values for edges in a cluster C_i. The cluster and edge with the largest *z-score*, $z_{\max}$ is identified. The variable e^i_{jk} denotes weight of an edge in this cluster C_i.

$$z_{max} = \underset{i}{Max} \underset{j,k \in C_i}{Max} \left(\frac{e^i_{jk} - \mu_i}{\sigma_i} \right) \tag{59.2}$$

- **Remove the Inconsistent Edge:** The edge identified in step above is further subject to the condition that its *z-score* be larger than pre-specified threshold. If this condition is satisfied, the inconsistent edge is removed, causing the cluster containing the inconsistent edge to be split into two child clusters[2] . However, if the edge's *z-score* falls below the

[2] Removing any edge of a tree (the MST in this case) causes the tree to be split into two trees (into two MSTs in this case).

threshold, the iterative subdivision process halts. It may be noted that the threshold for inconsistent edge removal is often specified in terms of the σ.

The categorization of the DNA pattern sequences into appropriate clusters is essential to train the pattern models as described in the following section. The quality of each cluster will be assessed to ensure that sufficient examples exist in the cluster to train a stochastic model. In the absence of sufficient examples, the patterns will be represented as boolean decision trees.

59.2.3 Learning Cluster Models

A DNA sequence matrix is a set of fixed-length DNA sequence segments aligned with respect to an experimentally determined biologically significant site. The columns of a DNA sequence matrix are numbered with respect to the biological site, usually starting with a negative number. A DNA sequence motif can be defined as a matrix of depth 4 utilizing a *cut-off* value. The 4-column/mononucleotide matrix description of a genetic signal is based on the assumptions that the motif is of fixed length, and that each nucleotide is independently recognized by a *trans*-acting mechanism. For example the following frequency matrix has been reported for the TATAA box.

Table 59.2. Weight Matrix for TATA Box

A	8	4	58	4	51	38	53	30
C	14	6	0	0	3	0	1	2
G	32	1	1	0	0	0	0	8
T	6	49	1	56	6	22	6	20

If a set of aligned signal sequences of length "L" corresponding to the functional signal under consideration, then $F = [f_{bi}], (b \in \Sigma), (j = 1..L)$ is the nucleotide frequency matrix, where f_{bi} is the absolute frequency of occurrence of the *b-th* type of the nucleotide out of the set $\Sigma = \{A, C, G, T\}$ at the *i-th* position along the functional site.

The frequency matrix may be utilized for developing an un-gapped score model when searching for the sites in a sequence. Typically a log-odds scoring scheme is utilized for this purpose of searching for pattern $x$of length Las shown in Eq. (63.3). The quantity $e_i(b)$ specifies the probability of observing the base b at position i is defined using the frequency matrix such as the one shown above. The quantity $q(b)$ represents the background probability for the base b.

$$S = \sum_{i=1}^{L} \log \frac{e_i(x_i)}{q(x_i)} \tag{59.3}$$

The elements of $\log \frac{e_i(x_i)}{q(x_i)}$ behave like a scoring matrix similar to the PAM and BLOSUM matrices. The term Position Specific Scoring Matrix (PSSM) is often used to define the pattern search with matrix. A PSSM can be used to search for a match in a longer sequence by evaluating a score S_j,for each starting point j in the sequence from position 1 to $(N-L+1)$ where $L$is the length of the PSSM. These optimized weight matrices can be used to search for functional signals in the nucleotide sequences. Any nucleotide fragment of length L is analyzed and tested for assignment to the proper functional signal. A matching score of $\sum_{i=1}^{L} W(b_i, i)$ is assigned to the nucleotide position being examined along the sequence. In the search formulation, $b_i$is the base at position i along the biological sequence, and $W(b_i, i)$ represents the corresponding weight matrix entry for symbol b_i occurring at position i along the motif. A more detailed example for learning the PSSM for a pattern cluster is shown in Figure 59.4(a).

A stochastic extension of the PSSM is based on a Markovian representation of biological sequence patterns. As a first step toward learning the pattern-HMM one of the two common HMM architectures must be selected to define the topology. These are the *fully connected ergodic architecture* and the *Left Right* (LR) architecture. The fully connected architecture offers a higher level modeling capability, but generally requires a larger set of training data. The left-right configuration on the other hand is powerful enough for modeling sequences, does not require a large training set, and facilitates model comprehension. Moderate level of available training data often dictates that the LR-HMM be utilized for representing the pattern clusters.

The initial parameters for the pattern-HMM are assigned heuristically. The number of states, N, denoted as, $S = \{S_1, S_2 ..., S_N\}$, in a pattern-HMM may set to as large a value as the total number of DNA symbols in the longest pattern in that cluster. Smaller number of states are heuristically chosen in practice. With each state S_i, an emission probability vector corresponding to the emission of each of the symbols, $\{A,C,T,G\}$, is associated. The process of generating each pattern is sequential such that the xth symbol generated, D_x, is a result of the HMM being in a hidden state $q_x = S_i$. The parameters of the HMM are denoted as λ=$\{A,B,\pi\}$ and defined as follows (Rabiner, 1989).

A: The $N \times N$ matrix $A = \{a_{i,j}\}$ representing the state transition probabilities.

$$a_{ij} = Pr[q_{x+1} = S_j | q_x = S_i] \qquad 1 \leq i,j \leq N \tag{59.4}$$

B: The $N \times k$ state dependent observation symbol probability matrix for each base n=$\{A,C,T,G\}$. The elements of this matrix, $B = \{b_j(n)\}$, are defined as follows:

$$b_j(n) = Pr[D_x = n | q_x = S_j] \qquad 1 \leq j \leq N, 1 \leq d \leq k \tag{59.5}$$

π: The initial state distribution probabilities, $\pi = \{\pi_i\}$.

$$\pi_i = Pr[q_1 = S_i] \qquad 1 \leq i \leq N \tag{59.6}$$

The Maximally Likelihood Estimation procedure suggested by BaumWelch is next utilized for training the each pattern-HMM such that the pattern sequences in a cluster would be the *maximally likely* set of samples generated by the underlying HMM. Figure 59.4(b) represents the training methodology applied for learning the HMM parameters based on the local alignment block used for training the PSSM in Figure 59.4(a).

Thus, pattern HMMs may be associated with clusters where the number of instances is large enough to allow us to adequately learn its parameters. In the case of smaller clusters, the pattern clusters will be represented as PSSMs, profiles or regular expressions. Profiles are similar to PSSMs (Gribskov *et al.*, 1990) and are generated using the sequences in a cluster when the alignment between the members of a cluster is strong. Regular expressions constitute the method of choice for smaller groups of shorter patterns where compositional statistics are hard to evaluate.

59.3 Searching for Meta-Patterns

The process of discovering hierarchical pattern associations is posed in terms of the relationships between models of a family of patterns, rather than between individual patterns. This

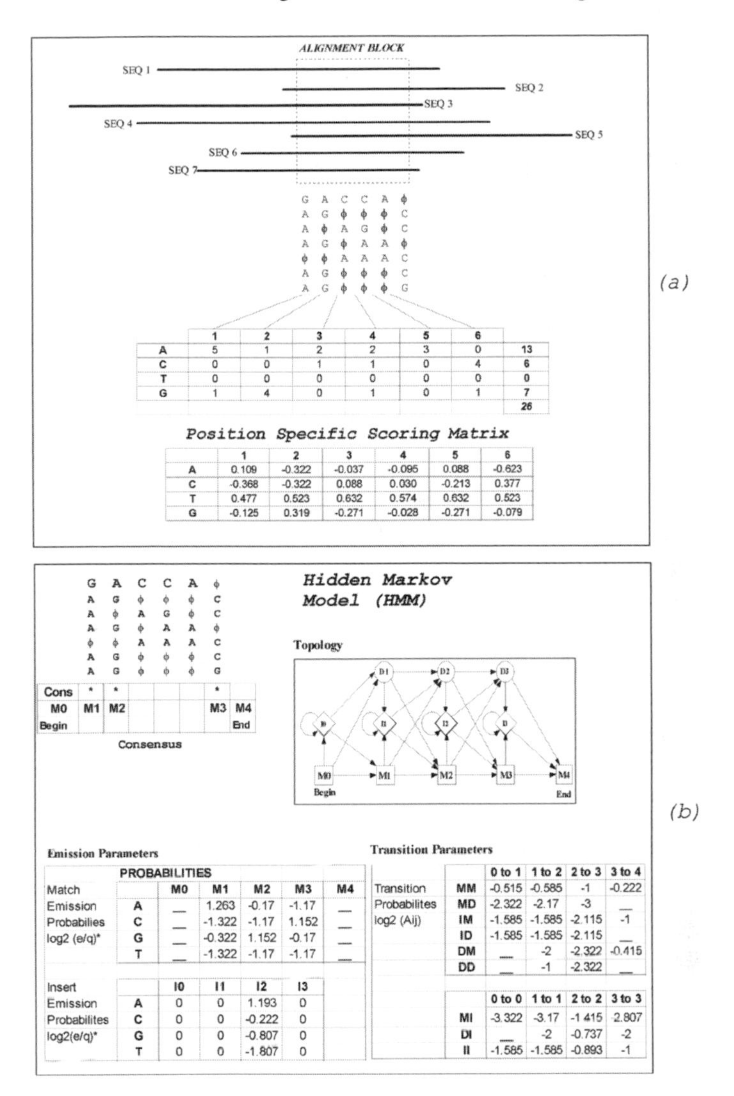

Fig. 59.4. (a) A PSSM based model induced from a Multiple Sequence Alignment (b) A HMM induced from the same alignment

will enable us to validate the meta-pattern hypotheses in a computationally tractable manner. The patterns are *associated* when they occur within a specific distance of each other, called their *association interval*. The association interval will be established using a split-and-merge procedure. Using a default association interval of 1000 bp, the overall significance of patterns found by splitting this intervals is assessed. Additionally, the neighboring windows are merged to assess the statistical significance of larger regions. In this manner, the region with the highest level of significance is considered as the association interval for a group of patterns.

The statistical significance is associated with each pattern-model pair detected within an association interval. This is achieved through two levels of searching the GenBank[3] . Level I search yields the regions that exhibit a high concentration of patterns. This is the first step toward generating pattern association hypotheses that are biologically significant, as patterns working in coordination are generally expected to be localized close to each other. In Level II search aims at building support and confidence where Level I hypotheses may be accepted or rejected based on pre-specified criteria. Consider, for example, two patterns A and B where there is a strong correlation between these two pattern HMMs in the Level I search. However, Level II search may reveal that there are a substantially large number of instances outside the *high pattern density* regions where their occurrence is independent of each other. This will lead to the rejection of the $A \equiv B$ hypothesis.

59.3.1 Level I Search: Locating High Pattern Density Region

High Pattern Density Regions or HPDRs aims at isolating the regions on the DNA sequence where the patterns modeled by the HMMs occur in a density that is higher than expected. The level I search is aimed at identifying HPDR as shown in Figure 59.5. These regions may be located by measuring the significance of patterns detected in a window of size W located at a given position on the sequence. A numerical value for *pattern-density* at location x on the DNA sequence is obtained by treating the pattern occurrences within a window centered at location x as trials from independent Poisson processes. The null hypothesis, H_0, tested in each window is essentially that the pattern frequencies observed in the window are no different from those expected in a random sequence.

Large deviation from the expected frequency of patterns in a window forces the rejection of H_0. The level of confidence with which H_0 is rejected is used to assign a *statistical* pattern-density metric to the window. Specifically, the pattern density in a window is defined to be, $\rho = -log(p)$, where the p is the probability of erroneously rejecting H_0. As a matter of detail it may be noted that the value of ρ is computed for both the forward and the reverse DNA strands and the average of the two is taken to be the true density estimate for that location.

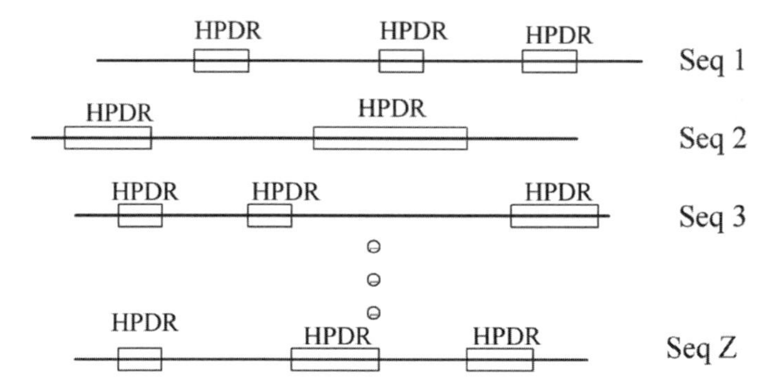

Fig. 59.5. High Pattern Density Regions or HPDRs are detected by statistical means for all sequences in the database.

[3] GenBank is the database of DNA sequences that is publicly accessible from the National Institute of Health, Bethesda, MD, USA.

In order to compute ρ, assume that we are searching for k distinct types of patterns within a given window of the sequence. In general, these patterns are defined as rules R_1, $R_2, \ldots,$ R_k. The probability of random occurrence of the various k patterns is calculated using the AND-OR relationships between the individual motifs. Assume that these probabilities for k patterns are p_1, $p_2, \ldots, p_k$. Next, a random vector of pattern frequencies, F, is constructed. F is a *k-dimensional* vector with components, $F = \{x_1, x_2, \ldots, x_k\}$, where each component x_i is a random variable representing the frequency of the pattern R_i in the W base-pair window. The component random variables x_i are assumed to be *independently* distributed Poisson processes, each with the parameter $\lambda_i = p_i \cdot W$. Thus, the joint probability of observing a frequency vector $F_{obs} = \{f_1, f_2, \ldots, f_k\}$ purely by chance is given by:

$$P(F_{obs}) = \prod_{i=1}^{k} \frac{e^{-\lambda_i} \lambda^{f_i}}{f_i!} \quad where\ \lambda_i = p_i.W \tag{59.7}$$

The steps required for computation of α, the cumulative probability that pattern frequencies equal to or greater than the vector F_{obs} occurs purely by chance is given by Eq. (59.8) below. This corresponds to the one-sided integral of the multivariate Poisson distribution and represents the probability that the H_0 is erroneously rejected.

$$\begin{aligned} \alpha &= \Pr(x_1 \geq f_1, x_2 \geq f_2, \ldots, x_k \geq f_k) \\ &= \Pr(x_1 \geq f_1) \wedge \Pr(x_2 \geq f_2) \wedge \ldots \wedge \Pr(x_k \geq f_k) \\ &= \sum_{x_1=f_1}^{\infty} \frac{\exp^{-\lambda_1} \lambda_1^{x_1}}{x_1!} \cdot \sum_{x_2=f_2}^{\infty} \frac{\exp^{-\lambda_2} \lambda_2^{x_2}}{x_2!} \cdots \sum_{x_K=f_K}^{\infty} \frac{\exp^{-\lambda_K} \lambda_k^{x_K}}{x_k!} \end{aligned} \tag{59.8}$$

The *p-value*, α, in Eq. (59.8) is utilized to compute the value of ρ or the *cluster-density* as specified in Eq. (59.9) below:

$$\begin{aligned} \rho &= \ln \tfrac{1.0}{\alpha} = -\ln(\alpha) \\ &= \sum_{i=1}^{k} \lambda_i + \sum_{i=1}^{k} \ln f_i! - \sum_{i=1}^{k} f_i \ln \lambda_i - \\ &\sum_{i=1}^{k} \ln\left(1 + \tfrac{\lambda_i}{f_i+1} + \ldots + \tfrac{\lambda_i^t}{(f_i+1)(f_i+2)\ldots(f_i+t)}\right) \end{aligned} \tag{59.9}$$

The infinite summation term in Eq. (59.9) quickly converges and thus can be adaptively calculated to the precision desired. For small values of λ_i, the series may be truncated such that the last term is smaller than an arbitrarily small constant, ε.

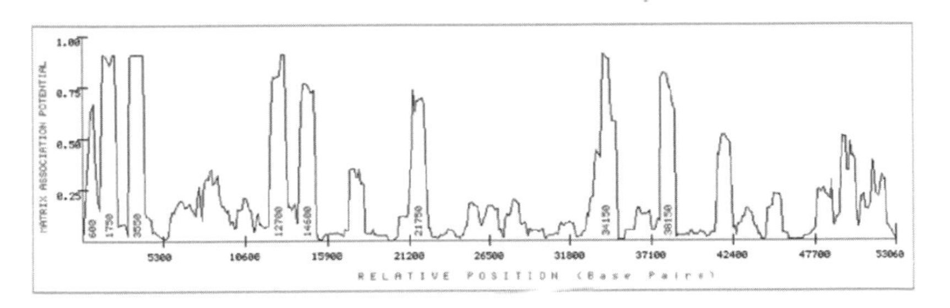

Fig. 59.6. The analysis of human protamine gene cluster using the MAR-Finder tool. Default analysis parameters were used.

Figure 59.6 presents the output from the analysis of the human protamine gene sequence. This statistical inference algorithm based on the association of patterns found within the close proximity of a DNA sequence region has been incorporated in the *MAR-Finder* tool. A java-enabled version of the tool described in (Singh *et al.*, 1997) is also available for public access from http://www.MarFinder.com.

We also need t take into consideration the interdependence of pattern occurrences. Let f_{ij} correspond to the observed frequency of pattern defined by pattern-HMM H_j in the ith. window sample. Using the frequency data from n window samples, and the mean frequency, $\overrightarrow{f} = (f_i)$, the correlation matrix, $R = (r_{ij})$ can be evaluated as follows:

$$r_{ij} = \frac{s_{ij}}{s_i s_j} = \frac{\sum_{r=1}^{n} (f_{ri} - f_i)(f_{rj} - f_j)}{\sqrt{\sum_{r=1}^{n} (f_{ri} - f_i)^2}\sqrt{\sum_{r=1}^{n} (f_{rj} - f_j)^2}} \tag{59.10}$$

If the *sample correlation matrix*, R, is equal to the identity matrix, the variables can be considered to be uncorrelated or independent. The hypothesis $r_{ij} = 0$ can be tested using the statistic t_{ij} defined in Eq.(59.11). t_{ij} follows a Student's distribution with $(n-2)$ degrees of freedom (Kachigan, 1986).

$$t_{ij} = \frac{r_{ij}\sqrt{n-2}}{\sqrt{1 - r_{ij}^2}} \tag{59.11}$$

If a pattern interdependence is detected, the pairwise correlation terms in R can be used to remove surrogate variables, i.e. one of the two patterns that exhibit a high degree of correlation. Removal of surrogate variables results in retaining a *core* subset of original patterns that account for the variability of the observed data (Hair *et al.*, 1987). Let there be k such *core patterns* that get retained for subsequent analysis stage. If the pairwise correlation terms of R_k are non-zero, the *Mahalanobis Transformation* can be applied to the vector $\overrightarrow{f_k}$ to transform it to a vector $\overrightarrow{z_k}$. The property of such a transformation is that the correlation matrix of the transformed variables is guaranteed to be the *identity* matrix I (Mardia *et al.*, 1979). The *Mahalanobis Transformation* for obtaining the uncorrelated vector $\overrightarrow{z_k}$ from the observed frequency of core vectors $\overrightarrow{f_k}$ is specified in Eq. (59.12), with the l_i denoting the eigenvalues.

$$\begin{aligned} \overrightarrow{z_k} &= S_k^{-\frac{1}{2}}(\overrightarrow{f_k} - \overrightarrow{f_k}) \\ S_k^{-\frac{1}{2}} &= \Gamma\Lambda^{-\frac{1}{2}}\Gamma \end{aligned} \tag{59.12}$$

where $\overrightarrow{f_k}$ is the observed frequency vector and $\Lambda^{-\frac{1}{2}} = diag(l_i^{-\frac{1}{2}})$

The value for α can next be computed based on the transformed vector $\overrightarrow{z_k}$ as shown in Eq. (59.13). The components of the transformed vector are independent, and thus the multiplication of individual probability terms is justifiable. Each component, z_i, represents a linear combination of the observed frequency values.

$$\begin{aligned} \alpha &= Pr(z_1 \geq z_{f_1}, z_2 \geq z_{f_2}, ..., z_c \geq z_{f_c}) \\ &= Pr(z_1 \geq z_{f_1}) \cdot Pr(z_2 \geq z_{f_2}) \cdot ... \cdot Pr(z_c \geq z_{f_c}) \\ &= \int_{z_1=\lfloor z_{f_1} \rfloor}^{\infty} \frac{e^{-1}}{z_1!} \cdot \int_{z_2=\lfloor z_{f_2} \rfloor}^{\infty} \frac{e^{-1}}{z_2!} \cdot ... \cdot \int_{z_c=\lfloor z_{f_k} \rfloor}^{\infty} \frac{e}{z_c!}{-1} \end{aligned} \tag{59.13}$$

59.3.2 Level II Search: Meta-Pattern Hypotheses

The meta- or higher level pattern hypotheses are generated and tested within the HPDRs. Specifically, the *Pattern Association* (PA) hypotheses are generated and verified within these HPDRs. These *PA* hypotheses are build in a bottom-up manner from the validation of pair-wise associations. For example, for two patterns A and B, a PA-hypothesis that we might validate is that $A \rightarrow B$, with the usual semantics that the occurrence of a pattern A implies that occurrence of pattern B within a pre-specified association distance. Furthermore, if the PA-hypothesis stating that $B \rightarrow A$ is also validated, the relationship between patterns A and B is promoted to that of *Pattern Equivalence* (PE), denoted as $A \leftrightarrow B$ or $A \equiv B$. Transitivity can be used to build larger groups of associations, such that if $A \rightarrow B$ and $B \rightarrow C$, then the implication $A \rightarrow BC$ may be concluded.

Similar statement can be made about the PE-hypotheses[4] . Meta pattern formation using transitivity rules will lead to the discovery of *mosaic* type meta-patterns. For the purpose of developing a methodology for systematically generating PA-hypotheses, the DNA sequence is represented as a *sequence* of a 2-elements. The first element in this sequence is the pattern match location, and the second element identifies the specific HMM(s) that matched. (It is possible for more than one pattern model to match the DNA sequence at a specific location). Such a representation shown in Eq.(59.14), is denoted as F_S, is the *pattern-sequence* corresponding to the biological sequence S.

$$F_S = \langle (x_1, P_a), (x_2, P_b), ..., (x_i, P_r), ..., (x_n, P_v) \rangle \tag{59.14}$$

Eq.(59.15) specifies the set of pattern hypotheses generated within each HPDR. The operator $cadr(L)$ is used to denote $car(cdr(L))$. The equation specifies that unique hypotheses are formed considering the closest pattern P_y instance to a given pattern P_x instance.

$$\begin{aligned} H_{AB} = & \\ & \{(A,B) | A = cadr(P_x) \wedge B = cadr(P_y) \wedge \\ \Delta_{A,B} = & \ ||car(P_x) - car(P_y)|| \wedge (\Delta_{A,B} < \theta) \wedge \\ & (\neg \exists P_z)(\ ||car(P_x) - car(P_z)|| < \Delta_{A,B})\} \end{aligned} \tag{59.15}$$

A $N \times N$ matrix C, similar to a contingency table (Gokhale, 1978, Brien, 1989) is used for recording significance of the each PA-hypotheses generated from the analysis of *all* HPDRs in the entire set of sequences. Recall that these regions were identified during the Level I search. The score for cell $C_{A,B}$ is updated according to Eq. (59.16) for every pattern pair (A,B) hypotheses H_{AB} generated in these regions. The probabilities of random occurrence of patterns A and B are p_A and p_B respectively, and $\Delta_{A,B}$ is the distance between them.

$$\begin{aligned} C_{AB} &= C_{AB} + \rho_{AB} \\ &= C_{AB} + (\lambda_A + \lambda_B - ln\lambda_A - ln\lambda_B) \\ where \quad & \lambda_A = p_A \Delta_{AB}, \\ and \quad & \lambda_B = p_B \Delta_{AB} \end{aligned} \tag{59.16}$$

Information theoretical approach based on mutual information content is next utilized for characterizing the strengths between pattern pairs. Contents of the contingency table (after all the sequences in the database have been processed) need to be converted to correspond to

[4] The functional significance of $A \rightarrow BC$ is that protein binding to site A will lead to the the binding of proteins at sites B and C. For a meta-pattern of the form $A \leftrightarrow B$, both the proteins must simultaneously bind to bring forth the necessary function.

probability density function. This is accomplished by associating the cell probability value, p_{ij} defined in Eq. (59.17).

$$p_{ij} = \frac{C_{\iota\varphi}}{\sum\limits_{\varphi=1,N} C_{\iota\varphi}} \tag{59.17}$$

In the final step, the *uncertainty of finding* a pattern *B*, given that a pattern *A* is present is defined by Eq. (59.18).

$$U(B|A) = \frac{H(B)-H(B|A)}{H(B)} = \frac{\sum\limits_{i} p_{B_i} \cdot lnp_{B_i} - p_{AB} \cdot lnp_{AB}}{\sum\limits_{i} p_{B_i} \cdot lnp_{B_i}} \tag{59.18}$$

If the presence of a pattern *A* results in a low value for the uncertainty that the pattern *B* is present, then we have a *meta-pattern*. Figure 59.7 shows the MAR and the transcription factor analysis of Chromosome I for *S. cerevisea*. A correlation between the high density of transcription factor binding sites and the matrix attachment regions is evident in this plot. This plot will assist in identifying regions further biological investigation.

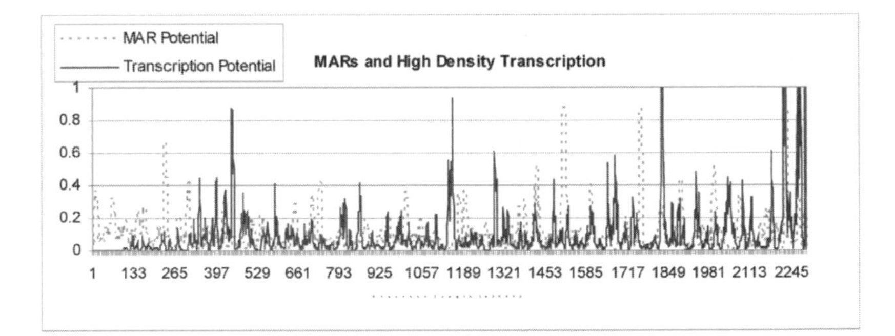

Fig. 59.7. A cumulative analysis of yeast Chromosome I using MAR detection algorithm and isolation of transcription density regions.

59.4 Conclusions

In this chapter we described the process for learning stochastic models of known lower-level patterns and using them in an inductive procedure to learn meta-pattern organization. The next logical step is to extend this unsupervised learning process to include lower level patterns that have not yet been discovered, and thus not included in the pattern sets available within the databases such as such as TFD, TRANSFAC, EPD. In this case our analogy is equivalent to *solving a jigsaw puzzle where we do not know what the solved puzzle will look like, and there may still be some pieces missing*. The process described in this chapter may in fact be applied to this problem if we first generate a hypothetical piece (pattern) and use it with all the known pieces (patterns) and create possible solution to the puzzle (generate a meta-pattern hypothesis). If there are abundant instances that indicate prevalence of our meta-pattern hypothesis in the database, we can associate a confidence and support to our discovery. Moreover, in this case, the newly found pattern as well as a meta-pattern will be added to the database of known patterns and used in the future discovery processes. In summary the potential for applying the algorithmic rich Data Mining and machine learning approaches to biological data has potential for discovery of novel concepts.

References

Berg, O. and Hippel, P. v., "Selection of DNA binding sites by regulatory proteins," *J.Mol.Biol.*, Vol. 193, 1987, pp. 723-750.

Bode, J., Stengert-Iber, M., Kay, V., Schlake, T., and Dietz-Pfeilstetter, A., "Scaffold/Matrix Attchment Regions: Topological Switches with Multiple Regulatory Functions," *Crit.Rev.in Eukaryot.Gene Expr.*, Vol. 6, 1996, pp. 115-138.

O'Brien, L., *The statistical analysis of contingency table designs*, no. 51 ed., Order from Environmental Publications, University of East Anglia, Norwich, 1989.

Bucher, P. and Trifonov, N., "CCAAT-box revisited: Bidirectionality, Location and Context," *J.Biomol.Struct.Dyn.*, Vol. 6, 1988, pp. 1231-1236.

Faisst, S. and Meyer, S., "Compilation of vertebrate encoded transcription factors," *Nucleic Acid Res.*, Vol. 20, 1992, pp. 1-26.

Ghosh, D., "A relational database of transcription factors," *Nucleic Acid Res.*, Vol. 18, 1990, pp. 1749-1756.

Ghosh, D., "OOTFD (Object-Oriented Transcription Factors Database): an object-oriented successor to TFD," *Nucleic Acid Res.*, Vol. 26, 1998, pp. 360-362.

Gokhale, D. V. and Kullback, S., *The information in contingency tables*, M. Dekker, New York, 1978.

Gribskov, M., Luethy, R., and Eisenberg, D., "Profile Analysis," *Methods in Enzymology*, Vol. 183, 1990, pp. 146-159.

Hair, J., Anderson, R., and Tatham, R., "Multivariate data analysis with readings," 1987.

Hartwell, L. and Kasten, M., "Cell cycle control and cancer," *Science*, Vol. 266, pp. 1821-1828, 1994.

Kachigan, S., "Statistical Analysis," 1986.

Kadonaga, J., "Eukaryotic transcription: An interlaced network of transcription factors and chromatin-modifying machines," *Cell*, Vol. 92, 1998, pp. 307-313.

Kliensmith, L. and Kish, V., *Principles of cell and molecular biology* 1995.

Liebich, I., Bode, J., Frisch, M., and Wingender, E., "S/MARt DB: a database on scaffold/matrix attached regions," *Nucleic Acids Res.*, Vol. 30, No. 1, 2002, pp. 372-374.

Mardia, K., Kent, J., and Bibby, J., "Multivariate Analysis," 1979.

Kel-Margoulis, O. V., Kel, A. E., Reuter, I., Deineko, I. V., and Wingender, E., "TRANSCompel: a database on composite regulatory elements in eukaryotic genes," *Nucleic Acids Res.*, Vol. 30, No. 1, 2002, pp. 332-334.

Matys, V., Fricke, E., Geffers, R., Gossling, E., Haubrock, M., Hehl, R., Hornischer, K., Karas, D., Kel, A. E., Kel-Margoulis, O. V., Kloos, D. U., Land, S., Lewicki-Potapov, B., Michael, H., Munch, R., Reuter, I., Rotert, S., Saxel, H., Scheer, M., Thiele, S., and Wingender, E., "TRANSFAC: transcriptional regulation, from patterns to profiles," *Nucleic Acids Res.*, Vol. 31, No. 1, 2003, pp. 374-378.

Nikolaev, L., Tsevegiyn, T., Akopov, S., Ashworth, L., and Sverdlov, E., "Construction of a chromosome specific library of MARs and mapping of matrix attachment regions on human chromosome 19," *Nucleic Acid Res.*, Vol. 24, 1996, pp. 1330-1336.

Nussinov, R., "Signals in DNA sequences and their potential properties," *Comput.Applic.Biosci.*, Vol. 7, 1991, pp. 295-299.

Page, R., "Minimal Spanning Tree Clustering Methods," *Comm.of the ACM*, Vol. 17, 1974, pp. 321-323.

Penotti, F., "Human DNA TATA boxes and transcription initiation sites. A Statistical Study," *J.Mol.Biol.*, Vol. 213, 1990, pp. 37-52.

Rabiner, L., "A tutorial on hidden Markov models and selected applications in speech recognition," *Proc.of the IEEE*, Vol. 77, 1989, pp. 257-286.

Roeder, R., "The role of general initiation factors in transcription by RNA Polymerase II," *Trends in Biochem.Sci.*, Vol. 21, 1996, pp. 327-335.

Singh, G., Kramer, J., and Krawetz, S., "Mathematical model to predict regions of chromatin attachment to the nuclear matrix," *Nucleic Acid Res.*, Vol. 25, 1997, pp. 1419-1425.

Wheeler, D. L., Church, D. M., Edgar, R., Federhen, S., Helmberg, W., Madden, T. L., Pontius, J. U., Schuler, G. D., Schriml, L. M., Sequeira, E., Suzek, T. O., Tatusova, T. A., and Wagner, L., "Database resources of the National Center for Biotechnology Information: update," *Nucleic Acids Res.*, Vol. 32 Database issue, 2004, pp. D35-D40.

Wingender, E., Chen, X., Fricke, E., Geffers, R., Hehl, R., Liebich, I., Krull, M., Matys, V., Michael, H., Ohnhauser, R., Pruss, M., Schacherer, F., Thiele, S., and Urbach, S., "The TRANSFAC system on gene expression regulation," *Nucleic Acids Res.*, Vol. 29, No. 1, 2001, pp. 281-283.

Zahn, C., "Graph-theoretical methods for detecting and describing Gestalt clusters," *IEEE Trans.Computers*, Vol. 20, 1971, pp. 68-86.

60

Data Mining for Financial Applications

Boris Kovalerchuk[1] and Evgenii Vityaev[2]

[1] Central Washington University, USA
[2] Institute of Mathematics, Russian Academy of Sciences, Russia

Summary. This chapter describes Data Mining in finance by discussing financial tasks, specifics of methodologies and techniques in this Data Mining area. It includes time dependence, data selection, forecast horizon, measures of success, quality of patterns, hypothesis evaluation, problem ID, method profile, attribute-based and relational methodologies. The second part of the chapter discusses Data Mining models and practice in finance. It covers use of neural networks in portfolio management, design of interpretable trading rules and discovering money laundering schemes using decision rules and relational Data Mining methodology.

Key words: finance time series, relational Data Mining, decision tree, neural network, success measure, portfolio management, stock market, trading rules.

> October. This is one of the peculiarly dangerous months to speculate in stocks in. The others are July, January, September, April, November, May, March, June, December, August and February. Mark Twain, 1894

60.1 Introduction: Financial Tasks

Forecasting stock market, currency exchange rate, bank bankruptcies, understanding and managing financial risk, trading futures, credit rating, loan management, bank customer profiling, and money laundering analyses are core financial tasks for Data Mining (Nakhaeizadeh *et al.*, 2002). Some of these tasks such as bank customer profiling (Berka, 2002) have many similarities with Data Mining for customer profiling in other fields.

Stock market forecasting includes uncovering market trends, planning investment strategies, identifying the best time to purchase the stocks and what stocks to purchase. Financial institutions produce huge datasets that build a foundation for approaching these enormously complex and dynamic problems with Data Mining tools. Potential significant benefits of solving these problems motivated extensive research for years.

O. Maimon, L. Rokach (eds.), *Data Mining and Knowledge Discovery Handbook*, 2nd ed.,
DOI 10.1007/978-0-387-09823-4_60,

Almost every computational method has been explored and used for *financial modeling*. We will name just a few recent studies: Monte-Carlo simulation of option pricing, finite-difference approach to interest rate derivatives, and fast Fourier transform for derivative pricing (Huang *et al.*, 2004, Zenios, 1999, Thulasiram and Thulasiraman, 2003). New developments augment traditional technical analysis of stock market curves (Murphy, 1999) that has been used extensively by financial institutions. Such stock charting helps to identify buy/sell signals (timing "flags") using graphical patterns.

Data Mining as a process of *discovering useful patterns, correlations* has its own niche in financial modeling. Similarly to other computational methods almost every Data Mining method and technique has been used in financial modeling. An incomplete list includes a variety of linear and non-linear models, multi-layer neural networks (Kingdon, 1997, Walczak, 2001, Thulasiram *et al.*, 2002, Huang *et al.*, 2004), k-means and hierarchical clustering; k-nearest neighbors, decision tree analysis, regression (logistic regression; general multiple regression), ARIMA, principal component analysis, and Bayesian learning.

Less traditional methods used include rough sets (Shen and Loh, 2004), relational Data Mining methods (deterministic inductive logic programming and newer probabilistic methods (Muggleton, 2002, Lachiche and Flach, 2002, Kovalerchuk and Vityaev, 2000), support vector machine, independent component analysis, Markov models and hidden Markov models.

Bootstrapping and other evaluation techniques have been extensively used for improving Data Mining results. Specifics of financial time series analyses with ARIMA, neural networks, relational methods, support vector machines and traditional technical analysis is discussed in (Back and Weigend, 1998, Kovalerchuk and Vityaev, 2000, Muller *et al.*, 1997, Murphy, 1999, Tsay, 2002).

The naïve approach to Data Mining in finance assumes that somebody can provide a cookbook instruction on "how to achieve the best result". Some publications continue to foster this unjustified belief. In fact, the only realistic approach proven to be successful is providing comparisons between different methods showing their strengths and weaknesses relative to problem characteristics (problem ID) conceptually and leaving for user the selection of the method that likely fits the specific user problem circumstances. In essence this means clear understanding that Data Mining in general, and in finance specifically, is still more art than hard science.

Fortunately now there is growing number of books that discuss issues of matching tasks and methods in a regular way (Dhar and Stein ,1997, Kovalerchuk and Vityaev, 2000, Wang, 2003). For instance, understanding the power of first-order If-Then rules over the decision trees can significantly change and improve Data Mining design. User's actual experiments with data provide a real judgment of Data Mining success in finance. In comparison with other fields such as geology or medicine, where test of the forecast is expensive, difficult, and even dangerous, a trading forecast can be tested next day in essence without cost and capital risk involved in real trading.

Attribute-based learning methods such as neural networks, the nearest neighbors method, and decision trees dominate in financial applications of Data Mining. These methods are relatively simple, efficient, and can handle noisy data. However, these methods have two serious drawbacks: a limited ability to represent background knowledge and the lack of complex relations. *Relational data mining* techniques that include Inductive Logic Programming (ILP) (Muggleton, 1999, Džeroski, 2002) intend to overcome these limitations.

Previously these methods have been relatively computationally inefficient (Thulasiram, 1999) and had rather limited facilities for handling numerical data (Bratko and Muggleton, 1995). Currently these methods are enhanced in both aspects (Kovalerchuk and Vityaev, 2000) and are especially actively used in bioinformatics (Turcotte *et al.*, 2001, Vityaev *et al.*, 2002).

We believe that now is the time for applying these methods to financial analyses more intensively especially to those analyses that deal with probabilistic relational reasoning.

Various publications have estimated the use of Data Mining methods like hybrid architectures of neural networks with genetic algorithms, chaos theory, and fuzzy logic in finance. "Conservative estimates place about $5 billion to $10 billion under the direct management of neural network trading models. This amount is growing steadily as more firms experiment with and gain confidence with neural networks techniques and methods" (Loofbourrow and Loofbourrow, 1995). Many other proprietary financial applications of Data Mining exist, but are not reported publicly as was stated in (Von Altrock, 1997, Groth, 1998).

60.2 Specifics of Data Mining in Finance

Specifics of Data Mining in finance are coming from the need to:

- forecast *multidimensional time series* with high level of *noise*
- accommodate specific efficiency criteria (e.g., the maximum of *trading profit*) in addition to prediction accuracy such as R^2;
- make coordinated *multiresolution forecast* (minutes, days, weeks, months, and years);
- incorporate a *stream of text signals* as input data for forecasting models (e.g., Enron case, September 11 and others);
- be able to *explain the forecast* and the *forecasting model* ("black box" models have limited interest and future for significant investment decisions);
- be able to benefit from very *subtle patterns* with a *short life time* and
- incorporate the impact of market players on market regularities.

The current *efficient market theory/hypothesis* discourages attempt to discover long-term stable trading rules/regularities with significant profit. This theory is based on the idea that if such regularities exist they would be discovered and used by the majority of the market players. This would make rules less profitable and eventfully useless or even damaging.

Greenstone and Oyer (2000) examine the month by month measures of return for the computer software and computer systems stock indexes to determine whether these indexes' price movements reflect genuine deviations from random chance using the standard t-test. They concluded that although Wall Street analysts recommended to use the "summer swoon" rule (sell computer stocks in May and buy them at the end of summer) this rule is not statistically significant. However they were able to confirm several previously known 'calendar effects" such as "January effect" noting meanwhile that they are not the first to warn of the dangers of easy Data Mining and unjustified claims of market inefficiency.

The market efficiency theory does not exclude that hidden *short-term local conditional regularities* may exist. These regularities can not work "forever," they should be corrected *frequently*.

It has been shown that the financial data are not random and that the efficient market hypothesis is merely a subset of a larger *chaotic market hypothesis* (Drake and Kim, 1997). This hypothesis does not exclude successful short term forecasting models for prediction of chaotic time series (Casdagli and Eubank, 1992).

Data Mining does not try to accept or reject the efficient market theory. Data Mining creates *tools*, which can be useful for discovering subtle short-term conditional patterns and trends in wide range of financial data. This means that retraining should be a permanent part of data mining in finance and any claim that a silver bullet trading has been found should be treated similarly to claims that a perpetuum mobile has been discovered.

The impact of market players on market regularities stimulated a surge of attempts to use ideas of *statistical physics* in finance (Bouchaud and Potters, 2000). If an observer is a large marketplace player then such observer can potentially change regularities of the marketplace dynamically. Attempts to forecast in such dynamic environment with thousands active agents leads to much more complex models than traditional Data Mining models designed for. This is one of the major reasons that such interactions are modeled using ideas from statistical physics rather than from statistical Data Mining. The *physics approach* in finance (Voit, 2003, Ilinski, 2001, Mantegna and Stanley, 2000, Mandelbrot, 1997) is also known as "econophysic" and "physics of finance". The major difference from Data Mining approach is coming from the fact that in essence the Data Mining approach is not about developing specific methods for financial tasks, but the physics approach is. It is deeper integrated into the finance subject mater. For instance, Mandelbrot (1997) (known for his famous work on fractals) worked also on proving that the price movement's distribution is scaling invariant.

Data Mining approach covers empirical models and regularities derived directly from data and almost only from data with little domain knowledge explicitly involved. Historically, in many domains, deep field-specific theories emerge after the field accumulates enough empirical regularities. We see that the future of Data Mining in finance would be to generate more empirical regularities and combine them with domain knowledge via generic analytical Data Mining approach (Mitchell, 1997). First attempts in this direction are presented in (Kovalerchuk and Vityaev, 2000) that exploit power of relational Data Mining as a mechanism that permits to encode domain knowledge in the first order logic language.

60.2.1 Time series analysis

A temporal dataset $\boldsymbol{T}$ called a *time series* is modeled in attempt to discover its main components such as *Long term trend, L(T)*, *Cyclic variation, C(T), Seasonal variation, S(T)* and *Irregular movements, I(T)*. Assume that $\boldsymbol{T}$ is a time series such as daily closing price of a share, or SP500 index from moment 0to current moment k, then the next value of the time series $T(k+n)$ is modeled by formula 63.1:

$$T(k+n) = L(T) + C(T) + S(T) + I(T) \tag{60.1}$$

Traditionally classical ARIMA models occupy this area for finding parameters of functions used in formula 63.1. ARIMA models are well developed but are difficult to use for highly non-stationary stochastic processes.

Potentially Data Mining methods can be used to build such models to overcome ARIMA limitations. The advantage of this four-component model in comparison with "black box" models such as neural networks is that components in formula 63.1 have an interpretation.

60.2.2 Data selection and forecast horizon

Data Mining in finance has the same challenge as general Data Mining in data selection for building models. In finance, this question is tightly connected to the selection of the target variable. There are several options for target variable y: $y=T(k+1)$, $y=T(k+2)$,...,$y=T(k+n)$, where $y=T(k+1)$ represents forecast for the next time moment, and $y=T(k+n)$ represents forecast for n moments ahead. Selection of dataset $\boldsymbol{T}$ and its size for a specific desired forecast horizon n is a significant challenge.

For stationary stochastic processes the answer is well-known a better model can be built for longer training duration. For financial time series such as SP500 index this is not the case (Mehta and Bhattacharyya, 2004). Longer training duration may produce many and contradictory profit patterns that reflect bear and bull market periods. Models built using too short durations may suffer from overfitting and hardly applicable to the situations where market is moving from the bull period to the bear period. Also in finance the long-horizon returns could be forecast better than short-horizon returns depending on the training data used and model parameters (Krolzig *et al.*, 2004).

In standard Data Mining it is typically assumed that the quality of the model does not depend on *frequency* of its use. In financial application the frequency of trading is one of the parameters that impact a quality of the model. This happens because in finance the *criterion of the model quality* is not limited by the accuracy of prediction, but is driven by profitability of the model. It is obvious that frequency of trading impacts the profit as well as the trading rules and strategy.

60.2.3 Measures of success

Traditionally the quality of financial Data Mining forecasting models is measured by the standard deviation between forecast and actual values on training and testing data. This approach works well in many domains, but this assumption should be revisited for trading tasks. Two models can have the same standard deviation but may provide very different trading return. The small R^2 is not sufficient to judge that the forecasting model will correctly forecast stock change direction (sign and magnitude). For more detail see (Kovalerchuk and Vityaev, 2000). More appropriate measures of success in financial Data Mining are measures such as Average Monthly Excess Return (AMER) and Potential trading profits (PTP) (Greenstone and Oyer, 2000):

$$AMER_j = R_{ij} - \beta_i R_{500j} - (\sum_{j=1}^{12} (R_{ij} - \beta_i R_{500j})/12)$$

where R_{ij} is the average return for the S&P500 index in industry i and month j and R_{500j} is the average return of the S&P 500 in month j. The β_i values adjust the AMER for the index's sensitivity to the overall market. A second measure of return is Potential Trading Profits (PTP):

$$PTP_{ij} =_{ij} - R_{500j}$$

PTP shows investor's trading profit versus the alternative investment based on the broader S&P 500 index.

60.2.4 QUALITY OF PATTERNS AND HYPOTHESIS EVALUATION

An important issue in Data Mining in general and in finance in particular is the evaluation of quality of discovered pattern P measured by its statistical significance. A typical approach assumes the testing of the null hypothesis H that pattern P is not statistically significant at level α. A meaningful statistical test requires that pattern parameters such as the month(s) of the year and the relevant sectoral index in a trading rule pattern P have been chosen *randomly* (Greenstone and Oyer, 2000). In many tasks this is not the case.

Greenstone and Oyer argue that in the summer "summer swoon" trading rule mentioned above, the parameters are not selected randomly, but are produced by data snooping – checking combination of industry sectors and months of return and then reporting only a few "significant" combinations. This means that rigorous test would require to test a different null hypothesis not only about one "significant" combination, but also about the *"family" of combinations.* Each combination is about an individual industry sector by month's return. In this setting the return for the "family" is tested versus the overall market return.

Several testing options are available. Sullivan *et al.* (1998, 1999) use a bootstrapping method to evaluate statistical significance of such hypotheses adjusted for the effects of data snooping in "trading rules" and calendar anomalies. Greenstone and Oyer (2000) suggest a simple computational method – combining individual *t-test* results by using the Bonferroni inequality that given any set of events $A_1, A_2, \ldots, A_n$, the probability of their union is smaller than or equal to the sum of their probabilities:

$$P(A_1 \,\&\, A_2 \&\ldots\&\, A_k) \quad \leq \quad \Sigma_{i=1:k} P(A_i)$$

where A_i denotes the false rejection of statement i, from a given family with k statements. One of the techniques to keep the family-wide error rate at reasonable levels is "Bonferroni correction" that sets a significance level of α/k for each of the k statements.

Another option would be to test whether the statements are jointly true using the traditional *F-test.* However if the null hypothesis about a joint statement is *rejected* it does not identify the profitable trading strategies (Greenstone and Oyer, 2000).

The sequential semantic probabilistic reasoning that uses F-test addresses this issue (Kovalerchuk and Vityaev, 2000). We were able to identify profitable and statistically significant patterns for SP500 index using this method. Informally the idea of semantic probabilistic reasoning is coming from the principle of *Occam's razor* (a law of simplicity) in science and philosophy. Informally for trading it was written by practical traders as follows:

- When you have two competing trading theories which make exactly *the same predictions*, the one that is simpler is the better & more profitable one.
- If you have two trading/investing theories which *both explain* the observed facts then you should use the simplest one until more evidence comes along.
- The *simplest explanation* for a commodity or stock price movement phenomenon is more likely to be accurate than more complicated explanations.
- If you have two *equally likely solutions* to a trading or day trading problem, pick the simplest.
- The price movement explanation requiring the *fewest assumptions* is most likely to be correct.

60.3 Aspects of Data Mining Methodology in Finance

Data Mining in finance typically follows a set of general for any Data Mining task steps such as problem understanding, data collection and refining, building a model, model evaluation and deployment (Klösgen and Zytkow, 2002). Some specifics of these steps for trading tasks are presented in (Zemke, 2002,Zemke, 2002) such as data enhancing techniques, predictability tests, performance improvements, and pitfalls to avoid.

Another important step in this process is adding expert-based rules in Data Mining loop when dealing with absent or insufficient data. "Expert mining" is a valuable additional source of regularities. However in finance, expert-based learning systems respond slowly to the to market changes (Cowan, 2002). A technique for efficiently mining regularities from an *expert's perspective* has been offered (Kovalerchuk and Vityaev, 2000). Such techniques need to be integrated into financial Data Mining loop similar to what was done for medical Data Mining applications (Kovalerchuk *et al.*, 2001).

60.3.1 Attribute-based and relational methodologies

Several parameters characterize data mining methodologies for financial forecasting. Data categories and mathematical algorithms are most important among them. The first data type is represented by *attributes* of objects, that is each object x is given by a set of values $A_1(x)$, $A_2(x),\ldots,A_n(x)$. The common Data Mining methodology assumes this type of data and it is known as an *attribute-based* or *attribute-value methodology*. It covers a wide range of statistical and connectionist (neural network) methods.

The *relational data type* is a second type, where objects are represented by their relations with other objects, for instance, x>y, y<z, x>z. In this example we may not know that x=3, y=1 and z=2. Thus attributes of objects are not known, but their relations are known. Objects may have different attributes (e.g., x=5, y=2, and z= 4), but still have the same relations. Less traditional *relational methodology* is based on the relational data type.

Another data characteristic important for financial modeling methodology is an actual *set of attributes* involved. A fundamental analysis approach incorporates all available attributes, but technical analysis approach is based only on a time series such as stock price and parameters derived from it. Most popular time series are index value at open, index value at close, highest index value, lowest index value and trading volume and lagged returns from the time series of interest. Fundamental factors include the price of gold, retail sales index, industrial production indices, and foreign currency exchange rates. Technical factors include variables that are derived from time series such as moving averages.

The next characteristic of a specific Data Mining methodology is a form of the relationship between objects. Many Data Mining methods assume a *functional form* of the relationship. For instance, the linear discriminant analysis assumes linearity of the border that discriminates between two classes in the space of attributes. Often it is hard to justify such functional form in advance. Relational Data Mining methodology in finance does not assume a functional form for the relationship. Its intention is *learning symbolic relations* on numerical data of financial time series.

60.3.2 Attribute-based relational methodologies

In this section we discuss a combination of both attribute-based and relational methodologies that permit to mitigate their difficulties. In most of the publications relational Data Mining was associated with *Inductive Logic Programming* (ILP) which is a deterministic technique in its purest form. The typical claim about relational data miming is that it can not handle large data sets (Thulasiram, 1999). This statement is based on the assumption that initial data are provided in the form of relations. For instance, to mine in a training data with m attributes for n data objects we need to store and operate with $n \times m$ data elements, but for m simplest binary relations (used to represent graphs) we need to store and operate with $n^2 \times m$ elements. This number is n times larger and for large training datasets the difference can be very significant. The *attribute-based relational Data Mining* does not need to store and operate with $n^2 \times m$ elements. It computes relations from attribute-based data sets on demand. For instance, to explore a relation, *Stock(t)>Stock(t+k)* for k days ahead we do not need to store this relation. It can be computed for every pair of stock data as needed to build a graph of stock relations. In finance with predominantly numeric input data, a dataset that should be represented in a relational form from the beginning can be relatively small.

We share Thuraisingham's (1999) vision that relational Data Mining is most suitable for applications where *structure can be extracted from the instances*. We also agree with her statement that Data Mining is now very much an art and to make it into a science, we need more work in areas like ILP that is a part of relational learning that includes probabilistic learning.

60.3.3 Problem ID and method profile

Selection of a method for discovering regularities in financial time series is a very complex task. Uncertainty of problem descriptions and method capabilities are among the most obvious difficulties in this process. Dhar and Stein (1997) introduced and applied a unified vocabulary for business computational intelligence problems and methods that provide a framework for matching problems and methods.

A problem is described using a set of *desirable values* (problem ID profile) and a method is described using its *capabilities* in the same terms. Use of unified terms (*dimensions*) for problems and methods enhances capabilities of comparing alternative methods. Introducing dimensions also accelerates their clarification. Next, users should not be forced to spend time determining a method's capabilities (values of dimensions for the method). This is a task for developers, but users should be able to identify desirable values of dimensions using natural language terms as suggested by (Dhar and Stein ,1997).

Along these lines Table 60.1 indicates three shortcomings of neural networks for stock price forecasting related to explainability, usage of logical relations and tolerance for sparse data.

The strength of neural networks is also indicated by lines where requested capabilities are satisfied by neural networks. The advantages of using neural network models include the ability to model highly complex functions and to use a high number of variables including both fundamental and technical factors.

Table 60.1. Comparison of model quality and resources

Dimension	*Desirable value for stock price forecast problem*	*Capability of a neural network method*
Accuracy	Moderate	High
Explainability	Moderate to High	Low to Moderate
Response speed	Moderate	High
Ease to use logical relations	High	Low
Ease to use numerical attributes	High	High
Tolerance for *noise* in data	High	Moderate to high
Tolerance for *sparse data*	High	Low
Tolerance for complexity	High	High
Independence from experts	Moderate	High

60.3.4 Relational Data Mining in finance

Decision tree methods are very popular in Data Mining applications in general and in finance specifically. They provide a set of human readable, consistent rules, but discovering small trees for complex problems can be a significant challenge in finance (Kovalerchuk and Vityaev, 2000). In addition, rules extracted from decision trees fail to compare two attribute values as it is possible with relational methods.

It seems that relational Data Mining methods also known as relational knowledge discovery methods are gaining momentum in different fields (Muggleton, 2002, Džeroski, 2002, Thulasiram, 1999, Neville and Jensen, 2002, Vityaev *et al.*, 2002).

Data Mining in finance not only follows this trend but also leads the application of relational Data Mining for multidimensional time series such as stock market time series. A. Cowan, a senior financial economist from US Department of the Treasury noticed that examples and arguments available in (Kovalerchuk and Vityaev, 2000) for the application of relational Data Mining to financial problems produce expectations of great advancements in this field in the near future for financial applications (Cowan, 2002).

It was strengthened in several publications that relational data mining area is moving toward probabilistic first-order rules to avoid the limitations of deterministic systems, e.g., (Muggleton, 2002). Relational methods in finance such as Machine Method for Discovering Regularities (MMDR) (Kovalerchuk and Vityaev, 2000) are equipped with probabilistic mechanism that is necessary for time series with high level of noise.

MMDR is well suited to financial applications given its ability to handle numerical data with high levels of noise (Cowan, 2002). In computational experiments, trading strategies developed based on MMDR consistently outperform trading strategies developed based on other data-mining methods and buy and hold strategy.

60.4 Data Mining Models and Practice in Finance

Prediction tasks in finance typically are posed in one of two forms: (1) straight prediction of the market numeric characteristic, e.g., stock return or exchange rate, and (2) the prediction whether the market characteristic will increase or decrease. Having in mind that we need to take into account the trading cost and significance of the trading return in the second case we need to forecast whether the market characteristic will increase or decrease no less than some threshold. Thus, the difference between data mining methods for (1) or (2) can be less obvious, because (2) may require some kind of numeric forecast.

Another type of task is presented in (Becerra-Fernandez *et al.*, 2002). This task is assessment of investing risk. It uses a decision tree technique C5.0 (Quinlan, 1993) and neural networks to a dataset of 52 countries whose investing risk category was assessed in a Wall Street Journal survey of international experts. The dataset included 27 variables (economic, stock market performance/risk and regulatory efficiencies).

60.4.1 Portfolio management and neural networks

The neural network most commonly used by financial institutions is a multi-layer perceptron (MLP) with a single hidden layer of nodes for time series prediction. The peak of research activities in finance based on neural networks was in mid 1990s (Trippi and Turban, 1996, Freedman *et al.*, 1995, Azoff, 1994) that covered MLP and recurrent NN (Refenes, 1995). Other neural networks used in prediction are time delay networks, Elman networks, Jordan networks, GMDH, milti-recurrent networks (Giles *et al.*, 1997).

Below we present typical steps of *portfolio management* using the neural network forecast of return values.

1. Collect 30- 40 historical fundamental and technical factors for stock S_1, say for 10-20 years.
2. Build a neural network NN_1 for predicting the return values for stock S_1.
3. Repeat steps 1 and 2 for every stock S_i, that is monitored by the investor. Say 3000 stocks are monitored and 3000 networks, NN_i are generated.
4. Forecast stock return $S_i(t+k)$ for each stock *i and k* days ahead (say a week, seven days) by computing $NN_i(S_i(t))=S(t+k)$.
5. Select *n* highest $S_i(t+k)$ values of predicted stock return.
6. Compute a total forecasted return of selected stocks, *T* and compute $S_i(t+k)/T$. Invest to each stock proportionally to $S_i(t+k)/T$.
7. Recompute NN_i model for each stock *i* every *k* days adding new arrived data to the training set. Repeat all steps for the next portfolio adjustment.

These steps show why neural networks became so popular in finance. Potentially all steps above can be done automatically including actual investment. Even institutional investors may have no resources to manually analyze 3000 stocks and their 3000 neural networks every week. If investment decisions are made more often, say every day, then the motivation to use neural networks with their high adaptability is even more evident.

This consideration also shows current challenges of Data Mining in finance – the need to build models that can be very quickly evaluated in both accuracy and interpretability. Because NN are difficult to interpret even without time limitation recently steps 1-6 have been adjusted by adding more steps after step 3 that include *extracting interpretable rules* from the trained neural networks and improving prediction accuracy using rules, e.g., (Giles *et al.*, 1997).

It is likely that extracting rules from the neural network is a *temporary solution*. It would be better to extract rules directly from data without introducing neural network artifacts to rules and potentially overlooking some better rules because of this. It is clear that it can happen from mathematical considerations. There is also a growing number of computational experiments that support this claim, e.g., see (Kovalerchuk and Vityaev, 2000) on experiments with SP500, where first order rules built directly from data outperformed backpropagation neural networks that are most common in financial applications. Moody and Saffell (2001) discuss advantages of incremental portfolio optimization and building trading models.

The logic of using Data Mining in trading futures is similar to portfolio management. The most significant difference is that it is possible to substitute numeric forecast of actual return to less difficult categorical forecast, will it be profitable buy or sell the stock at price S(t) on date t. This corresponds to long and short terms used in stock market, where *Long* stands for buying the stock and *Short* stands for sell the stock on date t.

60.4.2 Interpretable trading rules and relational Data Mining

The logic of portfolio management based on discovering interpretable trading rules is the same as for neural networks with the substitution of NN for rule discovering techniques. Depending on the rule discovering techniques produced rules can be quite different. Below we present categories of rules that can be discovered.

Categorical rules predict a categorical attribute, such as increase/decrease, buy/sell. A typical example of a *monadic categorical rule* is the following rule:

If $S_i(t)$ <Value1 and $S_i(t-2)$ <Value2 then $S_i(t+1)$ will increase.

In this example, $S_i(t)$ is a continuous variable, e.g., stock price at the moment *t*. If $S_i(t)$ *is* a discrete variable that Value1 and Value2 are taken from *m* discrete values. This rule is called monadic because it compared a *single attribute* value with a *constant*. Such rules can be discovered from a trained decision trees by tracing its branches to the terminal nodes. Unfortunately decision trees produce only such rules.

The following technical analysis rule is a *relational categorical rule*, because to derive a conclusion it compares values of two attributes such as 5 and 15 day moving averages (ME5 and ME15) and derivatives of moving averages for 10 and 30 days (DerivativeME10, DerivativeME30) :

If ME5(t)=ME15(t) & DerivativeME10(t)>0 DerivativeME30(t)>0 then Buy stock at moment (t+1).

This rule can be read as "If moving averages for 5 and 15 days are equal and derivatives for moving averages for 10 and 30 days are positive then buy stock on the next day". The statement ME5(t)=ME15(t) compares two attribute values. Thus, in this sense classical for stock market technical analysis is superior to decision trees. The presented rule is written in a first order logic form. Note that typically technical analysis rules are not discovered in this form, but relational Data Mining technique does.

Classical categorical rules assume crisp relations such as $S_i(t) <$Value1 and ME5(t)=ME15(t). More realistic would be to assume that ME5(t) and ME15(t) are equal only approximately and Value1 is not exact. Fuzzy logic and rough sets rules are used in finance to work with *"soft" relations* (Von Altrock, 1997, Kovalerchuk and Vityaev, 2000, Shen and Loh, 2004). The logic of using *"soft" trading rules* in finance includes the conversion of time series to soft objects, discovering temporal "soft" rule from stock market data, discovering temporal "soft" rule from experts ("expert mining"), testing consistency of expert rules and rules extracted from data, and finally using rules for forecasting and trading.

60.4.3 Discovering money laundering and attribute-based relational Data Mining

Problem statement

Forensic accounting is a field that deals with possible illegal and fraudulent financial transactions. One current focus in this field is the analysis of funding mechanisms for terrorism (Prentice, 2002) where *clean money (e.g., charity money) and laundered money are both used* for a variety of activities including acquisition and production of weapons and their precursors. In contrast, traditional illegal businesses and drug trafficking *make dirty money appear clean.*

The specific tasks in automated forensic accounting related to Data Mining are the identification of suspicious and unusual electronic transactions and the reduction in the number of 'false positive' suspicious transactions. Currently inexpensive, simple rule-based systems, customer profiling, statistical techniques, neural networks, fuzzy logic and genetic algorithms are considered as appropriate tools (Prentice, 2002).

There are many indicators of possible suspicious (abnormal) transactions in traditional illegal business. These include (1) the use of several related and/or unrelated accounts before money is moved offshore, (2) a lack of account holder concern with commissions and fees (Vangel and James, 2002), (3) correspondent banking transactions to offshore shell banks (Vangel and James, 2002), (4) transferor insolvency after the transfer or insolvency at the time of transfer, (5) wire transfers to new places (Chabrow, 2002), (6) transactions without identifiable business purposes, and (7) transfers for less than reasonably equivalent value.

Some of these indicators can be easily implemented as simple flags in software. However, indicators such as wire transfers to new places produce a large number of 'false positive' suspicious transactions. Thus, the goal is to develop more sophisticated mechanisms based on interrelations among many indicators. To meet these challenges link analysis software for forensic accountants, attorneys and fraud examiners such as NetMap, Analyst's Notebook and others (Chabrow, 2002, i2, Evett *et al.*, 2000) have been and are being developed.

Data Mining can assist in discovering patterns of fraudulent activities that are closely related to terrorism such as transactions without identifiable business purposes. The problem is that often an individual transaction does not reveal that it has no identifiable business purpose or that it was done for no reasonably equivalent value. Thus, Data Mining techniques can search for suspicious patterns in the form of more complex combinations of transactions and other evidence using background knowledge. This means that the training data are formed not by transactions themselves but combination of two, three or more transactions. This implies that the number of training objects exploded. The percentage of suspicion records in the set of all transactions is very small, but the percentage of suspicious combinations in the set of combinations is minuscule. This is a typical task of *discovering rare patterns*. Traditional Data Mining methods and approaches are ill-equipped to deal this such problems. Relational Data Mining methods open new opportunities for solving these tasks by discovering "negated patters" described below.

Approach and method

Consider a transactions dataset with attributes such as seller, buyer, item sold, item type, amount, cost, date, company name, type, company type. We will denote each record in this dataset as (<S>, <B>, <I>), where <S>, <B>, and <I> are sets of attributes about the seller, buyer, and item, respectively. We may have two linked records R1=(<S1>, <B1>, <I1>) and R2=(<S2>, <B2>, <I2>), such that the first buyer B1 is also a seller S2, B1=S2. It is also possible that the item sold in both records is the same I1=I2. We create a new dataset of pairs of linked records {<R1,R2>}. Data Mining methods will work in this dataset to discover suspicious records if samples or definitions of normal and suspicious patterns provided. Below we list such patterns:

- a normal pattern (NP) – a Manufacturer Buys a Precursor & Sells the Result of manufacturing (MBPSR);
- a suspicious (abnormal) pattern (SP) – a Manufacturer Buys a Precursor & Sells the same Precursor (MBPSP);
- a suspicious pattern (SP) – a Trading Co. Buys a Precursor and Sells the same Precursor Cheaper (TBPSPC);
- a normal pattern (NP) – a Conglomerate Buys a Precursor & Sells the Result of manufacturing (CBPSR).

A Data Mining algorithm A analyzes pairs of records {<R1,R2>} with say 18 attributes total and can match a pair (#5,#6) with a normal pattern MBPSR, A(#5,#6)= MBPSR, while another pair (#1,#3) can be matched with a suspicious pattern, *A*(#1,#3)= MBPSP.

If definitions of suspicious patterns are given then finding suspicious records is a matter of computationally efficient search is a database that can be distributed. This is not the major challenge. The *automatic generation of patterns/hypotheses descriptions* is a major challenge. One can ask: "Why do we need to discover these definitions (rules) automatically?" A manual way can work if the number of types of suspicious patterns is small and an expert is available. For multistage money-laundering transactions, this is difficult to accomplish manually. Creative criminals and terrorists permanently invent new and more sophisticated money laundering schemes. There is no statistics for such new schemes to learn as it is done in traditional Data Mining approaches.

An approach based on the idea of "negated patters" can uncover such unique schemes. According to this approach *highly probable patterns* are discovered and then *negated.* It is assumed that a highly probable pattern should be *normal*. In more formal terms, the *main hypothesis (MH)* of this approach is:

If Q is a highly probable pattern (>0.9) then Q constitutes a normal pattern and not(Q) can constitute a suspicious (abnormal) pattern

Below we outline an algorithm based on this hypothesis to find suspicious patterns. Computational experiments with two synthesized databases and few suspicious transactions schemes permitted us to discover such transactions. The actual relational data mining algorithm used was algorithm MMRD (Machine Method for Discovery Regularities). Previous research has shown that MMDR based on first-order logic and probabilistic semantic inference is computationally efficient and complete for statistically significant patterns (Kovalerchuk and Vityaev, 2000).

The algorithm finding suspicious patterns based on the main hypotheis (MH) consists of four steps:

1. *Discover* patterns, compute probability of each pattern, select patterns with probabilities above a threshold, say 0.9. To be able to compute conditional probabilities patterns should have a rule form: IF A then B. Such patterns can be extracted using decision tree methods for relatively simple rules and using relational Data Mining for discovering more complex rules. Neural Network (NN) and regression methods typically have no if-part. With additional effort rules can be extracted from NN and regression equations.
2. *Negate* patterns and compute *probability* of each negated pattern,
3. Find records database that satisfy negated patterns and analyze these records for possible *false alarm* (records maybe normal not suspicious).
4. Remove false alarm records and provide detailed analysis of suspicious records.

60.5 Conclusion

To be successful a Data Mining project should be driven by the application needs and results should be tested quickly. Financial applications provide a unique environment where efficiency of the methods can be tested instantly, not only by using traditional training and testing data but making real stock forecast and testing it the same day. This process can be repeated daily for several months collecting quality estimates.

This chapter highlighted problems of Data Mining in finance and specific requirements for Data Mining methods including in making interpretations, incorporating relations and probabilistic learning.

The relational Data Mining methods outlined in this chapter advances pattern discovery methods that deal with complex numeric and non-numeric data, involve structured objects, text and data in a variety of discrete and continuous scales (nominal, order, absolute and so on). The chapter shows benefits of using such methods for stock market forecast and forensic accounting that includes uncovering money laundering schemes. The technique combines first-order logic and probabilistic semantic inference. The approach has been illustrated with an example of discovery of suspicious patterns in forensic accounting.

Currently the success of Data Mining exercises has been reported in literature extensively. Typically it is done by comparing simulated trading and forecasting results with results of other methods and real gain/loss and stock. For instance, recently Huang *et al.* (2003) claimed that Data Mining methods achieved better performance than traditional statistical methods in predicting credit ratings. Much less has been reported publicly on success of Data Mining in real trading by financial institutions. It seems that the market efficiency theory is applicable to reporting success. If real success is reported then competitors can apply the same methods and the leverage will disappear because in essence all fundamental Data Mining methods are not proprietary.

Next future direction is developing practical decision support software tools that make easier to operate in Data Mining environment specific for financial tasks, where hundreds and thousands of models such as neural networks, and decision trees need to be analyzed and adjusted every day with a new data stream coming every minute. E.g., Tsang, Yung, Li (2003) reported an architecture for learning from and monitoring the stock market.

Inside of the field of Data Mining in finance we expect an extensive growth of *hybrid methods* that combine different models and provide a better performance than can be achieved by individuals. In such integrative approach individual models are interpreted as *trained artificial "experts"*. Therefore their combinations can be organized similar to a consultation of real *human experts*. Moreover, these artificial experts can be effectively combined with real experts. It is expected that these artificial experts will be built as autonomous *intelligent software agents*. Thus "experts" to be combined can be Data Mining models, real financial experts, trader and virtual *experts* that runs trading rules extracted from real experts. A virtual expert is a software intelligent agent that is in essence an expert system. We coined a new term ***"expert mining"*** as an umbrella term for extracting knowledge from real human experts that is needed to populate virtual experts.

We expect that in coming years Data Mining in finance will be shaped as a distinct field that blends knowledge from finance and Data Mining, similar to what we see now in bioinformatics where integration of field specifics and Data Mining is close to maturity. We also expect that the blending with ideas from the theory of dynamic systems, chaos theory, and physics of finance will deepen.

References

Azoff, E., Neural networks time series forecasting of financial markets, Wiley, 1994.

Back, A., Weigend, A., A first application of independent component analysis to extracting structure from stock returns. Int. J. on Neural Systems, 8(4):473–484, 1998.

Becerra-Fernandez, I., Zanakis, S. Walczak,S., Knowledge discovery techniques for predicting country investment risk, Computers and Industrial Engineering Vol. 43 , Issue 4:787 – 800, 2002.

Berka, P. PKDD Discovery Challenge on Financial Data, In: Proceedings of the First International Workshop on Data Mining Lessons Learned, (DMLL-2002), 8-12 July 2002, Sydney, Australia.

Bouchaud, J., Potters,M., Theory of Financial Risks: From Statistical Physics to Risk Management, 2000, Cambridge Univ. Press, Cambridge, UK.

Bratko, I., Muggleton, S., Applications of Inductive Logic Programming. Communications of ACM, 38(11): 65-70, 1995.

Casdagli, M., Eubank S., (Eds). Nonlinear modeling and forecasting, Addison Wesley, 1992.

Chabrow, E. Tracking the terrorists, Information week, Jan. 14, 2002, http://www.tpirsrelief.com/forensic_accounting.htm

Cowan, A., Book review: Data Mining in Finance, International journal of forecasting, Vol.18, Issue 1, 155-156, Jan-March 2002.

Dhar, V., Stein,R., Intelligent decision support methods, Prentice Hall, 1997.

Džeroski S., Inductive Logic programming Approaches, In: Klösgen W., Zytkow J. Handbook of Data Mining and knowledge discovery, Oxford Univ. Press, 2002, 348-353.

Drake, K., Kim Y., Abductive information modeling applied to financial time series forecasting, In: Nonlinear financial forecasting, Finance and Technology, 1997, 95-109.

Evett, IW., Jackson, G. Lambert, JA , McCrossan, S. The impact of the principles of evidence interpretation on the structure and content of statements. Science and Justice, 40, 2000, 233–239.

Freedman R., Klein R., Lederman J., Artificial intelligence in the capital markets, Irwin, Chicago, 1995.

Giles, G., Lawrence S., Tshoi, A. Rule inference for financial prediction using recurrent neural networks, In: Proc. Of IEEE/IAAFE Conference on Computational Intelligence for financial Engineering, IEEE, NJ, 1997, 253-259.

Groth, R., Data Mining, Prentice Hall, 1998.

Greenstone, M., Oyer, P., Are There Sectoral Anomalies Too? The Pitfalls of Unreported Multiple Hypothesis Testing and a Simple Solution, Review of Quantitative Finance and Accounting, 15, 2000: 37-55, http://faculty-gsb.stanford.edu/oyer/wp/tech.pdf

Haugh, M., Lo, A., Computational Challenges in Portfolio Management, Tomorrow's Hardest Problems, IEEE Computing in Science and Engineering, May/June 2001, 54-59.

Huang, Z, Chen H, Hsu C.-J., Chen W.-H., Wu S., Credit rating analysis with support vector machines and neural networks: a market comparative study, Decision support systems, Volume 37, Issue 4, pp. 543-558, 2004.

Ilinski, K., Physics of Finance: Gauge Modeling in Non-Equilibrium Pricing, Wiley, 2001

i2 Applications-Fraud Investigation Techniques, http://www.i2.co.uk/Products/

Kingdon, J., Intelligent systems and financial forecasting. Springer, 1997.

Klösgen W., Zytkow J. Handbook of Data Mining and knowledge discovery, Oxford Univ. Press, Oxford, 2002.

Kovalerchuk, B., Vityaev, E., Data Mining in Finance: Advances in Relational and Hybrid Methods, Kluwer, 2000.

Kovalerchuk, B., Vityaev E., Ruiz J.F., Consistent and Complete Data and "Expert Mining" in Medicine, In: Medical Data Mining and Knowledge Discovery, Springer, 2001, 238-280.

Krolzig, M., Toro, J., Multiperiod Forecasting in Stock Markets: A Paradox Solved, Decision Support Systems, Volume 37, Issue 4, pp. 531-542, 2004.

Lachiche, N., Flach, P.A True First-Order Bayesian Classifier. 12th International Conference, ILP 2002, Sydney, Australia, July 9-11, 2002. Lecture Notes in Computer Science 2583 Springer 2003,133-148.

Loofbourrow, J., Loofbourrow, T., What AI brings to trading and portfolio management, In: Freedman R., Klein R., Lederman J., Artificial intelligence in the capital markets, Irwin, Chicago, 1995, 3-28.

Mandelbrot, B., Fractals and scaling in finance, Springer, 1997
Mantegna, R., Stanley, H., An Introduction to Econophysics: Correlations and Complexity in Finance, Cambridge Univ. Press, Cambridge, UK, 2000
Mehta, K., Bhattacharyya S., Adequacy of Training Data for Evolutionary Mining of Trading Rules, Decision support systems, Volume 37, Issue 4, pp. 461-474, 2004.
Mitchell, T., Machine learning. 1997, McGraw Hill.
Moody, J. Saffell, M. Learning to trade via direct reinforcement, IEEE transactions on neural Networks, Vol. 12, No. 4, 2001, 875-889.
Muller, K.-R., Smola, A., Rtsch, G., Schlkopf, B., Kohlmorgen, J., & Vapnik, V., 1997. Using support vector machines for time series prediction, In: Advances in Kernel Methods – Support Vector Learning, MIT Press, 1997.
Murphy, J. Technical analysis of the financial markets: A comprehensive guide to trading methods and applications, Prentice Hall, 1999.
Muggleton, S., Learning Structure and Parameters of Stochastic Logic Programs, 12th International Conference, ILP 2002, Sydney, Australia, July 9-11, 2002. Lecture Notes in Computer Science 2583 Springer 2003, 198-206.
Muggleton S., Scientific Knowledge Discovery Using Inductive Logic Programming. Communications of ACM, 42(11), 1999, 42-46.
Nakhaeizadeh, G., Steurer, E., Bartmae, K., Banking and Finance, In: Klösgen W., Zytkow J. Handbook of Data Mining and knowledge discovery, Oxford Univ. Press, Oxford, 2002, 771-780.
Neville, J., Jensen, D. , Supporting relational knowledge discovery: Lessons in architecture and algorithm design, In: Proceedings of the First International Workshop on Data Mining Lessons Learned, (DMLL-2002), 8-12 July 2002, Sydney, Australia.
Prentice, M., Forensic Services-tracking terrorist networks,2002, Ernst & Young, UK.
Quinlan J.R., C4.5: programs for machine learning, Morgan Kaufmann Publishers Inc., San Francisco, CA, 1993.
Refenes A., (Ed.) Neural Networks in the Capital Markets, Wiley, 1995
Shen L., Loh, H., Applying rough sets to market timing decisions, Decision support systems, Volume 37, Issue 4, 583-597, 2004.
Sullivan, R., Timmermann, A., White, H., Dangers of Data-Driven Inference: The Case of Calendar Effects in Stock Returns. University of California. San Diego Department of Economics, Discussion Paper 98-16, 1998.
Sullivan, R., Timmermann, A., White, H., Data-Snooping, Technical Trading Rule Performance, and the Bootstrap. Journal of Finance 54, 1999, 1647-1691.
Thulasiram, R., Thulasiraman, P., Performance Evaluation of a Multithreaded Fast Fourier Transform Algorithm for Derivative Pricing, Journal of Supercomputing, Vol.26 No.1, 43-58, August 2003.
Thulasiram, R. Jayaraman, S. Sampath, S. Financial Forecasting using Neural Networks under Multithreaded Environment, IIIS Proc. of the 6th World Multiconference on Systems, Cybernetics and Informatics, SCI 2002 , Orlando, FL, USA, July 14-17, 2002, 147-152.
Thuraisingham, B, Data Mining: technologies, techniques, tools and trends. CRC Press, 1999
Trippi, R., Turban, E., Neural networks in finance and investing, Irwin, Chicago 1996.
Tsay, R. ,Analysis of financial time series. Wiley, 2002.
Turcotte, M., Muggleton, S., Sternberg, M., The Effect of Relational Background Knowledge on Learning of Protein Three-Dimensional Fold Signatures. Machine Learning, 43(1/2), 2001, 81-95.

Vangel, D., James A. Terrorist Financing: Cleaning Up a Dirty Business, the issue of Ernst & Young's financial services quarterly, Springer, 2002.

Vityaev E.E., Orlov Yu. L., Vishnevsky O.V., Kovalerchuk B.Ya., Belenok A.S., Podkolodnii N.L., Kolchanov N.A. Knowledge Discovery for Gene Regulatory Regions Analysis, In: Knowledge-Based Intelligent Information Engineering Systems and Allied Technologies, KES 2002. Eds. E. Damiani, R. Howlett, L.Jain, N. Ichalkaranje, IOS Press, Amsterdam, 2002, part 1, 487-491.

Voit, J., The Statistical Mechanics of Financial Markets, Vol. 2, Springer, 2003.

Von Altrock C. , Fuzzy Logic and NeuroFuzzy Applications in Business and Finance, Prentice Hall, 1997.

Walczak, S., An empirical analysis of data requirements for financial forecasting with neural networks, Journal of Management Information Systems, 17(4), 2001, 203-222, 2001.

Wang, H., Weigend A. (Eds), Data Mining for financial decision making, Special Issue, Decision support systems, Volume 37, Issue 4,2004.

Wang J., Data Mining; opportunities and challenges, Idea Group, London, 2003

Zemke, S. On Developing a Financial Prediction System: Pitfalls and Possibilities, In: Proceedings of the First International Workshop on Data Mining Lessons Learned (DMLL-2002), 8-12 July 2002, Sydney, Australia.

Zemke, S. , Data Mining for Prediction. Financial Series Case, Doctoral Thesis, The Royal Institute of Technology, Department of Computer and Systems Sciences, Sweden, December 2003.

Zenios, S. High Performance Computing in Finance - Last Ten Years and Next, Parallel Computing, Dec. 1999, 2149-2175.

61

Data Mining for Intrusion Detection

Anoop Singhal[1] and Sushil Jajodia[2]

[1] Center for Secure Information Systems,
George Mason University, Fairfax, VA 22030-4444
[2] Center for Secure Information Systems,
George Mason University, Fairfax, VA 22030-4444

Summary. Data Mining Techniques have been successfully applied in many different fields including marketing, manufacturing, fraud detection and network management. Over the past years there is a lot of interest in security technologies such as intrusion detection, cryptography, authentication and firewalls. This chapter discusses the application of Data Mining techniques to computer security. Conclusions are drawn and directions for future research are suggested.

Key words: computer security, intrusion detection, data warehouse, alert correlation

61.1 Introduction

Computer security is of importance to a wide variety of practical domains ranging from banking industry to multinational corporations, from space exploration to the intelligence community and so on. The following principles are generally accepted as the foundation of a good security solution:

- Authentication: The process of establishing the validity of a claimed identity.
- Authorization: The process of determining whether a validated entity is allowed access to a resource based on attributes, predicates, or context.
- Integrity: The prevention of modification or destruction of an asset by an unauthorized user.
- Availability: The protection of assets from denial-of-service threats that might impact system availability.
- Confidentiality: The property of non-disclosure of information to unauthorized users.
- Auditing: The property of logging all system activities at a sufficient level so that events can be reconstructed if it is required.

Intrusion detection is the process of monitoring and analyzing the events occurring in a computer system in order to detect signs of security problems. Over the past several years,

O. Maimon, L. Rokach (eds.), *Data Mining and Knowledge Discovery Handbook*, 2nd ed.,
DOI 10.1007/978-0-387-09823-4_61,

intrusion detection and other security technologies such as cryptography, authentication and firewalls have increasingly gained in importance. There is a lot of interest in applying Data Mining techniques to intrusion detection. This chapter gives a critical summary of Data Mining research for intrusion detection. We first give the basics of Data Mining techniques. We then survey a list of research projects that apply Data Mining techniques to intrusion detection. We then suggest new directions for research and then give our conclusions.

61.2 Data Mining Basics

Recent progress in scientific and engineering applications has accumulated huge volumes of data. The fast growing, tremendous amount of data, collected and stored in large databases has far exceeded our human ability to comprehend it without proper tools. It is estimated that the total database size for a retail store chain such as Walmart will exceed 1 Petabyte (1K Terabyte) by 2005. Similarly, the scope, coverage and volume of digital geographic data sets and multidimensional data have grown rapidly in recent years. These data sets include digital data of all sorts created and disseminated by government and private agencies on land use, climate data and vast amounts of data acquired through remote sensing systems and other monitoring devices. It is estimated that multimedia data is growing at about 70% per year. Therefore, there is a critical need of data analysis systems that can automatically analyze the data, to summarize it and predict future trends. Data Mining is a necessary technology for collecting information from distributed databases and then performing data analysis.

The process of knowledge discovery in databases is explained in Figure 61.1 and it consists of the following steps (Han and Kamber, 2000):

1. Data cleaning to remove noise and inconsistencies.
2. Data integration to get data from multiple sources.
3. Data selection step where data relevant for the task is retrieved.
4. Data transformation step where data is transformed into an appropriate form for data analysis.
5. Data Analysis where complex queries are executed for in depth analysis.

The following are different kinds of techniques and algorithms that data mining can provide:

Association Analysis: This involves discovery of *association rules* showing attribute-value conditions that occur frequently together in a given set of data. This is used frequently for market basket or transaction data analysis. For example, the following rule says that if a customer is in age group 20 to 29 years and income is greater than 40K/year then he or she is likely to buy a DVD player.

Age(X, “20-29”) & income(X, “>40K”) => buys (X, “DVD player”)

[support = 2% , confidence = 60%]

Rule *support* and *confidence* are two measures of rule interestingness. A support of 2% means that 2% of all transactions under analysis show that this rule is true. A confidence of 60% means that among all customers in the age group 20-29 and income greater than 40K, 60% of them bought DVD players.

A popular algorithm for discovering association rules is the **Apriori** method. This algorithm uses an iterative approach known as *level-wise* search where k-itemsets are used to explore (k+1) itemsets. Association rules are widely used for prediction.

Classification and Prediction: Classification and prediction are two forms of data analysis that can be used to extract models describing important data classes or to predict future

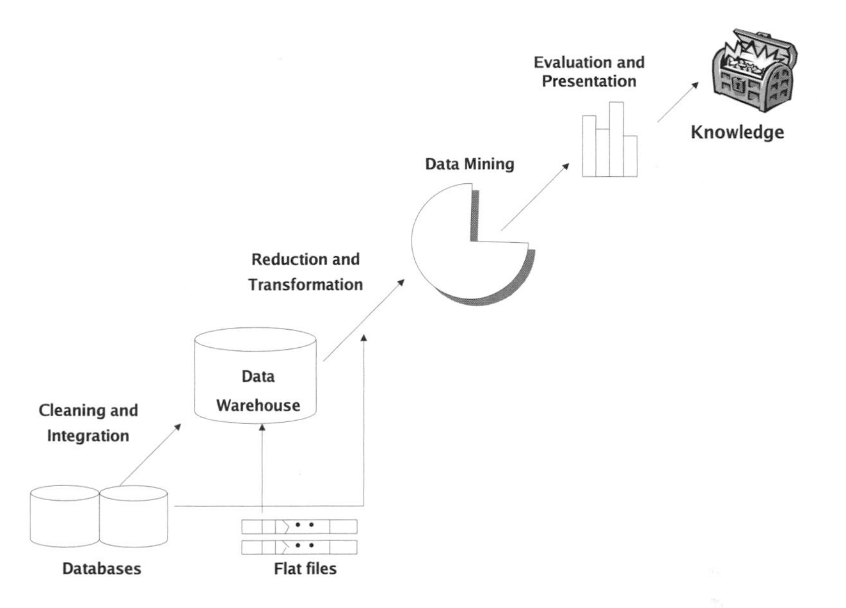

Fig. 61.1. The process of Knowledge Discovery

data trends. For example, a classification model can be built to categorize bank loan applications as either safe or risky. A prediction model can be built to predict the expenditures of potential customers on computer equipment given their income and occupation. Some of the basic techniques for data classification are decision tree induction, Bayesian classification and neural networks.

These techniques find a set of models that describe the different *classes* of objects. These models can be used to predict the class of an object for which the class is unknown. The derived model can be represented as rules (IF-THEN), decision trees or other formulae.

Clustering: This involves grouping objects so that objects within a cluster have high similarity but are very dissimilar to objects in other clusters. Clustering is based on the principle of *maximizing the intraclass similarity and minimizing the interclass similarity.*

In business, clustering can be used to identify customer groups based on their purchasing patterns. It can also be used to help classify documents on the web for information discovery. Due to the large amount of data collected, cluster analysis has recently become a highly active topic in Data Mining research. As a branch of statistics, cluster analysis has been extensively studied for many years, focusing primarily on *distance based cluster analysis.* These techniques have been built into statistical analysis packages such as S-PLUS and SAS. In machine learning, clustering is an example of *unsupervised learning*. For this reason clustering is an example of *learning by observation.*

Outlier Analysis: A database may contain data objects that do not comply with the general model or behavior of data. These data objects are called *outliers*. Most Data Mining methods discard outliers as noise or exceptions. These outliers are useful for applications such as fraud detection and network intrusion detection. The analysis of outlier data is referred to as *outlier mining.* We will describe some intrusion detection systems that use outlier analysis.

Outliers may be detected using statistical tests that assume a distribution or probability model for the data, or using distance measures where objects that are a substantial distance from other clusters are consider outliers.

61.3 Data Mining Meets Intrusion Detection

Since the cost of information processing and Internet accessibility is dropping, more and more organizations are becoming vulnerable to a wide variety of cyber threats. According to a recent survey by CERT, the rate of cyber attacks has been doubling every year in recent times. Therefore, it has become increasingly important to make our information systems, especially those used for critical functions such as military and commercial purpose, resistant to and tolerant of such attacks. Intrusion Detection Systems (IDS) are an integral part of any security package of a modern networked information system. An IDS detects intrusions by monitoring a network or system and analyzing an audit stream collected from the network or system to look for clues of malicious behavior.

Intrusion detection systems can be classified into the following two categories:

- Misuse Detection: This method finds intrusions by monitoring network traffic in search of direct matches to known patterns of attack (called signatures or rules). A disadvantage of this approach is that it can only detect intrusions that match a pre-defined rule. One advantage of these systems is that they have low false alarm rates.
- Anomaly Detection: In this approach, the system defines the expected behavior of the network in advance. The profile of normal behavior is built using techniques that include statistical methods, association rules and neural networks. Any significant deviations from this expected behavior are reported as possible attacks. In principle, the primary advantage of anomaly based detection is the ability to detect novel attacks for which signatures have not been defined yet. However, in practice, this is difficult to achieve because it is hard to obtain accurate and comprehensive profiles of normal behavior. This makes an anomaly detection system generate too many false alarms and it can be very time consuming and labor intensive to sift through this data.

Intrusion Detection Systems (IDS) can also be categorized according to the kind of information they analyze. This leads to the distinction between *host-based* and *network-based* IDSs. A host based IDS analyzes host-bound audit sources such as operating system audit trails, system logs or application logs. Since host based systems directly monitor the host data files and operating system processes, they can determine exactly which host resources are targets of a particular attack. Due to the rapid development of computer networks, traditional single host intrusion detection systems have been modified to monitor a number of hosts on a network. They transfer the monitored information from multiple monitored hosts to a central site for processing. These are termed as distributed intrusion detection systems.

A network based IDS analyzes network packets that are captured on a network. This involves placing a set of traffic sensors within the network. The sensors typically perform local analysis and detection and report suspicious events to a central location.

Recently, there is a great interest in application of Data Mining techniques to intrusion detection systems. The problem of intrusion detection can be reduced to a Data Mining task of classifying data. Briefly, one is given a set of data points belonging to different classes (normal activity, different attacks) and aims to separate them as accurately as possible by means of a model. This section gives a summary of the current research project in this area.

61.3.1 ADAM

The ADAM project at George Mason University (Barbara *et al.*, 2001, Barbara*et al.*, 2001) is a network-based anomaly detection system. ADAM learns normal network behavior from attack-free training data and represents it as a set of association rules, the so called profile. At run time, the connection records of past delta seconds are continuously mined for new association rules that are not contained in the profile.

ADAM is an anomaly detection system. It is composed of three modules: a preprocessing engine, a mining engine and a classification engine. The preprocessing engine sniffs TCP/IP traffic data and extracts information from the header of each connection according to a predefined schema. The mining engine applies mining association rules to the connection records. It works in two modes: training mode and detecting mode. In training mode, the mining engine builds a profile of the users and systems normal behavior and generates association rules that are used to train the classification engine. In detecting mode, the mining engine mines unexpected association rules that are different from the profile. The classification engine will classify the unexpected association rules into normal and abnormal events. Some abnormal events can be further classified as attacks. Although mining of association rules has used previously to detect intrusions in audit trail data, the ADAM system is unique in the following ways:

- It is on-line; it uses an incremental mining (on-line mining) which does not look at a batch of TCP connections, but rather uses a sliding window of time to find the suspicious rules within that window.
- It is an anomaly detection system that aims to categorize using Data Mining the rules that govern misuse of a system. For this, the technique builds, apriori, a profile of "normal" rules, obtained by mining past periods of time in which there were no attacks. Any rule discovered during the on-line mining that also belongs to this profile is ignored, assuming that it corresponds to a normal behavior.

Figures 61.2 and 61.3 show the basic architecture of ADAM. ADAM performs its task in two phases. In the training phase, ADAM uses a data stream for which it knows where the attacks are located. The attack free parts of the stream are fed into a module that performs off-line association rules discovery. The output of this module is a profile of rules that we call "normal" i.e. it provides the behavior during periods when there are no attacks. The profile along with the training data set is also fed into a module that uses a combination of dynamic, on line algorithm for association rules, whose output consists of frequent item sets that characterize attacks to the system. These item sets are used as a classifier or decision tree. This whole phase takes place off-line before we use the system to detect attacks.

The second phase of ADAM in which we actually detect attacks is shown in the figure below. Again, the on-line association rules mining algorithm is used to process a window of current connections. Suspicious connections are flagged and sent along with their feature vectors to the trained classifier, where they are labeled as attacks, false alarms or unknown. When, the classifier labels connections as false alarms, it is filtering them out of the attacks set and avoiding passing these alerts to the security officer. The last class, i.e. unknown is reserved for the events whose exact nature cannot be confirmed by the classifier. These events are also considered as attacks and they are included in the set of alerts that are passed to the security officer.

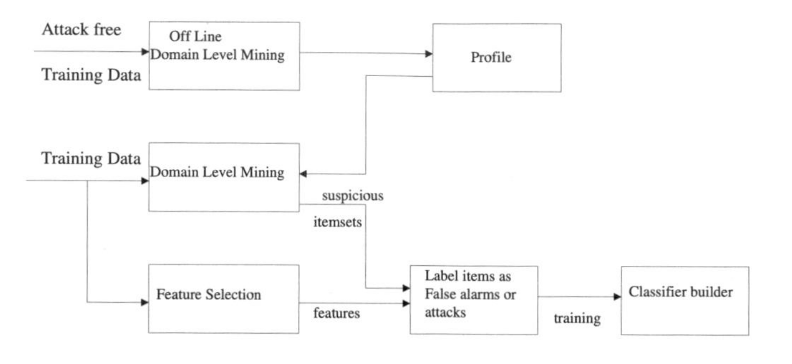

Fig. 61.2. The Training Phase of ADAM

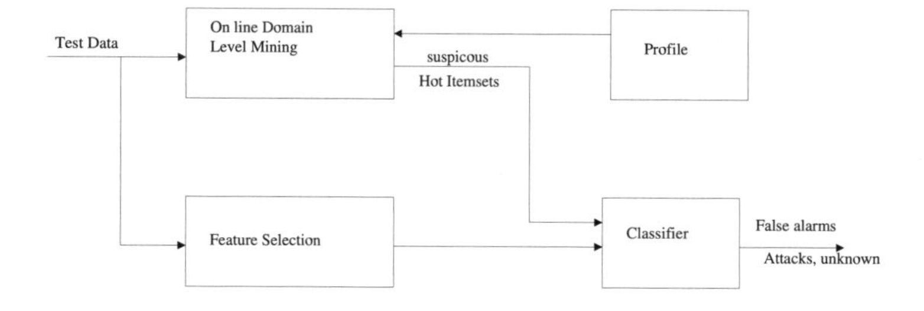

Fig. 61.3. The Intrusion Detection Phase of ADAM

61.3.2 MADAM ID

The MADAM ID project at Columbia University (Lee, 1998, Lee *et al.*, 1998) has shown how Data Mining techniques can be used to construct an IDS in a more systematic and automated manner. Specifically, the approach used by MADAM ID is to learn classifiers that distinguish between intrusions and normal activities. Unfortunately, classifiers can perform really poorly when they have to rely on attributes that are not predictive of the target concept. Therefore, MADAM ID proposes association rules and frequent episode rules as means to construct additional more predictive attributes. These attributes are termed as *features.*

We will describe briefly how MADAM ID is used to construct network based misuse detection systems. First all network traffic is preprocessed to create *connection records.* The attributes of connection records are intrinsic connection characteristics such as source host, the destination host, the source and destination posts, the start time, the duration, header flags and so on. In the case of TCP/IP networks, connection records summarize TCP sessions.

The most important characteristic of MADAM ID is that it *learns* a misuse detection model from examples. In order to use MADAM ID, one needs a large set of connection records that have already been classified into "normal records" or some kind of attacks. MADAM ID proceeds in two steps. In the first step it does *feature construction* in which some additional features are constructed that are considered useful for doing the analysis. One example for this step is to calculate the count of the number of connections that have been initiated during the last two seconds to the same destination host as the current host. The feature construction step is followed by the *classifier learning* step. It consists of the following process:

1. The training connection records are partitioned into two sets, namely *normal connection records* and *intrusion connection records.*
2. Association rules and frequent episode rules are mined separately from the normal connection records and from the intrusion connection records. The resulting patterns are compared and all patterns that are exclusively contained in the intrusion connection records are collected to form *the intrusion only* patterns.
3. The intrusion only patterns are used to derive additional attributes such as count or percentage of connection records that share some attribute values with the current connection records.
4. A classifier is learned that distinguishes normal connection records from intrusion connection records, This classifier is the end product of MADAM ID.

61.3.3 MINDS

The MINDS project (Ertoz *et al.*, 2003, Kumar *et al.*, 2003) at University of Minnesota uses a suite of Data Mining techniques to automatically detect attacks against computer networks and systems. Their system uses an anomaly detection technique to assign a score to each connection to determine how anomalous the connection is compared to normal network traffic. Their experiments have shown that anomaly detection algorithms can be successful in detecting numerous novel intrusions that could not be identified using widely popular tools such as SNORT.

Input to MINDS is Netflow data that is collected using Netflow tools. The netflow data contains packet header information i.e. they do not capture message contents. Netflow data for each 10 minute window which typically results in 1 to 2 million records is stored in a flat file. The analyst uses MINDS to analyze these 10 minute data files in a batch mode. The first step in MINDS involves constructing features that are used in the Data Mining analysis.

Basic features include source IP address and port, destination IP address and port, protocol, flags, number of bytes and number of packets. Derived features include time-window and connection window based features. After the feature construction step, the data is fed into the MINDS anomaly detection module that uses an outlier detection algorithm to assign an anomaly score to each network connection. A human analyst then has to look at only the most anomalous connections to determine if they are actual attacks or other interesting behavior.

MINDS uses a density based outlier detection scheme for anomaly detection. The reader is referred to (Ertoz *et al.*, 2003) for a more detailed overview of their research. MINDS assigns a degree of being an outlier to each data point which is called the local outlier factor (LOF). The output of the anomaly detector contains the original Netflow data with the addition of the anomaly score and relative contribution of the different attributes to that score. The analyst typically looks at only the top few connections that have the highest anomaly scores. The researchers of MINDS have their system to analyze the University of Minnesota network traffic. They have been successful in detecting scanning activities, worms and non standard behavior such as policy violations and insider attacks.

61.3.4 Clustering of Unlabeled ID

Traditional anomaly detection systems require "clean" training data in order to learn the model of normal behavior. A major drawback of these systems is that clean training data is not easily available. To overcome this weakness, recent research has investigated the possibility of training anomaly detection systems over noisy data (Portnoy *et al.*, 2001). Anomaly detection over noisy data makes two key assumptions about the training data. First, the number of normal elements in the training data is assumed to be significantly larger than the number of anomalous elements. Secondly, anomalous elements are assumed to be qualitatively different from normal ones. Then, given that anomalies are both rare and different, they are expected to appear as outliers that stand out from the normal baseline data. Portnoy et al. (Portnoy *et al.*, 2001) apply clustering to the training data. Here the hope is that intrusive elements will bundle with other intrusive elements whereas normal elements will bundle with other normal ones. Moreover, as intrusive elements are assumed to be rare, they should end up in small clusters. Thus, all small clusters are assumed to contain intrusions/anomalies, whereas large clusters are assumed to represent normal activities. At run time, new elements are compared against all clusters and the most similar cluster determines the new element's classification as either "normal" or "intrusive".

61.3.5 Alert Correlation

Correlation techniques from multiple sensors for large networks is described in (Ning *et al.*, 2002,Ning and Xu, 2003). A language for modeling alert correlation is described in (Cuppens and Miege, 2002). Traditional IDS systems focus on low level alerts and they raise alerts independently though there may be a logical connection between them. In case of attacks, the number of alerts that are generated become unmanageable. As a result, it is difficult for human users to understand the alerts and take appropriate actions. Ning et al. present a practical method for constructing attack scenarios through alert correlation, using prerequisites and consequences of intrusions. Their approach is based on the observation that in a series of attacks, alerts are not isolated, but related as different stages, with earlier stages preparing for the later ones. They proposed a formal framework to represent alerts with their prerequisites and consequences using the concept of *hyper-alerts*. They evaluated their approach using the 2000 DARPA intrusion detection scenario specific datasets.

61.4 Conclusions and Future Research Directions

In this chapter, we reviewed the application of Data Mining techniques to the area of computer security. Data Mining is primarily being used to detect intrusions rather than to discover new knowledge about the nature of attacks. Moreover, most research is based on strong assumptions that complicate building of practical applications. First, it is assumed that labeled training data is readily available, and second it is assumed that this data is of high quality. Different authors have remarked that in many cases, it is not easy to obtain labeled data. Even if one could obtain labeled training data by simulating intrusions, there are many problems with this approach. Additionally, attack simulation limits the approach to the set of known attacks. We think that the difficulties associated with the generation of high quality training data will make it difficult to apply Data Mining techniques that depend on availability of high quality labeled training data. Finally, Data Mining in intrusion detection focuses on a small subset of possible applications. Interesting future applications of Data Mining might include the discovery of new attacks, the development of better IDS signatures and the construction of alarm correlation systems.

For future research, it should be possible to focus more on the KDD process and detection of novel attacks. It is known that attackers use a similar strategy to attack in the future as what they used in the past. The current IDSs can only detect a fraction of these attacks. There are new attacks that are hidden in the audit logs, and it would be useful to see how Data Mining can be used to detect these attacks.

Data Mining can also be applied to improve IDS signatures. IDS vendors can run their systems in operational environment where all alarms and audit logs are collected. Then, Data Mining can be used to search for audit log patterns that are closely related with particular alarms. This might lead to new knowledge as to why false positives arise and how they can be avoided.

Finally, Data Mining projects should focus on the construction of alarm correlation systems. Traditional intrusion detection systems focus on low level alerts and they raise alerts independently even though there is a logical connection among them. More work needs to be done on alert correlation techniques that can construct "attack strategies" and facilitate intrusion analysis. One way is to store data from multiple sources in a data warehouse and then perform data analysis. Alert correlation techniques will have several advantages. First, it will provide a high level representation of the alerts along with a temporal relationship of the sequence in which these alerts occurred. Second, it will provide a way to distinguish a true alert from a false alert. We think that true alerts are likely to be correlated with other alerts whereas false alerts will tend to be random and, therefore, less likely to be related to other alerts. Third, it can be used to anticipate the future steps of an attack and, thereby, come up with a strategy to reduce the damage.

References

Barbara D., Wu N., and Jajodia S., Detecting novel network intrusions using bayes estimators. In Proc. First SIAM Conference on Data Mining, Chicago, IL, April 2001.

Barbara D., Couto J., Jajodia S., and Wu N., Adam: Detecting Intrusions by Data Mining, In Proc. 2nd Annual IEEE Information Assurance Workshop, West Point, NY, June 2001.

Cuppens F. and Miege A., Alert Correlation in a Cooperative Intrusion Detection Framework, Proc. IEEE Symposium on Security and Privacy, May 2002.

Ertoz L., Eilertson E., Lazarevic A., Tan P., Dokes P., Kumar V., Srivastava J., Detection of Novel Attacks using Data Mining, Proc. IEEE Workshop on Data Mining and Computer Security, November 2003.

Han J. and Kamber M., Data Mining: Concepts and Techniques, Morgan Kaufmann, August 2000.

Kumar V., Lazarevic A., Ertoz L., Ozgur A., Srivastava J., A Comparative Study of Anomaly Detection Schemes in Network Intrusion Detection, In Proc. Third SIAM International Conference on Data Mining, San Francisco, May 2003.

Lee W., Stolfo, S. J., and Kwok K. W. Mining audit data to build intrusion detection models. In Proc. Fourth International Conference on Knowledge Discovery and Data Mining, NewYork, 1998.

Lee W. and Stolfo S. J. Data Mining approaches for intrusion detection, In Proc. Seventh USENIX Security Symposium, San Antonio, TX, 1998.

Ning P., Cui Y., Reeves D. S., Constructing Attack Scenarios through Correlation of Intrusion Alerts, Proc. ACM Computer and Communications Security Conf., 2002.

Ning P., Xu D., earning Attack Strategies from Intrusion Alerts, Proc. ACM Computer and Communications Security Conf., 2003.

Portnoy L., Eskin E., Stolfo S. J., Intrusion Detection with unlabeled data using clustering. In Proceedings of ACM Workshop on Data Mining Applied to Security, 2001.

62

Data Mining for CRM

Kurt Thearling

Vertex Business Services

Summary. Data Mining technology allows marketing organizations to better understand their customers and respond to their needs. This chapter describes how Data Mining can be combined with customer relationship management to help drive improved interactions with customers. An example showing how to use Data Mining to drive customer acquisition activities is presented.

Key words: Customer Relationship Management (CRM), campaign management, customer acquisition, scoring

62.1 What is CRM?

It is now a cliché that in the days of the corner market, shopkeepers had no trouble understanding their customers and responding quickly to their needs. The shopkeepers would simply keep track of each customer in their heads, and would know what to do when a customer walked into the store. But today's shopkeepers face a much more complex situation. More customers, more products, more competitors, and less time to react means that understanding your customers is now much harder to do. This is where customer relationship management (CRM) comes in. CRM lets companies design, manage, and execute strategies for interacting with customers (and potential customers). CRM can be applied to the complete customer life-cycle, from acquisition, to ongoing account management, to cross-selling, to customer retention and attrition.

The goal of CRM is to allow marketing organizations to tune the customer interaction strategies to the specific needs of each individual, giving customers what they want, when they want it. Instead of interacting with large numbers of customers en masse (consider billboards or magazine advertisements), the new role of marketing is to interact with individual customers. This involves identifying and understanding unique customer patterns as well as being able to create customized offers for small customer groups that correspond to those patterns. For example, a pattern might be that 71% of cell-phone customers that make five or

O. Maimon, L. Rokach (eds.), *Data Mining and Knowledge Discovery Handbook*, 2nd ed., DOI 10.1007/978-0-387-09823-4_62,

more calls to customer support in the first month cancel their service. A marketing manger could use this pattern to identify dissatisfied customers and proactively respond to their needs before they cancel.

As a result of the complex interactions that are now possible, the function of marketing is increasingly becoming tied to technology, ranging from complex Data Mining algorithms to campaign management software applications. Campaign management software allows marketing professionals to segment groups of customers (and prospective customers) into smaller groups and then specify the interaction that should take place with those individuals. For example, consider a marketing manager for a cellular phone company that is focusing on customer retention. There might be a large number of reasons that a customer chooses to leave their cellular provider, and the marketing manager is responsible for identifying ways to reduce this problem. One group of customers might be leaving because they are experiencing technical problems (e.g., frequent dropped calls) while another group might be leaving because the plan they are signed up for does not match their current calling patterns (e.g., a local calling plan with a large number of national calls).

A user of campaign management software would define these segments by selecting customers in the database that have the desired characteristics. For the customers with technical problems, the marketer could create a customer segment that selects those customers who have had more than five dropped calls within the last month. Once the segment is defined, it needs to be associated with offers that will be communicated to the customers in order to improve retention. In the case of customers with technical problems, the offer might be a rebate of one month's charges and a promise to improve service quality. This offer could be communicated by a call from customer service or via a piece of direct mail (email might be a third option, if the customer's email address is available). The campaign management application would take the segment and split it into two groups, half receiving a phone call and the other half receiving a piece of direct mail (which half a customer fell into would be by random selection). Once the the segmentation is defined and the marketing manager is satisfied with the campaign, it would need to be executed. This would be handled by a scheduler that executes the campaign at regular intervals (e.g., every night at 2am). Upon execution of the campaign, the segments associated with the phone call would be passed to the call-center software system, which would queue up the customers who are supposed to receive the offer along with the specifics of the script that the operator is supposed to use (full or half month rebate). The direct mail segments would likely be handled differently, possibly by using an external vendor (a "mail shop") that would take a list of customers and produce the actual envelopes that would be mailed. In this case, the campaign management system would generate a file listing each of the customers, including their address and offer type.

62.2 Data Mining and Campaign Management

In the above discussion of campaign management, the selection criteria used to define customer segments ("five dropped calls within the last month") was static, based on historical values stored in a database. Alternatively, some decisions might be based on predicted values (scores) that are output by Data Mining models. Scores can take just about any form, from numbers to strings to entire data structures, but the most common scores are numbers (for example, the probability of responding to a particular promotional offer). These scores can be combined with static values to select the most appropriate prospects for a targeted marketing campaign.

The actual execution of a Data Mining model (scoring) is distinct from the process that creates the model. Typically, a model is used multiple times after it is created to score data in different marketing campaigns. For example, consider a model that has been created to predict the probability that a customer will respond to the cell-phone retention campaign. The model would be built by using historical data from customers and calls, as well as the responses those customers had to various retention offers. After the model has been created based on historical data, it can then be scored on new data in order to make predictions about unseen behavior. This is what Data Mining is all about.

Scoring is the unglamorous workhorse of Data Mining. It doesn't have the sexiness of a neural network or a genetic algorithm, but without it, data mining is pretty useless. At the end of the day, after your Data Mining tools have given you a great predictive model, there's still a lot of work to be done. Scoring models against a customer database can be a time-consuming, error-prone activity, so the key is to integrate it smoothly with the rest of the CRM process. In the past, when a marketer wanted to run a campaign based on model scores, he or she would call the model builder to have the model manually run against a database so that a score file could be created. The marketer then had to solicit the help of an IT staffer to merge the scores with the marketing database. This disjointed process was fraught with problems and errors and could take several weeks. Often, by the time the models were integrated with the database, either the models were outdated or the campaign opportunity had passed.

The current solution is to integrate tightly Data Mining and campaign management technologies. Under this scenario, marketers can invoke statistical models from within the campaign management application, score customer segments on the fly, and quickly create campaigns targeted at customer segments offering the greatest potential. The past few years have seen significant improvements by CRM vendors with respect to integrating Data Mining into the CRM process. This trend is expected to continue resulting in CRM applications driving more and more marketing activities based on Data Mining results.

62.3 An Example: Customer Acquisition

For most businesses, the primary means of growth involves acquiring new customers. This could involve finding customers who previously were not aware of your product, were not candidates for purchasing your product (for example, baby diapers for new parents), or customers who in the past have bought from your competitors. Some of these customers might have been your customers previously, which could be an advantage (more data might be available about them) or a disadvantage (they might have switched as a result of poor service). In any case, Data Mining can often help segment these prospective customers and increase the response rates that an acquisition marketing campaign can achieve.

The traditional approach to customer acquisition involved a marketing manager developing a combination of mass marketing (magazine advertisements, billboards, etc.) and direct marketing (telemarketing, mail, etc.) campaigns based on their knowledge of the particular customer base that was being targeted. In the case of a marketing campaign trying to influence new parents to purchase a particular brand of diapers, the mass marketing advertisements might be focused in parenting magazines (naturally). The ads could also be placed in more mainstream publications whose readership demographics (age, marital status, gender, etc.) were similar to those of new parents.

In direct marketing, a marketing manager would select the demographics that they are interested in (which could very well be the same characteristics used for mass market advertising), and then work with a data vendor (sometimes known as a service bureau) to obtain lists of customers meeting those criteria. Service bureaus have large databases containing millions of prospective customers that can be segmented based on specific demographic criteria (age, gender, interest in particular subjects, etc.). To prepare for the "diapers" direct mail campaign, the marketing manager might request a list of prospects from a service bureau. This list could contain people, aged 18 to 30, who have recently purchased a baby stroller or crib (this information might be collected from people who have returned warranty cards for strollers or cribs). The service bureau would then provide the marketer with a computer file containing the names and addresses for these customers so that the diaper company can contact these customers with their marketing message.

It should be noted that because of the number of possible customer characteristics, the concept of "similar demographics" has traditionally been an art rather than a science. There usually are no hard-and-fast rules about whether two groups of customers share the same characteristics. In the end, much of the segmentation that took place in traditional direct marketing involved hunches on the part of the marketing professional. In the case of 18-to-30 year old purchasers of baby strollers, the hunch might be that people who purchase a stroller in this age group are probably making the purchase before the arrival of their first child (because strollers are saved and used for additional children) and they also haven't yet decided which brand of diapers to use. Seasoned veterans of the marketing game know their customers well and are often quite successful in making these kinds of decisions.

62.3.1 How Data Mining and Statistical Modeling Changes Things

Although a marketer with a wealth of experience can often choose relevant demographic selection criteria, the process becomes more difficult as the amount of data increases. The complexities of the patterns increase, both with the number of customers being considered and the increasing detail known about each customer. The past few years have seen tremendous growth in consumer databases, so the job of segmenting prospective customers is becoming overwhelming.

Data Mining can help this process, but it is by no means a solution to all of the problems associated with customer acquisition. The marketer will need to combine the potential customer list that Data Mining generates with offers that people are interested in. Deciding what is an interesting offer is where the art of marketing comes in.

62.3.2 Defining Some Key Acquisition Concepts

Before the process of customer acquisition begins, it is important to think about the goals of the marketing campaign. In most situations, the goal of an acquisition marketing campaign is to turn a group of potential customers into actual customers of your product or service. This is where things can get a bit fuzzy. There are usually many kinds of customers, and it can often take a significant amount of time before someone becomes a valuable customer. When the results of an acquisition campaign are evaluated, there are often different kinds of responses requiring consideration.

The responses that come in as a result of a marketing campaign are called "response behaviors". The use of the word "behavior" is important because the way in which different people respond to a particular marketing message can vary. This variation needs to be taken into consideration by the campaign and will likely result in different follow-up actions. A response behavior defines a distinct kind of customer action and categorizes the different possibilities so that they can be further analyzed and reported on.

Binary response behaviors are the simplest kind of response. With a binary response behavior, the customer response is either a yes or no. If someone is sent a catalog, did they buy something from the catalog or not? At the highest level, this is often the kind of response that is talked about. Binary response behaviors do not convey any subtle distinctions between customer actions, and these distinctions are not always necessary for effective marketing campaigns.

Beyond binary response behaviors are categorical response behaviors. As you would expect, a categorical response behavior allows for multiple behaviors to be defined. The rules that define the behaviors are arbitrary and are based on the kind of business in which you are involved. Returning to the example of sending out catalogs, one response behavior might be defined to match if the customer purchased women's clothing from the catalog, whereas a different behavior might match when the customer purchased men's clothing. These behaviors can be refined a far as deemed necessary (for example, "purchased men's red polo shirt."

It should be noted that it is possible for different response behaviors to overlap. A behavior might be defined for customers that purchased over $100 from the catalog. This could overlap with the "purchased men's clothing" behavior if the clothing that was purchased cost more than $100. Overlap can also be triggered if the customer purchases more than one item (both men's and women's shirts, for example) as a result of a single offer. Although the use of overlapping behaviors can tend to complicate analysis and reporting, the use of overlapping categorical response behaviors tends to be richer and therefore will provide a better understanding of your customers in the future.

VALEX Response Analyzer: Response Counts by Behavior

Behavior	Measures	Day 12/12/98	12/19/98	12/26/98	1/2/99	TOTAL
Inquiry	Number of Responses	1,556	1,340	328	352	3,576
Purchase A	Number of Responses	210	599	128	167	1,104
Purchase B	Number of Responses	739	476	164	97	1,476
Purchase C	Number of Responses	639	647	113	105	1,504

Fig. 62.1. Example Response Analysis Broken Down by Behavior.

There are usually several different kinds of positive response behaviors that can be associated with an acquisition marketing campaign. This assumes that the goal of the campaign is to increase customer purchases, as opposed to an informational marketing campaign in which customers are simply told of your company's existence. Some of the general categories of response behaviors (see Figure 62.1) are the following:

- **Customer inquiry**: The customer asks for more information about your products or services. This is a good start. The customer is definitely interested in your products — it could signal the beginning of a long-term customer relationship. You might also want to track conversions, which are follow-ups to inquiries that result in the purchase of a product.

- **Purchase of the offered product or products**: This is the usual definition of success. You offered your products to someone, and they decided to buy one or more of them. Within this category of response behaviors, there can be many different kinds of responses. As mentioned earlier, both "purchased men's clothing" and "purchased women's clothing" fit within this category.
- **Purchase of a product different than the ones offered**: Despite the fact that the customer purchased one of your products, it wasn't the one you offered. You might have offered the deluxe product and they chose to purchase the standard model (or vice-versa). In some sense, this is very valuable response because you now have data on a customer/product combination that you would not otherwise have collected.

There are also typically two kinds of negative responses. The first is a non-response. This is not to be confused with a definite refusal of your offer. For example, if you contacted the customer via direct mail, there may be any number of reasons why there was no response (wrong address, offer misplaced, etc.). Other customer contact channels (outbound telemarketing, email, etc.) can also result in ambiguous non-responses. The fact there was no response does not necessarily mean that the offer was rejected. As a result, the way you interpret a non-response as part of additional data analysis will need further consideration(more on this later).

A rejection by the prospective customer is the other kind of negative response. Depending on the offer and the contact channel, you can often determine exactly whether or not the customer is interested in the offer (for example, an offer made via outbound telemarketing might result in a definitive "no, I'm not interested" response). Although it probably does not seem useful, the definitive "no" response is often as valuable as the positive response when it comes to further analysis of customer interests.

62.3.3 It All Begins with the Data

One of the differences between customer acquisition and most other marketing applications of Data Mining revolves around the data that is used to build predictive models. The amount of information that you have about people with whom you do not yet have a relationship is much more limited than the information you have about your existing customers. In some cases, the data might be limited to their address and/or phone number. The key to this process is finding a relationship between the information that you do have and the behaviors you want to model.

Most acquisition marketing campaigns begin with the prospect list. A prospect list is simply a list of customers that have been selected because they are likely to be interested in your products or services. There are numerous companies around the world that will sell lists of customers, often with a particular focus (for example, new parents, retired people, new car purchasers, etc.).

Sometimes, it is necessary to add additional information to a prospect list by overlaying data from other sources. For example, consider a prospect list that contains only names and addresses. In terms of a potential data mining analysis, the information contained in the prospect list is very weak. There might be some patterns in the city, state, or Zip code fields, but they would be limited in their predictive power. To augment the data, information about customers on the prospect list could be matched with external data. One simple overlay involves combining the customer's ZIP code with U.S. census data about average income, average age, and so on. This can be done manually or, as is often the case with overlays, your list provider can take care of this automatically.

More complicated overlays are also possible. Customers can be matched against purchase, response, and other detailed data that the data vendors collect and refine. This data comes from a variety of sources including retailers, state and local governments, and the customers themselves. If you are mailing out a car accessories catalog, it might be useful to overlay information (make, model, year) about any known cars that people on the prospect list might have registered with their department of motor vehicles.

62.3.4 Test Campaigns

Once you have a list of prospective customers, there is still some work that needs to be done before you can create predictive models for customer acquisition. Unless you have data available from previous acquisition campaigns, you will need to send out a test campaign in order to collect data for analysis. Besides the customers you have selected for your prospect list, it is important to include some other customers in the campaign, so that the data is as rich as possible for future analysis. For example, assume that your prospect list (that you purchased from a list broker) was composed of men over age 30 who recently purchased a new car. If you were to market to these prospective customers and then analyze the results, any patterns found by Data Mining would be limited to sub-segments of the group of men over 30 who bought a new car. What about women or people under age 30? By not including these people in your test campaign, it will be difficult to expand future campaigns to include segments of the population that are not in your initial prospect list. The solution is to include a small random selection of customers whose demographics differ from the initial prospect list. This random selection should constitute only a small percentage of the overall marketing campaign, but it will provide valuable information for data mining. You will need to work with your data vendor in order to add a random sample to the prospect list. More sophisticated techniques than random selection do exist, such as those found in statistical design of experiments (DoE).

Although this circular process (customer interaction $\rightarrow$ data collection $\rightarrow$ Data Mining $\rightarrow$ customer interaction) exists in almost every application of Data Mining to marketing, there is more room for refinement in customer acquisition campaigns. Not only do the customers that are included in the campaigns change over time, but the data itself can also change. Additional overlay information can be included in the analysis when it becomes available. Also, by using random selection in the test campaigns, new segments of people can be added to your customer pool.

Once you have started your test campaign, the job of collecting and categorizing the response behaviors begins. Immediately after the campaign offers go out, responses must be tracked. The nature of the response process is such that responses tend to trickle in over time, which means that the campaign can drag on forever. In most real-world situations, though, there is a threshold after which you no longer look for responses. At that time, any customers on the prospect list that have not responded are deemed “non-responses.” Before the threshold, customers who have not responded are in a state of limbo, somewhere between a response and a non-response.

62.3.5 Building Data Mining Models Using Response Behaviors

With the test campaign response data in hand, the actual mining of customer response behaviors can begin. The first part of this process requires you to choose which behaviors you are interested in predicting, and at what level of detail. The level at which the predictive models work should reflect the kinds of offers that you can make, not the kinds of responses that you can track. It might be useful (for reporting purposes) to track catalog clothing purchases down to the level of color and size. If all catalogs are the same, however, the specifics of a customer purchase don't really matter for the Data Mining analysis. In this case (all catalogs are the same), binary response prediction is the way to go. If separate men's and women's catalogs are available, analyzing response behaviors at the gender level would be appropriate. In either case, it is a straightforward process to turn the lower-level categorical behaviors into a set of responses at the desired level of granularity. If there are overlapping response behaviors, the duplicates should be removed prior to mining.

In some circumstances, predicting individual response behaviors might be an appropriate course of action. With the movement toward one-to-one customer marketing, the idea of catalogs that are custom-produced for each customer is moving closer to reality. Existing channels such as the Internet or outbound telemarketing also allow you to be more specific in the ways you target the exact wants and needs of your prospective customers. A significant drawback of the modeling of individual response behaviors is that the analytical processing power required can grow dramatically because the Data Mining process needs to be carried our multiple times, once for each response behavior that you are interested in.

How you handle negative responses also needs to be thought out prior to the data analysis phase. As discussed previously, there are two kinds of negative responses: rejections and non-responses. Rejections, by their nature, correspond to specific records in the database that indicate the negative customer response. Non-responses, on the other hand, typically do not represent records in the database. Non-responses usually correspond to the absence of a response behavior record in the database for customers who received the offer.

There are two ways in which to handle non-responses. The most common way is to translate all non-responses into rejections, either explicitly (by creating rejection records for the non-responding customers) or implicitly (usually a function of the Data Mining software used). This approach will create a data set comprised of all customers who have received offers, with each customer's response being positive (inquiry or purchase) or negative (rejections and non-responses).

The second approach is to leave non-responses out of the analysis data set. This approach is not typically used because it throws away so much data, but it might make sense if the number of actual rejections is large (relative to the number of non-responses); experience has shown that non-responses do not necessarily correspond to a rejection of your product or services offering.

Once the data has been prepared, the actual Data Mining can be performed. The target variable that the Data Mining software will predict is the response behavior type at the level you have chosen (binary or categorical). Because some Data Mining applications cannot predict non-binary variables, some finessing of the data will be required if you are modeling categorical responses using non-categorical software. The inputs to the Data Mining system are the input variables and all of the demographic characteristics that you might have available, especially any overlay data that you combined with your prospect list.

In the end, a model (or models, if you are predicting multiple categorical response behaviors) will be produced that will predict the response behaviors that you are interested in. The models can then be used to score lists of prospect customers in order to select only those who are likely to response to your offer. Depending on how the data vendors you work with operate, you might be able to provide them with the model, and have them send you only the best prospects. In the situation in which you are purchasing overlay data in order to aid in the selection of prospects, the output of the modeling process should be used to determine whether all of the overlay data is necessary. If a model does not use some of the overlay variables, in the interests of economy, you might consider leaving out these unused variables the next time you purchase a prospect list.

63

Data Mining for Target Marketing

Nissan Levin[1] and Jacob Zahavi[2]

[1] Q-Ware Software Company, Israel
[2] Tel-Aviv University

Summary. Targeting is the core of marketing management. It is concerned with offering the right product/service to the customer at the right time and using the proper channel. In this chapter we discuss how Data Mining modeling and analysis can support targeting applications. We focus on three types of targeting models: continuous-choice models, discrete-choice models and in-market timing models, discussing alternative modeling for each application and decision making. We also discuss a range of pitfalls that one needs to be aware of in implementing a data mining solution for a targeting problem.

Key words: Targeting, predictive modeling, decision trees, clustering, survival analysis, in-market timing

63.1 Introduction

Targeting is at the core of marketing management. It is concerned with offering the right product to the customer at the right time and using the proper channel. Indeed, marketing has gone a long way from the mass marketing era where everybody was exposed to the same product, to today's fragmented and diversified markets. The focus has changed from the product to the customer. Instead of increasing market share the objective has shifted to increasing customer share and enhancing customers' loyalty and satisfaction. Recent developments in computer and database technologies are helping these goals by harnessing database marketing, Data Mining and more recently CRM technologies to better understand the customer thus approach her only with products and services that are keen to her. Various marketing metrics have been developed to evaluate the effectiveness of marketing programs and keep track of the profit and costs of each individual customer.

From a Data Mining point of view, we classify the targeting problems into three main categories, according to the variable that we are attempting to predict (the dependent, the choice or the response variable) – discrete choice, continuous choice and in-market timing problems. Each type of problem requires a different type of model to solve.

O. Maimon, L. Rokach (eds.), *Data Mining and Knowledge Discovery Handbook*, 2nd ed.,
DOI 10.1007/978-0-387-09823-4_63, © Springer Science+Business Media, LLC 2010

Discrete choice problems are targeting problems where the response variable is discrete (integer value). The simplest is the binary choice model where the dependent variable assumes two values, usually 0 and 1, e.g.: 0 – do not buy, 1- buy (a product or service). A generalization is the multiple choice model where the dependent variable assumes more than 2 nominal values, e.g., 3 values: 0 – do not buy, 1 - buy a new car, 2 - buy a used car. A special case of a discrete choice is where the dependent variable assumes several discrete values which possess some type of order, or preference. An example in the automotive industry would be: 0- no buy, 1 – buy a compact car, 2 – buy an economy car, 3- buy a midsize car, 4 – buy a luxury car, where the order here is defined in terms of the car segment in increasing order of size.

Continuous choice problems are targeting problems where the choice variable is continuous. Examples are money spent on purchasing from a catalog, donations to charity, year-to-date interest paid on a loan/mortgage, and others. What makes continuous targeting problem in marketing special is the fact that the choice variable is non-negative, i.e., either the customer responds to the solicitation and purchases from the catalog or the customer declines the offer and spends nothing.

Mixed types of problems also exist. For example, continuous choice problems which are formulated as discrete choice models (binary or ordinal), and discrete choice models which are expressed as continuous choice problems (e.g., predicting the number of purchases, where the frequency of purchase assumes many discrete values 0,1,2,... and is thus approximated by a continuous choice).

In-Market timing problems are time-related targeting problems where the objective is to predict the time of next purchase of a product or service. For example, when the customer will be in the market to purchase a new car? When s/he is up to taking the next flight or next cruise trip? Etc.

In this chapter, we discuss how Data Mining modeling and analysis can support these targeting problems, ranging from segmentation-based targeting programs to detailed "one-to-one" programs. For each of models we also discuss the decision making process. Yet, this process is not risk free as there are many pitfalls that one needs to be aware of in building and implementing a targeting program based on Data Mining, which, if not cared for, could lead to erroneous results. So we devote a great deal of efforts to suggesting ways to identify these pitfalls and ways to fix them.

This chapter is organized as follows: In Section 63.2 we discuss the modeling process for a typical targeting application of a new product, followed by a brief review, in Section 63.3, of the common metrics used to evaluate the quality of targeting models. In sections 63.4,63.5,63.6, we discuss the three class of models to support targeting decisions - segmentation, predictive modeling and in-market timing models, respectively. In Section 63.7 we review a host of pitfalls and issues that one needs to be aware of when building and implementing a targeting application involving Data Mining. We conclude, in Section 63.8, with a short summary

63.2 Modeling Process

Figure 63.1 exhibits the decision process for a targeting application. In the case of a new product, the process is often initiated by a test mailing to a sample of customers in order to assess customers' response. Then people in the audience who "look like" the test buyers are selected for the promotion. For a previously promoted program, the modeling process is based on the results of the previous campaign for the same product. The left hand side of Figure 63.1

corresponds to the testing phase, the right hand side to the rollout phase. The target audience, often referred to as the universe, is typically, but not necessarily, a subset of the customer list containing only customers who, based upon some previous consideration, make up potential prospects for the current product (e.g., people who have been active in the last, say, three years). The test results are used to calibrate a response model to identify the characteristics of the likely buyers. The model results are then applied against the balance of the database to select customers for the promotion. As discussed below, it is a good practice to split the test audience into two mutually exclusive data sets, a training set to build the model with and a validation (or a holdout set) to validate the model with. The validation procedure is essential to avoid over fitting and make sure that the model produces stable results that could be applied to score a set of new observations. Often there is a time gap between the time of the test and the time of the rollout campaign because of the lead time to stock up on the product. Since the customer database is highly dynamic and changes by the minute, one has to make sure that the test universe and the rollout universe are compatible and contain the same "kind" of people. For example, if the test universe contains only people who have been active in the last three years prior to the test, the rollout universe should also include only the last three-year buyers. Otherwise we will be comparing apples to oranges thereby distorting the targeting results.

We note that the validation data set is used only to validate the model by comparing predicted to actual results. The actual decisions, however, are based only on the predicted profit/response for the training set. The decision process proceeds as follows:

- Build a model based on the training set
- Validate the model based on the validation set. Below we discuss a variety of metrics to evaluate and assess the quality of a predictive model.
- If the resulting model is not "good enough", build a new model by changing the parameters, adding observations, trying a different set of influential predictors, use a different type of model, new transformations, etc. Iterate, if necessary
- Once happy with the model, apply the model to predict the value of the dependent variable for each customer in the rollout universe. This process is often referred to as "scoring" and the resulting predicted values as "scores". These scores may vary from model to model. For example, in logistic regression, the resulting score is the purchase probability of the customer.
- Finally, use economic criterion to select the customers for targeting from the rollout universe. These economic criteria may vary between models, and so we discuss them below in the context of each class of models. Note that the rollout universe does not have any actual values for the current promotions. Hence decisions should be based solely on predicted values, i.e., the calculated scores.

63.3 Evaluation Metrics

Several metrics are used to evaluate the results of targeting models. These are divided into goodness-of-fit measures, prediction accuracy and profitability/ROI measures.

63.3.1 Gains Charts

Prediction models are evaluated based on some goodness-of-fit measures which assess how good the model fits the data. However, unlike the scalar values used to assess overall fit (e.g.,

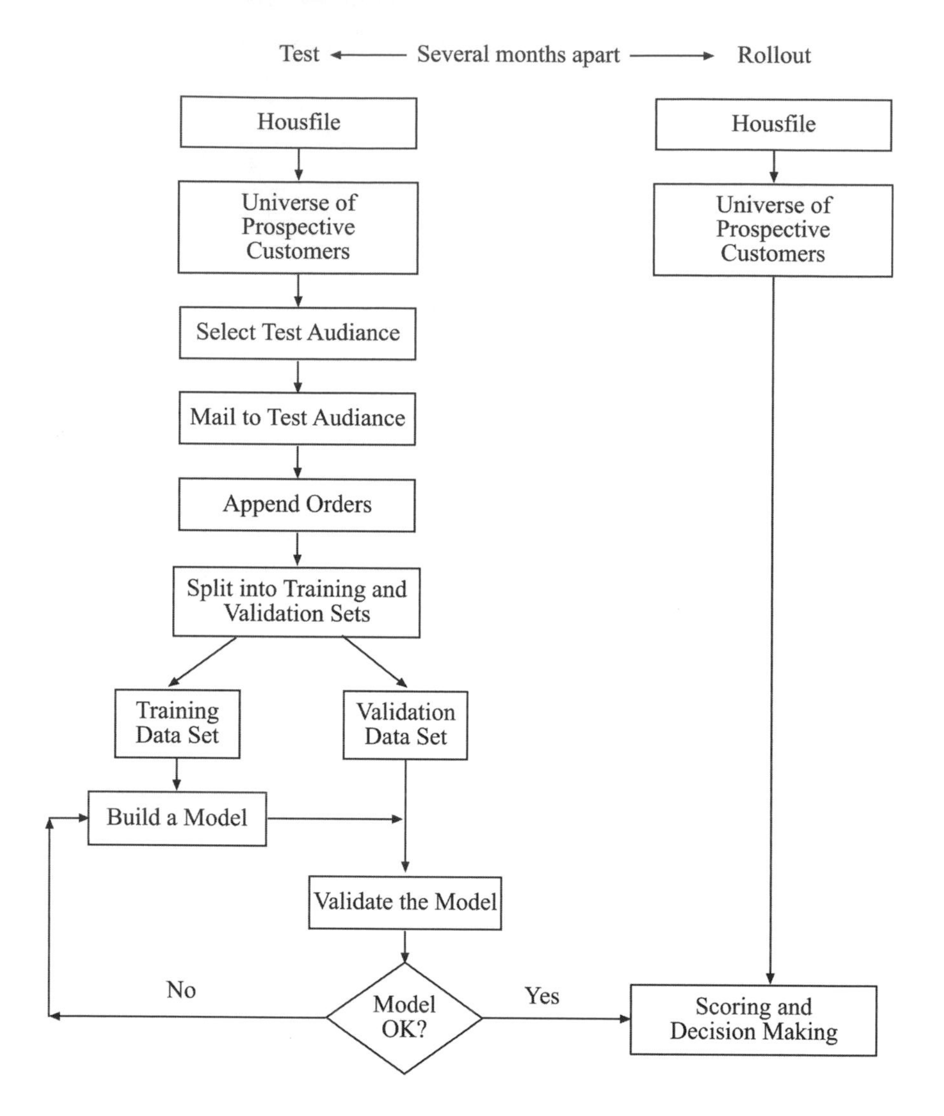

Fig. 63.1. The decision Making Process for New Promotions

the coefficient of determination in linear regression, misclassification rates in classification models, etc), in targeting applications we are interested in assessing the gains achieved by the model, or how well the model is capable of discriminating between buyers and non buyers. Thus the relevant goodness-of-fit measure is based on the distribution of the targeting results, known as gains chart.

Basically, gains-chart displays the added gains (for instance profitability or response) by using a predictive model versus a null model that assumes that all customers are the same. The X-axis represents the cumulative proportion of the population $X_i = 100 \cdot i/n$, (where n is the size of the audience, i - the customer index). The Y-axis represents the cumulative proportion of the actual response (e.g., proportion of buyers), $Y_i = 100 \cdot \frac{\sum_{j=1}^{i} y_j}{\sum_{j=1}^{n} y_j}$ where the observations are ordered in descending order of the predicted values of the dependent variable, i.e., $\hat{y}_i \leq \hat{y}_{i+1}$. A typical gains chart is exhibited in Figure 63.2. We note that gains charts are similar to Lorenz curves in economics.

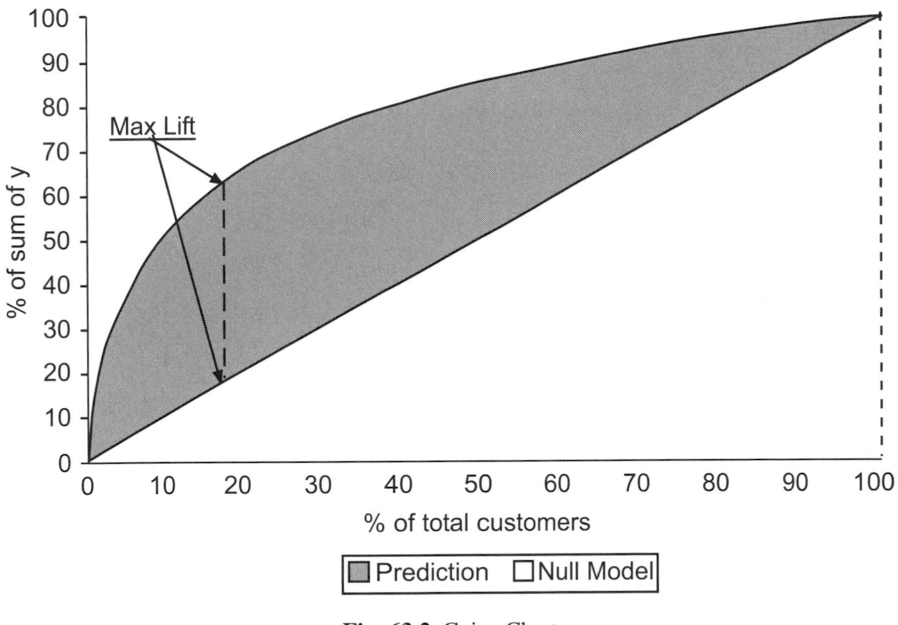

Fig. 63.2. Gains Chart

Two metrics, based on the gains chart, are typically used to assess how the model results differs from the null model:

- Maximum Lift (ML), more commonly known as the Kolmogorov Smirnov (K-S) criterion (Lambert, 1993), which is the maximum distance between the model curve and the null model. The K-S statistics has a distribution known as the D distribution (DeGroot, 1991). In most applications, a large ML indicates that the distribution of the model results is different from the null model. The D distribution can be approximated when the number of observation n is large. For large n, the null hypothesis that the two distributions are the same, is rejected with a significance level of 5% if $ML > D_{95} \approx \frac{1.36}{\sqrt{n}}$ (Gilbert, 1999, Hodges, 1957)
- The Gini Coefficient (Lambert, 1993) which is calculated as the area between the model curve and the null model (the gray area in Figure 63.2) divided by the area below the null model. In most applications, a large Gini coefficient indicates that the distribution of the model results is different from the null model.

Clearly, the closer the model curve to the upper left corner of the chart, the better the model is capable of distinguishing the buyers from the non buyers. Equivalently, the larger is the maximum lift or the larger the Gini coefficient, the better the model.

We note that the gains chart is the metrics which reflect the true prediction quality of the model. The lift and the Gini coefficients are summary measures that are often used to compare between several candidate models. Moreover, the maximum lift and the Gini coefficient may not be consistent with one other. For example, it is possible to find two alternative models, build off the same data set, where in one the Gini coefficient is larger, and in the other the *ML* is larger.

63.3.2 Prediction Accuracy

Prediction accuracy is measured by means of the difference in the predicted response versus the actual results, the closer the values the better. Again, it is convenient to view the prediction results at a percentile level, say deciles. Gains tables or bar charts are often used to exhibit the prediction results.

63.3.3 Profitability/ROI

Definitely, the ultimate merit of any model is given by profitability/ROI measures, such as the total profits/return for the target mailing audience and/or the average profits/return per mailed customer. A common ROI measure is given by the incremental profits of the targeted mailing versus using no model (null model) and mailing everybody in the list.

63.3.4 Gains Table

The tabular form of the gains chart is referred to as the gains table. Gains tables are exhibited at some percentiles level, often deciles, i.e. the predicted y-values (e.g., response probabilities) are arranged in decreasing order and the audience is divided into percentiles. The actual results are then summarized at the percentile level (see Table 63.1).

A "good" model is a model which satisfies the following criteria:

- The actual response rate (the ratio of the number of buyers "captured" by the model to the size of the corresponding audience) monotonically decreases as one traverses from the top to the bottom percentiles.
- A large difference in the response rate between the top and the bottom percentiles.

For example, in Table 63.1, the first decile captures 6 times as many buyers as the bottom decile (30 vs. 5) for the same audience size (708 customers). Except for minor fluctuations at the lower deciles because of few buyers, the number of buyers nicely declines as one traverse along deciles suggesting that the model is capable of distinguishing between the better and worse customers.

The economic cutoff rate for this problem (more on this below) falls in the fifth decile, at which point the profits attains a maximum value (highlighted in bold). The profits are calculated by multiplying the actual number of buyers by the profit per order and subtracting the mailing costs.

To assess the prediction accuracy of the model, one should compare the actual number of buyers to the predicted number of buyers (calculated by summing up the purchase probabilities) at the decile level, the closer the value, the better. For the interesting deciles at the top of the list, the prediction accuracy is indeed high (e.g., 30 actual buyers for the first decile vs. 32 predicted buyers, etc.)

Note that the example above represents the results of a discrete choice model where the model performance is measured by means of the response rates. These measures should be substituted by profit values in case of continuous choice modeling. In other words, one needs to arrange the observations in decreasing order of the predicted profits, divide up the list into deciles, or some other percentiles, and create the gains chart or gains table.

Finally we emphasize that whatever model is used, whether discrete or continuous, the validation process of a model should always be based on the validation data set and not on the training data set. Being an independent data set, the validation file is a representative of the audience at large and as such is the only file to assess the performance of the model when applied in the "real" world.

63.4 Segmentation Methods

Segmentation is a central concept in marketing. The concept was formally introduced by Smith (1956) and since then has become a core method for supporting targeting applications. Segmentation is concerned with splitting the market into groups or segments of "like" people with similar purchase characteristics. The key to successful segmentation is identifying a measure of "similarity" between customers with respect to purchasing pattern. The objective of segmentation is to partition the market into segments which are as homogenous as possible within the segments and as heterogeneous as possible in between segments. Then, one may offer each segment only the products/services which are of most interest to the members of the segment. Hence the decision process is conducted at the segmentation level, either the entire segment is targeted for the promotion or the entire segment is declined.

Segmentation methods which are used to address targeting decisions consist of unsupervised judgmentally-based RFM/FRAT methods, clustering methods and supervised classification methods. Supervised models are models where learning is based on some type of a dependent variable. In unsupervised learning, no dependent variable is given and the learning proccss is based on the attributes themselves.

63.4.1 Judgmentally-based RFM/FRAT methods

Judgmentally based or "manual" segmentation are still commonly used to partition a customer list into "homogenous" segments for targeting applications. Typical segmentation criteria include previous purchase behavior, demographics, geographic and psychographics. Previous purchase behavior is often considered to be the most powerful criterion in predicting likelihood of future response. This criterion is operationalized for the segmentation process by means of Recency, Frequency, Monetary (RFM) variables (Shepard, 1995). Recency corresponds to the number of weeks (or months) since the most recent purchase; frequency to the number of previous purchases or the proportion of mailings to which the customer responded; and monetary to the total amount of money spent on all purchases (or purchases within a product category), or the average amount of money per purchase. The general convention is that the more recently the customer has placed the last order, the more items s/he bought from the company in the past, and the more money s/he spent on the company's products, the higher is his/her likelihood of purchasing the next offering and the better target s/he is. This simple rule allows one to arrange the segments in decreasing likelihood of purchase.

Table 63.1. Gains Table by Deciles

Response Prob. (%)	Cum. Aud.	Actual No. of Responders Decile	Cum.	% Buyer	Actual Profit ($)	Pred. # of Responders Decile	Cum.
2.90	708	30	30	26.55	9168	32	32
2.15	1416	20	50	44.25	14436	18	50
1.55	2124	14	64	56.64	17104	14	64
1.17	2832	10	74	65.49	18272	9	73
0.85	3540	10	84	74.34	**19440**	7	80
0.73	4248	5	89	78.76	18608	6	86
0.57	4956	8	97	85.84	18976	5	91
0.46	5664	7	104	92.04	18056	4	95
0.27	6372	4	108	95.58	17712	3	98
0.02	7077	5	113	100.00	16892	1	99

The more sophisticated manual methods also make use of product/attribute proximity considerations in segmenting a file. By and large, the more similar the products bought in the past are to the current product offering, or the more related are the attributes (e.g., themes), the higher the likelihood of purchase. For example, when promoting a sporting good, it is plausible that a person who bought another sporting good in the past is more likely to respond to a new sporting good offer; next in line are probably people who like camping, followed by people who like outdoor activities, etc. In cases where males and females may react differently to the product offering, gender may also be used to partition customers into groups. By and large, the list is first partitioned by product/attribute type, then by RFM and then by gender (i.e., the segmentation process is hierarchical). This segmentation scheme is also known as FRAT - Frequency, Recency, Amount (of money) and Type (of product).

RFM and FRAT methods are subject to judgmental and subjective considerations. Also, the basic assumption behind the RFM method may not always hold. For example, in durable products, such as cars or refrigerators, recency may work in a reverse way - the longer the time since last purchase, the higher the likelihood of purchase. Finally, to meet segment size constraints, it may be necessary to run the RFM/FRAT iteratively, each time combining small segments and splitting up large segments, until a satisfactory solution is obtained. This may increase computation time significantly.

63.4.2 Clustering

Clustering are methods for grouping unlabeled observations. A data item is mapped into one of several clusters as determined from the data. Modern clustering algorithms group data elements based on the proximity (or similarity) of their attributes. The objective is to partition the observations into "homogeneous" clusters, or groups, such that all observations (e.g., customers) within a cluster are "alike", and those in between clusters are dissimilar. In the context of our targeting applications, the purpose of clustering is to partition the audience into clusters of people with similar purchasing characteristics. The attributes used in the clustering process are the same as those used by the RFM/FRAT method discussed above. In fact, clustering methods take away the judgmental considerations used by the subjective RFM/FRAT methods, thereby providing more "objective" segments. Once the audience is partitioned into clusters, the targeting decision process proceeds as above.

Let $X_i = (x_{i1}, x_{i2}, ..., x_{iJ})$ denotes the attribute vector of customer i with attributes $x_{ij}, j = 1,2,...,J$. To find which customers cluster together, one needs to define a similarity measure between customers. The most common one is the Euclidean distance. Given that ℓ and m are two customers from a list of n customers, the Euclidean distance is defined by:

$$\text{distance}\ (X_\ell, X_m) = \sqrt{\sum_j (x_{\ell j} - x_{mj})^2}$$

Clearly, the shorter the distance, the more similar are the customers. In the case of two identical customers, the Euclidean distance returns the value of zero.

An alternative distance measure, which may be more appropriate for binary or integer attributes, is the cosine distance (Herz *et al.*, 1997), defined by:

$$\text{distance}\ (X_\ell, X_m) = \frac{\sum_{j=1}^{J} x_{\ell j}\, x_{mj}}{\sqrt{\sum_{j=1}^{J} x_{\ell j}^2 \sum_{j=1}^{J} x_{mj}^2}}$$

Note that unlike in the Euclidean distance, here if the purchase profile of customers are identical, the cosine measure returns the value of 1; if orthogonal (i.e., totally dissimilar), it returns the value of 0.

Several clustering algorithms have been devised in the literature, ranging from K-Means algorithms (Fukunaga, 1990), Expectation Maximization (Lauritzen, 1995), Linkage-based methods (Bock, 1974), Kernel Density estimation (Silverman, 1986), even Neural Network based models (Kohonen *et al.*, 1991). Variations of these models that address the scalability issues are appearing recently in the literature, e.g., the BIRCH algorithm (Zhang *et al.*, 1996).

The K-Means algorithm is, undoubtedly, the most popular of all clustering algorithms. It partitions the observations (customers, in our case) into K clusters, where the number of clusters K is defined in advance, based on the proximity of the customer attributes from the center of the cluster (called the centroid).

Let S_k denote the centroid of cluster k. S_k is a vector with J dimensions (or coordinates), one coordinate per attribute j. Each coordinate of the centroid of a cluster is calculated as the mean value of the corresponding coordinates of all customers which belong to the cluster (hence the name K-Means).

To find which customers belong to which cluster, the algorithm proceeds iteratively, as follows:

Step 1 – Initialization: Determine the K centroids, S_k, $k = 1, ..., K$ - (e.g., randomly).

Step 2 – Loop on all customers: For each customer, find the distance of his/her profile to each of the centroids S_k, $k = 1, ..., K$, using the specified similarity measure, and assign the customer to the cluster corresponding to the nearest centroid.

Step 3 – Loop on all clusters: For each cluster $k = 1, ..., K$, recalculate the coordinates of its centroid by averaging out the coordinates of all customers currently belonging to the centriod.

Step 4 – Termination: Stop the process when the termination criteria are met – otherwise, return to Step 2.

63.4.3 Classification Methods

Classification models are segmentation methods which use previous observations with known class labels (i.e., whether the customer responded to the offer, or not) in order to classify the audience into one of several predefined classes. Hence these models belong to the realm of supervised learning. They too take away the judgmental bias that is inherent in the subjective RFM/FRAT methods.

The leading classification models are decision trees. In the binary response case, the purpose is to segment customers into one of two classes – likely buyers and likely non buyers. Some decision trees allow splitting an audience into more than 2 classes. Several automatic tree classifiers were discussed in the literature, among them AID - Automatic Interaction Detection (Sonquist *et al.*, 1971); CHAID - Chi square AID (Kass, 1983), CART - Classification and Regression Trees (Breiman *et al.*, 1984), ID3 (Quinlan, 1986), C4.5 (Quinlan, 1993), and others. A comprehensive survey of automatic construction of decision trees from data can be found in Chapter 8.8 of this volume.

Basically, all automatic tree classifiers share the same structure. Starting from a "root" node (the whole population), tree classifiers employ a systematic approach to grow a tree into "branches" and "leaves". In each stage, the algorithm looks for the "best" way to split a "father" node into several "children" nodes, based on some splitting criteria. Then, using a set of predefined termination rules, some nodes are declared as "undetermined" and become the father nodes in the next stages of the tree development process, some others are declared as "terminal" nodes. The process proceeds in this way until no more nodes are left in the tree which are worth splitting any further. The terminal nodes define the resulting segments. If each node in a tree is split into two children only, one of which is a terminal node, the tree is said to be "hierarchical".

Three main considerations are involved in developing automatic trees:
- Growing the tree
- Determining the best split
- Termination rules

Growing the Tree

One grows the tree by successively partitioning nodes based on the data. With so many variables involved, there is practically infinite number of ways to split a node. Several methods have been applied in practice to reduce the number of possible partitions of a node to a manageable number:

- All continuous variables are categorized prior to the tree development process into small number of ranges ("binning"). A similar procedure applies for integer variables which assume many values (such as the frequency of purchase).

- Nodes are partitioned only on one variable at a time ("univariate" algorithm).

- The number of splits per each "father" node is often restricted to two ("binary" trees).

- Splits are based on a "greedy" algorithm in which splitting decisions are made sequentially looking only on the impact of the split in the current stage, but never beyond (i.e., there is no "looking ahead").

Several algorithms exist that relax some of these restrictions. For example, CHAID is a non binary tree as it allows splitting a node into several descendants. By and large, Genetic algorithms (GAs) are non greedy methods and can also handle multiple variables to split a node.

Determining the Best Split

With so many possible partitions per node, the question is what is the best split? There is no unique answer to this question as one may use a variety of splitting criteria each may result in a different "best" split. We can classify the splitting criteria into two "families": node-value based criteria and partition-value based criteria.

- Node-value based criteria: seeking the split that yields the best improvement in the node value.

- Partition-value based criteria: seeking the split that separates the node into groups which are as different from each other as possible

Termination Rules

Theoretically, one can grow a tree indefinitely, until all terminal nodes contain very few customers, as low as one customer per segment. The resulting tree in this case is unbounded and unintelligible, having the effect of "can't see the forest because of too many trees". It misses the whole point of tree classifiers whose purpose is to divide the population into buckets of "like" people, where each bucket contains a meaningful number of people for statistical significance. Also, the larger the tree, the larger the risk of overfitting. Hence it is necessary to control the size of a tree by means of termination rules that determine when to stop growing the tree. These termination rules should be set to ensure statistical validity of the results and avoid overfitting.

63.4.4 Decision Making

There's quite a distinction between the decision process in unsupervised and supervised models.

In the unsupervised RFM/FRAT and clustering methods, one would normally contact the top segments in the list, i.e., the segments which are most likely to respond to the solicitation. In the RFM/FRAT approach, the position of the segments in the hierarchy of segments represents, more-or-less, its likelihood of purchase, in ordinal terms. Thus people at the segment occupying, say, the 10th position in the hierarchy of segments are usually (but not necessarily) more likely to buy the current product than people belonging to the succeeding segments, but are less likely to purchase the current product than people belonging to the preceding segments. In the clustering approach, there is no such clear-cut definition of the quality of segments, and one needs to assess how good are the resulting clusters by analyzing the leading attributes of the customers in each segment based on domain knowledge.

A more accurate approach for targeting is to conduct a live test mailing involving a sample of customers from the resulting segment to predict the response rate of each segment in the list for the current product offering. Several rules of thumb exist to determine the size of the sample to use for testing from each segment. The convention among practitioners is to randomly pick a proportional sample, typically 10% from each segment, to participate in the test mailing. Then, if the predicted response rate of the segment, based on the test result, is high enough the remainder of the segment is rolled out; otherwise, it is not. The threshold level separating the strong and the weak segments depend on economical considerations. In particular, a segment is worth promoting if the expected profit contribution for a customer exceeds the cost of contacting the customer. The expected profit per customer is obtained as the product of the customer purchase probability, estimated by the response rate of the segment that the customer belongs to, by the profit per sold item.

This decision process is subject to several inaccuracies because of large Type-I and Type-II errors, poor prediction accuracy and regression to the mean, which fall beyond the scope of this chapter. Further discussion of these issues can be found in (Levin and Zahavi, 1996).

The decision process may be simpler with supervised classification models as no test mailing is required here. The objective is to contact only the "profitable" segments whose response rate exceeds a certain cutoff response rate (CRR) based on economical considerations. As discussed in the next section, the CRR is given by the ratio of the contact cost to the profit per order, perhaps bumped up by a certain profit margin set by management.

63.5 Predictive Modeling

Predictive modeling is the work horse of targeting issues in marketing. Whether the model involved is discrete or continuous, the purpose of predictive modeling is to estimate the expected return per customer as a function of a host of explanatory variables (or predictors). Then, if the predicted response measure exceeds a given cutoff point, often calculated based on economical and financial parameters, the customer is targeted for the promotion; otherwise, the customer is rejected.

A typical predictive model has the general form:

$Y = f(x_1, x_2, ..., x_J, U)$

Where:

Y - the response (choice variable)

X - $(x_1, ..., x_J)$ - a vector of predictors "explaining" customers' choice

U - a random disturbance (error)

There are a variety of predictive models and it is beyond the scope of this chapter to discuss them all. So we will only review here the two most important regression models used for targeting decisions – linear regression and logistic regression, as well as the AI-based neural network model. More information about these and other predictive models can be found in the database marketing and econometric literature.

63.5.1 Linear Regression

The linear regression model is the most commonly used continuous choice model. The model has the general form:

$$Y_i = \beta' X_i + U_i$$

Where:

- Y_i - The continuous choice variable for observation i
- X_i - Vector of explanatory variables, or predictors, for observation i
- β - Vector of coefficients
- U_i - Random disturbance, or residual, of observation i, and there exist $E(U_i) = 0$

Denoting the coefficient estimate vector by $\hat{\beta}$, the predicted continuous choice value for each customer, given the attribute vector X_i, is given by:

$$E(Y_i\,|\,X_i) = \hat{\beta}'X_i$$

Since the linear regression model is not bounded from below, the predicted response may turn out negative, in contrast with the fact that actual response values in targeting applications are always non-negative (either the customer responds to the offer and incurs positive cost/revenues, or does not respond and incurs no cost/revenues). This may render the prediction results of a linear regression model somewhat inaccurate.

In addition, the linear regression model violates two of the basic assumptions underlying the linear model:

- Because the actual observed values of Y_i consists of many zeros (non responders) but only a few responders, there is a large probability mass at the origin which ordinary least squares methods are not "equipped" to deal with. Indeed, other methods have been devised to deal with this situation, the most prominent ones are the Tobit (Tobin, 1958), and the two-stage model (Heckman, 1979).

- Many of the predictors in database marketing, if not most of them, are dichotomous (i.e., 0/1 variables). This may affect the test of hypotheses process and the interpretability of the analysis results.

A variation of the linear regression model, in which the choice variable Y_i is defined as a binary variable which takes on the value of 1 if the event occurs (e.g., the customer buys the product), and the value of 0 if the event does not occur (the customer declines the product), is referred to as the linear probability model (LPM). The conditional expectation $E(Y_i/X_i)$ in this case may be interpreted as the probability that the event occurs, given the attribute vector X_i. However, because the linear regression model in unbounded, $E(Y_i/X_i)$ can lie outside the probability range (0,1).

63.5.2 Logistic Regression

Logistic regression models are at the forefront of predictive models for targeting decisions. Most common is the binary model, where the choice variable is a simple yes/no, which is coded as 0/1: 0 – for "no" (e.g., no purchase), 1 - for "yes" (purchase). The formulation of this model stems from the assumption that there is an underlying latent variable Y_i^* defined by the linear relationship:

$$Y_i^* = \beta' X_i + U_i \tag{63.1}$$

Y_i^* is often referred to as the "utility" that the customer derives by making the choice (e.g., purchasing a product). But in practice, Y_i^* is not observable. Instead, one observes the response variable Y_i, which is related to the latent variable Y_i^* by:

$$Y_i = \begin{cases} 1 & if\, Y_i^* > 0 \\ 0 & otherwise \end{cases} \tag{63.2}$$

From (63.1) and (63.2), we obtain:

$$\begin{aligned} Prob(Y_i = 1) &= Prob(Y_i^* = \beta' X_i + U_i > 0) \\ &= Prob(U_i > -\beta' X_i) = 1 - F(-\beta' X_i) \end{aligned} \tag{63.3}$$

Which yields, for symmetrical distribution of U_i around zero:

$$Prob\,(Y_i = 1) = F\left(\beta' X_i\right)$$

$$Prob(Y_i = 0) = F(-\beta' X_i)$$

Where F(·) denotes the CDF of the disturbance U_i.

The parameters $\beta's$ are estimated by the method of maximum likelihood. In case the distribution of U_i is logistic, we obtain the **logit** model with closed-form purchase probabilities (Ben Akiva and Lerman, 1987):

$$Prob(Y_i = 1) = \frac{1}{1+exp(-\hat{\beta}'X)}$$

$$Prob(Y_i = 0) = \frac{1}{1+exp(\hat{\beta}'X)}$$

Where $\hat{\beta}$, the MLE (Maximum likelihood estimate) of β

An alternative assumption is that U_i is normally distributed. The resulting model in this case is referred to as the **probit** model. This model is more complicated to estimate because the cumulative normal variable does not have a closed-form solution. But fortunately, the cumulative normal distribution and the logistic distribution are very close to each other. Consequently, the resulting probability estimates are similar. Thus, for all practical purposes, one can use the more convenient and more efficient logit model instead of the probit model.

Finally we mentioned two more models which belong to the family of discrete choice models - multinomial regression models and ordinal regression models (Long, 1997). In multinomial models, the choice variable may assume more than two values. Examples are a trinomial model with 3 choice values (e.g., 0 – no purchase, 1 – purchase a new car, 2 – purchase a used car), and a quadrinomial model with 4 choice values (e.g., 0 – no purchase, 1 – purchase a compact car, 2 – purchase a mid size car, 3 – purchase a full size luxury car). Higher order multinomial models are very hard to estimate and are therefore much less common.

In ordinal regression models the choice variable assumes several discrete values which possess some type of an order, or preference. The above example involving the compact, mid size and luxury car, can also be conceived as an ordinal regression model with the size of the car being the ranking measure. By and large, ordinal regression models are easier to solve than multinomial regression models.

63.5.3 Neural Networks

Neural Networks (NN) are AI-based predictive modeling method which has gained a lot of popularity recently. NN is a biologically inspired model which tries to mimic the performance of the network of neurons, or nerve cells, in the human brain. Mathematically, a NN is made up of a collection of processing units (neurons, cells), connected by means of branches, each characterized by a weight representing the strength of the connection between the neurons. These weights are determined by means of a learning process by repeatedly showing the NN with examples of past cases for which the actual output is known, thereby inducing the system to adjust the strength of the weight between neurons. On the first try, since the NN is still untrained, the input neuron will send a current of initial strength to the output neurons, as determined by the initial conditions. But as more and more cases are presented, the NN will eventually learn to weigh each signal appropriately. Then, given a set of new observations, these weights can be used to predict the resulting output.

Many types of NN have been devised in the literature. Perhaps the most common one, which forms the basis of most business applications of neural computing, is the supervised-learning, feed-forward networks, also referred to as backpropagation networks. In this model, which resulted from the seminal work of (Rumelhart and McClelland, 1986), and the PDP

Research Group (1986), the NN is represented by a weighted directed graph, with nodes representing neurons and links representing connections. A typical feedforward network contain three types of processing units: input units, output units and hidden units, organized in a hierarchy of layers, as demonstrated in Figure 63.3 for a three-layer network. The flow of information in the network is governed by the topology of the network. A unit receiving input signal from units in a previous layer aggregates those signals based on an input function I, and generates an output signal based on an output function O (sometimes called a transfer function). The output signal is then routed to other units as directed by the topology of the network. The input function I often used in practice is the linear one, and the transfer function O either the tangent hyperbolic or the sigmoid (logit) function.

The weight vector W is determined through a learning process to minimize the sum of squared deviations between the actual and the calculated output, where the sum is taken over all output nodes in the network. The backpropagation algorithm consists of two phases: feedforward propagation and backward propagation. In forward propagation, outputs are generated for each node on the basis of the current weight vector W and propagated to the output nodes to generate the total sum of squared deviations. In backward propagation, errors are propagated back, layer by layer, adjusting the weights of the connections between the nodes to minimize the total error. The forward and backward propagation are executed iteratively once for each number of iterations (called epoch) until convergence occurs.

The type and topology of the backpropagation network depends on the structure and dimension of the application problem involved, and could vary from one problem to the other.

In addition, there are other considerations in applying NN for target marketing, which are not usually encountered in other marketing applications of NN (see Levin and Zahavi, 1997b, for descriptions of these factors). Recent research also indicates that NN may not have any advantage over logistic models for supporting binary targeting applications (Levin and Zahavi, 1997a). All this suggest that one should apply NN to targeting applications with cautious.

63.5.4 Decision Making

From the marketer's point of view, it is worth mailing to a customer as long as the expected return from an order exceeds the cost invested in generating the order, i.e., the cost of promotion. The return per order depends on the economical/financial parameters of the current offering. The promotion cost usually includes the brochure and the postal costs. Denoting by:

g - the expected return from the customer (e.g., expected order size in a catalog promotion).

c - the promotion cost

M - the minimum required rate of return

Then, the rate of return per customer (mailing) is given by:

$$\frac{g-c}{c} = \frac{g}{c} - 1$$

And the customer is worth promoting to if his/her rate of return exceeds the minimal required rate of return, M, i.e.:

$$\frac{g}{c} - 1 \geq M \rightarrow g \geq c \bullet (M+1) \tag{63.4}$$

The quantity on the right-hand side of (63.4) is the cutoff point separating out between the promotable and the nonpromotable customers.

Alternatively, equation (63.4) can be expressed as:

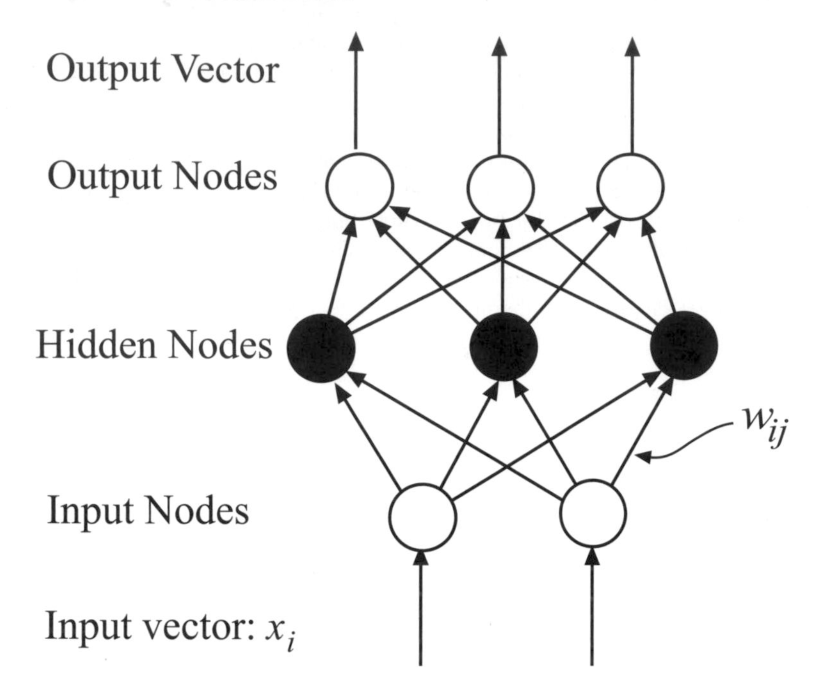

Fig. 63.3. A multi-layer Neural Network

$$g - c \bullet (M+1) \geq 0 \tag{63.5}$$

Where the quantity on the left-hand side denotes the net profit per order.

Then, if the net profit per order is non-negative, the customer is promoted; otherwise, s/he is not.

In practical applications, the quantity c is determined by the promotion cost; M is a threshold margin level set up by management. Hence the only unknown quantity is the value of g - the expected return from the customer, which is estimated by the predictive model. Two possibilities exist:

In a continuous response model, g is estimated directly by the model.

In a binary response model, the value of g is given by:

$$g = p \bullet R \tag{63.6}$$

Where:

p - The purchase probability estimated by the model, i.e. $p = Prob(Y = 1)$. Y is the purchase indicator - 1 for purchase, 0 for no purchase. R is the return/profit per responder.

In this case, it is customary to express the selection criterion by means of purchase probabilities. Plugging (63.6) in (63.4) we obtain:

$$p \geq \frac{c(M+1)}{R} \tag{63.7}$$

The right hand side of (63.7) is the cutoff response rate (CRR). If the customer's response probability exceeds CRR, s/he is promoted; otherwise, s/he is not.

Thus, the core of the decision process in targeting applications is to estimate the expected return per customer, g. Then, depending upon the model type, one may use either (63.4) or (63.7) to select customers for the campaign.

Finally we note that CRR calculation applies only to the case where the scores coming out from the model represent well-defined purchase probabilities. This is true of logistic regression, but less true for NN where the score is ordinal. But ordinal scores still allow the user to rank customers in decreasing order of their likelihood of purchase, placing the best customers at the top of the list and the worst customers at the bottom of the list. Then, in the absence of a well defined CRR, one can select customers for promotion based on "executive decision", say promote the top four deciles in the list.

63.6 In-Market Timing

For durable products such as cars or appliances, or events such as vacations, cruise trips, flights, bank loans, etc, the targeting problem boils down to the timing when the customer will be in the market "looking around" for these products/events. We refer to this problem as the in-market timing problem. The in-market timing depends on the customer's characteristics as well as the time that elapsed since last acquisition, e.g., the time since the last car purchase. Clearly, a customer that just purchased a new car is less likely to be in the market in the next, say, three months than a customer who bought his current car three years ago. Not only this, but the time until next car purchase is a random variable. We offer two approaches for addressing the in-market timing problem:

- Logistic regression – estimating the probability that the next event (car purchase, next flight, next vacation,...) takes place in the following time period, say next quarter.
- Survival analysis – estimating the probability distribution that the event will take place within the next time period t (called survival time), given that the last event took t_L units of time ago.

63.6.1 Logistic Regression

We demonstrate this process for estimating the probability that a customer will replace his/her old car in the next quarter. For this sake, we summarize the purchase information by, say, quarters, as demonstrated in Figure 63.4 below, and split the time axis into two mutually exclusive time periods – the "targeting period", to define the choice variable (e.g., 1 – if the customer bought a new car in the present quarter, 0 – if not), and the "history period" to define the independent variables (the predictors). In the example below, we define the present quarter as the target period and the previous four quarters as the history period. Then, in the modeling stage we build a logistic regression model expressing the choice probability as a function of the customer's behavior in the past quarters (the history period) and his/her demographics.

In the scoring stage, we apply the resulting model to score customers and estimate their probability of purchasing a car in the next quarter. Note the shift in the history period in the scoring process. This is because the model explains the purchase probability in terms of the customers' behavior in the previous four quarters. Consequently, and in order to be compatible with the model, one needs to shift the data for scoring by discarding the earliest quarter (the fourth quarter, in this example) and adding the present one.

We also note that the "target" period used to define the choice variable and the "history" period used to define the predictors, are not necessarily consecutive. This applies primarily

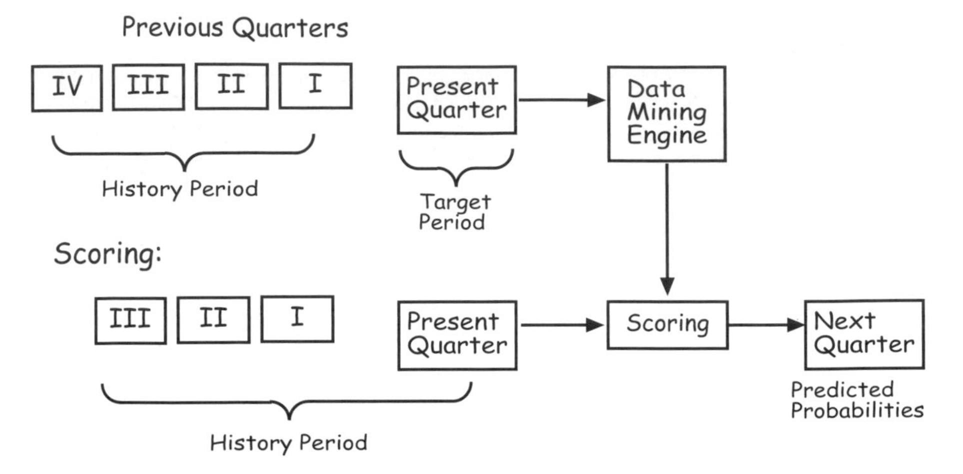

Fig. 63.4. In-Market Timing Using Logistic Regression

to purchase history and less to demographics. For example, in the automotive industry, since customers who bought a car recently are less likely to look around for a new car in the next quarter, one may discard customers who purchased a new car in the last, say, two years from the universe. So if the target period in the above example corresponds to the first quarter of 2004, the history period would correspond to the year 2001. There could also be some shift in the data because of the time lag that takes place between the actual transaction and the time the data becomes available for the analysis. Finally, we note that we used quarters in the above example just for demonstration purposes. In practice, one may use a different time period to summarize the data by, or a longer time period to express the history period. It all depends on the application. Certainly, in the automotive industry, because the purchase cycle to replace a car is rather long, the history period could extends over several years; moreover, this period should even vary from one country to the other, because the "typical" purchase cycle time for each country is not the same. In other industries, these time periods could be much shorter. So domain knowledge should play an important role in setting up the problem. Data availability may also dictate what time units to use to summarize the data and how long the history period and the targeting period should be.

63.6.2 Survival Analysis

Survival Analysis (SA) is concerned with estimating the duration time distribution until an event occurs (called the survival time). Given the probability distribution, one can estimate various measures of the survival time, primarily the expected time or the median time until an event occurs. The roots of survival analysis are in health and life sciences (Cox and Oakes, 1984). Targeting applications include purchasing a new vehicle, applying for a loan, taking a cruise trip, a flight, a vacation...

The survival analysis process is demonstrated in Figure 63.5 below. The period from the starting time to the ending time ("today") is the experimental or the analysis period. As alluded to earlier, each application may have its own "typical" analysis period (e.g., several years for the automotive industry). Now, because the time until an event occurs is a random variable, the observations may be left-censored or right-censored. In the former, the observation commences prior to the beginning of the analysis period (e.g., the analysis period for car purchases is three years and the customer purchased her current car more than three years ago); in the latter, the event occurs after the analysis period (e.g., the customer did not purchase a new car within the three-year analysis period). Of course, both types of censoring may occur. For example, a customer that has bought her car prior to the analysis period (left censoring) and replaced it after the end of the analysis period (right censoring).

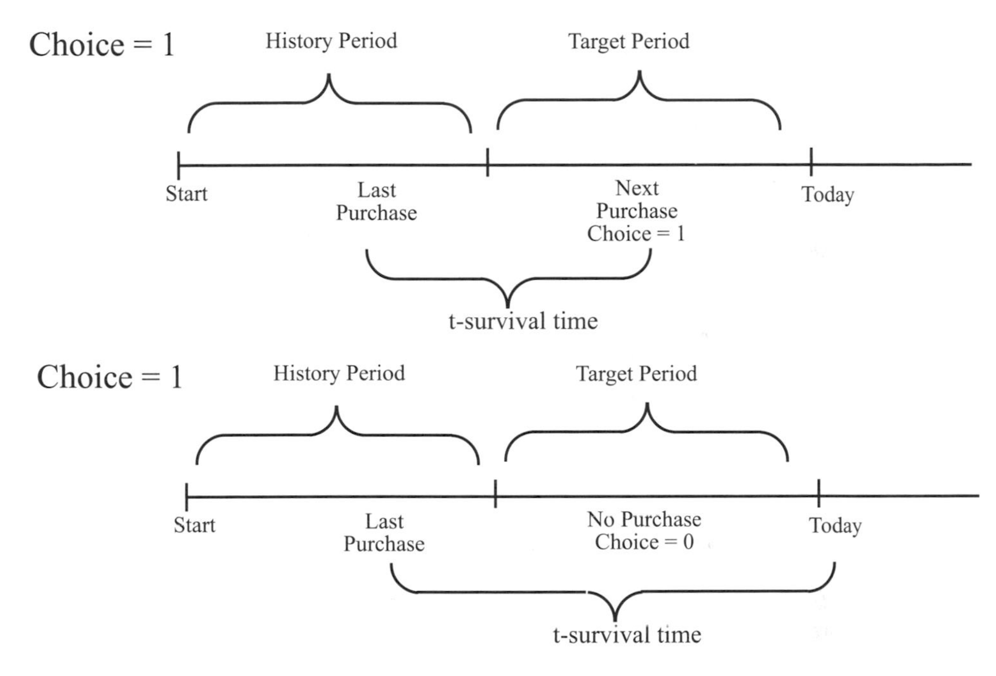

Fig. 63.5. In-Market Timing Using Survival Analysis

As in the logistic regression case, we divide the time axis into two mutually exclusive time periods – the target period, to define the choice variable, and the history period, to define the predictors. But in addition, we also define the survival time, i.e., the time between the last event in the history period and the time until the first event in the target period, as shown in Figure 63.5 (if no event took place in the history period, the survival time commences at the start of the analysis period). Clearly the survival time is a random variable expressed by means of a survival function S(t), which describes the probability that the time until the next event occurs exceed a given time t. The most commonly used distributions to express the survival process are the exponential, the Weibull, the log-logistic and the log-normal distributions. The type of the distribution to use in each occasion depends on the corresponding hazard function, which is defined as the instantaneous probability that the event occurs in an infinitesimally short period of time, given that the event has not occurred earlier. The hazard function is constant for the exponential distribution; It increases or decreases with time, for the other survival functions, depending upon the parameters of the distribution. For example, in the insurance industry, the exponential distribution is often used to represent the survival time, because the hazard function for filing a claim is likely to be constant as the probability of being involved in an accident is independent of the time that elapses since the preceding accident. In the car industry, for the same make, the hazard function is likely to assume an inverted U-shape function. This is because, right after the customer purchases a new car, the instantaneous probability that s/he buys a new car is almost zero, but it increases with time as the car gets older. Then, if after a while the customer still did not buy a new car, the instantaneous probability goes down, most likely because s/he bought a car from a different manufacturer. Note that in the case any car is involved (not a specific brand), the hazard function is likely to rise with time as the longer one keeps her car, the larger the probability she will replace the car in the next time period. In both cases, the log-logistic distribution could be a reasonable candidate to represent the survival process, with the parameter of the log-logistic distribution determining the shape of the hazard function.

Now, in marketing applications, the survival functions are expressed in terms of a linear function of the customer's attributes (the "utility") and the scaling factor (often denoted by σ). These parameters are estimated based on observations using the method of maximum likelihood.

Given the model, one can estimate the in-market timing probabilities for any new observation for any period Q from "today", using the formula:

$$P(t < t_L + Q | t > t_L) = 1 - \frac{S(t_L + Q)}{S(t_L)}$$

Where:

S(t) – The survival function estimated by the model

t – The time index

t_L – The time since last purchase

We note that the main difference between the logit and the survival analysis models is that prediction based on logit could only be made for a fixed period length (i.e., the period Q above) while in survival analysis Q could be of any length. Also, survival analysis is better "equipped" to handle censored data which is prevalent in time-related applications. This allows the marketer to target customers more accurately by going after them only at the time when their in-market timing probabilities are the highest.

Given the in-market probabilities, either using logistic regression or survival analysis, one may use a judgmentally-based cutoff rate, or a one based on economical considerations, to pick the customers to go after.

63.7 Pitfalls of Targeting

As alluded to earlier, the application of Data Mining to address targeting applications is not all that straightforward and definitely not automatic. Whether by overlooking, ignorance, carelessness, or whatever, it is very easy to abuse the results of Data Mining tools, especially predictive modeling and make wrong decisions. An example which is widely publicized is the 1998 KDD (Knowledge Discovery in Databases) CUP. The KDD-CUP is a Data Mining competition that provides a forum for comparing and evaluating the performance of Data Mining tools on a predefined business problem using real data. The competition in 1998 involved a charity application and the objective was to predict the donation amount for each customer in a validation sample, based on a model built using an independent training sample. Competitors were evaluated based on the net donation amount obtained by summing up the actual donation amount of all people in the validation set whose expected donation amount exceeded the contact cost ($0.68 per piece). All in all, 21 groups submitted their entry. The results show quite a variation. The first two winners were able to identify a subset of the validation audience to solicit that would increase the net donation by almost 40 percent as compared to mailing to everybody. However, the net donation amount of all other participants lagged far behind the first two. In all, 12 entrants did better than mailing to the whole list, 9 did worse than mailing to the entire list and the last group even lost money on the campaign! The variation in the competition results is indeed astonishing! It tells us that Data Mining is more than just applying modeling software. It is basically a blend of art and science. The scientific part involves applying an appropriate model for the occasion, whether regression model, clustering model, classification model, or whatever. The art part has to do with evaluating of the data that goes into the model and the knowledge that comes out from the modeling process. Our guess is that the dramatic variations in the results of the 1998 KDD-CUP competition is due to the fact that many groups were "trapped" into the mines of Data Mining. So in this section we discuss some of the pitfalls to beware of in building Data Mining models for targeting applications. Some of these are not necessarily pitfalls but issues that one needs to account for in order to render strong models. We divide these pitfalls into 3 main categories – modeling, data and implementation.

63.7.1 Modeling Pitfalls

Misspecified Models

Modern databases often contain tons of information about each customer, which may be translated into hundreds, if not more, of potential predictors. Usually only a handful of which suffices to explain response. The process of selecting the most influential predictors in predictive modeling affecting response from the much larger set of potential predictors is referred to in Data Mining as the feature selection problem. Statisticians refer to this problem as the specification problem. It is a hard combinatorial optimization problem which usually requires heuristic methods to solve, the most common of which is the stepwise regression method (SWR). It is beyond the scope of this chapter to review the feature selection problem in full. So we only demonstrate below the problems that may be introduced to the feature selection problem because of sampling error. For a more comprehensive review of feature selection methods see (Miller, 2002), (George, 2000), and others.

The sheer magnitude of today's databases makes it impossible to build models based on the entire audience. A compromise is to use sampling. The benefits of sampling is that it reduces processing time significantly, but on the other hand it reduces model accuracy by introducing to the model insignificant predictors while eliminating significant ones, both result in misspecified model. We demonstrate this with respect to the linear regression model.

Recalling, in linear regression the objective is to "explain" a continuous dependent variable, Y, in terms of a host of explanatory variables $X_j,\ j = 0, 1, 2, .., J$

$$Y = \sum_{j=0}^{J} \beta_j X_j + U$$

Where:

$\beta_j,\ j = 0, 1, 2, ..., J$ – The coefficients estimated based on real observations

U – A random disturbance

Assuming all other factors are equal, one can check whether a variable, say X_k, is significant by testing the hypothesis.

$$H_0: \quad \beta_k = 0$$
$$H_1: \quad \beta_k \neq 0$$

The test statistics for testing the hypothesis is given by:

$$t = \left| \hat{\beta}_k \big/ s(\hat{\beta}_k) \right|$$

Where:

$\hat{\beta}_k$ – the coefficient estimate of β_k

$s(\hat{\beta}_k)$ – the standard error of the coefficient estimate

In small samples, the test statistics t is distributed as the t (student) distribution with $n-J-1$ degrees of freedom. In Data Mining applications, where the sample size is very large, often containing as many as several hundred observations, or more, the t-distribution may be approximated by the normal distribution.

Given the test statistics and its sampling distribution, one calculates the minimum probability level to reject H_0 where it is true, $P-value$:

$$P-Value = 2P(T > |\hat{\beta}_k / s(\hat{\beta}_k))$$

And if the resulting $P-value$ is smaller than, or equal, to a predefined level of significance, often denoted by α, one rejects H_0; otherwise, one does not reject H_0.

The level of significance α is the upper bound on the probability of Type-I error (rejecting H_0 when true). It is the proportion of times that we reject H_0 when true, out of all possible samples of size n drawn from the population. In fact, the P-value is just one realization of this phenomenon. It is the actual Type-I error probability for the given sample statistics.

Now, suppose that X_k is an insignificant variable having no relation whatsoever to the dependent variable Y (i.e., the correlation coefficient between X_k and Y is zero). Then, if we build the regression model based on a sample of observations, there is a probability of α that X_k will turn out significant just by pure chance, thus making it into the model and resulting in Type-I error, in contradiction to the fact that X_k *and* Y are not correlated.

Extending the analysis to the case of multiple insignificant predictors, even a small Type-I error may result in several of those variables making it into the model as significant. Taking this to the extreme case where all predictors involved are insignificant, we are almost sure to find a significant model, indicating a true relationship between the dependent variable (e.g., response) and some of the regressors, where such a relationship does not exist! This phenomenon also extends for the more realistic case which involves both significant and insignificant predictors.

The converse is also true, i.e., that there is a fairly large probability for significant predictors in a population to come out insignificant in a sample, and thus wrongly excluded from the model (Type-II error).

In either case, the resulting predictive model is misspecified. In the context of targeting decisions in database marketing, a misspecified model may result in either some profitable people being excluded from the mailing (Type-I error) and some unprofitable people being included in the mailing (Type-II error), both incur some costs: Type-I error – forgone profits due to missing out good people from the mailing as well as lost of reputation; Type-II error – real losses for contacting unprofitable people.

Clearly, one cannot avoid Type-I and Type-II errors altogether, unless the model is built off the entire database, which is not feasible. But one can reduce the error probabilities by several means - controlling the sample size, controlling the Type-I and Type-II errors using Bonferonni coefficients, False Discovery Rates (FDR) (Benjamini and Hochberg, 1995), Akaike Information Criterion (AIC) (Akaike, 1973), Bayesian Information Criterion (BIC) (Schwarz, 1978), and others.

Detecting misspecified models is an essential component of the knowledge discovery process, because applying a wrong model to target audiences for promotion may incur substantial losses. This is why it is important that one validates the model on an independent data set, so that if a model is wrongly specified, this will show up in the validation results.

Over-Fitting

Over-fitting pertains to case where the model gives good results when applied on the data used to build the model, but yields poor results when applied against a set of new observations. An extreme case of overfitting is when the model is doing too good a job in discriminating between the buyers and the non buyers (e.g., "capturing" all the buyers in the first top percentiles of the audience ("too good to be true")). In either case, the model is not valid and definitely can not be used to support targeting decisions. Over-fitting is a problem that plagues large-scale predictive model, often as a result of a misspecified model, introducing insignificant predictors to a regression model (Type-I error) or eliminating significant predictors from a model (Type-II error).

To test for over-fitting, it is necessary to validate the model using a different set of observations than those used to build the model. The simplest way is to set aside a portion of the observations for building the model (the training set) and hold out the balance to validate the model (the holdout, or the validation data). After building the model based on the training set, the model is used to predict the value of the dependent variable (e.g., purchase probabilities) in predictive model or the class label in classification models, of the validation audience. Then, if the scores obtained for the training and validation data sets are more-or-less compatible, the model appears to be OK (no over-fitting). The best way to check for the compatibility is to summarize the scores in gains table at some percentile level and then compare the actual results between the two tables at each audience level. The more sophisticated validation involves n-fold cross-validation.

Over-fitting results when there is too little information to build the model upon. For example, there are too many predictors to estimate and only relatively few responders in the test data. The cure for this problem is to reduce the number of predictors in the model (parsimonious model). Recent research focuses on combining estimators from several models to decrease variability in predictions and yield more stable results. The leading approaches are bagging (Breiman, 1996) and boosting (Friedman *et al.*, 1998).

Under-Fitting

Under-fitting is the counterpart of over-fitting. Under fitting refers to a wrong model that is not fulfilling its mission. For example, in direct marketing applications, under-fitting results when the model is not capable of distinguishing well between the likely respondents and the likely non-respondents. A fluctuation of the response rate across a gains table may be an indication of a poor fit, or too small of a difference between the top and the bottom deciles. Reasons for under-fitting could vary: wrong model, wrong transformations, missing out the influential predictors in the feature selection process, and others. There is no clear prescription to resolve the under-fitting issue. Some possibilities are: trying different models, partitioning the audience into several key segments and building a separate model for each, enriching data, adding interaction terms, appending additional data from outside sources (e.g., demographic data, lifestyle indicators), using larger samples to build the model, introducing new transformations, and others. The process may require some creativity and ingenuity.

Non-Linearity/ Non-Monotonic Relationships

Regression-based models are linear-in-parameters models. In linear regression, the response is linearly related to the attributes; in logistic regression model, the utility is linearly related to the attributes. But more often than not, the relationship between the output variable and the attribute is not linear. In this case one needs to specify the non-linear relationships using a transformation of the attribute. A common transformation is a polynomial transformation of the form $y = x^a$ where $-2 < a < 2$. Depending upon the value of a, this function provide a variety of ways to express non-linear relationships between the input variable x and the output variable y. For example, if a<1, the transformation has the effect of moderating the impact of x on the choice variable. Conversely, if a>1, the transformation has the effect of magnifying the impact of x on the choice variable. For the special case of a=0, the transformation is defined as y=log(x). The disadvantage of the power transformation above is that it requires that the type of the non-linear relationship be defined in advance. A more preferable approach is to define the non-linear relationships based on the data. Candidate transformations of this type are the step function or the piecewise linear transformation. In step function, the attribute range is partitioned into several mutually exclusive and exhaustive intervals (say, by quartiles). Each interval is then represented by means of a categorical variable, assuming the value of 1 if the attribute value falls in the interval, 0 – otherwise. A piecewise transformation splits a variable into several non-overlapping and continuously-linked linear segments, each with a given slope. Then, the coefficient estimates of the categorical variables in the step function, and the estimate of the slopes of the linear segments in the piecewise function, actually determine the type of relationships that exist between the input and the output variables.

Variable Transformations

More often than not, the intrinsic prediction power resides not in the original variables themselves but on transformations of these variables. There are basically infinite number of ways to define transformations, and the "sky is the limit". We mention here only proportions and ratios, which are very powerful transformations in regression-based models. For example, the response rate, defined as the ratio of the number of responses to the number of promotions, is considered to be a more powerful predictor of response than either the number of responses or the number of promotions. Proportions are also used to scale variables. For example, instead of using the dollar amount in a given time segment as a predictor, one may use the proportion of the amount of money spent in the time segment relative to the total amount of money spent. Proportions possess the advantage of having a common reference point which makes them comparable. For example, in marketing applications it is more meaningful to compare the response rates of two people rather than their number of purchases, because the number of purchases does not make sense unless related to the number of opportunities (contacts) the customer has had to respond to the solicitation.

Space is too short to review the range of possible transformations to build a model. Suffice it to say that one needs to pay a serious consideration to defining transformations to obtain a good model. Using domain knowledge could be very helpful in defining the "right" transformation.

Choice-Based Sampling

Targeting applications are characterized by very low response rates, often less than 1%. As a result, one may have to draw a larger proportion of buyers than their proportion in the population, in order to build a significant model. It is not uncommon in targeting applications to draw a stratified sample for building a model which includes all of the buyers in the test audience and a sample of the non buyers. These types of samples are referred to as choice-based sample (Ben Akiva and Lerman, 1987). But choice-based samples yield results which are compatible with the sample, not the population. For example, a logistic regression model based on a choice-based sample that contains higher proportion of buyers than in the population will yield inflated probabilities of purchase. Consequently, one needs to update the purchase probabilities in the final stage of the analysis to reflect the true proportion of buyers and non buyers in the population in order to make the right selection decision. For discrete choice models, this can be done rather easily by simply updating the intercept of the regression equation (Ben Akiva and Lerman, 1987). In other models, this may be more complicated.

Observations Weights

Sampling may apply not just to the dependent variable but also to the independent variable. For example, one may select for the test audience only 50% of the female and 25% of the males. However, unlike the choice-based sampling which does not affect the model, proportion-based sampling affect the modeling results (e.g., the regression coefficients). To correct for this bias, one needs to inflate the number of males by a factor of 2 and the number of females by a factor of 4 to reflect their "true" numbers in the population. We refer to these factors as observations weights.

Of course, a combination of choice-based sampling and proportional sampling may also exist. For example, suppose we first create a universe which contains 50% of the females and 25% of the males and then pick all of the buyers and 10% of the non buyers for building the model. In this case, each female buyer represents 2 customers in the population whereas each female non-buyer represents 20 customers in the population. Likewise, each male buyer represents 4 customers in the population whereas each male non-buyer represents 40 customers in the population. Clearly, one needs to account for these proportions to yield unbiased targeting models.

63.7.2 Data Pitfalls

Data Bias

By data bias we mean that not all observations in the database have the same items of data, with certain segments of the population are having the full data whereas other segments containing only partial data. For example, new entrants usually contain only demographics information but no purchase history, automotive customers may have purchase history information only for the so-called unrestricted states and only demographic variables for the restricted sates, survey data may be available only for buyers and not for non buyers, some outfits may introduce certain type of data, say prices, only for buyers and not for non-buyers, etc. If not taken care of, this can distort the model results. For example, using data available only for buyers but not for non buyers, say the price, may yield "perfect" model in the sense that price is the perfect predictor of response, which is of course not true. Building one model for "old" customers and new entrants may underestimate the effect of certain predictors on response, while over estimating the effect of others. So one need to exercise caution in these cases, perhaps build a different model for each type of data, use adjustment factors to correct for the biased data, etc.

Missing values

Missing data is very typical of large realistic data sets. But unlike in the previous case, where the missing information was confined to certain segments of the population, in this case missing value could be everywhere, with some attributes having only a few observations with missing values with others having a large proportion of observations with missing values. Unless accounted for, missing values could definitely affect the model results. There's a trade off here. Dropping attributes with missing data from the modeling process results in loss of information; but including attributes with missing data in the modeling process may distort the model results. The compromise is to discard attributes for which the proportion of observations with missing value for that attribute exceeds a pre defined threshold level.

As to the others, one can "capture" the effect of missing value by defining an additional predictor for each attribute which will be "flagged" for each observation with a missing value, or impute a value for missing data. The value to impute depends on the type of the attribute involved. For interval and ratio variables, candidate values to impute are the mean value, the median value, the maximum value or the minimum value of the attribute across all observations; for ordinal variables – the median of the attribute is the likely candidate; and for nominal variables - the mode of the attribute. More sophisticated approaches to dealing with missing value exist, e.g., for numerical variables, imputing a value obtained by means of a regression model.

Outliers

Outliers are the other extreme of missing value. We define an outlier as an attribute value which is several standards deviations away from the mean value of the observations. As in the case of a missing value, there's also a tradeoff here. Dropping observations with outlier attributes may result in a loss of information, while including them in the modeling process may distort the modeling results. A reasonable compromise is to trim outlier value from above by setting the value of an outlier attribute at the mean value of the attribute plus a pre-defined number of standard deviations (say 5), and trim an outlier value from below by setting the value of an outlier at the mean value minus a certain number of standard deviations.

Noisy Data

We define by noisy data binary attributes which appear with very low frequency, e.g., the proportion of observations in the database having a value of 1 for the attribute is less than a small threshold level of the audience, say 0.5%. The mirror image are attributes for which the proportion of observations having a value of 1 for the attribute exceeds a large threshold level, say 99.5%. These types of attributes are not strong enough to be used as predictors of response and should either be eliminated from the model, or combined with related binary predictors (e.g., all the Caribbean islands may be combined into one predictor for model building, thereby mitigating the effect of noisy data).

Confounded Dependent Variables

By a confounded dependent variable we mean a dependent variable which is "contaminated" by one or more of the independent variables. This is quite a common mistake in building predictive models. For example, in a binary choice application the value of the current purchase in a test mailing is included in the predictor Money_Spent. Then, when one uses the test mailing to build a response model, the variable Money_Spent fully explains customer's choice, yielding a model which is "too good to be true". This is definitely wrong. The way to avoid this type of errors is to keep the dependent variable clean of any effect of the independent variables.

Incomplete Data

Data is never complete. Yet, one needs to make best use of the data, introducing adjustment and modification factors, as necessary, to compensate for the lack of data. Take for example the in-market timing problem in the automotive industry. Suppose we are interested in estimating the mean time or the median time until the next car replacement for any vehicle. But often, the data available for such an analysis contain, in the best case, only the purchase history for a given OEM (Original Equipment Manufacturer) which allows one to predict only the replacement time of an OEM vehicle. This time is likely to be much longer than the replacement time of any vehicle. One may therefore have to adjust the estimates to attain time estimates which are more compatible with the industry standards.

63.7.3 Implementation Pitfalls

Selection Bias

By selection bias we mean samples which are not randomly selected. In predictive modeling this type of sample is likely to render biased coefficient estimates. This situation may arise in several cases. We consider here the case of subsequent promotions with the "funnel effect", also referred to as rerolls. In this targeting application, the audience for each subsequent promotion is selected based on the results of the previous promotion in a kind of a "chain" mode. In the first time around, the chain is usually initiated by conducting a live market test to build a response model (as in Figure 63.1), involving a random sample of customers from the universe. The predictive model based on the test results is then used to select the audience for the first rollout campaign (the first-pass mailing). The reroll campaign (the second-pass mailing) is then selected using a response model which is calibrated based on the rollout campaign. But we note that the rollout audience was selected based on a response model and it is therefore not a random sample of the universe. This gives rise to a selection bias. Similarly, the second reroll (the third-pass campaign) is selected based on a response model built based upon the reroll audience, the third reroll is based on the second reroll, and so on. .

Now, consider the plausible purchase situation where once a customer purchases a product, h/se is not likely to purchase it again in the near future. Certainly, it makes no sense to approach these customers in the next campaign and they are usually removed from the universe for the next solicitation. In this case, the rollout audience, the first campaign in the sequence of campaigns, consists only of people who were never exposed to the product before. But moving on to the next campaign, the reroll, the audience here consists of both exposed and unexposed people.

The exposed people are people who were approached in the roll campaign, declined the product, but are promoted again in the reroll because they still meet the promotability criteria (e.g., they belong to the "right" segment)

The unexposed people are people contacted in the reroll for the first time. They consist of two types of people:

- New entrants to the database who have joined the list in the time period between the first rollout campaign and the reroll campaign.
- "Older" people who were not eligible for the rollout campaign, but have "graduated" since then and now meet the promotability criteria for the reroll campaign (e.g., people who have bought a product from the company in the time gap between the rollout and the reroll campaigns, and have thus been elevated into a status of "recent buyers" which qualifies them to take part in the reroll promotion).

Hence the reroll audience is not compatible with the rollout audience, i.e., it contains "different" type of people. The question then is how one can adjust the purchase probabilities of the exposed people in the reroll given that the model is calibrated based on the rollout audience which contains unexposed people only?

Now, going one step further, the second reroll audience is selected based on the results of the first reroll audience. But the first reroll audience consists only of unexposed and first-time exposed people, whereas the second reroll audience also contains twice-exposed people. The question, again, is how to adjust the probabilities of second-time exposures given the probabilities of the first-time exposures and the probabilities of the unexposed people? The problem extends in this way to all subsequent rerolls.

Empirical evidence show that the response rate of repeated campaigns for the same product drops down with each additional promotion. This decline in response is often referred to as the "list dropoff" phenomenon (Buchannan and Morrison 1988). The list falloff rate is not consistent across subsequent solicitations. It is usually the largest, as high as 50% or more, when going from the first rollout to the reroll campaigns and then more-or-less stabilizes at a more moderate level, often 20%, with each additional solicitation. Clearly, with the response rate of the list going down from one solicitation to the other, there comes a point where it is not worth promoting the list, or certain segments of the list, any more, because the response rate becomes too small to yield any meaningful expected net profits. Thus, it is very important to accurately model the list falloff phenomenon to ensure that the right people are promoted in any campaign, whether the first one or a subsequent one.

Regression to the Mean (RTM) Effect

Another type of selection bias, which applies primarily to segmentation-based models, is the regression to the mean (RTM) phenomenon. Recall that in the segmentation approach, either the entire segment is rolled out or the entire segment is excluded from the campaign. The RTM effect arises because only the segments that performed well in the test campaign, i.e., the "winners" are recommended for the roll. Now because of the random nature of the process, it is likely that several of the "good" segments that performed well in the test happened to do so just because of pure chance; as a result, when the remainder of the segment is promoted, its response rate drops back to the "true" response rate of the segment, which is lower than the response rate observed in the test mailing. Conversely, it is possible that some of the segments that performed poorly in the test campaign happened to do so also because of pure chance; as a result, if the remainder of the segment is rolled out, it is likely to perform above the test average response rate. These effects are commonly referred to as RTM (Shepard, 1995). When both the "good" and "bad" segments are rolled out, the over and under effects of RTM cancels out and the overall response rate in the rollout audience should be more-or-less equal to the response rate of the testing audience. But since only the "good" segments, or the "winners", are promoted, one usually witnesses a dropoff in the roll response rate as compared to the test response rate.

Since the RTM effect is not known in advance for any segment, one needs to estimate this effect based on the test results for better targeting decisions. This is a complicated problem because the RTM effect for any segment depends on the "true" response rate of the segment, which is not known in advance. Levin and Zahavi (1996) offer an approach to estimate the RTM effect for each segment which uses a prior knowledge on the "quality" of the segment (either "good", "medium" or "bad"). Practitioners use a knock down factor (often 20%-25%) to project the rollout response rate. While the latter is a crude approximation to the RTM effect, it is better than using no correction at all, as failure to account for the RTM may results in some "good" segments eliminated from the rollout campaign and some "bad" segments included in the campaign, both incur substantial costs.

As-of-Date

Because of the lead time to stock up on product, the promotion process could extend over time, with the time gap between the testing and the rollout campaign could extend over several months, sometime a year (see Figure 63.1). In case of subsequent rerolls, the time period between any two consecutive rerolls may be even longer. This introduces a time dimension into the modeling process.

Now, most predictors of response also have a time dimension. Certainly, this applies to the RFM variables which have proven to be the most important predictors of response in numerous applications. This goes without saying for recency which is a direct measure of time since last purchase. But frequency and monetary variables are also linked to time, because they often measure the number of previous purchases (frequency) and money spent (monetary) for a given time period, say a year. We note that some demographic variables such as age, number of children, etc., also change over time.

As a result, all data files for supporting targeting decisions ought to be created as of the date of the promotion. So if testing took place on January 1st, 2003 and the rollout campaign on July 30th, 2003, one needs to create a snap shot of the test audience as of January 1, 2003, for building the model and another snap shot of the universe as of July 30, 2003, for scoring the audience.

We note that if the time gap between two successive promotions (say the test and the rollout campaigns) is very long, several models may be needed to support a promotion. One model to predict the expected number of orders to be generated by the rollout campaign, based on the test audience reflecting customers' data as of the time of the test (January 1 2003, in the above example). Then, at the time of the roll, when one applies the model results for selecting customers for the rollout campaign, it might be necessary to recalibrate the model based on a snap shot of the test audience as of the rollout date (July 30th, 2003, in the above example).

63.8 Conclusions

In this chapter we have discussed the application of Data Mining models to support targeting decisions in direct marketing. We distinguished between three targeting categories – discrete choice problems, continuous choice problems and in-market timing problems, and reviewed a range of models for addressing each of these categories. We also discussed some pitfalls and issues that need to be taken care of in implementing a Data Mining solution for targeting applications.

But we note that the discussion in this chapter is somewhat simplified as it is confined mainly to targeting problem where each product/service is promoted on its own, by means of a single channel (mostly mail), independently of other products/services. But clearly, targeting problems can be much more complicated than that. We discuss below two extensions to the basic problem above – multiple offers and multiple products.

63.8.1 Multiple Offers

An "offer" is generalized here to include any combination of the marketing mix attributes, including price point, position, package, payment terms, incentive levels... For example, in the credit card industry, the two dominant offers are the line of credit to grant to a customer and the interest rate. In the collectible industry, the leading offers are price points, positioning of the product (i.e., as a gift or for own use), packaging,...

Incentive offers are gaining increasing popularity as more and more companies recognize the need to incorporate an incentive management program into the promotion campaigns to maximize customers' value chain. Clearly it does not make sense to offer any incentive to customers who are "captive audience" who are going to purchase the product no matter what. But it does make sense to offer an incentive to borderline customers "on the fence" for whom the incentive can make the difference between purchasing the product/service or declining it. This is true for each offer, not just for incentives. In general, the objective is to find the best offer to each customer to maximize expected net benefits. This gives rise to a very large constrained optimization problem containing hundreds of thousands, perhaps millions, of rows (each row corresponds to a customer) and multiple columns, one for each offer combination. The optimization problem may be hard to solve analytically, if any, and a resort to heuristic methods may be required.

From a Data Mining perspective, one needs to estimate the effect of each offer combination on the purchase probabilities, which typically requires that one designs an experiment whereby customers are randomly split into groups, each exposed to one offer combination. Then, based on the response results, one may estimate the offer effect. But, because the response rates in the direct marketing industry is very low, it is often necessary to test only part of the offer combinations (partial factorial design) and then deduct from the partial experiment onto the full factorial experiment. Further complication arises when optimizing the test design to maximize the information content of the test, using feedback from previous tests.

63.8.2 Multiple Products/Services

The case of multiple products adds another dimension of complexity to the targeting problem. Not only it is required to find the best offer for a given product to each customer, but it is also necessary to optimize the promotion stream to each customer over time, controlling the timing, number and mix of promotions to expose to each individual customer at each time window. This gives rise to even a bigger optimization problem which now contains many more columns, one column for each product/offer combination.

From a modeling standpoint, this requires that one estimate the cannibalization and saturation effects. The cannibalization effect is defined as the rate of the reduction in the purchase probability of the product as a result of over-promotion. Because of the RFM effect discussed above, it so happens that the "good" customers are often bombarded with too many mailings at any given time window. One of the well known effects of over-promotion is that it turns down customers, resulting in a decline in their likelihood of purchase of either product promoted to them. Experience shows that too many promotions may cause customers to discard the promotional material without even looking at them. The end result is often a loss in the number of active customers, not to mention the fact that over promotion results in misallocation of the promotion budget.

While the cannibalization effect is a result of over-promotion, the saturation effect is the result of over-purchase. Clearly, the more a customer buys from a given product category, the less likely s/he is to respond to a future solicitation for a product from the same product category. From a modeling perspective, the saturation effect is defined as the rate of reduction in the purchase probability of a product as a function of the number of products in the same product line that the customer has bought in the past. Since the saturation effect is not known in advance, it must be estimated based on past observations.

And these are not the only issues involved, and there are a myriad of others. Clearly, targeting applications in marketing are at the top of the analytical hierarchy, requiring a combination of tools from Data Mining, operations research, design of experiments, direct and database marketing, database technologies, and others. And we have not discussed here the organizational aspects involved in implementing a targeting system, and the integration with other operational units of the organization, such as inventory, logistics, financial, and others.

References

Akaike, H., Information Theory and an Extension of the Maximum Likelihood Principle, in 2nd International Symposium on Information Theory, B.N. Petrov and F. Csaki, eds, pp. 267-281, Budapest, 1973.

Ben-Akiva, M., and S.R. Lerman, Discrete Choice Analysis, the MIT Press, Cambridge, MA, 1987.

Benjamini, Y. and Hochberg, Y., Controlling the False Discovery Rate: a Practical and Powerful Approach to Multiple Testing, Journal Royal Statistical Society, Ser. B, 57, pp. 289-300, 1995.

Bock, H.H. Automatic Classification. Vandenhoeck and Ruprecht, Gottingen, 1974.

Breiman, L., Bagging Predictors, Machine Learning, Vol. 2, pp. 123-140, 1996.

Breiman, L., Friedman, J., Olshen, R. and Stone, C., Classification and Regression Trees, Belmont, CA., Wadsworth, 1984.

Buchanan, B. and Morrison, D.G., A Stochastic Model of List Falloff with Implications for Repeated Mailings", The Journal of Direct Marketing, Summer, 1988.

Cox, D.R. and Oakes, D., Analysis of Survival Data, Chapman and Hall, London, 1984.

DeGroot, M. H., Probability and Statistics 3rd edition. *Addison-Wesley*, 1991.

Friedman, J., Hastie, T. and Tibshirani, R., Additive Logistic Regression: a Statistical View of Boosting, Technical Report, Department of Statistics, Stanford University, 1998.

Fukunaga, K., Introduction to Statistical Pattern Recognition. San Diego, CA: Academic Press, 1990.

Heckman, J., Sample Selection Bias as a Specification Error, Econometrica, Vol. 47, No. 1, pp. 153-161, 1979.
Gilbert A. and Churchill, Jr., Marketing Research. *Seventh edition The Dryden Press*, 1999.
George, E.I., The Variable Selection Problem, University of Texas, Austin, 2000.
Herz, F., Ungar, L. and Labys, P., A Collaborative Filtering System for the Analysis of Consumer Data. Univ. of Pennsylvania, Philadelphia, 1997.
Hodges, J.L. Jr., "The Significance Probability of the Smirnov Two-Sample Test," Arkiv for Matematik, 3, 469 -486, 1957.
Kass, G., An Exploratory Technique for Investigating large Quantities of Categorical Data, Applied Statistics, 29, 1983.
Kohonen, K., Makisara, K., Simula, O. and Kangas, J., Artificial Networks. Amsterdam, 1991.
Lauritzen, S.L., The EM algorithm for Graphical Association Models with Missing Data. Computational Statistics and Data Analysis, 19, 191-201, 1995.
Long, S.J., Regression Models for Categorical and Limited Dependent Variables, Sage Publications, Thousand Oaks, CA, 1997.
Lambert P.J., The Distribution and Redistribution of Income. *Manchester University Press.*, 1993.
Levin, N. and Zahavi, J., Segmentation Analysis with Managerial Judgment, Journal of Direct Marketing, Vol. 10, pp. 28-47, 1996.
Levin, N. and Zahavi, J., Applying Neural Computing to Target Marketing, The Journal of Direct Marketing, Vol. 11, No. 1, pp. 5-22, 1997a.
Levin, N. and Zahavi, J., Issues and Problems in Applying Neural Computing to Target Marketing, The Journal of Direct marketing, Vol. 11, No. 4, pp. 63-75, 1997b.
Miller, A., Subset Selection in Regression, Chapman and Hall, London, 2002.
Quinlan, J.R., Induction of Decision Trees, Machine Learning, 1, pp. 81-106, 1986.
Quinlan, J.R., C4.5: Program for Machine Learning, CA., Morgan Kaufman Publishing, 1993.
Rumelhart, D.E., McClelland, J.L., and Williams, R.J., Learning Internal Representation by Error Propagation, in Parallel Distributed Processing: Exploring the Microstructure of Cognition, Rumelhart, D.E., McClelland, J.L. and the PDP Researcg Group, eds., MIT Press, Cambridge, MA, 1986.
Schwarz, G., Estimating the Dimension of a Model, Annals of Statistics, Vol. 6, pp. 486-494, 1978.
Shepard, D. (ed.), The New Direct Marketing, New York, Irwin, 1995.
Silverman, B.W., Density Estimation for Statistics and Data Analysis. Chapman and Hall, 1986.
Smith, W.R., Product Differentiation and Market Segmentation as Alternative Marketing Strategies, Journal of Marketing, 21, 3-8, 1956.
Sonquist, J., Baker, E. and Morgan, J.N., Searching for Structure, Ann Arbor, University of Michigan, Survey Research Center, 1971.
Tobin, J., Estimation of Relationships for Limited-Dependent Variables, Econometrica, Vol. 26, pp. 24-36, 1958.
Zhang, R., Ramakrishnan, R. and Livny, M., An Efficient Data Clustering Method for Very Large Databases. Proceedings ACM SIGKDD International Conference on Management of Data. 103-114, 1996.

NHECD - Nano Health and Environmental Commented Database

Oded Maimon[1] and Abel Browarnik[1]

Department of Industrial Engineering, Tel-Aviv University, Ramat-Aviv 69978, Israel, maimon@eng.tau.ac.il

Summary. The impact of nanoparticles on health and the environment is a significant research subject, driving increasing interest from the scientific community, regulatory bodies and the general public. We present a smart repository system with text and data mining for this domain. The growing body of knowledge in this area, consisting of scientific papers and other types of publications (such as surveys and whitepapers) emphasize the need for a methodology to alleviate the complexity of reviewing all the available information and discovering all the underlying facts, using data mining algorithms and methods.

The European Commission-funded project NHECD (whose full name is "Creation of a critical and commented database on the health, safety and environmental impact of nanoparticles") converts the unstructured body of knowledge produced by the different groups of users (such as researchers and regulators) into a repository of scientific papers and reviews *augmented* by layers of information extracted from the papers. Towards this end we use taxonomies built by domain experts and metadata, using advanced methodologies. We implement algorithms for textual information extraction, graph mining and table information extraction. Rating and relevance assessment of the papers are also part of the system. The project is composed of two major layers, a backend consisting of all the above taxonomies, algorithms and methods, and a frontend consisting of a query and navigation system. The frontend has web interface which address the needs (and knowledge) of the different user groups. Documentum, a content management system (CMS), is the backbone of the backend process component. The frontend is a customized application built using an open source CMS. It is designed to take advantage of the taxonomies and metadata for search and navigation, while allowing the user to query the system, taking advantage of the extracted information.

64.1 Introduction

Nanoparticles toxicity (or NanoTox) is currently one of the main concerns for the scientific community, for regulators and for the public. Nanoparticles impact on health and the environment is a research subject driving increasing interest. This fact is reflected by the number of papers published on the subject, both in scientific journals and on the press.

O. Maimon, L. Rokach (eds.), *Data Mining and Knowledge Discovery Handbook*, 2nd ed., DOI 10.1007/978-0-387-09823-4_64, © Springer Science+Business Media, LLC 2010

The published material (e.g., scientific papers) is essentially unstructured. It always uses natural language (in the form of text), sometimes accompanied by tables and/or graphs.

Usually, when searching a body of unstructured knowledge (such as a corpus of scientific papers) the "search engine" uses a method called "full text search". Full text search can be done either directly, by scanning all the available text or by using indexing mechanisms. Direct search is feasible only for small volumes of data. Index-based search applies when the amount of data rules out direct search. There are several indexing mechanisms, the most famous being Google's *Page Rank* (Brin, 1998). Indexing mechanisms are rated according to the results returned by the search engines using them. In either case, users interact with the search engine (and through it with the indexing mechanism) by means of queries.

Scientific papers are written in natural language. It could be easier for users to formulate queries using the same natural language. However, the understanding of natural language is an extremely non-trivial task. Therefore, using it for queries would add additional complexity to a problem with enough complexity by itself. To avoid it, search engines use different approaches to deal with queries:

1. Keywords: Document creators (or trained indexers) are asked to supply a list of words that describe the subject of the text, including synonyms of words that describe this subject. Keywords improve recall, particularly if the keyword list includes a search word that is not in the document text.
2. Boolean queries: Searches using Boolean operators can dramatically increase the precision of a free text search. The AND operator says, in effect, "Do not retrieve any document unless it contains both of these terms." The NOT operator says, in effect, "Do not retrieve any document that contains this word." If the retrieval list retrieves too few documents, the OR operator can be used to increase recall.
3. Phrase search: A phrase search matches only those documents that contain a specified phrase.
4. Concordance search: A concordance search produces an alphabetical list of all principal words that occur in a text with their immediate context.
5. Proximity search: A phrase search matches only those documents that contain two or more words that are separated by a specified number of words.
6. Regular expression: A regular expression employs a complex but powerful querying syntax that can be used to specify retrieval conditions with precision.
7. Wildcard search: A search that substitutes one or more characters in a search query for a wildcard character such as an asterisk.

"Skin Deep" [1], a product safety guide dealing with cosmetics, run by the Environmental Working Group, grants public access to a database containing more than 42,000 products with more than 8,300 ingredients from the U.S., nearly a quarter of all products on the market (figures updated to May 2009). The database is based on a link between a collection of personal care product ingredient listings with more than 50 toxicity and regulatory databases.

Skin Deep uses a restricted user interface for simple queries. The visitor is asked on a product, ingredient or company (see Figure 1).

A query for "vitamin a" returned 614 results, matching at least one word. The advanced query screen allows for a much more detailed search (see Figure 2). Visitors can ask to find products, ingredients or companies with higher granularity.

The results returned by Skin Deep consist of an exhaustive analysis of the substance, as shown in Figure 3.

[1] The Environmental Working Group Repository of Cosmetics, http://www.cosmeticsdatabase.com/about.php

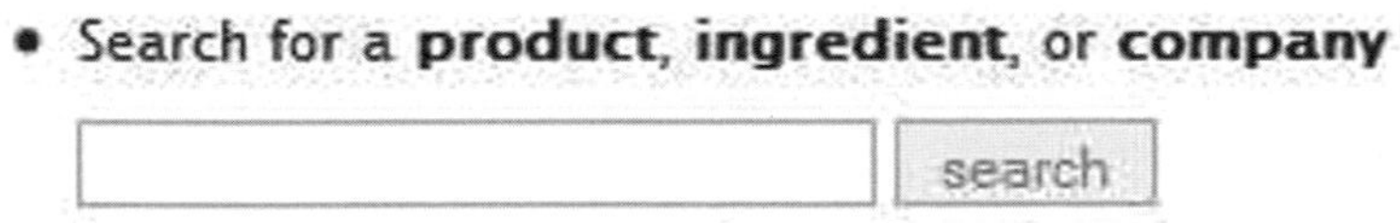

- Search for a **product**, **ingredient**, or **company**

search

Fig. 64.1. Skin Deep simple query.

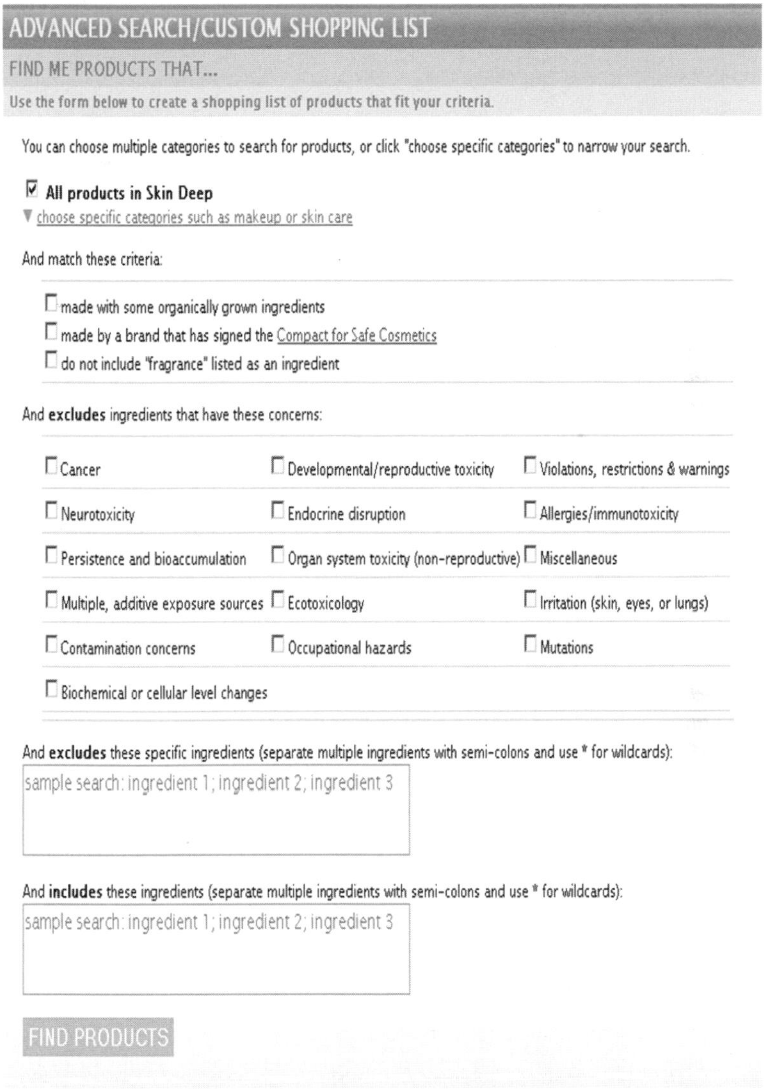

ADVANCED SEARCH/CUSTOM SHOPPING LIST

FIND ME PRODUCTS THAT...

Use the form below to create a shopping list of products that fit your criteria.

You can choose multiple categories to search for products, or click "choose specific categories" to narrow your search.

☑ **All products in Skin Deep**

▼ choose specific categories such as makeup or skin care

And match these criteria:

☐ made with some organically grown ingredients
☐ made by a brand that has signed the Compact for Safe Cosmetics
☐ do not include "fragrance" listed as an ingredient

And **excludes** ingredients that have these concerns:

☐ Cancer ☐ Developmental/reproductive toxicity ☐ Violations, restrictions & warnings
☐ Neurotoxicity ☐ Endocrine disruption ☐ Allergies/immunotoxicity
☐ Persistence and bioaccumulation ☐ Organ system toxicity (non-reproductive) ☐ Miscellaneous
☐ Multiple, additive exposure sources ☐ Ecotoxicology ☐ Irritation (skin, eyes, or lungs)
☐ Contamination concerns ☐ Occupational hazards ☐ Mutations
☐ Biochemical or cellular level changes

And **excludes** these specific ingredients (separate multiple ingredients with semi-colons and use * for wildcards):

sample search: ingredient 1; ingredient 2; ingredient 3

And **includes** these ingredients (separate multiple ingredients with semi-colons and use * for wildcards):

sample search: ingredient 1; ingredient 2; ingredient 3

FIND PRODUCTS

Fig. 64.2. Skin Deep advanced queries.

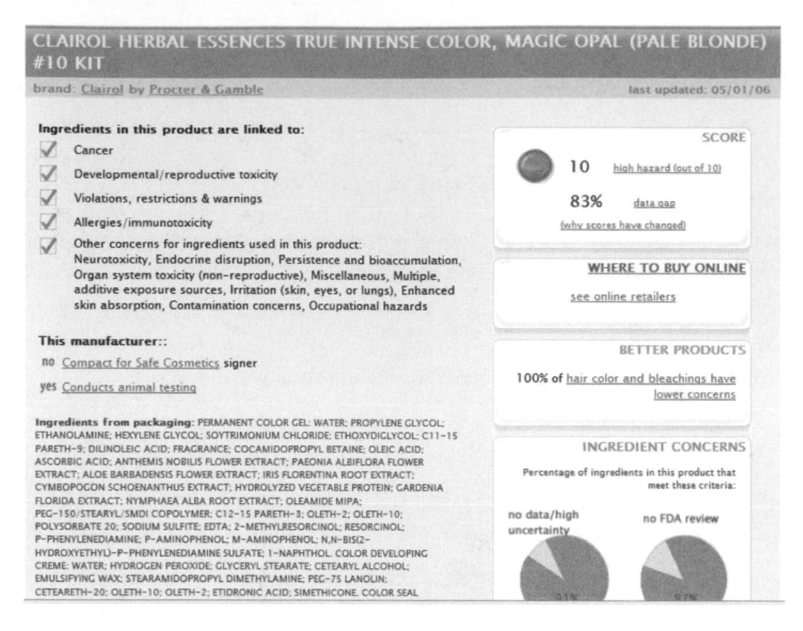

Fig. 64.3. Skin Deep result example.

ICON, the International Council on Nanotechnology, from RICE University, uses an approach that constrains the user to formulate a query within a restricted (although very rich) template, together with a "controlled vocabulary" (e.g., a list of predefined values) as shown in Figure 4.

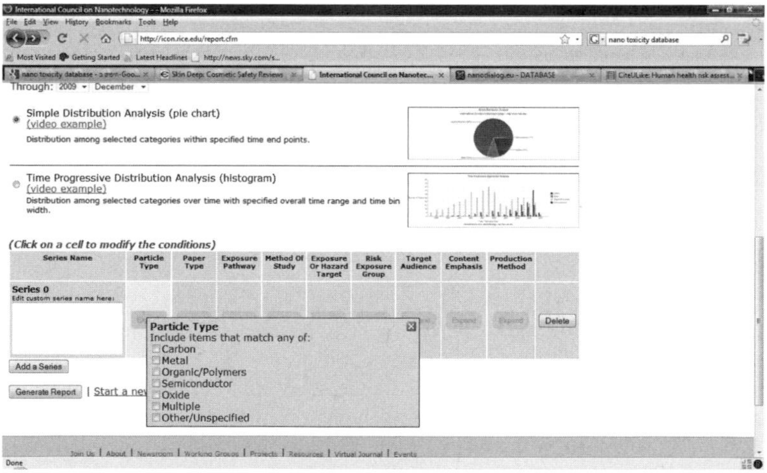

Fig. 64.4. ICON database.

Results obtained from ICON are, as stated on the ICON website:

"... a quick and thorough synopsis of our Environment, Health and Safety Database using two types of analyses. The first is a Simple Distribution Analysis (pie chart) which compares categories within a specified time range. The second type is a Time Progressive Distribution Analysis (histogram) which compares categories over a specified overall time range and data grouping period.

Other useful features include the ability to:

1. *Generate and export custom reports in pdf and xls formats.*
2. *Click on a report result to generate a list of publications meeting your criteria".*

TOXNET - Databases on toxicology, hazardous chemicals, environmental health, and toxic releases, an initiative by the US National Library of Medicine, lets visitors query its network of databases by using keywords, as shown in Figure 5.

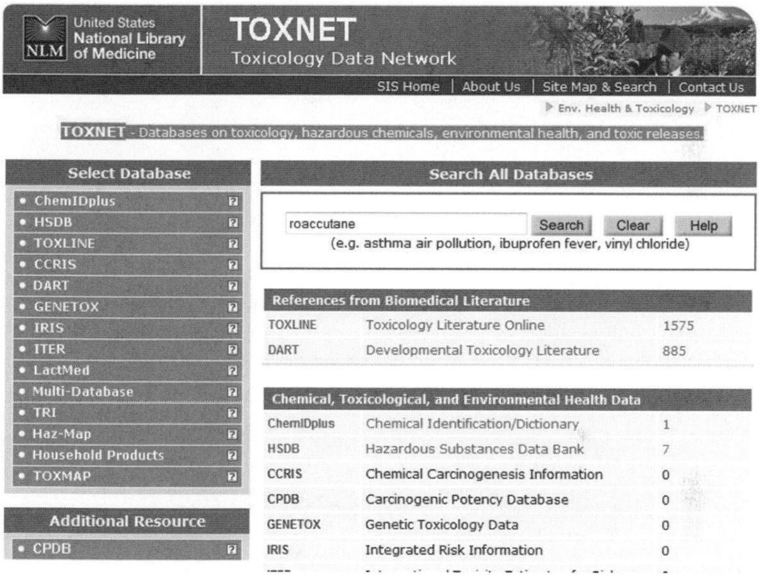

Fig. 64.5. Toxnet query.

There are several initiatives related to toxicity of nanoparticles, but to date none of it is a real alternative to the existing (and limited) databases. Examples of such initiatives are the Environmental Defense Fund Nanotech section [2] and the NANO Risk Framework.

There are also initiatives that aim at mapping current *nanotox* research.

The OECD (Organisation for Economic Co-operation and Development) runs a "Database on Research into the Safety of Manufactured Nanomaterials" [3]. As suggested by its name, the database maps research on the area. It uses extensive metadata[1], as seen in Figure 6.

---

[2] Environmental Defense Fund, http://www.edf.org/page.cfm?tagID=77

[3] http://webnet.oecd.org/NanoMaterials/Pagelet/Front/Default.aspx

[4] Metadata: "data about data" (see http://en.wikipedia.org/wiki/Metadata)

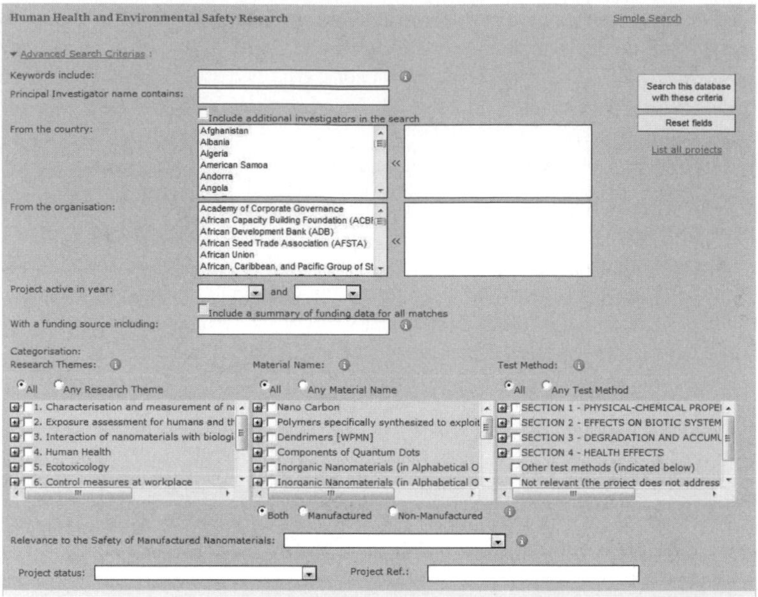

Human Health and Environmental Safety Research
Simple Search
Advanced Search Criterias :
Keywords include:
Principal Investigator name contains:
Include additional investigators in the search
Search this database with these criteria
Reset fields
List all projects
From the country:
Afghanistan
Albania
Algeria
American Samoa
Andorra
Angola
From the organisation:
Academy of Corporate Governance
African Capacity Building Foundation (ACBF
African Development Bank (ADB)
African Seed Trade Association (AFSTA)
African Union
African, Caribbean, and Pacific Group of St
Project active in year: and
Include a summary of funding data for all matches
With a funding source including:
Categorisation:
Research Themes: All Any Research Theme
1. Characterisation and measurement of n
2. Exposure assessment for humans and th
3. Interaction of nanomaterials with biologi
4. Human Health
5. Ecotoxicology
6. Control measures at workplace
Material Name: All Any Material Name
Nano Carbon
Polymers specifically synthesized to exploit
Dendrimers [WPMN]
Components of Quantum Dots
Inorganic Nanomaterials (in Alphabetical O
Inorganic Nanomaterials (in Alphabetical O
Test Method: All Any Test Method
SECTION 1 - PHYSICAL-CHEMICAL PROPE
SECTION 2 - EFFECTS ON BIOTIC SYSTEM
SECTION 3 - DEGRADATION AND ACCUMU
SECTION 4 - HEALTH EFFECTS
Other test methods (indicated below)
Not relevant (the project does not address
Both Manufactured Non-Manufactured
Relevance to the Safety of Manufactured Nanomaterials:
Project status:
Project Ref.:

Fig. 64.6. OECD NanoTox advanced search.

NIOSH, the U.S. National Institute for Occupational Safety and Health, runs a Nanoparticle Information Library (NIL) [5]

IMPART-Nanotox [6], an EU funded project ended in 2008, includes a public, web accessible database of *nanotox* publications. The search can be done by publications' metadata, as seen in Figure 7.

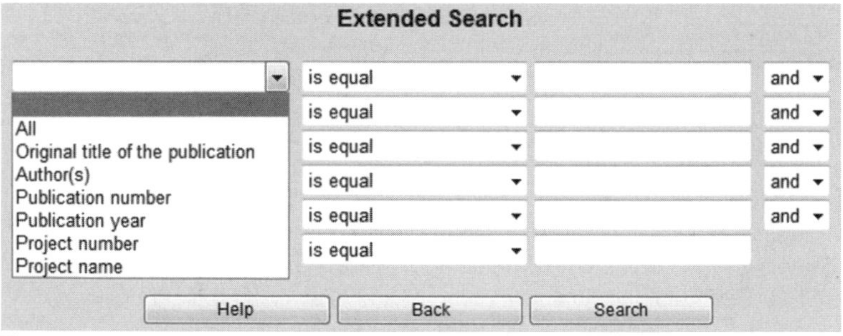

Extended Search
All
Original title of the publication
Author(s)
Publication number
Publication year
Project number
Project name
is equal — and
is equal — and
is equal — and
is equal — and
is equal — and
is equal
Help Back Search

Figure 7 - Impart-Nanotox extended search

Fig. 64.7. Impart-Nanotox extended search.

[5] The U.S. National Institute for Occupational Safety and Health Nanoparticle Information Library (NIL) – http://www.cdc.gov/search.do?action=search&subset=niosh

[6] http://www.impart-nanotox.org

SAFENANO [7], another EU funded project, contains a database of publications and metadata searchable on the web (see Figure 6).

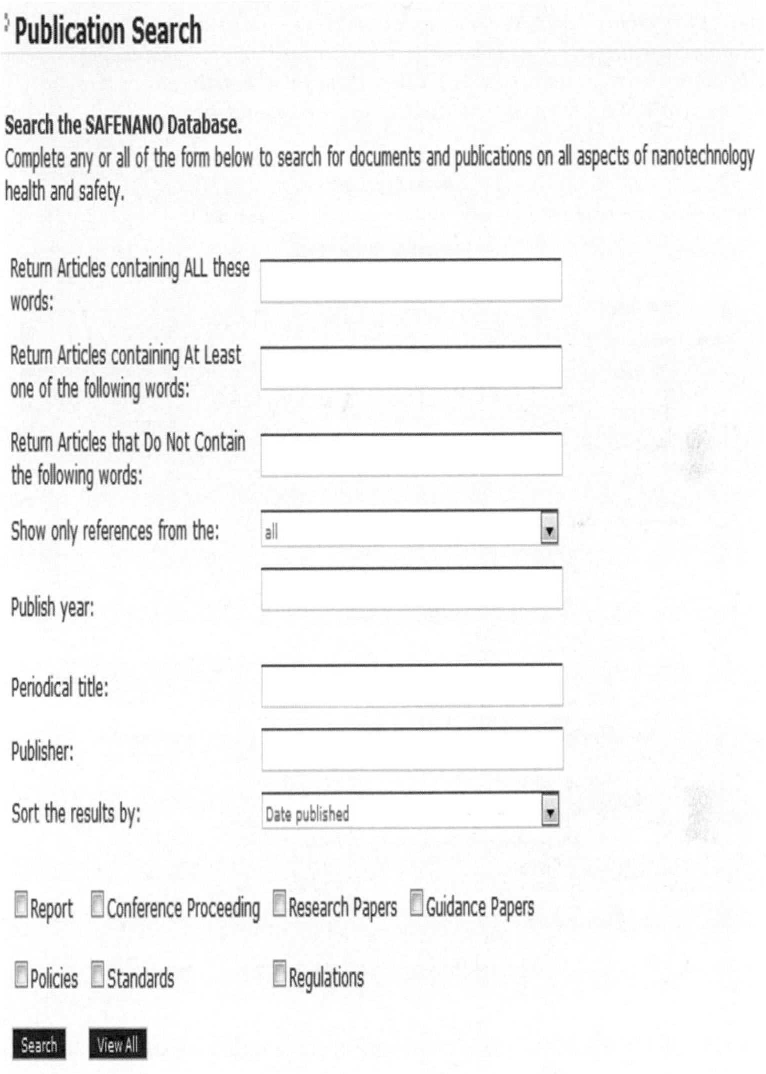

Publication Search

Search the SAFENANO Database.
Complete any or all of the form below to search for documents and publications on all aspects of nanotechnology health and safety.

Return Articles containing ALL these words:

Return Articles containing At Least one of the following words:

Return Articles that Do Not Contain the following words:

Show only references from the: all

Publish year:

Periodical title:

Publisher:

Sort the results by: Date published

Report Conference Proceeding Research Papers Guidance Papers

Policies Standards Regulations

Search View All

Fig. 64.8. SAFENANO Publication Search.

[7] http://www.safenano.org

Nano Archive [8], another EU FP7 project, has the objective of allowing researchers to share and search information, mainly through metadata exchange (see Figure 9). ObservatoryNano [9], yet another EU FP7 funded project has an ambitious target,

"*to create a European Observatory on Nanotechnologies to present reliable, complete and responsible science-based and economic expert analysis, across different technology sectors, establish dialogue with decision makers and others regarding the benefits and opportunities, balanced against barriers and risks, and allow them to take action to ensure that scientific and technological developments are realized as socio-economic benefits.*"

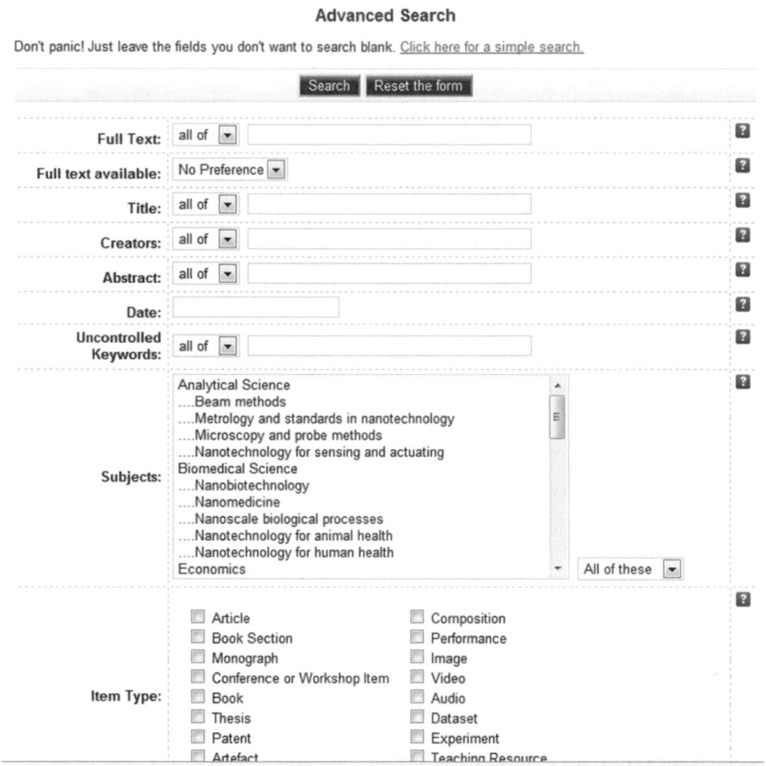

Advanced Search

Don't panic! Just leave the fields you don't want to search blank. Click here for a simple search.

Search | Reset the form

Full Text: all of

Full text available: No Preference

Title: all of

Creators: all of

Abstract: all of

Date:

Uncontrolled Keywords: all of

Subjects:
Analytical Science
....Beam methods
....Metrology and standards in nanotechnology
....Microscopy and probe methods
....Nanotechnology for sensing and actuating
Biomedical Science
....Nanobiotechnology
....Nanomedicine
....Nanoscale biological processes
....Nanotechnology for animal health
....Nanotechnology for human health
Economics
All of these

Item Type:
Article | Composition
Book Section | Performance
Monograph | Image
Conference or Workshop Item | Video
Book | Audio
Thesis | Dataset
Patent | Experiment
Artefact | Teaching Resource

Fig. 64.9. Nano Archive search.

The review above brought us to the conclusion that the following shortcomings should be dealt with:

1. Many efforts are being dedicated to creating repositories of raw metadata of nanotox publications. There is no evidence as to the contribution of such repositories to the advancement of nanotox research and implementation.
2. No significant searchable repository of nanotox data (as compared to metadata) exists currently.

[8] http://www.nanoarchive.org/information.html

[9] http://www.observatorynano.eu/project/search/extendedsearch/

3. The query capabilities of widely used search engines do not include the option to query the text for fact patterns (as well as more complex, derived patterns), such as "what conclusions were reached in scientific papers where <fact X> and <fact Y> occurred in that order?". Those are examples of queries that may help nanotox researchers and regulators, as well as the general public.
4. There is no tool capable of extracting information specific to nanotox.

NHECD[10], an EU FP7 funded project, is aiming at transforming the emerging body of unstructured knowledge (in the form of scientific papers and other publications) into structured data by means of textual information extraction, solves the above shortcomings by:

1. Developing taxonomies for the nanotox domain
2. Developing and implementing algorithms for information extraction from nanotox papers
3. Creating a repository of papers augmented by structured knowledge extracted from the papers
4. Allowing visitors (e.g., nanotox scientists, regulators, general public) to navigate the repository using the taxonomies
5. Letting visitors search the repository using complex patterns (such as facts)
6. Enabling data mining algorithms to predict toxicity based on characteristics extracted by text mining methods. Thus free text can be used for data mining inference.

64.2 The NHECD Model

NHECD is, as suggested by its full name (Nanotox Health and Environment Commented Database) an initiative to obtain a database (e.g., structured information that can be queried) from available unstructured information such as scientific papers and other publications.

The process of obtaining the structured data involves many resources, from the domain of Nanotox and from the areas of information sciences and technologies (IT).

The NHECD model is depicted in Figure 10.

The process starts with a collection of documents (e.g., scientific papers) gathered by means of a search using criteria given by Nanotox experts. The process used to *populate* the repository is called *crawling*. The documents are accompanied by the corresponding metadata (e.g., authors, publication dates, journals, keywords supplied by the authors, abstract and more). The process requires Nanotox taxonomies. Taxonomies are classification artifacts used at the information extraction stage (taxonomies are also used in NHECD for document navigation). Taxonomy building tasks are "located" at the boundary between the Nanotox experts and the IT experts (see Figure 10), due to its interdisciplinary nature.

Nanotox experts annotate papers to train the system towards the information extraction stage. This stage is implemented using text mining algorithms. Further to the information extraction process, a set of rating algorithms is applied on the documents to provide an additional layer of information (e.g., the rating).

The result of the process consists of:

1. A corpus of results, updated on an ongoing, asynchronous basis.
2. A *commented* collection of scientific papers. By commented we refer to the added layer of metadata, rating and other information extracted from the document.

The whole process can be represented with a block diagram as shown in figure 11.

[10] "http://www.nhecd-fp7.eu - Creation of a critical and commented database on the health, safety and environmental impact of nanoparticles"

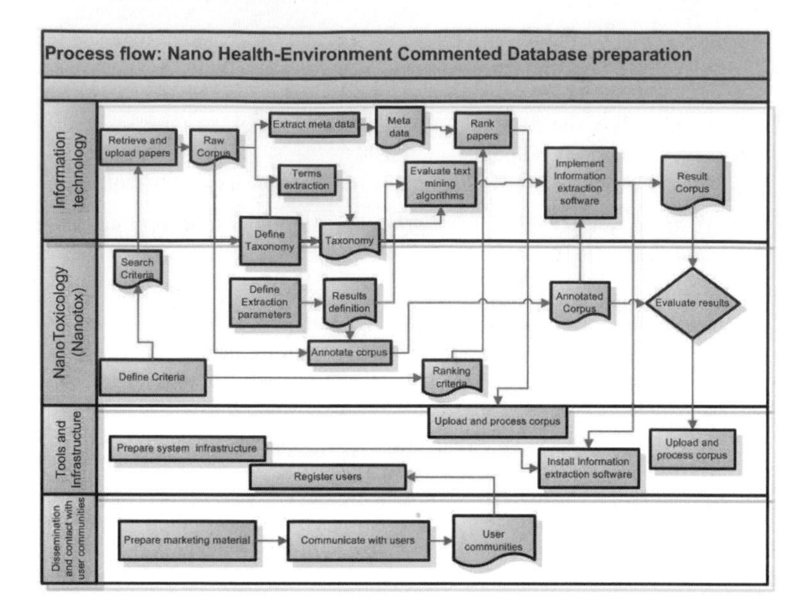

Fig. 64.10. NHECD model.

64.3 NHECD implementation

NHECD is built around Documentum [11], an enterprise content management system (ECM). Documentum acts as a central repository for documents (e.g., unstructured data), metadata (semi structured to structured data, mostly in XML) and extracted information (mostly structured, tabular data).

NHECD assumes that all target scientific papers to be included in the repository are in Adobe PDF format. A review of several websites hosting scientific papers shows that this is a safe choice. If other formats are found the crawler can be instructed to convert the format found to PDF almost seamlessly.

64.3.1 Taxonomies

Each taxonomy deals with a certain aspect of the Nanotox domain. They were built by teams of domain experts and information management experts. Taxonomies are not an expected output of NHECD. Yet, they are essential to the NHECD process. Hence it was one of the initial steps in NHECD implementation. The taxonomies are:

Following are the taxonomies describing the subject "commercial NP characterization".

64.3.2 Crawling

The process of automatically obtaining scientific papers and data about the paper (such as the name of the author or authors, the publication date, the name of the journal, keywords, abstract,

[11] http://www.emc.com/products/family/documentum-family.htm

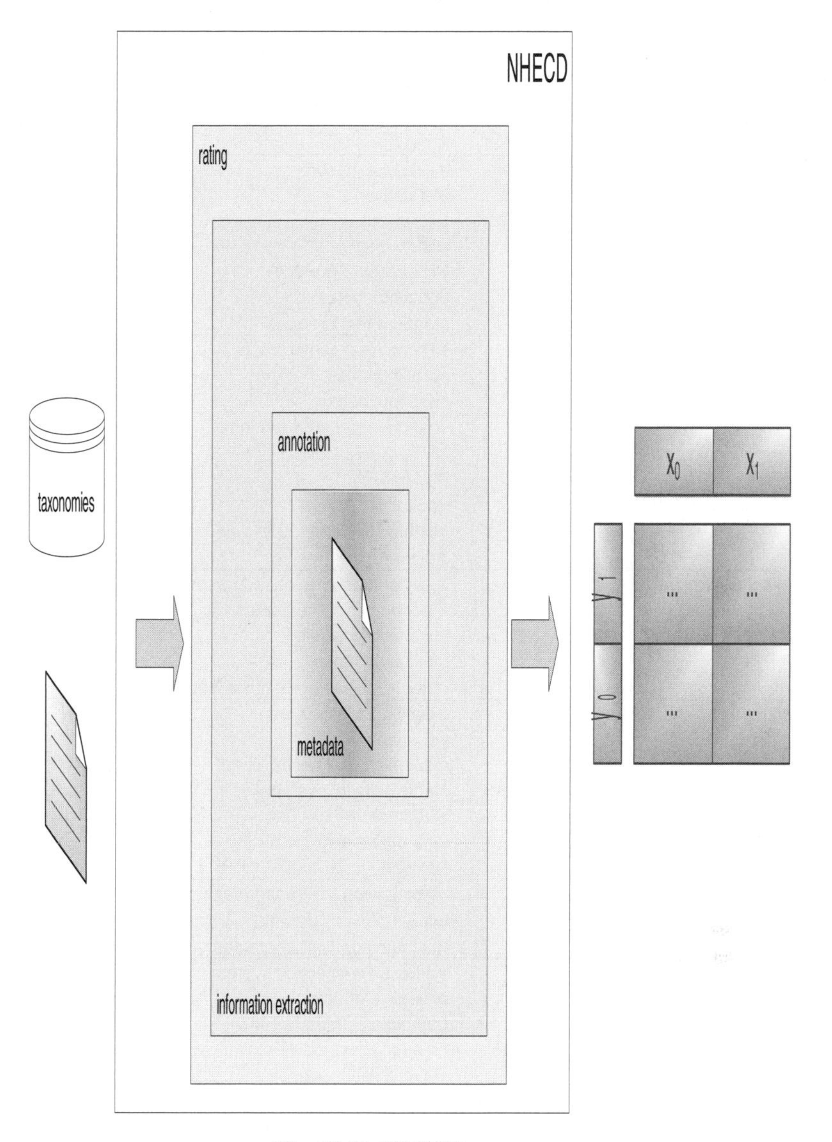

Fig. 64.11. NHECD process.

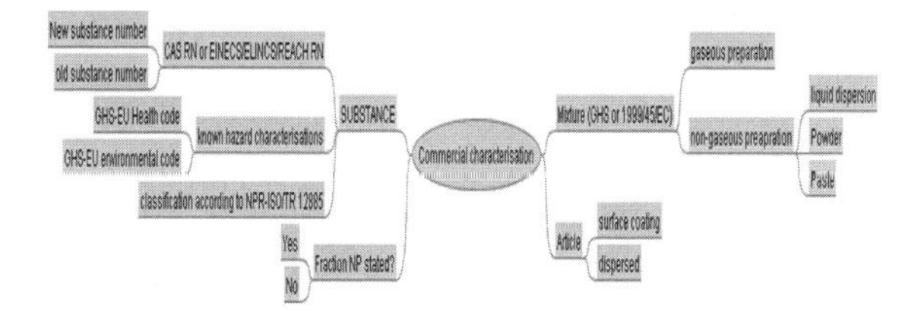

Fig. 64.12. Commercial characterization of NP.

Table 64.1. Taxonomies.

Subject	Taxonomies
animal model	animal gender Species
experimental exposure parameters	mode of exposure Metabolism Excretion Distribution NP exposure protocol effecting agents standard test protocol
NP chemical characterization	NP chemical composition core impurities coat impurities coat chemical composition NP carrier solubility
commercial NP characterization	Substance fraction NP stated Mixture (GHS or 1999/45/EC) Article
NP characterization methods	X-ray and neutron based instruments Electron beam methods Ion Beam analysis Optical methods other NP measurements methods
NP general characterization	Specific Surface Area NP shape dispersion and adsorption Nano Delivery system Structure zeta potential NP type
Overall NP	taxonomies of NP characterization taxonomies of NP chemical characterization taxonomies of NP characterization methods
Results	visible toxic effect by system biological effects pathologic effects time for effect manifestation Reversibility site of effect measurement methods for biological effect measurements methods for pathological effects

and in general any detail made available with the paper itself) by visiting scientific paper repositories available on the web (whether restricted to subscribers or available to everyone) and searching by keywords on the paper text is called *crawling*. NHECD developed a crawler for Pubmed [12]. Crawlers for other leading scientific sites, such as, ISIWEB [13] or SciFinder

[12] http://www.ncbi.nlm.nih.gov/pubmed/
[13] http://www.webofknowledge.com/

[14] are in a development stage. The main obstacles found often refer to intellectual property issues and the efforts by the publishers to enforce it.

The crawler is written in java. It takes as input a set of keywords. Using websites API[15] it obtains a list of *pointers* to the targeted scientific papers. Those pointers are processed to transform it into downloadable links. If a downloadable link is obtained, the scientific paper is downloaded, provided that NHECD has access to the paper (e.g., there is a subscription to the resource or it is publicly available). The paper (if available, otherwise its place holder), along with the metadata already converted to XML, are uploaded to the NHECD document repository.

64.3.3 Information extraction

The goals of information extraction in NHECD are:

1. To enable users to ask specific questions about specific attributes and receive answers. If possible, a link to the paper is given, along with a pointer to the location of the requested information within the document.
2. To enable, in the future, data mining on extracted data (e.g., patterns).

The process starts with a multistep ***preprocessing*** stage:

1. convert the input documents from PDF to text
2. perform parsing and stemming
3. perform zoning within the document
4. classify the document according to NHECD taxonomies

Next in the process is the ***tagging*** stage, used to recognize keywords, either by using the taxonomies or the values involved. As an example, the input phrase

> *"To determine the effect of particle size, labeled microspheres of 500 and 1000 nm in diameter were incubated with mouse melanoma B16 cells"*

would result in the tagged form

> *"To determine the effect of particle size, labeled microspheres of <NUMBER_1> and <NUMBER_2> <LENGTH-UNIT_3> in diameter were incubated with <SPECIES_4> <CELL-TYPE_5> <CELL-LINE_6> cells"*

The ***pattern matching*** stage is based on the output of previous stages and on the process of annotation, an auxiliary step performed by Nanotox domain experts to prepare a training set for this stage.

The tasks needed to obtain patterns are:

1. Define the list of features to be extracted (based on the taxonomy)

---

[14] http://pubs.acs.org/

[15] Application Program Interface

2. For each feature that needed to be extracted we define a list of extraction patterns
3. Each extraction pattern (p) consists of the following items:
   a) p.attributes – Associated attributes to be extracted. (note the same pattern can be used to extract several attributes concurrently)
   b) p.precondiction – A pre-condition
   c) p.match - A regular expression to be matched.
   d) p.extraction – A regular extraction expression to be used for extraction the values assuming that pattern p.t has been matched.
   e) p.scope – determine the scope of the extracted values in the text
   f) p.store – A SQL query for storing the results in the database

The closing stage of the process is the ***conflict resolution*** stage. It is required for cases where several possible contradicting patterns can be matched to the same text or the same pattern can be matched to different part of the text.

The information extraction process is depicted in Figure 13:

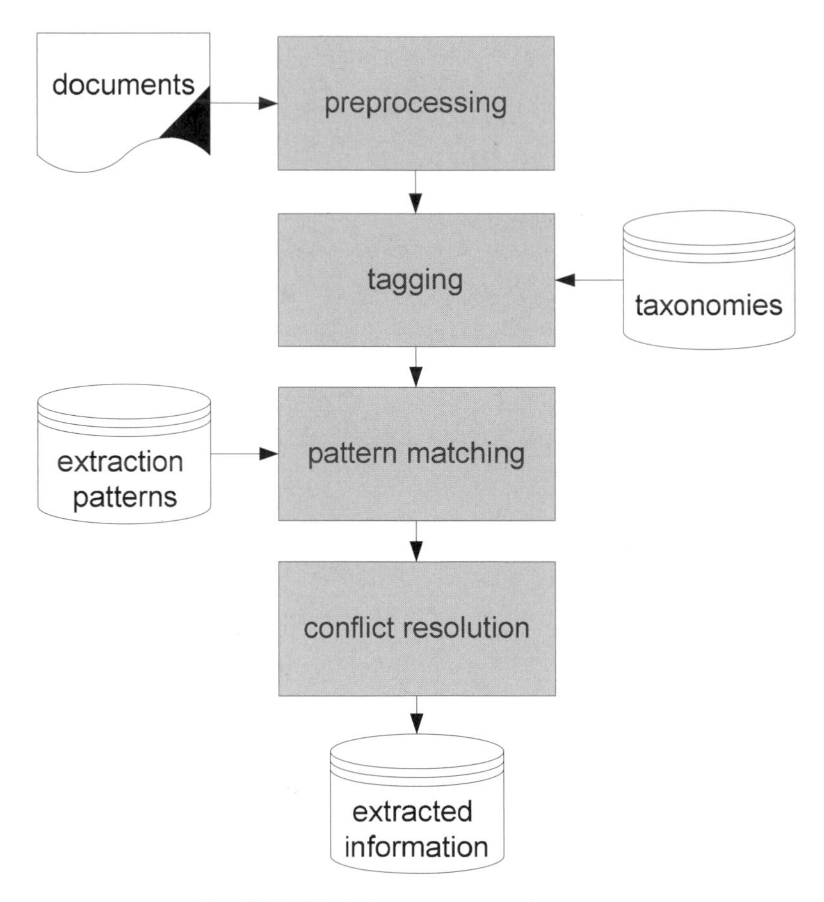

Fig. 64.13. The information extraction process.

64.3.4 NHECD products

The results of NHECD consist mainly of two *products*:

1. A repository of scientific papers related to Nanotox, augmented by metadata provided by authors and publishers, metadata extracted from the papers using text mining algorithms, and ratings for the articles based on methods adopted by NHECD. All the above, indexed using NHECD taxonomies. As a result, it is possible to retrieve scientific papers using sophisticated queries.
2. A set of structured *facts* extracted from the scientific papers in tabular format. The structured facts should make it possible to perform data mining to obtain new, unforeseen knowledge.

64.3.5 Scientific paper rating

A scientific paper has a well established life cycle. After the paper is written, refereed and eventually accepted, it is published. From this point in time the paper can be cited.

The rating of a paper depends on several variables:

1. Journal Name
2. Publication Year
3. Full Author Names
4. For each citing article:
5. Citing article name (and a unique identifier for the paper itself. NHECD decided to adopt SICI [16] for this purpose)
6. Citing journal name
7. From JCR (Journal Citation Report) , for journal name (including citing journals):
8. Impact Factor
9. Cited Half Life
10. H-Indices [17] per Author, From PoP

The rating algorithm is applied when the paper is loaded and then on a periodic basis, to reflect changes such as new citations, changes in impact factors, in “Cited Half Life”, in JCR data and more. The rating algorithm takes into account the publication date of newly published papers to avoid less-than-fair ratings for such papers.

The scientific paper rating devised by NHECD is composed a Journal Impact Factor and by H-indices. These components are defined below.

1. **Rating By Journal Impact Factor**

$$Rating_1(Article(i)) = 1 - 2^{-0.6 \bullet CitationScore_{Article(i)}}$$

where

$$CitationScore_{Article(i)} = \frac{\sum_{Article(j) \in citations(Article(i))} Map(impact(Journal(Article(j))))}{Age(Article(i))}$$

and

[16] http://en.wikipedia.org/wiki/SICI

[17] http://en.wikipedia.org/wiki/Hirsch_number

$$Map(impact(Journal(Article(j)))) = \begin{cases} 0.08 & 0 \leq impact(Journal(Article(j))) \leq 1.296 \\ 0.4 & 1.297 \leq impact(Journal(Article(j))) \leq 3.76 \\ 1 & 3.77 \leq impact(Journal(Article(j))) \leq \infty \end{cases}$$

2. **Rating By H-Indices**

$$Rating_2(Article(i)) = 1 - 1.05^{-HScore_{Article(i)}}$$

where

$$HScore_{Article(i)} = \frac{\sum_{Article(j) \in citations(Article(i))} Average_{Author(k) \in Article(j)}(H - Index_k)}{Age(Article(i))}$$

3. **Final Rating**

$$Rating = \frac{\alpha_1 Rating_1 + \alpha_2 Rating_2}{\alpha_1 + \alpha_2}$$

$$0 \leq \alpha_i \leq 1$$

64.3.6 NHECD Frontend

NHECD provides a free access website including information retrieval functionalities to facilitate the search on NHECD repository.

It includes the following components:

1. An open source content management system implemented on Drupal, which stores and manages the entire frontend database (including user information and usage patterns).
2. The user interface component that handles all the input or requests from the user.

The frontend interacts with the backend repository, stored and managed on Documentum. Figure 14 shows the architecture design of NHECD Frontend.

1. User communities and Characteristics – NHECD front end is designed to meet the different needs of three main communities and an additional group – the administrators.
2. Scientists – Users in this community will be scientists from academia and industry – the most expert users among all three communities. These users should have an extensive prior knowledge in the domain of nanotoxicology. The system assumes that these users are proficient in information searches.
3. Regulators – Users working for (or on behalf of) government institutes and regulatory agencies are part of the NHECD regulatory community. This community aims at providing legislation and regulation on the health, safety or environmental concerns regarding the use of nano-particles. Usage patterns of this group often overlap with those of the other communities.
4. General public – This community is composed of individuals and NGO's who are active in a wide range of fields where information provided by NHECD may be relevant. We assume that most of the general public users are NOT able to read/evaluate the scientific material NHECD provides. Therefore, the frontend provides - for this community - mainly answers to queries on general information/*light* reviews or news on the impact of exposure to nanoparticles.

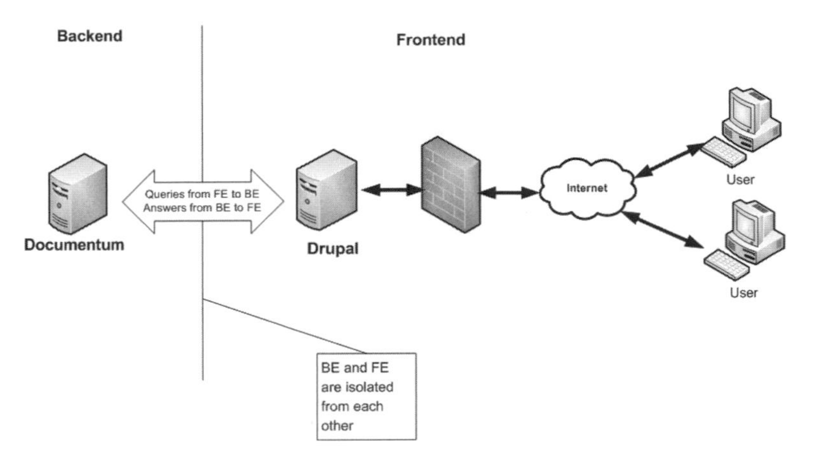

Fig. 64.14. Architecture.

5. Administrator – The administrator is in charge of managing the daily operation of the system. Administrators are responsible for managing user accounts, general settings and monitoring.

The NHECD frontend provides the following features:

1. Basic search
2. Advanced search
3. Intelligent search
4. Taxonomic navigation
5. Recommender results (i.e., recommendations based on the analysis of usage patterns of other users)
6. Option to resubmit queries, adding additional criteria for the refinement of results
7. Site registration
8. Personalization features
9. Displaying a list of most viewed papers
10. Links to other nanotox related sites
11. NHECD news, updates and FAQ's

64.4 Conclusions

NHECD provides two important *products*:

1. An extensive and commented repository of scientific papers and other publications in the Nanotox area, searchable using taxonomies and full text search. The scientific papers are rated according to published NHECD criteria, to help users to better estimate their findings. Such a repository significantly expand currently available repositories due to the fact that it goes beyond the mapping of existing research in Nanotox (as most current initiatives do). NHECD gives access to the research papers results, extracted from the sources using text mining algorithms. Access to scientific papers is granted to visitors following copyright and restrictions as imposed by publishers. This NHECD result is intended for Nanotox scientists, regulators and for the general public.

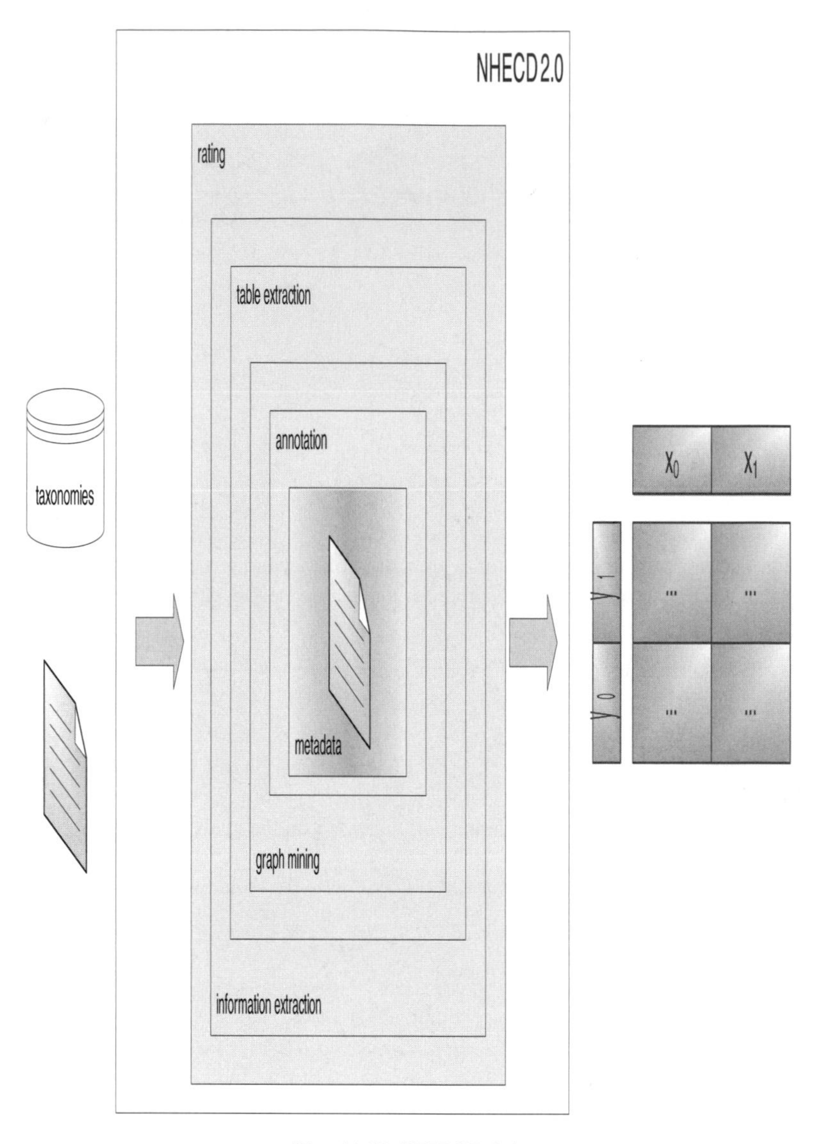

Fig. 64.15. NHECD 2.0.

2. A set of structured results extracted from the scientific papers populating the NHECD repository. Using these results it will be possible to perform data mining on the results. Data mining will result in validated results and further knowledge discovery. This part of NHECD results is targeted at Nanotox scientists and regulators.

64.5 Further research

Graph and table mining

NHECD makes resort to text mining algorithms, allowing for information extraction from textual data. It appears that scientific Nanotox papers (as in many other areas) often include other type of elements, such as graphs and tables. Moreover, the expressiveness of these elements is generally higher than that conveyed by text. Hence, expanding NHECD to include graph and table mining seems desirable. Preliminary research on these subjects made by the NHECD team shows that – at least for some types of graphs and tables – the task is feasible.

The concept of the future NHECD (touted NHECD 2.0) is shown in Figure 15.

Taxonomies and ontologies NHECD uses, at several stages, manually prepared taxonomies. It is arguable that using an ontology of the Nanotox domain could enhance the quality of information extraction (either textual, graphic or tabular). On the other hand, no Nanotox ontology exists. Research towards ontology learning could use NHECD results. In turn, the learned ontology could improve information extraction, implementing a kind of bootstrapping process. Data mining on the second NHECD product can have a strong influence on the ontology learning process. As a result, the ontology can be further enhanced.

References

Arbel, R. and Rokach, L., Classifier evaluation under limited resources, Pattern Recognition Letters, 27(14): 1619–1631, 2006, Elsevier.

Averbuch, M. and Karson, T. and Ben-Ami, B. and Maimon, O. and Rokach, L., Context-sensitive medical information retrieval, The 11th World Congress on Medical Informatics (MEDINFO 2004), San Francisco, CA, September 2004, IOS Press, pp. 282–286.

Brin, S. and Page, L. 1998. The anatomy of a large-scale hypertextual Web search engine. Comput. Netw. ISDN Syst. 30, 1-7 (Apr. 1998), 107-117.

Cohen S., Rokach L., Maimon O., Decision Tree Instance Space Decomposition with Grouped Gain-Ratio, Information Science, Volume 177, Issue 17, pp. 3592-3612, 2007.

Maimon O., and Rokach, L. Data Mining by Attribute Decomposition with semiconductors manufacturing case study, in Data Mining for Design and Manufacturing: Methods and Applications, D. Braha (ed.), Kluwer Academic Publishers, pp. 311–336, 2001.

Maimon O. and Rokach L., “Improving supervised learning by feature decomposition”, Proceedings of the Second International Symposium on Foundations of Information and Knowledge Systems, Lecture Notes in Computer Science, Springer, pp. 178-196, 2002.

Maimon, O. and Rokach, L., Decomposition Methodology for Knowledge Discovery and Data Mining: Theory and Applications, Series in Machine Perception and Artificial Intelligence - Vol. 61, World Scientific Publishing, ISBN:981-256-079-3, 2005.

Rokach, L., Decomposition methodology for classification tasks: a meta decomposer framework, Pattern Analysis and Applications, 9(2006):257–271.

Rokach L., Genetic algorithm-based feature set partitioning for classification problems,Pattern Recognition, 41(5):1676–1700, 2008.

Rokach L., Mining manufacturing data using genetic algorithm-based feature set decomposition, Int. J. Intelligent Systems Technologies and Applications, 4(1):57-78, 2008.

Rokach L., Maimon O. and Lavi I., Space Decomposition In Data Mining: A Clustering Approach, Proceedings of the 14th International Symposium On Methodologies For Intelligent Systems, Maebashi, Japan, Lecture Notes in Computer Science, Springer-Verlag, 2003, pp. 24–31.

Rokach, L. and Maimon, O. and Averbuch, M., Information Retrieval System for Medical Narrative Reports, Lecture Notes in Artificial intelligence 3055, page 217-228 Springer-Verlag, 2004.

Rokach, L. and Maimon, O. and Arbel, R., Selective voting-getting more for less in sensor fusion, International Journal of Pattern Recognition and Artificial Intelligence 20 (3) (2006), pp. 329–350.

Rokach, L. and Maimon, O., Theory and applications of attribute decomposition, IEEE International Conference on Data Mining, IEEE Computer Society Press, pp. 473–480, 2001.

Rokach L. and Maimon O., Feature Set Decomposition for Decision Trees, Journal of Intelligent Data Analysis, Volume 9, Number 2, 2005b, pp 131–158.

Rokach, L. and Maimon, O., Clustering methods, Data Mining and Knowledge Discovery Handbook, pp. 321–352, 2005, Springer.

Rokach, L. and Maimon, O., Data mining for improving the quality of manufacturing: a feature set decomposition approach, Journal of Intelligent Manufacturing, 17(3):285–299, 2006, Springer.

Rokach, L., Maimon, O., Data Mining with Decision Trees: Theory and Applications, World Scientific Publishing, 2008.

Part VIII

Software

65

Commercial Data Mining Software

Qingyu Zhang and Richard S. Segall

[1] Arkansas State University, Department of Computer and Info. Tech., Jonesboro, AR 72467-0130,USA. qzhang@astate.edu
[2] Arkansas State University, Department of Computer and Info. Tech., Jonesboro, AR 72467-0130,USA. rsegall@astate.edu

Summary. This chapter discusses selected commercial software for data mining, supercomputing data mining, text mining, and web mining. The selected software are compared with their features and also applied to available data sets. The software for data mining are SAS Enterprise Miner, Megaputer PolyAnalyst 5.0, PASW (formerly SPSS Clementine), IBM Intelligent Miner, and BioDiscovery GeneSight. The software for supercomputing are Avizo by Visualization Science Group and JMP Genomics from SAS Institute. The software for text mining are SAS Text Miner and Megaputer PolyAnalyst 5.0. The software for web mining are Megaputer PolyAnalyst and SPSS Clementine . Background on related literature and software are presented. Screen shots of each of the selected software are presented, as are conclusions and future directions.

65.1 Introduction

In the data mining community, there are three basic types of mining: data mining, web mining, and text mining (Zhang and Segall, 2008). In addition, there is a special category called supercomputing data mining, which is today used for high performance data mining and data intensive computing of large and distributed data sets. Much software has been developed for visualization of data intensive computing for use with supercomputers, including that for large-scale parallel data mining.

Data mining primarily deals with structured data. Text mining mostly handles unstructured data/text. Web mining lies in between and copes with semi-structured data and/or unstructured data. The mining process includes preprocessing, patterns analysis, and visualization. To effectively mine data, a software with sufficient functionalities should be used. Currently there are many different software, commercial or free, available on the market. A comprehensive list of mining software is available on web page of KDnuggets (http:// www.kdnuggets.com/software /index.html).

This chapter discusses selected software for data mining, supercomputing data mining, text mining, and web mining that are not available as free open source software. The selected

O. Maimon, L. Rokach (eds.), *Data Mining and Knowledge Discovery Handbook*, 2nd ed., DOI 10.1007/978-0-387-09823-4_65,

software for data mining are SAS Enterprise Miner, Megaputer PolyAnalyst 5.0, PASW (formerly SPSS Clementine), IBM Intelligent Miner, and BioDiscovery GeneSight. The selected software for text mining are SAS Text Miner and Megaputer PolyAnalyst 5.0. The selected software for web mining are Megaputer PolyAnalyst and SPSS Clementine. The software for supercomputing are Avizo by Visualization Science Group and JMP Genomics from SAS Institute. Avizo is 3-D visualization software for scientific and industrial data that can process very large datasets at interactive speed. JMP Genomics from SAS is used for discovering the biological patterns in genomics data.

These software are described and compared as to the existing features and algorithms for each and also applied to different available data sets. Background on related literature and software are also presented. Screen shots of each of the selected software are reported as are conclusions and future directions.

65.2 Literature Review

Data mining is defined by the Data Intelligence Group (1995) as the extraction of hidden predictive information form large databases. According to them, "data mining tools scour databases for hidden patterns, finding predictive information that experts may miss because it lies outside their expectations." According to StatSoft (2006), algorithms are operations or procedures that will produce a particular outcome with a completely defined set of steps or operations. This is opposed to heuristics that are general recommendations or guides based upon theoretical reasoning or statistical evidence such as "data mining can be a useful tool if used appropriately." Data mining and algorithms are widely implemented and rapidly developed (Kim et al., 2008; Nayak, 2008; Segall and Zhang, 2006).

According to Wikipedia (2009), supercomputers or HPC (High Performance Computing) are used for highly calculation-intensive tasks such as problems involving quantum mechanical physics, weather forecasting, global warming, molecular modeling, physical simulations (such as for simulation of airplanes in wind tunnels and simulation of detonation of nuclear weapons). Sanchez (1996) cited the importance of data mining using supercomputers by stating "Data mining with these big, superfast computers is a hot topic in business, medicine and research because data mining means creating new knowledge from vast quantities of information, just like searching for tiny bits of gold in a stream bed". According to Sanchez (1996), The Children's Hospital of Pennsylvania took MRI scans of a child's brain in 17 seconds using supercomputing for that which otherwise normally would require 17 minutes assuming no movement of the patient.

The increasing availability of textual knowledge applications and online textual sources has caused a boost in text mining and web mining research. Hearst (2003) defines text mining as "the discovery of new, previously unknown information, by automatically extracting information from different written sources." He distinguishes text mining from data mining by noting that "in text mining the patterns are extracted from natural language rather than from structured database of facts." Metz (2003) describes text mining as those for that "applications are clever enough to run conceptual searches, locating, say, all the phone numbers and places names buried in a collection of intelligence communiqus." More impressive, the software can identify relationships, patterns, and trends involving words, phrases, numbers, and other data.

Web mining is the application of data mining techniques to discover patterns from the Web and can be classified into three different types of web content mining, web usage mining, and web structure mining (Pabarskaite and Raudys, 2007; Sanchez et al., 2008). Web content mining is the process to discover useful information from the content of a web page that may consist of text, image, audio or video data in the web; web usage mining is the application that uses data mining to analyze and discover interesting patterns of user's usage of data on the web; and web structure mining is the process of using graph theory to analyze the node and connection structure of a web site (Wikipedia, 2007). An example of the latter would be discovering the authorities and hubs of any web document, e.g. identifying the most appropriate web links for a web page.

There is a wealth of software today for data, supercomputing, text and web mining such as presented in American Association for Artificial Intelligence (AAAI) (2002) and Ducatelle (2006) for teaching data mining, Nisbet (2006) for CRM (Customer Relationship Management) and software review of Deshmukah (1997). StatSoft (2006) presents screen shots of several softwares that are used for exploratory data analysis and various data mining techniques. Kim et al. (2008) classify software changes in data mining and Ceccato et al. (2006) combine three mining techniques. Nayak (2008) develops and applies data mining techniques in web services discovery and monitoring.

Davi et al. (2005) review two text mining packages of SAS text mining and Wordstat. Chou et al. (2008) apply text mining approach to Internet abuse detection and Lau et al. (2005) discuss text mining for the hotel industry. Lazarevic et al. (2006) discussed a software system for spatial data analysis and modeling. Leung (2004) compares microarray data mining software. National Center for Biotechnology Information (2006) referred to as NCBI provides tools for data mining including those specifically for each of the following categories of nucleotide sequence analysis, protein sequence analysis and proteomics, genome analysis, and gene expression.

Chang and Lee (2006) find frequent itemsets using online data streams. Pabarskaite and Raudys (2007) review the knowledge discovery process from web log data. Sanchez et al. (2008) integrate software engineering and web mining techniques in the development of an e-commerce recommender system capable of predicting the preferences of its users and present them a personalized catalogue. Ganapathy et al. (2004) discuss visualization strategies and tools for enhancing customer relationship management.

Some applications of supercomputers for data mining include that of Davies (2007) using Internet distributed supercomputers, Seigle (2002) for CIA/FBI, Mesrobian et al. (1995) for real time data mining, and Curry et al. (2007) for detecting changes in large data sets of payment card data. DMMGS06 conducted a workshop on data mining and management on the grid and supercomputers in Nottingham, UK. Grossman (2007) wrote a survey of high performance and distributed data mining. Sekijima (2007) studied the application of HPC to analysis of disease related protein.

65.3 Data Mining Software

The research is to compare the five selected software for data mining including SAS Enterprise Miner, Megaputer PolyAnalyst 5.0, PASW Modeler/ formerly SPSS Clementine, IBM Intelligent Miner, and BioDiscovery GeneSight. The data mining algorithms to be performed include those for neural networks, genetic algorithms, clustering, and decision trees. As can be visualized from Table 1, SAS Enterprise Miner , PolyAnalyst 5, PASW, and IBM Intelligent Miner offer more algorithms than GeneSight.

Table 65.1. Data Mining Software

ALGORITHMS	GeneSight	PolyAnalyst	SAS Enterprise Miner	PASW Modeler/ SPSS Clementine	IBM Intelligent Miner
Statistical Analysis	x	x	x	x	x
Neural Networks		x	x	x(add on)	x
Decision Trees		x	x		x
Regression Analysis		x	x	x	x
Cluster Analysis	x	x	x	x	x
Self-Organizing Map (SOM)			x	x	
Link/Association Analysis		x	x	x	x

65.3.1 BioDiscovery GeneSight

GeneSight is a product of BioDiscovery, Inc. of El Segundo, CA that focuses on cluster analysis using two main techniques of hierarchical and partitioning for data mining of microarray gene expressions.

Figure 1 shows the k-means clustering of global variations using the Pearson correlation. This can also be done by self-organizing map (SOM) clustering using the Euclidean distance metric for the first three variables of aspect, slope and elevation. Figure 2 shows the two-dimensional self-organizing map (SOM) for the eleven variables for all of the data using the Chebychev distance metric.

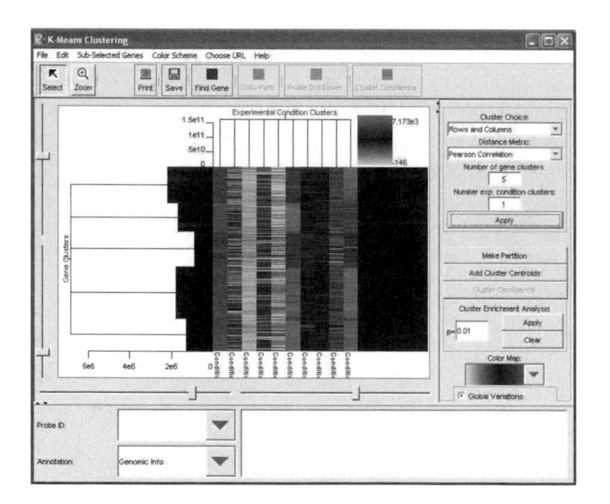

Fig. 65.1. K-means clustering of global variations with the Pearson correlation using GeneSight

65.3.2 Megaputer PolyAnalyst 5.0

PolyAnalyst 5 is a product of Megaputer Intelligence, Inc. of Bloomington, IN and contains sixteen (16) advanced knowledge discovery algorithms.

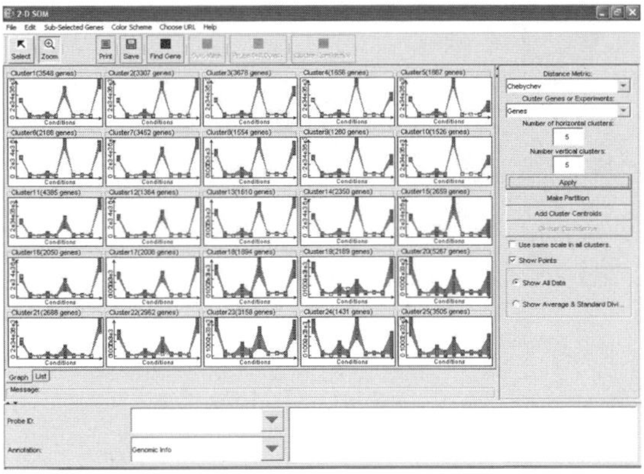

Fig. 65.2. Self-organizing map (SOM) with the Chebychev distance metric using GeneSight

Figure 3 shows input data window for the forest cover type data in PolyAnalyst 5.0. The link diagram given by Figure 4, illustrates for each of the six (6) forest cover types for each of the 5 elevations present for each of the 40 soil types. Figure 5 provides the bin selection rule for the variable of selection. The Decision Tree Report indicates a classification probability of 80.19% with a total classification error of 19.81%. Per PolyAnalyst output the decision tree has a tree depth of 100 with 210 leaves, and a depth of constructed tree of 16, and a classification efficiency of 47.52%.

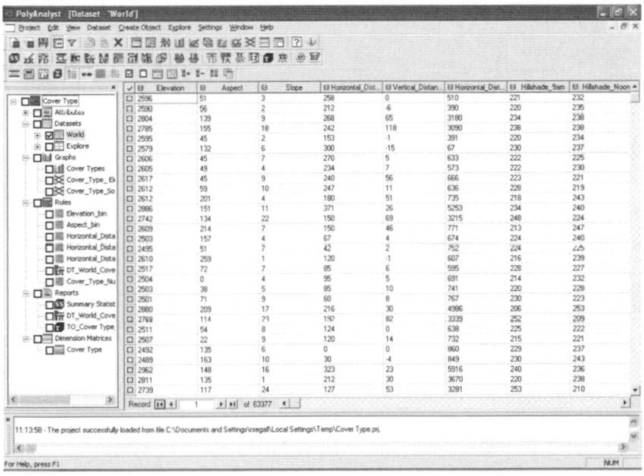

Fig. 65.3. Input data window for the forest cover type data in PolyAnalyst 5.0

65.3.3 SAS Enterprise Miner

SAS Enterprise Miner is a product of SAS Institute Inc. of Cary, NC and is based on the SEMMA approach that is the process of Sampling (S), Exploring (E), Modifying (M), Modeling (M), and Assessing (A) large amounts of data. SAS Enterprise Miner utilizes a workspace with a drop-and-drag of icons approach to constructing data mining models. SAS Enterprise Miner utilizes algorithms for decision trees, regression, neural networks, cluster analysis, and association and sequence analysis.

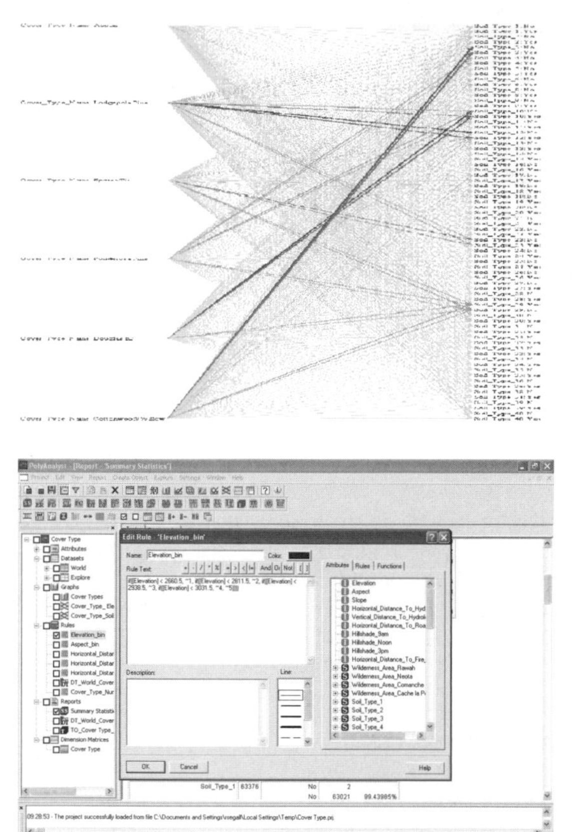

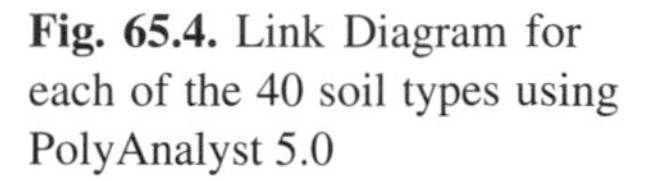

Fig. 65.4. Link Diagram for each of the 40 soil types using PolyAnalyst 5.0

Fig. 65.5. Bin selection rule for the forest cover type data using PolyAnalyst 5.0

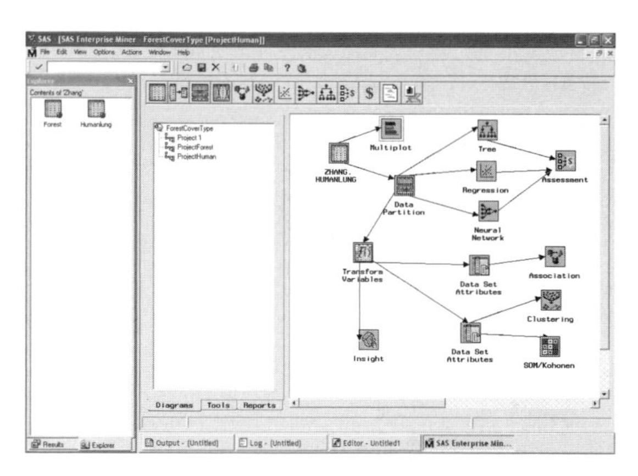

Fig. 65.6. Workspace of SAS Enterprise Miner for human lung project

Figure 6 shows the workspace of SAS Enterprise Miner that was used in the data mining of the human lung dataset. Figure 7 shows a partial view of the decision tree diagram obtained by data mining using SAS Enterprise Miner as specified for a depth of 6 from the initial node of NL279. Figure 8 shows a 2x3 Self-Organized Maps (SOM) that provides results in the form

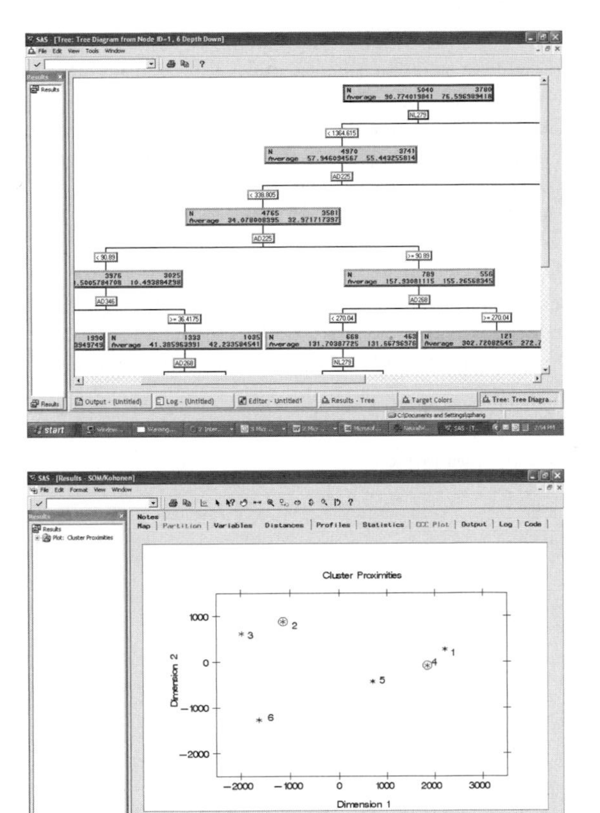

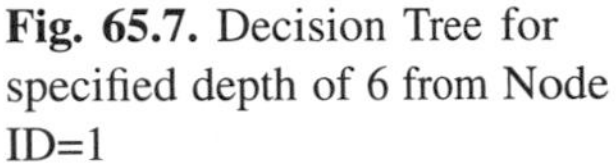

Fig. 65.7. Decision Tree for specified depth of 6 from Node ID=1

Fig. 65.8. SOM two-dimensional cluster proximities for the forest cover data

of an interactive map that illustrates the characteristics of the clusters and importance of each variable. Especially it shows the normalized means for the clusters of the variables and cluster proximities respectively. Figure 8 shows that the normalized means for the cluster proximities of the gene type variables are scattered and not uniform.

65.3.4 PASW Modeler/ Formerly SPSS Clementine

PASW (Predictive Analytics Software) Modeler (formerly Clementine) makes it easy to discover insights in your data with a simple graphical interface. The PASW base model performs decision lists, factor and principle component analysis, linear regression, CHAID (Chi-squared Automatic Interaction Detector) which is one of the oldest tree classification methods. Add-on modules to the PASW base model are available for neural networks, time series forecast models, clustering, and sequential association algorithm, binomial and multinomial logistic regression, and anomaly detection.

Some of the key features of PASW Modeler include the ability to incorporate all types of data including structured (tabular), unstructured (textual), web site, and survey data. PASW Modeler has several data-cleaning options and also visual link analysis for revealing meaningful association in the data.

PASW Modeler is able to perform data mining within existing databases and score millions of records in a matter of minutes without additional hardware requirements. Through multithreading, clustering, embedded algorithms, and other techniques, the user of PASW Modeler can conserve resources and control information technology costs while delivering results faster. PASW Modeler offers many features that provide faster and greater return on your analytical investment. Automated modeling, for example, helps you quickly identify the best performing models and combine multiple predictions for the most accurate results (SPSS 2009a).

Figure 9 shows the workspace of PASW Modeler with the result file that can be written to a database, an Excel spreadsheet, or other. Figure 10 is a screen of PASW Modeler for selection of the fields to be included and excluded for the cluster analysis. Figure 11 is a screen of PASW Modeler where on the left is the available clusters for selection for performing cluster comparisons as shown on the right.

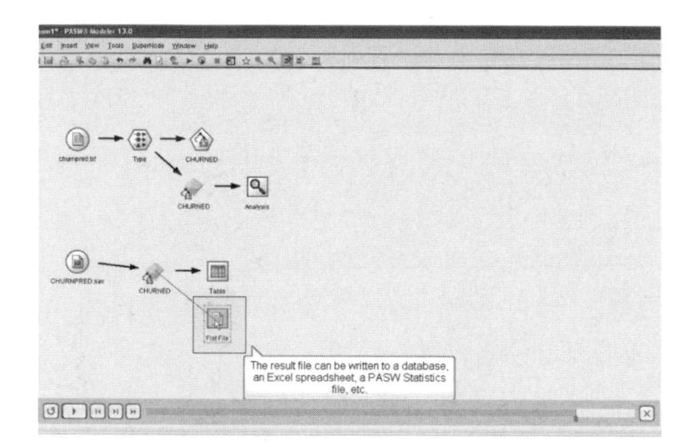

Fig. 65.9. Workspace of PASW Modeler (SPSS 2009a)

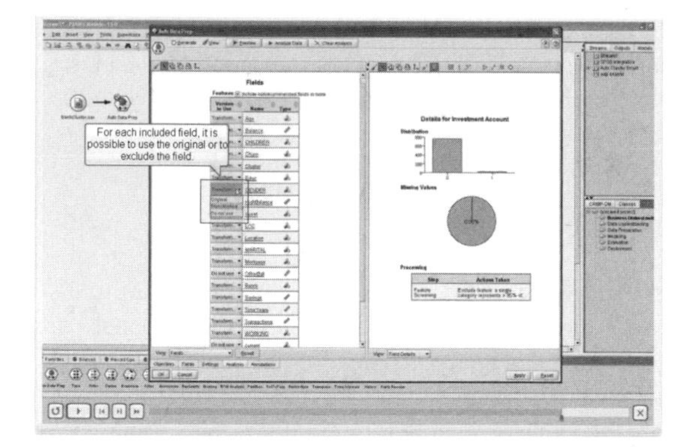

Fig. 65.10. Screen of PASW Modeler for selection of the fields to be included and excluded for the cluster analysis (SPSS 2009b)

65.3.5 IBM DB2 Intelligent Miner

IBM DB2 Intelligent Miner for Data performs mining functions against traditional DB2 databases or flat files. IBM's data mining capabilities help you detect fraud, segment your customers, and

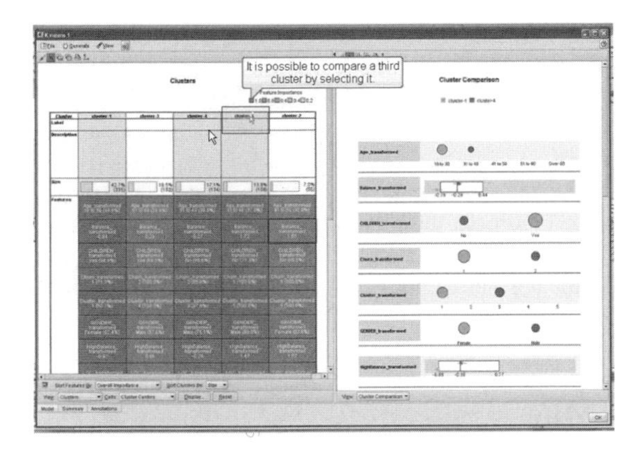

Fig. 65.11. Screen of PASW Modeler where on the left are the available clusters for selection for performing Cluster comparisons as shown on the right. (SPSS 2009b)

simplify market basket analysis. IBM's in-database mining capabilities integrate with your existing systems to provide scalable, high performing predictive analysis without moving your data into proprietary data mining platforms. It also has capabilities to access data in other re-

Fig. 65.12. The Graphics View of the Clustering Visualize (IBM 2004)

Fig. 65.13. The Tree Node Distribution View of the Classification Visualizer (IBM 2004)

lational Database Management Systems (DBMSs) using ODBC (Open Database Connectivity Standard). IBM Intelligent Miner performs functions of association rules, clustering, prediction, sequential patterns, and time series. IBM Intelligent Miner for Text performs mining activities against textual data, including e-mail and Web pages.

In April 2008, IBM D2 Intellgent Miner was replaced with IBM InfoSphere Warehouse 9.7 that incorporates Intelligent Miner Modeling and Text Analytics within the Departmental and Enterprise full versions.

Figure 12 shows the graphics view for an overview of the relevant clusters, their size, and their fields. Each row in the table of Figure 12 describes one cluster by showing how the field values are distributed for the records in this cluster. Figure 13 shows the Tree Node Distribution View that includes the tree, the appending node IDs, and the distribution of field values in the nodes.

65.4 Supercomputing Data Mining Software

Table 65.2. Supercomputing Data Mining Software

Features		Avizo	JMP
Data Acquisition	Data Import	x	x
	Image segmentation	x	x
Data Analysis	Slicing and clipping	x	
	Analyze large microarrays		x
	Surface rendering	x	x
	Volume rendering	x	x
	Scaler and vector visualization	x	x
	Molecular data support	x	x
	Matlab bridge	x	
	Geometric models	x	x
	Surface reconstruction	x	x
	Geometric models	x	x
Results Reporting	Visual presentation	x	x
	Scripting	x	
Unique features		Special editions for Avizo earth, wind, fire, green	For genetic data

Supercomputing data mining is used for highly calculation intensive tasks such as problems involving quantum mechanical physics, weather forecasting, molecular modeling, and physical simulations. The selected software for supercomputing are Avizo by Visualization Science Group and JMP Genomics from SAS Institute. Shown in Table 2, Avizo is a general supercomputing software with unique editions specifically for earth, wind, fire, and environmental data while JMP Genomics is specialized specifically for genetic data.

65.4.1 Data Visualization using Avizo

Avizo software is a powerful, multifaceted tool for visualizing, manipulating, and understanding scientific and industrial data. Wherever three-dimensional datasets need to be processed,

Avizo offers a comprehensive feature set within an intuitive workflow and easy-to-use graphical user interface (VSG, 2009).

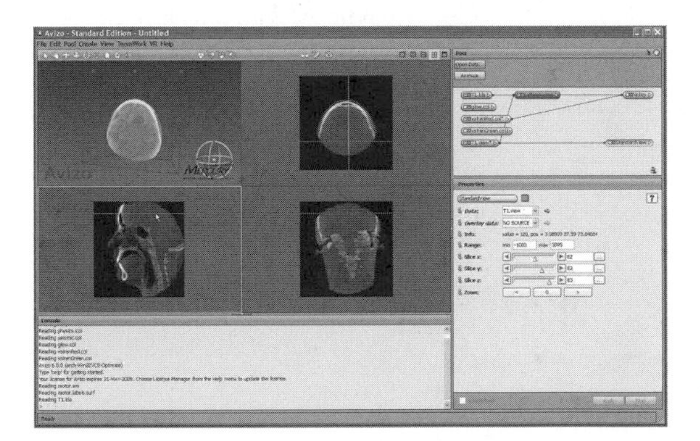

Fig. 65.14. Four views of a human skull in single screen of 3D visualization by Avizo

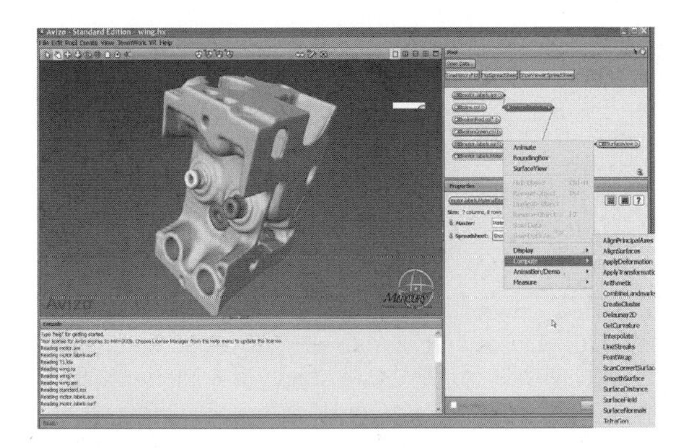

Fig. 65.15. Avizo workspace for the 3D visualization of a motor in a front view

Fig. 65.16. The atomic structure of 14 atoms having 13 bonds and 1 residue

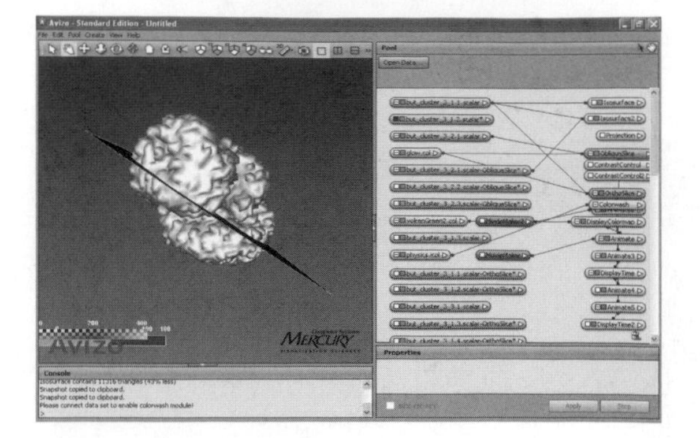

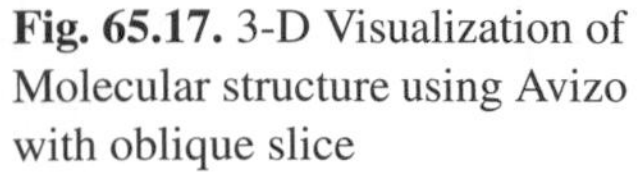

Fig. 65.17. 3-D Visualization of Molecular structure using Avizo with oblique slice

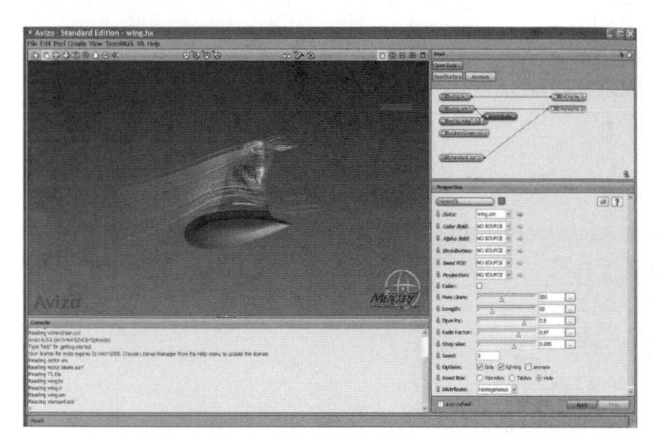

Fig. 65.18. The airflow around a wing component such as found on aircraft

Some of the core features of Avizo include advanced 3D visualization by surface and volume rendering, scientific visualization of flow data and processing very large datasets at interactive speed, and 3D data exploration and analysis by displaying single or multiple datasets in a single or multiple viewer window, and navigate freely or around or through these objects. Avizo can also perform 3D reconstruction by employing innovative and robust algorithms from image processing and computational geometry to reconstruct high resolution 3D images generated by CT or MRI scanners, 3D ultrasonic devices, or confocal microscopes (VSG, 2009).

Avizo software is used for supercomputing data mining as shown in this paper using the standard edition. Figure 14 shows four views of 3D visualization of a human skull in single screen by Avizo. Figure 15 show Avizo workspace for the 3D visualization of a component of a motor in a front view. Figure 16 shows the atomic structure of 14 atoms having 13 bonds and 1 residue. Fig. 17 shows 3-D Visualization of Molecular structure using Avizo with oblique slice. Figure 18 shows the airflow around a wing component such as found on aircraft or other.

65.4.2 Data Visualization using JMP Genomics

JMP Genomics is statistical discovery software that can uncover meaningful patterns in high throughput genomics and proteomics data. JMP Genomics is designed for biologists, biostatisticians, statistical geneticists, and those engaged in analyzing the vast stores of data that are common in genomic research (SAS, 2009).

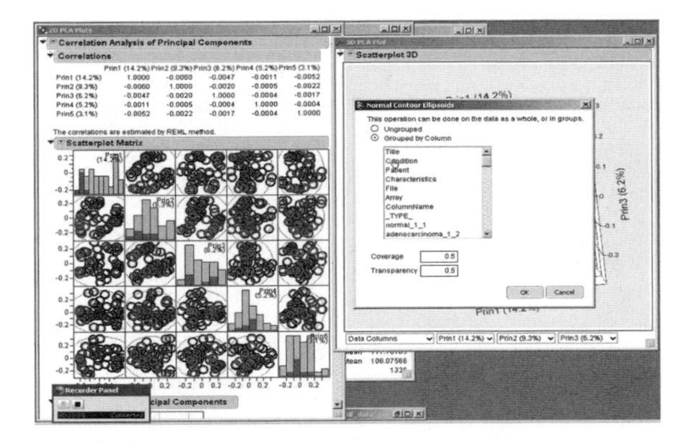

Fig. 65.19. Correlation of 5 principal components with respective scatterplot matrices

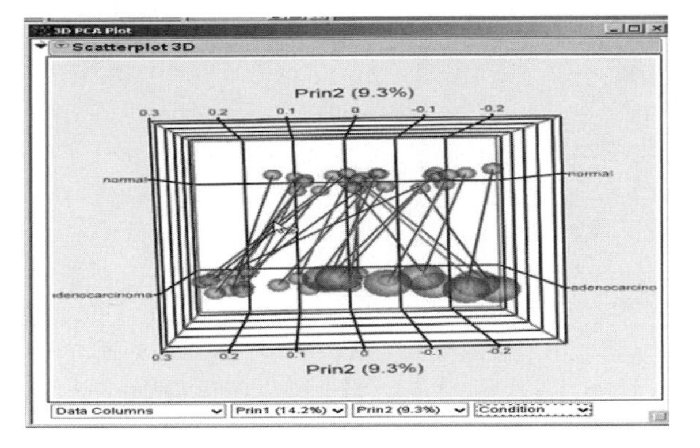

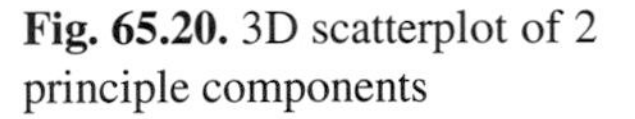
Fig. 65.20. 3D scatterplot of 2 principle components

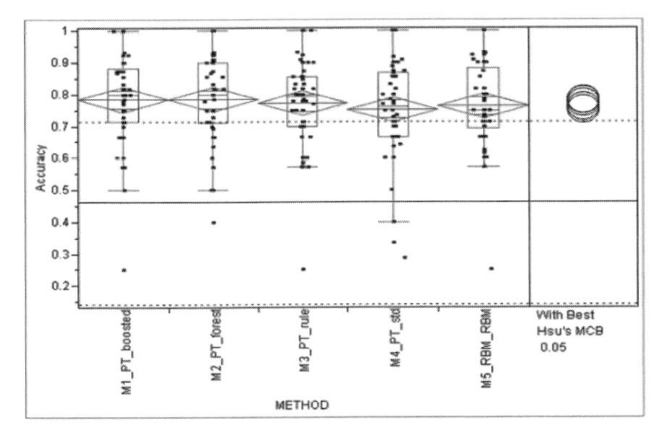

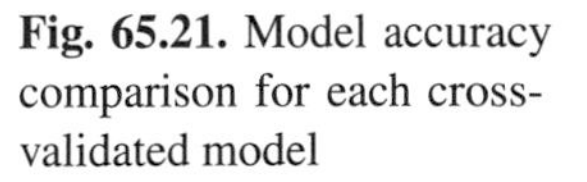
Fig. 65.21. Model accuracy comparison for each cross-validated model

Using data of characteristics for patients diagnosed with adenocarcinoma cancer, Figure 19 shows correlation analysis of 5 principal components for correlations with their respective scatterplot matrices presented by grouping by columns. Fig. 20 shows 3D scatterplot of 2 principle components. Figure 21 shows model accuracy comparison for each cross-validated model after 50 iterations for the dependent variable of grade. Figure 22 shows almost equal distributions of data type for training data for "true_grade" variable, with the corresponding actual probabilities, quartiles, and correct prediction frequencies. Figure 22 also shows the

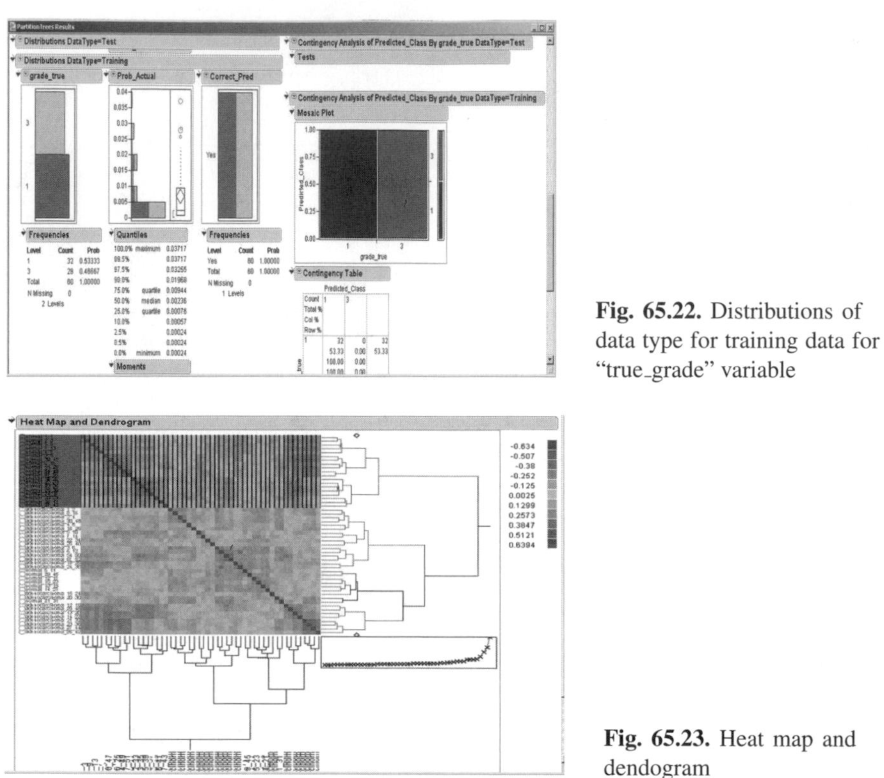

Fig. 65.22. Distributions of data type for training data for "true_grade" variable

Fig. 65.23. Heat map and dendogram

contingency analysis of predicted class by grade_true data type of training as shown in the mosaic plot and contingency table. Figure 23 shows a complete heat map and dendrogram.

65.5 Text Mining Software

Some of the popular software currently available for text mining include SAS Text Miner and Megaputer PolyAnalyst. Both software provide a variety of graphical views and analysis tools with powerful capabilities to discover knowledge from text databases (shown in Table 3). The main focus here is to compare, discuss, and provide sample output for each as visual comparisons. As a visual comparison of the features for both selected text mining software, the authors of this chapter constructed Table 3, where essential functions are indicated as being either present or absent with regard to data preparation, data analysis, results reporting, and unique features. As Table 3 shows, both Megaputer PolyAnalyst and SAS Text Miner have extensive text mining capabilities.

65.5.1 SAS Text Miner

SAS Text Miner is actually an "add-on" to SAS Enterprise Miner with the inclusion of an extra icon in the "Explore" section of the tool bar (Woodfield, 2004). SAS Text Miner performs

Table 65.3. Text Mining Software

Features		SAS Text Miner	Megaputer PolyAnalyst
Data Preparation Preparation	Text parsing and extraction	x	x
	Define dictionary		x
	Automatic Text Cleaning	x	
Data Analysis	Categorization		x
	Filtering		x
	Concept Linking	x	x
	Text Clustering	x	x
	Dimension reduction techniques	x	x
Results Reporting	Interactive Results Window	x	x
	Support for multiple languages	x	x

simple statistical analysis, exploratory analysis of textual data, clustering, and predictive modeling of textual data.

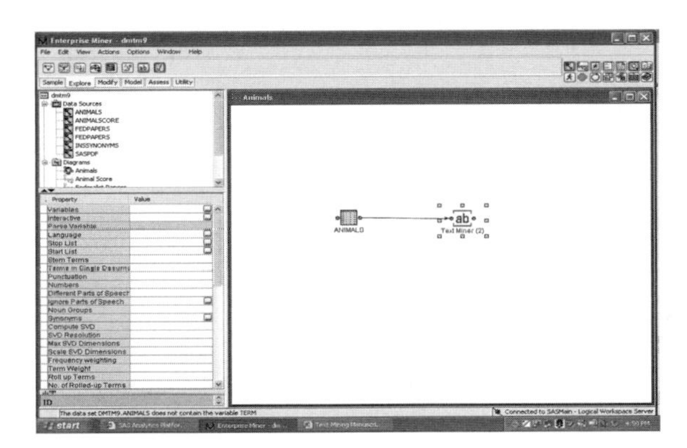

Fig. 65.24. Workspace of SAS Text Miner for Animal Text

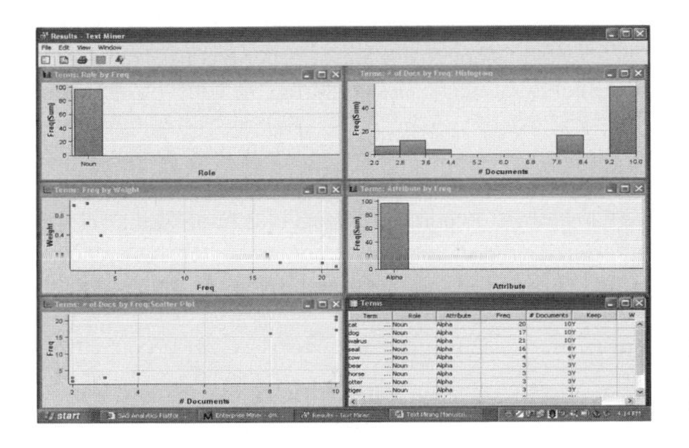

Fig. 65.25. Interactive Window of SAS Text Miner for Animal Text

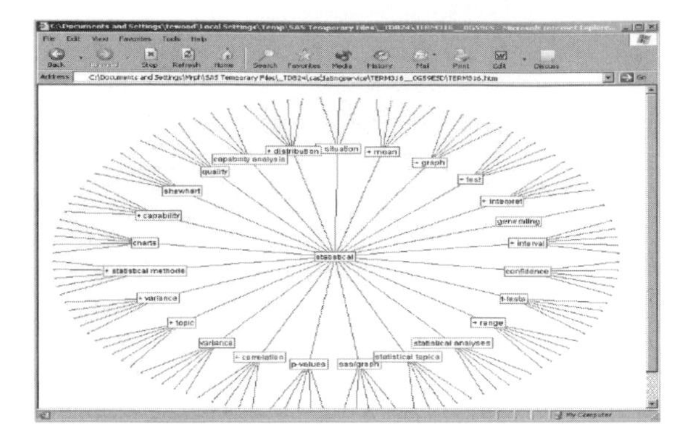

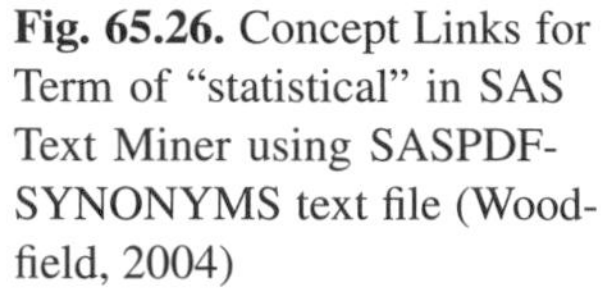

Fig. 65.26. Concept Links for Term of "statistical" in SAS Text Miner using SASPDF-SYNONYMS text file (Woodfield, 2004)

SAS Text Miner uses the "drag-and-drop" principle by dragging the selected icon in the tool set to dropping it into the workspace. The workspace of SAS Text Miner was constructed with a data icon of selected animal data that was provided by SAS in their Instructor's Trainer Kit as shown in Figure 24. Figure 25 shows the results of using SAS Text Miner with individual plots for "role by frequency", "number of documents by frequency", "frequency by weight", "attribute by frequency", and "number of documents by frequency scatter plot." Figure 26 shows "Concept Linking Figure" as generated by SAS Text Miner using SASPDF-SYNONYMS text file.

65.5.2 Megaputer PolyAnalyst

Previous work by the authors Segall and Zhang (2006) have utilized Megaputer PolyAnalyst for data mining. The new release of PolyAnalyst version 6.0 includes text mining and specifically new features for text OLAP (on-line analytical processing) and taxonomy based categorization which is useful for when dealing with large collections of unstructured documents as discussed in Megaputer Intelligence Inc. (2007). The latter cites that taxonomy based classifications are useful when dealing with large collections of unstructured documents such as tracking the number of known issues in product repair notes and customer support letters.

According to Megaputer Intelligence Inc. (2007), PolyAnalyst "provides simple means for creating, importing, and managing taxonomies, and carries out automated categorization of text records against existing taxonomies." Megaputer Intelligence Inc. (2007) provides examples of applications to executives, customer support specialists, and analysts. According to Megaputer Intelligence Inc. (2007), "executives are able to make better business decisions upon viewing a concise report on the distribution of tracked issues during the latest observation period".

This chapter provides several figures of actual screen shots of Megaputer PolyAnalyst version 6.0 for text mining. These are Figure 27 for workspace of text mining of Megaputer PolyAnalyst, Figure 28 is "Suffix Tree Clustering" Report for the text cluster of (desk; front), and Figure 29 is screen shot of "Link Term" Report of hotel customer survey text. Megaputer PolyAnalyst can also provide screen shots with drill-down text analysis and histogram plot of text analysis.

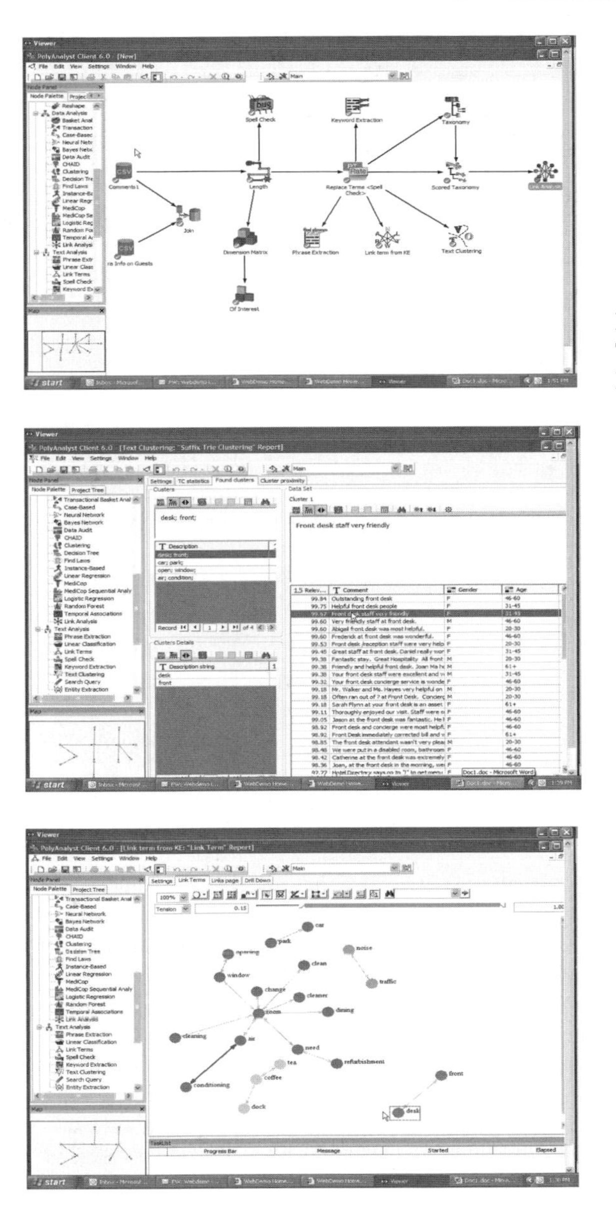

Fig. 65.27. Workspace for Text Mining in Megaputer PolyAnalyst

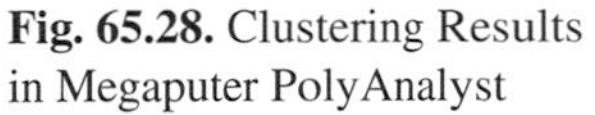

Fig. 65.28. Clustering Results in Megaputer PolyAnalyst

Fig. 65.29. Link Term Report using Text Analysis in Megaputer PolyAnalyst

65.6 Web Mining Software

Two selected software are reviewed and compared in terms of data preparation, data analysis, and results reporting (see Table 4). As shown in the table below, Megaputer PolyAnalyst has unique feature of data and text mining tool integrated with web site data source input, while SPSS Clementine has linguistic approach rather than statistics based approach, Table 4 gives a visual interpretation of the differences and similarities among both selected software as shown below.

Table 65.4. Web Mining Software

Features		Megaputer PolyAnalyst	SPSS Clementine
Data	Data extraction	x (web site as data source input)	Import server files
Preparation	Automatic Data Cleaning	x	x
	user segmentation	x	x
	Detect users' sequences		x
Data	Understand product and content affinities (link analysis)	x	x
Analysis	Predict user propensity to convert, buy, or churn		x
	Navigation report		x
	Keyword and Search Engine	x	x
Results	Interactive Results Window		x
Reporting	Support for multiple languages	x	x
	Visual presentation	x	x
Unique features		Data and text mining tool integrated with web site data source input	Linguistic approach rather than statistics based approach

65.6.1 Megaputer PolyAnalyst

Megaputer PolyAnalyst is an enterprise analytical system that integrates Web mining together with data and text mining because it does not have a separate module for Web mining. Web pages or sites can be inputted directly to Megaputer PolyAnlayst as data source nodes.

Megaputer PolyAnlayst has the standard data and text mining functionalities such as Categorization, Clustering, Prediction, Link Analysis, Keyword and entity extraction, Pattern discovery, and Anomaly detection. These different functional nodes can be directly connected to the web data source node for performing web mining analysis. Megaputer PolyAnalyst user interface allows the user to develop complex data analysis scenarios without loading data in the system, thus saving analyst's time. According to Megaputer (2007), whatever data sources are used, PolyAnalyst provides means for loading and integrating these data. PolyAnalyst can load data from disparate data sources including all popular databases, statistical, and spreadsheet systems. In addition, it can load collections of documents in html, doc, pdf and txt formats, as well as load data from an internet web source. PolyAnalyst offers visual "on-the-fly integration" and merging of data coming from disparate sources to create data marts for further analysis. It supports incremental data appending and referencing data sets in previously created PolyAnalyst projects.

Figures 30-32 are screen shots illustrating the applications of Megaputer PolyAnalyst for web mining to available data sets. Figure 30 shows an expanded view of PolyAnalyst workspace. Figure 31 shows screen shot of PolyAnalyst using website of Arkansas State University (ASU) as the web data source. Figure 32 shows a keyword extraction report from a web page of undergraduate admission of website of Arkansas State University (ASU).

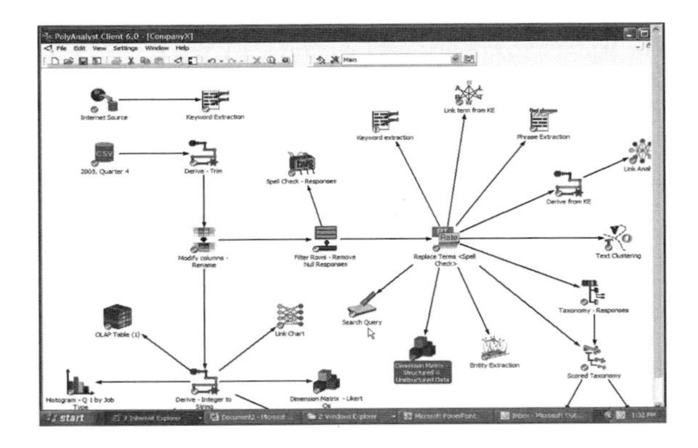

Fig. 65.30. PolyAnalyst workspace with Internet data source

PolyAnalyst Client 6.0 - [CompanyX: Internet Source]

File Edit View Settings Window Help

Data | Statistics | Distinct

<?xml version="1.0" encoding="ISO-8859-1"?>
Arkansas State University - Powering Minds
Mascot Review
R26;
Home
R26;
Review Committee
R26;
Materials
Mascot Information
Dr. Robert Potts, chancellor of Arkansas State University-Jonesboro has appointed a broadly-based committee of constituent representatives to study the Indian mascot and nickname issue. Jim Pickens, immediate past chair of the ASU Board of Trustees, is chairing the committee which will meet over the summer and develop a recommendation to submit to Dr. Potts. Materials submitted to the committee will be posted on this web page for the public to review. If you would like to make a comment to be presented to the committee please direct your message to mascot@astate.edu.
This page is maintained by: ITS Web Team
Please e-mail problems, comments, and suggestions to: mascot@astate.edu

URL	MIME Type	Date Modified	Language	Code Page	Text Content
http://admissions.astate.edu	text/html	6/8/2007 6:12:28 PM	English	Western European (	Arkansas State University - Undergraduat
http://www.astate.edu/./sections/artsandculture/tickets/	text/html	6/8/2007 6:12:28 PM	English	Western European (	<?xml version="1.0" encoding="ISO-8859
http://www.astate.edu/./sections/directories/	text/html	6/8/2007 6:12:28 PM	English	Western European (	<?xml version="1.0" encoding="ISO-8859
http://www.astate.edu/./sections/athletics/	text/html	6/8/2007 6:12:28 PM	English	Western European (	<?xml version="1.0" encoding="ISO-8859
http://www.astate.edu/./sections/employment/	text/html	6/8/2007 6:12:29 PM	English	Western European (	<?xml version="1.0" encoding="ISO-8859
http://admissions.astate.edu/application_options.shtml	text/html	6/8/2007 6:12:29 PM	English	Western European (	n>i Arkansas State University - Undergrad
http://asunews.astate.edu/SummerChildTheatre07.htm	text/html	6/8/2007 6:12:29 PM	English	Western European (	News from Arkansas State University Uni
http://apps.astate.edu/calendar/	text/html	6/8/2007 6:12:29 PM	English	Western European (	<?xml version="1.0" encoding="ISO-8859
http://www.astate.edu/./sections/sitemap/	text/html	6/8/2007 6:12:29 PM	English	Western European (	<?xml version="1.0" encoding="ISO-8859
http://asunews.astate.edu/newspage.htm	text/html	6/8/2007 6:12:30 PM	English	Western European (	
http://its.astate.edu/aboutits/computingresources.php	text/html	6/8/2007 6:12:31 PM	English	Western European (	ITS Mission Statement Students | Future
http://connectasu.astate.edu/pub-session.html?ocs=7806	text/html	6/8/2007 6:12:31 PM	English	Western European (	How to Enable Cookies This site requires
http://studentaffairs.astate.edu/nso/orientation.htm	text/html	6/8/2007 6:12:31 PM	English	Western European (	Arkansas State University - New Student
http://asunews.astate.edu/PresDeansLists07spring.htm	text/html	6/8/2007 6:12:31 PM	English	Western European (	
http://www.astate.edu/mascot/	text/html	6/8/2007 6:12:31 PM	English	Western European (	<?xml version="1.0" encoding="ISO-8859
http://server.iad.liveperson.net/hc/36307506/?cmd=file&f	text/html	6/8/2007 6:12:32 PM	English	Western European (	
http://www.astate.edu/sections/directories/	text/html	6/8/2007 6:12:32 PM	English	Western European (	<?xml version="1.0" encoding="ISO-8859
http://www.astate.edu/vcaarsearch/	text/html	6/8/2007 6:12:33 PM	English	Western European (	<?xml version="1.0" encoding="ISO-8859
http://www.astate.edu/sections/campusmap/	text/html	6/8/2007 6:12:33 PM	English	Western European (	<?xml version="1.0" encoding="ISO-8859
http://www.astate.edu/sections/admissions/financialaid/	text/html	6/8/2007 6:12:33 PM	English	Western European (	<?xml version="1.0" encoding="ISO-8859

Fig. 65.31. PolyAnalyst using www.astate.edu as web data source

PolyAnalyst Client 6.0 - [Keyword Extraction: "Keyword Extraction" Report]

File Edit View Settings Window Help

Settings | Keywords

Term

Term	1.5 Signi...	Suppo...	Fr...
d	81.1	1	1,045
dl	79.6	1	489
craighead	80.4	1	374
jonesboro	80.0	7	331
greene	78.5	1	155
paragould	78.2	2	136
asu	77.3	17	99
arkansas	74.8	28	87
university	72.9	28	79
student	72.1	17	73
mo	73.6	1	72
mississippi	73.2	2	71
home	67.8	24	67
state	67.1	28	61
pulaski	75.9	1	60
poinsett	75.6	1	55
city	63.9	3	52
crittenden	75.4	1	51
faculty	71.1	13	47
[illegible]	75.0	1	46
lawrence	74.6	1	45
pocahontas	72.0	1	43
information	68.7	16	42
college	69.2	10	41
center	70.4	9	41
rock	67.9	2	41
spring	67.1	4	39
concert	64.7	3	38
mountain	65.0	2	38
jackson	69.4	1	37
little	53.6	2	37
mc	71.2	1	36

Data set

1.5 Relev...	Text Content	Title
98.4	<?xml version="1.0" encoding="ISO-8859	
97.1		
95.1	<?xml version="1.0" encoding="ISO-8859	
94.9	<?xml version="1.0" encoding="ISO-8859	
94.9	<?xml version="1.0" encoding="ISO-8859	
92.6	<?xml version="1.0" encoding="ISO-8859	
91.6	<?xml version="1.0" encoding="ISO-8859	
91.6	<?xml version="1.0" encoding="ISO-8859	
90.2	ITS Mission Statement Students | Future	
87.1	<?xml version="1.0" encoding="ISO-8859	
74.6	News from Arkansas State University Uni	
74.2	<?xml version="1.0" encoding="ISO-8859	

Fig. 65.32. Keyword extraction report

65.6.2 SPSS Clementine

"Web Mining for Clementine is an add-on module that makes it easy for analysts to perform ad hoc predictive Web analysis within Clementine's intuitive visual workflow interface." Web Mining for Clementine combines both Web analytics and data mining with SPSS analytical capabilities to transform raw Web data into "actionable insights". It enables business decision makers to take more effective actions in real time. SPSS (2007) claims examples of automatically discovering user segments, detecting the most significant sequences, understanding product and content affinities, and predicting user intention to convert, buy, or churn.

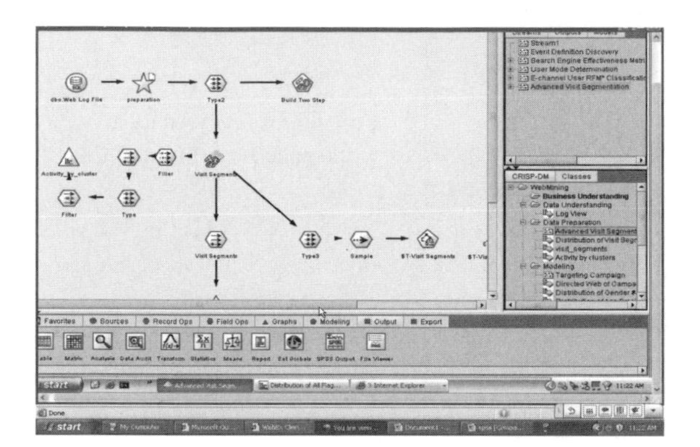

Fig. 65.33. SPSS Clementine workspace

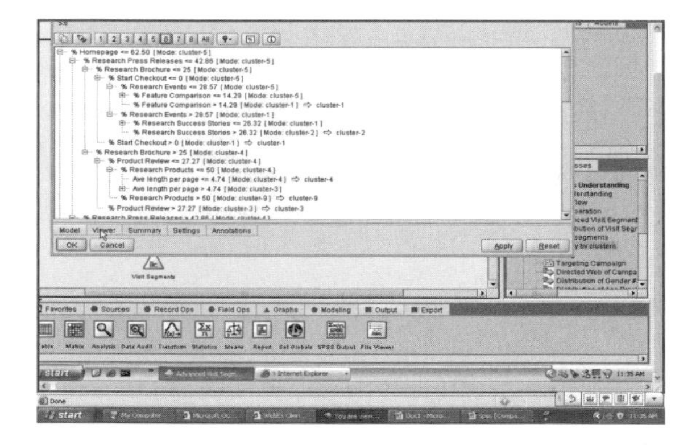

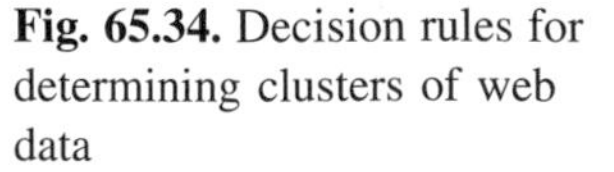

Fig. 65.34. Decision rules for determining clusters of web data

SPSS (2007) claims four key data mining capabilities: segmentation, sequence detection, affinity analysis, and propensity modeling. Specifically, SPSS (2007) indicates six Web analysis application modules within SPSS Clementine that are: search engine optimization, automated user and visit segmentation, Web site activity and user behavior analysis, home page activity, activity sequence analysis, and propensity analysis.

Unlike other platforms used for Web mining that provide only simple frequency counts (e.g., number of visits, ad hits, top pages, total purchase visits, and top click streams), SPSS (2007) Clementine provides more meaningful customer intelligence such as: likelihood to

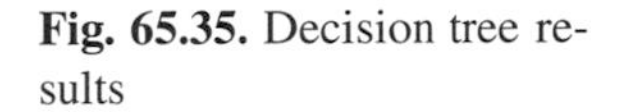

Fig. 65.35. Decision tree results

convert by individual visitor, likelihood to respond by individual prospect, content clusters by customer value, missed crossed-sell opportunities, and event sequences by outcome.

Figures 33-35 are screen shots illustrating the applications of SPSS Clementine for web mining to available data sets. Figure 33 shows the SPSS Clementine workspace. Different user modes can be defined including research mode, shopping mode, search mode, evaluation mode, and so on. Decision rules for determining clusters of web data are demonstrated in Figure 34. Figure 35 exhibits decision tree results with classifiers using different model types (e.g., CHAID, logistic, neural).

65.7 Conclusion and Future Research

The conclusions of this research include the fact that each of the software selected for this research has its own unique characteristics and properties that can be displayed when applied to the available data sets. As indicated, each software has it own set of algorithm types to which it can be applied.

Comparing five data mining software, Biodiscovery GeneSight focuses on cluster analysis and is able to provide a variety of data mining visualization charts and colors. BioDiscovery GeneSight have less data mining functions than the other four do. SAS Enterprise Miner, Megaputer PolyAnalyst, PASW, and IBM Intelligent Miner employ each of the same algorithms as illustrated in Table 1 except that SAS has a separate software SAS Text Miner for text analysis. The regression results are comparable for those obtained using these software. The cluster analysis results for SAS Enterprise Miner, Biodiscovery GeneSight, and Megaputer PolyAnalyst each are unique to each software as to how they represent their results. In conclusion, SAS Enterprise Miner, Megaputer PolyAnalyst, PASW, and IBM Intelligent Miner offer the greatest diversification of data mining algorithms.

This chapter has discussed commercial data mining software that is applicable to supercomputing for 3-D visualization and very large microarray databases. Specifically it illustrated the applications of supercomputing for data visualization using two selected software of Avizo and JMP Genomics. Avizo is a general supercomputing software and JMP Genomics is a special software for genetic data. Supercomputing data mining for 3-D visualization with Avizo is applied to diverse applications such as the human skull for medical research, and the atomic structure that can be used for multipurpose applications such as chemical or nuclear. We have also presented, using JMP Genomics, the data distributions of condition, patient, frequencies, and characteristics for patient data of adenocarcinoma cancer. The figures of this chapter illustrate the level of visualization that is able to be provided by these two softwares.

Comparing two text mining software, both Megaputer PolyAnalyst, and SAS Text Miner have extensive text mining capabilities. SAS Text Miner is an add-on to base SAS Enterprise Miner by inserting an additional Text Miner icon on the SAS Enterprise Miner workspace toolbar. SAS Text Miner tags parts of speech and performs transformations such as those using Singular Value Decompositions (SVD) to generate term-document frequency matrix for viewing in the Text Miner node. Megaputer PolyAnalyst similarly is a software that combines both data mining and text mining, but also includes web mining capabilities. Megaputer also has standalone Text Analyst software for text mining.

Regarding web mining software, PolyAnalyst can mine web data integrated within a data mining enterprise analytical system and provide visual tools such as link analysis of the critical terms of the text. SPSS Clementine can be used for graphical illustrations of customer web activities as well as also for link analysis of different data categories such as campaign, age, gender, and income. The selection of appropriate web mining software should be based on both its available web mining technologies and also the type of data to be encountered.

The future direction of the research is to investigate other data, text, web, and supercomputing mining software for analyzing various types of data and making comparisons of the capabilities of these software between and among each other. This future research would also include the acquisition of other data sets to perform these new analyses and comparisons.

Acknowledgement. The authors would like to acknowledge the support provided by a 2009 Summer Faculty Research Grant as awarded to them by the College of Business of Arkansas State University without whose program and support this work cannot be done. The authors also want to acknowledge each of the software manufactures for their support of this research.

References

AAAI (2002), American Association for Artificial Intelligence (AAAI) Spring Symposium on Information Refinement and Revision for Decision Making: Modeling for Diagnostics, Prognostics, and Prediction, Software and Data, retrieved from `http://www.cs.rpi.edu/~goebel/ss02/software-and-data.html`.

Ceccato, M., M. Marin, K. Mens, L. Moonen, et al., (2006), Applying and combining three different aspect Mining Techniques, Software Quality Journal. 14(3), 209-214.

Chang, J. and Lee, W. (2006), Finding frequent itemsets over online data streams, Information and Software Technology. 48(7), 606-619.

Chou, C., Sinha, A. and Zhao, H. (2008), A text mining approach to Internet abuse detection, Information Systems and eBusiness Management. 6(4), 419-440.

Curry, C., Grossman, R., Locke, D., Vejcik, S., and Bugajski, J. (2007), Detecting changes in large data sets of payment card data: A case study, KDD'07, August 12-15, San Jose, CA.

Data Intelligence Group (1995), An overview of data mining at Dun & Bradstreet, DIG White Paper 95/01, retrieved from http://www.thearling.com.text/wp9501/wp9501.htm.

Davi, A, Dominique Haughton, Nada Nasr, Gaurav Shah, et al (2005), A Review of Two Text-Mining Packages: SAS TextMining and WordStat. The American Statistician. 59(1), 89-104.

Davies, A. (2007), Identification of spurious results generated via data mining using an Internet distributed supercomputer grant, Duquesne University Donahue School of Business, http://www.business.duq.edu/Research/details.asp?id=83

Deshmukah, A. V. (1997), Software review: ModelQuest Expert 1.0, ORMS Today, December 1997, retrieved from http://www.lionhrtpub.com/orms/orms-12-97/software-review.html.

Ducatelle, F., (2006), Software for the data mining course, School of Informatics, The University of Edinburgh, Scotland, UK, retrieved from http://www.inf.ed.ac.uk/teaching/courses/dme/html/software2.html.

Ganapathy, S., Ranganathan, C. and Sankaranarayanan, B. (2004), Visualization strategies and tools for enhancing customer relationship management, Communications of the ACM. 47(11), 92-98.

Grossman, R. (2007), Data grids, data clouds and data webs: a survey of high performance and distributed data mining, HPC Workshop: Hardware and software for large-scale biological computing in the next decade, December 11-14, Okinawa, Japan, http://www.irp.oist.jp/hpc-workshop/slides.html

Hearst, M. A.(2003), What is Data Mining?, http://www.ischool.berkeley.edu/˜hearstr/text_mining.html

IBM DB2 Intelligent Miner Visualization: Using the Intelligent Miner Visualizers Version 8.2 SH12, Second Edition, August 2004

Kim, S., E James Whitehead Jr and Yi Zhang, (2008), Classifying Software Changes: Clean or Buggy? IEEE Transactions on Software Engineering. 34(2), 181-197.

Lau, K., Lee, K. and Ho, Y. (2005), Text Mining for the Hotel Industry, Cornell Hotel and Restaurant Administration Quarterly. 46(3), 344-363.

Lazarevic A., Fiea T., & Obradovic, Z., (2006), A software system for spatial data analysis and modeling, retrieved from http://www.ist.temple.edu?˜zoran/papers/lazarevic00.pdf.

Leung, Y. F. (2004), My microarray software comparison - Data mining software, September 2004, Chinese University of Hong Kong, retrieved from http://www.ihome.cuhk.edu.hk/˜b400559/arraysoft mining specific.html.

Megaputer Intelligence Inc.(2007), Data Mining, Text Mining, and Web Mining Software, http:///www.megaputer.com

Mesrobian, E. , Muntz, R., Shek,E., Mechoso,, C. R., Farrara, J.D., Spahr, J.A., Stolorz, P.(1995), Real time data mining, management, and visualization of GCM output, IEEE Computer Society, v.81, http://dml.cs.ucla.edu/˜shek/publications/sc_94.ps.gz

Metz. C.(2003), Software: Text Mining, PC Magazine, July 1, http://www.pcmag.com/print_article2/0,1217.a=43573,00.asp

National Center for Biotechnology Information (2006), National Library of Medicine, National Institutes of Health, NCBI tools for data mining, retrieved from http://www.ncbi.nlm,nih.gov/Tools/.

Nayak, R. (2008), Data Mining in Web Services Discovery and Monitoring, International Journal of Web Services Research. 5(1), 63-82.

Nisbet, R. A.(2006), Data mining tools: Which one is best for CRM? Part 3, DM Review, March 21, 2006, retrieved from http://www.dmreview.com/editorial/dmreview/print_action.cfm?articleId=1049954.

Pabarskaite, Z. and Raudys, A. (2007), A process of knowledge discovery from web log data: Systematization and critical review, Journal of Intelligent Information Systems. 28(1), 79-105.

Rokach L., Mining manufacturing data using genetic algorithm-based feature set decomposition, Int. J. Intelligent Systems Technologies and Applications, 4(1):57-78, 2008.

Rokach, L. and Maimon, O., Theory and applications of attribute decomposition, IEEE International Conference on Data Mining, IEEE Computer Society Press, pp. 473–480, 2001.

Rokach, L. and Maimon, O. and Averbuch, M., Information Retrieval System for Medical Narrative Reports, Lecture Notes in Artificial intelligence 3055, page 217-228 Springer-Verlag, 2004.

Sanchez, E. (1996), Speedier: Penn researchers to link supercomputers to community problems, The Compass, v. 43, n. 4, p. 14, September 17, http://www.upenn.edu/pennnews/features/1996/091796/research

Sanchez, M., Moreno, M., Segrera,S. and Lopez, V. (2008), Framework for the development of a personalised recommender system with integrated web-mining functionalities,International Journal of Computer Applications in Technology, 33(4), 312-327.

SAS (2009), JMP Genomics 4.0 Product Brief, http://www.jmp.com/software/genomics /pdf/103112_jmpg4_prodbrief.pdf

Segall, R. and Zhang, Q. (2006), Data visualization and data mining of continuous numerical and discrete nominal-valued microarray databases for biotechnology, Kybernetes: International Journal of Systems and Cybernetics, 35(9/10),1538-1566.

Seigle, G. (2002), CIA, FBI developing intelligence supercomputer, Global Security.

Sekijima, M. (2007), Application of HPC to the analysis of disease related protein and the design of novel proteins, HPC Workshop: "Hardware and software for large-scale biological computing in the next decade", December 11-14, Okinawa, Japan, http://www.irp.oist.jp/hpc-workshop/slides.html

SPPS (2009a): PASW Modeler 13: Overview Demo, http://www.spss.com/media/demos/modeler/ demo-modeler-overview/index.htm

SPPS (2009b): PAWS Modeler Auto Cluster and Cluster Viewer, http://www.spss.com/media/demos/modeler/demo-modeler-autocluster/index.htm

SPSS (2007), Web Mining for Clementine, http://www.spss.com/web_mining_for_clementine, viewed 16 May 2007.

StatSoft, Inc. (2006), Electronic textbook, retrieved from http://www.statsoft.com/textbook/glosa.html.

VSG Visualization Sciences Group (2009), Avizo The 3D visualization software for scientific and industrial data, http://www.vsg3d.com/vsg_prod_avizo_overview.php

Wikipedia (2006), Supercomputers, Retrieved May 19, 2009 from BookRags.com: http://www.bookrags.com/wiki/Supercomputer

Wikipedia (2007), Web mining, http://en.wikipedia.org/wiki/Web_mining

Woodfield, Terry (2004), Mining Textual Data Using SAS Text Miner for SAS9 Course Notes, SAS Institute, Inc., Cary, NC.

Zhang, Q. and Segall, R. (2008), Web mining: a survey of current research, techniques, and software, International Journal of Information Technology & Decision Making, 7(4), 683-720.

Weka-A Machine Learning Workbench for Data Mining

Eibe Frank[1], Mark Hall[1], Geoffrey Holmes[1], Richard Kirkby[1], Bernhard Pfahringer[1], Ian H. Witten[1], and Len Trigg[2]

[1] Department of Computer Science, University of Waikato, Hamilton, New Zealand
{eibe, mhall, geoff, rkirkby, bernhard, ihw}@cs.waikato.ac.nz

[2] Reel Two, P O Box 1538, Hamilton, New Zealand
len@reeltwo.com

Summary. The Weka workbench is an organized collection of state-of-the-art machine learning algorithms and data preprocessing tools. The basic way of interacting with these methods is by invoking them from the command line. However, convenient interactive graphical user interfaces are provided for data exploration, for setting up large-scale experiments on distributed computing platforms, and for designing configurations for streamed data processing. These interfaces constitute an advanced environment for experimental data mining. The system is written in Java and distributed under the terms of the GNU General Public License.

Key words: machine learning software, Data Mining, data preprocessing, data visualization, extensible workbench

66.1 Introduction

Experience shows that no single machine learning method is appropriate for all possible learning problems. The universal learner is an idealistic fantasy. Real datasets vary, and to obtain accurate models the bias of the learning algorithm must match the structure of the domain.

The Weka workbench is a collection of state-of-the-art machine learning algorithms and data preprocessing tools. It is designed so that users can quickly try out existing machine learning methods on new datasets in very flexible ways. It provides extensive support for the whole process of experimental Data Mining, including preparing the input data, evaluating learning schemes statistically, and visualizing both the input data and the result of learning. This has been accomplished by including a wide variety of algorithms for learning different types of concepts, as well as a wide range of preprocessing methods. This diverse and comprehensive set of tools can be invoked through a common interface, making it possible for users

O. Maimon, L. Rokach (eds.), *Data Mining and Knowledge Discovery Handbook*, 2nd ed., DOI 10.1007/978-0-387-09823-4_66,

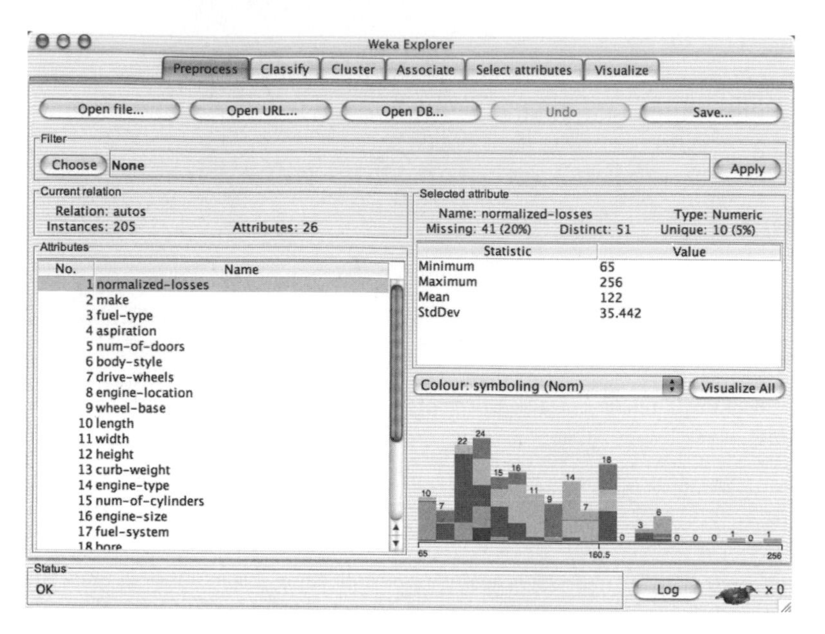

Fig. 66.1. The Explorer Interface.

to compare different methods and identify those that are most appropriate for the problem at hand.

The workbench includes methods for all the standard Data Mining problems: regression, classification, clustering, association rule mining, and attribute selection. Getting to know the data is is a very important part of Data Mining, and many data visualization facilities and data preprocessing tools are provided. All algorithms and methods take their input in the form of a single relational table, which can be read from a file or generated by a database query.

Exploring the Data

The main graphical user interface, the "Explorer," is shown in Figure 66.1. It has six different panels, accessed by the tabs at the top, that correspond to the various Data Mining tasks supported. In the "Preprocess" panel shown in Figure 66.1, data can be loaded from a file or extracted from a database using an SQL query. The file can be in CSV format, or in the system's native ARFF file format. Database access is provided through Java Database Connectivity, which allows SQL queries to be posed to any database for which a suitable driver exists. Once a dataset has been read, various data preprocessing tools, called "filters," can be applied—for example, numeric data can be discretized. In Figure 66.1 the user has loaded a data file and is focusing on a particular attribute, *normalized-losses*, examining its statistics and a histogram.

Through the Explorer's second panel, called "Classify," classification and regression algorithms can be applied to the preprocessed data. This panel also enables users to evaluate the resulting models, both numerically through statistical estimation and graphically through visualization of the data and examination of the model (if the model structure is amenable to visualization). Users can also load and save models.

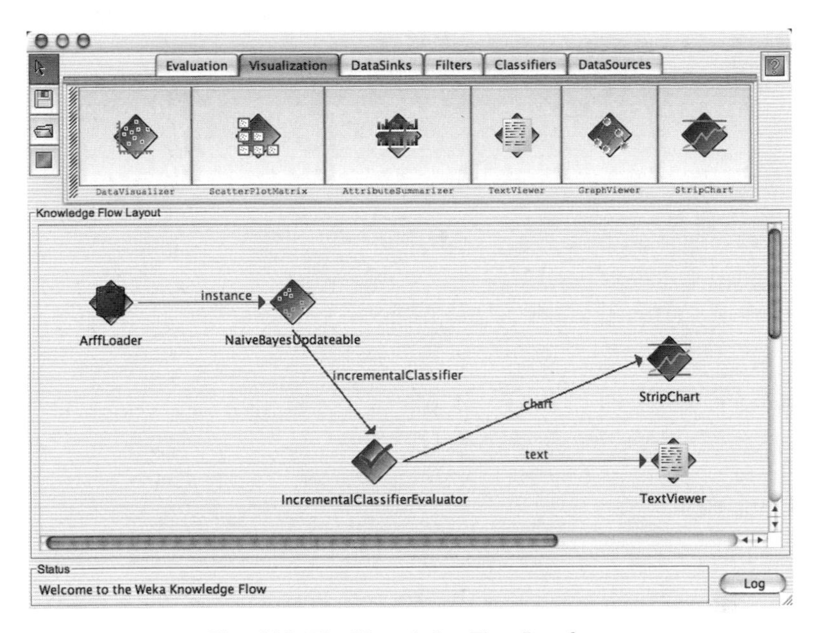

Fig. 66.2. The Knowledge Flow Interface.

The third panel, "Cluster," enables users to apply clustering algorithms to the dataset. Again the outcome can be visualized, and, if the clusters represent density estimates, evaluated based on the statistical likelihood of the data. Clustering is one of two methodologies for analyzing data without an explicit target attribute that must be predicted. The other one comprises association rules, which enable users to perform a market-basket type analysis of the data. The fourth panel, "Associate," provides access to algorithms for learning association rules.

Attribute selection, another important Data Mining task, is supported by the next panel. This provides access to various methods for measuring the utility of attributes, and for finding attribute subsets that are predictive of the data. Users who like to analyze the data visually are supported by the final panel, "Visualize." This presents a color-coded scatter plot matrix, and users can then select and enlarge individual plots. It is also possible to zoom in on portions of the data, to retrieve the exact record underlying a particular data point, and so on.

The Explorer interface does not allow for incremental learning, because the Preprocess panel loads the dataset into main memory in its entirety. That means that it can only be used for small to medium sized problems. However, some incremental algorithms are implemented that can be used to process very large datasets. One way to apply these is through the command-line interface, which gives access to all features of the system. An alternative, more convenient, approach is to use the second major graphical user interface, called "Knowledge Flow." Illustrated in Figure 66.2, this enables users to specify a data stream by graphically connecting components representing data sources, preprocessing tools, learning algorithms, evaluation methods, and visualization tools. Using it, data can be processed in batches as in the Explorer, or loaded and processed incrementally by those filters and learning algorithms that are capable of incremental learning.

An important practical question when applying classification and regression techniques is to determine which methods work best for a given problem. There is usually no way to answer

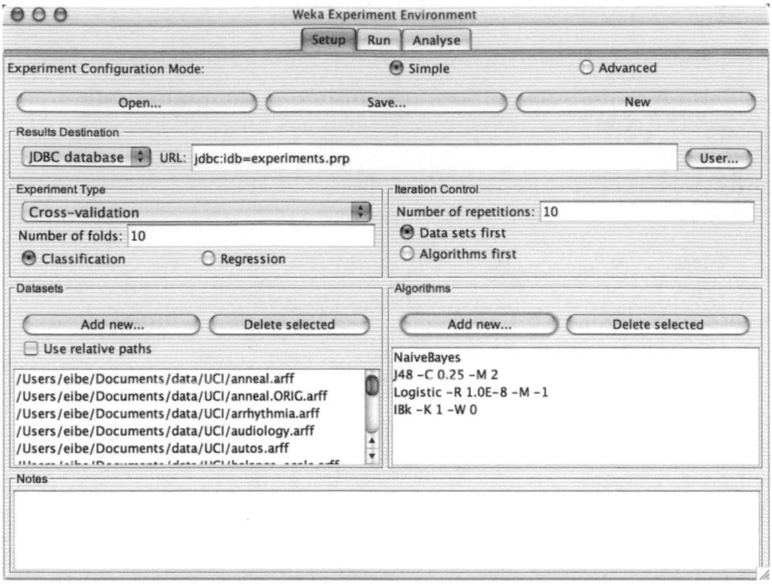

Fig. 66.3. The Experimenter Interface.

this question *a priori*, and one of the main motivations for the development of the workbench was to provide an environment that enables users to try a variety of learning techniques on a particular problem. This can be done interactively in the Explorer. However, to automate the process Weka includes a third interface, the "Experimenter," shown in Figure 66.3. This makes it easy to run the classification and regression algorithms with different parameter settings on a corpus of datasets, collect performance statistics, and perform significance tests on the results. Advanced users can also use the Experimenter to distribute the computing load across multiple machines using Java Remote Method Invocation.

Methods and Algorithms

Weka contains a comprehensive set of useful algorithms for a panoply of Data Mining tasks. These include tools for data engineering (called "filters"), algorithms for attribute selection, clustering, association rule learning, classification and regression. In the following subsections we list the most important algorithms in each category. Most well-known algorithms are included, along with a few less common ones that naturally reflect the interests of our research group.

An important aspect of the architecture is its modularity. This allows algorithms to be combined in many different ways. For example, one can combine bagging! boosting, decision tree learning and arbitrary filters directly from the graphical user interface, without having to write a single line of code. Most algorithms have one or more options that can be specified. Explanations of these options and their legal values are available as built-in help in the graphical user interfaces. They can also be listed from the command line. Additional information and pointers to research publications describing particular algorithms may be found in the internal Javadoc documentation.

Classification

Implementations of almost all main-stream classification algorithms are included. Bayesian methods include naive Bayes, complement naive Bayes, multinomial naive Bayes, Bayesian networks, and AODE. There are many decision tree learners: decision stumps, ID3, a C4.5 clone called "J48," trees generated by reduced error pruning, alternating decision trees, and random trees and forests thereof. Rule learners include OneR, an implementation of Ripper called "JRip," PART, decision tables, single conjunctive rules, and Prism. There are several separating hyperplane approaches like support vector machines with a variety of kernels, logistic regression, voted perceptrons, Winnow and a multi-layer perceptron. There are many lazy learning methods like IB1, IBk, lazy Bayesian rules, KStar, and locally-weighted learning.

As well as the basic classification learning methods, so-called "meta-learning" schemes enable users to combine instances of one or more of the basic algorithms in various ways: bagging! boosting (including the variants AdaboostM1 and LogitBoost), and stacking. A method called "FilteredClassifier" allows a filter to be paired up with a classifier. Classification can be made cost-sensitive, or multi-class, or ordinal-class. Parameter values can be selected using cross-validation.

Regression

There are implementations of many regression schemes. They include simple and multiple linear regression, pace regression, a multi-layer perceptron, support vector regression, locally-weighted learning, decision stumps, regression and model trees (M5) and rules (M5rules). The standard instance-based learning schemes IB1 and IBk can be applied to regression problems (as well as classification problems). Moreover, there are additional meta-learning schemes that apply to regression problems, such as additive regression and regression by discretization.

Clustering

At present, only a few standard clustering algorithms are included: KMeans, EM for naive Bayes models, farthest-first clustering, and Cobweb. This list is likely to grow in the near future.

Association rule learning

The standard algorithm for association rule induction is Apriori, which is implemented in the workbench. Two other algorithms implemented in Weka are Tertius, which can extract first-order rules, and Predictive Apriori, which combines the standard confidence and support statistics into a single measure.

Attribute selection

Both wrapper and filter approaches to attribute selection are supported. A wide range of filtering criteria are implemented, including correlation-based feature selection, the chi-square statistic, gain ratio, information gain, symmetric uncertainty, and a support vector machine-based criterion. There are also a variety of search methods: forward and backward selection, best-first search, genetic search, and random search. Additionally, principal components analysis can be used to reduce the dimensionality of a problem.

Filters

Processes that transform instances and sets of instances are called "filters," and they are classified according to whether they make sense only in a prediction context (called "supervised") or in any context (called "unsupervised"). We further split them into "attribute filters," which work on one or more attributes of an instance, and "instance filters," which manipulate sets of instances.

Unsupervised attribute filters include adding a new attribute, adding a cluster indicator, adding noise, copying an attribute, discretizing a numeric attribute, normalizing or standardizing a numeric attribute, making indicators, merging attribute values, transforming nominal to binary values, obfuscating values, swapping values, removing attributes, replacing missing values, turning string attributes into nominal ones or word vectors, computing random projections, and processing time series data. Unsupervised instance filters transform sparse instances into non-sparse instances and vice versa, randomize and resample sets of instances, and remove instances according to certain criteria.

Supervised attribute filters include support for attribute selection, discretization, nominal to binary transformation, and re-ordering the class values. Finally, supervised instance filters resample and subsample sets of instances to generate different class distributions—stratified, uniform, and arbitrary user-specified spreads.

System Architecture

In order to make its operation as flexible as possible, the workbench was designed with a modular, object-oriented architecture that allows new classifiers, filters, clustering algorithms and so on to be added easily. A set of abstract Java classes, one for each major type of component, were designed and placed in a corresponding top-level package.

All classifiers reside in subpackages of the top level "classifiers" package and extend a common base class called "Classifier." The Classifier class prescribes a public interface for classifiers and a set of conventions by which they should abide. Subpackages group components according to functionality or purpose. For example, filters are separated into those that are supervised or unsupervised, and then further by whether they operate on an attribute or instance basis. Classifiers are organized according to the general type of learning algorithm, so there are subpackages for Bayesian methods, tree inducers, rule learners, etc.

All components rely to a greater or lesser extent on supporting classes that reside in a top level package called "core." This package provides classes and data structures that read data sets, represent instances and attributes, and provide various common utility methods. The core package also contains additional interfaces that components may implement in order to indicate that they support various extra functionality. For example, a classifier can implement the "WeightedInstancesHandler" interface to indicate that it can take advantage of instance weights.

A major part of the appeal of the system for end users lies in its graphical user interfaces. In order to maintain flexibility it was necessary to engineer the interfaces to make it as painless as possible for developers to add new components into the workbench. To this end, the user interfaces capitalize upon Java's introspection mechanisms to provide the ability to configure each component's options dynamically at runtime. This frees the developer from having to consider user interface issues when developing a new component. For example, to enable a new classifier to be used with the Explorer (or either of the other two graphical user interfaces), all a developer need do is follow the Java Bean convention of supplying "get" and "set" methods for each of the classifier's public options.

Applications

Weka was originally developed for the purpose of processing agricultural data, motivated by the importance of this application area in New Zealand. However, the machine learning methods and data engineering capability it embodies have grown so quickly, and so radically, that the workbench is now commonly used in all forms of Data Mining applications—from bioinformatics to competition datasets issued by major conferences such as *Knowledge Discovery in Databases*.

New Zealand has several research centres dedicated to agriculture and horticulture, which provided the original impetus for our work, and many of our early applications. For example, we worked on predicting the internal bruising sustained by different varieties of apple as they make their way through a packing-house on a conveyor belt (Holmes et al., 1998); predicting, in real time, the quality of a mushroom from a photograph in order to provide automatic grading (Kusabs et al., 1998); and classifying kiwifruit vines into twelve classes, based on visible-NIR spectra, in order to determine which of twelve pre-harvest fruit management treatments has been applied to the vines (Holmes and Hall, 2002). The applicability of the workbench in agricultural domains was the subject of user studies (McQueen et al., 1998) that demonstrated a high level of satisfaction with the tool and gave some advice on improvements.

There are countless other applications, actual and potential. As just one example, Weka has been used extensively in the field of bioinformatics. Published studies include automated protein annotation (Bazzan et al., 2002), probe selection for gene expression arrays (Tobler et al., 2002), plant genotype discrimination (Taylor et al., 2002), and classifying gene expression profiles and extracting rules from them (Li et al., 2003). Text mining is another major field of application, and the workbench has been used to automatically extract key phrases from text (Frank et al., 1999), and for document categorization (Sauban and Pfahringer, 2003) and word sense disambiguation (Pedersen, 2002).

The workbench makes it very easy to perform interactive experiments, so it is not surprising that most work has been done with small to medium sized datasets. However, larger datasets have been successfully processed. Very large datasets are typically split into several training sets, and a voting-committee structure is used for prediction. The recent development of the knowledge flow interface should see larger scale application development, including online learning from streamed data.

Many future applications will be developed in an online setting. Recent work on data streams (Holmes et al., 2003) has enabled machine learning algorithms to be used in situations where a potentially infinite source of data is available. These are common in manufacturing industries with 24/7 processing. The challenge is to develop models that constantly monitor data in order to detect changes from the steady state. Such changes may indicate failure in the process, providing operators with warning signals that equipment needs re-calibrating or replacing.

Summing up the Workbench

Weka has three principal advantages over most other Data Mining software. First, it is open source, which not only means that it can be obtained free, but—more importantly—it is maintainable, and modifiable, without depending on the commitment, health, or longevity of any particular institution or company. Second, it provides a wealth of state-of-the-art machine learning algorithms that can be deployed on any given problem. Third, it is fully implemented in Java and runs on almost any platform—even a Personal Digital Assistant.

The main disadvantage is that most of the functionality is only applicable if all data is held in main memory. A few algorithms are included that are able to process data incrementally or in batches (Frank et al., 2002). However, for most of the methods the amount of available memory imposes a limit on the data size, which restricts application to small or medium-sized datasets. If larger datasets are to be processed, some form of subsampling is generally required. A second disadvantage is the flip side of portability: a Java implementation may be somewhat slower than an equivalent in C/C++.

Acknowledgments

Many thanks to past and present members of the Waikato machine learning group and the many external contributors for all the work they have put into Weka.

References

Bazzan, A. L., Engel, P. M., Schroeder, L. F., and da Silva, S. C. (2002). Automated annotation of keywords for proteins related to mycoplasmataceae using machine learning techniques. *Bioinformatics*, 18:35S–43S.

Frank, E., Holmes, G., Kirkby, R., and Hall, M. (2002). Racing committees for large datasets. In *Proceedings of the International Conference on Discovery Science*, pages 153–164. Springer-Verlag.

Frank, E., Paynter, G. W., Witten, I. H., Gutwin, C., and Nevill-Manning, C. G. (1999). Domain-specific keyphrase extraction. In *Proceedings of the 16th International Joint Conference on Artificial Intelligence*, pages 668–673. Morgan Kaufmann.

Holmes, G., Cunningham, S. J., Rue, B. D., and Bollen, F. (1998). Predicting apple bruising using machine learning. *Acta Hort*, 476:289–296.

Holmes, G. and Hall, M. (2002). A development environment for predictive modelling in foods. *International Journal of Food Microbiology*, 73:351–362.

Holmes, G., Kirkby, R., and Pfahringer, B. (2003). Mining data streams using option trees. Technical Report 08/03, Department of Computer Science, University of Waikato.

Kusabs, N., Bollen, F., Trigg, L., Holmes, G., and Inglis, S. (1998). Objective measurement of mushroom quality. In *Proc New Zealand Institute of Agricultural Science and the New Zealand Society for Horticultural Science Annual Convention*, page 51.

Li, J., Liu, H., Downing, J. R., Yeoh, A. E.-J., and Wong, L. (2003). Simple rules underlying gene expression profiles of more than six subtypes of acute lymphoblastic leukemia (all) patients. *Bioinformatics*, 19:71–78.

McQueen, R., Holmes, G., and Hunt, L. (1998). User satisfaction with machine learning as a data analysis method in agricultural research. *New Zealand Journal of Agricultural Research*, 41(4):577–584.

Pedersen, T. (2002). Evaluating the effectiveness of ensembles of decision trees in disambiguating Senseval lexical samples. In *Proceedings of the ACL-02 Workshop on Word Sense Disambiguation: Recent Successes and Future Directions*.

Sauban, M. and Pfahringer, B. (2003). Text categorisation using document profiling. In *Proceedings of the 7th European Conference on Principles and Practice of Knowledge Discovery in Databases*, pages 411–422. Springer.

Taylor, J., King, R. D., Altmann, T., and Fiehn, O. (2002). Application of metabolomics to plant genotype discrimination using statistics and machine learning. *Bioinformatics*, 18:241S–248S.

Tobler, J. B., Molla, M., Nuwaysir, E., Green, R., and Shavlik, J. (2002). Evaluating machine learning approaches for aiding probe selection for gene-expression arrays. *Bioinformatics*, 18:164S–171S.

Index

O. Maimon, L. Rokach (eds.), *Data Mining and Knowledge Discovery Handbook*, 2nd ed.,
DOI 10.1007/978-0-387-09823-4,

Printed in the United States of America